Meyler's Side Effects of Psychiatric Drugs

Meyler's Side Effects of Psychiatric Drugs

Editor

J K Aronson, MA, DPhil, MBChB, FRCP, FBPharmacolS, FFPM (Hon)
Oxford, United Kingdom

ELSEVIER

AMSTERDAM • BOSTON • HEIDELBERG • LONDON • NEW YORK • OXFORD
PARIS • SAN DIEGO • SAN FRANCISCO • SINGAPORE • SYDNEY • TOKYO

Elsevier
Radarweg 29, PO Box 211, 1000 AE Amsterdam, The Netherlands
The Boulevard, Langford Lane, Kidlington, Oxford OX5 1GB, UK
525 B Street, Suite 1900, San Diego, CA 92101-4495, USA

Notice
No responsibility is assumed by the publisher for any injury and/or damage to persons
or property as a matter of products liability, negligence or otherwise, or from any use or operation
of any methods, products, instructions or ideas contained in the material herein. Because of rapid
advances in the medical sciences, in particular, independent verification of diagnoses and drug
dosages should be made

Medicine is an ever-changing field. Standard safety precautions must be followed, but as new
research and clinical experience broaden our knowledge, changes in treatment and drug therapy
may become necessary or appropriate. Readers are advised to check the most current product
information provided by the manufacturer of each drug to be administered to verify the
recommended dose, the method and duration of administrations, and contraindications. It is the
responsibility of the treating physician, relying on experience and knowledge of the patient, to
determine dosages and the best treatment for each individual patient. Neither the publisher nor the
authors assume any liability for any injury and/or damage to persons or property arising from this
publication.

British Library Cataloguing in Publication Data
A catalogue record for this book is available from the British Library

Library of Congress Catalog Number: 2008933970

ISBN: 978-044-453266-4

For information on all Elsevier publications
visit our web site at http://www.elsevierdirect.com

Typeset by Integra Software Services Pvt. Ltd, Pondicherry, India www.integra-india.com
Printed and bound in the USA

08 09 10   10 9 8 7 6 5 4 3 2 1

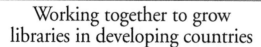

Working together to grow
libraries in developing countries

www.elsevier.com | www.bookaid.org | www.sabre.org

ELSEVIER   BOOK AID International   Sabre Foundation

Contents

Preface

This volume covers the adverse effects of medicines used in psychiatry and drugs of abuse. The material has been collected from *Meyler's Side Effects of Drugs: The International Encyclopedia of Adverse Drug Reactions and Interactions* (15th edition, 2006, in six volumes), which was itself based on previous editions of *Meyler's Side Effects of Drugs*, and from the *Side Effects of Drugs Annuals* (SEDA) 28, 29, and 30. The main contributors of this material were Jeffrey K Aronson, Stefan Borg, Andrew Byrne, Alfonso Carvajal, Philip J Cowen, Stephen Curran, DL Dunner, Everett H Ellingwood, Rif S El-Mallakh, Peter J Geerlings, Sarah Guzofski , Leslie L Iversen, Jeffrey W Jefferson, Natalia Jimeno, Tong H Lee, Luis H Martín Arias, David B Menkes, Shabir Musa, Inger Öhman, Jayendra K Patel, Edmond H Pi, TE Ralston, Reginald P Sequeira, George M Simpson, John J Sramek, and Eileen J Wong. For contributors to earlier editions of *Meyler's Side Effects of Drugs* and the *Side Effects of Drugs Annuals*, see http://www.elsevier.com/wps/find/bookseriesdescription.cws_home/BS_SED/description.

A brief history of the Meyler series

Leopold Meyler was a physician who was treated for tuberculosis after the end of the Nazi occupation of The Netherlands. According to Professor Wim Lammers, writing a tribute in Volume VIII (1975), Meyler got a fever from para-aminosalicylic acid, but elsewhere Graham Dukes has written, based on information from Meyler's widow, that it was deafness from dihydrostreptomycin; perhaps it was both. Meyler discovered that there was no single text to which medical practitioners could look for information about unwanted effects of drug therapy; Louis Lewin's text "Die Nebenwirkungen der Arzneimittel" ("The Untoward Effects of Drugs") of 1881 had long been out of print (SEDA-27, xxv–xxix). Meyler therefore determined to make such information available and persuaded the Netherlands publishing firm of Van Gorcum to publish a book, in Dutch, entirely devoted to descriptions of the adverse effects that drugs could cause. He went on to agree with the Elsevier Publishing Company, as it was then called, to prepare and issue an English translation. The first edition of 192 pages (*Schadelijke Nevenwerkingen van Geneesmiddelen*) appeared in 1951 and the English version (*Side Effects of Drugs*) a year later.

The book was a great success, and a few years later Meyler started to publish what he called surveys of unwanted effects of drugs. Each survey covered a period of two to four years. They were labelled as volumes rather than editions, and after Volume IV had been published Meyler could no longer handle the task alone. For subsequent volumes he recruited collaborators, such as Andrew Herxheimer. In September 1973 Meyler died unexpectedly, and Elsevier invited Graham Dukes to take over the editing of Volume VIII.

Dukes persuaded Elsevier that the published literature was too large to be comfortably encompassed in a four-yearly cycle, and he suggested that the volumes should be produced annually instead. The four-yearly volume could then concentrate on providing a complementary critical encyclopaedic survey of the entire field. The first *Side Effects of Drugs Annual* was published in 1977. The first encyclopaedic edition of *Meyler's Side Effects of Drugs*, which appeared in 1980, was labelled the ninth edition, and since then a new encyclopaedic edition has appeared every four years. The 15th edition was published in 2006, in both hard and electronic versions.

Monograph structure

This volume is in six sections:

- antidepressants—a general introduction to their adverse effects, followed by monographs on individual drugs and groups of drugs (including lithium);
- neuroleptic drugs—a general introduction to their adverse effects, followed by monographs on individual drugs;
- hypnosedatives—a general introduction to their adverse effects, followed by monographs on individual drugs;
- drugs of abuse;
- drugs used to treat Alzheimer's disease;
- psychological and psychiatric adverse effects of non-psychoactive drugs.

In each monograph in the Meyler series the information is organized into sections as shown below (although not all the sections are covered in each monograph).

DoTS classification of adverse drug reactions

A few adverse effects have been classified using the system known as DoTS. In this system adverse reactions are classified according to the *Dose* at which they usually occur, the *Time-course* over which they occur, and the *Susceptibility factors* that make them more likely, as follows:

- *Relation to Dose*

  - Toxic reactions—reactions that occur at supratherapeutic doses
  - Collateral reactions—reactions that occur at standard therapeutic doses
  - Hypersusceptibility reactions—reactions that occur at subtherapeutic doses in susceptible individuals

- *Time-course*

  - Time-independent reactions—reactions that occur at any time during a course of therapy
  - Time-dependent reactions

    - Immediate or rapid reactions—reactions that occur only when a drug is administered too rapidly
    - First-dose reactions—reactions that occur after the first dose of a course of treatment and not necessarily thereafter

- Early reactions—reactions that occur early in treatment then either abate with continuing treatment (owing to tolerance) or persist
- Intermediate reactions—reactions that occur after some delay but with less risk during longer term therapy, owing to the "healthy survivor" effect
- Late reactions—reactions the risk of which increases with continued or repeated exposure
- Withdrawal reactions—reactions that occur when, after prolonged treatment, a drug is withdrawn or its effective dose is reduced
- Delayed reactions—reactions that occur some time after exposure, even if the drug is withdrawn before the reaction appears

- *Susceptibility factors*

  - Genetic
  - Age
  - Sex
  - Physiological variation
  - Exogenous factors (for example drug–drug or drug–food interactions, smoking)
  - Diseases

Drug names

Drugs have usually been designated by their recommended or proposed International Non-proprietary Names (rINN or pINN); when these are not available, chemical names have been used. In some cases brand names have been used.

Spelling

For indexing purposes, American spelling has been used, e.g. anemia, estrogen rather than anaemia, oestrogen.

Cross-references

The various editions of *Meyler's Side Effects of Drugs* are cited in the text as SED-l3, SED-14, etc; the *Side Effects of Drugs Annuals* are cited as SEDA-1, SEDA-2, etc.

J K Aronson
Oxford, June 2008

Organization of material in monographs in the Meyler series (not all sections are included in each monograph)

General information
Drug studies
    Observational studies
    Comparative studies
    Drug-combination studies
    Placebo-controlled studies
    Systematic reviews
Organs and systems
    Cardiovascular
    Respiratory
    Ear, nose, throat
    Nervous system
    Neuromuscular function
    Sensory systems
    Psychological
    Psychiatric
    Endocrine
    Metabolism
    Nutrition
    Electrolyte balance
    Mineral balance
    Metal metabolism
    Acid-base balance
    Fluid balance
    Hematologic
    Mouth
    Teeth
    Salivary glands
    Gastrointestinal
    Liver
    Biliary tract
    Pancreas
    Urinary tract
    Skin
    Hair
    Nails
    Sweat glands
    Serosae
    Musculoskeletal
    Sexual function
    Reproductive system
    Breasts
    Immunologic
    Autacoids
    Infection risk

    Body temperature
    Multiorgan failure
    Trauma
    Death
Long-term effects
    Drug abuse
    Drug misuse
    Drug tolerance
    Drug resistance
    Drug dependence
    Drug withdrawal
    Genotoxicity
    Cytotoxicity
    Mutagenicity
    Tumorigenicity
Second-generation effects
    Fertility
    Pregnancy
    Teratogenicity
    Fetotoxicity
    Lactation
    Breast feeding
Susceptibility factors
    Genetic factors
    Age
    Sex
    Physiological factors
    Disease
    Other features of the patient
Drug administration
    Drug formulations
    Drug additives
    Drug contamination and adulteration
    Drug dosage regimens
    Drug administration route
    Drug overdose
Interactions
    Drug-drug interactions
    Food-drug interactions
    Drug-device interactions
    Smoking
    Other environmental interactions
Interference with diagnostic tests
Diagnosis of adverse drug reactions
Management of adverse drug reactions
Monitoring therapy
References

ANTIDEPRESSANTS

This section contains monographs on the following groups of antidepressants:

- tricyclic antidepressants;
- selective serotonin re-uptake inhibitors (SSRIs);
- monoamine oxidase inhibitors;
- other antidepressants (including serotonin and noradrenaline re-uptake inhibitors)
- lithium.

Here we cover some of the adverse effects that affect antidepressants as a class or for which there have been comparisons between different types of antidepressants.

Organs and Systems

Psychiatric

Mania is listed as a possible adverse effect of all antidepressant drugs, and there is a well recognized association between the use of antidepressants and reports of switches from depression into mania (SEDA-23, 17). There are three kinds of explanation for this phenomenon: (i) a spontaneous switch from depression into mania as part of a bipolar illness, which happens by chance to coincide with the use of antidepressant treatment; (ii) the mania is indeed triggered by the antidepressant drug but the depression is part of a bipolar illness and bipolar patients are unusually susceptible to antidepressant-induced hypomania; (iii) antidepressants can induce mania in patients with true unipolar depression who would never otherwise suffer from manic illness.

In a systematic review of antidepressant-induced hypomania and mania the rate of switching from depression into mania with antidepressant drug treatment in patients identified as having unipolar depression was quite low and generally under 5% (1). This rate is within that described for the spontaneous conversion from unipolar to bipolar disorder during longer-term follow-up in the absence of antidepressant treatment. In contrast, in patients with established bipolar illness, antidepressant drugs did increase the risk of mania, and the risk with tricyclic antidepressants was probably greater than that seen with selective serotonin re-uptake inhibitors (SSRIs). The authors concluded that the rate of antidepressant-induced hypomania and mania in major depression is within the rate of misdiagnosis of bipolar depression as unipolar depression, and that depressed patients who experience antidepressant-induced hypomania are truly bipolar. However, this conclusion is somewhat circular, and if accepted would make it impossible to diagnose antidepressant-induced hypomania without also diagnosing bipolar illness. Longer-term follow-up studies will help show whether patients with antidepressant-induced hypomania also develop manic illness in circumstances other than exposure to antidepressant medications. However, the best way of resolving this issue would be to identify a genetic or other biological marker that distinguished bipolar subjects from those with unipolar illness.

Evidence from 12 trials has been assessed in a systematic review, in which 1088 patients with bipolar depression were randomized to different kinds of antidepressant medication or placebo (2). A high proportion of patients (75%) were taking concomitant therapy with a mood stabilizer or an atypical antipsychotic drug. Overall, antidepressant treatment did not increase the risk of switching to mania relative to placebo (3.8% versus 4.7%, difference = 0.9%, CI= –2.0, 3.8%). However, patients who were taking tricyclic antidepressants had a higher rate of manic switching compared with other antidepressants (mainly SSRIs and MAOIs).The rate of manic switch on tricyclics was 10% and for other antidepressants 3.2% (absolute risk difference of 6.8%, CI= 1.7, 12%). These findings suggest that tricyclic antidepressants should not be used in depressed patients with bipolar disorder unless it is essential and that SSRIs should probably be first-line treatment. However, the data are derived from short-term studies (up to 10 weeks) and the course of the bipolar disorder during longer-term treatment with antidepressants is unclear. Also most of the patients in the systematic review were taking other drugs (mood stabilizers and antipsychotic drugs), which would be expected to lessen the risk of mania. Hence, the rate of manic switching in patients taking antidepressants as sole therapy could be higher.

Sexual function

Most classes of antidepressant drugs are associated with sexual dysfunction of various kinds, including reduced desire and arousal, erectile difficulties, and ejaculatory and orgasmic inhibition. A literature review has shown that rates of sexual dysfunction of all kinds were highest with selective serotonin re-uptake inhibitors (SSRIs) and venlafaxine and least with amfebutamone (bupropion) and the reversible inhibitor of monoamine oxidase type A, moclobemide (3). Switching to amfebutamone or mirtazapine from SSRIs often enabled patients to obtain relief from sexual dysfunction. However, commonly recommended antidotes to SSRI-induced sexual dysfunction, such as *Ginkgo biloba*, serotonin ($5HT_2$) receptor antagonists, such as cyproheptadine, and amfebutamone augmentation were not supported by placebo-controlled trials.

These data suggest that antidepressant-induced sexual dysfunction is more likely to be associated with agents that greatly potentiate $5HT$ neurotransmission. This notion is supported by the results of a 6-week double-blind study of 24 men with premature ejaculation, in which paroxetine (20 mg/day) increased latency to ejaculation six-fold while mirtazapine (30 mg/day) had minimal effect (4). In a randomized, 8-week, double-blind, placebo-controlled study in 450 patients with major depression, fluoxetine (20–40 mg/day) significantly impaired sexual function, while the noradrenaline re-uptake inhibitor reboxetine had no effect (5).

In a double-blind, placebo-controlled study in 90 patients with sexual dysfunction who were taking a variety of $5HT$ re-uptake inhibitor antidepressants, sildenafil (50–100 mg) produced improvement in all aspects of the sexual response in 54% of antidepressant-treated patients compared with a placebo response rate of 4.4% (NNT = 2) (6). This suggests that sildenafil is an effective treatment for antidepressant-induced sexual dysfunction.

Long-Term Effects

Tumorigenicity

Studies in rats have suggested that both tricyclic antidepressants and selective serotonin re-uptake inhibitors (SSRIs) promote the growth of mammary tumors. Rates of antidepressant prescribing have increased over the last 20 years, and breast cancer remains a leading cause of mortality in women. Two recent non-systematic reviews (7,8) have examined the question of whether antidepressant treatment might increase the risk of *breast cancer*. A variety of studies have been conducted, including hospital- and population-based case-control studies and prospective cohort studies. Overall the evidence for an association between any use of antidepressant medication and breast cancer is weak and inconclusive. However, it is possible that sustained use of certain drugs, for example, the SSRI paroxetine, might be associated with an increased risk. Problems with the studies reviewed include defining what kind of antidepressant treatment regimen represents significant exposure, and the unknown nature of the time-course of any effect of antidepressants on the development of breast cancer. The increasing use of SSRIs, which are known to be associated with breast enlargement (SEDA-22, 12), suggests that further prospective studies are needed.

Drug Administration

Drug overdose

Suicidal ideation of some kind almost invariably accompanies severe depression. Hence the relative toxicity of antidepressants in overdose can be important in determining treatment choice. It is accepted that SSRIs are less dangerous in overdose than tricyclic antidepressants, but there are fewer data on the toxicity of other antidepressants. The presentation and likely toxicity in overdose of several newer antidepressant drugs have been reviewed (9). Deaths in overdose have been most clearly associated with amfebutamone and venlafaxine.

Amfebutamone overdose typically presents with neurological symptoms, including delirium, agitation, and seizures; however, cardiac dysrhythmias, with QT interval prolongation and cardiac arrest, have occurred (10,11). Venlafaxine overdose is also associated with seizures and cardiac dysrhythmias (Nurnberg 56). Venlafaxine is a potent 5-HT re-uptake inhibitor, and signs of 5HT toxicity (agitation, myoclonus, hyperthermia) are common.

In a prospective cohort study of over 450 patients who had attempted suicide by antidepressant ingestion the risk of seizures after venlafaxine overdose (14%) was significantly greater than that of SSRIs (1.3%) and similar to that seen with dosulepin (11%) (12). Rates of 5HT toxicity did not differ significantly between venlafaxine and SSRIs (29% versus 19%) but were greater than with tricyclic antidepressants (1.2%). Unlike SSRIs, venlafaxine was associated with significant prolongation of the QT interval; tricyclic antidepressants had a similar effect.

Data on the consequences of overdose of other new antidepressant agents are limited, but current evidence suggests that reboxetine and mirtazapine have low toxicity in overdose (Buckley 539). Reboxetine, as would be expected, presents with signs of noradrenergic overactivity, such as sweating, tachycardia, hypertension, and anxiety. The characteristic feature of mirtazapine overdose is sedation (Buckley 539).

Overdose of moclobemide by itself rarely appears to give rise to serious problems. This is in contrast to overdose with conventional monoamine oxidase inhibitors, which can cause fatal 5HT toxicity. However, if patients take moclobemide together with serotonergic antidepressants, such as SSRIs or clomipramine, 5HT toxicity is common. 5HT toxicity occurred in 11 of 21 patients who took overdoses of moclobemide and serotonergic agents but in only one of 33 patients who took moclobemide alone (13). Consistent with this, four patients died, presumably of 5HT toxicity, after co-ingesting 3,4-methylenedioxymethamphetamine (MDMA, ecstasy) and moclobemide (14).

Overall the current data suggest that the safety advantage in overdose relative to tricyclic antidepressants enjoyed by SSRIs may extend to reboxetine and mirtazapine. Amfebutamone and venlafaxine are more toxic than SSRIs in overdose, but they are still likely to be safer than tricyclic antidepressants.

References

1. Chun BJ, Dunner DL. A review of antidepressant-induced hypomania in major depression: suggestions for DSM-V. Bipolar Disord 2004;6:32-42.2.
2. Gijsman HJ, Geddes JR, Rendell JM, Nolen WA, Goodwin GM. Antidepressants for bipolar depression: a systematic review of randomized, controlled trials. Am J Psychiatry 2004;161:1537–47.
3. Labbate LA, Croft HA, Oleshansky MA. Antidepressant-related erectile dysfunction: management via avoidance, switching antidepressants, antidotes and adaptation. J Clin Psychiatry 2003;64:11–19.
4. Waldinger MD, Zwinderman AH, Olivier B. Antidepressants and ejaculation: a double-blind, randomised, fixed-dose study with mirtazapine and paroxetine. J Clin Psychopharmacol 2003;23:467–70.
5. Clayton AH, Zajecka J, Ferguson JM, Filipiak-Reisner JK, Brown MT, Schwartz GE. Lack of sexual dysfunction with the selective noradrenaline reuptake inhibitor reboxetine during treatment for major depressive disorder. Int Clin Psychopharmacol 2003;18:151–6.
6. Nurnberg HG, Hensley PL, Gelenberg AJ, Fava M, Lauriello J, Paine S. Treatment of antidepressant-associated sexual dysfunction with sildenafil. J Am Med Assoc 2003;289:56–64.
7. Bahl S, Cotterchio M, Kreiger N. Use of antidepressant medications and the possible association with breast cancer risk. Psychother Psychosom 2003;72:185–94.
8. Lawlor DA, Juni P, Ebrahim S, Egger M. Systematic review of the epidemiologic and trial evidence of an association between antidepressant medication and breast cancer. J Clin Epidemiol 2003;56:155–63.

9. Buckley NA, Faunce TA. "Atypical" antidepressants in overdose. Clinical considerations with respect to safety. Drug Saf 2003;26:539–51.

10. Balit CR, Lynch CN, Isbister GK. Bupropion poisoning: a case series. Med J Aust 2003;178:61–3.

11. Isbister GK, Balit CR. Bupropion overdose: QTc prolongation and its clinical significance. Ann Pharmacother 2003;37:999–1002.

12. Whyte IM, Dawson AH, Buckley NA. Relative toxicity of venlafaxine and selective serotonin reuptake inhibitors in overdose compared to tricyclic antidepressants. Assoc Phys 2003;96:369–74.

13. Isbister GK, Hackett LP, Dawson AH, Whyte IM, Smith AJ. Moclobemide poisoning: toxicokinetics and occurrence of serotonin toxicity. Br J Clin Pharmacol 2003;56:441–50.

14. Vuori E, Henry JA, Ojanpera I, Nieminen R, Savolainen T, Wahlsten P, Jantti M. Death following ingestion of MDMA (ecstasy) and moclobemide. Addiction. Soc Study Addiction Alcohol Other Drugs 2003;98:365–8.

TRICYCLIC ANTIDEPRESSANTS

General Information

There are no accurate data on the worldwide use of the many tricyclic compounds listed in Table 1, and the availability of particular drugs varies from country to country. The dosage range for all these compounds is 50–300 mg/day, with the exception of nortriptyline, which has an upper limit of 200 mg, and protriptyline, which is more potent (range 10–60 mg/day). Well-controlled comparisons are few, but it is clear that these drugs resemble each other more than they differ. Their adverse effects will be discussed for the class as a whole, with distinguishing features of specific compounds mentioned when appropriate.

Tricyclic antidepressants interfere with the activity of at least five putative neurotransmitters by several different potential mechanisms at both central and peripheral sites. This gives rise to uncertainty in understanding the mechanisms that underlie a bewildering variety of untoward effects, which are in turn further modified by temporal factors, probably related to changes in receptor sensitivity. Differences between drugs are often inferred on the basis of selective effects on isolated organs in specific species, but their relevance to actions in man is largely unsubstantiated, owing to a lack of early clinical pharmacology studies. There is a high base rate for spontaneously occurring complaints or placebo-induced adverse effects in psychiatric populations, which complicates interpretation, even in controlled studies. There is also a wide range of interindividual sensitivity and susceptibility between patients, and very little consistent correlation between plasma concentrations and particular adverse effects.

All tricyclic compounds (with the possible exception of protriptyline) have sedative effects, and this may be desirable or undesirable, depending on a particular patient's state of apathy or agitation. They have a spectrum of anticholinergic activity, presenting as troublesome adverse effects, such as dry mouth, sweating, confusion, constipation, blurred vision, and urinary hesitancy, depending on individual patient susceptibility. Weight gain is a common and troublesome adverse effect; it is mediated in part by histamine H_1 receptor antagonism.

The adverse effects of most serious concern relate to the cardiovascular system and seizure threshold. Actions on the adrenergic and cholinergic systems probably contribute to both hypotensive and direct cardiac effects, including alterations in heart rate, quinidine-like delays in conduction, and reduced myocardial contractility. The seizure threshold is lowered, increasing the frequency of epileptic seizures. All of these adverse effects can occur at therapeutic dosages in susceptible populations, such as elderly people, children, and people with cardiac problems or epilepsy, but are also a major cause of morbidity and mortality in accidental or intentional overdosage. Doses in excess of 500 mg can be seriously toxic, and death is fairly common when doses of 2 g or more are taken.

Tricyclic antidepressants rarely cause cholestatic jaundice and agranulocytosis due to hypersensitivity reactions. The rare liver necrosis may reflect severe hypersensitivity. Two fatal cases of hypersensitivity myocarditis and hepatitis have been described (1). A variety of dermatological manifestations have been reported (rash, urticaria, vasculitis), but their relation to drug ingestion is often uncertain. A single case of pulmonary hypersensitivity with

Table 1 Tricyclic antidepressants that have been widely studied or are currently available for treating depression (all rINNs)

Compound	Structure	Comments
Imipramine	Dibenzapine; tertiary amine	Prototype compound
Desipramine	Secondary amine	First metabolite of imipramine
Amitriptyline	Dibenzocycloheptene; tertiary amine	Meta-analyses suggest the most effective
Nortriptyline	Secondary amine	First metabolite of amitriptyline
Protriptyline	Secondary amine	Most potent; least sedative
Doxepin	Dibenzoxepine ring	Sedative
Clomipramine	Halogenated ring	Available for intravenous use
Dimetacrine	Acridine ring	
Lofepramine	Propylamine side-chain	Relatively safe in overdose
Noxiptiline	Oxyimino side-chain	
Butriptyline	Isobutyl side-chain	More potent dopamine effects
Imipramine oxide	Oxygenated ring	Metabolite of imipramine
Amitriptyline oxide	Oxygenated ring	Metabolite of amitriptyline
Dibenzepin	Dibenzodiazepine ring	
Melitracen	Anthracene ring	
Amoxapine	Dibenzoxazepine ring; piperazine side-chain	Less potent than other tricyclics; dopamine D_2 receptor antagonist
Iprindole	6,5,8 ring structure	Weak action on amino pump mechanism
Dosulepin	Dibenzothiepine ring	
Trimipramine	Propyl side-chain	Little effect on monoamine re-uptake
Amineptine	Seven-carbon side-chain	Less sedative than other tricyclics

pleural effusions and eosinophilia has been reported with desipramine (2).

Tumor-inducing effects have not been reported.

Plasma concentrations and adverse effects

Plasma concentrations of antidepressants are influenced by pharmacogenetic factors, age, and drug interactions. Several studies have attempted to define the relation between plasma concentrations of tricyclic antidepressants and their therapeutic or adverse effects. However, the results are conflicting (SEDA-3, 12; SEDA-5, 15) (3). Sometimes there is a clear relation between cardiac toxicity and high plasma concentrations (4–6), but in some individuals cardiotoxic effects occur at presumed therapeutic concentrations. Sporadic reports of severe adverse effects, mostly of the anticholinergic type, have often been associated with high plasma concentrations of amitriptyline or nortriptyline (7,8), although these vary with each drug and from patient to patient. The risk of central nervous system toxicity in patients treated with tricyclic antidepressants may be correlated with plasma concentrations, age, and sex (SEDA-16, 8; SEDA-17, 17). It can be concluded that routine plasma concentration monitoring is generally of little practical value in managing patients with adverse effects, but in view of the great inter-individual variability in the pharmacokinetics of the tricyclic antidepressants, monitoring may be useful in those patients who report adverse effects at low doses.

Once-daily dosage

Most comparisons of once-daily with divided regimens for a variety of tricyclic compounds have shown equal efficacy and reduced adverse effects with once-daily regimens (SEDA-3, 10). However, the significance of such regimens for compliance may have been exaggerated. Multiple drugs have a much clearer impact on compliance than do multiple dosages of a single drug. Compliance is not usually impaired until more than three tablets a day are prescribed (9). A patient who forgets to take a single dose of a once-daily regimen loses more therapeutic effect than a patient who is equally forgetful about a divided regimen. Divided regimens may also be useful for patients who benefit from short-term sedation during the daytime.

Some adverse effects are more marked with single large doses, particularly in vulnerable patients. One study showed an increased frequency of frightening dreams when tricyclic antidepressants were given in a single bedtime dose (10). Elderly patients who take large single doses at bedtime may be at risk of dizziness, ataxia, and confusion caused by postural hypotension when they attempt to get out of bed in the dark (8).

Compatibility with electroconvulsive therapy

An important practical question concerns the compatibility of tricyclic antidepressants and electroconvulsive therapy (ECT). This has been studied in 15 patients taking 150–250 mg/day of imipramine or amitriptyline who received 4–16 ECT treatments (11). Continuous monitoring of cardiac function by oscilloscopy showed dysrhythmias in 40% of the ECT sessions. There were single extra

atrial beats in 46 of 151 sessions and 1–3 premature ventricular beats in 12 sessions. Transient ventricular tachycardia occurred in one 32-year-old woman taking amitriptyline 250 mg/day who received 12 ECT treatments. The authors concluded that the cardiac effects are similar to those observed in previous studies of electrocardiographic changes during ECT in patients not taking antidepressants, and expressed the opinion that combined therapy does not involve any increased risk of serious cardiac dysrhythmias. A controlled comparison of ECT given to 19 patients taking tricyclic antidepressants and 27 control patients showed no differences in heart rate, blood pressure, or ectopic heart beats (12).

Tricyclic antidepressants in the treatment of enuresis

Although pediatric psychopharmacology is much neglected, nocturnal enuresis is an area of extensive research. An earlier review catalogued almost 100 publications on the topic (13). The tricyclic antidepressants have been shown to be effective in well-controlled trials, and over 40 publications had appeared before 1970. At that time adverse effects in children appeared to be minimal and comparable to those in adults. Since then considerable concern has developed over cardiotoxic effects and the risks of accidental overdose in children. The earlier reports have been summarized (SEDA-1, 10); managing overdose in children has been reviewed (SEDA-2, 10); death in a 16-month-old infant has been reported (SEDA-3, 9).

Sudden death, possibly related to cardiac effects in children and adolescents, has been discussed (SEDA-15, 13; SEDA-16, 9; SEDA-18, 18); it was concluded that children taking tricyclic antidepressants require careful monitoring of the electrocardiogram, even when relatively low doses are used (SEDA-18, 18).

Symptoms of intoxication in children taking tricyclic antidepressants may fail to be recognized in time, as shown by a number of cases published in the former German Democratic Republic (14), which led to serious consideration of the abandonment of such treatment there.

The British Committee on Review of Medicines recommended that these drugs should not be used in children under 6 years of age and should be given for periods not exceeding 3 months, and then only after a full examination (including electrocardiography) and consideration of other treatment options.

Long-term adverse effects

Treatment guidelines for depression and anxiety increasingly emphasize the value of longer-term maintenance treatment with antidepressants in order to prevent recurrence of illness. It is therefore important to assess the adverse effects burden of longer-term medication. The change in adverse effects profile over 1 year of treatment has been studied in a double-blind, placebo-controlled study of maintenance treatment with imipramine (average daily dose 160 mg) in 53 patients with panic disorder (15). Adverse effects of imipramine, such as sweating, dry mouth, and increased heart rate, persisted over the year of treatment, while rates of sexual dysfunction fell. Weight gain became increasingly problematic, and by the end of the trial imipramine-treated patients had a mean increase in weight of 4.5 kg, significantly more than the placebo-treated subjects, who gained only 1.3 kg. Data of this kind are helpful when advising patients on which adverse effects may remit and which are likely to persist during longer-term treatment. These data suggest that relative to SSRIs, such as fluoxetine and sertraline, maintenance treatment with imipramine is less likely to cause sexual dysfunction but has a greater risk of weight gain.

Organs and Systems

Cardiovascular

The cardiac toxicity of tricyclic antidepressants in overdose has been a source of continued concern. Undesirable cardiovascular effects, besides representing a major therapeutic limitation for this category of drugs, delineate an area in which tricyclic compounds with novel structures, as well as second-generation antidepressants, may have significant advantages. The cardiovascular effects of tricyclic antidepressants and the new generation of antidepressants have been reviewed (SEDA-18, 16) (16).

Since the inception of SEDA in 1977, each volume has included a review of the evolving literature, focusing on specific aspects, including direct myocardial actions (SEDA-12, 13), hypotension (SEDA-12, 13), and the incidence, severity, and management of overdosage in adults and children (SEDA-10, 19) (SEDA-11, 16) (4,17–21).

Direct myocardial actions

Tricyclic antidepressants are highly concentrated in the myocardium; this may account for the vulnerability of the heart as a target organ as well as for inconsistently and inconclusively reported relations between plasma drug concentrations and specific manifestations of cardiac toxicity. These drugs interfere with the normal rate, rhythm, and contractility of the heart through actions on both nerve and muscle that are mediated by at least four different mechanisms (singly, in combination, or due to imbalance), including an anticholinergic action, interference with re-uptake of catecholamines, direct myocardial depression, and alterations in membrane permeability due to lipophilic and surfactant properties.

Acute experiments in dogs have shown a negative inotropic effect sufficient to cause congestive cardiac failure, but a carefully conducted long-term study in man did not show impaired left ventricular function in depressed patients with concurrent congestive failure (SEDA-9, 18). In another study (22) nortriptyline (mean dose 76 mg/day, mean plasma concentration 107 ng/ml) was given to 21 depressed patients with either congestive heart failure or enlarged hearts. In this study, nortriptyline was effective and well tolerated, producing only one episode of intolerable hypotension.

The most readily observable change in cardiac function is sinus tachycardia, which occurs to a greater or lesser extent in more patients and which correlates weakly or inconsistently with plasma concentrations (4,17,18). The mechanism may be related to both central and peripheral effects on cholinergic and adrenergic systems, but is not simply a reflex response to hypotension (4). The presence of tachycardia can serve as an indirect measure of compliance (4), but it is seldom a cause for concern, except in individuals who anxiously monitor their own physiological functions.

The complex changes that occur in cardiac rhythm have been intensively studied using 24-hour high-speed and high-fidelity cardiographic tracings, His bundle electrocardiography (23,24), and cardiac catheterization (25,26). Changes in conduction and repolarization cause prolongation of the PR, QRS, and QT intervals and flattening or inversion of T-waves on routine electrocardiograms; conduction delay occurs distal to the atrioventricular node and is apparent as a prolonged HV interval (the time from activation of the bundle of His to contraction of the ventricular muscle). This effect resembles that due to type I cardiac antidysrhythmic drugs, such as quinidine and procainamide. These conduction changes can cause atrioventricular or bundle branch block and can predispose to re-entrant excitation currents with ventricular extra beats, tachycardia, or fibrillation.

The implications and complications of these changes in cardiac function and rhythm are less clear; knowledge of their existence provoked concern about the incidence of sudden death in patients with cardiovascular disease, but the evidence from epidemiological sources is equivocal (19). General guidelines for the use of these drugs in the elderly have been discussed above, and a review of studies on the cardiovascular effects of therapeutic doses of tricyclic antidepressants has supported their use in elderly patients and those with pre-existing cardiovascular disease, provided precautions are taken. Atrial fibrillation has been reported in predisposed elderly subjects (SEDA-18, 19). In children taking desipramine there have been reports of sudden death (27,28) and tachycardia, and cardiographic evidence of an intraventricular conduction defect (SEDA-18, 18) (29).

The antidysrhythmic effect of imipramine was first reported in 1977 during treatment of two depressed patients whose ventricular extra beats improved during treatment (30). Tricyclic compounds can trigger serious dysrhythmias at high doses and perhaps also when the myocardium is sensitized.

Care should be taken in patients with a recent myocardial infarction who show evidence of impaired conduction (first-degree heart block, bundle-branch block, or prolongation of the QT$_c$ interval), since tricyclic antidepressants can theoretically add to the already increased risk of ventricular fibrillation in such patients (31). Reviews in earlier editions of Meyler's *Side Effects of Drugs* discussed these effects and gave practical guidelines on the use of tricyclic antidepressants in patients with heart disease (32).

There is sometimes a clear-cut correlation between cardiac toxicity and high plasma concentrations (SEDA-18, 18) (4–6), but this may not always be so in individuals who are highly sensitive to the drug or in whom prolonged treatment may have led to drug accumulation in the myocardium, despite plasma concentrations in the usual target range. Routine plasma concentration monitoring does not seem indicated, since plasma concentrations account for only a small part of the variance in cardiac effects (17). If an individual shows significant changes clinically or cardiographically, a spot measurement may show a high plasma concentration, requiring dosage reduction.

To date there have been no prospective studies that clearly show increased mortality in cardiac patients who use tricyclic antidepressants. It has been suggested that overall mortality due to cardiac disease may be higher in depressed patients who remain untreated than in those who receive either an antidepressant or electroconvulsive therapy (33). Even in patients who have chronic heart disease, the risks of effective treatment with a tricyclic appear to be minimal (34).

Hypotension

As many as 20% of patients taking adequate doses of a tricyclic antidepressant experience marked postural hypotension. This effect is not consistently correlated with plasma concentrations and tolerance does not develop during treatment (35–37). The mechanism for this effect is uncertain; it has been attributed to a peripheral antiadrenergic action, to a myocardial depressant effect, and to an action mediated by alpha-adrenoceptors in the central nervous system (38). Studies of left ventricular function in man are conflicting. One study of systolic time intervals showed a decrement in left ventricular function with therapeutic doses (39), while two in which cardiac function was observed directly during cardiac catheterization after overdosage showed no evidence of impaired myocardial efficiency, whereas the hypotension persisted after left ventricular filling pressures and cardiac output had returned to normal (40,41).

Postural hypotension can lead to falls. Its occurrence may be predictable, since patients who have raised systolic pressures and who have a pronounced postural drop before treatment are most likely to experience drug-induced hypotension (37). Such patients should be cautioned to rise slowly from sitting positions and, since elderly people are especially at risk of falling at night, single large bedtime doses of sedative tricyclic drugs should be avoided. These preventive measures are the most helpful, since the hypotensive effect is not directly related to plasma drug concentrations and may not improve with dosage reduction. The wisdom of using sympathomimetic drugs to counter this undesirable effect is questionable (42).

Use in patients with cardiac disease

The diagnosis of depression and the use of antidepressant medication are both associated with an increased risk of myocardial infarction. The relative contribution of these two factors is uncertain. In a case-control study of 2247 subjects, taking antidepressants was associated with a 2.2-fold (CI = 1.3, 3.7) increase in the risk of myocardial infarction (43). This increased risk seemed to be accounted for entirely by the use of tricyclic antidepressants, because selective serotonin re-uptake inhibitors were not associated with an increased risk, although the confidence intervals were wide (relative risk 0.8; CI = 0.2, 3.5). These findings support the usual clinical advice that

tricyclic antidepressants are best avoided in those with known cardiovascular disease or significant risk factors.

Cardiovascular complications of overdosage

The relation between the dosage of a tricyclic antidepressant and the development of life-threatening cardiovascular complications is unclear and individually variable (44,45), although plasma concentrations above 1000 ng/ml give cause for serious concern. Plasma concentrations vary widely, and absorption can be delayed or deceptive, owing to gastric stasis and enterohepatic recycling (46). Dysrhythmias can occur for the first time up to 36 hours after drug ingestion or admission to hospital (46). The frequency of serious cardiac conditions in one series of 68 cases (47) was 46% in patients who took over 2000 mg of imipramine or its equivalent, almost twice that of those who took less (25%). An intensive study of cardiovascular complications among 35 overdose patients showed that 51% had significant hypotension and 80% had abnormal electrocardiograms (40). The latter consisted of sinus tachycardia (71%) and various abnormalities that reflect impaired conduction, including prolongation of the QT_c interval (86%), QRS complex (29%), and PR interval (11%). The ST segments and T-waves were abnormal in 28% of patients. Despite these manifestations of disturbed conduction and repolarization, there were relatively few dysrhythmias; 13 patients (37%) had ventricular extra beats, which lasted up to 72 hours after admission and subsided within 36 hours in 10 cases. No patients developed sustained repeated ventricular tachydysrhythmias, and the authors speculated that bizarre wide QRS complexes seen in aberrantly conducted supraventricular tachycardia (present in some cases) may sometimes be misinterpreted as ventricular tachycardia.

The basic principles of intensive supportive care should be applied early, and artificial ventilation is often necessary, since respiratory depression is more frequent than is commonly supposed (48,49). Patients should be monitored for 24 hours if the initial (or subsequent) electrocardiogram shows a dysrhythmia. Among 75 patients with overdose, none who had a normal electrocardiogram and level of consciousness for 24 hours went on to develop any significant dysrhythmia (50). However, a case was subsequently reported of a patient who died an acute cardiac death 57 hours after admission and 33 hours after normalization of the electrocardiogram (51). The authors suggested that prolonged monitoring may be justified in individuals who have taken antidepressants for prolonged periods, compared with those who overdose early in treatment.

More specific treatment to combat cardiotoxic effects is usually necessary in only a minority of instances; in the series reported above (40), five patients (14%) had marked hypotension. Initial low left ventricular filling pressures were corrected within 3 hours by infusion of isotonic saline. Systemic hypotension persisted and was corrected by infusion of sympathomimetic amines. Routine insertion of a pulmonary artery catheter, with continuous monitoring of blood gases, pulmonary arterial pressure, left atrial wedge pressure, and cardiac output have been recommended (40). Volume expansion is suggested for low left atrial pressure, with dopamine infusion to improve myocardial contractility if cardiac output remains low.

The management of ventricular extra beats is based on recognition of the quinidine-like basis of the conduction defect. In the series reported above (40) all 13 patients with ventricular arrhythmias responded to intravenous infusion of lidocaine (mean dose 2.0 mg/minute). This alone might account for the absence of deaths in this series.

Another indirect method of benefiting the patient with cardiotoxic effects has been the alkalinization of plasma to a pH of 7.50–7.55 using sodium bicarbonate infusion (52). This enhances plasma protein binding, making the drug less available to the tissues. It is claimed that this technique reverses both hypotension and cardiac dysrhythmias without the risk of the undesirable effects of antidysrhythmic drugs (46). Recommendations for the management of poisoning with tricyclic antidepressants given in a recent review have been summarized (SEDA-16, 8).

Differences among tricyclic compounds

There is evidence that doxepin is significantly less cardiotoxic than other tricyclic compounds (53). However, a complete review of all the animal and clinical data has suggested that doxepin overdose can still cause lethal dysrhythmias in man, probably by producing more marked respiratory depression.

Three different studies have shown that there is less risk of hypotension in patients treated with nortriptyline (54–56) than with other tricyclic compounds. However, a similar claim has been made for doxepin (57).

Increased pulse rate and blood pressure have been associated with desipramine in the treatment of bulimia nervosa (SEDA-17, 18).

Respiratory

There has been one case report of reduced ventilatory response to hypercapnia after nortriptyline in a woman with chronic obstructive pulmonary disease (SEDA-18, 19).

Nervous system

Miscellaneous symptoms that have been attributed to the tricyclic antidepressants include fatigue, weakness, dizziness, headache, and tremor; patients are likely to fall because of these disturbances.

Seizures are serious adverse events associated with the use of antidepressants, and the relative frequencies for different antidepressants have for a long time been a matter of controversy. The literature has been critically evaluated, taking into account predisposing factors, drug doses, plasma drug concentrations, and the duration of treatment (58). A significant proportion of seizures occurred in predisposed individuals, and the risk of seizure for most antidepressants increased with dose or blood concentration. After overdose the risk was higher for amoxapine and the tetracyclic drug maprotiline than for other antidepressants, but for several drugs there are not enough data to estimate the risk. Imipramine was the most frequently studied tricyclic, with a seizure risk of 0.3–0.6% at effective doses. For several of the second-generation antidepressants, a lower seizure risk has been reported in large clinical trials. Caution with all antidepressants should certainly be exercised in people who are predisposed to seizure activity because of brain damage or alcohol or drug abuse.

Concurrent use of lithium may be a risk factor for neurological adverse effects.

- A 34-year-old woman took amitriptyline 300 mg each night for several years. Six days after starting lithium 300 mg tds she had several generalized tonic-clonic seizures. A second episode occurred on re-exposure to lithium (59).

Patients with phobias or panic disorders are extremely sensitive to the adverse effects of tricyclic drugs early in treatment. They often have a syndrome of fine tremor, insomnia, and anxiety, which can be characterized as "jitteriness" and which is sometimes attributed to adrenergic hypersensitivity (SEDA-13, 9). Serum iron concentrations were significantly lower in jittery patients than in those who were not affected (SEDA-17, 19) (60). The authors suggested that this may be related to the role of iron as a co-factor for tyrosine hydroxylase.

Because tricyclic antidepressants suppress REM sleep, they have been used in the management of narcolepsy and cataplexy when amphetamines fail or abuse potential is high. Clomipramine is the most effective, possibly because it has more pronounced actions on serotonergic mechanisms (61).

Three reports have referred to a type of difficulty in articulation described as "speech blockage" or "dysarthria" (62–64). The disturbance was described as a delay in thinking and speech, in which the patient has difficulty in conceptualizing or transferring the next logical thought into words. The effect resembles stammering.

There have been sporadic reports of bilateral foot-drop with peroneal nerve involvement (65). A major neuropathy, with high stepping gait and inability to dorsiflex the foot, occurred in an 84-year-old woman; this presumed adverse effect remitted 8 weeks after withdrawal (66).

Unusual neurological reactions have been observed in some patients, generally when combinations of drugs (often including maprotiline) have been used (67). The symptoms were ataxia, akathisia, hypokinetic disorders of speech and motion, a dream-like state, and transiently impaired memory.

Confusion was observed in 13% of 150 patients taking tricyclic drugs, and in as many as 35% of patients over 40 years of age. All responded rapidly to drug withdrawal (SEDA-9, 26).

Aggressiveness during treatment in patients taking imipramine and amitriptyline in relatively low doses has been described (68). Violent behavior has also been attributed to amitriptyline (SEDA-17, 18).

Anticholinergic actions

Several organs are the target for the anticholinergic (antimuscarinic) activity of the tricyclic antidepressants. They constitute the most common and troublesome adverse effects of the tricyclic antidepressants, but the peripheral anticholinergic actions can also be put to therapeutic use in conditions such as irritable bowel syndrome, premature ejaculation, and nocturnal enuresis.

Experiments on receptor binding in rat brain and guinea-pig ileum have shown a spectrum of activity at muscarinic acetylcholine receptors for different tricyclic antidepressants (69). A comparison of five compounds showed that amitriptyline and doxepin were the most and desipramine the least potent, imipramine and nortriptyline being intermediate. Single-dose experiments in volunteers (given up to 100 mg of each drug) have confirmed the same rank order for both the peripheral anticholinergic actions (on salivary flow) and the central effects (sedation and other measures on a mood scale) (70). The significance of these differences between relatively low single doses of these drugs in healthy volunteers should not be uncritically extrapolated to clinical practice. However, they do provide a rationale for selecting among the different drugs for patients in whom either a high degree of sedation is desirable or anticholinergic effects are likely to be troublesome. No tricyclic is entirely free of anticholinergic action, and individual differences in susceptibility or metabolism can still cause serious problems in some patients. It is also difficult to predict which particular organ will become the major target for anticholinergic activity; some patients complain bitterly of a dry mouth, others report blurred vision, and some develop bowel or bladder symptoms. Careful history taking and physical examination will often reveal a possible cause for concern, based on the patient's previous response to similar drugs, existing disease (such as narrow-angle glaucoma, enlarged prostate, constipation), or advancing age.

Older people are supposedly susceptible to delirium caused by centrally acting anticholinergic drugs. This can take the form of anxiety, agitation, or frank hypomania. Such difficulties are more likely if the patient is also taking antipsychotic drugs (many of which also have anticholinergic effects) or anticholinergic antiparkinsonian drugs. Although such effects have long been recognized as a risk associated with the anticholinergic properties of these drugs, it has been suggested that their incidence may be lower than is often assumed (71). In an epidemiological study from a West German multicenter drug surveillance system in almost 14 000 patients for 5 years, exposure-related incidence rates were 1.2% for tricyclic antidepressants compared with 0.8% for both tranylcypromine and neuroleptic drugs (72). The risk increased steadily with age in both sexes and all diagnostic subgroups, and was six-fold higher (3.4%) in those over 60 years. Also at greater risk were women and patients with affective psychosis.

Adverse anticholinergic effects can occur immediately after the first dose of a tricyclic antidepressant. They are a major cause of poor compliance in patients who expect immediate relief, but who are not properly prepared for the delay that can occur in the beneficial effects of these drugs on mood and energy.

Sensory systems
Loss of accommodation and blurred vision are common inconveniences that can usually be tolerated in the knowledge that they lessen with the duration of treatment. Exacerbation of narrow-angle glaucoma in the elderly can occur, but is not an absolute contraindication to treatment with a tricyclic antidepressant, since the anticholinergic effects can be balanced by judicious use of pilocarpine (73,74).

Damage can occur to the corneal epithelium, due to reduced lacrimation and relative accumulation of mucoid secretions in patients who wear contact lenses (75).

Mouth and teeth
Rampant dental caries due to xerostomia occurred in a patient who took doxepin up to 300 mg/day for over 2 years (76). The author warned of the need to counsel patients to carry out rigorous and regular dental hygiene, as well as simple measures to promote increased salivation, such as sugarless lemon drops or chewing gum. In another study there was an increase in the number of decayed teeth in 35 children treated with amitriptyline or nortriptyline for enuresis, compared with a smaller group of untreated children with enuresis and a larger matched population control group (77).

Gastrointestinal
Simple dietary advice about bulk foods can mitigate minor bowel disturbances. More serious complications that can arise include paralytic ileus, which can be life-threatening, especially in the elderly (78). A less well-known adverse effect is the potential for aggravating or even possibly inducing a hiatus hernia, presumably due to an anticholinergic effect on the cardiac sphincter (79).

Urinary tract
The tricyclic antidepressants increase bladder sphincter tone and the volume of fluid necessary to trigger detrusor contraction (80). Such effects may account for their efficacy in nocturnal enuresis, in which the benefit occurs early and at a low dosage, consistent with anticholinergic activity. However, this pharmacological action can cause hesitancy and urinary retention, especially in predisposed men who have prostatic hyperplasia.

Renal damage from tricyclic antidepressants has been suggested on only one occasion.

- A 65-year-old man taking imipramine 300 mg/day developed toxic psychosis, anorexia, and nausea after 24 days. He was mildly azotemic, but these changes quickly reverted after withdrawal. No biopsy or renal function tests were reported.

The findings are entirely compatible with prerenal azotemia associated with diminished fluid intake during a drug-induced psychosis (81).

Neuroleptic malignant syndrome
The neuroleptic malignant syndrome, which is classically associated with antipsychotic drugs and is usually attributed to excessive dopamine D2 receptor blockade, can rarely occur with other medications, including tricyclic antidepressants.

- A 62-year-old man was found unresponsive in his apartment. He had a past history of bipolar disorder, for which he was taking nortriptyline and sodium valproate (dosages not stated). On admission to hospital his rectal temperature was 107.1°F and he had increased muscle tone. He was intubated and cooled with ice packs. His creatine kinase activity was raised (1046 IU/l) but valproate and nortriptyline concentrations were within the target ranges. Extensive investigations, including biochemical screens and brain scans, showed no clear cause for his condition. Shortly afterwards he developed disseminated intravenous coagulation and died.

The authors concluded that the features of the illness were consistent with nortriptyline-induced neuroleptic malignant syndrome (82). However, a contributory role for sodium valproate was also possible.

The neuroleptic malignant syndrome has also been reported in association with the antidepressants trimipramine (SEDA-21, 11) (83), desipramine (SEDA-17, 18), and amoxapine (SEDA-16, 9; SEDA-17, 18) (84). Amoxapine in particular has significant dopamine D2 receptor antagonistic properties. In most cases the patients were taking several other drugs, but there have been reports of the syndrome in association with amoxapine or desipramine alone.

Serotonin syndrome

The serotonin syndrome is usually associated with the use of combinations of drugs that potentiate brain serotonin function. This syndrome has been associated with several tricyclic antidepressants when used in combination with monoamine oxidase (MAO) inhibitors (85). Rarely it can be associated with the use of a single agent, such as clomipramine, a tricyclic antidepressant with potent serotonin re-uptake inhibitor properties (86).

- A 60-year-old woman, with a history of hypertension, type 2 diabetes mellitus, and depression, took clomipramine 200 mg/day for 8 months and then 250 mg/day for 3 months. Her other medications were glibenclamide 7.5 mg/day, lisinopril 5 mg/day, and clonazepam 0.5 mg/day. Without any change in medications or other precipitants, she began to feel confused and weak. She became pyrexial (41.6°C), confused, and tremulous, with myoclonic jerking. The combined plasma concentrations of clomipramine and desmethylclomipramine were 2230 nmol/l, somewhat over the usual target range (below 1900 nmol/l). Within hours her condition deteriorated, with seizures, ventricular tachycardia, and disseminated intravascular coagulation. Rhabdomyolysis led to acute renal insufficiency, which required dialysis. She remained severely ill over the next 4 weeks and eventually died of opportunistic Gram-negative infections.

This case illustrates that the serotonin syndrome can occasionally occur apparently spontaneously in patients taking a single serotonergic drug. The authors were unable to find any reason why the syndrome developed so catastrophically when it did, except for the modestly increased concentrations of clomipramine and its metabolite.

Extrapyramidal symptoms

Tricyclic antidepressants are often listed among the many drugs that can produce buccofaciolingual or choreoathetoid movements (87). A putative mechanism is a central anticholinergic action, which upsets the balance between the dopaminergic and cholinergic systems. The spontaneous occurrence of this syndrome makes it difficult to establish a cause-and-effect relation, although both patients described in the above report had symptoms again when rechallenged.

There is a more clear-cut cause-and-effect relation in the parkinsonian symptoms that occasionally occur with high-dosage tricyclic therapy in susceptible individuals (particularly elderly women). Because of its piperazine side-chain and structural resemblance to the phenothiazines, amoxapine has antidopaminergic properties that appear to produce typical dystonic reactions (88), but other tricyclic antidepressants may also be implicated in producing the full range of so-called extrapyramidal syndromes, including akathisia, dystonic reactions, parkinsonism, and tardive dyskinesia. Case reports have been described before (SEDA-16, 9; SEDA-17, 18; SEDA-18, 18). In the reports of tardive dyskinesia the problem is often that these patients have taken many different drugs.

Sensory systems

Eyes

Loss of accommodation and blurred vision are common inconveniences that can usually be tolerated, in the knowledge that they lessen with the duration of treatment. Exacerbation of narrow-angle glaucoma in the elderly can occur, but is not an absolute contraindication to treatment with a tricyclic antidepressant, since the anticholinergic effects can be balanced by judicious use of pilocarpine (73,74).

Treatment with bright light is used for mood disorders, and it has been suggested that antidepressants, which may act as photosensitizers, could enhance the effect of bright light on the eye, giving rise to adverse effects (SEDA-18, 17).

Ears

Tinnitus can occur after prolonged treatment with tricyclic antidepressants. In an early trial of imipramine (89) there were two cases of transient deafness, but in the subsequent 20 years no auditory effects have been recorded. In 1980, a report appeared (90) concerning four patients, all taking imipramine in dosages below 150 mg/day. Each complained of buzzing or ringing in the ears. In each case the symptoms improved or disappeared on dosage reduction, with maintenance of therapeutic benefit, and in one case the patient was switched to an equivalent dosage of desipramine without recurrence. A further report concerned a patient taking protriptyline 45 mg/day who developed ringing in both ears after 12 days (91). Symptom severity fell with dosage reduction and disappeared entirely after desipramine 100 mg/day was substituted. The authors postulated that non-vibratory tinnitus had originated from either neurological factors or changes in blood flow. In a chart review of 475 patients treated with tricyclic antidepressants there were five patients with tinnitus. Each developed the symptom in the second or third week of treatment at dosages of imipramine of 150–250 mg/day, with plasma concentrations of 200–400 ng/ml (92). In every case tinnitus subsided spontaneously within a further 2–4 weeks after onset, even though dosage and plasma concentrations remained constant.

Another unusual disturbance of hearing has been reported in a child taking tricyclic antidepressants (93). Auditory acuity was normal, but discriminative ability for both clear and distorted speech was depressed. The disorder cleared within some days of withdrawal.

Psychological, psychiatric

Cognitive impairment has been associated with nortriptyline in elderly subjects (SEDA-17, 19).

Musical hallucinations have been reported in association with clomipramine (SEDA-17, 18).

Mania

It is widely believed that there is a significant risk that tricyclic antidepressants can precipitate mania or rapid cycling in up to 10% of patients, and that various factors increase this possibility, including being female or younger, having an earlier onset of illness, and having a positive first-degree family history (SED-11, 40). Biochemical risk factors have been alleged to include patients with a low urinary excretion of the noradrenaline metabolite methoxyhydroxyphenol glycol (MHPG); the risk is possibly greater in patients taking tricyclic antidepressants rather than MAO inhibitors, and particularly in the case of clomipramine. The data on which these conclusions were based have now been rigorously analysed (94) in a review of the controversy surrounding the alternative suggestion that the so-called switch effect is a random manifestation of bipolar illness. There is a paucity of both prospective and long-term placebo-controlled studies, and existing research has suffered from unrepresentative samples and poor definition of manic outcomes. The reviewers concluded that "⋯some bipolar patients and few, if any, unipolar patients become manic when they are treated with antidepressants. A small number of patients develop rapid cycling."

This more cautious conclusion is supported by the results of a prospective study of 230 carefully selected patients with recurrent depression (at least two episodes, with an average of six) who took imipramine (200 mg/day) for an average of over 46 weeks (95). Mania and hypomania were defined and measured by the Raskin rating scale. Only six patients (2.6%) developed hypomania, and four of these did so after withdrawal. Younger patients, women, and those with a previous history of hypomania (bipolar II) were no more likely to switch than unipolar patients.

These results suggest that the risk of mania or hypomania in the long-term treatment of recurrent unipolar depressed patients is relatively small. The 12 placebo-controlled studies of acute treatment in less carefully defined samples support higher incidence rates (around 6–7% for hypomania and 1–2% for mania), but these figures may be inflated owing to the inclusion of bipolar patients with a high risk of a spontaneous switch (94).

Metabolism

Weight gain has long been recognized as a concomitant of antidepressant and antipsychotic drug therapy. This may in part reflect improvement in mental state, but there also appears to be a physiological component, with an increased craving for sweets (96). No abnormalities have been found in fasting glucose and insulin concentrations or in intravenous insulin tolerance tests (96,97). Another possible suggestion for weight gain is that taste perception in depression improves after therapy with tricyclic antidepressants (98). A study of 50 depressed patients attempted to address some of these issues (99). Increased energy efficiency during antidepressant treatment has also been suggested as a reason for weight gain (SEDA-17, 8). A warning to patients with diabetes that hypoglycemia can be masked seems appropriate (100).

Interest in the metabolic effects of psychotropic drugs has been heightened by reported increases in serum cholesterol concentrations produced by atypical antipsychotic drugs, such as olanzapine and clozapine. Raised cholesterol concentrations have also been reported in conjunction with the antidepressant drug, mirtazapine, which, like olanzapine and clozapine, blocks histamine H_1 and $5HT_2$ receptors. A striking increase in serum cholesterol was reported in 32-year-old woman during treatment with the tricyclic antidepressant, doxepin (101). When reboxetine was substituted for doxepin the cholesterol concentration returned to normal. Doxepin has particularly potent H_1 receptor antagonist properties, which suggests that blockade of H_1 receptors may play a role in the cholesterol raising properties of some psychotropic drugs. This effect may have implications for the association between the use of tricyclic antidepressants and an increased risk of myocardial infarction (SEDA-24, 12).

Endocrine

The "Division of Drug Experience" of the US Department of Health and Welfare issued a note on five cases of the syndrome of inappropriate antidiuretic hormone secretion and drugs to which it has been attributed (102). All involved drugs with a tricyclic structure; one patient was taking imipramine, three carbamazepine, and the others the closely related muscle relaxant cyclobenzaprine. The dosage of imipramine was 50 mg/day for 3 weeks and the patient was a 72-year-old woman. Other cases have been reported, involving amitriptyline (102), imipramine, and protriptyline (SEDA-17, 17).

Tricyclic antidepressants, presumably through blocking the re-uptake of noradrenaline, can cause a crisis in a patient with a pheochromocytoma (SEDA-21, 11) (103).

Prolactin concentrations are very rarely altered during treatment with tricyclic antidepressants, but this is more likely to occur and to produce galactorrhea or amenorrhea with clomipramine and amoxapine and when there are other contributory factors that may stimulate prolactin secretion, such as stress or electroconvulsive therapy (104).

Hematologic

Occasional cases of blood dyscrasias with tricyclic antidepressants continue to be reported (SEDA-12, 46) (SEDA-18, 18) (SEDA-21, 10). Antidepressant-induced blood dyscrasias have recently been reviewed (105).

Agranulocytosis has been associated with tricyclic antidepressants (106). Of 20 cases, eight were fatal and 12 recovered after 3–20 days.

Two cases of non-thrombocytopenic purpura occurred in patients taking tricyclic antidepressants. Both improved on withdrawal (107). True thrombocytopenia, with platelets counts as low as $120 \times 10^{12}/l$, occurred in a 79-year-old woman taking doxepin. During later treatment with amitriptyline, thrombocytopenia recurred, but not when she took imipramine (108). Cross-sensitivity must be highly specific to chemical structure, doxepin and amitriptyline being more like each other than imipramine.

Liver

Cholestasis was among the first adverse effects reported with phenothiazines, tricyclic antidepressants, and MAO inhibitors. Its incidence appears to have fallen, for reasons that are not understood, but under-reporting may be a factor. Increases in liver enzymes, especially transaminases and alkaline phosphatase, are quite common during treatment with both phenothiazines and tricyclic antidepressants. Such effects are usually benign, but one careful study of a patient taking amitriptyline showed biopsy findings of mononuclear and eosinophilic infiltration; cholestasis and jaundice were absent (109). In a controlled comparison of lofepramine and fluoxetine, there were significant increases in alkaline phosphatase, alanine aminotransferase, and gammaglutamyl transferase in patients taking lofepramine but not fluoxetine (110).

More serious and sometimes fatal liver necrosis has been reported with a number of different tricyclic structures and probably represents an extreme form of hypersensitivity (111,112). Liver necrosis in a 33-year-old woman who took imipramine 300 mg/day for over 1 month led the authors to suggest that once-daily dosage may pose a special hazard, because peak concentrations can exceed the toxic concentration, even though steady-state plasma concentrations are in the usual target range (113). A particular hazard appears to be posed by amineptine (qv).

There is no indication for routine liver function tests in patients taking tricyclic antidepressants; raised transaminases and alkaline phosphatase within the limits of the reference ranges are not a cause for serious concern, unless they are accompanied by clinical signs or symptoms indicative of liver dysfunction.

Skin

Skin rashes are so common that it is difficult to determine a cause-and-effect relation. A choice must be made between waiting to see if the rash clears despite continued treatment or switching to a different compound and, if necessary, rechallenging at a later date. Serious reported skin reactions include cutaneous vasculitis, urticaria, and photosensitivity. A grey discoloration of the skin in light-exposed areas has been associated with long-term therapy with imipramine and desipramine (SEDA-17, 19) (SEDA-18, 18). Pigmentary changes in the iris were also reported in one of the cases. Amineptine has been reported (114) to cause a particularly active acne, occurring beyond the usual distribution on the body and beyond the usual age, especially in women. The remedy recommended is withdrawal of amineptine. A single case of a rosacea-like eruption on the face of a 76-year-old woman was also caused by this drug, and the association was confirmed by re-challenge (115).

Hyperpigmentation is a recognized adverse effect of the antipsychotic drug chlorpromazine. Four cases of hyperpigmentation have been described in patients taking the structurally related tricyclic antidepressant imipramine (116). All were women and had taken imipramine for at least 2 years. The hyperpigmentation occurred in a photodistribution on the face, arms, and the backs of the hands. In two patients who discontinued imipramine the hyperpigmentation resolved within 1 year. The authors speculated that the pigmentation might have been due to deposition of melanin in an unusual form, possibly in a complex with a metabolite of imipramine.

Musculoskeletal

Elderly people are at increased risk of fractures, and a case-control study of patients admitted to hospital suggested that both tricyclic antidepressants and SSRIs increased the probability of hip fracture about 2.5 times (SEDA-22, 11). In a prospective study of 8127 women aged 65 years and older who were followed for 4.8 years, the risk of a first hip fracture was 4% and the risk in women taking antidepressants was increased 1.7 times (95% CI = 1.05, 2.07) (117). The relative risk among women taking tricyclic antidepressants (RR = 1.83; CI = 1.08, 3.09) was slightly higher than that for SSRIs, which had wider confidence intervals (RR = 1.54; CI = 0.62, 3.08). Depression as an independent variable did not increase the risk of hip fracture. This study has confirmed that women taking antidepressants are at increased risk of hip fracture and has suggested that the effects of SSRIs and tricyclic antidepressants are similar.

Sexual function

A review of sexual dysfunction due to antidepressant drugs in men, citing both published findings and reports provided by the manufacturers, drew a distinction between erectile dysfunction, ejaculatory problems, and changes in libido (118). A complicating factor is the lack of information concerning the base rate of these problems in depression itself. Erectile impotence has been reported with all of the commonly used tricyclic compounds in low normal dosages. Delayed and occasionally painful ejaculation occurs, and four cases of painful ejaculation in association with imipramine and clomipramine have been reported (SEDA-17, 18). Priapism can occur (118). Both increased and decreased libido have been reported, but it is virtually impossible to distinguish drug relatedness from the natural history of the condition. A small number of uncontrolled studies have shown therapeutic benefits in patients with premature ejaculation or disturbed sexual function accompanied by depression (119).

Delayed orgasm or loss of ability to obtain orgasm has been reported in women taking desipramine (120).

The sexual dysfunction is probably due to 5-HT reuptake blockade, because similar effects are seen in patients taking SSRIs.

Long-Term Effects

Drug withdrawal

There is compelling evidence for a withdrawal syndrome due to abrupt discontinuation of tricyclic antidepressants (SEDA-5, 16), and the literature has been reviewed (121). Reports have involved both imipramine and doxepin (122). Symptoms occur as early as the morning after a missed dose (123), but more often after 48 hours and up to 2 weeks after withdrawal. They include anxiety, restlessness, sweating, diarrhea, hot or cold flushes, and piloerection. Amitriptyline withdrawal was followed by similar physical symptoms 36 hours after the last dose, followed by severe depressive illness (SEDA-17, 18).

Delirium after withdrawal of doxepin has been reported (124), as well as instances of mania (125). Pronounced neurological symptoms have been described after sudden withdrawal of amitriptyline (SEDA-18, 18).

The existence of a withdrawal syndrome was subjected to a controlled test (122) in seven patients who had been taking long-term amitriptyline up to 250 mg/day, imipramine 200 mg/day, or desipramine 250 mg/day. After 4 weeks placebo was substituted double-blind for 10–21 days. Both plasma and urine MHPG concentrations increased by an average of 74% above baseline, starting within 36 hours and reaching a peak 3 weeks after withdrawal. Despite these pronounced neurochemical changes there were no alterations in pulse rate or heart beat, and only two patients had definitely worse anxiety.

There is clearly a need for larger controlled studies to determine the incidence and severity of withdrawal effects after withdrawal of tricyclic antidepressants. Based on uncontrolled observations, it has been suggested that the incidence varies from under a quarter to over half of all patients.

A report of three cases (126) has suggested that central cholinergic overactivity is implicated, and that atropine sulfate 3 mg/day or synthetic anticholinergic agents (such as benzatropine mesylate 4 mg/day) ameliorate withdrawal symptoms within a few hours. The authors suggested that this technique may be especially useful in patients in whom tricyclic antidepressants must be abruptly withdrawn because of allergic or idiosyncratic reactions.

Second-Generation Effects

Teratogenicity

In chick embryos there was a high prevalence of abnormalities, including microphthalmia, micromelia, and reduced body size, after the administration of imipramine, but the dosages of imipramine were close to lethal (127).

The issue of dysmorphogenesis due to these drugs was first seriously discussed after a report on three possible cases from Australia in 1972 (128). Although the data underlying this report were later discredited, it led to a careful study of the case records of women who had taken tricyclic antidepressants in pregnancy, and more than 300 cases were rapidly identified in which such treatment had been followed by the birth of a normal infant. Other negative reports exonerating the drugs have appeared since (129–132), although sporadic case reports continue to appear (133).

In spite of these reports, which negate a possible association between tricyclic antidepressants and teratogenicity, it may be advisable to avoid these drugs during pregnancy, especially in the early stages, unless there is a compelling need. In many of the reports the drugs were given in low doses or for indications not justifying their use.

Fetotoxicity

Instances of distress in the newborn have been reported after treatment of their mothers with tricyclic antidepressants in the period before delivery (134). In one case, a neonate had signs of congestive heart failure without cardiac abnormality; another had tachycardia and myoclonus; a third had respiratory distress and neuromuscular spasms. These effects were thought to have resulted from both the adrenergic and anticholinergic effects of the tricyclic antidepressants, which readily pass the placenta and should be avoided during the perinatal period.

Infants born to mothers taking tricyclic antidepressants can become jittery in the first few days of life (135).

- A healthy boy (weight 3370 g) was born to a mother who had taken clomipramine 100 mg/day throughout pregnancy. On the second day after delivery, the child was jittery and on the fifth day he developed myoclonic jerking of his arms and legs. There was no epileptic activity on a 10-channel electroencephalogram, and clonazepam and phenobarbital did not suppress the movements. However, the movements were suppressed by a single dose of intravenous clomipramine 0.5 mg. The myoclonus recurred only occasionally over the next 4 days. Examination at one month later was normal, apart from mild jitteriness in response to touch.

Maternal clomipramine use is associated with seizures in neonates but this case appears to have been caused by a withdrawal state that was relieved by clomipramine administration.

Another report has suggested that effects in the neonate may be due to withdrawal from maternal antidepressants after birth (136). Two cases of neonatal convulsions have been reported in infants whose mothers had been treated with clomipramine. In both cases the seizures occurred on the first day of life coincident with a fall in plasma clomipramine concentrations. In one case the convulsions were controlled by administration of clomipramine followed by tapered withdrawal.

Hypotonia has been described in four children, one of whom also developed jitteriness (SEDA-17, 18).

Lactation

There are few reports on the excretion of antidepressant drugs in breast milk, even though postpartum depression is relatively common. In a 32-year-old woman who took imipramine 200 mg/day from 1 month postpartum imipramine and desipramine were detectable in breast milk (137). There have also been reports that amitriptyline (138), desipramine (139), and nortriptyline (140,141) were detectable in the milk of nursing mothers and in the plasma of the mothers and infants. Neither parent compound nor metabolite were detected in infants' serum, except for two infants who had low concentrations of 10-hydroxynortriptyline. There were no adverse effects in any of the infants. The use of antidepressants during lactation has been reviewed, including 15 studies in which serum concentrations of antidepressants were obtained from nursing infants (142).

There has been a report of an 8-week-old breast-fed infant whose mother was taking doxepin (143). Four days after an increase in the daily dosage from 10 to 75 mg the infant developed respiratory depression; desmethyldoxepin was detected in the baby's plasma.

Susceptibility Factors

Age

Elderly people
Elderly people have high rates of depression but tend to be excluded from randomized trials of antidepressant treatment. In general, older people metabolize tricyclic antidepressants more slowly and have higher steady-state plasma concentrations. There is an increased risk of adverse effects of tricyclic antidepressants in elderly patients (SEDA-18, 17). Elderly people often take other drugs that can cause depression or interact with its treatment (144). These potential sources of variation are superimposed on the wide range of plasma concentrations reported among individuals, and on the differences between drugs in dose–response profiles. The dictates of safe practice are that treatment in elderly people be begun with low dosages (50 mg/day of imipramine or equivalent) in divided amounts, with dosage increases in small increments (50 mg/day of imipramine each week) and a reduced total dose range (75–150 mg/day, except in exceptional circumstances) (145). Close attention should be paid to the potential anticholinergic, neurological, or cardiovascular complications to which elderly people are especially vulnerable (31).

The adverse effects profile of fixed-dose clomipramine (150 mg/day) has been assessed in 112 hospitalized depressed patients (aged 22–70 years), of whom 38 were over 55 years of age (146). The only adverse effect that distinguished patients over 55 years was orthostatic hypotension: older subjects had a significantly greater fall in systolic blood pressure on standing. Orthostatic hypotension can lead to falls and injuries, particularly in patients being treated at home, and this suggests that blood pressure should be monitored in older patients who are taking psychiatric doses of tricyclic antidepressants. It is also worth noting that the upper age limit of patients in this study was only 70 years, so it is unclear how more elderly patients would fare with this rather substantial dose of clomipramine.

Drug Administration

Drug overdose

Both accidental and intentional overdose are relatively frequent and pose difficult management problems. Particular concern has been expressed for children, either because they gain access to parents' tablets or have been treated for enuresis. During one year a Melbourne hospital admitted 35 children poisoned with tricyclic antidepressants (147). In 1979 it was reported that tricyclic antidepressants had replaced salicylates as the most common cause of accidental death in English children under the age of five. Concern was expressed about this (148), and Swiss federal statistics raised similar worries (149).

The majority of all deaths from antidepressant poisoning in Scotland, England, and Wales, during 1975–1984 and 1985–1989 were due to two tricyclic drugs, amitriptyline and dosulepin, and they, as well as the entire group of older tricyclic antidepressants, had a fatality index (deaths per million prescriptions) significantly higher than the mean (150).

A nation-wide analysis of suicide mortality in Finland showed that between 1990 and 1995 the overall suicide mortality fell significantly from 30 to 27 per 100 000 (151). However, the use of antidepressants in completed suicide showed an upward trend, while the use of more violent methods (gassing, hanging) fell. During this time prescription of moclobemide and two SSRIs (citalopram and fluoxetine) increased, while that of tricyclics (mainly doxepin and amitriptyline) remained steady. The mean annual fatal toxicity index was highest for tricyclics, such as doxepin, trimipramine, and amitripyline, and lowest for SSRIs.

Educational policies and commercial marketing of antidepressant drugs have led to an increase in the detection and treatment of depression. Conceivably this may be associated with the fall in suicide rates noted in Finland. However, overdosage of tricyclic antidepressants continues to contribute to deaths from suicide. Whether completely replacing tricyclics with less toxic compounds would lower overall suicide rate remains controversial.

The results of a British survey of all deaths due to acute poisoning due to antidepressants are shown in Table 2 (152).

A major problem in evaluating the incidence and severity of complications due to overdosage in adults has been the reporting of individual cases or selected samples, often with the inclusion of patients who have taken several drugs or who have concurrent physical disease. More reliable epidemiological surveys from different countries (45,153,154) have suggested that the mean ingested overdose of a tricyclic antidepressant in adults is around

Table 2 Deaths per million prescriptions for antidepressants in the UK

Drug	No. of prescriptions	Total deaths	Deaths per million prescriptions (95% CI)
Tricyclic antidepressants and related drugs	74 598 000	2598	35 (34, 36)
Desipramine	45 000	9	201 (92, 382)
Amoxapine	107 000	10	94 (45, 172)
Dosulepin	26 210 000	1398	53 (51, 56)
Amitriptyline	23 844 000	906	38 (36, 41)
Imipramine	3 354 000	110	33 (27, 40)
Doxepin	1 587 000	40	25 (18, 34)
Trimipramine	2 370 000	39	17 (12, 23)
Clomipramine	4 315 000	54	13 (9.4, 16)
Nortriptyline	1 269 000	7	5.5 (2.2, 11)
Maprotiline	201 000	1	5.0 (0.1, 28)
Trazodone	2 753 000	11	4.0 (2.0, 7.1)
Mianserin	922 000	3	3.3 (0.7, 9.5)
Mirtazapine	324 000	1	3.1 (0.1, 17)
Lofepramine	7 189 000	9	1.3 (0.6, 2.4)
Butriptyline	1000	0	0 (0, 3372)
Iprindole	3000	0	0 (0, 1218)
Viloxazine	10 000	0	0 (0, 357)
Protriptyline	94 000	0	0 (0, 39)
Serotonin re-uptake inhibitors	47 329 000	77	1.6 (1.3, 2.0)
Venlafaxine	2 570 000	34	13 (9.2, 19)
Fluvoxamine	660 000	2	3.0 (0.3, 11)
Citalopram	2 603 000	5	1.9 (0.6, 4.5)
Sertraline	5 964 000	7	1.2 (0.5, 2.4)
Fluoxetine	19 926 000	18	0.9 (0.5, 1.4)
Paroxetine	15 031 000	11	0.7 (0.4, 1.3)
Nefazodone	576 000	0	0 (0, 6.4)
Monoamine oxidase inhibitors	1 203 000	24	20 (13, 30)
Tranylcypromine	367 000	16	44 (25, 71)
Phenelzine	404 000	6	15 (5.5, 32)
Moclobemide	365 000	2	5.5 (0.6, 20)
Iproniazid	200	0	0 (0, 18 444)
Isocarboxazid	68 000	0	0 (0, 55)
Other antidepressants	2 523 000	1	0.4 (0, 2.2)
Flupentixol	28 000	0	0 (0, 2.4)
Tryptophan	28 000	0	0 (0, 133)
Reboxetine	175 000	0	0 (0, 21)
Lithium	5 106 000	37	7.2 (5.1, 10)

Numbers may not add up to the total because of rounding.

1000 mg, and that only about 3% of patients take enough to cause fatal complications. There is considerable individual variability in response to overdosage, and there are conflicting reports concerning the degree of correlation between plasma concentrations and complications. One large study showed that plasma concentrations in excess of 1000 ng/ml were associated with coma, convulsions, cardiac dysrhythmias, and a need for supportive measures (21). In other studies there was no clear-cut correlation between plasma concentration and the incidence of toxic complications (155). Cardiac complications have been reviewed above and are serious, but should not deflect attention from other serious effects. A review of all deaths due to these drugs in Britain during 1976 showed that dysrhythmias were less common than supposed (11 of 113 patients who died in hospitals), but that respiratory depression was more frequent (54 of 113 patients) (48). Some such deaths might have been avoided by more frequent artificial ventilation and better attention to the principles of supportive care.

Management of overdose
The literature on poisoning with tricyclic antidepressants has been reviewed and recommendations for the management of overdose proposed (SEDA-16, 8; 156). In overdose, physostigmine can reverse life-threatening dysrhythmias, but its effect is short-lasting and it has adverse effects of its own; it should be avoided (156).

Even in an intensive care unit, treatment of overdose with tricyclic antidepressants can be very challenging (157).

- A 17-month-old girl took about 750 mg of amitriptyline (usual adult daily dose 150 mg). Two hours later she was comatose with minimal response to painful stimuli. She also had multifocal clonic seizures. Her blood pressure was 70/40 mmHg and heart rate 140/minute. Arterial pH was 7.24. Electrocardiography showed a ventricular tachycardia and wide QRS complexes. She was treated with intravenous fluids for circulatory support, diazepam (0.3 mg/kg/hour) for seizures, intravenous lidocaine (20 micrograms/kg/minute), and sodium bicarbonate (2 mmol/kg). There was no response to this therapy; and she remained in a deep coma with persistent cardiac dysrhythmias and seizures. Ten hours after admission, hemoperfusion was started and continued for 2 hours. Just before and during hemoperfusion, she had several episodes of ventricular fibrillation, which were treated successfully with cardioversion. After hemoperfusion her cardiac rhythm returned to normal and the seizures stopped. Her serum amitriptyline concentration fell from 1299 µg/l to 849 µg/l. Her cardiovascular and neurological status returned to normal the next day.

This case vividly illustrates the life-threatening consequences of tricyclic antidepressant overdose, in which serum concentrations of more than 1000 µg/l carry a high risk of mortality. Because of the high toxicity of tricyclic antidepressants, children are particularly at risk from accidental ingestion of tablets from inadequately secured containers, a point worth making when prescribing tricyclic antidepressant for patients who have young children. The value of hemoperfusion in tricyclic antidepressant overdose has been debated. General opinion is that it is unlikely to be helpful, because the large volume of distribution of tricyclic antidepressants means that relatively little drug will be removed.

Drug–Drug Interactions

General

The large number of drugs with which tricyclic antidepressants interact are summarized in Table 3; those of major concern are discussed in more detail below.

Alcohol

Alcohol in combination with tricyclic antidepressants impairs performance, and it is customary to caution patients about the risk of impairment in their ability to drive or handle machinery. Not only can additive impairment occur (mediated at the receptor level), but alcohol can also alter the metabolism of the antidepressant (SEDA-8, 24).

Amines

Tricyclic antidepressants act on both presynaptic and postsynaptic neurons, as well as on alpha- and beta-adrenoceptors. Because their principal action is to block the re-uptake of noradrenaline at the presynaptic neuron, they potentiate the hypertensive effects of both directly acting and indirectly acting amines (158,159). The hypertensive effects of phenylephrine are increased by a factor of 2–3, and of noradrenaline by a factor of 4–8. Even the administration of local anesthetics containing noradrenaline as a vasoconstrictor has proven fatal. The types of

compound that can produce this interaction and the symptoms that result have been previously reviewed in detail (SEDA-1, 11).

Anesthetics

In 190 patients taking tricyclic antidepressants that could not be discontinued before surgery, who underwent general and 61 local or regional anesthesia, there were no changes in the cardiovascular effect of halothane, induction time with pentobarbital, propanidid, or ketamine, or the duration of depolarization or recovery time (160). The general conclusion was that it is safer to continue treatment with tricyclic antidepressants than to risk potential disruption from withdrawal before surgery.

Anticoagulants

Tricyclic antidepressants can interfere with the metabolism of oral anticoagulants, increase their serum concentrations, and prolong their half-lives by as much as 300% (SEDA-21, 10) (161). Prothrombin activity should be carefully monitored in patients taking oral anticoagulants.

Antidysrhythmic drugs

Because of their similar lipophilic and surfactant properties, tricyclic antidepressants interact with antidysrhythmic drugs of the quinidine type, interfering with the voltage-dependent stimulus and producing dose-related synergy (162). Cardiac glycosides and beta-blockers are free of this interaction, although animal studies have suggested increased lethality of digoxin in rats pretreated with tricyclic antidepressants (163), while propranolol may potentiate direct depression of myocardial contractility due to tricyclic antidepressants. For all of these reasons, the preferred treatment for tricyclic-induced dysrhythmias is lidocaine, but even this is reported to be only variably effective and possibly to potentiate the hypotensive effects of tricyclic drugs (46).

Antipsychotic drugs

Tricyclic antidepressants lower the seizure threshold and should therefore be used with caution with other agents that can also lower seizure threshold, such as antipsychotic drugs (164).

- A 34-year-old man with schizophrenia responded well to olanzapine 20 mg/day. However, he then had obsessional hand washing and was given clomipramine, which was increased to a dosage of 250 mg/day. He then reported myoclonic jerks with some dizziness, and 10 days later had a generalized tonic-clonic seizure. The combined clomipramine + desmethylclomipramine concentration was 2212 nmol/l, higher than the upper end of the usual target range (1300 nmol/l). An electroencephalogram showed paroxysmal slowing and spike-and-wave activity. Both olanzapine and clomipramine were withdrawn. Later the clomipramine was restarted as monotherapy and a dose of 300 mg/day was reached, which led to an even higher

Table 3 Interactions of tricyclic antidepressants with other substances

Interacting substance(s)	Type of interaction(s)	Comments
Alcohol	Additive sedative effects	May be more pronounced with more sedative tricyclic compounds
Anticholinergic drugs	Additive effects on pupils, nervous system, bowel, and bladder	Particularly likely to occur when a tricyclic drug, a phenothiazine, and a antiparkinsonian drug are prescribed concurrently
Anticoagulants	Increased serum concentrations of oral anticoagulant	Tricyclic drugs can interfere with the metabolism of oral anticoagulants
Antiepileptic drugs	Antagonism	Convulsive threshold lowered
	Decreased plasma concentrations of some tricyclic drugs	Hepatic microsomal enzyme induction
Antihypertensive agents	Reversal of hypotensive effects of clonidine, reserpine, α-methyldopa, and guanethidine	Hypertension should be controlled with diuretics, β-blockers, or vasodilators before treatment of depression
Barbiturates	Reduced plasma concentrations of tricyclic drugs	Hepatic microsomal enzyme induction
Calcium channel blockers	Increased plasma concentrations of imipramine and possibly other drugs	
Cigarette smoking	Decreased plasma concentrations of tricyclic drugs	Probably increased metabolism
Cimetidine	Increased systemic availability of tricyclic drugs	Probably due to impaired hepatic extraction
Disulfiram	Increased plasma concentrations of tricyclic drugs	Probably inhibition of metabolism
Diuretics	Increased risk of postural hypotension	Pharmacodynamic interaction
Halofantrine	Increased risk of ventricular dysrhythmias	Prolonged QT interval
Levodopa	Impaired gastrointestinal absorption	May apply to other drugs metabolized in the gut
Monoamine oxidase inhibitors	Increased incidence of weight gain	See text; see also Monoamine oxidase inhibitors
	Increased anticholinergic effects	
	Hyperthermia, hyper-reflexia, convulsions, death (very rare)	
Membrane-stabilizing drugs (quinidine type)	Synergism	
Methylphenidate	Increased plasma concentrations of tricyclic drugs	Probably inhibition of metabolism
Morphine	Potentiation of analgesic effect of morphine	May be partly due to increased systemic availability of morphine
Nitrates	Reduced effect of sublingual nitrates	Owing to dry mouth
Phenothiazines	Increased plasma concentrations of tricyclic drugs	Owing to inhibition of metabolism
	Possible potentiation and/or increased speed of onset of antidepressant effect	
	Enhanced cardiotoxic effects with thioridazine	Presynaptic α_2-adrenoceptor blockade (animal studies)
		Ventricular dysrhythmias can occur
Rifampicin	Decreased plasma concentrations of some tricyclics	Hepatic microsomal enzyme induction
Ritonavir	Possibly increased plasma concentrations of tricyclic drugs	Inhibition of metabolism
SSRIs	Increased plasma concentrations of tricyclic drugs	Inhibition of metabolism
Corticosteroids	Altered systemic availability of tricyclic drugs	Inhibition of metabolism
	Akathisia	Receptor interaction
	Exacerbation of psychosis	
Sympathomimetic agents	Potentiation of hypertensive effects with phenylephrine, noradrenaline, methylphenidate	
	Potentiation of effects of pheochromocytoma	
Thyroid hormones	Can manifest as spurious hyperthyroidism, cardiac dysrhythmias, or, in some cases, enhanced therapeutic actions of tricyclics	Increased receptor sensitivity to catecholamines

clomipramine + desmethylclomipramine concentration (3234 nmol/l), but there was no clinical or electroencephalographic evidence of seizure activity. However, when olanzapine was added in a dosage of 15 mg/day, myoclonic jerking and abnormal electroencephalographic activity recurred within 7 days.

The seizures occurred in the presence of high concentrations of clomipramine in this case, suggesting a pharmacodynamic drug interaction, since neither agent given alone provoked seizure activity, whereas the combination did. It is, however, possible that clomipramine might have caused a rise in olanzapine blood concentrations, which were not measured.

Baclofen

Tricyclic antidepressants can potentiate the muscle relaxant effects of baclofen, resulting in severe hypotonic weakness (165); the combination has also been incriminated in the causation of short-term memory impairment (SEDA-11, 127) (166).

Benzodiazepines

Four patients developed adverse effects attributable to combinations of benzodiazepines with tricyclic antidepressants, including exacerbations of delusional thought disorders (167).

Clozapine

Nefazodone had minimal effects on clozapine metabolism when co-administered in six patients: mean clozapine concentrations rose by 4% and norclozapine concentrations by 16% (168).

Corticosteroids

Patients taking prednisone have experienced psychosis, perhaps due to an interaction at dopamine receptor sites (169,170).

Desmopressin

Tricyclic antidepressants increase the release of endogenous antidiuretic hormone and can therefore potentiate the antidiuretic effect of desmopressin. A hyponatremic convulsion occurred in a child who was given desmopressin and imipramine (171).

Fluconazole

Fluconazole can increase plasma concentrations of amitriptyline, presumably by inhibiting cytochrome P450 isozymes (CYP3A4 and CYP2C19), preventing its demethylation. Two patients taking amitriptyline and fluconazole developed syncope and in one concomitant electrocardiographic monitoring showed a prolonged QT interval and torsade de pointes (172,173). In neither case were serum amitriptyline concentrations measured, but the symptoms were consistent with tricyclic antidepressant toxicity. These case reports suggest that this combination should be used with caution and probably with monitoring of amitriptyline concentrations.

Fluvoxamine

Fluvoxamine is a potent inhibitor of cytochrome CYP1A2, which may lead to interactions with several tricyclic antidepressants and theophylline (174).

Hormonal contraceptives—oral

Oral contraceptives have a complex effect on the metabolism of tricyclic drugs, resulting in reduced systemic availability (175). Oral contraceptives reduce the clearance of imipramine, probably by reducing hepatic oxidation, and thus increase its half-life. Hydroxylation of amitriptyline is inhibited by contraceptive steroids. The clinical significance is uncertain, but there is at least anecdotal evidence of an increase in antidepressant adverse effects (176). Caution should be exercised when tricyclic antidepressants are used long term in women taking oral contraceptives.

Levodopa

The anticholinergic effects of imipramine and other tricyclic antidepressants can delay gastrointestinal motility enough to interfere with the absorption of various other drugs. Such was the case in an experimental study of the absorption of levodopa in four healthy subjects (177). It is likely that this effect may interfere with absorption of other drugs, especially those, such as chlorpromazine, that are extensively metabolized in the gut.

Lithium

Amitriptyline

Serum lithium concentrations were unchanged when breakthrough depression was treated double blind by the addition of amitriptyline (75–150 mg/day, $n = 23$) and the combination was generally well tolerated (178).

- A 34-year-old woman took amitriptyline 300 mg each night for several years (179). Six days after starting to take lithium 300 mg tds, she had several generalized tonic-clonic seizures. A second episode occurred on re-exposure to lithium.

Doxepin

An interaction of lithium with doxepin has been described.

- A 65-year-old man who took lithium and doxepin for 13 years presented with a 6-month history of myoclonic jerking of both arms, which resolved when both drugs were stopped (180).

Whether this represented a drug interaction or a single drug effect is unclear.

Six patients taking a stable dose of fluvoxamine had a minor increase in plasma fluvoxamine concentration (from 67 to 76 ng/ml) 2 weeks after starting to take unspecified doses of lithium; this is unlikely to be of clinical significance (181).

Methylphenidate

The interaction with methylphenidate may be of particular significance, because of claims that tricyclic antidepressants and methylphenidate have a synergistic effect on mood, owing to interference by methylphenidate with the metabolism of imipramine, resulting in increased blood imipramine concentrations (182). The occurrence of this hypertensive interaction calls for caution in the use of such combinations, for which there is no established evidence.

Monoamine oxidase inhibitors

The interactions of tricyclic antidepressants with MAO inhibitors are so dangerous that they constitute a formidable barrier to their combined clinical use. The temptation to treat refractory patients with this combination is great, but considerable care should be taken (183,184), and it should be remembered that there is no firm proof of the superiority of such combinations. Several reviewers (185–187) have concluded that the dangers of these interactions have been overstated and the potential therapeutic advantages underestimated. On the other hand, reports of serious and fatal complications continue to appear (188,189). Specialists have advised that the combination of an MAO inhibitor with trimipramine or amitriptyline is usually safe, but that imipramine and clomipramine must be avoided.

Morphine

Clomipramine, amitriptyline, and probably other antidepressants that potentiate the serotonergic system, enhance the analgesic effect of morphine.

Olanzapine

Clomipramine and olanzapine are both metabolized by CYP1A2 and CYP2D6, and it is therefore possible that raised concentrations of both compounds can result from co-administration.

- Seizures in a 34-year-old man were attributed to an interaction of olanzapine with clomipramine (190).

A pharmacodynamic interaction occurred between olanzapine and imipramine 191).

Pancuronium bromide

The use of both halothane and pancuronium in patients taking tricyclic antidepressant has been reported as resulting in severe tachydysrhythmias. Experiments in dogs have shown that this combination can produce ventricular fibrillation and cardiac arrest (192). Enflurane also resulted in tachycardias in dogs given both imipramine and pancuronium acutely, but not when the imipramine was given chronically for 15 days beforehand. Pancuronium should not be used in patients taking tricyclic antidepressants.

Paroxetine

Paroxetine can cause clinically important increases in plasma concentrations of tricyclic antidepressants (193).

Phenothiazines

Two Scandinavian patients taking combinations of neuroleptic drugs and tricyclic antidepressants developed epileptic seizures (194). The risk of seizures is greater in patients with brain damage or epilepsy and with high dosages, sudden increases in dosage, or shortly after the introduction of a second compound.

The possible mechanism for these interactions with phenothiazines is raised plasma concentrations of the tricyclic compounds due to competition for the hepatic cytochrome P450 that metabolizes both types of compound (SEDA-9, 30) (195). Antidepressant plasma concentrations may increase by up to 70% (196).

Postganglionic blockers

Tricyclic antidepressants reverse the hypotensive effects of postganglionic blocking agents, guanethidine, reserpine, clonidine, and alpha-methyldopa, and the addition of a tricyclic can result in loss of blood pressure control (159,197). Sudden withdrawal of a tricyclic compound from a patient stabilized with these compounds can also result in serious hypotension. An additional reason for avoiding drugs such as reserpine, methyldopa, and propranolol in depressed cardiovascular patients is that these drugs can also cause or aggravate depression.

Psychotropic drugs

The evidence for interactions with other commonly prescribed psychotropic drugs has been reviewed extensively (SEDA-8, 25).

Salicylates

Tricyclic antidepressants are highly bound to plasma proteins. Theoretically, other drugs that are highly protein bound could displace tricyclic antidepressants from protein binding sites; this would be expected to lead to increased plasma concentrations of unbound antidepressant and a greater likelihood of adverse effects. In practice, however, important interactions of this kind are unusual, probably because these drugs have high apparent volumes of distribution, and the amount of drug that can be displaced from plasma proteins is very small compared with tissue concentrations. Salicylates are highly protein bound, and the effect of adding acetylsalicylic acid 1 g/day for 2 days on the plasma availability of imipramine (150 mg/day) has been studied in 20 depressed patients (198). Acetylsalicylic acid reduced the protein binding of imipramine and while the total plasma concentration of imipramine was not altered, the unbound concentration rose almost four-fold. The number of characteristic adverse effects of imipramine also increased significantly. However, this study was not blind and the results must be in doubt, since it is unlikely that brain concentrations of the drug changed significantly.

Sex hormones

Conjugated estrogens cause akathisia in some patients taking tricyclic drugs, perhaps due to an interaction at dopamine receptor sites (199).

Sodium valproate

Sodium valproate can increase serum tricyclic drug concentrations (SEDA-21, 22).

- In a 46-year-old woman, when valproate 1 g/day was added to clomipramine 150 mg/day, there was a substantial increase in serum clomipramine concentrations (from 185 ng/ml to 447 ng/ml); she had feelings of numbness and sleep disturbance, which disappeared when the dose of clomipramine was reduced (200).

The increase in tricyclic concentrations following valproate is probably partly due to inhibition of CYP2C isozymes, preventing demethylation of tertiary tricyclic antidepressants to the corresponding secondary amines (desmethylimipramine in the case of clomipramine). However, valproate can also increase the plasma concentrations of secondary amine tricyclics, such as nortriptyline (200). The current data suggest that the combined use of tricyclic antidepressants and valproate should be undertaken with caution.

SSRIs

Fluoxetine can increase blood concentrations of desipramine and nortriptyline (201–204).

Interactions of SSRIs with tricyclic antidepressants have been reported. For example, plasma concentrations of tricyclic antidepressants rise after the addition of fluoxetine (205,206), fluvoxamine (207,208), and sertraline (209).

St John's wort

St John's wort can cause drug interactions by inducing hepatic microsomal drug-metabolizing enzymes or the drug transporter P-glycoprotein, which causes a net efflux of substrates, such as amitriptyline, from intestinal epithelial cells into the gut lumen (SEDA-24, 12). In 12 patients (9 women, 3 men) the addition of St John's wort 900 mg/day to amitriptyline 150 mg/day led to a 20% reduction in plasma amitriptyline concentrations, while nortriptyline concentrations were almost halved (210).

Terbinafine

Desipramine

Inhibition of CYP2D6 by terbinafine has been evaluated by assessing 48-hour concentration-time profiles of the tricyclic antidepressant desipramine in 12 healthy volunteers identified as extensive CYP2D6 metabolizers by genotyping and phenotyping (211). The pharmacokinetics were evaluated at baseline (50 mg oral desipramine given alone), steady state (after 250 mg oral terbinafine for 21 days), and 2 and 4 weeks after terbinafine withdrawal. The pharmacodynamics were evaluated before and 2 hours after each dose of desipramine, using Mini-Mental Status Examination and electroencephalography. Terbinafine inhibited CYP2D6 metabolism, as indicated by significant increases in desipramine C_{max} and AUC and reductions in the C_{max} and AUC of the CYP2D6-mediated metabolite, 2-hydroxydesipramine, both of which were still altered 4 weeks after terbinafine withdrawal. Caution should be exercised when co-prescribing terbinafine and drugs that are metabolized by CYP2D6, particularly those with a narrow therapeutic index.

Imipramine

- A 51-year-old patient developed imipramine toxicity and increased plasma concentrations associated with the introduction of terbinafine, possibly due to inhibition of CYP2D6 (212).

Nortriptyline

Metabolism by CYP2D6 is of major importance for the hydroxylation of nortriptyline, making it susceptible to competitive inhibition by terbinafine. Nortriptyline intoxication provoked by terbinafine has been reported (213).

- A 74-year-old man taking a stable dose of nortriptyline for depression developed signs of nortriptyline intoxication 14 days after he started to take terbinafine (210). Nortriptyline serum concentrations were several times higher than the usual target range and fell to baseline after withdrawal of terbinafine. Re-challenge led to the same clinical and laboratory findings.

Nortriptyline intoxication secondary to terbinafine has been observed in a woman with a major depressive disorder (214). After rechallenge her serum nortriptyline concentration rose and the serum concentrations of its two hydroxylated metabolites fell. She had a normal genotype for CYP2D6, suggesting that this interaction can occur even in people without reduced CYP2D6 activity.

Thioridazine

The interaction of thioridazine with a tricyclic antidepressant is particularly dangerous, since both cause cardiac toxicity. One report concerned two young patients who developed ventricular dysrhythmias, from which they recovered (215).

Thyroid hormone

Spurious hyperthyroidism occurred in a child taking thyroid hormone and imipramine for enuresis (216). The ability of thyroid hormone to increase receptor sensitivity to catecholamines has long been known, and has been used to enhance the clinical response in some refractory patients, especially women.

Yohimbine

Yohimbine has a sialogenic effect in depressed patients with a dry mouth due to tricyclic antidepressants or neuroleptic drugs (217).

However, there is an increased risk of hypertension when tricyclic antidepressants are combined with yohimbine (218).

Yohimbine can cause hypertension in patients taking tricyclic antidepressants. A drug history should include the use of herbal remedies before conventional treatments are prescribed.

References

1. Morrow PL, Hardin NJ, Bonadies J. Hypersensitivity myocarditis and hepatitis associated with imipramine and its metabolite, desipramine. J Forensic Sci 1989;34(4):1016–20.
2. Carlson DH, Healy J. Pulmonary hypersensitivity to imipramine. South Med J 1982;75(4):514.
3. Anonymous. Tricyclic antidepressants—blood level measurements and clinical outcome: an APA Task Force report. Task Force on the Use of Laboratory Tests in Psychiatry. Am J Psychiatry 1985;142(2):155–62.
4. Ziegler VE, Co BT, Biggs JT. Plasma nortriptyline levels and ECG findings. Am J Psychiatry 1977;134(4):441–3.

5. Kantor SJ, Glassman AH, Bigger JT Jr, Perel JM, Giardina EV. The cardiac effects of therapeutic plasma concentrations of imipramine. Am J Psychiatry 1978;135(5):534–8.

6. Kantor SJ, Bigger JT Jr, Glassman AH, Macken DL, Perel JM. Imipramine-induced heart block. A longitudinal case study. JAMA 1975;231(13):1364–6.

7. Preskorn SH, Biggs JT. Use of tricyclic antidepressant blood levels. N Engl J Med 1978;298(3):166.

8. Carr AC, Hobson RP. High serum concentrations of antidepressants in elderly patients. BMJ 1977;2(6095):1151.

9. Blackwell B. The drug regimen and treatment compliance. In: Haynes RB, Taylor DW, Sacket DL, editors. Compliance in Health Care. Baltimore: Johns Hopkins University Press, 1979:144.

10. Flemenbaum A. Pavor nocturnus: a complication of single daily tricyclic or neuroleptic dosage. Am J Psychiatry 1976;133(5):570–2.

11. Hoppe E, Kramp P, Sandoe E, Bolwig TG. Elektrokonvulsiv behandling og tricykliske antidepressiva: risiko for udvikling af hjertearitmi. [Electroconvulsive therapy and tricyclic antidepressants: risk of development of cardiac arrythmia.] Ugeskr Laeger 1977;139(44):2636–8.

12. Azar I, Lear E. Cardiovascular effects of electroconvulsive therapy in patients taking tricyclic antidepressants. Anesth Analg 1984;63(12):1140.

13. Blackwell B, Currah J. The psychopharmacology of nocturnal enuresis. In: Kolvin IL, MacKeith RC, Meadow SR, editors. Bladder Control and Enuresis. London: William Heinemann Medical Books, 1973:231.

14. Ratzman GW, Seer OR. Zur Symptomatologie toxischer Nebenwirkungen bei der Therapie der Enuresis im Kindesalter mit Imipramin (Pryleugan). [On the symptomatology of toxic adverse effects of imipramine (Pryleugan) in the treatment of enuresis in children.] Dtsch Gesundheitswes 1978;33:XII.

15. Mavissakalian M, Perel J, Guo S. Specific side effects of long-term imipramine management of panic disorder. J Clin Psychopharmacol 2002;22(2):155–61.

16. Glassman AH, Preud'homme XA. Review of the cardiovascular effects of heterocyclic antidepressants. J Clin Psychiatry 1993;54(Suppl):16–22.

17. Veith RC, Friedel RO, Bloom V, Bielski R. Electrocardiogram changes and plasma desipramine levels during treatment of depression. Clin Pharmacol Ther 1980;27(6):796–802.

18. Spiker DG, Weiss AN, Chang SS, Ruwitch JF Jr, Biggs JT. Tricyclic antidepressant overdose: clinical presentation and plasma levels. Clin Pharmacol Ther 1975;18(5 Part 1):539–46.

19. Burrows GD, Vohra J, Hunt D, Sloman JG, Scoggins BA, Davies B. Cardiac effects of different tricyclic antidepressant drugs. Br J Psychiatry 1976;129:335–41.

20. Hallstrom C, Gifford L. Antidepressant blood levels in acute overdose. Postgrad Med J 1976;52(613):687–8.

21. Petit JM, Spiker DG, Ruwitch JF, Ziegler VE, Weiss AN, Biggs JT. Tricyclic antidepressant plasma levels and adverse effects after overdose. Clin Pharmacol Ther 1977;21(1):47–51.

22. Roose SP, Glassman AH, Giardina EG, Johnson LL, Walsh BT, Woodring S, Bigger JT Jr. Nortriptyline in depressed patients with left ventricular impairment. JAMA 1986;256(23):3253–7.

23. Bigger JT, Kantor SJ, Glassman AH, et al. Cardiovascular effects of tricyclic antidepressant drugs. In: Lipton MA, Dimascio A, Killam KF, editors. Psychopharmacology: a Generation of Progress. New York: Raven Press, 1978:1033.

24. Burrows GD, Vohra J, Dumovic P, et al. Tricyclic antidepressant drugs and cardiac conduction. Prog Neuropsychopharmacol 1977;1:329.

25. Brorson L, Wennerblom B. Electrophysiological methods in assessing cardiac effects of the tricyclic antidepressant imipramine. Acta Med Scand 1978;203(5):429–32.

26. Scherlag BJ, Lau SH, Helfant RH, Berkowitz WD, Stein E, Damato AN. Catheter technique for recording His bundle activity in man. Circulation 1969;39(1):13–8.

27. Anonymous. Sudden death in children treated with a tricyclic antidepressant. Med Lett Drugs Ther 1990;32(819):53.

28. Riddle MA, Nelson JC, Kleinman CS, Rasmusson A, Leckman JF, King RA, Cohen DJ. Sudden death in children receiving Norpramin: a review of three reported cases and commentary. J Am Acad Child Adolesc Psychiatry 1991;30(1):104–8.

29. Biederman J, Baldessarini RJ, Wright V, Knee D, Harmatz JS, Goldblatt A. A double-blind placebo controlled study of desipramine in the treatment ADD. II. Serum drug levels and cardiovascular findings. J Am Acad Child Adolesc Psychiatry 1989;28(6):903–11.

30. Bigger JT, Giardina EG, Perel JM, Kantor SJ, Glassman AH. Cardiac antiarrhythmic effect of imipramine hydrochloride. N Engl J Med 1977;296(4):206–8.

31. Wasylenki D. Depression in the elderly. Can Med Assoc J 1980;122(5):525–32.

32. Todd RD, Faber R. Ventricular arrhythmias induced by doxepin and amitriptyline: case report. J Clin Psychiatry 1983;44(11):423–5.

33. Avery D, Winokur G. Mortality in depressed patients treated with electroconvulsive therapy and antidepressants. Arch Gen Psychiatry 1976;33(9):1029–37.

34. Veith RC, Raskind MA, Caldwell JH, Barnes RF, Gumbrecht G, Ritchie JL. Cardiovascular effects of tricyclic antidepressants in depressed patients with chronic heart disease. N Engl J Med 1982;306(16):954–9.

35. Ziegler VE, Taylor JR, Wetzel RD, Biggs JT. Nortriptyline plasma levels and subjective side effects. Br J Psychiatry 1978;132:55.

36. Reisby N, Gram LF, Bech P, Nagy A, Petersen GO, Ortmann J, Ibsen I, Dencker SJ, Jacobsen O, Krautwald O, Sondergaard I, Christiansen J. Imipramine: clinical effects and pharmacokinetic variability. Psychopharmacology (Berl) 1977;54(3):263–72.

37. Glassman AH, Bigger JT Jr, Giardina EV, Kantor SJ, Perel JM, Davies M. Clinical characteristics of imipramine-induced orthostatic hypotension. Lancet 1979;1(8114):468–72.

38. van Zwieten PA. The central action of antihypertensive drugs, mediated via central alpha-receptors. J Pharm Pharmacol 1973;25(2):89–95.

39. Taylor DJ, Braithwaite RA. Cardiac effects of tricyclic antidepressant medication. A preliminary study of nortriptyline. Br Heart J 1978;40(9):1005–9.

40. Langou RA, Van Dyke C, Tahan SR, Cohen LS. Cardiovascular manifestations of tricyclic antidepressant overdose. Am Heart J 1980;100(4):458–64.

41. Thorstrand C. Cardiovascular effects of poisoning with tricyclic antidepressants. Acta Med Scand 1974;195(6):505–14.

42. Sternon J, Owieczka J. La prophylaxie de l'hypotension orthostatique induite par les antidepresseurs tricycliques.

[The prophylaxis of orthostatic hypotension caused by tricyclic antidepressants.] Ars Med (Brux) 1979;34:641.

43. Cohen HW, Gibson G, Alderman MH. Excess risk of myocardial infarction in patients treated with antidepressant medications: association with use of tricyclic agents. Am J Med 2000;108(1):2–8.

44. Siddiqui JH, Vakassi MM, Ghani MF. Cardiac effects of amitriptyline overdose. Curr Ther Res Clin Exp 1977;22:321.

45. O'Brien JP. A study of low-dose amitriptyline overdoses. Am J Psychiatry 1977;134(1):66–8.

46. Hoffman JR, McElroy CR. Bicarbonate therapy for dysrhythmia hypotension in tricyclic antidepressant overdose. West J Med 1981;134(1):60–4.

47. Serafimovski N, Thorball N, Asmussen I, Lunding M. Tricyclic antidepressive poisoning with special reference to cardiac complications. Acta Anaesthesiol Scand Suppl 1975;57:55–63.

48. Crome P, Newman B. Fatal tricyclic antidepressant poisoning. J R Soc Med 1979;72(9):649–53.

49. Nogue Xarau S, Nadal Trias P, Bertran Georges A, Mas Ordeig A, Munne Mas P, Milla Santos J. Intoxicacion agoda grave por antidepresivos triciclcos. Estudio retrospective de is casos. [Severe acute poisoning following the ingestion of tricyclic antidepressants.] Med Clin (Barc) 1980;74(7):257–62.

50. Goldberg RJ, Capone RJ, Hunt JD. Cardiac complications following tricyclic antidepressant overdose. Issues for monitoring policy. JAMA 1985;254(13):1772–5.

51. McAlpine SB, Calabro JJ, Robinson MD, Burkle FM Jr. Late death in tricyclic antidepressant overdose revisited. Ann Emerg Med 1986;15(11):1349–52.

52. Brown TC, Barker GA, Dunlop ME, Loughnan PM. The use of sodium bicarbonate in the treatment of tricyclic antidepressant-induced arrhythmias. Anaesth Intensive Care 1973;1(3):203–10.

53. Pinder RM, Brogden RN, Speight TM, Avery GS. Doxepin up-to-date: a review of its pharmacological properties and therapeutic efficacy with particular reference to depression. Drugs 1977;13(3):161–218.

54. Reed K, Smith RC, Schoolar JC, Hu R, Leelavathi DE, Mann E, Lippman L. Cardiovascular effects of nortriptyline in geriatric patients. Am J Psychiatry 1980;137(8):986–9.

55. Roose SP, Glassman AH, Siris SG, Walsh BT, Bruno RL, Wright LB. Comparison of imipramine- and nortriptyline-induced orthostatic hypotension: a meaningful difference. J Clin Psychopharmacol 1981;1(5):316–9.

56. Thayssen P, Bjerre M, Kragh-Sorensen P, Moller M, Petersen OL, Kristensen CB, Gram LF. Cardiovascular effect of imipramine and nortriptyline in elderly patients. Psychopharmacology (Berl) 1981;74(4):360–4.

57. Neshkes RE, Gerner R, Jarvik LF, Mintz J, Joseph J, Linde S, Aldrich J, Conolly ME, Rosen R, Hill M. Orthostatic effect of imipramine and doxepin in depressed geriatric outpatients. J Clin Psychopharmacol 1985;5(2):102–6.

58. Rosenstein DL, Nelson JC, Jacobs SC. Seizures associated with antidepressants: a review. J Clin Psychiatry 1993;54(8):289–99.

59. Solomon JG. Seizures during lithium-amitriptyline therapy. Postgrad Med 1979;66(3):145–6.

60. Yeragani VK, Pohl R, Balon R, Kulkarni A, Keshavan M. Low serum iron levels and tricyclic antidepressant-induced jitteriness. J Clin Psychopharmacol 1989;9(6):447–8.

61. Bental E, Lavie P, Sharf B. Severe hypermotility during sleep in treatment of cataplexy with clomipramine. Isr J Med Sci 1979;15(7):607–9.

62. Schatzberg AF, Cole JO, Blumer DP. Speech blockage: a tricyclic side effect. Am J Psychiatry 1978;135(5):600–1.

63. Quader SE. Dysarthria: an unusual side effect of tricyclic antidepressants. BMJ 1977;2(6079):97.

64. Saunders M. Dysarthria: an unusual side effect of tricyclic antidepressants. BMJ 1977;2:317.

65. Casarino JP. Neuropathy associated with amitriptyline. Bilateral footdrop. NY State J Med 1977;77(13):2124–6.

66. Yeragani VK, Meiri P, Balon R, Pohl R, Golec S. Effect of imipramine treatment on changes in heart rate and blood pressure during postural and isometric handgrip tests. Eur J Clin Pharmacol 1990;38(2):139–44.

67. Davies RK, Tucker GJ, Harrow M, Detre TP. Confusional episodes and antidepressant medication. Am J Psychiatry 1971;128(1):95–9.

68. Rampling D. Aggression: a paradoxical response to tricyclic antidepressants. Am J Psychiatry 1978;135(1):117–8.

69. Snyder SH, Yamamura HI. Antidepressants and the muscarinic acetylcholine receptor. Arch Gen Psychiatry 1977;34(2):236–9.

70. Blackwell B, Peterson GR, Kuzma RJ, Hostetler RM, Adolphe AB. The effect of five tricyclic antidepressants on salivary flow and mood in healthy volunteers. Commun Psychopharmacol 1980;4(4):255–61.

71. Meyers BS, Mei-Tal V. Psychiatric reactions during tricyclic treatment of the elderly reconsidered. J Clin Psychopharmacol 1983;3(1):2–6.

72. Schmidt LG, Grohmann R, Strauss A, Spiess-Kiefer C, Lindmeier D, Muller-Oerlinghausen B. Epidemiology of toxic delirium due to psychotropic drugs in psychiatric hospitals. Compr Psychiatry 1987;28(3):242–9.

73. Nouri A, Cuendet JF. Atteintes oculaires au cours des traitements aux thymoleptiques. [Ocular disturbances during treatment with thymoleptics.] Schweiz Med Wochenschr 1971;101(32):1178–80.

74. Reid WH, Blouin P, Schermer M. A review of psychotropic medications and the glaucomas. Int Pharmacopsychiatry 1976;11(3):163–74.

75. Litovitz GL. Amitriptyline and contact lenses. J Clin Psychiatry 1984;45(4):188.

76. Slome BA. Rampant caries: a side effect of tricyclic antidepressant therapy. Gen Dent 1984;32(6):494–6.

77. von Knorring AL, Wahlin YB. Tricyclic antidepressants and dental caries in children. Neuropsychobiology 1986;15(3–4):143–5.

78. Clarke IM. Adynamic ileus and amitriptyline. BMJ 1971;2(760):531.

79. Tyber MA. The relationship between hiatus hernia and tricyclic antidepressants: a report of five cases. Am J Psychiatry 1975;132(6):652–3.

80. Appel P, Eckel K, Harrer G. Veränderungen des Blasen- und Blasensphinktertonus durch Thymoleptika. [Changes in tonus of the bladder and its sphincter under thymoleptic treatment. Cystomanometric measurements in man.] Int Pharmacopsychiatry 1971;6(1):15–22.

81. Sathananthan G, Gershon S. Renal damage due to imipramine. Lancet 1973;1(7807):833–4.

82. June R, Yunus M, Gossman W. Neuroleptic malignant syndrome associated with nortriptyline. Am J Emerg Med 1999;17(7):736–7.

83. Langlow JR, Alarcon RD. Trimipramine–induced neuroleptic malignant syndrome after transient psychogenic polydipsia in one patient. J Clin Psychiatry 1989;50(4):144–5.

84. Washington C, Haines KA, Tam CW. Amoxapine-induced neuroleptic malignant syndrome. DICP 1989;23(9):713.

85. Sporer KA. The serotonin syndrome. Implicated drugs, pathophysiology and management. Drug Saf 1995;13(2):94–104.

86. Rosebush PI, Margetts P, Mazurek MF. Serotonin syndrome as a result of clomipramine monotherapy. J Clin Psychopharmacol 1999;19(3):285–7.

87. Fann WE, Sullivan JL, Richman BW. Dyskinesias associated with tricyclic antidepressants. Br J Psychiatry 1976;128:490–3.

88. Steele TE. Adverse reactions suggesting amoxapine-induced dopamine blockade. Am J Psychiatry 1982;139(11):1500–1.

89. Barker PA, Ashcroft GW, Binns JK. Imipramine in chronic depression. J Ment Sci 1960;106:1447–51.

90. Racy J, Ward-Racy EA. Tinnitus in imipramine therapy. Am J Psychiatry 1980;137(7):854–5.

91. Evans DL, Golden RN. Protriptyline and tinnitus. J Clin Psychopharmacol 1981;1(6):404–6.

92. Tandon R, Grunhaus L, Greden JF. Imipramine and tinnitus. J Clin Psychiatry 1987;48(3):109–11.

93. Smith EE, Reece CA, Kauffman R. Ototoxic reaction associated with use of nortriptyline hydrochloride: case report. J Pediatr 1972;80(6):1046–8.

94. Wehr TA, Goodwin FK. Can antidepressants cause mania and worsen the course of affective illness? Am J Psychiatry 1987;144(11):1403–11.

95. Kupfer DJ, Carpenter LL, Frank E. Possible role of antidepressants in precipitating mania and hypomania in recurrent depression. Am J Psychiatry 1988;145(7):804–8.

96. Paykel ES, Mueller PS, De la Vergne PM. Amitriptyline, weight gain and carbohydrate craving: a side effect. Br J Psychiatry 1973;123(576):501–7.

97. Nakra BR, Rutland P, Verma S, Gaind R. Amitriptyline and weight gain: a biochemical and endocrinological study. Curr Med Res Opin 1977;4(8):602–6.

98. Steiner JE, Rosenthal-Zifroni A, Edelstein EL. Taste perception in depressive illness. Isr Ann Psychiatr Relat Discip 1969;7(2):223–32.

99. Fernstrom MH, Krowinski RL, Kupfer DJ. Appetite and food preference in depression: effects of imipramine treatment. Biol Psychiatry 1987;22(5):529–39.

100. Sherman KE, Bornemann M. Amitriptyline and asymptomatic hypoglycemia. Ann Intern Med 1988;109(8):683–4.

101. Roessner MD, Demling J, Bleich S. Doxepin increases serum cholesterol levels. Can J Psychiatry 2004;49:74–5.

102. Luzecky MH, Burman KD, Schultz ER. The syndrome of inappropriate secretion of antidiuretic hormone associated with amitriptyline administration. South Med J 1974;67(4):495–7.

103. Kuhs H. Demaskierung eines Phäochromozytoms durch Amitriptylin. [Unmasking pheochromocytoma by amitriptyline.] Nervenarzt 1998;69(1):76–7.

104. Anand VS. Clomipramine–induced galactorrhoea and amenorrhoea. Br J Psychiatry 1985;147:87–8.

105. Levin GM, DeVane CL. A review of cyclic antidepressant-induced blood dyscrasias. Ann Pharmacother 1992;26(3):378–83.

106. Albertini RS, Penders TM. Agranulocytosis associated with tricyclics. J Clin Psychiatry 1978;39(5):483–5.

107. Kozakova M. Liekova purpura po antidepresivach. [Drug-induced purpura due to antidepressive drugs.] Cesk Dermatol 1971;46(4):158–60.

108. Nixon DD. Thrombocytopenia following doxepin treatment. JAMA 1972;220(3):418.

109. Yon J, Anuras S. Hepatitis caused by amitriptyline therapy. JAMA 1975;232(8):833–4.

110. Robertson MM, Abou Saleh MR, Harrison DA, et al. A double blind controlled comparison of fluoxetine and lofepramine in major depressive illness. J Psychopharmacol 1994;8:98–103.

111. In: Schiff L, editor. Diseases of the Liver. Philadelphia: Lippinc-ott, 1982:604.

112. van Vliet AC, Frenkel M, Wilson JH. Acute leverinsufficientie na opipramolgebruik. [Acute liver necrosis after treatment with opipramol.] Ned Tijdschr Geneeskd 1977;121(34):1325–7.

113. Moskovitz R, DeVane CL, Harris R, Stewart RB. Toxic hepatitis and single daily dosage imipramine therapy. J Clin Psychiatry 1982;43(4):165–6.

114. Thioly-Bensoussan D, Charpentier A, Triller R, Thioly F, Blanchet P, Tricoire N, Noury JY, Grupper C. Acne iatrogène à l'amineptine (Survector): à propos de 8 cas. [Iatrogenic acne caused by amineptin (Survector): apropos 8 cases.] Ann Dermatol Venereol 1988;115(11):1177–80.

115. Jeanmougin M, Civatte J, Cavelier-Balloy B. Toxidermie rosacéiforme à l'amineptine (Survector). [Rosaceous drug eruption caused by amineptin (Survector).] Ann Dermatol Venereol 1988;115(11):1185–6.

116. Ming ME, Bhawan J, Stefanato CM, McCalmont TH, Cohen LM. Imipramine-induced hyperpigmentation: four cases and a review of the literature. J Am Acad Dermatol 1999;40(2 Part 1):159–66.

117. Ensrud KE, Blackwell T, Mangione CM, Bowman PJ, Bauer DC, Schwartz A, Hanlon JT, Nevitt MC, Whooley MA. Central nervous system active medications and risk for fractures in older women. Arch Intern Med 2003;163:949–57.

118. Mitchell JE, Popkin MK. Antidepressant drug therapy and sexual dysfunction in men: a review. J Clin Psychopharmacol 1983;3(2):76–9.

119. Renshaw DC. Doxepin treatment of sexual dysfunction associated with depression. Sinequan: a monograph of clinical studiesAmsterdam: Excerpta Medica;. 1975.

120. Yeragani VK. Anorgasmia associated with desipramine. Can J Psychiatry 1988;33(1):76.

121. Dilsaver SC. Antidepressant withdrawal syndromes: phenomenology and pathophysiology. Acta Psychiatr Scand 1989;79(2):113–7.

122. Charney DS, Heninger GR, Sternberg DE, Landis H. Abrupt discontinuation of tricyclic antidepressant drugs: evidence for noradrenergic hyperactivity. Br J Psychiatry 1982;141:377–86.

123. Stern SL, Mendels J. Withdrawal symptoms during the course of imipramine therapy. J Clin Psychiatry 1980;41(2):66–7.

124. Santos AB Jr, Mccurdy L. Delirium after abrupt withdrawal from doxepin: case report. Am J Psychiatry 1980;137(2):239–40.

125. Mirin SM, Schatzberg AF, Creasey DE. Hypomania and mania after withdrawal of tricyclic antidepressants. Am J Psychiatry 1981;138(1):87–9.

126. Dilsaver SC, Feinberg M, Greden JF. Antidepressant withdrawal symptoms treated with anticholinergic agents. Am J Psychiatry 1983;140(2):249–51.

127. Gilani SH. Imipramine and congenital abnormalities. Pathol Microbiol (Basel) 1974;40(1):37–42.

128. McBride WG. Limb deformities associated with iminodibenzyl hydrochloride. Med J Aust 1972;1(10):492.

129. Crombie DL, Pinsent RJ, Fleming D. Imipramine in pregnancy. BMJ 1972;1(5802):745.

130. Kuenssberg EV, Knox JD. Imipramine in pregnancy. BMJ 1972;2(808):292.

131. Idanpaan-Heikkila J, Saxen L. Possible teratogenicity of imipramine–chloropyramine. Lancet 1973;2(7824):282–4.
132. Fu TK, Jarvik LF, Yen FS, Matsuyama SS, Glassman AH, Perel JM, Maltiz S. Effects of imipramine on chromosomes in psychiatric patients. Neuropsychobiology 1978;4(2):113–20.
133. Anonymous. Report on clomipramine H. Kuwait University Drug Adverse React Series 1983;2:14.
134. Eggermont E, Raveschot J, Deneve V, Casteels-van Daele M. The adverse influence of imipramine on the adaptation of the newborn infant to extrauterine life. Acta Paediatr Belg 1972;26(4):197–204.
135. Bloem BR, Lammers GJ, Roofthooft DW, De Beaufort AJ, Brouwer OF. Clomipramine withdrawal in newborns. Arch Dis Child Fetal Neonatal Ed 1999;81(1):F77.
136. Cowe L, Lloyd DJ, Dawling S. Neonatal convulsions caused by withdrawal from maternal clomipramine. BMJ (Clin Res Ed) 1982;284(6332):1837–8.
137. Sovner R, Orsulak PJ. Excretion of imipramine and desipramine in human breast milk. Am J Psychiatry 1979;136(4A):451–2.
138. Pittard WB 3rd, O'Neal W Jr. Amitriptyline excretion in human milk. J Clin Psychopharmacol 1986;6(6):383–4.
139. Stancer HC, Reed KC. Desipramine and 2-hydroxydesipramine in human breast milk and the nursing infants serum. Am J Psychiatry 1986;143(12):1597–600.
140. Wisner KL, Perel JM, Findling RL, Hinnes RL. Nortriptyline and its hydroxymetabolites in breastfeeding mothers and newborns. Psychopharmacol Bull 1997;33(2):249–51.
141. Wisner KL, Perel JM. Serum nortriptyline levels in nursing mothers and their infants. Am J Psychiatry 1991;148(9):1234–6.
142. Wisner KL, Perel JM, Findling RL. Antidepressant treatment during breast-feeding. Am J Psychiatry 1996;153(9):1132–7.
143. Matheson I, Pande H, Alertsen AR. Respiratory depression caused by N-desmethyldoxepin in breast milk. Lancet 1985;2(8464):1124.
144. Salzman C, Shader RI. Drugs that may contribute to depression in the elderly. In: Raskin A, Jarvik LF, editors. Psychiatric Symptoms and Cognitive Loss in the Elderly. New York: Wiley and Sons, 1979:57.
145. Gulevich G. Psychopharmacological treatment of the aged. In: Barelos ID, et al., editors. Psychopharmacology. Oxford: Oxford University Press, 1977:448.
146. Stage KB, Kragh-Sorensen PBDanish University Antidepressant Group. Age-related adverse drug reactions to clomipramine. Acta Psychiatr Scand 2002;105(1):55–9.
147. Brown TC, Dwyer ME, Stocks JG. Antidepressant overdosage in children—a new menace. Med J Aust 1971;2(17):848–51.
148. Anonymous. Tricyclic antidepressant poisoning in children. Lancet 1979;2(8141):511.
149. Haner H, Brandenberger H, Pasi A, Moccetti T, Hartmann H. Ausserg ewohnliche Todesfalle durch trizyclische Antidepressivo. [Tricyclic antidepressants (TAD) as a cause of unexpected death.] Z Rechtsmed 1980;84(4):255–62.
150. Henry JA, Antao CA. Suicide and fatal antidepressant poisoning. Eur J Med 1992;1(6):343–8.
151. Ohberg A, Vuori E, Klaukka T, Lonnqvist J. Antidepressants and suicide mortality. J Affect Disord 1998;50(2–3):225–33.
152. Buckley NA, McManus PR. Fatal toxicity of serotoninergic and other antidepressant drugs: analysis of United Kingdom mortality data. BMJ 2002;325(7376):1332–3.
153. Biggs JT. Clinical pharmacology and toxicology of antidepressants. Hosp Pract 1978;13(2):79–84.
154. van de Ree JK, Zimmerman AN, van Heijst AN. Intoxication by tricyclic antidepressant drugs: experimental study and therapeutic considerations. Neth J Med 1977;20(4–5):149–55.
155. Hulten BA, Adams R, Askenasi R, Dallos V, Dawling S, Volans G, Heath A. Predicting severity of tricyclic antidepressant overdose. J Toxicol Clin Toxicol 1992;30(2):161–70.
156. Dziukas LJ, Vohra J. Tricyclic antidepressant poisoning. Med J Aust 1991;154(5):344–50.
157. Donmez O, Cetinkaya M, Canbek R. Hemoperfusion in a child with amitriptyline intoxication. Pediatr Nephrol 2005;20:105–7.
158. Rumack BH, Anderson RJ, Wolfe R, Fletcher EC, Vestal BK. Ornade and anticholinergic toxicity. Hypertension, hallucinations, arrhythmias. Clin Toxicol 1974;7(6):573–81.
159. Kadar D. Letter: Amitriptyline and isoproterenol: fatal drug combination. Can Med Assoc J 1975;112(5):556–7.
160. Meignan L. Anesthésie et anti-dépresseurs tricycliques. [Anesthesia and tricyclic antidepressants.] Cah Anesthesiol 1977;25:735.
161. Vesell ES, Passananti GT, Greene FE. Impairment of drug metabolism in man by allopurinol and nortriptyline. N Engl J Med 1970;283(27):1484–8.
162. Cocco G, Ague C. Interactions between cardioactive drugs and antidepressants. Eur J Clin Pharmacol 1977;11(5):389–93.
163. Attree T, Sawyer P, Turnbull MJ. Interaction between digoxin and tricyclic antidepressants in the rat. Eur J Pharmacol 1972;19(2):294–6.
164. Deshauer D, Albuquerque J, Alda M, Grof P. Seizures caused by possible interaction between olanzapine and clomipramine. J Clin Psychopharmacol 2000;20(2):283–4.
165. Silverglat MJ. Baclofen and tricyclic antidepressants: possible interaction. JAMA 1981;246(15):1659.
166. Sandyk R, Gillman MA. Baclofen-induced memory impairment. Clin Neuropharmacol 1985;8(3):294–5.
167. Messiha FS, Morgan JP. Imipramine-mediated effects on levodopa metabolism in man. Biochem Pharmacol 1974;23(10):1503–7.
168. Taylor D, Bodani M, Hubbeling A, Murray R. The effect of nefazodone on clozapine plasma concentrations. Int Clin Psychopharmacol 1999;14(3):185–7.
169. Hall RC, Popkin MK, Kirkpatrick B. Tricyclic exacerbation of steroid psychosis. J Nerv Ment Dis 1978;166(10):738–42.
170. Malinow KL, Dorsch C. Tricyclic precipitation of steroid psychosis. Psychiatr Med 1984;2(4):351–4.
171. Hamed M, Mitchell H, Clow DJ. Hyponatraemic convulsion associated with desmopressin and imipramine treatment. BMJ 1993;306(6886):1169.
172. Dorsey ST, Biblo LA. Prolonged QT interval and torsades de pointes caused by the combination of fluconazole and amitriptyline. Am J Emerg Med 2000;18(2):227–9.
173. Robinson RF, Nahata MC, Olshefski RS. Syncope associated with concurrent amitriptyline and fluconazole therapy. Ann Pharmacother 2000;34(12):1406–9.
174. Barnes TR, Kidger T, Greenwood DT. Viloxazine and migraine. Lancet 1979;2(8156–8157):1368.
175. Abernethy DR, Greenblatt DJ, Shader RI. Imipramine disposition in users of oral contraceptive steroids. Clin Pharmacol Ther 1984;35(6):792–7.

176. Waehrens J. Krampeanfald ved samtidig behandling med neuroleptika og antidepressiva. [Convulsions during simultaneous treatment with neuroleptics and antidepressive agents.] Ugeskr Laeger 1982;144(2):106–8.

177. Beresford TP, Feinsilver DL, Hall RC. Adverse reactions to benzodiazepine–tricyclic antidepressant compound. J Clin Psychopharmacol 1981;1(6):392–4.

178. Bauer M, Zaninelli R, Muller-Oerlinghausen B, Meister W. Paroxetine and amitriptyline augmentation of lithium in the treatment of major depression: a double-blind study. J Clin Psychopharmacol 1999;19(2):164–71.

179. Solomon JG. Seizures during lithium-amitriptyline therapy. Postgrad Med 1979;66(3):145–6148.

180. Evidente VG, Caviness JN. Focal cortical transient preceding myoclonus during lithium and tricyclic antidepressant therapy. Neurology 1999;52(1):211–3.

181. Takano A, Suhara T, Yasuno F, Ichimiya T, Inoue M, Sudo Y, Suzuki K. Characteristics of clomipramine and fluvoxamine on serotonin transporter evaluated by PET. Int Clin Psychopharmacol 2002;17(Suppl 2):S84–5.

182. Flemenbaum A. Hypertensive episodes after adding methylphenidate (Ritalin) to tricyclic antidepressants. [Report of three cases and review of clinical advantages.] Psychosomatics 1972;13(4):265–8.

183. Frejaville JP. De mauvais mélanges d'antidépresseurs. [Bad combinations of antidepressant drugs.] Concours Med 1972;94:8543.

184. Ponto LB, Perry PJ, Liskow BI, Seaba HH. Drug therapy reviews: tricyclic antidepressant and monoamine oxidase inhibitor combination therapy. Am J Hosp Pharm 1977;34(9):954–61.

185. Goldberg RS, Thornton WE. Combined tricyclic–MAOI therapy for refractory depression: a review, with guidelines for appropriate usage. J Clin Pharmacol 1978;18(2–3):143–7.

186. Anath J, Luchins DJ. Combined MAOI–tricyclic therapy (a critical review). Indian J Psychiatry 1976;18:26.

187. White K, Simpson G. Combined MAOI–tricyclic antidepressant treatment: a reevaluation. J Clin Psychopharmacol 1981;1(5):264–82.

188. Johne A, Schmider J, Brockmoller J, Stadelmann AM, Stormer E, Bauer S, Scholler G, Langheinrich M, Roots I. Decreased plasma levels of amitriptyline and its metabolites on comedication with an extract from St. John's wort (Hypericum perforatum). J Clin Psychopharmacol 2002;22(1):46–54.

189. Graham PM, Potter JM, Paterson J. Combination monoamine oxidase inhibitor/tricyclic antidepressants interaction. Lancet 1982;2(8295):440.

190. Zullino DF, Delessert D, Eap CB, Preisig M, Baumann P. Tobacco and cannabis smoking cessation can lead to intoxication with clozapine or olanzapine. Int Clin Psychopharmacol 2002;17(3):141–3.

191. Callaghan JT, Bergstrom RF, Ptak LR, Beasley CM. Olanzapine. Pharmacokinetic and pharmacodynamic profile. Clin Pharmacokinet 1999;37(3):177–93.

192. Edwards RP, Miller RD, Roizen MF, Ham J, Way WL, Lake CR, Roderick L. Cardiac responses to imipramine and pancuronium during anesthesia with halothane or enflurane. Anesthesiology 1979;50(5):421–5.

193. Leucht S, Hackl HJ, Steimer W, Angersbach D, Zimmer R. Effect of adjunctive paroxetine on serum levels and side-effects of tricyclic antidepressants in depressive inpatients. Psychopharmacology (Berl) 2000;147(4):378–83.

194. Krishnan KR, France RD, Ellinwood EH Jr. Tricyclic-induced akathisia in patients taking conjugated estrogens. Am J Psychiatry 1984;141(5):696–7.

195. Siris SG, Cooper TB, Rifkin AE, Brenner R, Lieberman JA. Plasma imipramine concentrations in patients receiving concomitant fluphenazine decanoate. Am J Psychiatry 1982;139(1):104–6.

196. Linnoila M, George L, Guthrie S. Interaction between antidepressants and perphenazine in psychiatric inpatients. Am J Psychiatry 1982;139(10):1329–31.

197. Reda G, Lacerna F, Reda M, Lauro R. Interazioni tra farmaci antidepressi ed antipertensivi. Riv Psichiatr 1977;XII:309.

198. Juarez-Olguin H, Jung-Cook H, Flores-Perez J, Asseff IL. Clinical evidence of an interaction between imipramine and acetylsalicylic acid on protein binding in depressed patients. Clin Neuropharmacol 2002;25(1):32–6.

199. Krishnan KR, France RD, Ellinwood EH Jr. Tricyclic-induced akathisia in patients taking conjugated estrogens. Am J Psychiatry 1984;141(5):696–7.

200. Fehr C, Grunder G, Hiemke C, Dahmen N. Increase in serum clomipramine concentrations caused by valproate. J Clin Psychopharmacol 2000;20(4):493–4.

201. Goodnick PJ. Influence of fluoxetine on plasma levels of desipramine. Am J Psychiatry 1989;146(4):552.

202. Bell IR, Cole JO. Fluoxetine induces elevation of desipramine level and exacerbation of geriatric nonpsychotic depression. J Clin Psychopharmacol 1988;8(6):447–8.

203. Vaughan DA. Interaction of fluoxetine with tricyclic antidepressants. Am J Psychiatry 1988;145(11):1478.

204. von Ammon Cavanaugh S. Drug–drug interactions of fluoxetine with tricyclics. Psychosomatics 1990;31(3):273–6.

205. Aranow AB, Hudson JI, Pope HG Jr, Grady TA, Laage TA, Bell IR, Cole JO. Elevated antidepressant plasma levels after addition of fluoxetine. Am J Psychiatry 1989;146(7):911–3.

206. Preskorn SH, Beber JH, Faul JC, Hirschfeld RM. Serious adverse effects of combining fluoxetine and tricyclic antidepressants. Am J Psychiatry 1990;147(4):532.

207. Spina E, Campo GM, Avenoso A, Pollicino MA, Caputi AP. Interaction between Fluvoxamine and imipramine/desipramine in four patients. Ther Drug Monit 1992;14(3):194–6.

208. Härtter S, Wetzel H, Hammes E, Hiemke C. Inhibition of antidepressant demethylation and hydroxylation by fluvoxamine in depressed patients. Psychopharmacology (Berl) 1993;110(3):302–8.

209. Lydiard RB, Anton RF, Cunningham I. Interaction between sertraline and tricyclic antidepressants. Am J Psychiatry 1993;150(7):1125–6.

210. Anonymous. Antidepressant interaction led to death. Pharm J 1982;13:191.

211. Madani S, Barilla D, Cramer J, Wang Y, Paul C. Effect of terbinafine on the pharmacokinetics and pharmacodynamics of desipramine in healthy volunteers identified as cytochrome P450 2D6 (CYP2D6) extensive metabolizers. J Clin Pharmacol 2002;42(11):1211–8.

212. Teitelbaum ML, Pearson VE. Imipramine toxicity and terbinafine. Am J Psychiatry 2001;158(12):2086.

213. van der Kuy PH, Hooymans PM. Nortriptyline intoxication induced by terbinafine. BMJ 1998;316(7129):441.

214. Bagheri H, Schmitt L, Berlan M, Montastruc JL. A comparative study of the effects of yohimbine and anethol-trithione on salivary secretion in depressed patients

treated with psychotropic drugs. Eur J Clin Pharmacol 1997;52(5):339–42.

215. Heiman EM. Cardiac toxicity with thioridazine–tricyclic antidepressant combination. J Nerv Ment Dis 1977;165(2):139–43.

216. Colantonio L, Orson J. Hyperthyroidism with normal T4-induction by imipramine. Clin Pharmacol Ther 1975;15:203.

217. Van Der Kuy PH, Van Den Heuvel HA, Kempen RW, Vanmolkot LM. Pharmacokinetic interaction between nortriptyline and terbinafine. Ann Pharmacother 2002;36(11):1712–4.

218. Fugh-Berman A. Herb–drug interactions. Lancet 2000;355(9198):134–8.

Amineptine

General Information

Amineptine is a tricyclic antidepressant that selectively reduces the dopamine reuptake without affecting the uptake of noradrenaline or serotonin (5-HT) (1). In vivo, it increases striatal homovanillic acid concentrations without affecting the concentrations of other metabolites of dopamine, 3,4,dihydroxyphenylacetic acid (DOPAC) and 3-methoxytyramine. However, high doses of amineptine reduced the extracellular DOPAC concentration in the nucleus accumbens but not in the striatum. Long-term treatment with amineptine causes down-regulation of beta-adrenoceptors.

Amineptine has a seven-membered carbon side-chain and is reported to have more stimulant and fewer sedative effects than other tricyclic compounds, possibly owing to differential actions on dopaminergic rather than serotonergic mechanisms. Amineptine appears to have an unusual propensity for causing hepatocellular damage, which may limit its clinical use (2).

Amineptine increases the release and reduces the reuptake of dopamine, and it is therefore not surprising that an amphetamine-like drug dependence has been reported (3–5). A withdrawal syndrome occurs and can be improved by clonidine (SEDA-16, 8).

Organs and Systems

Liver

Of all the tricyclic antidepressants, amineptine appears to be the most likely to cause liver damage. More than 26 cases of toxic hepatitis have been reported in France (6). In most cases, hepatitis occurred within the usual dosage range and recurred on rechallenge. However, in several instances it was reported after overdosage. There are other reports of hepatotoxicity associated with amineptine (7,8).

Skin

Amineptine has been reported (9) to cause particularly active acne, occurring beyond the usual distribution on the body and beyond the usual age, especially in women. The remedy recommended is withdrawal of amineptine.

A rosacea-like eruption on the face of a 76-year-old woman was associated with amineptine; the association was confirmed by re-challenge (10).

Long-Term Effects

Drug dependence

Dependence on amineptine in a patient also taking midazolam has been reported (11).

References

1. Garattini S. Pharmacology of amineptine, an antidepressant agent acting on the dopaminergic system: a review. Int Clin Psychopharmacol 1997;12(Suppl 3):S15–9.

2. Lefebure B, Castot A, Danan G, Elmalem J, Jean-Pastor MJ, Efthymiou ML. Hépatites aux antidépresseurs. (Hepatitis from antidepressants. Evaluation of cases from French Assoc Drug Surveillance Centers Techn Committee. Thérapie) 1984;39(5):509–16.

3. Ginestet D, Cazas O, Branciard M. Deux cas de dependance à l'amineptine. [2 cases of amineptine dependence.] Encephale 1984;10(4):189–91.

4. Castot A, Benzaken C, Wagniart F, Efthymiou ML. Surconsommation d'amineptine. [Amineptin abuse. Analysis of 155 cases. An evaluation of the official cooperative survey of Regional Centers of Pharmacovigilance.] Therapie 1990;45(5):399–405.

5. Bertschy G, Luxembourger I, Bizouard P, Vandel S, Allers G, Volmat R. Dépendance à l'amineptine. [Amineptin dependence. Detection of patients at risk. Report 8 Cases.] Encephale 1990;16(5):405–9.

6. Andrieu J, Doll J, Coffinier C. Hépatite due à l'amineptine: quatre observations. [Hepatitis due to amineptin. 4 Cases.] Gastroenterol Clin Biol 1982;6(11):915–8.

7. Lazaros GA, Stavrinos C, Papatheodoridis GV, Delladetsima JK, Toliopoulos A, Tassopoulos NC. Amineptine induced liver injury. Report of two cases and brief review of the literature. Hepatogastroenterology 1996;43(10):1015–9.

8. Lonardo A, Grisendi A, Mazzone E. One more case of hepatic injury due to amineptine. Ital J Gastroenterol Hepatol 1997;29(6):580–1.

9. Thioly-Bensoussan D, Charpentier A, Triller R, Thioly F, Blanchet P, Tricoire N, Noury JY, Grupper C. Acne iatrogène à l'amineptine (Survector): à propos de 8 cases. [Iatrogenic acne caused by amineptin (Survector). Report of 8 cases.] Ann Dermatol Venereol 1988;115(11):1177–80.

10. Jeanmougin M, Civatte J, Cavelier-Balloy B. Toxidermie rosacéiforme à l'amineptine (Survector). [Rosaceous Drug Eruption Caused by Amineptin (Survector).] Ann Dermatol Venereol 1988;115(11):1185–6.

11. Perera I, Lim L. Amineptine and midazolam dependence. Singapore Med J 1998;39(3):129–31.

Amitriptylinoxide

General Information

Amitriptylinoxide, a metabolite of amitriptyline, has been compared with the parent drug (1). The antidepressant effects were comparable, but the metabolite was thought to have fewer adverse effects, especially cardiotoxic ones. This conclusion was based on the absence of cardiographic abnormalities in 15 patients, but this is a very small series on which to base such a conclusion.

Reference

1. Godt HH, Fredslund-Andersen K, Edlund AH. Amitriptyline N-oxid, et nyt antidepressivum: sammenlignende klinisk vurdering i forhold til amitriptylin. [Amitriptyline N-oxide. A new antidepressant. A clinical double-blind comparison with amitriptyline.] Nord Psykiatr Tidsskr 1971;25(3):237–46.

Amoxapine

General Information

In a major review of amoxapine and its pharmacology it was concluded that it is similar in efficacy and potency to standard tricyclic compounds, with a sedative action intermediate between amitriptyline and imipramine (1).

As previously noted (SEDA-4, 21; SEDA-5, 17), amoxapine has a tricyclic nucleus with a modified piperazine side chain and is closely related to the neuroleptic drug loxapine. In animals, amoxapine has no antiserotonergic properties and less anticholinergic activity than prototype drugs. Peak plasma concentrations are achieved in less than 2 hours, and the half-lives of the parent drug and its two active metabolites are 8, 6, and 30 hours respectively.

Overall, amoxapine appears to have some advantage over other tricyclic antidepressants: possible earlier onset of action and relative freedom from serious cardiotoxic effects. Its major drawbacks are the potential for neuroleptic adverse effects, a high incidence of seizures, deaths in overdose (2), and the possibility of long-term neurological damage.

Amoxapine is less potent than other tricyclic antidepressants, with a therapeutic dosage range of 75–600 mg/day (usually 200–400 mg/day). Clinical effects have not been consistently correlated with plasma concentrations, but amoxapine has similar efficacy to other tricyclic antidepressants in heterogeneous populations of depressed patients. Controlled comparisons have shown that its clinical profile is very similar to that of imipramine (3) and that it is somewhat less sedative than amitriptyline (4–6). In two of these studies (4,6) the

results confirmed the suggestion of a somewhat earlier onset of action.

Amoxapine appears to have the same common adverse effects as other tricyclic compounds, including those attributable to anticholinergic activity. Its structural similarity to the classic neuroleptic drugs appears to confer an additional hazard of adverse effects usually found in that category of drugs, such as galactorrhea and extrapyramidal disorders (SEDA-9, 20). In a study of its potential neuroleptic properties (7), using a radioreceptor assay in vivo and using plasma drawn from patients taking neuroleptic drugs or antidepressants, amoxapine and its metabolite, 7-hydroxyamoxapine, had potent neuroleptic activity. In patients taking amoxapine, there was neuroleptic activity comparable to that in patients taking loxapine, while none of the other tricyclic antidepressants (amitriptyline, nortriptyline, imipramine, and desipramine) had this property. Further reports of adverse effects attributable to its neuroleptic profile continue to appear, including tardive dyskinesia (8,9), acute torticollis (10), and malignant neuroleptic syndrome (SEDA-16, 9; SEDA-17, 18; 11).

Organs and Systems

Metabolism

A further reminder of the structural resemblance of amoxapine to the neuroleptic drugs has been provided by an adverse effect reported with both amoxapine and its close congener, loxapine (12).

- A 49-year-old woman with no history of diabetes was admitted in unexplained hyperglycemic coma (blood glucose 26 mmol/l) while taking lithium 1500 mg/day and loxapine 150 mg/day. She responded to insulin, but insulin responses on testing were not delayed and suggested an iatrogenic rather than a diabetic cause. The fasting glucose fell to 4.2 mmol/l after withdrawal of loxapine but continuing lithium. Two weeks later amoxapine 150 mg/day was started and she became acutely confused, with a serum glucose of 5.7 mmol/l. Two weeks after stopping amoxapine the serum glucose returned to normal.

The authors speculated that a common metabolite of both drugs, 7-hydroxyamoxapine, was responsible for the hyperglycemia, owing to its antidopaminergic properties.

Drug Administration

Drug overdose

A bleak picture has rapidly evolved for amoxapine with regard to its toxicity in overdosage. Over 18 months, 33 cases were reported from Washington, DC, and New Mexico Poison Centers (2). These cases included 5 patients who died and 12 who developed seizures. Thus, the mortality rate of 15% greatly exceeds that of 0.7% for other antidepressants in the same centers, and the seizure rate is 36% compared to 4.3%. The authors noted that "the striking CNS toxicity of amoxapine overdose with frequent, persistent, and poorly controlled seizure activity is disconcerting." In a retrospective comparison of deaths from antidepressant overdosage in Scotland, England, and Wales between 1987 and 1992, the number of deaths per million prescriptions of amoxapine was significantly higher than expected (13).

Renal damage has occurred in cases of amoxapine overdose.

● Acute renal insufficiency and rhabdomyolysis were reported in a 27-year-old man who took 1–2 g of amoxapine (14). The authors recommended aggressive volume expansion and diuresis with loop diuretics, because of the futility of hemodialysis.

A 24-year-old man took 4 g of amoxapine and developed gross hematuria and a high serum uric acid concentration on the second day of hospitalization (15). As in previously reported cases, serum creatine phosphokinase was grossly raised. The patient remained obtunded and stuporose for 7 days but eventually recovered.

References

1. Jue SG, Dawson GW, Brogden RN. Amoxapine: a review of its pharmacology and efficacy in depressed states. Drugs 1982;24(1):1–23.
2. Litovitz TL, Troutman WG. Amoxapine overdose. Seizures and fatalities. JAMA 1983;250(8):1069–71.
3. Bagodia VN, Shah LP, Pradan PV, Gada MT. A double-blind controlled study of amoxapine and imipramine in cases of depression. Curr Ther Res Clin Exp 1979;26:417.
4. Sethi BB, Sharma I, Singh H, Metha VK. Amoxapine and amitriptyline: a double-blind study in depressed patients. Curr Ther Res Clin Exp 1979;25:726.
5. Fruensgaard K, Hansen CE, Korsgaard S, Nymgaard K, Vaag UH. Amoxapine versus amitriptyline in endogenous depression. A double-blind study. Acta Psychiatr Scand 1979;59(5):502–8.
6. Kaumeier HS, Haase HJ. A double-blind comparison between amoxapine and amitriptyline in depressed in-patients. Int J Clin Pharmacol Ther Toxicol 1980;18(4):177–84.
7. Cohen BM, Harris PQ, Altesman RI, Cole JO. Amoxapine: neuroleptic as well as antidepressant? Am J Psychiatry 1982;139(9):1165–7.
8. Huang CC. Persistent tardive dyskinesia associated with amoxapine therapy. Am J Psychiatry 1986;143(8):1069–70.
9. Price WA, Giannini AJ. Withdrawal dyskinesia following amoxapine therapy. J Clin Psychiatry 1986;47(6):329–30.
10. Matot JP, Ziegler M, Olie JP, Rondot P. Amoxapine. Un antidepresseur responsable d'effets secondaires extrapyramidaux?. [Amoxapine. An antidepressant responsible for extrapyramidal side effects?.] Therapie 1985;40(3):187–90.
11. Burch EA Jr, Downs J. Development of neuroleptic malignant syndrome during simultaneous amoxapine treatment and alprazolam discontinuation. J Clin Psychopharmacol 1987;7(1):55–6.
12. Tollefson G, Lesar T. Nonketotic hyperglycemia associated with loxapine and amoxapine: case report. J Clin Psychiatry 1983;44(9):347–8.
13. Henry JA, Alexander CA, Sener EK. Relative mortality from overdose of antidepressants. BMJ 1995;310(6974):221–4.
14. Jennings AE, Levey AS, Harrington JT. Amoxapine-associated acute renal failure. Arch Intern Med 1983;143(8):1525–7.
15. Thompson M, Dempsey W. Hyperuricemia, renal failure, and elevated creatine phosphokinase after amoxapine overdose. Clin Pharm 1983;2(6):579–81.

Butriptyline

General Information

Butriptyline is the isobutyl side-chain homologue of amitriptyline. Its adverse effects, reported in two open studies, were no different from those of other tricyclic antidepressants (1,2).

References

1. Madalena JC, De Matos HG. Preliminary clinical observations with butriptyline. J Med 1971;2(5):322–6.
2. Grivois H. Butriptyline. A new antidepressant compound. J Med 1971;2(5):276–89.

Clomipramine

General Information

Clomipramine is the imipramine analogue of chlorpromazine. However, while the difference between chlorpromazine and promazine is large, adding a chloride atom to imipramine hardly affects its actions. Most trials have failed to show any superiority of the chlorinated compound over imipramine. The adverse effects profile is similar (1), but drowsiness, confusion, and "feeling awful" are commonly reported (2).

In a controlled comparison between clomipramine and amitriptyline, the former caused adverse effects more often, especially drowsiness (3). Overdose toxicity is the same as with other tricyclic antidepressants (4); fatal interactions with monoamine oxidase (MAO) inhibitors have been reported (SEDA-18, 16; 5). Altogether, toxic effects are not substantially different.

Besides depression, clomipramine is also widely used in the treatment of phobic and obsessive-compulsive disorders (6–8) and in panic disorders (9).

Organs and Systems

Cardiovascular

Venous thrombosis is a recognized complication of tricyclic antidepressants. Thrombosis of the cerebral veins occurred in a 61-year-old woman after intravenous clomipramine, and the authors suggested that the risk may be greater when the intravenous route is used (10).

Intravenous administration invariably produced electrocardiographic changes, sometimes slow to reverse, in elderly patients (11).

At therapeutic doses, tricyclic antidepressants can cause postural hypotension, but they are regarded as being safe in patients who require general anesthesia. However, hypotension during surgery has been associated with clomipramine (12).

- A 57-year-old man due to undergo mitral valve surgery took clomipramine (150 mg at night) up to the night before surgery. His blood pressure before induction with thiopental (250 mg) and fentanyl (250 micrograms) was 105/65 mmHg, with a heart rate of 70 beats/minute. Anesthesia was maintained with isofluorothane, and 45 minutes after induction, his systolic blood pressure fell to 90 mmHg. Ephedrine (30 mg total), phenylephrine (500 micrograms total), or dopamine (10 micrograms/kg/minute) did not increase the blood pressure. After sternotomy, his systolic blood pressure fell to 55 mmHg and his pulse rate to 60 beats/minute, and he had third-degree atrioventricular block. Further ephedrine, phenylephrine, and adrenaline were without effect. During cardiopulmonary bypass, a noradrenaline infusion was started (0.2 micrograms/kg/minute) and isofluorothane was withdrawn. After he had been weaned from bypass the noradrenaline infusion was continued at a dose of 0.2–0.8 micrograms/kg/minute, sufficient to maintain the systolic blood pressure at 90–100 mmHg. After the operation, clomipramine was withheld and the noradrenaline infusion tapered off, and 3 days later the hypotension had resolved.

The hypotension in this case was severe and refractory to noradrenergic stimulation, perhaps because of the alpha$_1$-adrenoceptor antagonist properties of clomipramine. The fall in systolic blood pressure was accompanied by a paradoxical fall in heart rate, perhaps because the anticholinergic effect of clomipramine removed the effect of vagal tone on the resting heart rate. It seems likely that the hypotensive effect of clomipramine was potentiated by general anesthesia; however, such a reaction is rare and the underlying cardiac problem may have contributed to this severe adverse reaction. This case reinforces current advice that tricyclic antidepressants are best avoided in patients with significant cardiac disease.

Nervous system

Use of the intravenous route is fraught with danger and without any demonstrable advantages.

- A 31-year-old woman who received clomipramine intravenously 300 mg/day developed seizures and cardiac arrest on the 15th day; she was successfully resuscitated (13).

Seizures occurred in four of 50 patients receiving intravenous treatment. Those vulnerable to this complication were not identified in advance by prescreening electroencephalography (14).

Endocrine

Clomipramine has also aroused interest because of an action on prolactin release, which occurs with major tranquillizers but not with other tricyclic antidepressants (15). This action of clomipramine is related to its chemical structure and reflects a greater effect on dopamine metabolism and serotonin uptake compared with other antidepressants.

Sexual function

Delayed or complete abolition of ejaculation is attributed to strong 5-HT re-uptake blockade but perhaps with additional alpha-adrenoceptor-blocking activity (16). Impotence may be due to a ganglionic blocking action.

Four patients taking up to 100 mg/day developed uncontrollable yawning, in three cases (two men and one woman) associated with sexual arousal, in two instances with spontaneous orgasm (17). In all four patients the symptoms disappeared after withdrawal.

References

1. Collins GH. The use of parenteral and oral chlorimipramine (Anafranil) in the treatment of depressive states. Br J Psychiatry 1973;122(567):189–90.
2. Capstick N. Psychiatric side-effects of clomipramine (Anafranil). J Int Med Res 1973;1:444.
3. Rickels K, Weise CC, Csanalosi I, Chung HR, Feldman HS, Rosenfeld H, Whalen EM. Clomipramine and amitriptyline in depressed outpatients. A controlled study. Psychopharmacologia 1974;34(4):361–76.
4. Haqqani MT, Gutteridge DR. Two cases of clomipramine hydrochloride (Anafranil) poisoning. Forensic Sci 1974;3(1):83–7.
5. Beaumont G. Drug interactions with clomipramine (Anafranil). J Int Med Res 1973;1:480.
6. Yaryura-Tobias JA, Neziroglu F, Bergman L. Chlorimipramine for obsessive-compulsive neurosis: an organic approach. Curr Ther Res 1976;20:541.
7. Silva FR, Wijewickrama HS. Clomipramine in phobic and obsessional states: preliminary report. NZ Med J 1976;84(567):4–6.
8. Kelly MW, Myers CW. Clomipramine: a tricyclic antidepressant effective in obsessive compulsive disorder. DICP 1990;24(7–8):739–44.
9. Modigh K, Westberg P, Eriksson E. Superiority of clomipramine over imipramine in the treatment of panic disorder: a placebo-controlled trial. J Clin Psychopharmacol 1992;12(4):251–61.
10. Eikmeier G, Kuhlmann R, Gastpar M. Thrombosis of cerebral veins following intravenous application of clomipramine. J Neurol Neurosurg Psychiatry 1988;51(11):1461.
11. Symes MH. Cardiovascular effects of clomipramine (Anafranil). J Int Med Res 1973;1:460.
12. Malan TP Jr, Nolan PE, Lichtenthal PR, Polson JS, Tebich SL, Bose RK, Copeland JG 3rd. Severe, refractory hypotension during anesthesia in a patient on

chronic clomipramine therapy. Anesthesiology 2001;95(1):264–6.

13. Singh G. Cardiac arrest with clomipramine. BMJ 1972;3(828):698.
14. Dickson J. Neurological and EEG effects of clomipramine (Anafranil). J Int Med Res 1973;1:449.
15. Jones RB, Luscombe DK, Groom GV. Plasma prolactin concentrations in normal subjects and depressive patients following oral clomipramine. Postgrad Med J 1977;53(Suppl. 4):166–71.
16. Beaumont G. Sexual side-effects of clomipramine (Anafranil). J Int Med Res 1973;1:469.
17. McLean JD, Forsythe RG, Kapkin IA. Unusual side effects of clomipramine associated with yawning. Can J Psychiatry 1983;28(7):569–70.

Dibenzepin

General Information

Dibenzepin is a 6,7,6 tricyclic compound of the dibenzodiazepine type. A comparison with imipramine was said to show equal efficacy; adverse effects were comparable in type and degree (1).

Reference

1. Sim M, Armitage GH, Davies WH, Gordon EB. The treatment of depressive states: a comparative trial of dibenzepin (Noveril) with imipramine (Tofranil). Clin Trial J 1971;1:29.

Dimetacrine

General Information

Dimetacrine is a 6,6,6 tricyclic acridine derivative. In a double-blind study, dimetacrine was less effective than imipramine and produced more weight loss and abnormal liver function tests more often (1).

Reference

1. Abuzzahab FS Sr. A double-blind investigation of dimethacrine versus imipramine in hospitalized depressive states. Int J Clin Pharmacol 1973;8(3):244–53.

Dosulepin

General Information

Dosulepin (dothiepin) is a tricyclic antidepressant that has been available in Europe for over 30 years and is particularly popular in the UK. Its animal and clinical pharmacology has been described in an extensive review (1). It appears to be equivalent to amitriptyline, although few studies have reported dosages above 225 mg/day. Dosulepin is as effective and sedative as amitriptyline, with somewhat fewer anticholinergic adverse effects in several studies. However, it does not appear to have been compared with other less sedative or anticholinergic tricyclic compounds or second-generation drugs. Although it is claimed to have fewer cardiovascular effects, this has not been well substantiated in controlled comparative studies, and the cardiovascular and other effects of overdosage appear to be identical for all tricyclic compounds. Fetal tachydysrhythmias were believed to be caused by maternal ingestion of dosulepin (2).

References

1 Goldstein BJ, Claghorn JL. An overview of seventeen years of experience with dothiepin in the treatment of depression in Europe. J Clin Psychiatry 1980;41(12 Pt 2):64–70.
2 Prentice A, Brown R. Fetal tachyarrhythmia and maternal antidepressant treatment. BMJ 1989;298(6667):190.

Doxepin

General Information

Doxepin is a tricyclic antidepressant that has also been used topically in the treatment of atopic dermatitis and other forms of eczematous dermatitis. It causes the adverse effects that one would expect, and systemic effects can result from absorption after topical administration.

Organs and Systems

Nervous system

Short-term treatment with 5% doxepin cream caused mild transient drowsiness in 16–28% of patients, severe enough to necessitate withdrawal in about 2% (1). In addition, 5% of treated patients noted a dry mouth. Excessive use in children can result in intoxication, with seizures, respiratory depression, electrocardiographic abnormalities, and coma (2).

References

1. Drake LA, Millikan LEDoxepin Study Group. The antipruritic effect of 5% doxepin cream in patients with eczematous dermatitis. Arch Dermatol 1995;131(12):1403–8.
2. Vo MY, Williamsen AR, Wasserman GS, Duthie SE. Toxic reaction from topically applied doxepin in a child with eczema. Arch Dermatol 1995;131(12):1467–8.

Imipraminoxide

General Information

Imipraminoxide is an imipramine metabolite. In a comparison with the parent drug, efficacy was identical; adverse effects were of the same type but possibly less frequent (1).

Reference

1. Rapp W, Noren MB, Pedersen F. Comparative trial of imipramine N-oxide and imipramine in the treatment of outpatients with depressive syndromes. Acta Psychiatr Scand 1973;49(1):77–90.

Lofepramine

General Information

Lofepramine is an imipramine analogue whose animal pharmacology suggests low toxicity (1). Its clinical efficacy is similar to that of imipramine and amitriptyline. Patients on lofepramine report less dryness of the mouth, fewer disturbances of accommodation, and less drowsiness, but a greater incidence of tremor (2).

Lofepramine has apparently not been compared with desipramine, to which it is mainly metabolized. The claim that lofepramine has fewer adverse effects than traditional tricyclic compounds must be viewed with skepticism (SEDA-11, 14). However, lofepramine is safer than conventional tricyclic antidepressants in overdose.

References

1. d'Elia G, Borg S, Hermann L, Lundin G, Perris C, Raotma H, Roman G, Siwers B. Comparative clinical evaluation of lofepramine and imipramine. Psychiatric aspects. Acta Psychiatr Scand 1977;55(1):10–20.
2. Bernik V, Maia E. Therapeutical and clinical evaluation of a new antidepressant drug, lofepramin (EMD 31.802), in comparison to amitriptyline in the treatment of depression Rev Bras Clin Ter 1978;7:43.

Melitracen

General Information

Melitracen is structurally and pharmacologically related to imipramine, with two methyl groups attached to the central ring. It has similar efficacy to amitriptyline, with a somewhat more rapid effect and similar adverse effects (1).

Reference

1. Francesconi G, LoCascio A, Mellina S, et al. Controlled comparison of melitracen and amitriptyline in depressed patients. Curr Ther Res Clin Exp 1976;20:529.

Nortriptyline

General Information

Nortriptyline is a tricyclic compound which differs from amitriptyline in being monodemethylated (1).

Reference

1. Jerling M, Bertilsson L, Sjoqvist F. The use of therapeutic drug monitoring data to document kinetic drug interactions: an example with amitriptyline and nortriptyline. Ther Drug Monit 1994;16(1):1–12.

Noxiptiline

General Information

Noxiptiline is a tricyclic compound that differs from amitriptyline in having a longer side-chain, containing an oxyimino-group.

In a double-blind comparison of noxiptiline ($n = 30$) and amitriptyline ($n = 32$) in hospitalized patients with primary depressive illness for 3–6 weeks there were no significant differences, but noxiptiline had a faster onset of action (1). A comparison with imipramine also showed no difference in therapeutic efficacy (2).

As regards adverse effects, noxiptiline seemed to cause more mental symptoms, such as delirium (3). In one study in 44 patients, 9 developed delirium, there were mild manic symptoms in 6, dry mouth in 21, disturbed micturition in 2, and difficulty in accommodation in 3 (2), the last 3 being evidence of anticholinergic effects.

References

1. Lingjaerde O, Asker T, Bugge A, Engstrand E, Eide A, Grinaker H, Herlofsen H, Ose E, Ofsti E. Noxiptilin (Agedal)—a new tricyclic antidepressant with a faster onset of action? A double-blind, multicentre comparison with amitriptyline. Pharmakopsychiatr Neuropsychopharmakol 1975;8(1):26–35.

2. Petrilowitsch N, Boeters U, Grahmann H, Reckel K, Spiegelberg U. Multiclinic trials of antidepressant BAY 1521. Int Pharmacopsychiatry (Basel) 1969;3:42.

3. Berner P, Guss H, Hofmann G, Kryspin-Exner K, Kufferle B. Doppelblindprüfung von Antidepressiva. [Double-blind study of the antidepressive drugs (noxiptilin–imipramine).] Arzneimittelforschung 1971;21(5):638–40.

Trimipramine

General Information

Trimipramine is a sedating tricyclic antidepressant that has been used as a hypnotic (1); it shares this activity with other drugs of its class, notably amitriptyline, dosulepin, doxepin, and trazodone, and with the tetracyclics mianserin and mirtazapine. Trimipramine may be preferred for this purpose, since it has less effect on sleep architecture, including REM sleep (2), and has only a modest propensity to produce rebound insomnia in a subset of patients (3). Sedative antidepressants may be particularly appropriate for individuals at risk of benzodiazepine abuse and patients with chronic pain (4). The usual pattern of tricyclic adverse effects, especially antimuscarinic and hypotensive effects and weight gain, can be expected. Some authors, enthusiastic about GABA enhancers, contend that antidepressants are not useful hypnotic alternatives (5).

Antidepressant drugs of various classes (tricyclics, monoamine oxidase inhibitors, SSRIs) have broad efficacy in generalized anxiety and in panic disorder, for which they are the treatments of choice (6,7). While not likely to cause benzodiazepine-like dependence or abuse, they do have a significant therapeutic latency, and the older drugs are very toxic in overdose.

References

1. Settle EC Jr, Ayd FJ Jr. Trimipramine: twenty years' worldwide clinical experience. J Clin Psychiatry 1980;41(8):266–74.

2. Dunleavy DL, Brezinova V, Oswald I, Maclean AW, Tinker M. Changes during weeks in effects of tricyclic drugs on the human sleeping brain. Br J Psychiatry 1972;120(559):663–72.

3. Hohagen F, Montero RF, Weiss E, Lis S, Schonbrunn E, Dressing H, Riemann D, Berger M. Treatment of primary insomnia with trimipramine: an alternative to benzodiazepine hypnotics? Eur Arch Psychiatry Clin Neurosci 1994;244(2):65–72.

4. Stiefel F, Stagno D. Management of insomnia in patients with chronic pain conditions. CNS Drugs 2004;18(5):285–96.

5. Laux G. Current status of treatment with benzodiazepines. Nervenarzt 1995;66(5):311–22.

6. Lader M. Psychiatric disorders. In: Speight T, Holford N, editors. Avery's Drug Treatment. 4th ed.. Auckland: ADIS International Press, 1997:1437.

7. Menkes DB. Antidepressant drugs. New Ethicals 1994;31:101.

SELECTIVE SEROTONIN RE-UPTAKE INHIBITORS (SSRIs)

General Information

The selective serotonin re-uptake inhibitors (SSRIs) that are currently available are fluoxetine, fluvoxamine, paroxetine, sertraline, citalopram, and escitalopram. They are widely marketed and are in many countries a major alternative to tricyclic antidepressants in the treatment of depression. The SSRIs are structurally diverse, but they are all inhibitors of serotonin uptake, with much less effect on noradrenaline. They have slight or no inhibitory effect on histaminergic, adrenergic, serotonergic, dopaminergic, and cholinergic receptors (1).

The SSRIs are eliminated mainly through hepatic metabolism. Their half-lives are about 1 day for fluvoxamine, paroxetine, and sertraline (2–5), 1.5 days for citalopram (6), and 2–3 days for fluoxetine (7). Norfluoxetine, the main metabolite of fluoxetine, is a potent and selective serotonin re-uptake inhibitor, and since this metabolite also has a very long half-life (7–15 days) it contributes significantly to the clinical effect. Norsertraline, the desmethylated metabolite of sertraline, is also an inhibitor of serotonin re-uptake, but with much lower potency than sertraline; although its half-life is about 2.5 times longer than that of sertraline, it is not considered to contribute to the clinical effect (3). The main metabolite of citalopram, desmethylcitalopram, is pharmacologically active but has much lower potency than the parent compound (1). The metabolites of fluvoxamine and paroxetine are inactive with respect to monoamine uptake (2). Escitalopram oxalate is the *S*-enantiomer of citalopram (8). The therapeutic activity of citalopram resides in the *S*-isomer and escitalopram binds with high affinity to the human serotonin transporter; *R*-citalopram is about 30-fold less potent. The half-life of escitalopram is 27–32 hours. Escitalopram and citalopram have negligible effects on CYP isozymes.

The major advantages of SSRIs over the tricyclic antidepressants are their less pronounced anticholinergic adverse effects and lack of severe cardiotoxicity. However, some studies have shown some degree of nervousness or agitation, sleep disturbances, gastrointestinal symptoms, and perhaps sexual adverse effects more commonly in patients treated with SSRIs than in those treated with tricyclic antidepressants. SSRIs may also be associated with an increased risk of suicide, particularly in children under 16 (9).

There seems to be little difference between SSRIs with respect to frequency and severity of adverse effects. The most common adverse effects are gastrointestinal disturbances (nausea, diarrhea/loose stools, constipation; incidence 6–37%), nervous system effects (insomnia, somnolence, tremor, dizziness and headache; 11–26%), and effects on the autonomic nervous system (dryness of the mouth and sweating; 9–30%) (2,10). Weight gain or weight loss have been documented relatively infrequently (2). A high frequency of sexual disturbances has been reported. SSRIs may selectively inhibit hepatic enzymes, causing pharmacokinetic interactions with other drugs that are metabolized by these enzymes; a pharmacodynamic interaction can also occur when SSRIs are given in combination with other serotonergic drugs, which may give rise to serotonergic hyperstimulation, the "serotonin syndrome." The SSRIs are safer than the tricyclic antidepressants in overdose.

A meta-analysis of 20 short-term studies of five SSRIs (citalopram, fluoxetine, fluvoxamine, paroxetine, and sertraline) has been published (11). There were no overall differences in efficacy, but fluoxetine had a slower onset of action. Analysis of tolerability showed the expected adverse effects profile, the most common adverse event being nausea, followed by headache, dizziness, and tremor. The rate of withdrawal from treatment because of adverse effects was significantly greater with fluvoxamine (RR = 1.9; CI = 1.2, 3.0). Data available from the UK Committee on Safety of Medicines suggested that withdrawal reactions were most common with paroxetine and least common with fluoxetine (presumably because of the long half-life of its active metabolite norfluoxetine). There were more gastrointestinal reactions to fluvoxamine and paroxetine. This pattern was reflected in prescription-event monitoring data. Citalopram and sertraline were least likely to cause drug interactions, but citalopram was implicated more often in fatal overdoses.

Organs and Systems

Nervous system

Movement disorders

Extrapyramidal symptoms (including akathisia, dystonia, dyskinesia, tardive dyskinesia, parkinsonism, and bruxism) have been reported in association with SSRIs, especially in the presence of predisposing factors (SEDA-14, 14; 12). Current data suggest that SSRIs should be used with caution in patients with parkinsonism (see the monograph on fluoxetine). Concomitant treatment with neuroleptic drugs and high concentrations of SSRIs seems to predispose to extrapyramidal symptoms. Elderly patients and women are also at increased risk.

It is believed that SSRIs produce movement disorders by facilitating inhibitory serotonin interactions with dopamine pathways. While all SSRIs are potent inhibitors of serotonin re-uptake, they have other pharmacological actions that might contribute to their clinical profile. Sertraline has an appreciable affinity for the dopamine re-uptake site, and for this reason might be presumed less likely to cause movement disorders than other SSRIs. However, there is little clinical evidence to support this suggestion and a case of sertraline-induced parkinsonism has been reported (13).

SSRI-induced movement disorders have been comprehensively reviewed (14). The use of SSRIs was associated with a range of movement disorders, the most frequent of which was akathisia. However, clinician-based reports of adverse events solicited from SSRI manufacturers suggested that parkinsonism might occur with an equal frequency but with a later onset during treatment. As suggested before (SEDA-22, 12), concomitant treatment with antipsychotic drugs and lithium, as well as pre-existing brain damage, predisposes to the development of movement disorders with SSRIs. Case reports have suggested that akathisia can be associated with suicidal impulses.

Spontaneous reports received by the Netherlands Pharmacovigilance Foundation between 1985 and 1999 have been analysed in a case-control study (15). Relative to other antidepressants, SSRIs were about twice as likely to be implicated in spontaneous reports of extrapyramidal reactions (OR = 2.2; 95% CI = 1.2, 3.9). The risk was greater in patients who were also taking neuroleptic drugs. This result suggests that SSRIs have a modestly increased risk of producing extrapyramidal reactions compared with other antidepressants. However, increased reporting can be influenced by increased awareness. In addition, no account was apparently taken in this study of relative prescription rates of different antidepressants.

Withdrawal of SSRIs usually results in remission of symptoms of extrapyramidal movement disorders. However, occasionally, SSRIs can unmask a vulnerability to Parkinson's disease, and can also worsen established Parkinson's disease (SEDA-22, 23) (see monographs on citalopram and sertraline) (16).

Serotonin syndrome

The serotonin syndrome is a well-established complication of SSRI treatment. It is usually associated with high doses of SSRIs or the use of SSRIs in combination with other serotonin-potentiating agents, such as monoamine oxidase inhibitors (SEDA-22, 24).

- A 45-year-old man had definite symptoms of serotonin toxicity (hypomania, myoclonus, sweating, and shivering), first when taking a low therapeutic dose of citalopram (20 mg/day) and then with low-dose sertraline (25 mg/day); he was also taking zolpidem (17).

The authors speculated that the combination of zolpidem with an SSRI might have predisposed to the serotonin syndrome. It is also possible that some people (for example poor metabolizers) are idiosyncratically vulnerable to serotonin toxicity at low doses of SSRIs.

Sleep

SSRIs can cause insomnia and daytime somnolence; however, the symptoms seem to reflect a sleep–wake cycle disorder. It is conceivable that disruptions in the normal pattern of melatonin secretion, particularly a delay in the normal early morning fall in plasma concentrations could be involved in the pathophysiology of these symptoms. The fact that fluvoxamine is the SSRI that is most likely to cause sleep disorders supports a role of melatonin, which is increased by fluvoxamine but not by other SSRIs (18).

Paroxetine has been associated with sleep-walking (somnambulism) (19).

- A 61-year-old woman taking paroxetine 10 mg/day for depression had difficulty falling asleep, but there was no personal or family history of parasomnia. After 2 weeks the paroxetine was increased to 20 mg/day and 1 week later she was noted by her husband to be sleep-walking and trying at times to leave the house. When wakened she was confused and had no memory of the event. The paroxetine was withdrawn; the sleep-walking stopped and did not recur.

Other classes of antidepressant drugs can sometimes cause somnambulism, and it seems likely that paroxetine provoked this rare adverse effect. Sleep-walking is thought to be initiated during slow-wave sleep, after which partial arousals activate motor behaviors in the absence of full consciousness. The disrupting effects of antidepressants on sleep architecture might lead to somnambulism in pre-disposed individuals.

Sensory systems

Tricyclic antidepressants can precipitate acute glaucoma through their anticholinergic effects. There are also reports that SSRIs can cause acute glaucoma, presumably by pupillary dilatation (see the monograph on paroxetine and fluoxetine).

Psychological, psychiatric

Like other antidepressants, SSRIs are occasionally associated with manic episodes, even in patients with no history of bipolar disorder. Some argue that the affected patients may have had an underlying predisposition to bipolar illness.

In a retrospective chart review of 167 patients with a variety of anxiety disorders, excluding patients with evidence of current or previous mood disorder, manic episodes were recorded in five patients, a rate of 3% (20). While this might suggest a clear effect of SSRIs to induce mania, two of the patients were taking clomipramine, a tricyclic antidepressant, albeit a potent serotonin re-uptake inhibitor. In addition, all the affected patients had additional diagnoses of histrionic or borderline personality disorder, known to be associated with mood instability. It is still therefore plausible that SSRIs cause mania only in patients with an underlying predisposition, although this may be more subtle than a personal or family history of bipolar illness.

Visual hallucinations are rare during antidepressant drug treatment, except after overdosage.

- A 38-year-old man with a history of chronic depression developed on waking visual hallucinations of different geometric shapes after treatment with both sertraline and fluoxetine. Eventually he responded to treatment with nefazodone, which did not cause hallucinations (21).

Many drugs that cause visual hallucinations (for example lysergic acid diethylamide) have agonist activity at 5-HT$_2$ receptors, and it is conceivable that in these patients both sertraline and fluoxetine caused sufficient activation of post-synaptic 5-HT$_2$ receptors to produce this visual disturbance. Although nefazodone can increase 5-HT neurotransmission, it is also a 5-HT$_2$-receptor antagonist, so presumably does not activate 5-HT$_2$ receptors.

Suicidal behavior

DoTS classification:
Reaction: Suicide risk due to SSRIs
Dose-relation: Collateral
Time-course: Intermediate
Susceptibility factors: Age (adolescents and children)

Over the last decade there has been a debate as to whether SSRIs might increase the risk of suicide in certain individuals. Some patients can respond to SSRIs by becoming agitated and restless and developing symptoms that resemble akathisia. Case reports have suggested that adverse effects of this type could underlie an increased risk of self-harm and aggression. However, results from the placebo-controlled randomized trials carried out for regulatory purposes have not supported the proposal that SSRIs increase the risk of suicide of suicidal behavior.

A review of the database of the Food and Drug Administration showed 77 suicides among 48 277 patients who had participated in placebo-controlled trials of antidepressants (22). The rate of suicide with SSRIs (0.59%; CI = 0.31, 0.87) was similar to that of other antidepressants (0.76%; CI = 0.49, 1.03) and placebo (0.45%; CI = 0.01, 0.89). While these data are reassuring, patients considered clinically at high risk of suicide were excluded from the trials and the patients received a greater degree of supervision than would occur in routine practice.

In a re-analysis of antidepressant trial data there was an odds ratio for a suicidal act while taking SSRIs relative to placebo of 2.0 (CI = 1.2, 3.3), while the risk for completed suicide on SSRIs, although raised, had wide confidence intervals (RR = 3.1; CI = 0.4, 23.1) (23). One of the author's arguments was that in previous analyses suicidal behaviors occurring during placebo wash-out have been misclassified as happening during placebo treatment. Separation of these two classes of event allows a pro-suicidal effect of SSRIs to be revealed.

In a matched case-control, primary-care study of over 150 000 patients who received at least one prescription for an antidepressant between 1993 and 1999, in which dosulepin was used as the reference standard, there was no relative increased risk of non-fatal self-harm for fluoxetine (OR = 1.16; 95% CI = 0.90, 1.50), while the risk for paroxetine approached significance (OR = 29; 95% CI = 0.97, 1.70) (24). The authors suggested that the latter finding might have been due to uncontrolled confounding by severity of depression or apparent suicide risk. In the small number of cases of fatal suicide there was no relation with any particular class of medication. However, the strong relation between the risk of suicidal behavior and the initiation of treatment was noteworthy; thus, in the first 9 days of treatment the risk of non-fatal suicidal behavior was increased four-fold over later weeks of therapy (OR = 4.07; 95%CI = 2.89, 5.74), while for fatal suicide the risk was over 30 times greater (OR = 38; 95%CI = 6, 231). This relation was independent of the class of medication prescribed.

These findings suggest that in the routine clinical management of depression in primary care, the risk of self-harm is not strongly associated with any particular class of antidepressant. In the absence of a placebo group we cannot say whether antidepressants might increase the risk of self-harm relative to placebo; conceivably, of course, they might reduce it. However, it does seem clear that there is a strong link between initiation of antidepressant drug treatment and both fatal and non-fatal self-harm. This fits in with the well-known clinical impression that the start of antidepressant treatment is a time of increased risk of suicidal behavior (25). There are several possible explanations for this phenomenon; for example, (i) antidepressant treatment increases the risk of suicidal behavior directly; (ii) antidepressants improve motor retardation before alleviating depressed mood, thereby increasing the probability that patients will act on their pre-existing suicidal thinking; (iii) patients only present for treatment when they are at a particularly low ebb, and the failure of antidepressant medication to have a quick onset of action adds to feelings of hopelessness and despair. Whatever the explanation, it is clearly prudent to monitor depressed patients closely during the first few weeks of treatment, while mood is still low and suicidal feelings have not resolved.

The UK Committee of Safety of Medicines has previously warned that paroxetine appeared to be no more effective than placebo in the treatment of depression in adolescents and might be associated with a greater risk of self harm (SEDA-28, 16). In a meta-analysis of both published and unpublished placebo-controlled trials of SSRIs in childhood and adolescent depression, only fluoxetine seemed clearly to be associated with a positive benefit–harm balance (26). The evidence of efficacy for sertraline and citalopram was doubtful, while the risk of serious adverse events was significantly increased. Additionally, for both drugs the risk of suicidal behavior was numerically increased. In regard to venlafaxine, the risk of suicidal behavior was significantly greater than placebo.

Subsequently, the UK Committee on Safety of Medicines issued guidance for practitioners, indicating that with the exception of fluoxetine, the benefit–harm balance of SSRIs and venlafaxine in depressed children and adolescents was unfavorable (27). A similar conclusion was drawn concerning mirtazapine. It is, however, puzzling that the therapeutic effects of fluoxetine should be quite so different from compounds that are similar pharmacologically. One important issue might be the relevance of regulatory trials of antidepressants to real-world treatment. Placebo-controlled trials in depression often exclude the more depressed patients as well as those with worrying suicidal ideation. Hence, it may be more difficult to show a specific benefit of a drug. However, the meta-analysis cited above (Whittington 1341) also suggested that young people might be more susceptible than adults to any effect of antidepressants to induce suicidal ideation. SSRIs are also used in adolescents to treat anxiety disorders, such as obsessive–compulsive disorder. Whether the risk of suicidal behavior with SSRIs is also increased in this group is uncertain, but clearly caution is indicated.

Another large systematic review was carried out by the UK Medicines Control Agency (MCA), focusing on all published and unpublished regulatory trials of SSRIs in depressed patients aged 18 and over (28). The MCA found no evidence that, relative to placebo, any individual SSRI increased the risk of fatal or non-fatal self-harm, but they concluded that "a modest increase in the risk of suicidal thoughts and self-harm for SSRIs compared with placebo cannot be ruled out". There was also no evidence from epidemiological studies that the prescribing of SSRIs had led to an increase in suicide rate, although such studies are difficult to interpret, because suicide rates are affected by many factors that are likely to have a greater impact than antidepressants.

Overall the randomized studies suggest that the risk of suicide and suicidal behavior in patients taking placebo is numerically less than in patients taking active antidepressants. Since on clinical rating scales, antidepressant drugs usually reduce suicidal ideation more than placebo, this is a paradoxical observation. It might, however, reflect the old adage that the risk of suicide is greatest during the early stages of antidepressant treatment, when motor retardation has remitted but depressed mood and suicidal thinking persist. Such an effect might be exaggerated with modern antidepressants, which are more activating than tricyclic antidepressants. The need for caution in both clinical practice and trial evaluation is underlined by a warning from the Committee of Safety of Medicines that paroxetine is no more effective than placebo in the treatment of depression in adolescence, and may also be associated with a greater risk of self-harm and suicidal behavior (29).

Furthermore, the data provided by the MCA and published subsequently suggest that in people aged 18 years and older, SSRIs do not increase the risk of completed suicide relative to placebo. In addition, there is no difference in the risk of suicide or self-harm in patients taking SSRIs compared with those taking tricyclic antidepressants (30). It is possible, however, that when all the data for individual SSRIs are combined, SSRIs as a group might produce a small increase in the risk of self-harm relative to placebo (31,32). It has been calculated that, if this is the case, on the present evidence the number needed to treat for harm (NNT$_H$, i.e. to produce one suicide) is 759; this contrasts with the much more favorable number need to treat for benefit (NNT$_B$) in depression, which is 4–7 (Gunnell 385). Thus, particularly in patients with significant depressive symptoms and functional disability, the benefit–harm balance of SSRIs appears to be favorable. It is also worth noting that the possible increase in the risk of self-harm appears to be similar between tricyclic antidepressants and SSRIs, suggesting that any putative mechanism is likely to be common to all antidepressant medications.

Endocrine

Serotonin pathways are involved in the regulation of prolactin secretion. Amenorrhea, galactorrhea, and hyperprolactinemia have been reported in patients taking SSRIs (see the monographs on fluoxetine, fluvoxamine, paroxetine, and sertraline).

Electrolyte balance

SSRIs can reportedly cause hyponatremia (SEDA-18, 20; SEDA-26, 13). In a case-control study of hyponatremia in 39 071 psychiatric inpatients and outpatients, the incidence of antidepressant-induced hyponatremia was 2.1% (33). SSRI users had a three times higher risk of developing hyponatremia relative to users of other antidepressant drugs (OR = 3.1; 95% CI = 1.3, 8.6). Additional risk factors included older age and concomitant treatment with diuretics.

SSRIs can cause hyponatremia in elderly patients (SEDA-18, 20, 21).

- A 45-year-old woman developed hyponatremia complicated by rhabdomyolysis while taking citalopram and the antipsychotic drug chlorprothixene for depressive psychosis (34). The hyponatremia became apparent 2 weeks after the dose of citalopram was increased to 40 mg/day, when she complained of weakness and lethargy.

SSRI-induced hyponatremia is unusual in non-geriatric populations, but the chlorprothixene may have played a role in this case.

- An 89-year-old man, who was taking aspirin 325 mg/day, clopidogrel 75 mg/day, atenolol 25 mg/day, and lansoprazole 30 mg/day, developed depression and was given citalopram 20 mg/day (35). His depressive symptoms started to improve, but after 12 days he developed malaise, nausea, and headache. His serum sodium concentration had fallen from 138 mmol/l before citalopram to 117 mmol/l. This was due to inappropriate secretion of antidiuretic hormone. Citalopram was withdrawn, he was given intravenous saline, and 72 hours later his serum sodium was normal and his symptoms of malaise, nausea, and headache had resolved. However, his depressive symptoms worsened, so citalopram (20 mg/day) was restarted with close monitoring of the serum sodium. Within 2 days his serum sodium had fallen to 126 mmol/l. Citalopram was withdrawn and mirtazapine was used as an alternative antidepressant with good effect and without any reduction in serum sodium.

This is an unusually well documented case, in which rechallenge clearly established the role of citalopram in causing hyponatremia in this patient.

In a prospective study of serum sodium concentrations in 75 men and women (aged 36–90 years) who received paroxetine for the treatment of depression, hyponatremia (defined as a serum sodium concentration of less than 135 mmol/l) developed in 12% within a median of 9 days of paroxetine treatment (36). The effect did not seem to be related to plasma concentrations of paroxetine, but lower baseline serum sodium concentrations and low body weight were risk factors for hyponatremia. In fact the authors' data showed that in most patients, paroxetine lowered serum sodium concentrations to some extent, but in a subgroup the effect was greater. Presumably patients who start treatment with lower baseline values of serum sodium are at more risk of reaching a sodium concentration at which symptoms will occur. In animals, serotonin agents can release ADH (37), which might provide an explanation for the ability of SSRIs to lower plasma sodium in humans.

There have been other reports of hyponatremia with SSRIs (38,39). Hyponatremia is probably more common with SSRIs than with tricyclic antidepressants and predominantly but not exclusively affects older patients. Most reports involve fluoxetine, but this might represent greater patient exposure. All SSRIs and venlafaxine can produce this adverse effect (SEDA-23, 21; SEDA-25, 14). According to published reports, the median time to the onset of hyponatremia is 13 days (range 3–120) and the presentation is of inappropriate secretion of antidiuretic hormone (38). Symptoms, such as lethargy and confusion, can be non-specific, so awareness of the possibility of SSRI-induced hyponatremia, particularly in elderly people, is needed.

Fluid balance

The syndrome of inappropriate secretion of antidiuretic hormone (SIADH) is a possible adverse effect of the SSRIs (SEDA-14, 14) (SEDA-18, 20) (SEDA 21, 11) (40). The mechanism is not known. Several of the affected patients have been elderly, and old people may be at greater risk.

Hematologic

SSRIs have in rare cases been reported to produce bruising, bleeding, prolonged bleeding time, increased prothrombin time, and other hematological disturbances (SEDA-24, 15) (see the monographs on fluoxetine, fluvoxamine, paroxetine, and sertraline). The suggested mechanism of these adverse effects is reduced granular storage of serotonin in platelets, leading to disturbances of platelet function, especially in predisposed patients with mild underlying platelet disorders (41,42). An alternative mechanism is increased capillary fragility. Some patients appear to have a pre-existing susceptibility, for example, by virtue of treatment with other medications that might predispose to bleeding.

Five children, aged 8–15 years, developed bruising or epistaxis 1–12 weeks after starting SSRI treatment (43). In all cases the bleeding problem resolved when the SSRI was withdrawn or the dose lowered. In a review of 30 cases of SSRI-induced bleeding disorders the most common events were bruising, petechiae, purpura, and

epistaxis, though gastrointestinal hemorrhage was also reported (44). The mean age of the affected patients was 42 years and the female:male ratio was 3:4. Symptoms were sometimes associated with prolonged bleeding time or platelet aggregation disorders, but often these indices were normal.

Gastrointestinal

Gastrointestinal adverse effects are one of the major disadvantages of SSRIs. The most common is nausea, and the incidence is said to be 20% or more for paroxetine (45,46), sertraline (47), fluvoxamine (5), fluoxetine (48), and citalopram (10,49). Although nausea can lead to drug withdrawal, it usually disappears after a few weeks. Other gastrointestinal symptoms that occur commonly with fluoxetine and sertraline are loose stools and diarrhea (47,48,50), while constipation has been more often reported with fluvoxamine (5) and paroxetine (45,46).

In a case-control study, current exposure to SSRIs was associated with an increased risk of upper gastrointestinal bleeding (RR = 3.0; CI = 2.1, 4.4) (51). The risk was substantially greater in subjects taking both SSRIs and non-steroidal anti-inflammatory drugs (NSAIDs) (RR = 16; CI = 6.6, 37). However, there was also an increased risk of bleeding with trazodone (RR = 8.6; CI = 2.1, 35), which does not block the uptake of 5-HT in therapeutic doses. Thus, the association between certain kinds of antidepressant drugs and upper gastrointestinal bleeding appears to be real.

This association has been confirmed in a cohort study in 26 005 antidepressant users (52). The risk of being hospitalized for upper gastrointestinal bleeding within 90 days of antidepressant prescription was increased in people taking SSRIs (OR = 3.6; CI = 2.7, 4.7). Combined use of SSRIs and aspirin increased the risk 5.2 times (CI = 3.2, 8.0), while combined use with an NSAID increased the risk 12 times (CI = 7.1, 20). There was a lower risk of gastrointestinal bleeding (OR = 2.3; CI = 1.5, 3.4) with antidepressants with less potency to inhibit 5HT re-uptake (amitriptyline, dosulepin, imipramine), while non-serotonergic antidepressants (desipramine, nortriptyline, mianserin) did not increase the risk. These findings suggest that SSRIs should be used with caution in patients taking aspirin or NSAIDs. The risk of upper gastrointestinal bleeding with SSRIs used by themselves appears to be of the same order as that with aspirin. This observation is of interest, in view of a case-control study suggesting a lower risk of first myocardial infarction in patients taking SSRIs (OR = 0.59; CI = 0.39, 0.91) (53). This effect was not seen with non-SSRI antidepressants.

Liver

SSRIs are rarely hepatotoxic and have only occasionally been reported (see the monographs on fluoxetine, paroxetine, and sertraline).

- A 23-year-old woman taking sertraline (dose and length of time not stated) developed a mild pyrexia, right-sided abdominal pain, nausea, vomiting, and chills. She had a raised serum alanine transaminase activity but no other abnormality of liver function tests. A liver biopsy showed changes consistent with either a drug-induced reaction or autoimmune disease. Sertraline was withdrawn and prednisone 20 mg/day started. Four weeks later her alanine transaminase activity had fallen significantly, but she then restarted sertraline because of increasing depression. The alanine transaminase activity rose again, despite continuing prednisone, and only fell when sertraline was withdrawn.

while there have been other cases of hepatitis in association with sertraline this is the first case to have been confirmed by re-challenge.

A similar case has been reported in association with citalopram (54).

- A 44-year-old man with depression was given hydroxyzine hydrochloride 100 mg/day, clonazepam 2 mg/day, and citalopram 20 mg/day. Two years before he had tolerated fluoxetine for 6 months for an episode of depression, with good effect. After 7 weeks he developed weakness and weight loss. Physical examination was normal but the serum aspartate transaminase (AsT) was raised at 277 IU/L (reference range < 36 IU/l). His bilirubin was normal. Citalopram was withdrawn and the other drugs were continued intermittently; 5 days later the serum aspartate transaminase had fallen by a half, and within 2 months it had returned to normal.

The time course of the raised aspartate transaminase and its resolution in response to citalopram withdrawal suggests that citalopram was responsible for the hepatic damage in this case. The earlier exposure to fluoxetine did not apparently cause liver damage, suggesting that the response to citalopram did not involve blockade of serotonin re-uptake.

- A 39-year-old woman with a severe depressive illness took fluvoxamine in increasing doses, but after 9 days reported upper quadrant pain and vomiting (55). Her liver enzymes were raised (aspartate transaminase 609 µmol/l, previously 11 µmol/l). Tests for hepatitis and HIV were negative, but hepatic biopsy showed cholestasis and hepatocytolysis with possible duct damage. The liver enzymes fell when fluvoxamine was withdrawn but increased again 4 days after fluvoxamine re-challenge. They continued to rise when citalopram was substituted for fluvoxamine but fell when citalopram was withdrawn. She eventually responded to ECT and was discharged with normal liver function tests. However, 10 months later she was again admitted with a depressive psychosis. Olanzapine was started and 7 days later citalopram added; 4 days later this she again developed upper quadrant pain and raised aspartate transaminase activity (251 µmol/l). Citalopram was withdrawn and the liver enzymes gradually returned to normal over the next 3 weeks. She eventually responded well to olanzapine alone.

In this case the hepatotoxicity with citalopram and fluvoxamine appears to have been confirmed by re-challenge. Sometimes patients who have experienced hepatotoxicity with one SSRI can apparently be treated with a different SSRI without recurrence. However, this did not seem to be the case here because the patient had the same adverse hepatic reaction to two structurally unrelated SSRIs.

Skin

Skin reactions to SSRIs have been reported (SEDA-17, 20) (SEDA-18, 20), including Stevens–Johnson syndrome (56), and rash due to fluoxetine occurs in a few percent of patients (57) (see also the monographs on fluoxetine, fluvoxamine, and sertraline).

Sexual function

Impaired sexual function

The adverse sexual effects of SSRIs have been reviewed (58). The use of SSRIs is most often associated with delayed ejaculation and absent or delayed orgasm, but reduced desire and arousal have also been reported. Estimates of the prevalence of sexual dysfunction with SSRIs vary from a small percentage to over 80%. Prospective studies that enquire specifically about sexual function have reported the highest figures. Similar sexual disturbances are seen in patients taking SSRIs for the treatment of anxiety disorders (59), showing that SSRI-induced sexual dysfunction is not limited to patients with depression. It is not clear whether the relative incidence of sexual dysfunction differs between the SSRIs, but it is possible that paroxetine carries the highest risk (58).

In a chart review of 22 adolescents who were taking SSRIs for a variety of indications and who had been systematically questioned about the effect of their illness and its treatment on their sexual function, five reported significant sexual dysfunction (three cases of anorgasmia, and two of reduced libido), probably attributable to the SSRI (60). This small sample suggests that the rate of sexual dysfunction associated with SSRI treatment in adolescents is very similar to that seen in adults. Issues of sexuality are particularly important in adolescence, but may be difficult to discuss. This study shows the importance of tactful and sensitive enquiry and the probable benefit of clear information about the sexual adverse effects of SSRIs.

The ability of SSRIs to cause delayed ejaculation has been used in controlled trials of men with premature ejaculation (61,62). Of the SSRIs, paroxetine and sertraline produced the most benefit in terms of increase in time to ejaculation, but fluvoxamine did not differ from placebo. Clomipramine was more effective than the SSRIs but caused most adverse effects. From a practical point of view many patients might prefer to take medication for sexual dysfunction when needed rather than on a regular daily basis, and it would be of interest to study the beneficial effects of SSRIs on premature ejaculation when used in this way.

Laboratory studies have shown that fluvoxamine differs from paroxetine, sertraline, and fluoxetine in not delaying the time to ejaculation. The effect of citalopram to delay ejaculation is also relatively modest (63). There are, however, many other ways in which SSRIs can interfere with sexual function, for example by causing loss of sexual interest and erectile difficulties. In an open, prospective study of 1000 Spanish patients taking a variety of antidepressants, there was an overall incidence of sexual dysfunction of 59% (64). The highest rates, 60–70%, were found with SSRIs (including fluvoxamine) and venlafaxine. The lowest rates were found with mirtazapine (24%), nefazodone (8%), and moclobemide (4%). Spontaneous resolution of this adverse effect was uncommon—80% of subjects had no improvement in sexual function over 6 months of treatment. This study suggests very high rates of sexual dysfunction in patients taking SSRIs and venlafaxine. However, in investigations of this nature, it can be difficult to tease out the effect of the drug from that of the underlying disorder. Nevertheless, while depressive symptoms should improve in most patients over 6 months of treatment, the sexual dysfunction in these subjects tended not to remit, suggesting that the antidepressant was the main culprit.

Increased sexual function

Occasionally, SSRIs are associated with increased sexual desire and behavior.

- A 27-year-old married woman with a borderline personality disorder was admitted to hospital with depression and suicidal ideation (65). Over 3 weeks she was given fluvoxamine in doses up to 150 mg/day, but because of lack of response the dosage was increased to 200 mg/day; 3 days later she reported that her sex drive was greater than it had ever been before and that she felt she could not control it. There was no evidence of mania. Within a week of withdrawal of fluvoxamine her sexual desire had returned to its previous level.

Patients with borderline personality disorder may behave in a sexually disinhibited manner and have mood swings. In this patient, however, it did not appear as though the hypersexuality was part of a manic syndrome, and she was clear that the sexual feelings were unusually great for her. Support for a role of the SSRI in this adverse effect comes from a series of five patients (four taking citalopram and one paroxetine) who had an unusual increase in sexual interest, with preoccupation with sexual thoughts, promiscuity, and excessive interest in pornography (66). In some of the cases symptoms such as diminished need for sleep suggested the possibility of a manic syndrome.

These reports suggest that occasionally SSRIs can be associated with increased sexual desire and behavior. This might be associated with mood instability, for example in a manic or mixed affective state, but in some people personality factors are likely to be important.

Treatment of sexual dysfunction due to SSRIs

Various treatments have been advocated to ameliorate sexual dysfunction in SSRI-treated patients, including 5-HT$_2$-receptor antagonists (cyproheptadine, mianserin, nefazodone) and 5-HT$_3$-receptor antagonists

such as granisetron (59). One of the most popular remedies is the use of dopaminergic agents, such as amfebutamone.

In a prospective study, 47 patients who complained of SSRI-induced sexual dysfunction took amfebutamone (bupropion) 75–150 mg 1–2 hours before sexual activity (67). If this was unsuccessful they were titrated to a dosage of 75 mg tds on a regular basis. Amfebutamone improved sexual function in 31 patients (66%). Anxiety and tremor were the most frequently reported adverse events, and seven patients discontinued for this reason. However, it should be noted that more serious adverse events (panic attacks, delirium, and seizures) have been reported when amfebutamone (bupropion) and SSRIs are combined (59).

In a placebo-controlled, parallel-group study of buspirone and amantadine in 57 patients with fluoxetine-induced sexual dysfunction, there was an overall improvement in all three treatment groups and no significant differences between them (68). These data suggest that anecdotal reports of treatment benefit in SSRI-induced sexual dysfunction should be regarded with caution.

Reproductive system

Vaginal bleeding and menorrhagia have been rarely reported with SSRI treatment and the mechanism is uncertain. SSRIs have been associated with bleeding diatheses, which might explain these observations (SEDA-24, 15) (SEDA-25, 14). Another possibility is an action of serotonin pathways on the neural regulation of gonadotropin release. More cases have been reported.

- A 67-year-old woman developed vaginal bleeding after taking sertraline 25 mg/day for 3 days (69). The sertraline was withdrawn and the bleeding stopped 48 hours later.
- A 41-year-old woman developed vaginal bleeding after taking venlafaxine 75 mg/day, a dose at which selective serotonin blockade is the predominant pharmacological action (70). The vaginal bleeding stopped 24 hours after venlafaxine withdrawal and occurred again on rechallenge.

SSRIs can be associated with breast enlargement in women and gynecomastia has been reported in men.

- Gynecomastia occurred in a 30-year-old man who had taken paroxetine 40 mg/day for 5 years (71). Over this time gynecomastia of the left breast became increasingly marked. There was no evidence of metabolic or hormonal abnormalities and plasma prolactin was within the reference range. Biopsy showed no evidence of malignancy. The gynecomastia was reduced by surgery and paroxetine was replaced with mirtazapine. At 2-year follow-up there was no evidence of recurrence.
- A similar but more acute case of gynecomastia in association with fluoxetine has been reported in a 21 year old man (72). The gynecomastia resolved after fluoxetine withdrawal.

Immunologic

Infection risk: three cases of *Herpes simplex* reactivation associated with fluoxetine have been described (SEDA-16, 10).

Death

Severe adverse reactions to SSRIs that were reported to Health Canada's database in 1986–1996 have been reviewed (73). There were 295 severe adverse reactions with 87 deaths. Of the fatal cases, 65 were due to intentional overdose. The other 22 deaths were due chiefly to other forms of suicide or were accidental or indeterminate (12 cases). Of the rest there were three cases of neuroleptic malignant syndrome and individual cases of cardiac or respiratory disease in which the role of SSRIs was less clear.

This report shows that the major causes of death in patients taking SSRIs are related to the risks of depression itself, particularly self-harm. SSRIs themselves appear to be relatively safe. It is possible, however, that the cases of neuroleptic malignant syndrome could have been misdiagnosed forms of the serotonin syndrome, which, like the neuroleptic malignant syndrome, can present with hyperthermia and changes in consciousness and is usually produced by pharmacodynamic interactions between SSRIs and other serotonin-potentiating compounds (SEDA-25, 24). In general, the risks of SSRIs are increased by co-prescription, and fatal overdose with SSRIs usually involves a mixture of psychotropic drugs and/or alcohol.

Sudden death has been reported in three elderly women with pulmonary disease and atrial fibrillation who had recently started taking fluoxetine. The authors recommended caution in such patients, although no causal link with fluoxetine was established (SEDA-17, 19).

Long-Term Effects

Drug withdrawal

Reports of withdrawal reactions after abrupt withdrawal of SSRIs continue to accumulate, and several reviews have summarized case reports of withdrawal symptoms (74,75).

SSRIs with shorter half-lives, such as fluvoxamine and paroxetine (qv), have a higher incidence of withdrawal symptoms than long-acting ones, but withdrawal effects have been reported, albeit rarely, after withdrawal of fluoxetine (qv) (SEDA-20, 8) (76) and citalopram (77). Patients taking high doses, long-term treatment, or both are at increased risk of developing withdrawal symptoms (76,78).

In a randomized, placebo-controlled trial, sudden withdrawal of paroxetine produced significant withdrawal symptoms by as early as the second day, while patients taking fluoxetine remained asymptomatic for the five-day withdrawal period (79). Patients taking sertraline had an intermediate level of abstinence symptoms. Both paroxetine- and sertraline-treated patients reported impaired

functioning during the withdrawal period, while those taking fluoxetine did not. These findings are consistent with earlier reports that suggested that acute withdrawal symptoms after fluoxetine withdrawal are unusual, presumably because of the long half-life of its active metabolite, norfluoxetine.

Second-Generation Effects

Teratogenicity

Tricyclic antidepressants appear to be generally free of teratogenic effects, but the status of newer antidepressants is unclear. Since 1993 four controlled prospective studies of antidepressant exposure during pregnancy have been published, involving about 400 women, most of whom took fluoxetine at various stages during pregnancy (80). There was no evidence that fluoxetine or other SSRIs caused an increase in intrauterine death, significant birth defects, or growth impairment. Follow-up behavioral studies of infants exposed to SSRIs during pregnancy also showed no difference from controls. In a summary of systematic reviews of the safety of SSRIs in pregnancy and lactation the authors concluded that thus far there is little evidence that SSRIs cause birth defects (81) (but see Paroxetine).

Fetotoxicity

An increasing number of women are being exposed to SSRIs during pregnancy and postanatany. It is therefore important to establish as carefully as possible whether SSRIs have teratogenic effects (SEDA-24, 15). In a retrospective study of the impact on birth outcome of the timing of fluoxetine exposure in 64 pregnant women there were no major differences between infants exposed to fluoxetine early in pregnancy (the first or second trimesters only) and those exposed to fluoxetine throughout the third trimester and delivery (82). However, the infants in the late-exposed group were about twice as likely to be admitted to a special-care nursery. No specific pattern of neonatal difficulties could be found to account for this difference, and it is possible that the excess of neonatal problems in the late-onset treatment group was due to worse depression in the women who took antidepressants at around the time of delivery.

Withdrawal effects of SSRI treatment may be apparent in neonates shortly after delivery. These include jitteriness, hypoglycemia, hypothermia, and respiratory distress (83).

- A boy (3.4 kg) was born by cesarean section, because of fetal distress, to a mother who had taken paroxetine (40 mg/day) and olanzapine (10 mg/day) until 48 hours before delivery (84). After delivery, the baby had increased tone and was hypoglycemic. The next day he was increasingly jittery, with profuse salivation. At 44 hours there was no detectable paroxetine or olanzapine in the neonatal plasma and CSF concentrations of the 5HT metabolite, 5-hydroxyindoleacetic acid (5HIAA) were normal. The baby's condition improved over the next few days, and 1 week after birth the only abnormality was slight jitteriness. At 6 months the baby was considered normal.

This infant's symptoms were similar to those described in other neonates whose mothers took SSRIs shortly before delivery, although in this case a contributory effect from olanzapine was also possible. The authors made the point that it can be difficult to decide from the clinical presentation whether neonatal problems, such as those described above, represent SSRI withdrawal (5HT deficiency) or SSRI toxicity (5HT excess). The fact that no paroxetine was detected in the infant's plasma led them to conclude that SSRI withdrawal was responsible for the symptoms in this case.

In a follow-up study in 20 infants born to mothers who had taken either fluoxetine (20–40 mg/day) or citalopram (20–40 mg/day), the children of mothers who had taken SSRIs had raised scores in the first 4 days of life on scales rating symptoms of 5HT excess, including tremor, rigidity, and restlessness (85). Children who had been exposed to SSRIs had lower concentrations of 5HIAA in cord blood (consistent with persistent 5HT re-uptake blockade) and there was a significant negative correlation between the 5HT symptom score and cord blood 5HIAA. At 4 days citalopram and fluoxetine were detectable in plasma samples from the infants. By 2 weeks there was no significant difference in 5HT symptom scores between exposed infants and controls. Weight gain over this time was similar in the two groups. These findings support the view that the neonatal syndrome associated with SSRI treatment of mothers around the time of delivery can sometimes be due to 5HT toxicity rather than SSRI withdrawal.

There are concerns that the use of SSRIs in later pregnancy may be associated with persistent pulmonary hypertension in the newborn (86).

Of 93 cases of neonatal symptoms associated with the use of SSRIs in mothers around the time of delivery 64 were associated with paroxetine but reactions were also reported in infants whose mothers had taken citalopram, fluoxetine, and sertraline (87). It is unclear from these data whether paroxetine is actually most likely to provoke the neonatal syndrome, but in adults its use is associated with more severe withdrawal reactions than other SSRIs. It should also be noted that it is not clear whether the syndrome described in neonates is due to SSRI withdrawal or a form of serotonin toxicity.

Lactation

The benefit : harm balance of breast-feeding during antidepressant treatment is difficult to compute and must be done on an individual basis. In general, breast-feeding with SSRIs is regarded as safe, as the amount of drug ingested by the infant is very low, and although all SSRIs are excreted in breast milk to some extent, there is little evidence that infants who are breast-fed by mothers who are taking SSRIs suffer acute adverse effects (4c). However, adverse effects in the child are reported occasionally (SEDA-25, 15) and it is difficult to exclude completely the possibility of long-term effects on brain development in the infant. Clearly, the lower the concentration of SSRI in the infant the less likely are problems of acute and longer-term toxicity. In two cases treatment of breast-feeding mothers with fluvoxamine (300 mg/day) was associated with undetectable concentrations of fluvoxamine (below 2.5 ng/ml) in the plasma of both infants (88). These results are encouraging, but further data will be needed before it can be concluded that fluvoxamine has an advantage over other SSRIs in this respect. Because fluoxetine has a long half-life it may be advisable not to use it in breast-feeding women.

When blood and milk were sampled in 22 nursing women taking sertraline (25–200 mg/day), sertraline and its metabolite, desmethylsertraline, were found in all the milk samples (89). The maximum concentration of sertraline and desmethylsertraline in the milk occurred 8–9 hours after maternal ingestion of the daily dose of sertraline. Eleven infants had detectable desmethylsertraline; of these, four also had detectable sertraline. No adverse effects were noted in any of the children. The authors calculated that the infants had received on average about 0.5% of the maternal sertraline dose.

These data confirm that the infants of nursing mothers taking SSRIs are likely to be exposed to low doses of antidepressants. While this rarely causes overt effects, the possibility of subtle long-term behavioral consequences cannot be excluded. In the case of sertraline it would be wise to discard breast milk accumulated 8–9 hours after dosing as this will reduce the daily dose that the infant receives.

In a prospective study of 31 nursing mothers who were taking citalopram for the treatment of depression, and 31 matched, healthy breast-feeding women, the mothers taking citalopram treatment did not report a significant excess of adverse effects in their infants (90). A further prospective study in 27 mothers who took paroxetine while breast feeding included two control groups: (i) 19 women who did not take any medication and did not breast feed; (ii) 27 mothers who breast-fed their infants but did not take any medications (91). Information about the infants was gathered from the mothers, who completed a detailed questionnaire at 3 months and 1 year post-partum. There were no significant differences between the infants in terms of weight gain or developmental milestones.

These data are reassuring, but the studies, although prospective, were not large. The current data suggest that SSRIs are unlikely to produce overt effects on infant development; however, subtle effects on neurological and psychological development are hard to exclude (92).

Drug Administration

Drug overdose

The SSRIs are in general considered to be less toxic in overdose than tricyclic antidepressants. From a review of cases of overdose with SSRI during 10 years it was concluded that SSRIs are rarely fatal in overdose when taken alone (93). In moderate overdoses of up to 30 times the daily dose, symptoms were either minor or absent. In higher doses, typical symptoms included drowsiness, tremor, nausea, and vomiting. At extremely high doses (over 75 times the usual daily dose), there were more serious toxic effects, including seizures, electrocardiographic changes, and disturbances of consciousness. When an SSRI overdose was taken in combination with alcohol or other drugs, toxicity increased, and almost all deaths involving SSRIs were in combination with other drugs.

In 225 patients who had taken overdoses of antidepressant drugs in suicide attempts, venlafaxine and citalopram were more likely to be associated with seizures than mirtazapine and nefazodone and 5HT toxicity was more common after overdose of venlafaxine (94). These findings confirm the potential toxicity of venlafaxine in overdose and also suggest a pro-convulsant effect of large doses of citalopram.

Drug–Drug Interactions

General

For the SSRIs there are two types of interaction of major concern: SSRIs can selectively inhibit hepatic drug metabolizing enzymes, giving rise to pharmacokinetic interactions (SEDA-22, 13), and pharmacodynamic interactions can occur when SSRIs are given in combination with other serotonergic drugs, which can cause serotonergic hyperstimulation: the "serotonin syndrome." Fluoxetine gives rise to special concern, since reactions can occur weeks after withdrawal, owing to the long half-life of both parent compound and metabolite. Of the various SSRIs, citalopram seems the least likely to produce this effect. Paroxetine and fluoxetine can cause interactions with tricyclic antidepressants, neuroleptic drugs, and some antidysrhythmic drugs that are metabolized by CYP2D6.

In 31 healthy men and women (mean age 28 years) the ability of four SSRIs (fluoxetine, fluvoxamine, paroxetine, and sertraline) to inhibit CYP2D6 activity was assessed in vivo, as judged by the dextromethorphan test (95). All were extensive metabolizers of dextromethorphan. After 8 days treatment at therapeutic doses, four of eight paroxetine-treated and five of eight fluoxetine-treated subjects had become poor metabolizers, presumably because of the inhibitory effect of the two SSRIs on CYP2D6 activity. In contrast, none of the eight fluvoxamine-treated or seven sertraline-treated subjects showed this effect. This in vivo study has confirmed the potent inhibitory effect of fluoxetine and paroxetine on CYP2D6 and the relative sparing by fluvoxamine and sertraline. However, fluvoxamine is a potent inhibitor of CYP1A2 and CYP3A3/4 (SEDA-22, 13). In addition, at higher doses (over 100 mg/day at steady state) sertraline can inhibit CYP2D6.

The drugs that can cause a serotonin syndrome when they are combined with SSRIs include monoamine oxidase inhibitors (including reversible inhibitors of monoamine oxidase types A and B), dextromethorphan,

tryptophan, lithium, pentazocine, and carbamazepine (SEDA-17, 23) (SEDA-18, 22) (96).

Alprazolam

SSRIs and the benzodiazepine alprazolam are often used to treat panic disorder. Pharmacokinetic reactions between them could therefore be important. Alprazolam is metabolized by CYP3A4, which fluvoxamine inhibits (SEDA-22, 13). In 23 out-patients (11 men, 12 women, mean age 39 years) who took alprazolam both as monotherapy (mean dose 1.0 mg/day) and in combination with fluvoxamine (mean dose 34 mg/day), fluvoxamine increased plasma alprazolam concentrations by 58% (97). This was not associated with increased sleepiness, measured by a subjective rating scale, but objective measures of psychomotor function were not carried out and these could have been impaired by raised alprazolam concentrations.

The effects of SSRIs on alprazolam pharmacokinetics have been studied in 21 healthy volunteers (age and sex not given) who were pre-treated with either fluoxetine or citalopram (20 mg/day for 21 days) (98). Fluoxetine increased the AUC of a single 1.0 mg dose of alprazolam by 32%; citalopram had no effect. These findings are consistent with previous reports that fluoxetine and its active metabolite, norfluoxetine, produce moderate inhibition of CYP3A4 while citalopram does not.

Buspirone

The serotonin syndrome can occur with therapeutic doses of SSRIs (see above), but it occurs most commonly when SSRIs are co-administered with other drugs that also potentiate serotonin function. Recent case reports have suggested that there is a risk of the serotonin syndrome when SSRIs are combined with buspirone (99).

Coumarin anticoagulants

Concomitant treatment with aspirin or NSAIDs increases the risk of gastrointestinal bleeding five and twelve times respectively. An interaction might also be expected with *warfarin*, not only because SSRIs and warfarin can cause bleeding, but also because SSRIs that inhibit CYP2C9 (fluoxetine and fluvoxamine) might inhibit the metabolism of warfarin and prolong its activity. The risk of gastrointestinal bleeding during combined treatment of SSRIs with warfarin has been studied in 98 784 patients aged 65 years or older who had used warfarin for at least a year (100). Of these patients 1538 were admitted to hospital with gastrointestinal bleeding. The risk of exposure to fluoxetine or fluvoxamine did not differ from that of a nested case–control group (1.2, CI = 0.8, 1.7) and the risk with other SSRIs was similar (1.1, CI = 0.9, 1.4). These data are reassuring, but people who take combined treatment with SSRIs and anticoagulants need careful monitoring (SEDA-27, 14).

Dextromethorphan

SSRIs inhibit hepatic CYP isozymes and can thereby increase the activity of co-administered drugs that are metabolized by this route (SEDA-22, 13) (SEDA-24, 15). In healthy volunteers randomly allocated to fluoxetine (20 mg/day), sertraline (100 mg/day), or paroxetine (20 mg/day) the activity of CYP2D6 was measured by dextromethorphan testing once steady state had been achieved and the medication was withdrawn (101). Extrapolated calculations showed that the mean time for full CYP2D6 recovery after fluoxetine (63 days) was significantly longer than that for sertraline (25 days) or paroxetine (20 days). Accordingly, even after SSRIs have been withdrawn, the potential for drug interactions persists for substantial periods of time, particularly in the case of fluoxetine.

Erythromycin

- A 12-year-old boy developed the serotonin syndrome, which is normally associated with the interaction of two or more serotonergic agents, after the co-administration of erythromycin and sertraline.

This could have been due to erythromycin-induced inhibition of sertraline metabolism by CYP3A (102).

Grapefruit juice

Grapefruit juice is also a modest inhibitor of CYP3A4. Grapefruit juice (250 ml tds) for 5 days produced a 1.6-fold increase in the AUC of a single dose of fluvoxamine 75 mg in 10 healthy men (aged 19–30 years) (103). Fluvoxamine is metabolized by CYP2D6 and CYP1A2, but this study suggests that there may also be a modest contribution from CYP3A4.

Haloperidol

Fluvoxamine has little inhibitory effect on CYP2D6 but is a potent inhibitor of CYP1A2 and CYP2C19. It also is a moderate inhibitor of CYP3A4, which is involved in the metabolism of haloperidol. When fluvoxamine (25, 75, and 150 mg/day, each for 2 weeks) was added to haloperidol in 12 patients with schizophrenia aged 22–59 years, plasma haloperidol concentrations rose dose-relatedly; after fluvoxamine 25 mg/day haloperidol concentrations rose by about 20%, with additional 20% increases with each increment in fluvoxamine dose; however, this was not associated with overt clinical toxicity (104).

Linezolid

SSRIs can provoke 5HT neurotoxicity (the 5HT syndrome) through pharmacodynamic interactions with other drugs that also potentiate 5HT function. Often the ability of the interacting drug to facilitate 5HT function is well known, as is the case, for example, when SSRIs are combined with monoamine oxidase inhibitors or lithium. In other cases, however, the potential 5HT activity of the co-administered drug is not widely known. The ability of the antibiotic linezolid to inhibit MAO and thereby to cause 5HT neurotoxicity in combination with SSRIs has been noted previously (SEDA-27, 14), and further cases have now been reported.

- An 85-year-old woman with an oxacillin-resistant *Staphylococcus aureus* infection took oral linezolid (105). She had been taking long-term citalopram maintenance treatment for depression. Shortly afterwards she developed a worsening tremor and became restless and confused. She had dysarthria and hyper-reflexia, with impaired gait. Citalopram was withdrawn, and her mental status and neurological signs returned to baseline within two days.

- A four-year-old girl took fluoxetine for symptoms of post-traumatic stress disorder after a severe burn injury and 2 days after the addition of linezolid developed agitation, mydriasis, and abnormal movements in her limbs (106). Linezolid was withdrawn, and the symptoms resolved after 2 days.

The signs of 5HT neurotoxicity can be sometimes rather non-specific, as in this case, and a high index of suspicion may be needed when SSRIs are used in drug combinations of which there is little previous experience.

Methylenedioxymetamfetamine (MDMA)

An MDMA-related psychiatric adverse effect ecstasy been enhanced by an SSRI (107).

- A 52-year-old prisoner, who was taking the SSRI citalopram 60 mg/day, suddenly became aggressive, agitated, and grandiose after using ecstasy. He carried out peculiar compulsive movements and had extreme motor restlessness, but no fever or rigidity. He was given chlordiazepoxide and 2 days later was asymptomatic. Citalopram was reintroduced, and 2 days later he reported visual hallucinations of little bugs in the cell. Promazine was substituted for citalopram and his condition improved 2 days later.

The authors suggested that SSRIs such as citalopram can potentiate the neurochemical and behavioral effects of MDMA.

Monoamine oxidase inhibitors

Monoamine oxidase inhibitors, including reversible inhibitors of monoamine oxidase types A and B, can cause a serotonin syndrome when they are combined with SSRIs (SEDA-17, 23) (SEDA-18, 22) (64).

Nefazodone

The serotonin syndrome can occur with therapeutic doses of SSRIs (see above), but it occurs most commonly when SSRIs are co-administered with other drugs that also potentiate serotonin function. Recent case reports have suggested that there is a risk of the serotonin syndrome when SSRIs are combined with nefazodone (108).

Concentrations of nefazodone and its metabolites can be increased by fluoxetine and paroxetine (SEDA-20, 9). Combinations of serotonin agents produce serotonin toxicity, and a case of serotonin syndrome occurred when nefazodone (200 mg/day) was combined with fluoxetine (40 mg/day) in a 50-year-old man (109). The toxic symptoms settled 3 days after withdrawal of both antidepressants.

Neuroleptic drugs

Interactions of SSRIs with neuroleptic drugs have been reported (110–112).

Olanzapine

Olanzapine is metabolized by CYP1A2, and fluvoxamine, which inhibits CYP1A2, increases plasma olanzapine concentrations (SEDA-24, 71).

- A 21-year-old woman had a very high olanzapine plasma concentration (120 ng/ml) during co-administration of fluvoxamine 150 mg/day and olanzapine 15 mg/day; her symptoms included slight tremor, rigid movements, and general discomfort (113).

Pharmacokinetic and pharmacodynamic profiles of olanzapine have been extensively reviewed (114). Olanzapine does not inhibit CYP isozymes, and no clinically significant metabolic interaction was found with fluoxetine.

Oxazolidinones

Serotonin syndrome was reported in a 56-year-old white woman who received intravenous linezolid shortly after withdrawal of a selective serotonin reuptake inhibitor, paroxetine (115).

The serotonin syndrome was reported in a 45-year-old white man who received intravenous linezolid (600 mg 12 hourly) and sertraline (116).

Pethidine

Pethidine plus an SSRI may have caused a 5HT syndrome (117).

- A 43 year old man was premedicated for endoscopy with intravenous midazolam 2 mg and pethidine 50 mg). He immediately became agitated and restless. His blood pressure rose to 180/100 mmHg, he sweated, had widely dilated pupils, and had diarrhea. Over the next 90 minutes his condition remitted without specific treatment. He then reported that he had been taking fluoxetine (20 mg every other day), which he had stopped taken about 2 weeks before.

The symptoms described here resemble the 5HT toxicity syndrome (SEDA-25, 16) and suggest that pethidine can provoke this reaction when combined with SSRIs, as it can with monoamine oxidase inhibitors. The report is also a useful reminder that the active metabolite of fluoxetine, norfluoxetine, has a half-life of about 1 week and would still have been present at the time of endoscopy, even though fluoxetine had been stopped 14 days before.

Risperidone

SSRIs are often prescribed with antipsychotic drugs, and some SSRIs inhibit CYP2D6, which can lead to increased blood concentrations of the antipsychotic drug. In 13 patients aged 26–56 years with schizophrenia, stabilized on risperidone 4–6 mg/day who took sertraline 50 mg/day for 4 weeks, plasma concentrations of risperidone and its major metabolite, 9-OH-risperidone, did not change

(118). Over the next 4 week, four patients continued to take sertraline 50 mg/day, while in the others the dose was increased at the discretion of the treating clinician. In patients taking sertraline 100 mg/day there was a small but non-significant increase in plasma risperidone concentrations. However, in two patients who took sertraline 150 mg/day, plasma risperidone concentrations increased by about 50%. Sertraline is a modest inhibitor of CYP2D6 compared with fluoxetine or paroxetine. This study supports the view that at doses under 100 mg/day sertraline is unlikely to produce significant pharmacokinetic interactions through CYP2D6 inhibition; however, at higher doses such interactions become increasingly likely.

Sibutramine

The serotonin syndrome has been attributed to a combination of citalopram and sibutramine.

- A 43-year-old woman who had taken citalopram 40 mg/day for 2 years was given sibutramine for obesity and a few hours after the first dose of 10 mg had irritability, racing thoughts, pressure of speech, agitation, shivering, and sweating (119). These symptoms persisted for 3 days, during which time she continued to take sibutramine. The day after sibutramine was withdrawn the symptoms resolved.

The serotonin syndrome can present with hypomanic features, and the clinical picture and rapid onset in this case suggested that the addition of sibutramine to citalopram provoked serotonin toxicity. Sibutramine blocks the reuptake of serotonin, dopamine, and noradrenaline. Whether the serotonin toxicity seen here resulted purely from the combined effects of both citalopram and sibutramine in blocking serotonin re-uptake is unclear. It is possible, for example, that potentiation of dopamine and noradrenaline activity by sibutramine might also have been involved.

Tizanidine

Fluvoxamine 100 mg/day resulted in a large (about 30-fold) increase in plasma concentrations of tizanidine in 10 healthy volunteers (120). This interaction caused significant physiological consequences, including a substantial fall in systolic blood pressure, perhaps because tizanidine is an α_2-adrenoceptor agonist. The authors suggested that the interaction was likely to be due to fluvoxamine-mediated inhibition of CYP3A4, which is involved in the metabolism of tizanidine.

Tramadol

Tramadol has some activity in blocking the re-uptake of 5HT and can also cause 5HT neurotoxicity in combination with SSRIs.

Citalopram

- A 70-year-old woman, who had been taking citalopram 10 mg/day for 3 years for depression, began taking tramadol 50 mg/day) for pain relief and rapidly developed tremor, restlessness, fever, and confusion (121).

The symptoms settled after tramadol was withdrawn. The same symptoms recurred 1 year later, when tramadol 20 mg/day was added to the citalopram. Genotyping for functional polymorphisms in CYP2D6 and CYP2C19 showed that she was heterozygous for alleles causing deficient activity in both these metabolizing enzymes. Consistent with this she had reduced clearance of citalopram, presumably due to CYP2C19 deficiency.

The simultaneous use of tramadol and SSRIs is probably not uncommon in clinical practice, and it is therefore likely that some patients can take this combination without developing 5HT toxicity. When symptoms of the 5HT syndrome develop it may be that, as in this case, that there is a factor in the individual patient that increases the risk.

Fluoxetine

Serotonin syndrome and mania occurred in a 72-year-old woman taking fluoxetine 20 mg/day and tramadol 150 mg/day 18 days after she started to take the combination (122). Inhibition of CYP2D6 may have played a part (123).

Sertraline

The serotonin syndrome has been reported after concurrent use of tramadol and sertraline (SEDA-22, 103).

- Serotonin syndrome occurred in an 88-year-old woman who took sertraline 50–100 mg/day and tramadol 200–400 mg/day for 10 days; the symptoms subsided 15 days after withdrawal of tramadol (124).

Venlafaxine

The death of a 36-year-old patient with a history of alcohol dependence who was taking tramadol, venlafaxine, trazodone, and quetiapine has highlighted the increased risk of seizures with concomitant use of tramadol and selective serotonin re-uptake inhibitors (125).

Tricyclic antidepressants

Interactions of SSRIs with tricyclic antidepressants have been reported. For example, plasma concentrations of tricyclic antidepressants rise after the addition of fluoxetine (126,127), fluvoxamine (128,129), and sertraline (130).

Triptans

Drugs used in the treatment of acute migraine, such as sumatriptan and rizatriptan, are 5-HT$_{1B/1D}$-receptor agonists and could theoretically interact pharmacodynamically with SSRIs to cause serotonin toxicity. Triptans are metabolized mainly by monoamine oxidase, which makes pharmacokinetic interactions with SSRIs unlikely. Although case series have suggested that sumatriptan can be safely combined with SSRIs (SEDA-22, 14), there are occasional reports of toxicity.

References

1. Hyttel J. Pharmacological characterization of selective serotonin reuptake inhibitors (SSRIs). Int Clin Psychopharmacol 1994;9(Suppl. 1):19–26.
2. Leonard BE. The comparative pharmacology of new antidepressants. J Clin Psychiatry 1993;54(Suppl.):3–15.
3. Doogan DP, Caillard V. Sertraline: a new antidepressant. J Clin Psychiatry 1988;49(Suppl.):46–51.
4. Kaye CM, Haddock RE, Langley PF, Mellows G, Tasker TC, Zussman BD, Greb WH. A review of the metabolism and pharmacokinetics of paroxetine in man. Acta Psychiatr Scand Suppl 1989;350:60–75.
5. Benfield P, Ward A. Fluvoxamine: a review of its pharmacodynamic and pharmacokinetic properties, and therapeutic efficacy in depressive illness. Drugs 1986;32(4):313–34.
6. Kragh-Sorensen P, Overo KF, Petersen OL, Jensen K, Parnas W. The kinetics of citalopram: single and multiple dose studies in man. Acta Pharmacol Toxicol (Copenh) 1981;48(1):53–60.
7. Bergstrom RF, Lemberger L, Farid NA, Wolen RL. Clinical pharmacology and pharmacokinetics of fluoxetine: a review. Br J Psychiatry Suppl 1988;3:47–50.
8. Burke WJ. Escitalopram. Expert Opin Investig Drugs 2002;11(10):1477–86.
9. Whittington CJ, Kendall T, Fonagy P, Cottrell D, Cotgrove A, Boddington E. Selective serotonin reuptake inhibitors in childhood depression: systematic review of published versus unpublished data. Lancet 2004;363(9418):1341–5.
10. Dencker SJ, Hopfner Petersen HE. Side effect profile of citalopram and reference antidepressants in depression. In: Montgomery SA, editor. Citalopram—The New Antidepressant from Lundbeck Research. Amsterdam: Excerpta Medica, 1989:31.
11. Edwards JG, Anderson I. Systematic review and guide to selection of selective serotonin reuptake inhibitors. Drugs 1999;57(4):507–33.
12. Gill HS, DeVane CL, Risch SC. Extrapyramidal symptoms associated with cyclic antidepressant treatment: a review of the literature and consolidating hypotheses. J Clin Psychopharmacol 1997;17(5):377–89.
13. Di Rocco A, Brannan T, Prikhojan A, Yahr MD. Sertraline induced parkinsonism: a case report and an in-vivo study of the effect of sertraline on dopamine metabolism. J Neural Transm 1998;105(2–3):247–51.
14. Gerber PE, Lynd LD, Leo RJ. Selective serotonin-reuptake inhibitor-induced movement disorders. Ann Pharmacother 1998;32(6):692–8.
15. Gray NA, Zhou R, Du J, Moore GJ, Manji HK. The use of mood stabilizers as plasticity enhancers in the treatment of neuropsychiatric disorders. J Clin Psychiatry 2003;64(Suppl. 5):3–17.
16. Stadtland C, Erfurth A, Arolt V. De novo onset of Parkinson's disease after antidepressant treatment with citalopram. Pharmacopsychiatry 2000;33(5):194–5.
17. Voirol P, Hodel PF, Zullino D, Baumann P. Serotonin syndrome after small doses of citalopram or sertraline. J Clin Psychopharmacol 2000;20(6):713–4.
18. Hartter S, Wang X, Weigmann H, Friedberg T, Arand M, Oesch F, Hiemke C. Differential effects of fluvoxamine other antidepressants on the biotransformation of melatonin. J Clin Psychopharmacol 2001;21(2):167–74.
19. Kawashima T, Yamada S. Paroxetine-induced somnambulism. J Clin Psychiatry 2003;64:483.
20. Levy D, Kimhi R, Barak Y, Aviv A, Elizur A. Antidepressant-associated mania: a study of anxiety

disorders patients. Psychopharmacology (Berl) 1998;136(3):243–6.

21. Bourgeois JA, Thomas D, Johansen T, Walker DM. Visual hallucinations associated with fluoxetine and sertraline. J Clin Psychopharmacol 1998;18(6):482–3.

22. Khan A, Kahn S, Kolts R, Brown WA. Suicide rates in clinical trials of SSRIs, other antidepressants and placebo: analysis of FDA reports. Am J Psychiatry 2003;160:790–2.

23. Healy D. Lines of evidence on the risks of suicide with selective serotonin reuptake inhibitors. Psychother Psychosom 2003;72:71–9.

24. Jick H, Kaye JA, Jick SS. Antidepressants and the risk of suicidal behaviors. JAMA 2004;292:338–43.

25. Slater E, Roth M. Clinical Psychiatry, 3rd edition. London: Baillère Tindall, 1977.

26. Whittington CJ, Kendall T, Cottrell D, Cotgrove, A, Boddington E. Selective serotonin reuptake inhibitors in childhood depression: systematic review of published versus unpublished data. Lancet 2004;363:1341–5.

27. Committee on Safety of Medicines. Selective Serotonin Reuptake Inhibitors (SSRIs): overview of regulatory status and CSM advice relating to major depressive disorder (MDD) in children and adolescents including a summary of available safety and efficacy data. http:/medicines.mh-ra.gov.uk/ourwork/monitorsafequalmed/safetymessages/ssrioverview_101203.htm, updated 8.10.2004.

28. Medicines Control Agency. Report of the CSM Expert Working Group on the Safety of Selective Serotonin Reuptake Inhibitor Antidepressants. http//medicines.mh-ra.gov.uk/ourwork/monitorsafequalmed/safetymessages//ssrifinal.pfd. 2004:December 31.

29. Committee on Safety of Medicines 2003. Paroxetine (Seroxat). Safety in children and adolescents. SSRI and venlafaxine use in children. 2003 September;29:4.

30. Martinez C, Rietbrock S, Wise L, Ashby D, Chick J, Moseley J, Evans S, Gunnell D. Antidepressant treatment and the risk of fatal and non-fatal self harm in first episode depression: nested case-control study. BMJ 2005;330:389–95.

31. Gunnell DJ, Saperia J, Ashby D. Selective serotonin reuptake inhibitors (SSRIs) and suicide in adults: meta-analysis of drug company data from placebo controlled, randomised controlled trials submitted to the MHRA's safety review. BMJ 2005;330:385–9.

32. Fergusson D, Doucette C, Glass KC, Shapiro S, Healy D, Hebert P, Hutton B. Association between suicide attempts and selective serotonin reuptake inhibitors: systematic and randomised controlled trials. BMJ 2005;330:396–402.

33. Movig KLL, Leufkens HGM, Lenderink AW, van den Akker VGA, Hodiamont PPG, Goldschmidt HMJ, Egberts ACG. Association between antidepressant drug use and hyponatremia: a case–control study. Br J Clin Pharmacol 2002;53(4):363–9.

34. Zullino D, Brauchli S, Horvath A, Baumann P. Inappropriate antidiuretic hormone secretion and rhabdomyolysis associated with citalopram. Thérapie 2000;55(5):651–2.

35. Iraqi A, Baickle E. A case report of hyponatremia with citalopram use. J Am Med Directors Assoc 2004;5:64–5.

36. Fabian TJ, Amico JA, Kroboth PD, Mulsant BH, Corey SE, Begley AE, Bensasi SG, Weber E, Dew MA, Reynolds CF, Pollock BG. Paroxetine induced hyponatremia in older adults. Arch Intern Med 2004;164:327–32.

37. Carrasco GA, Van De Kar LD. Neuroendocrine pharmacology of stress. Eur J Pharmacol 2003;463:275–2.

38. Odeh M, Beny A, Oliven A. Severe symptomatic hyponatremia during citalopram therapy. Am J Med Sci 2001;321(2):159–60.

39. Movig KLL, Egberts ACG, Van den Akker VGA, Goldschmidt HMJ, Leufkens HGM, Lenderink AW. SSRIs are different than other antidepressant agents: a larger risk of hyponatremia. Pharm Weekbl 2001;136:461–3.

40. Staab JP, Yerkes SA, Cheney EM, Clayton AH. Transient SIADH associated with fluoxetine. Am J Psychiatry 1990;147(11):1569–70.

41. Skop BP, Brown TM. Potential vascular and bleeding complications of treatment with selective serotonin reuptake inhibitors. Psychosomatics 1996;37(1):12–6.

42. Pai VB, Kelly MW. Bruising associated with the use of fluoxetine. Ann Pharmacother 1996;30(7–8):786–8.

43. Lake MB, Birmaher B, Wassick S, Mathos K, Yelovich AK. Bleeding and selective serotonin reuptake inhibitors in childhood and adolescence. J Child Adolesc Psychopharmacol 2000;10(1):35–8.

44. Nelva A, Guy C, Tardy-Poncet B, Beyens MN, Ratrema M, Benedetti C, Ollagnier M. Syndromes hemorragiques sons antidepresseurs inhibiteurs selectifs de la recapture de la serotonine (ISRS). A propos de sept cas et revue de la literature. [Hemorrhagic syndromes related to selective serotonin reuptake inhibitor (SSRI) antidepressants Seven case reports and review of the literature.] Rev Méd Interne 2000;21(2):152–60.

45. Dunbar GC. An interim overview of the safety and tolerability of paroxetine. Acta Psychiatr Scand Suppl 1989;350:135–7.

46. Boyer WF, Blumhardt CL. The safety profile of paroxetine. J Clin Psychiatry 1992;53(Suppl.):61–6.

47. Murdoch D, McTavish D. Sertraline: a review of its pharmacodynamic and pharmacokinetic properties, and therapeutic potential in depression and obsessive-compulsive disorder. Drugs 1992;44(4):604–24.

48. Stokes PE. Fluoxetine: a five-year review. Clin Ther 1993;15(2):216–43.

49. Shaw DM, Crimmins R. A multicenter trial of citalopram and amitriptyline in major depressive illness. In: Montgomery SA, editor. Citalopram—The New Antidepressant from Lundbeck Research. Amsterdam: Excerpta Medica, 1989:43.

50. Nemeroff CB. Evolutionary trends in the pharmacotherapeutic management of depression. J Clin Psychiatry 1994;55(Suppl.):3–15.

51. de Abajo FJ, Rodriguez LA, Montero D. Association between selective serotonin reuptake inhibitors and upper gastrointestinal bleeding: population based case-control study. BMJ 1999;319(7217):1106–9.

52. Dalton SO, Johansen C, Mellemkjaer L, Norgard B, Sorensen HT, Olsen JH. Use of selective serotonin reuptake inhibitors and risk of upper gastrointestinal tract bleeding: a population-based cohort study. Arch Intern Med 2003;163:59–64.

53. Sauer WH, Berlin JA, Kimmel SE. Effect of antidepressants and their relative affinity for the serotonin transporter on the risk of myocardial infarction. Circulation 2003;108:32.

54. Lopez-Torres E, Lucena MI, Seoane J, Verge C, Andrade RJ. Hepatoxicity related to citalopram. Am J Psychiatry 2004;161:923–4.

55. Solomons K, Gooch S, Wong A. Toxicity with selective serotonin re-uptake inhibitors. Am J Psychiatry 2005;162;1225–6.

56. Jan V, Toledano C, Machet L, Machet MC, Vaillant L, Lorette G. Stevens–Johnson syndrome after sertraline. Acta Dermatol Venereol 1999;79(5):401.

57. Cooper GL. The safety of fluoxetine—an update. Br J Psychiatry Suppl 1988;3:77–86.

58. Rosen RC, Lane RM, Menza M. Effects of SSRIs on sexual function: a critical review. J Clin Psychopharmacol 1999;19(1):67–85.

59. Labbate LA, Grimes JB, Arana GW. Serotonin reuptake antidepressant effects on sexual function in patients with anxiety disorders. Biol Psychiatry 1998;43(12):904–7.

60. Scharko AM, Reiner S. SSRI-induced sexual dysfunction in adolescents. J Am Acad Child Adolesc Psychiatry 2004;43:1067–8.

61. Kim SC, Seo KK. Efficacy and safety of fluoxetine, sertraline and clomipramine in patients with premature ejaculation: a double-blind, placebo controlled study. J Urol 1998;159(2):425–7.

62. Waldinger MD, Hengeveld MW, Zwinderman AH, Olivier B. Effect of SSRI antidepressants on ejaculation: a double-blind, randomized, placebo-controlled study with fluoxetine, fluvoxamine, paroxetine, and sertraline. J Clin Psychopharmacol 1998;18(4):274–81.

63. Waldinger MD, Olivier B. Sexual dysfunction and fluvoxamine therapy. J Clin Psychiatry 2001;62(2):126–7.

64. Montejo AL, Llorca G, Izquierdo JA, Rico-Villademoros F. Incidence of sexual dysfunction associated with antidepressant agents: a prospective multicenter study of 1022 outpatients. Spanish Working Group for the Study of Psychotropic-Related Sexual Dysfunction. J Clin Psychiatry 2001;62(Suppl. 3):10–21.

65. Hori H, Yoshimura R, Nakamura J. Increased libido in a woman treated with fluvoxamine: a case report. Acta Psychiatr Scand 2001;103(4):312–4.

66. Greil W, Horvath A, Sassim N, Erazo N, Grohmann R. Disinhibition of libido: an adverse effect of SSRI? J Affect Disord 2001;62(3):225–8.

67. Ashton AK, Rosen RC. Bupropion as an antidote for serotonin reuptake inhibitor-induced sexual dysfunction. J Clin Psychiatry 1998;59(3):112–5.

68. Michelson D, Bancroft J, Targum S, Kim Y, Tepner R. Female sexual dysfunction associated with antidepressant administration: a randomized, placebo-controlled study of pharmacologic intervention. Am J Psychiatry 2000;157(2):239–43.

69. Smith M, Robinson D. Sertraline and vaginal bleeding—a possible association. J Am Geriatr Soc 2002;50(1):200–1.

70. Linnebur SA, Saseen JJ, Pace WD. Venlafaxine-associated vaginal bleeding. Pharmacotherapy 2002;22(5):652–5.

71. Damsa C, Sterck R, Schulz P. Case of gynecomastia during paroxetine therapy. J Clin Psychiatry 2003;64:971.

72. Boulenger A, Viseux V, Plantin-Eon I, Commegeille P, Plantin P. Gynaecomastia following treatment by fluoxetine. J Eur Acad Dermatol Venereol 2003;17:97–116.

73. Dalfen AK, Stewart DE. Who develops severe or fatal adverse drug reactions to selective serotonin reuptake inhibitors? Can J Psychiatry 2001;46(3):258–63.

74. Haddad P. Newer antidepressants and the discontinuation syndrome. J Clin Psychiatry 1997;58(Suppl. 7):17–21discussion 22.

75. Coupland NJ, Bell CJ, Potokar JP. Serotonin reuptake inhibitor withdrawal. J Clin Psychopharmacol 1996;16(5):356–62.

76. Zajecka J, Tracy KA, Mitchell S. Discontinuation symptoms after treatment with serotonin reuptake inhibitors: a literature review. J Clin Psychiatry 1997;58(7):291–7.

77. Fernando AT 3rd, Schwader P. A case of citalopram withdrawal. J Clin Psychopharmacol 2000;20(5):581–2.

78. Lejoyeux M, Ades J. Antidepressant discontinuation: a review of the literature. J Clin Psychiatry 1997;58(Suppl. 7):11–5.

79. Michelson D, Fava M, Amsterdam J, Apter J, Londborg P, Tamura R, Tepner RG. Interruption of selective serotonin reuptake inhibitor treatment. Double-blind, placebo-controlled trial. Br J Psychiatry 2000;176:363–8.

80. Wisner KL, Gelenberg AJ, Leonard H, Zarin D, Frank E. Pharmacologic treatment of depression during pregnancy. JAMA 1999;282(13):1264–9.

81. Hallberg P, Sjoblom V. The use of selective serotonin reuptake inhibitors during pregnancy and breast-feeding: a review and clinical aspects. J Clin Psychopharmacol 2005;25:59–73.

82. Cohen LS, Heller VL, Bailey JW, Grush L, Ablon JS, Bouffard SM. Birth outcomes following prenatal exposure to fluoxetine. Biol Psychiatry 2000;48(10):996–1000.

83. Gerola O, Fiocchi S, Rondini G. Rischi da farmaci antidepressivi in gravidanza: revisione della letteratura e presentazione di un caso di sospetta sindrome da astinenza da paroxetina in neonato. [Risk of antidepressant drugs in pregnancy: review of the literature and presentation of a case of suspected sevotonin syndrome caused by abstinence of paroxetine in neonate.] Riv Ital Pediatr 1999;25:216–8.

84. Jaiswal S, Coombs RC, Isbister GK. Paroxetine withdrawal in a neonate with historical and laboratory confirmation. Eur J Pediatr 2003;162:723–4.

85. Laine Kari, Heikkinen T, Ekblad U, Kero P. Effects of exposure to selective serotonin reuptake inhibitors during pregnancy on serotonergic symptoms in newborns and cord blood monoamine and prolactin concentrations. Arch Gen Psychiatry 2003;60:720–6.

86. National Institute for Health and Clinical Excellence. Antenatal and postnatal mental health: clinical management and service guidance—NICE Guidance, February 2007: http://guidance.nice.org.uk/CG45.

87. Sanz EJ, De Las-Cuevas C, Bate A, Edwards R. Selective serotonin reuptake inhibitors in pregnant women and neonatal withdrawal syndrome: a database analysis. Lancet 2005;365:482–87.

88. Piontek CM, Wisner KL, Perel JM, Peindl KS. Serum fluvoxamine levels in breastfed infants. J Clin Psychiatry 2001;62(2):111–3.

89. Stowe, ZN, Hostetter AL, Owens MJ, Ritchie JC, Sternberg K, Cohen LS, Nemeroff CB. The pharmacokinetics of sertraline excretion into human breast milk: determinants of infant serum concentrations. J Clin Psychiatry 2003;64:73–80.

90. Lee A, Woo J, Ito S. Frequency of infant adverse events that are associated with citalopram use during breast-feeding. Am J Obstet Gynecol 2003;190:218–21.

91. Merlob P, Stahl B, Sulkes J. Paroxetine during breast-feeding: infant weight gain and maternal adherence to counsel. Eur J Pediatr 2004;163:135–9.

92. de Vries TW, de Jong-van de Berg LTW, Hadders-Algra M., Paroxetine during lactation: is it really safe for the infant? Acta Paediatr 2004;93:1406–7.

93. Barbey JT, Roose SP. SSRI safety in overdose. J Clin Psychiatry 1998;59(Suppl. 15):42–8.

94. Kelly CA, Dhaun N, laing WJ, Strachan FE, Good AM, Bateman DN. Comparative toxicity of citalopram and the newer antidepressants after overdose. J Toxicol 2004;42:67–71.

95. Alfaro CL, Lam YW, Simpson J, Ereshefsky L. CYP2D6 status of extensive metabolizers after multiple-dose fluoxetine, fluvoxamine, paroxetine, or sertraline. J Clin Psychopharmacol 1999;19(2):155–63.

96. Sporer KA. The serotonin syndrome. Implicated drugs, pathophysiology and management. Drug Saf 1995;13(2):94–104.

97. Suzuki Y, Shioiri T, Muratake T, Kawashima Y, Sato S, Hagiwara M, Inoue Y, Shimoda K, Someya T. Effects of concomitant fluvoxamine on the metabolism of alprazolam in Japanese psychiatric patients: interaction with CYP2C19 mutated alleles. Eur J Clin Pharmacol 2003;58:829–33.

98. Hall J, Naranjo CA, Sproule BA, Herrmann N. Pharmacokinetic and pharmacodynamic evaluation of the inhibition of alprazolam by citalopram and fluoxetine. J Clin Psychopharmacol 2003;23:349–57.

99. Manos GH. Possible serotonin syndrome associated with buspirone added to fluoxetine. Ann Pharmacother 2000;34(7–8):871–4.

100. Kurdyak PA, Juurlink DN, Kopp A, Hermann N, Mamdani MM. Antidepressants, warfarin and risk of hemorrhage. J Clin Psychopharmacol 2005;25:561–4.

101. Liston HL, DeVane CL, Boulton DW, Risch SC, Markowitz JS, Goldman J. Differential time course of cytochrome P450 2D6 enzyme inhibition by fluoxetine, sertraline, and paroxetine in healthy volunteers. J Clin Psychopharmacol 2002;22(2):169–73.

102. Lee DO, Lee CD. Serotonin syndrome in a child associated with erythromycin and sertraline. Pharmacotherapy 1999;19(7):894–6.

103. Hori H, Yoshimura R, Yeda N, Eto S, Shinkai K, Sakata S, Ohmori O, Terao T, Nakamura J. Grapefruit juice-fluvoxamine interaction. Is it risky or not? J Clin Psychopharmacol 2003;23:422–4.

104. Yasui-Furukori N, Kondo T, Mihara K, Inoue Y, Kaneko S. Fluvoxamine dose-dependent interaction and the effects on negative symptoms in schizophrenia. Psychopharmacology 2004;171:223–7.

105. Tahir N. Serotonin syndrome as consequence drug resistant infections: an interaction between linezolid and citalopram. J Am Med Directors Assoc 2004;5:111–3.

106. Thomas CR, Rosenberg M, Blythe V, Meyer WJ. Serotonin syndrome and linezolid. J Am Acad Child Adolesc Psychiatry 2004;43:790.

107. Hernandez-Lopez C, Farre M, Roset PN, Menoyo E, Pizarro N, Ortuno J, Torrens M, Cami J, de La Torre R. 3,4-Methylenedioxymethamphetamine (ecstasy) and alcohol interactions in humans: psychomotor performance, subjective effects, and pharmacokinetics. J Pharmacol Exp Ther 2002;300(1):236–44.

108. Smith DL, Wenegrat BG. A case report of serotonin syndrome associated with combined nefazodone and fluoxetine. J Clin Psychiatry 2000;61(2):146.

109. Smith DL, Wenegrat BG. A case report of serotonin syndrome associated with combined nefazodone and fluoxetine. J Clin Psychiatry 2000;61(2):146.

110. Hamilton S, Malone K. Serotonin syndrome during treatment with paroxetine and risperidone. J Clin Psychopharmacol 2000;20(1):103–5.

111. Spina E, Avenoso A, Facciola G, Scordo MG, Ancione M, Madia A. Plasma concentrations of risperidone and 9-hydroxyrisperidone during combined treatment with paroxetine. Ther Drug Monit 2001;23(3):223–7.

112. de Jong J, Hoogenboom B, van Troostwijk LD, de Haan L. Interaction of olanzapine with fluvoxamine. Psychopharmacology (Berl) 2001;155(2):219–20.

113. de Jong J, Hoogenboom B, van Troostwijk LD, de Haan L. Interaction of olanzapine with fluvoxamine. Psychopharmacology (Berl) 2001;155(2):219–20.

114. Callaghan JT, Bergstrom RF, Ptak LR, Beasley CM. Olanzapine. Pharmacokinetic and pharmacodynamic profile. Clin Pharmacokinet 1999;37(3):177–93.

115. Wigen CL, Goetz MB. Serotonin syndrome and linezolid. Clin Infect Dis 2002;34(12):1651–2.

116. Lavery S, Ravi H, McDaniel WW, Pushkin YR. Linezolid and serotonin syndrome. Psychosomatics 2001;42(5):432–4.

117. Tissot TA. Probable meperidine-induced serotonin syndrome in a patient with a history of fluoxetine use. Anesthesiol (Phil) 2003;98:1511–12.

118. Spina E, D'Arrigo C, Miglardi G, Morgante L, Zoccali R, Ancione M, Madia A. Plasma risperidone concentrations during combined treatment with sertraline. Ther Drug Monit 2004;26:386–90.

119. Benazzi F. Organic hypomania secondary to sibutramine–citalopram interaction. J Clin Psychiatry 2002;63(2):165.

120. Granfors MT, Backmann JT, Neuvonen M, Ahonen J, Neuvonen PJ. Fluvoxamine drastically increases concentrations and effects of tizanidine: a potentially hazardous interaction. Am Soc Clin Pharmacol Ther 2004;75:331–41.

121. Mahlberg R, Kunz D, Sasse J, Kirchheiner J. Serotonin syndrome with tramadol and citalopram. Am J Psychiatry 2004;161:1129.

122. Gonzalez-Pinto A, Imaz H, De Heredia JL, Gutierrez M, Mico JA. Mania and tramadol–fluoxetine combination. Am J Psychiatry 2001;158(6):964–5.

123. Ingelman-Sundberg M. Genetic susceptibility to adverse effects of drugs and environmental toxicants. The role of the CYP family of enzymes. Mutat Res 2001;482(1–2):11–9.

124. Sauget D, Franco PS, Amaniou M, Mazere J, Dantoine T. Possible syndrome sérotoninergiques induit par l'association de tramadol à de la sertraline chez une femme agée. [Possible serotonergic syndrome caused by combination of tramadol and sertraline in an elderly woman.] Thérapie. 57(3):309–10.

125. Ripple MG, Pestaner JP, Levine BS, Smialek JE. Lethal combination of tramadol and multiple drugs affecting serotonin. Am J Forensic Med Pathol 2000;21(4):370–4.

126. Aranow AB, Hudson JI, Pope HG Jr, Grady TA, Laage TA, Bell IR, Cole JO. Elevated antidepressant plasma levels after addition of fluoxetine. Am J Psychiatry 1989;146(7):911–3.

127. Preskorn SH, Beber JH, Faul JC, Hirschfeld RM. Serious adverse effects of combining fluoxetine and tricyclic antidepressants. Am J Psychiatry 1990;147(4):532.

128. Spina E, Campo GM, Avenoso A, Pollicino MA, Caputi AP. Interaction between Fluvoxamine and imipramine/desipramine in four patients. Ther Drug Monit 1992;14(3):194–6.

129. Härtter S, Wetzel H, Hammes E, Hiemke C. Inhibition of antidepressant demethylation and hydroxylation by fluvoxamine in depressed patients. Psychopharmacology (Berl) 1993;110(3):302–8.

130. Lydiard RB, Anton RF, Cunningham I. Interaction between sertraline and tricyclic antidepressants. Am J Psychiatry 1993;150(7):1125–6.

Citalopram and escitalopram

General Information

Citalopram is a racemic bicyclic phthalane derivative and is a highly selective serotonin re-uptake inhibitor with minimal effects on noradrenaline and dopamine neuronal reuptake. Inhibition of 5-HT re-uptake by citalopram is primarily due to escitalopram, the active S-enantiomer of citalopram (1). One would expect escitalopram to be twice as potent as citalopram but otherwise not to differ significantly from the racemic mixture. However, escitalopram is marketed as being more efficacious than citalopram because, it is argued, the inactive R-isomer present in the racemate actually inhibits binding of the S-enantiomer to its site of action, the serotonin transporter. In some, but not all, clinical trials escitalopram has been statistically superior to citalopram in terms of speed of onset of therapeutic action and improvement on depression rating scales. The clinical significance of these differences is debatable (2).

Their most frequent adverse events (nausea, somnolence, dry mouth, increased sweating) are mainly transient and mostly mild to moderate (3). In terms of adverse effects escitalopram appears to be equivalent to citalopram. For example, in placebo-controlled trials, escitalopram produced unwanted effects typical of the SSRI class, including nausea (15%), ejaculation disorders (9%), insomnia (9%), diarrhea (8%), somnolence (7%), dry mouth (6%), and dizziness (6%).

The single and multiple-dose pharmacokinetics of citalopram are linear and dose-proportional in the range 10–60 mg/day. Citalopram is metabolized to demethylcitalopram, didemethylcitalopram, citalopram-N-oxide, and a deaminated propionic acid derivative. Citalopram has a mean half-life of about 35 hours (4). Racemic citalopram is several times more potent than its metabolites in inhibiting serotonin reuptake (5).

In a systematic review of clinical trials the therapeutic efficacy of citalopram was significantly greater than that of placebo and comparable with that of other antidepressants (6).

Escitalopram oxalate is the S-enantiomer of citalopram (7). The therapeutic activity of citalopram resides in the S-isomer and escitalopram binds with high affinity to the human serotonin transporter; R-citalopram is about 30-fold less potent. Escitalopram is extensively metabolized in the liver by CYP2C19, CYP3A4, and CYP2D6, and its blood concentrations are increased by drugs that inhibit one or more of these enzymes. The half-life of escitalopram is 27–32 hours. Citalopram has negligible effects on CYP isozymes. In vitro, escitalopram is a weak inhibitor of CYP2D6. Drugs that are substrates for CYP2D6 and that have a narrow therapeutic index (for example, flecainide and metoprolol) should be prescribed with caution in conjunction with escitalopram. As with other SSRIs, escitalopram should not be co-administered with monoamine oxidase inhibitors.

Escitalopram was efficacious in patients with major depressive disorder in short-term, placebo-controlled trials, three of which included citalopram as an active control, and in a 36-week study in the prevention of relapse in depression (7). It has also been used to treat generalized anxiety disorder, panic disorder, and social anxiety disorder. Results also suggest that, at comparable doses, escitalopram demonstrates clinically relevant and statistically significant superiority to placebo treatment earlier than citalopram. The most common adverse events associated with escitalopram include nausea, insomnia, disorders of ejaculation, diarrhea, dry mouth, and somnolence. Only nausea occurred in more than 10% of patients taking escitalopram.

A meta-analysis of 20 short-term studies of five SSRIs (citalopram, fluoxetine, fluvoxamine, paroxetine, and sertraline) has been published (8). There were no overall differences in efficacy, but fluoxetine had a slower onset of action. Citalopram and sertraline were least likely to cause drug interactions, but citalopram was implicated more often in fatal overdoses.

Organs and Systems

Cardiovascular

There has been some concern about the cardiovascular safety of citalopram, mainly because of animal studies showing effects on cardiac conduction. These most commonly occur in large overdoses, in which a variety of cardiac abnormalities, including QT_c prolongation, have been noted. However, this can occur with therapeutic doses too.

- Bradycardia (34/minute) with a prolonged QT_c interval of 463 ms occurred in a patient taking citalopram 40 mg/day (9). The bradycardia resolved when citalopram was withdrawn. The patient also had alcohol dependence and evidence of cardiomyopathy; presumably this may have potentiated the effect of citalopram on cardiac conduction.
- A 21-year-old woman developed QT_c prolongation (457 ms) after taking a fairly modest overdose (400 mg) of citalopram (usual daily dose 20–60 mg) (10). The QT_c prolongation resolved uneventfully over the next 30 hours.

This suggests that even modest overdoses of citalopram can cause QT_c prolongation and that cardiac monitoring should be considered. Based on the pharmacokinetic profile of citalopram and the temporal pattern of QT_c change, the authors suggested that the effect of citalopram on the QT_c interval was mediated by one of its metabolites, dimethylcitalopram.

Prolongation of the QT_c interval has been reported in five patients who made non-fatal suicide attempts by taking large amounts of citalopram. Their electrocardiograms showed other conduction disorders, including sinus tachycardia and inferolateral repolarization disturbances (SEDA-21, 12).

Nervous system

SSRIs can infrequently cause extrapyramidal movement disorders and can also worsen established Parkinson's disease (SEDA-22, 23) and another case has been reported (11).

- A 68-year-old woman developed major depression. A neurological assessment excluded neurological diseases, including Parkinson's disease. After treatment with citalopram, 20 mg/day for 7 days, she developed severe parkinsonism, with rigidity, tremor, and bradykinesia, and became unable to walk. The citalopram was withdrawn after a further week and nortriptyline was substituted; however, 10 days later parkinsonism was still present. Her symptoms eventually responded to cobeneldopa.

The authors concluded that the citalopram had probably precipitated latent Parkinson's disease. Citalopram is the most highly selective SSRI and, in anecdotal accounts, has been implicated somewhat less often than other SSRIs in extrapyramidal movement disorders. However, this case, together with another report of citalopram-induced worsening of pre-existing Parkinson's disease (12), suggests that it should be used with caution in patients with this disorder.

The rabbit syndrome is a movement disorder characterized by involuntary perioral movements that mimic the chewing movements of a rabbit. The condition is distinguished from tardive dyskinesia by the lack of tongue involvement. Rabbit syndrome is usually associated with antipsychotic drug treatment, but two cases have been associated with escitalopram and citalopram; both resolved when the antidepressant was withdrawn (13). SSRIs can rarely cause extrapyramidal movement disorders (SEDA-22, 23), probably through indirect interaction with dopaminergic pathways, and this is presumably the mechanism here. Venlafaxine did not produce the rabbit syndrome in either patient, even though venlafaxine is a potent serotonin re-uptake inhibitor. This suggests that rabbit syndrome may be more specifically associated with citalopram and escitalopram than with SSRIs in general. Alternatively, the concomitant noradrenergic potentiation produced by venlafaxine may have prevented expression of the movement disorder.

Sensory systems

Sudden-onset diplopia can be an alarming symptom and can be associated with serious underlying disorders, such as cranial nerve lesions, orbital disease, intranuclear ophthalmoplegia, and vertebrobasilar insufficiency.

- A 28-year-old medical student developed major depression and was given citalopram (20 mg/day) (14). After 12 days he described incapacitating diplopia which resolved when he closed one eye. Neurological and ophthalmic examination was normal and no structural lesion was detected with brain magnetic resonance imaging. The citalopram was withdrawn and the diplopia resolved within 3 days.

The fact that withdrawal of citalopram led to rapid resolution of the diplopia suggests that it was due to the citalopram, but the mechanism of this rare adverse effect is unclear.

Psychological, psychiatric

Mania has been reported in six patients, five of whom were taking citalopram and one paroxetine (SEDA-22, 12).

SSRIs are generally thought to have relatively little effect on tasks of psychological performance in comparison to tricyclic antidepressants and agents such as mirtazapine (SEDA-28, 19). In 24 healthy men and women aged 30–50 years who were randomized to receive citalopram (40 mg/day), sertraline (100 mg/day), and placebo for 14 days in a crossover, within-subject design, citalopram (but not sertraline) caused impaired vigilance on the Mackworth clock task (15). The same authors have previously reported similar deficits after fluoxetine, paroxetine, and venlafaxine, suggesting that impairment of vigilance might be a general consequence of drugs that potently block serotonin re-uptake. In contrast to other SSRIs, sertraline did not apparently impair vigilance in the Mackworth clock task. The authors speculated that this might be due to its concomitant dopamine re-uptake blocking properties. This study has a certain ecological validity because it investigated the effects of subchronic treatment with SSRIs rather than the more common approach of using single doses. However, it is not clear how far the reduction in vigilance that the authors detected would lead to deficits in performance of real-world tasks such as driving (see below). Also we do not know how SSRI treatment might alter psychological performance in depressed patients, many of whom have pre-existing cognitive deficits due to the depressive disorder.

The effects of both acute and subchronic mirtazapine and escitalopram on driving performance in a specially adapted vehicle have been studied in 18 healthy participants (9 men and 9 women) mean age 31.4 years (16). They were randomly assigned to escitalopram (increasing to 20 mg/day over 15 days), mirtazapine (increasing to 45 mg/day over 15 days), and placebo in a double-blind crossover design. Escitalopram did not alter driving performance at any time, whereas mirtazapine impaired driving after 2 days but not after 9 and 16 days. These data suggest that escitalopram in standard clinical doses does not alter driving performance in healthy volunteers. The effects of mirtazapine, even at the highest dose, were fairly transient; however, it seems sensible to warn patients starting mirtazapine treatment to be cautious about driving until they have adapted to the sedative effects of the drug.

Electrolyte balance

Hyponatremia can sometimes cause severe disturbances of consciousness.

- A 47-year-old woman with multiple sclerosis took citalopram 20 mg/day for 4 weeks and was found unconscious in her apartment (17). The main finding was a low plasma sodium (108 mmol/l). As a result of prolonged coma, she had rhabdomyolysis and required intubation for 3 days as well as sodium replacement therapy. She eventually made a full recovery.

It is possible in this case that the underlying demyelinating disease may have made the patient more susceptible to the sodium-lowering effects of the SSRI.

- A 45-year-old woman developed hyponatremia complicated by rhabdomyolysis while taking citalopram and the antipsychotic drug chlorprothixene for depressive psychosis (18). The hyponatremia became apparent 2 weeks after the dose of citalopram was increased to 40 mg/day, when she complained of weakness and lethargy.

SSRI-induced hyponatremia is unusual in non-geriatric populations, but the chlorprothixene may have played a role in this case.

Gastrointestinal

Gastrointestinal adverse effects are one of the major disadvantages of SSRIs. The most common is nausea, and the incidence is said to be 20% or more for citalopram (19,20).

Sexual function

Citalopram has a relatively modest effect in delaying ejaculation (21).

Sexual disinhibition has been reported in five patients, four of whom were taking citalopram; they had an unusual increase in sexual interest, with preoccupation with sexual thoughts, promiscuity, and excessive interest in pornography (22). In some of the cases symptoms such as diminished need for sleep suggested the possibility of a manic syndrome.

Long-Term Effects

Drug withdrawal

Reports of withdrawal symptoms after citalopram withdrawal are rare, but it is uncertain whether this reflects a truly lower propensity to cause withdrawal symptoms.

- A 30-year-old man with a history of major depression and panic disorder had been in remission for a year with citalopram 20 mg/day, valproate 600 mg/day, and alprazolam 3 mg/day (23). The citalopram was tapered over 3 weeks to 5 mg/day and then withdrawn. The day after the last dose he experienced anxiety and irritability together with frequent short-lasting bursts of dizziness, not having had the latter previously panic and depression did not recur and after a week the symptoms resolved spontaneously.
- A 45-year-old woman achieved remission from an episode of major depression within 2 weeks of taking citalopram (40 mg/day). After about 3 months of treatment she missed her daily dose of citalopram, and 3 hours later had a sudden episode of dizziness while driving. A similar episode occurred 2 weeks later

again after a missed dose of citalopram. The dizziness remitted about 1 hour after the citalopram was taken.

These symptoms, particularly dizziness, are characteristic of SSRI withdrawal, and suggest that citalopram, like other SSRIs, can cause a withdrawal syndrome in some patients, despite slow tapering of the dose.

Withdrawal symptoms in the 2 weeks after sudden discontinuation of citalopram have been examined in a double-blind, placebo-controlled study (24). Withdrawal symptoms were overall mild, but neurological and psychiatric disturbances were 2–3 times as common in patients randomized to placebo than in those randomized to continue with citalopram. The authors pointed out that withdrawal symptoms were particularly common in patients who were randomized to placebo who also had depressive relapses. This shows the difficulty of disentangling the effects of depressive relapse from those of pure treatment withdrawal. However, it is also possible that acute withdrawal of medication induces an abnormal neurobiological state, in which both depression and abstinence symptoms are more likely to occur. It would be wise to warn patients about the possible effects of missing doses of the shorter-acting SSRIs.

Second-Generation Effects

Lactation

Citalopram has been reported to cause sleep disturbance in a breast-fed infant (25).

- A 29-year-old woman took citalopram (40 mg/day) while breast feeding her 5-week-old daughter. The maternal citalopram concentrations were 99 ng/ml in the serum and 205 ng/ml in the breast milk. The serum concentration in the infant was 13 ng/ml, and the child's sleep was fitful and disturbed. The dosage of citalopram was reduced to 20 mg/day and the two feeds after each daily dose were replaced by artificial nutrition. One week later the infant was sleeping normally, and the serum citalopram concentrations in mother and infant had fallen to 35 ng/ml and 2 ng/ml, respectively.

These data suggest that although breast feeding during citalopram treatment is possible, careful dosing and close observation of mother and infant are necessary.

Drug Administration

Drug overdose

Six deaths have been reported after overdosage of citalopram. Although five of the six had also taken other substances, these were not thought to have contributed significantly (SEDA-20, 8). Of five patients who made non-fatal suicide attempts by taking large amounts of citalopram (up to 5200 mg), four developed generalized seizures and all had prolonged QT_c intervals. Other conduction disorders included sinus tachycardia and inferolateral repolarization disturbances. Two patients developed rhabdomyolysis and one hypokalemia. These data suggest that citalopram overdose can cause seizures and disturbances of cardiac conduction that might predispose to fatal dysrhythmias (SEDA-21, 12) (26,27).

In another case prolonged sinus bradycardia occurred (28).

- A 32-year-old woman took 800 mg of citalopram, 20 times her usual daily dose, in a suicide attempt. On admission to hospital she had a sinus bradycardia (41/minute) but the electrocardiogram was otherwise normal, with a QT interval of 430 ms. Treatment with atropine failed to increase her heart rate and she had hypotension and syncope. A temporary pacemaker was inserted and was required for the next 6 days before it could be safely removed.

There are concerns that citalopram may be less safe in acute overdose than other SSRIs (SEDA-21, 12). Among all fatal poisonings in one forensic district of Sweden, citalopram was the fourth most commonly used drug (22 of 358 cases) (29). However, when correction was made for prescription rate, citalopram was less toxic than amitriptyline, dextropropoxyphene, or nitrazepam. This study has confirmed that citalopram is less toxic than tricyclic antidepressants such as amitriptyline. However, whether it is more toxic than other SSRIs is still uncertain.

Drug–Drug Interactions

General

Of all the SSRIs citalopram has the least inhibitory effect on cytochrome P450 enzymes and has not been associated with clinically significant interactions with other CNS drugs.

It has been used successfully in combination with the tricyclic antidepressant desipramine in a 45-year-old woman who had previously suffered tricyclic toxicity when desipramine had been combined with paroxetine (30).

Acenocoumarol

When a drug has a relatively narrow therapeutic index, such as acenocoumarol, pharmacokinetic interactions can have serious clinical consequences.

- A 63-year-old woman taking acenocoumarol 18 mg/week (INR 1.8) started to take citalopram 20 mg/day and 10 days later noted spontaneous bleeding from her gums; the INR had risen to more than 15 (31). She was treated with two units of whole blood and the citalopram was withdrawn. Five days later the INR had returned to the target range.

Citalopram is said to be less likely than other SSRIs to cause drug interactions, because it is a relatively weak inhibitor of CYP isozymes. However, even slight inhibition may have produced serious consequences in this case.

Clozapine

The effect of citalopram on plasma concentrations of clozapine have been prospectively studied in 15 patients with schizophrenia taking clozapine 200–400 mg/day (32). The addition of citalopram 40 mg/day did not alter plasma clozapine concentrations.

However, in a 39-year-old man with a schizoaffective disorder, citalopram 40 mg/day appeared to increase plasma clozapine concentrations and increased adverse effects (33). The adverse effects settled within 2 weeks of a reduction in citalopram dosage to 20 mg/day, with a corresponding 25% fall in clozapine concentrations. It is possible that at higher doses, citalopram can increase clozapine concentrations, perhaps through inhibition of CYP1A2 or CYP3A4.

Digoxin

Citalopram 40 mg/day for 4 weeks did not alter the pharmacokinetics of digoxin 1 mg orally (34). Digoxin is not a CYP substrate, so an interaction with SSRIs is unlikely, but the authors cited a report that fluoxetine increased plasma digoxin concentrations (35).

Risperidone

The effect of citalopram on plasma concentrations of risperidone has been prospectively studied in 15 patients with schizophrenia (32). The addition of citalopram did not alter plasma risperidone concentrations.

References

1 Murdoch D, Keam SJ. Escitalopram: a review of its use in the management of major depressive disorder. Drugs 2005;65:2379–404.

2 National Institute for Health and Clinical Excellence. CG23. Depression: management of depression in primary and secondary care—NICE Guidance, December 2005: http://guidance.nice.org.uk/CG23.

3 Nemeroff CB. Overview of the safety of citalopram. Psychopharmacol Bull 2003;37(1):96–121.

4 Kragh-Sorensen P, Overo KF, Petersen OL, Jensen K, Parnas W. The kinetics of citalopram: single and multiple dose studies in man. Acta Pharmacol Toxicol (Copenh) 1981;48(1):53–60.

5 Sanchez C, Hyttel J. Comparison of the effects of antidepressants and their metabolites on reuptake of biogenic amines and on receptor binding. Cell Mol Neurobiol 1999;19(4):467–89.

6 Parker NG, Brown CS. Citalopram in the treatment of depression. Ann Pharmacother 2000;34(6):761–71.

7 Burke WJ. Escitalopram. Expert Opin Invest Drugs 2002;11(10):1477–86.

8 Edwards JG, Anderson I. Systematic review and guide to selection of selective serotonin reuptake inhibitors. Drugs 1999;57(4):507–33.

9 Favre MP, Sztajzel J, Bertschy G. Bradycardia during citalopram treatment: a case report. Pharmacol Res 1999;39(2):149–50.

10 Catalano G, Catalano MC, Epstein MA, Tsambiras PE. QTc interval prolongation associated with citalopram overdose: a case report literature review. Clin Neuropharmacol 2001;24(3):158–62.

11 Stadtland C, Erfurth A, Arolt V. De novo onset of Parkinson's disease after antidepressant treatment with citalopram. Pharmacopsychiatry 2000;33(5):194–5.

12 Linazasoro G. Worsening of Parkinson's disease by citalopram. Parkinsonism Relat Disord 2000;6(2):111–3.

13 Parvin M, Swatz CM. Dystonic rabbit syndrome from citalopram. Clin Neuropharmacol 2005;28:289–91.

14 Mowla A, Ghanizadeh AG, Ashkani HA. Diplopia with citalopram. J Clin Psychopharmacol 2005;25:623–4.

15 Riedel WJ, Eikmans K, Heldens A, Schmitt JAJ. Specific serotonergic reuptake inhibition impairs vigilance performance acutely and after subchronic treatment. J Psychopharmacol 2005;19:12–20.

16 Wingen M, Bothmer J, Langer S, Ramaekers G. Actual driving performance and psychomotor function in healthy subjects after acute and subchronic treatment with escitalopram, mirtazapine, and placebo: a crossover trial. J Clin Psychiatry 2005;66:436–43.

17 Hull M, Kottlors M, Braune S. Prolonged coma caused by low sodium and hypo-osmolarity during treatment with citalopram. J Clin Psychopharmacol 2002;22(3):337–8.

18 Zullino D, Brauchli S, Horvath A, Baumann P. Inappropriate antidiuretic hormone secretion and rhabdomyolysis associated with citalopram. Thérapie 2000;55(5):651–2.

19 Dencker SJ, Hopfner Petersen HE. Side effect profile of citalopram and reference antidepressants in depression. In: Montgomery SA, editor. Citalopram: The New Antidepressant from Lundbeck Research. Amsterdam: Excerpta Medica, 1989:31.

20 Shaw DM, Crimmins R. A multicenter trial of citalopram and amitriptyline in major depressive illness. In: Montgomery SA, editor. Citalopram: The New Antidepressant from Lundbeck research. Amsterdam: Excerpta Medica, 1989:43–9.

21 Waldinger MD, Olivier B, Nafziger AN, Bertino JS Jr, Goss-Bley AI, Kashuba ADM. Sexual dysfunction and fluvoxamine therapy. J Clin Psychiatry 2001;62(2):126–7.

22 Greil W, Horvath A, Sassim N, Erazo N, Grohmann R. Disinhibition of libido: an adverse effect of SSRI? J Affect Disord 2001;62(3):225–8.

23 Benazzi F. Citalopram withdrawal symptoms. Eur Psychiatry 1998;13:219.

24 Markowitz JS, DeVane CL, Liston HL, Montgomery SA. An assessment of selective serotonin reuptake inhibitor discontinuation symptoms with citalopram. Int Clin Psychopharmacol 2000;15(6):329–33.

25 Schmidt K, Olesen OV, Jensen PN. Citalopram and breast-feeding: serum concentration and side effects in the infant. Biol Psychiatry 2000;47(2):164–5.

26 Personne M, Persson H, Sjoberg E. Citalopram toxicity. Lancet 1997;350(9076):518–9.

27 Power A. Drug treatment of depression. Citalopram in overdose may result in serious morbidity and death. BMJ 1998;316(7127):307–8.

28 Rothenhausler HB, Hoberl C, Ehrentrout S, Kapfhammer HP, Weber MM. Suicide attempt by pure citalopram overdose causing long-lasting severe sinus bradycardia, hypotension and syncopes: successful therapy with a temporary pacemaker. Pharmacopsychiatry 2000;33(4):150–2.

29 Jonasson B, Saldeen T. Citalopram in fatal poisoning cases. Forensic Sci Int 2002;126(1):1–6.

30 Ashton AK. Lack of desipramine toxicity with citalopram. J Clin Psychiatry 2000;61(2):144.

31 Borras-Blasco J, Marco-Garbayo JL, Bosca-Sanleon B, Navarro-Ruiz A. Probable interaction between citalopram and acenocoumarol. Ann Pharmacother 2002;36(2):345.

32 Avenoso A, Facciola G, Scordo MG, Gitto C, Ferrante GD. No effect of citalopram on plasma levels of clozapine, risperidone and their active metabolites in patients with chronic schizophrenia. Clin Drug Invest 1998;16:393–8.

33 Borba CP, Henderson DC. Citalopram and clozapine: potential drug interaction. J Clin Psychiatry 2000;61(4):301–2.

34 Larsen F, Priskorn M, Overo KF. Lack of citalopram effect on oral digoxin pharmacokinetics. J Clin Pharmacol 2001;41(3):340–6.

35 Leibovitz A, Bilchinsky T, Gil I, Habot B. Elevated serum digoxin level associated with coadministered fluoxetine. Arch Intern Med 1998;158(10):1152–3.

Fluoxetine

General Information

Fluoxetine is a selective serotonin re-uptake inhibitor (SSRI). The manufacturers of fluoxetine have published a review of the adverse effects that were noted in 1378 patients who took it for up to 2 years (1).

In a meta-analysis based on 9087 patients in 87 different randomized clinical trials fluoxetine was more effective than placebo from the first week of therapy (2). In bulimia nervosa, fluoxetine was as effective as other agents. It was as effective as clomipramine in the treatment of obsessive-compulsive disorder.

The major adverse effects of fluoxetine confirm its stimulant profile and its relative lack of anticholinergic actions. The most frequent adverse effects, which occurred in 10–25% of patients, were nausea (25%), nervousness, insomnia, headache, tremor, anxiety, drowsiness, dry mouth, sweating, and diarrhea (10%). Most of these adverse effects occurred early in treatment and seldom led to drug withdrawal.

Organs and Systems

Cardiovascular

Fluoxetine appears not to have the cardiovascular effects associated with tricyclic compounds, but 10 patients did discontinue treatment because of tachycardia, palpitation, and dyspnea (3). Two older women each had a myocardial infarction and subsequently died, although these events may not have been drug-related.

In general, SSRIs are assumed to be safe in patients with cardiovascular disease, although there have been few systematic investigations in these patients. In a prospective study of 27 depressed patients with established cardiac disease, fluoxetine (up to 60 mg/day for 7 weeks) produced a statistically significant reduction in heart rate (6%) and an increase in supine systolic blood pressure (2%) (4). One patient had worsening of a pre-existing dysrhythmia and this persisted after fluoxetine withdrawal. These findings suggest that, relative to tricyclic antidepressants, fluoxetine may have a relatively benign profile in patients with cardiovascular disease. However, the authors cautioned that in view of the small number of patients studied, these findings cannot be widely generalized.

The effects of fluoxetine (20 mg/day for 12 weeks) on sitting and standing blood pressures have been reported (5). Fluoxetine modestly but significantly lowered sitting and standing systolic and diastolic blood pressures by about 2 mmHg. Patients with pre-existing cardiovascular disease showed no change. This study confirms that fluoxetine has little effect on blood pressure in physically healthy depressed patients and in those with moderate cardiovascular disease.

Fluoxetine-induced remission of Raynaud's phenomenon has been reported (SEDA-18, 20) (6).

Cardiac dysrhythmias

Fluoxetine has reportedly caused prolongation of the QT_c interval (7).

- A 52-year-old man had an abnormally prolonged QT_c interval of 560 ms, with broad-based T-waves. He had taken fluoxetine 40 mg/day over the previous 3 months, before which an electrocardiogram had shown a normal QT_c interval (380 ms). The fluoxetine was withdrawn, and 10 days later the QT_c interval was 380 ms. His only other medication was verapamil which he had taken for 3 years for hypertension.

Systematic studies of fluoxetine as monotherapy have not shown evidence of QT_c prolongation. It is possible in this case that fluoxetine interacted with verapamil to produce a conduction disorder.

An elderly man developed atrial fibrillation and bradycardia shortly after starting fluoxetine, and again on rechallenge (SEDA-16, 9). Dose-dependent bradycardia with dizziness and syncope has also been reported in a few patients taking fluoxetine (SEDA-16, 9) and in a presenile patient (8).

Nervous system

Several cases in which fluoxetine worsened parkinsonian disability have been described, and the problem of exacerbation of Parkinson's disease by fluoxetine has been reviewed (SEDA-18, 19).

Five patients taking fluoxetine developed akathisia, perhaps due to enhanced serotonergic inhibition of dopamine neurons (9). A causal link between fluoxetine-induced akathisia and suicidal behavior has been suggested (SEDA-17, 19) (SEDA-18, 19), and akathisia has also been associated with "indifference" (SEDA-18, 19).

One case of neuroleptic malignant syndrome has been described with fluoxetine (10).

One case of tics after long-term fluoxetine therapy has been described; the symptoms subsided several months after withdrawal (SEDA-18, 19).

One case of migraine associated with fluoxetine has been reported, with no further attacks when the drug was withdrawn (SEDA-18, 20).

Stuttering has been reported with fluoxetine (11).

Two patients who had been maintained successfully on fluoxetine (20 mg/day) for 6 and 10 years respectively began to have agitation, tension, and sleep disturbance (12). There had been no recent changes in medications or life events to explain these symptoms, which closely resembled the kind of adverse effects that can occur shortly after the start of SSRI treatment. Both patients improved after downward titration of the dose of fluoxetine. Blood concentrations of fluoxetine were not reported, so it is possible that for some reason (for example a change in diet or activity) plasma fluoxetine concentrations had recently increased in these subjects. However, the development of characteristic adverse effects after such a long trouble-free period suggests that patients taking maintenance medication need long-term follow-up, or at least ready access to specialist advice.

Seizures

Fluoxetine has been associated with seizures, ((3,13) and in overdose (3,14). It has also been shown to lengthen seizure duration during electroconvulsive therapy (SEDA-17, 20). Four patients had suspected seizures during studies (3) and one who took a 3000 mg overdose (3) had unequivocal convulsions but recovered.

Sensory systems

Tricyclic antidepressants can precipitate acute glaucoma through their anticholinergic effects. There are also reports that SSRIs can cause acute glaucoma, presumably by pupillary dilatation (see the monograph on Paroxetine).

In a placebo-controlled study in depressed patients a single dose of fluoxetine (20 mg) increased intraocular pressure by 4 mmHg (15). This increase is within the normal diurnal range, but could be of clinical consequence in individuals predisposed to glaucoma. However, post-marketing surveillance has not suggested an association between the use of fluoxetine and glaucoma (16).

Blurred vision has occasionally required withdrawal of fluoxetine (3).

Psychological, psychiatric

An analysis of severe adverse effects that caused drug withdrawal showed that psychotic reactions occurred in nine of 1378 patients; in four cases this appeared to be a stimulant psychosis and in three a conversion to mania (3).

Cognitive function can be impaired by fluoxetine; a negative effect on learning and memory has been described (SEDA-17, 20).

Reports of acute mania and manic-like behavior after treatment with fluoxetine or fluvoxamine have appeared (SEDA-13, 12; SEDA-17, 20; SEDA-18, 20), but there are not enough data to estimate the incidence.

Suicidal ideation has been described after 2–7 weeks of fluoxetine (17) and other case reports (SEDA-16, 9; SEDA-17, 19). A causal link was initially questioned (SED-12, 57; SEDA-15, 15; SEDA-17, 19), and in one controlled trial there was no increase (SEDA-16, 9).

Furthermore, a meta-analysis of controlled trials did not point to a greater risk of suicide attempts or suicidal ideation with fluoxetine than with tricyclic antidepressants (1).

In a meta-analysis based on 9087 patients in 87 different randomized clinical trials with fluoxetine there was no increased risk of suicide (2).

In an analysis of data from the National Institute of Mental Health Collaborative Depression Study in 643 patients with affective disorders who were followed up after fluoxetine was approved by the FDA in December 1987 for the treatment of depression, nearly 30% (n = 185) took fluoxetine at some point (18). There was an increased rate of suicide attempts before fluoxetine treatment in those who subsequently took fluoxetine. Relative to no treatment, fluoxetine and other antidepressants were associated with non-significant reductions in the likelihood of suicide attempts or completions. Severity of psychopathology was strongly associated with increased risk, and each suicide attempt after admission to the study was associated with a marginally significant increase in the risk of suicidal behavior. The authors concluded that the results did not support the speculation that fluoxetine increases the risk of suicide.

Endocrine

Fluoxetine causes weight loss, in contrast to tricyclic antidepressants (19). In one study there was a mean fall in weight of 3.88 pounds over 6 weeks compared with a gain of 4.6 pounds with amitriptyline (20).

Serotonin pathways are involved in the regulation of prolactin secretion. Amenorrhea, galactorrhea, and hyperprolactinemia have been reported in a patient taking SSRIs.

- A 71-year-old woman who had taken fluoxetine (dose unspecified) for a number of weeks noted unilateral galactorrhea and had a raised prolactin of 37 ng/ml (reference range 1.2–24 ng/ml) (21). She was also taking estrogen hormonal replacement therapy, benazapril, and occasional alprazolam. Withdrawal of the fluoxetine led to normalization of the prolactin concentration and resolution of the galactorrhea.

Estrogens also facilitate prolactin release, and so hormone replacement therapy may have played a part in this case.

Hematologic

Petechiae and prolongation of bleeding time have been reported in association with fluoxetine (SEDA-16, 10). In one case a 31-year-old woman taking fluoxetine developed bruising (22).

Aplastic anemia with fever, pleuritic chest pain, and pancytopenia developed in a 28-year-old woman (23) after 6 weeks of fluoxetine therapy. Bone-marrow examination confirmed acute marrow aplasia. She recovered 19 days after withdrawal. Neutropenia occurred rapidly after rechallenge. Another report described severe neutropenia during fluoxetine treatment (24).

Gastrointestinal

Gastrointestinal adverse effects are one of the major disadvantages of SSRIs (see General section). Two women developed stomatitis while taking fluoxetine; in one it recurred on rechallenge (25).

Liver

Chronic hepatitis associated with fluoxetine has been reported in a 35-year-old man (26).

In a post-marketing surveillance study there were some cases in which fluoxetine alone appeared to have precipitated hepatitis, which remitted when treatment was withdrawn (27). Fluoxetine can cause mild increases in liver enzymes, with a rate in clinical trials of about 0.5%. Rarely this can progress to hepatitis.

Skin

Rashes due to fluoxetine occur in a few percent of patients (28). Fluoxetine has been implicated in two cases of psoriasis (SEDA-17, 20). A woman taking fluoxetine monotherapy had painful burning, persistent erythema, and blisters on sun-exposed areas (SEDA-20, 8).

Hair

Reversible hair loss has been reported in patients taking fluoxetine (SEDA-16, 10) (SEDA-17, 20).

Sexual function

Fluoxetine can impair sexual function in both sexes, and particularly causes delayed orgasm or anorgasmia (SEDA-14, 14; SEDA-17, 21; SEDA-18, 21), in 5–10% of patients (29,30).

Fluoxetine has been implicated in one case of prolonged erection (SEDA-18, 21).

Yawning, clitoral engorgement, and spontaneous orgasm have been associated with fluoxetine (31).

Penile anesthesia has been reported in association with fluoxetine and sertraline (SEDA-17, 20).

Immunologic

Vasculitis has been attributed to fluoxetine (32,33).

- A patient who took fluoxetine for a manic-depressive disorder developed pulmonary inflammatory nodules with non-caseating giant cell granulomas, interstitial pneumonia, and non-necrotizing vasculitis, but remained asymptomatic (34). The diagnosis was made by open lung biopsy. The pulmonary nodules progressively resolved after withdrawal and the chest X-ray returned to normal in 9 months.

Second-Generation Effects

Fetotoxicity

Fluoxetine can occasionally cause cardiac dysrhythmias in adults, and may have done so in a fetus, whose mother took fluoxetine during pregnancy.

- A woman took fluoxetine (20–30 mg/day) from the 28th week of pregnancy (35). The fluoxetine was withdrawn during the 37th week, and at 38 weeks a male infant (2700 g) was born by spontaneous vaginal delivery. Both before and after delivery the baby was noted to have multiple atrial and ventricular extra beats. Echocardiography showed a normal heart and the baby was otherwise well. By discharge on day 5 the frequency of extra beats had fallen, and on follow-up 1 month later they were no longer present.

The long half-life of fluoxetine and its active metabolite, norfluoxetine, means that the active drug would have been present in the mother and baby for some weeks after withdrawal. However, as the authors pointed out, the baseline incidence of atrial extra beats in a fetus in utero is about 1–2%. It is therefore possible that treatment with fluoxetine in this case was coincidental.

Fluoxetine-related withdrawal effects have been reported in a neonate whose mother had taken fluoxetine during pregnancy (SEDA-18, 21).

Drug Administration

Drug overdose

Nine patients took overdoses of fluoxetine in amounts up to 3000 mg (37 times the recommended dose) (3). One, who also took several other drugs, including amitriptyline, died, but the other eight all recovered with relatively minor symptoms in most cases. Four patients had suspected seizures during studies (3) and one who took a 3000 mg overdose had unequivocal convulsions but recovered.

Overdosage of fluoxetine has been implicated in the development of seizures, but usually only when taken with other substances.

- A 15-year-old girl had a tonic-clonic seizure after overdosage of fluoxetine alone; she recovered uneventfully (36).

Drug–Drug Interactions

Beta-blockers

An interaction between metoprolol and fluoxetine with severe bradycardia has been described (SEDA-18, 23).

Carbamazepine

Fluoxetine can increase plasma concentrations of carbamazepine (SEDA-17, 21) (SEDA-18, 23).

Clozapine

In 10 patients stabilized on clozapine (200–450 mg/day) who took fluoxetine (20 mg/day) for 8 weeks, mean plasma concentrations of clozapine, norclozapine, and clozapine N-oxide increased significantly by 58%, 36%, and 38% respectively (44).

In two cases ingestion of clozapine and fluoxetine had a fatal outcome (45). The blood fluoxetine concentration was 0.7 µg/ml, which would be considered a high therapeutic concentration (usual target range 0.03–0.5 µg/ml). The blood clozapine concentration was 4900 ng/ml, which is within the lethal concentration range (1600–7100 ng/ml).

Dual effects were observed in a 44-year-old schizophrenic patient taking clozapine with both fluoxetine and sertraline for mood stabilization (46). Clinical and motor status improved with both fluoxetine and sertraline; cognitive function improved with clozapine and fluoxetine, but was not sustained with sertraline.

Digoxin

Fluoxetine has been reported to increase serum digoxin concentrations (37).

- A 93-year-old woman with congestive cardiac failure, paroxysmal atrial fibrillation, and hypertension developed depression after the death of her daughter. For several months she had been taking captopril 25 mg bd, furosemide 40 mg/day, digoxin 0.125 mg/day, and ranitidine 150 mg bd. She was in sinus rhythm and her serum digoxin concentration was 1.0–1.4 nmol/l. She was treated with fluoxetine (10 mg/day), but a week later she developed anorexia. At that time the serum digoxin concentration was 4.2 nmol/l. Both digoxin and fluoxetine were withdrawn and the digoxin concentration returned to the target range within 5 days, with resolution of the anorexia. Digoxin was restarted and concentrations in the usual target range were achieved during the next 3 weeks (0.9–1.4 nmol/l). Because of persisting depressive symptoms, fluoxetine (10 mg/day) was given again, but the digoxin concentrations rose and after 4 days were 2.8 nmol/l. Both digoxin and fluoxetine were then withdrawn.

Because digoxin has a narrow therapeutic range, this interaction, if confirmed, may be of clinical significance. The mechanism is not clear, because digoxin is not a substrate for the cytochrome P450 enzymes that are inhibited by fluoxetine. The authors speculated that fluoxetine may reduce the renal clearance of digoxin; if so, it might do that by inhibiting the P-glycoprotein that is responsible for the active tubular secretion of digoxin.

Lithium

A report has been published of lithium toxicity induced by combined treatment with fluoxetine (38).

Lithium toxicity has also been reported during co-administration with fluoxetine (47).

- After 4 hours of mild, intermittent, hot-weather work, a 45-year-old man taking fluoxetine and lithium (serum concentration not mentioned) collapsed, became comatose, convulsed, and was febrile (42°C); consciousness returned after 6 days but cerebellar symptoms and atrophy persisted (48). It was suggested that disruption of temperature regulation had been caused by a synergistic effect of the two drugs (although he had taken neither drug for 36 hours before the episode).

Mirtazapine

In patients who do not respond to SSRI treatment, the addition of mirtazapine is an increasingly popular option. Mirtazapine is an α_2-adrenoceptor antagonist, so its acute pharmacological effects predominantly involve potentiation of noradrenergic function. This might be expected to enhance the serotonergic actions of SSRIs.

- A 48-year-old woman unresponsive to fluoxetine (20–40 mg/day for 8 weeks) was given additional mirtazapine (30 mg/day) (49). Over the next month her depression remitted, but after taking the combined treatment for 7 weeks she slept less, spent more, and was more sociable than usual. She was also argumentative and agitated. Mirtazapine and fluoxetine were withdrawn and valproate introduced as a mood stabilizer. Within 2 weeks her mood had settled but she subsequently became depressed again.

The manic episode here could have been due to mirtazapine alone or to the combination of fluoxetine and mirtazapine. The case also demonstrates the common clinical observation that when an antidepressant treatment has apparently caused mania, withdrawal of antidepressant therapy often results in re-emergence of depressive symptoms, even when mood stabilizer treatment has been introduced.

Monoamine oxidase inhibitors

The problem of the long half-life of fluoxetine, leading to interactions with monoamine oxidase inhibitors, even after withdrawal, has been discussed previously (SEDA-13, 12), and caused the manufacturer to circulate a warning to that effect.

Pseudopheochromocytoma, with hypertension, palpitation, and headache, has been reported in a patient taking fluoxetine and selegiline (SEDA-18, 23).

Phenytoin

Several case reports and one in vitro study have suggested that combined administration of fluoxetine with phenytoin can significantly increase phenytoin serum concentrations, leading to toxicity (19,39,40).

Reboxetine

In the pharmacological management of patients with treatment-resistant depression, it is a common strategy to combine a drug that selectively inhibits noradrenaline re-uptake (for example reboxetine) with one that selectively inhibits the re-uptake of serotonin (for example an SSRI). As well as the hoped-for pharmacodynamic interaction, a kinetic interaction can also occur, because of the effect of SSRIs on CYP450 enzymes. The effect of combined treatment with reboxetine (8 mg/day) and fluoxetine (20 mg/day) has been compared with each treatment given alone for 8 days in 30 healthy volunteers in a parallel design (50). There was no potentiation of adverse effects by the combination. Fluoxetine increased the plasma AUC of reboxetine by 20%, but this was not statistically significant. The authors suggested that the

combination of fluoxetine and reboxetine should have minimal adverse impact in depressed patients. However, the major metabolite of fluoxetine, norfluoxetine, is also an inhibitor of CYP3A4 and would not have reached steady-state concentrations during the time of the study. This suggests that caution might still be needed during longer-term use of this combination in depressed patients.

Risperidone

Combined treatment with atypical neuroleptic drugs and SSRIs is common and case reports have suggested that SSRIs can increase risperidone concentrations and increase the risk of extrapyramidal disorders (SEDA-23, 18).

- Severe parkinsonism with urinary retention occurred when fluoxetine 20 mg/day was added to risperidone 2 mg/day in a 46-year-old man with schizophrenia. Risperidone had been prescribed at this dose for 1 month without any adverse effects, and the authors considered that a pharmacokinetic interaction between fluoxetine and risperidone was the most likely mechanism (41).

In a systematic open study in 11 hospitalized patients taking a steady dose of risperidone (4–6 mg/day), fluoxetine (20 mg/day) increased plasma concentrations of active antipsychotic medication (combined concentrations of risperidone and 9-hydroxyrisperidone) by 50% after treatment for 25 days (42). Despite this, the treatment was well tolerated and there were improvements in rating scales of psychosis and depressed mood. Whether this was due to the introduction of fluoxetine or the higher plasma concentrations of risperidone is not clear. Fluoxetine and norfluoxetine would require at least a further 2 weeks to reach steady state, so additional increases in risperidone concentrations might be anticipated over this time.

A pharmacokinetic interaction of risperidone with fluoxetine has been reported (SEDA-22, 71). When 10 schizophrenic patients stabilized on risperidone 4–6 mg/day took fluoxetine 20 mg/day for concomitant depression the mean plasma risperidone concentration increased from 12 to 56 ng/ml at week 4; the concentration of 9-hydroxyrisperidone was not significantly affected (51). One patient dropped out after 1 week because of akathisia associated with a markedly increased plasma risperidone concentration.

In an open, 30-day trial, the pharmacokinetics, safety, and tolerability of a combination of risperidone 4 or 6 mg/day with fluoxetine 20 mg/day were evaluated in 11 psychotic inpatients (52). CYP2D6 genotyping showed that three were poor metabolizers and eight were extensive metabolizers. The mean AUC of risperidone increased from 83 and 398 h.ng/ml to 341 and 514 h.ng/ml when risperidone was co-administered with fluoxetine in extensive and poor metabolizers respectively. However, despite this pharmacokinetic interaction, the severity and incidence of extrapyramidal symptoms and adverse events did not increase significantly when fluoxetine was added; 10 of the 11 patients improved clinically.

• Catastrophic deterioration, with the severity of obsessive-compulsive symptoms returning to pretreatment levels, was observed in a 21-year-old man when risperidone was added to fluoxetine in a dosage that was stepped up to 3 mg/day (53).

Tramadol

Serotonin syndrome and mania occurred in a 72-year-old woman taking fluoxetine 20 mg/day and tramadol 150 mg/day 18 days after she started to take the combination (54). Inhibition of CYP2D6 may have played a part (55).

Triptans

Caution has been advocated when SSRIs such as fluoxetine are combined with the triptans that are used to treat acute episodes of migraine (SEDA-24, 16). There are case reports of symptoms suggestive of serotonin toxicity when fluoxetine has been combined with sumatriptan, perhaps because the SSRI can potentiate the $5\text{-HT}_{1B/1D}$ agonist effects of the triptan (SEDA-22, 14).

The effect of fluoxetine 60 mg/day for 8 days on the pharmacokinetics of almotriptan has been studied in 14 healthy volunteers (43). Fluoxetine produced a significant increase in the peak concentration of almotriptan, but the AUC was not altered. These results suggest that CYP2D6 plays a minor role in the metabolism of almotriptan. The combined treatment was reported to be well tolerated, but this does not exclude the possibility of occasional cases of serotonin toxicity in some individuals.

Venlafaxine

Four depressed patients (age range 21–73 years) had anticholinergic adverse effects when venlafaxine 37.5 mg/day was added to fluoxetine 20 mg/day (56). Venlafaxine does not have direct anticholinergic effects, but could cause them indirectly as a result of its ability to increase noradrenergic neurotransmission. This effect would be expected only at much higher doses of venlafaxine than those used here, which led the author to suggest that in the presence of CYP2D6 inhibition by fluoxetine, venlafaxine concentrations could have been substantially higher than expected.

Zolpidem

The possible pharmacokinetic and pharmacodynamic interactions of repeated nightly zolpidem dosing with fluoxetine were evaluated in 29 healthy women. There were no clinically significant pharmacokinetic or pharmacodynamic interactions (57).

References

1. Beasley CM Jr, Dornseif BE, Bosomworth JC, Sayler ME, Rampey AH Jr, Heiligenstein JH, Thompson VL, Murphy DJ, Masica DN. Fluoxetine and suicide: a meta-analysis of controlled trials of treatment for depression. BMJ 1991;303(6804):685–92.

2. Rossi A, Barraco A, Donda P. Fluoxetine: a review on evidence based medicine. Ann Gen Hosp Psychiatry 2004;3(1):2.

3. Wernicke JF. The side effect profile and safety of fluoxetine. J Clin Psychiatry 1985;46(3 Part 2):59–67.

4. Roose SP, Glassman AH, Attia E, Woodring S, Giardina EG, Bigger JT Jr. Cardiovascular effects of fluoxetine in depressed patients with heart disease. Am J Psychiatry 1998;155(5):660–5.

5. Amsterdam JD, Garcia-Espana F, Fawcett J, Quitkin FM, Reimherr FW, Rosenbaum JF, Beasley C. Blood pressure changes during short-term fluoxetine treatment. J Clin Psychopharmacol 1999;19(1):9–14.

6. Rudnick A, Modai I, Zelikovski A. Fluoxetine-induced Raynaud's phenomenon. Biol Psychiatry 1997;41(12):1218–21.

7. Varriale P. Fluoxetine (Prozac) as a cause of QT prolongation. Arch Intern Med 2001;161(4):12.

8. Anderson J, Compton SA. Fluoxetine induced bradycardia in presenile dementia. Ulster Med J 1997;66(2):144–5.

9. Lipinski JF Jr, Mallya G, Zimmerman P, Pope HG Jr. Fluoxetine-induced akathisia: clinical and theoretical implications. J Clin Psychiatry 1989;50(9):339–42.

10. Halman M, Goldbloom DS. Fluoxetine and neuroleptic malignant syndrome. Biol Psychiatry 1990;28(6):518–21.

11. Guthrie S, Grunhaus L. Fluoxetine-induced stuttering. J Clin Psychiatry 1990;51(2):85.

12. Buchman N, Strous RD, Baruch Y. Side effects of long-term treatment with fluoxetine. Clin Neuropharmacol 2002;25(1):55–7.

13. Weber JJ. Seizure activity associated with fluoxetine therapy. Clin Pharm 1989;8(4):296–8.

14. Riddle MA, Brown N, Dzubinski D, Jetmalani AN, Law Y, Woolston JL. Fluoxetine overdose in an adolescent. J Am Acad Child Adolesc Psychiatry 1989;28(4):587–8.

15. Costagliola C, Mastropasqua L, Steardo L, Testa N. Fluoxetine oral administration increases intraocular pressure. Br J Ophthalmol 1996;80(7):678.

16. Eke T, Carr S, Costagliola C, Mastropasqua L, Steardo L. Acute glaucoma, chronic glaucoma, and serotoninergic drugs. Br J Ophthalmol 1998;82(8):976–8.

17. Teicher MH, Glod C, Cole JO. Emergence of intense suicidal preoccupation during fluoxetine treatment. Am J Psychiatry 1990;147(2):207–10.

18. Leon AC, Keller MB, Warshaw MG, Mueller TI, Solomon DA, Coryell W, Endicott J. Prospective study of fluoxetine treatment and suicidal behavior in affectively ill subjects. Am J Psychiatry 1999;156(2):195–201.

19. Jalil P. Toxic reaction following the combined administration of fluoxetine and phenytoin: two case reports. J Neurol Neurosurg Psychiatry 1992;55(5):412–3.

20. Fawcett J, Zajecka JM, Kravitz HM, et al. Fluoxetine versus amitriptyline in adult outpatients with major depression. Psychiatry Res 1989;45:821.

21. Peterson MC. Reversible galactorrhea and prolactin elevation related to fluoxetine use. Mayo Clin Proc 2001;76(2):215–6.

22. Bottlender R, Dobmeier P, Moller HJ. Der Einfluss von selektiven Serotonin-Wiederaufnahmeinhibitoren auf die Blutgerinnung. [The effect of selective serotonin-reuptake inhibitors in blood coagulation.] Fortschr Neurol Psychiatr 1998;66(1):32–5.

23. Calhoun JW, Calhoun DD. Prolonged bleeding time in a patient treated with setraline. Am J Psychiatry 1996;153:443.

24. Vilinsky FD, Lubin A. Severe neutropenia associated with fluoxetine hydrochloride. Ann Intern Med 1997;127(7):573–4.

25. Palop V, Sancho A, Morales-Olivas FJ, Martinez-Mir I. Fluoxetine-associated stomatitis. Ann Pharmacother 1997;31(12):1478–80.
26. Johnston DE, Wheeler DE. Chronic hepatitis related to use of fluoxetine. Am J Gastroenterol 1997;92(7):1225–6.
27. Capella D, Bruguera M, Figueras A, Laporte J. Fluoxetine-induced hepatitis: why is postmarketing surveillance needed? Eur J Clin Pharmacol 1999;55(7):545–6.
28. Cooper GL. The safety of fluoxetine – an update. Br J Psychiatry Suppl 1988;3:77–86.
29. Stark P, Hardison CD. A review of multicenter controlled studies of fluoxetine vs. imipramine and placebo in outpatients with major depressive disorder. J Clin Psychiatry 1985;46(3 Part 2):53–8.
30. Herman JB, Brotman AW, Pollack MH, Falk WE, Biederman J, Rosenbaum JF. Fluoxetine-induced sexual dysfunction. J Clin Psychiatry 1990;51(1):25–7.
31. Modell JG. Repeated observations of yawning, clitoral engorgement, and orgasm associated with fluoxetine administration. J Clin Psychopharmacol 1989;9(1):63–5.
32. Roger D, Rolle F, Mausset J, Lavignac C, Bonnetblanc JM. Urticarial vasculitis induced by fluoxetine. Dermatology 1995;191(2):164.
33. Fisher A, McLean AJ, Purcell P, Herdson PB, Dahlstrom JE, Le Couteur DG. Focal necrotising vasculitis with secondary myositis following fluoxetine administration. Aust NZ J Med 1999;29(3):375–6.
34. de Kerviler E, Tredaniel J, Revlon G, Groussard O, Zalcman G, Ortoli JM, Espie M, Hirsch A, Frija J. Fluoxetin-induced pulmonary granulomatosis. Eur Respir J 1996;9(3):615–7.
35. Abebe-Campino G, Offer D, Stahl B, Merlob P. Cardiac arrhythmia in a newborn infant associated with fluoxetine use during pregnancy. Ann Pharmacother 2002;36(3):533–4.
36. Brosen K. The pharmacogenetics of the selective serotonin reuptake inhibitors. Clin Investig 1993;71(12):1002–9.
37. Leibovitz A, Bilchinsky T, Gil I, Habot B. Elevated serum digoxin level associated with coadministered fluoxetine. Arch Intern Med 1998;158(10):1152–3.
38. Salama AA, Shafey M. A case of severe lithium toxicity induced by combined fluoxetine and lithium carbonate. Am J Psychiatry 1989;146(2):278.
39. Woods DJ, Coulter DM, Pillans P. Interaction of phenytoin and fluoxetine. NZ Med J 1994;107(970):19.
40. Schmider J, Greenblatt DJ, von Moltke LL, Karsov D, Shader RI. Inhibition of CYP2C9 by selective serotonin reuptake inhibitors in vitro: studies of phenytoin p-hydroxylation. Br J Clin Pharmacol 1997;44(5):495–8.
41. Bozikas V, Petrikis P, Karavatos A. Urinary retention caused after fluoxetine–risperidone combination. J Psychopharmacol 2001;15(2):142–3.
42. Bondolfi G, Eap CB, Bertschy G, Zullino D, Vermeulen A, Baumann P. The effect of fluoxetine on the pharmacokinetics and safety of risperidone in psychiatric patients. Pharmacopsychiatry 2002;35(2):50–6.
43. Fleishaker JC, Ryan KK, Carel BJ, Azie NE. Evaluation of the potential pharmacokinetic interaction between almotriptan and fluoxetine in healthy volunteers. J Clin Pharmacol 2001;41(2):217–23.
44. Spina E, Avenoso A, Facciola G, Fabrazzo M, Monteleone P, Maj M, Perucca E, Caputi AP. Effect of fluoxetine on the plasma concentrations of clozapine and its major metabolites in patients with schizophrenia. Int Clin Psychopharmacol 1998;13(3):141–5.
45. Ferslew KE, Hagardorn AN, Harlan GC, McCormick WF. A fatal drug interaction between clozapine and fluoxetine. J Forensic Sci 1998;43(5):1082–5.
46. Purdon SE, Snaterse M. Selective serotonin reuptake inhibitor modulation of clozapine effects on cognition in schizophrenia. Can J Psychiatry 1998;43(1):84–5.
47. Salama AA, Shafey M. A case of severe lithium toxicity induced by combined fluoxetine and lithium carbonate. Am J Psychiatry 1989;146(2):278.
48. Epstein Y, Albukrek D, Kalmovitc B, Moran DS, Shapiro Y. Heat intolerance induced by antidepressants. Ann N Y Acad Sci 1997;813:553–8.
49. Ng B. Mania associated with mirtazapine augmentation of fluoxetine. Depress Anxiety 2002;15(1):46–7.
50. Fleishaker JC, Herman BD, Pearson LK, Ionita A, Mucci M. Evaluation of the potential pharmacokinetic/pharmacodynamic interaction between fluoxetine and reboxetine in healthy volunteers. Clin Drug Invest 1999;18:141–50.
51. Spina E, Avenoso A, Scordo MG, Ancione M, Madia A, Gatti G, Perucca E. Inhibition of risperidone metabolism by fluoxetine in patients with schizophrenia: a clinically relevant pharmacokinetic drug interaction. J Clin Psychopharmacol 2002;22(4):419–23.
52. Bondolfi G, Eap CB, Bertschy G, Zullino D, Vermeulen A, Baumann P. The effect of fluoxetine on the pharmacokinetics and safety of risperidone in psychiatric patients. Pharmacopsychiatry 2002;35(2):50–6.
53. Andrade C. Risperidone may worsen fluoxetine-treated OCD. J Clin Psychiatry 1998;59(5):255–6.
54. Gonzalez-Pinto A, Imaz H, De Heredia JL, Gutierrez M, Mico JA. Mania and tramadol–fluoxetine combination. Am J Psychiatry 2001;158(6):964–5.
55. Ingelman-Sundberg M. Genetic susceptibility to adverse effects of drugs and environmental toxicants. The role of the CYP family of enzymes. Mutat Res 2001;482(1–2):11–9.
56. Benazzi F. Venlafaxine–fluoxetine interaction. J Clin Psychopharmacol 1999;19(1):96–8.
57. Allard S, Sainati S, Roth-Schechter B, MacIntyre J. Minimal interaction between fluoxetine and multiple-dose zolpidem in healthy women. Drug Metab Dispos 1998;26(7):617–22.

Fluvoxamine

General Information

Fluvoxamine is a non-sedating antidepressant with fewer anticholinergic adverse effects than clomipramine or imipramine (1–3). Its major adverse effects, like those of other SSRIs, include nausea and vomiting. It has a half-life of 15 hours, and peak plasma concentrations occur at 1–8 hours after oral administration (4). It is metabolized (by oxidation, oxidative deamination, and hydrolysis) to nine metabolites, none of which is pharmacologically active (5). Studies of single versus multiple dosing have shown no significant differences (6,7). Usual doses are in the range of 150–300 mg/day, and once-a-day dosing is possible.

Fluvoxamine appears to have no specific effects on laboratory tests (4,8,9); although some have reported a significant fall in platelet count and an increase in serum creatinine, most of the values remained well within the reference ranges (10).

Organs and Systems

Cardiovascular

A slight, clinically unimportant reduction in heart rate has been reported with fluvoxamine (11,12). There has been one report of supraventricular tachycardia in a woman with no previous cardiovascular disease, but the association with fluvoxamine was unclear since there was no rechallenge (SEDA-16, 9).

Nervous system

In a yearlong study of 31 patients there was agitation, which required withdrawal in two patients early in treatment, dry mouth, tremor, and insomnia (13). Increased agitation and insomnia may require the addition of a sedative or hypnotic.

Acute dystonia has been described in association with fluvoxamine (SEDA-18, 20).

Seizures have been reported in a predisposed subject given fluvoxamine (SEDA-17, 20).

Treatment with serotonin-potentiating drugs in usual therapeutic doses can sometimes produce the serotonin syndrome. There are also case reports of this reaction with single doses of SSRIs (see also the monograph on Sertraline) (14).

- An 11-year-old boy was brought to the emergency room about 2 hours after taking a single tablet of fluvoxamine (50 mg) prescribed for treatment of attention-deficit disorder. He was also taking benzatropine and perphenazine (dosages not stated). On arrival, he was agitated and unresponsive, with bilateral ankle clonus, muscle rigidity, fasciculations, and profuse sweating. His temperature rose to 39.7°C. He was paralysed, intubated, and ventilated, after which his condition improved. Two days after admission, he had fully recovered.

The dopamine receptor antagonist properties of perphenazine may have played a part in producing the syndrome in this case.

SSRIs can cause insomnia and daytime somnolence; however, the symptoms seem to reflect a sleep–wake cycle disorder. It is conceivable that disruptions in the normal pattern of melatonin secretion, particularly a delay in the normal early morning fall in plasma concentrations, could be involved in the pathophysiology of these symptoms. The fact that these sleep disorders were seen only with fluvoxamine would also support a role of melatonin (see the section on Endocrine in this monograph).

Psychological, psychiatric

Reports of acute mania and manic-like behavior after treatment with fluoxetine or fluvoxamine have appeared

(SEDA-13, 12) (SEDA-17, 20) (SEDA-18, 20), but there are not enough data to estimate the incidence.

Endocrine

Fluvoxamine causes increased plasma melatonin concentrations. In an in vitro preparation, melatonin was metabolized to 6-hydroxymelatonin by CYP1A2, which was inhibited by fluvoxamine at concentrations similar to those found in the plasma during therapy (15). This effect was not shared by other SSRIs or by tricyclic antidepressants, which do not have prominent effects on melatonin secretion. Whether increased concentrations of melatonin and loss of its normal circadian rhythm might cause symptoms is unclear. However, melatonin is believed to play a role in the regulation of circadian rhythms, including entrainment of the sleep–wake cycle. There have been 10 cases of circadian rhythm sleep disorder associated with fluvoxamine (16). All the patients had delayed sleep-phase syndrome, which is characterized by delayed-sleep onset and late awakening. The delay in falling asleep and waking up in the morning was 2.5–4 hours. In nine of the cases withdrawal of fluvoxamine or a reduced dosage led to resolution of the sleep disorder. When the patients were given alternative serotonin potentiating agents, such as clomipramine or fluoxetine, the sleep disorder did not recur.

Serotonin pathways are involved in the regulation of prolactin secretion. Amenorrhea, galactorrhea, and hyperprolactinemia have been reported in a patient who was already taking an antipsychotic drug after starting treatment with fluvoxamine (SEDA-17, 20).

Three cases of fluvoxamine-induced polydipsia, attributed to the syndrome of inappropriate ADH secretion (SIADH), have been reported (SEDA-18, 20).

Hematologic

Fluvoxamine-associated bleeding has been described (17–19).

Skin

A case of toxic epidermal necrolysis after fluvoxamine has been described; although the patient was taking other drugs, the authors concluded that the skin reaction was probably due to fluvoxamine (SEDA-18, 20).

Hair

Hair loss has been associated with fluvoxamine (20).

Long-Term Effects

Drug withdrawal

Withdrawal symptoms have been reported with fluvoxamine (SEDA-17, 20; SEDA-18, 21).

Second-Generation Effects

Lactation

In two cases, treatment of breast-feeding mothers with fluvoxamine (300 mg/day) was associated with undetectable concentrations of fluvoxamine (below 2.5 ng/ml) in the plasma of both infants (21). These results are encouraging, but further data will be needed before it can be concluded that fluvoxamine has an advantage over other SSRIs in this respect.

Drug Administrations

Drug overdose

There have been four cases of overdosage with fluvoxamine in amounts ranging from 600 to 2500 mg (22).

Drug–Drug Interactions

Buspirone

In 10 healthy volunteers, fluvoxamine (100 mg/day for 5 days) significantly increased the peak concentrations of the anxiolytic drug buspirone. Concentrations of the active metabolite, 1-(2-pyrimidinyl)-piperazine, were reduced (23). These effects were probably mediated through inhibition of CYP3A4 by fluvoxamine.

The effects of fluvoxamine on the pharmacokinetics and pharmacodynamics of buspirone have been investigated in 10 healthy volunteers. Fluvoxamine moderately increased plasma buspirone concentrations and reduced the production of the active metabolite of buspirone. The mechanism of this interaction is probably inhibition of CYP3A4. However, this pharmacokinetic interaction was not associated with impaired psychomotor performance and is probably of limited clinical significance (33).

Clozapine

In a prospective study fluvoxamine, in a low dosage of 50 mg/day, produced a threefold increase in plasma clozapine concentrations ($n = 16$) (24).

Other reports have confirmed that fluvoxamine increases plasma concentrations of clozapine and its metabolites 25,26 (SEDA-21, 12). The mechanism is probably inhibition of CYP1A2.

Fluvoxamine increases clozapine plasma concentrations (34,35). In 16 patients taking clozapine monotherapy, fluvoxamine 50 mg was added in the hope of ameliorating the negative symptoms of schizophrenia (36). At steady state the serum concentrations of clozapine and its metabolites increased up to five-fold (average two- to three-fold). However, adverse effects were almost unchanged in frequency and severity, in spite of the pharmacokinetic interaction.

In another study there were similar increases in plasma clozapine concentrations and adverse reactions in 18 patients taking fluvoxamine 50 mg (week 5, mean dose 97 mg) (37).

In nine men who were given a single dose of clozapine 50 mg on two separate occasions with a 2-week interval, fluvoxamine increased clozapine plasma concentrations, and the total mean clozapine AUC was increased by a factor of 2.6; all the patients were sedated during combined therapy (38).

Combined therapy with clozapine and fluvoxamine ($n = 11$) and clozapine monotherapy ($n = 12$) have been monitored before and during the first 6 weeks of medication (39). The co-administration of fluvoxamine attenuated and delayed the clozapine-induced increase in plasma concentrations of tumor necrosis factor-alpha, enhanced and accelerated the clozapine-induced increase in leptin plasma concentrations without a significant effect on clozapine-induced weight gain, and reduced granulocyte counts.

In two studies of short duration (18 patients each) there were benefits of using low doses of clozapine plus fluvoxamine, and the authors suggested taking advantage of this interaction (40). Patients taking fluvoxamine required relatively low doses of clozapine and had clinically significant reductions in the symptoms of their illness while avoiding the sedative adverse effects associated with the usual doses of clozapine.

Lithium

Six patients taking a stable dose of fluvoxamine had a minor increase in plasma fluvoxamine concentration (from 67 to 76 ng/ml) 2 weeks after starting to take unspecified doses of lithium; this is unlikely to be of clinical significance (41).

Methadone

Fluvoxamine increases methadone concentrations in patients taking methadone maintenance treatment for the management of opioid dependence (SEDA-19, 11). The addition of sertraline (200 mg/day) produced a modest (16%) increase in methadone concentrations in 31 depressed opioid-dependent subjects after 6 but not 12 weeks of combined treatment (27). The increase in methadone concentrations was more modest than that reported with fluvoxamine, presumably because sertraline is a less potent inhibitor of CYP1A2 and CYP3A4, both of which are involved in methadone metabolism.

Fluvoxamine increases the effects of methadone, probably by inhibition of methadone metabolism (42).

Mexiletine

Mexiletine is metabolized by CYP2D6, CYP1A2, and CYP3A4; fluvoxamine inhibits CYP1A2. It is not surprising therefore that fluvoxamine 50 mg bd for 7 days increased the $C_{\max}$ and AUC of a single oral dose of mexiletine 200 mg in six healthy Japanese men (43).

Olanzapine

The atypical antipsychotic drug olanzapine is also metabolized by CYP1A2.

- A 21-year-old woman with schizophrenia and depression, who had been taking fluvoxamine (150 mg/day) and olanzapine (15 mg/day) for several months, developed an extrapyramidal movement disorder, including rigidity and tremor (28). The plasma fluvoxamine concentration was 70 ng/ml (usual target range is 20–500 ng/ml), while that of olanzapine was 120 ng/ml (usual target range is 9–25 ng/ml). The dosage of olanzapine was reduced to 5 mg/day and the plasma olanzapine concentration fell to 38 ng/ml, with resolution of the tremor and rigidity. When fluvoxamine was replaced with paroxetine (20 mg/day) the olanzapine concentration fell further to 22 ng/ml.

Of the SSRIs, fluvoxamine is the most potent inhibitor of CYP1A2 and is therefore likely to increase plasma olanzapine concentrations. The extrapyramidal effects in this case were presumably due to excessive blockade of dopamine D_2 receptors by raised olanzapine concentrations.

Phenytoin

Fluvoxamine inhibits CYP2C9 and CYP2C19, the enzymes responsible for the metabolism of phenytoin.

- A 45-year-old woman taking phenytoin 300 mg/day had a plasma phenytoin concentration of 66 µmol/l (29). When she became depressed fluvoxamine 50 mg/day was added. A month later her depressive symptoms had improved, but she was ataxic and the plasma phenytoin concentration was 196 µmol/l. The fluvoxamine was withdrawn and the phenytoin dose reduced to 150 mg/day. Her plasma phenytoin concentration fell to 99 µmol/l, with resolution of the ataxia.

In vitro studies suggest that fluvoxamine is the most potent SSRI in terms of its ability to inhibit phenytoin metabolism. Inhibition of CYP2C19 or CYP2C9 could be responsible, although fluvoxamine is a relatively weak in vitro inhibitor of CYP2C9. However, in 14 healthy volunteers fluvoxamine (150–300 mg/day for 5 days) significantly reduced the clearance of tolbutamide (30). This suggests that fluvoxamine should be used with caution when it is co-administered with drugs such as tolbutamide, phenytoin, and warfarin, which are substrates for CYP2C9.

Risperidone

- A 24-year-old woman with auditory hallucinations taking risperidone 6 mg/day developed neuroleptic malignant syndrome after adding fluvoxamine 50 mg/day (44).

Sildenafil

Sexual dysfunction is a common adverse effect of SSRIs and various treatments have been proposed, of which sildenafil is the only strategy with consistent support from controlled trials. Sildenafil is metabolized by CYP3A4, which is inhibited by fluvoxamine. The effects of fluvoxamine (100 mg/day for 10 days) on the pharmacokinetics of sildenafil (50 mg orally) has been evaluated in 12 healthy men (mean age 25 years) using a double-blind, placebo-controlled, crossover design (46). Fluvoxamine increased the AUC of sildenafil by about 40% and prolonged its half-life by about 20%. This suggests that patients taking fluvoxamine should use lower doses of sildenafil for the treatment of SSRI-induced sexual dysfunction. A similar but smaller effect might occur with co-prescription of sildenafil and fluoxetine, which is a weaker inhibitor of CYP3A4 than fluvoxamine.

Thioridazine

Fluvoxamine inhibits CYP1A2, and a low dosage (50 mg/day) produced a 225% increase in plasma thioridazine concentrations in 10 patients with schizophrenia (31). This was not reflected in an increased incidence of clinical adverse events; however, thioridazine prolongs the QT_c interval, which was not measured.

Fluvoxamine increased thioridazine concentrations three-fold (45) in 10 patients with schizophrenia taking steady-state thioridazine who were given fluvoxamine (25 mg bd) for 1 week. Fluvoxamine interferes with the metabolism of thioridazine, probably via CYP2C19 and/or CYP1A2.

Tolbutamide

In 14 healthy volunteers fluvoxamine (150–300 mg/day for 5 days) significantly reduced the clearance of tolbutamide (30). This suggests that fluvoxamine should be used with caution when it is co-administered with drugs that are substrates for CYP2C9, such as tolbutamide, phenytoin, and warfarin.

Tricyclic antidepressants

Fluvoxamine is a potent inhibitor of cytochrome CYP1A2, which may lead to interactions with several tricyclic antidepressants and theophylline (32).

References

1. De Wilde JE, Mertens C, Wakelin JS. Clinical trials of fluvoxamine vs chlorimipramine with single and three times daily dosing. Br J Clin Pharmacol 1983;15(Suppl. 3):427S–31S.
2. Itil TM, Shrivastava RK, Mukherjee S, Coleman BS, Michael ST. A double-blind placebo-controlled study of fluvoxamine and imipramine in out-patients with primary depression. Br J Clin Pharmacol 1983;15(Suppl. 3):433S–8S.
3. Klok CJ, Brouwer GJ, van Praag HM, Doogan D. Fluvoxamine and clomipramine in depressed patients. A double-blind clinical study. Acta Psychiatr Scand 1981;64(1):1–11.
4. Claassen V. Review of the animal pharmacology and pharmacokinetics of fluvoxamine. Br J Clin Pharmacol 1983;15(Suppl. 3):349S–55S.
5. Kinney-Parker JL, Smith D, Ingle SF. Fluoxetine and weight: something lost and something gained? Clin Pharm 1989;8(10):727–33.
6. Doogan DP. Fluvoxamine as an antidepressant drug. Neuropharmacology 1980;19(12):1215–6.
7. Siddiqui UA, Chakravarti SK, Jesinger DK. The tolerance and antidepressive activity of fluvoxamine as a single dose compared to a twice daily dose. Curr Med Res Opin 1985;9(10):681–90.

8. Coleman BS, Block BA. Fluvoxamine maleate, a serotonergic antidepressant; a comparison with chlorimipramine. Prog Neuropsychopharmacol Biol Psychiatry 1982;6(4–6):475–8.

9. De Wilde JE, Doogan DP. Fluvoxamine and chlorimipramine in endogenous depression. J Affect Disord 1982;4(3):249–59.

10. Guelfi JD, Dreyfus JF, Pichot P. A double-blind controlled clinical trial comparing fluvoxamine with imipramine. Br J Clin Pharmacol 1983;15(Suppl. 3):411S–7S.

11. Roos JC. Cardiac effects of antidepressant drugs. A comparison of the tricyclic antidepressants and fluvoxamine. Br J Clin Pharmacol 1983;15(Suppl. 3):439S–45S.

12. Robinson JF, Doogan DP. A placebo controlled study of the cardiovascular effects of fluvoxamine and clovoxamine in human volunteers. Br J Clin Pharmacol 1982;4(6):805–8.

13. Feldmann HS, Denber HC. Long-term study of fluvoxamine: a new rapid-acting antidepressant. Int Pharmacopsychiatry 1982;17(2):114–22.

14. Gill M, LoVecchio F, Selden B. Serotonin syndrome in a child after a single dose of fluvoxamine. Ann Emerg Med 1999;33(4):457–9.

15. Peterson MC. Reversible galactorrhea and prolactin elevation related to fluoxetine use. Mayo Clin Proc 2001;76(2):215–6.

16. Morrison J, Remick RA, Leung M, Wrixon KJ, Bebb RA. Galactorrhea induced by paroxetine. Can J Psychiatry 2001;46(1):88–9.

17. Calhoun JW, Calhoun DD. Two other published reports describing fluvoxamine-associated bleeding

18. Leung M, Shore R. Fluvoxamine-associated bleeding. Can J Psychiatry 1996;41(9):604–5.

19. Wilmshurst PT, Kumar AV. Subhyaloid haemorrhage with fluoxetine. Eye 1996;10(Pt 1):141.

20. Parameshwar E. Hair loss associated with fluvoxamine use. Am J Psychiatry 1996;153(4):581–2.

21. Piontek CM, Wisner KL, Perel JM, Peindl KS. Serum fluvoxamine levels in breastfed infants. J Clin Psychiatry 2001;62(2):111–3.

22. Bradford LD, Coleman BS, Hoeve L. Summary of the properties of fluvoxamine maleate. Duphar Report no. H114.058. 1984.

23. Lamberg TS, Kivisto KT, Laitila J, Martensson K, Neuvonen PJ. The effect of fluvoxamine on the pharmacokinetics and pharmacodynamics of buspirone. Eur J Clin Pharmacol 1998;54(9–10):761–6.

24. Lamberg TS, Kivisto KT, Laitila J, Martensson K, Neuvonen PJ. The effect of fluvoxamine on the pharmacokinetics and pharmacodynamics of buspirone. Eur J Clin Pharmacol 1998;54(9–10):761–6.

25. Wetzel H, Anghelescu I, Szegedi A, Wiesner J, Weigmann H, Harter S, Hiemke C. Pharmacokinetic interactions of clozapine with selective serotonin reuptake inhibitors: differential effects of fluvoxamine and paroxetine in a prospective study. J Clin Psychopharmacol 1998;18(1):2–9.

26. Fabrazzo M, La Pia S, Monteleone P, Mennella R, Esposito G, Pinto A, Maj M. Fluvoxamine increases plasma and urinary levels of clozapine and its major metabolites in a time- and dose-dependent manner. J Clin Psychopharmacol 2000;20(6):708–10.

27. Lu ML, Lane HY, Chen KP, Jann MW, Su MH, Chang WH. Fluvoxamine reduces the clozapine dosage needed in refractory schizophrenic patients. J Clin Psychiatry 2000;61(8):594–9.

28. Spina E, Avenoso A, Facciola G, Fabrazzo M, Monteleone P, Maj M, Perucca E, Caputi AP. Effect of

29. Markowitz JS, Gill HS, Lavia M, Brewerton TD, DeVane CL. Fluvoxamine–clozapine dose-dependent interaction. Can J Psychiatry 1996;41(10):670–1.

fluoxetine on the plasma concentrations of clozapine and its major metabolites in patients with schizophrenia. Int Clin Psychopharmacol 1998;13(3):141–5.

30. Szegedi A, Anghelescu I, Wiesner J, Schlegel S, Weigmann H, Hartter S, Hiemke C, Wetzel H. Addition of low-dose fluvoxamine to low-dose clozapine monotherapy in schizophrenia: drug monitoring and tolerability data from a prospective clinical trial. Pharmacopsychiatry 1999;32(4):148–53.

31. Lammers CH, Deuschle M, Weigmann H, Hartter S, Hiemke C, Heese C, Heuser I. Coadministration of clozapine and fluvoxamine in psychotic patients—clinical experience. Pharmacopsychiatry 1999;32(2):76–7.

32. Chang WH, Augustin B, Lane HY, ZumBrunnen T, Liu HC, Kazmi Y, Jann MW. In-vitro and in-vivo evaluation of the drug–drug interaction between fluvoxamine and clozapine. Psychopharmacology (Berl) 1999;145(1):91–8.

33. Hinze-Selch D, Deuschle M, Weber B, Heuser I, Pollmacher T. Effect of coadministration of clozapine and fluvoxamine versus clozapine monotherapy on blood cell counts, plasma levels of cytokines and body weight. Psychopharmacology (Berl) 2000;149(2):163–9.

34. Prior TI. Is there a way to overcome over-sedation in a patient being treated with clozapine? J Psychiatry Neurosci 2002;27:224.

35. Takano A, Suhara T, Yasuno F, Ichimiya T, Inoue M, Sudo Y, Suzuki K. Characteristics of clomipramine and fluvoxamine on serotonin transporter evaluated by PET. Int Clin Psychopharmacol 2002;17(Suppl 2):S84–5.

36. Hamilton SP, Nunes EV, Janal M, Weber L. The effect of sertraline on methadone plasma levels in methadone-maintenance patients. Am J Addict 2000;9(1):63–9.

37. Iribarne C, Picart D, Dreano Y, Berthou F. In vitro interactions between fluoxetine or fluvoxamine and methadone or buprenorphine. Fundam Clin Pharmacol 1998;12(2):194–9.

38. Kusumoto M, Ueno K, Oda A, Takeda K, Mashimo K, Takaya K, Fujimura Y, Nishihori T, Tanaka K. Effect of fluvoxamine on the pharmacokinetics of mexiletine in healthy Japanese men. Clin Pharmacol Ther 2001;69(3):104–7.

39. de Jong J, Hoogenboom B, van Troostwijk LD, de Haan L. Interaction of olanzapine with fluvoxamine. Psychopharmacology (Berl) 2001;155(2):219–20.

40. Mamiya K, Kojima K, Yukawa E, Higuchi S, Ieiri I, Ninomiya H, Tashiro N. Phenytoin intoxication induced by fluvoxamine. Ther Drug Monit 2001;23(1):75–7.

41. Madsen H, Enggaard TP, Hansen LL, Klitgaard NA, Brosen K. Fluvoxamine inhibits the CYP2C9 catalysed biotransformation of tolbutamide. Clin Pharmacol Ther 2001;69(1):41–7.

42. Reeves RR, Mack JE, Beddingfield JJ. Neurotoxic syndrome associated with risperidone and fluvoxamine. Ann Pharmacother 2002;36(3):440–3.

43 Hesse C, Siedler H, Burhenne J, Riedel K-D, Haefeli WE. Fluvoxamine affects sildenafil kinetics and dynamics. J Clin Psychopharmacol 2005;25:589–92.

44. Carrillo JA, Ramos SI, Herraiz AG, Llerena A, Agundez JA, Berecz R, Duran M, Benitez J. Pharmacokinetic interaction of fluvoxamine and thioridazine in schizophrenic patients. J Clin Psychopharmacol 1999;19(6):494–9.

45.   Carrillo JA, Ramos SI, Herraiz AG, Llerena A, Agundez JA, Berecz R, Duran M, Benitez J. Pharmacokinetic interaction of fluvoxamine and thioridazine in schizophrenic patients. J Clin Psychopharmacol 1999;19(6):494–9.
46.   Barnes TR, Kidger T, Greenwood DT. Viloxazine and migraine. Lancet 1979;2(8156–8157):1368.

Paroxetine

General Information

Paroxetine is a phenylpiperidine derivative. Its half-life is about 17–22 hours and about 95% of it is bound to plasma proteins. Its metabolites have no more than 1/50 of the potency of the parent compound in inhibiting serotonin re-uptake. The metabolism of paroxetine is accomplished in part by CYP2D6, saturation of which at therapeutic doses appears to account for the non-linearity of paroxetine kinetics at higher doses and increasing durations of treatment. The adverse effects of paroxetine are those of the SSRIs in general. Commonly observed adverse events in placebo-controlled clinical trials were weakness, sweating, nausea, reduced appetite, somnolence, dizziness, insomnia, tremor, nervousness, ejaculatory disturbance, and other male genital disorders. Paroxetine seems to have a higher incidence of withdrawal symptoms than other SSRIs.

Organs and Systems

Cardiovascular

Electrocardiographic changes, with a prolonged QT_c interval and bradycardia, have been reported with paroxetine (1).

Nervous system

Serotonin syndrome
The serotonin syndrome is a recognized complication of SSRI treatment. Usually it occurs as part of a drug interaction, when the serotonergic effects of SSRIs are augmented by medications that also have serotonin-potentiating properties (SEDA-22, 14) (SEDA-25, 16). Occasionally, however, the serotonin syndrome can occur after SSRI monotherapy.

- A 23-year-old Japanese woman with major depression took a single dose of paroxetine (20 mg) and 1 hour later had agitation, myoclonus, mild hyperthermia (37.5°C), sweating, and diarrhea, symptoms that meet the criteria for the serotonin syndrome; she recovered with supportive treatment over 3 days (2).

Blood concentrations of paroxetine were not obtained, but the authors reported that the patient was homozygous for the 10* form of the CYP2D6 allele, which is associated with low CYP2D6 activity in vivo. While this is an interesting observation, it is unlikely by itself to explain the patient's sensitivity to paroxetine, because this genotype is not uncommon in the Japanese population, and if a lot of Japanese have this genotype, the serotonin syndrome with SSRI monotherapy would be quite common, given the widespread prescription of SSRIs. However, it reinforces clinical advice that in patients new to SSRI treatment it is advisable to start therapy with half the standard dose.

Dystonias
Acute dystonia has been described during the first days of paroxetine treatment (SEDA-17, 19).

Paroxetine-induced akathisia has been described in an 81-year-old man with bipolar depression. The akathisia began one week after paroxetine treatment (20 mg/day) and remitted within 6 days of withdrawal (3).

The authors pointed out that it is important to recognize SSRI-induced akathisia, because increasing agitation and restlessness early in treatment can be mistaken for worsening depression. In addition, case reports have suggested that akathisia can be associated with suicidal impulses.

Sensory systems
Tricyclic antidepressants can precipitate acute glaucoma through their anticholinergic effects. There are also reports that paroxetine can cause acute glaucoma, presumably by pupillary dilatation.

- An 84-year-old woman developed acute closed-angle glaucoma after taking paroxetine for 6 days (SEDA-21, 13).
- A 70-year-old woman taking paroxetine developed acute closed-angle glaucoma (4).
- A 91-year-old developed bilateral acute closed-angle glaucoma after taking paroxetine (5).

Three patients taking paroxetine for interferon-alpha-induced depression developed retinal hemorrhages, including one with irreversible loss of vision (6).

Psychological, psychiatric

In three children (two aged 9 years and one aged 10 years) who took paroxetine 10–20 mg/day for the treatment of childhood obsessive-compulsive disorder, symptoms of mania, including overactivity, pressure of speech, irritability, and antisocial behavior, occurred within 3 weeks of starting paroxetine and remitted after paroxetine withdrawal or dosage reduction (7). Symptoms of mania are rare in childhood, suggesting that the elevated mood in these cases was a direct effect of the paroxetine.

Psychomotor retardation with semistupor has been reported in one patient taking paroxetine; however, she was also taking antipsychotic drugs, which may have contributed (SEDA-18, 20).

Endocrine

Serotonin pathways are involved in the regulation of prolactin secretion. Galactorrhea has been associated with paroxetine (8).

Serotonin pathways are involved in the regulation of prolactin secretion. Amenorrhea, galactorrhea, and hyperprolactinemia have been reported in a patient taking SSRIs.

- A 32-year-old woman taking paroxetine 40 mg/day had a raised prolactin concentration (46 ng/ml) and galactorrhea, both of which resolved a few days after paroxetine withdrawal (9).

Metabolism

Increases in cholesterol are most commonly reported in association with atypical antipsychotic drugs, such as olanzapine; however, similar reactions have been reported with some antidepressants, including mirtazapine and doxepin, which cause significant weight gain, probably through histamine H_1 receptor blockade. Serum cholesterol concentrations have been measured in 38 patients (23 men and 15 women) suffering from panic disorder before and after 3 months of treatment with paroxetine (20–40 mg/day) (10). At baseline the mean total cholesterol concentrations of the patients did not differ from those of controls (4.06 versus 4.29 mmol/l; 156 versus 165 mg/dl). However, after paroxetine the cholesterol concentrations rose significantly to 4.55 mmol/l (175 mg/dl).

Of the SSRIs, paroxetine is the agent most likely to cause *weight gain*; however, in this study the authors reported no change in body mass index during the 3 months of treatment, suggesting a more direct effect of paroxetine on metabolism. Further work will be needed to see if similar metabolic effects are associated with other SSRIs and in patients with other treatment indications.

Hematologic

Clozapine and SSRIs are often used together, because depressive syndromes are common in patients with schizophrenia. Clozapine carries a relatively high risk of agranulocytosis, but this adverse effect is very rarely seen with SSRIs, although a case of possible fluoxetine-induced neutropenia has been described (SEDA-22, 15). Two cases in which the addition of paroxetine to clozapine was associated with neutropenia have been reported (11). The patients had been taking stable doses of clozapine for 6–12 months and had previously tolerated other SSRIs without adverse hematological consequences. In both cases the white cell count recovered when clozapine was withdrawn, although paroxetine was continued.

Fluoxetine has infrequently been associated with abnormal bleeding, including ecchymoses, melena, and hematuria. Spontaneous ecchymoses have been reported in a woman taking paroxetine (12).

- A 47-year-old woman who had had bilateral mastectomies for breast cancer became depressed and was given paroxetine 20 mg/day. After 15 days she developed widespread multiple ecchymoses over the arms, legs, and abdomen. Her platelet count, prothrombin time, partial thromboplastin time, and bleeding time were normal. Paroxetine was withdrawn, and 5 days later, the bruising had markedly abated and no new

lesions were identified. She was subsequently treated with a tricyclic antidepressant without recurrence of the ecchymoses.

The authors noted two earlier reports of ecchymoses with paroxetine, with normal laboratory values. They speculated that an indirect effect on platelet function through inhibition of platelet 5-HT uptake may be involved.

Paroxetine caused neutropenia in a patient whose white cell count fell to 2.9×10^9/l (neutrophils 1.37×10^9/l) (13). The white cell count gradually recovered over 6 weeks after paroxetine withdrawal.

Liver

Hepatotoxicity associated with paroxetine has been documented in two case reports (SEDA-21, 12). Severe hepatitis has been reported in two young women who took Atrium (febarbamate + difebarbamate + phenobarbital) and paroxetine (14). Liver biopsy showed lesions compatible with drug-related injury. Both recovered completely. The authors suggested that simultaneous treatment with Atrium and paroxetine increased each drug's hepatotoxicity.

Sexual function

Delayed ejaculation associated with paroxetine has been reported (SEDA-18, 21).

Immunologic

A skin reaction consistent with a vasculitis has been reported.

- A 20-year-old woman taking paroxetine 10 mg/day for obsessive-compulsive disorder developed multiple purple lesions on the fingers of both hands after 15 weeks (15). The lesions disappeared after 1 week but returned in 2 days after rechallenge with paroxetine.

Long-Term Effects

Drug withdrawal

Sudden withdrawal of paroxetine has been associated with nausea, dizziness, tremor, insomnia, irritation, and agitation (SEDA-17, 20). In three cases, a withdrawal reaction occurred with paroxetine, despite tapering of the dose for 7–14 days before discontinuation; the symptoms were the same as in earlier reports, but also included in one case myalgia, and in another rhinorrhea and visual phenomena similar to those associated with migraine (16). The authors suggested that cholinergic mechanisms and functional changes in 5-HT may play a role in the mediation of withdrawal symptoms.

Second-Generation Effects

Teratogenicity

In a recent guideline the National Institute for Health and Clinical Excellence has suggested that the use of paroxetine in the first trimester of pregnancy could be associated with an increased risk of fetal heart defects (17).

Drug Administration

Drug overdose

There were gastrointestinal symptoms and CNS disturbances in patients who took overdoses of paroxetine, the largest amount being 850 mg (18).

Drug–Drug Interactions

Beta-blockers

Paroxetine can cause clinically important increases in plasma concentrations of metoprolol (19).

Clozapine

In a prospective study, paroxetine (20 mg/day), a potent inhibitor of CYP2D6, did not increase clozapine concentrations ($n = 14$), and the authors therefore suggested that CYP2D6 is not an important pathway of metabolism of clozapine (20). However, other studies (SEDA-21, 22) have shown that paroxetine can increase clozapine concentrations, suggesting that this combination should be used with caution.

The effect of paroxetine on steady-state plasma concentrations of clozapine and its metabolites has been studied in 17 patients taking clozapine (200–400 mg/day), nine of whom took additional paroxetine (20–40 mg/day) (21). Paroxetine, a potent inhibitor of CYP2D6, inhibited the metabolism of clozapine, possibly by affecting a pathway other than N-desmethylation and N-oxidation. After 3 weeks of paroxetine, mean plasma concentrations of clozapine and norclozapine increased significantly by 31% and 20% respectively, while concentrations of clozapine N-oxide were unchanged.

Linezolid

SSRIs can also cause pharmacodynamic drug interactions for some time after withdrawal, through residual serotonin reuptake blocking activity.

- A 56-year-old woman with a postoperative wound infection developed the serotonin syndrome when she was given the antibiotic linezolid intravenously (22). The dose of paroxetine had been tapered and it had been withdrawn 5 days before.

Linezolid inhibits monoamine oxidase activity and has been reported to cause serotonin toxicity in combination with paroxetine. While some patients have apparently taken the combination of linezolid and an SSRI safely, this report suggests that patients taking combined treatment should be monitored for serotonin toxicity.

Lithium

Serum lithium concentrations were unchanged when breakthrough depression was treated double blind by the addition of paroxetine (20–40 mg/day, $n = 19$) and the combination was generally well tolerated (23).

- A 20-year-old woman taking lithium and risperidone became catatonic 5–7 days after the addition of paroxetine, leading to speculation that this was due to an interaction between the three drugs (24). Of 17 patients 4 who had paroxetine added to lithium as an adjunctive antidepressant developed symptoms suggestive of emerging serotonin syndrome (for example nausea, vomiting, diarrhea, sweating, anxiety, oversleeping) (25).

Methadone

Methadone-maintenance treatment is now established by controlled trials as effective in managing patients with opioid dependence. SSRIs are often co-prescribed for such patients, and there have been reports that SSRIs can increase methadone concentrations, presumably by inhibition of CYP2D6 (SEDA-25, 15). The effect of adding paroxetine (20 mg/day) to the treatment regimen has been studied in 10 opiate-dependent patients taking methadone maintenance (26). Methadone concentrations increased on an average by about one-third, although there was much individual variation. There were no obvious clinical consequences, presumably because the patients were fairly tolerant to the effects of methadone. However, the authors cautioned that sudden withdrawal of an SSRI in methadone users has the potential to trigger opioid-withdrawal symptoms.

Paroxetine 20 mg/day, a selective CYP2D6 inhibitor, was given for 12 days to 10 patients on methadone maintenance (27). Eight were genotyped as CYP2D6 homozygous extensive metabolizers and two as poor metabolizers. Paroxetine increased the steady-state concentrations of R-methadone and S-methadone, especially in the extensive metabolizers.

Tricyclic antidepressants

Paroxetine can cause clinically important increases in plasma concentrations of tricyclic antidepressants (28).

Triptans

Treatment of 12 healthy volunteers with paroxetine (20 mg/day for 14 days) did not alter the pharmacokinetics or pharmacodynamic effects of an acute dose of rizatriptan (10 mg orally) (29). These data are reassuring, but (as with sumatriptan) it is possible that sporadic cases of 5-HT neurotoxicity could still occur when rizatriptan is combined with an SSRI.

Risperidone

The addition of paroxetine 20 mg/day to risperidone 4–8 mg/day in 10 patients with schizophrenia produced a 45% increase in plasma concentrations of risperidone and its active metabolite, 9-hydroxyrisperidone (30).

One of the patients developed signs of drug-induced Parkinson's disease following the addition of paroxetine.

The serotonin syndrome occurred in a patient taking paroxetine plus an atypical antipsychotic drug, risperidone (31).

- A 53-year-old man took paroxetine 40 mg/day and risperidone 6 mg/day, having previously taken lower doses of both. Within 2 hours he developed ataxia, shivering, and tremor. He had profound sweating but was apyrexial, and was confused, with involuntary jerking movements of his limbs. He recovered without specific treatment over the next 2 days.

This was the first report of the serotonin syndrome in a patient taking an SSRI and an atypical antipsychotic drug. The reaction was unexpected because risperidone, in addition to being a potent dopamine receptor antagonist, is also a 5-HT$_2$-receptor antagonist. Subsequent animal studies suggested that 5-HT$_2$-receptor antagonists increase the firing of serotonergic neurons, perhaps through a postsynaptic feedback loop. This could account for potentiation of the effects of SSRIs by 5-HT$_2$-receptor antagonists, such as risperidone.

The serotonin syndrome has been reported during treatment with paroxetine and risperidone (32). A case of edema in a patient taking risperidone and paroxetine has also been reported (33).

The effects of paroxetine 20 mg/day for 4 weeks on steady-state plasma concentrations of risperidone and its active metabolite 9-hydroxyrisperidone have been studied in 10 patients taking risperidone 4–8 mg/day (34). During paroxetine administration, mean plasma risperidone concentrations increased significantly, while 9-hydroxyrisperidone concentrations fell slightly but not significantly; after 4 weeks, the sum of the risperidone and 9-hydroxyrisperidone concentrations increased significantly by 45% over baseline, and the mean plasma risperidone/9-hydroxyrisperidone concentration ratio was also significantly changed. However, the drug combination was generally well tolerated, with the exception of one patient who developed parkinsonian symptoms during the second week, and whose total plasma risperidone and 9-hydroxyrisperidone concentrations increased by 62%.

References

1. Erfurth A, Loew M, Dobmeier P, Wendler G. EKG – Veran derungen nach faroxetineo Drei Fallberichte. [ECG changes after paroxetine. 3 case reports.] Nervenarzt 1998;69(7):629–31.
2. Kaneda Y, Kawamura I, Fujii A, Ohmori T. Serotonin syndrome—"potential" role of the CYP2D6 genetic polymorphism in Asians. Int J Neuropsychopharmacol 2002;5(1):105–6.
3. Bonnet-Brilhault F, Thibaut F, Leprieur A, Petit M. A case of paroxetine-induced akathisia and a review of SSRI-induced akathisia. Eur Psychiatry 1998;13:109–11.
4. Lewis CF, DeQuardo JR, DuBose C, Tandon R. Acute angle-closure glaucoma and paroxetine. J Clin Psychiatry 1997;58(3):123–4.
5. Kirwan JF, Subak-Sharpe I, Teimory M. Bilateral acute angle closure glaucoma after administration of paroxetine. Br J Ophthalmol 1997;81(3):252.
6. Musselman DL, Lawson DH, Gumnick JF, Manatunga AK, Penna S, Goodkin RS, Greiner K, Nemeroff CB, Miller AH. Paroxetine for the prevention of depression induced by high-dose interferon alfa. N Engl J Med 2001;344(13):961–6.
7. Diler RS, Avci A. SSRI-induced mania in obsessive-compulsive disorder. J Am Acad Child Adolesc Psychiatry 1999;38(1):6–7.
8. Bonin B, Vandel P, Sechter D, Bizouard P. Paroxetine and galactorrhea. Pharmacopsychiatry 1997;30(4):133–4.
9. Morrison J, Remick RA, Leung M, Wrixon KJ, Bebb RA. Galactorrhea induced by paroxetine. Can J Psychiatry 2001;46(1):88–9.
10. Kim EJ, Yu B-H. Increased cholesterol levels after paroxetine treatment in patients with panic disorder. J Clin Psychopharmacol 2005;25:597–9.
11. George TP, Innamorato L, Sernyak MJ, Baldessarini RJ, Centorrino F. Leukopenia associated with addition of paroxetine to clozapine. J Clin Psychiatry 1998;59(1):31.
12. Cooper TA, Valcour VG, Gibbons RB, O'Brien-Falls K. Spontaneous ecchymoses due to paroxetine administration. Am J Med 1998;104(2):197–8.
13. Moselhy HF, Conlon W. Neutropenia associated with paroxetine. Ir J Psychol Med 1999;16:75.
14. Cadranel JF, Di Martino V, Cazier A, Pras V, Bachmeyer C, Olympio P, Gonzenbach A, Mofredj A, Coutarel P, Devergie B, Biour M. Atrium and paroxetine-related severe hepatitis. J Clin Gastroenterol 1999;28(1):52–5.
15. Margolese HC, Chouinard G, Beauclair L, Rubino M. Cutaneous vasculitis induced by paroxetine. Am J Psychiatry 2001;158(3):497.
16. Barr LC, Goodman WK, Price LH. Physical symptoms associated with paroxetine discontinuation. Am J Psychiatry 1994;151(2):289.
17. National Institute for Health and Clinical Excellence. Antenatal and postnatal mental health: clinical management and service guidance—NICE Guidance, February 2007: http://guidance.nice.org.uk/CG45.
18. Grimsley SR, Jann MW. Paroxetine, sertraline, and fluvoxamine: new selective serotonin reuptake inhibitors. Clin Pharm 1992;11(11):930–57.
19. Wigen CL, Goetz MB. Serotonin syndrome and linezolid. Clin Infect Dis 2002;34(12):1651–2.
20. Wetzel H, Anghelescu I, Szegedi A, Wiesner J, Weigmann H, Harter S, Hiemke C. Pharmacokinetic interactions of clozapine with selective serotonin reuptake inhibitors: differential effects of fluvoxamine and paroxetine in a prospective study. J Clin Psychopharmacol 1998;18(1):2–9.
21. Leucht S, Hackl HJ, Steimer W, Angersbach D, Zimmer R. Effect of adjunctive paroxetine on serum levels and side-effects of tricyclic antidepressants in depressive inpatients. Psychopharmacology (Berl) 2000;147(4):378–83.
22. Hemeryck A, Lefebvre RA, De Vriendt C, Belpaire FM. Paroxetine affects metoprolol pharmacokinetics and pharmacodynamics in healthy volunteers. Clin Pharmacol Ther 2000;67(3):283–91.
23. Bauer M, Zaninelli R, Muller-Oerlinghausen B, Meister W. Paroxetine and amitriptyline augmentation of lithium in the treatment of major depression: a double-blind study. J Clin Psychopharmacol 1999;19(2):164–71.
24. Shad U, Preskorn SH, Izgur Z. Failure to consider drug–drug interactions as a likely cause of behavioral

deterioration in a patient with bipolar disorder. J Clin Psychopharmacol 2000;20(3):390–2.

25. Fagiolini A, Buysse DJ, Frank E, Houck PR, Luther JF, Kupfer DJ. Tolerability of combined treatment with lithium and paroxetine in patients with bipolar disorder and depression. J Clin Psychopharmacol 2001;21(5):474–8.

26. Begre S, von Bardeleben U, Ladewig D, Jaquet-Rochat S, Cosendai-Savary L, Golay KP, Kosel M, Baumann P, Eap CB. Paroxetine increases steady-state concentrations of (R)-methadone in CYP2D6 extensive but not poor metabolizers. J Clin Psychopharmacol 2002;22(2):211–5.

27. Begre S, von Bardeleben U, Ladewig D, Jaquet-Rochat S, Cosendai-Savary L, Golay KP, Kosel M, Baumann P, Eap CB. Paroxetine increases steady-state concentrations of (R)-methadone in CYP2D6 extensive but not poor metabolizers. J Clin Psychopharmacol 2002;22(2):211–5.

28. Spina E, Avenoso A, Salemi M, Facciola G, Scordo MG, Ancione M, Madia A. Plasma concentrations of clozapine and its major metabolites during combined treatment with paroxetine or sertraline. Pharmacopsychiatry 2000;33(6):213–7.

29. Goldberg MR, Lowry RC, Musson DG, Birk KL, Fisher A, De Puy ME, Shadle CR. Lack of pharmacokinetic and pharmacodynamic interaction between rizatriptan and paroxetine. J Clin Pharmacol 1999;39(2):192–9.

30. Spina E, Avenoso A, Facciola G, Scordo MG, Ancione M, Madia A. Plasma concentrations of risperidone and 9-hydroxyrisperidone during combined treatment with paroxetine. Ther Drug Monit 2001;23(3):223–7.

31. Hamilton S, Malone K. Serotonin syndrome during treatment with paroxetine and risperidone. J Clin Psychopharmacol 2000;20(1):103–5.

32. Hamilton S, Malone K. Serotonin syndrome during treatment with paroxetine and risperidone. J Clin Psychopharmacol 2000;20(1):103–5.

33. Masson M, Elayli R, Verdoux H. Rispéridone et edème: à propos d'un cas. [Risperidone and edema: apropos of a case.] Encephale 2000;26(3):91–2.

34. Spina E, Avenoso A, Facciola G, Scordo MG, Ancione M, Madia A. Plasma concentrations of risperidone and 9-hydroxyrisperidone during combined treatment with paroxetine. Ther Drug Monit 2001;23(3):223–7.

Sertraline

General Information

Sertraline hydrochloride has an average half-life of about 26 hours, and mean peak plasma concentrations occur at 4.5–8.4 hours. The dosage is 50–200 mg/day orally. Its major metabolite, N-desmethylsertraline, is less active than sertraline. Adverse effects are as for the SSRIs in general.

Organs and Systems

Cardiovascular

Angina occurred in an elderly woman shortly after she started to take sertraline and on rechallenge (1).

Nervous system

Parkinsonism

Sertraline has an appreciable affinity for the dopamine re-uptake site, and for this reason might be presumed less likely to cause movement disorders than other SSRIs. However, there is little clinical evidence to support this suggestion and cases of sertraline-induced parkinsonism have been reported (2).

- A 70-year-old woman, who had been taking sertraline 100 mg/day for 7 months, gave a 6-month history of resting tremor and loss of dexterity in the right hand. She had mild bradykinesia and cogwheel rigidity in the right arm and leg. She had not taken any other medications. A brain MRI scan was normal. The sertraline was withdrawn and within 1 month all the neurological symptoms and signs had remitted.
- An 81-year-old woman took sertraline 100 mg/day for depression and 6 months later presented with tremor and difficulty in moving her right arm and leg (3). A diagnosis of right hemiparkinsonism was made and the sertraline was withdrawn. Her extrapyramidal symptoms resolved within 3 months, but 14 months later she developed parkinsonism and was treated with levodopa and carbidopa.
- A 70-year-old man developed parkinsonian symptoms 1 month after starting to take sertraline 100 mg/day (4). Withdrawal of sertraline resulted in amelioration but not complete remission of his symptoms which then required treatment with carbidopa and levodopa.

In these cases presumably the sertraline prematurely precipitated Parkinson's disease. Two cases of sertraline-induced akathisia have been reported (SEDA-18, 19).

Serotonin syndrome

Treatment with serotonin-potentiating drugs in usual therapeutic doses can sometimes produce the serotonin syndrome. There are also case reports of this reaction with single doses of SSRIs (see also the monograph on fluvoxamine) (5).

Serotonin syndrome after a single dose of sertraline 100 mg has been reported in a 16-year-old girl (6). It responded to a single 4 mg dose of the serotonin antagonist cyproheptadine.

There was therapeutic benefit from cyproheptadine in a case of serotonin syndrome in a 2-year-old girl who accidentally swallowed ten 50 mg tablets of sertraline (7).

Endocrine

Serotonin pathways are involved in the regulation of prolactin secretion. Galactorrhea has also been reported in a patient taking sertraline, in whom lactation ceased after withdrawal (SEDA-18, 20) and in another case (8).

Hematologic

Prolonged bleeding time has been reported with sertraline (9).

Sertraline has been associated with agranulocytosis (10).

Liver

Hepatitis has been associated with sertraline (11).

Skin

Stevens–Johnson syndrome has been reported in a 96-year-old woman who had taken sertraline (dosage not stated) for 3 weeks (12). The lesions involved the skin, oral mucosa, and conjunctiva. The eruption disappeared within 7 days of sertraline withdrawal.

Hair

Hair loss has been associated with sertraline (13).

Sexual function

Sexual disturbance has also been associated with sertraline (14), and a high frequency of such adverse effects has been reported in studies in which high doses were used. In a double-blind, placebo-controlled study of sertraline and amitriptyline in patients with major depression, male sexual dysfunction, mainly ejaculatory disturbance, was reported significantly more often with sertraline (in 21% of the patients) (15). Male sexual dysfunction in 15% of sertraline-treated patients has also been reported (16).

Priapism is occasionally associated with the use of psychotropic drugs, such as trazodone, that are α_1-adrenoceptor antagonists. It has also been reported in a man taking sertraline (17).

- A 47-year-old man presented with a 4-day history of priapism and moderate pain. Several brief but otherwise similar episodes had occurred during the previous month. He had a history of depression and had been taking sertraline 200 mg/day and dexamfetamine 10 mg/day. He received intracorporeal methoxamine, but when this proved ineffective he was treated with intracorporeal adrenaline and a shunt procedure. However, detumescence was incomplete. At follow-up after several weeks the priapism had resolved and he had not become impotent (a significant risk in cases of prolonged priapism). He was given nefazodone with no recurrence of erectile dysfunction.

The dosage of sertraline used in this case was high and the combined use of dexamfetamine may also have been relevant.

Long-Term Effects

Drug withdrawal

Withdrawal symptoms have been reported with sertraline (18). All the symptoms, including gastrointestinal discomfort, insomnia, and influenza-like symptoms, remitted when sertraline was reinstituted.

Drug Administration

Drug overdose

In 40 patients who took up to 8 g of sertraline there were no serious sequelae (19).

One of the largest overdoses of sertraline has been reported (20).

- A 51-year-old woman took about 8 g of sertraline, about 80 times the usual daily dose. On admission to hospital she was somnolent but rousable. Her electrocardiogram showed a transiently prolonged QT$_c$ interval (510 ms falling to 470 ms). On the third day she developed agitation, disorientation, myoclonus, and pyrexia (38.5°C), was treated with supportive measures, and recovered over the next 3 days.

While cardiac toxicity was not prominent in this case, the patient developed clear evidence of the serotonin syndrome, which proved self-limiting.

Drug–Drug Interactions

Alprazolam

Sertraline (50–150 mg/day) had no effects on alprazolam metabolism in a randomized, double-blind, placebo-controlled study in 10 healthy volunteers (24).

Buspirone

Serotonin syndrome has been attributed to the combination of buspirone and sertraline.

- A 49-year-old man had major adverse effects 11 days after taking a combination of sertraline, buspirone, and loxapine (25). The adverse effects were characteristic of the serotonin syndrome, which is characterized by a constellation of symptoms, including hypomania, agitation, seizures, confusion, restlessness, hyper-reflexia, tremor, myoclonus, ataxia, incoordination, anxiety, double vision, fever, shivering, variable effects on blood pressure, nausea and vomiting, sweating, and diarrhea.

Carbamazepine

Sertraline is a substrate for a number of CYP450 isozymes, including CYP2C9, CYP2C19, and CYP3A4. Several case reports have shown loss of antidepressant activity of sertraline at usual therapeutic doses when depressed patients have also taken drugs that induce CYP3A4, including carbamazepine (21).

Clonazepam

In a randomized, double-blind, placebo-controlled, crossover study in 13 subjects, sertraline did not affect the pharmacokinetics or pharmacodynamics of clonazepam (26).

Clozapine

Dual effects were observed in a 44-year-old schizophrenic patient taking clozapine with both fluoxetine and sertraline for mood stabilization (27). Clinical and motor status improved with both fluoxetine and sertraline; cognitive function improved with clozapine and fluoxetine, but was not sustained with sertraline. However, sertraline did not affect steady-state plasma concentrations of clozapine and its metabolites in 17 patients taking clozapine (200–400 mg/day), 8 of whom took additional sertraline (50–100 mg/day) (28).

Erythromycin

Drug interactions leading to the serotonin syndrome usually result from pharmacodynamic mechanisms. However, the antibiotic erythromycin may have precipitated the serotonin syndrome in a patient taking sertraline by a pharmacokinetic mechanism (22).

- A 12-year-old boy, who had taken sertraline 37.5 mg/day for 5 weeks for obsessive-compulsive disorder, started to take erythromycin 200 mg bd. Within 4 days he began to feel anxious; this was followed over the next 10 days by panic, restlessness, irritability, tremulousness, and confusion. These symptoms resolved within 72 hours of withdrawal of sertraline and erythromycin.

The authors proposed that in this case erythromycin had inhibited sertraline metabolism by inhibiting CYP3A. This could have led to increased concentrations of sertraline and signs of serotonin toxicity. Unfortunately sertraline concentrations were not measured to confirm this suggestion.

Rifampicin

Sertraline is a substrate for a number of CYP450 isozymes, including CYP2C9, CYP2C19, and CYP3A4. Several case reports have shown loss of antidepressant activity of sertraline at usual therapeutic doses when depressed patients have also taken drugs that induce CYP3A4, including rifampicin (23).

Tramadol

The serotonin syndrome has been reported after concurrent use of tramadol and sertraline (SEDA-22, 103).

- Serotonin syndrome occurred in an 88-year-old woman who took sertraline 50–100 mg/day and tramadol 200–400 mg/day for 10 days; the symptoms subsided 15 days after withdrawal of tramadol (29).

Zolpidem

Interactions between zolpidem and sertraline have been studied in 28 healthy women, who took a single dose of zolpidem alone and five consecutive doses of zolpidem 10 mg while taking chronic doses of sertraline 50 mg (30). Co-administration of sertraline 50 mg and zolpidem 10 mg was safe but could result in a shortened onset of action and increased effect of zolpidem.

References

1. Sunderji R, Press N, Amin H, Gin K. Unstable angina associated with sertraline. Can J Cardiol 1997;13(9):849–51.
2. Di Rocco A, Brannan T, Prikhojan A, Yahr MD. Sertraline induced parkinsonism. A case report and an in-vivo study of the effect of sertraline on dopamine metabolism. J Neural Transm 1998;105(2–3):247–51.
3. Pina Latorre MA, Modrego PJ, Rodilla F, Catalan C, Calvo M. Parkinsonism and Parkinson's disease associated with long-term administration of sertraline. J Clin Pharm Ther 2001;26(2):111–2.
4. Gregory RJ, White JF. Can sertraline induce parkinson's disease? Psychosomatics 2001;42(2):163–4.
5. Gill M, LoVecchio F, Selden B. Serotonin syndrome in a child after a single dose of fluvoxamine. Ann Emerg Med 1999;33(4):457–9.
6. Mullins ME, Horowitz BZ. Serotonin syndrome after a single dose of fluvoxamine. Ann Emerg Med 1999;34(6):806–7.
7. Horowitz BZ, Mullins ME. Cyproheptadine for serotonin syndrome in an accidental pediatric sertraline ingestion. Pediatr Emerg Care 1999;15(5):325–7.
8. Lesaca TG. Sertraline and galactorrhea. J Clin Psychopharmacol 1996;16(4):333–4.
9. Calhoun JW, Calhoun DD. Prolonged bleeding time in a patient treated with sertraline. Am J Psychiatry 1996;153(3):443.
10. Trescoli-Serrano C, Smith NK. Sertraline-induced agranulocytosis. Postgrad Med J 1996;72(849):446.
11. Hautekeete ML, Colle I, van Vlierberghe H, Elewaut A. Symptomatic liver injury probably related to sertraline. Gastroenterol Clin Biol 1998;22(3):364–5.
12. Jan V, Toledano C, Machet L, Machet MC, Vaillant L, Lorette G. Stevens-Johnson syndrome after sertraline. Acta Derm Venereol 1999;79(5):401.
13. Bourgeois JA. Two cases of hair loss after sertraline use. J Clin Psychopharmacol 1996;16(1):91–2.
14. Cooper GL. The safety of fluoxetine – an update. Br J Psychiatry Suppl 1988;3:77–86.
15. Reimherr FW, Chouinard G, Cohn CK, Cole JO, Itil TM, LaPierre YD, Masco HL, Mendels J. Antidepressant efficacy of sertraline: a double-blind, placebo- and amitriptyline-controlled, multicenter comparison study in outpatients with major depression. J Clin Psychiatry 1990;51(Suppl. B):18–27.
16. Doogan DP. Toleration and safety of sertraline: experience worldwide. Int Clin Psychopharmacol 1991;6(Suppl. 2):47–56.
17. Rand EH. Priapism in a patient taking sertraline. J Clin Psychiatry 1998;59(10):538.
18. Louie AK, Lannon RA, Ajari LJ. Withdrawal reaction after sertraline discontinuation. Am J Psychiatry 1994;151(3):450–1.
19. Lau GT, Horowitz BZ. Sertraline overdose. Acad Emerg Med 1996;3(2):132–6.
20. Brendel DH, Bodkin JA, Yang JM. Massive sertraline overdose. Ann Emerg Med 2000;36(5):524–6.
21. Khan A, Shad MU, Preskorn SH. Lack of sertraline efficacy probably due to an interaction with carbamazepine. J Clin Psychiatry 2000;61(7):526–7.
22. Lee DO, Lee CD. Serotonin syndrome in a child associated with erythromycin and sertraline. Pharmacotherapy 1999;19(7):894–6.
23. Markowitz JS, DeVane CL. Rifampin-induced selective serotonin reuptake inhibitor withdrawal syndrome in a patient treated with sertraline. J Clin Psychopharmacol 2000;20(1):109–10.
24. Hassan PC, Sproule BA, Naranjo CA, Herrmann N. Dose-response evaluation of the interaction between sertraline and alprazolam in vivo. J Clin Psychopharmacol 2000;20(2):150–8.
25. Bonin B, Vandel P, Vandel S, Sechter D, Bizouard P. Serotonin syndrome after sertraline, buspirone and loxapine? Thérapie 1999;54(2):269–71.

26. Bonate PL, Kroboth PD, Smith RB, Suarez E, Oo C. Clonazepam and sertraline: absence of drug interaction in a multiple-dose study. J Clin Psychopharmacol 2000;20(1):19–27.

27. Purdon SE, Snaterse M. Selective serotonin reuptake inhibitor modulation of clozapine effects on cognition in schizophrenia. Can J Psychiatry 1998;43(1):84–5.

28. Spina E, Avenoso A, Salemi M, Facciola G, Scordo MG, Ancione M, Madia A. Plasma concentrations of clozapine and its major metabolites during combined treatment with paroxetine or sertraline. Pharmacopsychiatry 2000;33(6):213–7.

29. Sauget D, Franco PS, Amaniou M, Mazere J, Dantoine T. Possible syndrome sérotoninergiques induit par l'association de tramadol à de la sertraline chez une femme agée. [Possible serotonergic syndrome caused by combination of tramadol and sertraline in an elderly woman.] Thérapie. 57(3):309–10.

30. Allard S, Sainati SM, Roth-Schechter BF. Coadministration of short-term zolpidem with sertraline in healthy women. J Clin Pharmacol 1999;39(2):184–91.

MONOAMINE OXIDASE INHIBITORS

General Information

All available monoamine oxidase (MAO) inhibitors (excepting moclobemide, toloxatone, brofaromine, and selegiline) act via a "suicide" mechanism, by causing long-lasting, irreversible, competitive inhibition of mitochondrial MAO, which persists until new enzyme is manufactured (1). Most of these drugs also produce a non-specific reduction in the activity of hepatic drug-metabolizing enzymes.

Table 1 lists the MAO inhibitors that are already available or under investigation. These compounds fall into different chemical categories, including compounds that have the following properties: antidepressant (moclobemide), antihypertensive (pargyline), antineoplastic (procarbazine), and antimicrobial (furazolidone).

Antidepressant drugs of various classes (tricyclics, MAO inhibitors, SSRIs) have broad efficacy in generalized anxiety and in panic disorder, but SSRIs are now the treatments of choice (2,3). Selegiline has been used to treat Parkinson's disease. Other drugs that have MAO inhibitory activity, but are not used as such, include debrisoquine, linezolid, and isoniazid.

The monoamine oxidase inhibitors epitomize cyclical fashions in drug use and the impact of adverse effects. They were the first psychotropic drugs for which a clear biochemical action was defined. Early excitement was quickly tempered by reports of liver toxicity with the hydrazine derivatives, leading to synthesis of the cyclopropylamine drug, tranylcypromine, which in turn elicited the food and drug interactions that led to an overall decline in popularity.

Then in the 1980s there was a reappraisal of the benefit-to-harm balance of the MAO inhibitors. This spawned both a search for safer and more selective or rapidly reversible enzyme inhibitors (including moclobemide, toloxatone, and brofaromine), as well as a review and retrial of the older compounds.

The scientific underpinnings of this renaissance have been reviewed (4). Much of the earlier work was conducted using inadequate doses of phenelzine, whose efficacy was later validated using adequate drug concentrations (to produce 85% or more enzyme inhibition) (5). A review of eleven studies conducted in 1963–1982 that compared MAO inhibitors and tricyclic compounds showed that in three studies there was no difference, four favored tricyclic antidepressants, and in three MAO inhibitors were superior (6). The three studies that favored MAO inhibitors were among the four most recently conducted (all since 1979). An article in 1985 entitled "Should the use of MAO be abandoned?" (7) was accompanied by commentaries from six British and US experts in psychopharmacology. The consensus was clearly in favor of continued use, with the recognition that even if a specific responder is difficult to define there are individuals who respond when all other drugs have failed. No clear-cut clinical or metabolic features distinguish such individuals; an earlier claim that the clinical response and susceptibility to adverse effects might be influenced by genetically determined rapid or slow acetylation of phenelzine was not confirmed by a review of seven studies (8).

There have been many studies of the efficacy and toxicity of selective inhibitors of MAO type A (SEDA-16, 7; SEDA-17, 16; SEDA-18, 16).

The adverse effects of the MAO inhibitors include hepatocellular damage, similar to that which led to the withdrawal of the earlier hydrazine derivatives, hypotension, often a pronounced adverse effect (possibly due to

Table 1 Monoamine oxidase (MAO) inhibitors that have been studied or are currently commercially available for treating depression or for other purposes (rINNs except where stated)

Compound	Structure	Comments
Fenoxypropazine Iproniazid Mebanazine Pheniprazine Pivhydrazine	Hydrazines	These drugs are earlier compounds that caused liver damage; they are obsolete
Nialamide	Hydrazine	Little used today
Phenelzine (pINN)	Hydrazine	Widely used
Isocarboxazid	Hydrazine	Similar to phenelzine
Tranylcypromine	Cyclopropylamine	Most amphetamine-like High propensity for interactions with foods and drugs
Pargyline	Propinylbenzylamine	Used for hypertension
Furazolidone	Nitrofuran	Antimicrobial (giardiasis)
Procarbazine	Methylhydrazine	Antineoplastic
Clorgiline	Propylamine	Type A MAO inhibitor (serotonin/noradrenaline)
Selegiline	Propinylamine	Type B MAO inhibitor (phenylethylamine)
Moclobemide	Benzamide	Type A MAO inhibitor
Toloxatone	Oxazolidinone	Type A MAO inhibitor
Brofaromine	Piperidine	Type A MAO inhibitor

the accumulation of a pseudotransmitter normally metabolized by MAO), and autonomic disturbances, such as dryness of the mouth, sweating, constipation, and weight gain.

Differences between MAO inhibitors in adverse effect profiles are poorly substantiated, and are confounded by different usage patterns and drug potencies. The selective MAO inhibitors are considered in other monographs.

Comparative risks of monoamine oxidase inhibitors and tricyclic antidepressants

MAO inhibitors suffered a long period of disfavor, largely because of their troublesome interactions. Their risks still exceed those of the tricyclic compounds, but some experienced clinicians feel that they can be used safely, provided appropriate precautions are taken.

Of particular interest is a chart review of 198 patients, aged 19–64, treated in a university research clinic by psychiatrists who specialized in the pharmacological treatment of affective disorders (9). Most of the patients (130) had atypical depression and participated in a 6- or 12-week double-blind comparison of phenelzine, imipramine, and placebo, with further follow-up. The authors selected 14 adverse effects generally considered to cause "serious medical risk or subjective discomfort great enough to require drug discontinuation." There were serious adverse effects in 14% of the patients taking placebo, 27% of those taking imipramine, 43% of those taking tranylcypromine, and 64% of those taking phenelzine; 38% had two or more serious adverse effects and all but one was taking phenelzine. The differences between imipramine and phenelzine were highly significant. By 33 weeks under half of the patients taking imipramine had had a serious adverse effect, compared with over 90% of those taking phenelzine. Adverse effects were sufficiently severe to require withdrawal in 45%. None of the 14 selected adverse effects was more common in the patients taking imipramine. The serious adverse effects that were more common in patients taking phenelzine included hypomania (10%), hypertensive crisis (8%), weight gain over 15 pounds (8%), and anorgasmia and impotence (22%). In a second article the authors discussed these adverse effects in detail (10). Of the eleven patients who had hypertensive crises, six had eaten tyramine-containing foods "despite meticulous dietary review and cautioning," and three had taken ephedrine-containing medications; four obtained emergency medical treatment in local hospitals and a fifth developed coma with intracranial bleeding due to an unsuspected aneurysm.

Organs and Systems

Cardiovascular

When MAO is inhibited, the concentrations of noradrenaline, dopamine, and serotonin increase in the central nervous system and heart. The concentrations of the precursors of these amines (dihydroxyphenylalanine and 5-hydroxytryptophan) are greatly increased. The effects are similar to those of the postganglionic blocking amines:

postural hypotension occurs and cardiac output is reduced, possibly due to the accumulation of a pseudotransmitter.

Hypotension

The hypotensive effects of the MAO inhibitors differ from those of the tricyclic antidepressants, inasmuch as the former affect supine as well as orthostatic blood pressure. This was confirmed both in a study (11) involving tranylcypromine and in an evaluation of blood pressure changes in 14 patients taking phenelzine averaging 65 mg/day for at least 3 weeks (12). The drug-free mean supine systolic blood pressure was 127 mmHg and it fell significantly (mean drop 5 mmHg) by the end of the first week. By contrast, an increase in the orthostatic drop from the mean predrug baseline of 2 mmHg did not reach significance until the end of the second week of treatment, after which the blood pressure continued to fall. Two patients developed profound orthostatic falls of up to 50 mmHg after more than 2 weeks of treatment. In one patient the hypotension and related symptoms (light-headedness and ataxia) improved with fludrocortisone, but not in the other, who required drug withdrawal. The authors commented that because orthostatic hypotension can develop late, cautious long-term monitoring is required.

Hypertension

Hypertensive crises usually occur when MAO inhibitors are combined with other drugs or foods that cause interactions (see next subsection).

Autopotentiation of hypertensive effects

In a few cases, a hypertensive crisis appears not to have been provoked by known drug or dietary precipitants (13). From a review of 12 reports of spontaneous hypertension in patients taking tranylcypromine or phenelzine, it has emerged that a family history of hypertension may be a risk factor (SEDA-18, 14). A significant increase in supine blood pressure without similar changes in standing blood pressure after the administration of tranylcypromine has also been described (SEDA-17, 17).

Cardiac arrhythmias

MAO inhibitors can cause bradycardia. A report of two cases of interactions of monoamine oxidase inhibitors with beta-blockers (nadolol and metoprolol) is of interest, and several possible mechanisms were discussed (14).

Nervous system

Autonomic nervous system

A carefully controlled 3-week comparison of phenelzine (up to 90 mg/day, mean 77 mg) and imipramine (up to 150 mg/day, mean 139 mg) showed no significant difference between the drugs in the incidence of dry mouth, blurred vision, constipation, and urinary hesitancy (15).

Central nervous system

Insomnia and daytime somnolence and fatigue have been reported with tranylcypromine, phenelzine, and isocarboxazid (SEDA-16, 8; SEDA-17, 16). In a comparison of phenelzine and imipramine (15) there were important and significant differences between the drugs: 19% of the patients taking phenelzine reported drowsiness compared with none of those taking imipramine; moreover, 18 patients taking phenelzine had to be withdrawn, compared with one taking imipramine (who developed urinary retention). Central nervous system toxicity was also reported after an abrupt switch from phenelzine to isocarboxazid (SEDA-17, 17).

Parkinsonism

The MAO inhibitors can be used in patients who have Parkinson's disease, for two reasons. First, this is a disorder in which depression is common. Secondly, the selective type B inhibitor selegiline was originally thought to benefit patients with parkinsonism, possibly by increasing brain dopamine concentrations (16).

However, a recent long-term study has shown higher death rates in patients with early mild Parkinson's disease taking combined selegiline and levodopa compared with those taking levodopa alone. The authors concluded that the combination of selegiline with levodopa offers no advantage to patients with early mild Parkinson's disease, and that although selegiline may help symptoms in advanced Parkinson's disease it is best avoided in patients with postural hypotension, frequent falls, confusion, and dementia (17).

Dystonias

Two Italian neurologists have described current theories about the pathophysiology of levodopa-associated motor fluctuations and the use of inhibitors of MAO type B, inhibitors of catechol-*O*-methyl transferase, and modified-release levodopa in minimizing this problem (18).

- Dystonia has been described in a 65-year-old woman who had taken tranylcypromine 10 mg bd for 3 days. She had major long-term depression that had failed to respond to other agents. The dystonia abated 48 hours after withdrawal but recurred on rechallenge (19).

Neuromuscular function

The tendency for MAO inhibitors to produce symptoms related to neuromuscular excitability, the serotonin syndrome, has been recognized in cases of overdose (SEDA-10, 18) and in interactions with other antidepressants or tryptophan (SEDA-10, 16, 17) (20). The authors of a thorough review of the preclinical and clinical literature have drawn attention to these phenomena, which occur at therapeutic doses with a MAO inhibitor alone, and have speculated that the mechanism is related to a combination of increased serotonergic tone and central disinhibition of alpha motor neuron-mediated spinal activity (21). They discussed ten previous reports of myoclonus, hyper-reflexia, muscle twitching, and increased muscle tone in patients taking MAO inhibitors. These neuromuscular effects appear to occur in up to 15% or more of patients, and are more likely when tryptophan is given in combination. They usually appear after 10–14 days. Tolerance does not occur, but the effects may abate or

disappear with dosage reduction. Symptoms predominate during sleep and are often associated with muscle or joint pains.

Endocrine

The syndrome of inappropriate secretion of antidiuretic hormone (SIADH) (22) may be the mechanism of action underlying cases of peripheral edema that have been described (SEDA-2, 12) (SEDA-6, 28). Diuretics are not helpful, but dosage reduction produces relief (23).

A number of antidepressant drugs, particularly selective serotonin re-uptake inhibitors (SSRIs), can increase plasma prolactin concentrations, although galactorrhea is uncommon. In a prescription-event monitoring survey of about 65 000 patients, compared with SSRIs, moclobemide was associated with a relative risk of galactorrhea of 6.7 (95% CI = 2.7, 15) (24). However, this was substantially less than the risk associated with the dopamine receptor antagonist risperidone (relative risk compared with SSRIs 32; 95% CI = 14, 70). Nevertheless, the data suggest that moclobemide may be more likely to cause galactorrhea than other antidepressant drugs.

Hematologic

Leukopenia and agranulocytosis are well-recognized complications of treatment with tricyclic antidepressants and have been reported with some second-generation compounds. In a report of leukopenia in a patient taking phenelzine attention was drawn to five other unpublished cases and to previous published reports involving isocarboxazid, tranylcypromine, and tryptamine (25).

Liver

The early reports of hepatotoxicity with iproniazid led to the synthesis of non-hydrazine inhibitors, and a report from 19 French drug surveillance centers (26) has reaffirmed the frequency and severity of this adverse effect. There were 91 cases of hepatitis due to all antidepressants (11 with iproniazid) over 5 years. Cytolytic reactions occurred in 11 patients treated with iproniazid; 5 died. There was a high titer of antimitochondrial antibodies (M6) in five patients.

Skin

Photosensitivity to MAO inhibitors must be most unusual, but one well-authenticated report has been published (27).

Sexual function

Impotence and delayed ejaculation in men and difficulty in achieving orgasm in two women have been reported with a variety of MAO inhibitors used to treat narcolepsy (28), phobic anxiety (29), and depression (30–32). Sexual symptoms are often dose-related and there is a delicate interplay between psychic and pathophysiological influences. In men with premature ejaculation this effect may even be considered therapeutic.

Immunologic

Rashes have been reported, but their relation to drug ingestion is poorly substantiated. The hepatocellular damage caused by hydrazine derivatives is probably mediated by an immunological mechanism.

Body temperature

Hyperthermia and labile blood pressure, perhaps due to an interaction of clomipramine, phenelzine, and chlorpromazine has been described in one patient (SEDA-17, 17).

Long-Term Effects

Drug withdrawal

The literature on a withdrawal syndrome (SEDA-10, 17) has been expanded by further reports. One of these (33) involved the development of an acute toxic delirium 3 days after withdrawal of phenelzine, and another (34) concerned patients who became manic after withdrawal of isocarboxazid. A withdrawal state similar to that caused by withdrawal from amphetamines has been described after withdrawal of tranylcypromine (SEDA-16, 8) (SEDA-18, 14).

In principle it should be possible to switch from one conventional MAO inhibitor to another without a washout period. However, there have been reports that patients who switched from phenelzine to tranylcypromine had hypertensive reactions, with disastrous consequences (35). Whenever possible, a 2-week washout period when switching from a conventional MAO inhibitor to tranylcypromine seems advisable.

Second-Generation Effects

Pregnancy

Little is known about the use of MAO inhibitors during pregnancy and the postpartum period. There is a literature review along with a case report of phenelzine exposure throughout pregnancy and the postpartum period (36).

Lactation

There are some reports of the excretion of MAO inhibitors in breast milk (37,38).

Susceptibility Factors

Age

Because of the concern expressed about the use of tricyclic antidepressants in elderly people, MAO inhibitors have been studied in this population (39). Patients with dementia benefited in mood (but not cognition), and some non-demented patients also improved. Adverse effects were considered less frequent or troublesome than those due to tricyclic compounds, although one patient taking tranylcypromine became paranoid and

another taking phenelzine developed a choreiform movement disorder. The fact that MAO concentrations increase with age supports the use of these drugs in the elderly, but their hypotensive effects and interactions with other drugs and foods suggest that they should be used with extreme caution in people who are exposed to polypharmacy and whose comprehension of and attention to warnings may be impaired.

Other features of the patient

It is inadvisable to give MAO inhibitors to schizophrenic patients, even when they seem to be anergic or depressed, since they may precipitate a psychotic crisis. It is also difficult to ensure that such patients respect the necessary dietary restrictions, that is, avoidance of tyramine-containing foods.

Drug Administration

Drug overdose

The fatal effects of antidepressants in England, Scotland, and Wales during 1975–1984 and 1985–1989 have been compared in retrospective epidemiological studies (40,41). The fatality index (deaths per million prescriptions) for individual drugs or groups of drugs was used to give an indication of the relative effects of the different drugs. There were 24 deaths due to tranylcypromine, 33 due to phenelzine, and 3 due to isocarboxazid. The fatality index for tranylcypromine was significantly higher than the mean for all antidepressants, while that for the other MAO inhibitors was lower. The majority of all deaths from antidepressant poisoning were due to two tricyclic drugs (amitriptyline and dosulepin), and they, as well as the entire group of older tricyclic antidepressants, had a fatality index significantly higher than the mean (41). It must of course be kept in mind that the fatality index of a drug is influenced by many factors other than its inherent toxicity, including dosage, differences in severity of illness and suicidal tendencies, and prescription for indications other than depression.

The minimum fatal dose of non-selective MAO inhibitors is about 5–10 times the maximum daily dose, although individuals have been reported to survive such amounts. Symptoms of overdosage can be initially mild and deceptive, but can progress over 24 hours to include agitation, tremor, alternating low and high blood pressure, severe muscle spasms, hyperpyrexia, and convulsions. Symptomatic treatment has included the beta-blocker practolol (10 mg intravenously, repeated after 2 hours) and muscle relaxants with assisted respiration (42), which may help reduce pyrexia by abolishing excessive muscle activity. Active elimination techniques are unhelpful, according to a comprehensive review of antidepressant overdosage (43).

In one patient who took 900 mg of phenelzine, there was a marked excess of urinary and plasma catecholamines, analogous to pheochromocytoma, and the patient was successfully managed with alpha-adrenoceptor antagonists (44). In another patient, who probably took about 2000 mg of phenelzine, hyperpyrexia was prominent and responded to dantrolene sodium (45). The authors noted clinical similarities to malignant hyperpyrexia and neuroleptic malignant syndrome.

Overdose of the selective monoamine oxidase A inhibitor moclobemide by itself rarely appears to give rise to serious problems, in contrast to overdose with conventional monoamine oxidase inhibitors, which can cause fatal 5HT toxicity. However, if patients take moclobemide together with serotonergic antidepressants, such as SSRIs or clomipramine, 5-HT toxicity is common. 5HT toxicity occurred in 11 of 21 patients who took overdoses of moclobemide and serotonergic agents but in only one of 33 patients who took moclobemide alone (46). Consistent with this, four patients died, presumably of 5HT toxicity, after co-ingesting 3,4-methylenedioxymethamphetamine (MDMA, ecstasy) and moclobemide (47). Massive ingestion of moclobemide (over 30 g) also appeared to be responsible for death in a suicide attempt in a 48-year-old man (48). The blood concentration of moclobemide measured at post mortem (498 µg/ml) was about 500 times the usual target concentration (0.2–2.1 µg/ml). Post-mortem findings were non-specific, but the authors concluded that death might have been due to the serotonin syndrome.

Drug–Drug Interactions

General

Problems with non-selective irreversible MAO inhibitors arise for two reasons: the large number of amines that are substrates for the enzyme and the fact that these drugs also inhibit non-specific drug-metabolizing enzymes in the liver. Together, these two actions produce a lengthy list of interactions with other substances, the effects of which can be enhanced or prolonged, both during treatment with a MAO inhibitor and for up to 2 weeks afterwards, owing to the irreversible nature of the enzyme inhibition (Table 2).

Amines that serve as substrates include both natural and synthetic ephedrine congeners contained in prescribed and proprietary formulations, and amines such as tyramine, which occur naturally in a variety of

Table 2 Drug and food interactions with MAO inhibitors

Hypertensive crisis: potentiation of central effects of drugs or food components due to inhibition of monoamine oxidase

Drugs	Amines, prescribed or proprietary
	Amphetamine and congeners
	Ephedrine and congeners
	Levodopa
	Reserpine
	Tryptophan
Foods	Alcohol (wine, beer)
	Amine-containing foodstuffs
	Banana (peel)
	Broad beans (including pods)
	Caviar
	Cheese
	Chicken livers
	Chocolate
	Hung meats
	Pickled herring
	Yeast extracts (Marmite, packeted soups)
Serotonin syndrome	Dextromethorphan
	Pethidine (meperidine)
	Selective serotonin re-uptake inhibitors
	Trazodone and dextropropoxyphene (combined with phenelzine)
	Tricyclic antidepressants
	Tryptophan
Potentiation or prolongation of other drug effects due to inhibition of hepatic drug-metabolizing enzymes	Alcohol (all types)
	Anesthetics
	Antihistamines
	Antiparkinsonian agents
	Barbiturates
	Benzodiazepines
	Chloral hydrate
	Hypoglycemic agents
	Opiate analgesics
	Thyroid extract
	Tricyclic antidepressants

protein-containing foods, in which amines are produced by bacterial decarboxylation of amino acids. When these "indirectly acting" amines are ingested, they are normally destroyed by MAO in the intestine or liver; however, after treatment with a MAO inhibitor they gain access to the systemic circulation, resulting in release of noradrenaline from sympathetic nerve terminals. If the amount ingested is sufficient, symptoms similar to those of pheochromocytoma occur, with a paroxysmal increase in blood pressure, sudden severe occipital headache, and cardiac irregularities. In rare instances, fatal cerebral hemorrhage or cardiac failure have occurred. The actions of MAO inhibitors on non-specific hepatic drug-metabolizing enzymes potentiate the effects of a large number of drugs, including narcotic analgesics, barbiturates, and anesthetics. This can complicate elective or emergency surgery in patients taking these antidepressants, with occasional fatal results.

A cause-and-effect relation for interactions between non-selective MAO inhibitors and other substances is often difficult to establish, for two reasons (49). First, MAO inhibitors act by forming an irreversible bond with an enzyme that is continuously being resynthesized. The degree of enzyme inhibition therefore varies with the dosage and duration of treatment, and can persist to some degree for up to 2 weeks after withdrawal. Secondly, any protein food that contains decarboxylating bacteria can convert amino acids to amines such as tyramine, phenylethylamine, and histamine, and the amine composition of foodstuffs is variable and unpredictable; for example, identical-looking pieces of cheese can vary 100-fold in tyramine content. The diversity and number of substances (listed in Table 2) and unpredictability in the occurrence of adverse effects have contributed to the unpopularity of the MAO inhibitors as therapeutic agents.

Two facts simplify the understanding and use of these drugs.

(1) The interactions listed in Table 2 fall into three categories:

  (a) hypertensive crisis due to the release and potentiation of catecholamines similar to that experienced in pheochromocytoma;
  (b) a serotonin syndrome, caused by excess serotonin availability in the nervous system;
  (c) exacerbation or prolongation of the normal actions of the object drug (sedation or coma due to alcohol, anesthetics, or opioid analgesics; atropine-like central toxicity due to tricyclic antidepressants).

The consequences of the hypertensive interaction are variable. Many individuals remain unaware of relatively minor increases in blood pressure. However, if the rise is large and rapid (an increase of 30 mmHg or more in systolic pressure within 20 minutes), the patient experiences a sudden severe occipital headache and palpitation, which may be associated in rare instances with subarachnoid hemorrhage or cardiac failure, if the cerebral vasculature or cardiac musculature are already weakened.

The serotonin syndrome is characterized by three or more of the following symptoms: confusion, hypomania, agitation, myoclonus, hyper-reflexia, hyperthermia, shivering, sweating, ataxia, and diarrhea (50). It is most often seen in patients taking a combination of drugs that increase serotonin availability by different mechanisms. Drug combinations that have been reported to cause the serotonin syndrome have been reviewed and the pathophysiology of the syndrome discussed (45). It is probably mediated by stimulation of 5-HT$_2$ receptors. In most reported cases the drug combination associated with the syndrome included a MAO inhibitor (see Table 2). The syndrome is rare, although it has been reported more often during the past few years, and is often due to a combination of a MAO inhibitor and a selective serotonin re-uptake inhibitor. In most cases the symptoms are mild and resolve rapidly after drug withdrawal and supportive therapy. It must, however, be borne in mind that this syndrome can be fatal, and if combinations of serotonergic drugs are used they should be used with great caution.

(2) A second factor that can improve the use of these drugs is adequate patient education. Patients must be aware of the name and nature of the drug they are taking and its potential to interact with proscribed foods and prescribed or over-the-counter medications. In addition, patients should understand what symptoms to expect when an interaction occurs, such as unexpected drowsiness (after alcohol or other drugs) or a sudden severe headache (within 2 hours of a meal or medication). Because of the many variables involved, foods or medications that are taken with impunity on one occasion can interact dangerously on another. The director of a British counselling service has reported that of 119 patients taking MAO inhibitors who experienced problems, 35 reported hypertensive crises, 4 fatal (51). Despite warnings, these patients had eaten amine-containing foodstuffs or taken over-the-counter cold remedies. MAO inhibitors should not be prescribed unless the patient is able to understand such instructions and repeat them after explanation; compliance can be confirmed at subsequent inquiry. Those who take multiple medications, who have difficulty with comprehension or compliance (such as the elderly), or who are frightened by such explanations should not be given these drugs. The management of a patient who experiences these drug interactions depends on their nature. For symptoms due to exacerbation or prolongation of drug effects, specific or supportive measures (cardiovascular or respiratory) may be indicated. In the case of a hypertensive crisis, symptoms usually abate within 1–2 hours, as the blood pressure falls. If hypertension is seen early, or if it persists or recurs, it can be treated with parenteral phentolamine or chlorpromazine, but the danger of causing hypotension should be weighed carefully. The combination of central nervous system stimulants with MAO inhibitors in treatment-resistant depressed patients has

been reported (52). However, this use should be restricted to patients in whom there is careful monitoring by specialists, because of the potential for hypertensive crisis.

Araliaceae

Interactions of antidepressants with herbal medicines have been reported (53). Ginseng has been reported to cause mania, tremor, and headache when used in combination with conventional monoamine oxidase inhibitors.

Buspirone

The combination of buspirone with a MAO inhibitor can cause the serotonin syndrome (54).

Buspirone interacts with monoamine inhibitors and can cause the serotonin syndrome (55).

Chlorpromazine

Hyperthermia and labile blood pressure occurred in a patient taking chlorpromazine, phenelzine, and clomipramine (56).

Cocaine

The combination of monoamine oxidase inhibitors with cocaine can cause hyperpyrexia (57).

Dextromethorphan

A possible interaction between dextromethorphan and the monoamine oxidase inhibitor isocarboxazid has been described, with myoclonic jerks, choreoathetoid movements, and marked urinary retention (58).

Individuals who take monoamine oxidase inhibitors are at increased risk of serotonergic adverse effects, as well as the narcotic adverse effects of coma and respiratory depression (58,59)

Four subjects had markedly reduced O-demethylation of dextromethorphan after they had taken moclobemide 300 mg bd for 9 days (60). N-demethylation was not affected. This result supports the hypothesis that moclobemide or a metabolite reduces the activity of the cytochrome enzyme CYP2D6. The clinical implications of this particular interaction remain to be clarified.

Etamivan

Etamivan is contraindicated in patients taking MAO inhibitors (61).

Fenfluramine

Fenfluramine can produce an acute confusional state if it is given together with a MAO inhibitor (SED-9, 9).

Fentanyl

There is a risk of hypertension, tachycardia, hyperpyrexia, and coma with the concurrent administration of opioids and monoamine oxidase inhibitors (SEDA-19, 83).

Fluoxetine

The problem of the long half-life of fluoxetine, which leads to interactions with MAO inhibitors, even after withdrawal, has been discussed previously (SEDA-13, 12), and caused the manufacturer to circulate a warning to that effect.

Pseudopheochromocytoma, with hypertension, palpitation, and headache, has been reported in a patient taking fluoxetine and selegiline (SEDA-18, 23).

Ginseng

Interactions of antidepressants with herbal medicines have been reported (62). Herbal medicines are widely used for their psychotropic properties and patients sometimes take herbal formulations in addition to conventional psychotropic drugs, such as antidepressants. Ginseng has been reported to cause mania, tremor, and headache in combination with conventional MAO inhibitors.

Isoniazid

Isoniazid inhibits MAO, and during isoniazid treatment the ingestion of several kinds of cheese can cause flushing, palpitation, tachycardia, and increased blood pressure (63). Similar symptoms occur with isoniazid after the ingestion of skipjack fish (*Thunnidae*) (63). The symptoms, notably headache, palpitation, erythema, redness of the eyes, itching, diarrhea, and wheezing, are thought to be caused mainly by the high histamine content of this fish. The undesirable effects that occur when MAO inhibitors are taken together with isoniazid resemble the symptoms seen after the simultaneous ingestion of these foods.

Isoniazid inhibits monoamine oxidase, and hence reduces tyramine metabolism; this effect is enhanced by co-administration of other monoamine oxidase inhibitors (64).

Lithium

Interactions of lithium with antidepressants have been reviewed: tricyclic antidepressants and MAO inhibitors—no serious problems; SSRIs—a few reports of the serotonin syndrome (65).

Lysergide

Chronic administration of a monoamine oxidase inhibitor causes a subjective reduction in the effects of LSD, perhaps due to differential changes in central serotonin and dopamine receptor systems (66).

Nefazodone

Co-administration of nefazodone with a MAO inhibitor or SSRI can cause the serotonin syndrome (67).

Neuroleptic drugs

Additive hypotensive effects can occur when a MAO inhibitor is combined with an antipsychotic drug (68).

However, the combination can be beneficial in treating the negative symptoms of schizophrenia (69).

Opioid analgesics

Opioid analgesics interact with non-selective monoamine oxidase inhibitors, causing nervous system excitation and hypertension (70). These interactions have been reviewed (SEDA-18, 14).

Concomitant use of some opioid analgesics, such as pethidine or dextropropoxyphene, with the selective monoamine oxidase inhibitors selegiline and moclobemide can enhance their nervous system toxicity (71).

Dextromethorphan is metabolized by CYP2D6 to dextrorphan, which binds to phencyclidine receptors and is thought to account for the toxic effects of hallucinations, tachycardia, hypertension, ataxia, and nystagmus. Individuals who are slow metabolizers, those who take long-acting dextromethorphan formulations, and those who take serotonin re-uptake inhibitors or MAO inhibitors are at increased risk of adverse effects.

Opioids interact with monoamine oxidase inhibitors, causing CNS excitation and hypertension (72).

Phenylephrine

Patients taking medications with pressor effects, such as MAO inhibitors, tricyclic antidepressants, and anticholinergic agents, should be monitored closely if phenylephrine is used (SEDA-16, 542).

Selective serotonin re-uptake inhibitors (SSRIs)

Monoamine oxidase inhibitors, including reversible inhibitors of monoamine oxidase types A and B, can cause a serotonin syndrome when they are combined with SSRIs (SEDA-17, 23) (SEDA-18, 22) (73).

Sumatriptan

On theoretical grounds sumatriptan should not be used with a MAO inhibitor (74).

Trazodone

Trazodone is related to nefazodone but probably potentiates 5-HT neurotransmission less. Trazodone is often added to MAO inhibitors and SSRIs at low doses (50–150 mg/day) as a hypnotic, but such combinations can rarely provoke the serotonin syndrome.

- The serotonin syndrome occurred in a 60-year-old woman when the addition of trazodone 50 mg/day to nefazodone 500 mg/day caused confusion, restlessness, sweating, and nausea after the third day of treatment (75). The symptoms settled quickly on withdrawal of both antidepressants.

Tricyclic antidepressants

The interactions of tricyclic antidepressants with MAO inhibitors are so dangerous that they constitute a formidable barrier to their combined clinical use. The temptation to treat refractory patients with this combination is great, but considerable care should be taken (76,77), and it should be remembered that there is no firm proof of the superiority of such combinations. Several reviewers (78–80) have concluded that the dangers of these interactions have been overstated and the potential therapeutic advantages underestimated. On the other hand, reports of serious and fatal complications continue to appear (81,82). Specialists have advised that the combination of an MAO inhibitor with trimipramine or amitriptyline is usually safe, but that imipramine and clomipramine must be avoided.

Venlafaxine

There have been previous reports of serotonin toxicity when venlafaxine was combined with therapeutic doses of conventional MAO inhibitors (SEDA 20, 21). The serotonin syndrome has been reported in four patients who were switched from the MAO inhibitor phenelzine to venlafaxine (83). In two of the subjects, the 14-day washout period recommended when switching from phenelzine to other antidepressant drugs had elapsed.

- A 25-year-old woman had been taking phenelzine 45 mg/day for refractory migraine and tension headache, but suffered intolerable adverse effects (weight gain, edema, and insomnia). Phenelzine was withdrawn and 15 days elapsed before she took a single dose of venlafaxine 37.5 mg. Within 1 hour she developed agitation, twitching, shakiness, sweating, and generalized erythema with hyperthermia (38°C). Her symptoms resolved within 3 hours with no sequelae.

These cases suggest that even after the recommended 2-week washout from MAO inhibitors, venlafaxine can provoke serotonin toxicity in some patients. In principle it should be possible to switch from one conventional MAO inhibitor to another without a washout period. However, there have been reports that patients who switched from phenelzine to tranylcypromine had hypertensive reactions, with disastrous consequences (84). Whenever possible, a 2-week washout period when switching MAO inhibitors or when introducing serotonin re-uptake inhibitors seems advisable.

References

1. Youdim MBH, Finberg JPM. Monoamine oxidase inhibitor antidepressants. In: Grahame-Smith DG, Cohen PJ, editors. Preclinical Psychopharmacology. Amsterdam: Excerpta Medica, 1983:38.
2. Lader M. Psychiatric disorders. In: Speight T, Holford N, editors. Avery's Drug Treatment. 4th ed.. Auckland: ADIS Internation Press, 1997:1437.
3. Menkes DB. Antidepressant drugs. New Ethicals 1994;31:101.
4. Murphy DL, Sunderland T, Cohen RM. Monoamine oxidase-inhibiting antidepressants. A clinical update. Psychiatr Clin North Am 1984;7(3):549–62.
5. Paykel ES, Parker RR, Penrose RJ, Rassaby ER. Depressive classification and prediction of response to phenelzine. Br J Psychiatry 1979;134:572–81.

6. Liebowitz MR, Quitkin FM, Stewart JW, McGrath PJ, Harrison W, Rabkin J, Tricamo E, Markowitz JS, Klein DF. Phenelzine vs imipramine in atypical depression. A preliminary report. Arch Gen Psychiatry 1984;41(7):669–77.

7. White K, Simpson G. Should the use of MAO inhibitors be abandoned? Integrat Psychiatry 1985;3:34.

8. Rose S. The relationship of acetylation phenotype to treatment with MAOIs: a review. J Clin Psychopharmacol 1982;2(3):161–4.

9. Rabkin J, Quitkin F, Harrison W, Tricamo E, McGrath P. Adverse reactions to monoamine oxidase inhibitors. Part I. A comparative study. J Clin Psychopharmacol 1984;4(5):270–8.

10. Rabkin JG, Quitkin FM, McGrath P, Harrison W, Tricamo E. Adverse reactions to monoamine oxidase inhibitors. Part II. Treatment correlates and clinical management. J Clin Psychopharmacol 1985;5(1):2–9.

11. Razani J, White KL, White J, Simpson G, Sloane RB, Rebal R, Palmer R. The safety and efficacy of combined amitriptyline and tranylcypromine antidepressant treatment. A controlled trial. Arch Gen Psychiatry 1983;40(6):657–61.

12. Kronig MH, Roose SP, Walsh BT, Woodring S, Glassman AH. Blood pressure effects of phenelzine. J Clin Psychopharmacol 1983;3(5):307–10.

13. Linet LS. Mysterious MAOI hypertensive episodes. J Clin Psychiatry 1986;47(11):563–5.

14. Blackwell B. Clinical and pharmacological observations of the interactions of monoamine oxidase inhibitors, amines and foodstuffs. Cambridge University: MD Thesis, 1966.

15. Evans DL, Davidson J, Raft D. Early and late side effects of phenelzine. J Clin Psychopharmacol 1982;2(3):208–10.

16. Reiderer P, Reynolds GP. Deprenyl is a selective inhibitor of brain MAO-B in the long-term treatment of Parkinsons's disease. Br J Clin Pharmacol 1980;9(1):98–9.

17. Ben-Shlomo Y, Churchyard A, Head J, Hurwitz B, Overstall P, Ockelford J, Lees AJ. Investigation by Parkinson's Disease Research Group of United Kingdom into excess mortality seen with combined levodopa and selegiline treatment in patients with early, mild Parkinson's disease: further results of randomised trial and confidential inquiry. BMJ 1998;316(7139):1191–6.

18. Colosimo C, De Michele M. Motor fluctuations in Parkinson's disease: pathophysiology and treatment. Eur J Neurol 1999;6(1):1–21.

19. Pande AC, Max P. A dystonic reaction occurring during treatment with tranylcypromine. J Clin Psychopharmacol 1989;9(3):229–30.

20. Jack RA. Myoclonus, hyperreflexia, diaphoresis. Can J Psychiatry 1986;31:178.

21. Lieberman JA, Kane JM, Reife R. Neuromuscular effects of monoamine oxidase inhibitors. J Clin Psychopharmacol 1985;5(4):221–8.

22. Peterson JC, Pollack RW, Mahoney JJ, Fuller TJ. Inappropriate antidiuretic hormone secondary to a monoamine oxidase inhibitor. JAMA 1978;239(14):1422–3.

23. Dunleavy DL. Phenelzine and oedema. BMJ 1977;1(6072):1353.

24. Dunn NR, Freemantle SN, Pearce GL, Mann RD. Galactorrhoea with moclobemide. Lancet 1998;351(9105):802.

25. Tipermas A, Gilman HE, Russakoff LM. A case report of leukopenia associated with phenelzine. Am J Psychiatry 1984;141(6):806–7.

26. Lefebure B, Castot A, Danan G, Elmalem J, Jean-Pastor MJ, Efthymiou ML. Hépatites aux antidépresseurs. [Hepatitis from antidepressants. Evaluation of cases from the French Association of Drug Surveillance Centers and the Technical Committee.] Thérapie 1984;39(5):509–16.

27. Bonkovsky HL, Blanchette PL, Schned AR. Severe liver injury due to phenelzine with unique hepatic deposition of extracellular material. Am J Med 1986;80(4):689–92.

28. Wyatt RJ, Fram DH, Buchbinder R, Snyder F. Treatment of intractable narcolepsy with a monoamine oxidase inhibitor. N Engl J Med 1971;285(18):987–91.

29. Hollender MH, Ban TA. Ejaculatio retarda due to phenelzine. Psychiatr J University Ottawa 1979;IV:233.

30. Rapp MS. Two cases of ejaculatory impairment related to phenelzine. Am J Psychiatry 1979;136(9):1200–1.

31. Barton JL. Orgasmic inhibition by phenelzine. Am J Psychiatry 1979;136(12):1616–7.

32. Lesko LM, Stotland NL, Segraves RT. Three cases of female anorgasmia associated with MAOIs. Am J Psychiatry 1982;139(10):1353–4.

33. Modai I, Beigel Y, Cygielman G. Urinary amine metabolite excretion in a patient with adrenergic hyperactivity state: reaction to phenelzine withdrawal and combined treatment. J Clin Psychiatry 1986;47(2):92–3.

34. Rothschild AJ. Mania after withdrawal of isocarboxazid. J Clin Psychopharmacol 1985;5(6):340–2.

35. Mattes JA. Stroke resulting from a rapid switch from phenelzine to tranylcypromine. J Clin Psychiatry 1998;59(7):382.

36. Gracious BL, Wisner KL. Phenelzine use throughout pregnancy and the puerperium: case report, review of the literature, and management recommendations. Depress Anxiety 1997;6(3):124–8.

37. Pons G, Schoerlin MP, Tam YK, Moran C, Pfefen JP, Francoual C, Pedarriosse AM, Chavinie J, Olive G. Moclobemide excretion in human breast milk. Br J Clin Pharmacol 1990;29(1):27–31.

38. Fitton A, Faulds D, Goa KL. Moclobemide. A review of its pharmacological properties and therapeutic use in depressive illness. Drugs 1992;43(4):561–96.

39. Ashford JW, Ford CV. Use of MAO inhibitors in elderly patients. Am J Psychiatry 1979;136(11):1466–7.

40. Cassidy S, Henry J. Fatal toxicity of antidepressant drugs in overdose. BMJ (Clin Res Ed) 1987;295(6605):1021–4.

41. Henry JA, Antao CA. Suicide and fatal antidepressant poisoning. Eur J Med 1992;1(6):343–8.

42. Shepherd JT, Whiting B. Letter. Beta-adrenergic blockade in the treatment of M.A.O.-I. self-poisoning Lancet 1974;2(7887):1021.

43. Crome P. Antidepressant overdosage. Drugs 1982;23(6):431–61.

44. Breheny FX, Dobb GJ, Clarke GM. Phenelzine poisoning. Anaesthesia 1986;41(1):53–6.

45. Kaplan RF, Feinglass NG, Webster W, Mudra S. Phenelzine overdose treated with dantrolene sodium. JAMA 1986;255(5):642–4.

46. Isbister GK, Hackett LP, Dawson AH, Whyte IM, Smith AJ. Moclobemide poisoning: toxicokinetics and occurrence of serotonin toxicity. Br J Clin Pharmacol 2003;56:441–50.

47. Vuori E, Henry JA, Ojanpera I, Nieminen R, Savolainen T, Wahlsten P, Jantti M. Death following ingestion of MDMA (ecstasy) and moclobemide. Addiction. Soc Study Addiction Alcohol Other Drugs 2003;98:365–8.

48. Giroud C, Horisberger B, Eap C, Augsburger M, Menetrey A, Baumann P, Mangin P. Death following acute poisoning by moclobemide. Forensic Sci Int 2004;140:101–7.

49. Blackwell B, Marley E, Price J, Taylor D. Hypertensive interactions between monoamine oxidase inhibitors and foodstuffs. Br J Psychiatry 1967;113(497):349–65.

50. Sternbach H. The serotonin syndrome. Am J Psychiatry 1991;148(6):705–13.

51. Wright SP. Hazards with monoamine-oxidase inhibitors: a persistent problem. Lancet 1978;1(8058):284–5.

52. Fawcett J, Kravitz HM, Zajecka JM, Schaff MR. CNS stimulant potentiation of monoamine oxidase inhibitors in treatment-refractory depression. J Clin Psychopharmacol 1991;11(2):127–32.

53. Napoliello MJ, Domantay AG. Buspirone: a worldwide update. Br J Psychiatry Suppl 1991;12:40–4.

54. Couch RB, Cate TR, Chanock RM. Evaluation of a new analeptic: ethamivan (Emivan). JAMA 1964;187:448–9.

55. Fugh-Berman A. Herb–drug interactions. Lancet 2000;355(9198):134–8.

56. Hauser MJ, Baier H. Interactions of isoniazid with foods. Drug Intell Clin Pharm 1982;16(7–8):617–8.

57. Schweitzer I, Tuckwell V. Risk of adverse events with the use of augmentation therapy for the treatment of resistant depression. Drug Saf 1998;19(6):455–64.

58. John L, Perreault MM, Tao T, Blew PG. Serotonin syndrome associated with nefazodone and paroxetine. Ann Emerg Med 1997;29(2):287–9.

59. Burggraf GW. Are psychotropic drugs at therapeutic levels a concern for cardiologists? Can J Cardiol 1997;13(1):75–80.

60. Bucci L. The negative symptoms of schizophrenia and the monoamine oxidase inhibitors. Psychopharmacology (Berl) 1987;91(1):104–8.

61. Hsu VD. Criteria for use of sumatriptan in adult inpatients and outpatients. Clin Pharm 1992;11(11):972–4.

62. Margolese HC, Chouinard G. Serotonin syndrome from addition of low-dose trazodone to nefazodone. Am J Psychiatry 2000;157(6):1022.

63. Fugh-Berman A. Herb–drug interactions. Lancet 2000;355(9198):134–8.

64. Napoliello MJ, Domantay AG. Buspirone: a worldwide update. Br J Psychiatry 1991;159(Suppl 12):40–4.

65. Stern TA, Schwartz JH, Shuster JL. Catastrophic illness associated with the combination of clomipramine, phenelzine and chlorpromazine. Ann Clin Psychiatry 1992;4:81–5.

66. Hurtova M, Duclos-Vallee JC, Saliba F, Emile JF, Bemelmans M, Castaing D, Samuel D. Liver transplantation for fulminant hepatic failure due to cocaine intoxication in an alcoholic hepatitis C virus-infected patient. Transplantation 2002;73(1):157–8.

67. Sovner R, Wolfe J. Interaction between dextromethorphan and monoamine oxidase inhibitor therapy with isocarboxazid. N Engl J Med 1988;319(25):1671.

68. Anonymous. High-dose fentanyl. Lancet 1979;1(8107):81–2.

69. Hartter S, Dingemanse J, Baier D, Ziegler G, Hiemke C. Inhibition of dextromethorphan metabolism by moclobemide. Psychopharmacology (Berl) 1998;135(1):22–6.

70. DiMartini A. Isoniazid, tricyclics and the "cheese reaction". Int Clin Psychopharmacol 1995;10(3):197–8.

71. Bonson KR, Murphy DL. Alterations in responses to LSD in humans associated with chronic administration of tricyclic antidepressants, monoamine oxidase inhibitors or lithium. Behav Brain Res 1996;73(1–2):229–33.

72. British Medical Association, Royal Pharmaceutical Society of Great Britain. In: British National Formulary. 22nd ed. 1991:458.

73. Bonnet U. Moclobemide: therapeutic use and clinical studies. CNS Drug Rev 2003;9(1):97–140.

74. Rossiter A, Souney PF. Interaction between MAOIs and opioids: pharmacologic and clinical considerations. Hosp Formul 1993;28(8):692–8.

75. Sporer KA. The serotonin syndrome. Implicated drugs, pathophysiology and management. Drug Saf 1995;13(2):94–104.

76. Frejaville JP. De mauvais mélanges d'antidépresseurs. [Bad combinations of antidepressant drugs.] Concours Med 1972;94:8543.

77. Ponto LB, Perry PJ, Liskow BI, Seaba HH. Drug therapy reviews: tricyclic antidepressant and monoamine oxidase inhibitor combination therapy. Am J Hosp Pharm 1977;34(9):954–61.

78. Goldberg RS, Thornton WE. Combined tricyclic–MAOI therapy for refractory depression: a review, with guidelines for appropriate usage. J Clin Pharmacol 1978;18(2–3):143–7.

79. Anath J, Luchins DJ. Combined MAOI–tricyclic therapy (a critical review). Indian J Psychiatry 1976;18:26.

80. White K, Simpson G. Combined MAOI–tricyclic antidepressant treatment: a reevaluation. J Clin Psychopharmacol 1981;1(5):264–82.

81. Anonymous. Antidepressant interaction led to death. Pharm J 1982;13:191.

82. Graham PM, Potter JM, Paterson J. Combination monoamine oxidase inhibitor/tricyclic antidepressants interaction. Lancet 1982;2(8295):440.

83. Diamond S, Pepper BJ, Diamond ML, Freitag FG, Urban GJ, Erdemoglu AK. Serotonin syndrome induced by transitioning from phenelzine to venlafaxine: four patient reports. Neurology 1998;51(1):274–6.

84. Mattes JA. Stroke resulting from a rapid switch from phenelzine to tranylcypromine. J Clin Psychiatry 1998;59(7):382.

Brofaromine

General Information

Brofaromine is a selective inhibitor of monoamine oxidase (MAO) type A. In an open study in endogenous depression, adverse effects were reported in nine of 51 patients; these included dry mouth, dizziness, tremor, hypomania, anxiety, and memory problems (SEDA-17, 16).

A randomized comparison of brofaromine with imipramine in inpatients with major depression showed that brofaromine was as effective as imipramine in the treatment of major depression but had a different adverse effects profile (1). Brofaromine was more likely to cause sleep disturbances but lacked the anticholinergic and certain cardiovascular adverse effects of imipramine.

Reference

1. Volz HP, Gleiter CH, Moller HJ. Brofaromine versus imipramine in in-patients with major depression—a controlled trial. J Affect Disord 1997;44(2–3):91–9.

Iprindole

General Information

Iprindole is a 6,5,8 tricyclic compound, whose first two rings form an indole nucleus (SED-7, 27). It is of some theoretical interest, in that it is weak in its action on the amine pump mechanism. Adverse effects are reported as being similar to those of other tricyclic antidepressants (1,2).

Organs and Systems

Liver

The major adverse effect of serious concern has been jaundice. A review of 21 cases of liver damage during iprindole treatment showed that the onset was at 4–21 days after starting treatment (3). Recovery was rapid after withdrawal. Liver biopsy showed predominantly cholestasis. This complication seems to be similar to that seen after chlorpromazine. In one instance (4), jaundice was accompanied by a marked eosinophilia of 30% and laboratory signs suggesting some hepatocellular damage as well. Jaundice promptly recurred when the drug was re-administered. All the features of this case suggested an allergic reaction.

References

1. Martin ICA, Hossain M, Hart J. Treatment of marked, persistent mood depression: a double-blind comparison of iprindole (Prondol) with nortriptyline. Clin Trials J 1972;3:39.
2. Narayanan HS, Reddy GN, Rao BS. A comparative double-blind evaluation of iprindole and imipramine in the treatment of depressive states. Indian J Med Sci 1973;27(1):1–7.
3. Ajdukiewicz AB, Grainger J, Scheuer PJ, Sherlock S. Jaundice due to iprindole. Gut 1971;12(9):705–8.
4. Aylett MJ. Allergy to iprindol (Prondol) with hepatotoxicity. BMJ 1971;1(740):112.

Moclobemide

General Information

Moclobemide was the first selective inhibitor of monoamine oxidase (MAO) type A to become commercially available. It binds reversibly to the enzyme, considerably shortening the duration of its effect. Both of these characteristics reduce the risk of hypertensive crisis (1,2).

The adverse effects of moclobemide have been well reported in several studies, mainly comparisons of moclobemide with standard antidepressants. The consensus has been that moclobemide produces fewer anticholinergic effects and less orthostatic hypotension and dizziness than clomipramine or imipramine. The main problems reported early on were insomnia, agitation, and paresthesia (3–9). The most common adverse effects in controlled trials have been insomnia and nausea (SEDA-18, 15).

In a multicenter comparison of moclobemide, amitriptyline, and placebo, gastrointestinal discomfort, headache, and dizziness occurred in over 20% of moclobemide-treated patients; insomnia was also common (10). In healthy volunteers there were no effects of moclobemide on psychomotor performance. Acute confusion and agitation was reported in one patient who dropped out of a clinical trial owing to this adverse effect (SEDA-16, 7). Hypomania was attributed to moclobemide in two cases (SEDA-17, 16). Aggressive behavior and mild manic symptoms were described in severely depressed patients, refractory to other treatments, who took moclobemide (SEDA-18, 15).

In a controlled multicenter study of 115 in-patients with relatively severe depression, moclobemide was better tolerated but had a weaker therapeutic effect than clomipramine. Nine of the patients taking moclobemide withdrew owing to worsening of depression and suicidality, although it is possible that the dosage of moclobemide (400 mg/day) was inadequate (SEDA-18, 15).

Organs and Systems

Cardiovascular

There have been two reports of single cases of a rise in blood pressure after moclobemide: in one there was a 40 mmHg increase in systolic pressure after one dose only (8), while the other report referred only to "episodes of hypertension" without further comment (4). Interactions were not suggested in either case.

In a prospective drug utilization study in 13 741 patients who received the reversible type A MAO inhibitor moclobemide in a variety of settings, including general practice, psychiatric out-patients, and psychiatric in-patients, there were few episodes of hypertension (0.11%; 95% CI = 0.06, 0.18%) or hypotension (0.04%; 95% CI = 0.02, 0.10%) and episodes were mostly associated with underlying cardiovascular disease (11). These findings suggest that moclobemide in usual therapeutic doses carries a low risk of producing significant changes in blood pressure relative to conventional MAO inhibitors.

Endocrine

A number of antidepressant drugs, particularly SSRIs, can increase plasma prolactin concentrations, although galactorrhea is uncommon. In a prescription event monitoring survey of about 65 000 patients, compared with SSRIs, moclobemide was associated with a relative risk of galactorrhea of 6.7 (95% CI = 2.7, 15) (12). However, this was substantially less than the risk associated with the dopamine receptor antagonist risperidone (relative risk compared with SSRIs 32; 95% CI = 14, 70). Nevertheless, the data suggest that moclobemide may be more likely to cause galactorrhea than other antidepressant drugs.

Liver

Hepatotoxicity has been associated with moclobemide, including fatal intrahepatic cholestasis in an 85-year-old woman 7 days after switching to moclobemide from fluoxetine (13).

Sexual function

Most conventional antidepressants lower sexual desire and performance. However, the reversible type A selective MAO inhibitor moclobemide produced intense pathological sexual desire in three men with organic brain disease (two with strokes and one with idiopathic Parkinson's disease) (14). In one man, the hypersexuality was associated with features of pathological jealousy in a paranoid state, but in the other two increased sexuality was an isolated symptom. One of these patients, who had been impotent after the stroke, resorted to telephone sex services, a most uncharacteristic behavior for him. In all cases, the hypersexuality remitted when moclobemide was withdrawn. There have been two previous case reports of moclobemide-induced hypersexuality in women without organic brain disease. This must be a rare adverse effect, but it is possible that in the cases reported here the organic brain disease may have contributed.

There are many other ways in which SSRIs can interfere with sexual function, for example by causing loss of sexual interest and erectile difficulties. In an open, prospective study of 1000 Spanish patients taking a variety of antidepressants, there was an overall incidence of sexual dysfunction of 59% (15). The highest rates, 60–70%, were found with SSRIs (including fluvoxamine) and venlafaxine. The lowest rates were found with mirtazepine (24%), nefazodone (8%), and moclobemide (4%). Spontaneous resolution of this adverse effect was uncommon – 80% of subjects had no improvement in sexual function over 6 months of treatment.

Second-Generation Effects

Lactation

Moclobemide was measured in the milk and plasma of six nursing mothers after single oral doses of moclobemide 300 mg, 3–5 days after delivery (16). The milk plasma ratio was 0.72 and the estimated dose of moclobemide to the infant was at most 1.2% of the weight-adjusted maternal dose.

Drug Administration

Drug overdose

Overdoses of moclobemide alone have been considered less toxic than overdoses of tricyclic antidepressants. Intoxication with moclobemide (0.95–2 g) produced symptoms of drowsiness, disorientation, and hyporeflexia (17,18). Fatal moclobemide overdose when the compound is taken alone is extremely rare (19).

However, if patients take moclobemide together with serotonergic antidepressants, such as SSRIs or clomipramine, 5-HT toxicity is common. 5HT toxicity occurred in 11 of 21 patients who took overdoses of moclobemide and serotonergic agents but in only one of 33 patients who took moclobemide alone (20). Consistent with this, four patients died, presumably of 5HT toxicity, after co-ingesting 3,4-methylenedioxymethamphetamine (MDMA, ecstasy) and moclobemide (21). Massive ingestion of moclobemide (over 30 g) also appeared to be responsible for death in a suicide attempt in a 48-year-old man (22). The blood concentration of moclobemide measured at post mortem (498 µg/ml) was about 500 times the usual target concentration (0.2–2.1 µg/ml). Post-mortem findings were non-specific, but the authors concluded that death might have been due to the serotonin syndrome.

Drug–Drug Interactions

Alprazolam

A 44-year-old man developed the serotonin syndrome after taking moclobemide and alprazolam for 1 year (23). The symptoms developed after 4 days of extreme heat, which was thought to have contributed.

Citalopram and clomipramine

Five fatal cases of serotonin syndrome after combined overdoses of moclobemide with citalopram and moclobemide with clomipramine have been reported (SEDA-18, 16). Several cases of this syndrome have been reported in patients taking moclobemide with drugs that potentiate serotonin function (24–31).

Co-beneldopa

In healthy subjects given co-beneldopa and moclobemide, no hypertension was noted, but the combination was poorly tolerated, headache and insomnia commonly being reported (SEDA-18, 16).

Dextromethorphan

Four healthy volunteers taking moclobemide (600 mg/day for 9 days) had reduced clearance of dextromethorphan, a marker of hepatic CYP2D6 activity (32). These findings suggest that at the higher end of its therapeutic dosage range, moclobemide inhibits the metabolism of drugs that are substrates for CYP2D6, for example antipsychotic drugs and tricyclic antidepressants (SEDA-14, 22).

Indirect sympathomimetics

Moclobemide increased the pressor effect of ephedrine in healthy volunteers, who also had more adverse effects including light-headedness and palpitation (SEDA-18, 16). Thus, moclobemide can potentiate the effect of indirect sympathomimetics; such combinations should be used with caution.

Rizatriptan

Rizatriptan is a 5-HT$_{1B/1D}$ receptor agonist that is metabolized by MAO type A. In 12 young healthy volunteers, moclobemide (300 mg/day for 4 days) potentiated the hypertensive effect of acute rizatriptan (10 mg orally) (33). This was associated with a significant increase in the AUC of rizatriptan and its N-monodesmethyl metabolite (by 2.2 and 5.3 times respectively). The authors suggested that moclobemide and rizatriptan should not be given together.

Selegiline

Patients taking moclobemide can usually eat a normal diet, since there is no significant potentiation of tyramine. However, the combination in high doses (20 mg or more) of selegiline and moclobemide potentiates the pressor effects of tyramine (presumably because of simultaneous inhibition of both MAO type A and MAO type B), and a tyramine-free diet is needed if the two drugs are used together (SEDA-20, 6).

SSRIs

The combination of conventional MAO inhibitors with serotonin-potentiating agents, such as SSRIs, is contraindicated, because of the risk of the serotonin syndrome. Moclobemide may be less likely to cause this interaction, although both case reports and clinical series have suggested that some patients have suffered significant adverse effects consistent with serotonin toxicity (SEDA-20, 6) (SEDA-21, 10). After steady-state therapy with the SSRI fluoxetine (20–40 mg/day for 23 days), 12 subjects took fluoxetine plus moclobemide (600 mg/day) and 6 took fluoxetine and placebo (34). There was no difference in the rates of adverse effects between the two groups and no evidence of serotonin toxicity, although fluoxetine inhibited the metabolism of moclobemide. The authors suggested that patients can be safely switched from fluoxetine to moclobemide without the need for the currently advised 5-week washout (which is necessary to allow the long-acting fluoxetine metabolite, norfluoxetine, to be eliminated). However, case reports have suggested that some patients can develop serotonin toxicity when moclobemide is combined with SSRIs (26,27,31) or clomipramine (29), so great caution is still needed in clinical practice. Overdoses of moclobemide and SSRIs can cause serious and sometimes fatal serotonin toxicity (SEDA-18, 16).

Tyramine

Moclobemide in conventional doses (300–600 mg/day) does not significantly potentiate the pressor effects of oral tyramine, and special dietary restrictions are therefore not usually necessary. However, higher doses of moclobemide are sometimes used to treat patients with resistant depression. Moclobemide 900 mg/day and 1200 mg/day have been compared with placebo in 12 healthy volunteers (35). Neither dose of moclobemide significantly potentiated the pressor effect of 50 mg of tyramine, which has been estimated to be the upper limit of dietary tyramine content, even of large meals containing substantial amounts of cheese. However, one subject taking moclobemide 1200 mg/day had an increase in systolic blood pressure of over 30 mmHg. Individuals have rather different sensitivities to the pressor effects of oral tyramine challenge, and the number studied in this series was small. Accordingly, clinical extrapolation might not be straightforward. Patients taking conventional doses of moclobemide are sometimes advised to restrict their intake of cheese or other tyramine-containing foods, and this advice is clearly prudent if subjects take higher doses.

Venlafaxine

Interactions of moclobemide with venlafaxine, a potent serotonin re-uptake inhibitor, have been reported (27,36).

- A 34-year-old man took 2.625 g of venlafaxine (therapeutic dose 75–375 mg/day) and 3 g of moclobemide, plus an unknown amount of alcohol 1 hour before being admitted to hospital. Within 20 minutes of arrival his conscious level deteriorated and he had increased muscle tone, with clonus in all limbs. He was treated with intubation, paralysis, and ventilation, and sedated with midazolam and morphine. He regained consciousness after 2 days.

This case report confirms the serious consequences of combined overdosage of moclobemide with drugs that potentiate brain serotonin function (see also the monograph on SSRIs).

References

1. Amrein R, Allen SR, Guentert TW, Hartmann D, Lorscheid T, Schoerlin MP, Vranesic D. The pharmacology of reversible monoamine oxidase inhibitors. Br J Psychiatry Suppl 1989;6:66–71.
2. Simpson GM, de Leon J. Tyramine and new monoamine oxidase inhibitor drugs. Br J Psychiatry Suppl 1989;6:32–7.
3. Versiani M, Oggero U, Alterwain P, Capponi R, Dajas F, Heinze-Martin G, Marquez CA, Poleo MA, Rivero-Almanzor LE, Rossel L, et al. A double-blind comparative trial of moclobemide v. imipramine and placebo in major depressive episodes. Br J Psychiatry Suppl 1989;6:72–7.
4. Realini R, Mascetti R, Calanchini C. Efficacité et tolerance du moclobemide (Ro 11-1163 *Aurorix*) en comparaison avec la maprotiline chez des patients ambulatoires présentant un épisode dépressif majeur. [Effectiveness and tolerance of moclobemide (Ro 11-1163 Aurorix) in comparison with maprotiline in ambulatory patients presenting with a major depressive episode.] Psychol Med 1989;21:1689.
5. Laux G. Moclobemid in der Depressionsbehandlung – eine Übersicht. [Moclobemide in the treatment of depression – an overview.] Psychiatr Prax 1989;16(Suppl 1):37–40.
6. Koczkas C, Holm P, Karlsson A, Nagy A, Ose E, Petursson H, Ulveras L, Wenedikter O. Moclobemide and clomipramine in endogenous depression. A randomized clinical trial. Acta Psychiatr Scand 1989;79(6):523–9.
7. Larsen JK, Holm P, Hoyer E, Mejlhede A, Mikkelsen PL, Olesen A, Schaumburg E. Moclobemide and clomipramine

in reactive depression. A placebo-controlled randomized clinical trial. Acta Psychiatr Scand 1989;79(6):530–6.

8. Baumhackl U, Biziere K, Fischbach R, Geretsegger C, Hebenstreit G, Radmayr E, Stabl M. Efficacy and tolerability of moclobemide compared with imipramine in depressive disorder (DSM-III): an Austrian double-blind, multicentre study. Br J Psychiatry Suppl 1989;6:78–83.

9. Burner M. Antidépresseur inhibiteur reversible et sélectif de la MAO-A. [Reversible Inhibitory antidepressant, selective for MAO-A.] Med Hyg 1990;48:2245.

10. Bakish D, Bradwejn J, Nair N, McClure J, Remick R, Bulger L. A comparison of moclobemide, amitriptyline and placebo in depression: a Canadian multicentre study. Psychopharmacology (Berl) 1992;106(Suppl):S98–S101.

11. Delini-Stula A, Baier D, Kohnen R, Laux G, Philipp M, Scholz HJ. Undesirable blood pressure changes under naturalistic treatment with moclobemide, a reversible MAO-A inhibitor – results of the drug utilization observation studies. Pharmacopsychiatry 1999;32(2):61–7.

12. Dunn NR, Freemantle SN, Pearce GL, Mann RD. Galactorrhoea with moclobemide. Lancet 1998;351(9105):802.

13. Timmings P, Lamont D. Intrahepatic cholestasis associated with moclobemide leading to death. Lancet 1996;347(9003):762–3.

14. Korpelainen JT, Hiltunen P, Myllyla VV. Moclobemide-induced hypersexuality in patients with stroke and Parkinson's disease. Clin Neuropharmacol 1998;21(4):251–4.

15. Montejo AL, Llorca G, Izquierdo JA, Rico-Villademoros F. Incidence of sexual dysfunction associated with antidepressant agents: a prospective multicenter study of 1022 outpatients. Spanish Working Group for the Study of Psychotropic-Related Sexual Dysfunction. J Clin Psychiatry 2001;62(Suppl 3):10–21.

16. Pons G, Schoerlin MP, Tam YK, Moran C, Pfefen JP, Francoual C, Pedarriosse AM, Chavinie J, Olive G. Moclobemide excretion in human breast milk. Br J Clin Pharmacol 1990;29(1):27–31.

17. Vine R, Norman TR, Burrows GD. A case of moclobemide overdose. Int Clin Psychopharmacol 1988;3(4):325–6.

18. Heinze G, Sanchez A. Overdose with moclobemide. J Clin Psychiatry 1986;47(8):438.

19. Camaris C, Little D. A fatality due to moclobemide. J Forensic Sci 1997;42(5):954–5.

20. Isbister GK, Hackett LP, Dawson AH, Whyte IM, Smith AJ. Moclobemide poisoning: toxicokinetics and occurrence of serotonin toxicity. Br J Clin Pharmacol 2003;56:441-50.

21. Vuori E, Henry JA, Ojanpera I, Nieminen R, Savolainen T, Wahlsten P, Jantti M. Death following ingestion of MDMA (ecstasy) and moclobemide. Addiction. Soc Study Addiction Alcohol Other Drugs 2003;98:365-8.

22. Giroud C, Horisberger B, Eap C, Augsburger M, Menetrey A, Baumann P, Mangin P. Death following acute poisoning by moclobemide. Forensic Sci Int 2004;140:101-7.

23. Butzkueven H. A case of serotonin syndrome induced by moclobemide during an extreme heatwave. Aust NZ J Med 1997;27(5):603–4.

24. Ferrer-Dufol A, Perez-Aradros C, Murillo EC, Marques-Alamo JM. Fatal serotonin syndrome caused by moclobemide–clomipramine overdose. J Toxicol Clin Toxicol 1998;36(1–2):31–2.

25. Gillman PK. Serotonin syndrome—clomipramine too soon after moclobemide? Int Clin Psychopharmacol 1997;12(6):339–42.

26. Singer PP, Jones GR. An uncommon fatality due to moclobemide and paroxetine. J Anal Toxicol 1997;21(6):518–20.

27. Francois B, Marquet P, Desachy A, Roustan J, Lachatre G, Gastinne H. Serotonin syndrome due to an overdose of moclobemide and clomipramine. A potentially life–threatening association. Intensive Care Med 1997;23(1):122–4.

28. Roxanas MG, Machado JF. Serotonin syndrome in combined moclobemide and venlafaxine ingestion. Med J Aust 1998;168(10):523–4.

29. Chan BS, Graudins A, Whyte IM, Dawson AH, Braitberg G, Duggin GG. Serotonin syndrome resulting from drug interactions. Med J Aust 1998;169(10):523–5.

30. Dardennes RM, Even C, Ballon N, Bange F. Serotonin syndrome caused by a clomipramine–moclobemide interaction. J Clin Psychiatry 1998;59(7):382–3.

31. Benazzi F. Serotonin syndrome with moclobemide–fluoxetine combination. Pharmacopsychiatry 1996;29(4):162.

32. Hartter S, Dingemanse J, Baier D, Ziegler G, Hiemke C. Inhibition of dextromethorphan metabolism by moclobemide. Psychopharmacology (Berl) 1998;135(1):22–6.

33. Van Haarst AD, Van Gerven JM, Cohen AF, De Smet M, Sterrett A, Birk KL, Fisher AL, De Puy ME, Goldberg MR, Musson DG. The effects of moclobemide on the pharmacokinetics of the 5-HT$_{1B/1D}$ agonist rizatriptan in healthy volunteers. Br J Clin Pharmacol 1999;48(2):190–6.

34. Dingemanse J, Wallnofer A, Gieschke R, Guentert T, Amrein R. Pharmacokinetic and pharmacodynamic interactions between fluoxetine and moclobemide in the investigation of development of the "serotonin syndrome". Clin Pharmacol Ther 1998;63(4):403–13.

35. Dingemanse J, Wood N, Guentert T, Oie S, Ouwerkerk M, Amrein R. Clinical pharmacology of moclobemide during chronic administration of high doses to healthy subjects. Psychopharmacology (Berl) 1998;140(2):164–72.

36. Coorey AN, Wenck DJ. Venlafaxine overdose. Med J Aust 1998;168(10):523.

Phenelzine

General Information

Phenelzine is a non-selective monoamine oxidase (MAO) inhibitor.

Organs and Systems

Nervous system

A potential risk of using a non-selective inhibitor in patients with Parkinson's disease is illustrated by separate reports of the appearance of parkinsonism in patients taking phenelzine (1,2).

- Speech blockage, so called, has been reported in a 34-year-old woman who had taken phenelzine 45 mg/day for 2 months (3). The adverse effect disappeared on withdrawal and did not recur when her depression was successfully treated with maprotiline 175 mg/day.

Psychological, psychiatric

In a carefully controlled 3-week comparison of phenelzine (up to 90 mg/day, mean 77 mg) and imipramine (up to

150 mg/day, mean 139 mg), four patients developed anti-social behavior, three overt paranoid psychosis, and one a hypertensive crisis, despite all precautions to avoid interacting foods and drugs (4).

There has also been a report of delusional parasitosis with phenelzine (SEDA-17, 14).

Dose-related visual hallucinations have been reported in a patient with macular degeneration taking phenelzine (the Charles Bonnet syndrome); the authors discussed the possibility that deprivation-induced visual phenomena had been intensified by increased central monoamine concentrations (5).

Nutrition

Authors who reported a case of carpal tunnel syndrome due to pyridoxine deficiency in a patient taking tranylcypromine (SEDA-9, 21) later collected data (6) on six patients taking phenelzine (up to 75 mg/day for up to 4 months). All developed low concentrations of pyridoxine and a variety of symptoms, including numbness, paresthesia, and edema of the hands, as well as an "electric shock" sensation in the head, neck, and arms. The symptoms resolved completely after the addition of pyridoxine 150–300 mg/day to the treatment regimen.

Hematologic

Leukopenia and agranulocytosis are well-recognized complications of treatment with tricyclic antidepressants and have been reported with some second-generation compounds. In a report of leukopenia in a patient taking phenelzine, attention was drawn to five other unpublished cases and to previous published reports involving isocarboxazid, tranylcypromine, and tryptamine (7).

Long-Term Effects

Drug withdrawal

In a study of the use of phenelzine in continuation therapy after recovery from an acute episode of depression, relapse rates were higher in patients subjected to tapered withdrawal than in those who continued taking the therapeutic dose (8).

In a study of the effects of sudden drug withdrawal in 34 patients taking phenelzine and 17 taking tricyclic antidepressants who had been treated for a mean duration of over 9 months, depressed patients taking phenelzine had significantly more symptoms than depressed patients taking tricyclic antidepressants, and a third of them relapsed, compared with a quarter taking the tricyclic (9). At 3 months follow-up 47% of the patients taking phenelzine had resumed treatment compared with 23% taking the tricyclic. An attempt to distinguish between withdrawal symptoms and relapse on the basis of the rapidity and severity of symptoms was unsuccessful, but about a third of the patients in both groups developed new symptoms of adrenergic hyperactivity, including anxiety and perceptual disturbances.

Acute psychotic symptoms have been reported in two young women shortly after withdrawal of long-term phenelzine 90 mg/day (10).

Tumorigenicity

A single case of angiosarcoma in the liver has been reported in a patient taking phenelzine (11); similar tumors have occurred in mice treated with phenelzine.

Drug–Drug Interactions

Amantadine

A possible hypertensive interaction of phenelzine with amantadine in Parkinson's disease has been reported (SEDA-10, 17).

Clonazepam

A flushing reaction has been associated with an interaction of phenelzine with clonazepam (SEDA-17, 17).

Suxamethonium

Phenelzine, a monoamine oxidase inhibitor, has been reported to cause significant prolongation of suxamethonium paralysis due to depressed plasma cholinesterase levels (to about 10% of normal). Recovery of plasma cholinesterase activity took 2 weeks (13).

Venlafaxine

There have been reports of serotonin toxicity when venlafaxine was combined with therapeutic doses of conventional MAO inhibitors (SEDA-20, 21). The serotonin syndrome has been reported in four patients who were switched from the MAO inhibitor phenelzine to venlafaxine (12). In two of them, the 14-day washout period recommended when switching from phenelzine to other antidepressant drugs had elapsed.

- A 25-year-old woman, who had taken phenelzine (45 mg/day) for refractory migraine and tension headache, suffered intolerable adverse effects (weight gain, edema, and insomnia). Phenelzine was withdrawn and 15 days elapsed before she took a single dose of 37.5 mg of venlafaxine. Within 1 hour she developed agitation, twitching, shakiness, sweating, and generalized erythema with hyperthermia (38°C). Her symptoms resolved within 3 hours with no sequelae.

This suggests that even after the recommended 2-week washout from MAO inhibitors, venlafaxine can provoke serotonin toxicity in some patients.

References

1. Teusink JP, Alexopoulos GS, Shamoian CA. Parkinsonian side effects induced by a monoamine oxidase inhibitor. Am J Psychiatry 1984;141(1):118–9.
2. Gillman MA, Sandyk R. Parkinsonism induced by a monoamine oxidase inhibitor. Postgrad Med J 1986;62(725):235–6.
3. Goldstein DM, Goldberg RL. Monoamine oxidase inhibitor-induced speech blockage. J Clin Psychiatry 1986;47(12):604.
4. Evans DL, Davidson J, Raft D. Early and late side effects of phenelzine. J Clin Psychopharmacol 1982;2(3):208–10.

5. Galynker I, Kampf R, Rosenthal R. Dose-related visual hallucinations in macular degeneration patients receiving phenelzine. Am J Psychiatry 1994;151(3):450.
6. Stewart JW, Harrison W, Quitkin F, Liebowitz MR. Phenelzine-induced pyridoxine deficiency. J Clin Psychopharmacol 1984;4(4):225–6.
7. Tipermas A, Gilman HE, Russakoff LM. A case report of leukopenia associated with phenelzine. Am J Psychiatry 1984;141(6):806–7.
8. Davidson J, Raft D. Use of phenelzine in continuation therapy. Neuropsychobiology 1984;11(3):191–4.
9. Tyrer P. Clinical effects of abrupt withdrawal from tricyclic antidepressants and monoamine oxidase inhibitors after long-term treatment. J Affect Disord 1984;6(1):1–7.
10. Liskin B, Roose SP, Walsh BT, Jackson WK. Acute psychosis following phenelzine discontinuation. J Clin Psychopharmacol 1985;5(1):46–7.
11. Daneshmend TK, Scott GL, Bradfield JW. Angiosarcoma of liver associated with phenelzine. BMJ 1979;1(6179):1679.
12. Diamond S, Pepper BJ, Diamond ML, Freitag FG, Urban GJ, Erdemoglu AK. Serotonin syndrome induced by transitioning from phenelzine to venlafaxine: four patient reports. Neurology 1998;51(1):274–6.
13. Bodley PO, Halwax K, Potts L. Low serum pseudocholinesterase levels complicating treatment with phenelzine. BMJ 1969;3(669):510–2.

Toloxatone

General Information

Like moclobemide, toloxatone, a selective and reversible inhibitor of monoamine oxidase type A, is thought to be relatively safe in combination with sympathomimetics (SEDA-18, 16). However, sweating, tachycardia, and headache have been reported when terbutaline was added to toloxatone and phenylephrine (SEDA-18, 16). In healthy volunteers, doses up to 600 mg/day did not produce hypertensive reactions on challenge with oral tyramine (SEDA-17, 17). Two fatal cases of fulminant hepatitis have been reported (SEDA-16, 7).

Interactions with toloxatone are similar to those with moclobemide (1).

Reference

1. Livingston MG, Livingston HM. Monoamine oxidase inhibitors. An update on drug interaction. Drug Saf 1996;14(4):219–27.

Tranylcypromine

General Information

Tranylcypromine is a non-hydrazine monoamine oxidase (MAO) inhibitor with actions and uses similar to those of phenelzine, but with less prolonged inhibition. Its half-life is 90–190 minutes. It is structurally related to amfetamine, to which it is metabolized in overdose (1).

Long-Term Effects

Drug abuse

Four cases of addiction to tranylcypromine have been described, in addition to the three reported since 1965 (2). The dosage was 150–300 mg/day. The mild euphoriant properties of tranylcypromine reflect its structural resemblance to amfetamine, and probably account for tolerance and addiction in predisposed individuals. Tranylcypromine abuse in 18 patients has been reviewed (3), and two further reports have appeared (SEDA-17, 17) (4). In one case (5), the patient took 440 mg/day without any adverse effects. The patient reported that she was longing for the "energizing" effect of the drug and for the feeling of "freedom and power." Withdrawal resulted in repeated generalized seizures and status epilepticus.

Susceptibility Factors

Hepatic disease

A carefully controlled study showed that patients with impaired liver function are especially sensitive to tranylcypromine, sometimes developing obtunded consciousness and slow electroencephalograms similar to those found in hepatic encephalopathy (6).

Drug Administration

Drug overdose

The fatal effects of antidepressants in England, Scotland, and Wales during 1975–1984 and 1985–1989 have been compared in retrospective epidemiological studies (7,8). There were 24 deaths due to tranylcypromine, the fatality index for which was significantly higher than the mean for all antidepressants, while that for the other MAO inhibitors was lower.

References

1. Youdim MB, Aronson JK, Blau K, Green AR, Grahame-Smith DG. Tranylcypromine ("Parnate") overdose: measurement of tranylcypromine concentrations and MAO inhibitory activity and identification of amfetamines in plasma. Psychol Med 1979;9(2):377–82.
2. Griffin N, Draper RJ, Webb MG. Addiction to tranylcypromine. BMJ (Clin Res Ed) 1981;283(6287):346.
3. Briggs NC, Jefferson JW, Koenecke FH. Tranylcypromine addiction: a case report and review. J Clin Psychiatry 1990;51(10):426–9.
4. Shepherd JT, Whiting B. Beta-adrenergic blockade in the treatment of M.A.O.I. self-poisoning Lancet 1974;2(7887):1021.
5. Vartzopoulos D, Krull F. Dependence on monoamine oxidase inhibitors in high dose. Br J Psychiatry 1991;158:856–7.
6. Morgan MH, Read AE. Antidepressants and liver disease. Gut 1972;13(9):697–701.
7. Cassidy S, Henry J. Fatal toxicity of antidepressant drugs in overdose. BMJ (Clin Res Ed) 1987;295(6605):1021–4.
8. Henry JA, Antao CA. Suicide and fatal antidepressant poisoning. Eur J Med 1992;1(6):343–8.

OTHER ANTIDEPRESSANTS

General Information

The newer antidepressants that have followed the mono-amine oxidase inhibitors and tricyclic antidepressants are listed in Table 1. Most of them are covered in separate monographs. They have a wide variety of chemical structures and pharmacological profiles, and are categorized as "second generation" antidepressants purely for convenience. Although these drugs are widely considered to be as effective as each other and as any of the older compounds, they have different adverse effects profiles. No new antidepressant has proven to be sufficiently free of adverse effects to establish itself as a routine first line compound; some share similar adverse effects profiles with the tricyclic compounds, while others have novel or unexpected adverse effects. Complete categorization of each compound will rest on wide-scale general use beyond the artificial confines of clinical trials. This also includes the experience that accumulates from cases of overdosage, which cannot be anticipated before a new drug is released. The selective serotonin reuptake inhibitors are dealt with as a separate group, since they have many class-specific adverse effects.

Amfebutamone (bupropion)

General Information

Amfebutamone is an amphetamine-like drug. It is structurally and pharmacologically distinct from other antidepressants, and apparently enhances both dopamine and noradrenaline function in the brain (SEDA-8, 18) (SEDA-10, 20).

Amfebutamone is used to encourage smoking cessation and in some countries as an antidepressant. In some respects its adverse effects profile is similar to that of the SSRIs: insomnia, agitation, tremor, and nausea are most often reported.

Randomized controlled trials have shown that amfebutamone is less likely to cause sexual dysfunction than SSRIs (1,2), unlike SSRIs, which gives it an important advantage in some patients.

Amfebutamone has been used as an antidepressant for many years in the USA and Canada and is now licensed in Europe for smoking cessation. There has been concern about the number of serious adverse reactions associated

Table 1 Second-generation antidepressants (rINNs except where stated)

Compound	Structure	Comments
Amfebutamone (bupropion)	Aminoketone	Modulates dopaminergic function Increased risk of seizures in high doses
Duloxetine	Aryloxypropremine	Inhibitor of serotouin and noradrenaline reuptake
Maprotiline	Tetracyclic	Strong inhibitory effect on noradrenaline uptake Skin rashes (3%) Increased incidence of seizures in overdose Similar adverse effects profile to tricyclic compounds
Mianserin (pINN)	Tetracyclic	Sedative profile Increased incidence of agranulocytosis Possibly safer in overdose Fewer cardiac effects
Milnacipran	Tetracyclic	Inhibitor of serotonin and noradrenaline reuptake
Mirtazapine	Piperazinoazepine	Noradrenergic and specific serotonergic antidepressant (NaSSA); similar to mianserin
Nefazodone	Phenylpiperazine	Weak serotonin reuptake inhibitor Blocks 5-HT$_2$ receptors Chemically related to trazodone
Reboxetine	Morpholine	Selective noradrenaline reuptake inhibitor (NRI or NARI)
Trazodone	Triazolopyridine	Weak effect on serotonin uptake Blocks 5-HT$_2$ receptors Fewer peripheral anticholinergic properties Sedative profile
Tryptophan	Amino acid	Precursor of serotonin Eosinophilia-myalgia syndrome
Venlafaxine	Bicyclic; cyclohexanol	Serotonin and noradrenaline uptake inhibitor Nausea, sexual dysfunction, and cardiovascular adverse effects
Viloxazine	Bicyclic	Fewer anticholinergic or sedative effects and weight gain Causes nausea, vomiting, and weight loss Can precipitate migraine

with amfebutamone, with particular emphasis on the risk of seizures. Deaths have also been reported from myocarditis and liver failure (3). A problem in determining the role of amfebutamone in these adverse effects is that it is prescribed for heavy smokers who are at risk of serious co-morbid disorders, particularly cardiovascular and respiratory disease. However, it has been clear for some time that the use of amfebutamone is associated with an increased risk of seizures, though this is less with the modified-release formulation (SEDA-23, 20).

Organs and Systems

Cardiovascular

The UK Committee on Safety of Medicines has received over 200 reports of chest pain in patients taking amfebutamone, and a case of myocardial infarction has been reported.

- A 43-year-old man who had smoked up to 20 cigarettes daily for several years and had a family history of heart disease was given amfebutamone and stopped smoking after reaching the recommended dose (300 mg/day) (4). Three days later he developed central chest and arm pain and 1 day later stopped taking amfebutamone. Three days after this he developed classical symptoms of acute inferoposterolateral myocardial infarction. He was treated with thrombolytic therapy and discharged taking secondary prevention therapy.

It is difficult to know how far amfebutamone might have contributed to this acute cardiac event. However, continuing vigilance and case-control studies are warranted.

Nervous system

The main concern with the use of amfebutamone is that in higher dosages it is associated with a risk of seizures (0.4%). Early in 1986, advertising began to appear in advance of the expected release of amfebutamone in the USA. Abruptly, in March, distribution was halted because of reports of seizures in patients with bulimia. Although the risk is greater than that seen with SSRIs (about 0.1%), data have suggested that amfebutamone has approximately the same seizure potential as the tricyclic compounds (SEDA-8, 30). However, a higher seizure risk for amfebutamone than for tricyclic antidepressants has been reported in patients taking amfebutamone in doses over 450 mg/day (SEDA 17, 21). Furthermore, the manufacturers have noted an increased risk of seizures in patients taking over 600 mg/day in combination with lithium or antipsychotic drugs (SEDA-10, 20).

Of 279 patients who presented to a hospital emergency service between 1994 and 1998 with a first tonic-clonic seizure, 17 (6.1%) had seizures that were thought to be drug-related (5). The most common drug-induced causes were cocaine intoxication (6/17) and benzodiazepine withdrawal (5/17) followed by amfebutamone use (4/17). While one amfebutamone-associated seizure occurred in a 26-year-old woman without any other risk factors, the

three other patients (all women) had additional risk factors, such as concomitant treatment with antidepressants that also lower seizure threshold and a history of bulimia nervosa. These results suggest that amfebutamone is not an infrequent cause of de novo seizures. However, because of the time frame of the study, many of the patients would have been taking standard-release amfebutamone. It would be of interest to repeat the study now that modified-release amfebutamone is available.

Amfebutamone lowers the seizure threshold and can cause paresthesia (6).

- A 38-year-old woman who was taking olanzapine 10 mg/day and lamotrigine 200 mg/day for a schizoaffective disorder started to take amfebutamone for a depressive mood swing. After 4 weeks the dose of amfebutamone was increased to 300 mg/day, at which point she complained of a twitching pain on the left side of her face. There was hypesthesia of two branches of the left trigeminal nerve, the ophthalmic and maxillary branches, and a reduced left corneal reflex. The amfebutamone was withdrawn and the neurological signs and symptoms disappeared within 8 days. Four weeks later, because of persisting depression, the amfebutamone was reintroduced; at a dosage of 300 mg/day identical neurological symptoms recurred.

The reason for this unusual reaction is not clear. Amfebutamone may potentiate dopamine neurotransmission, and dopamine D_2 receptors modulate trigeminal nerve function (7).

Amfebutamone has been reported to mimic transient ischemic attacks (8).

- A 67-year-old man with a strong history of ischemic heart disease, who had been smoking 20 cigarettes a day for many years, started to take amfebutamone (100 mg tds) as an aid to smoking cessation. One week later he had episodes of disorientation, a tingling sensation over his body, and roaring sounds in his ears. MRI scanning and angiography showed evidence of previous strokes. His current episodes were ascribed to transient ischemic attacks. He had stopped taking amfebutamone while in hospital and then restarted it 2 days after discharge, when the same symptoms recurred. He then stopped taking amfebutamone completely and over the next 9 months had no further neurological episodes.

It is possible that these symptoms were entirely due to amfebutamone, but more probably the amfebutamone interacted in some way with the patient's established cerebrovascular disease.

Amfebutamone has been associated with extrapyramidal movement disorders (SEDA-27, 15) and ballismus has been reported (22).

- A 42 year old woman developed ballismus 4 days after increasing her dose of amfebutamone to 150 mg bd. She had an involuntary urge to move, gross flexion movements of the torso, and slapping movements of her arms. She had no previous history of movement disorders and was taking only occasional sumatriptan for migraine. Amfebutamone was withdrawn and she was given haloperidol and oxazepam. The movements abated.

It seems likely from this that the dopaminergic effects of amfebutamone can provoke movement disorders in therapeutic doses in some individuals.

Acute dystonia is a recognized complication of treatment with antipsychotic drugs and it can also occur with SSRIs and the anxiolytic drug buspirone.

- A 44-year-old man took amfebutamone 300 mg/day and buspirone 45 mg/day for depression (9). Over 2–3 weeks he developed increasing neck stiffness, with tightening and spasm of the jaw muscles and pain in the left temporomandibular joint. He stopped taking his medications, and all his symptoms resolved over the next 3 weeks. Rechallenge with buspirone (45 mg/day) failed to reproduce the dystonic symptoms. The buspirone was withdrawn and amfebutamone 150 mg/day was started; the dystonic symptoms did not recur. However, 24 hours after the dose of amfebutamone was increased to 300 mg/day there was a return of neck stiffness and jaw spasm.

Amfebutamone has dopaminergic properties, which seem to have played a part in this dystonia. Whether the effect was initiated by concomitant buspirone therapy is unclear, but subsequently amfebutamone alone was sufficient to produce the dystonic symptoms.

Visual hallucinations and tinnitus have been reported in patients taking amfebutamone (SEDA-17, 21).

Somnambulism has been reported with amfebutamone (23).

- A 35-year-old man took amfebutamone 150 mg bd as part of a smoking cessation program. After 14 days he stopped smoking, after which he noted some increase in mood and appetite. Three days later a friend reported that the patient had telephoned him at 1.00 am, about 2 hours after he had gone to sleep, but the patient had no memory of this. Over the next week he discovered evidence of several episodes of nocturnal eating but again had no recall of getting up at night to obtain food. He stopped taking amfebutamone and the sleep-walking episodes disappeared and did not recur.

Amfebutamone promotes slow wave sleep, the sleep stage during which somnambulism is initiated. However, this case was complex, because of the possible interaction with nicotine withdrawal. Nicotine withdrawal is associated with increased appetite and weight gain, and sleep walking can be associated with nocturnal eating

Psychological, psychiatric

Amfebutamone potentiates dopamine neurotransmission and at higher doses can cause toxic delirium and psychosis, particularly in patients with a history of psychosis and those taking other dopaminergic medications (SEDA-8, 31).

- A 79-year-old man developed a paranoid psychosis with auditory hallucinations during treatment with amfebutamone (100 mg tds) (10). The symptoms remitted with reduction of the dose of amfebutamone and the introduction of haloperidol 5 mg/day.

Paranoid ideas and hallucinations can occur during the course of a severe depressive episode, but in this case the patient's mood was gradually improving, so a link with amfebutamone seems more likely. It seems prudent to dose amfebutamone conservatively in elderly patients.

- A 35-year-old woman with no history of a psychiatric disorder was given bupropion 300 mg/day for smoking cessation (24). After 5 days she developed an acute paranoid state with ideas of reference and fixed convictions concerning her partner's fidelity. She was also irritable and slightly grandiose. Bupropion was withdrawn and she was given benzodiazepines. She recovered over the next 2 days.

This case suggests that bupropion can rarely cause psychotic symptoms even in people without susceptibility factors, although it is conceivable that this patient might, for example, have had a family history of psychiatric disorder. Amfebutamone is believed to enhance dopaminergic function, and the psychiatric phenomena experienced by the patient, principally paranoid delusions and elevated mood, are consistent with a hyperdopaminergic state.

In an open trial, two of 16 patients became psychotic and were withdrawn from the study (11). A more detailed report of this adverse effect described four of 13 patients who became psychotic during a trial of amfebutamone; three had a history of psychosis, but none had previously had this response to other antidepressants (12). The usual dosage range of amfebutamone is 300–750 mg/day, and the psychotic symptoms occurred at dosages of 300–500 mg/day. In two cases psychotic symptoms were absent at lower dosages, but in one case the dose was therapeutically inadequate.

Other reports of psychotic reactions and a manic syndrome associated with amfebutamone have appeared (SEDA-16, 10) (SEDA-17, 21). In two of these cases a possible drug interaction with fluoxetine, with inhibition of the metabolism of amfebutamone, could not be excluded.

Electrolyte balance

Hyponatremia with SSRIs is well described (SEDA-27, 12) but has also been reported with other antidepressants, including the selective noradrenaline re-uptake inhibitor reboxetine.

- A 49-year-old woman, who was taking oxcarbazepine for bipolar disorder, developed hyponatremia after also taking bupropion (25).

Oxcarbazepine can also cause hyponatremia, but in this case the hyponatremia appeared to correlate with the prescription of bupropion rather than oxcarbazepine and was also reproduced by amfebutamone re-challenge. This case suggests that amfebutamone can also cause hyponatremia, although it is possible that the presence of oxcarbazepine was necessary for the adverse effect to occur.

Hematologic

Eosinophilia reportedly developed in a 72-year-old woman shortly after she was given amfebutamone. The eosinophil count fell rapidly after all medications (including glibenclamide and tolmetin) were withdrawn (13).

Skin

Adverse skin reactions, such as rash, urticaria, and pruritus, have been reported in 1–4% of patients taking amfebutamone (14).

Sexual function

In contrast to SSRIs, amfebutamone is believed to have minimal effect on sexual function (SEDA-23, 20).

- A 36-year-old man took amfebutamone 150 mg/day as part of a smoking cessation program and soon complained of complete anorgasmia; sexual desire and arousal were preserved (26). Amfebutamone was withdrawn and the sexual dysfunction resolved within 3 days.

Anorgasmia is a sexual adverse effect typically seen with SSRIs. However, this case report suggests that rarely amfebutamone can produce a similar effect.

Immunologic

Amfebutamone has been associated with a variety of generalized sensitivity reactions (SEDA-25, 17; SEDA-27, 28), including urticaria and serum sickness-like reactions.

- A 17-year-old boy took amfebutamone (dose unstated) for attention deficit disorder and 1 week later developed a generalized pruritic rash, but continued to take amfebutamone (27). After a further week he presented as an emergency with large joint tenderness and joint swelling. A punch biopsy of a skin lesion showed urticaria with vasculitis. Amfebutamone was withdrawn and a single dose of methylprednisolone sodium succinate was given. His symptoms resolved completely within 36 hours.
- A 24 year old man developed a fever and a generalized maculopapular rash after taking amfebutamone 150 mg bd for 3 weeks (28). Amfebutamone was withdrawn but he went on to develop angioedema, eosinophilia, and a systemic syndrome with hepatitis, myositis, and obstructive lung disease. His symptoms resolved over several weeks with a glucocorticoid.

Two cases of serum sickness-like reactions have been reported in association with amfebutamone when used as an aid to smoking cessation (15,16). Both patients developed localized swellings of the fingers and hands, urticaria, and arthralgia. In both cases treatment with antihistamines and corticosteroids produced rapid relief of symptoms.

In three other patients (two women, one man) a serum sickness-like reaction developed 6–21 days after the start of amfebutamone treatment (14). The symptoms, arthralgia, pruritus, and tongue swelling, abated within 2 weeks of treatment with oral corticosteroids. Serum sickness reactions to drugs are rare, and it will be important to find out whether amfebutamone carries an increased risk of this unusual reaction.

Death

Amfebutamone has been linked to 41 deaths (17). From the reports of suspected adverse events received by the Netherlands Pharmacovigilance Foundation, it appears that more than half concerned patients at risk of smoking-related diseases. In 15 cases there had been simultaneous use of amfebutamone with another antidepressant (10 patients), theophylline (1 patient), or insulin (4 patients). These combinations may lead to an increase in the risk of seizures. Furthermore, two patients reported having taken antiepileptic drugs, despite the fact that amfebutamone is contraindicated in patients with seizure disorders. These results suggest that the guidelines described in the product information are not being adhered to in some cases.

Drug Administration

Drug formulations

A modified-release formulation of amfebutamone has been marketed (18). There are insufficient data yet on the risk of seizures with higher doses of the modified-release formulation, but preliminary indications suggest that it is likely to be lower than that seen with the immediate-release formulation (18).

Drug overdose

Seizure is a significant risk in amfebutamone overdose.

- An 18-year-old man attempted suicide by taking amfebutamone 7.5 g (50 × 150 mg tablets) (19). On assessment 90 minutes later he was agitated and aggressive, with a resting pulse of 160 beats/minute and a blood pressure of 142/63 mmHg. He quickly developed a persistently low blood pressure (65/40 mmHg), followed by three generalized tonic-clonic seizures, which were controlled by diazepam. He was ventilated and a metabolic acidosis was corrected with sodium bicarbonate. Despite this, his blood pressure remained persistently low (70/40 mmHg) with a sinus tachycardia (150 beats/minute). Dopamine was ineffective and adrenaline was required. After 24 hours he was extubated and made a full recovery over the next 3 days.
- A 14-year-old boy took 1.5–3.0 g of amfebutamone and had a persistent tachycardia, seizures, and brief agitation and aggression (20). He also had visual hallucinations, disorientation, and confusion but recovered about 24 hours after ingestion.

These cases show that overdose of amfebutamone can be serious.

Drug–Drug Interactions

Antidepressants

The immediate-release form of amfebutamone may be associated with a higher risk of seizures than other antidepressant drugs, but the risk with the modified-release formulation is believed to be about the same as SSRIs (SEDA-23, 27).

- A 28-year-old woman with schizophrenia taking risperidone 4 mg/day and clomipramine 25 mg/day started to take modified-release amfebutamone for depressive symptoms 150 mg/day initially and 1 month later 300 mg/day; 1 week after this she had a generalized tonic–clonic seizure with unconsciousness (29). The amfebutamone was withdrawn, but she had a similar seizure 3 days later, after which the clomipramine was changed to sertraline. Following this no further seizures occurred. However, epileptic discharges on the electroencephalogram, which had previously been normal, continued for another month, after which she was given valproate.

As in many cases of seizures associated with amfebutamone, there were other risk factors in this case, in particular the use of other psychotropic drugs known to lower the seizure threshold. The dose of clomipramine was low, but amfebutamone may have some inhibitory effect on CYP2D6, which is involved in the metabolism of tertiary tricyclics such as clomipramine. The persistent electroencephalographic abnormalities also suggested that the patient had an underlying vulnerability to seizure disorder although an earlier electroencephalogram had been normal. In general, it seems prudent to avoid the use of even low doses of tricyclic antidepressants in combination with amfebutamone. However, the combination of SSRIs and amfebutamone is quite widely used in the USA in the treatment of resistant depression (30).

Amfebutamone is sometimes used to augment the action of other antidepressant drugs, but there are few data on the pharmacokinetic effects of this strategy. In an open study in 19 consecutive patients (7 men and 12 women), amfebutamone (150 mg/day for 8 weeks) did not alter plasma concentrations of paroxetine or fluoxetine (21). Plasma concentrations of venlafaxine increased, but concentrations of its active metabolite, O-desmethylvenlafaxine, fell. Overall tolerability was good and there was a significant reduction in depression rating scales. Sexual function, particularly orgasmic function in women, improved. These data suggest that adding amfebutamone to SSRI treatment does not alter plasma concentrations of SSRIs. The addition of amfebutamone to venlafaxine seems unlikely to cause pharmacokinetic problems in practice, because the increase in plasma concentration of the parent compound was offset by a reduction in the concentration of its major active metabolite. However, there is scope for an interaction of venlafaxine with amfebutamone and the effects in individual patients might be hard to predict. This study also supports clinical suggestions that the addition of amfebutamone to SSRI treatment can enhance antidepressant efficacy and

improve sexual function. However, controlled trials are needed to confirm this view.

Dextromethorphan

While in vitro studies have suggested that amfebutamone (bupropion) is a weak inhibitor of CYP2D6, case reports suggest that it may in fact have cause significant drug interactions though inhibition of CYP2D6 (SEDA-29, 25). In 21 subjects, mean age 40 years, 48% female, who were randomly assigned to receive either amfebutamone 300 mg/day for 17 days or placebo in a parallel group design, amfebutamone, but not placebo, produced a substantial increase in the dextromethorphan/dextrorphan ratio, showing that it is an effective inhibitor of CYP2D6 in vivo (31).

Metamfetamine

When metamfetamine and amfebutamone (bupropion) were co-administered to 26 subjects, 20 of whom completed the protocol, there was no evidence of additive cardiovascular effects [32]. The subjects received metamfetamine 0, 15, and 30 mg intravenously before and after randomization to bupropion 150 mg bd in a modified-release formulation or matched placebo. There was a non-significant trend for amfebutamone to reduce metamfetamine-associated increases in blood pressure and a significant reduction in the metamfetamine-associated increase in heart rate. Amfebutamone reduced the plasma clearance of metamfetamine and the appearance of amfetamine in the plasma. Metamfetamine did not alter the peak and trough concentrations of amfebutamone or its metabolites. These findings are relevant to the potential use of amfebutamone in ameliorating acute abstinence in metamfetamine users. However, the risk of seizures during amfebutamone treatment for metamfetamine abuse has not been estimated.

Metoprolol

A possible inhibitory effect of amfebutamone on CYP2D6 was used to explain an episode of severe bradycardia in a man taking the β-adrenoceptor antagonist, metoprolol (33).

- A 56-year-old man taking metoprolol 75 mg/day for hypertension developed fatigue and dyspnea 12 days after starting to take amfebutamone 300 mg/day to help smoking cessation. His heart rate was 43/minute and his blood pressure 102/65 mmHg. His chest X-ray showed evidence of mild congestive cardiac failure and an electrocardiogram showed an atrial rate of 40/minute together with a junctional escape rhythm of 43/minute. All medications were withheld and the next morning he was asymptomatic and in sinus rhythm. Metoprolol was then restarted and he remained asymptomatic after 1 month of follow up.

Although metoprolol concentrations were not measured in this patient it seems likely that they were increased by co-administration of amfebutamone.

References

1. Coleman CC, Cunningham LA, Foster VJ, Batey SR, Donahue RM, Houser TL, Ascher JA. Sexual dysfunction associated with the treatment of depression: a placebo-controlled comparison of bupropion sustained release and sertraline treatment. Ann Clin Psychiatry 1999;11(4):205–15.

2. Croft H, Settle E Jr, Houser T, Batey SR, Donahue RM, Ascher JA. A placebo-controlled comparison of the antidepressant efficacy and effects on sexual functioning of sustained-release bupropion and sertraline. Clin Ther 1999;21(4):643–58.

3. Wooltorton E. Bupropion (Zyban, Wellbutrin SR): reports of deaths, seizures, serum sickness. CMAJ 2002;166(1):68.

4. Patterson RN, Herity NA. Acute myocardial infarction following bupropion (Zyban). Q J Med 2002;95(1):58–9.

5. Pesola GR, Avasarala J. Bupropion seizure proportion among new-onset generalized seizures and drug related seizures presenting to an emergency department. J Emerg Med 2002;22(3):235–9.

6. Amann B, Hummel B, Rall-Autenrieth H, Walden J, Grunze H. Bupropion-induced isolated impairment of sensory trigeminal nerve function. Int Clin Psychopharmacol 2000;15(2):115–6.

7. Peterfreund RA, Kosofsky BE, Fink JS. Cellular localization of dopamine D2 receptor messenger RNA in the rat trigeminal ganglion. Anesth Analg 1995;81(6):1181–5.

8. Humma LM, Swims MP. Bupropion mimics a transient ischemic attack. Ann Pharmacother 1999;33(3):305–7.

9. Detweiler MB, Harpold GJ. Bupropion-induced acute dystonia. Ann Pharmacother 2002;36(2):251–4.

10. Howard WT, Warnock JK. Bupropion-induced psychosis. Am J Psychiatry 1999;156(12):2017–8.

11. Dufresne R, Becker R, Blitzer R, et al. Safety and efficacy of bupropion. Drug Dev Res 1985;6:39.

12. Golden RN, James SP, Sherer MA, Rudorfer MV, Sack DA, Potter WZ. Psychoses associated with bupropion treatment. Am J Psychiatry 1985;142(12):1459–62.

13. Malesker MA, Soori GS, Malone PM, Mahowald JA, Housel GJ. Eosinophilia associated with bupropion. Ann Pharmacother 1995;29(9):867–9.

14. McCollom RA, Elbe DH, Ritchie AH. Bupropion-induced serum sickness-like reaction. Ann Pharmacother 2000;34(4):471–3.

15. Peloso PM, Baillie C. Serum sickness-like reaction with bupropion. JAMA 1999;282(19):1817.

16. Tripathi A, Greenberger PA. Bupropion hydrochloride induced serum sickness-like reaction. Ann Allergy Asthma Immunol 1999;83(2):165–6.

17. Bhattacharjee C, Smith M, Todd F, Gillespie M. Bupropion overdose: a potential problem with the new "miracle" anti-smoking drug. Int J Clin Pract 2001;55(3):221–2.

18. Settle EC Jr. Bupropion sustained release: side effect profile. J Clin Psychiatry 1998;59(Suppl 4):32–6.

19. Ayers S, Tobias JD. Bupropion overdose in an adolescent. Pediatr Emerg Care 2001;17(2):104–6.

20. Carvajal Garcia-Pando A, Garcia del Pozo J, Sanchez AS, Velasco MA, Rueda de Castro AM, Lucena MI. Hepatotoxicity associated with the new antidepressants. J Clin Psychiatry 2002;63(2):135–7.

21. Kennedy SH, McCann SM, Masellis M, McIntyre RS, Raskin J, McKay G, Baker GB. Combining bupropion SR with venlafaxine, paroxetine, or fluoxetine: a preliminary report on pharmacokinetic, therapeutic, and sexual dysfunction effects. J Clin Psychiatry 2002;63(3):181–6.

22. De Graaf L, Admiraal P, Van Puijenbroek EP. Ballism associated with bupropion use. Ann Pharmacother 2003;37:302–3.

23. Khazaal Y, Krenz S, Zullino DF. Bupropion-induced somnambulism. Addict Biol 2003;8:359–62.

24. Khazaal Y, Kolly S, Zullino DF. Psychotic symptoms associated with bupropion treatment for smoking cessation. J Subst Use 2005;10:62–4.

25. Bagley SC, Yaeger D. Hyponatremia associated with bupropion, a case verified by rechallenge. J Clin Psychopharmacol 2005;1:98–9.

26. Martinez-Raga J, Sabater A, Cervera G. Anorgasmia in a patient treated with bupropion SR. J Clin Psychopharmacol 2004;24:460–1.

27. Waibel KH, Katial RR. Serum sickness- like reaction and bupropion. J Am Acad Child Adolesc 2004;43:509.

28. Bagshaw SM, Cload B, Gilmour J, Leung ST, Bowen TJ. Drug-induced rash with eosinophilia and systemic symptoms syndrome with bupropion administration. Ann Allergy Asthma Immunol 2003;90:572–5.

29. Shin YW, Erm TM, Choi EJ, Kim SY. A case of prolonged seizure activity after combined use of bupropion and clomipramine. Clin Neuropharmacol 2004;27:192–4.

30. Lam RW, Hossie H, Solomons K, Yatham LN. Citalopram and bupropion-SR: combining versus switching in patients with treatment-resistant depression. J Clin Psychiatry 2004;65:337–40.

31. Kotlyar M, Brauer LH, Tracey TS, Hatsukami DK, Harris J, Bronanrs CA, Adson DE. Inhibition of CYP2D6 activity by bupropion. J Clin Psychopharmacol 2005;25:226–9.

32. Newton TF, Roache JD, De la Garza R II, Fong T, Wallace CL, Li S-H, Elkashef A, Chiang N, Kahn R. Safety of intravenous methamphetamine administration during treatment with bupropion. Psychopharmacology 2005;182:426–35.

33. McCollum DL, Greene JL, McGuire DK. Severe sinus bradycardia after initiation of bupropion therapy: a probable drug–drug interaction with metoprolol. Cardiovasc Drugs Ther 2004;18:329–30.

Duloxetine

Duloxetine has recently been marketed as an antidepressant in Europe. It inhibits the re-uptake of serotonin and noradrenaline, with minimal effects on other neurotransmitter mechanisms. It is therefore classified as a serotonin and noradrenaline re-uptake inhibitor (SNRI) and is grouped with venlafaxine. The adverse effect profile of duloxetine appears to be similar to that of the SSRIs and venlafaxine. In placebo-controlled trials the most common adverse effects were nausea (37%), dry mouth (32%), dizziness (22%), somnolence (20%), insomnia (20%), and diarrhea (14%). Sexual dysfunction has also been reported. Current data suggest that, unlike venlafaxine, duloxetine does not increase the blood pressure, but further post-marketing surveillance studies will be needed to confirm this (1).

Organs and Systems

Cardiovascular

Early descriptions of duloxetine suggested that it might be less likely than venlafaxine to cause increased blood pressure. The cardiovascular profile of duloxetine has been reviewed from a database of eight double-blind randomized trials in depression, in which 1139 patients took duloxetine (40–120 mg/day) and 777 took placebo (2). Relative to placebo, duloxetine produced a small but significant increase in heart rate (about 2/minute). Duloxetine produced a greater rate of sustained increase in systolic blood pressure than placebo (1% versus 0.4%, relative risk 2.1) but no difference in diastolic blood pressure. There were also significantly more instances of significantly increased systolic blood pressure in patients taking duloxetine than placebo at any clinic visit (19% versus 13%). These findings suggest that duloxetine can produce increases in blood pressure in some people, presumably through its potentiation of noradrenaline function. How the effect of duloxetine compares with that of venlafaxine remains to be established.

Long-Term Effects

Drug withdrawal

Like venlafaxine and the SSRIs, acute withdrawal of duloxetine causes a characteristic abstinence syndrome, with tachycardia, dizziness, insomnia, headache, and anxiety.

Drug-drug interactions

Duloxetine is metabolized by CYP2D6, the activity of which it inhibits to a moderate extent. Caution is therefore needed when duloxetine is co-prescribed with other drugs that are substrates of this enzyme. Duloxetine is also partly metabolized by CYP1A2, so blood concentrations of duloxetine might be increased by co-prescription of potent inhibitors of CYP1A2 (1).

References

1. Cowen PJ, Ogilvie AD, Gama J. Efficacy, safety and tolerability of duloxetine 60 mg once daily in major depression. Curr Med Res Opin 2005;3:345–55.
2. Thase ME, Tran PV, Curtis W, Pangallo B, Mallinckrodt C, Detke MJ. Cardiovascular profile of duloxetine, a dual reuptake inhibitor of serotonin and norepinephrine. J Clin Psychopharmacol 2005;25:132–40.

Maprotiline

General Information

Maprotiline has a tetracyclic structure, in which the tricyclic nucleus adjoins a fourth ring formed by an ethylene bridge vertical to the major plane of the molecule (SEDA-5, 34). It has a strong inhibitory effect on noradrenaline uptake across cell membranes, and relatively weak effects on serotonergic mechanisms. The half-life averages 43 hours (1), allowing once-daily dosing.

After the release of maprotiline in Britain, information was collected from 10 000 patients treated in general practice during 9 months (2). Patients were given 75 mg/day, preferably at night, and were followed for 3 weeks. By the end of that time 1343 patients had dropped out owing to adverse effects, of which drowsiness was the most common, followed by dizziness and headache. Skin rashes reached a constant peak incidence of about 3% after 2 weeks; rashes were reported in some earlier clinical trials (SEDA-2, 12).

Blood concentrations were measured in two trials in general practice (3). On both regimens (single and divided doses), a steady state was reached after 1 week, but there was considerable individual variability (up to ten-fold) in plasma concentrations. In a multicenter study of over 2000 records from more than 500 general practitioners (4) the onset of action usually occurred within the first week of treatment and the effect was complete after 3–4 weeks; roughly half of the total improvement occurred by the end of the first week.

There have been several comparisons of maprotiline with other antidepressants. There was no significant difference in adverse effects compared with doxepin (5), amitriptyline (6), or imipramine (7).

Organs and Systems

Cardiovascular

Maprotiline is among the antidepressants with the lowest reported incidence of tachycardia and postural hypotension.

Respiratory

Exercise-induced bronchospasm has been reported (8).

Nervous system

Information released by the UK Committee on Safety of Medicines showed that maprotiline was associated with a disproportionate risk of seizures (9). Among 186 patients taking antidepressants who were admitted to a medical center during one year only one of 45 patients taking tricyclic compounds and none of the 109 patients not taking antidepressants had a seizure; in contrast, five of 32 patients taking maprotiline had seizures, a significant difference (10). The patients who had seizures were aged 20–66 years and none had a previous history. Dosages of maprotiline were 75–300 mg/day, and each patient had undergone a dosage increase in the week before the seizure. A report based on experience in one hospital, in addition to a review of 87 cases reported to the manufacturer, showed that seizures tended to occur at high doses (regardless of dose escalation or plasma concentrations), and that seizures presented after patients had been taking maprotiline for several weeks (11). Post-marketing surveillance has also shown a higher seizure risk with maprotiline than with tricyclic antidepressants, but since the maximum dose of maprotiline was reduced from 300 to 225 mg, few seizures have been reported (SEDA-17, 22).

So-called speech blockage has been reported with other antidepressants.

- A 54-year-old man developed a stammer and speech blockage while taking maprotiline 75 mg/day (12). It responded to a reduction in dose to 50 mg/day, reappeared with another challenge of 75 mg/day, and did not respond to physostigmine intramuscularly on two occasions. It did not occur when he took desipramine 50 mg/day.

Myoclonus due to maprotiline has been reported (13). Further neuromuscular symptoms that have been reported with maprotiline include cerebellar ataxia in a 54-year-old man with a history of unipolar depression and chronic alcohol abuse who was taking maprotiline 200 mg/day (14). The question of whether his history of alcohol abuse contributed by sensitizing his cerebellum to maprotiline-induced ataxia was unresolved.

Non-convulsive status epilepticus occurred in a 27-year-old woman shortly after starting treatment with clomipramine and maprotiline (15). On withdrawal of medication she gradually improved.

Endocrine

Galactorrhea occurred in a 23-year-old woman 2 weeks after she started to take maprotiline 50 mg/day (increased to 75 mg after 10 days) and resolved after withdrawal (16).

Metabolism

An often troublesome adverse effect of antidepressant medication is weight gain. Two cases of this adverse effect have been reported in patients taking low doses of maprotiline (17).

Hematologic

Neutropenia has been reported in a 51-year-old woman who took 150 mg/day for several months and developed myeloid hyperplasia with a virtual absence of mature granulocytes; she recovered fully with conservative management (18).

Liver

As noted previously (SEDA-10, 21), hepatotoxicity has been described with maprotiline, and two further reports have been published. In one of these, hepatotoxicity was observed in a 54-year-old man treated with maprotiline for chronic head and neck pain (19). The other report described a patient who developed a marked increase in liver enzymes and symptoms during therapy with maprotiline and opipramol; after withdrawal the patient recovered completely (20).

Skin

Ichthyosiform desquamation of the skin and alopecia have been reported (SEDA-17, 22).

Drug Administration

Drug overdose

Reports of overdose with maprotiline relate to doses of 750–3200 mg (21,22). The symptoms included impaired consciousness, convulsions, confusion, disorientation, visual hallucinations, and electrocardiographic changes similar to those seen with tricyclic compounds. Among 41 patients who had taken overdoses of maprotiline cardiotoxicity was equal to or greater than that of tricyclic drugs (23). Mania occurred in one case of maprotiline overdose (SEDA-17, 22).

Fatal overdose has been reported in detail in a 13-year-old girl who took 3000 mg of maprotiline and developed lactic acidosis, which the authors believed was responsible for her death (24). Another death due to maprotiline was reported in the USA in a 23-year-old woman who took 4.5–6.0 g (25).

Drug–Drug Interactions

Insulin

- Maprotiline, a tetracyclic antidepressant, repeatedly induced hypoglycemia in a 39-year-old woman with type 1 diabetes, even when the insulin dosage was reduced from 20 U/day to 4–10 U/day. Maprotiline seems to prolong the half-life of insulin. A glucagon stimulation test showed a maximum C-peptide concentration of only 0.22 nmol/l (27).

Risperidone

In three patients, plasma concentrations of maprotiline were increased by concomitant treatment with the atypical antipsychotic drug, risperidone (26). The data suggest that risperidone is a modest inhibitor of CYP2D6 and should therefore be used with caution in combination with drugs such as maprotiline, where increased plasma concentrations have the potential to cause serious toxicity.

In three patients, maprotiline plasma concentrations increased when risperidone was added (28). The rise was explained by inhibition of CYP2D6, by which maprotiline is mainly metabolized.

References

1. Carpenter EH, Plant MJ, Hassell AB, Shadforth MF, Fisher J, Clarke S, Hothersall TE, Dawes PT. Management of oral complications of disease-modifying drugs in rheumatoid arthritis. Br J Rheumatol 1997;36(4):473–8.
2. Forrest WA. Maprotiline (Ludiomil) in depression: a report of a monitored release study in general practice. J Int Med Res 1977;5(Suppl 4):112–5.
3. Miller PI, Beaumont G, Seldrup J, John V, Luscombe DK, Jones R. Efficacy, side-effects, plasma and blood levels of maprotiline (Ludiomil). J Int Med Res 1977;5(Suppl 4):101–11.
4. Forrest WA. Maprotiline (Ludiomil) in depression: a multi-centre assessment of onset of action, efficacy and tolerability. J Int Med Res 1977;5(Suppl 4):116–21.

5. Vaisanen E, Naarala M, Kontiainen H, Merilainen V, Heikkila L, Malinen L. Maprotiline and doxepin in the treatment of depression. A double-blind multicentre comparison. Acta Psychiatr Scand 1978;57(3):193–201.

6. Dell AJ. A comparison of maprotiline (Ludiomil) and amitriptyline. J Int Med Res 1977;5(Suppl 4):22–4.

7. Claghorn JL. A double-blind study of maprotiline (Ludiomil) and imipramine in depressed outpatients. Curr Ther Res Clin Exp 1977;22:446.

8. Dubovsky SL, Freed C. Exercise-induced bronchospasm caused by maprotiline. Psychosomatics 1988;29(1):104–6.

9. Edwards JG. Antidepressants and convulsions. Lancet 1979;2(8156–8157):1368–9.

10. Jabbari B, Bryan GE, Marsh EE, Gunderson CH. Incidence of seizures with tricyclic and tetracyclic antidepressants. Arch Neurol 1985;42(5):480–1.

11. Dessain EC, Schatzberg AF, Woods BT, Cole JO. Maprotiline treatment in depression. A perspective on seizures. Arch General Psychiatry 1986;43(1):86–90.

12. Sandyk R. Speech blockage induced by maprotiline. Am J Psychiatry 1986;143(3):391–2.

13. DeCastro RM. Antidepressants and myoclonus: case report. J Clin Psychiatry 1985;46(7):284–7.

14. Buckler RA, Friedman JH. Maprotiline-induced ataxia. J Clin Psychopharmacol 1986;6(6):382–3.

15. Miyata H, Kubota F, Shibata N, Kifune A. Non-convulsive status epilepticus induced by antidepressants. Seizure 1997;6(5):405–7.

16. Perez OE, Henriquez N. Galactorrhea associated with maprotiline HCl. Am J Psychiatry 1983;140(5):641.

17. Nakra BR, Grossberg GT. Carbohydrate craving weight gain with maprotiline. Psychosomatics 1986;27(5):376381.

18. Ream RS, Kerr RO. Neutropenia associated with maprotiline. JAMA 1982;248(7):871.

19. Aleem A, Lingam V. Hepatotoxicity following treatment with maprotiline. J Clin Psychopharmacol 1987;7(1):54–5.

20. Braun JS, Geiger R, Wehner H, Schaffer S, Berger M. Hepatitis caused by antidepressive therapy with maprotiline and opipramol. Pharmacopsychiatry 1998;31(4):152–5.

21. Lutier F, Lefebvre JP, Stephan E. Intoxication aiguë par le Ludiomil. [Acute intoxication caused by Ludiomil.] Arch Med Normandie 1977;8:539.

22. Park J, Proudfoot AT. Acute poisoning with maprotiline hydrochloride. BMJ 1977;1(6076):1573.

23. Knudsen K, Heath A. Effects of self poisoning with maprotiline. BMJ (Clin Res Ed) 1984;288(6417):601–3.

24. Alten HE, Koppel C, Ibe K. Akute Vergiftung mit Maprotiline mit lethalem Ausgang. [Acute poisoning by maprotiline with fatal outcome.] Intensivmedizin 1980;17:71.

25. Rejent TA, Doyle RE. Maprotiline fatality: case report and analytical determinations. J Anal Toxicol 1982;6(4):199–201.

26. Normann C, Lieb K, Walden J. Increased plasma concentration of maprotiline by coadministration of risperidone. J Clin Psychopharmacol 2002;22(1):92–4.

27. Isotani H, Kameoka K. Hypoglycemia associated with maprotiline in a patient with type 1 diabetes. Diabetes Care 1999;22(5):862–3.

28. Normann C, Lieb K, Walden J. Increased plasma concentration of maprotiline by coadministration of risperidone. J Clin Psychopharmacol 2002;22(1):92–4.

Mianserin

General Information

Mianserin is a tetracyclic compound, related to cyproheptadine (SEDA-5, 18). It is an effective antidepressant with antiserotonergic properties and a sedative profile, but minimal anticholinergic effect and less action on the cardiovascular system than tricyclic antidepressants (1).

Organs and Systems

Cardiovascular

Cardiotoxic effects are relatively uncommon with mianserin (2). In a placebo-controlled study in 50 patients with a variety of cardiac conditions who were taking anticoagulants, mianserin (up to 30 or 60 mg) had no effects on electrocardiography, blood pressure, or pulse rate after 3 weeks. In a second phase, mianserin (up to 60 mg/day) was compared with amitriptyline (up to 150 mg/day) and placebo in 18 healthy volunteers. Measurements included systolic time intervals, electrocardiography at rest and during exercise, echocardiography, and blood pressure. Amitriptyline had a negative inotropic effect; mianserin increased ejection fraction. The results of both these experiments led the authors to conclude that mianserin is an antidepressant with very low cardiac toxicity.

In another study of the electrocardiographic effects of mianserin and a review of previous experiments with doses up to 120 mg/day, there were no consistent cardiovascular effects, although four patients who took 40 mg/day for 13 months had significant increases in pulse rate, with prolonged PR intervals and reduced T-wave amplitude (3). Fainting and persistent bradycardia were noted in one woman after a single dose of mianserin 60 mg, and her symptoms recurred on rechallenge (SEDA-17, 21).

There have been two instances of possible cardiac effects due to mianserin in elderly patients with pre-existing cardiovascular disorders (4).

- A 71-year-old man with hypertension who took 30 mg/day developed cardiac failure.
- A 66-year-old woman with mitral regurgitation and atrial fibrillation developed hypokalemia and a variety of rhythm disturbances at dosages of 40 mg or more per day.

Nervous system

In 40 patients who had seizures while taking therapeutic dosages of mianserin it was concluded that mianserin is probably no more likely to produce convulsions in therapeutic dosages than tricyclic compounds (5). When it does so, seizures are more likely to occur in the first 2 weeks of treatment in patients with a family or personal history indicative of risk, and after a dosage increase.

Mianserin can cause restless legs (SEDA-14, 15) (SEDA-17, 21).

Hematologic

In 1979 the first case of leukopenia due to mianserin was reported (6), followed by another soon after (7). In July 1980, the Australian Drug Evaluation Committee issued a preliminary caution concerning further cases that had been reported to its National Monitoring Centres and 6 months later published detailed reports of four other cases (8), all of which had occurred within 1 year of marketing in Australia, despite 6 years of use in Europe. Inquiries unearthed reports of seven similar episodes held by drug surveillance organizations in other countries, and the Committee expressed its concern that "if a causal relationship between mianserin and these blood disorders is subsequently confirmed, the incidence of disorders due to mianserin may be significantly greater than that of disorders due either to the phenothiazines or the tricyclic antidepressants" (9). In 1985 the UK Committee on Safety of Medicines issued a statement concerning 15 reports that the Committee had received during 5 years (10). Three of the patients subsequently died, one with aplastic anemia, one with granulocytopenia, and one with both erythrocyte and leukocyte hypoplasia. The New Zealand Department of Health's Clinical Services Letter mentioned three cases of agranulocytosis that occurred in New Zealand in 1982 (11). Another report (12) described thrombocytopenia and leukopenia based on an immune mechanism without generalized marrow depression. A high rate of agranulocytosis has been reported by the IMMP in New Zealand: one case out of 1822 and one fatal case out of 11 537 (12).

Mouth and teeth

The lack of anticholinergic effects reported in clinical studies has been supported by a double blind, placebo-controlled comparison of mianserin (10–60 mg/day) and amitriptyline (25–150 mg/day) in healthy volunteers (13). Mianserin tended to increase saliva flow, while amitriptyline significantly reduced it. However, a further experiment in healthy volunteers showed that single doses of mianserin (50–70 mg) produced a significant 29% reduction of salivation compared with placebo (14). This may account for the glossitis that has previously been reported (SEDA-9, 23).

Liver

Hepatotoxicity from mianserin has been reported in eight cases (SEDA-17, 22).

Musculoskeletal

Mianserin can cause joint symptoms and arthritis (SEDA-16, 11).

Drug Administration

Drug overdose

One of the putative benefits of mianserin is its alleged safety in overdose, which may be related to a reduced risk of cardiovascular adverse effects and convulsions. Data from the UK Committee on Safety of Medicines suggest that mianserin accounts for 11% of reported convulsions and 5.8% of use, putting it intermediate between amitriptyline and maprotiline (15). On the other hand, in the London Poisons Unit survey, involving 84 patients who took mianserin alone (up to 1000 mg), there were no deaths and no patients with convulsions, although this could represent a frequency of up to 3.6% (12).

A survey of 100 cases of overdose with mianserin reported to the London Centre of the UK Poisons Information Service included 54 patients who took mianserin alone in amounts up to and in three cases in excess of 1000 mg (12). Plasma mianserin concentrations were 70–665 ng/l (the usual mean target concentration is 50 ng/l). There were no reports of convulsions, cardiac dysrhythmias, or profound coma in any patient taking mianserin alone, but there were two deaths in patients who took multiple drugs. The authors concluded that in acute overdosage mianserin is less toxic than tricyclic antidepressants.

Drug–Drug Interactions

Clonidine and apraclonidine

Mianserin has alpha-adrenoceptor activity and so might interact with clonidine (17). In healthy volunteers, pretreatment with mianserin 60 mg/day for 3 days did not modify the hypotensive effects of a single 300 mg dose of clonidine. In 11 patients with essential hypertension, the addition of mianserin 60 mg/day (in divided doses) for 2 weeks did not reduce the hypotensive effect of clonidine. The results of this study appear to have justified the authors' conclusion that adding mianserin to treatment with clonidine will not result in loss of blood pressure control.

Clonidine and methyldopa

Mianserin lacks potential for peripheral adrenergic interactions, but since it has α-adrenoceptor activity, it might interact with the centrally acting α-adrenoceptor agonists clonidine and methyldopa. In healthy volunteers, pretreatment with mianserin 60 mg/day for 3 days did not modify the hypotensive effects of a single 300 mg dose of clonidine, and in 11 patients with essential hypertension, the addition of mianserin 60 mg/day (in divided doses) for 2 weeks did not reduce the hypotensive effects of clonidine or methyldopa (16). In patients taking methyldopa, there were additive hypotensive effects after the first dose of mianserin, but these were not significant after 1 or 2 weeks of combined treatment. The results of this study appear to have justified the authors' conclusion that combining mianserin with centrally-acting hypotensive agents will not result in loss of blood pressure control.

Methyldopa

Mianserin has alpha-adrenoceptor activity and so might interact with methyldopa (17). In 11 patients with essential hypertension, the addition of mianserin 60 mg/day (in divided doses) for 2 weeks did not reduce the hypotensive effect of methyldopa. In patients treated with methyldopa, there were additive hypotensive effects after the first dose of mianserin, but these were not significant after 1 or 2 weeks of combined treatment. The results of this study appear to have justified the authors' conclusion that adding mianserin to treatment with methyldopa will not result in loss of blood pressure control.

References

1. Brogden RN, Heel RC, Speight TM, Avery GS. Mianserin: a review of its pharmacological properties and therapeutic efficacy in depressive illness. Drugs 1978;16(4):273–301.
2. Kopera H, Klein W, Schenk H. Psychotropic drugs and the heart: clinical implications. Prog Neuropsychopharmacol 1980;4(4–5):527–35.
3. Goldie A, Edwards JG. Electrocardiographic changes during treatment with maprotiline and mianserin. Neuropharmacology 1984;23(2B):273–5.
4. Whiteford H, Klug P, Evans L. Disturbed cardiac function possibly associated with mianserin therapy. Med J Aust 1984;140(3):166–7.
5. Edwards JG, Glen-Bott M. Mianserin and convulsive seizures. Br J Clin Pharmacol 1983;15(Suppl 2):299S–311S.
6. Curson DA, Hale AS. Mianserin and agranulocytosis. BMJ 1979;1(6160):378–9.
7. McHarg AM, McHarg JF. Leucopenia in association with mianserin treatment. BMJ 1979;1(6163):623–4.
8. Anonymous. Mianserin: a possible cause of neutropenia and agranulocytosis. Adverse Drug Reactions Advisory Committee. Med J Aust 1980;2(12):673–4.
9. Anonymous. Side effects associated with mianserin. Scrip 1982;707:11.
10. Stricker BH, Barendregt JN, Claas FH. Thrombocytopenia and leucopenia with mianserin-dependent antibodies. Br J Clin Pharmacol 1985;19(1):102–4.
11. Coulter DM, Edwards IR. Mianserin and agranulocytosis in New Zealand. Lancet 1990;336(8718):785–7.
12. Shaw WL. The comparative safety of mianserin in overdose. Curr Med Res Opin 1980;6(Suppl 7):44.
13. Kopera H. Anticholinergic effects of mianserin. Curr Med Res Opin 1980;6(Suppl 7):132.
14. Clemmesen L, Jensen E, Min SK, Bolwig TG, Rafaelsen OJ. Salivation after single-doses of the new antidepressants femoxetine, mianserin and citalopram. A crossover study. Pharmacopsychiatry 1984;17(4):126–32.
15. Edwards JG. Antidepressants and convulsions. Lancet 1979;2(8156–8157):1368–9.
16. Elliott HL, Whiting B, Reid JL. Assessment of the interaction between mianserin and centrally-acting antihypertensive drugs. Br J Clin Pharmacol 1983;15(Suppl 2):323S–8S.
17. Elliott HL, Whiting B, Reid JL. Assessment of the interaction between mianserin and centrally-acting antihypertensive drugs. Br J Clin Pharmacol 1983;15(Suppl 2):S323–8.

Mirtazapine

General Information

The antidepressant mirtazapine is pharmacologically similar to mianserin. It has a slightly weaker blocking action at α_1-adrenoceptors, and this is claimed to give it a dual mode of action, increasing the release of both noradrenaline and serotonin (1). It is claimed to have a more rapid onset of action than some SSRIs, but this may be due to its prominent sedative effect, which improves insomnia from the start of treatment.

Mirtazapine is extensively metabolized in the liver, mainly by CYP1A2, CYP2D6, and CYP3A4. With once-daily dosing, steady-state concentrations are reached after 4 days in adults and 6 days in elderly people.

In major depression, the efficacy of mirtazapine is comparable to that of amitriptyline, clomipramine, doxepin, fluoxetine, paroxetine, citalopram, and venlafaxine.

In placebo-controlled trials the most common adverse effects of mirtazapine were dry mouth (25%), drowsiness (23%), increased appetite (11%), and weight gain (10%) (SEDA 21, 13).

Organs and Systems

Nervous system

Mirtazapine is a potent histamine H_1 receptor antagonist and is perceived by patients as being sedative. While this may improve disturbed sleep, it also has implications for daytime psychomotor performance. The effect of mirtazapine (15 mg at night for 2 days, then 30 mg at night for 2 days) and paroxetine (20 mg in the morning for 5 days) has been studied in a placebo-controlled, crossover design in 12 healthy volunteers (10 women, 2 men, median age 26 years) (2). On the fifth day the subjects undertook a battery of psychomotor tests. Mirtazapine significantly impaired several aspects of psychomotor performance and increased daytime sleepiness, while paroxetine did not. These results suggest that patients who take mirtazapine should be warned about its deleterious effects on psychomotor performance, particularly driving. It is likely that tolerance to this effect will occur, but the time course of adaptation is uncertain. Also uncertain is how far these results in healthy subjects can be extrapolated to depressed patients, many of whom will have performance deficits due to poor sleep, which could conceivably be helped by mirtazapine.

Psychological, psychiatric

Some patients appear to develop manic symptoms in response to some antidepressants but not to others. Psychosis was reportedly triggered by mirtazapine in a patient taking long-term levodopa (3) and in another patient hypomania occurred.

- A 45-year-old woman had a long history of dysthymia and depression. She had taken many antidepressants, including tricyclics, SSRIs, amfebutamone, and venla-faxine. She had no history of mania or hypomania. She took sertraline 250 mg/day, with only a transient response, and mirtazapine 15 mg/day was added. Within 4 days she developed clear symptoms of hypomania, with euphoric mood, mild grandiosity, pressure of speech, increased energy, and a reduced need for sleep. Mirtazapine was withdrawn and sertraline continued; within 3 days the hypomanic symptoms had remitted. The depressive disorder then re-emerged (4).

Unlike most other antidepressants, mirtazapine does not inhibit the reuptake of monoamines, but instead blocks inhibitory α_2-adrenoceptors. It is conceivable that this pharmacological mechanism led to hypomanic symptoms in this patient.

Hematologic

Mianserin, an analogue of mirtazapine, has been associated with an increased risk of agranulocytosis.

- A 44-year-old woman with major depression was given mirtazapine 30 mg/day. She had a normal white blood cell count (6.8×10^9/l) (5). After 3 weeks she complained of a sore throat, difficulty in swallowing, and an aphthous ulcer in the mouth. Her temperature was slightly raised ($37.7°C$) and her white cell count was 2.2×10^9/l, with a granulocyte count of 1.1×10^9/l. Mirtazapine was withdrawn and she was given sultamicillin 375 mg bd. Within 2 weeks her white cell count had risen to 3.8×10^9/l, with a granulocyte count of 3.2×10^9/l. Four weeks later she was treated successfully with sertraline while her white count continued to rise.

The neutropenia in this case was consistent with an effect of mirtazapine. This is presumably an uncommon adverse effect, but this case reinforces advice given in the *British National Formulary* that patients taking mirtazapine should be advised to report any sore throat, fever, or stomatitis.

During clinical trials two of 2796 patients (0.08%) developed neutropenia, but both recovered after withdrawal. No other cases were detected in postmarketing surveillance of 13 500 patients in the Netherlands (SEDA-21, 14).

Liver

Mirtazapine has been associated with increases in liver enzymes (SEDA-21, 14).

Drug–Drug Interactions

Clonidine and apraclonidine

Hypertension has been reported with mirtazapine plus clonidine (6).

- A 20-year-old man had had Goodpasture's syndrome for 2.5 years, end-stage renal disease on chronic hemodialysis for 15 months, and hypertension controlled with metoprolol, losartan, and clonidine. He developed dyspnea and hypertension (blood pressure 178/115 mmHg) 2 weeks after his psychiatrist first gave him mirtazapine 15 mg at bedtime to treat depression. His blood pressure did not fall significantly, despite the addition of losartan and minoxidil and the use of intravenous glyceryl trinitrate and labetalol. Only after emergency dialysis and intravenous nitroprusside did his blood pressure fall to 150–180/80–100 mmHg. When mirtazapine was withdrawn, his blood pressure was controlled with minoxidil 5 mg, clonidine 0.1 mg, and metoprolol 10 mg, all bd.

The authors recognized that mirtazapine alone could have caused the hypertensive event. In postmarketing surveillance of mirtazapine, hypertension occurred in at least 1% of patients. However, it is likely that the patient lost antihypertensive control because mirtazapine antagonized the antihypertensive effect of clonidine. Mirtazapine, a tetracyclic antidepressant, stimulates the noradrenergic system through antagonism at central alpha$_2$ inhibitory receptors, which is precisely opposite to the effect of clonidine.

Fluoxetine

In patients who do not respond to SSRI treatment, the addition of mirtazapine is an increasingly popular option. Mirtazapine is an α_2-adrenoceptor antagonist, so its acute pharmacological effects predominantly involve potentiation of noradrenergic function. This might be expected to enhance the serotonergic actions of SSRIs.

- A 48-year-old woman unresponsive to fluoxetine (20–40 mg/day for 8 weeks) was given additional mirtazapine (30 mg/day) (7). Over the next month her depression remitted, but after taking the combined treatment for 7 weeks she slept less, spent more, and was more sociable than usual. She was also argumentative and agitated. Mirtazapine and fluoxetine were withdrawn and valproate introduced as a mood stabilizer. Within 2 weeks her mood had settled but she subsequently became depressed again.

The manic episode here could have been due to mirtazapine alone or to the combination of fluoxetine and mirtazapine. The case also demonstrates the common clinical observation that when an antidepressant treatment has apparently caused mania, withdrawal of antidepressant therapy often results in re-emergence of depressive symptoms, even when mood stabilizer treatment has been introduced.

Lithium

In a placebo-controlled, crossover study in 12 healthy men, lithium and mirtazapine had no effect on the pharmacokinetics of each other and there was no difference in psychometric testing between the addition of lithium and placebo (8).

References

1. Anttila SA, Leinonen EV. A review of the pharmacological and clinical profile of mirtazapine. CNS Drug Rev 2001;7(3):249–64.
2. Ridout F, Meadows R, Johnsen S, Hindmarch I. A placebo controlled investigation in to the effects of paroxetine and mirtazapine on measures related to a car driving performance. Human Psychopharmacol 2003;18:261–9.
3. Normann C, Hesslinger B, Frauenknecht S, Berger M, Walden J. Psychosis during chronic levodopa therapy triggered by the new antidepressive drug mirtazapine. Pharmacopsychiatry 1997;30(6):263–5.
4. Soutullo CA, McElroy SL, Keck PE Jr. Hypomania associated with mirtazapine augmentation of sertraline. J Clin Psychiatry 1998;59(6):320.
5. Ozcanli T, Unsalver B, Ozdemir S, Ozmen M, Sertraline and mirtazapine-induced severe neutropenia, Am J Psychiatry 2005;162:1386.
6. Abo-Zena RA, Bobek MB, Dweik RA. Hypertensive urgency induced by an interaction of mirtazapine and clonidine. Pharmacotherapy 2000;20(4):476–8.
7. Ng B. Mania associated with mirtazapine augmentation of fluoxetine. Depress Anxiety 2002;15(1):46–7.
8. Sitsen JM, Voortman G, Timmer CJ. Pharmacokinetics of mirtazapine and lithium in healthy male subjects. J Psychopharmacol 2000;14(2):172–6.

Nefazodone

General Information

Nefazodone is a phenylpiperazine antidepressant, structurally related to trazodone, but it has less alpha$_1$-adrenoceptor antagonist activity and is therefore less sedative. It is a potent 5-HT$_2$ receptor antagonist at postsynaptic receptors, and a weak serotonin reuptake inhibitor. In trials it was more effective than placebo and comparable to imipramine.

Commonly reported adverse effects are dry mouth, weakness, somnolence, nausea, dizziness, constipation, light-headedness, amblyopia, and blurred vision (1,2).

Organs and Systems

Nervous system

Nefazodone is unusual amongst antidepressants in that it does not reduce and in fact may increase rapid eye movement (REM) sleep or dream sleep. In a controlled study in depressed patients nefazodone 400 mg/day increased REM sleep, while fluoxetine 20 mg/day produced the opposite effect (3). In addition, nefazodone increased sleep efficiency, while fluoxetine reduced it. The patients' subjective assessments of sleep improved more with nefazodone than with fluoxetine. However, the overall antidepressant effects of the drugs were similar. The function of REM sleep and its relation to depressive disorders is not clear, so it is uncertain whether preservation of REM sleep by nefazodone is likely to be of clinical consequence.

Three men (aged 28–63 years) had troublesome burning sensations, not clearly localized to any one area of the body, 3 days to 3 weeks after starting to take nefazodone (4). The episodes lasted for about 30 minutes and recurred several times each day. The symptoms responded to nefazodone withdrawal or dosage reduction. These unpleasant sensations appeared to be linked to nefazodone, but the mechanism was obscure.

Psychological, psychiatric

- A 55-year-old woman with no previous history of bipolar disorder became manic while taking nefazodone (5).

Metabolism

Hypoglycemia occurred in a 54-year-old woman with diabetes mellitus after she started to take nefazodone (6). She also reported weight loss of 5 pounds.

Liver

There have been several reports of serious hepatotoxicity associated with nefazodone, in some cases leading to liver failure.

- A 27-year-old man developed hepatitis (with raised bilirubin and liver enzymes) without overt jaundice after taking nefazodone 200 mg/day for 12 weeks (7). No other cause for the hepatitis could be established and the abnormal liver function tests settled 4 weeks after nefazodone withdrawal. They became abnormal again 1 week after nefazodone rechallenge and settled once again on withdrawal.
- Four women, aged 16–73 years, developed catastrophic liver failure while taking nefazodone (8,9). Nefazodone was given for 14–28 weeks before the onset of symptoms. The cases had similar histological appearances, with prominent centrilobular necrosis. Three died, one survived after liver transplantation, and one improved sufficiently to obviate the need for transplantation. The duration of nefazodone treatment before the onset of symptoms was 7–28 weeks. The patients had no history of liver disease and other causes were excluded as far as possible.
- An 80-year-old man developed acute fatal liver failure while taking nefazodone 50 mg/day (8).
- A 52-year-old man with a 10-day history of fatigue and jaundice had been taking nefazodone (300 mg/day) for depression for about 6 weeks. Biochemical investigations showed acute liver failure. Infective hepatitis and immune disorders were excluded. He failed to respond to medical treatment, and hepatic transplantation was performed. Histological examination of the liver showed parenchymal necrosis, particularly in centrilobular areas, together with lymphocytic infiltration (10).
- A 46-year-old woman developed fatigue and jaundice about 20 weeks after she started to take nefazodone (300 mg/day). She had raised liver enzymes and bilirubin concentrations. There was no evidence of infectious hepatitis or immune disorders. Liver biopsy showed ballooning degeneration and necrosis of

hepatocytes with mixed inflammatory infiltrates. The nefazodone was withdrawn and corticosteroid treatment started. Within 4 months she recovered clinically and her liver function tests returned to normal (11).

These severe reactions, although rare, cause concern.

The cumulative incidence of hepatic adverse reactions associated with antidepressant treatment has been estimated through spontaneous reports to the Spanish Pharmacovigilance System (12). For classical tricyclic antidepressants and SSRIs the estimated rate of adverse hepatic reactions was 1.28–4.00 per 100 000 patient years. However, the rate with nefazodone was much higher (29 per 100 000 patient years). This report supports concerns that nefazodone may be more hepatotoxic than other antidepressants. Significant hepatic reactions to nefazodone are relatively rare but can be serious.

Hair

Nefazodone was associated with hair loss in a 40-year-old woman after dose escalation from 100 to 200 mg (13).

Sexual function

Sexual adverse effects seem to occur less often with nefazodone than with SSRIs (SEDA 20, 9).

- A 48-year-old woman developed clitoral priapism on the first day of nefazodone treatment (100 mg) (14). The problem persisted for another day, after which she discontinued the nefazodone and the priapism subsided.

Clitoral priapism has been reported with other antidepressants, including SSRIs both alone and in combination with trazodone. In men, trazodone rather than nefazodone is occasionally associated with priapism (SEDA-21, 14).

Long-Term Effects

Drug withdrawal

Discontinuation of SSRIs causes significant withdrawal symptoms (SEDA-22, 12). Withdrawal reactions have been less often described after discontinuation of nefazodone.

- A 27-year-old man developed withdrawal symptoms (dizziness, nausea, vomiting, sweating, anxiety, insomnia, and restlessness) after nefazodone was abruptly withdrawn after 1 month (15).
- A 22-year-old woman gradually increased her dose of nefazodone to 400 mg/day over 3 weeks (16). One week later she abruptly discontinued treatment. About 48 hours later she began to have nausea, vomiting, fatigue, headache, myalgia, restlessness, dizziness, nightmares, emotional lability, and anxiety. These symptoms continued unabated for a further 48 hours, after which they remitted a few hours after nefazodone was restarted.

These symptoms are very similar to those of severe SSRI-induced withdrawal, as were those described in two further case reports (17,18). As with SSRIs, it seems prudent to withdraw nefazodone slowly and to be aware of the possibility of SSRI-like withdrawal symptoms.

Second-Generation Effects

Lactation

Nefazodone passes into the breast milk and has been reported to cause drowsiness and feeding problems in a breast-fed infant (19).

- The mother of a 7-week-old premature girl became depressed and was given nefazodone 300 mg/day. Two weeks later the infant became drowsy and lethargic and did not feed well. No medical condition was found to account for these symptoms and breast-feeding was stopped. The symptoms improved over the next 3 days. The concentrations of nefazodone in breast milk were about 10 times less than those in maternal plasma, and the total infant dose of nefazodone was calculated to be only 0.45% of the maternal dose.

In this case, nefazodone in the breast milk may have caused the infant's drowsiness and feeding problems, despite the fact that the transferred dose of nefazodone would have been very low, since the low gestational age of the infant might have been associated with impaired hepatic and renal clearance, making her susceptible to even very small quantities of nefazodone.

Drug Administration

Drug overdose

The outcomes of over 1300 cases of poisoning with nefazodone as a sole agent have been reported (20). Generally, the toxic effects were mild, consisting mainly of drowsiness, nausea, and dizziness. Clinical signs developed within 1–4 hours of ingestion and dissipated within the following 24 hours. The most serious toxic effect was hypotension, which occurred in 1.6% of cases. Bradycardia occurred in 1.4% of cases. Only one patient had a seizure and none required intubation. These data suggest that the toxicity of nefazodone in overdose is low. However, in suicide attempts, antidepressants are often ingested with other agents, particularly alcohol and other sedating drugs. This might increase the toxic effects of nefazodone, particularly with regard to respiratory depression.

Drug–Drug Interactions

Alprazolam

Nefazodone is a weak inhibitor of CYP2D6 but a potent inhibitor of CYP3A4 and it increases plasma concentrations of drugs that are substrates of CYP3A4, such as alprazolam, astemizole, carbamazepine, ciclosporin, cisapride, terfenadine, and triazolam.

Carbamazepine

In a controlled study in 12 healthy men the combination of carbamazepine and nefazodone led to increased plasma carbamazepine concentrations and substantially lowered plasma nefazodone concentrations (21). This interaction, which is due to inhibition of CYP3A4 by nefazodone, could lead to difficulties during treatment, since the risk of carbamazepine toxicity would have to be balanced against the loss of therapeutic effect of nefazodone. Except in special circumstances it is probably better to avoid the combination.

Clozapine

When nefazodone (400 mg/day for 3 weeks) was added to clozapine treatment in six patients with schizophrenia there was no significant change in plasma clozapine concentrations (22). This finding suggests that CYP3A4 does not play a major role in the metabolism of clozapine and that nefazodone may be a suitable adjunctive treatment for depression in clozapine-treated patients.

However, in another case the addition of nefazodone appeared to produce clozapine toxicity (23).

- A 40-year-old man had taken clozapine 450 mg/day and risperidone 6 mg/day for several years. Nefazodone 200 mg/day was added in an attempt to improve persistent negative symptoms, and after a week the dosage was increased to 300 mg/day. One week later, he reported anxiety and dizziness and was hypotensive. The combined concentrations of clozapine and its active metabolite norclozapine had increased from 309 ng/ml before nefazodone to 566 ng/ml. The nefazodone dosage was reduced to 200 mg/day and the anxiety, dizziness, and hypotension resolved over the next 7 days. At the same time plasma concentrations of clozapine and norclozapine fell to 370 ng/ml.

These results suggest that in some individuals CYP3A4 plays a significant role in the metabolism of clozapine and that the combination of nefazodone and clozapine should therefore be used with caution. It is possible that in this case concomitant treatment with risperidone may have increased the effect of nefazodone to reduce the clearance of clozapine.

HMG coenzyme-A reductase inhibitors

The antidepressant nefazodone increases the risk of rhabdomyolysis from statins (28).

Haloperidol

Hepatic enzyme inhibition by nefazodone can significantly raise concentrations and toxicity of haloperidol (29).

Lithium

In 12 healthy volunteers, there were no clinically significant alterations in blood concentrations of lithium or nefazodone and its metabolites when the drugs were co-administered (30). The addition of lithium for 6 weeks to nefazodone in 14 treatment-resistant patients produced no serious adverse effects and no dropouts (31). Lithium augmentation of nefazodone in 13 treatment-resistant depressed patients was associated with a variety of annoying adverse effects, but none led to treatment withdrawal (32).

Monoamine oxidase inhibitors

Co-administration of nefazodone with a MAO inhibitor or SSRI can cause the serotonin syndrome (33).

Selective serotonin re-uptake inhibitors (SSRIs)

Concentrations of nefazodone and its metabolites can be increased by fluoxetine and paroxetine (SEDA-20, 9). Combinations of serotonin agents produce serotonin toxicity, and a case of serotonin syndrome occurred when nefazodone (200 mg/day) was combined with fluoxetine (40 mg/day) in a 50-year-old man (24). The toxic symptoms settled 3 days after withdrawal of both antidepressants.

The serotonin syndrome can occur with therapeutic doses of SSRIs (see above), but it occurs most commonly when SSRIs are co-administered with other drugs that also potentiate serotonin function. Recent case reports have suggested that there is a risk of the serotonin syndrome when SSRIs are combined with nefazodone (34).

Statins

Nefazodone can cause myositis and rhabdomyolysis in patients taking pravastatin and simvastatin (25). The postulated mechanism involves inhibition of CYP3A4, leading to reduced clearance of the HMG-CoA reductase inhibitors and muscle toxicity.

Substrates of CYP3A4

Nefazodone is a weak inhibitor of CYP2D6, but a potent inhibitor of CYP3A4, and increases plasma concentrations of drugs that are substrates of CYP3A4, such as triazolam, alprazolam, ciclosporin, astemizole, cisapride, terfenadine, and carbamazepine (26). Co-administration with terfenadine, astemizole, or cisapride should be avoided, because of the risk of cardiac dysrhythmias (SEDA-20, 9).

Terfenadine

Drugs that inhibit CYP3A4 inhibit the clearance of terfenadine, an antihistamine that can prolong the QT_c interval. This can cause potentially dangerous interactions. In a double-blind, placebo-controlled study of the effect of nefazodone (600 mg/day for 1 week) on the pharmacokinetics of terfenadine (120 mg/day for 14 days) and another antihistamine, loratadine (20 mg/day for 14 days), in 67 healthy volunteers, nefazodone significantly reduced the clearance of terfenadine and prolonged the mean QT_c interval (27). In addition, nefazodone produced a similar but smaller decrease in the clearance of loratadine and combined treatment also significantly increased the QT_c interval. This effect of nefazodone on the clearance of terfenadine is predictable, as is the increase in QT_c interval. Loratadine is also partly metabolized by CYP2D6, which probably explains the lesser effect of nefazodone on loratadine clearance. Loratadine by itself does not increase the QT_c interval significantly, but cardiotoxicity may be a possibility when it is combined with nefazodone.

Triazolam

Nefazodone is a weak inhibitor of CYP2D6, but a potent inhibitor of CYP3A4, and increases plasma concentrations of drugs that are substrates of CYP3A4, such as triazolam (35,36).

Zopiclone

- An 86-year-old white woman taking nefazodone for depression started to take zopiclone for insomnia, but subsequently had morning drowsiness (37). The plasma concentration of zopiclone was measured 8 hours after administration on two occasions, during and after nefazodone therapy. After withdrawal of nefazodone, the plasma concentration of the S-enantiomer of zopiclone fell from 107 to 17 ng/ml, while the plasma concentration of the R-enantiomer fell from 21 to 1.5 ng/ml.

The substantial fall in plasma zopiclone concentration after withdrawal of nefazodone probably reflected a drug interaction due to inhibition of CYP3A4 by nefazodone (38). Despite the normally short half-life of zopiclone, the residual sedation initially observed in this case suggests that the interaction had clinical significance.

References

1. Fontaine R. Novel serotonergic mechanisms and clinical experience with nefazodone. Clin Neuropharmacol 1992;15(Suppl 1:Part A):99A.
2. Nemeroff CB. Evolutionary trends in the pharmacotherapeutic management of depression. J Clin Psychiatry 1994;55(Suppl):3–15.
3. Rush AJ, Armitage R, Gillin JC, Yonkers KA, Winokur A, Moldofsky H, Vogel GW, Kaplita SB, Fleming JB, Montplaisir J, Erman MK, Albala BJ, McQuade RD. Comparative effects of nefazodone and fluoxetine on sleep in outpatients with major depressive disorder. Biol Psychiatry 1998;44(1):3–14.
4. Lerner V, Matar MA, Polyakova I. Nefazodone-associated subjective complaints of burning sensations. J Clin Psychiatry 2000;61(3):216–7.
5. Dubin H, Spier S, Giannandrea P. Nefazodone-induced mania. Am J Psychiatry 1997;154(4):578–9.
6. Warnock JK, Biggs F. Nefazodone-induced hypoglycemia in a diabetic patient with major depression. Am J Psychiatry 1997;154(2):288–9.
7. Schrader GD, Roberts-Thompson IC. Adverse effect of nefazodone: hepatitis. Med J Aust 1999;170(9):452.
8. Lucena MI, Andrade RJ, Gomez-Outes A, Rubio M, Cabello MR. Acute liver failure after treatment with nefazodone. Dig Dis Sci 1999;44(12):2577–9.
9. Aranda-Michel J, Koehler A, Bejarano PA, Poulos JE, Luxon BA, Khan CM, Ee LC, Balistreri WF, Weber FL
Jr. Nefazodone-induced liver failure: report of three cases. Ann Intern Med 1999;130(4 Part 1):285–8.
10. Schirren CA, Baretton G. Nefazodone-induced acute liver failure. Am J Gastroenterol 2000;95(6):1596–7.
11. Eloubeidi MA, Gaede JT, Swaim MW. Reversible nefazodone-induced liver failure. Dig Dis Sci 2000;45(5):1036–8.
12. Khan AY, Preskorn SH. Increase in plasma levels of clozapine and norclozapine after administration of nefazodone. J Clin Psychiatry 2001;62(5):375–6.
13. Gupta S, Gilroy WR. Hair loss associated with nefazodone. J Fam Pract 1997;44:20–1.
14. Brodie-Meijer CC, Diemont WL, Buijs PJ. Nefazodone-induced clitoral priapism. Int Clin Psychopharmacol 1999;14(4):257–8.
15. Benazzi F. Nefazodone withdrawal symptoms. Can J Psychiatry 1998;43(2):194–5.
16. Rajagopalan M, Little J. Discontinuation symptoms with nefazodone. Aust NZ J Psychiatry 1999;33(4):594–7.
17. Lauber C. Nefazodone withdrawal syndrome. Can J Psychiatry 1999;44(3):285–6.
18. Kotlyar M, Golding M, Brewer ER, Carson SW. Possible nefazodone withdrawal syndrome. Am J Psychiatry 1999;156(7):1117.
19. Yapp P, Ilett KF, Kristensen JH, Hackett LP, Paech MJ, Rampono J. Drowsiness and poor feeding in a breast-fed infant: association with nefazodone and its metabolites. Ann Pharmacother 2000;34(11):1269–72.
20. Benson BE, Mathiason M, Dahl B, Smith K, Foley MM, Easom LA, Butler AY. Toxicities and outcomes associated with nefazodone poisoning. An analysis of 1,338 exposures. Am J Emerg Med 2000;18(5):587–92.
21. Laroudie C, Salazar DE, Cosson JP, Cheuvart B, Istin B, Girault J, Ingrand I, Decourt JP. Carbamazepine–nefazodone interaction in healthy subjects. J Clin Psychopharmacol 2000;20(1):46–53.
22. Taylor D, Bodani M, Hubbeling A, Murray R. The effect of nefazodone on clozapine plasma concentrations. Int Clin Psychopharmacol 1999;14(3):185–7.
23. Khan AY, Preskorn SH. Increase in plasma levels of clozapine and norclozapine after administration of nefazodone. J Clin Psychiatry 2001;62:375–6.
24. Smith DL, Wenegrat BG. A case report of serotonin syndrome associated with combined nefazodone and fluoxetine. J Clin Psychiatry 2000;61(2):146.
25. Alderman CP. Possible interaction between nefazodone and pravastatin. Ann Pharmacother 1999;33(7–8):871.
26. Wright DH, Lake KD, Bruhn PS, Emery RW Jr. Nefazodone and cyclosporine drug–drug interaction. J Heart Lung Transplant 1999;18(9):913–5.
27. Abernethy DR, Barbey JT, Franc J, Brown KS, Feirrera I, Ford N, Salazar DE. Loratadine and terfenadine interaction with nefazodone: both antihistamines are associated with QTc prolongation. Clin Pharmacol Ther 2001;69(3):96–103.
28. Jacobson RH, Wang P, Glueck CJ. Myositis and rhabdomyolysis associated with concurrent use of simvastatin and nefazodone. JAMA 1997;277(4):296–7.
29. Kudo S, Ishizaki T. Pharmacokinetics of haloperidol: an update. Clin Pharmacokinet 1999;37(6):435–56.
30. Laroudie C, Salazar DE, Cosson JP, Cheuvart B, Istin B, Girault J, Ingrand I, Decourt JP. Pharmacokinetic evaluation of co-administration of nefazodone and lithium in healthy subjects. Eur J Clin Pharmacol 1999;54(12):923–8.
31. Hawley C, Sivakumaran T, Huber TJ, Ige AK. Combination therapy with nefazodone and lithium: safety and tolerability in fourteen patients. Int J Psychiatry Clin Pract 1998;2:251–4.

32. Hawley C, Sivakumaran T, Ochocki M, Ratnam S, Huber T. A preliminary safety study of combined therapy with nefazodone and lithium. Int J Neuropsychopharmacol 1999;2(Suppl 1):S30–1.

33. John L, Perreault MM, Tao T, Blew PG. Serotonin syndrome associated with nefazodone and paroxetine. Ann Emerg Med 1997;29(2):287–9.

34. Smith DL, Wenegrat BG. A case report of serotonin syndrome associated with combined nefazodone and fluoxetine. J Clin Psychiatry 2000;61(2):146.

35. Ereshefsky L, Riesenman C, Lam YW. Serotonin selective reuptake inhibitor drug interactions and the cytochrome P450 system. J Clin Psychiatry 1996;57(Suppl 8):17–25.

36. Nemeroff CB, DeVane CL, Pollock BG. Newer antidepressants and the cytochrome P450 system. Am J Psychiatry 1996;153(3):311–20.

37. Alderman CP, Gebauer MG, Gilbert AL, Condon JT. Possible interaction of zopiclone and nefazodone. Ann Pharmacother 2001;35(11):1378–80.

38. Nemeroff CB, DeVane CL, Pollock BG. Newer antidepressants and the cytochrome P450 system. Am J Psychiatry 1996;153(3):311–20.

Nomifensine

General Information

Nomifensine was first introduced in 1977 and eventually became available in over 70 countries. There has been increasing evidence of problems with adverse effects since 1981 (SEDA-5, 19), when cases of "drug fever" were recorded (SEDA-10, 22); this vague term covers a multitude of sins, the origins of which were finally elucidated at almost the exact moment when nomifensine was approved for release in the USA. Finally, in January 1986, the manufacturers withdrew nomifensine worldwide, citing grave concern elicited particularly by cases of rapid acute hemolytic anemia with intravascular hemolysis (SEDA-10, 22) (1).

Reference

1. Hayes PE, Kristoff CA. Adverse reactions to five new antidepressants. Clin Pharm 1986;5(6):471–80.

Reboxetine

General Information

Reboxetine is a selective noradrenaline re-uptake inhibitor with low affinity for α-adrenoceptors and muscarinic receptors. In controlled trials the following adverse events occurred significantly more often with reboxetine than with placebo: dry mouth (27%), constipation (17%), increased sweating (14%), insomnia (14%), urinary hesitancy (5%), impotence (5%), tachycardia (5%), and vertigo (2%) (SEDA-21, 13).

Organs and Systems

Metabolism

Reboxetine does not cause weight gain during routine clinical use and indeed has been advocated as an adjunctive treatment in the management of olanzapine-induced weight gain.

- A 44-year-old woman with bipolar disorder took lamotrigine 100 mg/day and reboxetine 12 mg/day (1). She noted a loss of appetite but continued to eat three meals a day. However, over the next year her weight fell from 55 kg to 43 kg. There seemed to be no psychiatric explanation for the weight loss and no medical cause could be discovered. The reboxetine and lamotrigine were withdrawn and her weight returned to baseline over the next 3 months. Around that time she again took reboxetine, this time as a sole agent, and once again her weight started to fall.

The loss of weight in this patient did not seem to be due her to psychiatric condition. The authors speculated that reboxetine might have some serotonergic activity, which could have accounted for a reduction in appetite and concomitant weight loss. However, drugs that potentiate noradrenaline activity can also reduce hunger, so this adverse effect could be due to the well characterized noradrenergic effects of reboxetine.

Electrolyte balance

Hyponatremia has been reported with reboxetine (2,3).

- A 72-year-old man with diabetes mellitus and cardiovascular disease developed major depression. He was taking aspirin (100 mg/day), enalapril (20 mg/day), and glibenclamide (5 mg/day). His serum sodium was 133 mmol/l (reference range 134–146 mmol/l). He started to take reboxetine (4 mg/day) and after 8 days experienced malaise and nausea, at which time his serum sodium had fallen to 118 mmol/l. The reboxetine was withdrawn, and both his symptoms and the low serum sodium remitted over the next 6 days. Rechallenge with reboxetine produced a recurrence of both the low sodium and the accompanying symptoms.

It appears that, like SSRIs, reboxetine can cause sodium depletion in elderly people. However, in this case the contributions of concomitant general medical illness and its treatment were uncertain.

Gastrointestinal

In six children aged 6–15 years with co-morbid enuresis and attention deficit/hyperactivity disorder (ADHD), which had failed to respond to methylphenidate, reboxetine 4–8 mg/day for 6 weeks reduced the frequency of bedwetting from an average of five times a week to once a week (4). Reboxetine was generally well-tolerated, although three of the children reported *anorexia*. It is difficult to draw firm conclusions from this small open study. However, the effect of reboxetine on enuresis would be consistent with its inhibitory effect on micturition. If reboxetine does prove effective for this indication it would probably be safer than the standard drug management, which is imipramine. However, psychological approaches, such as behavior therapy, are first-line treatments.

Urinary tract

One of the adverse effects of reboxetine is difficulty in passing urine (SEDA-21, 13). Eight patients taking reboxetine (4–8 mg/day) had troublesome urinary hesitancy (5). They were successfully treated with tamsulosin (0.4 mg/day), and in two patients tamsulosin was withdrawn after 2 weeks without recurrence of the urinary symptoms. Reboxetine is a selective noradrenaline re-uptake inhibitor and may therefore produce urinary symptoms by activating α_1-adrenoceptors in the bladder, which tamsulosin would be expected to reverse. However, tamsulosin is also effective for urinary symptoms caused by other mechanisms, for example benign prostatic hyperplasia. Whether its apparent effectiveness in reboxetine-induced dysuria represents specific pharmacological antagonism is therefore uncertain.

Drug–Drug Interactions

Fluoxetine

In the pharmacological management of patients with treatment-resistant depression, it is a common strategy to combine a drug that selectively inhibits noradrenaline re-uptake (for example reboxetine) with one that selectively inhibits the re-uptake of serotonin (for example an SSRI). As well as the hoped-for pharmacodynamic interaction, a kinetic interaction can also occur, because of the effect of SSRIs on CYP450 enzymes. The effect of combined treatment with reboxetine (8 mg/day) and fluoxetine (20 mg/day) has been compared with each treatment given alone for 8 days in 30 healthy volunteers in a parallel design (6). There was no potentiation of adverse effects by the combination. Fluoxetine increased the plasma AUC of reboxetine by 20%, but this was not statistically significant. The authors suggested that the combination of fluoxetine and reboxetine should have minimal adverse impact in depressed patients. However, the major metabolite of fluoxetine, norfluoxetine, is also an inhibitor of CYP3A4 and would not have reached steady-state concentrations during the time of the study. This suggests that caution might still be needed during longer-term use of this combination in depressed patients.

Ketoconazole

Reboxetine is metabolized by CYP3A4. In 11 healthy volunteers ketoconazole, an inhibitor of CYP3A4, increased the plasma AUC of reboxetine by about 50% and prolonged the half-life (7). The adverse effects profile of reboxetine was not altered by ketoconazole, but the finding suggests that reboxetine should be used with caution in combination with drugs that inhibit CYP3A4, for example nefazodone and fluvoxamine.

References

1. Lu T Y-T, Kupa A, Easterbrook G, Mangoni AA. Profound weight loss associated with reboxetine use in a 44 year old woman. Br J Clin Pharmacol 2005;60:218–20.
2. Ranieri P, Franzoni S, Trabucchi M. Reboxetine and hyponatremia. N Engl J Med 2000;342(3):215–6.
3. Schwartz GE, Veith J. Reboxetine and hyponatremia. N Engl J Med 2000;342:216.
4. Toren P, Ratner S, Laor N, Lerer-Amisar D, Weizman A. A possible antienuretic effect of reboxetine in children and adolescents with attention deficit/hyperactivity disorder: case series. Neuropsychobiol 2005;51:239–42.
5. Kasper S, Wolf R. Successful treatment of reboxetine-induced urinary hesitancy with tamsulosin. Eur Neuropsychopharmacol 2002;12(2):119–22.
6. Fleishaker JC, Herman BD, Pearson LK, Ionita A, Mucci M. Evaluation of the potential pharmacokinetic/pharmacodynamic interaction between fluoxetine and reboxetine in healthy volunteers. Clin Drug Invest 1999;18:141–50.
7. Herman BD, Fleishaker JC, Brown MT. Ketoconazole inhibits the clearance of the enantiomers of the antidepressant reboxetine in humans. Clin Pharmacol Ther 1999;66(4):374–9.

Trazodone

General Information

Trazodone (SEDA-7, 19–21) is a triazolopyridine derivative that selectively but weakly inhibits 5-hydroxytryptamine (5-HT) re-uptake and is an alpha-adrenoceptor antagonist at both presynaptic and postsynaptic receptors (1). During long-term administration it down-regulates serotonin receptors (2).

Experience with therapeutic and toxic dose ranges of trazodone has thrown much doubt on earlier claims for its greater safety, and some unanticipated problems have surfaced.

Trazodone is rapidly absorbed, with peak plasma concentrations at 0.5–2 hours. It is extensively metabolized and eliminated mainly via the kidneys, with a half-life of 13 hours for total drug and metabolites.

A review of controlled clinical studies in depressed patients in Europe and the USA showed that trazodone is superior to placebo (3) and can be as effective as imipramine or amitriptyline (3,4). The doses needed to secure an equivalent effect (150–800 mg/day) were generally double those of the tricyclic antidepressants.

In 28 depressed patients treated with up to 600 mg/day for 4 weeks, plasma concentrations varied eightfold (5). There was a significant correlation between dosage and plasma concentration, but no association with outcome and no significant difference between the mean plasma concentrations in responders (1.51 µg/ml) and non-responders (1.64 µg/ml).

The short half-life of trazodone naturally raises the question of the optimal dosage frequency. The General Practitioner Research Group in Britain (6) has conducted a series of comparisons of trazodone given once, twice, and thrice a day to anxious and depressed patients taking up to 200 mg/day. Efficacy was the same with all three regimens, although there were more complaints of dizziness and fewer of drowsiness with the thrice-daily regimen.

Organs and Systems

Cardiovascular

Based on animal research and restricted experience in overdosage (SEDA-7, 19–21), early attempts to differentiate trazodone from tricyclic antidepressants suggested that it might be relatively free of cardiotoxic effects. However, a preliminary report of a study of the effects of trazodone on the cardiovascular system in 20 subjects mentioned two patients who had ventricular dysrhythmias (7). Others have reported ventricular tachycardia (8–10), atrial fibrillation (11), and complete heart block (12).

Additive hypotensive effects of trazodone and phenothiazines have been reported (13).

Trazodone can cause peripheral edema, as outlined in a report of 10 cases (14).

Nervous system

Trazodone is markedly sedative, very similar in profile to amitriptyline. It produces marked sedation and lethargy within an hour, lasting up to 6 hours, and is about half as potent as imipramine (4).

Although it is claimed to be relatively free of effects usually attributed to anticholinergic activity, trazodone does cause complaints of dry mouth and blurred vision, which may be mediated through its actions on alpha-adrenoceptors. Urinary retention has been reported in a 69-year-old woman taking a combination of trazodone and an anticholinergic drug (isopropamide iodide) (15). Cholinergic overactivity has also been described in two patients after withdrawal (16).

The manufacturers of trazodone received 30 unpublished reports of seizures in patients, most of whom had evidence of seizure predisposition or concurrent contributory conditions. Two reports are detailed here (17,18).

- A 50-year-old woman with an electroencephalographic abnormality but no history of epilepsy, who had taken amitriptyline and perphenazine uneventfully for years, suffered two seizures 18 days after switching to trazodone 50 mg/day.
- A 47-year-old man developed 30-second episodes of facial contortions, aphasia, garbled speech, neologism, nocturnal episodes of deep breathing, swallowing, and incomprehensible speech after taking trazodone 150 mg/day for 3 weeks. On withdrawal of trazodone, his symptoms abated, but an electroencephalogram showed a left anterior lobe spike. He was treated

with carbamazepine and within 6 months was not depressed and had no further convulsive symptoms.

Sensory systems

Three cases of palinopsia (the persistence or reappearance of an image of a recently viewed object) have been reported (SEDA-17, 22); the authors speculated that this may have been due to pharmacological effects resembling those of LSD and mescaline.

Psychological, psychiatric

Conversion to mania has been reported in patients with unipolar depression (19,20) and bipolar illness (20).

Delirium, which also occurs with tricyclic antidepressants, has been reported with trazodone (21).

Liver

Hepatotoxicity, a known hazard of tricyclic compounds, has been reported with trazodone (22,23).

Skin

Trazodone has caused generalized erythematous maculo-papular eruptions (24), erythema multiforme (although the patient was also taking lithium) (25), and generalized pustular psoriasis in a patient who had had stable plaque psoriasis for 19 years (26).

Sexual function

Inhibition of ejaculation was reported in a middle-aged man 1 week after he started taking trazodone 100 mg at night (27). The symptoms abated 3 days after withdrawal and did not return when treatment was changed to doxepin 50 mg/day. Because trazodone has relatively more alpha-adrenoceptor blocking action and less anticholinergic activity than doxepin, the author speculated that this was the mechanism.

Trazodone can cause severe and persistent priapism. In a communication from the manufacturers in 1987, it was noted that there had been 136 reports of "increased penile tumescence" in patients taking trazodone. These included all reports of abnormal erectile activity. The company reported that the incidence of all abnormal erectile activity was about one in 6000 men, that in 36 of the 136 patients surgical interventions had been performed, and that at least nine of these patients were impotent. They suggested that early intervention is the treatment of choice, and that trazodone should be withdrawn immediately at the first signs of prolonged erections. They recommended that if conservative measures fail, intracavernosal irrigation with a weak solution of adrenaline should be considered (28). This rare adverse effect has been reported previously with a number of psychoactive and antihypertensive drugs (phenothiazines, guanethidine, prazosin, hydralazine) and has been ascribed to their alpha-adrenoceptor antagonist activity. The same mechanism probably explains the effect of trazodone.

Increased libido (29) and anorgasmia (30) have been reported with trazodone.

Second-Generation Effects

Teratogenicity

The use of antidepressants in pregnancy is a difficult area, because the teratogenic effects of many antidepressants are not known. While tricyclic antidepressants and SSRIs are believed to present a low risk of fetal abnormality (SEDA-24, 15), the teratogenic potential of trazodone has been unclear. In 147 women who took either trazodone or nefazodone during the first trimester of pregnancy, the rate of major malformations in the trazodone/nefazodone group (1.6%) did not differ statistically from that in a control group (3.0%) (31). Similarly, there were no statistically significant differences in rates of spontaneous abortion or stillbirth or in birth weight in the babies of women who took trazodone or nefazodone. These findings are reassuring. However, as the authors pointed out, the study, while prospective, was not randomized. In addition they did not examine the effects of trazodone and nefazodone separately, and did not state how many women were taking each drug. While trazodone and nefazodone are pharmacologically similar, they are not identical, and nefazodone has recently been withdrawn in the UK because of a rare association with severe liver disease (SEDA-26, 16).

Drug Administration

Drug overdose

In a 1980 review there were 68 cases of trazodone overdosage in amounts ranging up to 5 g (12 times the maximum therapeutic dose) (32). The predominant symptoms were drowsiness, dizziness, and rarely coma. There were no deaths in patients who took trazodone alone, and only two in those who took it in combination with other potentially lethal drugs. A brief review of data obtained from 46 cases of trazodone overdose reported to the National Poisons Information Service in London showed that 25 of the patients took overdoses of trazodone alone, while 21 took trazodone in combination with other drugs (33). This retrospective questionnaire study had inherent methodological flaws, but it did lend credence to the belief that trazodone is safe in overdose.

Trazodone is less cardiotoxic than tricyclic antidepressants, although it has rarely been reported to cause ventricular tachycardia. QT interval prolongation has been reported in overdose (34).

- A patient who took an overdose of trazodone (3 g) had sinus bradycardia (57 beats/minute) and a prolonged QT_c interval (60 msec). The abnormal QT_c interval gradually normalized over the next 3 days with supportive hospital treatment.

Drug–Drug Interactions

Amiodarone

A patient taking both trazodone and amiodarone developed prolongation of the QT interval and a polymorphous ventricular tachycardia, perhaps by mutual potentiation (35).

Carbamazepine

In a 53-year-old man, trazodone produced a clinically significant increase in carbamazepine concentrations (36). This suggests that trazodone inhibits CYP3A4. It should be used with caution in combination with carbamazepine.

MAOIs and SSRIs

Although trazodone is related to nefazodone it probably potentiates 5-HT neurotransmission less. Trazodone is often added to monoamine oxidase inhibitors and serotonin re-uptake inhibitors at low doses (50–150 mg/day) as a hypnotic. In one case the combination of trazodone with nefazodone provoked the serotonin syndrome (37).

- When a 60-year-old woman taking nefazodone 500 mg/day also took trazodone 50 mg/day she developed confusion, restlessness, sweating, and nausea after 3 days. Her symptoms settled quickly on withdrawal of both antidepressants.

Trazodone is commonly used as a hypnotic in patients taking non-sedating antidepressants, particularly SSRIs (38). It is believed to have a fairly wide safety margin, but the effects of SSRIs on its pharmacokinetics have not been widely studied. In 97 patients, mean age 40 years, 40 of whom took trazodone as monotherapy, 41 trazodone + citalopram, and 16 trazodone + fluoxetine, there were no differences in plasma concentrations of trazodone between the three treatment groups and no significant adverse effects as a result of the combinations (39). The authors concluded that trazodone has a wide safety margin in combination with SSRIs, perhaps because it is metabolized mainly by CYP3A4. However, if that is the case, combination with fluvoxamine may be more problematic (see SSRIs). In addition, the active metabolite of trazodone, m-chlorophenylpiperazine, a 5HT receptor agonist, is a substrate for CYP2D6, which raises the possibility that its concentrations may be increased when trazodone is combined with fluoxetine or paroxetine. In addition, there have been occasional case reports of signs of 5HT toxicity when trazodone has been combined with SSRIs.

References

1. Riblet LA, Taylor DP. Pharmacology and neurochemistry of trazodone. J Clin Psychopharmacol 1981;1(Suppl):175.
2. Subhash MN, Srinivas BN, Vinod KY. Alterations in 5-HT (1A) receptors and adenylyl cyclase response by trazodone in regions of rat brain. Life Sci 2002;71(13):1559–67.
3. Davis JM, Vogel C. Efficacy of trazodone: data from European and United States studies. J Clin Psychopharmacol 1981;1(Suppl.):275.
4. Brogden RN, Heel RC, Speight TM, Avery GS. Trazodone: a review of its pharmacological properties and therapeutic use in depression and anxiety. Drugs 1981;21(6):401–29.

5. Mann JJ, Georgotas A, Newton R, Gershon S. A controlled study of trazodone, imipramine, and placebo in outpatients with endogenous depression. J Clin Psychopharmacol 1981;1(2):75–80.

6. Wheatley D. Trazodone in depression. Int Pharmacopsychiatry 1980;15(4):240–6.

7. Janowsky D, Curtis G, Zisook S, Kuhn K, Resovsky K, Le Winter M. Trazodone-aggravated ventricular arrhythmias. J Clin Psychopharmacol 1983;3(6):372–6.

8. Vlay SC, Friedling S. Trazodone exacerbation of VT. Am Heart J 1983;106(3):604.

9. Vitullo RN, Wharton JM, Allen NB, Pritchett EL. Trazodone-related exercise-induced nonsustained ventricular tachycardia. Chest 1990;98(1):247–8.

10. Aronson MD, Hafez H. A case of trazodone-induced ventricular tachycardia. J Clin Psychiatry 1986;47(7):388–9.

11. White WB, Wong SH. Rapid atrial fibrillation associated with trazodone hydrochloride. Arch Gen Psychiatry 1985;42(4):424.

12. Rausch JL, Pavlinac DM, Newman PE. Complete heart block following a single dose of trazodone. Am J Psychiatry 1984;141(11):1472–3.

13. Asayesh K. Combination of trazodone and phenothiazines: a possible additive hypotensive effect. Can J Psychiatry 1986;31(9):857–8.

14. Barrnett J, Frances A, Kocsis J, Brown R, Mann JJ. Peripheral edema associated with trazodone: a report of ten cases. J Clin Psychopharmacol 1985;5(3):161–4.

15. Chan CH, Ruskiewicz RJ. Anticholinergic side effects of trazodone combined with another pharmacologic agent. Am J Psychiatry 1990;147(4):533.

16. Montalbetti DJ, Zis AP. Cholinergic rebound following trazodone withdrawal? J Clin Psychopharmacol 1988;8(1):73.

17. Lefkowitz D, Kilgo G, Lee S. Seizures and trazodone therapy. Arch Gen Psychiatry 1985;42(5):523.

18. Tasini M. Complex partial seizures in a patient receiving trazodone. J Clin Psychiatry 1986;47(6):318–9.

19. Warren M, Bick PA. Two case reports of trazodone-induced mania. Am J Psychiatry 1984;141(9):1103–4.

20. Knobler HY, Itzchaky S, Emanuel D, Mester R, Maizel S. Trazodone-induced mania. Br J Psychiatry 1986;149:787–9.

21. Damlouji NF, Ferguson JM. Trazodone-induced delirium in bulimic patients. Am J Psychiatry 1984;141(3):434–5.

22. Sheikh KH, Nies AS. Trazodone and intrahepatic cholestasis. Ann Intern Med 1983;99(4):572.

23. Chu AG, Gunsolly BL, Summers RW, Alexander B, McChesney C, Tanna VL. Trazodone and liver toxicity. Ann Intern Med 1983;99(1):128–9.

24. Rongioletti F, Rebora A. Drug eruption from trazodone. J Am Acad Dermatol 1986;14(2 Part 1):274–5.

25. Ford HE, Jenike MA. Erythema multiforme associated with trazodone therapy: case report. J Clin Psychiatry 1985;46(7):294–5.

26. Barth JH, Baker H. Generalized pustular psoriasis precipitated by trazodone in the treatment of depression. Br J Dermatol 1986;115(5):629–30.

27. Jones SD. Ejaculatory inhibition with trazodone. J Clin Psychopharmacol 1984;4(5):279–81.

28. Goldstein I, Payton TR. Pharmacologic detumescence – the alternative to surgical shunting. J Urol 1986;135:308A.

29. Gartrell N. Increased libido in women receiving trazodone. Am J Psychiatry 1986;143(6):781–2.

30. Garvey MJ, Tollefson GD. Occurrence of myoclonus in patients treated with cyclic antidepressants. Arch Gen Psychiatry 1987;44(3):269–72.

31. Einarson A, Bonari L, Voyer-Lavigne S, Addis A, Matsui D, Johnson Y, Koren G. A multicentre prospective controlled study to determine the safety of trazodone and nefazodone use during pregnancy. Can J Psychiatry 2003;48:106–10.

32. Faillace LA. Antidepressant therapy: risks and alternatives. Emerg Med Spec 1983;(Suppl):20.

33. Ali CJ, Henry JA. Trazodone overdosage: experience over 5 years. Neuropsychobiology 1986;15(Suppl 1):44–5.

34. Levenson JL. Prolonged QT interval after trazodone overdose. Am J Psychiatry 1999;156(6):969–70.

35. Mazur A, Strasberg B, Kusniec J, Sclarovsky S. QT prolongation and polymorphous ventricular tachycardia associated with trasodone-amiodarone combination. Int J Cardiol 1995;52(1):27–9.

36. Romero AS, Delgado RG, Pena MF. Interaction between trazodone and carbamazepine. Ann Pharmacother 1999;33(12):1370.

37. Margolese HC, Chouinard G. Serotonin syndrome from addition of low-dose trazodone to nefazodone. Am J Psychiatry 2000;157(6):1022.

38. Kaynak H, Kaynack D, Gozukirmizi E, Guilleminault C. The effects of trazodone on sleep in patients treated with stimulant antidepressants. Sleep Med 2004;5:15–20.

39. Propotnik M, Waschgler R, Konig P, Moll W, Conca A. Therapeutic drug monitoring of trazodone: are there pharmacokinetic interactions involving citalopram and fluoxetine? Int J Clin Pharmacol Ther 2004;42:120–4.

Tryptophan

General Information

The possibility that mental illness may be alleviated by biogenic amine precursors is an appealing one (SEDA-4, 18). Tryptophan is a naturally-occurring essential amino acid, which has been advocated as an innocuous health food for the treatment of depression, insomnia, stress, behavioral disorders, and premenstrual syndrome. The availability of amino acids in health food stores and a contemporary interest in natural remedies led to reported widespread use of tryptophan to treat depression. It was estimated in 1976 that up to that time several hundred patients with affective disorders had been studied, with results reported in at least 21 papers (1). However, the results of clinical trials with L-tryptophan in the treatment of depressive disorders are inconsistent (2).

It has been suggested that there may be some benefit of using tryptophan in selected patients, particularly those with psychomotor retardation (3). Unfortunately, most of these reports have appeared as letters to the editors of journals (4–6) or as preliminary communications (7). In addition to the possible absence of any consistent effect, there are many plausible reasons to explain the variability in response. Tryptophan has been given in both the racemic and monomeric (levorotatory) forms, both alone and together with a number of substances intended to increase the synthesis or availability of serotonin, including monoamine oxidase (MAO) inhibitors (8), potassium or carbohydrate supplements (9), and co-enzymes such as pyridoxine or ascorbic acid (10). It has also been suggested that tryptophan plasma concentrations have a therapeutic window (4), and that repeated administration induces hepatic tryptophan pyrrolase, resulting in lowered plasma concentrations and loss of therapeutic effect after 2 weeks of treatment (8). Attempts have been made to ameliorate this problem by coadministration of nicotinamide (4).

In addition to the difficulty of interpreting possible benefits due to tryptophan, there is a paucity of information on its adverse effects. This may be partly accounted for by the assumed safety of a natural substance, but it is also contributed to by the preliminary nature of many communications. In at least two studies (5,7) in which tryptophan was compared with a tricyclic antidepressant, inquiry about adverse effects was deliberately avoided, in order to protect the double-blind integrity of the study. Two studies have reported the lack of any consistent or definite changes in hematological values, serum electrolytes, plasma proteins, or liver function tests after 4 weeks of treatment with L-tryptophan up to 8 g/day (7,10). Nausea early in treatment (11), light-headedness, which does not appear to be related to postural hypotension (12), and deterioration in mental status (13,14) have been reported. Hypomania on combining tryptophan with a MAO inhibitor has been reported (14) and toxic effects, including muscle tremor, hypomanic mood, hyper-reflexia, and bilateral Babinski signs, were seen in a patient taking phenelzine and tryptophan.

Eosinophilia–myalgia syndrome

No severe or irreversible adverse effects of tryptophan were reported until 1989, when an eosinophilia–myalgia syndrome was described, and L-tryptophan-containing products were withdrawn from the market (SEDA-15, 518). This syndrome was characterized by an eosinophil count of at least $1 \times 10^9/l$ and intense generalized myalgia. Other relatively frequent signs and symptoms were fatigue, arthralgia, skin rash, cough and dyspnea, edema of the limbs, fever, scleroderma-like skin abnormalities, increased hair loss, xerostomia, neuropathy, and pneumonia or pneumonitis with or without pulmonary vasculitis. About one-third of the cases required hospitalization, and a substantial number of patients died. The syndrome is suspected to have been due to an unidentified impurity in products from one manufacturer (SEDA-18, 22).

The syndrome appears to be only part of a spectrum of adverse effects associated with tryptophan (15). There has been much discussion, but finally it appears that the links are causal, as consistent findings were found in multiple independently conducted studies and the incidence of eosinophilia–myalgia syndrome in the USA fell abruptly once tryptophan-containing products were recalled (16).

Although L-tryptophan was withdrawn in many countries, in 1994 it became available again in the UK for combination treatment of patients with long-standing refractory depression, on the strict condition that it should only be prescribed by hospital specialists for patients with long-standing resistant depression (SEDA-18, 22). It is also still in use in Canada but not in the USA.

Organs and Systems

Respiratory

A case of interstitial pneumonitis and pulmonary vasculitis was ascribed to L-tryptophan, and unintended rechallenge supported a causal relation (17).

References

1. Farkas T, Dunner DL, Fieve RR. L-tryptophan in depression. Biol Psychiatry 1976;11(3):95–302.
2. Mendels J, Stinnett JL, Burns D, Frazer A. Amine precursors and depression. Arch Gen Psychiatry 1975;32(1):22–30.
3. Cooper AJ. Tryptophan antidepressant "physiological sedative": fact or fancy? Psychopharmacology (Berl) 1979;61(1):97–102.
4. Chouinard G, Young SN, Annable L, Sourkes TL. Tryptophan dosage critical for its antidepressant effect. BMJ 1978;1(6124):1422.
5. Rao B, Broadhurst AD. Letter: Tryptophan and depression. BMJ 1976;1(6007):460.
6. Jensen K, Fruensgaard K, Ahlfors UG, Pihkanen TA, Tuomikoski S, Ose E, Dencker SJ, Lindberg D, Nagy A. Letter: Tryptophan/imipramine in depression. Lancet 1975;2(7941):920.
7. Herrington RN, Bruce A, Johnstone EC, Lader MH. Comparative trial of L-tryptophan and amitriptyline in depressive illness. Psychol Med 1976;6(4):673–8.
8. Coppen A, Shaw DM, Farrell JP. Potentiation of the antidepressive effect of a monoamine-oxidase inhibitor by tryptophan. Lancet 1963;1:79–81.
9. Coppen A, Shaw DM, Herzberg B, Maggs R. Tryptophan in the treatment of depression. Lancet 1967;2(7527):1178–80.
10. Herrington RN, Bruce A, Johnstone EC. Comparative trial of L-tryptophan and E.C.T. in severe depressive illness Lancet 1974;2(7883):731–4.
11. Broadhurst AD. L-tryptophan verses E.C.T Lancet 1970;1(7661):1392–3.
12. Carroll BJ, Mowbray RM, Davies B. Sequential comparison of L-tryptophan with E.C.T. in severe depression Lancet 1970;1(7654):967–9.
13. Murphy DL, Baker M, Goodwin FK, Miller H, Kotin J, Bunney WE Jr. L-tryptophan in affective disorders: indoleamine changes and differential clinical effects. Psychopharmacologia 1974;34(1):11–20.
14. Gayford JJ, Parker AL, Phillips EM, Rowsell AR. Whole blood 5-hydroxytryptamine during treatment of endogenous depressive illness. Br J Psychiatry 1973;122(570):597–8.
15. Varga J, Uitto J, Jimenez SA. The cause and pathogenesis of the eosinophilia-myalgia syndrome. Ann Intern Med 1992;116(2):140–7.
16. Kilbourne EM, Philen RM, Kamb ML, Falk H. Tryptophan produced by Showa Denko and epidemic eosinophilia-myalgia syndrome. J Rheumatol Suppl 1996;46:81–8.
17. Bogaerts Y, Van Renterghem D, Vanvuchelen J, Praet M, Michielssen P, Blaton V, Willemot JP. Interstitial pneumonitis and pulmonary vasculitis in a patient taking an L-tryptophan preparation. Eur Respir J 1991;4(8):1033–6.

Venlafaxine

General Information

Venlafaxine inhibits the re-uptake of both serotonin and noradrenaline. Trials have shown efficacy comparable to that of tricyclic antidepressants, and it may be particularly effective in refractory depression (1)(2).

Reported adverse effects were nausea, somnolence, dry mouth, insomnia, dizziness, constipation, weakness, nervousness, and sweating. Other adverse effects include nervousness, weight loss, and a dose-dependent increase in blood pressure.

Because of its serotonin-potentiating effect, venlafaxine can cause a serotonin syndrome when combined with monoamine oxidase (MAO) inhibitors (SEDA-20, 9) (SEDA-21, 13).

Adverse sexual effects are reported in a frequency similar to that of SSRIs (SEDA-20, 9).

Withdrawal reactions have been reported in several cases after abrupt discontinuation (SEDA-21, 13) (3)(4).

Preliminary data suggest that venlafaxine is safer in acute overdose than tricyclic antidepressants but more dangerous than SSRIs. Seizures have been reported in several cases of overdose (5–7).

Organs and Systems

Cardiovascular

A meta-analysis of the effect of venlafaxine on blood pressure in patients studied in randomized placebo-controlled trials of venlafaxine, imipramine, and placebo showed that at the end of the acute phase (6 weeks) the incidence of a sustained rise in supine diastolic blood pressure (over 90 mmHg) was significantly higher in both active treatment groups: venlafaxine 4.8% (135/2817), imipramine 4.7% (15/319), and placebo 2.1% (13/605) (8). The effect of venlafaxine in causing a rise in diastolic blood pressure appeared to be dose related, with an incidence of 1.7% in patients taking under 100 mg/day and 9.1% in those taking over 300 mg/day.

These data confirm that venlafaxine, particularly in higher dosages, can significantly increase blood pressure. At high doses, venlafaxine inhibits the re-uptake of noradrenaline as well as that of serotonin, which probably accounts for the pressor effect.

In physically healthy subjects venlafaxine has a generally benign cardiovascular profile, although hypotension and dose-related hypertension have been reported (SEDA-23, 20).

Venlafaxine is often used in high doses in patients with treatment-resistant depression. If there is continuing failure to respond, electroconvulsive therapy might be used, often in combination with venlafaxine. A 73-year-old woman taking venlafaxine (112.5 mg/day) had sustained hypertension for several hours after her first treatment (9). However, electroconvulsive therapy can cause transient hypertension, and the patient had essential hypertension controlled by bendroflumethiazide. It is therefore possible that the reaction might have occurred had she not been taking venlafaxine. Nevertheless, the fact that venlafaxine can cause hypertension when used as sole treatment suggests that blood pressure should be monitored carefully in patients receiving electroconvulsive therapy and venlafaxine together, particularly if there is a history or current evidence of hypertension.

Venlafaxine has not been studied systematically in patients with cardiovascular disease, although there are reports that older patients can have clinically significant disturbances of cardiac rhythm (10).

- A 69-year-old woman with stable angina and mild single-vessel coronary artery disease developed acute myocardial ischemia within a week of starting venlafaxine (75 mg/day).

Taken together with the information that the authors cited in their review, the current data suggest that venlafaxine should be used with caution in patients with established cardiovascular disease.

Respiratory

There has been a previous report linking venlafaxine to pneumonia associated with eosinophilic infiltration, and SSRIs have been associated with drug-induced infiltrative lung disease.

- A 21-year-old woman developed progressive dyspnea, a non-productive cough, weight loss, and syncope (11). She had been taking venlafaxine for depression for 2 months (75 mg/day for 1 month, then 37.5 mg/day). A chest X-ray showed diffuse reticulonodular opacities throughout both lung fields, and a CT scan showed numerous diffuse, ill-defined pulmonary nodules. Lung function tests showed a restrictive ventilatory defect, depression of gas transfer, and resting hypoxia. Histological examination showed a lymphocytic interstitial infiltrate. Venlafaxine was withdrawn and glucocorticoid treatment started. Her clinical condition improved rapidly over the next 2 weeks after which the glucocorticoid treatment was stopped. At 3 years she remained well.

It is difficult to be sure how far venlafaxine contributed to this presentation, but no other cause could be established and the patient improved quickly when venlafaxine was withdrawn. Venlafaxine is a potent 5HT re-uptake inhibitor and 5HT has been implicated in fibrotic reactions in a variety of tissues.

Nervous system

Antidepressant drugs, such as tricyclic antidepressants and SSRIs, suppress rapid eye movement (REM) sleep, the period of sleep during which dreaming occurs. Despite this, antidepressant treatment is sometimes associated with recurrent nightmares. Venlafaxine also suppresses REM sleep and was associated with nightmares in a 35-year-old woman with body image disturbance; the nightmares remitted when the venlafaxine was withdrawn (12). The authors speculated that indirect activation of 5-

HT_2 receptors might have played a causal role, because 5-HT_2 receptor antagonists, such as nefazodone, can be helpful in the treatment of nightmares, for example in patients with post-traumatic stress disorder.

The 5HT toxicity (serotonin) syndrome usually results from the combination of two drugs with potentiating effects on 5HT neurotransmission. However, it can occasionally result from therapeutic doses of a conventional 5HT re-uptake inhibitor antidepressant.

- A 29 year old woman developed anxiety, restlessness, shivering, diarrhea, nausea, and vomiting after taking venlafaxine 37.5 mg/day for 3 days (13). She also had ataxia and myoclonus. Her symptoms resolved after a few hours and treatment with prochlorperazine and lorazepam. Two weeks later she took fluoxetine 20 mg/day without adverse effects. This case shows how sensitive some patients can be to even low doses of 5HT potentiating drugs and also the rather puzzling fact that another drug, equally potent at facilitating 5HT neurotransmission, can then be taken without apparent adverse consequences.

Venlafaxine as monotherapy has been associated with both the serotonin syndrome and the neuroleptic malignant syndrome, which can be difficult to distinguish.

- A 40-year-old woman took venlafaxine 150 mg bd for a severe depressive disorder and after 12 days developed a tremor, posturing, confusion, muscular rigidity, a mask-like facies, dysphagia, hypertension, and tachycardia (14). The venlafaxine was withdrawn and she improved. The next day the venlafaxine was re-started and the symptoms returned more severely, this time accompanied by a pyrexia of 38°C and a dramatically raised creatine kinase activity (2097 U/l, reference range 0–190 U/l). A brain scan and cerebrospinal fluid were normal. There were no myoclonic movements and no sweating. Venlafaxine was again withdrawn and she recovered over the following week. She was then given mirtazapine without recurrence of the abnormal neurological state and her depression remitted within the next month

The authors described this as a case of serotonin syndrome rather than neuroleptic malignant syndrome, because no dopamine receptor blocking drug was involved. However, drugs other than antipsychotic agents have sometimes been implicated in neuroleptic malignant syndrome, and it is well established that drugs that potentiate serotonin can indirectly affect dopamine pathways. In this case the reported symptoms, including muscular rigidity, change in conscious state, hyperthermia, autonomic instability, and increased creatine kinase activity are more consistent with neuroleptic malignant syndrome than the serotonin syndrome.

Sensory systems

Tricyclic antidepressants predispose to acute angle closure glaucoma, probably through an anticholinergic action. However, acute angle closure has also been reported with SSRIs, suggesting that all serotonin-potentiating drugs can cause glaucoma (SEDA-21, 13) (SEDA-23, 17).

- Acute angle closure glaucoma has been reported in a 45-year-old woman taking venlafaxine (75 mg/day) and chlorpromazine (150 mg/day) (15). She had previously taken low-dose dosulepin (75 mg/day), chlorpromazine, and SSRIs without ophthalmic problems.

Chlorpromazine has mild anticholinergic properties and may have contributed to the angle closure in this case. She also had hypermetropia which is an another risk factor for glaucoma. However, this case supports the view that potent blockade of serotonin re-uptake can cause glaucoma in predisposed subjects.

Psychological, psychiatric

The use of antidepressants in patients with documented bipolar illness is often associated with a risk of manic illness (16).

- A 63-year-old man with a long history of bipolar illness and alcohol dependence took nefazodone 400 mg/day for 8 months, with no relief of a depressive episode. He was also taking valproate 1500 mg/day as a mood stabilizer, aspirin 81 mg/day, ranitidine 300 mg/day, and docusate calcium 240 mg/day. Nefazodone was tapered over 4 days and venlafaxine begun in a dosage of 37.5 mg/day and gradually increased to 150 mg/day over 3 weeks. Venlafaxine led to improvement in depressive symptoms, but 6 days after the dose reached 150 mg/day he became agitated, verbally and physically threatening, grandiose, and sexually disinhibited. He also had paranoid thinking and appeared to be hallucinating. The venlafaxine was withdrawn but the manic symptoms persisted and eventually required increases in the dosages of valproate and antipsychotic drugs.

This report illustrates the difficulty of treating depression in bipolar disorder with antidepressants. The presence of a mood stabilizer did not prevent the manic episode that emerged during venlafaxine treatment, and it is in any case difficult to know whether the mania was in fact due to the venlafaxine or instead represented a spontaneous mood swing.

Delusions of love (erotomania or De Clérambault's syndrome) is a rare but striking disorder.

- A 39-year-old woman with a history of treatment-resistant depression developed delusions that her medical attendants were in love with her on two separate occasions when taking venlafaxine in doses of 225 mg and more (17). There was no evidence of mania and no other psychotic symptoms. On both occasions the delusional beliefs subsided when venlafaxine was withdrawn. She was subsequently treated with another antidepressant and made a good recovery.

The absence of a history of psychosis and the re-emergence of delusional thinking when venlafaxine was prescribed again suggest that venlafaxine played a role in producing this psychotic state. At high doses venlafaxine potentiates dopamine activity, which could lead to psychotic reactions in predisposed individuals.

Endocrine

SSRIs can facilitate prolactin release and have been associated with galactorrhea (SEDA-26, 13). A similar effect would be expected with venlafaxine.

- A 38-year-old woman developed galactorrhea on two separate occasions while taking venlafaxine 225 mg/day and 75 mg/day (18). On the first occasion prolactin concentrations were modestly raised but on the second they were not.

This report confirms that, like SSRIs, venlafaxine can cause galactorrhea and also suggests, as has been observed with other drugs, that the symptom of lactation is not necessarily closely linked to plasma prolactin concentrations. This suggests that other mechanisms could be involved in the production of drug-induced galactorrhea.

Electrolyte balance

It is well established that SSRIs can cause hyponatremia. Similar cases have now been reported with venlafaxine.

- A 90-year-old woman had a depressive disorder in addition to other medical problems, including congestive cardiac failure, seizures, dementia, and osteoporosis (19). She was also taking phenobarbital 120 mg/day, enalapril 40 mg/day, furosemide 80 mg/day, calcium carbonate 1300 mg/day, and nortriptyline 50 mg at night. She was given additional paroxetine 20 mg/day, but there was no clinical response and she was changed to venlafaxine, increasing to 75 mg/day. Two months later her sodium concentration had fallen to 130 mmol/l and the furosemide was withdrawn. However, 4 months later the sodium concentration had fallen to 124 mmol/l. The venlafaxine was then withdrawn and the sodium normalized within a week and remained within the reference range for the following year.

This was a complex case, because of the multiple medical problems and treatments. However, it has been reported that since the launch of venlafaxine in Australia in 1996, the Adverse Drug Reactions Advisory Committee there has received 15 reports of hyponatremia, suggesting that the association is real (20).

Liver

Raised liver enzymes have been rarely found in patients taking venlafaxine, and it appears that hepatitis may also be rare adverse effect (21).

- A 44-year-old woman developed weakness with abnormal liver function tests (aspartate transaminase 661 IU/l) about 6 months after starting to take venlafaxine 150 mg/day. Biopsy showed confluent necrosis in zone 3, with unaffected portal tracts. No other cause for the hepatitis could be found. The clinical and biochemical features resolved within 12 weeks of withdrawal of venlafaxine.

Hair

SSRIs can cause occasional, idiosyncratic hair loss in some patients (SEDA-19, 10), and this has also been attributed to venlafaxine (22).

- A 50-year-old woman took venlafaxine 75 mg/day for depression, and the dosage was increased to 150 mg/day after 2 weeks. After 4 weeks she noted increased hair loss when brushing or washing her hair. After 3 months she stopped taking venlafaxine and 1 month later her hair loss stopped completely. However, 10 months later she had another depressive episode and was again successfully treated with venlafaxine. Once again she noted excessive hair loss. She was subsequently treated with sertraline 50 mg/day, which helped her depression without causing hair loss.

The fact that sertraline did not cause hair loss in this patient suggests that the mechanism was not related to blockade of serotonin re-uptake.

Sexual function

Antidepressant drugs can rarely cause priapism. The agent most often implicated has been trazodone, perhaps because of its α_1-adrenoceptor antagonist properties. In general venlafaxine has an inhibitory effect on sexual function, but perhaps, like SSRIs, it can rarely cause priapism (23).

- A 16-year-old youth taking venlafaxine (150 mg/day) had several episodes of prolonged erections, which persisted for several hours after intercourse. The episodes remitted after the venlafaxine was withdrawn.

Pain on ejaculation is also a rare adverse effect of tricyclic antidepressants and SSRIs and has been reported with venlafaxine (24).

- Painful ejaculation occurred in a 59-year-old man taking venlafaxine 150 mg/day. It remitted when the venlafaxine was withdrawn and did not recur when citalopram 40 mg/day was used instead.

Drugs that potentiate serotonin function can cause ejaculatory delay in men, and this has led to the use of SSRIs to treat premature ejaculation (SEDA-26, 13). Venlafaxine is also reported to cause problems with ejaculation during routine use and its efficacy has been studied in a placebo-controlled, crossover study in 31 men with ejaculation latencies of less than 2 minutes (25). Both placebo and venlafaxine (75 mg/day of the XL formulation) significantly increased latency to ejaculation over baseline, placebo by 2 minutes and venlafaxine by 3 minutes; there was no difference between the two treatments. The authors concluded that venlafaxine is not effective for the management of premature ejaculation. However, the small number of subjects studied and the large placebo effect makes this conclusion tentative. It does appear, however, that the effect of venlafaxine on ejaculation delay is probably less striking than, for example, that of paroxetine.

Long-Term Effects

Drug abuse

Misuse of antidepressants is unusual.

- A 38-year-old man without a previous history of substance misuse took up to 3600 mg of venlafaxine daily (about 10 times the maximum therapeutic dose) because it caused a subjective "high" (26). He maintained his high doses by obtaining illicit supplies of venlafaxine until a dose of 4050 mg produced acute central chest pain. He was then admitted to be treated for depression and substance misuse.

This patient developed features of drug dependence. Such reactions are unusual but might be related to the ability of venlafaxine to potentiate dopamine function at high doses. This pharmacological property characterizes many drugs that are misused for their euphoriant properties.

Drug withdrawal

Both SSRIs and venlafaxine can cause troublesome withdrawal symptoms (SEDA-22, 12) (SEDA-22, 17). Because venlafaxine blocks the re-uptake of both serotonin and noradrenaline the mechanism of venlafaxine-induced abstinence symptoms is not clear.

- A 72-year-old woman taking venlafaxine 150 mg/day for depression was abruptly switched to the noradrenaline re-uptake inhibitor maprotiline 75 mg/day; 1 day later she developed agitation, sweating, nausea, vomiting, tinnitus, and insomnia (27). These symptoms continued for another week, but disappeared on the second day of sertraline treatment 50 mg/day.

The symptoms experienced by this patient were typical of venlafaxine and SSRI withdrawal (although they could also be experienced after sudden withdrawal of tricyclic antidepressants). The fact that they were relieved by a serotonin but not a noradrenaline re-uptake inhibitor suggests that venlafaxine-induced withdrawal symptoms are mediated by serotonergic mechanisms.

Drug Administration

Drug overdose

Venlafaxine is generally regarded as being reasonably safe in overdose relative to conventional tricyclic antidepressants, but may be more toxic in overdose than SSRIs. In a survey of Coroners in England and Wales of antidepressant-related deaths between 1998 and 2000, the numbers of expected and observed deaths were computed for each antidepressant using prescription frequency data to yield a standardized mortality ratio (28). The highest mortality ratios, as expected, were found with the conventional tricyclic antidepressants, amitriptyline and dosulepin (1.8 and 1.7 respectively). The corresponding ratios for SSRIs were substantially lower (between 0.1 and 0.3). However, the value for venlafaxine was 1.6,

very similar to that of tricyclics. This study supports previous findings that suggested that venlafaxine may be significantly more toxic in overdose than SSRIs. However, these data need to be interpreted cautiously. Venlafaxine tends to be prescribed for patients with treatment-resistant depression. Such patients may be more likely to make suicide attempts and also to take their antidepressant medication in combination with other psychotropic drugs (for example, antipsychotic drugs), which may themselves be toxic in overdose. Further epidemiological data will be needed to resolve these uncertainties. In the meantime, the UK Committee on Safety of Medicines has restricted the prescription of venlafaxine to patients who have failed to respond to at least two other antidepressant agents and has required that a baseline electrocardiogram be carried out before treatment is begun. In addition, venlafaxine is now contraindicated in patients with evidence of cardiovascular disease or electrolyte imbalance (29), although the matter is still under review.

Venlafaxine has been associated with occasional reports of cardiac conduction disturbances at both therapeutic doses and in overdose (SEDA-24, 19).

Deaths have been described after venlafaxine overdose, but in combination with other agents and alcohol. However, there have been two fatal cases of overdosage in which venlafaxine was the only agent detected postmortem (30). It therefore appears that venlafaxine can occasionally prove fatal in overdosage, probably through cardiac conduction abnormalities and seizures (30,31). It is possible that poor metabolizers may be especially liable to develop toxic effects.

- A 44-year-old woman took an overdose of venlafaxine 3 g. An electrocardiogram showed sinus rhythm and incomplete right bundle branch block (32). She was monitored in an intensive care unit and 10 hours later a further electrocardiogram showed atrial fibrillation with a wide QRS complex. Both of these abnormalities resolved with sodium bicarbonate (100 ml of a 1 M solution). No further conduction disturbances were noted over the following days.

The authors suggested that the effect of venlafaxine on cardiac conduction is mediated by its ability to block the fast inward sodium current in cardiac myocytes. This might promote membrane stabilizing effects in a similar way to tricyclic antidepressants. They recommended that the management of venlafaxine overdose should include cardiac monitoring.

Drug–Drug Interactions

Alprazolam

The effects of venlafaxine on the pharmacokinetics of alprazolam have been investigated in 16 healthy volunteers. Steady-state venlafaxine 75 mg bd did not inhibit CYP3A4 metabolism of a single dose of alprazolam 2 mg (33).

Caffeine

Venlafaxine 150 mg/day did not alter the disposition of caffeine, a probe of CYP1A2 activity (34). This suggests that at therapeutic doses venlafaxine, in contrast to fluvoxamine (SEDA-22, 13), does not inhibit CYP1A2.

Clozapine

The effect of venlafaxine on clozapine concentrations has been studied in 11 in-patients with chronic schizophrenia (35). Clozapine concentrations were not altered by venlafaxine. This is consistent with relative of lack inhibition of CYP enzymes by venlafaxine; however the doses of venlafaxine were modest (maximum 150 mg/day) and it is possible that higher doses might alter clozapine concentrations.

Dexamfetamine

Psychostimulants, such as dexamfetamine, are being increasingly prescribed for patients who meet the criteria for adult attention deficit disorder. Many such patients also have co-morbid major depression.

- A 32-year-old man taking dexamfetamine 15 mg/day developed the serotonin syndrome, with shivering, sweating, myoclonus, and pyrexia, following the addition of venlafaxine 150 mg/day (36). Venlafaxine and dexamfetamine were withdrawn, and he was given the serotonin receptor antagonist, cyproheptadine (32 mg over 3 hours). His symptoms remitted a few hours later and he subsequently restarted dexamfetamine without incident. A later trial of citalopram, while he was still taking dexamfetamine, also led to symptoms of serotonin toxicity.

The combination of stimulants and SSRIs is not uncommon in clinical practice, but reports of serotonin toxicity are unusual, perhaps because drugs such as dexamfetamine and methylphenidate predominantly release dopamine and noradrenaline rather than serotonin. However, psychostimulants do cause some degree of serotonin release, which might have been sufficient to cause serotonin toxicity in this case.

Fluoxetine

Four depressed patients (age range 21–73 years) had anticholinergic adverse effects when venlafaxine 37.5 mg/day was added to fluoxetine 20 mg/day (37). Venlafaxine does not have direct anticholinergic effects, but could cause them indirectly as a result of its ability to increase noradrenergic neurotransmission. This effect would be expected only at much higher doses of venlafaxine than those used here, which led the author to suggest that in the presence of CYP2D6 inhibition by fluoxetine, venlafaxine concentrations could have been substantially higher than expected.

Imipramine

In six healthy men, venlafaxine (150 mg/day for 10 days) produced an increase of 28% in the systemic availability of a single dose of imipramine (100 mg); the availability of the active metabolite of imipramine, desipramine, was also increased (38). These data suggest that venlafaxine inhibits CYP2D6; the effects appear to be modest relative to those of fluoxetine and are similar to those produced by sertraline.

Indinavir

Venlafaxine 50 mg 8-hourly reduced the AUC of a single dose of indinavir 800 mg by 28% in nine healthy subjects (39).

Moclobemide

Interactions of moclobemide with venlafaxine, a potent serotonin re-uptake inhibitor, have been reported (40,41).

- A 34-year-old man took 2.625 g of venlafaxine (therapeutic dose 75–375 mg/day) and 3 g of moclobemide, plus an unknown amount of alcohol 1 hour before being admitted to hospital. Within 20 minutes of arrival his conscious level deteriorated and he had increased muscle tone, with clonus in all limbs. He was treated with intubation, paralysis, and ventilation, and sedated with midazolam and morphine. He regained consciousness after 2 days.

This case report confirms the serious consequences of combined overdosage of moclobemide with drugs that potentiate brain serotonin function (see also the monograph on SSRIs).

Monoamine oxidase inhibitors

There have been previous reports of serotonin toxicity when venlafaxine was combined with therapeutic doses of conventional MAO inhibitors (SEDA 20, 21). The serotonin syndrome has been reported in four patients who were switched from the MAO inhibitor phenelzine to venlafaxine (42). In two of the subjects, the 14-day washout period recommended when switching from phenelzine to other antidepressant drugs had elapsed.

- A 25-year-old woman had been taking phenelzine 45 mg/day for refractory migraine and tension headache, but suffered intolerable adverse effects (weight gain, edema, and insomnia). Phenelzine was withdrawn and 15 days elapsed before she took a single dose of venlafaxine 37.5 mg. Within 1 hour she developed agitation, twitching, shakiness, sweating, and generalized erythema with hyperthermia (38°C). Her symptoms resolved within 3 hours with no sequelae.

These cases suggest that even after the recommended 2-week washout from MAO inhibitors, venlafaxine can provoke serotonin toxicity in some patients. In principle it should be possible to switch from one conventional MAO inhibitor to another without a washout period. However, there have been reports that patients who switched from phenelzine to tranylcypromine had hypertensive reactions, with disastrous consequences (43). Whenever possible, a 2-week washout period when switching MAO inhibitors or when introducing serotonin re-uptake inhibitors seems advisable.

Phenelzine

There have been reports of serotonin toxicity when venlafaxine was combined with therapeutic doses of conventional MAO inhibitors (SEDA-20, 21). The serotonin syndrome has been reported in four patients who were switched from the MAO inhibitor phenelzine to venlafaxine (44). In two of them, the 14-day washout period recommended when switching from phenelzine to other antidepressant drugs had elapsed.

- A 25-year-old woman, who had taken phenelzine (45 mg/day) for refractory migraine and tension headache, suffered intolerable adverse effects (weight gain, edema, and insomnia). Phenelzine was withdrawn and 15 days elapsed before she took a single dose of 37.5 mg of venlafaxine. Within 1 hour she developed agitation, twitching, shakiness, sweating, and generalized erythema with hyperthermia (38°C). Her symptoms resolved within 3 hours with no sequelae.

This suggests that even after the recommended 2-week washout from MAO inhibitors, venlafaxine can provoke serotonin toxicity in some patients.

Quinidine

Venlafaxine is metabolized by CYP2D6. In healthy volunteers the oral clearance of venlafaxine (37.5 mg/day for 2 days) was fourfold less in poor metabolizers ($n = 6$) than extensive metabolizers ($n = 8$) (45). Administration of the CYP2D6 inhibitor quinidine, 200 mg/day for 2 days, to the extensive metabolizers reduced the oral clearance of venlafaxine to the level seen in poor metabolizers. Quinidine had no effect on venlafaxine clearance in subjects who were poor metabolizers before treatment. The authors suggested that poor metabolizers may be at particular risk of venlafaxine toxicity, as could subjects who take inhibitors of CYP2D6.

Risperidone

Unlike SSRIs, venlafaxine only weakly inhibits CYP2D6. Venlafaxine (up to 150 mg/day for 10 days) modestly but significantly increased the plasma AUC of a single dose of risperidone (1 mg) in 38 healthy volunteers in an open longitudinal design, but did not affect its active metabolite, 9-hydroxyrisperidone (46). The authors suggested that their study had shown a small potentiating effect of venlafaxine on risperidone availability, probably through weak inhibition by venlafaxine of CYP2D6. Venlafaxine is therefore less likely to increase risperidone concentrations than are fluoxetine and paroxetine.

In eight patients with major depressive disorder without psychotic features, who did not respond to serotonin re-uptake inhibitors therapy when risperidone was added, all improved within 1 week. Furthermore, risperidone also seemed to have beneficial effects on sleep disturbance and sexual dysfunction (47). In an open study in 30 healthy subjects who took risperidone 1 mg orally

before and after venlafaxine dosing to steady state, the oral clearance of risperidone fell by 38% and the volume of distribution by 17%, resulting in a 32% increase in AUC; renal clearance of 9-hydroxyrisperidone also fell by 20% (48). The authors concluded that these small effects were consistent with the fact that venlafaxine is unlikely to alter the clearance of risperidone, which is mainly by CYP2D6.

Tramadol

The analgesic drug tramadol can cause 5HT toxicity in association with SSRIs. Venlafaxine is also a potent re-uptake inhibitor, and perhaps not surprisingly has been reported to cause features of the 5HT syndrome in a patient taking tramadol (49).

- A 47-year-old man with a long history of depression had been stable on a combination of venlafaxine 300 mg/day and mirtazapine 30 mg/day for 3 months. He started to take tramadol for a chronic pain syndrome and the dose was titrated up to 300 mg/day over the next 4 weeks. The dose was then increased to 400 mg/day, and 8 days later he developed shivering, sweating, myoclonus, hyper-reflexia, and mydriasis. His medications were withdrawn, but over the next 4 hours he developed a fever (39.2°C) and a tachycardia. He was given intravenous hydration and closely monitored, and the symptoms resolved over the next 36 hours. Venlafaxine and mirtazapine were restarted and he remained symptom free.

In this case the 5HT toxicity produced by tramadol may have been dose-related within the therapeutic range, because it did not become apparent until the dose had reached 400 mg/day. It is of interest that the 5HT syndrome developed despite the fact that the patient was taking mirtazapine, which is a potent 5HT$_2$ receptor antagonist. 5HT$_2$ receptor antagonists have been suggested for the treatment of drug-induced 5HT toxicity. However, activation of postsynaptic 5HT$_{1A}$ receptors has also been implicated in the development of the 5HT syndrome, which presumably may have been the mechanism here.

The death of a 36-year-old patient with a history of alcohol dependence who was taking tramadol, venlafaxine, trazodone, and quetiapine has highlighted the increased risk of seizures with concomitant use of tramadol and selective serotonin re-uptake inhibitors (50).

Trifluoperazine

Drugs that potentiate the function of serotonin, such as venlafaxine, can produce pharmacodynamic interactions with dopamine receptor antagonists, perhaps because serotonin pathways can inhibit dopamine neurotransmission.

- A 44-year-old man with major depression had been taking the antipsychotic drug trifluoperazine (3 mg/day) for anxiety for several years (51). He was given venlafaxine 75 mg/day, and 12 hours after the first dose developed anxiety and malaise. He was sweating and had tremor and rigidity. His blood pressure fluctuated and his creatine phosphokinase activity was raised at 11 320 IU/l. A diagnosis of neuroleptic malignant syndrome was made and he was given dantrolene and bromocriptine. His symptoms settled within 24 hours and trifluoperazine was reintroduced uneventfully.

This case suggests that venlafaxine can rarely increase the risk of neuroleptic malignant syndrome in patients taking a dopamine receptor antagonist. However, this combination of drugs is often used, and it is not clear why only very few patients appear to be susceptible to this reaction, and then only on some occasions.

References

1. Nemeroff CB. Evolutionary trends in the pharmacotherapeutic management of depression. J Clin Psychiatry 1994;55(Suppl):3–15.

2. Montgomery SA. Venlafaxine: a new dimension in antidepressant pharmacotherapy. J Clin Psychiatry 1993;54(3):119–26.

3. Dallal A, Chouinard G. Withdrawal and rebound symptoms associated with abrupt discontinuation of venlafaxine. J Clin Psychopharmacol 1998;18(4):343–4.

4. Parker G, Blennerhassett J. Withdrawal reactions associated with venlafaxine. Aust NZ J Psychiatry 1998;32(2):291–4.

5. White CM, Gailey RA, Levin GM, Smith T. Seizure resulting from a venlafaxine overdose. Ann Pharmacother 1997;31(2):178–80.

6. Zhalkovsky B, Walker D, Bourgeois JA. Seizure activity and enzyme elevations after venlafaxine overdose. J Clin Psychopharmacol 1997;17(6):490–1.

7. Leaf EV, Peano C, Leikin JB, Hanashiro PK. Seizures, ventricular tachycardia, and rhabdomyolysis as a result of ingestion of venlafaxine and lamotrigine. Ann Emerg Med 1997;30(5):704–8.

8. Thase ME. Effects of venlafaxine on blood pressure: a meta-analysis of original data from 3744 depressed patients. J Clin Psychiatry 1998;59(10):502–8.

9. West S, Hewitt J. Prolonged hypertension: a case report of a potential interaction between electroconvulsive therapy and venlafaxine. Int J Psychiatr Clin Pract 1999;3:55–7.

10. Reznik I, Rosen Y, Rosen B. An acute ischaemic event associated with the use of venlafaxine: a case report and proposed pathophysiological mechanisms. J Psychopharmacol 1999;13(2):193–5.

11. Drent M, Singh S, Gorgels APM, Hansell DM, Bekers O, Nicholson AG, Van Suylen RJ, Du Bois RM. Drug-induced pneumonitis and heart failure simultaneously associated with venlafaxine. Am J Resp Crit Care 2003;167: 958–61.

12. Zullino DF, Riquier F. Venlafaxine and vivid dreaming. J Clin Psychiatry 2000;61(8):600.

13. Pan JJ, Shen WW. Serotonin syndrome induced by low-dose venlafaxine. Ann Pharmacother 2003;37: 209–11.

14. Montanes-Rada F, Bilbao-Garay J, de Lucas-Taracena MT, Ortzi-Ortiz ME. Venlafaxine, serotonin syndrome and differential diagnosis. J Clin Psychopharmacol 2005;25: 101–2.

15. Ng B, Sanbrook GM, Malouf AJ, Agarwal SA. Venlafaxine and bilateral acute angle closure glaucoma. Med J Aust 2002;176(5):241.

16. Stoner SC, Williams RJ, Worrel J, Ramlatchman L. Possible venlafaxine-induced mania. J Clin Psychopharmacol 1999;19(2):184–5.

17. Adamou M, Hale AS. Erotomania induced by venlafaxine: a case study. Acta Psychiatr Scand 2003;107: 314–17.

18. Sternbach H. Venlafaxine-induced galactorrhea. J Clin Psychopharmacol 2003;23: 109.

19. Masood GR, Karki SD, Patterson WR. Hyponatremia with venlafaxine. Ann Pharmacother 1998;32(1):49–51.

20. Boyd IW, Karki SD, Masood GR. Comment: hyponatremia with venlafaxine. Ann Pharmacother 1998;32(9):981–2.

21. Horsmans Y, De Clercq M, Sempoux C. Venlafaxine-associated hepatitis. Ann Intern Med 1999;130(11):944.

22. Pitchot W, Ansseau M. Venlafaxine-induced hair loss. Am J Psychiatry 2001;158(7):159–60.

23. Samuel RZ, Horrigan JP, Barnhill LJ. Priapism associated with venlafaxine use. J Am Acad Child Adolesc Psychiatry 2000;39(1):16–7.

24. Michael A. Venlafaxine-induced painful ejaculation. Br J Psychiatry 2000;177:282.

25. Kilic S, Ergin H, Baydinc YC. Venlafaxine extended release for the treatment of patients with premature ejaculation: a pilot, single-blind, placebo-controlled, fixed dose crossover study on short-term administration of an antidepressant drug. Int J Androl 2005;28: 47–52.

26. Sattar SP, Grant KM, Bhatia SC. A case of venlafaxine abuse. New Engl J Med 2003;348(8):764–5.

27. Luckhaus C, Jacob C. Venlafaxine withdrawal syndrome not prevented by maprotiline, but resolved by sertraline. Int J Neuropsychopharmacol 2001;4(1):43–4.

28. Cheeta S, Schifano F, Oyefeso A, Webb L, Ghodse AH. Antidepressant-related deaths and antidepressant prescriptions in England and Wales, 1998–2000. Br J Psychiatry 2004;184:41–7.

29. Chief Medical Officer. Safety of Selective Serotonin Reuptake Inhibitor Antidepressants. CEM/CMO/2004/11.

30. Jaffe PD, Batziris HP, van der Hoeven P, DeSilva D, McIntyre IM. A study involving venlafaxine overdoses: comparison of fatal and therapeutic concentrations in post-mortem specimens. J Forensic Sci 1999;44(1):193–6.

31. Blythe D, Hackett LP. Cardiovascular and neurological toxicity of venlafaxine. Hum Exp Toxicol 1999;18(5):309–13.

32. Combes A, Peytavin G, Theron D. Conduction disturbances associated with venlafaxine. Ann Intern Med 2001;134(2):166–7.

33. Amchin J, Zarycranski W, Taylor KP, Albano D, Klockowski PM. Effect of venlafaxine on the pharmacokinetics of alprazolam. Psychopharmacol Bull 1998;34(2): 211–9.

34. Amchin J, Zarycranski W, Taylor KP, Albano D, Klockowski PM. Effect of venlafaxine on CYP1A2-dependent pharmacokinetics and metabolism of caffeine. J Clin Pharmacol 1999;39(3):252–9.

35. Repo-Tiihonen E, Eloranta A, Hallikainen T, Tiihonen J. Effects of venlafaxine treatment on clozapine plasma levels in schizophrenia patients. Neuropsychobiology 2005;51: 173–6.

36. Prior FH, Isbister GK, Dawson AH, Whyte IM. Serotonin toxicity with therapeutic doses of dexamphetamine and venlafaxine. Med J Aust 2002;176(5):240–1.

37. Benazzi F. Venlafaxine–fluoxetine interaction. J Clin Psychopharmacol 1999;19(1):96–8.

38. Albers LJ, Reist C, Vu RL, Fujimoto K, Ozdemir V, Helmeste D, Poland R, Tang SW. Effect of venlafaxine

on imipramine metabolism. Psychiatry Res 2000;96(3): 235–43.

39. Levin GM, Nelson LA, DeVane CL, Preston SL, Eisele G, Carson SW. A pharmacokinetic drug–drug interaction study of venlafaxine and indinavir. Psychopharmacol Bull 2001;35(2):62–71.

40. Roxanas MG, Machado JF. Serotonin syndrome in combined moclobemide and venlafaxine ingestion. Med J Aust 1998;168(10):523–4.

41. Coorey AN, Wenck DJ. Venlafaxine overdose. Med J Aust 1998;168(10):523.

42. Diamond S, Pepper BJ, Diamond ML, Freitag FG, Urban GJ, Erdemoglu AK. Serotonin syndrome induced by transitioning from phenelzine to venlafaxine: four patient reports. Neurology 1998;51(1):274–6.

43. Mattes JA. Stroke resulting from a rapid switch from phenelzine to tranylcypromine. J Clin Psychiatry 1998;59(7):382.

44. Diamond S, Pepper BJ, Diamond ML, Freitag FG, Urban GJ, Erdemoglu AK. Serotonin syndrome induced by transitioning from phenelzine to venlafaxine: four patient reports. Neurology 1998;51(1):274–6.

45. Lessard E, Yessine MA, Hamelin BA, O'Hara G, LeBlanc J, Turgeon J. Influence of CYP2D6 activity on the disposition and cardiovascular toxicity of the antidepressant agent venlafaxine in humans. Pharmacogenetics 1999;9(4):435–43.

46. Amchin J, Zarycranski W, Taylor KP, Albano D, Klockowski PM. Effect of venlafaxine on the pharmacokinetics of risperidone. J Clin Pharmacol 1999;39(3):297–309.

47. Ostroff RB, Nelson JC. Risperidone augmentation of selective serotonin reuptake inhibitors in major depression. J Clin Psychiatry 1999;60(4):256–9.

48. Amchin J, Zarycranski W, Taylor KP, Albano D, Klockowski PM. Effect of venlafaxine on the pharmacokinetics of risperidone. J Clin Pharmacol 1999;39(3):297–309.

49. Houlihan DJ. Serotonin syndrome resulting from coadministration of tramadol, venlafaxine and mirtazapine. Ann Pharmacother 2004;38: 411–3.

50. Ripple MG, Pestaner JP, Levine BS, Smialek JE. Lethal combination of tramadol and multiple drugs affecting serotonin. Am J Forensic Med Pathol 2000;21(4):370–4.

51. Nimmagadda SR, Ryan DH, Atkin SL. Neuroleptic malignant syndrome after venlafaxine. Lancet 2000;355(9200):289–90.

Viloxazine

General Information

Viloxazine, a bicyclic compound, is related structurally (but not pharmacologically) to the beta-adrenoceptor antagonists. In animal tests, its profile shows properties of both the imipramine-like compounds (reversal of reserpine-induced hypothermia) and amphetamine (stimulation of the electroencephalogram). A review of animal and clinical data confirmed the impression that viloxazine has efficacy comparable to that of imipramine but with a different adverse effects profile (1). There is a reduced frequency of anticholinergic and sedative effects and a tendency to lose rather than to gain weight. However, viloxazine has some limiting adverse effects of its own. These include nausea, vomiting, and gastrointestinal distress, which may be reduced by the use of an enteric-coated formulation (2,3). Viloxazine has also been implicated in migraine, even in patients with no previous history (4).

Organs and Systems

Cardiovascular

Although hypotension and tachycardia can occur, an extensive review of its animal and clinical pharmacology suggested that viloxazine is relatively free of direct cardiotoxic effects (1). In a controlled comparison of doxepin and viloxazine (150–450 mg/day), one patient taking viloxazine developed chest pain after 26 days; an electrocardiogram confirmed changes compatible with ischemia but there were no progressive electrocardiographic or enzyme changes, and the patient recovered fully after being dropped from the study (5).

Nervous system

Viloxazine has previously been reported to lack epileptogenic properties (SEDA-10, 52). In a review of eight patients (six reported to the UK Committee on Safety of Medicines and two in Japan) it was concluded that there was a possible causal connection with seizures in only two cases, and that such an association was inconsistent with the results of animal studies and with a worldwide review of clinical trials (6). The reviewers concluded that if there is a risk of inducing epilepsy with viloxazine, it is probably only one-tenth that of tricyclic compounds.

Drug Administration

Drug overdose

It has been claimed that viloxazine is relatively free of cardiotoxic effects and therefore safe in patients who take an overdose. Indeed, 12 cases of overdose have been reported, with complete recovery and no electrocardiographic changes (SEDA-1, 9). In an authoritative review it was concluded that although coma, hypotension, and loss of tendon reflexes have been reported, serious complications do not develop in most cases (7).

Drug–Drug Interactions

Carbamazepine

In seven patients with epilepsy taking carbamazepine the addition of viloxazine 300 mg/day resulted in a pronounced increase in plasma carbamazepine concentrations (8.1–12.1 ng/ml), with symptoms of intoxication (dizziness, fatigue, ataxia, and drowsiness), resolving on withdrawal of viloxazine (8). The mechanism was unclear, but serum concentrations of phenytoin in two patients were unchanged.

References

1. Pinder RM, Brogden RN, Speight TM, Avery GS. Viloxazine: a review of its pharmacological properties and therapeutic efficacy in depressive illness. Drugs 1977;13(6):401–21.

2. Pichot P, Guelfi J, Dreyfus JF. A controlled multicentre therapeutic trial of viloxazine. J Int Med Res 1975;3:80.

3. Ichimaru I. Evaluation of Vivalan: additional studies (Japan). J Int Med Res 1975;3:97.

4. Barnes TR, Kidger T, Greenwood DT. Viloxazine and migraine. Lancet 1979;2(8156–8157):1368.

5. Pinder RM, Brogden RN, Speight TM, Avery GS. Doxepin up-to-date: a review of its pharmacological properties and therapeutic efficacy with particular reference to depression. Drugs 1977;13(3):161–218.

6. Edwards JG, Glen-Bott M. Does viloxazine have epileptogenic properties? J Neurol Neurosurg Psychiatry 1984;47(9):960–4.

7. Crome P. Antidepressant overdosage. Drugs 1982;23(6):431–61.

8. Pisani F, Narbone MC, Fazio A, Crisafulli P, Primerano G, D'Agostino AA, Oteri G, Di Perri R. Effect of viloxazine on serum carbamazepine levels in epileptic patients. Epilepsia 1984;25(4):482–5.

LITHIUM

General Information

Lithium is an alkaline earth element that is used medicinally in the form of salts such as lithium chloride and lithium carbonate. Its main use is in the prevention or attenuation of recurrent episodes of mania and depression in individuals with bipolar mood disorder (manic depression). Lithium also has clearly established antimanic activity, although its relatively slow onset of action often necessitates the use of ancillary drugs, such as antipsychotic drugs and/or benzodiazepines, at the start of therapy. If lithium alone is ineffective for recurrent bipolar mood disorder, combining it or replacing it with carbamazepine or valproate may be of value; reports with lamotrigine and olanzapine are also encouraging.

Lithium also has antidepressant activity in bipolar disorder, has prophylactic value in recurrent major depression, and is a useful augmenting agent for antidepressant-resistant depression. Other uses in psychiatry include schizoaffective disorder, emotional instability, and pathological aggression. The point prevalence of lithium use has been estimated to be as high as 1 in 1000 people in populations in industrial countries. The complex relation between sub-syndromal manifestations of bipolar disorder, particularly cognitive dysfunction, and the role that lithium might play in alleviating or aggravating this problem have been discussed in a thoughtful review (1).

The well-established effectiveness of long-term lithium in reducing manic-depressive morbidity includes a reduced risk of suicide and suicidal behavior. For example, in one study, suicidal acts per 1000 patient-years were 23 before lithium, 3.6 during lithium, 71 in the first year after withdrawal, and 23 in subsequent years (2). In another study, suicide rates during treatment with lithium were 31 per 1000 person-years (emergency department suicide attempts), 11 per 1000 person-years (suicide attempts resulting in hospitalization), and 1.7 per 1000 person-years (suicide deaths); the risk of suicide death was 2.7 times higher during treatment with divalproex than during treatment with lithium (3). A retrospective study divided high-risk patients into excellent, moderate, and poor responders to lithium and showed that no further suicide attempts occurred in 93%, 83%, and 49% respectively (4). The substantial reduction in suicidal tendency in the poor responder group suggested an antisuicidal effect of lithium beyond its mood-stabilizing properties, although the psychosocial benefits of lithium clinic treatment could have been contributing factors.

Since bipolar disorder is a condition for which long-term treatment is usually necessary, both acute and long-term adverse effects are important, especially since patients in remission are often less likely to tolerate them (5). With this in mind, one might consider some speculatively positive findings involving the neurotropic and neuroprotective effects of lithium (6). The concentration of bcl-2, a cytoprotective protein, was upregulated by lithium in both rodent brains and human neuronal cells, as was the concentration of N-acetylaspartate, a marker of neuronal viability and function, in human gray matter (7). In addition, a 3-dimensional magnetic resonance imaging study with quantitative brain-tissue segmentation showed that treatment with lithium for 4 weeks increased the total volume of gray matter by about 3% in eight of 10 patients in the depressed phase of bipolar I disorder.

Guidelines for treating bipolar disorder, developed in a consensus meeting of experts, have been published by the British Association of Psychopharmacology (8). Lithium was recommended for many of the phases of bipolar disorder, often in combination with other treatments. Although the efficacy of lithium salts in the treatment of bipolar disorders, particularly in the prevention of recurrence of manic, hypomanic, mixed, and depressive episodes is well established, recent data suggest that it also has a prominent role in reducing the rate of suicide among patients with bipolar mood disorders.

The molecular effects of lithium have been reviewed (9). Its direct targets include inositol monophosphatase, inositol polyphosphate 1-phosphatase, biophosphate nucleotidase, fructose 1, 6)biophosphatase, phosphoglucomutase, and glycogen synthase kinase-3. These enzymes are largely phosphomonoesterases, which are magnesium-dependent. Lithium also has effects on adenylate cyclase, arachidonic acid, and myristoylated alanine-rich C kinase substrate (MARCKS). MARCKS is a presynaptic and postsynaptic protein that affects cellular signalling and cytoskeletal plasticity, and its expression is regulated by lithium (10).

During 2004 a number of clinical trials were reported involving acute and maintenance studies of lithium, mostly either comparing new atypical antipsychotic drugs with lithium in bipolar disorder or in combined treatment studies. Of the relatively few studies of the adverse effects of lithium, most clustered in the areas of cardiovascular effects and issues regarding lithium toxicity.

Efficacy of lithium

Bipolar illness appears to be a changing illness. A comparison of psychiatric services in North-West Wales in the 1890s and the 1990s has shown that the rate of admissions increased from 4.0 every 10 years to 6.3 every 10 years (11). Similarly, the daily hospital occupancy rate for patients with bipolar affective disorder rose from 16 per million to 24 per million. While acknowledging that there have been many social changes that may have contributed to these differences, the authors suggested that current treatments leave much to be desired. Reviews of lithium treatment have reached similar conclusions, particularly regarding the effect of lithium in acute episodes (12).

A review of treatment guidelines has shown that there is great variability in the various recommendations, despite the claim that they are all evidence-based (13). Nevertheless, there are foundational recommendations that seem to be consistent, which include the use of lithium and valproate for most patients with bipolar affective disorder.

There are several reasons for the perception that lithium may not be as effective as initially believed. Among these are questions about study design and diagnostic drift (i.e. changes in how bipolar illness is diagnosed over time) (14). Many maintenance studies involve enrichment of the sample for alternative agents, which reduces the apparent effect of lithium. In addition, many factors are associated with a good response to lithium. These include a family history of a response to lithium (15), higher social status, social support, and compliance with medication (16). Predictors of a poor response to lithium include stress, high expressed emotion, neurotic personality traits, unemployment, and a high number of previous episodes (Harris 423). It is likely that most of these factors are simply predictors of a good prognosis rather than lithium-specific factors.

Beneficial non-psychiatric uses of lithium

Some collateral drug effects that are not related to the intended therapeutic effect are potentially beneficial. Several beneficial effects of lithium, besides its action in bipolar disorder, have been described.

Nervous system

There has been a spate of publications describing the neuroprotective effects of lithium. By inhibiting glycogen synthase kinase-3, lithium inhibits tau hyperphosphorylation and protects against β-amyloid-induced cell death, suggesting a possible role in the treatment of Alzheimer's disease (17,18). Studies in mice and cell lines show that lithium reduces gp120-associated neurotoxicity, suggesting that it may be useful in preventing progression of HIV-associated cognitive deterioration (19). Low-dose lithium reduced infarct volume and neurological deficits in a rat model of transient focal cerebral ischemia (20).

Overall, the evidence that lithium has neuroprotective and neurotropic effects through a variety of mechanisms is striking (21), although whether those findings will evolve into therapies of practical clinical value remains to be seen.

Chronic cluster headache has responded to lithium (22).

Sensory systems

In cultured mouse retinal ganglion cells, lithium supported the survival and regeneration of axons (23). This led the authors to the very speculative suggestion that lithium might be useful in treating conditions such as glaucoma, optic neuritis, and other neuron loss disorders.

Endocrine

Lithium blocks the release of iodine and thyroid hormones from the thyroid and has been used to treat hyperthyroidism, as an adjunct to radioiodine therapy

(24–27) and in metastatic thyroid carcinoma (28). However, it can also cause hyperthyroidism. Lithium enhanced the efficacy of radioiodine in 23 patients (29), but was ineffective in a larger comparison of lithium ($n = 175$) or radioiodine alone ($n = 175$) (30). In 24 patients with Graves' disease, lithium attenuated or prevented increases in thyroid hormone concentration after methimazole withdrawal and radioiodine treatment (24,31).

In a case of amiodarone-induced thyrotoxicosis that did not respond to antithyroid drugs and glucocorticoids, low-dose lithium normalized thyroid function (32).

Lithium has been used, with several other drugs, to treat four patients with amiodarone-associated thyrotoxicosis, but the drugs were ineffective in two patients, who required thyroidectomy (33). One hopes that the authors actually used milligram amounts of lithium carbonate rather than the microgram amounts listed in the article.

Metabolism

Lithium therapy in a 17-year-old man with Kleine–Levin syndrome led to remission of the characteristic manifestations, including hyperphagia (34).

Hematologic

Lithium has beneficial granulocytopoietic effects (35–37). For example, lithium carbonate (800–900 mg/day) effectively corrected neutropenia due to chemotherapy or radiotherapy in over 85% of 100 cancer patients (38). The potential benefit and possible risks of using lithium to treat clozapine-induced neutropenia/agranulocytosis have been reviewed (39).

- A 29-year-old man with agranulocytosis, who could not tolerate granulocyte-colony stimulating factor, had normalization of peripheral granulocyte counts when he took lithium carbonate 800 mg/day (40).
- A 16-year-old with severe aplastic anemia failed to respond to treatment with corticosteroids plus an androgen and to antilymphocyte globulin, but had a strikingly positive response to the combination of lithium and an androgen derivative (41). Leukopenia and thrombocytopenia recurred 2 months after lithium was withdrawn and responded to reintroduction of the drug.

The potential of lithium to prevent or treat clozapine-induced granulocytopenia has been reviewed (39). In a study of 38 patients on clozapine for schizophrenia or schizoaffective disorder, the addition of lithium increased the leukocyte count (42). A 20-year-old man with olanzapine-induced neutropenia 5 mg/day was able to tolerate 20 mg/day while taking lithium (43).

Skin

Lithium succinate is used topically to treat seborrheic dermatitis (44). A topical 8% lithium gluconate ointment was more effective than a placebo ointment in treating 129 patients with facial seborrheic dermatitis (complete remission in 29 versus 3.8%) (45).

Immunologic and infections

Comments on the generally favorable effects of lithium on immune function have been summarized (46). The antiviral and neuroprotective properties of lithium were mentioned in a review of the immune system and bipolar disorder (47). The potential benefit of lithium in treating AIDS and AIDS-related dementia, owing in part to its cytokine-regulating and neuroprotective effects, has been reviewed (48). Genital *Herpes simplex* infection has responded to lithium (49).

- Lithium had an antiviral effect in a 44-year-old woman whose psychiatric symptoms did not improve but who had complete suppression of *Herpes labialis* for 2 years (after a 30-year history of at least twice-yearly episodes) followed by a recurrence, only 5 days after stopping the drug (50).

Cancers

In one study, there was a lower risk of cancer in both 609 lithium patients and 2396 psychiatric controls compared with the general population, and in the lithium group, there was a nonsignificant trend toward an even lower risk of nonepithelial tumors (51).

Lithium gamolenate, a compound with in vitro antitumor activity, given intravenously or orally, was ineffective in treating advanced pancreatic adenocarcinoma (*n* = 278) (52). Adverse effects attributed to lithium (type unspecified) were reported in two of 93 in the oral group (mean serum lithium 0.15 mmol/l), five of 90 with low-dose intravenous administration (mean serum lithium 0.4 mmol/l), and seven of 95 with the high-dose intravenous administration (mean serum lithium 0.8 mmol/l).

Pharmacokinetics

Ionized lithium is readily absorbed from the gastrointestinal tract and is excreted almost entirely by the kidney, which ordinarily clears it at a rate of about one-quarter to that of creatinine clearance (53).

A reduction in glomerular filtration rate (GFR) will reduce lithium clearance, as will a negative sodium balance. Lithium is not metabolized and is not bound to plasma proteins.

Lithium is easily, inexpensively, and accurately measurable in the serum, and serum concentration determination is a useful adjunct to monitoring its therapeutic efficacy and avoiding toxicity (54).

Both immediate-release and modified-release formulations of lithium carbonate are available. Peak blood concentrations are lower and occur more slowly with modified-release formulations than with immediate-release formulations, but all formulations are supposed to deliver equivalent amounts of lithium per millimole. The effectiveness of lithium should not be altered by the formulation used or the number of daily doses (assuming full adherence to therapy), but if it is given once a day the 12-hours serum lithium concentration will be somewhat higher than if the same amount is given in divided doses.

Standardizing the timing of blood sampling to about 12 hours after the last dose (which is preferably taken in the morning) will do much to avoid the vagaries of absorption and should provide a consistency that will allow accurate comparisons across samples (54). Blood concentrations are most useful if a steady state has been reached before sampling (4–5 days after starting treatment or after a dosage change).

Recommended serum lithium concentrations for the treatment of mania and for maintenance treatment are not uniformly agreed on, although differences of opinion are not large. For example, a concentration of 0.5–0.8 mmol/l has been advised for patients starting treatment, with the recognition that both lower and higher concentrations may be necessary. In US product monographs, concentration ranges for acute mania (1.0–1.5 mmol/l) and long-term treatment (0.6–1.2 mmol/l) tend to be higher than currently practiced (mania 0.8–1.2 mmol/l; long-term 0.6–1.0 mmol/l) and considerably higher than concentrations recommended in Europe. While there is uniform agreement that the serum concentration should be kept as low as is compatible with therapeutic efficacy, there is as yet no accurate way to predict this concentration in an individual. Within the target range, one can generally expect efficacy to increase with blood concentration, but at a price—adverse effects are likely to increase. Elderly people will require a lower dosage of lithium to achieve a given serum concentration, and they tend to be more sensitive to adverse effects than younger individuals, but whether they respond better to lower concentrations is unclear.

The frequency with which blood concentrations should be measured varies with the stage of the illness and with a number of patient factors (including age, associated illnesses, concomitant medications, and diet). Generally, concentrations are measured every 5–7 days during the start of therapy, and then less and less often as stability occurs and persists. In reliable patients taking long-term treatment, concentrations may be measured every 3–4 or even 6 months.

Pharmacogenetics

Individuals with the short form of the serotonin transporter have a poorer response to lithium (55). The short form of the serotonin transporter is a genetic polymorphism that increases the risk of depression in the setting of adversity (56,57,58) and reduces the likelihood of a response to antidepressant treatment, and is therefore associated with a poorer outcome (59,60,61,62). Similarly, a single nucleotide polymorphism (SNP) in the gene that encodes brain-derived neurotrophic factor (the val66met SNP of BDNF) has been associated with a poor response to lithium (63). This SNP is over-represented among patients with rapid cycling (64), who are less likely to respond well to lithium.

Clinicians have some control of a few factors that are associated with a poor response to lithium. For example, a psychoeducational program can have a dramatic effect on compliance with lithium, reflected in more stable lithium concentrations (65).

Instruction and information

The safe and effective use of lithium is best ensured by close collaboration among patients, physicians, and significant others, all of whom must remain well-informed and up to date about treatment guidelines, benefits, adverse effects, risks, and precautions. While verbal communication and education is invaluable, it should be constructively supplemented with written information (66,67).

Observational studies Beneficial effects have been reported in patients with schizophrenia (n = 10) and schizoaffective disorder (n = 10) who were treated with a combination of clozapine and lithium (68). When lithium (serum lithium concentration titrated to at least 0.5 mmol/l) was added to clozapine 100–800 mg/day there was a positive effect in the patients with treatment-resistant schizoaffective disorder, but not in the patients with schizophrenia.

In a retrospective chart analysis of patients with depression, 23 took added lithium (69). There was improvement in 13 and 11 improved over 4 weeks. Patients who continued to take lithium continued to improve, compared with patients who stopped taking lithium.

In five adolescents with somnolence from Kleine–Levin syndrome lithium (serum concentrations 0.6–0.9 mmol/l) reduced the frequency and severity of the somnolence, without severe adverse effects (70).

The addition of lithium in treating major depressive disorder in patients unresponsive to antidepressant drugs has been discussed, and it has been noted that about 50% of patients respond to lithium augmentation in 2–4 weeks (71), while others have pointed to the absence of controlled data for this treatment in bipolar depression, while nevertheless recommending its use (72). In summary, there are data that support the use of lithium augmentation for treatment-resistant unipolar major depression. However, the data are not robust and are based on only a few hundred patients. Placebo-controlled studies of lithium augmentation for treatment-resistant bipolar depression are lacking (73).

The diagnosis of bipolar affective disorder, including the bipolar spectrum, co-morbidity of bipolar disorder, issues of bipolar disorder in children and adolescence, and the pathophysiology of bipolar disorder have been reviewed in relation to neuroimaging studies (74). These imaging studies include Positron Emission Tomography (PET) studies, functional MRI, Single Photon Emission Computed Tomography (SPECT), and Magnetic Resonance Spectroscopy (MRS). They point to abnormalities in the brains of patients with bipolar disorder. The reviewer noted the effects of mood stabilizers, including lithium, and hypothesized that the lines of evidence regarding efficacy of lithium and other mood stabilizers, along with biological evidence, point to defects in the cell related to memory dysfunction. Various "alternative" treatments and their possible effects on stabilization of bipolar disorder were also reviewed.

Comparative studies

The availability of anticonvulsant mood stabilizers has led to comparative studies with lithium. A review of the comparative efficacy and tolerability of drug treatments for bipolar disorder included tolerability comparisons of lithium versus carbamazepine, lithium versus valproate semisodium, and lithium versus other medications (75).

In a 4-week, placebo-controlled study of lithium in 40 hospitalized children and adolescents (mean age 12.5 years) with aggression related to conduct disorder, lithium was "statistically and clinically superior to placebo" (76). Although there were no dropouts related to adverse events, nausea, vomiting, and increased urinary frequency occurred significantly more often in the lithium group. The 55% incidence of vomiting with lithium (versus 20% with placebo) may have been related to the relatively high mean serum lithium concentration of 1.07 mmol/l (range 0.78–1.55 mmol/l).

While the efficacy of lithium alone or in combination continues to be reconfirmed, drawbacks related to adherence to therapy or genetic links to poorer outcomes have also been highlighted. For those reasons, alternatives to lithium, such as anticonvulsants and antipsychotic drugs, have often been discussed (77). However, several studies have reconfirmed the efficacy of lithium in acute mania and its equivalence to some of the newer options.

Lithium versus carbamazepine

In a 2-year, double-blind study, lithium was superior to carbamazepine in prophylactic efficacy, although it caused more adverse effects (Table 1) (78).

Lithium versus divalproex

In a randomized, placebo-controlled, 12-month maintenance comparison of lithium and divalproex in 372 bipolar I outpatients, neither active drug was more effective than placebo on the primary outcome measure—the time to recurrence of any mood episode (79). While a history of intolerance to either lithium or divalproex was an exclusion criterion, it was not stated whether or not prior nonresponders were entered and, if so, how many. The following adverse effects were significantly more frequent:

(a) with lithium than with placebo: nausea, diarrhea, and tremor;
(b) with divalproex than with placebo: tremor, weight gain, and alopecia;
(c) with lithium than with divalproex: polyuria, thirst, tachycardia, akathisia, and dry eyes;

Table 1 Adverse effects of lithium and carbamazepine in a double-blind trial (in %)

Adverse effect	Lithium (n = 42)	Carbamazepine (n = 46)
Difficulty concentrating	45	33
Thirst	41	22
Hand tremor	31	4
Blurred vision	26	11
Reduced appetite	21	9
Increased appetite	17	33
Weakness	14	4

(d) with divalproex than with lithium: sedation, infection, and tinnitus.

Unfortunately, all dropouts were pooled, whether due to adverse events or noncompliance, making overall tolerability comparisons impossible.

In an open study of 37 patients aged 5–18 years with a current manic or mixed episode, who were treated for 6 months with either divalproex sodium plus risperidone or lithium plus risperidone, lithium was given in a dose of 10–30 mg/kg/day, beginning with a single dose of 150 mg or 300 mg, and gradually increasing the dose to produce a plasma concentration in the usual target range (80). About 80% of patients responded in each group and over 60% of patients "remitted." Adverse effects were similar in the two groups over the 6-month treatment period, and included weight gain, sedation, nausea, increased appetite, stomach pain, tremor, and cognitive dulling. Three of 17 patients taking lithium plus risperidone developed polyuria compared with none of those taking divalproex sodium plus risperidone. Two of 20 patients taking divalproex sodium plus risperidone developed galactorrhea compared with none of those taking lithium plus risperidone.

Lithium versus lamotrigine

The role of lamotrigine in the treatment of bipolar disorder has been reviewed, and combination therapy with lamotrigine plus other mood stabilizers, including lithium, has been particularly discussed (81). Lamotrigine has a favorable tolerability profile compared with lithium, but lithium has better antimanic effects than lamotrigine, which exerts its antidepressant effects sooner than lithium.

In a subanalysis of two 18-month maintenance studies of the use of lithium, lamotrigine, or placebo in delaying relapse in subjects with type I bipolar illness, 98 subjects 55 years of age or older were identified (82). Lithium delayed the time to mania compared with placebo, but lamotrigine also delayed the time to either mania or depression compared with placebo.

Lithium versus olanzapine

In a 12-month relapse prevention comparison of olanzapine and lithium in subjects initially stabilized on the combination, each was efficacious in preventing relapse into either mania or depression (83[C]). Olanzapine was significantly better than lithium in preventing relapse into mania or mixed mania, but lithium was better than olanzapine in preventing relapse into depression.

Lithium versus quetiapine

Quetiapine has been approved by the FDA as monotherapy for the treatment of acute mania. Quetiapine has been evaluated in combination with lithium or divalproex in 191 patients who had been recently manic (84). After treatment with quetiapine plus lithium or divalproex for 7–28 days, the patients were randomized to either additional quetiapine or placebo and followed for 3 weeks more. Early discontinuation was more frequent in the placebo group than in the quetiapine group. The intention-to-treat population included 81 taking quetiapine and 89 taking placebo. The mean dose during the last week in patients taking quetiapine was 504 mg/day. Patients taking quetiapine had a greater improvement in their Young Mania Rating Scale score (YMRS) than patients taking placebo. The response rate (50% or greater improvement from baseline using the YMRS) was significantly higher in the group with added quetiapine than added placebo. Common adverse events included somnolence, dry mouth, weakness, and postural hypotension. The authors concluded that quetiapine was a useful adjunct to treat mania in combination with standard measures and that it was well tolerated.

In acute non-mixed mania, lithium was significantly more effective than placebo and equivalent to quetiapine in reducing manic symptoms, as measured by the Young Mania Rating Scale (85). In addition, a larger fraction of subjects taking lithium or quetiapine remained in the study compared with those taking placebo. The effect was evident in the first week of treatment and was maintained throughout the 3 months of the study.

Lithium versus valproate

In 29 patients, the burden of taking lithium ($n = 17$) was compared with that of valproate ($n = 12$) using a visual analogue scale. Adverse effects were common but not significantly different between drugs (86), a finding that contrasts with the common impression that valproate is better tolerated. Indeed, a telephone interview of 11 adolescents taking lithium and 32 taking valproate found more adverse effects, poorer compliance, and greater perceived burden in the lithium group. There was a nonsignificant trend toward more weight gain with valproate (mean 12 kg) than with lithium (mean 9 kg) (87).

Lithium versus valproate and divalproex

Lithium plus risperidone (n = 33) has been compared with valproate plus risperidone (n = 46) in the acute and continuation treatment of mania (88). Both groups were initially studied during a bout of mania, and both groups improved without a significant difference in response rate between the two groups. At week 12, 88% of the patients taking lithium plus risperidone and 80% of those taking valproate plus risperidone were in remission. There were no differences in adverse effects. The major findings of this study suggested that risperidone could be combined with either lithium or valproate and that efficacy was similar, independent of which mood stabilizer was used. The dose of risperidone was 0.5–6.0 mg/day, with a mean of 1.7 mg/day in the lithium group and 2.2 mg/day in the valproate group.

The Bipolar Affective Disorder: Lithium/Anticonvulsant Evaluation (BALANCE) Study is now under way. It involves combination therapy with either lithium 400 mg/day plus valproate 500 mg/day or double those doses and then randomization to double-blind treatment with lithium plus valproate placebo, lithium plus valproate, or valproate plus lithium placebo. The study involves patients with bipolar I affective disorder who require maintenance treatment, and the randomization phase is scheduled to last 2 years, with the initial intention of randomizing 3000 patients. The start-up phase, which was designed to refine the trial design and procedures, has been reported as a pilot study in 30 patients (89). The main findings were that the combination of lithium plus valproate was tolerable and the results suggested that an initial open design would enhance patient flow. Also, the targeted sample size was revised to about 1068 patients based on the data from the pilot study.

In relapse prevention studies, valproate was superior to lithium in the prevention of recurrences over 1 year in subjects who initially presented with a dysphoric manic episode (90).

However, lithium was equivalent to divalproex in the prevention of relapse in rapid cycling subjects over 20 months (91). It was also equivalent to divalproex in youths aged 5–17 years with type I or II bipolar illness, randomly assigned to either agent and followed for 76 weeks (92).

Lithium versus psychotherapy

Psychodynamic supportive psychotherapy (n = 107) has been compared with psychotherapy plus medication (n = 101) in patients with major depressive disorder (93). The medications included venlafaxine, selective serotonin reuptake inhibitors, nortriptyline, and nortriptyline plus lithium. Lithium was used as an augmentation strategy in the patients who took lithium and nortriptyline (number not given). There were no differences in outcomes between the two treatment groups. No adverse effects specific to lithium were reported.

Placebo-controlled studies

In a double-blind, placebo-controlled study, 175 manic or recently manic patients were stabilized over 8–16 weeks with lamotrigine 100–400 mg/day (n = 59), lithium in a dose sufficient to produce a serum concentration of 0.8–1.1 mmol/l (n = 46), or placebo and were then randomized to continued treatment (94). Both lamotrigine and lithium were superior to placebo in prolonging the time to the next episode of any mood disturbance. Lamotrigine, but not lithium, was superior to placebo in prolonging the time to a depressive episode. Lithium, but not lamotrigine, was superior to placebo in prolonging the time to a manic, hypomanic, or mixed episode.

A similarly designed study in 463 bipolar I patients whose most recent episode was depression gave similar results (95).

A combined analysis of the data from these two studies, involving over 1300 bipolar I patients, showed that in the 638 randomized patients lamotrigine and lithium were superior to placebo regarding time to intervention to the next mood episode (96).

The addition of lithium to other drug therapy has been studied in 92 patients with treatment-resistant major depression taking nortriptyline (97). Non-responders to nortriptyline (n = 35) were randomized to added lithium or placebo; there was no significant difference.

In a review of five randomized controlled trials of prevention of relapse in 770 patients with bipolar affective disorder, lithium has been compared with placebo (98). Lithium was more effective than placebo in preventing all relapses and manic relapses, but the effect on depressive relapses was not as impressive and was termed "equivocal" by the authors. This is not particularly new information, although several of the studies that were included in this meta-analysis were more recent and the analysis was presented as odds ratios rather than episode frequency.

A systematic review of controlled trials has shown that lithium + haloperidol is superior to placebo in the control of manic symptoms (99).

Lithium is also effective in individuals with co-morbid pathological gambling and a mood disturbance. In a randomized, 10-week, placebo-controlled study in 40 subjects randomly assigned to modified-release lithium or placebo, 83% of those who took lithium responded compared with only 29% of those who took placebo (100).

Systematic reviews

A meta-analysis showed a considerable reduction in the suicide rate in bipolar patients taking lithium carbonate compared with patients who were not taking lithium or patients who had stopped taking lithium (101), and this finding has been confirmed in an evaluation of patients who were being treated in two large health maintenance organizations between 1994 and 2001 (102). The subjects were patients with bipolar disorders taking maintenance lithium, divalproex, or carbamazepine. There was a significant reduction in the rate of suicidal behaviors among those taking lithium compared with those taking divalproex or carbamazepine. The suicide attempt rate resulting in hospitalization (events per 1000 patient years) was 4.2 with lithium, 10.5 with divalproex, and 15.5 with carbamazepine. The rate of suicide deaths per 1000 patient years was 0.7 with lithium, 1.7 with divalproex, and 1.0 with carbamazepine.

General adverse effects

The adverse effects and interactions of lithium have been reviewed (5,103–105, 106). One review focused on lithium toxicity in elderly people (107). The adverse effects of lithium range widely in intensity and can be a major cause of nonadherence to therapy. However, with proper attention to the prevention and management of adverse effects, most patients can be treated effectively and safely. Withdrawal of lithium is almost always followed by resolution of adverse effects, although certain problems can sometimes persist (for example renal).

Adverse effects that can occur at concentrations in the target range, especially in the upper part of the range, include mild cognitive complaints, postural tremor, hypothyroidism, weight gain, leukocytosis, hypercalcemia, loss of appetite, nausea, loose stools, acne, and psoriasis. Renal adverse effects include impaired concentrating ability and polyuria with secondary polydipsia. Cardiac adverse effects are rarely symptomatic and are usually reversible. Lithium does not cause physiological dependence, although there may be an increased risk of early recurrence if it is withdrawn rapidly.

When 60 patients (22 men, 38 women) who had taken lithium for 1 year or more (mean 6.9 years; mean serum concentration 0.74 mmol/l) were interviewed about adverse effects, 60% complained of polyuria–polydipsia syndrome (serum creatinine concentrations were normal) and 27% had hypothyroidism requiring treatment (108). Weight gain was more common in women (47 versus 18%) as were hypothyroidism (37 versus 9%) and skin problems (16 versus 9%), while tremor was more common in men (54 versus 26%). Weight gain of over 5 kg in the first year of treatment was the only independent variable predictive of hypothyroidism.

The severity of lithium toxicity depends on the magnitude and duration of exposure and idiosyncratic factors. Manifestations of acute toxicity range in intensity from mild (tremor, unsteadiness, ataxia, dysarthria) to severe (impaired consciousness, neuromuscular irritability, seizures, heart block, and renal insufficiency), and the sequels of toxicity range from none at all to permanent neurological damage (often cerebellar) to death. Causes of raised serum lithium concentrations include increased intake and reduced excretion (due to kidney disease, low sodium intake, drug interactions). Whether long-term lithium use carries a risk of progressive renal insufficiency in a few patients continues to be debated. Reviews have addressed lithium toxicity in the elderly (109) and lithium intoxication with an emphasis on the kidney (110).

Dose- and time-related adverse effects

During initiation and stabilization of treatment, adverse effects, such as gastrointestinal upset, tremor, dysphoria, fatigue, muscle weakness, unsteadiness, thirst, and excessive urination, are not uncommon, but they usually abate with time. Such adverse effects are more likely to occur at higher dosages or higher serum concentrations and can usually be avoided or attenuated by proper attention to dosage, clinical and laboratory monitoring, and, if necessary, the use of adjunctive medication.

Long-term use of lithium is sometimes associated with weight gain, polyuria and polydipsia, and thyroid dysfunction (see below), but many patients have been treated successfully for several decades without developing treatment-limiting adverse effects. However, long-term success should not breed complacency, since there is an ever-present risk of recurrence (if concentrations are too low) and toxicity (if concentrations are too high).

Treatment of toxicity

Two patients with lithium toxicity (serum concentrations 3.5 and 4.2 mmol/l) had the well-recognized rebound increase in serum concentrations after the end of hemodialysis. Both died during hospitalization from what was cryptically described as "unrelated events" (111).

- A 49-year-old with severe lithium toxicity was treated successfully with continuous veno-venous hemodiafiltration; the serum lithium concentration fell from 3.0

to 0.93 mmol/l after 7 hours and there was no rebound increase after the end of the procedure (112). The maximum lithium clearance was 28 ml/minute which is considerably lower than usually attained with hemodialysis.

An all too common treatment error in the face of severe toxicity is "watchful waiting," during which the patient's condition is more likely to worsen than improve.

Organs and Systems

Cardiovascular

Cardiovascular disease is not a contraindication to lithium, but the risks may be greater, in view of factors such as fluid and electrolyte imbalance and the use of concomitant medications. Close clinical and laboratory monitoring is necessary, and an alternative mood stabilizer may be preferred. While long-term tricyclic antidepressant therapy may be more cardiotoxic than lithium, the newer antidepressants (SSRIs and others) seem to be safe.

In two studies of 277 and 133 patients taking long-term lithium, there was no evidence of increased cardiovascular mortality compared with the general population (113,114). While the latter study reported on 16-year mortality, it did not provide information about which patients continued to take lithium after the first 2 years.

Blood pressure

There is a higher mortality from cardiovascular diseases among patients with bipolar affective disorder than in the general population. In a study of 81 patients taking lithium monotherapy, 40 were studied in detail; one had hypothyroidism and six had hypertension (115). Of the 81 patients 13 were taking antihypertensive drugs, suggesting a high prevalence of hypertension. One of the points of the study was to assess if lithium was a factor in cardiovascular risk in these patients, but there was no correlation between the duration of lithium treatment or the duration of bipolar disorder and the presence of hypertension.

Two patients who were taking lithium carbonate for mood disorders and who underwent coronary artery bypass grafting developed refractory hypotension during cardiac surgery, which responded to methylthioninium chloride (116). The authors suspected that chronic lithium therapy had caused cardiac embarrassment and recommended that lithium be withdrawn before cardiac surgery.

Cardiac dysrhythmias

In 30 patients there were only minimal electrocardiographic changes during long-term treatment with lithium using the method of body surface electrocardiographic mapping (117). In contrast, a tricyclic antidepressant showed dose-related effects. Nonspecific, benign ST-T wave electrocardiographic changes are the most common cardiovascular effects of lithium.

- A 13-year-old boy taking lithium developed a "pseudo-myocardial infarct pattern" on the electrocardiogram; this may have been an overinterpretation of nonspecific T-wave changes (118).

A very uncommon adverse effect involves sinus node dysfunction (extreme bradycardia, sinus arrest, sinoatrial block), which can be associated with syncopal episodes, perhaps due to hypothyroidism (119,120). In such cases, lithium must either be withdrawn or continued in the presence of a pacemaker. At therapeutic concentrations, other cardiac conduction disturbances have been reported, sometimes in conjunction with hypercalcemia (121), but are uncommon.

Two reviews of the cardiac effects of psychotropic drugs briefly mentioned lithium and dysrhythmias, with a focus on sinus node dysfunction (122,123), reports of which, as manifested by bradycardia, sinoatrial block, and sinus arrest, continue to accumulate in association with both toxic (124) and therapeutic (125,126) serum lithium concentrations. The rhythm disturbance normalized in some cases when lithium was stopped (124,126), persisted despite discontinuation (125), or was treated with a permanent cardiac pacemaker (126). Of historical interest is the observation that the first patient treated with lithium by Cade developed manifestations of toxicity in 1950, including bradycardia (127).

There have been several reports of bradycardia and sinus node dysfunction.

- During an episode of lithium toxicity (serum concentration 3.86 mmol/l), a 42-year-old woman developed sinus bradycardia that required a temporary pacemaker (124). There was marked prolongation of sinus node recovery time. Lithium was withdrawn and the patient underwent hemodialysis once daily for 3 days; sinus node recovery time normalized. The presence of nontoxic concentrations of carbamazepine may have contributed to the condition.

- A 65-year-old man taking lithium for 2 years, with therapeutic concentrations, developed sinus bradycardia (30 beats/minute), which remitted when the drug was stopped and recurred when it was restarted (128). Implantation of a permanent pacemaker allowed lithium to be continued.

- Asymptomatic bradycardia occurred in three of 15 patients treated for mania with a 20 mg/kg oral loading dose of slow-release lithium carbonate (129).

- A 9-year-old boy whose serum lithium concentration was 1.29 mmol/l had a sinus bradycardia with a junctional escape rhythm (40 beats/minute), which normalized at a lower lithium concentration (130).

- A 58-year-old woman with lithium toxicity developed an irregular bradycardia (as low as 20 beats/minute), which resolved during hemodialysis; persistent sinoatrial conduction delay suggested that she was predisposed to the bradydysrhythmia (131).

- A 52-year-old man took an overdose of lithium (serum concentration 4.58 mmol/l) and developed asymptomatic sinus bradycardia with sinus node dysfunction and multiple atrial extra beats, which resolved after hemodialysis (132).

- A 66-year-old woman with pre-existing first-degree AV block, developed sinus bradycardia, a junctional rhythm, a prolonged QT interval, and syncopal episodes (serum lithium concentration 1.4 mmol/l in a 40-hours sample) about 2 weeks after beginning lithium therapy. She was treated successfully with a pacemaker and a lower dose of lithium (133).

- A 36-year-old man became hypomanic after lithium was withdrawn because of symptomatic first-degree atrioventricular block (although, how first-degree block could have caused symptoms is unclear) (134).

- A 44-year-old woman developed atropine-resistant but isoprenaline-sensitive bradycardia (36 beats/minute), thought to be due to sinus node dysfunction related to lithium, fentanyl, and propofol (120).

- A 52-year-old man with a serum lithium concentration of 4.58 mmol/l had sinus node dysfunction with multiple atrial extra beats and an intraventricular conduction delay, which normalized following hemodialysis (132). Two patients, a 58-year-old woman and a 74-year-old woman, developed sick sinus syndrome while taking lithium but were able to continue taking it after pacemaker implantation (135,136).

- A 59-year-old woman with syncope and sick sinus syndrome, which remitted when lithium was withdrawn, recurred when lithium was restarted, and then persisted despite lithium withdrawal; after a pacemaker was implanted she was treated successfully with lithium for 7 years (135).

Lithium can also occasionally cause tachycardia.

- A 59-year-old man was noted to have tachycardia, a shortened QT interval, and nonspecific ST-T changes, while hospitalized with lithium-associated hypercalcemia (137).

An extension of a previously published study (121) added a third comparator group of 18 hypercalcemic non-lithium treated patients and compared them with 12 hypercalcemic lithium patients, 40 normocalcemic lithium patients, and 20 normocalcemic bipolar patients taking anticonvulsant mood stabilizers (138). Both hypercalcemic groups had more conduction abnormalities than the other two groups, but did not differ from each other in this regard. While the authors concluded that both lithium and calcium played important roles in the dysrhythmias, their data suggested that hypercalcemia alone was the critical factor.

Cardiac dysrhythmias associated with lithium intoxication in the elderly included sinus node dysfunction and junctional bradycardia (109). A retrospective chart review of patients on lithium who had mild but persistent hypercalcemia ($n = 12$) showed a greater frequency of cardiographic conduction disturbances compared with normocalcemic patients taking lithium ($n = 40$) and normocalcemic bipolar patients taking anticonvulsant mood stabilizers ($n = 20$), although the overall frequency of cardiographic abnormalities did not differ significantly among the groups (121). When 21 patients without cardiovascular disease (mean serum lithium 0.66 mmol/l) were compared with healthy controls using standard electrocardiography, vector cardiography, and electrocardiographic body surface potential mapping, the only abnormality was a reduction in the initial phase of depolarization, a finding of questionable clinical significance (139).

Abnormalities of the QT$_c$ interval have been explored in 495 psychiatric patients (87 taking lithium, but many of them also taking other drugs) and 101 healthy controls (140). There was no association of lithium with QT$_c$ prolongation but it was associated with nonspecific T-wave abnormalities (odds ratio 1.9) and increased QT dispersion (odds ratio 2.9). Caution was suggested if lithium is used with drugs associated with QT$_c$ prolongation, such as tricyclic antidepressants, droperidol, and thioridazine.

Sudden death has been reported in 14 psychiatric patients and the literature has been reviewed regarding occult cardiac problems, psychotropic drugs, and sudden death (141).

- A 57-year-old man with bipolar disorder taking olanzapine, lithium, and other drugs had underlying mitral valve prolapse, left ventricular hypertrophy, and His bundle anomalies; he died suddenly, probably because of a cardiac dysrhythmia.

The authors suggested that cardiac pathology should be systematically evaluated in patients who take psychotropic drugs.

Cases of lithium toxicity, its cardiac effects, and issues of cardiac dysfunction in children have been reviewed in the light of a cardiac dysrhythmia in a child.

- A 10-year-old boy developed abdominal pain, diarrhea, and vomiting over 2 days (142). He had a history of bipolar disorder, with psychotic features, a schizoaffective disorder, an intermittent explosive disorder, and attention deficit hyperactivity disorder. He had several other medical problems, including hypothyroidism, asthma, and seizures. He was taking many drugs, including methylphenidate, escitalopram, oxcarbazepine, clonidine, Depakote, thyroid hormone, and lithium. The serum lithium concentration was 3.1 mmol/l. Electrocardiography showed a broad-complex tachydysrhythmia, which persisted despite treatment with intravenous adenosine and lidocaine. The cardiac rhythm was interpreted as a ventricular tachycardia. He was given intravenous procainamide, resulting in temporary slowing of his cardiac rhythm, and a continuous procainamide infusion produced stable sinus rhythm. Over the next 36 hours, he continued to have treatment for his lithium toxicity and procainamide for his ventricular dysrhythmia, and improved. At follow-up a 24-hour Holter monitor showed first-degree atrioventricular block.

I wonder if the diagnosis in this patient was correct. He obviously had a severe behavioral disturbance, which required treatment; however, it is not clear if his polypharmacy was appropriate for his condition.

In two studies lithium treatment was associated with prolongation of the corrected QT interval (QT$_c$). A retrospective analysis of the records of 76 patients taking lithium showed that intervals of over 440 msec were significantly more common in subjects with lithium concentrations over 1.2 mmol/l than in those with concentrations in the usual target range (55% versus 8%); T wave inversion was also more common in subjects with high lithium concentrations (73% versus 17%) (143).

Similar results have been reported in 39 in-patients with either bipolar illness or schizophrenia; the duration of the QT$_c$ interval correlated significantly with lithium concentrations and those with the longest QT$_c$ intervals had the highest lithium concentrations (144).

Cardiomyopathies
A study of cardiomyopathies found a specific cause in 614 of 1230 patients (the remainder were diagnosed as idiopathic). One was attributed to lithium but no details were provided (145).

In a study of 1230 patients with initially unexplained cardiomyopathies, lithium was implicated in one case (114). Using a data-based mining Bayesian statistical approach to the WHO database of adverse reactions to examine antipsychotic drugs and heart muscle disorders, a significant association was found between lithium and cardiomyopathy, but not myocarditis (146). The authors acknowledged that further study is needed to determine if the association is causal.

- A 78-year-old woman developed a cardiomyopathy while taking lithium, imipramine, amineptine, levomepromazine, and lorazepam; it resolved when the medications were withdrawn (147). Whether lithium was causally involved is not known.

Respiratory

A 60-year-old woman with bipolar disorder since the age of 29 developed idiopathic pulmonary fibrosis (cryptogenic fibrosing alveolitis) after having taken lithium for 9 years (148). Whether lithium played a causal role is at best highly speculative.

Ear, nose, throat

In two patients, profuse paroxysmal rhinorrhea improved when lithium was withdrawn (149). The rhinorrhea was thought to be a manifestation of mesotemporal lobe epilepsy that had been worsened by lithium.

Nervous system

Reports of overdose-related neurological symptoms abound (150–160). Among these reports, are cases of neurotoxicity at therapeutic serum lithium concentrations (150,158,160) and neurotoxicity associated with nonconvulsive status epilepticus (154,156).

There have been scattered reports of lithium-associated myasthenia gravis, paresthesia, somnambulism (161), seizures and electroencephalographic abnormalities, confusional states, and a reversible Creutzfeldt–Jakob-like syndrome (162). One report rather inconclusively suggested that lithium caused periodic alternating nystagmus in a 61-year-old man (163).

Electroneurographic studies in 34 lithium maintenance patients and controls (both healthy subjects and mood-disorder patients never on lithium) showed statistically significant reductions in sensory and motor conduction in the lithium group (164). None of the decrements was severe enough to be abnormal, and whether these findings have clinical implications is unknown.

Non-reversible lithium neurotoxicity continues to be reported (165,166), including a case of lithium overdose (serum lithium concentration 3.9 mmol/l) with persistent severe ataxia for 9 months that improved markedly when inadvertently treated with high-dose buspirone (120–160 mg/day) (167).

- An 85-year-old woman became gradually toxic (serum lithium 2.9 mmol/l) in a nursing home (168). Despite only conservative management, there was slow but complete resolution of severe neurological symptoms, including coma, fixed pupils, and Cheyne–Stokes respiration.

Twelve patients with mood disorder had electroencephalography before and after taking lithium for an average of 4.4 months (169). Lithium-related changes included:

(a) increased relative power in the delta- and theta-band frequencies;
(b) decreased relative alpha power;
(c) decreased dominant alpha frequency.

The clinical implications of these observations, if any, are unclear.

- A 73-year-old patient had toxic symptoms and moderately severe, generalized slowing on the electroencephalogram at a therapeutic serum concentration (170).

There has been a report of ataxia and dysarthria/choreoathetosis in a patient taking lithium.

- Ataxia and dysarthria/choreoathetosis occurred in a 76-year-old woman who was taking lithium and had a serum lithium concentration of 2.43 mmol/l (171). She underwent hemodialysis and the choreoathetotic movements disappeared. She also had repeated episodes of a tachydysrhythmia and atrial fibrillation, but returned to sinus rhythm after the lithium toxicity had resolved.

In a Canadian study of 200 000 automobile drivers, aged 67–84, who were followed from 1990 until they reached age 85, or until they emigrated from Quebec, or until 31 May 1993, those who had been involved in an automobile accident in which one person sustained a physical injury were assessed; the controls were a 6% random sample of the others (172). Of 5579 patients 20 had been taking lithium within the year before the index date, 19 of whom had been taking it within 16 days before the accident. This compared with 27 of 13 300 patients in the control group (OR = 1.8); for current lithium use, the odds ratio was 2.08. The data on carbamazepine did not show a raised odds ratio. The authors concluded that elderly patients taking lithium have a two-fold increase

in the risk of an injurious motor vehicle accident while driving. Whether this was due to lithium, other medications, or the severity of illness factors could not be determined.

Of 44 patients who used a combination of lithium plus clozapine for a mean of 23.5 months, 37 were rated as having responded (173). Most had schizophrenia or schizoaffective disorder, and only two had bipolar I disorder. There were adverse events, mostly benign and transient, in 28; however, eight patients developed transient new neurological adverse events, including two episodes of myoclonus and one generalized tonic-clonic seizure. In 23 patients who agreed to a reassessment there were no neurological or neurotoxic events. During treatment there were three neurological adverse events that had not been present before treatment; these included three instances of myoclonus and one generalized tonic-clonic seizure. The authors concluded that the combination of lithium plus clozapine does not result in an increased risk of neurological adverse effects compared with either drug alone. Furthermore, the combination produced improved efficacy over clozapine alone. These data contrast with data from previous studies, which suggested that the combination of lithium plus clozapine produced an increased risk of adverse neurological events due to a serotonergic interaction. However, the authors noted that some of the neurological events occurred during co-treatment with serotonergic compounds, such as paroxetine or trimipramine, and also that paroxetine could increase clozapine plasma concentrations and therefore increase the likelihood of a neurological adverse event. They suggested that if lithium and clozapine are combined, co-medication with serotonergic antidepressant drugs or drugs that interfere with clozapine metabolism or renal clearance should be avoided.

- A 70-year-old man developed lithium intoxication after a transient ischemic attack (174). He had a bradycardia (35–40/minute) and episodes of sinoatrial block. The clinical presentation was suggestive of a stroke.

The authors discussed the difficulty of the differential diagnosis between lithium intoxication and other neurological disorders, such as strokes. What they did not discuss was the possibility that the presentation was caused by sinus node dysfunction, which has been reported as a complication of lithium treatment.

Cerebellar syndromes

Chronic neurological sequelae of intoxication included two patients with a persistent cerebellar syndrome and severe cerebellar atrophy (175), one with subcortical dementia (176), and one with a diffuse sensorimotor peripheral neuropathy (177).

- An 86-year-old man with a serum lithium concentration of 0.7 mmol/l presented with a several month history of asterixis which resolved fully within 2 weeks of stopping lithium (178).

- Symptoms suggestive of toxicity at therapeutic serum concentrations also occurred in a 49-year-old man taking lithium (0.7 mmol/l), carbamazepine, and trifluperidol, who developed persistent cerebellar deterioration during a febrile episode of lobar pneumonia (179).
- Another patient who developed cerebellar symptoms consistent with lithium neurotoxicity despite a low therapeutic serum concentration (0.5 mmol/l) was more fortunate, as the symptoms resolved promptly when lithium was withdrawn (180).
- Permanent cerebellar sequelae occurred in a 36-year-old man after intoxication at therapeutic lithium concentrations (181).

A literature review of permanent neurological complications of acute lithium toxicity noted that a cerebellar syndrome is quite common and that neuroleptic drugs can worsen toxicity, as might rapid reduction of raised serum lithium concentrations (182). However, the latter point is speculative at best; hemodialysis remains the treatment of choice for severe lithium poisoning.

In a review of published cases of the syndrome of irreversible lithium-induced neurotoxicity (SILENT), cerebellar dysfunction was the most commonly reported long-lived outcome (183).

Convulsions
A proconvulsive effect of lithium was also suggested in two patients purported to have temporal lobe epilepsy, who improved when carbamazepine replaced lithium (149). Twelve cases of status dystonicus of varying causes included a woman with post-traumatic dystonia, who was treated unsuccessfully with lithium. Despite the lack of response, her muscular spasms worsened when lithium was stopped (184).

Five patients taking lithium for unspecified durations developed multifocal action myoclonus without reflex activation, which resolved fully when the drug was withdrawn (185).

- A patient presented with nonconvulsive status epilepticus and a serum lithium concentration of 1.9 mmol/l (186).
- A prolonged seizure occurred after electroconvulsive therapy (ECT) in a 45-year-old man taking lithium, amfebutamone, and venlafaxine (187).

Creutzfeldt–Jakob-like syndrome
A reversible Creutzfeldt–Jakob-like syndrome has been described in patients taking lithium.

- A 65-year-old woman taking lithium, levomepromazine, and phenobarbital developed a Creutzfeldt–Jakob-like syndrome after she had mistakenly increased her lithium dosage (162).
- A 66-year-old man presented comatose with an EEG suggestive of Creutzfeldt–Jakob encephalopathy after an 11-month history of progressive dementia and parkinsonism (188). Lithium (serum concentration 1.3 mmol/l), which he had been taking for 13 years,

was withdrawn, and by day 78, clinical examination showed only mild neurological impairment.

Extrapyramidal effects
Parkinsonism
Occasionally, long-term use of lithium is associated with cogwheel rigidity and a parkinsonian tremor (189). More often than not, concurrent or past treatment with an antipsychotic drug is involved. In a review of SSRI-induced extrapyramidal adverse effects, lithium was listed, but not discussed, as a possible risk factor (190). A review of drug-induced parkinsonism provided references to case reports of lithium's "occasionally inducing or exacerbating parkinsonism" (191).

Two patients developed parkinsonism after taking lithium for many years, but did not improve when the drug was withdrawn and a causal relation could not be established (136).

- An 86-year-old man taking lithium monotherapy (serum concentration 0.7 mmol/l) had asterixis for several months; it resolved fully within 2 weeks of stopping lithium (178).
- A 65-year-old woman, after a brief exposure to lithium, developed severe akathisia and disabling parkinsonism; within 4 days of stopping lithium, the symptoms improved, but orofacial dyskinesia appeared; 10 months later both the parkinsonism and the dyskinesia had improved, but were still present (189).

Dystonias
Reversible tardive dystonia has been attributed to lithium (192).

Neuroleptic malignant syndrome
There have been reports of neuroleptic malignant syndrome in patients taking lithium and a neuroleptic drug (193–195). However, while previous reports have suggested an association between lithium and neuroleptic malignant syndrome (196), a case-control study ($n = 12$ with the syndrome, $n = 24$ controls) found no such association (197). Since none of the patients with neuroleptic malignant syndrome was taking lithium, little can be concluded from such a finding.

However, lithium can be associated with the neuroleptic malignant syndrome in patients who are taking neuroleptic drugs, as in the following cases:

- A 63-year-old man taking lithium and amoxapine (193).
- A 23-year-old woman taking lithium, olanzapine, and fluoxetine (194).
- A 39-year-old woman who had overdosed with lithium and who took a single dose of haloperidol (195).

Whether lithium alone can cause neuroleptic malignant syndrome is controversial (SEDA-23, 24).

- A 45-year-old man took an intentional overdose of lithium (198). He was dialysed and stabilized, but on day 10 developed neuroleptic malignant syndrome. He died after developing acute renal insufficiency and acute respiratory distress syndrome.

The authors concluded that lithium alone can cause neuroleptic malignant syndrome, since no neuroleptic drugs had been co-administered in this case.

Headache

- A 64-year-old woman had a 2-week history of daily bilateral holocranial headache as the presenting complaint of lithium toxicity (serum concentration 2.5 mmol/l); dosage reduction resolved the headache and the extrapyramidal and cerebellar effects (199).

Myasthenia

- A 51-year-old man developed a myasthenic syndrome which resolved when lithium was withdrawn (200). The authors referred to three other previously reported cases of lithium-induced myasthenia.

Pseudotumor cerebri

Pseudotumor cerebri (benign intracranial hypertension) has been linked to lithium in over 30 cases, with headache, papilledema, increased intracranial pressure, reduced vision, and a risk of blindness (201). The condition tends to improve on withdrawal, but surgical intervention may sometimes be necessary. A review of pseudotumor cerebri devoted one paragraph to induction of this condition by lithium and provided six references but no new information (202).

- A 17-year-old woman developed pseudotumor cerebri with headache after she had taken lithium for 6.5 weeks (203). Papilledema and increased intracranial pressure resolved fully when lithium was withdrawn, and she was given acetazolamide.

Serotonin syndrome

Serotonin syndrome has been reviewed twice, but mention of lithium as a possible contributing factor was scanty (204,205).

Somnambulism

After almost a two-decade nap, interest in lithium-induced somnambulism was reawakened by a questionnaire survey of 389 clinic patients that found sleepwalking in 6.9% of those taking lithium alone or with other drugs compared with a 2.5% prevalence in the general population (161).

- A 52-year-old man had a 5-year history of sleepwalking 2–3 times a week beginning 3 months after starting lithium (206). On one occasion, he was injured in a fall when sleepwalking through a second-story window. When lithium was stopped for 3 months, the problem resolved only to recur when it was restarted.

Stuttering

Stuttering worsened in a 48-year-old man while he was taking lithium, with improvement when he was switched to valproate (207).

Tremor

Lithium commonly causes a benign postural tremor, especially early in the course of treatment (208,209) and can enhance physiological tremor (209). Risk factors include higher serum lithium concentrations, a family history of tremor, high caffeine intake, and the use of other drugs that cause tremor (for example antidepressants, valproate). The tremor is often well tolerated, but in some cases, it can be socially or occupationally problematic. In such cases, treatment possibilities include reducing the dose of lithium, the use of a modified-release formulation, reducing caffeine intake, or the addition of a tremor-reducing drug, such as a beta-blocker (210), primidone (211), or possibly gabapentin (212).

Tremor was reported as an adverse effect of lithium in 12 of 22 men and 10 of 38 women who had taken it for at least 1 year (108). In an open study, there was full remission of lithium tremor in four of five patients treated with vitamin B6 (213).

A double-blind study in which 31 patients with breakthrough depression taking lithium received augmentation with either paroxetine or amitriptyline and showed a quantitative increase in tremor activity with combined therapy, but no significant change in tremor frequency (214).

- Severe essential tremor, which was at first mistaken for tardive dyskinesia, occurred in an 80-year-old woman taking lithium; it resolved almost completely when lithium was withdrawn (215).

Neuromuscular function

Lithium has been implicated in impaired athletic prowess in two cases (216).

- A 21-year-old man had muscular incoordination while fast bowling (cricket), which improved when he switched to valproate.
- A 71-year-old woman was unable to serve properly at tennis until she stopped taking lithium.

Sensory systems

Eyes

Lithium can alter retinal sensitivity to light, and concern has been expressed that bright light therapy for seasonal mood disorder can increase the risk of photoreceptor cell damage, although such concerns have not been confirmed clinically (217,218).

Other effects on the visual system attributed to lithium include reduced accommodation, exophthalmos, extraocular muscle abnormalities, nystagmus (most characteristically downbeat), oscillopsia, photophobia, and papilledema with visual impairment (pseudotumor cerebri) (201).

- A 21-year-old woman developed temporary blindness possibly related to lithium toxicity (159).

Ears

The finding of reversible lithium-induced moderate hearing loss (especially low frequency) in guinea pigs has led to speculation that similar findings could occur in humans (219).

Psychological, psychiatric

Neuropsychological testing in a 51-year-old woman with a serum lithium concentration of 2.4 mmol/l showed striking cortical and subcortical deficits, which had only partially resolved when she was retested 4 and 14 weeks later (157).

Cognitive effects

Most people taking lithium have no complaints about its effects on mental processes; they feel normal and function normally. Mild cognitive effects can occur, although it is sometimes difficult to determine whether they are due to the lithium, the resolution of manic euphoria, the presence of mild depression or hypothyroidism, or the effects of other medications. Complaints can include reduced reactivity, lack of spontaneity, loss of emotional tone, and patchy memory impairment. Lithium reportedly has adverse effects on memory, speed of information processing, and reaction time (220,221). It has been suggested that the risk of driving accidents may be increased in patients taking lithium, but there is no evidence that this is so. As with any psychotropic drug, patients should be cautious about driving or operating dangerous machinery until they know how they will react to lithium. These mild cognitive adverse effects can be underappreciated by clinicians, yet be a cause of nonadherence to therapy. The action of lithium on the hypothalamic–pituitary–thyroid axis has been discussed in relation to its cognitive effects (222). Although, the English abstract of a Polish review concluded that there is no evidence of significant cognitive deficits caused by lithium (223), others dispute this conclusion (221).

Assessment of cognitive performance in 43 patients with bipolar I disorder showed no correlation between the use of lithium, carbamazepine, or valproate, and full-scale IQ or general and working memory (all of which were impaired by antipsychotic drugs) (224). While a deficit in sustained attention was noted in 19 euthymic patients with bipolar disorder (17 taking medications, 11 taking lithium alone or in combination), there was no difference between those taking and those not taking lithium (225).

In a 3-week, double-blind study of the cognitive effects of lithium (serum concentration about 0.8 mmol/l; $n = 15$) versus placebo ($n = 15$) in healthy subjects, lithium did not impair implicit recall, ability to process two tasks concurrently and simultaneously, short-term memory, or selective attention, but caused impaired learning during repeated administration of memory tests (226). Since neuropsychological testing could not distinguish lithium treatment from pre- and post-treatment in the lithium group, any lithium effects must be considered subtle at best.

It has been suggested that thyroid hormone may minimize the cognitive effects of lithium (227) and a review of this suggested benefit in those patients taking lithium who had subclinical hypothyroidism (222).

Lithium has diverse effects on creativity. While creativity can suffer from the absence of mania or from lithium-induced adverse effects, it can also be improved by the mood stabilization that lithium produces, or may be unaffected by treatment (227).

Four of seventeen studies of the cognitive effects of lithium were deemed to have acceptable methods. Reported adverse effects included effects on memory, speed of information processing, and reaction time (often in the absence of subjective complaints), suggesting that the risk of driving accidents might be increased when patients are taking lithium (220,221). Interventions for lithium-induced cognitive impairment include dosage reduction, use of a modified-release formulation, treatment of thyroid dysfunction, and assessing the role of concomitant illness and medication. When 67 patients who complained of cognitive deterioration while taking lithium were treated with aniracetam, 97% reported subjective improvement (228).

Delirium and demention

In a case-control study, lithium was found to be one of several risk factors for delirium in 91 psychiatric inpatients (odds ratio 2.23), although the authors concluded that this observation may have been confounded by an association with manic episodes (229).

- A 38-year-old woman who had tolerated lithium for 20 years developed delirium following a manic episode, despite therapeutic concentrations of lithium (0.7–1.0 mmol/l); the episode remitted fully after lithium was withdrawn (230).
- A 36-year-old developed a febrile confusional state in the absence of infection while taking lithium (serum concentrations 0.37 and 0.58 mmol/l) and cyamemazine, resulting in a persistent cerebellar syndrome (231).
- A 62-year-old woman became delirious when lithium at therapeutic concentrations was added to valproate, haloperidol, and biperiden (232). The delirium resolved after all drugs were withdrawn, but 6 months later she still had choreoathetoid movements.

The incidence of delirium in elderly subjects (over 65 years old) taking lithium does not appear to be higher than in those who are taking valproate (233). Among 5360 subjects with a mood disorder who had taken lithium or valproate in the previous year, the incidence of delirium with valproate was very similar to that of lithium (4.1 versus 2.8 cases/100 person years; HR = 1.36; 95% CI = 0.94, 1.97). Both of these rates were significantly lower than the rates observed with the anticholinergic drug benzatropine.

Actually, lithium has been associated with many effects that are believed to be neuroprotective (234). Most importantly, it reduces the activity of glycogen synthase kinase-3 (GSK-3), which leads to reduced production of the tau protein (235,236). However, lithium may actually increase the production of amyloid beta (237), although previous reports have suggested that lithium reduces amyloid beta and its consequent toxicity (238,239).

Clinically, lithium may protect patients against dementia, although it has been reported that patients who take lithium are actually at a higher risk of developing dementia (240). However, in truth, patients with bipolar affective disorder are at a higher risk of dementia, and lithium treatment reduces the relative risk nearly to that of the general population (241).

Suicide

In a systematic review of 32 randomized trials in which 1389 patients took lithium and 2069 took another agent (carbamazepine, divalproex, lamotrigine, or the antidepressants amitriptyline, fluvoxamine, mianserin, and maprotiline), among the seven studies that reported suicides, lithium-treated patients had significantly fewer completed events (242). These included two suicides on lithium (out of 503, 0.4%) and 11 suicides on other agents (two placebo, two amitriptyline, six carbamazepine, and one lamotrigine, out of a total of 601, 1.8%) (OR = 0.26; 95% CI = 0.09, 0.77).

The lower rate of completed suicides in those taking lithium is of particular note, since suicidal ideation may actually be more common in lithium-treated patients. In 128 patients followed prospectively for an average of 13 years suicidal ideation was non-significantly higher in lithium-treated patients than in those taking valproate or carbamazepine (243). This may be related to clinician preference, since among patients with bipolar affective disorder and suicidal ideation in the Systematic Treatment Enhancement Program for Bipolar Disorder (STEP-BD) clinicians were more likely to prescribe antidepressants and second-generation antipsychotic drugs, while they reserved lithium for more severely ill individuals (244).

There was a similar pattern in an observational study of all lithium prescriptions and recorded suicides in Demark from 1995 to1999 inclusive (245). Purchasing lithium was associated with a higher rate of suicide, but purchasing lithium at least twice was associated with a significantly lower risk (0.44; 95% CI = 0.28, 0.70). In other words, lithium is prescribed for patients who are at high risk of suicide, but continuing to take lithium appears to be protective. These studies have laid the foundations for a current adequately powered, prospective study of the purported anti-suicide effect of lithium (246).

Endocrine

Corticotropin (ACTH)

Calcium infusion in lithium patients ($n = 7$) and controls ($n = 7$) caused similar increases in ACTH concentrations across a physiological range of calcium (247).

Prolactin

Serum prolactin concentrations in patients taking long-term ($n = 15$) or short-term ($n = 15$) lithium did not differ from controls (248). In another study, when compared with 17 healthy controls, 20 euthymic bipolar patients who had taken lithium for more than 6 months had significantly lower serum prolactin concentrations (9.72 ng/ml versus 15.56 ng/ml), but prolactin concentrations in short-term lithium users ($n = 15$) did not differ from controls (249). Antipsychotic drugs were not involved.

Diabetes insipidus

Two of ten patients taking long-term lithium therapy were thought to have hypothalamic diabetes insipidus, because of a positive response to desmopressin (250).

For nephrogenic diabetes insipidus, see under Urinary tract below.

Hypothalamic–pituitary–adrenal axis

Hypothalamic–pituitary–adrenal axis function in bipolar disorder has been reviewed, but lithium was mentioned only in passing (251). Two studies ($n = 25$, $n = 24$), possibly reporting many of the same patients, showed that lithium augmentation of antidepressant-resistant unipolar depression increased hypothalamic–pituitary–adrenal axis activity, measured by the dexamethasone suppression test, either alone or combined with the corticotropin releasing hormone test (252,253). However, the tests did not distinguish between lithium responders and nonresponders.

Thyroid

The many effects of lithium on thyroid physiology and on the hypothalamic–pituitary axis and their clinical impact (goiter, hypothyroidism, and hyperthyroidism) have been reviewed (254). Lithium has a variety of effects on the hypothalamic–pituitary–thyroid axis, but it predominantly inhibits the release of thyroid hormone. It can also block the action of thyroid stimulating hormone (TSH) and enhance the peripheral degradation of thyroxine (254). Most patients have enough thyroid reserve to remain euthyroid during treatment, although some initially have modest rises in serum TSH that normalize over time.

Both goiter and hypothyroidism continue to be reported as complications of lithium therapy (136,255,256).

In a cross-sectional study of 121 patients taking lithium, there was no difference in thyroid function tests among those taking treatment for 0.7–6 months, 7–10 months, or 61–240 months. However, when compared with healthy volunteers ($n = 24$) and prelithium controls ($n = 11$), there was a significant increase in radioiodine uptake in all lithium groups. Serum TSH concentrations were higher in prelithium patients than controls and highest in those taking lithium. Being from an iodine-deficient area appeared to predispose lithium patients to abnormally high TSH values and clinical hypothyroidism (257).

In 1989, in 150 patients at different stages of lithium therapy, thyroid function was assessed and subsequently 118 were reassessed at least once and 54 completed a 10-year follow-up (258). The annual rates of new cases of thyroid dysfunction were subclinical hypothyroidism 1.7%, goiter 2.1%, and autoimmunity 1.4%. While these figures were little different from those found in the general population, the authors acknowledged that lithium was a potential cause of thyroid dysfunction.

Of 42 bipolar patients who had taken lithium for 4–156 months, three had subclinical hypothyroidism, three had subclinical hyperthyroidism, and one was overtly hyperthyroid (256). Ultrasonography showed that goiter was present in 38% and mild thyroid dysfunction was suggested in 48% because of an apparent increased conversion of free T4 to free T3. There was no correlation between the duration of lithium therapy and thyroid abnormalities.

In a controlled, cross-sectional comparison of 100 patients with mood disturbance who had taken lithium for at least 6 months and 100 psychiatrically normal controls, lithium did not increase the prevalence of thyroid autoimmunity; a minimally larger number of control subjects had antithyroid peroxidase antibodies (11 controls versus 7 patients with mood disorders) and anti-thyroglobulin antibodies (15 versus 8) (259).

Hypothyroidism

Lithium-induced hypothyroidism has been briefly reviewed (260). Some patients develop more persistent subclinical hypothyroidism (TSH over 5 mU/l, free thyroxine normal) and others overt hypothyroidism (higher risk in women, in those with pre-existing thyroid dysfunction, and those with a family history of hypothyroidism). Since subclinical hypothyroidism is not necessarily asymptomatic, treatment with thyroxine may be necessary in this group (261), as well as in those with more obvious hypothyroidism (262).

The prevalence of thyroperoxidase antibodies was higher in 226 bipolar patients (28%) than in population- and psychiatric-control groups (3–18%). While there was no association with lithium exposure, the presence of antibodies increased the risk of lithium-induced hypothyroidism (263).

In 1705 patients, aged 65 years or over, who had recently started to take lithium, identified from the 1.3 million adults in Ontario receiving universal health care coverage, the rate of treatment with thyroxine was 5.65 per 100 person-years, significantly higher that the rate of 2.70/100 person-years found in 2406 new users of valproate (264). Of 46 adults taking lithium in a psychiatric clinic, 17% developed overt hypothyroidism while 35% had subclinical hypothyroidism (raised concentrations of thyroid stimulating hormone, TSH) (265).

Thyroid function tests in 101 lithium maintenance patients were compared with their baseline values and with results in 82 controls without psychiatric or endocrine diagnoses. With hypothyroidism defined as a serum TSH above the reference range, 8 patients were hypothyroid at baseline, and another 40 became so during treatment. Women over 60 years of age were at slightly higher risk and had higher TSH values. Patients with a positive family history of hypothyroidism had raised TSH concentrations sooner after starting lithium (3.7 versus 8.7 years). Whether any patients became clinically hypothyroid was not noted (it was stated that those with grade II hypothyroidism were almost free of symptoms) (266).

Serum TSH concentrations were raised (10 mU/l or more) in 13 of 61 children aged 5–17 years taking lithium and valproate for up to 20 weeks (267).

In a review of lithium-induced subclinical hypothyroidism (TSH over 5 mU/l, free thyroxine normal), a prevalence of up to 23% in lithium patients was contrasted with up to 10% in the general population. It was stressed that subclinical hypothyroidism from any cause can be associated with subtle neuropsychiatric symptoms, such as depression, impaired memory and concentration, and mental slowing and lethargy, as well as with other somatic symptoms. Management guidelines were discussed (261).

An abstract reported that 23% of 61 children and adolescents taking lithium and divalproex sodium for up to 20 weeks had a TSH concentration over 10 mU/l (reference range 0.2–6.0); however, no clinical information was provided (268). Another abstract reported that the prevalence of thyroperoxidase antibodies was higher in bipolar outpatients (28% of 226) than in psychiatric inpatients with any diagnosis (10% of 2782) or healthy controls (14% of 225), but this was not related to lithium exposure; on the other hand, hypothyroidism was associated with lithium exposure, especially in the presence of antithyroid antibodies (269).

When 22 men and 38 women who had taken lithium for at least a year (mean 6.9 years) for bipolar disorder were evaluated for adverse effects, hypothyroidism requiring thyroid supplementation was found in 16 (14 women and 2 men); 9 had a goiter (108). The area from which some of the patients came was known to have a high background incidence of thyroid dysfunction.

The observation that Canada, with ample nutritional iodine, has a relatively high rate of lithium-related hypothyroidism compared with relatively low rates in iodine-deficient countries such as Italy, Spain, and Germany led to the suggestion that ambient iodine may play a role in the genesis of this condition (270). This is reminiscent of the association of amiodarone with hypothyroidism or hyperthyroidism in iodine-replete and iodine-deficient areas respectively (SEDA-10, 148).

Case reports of adverse thyroid effects of lithium have included the following:

- A 56-year-old man taking lithium whose TSH concentration was abnormally high (50 mU/l) (271).
- A 44-year-old woman who had taken lithium for 10 years and who developed swelling of the right lobe of the thyroid and hypothyroidism (120).
- A 63-year-old woman taking long-term lithium who developed subclinical hypothyroidism and primary hyperparathyroidism (272).

Hyperthyroidism

Despite the predominantly antithyroid effects of lithium, thyrotoxicosis continues to be described during treatment and after withdrawal (273–275). In a retrospective review of 201 patients taking lithium (mean duration 6.4 years), hypothyroidism requiring supplemental thyroxine developed in 10% (3.4% of men, 15% of women) after a mean duration of 56 months. Women over 50 years of age tended to have an earlier onset. Two patients developed goiter requiring surgery and two others developed thyrotoxicosis (262).

Reports of hyperthyroidism associated with lithium include one in a woman who was also hypercalcemic with a normal parathyroid hormone (PTH) concentration (276) and two discovered while treating lithium toxicity (277).

- A 27-year-old man developed thyrotoxicosis while taking lithium (275).
- A 52-year-old woman became thyrotoxic 2 months after stopping long-term lithium therapy; the authors briefly reviewed 10 previous reports (278).
- A woman with lithium-associated hyperthyroidism lost 2 kg over 3 months, suggesting that lithium may have indirectly caused the weight loss (279).

Thyroiditis

A retrospective record review of 300 patients with Graves' disease and 100 with silent thyroiditis who had undergone thyroid scans showed that the likelihood of lithium exposure was 4.7 times higher in the latter, suggesting a link between lithium and thyrotoxicosis caused by silent thyroiditis (280).

- A 30-year-old man, who had taken lithium for 16 years for bipolar disorder and long-term ciclosporin and prednisolone after a bone-marrow transplant, developed subacute thyroiditis associated with a diffusely enlarged gland that showed heterogeneous echogenicity, but without a clear relation to lithium (281).

Goiter

Euthyroid or hypothyroid goiter can also complicate lithium therapy, although the goiter is seldom of clinical importance and tends to resolve on withdrawal or with thyroxine treatment. In one ultrasound study, there was a 44% incidence of goiter in patients who had taken lithium for 1–5 years compared with 16% in a control group; cigarette smoking was associated with a greater size and frequency of goiter in both groups (282).

Hyperthyroidism has also been associated with lithium use and withdrawal, although a cause-and-effect relation has been more difficult to establish. In fact, lithium has been used with some success to treat hyperthyroidism, particularly in conjunction with propylthiouracil (283) and [131]I (25).

Parathyroid and calcium

Lithium toxicity has been reported in a patient with hypercalcemia (284).

- A 54-year-old man, who had taken lithium for 15 years without problems, suddenly developed food and water aversion, hypercalcemia (2.75 mmol/l), and lithium toxicity, with a serum lithium concentration of 4.3 mmol/l. He was confused, delirious, and irritable. Hemodialysis produced a marked improvement in laboratory tests, which became normal after 9 days.

The authors concluded that the hypercalcemia was due to long-term lithium treatment. The patient's chief complaint included nausea when he was exposed to food and water, and he therefore refused food and water for 2–3 days before admission. He also had acute renal insufficiency, which was thought to be due to the hypercalcemia, water aversion, and perhaps "idiosyncratic reasons." The

renal insufficiency and water aversion resulted in lithium toxicity. The studies cited in this paper showned that hyperparathyroidism occurs in 5–40% of patients taking long-term lithium, compared with a population frequency of less than 4%, and in areview, 27 reports of parathyroid adenoma and 11 of hyperplasia were mentioned (285). The hypercalcemia and raised PTH concentrations are often reversible on withdrawal, but surgical intervention may be necessary. So far, long-term lithium therapy has not emerged as a risk factor for reduced bone mineral density or osteoporosis (286). In one study, there was a greater frequency of electrocardiographic conduction defects in hypercalcemic patients taking lithium than in normocalcemic patients taking lithium (121).

Of 15 patients who had taken long-term lithium and who had also had surgery for primary hyperparathyroidism, one had recurrent hyperparathyroidism 2 years after the first operation (287). The authors noted that in their experience hyperparathyroidism during lithium treatment was associated with a high incidence of parathyroid adenomas rather than parathyroid gland hyperplasia, and they suggested that lithium might selectively stimulate the growth of parathyroid adenomas in individuals who are susceptible to developing parathyroid adenomas. Furthermore, such adenomas were best treated by excision rather than subtotal parathyroidectomy.

Of 12 patients who underwent parathyroid gland resection while maintaining their intake of lithium, only eight remained normocalcemic (288). Nevertheless, the authors recommended that surgery should be considered if there is hyperparathyroidism.

Of 537 patients who had parathyroid glands excised for hyperparathyroidism, 12 (2.2%) had been taking lithium and 11 (2.0%) had been taking it long-term (mean 15.3 years, range 2–30). Manifestations included fatigue, bone pain and fracture, and abdominal pain and constipation. Six had a single adenoma and five had multigland hyperplasia. All resumed lithium, but one had a recurrence after 3 years and one had increased PTH concentrations, but a normal serum calcium. A literature review detected 27 prior reports of parathyroid adenoma and 11 of hyperplasia associated with lithium (285).

When 15 euthymic bipolar patients who had taken lithium for a mean of 49 months were compared with 10 nonlithium euthymic bipolar controls, the former had significantly higher total serum calcium concentrations and intact PTH (iPTH) concentrations (289). The authors advised baseline and periodic serum calcium and iPTH concentrations and bone density measurements in all lithium patients, although whether the benefit outweighs cost is open to question.

Ten patients who had taken lithium for less than 1 year and 13 who had taken it for more than 3 years were assessed for alterations in bone metabolism and parathyroid function (286). There were no differences in bone mineral density, serum calcium concentration, or PTH concentration, but both groups had increased bone turnover and the long-term group had nonsignificantly higher calcium and PTH concentrations (including one hyperparathyroid patient who had an adenoma excised). The authors' conclusion that lithium therapy is not a risk factor for osteoporosis needs to be tempered by the small sample size, the case of adenoma, and the blood concentration trends.

Total serum calcium and iPTH concentrations were measured in 15 patients taking long-term lithium and 10 lithium-naïve patients; both were significantly higher in the lithium group (290). While the number of lithium patients with abnormally high concentrations was not stated, mean iPTH concentrations were almost twice the upper limit of the reference range (102 versus 55 pg/ml).

Parathyroid tumors from nine patients with lithium-associated hyperparathyroidism (six multiglandular, three uniglandular) were compared with 13 nonlithium-associated sporadic parathyroid tumors with regard to gross genomic alterations (291). Gross chromosomal alterations were absent in most of the lithium group and were more common in the sporadic group.

In 53 patients studied prospectively at 1, 6, 12, and 24 months, lithium increased serum PTH concentrations (apparent by 6 months) and increased renal reabsorption of calcium in the absence of a significant change in serum calcium (292). A prospective study of 101 lithium maintenance patients and 82 healthy controls showed higher serum calcium concentrations during lithium treatment than at baseline or in the controls, and higher calcium serum concentrations in those lithium patients over 60 years of age (266).

When compared with 12 healthy matched controls, 13 women who had taken lithium for a mean of 8 (range 3–16) years had higher mean ionized and total calcium concentrations, but mean plasma parathormone concentrations did not differ. In eight of the women taking lithium, the calcium concentration was above the upper end of the reference range, and in one the parathormone concentration was abnormally high (293).

Of 15 patients taking long-term lithium who had surgery for primary hyperparathyroidism, 14 had adenomas (11 single, 3 double) and one had four-gland hyperplasia. All restarted lithium successfully after surgery, except one who again developed hyperparathyroidism, resulting in removal of another adenoma (294).

Hyperparathyroidism was considered a possible cause of treatment-resistant manic psychosis in a patient taking lithium (295).

- Hypercalcemia and raised PTH concentrations improved in a woman who had taken lithium for over 20 years after she was switched to divalproex (296).
- A 64-year-old woman who had taken lithium for over 10 years was admitted with altered consciousness, agitation, and disorientation. The serum calcium was 3.35 mmol/l (reference range 2.1–2.6 mmol/l) and the PTH concentration was raised. With hydration and conversion from lithium to valproate, the serum calcium concentration normalized, but 2 years later disorientation and hypercalcemia recurred and a 150 mg parathyroid adenoma was removed surgically (297).
- A 53-year-old woman who had taken lithium for 9 years and carbamazepine for 3 years and had a 3-month history of lethargy was found to be hypercalcemic with a raised concentration of iPTH. She was

saved from parathyroid surgery when withdrawal of lithium resolved the hypercalcemia (298).
- A 51-year-old man who had taken lithium for over 10 years presented with nausea, vomiting, anorexia, hypercalcemia (3.1 mmol/l), and increased PTH concentration (iPTH 110 ng/l). Abnormalities resolved after an oxyphilic parathyroid adenoma was excised (299).

Other reports of hyperparathyroidism in patients taking lithium have included the following:

- Three cases among 26 cases of chronic lithium poisoning (300).
- A 78-year-old man who had taken lithium for 30 years who presented with dehydration, azotemia, hypernatremia, hypercalcemia, and increased PTH concentrations (301).
- A 63-year-old woman taking long-term lithium therapy (272).
- A woman who had taken lithium for 15 years who became hypercalcemic and stopped taking lithium, but 2 years later had two parathyroid adenomas removed surgically (302).
- A 42-year-old man who had taken lithium for 17 years and who had raised serum calcium and PTH concentrations which normalized after removal of a parathyroid adenoma (303).
- A 59-year-old man with hypercalcemia and increased PTH concentrations 3 months after starting lithium, which normalized after lithium was withdrawn (137).
- Three cases from Denmark (304) and one from Spain (305);
- A 78-year-old woman who had taken lithium for 25 years (306).
- A 74-year-old man who had an adenoma resected (136).
- Two 77-year-old women who developed hyperparathyroidism which was managed medically (136).
- A 39-year-old (sex unspecified) whose adenoma was resected after taking lithium for 10 years (294).
- A 59-year-old woman with hyperparathyroidism (307).

A lithium chloride solution caused changes in gravicurvature, statocyte ultrastructure, and calcium balance in pea root, believed to be due to effects of lithium on the phosphoinositide second messenger system (308). The implications with regard to human parathyroid function are obscure.

Metabolism

Diabetes mellitus

It has been reported that diabetes mellitus is three times more common in bipolar patients than in the general population (309). However, lithium does not appear to increase the risk of diabetes mellitus, and its use in patients with pre-existing diabetes is generally safe, assuming that the diabetes is well controlled.

When lithium toxicity has been reported in patients with diabetes mellitus, it has been attributed to impaired glucose intolerance (310).

- An increased lithium dosage requirement in a hyperglycemic 40-year-old woman was attributed to the osmotic diuretic effect of glycosuria, increasing lithium excretion (311).

- Two patients with diabetes mellitus developed lithium toxicity (serum concentrations 3.3 and 3.0 mmol/l) in association with impaired consciousness, and hyperglycemia that resolved after intravenous insulin and fluids (312).

While the authors of the second report concluded that impaired glucose tolerance had predisposed to lithium intoxication, the opposite is also possible.

When a 45-year-old man with severe lithium-induced diabetes insipidus developed hyperosmolar, nonketotic hyperglycemia, it was suggested that poorly controlled diabetes mellitus may have contributed to the polyuria (313). Prior contact with a female patient who had developed hyperosmolar coma secondary to lithium-induced diabetes insipidus (314) allowed physicians 4 years later to treat her safely after a drug overdose and a surgical procedure, by avoiding intravenous replacement fluids with a high dextrose content (despite stopping lithium several years earlier, the patient continued to put out 10 liters of urine daily) (315).

Difficulty in attaining a therapeutic serum concentration of lithium despite increased doses was attributed to increased renal clearance due to the osmotic effect of glycosuria in a 44-year-old man with poorly controlled diabetes mellitus (311).

Weight gain

Weight gain, a well-recognized adverse effect of lithium, occurs in one-third to two-thirds of patients (316). It is more common in those with prior weight problems and at higher dosages of lithium. Possible mechanisms include complex effects on carbohydrate and lipid metabolism, mood stabilization itself, lithium-induced hypothyroidism, the use of high-calorie beverages to treat lithium-induced polydipsia, and the concomitant use of other weight-gaining drugs (for example valproate, olanzapine, mirtazapine). Recognizing and managing weight gain early in the course of treatment can do much to ensure continued adherence to lithium regimens. Two reviews of weight gain with psychotropic drugs mentioned lithium (317,318).

Risk and magnitude

In a review of psychotropic drug-induced weight gain, the prevalence and magnitude of the problem with lithium was discussed together with risk factors, mechanisms, and management (316). Adolescent inpatients treated with risperidone (n = 18) or conventional antipsychotic drugs (n = 19) over 6 months gained more weight than a control group but concomitant treatment with lithium was not a contributing factor (319).

A review of psychotropic drugs and weight gain included a brief summary of lithium-related weight gain (320). A retrospective evaluation of 176 patients taking long-term lithium showed that weight gain was an adverse effect in 18%. While 34% of the total did not adhere to treatment because of somatic adverse effects, no specific adverse effect (including weight gain) was associated with nonadherence (321).

The prevalence of overweight (BMI 25–29) and obesity (BMI 30 or more) was determined in 89 euthyroid bipolar patients and 445 reference subjects. The rate of obesity in patients taking only lithium was 1.5 times greater than in the reference population (a nonsignificant difference), compared with a statistically significant 2.5 times greater rate associated with antipsychotic drugs (322).

The prevalence of overweight (BMI 20 or more) and obesity (BMI 30 or more) has been evaluated in 89 euthymic bipolar patients and 445 age- and sex-matched controls (322). The bipolar women were more overweight and more obese than the controls and the bipolar men were more obese but not more overweight. Obesity was clearly related to antipsychotic drug use and less so to lithium and anticonvulsants (but patients taking lithium alone had an obesity rate 1.5 times that of the general population).

A review of the effects of mood stabilizers on weight included a section on lithium in which the authors concluded that lithium-related weight gain occurs in one-third to two-thirds of patients, with a mean increase of 4–7 kg; possible mechanisms were discussed (323).

An open chart review of 74 hospitalized patients showed a mean weight gain of 6.3 kg and an increase in BMI of 2.1 kg/m^2 after they had taken lithium for a mean of 89 days (324). Of 47 lithium-treated patients, 14 gained at least 5% of their baseline BMI, 6 gained over 10%, and 2 gained over 15% during an acute treatment phase of unspecified duration, while during the 1-year maintenance phase 11 gained over 5% and 2 gained over 10% (325).

Comparative studies

In a 1-year, placebo-controlled study of bipolar I prophylaxis (n = 372), weight gain with divalproex, but not with lithium, was significantly more common than with placebo (81). A patient who gained 18 kg over 18 months while taking lithium and perphenazine lost 16 kg when the latter was changed to loxapine (she also participated in a weight loss program) (326). Whether lithium played a role in the weight gain was unclear.

In a 12-month maintenance study, weight gain was an adverse event in 21% of patients taking divalproex, 13% of those taking lithium, and 7% of those taking placebo (81). The divalproex/placebo difference was statistically significant, but the lithium/placebo difference was not.

Mechanism

In 15 consecutive patients, serum leptin concentrations were measured at baseline and after 8 weeks of lithium treatment. There was a significant mean increase of 3.5 ng/ml and serum leptin correlated positively with weight gain (5.9 kg), increased BMI (24–27), and clinical efficacy (327). The authors suggested that leptin might play a role in lithium-induced weight gain.

Management

The Expert Consensus Guideline Series, Medication Treatment of Bipolar Disorder 2000, has recommended to "continue present medication, focus on diet and exercise" as the preferred first-line treatment for managing weight gain in patients taking lithium or divalproex. The next approach was to continue medication and add topiramate. Second-line treatments included switching from divalproex to lithium or vice versa, reducing the dosage, and switching to another drug. The addition of an appetite suppressant was a lower second-line recommendation (328).

Nutrition

In a review of naturally occurring dietary lithium (food and water sources), the author acknowledged that human lithium deficiency states have not been identified, but concluded that lithium should be considered an essential element, a conclusion reached on rather shaky grounds (329).

Electrolyte balance

Potassium

Episodes of acute hypokalemic paralysis have been associated with long-term lithium therapy (330).

- A 25-year-old man, who had taken lithium for 5 years, awakened from sleep unable to move his limbs and had a generalized flaccid paralysis with a serum potassium concentration of 2.1 mmol/l (330). Lithium was withdrawn and he responded to treatment with intravenous potassium chloride.

The diagnosis of acute hypokalemic paralysis was attributed to lithium but without confirmation by rechallenge it was unclear whether this was causal or coincidental.

Sodium

Hypernatremia can occur secondary to dehydration in patients taking lithium and is not uncommon in association with lithium poisoning. Lithium-induced diabetes insipidus is often a contributing factor.

Mineral balance

Fluid and sodium balance are important to the safe use of lithium. Both dehydration and a negative sodium balance (for example a low salt intake, diuretic-induced sodium loss) will reduce renal lithium clearance and predispose to toxicity (331). Hyponatremia (for example, secondary to polydipsia or SIADH) may also increase the risk of lithium toxicity (332).

When nine trace elements were measured in whole blood (oven dried, moisture-free) from controls, prelithium, and lithium patients, there were many changes related to lithium, but none appeared to be clinically important (333).

Fluid balance

Edema associated with lithium is uncommon (334,335). It is usually restricted to the legs, and is usually transient or intermittent. If treatment is necessary, the intermittent and cautious use of a loop diuretic may be helpful (but see drug–drug interactions).

Dehydration, secondary to lithium-induced nephrogenic diabetes insipidus, was thought to be the cause of a superior sagittal sinus thrombosis in a 30-year-old woman who presented with confusion, papilledema, and a left hemiparesis (336).

Hematologic

The most common hematological effect of lithium is a benign leukocytosis (337), consisting primarily of mature granulocytes, which is reversible on withdrawal. An increase in platelet count is a much less consistent finding (338), and there are no clinically important effects on erythrocytes. Despite anecdotal reports to the contrary, epidemiological studies have shown no increased risk of leukemia with lithium (339). The leukocyte-inducing effects of lithium have been used with some success to treat granulocytopenia (35,36).

The effects of lithium on hemopoiesis have been studied in 100 patients who had developed chronic granulocytopenia after cancer chemotherapy or radiotherapy (240). The mean leukocyte count rose by 46%, but there were no changes in platelet or erythrocyte counts. However, there was a significant increase in platelet count in those whose baseline values were below 150×10^9/l. Lithium was well tolerated (mean serum concentration 0.59 mmol/l).

Granulocyte counts and granulocyte colony-stimulating factor (G-CSF) concentrations were measured in 18 patients before and after 1 and 4 weeks of lithium treatment, and compared with values in 20 patients taking long-term lithium (241). At week 4, the granulocyte count was significantly higher than at baseline or at week 1, or in the long-term group. There was only a nonsignificant increase in G-CSF concentration at weeks 1 and 4. The granulocyte count in those taking long-term lithium did not differ significantly from the baseline values in the other group.

A retrospective review of inpatients showed higher leukocyte and granulocyte counts in those taking lithium alone ($n = 38$) compared with those taking antipsychotic drugs alone ($n = 207$); lymphocyte counts were not affected. Rises in leukocyte counts above normal occurred in 18% of those taking lithium and 6% of those taking antipsychotic drugs (337). The neutrophil-stimulating effect of lithium was used to advantage to successfully re-treat a patient with clozapine several years after stopping it because of neutropenia (342). Likewise, lithium was used to successfully stimulate neutrophil production in a patient with clozapine-induced neutropenia and in another with clozapine-induced agranulocytosis (recovery was as fast as seen with colony-stimulating factor and twice as rapid as expected spontaneously) (35).

In eight patients with bipolar disorder, lithium for 3–4 weeks increased neutrophil count by 88% and also caused a significant increase in CD34+ cells (although three patients had no increase in either) (343).

- A lithium-treated bipolar patient with acute myeloid leukemia had an unusually great increase in CD34+ cells following administration of G-CSF, suggesting a boosting effect from lithium (344).
- After he had failed to respond to combined treatment with corticosteroids and androgens and to antilymphocyte globulin, a 16-year-old with aplastic anemia responded to lithium combined with an androgen derivative, relapsed when lithium was stopped, and responded again when it was restarted (41).

Of 39 patients taking lithium, 18% had neutrophilia and 15% had raised activity of polymorphonuclear elastase (a marker of granulocyte activation) (345). In keeping with these observations, a chart review of 38 patients taking clozapine showed an increase in leukocyte count when lithium was added (42). A man with olanzapine-induced neutropenia (with a prior history of risperidone-induced neutropenia), which normalized with drug withdrawal, had no difficulty when the drug was reintroduced after the patient had been treated with lithium (43).

When 50 bipolar lithium patients were compared with 30 healthy controls, platelet counts were similar, but the lithium group had higher concentrations of plasma beta-thromboglobulin and platelet factor 4, suggesting lithium-induced platelet activation (338).

A cross-sectional study showed a 20% lower serum vitamin B12 concentration in patients taking lithium ($n = 81$) than in controls ($n = 14$) (serum and erythrocyte folate concentrations were normal) (346).

Mouth and teeth

A comprehensive review of psychoactive drug-induced hyposalivation and hypersalivation included a discussion of lithium-induced dry mouth (common) and sialorrhea (347). A review of dental findings and their management in patients with bipolar disorder briefly mentioned that xerostomia, sialadenitis, dysgeusia, and stomatitis have been attributed to lithium (348). Dry mouth is common and can be due to lithium-induced polyuria, a direct effect on thirst, salivary gland hypofunction, or other drugs (347). Hypersalivation attributable to lithium is rare (349).

A review of drug-induced oral ulceration mentioned lithium as a possible cause, based on two older references (350,351).

- An 8-year-old taking lithium citrate for 6 months developed waxing and waning areas of denuded papillae on her tongue, diagnosed as benign migratory glossitis (geographic tongue) and attributed to lithium (352).

The issue has been raised of whether oral lithium therapy was responsible for failure of titanium dental implants in a 62-year-old man (353).

Gastrointestinal

Lithium can cause loss of appetite, nausea, and at times vomiting and loose stools, especially early in therapy, but these can be minimized by the passage of time and by making dosage increases gradually (81). Gastrointestinal symptoms can also be an early warning sign of lithium intoxication.

In a 12-month maintenance study, lithium ($n = 94$) was not unexpectedly associated with more nausea (45 versus 31%) and diarrhea (46 versus 30%) than placebo ($n = 94$) (81).

- A 72-year-old man who had recently started to take lithium developed severe nausea, vomiting, and oliguric renal insufficiency which was initially attributed to lithium toxicity, until a serum lithium concentration of only 0.35 mmol/l directed evaluation to the correct diagnosis of acute gastric volvulus (354).
- An 80-year-old woman taking lithium developed constipation, nausea, vomiting, and abdominal pain after starting to take amfebutamone. A diagnosis of acute paralytic ileus was made and attributed to amfebutamone, although an amfebutamone–lithium interaction could not be excluded (355).
- In an 80-year-old woman a 2-month history of diarrhea, nausea, and abdominal distress attributed to irritable bowel syndrome was ultimately determined to be due to early lithium intoxication (356). Her lithium concentration when she was hospitalized was 1.2 mmol/l, although she had taken no lithium for the previous 10 days. Treatment with a thiazide diuretic contributed to the toxicity.

Pancreas

Of 47 cases of drug-induced pancreatitis reported to the Danish Committee on Adverse Drug Reactions between 1968 and 1999, one involved lithium (plus a neuroleptic drug) (357). Whether lithium was causally involved is not known.

- A 78-year-old woman taking lithium had hyperamylasemia and hyperlipasemia in the absence of gastrointestinal symptoms. Ultrasound examination showed the pancreas and liver to be normal. She also had hyperparathyroidism and renal dysfunction (306).

Urinary tract

There have been several reviews of the effects of lithium on the kidney (358–361).

In a retrospective study, 114 patients who had taken lithium for 4–30 years were compared with 94 unmedicated age- and sex-matched controls with regard to changes in creatinine concentrations (362). Of the patients taking lithium, 21% had blood creatinine concentrations that had increased gradually and were now over the top of the reference range. This finding was associated with episodes of lithium intoxication and with diseases and other medications that could also affect glomerular function. Sex, psychiatric diagnosis, duration of treatment, cumulative dose, and serum lithium concentrations did not predict an abnormal creatinine concentration.

Renal function was assessed in 10 patients taking long-term lithium (over 3 years, mean 80 months), 10 taking short-term lithium (3 years or less, mean 16 months), and 10 lithium-naïve patients (250). Blood urea nitrogen and serum creatinine concentrations were within the reference ranges and did not differ among the groups, but 24-hour creatinine clearance was significantly lower in those taking long-term lithium (73 versus 125 and 150 ml/minute). There were no significant differences among the groups in urine osmolality after 8-hour water deprivation and desmopressin, but partial nephrogenic diabetes insipidus was diagnosed in four long-term and two short-term patients and hypothalamic diabetes insipidus in two long-term patients. The authors concluded that long-term lithium therapy is a risk factor for renal impairment.

In a retrospective review of lithium concentrations in 2210 psychiatric hospital patients, 151 (6.8%) had serum lithium concentrations of 1.5 mmol/l or more. Of those with high serum concentrations, 10 (6.6%) had a raised blood urea nitrogen or serum creatinine concentration (363). In a retrospective study of 114 patients who had taken lithium for 4–30 years and 94 matched unmedicated subjects, 21% of those taking lithium had blood creatinine concentrations greater than 1.5 mg/ml [sic]; comparative figures were not given for the controls (362). Raised creatinine concentrations tended to be associated with episodes of lithium toxicity and drugs or diseases that could alter glomerular function.

Renal disease is not a contraindication to lithium, but it does complicate its use, and alternative mood stabilizers should be considered. Despite the diverse effects of lithium on the kidney, most patients find it to be generally well tolerated; nevertheless, monitoring should include periodic testing of renal function. In one case, a psychiatrist and family physician were sued for failing to monitor renal function in a patient who developed renal insufficiency (364). The distribution of monitoring guidelines in the area of Aberdeen, Scotland, led to an increase in the number of lithium patients who had at least once-yearly serum creatinine concentration measurements from 71% to 78%, still leaving 22% without adequate renal function monitoring (365).

To what extent long-term treatment with lithium impairs GFR is a matter of continued study (360). Lithium does not appear to impair GFR consistently, especially if correction is made for age-related changes in kidney function, although in one study there was an age-related reduction in 21% of 142 patients who had taken lithium for at least 15 years (366). There have been a few case reports of progressive renal insufficiency attributed to lithium, but it has not been possible to establish a cause-and-effect relation with absolute certainty.

In a review of the renal and metabolic complications of lithium, the example of a 78-year-old woman taking long-term lithium who had urinary incontinence, moderate renal insufficiency, a 5–7 litres 24-hour urine volume, and thyroid and parathyroid abnormalities was used to set the scene (306).

In a historical cohort study, changes in renal function in 86 patients taking lithium were evaluated first after a median treatment duration of 5.8 years and again after 16 years (367). Maximum plasma osmolality was reduced in nine of 63 patients in the initial study and in 24 of 63 at follow-up. Other findings included increased serum creatinine (in one of 76 patients initially and eight of 76 at follow-up) and reduced GFR (in three of 29 patients initially and six of 29 at follow-up); only the latter of these changes was not significant. The authors noted that this progressive impairment in renal dysfunction was greater than expected for age and advised strict surveillance of renal function in patients taking long-term lithium.

In a retrospective review of 6514 renal biopsies, there were 24 patients with renal insufficiency who had taken lithium for a mean duration of 13.6 years (range 2–25 years); the histological changes included chronic tubulointerstitial nephropathy (100%), cortical and medullary tubular cysts (63%) or tubular dilatation (33%), global glomerulosclerosis (100%), and focal segmental glomerulosclerosis (50%) (368). Only two had a history of acute lithium toxicity. Clinical findings included proteinuria (42%), nephrotic syndrome (25%), nephrogenic diabetes insipidus (87%), and hypertension (33%). Despite lithium withdrawal, either seven (abstract) or eight (text) of nine patients with an initial serum creatinine of over 221 μmol/l (2.5 mg/dl) progressed to end-stage renal insufficiency, whereas this occurred in only one of ten with lower creatinine concentrations. The study design was such that the risk of renal insufficiency with long-term lithium therapy could not be established and the possibility of alternative causes could not be excluded.

Two studies in rats have potential implications for humans. In rats with mild to severe lithium-induced nephropathy, urine N-acetyl-β-D-glucosaminidase was an early indicator of renal insufficiency (369). Both ^{6}Li and ^{7}Li caused reduced urine concentrating ability and increased urine volume and renal tubular lesions, but ^{6}Li was more nephrotoxic (370). The authors suggested that eliminating ^{6}Li from pharmaceutical products might reduce nephrotoxicity (although ^{6}Li accounts for only about 7% of the lithium in such products).

Concentrating ability

Lithium often impairs renal concentrating ability, an effect that is related in part to both dosage and duration of treatment (358,360). While it is initially reversible on withdrawal, it may eventually become irreversible and indicative of structural tubular damage. Impaired concentration is of no clinical consequence in itself, but it may go hand in hand with polyuria (lithium-induced nephrogenic diabetes insipidus), which can sometimes be a social and occupational nuisance and can increase the risk of toxicity secondary to dehydration. These problems can be minimized if the maintenance serum concentration is low, but whether single daily dosing is kinder to the kidney has yet to be resolved. Since the defect is at the level of the kidney (both before and after the site of cyclic AMP generation and probably involving the water channel protein aquaporin-2, a vasopressin-regulated water channel protein (371)), vasopressin (ADH) is unlikely to be effective. A thiazide and/or potassium-sparing diuretic (especially amiloride) can reduce the urine volume, although serum lithium concentrations can rise at the same time. Avoiding hypokalemia is essential, but whether dietary potassium supplementation can also be helpful is not known. Inositol is no longer considered promising.

In a comparison of patients taking long-term lithium ($n = 10$) or short-term lithium ($n = 9$) and bipolar patients not taking lithium ($n = 10$), there was significantly lower creatinine clearance and renal concentrating ability in the long-term group (372).

Acute and chronic renal insufficiency

Renal insufficiency was attributed to lithium in a 40-year-old who had had nontoxic concentrations for 15 years (interstitial nephropathy on biopsy) (373) and in a 55-year-old woman who had taken lithium for 6 years (serum creatinine 141 µmol/l (374). A few patients develop progressive renal insufficiency that can best be attributed to lithium (360).

Chronic renal insufficiency (creatinine clearance under 80 ml/minute), for which there was no apparent alternative explanation, developed in 53 patients taking long-term lithium (mean 17.7 years); 7 required periodic dialysis (375).

After Taking lithim for 6 years, a 55-year-old woman development mild rental insuffiency (serum creatinine 1.6 mg/dl) and lithium was withdrawn (298)

In a follow-up study of 54 patients with lithium-induced renal insufficiency and 20 patients who had been referred for renal biopsy, the authors concluded that lithium-induced chronic renal disease is slowly progressive (376). The rate of progression was related to the duration of lithium treatment, and lithium-related end-stage renal disease accounted for a small percentage (0.22%) of all cases of end-stage renal disease in France. The authors strongly suggested that regular monitoring of creatinine clearance is important during long-term lithium management.

Renal size and structure have been evaluated by MRI in 16 patients with renal insufficiency and nephropathy thought to be secondary to lithium (377). There were renal microcysts in all patients. All the patients had nephrogenic diabetes insipidus, in which antidiuretic hormone concentrations are raised, and there is evidence that antidiuretic hormone can stimulate the production of renal cysts, by an action mediated via cyclic AMP (378).

Interstitial nephritis

Lithium-associated changes in kidney morphology include an acute, reversible, and possibly lithium-specific distal tubular lesion and a chronic, nonspecific, and tubulointerstitial nephritis (379). The differential diagnosis of the latter is extensive, and it is not clear if lithium is causative. Lithium received a brief mention in a review of tubulointerstitial nephritis (379).

- A 48-year-old man taking lithium and chlorprothixene had a creatinine clearance of 60 ml/minute and a renal biopsy showing chronic interstitial nephritis (380).
- Lithium-induced interstitial nephritis (serum creatinine 2.3 mg/dl) occurred in an 89-year-old woman who had taken lithium for 29 years (136).

Nephrogenic diabetes insipidus

There have been several case reports of lithium-related nephrogenic diabetes insipidus, sometimes associated with dehydration and lithium intoxication (301,381–384, 385,386). This effect is clearly distinct from bipolar illness, as non-psychiatric subjects taking lithium develop a reduced response to desmopressin and a reduced ability to concentrate their urine during water deprivation (387). Lithium is a competitive inhibitor of the action of antidiuretic hormone and diabetes insipidus in lithium-treated subjects appears to be mediated by second messengers (386) with consequent effects on cytoplasmic aquaporin 2 (386,388) or cyclo-oxygenases 1 and 2 (COX-1, COX-2) (389,390).

Nephrogenic diabetes insipidus has been specifically reviewed in the context of a case in which resolution did not occur despite withdrawal of lithium (391).

Nephrogenic diabetes insipidus secondary to lithium led to severe dehydration in two patients who required intravenous rehydration followed by a thiazide diuretic to reduce urine volume (382). One patient had persistent polyuria (6.7 l/day) 57 months after stopping lithium (296).

- A 77-year-old woman who had taken lithium for 10 years developed delirium, hypernatremia, prerenal azotemia, and a serum lithium concentration of 1.4 mmol/l; her condition was attributed to dehydration related to partial nephrogenic diabetes insipidus (381).
- Eight years after stopping lithium because of polydipsia and polyuria, a 55-year-old woman was hospitalized with lethargy, coma, and hypernatremia (sodium concentration 156 mmol/l) after her fluid intake had been restricted (391).
- A 30-year-old woman who became dehydrated secondary to lithium-induced nephrogenic diabetes insipidus developed a superior sagittal sinus thrombosis (336).
- A 78-year-old woman who had taken lithium for 25 years had hypotonic polyuria (4.7 l/day), mild renal insufficiency, and hyperparathyroidism attributed to lithium (306).
- Nephrogenic diabetes insipidus resulted in dehydration and hypernatremia in a 78-year-old man who had taken lithium for 30 years (301).
- Nephrogenic diabetes insipidus in a 63-year-old woman was treated successfully with lithium withdrawal and amiloride (395).
- A 33-year-old man who had taken multiple medications, including valproic acid and lithium carbonate, started to take olanzapine instead of haloperidol (393). Five days later he reported malaise, polyuria, and polydipsia. He had a raised blood glucose, an increased serum osmolality, ketonemia, and glycosuria.

The authors of the last case suggested that the patient had insulin resistance (probably due to olanzapine), which was exacerbated by valproic acid and lithium.

- A 47-year-old woman, who was taking lithium (serum concentration 0.7 mmol/l) for bipolar I disorder, developed an acute abdominal syndrome (394). She had a recent history of drinking about 4 l of fluid a day. After surgery, she developed nephrogenic diabetes insipidus, with 19 l/day intake and 15 l/day output. The diuresis fell to 8 l/day over the next 10 days.

The authors wondered whether physiological stress could trigger diabetes insipidus and cited other articles showing that diabetes insipidus due to lithium occurred after surgery or brain injury, during pregnancy, or while fasting. They also suggested that patients who have been stable for a long time might have their lithium withdrawn. Withdrawal of lithium is thought to reverse lithium-induced nephrogenic diabetes insipidus, but the authors reviewed the literature and cited cases of nephrogenic diabetes insipidus that did not reverse after withdrawal of lithium.

- A 76-year-old man developed severe intractable diabetes insipidus which was attributed to lithium (395). He was hospitalized for over 2 weeks and eventually died from intestinal hemorrhage. Vigorous efforts were made to treat his polyuria, electrolyte disturbances, hypernatremia, and dehydration. He had been taking chlorpromazine, lithium, and furosemide, along with other medications, and the diagnosis of lithium-induced nephrogenic diabetes insipidus was considered because of a lack of alternative explanations.

The authors reviewed the causes and pathophysiological mechanisms of nephrogenic diabetes insipidus. They also discussed the metabolic effects of lithium, including renal and thyroid effects, hypercalcemia, leukocytosis, and weight gain.

- A 63-year-old man taking long-term lithium for a schizoaffective disorder developed a dural sinus thrombosis and severe hypernatremia and died (396).

The authors suggested that the sequence of events was lithium-induced nephrogenic diabetes insipidus resulting in hypernatremia followed by the dural sinus thrombosis.

Nephrotic syndrome

Nephrotic syndrome (proteinuria, edema, hypoalbuminemia, hyperlipidemia) is a rare and idiosyncratic complication of lithium therapy; it usually resolves on withdrawal, and can recur on rechallenge (397,398). Lithium-associated nephrotic syndrome occurred in a 59-year-old woman with lithium toxicity (serum concentration 1.9 mmol/l) whose renal biopsy showed focal segmental glomerulosclerosis. Lithium withdrawal led to resolution of edema and marked improvement in proteinuria and albuminemia (398).

- An 83-year-old man developed nephrotic syndrome while taking lithium (399).
- An 11-year-old boy who had taken lithium for an unstated duration developed nephrotic syndrome with focal glomerulosclerosis which remitted fully after lithium was withdrawn (397).
- A 59-year-old woman with lithium-associated nephrotic syndrome (focal segmental glomerulosclerosis on biopsy) had resolution of edema and pleural effusions and marked improvement in albuminemia and proteinuria after withdrawal of lithium (398).

Renal tubular acidosis

Incomplete distal renal tubular acidosis has been attributed to lithium, but appears to be of no clinical significance (400).

Skin

The cutaneous adverse effects of lithium have been reviewed (401,402). Lithium can cause aggravation of psoriasis. Other dermatological problems related to lithium treatment include acne, folliculitis, and maculopapular eruptions. The prevalence of dermatological difficulties is up to 45%, although many have reported a much lower rate, less than 4%. Men are more susceptible to than women. Most patients can be managed without withdrawing lithium, but aggravation of psoriasis may make it necessary.

- A 19-year-old patient taking lithium therapy for bipolar I disorder developed a pityriasis rosea-like dermatitis, which resolved when valproic acid was substituted for lithium (403). The rash was defined as "erythematous, sharply bordered oval and finely scaling lesions along skin cleavages on the trunk and proximal parts of the extremities." There was no evidence of a herald patch.

In a review of acute skin reactions to psychotropic drugs, alopecia, psoriasiform, acneiform, and lichenoid eruptions, and drug-induced lupus were attributed to lithium, but critical comment was not provided (404).

There have been anecdotal reports of papular and non-papular rashes, pityriasis versicolor, hidradenitis suppurativa, and nail dystrophy (405,406), although causal relations are difficult to establish.

In a review of lichenoid drug eruptions, lithium was implicated in one patient with ulcerative oral lesions and in another with ulcerative genital lesions (407).

- A 60-year-old man developed systemic swelling, redness, and pruritus 1 month after starting to take lithium; it was attributed to the drug based on positive patch and challenge tests (408).
- A 55-year-old man who had taken lithium and haloperidol for 11 years developed hyperkeratotic follicular papules on his scalp, extremities, and trunk, which on biopsy were suggestive of follicular mycoses fungoides (409). He also had a 1-year history of scalp, axillary, and pubic hair loss. Following replacement of lithium with valproate, his hair regrew and the papules cleared almost completely in 3 months.

A man and a woman developed vegetating plaques with peripheral pustules (halogenoderma) after taking lithium for 6 and 8 years respectively (410). No follow-up information was provided and it could not be established whether the lesions were caused by, worsened by, or unrelated to lithium.

Acne

Acne can occur or worsen during lithium treatment (411). In a comparison of 51 patients taking lithium with 57 patients taking other psychotropic drugs, there were secondary skin reactions in 45% of the former and 25 of the latter; while acne (33 versus 9%) and psoriasis (6 versus 0%)

were more common in the lithium group, the only statistically significant association was acne in males (412).

- Six months after beginning lithium, a man in his late twenties developed severe truncal acne which worsened over 5 years, at which time lithium was withdrawn (413). Nevertheless, the lesions were still present 4 years later, leading to the conclusion that lithium had caused irreversible acne.

In this case, the association could have been coincidental.

Alopecia

Hair loss in psychopharmacology has been reviewed (414). It appears to be an uncommon and unpredictable adverse effect of lithium. Hypothyroidism is occasionally involved. Hair texture can also be altered (415).

- A 36-year-old woman had taken lithium for 4 months when her scalp hair became thinner and stopped growing. She continued to take lithium, and 2 months later was diagnosed and treated for hypothyroidism, after which her hair became curlier but did not grow longer or fuller.

Darier's disease

Darier's disease (follicular keratosis) can be exacerbated by lithium (416,417).

- In a woman who had had keratosis follicularis (Darier's disease) since she was 16, the condition worsened when she was given lithium at age 50 (and improved when she switched to valproate) (417).

However, despite reports that Darier's disease is worsened by lithium, a 52-year-old man noted no exacerbation despite taking lithium for many years (418).

Psoriasis

Psoriasis can occur or worsen during lithium treatment (411). Before psoriasis can be treated effectively, lithium may have to be withdrawn. In a comparison of 51 patients taking lithium with 57 patients taking other psychotropic drugs, there were secondary skin reactions in 45% of the former and 25% of the latter; while acne (33 versus 9%) and psoriasis (6 versus 0%) were more common in the lithium group, the only statistically significant association was acne in males (412).

- A 42-year-old woman taking lithium developed psoriasis which resolved when the drug was withdrawn (419). Brief mention was made of two patients (age and sex unstated) taking lithium whose psoriasis improved with oral omega-3 fatty acids (420).
- Lithium worsened psoriasis in a 54-year-old woman; improvement followed withdrawal (421).

Hair

Hair loss in psychopharmacology has been reviewed (414). It appears to be an uncommon and unpredictable adverse effect of lithium. Hypothyroidism is occasionally involved. Hair texture can also be altered (415).

- A 36-year-old woman had taken lithium for 4 months when her scalp hair became thinner and stopped growing. She continued to take lithium, and 2 months later was diagnosed and treated for hypothyroidism, after which her hair became curlier but did not grow longer or fuller.

Musculoskeletal

Despite lithium-induced increases in serum calcium and PTH concentrations, most recent studies have not shown a reduction in bone mineral density or increased risks of osteoporosis or bone fractures (422). In 26 patients who had taken lithium for at least 10 years, bone mineral density did not differ from controls (423).

Indeed, in a case-control study of 124 655 individuals who suffered a bone fracture matched with 373 962 subjects who had not had any bone injury, the relative risk of a bone fracture was significantly lower in those who took lithium, both before and after correction for psychotropic drug use (OR = 0.67; 95% CI = 0.55, 0.81) (424).

A patient with a diffuse sensorimotor peripheral neuropathy also developed rhabdomyolysis during the acute episode in association with a serum lithium concentration of 3.1 mmol/l and a serum sodium of 163 mmol/l (177).

Sexual function

Despite occasional reports of reduced libido and erectile dysfunction, lithium causes little in the way of sexual dysfunction in men or women (425). In a brief review of the sexual adverse effects of psychotropic drugs, reduced libido and arousal with lithium were mentioned in passing, particularly when it was combined with other drugs (which, of course, makes it difficult to implicate lithium) (426).

Reproductive system

In a cross-sectional pilot study of 22 bipolar women taking lithium (n = 10), divalproex (n = 10), or both (n = 10), there was an increased number of ovarian follicles in one woman taking lithium, but no evidence of hormonal changes suggestive of polycystic ovary syndrome in any patient (427). The small sample size was a limiting factor.

When 22 women with bipolar disorder (10 taking lithium alone, 10 taking divalproex alone, and 2 taking both) were evaluated for polycystic ovary syndrome, none had typical hormonal screening abnormalities (427). Some type of menstrual dysfunction was present in all ten women taking lithium alone, but it predated use of the drug in all but one.

Compared with 13 women taking placebo, 10 women taking lithium carbonate 900 mg/day for one menstrual cycle had no significant alterations in reproductive hormone concentrations (428).

In an in vitro study, LY294002, a phosphatidylinositol-3-kinase inhibitor, overcame impaired human sperm motility induced by lithium chloride (429).

At blood concentrations within the human target range, oral lithium caused degenerative changes in testicular morphology in spotted munia (*Lonchura punctulata*), a seasonally breeding subtropical finch (430). Lithium at a serum concentration of about 0.6 mmol/l reduced sperm motility, number, and viability, and markedly altered testicular histopathology in *Viscacha*, a nocturnal rodent from the pampas of Argentina (431). How these findings might relate to effects in men is open to question.

Immunologic

Lithium is not an allergen. Allergic reactions that have been reported in patients taking lithium have been attributed to excipients in the formulation (432), as in a case of leukocytoclastic vasculitis (433).

In a brief report, evidence has been presented that short-term exposure to lithium (less than 2 months) caused alterations in the expression of histocompatibility antigens (434).

In 10 healthy volunteers, lithium caused increases in interleukin-4 and interleukin-10 concentrations and falls in interleukin-2 and interferon concentrations (435). In in vitro studies of monocytes from women with breast cancer, lithium chloride suppressed production of interleukin-8 and induced production of interleukin-15 (436,437). The clinical implications of these findings are unclear.

The immunomodulatory effects of lithium have been reviewed (438,439). Lithium

(a) stimulated the production of pro-inflammatory cytokines and negative immunoregulatory cytokines or proteins in nine healthy subjects (440);
(b) altered the expression of human leukocyte antigens (HLA) in 11 of 15 subjects (434);
(c) normalized manifestations of mild immune activation in 17 rapid cycling bipolar patients (441).

The clinical implications of these findings are unclear.

The very complex antiviral and immunomodulatory effects of lithium have been reviewed (442). In 15 inpatients, lithium produced changes in a number of histocompatibility antigens, but whether these have any clinical implications is unknown (443).

Death

The 1997 Annual Report of the American Association of Poison Control Centers Toxic Exposure Surveillance System listed nine lithium-associated deaths and provided some clinical details in seven cases, including serum concentrations of 2.4–7.8 mmol/l (444).

However, in a systematic review there were significantly fewer deaths from all causes in lithium-treated patients (9 out of 696, 1.3%) compared with those who took other agents (22 out of 788, 2.8%; OR = 0.42; 95% CI = 0.21, 0.87) (241). Individuals with bipolar illness have a doubled standardized mortality ratio compared with the general population (445). This difference appears to be due to factors such as frequent depressive episodes, co-morbidity with substance abuse, and lifestyle problems, such as reduced exercise, poor diet, and a lower quality of medical care. The use of lithium may reduce overall mortality.

Long-Term Effects

Drug abuse

Lithium is not a drug of abuse or dependence, although when one bipolar alcohol abuser was prevented from drinking, he tried to get a "buzz" by increasing his lithium dose to the point of toxicity (serum concentration 3.0 mmol/l) (446). The only other suggestion of abuse appeared in 1977 when passing mention to the "fairly recent (over the past 2 years or so) abuse of lithium" by poly-drug abusers (447).

Drug tolerance

In a review of whether the prophylactic efficacy of lithium was transient or persistent, the authors concluded that "the balance of evidence does not indicate a general loss of lithium efficacy" (448). A similar conclusion was reached in a study of 22 patients who had taken lithium for at least 20 years (449). There was no change in affective morbidity over the second 10 years compared with the first 10 years. However, individual exceptions could not be excluded.

Drug withdrawal

Three issues have been addressed:

(a) does rapid withdrawal increase recurrence risk?
(b) how common is post-withdrawal refractoriness to the reinstitution of lithium?
(c) does withdrawal increase the risk of thyrotoxicosis?

Recurrence risk
Abrupt or rapid withdrawal of lithium is not associated with a physical withdrawal reaction, but there does appear to be an increased risk of early recurrence of mania and depression compared with more gradual withdrawal (450,451), although this conclusion has been questioned (452). Data from Italy have suggested that gradual withdrawal of lithium (over 15–30 days) was associated with a markedly reduced risk of early recurrence of mania and depression and a much greater likelihood of prolonged stability compared with rapid discontinuation (over 1–14 days) (450). There was a marked increase in suicidal acts during the first year after lithium withdrawal; after that the risk returned to what it had been before the start of lithium treatment. The increased risk in the first year exceeded that expected from increased affective morbidity alone. There was a 1.95-fold greater risk, during this first year in those who discontinued lithium rapidly, although this was not statistically significant (453).

Of 30 patients with major depressive disorder who had responded to lithium augmentation for antidepressant-resistant depression, 15 were switched to placebo over 1–7 days (454). Two became manic, and it was suggested that lithium withdrawal may have uncovered latent bipolar disorder (455).

When 21 elderly patients with a major depressive episode who had responded to lithium augmentation had lithium withdrawn gradually (over 2–12 weeks), 9 relapsed but none became manic (456). Whether gradual withdrawal protected against withdrawal mania or whether there were no latent bipolar patients in the study is unknown.

A retrospective study of lithium withdrawal in pregnant and nonpregnant women showed similar rates and times of recurrence, including a higher risk of early occurrence with rapid withdrawal (1–14 days) versus gradual withdrawal (15–30 days) (457). How to balance the first trimester fetal risk of greater lithium exposure during gradual withdrawal with the greater maternal risk of potentially devastating relapse after rapid withdrawal remains a challenge.

There is a higher risk of postpartum recurrence of bipolar disorder in women who have discontinued lithium; the protective effect of restarting lithium late in pregnancy or soon after delivery has been emphasized (457).

The observation that rapid or abrupt lithium withdrawal might be associated with a more immediate or higher likelihood of recurrence has gathered further support from a reanalysis of data from a double-blind lithium maintenance study, in which the benefits of low serum concentrations (0.4–0.6 mmol/l) and standard serum concentrations (0.8–1.0 mmol/l) were compared (458). Recurrence rates were greater only in those whose concentrations were abruptly reduced from standard to low at the start of the study. The authors suggested that rapid dosage reduction, rather than a low maintenance concentration itself, accounted for their initial conclusion that standard concentrations were more effective than low concentrations.

Others have concluded that withdrawal mania is "a major and sinister complication of the everyday use of lithium" (459). Withdrawal of effective lithium therapy was associated with an increased risk of suicide and suicidal acts, especially during the first 12 months. Gradual withdrawal (over 15–30 days) was associated with half the rate of suicidal acts compared with more rapid withdrawal (strong trend toward statistical significance) (460).

Refractoriness to retreatment

In 28 patients who had responded to lithium treatment of mania or schizoaffective mania and who had recurrences after withdrawal, there were equally good responses to retreatment with lithium (461). These findings add to the evidence that lithium discontinuation-induced refractoriness is the exception rather than the rule. However, the issue of whether post-withdrawal refractoriness to reintroduction of lithium is a real phenomenon and, if so, how often it occurs continues to be debated (462). Three patients failed to respond to the reintroduction of lithium, despite having had sustained beneficial responses before withdrawal (463).

Risk of thyrotoxicosis

Lithium withdrawal may be associated with an outpouring of thyroid hormone in predisposed individuals; there have been reports of thyrotoxicosis (274) and a thyrotoxic crisis (274) after lithium withdrawal.

- A 37-year-old man who had taken lithium and sulpiride for 14 years and who was a long-time smoker without respiratory symptoms or a history of asthma had lithium withdrawn because of an asymptomatic bradycardia (44 beats/minute) (125). Six weeks later, he developed symptoms of asthma, including nocturnal cough, exertional wheezing, increased airway resistance, and a low FEV1, attributed to lithium withdrawal.

Mutagenicity

In 18 patients taking benzodiazepines and/or neuroleptic drugs, there were increased chromosomal aberrations and increased sister chromatid exchange, but there were no significant differences between this group and another group of 18 patients taking lithium in addition to benzodiazepines and/or antipsychotic drugs (464).

Tumorigenicity

There is no evidence that lithium causes or promotes the growth of tumors. Since tumors are not rare events over a lifetime, and since lithium is taken for long periods of time, any apparent association is likely to be coincidental.

Second-Generation Effects

Fertility

Intramuscular lithium chloride produced subtherapeutic blood concentrations (by human standards) in male rose-ringed parakeets, but significantly reduced testicular weight and caused widespread degenerative changes in the testes (465). Fertility was not assessed directly.

Pregnancy

The treatment of bipolar disorder during pregnancy and lactation has been reviewed, with reference to lithium-related maternal, fetal, and neonatal toxicity, morphological and behavioral teratogenicity, carcinogenicity and mutagenicity, and miscellaneous effects (466–478). Elsewhere, the effects of lithium, valproic acid, and carbamazepine during pregnancy (479) and drug-induced congenital defects (with only a brief mention of lithium) (480) have been reviewed. A more specific review dealing with the use of drugs during pregnancy in women with renal disease mentioned the need for lithium-dosage reduction in such cases (481).

Pregnancy in bipolar women was found to be a "risk-neutral" condition, in that it neither protected against nor increased episode risk in a comparison of 42 pregnant with 42 nonpregnant women who stopped lithium either rapidly (over 1–14 days) or gradually (over 15–30 days). Stopping lithium was not "risk-neutral," and the risk was especially high in those who stopped rapidly (482) (see also Withdrawal effects). These observations must be balanced against the low but real risk of teratogenesis from first trimester lithium exposure (483).

- A 37-year-old woman with severe bipolar disorder, who continued to take lithium throughout pregnancy, had a normal delivery (484).

Teratogenicity

While there is an increased risk of teratogenesis (particularly cardiovascular) from exposure to lithium in the first trimester, this risk has been overstated initially because of selection bias (482,483,485,486). Subsequently, case-control and cohort studies have substantially reduced, but not eliminated, such concerns (487). The risk of major malformations from exposure to lithium in the first trimester appears to be lower than from exposure to carbamazepine or valproate.

The teratogenic effects of lithium have been reviewed in several articles (466,467,488); two reviews of drug-related congenital malformation briefly mentioned lithium and cardiovascular teratogenesis (480,489). The authors concluded that while the risk of cardiovascular malformation is lower than once believed, it is nevertheless increased.

The cardiovascular teratogenicity of lithium has been summarized in a review of managing bipolar disorder during pregnancy and postpartum (473). While the risk of Ebstein's anomaly is increased, likely 10–20 times more than in the general population, the absolute risk (0.05–0.10%) is small. Fetal ultrasonography was advised at 18–20 weeks of gestation in cases of first trimester lithium exposure (488).

In a review of all pregnancies in Leiden in The Netherlands between 1994 and 2002, none of the 20 children of mothers who had taken lithium had major problems or congenital anomalies (490).

There is no evidence that children exposed to lithium during pregnancy who are born without malformations develop other than normally.

Anencephaly in the child of a woman taking lithium appears to have been coincidental (491).

Fetotoxicity

In a prospective study of 10 pregnant women who were taking lithium before delivery lithium equilibrated completely across the placenta (ratio of umbilical cord lithium concentration to maternal blood = 1.05) across a wide range of maternal lithium serum concentrations (0.2–2.6 mmol/l); infant serum concentrations exceeding 0.64 mmol/l were associated with lower Apgar scores, longer hospital stays, and higher rates of neuromuscular complications (492).

- A 36-year-old woman with bipolar disorder who had taken lithium for 17 years continued to take it and other medications during pregnancy (493). At 35 weeks she developed signs of lithium toxicity, with nausea, diarrhea, and a concentration of 1.25 mmol/l. She delivered a lethargic infant with poor muscle tone, who showed signs of respiratory distress and hypopnea.

The infant's lithium concentration was 3.58 mmol/l. The infant was treated and responded, but at day 6 the lithium concentration increased to 2.6 mmol/L after dropping to 2.4 mmol/L on day 4.

However, the authors identified a laboratory error, in that the blood tube for analysis for lithium contained lithium heparin rather than sodium heparin. When the infant's serum lithium concentration was checked using a proper tube on day 6 it was 0.06 mmol/l, not 2.6 mmol/l, so the fetus may not after all have had lithium toxicity.

Hypoglycemia, requiring temporary glucagon and glucose supplementation, was noted in a neonate whose mother had taken lithium throughout pregnancy (cord lithium concentration 1.73 mmol/l) (472).

Reports of fetal goiter and a variety of other lithium-related adverse events in newborns have been reviewed (488).

In 20 infants exposed to lithium during labor and delivery, there were higher rates of perinatal complications (65%) and special care nursery admissions (45%) than in nonexposed infants, although most complications were transient (494). An infant who died shortly after birth had oromandibular-limb hypogenesis spectrum which was speculatively attributed to lithium that the mother had taken during most of her pregnancy (495).

A neonate whose mother had taken lithium throughout pregnancy developed a supraventricular tachycardia (240/minute) which was treated successfully with adenosine (496). Another newborn girl exposed to lithium in utero had asymptomatic cardiomegaly that resolved by 1 month of age (311).

Transient polyuria in two newborns was attributed to the lithium their mothers had taken throughout pregnancy (472,496).

Neonates born to mothers taking lithium included a boy with a goiter and chemical hypothyroidism who required temporary treatment with oral thyroxine for 11 weeks (497), a girl with respiratory distress, cardiomegaly, hyperbilirubinemia, nephrogenic diabetes insipidus, and hypoglycemia who responded to various treatments and eventually remitted fully (472), and a preterm infant with a supraventricular tachycardia and temporary polyuria associated with 20% weight loss (496). The mother of the preterm infant developed polyhydramnios during pregnancy, which was attributed to lithium-induced fetal polyuria.

Lactation

Lithium concentrations in breast milk are about 40% of maternal serum concentrations, although the range is wide (471,498,499). Because of this, breastfeeding is often discouraged (499), although the authors of a comprehensive review of the use of mood stabilizers during breastfeeding have pointed out that there is a paucity of information to support adverse effects of lithium in breastfeeding infants—one case of toxicity in a 2-month-old with an intercurrent infection (serum lithium 1.4 mmol/l) and one report of lethargy and cyanosis at days 5 and 6 (infant serum concentration 0.6 mmol/l on day 5) secondary to fetal and breast milk exposure. However, there have been only two reports of lithium toxicity attributed to breastfeeding (500,501). Others, however, have stated that "lithium should only be used with great caution ···" (502); "[breastfeeding by women taking lithium] has been repeatedly discouraged in the literature" (503); "··· it also seems unwise to expose infants unnecessarily to lithium" (499). Lithium has also been stated to be "an excellent example of a drug that requires monitoring and case-by-case assessment so that nursing mothers can be successfully treated" (498).

In a review of the use of psychotropic drugs during breastfeeding, it was briefly mentioned that lithium was not advisable, but was justified under certain circumstances (504). Lithium was also briefly discussed in a review of xenobiotics and breastfeeding (505).

Breast-milk lithium concentrations were measured in 11 women taking lithium carbonate 600–1,500 mg/day (506). Maternal serum concentrations were available in only three and infant concentrations in two. No infants had adverse effects that could be attributed to lithium, and the authors calculated that infant lithium exposure was low, leading them to challenge the general contraindication to breastfeeding under such circumstances.

In its 2001 Policy Statement, the American Academy of Pediatrics Committee on Drugs modified its earlier contraindication to lithium during breastfeeding by listing it with "drugs that have been associated with significant effects on some nursing infants, and should be given to nursing mothers with caution" (507).

The Motherisk Team in Toronto has recommended use of the Exposure Index (the maternal milk/plasma ratio times 100 divided by the infants' clearance in ml/kg/minute) to determine the advisability of breastfeeding when a mother is taking lithium (501). However, this approach does not seem very practical.

Given the extremely high risk of postpartum recurrence in bipolar patients, the issue is not whether the mother should take a mood stabilizer (she should), but whether the baby should be breastfed. A review of mood stabilizers during breastfeeding reaffirmed that there are only two reported cases of infant lithium toxicity associated with breastfeeding (one of which involved both fetal and breast milk exposure) (500). The following recommendations were made in the case of a mother taking lithium who chooses to breastfeed:

(a) educate her about the manifestations of toxicity;
(b) explain the risks of dehydration;
(c) consider partial or total formula supplements during episodes of illness or dehydration;
(d) suspend breastfeeding if toxicity is suspected;
(e) check infant and maternal serum lithium concentrations.

Susceptibility Factors

Age

Whether elderly patients taking lithium received proper monitoring was questioned in a case note audit of 91 patients, over 40% of whom had deviations from practice standards. These included absence of pretreatment laboratory tests, infrequent monitoring of serum lithium concentrations, lack of adequate adverse effects documentation, and the use of risky concomitant drugs (508). In a placebo-controlled study, there was poor tolerance of lithium augmentation of antidepressants in 76% (13/17) of elderly (mean age 70 years) patients at a mean serum concentration of 0.63 mmol/l, due to tremor and muscle twitches, cognitive disturbance, tiredness and sedation, and gastrointestinal upsets (509).

In a cross-sectional study of 12 octogenarians (average age 84 years) who had taken lithium for an average of 54 months (mean serum concentration 0.42 mmol/l), none became toxic and none had to stop treatment because of adverse effects. Transient renal function abnormalities were noted: one patient developed nephrogenic diabetes insipidus and one became hypothyroidic (510). For lithium therapy in very old people, the authors advised close monitoring in a specialized setting.

The database for pharmacological treatment of mania and bipolar depression in late life has been reviewed, focusing on four studies from 1970–99 (511). The studies were fairly small (n = 12–81). One study included a small number of patients with bipolar II affective disorder. Five more recent studies (1995–9) involving divalproex in elderly patients with mania or bipolar disorder were also reported; sample sizes again were modest (n = 13–39). Both short-term and long-term outcomes were assessed and the issue of lithium toxicity was also discussed. The authors concluded that lithium remains the treatment of choice, but clinicians should target lower serum concentrations (0.4–0.8 mmol/l), although occasional patients may require higher concentrations.

The adverse effects of lithium in elderly patients include cognitive status worsening, tremor, and hypothyroidism. The authors suggested that divalproex is also useful in elderly patients with mania and that concentrations of divalproex in the elderly are similar to those useful for the treatment of mania in younger patients. They noted that carbamazepine should be considered a second-line treatment for mania in the elderly. A partial response would warrant the addition of an atypical antipsychotic drug. For bipolar depression, they recommended lithium in combination with an antidepressant, such as an SSRI. They also noted that lamotrigine may be useful for bipolar depression. Electroconvulsive therapy (ECT) may also be useful, but there have been no comparisons of ECT and pharmacotherapy in elderly patients with bipolar depression.

The risk of hospital admission related to lithium toxicity has been estimated in a case-control study of 10 615 elderly patients over 9 years (512). Lithium toxicity occurred at least once in 413 of the patients who were taking lithium. Factors that increase the likelihood of hospital admission included starting treatment with a loop diuretic or ACE inhibitors during the month before hospitalization. Although furosemide has been suggested as the diuretic of choice for patients taking lithium, the authors suggested that furosemide may cause lithium toxicity in elderly patients. Non-steroidal anti-inflammatory drugs (NSAIDs) and thiazide diuretics, which are commonly associated with raised lithium concentrations, were not associated with lithium toxicity in this population.

Most of 78 patients who had lithium intoxication and a serum lithium concentration equal to or greater than 1.2 mmol/l had mild symptoms (513). The symptoms were more severe in patients with chronic lithium intoxication than in those with acute intoxication. None of the patients died or had permanent neurological damage as a sequel of lithium intoxication. The authors concluded that concomitant medications, older age, and pre-existing neurological illness could increase the susceptibility to lithium toxicity. Of the 43 patients with acute lithium intoxication, 34 had taken lithium in a suicide attempt and nine had taken an accidental overdose. Thirty-five patients had chronic lithium intoxication without any change in lithium dosing. Lithium concentrations in this particular study were mostly clustered in the 1.2–1.5 mmol/l range, but eight of the patients in the acute intoxication group and two in the chronic intoxication group had serum lithium concentrations greater than 2.5 mmol/l.

Renal disease

While there are no absolute contraindications to lithium, patients with advanced kidney disease or unstable fluid/electrolyte balance may be more safely treated with an alternative mood stabilizer, such as carbamazepine, valproate, lamotrigine, or olanzapine.

Other features of the patient

Factors that put patients at risk of lithium intoxication are those that increase intake (deliberately or accidentally), reduce excretion (kidney disease, dehydration, low sodium intake, drug interactions), or reduce body water (dehydration secondary to fluid restriction, vomiting, diarrhea, or polyuria) (66). Patients with lithium-induced polyuria are at a particular risk of toxicity if their ability to replace fluids is compromised (for example by anesthesia, over-sedation, CNS trauma).

Thyrotoxicosis was considered a possible contributor to lithium toxicity in two patients, possibly by increasing tubular lithium reabsorption through induction of the sodium-hydrogen antiporter (227).

Two patients developed lithium intoxication (serum concentrations 3.3 and 3.0 mmol/l) in association with poorly controlled diabetes mellitus, suggesting that the latter is a risk factor (312).

A review of the effects of obesity on drug pharmacokinetics briefly mentioned that the steady-state volume of distribution of lithium correlated with ideal body weight and fat-free mass but not with total body weight (514). Lithium clearance was greater in those with obesity than in lean controls, suggesting that obese patients may require larger maintenance doses to maintain target serum concentrations.

- Lithium toxicity occurred in an elderly patient who had a Norwalk virus-like infection 6 days before presentation (515). The serum lithium concentration 48 hours before presentation was 1.85 mmol/l. She had hypernatremia and abnormal liver function tests. She was hydrated and treated aggressively for lithium toxicity and recovered fully after 9 days.

The point of this case is that lithium toxicity can occur in a patient who has been taking a stable lithium dose during an episode that can be associated with nausea, vomiting, and diarrhea, which could reduce lithium excretion.

Drug Administration

Drug formulations

Despite the availability of modified-release lithium formulations for several decades, there continues to be a paucity of information about their efficacy and tolerability compared with less expensive immediate-release formulations (516).

Brain lithium concentrations (measured by magnetic resonance spectroscopy) after the use of a modified-release formulation (Lithobid SR) or an immediate-release formulation have been compared in a crossover design in 12 patients with bipolar disorder (517). There were higher brain concentrations with the modified-release formulation, but whether this has clinical implications requires further study.

Formulations of lithium carbonate tablets with various binding substances have been discussed (518).

A modified-release formulation (Carbolithium Once-A-Day) produced a reduction in peak/trough lithium ratio compared with a standard formulation (519), and an interim analysis of an open switch to the modified-release formulation suggested better tolerability and efficacy at 4 weeks ($n = 27$) and 6 weeks ($n = 15$) (520). A paucity of detail, however, prevents firm conclusions.

A modified-release multi-particulate lithium capsule has been described consisting of five copolymer-coated prolonged-release tablets and one standard tablet, each 6 mm in size (521).

Following an overdose with a sustained-release formulation (8,000 mg of Teralithe 400 LP), the appearance of clinical symptoms (vomiting and dizziness) was delayed for 35 hours, despite a serum lithium concentration of 2.38 mmol/l at 15 hours and 3.12 mmol/l at 25 hours (522).

A woman taking conventional lithium developed lithium toxicity after two doses of a homeopathic formulation, "Lithium carb. 30"; a paucity of detail allows no conclusions to be drawn from this observation (523).

Drug additives

Gelatin is derived from natural pork and beef products and is present in some lithium formulations. Since certain religions forbid the consumption of gelatin, knowing that it is present in Eskalith capsules and Eskalith CR and absent in Eskalith tablets (not available in the USA) and Lithobid SR might influence prescribing practices under certain circumstances (524). The same would apply to other lithium products.

Drug dosage regimens

In an open-label pilot study of rapid administration of slow-release lithium (20 mg/kg/day in two divided doses) for acute mania, five of 15 patients completed 10 days of treatment, seven improved sooner and were discharged, two withdrew because of adverse effects (bradycardia in one and tremor, fatigue, and diarrhea in the other; and one patient appears not to have been accounted for) two other patients also had asymptomatic bradycardia (129).

A review of loading strategies in acute mania included a section on lithium (525).

A small study of brain lithium concentrations measured by magnetic resonance spectroscopy showed higher brain:serum lithium concentration ratios in subjects taking a single daily dose ($n = 5$) than in those taking a twice-daily regimen ($n = 3$) (526). Even to speculate about the possible clinical implications of this finding would be premature.

Drug administration route

Intravenous
To determine the safety of using lithium chloride dilution to measure cardiac output, the pharmacokinetic and toxic effects of intravenous lithium chloride have been studied in six conscious healthy Standardbred horses (527). The mean peak serum concentration was 0.56 mmol/l. There were neither toxic effects nor significant changes in laboratory studies, electrocardiograms, or gastrointestinal motility. Three horses had increased urine output.

A similar study was performed in patients undergoing cardiac surgery and healthy volunteers; the highest dose of lithium chloride was 0.6 mmol given intravenously five times at 2-minute intervals (528). Unfortunately, no mention was made of tolerability or adverse effects.

When 10 volunteers were given 500 ml of a 0.1% lithium carbonate solution (13.5 mmol of lithium) intravenously over 1 hour, the peak serum concentration was 0.93 mmol/l, the elimination half-life was 7.8 hours, and there were no adverse effects (529).

Drug overdose

There are three forms of lithium overdose:

(a) acute (abrupt overdose in a drug-naïve person);
(b) chronic (gradual accumulation, reaching toxic concentrations);
(c) acute-on-chronic (abrupt overdose by a person already taking lithium), which can also be due to a drug interaction.

Thus, the term overdose (109,530,531) can be misleading, because poisoning can develop not only as a result of overdosage but also by a fall in lithium clearance. Of 205 cases of lithium poisoning reported to the Ontario Canada Regional Poison Information Centre in 1996, 12 were acute overdoses (someone else's tablets), 19 were chronic poisonings, and 174 were acute-on-chronic poisonings. Over 80% had no or minimal symptoms, two patients died, and one had persistent renal sequelae (532). A retrospective study of 97 cases of lithium poisoning treated at a regional center in Australia over 13 years found severe neurotoxicity in 28 cases (26 were cases of chronic and two of acute-on-chronic poisonings) (300). Risk factors were nephrogenic diabetes insipidus, older age, abnormal thyroid function, and impaired renal function.

Early signs of intoxication include ataxia, dysarthria, coarse tremor, weakness, and drowsiness. More advanced toxicity can involve progressively impaired consciousness, neuromuscular irritability (myoclonic jerks), seizures, cardiac dysrhythmias, and renal insufficiency. A reversible Creutzfeldt–Jakob-like syndrome has been described (162). The severity of intoxication depends on both the extent and duration of exposure to raised lithium concentrations, as well as idiosyncratic factors.

When the Marseilles Poisons Centre analysed information on lithium overdose between 1991 and 2000, in addition to an unspecified number of suicide attempts and accidental poisonings in children, the next most frequent reports were prescription misinterpretation ($n = 43$), dehydration in the elderly ($n = 35$), renal insufficiency ($n = 15$), and diuretic interactions ($n = 8$) (533).

The 2000 Annual Report of the American Association of Poison Control Centers Toxic Exposure Surveillance System listed six lithium-related deaths (four cases of intentional suicide and two of therapeutic error) and two other deaths in which lithium was not listed as the primary cause (534). A total of 4663 lithium-related exposures were reported, in which death was the outcome in 13 and a major life-threatening event or cause of significant disability in 267.

A chart review of psychiatric hospital admissions between 1990 and 1996 showed that 6.8% of 2210 patients who were given lithium had at least one serum concentration of 1.5 mmol/l or over (43% of these were increased at admission) and of those only 28% had signs and symptoms of toxicity (363).

In another case, following an overdose with a modified-release formulation, the appearance of clinical symptoms (vomiting and dizziness) was delayed for 35 hours, despite a serum lithium concentration of 2.38 mmol/l at 15 hours and 3.12 mmol/l at 25 hours (522).

The distribution of clinical practice guidelines in Northeast Scotland had no impact on whether appropriate action was taken for high serum lithium concentrations (80% before the guidelines, 82% after). There was no significant difference in proper attention to high concentrations between those in primary care alone (77%) and those in shared care (85%) (365).

In a retrospective study of 114 patients admitted to a toxicological ICU with suspected lithium intoxication, 81 had definite intoxication; 78% were deliberate overdoses, and 22% were accidental (due, for example, to renal insufficiency, dehydration, drug interactions, poor compliance, drunkenness). Most were treated conservatively with gastric lavage and forced diuresis; hemodialysis was used only in 3–6%. Two of those who took a deliberate overdose and one of those who took an accidental overdose died (535).

Cases of lithium toxicity in a municipal hospital over a 10-year period involved eight women (mean age 66 years); neurological symptoms were the most common presentations (151). Two were acute overdoses and the rest were chronic intoxications. There was one death (group not specified).

Convincing manifestations of toxicity have occasionally been reported in patients with concentrations within the preferred target range (sometimes called "therapeutic" concentrations) (536). While full recovery is often the case, improvement may lag many days behind the fall in serum lithium concentration. Unfortunately, both persistent neurological damage (characteristically cerebellar) and deaths have occurred. The risk of an adverse outcome from toxicity can be minimized by prompt and comprehensive treatment (537–539), which should include gastric lavage and ion exchange resins for acute overdose, volume expansion to restore fluid and electrolyte balance and improve kidney function, and in severe cases, hemodialysis. Hemodiafiltration has also been used, and while this technique eliminates rebound increases in lithium concentration after dialysis, it is also less efficient than hemodialysis.

The 1998 Annual Report of the American Association of Poison Control Centers Toxic Exposure Surveillance System listed three lithium-related fatal exposures (two intentional and one a therapeutic error) and three other fatalities in which lithium was not the primary cause of death. A total of 4486 lithium-related poison exposure cases were reported, in which the outcome was death in five and a major life-threatening event or cause-significant residual disability in 212 (540).

A study of drug intoxication in the south of Brazil reported 2938 cases of drug ingestion, 25 of which involved lithium (including 14 suicide attempts) (541).

Of 133 patients 40 who had begun treatment with lithium died over an observation period of 16 years. Suicide (in 11 cases) was twice as common as in the general population, but it was more likely to occur in lithium noncompliant patients (542). It is important to be aware that suicidal behavior is actually reduced in patients who are compliant with long-term lithium therapy (although still somewhat higher than in the general population).

The 2001 Annual Report of the American Association of Poison Control Centers Toxic Exposure Surveillance System included six fatal exposures to lithium, three of which were intentional suicides. A total of 4607 exposures to lithium were reported, and death was the outcome in eight (543). A retrospective review of eight cases of lithium poisoning included one death (151).

There has been a 10-year review of lithium overdose in 304 patients (544). The circumstances were accidental ingestion, mistakes in the quantity of ingested tablets, raised lithium concentrations due to diuretic therapy, renal insufficiency or dehydration, and suicide attempts. About half the patients required management in an intensive care unit, 5% needed hemodialysis, 10% had cardiac disturbances or neurological complications, and 2% died. The authors concluded that modified-release formulations when taken in large amounts present the greatest danger.

Of 56 patients with lithium toxicity, 42 had initially overdosed and they were compared with those who had toxicity that was described as inadvertent and associated with volume depletion (545). The initial lithium concentration was lower in the cases of intentional overdose than in the cases of inadvertent intoxication (2.4 mmol/l versus 3.4 mmol/l). Hemodialysis for lithium toxicity was required in 9% of those who had taken an intentional overdose compared with 50% of those who had inadvertent intoxication. These findings were in contrast to the amount of lithium taken during the 24 hours before hospitalization, which was much higher in those who had taken an intentional overdose, because of the large inhibitory effect of dehydration on lithium excretion.

Lithium toxicity due to accidental ingestion of a lithium-containing battery has been reported.

- A five-year-old boy swallowed a button battery containing lithium (546). During the 4 days after ingestion, he developed a serum lithium concentration of 0.71 mmol/l without signs of lithium toxicity and with normal renal function. The battery was eventually retrieved by gastrotomy.

The authors warned that button battery ingestion can be a source of lithium poisoning in youngsters.

Other reports include the following:

- A 71-year-old woman developed lithium toxicity (serum concentration 2.1 mmol/l) because of increased absorption of urinary lithium from the bowel following urinary diversion with ileal conduits for stress incontinence (547).
- A 29-year-old man who overdosed on 8,000 mg of a sustained-release lithium formulation had a serum concentration of 3.12 mmol/l 25 hours later, but only became symptomatic (with vomiting and dizziness) 35 hours later; his symptoms resolved with hemodialysis (522).
- A 72-year-old woman developed an acute confusional state, gait instability, and blepharospasm and apraxia of eyelid opening (24 hours serum concentration 1.8 mmol/l), which resolved after withdrawal (548).
- In the presence of high lithium concentrations (2.6 and 1.6 mmol/l), two patients had high amplitude of the primary complex in median nerve somatosensory evoked potentials, which normalized as concentrations fell (549).
- Two women with lithium toxicity and stormy clinical courses were found to be hyperthyroid (277).
- A 32-month-old boy developed lithium toxicity after ingesting a relative's tablets (550).
- A 46-year-old man became toxic after an initially unrecognized pontine hemorrhage (551).
- A 39-year-old woman took an overdose of lithium tablets, was given a single dose of haloperidol, and developed neuroleptic malignant syndrome (195).
- A 62-year-old woman developed persistent cerebellar and extrapyramidal sequelae at a serum concentration of 3.61 mmol/l (166).

Management

An all too common treatment error in the face of severe toxicity is "watchful waiting," during which the patient's condition is more likely to worsen than improve.

Hemodialysis (383,552,553), sometimes with additional continuous venovenous hemofiltration dialysis (554,555), continues to be described as a successful intervention for lithium poisoning. Peritoneal dialysis is a far less efficient way to clear lithium from the body. One patient treated in this way had permanent neurological abnormalities and another died; a third toxic patient who also had diabetic ketoacidosis died after treatment with hydration and insulin (556). On the other hand, a 51-year-old woman who took 50 slow-release lithium carbonate tablets (450 mg) had a serum lithium concentration of 10.6 mmol/l 13 hours later, but no evidence of neurotoxicity or nephrotoxicity. She was treated conservatively with intravenous fluids and recovered fully (557). Acute lithium overdose is often better tolerated than chronic intoxication.

Several reports have described severe poisoning responding to treatment that included hemodialysis or hemodiafiltration:

- A 57-year-old man, serum lithium concentration 3.1 mmol/l (558).
- A 52-year-old woman, serum lithium concentration 3.2 mmol/l (558).
- A 58-year-old woman, serum lithium concentration 4.0 mmol/l (131).
- A 52-year-old man, serum lithium concentration 4.6 mmol/l (132).
- A 40-year-old man, serum lithium concentration 5.4 mmol/l (559).
- A 39-year-old man, serum lithium concentration 5.9 mmol/l, with renal insufficiency associated with a polydrug overdose (560).

Some patients recovered from severe intoxication without dialysis:

- A 24-year-old woman, who survived a lithium carbonate overdose (5600 mg; serum concentration 4.0 mmol/l) with conservative treatment (561).
- A 55-year-old woman, serum lithium concentration 4.5 mmol/l, who was left with residual slurred speech (562).
- A 52-year-old woman, serum lithium concentration 10.6 mmol/l after an acute overdose (557).

In a small number of patients for whom hemodialysis was recommended, the outcomes were similar in those who were actually dialysed and those who were not, leading the authors to conclude that dialysis should be reserved for the more severe cases (563).

- Two teenagers with neurological toxicity (serum concentrations 5.4 mmol/l and 4.81 mmol/l) were treated successfully with hemodialysis followed by continuous venovenous hemofiltration, which prevented a post-dialysis rebound in serum lithium concentrations (554).
- An agitated, confused, disoriented 52-year-old woman who took an overdose of lithium recovered fully after high-volume continuous venovenous hemofiltration (564).

A brief review of cases of lithium toxicity has suggested that there is no clinical evidence that gastric lavage is useful in lithium overdose (565). However, an anecdote is of interest, since it suggests that gastric lavage soon after ingestion may be beneficial (566).

- A 32-year-old man took 50 modified-release tablets of lithium carbonate and arrived at the emergency services soon after ingestion. Gastric lavage yielded several tablet fragments. Associated with other supportive measures, his serum lithium concentration never exceeded 0.75 mmol/l.

However, the more problematic cases of lithium poisoning are those in which the lithium concentration increases slowly.

- A 42-year-old woman who presented with a change in mental status and rapidly decompensated into respiratory failure and required ventilatory assistance for 2 months had not taken an overdose—her lithium concentration had increased slowly (567).

In an in vitro study, bentonite was an effective adsorbent of lithium; the authors suggested that it be explored as an overdose treatment (562).

Product monographs written by pharmaceutical companies and published by the Canadian Pharmacists Association in the Compendium of Pharmaceutical Specialties have been reviewed with regard to the adequacy of lithium overdose management advice (568). All five were rated "fair" for listing essential interventions for managing overdose but "poor" for warning against contraindicated interventions, and all contained misleading or dangerous information. All in all, a dismal showing.

Drug–Drug Interactions

General

Drug interactions with lithium have been reviewed (569–573); another review focused on interactions in the elderly (573). A review of drug interactions with lithium considered both pharmacokinetic interactions [for example diuretics, nonsteroidal anti-inflammatory drugs (NSAIDs)] and pharmacodynamic interactions (for example antipsychotic drugs, SSRIs) and summarized the most important ones in tabular form (569).

Alcohol

Excessive use of alcohol can interfere with adherence to lithium therapy. Alcohol does not itself appear to alter lithium pharmacokinetics (574).

Amiloride

Although amiloride may reduce the renal clearance of lithium, it appears to be free of the troublesome interaction with lithium that complicates the use of thiazides and loop diuretics.

Anesthetics

Sinus bradycardia (36/minute) developed in a 44-year-old woman taking lithium who received fentanyl and propofol (120).

Angiotensin converting enzyme (ACE) inhibitors

There have been scattered reports of lithium toxicity associated with the use of ACE inhibitors and attributed to reduced lithium excretion (573,575). This is not a predictable interaction.

- A 57-year-old man developed confusion, lethargy, ataxia, and myoclonus in conjunction with a serum lithium concentration of 2.6 mmol/l 4 days after starting to take captopril 50 mg tds (575).

  - Lithium toxicity occurred in a 46-year-old man when his antihypertensive agent was changed from fosinopril to lisinopril (576).

In rats, ramipril reduced renal lithium clearance and increased fractional lithium reabsorption in association with decreased systolic blood pressure and decreased sodium excretion. These effects were attenuated by icatibant, a specific bradykinin B2 receptor antagonist (577).

Angiotensin-2 receptor antagonists

There have been occasional reports of lithium toxicity in patients taking angiotensin receptor blockers.

- A 58-year-old bipolar woman with previously stable therapeutic lithium concentrations was hospitalized with a 10-day history of confusion, disorientation, and agitation 8 weeks after starting to take candesartan 16 mg/day. Both drugs were withdrawn, the serum lithium concentration fell from a high of 3.25 mmol/l, and she was again maintained on her usual therapeutic concentration of lithium (578).
- Lithium toxicity occurred in an elderly patient after the addition of losartan (579).
- A 51-year-old woman developed symptoms of lithium toxicity (serum concentration 1.4 mmol/l) while taking valsartan, which resolved when the valsartan was replaced by diltiazem (serum lithium concentration 0.8 mmol/l) (580).

Antidepressants

Adverse interactions of lithium with tricyclic antidepressants, SSRIs, and monoamine oxidase inhibitors have been reviewed (581). In reviews of antidepressants and the serotonin syndrome, a possible contributory role has been suggested for lithium, based on case reports with tricyclic antidepressants, SSRIs, trazodone, and venlafaxine (204,582).

Lithium augmentation of antidepressants is a well-established treatment for resistant depression and is usually well tolerated with all classes of antidepressants, although there have been a few reports of the serotonin syndrome with SSRIs (581). It is possible that shared adverse effects could be magnified by combining lithium with various antidepressants (for example tremor, weight gain, gastrointestinal upset). Hyponatremia secondary to the SIADH has been linked to SSRIs and tricyclic antidepressants, especially in elderly patients, and could predispose to lithium toxicity.

In 28 of 75 patients taking lithium, 24-hour urine volumes were over 3 l/day, and this group had a greater duration of lithium exposure (6.0 versus 3.9 years) (583). There was no relation between polyuria and serum lithium concentrations or dosing regimens, but there was an association with the concurrent use of serotonergic antidepressants (odds ratio 4.25).

Amfebutamone

Although data have suggested that amfebutamone has approximately the same seizure potential as the tricyclic compounds (SEDA-8, 30) (584), the manufacturers reported an increased risk of seizures in patients taking over 600 mg/day in combination with lithium or antipsychotic drugs (SEDA-10, 20) (585).

- A 45-year-old man taking lithium, amfebutamone, and venlafaxine developed a prolonged seizure after ECT, thought to have been caused by a lowering of the seizure threshold due to amfebutamone (although a role of the other two drugs could not be excluded) (187).

Mirtazapine

In a placebo-controlled, crossover study in 12 healthy men, lithium and mirtazapine had no effect on the pharmacokinetics of each other and there was no difference in psychometric testing between the addition of lithium and placebo (586).

Nefazodone

In 12 healthy volunteers, there were no clinically significant alterations in blood concentrations of lithium or nefazodone and its metabolites when the drugs were co-administered (587). The addition of lithium for 6 weeks to nefazodone in 14 treatment-resistant patients produced no serious adverse effects and no dropouts (588). Lithium augmentation of nefazodone in 13 treatment-resistant depressed patients was associated with a variety of annoying adverse effects, but none led to treatment withdrawal (589).

SSRIs

The authors of a thorough literature review of 503 patients treated with lithium and SSRIs (590) acknowledged that conclusions would be hedged with qualifications and equivocations but suggested the following:

(a) "when lithium is added to SSRIs new, nonserious, events occur frequently";
(b) "serotonin syndrome is associated with combined lithium/SSRI therapy, but is rare";
(c) "the evidence for the efficacy of lithium add-on to SSRIs is at best provisional".

There was no systematic evidence that SSRIs alter serum lithium concentrations.

Fluoxetine

Lithium toxicity has also been reported during co-administration with fluoxetine (591).

- After 4 hours of mild, intermittent, hot-weather work, a 45-year-old man taking fluoxetine and lithium (serum concentration not mentioned) collapsed, became comatose, convulsed, and was febrile (42°C); consciousness returned after 6 days but cerebellar symptoms and atrophy persisted (592). It was suggested that disruption of temperature regulation had been caused by a synergistic effect of the two drugs (although he had taken neither drug for 36 hours before the episode).

Fluvoxamine

Six patients taking a stable dose of fluvoxamine had a minor increase in plasma fluvoxamine concentration (from 67 to 76 ng/ml) 2 weeks after starting to take unspecified doses of lithium; this is unlikely to be of clinical significance (593).

Paroxetine

Serum lithium concentrations were unchanged when breakthrough depression was treated double blind by the addition of paroxetine (20–40 mg/day, $n = 19$) and the combination was generally well tolerated (954).

- A 20-year-old woman taking lithium and risperidone became catatonic 5–7 days after the addition of paroxetine, leading to speculation that this was due to an interaction between the three drugs (595). Of 17 patients 4 who had paroxetine added to lithium as an adjunctive antidepressant developed symptoms suggestive of emerging serotonin syndrome (for example nausea, vomiting, diarrhea, sweating, anxiety, oversleeping) (596).

Tricyclic antidepressants
Amitriptyline

Serum lithium concentrations were unchanged when breakthrough depression was treated double blind by the addition of amitriptyline (75–150 mg/day, $n = 23$) and the combination was generally well tolerated (594).

- A 34-year-old woman took amitriptyline 300 mg each night for several years (597). Six days after starting to take lithium 300 mg tds, she had several generalized tonic-clonic seizures. A second episode occurred on re-exposure to lithium.

Doxepin

An interaction of lithium with doxepin has been described.

- A 65-year-old man who took lithium and doxepin for 13 years presented with a 6-month history of myoclonic jerking of both arms, which resolved when both drugs were stopped (598).

Whether this represented a drug interaction or a single drug effect is unclear.

Six patients taking a stable dose of fluvoxamine had a minor increase in plasma fluvoxamine concentration (from 67 to 76 ng/ml) 2 weeks after starting to take unspecified doses of lithium; this is unlikely to be of clinical significance (593).

Antiepileptic drugs

The combination of lithium with an anticonvulsant mood stabilizer can be beneficial. There have been reports of lithium/carbamazepine neurotoxicity, but on the other hand, lithium can benefit carbamazepine-induced leukopenia. There do not appear to be clinically important pharmacokinetic interactions of lithium with gabapentin (599), lamotrigine, valproate, or topiramate (although one subject did have a 70% fall in lithium concentration) (600). In a review of pharmacokinetic interactions between antiepileptic drugs and psychotropic drugs, there were no clinically significant interactions of lithium with gabapentin, lamotrigine, valproate, or topiramate, although serum lithium concentrations were reduced slightly by topiramate (600).

Although reviews have generally been favorable regarding the combination of lithium with anticonvulsants (600–602), there have been occasional anecdotal reports of possible interactions.

Carbamazepine

Lithium intoxication in a 33-year-old man was attributed to carbamazepine-induced renal insufficiency (603).

A case report suggested an association between lithium and carbamazepine in causing sinus node dysfunction (124).

Lamotrigine

In an open crossover study in 20 healthy men, the serum lithium concentration was slightly lower (0.65 versus 0.71 mmol/l) when lamotrigine 100 mg/day was added for 6 days, but the difference was not statistically significant (604).

Oxcarbazepine

Priapism was associated with co-administration of lithium, oxcarbazepine, and aripiprazole in a 16 year-old boy (605). It started soon after oxcarbazepine 300 mg bd had been added to lithium 1200 mg/day and resolved without recurrence after oxcarbazepine was withdrawn.

Topiramate

In a 42-year-old woman, the serum lithium concentration rose from 0.5 to 1.4 mmol/l after she increased her topiramate dose from 500 to 800 mg/day (606). The authors speculated that topiramate had interfered with lithium excretion. On the other hand, in a crossover study in healthy volunteers, 6 days of treatment with topiramate did not significantly alter serum lithium concentrations; however, the maximum topiramate dose was only 200 mg/day and one subject did have about a 70% fall in lithium C_{max} and AUC (607).

Lithium toxicity occurred in a patient who was taking lithium and had topiramate added (608).

- A 26-year-old woman with bipolar I disorder took lithium and valproate, and sometimes additional risperidone and lamotrigine. Both risperidone and lamotrigine produced dermatological adverse effects. Her serum lithium concentration was 0.82 mmol/l. Topiramate 75 mg/day was added. A week later, she continued to show a mixed state with mostly manic features and a raised lithium concentration of 1.24 mmol/l. The lithium concentration continued to increase over the next 4 days to 1.97 mmol/l even though the lithium dosage was reduced from 900 to 750 mg/day. Lithium was withdrawn and the lithium concentration fell. Lithium was then restarted at half the admission dose to achieve a blood concentration of 0.67 mmol/l. Subsequent increases in the dose of topiramate resulted in further increases in the lithium concentration.

The authors suggested that topiramate reduced renal lithium excretion through several mechanisms, possibly as a carbonic anhydrase inhibitor coupled with sodium depletion. They suggested that patients taking lithium and topiramate be carefully monitored for lithium concentrations and hydration.

Valproate
In rats, lithium pretreatment reduced the plasma half-life of valproate by 25% and increased urinary excretion of valproate glucuronide (609).

Ziprasidone
Two patients with schizoaffective illness developed new signs of mild lithium toxicity after intramuscular injections of ziprasidone (610).

Antimicrobial drugs

Two reviews of drug interactions with antibiotics briefly and incompletely discussed lithium (611,612).

In a case-control study in subjects whose lithium concentration exceeded 1.3 mmol/l, initiation of a concomitant medication (OR = 2.70, 95% CI = 0.78, 9.31) and particularly antibiotics (OR = 3.14, 95% CI = 1.15, 8.61) were associated with potential toxicity (613). Parenthetically, high temperature can precipitate lithium toxicity (614).

Quinolones
- A 56-year-old man with normal renal function and therapeutic lithium concentrations became toxic (serum concentration 2.53 mmol/l 24 hours after the last dose) with renal impairment (serum creatinine 141 µmol/l; 1.6 mg/dl) within days of starting levofloxacin. Both symptoms and laboratory abnormalities resolved with withdrawal of both lithium and levofloxacin (615).

Trimethoprim
- A 40-year-old woman developed nausea, malaise, impaired concentration, trembling, unsteadiness, diarrhea, and muscle spasm in association with a serum lithium concentration of 2.1 mmol/l while taking trimethoprim 300 mg/day (616).
- A 42-year-old woman developed symptoms of lithium toxicity and a raised serum concentration (2.1 mmol/l) while taking trimethoprim (616).

This interaction may be due to an amiloride-like diuretic effect of trimethoprim, causing lithium retention.

Antipsychotic drugs

In a discussion of drug interactions with antipsychotic drugs, the literature on lithium was reviewed (617). Caution was advised when lithium is combined with antipsychotic drugs, especially with high dosages of high-potency drugs. In a review of acute, life-threatening, drug-induced neurological syndromes, the controversy of whether lithium increases the risk of neuroleptic malignant syndrome was mentioned briefly (618) (see also Nervous system).

While it is still not clear whether there is a unique encephalopathic interaction of lithium with haloperidol, there is a consensus that the judicious use of these two drugs in combination should be safe. In general, caution is advised if lithium is combined with antipsychotic drugs, especially with high dosages of high-potency drugs (617). There have been reports of neuroleptic malignant syndrome in patients taking lithium plus antipsychotic drugs, but a causal relation has not been established (619).

The risk of extrapyramidal adverse effects may be increased when lithium is combined with antipsychotic drugs.

Erythrocyte/plasma lithium concentration ratios were lower in patients taking phenothiazines or haloperidol than in those taking lithium alone (620,621), and the former group had a higher incidence of neurological and renal adverse effects (621).

- A 59-year-old man taking lithium, haloperidol, and carbamazepine had impaired memory, impaired attention, and an encephalopathy-like pattern on the electroencephalogram that normalized when haloperidol was withdrawn (622). Olanzapine 5 mg/day was added, and 3 weeks later he became disoriented. Surprisingly, the olanzapine was continued and he remained disoriented.

Amisulpride
In a placebo-controlled, parallel-design, double-blind study in 24 male volunteers, amisulpride 100 mg bd for 7 days did not alter lithium pharmacokinetics (623).

Amoxapine
The neuroleptic malignant syndrome occurred in a 63-year-old man when lithium was added to amoxapine (193).

Aripiprazole
A review of aripiprazole included a brief mention of no apparent pharmacokinetic interaction with lithium (624).

Chlorpromazine

The neuroleptic malignant syndrome in a 49-year-old man was attributed to a combination of lithium and chlorpromazine (625).

Clozapine

Seizures and other neurological effects have been described in a few cases when lithium was added to clozapine (626), but in other instances the combination was beneficial in overcoming treatment resistance or attenuating clozapine-induced leukopenia. Five treatment-resistant patients were treated successfully with a combination of clozapine and lithium with no clinically significant adverse events (627). However, a 59-year-old woman developed neurotoxic symptoms 3 days after lithium was added to clozapine; the symptoms resolved when both drugs were stopped and recurred with rechallenge (628).

- Multisystem organ failure occurred shortly after clozapine was added to a therapeutic dose of lithium in a 23-year-old woman (629). Improvement occurred when clozapine was stopped and the toxicity was attributed to clozapine.

Haloperidol

The neuroleptic malignant syndrome has been described when lithium was added to haloperidol (195).

- A pharmacodynamic drug interaction could not be excluded when a 60-year-old man developed delirium at a serum lithium concentration of 0.97 mmol/l when taking lithium and haloperidol (158).

Persistent dysarthria with apraxia has been reported with a combination of lithium carbonate and haloperidol (717).

Olanzapine

The neuroleptic malignant syndrome occurred when lithium was added to olanzapine (194).

- A 13-year-old boy with rhabdomyolysis ascribed to olanzapine was also taking lithium, so that a drug interaction could not be excluded (118).
- A 16-year-old boy developed the neuroleptic malignant syndrome when his olanzapine dose was increased (630).

Lithium toxicity in a 62-year-old woman, in whom lithium concentrations peaked at 3.0 mmol/l and who presented with delirium and extrapyramidal symptoms, was attributed to the combination of lithium with olanzapine, but the course was what one would expect with lithium toxicity alone (631).

In a double-blind study of 344 patients inadequately responsive to lithium or valproate who were randomized to olanzapine or lithium for 6 weeks, 21% gained weight on lithium plus olanzapine compared with 4.9% taking lithium and placebo (632). Whether lithium contributed to weight gain in the olanzapine group is unclear.

Quetiapine

In an open study in 10 patients, the addition of quetiapine 250 mg tds did not significantly alter serum lithium concentrations (633).

Risperidone

There were no changes in lithium pharmacokinetics when risperidone was substituted open-label for another neuroleptic drug in 13 patients (634). On the other hand, an 81-year-old man had an acute dystonic reaction 4 days after lithium was added to a regimen of risperidone, valproic acid, and benzatropine (635).

- A 17-year-old man had taken risperidone for 2 years without adverse effects, but 12 weeks after lithium was added, he reported prolonged erections (lasting 1–3 hours) 2–5 times daily; risperidone was tapered and withdrawn and the problem resolved (636).
- A 30-year-old man developed neuroleptic malignant syndrome after taking risperidone and lithium carbonate for 1 week, having previously been taking olanzapine and divalproex (637).

Although the neuroleptic malignant syndrome has been reported in patients taking these atypical neuroleptic drugs, it is less common than in patients taking typical neuroleptic drugs, and lithium may have increased the risk in these cases. Co-administration of lithium and risperidone has been associated with the rabbit syndrome (638), but this reaction was probably caused by the risperidone, and the role of lithium was not clear.

In an in vitro study, there was no visible precipitate formation when lithium citrate syrup was mixed with risperidone solution (639).

Ziprasidone

In 34 healthy men, ziprasidone did not alter serum lithium concentrations or renal lithium clearance (640).

In a placebo-controlled, open-label study in 25 healthy subjects there were no changes in serum lithium concentration or renal lithium clearance when ziprasidone (40–80 mg/day) was added for 7 days (640).

Anxiolytics

There has been a well-documented case of profound hypothermia in a patient taking lithium and diazepam; it did not occur with either drug alone (641). Otherwise, benzodiazepines and lithium have proven to be compatible.

Calcitonin

Serum lithium concentration should be monitored at the start of calcitonin therapy.

Serum lithium concentrations fell significantly within 3 days of starting calcitonin in four women (642), due to increased renal clearance of lithium (642,643).

After they had received 100 units of salmon calcitonin subcutaneously for 3 days, four patients had a 30% mean reduction in serum lithium concentration, which was attributed to reduced absorption and/or increased renal excretion (642).

Calcium channel blockers

Lithium clearance is reduced by about 30% by nifedipine (644).

- A 30-year-old man required a reduction in lithium dosage from 1500 to 900 mg/day to maintain his serum lithium concentration in the target range shortly after he started to take nifedipine 60 mg/day (645).

There have been reports of neurotoxicity, bradycardia, and reduced lithium concentrations associated with verapamil (646–648).

Ciclosporin

In rats, lithium chloride alone had no significant renal toxicity, but when it was combined with ciclosporin, the renal toxicity of the latter was worsened (649). There was also a strong ciclosporin dose-dependent increase in serum lithium concentrations (tenfold at the highest dose).

Diazepam

There has been a well-documented case of profound hypothermia in a patient taking lithium and diazepam; it did not occur with either drug alone (710). Otherwise, benzodiazepines and lithium are compatible.

Diuretics

The effects of lithium can be enhanced by sodium depletion caused by any diuretic.

Osmotic diuretics increase lithium clearance, the change being proportional to the increased rate of urine flow.

Acetazolamide increases lithium renal clearance (650).

Furosemide can cause lithium toxicity by inhibiting the tubular excretion of lithium ions (651).

Thiazide diuretics predictably reduce renal lithium clearance and increase the risk of toxicity (652). The same may be true of potassium-sparing diuretics, although this is less well established.

- A 26-year-old woman had been stable on lithium (serum concentration 1.1 mmol/l), but after taking herbal diuretics for 2–3 weeks developed manifestations of lithium toxicity (serum concentration 4.5 mmol/l) (653).

Etacrynic acid

The loop diuretics increase the renal excretion of lithium after single-dose intravenous administration in both animals (711) and man (712). Furosemide has been used to treat lithium intoxication (713). The effect of etacrynic acid is larger than those of furosemide and bumetanide (712). However, long-term treatment with furosemide and bumetanide can cause lithium intoxication in some patients (714,715), perhaps by causing sodium depletion and a secondary increase in lithium reabsorption. An adverse interaction of lithium during long-term therapy with etacrynic acid is therefore theoretically likely.

Furosemide

Furosemide can cause lithium toxicity by inhibiting the renal tubular excretion of lithium ions (716).

Laxatives

Certain bulk-forming laxatives, such as ispaghula husk and psyllium, can impair the absorption of lithium (654).

Levofloxacin

Co-administration with levofloxacin can cause severe lithium toxicity; the authors did not discuss the mechanism (718).

Losartan

An interaction of losartan with lithium has been reported (719).

- A 77-year-old woman who had been taken lithium carbonate 625 mg/day with a stable lithium concentration of around 0.6 mmol/l started to take losartan for hypertension. Within 4 weeks she developed ataxia, dysarthria, and confusion, and her serum lithium concentration was 2.0 mmol/l. Her symptoms resolved on withdrawal of losartan.

The authors proposed that this effect had occurred by reduced aldosterone secretion, an effect which is greater with ACE inhibitors than with angiotensin II receptor antagonists. In healthy volunteers losartan did not alter the fractional secretion of lithium (720), so presumably this patient had some susceptibility that caused the interaction.

Mazindol

There have been reports of lithium toxicity associated with the use of the appetite suppressant mazindol.

- A woman stabilized on lithium developed toxic symptoms 3 days after starting the appetite suppressant mazindol. After 9 days, her serum lithium concentration was 3.2 mmol/l (655).
- A 58-year-old woman developed lithium toxicity (serum concentration 3.2 mmol/l) after the addition of mazindol (656).

Methyldopa

There have been occasional reports of neurotoxic symptoms when methyldopa was combined with lithium, both with and without an increase in serum lithium concentration (657).

Metronidazole

Nephrotoxicity due to lithium was reportedly exacerbated by metronidazole (658).

Minocycline

The co-administration of lithium and minocycline in an adolescent was associated with pseudotumor cerebri (659), although minocycline can do this on its own (SED-15, 2349).

Monoamine oxidase inhibitors

Interactions of lithium with antidepressants have been reviewed: tricyclic antidepressants and MAO inhibitors—no serious problems; SSRIs—a few reports of the serotonin syndrome (721).

Neuromuscular blocking drugs

A few cases of potentiation of the neuromuscular blocking effects of suxamethonium and pancuronium were reported about 30 years ago (660,661) and have been reviewed (662).

Nifedipine

Lithium clearance is reduced by about 30% by nifedipine (722).

Non-steroidal anti-inflammatory drugs

The interaction of NSAIDs with lithium has been reviewed briefly (663). Most NSAIDs, although perhaps not all (for example aspirin, sulindac), if given in sufficient dosages for sufficient time, can increase the serum lithium concentration, sometimes to the point of toxicity (664,665).

A review of drug interactions with analgesics in the dental literature concluded that NSAIDs should be used briefly, if at all, in patients taking lithium, especially in the elderly (663).

A review of the psychiatric effects of NSAIDs included a section on renal function and lithium clearance (666).

COX-2 inhibitors

In one study, there was a mean increase of only 17% in healthy volunteers taking celecoxib 200 mg bd (667). When celecoxib was co-administered with lithium, celecoxib concentrations were higher for the first 6 hours after the dose but the AUC was not altered significantly (668). In another review it was mentioned that clinically significant interactions with lithium (increased lithium concentrations) had been identified, but no detail was presented (669). Both celecoxib and naproxen reduced the renal clearance of lithium (used as a measure of proximal tubular sodium reabsorption) (670).

There have been several reports of raised serum lithium concentrations and neurotoxic symptoms when COX-2 inhibitors (celecoxib and rofecoxib) were added to an otherwise stable lithium regimen (130)(672,673). In 10 patients taking lithium who took rofecoxib 50 mg/day for 5 days serum lithium concentrations increased in 9 and reached 1.26, 1.47, and 1.63 mmol/l in three (details not provided) (674).

Serum lithium concentrations increased in 18 patients taking lithium who started to take a COX-2 inhibitor (rofecoxib or celecoxib) (671). The authors stressed the need for lithium monitoring when COX-2 inhibitors are used concomitantly.

- A 44-year-old woman taking nimesulide and ciprofloxacin developed lithium intoxication (serum concentration 3.23 mmol/l) complicated by renal insufficiency; the interaction was attributed to nimesulide (675).

All in all, the COX-2 inhibitors appear to be similar to non-selective NSAIDs with regard to the likelihood of increasing lithium concentrations.

A possible interaction of rofecoxib with lithium has been reported (709).

- A 73-year-old man with manic-depressive illness, who had taken lithium for 40 years, underwent coronary bypass surgery and was given long-term warfarin. He developed signs of lithium intoxication (confusion, irritability, tremor, and gait disturbance) after having taken rofecoxib 12.5 mg/day for 9 days for arthritis. Rofecoxib had been chosen in order to avoid a possible drug interaction with warfarin. Lithium and rofecoxib were withdrawn and the signs resolved within 1 week. His serum lithium concentration was 1.5 mmol/l and his serum creatinine 1430 µmol/l.

Both lithium and rofecoxib have been associated with nephrotoxicity, and it is likely that lithium intoxication was caused by concomitant administration of rofecoxib, causing a reversible reduction in renal function.

Diclofenac

Diclofenac increases serum lithium concentrations by impairing its renal excretion (676).

- The serum lithium concentration nearly doubled to 1.3 mmol/l after a 57-year-old woman started to take diclofenac (677).

Ibuprofen

Ibuprofen can increase the serum lithium concentration (SEDA-13, 81) (678,679).

- A man of unspecified age developed cognitive impairment and a serum lithium concentration of 2.4 mmol/l after taking ibuprofen for shoulder pain (680).

Ketorolac

Reports of lithium neurotoxicity resulting from interaction with ketorolac have been published (SEDA-22, 117) (681). In five male volunteers, ketorolac 10 mg qds for 5 days increased lithium AUC by 24% and increased the incidence and severity of lithium-related adverse effects (682).

In a pharmacokinetic study in healthy volunteers ketorolac increased the concentration of lithium in both serum and erythrocytes, which may reflect concentration of the drug in the nervous system more accurately. Ketorolac can therefore increase the risk of adverse reactions of lithium (683), as do many other NSAIDs.

Mefenamic acid

Mefenamic acid may have interacted with lithium in a patient with reduced renal function (SEDA-13, 83) (684).

Meloxicam

In 16 subjects, meloxicam 15 mg increased plasma lithium concentrations by 21% (range −9 to 59%) and reduced total plasma lithium clearance by 18% (685).

Naproxen

There were no significant changes in serum lithium concentrations in 12 men taking over-the-counter doses of naproxen (220 mg tds) or paracetamol (650 mg qds) for 5 days (685).

Nimesulide

Nimesulide increased lithium concentrations in a crossover study (686).

Piroxicam

Piroxicam may interact with lithium (687–689).

Sulindac

Unlike other NSAIDs, sulindac supposedly does not interact with lithium (SEDA-10, 82). However, there has been a report of a toxic increase in serum lithium concentration in a 23-year-old man and a 27-year-old woman (to 2.0 and 1.7 mmol/l respectively) (690).

Tenidap

Tenidap increased serum lithium concentrations by reducing its renal clearance and increasing steady-state lithium concentrations; the dosage of lithium should be reduced to avoid toxicity (691,724).

Plantago species

Plantago seeds are widely used as bulk laxatives under the names of psyllium (from *P. psyllium* or *P. indica*) and ispaghula (from *P. ovata*). The possibility that the intestinal absorption of lithium and other drugs may be inhibited should also be considered (658).

Platinum-containing cytostatic drugs

Some data have suggested that cisplatin-containing chemotherapy can alter lithium clearance through impaired renal function, and lithium therapy should be closely monitored during treatment with cisplatin-containing regimens (723).

Sumatriptan

The 1996 Canadian Product Monograph for sumatriptan (but not the 1998 US package insert) listed the combination of sumatriptan and lithium as contraindicated. However, in a review, there was little evidence of a severe interaction of sumatriptan with lithium (692).

Thioridazine

Several cases of neurotoxicity in patients taking lithium and thioridazine have been reported. The cause of this interaction has not been resolved, but lithium seems compatible with all neuroleptic drugs, although patients should be carefully monitored (726–728).

Trimethoprim and co-trimoxazole

Trimethoprim has the same effect on the kidney as amiloride, whose combined use with lithium can cause a raised serum lithium concentration.

- The addition of trimethoprim caused severe lithium toxicity in a 40-year-old woman with a schizoaffective disorder; following rehydration, she made a good recovery (729).

Xanthines

Renal lithium clearance is increased and serum lithium concentrations are reduced by theophylline and aminophylline (496).

Caffeine increases renal lithium clearance (693), and there have been case reports of caffeine withdrawal leading to increased serum lithium concentrations, assumed to be due to reduced renal lithium clearance (693).

Food–Drug Interactions

In 12 healthy men, there was no food-induced change in the systemic availability of a sustained-release lithium formulation that used an acrylic matrix of Eudragit RSPM as a sustaining agent (694).

Dietary restriction of sodium causes lithium retention and an increased risk of toxicity. The same would be true of a diet that markedly restricted fluid intake. In brief, dietary extremes should be avoided.

Drug–procedure interactions

Electroconvulsive therapy

While electroconvulsive therapy (ECT) does not alter serum lithium concentrations, the risk of prolonged confusion after ECT can be increased in the presence of lithium (695). However, there is no universal agreement about this association (696), and there may be circumstances that justify the continuation of lithium during ECT (697,698).

A case series of patients who underwent ECT while taking lithium, suggested that concern about the concomitant use of lithium and may be exaggerated (730).

Transcranial magnetic stimulation

A seizure induced by transcranial magnetic stimulation (TMS) in a patient with bipolar affective disorder taking maintenance lithium (731), may not have been related to lithium, since therapeutic concentrations of lithium can actually increase the seizure threshold (732).

Interference with Diagnostic Tests

Serum lithium measurement

The inadvertent use of lithium heparin as an anticoagulant in the collection tube will lead to a spuriously high serum lithium determination.

- A 33-year-old woman was admitted to an intensive care unit after ingesting unknown quantities of a variety of medications, including lithium. She remained asymptomatic, except for drowsiness, despite serum lithium concentrations of 2.7, 3.1, 3.6, and 5.6 mmol/l. The latter concentration was discovered to be spuriously high when it was realized that blood had been collected in a lithium-heparin anticoagulant tube. When the test was repeated using the proper collection tube, the concentration was 2.2 mmol/l (699).

- A similar problem occurred in a 20-month-old with recurrent convulsions, who had an apparent serum lithium concentration of 3.2 mmol/l. The absence of lithium in the urine prompted further investigation and the source of the lithium was found to be from the use of a lithium-heparin collection tube (700).

Ageing of ion-selective electrodes has been reported to give inaccurately high serum lithium concentration. In one case, a concentration of 0.4 mmol/l was reported in a patient who was not taking lithium (701). The possibility that other substances could interfere with ion-selective electrode lithium analysis has been briefly reviewed (702).

Contamination of lithium heparin blood culture bottles with *Pseudomonas fluorescens* led to an outbreak of pseudobacteremia and the unnecessary treatment of a number of children with antibiotics. Lithium had no direct role in this misadventure (703).

In blood samples from 32 subjects, TSH and free T4 concentrations were no different in collection tubes that contained a lithium heparin anticoagulant compared with dry tubes, but free T3 concentrations were significantly lower (704).

Lithium treatment of adolescents with bipolar disorder and secondary substance abuse improved both conditions (705). It was suggested, however, that lithium-induced polyuria may have diluted urine to the extent that false negative test results were obtained when screening for drugs of abuse (706). While the number of positive urine tests was actually higher (but not significantly so) in those with polyuria (707), there is still a possibility that lithium-induced diabetes insipidus could increase the frequency of false negative urine drug assays, because of urine dilution.

Diagnosis of Adverse Drug Reactions

When a split hair sample from a healthy volunteer was sent to six commercial laboratories in the USA for trace mineral analysis, marked variations in results were found (including lithium concentrations), leading to the conclusion that such analyses were unreliable (708).

Monitoring therapy

Monitroing serum lithium concentrations has been reviewed (733). Lithium concentrations are important predictors of outcome, but can also be associated with the type of outcome. Low lithium concentrations (< 0.6 mmol/l) are associated with a low likelihood of depressive relapse (12%), while higher concentrations (> 0.8 mmol/l) are associated with a higher likelihood of depressive relapse (64%) (734). This has been interpreted as showing that low concentrations, rather than higher concentrations, of lithium are more effective in preventing depressive relapse; however, the difference in concentrations may be reflections of more problematic illness in which the clinician has maximized lithium treatment.

In a comparison of lithium concentrations in erythrocytes and plasma during acute or chronic lithium intoxication (309 samples in 165 patients) good general correlation between erythrocyte and plasma lithium concentrations was confirmed (735). There were higher plasma lithium concentrations in acute intoxication and higher erythrocyte lithium concentrations in chronic intoxication; the lithium erythrocyte:plasma concentration ratio was highest in those with chronic intoxication.

Biochemical variables in erythrocytes, mood states, and adverse effects of lithium were measured in 30 patients, mostly men, who had bipolar disorder and were undergoing lithium treatment (736). Most (87%) had bipolar I affective disorder. The major finding was that when the serum lithium concentration was in the 0.93–1.42 mmol/l range, there was a full response without toxicity. Higher values predicted toxicity and lower values predicted partial response.

References

1. Muller-Oerlinghausen B. Does effective lithium prophylaxis result in a symptom-free state of manic-depressive illness? Some thoughts on the fine-tuning of mood stabilization. Compr Psychiatry 2000;41(2):26–31(Suppl. 1).
2. Tondo L, Baldessarini RJ, Hennen J, Floris G, Silvetti F, Tohen M. Lithium treatment and risk of suicidal behavior in bipolar disorder patients. J Clin Psychiatry 1998;59(8):405–14.
3. Goodwin FK, Fireman B, Simon GE, Hunkeler EM, Lee J, Revicki D. Suicide risk in bipolar disorder during treatment with lithium and divalproex. JAMA 2003;290(11):1467–73.
4. Ahrens B, Muller-Oerlinghausen B. Does lithium exert an independent antisuicidal effect? Pharmacopsychiatry 2001;34(4):132–6.
5. Dunner DL. Optimizing lithium treatment. J Clin Psychiatry 2000;61(Suppl 9):76–81.
6. Manji HK, Moore GJ, Chen G. Clinical and preclinical evidence for the neurotrophic effects of mood stabilizers: implications for the pathophysiology and treatment of manic-depressive illness. Biol Psychiatry 2000;48(8):740–54.
7. Moore GJ, Bebchuk JM, Wilds IB, Chen G, Manji HK. Lithium-induced increase in human brain grey matter. Lancet 2000;356(9237):1241–2.
8. Goodwin GM. Evidence-based guidelines for treating bipolar disorder: recommendation from the British Association of Psychopharmacology. J Psychopharmacol 2003;17:149–73.
9. Quiroz JA, Gould TD, Manji HK. Molecular effects of lithium. Mol Interventions 2004;4:259–72.
10. McNamara RK, Lenox RH. The myristoylated alanine-rich C kinase substrate: a lithium-regulated protein linking cellular signaling and cytoskeletal plasticity. Clin. Neurosci Res 2004;4:155–69.
11. Harris M, Chandran S, Chakraborty N, Healy D. The impact of mood stabilizers on bipolar disorder: the 1890s and 1990s compared. Hist Psychiatry 2005;16:423–34.

12. Carney SM, Goodwin GM. Lithium—a continuing story in the treatment of bipolar disorder. Acta Psychiatr Scand Suppl 2005;426:7–12.

13. Fountoulakis KN, Vieta E, Sanchez-Moreno J, Kaprinis SG, Goikolea JM, Kaprinis GS. Treatment guidelines for bipolar disorder: a critical review. J Affect Disord 2005;86:1–10.

14. Deshauer D, Fergusson D, Duffy A, Albuquerque J, Grof P. Re-evaluation of randomized control trials of lithium monotherapy: a cohort effect. Bipolar Disord 2005;7:382–7.

15. Alda M, Grof P, Rouleau GA, Turecki G, Young LT. Investigating responders to lithium prophylaxis as a strategy for mapping susceptibility genes for bipolar disorder. Prog Neuropsychopharmacol Biol Psychiatry 2005;29:1038–45.

16. Kleindienst N, Engel RR, Greil W. Psychosocial and demographic factors associated with response to prophylactic lithium: a systematic review for bipolar disorders. Psychol Med 2005;35:1685–94.

17. Alvarez G, Munoz-Montano JR, Satrustegui J, Avila J, Bogonez E, Diaz-Nido J. Regulation of tau phosphorylation and protection against beta-amyloid-induced neurodegeneration by lithium. Possible implications for Alzheimer's disease. Bipolar Disord 2002;4(3):153–65.

18. Phiel CJ, Wilson CA, Lee VM, Klein PS. GSK-3alpha regulates production of Alzheimer's disease amyloid-beta peptides. Nature 2003;423(6938):435–9.

19. Everall IP, Bell C, Mallory M, Langford D, Adame A, Rockestein E, Masliah E. Lithium ameliorates HIV-gp120-mediated neurotoxicity. Mol Cell Neurosci 2002;21(3):493–501.

20. Xu J, Culman J, Blume A, Brecht S, Gohlke P. Chronic treatment with a low dose of lithium protects the brain against ischemic injury by reducing apoptotic death. Stroke 2003;34(5):1287–92.

21. Gray NA, Du Zhou RJ, Moore GJ, Manji HK. The use of mood stabilizers as plasticity enhancers in the treatment of neuropsychiatric disorders. J Clin Psychiatry 2003;64(Suppl 5):3–17.

22. Ducros A, Bousser MG. L'algie vasculaire de la face. [Cluster headache.] Ann Medical Interne (Paris) 2003;154(7):468–74.

23. Huang X, Wu DY, Chen G, Manji H, Chen DF. Support of retinal ganglion cell survival and axon regeneration by lithium through a Bcl-2-dependent mechanism. Invest Ophthalmol Vis Sci 2003;44(1):347–54.

24. Bogazzi F, Bartalena L, Campomori A, Brogioni S, Traino C, De Martino F, Rossi G, Lippi F, Pinchera A, Martino E. Treatment with lithium prevents serum thyroid hormone increase after thionamide withdrawal and radio-iodine therapy in patients with Graves' disease. J Clin Endocrinol Metab 2002;87(10):4490–5.

25. Bogazzi F, Bartalena L, Brogioni S, Scarcello G, Burelli A, Campomori A, Manetti L, Rossi G, Pinchera A, Martino E. Comparison of radioiodine with radioiodine plus lithium in the treatment of Graves' hyperthyroidism. J Clin Endocrinol Metab 1999;84(2):499–503.

26. Benbassat CA, Molitch ME. The use of lithium in the treatment of hyperthyroidism. Endocrinologist 1998;8:383–7.

27. Hoogenberg K, Beentjes JA, Piers DA. Lithium as an adjunct to radioactive iodine in treatment-resistant Graves' thyrotoxicosis. Ann Intern Med 1998;129(8):670.

28. Koong SS, Reynolds JC, Movius EG, Keenan AM, Ain KB, Lakshmanan MC, Robbins J. Lithium as a potential adjuvant to 131I therapy of metastatic, well differentiated thyroid carcinoma. J Clin Endocrinol Metab 1999;84(3):912–6.

29. Murphy E, Bassett JD, Meeran K, Frank JW. The efficacy of radioiodine in thyrotoxicosis is enhanced by lithium carbonate. J Nucl Med 2002;43:1280.

30. Bal CS, Kumar A, Pandey RM. A randomized controlled trial to evaluate the adjuvant effect of lithium on radioiodine treatment of hyperthyroidism. Thyroid 2002;12(5):399–405.

31. Bogazzi F, Bartalena L, Pinchera A, Martino E. Adjuvant effect of lithium on radioiodine treatment of hyperthyroidism. Thyroid 2002;12(12):1153–4.

32. Boeving A, Cubas ER, Santos CM, Carvalho GA, Graf H. O uso de carbonato de lítio no tratamento da tireotoxicose induzida por amiodarona. [Use of lithium carbonate for the treatment of amiodarone-induced thyrotoxicosis.] Arq Bras Endocrinol Metabol 2005;49:991–5.

33. Claxton S, Sinha SN, Donovan S, Greenaway TM, Hoffman L, Loughhead M, Burgess JR. Refractory amiodarone-associated thyrotoxicosis: an indication for thyroidectomy. Aust NZ J Surg 2000;70(3):174–8.

34. Muratori F, Bertini N, Masi G. Efficacy of lithium treatment in Kleine–Levin syndrome. Eur Psychiatry 2002;17(4):232–3.

35. Blier P, Slater S, Measham T, Koch M, Wiviott G. Lithium and clozapine-induced neutropenia/agranulocytosis. Int Clin Psychopharmacol 1998;13(3):137–40.

36. Barrett AJ. Haematological effects of lithium and its use in treatment of neutropenia. Blut 1980;40(1):1–6.

37. Yoshiaki S, Naoko M, Kaoru S. [A case of bipolar disorder which responded to lithium in the treatment of drug-induced granulocytopenia.] Seishin Igaku. Clin Psychiatry 2002;44:1101–5.

38. Hager ED, Dziambor H, Winkler P, Hohmann D, Macholdt K. Effects of lithium carbonate on hematopoietic cells in patients with persistent neutropenia following chemotherapy or radiotherapy. J Trace Elem Med Biol 2002;16(2):91–7.

39. Oyewumi LK. Does lithium have a role in the prevention and management of clozapine-induced granulocytopenia? Psychiatr Ann 1999;29:597–603.

40. Iwamoto J, Hakozaki Y, Sakuta H, Kobari S, Kayashima S, Fujioka T, Ooba K, Shirahama T. A case of agranulocytosis associated with severe acute hepatitis B. Hepatol Res 2001;21(2):181–5.

41. Amano I, Morii T, Yamanaka T, Tsukaguchi N, Nishikawa K, Narita N, Shimoyama T. [Successful lithium carbonate therapy for a patient with intractable and severe aplastic anemia.]Rinsho Ketsueki 1999;40(1):46–50.

42. Wolstein J, Bender S, Hesse A, Jura S, Dittmann-Balcar A. Leukocyte count increases with the addition of lithium to concurrent clozapine treatment. Schizophr Res 2000;41(NSI):B25.

43. Gajwani P, Tesar GE. Olanzapine-induced neutropenia. Psychosomatics 2000;41(2):150–1.

44. Dreno B, Chosidow O, Revuz J, Moyse D. Lithium gluconate 8% vs ketoconazole 2% in the treatment of seborrhoeic dermatitis: a multicentre, randomized study. Br J Dermatol 2003;148(6):1230–6.

45. Dreno B, Moyse D. Lithium gluconate in the treatment of seborrhoeic dermatitis: a multicenter, randomised, double-blind study versus placebo. Eur J Dermatol 2002;12(6):549–52.

46. Nishida A, Hisaoka K, Zensho H, Uchitomi Y, Morinobu S, Yamawaki S. Antidepressant drugs and cytokines in mood disorders. Int Immunopharmacol 2002;2(12):1619–26.

47. Soto O, Murphy TK. The immune system and bipolar affective disorder. In: De Geller B, DelBello MP, editors. Bipolar Disorder in Childhood and Early Adolescence. New York: Gilford Press, 2003:193–214.

48. Harvey BH, Meyer CL, Gallichio VS, Manji HK. Lithium salts in AIDS and AIDS-related dementia. Psychopharmacol Bull 2002;36(1):5–26.

49. Amsterdam JD, Maislin G, Potter L, Giuntoli R. Reduced rate of recurrent genital herpes infections with lithium carbonate. Psychopharmacol Bull 1990;26(3):343–7.

50. Bschor T. Complete suppression of recurrent herpes labialis with lithium carbonate. Pharmacopsychiatry 1999;32(4):158.

51. Cohen Y, Chetrit A, Cohen Y, Sirota P, Modan B. Cancer morbidity in psychiatric patients: influence of lithium carbonate treatment. Med Oncol 1998;15(1):32–6.

52. Johnson CD, Puntis M, Davidson N, Todd S, Bryce R. Randomized, dose-finding phase III study of lithium gamolenate in patients with advanced pancreatic adenocarcinoma. Br J Surg 2001;88(5):662–8.

53. Sansoe G, Ferrari A, Castellana CN, Bonardi L, Villa E, Manenti F. Cimetidine administration and tubular creatinine secretion in patients with compensated cirrhosis. Clin Sci (Lond) 2002;102(1):91–8.

54. Aronson JK, Reynolds DJ. ABC of monitoring drug therapy. Lithium. BMJ 1992;305(6864):1273–4.

55. Rybakowski JK, Suwalska A, Czerski PM, Dmitrzak-Weglarz M, Leszczniska-Rodziewicz A, Hauser J. Prophylactic effect of lithium in bipolar affective illness may be related to serotonin transporter genotype. Pharmacol Rep 2005;57:124–7.

56. Caspi A, Sugden K, Moffitt TE, Taylor A, Craig IW, Harrington H, McClay J, Mill J, Martin J, Braithwaite A, Poulton R. Influence of life stress on depression: moderation by a polymorphism in the 5-HTT gene. Science 2003;301:386–9.

57. Jacobs N, Kenis G, Peeters F, Deron C, Vlietinck R, van Os J. Stress-related negative affectivity and genetically altered serotonin transporter function: evidence of synergism in shaping risk of depression. Arch Gen Psychiatry 2006;63:989–96.

58. Wilhelm K, Mitchell PB, Niven H, Finch A, Wedgwood L, Scimone A, Blair IP, Parker G, Schofield PR. Life events, first depression onset and the serotonin transporter gene. Br J Psychiatry 2006;188:210–5.

59. Murphy GM, Hollander SB, Rodrigues HE, Kremer C, Schatzberg AF. Effects of the serotonin trasporter gene promoter polymorphism on mirtrazapine and paroxetine efficacy and adverse events in geriatric major depression. Arch Gen Psychiatry 2004;61:1163–9.

60. Pollock BG, Ferrell RE, Mulsant BH, Mazumdar S, Miller M, Sweet RA, Davis S, Kirshner MA, Houck PR, Stack JA, Reynolds CF III, Kupfer DJ. Allelic variation in the serotonin transporter promoter affects onset of paroxetine treatment response in late-life depression. Neuropsychopharmacology 2000;23:587–90.

61. Durham LK, Webb SM, Milos PM, Clary CM, Seymour AB. The serotonin transporter polymorphism, 5HTTLPR, is associated with a faster response time to sertraline in an elderly population with major depressive disorder. Psychopharmacology (Berl) 2004;174(4):525–9.

62. Yu YW, Tsai SJ, Chen TJ, Lin CH, Hong CJ. Association study of the serotonin transporter promoter polymorphism and symptomatology and antidepressant response in major depressive disorders. Mol Psychiatry 2002;7:1115–9.

63. Rybakowski JK, Suwalska A, Skibinska M, Szczepankiewicz A, Leszczniska-Rodziewicz A, Permoda A, Czerski PM, Hauser J. Prophylactic lithium response and polymorphism of the brain-derived neurotrophic factor gene. Pharmacopsychiatry 2005;38:166–70.

64. Muller DJ, de Luca V, Sicard T, King N, Strauss J, Kennedy JL. Brain-derived neurotrophic factor (BDNF) gene and rapid-cycling bipolar disorder: family-based association study. Br J Psychiatry 2006;189:317–23.

65. Colom F, Vieta E, Sánchez-Moreno J, Martínez-Arán A, Reinares M, Gokolea JM, Scott J. Stabilizing the stabilizer: group psychoeducation enhances the stability of serum lithium levels. Bipolar Disord 2005;7 (Suppl 5):32–6.

66. Schou M. Lithium treatment of manic-depressive illness: a practical guide. 5th revised edn.. Basel: Karger;. 1993.

67. Jefferson JW, Bohn J. Lithium and Manic Depression: a GuideMadison, WI: Lithium Information Center. Madison Institute of Medicine;. 1999.

68. Small JG, Klapper MH, Malloy FW, Steadman TM. Tolerability and efficacy of clozapine combined with lithium in schizophrenia and schizoaffective disorder. J Clin Psychopharmacol 2003;23:223–8.

69. Morishita S, Arita S. Lithium augmentation of antidepressants in the treatment of protracted depression. Int Med J 2003;10:29–32.

70. Poppe M, Friebel D, Reuner U, Todt H, Koch R, Heubnerb G. The Kleine-Levin Syndrome—effects of treatment with lithium. Neuropediatrics 2003;34:113–19.

71. Bauer M, Adli M, Baethge C, Berghofer A, Sasse J, Heinz A, Bschor J. Lithium augmentation therapy in refractory depression: clinical evidence and neurological mechanisms. Can J Psychiatry 2003;48:440–8.

72. Bschor T, Lewitzka U, Bauer M. Lithiumaugmentation zur Behandlung der bipolaren Depression. PsychoNeurology 2003;29:392–9.

73. Lee W, Cleare A. Lithium augmentation in treatment-refractory unipolar depression. Br J Psychiatry 2003;182:456–7.

74. Kidd PM. Bipolar disorder as cell membrane dysfunction. Progress toward integrative management. Alternative Med Rev 2004;9:107–35.

75. Strakowski SMI, DelBello MP, Adler CM. Comparative efficacy and tolerability of drug treatments for bipolar disorder. CNS Drugs 2001;15(9):701–18.

76. Malone RP, Delaney MA, Luebbert JF, Cater J, Campbell M. A double-blind placebo-controlled study of lithium in hospitalized aggressive children and adolescents with conduct disorder. Arch Gen Psychiatry 2000;57(7):649–54.

77. Dunner DL. Safety and tolerability of emerging pharmacological treatments for bipolar disorder. Bipolar Disord 2005;7:307–25.

78. Hartong EG, Moleman P, Hoogduin CA, Broekman TG, Nolen WA. LitCar Group. Prophylactic efficacy of lithium versus carbamazepine in treatment-naive bipolar patients. J Clin Psychiatry 2003;64(2):144–51.

79. Bowden CL, Calabrese JR, McElroy SL, Gyulai L, Wassef A, Petty F, Pope HG Jr, Chou JC, Keck PE Jr, Rhodes LJ, Swann AC, Hirschfeld RM, Wozniak PJDivalproex Maintenance Study Group. A randomized, placebo-controlled 12-month trial of divalproex

and lithium in treatment of outpatients with bipolar I disorder. Arch Gen Psychiatry 2000;57(5):481–9.

80. Pavuluri MN, Henry DB, Carbray JA, Sampson G, Naylor MW, Janicak PG. Open-label prospective trial of risperidone in combination with lithium or divalproex sodium in pediatric mania. J Affect Dis 2004;82S:S103–11.

81. Calabrese J. Depressive mood stabilization: novel concepts and clinical management. Eur Neuropsychopharmacol 2004;S4:S100–5.

82. Sajatovic M, Gyulai L, Calabrese JR, Thompson TR, Wilson BG, White R, Evoniuk G. Maintenance treatment outcomes in older patients with bipolar I disorder. Am J Geriatr Psychiatry 2005;13:305–11.

83. Tohen M, Greil W, Calabrese JR, Sachs GS, Yatham LN, Oerlinghausen BM, Koukopoulos A, Cassano GB, Grunze H, Licht RW, Dell'Osso L, Evans AR, Risser R, Baker RW, Crane H, Dossenback MR, Bowden CL. Olanzapine versus lithium in the maintenance treatment of bipolar disorder: a 12-month, randomized double-blind, controlled clinical trial. Am J Psychiatry 2005;162:1281–90.

84. Sachs G, Chengappa KNR, Suppes T, Mullen JA, Brecher M, Devine NA, Sweitzer DE. Quetiapine with lithium or divalproex for the treatment of bipolar mania: a randomized, double-blind, placebo-controlled study. Bipolar Disord 2004;6:213–23.

85. Bowden CL, Grunze H, Mullen J, Brecher M, Paulsson B, Jones M, Vågerö M, Svensson K. A randomized, double-blind, placebo-controlled efficacy and safety study of quetiapine or lithium as monotherapy for mania in bipolar disorder. J Clin Psychiatry 2005;66:111–21.

86. Levin GM. A comparison of patient-rated burden and incidence of side-effects: lithium versus valproate. Int J Psychiatry Clin Pract 1997;1:89–93.

87. McConville BJ, Sorter MT, Foster K, Barken A, Browne K, Chaney R. In: Lithium versus Valproate Side Effects in Adolescents with Bipolar Disorder. New Clinical Drug Evaluation Unit ProgamPresented at the NCDEU 38th Annual Meeting, 10–13 June, Boca Raton, FL 1998:144 poster no. 74.

88. Yatham LN, Binder C, Kusumakar V, Riccardelli R. Risperidone plus lithium versus risperidone plus valproate in acute and continuation treatment of mania. Int Clin Psychopharmacol 2004;19:103–9.

89. Rendell JM, Juszczak E, Hainsworth J, Van der Gucht E, Healey C, Morriss R, Ferrier N, Young AH, Young H, Goodwin GM, Geddes JR. Developing the BALANCE trial–the role of the pilot study and start-up phase. Bipolar Disord 2004;6:26–31.

90. Bowden CL, Collins MA. McElroy SL, Calabrese JR, Swann AC, Weisler RH, Wozniak PJ. Relationship of mania symptomatology to maintenance treatment response with divalproex, lithium, or placebo. Neuropsychopharmacology 2005;30:1932–9.

91. Calabrese JR, Shelton MD, rapport DJ, Youngstrom EA, Jackson K, Bilali S, Ganocy SJ, Findling RL. A 20-month, double-blind, maintenance trial of lithium versus divalproex in rapid cycling bipolar disorder. Am J Psychiatry 2005;162:2152–61.

92. Findling RL, McNamara NK, Youngstrom EA, Stansbrey R, Gracious BL, Reed MD, Calabrese JR. Double-blind 18-month trial of lithium versus divalproex maintenance treatment in pediatric bipolar disorder. J Am Acad Child Adolesc Psychiatry 2005;44:409–17.

93. De Jonghe F, Hendriksen M, Van Aalst G, Kool S, Peen J, Van R, Van den Eijnden E, Dekker J. Psychotherapy alone and combined with pharmacotherapy in the treatment of depression. Br J Psychiatry 2004; 185:37–45.

94. Bowden, CL, Calabrese JR, Sachs G, Yatham LM, Asghar SA, Hompland M, Montgomery P, Earl N, Smoot TM, De Vaugh-Geiss J. A placebo-controlled 18-month trial of lamotrigine and lithium maintenance treatment in recently manic or hypomanic patients with bipolar I disorder. Arch Gen Psychiatry 2003;60:392–400.

95. Calabrese JR, Bowden CL, Sachs G, Yatham LM, Behnke K, Mehtonen OP, Montgomery P, Ascher J, Paska W, De Vaugh Geiss J. Lamictal 605 Study Group. A placebo-controlled 18-month trial of lamotrigine and lithium maintenance treatment in recently depressed patients with bipolar I disorder. J Clin Psychiatry 2003;64:1013–24.

96. Goodwin GM, Bowden CL, Calabrese JR, Grunze H, Kasper S, White R, Greene P, Leadbetter R. A pooled analysis of two placebo-controlled 18-month trials of lamotrigine and lithium maintenance in bipolar I disorder. J Clin Psychiatry 2004;65:432–41.

97. Nierenberg AA, Papakostas GI, Petersen J, Montoya HD, Worthington JJ, Tedlow J, Alpert JE, Fava M. Lithium augmentation of nortriptyline for subjects resistant to multiple antidepressants. J Clin Psychopharmacol 2003;23:92–5.

98. Geddes JR, Burgess S, Hawton K, Jamison K, Goodwin GM. Long-term lithium therapy for bipolar disorder: systematic review and meta-analysis of randomized controlled trials. Am J Psychiatry 2004;161:217–22.

99. Cipriani A, Rendell JM, Geddes JR. Haloperidol alone or in combination in acute mania. Cochrane Database Syst Rev 2006;3:CD004362.

100. Hollander E, Pallanti S, Allen A, Sood E, Baldini Rossi N. Does sustained-release lithium reduce impulsive gambling and affective instability versus placebo in pathological gamblers with bipolar spectrum disorders? Am J Psychiatry 2005;162:137–45.

101. Baldessarini RJ, Tondo L, Hennen J. Treating the suicidal patient with bipolar disorder: reducing suicide risk with lithium. Ann NY Acad Sci 2001;932:24–38.

102. Goodwin FK, Fireman B, Simon GE, Hunkeler EM, Lee J, Revicki D. Suicide risk in bipolar disorder during treatment with lithium and divalproex. J Am Med Assoc 2003;290:1467–13.

103. Birch NJ. Inorganic pharmacology of lithium. Chem Rev 1999;99(9):2659–82.

104. Jefferson JW, Greist JH. Lithium. In: Sadock BJ, Sadock VA, editors. 7th ed.Kaplan & Sadock's Comprehensive Textbook of Psychiatry 2. Philadelphia: Lippincott, William & Wilkins, 2000:2377–90.

105. Schopf J. In: Lithium [in German]. Darmstadt: Steinkopff, 1999:1–89.

106. Dunner DL. Drug interactions of lithium and other antimanic/mood-stabilizing medications. J Clin Psychiatry 2003;64 Suppl 5:38–43.

107. Alderman CP. Developments in psychiatry–2004. J Pharm Pract Res 2004;34:149–51.

108. Henry C. Lithium side-effects and predictors of hypothyroidism in patients with bipolar disorder: sex differences. J Psychiatry Neurosci 2002;27(2):104–7.

109. Jefferson JW. Lithium toxicity in the elderly. In: Nelson JC, editor. Geriatric Psychopharmacology. New York: Marcel Dekker, 1998:273–83.

110. Timmer RT, Sands JM. Lithium intoxication. J Am Soc Nephrol 1999;10(3):666–74.

111. Bosinski T, Bailie GR, Eisele G. Massive and extended rebound of serum lithium concentrations following

hemodialysis in two chronic overdose cases. Am J Emerg Med 1998;16(1):98–100.

112. Hazouard E, Ferrandiere M, Rateau H, Doucet O, Perrotin D, Legras A. Continuous veno-venous haemofiltration versus continuous veno-venous haemodialysis in severe lithium self-poisoning: a toxicokinetics study in an intensive care unit. Nephrol Dial Transplant 1999;14(6):1605–6.

113. Kallner G, Lindelius R, Petterson U, Stockman O, Tham A. Mortality in 497 patients with affective disorders attending a lithium clinic or after having left it. Pharmacopsychiatry 2000;33(1):8–13.

114. Brodersen A, Licht RW, Vestergaard P, Olesen AV, Mortensen PB. Sixteen-year mortality in patients with affective disorder commenced on lithium. Br J Psychiatry 2000;176:429–33.

115. Klumpers UMH, Boom K, Janssen FMG, Tulen JHM, Loonen AJM. Cardiovascular risk factors in outpatients with bipolar disorder. Pharmacopsychiatry 2004;32:211–16.

116. Sparicio D, Landoni G, Pappalardo F, Crivellari M, Cerchierini E, Marino G, Zangrillo A. Methyline blue for lithium-induced refractory hypotension in off-pump coronary artery bypass graft: report of two cases. J Thorac Cardiovasc Surg 2004;127:592–3.

117. Paclt I, Slavicek J, Dohnalova A, Kitzlerova E, Pisvejcova K. Electrocardiographic dose-dependent changes in prophylactic doses of dosulepine, lithium and citalopram. Physiol Res 2003;52:311–17.

118. Rosebraugh CJ, Flockhart DA, Yasuda SU, Woosley RL. Olanzapine-induced rhabdomyolysis. Ann Pharmacother 2001;35(9):1020–3.

119. Terao T, Abe H, Abe K. Irreversible sinus node dysfunction induced by resumption of lithium therapy. Acta Psychiatr Scand 1996;93(5):407–8.

120. Uchiyama Y, Nakao S, Asai T, Shingu K. [A case of atropine-resistant bradycardia in a patient on long-term lithium medication.]Masui 2001;50(11):1229–31.

121. Wolf ME, Moffat M, Ranade V, Somberg JC, Lehrer E, Mosnaim AD. Lithium, hypercalcemia, and arrhythmia. J Clin Psychopharmacol 1998;18(5):420–3.

122. Goodnick PJ, Jerry J, Parra F. Psychotropic drugs and the ECG: focus on the QTc interval. Expert Opin Pharmacother 2002;3(5):479–98.

123. Chong SA, Mythily, Mahendran R. Cardiac effects of psychotropic drugs. Ann Acad Med Singapore 2001;30(6):625–31.

124. Lai CL, Chen WJ, Huang CH, Lin FY, Lee YT. Sinus node dysfunction in a patient with lithium intoxication. J Formos Med Assoc 2000;99(1):66–8.

125. Convery RP, Hendrick DJ, Bourke SJ. Asthma precipitated by cessation of lithium treatment. Postgrad Med J 1999;75(888):637–8.

126. Numata T, Abe H, Terao T, Nakashima Y. Possible involvement of hypothyroidism as a cause of lithium-induced sinus node dysfunction. Pacing Clin Electrophysiol 1999;22(6 Part 1):954–7.

127. Davies B. The first patient to receive lithium. Aust NZ J Psychiatry 1999 1983;33:s32–4.

128. Kahkonen S, Kaartinen M, Juhela P. Permanent pacing-aid to carry out long-term lithium therapy in manic patient with symptomatic bradycardia. Pharmacopsychiatry 2000;33(4):157.

129. Keck PE Jr, Strakowski SMI, Hawkins JM, Dunayevich E, Tugrul KC, Bennett JA, McElroy SL. A pilot study of rapid lithium administration in the treatment of acute mania. Bipolar Disord 2001;3(2):68–72.

130. Moltedo JM, Porter GA, State MW, Snyder CS. Sinus node dysfunction associated with lithium therapy in a child. Tex Heart Inst J 2002;29(3):200–2.

131. Slordal L, Samstad S, Bathen J, Spigset O. A life-threatening interaction between lithium and celecoxib. Br J Clin Pharmacol 2003;55(4):413–4.

132. Newland KD, Mycyk MB. Hemodialysis reversal of lithium overdose cardiotoxicity. Am J Emerg Med 2002;20(1):67–8.

133. Delva NJ, Hawken ER. Preventing lithium intoxication. Guide for physicians. Can Fam Physician 2001;47:1595–600.

134. Montes JM, Ferrando L. Gabapentin-induced anorgasmia as a cause of noncompliance in a bipolar patient. Bipolar Disord 2001;3(1):52.

135. Terao T. Lithium therapy with pacemaker. Pharmacopsychiatry 2002;35(1):35.

136. Luby ED, Singareddy RK. Long-term therapy with lithium in a private practice clinic: a naturalistic study. Bipolar Disord 2003;5(1):62–8.

137. Rifai MA, Moles JK, Harrington DP. Lithium-induced hypercalcemia and parathyroid dysfunction. Psychosomatics 2001;42(4):359–61.

138. Wolf ME, Ranade V, Molnar J, Somberg J, Mosnaim AD. Hypercalcemia, arrhythmia, and mood stabilizers. J Clin Psychopharmacol 2000;20(2):260–4.

139. Slavicek J, Paclt I, Hamplova J, Kittnar O, Trefny Z, Horacek BM. Antidepressant drugs and heart electrical field. Physiol Res 1998;47(4):297–300.

140. Reilly JG, Ayis SA, Ferrier IN, Jones SJ, Thomas SH. QTc-interval abnormalities and psychotropic drug therapy in psychiatric patients. Lancet 2000;355(9209):1048–52.

141. Frassati D, Tabib A, Lachaux B, Giloux N, Daléry J, Vittori F, Charvet D, Barel C, Bui-Xuan B, Mégard R, Jenoudet LP, Descotes J, Vial T, Timour Q. Hidden cardiac lesions and psychotropic drugs as a possible cause of sudden death in psychiatric patients: a report of 14 cases and review of the literature. Can J Psychiatry 2004;491:100–5.

142. Francis J, Hamzeh RK, Cantin-Hermoso MR. Lithium toxicity-induced wide-complex tachycardia in a pediatric patient. J Pediatr 2004;145:235–40.

143. Hsu CH, Liu PY, Chen JH, Yeh TL, Tsai HY, Lin LJ. Electrocardiographic abnormalities as predictors for over-range lithium levels. Cardiology 2005;103:101–6.

144. Mamiya K, Sadanaga T, Sekita A, Nebeyama Y, Yao H, Yukawa E. Lithium concentration correlates with QTc in patients with psychosis. J Electrocardiol 2005;38:148–51.

145. Felker GM, Thompson RE, Hare JM, Hruban RH, Clemetson DE, Howard DL, Baughman KL, Kasper EK. Underlying causes and long-term survival in patients with initially unexplained cardiomyopathy. N Engl J Med 2000;342(15):1077–84.

146. Coulter DM, Bate A, Meyboom RH, Lindquist M, Edwards IR. Antipsychotic drugs and heart muscle disorder in international pharmacovigilance: data mining study. BMJ 2001;322(7296):1207–9.

147. Cruchaudet B, Eicher JC, Sgro C, Wolf JE. [Reversible cardiomyopathy induced by psychotropic drugs: case report and literature overview.]Ann Cardiol Angéiol (Paris) 2002;51(6):386–90.

148. Bhandari S, Samellas D. Bipolar affective disorder and idiopathic pulmonary fibrosis. J Clin Psychiatry 2001;62(7):574–5.

149. Berner J. Lithium-treated mood disorders, paroxysmal rhi-norrhea, and mesial temporal lobe epilepsy. J Neuropsychiatry Clin Neurosci 1999;11(3):414–5.

150. Miao YK. Lithium neurotoxicity within the therapeutic serum range. Hong Kong J Psychiatry 2002;12:19–22.

151. Meltzer E, Steinlauf S. The clinical manifestations of lithium intoxication. Isr Med Assoc J 2002;4(4):265–7.

152. Dallocchio C, Mazzarello P. A case of parkinsonism due to lithium intoxication: treatment with pramipexole. J Clin Neurosci 2002;9(3):310–1.

153. Lang EJ, Davis SM. Lithium neurotoxicity: the develop-ment of irreversible neurological impairment despite stan-dard monitoring of serum lithium levels. J Clin Neurosci 2002;9(3):308–9.

154. Roccatagliata L, Audenino D, Primavera A, Cocito L. Nonconvulsive status epilepticus from accidental lithium ingestion. Am J Emerg Med 2002;20(6):570–2.

155. O'Brien B, Crowley K. Protracted neurological recovery after chronic lithium intoxication. Ir Med J 2002;95(9):278.

156. Gansaeuer M, Alsaadi TM. Lithium intoxication mimick-ing clinical and electrographic features of status epilepti-cus. a case report and review of the literature. Clin Electroencephalogr 2003;34(1):28–31.

157. Bartha L, Marksteiner J, Bauer G, Benke T. Persistent cognitive deficits associated with lithium intoxication: a neuropsychological case description. Cortex 2002;38(5):743–52.

158. Omata N, Murata T, Omori M, Wada Y. A patient with lithium intoxication developing at therapeutic serum lithium levels and persistent delirium after discontinuation of its administration. Gen Hosp Psychiatry 2003;25(1):53–5.

159. Fabisiak DB, Murray GB, Stern TA. Central pontine mye-linolysis manifested by temporary blindness: a possible complication of lithium toxicity. Ann Clin Psychiatry 2002;14(4):247–51.

160. Dembowski C, Rechlin T. Successful antimanic treatment and mood stabilization with lamotrigine, clozapine, and valproate in a bipolar patient after lithium-induced cere-bellar deterioration. A case report. Pharmacopsychiatry 2003;36(2):83–6.

161. Landry P, Warnes H, Nielsen T, Montplaisir J. Somnambulistic-like behaviour in patients attending a lithium clinic. Int Clin Psychopharmacol 1999;14(3):173–5.

162. Kikyo H, Furukawa T. Creutzfeldt–Jakob-like syndrome induced by lithium, levomepromazine, and phenobarbi-tone. J Neurol Neurosurg Psychiatry 1999;66(6):802–3.

163. Lee MS, Lessell S. Lithium-induced periodic alternating nystagmus. Neurology 2003;60(2):344.

164. Faravelli C, Di Bernardo M, Ricca V, Benvenuti P, Bartelli M, Ronchi O. Effects of chronic lithium treatment on the peripheral nervous system. J Clin Psychiatry 1999;60(5):306–10.

165. Garcia-Resa E, Blasco Fontecilla H, Valbuena Briones A. Sindrome de neurotoxicidad irreversible por litio. [Non reversible lithium neurotoxicity: a case report.] Med Clin (Barc) 2001;116(9):357.

166. Roy M, Fond L, Ratrema M, Convers P, Lutz MF, Cathebras P. Intoxication au lithium: complications neuro-logiques sévères. [Lithium poisoning: severe neurologic complications.] Presse Méd 2001;30(18):900–1.

167. Megna J, O'dell M. Ataxia from lithium toxicity success-fully treated with high-dose buspirone: a single-case experimental design. Arch Phys Med Rehabil 2001;82(8):1145–8.

168. Dolamore MJ. Case report: lithium toxicity in a nursing home patient. Ann Long-Term Care 2001;9:56–61.

169. Schulz C, Mavrogiorgou P, Schroter A, Hegerl U, Juckel G. Lithium-induced EEG changes in patients with affective disorders. Neuropsychobiology 2000;42(Suppl 1):33–7.

170. Gallinat J, Boetsch T, Padberg F, Hampel H, Herrmann WM, Hegerl U. Is the EEG helpful in diagnosing and monitoring lithium intoxication? A case report and review of the litera-ture. Pharmacopsychiatry 2000;33(5):169–73.

171. Stemper B, Thurauf N, Neundorfer B, Heckmann JG. Choreoathetosis related to lithium intoxication. Eur J Neurol 2003;10:743–4.

172. Etminan M, Hemmelgaru B, Delaney JAC, Suissa S. Use of lithium and the risk of injurous motor vehicle crash in elderly adults: case-control study nested within a cohort. Br Med J 2004;328:558–9.

173. Bender S, Linka T, Wolstein J, Gehendges S, Paulus H-J, Schall U, Gastpar M. Safety and efficacy of combined clozapine-lithium pharmacotherapy. Int J Neuropsychopharmacol 2004;7:59–63.

174. Marque N, Mansencal N, Morrison-Castaqhet JF, Dubourg O. Intoxication au lithium. Arch Mal Coeur Vaiss 2004;97:271–4.

175. Roy M, Stip E, Black D, Lew V, Langlois R. Dégénérescence cérébelleuse secondaire à une intoxica-tion aiguë au lithium. [Cerebellar degeneration following acute lithium intoxication.] Rev Neurol (Paris) 1998;154(6–7):546–8.

176. Brumm VL, van Gorp WG, Wirshing W. Chronic neurop-sychological sequelae in a case of severe lithium intoxica-tion. Neuropsychiatry Neuropsychol Behav Neurol 1998;11(4):245–9.

177. Su KP, Lee YJ, Lee MB. Severe peripheral polyneuropa-thy and rhabdomyolysis in lithium intoxication: a case report. Gen Hosp Psychiatry 1999;21(2):136–7.

178. Stewart JT, Williams LS. A case of lithium-induced aster-ixis. J Am Geriatr Soc 2000;48(4):457.

179. Bischof F, Melms A, Fetter M. Persistent cerebellar dete-rioration in a patient with lobar pneumonia under lithium, carbamazepine, and trifluperidol treatment. Eur Psychiatry 1999;14(3):175–6.

180. Kumar R, Deb JK, Sinha VK. Lithium neurotoxicity at therapeutic level—a case report. J Indian Med Assoc 1999;97(11):473–4.

181. Van der Steenstraten IM, Achilles RA. Irreversibele neu-rologische schade na een lithiumintoxicatie bij therapeu-tische concentraties. [Irreversible neurological damage after lithium intoxication at therapeutical concentrations.] Tijdschr Psychiatr 2001;43:271–5.

182. Roy M, Stip E, Black DN, Lew V, Langlois R. Séquelles neurologiques secondaires à une intoxication aiguë au lithium. [Neurologic sequelae secondary to acute lithium poisoning.] Can J Psychiatry 1999;44(7):671–9.

183. Adityanjee, Munshi KR, Thampy A. The syndrome of irreversible lithium-effectuated neurotoxicity. Clin Neuropharmacol 2005;28:38–49.

184. Manji H, Howard RS, Miller DH, Hirsch NP, Carr L, Bhatia K, Quinn N, Marsden CD. Status dystonicus: the syndrome and its management. Brain 1998;121(2):243–52.

185. Caviness JN, Evidente VG. Cortical myoclonus during lithium exposure. Arch Neurol 2003;60(3):401–4.

186. Kuruvilla PK, Alexander J. Lithium toxicity presenting as non-convulsive status epilepticus (NCSE). Aust NZ J Psychiatry 2001;35(6):852.

187. Conway CR, Nelson LA. The combined use of bupropion, lithium, and venlafaxine during ECT: a case of prolonged seizure activity. J ECT 2001;17(3):216–8.

188. Slama M, Masmoudi K, Blanchard N, Andrejak M. A possible case of lithium intoxication mimicking Creutzfeldt–Jakob syndrome. Pharmacopsychiatry 2000;33(4):145–6.

189. Muthane UB, Prasad BN, Vasanth A, Satishchandra P. Tardive parkinsonism, orofacial dyskinesia and akathisia following brief exposure to lithium carbonate. J Neurol Sci 2000;176(1):78–9.

190. Lane RM. SSRI-induced extrapyramidal side-effects and akathisia: implications for treatment. J Psychopharmacol 1998;12(2):192–214.

191. Van Gerpen JA. Drug-induced parkinsonism. Neurologist 2002;8(6):363–70.

192. Chakrabarti S, Chand PK. Lithium-induced tardive dystonia. Neurol India 2002;50(4):473–5.

193. Gupta S, Racaniello AA. Neuroleptic malignant syndrome associated with amoxapine and lithium in an older adult. Ann Clin Psychiatry 2000;12(2):107–9.

194. Sierra-Biddle D, Herran A, Diez-Aja S, Gonzalez-Mata JM, Vidal E, Diez-Manrique F, Vazquez-Barquero JL. Neuropletic malignant syndrome and olanzapine. J Clin Psychopharmacol 2000;20(6):704–5.

195. Lin PY, Wu CK, Sun TF. Concomitant neuroleptic malignant syndrome and lithium intoxication in a patient with bipolar I disorder: case report. Chang Gung Med J 2000;23(10):624–9.

196. Gill J, Singh H, Nugent K. Acute lithium intoxication and neuroleptic malignant syndrome. Pharmacotherapy 2003;23:811–5.

197. Pelonero AL, Levenson JL, Pandurangi AK. Neuroleptic malignant syndrome: a review. Psychiatr Serv 1998;49(9):1163–72.

198. Berardi D, Amore M, Keck PE Jr, Troia M, Dell'Atti M. Clinical and pharmacologic risk factors for neuroleptic malignant syndrome: a case-control study. Biol Psychiatry 1998;44(8):748–54.

199. Bigal ME, Bordini CA, Speciali JG. Daily headache as a manifestation of lithium intoxication. Neurology 2001;57(9):1733–4.

200. Ronziere T, Auzou P, Ozsancak C, Magnier P, Senant J, Hannequin D. Syndrome myasthénique induit par le lithium. [Myasthenic syndrome induced by lithium.] Presse Méd 2000;29(19):1043–4.

201. Fraunfelder FT, Fraunfelder FW, Jefferson JW. The effects of lithium on the human visual system. J Toxicol Cutaneous Ocul Toxicol 1992;11:97–169.

202. Go KG. Pseudotumour cerebri. Incidence, management and prevention. CNS Drugs 2000;14:33–49.

203. Mackirdy C, Abass H. Lithium induced raised intracranial pressure. Aust NZ J Psychiatry 2002;36(3):426.

204. Keck PE, Arnold LM. The serotonin syndrome. Psychiatr Ann 2000;30:333–43.

205. Mason PJ, Morris VA, Balcezak TJ. Serotonin syndrome. Presentation of 2 cases and review of the literature. Medicine (Baltimore) 2000;79(4):201–9.

206. Landry P, Montplaisir J. Lithium-induced somnambulism. Can J Psychiatry 1998;43(9):957–8.

207. Netski AL, Piasecki M. Lithium-induced exacerbation of stutter. Ann Pharmacother 2001;35(7–8):961.

208. Sansone ME, Ziegler K. Brain and nervous system. In: Johnson FN, editor. Depression and Mania: Modern Lithium Therapy. Oxford: IRL Press, 1987:240.

209. O'Sullivan JD, Lees AJ. Nonparkinsonian tremors. Clin Neuropharmacol 2000;23(5):233–8.

210. Floru L. Die Anwendung beta-blockierender Substanzen in der Psychiatrie und Neurologie. [The use of beta-blocking agents in psychiatry and neurology.] Fortschr Neurol Psychiatr Grenzgeb 1977;45(2):112–27.

211. Gelenberg AJ, Jefferson JW. Lithium tremor. J Clin Psychiatry 1995;56(7):283–7.

212. Faulkner MA, Bertoni JM, Lenz TL. Gabapentin for the treatment of tremor. Ann Pharmacother 2003;37(2):282–6.

213. Miodownik C, Witztum E, Lerner V. Lithium-induced tremor treated with vitamin B6: a preliminary case series. Int J Psychiatry Med 2002;32(1):103–8.

214. Zaninelli R, Bauer M, Jobert M, Muller-Oerlinghausen B. Changes in quantitatively assessed tremor during treatment of major depression with lithium augmented by paroxetine or amitriptyline. J Clin Psychopharmacol 2001;21(2):190–8.

215. Davis JM. Lithium intoxication and pontine haemorrhage. Acta Psychiatr Scand 2001;103:401.

216. Grounds D. Connection between lithium and muscular incoordination. Aust NZ J Psychiatry 2002;36(1):142–3.

217. Lam RW, Allain S, Sullivan K, Beattie CW, Remick RA, Zis AP. Effects of chronic lithium treatment on retinal electrophysiologic function. Biol Psychiatry 1997;41(6):737–42.

218. Wirz-Justice A, Reme C, Prunte A, Heinen U, Graw P, Urner U. Lithium decreases retinal sensitivity, but this is not cumulative with years of treatment. Biol Psychiatry 1997;41(6):743–6.

219. Horner KC, Huang ZW, Higuerie D, Cazals Y. Reversible hearing impairment induced by lithium in the guinea pig. Neuroreport 1997;8(6):1341–5.

220. Honig A, Arts BM, Ponds RW, Riedel WJ. Lithium induced cognitive side-effects in bipolar disorder: a qualitative analysis and implications for daily practice. Int Clin Psychopharmacol 1999;14(3):167–71.

221. Arts BMG, Honig A, Riedel WJ, Ponds RWHM. Cognitive side effects of lithium; a meta-analysis and proposal for screening. Tijdschr Psychiatr 1998;40:460–8.

222. Tremont G, Stern RA. Minimizing the cognitive effects of lithium therapy and electroconvulsive therapy using thyroid hormone. Int J Neuropsychopharmacol 2000;3(2):175–86.

223. Suwalska A, Lojko D, Rybakowski J. Wplyw lekow normotymicznych na czynnosci poznawcze. [The influence of mood-normalizing agents on cognitive functions.] Psychiatr Pol 2001;35(2):245–56.

224. Donaldson S, Goldstein LH, Landau S, Raymont V, Frangou S. The Maudsley Bipolar Disorder Project: the effect of medication, family history, and duration of illness on IQ and memory in bipolar I disorder. J Clin Psychiatry 2003;64(1):86–93.

225. Harmer CJ, Clark L, Grayson L, Goodwin GM. Sustained attention deficit in bipolar disorder is not a working memory impairment in disguise. Neuropsychologia 2002;40(9):1586–90.

226. Stip E, Dufresne J, Lussier I, Yatham L. A double-blind, placebo-controlled study of the effects of lithium on cognition in healthy subjects: mild and selective effects on learning. J Affect Disord 2000;60(3):147–57.

227. Schou M. Artistic productivity and lithium prophylaxis in manic-depressive illness. Br J Psychiatry 1979;135:97–103.

228. Malitas PN, Alevizos B, Christodoulou GN. Aniracetam treatment for lithium-produced cognitive deficits of

bipolar patients. Eur Neuropsychopharmacol 1999;9(Suppl 5):S240.

229. Patten SB, Williams JV, Petcu R, Oldfield R. Delirium in psychiatric inpatients: a case-control study. Can J Psychiatry 2001;46(2):162–6.

230. Niethammer R, Keller A, Weisbrod M. Delirantes Syndrom als Lithium-nebenwirkung bei normalen Lithiumspiegeln. [Delirium syndrome as a side-effect of lithium in normal lithium levels.] Psychiatr Prax 2000;27(6):296–7.

231. Merle C, Sotto A, Galland MC, Jourdan E, Jourdan J. Syndrome cérébelleux persistant après traitement par lithium et neuroleptique. [Persistent cerebellar syndrome after treatment with lithium and a neuroleptic.] Therapie 1998;53(5):511–3.

232. Normann C, Brandt C, Berger M, Walden J. Delirium and persistent dyskinesia induced by a lithium–neuroleptic interaction. Pharmacopsychiatry 1998;31(5):201–4.

233. Shulman KI, Sykora K, Gill SS, Mamdani M, Bronskill S, Wodchis WP, Anderson G, Rochon P. Incidence of delirium in older adults newly prescribed lithium or valproate: a population-based cohort study. J Clin Psychiatry 2005;66:424–7.

234. Wada A, Yokoo H, Yanagita T, Kobayashi H. Lithium: potential therapeutics against acute brain injuries and chronic neurodegenerative diseases. J Pharmacol Sci 2005;99(4):307–21.

235. Nakashima H, Ishihara T, Suguimoto P, Yokota O, Oshima E, Kugo A, Terada S, Hamamura T, Trojanowski JQ, Lee VM, Kuroda S. Chronic lithium treatment decreases tau lesions by promoting ubiquitination in a mouse model of tauopathies. Acta Neuropathol (Berl) 2005;110(6):547–56.

236. Noble W, Planel E, Zehr C, Olm V, Meyerson J, Suleman F, Gaynor K, Wang L, LaFrancois J, Feinstein B, Burns M, Krishnamurthy P, Wen Y, Bhat R, Lewis J, Dickson D, Duff K. Inhibition of glycogen synthase kinase-3 by lithium correlates with reduced tauopathy and degeneration in vivo. Proc Natl Acad Sci USA 2005;102:6990–5.

237. Feyt C, Kienlen-Campard P, Leroy K, N'Kuli F, Courtoy PJ, Brion JP, Octave JN. Lithium chloride increases the production of amyloid-beta peptide independently from its inhibition of glycogen synthase kinase 3. J Biol Chem 2005;280:33220–7.

238. Su Y, Ryder J, Li B, Wu X, Fox N, Solenberg P, Brune K, Paul S, Zhou Y, Liu F, Ni B. Lithium, a common drug for bipolar disorder treatment, regulates amyloid-beta precursor protein processing. Biochemistry 2004;43:6899–908.

239. Wei H, Leeds PR, Qian Y, Wei W, Chen R, Chuang D. Beta-amyloid peptide-induce death of PC12 cells and cerebellar granule cell neurons is inhibited by long-term lithium treatment. Eur J Pharmacol 2000;392:117–23.

240. Dunn N, Holmes C, Mullee M. Does lithium therapy protect against the onset of dementia? Alzheimer Dis Assoc Disord 2005;19(1):20–2.

241. Nunes PV, Forlenza OV, Gattaz WF. Lithium and risk for Alzheimer's disease in elderly patients with bipolar disorder. Br J Psychiatry 2007;190:359–60.

242. Cipriani A, Pretty H, Hawton K, Geddes JR. Lithium in the prevention of suicidal behavior and all-cause mortality in patients with mood disorders: a systematic review of randomized trials. Am J Psychiatry 2005;162(10):1805–19.

243. Born C, Dittmann S, Post RM, Grunze H. New prophylactic agents for bipolar disorder and their influence on suicidality. Arch Suicide Res 2005;9(3):301–6.

244. Goldberg JF, Allen MH, Miklowitz DA, Bowden CL, Endick CJ, Chessick CA, Wisniewski SR, Miyahara S, Sagduyu K, Thase ME, Calabrese JR, Sachs GS. Suicidal ideation and pharmacology among STEP-BD patients. Psychiatr Serv 2005;56(12):1534–40.

245. Kessing LV, Søndergård L, Kvist K, Andersen PK. Suicide risk in patients treated with lithium. Arch Gen Psychiatry 2005;62(8):860–6.

246. Lauterbach E, Ahrens B, Felber W, Oerlinghausen BM, Kilb B, Bischof G, Heuser I, Werner P, Hawellek B, Maier W, Lewitzka U, Pogarell O, Hegerl U, Bronisch T, Richter K, Niklewski G, Broocks A, Hohagen F. Suicide prevention by lithium SUPLI—challenges of a multicenter prospective study. Arch Suicide Res 2005;9(1):27–34.

247. Haden ST, Brown EM, Stoll AL, Scott J, Fuleihan GE. The effect of lithium on calcium-induced changes in adrenocorticotrophin levels. J Clin Endocrinol Metab 1999;84(1):198–200.

248. Sofuoglu S, Karaaslan F, Tutus A, Basturk M, Yabanoglu I, Esel E. Effects of short and long-term lithium treatment on serum prolactin levels in patients with bipolar affective disorder. Int J Neuropsychopharmacol 1999;2:S56.

249. Basturk M, Karaaslan F, Esel E, Sofuoglu S, Tutus A, Yabanoglu I. Effects of short and long-term lithium treatment on serum prolactin levels in patients with bipolar affective disorder. Prog Neuropsychopharmacol Biol Psychiatry 2001;25(2):315–22.

250. Turan T, Esel E, Tokgoz B, Aslan S, Sofuoglu S, Utas C, Kelestimur F. Effects of short- and long-term lithium treatment on kidney functioning in patients with bipolar mood disorder. Prog Neuropsychopharmacol Biol Psychiatry 2002;26(3):561–5.

251. Watson S, Young AH. Hypothalamic-pituitary-adrenal-axis function in bipolar disorder. Clin Approaches Bipolar Disorders 2002;2:57–64.

252. Bschor T, Baethge C, Adli M, Eichmann U, Ising M, Uhr M, Muller-Oerlinghausen B, Bauer M. Lithium augmentation increases post-dexamethasone cortisol in the dexamethasone suppression test in unipolar major depression. Depress Anxiety 2003;17(1):43–8.

253. Bschor T, Adli M, Baethge C, Eichmann U, Ising M, Uhr M, Modell S, Kunzel H, Muller-Oerlinghausen B, Bauer M. Lithium augmentation increases the ACTH and cortisol response in the combined DEX/CRH test in unipolar major depression. Neuropsychopharmacology 2002;27(3):470–8.

254. Lazarus JH. The effects of lithium therapy on thyroid and thyrotropin-releasing hormone. Thyroid 1998;8(10):909–13.

255. Schiemann U, Hengst K. Thyroid echogenicity in manic-depressive patients receiving lithium therapy. J Affect Disord 2002;70(1):85–90.

256. Caykoylu A, Capoglu I, Unuvar N, Erdem F, Cetinkaya R. Thyroid abnormalities in lithium-treated patients with bipolar affective disorder. J Int Med Res 2002;30(1):80–4.

257. Deodhar SD, Singh B, Pathak CM, Sharan P, Kulhara P. Thyroid functions in lithium-treated psychiatric patients: a cross-sectional study. Biol Trace Elem Res 1999;67(2):151–63.

258. Bocchetta A, Mossa P, Velluzzi F, Mariotti S, Zompo MD, Loviselli A. Ten-year follow-up of thyroid function in lithium patients. J Clin Psychopharmacol 2001;21(6):594–8.

259. Baethge C, Blumentritt H, Berghöfer A, Bschor T, Gleen T, Adli M, Schlattmann P, Bauer M, Finke R. Long-term

lithium treatment and thyroid antibodies: a controlled study. J Psychiatry Neurosci 2005;30:423–7.

260. Jefferson JW. Lithium-associated clinical hypothyroidism. Int Drug Ther Newsl 2000;35:84–6.

261. Kleiner J, Altshuler L, Hendrick V, Hershman JM. Lithium-induced subclinical hypothyroidism: review of the literature and guidelines for treatment. J Clin Psychiatry 1999;60(4):249–55.

262. Kirov G. Thyroid disorders in lithium-treated patients. J Affect Disord 1998;50(1):33–40.

263. Kupka RW, Nolen WA, Post RM, McElroy SL, Altshuler LL, Denicoff KD, Frye MA, Keck PE Jr, Leverich GS, Rush AJ, Suppes T, Pollio C, Drexhage HA. High rate of autoimmune thyroiditis in bipolar disorder: lack of association with lithium exposure. Biol Psychiatry 2002;51(4):305–11.

264. Shulman KI, Sykora K, Gill Ss, Mamdani M, Anderson G, Marras C, Wodchis WP, Lee PE, Rochon P. New thyroxine treatment in older adults beginning lithium therapy: implications for clinical practice. Am J Geriatr Psychiatry 2005;13:299–304.

265. Aliasgharpour M, Abbassi M, Shafaroodi H, Razi F. Subclinical hypothyroidism in lithium-treated psychiatric patients in Tehran, Islamic Republic of Iran. East Mediterr Health J 2005;11:329–33.

266. Kusalic M, Engelsmann F. Effect of lithium maintenance therapy on thyroid and parathyroid function. J Psychiatry Neurosci 1999;24(3):227–33.

267. Gracious BL. Elevated TSH in bipolar youth prescribed both lithium and divalproex sodium. Int Drug Ther Newsl 2001;36:94–5.

268. Gracious BL, Findling RL, McNamara NK, Youngstrom EA, Calabrese JR. Elevated TSH in bipolar youth prescribed both lithium and divalproex sodium. Bipolar Disord 2001;3(Suppl 1):38–9.

269. Kupka RW, Nolen WA, Drexhage HA, McElroy SL, Altshuler LL, Denicoff KD, Frye MA, Keck PE, Leverich GS, Rush AJ, Suppes T, Pollio C, Post RM. High rate of autoimmune thyroiditis in bipolar disorder is not associated with lithium. Bipolar Disord 2001;3(Suppl 1):44–5.

270. Leutgeb U. Ambient iodine and lithium associated with clinical hypothyroidism. Br J Psychiatry 2000;176:495–6.

271. Bermudes RA. Psychiatric illness or thyroid disease? Don't be misled by false lab tests. Curr Psychiatry 2002;1(51–52):57–61.

272. Mira SA, Gimeno EJ, Diaz-Guerra GM, Carranza YFH. Alteraciones tiroideas y paratiroideas asociadas al tratamiento crónico con litio. A propósito de un caso. [Thyroid and parathyroid alterations associated with chronic lithium treatment. A case report.] Rev Esp Enferm Metab Oseas 2001;10:153–6.

273. Depoot I, Van Imschoot S, Lamberigts G. Lithium-geassocieerde hyperthyroïdie. [Lithium-associated hyperthyroidism.] Tijdschr Geneeskd 1998;54:413–6.

274. Calvo Romero JM, Puerto Pica JM. Crisis tirotóxica tras la retirada de tratamiento con litio. [A thyrotoxic crisis following the withdrawal of lithium treatment.] Rev Clin Esp 1998;198(11):782–3.

275. Scanelli G. Tireotossicosi da litio. Descrizione di un caso e revisione della letteratura. [Lithium thyrotoxicosis. Report of a case and review of the literature.] Recenti Prog Med 2002;93(2):100–3.

276. Ripoll Mairal M, Len Abad O, Falco Ferrer V, Fernandez de Sevilla Ribosa T. Hipertiroidismo e hypercalcemia asociados al tratamiento con litio. [Hyperthyroidism and

277. hypercalcemia associated with lithium treatment.] Rev Clin Esp 2000;200(1):48–9.

277. Oakley PW, Dawson AH, Whyte IM. Lithium: thyroid effects and altered renal handling. J Toxicol Clin Toxicol 2000;38(3):333–7.

278. Dang AH, Hershman JM. Lithium-associated thyroiditis. Endocr Pract 2002;8(3):232–6.

279. Yamagishi S, Yokoyama-ohta M. A case of lithium-associated hyperthyroidism. Postgrad Med J 1999;75(881):188–9.

280. Miller KK, Daniels GH. Association between lithium use and thyrotoxicosis caused by silent thyroiditis. Clin Endocrinol (Oxf) 2001;55(4):501–8.

281. Obuobie K, Al-Sabah A, Lazarus JH. Subacute thyroiditis in an immunosuppressed patient. J Endocrinol Invest 2002;25(2):169–71.

282. Perrild H, Hegedus L, Baastrup PC, Kayser L, Kastberg S. Thyroid function and ultrasonically determined thyroid size in patients receiving long-term lithium treatment. Am J Psychiatry 1990;147(11):1518–21.

283. Dickstein G, Shechner C, Adawi F, Kaplan J, Baron E, Ish-Shalom S. Lithium treatment in amiodarone-induced thyrotoxicosis. Am J Med 1997;102(5):454–8.

284. Bilanakis N, Gibiriti M. Lithium intoxication, hypercalcemia and "accidently" induced food and water inversion: a case report. Prog Neuropsychopharmacol Biol Psychiatry 2004;28:201–3.

285. Abdullah H, Bliss R, Guinea AI, Delbridge L. Pathology and outcome of surgical treatment for lithium-associated hyperparathyroidism. Br J Surg 1999;86(1):91–3.

286. Cohen O, Rais T, Lepkifker E, Vered I. Lithium carbonate therapy is not a risk factor for osteoporosis. Horm Metab Res 1998;30(9):594–7.

287. Awad SS, Miskulin J, Thompson N. Parathyroid adenomas versus four-gland hyperplasia as the cause of primary hyperparathyroidism in patients with prolonged lithium therapy World J Surg 2003;27:486–8.

288. Hundley JC, Woodrum DT, Saunders BD, Doherty GM, gauger PG. Revisiting lithium-associated hyperparathyroidism in the era of intraoperative parathyroid hormone monitoring. Surgery 2005;138:1027–31.

289. Sofuoglu S, Basturk M, Tutus A, Karaaslan F, Aslan SS, Gonuk AS. Lithium-induced alterations in parathormone function in patients with bipolar affective disorder. Int J Neuropsychopharmacol 1999;2:S56.

290. Turan MT, Esel E, Tutus A, Sofuoglu S, Gonuk AS. Lithium-induced alterations in parathormone function in patients with bipolar disorder. Bull Clin Psychopharmacol 2001;11:96–100.

291. Dwight T, Kytola S, Teh BT, Theodosopoulos G, Richardson AL, Philips J, Twigg S, Delbridge L, Marsh DJ, Nelson AE, Larsson C, Robinson BG. Genetic analysis of lithium-associated parathyroid tumors. Eur J Endocrinol 2002;146(5):619–27.

292. Mak TW, Shek CC, Chow CC, Wing YK, Lee S. Effects of lithium therapy on bone mineral metabolism: a two-year prospective longitudinal study. J Clin Endocrinol Metab 1998;83(11):3857–9.

293. El Khoury A, Petterson U, Kallner G, Aberg-Wistedt A, Stain-Malmgren R. Calcium homeostasis in long-term lithium-treated women with bipolar affective disorder. Prog Neuropsychopharmacol Biol Psychiatry 2002;26(6):1063–9.

294. Morillas Arino C, Jordan Lluch M, Sola Izquierdo E, Serra Cerda M, Garzon Pastor S, Gomez Balaguer M, Hernandez Mijares YA. [Parathyroid adenoma and lithium therapy.]Endocrinol Nutr 2002;49:56–7.

295. Collumbien ECA. Een geval van therapieresistente manie. Lithium en hyperparathyreoïdie. [A case of therapy-resistant mania. Lithium and hyperparathyroidism.] Tijdschr Psychiatrie 2000;42:851–5.

296. Guirguis AF, Taylor HC. Nephrogenic diabetes insipidus persisting 57 months after cessation of lithium carbonate therapy: report of a case and review of the literature. Endocr Pract 2000;6(4):324–8.

297. Lam P, Tsai S-J, Chou Y-C. Lithium associated hyperparathyroidism with adenoma: a case report. Int Med J 2000;7:283–5.

298. Gama R, Wright J, Ferns G. An unusual case of hypercalcaemia. Postgrad Med J 1999;75(890):769–70.

299. de Celis G, Fiter M, Latorre X, Llebaria C. Oxyphilic parathyroid adenoma and lithium therapy. Lancet 1998;352(9133):1070.

300. Oakley PW, Whyte IM, Carter GL. Lithium toxicity: an iatrogenic problem in susceptible individuals. Aust NZ J Psychiatry 2001;35(6):833–40.

301. Krastins MG, Phelps KR. Nephrogenic diabetes insipidus and hyperparathyroidism in a patient receiving chronic lithium therapy. J Am Geriatr Soc 2002;50:S140.

302. Dieserud F, Brun AC, Lahne PE, Normann E. Litiumbehandling og hyperparatyreoidisme. [Lithium treatment and hyperparathyroidism.] Tidsskr Nor Laegeforen 2001;121(22):2602–3.

303. Pieri-Balandraud N, Hugueny P, Henry JF, Tournebise H, Dupont C. Hyperparathyroïdie induite par le lithium. Un nouveau cas. [Hyperparathyroidism induced by lithium. A new case.] Rev Medical Interne 2001;22(5):460–4.

304. Valeur N, Andersen RS. Lithium induceret parathyreoideahormon dysfunktion. [Lithium induced dysfunction of the parathyroid hormone.] Ugeskr Laeger 2002;164(5):639–40.

305. Catala JC, Rubio AB, Fernandez CC, Ballester YAH. Hiperparatiroidismo asociada al tratamiento con litio. Rev Esp Enferm Metab Oseas 2001;10:157–8.

306. Kanfer A, Blondiaux I. Complications rénales et métaboliques du lithium. [Renal and metabolic complications of lithium.] Nephrologie 2000;21(2):65–70.

307. Gomez Moreno R, Lobo Fresnillo T, Calvo Cebrian A, Monge Ropero N. Hiperparatiroidismo y litio. [Hyperparathyroidism and lithium.] Aten Primaria 2003;31(5):337.

308. Belyavskaya NA. Lithium-induced changes in gravicurvature, statocyte ultrastructure and calcium balance of pea roots. Adv Space Res 2001;27(5):961–6.

309. Cassidy F. Diabetes mellitus in manic-depressive patients. Essent Psychopharmacol 2001;4:49–57.

310. Uzu T, Ichida K, Ko M, Tsukurimichi S, Yamato M, Takahara K, Ohashi M, Yamauchi A, Nomura M. [Two cases of lithium intoxication complicated by type 2 diabetes mellitus.]J Jpn Diabetes Soc 2001;44:767–70.

311. Cyr M, Guia MA, Laizure SC. Increased lithium dose requirement in a hyperglycemic patient. Ann Pharmacother 2002;36(3):427–9.

312. Oomura S, Mukasa H, Ooji T, Mukasa H, Mukasa H, Satomura T, Tatsumoto Y. Does impaired glucose tolerance predispose to lithium intoxication in treatment of MDI? Int J Neuropsychopharmacol 1999;2:S63.

313. Azam H, Newton RW, Morris AD, Thompson CJ. Hyperosmolar nonketotic coma precipitated by lithium-induced nephrogenic diabetes insipidus. Postgrad Med J 1998;74(867):39–41.

314. MacGregor DA, Baker AM, Appel RG, Ober KP, Zaloga GP. Hyperosmolar coma due to lithium-induced diabetes insipidus. Lancet 1995;346(8972):413–7.

315. MacGregor DA, Dolinski SY. Hyperosmolar coma. Lancet 1999;353(9159):1189.

316. Ackerman S, Nolan LJ. Bodyweight gain induced by psychotropic drugs. Incidence, mechanisms and management. CNS Drugs 1998;9:135–51.

317. Malhotra S, McElroy SL. Medical management of obesity associated with mental disorders. J Clin Psychiatry 2002;63(Suppl 4):24–32.

318. Vanina Y, Podolskaya A, Sedky K, Shahab H, Siddiqui A, Munshi F, Lippmann S. Body weight changes associated with psychopharmacology. Psychiatr Serv 2002;53(7):842–7.

319. Kelly DL, Conley RR, Love RC, Horn DS, Ushchak CM. Weight gain in adolescents treated with risperidone and conventional antipsychotics over six months. J Child Adolesc Psychopharmacol 1998;8(3):151–9.

320. Sussman N, Ginsberg D. Effects of psychotropic drugs on weight. Psychiatr Ann 1999;29:580–94.

321. Schumann C, Lenz G, Berghofer A, Muller-Oerlinghausen B. Non-adherence with long-term prophylaxis: a 6-year naturalistic follow-up study of affectively ill patients. Psychiatry Res 1999;89(3):247–57.

322. Elmslie JL, Silverstone JT, Mann JI, Williams SM, Romans SE. Prevalence of overweight and obesity in bipolar patients. J Clin Psychiatry 2000;61(3):179–84.

323. Ginsberg DL, Sussman N. Effects of mood stabilizers on weight. Primary Psychiatry 2000;7:49–58.

324. Chengappa KN, Chalasani L, Brar JS, Parepally H, Houck P, Levine J. Changes in body weight and body mass index among psychiatric patients receiving lithium, valproate, or topiramate: an open-label, nonrandomized chart review. Clin Ther 2002;24(10):1576–84.

325. Fagiolini A, Frank E, Houck PR, Mallinger AG, Swartz HA, Buysse DJ, Ombao H, Kupfer DJ. Prevalence of obesity and weight change during treatment in patients with bipolar I disorder. J Clin Psychiatry 2002;63(6):528–33.

326. Cheskin LJ, Bartlett SJ, Zayas R, Twilley CH, Allison DB, Contoreggi C. Prescription medications: a modifiable contributor to obesity. South Med J 1999;92(9):898–904.

327. Atmaca M, Kuloglu M, Tezcan E, Ustundag B. Weight gain and serum leptin levels in patients on lithium treatment. Neuropsychobiology 2002;46(2):67–9.

328. Sachs GS, Printz DJ, Kahn DA, Carpenter D, Docherty JP. The Expert Consensus Guideline Series. Medication Treatment of Bipolar Disorder 2000. Postgrad Med 2000;Special Number:1–104.

329. Schrauzer GN. Lithium: occurrence, dietary intakes, nutritional essentiality. J Am Coll Nutr 2002;21(1):14–21.

330. Chemali KR, Suarez JI, Katirji B. Acute hypokalemic paralysis associated with long-term lithium therapy. Muscle Nerve 2001;24(2):297–8.

331. Thomsen K, Schou M. Avoidance of lithium intoxication: advice based on knowledge about the renal lithium clearance under various circumstances. Pharmacopsychiatry 1999;32(3):83–6.

332. Looi JC, Cubis JC, Saboisky J. Hyponatremia, convulsions and neuroleptic malignant syndrome in a male with schizoaffective disorder. Aust NZ J Psychiatry 1995;29(4):683–7.

333. Singh B, Bandhu HK, Pathak CM, Garg ML, Mittal BR, Kulhara P, Singh N, Deodhar SD. Effect of lithium therapy

on trace elements in blood of psychiatric patients. Trace Elem Electrolytes 1998;15:94–100.

334. Danielson DA, Jick H, Porter JB, Perera DR, Hunter JR, Werrbach JH. Drug toxicity and hospitalization among lithium users. J Clin Psychopharmacol 1984;4(2):108–10.

335. Stancer HC, Kivi R. Lithium carbonate and oedema. Lancet 1971;2(7731):985.

336. Wasay M, Bakshi R, Kojan S, Bobustuc G, Dubey N. Superior sagittal sinus thrombosis due to lithium: local urokinase thrombolysis treatment. Neurology 2000;54(2):532–3.

337. Oyewumi LK, McKnight M, Cernovsky ZZ. Lithium dosage and leukocyte counts in psychiatric patients. J Psychiatry Neurosci 1999;24(3):215–21.

338. Celik C, Konukoglu D, Ozmen M, Akcay T. Platelet-specific proteins in lithium carbonate treatment. Med Sci Res 1998;26:417–8.

339. Volf N, Crismon ML. Leukemia in bipolar mood disorder: is lithium contraindicated? DICP 1991;25(9):948–51.

340. Hager ED, Dziambor H, Hohmann D, Winkler P, Strama H. Effects of lithium on thrombopoiesis in patients with low platelet cell counts following chemotherapy or radiotherapy. Biol Trace Elem Res 2001;83(2):139–48.

341. Esel E, Ozdemir MA, Turan MT, Basturk M, Kilic H, Kose K, Gonuk AS, Sofuoglu S. Effects of lithium treatment on granulocytes and granulocyte colony-stimulating factor in patients with bipolar affective disorder. Bull Clin Psychopharmacol 2001;11:28–32.

342. Silverstone PH. Prevention of clozapine-induced neutropenia by pretreatment with lithium. J Clin Psychopharmacol 1998;18(1):86–8.

343. Ballin A, Lehman D, Sirota P, Litvinjuk U, Meytes D. Increased number of peripheral blood CD34+ cells in lithium-treated patients. Br J Haematol 1998;100(1):219–21.

344. Canales MA, Arrieta R, Hernandez-Garcia C, Bustos JG, Aguado MJ, Hernandez-Navarro F. A single apheresis to achieve a high number of peripheral blood CD34+ cells in a lithium-treated patient with acute myeloid leukaemia. Bone Marrow Transplant 1999;23(3):305.

345. Capodicasa E, Russano AM, Ciurnella E, De Bellis F, Rossi R, Scuteri A, Biondi R. Neutrophil peripheral count and human leukocyte elastase during chronic lithium carbonate therapy. Immunopharmacol Immunotoxicol 2000;22(4):671–83.

346. Cervantes P, Ghadirian AM, Vida S. Vitamin B12 and folate levels and lithium administration in patients with affective disorders. Biol Psychiatry 1999;45(2):214–21.

347. Szabadi E, Tavernor S. Hypo- and hypersalivation induced by psychoactive drugs. Incidence, mechanisms and therapeutic implications. CNS Drugs 1999;11:449–66.

348. Friedlander AH, Friedlander IK, Marder SR. Bipolar I disorder: psychopathology, medical management and dental implications. J Am Dent Assoc 2002;133(9):1209–17.

349. Rampes H, Bhandari S. Hypersalivation and lithium toxicity. J Psychopharmacol 1994;14:436.

350. Muniz CE, Berghman DH. Contact stomatitis and lithium carbonate tablets. JAMA 1978;239(26):2759.

351. Nathan KI. Development of mucosal ulcerations with lithium carbonate therapy. Am J Psychiatry 1995;152(6):956–7.

352. Gracious BL, Llana M, Barton DD. Lithium and geographic tongue. J Am Acad Child Adolesc Psychiatry 1999;38(9):1069–70.

353. Corica M, Borcese R, Savoldi E. Can the lithium therapy cause titanium implant failure? J Dent Res 2001;80:1246.

354. Ala A, Arnold N, Khin CC, van Someren N. Nausea and vomiting, a cause for concern? Postgrad Med J 2000;76(896):375379–380.

355. Kales HC, Mellow AM. Ileus as a possible result of bupropion in an elderly woman. J Clin Psychiatry 1999;60(5):337.

356. Haude V, Kretschmer H. Chronic diarrhea and increasing disability in an older woman due to an unusual cause. J Am Geriatr Soc 1999;47(2):261–2.

357. Andersen V, Sonne J, Andersen M. Spontaneous reports on drug-induced pancreatitis in Denmark from 1968 to 1999. Eur J Clin Pharmacol 2001;57(6–7):517–21.

358. Batlle D, Dorhout-Mees EJ. Lithium and the kidney. In: De Broe ME, Porter GA, Bennett WM, Verpooten GA, editors. Clinical Nephrotoxins: Renal Injury from Drugs and Chemicals. Dordrecht, The Netherlands, Boston: Kluwer Academic Publishers, 1998:383–95.

359. Braden GL. Lithium-induced renal disease. In: Greenberg A, editor. Primer on Kidney Diseases. San Diego: Academic Press, 1998:332–4.

360. Gitlin M. Lithium and the kidney: an updated review. Drug Saf 1999;20(3):231–43.

361. Johnson G. Lithium—early development, toxicity, and renal function. Neuropsychopharmacology 1998;19(3):200–5.

362. Lepkifker E, Sverdlik A, Iancu I, Ziv R. Renal failure in long-term lithium treatment. Bipolar Disord 2001;3(Suppl 1):45.

363. Webb AL, Solomon DA, Ryan CE. Lithium levels and toxicity among hospitalized patients. Psychiatr Serv 2001;52(2):229–31.

364. DiGiacomo JN, Sadoff RL. Managing malpractice risks during psychopharmacologic treatment. Essent Psychopharmacol 1999;3:65–89.

365. Eagles JM, McCann I, MacLeod TN, Paterson N. Lithium monitoring before and after the distribution of clinical practice guidelines. Acta Psychiatr Scand 2000;101(5):349–53.

366. Bendz H, Aurell M, Balldin J, Mathe AA, Sjodin I. Kidney damage in long-term lithium patients: a cross-sectional study of patients with 15 years or more on lithium. Nephrol Dial Transplant 1994;9(9):1250–4.

367. Bendz H, Aurell M, Lanke J. A historical cohort study of kidney damage in long-term lithium patients: continued surveillance needed. Eur Psychiatry 2001;16(4):199–206.

368. Markowitz GS, Radhakrishnan J, Kambham N, Valeri AM, Hines WH, D'Agati VD. Lithium nephrotoxicity: a progressive combined glomerular and tubulointerstitial nephropathy. J Am Soc Nephrol 2000;11(8):1439–48.

369. Ida S, Yokota M, Ueoka M, Kiyoi K, Takiguchi Y. Mild to severe lithium-induced nephropathy models and urine N-acetyl-beta-D-glucosaminidase in rats. Meth Find Exp Clin Pharmacol 2001;23(8):445–8.

370. Stoll PM, Stokes PE, Okamoto M. Lithium isotopes: differential effects on renal function and histology. Bipolar Disord 2001;3(4):174–80.

371. Frokiaer J, Marples D, Knepper MA, Nielsen S. Pathophysiology of aquaporin-2 in water balance disorders. Am J Med Sci 1998;316(5):291–9.

372. Sofuoglu S, Utas C, Aslan SS, Yalcindag C, Basturk M, Gonuk AS. Kidney functioning during short- and long-term lithium treatment in bipolar patients. Eur Neuropsychopharmacol 1999;9(Suppl 5):S215–6.

373. Autret L, Meynard JA, Sanchez MF, Bernadet R, Charpin C. Entre efficacité et toxicité: histoire d'un patient

guéri par le lithium chez qui l'on découvre une néphropathie iatrogène. [Between efficiency and toxicity: the case of a patient improved by lithium who developed iatrogenic nephropathy.] Encephale 1999;25(5):485–7.

374. Lauterbach EC, Abdelhamid A, Annandale JB. Posthallucinogen-like visual illusions (palinopsia) with risperidone in a patient without previous hallucinogen exposure: possible relation to serotonin 5HT2a receptor blockade. Pharmacopsychiatry 2000;33(1):38–41.

375. Presne C, Fakhouri F, Kenouch S, Stengel B, Kreis H, Grunfeld JP. Insuffisance renale evolutive par nephropathie au lithium. [Progressive renal failure caused by lithium nephropathy.] Presse Méd 2002;31(18):828–33.

376. Presne C, Fakhouri F, Noel L-H, Stengel B, Even C, Kreis H, Mignon F, Grunfeld J-P. Lithium-induced nephropathy: rate of progression and prognostic factors. Kidney Int 2003;64:585–92.

377. Farres MT, Ronco P, Saadoun D, Remy P, Vincent F, Khalil A, Le Blanche AF. Chronic lithium nephropathy: MR imaging for diagnosis. Radiology 2003;229:570–4.

378. Gattone VH 2nd, Wang X, Harris PC, Torres VE. Inhibition of renal cystic disease development and progression by a vasopressin V2 receptor antagonist. Nature Med 2003;9:1323–6.

379. Rastegar A, Kashgarian M. The clinical spectrum of tubulointerstitial nephritis. Kidney Int 1998;54(2):313–27.

380. Gabutti L, Gugger M, Marti HP. Eingeschränkte Nierenfunktion bei Lithium-therapie. [Impaired kidney function in lithium therapy.] Ther Umsch 1998;55(9):562–4.

381. Mukhopadhyay D, Gokulkrishnan L, Mohanaruban K. Lithium-induced nephrogenic diabetes insipidus in older people. Age Ageing 2001;30(4):347–50.

382. Eustatia-Rutten CF, Tamsma JT, Meinders AE. Lithium-induced nephrogenic diabetes insipidus. Neth J Med 2001;58(3):137–42.

383. Lassnig E, Berent R, Wallner M, Jagsch C, Auer J, Eber B. Manisch-depressive Patientin mit Polyurie unter Lithiumtherapie. [Manic-depressive female patient with polyuria during lithium therapy.] Intensivmedizin 2001;38:26–30.

384. Waise A, Fisken RA. Unsuspected nephrogenic diabetes insipidus. BMJ 2001;323(7304):96–7.

385. Di Paolo A. Complications during lithium maintenance therapy. Rev Med Brux 2005;26:173–7.

386. Imam SK, Hasan A, Shahid SK. Lithium-induced nephrogenic diabetes insipidus. J Pak Med Assoc 2005;55:125–7.

387. Walker RJ, Weggery S, Bedford JJ, McDonald FJ, Ellis G, Leader JP. Lithium-induced reduction in urinary concentrating ability and urinary aquaporin 2 (AQP2) excretion in healthy volunteers. Kidney Int 2005;67:291–4.

388. Rojek A, Nielsen J, Brooks HL, Gong H, Kim YH, Kwon TH, Frokiaer J, Nielsen S. Altered expression of selected genes in kidney of rats with lithium-induced NDI. Am J Physiol Renal Physiol 2005;288(6):F1276-F1289.

389. Kotnik P, Nielsen J, Kwon TH, Krzisnik C, Frøkiaer J, Nielsen S. Altered expression of COX-1, COX-2, and mPGES in rats with nephrogenic and central diabetes insipidus. Am J Physiol Renal Physiol 2005;288: F1053–F1068.

390. Rao R, Zhang MZ, Zhao M, Cai H, Harris RC, Breyer MD, Hao CM. Lithium treatment inhibits renal GSK-3 activity and promotes cyclooxygenase 2-dependent polyuria. Am J Physiol Renal Physiol 2005;288:F642-F649.

391. Stone KA. Lithium-induced nephrogenic diabetes insipidus. J Am Board Fam Pract 1999;12(1):43–7.

392. Finch CK, Kelley KW, Williams RB. Treatment of lithium-induced diabetes insipidus with amiloride. Pharmacotherapy 2003;23(4):546–50.

393. Chen PS, Yang YK, Yeh TL, Lo YC, Wang YT. Nonketotic hyperosmolar syndrome from olanzapine, lithium, and valproic acid cotreatment. Ann Pharmacother 2003;37:919–20.

394. Sirois F. Lithium-induced neurogenic diabetes insipidus in a surgical patient. Psychosomatics 2004;45:82–3.

395. Ashrafian H, Bogle RG. Sensitizing the insensitive. Clin Nephrol 2004;61:440–3.

396. Kamijo Y, Soma K, Hamanaka S, Nagai T, Kurihara K. Dural sinus thrombosis with severe hypernatremia developing in a patient on long-term lithium therapy. J Toxicol Clin Toxicol 2003;41:359–62.

397. Sakarcan A, Thomas DB, O'Reilly KP, Richards RW. Lithium-induced nephrotic syndrome in a young pediatric patient. Pediatr Nephrol 2002;17(4):290–2.

398. Schreiner A, Waldherr R, Rohmeiss P, Hewer W. Focal segmental glomerulosclerosis and lithium treatment. Am J Psychiatry 2000;157(5):834.

399. Herrero Mendoza MD, Caramelo C, Bellver Alvarez TM, Lopez Cubero L. Sindrome nefrotico y tratamiento con litio. [Nephrotic syndrome and lithium therapy.] Med Clin (Barc) 2001;116(19):758–9.

400. Perez GO, Oster JR, Vaamonde CA. Incomplete syndrome of renal tubular acidosis induced by lithium carbonate. J Laboratory Clin Med 1975;86(3):386–94.

401. Warnock JK, Morris DW. Adverse cutaneous reactions to mood stabilizers. Am J Clin Dermatol 2003;4(1):21–30.

402. Yeung CK, Chan HHL. Cutaneous adverse effects of lithium. Am J Clin Dermatol 2004;5:3–8.

403. Senol M, Ozcan A, Ozcan EM, Aydin EN. Pityriasis rosea-like eruption due to lithium. Clin Drug Invest 2004;24:493–4.

404. Kimyai-Asadi A, Harris JC, Nousari HC. Critical overview: adverse cutaneous reactions to psychotropic medications. J Clin Psychiatry 1999;60(10):714–25.

405. Albrecht G. Cutaneous side effects in patients on long-term lithium therapy. In: Birch NJ, editor. Lithium: Inorganic Pharmacology and Psychiatric Use. Oxford: IRL Press, 1988:144.

406. Krahn LE, Goldberg RL. Psychotropic medications and the skin. Adv Psychosom Med 1994;21:90–106.

407. Ellgehausen P, Elsner P, Burg G. Drug-induced lichen planus. Clin Dermatol 1998;16(3):325–32.

408. Arakaki T, Miyahira Y, Miyazato H, Arakaki H, Asato Y. A case of serious systemic swelling and erythema following lithium carbonate treatment. Int Clin Psychopharmacol 2002;17(Suppl 2):S69.

409. Francis GJ, Silverman AR, Saleh O, Lee GJ. Follicular mycosis fungoides associated with lithium. J Am Acad Dermatol 2001;44(2):308–9.

410. Alagheband M, Engineer L. Lithium and halogenoderma. Arch Dermatol 2000;136(1):126–7.

411. Mantelet S, Feline A. Effets cutanés du lithium: revue de la littérature à propos d'un cas. [Cutaneous effects of lithium: review of the literature and a case report.] Ann Med Psychol 1997;155:664–8.

412. Chan HH, Wing Y, Su R, Van Krevel C, Lee S. A control study of the cutaneous side effects of chronic lithium therapy. J Affect Disord 2000;57(1–3):107–13.

413. Oztas P, Aksakal AB, Oztas MO, Onder M. Severe acne with lithium. Ann Pharmacother 2001;35(7–8):961–2.

414. Mercke Y, Sheng H, Khan T, Lippmann S. Hair loss in psychopharmacology. Ann Clin Psychiatry 2000;12(1):35–42.

415. McCreadie RG, Farmer JG. Lithium and hair texture. Acta Psychiatr Scand 1985;72(4):387–8.

416. Jones AV, Grabczynska SA, Vijayasingham S, Eady RAJ. Exacerbation of Darier's disease with oral lithium carbonate therapy. Eur J Dermatol 1996;6:527–8.

417. Ehrt U, Brieger P. Comorbidity of keratosis follicularis (Darier's disease) and bipolar affective disorder: an indication for valproate instead of lithium. Gen Hosp Psychiatry 2000;22(2):128–9.

418. Wang SL, Yang SF, Chen CC, Tsai PT, Chai CY. Darier's disease associated with bipolar affective disorder: a case report. Kaohsiung J Med Sci 2002;18(12):622–6.

419. Lozano Garcia MC, Baca Garcia E. Psoriasis y tratamento conlitio: un mecanismo fisiopatologico commun?. [Psoriasis and lithium treatment: a common pathophysiology?.] Actas Esp Psiquiatr 2002;30(6):400–3.

420. Akkerhuis GW, Nolen WA. Lithium-associated psoriasis and omega-3 fatty acids: two case reports. Bipolar Disord 2002;4:117.

421. Miyagawa M, Shimoda K, Danno K, Kato N. Exacerbation of psoriasis during lithium treatment in a patient with bipolar I disorder. Int Clin Psychopharmacol 2000;15:368.

422. Birch NJ. Bone. In: Johnson FN, editor. Depression and Mania: Modern Lithium Therapy. Oxford: IRL Press, 1987:158.

423. Nordenstrom J, Elvius M, Bagedahl-Strindlund M, Zhao B, Torring O. Biochemical hyperparathyroidism and bone mineral status in patients treated long-term with lithium. Metabolism 1994;43(12):1563–7.

424. Vestergaard P, Rejnmark L, Mosekilde L. Reduced relative risk of fractures among users of lithium. Calcif Tissue Int 2005;77:1–8.

425. Demyttenaere K, De Fruyt J, Sienaert P. Psychotropics and sexuality. Int Clin Psychopharmacol 1998;13(Suppl 6):S35–41.

426. Kristensen E. Sexual side effects induced by psychotropic drugs. Dan Med Bull 2002;49(4):349–52.

427. Rasgon NL, Altshuler LL, Gudeman D, Burt VK, Tanavoli S, Hendrick V, Korenman S. Medication status and polycystic ovary syndrome in women with bipolar disorder: a preliminary report. J Clin Psychiatry 2000;61(3):173–8.

428. Baptista T, Lacruz A, de Mendoza S, Guillen MM, Burguera JL, de Burguera M, Hernandez L. Endocrine effects of lithium carbonate in healthy premenopausal women: relationship with body weight regulation. Prog Neuropsychopharmacol Biol Psychiatry 2000;24(1):1–16.

429. Luconi M, Marra F, Gandini L, Filimberti E, Lenzi A, Forti G, Baldi E. Phosphatidylinositol 3-kinase inhibition enhances human sperm motility. Hum Reprod 2001;16(9):1931–7.

430. Banerji TK, Maitra SK, Dey M, Hawkins HK. Gametogenic responses of the testis in spotted munia (*Lonchura punctulata*; Aves) to oral administration of lithium chloride. Endocr Res 2001;27(3):345–56.

431. Perez Romera E, Munoz E, Mohamed F, Dominguez S, Scardapane L, Villegas O, Garcia Aseff S, Guzman JA. Lithium effect on testicular tissue and spermatozoa of viscacha (*Lagostomus maximus maximus*). A comparative study with rats. J Trace Elem Med Biol 2000;14(2):81–3.

432. Clark KJ, Jefferson JW. Lithium allergy. J Clin Psychopharmacol 1987;7(4):287–9.

433. Lowry MD, Hudson CF, Callen JP. Leukocytoclastic vasculitis caused by drug additives. J Am Acad Dermatol 1994;30(5 Part 2):854–5.

434. Kang BJ, Park SW, Chung TH. Can the expression of histocompatibility antigen be changed by lithium? Int J Neuropsychopharmacol 1999;2(Suppl 1):S55.

435. Rapaport MH, Manji HK. The effects of lithium on ex vivo cytokine production. Biol Psychiatry 2001;50(3):217–24.

436. Merendino RA, Arena A, Gangemi S, Ruello A, Losi E, Bene A, Valenti A, D'Ambrosio FP. In vitro effect of lithium chloride on interleukin-15 production by monocytes from IL-breast cancer patients. J Chemother 2000;12(3):252–7.

437. Merendino RA, Arena A, Gangemi S, Ruello A, Losi E, Bene A, D'Ambrosio FP. In vitro interleukin-8 production by monocytes treated with lithium chloride from breast cancer patients. Tumori 2000;86(2):149–52.

438. Rybakowski JK. The effect of lithium on the immune system. Hum Psychopharmacol 1999;14:345–53.

439. Harvey BH, Meyer CL, Gallicchio VS. The hemopoietic and immuno-modulating action of lithium salts: an investigation into the chemotherapy of HIV infection in South Africa. In: Lucas KC, Becker RW, Gallicchio VS, editors. Lithium-50 Years: Recent Advances in Biology and Medicine. Cheshire, Connecticut: Weidner Publishing, 1999:137–52.

440. Maes M, Song C, Lin AH, Pioli R, Kenis G, Kubera M, Bosmans E. In vitro immunoregulatory effects of lithium in healthy volunteers. Psychopharmacology (Berl) 1999;143(4):401–7.

441. Rapaport MH, Guylai L, Whybrow P. Immune parameters in rapid cycling bipolar patients before and after lithium treatment. J Psychiatr Res 1999;33(4):335–40.

442. Rybakowski JK. Antiviral and immunomodulatory effect of lithium. Pharmacopsychiatry 2000;33(5):159–64.

443. Kang BJ, Park SW, Chung TH. Can the expression of histocompatibility antigen be changed by lithium? Bipolar Disord 2000;2(2):140–4.

444. Litovitz TL, Klein-Schwartz W, Dyer KS, Shannon M, Lee S, Powers M. 1997 Annual Report of the American Association of Poison Control Centers Toxic Exposure Surveillance System. Am J Emerg Med 1998;16(5):443–97.

445. Morriss R, Mohammed FA. Metabolism, lifestyle and bipolar affective disorder. J Psychopharmacol 2005;19(6 Suppl):94–101.

446. O'Boyle M, Emory E. A case of lithium carbonate abuse. Am J Psychiatry 1988;145(8):1036.

447. Lipkin B. Lithium as a drug of abuse. BMJ 1977;1(6073):1411–2.

448. Kleindienst N, Greil W, Ruger B, Moller HJ. The prophylactic efficacy of lithium—transient or persistent? Eur Arch Psychiatry Clin Neurosci 1999;249(3):144–9.

449. Berghofer A, Muller-Oerlinghausen B. Is there a loss of efficacy of lithium in patients treated for over 20 years? Neuropsychobiology 2000;42(Suppl 1):46–9.

450. Baldessarini RJ, Tondo L. Recurrence risk in bipolar manic-depressive disorders after discontinuing lithium maintenance treatment: an overview. Clin Drug Invest 1998;15:337–51.

451. Baldessarini RJ, Tondo L, Floris G, Rudas N. Reduced morbidity after gradual discontinuation of lithium treatment for bipolar I and II disorders: a replication study. Am J Psychiatry 1997;154(4):551–3.

452. Schou M. Is there a lithium withdrawal syndrome? An examination of the evidence. Br J Psychiatry 1993;163:514–8.

453. Baldessarini RJ, Tondo L, Hennen J. Effects of lithium treatment and its discontinuation on suicidal behavior in

bipolar manic-depressive disorders. J Clin Psychiatry 1999;60(Suppl 2):77–84.

454. Bauer M, Bschor T, Kunz D, Berghofer A, Strohle A, Muller-Oerlinghausen B. Double-blind, placebo-controlled trial of the use of lithium to augment antidepressant medication in continuation treatment of unipolar major depression. Am J Psychiatry 2000;157(9):1429–35.

455. Faedda GL, Tondo L, Baldessarini RJ. Lithium discontinuation: uncovering latent bipolar disorder? Am J Psychiatry 2001;158(8):1337–9.

456. Fahy S, Lawlor BA. Discontinuation of lithium augmentation in an elderly cohort. Int J Geriatr Psychiatry 2001;16(10):1004–9.

457. Viguera AC, Nonacs R, Cohen LS, Tondo L, Murray A, Baldessarini RJ. Risk of recurrence of bipolar disorder in pregnant and nonpregnant women after discontinuing lithium maintenance. Am J Psychiatry 2000;157(2):179–84.

458. Perlis RH, Sachs GS, Lafer B, Otto MW, Faraone SV, Kane JM, Rosenbaum JF. Effect of abrupt change from standard to low serum levels of lithium: a reanalysis of double-blind lithium maintenance data. Am J Psychiatry 2002;159(7):1155–9.

459. Goodwin GM, Phil D. Clinical and biological investigation of mania following lithium withdrawal. In: Manji HK, Bowden CL, Belmaker RH, editors. Bipolar Medications: Mechanisms of Action. Washington DC: American Psychiatric Press, 2000:347–56.

460. Tondo L, Baldessarini RJ. Reduced suicide risk during lithium maintenance treatment. J Clin Psychiatry 2000;61(Suppl 9):97–104.

461. Coryell W, Solomon D, Leon AC, Akiskal HS, Keller MB, Scheftner WA, Mueller T. Lithium discontinuation and subsequent effectiveness. Am J Psychiatry 1998;155(7):895–8.

462. Coryell WH, Leon AC, Scheftner W. Lithium discontinuation. Dr Coryell and colleagues reply. Am J Psychiatry 1999;156:1130.

463. Oostervink F, Nolen WA, Hoenderboom AC, Kupka RW. Het risico van lithiumresistentie na stoppen en herstart na langdurig gebruik. [Risk of inducing resistance upon stopping and restarting lithium after long-term usage.] Ned Tijdschr Geneeskd 2000;144(9):401–4.

464. Bigatti MP, Corona D, Munizza C. Increased sister chromatid exchange and chromosomal aberration frequencies in psychiatric patients receiving psychopharmacological therapy. Mutat Res 1998;413(2):169–75.

465. Banerji TK, Maitra SK, Basu A, Hawkins HK. Lithium-induced alterations in the testis of the male roseringed parakeet (*Psittacula krameri*): evidence for significant structural changes disruption spermatogenetic activity. Endocr Res 1999;25(1):35–49.

466. Iqbal MM. The effects of lithium on fetuses, neonates, and nursing infants. Psychiatr Ann 2000;30:159–64.

467. Warner JP. Evidence-based psychopharmacology 3. Assessing evidence of harm: what are the teratogenic effects of lithium carbonate? J Psychopharmacol 2000;14(1):77–80.

468. Stowe ZN, Calhoun K, Ramsey C, Sadek N, Newport J. Mood disorders during pregnancy and lactation: defining issues of exposure and treatment. CNS Spectrums 2001;6:150–66.

469. Iqbal MM, Gundlapalli SP, Ryan WG, Ryals T, Passman TE. Effects of antimanic mood-stabilizing drugs on fetuses, neonates, and nursing infants. South Med J 2001;94(3):304–22.

470. Williams K, Oke S. Lithium and pregnancy. Psychiatr Bull 2000;24:229–31.

471. Davis LL, Shannon S, Drake RG, Petty F. The treatment of bipolar disorder during pregnancy. In: Yonkers KA, Little BB, editors. Management of Psychiatric Disorders in Pregnancy. London: Arnold, 2001:122–33.

472. Pinelli JM, Symington AJ, Cunningham KA, Paes BA. Case report and review of the perinatal implications of maternal lithium use. Am J Obstet Gynecol 2002;187(1):245–9.

473. Viguera AC, Cohen LS, Baldessarini RJ, Nonacs R. Managing bipolar disorder during pregnancy: weighing the risks and benefits. Can J Psychiatry 2002;47(5):426–36.

474. Amsterdam JD, Brunswick DJ, O'Reardon J. Bipolar disorder in women. Psychiatr Ann 2002;32:397–404.

475. Harris B. Postpartum depression. Psychiatr Ann 2002;32:405–15.

476. Kruger S, Braunig P. Clinical issues in bipolar disorder during pregnancy and the postpartum period. Clin Approaches Bipolar Disord 2002;1:65–71.

477. Kusumaker V, MacMaster FP, Kutcher SP, Shulman KI. Bipolar disorder in young people, the elderly and pregnant women. In: Yatham LN, Kusumakar V, Kutcher SP, editors. Bipolar Disorder: A Clinician's Guide to Biological Treatments. New York: Brunner-Routledge, 2002:85–113.

478. Ward RK, Zamorski MA. Benefits and risks of psychiatric medications during pregnancy. Am Fam Physician 2002;66(4):629–36.

479. Iqbal MM, Sohhan T, Mahmud SZ. The effects of lithium, valproic acid, and carbamazepine during pregnancy and lactation. J Toxicol Clin Toxicol 2001;39(4):381–92.

480. De Santis M, Carducci B, Cavaliere AF, De Santis L, Straface G, Caruso A. Drug-induced congenital defects: strategies to reduce the incidence. Drug Saf 2001;24(12):889–901.

481. Keller F, Griesshammer M, Haussler U, Paulus W, Schwarz A. Pregnancy and renal failure: the case for application of dosage guidelines. Drugs 2001;61(13):1901–20.

482. Viguera AC, Cohen LS. The course and management of bipolar disorder during pregnancy. Psychopharmacol Bull 1998;34(3):339–46.

483. Cohen LS, Rosenbaum JF. Psychotropic drug use during pregnancy: weighing the risks. J Clin Psychiatry 1998;59(Suppl 2):18–28.

484. Retamal P, Cantillano V. Tratamiento de la enfermedad bipolar durante el embarazo y puerperio. [Treatment of bipolar disorder during pregnancy and puerperium period. A case report.] Rev Medical Chil 2001;129(5):556–60.

485. Yonkers KA, Little BB, March D. Lithium during pregnancy. Drug effects and their therapeutic implications. CNS Drugs 1998;9:261–9.

486. Schou M. Treating recurrent affective disorders during and after pregnancy. What can be taken safely? Drug Saf 1998;18(2):143–52.

487. Cohen LS, Friedman JM, Jefferson JW, Johnson EM, Weiner ML. A reevaluation of risk of in utero exposure to lithium. JAMA 1994;271(2):146–50.

488. American Academy of Pediatrics—Committee on Drugs. Use of psychoactive medication during pregnancy and possible effects on the fetus and newborn. Pediatrics 2000;105(4 Part 1):880–7.

489. Ernst CL, Goldberg JF. The reproductive safety profile of mood stabilizers, atypical antipsychotics, and broad-spectrum psychotropics. J Clin Psychiatry 2002;63(Suppl 4):42–55.

490. Knoppert van der Klein EAM, Van Kamp IL. A pilot risk evaluation of lithium in pregnancy. Bipolar Disord 2002;4(Suppl 1):127.

491. Grover S, Gupta N. Lithium-associated anencephaly. Can J Psychiatry 2005;50(3):185–6.

492. Newport DJ, Viguera AC, Beach AJ, Ritchie JC, Cohen LS, Stowe AN. Lithium and placental passage and obstetrical outcome: implications for clinical management during late pregnancy. Am J Psychiatry 2005;162(11):2162–70.

493. Malzacher A, Engler H, Drack G, Kind C. Lethargy in a newborn: lithium toxicity or lab error? J Perinat Med 2003;31:340–2.

494. Viguera AC, Howlett SA, Cohen LS, Nonacs RM, Stoller J. In: Neonatal outcome associated with lithium use during pregnancyPresented at the NCDEU 40th Annual Meeting, May 30–June 2, Boca Raton, FL 2000:144 New Clinical Drug Evaluation Unit Program: Poster number: 49.

495. Tekin M, Ellison J. Oromandibular-limb hypogenesis spectrum and maternal lithium use. Clin Dysmorphol 2000;9(2):139–41.

496. Zegers B, Andriessen P. Maternal lithium therapy and neonatal morbidity. Eur J Pediatr 2003;162(5):348–9.

497. Frassetto F, Tourneur Martel F, Barjhoux CE, Villier C, Bot BL, Vincent F. Goiter in a newborn exposed to lithium in utero. Ann Pharmacother 2002;36(11):1745–8.

498. Moretti ME, Lee A, Ito S. Which drugs are contraindicated during breastfeeding? Practice guidelines. Can Fam Physician 2000;46:1753–7.

499. Yoshida K, Smith B, Kumar R. Psychotropic drugs in mothers' milk: a comprehensive review of assay methods, pharmacokinetics and of safety of breast-feeding. J Psychopharmacol 1999;13(1):64–80.

500. Chaudron LH, Jefferson JW. Mood stabilizers during breastfeeding: a review. J Clin Psychiatry 2000;61(2):79–90.

501. Koren G, Moretti M, Ito S. Continuing drug therapy while breastfeeding. Part 2. Common misconceptions of physicians. Can Fam Physician 1999;45:1173–5.

502. Austin MP, Mitchell PB. Use of psychotropic medications in breast-feeding women: acute and prophylactic treatment. Aust NZ J Psychiatry 1998;32(6):778–84.

503. Llewellyn A, Stowe ZN. Psychotropic medications in lactation. J Clin Psychiatry 1998;59(Suppl 2):41–52.

504. Burt VK, Suri R, Altshuler L, Stowe Z, Hendrick VC, Muntean E. The use of psychotropic medications during breast-feeding. Am J Psychiatry 2001;158(7):1001–9.

505. Howard CR, Lawrence RA. Xenobiotics and breastfeeding. Pediatr Clin North Am 2001;48(2):485–504.

506. Moretti ME, Koren G, Verjee Z, Ito S. Monitoring lithium in breast milk: an individualized approach for breast-feeding mothers. Ther Drug Monit 2003;25(3):364–6.

507. Ward RM, Bates BA, Benitz WE, Burchfield DJ, Ring JC, Walls RP, Walson PD. Transfer of drugs and other chemicals into human milk. Pediatrics 2001;108(3):776–89.

508. Olugbemi E, Katona C. Case note audit of lithium use in the elderly. Aging Ment Health 1998;2:151–4.

509. Stoudemire A, Hill CD, Lewison BJ, Marquardt M, Dalton S. Lithium intolerance in a medical-psychiatric population. Gen Hosp Psychiatry 1998;20(2):85–90.

510. Fahy S, Lawlor BA. Lithium use in octogenarians. Int J Geriatr Psychiatry 2001;16(10):1000–3.

511. Young RC, Gyulai L, Mulsant BH, Flint A, Beyer JI, Shulman KI, Reynolds CF III. Pharmacotherapy of bipolar disorder in old age. Am J Geriatr Psychiatry 2004;12:342–57.

512. Juurlink DN, Mamdani MM, Kopp A, Rochon PA, Shulman KI, Redelmeier DA. Drug-induced lithium toxicity in the elderly: a population-based study. J Am Geriatr Soc 2004;52:794–8.

513. Chen K-P, Shen W, Lu M-L. Implication of serum concentration monitoring in patients with lithium intoxication. Psychiatry Clin Neurosci 2004;58:25–9.

514. Cheymol G. Effects of obesity on pharmacokinetics implications for drug therapy. Clin Pharmacokinet 2000;39(3):215–31.

515. Abraham G, Voutsilakos F. Norwalk precipitates severe lithium toxicity. Can J Psychiatry 2004;49:215–16.

516. Kilts CD. The ups and downs of oral lithium dosing. J Clin Psychiatry 1998;59(Suppl 6):21–6.

517. Henry ME, Moore CM, Demopolas C, Cote J, Renshaw PF. A comparison of brain lithium levels attained with immediate and sustained release lithium. Biol Psychiatry 2001;49(Suppl 8):119S.

518. Gazikolovic E, Obrenovic D, Nicovic Z. Formulaciji tableta litijom-karbonata razlicitim sredstvima za vezivanje. [Formulation of lithium carbonate tablets with various binding substances.] Vojnosanit Pregl 2001;58(6):641–4.

519. Castrogiovanni P. A novel slow-release formulation of lithium carbonate (Carbolithium Once-A-Day) vs standard Carbolithium: a comparative pharmacokinetic study. Clin Ter 2002;153(2):107–15.

520. Durbano F, Mencacci C, Dorigo D, Riva M, Buffa G. The long-term efficacy and tolerability of carbolithium once a day: an interim analysis at 6 months. Clin Ter 2002;153(3):161–6.

521. Pietkiewicz P, Sznitowska M, Dorosz A, Lukasiak J. Lithium carbonate 24-hours extended-release capsule filled with 6 mm tablets Boll Chim Farm 2003;142(2):69–71.

522. Astruc B, Petit P, Abbar M. Overdose with sustained-release lithium preparations. Eur Psychiatry 1999;14(3):172–4.

523. Owen D. Interactions between homeopathy and drug treatment. Br Homeopath J 2000;89(1):60.

524. Sattar SP. Pinals DA. When taking medications is a sin. Psychiatr Serv 2002;53(2):213–4.

525. Carroll BT, Thalassinos A, Fawver JD. Loading strategies in acute mania. CNS Spectr 2001;6(11):919–22930.

526. Soares JC, Boada F, Spencer S, Mallinger AG, Dippold CS, Wells KF, Frank E, Keshavan MS, Gershon S, Kupfer DJ. Brain lithium concentrations in bipolar disorder patients: preliminary (7)Li magnetic resonance studies at 3 T Biol Psychiatry 2001;49(5):437–43.

527. Hatfield CL, McDonell WN, Lemke KA, Black WD. Pharmacokinetics and toxic effects of lithium chloride after intravenous administration in conscious horses. Am J Vet Res 2001;62(9):1387–92.

528. Jonas MM, Linton RAF, O'Brien TK, Band DM, Linton NWF, Kelly F, Burden TJ, Chevalier SFA, Thompson RPH, Birch NJ, Powell JJ. The pharmacokinetics of intravenous lithium chloride in patients and normal volunteers. J Trace Microprobe Techn 2001;19:313–20.

529. Waring WS, Webb DJ, Maxwell SR. Lithium carbonate as a potential pharmacological vehicle. Intravenous kinetics of single-dose administration in healthy subjects. Eur J Clin Pharmacol 2002;58(6):431–4.

530. Scharman EJ. Methods used to decrease lithium absorption or enhance elimination. J Toxicol Clin Toxicol 1997;35(6):601–8.

531. Kores B, Lader MH. Irreversible lithium neurotoxicity: an overview. Clin Neuropharmacol 1997;20(4):283–99.

532. Bailey B, McGuigan M. Lithium poisoning from a poison control center perspective. Ther Drug Monit 2000;22(6):650–5.

533. Anonymous. Lithium overdose. Prescrire Int 2003;12(63):19.

534. Litovitz TL, Klein-Schwartz W, White S, Cobaugh DJ, Youniss J, Omslaer JC, Drab A, Benson BE. 2000 Annual report of the American Association of Poison Control Centers Toxic Exposure Surveillance System. Am J Emerg Med 2001;19(5):337–95.

535. Montagnon F, Said S, Lepine JP. Lithium: poisonings and suicide prevention. Eur Psychiatry 2002;17(2):92–5.

536. Bell AJ, Ferrier IN. Lithium induced neurotoxicity at therapeutic levels: an aetiological review. Lithium 1994;5:181.

537. Groleau G. Lithium toxicity. Emerg Med Clin North Am 1994;12(2):511–31.

538. Kasahara H, Shinozaki T, Nukariya K, Nishimura H, Nakano H, Nakagawa T, Ushijima S. Hemodialysis for lithium intoxication: preliminary guidelines for emergency. Jpn J Psychiatry Neurol 1994;48(1):1–12.

539. Okusa MD, Crystal LJ. Clinical manifestations and management of acute lithium intoxication. Am J Med 1994;97(4):383–9.

540. Litovitz TL, Klein-Schwartz W, Caravati EM, Youniss J, Crouch B, Lee S. 1998 Annual Report of the American Association of Poison Control Centers Toxic Exposure Surveillance System. Am J Emerg Med 1999;17(5):435–87.

541. De Almeida Teixeira A, Machado MF, Ferreira WM, Torres JB, Brunstein MG, Barros HMT, Barros E. Lithium acute intoxication: epidemiologic study of the causes. J Bras Psiquiatr 1999;48:399–403.

542. Brodersen A, Licht RW, Vestergaard P, Olesen AV, Mortensen PB. Dodeligheden blandt patienter med affektiv sygdom der pategynder lithium behandling. En 16-ar s opfolgning. [Mortality in patients with affective disorder who commenced treatment with lithium. A 16-year follow-up.] Ugeskr Laeger 2001;163(46):6428–32.

543. Litovitz TL, Klein-Schwartz W, Rodgers GC Jr, Cobaugh DJ, Youniss J, Omslaer JC, May ME, Woolf AD, Benson BE. 2001 Annual Report of the American Association of Poison Control Centers Toxic Exposure Surveillance System. Am J Emerg Med 2002;20(5):391–452.

544. De Haro L, Roelandt J, Pommier P, Prost N, Arditti J, Hayek-Lanthois M, Valli M. Circconstances d'intoxication par sels de lithium: experience du centre antipoison de Marseille sur 10 ans. Ann Fr Anesth Reanim 2003;22:514–19.

545. Tuohy K, Shemin D. Acute lithium intoxication. Dial Transplant 2003;32:478–81.

546. Mallon PT, White JS, Thompson RLE. Systemic absorption of lithium following ingestion of a lithium button battery. Hum Exp Toxicol 2004;23:192–5.

547. Alhasso A, Bryden AA, Neilson D. Lithium toxicity after urinary diversion with ileal conduit. BMJ 2000;320(7241):1037.

548. Micheli F, Cersosimo G, Scorticati MC, Ledesma D, Molinos J. Blepharospasm and apraxia of eyelid opening in lithium intoxication. Clin Neuropharmacol 1999;22(3):176–9.

549. Vollhardt M, Ferbert A. [Influence of hypocalcemia and high serum levels of lithium on the amplitude of N20/P25 components of median nerve SEP.]EEG-Labor 1999;21:65–70.

550. Ochoa ER, Farrar HC, Shirm SW. Lithium poisoning in a toddler with fever and altered consciousness. Case presentation and discussion. J Invest Med 2000;48:612.

551. Novak-Grubic V, Tavcar R. Lithium intoxication secondary to unrecognized pontine haemorrhage. Acta Psychiatr Scand 2001;103(5):400–1.

552. Peces R, Pobes A. Effectiveness of haemodialysis with high-flux membranes in the extracorporeal therapy of life-threatening acute lithium intoxication. Nephrol Dial Transplant 2001;16(6):1301–3.

553. Danel V, Rhodes AS, Saviuc P, Hanna J. Intoxication grave par le lithium: à propos de deux cas. JEUR 2001;14:134–6.

554. Meyer RJ, Flynn JT, Brophy PD, Smoyer WE, Kershaw DB, Custer JR, Bunchman TE. Hemodialysis followed by continuous hemofiltration for treatment of lithium intoxication in children. Am J Kidney Dis 2001;37(5):1044–7.

555. Beckmann U, Oakley PW, Dawson AH, Byth PL. Efficacy of continuous venovenous hemodialysis in the treatment of severe lithium toxicity. J Toxicol Clin Toxicol 2001;39(4):393–7.

556. Suraya Y, Yoong KY. Lithium neurotoxicity. Med J Malaysia 2001;56(3):378–81.

557. Nagappan R, Parkin WG, Holdsworth SR. Acute lithium intoxication. Anaesth Intensive Care 2002;30(1):90–2.

558. Ilagan MC, Carlson D, Madden JF. Lithium toxicity: two case reports. Del Med J 2002;74(6):263–70.

559. De Ridder K, De Meester J, Demeyer I, Verbeke J, Nollet G. [Management of a case of lithium intoxication.]Tijdschr Geneeskd 2002;58:769–72.

560. Kerbusch T, Mathot RA, Otten HM, Meesters EW, van Kan HJ, Schellens JH, Beijnen JH. Bayesian pharmacokinetics of lithium after an acute self-intoxication and subsequent haemodialysis: a case report. Pharmacol Toxicol 2002;90(5):243–5.

561. Yoshimura R, Yamada Y, Ueda N, Nakamura J. Changes in plasma monoamine metabolites during acute lithium intoxication. Hum Psychopharmacol 2000;15(5):357–60.

562. Ponampalam R, Otten EJ. In vitro adsorption of lithium by bentonite. Singapore Med J 2002;43(2):86–9.

563. Bailey B, McGuigan M. Comparison of patients hemodialyzed for lithium poisoning and those for whom dialysis was recommended by PCC but not done: what lesson can we learn? Clin Nephrol 2000;54(5):388–92.

564. van Bommel EF, Kalmeijer MD, Ponssen HH. Treatment of life-threatening lithium toxicity with high-volume continuous venovenous hemofiltration. Am J Nephrol 2000;20(5):408–11.

565. Teece S, Crawford I. Best evidence topic report: no clinical evidence for gastric lavage in lithium overdose. Emerg Med J 2005;22:43–4.

566. Borrás Blasco J, Murcia López A, Romero Crespo I, sirvent Pedreño A, Navarro Ruiz A. Intoxicación aguda por comprimidos de liberación sostenida decarbonato de litio. A propósito de un caso clínico. [Acute intoxication with sustained-release lithium carbonate tablets.] A propos of a case. Farm Hosp 2005;29:140–3.

567. Toronjadze T, Polena S, Santucci T, Naik S, Watson C, Lakovou C, Babury MA, Gintautas J. Prolonged requirement for ventilatory support in a patient with Eskalith overdose. Proc West Pharmacol Soc 2005;48:148–9.

568. Brubacher JR, Purssell R, Kent DA. Salty broth for salicylate poisoning? Adequacy of overdose management

advice in the 2001 Compendium Pharmaceuticals Specialties. CMAJ 2002;167(9):992–6.

569. Muller-Oerlinghausen B. Drug interactions with lithium. A guide for clinicians. CNS Drugs 1999;11:41–8.

570. Jefferson JW, Greist JH, Ackerman DL, Carroll JA. Lithium Encyclopedia for Clinical Practice. 2nd edn. Washington DC: American Psychiatric Press;. 1987.

571. Janicak PG, Davis JM. Pharmacokinetics and drug interactions. In: Sadock BJ, Sadock VA, editors. 7th ed.Kaplan & Sadock's Comprehensive Textbook of Psychiatry 2. Philadelphia: Lippincott, Williams & Wilkins, 2000:2250–9.

572. DeVane CL, Nemeroff CB. 2000 Guide to psychotropic drug interactions. Primary Psychiatry 2000;7:40–68.

573. Sproule BA, Hardy BG, Shulman KI. Differential pharmacokinetics of lithium in elderly patients. Drugs Aging 2000;16(3):165–77.

574. Schou M. Treatment of Manic-Depressive Illness: A Practical Guide. 3rd edn.. Basel: Karger;. 1989.

575. Ventura JM, Igual MJ, Borrell C, Lozano MD, Maiques FJ, Alos M. Toxicidad de litio inducida por captoprilo. A propósito de un caso. [Lithium toxicity induced by captopril. A case study.] Farm Hosp 2000;24:166–9.

576. Meyer JM, Dollarhide A, Tuan IL. Lithium toxicity after switch from fosinopril to lisinopril. Int Clin Psychopharmacol 2005;20:115–8.

577. Bagate K, Grima M, De Jong W, Imbs JL, Barthelmebs M. Effects of icatibant on the ramipril-induced decreased in renal lithium clearance in the rat. Naunyn Schmiedebergs Arch Pharmacol 2001;363(3):281–7.

578. Zwanzger P, Marcuse A, Boerner RJ, Walther A, Rupprecht R. Lithium intoxication after administration of AT1 blockers. J Clin Psychiatry 2001;62(3):208–9.

579. Blanche P, Raynaud E, Kerob D, Galezowski N. Lithium intoxication in an elderly patient after combined treatment with losartan. Eur J Clin Pharmacol 1997;52(6):501.

580. Leung M, Remick RA. Potential drug interaction between lithium and valsartan. J Clin Psychopharmacol 2000;20(3):392–3.

581. Schweitzer I, Tuckwell V. Risk of adverse events with the use of augmentation therapy for the treatment of resistant depression. Drug Saf 1998;19(6):455–64.

582. Sternbach H. Serotonin syndrome: how to avoid, identify & treat dangerous drug interactions. Current Psychiatry 2003;2:1516,19,24.

583. Movig KL, Baumgarten R, Leufkens HG, van Laarhoven JH, Egberts AC. Risk factors for the development of lithium-induced polyuria. Br J Psychiatry 2003;182:319–23.

584. Peck AW, Stern WC, Watkinson C. Incidence of seizures during treatment with tricyclic antidepressant drugs and bupropion. J Clin Psychiatry 1983;44(5 Part 2):197–201.

585. Dufresne RL, Weber SS, Becker RE. Bupropion hydrochloride. Drug Intell Clin Pharm 1984;18(12):957–64.

586. Sitsen JM, Voortman G, Timmer CJ. Pharmacokinetics of mirtazapine and lithium in healthy male subjects. J Psychopharmacol 2000;14(2):172–6.

587. Laroudie C, Salazar DE, Cosson JP, Cheuvart B, Istin B, Girault J, Ingrand I, Decourt JP. Pharmacokinetic evaluation of co-administration of nefazodone and lithium in healthy subjects. Eur J Clin Pharmacol 1999;54(12):923–8.

588. Hawley C, Sivakumaran T, Huber TJ, Ige AK. Combination therapy with nefazodone and lithium: safety and tolerability in fourteen patients. Int J Psychiatry Clin Pract 1998;2:251–4.

589. Hawley C, Sivakumaran T, Ochocki M, Ratnam S, Huber T. A preliminary safety study of combined therapy with nefazodone and lithium. Int J Neuropsychopharmacol 1999;2(Suppl 1):S30–1.

590. Hawley CJ, Loughlin PJ, Quick SJ, Gale TM, Sivakumaran T, Hayes J, McPhee S. Efficacy, safety and tolerability of combined administration of lithium and selective serotonin reuptake inhibitors: a review of the current evidence. Hertfordshire Neuroscience Research Group. Int Clin Psychopharmacol 2000;15(4):197–206.

591. Salama AA, Shafey M. A case of severe lithium toxicity induced by combined fluoxetine and lithium carbonate. Am J Psychiatry 1989;146(2):278.

592. Epstein Y, Albukrek D, Kalmovitc B, Moran DS, Shapiro Y. Heat intolerance induced by antidepressants. Ann N Y Acad Sci 1997;813:553–8.

593. Takano A, Suhara T, Yasuno F, Ichimiya T, Inoue M, Sudo Y, Suzuki K. Characteristics of clomipramine and fluvoxamine on serotonin transporter evaluated by PET. Int Clin Psychopharmacol 2002;17(Suppl 2):S84–5.

594. Bauer M, Zaninelli R, Muller-Oerlinghausen B, Meister W. Paroxetine and amitriptyline augmentation of lithium in the treatment of major depression: a double-blind study. J Clin Psychopharmacol 1999;19(2):164–71.

595. Shad U, Preskorn SH, Izgur Z. Failure to consider drug–drug interactions as a likely cause of behavioral deterioration in a patient with bipolar disorder. J Clin Psychopharmacol 2000;20(3):390–2.

596. Fagiolini A, Buysse DJ, Frank E, Houck PR, Luther JF, Kupfer DJ. Tolerability of combined treatment with lithium and paroxetine in patients with bipolar disorder and depression. J Clin Psychopharmacol 2001;21(5):474–8.

597. Solomon JG. Seizures during lithium-amitriptyline therapy. Postgrad Med 1979;66(3):145–6148.

598. Evidente VG, Caviness JN. Focal cortical transient preceding myoclonus during lithium and tricyclic antidepressant therapy. Neurology 1999;52(1):211–3.

599. Frye MA, Kimbrell TA, Dunn RT, Piscitelli S, Grothe D, Vanderham E, Cora-Locatelli G, Post RM, Ketter TA. Gabapentin does not alter single-dose lithium pharmacokinetics. J Clin Psychopharmacol 1998;18(6):461–4.

600. Spina E, Perucca E. Clinical significance of pharmacokinetic interactions between antiepileptic and psychotropic drugs. Epilepsia 2002;43(Suppl 2):37–44.

601. Wang PW, Ketter TA. Pharmacokinetics of mood stabilizers and new anticonvulsants. Psychopharmacol Bull 2002;36(1):44–66.

602. Pies R. Combining lithium and anticonvulsants in bipolar disorder: a review. Ann Clin Psychiatry 2002;14(4):223–32.

603. Mayan H, Golubev N, Dinour D, Farfel Z. Lithium intoxication due to carbamazepine-induced renal failure. Ann Pharmacother 2001;35(5):560–2.

604. Chen C, Veronese L, Yin Y. The effects of lamotrigine on the pharmacokinetics of lithium. Br J Clin Pharmacol 2000;50(3):193–5.

605. Negin B, Murphy TK. Priapism associated with oxcarbazepine, aripiprazole, and lithium. J Am Acad Child Adolesc Psychiatry 2005;44:1223–4.

606. Pinninti NR, Zelinski G. Does topiramate elevate serum lithium levels? J Clin Psychopharmacol 2002;22(3):340.

607. Doose DR, Kohl KA, Desai-Krieger D, Natarajan J, Van Kammen DP. The effect of topiramate of lithium serum concentration. In: Presented at the 37th Annual Meeting of the American College of Neuropsychopharmacology14–18 December, Las Croabas, Puerto Rico 1998:144.

608. Abraham G, Owen J. Topiramate can cause lithium toxicity. J Clin Psychopharmacol 2004;24:565–7.

609. Yoshioka H, Ida S, Yokota M, Nishimoto A, Shibata S, Sugawara A, Takiguchi Y. Effects of lithium on the pharmacokinetics of valproate in rats. J Pharm Pharmacol 2000;52(3):297–301.

610. Miodownik C, Hausmann M, Frolova K, Lerner V. Lithium intoxication associated with intramuscular ziprasidone in schizoaffective patients. Clin Neuropharmacol 2005;28:295–7.

611. Joos AA. Pharmakologische Interaktionen von Antibiotika und psychopharmaka. [Pharmacologic interactions of antibiotics and Psychotropic drugs.] Psychiatr Prax 1998;25(2):57–60.

612. Hersh EV. Adverse drug interactions in dental practice: interactions involving antibiotics. Part II of a series. J Am Dent Assoc 1999;130(2):236–51.

613. Wilting I, Movig KL, Moolenaar M, Hekster YA, Brouwers JR, Heerdink ER, Nolen WA, Egberts AC. Drug–drug interactions as a determinant of elevated lithium serum levels in daily clinical practice. Bipolar Disord 2005;7:274–80.

614. Cebezón Pérez N, García Lloret T, Redondo de Pedro M. Intoxicación por litio desencadenada por un proceso febril. A propósito de un caso. [Lithium intoxication triggered by a process of high temperature. Concerning a case.] Aten Primaria 2005;36:344–5.

615. Takahashi H, Higuchi H, Shimizu T. Severe lithium toxicity induced by combined levofloxacin administration. J Clin Psychiatry 2000;61(12):949–50.

616. de Vries PL. Lithiumintoxicatie bij gelijktijdig gebruik van trimethoprim. [Lithium intoxication due to simultaneous use of trimethoprim.] Ned Tijdschr Geneeskd 2001;145(11):539–40.

617. ZumBrunnen TL, Jann MW. Drug interactions with antipsychotic agents. Incidence and therapeutic implications. CNS Drugs 1998;9:381–401.

618. Richard IH. Acute, drug-induced, life-threatening neurological syndromes. Neurologist 1998;4:196–210.

619. Schou M. Adverse lithium–neuroleptic interactions: are there permanent effects? Hum Psychopharmacol 1990;5:263.

620. Ahmadi-Abhari SA, Dehpour AR, Emamian ES, Azizabadi-Farahani M, Farsam H, Samini M, Shokri J. The effect of concurrent administration of psychotropic drugs and lithium on lithium ratio in bipolar patients. Hum Psychopharmacol 1998;13:29–34.

621. Dehpour AR, Emamian ES, Ahmadi-Abhari SA, Azizabadi-Farahani M. The lithium ratio and the incidence of side effects. Prog Neuropsychopharmacol Biol Psychiatry 1998;22(6):959–70.

622. Swartz CM. Olanzapine-lithium encephalopathy. Psychosomatics 2001;42(4):370.

623. Chaufour S, Borgstein NG, Van Den Eynde W, Bernard F, Canal M, Zieleniuk I, Pinquier JL. Repeated administrations of amisulpride (A) do not modify lithium carbonate (L) pharmacokinetics in healthy volunteers. Clin Pharmacol Ther 1999;65:143.

624. Taylor DM. Aripiprazole: a review of its pharmacology and clinical use. Int J Clin Pract 2003;57(1):49–54.

625. Leber K, Malek A, D'Agostino A, Adelman HM. A veteran with acute mental changes years after combat. Hosp Pract (Off Ed) 1999;34(6):21–2.

626. Edge SC, Markowitz JS, DeVane CL. Clozapine drug–drug interactions: a review of the literature. Hum Psychopharmacol 1997;12:5–20.

627. Moldavsky M, Stein D, Benatov R, Sirota P, Elizur A, Matzner Y, Weizman A. Combined clozapine-lithium treatment for schizophrenia and schizoaffective disorder. Eur Psychiatry 1998;13:104–6.

628. Lee SH, Yang YY. Reversible neurotoxicity induced by a combination of clozapine and lithium: a case report. Zhonghua Yi Xue Za Zhi (Taipei) 1999;62(3):184–7.

629. Patton S, Remick RA, Isomura T. Clozapine—an atypical reaction. Can J Psychiatry 2000;45(4):393–4.

630. Berry N, Pradhan S, Sagar R, Gupta SK. Neuroleptic malignant syndrome in an adolescent receiving olanzapine–lithium combination therapy. Pharmacotherapy 2003;23(2):255–9.

631. Tuglu C, Erdogan E, Abay E. Delirium and extrapyramidal symptoms due to a lithium-olanzapine combination therapy: a case report. J Korean Med Sci 2005;20:691–4.

632. Tohen M, Chengappa KN, Suppes T, Zarate CA Jr, Calabrese JR, Bowden CL, Sachs GS, Kupfer DJ, Baker RW, Risser RC, Keeter EL, Feldman PD, Tollefson GD, Breier A. Efficacy of olanzapine in combination with valproate or lithium in the treatment of mania in patients partially nonresponsive to valproate or lithium monotherapy. Arch Gen Psychiatry 2002;59(1):62–9.

633. Munera PA, Perel JM, Asato M. Medication interaction causing seizures in a patient with bipolar disorder and cystic fibrosis. J Child Adolesc Psychopharmacol 2002;12(3):275–6.

634. Demling J, Huang ML, De Smedt G. Pharmacokinetics and safety of combination therapy with lithium and risperidone in adult patients with psychosis. Int J Neuropsychopharmacol 1999;2:S63.

635. Durrenberger S, de Leon J. Acute dystonic reaction to lithium and risperidone. J Neuropsychiatry Clin Neurosci 1999;11(4):518–9.

636. Owley T, Leventhal B, Cook EH Jr. Risperidone-induced prolonged erections following the addition of lithium. J Child Adolesc Psychopharmacol 2001;11(4):441–2.

637. Bourgeois JA, Kahn DR. Neuroleptic malignant syndrome following administration of risperidone and lithium. J Clin Psychopharmacol 2003;23:315–17.

638. Mendhekar DN. Rabbit syndrome induced by combined lithium and risperidone. Can J Psychiatry 2005;50:369.

639. Park SH, Gill MA, Dopheide JA. Visual compatibility of risperidone solution and lithium citrate syrup. Am J Health Syst Pharm 2003;60(6):612–3.

640. Apseloff G, Mullet D, Wilner KD, Anziano RJ, Tensfeldt TG, Pelletier SM, Gerber N. The effects of ziprasidone on steady-state lithium levels and renal clearance of lithium. Br J Clin Pharmacol 2000;49(Suppl 1):61S–4S.

641. Naylor GJ, McHarg A. Profound hypothermia on combined lithium carbonate and diazepam treatment. BMJ 1977;2(6078):22.

642. Passiu G, Bocchetta A, Martinelli V, Garau P, Del Zompo M, Mathieu A. Calcitonin decreases lithium plasma levels in man. Preliminary report. Int J Clin Pharmacol Res 1998;18(4):179–81.

643. Bachofen M, Bock H, Beglinger C, Fischer JA, Thiel G. [Calcitonin, a proximal-tubular-acting diuretic: lithium clearance measurements in humans.]Schweiz Med Wochenschr 1997;127(18):747–52.

644. Bruun NE, Ibsen H, Skott P, Toftdahl D, Giese J, Holstein-Rathlou NH. Lithium clearance and renal tubular sodium handling during acute and long-term nifedipine treatment in essential hypertension. Clin Sci (Lond) 1988;75(6):609–13.

645. Pinkofsky HB, Sabu R, Reeves RR. A nifedipine-induced inhibition of lithium clearance. Psychosomatics 1997;38(4):400–1.

646. Price WA, Giannini AJ. Neurotoxicity caused by lithium-verapamil synergism. J Clin Pharmacol 1986;26(8):717–9.

647. Price WA, Shalley JE. Lithium–verapamil toxicity in the elderly. J Am Geriatr Soc 1987;35(2):177–8.

648. Dubovsky SL, Franks RD, Allen S. Verapamil: a new antimanic drug with potential interactions with lithium. J Clin Psychiatry 1987;48(9):371–2.

649. Tariq M, Morais C, Sobki S, Al Sulaiman M, Al Khader A. Effect of lithium on cyclosporin induced nephrotoxicity in rats. Ren Fail 2000;22(5):545–60.

650. Colussi G, Rombola G, Surian M, De Ferrari ME, Airaghi C, Benazzi E, Malberti F, Minetti L. Lithium clearance in humans: effects of acute administration of acetazolamide and furosemide. Kidney Int Suppl 1990;28:S63–6.

651. Hurtig HI, Dyson WL. Lithium toxicity enhanced by diuresis. N Engl J Med 1974;290(13):748–9.

652. Finley PR, Warner MD, Peabody CA. Clinical relevance of drug interactions with lithium. Clin Pharmacokinet 1995;29(3):172–91.

653. Pyevich D, Bogenschutz MP. Herbal diuretics and lithium toxicity. Am J Psychiatry 2001;158(8):1329.

654. Perlman BB. Interaction between lithium salts and ispaghula husk. Lancet 1990;335(8686):416.

655. Hendy MS, Dove AF, Arblaster PG. Mazindol-induced lithium toxicity. BMJ 1980;280(6215):684–5.

656. Verduijn M. Lithiumtoxiciteit door mazindol: Patiënt sprak niet te volgen taal. [Lithium toxicity caused by mazindol: the patient spoke incomprehensible language.] Pharm Weekbl 1998;133:1901.

657. O'Regan JB. Letter: Adverse interaction of lithium carbonate and methyldopa. Can Med Assoc J 1976;115(5):385–6.

658. Teicher MH, Altesman RI, Cole JO, Schatzberg AF. Possible nephrotoxic interaction of lithium and metronidazole. JAMA 1987;257(24):3365–6.

659. Jonnalagadda J, Saito E, Kafantaris V. Lithium, minocycline, and pseudotumor cerebri. J Am Acad Child Adolesc Psychiatry 2005;44:209.

660. Hill GE, Wong KC, Hodges MR. Potentiation of succinylcholine neuromuscular blockade by lithium carbonate. Anesthesiology 1976;44(5):439–42.

661. Hill GE, Wong KC, Hodges MR. Lithium carbonate and neuromuscular blocking agents. Anesthesiology 1977;46(2):122–6.

662. Naguib M, Koorn R. Interactions between psychotropics, anaesthetics and electroconvulsive therapy: implications for drug choice and patient management. CNS Drugs 2002;16(4):229–47.

663. Haas DA. Adverse drug interactions in dental practice: interactions associated with analgesics. Part III in a series. J Am Dent Assoc 1999;130(3):397–407.

664. Ragheb M. The clinical significance of lithium–nonsteroidal anti-inflammatory drug interactions. J Clin Psychopharmacol 1990;10(5):350–4.

665. Phelan KM, Mosholder AD, Lu S. Lithium interaction with the cyclooxygenase 2 inhibitors rofecoxib and celecoxib and other nonsteroidal anti-inflammatory drugs. J Clin Psychiatry 2003;64:1328–34.

666. Sussman N, Magid S. Psychiatric manifestations of nonsteroidal anti-inflammatory drugs. Prim Psychiatry 2000;7:26–30.

667. Montvale NJ, Physicians' Desk Reference. Medical Economics Company, Inc; 2001:2484.

668. Davies NM, McLachlan AJ, Day RO, Williams KM. Clinical pharmacokinetics and pharmacodynamics of celecoxib: a selective cyclo-oxygenase-2 inhibitor. Clin Pharmacokinet 2000;38(3):225–42.

669. Davies NM, Gudde TW, de Leeuw MA. Celecoxib: a new option in the treatment of arthropathies and familial adenomatous polyposis. Expert Opin Pharmacother 2001;2(1):139–52.

670. Rossat J, Maillard M, Nussberger J, Brunner HR, Burnier M. Renal effects of selective cyclooxygenase-2 inhibition in normotensive salt-depleted subjects. Clin Pharmacol Ther 1999;66(1):76–84.

671. Gunja N, Graudins A, Dowsett R. Lithium toxicity: a potential interaction with celecoxib. Intern Med J 2002;32(9–10):494.

672. Lundmark J, Gunnarsson T, Bengtsson F. A possible interaction between lithium and rofecoxib. Br J Clin Pharmacol 2002;53(4):403–4.

673. Sajbel TA, Carter GW, Wiley RB. Pharmacokinetics/pharmacodynamics/pharmacometrics/drug metabolism. Pharmacotherapy 2001;21:380.

674. Phelan KM, Mosholder AD, Lu S. Lithium interaction with cyclooxygenase 2 inhibitors refocoxib and celecoxib and other nonsteroidal anti-inflammatory drugs. J Clin Psychiatry 2003;64:1328–34.

675. Bocchia M, Bertola G, Morganti D, Toscano M, Colombo E. Intossicazione da litio e uso di nimesulide. [Lithium poisoning and the use of nimesulide.] Recenti Prog Med 2001;92(7–8):462.

676. Reimann IW, Frolich JC. Effects of diclofenac on lithium kinetics. Clin Pharmacol Ther 1981;30(3):348–52.

677. Monji A, Maekawa T, Miura T, Nishi D, Horikawa H, Nakagawa Y, Tashiro N. Interactions between lithium and non-steroidal antiinflammatory drugs. Clin Neuropharmacol 2002;25(5):241–2.

678. Ragheb M. Ibuprofen can increase serum lithium level in lithium-treated patients. J Clin Psychiatry 1987;48(4):161–3.

679. Bailey CE, Stewart JT, McElroy RA. Ibuprofen-induced lithium toxicity. South Med J 1989;82(9):1197.

680. Joseph DiGiacomo. Interview with F Flach. Risk management issues associated with psychopharmacological treatment. Essent Psychopharmacol 2001;4:137–50.

681. Iyer V. Ketorolac (Toradol) induced lithium toxicity. Headache 1994;34(7):442–4.

682. Cold JA, ZumBrunnen TL, Simpson MA, Augustin BG, Awad E, Jann MW. Increased lithium serum and red blood cell concentrations during ketorolac coadministration. J Clin Psychopharmacol 1998;18(1):33–7.

683. Danion JM, Schmidt M, Welsch M, Imbs JL, Singer L. Interaction entre les anti-inflammatoires non steroidiens et les sels de lithium. [Interaction between non-steroidal anti-inflammatory agents and lithium salts.] Encephale 1987;13(4):255–60.

684. Turck D, Heinzel G, Luik G. Steady-state pharmacokinetics of lithium in healthy volunteers receiving concomitant meloxicam. Br J Clin Pharmacol 2000;50(3):197–204.

685. Levin GM, Grum C, Eisele G. Effect of over-the-counter dosages of naproxen sodium and acetaminophen on plasma lithium concentrations in normal volunteers. J Clin Psychopharmacol 1998;18(3):237–40.

686. Sidhu S, Kondal A, Malhotra S, Garg SK, Pandhi P. Effect of nimesulide co-administration on pharmacokinetics of lithium. Indian J Exp Biol 2004;42:1248–50.

687. Kerry RJ, Owen G, Michaelson S. Possible toxic interaction between lithium and piroxicam. Lancet 1983;1(8321):418–9.

688. Nadarajah J, Stein GS. Piroxicam induced lithium toxicity. Ann Rheum Dis 1985;44(7):502.

689. Walbridge DG, Bazire SR. An interaction between lithium carbonate and piroxicam presenting as lithium toxicity. Br J Psychiatry 1985;147:206–7.

690. Jones MT, Stoner SC. Increased lithium concentrations reported in patients treated with sulindac. J Clin Psychiatry 2000;61(7):527–8.

691. Apseloff G, Wilner KD, von Deutsch DA, Gerber N. Tenidap sodium decreases renal clearance and increases steady-state concentrations of lithium in healthy volunteers. Br J Clin Pharmacol 1995;39(Suppl 1):25S–8S.

692. Gardner DM, Lynd LD. Sumatriptan contraindications and the serotonin syndrome. Ann Pharmacother 1998;32(1):33–8.

693. Donovan JL, DeVane CL. A primer on caffeine pharmacology and its drug interactions in clinical psychopharmacology. Psychopharmacol Bull 2001;35(3):30–48.

694. Gai MN, Thielemann AM, Arancibia A. Effect of three different diets on the bioavailability of a sustained release lithium carbonate matrix tablet. Int J Clin Pharmacol Ther 2000;38(6):320–6.

695. Gupta S, Austin R, Devanand DP. Lithium and maintenance electroconvulsive therapy. J ECT 1998;14(4):241–4.

696. Jha AK, Stein GS, Fenwick P. Negative interaction between lithium and electroconvulsive therapy—a case-control study. Br J Psychiatry 1996;168(2):241–3.

697. Schou M. Lithium and electroconvulsive therapy: adversaries, competitors, allies? Acta Psychiatr Scand 1991;84(5):435–8.

698. Lippmann SB, El-Mallakh R. Can electroconvulsive therapy be given during lithium treatment? Lithium 1994;5:205–9.

699. Dolenc TJ, Rasmussen KG. The safety of electroconvulsive therapy and lithium in combination: a case series and review of the literature. JECT 2005;21:165–70.

700. Tharayil BS, Gangadhar BN, Thirthalli J, Anand J. Seizure with single pulse transcranial magnetic stimulation in a 35-year-old otherwise-healthy patient with bipolar disorder. JECT 2005;21:188–9.

701. El-Mallakh RS. Acute lithium neurotoxicity. Psychiatr Develop 1986;4:311–28.

702. Nordt SP, Cantrell FL. Elevated lithium level: a case and brief overview of lithium poisoning. Psychosom Med 1999;61(4):564–5.

703. Van Osch-Gevers M, Draaisma JMTh, Verzijl JM. Een 20 maanden oude peuter met onbegrepen convulsies. [A 20-month-old toddler suffering from unexplained convulsions.] Pharm Weekbl 1999;134:1163–4.

704. Lovell RW, Bunker WW. Lithium assay errors. Am J Psychiatry 1997;154(10):1477.

705. Linder MW, Keck PE Jr. Standards of laboratory practice: antidepressant drug monitoring. National Academy of Clinical Biochemistry. Clin Chem 1998;44(5):1073–84.

706. Namnyak S, Hussain S, Davalle J, Roker K, Strickland M. Contaminated lithium heparin bottles as a source of pseudobacteraemia due to *Pseudomonas fluorescens*. J Hosp Infect 1999;41(1):23–8.

707. He GR, Cheng W, Huang YY. [Effect of heparin lithium as anticoagulant in assay of FT3, FT4 and TSH.]Di Yi Jun Yi Da Xue Xue Bao 2002;22(8):721–3.

708. Geller B, Cooper TB, Sun K, Zimerman B, Frazier J, Williams M, Heath J. Double-blind and placebo-controlled study of lithium for adolescent bipolar disorders with secondary substance dependency. J Am Acad Child Adolesc Psychiatry 1998;37(2):171–8.

709. Rohde LA, Szobot C. Lithium in bipolar adolescents with secondary substance dependency. J Am Acad Child Adolesc Psychiatry 1999;38(1):4.

710. Geller B. Lithium in bipolar adolescents with secondary substance dependency. J Am Acad Child Adolesc Psychiatry 1999;38:4.

711. Seidel S, Kreutzer R, Smith D, McNeel S, Gilliss D. Assessment of commercial laboratories performing hair mineral analysis. JAMA 2001;285(1):67–72.

712. Lundmark J, Gunnarsson T, Bengtsson F. A possible interaction between lithium and rofecoxib. Br J Clin Pharmacol 2002;53(4):403–4.

713. Naylor GJ, McHarg A. Profound hypothermia on combined lithium carbonate and diazepam treatment. BMJ 1977;2(6078):22.

714. Stokke ES, Ostensen J, Hartmann A, Kiil F. Loop diuretics reduce lithium reabsorption without affecting bicarbonate and phosphate reabsorption. Acta Physiol Scand 1990;140(1):111–8.

715. Beutler JJ, Boer WH, Koomans HA, Dorhout Mees EJ. Comparative study of the effects of furosemide, ethacrynic acid and bumetanide on the lithium clearance and diluting segment reabsorption in humans. J Pharmacol Exp Ther 1992;260(2):768–72.

716. Hansen HE, Amdisen A. Lithium intoxication. (Report of 23 cases and review of 100 cases from the literature). Q J Med 1978;47(186):123–44.

717. Jefferson JW, Kalin NH. Serum lithium levels and long-term diuretic use. JAMA 1979;241(11):1134–6.

718. Huang LG. Lithium intoxication with coadministration of a loop-diuretic J Clin Psychopharmacol 1990;10(3):228.

719. Hurtig HI, Dyson WL. Letter: Lithium toxicity enhanced by diuresis. N Engl J Med 1974;290(13):748–9.

720. Bond WS, Carvalho M, Foulks EF. Persistent dysarthria with apraxia associated with a combination of lithium carbonate and haloperidol. J Clin Psychiatry 1982;43(6):256–7.

721. Takahashi H, Higuchi H, Shimizu T. Severe lithium toxicity induced by combined levofloxacin administration. J Clin Psychiatry 2000;61(12):949–50.

722. Blanche P, Raynaud E, Kerob D, Galezowski N. Lithium intoxication in an elderly patient after combined treatment with losartan. Eur J Clin Pharmacol 1997;52(6):501.

723. Burnier M, Rutschmann B, Nussberger J, Versaggi J, Shahinfar S, Waeber B, Brunner HR. Salt-dependent renal effects of an angiotensin II antagonist in healthy subjects. Hypertension 1993;22(3):339–47.

724. Schweitzer I, Tuckwell V. Risk of adverse events with the use of augmentation therapy for the treatment of resistant depression. Drug Saf 1998;19(6):455–64.

725. Bruun NE, Ibsen H, Skott P, Toftdahl D, Giese J, Holstein-Rathlou NH. Lithium clearance and renal tubular sodium handling during acute and long-term nifedipine

treatment in essential hypertension. Clin Sci (Lond) 1988;75(6):609–13.

726. Beijnen JH, Vlasveld LT, Wanders J, ten Bokkel Huinink WW, Rodenhuis S. Effect of cisplatin-containing chemotherapy on lithium serum concentrations. Ann Pharmacother 1992;26(4):488–90.

727. Pullar T. The pharmacokinetics of tenidap sodium: introduction. Br J Clin Pharmacol 1995;39(Suppl 1):S1–2.

728. Apseloff G, Wilner KD, von Deutsch DA, Gerber N. Tenidap sodium decreases renal clearance and increases steady-state concentrations of lithium in healthy volunteers. Br J Clin Pharmacol 1995;39(Suppl 1):S25–8.

729. Jefferson JW, Greist JH, Baudhuin M. Lithium: interactions with other drugs. J Clin Psychopharmacol 1981;1(3):124–34.

730. Waddington JL. Some pharmacological aspects relating to the issue of possible neurotoxic interactions during combined lithium-neuroleptic therapy. Hum Psychopharmacol 1990;5:293–7.

731. Batchelor DH, Lowe MR. Reported neurotoxicity with the lithium/haloperidol combination and other neuroleptics. A literature review. Hum Psychopharmacol 1990;5:275–80.

732. de Vries PL. Lithiumintoxicatie bij gelijktijdig gebruik van trimethoprim. [Lithium intoxication due to simultaneous use of trimethoprim.] Ned Tijdschr Geneeskd 2001;145(11):539–40.

733. Aronson JK, Reynolds DJM. ABC of monitoring drug therapy. Lithium. BMJ 1992;305:1273–6.

734. Kleindienst N, Severus WE, Möller HJ, Greil W. Is polarity of recurrence related to serum lithium level in patients with bipolar disorder? Eur Arch Psychiatry Clin Neurosci 2005;255:72–4.

735. Camus M, Hennere G, Baron G, Peytavin G, Massias L, Mentre F, Farinotti R. Comparison of lithium concentrations in red blood cells and plasma in samples collected for TDM, acute toxicity, or acute-or-chronic toxicity. Eur J Clin Pharmacol 2003;59:583–7.

736. Layden BT, Minadeo N. Suhy J, Abukhdeir AM, Metreger T, Foley K, Borge G, Crayton JW, Bryant FB, De Freitas DM. Biochemical and psychiatric predictors of Li$^+$ response and toxicity in Li$^+$-treated bipolar patients. Bipolar Disord 2004;6:53–61.

NEUROLEPTIC DRUGS

General Information

Since the introduction of chlorpromazine in the early 1950s, a large number of phenothiazines with neuroleptic properties have been discovered; they include fluphenazine, perphenazine, prochlorperazine, and thioridazine. Several other chemical structures with similar therapeutic properties have also been introduced, including the butyrophenones (such as haloperidol and droperidol) and the thioxanthenes (such as flupenthixol). In recent years the so-called atypical neuroleptic drugs (such as clozapine, olanzapine, and risperidone) have also become available. They have less affinity for dopamine receptors in the basal ganglia than the typical neuroleptic drugs, and therefore cause fewer extrapyramidal adverse effects. The range of usefulness of these agents includes the treatment of schizophrenia (to treat acute episodes and in long-term maintenance treatment), mania, certain organic psychoses, certain depressive states, and a variety of lesser indications. They have also been used to treat autism (1) and psychosis in patients with dementia (2).

Equipotent doses of neuroleptic drugs are listed in Table 1.

New updated treatment recommendations have been issued by the Schizophrenia Patient Outcomes Research Team (3). There is no definitive evidence that the atypical drugs differ from the typical ones for the treatment of acute positive symptoms; there are no data to support a change to a second-generation neuroleptic drug for patients who have adequate symptom control and minimal adverse effects with first-generation neuroleptic drug therapy. For haloperidol, a maintenance dosage of 6–12 mg is recommended; this dosage is lower than that previously recommended by the same panel (6–20 mg)

(4). In fact, the optimal daily dose remains a topic of controversy.

General adverse reactions

Neuroleptic drugs can produce a variety of adverse effects in several organ systems. Extrapyramidal reactions and sedation are common; less common are seizures, unwanted behavioral effects, and tardive dyskinesia. Most neuroleptic drugs have anticholinergic effects and commonly produce dry mouth, blurred vision, and constipation. Postural hypotension is common. These effects usually disappear when the drug is stopped or the dosage is reduced.

Non-specific, usually reversible, cardiographic changes have been reported, but their relation to myocardial toxicity has not been confirmed. Sudden death related to cardiac arrest cannot be fully explained on the basis of the administration of neuroleptic drugs. Weight gain is a common adverse effect. Breast engorgement and galactorrhea can occur in women and even in men. Amenorrhea, gynecomastia, hyperglycemia, hypoglycemia, raised growth hormone, inappropriate ADH secretion, and disturbance of sex hormones have been reliably documented, although they are unusual.

The contraindications to neuroleptic drug therapy include coma, the presence or withdrawal of high doses of other CNS depressants (alcohol, barbiturates, narcotics, etc.), serious hematological conditions (for example bone-marrow suppression), and a previous history of hypersensitivity reactions for example jaundice or severe photosensitivity. Since neuroleptic drugs cause sedation, they can impair mental or physical abilities (including reaction times), especially during the first few days of therapy. A large number of substances can interact with neuroleptic drugs (SED-14, 153) (5).

Hypersusceptibility reactions

Neuroleptic drugs infrequently produce allergic reactions. There are no reports of anaphylactic reactions, but various skin reactions, for example rashes, photosensitivity, and dermatitis, can be viewed as delayed forms of hypersensitivity. Jaundice and blood dyscrasias (hemolytic anemia, agranulocytosis) are rare and may be types of allergic reactions.

Tumorigenicity

Since neuroleptic drugs raise prolactin concentrations, there is concern that this may increase the risk of breast cancer. Although studies have failed to show an association, it would be best to avoid neuroleptic drugs in a patient with a hormone-dependent breast tumor.

Effects on fertility

Amenorrhea and infertility were consequences of the effects of typical neuroleptic drugs and of risperidone. Clinicians should be aware that patients changed from these agents to drugs like olanzapine, quetiapine, or clozapine are therefore at risk of pregnancy (6).

Table 1 Equipotent doses of neuroleptic drugs (all rINNs)

Drug	Dose (mg)
Typical neuroleptic drugs	
Chlorpromazine	100
Chlorprothixene	100
Fluphenazine hydrochloride	2
Haloperidol	2
Loxapine	10
Mesoridazine	50
Molindone	10
Perphenazine	10
Prochlorperazine	15
Pimozide	1
Thioridazine	100
Tiotixene	5
Trifluoperazine	5
Trifluoperazine	25
Atypical neuroleptic drugs	
Aripiprazole	6
Clozapine	50
Olanzapine	4
Quetiapine	80
Risperidone	1
Ziprasidone	20

A project group at the Swedish Council on Technology Assessment in Health Care has analysed more than 2000 published manuscripts about neuroleptic drugs (7). They concluded that neuroleptic drug therapy is often accompanied by serious, sometimes permanent, adverse effects. Hence, neuroleptic drugs should be reserved for patients with severe psychoses. Agitated and demented elderly patients should not be treated with neuroleptic drugs unless they have pronounced psychotic symptoms. Nor should neuroleptic drugs be used in young mentally retarded patients and other children and adolescents, except in those with severe autism, Tourette's syndrome, or schizophrenia. In fact, in patients with schizophrenia there is increased mortality, and the involvement of neuroleptic drug treatment has been investigated (8).

Severe adverse events associated with neuroleptic drug treatment are epileptic seizures, QT interval prolongation, myocarditis associated with clozapine, neuroleptic malignant syndrome, hypothermia, respiratory arrest, and pulmonary embolism associated with clozapine. To minimize these potential risks, practical prescribing guidelines have recently been proposed (SEDA-21, 42); they recommend careful titration of therapy, checking for a history of cardiac disorders, seizures, neuroleptic malignant syndrome, and hypotension, and regular monitoring for known adverse effects.

Further comparisons of the main features of typical and atypical neuroleptic drugs have emerged (SEDA-22, 45) (9). The review by the Collaborative Working Group on Clinical Trial Evaluations has addressed adverse effects extensively (10). The authors stressed that atypical neuroleptic drugs cause fewer extrapyramidal signs and may have a lower risk of causing tardive dyskinesia than typical neuroleptic drugs. The adverse effects of the atypical drugs of which one should be aware are the following (listed with the drug(s) most likely to cause them):

(a) orthostatic hypotension (clozapine, olanzapine, quetiapine);
(b) myocarditis (clozapine);
(c) pulmonary embolism (clozapine);
(d) seizures (clozapine);
(e) anticholinergic effects (clozapine, olanzapine);
(f) increased prolactin (risperidone);
(g) weight gain (clozapine, olanzapine);
(h) hepatic changes (clozapine, risperidone);
(i) agranulocytosis (clozapine).

Drug interactions can be dangerous or fatal and should be avoided. Patients' individual concerns and health needs must be taken into account when selecting a drug. The atypical drugs, it is said, would be better first-line drugs for patients with specific health concerns.

The Collaborative Working Group has also drawn attention to the interpretation of certain results from clinical trials with novel neuroleptic drugs (10). It has been stated, for instance, that when extrapyramidal signs are not significantly different from those with placebo, it does not necessarily mean that a new neuroleptic drug absolutely lacks extrapyramidal effects. Patients who enter studies of new neuroleptic drugs may have been previously treated with traditional neuroleptic drugs, and extrapyramidal symptoms may have persisted from this prior drug treatment.

A comparison of patients' and prescribers' beliefs about the adverse effects of neuroleptic drugs has been carried out (11). Psychiatrists' estimates of prevalence, but not of distress, correlated significantly with patients' reports. The authors concluded that the apparent lack of understanding of which adverse effects are most likely to cause distress to patients can adversely affect the therapeutic alliance between prescribers and patients.

Patients with good adherence to therapy have a higher incidence of adverse effects (12). Logistic regression analysis identified four factors that discriminate adherent ($n = 48$) from non-adherent ($n = 30$) patients: the course of the illness, the employment status of a key relative, age at onset of the illness, and the presence or absence of adverse effects.

Several atypical neuroleptic drugs are currently considered to be first-line therapies for schizophrenia. However, choosing one can be difficult, because head-to-head comparisons are just beginning to appear and most published trials are comparisons with typical neuroleptic drugs, particularly haloperidol. Several reviews (13,14) and clinical comparisons (15,16) of atypical neuroleptic drugs have been published; a meta-analysis of the efficacy and extrapyramidal effects of the new neuroleptic drugs olanzapine, quetiapine, and risperidone, compared with typical neuroleptic drugs and placebo, deserves particular attention (17). Only randomized, double-blind, controlled trials in patients with schizophrenia or schizophrenia-like psychoses were selected. Combining all Brief Psychiatric Rating Scale (BPRS) comparisons between all neuroleptic drugs and placebo resulted in a mean effect size of 0.25 (CI = 0.22, 0.28; $n = 2477$), which is only a moderate treatment effect. Quetiapine (six studies; $n = 1414$) was as effective as haloperidol. Risperidone (nine studies; $n = 2215$) and olanzapine (four studies; $n = 2914$) were statistically superior to haloperidol. However, the latter two effect sizes (0.06 and 0.07) were very small. The new neuroleptic drugs and placebo were associated with similar use of antiparkinsonian medications, and all were clearly superior to haloperidol in this respect; risperidone had the weakest effect (effect size 0.09 versus placebo and 0.12 versus haloperidol). Many of the trials used relatively high doses of haloperidol, favoring the occurrence of extrapyramidal signs.

Special subgroups have been considered in different studies. Agitated and demented elderly patients should not be treated with neuroleptic drugs, unless they have pronounced psychotic symptoms (SEDA-23, 48). Since some studies have reported the lack of a documented diagnosis associated with the use of neuroleptic drugs, a drug utilization study using a chart review has been carried out in Canada to discover the pattern of use of neuroleptic drugs in a long-term care facility (18). Neuroleptic drugs were prescribed for 86 (19%) of 446 patients from four units of a 360-bed nursing home, mainly for agitation ($n = 21$); their mean age was 84 years and the mean length of time they had been receiving neuroleptic drugs was 37 (7–263) months. The authors thought that this was of definite concern, given the modest benefit that has been observed during long periods of treatment in such patients and the risk of tardive dyskinesia and other adverse effects.

In the light of evidence from large, randomized, double-blind trials, the authors of a thorough review of the role of atypical neuroleptic drugs in the treatment of psychosis and agitation associated with dementia have concluded that low-dose risperidone (0.25–1.5 mg/day) can be used as first-line treatment (19).

A further comprehensive review on the use of atypical neuroleptic drugs in older adults has also been published (20).

Observational studies

Of 300 hospitalized Asian patients with schizophrenia, 215 were being given polytherapy, defined as the use of more than one neuroleptic drug at one time (21). The average number of drugs, including depot formulations, was 1.8 neuroleptic drugs (range 1–4) and the mean daily dose in chlorpromazine equivalents was 612 (median 464; range: 25–2500) mg. Atypical neuroleptic drugs were scarcely used (n = 20). Anticholinergic antiparkinsonian drugs were prescribed in 82%. The frequency of adverse events increased with total dose, and dry mouth (19%), tremor (8.7%), and constipation (5.3%) were the most common.

Combinations are often used in patients with treatment-resistant schizophrenia and schizoaffective disorders, and the published data (29 case reports and case series, n = 172, and one double-blind placebo-controlled trial, n = 28) between 1985 and 2003 has been reviewed (22). Significant adverse effects were rarely reported (by only 36 of the 200 subjects) and did not appear to be different in nature from those of monotherapy regimens: hypersalivation (n = 6), mild akathisia (n = 4), exacerbation of hoarding behavior (n = 2), neuroleptic malignant syndrome (n = 2), and neutropenia, agranulocytosis, oculogyric crises, atrial extra beats, and aggravation of previous tardive dyskinesia (n = 1 each).

Comparisons of neuroleptic drugs

The impact of new neuroleptic drugs on the pattern of neuroleptic drug use has been studied in Spain (23). The use of neuroleptic drugs rose by 146% from 1990 to 2001; the atypical neuroleptic drugs accounted for 49% of the total consumption of neuroleptic drugs in 2001 and 90% of the costs. There is a similar pattern worldwide. This is surprising, since there is no clear evidence that atypical neuroleptic drugs are more effective or better tolerated than conventional neuroleptic drugs (SEDA-25, 53). Moreover, a recent meta-analysis of typical and atypical neuroleptic drugs (31 studies, 2320 participants) showed that optimum doses of low-potency conventional neuroleptic drugs might not induce more extrapyramidal signs than newer drugs (24); mean doses less than 600 mg/day of chlorpromazine equivalents had no higher risk of extrapyramidal signs than newer neuroleptic drugs.

In contrast to current beliefs, neuroleptic drugs caused more severe adverse effects than typical neuroleptic drugs did in a program intended to monitor the adverse effects

of neuroleptic drugs in routine clinical practice (25). From 1993 to 2000, 86 439 patients who took at least one neuroleptic drug were monitored in 35 psychiatric hospitals in Germany and Switzerland. Overall, 975 clinically severe adverse effects were identified (1.1%). As to individual drugs in monotherapy, the incidences were:

- haloperidol, 0.2%;
- perazine, 0.4%;
- olanzapine, 0.5%;
- risperidone, 0.5%;
- clozapine, 0.9%.

The authors stated that their results were in accordance with those of a meta-analysis of 52 randomized trials (SEDA-25, 53).

Coming from the same program, separate estimates have been made for different reactions:

- severe galactorrhea occurred in 27 cases (0.03%); haloperidol, 0%; clozapine, 0%; perazine, 0.01%; olanzapine, 0.03%; risperidone, 0.09% (26).
- severe neutropenia (neutrophils $< 1.5 \times 10^9$/l) occurred in 43 cases (0.05%); haloperidol, 0.01%; risperidone, 0.01%; olanzapine, 0.05%; perazine, 0.07%; clozapine, 0.16% (27).
- severe and uncommon involuntary movement disorders occurred in 115 cases (0.13%); clozapine, 0.01%; perazine, 0.02%; olanzapine, 0.04 %; risperidone, 0.09%; haloperidol, 0.18% (28).

These estimates are very low, since only those reactions rated as probably or definitely drug related were considered.

The primary and main outcomes of the Clinical Antipsychotic Trials of Intervention Effectiveness both in schizophrenic patients and in patients with dementia have been released. In the first study, 1493 patients with schizophrenia recruited at 57 US sites were randomly allocated to olanzapine (7.5–30 mg/day), perphenazine (8–32 mg/day), quetiapine (200–800 mg/day) and ziprasidone (40–160 mg/day) for up to 18 months (29). The main conclusion of the study was that patients with chronic schizophrenia discontinued their neuroleptic drug medication at a high rate, indicating substantial limitations in the effectiveness of the drugs. In fact, overall 74% of patients withdrew within 18 months; the time to withdrawal for any cause was significantly longer with olanzapine than quetiapine or risperidone but not perphenazine or ziprasidone. The time to withdrawal because of intolerable adverse effects was similar across the groups (olanzapine, 19%; quetiapine, 15%; risperidone, 10%; perphenazine, 16%; ziprasidone, 15%). The authors suggested that the ways in which clinicians, patients, families, and policymakers evaluate the trade-offs between efficacy and adverse effects, as well as drug prices, will determine future patterns of use.

Comparisons of different typical drugs
Amisulpride versus typical neuroleptic drugs
In an extensive review of 19 randomized studies for the Cochrane Library (*n* = 2443), most of the trials were

small and of short duration (30). Compared with typical neuroleptic drugs, the pooled results of a total of 14 trials suggested that amisulpride was more effective in improving global state ($n = 651$), general mental state ($n = 695$), and the negative symptoms of schizophrenia ($n = 506$). Amisulpride was as effective as typical neuroleptic drugs in relieving positive symptoms. It was less likely to cause at least one general adverse event ($n = 751$), to cause one extrapyramidal symptom ($n = 771$), or to require the use of antiparkinsonian drugs ($n = 851$). There were no clear differences in other adverse events compared with typical drugs. Amisulpride also seemed to be more acceptable than typical drugs, as measured by early withdrawal ($n = 1512$) than typical drugs, but this result may have been overestimated owing to publication bias, which could not be excluded with certainty.

A further meta-analysis of 10 randomized controlled clinical trials of amisulpride in "acutely ill patients" ($n = 1654$) was supported in part by a grant from Sanofi-Synthélabo, the marketing authorization holder (31). Amisulpride was significantly better than typical neuroleptic drugs by about 11% on the BPRS. In four studies in patients with "persistent negative symptoms," amisulpride was significantly better than placebo ($n = 514$), but there was no significant difference between amisulpride and typical drugs (only three studies; $n = 130$). Low doses of amisulpride (50–300 mg/day) were not associated with significantly more use of antiparkinsonian drugs than placebo ($n = 507$), and usual doses caused fewer extrapyramidal adverse effects than typical neuroleptic drugs ($n = 1599$). In four studies in acutely ill patients, significantly fewer patients taking amisulpride dropped out compared with patients taking typical drugs, mainly because of fewer adverse events; in three small studies with conventional neuroleptic drugs as comparators there was only a trend in favor of amisulpride in this regard.

Amisulpride versus flupenthixol

In a randomized, double-blind, multicenter comparison of amisulpride 1000 mg/day ($n = 70$) and flupenthixol 25 mg/day ($n = 62$) for 6 weeks, the two drugs significantly improved the acute psychotic symptoms to a similar extent (32). The total numbers of dropouts were 19 with amisulpride and 25 with flupenthixol. Adverse effects accounted for 8.6 and 18% respectively of the totals. Amisulpride caused significantly fewer extrapyramidal adverse effects. Apart from the extrapyramidal adverse effects, there were treatment-related adverse events in 87% of the patients given amisulpride and 92% of those given flupenthixol. Prolactin concentrations were higher with amisulpride.

Amisulpride versus haloperidol

Fixed doses of amisulpride (100, 400, 800, and 1200 mg/day) and haloperidol (16 mg/day) have been compared in a 4-week, double-blind, randomized trial in 319 patients with acute exacerbations of schizophrenia (33). Amisulpride 400 mg/day and 800 mg/day was effective in treating the positive symptoms of schizophrenia, with fewer extrapyramidal adverse effects than haloperidol,

which was associated with the highest proportion of extrapyramidal symptoms. The incidence of extrapyramidal symptoms in patients treated with amisulpride increased with increasing dose (31%, 42%, 45%, and 55% for 100, 400, 800, and 1200 mg/day respectively). The rate of withdrawals due to adverse events was higher with haloperidol (16%) than with amisulpride (0%, 5%, 3%, and 5% respectively).

The long-term safety and efficacy of amisulpride in subchronic and chronic schizophrenia have been assessed in an open, multicenter study in 489 patients randomly allocated to amisulpride (mean dose 605 mg/day; $n = 370$) or haloperidol (mean dose 14.6 mg/day; $n = 119$) for 12 months (34). Improvement in mean total score on the BPRS was significantly greater with amisulpride than with haloperidol. The proportion of patients with at least one treatment-emergent adverse event was similar in the two groups, 69% with amisulpride and 70% with haloperidol, but extrapyramidal symptoms occurred more often with haloperidol (41%) than with amisulpride (26%); endocrine disorders occurred in 4% of those taking amisulpride and 3% of those taking haloperidol. Amenorrhea (6 versus 0%) and weight increase (11 versus 4%) were more frequent with amisulpride. There were serious adverse events in 10% of the patients taking amisulpride and 7% of those taking haloperidol.

Atypical versus typical drugs

There is no clear evidence that atypical neuroleptic drugs are more effective or better tolerated than typical neuroleptic drugs. It has therefore been suggested that typical neuroleptic drugs can be used in the initial treatment of schizophrenia, unless the patient has previously not responded to these drugs or has unacceptable extrapyramidal adverse effects, based on a meta-analysis of 52 randomized comparisons of atypical neuroleptic drugs with typical neuroleptic drugs (12 649 patients) or alternative atypical neuroleptic drugs (35). After correction for the higher than recommended doses of typical neuroleptic drugs that are used in some trials, there was a modest advantage of atypical neuroleptic drugs in terms of extrapyramidal adverse effects, but the differences in efficacy and overall tolerability disappeared, suggesting that many of the perceived benefits of atypical neuroleptic drugs are really due to excessive doses of the comparator drugs used in trials. Other reviews of atypical neuroleptic drugs have not added anything new (36,37). Both types of neuroleptic drugs have serious shortcomings, particularly their adverse effects on the extrapyramidal and endocrine systems. However, quality of life is said to be superior with atypical neuroleptic drugs, from an analysis of seven controlled trials and eight open trials (38).

In a meta-analysis the effect sizes of olanzapine, risperidone, amisulpride, and clozapine were respectively 0.21, 0.25, 0.29, and 0.49 times greater than those of first-generation neuroleptic drugs; other atypical neuroleptic drugs were not significantly different from typical neuroleptic drugs, although zotepine was marginally different (39). No efficacy difference was detected among amisulpride, risperidone, and olanzapine. There was no evidence that the dosage of haloperidol (or of any first-generation neuroleptic drug converted to haloperidol-equivalent doses) affected these results when effects by drug were examined or in a 2-way analysis of variance model, in which the effectiveness of the atypical neuroleptic drug was entered as a second factor. Further information has been provided in later comparisons of neuroleptic drugs (SEDA-27, 50)(40,41).

An independent cross-sectional survey, not sponsored by the pharmaceutical industry, in schizophrenic outpatients clinically stabilized on a neuroleptic drug for a period of 6 months showed that quality-of-life measures and Global Assessment of Functioning did not differ significantly in patients taking typical neuroleptic drugs ($n = 44$) and novel ones (risperidone, $n = 50$; olanzapine, $n = 48$; quetiapine, $n = 42$; clozapine, $n = 46$) (42).

Adverse effects have been studied in people with mental retardation treated with atypical neuroleptic drugs ($n = 17$), typical neuroleptic drugs ($n = 17$), or no drugs ($n = 17$) (43). The patients taking atypical neuroleptic drugs did not have different overall adverse events from those taking no medications, and both had significantly fewer overall adverse effects than those taking typical neuroleptic drugs. However, the study had some important flaws: patients taking typical neuroleptic drugs were on average 7 years older than those taking atypical drugs; they had also taken medication for longer and had more stereotypic movement disorders at baseline. This jeopardizes the conclusions.

Atypical drugs versus haloperidol

In a randomized double-blind trial in 157 inpatients with chronic schizophrenia, clozapine ($n = 40$), olanzapine ($n = 39$), and risperidone ($n = 41$), but not haloperidol ($n = 37$), produced statistically significant improvements in total scores on the Positive and Negative Syndrome Scale after 14 weeks (44). Patients who had failed to respond to any typical neuroleptic drug were eligible, including patients who had previously failed to respond to haloperidol. High doses were used (the mean daily doses achieved during the last period of study were: clozapine 525 mg, olanzapine 30 mg, risperidone 12 mg, and haloperidol 26 mg). There was a significant fall in the Extrapyramidal Symptom Rating Score with the three atypical drugs at the end of the study and no change with haloperidol. One patient developed agranulocytosis, two had hypertensive episodes, and four had seizures while taking clozapine.

The results of 11 studies in 1933 patients with schizophrenia, who were randomly assigned to amisulpride ($n = 1247$), haloperidol ($n = 309$), risperidone ($n = 113$), flupenthixol ($n = 62$), or placebo ($n = 202$) have been reviewed (45). Extrapyramidal signs occurred in 15% of those given amisulpride ($n = 579$), 12% of those given risperidone ($n = 113$), and 31% of those given haloperidol ($n = 214$). In contrast, endocrine disorders were more frequent with amisulpride (4%) and risperidone (6%) than with haloperidol (1%). In a subgroup of patients with predominant negative schizophrenia who had at least one electrocardiogram recorded during treatment, there was a relative prolongation of the QT_c interval of at least 60 ms in three of 296 patients treated with amisulpride compared with no cases with both haloperidol ($n = 80$) and risperidone ($n = 91$); however, there were no ventricular dysrhythmias.

Clozapine versus typical neuroleptic drugs

A meta-analysis of 30 randomized controlled comparisons of clozapine with typical neuroleptic drugs ($n = 2530$) has been published (46). Clozapine was more effective in reducing symptoms in patients with both treatment-resistant and non-resistant schizophrenia. In a subset of 13 trials hematological problems tended to be more frequent in patients taking clozapine (OR = 1.93, CI = 0.96, 3.87); hypersalivation was also more frequent (OR = 5.50, CI = 4.26, 7.10), as were fever (OR = 1.89, CI = 1.38, 2.60) and sedation (OR = 1.94, CI = 1.50, 2.50). Xerostomia (OR = 0.29, CI = 0.20, 0.42) and extrapyramidal symptoms (OR = 0.46, CI = 0.28–0.75) were more frequent in the patients treated with typical neuroleptic drugs. There was no difference between the two groups in weight gain (OR = 1.07, CI = 0.37, 3.10), hypotension or dizziness (OR = 1.66, CI = 0.74, 3.71), or seizures (OR = 1.60, CI = 0.84, 3.04). Although there were fewer deaths in the clozapine group, statistical significance was not reached (OR = 0.50, CI = 0.11, 2.30). One of the flaws of this meta-analysis was that there was no information on dropouts.

Clozapine versus chlorpromazine

In 42 elderly patients (mean age 67 years) with schizophrenia randomly assigned to clozapine, titrated to 300 mg/day ($n = 24$), or chlorpromazine, titrated to a maximum of 600 mg/day ($n = 18$), the two medications were equally effective at 5 weeks (47). In each group there was one patient with a serious and potentially fatal adverse effect: agranulocytosis in the clozapine group and paralytic ileus in the chlorpromazine group; both drugs significantly lowered the white cell count.

Clozapine versus chlorpromazine and haloperidol

Clozapine has been compared with typical neuroleptic drugs in seven studies in patients with chronic refractory schizophrenia (48) ($n = 1124$). Of the 10 comparisons of second-generation versus typical neuroleptic drugs ($n = 1801$), there was a significant difference in six favoring second-generation neuroleptic drugs on measures of treatment efficacy; in the other four there was no significant difference between treatments. ANCOVA with the baseline score as a co-variate was performed to compare the efficacy of clozapine with that of typical neuroleptic drugs in terms of BPRS total score; there was a significant reduction in psychopathology in those who took clozapine, and the reduction was greater among those with higher BPRS scores. When the assessment was performed with other scales (BPRS positive symptom subscale, SANS) there were no significant treatment effects for clozapine over typical neuroleptic drugs. The subjects who took clozapine had significantly fewer extrapyramidal effects; tardive dyskinesia occurred equally in the two groups. Weight gain was reported in patients who took chlorpromazine (1%) and clozapine (7.1%), but in none of those who took haloperidol. Patients taking chlorpromazine (3.4%) or clozapine (2.9%) had difficulty in concentrating, but none of those taking haloperidol did. Other adverse events reported in patients taking clozapine included neutropenia (2.0%), enuresis (1.0%), and seizures (0.4%); neuroleptic malignant syndrome developed in 0.5% of the patients who took chlorpromazine. Completion rates were higher for clozapine (70%; $n = 400$) than for the typical neuroleptic drugs (56%; $n = 398$). Despite the superior efficacy of clozapine in the treatment of resistant patients, the extent of this in terms of scope (symptoms improved) and magnitude (effect size) was variable. Using what might be regarded as a non-stringent criterion of a 20–30% reduction in total psychopathology scores, under half of the patients responded in most of the studies.

Clozapine versus haloperidol

Clozapine and haloperidol have been compared in 75 schizophrenic outpatients, who met criteria for residual positive or negative symptoms after being treated for at least 6 weeks with typical neuroleptic drugs, in a 10-week randomized, double-blind, parallel-group comparison (49). There was no evidence of any superior efficacy or long-term effect of clozapine ($n = 38$) on primary or secondary negative symptoms compared with haloperidol ($n = 37$). Long-term clozapine was associated with significant improvements in social and occupational functioning, but not in overall quality of life. There were no significant differences between the two groups in adverse effects from previous neuroleptic drug treatment, and dizziness (50 versus 19%), salivation (82 versus 19%), and nausea (37 versus 11%) were significantly more common in patients treated with clozapine. In contrast, dry mouth was significantly more common with haloperidol (62 versus 18%). Patients who completed the double-blind study ($n = 58$) entered a 1-year open clozapine study. Over the course of that year, there was a small reduction in adverse effects, apart from hypersalivation, which increased significantly. Over the first 6 months there was weight gain followed by a plateau.

Clinical predictors of response to clozapine have been examined in 37 partially treatment-refractory outpatients who had been assigned to clozapine in a double-blind, haloperidol-controlled, 29-week study (50). Clozapine responders were rated as less severely ill, had fewer negative symptoms, and had fewer extrapyramidal adverse effects at baseline compared with non-responders.

Olanzapine versus chlorpromazine

In 103 previously treatment-resistant patients with schizophrenia were given a prospective 6-week trial of 10–40 mg/day of haloperidol; 84 failed to respond and were randomly assigned to a double-blind, 8-week, fixed-dose trial of either olanzapine 25 mg/day alone ($n = 42$) or chlorpromazine 1200 mg/day plus benzatropine mesylate 4 mg/day ($n = 39$) (51). There was no significant

difference in completion rates. Neither drug produced a substantial change in the level of psychosis from prerandomization baseline, and there were no differences between the groups in efficacy. However, olanzapine had a better adverse effects profile.

Olanzapine versus haloperidol

The global efficacy of olanzapine was not substantially different from that of haloperidol in two of three comparative trials involving 2500 patients, according to a comprehensive review of the safety and efficacy of olanzapine; in addition, the only relevant comparative trial failed to demonstrate superiority of olanzapine over risperidone (52). Olanzapine has fewer adverse neurological effects than haloperidol, but there is no evidence that it differs from other atypical neuroleptic drugs in this respect.

A meta-analysis of four comparisons of olanzapine and haloperidol showed similar efficacy, with fewer extrapyramidal effects with olanzapine (17). Similar conclusions were reached in another meta-analysis of three randomized, double-blind, controlled comparisons of olanzapine and haloperidol (53). Only 15% of olanzapine-treated patients ($n = 1620$, dosage 5–20 mg/day) needed anticholinergic drugs compared with 49% of those treated with haloperidol ($n = 786$, dosage 5–20 mg/day).

Olanzapine has been compared with haloperidol in cannabis-induced psychosis (54), schizoaffective disorder (55), first-episode psychosis (56), and treatment-resistant schizophrenia (57); the two last studies were reanalyses of data from large clinical trials promoted by the manufacturers, Eli Lilly. In all cases olanzapine was better than haloperidol at reducing BPRS scores, but in patients with cannabis-induced psychotic disorders they were similar. Increased appetite was consistently reported more often in olanzapine-treated patients and extrapyramidal signs more often in those treated with haloperidol.

A re-analysis of data from a large double-blind comparison of olanzapine with haloperidol ($n = 1996$) showed that in patients who had an initial response, there was no significant difference between olanzapine and haloperidol when outcome was measured using either 52-week relapse rates or the time to first non-compliance; after 12 months, the estimated mean times to discontinuation were 271 days and 241 days respectively (58). However, while the dose of olanzapine was well within the recommended range, the dose of haloperidol was too high (modal doses 13 and 12 mg/day respectively).

In a double-blind trial, 28 patients with paranoid schizophrenia who had had a partial response to typical neuroleptic drugs were randomized to receive either olanzapine or haloperidol in flexible doses for 14 weeks (59). The two groups showed similar improvement on the Brief Psychiatric Rating Scale positive symptoms subscale, while improvement in the same negative symptoms subscale was significant only with olanzapine. There were no significant differences between the two groups on the Simpson and Angus rating Scale scores, but there was a significant difference in the Barnes Akathisia Rating

Scale. No patient taking olanzapine had akathisia, while a few patients taking haloperidol did. Apart from the small sample, patients with a partial response to typical neuroleptic drugs were included, which precludes firm conclusions.

In a double-blind, placebo-controlled, dose-response trial, 270 acutely agitated patients were randomized to receive 1–3 intramuscular injections of olanzapine (2.5, 5, 7.5, or 10 mg), haloperidol (7.5 mg), or placebo within 24 hours (60). Olanzapine had a dose-related effect in reducing agitation; olanzapine was better than placebo but not better than haloperidol; the most frequently reported adverse event was hypotension, which occurred with olanzapine ($n = 7$) but not haloperidol or placebo. Acute dystonias did not occur in patients given olanzapine or placebo but occurred in two patients given haloperidol.

Quality of life is said to be superior with atypical neuroleptic drugs, from an analysis of seven controlled trials and eight open trials (38). However, in a comparison of olanzapine and haloperidol in 335 patients there was no significant change in quality of life (61). Moreover, in all the studies, mean doses of haloperidol were higher than 12 mg/day, and in some studies as high as 28 mg/day. This is in accordance with the results of a double-blind, randomized, comparison of olanzapine (mean dosage 11 mg/day; $n = 159$) and haloperidol (mean dosage 11 mg/day; $n = 150$) in 309 schizophrenic patients (62). There were no significant differences between the groups in positive, negative, or total symptoms of schizophrenia, quality of life, extrapyramidal symptoms, or withdrawals. Olanzapine was associated with less akathisia but there were more reports of weight gain and significantly higher costs.

Olanzapine versus pimozide

The use of olanzapine in patients with Gilles de la Tourette syndrome has been explored in a 52-week, double-blind, crossover comparison of olanzapine (5 and 10 mg/day) and low-dose pimozide (2 and 4 mg/day) in four patients (aged 19–40 years) with a high frequency of tics (2–10/minute), vocalizations, and lack of co-morbidity (63). There was a highly significant reduction in rating scale scores for the syndrome with olanzapine 10 mg versus baseline and versus pimozide 2 mg, and a significant reduction with olanzapine 5 mg versus pimozide 4 mg; only moderate sedation was reported by one patient during olanzapine treatment while three complained of minor motor adverse effects and sedation during pimozide treatment.

Olanzapine and risperidone versus haloperidol

A prospective 6-month open study, promoted by Lilly, the market authorization holder of olanzapine, has been conducted, in which olanzapine ($n = 2128$), risperidone ($n = 417$), and haloperidol ($n = 112$) were compared (64). Age, sex, and duration of disease did not differ among the groups. The initial and overall mean daily doses were respectively: olanzapine 12.2 and 13.0 mg; risperidone 5.2 and 5.4 mg; haloperidol 13.9 and

13.6 mg. The improvements were similar in the three groups. A lower proportion of patients taking olanzapine had extrapyramidal symptoms (37%) compared with risperidone (50%) and haloperidol (76%). Weight gain was significantly more common with olanzapine (6.9%) than risperidone (1.9%) or haloperidol (0.9%).

In other studies, outcome variables have been discontinuation, relapse, and compliance (58) or quality of life (38). A re-analysis of data from two large double-blind comparisons of olanzapine with haloperidol ($n = 1996$) and of olanzapine with risperidone ($n = 336$) showed that in patients who had an initial response, there was no significant difference between olanzapine and haloperidol when outcome was measured using either 52-week relapse rates or the time to first non-compliance; after 12 months, the estimated mean times to discontinuation were 271 days and 241 days respectively (58). There were no differences between olanzapine and risperidone. However, while the dose of olanzapine was well within the recommended range, the dose of haloperidol was too high (modal doses, 13 and 12 mg/day respectively).

Olanzapine and risperidone versus promazine

Risperidone, olanzapine, and promazine have been compared in the treatment of the behavioral and psychological symptoms of dementia (65). At the end of the eighth week, there was complete regression of symptoms in 70% of those who took risperidone (mean age 77 years; dose 1–2 mg/day; n = 20), in 80% of those who took olanzapine (mean age 83 years; dose 5–10 mg/day; n = 20), and in 65% of those who took promazine (mean age 78 years; dose 50–100 mg/day; n = 20). The main adverse effects of risperidone were hypotension and somnolence (20%), dyspepsia (12%), sinus tachycardia, weakness, constipation, and extrapyramidal symptoms (8%), increased libido and disinhibition, abdominal pain and insomnia (4%). The main adverse effects of olanzapine were somnolence and weight gain (32%), dizziness and constipation (16%), postural hypotension (8%), akathisia (4%), and worsening of the blood sugar in one patient with diabetes (4%). The main adverse effects of promazine were constipation and hypotension (35%), dry mouth (30%), sinus tachycardia (25%), cognitive impairment and extrapyramidal symptoms (20%), confusion (15%), somnolence (10%), and nausea (5%). The small sample size, the short duration, the lack of blinding, and the lack of data on comparability were some of the flaws that preclude firm conclusions from this study.

Quetiapine versus haloperidol

A 6-week, multicenter, double blind, randomized comparison of quetiapine (mean dose 455 mg/day) and haloperidol (8 mg/day) has been carried out in 448 patients with schizophrenia (66). By day 42, the total score on the Positive and Negative Symptom Scale (PNSS) was equally reduced in the two groups. There were 69 withdrawals among the 221 patients who took quetiapine and 80 among the 227 who took haloperidol. Of the patients who took quetiapine, 154 (70%) had one or more spontaneously reported adverse events, whereas 171 (75%) of the haloperidol-treated patients reported adverse events. Motor and extrapyramidal effects were much more frequent among haloperidol users. In the quetiapine group the most common adverse events were somnolence (20%), insomnia (13%), and dry mouth (10%); in the haloperidol group the most frequent adverse events were akathisia (20%), insomnia (15%), and hypertonia (13%). The overall mean weight gain was 1.9 kg in those who took quetiapine and 0.3 kg in those who took haloperidol. With quetiapine, there was a fall in prolactin concentration from baseline by more than 16 µg/l and with haloperidol a rise of just under 6 µg/l during the same period. There was one death from acute heart failure in the haloperidol group.

Another similarly designed comparison of quetiapine and haloperidol in patients with schizophrenia with a partial response to other neuroleptic drugs gave similar results (67). Patients with a history of partial response to typical neuroleptic drugs and a partial or no response to fluphenazine (20 mg/day) for 1 month were randomly allocated to haloperidol 20 mg/day (*n* = 145) or quetiapine 600 mg/day (*n* = 143). By 8 weeks the total score on the PNSS was reduced to the same extent in the two groups. Similar numbers of patients withdrew during the randomized phase of the trial in the quetiapine (*n* = 32) and haloperidol (*n* = 28) treatment arms. The proportions of patients who were taking anticholinergic drugs at the end of the study were 44% and 60% for quetiapine and haloperidol respectively. The most frequently reported adverse events with quetiapine were somnolence (9.8%), postural hypotension/dizziness (7.7%), dry mouth (5.6%), increased muscle tone (5.6%), and akathisia (5.6%). In contrast, the most common adverse events with haloperidol were related to extrapyramidal signs: tremor (12%), akathisia (9.0%), hypertonia (6.9%), extrapyramidal syndrome (6.3%), and insomnia (6.2%). The mean increase in body weight was 1.4 kg with quetiapine and 0.7 kg with haloperidol.

In a study funded by Astra Zeneca, the marketing authorization holder of quetiapine, 25 patients with schizophrenia were randomly assigned double-blind to quetiapine 300–600 mg/day (*n* = 13) or haloperidol 10–20 mg/day (*n* = 12) for 6 months; 11 completed the study, 8 with quetiapine and 3 with haloperidol (68). After Bonferroni correction for 34 comparisons, only the beneficial effect of quetiapine on the revised Wechsler Adult Intelligence Scale (WAIS-R) was significant. Three patients had weight gain with quetiapine versus one with haloperidol. Two patients had raised hepatic enzymes with quetiapine. One patient taking quetiapine had tiny subcapsular blebs, which were apparent at baseline as small dots and appeared larger after 8 weeks of treatment.

Risperidone versus haloperidol
In 62 patients with co-existing psychotic and depressive symptoms, risperidone (mean dose 6.9 mg/day) was compared with a combination of haloperidol 9 mg/day and amitriptyline 180 mg/day (*n* = 61) in a multicenter, randomized, double-blind study over 6 weeks (69). The results suggested that haloperidol plus amitriptyline was superior to risperidone alone. The incidence of extrapyramidal adverse effects was slightly higher with risperidone (37%) than haloperidol plus amitriptyline (31%); the use of concurrent anticholinergic drugs was significantly higher with risperidone (37%) than haloperidol plus amitriptyline (20%).

In patients aged 18–73 years with disturbing extrapyramidal symptoms during previous neuroleptic drug treatment (*n* = 77), there was a greater reduction of parkinsonism over 8 weeks with risperidone (*n* = 40; average dose 7.4 mg/day) than with haloperidol (*n* = 37; average dose 9.9 mg/day) (70). With risperidone the most frequently mentioned adverse effects were headache (*n* = 4), oculogyric crisis (*n* = 3), and hypersalivation (*n* = 3); with haloperidol the adverse effects were sleep disorders (*n* = 4), tremor (*n* = 4), and vomiting (*n* = 3).

The effects of risperidone and haloperidol in preventing relapse in 365 patients with schizophrenia have been compared (71). The patients were randomly assigned to receive risperidone (*n* = 177; median duration 364 days; mean dose 4.9 mg/day) or haloperidol (*n* = 188; median duration 238 days; mean dose 11.7 mg/day). Primary analyses were performed on all subjects who underwent randomization and were assessed at least once during drug treatment. The risk ratio for relapse with haloperidol was 1.93 (95% CI = 1.33, 2.80), and early withdrawal for any reason was more frequent among haloperidol-treated patients (RR = 1.52; 95% CI = 1.18, 1.96). Patients who took risperidone had greater reductions in mean severities of both psychotic symptoms and extrapyramidal adverse effects than those who took haloperidol. There were adverse events in 90% of those who took risperidone and 91% of those who took haloperidol. Events that were reported in more than 10% of subjects in at least one group were somnolence (14% with risperidone; 25% with haloperidol), agitation (10% and 18%), and hyperkinesia (5% and 20%). There was a mean weight gain of 2.3 kg in those who took risperidone—similar in magnitude to the weight gain seen in short-term studies—and a mean loss of 0.73 kg in those who took haloperidol.

A qualitative analysis of extrapyramidal effects has been performed using Extrapyramidal Symptom Rating Scale (ESRC) data reported in 11 double-blind risperidone comparisons with haloperidol or placebo (72). Between-group comparisons showed no differences between placebo and risperidone 1–2 mg/day, but parkinsonism, tremor, akathisia, and sialorrhea were more likely to occur with haloperidol 1–6 mg/day than with placebo or risperidone. At risperidone dosages of more than 8 mg/day, the severity of acute extrapyramidal effects lay between those of placebo and haloperidol; the severity of tardive dyskinesia was greater with placebo than with either drug. However, others emphasized that those who received haloperidol in this study took a relatively high mean modal dose (12 mg per day) and 81% of the patients taking haloperidol took a daily dose of 7.5–20 mg (73–75). Many studies have shown that doses of haloperidol over 3.75–7.5 mg/day have no increased clinical efficacy but are associated with a significantly increased risk of extrapyramidal effects.

In a comparison of risperidone and haloperidol in patients with chronic schizophrenia, those taking risperidone had greater reductions in the mean severity of both psychotic symptoms and extrapyramidal adverse effects than those taking haloperidol (71) (SEDA-26, 53). The conclusion was that "the preponderance of the evidence now supports the use of risperidone as a first-line treatment for patients with schizophrenia" (76). Nevertheless, the use of "higher-than-optimal doses of typical drugs" was also mentioned as a caveat of the study.

A meta-analysis has compared risperidone with haloperidol (77). Six of the nine trials met criteria for inclusion, that is they were randomized and double-blind, had a duration of at least 4 weeks, and used risperidone in schizophrenic patients in a dosage range of 4–8 mg/day or in a flexible dose regimen. Risperidone was associated with higher clinical response rates (mean difference 14%; 95% CI = 5.6, 22), less prescribing of anticholinergic drugs (mean difference 18%; CI = 9.4, 26), and fewer treatment dropouts (mean difference 13%; CI = 9.4, 26). It was concluded that risperidone was more efficacious than haloperidol, suggesting both a lower incidence of extrapyramidal symptoms and improved treatment compliance.

Another meta-analysis has further compared risperidone and haloperidol (78). Almost all the randomized controlled clinical trials of risperidone versus haloperidol (number of studies = 18; number of patients = 3591) used a 20% improvement on the BPRS and PNSS to define a clinical response. However, an overall analysis using a 50% improvement showed that risperidone was better than haloperidol (OR = 1.6; 95% CI = 1.4, 1.8). Meta-regression did not show that the dose of the haloperidol comparator had a significant effect on positive or negative symptoms. The authors stated that previous meta-analyses had underestimated the risperidone-comparator difference, because they included the suboptimal 1 mg or 2 mg dose in their comparisons, which were not included in this one. Others have suggested that the apparent advantage of risperidone in some trials may be explained by the use of doses of typical drugs that are less than optimal, leading to poorer tolerability and effectiveness (76).

Two other studies have addressed the effects of neuroleptic drugs in the treatment of delirium, defined as an alteration in mental status that is characterized by disturbance of consciousness and attention, cognition, and perception for a brief period of time. In 28 patients with delirium, randomly assigned to a flexible-dose regimen of haloperidol (mean age 67; n = 12; starting dose 0.75 mg) or risperidone (mean age 66; n = 12; starting dose 0.5 mg) for 7 days, the response to haloperidol was 75% and to risperidone 42%, although the difference was not significant, probably owing to the small sample size and short duration of the study (79). One patient taking haloperidol had mild akathisia. In the second randomized study, there was similar clinical improvement with haloperidol (mean age 63 years; mean dose 6.5 mg/day; n = 45) and olanzapine (mean age 68 years; mean dose 4.5 mg/day; n = 28) over 5 days (80). Among the patients who received haloperidol, six rated low scores on extra-pyramidal symptoms; there were no adverse effects attributable to olanzapine.

Risperidone versus pimozide
Risperidone (n = 26, mean dosage 3.8 mg/day) has been compared with pimozide (n = 24, mean dosage 2.9 mg/day) in a 12-week, multicenter, double-blind, parallel-group study in patients with Tourette's disorder (81). Tics significantly improved in both groups, as measured by the Tourette's Symptom Severity Scale (TSSS) and there was also significant improvement in scales of Global Assessment of Functioning and Clinical Global Impressions, and in symptoms of anxiety and depressive mood; however, obsessive-compulsive behavior was significantly improved only by risperidone. Fewer patients who took risperidone reported extrapyramidal adverse effects (n = 4) compared with those who took pimozide (n = 8), and depression, fatigue, and somnolence were the most prominent adverse effects in both groups.

Risperidone versus typical drugs
When compared with typical neuroleptic drugs, risperidone may be more acceptable to those with schizophrenia. It may also have marginal benefits in terms of clinical improvement and its adverse effects profile. The superiority of risperidone in these respects may have been overestimated, owing to possible publication bias in favor of risperidone. Any marginal benefit has to be balanced against the greater cost of risperidone and its increased tendency to cause other adverse effects, such as weight gain.

Tiapride versus haloperidol
Tiapride has been assessed for the treatment of agitation and aggressiveness in elderly patients with cognitive impairment in a multicenter, double-blind study, in which patients were randomly allocated to tiapride 100 mg/day (n = 102), haloperidol 2 mg/day (n = 101), or placebo (n = 103) (82). The percentage of responders after 21 days, according to the Multidimensional Observation Scale for Elderly Subjects (MOSES) irritability/aggressiveness subscale, was significantly greater in both of the active treatment groups (haloperidol 63%, tiapride 69%) than in the placebo group (49%). There were 10 dropouts in the tiapride group, 21 in the haloperidol group, and 16 in the placebo group. The number of patients with at least one extrapyramidal symptom was significantly smaller with tiapride (16%) than with haloperidol (34%) and identical to that with placebo (17%); there was no significant difference across the groups in the numbers of patients with endocrinological adverse events. Four deaths were reported: one with placebo (stroke), one with tiapride (pneumonia), and two with haloperidol (stroke and heart failure).

Ziprasidone versus haloperidol
Intramuscular ziprasidone has recently been compared with intramuscular haloperidol in the treatment of acute psychosis for a very short period (83). Patients were randomly allocated to intramuscular ziprasidone for up to 3

days of flexible dosing (n = 90; last oral daily dose 91 mg) or haloperidol (n = 42; last oral daily dose 14 mg) followed by oral treatment to day 7. Mean reductions from baseline in all efficacy variables were significantly greater with ziprasidone than with haloperidol at the end of the study. The percentage of patients who had any adverse event was lower with ziprasidone (46%) than with haloperidol (60%); most of the adverse effects were mild or moderate. Four patients discontinued ziprasidone owing to adverse events compared with one in the haloperidol group.

Both ziprasidone and haloperidol reduced overall psychopathology in 301 schizophrenic outpatients aged 18–64 years who were randomized to flexible-dose oral ziprasidone 80–160 mg/day (n = 148) or haloperidol 5–15 mg/day (n = 153) for 28 weeks in a double-blind, multicenter study (84). The median duration of treatment was 113 days with ziprasidone and 139 days with haloperidol; the rates of withdrawal due to insufficient clinical response were the same in the two groups (18%), but twice as many patients discontinued haloperidol (16%) as ziprasidone (8%) because of treatment-related adverse effects; the percentage of patients in whom any movement disorder emerged was markedly higher with haloperidol (41%) than with ziprasidone (15%). Mean changes in body weight from baseline to end-point were small and similar with ziprasidone (+0.31 kg) and haloperidol (+0.22 kg). With regard to electrocardiographic changes, the mean baseline and endpoint QT_c intervals were 398 and 404 ms with ziprasidone and 389 and 387 ms with haloperidol; no patient had a QT_c interval over 500 ms at any time. It must be stated that some patients included in the study had already been taking haloperidol, 26% in the ziprasidone group and 25% in the haloperidol group. The study was funded by Pfizer, the marketing authorization holder for ziprasidone.

Ziprasidone 160 mg/day is as effective as haloperidol 15 mg/day, and is less likely to cause extrapyramidal symptoms. Such was the conclusion of a study in which, after a single-blind washout period of 4–7 days, patients were randomly assigned to one of four dosages of ziprasidone—4 mg/day (n = 19), 10 mg/day (n = 17), 40 mg/day (n = 17), or 160 mg/day (n = 20)—or haloperidol 15 mg/day (n = 17) for 4 weeks (85). There was less use of benzatropine at any time during the study with ziprasidone 160 mg/day (15%) than with haloperidol (53%). One patient taking ziprasidone 4 mg/day had confusion and hyponatremia, and another taking 40 mg/day had seizures.

Zotepine versus typical drugs
The use of zotepine in short-term trials (4–8 weeks) has been reviewed (86). Comparisons have been carried out with haloperidol (three trials; n = 212), chlorpromazine (two trials; n = 328), perazine (two trials; n = 81), and tiotixene (one trial; n = 94). In these double-blind trials, zotepine 150–300 mg/day was as effective as typical neuroleptic drugs in controlling the symptoms of schizophrenia. The results suggested that zotepine may be effective in the management of patients with negative symptoms

and in those with treatment-resistant schizophrenia. The most common adverse effects of zotepine were constipation, dry mouth, insomnia, sleepiness, weakness, and weight gain. The incidence of extrapyramidal symptoms was 8–29%, significantly less than with haloperidol and chlorpromazine, but there were no differences between zotepine and some other typical neuroleptic drugs in this respect. There was an increase in the risk of generalized seizures with zotepine at dosages over 300 mg/day.

Zotepine versus chlorpromazine
In a trial funded by Knoll Pharmaceuticals, the market authorization holder, zotepine was compared with chlorpromazine and placebo (87). Patients with exacerbation of schizophrenia were randomly allocated to zotepine 150–300 mg/day (n = 53), chlorpromazine 300–600 mg/day (n = 53), or placebo (n = 53) for 8 weeks. Mean BPRS scores improved statistically significantly more with zotepine than chlorpromazine or placebo. During the study, 14 patients reported extrapyramidal symptoms, 5 taking zotepine, 4 taking chlorpromazine, and 5 taking placebo. In all, 99 patients (zotepine 43; chlorpromazine 33; placebo 23) reported a total of 257 adverse events (zotepine 120; chlorpromazine 85; placebo 52) during the study. The mean weight reduction with placebo (1.4 kg) was significantly different from the mean weight gain with zotepine (2.4 kg) and chlorpromazine (1.4 kg). Two patients (one taking zotepine, one taking chlorpromazine) with suspected myocardial infarction required hospitalization, but both subsequently recovered.

Zuclopenthixol versus typical drugs
Case series and reviews have suggested superior effectiveness of zuclopenthixol acetate in the acute management of disturbed behavior caused by serious mental illness. However, this seems not to have been supported by the evidence from an analysis of randomized controlled trials (88). A meta-analysis of five randomized comparisons of zuclopenthixol acetate with other neuroleptic drugs in patients with considerable behavioral disturbance showed that in all studies there was some improvement in mental state scores (BPRS; CGI), but none showed statistically significant differences between zuclopenthixol acetate and "standard treatment." In three studies there was more sedation in those who took zuclopenthixol acetate than in those allocated to haloperidol. With regard to adverse effects, the studies were not homogeneous: one study showed that people who took zuclopenthixol acetate were more likely to need antiparkinsonian drugs (OR = 6.4; CI = 1.5, 17); other studies did not show any differences in this particular outcome.

Comparisons of different atypical drugs

There have been systematic reviews from the Cochrane Collaboration (www.cochrane.org.uk) of head-to-head comparisons of atypical neuroleptic drugs in non-treatment-resistant schizophrenia.

The benefits of the atypical neuroleptic drugs in patients with bipolar disorder and the possibility that risperidone, quetiapine, olanzapine, and clozapine cause tardive dyskinesia have been studied (89), and there has been a retrospective comparison of clozapine (5 trials), risperidone (25 trials), and olanzapine (20 trials) using data from consecutive treatment trials in the Massachusetts General Hospital (90). The overall results suggested equivalent efficacy of the novel neuroleptic drugs according to changes in Clinical Global Impressions Scale scores. Extrapyramidal symptoms occurred in about 29% of cases in all groups. There were no prolactin-related adverse effects. Substantial weight gain of more than 4.5 kg was significantly more frequent in patients taking olanzapine.

Amisulpride versus risperidone
Amisulpride has been compared with risperidone in patients aged 18–65 years with an acute exacerbation of schizophrenia who were randomized double-blind to receive amisulpride 800 mg/day ($n = 115$) or risperidone 8 mg/day ($n = 113$) for 8 weeks (91). The two treatments were equally effective. Antiparkinsonian medication was begun in 23% of patients taking risperidone and in 30% of those taking amisulpride. The only notable difference was that patients taking risperidone had significant weight gain (1.4 kg) compared with those taking amisulpride (0.4 kg).

In another comparison of amisulpride and risperidone ($n = 228$) there were no significant differences in efficacy or acceptability, with the exception of agitation, which was more frequent with amisulpride (30).

Dose-ranging studies
In a double-blind, randomized, controlled study, comparison of haloperidol 2 mg/day ($n = 20$) and 8 mg/day ($n = 20$) in subjects with first-episode psychosis over 6 weeks the two doses had similar efficacy (92). Two patients developed dystonic reactions in the low-dose group and six in the high-dose group; the figures for akathisia were three and eight respectively; two developed dyskinesia, both in the high-dose group. Eleven patients did not complete the study, three in the low-dose group and eight in the high-dose group. Although the small sample size precludes firm conclusions, the results of this study are in accordance with those of a meta-analysis (SEDA-25, 53).

Clozapine versus olanzapine
Olanzapine and clozapine have been compared in a double-blind study by Lilly Research Laboratories for 18 weeks in 220 patients with schizophrenia; conclusions were based on the one-sided lower 95% confidence limit of the treatment effect observed from the primary efficacy variable (Positive and Negative Syndrome Scale (PANSS) total score) (93). The two drugs were comparably effective in neuroleptic drug-resistant patients. Significantly fewer olanzapine-treated patients (4%) withdrew owing to adverse effects than their clozapine-treated counterparts (14%). Among spontaneously reported adverse effects, increased salivation, constipation, dizziness, and nausea were reported significantly more often by patients taking clozapine, whereas only dry mouth was reported more often by patients taking olanzapine.

The efficacy and safety of olanzapine and clozapine have been studied in an open 12-week study in 18 patients with Parkinson's disease and dopaminergic drug-induced psychosis (94). All the patients who took clozapine (mean dosage 13 mg/day) completed the study, despite reporting a number of adverse effects, including somnolence, falls, orthostatic hypotension, and syncope. In contrast, early withdrawal was required in three of the nine patients who took olanzapine, owing to severe gait deterioration and drowsiness (mean dosage 3.9 mg/day). Psychotic symptoms improved in both groups. However, parkinsonism improved in the clozapine group according to the Cornell University Rating Scale (with a 20% fall in raw score and a 7.9% fall in weighted score), while the six patients who took olanzapine and who finished the study had worse parkinsonian symptoms (with a 26% worsening in raw score and a 25% worsening in weighted score).

Clozapine versus risperidone
After a baseline period of treatment with fluphenazine for a minimum of 2 weeks, 29 patients with chronic schizophrenia participated in a randomized, double-blind, 6-week comparison of clozapine and risperidone (15). Clozapine was superior to risperidone for positive symptoms and parkinsonian adverse effects. In addition, clozapine produced fewer effects on plasma prolactin than risperidone. The mean daily doses during week 6 of the trial were 404 mg of clozapine and 5.9 mg of risperidone.

Clozapine has also been compared with risperidone in 60 treatment-resistant patients with schizophrenia in India (16). There was clinical improvement (a more than 20% reduction from baseline PANSS scale scores) in 80% of the clozapine-treated patients and 67% of the risperidone-treated patients. The predominant adverse effects with clozapine ($n = 30$) were tachycardia (77%), hypersalivation (60%), sedation (60%), weight gain (43%), and constipation (30%); one patient had a seizure. The adverse effects of risperidone ($n = 30$) were constipation (50%), dry mouth (47%), weight gain (43%), akathisia (37%), insomnia (33%), tachycardia (30%), and impotence (27%). The final mean daily doses after 16 weeks of treatment were 343 mg for clozapine and 5.8 mg for risperidone.

Clozapine and risperidone have been compared in 10 patients with psychosis in Parkinson's disease, who were randomized to risperidone or clozapine for 12 weeks (95). The mean improvement in the total BPRS score was 3.0 with clozapine (mean dose 62.5 mg/day) and 6.0 with risperidone (mean dose 1.2 mg/day). The white blood cell count fell below $3.0 \times 10^9/l$ after 10 weeks in one subject taking clozapine and rose to $5.0 \times 10^9/l$ after withdrawal. One subject taking clozapine had a marked increase in rigidity and incontinence of urine after 4 weeks, and there were similar effects in a patient who took risperidone for 10 weeks. All three adverse events resolved on withdrawal.

Clozapine and risperidone have been compared in a randomized, open study in patients with schizophrenia for 10 weeks; treatment outcomes were assessed blindly and 19 patients entered the randomized phase (96). There were no significant differences between the groups in baseline or end-point positive or negative symptoms, disease severity, or global or social functioning scores, and patients' opinion on the two drugs did not differ. These results have corroborated previous evidence that risperidone may be as effective as clozapine, but it is probable that this study does not have enough power to detect a difference.

The relation between the anticholinergic effects of clozapine and risperidone and their cognitive adverse effects has been studied in 22 patients (97). Anticholinergic potency was indexed by a reduction in the receptor occupancy rates of quinuclidinyl benzylate. Patients who took clozapine ($n = 15$) had significantly higher anticholinergic concentrations than those who took risperidone ($n = 7$). However, they all had essentially equivalent scores on cognitive measures. These data suggest that anticholinergic activity distinguishes clozapine and risperidone in vivo, but that this effect is not associated with differences in global cognitive functioning.

Olanzapine versus risperidone

Olanzapine and risperidone seem to be broadly similar, according to the numbers of patients who respond to treatment (40% reduction in PANSS scores: $n = 339$, RR = 1.14, 95% CI = 0.99, 1.32) (www.cochrane.org.uk). More of those who took risperidone withdrew early ($n = 404$, RR = 1.31, 95% CI = 1.06, 1.60; NNT = 8, 95% CI = 4, 32) and they had more extrapyramidal adverse effects ($n = 339$, RR = 1.67, 95% CI = 1.14, 2.46; NNH = 8, 95% CI = 5, 33), although comparative doses of risperidone were higher than those recommended in practice.

In an open comparison of risperidone and olanzapine (mean daily doses at follow-up 4.5 and 13.8 mg respectively) in 42 schizophrenic patients, there was a greater reduction in psychotic symptoms after 6 months of treatment with risperidone, although akathisia was more frequent (98).

A re-analysis of data from a large double-blind comparison of olanzapine with risperidone ($n = 336$) showed that there were no differences between the two (58).

Olanzapine ($n = 172$) has been compared with risperidone ($n = 167$) in an international, multicenter, double-blind, parallel-group, 28-week prospective study in 339 patients who met DSM-IV criteria for schizophrenia, schizophreniform disorder, or schizoaffective disorder (99). Both olanzapine (starting dosage 15 mg/day) and risperidone (starting dosage 1 mg bd) were effective in the management of psychotic symptoms. However, olanzapine had greater efficacy for negative symptoms and overall response rate. The most common adverse events were somnolence, headache, insomnia, rhinitis, depression, and nausea. Weight gain was reported significantly more often with olanzapine, whereas nausea, amblyopia, extrapyramidal syndrome, increased salivation, attempted suicide, abnormal ejaculation, back pain, increased creatine kinase, and urinary tract infection were reported statistically significantly more often with risperidone. A significantly greater proportion of patients taking olanzapine had raised alanine transaminase activity at any time than

those taking risperidone. Similarly, significantly more patients taking olanzapine had low neutrophil counts at any time (olanzapine 4.3%, risperidone 0.6%).

The results of one of the clinical trials in which olanzapine (Eli Lilly) was compared with risperidone (Janssen Pharmaceutica) (SEDA-22, 64) gave rise to a debate between researchers of the two pharmaceuticals companies on some of the possible flaws (100–105). Since the modal dosage over the 28-week trial was 7.2 mg/day, significantly higher than that used in actual clinical practice (average dose 4.6 mg/day), the higher incidence of risperidone-associated adverse effects could have been explained by this dosage difference.

In other studies, outcome variables have been discontinuation, relapse, and compliance (58) or quality of life (38). A re-analysis of data from two large double-blind comparisons of olanzapine with haloperidol ($n = 1996$) and of olanzapine with risperidone ($n = 336$) showed that in patients who had an initial response, there was no significant difference between olanzapine and haloperidol when outcome was measured using either 52-week relapse rates or the time to first non-compliance; after 12 months, the estimated mean times to discontinuation were 271 and 241 days respectively (58). There were no differences between olanzapine and risperidone. However, while the dose of olanzapine was well within the recommended range, the dose of haloperidol was too high (modal doses 13 and 12 mg/day respectively).

The use of neuroleptic drugs in conditions other than schizophrenia

There is enormous interest in the effectiveness of neuroleptic drugs in conditions other than schizophrenia, such as autism (1) and psychosis in patients with dementia (2). Neuroleptic medication for treatment of psychosis and agitation in patients with dementia was generally effective; in double-blind, placebo-controlled trials; mean improvement rates were 61% with neuroleptic drugs and 35% with placebo. However, the number of well-designed studies in this area has been small so far.

Amisulpride

In a randomized, double-blind, multicenter trial for 8 weeks in 278 patients with depression there were no differences in efficacy or tolerability between amisulpride 50 mg and SSRIs (106).

Clozapine

Special subgroups of patients can benefit from clozapine (SEDA 24, 61). Of 10 adolescent inpatients (aged 12–17 years) with severe acute manic or mixed episodes, who did not improve after treatment with typical drugs and who were given clozapine (mean dose 143 mg/day), all responded positively after 15–28 days and adverse effects (increased appetite, sedation, enuresis, sialorrhea) were frequent but not severe enough to require reduced dosages (107). Mean weight gain after 6 months was 7 kg (11%), and neither reduced white cell counts nor epileptic seizures were reported during follow-up for 12–24 months.

In a case series in which clozapine was used as add-on medication, two patients with bipolar disorder and one with schizoaffective disorder had marked reductions in affective symptoms after clozapine had been added to pretreatment with a mood stabilizer; transient and moderate weight gain and fatigue were the only adverse effects (108).

Olanzapine

The efficacy and safety of olanzapine in disorders other than schizophrenia have been studied (SEDA-24, 67) (SEDA-25, 64) (SEDA-26, 61). In a 3-week, randomized, double-blind clinical trial, the effects of a flexible dose of olanzapine (5–20 mg/day) and divalproex (500–2500 mg/day) for the treatment of patients with acute bipolar manic or mixed episodes have been compared (109). The olanzapine treatment group ($n = 125$) had significantly greater mean improvements in mania rating and a significantly greater proportion of patients achieved protocol-defined remission. With olanzapine there was significantly more weight gain (12 versus 7.9%), dry mouth (34 versus 6.3%), increased appetite (12 versus 2.4%), and somnolence (39 versus 21%), while more cases of nausea (29 versus 10%) were reported with divalproex.

The efficacy of adding olanzapine to either valproate or lithium alone in acute manic or mixed bipolar episodes has been studied in a 6-week, double-blind, randomized, placebo-controlled trial (110). Compared with valproate or lithium alone, the addition of olanzapine provided better efficacy. Olanzapine was associated with somnolence, dry mouth, weight gain, increased appetite, tremor, and slurred speech.

Olanzapine, mean dose 5.4 mg/day, has been given to 21 patients with apathy in the absence of depression after long-term treatment with SSRIs for non-psychotic depression in an open, flexible-dose study (111). The more frequent adverse effects were sedation ($n = 12$), increased appetite ($n = 8$), stiffness ($n = 7$), edema ($n = 6$), and dry mouth ($n = 5$).

The medical records of 10 patients with a DSM-IV diagnostic of cluster B personality disorder (narcissistic personality disorder) who had received olanzapine 2.5–20 mg/day for 8 weeks have been reviewed (112). The mean Social Dysfunction and Aggression Scale score was 29 for the 8 weeks before olanzapine therapy and improved to 14 after 8 weeks of treatment. Five of the ten patients developed severe weight gain.

Perphenazine

Immediately after remission of an episode of mania treated with perphenazine + a mood stabilizer (lithium, carbamazepine, or valproate), patients were randomly assigned to 6 months of double-blind treatment in which, in addition to the mood stabilizer, they received either continued perphenazine (n = 19) or placebo (n = 18) (113). There were no between-group differences in various important demographic and clinical characteristics. Those given placebo were more likely than those who

continued to take perphenazine to complete the study (83% versus 47%), to take longer to have a depressive relapse, to remain in the study for longer, and to have akinesia, dysphoria, and parkinsonism less often The authors concluded that to continue a typical neuroleptic drug after remission from mania for an extended time may be detrimental for some patients.

Risperidone

Risperidone has been used in bipolar disorder, dementia, disruptive behavior disorder with subaverage intelligence, Tourette's syndrome, and autism (SEDA-23, 69; SEDA-25, 67; SEDA-26, 64). The efficacy and safety of low doses of risperidone in the treatment of autism and serious behavioral problems have been studied in 101 children aged 5–17 years with autistic disorder accompanied by severe tantrums, aggression, or self-injurious behavior, who were randomly assigned to risperidone for 8 weeks ($n = 49$; dosage 0.5–3.5 mg/day) or placebo ($n = 52$) (SEDA-27,5; 114). Risperidone produced a 57% reduction in the Irritability Score, compared with a 14% reduction in the placebo group; all other parameters were also significantly improved. Risperidone was associated with an average weight gain of 2.7 kg, compared with 0.8 kg with placebo; increased appetite, fatigue, drowsiness, dizziness, and drooling were more common with risperidone. In two-thirds of the children with a positive response to risperidone at 8 weeks, the benefit was maintained at 6 months.

There is no clear evidence that typical neuroleptic drugs are effective in the management of the behavioral and psychological symptoms of dementia, and a systematic review has been conducted to assess the role of atypical drugs (115). Five good-quality randomized trials (1570 elderly patients with dementia; mean age 82 years), four evaluating risperidone and one olanzapine, were identified; all had been sponsored by a pharmaceutical company. In the short term (6–12 weeks) treatment with atypical neuroleptic drugs was superior to placebo for the primary endpoint in only three of the five trials. One of the studies reported serious adverse events in 9% of participants taking placebo and in 17% of those taking risperidone; in the risperidone group, there were six cerebrovascular adverse events and none in the placebo group. Despite their short duration, most trials reported high withdrawal rates in both treatment and placebo groups. The conclusion was that, although atypical neuroleptic drugs are being used with increasing frequency, limited evidence supports the perception of improved efficacy and adverse event profiles compared with typical neuroleptic drugs.

Organs and Systems

Cardiovascular

Of 86 439 patients who had been exposed to neuroleptic drugs, 59 developed a cardiovascular adverse effect (116). Among the commonly used neuroleptic drugs, the highest rate of cardiovascular adverse effects was found for clozapine (4.5 cases per 10 000 patients) including a case of myocarditis. The study was supported by a pharmaceutical company, and since no exposure times were provided, comparative estimates cannot be calculated.

Neuroleptic drugs can reduce exercise-induced cardiac output as a result of drug-induced increases in plasma catecholamine concentrations and concurrent alpha-adrenoceptor blockade (117). Cardiomyopathy has been associated with neuroleptic drugs, including clozapine (118–120).

Hypotension

Neuroleptic drugs with alpha-blocking activity can cause hypotension. Those of high and intermediate potency, such as haloperidol and loxitane, have minimal alpha-blocking effects and should be less likely to cause hypotension, although in one report orthostatic changes (a fall of 30 mmHg) were reported with these drugs in 27% and 22% of cases respectively (SED-11, 106) (121). An exception to the relatively safe use of high-potency agents has been noted in the combination of droperidol with the narcotic fentanyl, which can cause marked hypotension (122).

Postural hypotension is particularly hazardous in susceptible patients, such as the elderly and those with depleted intravascular volume or reduced cardiovascular output. The risk of orthostatic hypotension is markedly increased after parenteral administration. The combination of alpha-adrenoceptor blockade and sedative effects may explain the increased risk of falling when taking neuroleptic drugs (SEDA-12, 52).

Significantly more low blood pressures were documented by 24-hour ambulatory blood pressure monitoring compared with typical blood pressure measurement obtained an average of 3.6 times a day in patients treated with psychotropic drugs ($n = 12$), most of which were neuroleptic drugs (123). This finding may be of clinical relevance, in view of the potential hemodynamic consequences of hypotension, especially in older patients taking more than one psychotropic drug.

Cardiac dysrhythmias

Neuroleptic drugs are vagolytic and can increase resting and exercise heart rates. Cardiac dysrhythmias have been reported, and include atrial dysrhythmias, ventricular tachycardia, and ventricular fibrillation. Bradycardia is unusual. The risk of cardiac dysrhythmias is dose-related and is increased by pre-existing cardiovascular pathology (SEDA-2, 48), interactions with other cardiovascular or psychotropic drugs (particularly the highly anticholinergic tricyclic antidepressants), increased cardiac sensitivity in the elderly, hypokalemia, and vigorous exercise.

In elderly people it is advisable to avoid low-potency neuroleptic drugs, such as thioridazine, which produce significantly more cardiographic changes than high-potency agents, such as fluphenazine (124). In any patient with pre-existing heart disease, a pretreatment electrocardiogram with routine follow-up is recommended.

QT interval prolongation due to neuroleptic drugs has been reviewed (125). It is not a class effect: among currently available agents, thioridazine and ziprasidone are associated with the greatest prolongation. Dysrhythmias are more likely to occur if drug-induced QT prolongation co-exists with other risk factors, such as individual susceptibility, congenital long QT syndromes, heart failure, bradycardia, electrolyte imbalance, overdose of a QT interval-prolonging drug, female sex, restraint, old age, hepatic or renal impairment, and slow metabolizer status; pharmacokinetic or pharmacodynamic interactions can also increase the risk of dysrhythmias. Prolongation of the QT interval occurs more often in patients taking more than 2000 mg of chlorpromazine equivalents daily (126).

In an open study in 164 patients with schizophrenia of the effect of several neuroleptic drugs on the QT interval, the study drugs were given for 21–29 days and three separate electrocardiograms were obtained after steady state had been achieved and drug concentrations were at their maximum (127). The mean changes in the QT_c interval were:

(a) thioridazine 36 ms (95% CI = 31, 41; $n = 30$);
(b) ziprasidone 20 ms (95% CI = 14, 26; $n = 31$);
(c) quetiapine 15 ms (95% CI = 9.5, 20; $n = 27$);
(d) risperidone 12 ms (95% CI = 7.4, 16; $n = 20$);
(e) olanzapine 6.8 ms (95% CI = 0.8, 13; $n = 24$);
(f) haloperidol 4.7 ms (95% CI = −2, 11; $n = 20$).

In a population-based retrospective case-control study performed in the Integrated Primary Care Information project, a longitudinal observational database, the use of neuroleptic drugs, particularly haloperidol, was associated with a significant increase in the risk of sudden death (adjusted OR = 5.0; 95% CI = 1.6, 15 for neuroleptic drugs as a whole; adjusted OR = 5.6; 95% CI = 1.6, 19 for haloperidol; number of cases and controls 775 and 6297 respectively; mean ages 71 and 69 years respectively) (128). Sudden death was defined as a natural death due to cardiac causes heralded by abrupt loss of consciousness within 1 hour after the onset of acute symptoms or unwitnessed, or an unexpected death of someone seen in a stable medical condition under 24 hours before with no evidence of a non-cardiac cause. For each case of sudden death, up to 10 controls were randomly drawn from the source population matched for age, sex, and practice.

Conflicting results have been found in two crossover studies with regard to haloperidol-induced QT interval prolongation. In the first study, QT interval prolongation was associated with sulpiride but not haloperidol (129). Eight schizophrenic patients who had been free of medication for at least 2 weeks took sulpiride 15 mg/kg for 2 weeks and then haloperidol 0.25 mg/kg for another 2 weeks. QT_c intervals during sulpiride treatment were significantly prolonged by 5.1% and 8.5% compared with haloperidol and no treatment. Conversely, in the second study there was a statistically longer mean QT_c interval with haloperidol (422 ms) than placebo (408 ms) 10 hours after haloperidol or placebo administration (130). The subjects of this study were 16 healthy volunteers who randomly took haloperidol (a single dose of 10 mg) or placebo during the first study period (4 days) and the alternative during the second period (4 days). Despite a statistically significant longer mean half-life of haloperidol (19 versus 13 hours) in poor metabolizers of CYP2D6 than in extensive metabolizers, this exposure change did not translate into marked changes in the QT_c interval.

- A 34-year-old alcoholic man with acute pancreatitis was given continuous intravenous infusion of haloperidol (2 mg/hour) for agitation; after 7 hours he received a bolus dose of haloperidol 10 mg for worsening agitation and 20 minutes later, QT interval was 560 ms (420 ms before treatment) (131). He developed torsade de pointes and ventricular fibrillation, which resolved with electric defibrillation. He was a smoker and was also taking tiapride and alprazolam for depression, in addition to pantoprazole, piperazilline + tazobactam, paracetamol, and vitamins B_1, B_6, and B_{12}.

Mechanism

Torsade de pointes, first described in 1966 by Dessertenne (132), is a potentially fatal ventricular tachydysrhythmia with a characteristic pattern of polymorphous QRS complexes, which appear to twist around the isoelectric line. It is often associated with ventricular extra beats immediately before the dysrhythmia. Torsade de pointes typically occurs in the setting of a prolonged QT interval, which includes depolarization and repolarization times. Conditions or agents that delay ventricular repolarization, causing long QT interval syndromes, can trigger the dysrhythmia. It has been hypothesized that prolongation of the QT interval is caused by early after-depolarization, which in turn may develop in response to abnormal ventricular repolarization (133–135). Evidence linking QT interval prolongation, potassium channel function, and torsade de pointes to neuroleptic drugs has been reviewed (136).

Among the non-cardiac drugs that can cause QT interval prolongation, torsade de pointes, and sudden death (137) the neuroleptic drugs have particularly been associated with conduction disturbances, torsade de pointes being one of the most worrisome (SED-14, 141; SEDA-20, 36; SEDA-21, 43; SEDA-22, 45; SEDA-23, 49; (138–141). Although the role of neuroleptic drugs in sudden death is controversial (SEDA-18, 47; SEDA-20, 36), and although there are other non-cardiac causes of this syndrome, including asphyxia, convulsions, or hyperpyrexia, QT interval prolongation and torsade de pointes provide a plausible mechanism of sudden death.

QT_c prolongation has been proposed as a predictor of sudden death, and psychiatrists are encouraged to perform electrocardiograms in patients taking high-dose neuroleptic drugs to detect conduction abnormalities, especially QT_c prolongation. It has been suggested that QT_c prolongation in itself is not necessarily an indicator of the risk of sudden death (142). Instead, QT_c dispersion, the difference between the longest and the shortest QT_c interval on the 12-lead electrocardiogram, is an indication of more extreme variability in ventricular repolarization,

which could be regarded as a better predictor of the risk of dysrhythmias.

Epidemiology

Electrocardiographic changes are relatively common during treatment with neuroleptic drugs, but there is a lack of unanimity regarding the clinical significance of these findings. The changes that are generally considered benign and non-specific are reversible after withdrawal. Potentially more serious changes include prolongation of the QT interval, depression of the ST segment, flattened T waves, and the appearance of U waves. Non-specific T wave changes are commonly seen during the mid-afternoon, and may be related to the potassium shift and other changes that result after meals, so that a prebreakfast cardiogram may be more desirable.

The prevalence of QT_c prolongation in psychiatric patients has been estimated, to assess whether it is associated with any particular neuroleptic drug (143). Electrocardiograms were obtained from 101 healthy control individuals and 495 psychiatric patients (aged 18–74 years) in various inpatient and community settings in North-East England. Exclusion criteria were atrial fibrillation, bundle-branch block, and a change in drug therapy within the previous 2 weeks (3 months for depot formulations). The threshold for QT_c prolongation (456 ms) was defined as 2 standard deviations above the mean value in the healthy controls. Values were abnormal in 40 patients. Significant independent predictors of QT_c prolongation in psychiatric patients after adjustment for potential confounding effects were aged over 65 years and the use of tricyclic antidepressants, droperidol (RR = 6.7; CI = 1.8, 25), or thioridazine (RR = 5.3; CI = 2.0, 14). Increasing neuroleptic drug dosage was also associated with an increased risk of QT_c prolongation. Abnormal QT dispersion or T wave abnormalities were not significantly associated with neuroleptic drug treatment. Based on these results, the authors recommended electrocardiographic screening, not only in patients taking high doses of any neuroleptic drug but also in those taking droperidol or thioridazine, even at low doses. In other studies, dose and age were similarly found to be predictive factors (126), as was co-administration of carbamazepine (144).

Susceptibility factors

The susceptibility factors for drug-induced torsade de pointes are: female sex, hypokalemia, bradycardia, recent conversion from atrial fibrillation (especially with a QT interval-prolonging drug), congestive heart failure, digitalis therapy, a high drug concentration, baseline QT interval prolongation, subclinical long QT syndrome, ion channel polymorphisms, and severe hypomagnesemia (145). Information on the strength of evidence linking some neuroleptic drugs to torsade de pointes has been reviewed (146).

Critically ill patients are particularly susceptible to torsade de pointes, owing to various co-morbidities, electrolyte disturbance, and many drugs, as a recent case shows (147).

- A 58-year-old woman with pneumonia and multiple co-morbidities developed disorientation, hypoxia, and respiratory failure. She was given intravenous haloperidol 70, 30, 300, 270, and 340 mg on days 1, 2, 3, 4, and 5 respectively; concomitant drugs were intravenous levofloxacin 500 mg/day for 14 days, piperacillin + tazobactam, doxycycline, midazolam, morphine, diltiazem, enoxaparin, famotidine, metoclopramide, hydroxychloroquine, and transdermal nicotine. On day 5, she developed non-sustained runs of ventricular tachycardia and torsade de pointes (QT interval 533 ms), which resolved after haloperidol was withdrawn and magnesium sulfate 4 g was given.

The authors of a review identified 45 reports containing 70 cases of torsade de pointes, most associated with neuroleptic drugs (148). Female sex, heart disease, hypokalemia, high doses of the offending agent, concomitant use of a QT interval prolonging agent, and a history of long QT syndrome were identified as susceptibility factors

Management

Adverse effects related to QT interval prolongation can be prevented by avoiding higher doses and by avoiding the following patients at risk: those with organic heart disease, particularly congestive heart failure; those with metabolic abnormalities (such as hypokalemia and hypomagnesemia); and those with sinus bradycardia or heart block. Concomitant administration of drugs that inhibit drug metabolism should also be avoided, and the potassium concentration should be controlled. If torsade de pointes is suspected, neuroleptic drugs must be withdrawn.

Magnesium sulfate 2 g (20 ml of a 10% solution intravenously) suppresses torsade de pointes (149). Most patients report flushing during this injection. A case of torsade de pointes related to high dose haloperidol and treated with magnesium has been reported (150).

- A 41-year-old woman, with liver lacerations, rib fractures, and pneumothorax after a motor vehicle accident, was given haloperidol for agitation on day 7. During the first 24 hours she received a cumulative intravenous dose of 15 mg, 70 mg on day 2, 190 mg on day 3, 160 mg on days 4 and 5, and 320 mg on day 6. An hour after the first dose of 80 mg on day 7, she had ventricular extra beats followed by 5-beat and 22-beat runs of ventricular tachycardia. The rhythm strips were consistent with polymorphous ventricular tachycardia or torsade de pointes and the QT_c interval was 610 ms (normally under 450 in women). She received intravenous magnesium sulfate 2 g. Concurrent medications included enoxaparin, famotidine, magnesium hydroxide, ampicillin/sulbactam, nystatin suspension, midazolam, and 0.45% saline with 20 mmol/l of potassium chloride. She had no further dysrhythmias after haloperidol was withdrawn. Eight days after the episode of torsade de pointes she had a QT_c interval of 426 ms.

A thorough review of neuroleptic drugs and QT prolongation included some practical observations and suggestions (151):

- there are almost no neuroleptic drugs available that do not prolong the QT interval;

- thioridazine, mesoridazine, and pimozide should not be prescribed for patients with known heart disease, a personal history of syncope, a family history of sudden death under age 40 years, or congenital long QT syndrome;

- a baseline electrocardiogram should be obtained in all patients to determine the QT_c interval as well as the presence of other abnormalities suggesting a cardiac disorder.

Cardiomyopathy

The relation between neuroleptic drug therapy and myocarditis and cardiomyopathy has been examined using the international database on adverse drug reactions run by the World Health Organization (152). Myocarditis and cardiomyopathy were reported rarely as suspected adverse drug reactions, and accounted for under 0.1% ($n = 2121$) of almost 2.5 million reports. The association of clozapine with those adverse reactions was statistically significant (231 reports out of 24 730), as was the association with "other antipsychotics" (89 of 60 775).

Venous thromboembolism

Typical neuroleptic drugs have been associated with an increased risk of venous thromboembolism; the mechanism may be related to increased platelet aggregation (153).

This association was first suggested in the 1950s after the introduction of the phenothiazines (154). Later, a 7-fold increase in the risk of idiopathic venous thromboembolism was found among users of conventional neuroleptic drugs who were under 60 years of age and had no major risk factors (155). More recently, a 6-month retrospective cohort study of residents of US nursing homes aged 65 years and over has shown that users of atypical but not typical neuroleptic drugs had an increased risk of hospitalization for venous thromboembolism compared with non-users (156). The adjusted hazard ratio was 2.0 (95% CI = 1.4, 2.8) for risperidone (43 events; 3451 person-years); 1.9 (1.1, 3.3) for olanzapine (15 events; 1279 person-years); and 2.7 (1.1, 6.3) for clozapine and quetiapine (10 events; 443 person-years); there were 439 events in non-users (50 604 person-years). Since dementia was much more prevalent among users of atypical neuroleptic drugs, confounding by indication was possible; however, the findings were confirmed after excluding residents with severe cognitive decline.

Respiratory

Gag and cough reflexes can be suppressed by neuroleptic drugs (SED-11, 107). Periodic examination of the gag reflex, particularly in patients with tardive dyskinesia, has been recommended.

Acute respiratory failure, which can be complicated by pneumonia, has been reported in psychiatric patients receiving long-term neuroleptic drugs (SED-11, 107) (157).

Pulmonary embolism without a primary focus was surprisingly frequent in cases of sudden death.

Aspiration asphyxia in patients treated with neuroleptic drugs has been described (SEDA-4, 40; SEDA-5, 51), and it has been suggested that this could have been due to laryngeal-pharyngeal dystonia (158). Patients with asthma treated with neuroleptic drugs may be at increased risk of serious complications of asthma (159).

Diaphragmatic, laryngeal, and glottal dyskinesias have been described as part of the tardive dyskinesia syndrome and can cause respiratory complications (160,161).

Nervous system

Information on neurological adverse effects in particular groups of patients, such as young patients (162) and those with Alzheimer's disease (163,164), AIDS (165), or Gilles de la Tourette syndrome (166), have been published. The use of neuroleptic drugs in patients with Alzheimer's disease is controversial because of the significant adverse effects profile associated with these drugs. Therefore, other pharmacological and psychological methods should be explored before using neuroleptic drugs in dementia. This was the conclusion of a retrospective study in which 80 patients, 40 with confirmed Alzheimer's disease and 40 with confirmed Lewy body dementia, were assessed for neuroleptic drug use and adverse effects (163). Neuroleptic drugs were used in 15 of the patients with Alzheimer's disease and 21 of those with Lewy body dementia. Only six of the latter (29%) had a definite severe sensitivity reaction to neuroleptic drugs, which included cognitive impairment, parkinsonism, drowsiness, and features of the neuroleptic malignant syndrome. All the reactions occurred within 2 weeks of new neuroleptic drug prescription or a dosage change and were associated with a reduction in survival. Certain motor disturbances, measured before neuroleptic drug treatment was begun, could be used to predict the development and severity of neuroleptic drug-induced parkinsonism in patients with Alzheimer's disease treated with very low-dose neuroleptic drugs (164). Parkinsonism occurred in 67% of the patients with Alzheimer's disease. Pretreatment instrumental, but not clinical, measurement of bradykinesia was a predictor of post-treatment parkinsonism.

The incidence and severity of neurological soft signs have been assessed in schizophrenic patients taking haloperidol (n = 37), risperidone (n = 19), clozapine (n = 34), and olanzapine (n = 18) (167). There were no significant differences across the four groups. The authors therefore suggested that neurological soft signs are independent of neuroleptic drug treatment. However, the cross-sectional character of the study, along with the small sample size, limited this conclusion.

Seizures

Neuroleptic drugs cause slowing of alpha rhythm and increased synchronization and amplitude with superimposed sharp fast activity. They also induce discharge patterns in the electroencephalogram similar to those associated with epileptic seizures of the tonic-clonic generalized or focal types (SED-11, 108; 168).

Convulsions associated with typical neuroleptic drugs are relatively rare (probably less than 1%) (SED-9, 81; 169).

Predisposing factors to neuroleptic drug-induced seizures include an abnormal electroencephalogram, pre-existing CNS abnormalities, parenteral administration of high doses, and a family history of seizures or febrile convulsions (168).

It has been suggested that the less potent sedative neuroleptic drugs (aliphatic or piperidine phenothiazines) lower the convulsive threshold more than the potent neuroleptic drugs (piperazine phenothiazines) (170). Variable and unpredictable effects on seizure activity related to butyrophenones have been reported (171).

An in vitro technique, claimed to assess the relative risks of neuroleptic drug-induced seizures, was reported to produce striking differences between neuroleptic drugs in spike activity in hippocampal slices. Tentatively, molindone, pimozide, and butaclamol were the safest compounds, based on these in vitro experiments (172).

However, they are more common with clozapine (SEDA-22, 57) and are said to occur in 0.9% of patients taking olanzapine (SEDA-24, 67). Olanzapine has been reported to cause a lowered seizure threshold (173).

- A 30-year-old man with paranoid psychosis for 5 years and seizures for 12 years (two generalized seizures a year) was switched to olanzapine 10 mg/day; he was also taking zuclopenthixol and valproate. He then had more frequent seizures, culminating in a generalized tonic-clonic seizure, which resulted in bilateral humeral head fractures. There were no metabolic or electrolyte disturbances. An electroencephalogram showed multifocal generalized epileptiform discharges similar to those seen with clozapine. They resolved on withdrawal of olanzapine and reintroduction of zuclopenthixol.

Extrapyramidal effects

Neuroleptic drugs cause several different types of extrapyramidal adverse effects:

(a) acute dystonic reactions;
(b) akathisia;
(c) parkinsonism and pseudoparkinsonism;
(d) tardive dyskinesia;
(e) tardive akathisia;
(f) tardive Tourette's syndrome;
(g) tardive dystonia;
(h) rabbit syndrome.

Except for tardive dyskinesia, the extrapyramidal adverse effects are largely reversible by giving anticholinergic drugs and withdrawing or lowering the dosage of the neuroleptic drug. These effects have been reviewed (SED-9, 78; SEDA-7, 61; SEDA-16, 40; SEDA-18, 48) (174).

The dose-response relations for a variety of adverse effects of neuroleptic drugs have been reviewed (175). Although the dose-response relations for extrapyramidal effects are not fully understood, the evidence supports a dose-related effect. The relation is probably not linear and is influenced by several factors, but it is reasonable to conclude that systematic attempts to use the lowest possible clinically effective dosage should be strongly encouraged. It is possible that there are susceptible patients in whom tardive dyskinesia can occur at a relatively low cumulative or average daily dose, while in others increasing doses beyond this range may not lead to a substantially increased risk. Since in most patients high dosages do not lead to better therapeutic effects, but are associated with more frequent and severe adverse effects, the use of the smallest dosage necessary to produce neuroleptic benefit is recommended. When sedation is required, the concurrent use of moderate doses of a neuroleptic drug with a benzodiazepine may be preferable to the use of high dosages of a neuroleptic drug alone.

The CYP2D6 genotype is not a determinant of susceptibility to acute dystonic reactions, but may be a contributory factor in neuroleptic drug-induced movement disorders, including tardive dyskinesia (176).

The hypothesis that extrapyramidal adverse effects result from neurotoxicity due to oxidative damage by neuroleptic drugs has been reviewed (SEDA-20, 39; 177).

Patients with AIDS are sensitive to the extrapyramidal adverse effects of neuroleptic drugs and have evidence of depletion of dopamine in the cerebrospinal fluid (SEDA-22, 52). Of 115 consecutive HIV-infected patients, six developed parkinsonism and three of the cases were precipitated by the use of neuroleptic drugs (165).

Since the gamut of the clinical pharmacology of tics is broad, it is often difficult to differentiate tics from other hyperkinetic movement disorders. Of 373 cases of Gilles de la Tourette syndrome, 18 had both tics and other abnormal movements; 12 were secondary to neuroleptic drug treatment (167). Akathisia was the most common movement disorder.

In 15 schizophrenic inpatients aged 16–55 years, there was a 50% probability that a patient would have a tremor when the plasma concentration of chlorpromazine was 46 ng/ml or more, corresponding to the minimum that has been associated with a good clinical response (178). The use of objective accelerometric recordings was said to improve the accuracy of diagnosis of neuroleptic drug-induced tremor. This conclusion was reached in a study in which repeated accelerometric recordings showed constant and regular waveforms and frequencies (4–7 Hz) in each of 14 patients treated with neuroleptic drugs and diagnosed as having neuroleptic drug-induced tremor (179).

In an Australian psychiatric hospital, a rating of movement disturbances was compared with standardized scales (the Abnormal Involuntary Movement Scale (AIMS), Simpson-Angus, and BARS) and the psychiatrists' clinical assessments of such disturbances in 38 consecutive patients (180). The patients were taking neuroleptic drugs and 47% fulfilled the Simpson-Angus criteria for the presence of parkinsonian symptoms compared with 39% who were rated as having mild, moderate, or severe parkinsonian symptoms by their psychiatrists. When BARS was used, 11% of the patients scored some degree of akathisia versus 17% identified by the psychiatrists. The corresponding figures for dyskinesia when using AIMS were 11 versus 14%. In the light of these results,

the authors suggested that standardized measures should be introduced into routine practice.

Several studies have shown a relation between neuroleptic drug dosages, extrapyramidal adverse effects, and the degree of dopamine D_2 receptor occupancy (SEDA-18, 48) (181,182). Atypical neuroleptic drugs, such as olanzapine, quetiapine, risperidone, and sertindole, which have lower affinities for D_2 receptors, cause fewer extrapyramidal effects than typical neuroleptic drugs (183,185,186). However, there are reports of extrapyramidal effects associated with these atypical neuroleptic drugs (187–189).

A comprehensive review of neuroleptic drug-induced abnormal movements focused on older patients (190). Since there is no effective treatment for patients with tardive dyskinesia once it develops, attention should be paid to its prevention and close monitoring. It was confirmed that striatal D_2 receptor occupancy is an important mediator of response and adverse effects in neuroleptic drug treatment (191). In a double-blind study, 22 patients with first-episode schizophrenia were randomly assigned to a starting dose of haloperidol 1 or 2.5 mg/day. After 2 weeks, D_2 receptor occupancy was determined with raclopride and positron emission tomography; the clinical response, extrapyramidal adverse effects, and prolactin concentrations were also measured. The patients had a wide range of D_2 receptor occupancy (38–87%). The likelihoods of clinical response, hyperprolactinemia, and extrapyramidal adverse effects and akathisia increased significantly as D_2 receptor occupancy exceeded 65%, 72%, and 78% respectively. Since 65–70% D_2 receptor occupancy was obtained with haloperidol 2.5 mg/day in most of the patients, the authors suggested that a dose of 2–3 mg/day should be the optimal starting dose for first-episode patients, which contrasts with the 10–20 mg/day reported in other studies.

Patients with catatonic schizophrenia are highly vulnerable to negative symptoms related to neuroleptic drugs, according to the results of a study in 1528 schizophrenic patients, of whom 51 had catatonic schizophrenia (192). Similarly, patients with frontotemporal lobar degeneration, commonly associated with behavioral disturbances, may be particularly sensitive to extrapyramidal adverse effects. This was observed in 100 patients with such degeneration (193). In 61 patients there were significant behavioral disturbances; of those, 24 were taking neuroleptic drugs and eight of the 24 developed extrapyramidal adverse effects; the effects were severe in five cases and in one case resulted in impaired consciousness.

Extrapyramidal symptoms have been identified at much higher rates in psychotic youths than in comparable adult populations (194). Subjects who selected because of prominent positive psychotic symptoms were randomly assigned to double-blind, parallel treatment with risperidone (mean age 15 years; mean dose at termination 4 mg; n = 19), olanzapine (mean age, 14.6 years; mean dose at termination, 12.3 mg; n = 16) or haloperidol (mean age 15 years; mean dose at termination 5 mg; n = 15) for 8 weeks; in all, 88% of those who took olanzapine, 74% of those who took risperidone, and 53% of those who took haloperidol met the response criteria. A large proportion of those in each treatment group required low-dose anticholinergic drugs to control their extrapyramidal symptoms (risperidone 53%; olanzapine 56%; haloperidol 67%). There was significant weight gain in all treatment groups (mean increases: risperidone 4.9 kg; olanzapine 7.1 kg; haloperidol 3.5 kg). Withdrawals were 47% with risperidone, 13% with olanzapine, and 47% with haloperidol.

In a prospective study, 111 patients with schizophrenia and schizoaffective disorders were evaluated by means of a semistructured interview, providing information about childhood (up to 11 years), early adolescence (12–15 years), late adolescence (16–18 years), and adulthood (19 years and above) (195). There were no differences in the incidences of parkinsonism, akathisia, or dystonia among the three different premorbid functioning groups (stable good, deteriorating, and chronically poor premorbid functioning), but there were differences in tardive dyskinesia. Moreover, Premorbid Adjustment Scale scores predicted susceptibility to tardive dyskinesia but not to other extrapyramidal symptoms. Possible explanations for these findings are that tardive dyskinesia has a different pathophysiology from other extrapyramidal symptoms or that there is a more severe illness subtype characterized by more negative symptoms, which might account for the observation of or0facial dyskinesias before medication.

As early as 1989, the World Health Organization recommended that neuroleptic drugs should be initiated without routinely adding anticholinergic drugs prophylactically (SEDA-21, 44). However, since anticholinergic drugs are used in this way in India, a study has been conducted to investigate whether Indians are more susceptible to extrapyramidal adverse effects (196). Of 71 consecutive patients who were taking conventional neuroleptic drugs and who were repeatedly evaluated over 2 months, 68 had extrapyramidal symptoms while taking 2–13 mg/day of haloperidol equivalents, the most common being tremor (49%), cogwheel rigidity (40%), and acute dystonias (34%). The authors concluded that their findings supported consideration of routine prophylactic use of antiparkinsonian drugs in this population.

A therapeutic role for nefazodone 100 mg bd has been suggested in the treatment of neuroleptic drug-induced extrapyramidal signs, based on the results of a placebo-controlled, randomized study in 49 patients (197). There were no differences in akathisia or tardive dyskinesia between the two groups. However; it should be noted that nefazodone has been withdrawn in most countries owing to the risk of severe liver damage.

Acute dystonic reactions

Acute dystonic reactions are dramatic, acute-onset muscular spasms that occur within the first 24–48 hours after starting therapy, or in a few cases when the dosage is increased. A circadian pattern of acute dystonic reactions has been described (198). Men are more susceptible than women to this reaction, and the young more so than the elderly (199). Drug-induced dystonia can also be precipitated by emotional arousal (200,201). The muscles of the head and neck are mainly affected: opisthotonos, torticollis, oculogyric crisis, and macroglossia (all of which can occur together) are dramatic effects relieved by the use of intramuscular antiparkinsonian or antihistaminic drugs.

- A man in his late twenties, with a several-year history of intravenous heroin use, developed diplopia after he had received single doses of chlorpromazine 100 mg and ibuprofen 400 mg for anxiety (202). There was no extraocular muscle paresis and neurological examination was unremarkable. The diplopia resolved after 6–8 hours.

Temporomandibular joint dislocation has also been reported (203).

Acute laryngeal dystonia is probably under-reported but it is potentially lethal and can mimic anaphylaxis (204). It can occur within hours to days of starting therapy with neuroleptic drugs and is characterized by sudden onset of difficulty in breathing and swallowing (SEDA-22, 51).

- A 26-year-old schizophrenic woman developed throat pain and dyspnea while taking haloperidol (205).
- A 76-year-old man reported difficulty in starting to swallow after taking haloperidol (6 mg/day), sertraline (200 mg/day), lithium, and temazepam for 6 weeks for recurrent depression (206).

- A 28-year-old man with auditory hallucinations was given haloperidol 5 mg/day, biperiden 4 mg/day and diazepam 6 mg/day (207). After 9 days, the dose of haloperidol was increased to 15 mg/day and biperiden to 6 mg/day. A day later he developed severe extrapyramidal signs. The haloperidol was withdrawn and switched to thioridazine 200 mg/day, which was followed by marked improvement. Five months later he relapsed and the dosage of thioridazine was increased to 300 mg/day. A month later, 4 hours after taking thioridazine 800 mg in an attempt to relieve anxiety, he developed severe respiratory distress; he was grasping his throat with his hands and his lips were slightly cyanotic. Laryngeal dystonia was suspected, based on the clinical presentation and on the previous history. He was given biperiden lactate 5 mg intramuscularly, which produced improvement; 10 minutes later he was given another dose and the dystonia resolved.

- A 35-year-old woman with schizophrenia developed acute respiratory distress and laryngeal stridor (196). Neuroleptic drug-induced laryngeal dystonia was diagnosed. She was given biperiden lactate 10 mg intramuscularly and her symptoms resolved fully within 30 minutes. On the advice of a psychiatrist, and to sedate her for admission to a psychiatric hospital, her mother had secretly put 50 drops of haloperidol (10 mg/ml) in her food 1 hour before the symptoms appeared.

In addition, the authors reviewed 26 previously published cases of neuroleptic drug-induced laryngeal dystonia. They suggested that this condition could be the cause of unexpected deaths related to neuroleptic drug use. The differential diagnosis includes acute anaphylaxis, tardive laryngeal dystonia, airway obstruction, and respiratory dyskinesia. The diagnosis can be confirmed by laryngoscopy, which shows intermittent dystonic movements of the laryngeal musculature, with no edema. Anticholinergic drugs appear to be effective and, unlike benzodiazepines, do not compromise respiration.

A prospective study has identified cocaine as a risk factor for neuroleptic drug-induced acute dystonia (208). The study sample consisted of a high-risk group for neuroleptic-induced acute dystonia: 29 men aged 17–45 years who had used high-potency neuroleptic drugs within 24 hour of admission and had not taken neuroleptic drugs in the last month. Nine men developed acute dystonia: six of nine cocaine users and three of twenty non-users, a relative risk of 4.4 (95% CI = 1.4, 14). Cocaine users did not differ significantly in age, mean daily neuroleptic drug dose, or peak dose.

A comprehensive review has strongly advocated immediate intravenous administration of anticholinergic drugs to relieve acute dystonia.

Akathisia

Akathisia is a variant of the restless legs syndrome associated with anxiety and/or dysphoria (SEDA-19, 44) (SEDA-20, 36) (209–211). It can be confused with an exacerbation of the disorder being treated. Suicidal tendencies can occur in psychotic patients who developed neuroleptic-induced akathisia (212). On the other hand, recent evidence suggests that depressive symptoms may not be induced or worsened, and may even be reduced, by prescribing atypical neuroleptic drugs (213–215). Subjects show various degrees of restlessness and an inability to sit or stand still; in severe cases, the presentation can merge with behavioral disorders.

Akathisia has been associated with strong effects of terror, anger, and extreme anxiety, the most serious complication being a feeling of helplessness or being out of control, which might lead to suicidal ideation or attempted suicide. Five reported cases of neuroleptic drug-induced akathisia and suicidal tendencies in psychotic patients have further emphasized these risks (212). Doctors ought to pay attention to these symptoms, as suicidal tendencies disappear when symptoms of akathisia are relieved. Patients should also be told that akathisia is a treatable adverse effect of neuroleptic drugs.

Incidence rates of akathisia are reported to be 25–75%. Most cases occur within the first few days of neuroleptic drug treatment, and dosage increase has been identified as a risk factor (216).

Of 23 patients with bipolar affective disorder taking neuroleptic drugs, 15 developed akathisia and 18 had parkinsonism during their entire inpatient treatment (217). The patients initially received high-potency neuroleptic drugs (maximum 2000 mg/day of chlorpromazine equivalents) and were assessed weekly. Akathisia developed in 44% of the patients with severe headache or nausea who received prochlorperazine 10 mg intravenously ($n = 100$) within 1 hour (218). None of the 40

patients receiving other medications, who served as controls, developed akathisia.

Two cases of acute neuroleptic drug-induced akathisia in patients with traumatic paraplegia have been published (219). The authors emphasized the possibility of drug-induced akathisia when patients with traumatic paraplegia or other physical disabilities develop increasing restlessness and an inability to sit or lie still.

Akathisia observed at any time, whether treated or not, has been associated with a poor outcome. Antiparkinsonian agents are sometimes helpful, beta-blockers often more so (SEDA-19, 43) (220–222). Low-dosage mianserin has also been used in the treatment of akathisia (223).

A review of several case reports and three studies relating to akathisia and violence has been published (224). Three cases involved homicidal behavior and two suicidal behavior linked to haloperidol-induced akathisia.

- A 29-year-old man with a sociopathic personality and transvestism was treated with haloperidol after developing psychotic symptoms; he developed violent behavior, which consisted of assaulting his dog.
- A 47-year-old man with bipolar disorder had haloperidol-induced akathisia thought to be associated with an attack on an emergency room staff member and subsequent destruction of ward property after he had been admitted as an in-patient.

Among 16 patients with no history of violence, there were significantly more violent episodes during haloperidol use than during placebo or chlorpromazine. Also, in a series of patients admitted to hospital (n = 313), there were higher scores for akathisia in the violent group. There was also a non-significant increase in akathisia among 31 physically aggressive schizophrenics compared with 31 non-aggressive ones. The authors concluded that clinicians need to differentiate between akathisia manifesting as violence and generalized psychotic agitation. From the forensic point of view, aggression associated with neuroleptic drug-induced akathisia is relevant when considering a criminal defendant's mental state at the time of an alleged crime.

Prevention and treatment

Following the development of akathisia, biperiden 5 mg was given intravenously to 17 patients and intramuscularly to 6 (225). The mean times to onset of effect were 1.6 and 31 minutes, respectively, and the maximum effects occurred at 9.2 and 50 minutes. Adverse effects occurred in six patients after intravenous biperiden (slight or mild confusion, drowsiness, dizziness, palpitation, and dry mouth) and in two after intramuscular biperiden (drowsiness and dry mouth).

In 100 adults who were randomly assigned to receive a 3-minute infusion of diphenhydramine or placebo after receiving intravenous prochlorperazine 10 mg for nausea/vomiting or headache, akathisia developed in 18 of 50 subjects in the control group and in seven of 50 subjects in the diphenhydramine group (226). The only adverse effect of diphenhydramine was sedation. Diphenhydramine may be effective in akathisia by more than just an anticholinergic mechanism, since it crosses the blood–brain barrier and blocks central alpha-adreno-ceptors and muscarinic, 5-HT, and H_1 histamine receptors. This is claimed to be the first randomized controlled study of the efficacy of single-dose adjuvant diphenhydramine in preventing intravenous drug-induced akathisia.

Mirtazapine is a tetracyclic antidepressant, similar to mianserin, and is a potent $5\text{-HT}_{2A/2C}$ receptor antagonist. It has been successfully used to treat akathisia. In a double-blind, placebo-controlled study in 26 patients with schizophrenia who were receiving neuroleptic drugs, mirtazapine 15 mg/day for 5 days was associated with less akathisia (227).

- A 28-year-old man with akathisia who was taking haloperidol 10 mg/day for schizophrenia complained of leg restlessness, an inability to sit still, and a constant urge to move (228). Biperiden (4 mg bd for 5 days) with the subsequent addition of diazepam (10 mg/day for 3 days) had no effect. Mirtazapine (15 mg/day for 5 days) produced substantial relief of the subjective component of the akathisia, and the abnormal movements disappeared.

Apart from mianserin, other $5\text{-HT}_{2A/2C}$ receptor antagonists, such as ritanserin and cyproheptadine, have also been used to treat akathisia (227).

In one case, severe akathisia was abolished by passive motion when travelling as a car passenger (229).

- A 48-year-old woman developed akathisia soon after having been treated with chlorpromazine 400 mg/day for a bipolar affective disorder. Although the medication was changed a number of times over the next 4 years, there was no improvement. Fluvoxamine 150 mg/day and fluoxetine 20 mg/day seemed to aggravate the movement disorder; neither orphenadrine 200 mg/day nor procyclidine 15 mg/day affected her akathisia; nor did she experience relief from propranolol up to 320 mg/day. Both the subjective feeling of a need to move and the fidgeting in her legs were abolished within seconds of the car's beginning to move. Other movements (for example rocking in a chair or using an exercise bike) were ineffective, as were relaxation techniques and hypnosis.

Since this case emphasizes the importance of sensory input in akathisia, the authors suggested that the recurrent pacing observed in akathisia may be an attempt to alleviate the condition through sensory stimulation, rather than through motor activity.

Akathisia has been associated with iron deficiency (SEDA-17, 49). However, the rationale for iron supplementation in the treatment of akathisia is poor, and there are potential long-term adverse consequences (SEDA-20, 38). This issue has been addressed in patients with acute psychotic disorders who received neuroleptic drugs; 33 patients who developed akathisia were compared with 23 who did not (230). Serum iron was similar in the two groups but ferritin concentrations were significantly lower in akathisia; nevertheless, iron and ferritin concentrations were within the reference ranges.

Parkinsonism and pseudoparkinsonism
Parkinsonism or pseudoparkinsonism induced by neuroleptic drugs is clinically indistinguishable from postencephalitic or classical parkinsonism. It begins in the head and neck and causes loss of movement in the muscles, which spreads to the arms, producing varying degrees of akinesia and rigidity. Cases of neuroleptic drug-induced parkinsonism with dysphagia as one of the main features have been reported (SEDA-19, 44; 231–233).

The neuroleptic drugs that elicit parkinsonism are those that bind with higher affinity than dopamine to D_2 receptors, while those that cause little or no parkinsonism (clozapine, melperone, quetiapine) bind with lower affinity (234). Antiparkinsonian drugs are more often required in patients taking typical neuroleptic drugs, but they can cause objective and subjective deficits (235,236).

Parkinsonism has been studied in special groups, such as elderly patients (237), patients with bipolar disorders (217), and patients who have received acute medication (218). The incidence of neuroleptic drug-induced parkinsonism has been examined in 50 older newly medicated psychiatric patients (237). After controlling for spontaneous parkinsonism with two groups of 15 unmedicated patients and 49 healthy elderly individuals, 32% of the patients, who were taking an average of 43 mg/day chlorpromazine equivalents of a typical neuroleptic drug, met strict criteria for neuroleptic drug-induced parkinsonism.

Different strategies for the treatment of neuroleptic drug-induced parkinsonism have been reviewed (SEDA-18, 48) (SEDA-20, 40) (238,239). The WHO has recommended that anticholinergic drugs should not be given routinely to patients who are starting to take neuroleptic drugs.

Olanzapine was effective in extinguishing typical neuroleptic drug-induced tremor (240).

When the potency and doses of neuroleptic drugs are considered, atypical neuroleptic drugs are not necessarily safer than typical neuroleptic drugs in relation to the development of parkinsonism, according to the results of a population-based retrospective cohort study in 57 838 adults with dementia aged 66 years and older (241). There was incident parkinsonism (a new diagnosis of Parkinson's disease or the dispensing of an antiparkinsonian drug) in 4.3 per 100 person-years of follow-up in those taking typical neuroleptic drugs (n = 14 198; adjusted HR = 1.3; 95% CI = 1.0, 1.6), in 3.5 per 100 person-years in those taking atypical neuroleptic drugs (n = 11 571; adjusted HR = 1.0), and in 1.3 per 100 person-years in a control group taking non-neuroleptic drugs (n = 32 069; adjusted HR = 0.4; 95% CI = 0.3, 0.4). Medium and high doses of higher-potency typical neuroleptic drugs accounted for adjusted HRs of 2.2 (95% CI = 1.5, 3.1) and 2.3 (95% CI = 1.4, 3.5) respectively; similarly, medium and high doses of atypical neuroleptic drugs accounted for adjusted HRs of 1.3 (95% CI = 0.9, 1.7) and 2.1 (95% CI = 1.4, 3.0) respectively. The authors stressed the importance of these findings, given that a quarter of older adults in the study were given a high-dose agent. Since risperidone was the most commonly used atypical drug, the results might not be generalizable to all atypical drugs. These results are similar to those of a meta-analysis in which there was no evidence that atypical neuroleptic drugs were less likely to produce extrapyramidal effects than low-potency conventional neuroleptic drugs (242).

Tardive dyskinesia

DoTS classification (BMJ 2003;327:1222–5)
Dose-relation: collateral effect
Time-course: intermediate or delayed
Susceptibility factors: age; sex

Originally, tardive dyskinesia was described as comprising spontaneous irregular movements, mainly affecting the mouth and tongue; chewing, licking, and smacking movements of the tongue and lips were involved, including protrusion of the tongue outside the buccal cavity and various abnormal movements of the tongue within the buccal cavity. This condition is now also known to include choreoathetoid movements of the fingers and toes, sometimes associated with rocking movements, and akathisia; truncal muscles and respiratory muscles can be involved. Tardive dyskinesia can affect the neural control of any voluntary muscle (SEDA-22, 50). However, smooth pursuit eye movements were not related to tardive dyskinesia in patients with schizophrenia with ($n = 40$) and without ($n = 25$) the condition (243).

The tongue protrusion test, that is an inability to maintain tongue protrusion, is a measure of severity of tardive dyskinesia, and is present mainly in advanced cases; abnormal movements within the buccal cavity occur over a much wider range of severity, suggesting that abnormal tongue movements may be more helpful in the early detection of tardive dyskinesia (244).

There are insufficient data to support the use of drug holidays in detecting the risk of tardive dyskinesia, but they may help to diagnose the covert type by unmasking dyskinesia (245).

When abnormal involuntary movements occur in a patient taking a neuroleptic drug they are not always an adverse effect of the drug, but may be, at least partly, an inherent part of some psychotic illness. Five of forty-nine neuroleptic-naive patients with a first episode of schizophrenia had spontaneous dyskinesias (246) and they were found in 5.8% of individuals in 18 population studies (SED-11, 109) (247). Even higher rates have been recorded, especially among elderly patients; senile dyskinesias in two studies of drug-free elderly subjects occurred in 9% and 37% of individuals (248,249). There is also evidence that dyskinetic movements can be a feature of severe chronic schizophrenia unmodified by neuroleptic drugs (SED-11, 109; 246,250–254), which could mean that neuroleptic drugs merely trigger already latent dyskinesias in patients with schizophrenia.

Whether certain typical neuroleptic drugs are more likely than others to cause tardive dyskinesia is unknown (245), but virtually all typical neuroleptic drugs have been associated with it (255). Although there is evidence that atypical neuroleptic drugs are associated with a low risk (256,257), this needs to be evaluated in more prospective studies (256,258).

A review of the literature did not support the notion that neuroleptic drugs with central anticholinergic properties are particularly likely to cause tardive dyskinesia. However, anticholinergic antiparkinsonian drugs tend to produce reversible increases in the severity of dyskinetic movements, and the authors suggested that antiparkinsonian agents can be used as pharmacological probes in the evaluation of neuroleptic drug-induced movement disorders (259–261).

Discussion about tardive dyskinesia is necessary in the process of obtaining informed consent to treatment with neuroleptic drugs (SEDA-15, 46; SEDA-21, 42; SEDA-21, 45). The effect of education about tardive dyskinesia has been evaluated in 56 patients taking maintenance neuroleptic drugs, who completed a questionnaire assessing their knowledge of the condition (262). Education made patients more knowledgeable at 6 months, but had no effect on the clinical outcome.

In a consensus meeting the risk of tardive dyskinesia with typical and atypical neuroleptic drugs was addressed (263). There is sufficient evidence to conclude that atypical neuroleptic drugs are less likely to cause tardive dyskinesia than typical drugs, and that atypical neuroleptic drugs, apart from clozapine or ziprasidone, should be selected before typical drugs for patients with a first episode of schizophrenia or for patients whose history of response to neuroleptic drugs is not available. However, a meta-analysis has shown that atypical neuroleptic drugs have advantages over typical drugs only in studies in which haloperidol in relatively high doses was used as the comparison drug; when doses of haloperidol were below 12 mg/day, there were no advantages of the newer agents (35). Furthermore, clinical studies have not shown an advantage of haloperidol in dosages over 5 mg/day, which leads to a high occupancy of dopamine D_2 receptors, as demonstrated by positron emission tomography and single-photon emission computed tomography (186).

- A 48-year-old man with a 25-year history of paranoid schizophrenia previously treated with neuroleptic drugs, took thioproperazine 40 mg/day, haloperidol 30 mg/day, penfluridol 20 mg/week, and biperiden 8 mg/day for 10 days, and developed dysphagia, voice problems, and tremor, which increased progressively despite dosage reductions (264). He could hardly speak and often choked on his food. Neuroleptic drug-induced tardive dystonia was diagnosed, and he was given biperiden up to 12 mg/day without significant improvement. Because of a psychotic relapse, olanzapine monotherapy 5 mg/day was introduced and titrated up to 20 mg/day. During the next 20 days both the dystonia and the psychotic symptoms improved dramatically; 1 year later he was taking olanzapine 15 mg/day and was free of dystonic movements and other adverse effects.

Incidence

Tardive dyskinesia usually occurs after long-term use of neuroleptic drugs, but some cases of early onset (less than 1 year) have been reported (SEDA-6, 47; SEDA-7, 61). The incidence among patients with extrapyramidal effects is 2.32 times higher than among those without, and the risk could be even higher during the first 2 years of exposure (265). In a prospective study, the cumulative incidence of tardive dyskinesia was 5% after 1 year, 10% after 2 years, 15% after 3 years, and 19% after 4 years. The authors suggested that prevalence increases with increasing duration of neuroleptic drug exposure and that the increase is linear for at least the first 4–5 years (265). The average incidence rate of tardive dyskinesia has been estimated at 0.053 per year and the 5-year risk was 20% (SEDA-18, 49; 266). In another study, the cumulative incidence in patients over age 45 years was 26, 52, and 60% after 1, 2, and 3 years, respectively (267). Figures vary from center to center. In one study, there was a cumulative incidence of presumptive tardive dyskinesia of 6.3% after 1 year, 12% after 2 years, 14% after 3 years, and 18% after 4 years (268). In patients receiving neuroleptic drugs, the prevalence has been reported as 0.5–65% (269). One reviewer has gone so far as to suggest that the prevalence of tardive dyskinesia among all patients taking neuroleptic drugs is so low as not to justify any alarm (270). No cases of the severe form were found in a 12-year follow-up study of 99 chronically hospitalized patients who had received extensive neuroleptic drug treatment (271). Absence of severe tardive dyskinesia was reported in a group of Hungarian schizophrenic outpatients (259). Prevalences of 8.4–39% were reported in other cross-cultural studies of different ethnic groups (272–275).

A group of 261 neuroleptic drug-naive patients aged 55 years or over (mean 77 years) were identified at the time they were starting to take neuroleptic drugs (276). The length of follow-up was 3–393 (mean 115). weeks; 60 developed dyskinesias. The cumulative incidences were 25%, 34%, and 53% after 1, 2, and 3 years respectively of cumulative neuroleptic drug treatment. A greater risk of tardive dyskinesia was associated with a history of electroconvulsive therapy (ECT), higher mean daily and cumulative neuroleptic drug doses, and the presence of extrapyramidal signs early in treatment.

Although there are difficulties in establishing a reference figure for the incidence or prevalence of dyskinesia, Australian psychiatrists seem to underestimate the prevalence of tardive dyskinesia. According to a survey of 139 psychiatrists, 80% estimated the prevalence of "mild reversible" tardive dyskinesia as being 5% of those treated with neuroleptic drugs (277).

These wide variations in incidence are perhaps due to the lack of a precise objective definition of the syndrome. The availability of standardized rating scales (278) and research diagnostic criteria represents a significant advance in resolving these problems (279). A multinational study of tardive dyskinesia in Asians ($n = 982$) found that the overall prevalence of tardive dyskinesia was 17% (range 8.2–22%). There was a significant difference in tardive dyskinesia prevalence from center to center within the same ethnic group. The reasons for such a difference are unclear. Two studies showed that Afro-Americans have a greater incidence of tardive dyskinesia than whites (266,267). However, there is still a lack of convincing evidence that there are true interethnic differences in the prevalence of tardive dyskinesia (280,281).

The risk of tardive dyskinesia with typical and atypical neuroleptic drugs has been addressed in a consensus meeting (263). There is sufficient evidence to conclude that atypical neuroleptic drugs are less likely to cause tardive dyskinesia than typical drugs, and that atypical neuroleptic drugs, apart from clozapine or ziprasidone, should be selected before typical drugs for patients with a first episode of schizophrenia or for patients whose history of response to neuroleptic drugs is not available. However, a meta-analysis has shown that atypical neuroleptic drugs have advantages over typical drugs only in studies in which haloperidol in relatively high doses was used as the comparison drug; when doses of haloperidol were below 12 mg/day, there were no advantages of the newer agents (35). Furthermore, clinical studies have not shown an advantage of haloperidol in dosages over 5 mg/day, which leads to a high occupancy of dopamine D_2 receptors, as demonstrated by positron emission tomography and single-photon emission computed tomography (186).

The prevalence of tardive dyskinesia in a Xhosa population in South Africa was within the range reported in other studies, 29 of 102 patients using typical neuroleptic drugs (282) (see SEDA-20, 38 for details of the prevalences of tardive dyskinesia in different populations). Inpatients and out-patients from different clinics who had been exposed to typical neuroleptic drugs for at least 6 months and were currently taking an neuroleptic drug were screened for abnormal movements using the Abnormal Involuntary Movement Scale; other data were gathered from the patients and from a chart review. In a logistic regression model, years of treatment and total cumulative neuroleptic drug dose were significant predictors of tardive dyskinesia. Subjects with higher total consumption of foods containing antioxidants had lower rates of tardive dyskinesia; only consumption of onions was significantly associated with a reduced prevalence of tardive dyskinesia.

Susceptibility factors
Considerable information, although not entirely consistent, is available on predisposing factors (SED-11, 108) (245,269,283). The condition is apparently more common in women and the elderly; and elderly women in particular seem susceptible to a relatively severe form of the condition (SED-11, 108) (269,284). It has been suggested that estrogen plays a role (285,286). Here too, however, the epidemiology has been challenged; not all studies have confirmed that women are more vulnerable (287), although the consensus supports this conclusion.

Chronically hospitalized elderly inpatients with schizophrenia ($n = 121$; mean age 74 years) were rated for tardive dyskinesia and cognition (288). Subjects with tardive dyskinesia (60%) were older than those without. In subjects who were taking typical neuroleptic drugs ($n = 119$) there was no difference in dosage between those with and without tardive dyskinesia. Cognitive scores (Mini-Mental Status Examination) were significantly lower in the subjects with tardive dyskinesia affecting the orofacial regions. This raises the question of whether neuroleptic drug-induced brain changes underlie both the dyskinesia and cognitive impairment or whether the cognitive impairment represents a degenerative process that is itself a vulnerability factor for the emergence of tardive dyskinesia.

Conversely, tardive dyskinesia is said to be less common in young than in old patients, and a lower incidence has been observed in a retrospective chart review in 40 adolescents taking neuroleptic drugs (289). After 2 years, the figure was 18%; although comparability of those studies is far from optimal. Average daily dose, non-adherence to therapy, early age of illness, and concomitant use of antiparkinsonian drugs were associated with increased susceptibility.

Of 34 schizophrenic children and adolescents followed after a drug-free period that lasted up to 4 weeks at 2-year intervals, 17 had either withdrawal dyskinesia or tardive dyskinesia at some time. Patients who developed dyskinesia had greater premorbid impairment and a greater severity of positive symptoms at baseline, and there was a trend toward more months of neuroleptic drug exposure.

Negative symptoms of schizophrenia have been associated with the severity of orofacial tardive dyskinesia, and positive symptoms were associated with limb dyskinesia (290).

In a retrospective study in a psychiatric hospital in Curaçao, Netherlands Antilles, 133 Afro-Caribbean inpatients (mean age 52 years), with no organic disorders and a history of current use of neuroleptic drugs for at least 3 months, were assessed for tardive dyskinesia (291). The prevalence was 36%. When the number of interruptions to neuroleptic drug therapy was split into up to two and more than two, the resulting adjusted odds ratio was 3.29 (95% CI = 1.27, 8.49). Thus, the number of interruptions turned out to be the second risk factor after age. Cumulative dosages of neuroleptic or anticholinergic drugs were not risk factors.

Another factor that increases the risk of tardive dyskinesia is prolonged treatment with neuroleptic drugs (287). This has been supported by the preliminary results of a prospective study, but there are conflicting results (245,283); one outpatient study showed that the duration of treatment with neuroleptic drugs did not explain the differences in severity of tardive dyskinesia (284). However, dose may be important; one study showed that while age, sex, diagnosis, and race had no significant effects on presumptive tardive dyskinesia, it was predicted by greater neuroleptic drug exposure: each increase in neuroleptic drug dose of 100 mg chlorpromazine equivalents resulted in a 6% increase in the hazard of presumptive tardive dyskinesia (268). Furthermore, a positive correlation between tardive dyskinesia and circulating neuroleptic drug concentrations has been reported (292); however, in one study there was no significant difference in serum concentrations of thioridazine, its metabolites, or radio-receptor activity between patients with and without tardive dyskinesia (293). Histories of more and longer drug-free periods were more common in moderate and severe tardive dyskinesia than in mild

forms (294). There was a positive association between neuroleptic drug-free periods and persistent tardive dyskinesia (295).

In one study, previous neuroleptic drug use at study entry was the only significant predictor of tardive dyskinesia (296). Psychiatric outpatients aged over 45 years, who had taken neuroleptic drugs for 0–30 days ($n = 176$), were compared with 131 who had taken neuroleptic drugs for more than 30 days. The cumulative incidences of tardive dyskinesia were 23% and 37% at the end of 12 months. In the patients who had never used neuroleptic drugs before ($n = 87$), the mean cumulative incidence of tardive dyskinesia after the use of typical neuroleptic drugs (median dose 68 mg/day of chlorpromazine equivalents) was 3.4% at baseline and 5.9% at 1 and 3 months.

The relation between handedness and tardive dyskinesia has been studied. The estimated rate ratio, comparing left-handers and mixed-handers with purehanders, adjusted for confounders, was 0.25. The handedness effect was stronger for men than for women (297).

There are various other possible risk factors, for example diabetes mellitus (SEDA-16, 47) (298), organicity, affective disorder, and a history of ECT or alcohol abuse (299).

Mechanism

Pharmacogenetic assessments of neuroleptic drug-induced tardive dyskinesia have been reviewed (300). The dopamine D_3 receptor (DRD3) gene has a single nucleotide polymorphism that results in a serine to glycine amino acid substitution (Ser9Gly) in the N terminal and gives rise to allelic differences in dopamine affinity; autosomal inheritance of two polymorphic Ser9Gly alleles (2–2 genotype), but not homozygosity for the wild-type allele (1–1 genotype), was a susceptibility factor (301). The severity of tardive dyskinesia was greater in homozygotes for the glycine variant of DRD3 than in serine/serine homozygotes or serine/glycine heterozygotes.

Another polymorphism has been identified in intron 1 of the CYP1A2 gene; similarly, there is an association between the severity of tardive dyskinesia and one of the corresponding genotypes. It is said that CYP1A2 may be important in neuroleptic drug metabolism after CYP2D6 saturation during long-term treatment.

The hypothesis of oxidative damage to striatal neurons mediated by neuroleptic drug enhancement of glutamatergic neurotransmission has been tested in a case-control study (302). Several markers of excitatory neurotransmission (N-acetylaspartylglutamate, N-acetylaspartate, aspartate, and glutamate) and of oxidative damage (superoxide dismutase, protein carbonyl content, and lipid hydroperoxides) were measured in the CSF of patients with schizophrenia who had taken neuroleptic drugs chronically, and who had ($n = 11$) or had not ($n = 9$) developed tardive dyskinesia. There was an inverse correlation between CSF concentrations of aspartate and superoxide dismutase activity in tardive dyskinesia, what suggests a causative relation between enhanced excitatory amino acid neurotransmission and the oxidative damage associated with tardive dyskinesia. Another plausible model for the development of tardive dyskinesia is that lower activity of superoxide dismutase renders the striatal neurons more vulnerable to excitatory neurotransmission that is exacerbated by neuroleptic drugs, as shown in a subgroup of patients with schizophrenia.

There are conflicting results on the possible relation between plasma iron concentrations and movement disorders (SEDA-19, 45). A significant correlation between serum ferritin concentrations and the severity of choreoathetoid movements has been observed (303). All 30 subjects had a minimum lifetime cumulative exposure to typical neuroleptic drugs of 3 years. Nevertheless, and as was stated by the authors, it is unclear whether higher body iron stores exacerbate the symptoms of tardive dyskinesia or predispose to its development.

Reversibility
The rate of reversibility of tardive dyskinesia after drug withdrawal is 0–90% (269). Since patients with tardive dyskinesia rarely have subjective complaints (304), periodic assessment of dyskinetic movements is essential in making an early diagnosis and can increase the chance of reversing the disorder. Some reports are relatively encouraging regarding reversibility (305,306); the characteristics of reversible and irreversible forms have been reviewed, but no firm conclusion can be drawn (307). However, the prognosis of tardive dyskinesia was better in patients treated for a shorter duration and in those treated with lower doses (308).

Prevention and treatment
In dealing with the whole problem of dyskinesia, the pre-eminent role of prevention must be emphasized (299), particularly because treatment is so unrewarding. Various agents have been studied, including agonists and antagonists at various CNS neurotransmitters, and newer dopamine receptor antagonists, which supposedly act only at dopamine D_2 receptors (sites that are not linked to adenyl cyclase). The few supposedly positive results that have been claimed for a number of drugs must be interpreted with great caution (SEDA-18, 49) (309).

Treatment of tardive dyskinesia is often unsatisfactory, especially in severe cases. A large number of treatments have been proposed (SEDA-20, 40), including antiparkinsonian drugs, benzodiazepines, baclofen, hormones, calcium channel blockers, valproate, propranolol, opiates, cyproheptadine, tryptophan, lithium, manganese, niacin, botulinum toxin, ECT, dietary control, and biofeedback training. In an open study, 20 patients (mean age 65 years) with severe unresponsive tardive dyskinesia (mean duration 44 months, mean exposure 52 months) were treated with tetrabenazine (mean dose 58 mg/day) (310). The mean score on the AIMS motor subset, determined from videotapes, improved by 54%. Sedation was the only subjective complaint.

Abnormal lipid peroxidation and low lipid-corrected plasma concentrations of vitamin E are associated with tardive dyskinesia (311).Vitamin E has therefore been used for its antioxidant properties (SEDA-21, 47) (312–315). A meta-analysis has summarized eight double-blind, placebo-controlled studies of vitamin E in the treatment of tardive dyskinesia in 221 patients (316). Overall, vitamin E had a better effect than placebo; 28% of those who took vitamin E had a 33% or greater reduction in AIMS scores, compared with only 4.6% in the placebo arm. The number of patients included in any study was very small, never more than 28. The rationale for using vitamin E and for finding a better indicator of vitamin E deficiency has been explored (311). Since much of the vitamin E in plasma is carried in the low-density lipoprotein fraction, relating vitamin E content to the sum of cholesterol and triglycerides in the plasma could produce a high degree of specificity and sensitivity in defining vitamin E deficiency. Patients with tardive dyskinesia had lower concentrations of lipid-corrected vitamin E. However, the authors admitted that the lower concentration of vitamin E could have resulted from, predisposed to, or served as a marker of the susceptibility to tardive dyskinesia. The possible involvement of free radicals and treatment with vitamin E has been extensively reviewed (317). A further study has concluded that the addition of vitamin E to neuroleptic drug medication at the start of treatment can reduce the severity of acute neuroleptic drug-induced parkinsonism (318). This has been observed by comparing two groups of randomly allocated patients treated with neuroleptic drugs ($n = 20$) or with neuroleptic drugs plus vitamin E 600 IU/day ($n = 19$). From days 0–14, there was a mean reduction in SARS scores of 10% in those treated with vitamin E versus an increase of 78% in the comparison group. More recently, vitamin E (alpha-tocopherol), with its antioxidant properties, has been reported to be effective.

A beneficial effect of pyridoxine has been reported (319).

- A 22-year-old man with chronic organic persecutory paranoid ideation and recurrent explosive attacks had received neuroleptic drugs since the age of 7. While taking haloperidol, up to 25 mg/day, and trihexyphenidyl, up to 4 mg/day, he developed involuntary movements that were diagnosed as tardive dyskinesia: blinking, movements of the forehead and eyebrows, tongue-thrusting, licking of the lips, smacking, and chewing. He scored 27 on items 1–7 of the AIMS. He began to take pyridoxine, 200 mg/day, and after 5 days had a drastic reduction in the severity of all his movement disorders.

- A 41-year-old man had dramatic relief from tardive dyskinesia and akathisia with high-dose piracetam (320).

Reserpine has been used with apparent improvement in symptoms, but deterioration followed withdrawal (321), and reserpine has also been reported to cause the condition.

A double-blind study of propranolol showed short-term improvement, and two of four subjects responded to long-term propranolol (322); unfortunately this study has not been replicated.

Tetrahydroisoxazoylpyridinol (THIP), an analogue of gamma-aminobutyric acid (GABA), which is a GABA receptor antagonist, produced no change in tardive dyskinesia, in either a dose-finding study or a 4-week placebo-controlled study, but pre-existing parkinsonism increased significantly and eye-blinking rates fell (323); these are preliminary findings, more than a decade old, and hard to interpret.

Tardive akathisia

Akathisia can occur relatively early, within days or weeks, but sometimes occurs later. A 1986 review (324) considered 24 cases of tardive akathisia seen over a long period; in three of these the condition had appeared after only 1 or 2 months of neuroleptic drug therapy, but the remainder had been treated for at least 1 year, and seven had been treated with neuroleptic drugs for at least 5 years. Whereas tardive dyskinesia most often starts in the bucco-oral region and extends to the fingers and occasionally to the lower limbs and trunk, tardive akathisia most often affects the legs or is described as a generalized sensation throughout the limbs and trunk. Moreover, while a reduction in dosage can induce a temporary worsening of masked tardive dyskinesia and an increased dosage an improvement, the effects of dosage changes on akathisia are less certain.

Tardive Tourette's syndrome

There have been at least seven published cases of Tourette's syndrome ascribed to neuroleptic drugs and emerging either during treatment or after withdrawal. Some authors believe that a tardive Tourette-like syndrome may be a subtype of the more frequent tardive dyskinesia, because it can be masked by an increase in neuroleptic drug dosage and exacerbated by withdrawal. However, the symptoms can readily be confused with exacerbation of the underlying psychosis; misdiagnosis of the condition, at least in some of the published case reports, cannot be completely ruled out.

Tardive dystonia

Tardive dystonia is a rare, late-onset, persistent dystonia associated with neuroleptic drugs, which usually affects young men. It tends to affect the muscles of the neck, shoulder girdle, and trunk, causing opisthotonos. Sometimes, patients can become incapacitated (325,226). The incidence is 1–2% (327). Once developed, it is a very persistent disorder, with a low remission rate of only 14%; withdrawal of neuroleptic drugs increases the chances of remission.

Tardive dystonia has sometimes been thought to be a subtype of tardive dyskinesia (SED-13, 123). However, some features of this condition are clearly different from those of tardive dyskinesia (SEDA-20, 41). The diagnosis of tardive dystonia should meet the following criteria: (a) the presence of chronic dystonia; (b) a history of neuroleptic drug treatment preceding (less than 2 months) or

concurrent with the onset of dystonia; and (c) exclusion of known causes of secondary dystonia (328).

Tardive dystonia developed at any time between 4 days and 23 years after exposure to a neuroleptic drug in 107 patients who fulfilled these diagnostic criteria (329). Although the majority had a focal onset involving the craniocervical region, tardive dystonia tended to spread over the next 1–2 years and resulted in segmental or generalized dystonia in most cases.

Pleurothotonus (Pisa syndrome) is a special form of tardive dystonia that involves tonic flexion of the trunk to one side accompanied by slight backward rotation, in the absence of other dystonic symptoms. However, it can also occur immediately after the administration of neuroleptic drugs (SED-13, 123). Nine of twenty patients (mean age 40 years) who developed Pisa syndrome while taking neuroleptic drugs (mean duration 12 years) improved within 3 weeks of treatment with trihexyphenidyl 12 mg/day (330). Reduction or withdrawal of the daily dose of the neuroleptic drugs was beneficial to the remaining patients.

Antecollis is a rare form of tardive dystonia, in which there is forward bending of the neck (331). Patients with this disorder are usually quite disabled and distressed, have severe difficulties with vision, speech, swallowing, and inspiratory obstruction, and are unable to lie supine. In three reported cases (331), the patients developed the symptoms after receiving a number of neuroleptic drugs for 4 months to 14 years. Neither withdrawal of neuroleptic drugs nor the administration of anticholinergic agents affected their symptoms.

The differential diagnosis includes idiopathic torsion dystonia, parkinsonism, idiopathic torticollis, Huntington's disease, Wilson's disease, and Meige's syndrome (blepharospasm, oromandibular dystonia). A retrospective evaluation of the records of patients with idiopathic cervical dystonias ($n = 82$) and tardive cervical dystonias ($n = 20$) has been performed, in a search for clinical features that could help separate these closely related disorders (332). Despite the overall similarity, the presence of a dystonic head tremor was strongly suggestive of the idiopathic form, which was present in 42% and did not occur at all in the tardive group. A family history of dystonia (10%) was also exclusive to the idiopathic group.

A thorough and extensive review has addressed the treatment of tardive dystonia (333). For patients taking typical neuroleptic drugs, a switch to clozapine is suggested, or, if clozapine is contraindicated, to risperidone, olanzapine, or sertindole. Other alternatives are anticholinergic drugs, particularly trihexyphenidyl, benzodiazepines, tetrabenazine, reserpine, tocopherol, bromocriptine, and (in children and adolescents) baclofen. Local injection of botulinum toxin, which blocks acetylcholine release at the neuromuscular junction, is effective in the treatment of focal dystonias (SEDA-19, 45) (SEDA-22, 52). The therapeutic effects of botulinum toxin last on average for 2–6 months. In one study, botulinum toxin injections produced symptomatic relief, but benzodiazepines were associated with a poorer outcome, probably because of reverse causality, since non-remitting patients were more likely to use benzodiazepines as a second-line treatment (329). In another study there were no differences in responses to injections of botulinum toxin in patients with tardive oromandibular dystonia ($n = 24$) or idiopathic oromandibular dystonia ($n = 92$) (334). Surgical treatment, such as local denervation myectomy, thalamotomy, and pallidotomy, and deep brain stimulation can be considered only for patients with disabling dystonia and in those in whom medical treatment has failed to provide improvement. The successful treatment of tardive dystonia with low-dose levodopa plus benserazide, 50/12.5 mg tds, has been reported (335).

About 50% of patients with tardive dystonia have retrocollis; if the trunk is involved, most of the patients have back-arching opisthotonos. Five such patients experienced relief with a custom-made mechanical device that delivers constant contact to the occiput and shoulders (336).

Misinterpretation of symptoms has led to ineffective management of dystonia (337).

- A 14-year-old boy with Tourette's syndrome developed withdrawal dystonia while being treated with pimozide 12 mg/day. Increased blinking, facial pain, dystonic movements, and other facial movements at each dose reduction pointed toward withdrawal dystonia rather than toward a worsening of Tourette's syndrome.

Rabbit syndrome

Rabbit syndrome is a late-onset extrapyramidal adverse effect associated with neuroleptic drugs. It is characterized by a rapid tremor of the lips and occasionally the jaw. The movements usually respond well to antiparkinsonian agents and withdrawal of neuroleptic drugs. It has been suggested that the rabbit syndrome is the clinical converse of tardive dyskinesia (338).

Neuroleptic malignant syndrome

DoTS classification (BMJ 2003; 327:1222–5)
Dose-relation: toxic effect
Time-course: intermediate or delayed
Susceptibility factors: high doses or intravenous administration; psychiatric disorders; dehydration; postpartum

The neuroleptic malignant syndrome (SEDA-11, 47) (SEDA-14, 50) (339,340) is a rare but potentially fatal disorder characterized by muscle rigidity, hyperthermia, altered consciousness, and autonomic dysfunction (341). Severe rigidity may explain the increased body temperature, because of heat generated by the muscles, and it may contribute to a rise in serum creatine kinase activity, reflecting a risk of myoglobinuria and acute renal insufficiency. Hyperthermia can lead to dehydration and electrolyte imbalance, leaving the patient exposed to infections and other consequences. Patients with neuroleptic malignant syndrome are more likely to be agitated or dehydrated before the syndrome develops, they often need restraint or seclusion, and they have often received larger doses of neuroleptic drugs soon after hospitalization. Previous treatment with ECT increases vulnerability (342).

This description applies to most of the cases reported since 1960, but the syndrome is still poorly defined and overlaps to some extent with lethal catatonia, neuroleptic drug-induced hyperpyrexia, and sudden death due to cardiac dysrhythmias. Similar symptoms have also been reported in non-schizophrenic patients after exposure to dopamine-depleting drugs (343) and after withdrawal of indirect dopamine receptor agonists (344). There is no consensus about the boundaries, causes, or management of the syndrome (SEDA-18, 50; 345,346). Neuroleptic malignant syndrome without pyrexia has been reported (SEDA-19, 4; 347). The syndrome shares clinical features with malignant hyperthermia, which is genetically acquired, but there does not appear to be a common pathophysiological link between the two (SED-11, 111).

The occurrence of neuroleptic malignant syndrome in a patient with cancer after the use of haloperidol has given rise to some comments: (a) clinical oncologists are not familiar with the neuroleptic malignant syndrome; (b) neuroleptic malignant syndrome is difficult to diagnose, because its presentation resembles that of cancer itself, and sometimes other treatment-related complications (348).

There have been reports of neuroleptic malignant syndrome precipitated by promethazine 100 mg/day to treat neuroleptic drug-induced extrapyramidal symptoms and lorazepam 6 mg/day to treat agitation (349), after the addition of intramuscular haloperidol 23 mg to atypical neuroleptic drugs (350), and in other instances in children and adolescents (351).

Cases of neuroleptic malignant syndrome with different features have been published (352–356).

- A 21-year-old woman had bilateral dislocations after being struck by an automobile (352). Postoperatively, she had signs of delirium and agitated behavior, and was given haloperidol 3–5 mg intravenously or intramuscularly and lorazepam 2 mg intravenously. Her general muscular tone increased during the next several days, leading to spontaneous dislocations/subluxations of multiple joints; there was mild hyperthermia (38.4° C), a raised neutrophil count (12×10^9/l) in the absence of an identifiable source of infection, and a raised creatine kinase activity (361 U/l; reference range 0–150). Neuroleptic malignant syndrome was diagnosed. Haloperidol and droperidol were withdrawn, and she was given dantrolene sodium 1 mg/kg orally every 8 hours for 14 days. Within 24 hours her temperature had fallen and during the next week her spasticity progressively improved.

In two other cases there were features of neuroleptic malignant syndrome within 10 days of autotransplantation, with rapid improvement after withdrawal of the neuroleptic drugs (353). Since neuroleptic drugs are often used during transplantation, it is important to recognize that the use of these drugs in concert with the physiological stress of the operation can result in this drug-related complication. In two other cases of neuroleptic malignant syndrome, computerized tomography showed evidence of cerebral edema, which is claimed not to have been reported before (354).

Acute renal insufficiency has been reported as a severe complication of neuroleptic malignant syndrome (SED-13, 124) (SEDA-21, 49), and a patient who developed severe hyponatremia and progressed to neuroleptic malignant syndrome, myoglobinuria, and acute renal insufficiency deserves attention (355).

- A 42-year-old man with a history of paranoid schizophrenia developed confusion and vomiting. His medications included buspirone and tiotixene. He also had a generalized seizure that resolved spontaneously. His serum sodium concentration was 114 mmol/l, which was thought to be secondary to psychogenic polydipsia and which was corrected. The next day his temperature rose to 40° C and the day after that he became lethargic and non-verbal and developed generalized muscle rigidity. A diagnosis of probable neuroleptic malignant syndrome was made, and he was given dantrolene sodium 1 mg/kg intravenously; neuroleptic drugs were withdrawn. The serum creatine kinase activity was over 234 500 U/l. Urinalysis showed more than 3 g/l of protein, a trace of ketones, and hemoglobin and leukocytes. By the following day, his muscle rigidity and fever had resolved and his mental status had markedly improved.

Two patients developed non-convulsive status epilepticus and neuroleptic malignant syndrome (356). Whether this constitutes one syndrome or merely the co-existence of the two is academic, because it makes no difference to therapy.

The syndrome is reported most often in young men. It can occur suddenly days or weeks after initiating or intensifying drug therapy, and usually lasts 5–10 days after withdrawal. Although it is unusual, the syndrome can occur while neuroleptic drug dosage is being reduced (SEDA-19, 46).

The prevalence has been suggested as being less than 1% (357), but the precise frequency is unknown, and a trend toward fewer reports in recent years suggests that factors such as the use of lower dosages of neuroleptic drugs may be important. In a 1986 review it was argued that the neuroleptic malignant syndrome represents a heterogeneous group of neuroleptic drug-induced extrapyramidal syndromes with concurrent fever, and the existence of neuroleptic malignant syndrome as a discrete syndrome was questioned (358).

There is no consistent evidence that any one neuroleptic drug or class of drugs is more or less likely to produce the syndrome. It has been reported with atypical neuroleptic drugs, such as clozapine (359–365) and risperidone (366–371), although less often than with typical neuroleptic drugs. Neuroleptic malignant syndrome attributed to clozapine (19 cases) and risperidone (13 cases) has been reviewed (372). However, relatively more cases have been associated with high-potency neuroleptic drugs, and the course of the syndrome may be particularly prolonged and difficult to treat when depot forms of these drugs have been used. The authors of one review found insufficient evidence to support the concept of an atypical neuroleptic malignant syndrome with novel neuroleptic drugs (372).

Mechanism

It has been suggested that neurotransmitter abnormalities other than reduced dopaminergic function may be responsible for the neuroleptic malignant syndrome, in the light of three cases in which the syndrome occurred after neuroleptic drug administration during benzodiazepine withdrawal (373). Benzodiazepines potentiate GABA transmission and their long-term use is associated with a compensatory reduction in GABA activity; since nigral dopaminergic neurons are modulated through the action of GABAergic projection neurons, reduced GABA could facilitate the occurrence of the syndrome.

Susceptibility factors

Susceptibility factors that have been proposed for the development of neuroleptic malignant syndrome include a history of prior episodes of neuroleptic malignant syndrome, the use of high potency drugs, rapid escalation in dosage or parenteral administration of neuroleptic drugs, dehydration, agitation, catatonia, and certain conditions, including mood disorders, schizophrenia, and mental disorders due to medical conditions (SEDA-22, 52). Acute encephalitis has been suggested to be another susceptibility factor (374). It has been reported in five patients who developed neuroleptic malignant syndrome after being treated with neuroleptic drugs for the psychiatric symptoms associated with encephalitis. HIV encephalitis has in particular been associated with neuroleptic malignant syndrome (SEDA-19, 47).

The postpartum period is said to be a susceptibility factor for neuroleptic malignant syndrome. During 30 months, 11 cases of neuroleptic malignant syndrome were detected in a university general hospital in India (375). Five patients were women and in three the onset was in the postpartum period. One was taking a small dose of oral chlorpromazine (20 mg). In contrast, agitation and ECT, which have been proposed as susceptibility factors (SEDA-22, 52), have been questioned as independent susceptibility factors (376,377). Since catatonia was mentioned without reporting the number of affected patients, the authors suggested that some of the agitated patients would actually correspond to catatonic patients. However, catatonia was present before the onset of neuroleptic malignant syndrome in only one subject from the neuroleptic malignant syndrome group (378). On the other hand, ECT remains, at best, an association that possibly flags the characteristics of the primary psychiatric disorder that prompted the treatment.

Malignant catatonia associated with a low serum iron concentration carries a high risk of evolving into neuroleptic malignant syndrome, going by the results of a retrospective study, in which 39 catatonic episodes in patients with low ($n = 17$) or normal ($n = 22$) serum iron concentrations were compared (379). All had been exposed to neuroleptic drugs. Hypoferremia has previously been related to the neuroleptic malignant syndrome.

Susceptibility factors that have been proposed for the development of neuroleptic malignant syndrome include, among others, a history of prior episodes (SEDA-22, 52). A case report has further illustrated that possibility (380).

- A 19-year-old man with bipolar disorder received intramuscular haloperidol 30 mg/day and chlorpromazine 300 mg/day and developed neuroleptic malignant syndrome; the neuroleptic drugs were withdrawn. One month later he had a recurrence. It transpired that he had discontinued his medication 2 weeks after discharge, but because his manic symptoms recurred his relatives had started to give him haloperidol 10 mg/day again, which led to the recurrence.

A case-control study in a psychiatric institution has shown an association between different susceptibility factors and the appearance of neuroleptic malignant syndrome (381). In a multivariate analysis of 15 patients who developed the reaction (cases) and 45 patients who did not (controls), increasing doses of neuroleptic drugs were significantly associated with an increased risk of neuroleptic malignant syndrome (OR = 44), which is in accordance with the results of other case-control studies (SEDA-22, 52). Intramuscular administration, but not environmental higher temperatures, was also a significant risk factor (OR = 36). There were no deaths. Caution is recommended in the use of intramuscular neuroleptic drugs and the use of rapidly increasing high doses, particularly in mentally retarded or agitated patients.

Prevention

Precautionary measures seem to reduce the risk of definite neuroleptic malignant syndrome in schizophrenia inpatients treated with typical neuroleptic drugs, according to the results of a study in which a group of consecutive drug-free in-patients with schizophrenia, who received various typical neuroleptic drugs for 28 days, were compared with a historical group of 192 similarly treated patients but in whom such precautionary measures were not adopted (382); the study group had a significantly lower incidence of definite neuroleptic malignant syndrome (1/657 versus 4/192). The protocol of measures included strict monitoring of signs that often precede neuroleptic malignant syndrome (i.e. altered level of consciousness, raised serum creatine kinase activity, white blood cell count, and body temperature, and altered autonomic function).

Treatment

Early recognition and prompt treatment of the neuroleptic malignant syndrome, in particular immediate recognition of new rigidity, may be the best means of arresting its progress and preventing further complications (345). If there is fever, although other possible causes should be investigated, there should be no delay in instituting appropriate treatment, and in particular anticholinergic treatment of severe rigidity. There is no proven specific treatment, but immediate withdrawal of neuroleptic drugs is essential, followed by supportive therapy and intensive monitoring of respiratory, renal, and cardiac function. Anticholinergic drugs are often used, but when the temperature exceeds 101° F, they can exacerbate fever, and other treatments may be preferred. It should be remembered that the simultaneous withdrawal of neuroleptic and anticholinergic drugs can exacerbate extrapyramidal

features and that anticholinergic drugs should if possible be continued for 1 week after the withdrawal of the neuroleptic drug. Carbamazepine (383), amantadine (384), and bromocriptine (385) have been successfully used as empirical treatments in isolated cases. However, dopamine receptor agonists are preferred when the temperature exceeds 101–103° F, and muscle contraction can be further alleviated with dantrolene or benzodiazepines (386). Botulinum toxin was used for preventing muscle contractions in one case (387).

Dantrolene may be effective in reducing muscle rigidity in neuroleptic malignant syndrome (388,389) (SEDA-20, 42), but a reliable regimen has not been established (390).

- A 39-year-old man receiving haloperidol 4 mg and trihexyphenidyl 2 mg for schizophrenia developed a fever of over 40° C and clouding of consciousness 19 days after the start of medication. Neuroleptic malignant syndrome was suspected and dantrolene 40–80 mg/day was given intravenously for 5 days, with no improvement; disseminated intravascular coagulation and acute renal insufficiency developed. Dantrolene 200 mg was given intravenously over 10 minutes. One hour later, his temperature fell to 38.4° C. Continuous intravenous infusion of dantrolene 400 mg/day was started in combination with oral bromocriptine 15 mg/day. Dantrolene was changed to the oral route after 14 days and was withdrawn after 28 days.

Six cases successfully treated with ECT have been reported (391,392).

Five patients had residual catatonia after withdrawal; two recovered gradually with supportive treatment (393). Three patients were treated with ECT; two had an initial positive response, but one died later of intercurrent pneumonia; the third did not respond.

After resolution of symptoms, neuroleptic drugs can be reintroduced safely in most patients (394). In all cases the lowest effective dose of neuroleptic drug should be used, along with anticholinergic therapy or thioridazine, which has anticholinergic effects.

It is said that the risk of recurrence on re-exposure to neuroleptic drugs (about 15–30%) can be minimized by delaying rechallenge by 2 weeks or by using a neuroleptic drug of an alternative class.

Meige's syndrome

Meige's syndrome is characterized by blepharospasm and spasm of the lower facial or oromandibular muscles (395).

- A 52-year-old woman developed Meige's syndrome 2 days after the appearance of akathisia. She had taken neuroleptic drugs for years, but her current medication had been changed to bromperidol 18 mg/day and trihexyphenidyl (benzhexol) 6 mg/day; 2 days later she developed akathisia. Oral perphenazine 12 mg and trihexyphenidyl 6 mg dose-dependently reduced the frequency of blepharospasm; the dosages of bromperidol and trihexyphenidyl were gradually reduced to 8 mg/day and 3 mg/day respectively over 3 months, by which time her symptoms had completely disappeared.

Sleep

Oversedation and/or drowsiness is the most common adverse effect of neuroleptic drugs. However, most patients develop tolerance to this effect. Patients with Alzheimer's disease, as well as having impaired cognitive function and a change in personality, have a tendency to develop sleep–wake cycle disturbances that may be aggravated by classical neuroleptic drugs.

- Haloperidol was given to a 54-year-old patient with dementia; 10 days later, the circadian rest–activity cycle began to disintegrate and there was total dysrhythmicity for over 2 months; deterioration was progressive (396). Because of extrapyramidal symptoms, haloperidol 40 mg/day was changed to clozapine 50 mg/day and treatment with donepezil was begun. Two weeks later there was sudden and rapid normalization of the rest-activity cycle, started by an apparent shift of wake-up time to earlier each day, until it attained a new stable timing; orientation and memory also improved.

Other neurological effects

Myasthenia gravis caused by neuroleptic drugs has been reported (SED-11, 110). Such cases may be due to impairment of neuromuscular transmission.

Different degrees of frontal atrophy have been observed in 31 psychotic patients who had taken neuroleptic drugs for 5 years (397). All underwent computed tomography when they were drug-naive and 5 years later. Logistic regression analysis identified neuroleptic drugs as having a significant impact on the development of frontal atrophy, and the estimated risk of atrophy increased by 6.4% for each additional 10 mg of chlorpromazine equivalents.

Sensory systems

Various neuroleptic drugs, particularly low-dose phenothiazines and thioxanthenes, commonly cause blurred vision secondary to their anticholinergic activity. This is primarily a nuisance, except in the rare patient with closed-angle glaucoma.

Of more concern are two distinct types of adverse effects in the eye, which can be produced by various neuroleptic drugs: lenticular and corneal deposits, cataract, and pigmentary retinopathy (174,398,399). Deposits in the lens or cornea probably result from deposition of melanin-drug complexes and are best detected by slit-lamp examination. These deposits are probably dose- and time-related, since they generally occur only after years of treatment. Fortunately, they are in large part benign and reversible, but if undetected they can progress to interfere with vision. They are most often reported with chlorpromazine or thioridazine and can occur in association with pigmentary changes in the skin. Non-phenothiazines appear to have minimal propensity to cause oculocutaneous reactions and may be preferred when these problems have occurred during treatment with a phenothiazine (SED-11, 113), although the patient should still be closely monitored.

- A 46-year-old schizophrenic patient, who had taken chlorpromazine 200 mg/day for 9 years and then 400 mg/day for 7 years, developed ocular pigmentation (400).

The interesting thing about this report was its striking photographs, in which multiple, small, dark pigment deposits in the conjunctiva were seen; in addition, the cornea had multiple, small, discrete, yellowish specks, and both lenses had anterior capsular opacities.

Various rarer ocular effects have been reported, including oculomotor palsies, transient myopia, optic atrophy, blue-green blindness, and night-blindness. For a long time it has been suspected that neuroleptic drugs may increase the risk of cataract (399).

A further case of cataract has been reported (401).

- A 50-year-old woman with schizophrenia had a 1-year history of gradual deterioration of vision in both eyes. For several years she had been taking chlorpromazine 300 mg/day, trifluoperazine 10 mg/day, and trihexyphenidyl 4 mg/day. Slit lamp examination showed fine, discrete, brown refractile deposits on the corneal endothelium in both eyes, and characteristic bilateral stellate cataracts with dense, dust-like brown-yellow granular deposits were noted along the suture lines in the anterior pole of the lens and obscured the visual axis.

Psychological, psychiatric

Although neuroleptic drugs are relatively effective in treating psychiatric symptoms, their mental adverse effects can impair the quality of life of some individuals. The effects of newer neuroleptic drugs on cognitive functioning and the implications for functional outcome have been summarized and methodological and conceptional issues addressed (402).

In 44 stable schizophrenic outpatients, a significant proportion of the variance was explained by a combination of protracted duration of illness and dysphoric subjective responses to neuroleptic drugs (403). In terms of the individual items on the Negative Subjective Response subscale, only the statement "I feel weird, like a zombie on medication" was associated with statistically significant differences in quality of life. Patients who agreed with this statement ($n = 10$) had a significantly poorer quality of life than those who disagreed ($n = 34$). Related to this possible impairment of the quality of life is the observation, in 45 patients with chronic schizophrenia, that neuroleptic drugs playing a role in the development of depressive symptoms and negative symptoms (214). Duration of treatment correlated positively with the depressive and the negative symptoms.

Notwithstanding this finding, the differentiation between disease-related and neuroleptic drug-induced negative symptoms, such as flat affect, can be especially difficult in some patients. To investigate the interactions among psychomotor performance, negative symptoms, and neuroleptic drug effects, dopamine D_2 receptor availability (measured by the binding of ^{123}I-iodobenzamide) psychopathology, and psychomotor reaction time have been assessed in eight drug-free and eight neuroleptic drug-treated patients with schizophrenia (404). Negative symptoms increased in the patients taking neuroleptic drugs compared with drug-free patients and correlated positively with neuroleptic blockade of dopamine D_2/D_3 receptors. Furthermore, parkinsonism correlated with a flat affect and psychomotor retardation. There were some limitations to this observational study: the sample size was small; different neuroleptic drugs were used, including risperidone (in seven patients), which may not behave in the same way than others neuroleptic drug owing to its antagonistic effect on 5-HT_2 receptors, although it has been observed that dopamine D_2 receptor occupancy and extrapyramidal adverse effects with risperidone do not differ significantly from those observed with traditional neuroleptic drugs; and the duration of disease was longer in the patients who took neuroleptic drugs.

Psychiatric abnormalities

Neuroleptic drugs can cause a depression-like syndrome. This should be differentiated from neuroleptic-induced akinesia. Long-acting drugs have been particularly implicated (SED-11, 107) (174). However, a double-blind, placebo-controlled, randomized study of long-acting neuroleptic drugs did not support this conclusion (405).

Toxic delirium caused by neuroleptic drugs with potent anticholinergic properties has been widely reported (SED-11, 107), and has been reported with low-dose clozapine (406).

Obsessive-compulsive symptoms have been described with both typical neuroleptic drugs (for example chlorpromazine) and atypical drugs (for example clozapine and risperidone) (SEDA-19, 51; SEDA-20, 47; 407,410).

Other uncommon effects include psychotic exacerbation, catatonia-like states, and Klüver–Bucy-like syndrome (SED-9, 81; SEDA-7, 67; 174).

Dysphoria is said to be an under-recognized adverse effect of neuroleptic drugs; it encompasses a variety of unpleasant subjective changes in arousal, mood, thinking, and motivation, which occur early during treatment and typically manifest as a dislike for the medication. The authors of a thorough review have suggested that dysphoria that persists over time can lead to adverse consequences, such as treatment non-adherence, substance abuse, a poor clinical outcome, increased suicidality, and compromised quality of life (408). The authors of another review recommended explicitly asking about dysphoria and observing its outcome in order to improve adherence (409); their estimate was that a mean of 30% of treated patients have dysphoria.

Cognitive effects

The effects of atypical neuroleptic drugs on cognitive functioning and the implications for functional outcome have been summarized and methodological and conceptional issues addressed (402).

Cognitive function is reduced in schizophrenia, and neuroleptic drugs can cause further impairment (411,412). However, the extent to which they do so is

disputed. In one study, the more anticholinergic drugs impaired short-term verbal memory (SEDA-18, 48) (413). Another study showed that patients with schizophrenia who took neuroleptic drugs had superior information processing compared with unmedicated patients with schizophrenia; the authors claimed that neuroleptic drugs probably do not cause, and may actually reverse, slow information processing in patients with schizophrenia (414). However, a substantial number of patients with schizophrenia declare that neuroleptic drugs slow their thinking, cause them to forget, and remove interest and motivation. These responses to neuroleptic drugs are claimed to be dysphoric and are often associated with drug-induced extrapyramidal symptoms, particularly akathisia (415). Akinesia can also contribute to feelings of apathy and diminished spontaneity, and can be difficult to distinguish from the negative symptoms of schizophrenia or the psychomotor retardation of post-psychotic depression (416).

Visuomotor testing has been performed in 76 patients with schizophrenia or a schizophreniform disorder receiving haloperidol ($n = 23$; mean dose 10 mg/day), olanzapine ($n = 26$; 10.6 mg/day), or risperidone ($n = 27$; 4.4 mg/day) (412). Cognitive function was better in patients receiving risperidone or olanzapine compared with those receiving haloperidol; patients receiving haloperidol or risperidone had more severe extrapyramidal signs than those receiving olanzapine. The authors concluded that the benefits on cognitive functioning had resulted from a direct effect and were not related to reduced extrapyramidal effects. However, the patients receiving haloperidol were older and had a longer duration of illness than those receiving olanzapine or risperidone; this precludes any firm conclusions.

A comparison between haloperidol ($n = 25$) and atypical neuroleptic drugs (clozapine, $n = 24$; olanzapine, $n = 26$; risperidone, $n = 26$) has been conducted in patients with treatment-resistant schizophrenia in a 14-week, randomized, double-blind trial (417). Global neurocognitive function improved significantly with olanzapine and risperidone, by about 8–9 "IQ-equivalent" points. According to the authors, the finding was not mediated by changes in symptoms, adverse effects, or blood drug concentrations. In the haloperidol group, higher blood concentrations were associated with less improvement in motor function.

In another study, there was a small but significant negative correlation between the dosages of typical neuroleptic drugs and "mental functioning" when the data were controlled for psychopathology; this correlation was also seen for risperidone and clozapine, but there were no significant differences between patients taking low or high dosages of atypical agents. The authors pointed out that the dosage range in patients taking atypical agents was rather narrow and it is possible that higher doses of atypical neuroleptic drugs may give different results (418).

In 207 schizophrenic patients who answered the short form of a questionnaire entitled "Subjective Well-Being under Neuroleptic Treatment" (SWN-S), those taking risperidone ($n = 26$) and olanzapine ($n = 40$) took part in a study with random assignment after a washout period of at least 3 days, whereas patients taking typical neuroleptic drugs (n = 106; predominantly haloperidol and flupenthixol) were assessed during treatment (419).

Endocrine

The effects of neuroleptic drugs on menstrual status and the relation between menstrual status and neuroleptic drug efficacy and adverse effects have been explored (420). In contrast to prior reports (SEDA-18, 50), there was not a high prevalence of menstrual irregularities or amenorrhea in 27 premenopausal women with chronic schizophrenia treated with typical neuroleptic drugs.

Prolactin

Neuroendocrine effects of neuroleptic drugs include a rise in growth hormone, inappropriate ADH and prolactin secretion, and disturbances of sex hormones (SEDA-7, 67). Galactorrhea (SEDA-20, 43) and gynecomastia can follow the rise in prolactin.

A correlation between serum concentrations of neuroleptic drugs and prolactin has been claimed (421,422).

However, no further rise in plasma prolactin concentration was observed with higher dosages of haloperidol (over 100 mg/day), which was explained as being related to saturation of the pituitary dopamine receptors by a modest amount of haloperidol (422). A low prolactin concentration during maintenance neuroleptic drug treatment predicted relapse after withdrawal, and it was suggested that serum prolactin concentration may be helpful in monitoring treatment (423). Hirsutism, amenorrhea, and a false-positive pregnancy test associated with neuroleptic treatment have also been reported (SED-11, 111) (420,424).

The effects of haloperidol and quetiapine on serum prolactin concentrations have been compared in 35 patients with schizophrenia after a 2-week washout period in a randomized study (425). There was no significant difference in prolactin concentration between the groups at the start of the study, although prolactin concentrations were significantly lower with quetiapine than haloperidol. After 6 weeks, the mean prolactin concentration was significantly increased in those taking haloperidol but not in those taking quetiapine. Two patients taking haloperidol had galactorrhea related to hyperprolactinemia.

There is concern that neuroleptic drugs may increase the risk of breast cancer because of raised prolactin concentrations. For a long time, findings did not confirm this association (426), but a Danish cohort study of 6152 patients showed a slight increase in the risk of breast cancer among schizophrenic women (427).

In eight patients receiving neuroleptic drugs, serum prolactin concentrations were grossly raised (428). The time-course of the prolactin increase was examined in 17 subjects whose prolactin concentrations rose during the first 6–9 days of treatment with haloperidol (429). The increase was followed by a plateau that persisted, with minor fluctuations, throughout the 18 days of observation.

Patients whose prolactin concentrations increased above 77 ng/ml (n = 2) had hypothyroidism, and it is known that TRH (thyrotropin) stimulates the release of prolactin (430). It was concluded that all patients should have had TSH determinations at the start of therapy with neuroleptic drugs.

Sex differences in hormone concentrations have been investigated in 47 patients (21 men and 26 women) with schizophrenia or related psychoses who were using different neuroleptic drugs (431). The median daily dose and the median body weight-adjusted daily dose were twice as high in men as in women. However, neuroleptic drug-induced hyperprolactinemia was more frequent and occurred at a lower daily dose in women. The growth hormone concentration was normal in all patients.

Some patients who take clozapine take another neuroleptic drug, and the consequences of this practice in terms of prolactin have been studied in five patients (437). After the addition of haloperidol (4 mg/day) to clozapine the mean prolactin concentration increased from 9.7 ng/ml to 16 ng/ml at week 4 and 19 ng/ml at week 6. Each subject had an increase in the percentage of D_2 receptor occupancy, and the group mean increased from 55% at baseline to 79% at week 4; the increased prolactin concentrations correlated with receptor occupancy.

The effects of haloperidol and quetiapine on serum prolactin concentrations have been compared in 35 schizophrenic patients after a 2-week washout period in a randomized study (425). There was no significant difference in prolactin concentration between the groups at the start of the study, although prolactin concentrations were significantly lower with quetiapine than haloperidol. After 6 weeks, the mean prolactin concentration was significantly increased in those taking haloperidol but not in those taking quetiapine. Two patients taking haloperidol had galactorrhea related to hyperprolactinemia.

Hyperprolactinemia with osteopenia has been attributed to haloperidol (432).

- A 58-year-old schizophrenic woman who had taken haloperidol decanoate 125 mg every 2 weeks had a mildly raised prolactin concentration (505 mIU/l; upper limit of the reference range 450 mIU/liter). Dual X-ray absorptiometry showed osteopenia in her spine and hips. She began taking alendronic acid 5 mg/day and absorptiometry at 1 year showed that her spine and hip had improved by 7% and 9%, although her prolactin concentrations remained mildly raised.

In 55 patients who had been taking neuroleptic drugs for more than 10 years, higher doses of medication were associated with increased rates of both hyperprolactinemia and bone mineral density loss, as shown by dual X-ray absortiometry of their lumbar and hip bones (433).

Three women, aged 61, 53, and 21 years, developed delusions of pregnancy while taking risperidone; their blood prolactin concentrations were 49, 78, and 52 ng/ml, respectively (reference range 2–26) (434). The relation between prolactin concentrations and osteoporosis has been studied by measuring prolactin concentrations and lumbar spine and hip bone mineral densities in women with schizophrenia taking so-called "prolactin-raising neuroleptic drugs" (n = 26; mean treatment duration, 8.4 years) or olanzapine (n = 12; mean treatment duration, 6.3 years) (435). Prolactin concentrations were 1692 IU/l and 446 IU/l respectively (reference range 50–350 IU/l). Hyperprolactinemia was associated with low bone mineral density: 95% of women with either osteopenia or osteoporosis had hyperprolactinemia, whereas only 11% of those with normal prolactin concentrations had abnormal bone mineral density.

The relation between neuroleptic drug-induced hyperprolactinemia and hypoestrogenism has been studied in 75 women with schizophrenia (436). Serum estradiol concentrations were generally reduced during the entire menstrual cycle compared with reference values. There was hypoestrogenism, defined as serum estradiol concentrations below 30 pg/ml in the follicular phase and below 100 pg/ml in the periovulatory phase, in about 60%.

Inappropriate ADH secretion

Polyuria and polydipsia have long been associated with schizophrenia, and neuroleptic drugs appear to aggravate these symptoms, sometimes with an accompanying syndrome of inappropriate ADH secretion. However, water retention and edema occur very rarely during treatment with neuroleptic drugs. Water intoxication has been reported during treatment with thioridazine and may be due to its pronounced anticholinergic properties and/or direct stimulation of the hypothalamic thirst center (438).

Metabolism

The charts of 94 long-term inpatients have been reviewed retrospectively to examine the changes in weight, fasting glucose, and fasting lipids in those taking either risperidone (n = 47) or olanzapine (n = 47) (441). The patients had increased weight, triglycerides, and cholesterol, and the changes were significantly higher with olanzapine; olanzapine but not risperidone considerably increased glucose concentrations. One case of new-onset diabetes mellitus occurred in a patient taking olanzapine.

Lipid concentrations

The effects of neuroleptic drugs on serum lipids in adults have been reviewed (439). Haloperidol and the atypical neuroleptic drugs ziprasidone, risperidone, and aripiprazole, are associated with lower risks of hyperlipidemia, whereas chlorpromazine, thioridazine, and the atypical drugs quetiapine, olanzapine, and clozapine are associated with higher risks. Treatment of the metabolic disturbances caused by neuroleptic drugs has also been reviewed (440).

In a cross-sectional study in 44 men, olanzapine had a worse metabolic risk profile than risperidone (444). The men (mean age 29 years) took olanzapine (n = 22; mean duration 18 months; mean dose 13 mg/day) or risperidone (n = 22; mean duration 17 months; mean dose 2.8 mg/day). Those who took olanzapine had significantly higher plasma triglyceride concentrations, significantly higher very low density lipoprotein cholesterol concentrations, a trend to a lower HDL cholesterol concentration, and a

trend to a higher cholesterol/HDL cholesterol ratio. Despite similar mean body weights (olanzapine 84 kg versus risperidone 81 kg), 32% of those who took olanzapine were characterized by the atherogenic metabolic triad (hyperinsulinemia and raised apolipoprotein B and small-density LDL concentrations) compared with only 5% of those who took risperidone.

Diabetes mellitus

DoTS classification (BMJ 2003; 327:1222–5)
Dose-relation: collateral reaction
Time-course: intermediate
Susceptibility factors: male sex, African–American Origin

Neuroleptic drugs, particularly atypical ones, have been associated with adverse effects on glucose regulation, including diabetes (SEDA-27, 53), and new-onset diabetes mellitus in such patients is of particular concern, owing to the associated cardiovascular morbidity and the difficulty in managing diabetes in psychiatric patients. Soon after the first neuroleptic drugs were used, associations with weight gain and diabetes were reported (442,443), including eight patients treated with clozapine (445,446) and two patients treated with olanzapine (446).

Nevertheless, even before neuroleptic drugs appeared, diabetes was observed to be more common in patients with schizophrenia (447). The rate of diabetes in patients with schizophrenia has been estimated at 6.2–8.7% (448) and at 0.8% in the general population in the USA (449).

Neuroleptic drug treatment of non-diabetic patients with schizophrenia can be associated with adverse effects on glucose regulation, as has been suggested by the results of a study in which modified oral glucose tolerance tests were performed in 48 patients with schizophrenia taking clozapine, olanzapine, risperidone, or typical neuroleptic drugs, and 31 untreated healthy control subjects (450). Newer neuroleptic drugs, such as clozapine and olanzapine, compared with typical agents, were associated with adverse effects on blood glucose regulation, which can vary in severity regardless of adiposity and age.

In 2004 the Food and Drug Administration (FDA) asked all manufacturers of atypical neuroleptic drugs to add the following warning statement describing the increased risk of hyperglycemia and diabetes in patients taking these medications (451): "Hyperglycemia, in some cases extreme and associated with ketoacidosis or hyperosmolar coma or death, has been reported in patients treated with atypical neuroleptic drugs including... [long list]. Assessment of the relationship between atypical neuroleptic use and glucose abnormalities is complicated by the possibility of an increased background risk of diabetes mellitus in patients with schizophrenia and the increasing incidence of diabetes mellitus in the general population. Given these confounders, the relationship between atypical neuroleptic use and hyperglycemia-related adverse events is not completely understood. However, epidemiological studies suggest an increased risk of treatment-emergent hyperglycemia-related adverse events in patients treated with atypical neuroleptic drugs. Precise risk estimates for hyperglycemia-related adverse events in patients treated with atypical neuroleptic drugs are not available.

Patients with an established diagnosis of diabetes mellitus who are started on atypical neuroleptic drugs should be monitored regularly for worsening of glucose control. Patients with risk factors for diabetes mellitus (e.g. obesity, family history of diabetes) who are starting treatment with atypical neuroleptic drugs should undergo fasting blood glucose testing at the beginning of treatment and periodically during treatment. Any patient treated with atypical neuroleptic drugs should be monitored for symptoms of hyperglycemia including polydipsia, polyuria, polyphagia, and weakness. Patients who develop symptoms of hyperglycemia during treatment with atypical neuroleptic drugs should undergo fasting blood glucose testing. In some cases, hyperglycemia has resolved when the atypical neuroleptic was discontinued; however, some patients required continuation of anti-diabetic treatment despite discontinuation of the suspect drug."

Whether diabetes is associated, and to what extent, with a particular neuroleptic drug or a particular type of neuroleptic drug is a matter of debate; since the two most commonly used atypical neuroleptic drugs are risperidone and olanzapine, comparison of these two drugs has been a recent focus of attention. There are more case reports for clozapine and olanzapine, but the studies discussed below have reached conflicting results; further information from clinical trials is needed.

Observational studies

Claims data for the period January 1996 to December 1997 were analysed for patients with mood disorders in two large US health plans (452). In all, 849 patients had been exposed to risperidone, 656 to olanzapine, 785 to high-potency conventional neuroleptic drugs, and 302 to low-potency conventional drugs; 2644 were untreated. Adjusted odds ratios of newly reported type 2 diabetes in patients who took risperidone or high-potency conventional neuroleptic drugs were not significantly different from those in untreated patients at 12 months; on the contrary, patients who took olanzapine or low-potency conventional neuroleptic drugs had a significantly higher risk of type 2 diabetes compared with untreated patients; the 12-month adjusted odds ratios compared with untreated patients were 4.3 (95%CI = 2.1, 8.8) in those who took olanzapine and 4.9 (95%CI = 1.9, 13) in those who took low-potency conventional neuroleptic drugs.

Olanzapine had significantly positive diabetic effects, based on both duration of treatment and dosage. All patients had been exposed to neuroleptic drugs for more than 60 days; because there was less awareness of the diabetic effects of atypical neuroleptic drugs during that period, the use of these data reduced the possibility of selection bias.

Additionally, three other epidemiological studies have identified a higher risk of diabetes associated with olanzapine (Garcia 725,Sernyak 561,623). The first of these studies included out-patients with schizophrenia treated over 4 months in 1999 (Sernyak 561). When patients who had taken atypical drugs (n = 22 648) were compared with those who had taken typical neuroleptic drugs (n = 15 984), the adjusted odds ratio of new diagnoses of diabetes was 1.09 (CI = 1.03, 1.15); and was even higher in patients under 40 years of age (OR = 1.63; CI = 1.23, 2.16; n = 3076 and n = 1105). The odds ratios for individual atypical drugs were:

- quetiapine 1.31 (1.11, 1.55; n = 955);
- clozapine 1.25 (1.07, 1.46; n = 1207);
- olanzapine 1.11 (1.04, 1.18; n = 10 970);
- risperidone 1.05 (0.98, 1.12; n = 9903).

In a nested case-control study using information from the UK General Practice Research Database 451 patients with diabetes out of 19 637 patients treated for schizophrenia were matched with 2696 controls (Koro 243). There was a significantly increased risk of diabetes in patients taking olanzapine compared with non-users of neuroleptic drugs (adjusted OR = 5.8; 95% CI = 2, 17) and compared with those taking conventional neuroleptic drugs (adjusted OR = 4.2, 95% CI = 1.5–12); there was a non-significantly increased risk in those taking risperidone compared with non-users of neuroleptic drugs (adjusted OR = 2.2, 95% CI = 0.9, 5.2) and compared with those taking conventional neuroleptic drugs (adjusted OR = 1.6; 95% CI = 0.7, 3.8).

In a comparison of olanzapine and risperidone, 319 patients out of 19 153 who were given a prescription for olanzapine between 1January 1997 and 31 December 1999 developed diabetes, compared with 217 who were given a prescription for risperidone (n = 14 793) (453). Proportional hazards analysis showed a 20% increased risk of diabetes with olanzapine relative to risperidone (RR = 1.20; CI = 1.00, 1.43).

Finally, a retrospective cohort study carried out with a database that included information from different care plans (Buse 164) showed that the incidence of diabetes mellitus in patients taking any conventional neuroleptic drug was 84 per 1000 patient-years (crude incidence 307 in 19 782) compared with 67 per 1000 patient-years (crude incidence 641 in 38 969) in patients taking any atypical neuroleptic drugs; the estimated incidence in the reference general population was 15.7 per 1000 patients-years (crude incidence 45 513 in 5 816 473). Cox proportional hazards regression, adjusted for age, sex, and duration of neuroleptic drug exposure, showed that the risk of diabetes mellitus was significantly higher for whatever neuroleptic drug than in the general population but not different between the conventional and atypical neuroleptic drugs (HR = 0.97; 95% CI = 0.84, 1.11). The hazard ratios for individual atypical neuroleptic drugs compared with haloperidol were:

- clozapine 1.31 (95% CI = 0.60, 2.86);
- risperidone 1.23 (95% CI = 1.01, 1.50);
- olanzapine 1.09 (95% CI = 0.86, 1.37);
- quetiapine 0.67 (95% CI = 0.46, 0.97).

The data cut-off point for this study was 31 August 2000. Only incident cases of diabetes mellitus that resulted in intervention with antidiabetic drugs were selected and only patients taking neuroleptic drug monotherapy were included; the average duration of neuroleptic drug treatment was not long (68–137 days). One of the main flaws of this study was the possibility of selection bias, since certain patient attributes that influence treatment selection might also affect the likelihood of diabetes mellitus.

In a recent open non-randomized study of the frequency of undiagnosed impaired fasting glucose and diabetes mellitus in 168 patients taking clozapine, 47 patients were discarded because a fasting plasma glucose was not successfully obtained (n = 20) or because they were identified as diabetic before the fasting plasma glucose screening (n = 27) (454). Of 121 patients not previously diagnosed as diabetic, 93 had normal fasting blood sugar concentrations (below 6 mmol/l or 110 mg/dl), and 2) had a raised plasma glucose concentration (fasting plasma glucose over 6 mmol/l), including seven with type 2 diabetes. Patients with hyperglycemia were significantly older and more commonly had concurrent bipolar affective disorder.

Glucose metabolism in 36 out-patients with schizophrenia aged 18–65 years taking clozapine (n = 12), olanzapine (n = 12), or risperidone (n = 12) has been examined in a cross-sectional study (455). There was no significant difference in fasting baseline plasma glucose concentrations. Those taking clozapine or olanzapine had significant insulin resistance compared with those taking risperidone. There were no significant differences in total cholesterol, high-density lipoprotein cholesterol, low-density lipoprotein cholesterol, or serum triglyceride concentrations; however, controlling for sex, there was a significant difference (clozapine > olanzapine > risperidone). Insulin resistance is a major but not necessary risk factor for type 2 diabetes; leptin, in turn, is important for the control of body weight and the authors proposed that leptin could be considered to be a link between obesity, insulin resistance syndrome, and treatment with some neuroleptic drugs. However, consistent with other results, the small sample size, the cross-sectional design, and the exclusion of obese subjects may limit the generalizability of these findings.

Randomized studies

In a randomized double-blind study, in-patients with schizophrenia were assessed for fasting glucose, cholesterol, and weight gain after 14 weeks of treatment with clozapine (n = 28), haloperidol (n = 25), olanzapine (n = 26), or risperidone (n = 22) (456). There was a statistically significant increase in mean blood glucose concentrations only in the olanzapine group; 14 out of 101 patients (14%) developed raised blood glucose concentrations (over 7 mmol/l or 125 mg/dl) at some time during the study: six while taking clozapine, one haloperidol, four olanzapine, and three risperidone. Similarly, there was a significant increase in cholesterol concentration in the olanzapine group. The largest weight gain was also observed with olanzapine (mean change 7.3 kg), followed by clozapine (4.8 kg) and risperidone (2.4 kg); there was minimal weight gain with haloperidol (0.9 kg). ANCOVA

showed no main effect or treatment interaction for the relation between change in blood glucose and weight gain at end-point; however, there was a significant main effect for the association between change in cholesterol and weight gain in the four groups. The rate of hyperglycemia (14%) in this study was about twice the rate of diabetes reported in a large survey of the US population (6–8%) (457) and somewhat higher than the current prevalence rate of 10% found in extensive samples of patients with schizophrenia (458).

Susceptibility factors and clinical features

Most of the patients who develop diabetes while taking neuroleptic drugs do so in under 6 months; men are at greater risk, and African–Americans are more susceptible than other ethnic groups (459). Weight gain can be present or not; in a retrospective series of 76 patients who developed diabetes while taking olanzapine or risperidone, the increase in fasting blood glucose concentration in patients taking olanzapine did not correlate with changes in weight (460).

Diabetic ketoacidosis is a common presentation of diabetes in patients taking atypical neuroleptic drugs. In 126 patients taking atypical neuroleptic drugs, new-onset, acute, marked glucose intolerance developed in 11 (8.7%) after treatment with clozapine, olanzapine, or quetiapine (461). Of these, six patients required insulin (four only transiently) and five developed diabetic ketoacidosis. The mean and median times to the onset of diabetic ketoacidosis after the start of treatment with an atypical neuroleptic drug were 81 and 33 days respectively. In 45 patients, 19 who presented with ketoacidosis were significantly younger, more often women, and less overweight at baseline than 26 patients who developed diabetes without ketoacidosis (462).

Death due to olanzapine-induced hyperglycemia has been reported (SEDA-27, 61).

Mechanisms

There has been some discussion about whether neuroleptic drug-induced hyperglycemia and diabetes are associated with weight gain (SEDA-25, 65; SEDA-26, 62; SEDA-27, 61). Clozapine and olanzapine have a high propensity to cause both weight gain and diabetes (SEDA-26, 56); however, there have been cases in which diabetes appeared or worsened in the absence of weight gain or even with weight loss (463,465). Both weight gain and diabetes seem to be effects that occur simultaneously, rather than diabetes being an indirect effect of weight gain. Other mechanisms have therefore been suggested (SEDA-21, 67).

Ten patients with schizophrenia taking olanzapine 7.5–20 mg/day were compared with 10 healthy volunteers with regard to body weight, fat mass, and insulin resistance over 8 weeks (466). Fasting serum glucose and insulin increased significantly in the olanzapine group, as did weight gain. The index for insulin resistance increased only in the olanzapine group, whereas beta-cell function did not change significantly. Consistent with this, it has been observed that some neuroleptic drugs inhibit glucose transport into PC12 cells in culture and increase cellular concentrations of glucose, which would in turn cause a homeostatic increase in insulin release.

The hormone leptin is synthesized by adipocytes and is important in controlling body weight; plasma leptin concentrations correlate with fat mass and insulin resistance. Plasma leptin concentrations are raised, regardless of weight, in patients taking clozapine (467).

These observations suggest that insulin resistance might be responsible for the development of diabetes during treatment with atypical neuroleptic drugs, but it may not be the only factor, since the rapid development of diabetes in many of these patients suggests a direct and possibly toxic effect of neuroleptic drugs.

Management

The relation between the metabolic syndrome and neuroleptic drug therapy has been reviewed (468,469,470). In most patients, diabetes improves or resolves after withdrawing neuroleptic drugs or switching to less diabetogenic drugs. Simple interventions, such as changes in diet and exercise, are recommended for patients with schizophrenia or bipolar disease taking neuroleptic drugs. Patients who develop conditions such as hyperlipidemia or diabetes mellitus often respond to routine treatment. However, managing diabetes in patients with schizophrenia is complicated by their lack of insight, loss of initiative, and cognitive deficits, which are central features of the illness. For patients with type 2 diabetes, the major risk is accelerated coronary heart disease and stroke, greatly aggravated by smoking, which is common in patients with schizophrenia.

In addition to considerations of the diabetogenic potential of each agent, the presence of susceptibility factors, such as obesity, sedentary lifestyle, and a family history of type 2 diabetes, should be taken into account in the selection of a suitable regimen. It is sensible to monitor fasting plasma glucose concentrations, glycosylated hemoglobin, and fasting cholesterol and triglycerides in order to anticipate hyperglycemia and hyperlipidemia. Given the serious implications for morbidity and mortality attributable to diabetes and raised cholesterol, clinicians need to be aware of these risk factors when treating patients with chronic schizophrenia.

Weight gain

Shortly after the typical neuroleptic drugs were introduced in the 1950s, marked increases in body weight were observed, and excessive weight gain has been reported in up to 50% of patients receiving long-term neuroleptic drug treatment (471–477).

However, changes in weight during psychosis are also related to the condition; Kraepelin wrote, "The taking of food fluctuates from complete refusal to the greatest voracity ··· Sometimes, in quite short periods, very considerable differences in the body weight are noticed ···" (478). It was observed early on that food intake and weight often fell as psychosis worsened, but eating and weight returned to normal or increased when an acute psychotic episode abated. However, since the start of the

neuroleptic drug era in the 1950s, a new pattern of sustained increased weight has commonly been detected. The question of whether weight gain is associated with efficacy is important; in one study there was no obvious relation between the magnitude of weight gain and therapeutic efficacy (479).

The association between clozapine-related weight gain and increased mean arterial blood pressure has been examined in 61 patients who were randomly assigned to either clozapine or haloperidol in a 10-week parallel group, double-blind study, and in 55 patients who chose to continue to take clozapine in a subsequent 1-year, open, prospective study (480). Clozapine was associated with significant weight gain in both the double-blind trial (mean 4.2 kg) and the open trial (mean 5.8 kg), but haloperidol was not associated with significant weight gain (mean 0.4 kg). There were no significant correlations between change in weight and change in mean arterial blood pressure for clozapine or haloperidol.

Serum leptin, a peripheral hormone secreted by fat, correlates inversely with body weight; in 22 patients taking clozapine who gained weight, those who had the most pronounced 2-week increase in leptin had the least gain in body weight after 6 and 8 months (481).

The published data suggest that clozapine and olanzapine are associated with considerable weight gain, risperidone and quetiapine have a moderate risk, and ziprasidone and aripiprazole a low risk (482). Moreover, obese children have an increased propensity for orthopedic, neurological, pulmonary, and gastroenterological complications. A thorough search for studies that addressed obesity in children and adolescents in relation to the new neuroleptic drugs showed that risperidone is associated with less weight gain than olanzapine (483).

Epidemiology

Several reviews have addressed the issue of weight gain (484–488). Comparison of different studies of weight gain during treatment with atypical neuroleptic drugs is hampered by problems with study design, recruitment procedures, patient characteristics, measurement of body weight, co-medications, and duration of therapy. These problems have to be considered when assessing this type of information, particularly figures collected in accordance with the last-observation-carried-forward technique, which is one of the most common approaches taken; this type of analysis can produce marked underestimates of the magnitude of weight gain.

There has been one comprehensive meta-analysis including over 80 studies and over 30 000 patients (489). A meta-analysis of trials of neuroleptic drugs showed the following mean weight gains in kg after 10 weeks of treatment: clozapine, 4.5; olanzapine, 4.2; thioridazine, 3.2; sertindole, 2.9; chlorpromazine, 2.6; risperidone, 2.1; haloperidol, 1.1; fluphenazine, 0.43; ziprasidone 0.04; molindone, −0.39; placebo, −0.74 (490,491). In one study, excessive appetite was a more frequent adverse event in patients treated with olanzapine versus haloperidol (24 versus 12%) (185). Loss of weight has been observed after withdrawal of neuroleptic drugs (492).

The percentages of patients who gain more than 7% of their baseline body weight are highest for olanzapine (29%) and lowest for ziprasidone (9.8%). Although these figures are useful for comparing different drugs, they do not illustrate how weight increases over time, nor the total gain. Body weight tends to increase at first but reaches a plateau by about 1 year; with olanzapine 12.5–17.5 mg/day, patients gained, on average, 12 kg (493).

Risperidone and olanzapine, the two most commonly used atypical drugs, have been compared in trials; patients who took olanzapine gained almost twice as much weight (4.1 kg) as those who took risperidone (2.3 kg) (99). In two groups of inpatients, who took either risperidone ($n = 50$) or olanzapine ($n = 50$) for schizophrenia the mean body weight at baseline was 83 kg in the risperidone group and 85 kg in the olanzapine group; after 4 months of treatment, the mean body weights were 83 and 87 kg respectively (494). The increase in body weight with olanzapine was statistically significant.

In a retrospective chart review of 91 patients with schizophrenia (120 treatment episodes; mean age 38 years), there was weight gain with zotepine (4.3 kg), clozapine (3.1 kg), sulpiride (1.9 kg), and risperidone (1.5 kg), but not in patients treated with typical neuroleptic drugs (mean gain 0.0–0.5 kg) (495). The mean duration of treatment was 28–34 days, the maximum weight gain being in the first 3–5 weeks. The mean increases in weight were 4.3% in the patients with normal weight, 3.0% in those with mild obesity, and 1.9% in those with severe obesity (BMI over 30).

In another retrospective study, 65 patients with schizophrenia who gained weight while taking clozapine for 6 months were followed (497). After that they were switched to a combination of clozapine with quetiapine for 10 months: clozapine was tapered to 25% of the current dose and quetiapine was added proportionately. During clozapine monotherapy the mean weight gain was 6.5 kg and 13 patients developed diabetes. At the end of the combination period, the mean weight loss was 9.4 kg; patients who developed diabetes showed significant improvement, resulting in a rapid fall in insulin requirements and/or withdrawal of insulin and replacement with an oral hypoglycemic agent. The mechanism of clozapine-associated weight gain is uncertain.

Weight gain due to atypical neuroleptic drugs has been addressed in a cross-sectional study of the weight at the time of a single visit compared with that recorded in the clinical chart (496). The proportion of patients with clinically relevant weight gain ($\geq 7\%$ increase versus initial weight) was higher with olanzapine (46%; n = 228) than with risperidone (31%; n = 232) or haloperidol (22%; n = 130); no patient had clinically relevant weight gain while taking quetiapine (n = 43). The risk of weight gain was higher in women (OR = 4.4), overweight patients (OR = 3.0) and patients with at least 1 year of treatment (OR = 6.3) in the olanzapine group; these risk factors are not entirely coincidental with other findings (SEDA-26, 56).

Mechanism

The mechanisms of weight gain are not known. Olanzapine, for example, affects at least 19 different receptor sites, may have reuptake inhibition properties, and may affect hormones such as prolactin. Animal models do not help to elucidate mechanisms, since they have not shown clear results: some studies have shown weight gain with neuroleptic drugs in rats and others have not.

However, a serotonergic mechanism has been proposed (498). Most of the atypical drugs, which most commonly cause weight gain, interact with 5-HT$_{2C}$ receptors; however, ziprasidone also binds to 5-HT$_{2C}$ receptors with high affinity and does not cause weight gain or does so to a lesser extent. In 152 patients treated with clozapine, Cys23Ser polymorphism of the 5-HT$_{2C}$ receptor did not explain the weight gain that occurred (499), nor was there an association between specific alleles of the dopamine D$_4$ receptor gene (500). In contrast, in 19-year-old monozygotic twins who gained around 40 kg after taking mainly clozapine, weight gain was related to an unspecified genotype (501). An association between weight gain and a 5-HT$_{2C}$ receptor gene polymorphism has been identified in 123 Chinese Han patients with schizophrenia taking chlorpromazine ($n = 69$), risperidone ($n = 46$), clozapine ($n = 4$), fluphenazine ($n = 3$), or sulpiride ($n = 1$) (502). Weight gain was substantially greater in the patients with the wild-type genotype than in those with a variant genotype (–759C/T), at both 6 and 10 weeks; this effect was seen in men and women.

Susceptibility factors

In general, people with schizophrenia have a greater tendency to be overweight and obese than those who do not have schizophrenia (503). The evidence suggests that weight gain will progress most rapidly during the first 3–20 weeks of treatment with second-generation neuroleptic drugs; there is little evidence that dose affects weight gain. There are no sex differences, but there is a positive correlation between weight gain and age. Smokers treated with neuroleptic drugs may gain less weight than non-smokers (495,504). Patients in hospital are more likely to gain weight than those in the community, perhaps because they have an unrestricted diet and limited physical activity; there is some evidence that those with a lower baseline BMI are likely to gain more weight (485).

However, in the USA, data from a large health and nutrition survey ($n = 17\,689$) and from patients with schizophrenia to be enrolled in a clinical trial ($n = 420$) showed no differences; the mean BMIs were high and a substantial proportion of the population was obese (503).

The genetic basis of some reactions associated with neuroleptic drugs is of particular interest. An important association between weight gain and a 5-HT$_{2C}$ receptor gene polymorphism has now been identified in 123 Chinese Han schizophrenic patients taking chlorpromazine ($n = 69$), risperidone ($n = 46$), clozapine ($n = 4$), fluphenazine ($n = 3$), or sulpiride ($n = 1$) (502). Weight gain was substantially greater in the patients with the wild-type genotype than in those with a variant genotype (–759C/T), at both 6 and 10 weeks; this effect was seen in men and women. In addition, a homozygous non-functional genotype, CYP2D6*4, was found in a 17-year-old schizophrenic patient who developed severe akathisia, parkinsonism, and drowsiness after taking risperidone 6 mg/day for 3 months; he had a high plasma concentration of risperidone and an active metabolite (505).

Clinical features

Obesity is associated with increased risks of dyslipidemia, hypertension, type 2 diabetes mellitus, cardiovascular disease, osteoarthritis, sleep apnea, and numerous other disorders; all these conditions have been associated with increased mortality. In the EUFAMI (European Federation of Associations of Families of Mentally Ill People) Patient Survey undertaken in 2001 across four countries and involving 441 patients, treatment-induced adverse effects were a fundamental problem; of the 91% of patients who had adverse effects, 60% had weight gain, and of these more than half (54%) rated weight gain as the most difficult problem to cope with (506). Furthermore, weight gain can adversely affect patients' adherence to medication, undermining the success of drug treatment for schizophrenia.

A report has illustrated how extreme the problem of weight gain associated with neuroleptic drugs can be (507).

- A 32-year-old man with schizophrenia taking neuroleptic drugs was switched to olanzapine 20 mg/day for better control. He weighed 101 kg and his BMI was 31 kg/m^2. After 6 months he had gained 4.5 kg and after 1 year 17 kg. Because of poor control, risperidone 4 mg/day was added and 16 months later he had had a 34.6 kg increase from baseline and had a BMI of 42 kg/m^2 (that is in the range of severe obesity). He had both increased appetite and impaired satiety. His serum triglycerides were 4.0 mmol/l, and his fasting blood glucose was 6.7 mmol/l. He and his physician agreed on a goal of losing 12 kg over 12 months and he was referred to a dietician. Over the next 5 weeks, he lost 2.7 kg and after 2 years his BMI was 38 kg/m^2 and 1 year later 39 kg/m^2. He was given nizatidine 300 mg bd and 1 year later his weight was 95 kg (BMI 29 kg/m^2). His triglycerides and fasting blood glucose concentrations were within the reference ranges.

Comparative studies

The charts of 94 long-term inpatients have been reviewed retrospectively to examine the changes in weight, fasting glucose, and fasting lipids in those taking either risperidone ($n = 47$) or olanzapine ($n = 47$) (441). The patients had increased weight, triglycerides, and cholesterol, and the changes were significantly higher with olanzapine; olanzapine but not risperidone considerably increased glucose concentrations. One case of new-onset diabetes mellitus occurred in a patient taking olanzapine. Weight should be monitored in patients taking maintenance atypical neuroleptic drugs, especially dibenzodiazepines.

Management

Behavioral treatment of obesity has given good results in patients taking neuroleptic drugs (508). More dubious is the use of antiobesity drugs, as some of them can cause psychotic reactions.

Good results have been reported with amantadine, which increases dopamine release (509). In 12 patients with a mean weight gain of 7.3 kg during olanzapine treatment, amantadine 100–300 mg/day over 3–6 months produced an average weight loss of 3.5 kg without adverse effects. In contrast, calorie restriction did not lead to weight loss in 39 patients with mental retardation who were taking risperidone, 37 of whom gained weight, the mean gain being 8.3 kg over 26 months (510).

Electrolyte balance

Care should be taken when treating hyponatremic patients with neuroleptic drugs (511).

Mineral balance

In one controlled study, 12 patients who were receiving various neuroleptic drugs had significantly increased urinary calcium and hydroxyproline concentrations and reduced urinary alkaline phosphatase compared with five controls (512). The possibility of a reduction in bone mineralization (513) may contribute to the increased risk of hip fracture associated with neuroleptic drug treatment in the elderly.

Hematologic

Rarely reported hematological reactions to various neuroleptic drugs include agranulocytosis, thrombocytopenic purpura, hemolytic anemia, leukopenia, and eosinophilia. These are thought to represent allergic or hypersensitivity reactions, although this has been questioned in one detailed case report of chlorpromazine-induced agranulocytosis (514).

The reported incidence of agranulocytosis is variable, ranging from 1 in 3000 to 1 in 250 000. Most cases are seen within the first 2 months after the start of treatment, but there have been a few reports in which agranulocytosis occurred only after many years. Frequent white cell counts are of limited value in monitoring, since counts fall rapidly and abruptly. Careful attention should be given to possible early warning signs, such as fever, sore throat, and lymphadenopathy. Treatment requires immediate withdrawal and preventive measures against infection. Granulocyte colony-stimulating factor (GCSF) has been used to treat neuroleptic drug-induced agranulocytosis (515).

In view of the association of agranulocytosis with clozapine and of aplastic anemia with remoxipride, hemopoietic disorders have been studied using information from the UK monitoring system (516). The Committee on Safety of Medicines and the erstwhile Medicines Control Agency in the UK received 999 reports of hemopoietic disorders related to neuroleptic drugs between 1963 and 1996; there were 65 deaths. There were 182 reports of agranulocytosis; chlorpromazine and thioridazine were associated with the highest number of deaths—27 of 56 and 9 of 24 respectively. The much lower mortality with clozapine-induced agranulocytosis—two of 91

(2.2%)—was explained as a function of the stringent monitoring requirements for this drug, which allow early detection and treatment.

There is greater variability in the reference ranges for all white blood cell indices in patients with schizophrenia than in the healthy population (517). This suggests that abnormal hematological findings in patients with schizophrenia should be assessed in the context of a reference range specifically determined in patients with schizophrenia.

Gastrointestinal

The possibility of fatal intestinal dilatation, although very rare, warrants careful evaluation of persistent complaints of constipation, particularly in patients who also have vomiting and abdominal pain, distension, or tenderness (518). Acute intestinal pseudo-obstruction (Ogilvie's syndrome) has been reported in a patient taking haloperidol plus benzatropine (519).

- A 64-year-old woman started to take oral haloperidol 0.5 mg tds, and 3 days later was given intravenous benzatropine 2 mg for dystonia plus a second dose 1 hour later because she had not responded to the first dose. Her dystonia improved, but she started to develop abdominal distension and discomfort, and within the next 3–4 hours her whole abdomen had become significantly distended. Haloperidol and benzatropine were withdrawn and she was treated with hydration, nasogastric suction, a rectal tube, and frequent change of position. With this conservative therapy, her abdominal distension resolved completely in 24 hour.

Liver

When a neuroleptic drug causes jaundice it generally occurs within 2–4 weeks and has many characteristics of an allergic reaction, being accompanied by fever, rashes, and eosinophilia, although a direct toxic mechanism has also been implicated. Symptoms generally subside rapidly after withdrawal, but cholestasis may be prolonged.

Liver damage due to atypical neuroleptic drugs generally occurs within the first weeks of treatment, but the delay is highly variable, being 1–8 weeks for clozapine, 12 days to 5 months for olanzapine, and 1 day to 17 months for risperidone (520).

There is evidence of a significant hepatotoxic effect of the phenothiazines and in persons under age 50, but not over 50 (521).

Hepatotoxicity may be as frequent with piperidine and piperazine phenothiazines as with chlorpromazine, despite previous suggestions that the toxicity of these compounds is less.

More common are minor dose-related abnormalities in liver function tests with various neuroleptic drugs.

Urinary tract

Urinary retention, incontinence, or dysuria can occasionally occur with neuroleptic drugs that have marked anticholinergic effects.

Skin

Many skin reactions have been reported with neuroleptic drugs, including urticaria, abscesses after intramuscular injection, rashes, photosensitivity or exaggerated sunburn, contact dermatitis, and melanosis or blue-gray skin discoloration. Skin rashes are usually benign. Chlorpromazine is most often implicated (incidence 5–10%).

Non-phenothiazines, such as haloperidol and loxitane, cause fewer urticarial reactions. As with any other class of drug, patients may be allergic to excipients in various tablet or capsule forms, or to preservatives, for example methylparabens in liquid dosage forms (522).

Abnormal skin pigmentation (SEDA-19, 49) is more common in women and generally occurs on the exposed parts of the body. This reaction may be due to deposition of melanin-drug complexes. It was commonly seen in the decade after the introduction of the phenothiazines, but rarely today (522).

- A 45-year-old schizophrenic woman with blue eyes and blond hair who had received a lifetime exposure of at least 1748 g of chlorpromazine had blue discoloration of the skin by age 36 (523). Chlorpromazine was withdrawn and clozapine substituted (up to a maximum of 600 mg/day). The skin pigmentation resolved over 4 years.

Complications at the site of injection of depot neuroleptic drugs, including pain, bleeding or hematoma, leakage of drug from the injection site, acute inflammatory induration, and transient nodules, have been reported (SEDA-20, 43).

Seborrheic dermatitis has been observed in patients receiving long-term neuroleptic drugs (SEDA-17, 57), and this adverse effect appears to be highly associated with drug-induced parkinsonism.

More serious types of skin reactions are rare, but angioedema, non-thrombocytopenic purpura, exfoliative dermatitis, and Stevens–Johnson syndrome have been reported.

Potentially serious skin reactions are best treated by withdrawing the offending agent and switching to a structurally unrelated neuroleptic drug. When the offending agent is a phenothiazine, non-phenothiazines such as haloperidol or molindone may be preferable to the more closely related thioxanthenes.

Musculoskeletal

Rhabdomyolysis has been described in a handicapped child without other symptoms of neuroleptic malignant syndrome (524).

- A 6-year-old boy who was taking clonazepam 2.6 mg/day, diazepam 10 mg/day, and phenobarbital 50 mg/day was given oral haloperidol (0.3 mg/day) plus biperiden (0.3 mg/day) for choreoathetosis. After haloperidol had been introduced, his mother noticed that his urine sometimes became dark brown. He had myoglobinuria (660 ng/ml; reference range below 10 ng/ml) but no renal insufficiency. Haloperidol and biperiden were withdrawn and 2 days later his urine was normal.

A marked increase in creatine kinase activity without neuroleptic malignant syndrome has been previously described (SEDA-21, 48) and, another report has further emphasized this possibility (525).

- A 19-year-old schizophrenic patient taking risperidone 6 mg/day and olanzapine 20 mg/day had a creatine kinase activity of 6940 U/l without clinical manifestations of neuroleptic malignant syndrome; when he switched to clozapine (dose not stated) the creatine kinase fell to about 300 U/l. Because he developed granulocytopenia, he was given quetiapine instead, and the creatine kinase again rose to 3942 U/l but fell after 4 days to 389 U/l without withdrawal of quetiapine.

The authors concluded that the mechanism by which creatine kinase increases is not comparable for olanzapine, quetiapine, and clozapine, and that the increase can be self-limiting.

Sexual function

Sexual dysfunction associated with neuroleptic drugs has been reviewed (526). The authors concluded that sexual dysfunction is common in patients with schizophrenia, but often not reported to doctors nor explored, and may be an unrecognized cause of non-adherence to treatment.

The frequency and course of sexual disturbances associated with neuroleptic drugs have been studied in a prospective open study of clozapine 261 mg/day (in 75 men and 25 women, mean age 29 years) and haloperidol 16 mg/day (41 men and 12 women, mean age 26 years) (527). There were no statistically significant differences between those taking haloperidol and those taking clozapine. During 1–6 weeks of treatment with clozapine, the most frequent sexual disturbances among women were diminished sexual desire (28%) and amenorrhea (12%), while among men they were diminished sexual desire (57%), erectile dysfunction (24%), orgasmic dysfunction (23%), ejaculatory dysfunction (21%), and increased sexual desire (15%).

Neuroleptic drug-induced sexual dysfunction, including erectile and ejaculatory dysfunction and changes in libido and the quality of orgasm, appear to be reversible on withdrawal.

Male dysfunction

Although the mechanism involved in neuroleptic drug-induced male sexual dysfunction is not entirely understood, it can occur at several levels, including the cortex, hypothalamus, pituitary gland, and the gonads, involving, for example, gonadotrophins and testosterone. Another mechanism involves the sympathetic and parasympathetic nervous systems and may explain why thioridazine and other highly anticholinergic drugs are mainly responsible for male sexual dysfunction, including impotence and retrograde ejaculation (528,529).

The first report of spontaneous ejaculation associated with the therapeutic use of neuroleptic drugs was described in 1983 (530).

One study showed a reduction in the strength of erection in men with schizophrenia, further accentuated in those who are taking neuroleptic drugs (SEDA-20, 44).

In 12 men with schizophrenia (mean age 36 years) receiving neuroleptic drugs, amantadine 100 mg/day for 6 weeks improved sexual function (531). All 12 patients, who had a sustained relationship with a female partner, had reported sexual dysfunction. Four areas of sexual function were assessed: desire, erection, ejaculation, and satisfaction; there was an improvement in all but ejaculation. Amantadine had no effect on the symptoms of schizophrenia.

Priapism

Neurotropic drug-induced priapism has been reviewed (532). It is said that neuroleptic drugs are most commonly implicated. A prior history of prolonged erections can be identified in as many as 50% of patients presenting with priapism who are using neuroleptic drugs. According to another review, the frequency of priapism with neuroleptic drugs may be increased by the fact that schizophrenia is said to be accompanied by an increase in sexual activity (533).

In one case, priapism followed the use of first risperidone and then ziprasidone (534).

- A 22-year-old African-American with chronic undifferentiated schizophrenia developed priapism after taking risperidone 4 mg bd, clonazepam 0.5 mg bd, vitamin E 400 IU bd, and a multivitamin for over 6 months. He did not respond to subcutaneous terbutaline 0.25 mg. Irrigation of the corpora with phenylephrine 200 μ resulted in detumescence; risperidone was withdrawn. A few months later he took ziprasidone 20 mg bd for 1 week, clonazepam 1 mg bd, and vitamin E 400 IU bd. The ziprasidone dosage was increased to 40 mg bd, but early the next morning he developed a firm erection with some discomfort that lasted about 2 hour and resolved when he urinated; the next morning he had a similar erection that also lasted 2 hour and resolved.

Priapism in twins suggests a hereditary predisposition (535).

- Identical twin brothers, aged 37 years, had both suffered from bipolar disorder since their early twenties and had been treated with chlorpromazine, haloperidol, lithium, and carbamazepine before developing priapism. One of them developed priapism after taking trazodone 400 mg/day, and in the 2 years after the initial episode he suffered recurrent painless erections. Initially they occurred daily and lasted 4–5 hours. During a relapse of mania at age 37, he was given oral zuclopenthixol 40 mg/day. On the tenth day he presented with priapism of 4 days duration, which persisted despite zuclopenthixol withdrawal, needle aspiration, and phenylephrine instillation, but subsided 2 weeks later with conservative management. The other twin presented with priapism of 75 hour duration and hypomania at age 31 years. He had been taking lithium and chlorpromazine for the preceding 2 years.

Priapism necessitates prompt urological consultation and sometimes even surgical intervention (528,536,537).

Female dysfunction

Female orgasm is inhibited by some central depressant and psychotropic drugs, including neuroleptic drugs, antidepressants, and anxiolytic benzodiazepines (204).

Immunologic

Hypogammaglobulinemia in a 22-year-old woman with brief psychotic disorder has been attributed to neuroleptic drug therapy (538). About 4 months after she had started to receive neuroleptic drugs, her serum concentrations of total protein had fallen to 58 g/l, with an IgG concentration of 3.49 g/l, an IgA concentration of 0.54 g/l, and an IgM concentration of 0.34 g/l.

Antiphospholipid syndrome is a disorder of recurrent arterial or venous thrombosis, thrombocytopenia, hemolytic anemia, or a positive Coombs' test, and in women recurrent idiopathic fetal loss, associated with raised concentrations of antiphospholipid antibodies. In systemic lupus erythematosus, the risk of this syndrome is about 40%, compared with a risk of 15% in the absence of antiphospholipid antibodies (539). However, only half of those with antiphospholipid antibodies have systemic lupus erythematosus, and the overall risk of the syndrome is about 30%. In patients who have antiphospholipid antibodies associated with chlorpromazine, there appears to be no increased risk of the syndrome. In contrast, in the primary antiphospholipid syndrome, the only clinical manifestations are the features of this syndrome.

- Symptomatic antiphospholipid syndrome has been described in a 42-year-old woman treated with chlorpromazine 260 mg/day (540). She presented with suddensided weakness, numbness, and headache. Examination confirmed upper motor neuron signs affecting the right arm, leg, and face, with hemiplegia and hemiparesthesia. Autoantibody screening showed positive antinuclear antibodies, with an IgG titer of 50 and an IgM titer of 1600. Anticardiolipin antibody was positive with a raised IgM titer of 24 (normal less than 9). The symptoms and the serological findings resolved after withdrawal of the phenothiazine.

Drug-induced lupus erythematosus has been reviewed (541,542). Neuroleptic drugs, particularly chlorpromazine and chlorprothixene, have often been associated with this autoimmune disorder. It is recommended that several diagnostic criteria for this condition should be met: (1) exposure to a drug suspected to cause lupus erythematosus; (2) no previous history of the condition; (5) detection of positive antinuclear antibodies; and (6) rapid improvement and a gradual fall in the antinuclear antibodies and other serological findings on drug withdrawal. Rare

disorders of connective tissue resembling systemic lupus erythematosus have been reported with chlorpromazine, perphenazine, and chlorprothixene (543).

Body temperature

Neuroleptic drugs interfere with the temperature regulatory function of the hypothalamus and peripherally with the sweating mechanism, resulting in poikilothermy. This can result in either hyperthermia or hypothermia, depending on the environmental temperature.

Hyperthermia
A series of cases has been reported in which heat stroke occurred during hot weather, probably due to impaired heat adaptation in patients taking benzatropine and ethylbenzatropine (544); this can occur with other anticholinergic and neuroleptic drugs as well.

Hypothermia
Reduced body temperature has been observed in a study of 14 drug-free and 7 patients with schizophrenia taking different neuroleptic drugs (545). The temperature fell by about 0.36° C at 24 hours after the drug-free subjects started to take neuroleptic drugs.

Hypothermia, defined as a body temperature lower than 35° C, has been observed in individual cases.

- A 73-year-old woman with diabetes mellitus and schizophrenia was given haloperidol 8 mg/day instead of zuclopenthixol; she also took levomepromazine 75 mg at night; 10 days later she was found unconscious with a rectal temperature of 31.5° C (546). A few weeks later, she had two further episodes, first with olanzapine 10 mg/day (rectal temperature 31.7° C) and then with thioridazine 40 mg/day (rectal temperature 34° C). She was therefore given dixyrazine 150 mg/day and alimemazine 20 mg at night, and had no more episodes of hypothermia.
- An 83-year-old woman developed a rectal temperature of 33.1° C 3 weeks after starting to take olanzapine 5 mg/day (547).
- A 68-year-old schizophrenic woman with type 1 diabetes mellitus developed a body temperature of 33.4° C 1 week after starting to take quetiapine 100 mg bd (547).

In the two last cases, the most common causes of hypothermia, such as hypothyroidism, infection, and cold exposure, were ruled out.

Death

The role of neuroleptic drugs in sudden death is controversial (SEDA-18, 47) (SEDA-20, 26) (SEDA-20, 36). Cardiac dysrhythmias may be involved, but there may also be multiple non-cardiac causes, including asphyxia, convulsions, or hyperpyrexia.

Long-Term Effects

Drug withdrawal

Various somatic complaints have been reported in patients in whom neuroleptic drugs are abruptly withdrawn (SEDA-20, 44). The incidence of these complaints varies widely in different reports, from 0 to 75%. Common complaints include headache, vomiting, nausea, diarrhea, insomnia, abdominal pain, rhinorrhea, and muscle aches. On rare occasions, the symptoms resemble those of benzodiazepine withdrawal (appetite change, dizziness, tremulousness, numbness, nightmares, a bad taste in the mouth, fever, sweating, vertigo, tachycardia, and anxiety), but it is possible that in some of the reported cases there was actually benzodiazepine withdrawal. Some of these symptoms may also have been linked to the simultaneous withdrawal of anticholinergic drugs (SED-11, 113) (548). Parkinsonism, not explained by withdrawal of anticholinergic drugs, has also been reported as an unusual withdrawal effect of neuroleptic drugs (549).

Worsening of psychotic symptoms and/or dyskinetic movements can occur when dosages are lowered or a neuroleptic drug is withdrawn. A functional increase in mesolimbic and striatal dopaminergic sensitivity has been suggested as an explanation (550). Psychotic relapse is rarely seen in the first 2 weeks after withdrawal, but physical withdrawal symptoms generally begin within 48 hours of the last dose (SEDA-14, 54).

Withdrawal emergent syndrome has been described in children (551,552) and consists of nausea, vomiting, ataxia, and choreiform dyskinesia primarily affecting the extremities, trunk, and head after sudden withdrawal of neuroleptic drugs (553). In one study, there were withdrawal symptoms in 51% of children, twice as many being affected by the withdrawal of low-dose, high-potency compounds compared with other drugs. Symptoms usually appear within a few days to 2 weeks after drug withdrawal; spontaneous remission is likely within the next 8–12 weeks.

The Omnibus Budget Reconciliation Act of 1987 (OBRA '87) regulations (www.elderlibrary.org) specify when neuroleptic drugs can and cannot be used to treat behavioral disturbances in nursing home residents in the USA. Accordingly, neuroleptic drugs can legally be used in patients with delirium or dementia only if there are psychotic or agitated features that present a danger to the patient or others. Preventable causes of agitation must be excluded and the nature and frequency of these behaviors must be documented. Non-dangerous agitation, uncooperativeness, wandering, restlessness, insomnia, and impaired memory are insufficient in isolation to justify the use of neuroleptic drugs. With this in mind, the effects of withdrawing haloperidol, thioridazine, and lorazepam have been examined in a double-blind, crossover study in 58 nursing home residents (43 women and 15 men, mean age 86 years), half of whom continued to take the psychotropic drugs that had been prescribed, while the other half were tapered to placebo (554). After 6 weeks, the patients were tapered to the reverse schedule and took it for

another 6 weeks. There were no differences between drug and placebo in functioning, adverse effects, and global impression. Cognitive functioning improved during placebo. The authors concluded that gradual dosage reductions of psychoactive medications must be attempted, unless clinically contraindicated, when trying to withdraw these drugs. Similar conclusions have been reached in other studies (SEDA-22, 54).

There has been a randomized controlled study of the factors that affect neuroleptic drug withdrawal or dosage reduction among people with learning disabilities, who were being treated for behavioral problems with typical neuroleptic drugs (555). Of 36 patients, 12 completed full withdrawal and a further 7 achieved and maintained at least a 50% reduction. Drug withdrawal or dosage reduction was not associated with increased maladaptive behavior. This result reinforces concerns that neuroleptic drug treatment for maladaptive behavior reduction is often ineffective and inappropriate.

Mutagenicity

There was an increase in markers of genotoxicity in patients receiving long-term neuroleptic drugs in combination with other psychotropic drugs ($n = 36$) compared with controls ($n = 36$) (556). In another study there was an association between the frequency of chromosomal aberrations in lymphocytes and the probability of tumor induction (557).

Tumorigenicity

Little information is available on the relation between neuroleptic drug use and cancers of any type or in any site, although an increase in markers of genotoxicity has been described (SEDA-23, 57). In a recent case-control study there was an association between the use of neuroleptic drugs and endometrial cancer (558). Premenopausal women with histologically confirmed endometrial cancer (41 cases) were compared with 123 controls. Five cases and three controls had been previously exposed to neuroleptic drugs for 48–252 months; the adjusted OR was 5.4 (95% CI = 1.1, 26). The authors hypothesized that hyperprolactinemia, which usually reduces the pulsatile secretion of follicle-stimulating and luteinizing hormones, inhibits gonadal function and might stimulate mitotic activity in the endometrium.

There is concern that neuroleptic drugs may increase the risk of breast cancer because of raised prolactin concentrations. For a long time, findings did not confirm this association (426), but a Danish cohort study of 6152 patients showed a slight increase in the risk of breast cancer among schizophrenic women (427).

Second-Generation Effects

Fertility

The effects of neuroleptic drugs on fertility are not clear; many of the data are controversial, often being based on animal studies, for example reduction in male rat copulation by chlorpromazine. However, oligospermia, polyspermia, necrospermia, and reduced sperm motility have been reported with various phenothiazines and butyrophenones; these are likely to improve after withdrawal (473). Furthermore, fertility may be impaired by neuroleptic drugs, since they increase prolactin concentrations and can cause amenorrhea (6).

Teratogenicity

There have been reviews of neuroleptic drug use during pregnancy (560–564).

All neuroleptic drugs cross the placenta and reach the fetus in potentially significant amounts. However, most large-scale controlled studies have shown that they can be used safely during pregnancy (174). Nevertheless, there have been isolated case reports of malformations (SEDA-7, 68) (565).

Should a neuroleptic drug be required during pregnancy, a drug such as chlorpromazine or trifluoperazine would be preferred, as there is considerably more worldwide experience with these drugs than with newer drugs (174). However, clozapine caused no serious complications or developmental abnormalities during pregnancy in two cases (566).

Although studies have failed to identify an increased rate of malformations with haloperidol, isolated cases have been reported (567) (SEDA-22, 5; SEDA-23, 57; 563,564).

Phocomelia associated with neuroleptic drugs has been reported (568).

- A 38-year-old woman was admitted to hospital in the third week after conception and acute schizophrenia was diagnosed. She was given oral haloperidol 1.5 mg at 12-hourly intervals. In the fourth week after conception, intramuscular depot fluphenazine 12.5 was added every 15 days. At 8 weeks of gestation, after the pregnancy had been diagnosed, haloperidol and fluphenazine were withdrawn and trifluoperazine 5 mg was started and continued until 8 weeks before the baby was born. The child was born with phocomelia of the left arm, with an extremely short humerus and an absent forearm.

Fetotoxicity

A variety of pharmacological effects can occur in the infant after birth, particularly when the mother has received these agents in the weeks before delivery.

These include postnatal depression and acute dystonic reactions (which may interfere with normal delivery). Hypotonia can persist for months (569) and may respond to diphenhydramine 5 mg/kg/day. Severe rhinorrhea and respiratory distress in a neonate exposed to fluphenazine hydrochloride prenatally has been reported (484). Neonatal jaundice, hyperbilirubinemia, and melanin deposits in the eyes have occurred when neuroleptic drugs were given during the last trimester or longer during pregnancy.

Lactation

Neuroleptic drugs appear in breast milk in very low concentrations, related to maternal dosage. Typical regimens of neuroleptic drugs yield low or negligible concentrations (174,571). In one study, breast-fed infants ingested up to 3% of the maternal daily dose, and small amounts of the drugs were detected in their plasma and urine. In addition, three breast-fed infants had a fall in developmental scores from the first to the second assessment at 12–18 months (572). Until these issues are further clarified, it would be best to avoid breast-feeding by mothers who are receiving neuroleptic drugs.

Susceptibility Factors

Genetic

Genetic factors, about which more and more information is emerging (Table 1), could help to explain some of the susceptibility factors for severe adverse reactions to neuroleptic drugs.

The ryanodine receptors (RYRs) are a family of intracellular Ca^{2+} release channels that play a pivotal role in the regulation of intracellular Ca^{2+} homeostasis in muscle cells and neurons; there are three isoforms, RYR1 (skeletal muscle), RYR2 (cardiac muscle and brain), and RYR3 (brain and some smooth muscle). So far, 22 different mutations in the RYR1 gene have been identified to play a causative role in the pathogenesis of malignant hyperthermia, but none of these mutations has been found in neuroleptic malignant syndrome. A pilot study to screen for mutations in RYR coding sequences has been conducted in two patients with recurrent neuroleptic malignant syndrome induced by different neuroleptic drugs (573). The analysis did not reveal sequence variants

Table 1 Genetic factors and neuroleptic drug-induced adverse reactions

Adverse reaction	Genetic factor	Comments	Reference
Tardive dyskinesia	D$_3$ receptor (Ser9Gly)	Greater severity	SEDA-26, 55
Tardive dyskinesia	CYP1A2 (intron 1)	Greater severity	SEDA-26, 55
Tardive dyskinesia	CYP2D6 variants	Conflicting results	
Clozapine-induced agranulocytosis	Human leukocyte antigen variants	Association	SEDA-26, 60
Olanzapine-induced agranulocytosis	Human leukocyte antigen variants	Association	SEDA-26, 63
Weight gain	5-HT$_{2C}$ receptor variant	Conflicting results	SEDA-27, 53 and 57
Weight gain	CYP2D6 variants	Association	SEDA-27, 61
Neuroleptic malignant syndrome	CYP2D6 variants	Conflicting results	SEDA-27, 58
Olanzapine toxicity	Gilbert's syndrome	Association	SEDA-28, 73

that are likely to confer vulnerability to neuroleptic malignant syndrome.

In a prospective study the prevalence of poor CYP2D6 metabolizers was significantly higher in those who developed extrapyramidal adverse events (574).

However, in a retrospective case-control study, CYP2D6 genotype was not associated with tardive dyskinesia (575). Patients with tardive dyskinesia (50 cases) were compared with 59 control). Genotyping identified the functional allele CYP2D6*1, the known major defective alleles CYP2D6*3, CYP2D6*4, CYP2D6*5, CYP2D6*6, and gene duplication; there was no statistical differences in the proportion of the different functional categories between the cases and the controls.

In a study funded by Eli Lilly, the marketing authorization holder of olanzapine, claims data for schizophrenic patients taking olanzapine (n = 1875), risperidone (n = 982), or haloperidol (n = 726) between January 1997 and August 1998 were retrieved to determine the role of ethnicity in predicting adherence to neuroleptic drugs (576). After controlling for other factors, African–Americans and Mexican–Americans were significantly less adherent to therapy than white patients; for all ethnicities, olanzapine was associated with 23 more adherent days than risperidone and 55 more adherent days than haloperidol.

No polymorphism of neural nitric oxide synthase gene was found in 128 Chinese patients with neuroleptic drug-induced tardive dyskinesia compared with 123 controls (577). After adjusting the effects of confounding factors, there was no significant association between NOS1 3′-UTR C276T genotypes and tardive dyskinesia.

An association between weight gain and a 5-HT_{2C} receptor gene polymorphism has been identified (SEDA-28, 53). However, two genetic variants of the β_3 adrenoceptor (Trp64Arg) and the G-protein β_3 subunit (C825T) did not play a significant role in clozapine-induced weight gain in a study in 87 treatment-resistant patients with schizophrenia (578)

In a survey of data from several studies in 126 African–American patients and 574 white patients who were randomly selected and for whom neuroleptic drugs were prescribed, the African–American patients, after adjustments for clinical, sociodemographic, and health-system characteristics, were less likely than the white patients to receive second-generation neuroleptic drugs (49% compared with 66%) (579).

Age

Children

Neuroleptic drugs have been prescribed for children in the treatment of psychotic disorders, Tourette's syndrome, attention deficit disorder, hyperactivity, behavioral and psychiatric complications of mental retardation, and pervasive developmental disorders, for example infantile autism (580,581).

Untoward effects of neuroleptic drugs in children are said to be similar to those seen in adults (582).

However, adverse effects of neuroleptic drugs in children can be unpredictable and a suggestion that they can cause sudden infant death remains hypothetical (551). Significant weight gain has been reported in almost 100% of neuroleptic drug-treated children, and there seems to be a relatively high incidence of extrapyramidal adverse effects (552). Since there is little information regarding the pharmacokinetics and pharmacodynamics of neuroleptic drugs in children, careful supervision of treatment is vital; the use of high dosages is inadvisable.

Their use in children and adolescents has been extensively reviewed (583–585). Typical neuroleptic drugs have been assessed in three randomized, double-blind, placebo-controlled studies in 122 patients and atypical drugs in five (one clozapine, n = 21; two amisulpride, n = 36; and two tiapride, n = 59). The studies were of short durations, 4–10 weeks. Extrapyramidal signs occurred in 25–73% of those treated with the typical neuroleptic drugs.

Elderly people

There is wide individual variation in the extent to which older people tolerate these drugs, and adverse effects, such as postural hypotension and anticholinergic effects, can be more problematic (586). Caution with neuroleptic drug use in old people is further warranted because several of these drugs are metabolized by CYP2D6, which is inhibited by many commonly used drugs (SEDA-20, 44).

Neuroleptic drugs have often been associated with increased rates of falls (SEDA-19, 40) (587).

Older people are supposedly susceptible to delirium caused by centrally acting anticholinergic drugs. This can take the form of anxiety, agitation, or frank hypomania. Such difficulties are more likely if the patient is also taking neuroleptic drugs (most of which also have a weak anticholinergic effect) or anticholinergic antiparkinsonian drugs.

Use and effectiveness

In a study that was claimed to be the first published placebo-controlled dose comparison of neuroleptic drugs used to treat psychoses and disruptive behavior in dementia, haloperidol 2–3 mg/day was compared with 0.50–0.75 mg/day in 71 outpatients with Alzheimer's disease (588). After 12 weeks, there was a favorable therapeutic effect of haloperidol 2–3 mg/day, although 25% of the patients developed moderate to severe extrapyramidal signs. No patient developed tardive dyskinesia, but neuroleptic drug exposure needs to be considerably longer for that to occur.

Neuroleptic drugs are currently used to treat the psychiatric and behavioral symptoms that affect elderly patients with dementia (589). This is an unlicensed indication, but 25% of elderly patients in nursing homes receive these drugs (590). The main conclusions from a double-blind, placebo-controlled study (CATIE-AD) in 421 outpatients were that adverse effects offset the efficacy of atypical neuroleptic drugs in managing psychosis, aggression, or agitation in patients with Alzheimer's disease, who were randomly assigned to olanzapine (mean dose 5.5 mg/day), quetiapine (57 mg/day), risperidone (mean dose 1.0 mg/day), or placebo (591). At 12 weeks there were no significant differences with regard to the time to withdrawal for any reason: olanzapine (median 8.1 weeks), quetiapine (median 5.3 weeks), risperidone (median 7.4 weeks), and placebo (median 8.0 weeks). Although the median time to withdrawal because of lack of efficacy favored olanzapine (22 weeks) and risperidone (27 weeks) compared with quetiapine (9.1 weeks) and placebo (9.0 weeks), the time to withdrawal because of adverse events or intolerability clearly favored placebo. Overall, 24% of patients who took olanzapine, 16% of those who took quetiapine, 18% of those who took risperidone, and 5% of those who took placebo withdrew because of intolerability. There were no significant differences among the groups with regard to improvements on the Clinical Global Impression of Change scale. Moreover, neither quetiapine nor rivastigmine was effective in agitation in people with dementia in institutional care.

In a double-blind, randomized, placebo-controlled study quetiapine was associated with significant cognitive decline in 93 patients with Alzheimer's disease, dementia, and clinically significant agitation (592).

These results coincide with those of a recent review of evidence (593) and a recent meta-analysis (594), although early pivotal comparisons of risperidone, haloperidol, and placebo in agitated and demented patients did not find substantial differences when the evaluation was performed at 12 months (SEDA-25, 68).

Neuroleptic drugs and stroke in patients with dementia
Different warnings to clinicians have been issued on the link between atypical neuroleptic drugs and cerebrovascular adverse events. The US Food and Drug Administration issued a similar warning in April 2003 (595). These warnings have led to a controversy among doctors (596,597).

A meta-analysis of the effect of olanzapine for the behavioral and psychological symptoms of dementia has shown that olanzapine may also be associated with an increased risk of cerebrovascular adverse effects (598). Nevertheless, several studies have not found any association between the use of atypical neuroleptic drugs and cerebrovascular events (599) (600).

Neither of two observational studies of the relation between atypical neuroleptic drugs and the risk of ischemic stroke showed a similar significant risk. In the first, a population-based retrospective cohort study, patients over 65 years with dementia who took atypical neuroleptic drugs showed no significant increase in the risk of ischemic stroke compared with those who took typical neuroleptic drugs (adjusted hazard ratio = 1.0; 95% CI = 0.8, 1.3) (601). The numbers of new admissions for ischemic stroke were 284 in those taking atypical neuroleptic drugs (n = 17 845) and 227 in those taking typical neuroleptic drugs (n = 14 865). In the second study, data from prescription-event monitoring of olanzapine (n = 8826), risperidone (n = 7684), and quetiapine (n = 1726) were examined (602). The patients were mainly old (median ages 83, 81, and 69 years respectively; women 33%, 74%, and 70% respectively). Within 6 months of

starting treatment, 10 patients had a first occurrence of a stroke or a transitory ischemic attack with olanzapine (0.1%; five fatal), 23 with risperidone (0.3%; nine fatal), and six with quetiapine (0.3%; one fatal). After adjusting for three confounders (age, sex, and indication) there were no significant differences in the relative risks of stroke between olanzapine and either risperidone or quetiapine, or between risperidone and quetiapine (RR = 1.9; 95 % CI = 0.5, 2.9 and RR = 2.1; 95% CI = 0.6, 7.6).

In another trial olanzapine 2.5–7.5 mg/day was not associated with a higher risk of adverse cardiovascular events compared to typical neuroleptic drugs (haloperidol or promazine chlorhydrate) in 346 patients aged 71-92 years with vascular dementia and behavioral problems (603).

Neuroleptic drugs and ventricular dysrhythmias and cardiac arrest
In a retrospective case-control study in US residents of nursing homes aged 65 years and over, conventional (adjusted OR = 1.9; 95% CI = 1.3, 2.7) but not atypical neuroleptic drugs (adjusted OR = 0.9; 95% CI = 0.6, 1.3) were associated with an increased risk of hospitalization for ventricular dysrhythmias and cardiac arrest (604). Among residents who took conventional neuroleptic drugs, those with cardiac disease were 3.3 times (95% CI = 1.9, 5.5) more likely to be hospitalized for ventricular dysrhythmias and cardiac arrest than non-users without cardiac disease. The number of patients hospitalized for ventricular dysrhythmias and cardiac arrest (cases) was 649, and 2962 controls were selected among in-patients in the inception cohort whose primary diagnosis at discharge was septicemia, gastrointestinal hemorrhage, rectal bleeding, gastritis with bleeding, duodenitis with bleeding, or influenza.

Atypical neuroleptic drugs and the risk of death
A thorough meta-analysis of published (n = 6) and unpublished (n = 9) double-blind, parallel-group, randomized, placebo-controlled trials has shown an increased risk of death in patients with dementia taking atypical neuroleptic drugs (605). There were more deaths among patients randomized to drugs (118 out of 3353; 3.5%) than those who took placebo (40 out of 1757; 2.3%) (OR = 1.5; 95% CI = 1.1, 2.2). There were no differences in drop-outs.

In a recent retrospective cohort study conventional neuroleptic drugs (n = 9142) were at least as likely as atypical agents (n = 13 748) to increase the risk of death among elderly people; accordingly, conventional drugs should not be used to replace atypical agents withdrawn in response to the FDA warning (606). The adjusted relative risk was significantly higher with conventional neuroleptic drugs (RR = 1.4; 95% CI = 1.3, 1.5) and in all subgroups defined according to the presence or absence of dementia or nursing home residency. Hence, the US Food and Drug Administration has newly issued a warning: "All of the atypical neuroleptic drugs are approved for the treatment of schizophrenia. None, however, is approved for the treatment of behavioral disorders in patients with dementia. Because of these findings, the Agency will ask the manufacturers of these drugs to include a Boxed Warning in their labeling describing this risk and noting that these drugs are not approved for this indication. Zambia, a combination product containing olanzapine and fluoxetine, approved for the treatment of depressive episodes associated with bipolar disorder, will also be included in the request... The Agency is also considering adding a similar warning to the labeling for older neuroleptic medications because the limited data available suggest a similar increase in mortality for these drugs" (607). Similarly, the European Agency for the Evaluation of Medicinal Products has underlined these risks in a public statement: "Neuroleptic drugs are known to be used in patients with dementia who experience psychotic symptoms and disturbed behavior. There are insufficient data to confirm any difference in the risk of mortality or cerebrovascular accidents among atypical neuroleptic drugs, including olanzapine, or between atypical and conventional neuroleptic drugs" (608).

Drug Administration

Drug formulations

Depot formulations
Guidelines for depot neuroleptic drug treatment in schizophrenia were developed during a consensus conference held in 1995 in Siena in Italy, since when a thorough review devoted to depot neuroleptic drugs has been published (609). The authors reported that depot neuroleptic drugs are associated with significantly fewer relapses and rehospitalizations than oral neuroleptic drugs. The potential disadvantages of depot drugs include patient reluctance to accept injections or a sense of being too subject to control. In addition, clinicians, and sometimes patients, fear that if adverse effects occur they will be more difficult to manage because of an inability to withdraw treatment rapidly. However, these formulations are not associated with a significantly higher incidence of adverse effects than oral drugs. Some patients have pain or discomfort at the injection site. In some cases the injection site becomes edematous and tender or pruritic, with a palpable mass for up to 3 months. The recommendations are to rotate the injection site, to limit the volume of the injection, and to assure deep intramuscular injection.

Drug administration route

Intramuscular injection of neuroleptic drugs in depot formulations is important for compliance in outpatients with schizophrenia, and the benefits and risks have been emphasized (SEDA-23, 73). In a retrospective study in a tertiary center in Riyadh, of 69 patients (mean age 35 years; mean duration of illness 11 years; 38 men) who had been given depot neuroleptic drugs during 1991–2000, only 24 were still receiving their depot injections regularly in that clinic (610). The authors suggested that negative attitudes of families and society toward mental illness and interference with psychiatric treatment by traditional and faith healers could have accounted for this lack of compliance. Although the use of depot neuroleptic drugs is advocated as maintenance treatment for patients with bipolar disorders, no such patients were identified. The authors pointed out that in spite of the fact that patients had been clinically stable, 7% had been receiving their depot neuroleptic drugs for more than 8 years, with no attempts to reduce the dosage or to increase the interval of administration.

Drug overdose

Suicide is the leading cause of premature death among patients with schizophrenia. The lifetime incidence of completed suicide among patients with schizophrenia is about 10% (611) and 18–55% of patients with schizophrenia make suicide attempts (612), sometimes with neuroleptic drugs. In autopsies and toxicological analyses in Vienna from 1991 to 1997, 97 neuroleptic compounds were detected in 85 fatal intoxications (613). Of these, 17 fatal poisonings were attributed to a single compound, 3 were due to two neuroleptic drugs, 57 to a combination of one neuroleptic drug with other drugs, 7 to two neuroleptic drugs with other drugs, and in one case three neuroleptic drugs combined with other drugs. Relating the number of deaths by a drug to the number of Defined Daily Doses, a fatality index was estimated. At the top of the rating were levopromazine, prothipendyl, and chlorprothixene, which had f-values significantly higher than the mean. However, since the estimates were not adjusted for age or condition, these data should be taken with caution.

In another study there were large differences in fatality from overdosage between different neuroleptic drugs (614). Pimozide had the lowest fatality index (the number of deaths divided by the number of prescriptions) and loxapine the highest.

Although mortality is generally low and infrequently associated with residual impairment (SED-13, 129), important exceptions occur when there is co-ingestion with alcohol, tricyclic antidepressants, or antiparkinsonian agents. In acute overdosage of neuroleptic drugs alone, the most serious complications include shock, seizures, and cardiac dysrhythmias. These can be more problematic when the neuroleptic drugs are of low potency, for example thioridazine and chlorpromazine, but taken in high dosages.

Acute extrapyramidal reactions occur more often after ingestion of high-potency drugs, such as haloperidol and fluphenazine; these respond to parenteral benzatropine, but anticholinergic drugs should be used judiciously, so as not to worsen peripheral or central autonomic toxicity. Other serious, but less frequent, complications include paralytic ileus and hypothermia. Acute renal insufficiency has been very rarely reported, but is apparently reversible and can occur secondary to severe hypotension or other causes after acute ingestion (615).

Of 524 inquiries received by the National Poisons Information Service concerning new neuroleptic drugs over 9 months, only 45 cases involved overdose with a single agent (olanzapine, $n = 10$; clozapine, $n = 8$;

risperidone, *n* = 10; sulpiride, *n* = 16) (616). There were no deaths or cases of convulsions. Cardiac dysrhythmias occurred only with sulpiride. Symptoms were most marked with clozapine: most patients had agitation, dystonia, central nervous system depression, and tachycardia. Most of the patients who had taken risperidone were asymptomatic.

Neuroleptic drug poisoning in 86 children has been retrospectively studied in two pediatric hospitals in the USA (1987–1997), with about 9000 and 11 000 annual admissions (617). Most (70%) occurred in children under 6 years of age; over two-thirds of the cases (78%) were unintentional. The owner of the medication, when identified (85% of cases), was the grandmother (22%), another family member (21%), the patient (13%), or a non-family caregiver (8%); the most common places where ingestion occurred were the patient's home (64%) or a relative's home (22%). There was a depressed level of consciousness in 91% and a dystonic reaction in 51%; there were no deaths.

- A 46-year-old woman took amisulpride (12 g), alprazolam (40 mg), and sertraline (1 g) with suicidal intent, and 3 hours later was still unconscious and was intubated (618). Her body temperature rose from 37.1° C on admission to 38° C soon after. She was discharged on the third day.
- A 50-year-old man who took an overdose of ziprasidone 3120 mg (52 tablets) had no serious effects; he was a little drowsy and his speech was slightly slurred (619). Ziprasidone blood concentrations were not measured.

Fatal intoxication has been reported with melperone (620).

- A 36-year-old woman was found dead in her flat about 2 days after her last contact with one of her relatives. Besides melperone, she was known to take diazepam and carbamazepine. The police found four empty containers of 100 melperone tablets (100 mg per tablet). Post-mortem blood concentration analysis showed melperone, diazepam, nordiazepam, and carbamazepine; the melperone concentration in venous blood was very high (17.1 µg/ml).

Minimal myocardial damage after promazine overdose has been reported (621).

- A 31-year-old woman with a borderline personality disorder took 15 ml of promazine hydrochloride (total 600 mg), 25 capsules of Optalidon™ (butalbital 1250 mg, propyphenazone 3125 mg, caffeine 500 mg), and 20 capsules of flurazepam (total 600 mg). Her electrocardiogram showed T wave inversion in V3–6, with no significant changes in ST segments or QRS duration and no dysrhythmias. Transthoracic echocardiography showed hypokinesis of the distal segments of the anterior and lateral segments and of the apex of the left ventricle; there was mild reduction of global systolic function, with an ejection fraction of 54%. Creatine kinase and myocardial band releases peaked at 36 hours, reaching maximum values of 469 U/l and

10 µg/l respectively. Her electrocardiogram, recorded every 12 hours, began to normalize within 2 days and was definitely normal after 5 days.

Because she also took a large dose of caffeine, a combined effect cannot be excluded.

Drug–Drug Interactions

Adrenoceptor agonists

Neuroleptic drugs can reduce or block the pressor effects of alpha-adrenoceptor agonists. When using drugs with both alpha- and beta-adrenoceptor activity, neuroleptic drug blockade of alpha-adrenoceptors can lead to unopposed beta predominance, resulting in severe hypotension (622).

A neuroleptic malignant-like syndrome occurred when norephedrine was combined with neuroleptic drugs (385).

Alcohol

Alcohol-induced CNS and respiratory depression is enhanced by neuroleptic drugs (623), but enhancement can be slight if both are used in reasonable amounts (624).

Alpha-adrenoceptor antagonists

Neuroleptic drugs can intensify the effects of alpha-adrenoceptor antagonists, for example phentolamine, causing severe hypotension (552).

Amfebutamone (bupropion)

Bupropion has about the same seizure potential as tricyclic antidepressants (SEDA-8, 30), but the manufacturers noted an increased risk of seizures in patients taking over 600 mg/day in combination with lithium or neuroleptic drugs (SEDA-10, 20).

Antacids

Antacids containing aluminium and magnesium reduce the gastrointestinal absorption of chlorpromazine and other phenothiazines by forming complexes (625). The clinical significance of this is unknown.

Antihistamines

Drugs with a CNS depressant effect (antihistamines, hypnotics, sedatives, narcotic analgesics, alcohol, etc.) will have an increase in effect caused by interaction with neuroleptic drugs.

Benzatropine

Benzatropine and ethylbenzatropine are particularly likely to interact additively with other drugs with both anticholinergic and antihistaminic activity, such as neuroleptic drugs; complications such as hyperpyrexia, coma, and toxic psychosis have been reported several times when such combinations were used (626–628).

Benzodiazepines

Neuroleptic drugs can potentiate the sedative effects of benzodiazepines pharmacodynamically.

In a brief review, emphasis has been placed on pharmacokinetic interactions between neuroleptic drugs and benzodiazepines, as much information on their metabolic pathways is emerging (629). Thus, CYP3A4, which plays a dominant role in the metabolism of benzodiazepines, also contributes to the metabolism of clozapine, haloperidol, and quetiapine, and plasma neuroleptic drug concentrations can rise.

Intramuscular levomepromazine in combination with an intravenous benzodiazepine has been said to increase the risk of airways obstruction, on the basis of five cases of respiratory impairment in patients who received injections of psychotropic drugs (630). The doses of levomepromazine were higher in the five cases that had accompanying airways obstruction than in another 95 patients who did not.

Beta-adrenoceptor antagonists

The cardiac effects of neuroleptic drugs can be potentiated by propranolol (631).

In general, concurrent use of neuroleptic and antihypertensive drugs merits close patient monitoring (632).

Bromocriptine

The dopamine-blocking activity of neuroleptic drugs can antagonize the effects of dopamine receptor agonists, such as bromocriptine. Conversely, bromocriptine has been reported to cause exacerbation of schizophrenic symptoms (633).

Caffeine

Neuroleptic drugs can precipitate from solution when mixed with coffee or tea (634), but the clinical significance of this physicochemical interaction is unknown (635).

Caffeine can alter blood concentrations of neuroleptic drugs (SEDA-5, 6).

Excess caffeine can stimulate the CNS, which can worsen psychosis and thus interfere with the effects of neuroleptic drugs (636).

Cannabis

In 10 patients, schizophrenia was acutely worsened after cannabis use, despite verified adequate depot treatment with neuroleptic drugs (637).

Carbamazepine

Plasma concentrations of neuroleptic drugs can be lowered by carbamazepine, and patients should be monitored for reduced efficacy (638).

- A 54-year-old man who had been taking neuroleptic drugs for about 30 years developed neuroleptic malignant syndrome within 3 days of taking add-on carbamazepine (400 mg/day) (639).

This syndrome does not appear to have been described with carbamazepine alone, and it was speculated that its pathogenesis could involve rebound cholinergic activity after a reduction in plasma neuroleptic drug concentrations by carbamazepine.

Cocaine

In a retrospective study of 116 patients taking neuroleptic drugs, 42% of cocaine users versus 14% of non-users developed dystonic reactions (640). This suggests that the use of cocaine may be a major risk factor for acute dystonic reactions secondary to the use of neuroleptic drugs.

This has been confirmed in a 2-year study on the island of Curaçao in the Netherlands Antilles, where cocaine and cannabis are often abused (208). The sample consisted of 29 men with neuroleptic-induced acute dystonias aged 17–45 years who had received high potency neuroleptic drugs in the month before admission; 9 were cocaine users and 20 non-users. Cocaine use was a major risk factor for neuroleptic-induced acute dystonia and should be added to the list of well-known risk factors such as male sex, younger age, neuroleptic dose and potency, and a history of neuroleptic-induced acute dystonias. The authors suggested that high-risk cocaine-using psychiatric patients who start to take neuroleptic drugs should be provided with a prophylactic anticholinergic drug to prevent neuroleptic drug-induced acute dystonias.

Corticosteroids

By reducing gastrointestinal motility, neuroleptic drugs enhance the absorption of corticosteroids (623).

Digoxin

By reducing gastrointestinal motility, neuroleptic drugs increase the systemic availability of digoxin and other inotropic drugs and thereby increase the potential for toxicity (624).

Guanethidine

Phenothiazines may inhibit the hypotensive action of guanethidine (624). This antagonism does not occur with molindone (622).

Hypoglycemic drugs

Because phenothiazines affect carbohydrate metabolism, they can interfere with the control of blood glucose in diabetes mellitus (445,623).

Levodopa

Levodopa and neuroleptic drugs can interfere with the effects of each other at dopamine receptors; the patient should be monitored for deterioration in both parkinsonism and mental state. If an antiemetic is required in a patient taking levodopa, one that does not affect central dopamine receptors should be chosen.

Lithium

Neuroleptic drugs are often used in mood stabilizer combinations. However, there have been few controlled studies of the use of such combinations, and interactions are potentially dangerous. The advantages and disadvantages of all currently used mood stabilizer combinations have been extensively reviewed (641). Some effects are well known: neurotoxicity, hypotension, somnambulistic-like events, and cardiac and respiratory arrest associated with the combination of lithium and traditional neuroleptic drugs considered as a first-line treatment for classic euphoric mania with psychotic features.

Several cases of neurotoxicity have been reported when lithium was combined with thioridazine, and patients should be carefully monitored (642–644).

Occasionally, long-term use of lithium is associated with cogwheel rigidity and a parkinsonian tremor. More often than not, concurrent or past treatment with a neuroleptic drug is involved. Neuroleptic malignant syndrome has also been reported in patients taking lithium and a neuroleptic drug, such as amoxapine (645), clozapine (646), haloperidol (647), olanzapine, (648), and risperidone (649). However, a causal interaction has not been established (650), and it should also be remembered that lithium toxicity itself can cause neuroleptic malignant syndrome (651).

Lysergic acid diethylamide

Phenothiazines and butyrophenones can counteract the psychoactive effects of lysergic acid diethylamide (LSD) (652).

Metrizamide

The role of chlorpromazine in facilitating metrizamide-induced convulsions has been confirmed in animal studies (653); phenothiazines should be withdrawn at least 48 hours before intrathecal use of metrizamide.

Monoamine oxidase inhibitors

There can be additive hypotensive effects when monoamine oxidase inhibitors are combined with neuroleptic drugs.

Four healthy volunteers taking moclobemide (600 mg/day for 9 days) had reduced clearance of dextromethorphan, a marker of hepatic CYP2D6 activity (654). These findings suggest that at the higher end of its therapeutic dosage range, moclobemide may inhibit the metabolism of drugs that are substrates for CYP2D6, for example neuroleptic drugs and tricyclic antidepressants (SEDA-22, 14).

Narcotic analgesics

CNS and respiratory depression due to narcotic analgesics can be enhanced by neuroleptic drugs (623).

Oral contraceptives

Estrogen-containing formulations can further promote neuroleptic drug-induced prolactin stimulation (623).

Neuroleptic malignant syndrome occurred in a woman taking haloperidol and thioridazine 12 hours after she started to take an oral contraceptive. The authors suggested that this could have been a pharmacodynamic interaction involving dopaminergic neurotransmission (655).

Phenytoin

Chlorpromazine, and in some cases other phenothiazines, has been reported to increase plasma phenytoin concentrations (656–659), to reduce plasma phenytoin concentrations (657–661), or to have no effect (658). In one case co-administration of thioridazine caused phenytoin toxicity (662).

Phenytoin induces hepatic microsomal drug-metabolizing enzymes and thus reduces concentrations of haloperidol (663,664), thioridazine (664), and tiotixene (665). In two patients phenytoin reduced plasma clozapine concentrations and worsened psychosis (666).

Piperazine

High doses of piperazine can enhance the adverse effects of chlorpromazine and other phenothiazines (667).

Quinidine

Concurrent administration of quinidine with neuroleptic drugs, particularly thioridazine, can cause myocardial depression (622).

Sedatives and hypnotics

Neuroleptic drugs increase sedative- and hypnotic-induced sleep time and respiratory depression. Lower dosages of barbiturates or other hypnotics should be used, at least initially, in patients receiving neuroleptic drugs (622).

Sevoflurane

- A 19-year-old taking cyamemazine for schizophrenia had severe dystonia during inhalation of sevoflurane (668). One minute after taking four or five maximum breaths in a closed system filled with 8% sevoflurane in a 50% nitrous oxide/oxygen mixture, the patient being unconscious, torticollic posturing began to develop and the stiffness rapidly extended to the left trapezius and scalenus muscles. This was followed by severe rotation of the head, trismus, and opisthotonos. Muscle spasm resolved with an injection of atracurium 30 mg.

An interaction of sevoflurane with cyamemazine was suggested as a possible explanation, without precluding a role of nitrous oxide.

Spiramycin

Acute dystonia occurred during spiramycin therapy in a patient who was being treated with neuroleptic drugs (368).

SSRIs

Paroxetine and fluoxetine are potent inhibitors of CYP2D6, and can therefore interact with neuroleptic drugs that are metabolized by this enzyme.

- Urinary retention and severe constipation occurred in a patient who started to take sertraline when already taking haloperidol and clonazepam (669). After a week, she discontinued sertraline and her symptoms disappeared in a few days.

Three of fourteen adults with Gilles de la Tourette syndrome had acute severe drug-induced parkinsonism when neuroleptic drugs and SSRIs were used together (670). There were no adverse reactions in almost 75 children receiving similar treatment. The authors suggested that age-related changes in dopamine neurotransmitter systems could confer vulnerability. On the other hand, SSRIs themselves can cause parkinsonism and also inhibit P450 enzymes, leading to increases in plasma concentrations of neuroleptic drugs during combined treatment (SEDA-20, 50) (SEDA-21, 51) (SEDA-22, 55). Surprisingly, in a double-blind, randomized, placebo-controlled study, there were no significant differences in plasma haloperidol concentrations when sertraline ($n = 18$) or placebo ($n = 18$) were added at 2, 4, 6, and 8 weeks (671).

Thiazide diuretics

Combined use of neuroleptic drugs and thiazide diuretics has rarely resulted in severe hypotension (622).

Tricyclic antidepressants

Various neuroleptic drugs inhibit the metabolism of imipramine, nortriptyline, and amitriptyline (SED-11, 114) (672,673).

Valproate

The combination of valproate and traditional neuroleptic drugs, a first-line treatment for mixed or rapid-cycling episodes or dysphoric mania with psychotic features, is associated with altered mental status and electroencephalographic abnormalities (641).

Smoking

Smoking, common in schizophrenics (SEDA-20, 44), has important effects on plasma concentrations of neuroleptic drugs.

Interference with Diagnostic Tests

Pregnancy testing

Phenothiazines can cause a false-positive pregnancy test, but only with the virtually obsolete Ascheim–Zondek animal test method (SED-11, 115).

Urinary ketones

Phenothiazines can cause increased urinary ketones (ferric chloride test) (SED-11, 115).

Urinary steroids

Phenothiazines can interfere with the measurement of urinary steroids (colorimetry) or oxosteroids (Porter–Silber test) (SED-11, 115).

References

1. Barnard L, Young AH, Pearson J, Geddes J, O'Brien G. A systematic review of the use of atypical antipsychotics in autism. J Psychopharmacol 2002;16(1):93–101.
2. Kindermann SS, Dolder CR, Bailey A, Katz IR, Jeste DV. Pharmacological treatment of psychosis and agitation in elderly patients with dementia: four decades of experience. Drugs Aging 2002;19(4):257–76.
3. Lehman AF, Kreyenbuhl J, Buchanan RW, Dickerson FB, Dixon LB, Goldberg R, Green-Paden LD, Tenhula WN, Boerescu D, Tek C, Sandson N, Steinwachs DM. The Schizophrenia Patient Outcomes Research Team (PORT): updated treatment recommendations 2003. Schizophr Bull 2004;30: 193–217.
4. Lehman AF, Steinwachs DM. Translating research into practice: The Schizophrenia Patient Outcomes Research Team (PORT) treatment recommendations. Schizophr Bull 1998;24: 1–10.
5. Zumbrunnen TL, Jann MW. Drug interactions with antipsychotic agents. Incidence and therapeutics implications. CNS Drugs 1998;9:381–401.
6. Currier GW, Simpson GM. Antipsychotic medications and fertility. Psychiatr Serv 1998;49(2):175–6.
7. Perry S. Report from the Swedish Council on Technology Assessment in Health Care (SBU). Treatment with neuroleptics. Int J Technol Assess Health Care 1998;14:394–400.
8. Bandelow B, Fritze J, Ruther E. Increased mortality in schizophrenia and the possible influence of antipsychotic treatment. Int J Psychiatry Clin Pract 1998;2(Suppl 2):S49–57.
9. Jibson MD, Tandon R. New atypical antipsychotic medications. J Psychiatr Res 1998;32(3–4):215–28.
10. Meltzer HY, Casey DE, Garver DL, Lasagna L, Marder SR, Masand PS, Miller D, Pickar D, Tandon R. Adverse effects of the atypical antipsychotics. J Clin Psychiatry 1998;59(Suppl 12):17–22.
11. Day JC, Kinderman P, Bentall R. A comparison of patients' and prescribers' beliefs about neuroleptic side-effects: prevalence, distress and causation. Acta Psychiatr Scand 1998;97(1):93–7.
12. Agarwal MR, Sharma VK, Kishore Kumar KV, Lowe D. Non-compliance with treatment in patients suffering from schizophrenia: a study to evaluate possible contributing factors. Int J Soc Psychiatry 1998;44(2):92–106.
13. Stahl SM. Selecting an atypical antipsychotic by combining clinical experience with guidelines from clinical trials. J Clin Psychiatry 1999;60(Suppl 10):31–41.
14. Fleischhacker WW. Clozapine: a comparison with other novel antipsychotics. J Clin Psychiatry 1999;60(Suppl 12):30–4.
15. Breier AF, Malhotra AK, Su TP, Pinals DA, Elman I, Adler CM, Lafargue RT, Clifton A, Pickar D. Clozapine risperidone in chronic schizophrenia: effects on symptoms,

parkinsonian side effects, and neuroendocrine response. Am J Psychiatry 1999;156(2):294–8.

16. Chowdhury AN, Mukherjee A, Ghosh K, Chowdhury S, Das Sen K. Horizon of a new hope: recovery of schizophrenia in India. Int Med J 1999;6:181–5.

17. Leucht S, Pitschel-Walz G, Abraham D, Kissling W. Efficacy and extrapyramidal side-effects of the new antipsychotics olanzapine, quetiapine, risperidone, and sertindole compared to conventional antipsychotics and placebo. A meta-analysis of randomized controlled trials. Schizophr Res 1999;35(1):51–68.

18. Conn DK, Fansabedian N. Pattern of use of neuroleptics and sedative-hypnotic medication in a Canadian long-term care facility. Int J Geriatr Psychopharmacol 1999;2:18–22.

19. Madhusoodanan S, Brenner R, Cohen CI. Role of atypical antipsychotics in the treatment of psychosis and agitation associated with dementia. CNS Drugs 1999;12:135–50.

20. Chan YC, Pariser SF, Neufeld G. Atypical antipsychotics in older adults. Pharmacotherapy 1999;19(7):811–22.

21. Sim K, Su A, Chan YH, Shinfuku N, Kua EH, Tan CH. Clinical correlates of antipsychotic polytherapy in patients with schizophrenia in Singapore. Psychiatry Clin Neurosci 2004;58: 324–9.

22. Lerner V, Libov I, Kotler M, Strous RD. Combination of "atypical" antipsychotic medication in the management of treatment-resistant schizophrenia and schizoaffective disorder. Prog Neuropsychopharmacol Biol Psychiatry 2004;28: 89–98.

23. García del Pozo J, Isusi L, Carvajal A, Martín I, Sáinz M, García del Pozo V, Velasco A. Evolución del consumo de fármacos antipsicóticos en Castilla y León (1990-2001). Rev Esp Salud Pública 2003;77: 725–33.

24. Leucht S, Barnes TRE, Kissling W, Engel RR, Correll C, Kane JM. Relapse prevention in schizophrenia with new-generation antipsychotics: a systematic review and exploratory meta-analysis of randomised, controlled trials. Am J Psychiatry 2003;160: 1209–22.

25. Bender S, Grohmann R, Engel RR, Degner D, Dittmann-Balcar A, Rüther E. Severe adverse drug reactions in psychiatric inpatients treated with neuroleptics. Pharmacopsychiatry 2004;37 Suppl 1: S46–54.

26. Kropp S, Ziegenbein M, Grohmann R, Engel RR, Degner D. Galactorrhea due to psychotropic drugs. Pharmacopsychiatry 2004;37 Suppl 1: S84–8.

27. Stübner S, Grohmann R, Engel R, Bandelow B, Ludwig WD, Wagner G, Müller-Oerlinghausen B, Möller HJ, Hippius H, Rüther E. Blood dyscrasias induced by psychotropic drugs. Pharmacopsychiatry 2004;37 Suppl 1: S70–8.

28. Stübner S, Grohmann R, Engel R, Bandelow B, Ludwig WD, Wagner G, Müller-Oerlinghausen B, Möller HJ, Hippius H, Rüther E. Severe and uncommon involuntary movement disorders due to psychotropic drugs. Pharmacopsychiatry 2004;37 Suppl 1: S54–64.

29. Lieberman JA, Stroup TS, McEvoy JP, Swartz MS, Rosenheck RA, Perkins DO, Keefe RSE, Davis SM, Davis CE, Lebowitz BD. Severe J, Hsiao JK. Effectiveness of antipsychotic drugs in patients with chronic schizophrenia. N Engl J Med 2005;353: 1209–23.

30. Mota NE, Lima MS, Soares BG. Amisulpride for schizophrenia. Cochrane Database Syst Rev 2002;(2):CD001357 http://www.cochrane.org/cochrane/revabstr/ AB001357.htm.

31. Leucht S, Pitschel-Walz G, Engel RR, Kissling W. Amisulpride, an unusual "atypical" antipsychotic: a meta-analysis of randomized controlled trials. Am J Psychiatry 2002;159(2):180–90.

32. Chabannes JP, Pelissolo A, Farah S, Gerard D. Evaluation de l'efficacité et de la tolérance de l'ámisulpride dans le traitement des psychoses schizophréniques. [Evaluation of efficacy and tolerance of amisulpride in treatment of schizophrenic psychoses.] Encephale 1998;24(4):386–92.

33. Puech A, Fleurot O, Rein W. Amisulpride, and atypical antipsychotic, in the treatment of acute episodes of schizophrenia: a dose-ranging study vs. haloperidol. The Amisulpride Study Group. Acta Psychiatr Scand 1998;98(1):65–72.

34. Colonna L, Saleem P, Dondey-Nouvel L, Rein W. Long-term safety and efficacy of amisulpride in subchronic or chronic schizophrenia. The Amisulpride Study Group. Int Clin Psychopharmacol 2000;15(1):13–22.

35. Geddes J, Freemantle N, Harrison P, Bebbington P. Atypical antipsychotics in the treatment of schizophrenia: systematic overview and meta-regression analysis. BMJ 2000;321(7273):1371–6.

36. Worrel JA, Marken PA, Beckman SE, Ruehter VL. Atypical antipsychotic agents: a critical review. Am J Health Syst Pharm 2000;57(3):238–55.

37. Beaumont G. Antipsychotics—the future of schizophrenia treatment. Curr Med Res Opin 2000;16:37–42.

38. Karow A, Naber D. Subjective well-being and quality of life under atypical antipsychotic treatment. Psychopharmacology (Berl) 2002;162(1):3–10.

39. Opolka Davis JM, Chen N, Glick ID. A meta-analysis of the efficacy of second-generation antipsychotics. Arch Gen Psychiatry 2003;60: 553–64.

40. Mensink GJ, Slooff CJ. Novel antipsychotics in bipolar and schizoaffective mania. Acta Psychiatr Scand 2004;109: 405–19.

41. Flashman LA, Green MF. Review of cognition and brain structure in schizophrenia: profiles, longitudinal course, and effects of treatment. Psychiatr Clin North Am 2004;27: 1–18.

42. Voruganti L, Cortese L, Oyewumi L, Cernovsky Z, Zirul S, Awad A. Comparative evaluation of conventional and novel antipsychotic drugs with reference to their subjective tolerability, side-effect profile and impact on quality of life. Schizophr Res 2000;43(2–3):135–45.

43. Advokat CD, Mayville EA, Matson JL. Side effect profiles of atypical antipsychotics, typical antipsychotics, or no psychotropic medications in persons with mental retardation. Res Dev Disabil 2000;21(1):75–84.

44. Volavka J, Czobor P, Sheitman B, Lindenmayer JP, Citrome L, McEvoy JP, Cooper TB, Chakos M, Lieberman JA. Clozapine, olanzapine, risperidone, and haloperidol in the treatment of patients with chronic schizophrenia and schizoaffective disorder. Am J Psychiatry 2002;159(2):255–62.

45. Coulouvrat C, Dondey-Nouvel L. Safety of amisulpride (Solian): a review of 11 clinical studies. Int Clin Psychopharmacol 1999;14(4):209–18.

46. Wahlbeck K, Cheine M, Essali A, Adams C. Evidence of clozapine's effectiveness in schizophrenia: a systematic review and meta-analysis of randomized trials. Am J Psychiatry 1999;156(7):990–9.

47. Howanitz E, Pardo M, Smelson DA, Engelhart C, Eisenstein N, Stern RG, Losonczy MF. The efficacy and safety of clozapine versus chlorpromazine in geriatric schizophrenia. J Clin Psychiatry 1999;60(1):41–4.

48. Chakos M, Lieberman J, Hoffman E, Bradford D, Sheitman B. Effectiveness of second-generation antipsychotics in patients with treatment-resistant schizophrenia:

a review and meta-analysis of randomized trials. Am J Psychiatry 2001;158(4):518–26.

49. Buchanan RW, Breier A, Kirkpatrick B, Ball P, Carpenter WT Jr. Positive and negative symptom response to clozapine in schizophrenic patients with and without the deficit syndrome. Am J Psychiatry 1998;155(6):751–60.

50. Umbricht DS, Wirshing WC, Wirshing DA, McMeniman M, Schooler NR, Marder SR, Kane JM. Clinical predictors of response to clozapine treatment in ambulatory patients with schizophrenia. J Clin Psychiatry 2002;63(5):420–4.

51. Conley RR, Tamminga CA, Bartko JJ, Richardson C, Peszke M, Lingle J, Hegerty J, Love R, Gounaris C, Zaremba S. Olanzapine compared with chlorpromazine in treatment-resistant schizophrenia. Am J Psychiatry 1998;155(7):914–20.

52. Anonymous. Olanzapine. Keep an eye on this neuroleptic. Can Fam Physician 2000;46(322–326):330–6.

53. Lima FB, Cunha RS, Costa LM, Santos-Jess R, De Sena EP, De Miranda-Scippa A, Ribeiro MG, De Oliveira IR. Meta-analysis for evaluate the efficacy and safety of olanzapine compared to haloperidol in the treatment of schizophrenia: preliminary findings. J Bras Psiquiatr 1999;48:169–75.

54. Sanger TM, Lieberman JA, Tohen M, Grundy S, Beasley C Jr, Tollefson GD. Olanzapine versus haloperidol treatment in first-episode psychosis. Am J Psychiatry 1999;156(1):79–87.

55. Breier A, Hamilton SH. Comparative efficacy of olanzapine and haloperidol for patients with treatment-resistant schizophrenia. Biol Psychiatry 1999;45(4):403–11.

56. Tran PV, Tollefson GD, Sanger TM, Lu Y, Berg PH, Beasley CM Jr. Olanzapine versus haloperidol in the treatment of schizoaffective disorder. Acute and long-term therapy. Br J Psychiatry 1999;174:15–22.

57. Berk M, Brook S, Trandafir AI. A comparison of olanzapine with haloperidol in cannabis-induced psychotic disorder: a double-blind randomized controlled trial. Int Clin Psychopharmacol 1999;14(3):177–80.

58. Glick ID, Berg PH. Time to study discontinuation, relapse, and compliance with atypical or conventional antipsychotics in schizophrenia and related disorders. Int Clin Psychopharmacol 2002;17(2):65–8.

59. Altamura AC, Velona I, Curreli R, Mundo E, Bravi D. Is olanzapine better than haloperidol in resistant schizophrenia? A double-blind study in partial responders. Int J Psychiatry Clin Pract 2002;6:107–11.

60. Breier A, Meehan K, Birkett M, David S, Ferchland I, Sutton V, Taylor CC, Palmer R, Dossenbach M, Kiesler G, Brook S, Wright P. A double-blind, placebo-controlled dose-response comparison of intramuscular olanzapine and haloperidol in the treatment of acute agitation in schizophrenia. Arch Gen Psychiatry 2002;59(5):441–8.

61. Hamilton SH, Revicki DA, Edgell ET, Genduso LA, Tollefson G. Clinical and economic outcomes of olanzapine compared with haloperidol for schizophrenia. Results from a randomised clinical trial. Pharmacoeconomics 1999;15(5):469–80.

62. Rosenheck R, Perlick D, Bingham S, Liu-Mares W, Collins J, Warren S, Leslie D, Allan E, Campbell EC, Caroff S, Corwin J, Davis L, Douyon R, Dunn L, Evans D, Frecska E, Grabowski J, Graeber D, Herz L, Kwon K, Lawson W, Mena F, Sheikh J, Smelson D, Smith-Gamble V. Department of Veterans Affairs Cooperative Study Group on the Cost-Effectiveness of Olanzapine.

Effectiveness and cost of olanzapine and haloperidol in the treatment of schizophrenia: a randomized controlled trial. JAMA 2003;290(20):2693–702.

63. Onofrj M, Paci C, D'Andreamatteo G, Toma L. Olanzapine in severe Gilles de la Tourette syndrome: a 52-week double-blind cross-over study vs. low-dose pimozide. J Neurol 2000;247(6):443–6.

64. Sacristan JA, Gomez JC, Montejo AL, Vieta E, Gregor KJ. Doses of olanzapine, risperidone, and haloperidol used in clinical practice: results of a prospective pharmacoepidemiologic study. EFESO Study Group. Estudio Farmacoepidemiologico en la Esquizofrenia con Olanzapina. Clin Ther 2000;22(5):583–99.

65. Gareri P, Cotroneo A, Lacava R, Seminara G, Marigliano N, Loiacono A, De Sarro G. Comparison of the efficacy of new and conventional antipsychotic drugs in the treatment of behavioral and psychological symptoms of dementia (BPSD). Arch Gerontol Geriatr 2004;Suppl 9: 207–15.

66. Copolov DL, Link CG, Kowalcyk B. A multicentre, double-blind, randomized comparison of quetiapine (ICI 204,636, "Seroquel") and haloperidol in schizophrenia. Psychol Med 2000;30(1):95–105.

67. Emsley RA, Raniwalla J, Bailey PJ, Jones AM. A comparison of the effects of quetiapine ("Seroquel") and haloperidol in schizophrenic patients with a history of and a demonstrated, partial response to conventional antipsychotic treatment. PRIZE Study Group. Int Clin Psychopharmacol 2000;15(3):121–31.

68. Purdon SE, Malla A, Labelle A, Lit W. Neuropsychological change in patients with schizophrenia after treatment with quetiapine or haloperidol. J Psychiatry Neurosci 2001;26(2):137–49.

69. Muller-Siecheneder F, Muller MJ, Hillert A, Szegedi A, Wetzel H, Benkert O. Risperidone versus haloperidol and amitriptyline in the treatment of patients with a combined psychotic and depressive syndrome. J Clin Psychopharmacol 1998;18(2):111–20.

70. Heck AH, Haffmans PM, de Groot IW, Hoencamp E. Risperidone versus haloperidol in psychotic patients with disturbing neuroleptic-induced extrapyramidal symptoms: a double-blind, multi-center trial. Schizophr Res 2000;46(2–3):97–105.

71. Csernansky JG, Mahmoud R, Brenner R. Risperidone-USA-79 Study Group. A comparison of risperidone and haloperidol for the prevention of relapse in patients with schizophrenia. N Engl J Med 2002;346(1):16–22.

72. Fleischhacker WW, Lemmens P, van Baelen B. A qualitative assessment of the neurological safety of antipsychotic drugs: an analysis of a risperidone database. Pharmacopsychiatry 2001;34(3):104–10.

73. Curtin F. Prevention of relapse in schizophrenia. N Engl J Med 2002;346(18):1412.

74. Stalman SL. Prevention of relapse in schizophrenia. N Engl J Med 2002;346(18):1412.

75. Lieberman J, Stroup S, Schneider L. Prevention of relapse in schizophrenia. N Engl J Med 2002;346(18):1412.

76. Geddes J. Prevention of relapse in schizophrenia. N Engl J Med 2002;346(1):56–8.

77. Davies A, Adena MA, Keks NA, Catts SV, Lambert T, Schweitzer I. Risperidone versus haloperidol: I. Meta-analysis of efficacy and safety. Clin Ther 1998;20(1):58–71.

78. Davis JM, Chen N. Clinical profile of an atypical antipsychotic: risperidone. Schizophr Bull 2002;28(1):43–61.

79. Han CS, Kim YK. A double-blind trial of risperidone and haloperidol for the treatment of delirium. Psychosomatics 2004;45: 297–301.

80. Skrobik YK, Bergeron N, Dumont M, Gottfried SB. Olanzapine vs haloperidol: treating delirium in a critical care setting. Intensive Care Med 2004;30: 444–9.

81. Bruggeman R, van der Linden C, Buitelaar JK, Gericke GS, Hawkridge SM, Temlett JA. Risperidone versus pimozide in Tourette's disorder: a comparative double-blind parallel-group study. J Clin Psychiatry 2001;62(1):50–6.

82. Allain H, Dautzenberg PH, Maurer K, Schuck S, Bonhomme D, Gerard D. Double blind study of tiapride versus haloperidol and placebo in agitation and aggressiveness in elderly patients with cognitive impairment. Psychopharmacology (Berl) 2000;148(4):361–6.

83. Brook S, Lucey JV, Gunn KP. Intramuscular ziprasidone compared with intramuscular haloperidol in the treatment of acute psychosis. Ziprasidone I.M. Study Group. J Clin Psychiatry 2000;61(12):933–41.

84. Hirsch SR, Kissling W, Bauml J, Power A, O'Connor R. A 28-week comparison of ziprasidone and haloperidol in outpatients with stable schizophrenia. J Clin Psychiatry 2002;63(6):516–23.

85. Goff DC, Posever T, Herz L, Simmons J, Kletti N, Lapierre K, Wilner KD, Law CG, Ko GN. An exploratory haloperidol-controlled dose-finding study of ziprasidone in hospitalized patients with schizophrenia or schizoaffective disorder. J Clin Psychopharmacol 1998;18(4):296–304.

86. Prakash A, Lamb HM. Zotepine. A review of its pharmacodynamic and pharmacokinetic properties and therapeutic efficacy in the management of schizophrenia. CNS Drugs 1998;9:153–75.

87. Cooper SJ, Tweed J, Raniwalla J, Butler A, Welch C. A placebo-controlled comparison of zotepine versus chlorpromazine in patients with acute exacerbation of schizophrenia. Acta Psychiatr Scand 2000;101(3):218–25.

88. Al-Sughayir MA. Depot antipsychotics. Patient characteristics and prescribing pattern. Saudi Med J 2000;21(12):1178–81.

89. Dunayevich E, McElroy SL. Atypical antipsychotics in the treatment of bipolar disorder: pharmacological and clinical effects. CNS Drugs 2000;13:433–41.

90. Guille C, Sachs GS, Ghaemi SN. A naturalistic comparison of clozapine, risperidone, and olanzapine in the treatment of bipolar disorder. J Clin Psychiatry 2000;61(9):638–42.

91. Peuskens J, Bech P, Moller HJ, Bale R, Fleurot O, Rein W. Amisulpride vs. risperidone in the treatment of acute exacerbations of schizophrenia. Amisulpride Study Group. Psychiatry Res 1999;88(2):107–17.

92. Oosthuizen P, Emsley R, Turner HJ, Keyter N. A randomized, controlled comparison of the efficacy and tolerability of low and high doses of haloperidol in the treatment of first-episode psychosis. Int J Neuropsychopharmacol 2004;7: 125–31.

93. Tollefson GD, Birkett MA, Kiesler GM, Wood AJ. Lilly Resistant Schizophrenia Study Group. Double-blind comparison of olanzapine versus clozapine in schizophrenic patients clinically eligible for treatment with clozapine. Biol Psychiatry 2001;49(1):52–63.

94. Gimenez-Roldan S, Mateo D, Navarro E, Gines MM. Efficacy and safety of clozapine and olanzapine: an open-label study comparing two groups of Parkinson's disease patients with dopaminergic-induced psychosis. Parkinsonism Relat Disord 2001;7(2):121–7.

95. Ellis T, Cudkowicz ME, Sexton PM, Growdon JH. Clozapine and risperidone treatment of psychosis in Parkinson's disease. J Neuropsychiatry Clin Neurosci 2000;12(3):364–9.

96. Wahlbeck K, Cheine M, Tuisku K, Ahokas A, Joffe G, Rimon R. Risperidone versus clozapine in treatment-resistant schizophrenia: a randomized pilot study. Prog Neuropsychopharmacol Biol Psychiatry 2000;24(6):911–22.

97. Tracy JI, Monaco CA, Abraham G, Josiassen RC, Pollock BG. Relation of serum anticholinergicity to cognitive status in schizophrenia patients taking clozapine or risperidone. J Clin Psychiatry 1998;59(4):184–8.

98. Ho BC, Miller D, Nopoulos P, Andreasen NC. A comparative effectiveness study of risperidone and olanzapine in the treatment of schizophrenia. J Clin Psychiatry 1999;60(10):658–63.

99. Tran PV, Hamilton SH, Kuntz AJ, Potvin JH, Andersen SW, Beasley C Jr, Tollefson GD. Double-blind comparison of olanzapine versus risperidone in the treatment of schizophrenia and other psychotic disorders. J Clin Psychopharmacol 1997;17(5):407–18.

100. Schooler NR. Comments on article by Tran and colleagues, "Double-blind comparison of olanzapine versus risperidone in treatment of schizophrenia and other psychotic disorders". J Clin Psychopharmacol 1998;18(2):174–6.

101. Tollefson G, Tran PV. Reply. J Clin Psychopharmacol 1998;18:175–6.

102. Gheuens J, Grebb JA. Comments on article by Tran and associates, "Double-blind comparison of olanzapine versus risperidone in treatment of schizophrenia and other psychotic disorders". J Clin Psychopharmacol 1998;18(2):176–9.

103. Tollefson G, Tran PV. Reply. J Clin Psychopharmacol 1998;18:177–9.

104. Kasper S, Kufferle B. Comments on "Double-blind comparison of olanzapine versus risperidone in the treatment of schizophrenia and other psychotic disorders" by Tran and Associates. J Clin Psychopharmacol 1998;18(4):353–6.

105. Tollefson G, Tran PV. Reply to Kasper and Kufferle. J Clin Psychopharmacol 1998;18:354–5.

106. Cassano GB, Jori MC. AMIMAJOR Group. Efficacy and safety of amisulpride 50 mg versus paroxetine 20 mg in major depression: a randomized, double-blind, parallel group study. Int Clin Psychopharmacol 2002;17(1):27–32.

107. Masi G, Mucci M, Millepiedi S. Clozapine in adolescent inpatients with acute mania. J Child Adolesc Psychopharmacol 2002;12(2):93–9.

108. Hummel B, Dittmann S, Forsthoff A, Matzner N, Amann B, Grunze H. Clozapine as add-on medication in the maintenance treatment of bipolar and schizoaffective disorders. A case series. Neuropsychobiology 2002;45(Suppl 1):37–42.

109. Tohen M, Baker RW, Altshuler LL, Zarate CA, Suppes T, Ketter TA, Milton DR, Risser R, Gilmore JA, Breier A, Tollefson GA. Olanzapine versus divalproex in the treatment of acute mania. Am J Psychiatry 2002;159(6):1011–7.

110. Tohen M, Chengappa KN, Suppes T, Zarate CA Jr, Calabrese JR, Bowden CL, Sachs GS, Kupfer DJ, Baker RW, Risser RC, Keeter EL, Feldman PD, Tollefson GD, Breier A. Efficacy of olanzapine in combination with valproate or lithium in the treatment of mania in patients partially nonresponsive to valproate or lithium monotherapy. Arch Gen Psychiatry 2002;59(1):62–9.

111. Marangell LB, Johnson CR, Kertz B, Zboyan HA, Martinez JM. Olanzapine in the treatment of apathy in previously depressed participants maintained with selective serotonin reuptake inhibitors: an open-label, flexible-dose study. J Clin Psychiatry 2002;63(5):391–5.

112. Zullino DF, Quinche P, Hafliger T, Stigler M. Olanzapine improves social dysfunction in cluster B personality disorder. Hum Psychopharmacol 2002;17(5):247–51.

113. Zarate CA, Tohen M. Double-blind comparison of the continued use of antipsychotic treatment versus its discontinuation in remitted manic patients. Am J Psychiatry 2004;161: 169–71.

114. McCracken JT, McGough J, Shah B, Cronin P, Hong D, Aman MG, Arnold LE, Lindsay R, Nash P, Hollway J, McDougle CJ, Posey D, Swiezy N, Kohn A, Scahill L, Martin A, Koenig K, Volkmar F, Carroll D, Lancor A, Tierney E, Ghuman J, Gonzalez NM, Grados M, Vitiello B, Ritz L, Davies M, Robinson J, McMahon D. Research Units on Pediatric Psychopharmacology Autism Network. Risperidone in children with autism and serious behavioral problems. N Engl J Med 2002;347(5):314–21.

115. Lee PE, Gill SS, Freedman M, Bronskill SE, Hillmer MP, Rochon PA. Atypical antipsychotic drugs in the treatment of behavioural and psychological symptoms of dementia: systematic review. BMJ 2004;329: 75.

116. Schmid C, Grohmann R, Engel RR, Rüther E, Kropp S. Cardiac adverse effects associated with psychotropic drugs. Pharmacopsychiatry 2004;37 Suppl 1: S65–9.

117. Carlsson C, Dencker SJ, Grimby G, Häggendal J. Noradrenaline in blood-plasma and urine during chlorpromazine treatment. Lancet 1966;1:1208.

118. Chatterton R. Eosinophilia after commencement of clozapine treatment. Aust NZ J Psychiatry 1997;31(6):874–6.

119. Leo RJ, Kreeger JL, Kim KY. Cardiomyopathy associated with clozapine. Ann Pharmacother 1996;30(6):603–5.

120. Juul Povlsen U, Noring U, Fog R, Gerlach J. Tolerability and therapeutic effect of clozapine. A retrospective investigation of 216 patients treated with clozapine for up to 12 years. Acta Psychiatr Scand 1985;71(2):176–85.

121. Petrie WM, Ban TA, Berney S, Fujimori M, Guy W, Ragheb M, Wilson WH, Schaffer JD. Loxapine in psychogeriatrics: a placebo- and standard-controlled clinical investigation. J Clin Psychopharmacol 1982;2(2):122–6.

122. Mandelstam JP. An inquiry into the use of Innovar for pediatric premedication. Anesth Analg 1970;49(5):746–50.

123. Yanovski A, Kron RE, Townsend RR, Ford V. The clinical utility of ambulatory blood pressure and heart rate monitoring in psychiatric inpatients. Am J Hypertens 1998;11(3 Pt 1):309–15.

124. Branchey MH, Lee JH, Amin R, Simpson GM. High- and low-potency neuroleptics in elderly psychiatric patients. JAMA 1978;239(18):1860–2.

125. Haddad PM, Anderson IM. Antipsychotic-related QT$_c$ prolongation, torsade de pointes and sudden death. Drugs 2002;62(11):1649–71.

126. Warner JP, Barnes TR, Henry JA. Electrocardiographic changes in patients receiving neuroleptic medication. Acta Psychiatr Scand 1996;93(4):311–3.

127. FDA Psychopharmacological Drugs Advisory Committee. Briefing Document of Zeldox Capsules (Ziprasidone HCl) 19 July 2000; www.fda.gov/ohrms/dockets/ac00/backgrd/361b1a.pdf 19/07/2000.

128. Straus SM, Sturkenboom MC, Bleumink GS, Dieleman JP, van der Lei J, de Graeff PA, Kingma JH, Stricker BH. Non-cardiac QTc-prolonging drugs and the risk of sudden cardiac death. Eur Heart J 2005;26: 2007–12.

129. Su K-P, Shen WW, Chuang C-L, Chen K-P, Chen CC. A pilot cross-over design study on QTc interval prolongation associated with sulpiride and haloperidol. Schizophr Res 2002;59: 93–4.

130. Desai M, Tanus-Santos JE, Li L, Gorski JC, Arefayene M, Liu Y, Desta Z, Flockhart DA. Pharmacokinetics and QT interval pharmacodynamics of oral haloperidol in poor and extensive metabolizers of CYP2D6. Pharmacogenomics J 2003;3: 105–13.

131. Herrero Hernández R, Cidoncha Gallego M, Herrero de Lucas E, Jiménez Lendínez M. Haloperidol por vía intravenosa y torsade de pointes. Med Intensiva 2004;28: 89–90.

132. Dessertenne F. La tachycardie ventriculaire a deux foyers opposés variables. [Ventricular tachycardia with 2 variable opposing foci.] Arch Mal Coeur Vaiss 1966;59(2):263–72.

133. Buckley NA, Sanders P. Cardiovascular adverse effects of antipsychotic drugs. Drug Saf 2000;23(3):215–28.

134. Moss AJ. The QT interval and torsade de pointes. Drug Saf 1999;21(Suppl 1):5–10.

135. Viskin S. Long QT syndromes torsade de pointes. Lancet 1999;354(9190):1625–33.

136. Glassman AH, Bigger JT Jr. Antipsychotic drugs: prolonged QT$_c$ interval, torsade de pointes, and sudden death. Am J Psychiatry 2001;158(11):1774–82.

137. Yap YG, Camm J. Risk of torsades de pointes with noncardiac drugs. Doctors need to be aware that many drugs can cause qt prolongation. BMJ 2000;320(7243):1158–9.

138. Kiriike N, Maeda Y, Nishiwaki S, Izumiya Y, Katahara S, Mui K, Kawakita Y, Nishikimi T, Takeuchi K, Takeda T. Iatrogenic torsade de pointes induced by thioridazine. Biol Psychiatry 1987;22(1):99–103.

139. Connolly MJ, Evemy KL, Snow MH. Torsade de pointes ventricular tachycardia in association with thioridazine therapy: report of two cases. New Trends Arrhythmias 1985;1:157.

140. Sharma ND, Rosman HS, Padhi ID, Tisdale JE. Torsades de pointes associated with intravenous haloperidol in critically ill patients. Am J Cardiol 1998;81(2):238–40.

141. Michalets EL, Smith LK, Van Tassel ED. Torsade de pointes resulting from the addition of droperidol to an existing cytochrome P450 drug interaction. Ann Pharmacother 1998;32(7–8):761–5.

142. Barber JM. Risk of sudden death on high-dose antipsychotic medication: QT$_c$ dispersion. Br J Psychiatry 1998;173:86–7.

143. Reilly JG, Ayis SA, Ferrier IN, Jones SJ, Thomas SH. QT$_c$-interval abnormalities and psychotropic drug therapy in psychiatric patients. Lancet 2000;355(9209):1048–52.

144. Iwahashi K. Significantly higher plasma haloperidol level during cotreatment with carbamazepine may herald cardiac change. Clin Neuropharmacol 1996;19(3):267–70.

145. Roden DM. Drug-induced prolongation of the QT interval. N Engl J Med 2004;350: 1013–22.

146. Arizona CERT. Center for Education and Research in Therapeutics. http://www.torsades.org/medical-pros/drug-lists/drug-lists.htm

147. Akers WS, Flynn JD, Davis GA, Green AE, Winstead PS, Strobel G. Prolonged cardiac repolarization after tacrolimus and haloperidol administration in the critically ill patient. Pharmacotherapy 2004;24: 404–8.

148. Justo D, Prokhorov V, Heller K, Zeltser D. Torsade de pointes induced by psychotropic drugs and the prevalence of its risk factors. Acta Psychiatr Scand 2005;111: 171–6.

149. Tzivoni D, Banai S, Schuger C, Benhorin J, Keren A, Gottlieb S, Stern S. Treatment of torsade de pointes with magnesium sulfate. Circulation 1988;77(2):392–7.

150. O'Brien JM, Rockwood RP, Suh KI. Haloperidol-induced torsade de pointes. Ann Pharmacother 1999;33(10):1046–50.

151. Stöllberger C, Huber JO, Finsterer J. Antipsychotic drugs and QT prolongation. Int Clin Psychopharmacol 2005;20: 243–51.

152. Coulter DM, Bate A, Meyboom RH, Lindquist M, Edwards IR. Antipsychotic drugs and heart muscle disorder in international pharmacovigilance: data mining study. BMJ 2001;322(7296):1207–9.

153. Cronqvist M, Pierot L, Boulin A, Cognard C, Castaings L, Moret J. Local intraarterial fibrinolysis of thromboemboli occurring during endovascular treatment of intracerebral aneurysm: a comparison of anatomic results and clinical outcome. AJNR Am J Neuroradiol 1998;19(1):157–65.

154. Grahmann H, Suchenwirth R. Thrombosis hazard in chlorpromazine and reserpine therapy of endogenous psychoses. Nervenarzt 1959;30: 224–5.

155. Zornberg GL, Jick H. Antipsychotic drug use and risk of first-time idiopathic venous thromboembolism: a case-control study. Lancet 2000;356(9237):1219–23.

156. Liperoti R, Pedone C, Lapane KL, Mor V, Bernabei R, Gambassi G. Venous thromboembolism among elderly patients treated with atypical and conventional antipsychotic agents. Arch Intern Med 2005;165: 2677–82.

157. Hilpert F, Ricome JL, Auzepy P. Insuffisances respiratoires aiguës durant les traitements au long cours par les neuroleptiques. [Acute respiratory failure during long-term treatment with neuroleptic drugs.] Nouv Presse Méd 1980;9(39):2897–900.

158. Flaherty JA, Lahmeyer HW. Laryngeal–pharyngeal dystonia as a possible cause of asphyxia with haloperidol treatment. Am J Psychiatry 1978;135(11):1414–5.

159. Joseph KS. Asthma mortality and antipsychotic or sedative use. What is the link? Drug Saf 1997;16(6):351–4.

160. Faheem AD, Brightwell DR, Burton GC, Struss A. Respiratory dyskinesia and dysarthria from prolonged neuroleptic use: tardive dyskinesia? Am J Psychiatry 1982;139(4):517–8.

161. Portnoy RA. Hyperkinetic dysarthria as an early indicator of impending tardive dyskinesia. J Speech Hear Disord 1979;44(2):214–9.

162. Kumra S, Jacobsen LK, Lenane M, Smith A, Lee P, Malanga CJ, Karp BI, Hamburger S, Rapoport JL. Case series: spectrum of neuroleptic-induced movement disorders and extrapyramidal side effects in childhood-onset schizophrenia. J Am Acad Child Adolesc Psychiatry 1998;37(2):221–7.

163. Ballard C, Grace J, McKeith I, Holmes C. Neuroleptic sensitivity in dementia with Lewy bodies and Alzheimer's disease. Lancet 1998;351(9108):1032–3.

164. Caligiuri MP, Rockwell E, Jeste DV. Extrapyramidal side effects in patients with Alzheimer's disease treated with low-dose neuroleptic medication. Am J Geriatr Psychiatry 1998;6(1):75–82.

165. Mirsattari SM, Power C, Nath A. Parkinsonism with HIV infection. Mov Disord 1998;13(4):684–9.

166. Kompoliti K, Goetz CG. Hyperkinetic movement disorders misdiagnosed as tics in Gilles de la Tourette syndrome. Mov Disord 1998;13(3):477–80.

167. Bersani G, Gherardelli S, Clemente R, Di Giannantonio M, Grilli A, Conti CM, Exton MS, Conti P, Doyle R, Pancheri P. Neurologic soft signs in schizophrenic patients treated with conventional and atypical antipsychotics. J Clin Psychopharmacol 2005;25: 372–5.

168. Sriwatanakul K. Minimizing the risk of antipsychotic-associated seizures. Drug Ther 1982;12:65.

169. Simpson GM, Cooper TA. Clozapine plasma levels and convulsions. Am J Psychiatry 1978;135(1):99–100.

170. Itil TM, Polvan N, Ucok A, Eper E, Guven F, Hsu W. Comparison of the clinical and electroencephalographical effects of molindone and trifluoperazine in acute schizophrenic patients. Behav Neuropsychiatry 1971;3(5):25–32.

171. Baldessarini RJ. Drugs and the treatment of psychiatric disorders. In: Goodman LS, Gilman A, editors. The Pharmacological Basis of Therapeutics. 6th ed.. New York: MacMillan, 1980:391.

172. Oliver AP, Luchins DJ, Wyatt RJ. Neuroleptic-induced seizures. An in vitro technique for assessing relative risk. Arch Gen Psychiatry 1982;39(2):206–9.

173. Woolley J, Smith S. Lowered seizure threshold on olanzapine. Br J Psychiatry 2001;178(1):85–6.

174. Simpson GM, Pi EH, Sramek JJ Jr. Adverse effects of antipsychotic agents. Drugs 1981;21(2):138–51.

175. Kane JM. Antipsychotic drug side effects: their relationship to dose. J Clin Psychiatry 1985;46(5 Pt 2):16–21.

176. Armstrong M, Daly AK, Blennerhassett R, Ferrier N, Idle JR. Antipsychotic drug-induced movement disorders in schizophrenics in relation to CYP2D6 genotype. Br J Psychiatry 1997;170:23–6.

177. Maurer I, Zierz S, Moller HJ, Jerusalem F. Neuroleptic associated extrapyramidal symptoms. Br J Psychiatry 1996;167:551–2.

178. Chetty M, Gouws E, Miller R, Moodley SV. The use of a side effect as a qualitative indicator of plasma chlorpromazine levels. Eur Neuropsychopharmacol 1999;9(1–2):77–82.

179. Rapoport A, Stein D, Shamir E, Schwartz M, Levine J, Elizur A, Weizman A. Clinico-tremorgraphic features of neuroleptic-induced tremor. Int Clin Psychopharmacol 1998;13(3):115–20.

180. Van Den Bos H, Rosenbauer C, Goldney RD. The assessment of involuntary movement disturbances in a psychiatric hospital population: an Australian experience. Aust J Psychopharmacol 1999;9:42–3.

181. Farde L, Nordstrom AL, Wiesel FA, Pauli S, Halldin C, Sedvall G. Positron emission tomographic analysis of central D_1 and D_2 dopamine receptor occupancy in patients treated with classical neuroleptics and clozapine. Relation to extrapyramidal side effects. Arch Gen Psychiatry 1992;49(7):538–44.

182. Nordstrom AL, Farde L, Wiesel FA, Forslund K, Pauli S, Halldin C, Uppfeldt G. Central D_2-dopamine receptor occupancy in relation to antipsychotic drug effects: a double-blind PET study of schizophrenic patients. Biol Psychiatry 1993;33(4):227–35.

183. Peuskens J. Risperidone in the treatment of patients with chronic schizophrenia: a multi-national, multi-centre, double-blind, parallel-group study versus haloperidol. Risperidone Study Group. Br J Psychiatry 1995;166(6):712–26.

184. Small JG, Hirsch SR, Arvanitis LA, Miller BG, Link CG. Quetiapine in patients with schizophrenia. A high- and low-dose double-blind comparison with placebo. Seroquel Study Group. Arch Gen Psychiatry 1997;54(6):549–57.

185. Tollefson GD, Beasley CM Jr, Tran PV, Street JS, Krueger JA, Tamura RN, Graffeo KA, Thieme ME. Olanzapine versus haloperidol in the treatment of schizophrenia and schizoaffective and schizophreniform disorders: results of an international collaborative trial. Am J Psychiatry 1997;154(4):457–65.

186. Zimbroff DL, Kane JM, Tamminga CA, Daniel DG, Mack RJ, Wozniak PJ, Sebree TB, Wallin BA, Kashkin KB. Controlled, dose-response study of sertindole and haloperidol in the treatment of schizophrenia. Sertindole Study Group. Am J Psychiatry 1997;154(6):782–91.

187. Brody AL. Acute dystonia induced by rapid increase in risperidone dosage. J Clin Psychopharmacol 1996;16(6):461–2.

188. Jauss M, Schroder J, Pantel J, Bachmann S, Gerdsen I, Mundt C. Severe akathisia during olanzapine treatment of acute schizophrenia. Pharmacopsychiatry 1998;31(4):146–8.

189. Landry P, Cournoyer J. Acute dystonia with olanzapine. J Clin Psychiatry 1998;59(7):384.

190. Caligiuri MR, Jeste DV, Lacro JP. Antipsychotic-induced movement disorders in the elderly: epidemiology and treatment recommendations. Drugs Aging 2000;17(5):363–84.

191. Kapur S, Zipursky R, Jones C, Remington G, Houle S. Relationship between dopamine D(2) occupancy, clinical response, and side effects: a double-blind PET study of first-episode schizophrenia. Am J Psychiatry 2000;157(4):514–20.

192. Salokangas RKR, Honkonen T, Stengard E, Koivisto A-M, Hietala J. Negative symptoms and neuroleptics in catatonic schizophrenia. Schizophr Res 2002;59: 73–6.

193. Pijnenburg YAL, Sampson EL, Harvey RJ, Fox NC, Rossor MN. Vulnerability to neuroleptic side effects in frontotemporal lobar degeneration. Int Geriatr Psychiatry 2003;18: 67–72.

194. Sikich L, Hamer RM, Bashford RA, Sheitman BB, Lieberman JA. A pilot study of risperidone, olanzapine, and haloperidol in psychotic youth: a double-blind, randomized, 8-week trial. Neuropsychopharmacology 2004;29: 133–45.

195. Strous RD, Alvir JM, Robinson D, Gal G, Sheitman B, Chakos M, Lieberman JA. Premorbid functioning in schizophrenia: relation to baseline symptoms, treatment response, and medication side effects. Schizophr Bull 2004;30: 265–78.

196. Dhavale HS, Pinto C, Dass J, Nayak A, Kedare J, Kamat M, Dewan M. Prophylaxis of antipsychotic-induced extrapyramidal side effects in east Indians: cultural practice or biological necessity?. J Psychiatr Pract 2004;10: 200–2.

197. Wynchank D, Berk M. Efficacy of nefazodone in the treatment of neuroleptic induced extrapyramidal side effects: a double-blind randomised parallel group placebo-controlled trial. Hum Psychopharmacol 2003;18: 271–5.

198. Mazurek MF, Rosebush PI. Circadian pattern of acute, neuroleptic-induced dystonic reactions. Am J Psychiatry 1996;153(5):708–10.

199. Koek RJ, Pi EH. Acute laryngeal dystonic reactions to neuroleptics. Psychosomatics 1989;30(4):359–64.

200. Sovner R, McGorrill S. Stress as a precipitant of neuroleptic-induced dystonia. Psychosomatics 1982;23(7):707–9.

201. Angus JW, Simpson GM. Hysteria and drug-induced dystonia. Acta Psychiatr Scand Suppl 1970;212:52–8.

202. Iqbal N. Heroin use, diplopia, largactil. Saudi Med J 2000;21(12):1194.

203. Ibrahim ZY, Brooks EF. Neuroleptic-induced bilateral temporomandibular joint dislocation. Am J Psychiatry 1996;153(2):293–4.

204. Ilchef R. Neuroleptic-induced laryngeal dystonia can mimic anaphylaxis. Aust NZ J Psychiatry 1997;31(6):877–9.

205. Fines RE, Brady WJ Jr, Martin ML. Acute laryngeal dystonia related to neuroleptic agents. Am J Emerg Med 1999;17(3):319–20.

206. Bulling M. Drug-induced dysphagia. Aust NZ J Med 1999;29(5):748.

207. Christodoulou C, Kalaitzi C. Antipsychotic drug-induced acute laryngeal dystonia: two case reports and a mini review. J Psychopharmacol 2005;19: 307–12.

208. van Harten PN, van Trier JC, Horwitz EH, Matroos GE, Hoek HW. Cocaine as a risk factor for neuroleptic-induced acute dystonia. J Clin Psychiatry 1998;59(3):128–30.

209. Lynch G, Green JF, King DJ. Antipsychotic drug-induced dysphoria. Br J Psychiatry 1996;169(4):524.

210. Hollistar LE. Antipsychotic drug-induced dysphoria. Br J Psychiatry 1997;170:387.

211. King DJ. Antipsychotic drug-induced dysphoria. Br J Psychiatry 1996;168:656.

212. Kasantikul D. Drug-induced akathisia and suicidal tendencies in psychotic patients. J Med Assoc Thai 1998;81(7):551–4.

213. Ashleigh EA, Larsen PD. A syndrome of increased affect in response to risperidone among patients with schizophrenia. Psychiatr Serv 1998;49(4):526–8.

214. Perenyi A, Norman T, Hopwood M, Burrows G. Negative symptoms, depression, and parkinsonian symptoms in chronic, hospitalised schizophrenic patients. J Affect Disord 1998;48(2–3):163–9.

215. Tollefson GD, Sanger TM, Lu Y, Thieme ME. Depressive signs and symptoms in schizophrenia: a prospective blinded trial of olanzapine and haloperidol. Arch Gen Psychiatry 1998;55(3):250–8.

216. Miller CH, Hummer M, Oberbauer H, Kurzthaler I, DeCol C, Fleischhacker WW. Risk factors for the development of neuroleptic induced akathisia. Eur Neuropsychopharmacol 1997;7(1):51–5.

217. Brune M. The incidence of akathisia in bipolar affective disorder treated with neuroleptics—a preliminary report. J Affect Disord 1999;53(2):175–7.

218. Drotts DL, Vinson DR. Prochlorperazine induces akathisia in emergency patients. Ann Emerg Med 1999;34(4 Pt 1):469–75.

219. Brune M. Acute neuroleptic-induced akathisia in patients with traumatic paraplegia: two case reports. Gen Hosp Psychiatry 1999;21(5):386–8.

220. Kurzthaler I, Hummer M, Kohl C, Miller C, Fleischhacker WW. Propranolol treatment of olanzapine-induced akathisia. Am J Psychiatry 1997;154(9):1316.

221. Auzou P, Ozsancak C, Hannequin D, Augustin P. Akathisie déclenchée par de faibles doses de neuroleptiques après un infarctus pallidal. [Akathisia induced by low doses of neuroleptics after a pallidal infarction.] Presse Méd 1996;25(6):260.

222. Weiss D, Aizenberg D, Hermesh H, Zemishlany Z, Munitz H, Radwan M, Weizman A. Cyproheptadine treatment in neuroleptic-induced akathisia. Br J Psychiatry 1995;167(4):483–6.

223. Poyurovsky M, Fuchs C, Weizman A. Low-dose mianserin in treatment of acute neuroleptic-induced akathisia. J Clin Psychopharmacol 1998;18(3):253–4.

224. Leong G, Arturo Silva J. Neuroleptic-induced akathisia and violence: a review. J Forensic Sci 2003;48: 187–9.

225. Hirose S, Ashby CR. Intravenous biperiden in akathisia: an open pilot study. Int J Psychiatry Med 2000;30(2):185–94.

226. Vinson DR, Drotts DL. Diphenhydramine for the prevention of akathisia induced by prochlorperazine: a randomized, controlled trial. Ann Emerg Med 2001;37(2):125–31.

227. Poyurovsky M, Epshtein S, Fuchs C, Schneidman M, Weizman R, Weizman A. Efficacy of low-dose mirtazapine in neuroleptic-induced akathisia: a double-blind

randomized placebo-controlled pilot study. J Clin Psychopharmacol 2003;23(3):305–8.

228. Poyurovsky M, Weizman A. Mirtazapine for neuroleptic-induced akathisia. Am J Psychiatry 2001;158(5):819.

229. Smith M. Case report: akathisia abolished by passive movement. Acta Psychiatr Scand 1998;97(2):168–9.

230. Hofmann M, Seifritz E, Botschev C, Krauchi K, Muller-Spahn F. Serum iron and ferritin in acute neuroleptic akathisia. Psychiatry Res 2000;93(3):201–7.

231. Hayashi T, Nishikawa T, Koga I, Uchida Y, Yamawaki S. Life-threatening dysphagia following prolonged neuroleptic therapy. Clin Neuropharmacol 1997;20(1):77–81.

232. Bashford G, Bradd P. Drug-induced Parkinsonism associated with dysphagia and aspiration: a brief report. J Geriatr Psychiatry Neurol 1996;9(3):133–5.

233. Leopold NA. Dysphagia in drug-induced parkinsonism: a case report. Dysphagia 1996;11(2):151–3.

234. Seeman P, Tallerico T. Antipsychotic drugs which elicit little or no parkinsonism bind more loosely than dopamine to brain D_2 receptors, yet occupy high levels of these receptors. Mol Psychiatry 1998;3(2):123–34.

235. Sweeney JA, Keilp JG, Haas GL, Hill J, Weiden PJ. Relationships between medication treatments and neuropsychological test performance in schizophrenia. Psychiatry Res 1991;37(3):297–308.

236. Krausz M, Moritz SH, Naber D, Lambert M, Andresen B. Neuroleptic-induced extrapyramidal symptoms are accompanied by cognitive dysfunction in schizophrenia. Eur Psychiatry 1999;14(2):84–8.

237. Caligiuri MP, Lacro JP, Jeste DV. Incidence and predictors of drug-induced parkinsonism in older psychiatric patients treated with very low doses of neuroleptics. J Clin Psychopharmacol 1999;19(4):322–8.

238. Merello M, Starkstein S, Petracca G, Cataneo EA, Manes F, Leiguarda R. Drug-induced parkinsonism in schizophrenic patients: motor response and psychiatric changes after acute challenge with L-dopa and apomorphine. Clin Neuropharmacol 1996;19(5):439–43.

239. Konig P, Chwatal K, Havelec L, Riedl F, Schubert H, Schultes H. Amantadine versus biperiden: a double-blind study of treatment efficacy in neuroleptic extrapyramidal movement disorders. Neuropsychobiology 1996;33(2):80–4.

240. Strauss AJ, Bailey RK, Dralle PW, Eschmann AJ, Wagner RB. Conventional psychotropic-induced tremor extinguished by olanzapine. Am J Psychiatry 1998;155(8):1132.

241. Rochon PA, Stukel TA, Sykora K, Gill S, Garfinkel S, Anderson GM, Normand SL, Mamdani M, Lee PE, Li P, Bronskill SE, Marras C, Gurwitz JH. Atypical antipsychotics and parkinsonism. Arch Intern Med 2005;165: 1882–8.

242. Leucht S, Wahlbeck K, Hamann J, Kissling W. New generation antipsychotics versus low-potency conventional antipsychotics: a systematic review and meta-analysis. Lancet 2003;361: 1581–9.

243. Ross DE, Buchanan RW, Lahti AC, Medoff D, Bartko JJ, Compton AD, Thaker GK. The relationship between smooth pursuit eye movements and tardive dyskinesia in schizophrenia. Schizophr Res 1998;31(2–3):141–50.

244. Pi EH, Simpson GM. Tardive dyskinesia and abnormal tongue movements. Br J Psychiatry 1981;139:526–8.

245. Kane JM, Smith JM. Tardive dyskinesia: prevalence and risk factors, 1959–1979. Arch Gen Psychiatry 1982;39(4):473–81.

246. Gervin M, Browne S, Lane A, Clarke M, Waddington JL, Larkin C, O'Callaghan E. Spontaneous abnormal involuntary movements in first-episode schizophrenia and schizophreniform disorder: baseline rate in a group of patients from an Irish catchment area. Am J Psychiatry 1998;155(9):1202–6.

247. Baldessarini RJ. Clinical and epidemiologic aspects of tardive dyskinesia. J Clin Psychiatry 1985;46(4 Pt 2):8–13.

248. Delwaide PJ, Desseilles M. Spontaneous buccolinguofacial dyskinesia in the elderly. Acta Neurol Scand 1977;56(3):256–62.

249. Varga E, Sugarman AA, Varga V, Zomorodi A, et al. Prevalence of spontaneous oral dyskinesia in elderly persons. 12th CINP Congress, Gothenburg 1980;.

250. Owens DG, Johnstone EC, Frith CD. Spontaneous involuntary disorders of movement: their prevalence, severity, and distribution in chronic schizophrenics with and without treatment with neuroleptics. Arch Gen Psychiatry 1982;39(4):452–61.

251. Cassady SL, Adami H, Moran M, Kunkel R, Thaker GK. Spontaneous dyskinesia in subjects with schizophrenia spectrum personality. Am J Psychiatry 1998;155(1):70–5.

252. Chatterjee A, Chakos M, Koreen A, Geisler S, Sheitman B, Woerner M, Kane JM, Alvir J, Lieberman JA. Prevalence and clinical correlates of extrapyramidal signs and spontaneous dyskinesia in never-medicated schizophrenic patients. Am J Psychiatry 1995;152(12):1724–9.

253. Fenton WS, Blyler CR, Wyatt RJ, McGlashan TH. Prevalence of spontaneous dyskinesia in schizophrenic and non-schizophrenic psychiatric patients. Br J Psychiatry 1997;171:265–8.

254. Kopala LC. Spontaneous and drug-induced movement disorders in schizophrenia. Acta Psychiatr Scand Suppl 1996;389:12–7.

255. Kane JM, Woerner MG, et al. Does clozapine cause tardive dyskinesia? J Clin Psychiatry 1993;54:327–30.

256. Tollefson GD, Beasley CM Jr, Tamura RN, Tran PV, Potvin JH. Blind, controlled, long-term study of the comparative incidence of treatment-emergent tardive dyskinesia with olanzapine or haloperidol. Am J Psychiatry 1997;154(9):1248–54.

257. Beasley CM, Dellva MA, Tamura RN, Morgenstern H, Glazer WM, Ferguson K, Tollefson GD. Randomised double-blind comparison of the incidence of tardive dyskinesia in patients with schizophrenia during long-term treatment with olanzapine or haloperidol. Br J Psychiatry 1999;174:23–30.

258. Casey DE. Effects of clozapine therapy in schizophrenic individuals at risk for tardive dyskinesia. J Clin Psychiatry 1998;59(Suppl 3):31–7.

259. Gardos G, Samu I, Kallos M, Cole JO. Absence of severe tardive dyskinesia in Hungarian schizophrenic outpatients. Psychopharmacology (Berl) 1980;71(1):29–34.

260. Gardos G, Cole JO. Tardive dyskinesia and anticholinergic drugs. Am J Psychiatry 1983;140(2):200–2.

261. Chouinard G, Bradwejn J. Reversible and irreversible tardive dyskinesia: a case report. Am J Psychiatry 1982;139(3):360–2.

262. Chaplin R, Kent A. Informing patients about tardive dyskinesia. Controlled trial of patient education. Br J Psychiatry 1998;172:78–81.

263. Marder SR, Essock SM, Miller AL, Buchanan RW, Davis JM, Kane JM, Lieberman J, Schooler NR. The Mount Sinai conference on the pharmacotherapy of schizophrenia. Schizophr Bull 2002;28(1):5–16.

264. Havaki-Kontaxaki BJ, Kontaxakis VP, Christodoulou GN. Treatment of tardive pharyngolaryngeal dystonia with olanzapine. Schizophr Res 2004;66: 199–200.

265. Kane JM, Woerner M, Borenstein M, Wegner J, Lieberman J. Integrating incidence and prevalence of tardive dyskinesia. Psychopharmacol Bull 1986;22(1):254–8.

266. Morgenstern H, Glazer WM. Identifying risk factors for tardive dyskinesia among long-term outpatients maintained with neuroleptic medications. Results of the Yale Tardive Dyskinesia Study. Arch Gen Psychiatry 1993;50:723–33.

267. Jeste DV, Caligiri MP, Paulsen JS, Heaton RK, Lacro JP, Harris MJ, Bailey A, Fell RL, McAdams LA. Risk of tardive dyskinesia in older patients: a prospective longitudinal study of 266 outpatients. Arch Gen Psychiatry 1995;52:756–65.

268. Chakos MH, Alvir JM, Woerner MG, Koreen A, Geisler S, Mayerhoff D, Sobel S, Kane JM, Borenstein M, Lieberman JA. Incidence and correlates of tardive dyskinesia in first episode of schizophrenia. Arch Gen Psychiatry 1996;53(4):313–9.

269. Simpson GM, Pi EH, Sramek JJ Jr. Management of tardive dyskinesia: current update. Drugs 1982;23(5):381–93.

270. Ananth J. Tardive dyskinesia: myths and realities. Psychosomatics 1980;21(5):394–6(389–391).

271. Gardos G, Cole JO. Overview: public health issues in tardive dyskinesia. Am J Psychiatry 1980;137(7):776–81.

272. Sramek J, Roy S, Ahrens T, Pinanong P, Cutler NR, Pi E. Prevalence of tardive dyskinesia among three ethnic groups of chronic psychiatric patients. Hosp Community Psychiatry 1991;42(6):590–2.

273. Gray GE, Pi EH. Ethnicity and psychotropic medication related movement disorders. J Pract Psychiatry Behav Health 1998;4:259–64.

274. Pi EH, Gray GE. A cross-cultural perspective on psychopharmacology. Essent Psychopharmacol 1998;2:233–62.

275. Pi EH. Transcultural psychopharmacology: present and future. Psychiatry Clin Newosci 1998;52(Suppl):S185–7.

276. Woerner MG, Alvir JM, Saltz BL, Lieberman JA, Kane JM. Prospective study of tardive dyskinesia in the elderly: rates and risk factors. Am J Psychiatry 1998;155(11):1521–8.

277. Parker G, Lambert T, McGrath J, McGorry P, Tiller J. Neuroleptic management of schizophrenia: a survey and commentary on Australian psychiatric practice. Aust NZ J Psychiatry 1998;32(1):50–8.

278. Simpson GM, Lee JH, Zoubok B, Gardos G. A rating scale for tardive dyskinesia. Psychopharmacology 1979;64:171.

279. Schooler NR, Kane JM. Research diagnoses for tardive dyskinesia. Arch Gen Psychiatry 1982;39(4):486–7.

280. Pi EH, Gutierrez MA, Gray GE. Cross-cultural studies in tardive dyskinesia. Am J Psychiatry 1993;150(6):991.

281. Pi EH, Gutierrez MA, Gray GE. Tardive dyskinesia: cross-cultural perspective. In: Lin KM, Poland R, Nakasaki G, editors. Psychopharmacology and Psychobiology of Ethnicity. Washington DC: American Psychiatric Press Inc, 1993:153–67.

282. Patterson BD, Swingler D, Willows S. Prevalence of and risk factors for tardive dyskinesia in a xhosa population in the eastern cape of south Africa. Schizophr Res 2005;76: 8997.

283. Kane JM, Woerner M, Wernhold P, Wegner J. Results from a prospective study of tardive dyskinesia over a 2-year period development: preliminary findings. Psychopharmacol Bull 1982;18:82.

284. Johnson GF, Hunt GE, Rey JM. Incidence and severity of tardive dyskinesia increase with age. Arch Gen Psychiatry 1982;39(4):486.

285. Chouinard G, Jones BD, Annable L, Ross-Chouinard A. Sex differences and tardive dyskineasia. Am J Psychiatry 1980;137(4):507.

286. Gordon JH, Borison RL, Diamond BI. Modulation of dopamine receptor sensitivity by estrogen. Biol Psychiatry 1980;15(3):389–96.

287. Perenyi A, Arato M. Tardive dyskinesia on Hungarian psychiatric wards. Psychosomatics 1980;21(11):904–9.

288. Byne W, White L, Parella M, Adams R, Harvey PD, Davis KL. Tardive dyskinesia in a chronically institutionalized population of elderly schizophrenic patients: prevalence and association with cognitive impairment. Int J Geriatr Psychiatry 1998;13(7):473–9.

289. McDermid SA, Hood J, Bockus S, D'Alessandro E. Adolescents on neuroleptic medication: is this population at risk for tardive dyskinesia? Can J Psychiatry 1998;43(6):629–31.

290. Yuen O, Caligiuri MP, Williams R, Dickson RA. Tardive dyskinesia and positive and negative symptoms of schizophrenia. A study using instrumental measures. Br J Psychiatry 1996;168(6):702–8.

291. van Harten PN, Hoek HW, Matroos GE, Koeter M, Kahn RS. Intermittent neuroleptic treatment and risk for tardive dyskinesia. Curacao Extrapyramidal Syndromes Study III. Am J Psychiatry 1998;155(4):565–7.

292. Jeste DV, Linnoila M, Wagner RL, Wyatt RJ. Serum neuroleptic concentrations and tardive dyskinesia. Psychopharmacology (Berl) 1982;76(4):377–80.

293. Widerlov E, Haggstrom JE, Kilts CD, Andersson U, Breese GR, Mailman RB. Serum concentrations of thioridazine, its major metabolites and serum neuroleptic-like activities in schizophrenics with and without tardive dyskinesia. Acta Psychiatr Scand 1982;66(4):294–305.

294. Yassa R, Nair NP, Iskander H, Schwartz G. Factors in the development of severe forms of tardive dyskinesia. Am J Psychiatry 1990;147(9):1156–63.

295. Jeste DV, Potkin SG, Sinha S, Feder S, Wyatt RJ. Tardive dyskinesia—reversible and persistent. Arch Gen Psychiatry 1979;36(5):585–90.

296. Jeste DV, Lacro JP, Palmer B, Rockwell E, Harris MJ, Caligiuri MP. Incidence of tardive dyskinesia in early stages of low-dose treatment with typical neuroleptics in older patients. Am J Psychiatry 1999;156(2):309–11.

297. Morgenstern H, Glazer WM, Doucette JT. Handedness and the risk of tardive dyskinesia. Biol Psychiatry 1996;40(1):35–42.

298. Woerner MG, Saltz BL, Kane JM, Lieberman JA, Alvir JM. Diabetes and development of tardive dyskinesia. Am J Psychiatry 1993;150(6):966–8.

299. Pi EH, Simpson GM. Prevention of tardive dyskinesia. In: Shah NS, Donald AG, editors. Neurobehavioral Dysfunction Induced by Psychotherapeutic Agents. Neurophysical, Neuropharmacological Bases and Clinical Implications. New York: Plenum Publishing, 1983.

300. Ozdemir V, Basile VS, Masellis M, Kennedy JL. Pharmacogenetic assessment of antipsychotic-induced movement disorders: contribution of the dopamine D_3 receptor cytochrome P450 1A2 genes. J Biochem Biophys Methods 2001;47(1–2):151–7.

301. Steen VM, Lovlie R, MacEwan T, McCreadie RG. Dopamine D_3-receptor gene variant and susceptibility to tardive dyskinesia in schizophrenic patients. Mol Psychiatry 1997;2(2):139–45.

302. Tsai G, Goff DC, Chang RW, Flood J, Baer L, Coyle JT. Markers of glutamatergic neurotransmission and oxidative

stress associated with tardive dyskinesia. Am J Psychiatry 1998;155(9):1207–13.

303. Wirshing DA, Bartzokis G, Pierre JM, Wirshing WC, Sun A, Tishler TA, Marder SR. Tardive dyskinesia and serum iron indices. Biol Psychiatry 1998;44(6):493–8.

304. Rosen AM, Mukherjee S, Olarte S, Varia V, Cardenas C. Perception of tardive dyskinesia in outpatients receiving maintenance neuroleptics. Am J Psychiatry 1982;139(3):372–4.

305. Wegner JT, Kane JM. Follow-up study on the reversibility of tardive dyskinesia. Am J Psychiatry 1982;139(3):368–9.

306. Seeman MV. Tardive dyskinesia: two-year recovery. Compr Psychiatry 1981;22(2):189–92.

307. Jeste DV, Wyatt RJ. Changing epidemiology of tardive dyskinesia: an overview. Am J Psychiatry 1981;138(3):297–309.

308. Pi EH, Kusuda M, Gray GE, et al. Cross-cultural psychopharmacology: neuroleptic-induced movement disorders. In: Keyzer H, Eckert GM, Forrest IT, Gupta RR, Gutmann F, Molnar J, editors. Thiazines and Structurally Related Compounds. Melbourne, Florida: Krieger Publishing Co, 1992.

309. Jeste DV, Wyatt RJ. Therapeutic strategies against tardive dyskinesia. Two decades of experience. Arch Gen Psychiatry 1982;39(7):803–16.

310. Ondo WG, Hanna PA, Jankovic J. Tetrabenazine treatment for tardive dyskinesia: assessment by randomized videotape protocol. Am J Psychiatry 1999;156(8):1279–81.

311. Brown K, Reid A, White T, Henderson T, Hukin S, Johnstone C, Glen A. Vitamin E, lipids, and lipid peroxidation products in tardive dyskinesia. Biol Psychiatry 1998;43(12):863–7.

312. Elkashef AM, Ruskin PE, Bacher N, Barrett D. Vitamin E in the treatment of tardive dyskinesia. Am J Psychiatry 1990;147(4):505–6.

313. Lohr JB, Cadet JL, Lohr MA, Jeste DV, Wyatt RJ. Alpha-tocopherol in tardive dyskinesia. Lancet 1987;1(8538):913–4.

314. Adler LA, Edson R, Lavori P, Peselow E, Duncan E, Rosenthal M, Rotrosen J. Long-term treatment effects of vitamin E for tardive dyskinesia. Biol Psychiatry 1998;43(12):868–72.

315. Lohr JB, Caligiuri MP. A double-blind placebo-controlled study of vitamin E treatment of tardive dyskinesia. J Clin Psychiatry 1996;57(4):167–73.

316. Barak Y, Swartz M, Shamir E, Stein D, Weizman A. Vitamin E (alpha-tocopherol) in the treatment of tardive dyskinesia: a statistical meta-analysis. Ann Clin Psychiatry 1998;10(3):101–5.

317. Elkashef AM, Wyatt RJ. Tardive dyskinesia: possible involvement of free radicals and treatment with vitamin E. Schizophr Bull 1999;25(4):731–40.

318. Dorfman-Etrog P, Hermesh H, Prilipko L, Weizman A, Munitz H. The effect of vitamin E addition to acute neuroleptic treatment on the emergence of extrapyramidal side effects in schizophrenic patients: an open label study. Eur Neuropsychopharmacol 1999;9(6):475–7.

319. Lerner V, Liberman M. Movement disorders and psychotic symptoms treated with pyridoxine: a case report. J Clin Psychiatry 1998;59(11):623–4.

320. Fehr C, Dahmen N, Klawe C, Eicke M, Szegedi A. Piracetam in the treatment of tardive dyskinesia and akathisia: a case report. J Clin Psychopharmacol 2001;21(2):248–9.

321. Donatelli A, Geisen L, Feuer E. Case report of adverse effect of reserpine on tardive dyskinesia. Am J Psychiatry 1983;140(2):239–40.

322. Schrodt GR Jr, Wright JH, Simpson R, Moore DP, Chase S. Treatment of tardive dyskinesia with propranolol. J Clin Psychiatry 1982;43(8):328–31.

323. Korsgaard S, Casey DE, Gerlach J, Hetmar O, Kaldan B, Mikkelsen LB. The effect of tetrahydroisoxazolopyridinol (THIP) in tardive dyskinesia: a new gamma-aminobutyric acid agonist. Arch Gen Psychiatry 1982;39(9):1017–21.

324. Jeste DV, Wisniewski AA, Wyatt RJ. Neuroleptic-associated tardive syndromes. Psychiatr Clin North Am 1986;9(1):183–92.

325. Simpson GM, Pi EH, Sramek JJ. An update on tardive dyskinesia. Hosp Community Psychiatry 1986;37(4):362–9.

326. Yadalam KG, Korn ML, Simpson GM. Tardive dystonia: four case histories. J Clin Psychiatry 1990;51(1):17–20.

327. Yassa R, Nair V, Dimitry R. Prevalence of tardive dystonia. Acta Psychiatr Scand 1986;73(6):629–33.

328. Burke RE, Fahn S, Jankovic J, Marsden CD, Lang AE, Gollomp S, Ilson J. Tardive dystonia: late-onset and persistent dystonia caused by antipsychotic drugs. Neurology 1982;32(12):1335–46.

329. Kiriakakis V, Bhatia KP, Quinn NP, Marsden CD. The natural history of tardive dystonia. A long-term follow-up study of 107 cases. Brain 1998;121(11):2053–66.

330. Suzuki T, Hori T, Baba A, Abe S, Shiraishi H, Moroji T, Piletz JE. Effectiveness of anticholinergics and neuroleptic dose reduction on neuroleptic-induced pleurothotonus (the Pisa syndrome). J Clin Psychopharmacol 1999;19(3):277–80.

331. Maeda K, Ohsaki T, Kuki K, Kin K, Ikeda M, Matsumoto Y. Severe antecollis during antipsychotics treatment: a report of three cases. Prog Neuropsychopharmacol Biol Psychiatry 1998;22(5):749–59.

332. Molho ES, Feustel PJ, Factor SA. Clinical comparison of tardive and idiopathic cervical dystonia. Mov Disord 1998;13(3):486–9.

333. Raja M. Managing antipsychotic-induced acute and tardive dystonia. Drug Saf 1998;19(1):57–72.

334. Tan EK, Jankovic J. Tardive and idiopathic oromandibular dystonia: a clinical comparison. J Neurol Neurosurg Psychiatry 2000;68(2):186–90.

335. Looper KJ, Chouinard G. Beneficial effects of combined L-dopa and central anticholinergic in a patient with severe drug-induced parkinsonism and tardive dystonia. Can J Psychiatry 1998;43(6):646–7.

336. Krack P, Schneider S, Deuschl G. Geste device in tardive dystonia with retrocollis and opisthotonic posturing. Mov Disord 1998;13(1):155–7.

337. Mennesson M, Klink BA, Fortin AH 6th. Case study: worsening Tourette's disorder or withdrawal dystonia? J Am Acad Child Adolesc Psychiatry 1998;37(7):785–8.

338. Deshmukh DK, Joshi VS, Agarwal MR. Rabbit syndrome—a rare complication of long-term neuroleptic medication. Br J Psychiatry 1990;157:293.

339. Pelonero AL, Levenson JL, Pandurangi AK. Neuroleptic malignant syndrome: a review. Psychiatr Serv 1998;49(9):1163–72.

340. Velamoor VR. Neuroleptic malignant syndrome. Recognition, prevention and management. Drug Saf 1998;19(1):73–82.

341. Caroff SN. The neuroleptic malignant syndrome. J Clin Psychiatry 1980;41(3):79–83.

342. Sachdev P, Mason C, Hadzi-Pavlovic D. Case-control study of neuroleptic malignant syndrome. Am J Psychiatry 1997;154(8):1156–8.

343. Burke RE, Fahn S, Mayeux R, Weinberg H, Louis K, Willner JH. Neuroleptic malignant syndrome caused by

dopamine-depleting drugs in a patient with Huntington disease. Neurology 1981;31(8):1022–5.

344. Keyser DL, Rodnitzky RL. Neuroleptic malignant syndrome in Parkinson's disease after withdrawal or alteration of dopaminergic therapy. Arch Intern Med 1991;151(4):794–6.

345. Simpson GM, Pi EH. Whats happening to the neuroleptic malignant syndrome? Psychiatr Ann 1996;26:172–4.

346. Weller M, Kornhuber J. Clozapine and neuroleptic malignant syndrome: a never-ending story. J Clin Psychopharmacol 1997;17(3):233–4.

347. Hynes AF, Vickar EL. Case study: neuroleptic malignant syndrome without pyrexia. J Am Acad Child Adolesc Psychiatry 1996;35(7):959–62.

348. Tanaka K, Akechi T, Yamazaki M, Hayashi R, Nishiwaki Y, Uchitomi Y. Neuroleptic malignant syndrome during haloperidol treatment in a cancer patient. A case report. Support Care Cancer 1998;6(6):536–8.

349. Duggal HS, Nizamie SH. Neuroleptic malignant syndrome precipitated by promethazine and lorazepam. Aust NZ J Psychiatry 2001;35(2):250–1.

350. Mujica R, Weiden P. Neuroleptic malignant syndrome after addition of haloperidol to atypical antipsychotic. Am J Psychiatry 2001;158(4):650–1.

351. Ty EB, Rothner AD. Neuroleptic malignant syndrome in children and adolescents. J Child Neurol 2001;16(3):157–63.

352. Cullinane CA, Brumfield C, Flint LM, Ferrara JJ. Neuroleptic malignant syndrome associated with multiple joint dislocations in a trauma patient. J Trauma 1998;45(1):168–71.

353. Garrido SM, Chauncey TR. Neuroleptic malignant syndrome following autologous peripheral blood stem cell transplantation. Bone Marrow Transplant 1998;21(4):427–8.

354. Blasi C, D'Amore F, Levati M, Bandinelli MC. Sindrome maligna da neurolettici: una patologia neurologica di grande interesse per l'internista. [Malignant neuroleptic syndrome: a neurologic pathology of great interest to internal medicine.] Ann Ital Med Interna 1998;13:111–6.

355. Elizalde-Sciavolino C, Racco A, Proscia-Lieto T, Kleiner M. Severe hyponatremia, neuroleptic malignant syndrome, rhabdomyolysis and acute renal failure: a case report. Mt Sinai J Med 1998;65(4):284–8.

356. Yoshino A, Yoshimasu H, Tatsuzawa Y, Asakura T, Hara T. Nonconvulsive status epilepticus in two patients with neuroleptic malignant syndrome. J Clin Psychopharmacol 1998;18(4):347–9.

357. Keck PE Jr, Pope HG Jr, McElroy SL. Frequency and presentation of neuroleptic malignant syndrome: a prospective study. Am J Psychiatry 1987;144(10):1344–6.

358. Levinson DF, Simpson GM. Neuroleptic-induced extrapyramidal symptoms with fever. Heterogeneity of the "neuroleptic malignant syndrome". Arch Gen Psychiatry 1986;43(9):839–48.

359. Anderson ES, Powers PS. Neuroleptic malignant syndrome associated with clozapine use. J Clin Psychiatry 1991;52(3):102–4.

360. DasGupta K, Young A. Clozapine-induced neuroleptic malignant syndrome. J Clin Psychiatry 1991;52(3):105–7.

361. Reddig S, Minnema AM, Tandon R. Neuroleptic malignant syndrome and clozapine. Ann Clin Psychiatry 1993;5(1):25–7.

362. Amore M, Zazzeri N, Berardi D. Atypical neuroleptic malignant syndrome associated with clozapine treatment. Neuropsychobiology 1997;35(4):197–9.

363. Dalkilic A, Grosch WN. Neuroleptic malignant syndrome following initiation of clozapine therapy. Am J Psychiatry 1997;154(6):881–2.

364. Sachdev P, Kruk J, Kneebone M, Kissane D. Clozapine-induced neuroleptic malignant syndrome: review and report of new cases. J Clin Psychopharmacol 1995;15(5):365–71.

365. Trayer JS, Fidler DC. Neuroleptic malignant syndrome related to use of clozapine. J Am Osteopath Assoc 1998;98(3):168–9.

366. Bajjoka I, Patel T, O'Sullivan T. Risperidone-induced neuroleptic malignant syndrome. Ann Emerg Med 1997;30(5):698–700.

367. Dursun SM, Oluboka OJ, Devarajan S, Kutcher SP. High-dose vitamin E plus vitamin B6 treatment of risperidone-related neuroleptic malignant syndrome. J Psychopharmacol 1998;12(2):220–1.

368. Gleason PP, Conigliaro RL. Neuroleptic malignant syndrome with risperidone. Pharmacotherapy 1997;17(3):617–21.

369. Newman M, Adityanjee, Jampala C. Atypical neuroleptic malignant syndrome associated with risperidone treatment. Am J Psychiatry 1997;154(10):1475.

370. Tarsy D. Risperidone and neuroleptic malignant syndrome. JAMA 1996;275(6):446.

371. Webster P, Wijeratne C. Risperidone-induced neuroleptic malignant syndrome. Lancet 1994;344(8931):1228–9.

372. Hasan S, Buckley P. Novel antipsychotics and the neuroleptic malignant syndrome: a review and critique. Am J Psychiatry 1998;155(8):1113–6.

373. Bobolakis I. Neuroleptic malignant syndrome after antipsychotic drug administration during benzodiazepine withdrawal. J Clin Psychopharmacol 2000;20(2):281–3.

374. Caroff SN, Mann SC, McCarthy M, Naser J, Rynn M, Morrison M. Acute infectious encephalitis complicated by neuroleptic malignant syndrome. J Clin Psychopharmacol 1998;18(4):349–51.

375. Alexander PJ, Thomas RM, Das A. Is risk of neuroleptic malignant syndrome increased in the postpartum period? J Clin Psychiatry 1998;59(5):254–5.

376. Rasmussen KG. Risk factors for neuroleptic malignant syndrome. Am J Psychiatry 1998;155(11):1639author reply 1639–1640.

377. Francis A, Chandragiri S, Petrides G. Risk factors for neuroleptic malignant syndrome. Am J Psychiatry 1998;155(11):1639–40.

378. Sachdev P. Risk factors for neuroleptic malignant syndrome. Replies. Am J Psychiatry 1998;155:1639–40.

379. Lee JW. Serum iron in catatonia and neuroleptic malignant syndrome. Biol Psychiatry 1998;44(6):499–507.

380. Askin R, Herken H, Derman H. Recurrent neuroleptic malignant syndrome: a case report. Int Med J 2000;7:287–8.

381. Viejo LF, Morales V, Puñal P, Pérez JL, Sancho RA. Risk factors in neuroleptic malignant syndrome. A case-control study. Acta Psychiatr Scand 2003;107: 45–9.

382. Shiloh R, Valevski A, Bodinger L, Misgav S, Aizenberg D, Dorfman-Etrog P, Weizman A, Munitz H. Precautionary measures reduce risk of definite neuroleptic malignant syndrome in newly typical neuroleptic-treated schizophrenia inpatients. Int Clin Psychopharmacol 2003;18: 147–79.

383. Thomas P, Maron M, Rascle C, Cottencin O, Vaiva G, Goudemand M. Carbamazepine in the treatment of neuroleptic malignant syndrome. Biol Psychiatry 1998;43(4):303–5.

384. McCarron MM, Boettger ML, Peck JJ. A case of neuroleptic malignant syndrome successfully treated with amantadine. J Clin Psychiatry 1982;43(9):381–2.

385. Mueller PS, Vester JW, Fermaglich J. Neuroleptic malignant syndrome. Successful treatment with bromocriptine. JAMA 1983;249(3):386–8.

386. Miyaoka H, Shishikura K, Otsubo T, Muramatsu D, Kamijima K. Diazepam-responsive neuroleptic malignant syndrome: a diagnostic subtype? Am J Psychiatry 1997;154(6):882.

387. Black KJ, Racette B, Perlmutter JS. Preventing contractions in neuroleptic malignant syndrome and dystonia. Am J Psychiatry 1998;155(9):1298–9.

388. Coons DJ, Hillman FJ, Marshall RW. Treatment of neuroleptic malignant syndrome with dantrolene sodium: a case report. Am J Psychiatry 1982;139(7):944–5.

389. Goekoop JG, Carbaat PA. Treatment of neuroleptic malignant syndrome with dantrolene. Lancet 1982;2(8288):49–50.

390. Tsujimoto S, Maeda K, Sugiyama T, Yokochi A, Chikusa H, Maruyama K. Efficacy of prolonged large-dose dantrolene for severe neuroleptic malignant syndrome. Anesth Analg 1998;86(5):1143–4.

391. Nisijima K, Ishiguro T. Electroconvulsive therapy for the treatment of neuroleptic malignant syndrome with psychotic symptoms: a report of five cases. J ECT 1999;15(2):158–63.

392. Alao AO, Yolles JC, Armenta W, Michaels AT. Neuroleptic malignant syndrome spectrum: a case report. J Pharm Technol 1999;15:162–4.

393. Caroff SN, Mann SC, Keck PE Jr, Francis A. Residual catatonic state following neuroleptic malignant syndrome. J Clin Psychopharmacol 2000;20(2):257–9.

394. Rosebush PI, Stewart TD, Gelenberg AJ. Twenty neuroleptic rechallenges after neuroleptic malignant syndrome in 15 patients. J Clin Psychiatry 1989;50(8):295–8.

395. Hayashi T, Furutani M, Taniyama J, Kiyasu M, Hikasa S, Horiguchi J, Yamawaki S. Neuroleptic-induced Meige's syndrome following akathisia: pharmacologic characteristics. Psychiatry Clin Neurosci 1998;52(4):445–8.

396. Wirz-Justice A, Werth E, Savaskan E, Knoblauch V, Gasio PF, Muller-Spahn F. Haloperidol disrupts, clozapine reinstates the circadian rest-activity cycle in a patient with early-onset Alzheimer disease. Alzheimer Dis Assoc Disord 2000;14(4):212–5.

397. Madsen AL, Keidling N, Karle A, Esbjerg S, Hemmingsen R. Neuroleptics in progressive structural brain abnormalities in psychiatric illness. Lancet 1998;352(9130):784–5.

398. Shah GK, Auerbach DB, Augsburger JJ, Savino PJ. Acute thioridazine retinopathy. Arch Ophthalmol 1998;116(6):826–7.

399. Anonymous. Epidemiology of cataract. Lancet 1982;1(8286):1392–3.

400. Webber SK, Domniz Y, Sutton GL, Rogers CM, Lawless MA. Corneal deposition after high-dose chlorpromazine hydrochloride therapy. Cornea 2001;20(2):217–9.

401. Leung AT, Cheng AC, Chan WM, Lam DS. Chlorpromazine-induced refractile corneal deposits and cataract. Arch Ophthalmol 1999;117(12):1662–3.

402. Weiss EM, Bilder RM, Fleischhacker WW. The effects of second-generation antipsychotics on cognitive functioning and psychosocial outcome in schizophrenia. Psychopharmacology (Berl) 2002;162(1):11–7.

403. Browne S, Garavan J, Gervin M, Roe M, Larkin C, O'Callaghan E. Quality of life in schizophrenia: insight and subjective response to neuroleptics. J Nerv Ment Dis 1998;186(2):74–8.

404. Heinz A, Knable MB, Coppola R, Gorey JG, Jones DW, Lee KS, Weinberger DR. Psychomotor slowing, negative symptoms and dopamine receptor availability—an IBZM SPECT study in neuroleptic-treated and drug-free schizophrenic patients. Schizophr Res 1998;31(1):19–26.

405. Wistedt B. Neuroleptics and depression. Arch Gen Psychiatry 1982;39(6):745.

406. Wilkins-Ho M, Hollander Y. Toxic delirium with low-dose clozapine. Can J Psychiatry 1997;42(4):429–30.

407. Voruganti L, Awad AG. Neuroleptic dysphoria: towards a new synthesis. Psychopharmacology 2004;171: 121–32.

408. Schimmelmann BG, Schacht M, Perro C, Lambert M. Die initial dysphorische Reaktion (IDR) auf die Ersteinnahme von Neuroleptika. Nervenarzt 2004;75: 36–43.

409. Mahendran R. Obsessional symptoms associated with risperidone treatment. Aust NZ J Psychiatry 1998;32(2):299–301.

410. Howland RH. Chlorpromazine and obsessive-compulsive symptoms. Am J Psychiatry 1996;153(11):1503.

411. Stip E. Memory impairment in schizophrenia: perspectives from psychopathology and pharmacotherapy. Can J Psychiatry 1996;41(8 Suppl 2):S27–34.

412. Weiser M, Shneider-Beeri M, Nakash N, Brill N, Bawnik O, Reiss S, Hocherman S, Davidson M. Improvement in cognition associated with novel antipsychotic drugs: a direct drug effect or reduction of EPS? Schizophr Res 2000;46(2–3):81–9.

413. Eitan N, Levin Y, Ben-Artzi E, Levy A, Neumann M. Effects of antipsychotic drugs on memory functions of schizophrenic patients. Acta Psychiatr Scand 1992;85(1):74–6.

414. Braff DL, Saccuzzo DP. Effect of antipsychotic medication on speed of information processing in schizophrenic patients. Am J Psychiatry 1982;139(9):1127–30.

415. Van Putten T, May PR, Marder SR, Wittmann LA. Subjective response to antipsychotic drugs. Arch Gen Psychiatry 1981;38(2):187–90.

416. Rifkin A, Quitkin F, Klein DF. Akinesia: a poorly recognized drug-induced extrapyramidal behavior disorder. Arch Gen Psychiatry 1975;32:672.

417. Bilder RM, Goldman RS, Volavka J, Czobor P, Hoptman M, Sheitman B, Lindenmayer JP, Citrome L, McEvoy J, Kunz M, Chakos M, Cooper TB, Horowitz TL, Lieberman JA. Neurocognitive effects of clozapine, olanzapine, risperidone, and haloperidol in patients with chronic schizophrenia or schizoaffective disorder. Am J Psychiatry 2002;159(6):1018–28.

418. Moritz S, Woodward TS, Krausz M, Naber D. PERSIST Study Group. Relationship between neuroleptic dosage and subjective cognitive dysfunction in schizophrenic patients treated with either conventional or atypical neuroleptic medication. Int Clin Psychopharmacol 2002;17(1):41–4.

419. Naber D, Moritz S, Lambert M, Pajonk FG, Holzbach R, Mass R, Andresen B. Improvement of schizophrenic patients' subjective well-being under atypical antipsychotic drugs. Schizophr Res 2001;50(1–2):79–88.

420. Magharious W, Goff DC, Amico E. Relationship of gender and menstrual status to symptoms and medication side effects in patients with schizophrenia. Psychiatry Res 1998;77(3):159–66.

421. Moller HJ, Kissling W, Maurach R. Beziehungen zwischen Haloperidol-Serumspiegel, Prolactin-Serumspiegel, antipsychotischen Effekt und extrapyramidalen Begleitwirkungen. [Relationship between haloperidol blood level, prolactin blood level, antipsychotic effect and extrapyramidal side effects.] Pharmacopsychiatrica 1981;14:27.

422. Zarifian E, Scatton B, Bianchetti G, Cuche H, Loo H, Morselli PL. High doses of haloperidol in schizophrenia.

A clinical, biochemical, and pharmacokinetic study. Arch Gen Psychiatry 1982;39(2):212–5.

423. Brown WA, Laughren T. Low serum prolactin and early relapse following neuroleptic withdrawal. Am J Psychiatry 1981;138(2):237–9.

424. Phillips P, Shraberg D, Weitzel WD. Hirsutism associated with long-term phenothiazine neuroleptic therapy. JAMA 1979;241(9):920–1.

425. Atmaca M, Kuloglu M, Tezcan E, Canatan H, Gecici O. Quetiapine is not associated with increase in prolactin secretion in contrast to haloperidol. Arch Med Res 2002;33(6):562–5.

426. Schyve PM, Smithline F, Meltzer HY. Neuroleptic-induced prolactin level elevation and breast cancer: an emerging clinical issue. Arch Gen Psychiatry 1978;35(11):1291–301.

427. Mortensen PB. The incidence of cancer in schizophrenic patients. J Epidemiol Community Health 1989;43(1):43–7.

428. Pollock A, McLaren EH. Serum prolactin concentration in patients taking neuroleptic drugs. Clin Endocrinol (Oxf) 1998;49(4):513–6.

429. Spitzer M, Sajjad R, Benjamin F. Pattern of development of hyperprolactinemia after initiation of haloperidol therapy. Obstet Gynecol 1998;91(5 Pt 1):693–5.

430. Feek CM, Sawers JS, Brown NS, Seth J, Irvine WJ, Toft AD. Influence of thyroid status on dopaminergic inhibition of thyrotropin and prolactin secretion: evidence for an additional feedback mechanism in the control of thyroid hormone secretion. J Clin Endocrinol Metab 1980;51(3):585–9.

431. Melkersson KI, Hulting AL, Rane AJ. Dose requirement and prolactin elevation of antipsychotics in male and female patients with schizophrenia or related psychoses. Br J Clin Pharmacol 2001;51(4):317–24.

432. Kapur S, Roy P, Daskalakis J, Remington G, Zipursky R. Increased dopamine D(2) receptor occupancy and elevated prolactin level associated with addition of haloperidol to clozapine. Am J Psychiatry 2001;158(2):311–4.

433. Howes O, Smith S. Alendronic acid for antipsychotic-related osteopenia. Am J Psychiatry 2004;161: 756.

434. Meaney AM, Smith S, Howes OD, O'Brien M, Murray RM, O'Keane V. Effects of long-term prolactin-raising antipsychotic medication on bone mineral density in patients with schizophrenia. Br J Psychiatry 2004;184: 503–8.

435. Ali JA, Desai KD, Ali LJ. Delusions of pregnancy associated with increased prolactin concentrations produced by antipsychotic treatment. Int J Neuropsychopharmacol 2003;6: 111–15.

436. O'Keane V, Meaney AM. A new risk factor for osteoporosis in young women with schizophrenia? J Clin Psychopharmacol 2005;25: 26–31.

437. Bergemann N, Mundt C, Parzer P, Jannakos I Nagl I, Salbach B, Klinga K, Runnebaum B, Resch F. Plasma concentrations of estradiol in women suffering from schizophrenia treated with conventional versus atypical antipsychotics. Schizophr Res 2005;73:357–66.

438. Rao KJ, Miller M, Moses A. Water intoxication and thioridazine (Mellaril). Ann Intern Med 1975;82(1):61.

439. Meyer JM. A retrospective comparison of weight, lipid, and glucose changes between risperidone- and olanzapine-treated inpatients: metabolic outcomes after 1 year. J Clin Psychiatry 2002;63(5):425–33.

440. Meyer JM, Koro CE. The effects of antipsychotic therapy on serum lipids: a comprehensive review. Schizophr Res 2004;70:1–17.

441. Baptista T, Kin NM, Beaulieu S. Treatment of the metabolic disturbances caused by antipsychotic drugs. Clin Pharmacokinet 2004;43:1–15.

442. Bouchard RH, Demers MF, Simoneau I, Almeras N, Villeneuve J, Mottard JP, Cadrin C, Lemieux I, Despres JP. Atypical antipsychotics and cardiovascular risk in schizophrenic patients. J Clin Psychopharmacol 2001;21(1):110–1.

443. Charatan FBE, Barlett NG. The effect of chorpromazine ("Largactil") on glucose tolerance. J Mental Sci 1955;191:351–53.

444. Hiles B. Hyperglycaemia and glycosuria following chlorpromazine therapy. J Am Med Assoc 1956;162:1651.

445. Popli AP, Konicki PE, Jurjus GJ, Fuller MA, Jaskiw GE. Clozapine and associated diabetes mellitus. J Clin Psychiatry 1997;58(3):108–11.

446. Wirshing DA, Spellberg BJ, Erhart SM, Marder SR, Wirshing WC. Novel antipsychotics and new onset diabetes. Biol Psychiatry 1998;44(8):778–83.

447. Braceland FJ, Meduna LJ, Vaichulis JA. Delayed action of insulin in schizophrenia. Am J Psychiatry 1945;102:108–10.

448. Sernyak MJ, Leslie DL, Alarcón RD, Losonczy MF, Rosenheck R. Association of diabetes mellitus with use of atypical neuroleptics in the treatment of schizophrenia. Am J Psychiatry 2002;159:561–6.

449. Buse JB, Cavazzoni P, Hornbuckle K, Hutchins D, Breier A, Jovanovic L. A retrospectivce cohort study diabetes mellitus and antipsychotic treatment in the United States. J Clin Epidemiol 2003;56:164–70.

450. Newcomer JW, Haupt DW, Fucetola R, Melson AK, Schweiger JA, Cooper BP, Selke G. Abnormalities in glucose regulation during antipsychotic treatment of schizophrenia. Arch Gen Psychiatry 2002;59(4):337–45.

451. Food and Drug Administration. www.fda.gov/medwatch/ SAFETY/ 2004.

452. Gianfrancesco F, Grogg A, Mahmoud R, Wang R-H, Meletiche D. Differential effects of antipsychotic agents on the risk of development of type 2 diabetes mellitus in patients with mood disorders. Clin Ther 2003;25:1150–71.

453. Koro CE, Fedder DO, L'Italien GJ, Weiss SS, Magder LS, Kreyenbuhl J, Revicki DA, Buchanan RW. Assessment of independent effect of olanzapine and risperidone on risk of diabetes among patients with schizophrenia: population based nested case-control study. BMJ 2002;325:243–7.

454. Caro JJ, Ward A, Levinton C, Robinson K. The risk of diabetes during olanzapine use compared with risperidone use: a retrospective database analysis. J Clin Psychiatry 2002;1135–9.

455. Sernyak MJ, Gulanski B, Leslie DL, Rosenheck R. Undiagnosed hyperglycemia in clozapine-treated patients with schizophrenia. J Clin Psychiatry 2003;64:605–8.

456. Henderson DC, Cagliero E, Copeland PM, Borba CP, Evins E, Hayden D, Weber MT, Anderson EJ, Allison DB, Daley TB, Schoenfeld D, Goff DC. Glucose metabolism in patients with schizophrenia treated with atypical antipsychotic agents. Arch Gen Psychiatry 2005;62:19–28.

457. Lindenmayer J-P, Czobor P, Volavka J, Citrome L, Sheitman B, McEvoy JP, Cooper TB, Chakos M, Lieberman JA. Changes in glucose and cholesterol levels in patients with schizophrenia treated with typical or atypical antipsychotics. Am J Psychiatry 2003;160:290–6.

458. Harris MI, Flegal KM, Cowie CC, Eberhardt MS, Goldstein DE, Little RR, Wiedmeyer HM, Byrd-Holt DD. Prevalence of diabetes, impaired fasting glucose and impaired glucose tolerance in US adults. Diabetes Care 1998;21:518–24.

459. Dixon L, Weiden P, Delahanty J, Goldberg R, Postrado L, Lucksted A, Lehman A. Prevalence and correlates of diabetes in national schizophrenia samples. Schizophr Bull 2000;26:903–12.

460. Koller EA, Doraiswamy PM. Olanzapine-associated diabetes mellitus. Pharmacotherapy 2002;22:841–52.

461. Meyer JM. A retrospective comparison of weight, lipid, and glucose changes between risperidone- and olanzapine-treated inpatients metabolic outcomes after 1 year. J Clin Psychiatry 2002;63:425–33.

462. Wilson DR, D'Souza L, Sarkar N, Newton M, Hammond C. New-onset diabetes and ketoacidosis with atypical antipsychotics. Schizophr Res 2002;59:1–6.

463. Jin H, Meyer JM, Jeste DV. Phenomenology of and risk factors for new-onset diabetes mellitus and diabetes ketoacidosis associated with atypical antipsychotics: an analysis of 45 cases. Ann Clin Psychiatry 2002;14:59–64.

464. Ramankutty G. Olanzapine-induced destabilization of diabetes in the absence of weight gain. Acta Psychiatr Scand 2002;105:235–7.

465. Wirshing DA. Adverse effects of atypical antipsychotics. J Clin Psychiatry 2001;62:7–10.

466. Ebenbichler CF, Laimer M, Eder U, Mangweth B, Weiss E, Hofer A, Hummer M, Kemmler G, Lechleitner M, Patsch JR, Fleischhacker WW. Olanzapine induces insulin resistance: results from a prospective study. J Clin Psychiatry 2003;64:1436–9.

467. Hagg S, Soderberg S, Ahren B, Olsson T, Mjornfal T. Leptin concentrations are increased in subjects treated with clozapine or conventional antipsychotics. J Clin Psychiatry 2001;62:843–8.

468. Dufresne RL. Metabolic syndrome and antipsychotic therapy: a summary of the findings. Drug Benefit Trends 2003;Suppl B:12–17.

469. Lean MEJ, Pajonk F-G. Patients on atypical antipsychotic drugs. Diabetes Care 2003;26:1597–605.

470. Liberty IF, Todder D, Umansky R, Harman-Boehm I. Atypical antipsychotics and diabetes mellitus: an association. Isr Med Assoc J 2004;6:276–9.

471. Baptista T. Body weight gain induced by antipsychotic drugs: mechanisms and management. Acta Psychiatr Scand 1999;100(1):3–16.

472. Taylor DM, McAskill R. Atypical antipsychotics and weight gain—a systematic review. Acta Psychiatr Scand 2000;101(6):416–32.

473. Gupta S, Droney T, Al-Samarrai S, Keller P, Frank B. Olanzapine-induced weight gain. Ann Clin Psychiatry 1998;10(1):39.

474. Brecher M, Geller W. Weight gain with risperidone. J Clin Psychopharmacol 1997;17(5):435–6.

475. Penn JV, Martini J, Radka D. Weight gain associated with risperidone. J Clin Psychopharmacol 1996;16(3):259–60.

476. Frankenburg FR, Zanarini MC, Kando J, Centorrino F. Clozapine and body mass change. Biol Psychiatry 1998;43(7):520–4.

477. Bustillo JR, Buchanan RW, Irish D, Breier A. Differential effect of clozapine on weight: a controlled study. Am J Psychiatry 1996;153(6):817–9.

478. Kraepelin E. Dementia Praecox and Paraphrenia. Edinburgh: E & S Livingstone 1919;87.

479. Gupta S, Droney T, Al-Samarrai S, Keller P, Frank B. Olanzapine: weight gain therapeutic efficacy. J Clin Psychopharmacol 1999;19(3):273–5.

480. Baymiller SP, Ball P, McMahon RP, Buchanan RW. Weight and blood pressure change during clozapine treatment. Clin Neuropharmacol 2002;25(4):202–6.

481. Monteleone P, Fabrazzo M, Tortorella A, La Pia S, Maj M. Pronounced early increase in circulating leptin predicts a lower weight gain during clozapine treatment. J Clin Psychopharmacol 2002;22(4):424–6.

482. Stigler KA, Potenza MN, Posey DJ, McDougle CJ. Weight gain associated with atypical antipsychotic use in children and adolescents. Pediatr Drugs 2004;6:33–44.

483. Vieweg WV, Sood AB, Pandurangi A, Silverman JJ. Newer antipsychotic drugs and obesity in children and adolescents. How should we assess drug-associated weight gain? Acta Psychiatr Scand 2005;111:177–84.

484. Wetterling T. Bodyweight gain with atypical antipsychotics. A comparative review. Drug Saf 2001;24(1):59–73.

485. Blin O, Micallef J. Antipsychotic-associated weight gain and clinical outcome parameters. J Clin Psychiatry 2001;62(Suppl 7):11–21.

486. Allison DB, Casey DE. Antipsychotic-induced weight gain: a review of the literature. J Clin Psychiatry 2001;62(Suppl 7):22–31.

487. Kurzthaler I, Fleischhacker WW. The clinical implications of weight gain in schizophrenia. J Clin Psychiatry 2001;62(Suppl 7):32–7.

488. Casey DE, Zorn SH. The pharmacology of weight gain with antipsychotics. J Clin Psychiatry 2001;62(Suppl 7):4–10.

489. Allison DB, Mentore JL, Heo M, Chandler LP, Cappelleri JC, Infante MC, Weiden PJ. Antipsychotic-induced weight gain: a comprehensive research synthesis. Am J Psychiatry 1999;156(11):1686–96.

490. Allison DB, Mentore JL, Heo M, et al. Weight gain associated with conventional and newer antipsychotics: a meta-analysis. Presented at the New Clinical Drug Evaluation Unit 38th Annual Meeting, Boca Raton, Florida, June 10–13, 1998;.

491. Kelly DL, Conley RR, Lore RC, et al. Weight gain in adolescents treated with risperidone and conventional antipsychotics over six months. J Child Adolesc Psychopharmacol 1998;813:151–9.

492. Wistedt B. A depot neuroleptic withdrawal study. A controlled study of the clinical effects of the withdrawal of depot fluphenazine decanoate and depot flupenthixol decanoate in chronic schizophrenic patients. Acta Psychiatr Scand 1981;64(1):65–84.

493. Nemeroff CB. Dosing the antipsychotic medication olanzapine. J Clin Psychiatry 1997;58(Suppl 10):45–9.

494. Ganguli R, Brar JS, Ayrton Z. Weight gain over 4 months in schizophrenia patients: a comparison of olanzapine and risperidone. Schizophr Res 2001;49(3):261–7.

495. Wetterling T, Mussigbrodt HE. Weight gain: side effect of atypical neuroleptics? J Clin Psychopharmacol 1999;19(4):316–21.

496. Reinstein MJ, Sirotovskaya LA, Jones LE, Mohan S, Chasanov MA. Effect of clozapine-quetiapine combination therapy on weight and glycaemic control. Clin Drug Invest 1999;18:99–104.

497. Bobes J, Rejas J, García-García M, Rico-Villademoros F, García-Portilla MP, Fernández I, Hernández G, for the EIRE Study Group. Weight gain in patients with schizophrenia treated with risperidone, olanzapine, quetiapine or haloperidol: results of the EIRE study. Schizophr Res 2003;62:77–88.

498. Owens DG. Extrapyramidal side effects and tolerability of risperidone: a review. J Clin Psychiatry 1994;55(Suppl 5):29–35.

499. Rietschel M, Naber D, Fimmers R, Moller HJ, Propping P, Nothen MM. Efficacy and side-effects of clozapine not

associated with variation in the 5-HT$_{2C}$ receptor. Neuroreport 1997;8(8):1999–2003.

500. Rietschel M, Naber D, Oberlander H, Holzbach R, Fimmers R, Eggermann K, Moller HJ, Propping P, Nothen MM. Efficacy and side-effects of clozapine: testing for association with allelic variation in the dopamine D$_4$ receptor gene. Neuropsychopharmacology 1996;15(5):491–6.

501. Theisen FM, Cichon S, Linden A, Martin M, Remschmidt H, Hebebrand J. Clozapine and weight gain. Am J Psychiatry 2001;158(5):816.

502. Reynolds GP, Zhang ZJ, Zhang XB. Association of antipsychotic drug-induced weight gain with a 5-HT$_{2C}$ receptor gene polymorphism. Lancet 2002;359(9323):2086–7.

503. Allison DB, Fontaine KR, Heo M, Mentore JL, Cappelleri JC, Chandler LP, Weiden PJ, Cheskin LJ. The distribution of body mass index among individuals with and without schizophrenia. J Clin Psychiatry 1999;60(4):215–20.

504. Hummer M, Kemmler G, Kurz M, Kurzthaler I, Oberbauer H, Fleischhacker WW. Weight gain induced by clozapine. Eur Neuropsychopharmacol 1995;5(4):437–40.

505. Kohnke MD, Griese EU, Stosser D, Gaertner I, Barth G. Cytochrome P450 2D6 deficiency and its clinical relevance in a patient treated with risperidone. Pharmacopsychiatry 2002;35(3):116–8.

506. European Federation of Associations of Families of Mentally Ill People. www.eufami.org

507. O'Keefe C, Noordsy D. Prevention and reversal of weight gain associated with antipsychotic treatment. J Clin Outcomes Manage 2002;9:575–82.

508. Rotatori AF, Fox R, Wicks A. Weight loss with psychiatric residents in a behavioral self control program. Psychol Rep 1980;46(2):483–6.

509. Floris M, Lejeune J, Deberdt W. Effect of amantadine on weight gain during olanzapine treatment. Eur Neuropsychopharmacol 2001;11(2):181–2.

510. Cohen S, Glazewski R, Khan S, Khan A. Weight gain with risperidone among patients with mental retardation: effect of calorie restriction. J Clin Psychiatry 2001;62(2):114–6.

511. Lawson WB, Karson CN, Bigelow LB. Increased urine volume in chronic schizophrenic patients. Psychiatry Res 1985;14(4):323–31.

512. Higuchi T, Komoda T, Sugishita M, Yamazaki J, Miura M, Sakagishi Y, Yamauchi T. Certain neuroleptics reduce bone mineralization in schizophrenic patients. Neuropsychobiology 1987;18(4):185–8.

513. al-Adwani A. Neuroleptics and bone mineral density. Am J Psychiatry 1997;154(8):1173.

514. Marcus J, Mulvihill FJ. Agranulocytosis and chlorpromazine. J Clin Psychiatry 1978;39(10):784–6.

515. Kendra JR, Rugman FP, Flaherty TA, Myers A, Horsfield N, Barton A, Russell L. First use of G-CSF in chlorpromazine-induced agranulocytosis: a report of two cases. Postgrad Med J 1993;69(817):885–7.

516. King DJ, Wager E. Haematological safety of antipsychotic drugs. J Psychopharmacol 1998;12(3):283–8.

517. Voss SN, Sanger T, Beasley C. Hematologic reference ranges in a population of patients with schizophrenia. J Clin Psychopharmacol 2000;20(6):653–7.

518. Evans DL, Rogers JF, Peiper SC. Intestinal dilatation associated with phenothiazine therapy: a case report and literature review. Am J Psychiatry 1979;136(7):970–2.

519. Sheikh RA, Prindiville T, Yasmeen S. Haloperidol and benztropine interaction presenting as acute intestinal pseudo-obstruction. Am J Gastroenterol 2001;96(3):934–5.

520. Dumortier G, Cabaret W, Stamatiadis L, Saba G, Benadhira R, Rocamora JF, Aubriot-Delmas B, Glikman J, Januel D. Tolerance hépatique des antipsychotiques atypiques. [Hepatic tolerance of atypical antipsychotic drugs.] Encephale 2002;28(6 Pt 1):542–51.

521. Jones JK, Van de Carr SW, Zimmerman H, Leroy A. Hepatotoxicity associated with phenothiazines. Psychopharmacol Bull 1983;19(1):24–7.

522. Kaminer Y, Apter A, Tyano S, Livni E, Wijsenbeek H. Use of the microphage migration inhibition factor test to determine the cause of delayed hypersensitivity reaction to haloperidol syrup. Am J Psychiatry 1982;139(11):1503–4.

523. Lal S, Lal S. Chlorpromazine-induced cutaneous pigmentation—effect of replacement with clozapine. J Psychiatry Neurosci 2000;25(3):281.

524. Yoshikawa H, Watanabe T, Abe T, Oda Y, Ozawa K. Haloperidol-induced rhabdomyolysis without neuroleptic malignant syndrome in a handicapped child. Brain Dev 2000;22(4):256–8.

525. Boot E, de Haan L. Massive increase in serum creatine kinase during olanzapine and quetiapine treatment, not during treatment with clozapine. Psychopharmacology (Berl) 2000;150(3):347–8.

526. Baldwin D, Mayers A. Sexual side-effects of antidepressant and antipsychotic drugs. Adv Psychiatr Treat 2003;9:202–10.

527. Hummer M, Kemmler G, Kurz M, Kurzthaler I, Oberbauer H, Fleischhacker WW. Sexual disturbances during clozapine and haloperidol treatment for schizophrenia. Am J Psychiatry 1999;156(4):631–3.

528. Mitchell JE, Popkin MK. Antipsychotic drug therapy and sexual dysfunction in men. Am J Psychiatry 1982;139(5):633–7.

529. Siris SG, Siris ES, van Kammen DP, Docherty JP, Alexander PE, Bunney WE Jr. Effects of dopamine blockade on gonadotropins and testosterone in men. Am J Psychiatry 1980;137(2):211–4.

530. Keitner GI, Selub S. Spontaneous ejaculations and neuroleptics. J Clin Psychopharmacol 1983;3(1):34–6.

531. Valevski A, Modai I, Zbarski E, Zemishlany Z, Weizman A. Effect of amantadine on sexual dysfunction in neuroleptic-treated male schizophrenic patients. Clin Neuropharmacol 1998;21(6):355–7.

532. Weiner DM, Lowe FC. Psychotropic drug-induced priapism: incidence, mechanism and management. CNS Drugs 1998;9:371–9.

533. Demyttenaere K, De Fruyt J, Sienaert P. Psychotropics and sexuality. Int Clin Psychopharmacol 1998;13(Suppl 6):S35–41.

534. Reeves RR, Mack JE. Priapism associated with two atypical antipsychotic agents. Pharmacotherapy 2002;22(8):1070–3.

535. Michael A, Calloway SP. Priapism in twins. Pharmacopsychiatry 1999;32(4):157.

536. Patel AG, Mukherji K, Lee A. Priapism associated with psychotropic drugs. Br J Hosp Med 1996;55(6):315–9.

537. Salleh MR, Mohamad H, Zainol J. Unpredictable neuroleptics induced priapism: a case report. Eur Psychiatry 1996;11:419–20.

538. Abe S, Suzuki T, Hori T, Baba A, Shiraishi H. Hypogammaglobulinemia during antipsychotic therapy. Psychiatry Clin Neurosci 1998;52(1):115–7.

539. McNeil HP, Chesterman CN, Krilis SA. Immunology and clinical importance of antiphospholipid antibodies. Adv Immunol 1991;49:193–280.

540. Lillicrap MS, Wright G, Jones AC. Symptomatic antiphospholipid syndrome induced by chlorpromazine. Br J Rheumatol 1998;37(3):346–7.

541. Pramatarov KD. Drug-induced lupus erythematosus. Clin Dermatol 1998;16(3):367–77.

542. Krohn K, Bennett R. Drug-induced autoimmune disorders. Inmunol Allergy Clin North Am 1998;18:897–911.

543. McNevin S, MacKay M. Chlorprothixene-induced systemic lupus erythematosus. J Clin Psychopharmacol 1982;2(6):411–2.

544. Adams BE, Manoguerra AS, Lilja GP, Long RS, Ruiz E. Heat stroke associated with medications having anticholinergic effects. Minn Med 1977;60(2):103–6.

545. Shiloh R, Hermesh H, Weizer N, Dorfman-Etrog P, Weizman A, Munitz H. Acute antipsychotic drug administration lowers body temperature in drug-free male schizophrenic patients. Eur Neuropsychopharmacol 2000;10(6):443–5.

546. Hagg S, Mjorndal T, Lindqvist L. Repeated episodes of hypothermia in a subject treated with haloperidol, levomepromazine, olanzapine, and thioridazine. J Clin Psychopharmacol 2001;21(1):113–5.

547. Parris C, Mack JM, Cochiolo JA, Steinmann AF, Tietjen J. Hypothermia in 2 patients treated with atypical antipsychotic medication. J Clin Psychiatry 2001;62(1):61–3.

548. Mitchell JE. Discontinuation of antipsychotic drug therapy. Psychosomatics 1981;22(3):241–7.

549. Nelli AC, Yarden PE, Guazzelli M, Feinberg I. Parkinsonism following neuroleptic withdrawal. Arch Gen Psychiatry 1989;46(4):383–4.

550. Chouinard G, Jones BD. Neuroleptic-induced supersensitivity psychosis: clinical and pharmacologic characteristics. Am J Psychiatry 1980;137(1):16–21.

551. Polizos P, Engelhardt DM. Dyskinetic phenomena in children treated with psychotropic medications. Psychopharmacol Bull 1978;14(4):65–8.

552. Gualtieri TC, Barnhill J, McGinsey J, Schell D. Tardive dyskinesia and other movement disorders in children treated with psychotropic drugs. J Am Acad Child Psychiatry 1980;19(3):491–510.

553. Polizos P, Engelhardt DM, Hoffman SP, Waizer J. Neurological consequences of psychotropic drug withdrawal in schizophrenic children. J Autism Child Schizophr 1973;3(3):247–53.

554. Cohen-Mansfield J, Lipson S, Werner P, Billig N, Taylor L, Woosley R. Withdrawal of haloperidol, thioridazine, and lorazepam in the nursing home: a controlled, double-blind study. Arch Intern Med 1999;159(15):1733–40.

555. Ahmed Z, Fraser W, Kerr MP, Kiernan C, Emerson E, Robertson J, Felce D, Allen D, Baxter H, Thomas J. Reducing antipsychotic medication in people with a learning disability. Br J Psychiatry 2000;176:42–6.

556. Bigatti MP, Corona D, Munizza C. Increased sister chromatid exchange and chromosomal aberration frequencies in psychiatric patients receiving psychopharmacological therapy. Mutat Res 1998;413(2):169–75.

557. Bonassi S, Abbondandolo A, Camurri L, Dal Pra L, De Ferrari M, Degrassi F, Forni A, Lamberti L, Lando C, Padovani P, Sbrana I, Vecchio D, Puntoni R. Are chromosome aberrations in circulating lymphocytes predictive of future cancer onset in humans? Preliminary results of an Italian cohort study. Cancer Genet Cytogenet 1995;79(2):133–5.

558. Yamazawa K, Matsui H, Seki K, Sekiya S. A case-control study of endometrial cancer after antipsychotics exposure in premenopausal women. Oncology 2003;64:116–23.

559. Blair JH, Simpson GM. Effect of antipsychotic drugs on reproductive functions. Dis Nerv Syst 1966;27(10):645–7.

560. Cohen LS, Rosenbaum JF. Psychotropic drug use during pregnancy: weighing the risks. J Clin Psychiatry 1998;59(Suppl 2):18–28.

561. Pinkofsky HB. Psychosis during pregnancy: treatment considerations. Ann Clin Psychiatry 1997;9(3):175–9.

562. Trixler M, Tenyi T. Antipsychotic use in pregnancy. What are the best treatment options? Drug Saf 1997;16(6):403–10.

563. Austin MP, Mitchell PB. Psychotropic medications in pregnant women: treatment dilemmas. Med J Aust 1998;169(8):428–31.

564. Cohen LS, Rosenbaum JF. Psychotropic drug use during pregnancy: weighing the risks. J Clin Psychiatry 1998;59(Suppl 2):18–28.

565. Coyle I, Wayner MJ, Singer G. Behavioral teratogenesis: a critical evaluation. Pharmacol Biochem Behav 1976;4(2):191–200.

566. Stoner SC, Sommi RW Jr, Marken PA, Anya I, Vaughn J. Clozapine use in two full-term pregnancies. J Clin Psychiatry 1997;58(8):364–5.

567. Kopelman AE, McCullar FW, Heggeness L. Limb malformations following maternal use of haloperidol. JAMA 1975;231(1):62–4.

568. Sriram B, Kumar P, Rajadurai VS. A case of phocomelia-drug association revisited. Singapore Paediatr J 1999;41:81–3.

569. O'Connor M, Johnson GH, James DI. Intrauterine effect of phenothiazines. Med J Aust 1981;1(8):416–7.

570. Nath SP, Miller DA, Muraskas JK. Severe rhinorrhea and respiratory distress in a neonate exposed to fluphenazine hydrochloride prenatally. Ann Pharmacother 1996;30(1):35–7.

571. Stewart RB, Karas B, Springer PK. Haloperidol excretion in human milk. Am J Psychiatry 1980;137(7):849–50.

572. Yoshida K, Smith B, Craggs M, Kumar R. Neuroleptic drugs in breast-milk: a study of pharmacokinetics and of possible adverse effects in breast-fed infants. Psychol Med 1998;28(1):81–91.

573. Dettling M, Sander T, Weber M, Steinlein OK. Mutation analysis of the ryanodine receptor gene isoform 3 (RYR3) in recurrent neuroleptic malignant syndrome. J Clin Psychopharmacol 2004;24:471–3.

574. Vandel P, Haffen E, Vandel S, Bonin B, Nezelhof S, Sechter D, Broly F, Biouard P, Dalery J. Drug extrapyramidal side effects. CYP2D6 genotypes and phenotypes. Eur J Clin Pharmacol 1999;55:659–65.

575. Lohmann PL, Bagli M, Krauss H, Müller DJ, Schulze TG, Fangerau H, Ludwig M, Barkow K, Held T, Heun R, Maier W, Rietschel M, Rao ML. CYP2D6 Polymorphism and tardive dyskinesia in schizophrenic patients. Pharmacopsychiatry 2003;36:73–8.

576. Opolka JL, Rascati KL, Brown CM, Gibson PJ. Role of ethnicity in predicting antipsychotic medication adherence. Ann Pharmacother 2003;37:625–30.

577. Wang YC, Liou YJ, Liao DL, Bai YM, Lin CC, Yu SC, Chen JY. Association analysis of a neural nitric oxide synthase gene polymorphism and antipsychotics-induced tardive dyskinesia in Chinese schizophrenic patients. J Neural Transm 2004;111:623–9.

578. Tsai SJ, Yu YW, Lin CH, Wang YC, Chen JY, Hong CJ. Association study of adrenergic β3 receptor (Trp64Arg) and G-Protein β3 subunit gene (C825T) polymorphisms and weight change during clozapine treatment. Neuropsychobiology 2004;50:37–40.

579. Herbeck DM, West JC, Ruditis I, Duffy FF, Fitek DJ, Bell CC, Snowden LR. Variations in use of second-generation

antipsychotic medication by race among adult psychiatric patients. Psychiatr Serv 2004;55:677–84.

580. Biederman J. New directions in pediatric psychopharmacology. Drug Ther 1982;12:33.

581. Scahill L, Lynch KA. Atypical neuroleptics in children and adolescents. J Child Adolesc Psychiatr Nurs 1998;11(1):38–43.

582. Campbell M, Rapoport JL, Simpson GM. Antipsychotics in children and adolescents. J Am Acad Child Adolesc Psychiatry 1999;38(5):537–45.

583. Lewis R. Typical and atypical antipsychotics in adolescent schizophrenia: efficacy, tolerability, and differential sensitivity to extrapyramidal symptoms. Can J Psychiatry 1998;43(6):596–604.

584. Toren P, Laor N, Weizman A. Use of atypical neuroleptics in child and adolescent psychiatry. J Clin Psychiatry 1998;59(12):644–56.

585. Naja WJ, Reneric JP, Bouvard MP. Neuroleptiques atypiques chez l'enfant et l'adolescent. [Atypical neuroleptics in the child and adolescent.] Encephale 1998;24(4):378–85.

586. Anonymous. Atypical antipsychotic agents in the treatment of schizophrenia and other psychiatric disorders. Part I: Unique patient populations. J Clin Psychiatry 1998;59(5):259–65.

587. Mendelson WB. The use of sedative/hypnotic medication and its correlation with falling down in the hospital. Sleep 1996;19(9):698–701.

588. Devanand DP, Marder K, Michaels KS, Sackeim HA, Bell K, Sullivan MA, Cooper TB, Pelton GH, Mayeux R. A randomized, placebo-controlled dose-comparison trial of haloperidol for psychosis and disruptive behaviors in Alzheimer's disease. Am J Psychiatry 1998;155(11):1512–20.

589. American Psychiatric Association. Practice guideline for the treatment of patients with Alzheimer's disease and other dementias of late life. Am J Psychiatry 1997;154 (5 Suppl):1–39.

590. Katz IR, Rovner BW, Schneider L. Use of psychoactive drugs in nursing homes. N Engl J Med 1992;327:1392–3.

591. Schneider LS, Tariot PN, Dagerman KS, Davis SM, Hsiao JK, Ismail MS, Lebowitz BD, Lyketsos CG, Ryan JM, Stroup TS, Sultzer DL, Weintraub D, Lieberman JA, for the CATIE-AD Study Group. Effectiveness of atypical antipsychotic drugs in patients with Alzheimer's disease. N Engl J Med 2006;355:1525–38.

592. Ballard C, Margallo-Lana M, Juszczak E, Douglas S, Swann A, Thomas A, O'Brien J, Everratt A, Sadler S, Maddison C, Lee L, Bannister C, Elvish R, Jacoby R. Quetiapine and rivastigmine and cognitive decline in Alzheimer's disease: randomised double blind placebo controlled trial. BMJ 2005;330:874–7.

593. Sink KM, Holden KF, Yaffe K. Pharmacological treatment of neuropsychiatric symptoms of dementia. JAMA 2005;293:596–608.

594. Schneider LS, Dagerman KS, Insel P. Efficacy and adverse effects of atypical antipsychotics for dementia: meta-analysis of randomized, placebo controlled trials. Am J Geriatr Psychiatry 2006;14:191–210.

595. US Food and Drug Administration. 2003 Safety alert: RISPERDAL (risperidone). http://www.fda.gov./medwatch/SAFETY/2003/risperdal.htm

596. Mowat D, Fowlie D, MacEwan T. CSM warning on atypical antipsychotics and stroke may be detrimental for dementia. BMJ 2004;328:1262.

597. Smith DA, Beier MT. Association between risperidone treatment and cerebrovascular adverse events: examining the evidence and postulating hypotheses for an underlying mechanism. J Am Med Dir Assoc 2004;5:129–32.

598. Wooltorton E. Olanzapine (Zyprexa): increase incidence of cerebrovascular events in dementia patients. CMAJ 2004;170:1395.

599. Herrmann N, Mamdani M, Lanctôt KL. Atypical antipsychotics and risk of cerebrovascular accidents. Am J Psychiatry 2004;161:1113–5.

600. Liperoti R. Cerebrovascular events among elderly patients treated with conventional or atypical antipsychotics. 2004 Annual meeting of the American Geriatrics Society. http://www.americangeriatrics.org/news/meeting/schedule_events.pdf //

601. Gill SS, Rochon PA, Herrmann N, Lee PE, Sykora K, Gunraj N, Normand SLT, Gurwitz JH, Marras C, Wodchis WP, Mamdani M. Atypical antipsychotic drugs and risk of ischaemic stroke: population based retrospective cohort study. BMJ 2005;330:445.

602. Layton D, Harris S, Wilton LV, Shakir SA. Comparison of incidence rates of cerebrovascular accidents and transient ischaemic attacks in observational cohort studies of patients prescribed risperidone, quetiapine or olanzapine in general practice in England including patients with dementia. J Psychopharmacol 2005;19:473–82.

603. Moretti R, Torre P, Antonello RM, Cattaruzza T, Cazzato G. Olanzapine as a possible treatment of behavioral symptoms in vascular dementia: risks of cerebrovascular events. A controlled, open-label study. J Neurol 2005;252:1186–93.

604. Liperoti R, Gambassi G, Lapane KL, Chiang C, Pedone C, Mor V, Bernabei R. Conventional and atypical antipsychotics and the risk of hospitalization for ventricular arrhythmias or cardiac arrest. Arch Intern Med 2005;165:696–701.

605. Schneider LS, Dagerman KS, Insel P. Risk of death with atypical antipsychotic drug treatment for dementia. JAMA 2005;294:1934–42.

606. Wang PS, Schneeweiss S, Avorn J, Fischer MA, Mogun H, Solomon DH, Brookhart MA. Risk of death in elderly users of conventional vs. atypical antipsychotic medications. N Engl J Med. 2005;353:2335–41.

607. US Food and Drug Administration. FDA Public Health Advisory. Deaths with antipsychotics in elderly patients with behavioral disturbances. http://www.fda.gov/cder/drug/advisory/antipsychotics.htm

608. EMEA. EMEA public statement on the safety of olanzapine (Zyprexa, Zyprexa Velotab). http://www.emea.europa.eu/pdfs/human/press/pus/085604en.pdf

609. Kane JM, Aguglia E, Altamura AC, Ayuso Gutierrez JL, Brunello N, Fleischhacker WW, Gaebel W, Gerlach J, Guelfi JD, Kissling W, Lapierre YD, Lindstrom E, Mendlewicz J, Racagni G, Carulla LS, Schooler NR. Guidelines for depot antipsychotic treatment in schizophrenia. European Neuropsychopharmacology Consensus Conference in Siena, Italy. Eur Neuropsychopharmacol 1998;8(1):55–66.

610. Hirshberg B, Gural A, Caraco Y. Zuclopenthixol-associated neutropenia and thrombocytopenia. Ann Pharmacother 2000;34(6):740–2.

611. Caldwell CB, Gottesman II. Schizophrenics kill themselves too: a review of risk factors for suicide. Schizophr Bull 1990;16(4):571–89.

612. Roy A. Suicide in Schizophrenia. In: Roy A, editor. Suicide. Baltimore MD: Williams & Wilkins, 1986:97–112.

613. Schreinzer D, Frey R, Stimpfl T, Vycudilik W, Berzlanovich A, Kasper S. Different fatal toxicity of neuroleptics identified by autopsy. Eur Neuropsychopharmacol 2001;11(2):117–24.

614. Buckley N, McManus P. Fatal toxicity of drugs used in the treatment of psychotic illnesses. Br J Psychiatry 1998;172:461–4.

615. Rossen B, Steiness I. The pathophysiology of acute renal failure after chlorprothixene overdosage. Acta Med Scand 1981;209(6):525–7.

616. Capel MM, Colbridge MG, Henry JA. Overdose profiles of new antipsychotic agents. Int J Neuropsychopharmacol 2000;3(1):51–4.

617. James LP, Abel K, Wilkinson J, Simpson PM, Nichols MH. Phenothiazine, butyrophenone, and other psychotropic medication poisonings in children and adolescents. J Toxicol Clin Toxicol 2000;38(6):615–23.

618. Dorne R, Pommier C, Manchon M, Berny C. Intoxication par l'amisulpride (Soliané): à propos d'une observation avec documentation toxicologique. [Intoxication with amisulpride (Solian): a case with toxicologic documentations.] Therapie 2000;55(2):325–8.

619. Burton S, Heslop K, Harrison K, Barnes M. Ziprasidone overdose. Am J Psychiatry 2000;157(5):835.

620. Stein S, Schmoldt A, Schulz M. Fatal intoxication with melperone. Forensic Sci Int 2000;113(1–3):409–13.

621. Garroni A, Palloshi A, Fragasso G, Margonato A. Minimal myocardial damage after tricyclic neuroleptic overdose. Pharmacopsychiatry 2003;36:33–4.

622. Risch SC, Groom GP, Janowsky DS. The effects of psychotropic drugs on the cardiovascular system. J Clin Psychiatry 1982;43(5 Pt 2):16–31.

623. Griffin JP, D'Arcy PF. A Manual of Adverse Drug InteractionsBristol: John Wright and Sons Ltd;. 1979.

624. Shopsin B, Kline NS, Ayd F. In: Evaluation of drug interactions. 2nd ed.. Washington DC: American Pharmaceutical Association, 1976:25 81, 122, 397.

625. Shader RI, Ciraulo DA, Greenblatt DJ. Drug interactions involving psychotropic drugs. Psychosomatics 1978;19(11):671–3677–681.

626. Warnes H. Toxic psychosis due to antiparkinsonian drugs. Can Psychiatr Assoc J 1967;12(3):323–6.

627. Dunlap JC, Miller WC. Toxic psychosis following the use of benztropine methanesulfonate (Congentin). J S C Med Assoc 1969;65(6):203–4.

628. el-Yosef MK, Janowsky DS, Davis JM, Sekerke HJ. Reversal of benztropine toxicity by physostigmine. JAMA 1972;220(1):125.

629. Bourin M, Baker GB. Therapeutic and adverse effect considerations when using combinations of neuroleptics and benziodiazepines. Saudi Pharm J 1998;3–4:262–5.

630. Hatta K, Takahashi T, Nakamura H, Yamashiro H, Endo H, Kito K, Saeki T, Masui K, Yonezawa Y. A risk for obstruction of the airways in the parenteral use of levomepromazine with benzodiazepine. Pharmacopsychiatry 1998;31(4):126–30.

631. Ayd FJ Jr. Loxapine update: 1966–1976. Dis Nerv Syst 1977;38(11):883–7.

632. Markowitz JS, Wells BG, Carson WH. Interactions between antipsychotic and antihypertensive drugs. Ann Pharmacother 1995;29(6):603–9.

633. Frye PE, Pariser SF, Kim MH, O'Shaughnessy RW. Bromocriptine associated with symptom exacerbation during neuroleptic treatment of schizoaffective schizophrenia. J Clin Psychiatry 1982;43(6):252–3.

634. Kulhanek F, Linde OK, Meisenberg G. Precipitation of antipsychotic drugs in interaction with coffee or tea. Lancet 1979;2(8152):1130.

635. Bowen S, Taylor KM, Gibb IA. Effect of coffee and tea on blood levels and efficacy of antipsychotic drugs. Lancet 1981;1(8231):1217–8.

636. Bezchlinbnyk KZ, Jeffries JJ. Should psychiatric patients drink coffee? Can Med Assoc 1981;124:357.

637. Knudsen P, Vilmar T. Cannabis and neuroleptic agents in schizophrenia. Acta Psychiatr Scand 1984;69(2):162–74.

638. Fast DK, Jones BD, Kusalic M, Erickson M. Effect of carbamazepine on neuroleptic plasma levels and efficacy. Am J Psychiatry 1986;143(1):117–8.

639. Nisijima K, Kusakabe Y, Ohtuka K, Ishiguro T. Addition of carbamazepine to long-term treatment with neuroleptics may induce neuroleptic malignant syndrome. Biol Psychiatry 1998;44(9):930–1.

640. Hegarty AM, Lipton RG, Merriam AE, Freeman K. Cocaine as a risk factor for acute dystonic reactions. Neurology 1991;41:1670–2.

641. Freeman MP, Stoll AL. Mood stabilizer combinations: a review of safety and efficacy. Am J Psychiatry 1998;155(1):12–21.

642. Jefferson JW, Greist JH, Baudhuin M. Lithium: interactions with other drugs. J Clin Psychopharmacol 1981;1(3):124–34.

643. Waddington JL. Some pharmacological aspects relating to the issue of possible neurotoxic interactions during combined lithium-neuroleptic therapy. Hum Psychopharmacol 1990;5:293–7.

644. Batchelor DH, Lowe MR. Reported neurotoxicity with the lithium/haloperidol combination and other neuroleptics. A literature review. Hum Psychopharmacol 1990;5:275–80.

645. Gupta S, Racaniello AA. Neuroleptic malignant syndrome associated with amoxapine and lithium in an older adult. Ann Clin Psychiatry 2000;12(2):107–9.

646. Pope HG Jr, Cole JO, Choras PT, Fulwiler CE. Apparent neuroleptic malignant syndrome with clozapine and lithium. J Nerv Ment Dis 1986;174(8):493–5.

647. Spring G, Frankel M. New data on lithium and haloperidol incompatibility. Am J Psychiatry 1981;138(6):818–21.

648. Berry N, Pradhan S, Sagar R, Gupta SK. Neuroleptic malignant syndrome in an adolescent receiving olanzapine–lithium combination therapy. Pharmacotherapy 2003;23(2):255–9.

649. Bourgeois JA, Kahn DR. Neuroleptic malignant syndrome following administration of risperidone and lithium. J Clin Psychopharmacol 2003;23(3):315–7.

650. Deng MZ, Chen GQ, Phillips MR. Neuroleptic malignant syndrome in 12 of 9,792 Chinese inpatients exposed to neuroleptics: a prospective study. Am J Psychiatry 1990;147(9):1149–55.

651. Gill J, Singh H, Nugent K. Acute lithium intoxication and neuroleptic malignant syndrome. Pharmacotherapy 2003;23(6):811–5.

652. Vardy MM, Kay SR. LSD psychosis or LSD-induced schizophrenia? A multimethod inquiry. Arch Gen Psychiatry 1983;40(8):877–83.

653. Robertson GH, Taveras JM, Tadmor R, et al. Computed tomography in metrizamide cisternography: importance of coronal and axial views. J Comput Assist Tomogr 1977;1:241.

654. Hartter S, Dingemanse J, Baier D, Ziegler G, Hiemke C. Inhibition of dextromethorphan metabolism by moclobemide. Psychopharmacology (Berl) 1998;135(1):22–6.

655. Rivera JM, Iriarte LM, Lozano F, Garcia-Bragado F, Salgado V, Grilo A. Possible estrogen-induced NMS. DICP 1989;23(10):811.

656. Houghton GW, Richens A. Inhibition of phenytoin metabolism by other drugs used in epilepsy. Int J Clin Pharmacol Biopharm 1975;12(1–2):210–6.

657. Siris JH, Pippenger CE, Werner WL, Masland RL. Anticonvulsant drug-serum levels in psychiatric patients with seizure disorders. Effects of certain psychotropic drugs. NY State J Med 1974;74(9):1554–6.

658. Sands CD, Robinson JD, Salem RB, Stewart RB, Muniz C. Effect of thioridazine on phenytoin serum concentration: a retrospective study. Drug Intell Clin Pharm 1987;21(3):267–72.

659. Kutt H, McDowell F. Management of epilepsy with diphenylhydantoin sodium. Dosage regulation for problem patients. JAMA 1968;203(11):969–72.

660. Haidukewych D, Rodin EA. Effect of phenothiazines on serum antiepileptic drug concentrations in psychiatric patients with seizure disorder. Ther Drug Monit 1985;7(4):401–4.

661. Gram LF, Christiansen J, Overo KF. Interaction between neuroleptics and tricyclic antidepressants. In: Morselli PL, Garranttini S, Cohen SN, editors. Drug Interactions. New York: Raven Press, 1974:271.

662. Vincent FM. Phenothiazine-induced phenytoin intoxication. Ann Intern Med 1980;93(1):56–7.

663. Jann MW, Fidone GS, Hernandez JM, Amrung S, Davis CM. Clinical implications of increased antipsychotic plasma concentrations upon anticonvulsant cessation. Psychiatry Res 1989;28(2):153–9.

664. Linnoila M, Viukari M, Vaisanen K, Auvinen J. Effect of anticonvulsants on plasma haloperidol and thioridazine levels. Am J Psychiatry 1980;137(7):819–21.

665. Ereshefsky L, Saklad SR, Watanabe MD, Davis CM, Jann MW. Thiothixene pharmacokinetic interactions: a study of hepatic enzyme inducers, clearance inhibitors, and demographic variables. J Clin Psychopharmacol 1991;11(5):296–301.

666. Miller DD. Effect of phenytoin on plasma clozapine concentrations in two patients. J Clin Psychiatry 1991;52(1):23–5.

667. Sturman G. Interaction between piperazine and chlorpromazine. Br J Pharmacol 1974;50(1):153–5.

668. Bernard JM, Le Roux D, Pereon Y. Acute dystonia during sevoflurane induction. Anesthesiology 1999;90(4):1215–6.

669. Benazzi F. Urinary retention with sertraline, haloperidol, and clonazepam combination. Can J Psychiatry 1998;43(10):1051–2.

670. Kurlan R. Acute parkinsonism induced by the combination of a serotonin reuptake inhibitor and a neuroleptic in adults with Tourette's syndrome. Mov Disord 1998;13(1):178–9.

671. Lee MS, Kim YK, Lee SK, Suh KY. A double-blind study of adjunctive sertraline in haloperidol-stabilized patients with chronic schizophrenia. J Clin Psychopharmacol 1998;18(5):399–403.

672. Linnoila M, George L, Guthrie S. Interaction between antidepressants and perphenazine in psychiatric inpatients. Am J Psychiatry 1982;139(10):1329–31.

673. Maynard GL, Soni P. Thioridazine interferences with imipramine metabolism and measurement. Ther Drug Monit 1996;18(6):729–31.

Alimemazine

General Information

Alimemazine is a phenothiazine derivative with sedative antihistaminic and antimuscarinic effects.

Organs and Systems

Nervous system

Neuroleptic malignant syndrome has been reported with alimemazine (1).

- A 4-year-old girl with damage in the basal ganglia who was receiving increasing doses of alimemazine for sedative purposes developed neuroleptic malignant syndrome. The alimemazine was withdrawn and she received dantrolene and supportive measures, including ventilation under sedation and paralysis with midazolam and vecuronium. As her symptoms were unchanged, she was given increasing doses of bromocriptine and improved. A few days after bromocriptine withdrawal, the neuroleptic malignant syndrome recurred and was complicated by cardiorespiratory arrest.

Reference

1. van Maldegem BT, Smit LM, Touw DJ, Gemke RJ. Neuroleptic malignant syndrome in a 4-year-old girl associated with alimemazine. Eur J Pediatr 2002;161(5):259–61.

Amisulpride

General Information

Amisulpride is an atypical antipsychotic drug, a benzamide derivative, which may have a low propensity to cause extrapyramidal symptoms (SEDA-22, 55).

Amisulpride 600–1200 mg/day for 3 months was effective and well tolerated in 445 patients with schizophrenia aged 18–45 years (1). During this time, 124 patients (28%) dropped out of the study; 21% reported adverse events, neurological (35%), psychiatric (15%), or endocrine (9.1%). Seven adverse events were assessed as serious: two suicides, two suicide attempts, one neuroleptic malignant syndrome, one somnolence, and one worsening of arteritis.

A lower dose of amisulpride (50 mg) has been tested in 20 healthy elderly volunteers (aged 65–79 years) (2). There were no serious adverse events, but one subject reported a moderate headache for 18 hours, a second subject vomited 9 hours after dosing, and a further subject complained of mild somnolence for 12 hours starting 4 hours after dosing; however, there were no extrapyramidal symptoms, clinically significant hemodynamic variations, or electrocardiographic abnormalities.

Observational studies

The prescribing of amisulpride for 811 schizophrenic inpatients from 240 psychiatric hospitals was monitored for 8 weeks; prescribed dosages were in the lower range of what is recommended for acute cases: the mean dose on day 56 was on average 550 mg/day, range 100–1600 mg/day (3). For the most severe cases there was a tendency for higher doses to be related to better improvement in positive symptoms and in negative symptoms; this correlation was not found in the least severe cases. The authors pointed out that this supports the notion that in more severe cases higher doses should be prescribed, while in milder cases lower doses may be sufficient; this would be in line with existing prescribing recommendations (800 mg/day as a standard dose in severe and recurrent episodes and especially in hospital; in the case of an insufficient response, the dose can be increased to 1200 mg/day) (4).

Comparisons with placebo and other antipsychotic drugs

In a randomized double-blind study, there were no differences in the numbers of patients with at least one adverse effect with amisulpride 100 mg (24%; n = 18), amisulpride 50 mg (25%; n = 21), or placebo (33%; n = 27) (5). Few patients had endocrine symptoms (2 out of 160 in the amisulpride groups).

Two narrative reviews of amisulpride have been published (6,7). The authors emphasized that amisulpride in low dosages (below 300 mg/day) causes a similar incidence of adverse effects to placebo; nevertheless, at higher dosages (400–1200 mg/day), the overall incidence of adverse events in those taking amisulpride was similar to that in patients taking haloperidol, flupenthixol, or risperidone. The most commonly reported adverse events associated with higher dosages of amisulpride were extrapyramidal symptoms, insomnia, hyperkinesia, anxiety, increased body weight, and agitation. The incidence of extrapyramidal symptoms was dose-related. In elderly people, amisulpride can cause hypotension and sedation. There are no systematic published data on efficacy in children aged under 15 years.

In an extensive review of 19 randomized studies for the Cochrane Library (n = 2443), most of the trials were small and of short duration (8). The data from four trials with 514 participants with predominantly negative symptoms suggested that low-dose amisulpride (up to 300 mg/day) was more acceptable than placebo (n = 514; RR = 0.6; 95% CI = 0.5, 0.8).

A meta-analysis of 10 randomized controlled clinical trials of amisulpride in "acutely ill patients" (n = 1654) has been published, supported in part by a grant from Sanofi-Synthélabo, the marketing authorization holder (9). Amisulpride was significantly better than conventional antipsychotic drugs by about 11 percentage points on the Brief Psychiatric Rating Scale. In four studies in patients with "persistent negative symptoms," amisulpride was significantly better than placebo (n = 514), but there was no significant difference between amisulpride and conventional drugs (only three trials; n = 130). Low doses of amisulpride (50–300 mg/day) were not associated with significantly more use of antiparkinsonian drugs than placebo (n = 507), and usual doses caused fewer extrapyramidal adverse effects than conventional antipsychotic

drugs (n = 1599). In studies in acutely ill patients, significantly fewer patients taking amisulpride dropped out compared with patients taking conventional drugs, mainly owing to fewer adverse events; there were no significant differences in dropout rates between amisulpride and conventional antipsychotic drugs (three small studies).

Eighteen randomized controlled trials that compared amisulpride with conventional neuroleptic drugs or placebo in patients with schizophrenia have been combined in a meta-analysis (10). The differences in the mean effect size for efficacy clearly showed that all types of neuroleptic drugs were more effective than placebo, but the difference in mean effect size for all neuroleptic drugs versus placebo (n = 2000) was only 25%. Amisulpride was associated with fewer extrapyramidal adverse effects and fewer drop-outs because of adverse events than conventional neuroleptic drugs. More risperidone recipients than amisulpride recipients had endocrine problems, but the difference was not statistically significant. Clozapine and olanzapine had the greatest potential to cause *weight gain*; after 10 weeks of treatment, mean increases in weight were 4.4 kg with clozapine and 4.1 kg with olanzapine. Risperidone also caused weight gain (a mean increase of 2.1 kg after 10 weeks), while ziprasidone caused the least gain; amisulpride has since been found to cause minor weight gain, about 0.8 kg after 10 weeks.

Amisulpride versus olanzapine

In a randomized comparison of amisulpride with olanzapine in 377 patients with schizophrenia with predominantly positive symptoms, who were treated for 6 months with either amisulpride 200–800 mg or olanzapine 5–20 mg, positive and negative scores were similar for amisulpride and olanzapine; new weight gain was less in amisulpride-treated patients; by day 56, amisulpride recipients had gained 0.4 kg whereas olanzapine recipients had gained 2.7 kg (Mortimer S21).

There were moderate but significant improvements in neurocognition (including executive function, working memory, and declarative memory) in a randomized, double-blind, 8-week study in 52 patients with schizophrenia assigned either to olanzapine (10–20 mg/day; n = 18) or amisulpride (400–800 mg/day; n = 18) (11). Of 16 dropouts, six were due to adverse events: olanzapine—sedation (n = 2) and increased transaminases (n = 1); amisulpride—rash, extrapyramidal symptoms, and galactorrhea (n = 1 each).

Amisulpride versus risperidone

In a 6-month randomized, controlled trial in 304 patients with schizophrenia amisulpride was compared with risperidone (12). The percentage of patients that responded to treatment was significantly greater with amisulpride 200–800 mg than risperidone 2–8 mg; amisulpride was superior to risperidone with respect to weight gain, as only 18% of amisulpride-treated patients increased their weight by more than 7% after 6 months compared with 33% of risperidone-treated patients.

Comparisons with antidepressants

Amisulpride has been compared with amitriptyline ($n = 250$; 6 months) (13) and with amineptine ($n = 323$; 3 months) in randomized double-blind trials in the treatment of dysthymia (14). In both trials, amisulpride was more efficacious than placebo but equal to amineptine and amitriptyline. Endocrine symptoms (such as galactorrhea and menstrual disorders) and weight gain were more frequent with amisulpride. There was galactorrhea in 8.4% and 11% respectively. In one of the studies (13), serious adverse events occurred in 13 of 165 patients taking amisulpride (15 events). Sudden death, probably secondary to myocardial infarction, occurred in a patient taking amisulpride 7 days after withdrawal. In the other cases, admission to hospital was required for fracture (two cases) and gastric pain, neuralgia, hyperglycemia, eczema, an injury, and an erythematous rash (one case each).

In a randomized, double-blind, multicenter trial for 8 weeks in 278 patients with depression, there were no differences in efficacy or tolerability between amisulpride 50 mg and SSRIs (15).

Organs and Systems

Cardiovascular

QT interval prolongation has been attributed to amisulpride.

- Sinus bradycardia and QT interval prolongation occurred in a 25-year-old man taking amisulpride 800 mg/day (16). The dosage of amisulpride was reduced to 600 mg/day and the electrocardiogram normalized within a few days.

Psychological, psychiatric

In a randomized, double-blind, crossover study in 21 healthy volunteers who took amisulpride 50 mg/day, amisulpride 400 mg/day, haloperidol 4 mg/day, or placebo, amisulpride 400 mg had several adverse effects on psychomotor performance and cognitive performance, similar to those of haloperidol, at the end of the 5-day course of treatment; however, there were no signs of mental disturbances on clinical rating scales or during a structured psychiatric interview (17).

Endocrine

Five women with psychoses treated with amisulpride developed hyperprolactinemia and were treated with bromocriptine 10–40 mg/day (18). Prolactin concentrations were markedly reduced in only three of the five; menses recurred in one of four patients with amenorrhea; lactation decreased in one of three patients with galactorrhea, and in two patients with reduced prolactin concentrations the psychotic symptoms exacerbated but fully remitted after withdrawal of bromocriptine. A 40-year-old woman also developed amenorrhea while taking a very low dose of amisulpride for 3 months (100 mg/day) (19).

A prolactinoma has been attributed to amisulpride.

- A 38-year-old woman with a borderline personality disorder developed a prolactinoma, with hyperprolactinemia, amenorrhea, and galactorrhea, probably induced by amisulpride 300 mg/day, which she had taken for 4 months following a bout of delirium with impaired attention, cognitive alteration, anxiety, and agitation (20). She had a microadenoma (5 mm) on the right side of the pituitary gland. Amisulpride was withdrawn and replaced by quetiapine 100 mg/day. The symptoms of hyperprolactinemia resolved.

References

1. Wetzel H, Grunder G, Hillert A, Philipp M, Gattaz WF, Sauer H, Adler G, Schroder J, Rein W, Benkert OThe Amisulpride Study Group. Amisulpride versus flupentixol in schizophrenia with predominantly positive symptomatology—a double-blind controlled study comparing a selective D2-like antagonist to a mixed D1-/D2-like antagonist. Psychopharmacology (Berl) 1998;137(3):223–32.
2. Hamon-Vilcot B, Chaufour S, Deschamps C, Canal M, Zieleniuk I, Ahtoy P, Chretien P, Rosenzweig P, Nasr A, Piette F. Safety and pharmacokinetics of a single oral dose of amisulpride in healthy elderly volunteers. Eur J Clin Pharmacol 1998;54(5):405–9.
3. Linden M, Schee T, Eich F. Dosage finding and outcome in the treatment of schizophrenic inpatients with amisulpride. Results of a drug utilization observation study. Hum Psychopharmacol Clin Exp 2004;19:111–9.
4. Lecrubier Y, Azorin M, Bottai T, Dalery J, Garreau G, Lemperiere T, Lisoprawski A, Petitjean F, Vanelle JM. Consensus on the practical use of amisulpride, an atypical antipsychotic, in the treatment of schizophrenia. Neuropsychobiology 2001;44(1):41–6.
5. Danion JM, Rein W, Fleurot OAmisulpride Study Group. Improvement of schizophrenic patients with primary negative symptoms treated with amisulpride. Am J Psychiatry 1999;156(4):610–6.
6. Curran MP, Perry CM. Spotlight on amisulpride in schizophrenia. CNS Drugs 2002;16(3):207–11.
7. Green B. Focus on amisulpride. Curr Med Res Opin 2002;18(3):113–7.
8. Mota NE, Lima MS, Soares BG. Amisulpride for schizophrenia. Cochrane Database Syst Rev 2002;(2):CD001357.
9. Leucht S, Pitschel-Walz G, Engel RR, Kissling W. Amisulpride, an unusual "atypical" antipsychotic: a meta-analysis of randomized controlled trials. Am J Psychiatry 2002;159(2):180–90.
10. Leucht S. Amisulpride – a selective dopamine antagonist and atypical antipsychotic: results of a meta-analysis of randomized controlled trials. Int J Neuropsychopharmacol 2004;7 Suppl 1:S15-S20.
11. Wagner M, Quednow BB, Westheide J, Schlaepfer TE, Maier W, Kuhn KU. Cognitive improvement in schizophrenic patients does not require a serotonergic mechanism: randomized controlled trial of olanzapine vs amisulpride. Neuropsychopharmacology 2005;30:381–90.
12. Mortimer A. How do we choose between atypical antipsychotics? The advantages of amisulpride. Int J Neuropsychopharmacol 2004;7 Suppl 1:S21–5.
13. Ravizza L. Amisulpride in medium-term treatment of dysthymia: a six-month, double-blind safety study versus

amitriptyline. AMILONG investigators. J Psychopharmacol 1999;13(3):248–54.

14. Boyer P, Lecrubier Y, Stalla-Bourdillon A, Fleurot O. Amisulpride versus amineptine and placebo for the treatment of dysthymia. Neuropsychobiology 1999;39(1):25–32.

15. Cassano GB, Jori MC. AMIMAJOR Group. Efficacy and safety of amisulpride 50 mg versus paroxetine 20 mg in major depression: a randomized, double-blind, parallel group study Int Clin Psychopharmacol 2002;17(1):27–32.

16. Pedrosa Gil F, Grohmann R, Ruther E. Asymptomatic bradycardia associated with amisulpride. Pharmacopsychiatry 2001;34(6):259–61.

17. Ramaekers JG, Louwerens JW, Muntjewerff ND, Milius H, de Bie A, Rosenzweig P, Patat A, O'Hanlon JF. Psychomotor, cognitive, extrapyramidal, and affective functions of healthy volunteers during treatment with an atypical (amisulpride) and a classic (haloperidol) antipsychotic. J Clin Psychopharmacol 1999;19(3):209–21.

18. Bliesener N, Yokusoglu H, Quednow B, Klingmüller D, Kühn K. Usefulness of bromocriptine in the treatment of amisulpride-induced hyperprolactinemia. Pharmaco psychiatry 2004;37:189–91.

19. Fountoulakis KN, Iacovides A, Kaprinis GS. Successful treatment of Tourette's disorder with amisulpride. Ann Pharmacother 2004;38:901.

20. Perroud N, Huguelet P. A possible effect of amisulpride on a prolactinoma growth in a woman with borderline personality disorder. Pharmacol Res 2004;50:377–9.

Aripiprazole

General Information

Aripiprazole is a partial agonist at dopamine D_2 receptors. It therefore acts as an agonist when there is low dopaminergic neurotransmission and an antagonist when there is excess dopaminergic neurotransmission. It has therefore been called a dopamine stabilizer (1). It is also a partial agonist at $5HT_{1A}$ receptors and antagonist at $5HT_2$ receptors (2). Its most common adverse effects include restlessness and akathisia, somnolence, and nausea. It can also worsen extrapyramidal symptoms (3).

Observational studies

In 142 adult patients who took aripiprazole (mean final daily dose 16 mg, 0.20 mg/kg) for psychotic, major affective, or other disorders, adverse effects occurred in 16, were three times more likely among women, and most often involved moderate behavioral activation or nausea, with no new episodes of mania (4).

Placebo-controlled studies

In a 4-week, double-blind, randomized placebo-controlled study in 36 US centers 414 patients with schizophrenia or schizoaffective disorder were randomized to aripiprazole (15–30 mg/day) or haloperidol (10 mg/day) (5). Haloperidol and both doses of aripiprazole produced statistically significant improvements from baseline compared with placebo. Unlike haloperidol, aripiprazole was not associated with significant extrapyramidal symptoms or raised prolactin at the end-point. There were no statistically significant differences in mean changes in body weight and no patients who took aripiprazole had clinically significant increases in the QT_c interval.

Organs and Systems

Nervous system

Oral dyskinesia emerged after several months of treatment with haloperidol 7.5 mg/day and gradually disappeared within 2 months after therapy was changed to aripiprazole 10 mg/day (6).

Neuroleptic malignant syndrome (7) and rabbit syndrome (8) have also been described.

Aripiprazole causes mild somnolence in about 11% of patients and severe somnolence has also been reported in a 9-year-old girl 3.5 hours after a single dose of aripiprazole 15 mg (9).

Psychiatric

Five patients (three women aged 30, 32, and 41 years and two men aged 36 and 56 years), had serious adverse effects developed after starting to take aripiprazole. There was agitation, akathisia, insomnia, and dysphoria; three made suicide attempts and two had suicidal thoughts (10).

- Paradoxical worsening of a schizoaffective disorder occurred in a 50-year-old man taking quetiapine 400 mg bd and divalproex 1000 mg bd, who was also given aripiprazole 15 mg/day. He recovered after withdrawal of aripiprazole and responded best to divalproex and olanzapine.

This effect was attributed to an agonistic effect of aripiprazole at dopamine receptors in the presence of quetiapine, a dopamine receptor antagonist (11). This suggests that aripiprazole should not be used in combination with another dopamine receptor antagonist.

Worsening of a psychosis has also been reported in four patients who took aripiprazole, two during tapering reduction of a previous atypical antipsychotic drug and two when aripiprazole was added to an atypical antipsychotic (12).

Endocrine

Hyperglycemia and diabetic ketoacidosis has been attributed to aripiprazole (13).

- A 34-year-old African–American woman with schizophrenia had nausea, vomiting, and malaise for 3–4 days shortly after starting to take aripiprazole therapy. She had hyperglycaemia and a metabolic acidosis, which responded rapidly to standard treatment and did not recur when aripiprazole was withdrawn.

Galactorrhea has been attributed to aripiprazole after only 2 days of administration (14).

- A 29-year-old woman with a schizoaffective disorder took haloperidol 5 mg/day and then 9 mg/day because of acute psychotic episodes. She had no adverse effects such as amenorrhea or galactorrhea. Haloperidol was then replaced by aripiprazole 15 mg/day and on the evening of the second day she developed breast tenderness and marked galactorrhea. The serum prolactin concentration was 32 ng/ml (reference range 5–25 ng/ml). Aripiprazole was withdrawn and haloperidol restarted. The galactorrhea resolved in 1 week.

Metabolism

In a 26-week, multicenter, randomized, double-blind, study in 317 patients with schizophrenia, 156 were randomized to aripiprazole and 161 to olanzapine; more of those who took olanzapine had clinically significant weight gain during the trial (37% versus 14%) (15). At week 26, there was a mean weight loss of 1.37 kg with aripiprazole compared with a mean increase of 4.23 kg with olanzapine. Changes in fasting plasma concentrations of total cholesterol, HDL cholesterol, and triglycerides were significantly different, with worsening in the patients who took olanzapine.

Drug Administration

Drug formulations

In January 2005, an oral solution of aripiprazole was approved by the FDA, providing an option for adults with difficulty in swallowing (16,17).

Drug overdose

An overdose of 195 mg (17.1 mg/kg) of aripiprazole in a 2.5 year-old child caused nervous system depression that did not require respiratory support but persisted for almost 2 weeks, because of the long half-life of aripiprazole; there were no significant cardiovascular effects (18).

A 27-year-old woman who took 330 mg of aripiprazole in a suicide attempt only suffered mild sedation despite a serum concentration of 716 μg/l, nearly six times the upper limit of the accepted target range (19).

References

1. Hirose T, Kikuchi T. Aripiprazole, a novel antipsychotic agent:dopamine D_2 receptor partial agonist. J Med Invest 2005;52 Suppl:284–90.
2. Fleischhacker WW. Aripiprazole. Expert Opin Pharmacother 2005;6(12):2091–101.
3. Kinghorn WA, McEvoy JP. Aripiprazole: pharmacology, efficacy, safety and tolerability. Expert Rev Neurother 2005;5(3):297–307.
4. Centorrino F, Fogarty KV, Cimbolli P, Salvatore P, Thompson TA, Sani G, Cincotta SL, Baldessarini RJ. Aripiprazole: initial clinical experience with 142 hospitalized psychiatric patients. J Psychiatr Pract 2005;11(4):241–7.
5. Kane JM, Carson WH, Saha AR, McQuade RD, Ingenito GG, Zimbroff DL, Ali MW. Efficacy and safety of aripiprazole and haloperidol versus placebo in patients with
schizophrenia and schizoaffective disorder. J Clin Psychiatry 2002;63:763–71.
6. Grant MJ, Baldessarini RJ. Possible improvement of neuroleptic-associated tardive dyskinesia during treatment with aripiprazole. Ann Pharmacother 2005;39:1953.
7. Chakraborty N, Johnston T. Aripiprazole and neuroleptic malignant syndrome. Int Clin Psychopharmacol 2004;19(6):351–3.
8. Mendhekar DN. Aripiprazole-induced rabbit syndrome. Aust N Z J Psychiatry 2004;38(7):561.
9. Davenport JD, McCarthy MW, Buck ML. Excessive somnolence from aripiprazole in a child. Pharmacotherapy 2004;24(4):522–5.
10. Scholten MRM, Selten JP. Suïcidale ideaties en suicidepogingen na instelling op aripiprazol, een nieuw antipsychoticum. [Suicidal ideations and suicide attempts after starting on aripiprazole, a new antipsychotic drug.] Ned Tijdschr Geneeskd 2005;149(41):2296–8.
11. Reeves RR, Mack JE. Worsening schizoaffective disorder with aripiprazole. Am J Psychiatry 2004;161(7):1308.
12. Ramaswamy S, Vijay D, William M, Sattar SP, Praveen F, Petty F. Aripiprazole possibly worsens psychosis. Int Clin Psychopharmacol 2004;19(1):45–8.
13. Church CO, Stevens DL, Fugate SE. Diabetic ketoacidosis associated with aripiprazole. Diabet Med 2005; 22(10):1440–3.
14. Ruffatti A, Minervini L, Romano M, Sonino N. Galactorrhea with aripiprazole. Psychother Psychosom 2005;74:391–2.
15. McQuade RD, Stock E, Marcus R, Jody D, Gharbia NA, Vanveggel S, Archibald D, Carson WH. A comparison of weight change during treatment with olanzapine or aripiprazole: results from a randomized, double-blind study. J Clin Psychiatry 2004;65 Suppl 18:47–56.
16. Medical News. FDA approves oral solution formulation of Abilify (aripiprazole) for schizophrenia. http://www.medicalnewstoday.com/articles/18670.php.
17. Fleischhacker WW. Aripiprazole. Expert Opin Pharmacother 2005;6:2091–101.
18. Seifert SA, Schwartz MD, Thomas JD. Aripiprazole (abilify) overdose in a child. Clin Toxicol (Phila) 2005;43(3):193–5.
19. Carstairs SD, Williams SR. Overdose of aripiprazole, a new type of antipsychotic. J Emerg Med 2005;28(3):311–3.

Chlorpromazine

General Information

Chlorpromazine is a phenothiazine with a large range of pharmacological actions; it is a dopamine receptor antagonist, an alpha-adrenoceptor antagonist, a muscarinic antagonist, and an antihistamine.

Organs and Systems

Cardiovascular

The possibility that some of the cardiac effects of chlorpromazine may be related to metabolites as well as the parent compound has been explored (1,2).

Some cases of sudden death in apparently young healthy individuals may be directly attributable to cardiac dysrhythmias after treatment with thioridazine or chlorpromazine (3).

Nervous system

In 15 schizophrenic inpatients aged 16–55 years, there was a 50% probability that a patient would have a tremor when the plasma concentration of chlorpromazine was 46 ng/ml or more, corresponding to the minimum plasma concentration that has been associated with a good clinical response (4).

Sensory systems

Deposits in the cornea and lens can complicate long-term chlorpromazine therapy and in vivo confocal imaging of such deposits has now been reported, supposedly for the first time (5).

- A 59-year-old woman who for 20 years had taken chlorpromazine up to 1200 mg/day (mean dose 400 mg/day) gradually developed blurred vision in her left eye. Slit-lamp biomicroscopy showed multiple fine creamy-white deposits on her corneal endothelium and anterior crystalline lens capsule bilaterally. Microstructural analysis of the corneal endothelium showed that there were no abnormalities in cellular morphology resulting from these deposits.

Skin

Chlorpromazine most often causes photosensitivity reactions (incidence around 3%), which may result from formation of a cytotoxic by-product after exposure to ultraviolet light. Patients should be advised to avoid prolonged exposure to strong indoor light, to wear protective clothing, and to use a combined para-aminobenzoic acid plus benzophenone sunscreen when exposure to strong sunlight is unavoidable.

Toxic epidermal necrolysis has been reported in association with chlorpromazine (6).

Chlorpromazine is thought to have caused an immunologically mediated contact urticaria (7–11).

Sexual function

Two cases of priapism have been attributed to chlorpromazine: a 65-year-old man who took a single dose of chlorpromazine 25 mg and a 27-year-old man who took chlorpromazine 200 mg for agitation after a suicide attempt (12).

- A 30-year-old man with schizophrenia developed priapism for 8 hours after taking chlorpromazine for 3 years (13). He was taking no other medications and there were no other pathological findings; a complete blood count and sickle cell screen were normal. Aspiration of blood from the corpora cavernosa, followed by saline irrigation led to complete detumescence.

Drug–Drug Interactions

Amfetamine

Chlorpromazine has sometimes been used to treat amfetamine psychosis, for example due to acute poisoning in children who did not respond to barbiturates (14).

Beta-adrenoceptor antagonists

Lipophilic beta-adrenoceptor antagonists are metabolized to varying degrees by oxidation by liver microsomal cytochrome P450 (for example propranolol by CYP1A2 and CYP2D6 and metoprolol by CYP2D6). They can therefore reduce the clearance and increase the steady-state plasma concentrations of other drugs that undergo similar metabolism, potentiating their effects. Drugs that interact in this way include chlorpromazine (15).

- A schizophrenic patient experienced delirium, tonic-clonic seizures, and photosensitivity after the addition of propranolol to chlorpromazine, suggesting that chlorpromazine concentrations are increased by propranolol (16).

Although high dosages of propranolol (up to 2 g) have been used in combination with chlorpromazine to treat schizophrenia, the combination of propranolol or pindolol with chlorpromazine should be avoided if possible (17).

Desmopressin

Chlorpromazine increases the release of endogenous antidiuretic hormone and can therefore potentiate the antidiuretic effect of desmopressin (18).

Haloperidol

- A 40-year-old man with schizophrenia developed a raised plasma concentration of haloperidol in combination with chlorpromazine and during overlap treatment with clozapine (19).

Like haloperidol, chlorpromazine is a competitive inhibitor of CYP2D6; however, clozapine appears to be largely metabolized by CYP1A2.

Lithium

The neuroleptic malignant syndrome in a 49-year-old man was attributed to a combination of lithium and chlorpromazine (20).

Monoamine oxidase inhibitors

Hyperthermia and labile blood pressure occurred in a patient taking chlorpromazine, phenelzine, and clomipramine (21).

Piperazine

High doses of piperazine can enhance the adverse effects of chlorpromazine and other phenothiazines (22,23).

Smoking

Chlorpromazine concentrations were reduced by 36% in smokers (24). Of factors that can affect chlorpromazine concentrations, smoking may be second in importance only to dosage (25).

Interference with Diagnostic Tests

Cholesterol

Chlorpromazine can cause overestimation of cholesterol (Zlatkis–Zak reaction) (26).

CSF protein

Chlorpromazine can cause falsely increased CSF protein (Folin–Ciocalteau method) (26).

Haptoglobin

Chlorpromazine can cause falsely reduced serum haptoglobin concentrations (26).

References

1. Axelsson R, Aspenstrom G. Electrocardiographic changes and serum concentrations in thioridazine-treated patients. J Clin Psychiatry 1982;43(8):332–5.
2. Dahl SG. Active metabolites of neuroleptic drugs: possible contribution to therapeutic and toxic effects. Ther Drug Monit 1982;4(1):33–40.
3. Risch SC, Groom GP, Janowsky DS. The effects of psychotropic drugs on the cardiovascular system. J Clin Psychiatry 1982;43(5 Pt 2):16–31.
4. Chetty M, Gouws E, Miller R, Moodley SV. The use of a side effect as a qualitative indicator of plasma chlorpromazine levels. Eur Neuropsychopharmacol 1999;9(1–2):77–82.
5. Phua YS, Patel DV, McGhee CN. In vivo confocal microstructural analysis of corneal endothelial changes in a patient on long-term chlorpromazine therapy. Graefe's Arch Clin Exp Ophthalmol 2005;243:721–3.
6. Purcell P, Valmana A. Toxic epidermal necrolysis following chlorpromazine ingestion complicated by SIADH. Postgrad Med J 1996;72(845):186.
7. Leliever WC. Topical gentamicin-induced positional vertigo. Otolaryngol Head Neck Surg 1985;93(4):553–5.
8. De Groot AC, Weyland JW, Nater JP. Unwanted Effects of Cosmetics and Drugs used in Dermatology. 3rd ed.. Amsterdam: Elsevier;. 1994.
9. Hannuksela M. Mechanisms in contact urticaria. Clin Dermatol 1997;15(4):619–22.
10. Wistedt B. Neuroleptics and depression. Arch Gen Psychiatry 1982;39(6):745.
11. Wilkins-Ho M, Hollander Y. Toxic delirium with low-dose clozapine. Can J Psychiatry 1997;42(4):429–30.
12. Mutlu N, Ozkurkcugil C, Culha M, Turkan S, Gokalp A. Priapism induced by chlorpromazine. Int J Clin Pract 1999;53(2):152–3.
13. Kilciler M, Bedir S, Sümer F, Dayanç M, Peker AF. Priapism in a patient receiving long-term chlorpromazine therapy. Urol Int 2003;71:127–8.
14. Espelin DE, Done AK. Amphetamine poisoning. Effectiveness of chlorpromazine. N Engl J Med 1968;278(25):1361–5.
15. Peet M, Middlemiss DN, Yates RA. Pharmacokinetic interaction between propranolol and chlorpromazine in schizophrenic patients. Lancet 1980;2(8201):978.
16. Miller FA, Rampling D. Adverse effects of combined propranolol and chlorpromazine therapy. Am J Psychiatry 1982;139(9):1198–9.
17. Markowitz JS, Wells BG, Carson WH. Interactions between antipsychotic and antihypertensive drugs. Ann Pharmacother 1995;29(6):603–9.
18. Allen SA. Effect of chlorpromazine and clozapine on plasma concentrations of haloperidol in a patient with schizophrenia. J Clin Pharmacol 2000;40(11):1296–7.
19. Stern TA, Schwartz JH, Shuster JL. Catastrophic illness associated with the combination of clomipramine, phenelzine and chlorpromazine. Ann Clin Psychiatry 1992;4:81–5.
20. Pantuck EJ, Pantuck CB, Anderson KE, Conney AH, Kappas A. Cigarette smoking and chlorpromazine disposition and actions. Clin Pharmacol Ther 1982;31(4):533–8.
21. Sramek J, Herrera J, Roy S, Parent M, Hudgins R, Costa J, Alatorre E. An analysis of steady state chlorpromazine plasma levels in the clinical setting. J Clin Psychopharmacol 1987;7(2):117–8.
22. Sher PP. Drug interferences with clinical laboratory tests. Drugs 1982;24(1):24–63.
23. Wilke RA. Posterior pituitary sigma receptors and drug-induced syndrome of inappropriate antidiuretic hormone release. Ann Intern Med 1999;131(10):799.
24. Leber K, Malek A, D'Agostino A, Adelman HM. A veteran with acute mental changes years after combat. Hosp Pract (Off Ed) 1999;34(6):21–2.
25. Sturman G. Interaction between piperazine and chlorpromazine. Br J Pharmacol 1974;50(1):153–5.
26. Boulos BM, Davis LE. Hazard of simultaneous administration of phenothiazine and piperazine. N Engl J Med 1969;280(22):1245–6.

Chlorprothixene

General Information

Chlorprothixene is a thioxanthene neuroleptic drug.

Organs and Systems

Immunologic

Rare disorders of connective tissue resembling systemic lupus erythematosus have been reported with chlorpromazine, perphenazine, and chlorprothixene (1).

Reference

1. McNevin S, MacKay M. Chlorprothixene-induced systemic lupus erythematosus. J Clin Psychopharmacol 1982;2(6):411–2.

Clotiapine

General Information

Clotiapine is a dibenzothiazepine neuroleptic drug.

Organs and Systems

Pancreas

Pancreatitis has been attributed to clotiapine (1).

- A 49-year-old woman started to take clotiapine 200 mg/day for a severe psychotic episode. On the second day, she complained of abdominal pain and nausea, followed by vomiting. Increased activities of serum amylase (1490 U/ml), serum lipase (3855 U/ml), and urinary amylase (3417 U/ml) suggested pancreatitis. There was no evidence of gallstones or tumor. She was given perphenazine instead, and her amylase and lipase activities fell to normal. Six months later, when she had a psychotic relapse, she was again treated with clotiapine and 2 days later had a rapid rise in amylase activity (887 U/ml). When switched to perphenazine, she had new peaks in amylase and lipase. A year later, her amylase and lipase were normal and her psychiatric disorder was stabilized with pimozide.

Reference

1. Francobandiera G, Rondalli G, Telattin P. Acute pancreatitis associated with clothiapine use. Hum Psychopharmacol Clin Exp 1999;14:211–2.

Clozapine

General Information

Clozapine is a dibenzodiazepine, an atypical neuroleptic drug with a high affinity for dopamine D_4 receptors and a low affinity for other subtypes (1). It is also an antagonist at alpha-adrenoceptors, $5\text{-}HT_{2A}$ receptors, muscarinic receptors, and histamine H_1 receptors.

Observational studies

Rates of hospitalization with clozapine have been analysed by Novartis, the manufacturers, based on two retrospective studies of hospitalization among patients with treatment-resistant schizophrenia (2). All the patients who began clozapine treatment in Texas State psychiatric facilities during the early 1990s ($n = 299$) were compared with controls who received traditional neuroleptic drugs ($n = 223$), matched for severity, age, and sex. More patients in the latter group required continuous hospitalization: at 4 years, four times as many patients taking a

traditional medication had a 6-month period of continuous hospitalization.

In a study in Ohio, patients with chronic borderline personality disorder who were in hospital for an average of 110 days/year were given clozapine (mean daily dosage at time of discharge 334 mg; range 175–550 mg) (3). None stopped taking clozapine and few adverse effects were reported. Among the seven patients who were taking clozapine when they were discharged, hospitalization fell to a mean of 6.3 days per patient per year. There was no control group in this study.

- Of 10 adolescent inpatients (aged 12–17 years) with severe acute manic or mixed episodes, who did not improve after treatment with conventional drugs and who were given clozapine (mean dose 143 mg/day), all responded positively after 15–28 days and adverse effects (increased appetite, sedation, enuresis, sialorrhea) were frequent but not severe enough to require reduced dosages (4). Mean weight gain after 6 months was 7 kg (11%), and neither reduced white cell counts nor epileptic seizures were reported during follow-up for 12–24 months.

In a case series in which clozapine was used as add-on medication, two patients with bipolar disorder and one with schizoaffective disorder had marked reductions in affective symptoms after clozapine had been added to pretreatment with a mood stabilizer; transient and moderate weight gain and fatigue were the only adverse effects (5).

In a retrospective open study of 46 patients taking clozapine for 4 years, clozapine had to be discontinued in 10 patients (21%) and serious adverse effects were rare; no patient had agranulocytosis (6). The most troublesome adverse effects were drooling, sedation, and weight gain, and three patients had seizures.

Experiences in uncontrolled open studies in Chinese patients have been summarized (7). The most common adverse effect of clozapine was hypersalivation, followed by sedation. Mandatory blood monitoring is considered an obstacle in persuading some patients to undergo a trial of clozapine, mainly for cultural reasons, summed up by the Chinese proverb that "a hundred grains of rice make a drop of blood."

Clozapine has been used in some special groups of patients, including patients with severe borderline personality disorder (8), patients with aggressive schizophrenia (9), and mentally retarded adults (10).

Of 12 in-patients with borderline personality disorder treated with clozapine for 16 weeks, 10 developed sedation, which disappeared during the first month of treatment; 9 had hypersialorrhea, and 6 had falls in white blood cell counts, which never reached unsafe values.

Patients with aggressive schizophrenia ($n = 29$) improved when treated with clozapine; one was withdrawn after the development of leukopenia. In 10 mentally retarded patients taking clozapine for 15 days to 46 months improvement was observed. Half of the patients developed sedation and hypersalivation, and one discontinued the drug after 2 weeks because of neutropenia. The putative neurotoxicity of clozapine in moderately to profoundly retarded patients (that is, those with an accentuation of cognitive deficits due to the drug's anticholinergic and sedating properties) was not observed. Fifty special hospitalized patients with schizophrenia associated with serious violence were treated with clozapine (mean dose at 2 years 465 mg) (11). The most frequent adverse effects were hypersalivation ($n = 14$), sedation ($n = 10$), and weight gain ($n = 6$); two patients had tonic-clonic seizures, two others developed mild neutropenia, and in one case treatment was stopped owing to agranulocytosis.

In a retrospective review, 33 mentally retarded patients were evaluated; adverse effects were mild and transient, constipation being the most common ($n = 10$) (12). There were no significant cardiovascular adverse effects and no seizures; no patient discontinued treatment because of agranulocytosis. Small sample sizes, short durations of treatment, and lack of controls in these studies preclude definite conclusions.

A 37-item survey covering a variety of somatopsychic domains has been administered to 130 patients with schizophrenia taking a stable clozapine regimen (mean dose 464 mg/day; mean duration 34 months) (13). Most of them reported an improvement in their level of satisfaction, quality of life, compliance with treatment, thinking, mood, and alertness. Most reported worse nocturnal salivation (88%); weight gain (35%) came second; fewer patients reported a worsening of various gastrointestinal and urinary symptoms.

Clinical predictors of response have been examined in 37 partially treatment-refractory outpatients who had been assigned to clozapine in a double-blind, haloperidol-controlled, long-term (29-week) study (14). Clozapine responders were rated as less severely ill, had fewer negative symptoms, and had fewer extrapyramidal adverse effects at baseline compared with non-responders.

Since clozapine may be the gold standard and the last resort in the treatment of refractory schizophrenia, the authors of a review aimed to discover whether a trial with clozapine is adequate (15). The results favored the approach of increasing the clozapine plasma concentration in treatment-refractory schizophrenic patients who do not respond to an initial low-to-medium dose. Some patients, especially young male smokers, will need dosages over 900 mg/day, and the addition of low-dose fluvoxamine while closely monitoring clozapine concentrations can help to reduce the large number of tablets required, since fluvoxamine increases the clozapine plasma concentration 2- to 3-fold, maximally 5-fold, and reduces N-desmethylclozapine concentrations; the combination can lead to non-linear kinetics of clozapine.

The aim of another study was to evaluate the long-term efficacy of clozapine in patients with treatment-resistant schizophrenia ($n = 34$), schizoaffective disorder, bipolar type ($n = 30$), or bipolar disorder with psychotic features ($n = 37$), who were treated with clozapine in flexible doses over 48 months (16). After this time, Global Assessment of Functioning scores were improved in all three groups, with significantly greater improvement in the bipolar disorder group compared with the others;

however, 54 patients withdrew during the study, five because of adverse effects (sedation, sleep cycle inversion, weight gain, and leukopenia).

Changes in regional cerebral blood flow induced by clozapine or haloperidol have been compared using positron emission tomography (PET) with radiolabelled water ($H_2^{15}O$), first after withdrawal of all psychotropic drugs ($n = 6$), then after treatment with therapeutic doses of haloperidol ($n = 5$) or clozapine ($n = 5$) (17). Cerebral blood flow increased in the ventral striatum and fell in the hippocampus and ventrolateral frontal cortex. The authors suggested that these changes might mediate a common component of the antipsychotic action of both drugs; however, the increase in cerebral blood flow in the dorsal striatum caused by haloperidol could well be associated with its prominent motor adverse effects, whereas the increased cerebral blood flow in the anterior cingulate or dorsolateral frontal cortex may mediate the probably superior antipsychotic action of clozapine.

Catatonic schizophrenia is a controversial syndrome, and there is debate about its etiology and treatment. There has been a report of two cases of catatonic schizophrenia successfully treated with clozapine: a 49-year-old woman and a 19-year-old man (18). Both responded to clozapine despite being resistant to several conventional and atypical antipsychotic drugs and, in the second case, a course of electroconvulsive therapy. These two cases are intriguing, because the dose of clozapine required to improve catatonia was about double the dose required to improve psychosis significantly (600 mg/day and 750 mg/day). The two patients had common adverse effects of clozapine; the first had mild nocturnal hypersalivation and mild/moderate constipation, and the second had moderate nocturnal hypersalivation.

- A 38-year-old woman with a 22-year history of resistant rapid-cycling bipolar I disorder finally responded well to a combination of clozapine 350 mg/day + topiramate 300 mg/day (19). Over 3 years of treatment she had no adverse effects, such as agranulocytosis, hyperglycemia, hyperlipidemia, or weight gain; in fact she had weight loss of 12 kg. She then developed daytime fatigue and palpitation, which resolved after she was given atenolol 100 mg/day.

Comparative studies

Maintenance therapy with neuroleptic drugs is an important aspect of the management of schizophrenia; maintenance therapy is usually conducted by halving the drug dosage that has proven effective during the acute phase. In a non-randomized study, records from patients taking clozapine ($n = 181$; mean duration of maintenance treatment 12.2 years) were compared with those of a control group of patients taking haloperidol ($n = 152$; mean duration of treatment 3.8 years) (20). The relapse rate was similar in the two groups, but the authors pointed out that compliance and therapeutic efficacy were superior with clozapine. The incidences of drowsiness, somnolence, delirium, and postural hypotension were similar in the two groups; however, only 1% of clozapine-treated patients complained of dry mouth, compared with 27% of those taking haloperidol. Extrapyramidal symptoms occurred in about 80% of those taking haloperidol (71% had parkinsonism and 9.2% had tardive dyskinesia) and none of those taking clozapine. Drooling was the sole adverse effect with a lower incidence in those taking haloperidol, and no patient in either group had agranulocytosis or granulocytopenia.

Clozapine has been evaluated in six treatment-resistant abused adolescents with post-traumatic stress disorder and psychotic symptoms, who had received at least two trials of conventional neuroleptic drugs, with particular attention to adverse effects (21). Clozapine was introduced after a 6-month baseline period and titrated over 173 days before the criterion dose (400 mg/day) was reached; the dose of clozapine reached a maximum of 600 mg/day during the treatment phase. Specific adverse effects were rated on a 0–4 scale, 0 representing "no difficulty" and 4 representing "severe difficulty"; the ratings of adverse effects were as follows: excessive salivation (2.7), dizziness (2.2), weight gain (2.0), nausea (1.7), feeling sleepy (1.5), constipation (1.3), palpitation (0.5), and dry mouth (0).

The combination of clozapine and amisulpride for the treatment of refractory schizophrenia has been assessed in seven patients. The response to clozapine plus amisulpride was rated as at least good in six cases (Clinical Global Impression Scale ≥ 3), the average length of treatment being 30 weeks (22). There were no significant changes in electrocardiographic time intervals after the addition of amisulpride to clozapine, and both the mean resting heart rate and mean QT_c interval were unaffected; mean clozapine plasma concentrations did not differ significantly from baseline.

- In a 28-year-old woman with psychotic symptoms resistant to monotherapy with clozapine or ziprasidone, the combination produced marked improvement in both positive and negative symptoms along with a reduction in adverse effects: body weight fell, blood pressure, pulse, and the electrocardiogram remained normal, and valproic acid, which had been introduced for epileptic seizures during clozapine monotherapy, was successfully withdrawn (23).

Placebo-controlled studies

In a 4-week, randomized, double-blind, parallel comparison of clozapine (n = 32) and placebo (n = 28), followed by a 12-week clozapine open period, plus a 1-month period after drug withdrawal in patients with Parkinson's disease with drug-induced psychosis, clozapine at a mean dose below 50 mg/day produced improvement without significantly worsening motor function according to the Unified Parkinson's Disease Rating Scale; however, the effect wore off once treatment was withdrawn (24). Somnolence was more common with clozapine than with placebo; one patient taking clozapine had seizures; two patients taking clozapine had transient neutropenia; in one the blood count normalized despite clozapine continuation and in one withdrawal was required. There was worsening of parkinsonism in 14 patients taking clozapine.

Combination therapy

Of 656 Danish patients who were taking clozapine, 35% were taking concomitant neuroleptic drugs, 28% benzodiazepines, 19% anticholinergic drugs, 11% antidepressants, 8% antiepileptic drugs, and 2% lithium (25). The rationale for supplementing clozapine treatment in refractory schizophrenia in this way has been thoroughly reviewed following a bibliographic search covering 1978–1998 (26). In all, 70 articles were retrieved but only a few were controlled studies, most being case reports/series. Among the many possible drug combinations, the evidence suggests that clozapine plus sulpiride is the most efficacious combination. The combined use of benzodiazepines and clozapine can cause cardiorespiratory collapse; valproate can cause hepatic dysfunction and more so with clozapine; lithium can cause neurotoxicity and seizures and more so with clozapine; and at least some serotonin reuptake inhibitors (SSRIs) appear to raise plasma clozapine concentrations to above the usual target range.

Organs and Systems

Cardiovascular

Clozapine has been associated with cardiomyopathy (SED-14, 142; SEDA-21, 52; 27), changes in blood pressure (SEDA-21, 52; SEDA-21, 52; SEDA-22, 57; 28,29), electrocardiographic changes (SEDA-22, 57; 30–32), and venous thromboembolism (SEDA-20, 47).

Hypertension
Several cases of hypertension have been associated with clozapine (SEDA-22, 57), and alpha$_2$-adrenoceptor blockade has been proposed as a possible mechanism (28). Four patients developed pseudopheochromocytoma syndrome associated with clozapine (33); all had hypertension, profuse sweating, and obesity. The authors suggested that clozapine could increase plasma noradrenaline concentrations by inhibiting presynaptic reuptake mediated by alpha$_2$-adrenoceptors.

Hypotension
Hypotension is the most commonly observed cardiovascular adverse effect of neuroleptic drugs, particularly after administration of those that are also potent alpha-adrenoceptor antagonists, such as chlorpromazine, thioridazine, and clozapine (34). A central mechanism involving the vasomotor regulatory center may also contribute to the lowering of blood pressure.

- A 51-year-old man taking maintenance clozapine developed profound hypotension after cardiopulmonary bypass (29).

Cardiac dysrhythmias
A substantial portion of patients taking clozapine develop electrocardiographic abnormalities; the prevalence was originally estimated at 10% (SEDA-20, 47) (SEDA-22, 57). However, although the prevalence may be higher,

most of the effects are benign and do not need treatment. In 61 patients with schizophrenia taking clozapine, in whom a retrospective chart review was conducted to identify electrocardiographic abnormalities, the prevalence of electrocardiographic abnormalities in those who used neuroleptic drugs other than clozapine was 14% (6/44), while in the neuroleptic drug-free patients it was 12% (2/17); when treatment was switched to clozapine, the prevalence of electrocardiographic abnormalities rose to 31% (19/61) (35).

The correlation between plasma clozapine concentration and heart rate variability has been studied in 40 patients with schizophrenia treated with clozapine 50–600 mg/day (31). The patients had reduced heart rate variability parameters, which correlated negatively with plasma clozapine concentration.

Clozapine can cause prolongation of the QT interval (36).

- In a 30-year-old man taking clozapine there were minor electrocardiographic abnormalities, including a prolonged QT interval. A power spectrum analysis of heart rate variability showed marked abnormalities in autonomic nervous system activity. When olanzapine was substituted, power spectrum analysis studies showed that his heart rate had improved significantly and that his cardiovascular parameters had returned to normal. Serial electrocardiograms showed minimal prolongation of the QT interval.

Tachycardia is the most common cardiovascular adverse effect of clozapine, and atrial fibrillation has also been reported (SEDA-22, 57) (30).

The reports of sudden death associated with clozapine and the possibility that it may have direct prodysrhythmic properties have been reviewed (37).

A patient developed ventricular fibrillation and atrial fibrillation after taking clozapine for 2 weeks (32).

- A 44-year-old man with no significant cardiac history was given clozapine and 12 days later had bibasal crackles in the chest and ST segment elevation in leads V2 and V3 of the electrocardiogram. He then developed ventricular tachycardia and needed resuscitation. He also developed atrial fibrillation for 24 hours, which subsequently resolved.

Cardiomyopathy

Cardiomyopathy has been associated with neuroleptic drugs, including clozapine (38–40), and partial data initially suggested an incidence of 1 in 500 in the first month.

In a review of articles on adverse cardiac effects associated with clozapine, the estimated risk of potentially fatal myocarditis or cardiomyopathy was 0.01–0.19%; the authors suggested that this low risk of serious adverse cardiac events should be outweighed by a reduction in suicide risk in most patients (41).

A thorough study of the risk of myocarditis or cardiomyopathy in Australia detected 23 cases (mean age 36 years; 20 men) out of 8000 patients treated with clozapine from January 1993 to March 1999 (absolute risk 0.29%; relative risk about 1000–2000) (27). All the accumulated data on previous reports of sudden death, myocarditis, or cardiac disease noted in connection with clozapine treatment were requested from the Adverse Drug Reactions Advisory Committee (ADRAC); there were 15 cases of myocarditis (five fatal) and 8 of cardiomyopathy (one fatal) associated with clozapine. All cases of myocarditis occurred within 3 weeks of starting clozapine. Cardiomyopathy was diagnosed up to 36 months after clozapine had been started. There were no confounding factors to account for cardiac illness. Necropsy results showed mainly eosinophilic infiltrates with myocytolysis, consistent with an acute drug reaction.

The manufacturers analysed 125 reports of myocarditis with clozapine and found 35 cases with fatal outcomes (42). A total of 53% occurred in the first month of therapy, and a small number (4.8%) occurred more than 2 years after the start of treatment. In this series, 70% of the patients were men.

Taking into account the results from an epidemiological study of deaths in users and former users of clozapine (43), the cardiovascular mortality risk related to clozapine may be outweighed by the overall lower mortality risk associated with its beneficial effects, since the death rate was lower among current users (322 per 100 000 person years) than among past users (696 per 100 000 person years). The reduction in death rate during current use was largely accounted for by a reduction in the suicide rate compared with past use (RR = 0.25; CI = 0.10, 0.30).

Since cardiomyopathy is potentially fatal, some precautions must be taken. If patients taking clozapine present with flu-like symptoms, fever, myalgia, dizziness or faintness, chest pain, dyspnea, tachycardia or palpitation, and other signs or symptoms of heart failure, consideration should always be given to a diagnosis of myocarditis. Suspicion should be heightened if the symptoms develop during the first 6–8 weeks of therapy. It should be noted, however, that flu-like symptoms can also occur during the titration period, supposedly as a result of alpha-adrenoceptor antagonism by clozapine. Patients in whom myocarditis is suspected should be referred immediately to a cardiac unit for evaluation.

Clozapine rechallenge after myocarditis has been described (44).

- A 23-year-old man with no history of cardiac disease was given clozapine 12.5 mg/day, increasing to 200 mg/day over 3 weeks; 5 weeks later he complained of shortness of breath and non-specific aches and pains in his legs and body. There was marked ST-segment depression and T wave inversion in the lateral and inferior leads of the electrocardiogram. There was no eosinophilia, and creatine kinase activity was not raised. An echocardiogram showed a hyperdynamic heart and left ventricular size was at the upper limit of normal. The heart valves were normal. Clozapine was withdrawn, but his mental state and quality of life deteriorated, and 2 years later clozapine was restarted because other drugs had not produced improvement. The dose of clozapine was built up to 225 mg at night and he remained well and free from cardiac adverse effects.

In this case a consultant cardiologist diagnosed myocarditis secondary to clozapine, as no other confounding comorbidity was identified. However, the negative rechallenge suggests either that the clozapine was not responsible or that there was tolerance to the effect.

Since selenium is an essential antioxidant, and its deficiency has been implicated in myocarditis and cardiomyopathy, the aim of an observational study was to measure plasma and erythrocyte selenium concentrations in random venous blood samples from four groups: patients with mood disorders (n = 36), patients with schizophrenia taking clozapine (n = 54), patients with schizophrenia not taking clozapine (n = 41), and healthy controls (n = 56) (45). Selenium concentrations in plasma and erythrocytes were significantly lower in the patients taking clozapine compared with all the others. Thus, low selenium concentrations in patients taking clozapine may be important in the pathogenesis of life-threatening cardiac adverse effects associated with clozapine.

Pericarditis

Serositis (pericarditis and pericardial effusion, with or without pleural effusion) has been reported in patients taking clozapine (46–48).

- A 43-year-old man developed a pericardial effusion after taking clozapine for 7 years. The condition resolved when the drug was withdrawn.
- A 16-year-old girl developed pericarditis associated with clozapine. There were electrocardiographic changes and serial rises in serum troponin I, a highly sensitive and specific marker of myocardial injury.

The latter is said to be the first reported case of pericarditis due to clozapine demonstrating rises in troponin I, which resolved despite continuation of therapy. The authors suggested that troponin I is the preferred marker for monitoring the cardiac adverse effects of clozapine.

Pericardial effusion associated with clozapine has been reported (SEDA-27, 55) and can be accompanied by pleural effusions (49).

- A 21-year-old man with paranoid schizophrenia was treated with zuclopentixol, which was withdrawn because of extrapyramidal adverse effects, He was given clozapine 300 mg/day, and from day 43 developed breathlessness and complained of pain in his shoulders on deep inspiration. A chest X-ray showed an enlarged cardiac silhouette and bilateral pleural effusions. An echocardiogram showed pericardial and pleural effusions with no compromise of cardiac function. Clozapine was withdrawn and all the symptoms resolved within 2 weeks.

Venous thromboembolism

Typical neuroleptic drugs have been associated with an increased risk of venous thromboembolism (50). Data from the Swedish Reactions Advisory Committee suggested that clozapine is also associated with venous thromboembolic complications (51). Between 1 April 1989 and 1 March 2000, 12 cases of venous thromboembolism were collected; in 5 the outcome was fatal.

Symptoms occurred in the first 3 months of treatment in eight patients; the mean clozapine dose was 277 mg/day (75–500). Although during the study total neuroleptic drug sales, excluding clozapine, accounted for 96% of all neuroleptic drug sales, only three cases of thromboembolism associated with those neuroleptic drugs were reported. The reported risk of thromboembolism associated with clozapine is estimated to be 1 per 2000–6000 treated patients, the true risk being higher owing to under-reporting. These conclusions were consistent with those from an observational study (43).

The mortality rate associated with pulmonary embolism in patients taking clozapine has been estimated to be about 28 times higher than in the general population of similar age and sex; it is not clear whether pulmonary embolism can be attributed to clozapine or some characteristics of its users (SEDA-27, 56).

Between February 1990, when clozapine was first marketed in the USA, and December 1999 the FDA received 99 reports of venous thromboembolism (83 mentioned pulmonary embolism with or without deep vein thrombosis and 16 mentioned deep vein thrombosis alone) (52). In 63 cases death had resulted from pulmonary embolism; 32 were confirmed by necropsy. Of 36 non-fatal cases, only 7 had been documented objectively by such diagnostic techniques as perfusion-ventilation lung scanning and venography. Thus, in 39 of the 99 reports there was objective evidence of pulmonary embolism or deep vein thrombosis. The median age of the 39 individuals was 38 (range 17–70) years and 20 were women. The median daily dose was 400 (range 125–900) mg. The median duration of clozapine exposure before diagnosis was 3 months (range 2 days to 6 years). Information on risk factors for pulmonary embolism and deep vein thrombosis varied; however, 18 of the 39 patients were obese. The frequency of fatal pulmonary embolism in this study is consistent with that described in the labelling for clozapine in the USA.

As of 31 December 1993, there were 18 cases of fatal pulmonary embolism in association with clozapine therapy in users aged 10–54 years. Based on the extent of use recorded in the Clozapine National Registry, the mortality rate associated with pulmonary embolism was 1 death per 3450 person years of use. This rate was about 28 times higher than that in the general population of a similar age and sex (95% CI = 17, 42). Whether pulmonary embolism can be attributed to clozapine or some characteristic(s) of its users is not clear (53).

Fatal pulmonary embolism occurred in a 29-year-old man who was not obese, did not smoke, and had not had recent surgery, after he had taken clozapine 300 mg/day for 6 weeks (54).

- A 58-year-old white deaf man, with a history of pulmonary embolism, two first-degree relatives with a history of stroke and myocardial infarction, and one first-degree relative who died suddenly, developed a new episode of pulmonary embolism shortly after clozapine was begun (55).

The authors pointed out that this case suggests that susceptibility factors, such as a previous history of pulmonary embolism or venous thrombosis or a strong family history, could be viewed as relative contraindications to treatment with clozapine; however, in an invited comment it was pointed out that in this case it was not mentioned whether the patient was immobile or not, because a crucial risk factor for thrombosis in psychiatric patients is reduced physical activity (56).

In another case venous thromboembolism occurred on two occasions in a 22-year-old man taking clozapine (57). Resolution of the first episode was most probably due to treatment with anticoagulants, and the authors thought that withdrawal of anticoagulants could be regarded as equivalent to reintroduction of clozapine. They suggested that, although the interval between the withdrawal of anticoagulant therapy and the second episode was long (26 months), the sequence of events suggested a causal relation between clozapine and venous thromboembolism.

Nervous system

Clozapine has been used to treat benign essential tremor refractory to the usual drugs (propranolol, primidone, alprazolam, phenobarbital, and botulinum toxin) in a randomized, double-blind, crossover study in 15 patients with essential tremor (58). Responders with more than 50% improvement after a single dose of clozapine 12.5 mg, compared with placebo, subsequently received 39–50 mg unblinded for a mean of 16 months. Tremor was effectively reduced by a single dose of clozapine in 13 of 15 patients; sedation was the only adverse effect reported.

Clozapine can cause stuttering (SEDA-27, 56).

- A 57-year-old man developed severe stuttering after taking clozapine 300 mg/day and sodium valproate 600 mg/day for 8 months (59). It abated 1 week after clozapine was reduced and finally withdrawn.

Sleep

In a study of the effects of clozapine on the electroencephalogram, 13% of patients developed spikes with no relation to dose or serum concentration of clozapine; 53% developed electroencephalographic slowing. Compared with plasma concentrations below 300 ng/ml, a clozapine serum concentration of 350–450 ng/ml led to more frequent and more severe electroencephalographic slowing (60). There were considerable differences in the electroencephalographic patterns between classical neuroleptic drugs and clozapine (61). Clozapine-treated patients showed significantly more stage 2 sleep, more stable non-REM sleep (stages 2, 3, and 4), and less stage 1 than patients treated with haloperidol or flupentixol. In a longitudinal study, clozapine significantly improved sleep continuity and significantly increased REM density, but did not affect the amount of REM sleep (62).

Seizures

Clozapine has a proconvulsant effect. Factors that increase the likelihood of seizures include high doses of clozapine, rapid dose titration, the concurrent use of other epileptogenic agents (such as antidepressants, neuroleptic drugs, and mood stabilizers) and a previous history of neurological abnormalities (63).

The prevalence of seizures with clozapine is higher than average (about 5%) and is dose-dependent (SED-14, 142) (64,65).

- A tonic-clonic seizure occurred in a 30-year-old man 4 weeks after he started to take clozapine 400 mg/day (66). This was followed by a large increase in liver enzymes, which had been normal the week before.

Seizure characteristics and electroencephalographic abnormalities in 12 patients taking clozapine have been identified; there was a surprisingly high incidence of focal epileptiform abnormalities (67). Seizures associated with clozapine are dose-dependent (SEDA-21, 53) (SEDA-22, 57). However, there have been reports of seizure activity in patients taking therapeutic or subtherapeutic doses of clozapine (64,65). Seizures have occasionally been reported in patients taking low doses.

- A 28-year-old woman, with no history of prior seizures and not taking concomitant medication, had seizures while taking clozapine 200 mg/day (65).
- A 75-year-old patient developed seizures while taking clozapine 12.5 mg/day.

However, in the second case the seizure was unlikely to have been due to clozapine, given the very low dose and non-recurrence with rechallenge at higher dosages (68).

Despite the risk of seizures in patients without pre-existing epilepsy, six patients with epilepsy and severe psychosis taking clozapine had no increases in seizure frequency, and three had a substantial reduction (69).

Several anticonvulsants have been shown to be helpful in the prevention and treatment of clozapine-induced seizures.

- A 15-year-old boy with refractory schizophrenia had seizures with clozapine; he was given gabapentin, and several years later was free of seizures (70).

The addition of lamotrigine to clozapine therapy has been associated with rapid improvement of psychiatric symptoms (71); this has been observed in three cases of poor response or resistance to clozapine monotherapy.

Stuttering has been associated with clozapine (SEDA-22, 58). The pathogenesis of developmental stuttering, as well as acquired or neurogenic stuttering, is unclear. However, since clozapine-induced stuttering can precede a seizure (SEDA-25, 64) it may be related to an effect on the brain rather than to a dystonic syndrome, as previously suggested.

- A 28-year-old man taking clozapine 300 mg/day developed severe stuttering and subsequently had a generalized tonic-clonic seizure while taking 425 mg/day (72). There were electroencephalographic abnormalities, especially left-sided slowing.
- A 49-year-old woman had prominent stuttering before a generalized epileptic seizure and recovered after antiepileptic treatment (73).

Extrapyramidal effects

Clozapine has a more favorable extrapyramidal effects profile than other neuroleptic drugs (74) and little or no parkinsonian effect (75).

Akathisia

The low prevalence of akathisia in patients taking clozapine has led to the proposal that clozapine should be used to treat patients with neuroleptic drug-induced chronic akathisia (76,77).

Parkinsonism

The efficacy and safety of treatment with clozapine in patients with Parkinson's disease have been discussed (SEDA-22, 57), and a multicenter retrospective review of the effects of clozapine in 172 patients with Parkinson's disease has been published (78). The mean duration of clozapine treatment was 17 (range 1–76) months. Low-dose clozapine improved the symptoms of psychosis, anxiety, depression, hypersexuality, sleep disturbances, and akathisia. Of the 40 patients, 24% withdrew as a result of adverse events, mostly sedation (*n* = 19). Sedation was reported in 46%, sialorrhea in 11%, and postural hypotension in 9.9%. Neutropenia was detected in four patients (2.3%).

Six patients who met the criteria for a diagnosis of HIV-associated psychosis, and who had previously developed moderate parkinsonism as a result of typical neuroleptic drugs, were treated with clozapine (79). Parkinsonism improved by an average of 77%, but one patient did not complete the trial because of a progressive fall in leukocyte count.

Clozapine has been used to treat psychosis related to Parkinson's disease (SEDA-22, 57) (78). In a randomized, double-blind, placebo-controlled trial of low doses of clozapine (6.25–50 mg/day) in 60 patients (mean age 72 years) with idiopathic Parkinson's disease and drug-induced psychosis, the patients in the clozapine group had significantly more improvement after 14 months than those in the placebo group in all measures used to determine the severity of psychosis (80). Clozapine improved tremor and had no deleterious effect on the severity of parkinsonism, but in one patient it was withdrawn because of leukopenia.

In a randomized, double-blind, placebo-controlled, 4-week trial in 60 patients with similar drug-induced psychosis in Parkinson's disease, assigned to clozapine (*n* = 32) or placebo (*n* = 28), the initial clozapine dose of 6.25 mg/day was titrated over at least 10 days to a maximum of 50 mg/day and was rapidly effective (81). Somnolence and worsening of parkinsonism were significantly more frequent in the clozapine group, seven of whom reported worsening of Parkinson's disease, usually mild or transient, which was confirmed by aggravation of the Schwab and England score by 10–20% in three patients; however, no-one withdrew for this reason.

Of 32 patients with Parkinson's disease and psychosis (mean age 73 years; mean disease duration 12.3 years) who were followed for 5 years in a non-randomized, open study 19 (eight with dementia) continued to take clozapine (mean dose 50 mg/day) and 13 stopped taking it (82). The average duration of treatment in those in whom medication was stopped was 8.5 (range 1-24) months; the reasons for withdrawal were: symptoms improved and did not return after weaning off clozapine (n = 9), somnolence (n = 3), and personal reasons (n = 1). There was no correlation between age, sex, duration and severity of disease, the presence of dementia, and the response to clozapine. Also, the Parkinsonian Psychosis Rating Scale scoring did not influence clozapine response.

Tardive dyskinesia

It is said that clozapine causes less tardive dyskinesia than haloperidol and even that it can improve pre-existing tardive dyskinesia (83–87).

Patients with schizophrenia with (*n* = 15) and without (*n* = 11) tardive dyskinesia differed markedly in their dopaminergic response to haloperidol, assessed by means of plasma homovanillic acid variations, which increased, whereas this difference was not observed after clozapine (88).

Nevertheless, 46 patients taking clozapine had higher tardive dyskinesia scores compared with 127 taking typical neuroleptic drugs (89). In a multiple regression analysis, there was a significant relation between the total score on the Abnormal Involuntary Movement Scale (AIMS) as a dependent variable and current neuroleptic drug dose, duration of treatment, age, sex, diagnosis, current antiparkinsonian therapy, and illness duration. There was no beneficial effect of clozapine on the prevalence of tardive dyskinesia, and the authors' conclusion was that certain patients develop tardive dyskinesia despite long-term intensive clozapine treatment; however, since most clozapine users were past users of typical neuroleptic drugs, this conclusion must be regarded with caution.

- A 45-year-old woman developed tardive dyskinesia while taking clozapine (90). She had never had any of the symptoms before she started to take 223 mg/day and first experienced involuntary tongue movements and akathisia 5 months after the start of treatment.
- Tardive dyskinesia has been attributed to clozapine in a 44-year-old man, who had discontinued haloperidol 24 days before the event (91).

Clozapine has been evaluated in an open study in seven patients (mean age 29 years; mean dose 428 mg/day) with chronic exacerbated schizophrenia and severe tardive dyskinesia (92). Extrapyramidal Symptoms Rating Scale scores fell by 83% after 3 years and 88% after 5 years. None of the patients had adverse effects related to clozapine: their weight did not change significantly and their serum glucose, cholesterol, and triglyceride concentrations remained within the reference ranges.

Tardive dystonia

It is generally considered that clozapine has little or no potential to cause tardive dystonia; it has even been speculated that it may be an effective therapy for this adverse effect (SEDA-21, 53). The efficacy of clozapine in severe dystonia was therefore assessed in an open trial in five patients (93). All had significant improvement; nevertheless, all had adverse effects, such as sedation and orthostatic hypotension: in one case persistent symptomatic orthostatic hypotension and tachycardia limited treatment.

However, there was no evidence of a beneficial effect of clozapine in primary dystonia, the most common form of dystonia and a difficult disorder to treat, until the report of a 56-year-old woman with severe and persistent primary cranial dystonia (Meige's syndrome), who responded to clozapine (50–100 mg) (94).

Tardive dystonia, probably associated with clozapine, has been described (95).

- A 37-year-old man, who had taken numerous neuroleptic drugs from 1975 to 1990, was switched to clozapine because of breakthrough psychosis. Clozapine was effective and was his only neuroleptic drug treatment from that time. In 1996, when his clozapine dosage was 825 mg/day, he had left torticollis of 60–70°, mild left laterocollis, and superimposed spasmodic head movements jerking his head to the left. He had difficulty rotating his head to the right past the midline.

Two cases of severe oromandibular dystonia refractory to other antidystonic therapies, including botulinum toxin, which improved with clozapine have been reported (96).

Tardive tremor

Tardive tremor is a hyperkinetic movement disorder associated with chronic neuroleptic drug treatment. It was first described in 1991 as a symmetrical tremor, of low frequency, present at rest and during voluntary movements but most prominent during maintenance of posture, and often accompanied by tardive dyskinesia. Tetrabenazine is the current treatment. Sequential responsiveness to both tetrabenazine and clozapine has been reported (97).

- A 55-year-old man with a 15-year history of schizophrenia treated with various neuroleptic drugs developed a tremor and was given tetrabenazine 75 mg/day, with complete regression of the tremor. Three months later he developed depression, a known adverse effect of tetrabenazine, which was discontinued, with subsequent partial improvement of his depressive symptoms but reappearance of the tardive tremor. Clozapine 25 mg/day was started and increased to 75 mg/day; his tardive tremor again disappeared.

Neuroleptic malignant syndrome

Neuroleptic malignant syndrome has been associated with clozapine (SEDA-22, 58) (98), although some doubts were expressed about the features of earlier cases. The presentation can be different from that associated with traditional antipsychotic drugs; for example, the patient may not develop rigidity or a rise in creatine kinase activity (SEDA-25, 62). In the light of two cases, a 35-year-old man and a 62-year-old woman, the literature was comprehensively reviewed and the characteristics of neuroleptic malignant syndrome due to clozapine and typical neuroleptic drugs were compared (99). Causation with clozapine

was deemed highly probable in 14 cases, of medium probability in 5 cases, and of low probability in 8 cases. The most commonly reported clinical features were tachycardia, changes in mental status, and sweating. Fever, rigidity, and raised creatine kinase activity were less prominent than in the neuroleptic malignant syndrome associated with typical neuroleptic drugs. This suggests that the presentation of clozapine-induced neuroleptic malignant syndrome may be different from that of typical neuroleptic drugs. Two other cases have also illustrated that possibility (100,101).

Neuroleptic malignant syndrome and subsequent acute interstitial nephritis has been reported in a 44-year-old woman (102). This patient met the main criteria for neuroleptic malignant syndrome, although she did not develop rigidity or a rise in creatine kinase activity. On the other hand, abnormal creatine kinase activity and signs of myotoxicity were respectively found in 14% and 2.1% of patients who took clozapine for an average of 18 months ($n = 94$) (103).

- A 22-year-old man developed atypical neuroleptic malignant syndrome while taking clozapine (104). He vomited and was sweating and agitated but afebrile, with mild hypertension (maximum 156/96 mmHg) and a tachycardia, with marked increases in white blood cell count (32×10^9/l), neutrophils (25×10^9/l), and creatine kinase (1442 IU/l); a similar syndrome occurred while he was taking haloperidol.
- A 52-year-old man with risk factors, including a subdural hematoma and three prior episodes of neuroleptic malignant syndrome secondary to chlorpromazine, loxapine, and lithium, developed neuroleptic malignant syndrome while taking clozapine (105).

Delirium

Toxic delirium caused by neuroleptic drugs with potent anticholinergic properties has been widely reported (SED-11, 107), and has been reported with low-dose clozapine (106).

Psychological, psychiatric

Suicide, suicidality, and suicidal ideation are very serious problems in patients with schizophrenia. Based on general observations that 1–2% of patients with schizophrenia complete suicide within 1 year after initial attempts, the authors of a retrospective study of 295 neuroleptic drug-resistant patients with schizophrenia who had taken clozapine monotherapy for at least 6 months would have expected as many as 10 or 11 successful suicides or suicide attempts, but none was observed (107).

Obsessive-compulsive symptoms during clozapine therapy have been suggested to be more common than first reported (SEDA-21, 54). In a retrospective cohort study, new or worse obsessiveness has been analysed in 121 consecutive young patients with recent-onset schizophrenia or other psychotic disorders taking clozapine and other neuroleptic drugs (108). More clozapine-treated subjects (21%) had new or worse obsessiveness than subjects treated with other neuroleptic drugs (1.3%).

However, there was no information on comparability of the groups.

Panic disorder has been attributed to clozapine (109).

- A 34-year-old woman taking clozapine 400 mg/day for psychiatric symptoms had recurrent attacks of sudden chest pressure, dizziness, fear of dying, and intense anxiety; reducing the dose of clozapine to 250 mg/day led to modest improvement. Olanzapine 10 mg/day was then substituted, without recurrence, and her panic symptoms progressively improved.

Metabolism

Hyperlipidaemia

Severe clozapine-induced hypercholesterolemia and hypertriglyceridemia has been reported in a patient taking clozapine (110).

- A 42-year-old man with a schizoaffective disorder had new-onset hyperlipidemia while taking clozapine (after failing therapy with traditional antipsychotic drugs). Before taking clozapine his total cholesterol measurements were 2.9–5.5 mmol/l and there were no triglyceride measurements. Despite treatment with various antihyperlipidemic agents, his total cholesterol concentration reached 12 mmol/l and his triglyceride concentration reached 54 mmol/l. His antipsychotic drug therapy was switched to aripiprazole and his lipid concentrations improved dramatically, to the point that antihyperlipidemic treatment was withdrawn. When he was given clozapine again his lipid concentrations again worsened.

Diabetes mellitus

Several cases of de novo diabetes mellitus or exacerbation of existing diabetes in patients taking neuroleptic drugs have been reported, including patients taking clozapine (111–114). There was no significant relation to weight gain.

- A 49-year-old man taking olanzapine developed diabetes mellitus and recovered after withdrawal (115).
- Diabetic ketoacidosis occurred in a 31-year-old man who had taken clozapine 200 mg/day for 3 months for refractory schizophrenia (116). Clozapine was withdrawn and he remained metabolically stable. Two months later, clozapine was restarted, and only 72 hours after drug re-exposure he had increased fasting glycemia and insulinemia, suggesting insulin resistance as the underlying mechanism. Apart from slight obesity, he had no predisposing factors.

Hyperglycemia occurs at 2 weeks to 3 months after the start of clozapine treatment and occurs without predisposing factors. Clozapine-induced hyperglycemia can be serious, leading to coma, but it is reversible if clozapine is withdrawn. In some cases, continuation of clozapine is possible by controlling blood glucose concentrations with hypoglycemic drugs. This approach can be useful in refractory schizophrenia responsive to clozapine. All patients should be advised to report altered consciousness, polyuria, or increased thirst.

Glucose metabolism has been studied in 17 patients taking clozapine (117). Six had impaired glucose tolerance and eight had a glycemic peak delay.

Diabetes was also more common in 63 patients taking clozapine than in 67 receiving typical depot neuroleptic drugs (118). The percentages of type 2 diabetes mellitus were 12% and 6% respectively. Nevertheless, the mechanism is not known. In six patients with schizophrenia, clozapine increased mean concentrations of blood glucose, insulin, and C peptide (119). The authors concluded that the glucose intolerance was due to increased insulin resistance.

However, opposite data have been found in a case-control study in 7227 patients with new diabetes and 6780 controls, all with psychiatric disorders (120). Clozapine was not significantly associated with diabetes (adjusted OR = 0.98; 95% CI = 0.74, 1.31) and there was no suggestion of relations between larger dosages or longer durations of clozapine use and an increased risk of diabetes. Among individual non-clozapine neuroleptic drugs, there were significantly increased risks for two phenothiazines: chlorpromazine (OR = 1.31; 95% CI = 1.09, 1.56) and perphenazine (OR = 1.34; 95% CI = 1.11, 1.62). The authors suggested that, in contrast to earlier reports, these results provided some reassurance that clozapine does not increase the risk of diabetes. However, cases of diabetes were identified by the new use of antidiabetic drugs, and it is therefore possible that clozapine was associated with less pronounced glucose intolerance that did not require drug therapy.

The effect of clozapine on glucose control and insulin sensitivity has been studied prospectively in 9 women and 11 men with schizophrenia (mean age 31 years) (121). Insulin resistance at baseline was unaffected by clozapine, but 11 of the patients developed abnormal glucose control (mean age 30 years; five women). Mean fasting and 2-hour glucose concentrations increased significantly by 0.55 mmol/l. There was no correlation between change in body mass index and change in fasting glucose concentrations. Weight gain with clozapine compared with other neuroleptic drugs has been studied in in-patients who were randomly assigned to switch to open treatment with clozapine (n = 138) or to continue receiving conventional neuroleptic drugs (n = 89) (122). Patients gained weight at the end of 2 years whether they switched to clozapine (5.9 kg, 7%) or continued to take first-generation neuroleptic drugs (2.3 kg, 4%), but weight gain was significantly greater (1 body mass index unit) in those taking clozapine, particularly women.

Weight gain

Weight gain is often associated with clozapine (SEDA-21, 54). In 42 patients who took clozapine for at least 1 year, men and women gained both weight and body mass, which is more directly related to cardiovascular morbidity (123). Over 10 weeks, leptin concentrations, which correlate with body mass index, increased significantly from baseline in 12 patients taking clozapine (124).

The relation between genetic variants of the β_3 adrenoceptor and the G-protein β_3 subunit and clozapine-induced body weight change has been investigated in 87 treatment-resistant patients with schizophrenia (125). They gained an average of 2.6 kg. There was no statistically significant relation between weight gain and either the β_3 adrenoceptor Trp6Arg or the G-protein β_3 subunit C8257 polymorphisms.

In a long-term follow-up study (14 months) of 93 patients with schizophrenia the possible relation between clozapine-induced weight gain and a genetic polymorphism in the adrenoceptor alpha 2a receptor, -1291 C>G, has been examined (126). The GG genotype was associated with a significantly higher mean body weight gain (8.4 kg) than the CC genotype (2.8 kg).

The effect of clozapine on serum ghrelin concentrations has been investigated in 12 patients over 10 weeks after the start of treatment (127). In contrast to increased body mass indices and serum leptin concentrations, there were no significant changes in serum ghrelin concentrations. The authors claimed that these results do not support a causal involvement of ghrelin in clozapine-related weight gain.

The association between clozapine-related weight gain and increased mean arterial blood pressure has been examined in 61 patients who were randomly assigned to either clozapine or haloperidol in a 10-week parallel-group, double-blind study, and in 55 patients who chose to continue to take clozapine in a subsequent 1-year open study (128). Clozapine was associated with significant weight gain in both the double-blind trial (mean 4.2 kg) and the open trial (mean 5.8 kg). There was no significant correlation between change in weight and change in mean arterial blood pressure.

There were no significant associations between cycle length and weight change during clozapine treatment in 13 premenopausal women with psychoses (129).

Sleep apnea associated with clozapine-induced obesity has been reported (130).

- A 45-year-old woman with schizophrenia who took clozapine 300 mg/day for 16 months gained 18 kg and had hypertriglyceridemia and glucose intolerance. She had daytime sedation, difficulty in sleeping at night, loud snoring, and periods of apnea during sleep.

Nasal continuous positive airway pressure produced improvement.

Phenylpropanolamine 75 mg/day did not promote weight loss in a randomized, placebo-controlled study in 16 patients with schizophrenia who had gained at least 10% of their body weight while taking clozapine (131).

- A 29-year-old man taking clozapine 800 mg/day gained 46 kg in weight after 25 months, and had myoclonic jerks in the hands, arms, and shoulders on both sides (132). He was treated with topiramate (which causes weight loss). The myoclonic jerks disappeared completely. He lost 21 kg over 5 months, with no significant change in eating habits or food consumption, and felt more energetic, more active, and more motivated to exercise.

- The 22-year-old son of healthy parents, with a life-long history of galactosemia, developed weight gain while taking an effective dose of clozapine (133).

Because patients with galactosemia need to avoid weight loss as a result of restrictive dietary measures, the authors suggested that this is an interesting example of weight gain as a positive side effect of clozapine, not necessarily associated with increased appetite and higher caloric intake.

Hematologic

Neutropenia and agranulocytosis

> DoTS classification (BMJ 2003; 327:1222–1225)
> Dose-relation: hypersusceptibility effect
> Time-course: intermediate or delayed
> Susceptibility factors: genetic; age.

Incidence
Clozapine-induced agranulocytosis was originally determined to be 0.21% in a selected Finnish population (SED-9, 83) (134), and the drug was withdrawn, only to be cautiously reintroduced in some countries a decade later, with hematological monitoring. With mandatory hematological monitoring by the Clozaril Patient Management System in the USA, the cumulative incidence of agranulocytosis was 0.8% at 1 year and 0.9% at 1.5 years of treatment; the risk was not related to dosage (SEDA-18, 54) (135). In France, the incidences of agranulocytosis and neutropenia in clozapine-treated patients from December 1991 were 0.46 and 2.1% respectively (136). Some of the available postmarketing data on clozapine-induced agranulocytosis are presented in Table 1 (137).

Mechanism
The underlying mechanisms of agranulocytosis are unknown, but hemopoietic cytokines, such as granulocyte colony-stimulating factor (G-CSF), are likely to be involved (138).

- In a 26-year-old woman who developed granulocytopenia twice, first when taking clozapine and again when taking olanzapine, G-CSF concentrations, but not those of other cytokines, closely paralleled the granulocyte count.
- In a 73-year-old patient who developed granulocytopenia while taking clozapine, G-CSF and leukocyte counts were reliable indicators of the evolution of the condition, showing an abortive form of toxic bone-marrow damage with subsequent recovery (139).

Immune-mediated mechanisms of clozapine-induced agranulocytosis have been reviewed in the context of agranulocytosis in a 46-year-old woman (140). Immune and toxic mechanisms have also been explored in patients taking clozapine, three who developed agranulocytosis, seven who developed neutropenia, and five who were asymptomatic. There was no evidence of antineutrophil antibodies in the blood of patients shortly after an episode of clozapine-induced agranulocytosis, and an antibody mechanism

Table 1 Reported incidences of clozapine-induced agranulocytosis

Country	Period	Number of patients	Incidence (mortality) (%)	Reference
Finland	1975	2260	0.70 (0.35)	(SED-9, 83; 105)
USA	1990–1991	11382	0.80 (0.02)	(SEDA-18, 54)
France	1992	2834	0.46 (ND)	(SEDA-21, 54)
USA	1990–1994	99502	0.38 (0.01)	(SEDA-22, 59)
UK and Ireland	1990–1994	6316	0.80 (0.03)	(SEDA-22, 59)
New Zealand	1988–1995	963	1.15 (0.00)	(SEDA-22, 59)
Australia	1993–1996	4061	0.90 (0.00)	(99)
Spain	1993–1999	6354	0.16 (0.02)	Agencia Española del Medicamento (personal communication)
Total		133402	0.44 (0.018)	

seems unlikely, in view of the delay in onset of clozapine-induced agranulocytosis on re-exposure to the drug (141).

- Increased apoptosis of neutrophils has been reported in a 45-year-old woman with clozapine-induced agranulocytosis (142). Withdrawal of clozapine and treatment with granulocyte colony-stimulating factor led to normalization of the blood neutrophil count within 3 weeks.

The authors suggested that enhanced apoptosis of blood neutrophils during the acute phase of clozapine-induced agranulocytosis could have resulted from enhanced expression of the pro-apoptotic proteins Bax and Bilk and from a reduction in the anti-apoptotic proteins BCI-X_LmRNA. The time-course of the fall and recovery of the neutrophils, as well as the release pattern of endogenous G-CSF, resembled those of chemotherapy-induced neutropenia. The kinetics of CD 34-positive cells mimic those of cytotoxic progenitor cell mobilization, for example after cytostatic drug administration. They suggested that clozapine-mediated inhibition of release of G-CSF or granulocyte-macrophage colony-stimulation factor (GM-CSF) is involved in clozapine-induced agranulocytosis.

Susceptibility factors
Some of the genetic aspects of clozapine-induced agranulocytosis have been evaluated (143). Polymorphisms of specific clozapine metabolizing enzyme systems were determined in 31 patients with agranulocytosis and in 77 without. Genotyping of a recently discovered G-463 A polymorphism of the myeloperoxidase gene and CYP2D6 showed no evidence of an association.

Because of the unusually high incidence of agranulocytosis in Finnish and Jewish patients (SEDA-20, 49), an ethnic susceptibility factor for agranulocytosis has been suggested. Human leukocyte antigen (HLA) B38 phenotype was found in 83% of patients who developed agranulocytosis and in 20% of clozapine-treated patients who did not develop agranulocytosis (144). Gene products contained in the haplotype may be involved. In an open study in 31 German patients with clozapine-induced agranulocytosis and 77 controls with schizophrenia, agranulocytosis was significantly associated with HLA-Cw*7, DQB*0502, DRB1*0101, and DRB3*020 (145). No other antigens were associated with agranulocytosis, but age was another major susceptibility factor. In another study in two groups of Finnish patients (19 "clozapine responders" and 26 patients with a history of non-fatal clozapine-induced granulocytopenia or agranulocytosis), the frequency of the HLA-A1 allele in the latter was low (12%), whereas HLA-A1 was associated with a good therapeutic response at an allele frequency of 58% (the frequency of HLA-A1 being 20% in the Finnish population) (146).

Concordant clozapine-induced agranulocytosis in monozygotic twins also suggested a genetic susceptibility; in both twins there was a low leukocyte count after 9 weeks of treatment (147). Serological typing of the HLA system showed identical patterns in the twins: HLA-A: 28, 26; HLA-B: 49, 63; DR: 2 (versus 16), 12, 52; DQ: 1. The authors pointed out that these data suggest that genetic factors may participate not only in the time of onset of schizophrenia, but also in the emergence and timing of agranulocytosis in response to clozapine.

Clinical features
Careful attention should be paid to possible early warnings of agranulocytosis, such as fever, sore throat, and lymphadenopathy.

Circadian variation in white cell count, with a dip in the morning, has been misdiagnosed as clozapine-induced neutropenia (148).

- A 31-year-old man with resistant schizophrenia took clozapine 500 mg/day. Although this was effective, the granulocyte count fell to 1.2×10^9/l (total count not given) and clozapine was withdrawn. During the subsequent year, several neuroleptic drugs were used, with unsatisfactory results. Careful monitoring showed a pronounced diurnal variation in both total white cell count ($2.9–4.2 \times 10^9$/l in the morning and $3.6–7.1 \times 10^9$/l in the afternoon) and granulocytes ($0.8–1.4 \times 10^9$/l in the morning and $2.9–5.5 \times 10^9$/l in the afternoon).

Thus, an apparently low white cell count may simply reflect the nadir of the diurnal variation and may not indicate a need to withdraw clozapine.

In over 11 000 patients the risk of agranulocytosis was higher in the first 3 months of treatment and is greater among women and elderly patients (149).

Agranulocytosis after very long-term clozapine therapy has been reported (150).

- A 41-year-old man suddenly developed agranulocytosis after taking clozapine nearly continuously for 89 months. During this time, his white blood cell and granulocyte counts remained stable. The white blood cell and granulocyte counts returned to baseline shortly after withdrawal of clozapine and administration of sargramostim.

Rechallenge

Cases of negative or positive rechallenge in patients with agranulocytosis have been reported (SEDA-20, 54; SEDA-22, 59).

- A 58-year-old man developed agranulocytosis during a second trial of clozapine, despite a successful previous trial (151).
- A 17-year-old boy with severe clozapine-induced neutropenia had a negative rechallenge; because he had had an unsatisfactory response to traditional neuroleptic drugs, clozapine was continued despite a fall in white blood cell count, since concomitant treatment with granulocyte colony-stimulating factor was followed by rapid normalization of the white blood cell count (152).
- A 29-year-old woman developed agranulocytosis after taking clozapine 300 mg/day for 5 years; 4 months after withdrawal, the clozapine was reintroduced (500 mg/day), and after 8 months the leukocyte count was still within the reference range (153).
- A 45-year-old woman developed neutropenia after taking clozapine 500 mg/day for 6 years combined with other agents (olanzapine 10 mg/day, benzopril hydrochloride 20 mg/day, and haloperidol 150 mg/day) (154). Clozapine was withdrawn immediately and the granulocytes recovered within a few days. However, 10 weeks later clozapine was restarted and there was no recurrence over more than 3 years.

Monitoring therapy

Over 10 000 patients have been treated with clozapine in Australia since its introduction in 1993, and the Clozaril monitoring system has ensured that since that time there have been no deaths from agranulocytosis in patients taking clozapine (155).

An increase in white blood cell count of at least 15% above previous counts is a sensitive, although not specific, predictor for the development of agranulocytosis within 75 days (149). Clozapine dosage and baseline white cell count do not appear to predict agranulocytosis.

Monitoring G-CSF concentrations, if available, may be useful in following patients in whom clozapine-induced marrow damage is suspected.

An example of a false sense of security gained by relying on monitoring monthly blood counts in patients taking clozapine has been published (156).

- A 61-year-old man who had taken clozapine for 3 years had normal blood counts. However, one day, his hemoglobin was 8.5 g/dl, having previously been 13 g/dl, following a steady asymptomatic fall over 6 months that had been documented but had gone unnoticed. He subsequently underwent investigation and treatment for anemia.

However, it is not clear in this case that clozapine was responsible for the anemia.

In a cohort study, based on a prospective drug exposure database, the effectiveness of centralized routine monitoring of blood counts was evaluated in 1500 patients taking clozapine between March 2001 and December 2001 (157). Seven patients developed severe neutropenia while taking clozapine (neutrophil counts below $1.5 \times 10^9/$l). The mean time to withdrawal of therapy was 1.6 days (maximum 6 days), and neutrophil counts recovered to normal in all cases after 6.4 days (maximum 13 days). Based on an estimate of 500 patient-years of exposure, the frequency of severe neutropenia was one case per 71 patient-years of therapy or 1.4% per annum.

According to the recommended guidelines by Novartis, neutropenia (a white blood cell count below $3.0 \times 10^9/$l or an absolute neutrophil count below $1.5 \times 10^9/$l) during clozapine treatment is classified as being in the "red-alert zone"; immediate withdrawal of clozapine is recommended and reinstitution prohibited. However, in some patients, this is not feasible, because of lack of effective alternatives to clozapine. In five patients who were maintained on clozapine despite red-alert zone neutropenia and two control patients who discontinued clozapine because of neutropenia, hematological and clinical progress was followed for more than 600 days (158). In all five patients, there were no additional episodes of neutropenia despite continued clozapine treatment.

Treatment

Withdrawal of clozapine can lead to resolution of agranulocytosis, but not always. Granulocytopenia, presumably induced by clozapine, persisted in a 53-year-old woman after she switched from clozapine to quetiapine (159).

Treatment with granulocyte colony-stimulating factor and granulocyte macrophage colony stimulating factors was helpful in a case of sepsis and neutropenia induced by clozapine (160) and in a case of agranulocytosis in a 45-year-old man (161).

Lithium can be used in combination with clozapine, and in these patients the possibility of inducing leukocytosis and increasing the total leukocyte count and the granulocyte count has been considered (SEDA-20, 50). Lithium has even been used to prevent clozapine-induced neutropenia (SEDA-22, 59) (162). It has also been used in a patient with clozapine-induced neutropenia and in another with complete agranulocytosis: in both cases lithium increased the neutrophil count to within the reference range within 6 days (163). In the patient who had neutropenia, clozapine was restarted in the presence of lithium and the neutrophil count did not fall thereafter. Five other patients who took combined clozapine and lithium had a significant improvement with this combination and there were no cases of agranulocytosis, neuroleptic malignant syndrome, or other adverse effects (164).

Eosinophilia

Clozapine-induced eosinophilia and subsequent neutropenia has been reported (165). As the patient had a high IgE concentration, an allergic cause was proposed. In a previous study in 70 patients there was no predictive value of eosinophilia for clozapine-induced neutropenia (166). Eosinophilia associated with clozapine treatment has been reported in 13% of treated patients in a study in Australia (38).

Platelet count

Thrombocytosis and thrombocytopenia have both been reported.

- Thrombocytosis (774×10^9/l) occurred in a middle-aged man taking clozapine (167).
- A 43-year-old man developed thrombocytopenia (platelet count 60×10^9/l), which persisted for 40 months after clozapine treatment (168). There was increased in vitro platelet serotonin release in the presence of clozapine (169).

In the second case, the authors suggested an immune mechanism and pointed out that the manufacturers recommend withdrawing clozapine when the platelet count falls below 100×10^9/l.

Mouth and teeth

Hypersalivation is a common adverse effect of clozapine (170), which has been estimated to occur in 10–23% of patients (SEDA-20, 49). Salivary gland swelling has been reported in patients treated with clozapine (SEDA-20, 49; 171).

There is evidence implicating alpha-adrenoceptors in hypersalivation caused by clozapine, and it has been hypothesized that a biallelic polymorphism in the promoter region of the alpha$_2$-adrenoceptor gene confers susceptibility to schizophrenia, which is associated with a clozapine-induced favorable therapeutic response and/or clozapine-induced hypersalivation (172). However, the results in 97 patients showed that the alpha$_2$-adrenoceptor gene polymorphism did not play a major role in susceptibility to hypersalivation or the therapeutic response of patients with schizophrenia.

A comprehensive review has been published on the evidence of the benefit of using antimuscarinic agents, adrenoceptor antagonists, and adrenoceptor agonists in treating clozapine-induced hypersalivation (173). There is a lack of good-quality controlled trials, most papers having reported series of uncontrolled cases dependent on subjective measures of improvement reported by patients; however, the authors suggested that the most effective treatment may be a combination of terazosin and benzhexol (174). Ten patients with sialorrhea associated with clozapine, who did not respond to anticholinergic or adrenergic drugs, received intranasal ipratropium bromide; at 6 months, six patients maintained improvement (175).

Salivary glands

In nine patients with clozapine-induced sialorrhea sublingual ipratropium produced complete responses in two patients and partial responses in five; the effect wore off after a few hours (range 2–8 hours) (176).

- A 50-year-old man with clozapine-induced hypersalivation was given botulinum toxin injections into each parotid gland and had a marked reduction in hypersalivation, which lasted for more than 12 weeks (177).

In 12 patients clonidine 50–100 micrograms/day relieved clozapine-induced sialorrhea, with good results in three and partial results in eight (178). Theoretically, the reduction in sialorrhea with clonidine could have been due to reduced plasma noradrenaline concentrations, resulting in less stimulation of unopposed β-adrenoceptors in the salivary glands.

Gastrointestinal

Reflux esophagitis has been reported in patients taking clozapine (179). It has been speculated that reduced esophageal motility was the mechanism (180).

Constipation is an adverse effect that has often been associated with clozapine; it can be serious and even fatal (SEDA-22, 60).

- A 49-year-old man taking clozapine developed a perforated colon and peritonitis (181). He survived, albeit with a markedly reduced quality of life.

The authors suggested that diet modification and regular exercise should be encouraged in patients taking clozapine, in order to prevent constipation.

- A 43-year-old man took clozapine 750 mg/day for 6 years and developed vomiting and epigastric pain (182). He had ulcerative esophagitis, and a CT scan was reportedly normal "apart from constipation." He was given omeprazole 20 mg/day and twice-daily psyllium. Six months later he developed abdominal pain with feculent vomiting. Emergency laparotomy revealed large-bowel obstruction secondary to severe fecal impaction. He died 3 weeks later with septic shock and progressive multisystem organ failure.

Liver

Transient asymptomatic liver enzyme rises are common with clozapine (183).

Hepatitis associated with clozapine has been reported (SEDA-20, 49) (SEDA-22, 59).

- A 49-year-old woman (184) took clozapine 300 mg/day and developed lethargy, anorexia, fever, eosinophilia, leukocytosis, and abnormal liver function tests. The serum clozapine concentration was 8595 nmol/l. Clozapine was withdrawn and after 8 days her condition stabilized and low-dose clozapine treatment was successfully restarted with serum monitoring.

Obstructive jaundice has been described (185).

- A 48-year-old man with schizophrenia started taking clozapine 12.5 mg/day, increasing over the next 18 days to 150 mg/day. By that time he was icteric, with mild distress and fever and raised bilirubin 149 µmol/l (reference range 5–26), direct bilirubin 92 µmol/l, gamma-glutamyl transpeptidase 446 IU/l (<65), alanine transaminase 100 IU/l (<40), and aspartate transaminase 56 IU/l (<40). Hepatitis serology showed positive hepatitis B surface antibodies, and HBs antigen was negative, as were hepatitis A, hepatitis C, Epstein-Barr virus, and cytomegalovirus. He also had hyperglycemia, pleural effusion, eosinophilia, hematuria, and proteinuria, which also resolved on clozapine withdrawal.

A case of fatal liver failure has been reported (186).

Pancreas

Occasional cases of pancreatitis have been related to clozapine therapy (SEDA-17, 63).

- A 73-year-old woman with a 4-year history of Parkinson's disease developed hallucinations and delusions that were interpreted as secondary effects of levodopa (187). She was given clozapine 25 mg/day and continued to take levodopa. Four days later she complained of abdominal pain. She had raised activities of serum amylase 806 IU/l (reference range <220 IU/l), lipase 2598 IU/l (<190 IU/l), and creatine kinase 464 IU/l (<190 IU/l), and normal concentrations of total and direct bilirubin. Other causes of pancreatitis were ruled out.
- A 17-year-old man took clozapine in a dose that was gradually increased to 175 mg/day, and 23 days after the start of treatment developed mild epigastric pain (188). He had raised pancreas amylase (140 U/l) and lipase (463 U/l) activities, and four days later developed increasing pain in both shoulders and a large pericardial effusion.

Two other cases have been reported in patients with no prior history of alcohol abuse or gallstones (189,190). The authors recommended monitoring serum amylase activity during slow increases in the dosage of clozapine if there is leukocytosis or eosinophilia, which may be associated with asymptomatic pancreatitis.

Urinary tract

Enuresis has been rarely associated with clozapine (0.23% of patients) (SEDA-19, 54) (191), and has been successfully treated with benzatropine in patients taking a variety of psychotropic medications (192).

In a retrospective study, 27 of 61 Chinese patients who took clozapine for more than 3 months developed urinary incontinence, persistent in 15 cases (193). The reaction could not be related to age, sex, clozapine dosage, duration of clozapine use, duration of hospitalization, duration of illness, age at onset of schizophrenia, or concurrent treatment with other psychiatric drugs.

Polymorphisms of the alpha 1a adrenoceptor gene were found to play no major role in the pathogenesis of schizophrenia or in clozapine-induced urinary incontinence (194).

Acute interstitial nephritis has been attributed to clozapine (195).

- A 38-year-old woman developed anorexia, lethargy, and vomiting, and noticed a profound reduction in urine output about 11 days after starting clozapine (125 mg bd). Severe renal insufficiency was confirmed (blood urea 33 mmol/l and creatinine 1200 µmol/l). There was no history of pre-existing renal or other systemic disease.

Musculoskeletal

Clozapine-induced myokymia (weakness and reduced muscle tone, with undulating movements of the muscle and skin, accompanied by involuntary repetitive firing of grouped motor unit action potentials) has been reported (196).

- A 33-year-old woman developed muscle twitching and spasms of the legs and back after having taken clozapine for 3 years (197). Neurological examination showed myokymia in both thighs, calves, and the lower lip. The myokymia disappeared 1 week after withdrawal of clozapine.

In an open study in 41 patients, strength control was evaluated before the start of clozapine therapy and again at the end of the titration period (on average 9 weeks later). The results suggested that the strength deficit was primarily due to clozapine and that there were two distinct effects: an initial transient stage characterized by "drowsiness" and a subsequent stage with dose-dependent myoclonic features (198).

- Rhabdomyolysis occurred in two men, aged 21 and 42 years, taking clozapine (198,199). The first had no risk factors, but calcium-dependent potassium efflux, normally responsible for membrane hyperpolarization and muscle refractoriness, was severely impaired in his erythrocytes. The second had marked hyponatremia, due to psychogenic polydipsia, and developed a marked rise in creatine kinase activity (62 730 U/l) after correction of hyponatremia with hyperosmolar fluids.

Sexual function

The frequency and course of sexual disturbances associated with clozapine have been studied in a prospective open study in 75 men and 25 women, mean age 29 years, and compared with the effects of haloperidol in 41 men and 12 women, mean age 26 years (200). There were no statistically significant differences between the patients taking haloperidol and those taking clozapine. During 1–6 weeks of treatment with clozapine, the most frequent sexual disturbances among women were diminished sexual desire (28%) and amenorrhea (12%), while among men they were diminished sexual desire (57%), erectile dysfunction (24%), orgasmic dysfunction (23%), ejaculatory dysfunction (21%), and increased sexual desire (15%). The mean daily doses were haloperidol 16 mg and clozapine 261 mg.

Retrograde ejaculation has been associated with clozapine (201).

Immunologic

Allergic reactions associated with clozapine are uncommon; however, a case of rash (SEDA-21, 55) and a case of pleural effusion (SEDA-22, 60) have previously been reported. Both rash and pleural effusion have been reported in a 37-year-old woman about 1 week after starting clozapine (202).

Lupus erythematosus has been associated with clozapine (203).

- A psychotic 55-year-old woman took clozapine for 15 years because of increasing extrapyramidal adverse effects with other neuroleptic drugs. When she started to take it she did not have raised serum inflammatory markers. She reduced the dosage of clozapine from 400 to 200 mg/day and developed increasing delusions. In hospital she was discovered to have generalized arthralgia, myalgia in the proximal limb muscles, a reticular skin rash, and a raised erythrocyte sedimentation rate. The syndrome resolved within 2 weeks of withdrawal of clozapine.

The authors thought that clozapine had caused lupus erythematosus.

Body temperature

Clozapine often causes a benign transient increase in body temperature early in treatment (204).

The presentation of neuroleptic malignant syndrome with clozapine can be different from that associated with traditional neuroleptic drugs (SEDA-28, 66), and the authors of a recent report have pointed out the differential diagnosis with heat stroke, a medical emergency with the two cardinal features of raised core body temperature (40°C) and central nervous system dysfunction, which is fatal in up to 50% of cases (205).

- A 32-year-old man with paranoid schizophrenia taking clozapine 200 mg bd and in gainful employment as a builder's laborer was found unconscious and fitting on an open building site during a heat wave. He had a respiratory arrest, a fever of 42.9°C, was hypotensive (90/60 mmHg), and had the minimum Glasgow coma scale score of 3. His creatine kinase activity was 6000 (24-195) U/l.

This patient was originally considered to have neuroleptic malignant syndrome, but the clinical picture was very much against this and strongly in favor of heat stroke.

Death

The causes of sudden unexpected death in psychiatric patients are often unknown; however, sudden cardiac death has been associated with clozapine (SEDA-26, 59). On the other hand, suicide is a common cause of death among patients with schizophrenia. In a recent report, the authors claimed that in cases of sudden unexpected death, toxic post-mortem drug concentrations can lead to erroneous conclusions and reported a case of supposed post-mortem redistribution of clozapine (206).

- A 22 year-old obese woman was found unresponsive after taking clozapine for about 6 weeks, in an eventual dose of 350 mg/day. Resuscitation was unsuccessful, and an autopsy about 8 hours later showed no clozapine in her stomach, but a clozapine concentration in cardiac blood of 4500 ng/ml; a concentration over 1300 ng/ml is considered toxic. The coroner expressed concern over the possibility of suicide, but the authors of the report suspected that suicide was very unlikely, since there were no overt signs of toxicity and the staff reported no change in behavior.

Long-Term Effects

Drug withdrawal

Rebound psychosis or delirium or both have been reported after withdrawal of clozapine (207–212). Clozapine withdrawal has also been associated with nausea, vomiting, diarrhea, headache, restlessness, agitation, and sweating (213,214), which occur as the result of cholinergic rebound and which may respond to anticholinergic drugs (215), and with dystonias and dyskinesias. Delirium and the return of dyskinetic movements can occur within days after clozapine withdrawal.

Four patients had severe dystonias and dyskinesias on abrupt withdrawal of clozapine (216), and another two had obsessive-compulsive symptoms during withdrawal; resumption of clozapine led to the complete disappearance of the obsessive-compulsive symptoms (217).

Second-Generation Effects

Pregnancy

All antipsychotic drugs cross the placenta and reach the fetus in potentially significant amounts; however, clozapine caused no serious complications or developmental abnormalities during pregnancy in two cases (218).

Tertatogenicity

Case reports have shown no congenital anomalies with clozapine in animals or humans (219).

- A 22-year-old woman took clozapine 75 mg/day throughout pregnancy (220). Her pregnancy was not detected until the end of the first trimester. She and her husband were counselled on the possible risk to the fetus, but decided to continue taking clozapine. She was brought to the hospital in labor with vaginal leaking at 9 months and 9 days of gestation. Ultrasound examination confirmed a fetus of 32 weeks gestation, with intrauterine growth retardation, oligohydramnios, and absence of fetal heart sounds. She delivered a macerated still-born boy weighing 2.2 kg with no gross congenital anomalies.

Although it is not clear whether clozapine, the disease process, poor nutrition, lack of antenatal care, or a combination of all these factors, contributed to the outcome of this pregnancy. The authors suggested that one should be cautious about using clozapine during pregnancy.

Lactation

Neonatal safety may be compromised during breast feeding by a woman taking clozapine (Ernst 42). The authors also pointed out that the available data suggest that neonates are at risk during breast-feeding by a mother who is taking lithium, are at moderate risk from lamotrigine and clozapine, and are probably at little risk from valproate or carbamazepine; the risks from most other new anticonvulsants and atypical antipsychotic drugs are currently undetermined.

Susceptibility Factors

Ethnicity

The results of an observational, non-randomized, interethnic comparison of clozapine dosage, clinical response, plasma concentrations, and adverse effects profiles have been published (221). Compared with Caucasian patients (n = 20), Asian patients (n = 20) appeared to have lower dosage requirements for clinical efficacy. The Asian patients scored significantly lower than the Caucasian patients on the Simpson and Angus Scale for extrapyramidal adverse effects, but there were no significant differences on the Abnormal Involuntary Movement Scale or the Liverpool University Neuroleptic Side-effect Rating Scale. Since there was no difference between ethnic groups in clozapine concentrations, the authors concluded that the higher extrapyramidal adverse effects scores in the Caucasian group might be attributed to more chronic illness duration, more concomitant medications, and the lack of standardized rating training. Genetic differences between these ethnic groups were not analysed in this study. However, it should be emphasized that Asians are genetically heterogeneous and differences are found between various Asian ethnic groups.

Age

Children

Since the adverse effects of clozapine may be more common in children than adults, perhaps reflecting developmental pharmacokinetic differences, clozapine and its metabolites, norclozapine and clozapine-N-oxide, have been studied in six youths aged 9-16 years, with childhood onset schizophrenia (222). Dose-normalized concentrations of clozapine did not vary with age and were similar to reported adult values. Clinical improvement in five patients correlated with serum clozapine concentrations, and clinical response and total number of common adverse effects (sialorrhea, n = 5; tachycardia, n = 4; sedation, n = 1; enuresis, n = 1) correlated with norclozapine concentrations. One child had a reduced neutrophil count (1.1×10^9/l) and another child had increased hepatic transaminases.

Elderly people

Data from an open study (n = 329) have suggested that patients aged 55–64 years may have a better response to clozapine than those aged 65 and older, but there were no significant differences between the two age groups in the number of patients remaining on clozapine therapy and the number in whom therapy was discontinued (n = 134) (223). The mean duration of clozapine therapy was 278 days. The most common adverse effects that required withdrawal were sedation (n = 12), hematological adverse effects (n = 7), and cardiovascular adverse effects (n = 6).

In over 11 000 patients the risk of agranulocytosis was greater among elderly patients (149).

Sex

In over 11 000 patients the risk of agranulocytosis was greater among women (149).

Other features of the patient

The evidence on predictors and markers of clozapine response has been reviewed (224). Higher baseline clinical symptoms and functioning in the previous years and low cerebrospinal homovanillic acid/5-hydroxyindoleacetic acid concentrations were identified as reliable, and three potential measures were also identified: reduced frontal cortex metabolic activity, reduced caudate volume, and improvement in P50 sensory gating. The authors pointed out that none of these is specific to clozapine, but that this does not reduce their value, instead showing that they do not clarify why clozapine is different from other antipsychotic drugs, for example in terms of efficacy in treatment-resistant patients.

Cancer

There is an increased risk of myelosuppression in patients who develop cancer and require chemotherapeutic agents while taking a stable dose of clozapine.

- A 46-year-old woman with schizophrenia who was taking clozapine 700 mg/day developed breast cancer and underwent segmental mastectomy; she developed neutropenia, which persisted for more than 6 months after her last radiation treatment (225).

Infection

Clozapine toxicity has been attributed to the effects of cytokines during an acute infection (226).

- A 51-year-old woman developed toxic serum concentrations of clozapine and N-desmethylclozapine (norclozapine) during an acute urinary tract infection and had a short period of aphasia and akinesia, followed by incoherence of speech and a gait disturbance.

The authors suspected that, since the prescribed antibiotics (trimethoprim, sulfamethoxazole, and ampicillin) did not seem to be responsible for the large rise in serum clozapine concentrations, the rise might have been secondary to cytokine-mediated inhibition of cytochrome P450.

Lewy body dementia

Patients with Lewy body dementia may be more intolerant of neuroleptic drugs, including atypical drugs, than other patients with neurodegenerative dementia. However, because hallucinations are common in this form of dementia, it is likely that people with Lewy body dementia will be exposed to neuroleptic drugs. Two patients with Lewy body dementia taking clozapine developed confusion and behavioral symptoms (227).

Drug Administration

Drug formulations

Since clozapine is expensive, it is interesting that generic clozapine (given as 25 and 100 mg tablets) behaves like Clozaril, the branded formulation; bioequivalence has been observed in 30 patients with schizophrenia (228).

Since the patent for Clozaril® (Novartis) expired in 1998, three manufacturers of generic clozapine have submitted abbreviated new drug applications to the FDA for review and approval to market generic clozapine. In a recent review, the authors suggested that until further studies have been conducted, patients who are refractory to treatment and stabilized on Clozaril® should not be switched to a generic formulation; on the other hand, if a patient is stabilized on Clozaril® and not refractory to treatment, then cautious switching to a generic formulation may be reasonable; finally, initiating a generic formulation in a clozapine-naïve individual would be appropriate (229). There has been an open, non-blinded out-patient study in 20 patients with schizophrenia to ascertain the effects of switching from branded to generic clozapine; at the end-point, there were no significant differences in the total Positive and Negative Syndrome Scale (PANSS), the positive symptom subscale, the negative symptom subscale, or the general psychopathology subscale, or the Beck Anxiety Inventory (BAI), and there were no clinically significant changes in any measure; these results are consistent with data that suggest bioequivalence of these products (230).

Drug dosage regimens

Since up to 17% of patients must discontinue clozapine because of adverse effects, strategies for minimizing and managing the adverse effects of clozapine have been reviewed (231,232). Treatment should begin with a low dosage, 12.5–25 mg/day.

The optimal plasma concentration of clozapine is 200–350 ng/ml, which usually corresponds to a daily dose of 200–400 mg (233,234). A nomogram to predict clozapine steady-state plasma concentrations has been generated using data from 71 patients (235). Clozapine steady-state plasma concentrations and demographic variables were obtained. The model explained 47% of the variance in clozapine concentrations. Two equations were obtained to predict steady-state plasma concentrations, one for men and one for women:

$$\text{clozapine (ng/ml)} = 0.464D + 111S + 145 \text{ (men)}$$

$$\text{clozapine (ng/ml)} = 1.590D + 111S - 149 \text{ (women)}$$

where D = dosage (mg/day), S = 1 for smokers, and S = 0 for non-smokers.

A further model for optimizing individual dosage regimens using Bayesian methods has been proposed (236).

Drug overdose

Fatal overdoses have been reported with clozapine (237).

Seven patients who took large doses of clozapine (mean 3 g, range 0.4–16 g) have been reported (238). All made a full recovery and toxicokinetic modelling suggested that norclozapine was formed by a saturable process but that clozapine kinetics were linear over the estimated doses.

- Clozapine overdose (2.5 g) in a 67-year-old woman resulted in seizures, loss of consciousness, metabolic acidosis, prolonged sedation, and aspiration pneumonia (239). By 9 days after intoxication she had recovered completely.
- A 40-year-old man who took 3–4 g of clozapine became unconscious, with constricted pupils, sinus tachycardia, and twitching; peak clozapine and norclozapine concentrations were 3.5 mg/l and 0.7 mg/l respectively, with secondary peaks at about 36 hours (240). Recovery was uneventful, and he was well 2 days after admission.
- A 41-year-old woman took 12.5 g of clozapine (125 tablets of 100 mg each) in a suicide attempt (241). She developed agitation, hallucinations, diminished distrust, and lethargy; she was given physostigmine 2 mg and recovered completely within 1 week.
- A 20-year-old woman presented 6 hours after taking clozapine 3500 mg (242). She had unexpectedly prolonged tachycardia and somnolence, and recovered only after the serum clozapine concentration began to fall after a 4-day plateau.

Suspected cases of poisoning in which blood clozapine and N-desmethylclozapine (norclozapine) were measured have been reviewed (243). There were seven fatal and five non-fatal clozapine cases of overdose, and 54 other people died while taking clozapine. Clozapine poisoning could not be diagnosed on the basis of blood clozapine and norclozapine concentrations alone. Analysis of antemortem blood specimens collected for white cell count monitoring and the blood clozapine:norclozapine ratio may provide additional interpretative information.

Drug–Drug Interactions

Benzodiazepines

Caution has been recommended when starting clozapine in patients taking benzodiazepines (SEDA-19, 55). Three cases of delirium associated with clozapine and benzodiazepines (244) have been reported. There have been several reports of synergistic reactions, resulting in increased sedation and ataxia, when lorazepam was begun in patients already taking clozapine (245).

Hypotension, collapse, and respiratory arrest occurred when low doses (12.5–25 mg) of clozapine were added to a pre-existing diazepam regimen in several patients (SEDA-22, 41).

- A 50-year-old man developed syncope and electrocardiographic changes (sinus bradycardia with deep inverted anteroseptal T waves and minor ST changes in other leads) with the concurrent administration of clozapine (after the dosage was increased to 300 mg/day) and diazepam (30 mg/day) (246).

Caffeine

Caffeine has been associated with changes in the metabolism of clozapine (SEDA-20, 50; SEDA-22, 61). Seven schizophrenic patients taking clozapine monotherapy participated in a study of the effects of caffeine withdrawal from the diet (247). After a caffeine-free diet for 5 days, clozapine plasma concentrations fell by 50%. The authors suggested that schizophrenic patients treated with clozapine should have their caffeine intake medically supervised, and that monitoring of concentrations of clozapine and its metabolite may be warranted.

The effects of caffeine-containing versus decaffeinated coffee on serum clozapine concentrations have been examined in 12 patients in a randomized, placebo-controlled, crossover study (248). Serum clozapine concentration increased, probably because of inhibition of CYP1A2. However, the effect was minor in most of the patients, although some individuals may be more sensitive to this interaction, owing for example to genetic factors.

Ciprofloxacin

A possible pharmacokinetic interaction between ciprofloxacin, which inhibits CYP1A2, and clozapine, with moderately increased serum concentrations of clozapine, has been reported (249).

Ciprofloxacin can alter plasma clozapine concentrations, perhaps by inhibition of cytochrome P450 enzymes (283).

Cisplatin

There is a potentially dangerous interaction with cancer treatment in patients with schizophrenia taking clozapine, because of the unpredictable risk of myelotoxicity. However, a 37-year-old patient taking clozapine for schizophrenia was given full-dose cisplatin and concomitant radiotherapy for an undifferentiated nasopharyngeal carcinoma, without significant neutropenia (250).

Citalopram and escitalopram

The effect of citalopram on plasma concentrations of clozapine have been prospectively studied in 15 patients with schizophrenia taking clozapine 200–400 mg/day (284). The addition of citalopram 40 mg/day did not alter plasma clozapine concentrations.

However, in a 39-year-old man with a schizoaffective disorder, citalopram 40 mg/day appeared to increase plasma clozapine concentrations and increased adverse effects (285). The adverse effects settled within 2 weeks of a reduction in citalopram dosage to 20 mg/day, with a corresponding 25% fall in clozapine concentrations. It is possible that at higher doses, citalopram can increase clozapine concentrations, perhaps through inhibition of CYP1A2 or CYP3A4.

Cocaine

An interaction between clozapine and cocaine, causing near syncope, has been reported (251). Eight male cocaine addicts underwent four oral challenges with increasing doses of clozapine (12.5, 25, and 50 mg) and placebo, followed 2 hours later by cocaine 2 mg/kg intranasally (286). Subjective and physiological responses, and serum cocaine concentrations were measured over 4 hours. Clozapine pretreatment increased cocaine concentrations during the study and significantly increased the peak serum cocaine concentrations dose-dependently. Despite this rise in blood concentrations, clozapine pretreatment significantly reduced subjective responses to cocaine, including "expected high," "high," and "rush" effects, notably at the 50 mg dose. There were also significant effects on "sleepiness," "paranoia" and "nervousness." Clozapine caused a significant near-syncopal episode in one subject, requiring withdrawal. Clozapine had no significant effect on baseline pulse rate or systolic blood pressure, but it attenuated the significant pressor effects of a single dose of intranasal cocaine. These data suggested a possible therapeutic role for clozapine in the treatment of cocaine addiction in humans, but also suggest caution due to the near-syncopal event and the increase in serum cocaine concentrations.

Diazepam

Hypotension, collapse, and respiratory arrest occurred when low doses (12.5–25 mg) of clozapine were added to a pre-existing diazepam regimen (SEDA-22, 41).

- A 50-year-old man with symptoms of chronic paranoid schizophrenia resistant to typical neuroleptic drugs had a brief syncopal attack with significant electrocardiographic changes (sinus bradycardia and deep anteroseptal inverted T waves and minor ST segment changes) after the dosage of clozapine was increased to 300 mg/day while he was taking diazepam 30 mg/day (287).

The mechanism of this presumed interaction is unknown.

Erythromycin

Increased clozapine serum concentrations have been reported with erythromycin (252,253) and can cause adverse effects (SEDA-21, 55). However, in 12 healthy men who took a single dose of clozapine 12.5 mg alone or in combination with a daily dose of erythromycin 1.5 g, the metabolism of clozapine was not altered (254). This confirms that CYP3A4 is a relatively minor pathway for clozapine metabolism, in contrast to CYP1A2.

In a case of neutropenia the authors suggested that an interaction of clozapine with erythromycin had been the precipitating factor (255).

Fluvoxamine

Fluvoxamine increases clozapine plasma concentrations (SEDA-27, 58). and the differential effects of steady-state fluvoxamine on the pharmacokinetics of olanzapine and clozapine have been studied in healthy volunteers (256). Fluvoxamine significantly reduced the N-desmethylclozapine to clozapine ratio, implying reduced metabolism of clozapine.

- A 34-year-old man, who had taken clozapine for 3 years in a dose that had been gradually increased up to 900 mg/day, was also given fluvoxamine 100 mg/day and 6 days later developed dystonia, dysarthria, hypersalivation, and dizziness; fluvoxamine was withdrawn and the adverse effects abated completely within 1 week (257).

Insulin

Severe insulin resistance with ketoacidosis (pH 6.9) has been reported with clozapine (291). After withdrawal, insulin requirements fell. Reinstitution of clozapine induced an identical increase in insulin need.

Interferon alfa

Agranulocytosis was observed after patients taking clozapine were given interferon alfa (SEDA-22, 404; 258). In one case, agranulocytosis occurred after 7 weeks of combined therapy in a 29-year-old patient who had been taking clozapine for more than 5 years without developing hematological abnormalities (258). Even so, it was not clear in these cases whether the agranulocytosis was due to the combination of clozapine with interferon-alfa or the clozapine alone.

Itraconazole

Itraconazole 200 mg had no significant effect on serum concentrations of clozapine 200–550 mg/day or desmethylclozapine in 7 schizophrenic patients (292).

Ketoconazole

The interaction of ketoconazole (400 mg/day for 7 days) with clozapine has been evaluated in five patients with schizophrenia given a single dose of clozapine 50 mg at the end of ketoconazole therapy (293). Ketoconazole did not significantly change the disposition of clozapine or its metabolism to its principal metabolites, desmethylclozapine and clozapine-*N*-oxide.

Lisinopril

Raised clozapine blood concentrations have been reported after the introduction of lisinopril (294).

- A 39-year-old man with schizophrenia and diabetes, who had taken clozapine 300 mg/day and glipizide 10 mg/day for a year, took lisinopril 5 mg/day for newly diagnosed hypertension. On several occasions afterwards he had roughly a doubling of his blood concentrations of clozapine and norclozapine. He had typical effects of clozapine toxicity. After replacement of lisinopril by diltiazem, the blood concentrations of clozapine and norclozapine returned to the values that were present before lisinopril was introduced.

The information given here was sketchy and there was no information on the timing of blood samples relative to the dose of clozapine. Clozapine is metabolized by CYP1A2 and CYP3A4, but there is no evidence that lisinopril affects these pathways.

Lithium

Seizures and other neurological effects have been described in a few cases when lithium was added to clozapine (259), but in other instances the combination was beneficial in overcoming treatment resistance or attenuating clozapine-induced leukopenia. Five treatment-resistant patients were treated successfully with a combination of clozapine and lithium with no clinically significant adverse events (295). However, a 59-year-old woman developed neurotoxic symptoms 3 days after lithium was added to clozapine; the symptoms resolved when both drugs were stopped and recurred with rechallenge (296).

- Multisystem organ failure occurred shortly after clozapine was added to a therapeutic dose of lithium in a 23-year-old woman (297). Improvement occurred when clozapine was stopped and the toxicity was attributed to clozapine.

The combination of clozapine and lithium has been examined in 44 patients (298). Medical records were retrospectively audited and a subsample of 23 patients was reassessed. The mean total duration of combination treatment was 23 months, and the combination was rated as effective in 84%; however, there were adverse effects in 64%, most often fatigue (32%), hypersalivation (14%), and, in 7% of patients, orthostatic dysregulation, muscle fatigue, weight gain, and oral dyskinesias.

The combination of clozapine and lithium has been studied in a randomized controlled trial in 10 patients with schizophrenia and 10 with a schizoaffective disorder taking clozapine with either lithium or placebo for 4 weeks (299). The combination was well tolerated, except for reversible neurotoxic reactions in two patients with schizophrenia, and safety measures showed no significant variations. The authors concluded that the lithium added to clozapine appears to afford potential benefit in schizoaffective disorders without harmful effects; for schizophrenia, however, it did not afford improvement but posed a risk of lithium toxicity.

Lorazepam

Caution has been recommended when starting clozapine in patients taking benzodiazepines (SEDA-19, 55). Three cases of delirium associated with clozapine and benzodiazepines (300) have been reported. There have been several reports of synergistic reactions, resulting in increased sedation and ataxia, when lorazepam was begun in patients already taking clozapine (301).

Modafinil

Clozapine toxicity occurred in a 42-year-old man after he was given modafinil 300 mg/day, a central stimulant, to combat sedation associated with clozapine (450 mg/day) (260). He complained of dizziness, had an unsteady gait, and fell twice. His serum clozapine concentration was 1400 ng/ml, which suggested a metabolic interaction between clozapine and modafinil. The authors suspected that inhibition of CYP2C19 by modafinil had reduced clozapine clearance.

Nefazodone

When nefazodone (400 mg/day for 3 weeks) was added to clozapine treatment in six patients with schizophrenia there was no significant change in plasma clozapine concentrations (302). This finding suggests that CYP3A4 does not play a major role in the metabolism of clozapine and that nefazodone may be a suitable adjunctive treatment for depression in clozapine-treated patients.

However, in another case the addition of nefazodone appeared to produce clozapine toxicity (303).

- A 40-year-old man had taken clozapine 450 mg/day and risperidone 6 mg/day for several years. Nefazodone 200 mg/day was added in an attempt to improve persistent negative symptoms, and after a week the dosage was increased to 300 mg/day. One week later, he reported anxiety and dizziness and was hypotensive. The combined concentrations of clozapine and its active metabolite norclozapine had increased from 309 ng/ml before nefazodone to 566 ng/ml. The nefazodone dosage was reduced to 200 mg/day and the anxiety, dizziness, and hypotension resolved over the next 7 days. At the same time plasma concentrations of clozapine and norclozapine fell to 370 ng/ml.

These results suggest that in some individuals CYP3A4 plays a significant role in the metabolism of clozapine and that the combination of nefazodone and clozapine should therefore be used with caution. It is possible that in this case concomitant treatment with risperidone may have increased the effect of nefazodone to reduce the clearance of clozapine.

Omeprazole

Clozapine is mainly metabolized by CYP1A2, which is induced by the proton pump inhibitor omeprazole. In two patients co-medication with omeprazole and clozapine was associated with reduced plasma concentrations of clozapine of 42% and 45% (304).

Perphenazine

Perphenazine doubled clozapine concentrations in a 46-year-old male smoker, with paradoxical myoclonus, hypersalivation, and worsening of psychosis (261). The mechanism of this effect is not known; perphenazine is a substrate for CYP2D6, but that is not important in the metabolism of clozapine.

Phenobarbital

Phenobarbital can stimulate the metabolism of clozapine, probably by inducing its N-oxidation and demethylation. Seven patients taking clozapine in combination with phenobarbital had significantly lower plasma clozapine concentrations than 15 controls taking clozapine only (262).

Rifamycins

Rifampicin can reduce plasma concentrations of clozapine and exacerbate psychotic symptoms (306).

Risperidone

Clozapine plasma concentrations increase when risperidone is introduced. The effects of risperidone 3.25 mg/day on cytochrome P450 isozymes have therefore been assessed in eight patients by determination of the metabolism of caffeine (for CYP1A2), dextromethorphan (for CYP2D6), and mephenytoin (for CYP2C19) (275). The results suggested that risperidone is a weak in vivo inhibitor of CYP2D6, CYP2C19, and CYP1A2. The authors concluded that inhibition by risperidone of those isozymes is an unlikely mechanism to explain increased clozapine concentrations.

SSRIs

Some SSRIs increase clozapine plasma concentrations (SEDA-20, 50; SEDA-21, 55; SEDA-22, 62) by inhibiting its metabolism.

Citalopram

Citalopram had no effect on plasma concentrations of clozapine ($n = 8$), risperidone ($n = 7$), or their active metabolites over 8 weeks (263). However, a possible interaction of clozapine with citalopram has been reported, with increased serum clozapine concentrations, perhaps dose-related (264).

Fluoxetine

In 10 patients stabilized on clozapine (200–450 mg/day) who took fluoxetine (20 mg/day) for 8 weeks, mean plasma concentrations of clozapine, norclozapine, and clozapine N-oxide increased significantly by 58%, 36%, and 38% respectively (265).

In two cases ingestion of clozapine and fluoxetine had a fatal outcome (266). The blood fluoxetine concentration was 0.7 µg/ml, which would be considered a high therapeutic concentration (usual target range 0.03–0.5 µg/ml). The blood clozapine concentration was 4900 ng/ml, which is within the lethal concentration range (1600–7100 ng/ml).

Dual effects were observed in a 44-year-old schizophrenic patient taking clozapine with both fluoxetine and sertraline for mood stabilization (267). Clinical and motor status improved with both fluoxetine and sertraline; cognitive function improved with clozapine and fluoxetine, but was not sustained with sertraline.

Fluvoxamine
Fluvoxamine increases clozapine plasma concentrations (265,268). In 16 patients taking clozapine monotherapy, fluvoxamine 50 mg was added in the hope of ameliorating the negative symptoms of schizophrenia (269). At steady state the serum concentrations of clozapine and its metabolites increased up to five-fold (average two- to threefold). However, adverse effects were almost unchanged in frequency and severity, in spite of the pharmacokinetic interaction.

In another study there were similar increases in plasma clozapine concentrations and adverse reactions in 18 patients taking fluvoxamine 50 mg (week 5, mean dose 97 mg) (270).

In nine men who were given a single dose of clozapine 50 mg on two separate occasions with a 2-week interval, fluvoxamine increased clozapine plasma concentrations, and the total mean clozapine AUC was increased by a factor of 2.6; all the patients were sedated during combined therapy (271).

Combined therapy with clozapine and fluvoxamine ($n = 11$) and clozapine monotherapy ($n = 12$) have been monitored before and during the first 6 weeks of medication (272). The co-administration of fluvoxamine attenuated and delayed the clozapine-induced increase in plasma concentrations of tumor necrosis factor-alpha, enhanced and accelerated the clozapine-induced increase in leptin plasma concentrations without a significant effect on clozapine-induced weight gain, and reduced granulocyte counts.

In two studies of short duration (18 patients each) there were benefits of using low doses of clozapine plus fluvoxamine, and the authors suggested taking advantage of this interaction (273). Patients taking fluvoxamine required relatively low doses of clozapine and had clinically significant reductions in the symptoms of their illness while avoiding the sedative adverse effects associated with the usual doses of clozapine.

In a prospective study fluvoxamine, in a low dosage of 50 mg/day, produced a threefold increase in plasma clozapine concentrations ($n = 16$) (288).

Other reports have confirmed that fluvoxamine increases plasma concentrations of clozapine and its metabolites 289,290 (SEDA-21, 12). The mechanism is probably inhibition of CYP1A2.

Paroxetine
The effect of paroxetine on steady-state plasma concentrations of clozapine and its metabolites has been studied in 17 patients taking clozapine (200–400 mg/day), nine of whom took additional paroxetine (20–40 mg/day) (274). Paroxetine, a potent inhibitor of CYP2D6, inhibited the metabolism of clozapine, possibly by affecting a pathway other than N-desmethylation and N-oxidation. After 3 weeks of paroxetine, mean plasma concentrations of clozapine and norclozapine increased significantly by 31% and 20% respectively, while concentrations of clozapine N-oxide were unchanged.

In a prospective study, paroxetine (20 mg/day), a potent inhibitor of CYP2D6, did not increase clozapine concentrations ($n = 14$), and the authors therefore suggested that CYP2D6 is not an important pathway of metabolism of clozapine (305). However, other studies (SEDA-21, 22) have shown that paroxetine can increase clozapine concentrations, suggesting that this combination should be used with caution.

Sertraline
Dual effects were observed in a 44-year-old schizophrenic patient taking clozapine with both fluoxetine and sertraline for mood stabilization (267). Clinical and motor status improved with both fluoxetine and sertraline; cognitive function improved with clozapine and fluoxetine, but was not sustained with sertraline. However, sertraline did not affect steady-state plasma concentrations of clozapine and its metabolites in 17 patients taking clozapine (200–400 mg/day), 8 of whom took additional sertraline (50–100 mg/day) (274).

Topiramate

Leukopenia has been reported after combination treatment with two leukopenic agents, clozapine and topiramate (307).

- A 28-year-old man with type I bipolar manic disorder was treated with clozapine 550 mg /day; he gained 25 kg, and topiramate, a mood stabilizer increasingly used as a weight suppressor, was added (25 mg bd for 1 week increased to 200 mg bd for 5 weeks). His white blood cell count had been 9–10 $\times 10^9$/l for 2 years while he was taking clozapine, and his complete blood counts were normal for 4 weeks after the addition of topiramate; however, but at 5 weeks his white blood cell count was 3.7 $\times 10^9$/l.

Tricyclic antidepressants

Nefazodone had minimal effects on clozapine metabolism when co-administered in six patients: mean clozapine concentrations rose by 4% and norclozapine concentrations by 16% (276).

Valproate

Clozapine inhibits the metabolism of valproate (277). Valproic acid has been reported to increase the sedative effects of clozapine (SEDA-20, 50) and alter serum concentrations of clozapine.

- In a 33-year-old woman taking clozapine and valproic acid, the serum concentrations of clozapine fell significantly (278).

The authors suggested that valproic acid had induced the metabolism of clozapine.

However, one study showed only small effects on plasma clozapine concentrations, which were thought to be unlikely to be clinically significant (279).

Smoking

Smoking is highly prevalent among patients with schizophrenia, of whom 70–80% smoke tobacco. In a before-and-after study, 55 smokers smoked less when treatment was switched to clozapine than when they were taking typical neuroleptic drugs (280). Nevertheless, it is probable that heavy smoking can induce CYP1A2, the main enzyme involved in the metabolism of clozapine, and plasma concentrations of clozapine are lower in smokers than in non-smokers. Conversely, sudden cessation of smoking can cause a rise in plasma clozapine concentrations. In one case, seizures have been reported as a result (281).

- A 35-year-old schizophrenic man successfully treated with clozapine 700–725 mg/day for more than 7 consecutive years abruptly stopped chronic heavy cigarette smoking and 2 weeks later suddenly developed tonic-clonic seizures followed by stupor and coma. After recovery, he successfully reduced the daily dose by about 40% before he stopped smoking.

Diagnosis of Adverse Drug Reactions

In an open study in 37 patients (27 men and 10 women; mean age 35 years) with treatment-resistant schizophrenia treated with clozapine for 18 weeks, there was no correlation between plasma clozapine concentrations and percentage improvement on the Positive and Negative Syndrome Scale (282). Plasma clozapine concentrations were not significantly different between those who responded to clozapine ($n = 19$) and those who did not ($n = 18$), nor between patients who smoked ($n = 28$) and those who did not ($n = 9$). Dosages were adjusted according to clinical response, and plasma concentrations of clozapine and its metabolites were measured weekly. The mean end-point clozapine dosage was 487 mg/day and there was a significant correlation between the daily dosage of clozapine and the plasma concentrations of clozapine and its metabolites. Three patients dropped out of the study owing to adverse effects (two because of significant sedation and one because of hypersalivation); there were no cases of agranulocytosis.

Monitoring therapy

The plasma concentrations of clozapine and its metabolite norclozapine (N-desmethylclozapine) have been measured in samples from 3775 patients (2648 men, 1127 women) (308). Step-wise backward multiple regression analysis (37% of the total sample) of log plasma clozapine concentration against log clozapine dose (mg/day), age (years), sex (male = 0, female = 1), cigarette smoking (non-smokers = 0, smokers = 1), body weight (kg), and plasma clozapine/norclozapine ratio (MR) showed that these co-variates explained 48% of the observed variation in plasma clozapine concentration (C = ng/ml x 10^3) according to the following equation:

$$\log_{10}(C) = 0.811 \log_{10}(\text{dose}) + 0.332(\text{MR}) + 69.42 \\ \times 10^{-3}(\text{sex}) + 2.263 \times 10^{-3}(\text{age}) + 1.976 \\ \times 10^{-3}(\text{weight}) - 0.171(\text{smoking habit}) \\ -3.180$$

The range of serum concentrations that corresponds to toxicity is unclear, and in three patients taking clozapine, a high serum concentration was interpreted as showing that they had a significant risk of toxicity, although they patients appeared to be well, with no signs of toxicity or evidence of adverse effects. It has therefore been suggested (Thomas 61) that serum concentration monitoring of clozapine should be used only in specific circumstances:

- a poor clinical response to routine doses;
- signs of toxicity or adverse events that could be linked to the serum concentration, such as seizures;
- the use of concurrent medications that interact with CYP isozymes, in particular CYP1A2;
- a change in the usual consumption of nicotine or caffeine;
- suspected or known liver disease;
- concerns about non-compliance.

However, serum concentrations should be interpreted with caution in view of significant intra-individual variation (309).

In contrast, other authors have stated that serum concentration monitoring is recommended during maintenance treatment with clozapine, in the light of a retrospective, open, non-randomized study in 86 patients, in which 404 serum concentrations were measured (310). Eight cases of intoxication were identified, and the conclusions were that the risk of toxicity is increased at serum concentrations over 750 ng/ml, while the risk of relapse is low at serum concentrations over 250 ng/ml, irrespective of concurrent psychotropic drugs, although intoxication was defined only according to the reason for the request for serum concentrations assay, as outlined in the request form.

- Serum concentrations of clozapine were measured 29 times in a 24-year-old woman took clozapine for 3 years (311). Relapses occurred repeatedly at low serum concentrations of 48, 109, and 138 ng/ml, because of non-adherence by the patient during out-patient treatment, and also because the dose was lowered too hastily after partial response during in-patient treatment. However, intoxication was evident at a serum concentration of clozapine of 1158 ng/ml during a trial with a dose of 800 mg/day.

References

1. Kulkarni SK, Ninan I. Dopamine D4 receptors and development of newer antipsychotic drugs. Fundam Clin Pharmacol 2000;14(6):529–39.

2. Reid WH. New vs. old antipsychotics: the Texas experience. J Clin Psychiatry 1999;60(Suppl 1):23–5.

3. Parker GF. Clozapine and borderline personality disorder. Psychiatr Serv 2002;53(3):348–9.

4. Masi G, Mucci M, Millepiedi S. Clozapine in adolescent inpatients with acute mania. J Child Adolesc Psychopharmacol 2002;12(2):93–9.

5. Hummel B, Dittmann S, Forsthoff A, Matzner N, Amann B, Grunze H. Clozapine as add-on medication in the maintenance treatment of bipolar and schizoaffective disorders. A case series. Neuropsychobiology 2002;45(Suppl 1):37–42.

6. Connelly JC, Fullick J. Experience with clozapine in a community mental health care setting. South Med J 1998;91(9):838–41.

7. Chong SA, Mahendran R, Wong KE. Use of atypical neuroleptics in a state mental institute. Ann Acad Med Singapore 1998;27(4):547–51.

8. Benedetti F, Sforzini L, Colombo C, Maffei C, Smeraldi E. Low-dose clozapine in acute and continuation treatment of severe borderline personality disorder. J Clin Psychiatry 1998;59(3):103–7.

9. Hector RI. The use of clozapine in the treatment of aggressive schizophrenia. Can J Psychiatry 1998;43(5):466–72.

10. Buzan RD, Dubovsky SL, Firestone D, Dal Pozzo E. Use of clozapine in 10 mentally retarded adults. J Neuropsychiatry Clin Neurosci 1998;10(1):93–5.

11. Dalal B, Larkin E, Leese M, Taylor PJ. Clozapine treatment of long-standing schizophrenia and serious violence: a two-year follow-up study of the first 50 patients treated with clozapine in Rampton high security hospital. Crim Behav Ment Health 1999;9:168–78.

12. Antonacci DJ, de Groot CM. Clozapine treatment in a population of adults with mental retardation. J Clin Psychiatry 2000;61(1):22–5.

13. Waserman J, Criollo M. Subjective experiences of clozapine treatment by patients with chronic schizophrenia. Psychiatr Serv 2000;51(5):666–8.

14. Umbricht DS, Wirshing WC, Wirshing DA, McMeniman M, Schooler NR, Marder SR, Kane JM. Clinical predictors of response to clozapine treatment in ambulatory patients with schizophrenia. J Clin Psychiatry 2002;63(5):420–4.

15. Schulte PFJ. What is an adequate trial with clozapine? Clin Pharmacokinet 2003;42:607–18.

16. Ciapparelli A, Dell'Osso L, Bandettini di Poggio A, Carmassi C, Cecconi D, Fenzi M, Chiavacci MC, Bottai M, Ramacciotti CE, Cassano GB. Clozapine in treatment-resistant patients with schizophrenia, schizoaffective disorder, or psychotic bipolar disorder: a naturalistic 48-month follow-up study. J Clin Psychiatry 2003;64:451–8.

17. Lahti AC, Holcomb HH, Weiler MA, Medoff DR, Tamminga CA. Functional effects of antipsychotic drugs: comparing clozapine with haloperidol. Biol Psychiatry 2003;53:601–8.

18. Dursun SM, Hallak JE, Haddad P, Leahy A, Byrne A, Strickland PL, Anderson IM, Zuardi AW, Deakin JF. Clozapine monotherapy for catatonic schizophrenia: should clozapine be the treatment of choice, with catatonia rather than psychosis as the main therapeutic index? J Psychopharmacol 2005;19:432–3.

19. Chen CK, Shiah IS, Yeh CB, Mao WC, Chang CC. Combination treatment of clozapine and topiramate in resistant rapid-cycling bipolar disorder. Clin Neuropharmacol 2005;28:136–8.

20. Gaszner P, Makkos Z. Clozapine maintenance therapy in schizophrenia. Prog Neuropsychopharmacol Biol Psychiatry 2004;28:465–9.

21. Wheatley M, Plant J, Reader H, Brown G, Cahill C. Clozapine treatment of adolescents with posttraumatic stress disorder and psychotic symptoms. J Clin Psychopharmacol 2004;24:167–73.

22. Agelink MW, Kavuk I, Ak I. Clozapine with amisulpride for refractory schizophrenia. Am J Psychiatry 2004;161:924–5.

23. Zink M, Mase E, Dressing H. Combination of ziprasidone and clozapine in treatment-resistant schizophrenia. Hum Psychopharmacol 2004;19:271–3.

24. Pollak P, Tison F, Rascol O, Destée A, Péré JJ, Senard JM, Durif F, Bourdeix I. Clozapine in drug induced psychosis in Parkinson's disease: a randomised, placebo controlled study with open follow up. J Neurol Neurosurg Psychiatry 2004;75:689–95.

25. Peacock L, Gerlach J. Clozapine treatment in Denmark: concomitant psychotropic medication and hematologic monitoring in a system with liberal usage practices. J Clin Psychiatry 1994;55(2):44–9.

26. Chong SA, Remington G. Clozapine augmentation: safety and efficacy. Schizophr Bull 2000;26(2):421–40.

27. Killian JG, Kerr K, Lawrence C, Celermajer DS. Myocarditis and cardiomyopathy associated with clozapine. Lancet 1999;354(9193):1841–5.

28. Shiwach RS. Treatment of clozapine induced hypertension and possible mechanisms. Clin Neuropharmacol 1998;21(2):139–40.

29. Donnelly JG, MacLeod AD. Hypotension associated with clozapine after cardiopulmonary bypass. J Cardiothorac Vasc Anesth 1999;13(5):597–9.

30. Low RA Jr, Fuller MA, Popli A. Clozapine induced atrial fibrillation. J Clin Psychopharmacol 1998;18(2):170.

31. Rechlin T, Beck G, Weis M, Kaschka WP. Correlation between plasma clozapine concentration and heart rate variability in schizophrenic patients. Psychopharmacology (Berl) 1998;135(4):338–41.

32. Varma S, Achan K. Dysrhythmia associated with clozapine. Aust NZ J Psychiatry 1999;33(1):118–9.

33. Krentz AJ, Mikhail S, Cantrell P, Hill GM. Pseudophaeochromocytoma syndrome associated with clozapine. BMJ 2001;322(7296):1213.

34. Bredbacka PE, Paukkala E, Kinnunen E, Koponen H. Can severe cardiorespiratory dysregulation induced by clozapine monotherapy be predicted? Int Clin Psychopharmacol 1993;8(3):205–6.

35. Kang UG, Kwon JS, Ahn YM, Chung SJ, Ha JH, Koo YJ, Kim YS. Electrocardiographic abnormalities in patients treated with clozapine. J Clin Psychiatry 2000;61(6):441–6.

36. Cohen H, Loewenthal U, Matar MA, Kotler M. Reversal of pathologic cardiac parameters after transition from clozapine to olanzapine treatment: a case report. Clin Neuropharmacol 2001;24(2):106–8.

37. Tie H, Walker BD, Singleton CB, Bursill JA, Wyse KR, Campbell TJ, Valenzuela SM, Breit SN. Clozapine and sudden death. J Clin Psychopharmacol 2001;21(6):630–2.

38. Chatterton R. Eosinophilia after commencement of clozapine treatment. Aust NZ J Psychiatry 1997;31(6):874–6.

39. Leo RJ, Kreeger JL, Kim KY. Cardiomyopathy associated with clozapine. Ann Pharmacother 1996;30(6):603–5.

40. Juul Povlsen U, Noring U, Fog R, Gerlach J. Tolerability and therapeutic effect of clozapine. A retrospective investigation of 216 patients treated with clozapine for up to 12 years. Acta Psychiatr Scand 1985;71(2):176–85.

41. Merrill DB, Dec GW, Goff DC. Adverse cardiac effects associated with clozapine. J Clin Psychopharmacol 2005;25:32–41.

42. Warner B, Schadelin J. Clinical safety and epidemiology. Leponex/Clozaril and myocarditisBasel, Switzerland: Novartis Pharm AG;. 1999.

43. Walker AM, Lanza LL, Arellano F, Rothman KJ. Mortality in current and former users of clozapine. Epidemiology 1997;8(6):671–7.

44. Reid P, McArthur M, Pridmore S. Clozapine rechallenge after myocarditis. Aust NZ J Psychiatry 2001;35(2):249.

45. Vaddadi KS, Soosai E, Vaddadi G. Low blood selenium concentrations in schizophrenic patients on clozapine. Br J Clin Pharmacol 2003;55:307–9.

46. Catalano G, Catalano MC, Frankel Wetter RL. Clozapine induced polyserositis. Clin Neuropharmacol 1997;20(4):352–6.

47. Murko A, Clarke S, Black DW. Clozapine and pericarditis with pericardial effusion. Am J Psychiatry 2002;159(3):494.

48. Kay SE, Doery J, Sholl D. Clozapine associated pericarditis and elevated troponin I. Aust NZ J Psychiatry 2002;36(1):143–4.

49. Boot E, De Haan L, Guzelcan Y, Scholte WF, Assies H. Pericardial and bilateral pleural effusion associated with clozapine treatment. Eur Psychiatry 2004;19:65–6.

50. Zornberg GL, Jick H. Antipsychotic drug use and risk of first-time idiopathic venous thromboembolism: a case-control study. Lancet 2000;356(9237):1219–23.

51. Hagg S, Spigset O, Soderstrom TG. Association of venous thromboembolism and clozapine. Lancet 2000;355(9210):1155–6.

52. Knudson JF, Kortepeter C, Dubitsky GM, Ahmad SR, Chen M. Antipsychotic drugs and venous thromboembolism. Lancet 2000;356(9225):252–3.

53. Kortepeter C, Chen M, Knudsen JF, Dubitsky GM, Ahmad SR, Beitz J. Clozapine and venous thromboembolism. Am J Psychiatry 2002;159(5):876–7.

54. Ihde-Scholl T, Rolli ML, Jefferson JW. Clozapine and pulmonary embolus. Am J Psychiatry 2001;158(3):499–500.

55. Pan R, John V. Clozapine and pulmonary embolism. Acta Psychiatr Scand 2003;108:76–7.

56. Hem E. Clozapine and pulmonary embolism: invited comment to letter to the editor. Acta Psychiatr Scand 2003;108:77.

57. Selten J-P, Büller H. Clozapine and venous thromboembolism: further evidence. J Clin Psychiatry 2003;64:609.

58. Ceravolo R, Salvetti S, Piccini P, Lucetti C, Gambaccini G, Bonuccelli U. Acute and chronic effects of clozapine in essential tremor. Mov Disord 1999;14(3):468–72.

59. Bär KJ, Häger F, Sauer H. Olanzapine and clozapine induced stuttering. Pharmacopsychiatry 2004;37:131-4.

60. Freudenreich O, Weiner RD, McEvoy JP. Clozapine-induced electroencephalogram changes as a function of clozapine serum levels. Biol Psychiatry 1997;42(2):132–7.

61. Wetter TC, Lauer CJ, Gillich G, Pollmacher T. The electroencephalographic sleep pattern in schizophrenic patients treated with clozapine or classical antipsychotic drugs. J Psychiatr Res 1996;30(6):411–9.

62. Hinze-Selch D, Mullington J, Orth A, Lauer CJ, Pollmacher T. Effects of clozapine on sleep: a longitudinal study. Biol Psychiatry 1997;42(4):260–6.

63. Toth P, Frankenburg FR. Clozapine and seizures: a review. Can J Psychiatry 1994;39(4):236–8.

64. Haller E, Binder RL. Clozapine and seizures. Am J Psychiatry 1990;147(8):1069–71.

65. Ravasia S, Dickson RA. Seizure on low-dose clozapine. Can J Psychiatry 1998;43(4):420.

66. Panagiotis B. Grand mal seizures with liver toxicity in a case of clozapine treatment. J Neuropsychiatry Clin Neurosci 1999;11(1):117–8.

67. Silvestri RC, Bromfield EB, Khoshbin S. Clozapine-induced seizures and EEG abnormalities in ambulatory psychiatric patients. Ann Pharmacother 1998;32(11):1147–51.

68. Solomons K, Berman KG, Gibson BA. All that seizes is not clozapine. Can J Psychiatry 1998;43(3):306–7.

69. Langosch JM, Trimble MR. Epilepsy, psychosis and clozapine. Hum Psychopharmacol 2002;17(2):115–9.

70. Usiskin SI, Nicolson R, Lenane M, Rapoport JL. Gabapentin prophylaxis of clozapine-induced seizures. Am J Psychiatry 2000;157(3):482–3.

71. Saba G, Dumortier G, Kalalou K, Benadhira R, Degrassat K, Glikman J, Januel D. Lamotrigine–clozapine combination in refractory schizophrenia: three cases. J Neuropsychiatry Clin Neurosci 2002;14(1):86.

72. Duggal HS, Jagadheesan K, Nizamie SH. Clozapine-induced stuttering and seizures. Am J Psychiatry 2002;159(2):315.

73. Supprian T, Retz W, Deckert J. Clozapine-induced stuttering: epileptic brain activity? Am J Psychiatry 1999;156(10):1663–4.

74. Miller CH, Mohr F, Umbricht D, Woerner M, Fleischhacker WW, Lieberman JA. The prevalence of acute extrapyramidal signs and symptoms in patients treated with clozapine, risperidone, and conventional antipsychotics. J Clin Psychiatry 1998;59(2):69–75.

75. Pi EH, Simpson GM. Medication-induced movement disorder. In: Sadock BJ, Sadock VA, editors. Comprehensive Textbook of Psychiatry. 7th ed.. Philadelphia: Lippincott Williams and Wilkins, 2000:2265–71.

76. Spivak B, Mester R, Abesgaus J, Wittenberg N, Adlersberg S, Gonen N, Weizman A. Clozapine treatment for neuroleptic-induced tardive dyskinesia, parkinsonism, chronic akathisia in schizophrenic patients. J Clin Psychiatry 1997;58(7):318–22.

77. Levine J, Chengappa KN. Second thoughts about clozapine as a treatment for neuroleptic-induced akathisia. J Clin Psychiatry 1998;59(4):195.

78. Trosch RM, Friedman JH, Lannon MC, Pahwa R, Smith D, Seeberger LC, O'Brien CF, LeWitt PA, Koller WC. Clozapine use in Parkinson's disease: a retrospective analysis of a large multicentered clinical experience. Mov Disord 1998;13(3):377–82.

79. Lera G, Zirulnik J. Pilot study with clozapine in patients with HIV-associated psychosis and drug-induced parkinsonism. Mov Disord 1999;14(1):128–31.

80. Friedman J, Lannon M, Cornelia C, Factor S, Kurlan R, Richard I. Low-dose clozapine for the treatment of drug-induced psychosis in Parkinson's disease. New Engl J Med 1999;340:757–63.

81. Pollak P, Destee A, Tison F, Pere JJ, Bordiex I, Agid YThe French Clozapine Parkinson Study Group. Clozapine in drug-induced psychosis in Parkinson's disease. Lancet 1999;353(9169):2041–2.

82. Klein C, Gordon J, Pollak L, Rabey M. Clozapine in Parkinson's disease psychosis: 5-year follow-up review. Clin Neuropharmacol 2003;26:8–11.

83. Littrell KH, Johnson CG, Littrell S, Peabody CD. Marked reduction of tardive dyskinesia with olanzapine. Arch Gen Psychiatry 1998;55(3):279–80.

84. O'Brien J, Barber R. Marked improvement in tardive dyskinesia following treatment with olanzapine in an elderly subject. Br J Psychiatry 1998;172:186.

85. Lykouras L, Malliori M, Christodoulou GN. Improvement of tardive dyskinesia following treatment with olanzapine. Eur Neuropsychopharmacol 1999;9(4):367–8.

86. Casey DE. Effects of clozapine therapy in schizophrenic individuals at risk for tardive dyskinesia. J Clin Psychiatry 1998;59(Suppl 3):31–7.

87. Dalack GW, Becks L, Meador-Woodruff JH. Tardive dyskinesia, clozapine, and treatment response. Prog Neuropsychopharmacol Biol Psychiatry 1998;22(4):567–73.

88. Andia I, Zumarraga M, Zabalo MJ, Bulbena A, Davila R. Differential effect of haloperidol and clozapine on plasma homovanillic acid in elderly schizophrenic patients with or without tardive dyskinesia. Biol Psychiatry 1998;43(1):20–3.

89. Modestin J, Stephan PL, Erni T, Umari T. Prevalence of extrapyramidal syndromes in psychiatric inpatients and the relationship of clozapine treatment to tardive dyskinesia. Schizophr Res 2000;42(3):223–30.

90. Kumet R, Freeman MP. Clozapine and tardive dyskinesia. J Clin Psychiatry 2002;63(2):167–8.

91. Elliott ES, Marken PA, Ruehter VL. Clozapine-associated extrapyramidal reaction. Ann Pharmacother 2000;34(5):615–8.

92. Louzã MR, Bassitt DP. Maintenance treatment of severe tardive dyskinesia with clozapine: 5 years' follow-up. J Clin Psychopharmacol 2005;25:180–2.

93. Karp BI, Goldstein SR, Chen R, Samii A, Bara-Jimenez W, Hallett M. An open trial of clozapine for dystonia. Mov Disord 1999;14(4):652–7.

94. Sieche A, Giedke H. Treatment of primary cranial dystonia (Meige's syndrome) with clozapine. J Clin Psychiatry 2000;61(12):949.

95. Molho ES, Factor SA. Worsening of motor features of parkinsonism with olanzapine. Mov Disord 1999;14(6):1014–6.

96. Hanagasi HA, Bilgic B, Gurvit H, Emre M. Clozapine treatment in oromandibular dystonia. Clin Neuropharmacol 2004;27:84–6.

97. Delecluse F, Elosegi JA, Gerard JM. A case of tardive tremor successfully treated with clozapine. Mov Disord 1998;13(5):846–7.

98. Trayer JS, Fidler DC. Neuroleptic malignant syndrome related to use of clozapine. J Am Osteopath Assoc 1998;98(3):168–9.

99. Karagianis JL, Phillips LC, Hogan KP, LeDrew KK. Clozapine-associated neuroleptic malignant syndrome: two new cases and a review of the literature. Ann Pharmacother 1999;33(5):623–30.

100. Lara DR, Wolf AL, Lobato MI, Baroni G, Kapczinski F. Clozapine-induced neuroleptic malignant syndrome: an interaction between dopaminergic and purinergic systems? J Psychopharmacol 1999;13(3):318–9.

101. Benazzi F. Clozapine-induced neuroleptic malignant syndrome not recurring with olanzapine, a structurally and pharmacologically similar antipsychotic. Hum Psychopharmacol Clin Exp 1999;14:511–2.

102. Doan RJ, Callaghan WD. Clozapine treatment and neuroleptic malignant syndrome. Can J Psychiatry 2000;45(4):394–5.

103. Reznik I, Volchek L, Mester R, Kotler M, Sarova-Pinhas I, Spivak B, Weizman A. Myotoxicity and neurotoxicity during clozapine treatment. Clin Neuropharmacol 2000;23(5):276–80.

104. Spivak M, Adams B, Crockford D. Atypical neuroleptic malignant syndrome with clozapine and subsequent haloperidol treatment. Can J Psychiatry 2003;48:66.

105. Duggal HS. Clozapine-induced neuroleptic malignant syndrome and subdural hematoma. J Neuropsychiatry Clin Neurosci 2004;16:118–9.

106. Wilkins-Ho M, Hollander Y. Toxic delirium with low-dose clozapine. Can J Psychiatry 1997;42(4):429–30.

107. Reinstein MJ, Chasonov MA, Colombo KD, Jones LE, Sonnenberg JG. Reduction of suicidality in patients with schizophrenia receiving clozapine. Clin Drug Invest 2002;22:341–6.

108. de Haan L, Linszen DH, Gorsira R. Clozapine and obsessions in patients with recent-onset schizophrenia and other psychotic disorders. J Clin Psychiatry 1999;60(6):364–5.

109. Bressan RA, Monteiro VB, Dias CC. Panic disorder associated with clozapine. Am J Psychiatry 2000;157(12):2056.

110. Ball MP, Hooper ET, Skipwith DF, Cates ME. Clozapine-induced hyperlipidemia resolved after switch to aripiprazole therapy. Ann Pharmacother 2005;39:1570–2.

111. Popli AP, Konicki PE, Jurjus GJ, Fuller MA, Jaskiw GE. Clozapine and associated diabetes mellitus. J Clin Psychiatry 1997;58(3):108–11.

112. Wirshing DA, Spellberg BJ, Erhart SM, Marder SR, Wirshing WC. Novel antipsychotics and new onset diabetes. Biol Psychiatry 1998;44(8):778–83.

113. Rigalleau V, Gatta B, Bonnaud S, Masson M, Bourgeois ML, Vergnot V, Gin H. Diabetes as a result of atypical anti-psychotic drugs—a report of three cases. Diabet Med 2000;17(6):484–6.

114. Wehring H, Alexander B, Perry PJ. Diabetes mellitus associated with clozapine therapy. Pharmacotherapy 2000;20(7):844–7.

115. Melkersson K, Hulting AL. Recovery from new-onset diabetes in a schizophrenic man after withdrawal of olanzapine. Psychosomatics 2002;43(1):67–70.

116. Colli A, Cocciolo M, Francobandiera F, Rogantin F, Cattalini N. Diabetic ketoacidosis associated with clozapine treatment. Diabetes Care 1999;22(1):176–7.

117. Chae BJ, Kang BJ. The effect of clozapine on blood glucose metabolism. Hum Psychopharmacol 2001;16(3):265–71.

118. Hagg S, Joelsson L, Mjorndal T, Spigset O, Oja G, Dahlqvist R. Prevalence of diabetes and impaired glucose tolerance in patients treated with clozapine compared with patients treated with conventional depot neuroleptic medications. J Clin Psychiatry 1998;59(6):294–9.

119. Yazici KM, Erbas T, Yazici AH. The effect of clozapine on glucose metabolism. Exp Clin Endocrinol Diabetes 1998;106(6):475–7.

120. Wang PS, Glynn RJ, Ganz DA, Schneeweiss S, Levin R, Avorn J. Clozapine use and risk of diabetes mellitus. J Clin Psychopharmacol 2002;22(3):236–43.

121. Howes OD, Bhatnagar A, Gaughran FP, Amiel SA, Murray RM, Pilowsky LS. A prospective study of impairment in glucose control caused by clozapine without changes in insulin resistance. Am J Psychiatry 2004;161:361–3.

122. Covell NH, Weissman EM, Essock SM. Weight gain with clozapine compared to first generation antipsychotic medications. Schizophr Bull 2004;30:229–40.

123. Frankenburg FR, Zanarini MC, Kando J, Centorrino F. Clozapine and body mass change. Biol Psychiatry 1998;43(7):520–4.

124. Bromel T, Blum WF, Ziegler A, Schulz E, Bender M, Fleischhaker C, Remschmidt H, Krieg JC, Hebebrand J.

Serum leptin levels increase rapidly after initiation of clozapine therapy. Mol Psychiatry 1998;3(1):76–80.

125. Tsai SJ, Yu YW, Lin CH, Wang YC, Chen JY, Hong CJ. Association study of adrenergic β3 receptor (Trp64Arg) and G-Protein β3 subunit gene (C825T) polymorphisms and weight change during clozapine treatment. Neuropsychobiology 2004;50:37–40.

126. Wang YC, Bai YM, Chen JY, Lin CC, Lai IC, Liou YJ. Polymorphism of the adrenergic receptor alpha 2a - 1291C>G genetic variation and clozapine-induced weight gain. J Neural Transm 2005;112:1463–8.

127. Theisen FM, Gebhardt S, Bromel T, Otto B, Heldwein W, Heinzel-Gutenbrunner M, Krieg JC, Remschmidt H, Tschop M, Hebebrand J. A prospective study of serum ghrelin levels in patients treated with clozapine. J Neural Transm 2005;112:1411–6.

128. Baymiller SP, Ball P, McMahon RP, Buchanan RW. Weight and blood pressure change during clozapine treatment. Clin Neuropharmacol 2002;25(4):202–6.

129. Feldman D, Goldberg JF. A preliminary study of the relationship between clozapine-induced weight gain and menstrual irregularities in schizophrenic, schizoaffective, and bipolar women. Ann Clin Psychiatry 2002;14(1):17–21.

130. Wirshing DA, Pierre JM, Wirshing WC. Sleep apnea associated with antipsychotic-induced obesity. J Clin Psychiatry 2002;63(4):369–70.

131. Borovicka MC, Fuller MA, Konicki PE, White JC, Steele VM, Jaskiw GE. Phenylpropanolamine appears not to promote weight loss in patients with schizophrenia who have gained weight during clozapine treatment. J Clin Psychiatry 2002;63(4):345–8.

132. Dursun SM, Devarajan S. Clozapine weight gain, plus topiramate weight loss. Can J Psychiatry 2000;45(2):198.

133. Haasen C, Lambert M, Yagdiran O, Karow A, Krausz M, Naber D. Comorbidity of schizophrenia and galactosemia: effective clozapine treatment with weight gain. Int Clin Psychopharmacol 2003;18:113–15.

134. de la Chapelle A, Kari C, Nurminen M, Hernberg S. Clozapine-induced agranulocytosis. A genetic and epidemiologic study. Hum Genet 1977;37(2):183–94.

135. Alvir JM, Lieberman JA, Safferman AZ, Schwimmer JL, Schaaf JA. Clozapine-induced agranulocytosis. Incidence and risk factors in the United States. N Engl J Med 1993;329(3):162–7.

136. Lamarque V. Effets hématologiques de la clozapine: bilan de l'experience internationale. [Hematologic effects of clozapine: a review of the international experience.] Encephale 1996;22(Spec No 6):35–6.

137. Copolov DL, Bell WR, Benson WJ, Keks NA, Strazzeri DC, Johnson GF. Clozapine treatment in Australia: a review of haematological monitoring. Med J Aust 1998;168(10):495–7.

138. Schuld A, Kraus T, Hinze-Selch D, Haack M, Pollmacher T. Granulocyte colony-stimulating factor plasma levels during clozapine- and olanzapine-induced granulocytopenia. Acta Psychiatr Scand 2000;102(2):153–5.

139. Jauss M, Pantel J, Werle E, Schroder J. G-CSF plasma levels in clozapine-induced neutropenia. Biol Psychiatry 2000;48(11):1113–5.

140. van de Loosdrecht AA, Faber HJ, Hordijk P, Uges DR, Smit A. Clozapine-induced agranulocytosis: a case report. Immunopathophysiological considerations. Neth J Med 1998;52(1):26–9.

141. Guest I, Sokoluk B, MacCrimmon J, Uetrecht J. Examination of possible toxic and immune mechanisms of clozapine-induced agranulocytosis. Toxicology 1998;131(1):53–65.

142. Loeffler S, Fehsel K, Henning U, Fischer J, Agelink M, Kolb-Bachofen V, Klimke A. Increased apoptosis of neutrophils in a case of clozapine-induced agranulocytosis. Pharmacopsychiatry 2003;36:37–41.

143. Dettling M, Sachse C, Muller-Oerlinghausen B, Roots I, Brockmoller J, Rolfs A, Cascorbi I. Clozapine-induced agranulocytosis and hereditary polymorphisms of clozapine metabolizing enzymes: no association with myeloperoxidase and cytochrome P4502D6. Pharmacopsychiatry 2000;33(6):218–20.

144. Lieberman JA, Yunis J, Egea E, Canoso RT, Kane JM, Yunis EJ. HLA-B38, DR4, DQw3 and clozapine-induced agranulocytosis in Jewish patients with schizophrenia. Arch Gen Psychiatry 1990;47(10):945–8.

145. Dettling M, Schaub RT, Mueller-Oerlinghausen B, Roots I, Cascorbi I. Further evidence of human leukocyte antigen-encoded susceptibility to clozapine-induced agranulocytosis independent of ancestry. Pharmacogenetics 2001;11(2):135–41.

146. Lahdelma L, Ahokas A, Andersson LC, Suvisaari J, Hovatta I, Huttunen MO, Koskimies S, Mitchell B. Balter Award. Human leukocyte antigen-A1 predicts a good therapeutic response to clozapine with a low risk of agranulocytosis in patients with schizophrenia. J Clin Psychopharmacol 2001;21(1):4–7.

147. Horacek J, Libiger J, Hoschl C, Borzova K, Hendrychova I. Clozapine-induced concordant agranulocytosis in monozygotic twins. Int J Psychiatry Clin Pract 2001;5:71–3.

148. Ahokas A, Elonen E. Circadian rhythm of white blood cells during clozapine treatment. Psychopharmacology (Berl) 1999;144(3):301–2.

149. Hu RJ, Malhotra AK, Pickar D. Predicting response to clozapine: status of current research. CNS Drugs 1999;11:317–26.

150. Patel NC, Dorson PG, Bettinger TL. Sudden late onset of clozapine-induced agranulocytosis. Ann Pharmacother 2002;36(6):1012–5.

151. Gupta S, Noor-Khan N, Frank B. Agranulocytosis in a second clozapine trial. Psychiatr Serv 1998;49(8):1094.

152. Sperner-Unterweger B, Czeipek I, Gaggl S, Geissler D, Spiel G, Fleischhacker WW. Treatment of severe clozapine-induced neutropenia with granulocyte colony-stimulating factor (G-CSF). Remission despite continuous treatment with clozapine. Br J Psychiatry 1998;172:82–4.

153. Silvestrini C, Arcangeli T, Biondi M, Pancheri P. A second trial of clozapine in a case of granulocytopenia. Hum Psychopharmacol 2000;15(4):275–9.

154. Small JG, Weber MC, Klapper MH, Kellams JJ. Rechallenge of late-onset neutropenia with clozapine. J Clin Psychopharmacol 2005;25:185–6.

155. Stewart P, Ezzy J. CPMSPlus, an innovative, web-based patient monitoring system for Clozaril centres. Aust J Hosp Pharm 2001;31:56.

156. Davies RH. Late awareness of anaemia in a patient receiving clozapine. Psychiatr Bull 2001;25:194–5.

157. Pascoe ST. The adjunctive use of a centralised database in the monitoring of clozapine-related neutropenia. Pharmacoepidemiol Drug Saf 2003;12:395–8.

158. Ahn YM, Jeong SH, Jang HS, Koo YJ, Kang UG, Lee KY, Kim YS. Experience of maintaining clozapine medication in patients with "red-alert zone" neutropenia: long-term follow-up results. Int Clin Psychopharmacol 2004;19:97–101.

159. Diaz P, Hogan TP. Granulocytopenia with clozapine and quetiapine. Am J Psychiatry 2001;158(4):651.

160. Melzer M, Hassanyeh FK, Snow MH, Ong EL. Sepsis and neutropenia induced by clozapine. Clin Microbiol Infect 1998;4(10):604–5.

161. Marcos F, Solano F, Arbol F, Caballero L, Maldonado G, Lopez P, Duran A. Clozapine-induced agranulocytosis. SN 2000;5:27–9.

162. Papetti F, Dariourt G, Giordana J-Y, Spreux A, Thauby S, Feral F, Pringuey D. Correction par le lithium des neutropénies induites par la clozapine (deux cas). L'Encephale 2004;30:570–82.

163. Blier P, Slater S, Measham T, Koch M, Wiviott G. Lithium and clozapine-induced neutropenia/agranulocytosis. Int Clin Psychopharmacol 1998;13(3):137–40.

164. Moldavsky M, Stein D, Benatov R, Sirota P, Elizur A, Matzner Y, Weizman A. Combined clozapine–lithium treatment for schizophrenia and schizoaffective disorder. Eur Psychiatry 1998;13:104–6.

165. Lucht MJ, Rietschel M. Clozapine-induced eosinophilia: subsequent neutropenia and corresponding allergic mechanisms. J Clin Psychiatry 1998;59(4):195–7.

166. Ames D, Wirshing WC, Baker RW, Umbricht DS, Sun AB, Carter J, Schooler NR, Kane JM, Marder SR. Predictive value of eosinophilia for neutropenia during clozapine treatment. J Clin Psychiatry 1996;57(12):579–81.

167. Hampson ME. Clozapine-induced thrombocytosis. Br J Psychiatry 2000;176:400.

168. Gonzales MF, Elmore J, Luebbert C. Evidence for immune etiology in clozapine-induced thrombocytopenia of 40 months' duration: a case report. CNS Spectr 2000;5:17–8.

169. Dunayevich E, McElroy SL. Atypical antipsychotics in the treatment of bipolar disorder: pharmacological and clinical effects. CNS Drugs 2000;13:433–41.

170. Szabadi E. Clozapine-induced hypersalivation. Br J Psychiatry 1997;171:89.

171. Patkar AA, Alexander RC. Parotid gland swelling with clozapine. J Clin Psychiatry 1996;57(10):488.

172. Tsai SJ, Wang YC, Yu, Younger WY, Lin CH, Yang KH, Hong CJ. Association analysis of polymorphism in the promoter region of the alpha$_{2a}$-adrenoceptor gene with schizophrenia and clozapine response. Schizophr Res 2001;49(1–2):53–8.

173. Cree A, Mir S, Fahy T. A review of the treatment options for clozapine-induced hypersalivation. Psychiatr Bull 2001;25:114–6.

174. Reinstein MJ, Sirotovskaya LA, Chasanov MA, Jones LE, Mohan S. Comparative efficacy and tolerability of benzatropine and terazosin in the treatment of hypersalivation secondary to clozapine. Clin Drug Invest 1999;17:97–102.

175. Calderon J, Rubin E, Sobota WL. Potential use of ipratropium bromide for the treatment of clozapine-induced hypersalivation: a preliminary report. Int Clin Psychopharmacol 2000;15(1):49–52.

176. Freudenreich O, Beebe M, Goff DC. Clozapine-induced sialorrhea treated with sublingual ipratropium spray: a case series. J Clin Psychopharmacol 2004;24:98–100.

177. Kahl KG, Hagenah J, Zapf S, Trillenberg P, Klein C, Lencer R. Botulinum toxin as an effective treatment of clozapine-induced hypersalivation. Psychopharmacology 2004;173:229–30.

178. Praharaj SK, Verma P, Roy D, Singh A. Is clonidine useful for treatment of clozapine-induced sialorrhea? J Psychopharmacol 2005;19:426–8.

179. Laker MK, Cookson JC. Reflux oesophagitis and clozapine. Int Clin Psychopharmacol 1997;12(1):37–9.

180. Baker RW, Chengappa KN. Gastroesophageal reflux as a possible result of clozapine treatment. J Clin Psychiatry 1998;59(5):257.

181. Freudenreich O, Goff DC. Colon perforation and peritonitis associated with clozapine. J Clin Psychiatry 2000;61(12):950–1.

182. Levin TT, Barrett J, Mendelowitz A. Death from clozapine-induced constipation: case report and literature review. Psychosomatics 2002;43(1):71–3.

183. Hummer M, Kurz M, Kurzthaler I, Oberbauer H, Miller C, Fleischhacker WW. Hepatotoxicity of clozapine. J Clin Psychopharmacol 1997;17(4):314–7.

184. Larsen JT, Clemensen SV, Klitgaard NA, Nielsen B, Brosen K. Clozapin-udlost toksisk hepatitis. [Clozapine-induced toxic hepatitis.] Ugeskr Laeger 2001;163(14):2013–4.

185. Thompson J, Chengappa KN, Good CB, Baker RW, Kiewe RP, Bezner J, Schooler NR. Hepatitis, hyperglycemia, pleural effusion, eosinophilia, hematuria and proteinuria occurring early in clozapine treatment. Int Clin Psychopharmacol 1998;13(2):95–8.

186. Macfarlane B, Davies S, Mannan K, Sarsam R, Pariente D, Dooley J. Fatal acute fulminant liver failure due to clozapine: a case report and review of clozapine-induced hepatotoxicity. Gastroenterology 1997;112(5):1707–9.

187. Gatto EM, Castronuovo AP, Uribe Roca MC. Clozapine and pancreatitis. Clin Neuropharmacol 1998;21(3):203.

188. Wehmeier PM, Heiser P, Remschmidt H. Pancreatitis followed by pericardial effusion in an adolescent treated with clozapine. J Clin Psychopharmacol 2003;23:102–3.

189. Cerulli TR. Clozapine-associated pancreatitis. Harv Rev Psychiatry 1999;7(1):61–3.

190. Bergemann N, Ehrig C, Diebold K, Mundt C, von Einsiedel R. Asymptomatic pancreatitis associated with clozapine. Pharmacopsychiatry 1999;32(2):78–80.

191. Poyurovsky M, Modai I, Weizman A. Trihexyphenidyl as a possible therapeutic option in clozapine-induced nocturnal enuresis. Int Clin Psychopharmacol 1996;11(1):61–3.

192. Costa JF, Sramek J, Bera RB, Brenneman M, Cristobal M. Control of bed-wetting with benztropine. Am J Psychiatry 1990;147(5):674.

193. Lin CC, Bai YM, Chen JY, Lin CY, Lan TH. A retrospective study of clozapine and urinary incontinence in Chinese in-patients. Acta Psychiatr Scand 1999;100(2):158–61.

194. Hsu JW, Wang YC, Lin CC, Bai YM, Chen JY, Chiu HJ, Tsai SJ, Hong CJ. No evidence for association of alpha 1a adrenoceptor gene polymorphism and clozapine-induced urinary incontinence. Neuropsychobiology 2000;42(2):62–5.

195. Elias TJ, Bannister KM, Clarkson AR, Faull D, Faull RJ. Clozapine-induced acute interstitial nephritis. Lancet 1999;354(9185):1180–1.

196. David WS, Sharif AA. Clozapine-induced myokymia. Muscle Nerve 1998;21(6):827–8.

197. Vrtunski PB, Konicki PE, Jaskiw GE, Brescan DW, Kwon KY, Jurjus G. Clozapine effects on force control in schizophrenic patients. Schizophr Res 1998;34(1–2):39–48.

198. Koren W, Koren E, Nacasch N, Ehrenfeld M, Gur H. Rhabdomyolysis associated with clozapine treatment in a patient with decreased calcium-dependent potassium permeability of cell membranes. Clin Neuropharmacol 1998;21(4):262–4.

199. Wicki J, Rutschmann OT, Burri H, Vecchietti G, Desmeules J. Rhabdomyolysis after correction of hyponatremia due to psychogenic polydipsia possibly complicated by clozapine. Ann Pharmacother 1998;32(9):892–5.

200. Hummer M, Kemmler G, Kurz M, Kurzthaler I, Oberbauer H, Fleischhacker WW. Sexual disturbances during clozapine and haloperidol treatment for schizophrenia. Am J Psychiatry 1999;156(4):631–3.

201. Jeffries JJ, Vanderhaeghe L, Remington GJ, Al-Jeshi A. Clozapine-associated retrograde ejaculation. Can J Psychiatry 1996;41(1):62–3.

202. Stanislav SW, Gonzalez-Blanco M. Papular rash and bilateral pleural effusion associated with clozapine. Ann Pharmacother 1999;33(9):1008–9.

203. Wolf J, Sartorius A, Alm B, Henn FA. Clozapine-induced lupus erythematosus. J Clin Psychopharmacol 2004;24:236–8.

204. Tremeau F, Clark SC, Printz D, Kegeles LS, Malaspina D. Spiking fevers with clozapine treatment. Clin Neuropharmacol 1997;20(2):168–70.

205. Kerwin RW, Osborne S, Sainz-Fuertes R. Heat stroke in schizophrenia during clozapine treatment: rapid recognition and management. J Psychopharmacol 2004;18:121–3.

206. Kerswill RM, Vicente MR. Clozapine and postmortem redistribution. Am J Psychiatry 2003;160:184.

207. Shiovitz TM, Welke TL, Tigel PD, Anand R, Hartman RD, Sramek JJ, Kurtz NM, Cutler NR. Cholinergic rebound and rapid onset psychosis following abrupt clozapine withdrawal. Schizophr Bull 1996;22(4):591–5.

208. Verghese C, DeLeon J, Nair C, Simpson GM. Clozapine withdrawal effects and receptor profiles of typical and atypical neuroleptics. Biol Psychiatry 1996;39(2):135–8.

209. Ekblom B, Eriksson K, Lindstrom LH. Supersensitivity psychosis in schizophrenic patients after sudden clozapine withdrawal. Psychopharmacology (Berl) 1984;83(3):293–4.

210. Perenyi A, Kuncz E, Bagdy G. Early relapse after sudden withdrawal or dose reduction of clozapine. Psychopharmacology (Berl) 1985;86(1–2):244.

211. Eklund K. Supersensitivity and clozapine withdrawal. Psychopharmacology (Berl) 1987;91(1):135.

212. Goudie AJ. What is the clinical significance of the discontinuation syndrome seen with clozapine? J Psychopharmacol 2000;14(2):188–92.

213. Simpson GM, Varga E. Clozapine—a new antipsychotic agent. Curr Ther Res Clin Exp 1974;16(7):679–86.

214. Lieberman JA, Kane JM, Johns CA. Clozapine: guidelines for clinical management. J Clin Psychiatry 1989;50(9):329–38.

215. de Leon J, Stanilla JK, White AO, Simpson GM. Anticholinergics to treat clozapine withdrawal. J Clin Psychiatry 1994;55(3):119–20.

216. Ahmed S, Chengappa KN, Naidu VR, Baker RW, Parepally H, Schooler NR. Clozapine withdrawal-emergent dystonias and dyskinesias: a case series. J Clin Psychiatry 1998;59(9):472–7.

217. Poyurovsky M, Bergman Y, Shoshani D, Schneidman M, Weizman A. Emergence of obsessive–compulsive symptoms and tics during clozapine withdrawal. Clin Neuropharmacol 1998;21(2):97–100.

218. Stoner SC, Sommi RW Jr, Marken PA, Anya I, Vaughn J. Clozapine use in two full-term pregnancies. J Clin Psychiatry 1997;58(8):364–5.

219. Ernst CL, Goldberg JF. The reproductive safety profile of mood stabilizers, atypical antipsychotics, and broad-spectrum psychotropics. J Clin Psychiatry 2002;63 Suppl 4:42–55.

220. Mendhekar DN, Sharma JB, Srivastava PK, War L. Clozapine and pregnancy. J Clin Psychiatry 2003;64:850.

221. Ng CH, Chong SA, Lambert T, Fan A, Hackett LP, Mahendran R, Subramaniam M, Schweitzer I. An interethnic comparison study of clozapine dosage, clinical response and plasma levels. Int Clin Psychopharmacol 2005;20:163–8.

222. Frazier JA, Cohen LG, Jacobsen L, Grothe D, Flood J, Baldessarini RJ, Piscitelli S, Kim GS, Rapoport JL. Clozapine pharmacokinetics in children and adolescents with childhood-onset schizophrenia. J Clin Psychopharmacol 2003;23:87–91.

223. Sajatovic M, Ramirez LF, Garver D, Thompson P, Ripper G, Lehmann LS. Clozapine therapy for older veterans. Psychiatr Serv 1998;49(3):340–4.

224. Chung C, Remington G. Predictors and markers of clozapine response. Psychopharmacology 2005;179:317–35.

225. Rosenstock J. Clozapine therapy during cancer treatment. Am J Psychiatry 2004;161:175.

226. Jecel J, Michel TM, Gutknecht L, Schmidt D, Pfuhlmann B, Jabs BE. Toxic clozapine serum levels during acute urinary tract infection: a case report. Eur J Clin Pharmacol 2005;60:909–10.

227. Burke WJ, Pfeiffer RF, McComb RD. Neuroleptic sensitivity to clozapine in dementia with Lewy bodies. J Neuropsychiatry Clin Neurosci 1998;10(2):227–9.

228. Sramek JJ, Anand R, Hartman RD, Schran HF, Hourani J, Barto S, Wardle TS, Shiovitz TM, Cutler NR. A bioequivalence study of brand and generic clozapine in patients with schizophrenia. Clin Drug Invest 1999;17:51–8.

229. Tse G, Thompson D, Procyshyn RM. A cost-saving alternative to brand name clozapine? Pharmacoeconomics 2003;21:1–11.

230. Makela EH, Cutlip WD, Stevenson JM, Weimer JM, Abdallah ES, Akhtar RS, Aboraya AS, Gunel E. Branded versus generic clozapine for treatment of schizophrenia. Ann Pharmacother 2003;37:350–3.

231. Young CR, Bowers MB Jr, Mazure CM. Management of the adverse effects of clozapine. Schizophr Bull 1998;24(3):381–90.

232. Lieberman JA. Maximizing clozapine therapy: managing side effects. J Clin Psychiatry 1998;59(Suppl 3):38–43.

233. Conley RR. Optimizing treatment with clozapine. J Clin Psychiatry 1998;59(Suppl 3):44–8.

234. Olesen OV. Therapeutic drug monitoring of clozapine treatment. Therapeutic threshold value for serum clozapine concentrations. Clin Pharmacokinet 1998;34(6):497–502.

235. Perry PJ, Bever KA, Arndt S, Combs MD. Relationship between patient variables and plasma clozapine concentrations: a dosing nomogram. Biol Psychiatry 1998;44(8):733–8.

236. Guitton C, Kinowski JM, Gomeni R, Bressolle F. A kinetic model for simultaneous fit of clozapine and norclozapine concentrations in chronic schizophrenic patients during long-term treatment. Clin Drug Invest 1998;16:35–43.

237. Keller T, Miki A, Binda S, Dirnhofer R. Fatal overdose of clozapine. Forensic Sci Int 1997;86(1–2):119–25.

238. Reith D, Monteleone JP, Whyte IM, Ebelling W, Holford NH, Carter GL. Features and toxicokinetics of clozapine in overdose. Ther Drug Monit 1998;20(1):92–7.

239. Hagg S, Spigset O, Edwardsson H, Bjork H. Prolonged sedation and slowly decreasing clozapine serum concentrations after an overdose. J Clin Psychopharmacol 1999;19(3):282–4.

240. Renwick AC, Renwick AG, Flanagan RJ, Ferner RE. Monitoring of clozapine and norclozapine plasma concentration-time curves in acute overdose. J Toxicol Clin Toxicol 2000;38(3):325–8.

241. Sartorius A, Hewer W, Zink M, Henn FA. High-dose clozapine intoxication. J Clin Psychopharmacol 2002;22(1):91–2.

242. Thomas L, Pollak PT. Delayed recovery associated with persistent serum concentrations after clozapine overdose. J Emerg Med 2003;25:61–6.

243. Flanagan RJ, Spencer EP, Morgan PE, Barnes TR, Dunk L. Suspected clozapine poisoning in the UK/Eire, 1992-2003. Forensic Sci Int 2005;155:91–9.

244. Jackson CW, Markowitz JS, Brewerton TD. Delirium associated with clozapine and benzodiazepine combinations. Ann Clin Psychiatry 1995;7(3):139–41.

245. Cobb CD, Anderson CB, Seidel DR. Possible interaction between clozapine and lorazepam. Am J Psychiatry 1991;148(11):1606–7.

246. Tupala E, Niskanen L, Tiihonen J. Transient syncope and ECG changes associated with the concurrent administration of clozapine and diazepam. J Clin Psychiatry 1999;60(9):619–20.

247. Carrillo JA, Herraiz AG, Ramos SI, Benitez J. Effects of caffeine withdrawal from the diet on the metabolism of clozapine in schizophrenic patients. J Clin Psychopharmacol 1998;18(4):311–6.

248. Raaska K, Raitasuo V, Laitila J, Neuvonen PJ. Effect of caffeine-containing versus decaffeinated coffee on serum clozapine concentrations in hospitalised patients. Basic Clin Pharmacol Toxicol 2004;94:18–8.

249. Raaska K, Neuvonen PJ. Ciprofloxacin increases serum clozapine and N-desmethylclozapine: a study in patients with schizophrenia. Eur J Clin Pharmacol 2000;56(8):585–9.

250. Bareggi C, Palazzi M, Locati LD, Cerrotta A, Licitral L. Clozapine and full-dose concomitant chemoradiation therapy in a schizophrenic patient with nasopharyngeal cancer. Tumori 2002;88(1):59–60.

251. Hameedi FA, Sernyak MJ, Navui SA, Kosten TR. Near syncope associated with concomitant clozapine and cocaine use. J Clin Psychiatry 1996;57(8):371–2.

252. Taylor D. Pharmacokinetic interactions involving clozapine. Br J Psychiatry 1997;171:109–12.

253. Cohen LG, Chesley S, Eugenio L, Flood JG, Fisch J, Goff DC. Erythromycin-induced clozapine toxic reaction. Arch Intern Med 1996;156(6):675–7.

254. Hagg S, Spigset O, Mjorndal T, Granberg K, Persbo-Lundqvist G, Dahlqvist R. Absence of interaction between erythromycin and a single dose of clozapine. Eur J Clin Pharmacol 1999;55(3):221–6.

255. Usiskin SI, Nicolson R, Lenane M, Rapoport JL. Retreatment with clozapine after erythromycin-induced neutropenia. Am J Psychiatry 2000;157(6):1021.

256. Wang CY, Zhang ZJ, Li WB, Zhai YM, Cai ZJ, Weng YZ, Zhu RH, Zhao JP, Zhou HH. The differential effects of steady-state fluvoxamine on the pharmacokinetics of olanzapine and clozapine in healthy volunteers. J Clin Pharmacol 2004;44:785–92.

257. Peritogiannis V, Tsouli S, Pappas D, Mavreas V. Acute effects of clozapine–fluvoxamine combination. Schizophr Res 2005;79:345–6.

258. Hoffmann RM, Ott S, Parhofer KG, Bartl R, Pape GR. Interferon-alpha-induced agranulocytosis in a patient on long-term clozapine therapy. J Hepatol 1998;29(1):170.

259. Edge SC, Markowitz JS, DeVane CL. Clozapine drug–drug interactions: a review of the literature. Hum Psychopharmacol 1997;12:5–20.

260. Dequardo JR. Modafinil-associated clozapine toxicity. Am J Psychiatry 2002;159(7):1243–4.

261. Cooke C, de Leon J. Adding other antipsychotics to clozapine. J Clin Psychiatry 1999;60(10):710.

262. Facciola G, Avenoso A, Spina E, Perucca E. Inducing effect of phenobarbital on clozapine metabolism in patients with chronic schizophrenia. Ther Drug Monit 1998;20(6):628–30.

263. Avenoso A, Facciolà G, Scordo MG, Gitto C, Ferrante GD, Madia AG, Spina E. No effect of citalopram on plasma levels of clozapine, risperidone and their active metabolites in patients with chronic schizophrenia. Clin Drug Invest 1998;16:393–8.

264. Borba CP, Henderson DC. Citalopram and clozapine: potential drug interaction. J Clin Psychiatry 2000;61(4):301–2.

265. Spina E, Avenoso A, Facciola G, Fabrazzo M, Monteleone P, Maj M, Perucca E, Caputi AP. Effect of fluoxetine on the plasma concentrations of clozapine and its major metabolites in patients with schizophrenia. Int Clin Psychopharmacol 1998;13(3):141–5.

266. Ferslew KE, Hagardorn AN, Harlan GC, McCormick WF. A fatal drug interaction between clozapine and fluoxetine. J Forensic Sci 1998;43(5):1082–5.

267. Purdon SE, Snaterse M. Selective serotonin reuptake inhibitor modulation of clozapine effects on cognition in schizophrenia. Can J Psychiatry 1998;43(1):84–5.

268. Markowitz JS, Gill HS, Lavia M, Brewerton TD, DeVane CL. Fluvoxamine–clozapine dose-dependent interaction. Can J Psychiatry 1996;41(10):670–1.

269. Szegedi A, Anghelescu I, Wiesner J, Schlegel S, Weigmann H, Hartter S, Hiemke C, Wetzel H. Addition of low-dose fluvoxamine to low-dose clozapine monotherapy in schizophrenia: drug monitoring and tolerability data from a prospective clinical trial. Pharmacopsychiatry 1999;32(4):148–53.

270. Lammers CH, Deuschle M, Weigmann H, Hartter S, Hiemke C, Heese C, Heuser I. Coadministration of clozapine and fluvoxamine in psychotic patients—clinical experience. Pharmacopsychiatry 1999;32(2):76–7.

271. Chang WH, Augustin B, Lane HY, ZumBrunnen T, Liu HC, Kazmi Y, Jann MW. In-vitro and in-vivo evaluation of the drug–drug interaction between fluvoxamine and clozapine. Psychopharmacology (Berl) 1999;145(1):91–8.

272. Hinze-Selch D, Deuschle M, Weber B, Heuser I, Pollmacher T. Effect of coadministration of clozapine and fluvoxamine versus clozapine monotherapy on blood cell counts, plasma levels of cytokines and body weight. Psychopharmacology (Berl) 2000;149(2):163–9.

273. Prior TI. Is there a way to overcome over-sedation in a patient being treated with clozapine? J Psychiatry Neurosci 2002;27:224.

274. Spina E, Avenoso A, Salemi M, Facciola G, Scordo MG, Ancione M, Madia A. Plasma concentrations of clozapine and its major metabolites during combined treatment with paroxetine or sertraline. Pharmacopsychiatry 2000;33(6):213–7.

275. Eap CB, Bondolfi G, Zullino D, Bryois C, Fuciec M, Savary L, Jonzier-Perey M, Baumann P. Pharmacokinetic drug interaction potential of risperidone with cytochrome p450 isozymes as assessed by the dextromethorphan, the caffeine, and the mephenytoin test. Ther Drug Monit 2001;23(3):228–31.

276. Taylor D, Bodani M, Hubbeling A, Murray R. The effect of nefazodone on clozapine plasma concentrations. Int Clin Psychopharmacol 1999;14(3):185–7.

277. Costello LE, Suppes T. A clinically significant interaction between clozapine and valproate. J Clin Psychopharmacol 1995;15(2):139–41.

278. Conca A, Beraus W, Konig P, Waschgler R. A case of pharmacokinetic interference in comedication of clozapine and valproic acid. Pharmacopsychiatry 2000;33(6):234–5.

279. Facciola G, Avenoso A, Scordo MG, Madia AG, Ventimiglia A, Perucca E, Spina E. Small effects of valproic acid on the plasma concentrations of clozapine and its major metabolites in patients with schizophrenic or affective disorders. Ther Drug Monit 1999;21(3):341–5.

280. McEvoy JP, Freudenreich O, Wilson WH. Smoking and therapeutic response to clozapine in patients with schizophrenia. Biol Psychiatry 1999;46(1):125–9.

281. Skogh E, Bengtsson F, Nordin C. Could discontinuing smoking be hazardous for patients administered clozapine medication? A case report. Ther Drug Monit 1999;21(5):580–2.

282. Llorca PM, Lancon C, Disdier B, Farisse J, Sapin C, Auquier P. Effectiveness of clozapine in neuroleptic-resistant schizophrenia: clinical response and plasma concentrations. J Psychiatry Neurosci 2002;27(1):30–7.

283. Joos AA. Pharmakologische Interaktionen von Antibiotika und Psychopharmaka. [Pharmacologic interactions of antibiotics and psychotropic drugs.] Psychiatr Prax 1998;25(2):57–60.

284. Avenoso A, Facciola G, Scordo MG, Gitto C, Ferrante GD. No effect of citalopram on plasma levels of clozapine, risperidone and their active metabolites in patients with chronic schizophrenia. Clin Drug Invest 1998;16:393–8.

285. Borba CP, Henderson DC. Citalopram and clozapine: potential drug interaction. J Clin Psychiatry 2000;61(4):301–2.

286. Farren CK, Hameedi FA, Rosen MA, Woods S, Jatlow P, Kosten TR. Significant interaction between clozapine and cocaine in cocaine addicts. Drug Alcohol Depend 2000;59(2):153–63.

287. Tupala E, Niskanen L, Tiihonen J. Transient syncope and ECG changes associated with the concurrent administration of clozapine and diazepam. J Clin Psychiatry 1999;60(9):619–20.

288. Wetzel H, Anghelescu I, Szegedi A, Wiesner J, Weigmann H, Harter S, Hiemke C. Pharmacokinetic interactions of clozapine with selective serotonin reuptake inhibitors: differential effects of fluvoxamine and paroxetine in a prospective study. J Clin Psychopharmacol 1998;18(1):2–9.

289. Fabrazzo M, La Pia S, Monteleone P, Mennella R, Esposito G, Pinto A, Maj M. Fluvoxamine increases plasma and urinary levels of clozapine and its major metabolites in a time- and dose-dependent manner. J Clin Psychopharmacol 2000;20(6):708–10.

290. Lu ML, Lane HY, Chen KP, Jann MW, Su MH, Chang WH. Fluvoxamine reduces the clozapine dosage needed in refractory schizophrenic patients. J Clin Psychiatry 2000;61(8):594–9.

291. Colli A, Cocciolo M, Francobandiera F, Rogantin F, Cattalini N. Diabetic ketoacidosis associated with clozapine treatment. Diabetes Care 1999;22(1):176–7.

292. Raaska K, Neuvonen PJ. Serum concentrations of clozapine and N-desmethylclozapine are unaffected by the potent CYP3A4 inhibitor itraconazole. Eur J Clin Pharmacol 1998;54(2):167–70.

293. Lane HY, Chiu CC, Kazmi Y, Desai H, Lam YW, Jann MW, Chang WH. Lack of CYP3A4 inhibition by grapefruit juice and ketoconazole upon clozapine administration in vivo. Drug Metabol Drug Interact 2001;18(3–4):263–78.

294. Abraham G, Grunberg B, Gratz S. Possible interaction of clozapine and lisinopril. Am J Psychiatry 2001;158(6):969.

295. Moldavsky M, Stein D, Benatov R, Sirota P, Elizur A, Matzner Y, Weizman A. Combined clozapine-lithium treatment for schizophrenia and schizoaffective disorder. Eur Psychiatry 1998;13:104–6.

296. Lee SH, Yang YY. Reversible neurotoxicity induced by a combination of clozapine and lithium: a case report. Zhonghua Yi Xue Za Zhi (Taipei) 1999;62(3):184–7.

297. Patton S, Remick RA, Isomura T. Clozapine—an atypical reaction. Can J Psychiatry 2000;45(4):393–4.

298. Bender S, Linka T, Wolstein J, Gehendges S, Paulus HJ, Schall U, Gastpar M. Safety and efficacy of combined clozapine–lithium pharmacotherapy. Int J Neuropsychopharmacol 2004;7:59–63.

299. Small JG, Klapper MH, Malloy FW, Steadman TM. Tolerability and efficacy of clozapine combined with lithium in schizophrenia and schizoaffective disorder. J Clin Psychopharmacol 2003;23:223–8.

300. Jackson CW, Markowitz JS, Brewerton TD. Delirium associated with clozapine and benzodiazepine combinations. Ann Clin Psychiatry 1995;7(3):139–41.

301. Cobb CD, Anderson CB, Seidel DR. Possible interaction between clozapine and lorazepam. Am J Psychiatry 1991;148(11):1606–7.

302. Taylor D, Bodani M, Hubbeling A, Murray R. The effect of nefazodone on clozapine plasma concentrations. Int Clin Psychopharmacol 1999;14(3):185–7.

303. Khan AY, Preskorn SH. Increase in plasma levels of clozapine and norclozapine after administration of nefazodone. J Clin Psychiatry 2001;62:375–6.

304. Frick A, Kopitz J, Bergemann N. Omeprazole reduces clozapine plasma concentrations. Pharmacopsychiatry 2003;36:121–3.

305. Wetzel H, Anghelescu I, Szegedi A, Wiesner J, Weigmann H, Harter S, Hiemke C. Pharmacokinetic interactions of clozapine with selective serotonin reuptake inhibitors: differential effects of fluvoxamine and paroxetine in a prospective study. J Clin Psychopharmacol 1998;18(1):2–9.

306. Behar D, Schaller JL. Topiramate leukopenia on clozapine. Eur Child Adolesc Psychiatry 2004;13:51–2.

307. Joos AA, Frank UG, Kaschka WP. Pharmacokinetic interaction of clozapine and rifampicin in a forensic patient with an atypical mycobacterial infection. J Clin Psychopharmacol 1998;18(1):83–5.

308. Rostami-Hodjegan A, Amin AM, Spencer EP, Lennard MS, Tucker GT, Flanagan RJ. Influence of dose, cigarette smoking, age, sex, and metabolic activity on plasma clozapine concentrations: a predictive model and nomograms to aid clozapine dose adjustment and to assess compliance in individual patients. J Clin Psychopharmacol 2004;24:70–8.

309. Greenwood-Smith C, Lubman DI, Castle DJ. Serum clozapine levels: a review of their clinical utility. J Psychopharmacol 2003;17:234–8.

310. Ulrich S, Baumann B, Wolf R, Lehmann D, Peters B, Bogerts B, Meyer FP. Therapeutic drug monitoring of clozapine and relapse–a retrospective study of routine clinical data. Int J Clin Pharmacol Ther 2003;41:3–13.

311. Ulrich S, Wolf R, Staedt J. Serum level of clozapine and relapse. Ther Drug Monit 2003;25:252–5.

Droperidol

General Information

Droperidol is a butyrophenone with actions similar to those of haloperidol.

Of 20 volunteers who took droperidol 5 mg orally in orange juice, none had a neutral or pleasant experience (1). All reported restlessness, 17 felt sedated, and 11 reported dysphoria, the onset being relatively immediate; one subject broke down in tears within an hour of taking droperidol. Suicidal feelings emerged acutely in two subjects and were entertained in two more subjects. Among other adverse events were skin hypersensitivity ($n = 5$), aching in the muscles ($n = 6$), wheezing consistent with respiratory dyskinesia ($n = 1$), change in voice quality ($n = 1$), and marked rhinorrhea ($n = 1$). Mental effort was difficult, and all subjects reported some problems with concentration.

Droperidol 5–7.5 mg given during induction of anesthesia was associated with impaired well-being scores 6 hours postoperatively in a randomized double-blind comparison of similar doses of droperidol ($n = 78$) and midazolam ($n = 72$) for preventing postoperative nausea and vomiting (2).

With regard to the benefit to harm balance of the use of droperidol as an antiemetic, it has been suggested that the acceptable risk for antiemetic drug administration be established and that prospective data be collected to establish the risk associated not only with the administration of droperidol but also with other commonly used antiemetics (3). However, according to the FDA (4), a causal relation between the drug and an adverse effect need not be established, and reasonable evidence of an association requires including a warning in the drug labelling.

Comparative studies

Prophylactic intravenous droperidol (10, 20, 40, or 80 micrograms/kg) dose-dependently reduced postoperative nausea and vomiting without increasing the time to discharge in 82 children who underwent strabismus surgery (5). There were no particular adverse effects, but sedation scores were higher in those who received the higher doses.

In contrast, the addition of droperidol (2.5 mg bd for 5 days) to granisetron (3 mg bd on the first day) and dexamethasone (16 mg bd on the first day, 8 mg bd on days 2 and 3, and 4 mg bd on days 4 and 5) did not reduce the delayed emesis induced by high-dose cisplatin in a double-blind, randomized, parallel study in 180 patients with lung cancer receiving chemotherapy (6). The incidence of sleepiness with droperidol was higher (69% versus 30%). A meta-analysis of 33 trials, with data from 3447 patients, showed that when only two drugs were used (a 5-HT3 receptor antagonist plus either droperidol or dexamethasone), both were similar and significantly more effective than the 5-HT3 antagonist alone for prophylaxis of postoperative nausea and vomiting (7). Adverse effects were reported in seven studies involving the combination with droperidol and in 13 studies involving the combination with dexamethasone. The most commonly reported adverse effects were dizziness, headache, and drowsiness and the incidences were not different across the groups.

Placebo-controlled studies

In a randomized, double-blind, dose-ranging study in 305 adults receiving droperidol 0.1, 2.75, 5.5, and 8.25 mg for the acute treatment of migraine, the number of patients who achieved a pain-free response at 2 hours after treatment was significantly greater than with placebo for droperidol 2.75, 5.5, and 8.25 mg (8). The most frequent adverse events were akathisia and weakness, and adverse events were dose related. Anorexia, anxiety, somnolence, tremor, and confusion were also reported. No patient had QT interval prolongation.

In a randomized, placebo-controlled trial in 140 patients a combination of metoclopramide 10 mg and droperidol 1.25 mg or two doses of droperidol provided a more effective antiemetic effect than metoclopramide alone (9). The level of *sedation* was significantly greater in the patients who received two doses of droperidol (8/35) and metoclopramide followed by droperidol (7/35) than in those who received only placebo (0/35) or metoclopramide followed by placebo (0/35).

Organs and Systems

Cardiovascular

Droperidol has been associated with QT interval prolongation (SED-14, 141) (10–12) and torsade de pointes has been reported (13).

- A 59-year-old woman with no history of cardiac problems, except for hypertension, who was taking amlodipine 5 mg qds, cyclobenzaprine 10 mg qds, and cotriamterzide 37.5 + 25 mg qds, and who had a QT_c interval of 497 ms, was given intravenous droperidol 0.625 mg and metoclopramide 10 mg 45 minutes before surgery. About 1.75 hours after surgery she developed a polymorphic ventricular tachycardia with findings consistent with torsade de pointes, which resolved with defibrillation.

In late 2001 the FDA decided to introduce a "black box" warning regarding the use of droperidol because of its potential cardiac effects (QT prolongation leading to torsade de pointes and death). This decision, based mainly on post-marketing surveillance (MedWatch program and other relevant sources), has produced several reactions. The authors of a review (14) of three clinical studies, one published abstract, and seven case reports, as well as MedWatch reports, stressed the following points: there are several risk factors for dysrhythmias, including underlying illnesses, other drug exposures, different clinical settings (including surgery, emergencies, and psychiatry), and high doses of the drug (up to 100 mg intramuscularly). Long and widespread clinical use is also mentioned. They found no convincing evidence of a causal relation between droperidol and serious cardiac events. Others reached the same conclusion after reviewing 10 cases, reported in the FDA database, that were possibly related to the administration of droperidol in doses of 1.25 or less (15).

A review of the cases in which serious cardiovascular events were probably related to droperidol at doses of 1.25 mg/day or less showed, according to the authors, that there are many confounding factors, such as the concomitant use of drugs that can cause QT prolongation and other risk factors (16) (17). It has also been suggested that using the more expensive 5-HT$_3$ receptor antagonists as first-line agents for antiemetic prophylaxis has significant cost implications, with no evidence that they are any safer than droperidol (18). On the other hand, several issues have been raised in defence of the FDA decision (19); it is assumed, for instance, that cardiovascular events would be more likely to occur in those with other risk factors, since such factors would be expected co-variates of droperidol-induced dysrhythmias.

Nervous system

Intramuscular droperidol 2.5 mg was used to treat 23 consecutive patients with acute migraine who had not responded to other drugs (20). If no relief was achieved by 30–60 minutes after treatment, and no significant adverse effects were reported, a second dose of droperidol 2.5 mg was given. Varying degrees of akathisia after treatment were reported by six patients. Similarly, in a retrospective series of 37 patients who received droperidol 2.5 mg for migraine, 3 developed mild akathisia and 5 had drowsiness (21).

Balance disturbances have been described with droperidol in 120 women undergoing gynecological dilatation and curettage, who were randomly assigned to receive either 0.9% saline (placebo) or droperidol 0.625 mg intravenously before surgery (22). The change in body sway from the baseline before anesthesia was significantly greater after droperidol (61%) than after placebo (33%).

In a randomized, double-blind study in 96 patients with uncomplicated headache, intravenous droperidol 2.5 mg produced a similar reduction in headache as intravenous prochlorperazine 10 mg. There was a lower incidence of akathisia at 24 hours after discharge in those who were given droperidol (1/40 versus 6/43 of those given prochlorperazine), but this did not reach significance (23).

Pancreas

Droperidol is used as an adjunct in conscious sedation for endoscopic procedures. In one study, basal biliary sphincter pressures measured in 35 patients before and after droperidol were 56 and 48 mmHg; the basal pancreatic sphincter pressures measured in 22 patients before and after droperidol were 92 and 67 mmHg (24). However, in another study basal pressures of the biliary sphincter and of the pancreatic sphincter were not significantly altered by droperidol (25).

Death

There have been two further reports of sudden death after the use of droperidol to sedate agitation secondary to cocaine and phencyclidine intoxication (26). Both patients were restrained by the police and were then given droperidol, either 5 mg (a 33-year-old obese man) or 10 mg (a 22-year-old man). The first patient stopped breathing 10–15 minutes later, while being transported to the emergency department; he was pulseless and couldn't be resuscitated. The other patient was unresponsive on arrival at the emergency department, with agonal respirations and no detectable pulse; after 30 minutes of resuscitative efforts he was pronounced dead.

Drug–Drug Interactions

Fentanyl

An exception to the relatively safe use of high-potency agents has been noted in the combination of droperidol with the narcotic fentanyl, which can cause marked hypotension (27).

The addition of droperidol 2.5 mg to fentanyl 0.4 mg in 40 ml of 0.125% bupivacaine lowered the incidence of postoperative nausea and vomiting compared with a solution without droperidol or with butorphanol added instead in patients undergoing anorectal surgery in a prospective randomized, single-blind study (28).

Fentanyl plus droperidol (neuroleptanalgesia) was more effective than morphine in relieving anginal pain during unstable angina. However, the patients who received the neuroleptanalgesia also had longer hospital stays, because of significantly more cardiac instability and anginal episodes, and a higher total mortality (29).

Morphine

A prospective, randomized, double-blind study of 97 women investigated whether droperidol alleviated the adverse effects of epidural morphine after cesarean section (30). All groups received morphine 5 mg epidurally on delivery, accompanied by no droperidol, or droperidol 2.5 mg epidurally, or droperidol 2.5 mg intravenously. Pruritus occurred in 70% of patients, starting at 6 hours after epidural morphine, peaking at 17 hours, and with no significant difference between the different treatment regimens. Nausea and vomiting were significantly reduced by intravenous droperidol, but not by epidural droperidol. The authors concluded that droperidol acts systemically to counter the adverse effects of epidural morphine but is not entirely effective, and they suggested that its failure to alleviate pruritus may have been due to the fact that they used larger doses of morphine than some other investigators.

Droperidol 0.5 micrograms reduced the need for postoperative morphine delivered via a patient-controlled analgesia device (31). At these doses it was non-sedating and caused no dyskinetic movements.

Neuromuscular blocking agents

Animal studies suggest that large doses of pethidine and droperidol can augment the myoneural effects of neuromuscular blocking agents (32).

References

1. Healy D, Farquhar G. Immediate effects of droperidol. Human Psychopharmacol 1998;13:113–20.
2. Eberhart LH, Seeling W. Droperidol-supplemented anaesthesia decreases post-operative nausea and vomiting but impairs post-operative mood and well-being. Eur J Anaesthesiol 1999;16(5):290–7.
3. Scuderi PE. Droperidol: many questions, few answers. Anesthesiology 2003;98:289–90.
4. Meyer RJ. FDA "black box" labeling. [Comment on: Kao LW, Kirk MA, Evers SJ, Rosenfeld SH. Droperidol, QT prolongation, and sudden death: what is the evidence? Ann Emerg Med 2003;41:546-58.] Ann Emerg Med 2003;41:559–60.
5. Stead SW, Beatie CD, Keyes MA, Isenberg SJ. Effects of droperidol dosage on postoperative emetic symptoms following pediatric strabismus surgery. J Clin Anesth 2004;16:34–9.
6. Minegishi Y, Ohmatsu H, Miyamoto T, Niho S, Goto K, Kubota K, Kakinuma R, Kudoh S, Nishiwaki Y. Efficacy of droperidol in the prevention of cisplatin-induced delayed emesis: a double-blind, randomised parallel study. Eur J Cancer 2004;40:1188–92.
7. Habib AS, El-Moalem HE, Gan TJ. The efficacy of the 5-HT$_3$ receptor antagonists combined with droperidol for PONV prophylaxis is similar to their combination with dexamethasone. A meta-analysis of randomized controlled trials. Can J Anesth 2004;51:311–9.
8. Silberstein SD, Young WB, Mendizabal JE, Rothrock JF, Alam AS. Acute migraine treatment with droperidol. A randomized, double-blind, placebo-controlled trial. Neurology 2003;60:315–21.
9. Nesek-Adam V, Grizelj-Stojcic E, Mrsic V, Smiljanic A, Rasic Z, Cala Z. Prohylactic antiemetics for laparoscopic cholecystectomy, droperidol, metoclopramide, and droperidol plus metoclopramide. J Laparoendoscopic Advanced Surg Tech 14;4:212–18.
10. Warner JP, Barnes TR, Henry JA. Electrocardiographic changes in patients receiving neuroleptic medication. Acta Psychiatr Scand 1996;93(4):311–3.
11. Iwahashi K. Significantly higher plasma haloperidol level during cotreatment with carbamazepine may herald cardiac change. Clin Neuropharmacol 1996;19(3):267–70.
12. Reilly JG, Ayis SA, Ferrier IN, Jones SJ, Thomas SH. QTc-interval abnormalities and psychotropic drug therapy in psychiatric patients. Lancet 2000;355(9209):1048–52.
13. Michalets EL, Smith LK, Van Tassel ED. Torsade de pointes resulting from the addition of droperidol to an existing cytochrome P450 drug interaction. Ann Pharmacother 1998;32(7–8):761–5.
14. Kao LW, Kirk MA, Evers SJ, Rosenfeld SH. Droperidol, QT prolongation, and sudden death: what is the evidence? Ann Emerg Med 2003;41:546–58.
15. Habib AS, Gan TJ. Food and drug administration black box warning on the perioperative use of droperidol: a review of the cases. Anesth Analg 2003;96:1377–9.
16. Gan TJ. "Black box" warning on droperidol: a report of the FDA convened expert panel. Anesth Analg 2004;98:1809.
17. van Zwieten K, Mullins ME, Jang T. Droperidol and the black box warning. Ann Emerg Med 2004;43:139–40.
18. Habib AS, Gan TJ. Safety of patients reason for FDA black box warning on droperidol. Anesth Analg 2004;98:551–2.
19. Shafer SL. Safety of patients reason for FDA black box warning on droperidol. Anesth Analg 2004;98:551–2.
20. Mendizabal JE, Watts JM, Riaz S, Rothrock JF. Open-label intramuscular droperidol for the treatment of refractory headache: a pilot study. Headache 1999;10:55–7.
21. Richman PB, Reischel U, Ostrow A, Irving C, Ritter A, Allegra J, Eskin B, Szucs P, Nashed AH. Droperidol for acute migraine headache. Am J Emerg Med 1999;17(4):398–400.
22. Song D, Chung F, Yogendran S, Wong J. Evaluation of postural stability after low-dose droperidol in outpatients undergoing gynaecological dilatation and curettage procedure. Br J Anaesth 2002;88(6):819–23.
23. Weaver CS, Jones JB, Chisholm CD, Foley MJ, Giles BK, Somerville GG, Brizendine EJ, Cordell WH. Droperidol vs prochlorperazine for the treatment of acute headache. J Emerg Med 2004;26:145–50.
24. Wilcox CM, Linder J. Prospective evaluation of droperidol on sphincter of Oddi motility. Gastrointest Endosc 2003;58:483–7.
25. Fogel EL, Sherman S, Bucksot L, Shelly L, Lehman GA. Effects of droperidol on the pancreatic and biliary sphincters. Gastrointest Endosc 2003;58:488–92.
26. Cox RD, Koelliker DE, Bradley KG. Association between droperidol use and sudden death in two patients intoxicated with illicit stimulant drugs. Vet Hum Toxicol 2004;46:21–3.
27. Mandelstam JP. An inquiry into the use of Innovar for pediatric premedication. Anesth Analg 1970;49(5):746–50.
28. Boros M, Chaudhry IA, Nagashima H, Duncalf RM, Sherman EH, Foldes FF. Myoneural effects of pethidine and droperidol. Br J Anaesth 1984;56(2):195–202.
29. Kotake Y, Matsumoto M, Ai K, Morisaki H, Takeda J. Additional droperidol, not butorphanol, augments epidural fentanyl analgesia following anorectal surgery. J Clin Anesth 2000;12(1):9–13.
30. Burduk P, Guzik P, Piechocka M, Bronisz M, Rozek A, Jazdon M, Jordan MR. Comparison of fentanyl and droperidol mixture (neuroleptanalgesia II) with morphine on clinical outcomes in unstable angina patients. Cardiovasc Drugs Ther 2000;14(3):259–69.
31. Sanansilp V, Areewatana S, Tonsukchai N. Droperidol and the side effects of epidural morphine after cesarean section. Anesth Analg 1998;86(3):532–7.
32. Lo Y, Chia Y-Y, Liu K, Ko N-H. Morphine sparing with droperidol in patient controlled analgesia. J Clin Anaesth 2005;17:271–2.

Flupentixol

General Information

Flupentixol is a thioxanthene neuroleptic drug.

Organs and Systems

Nervous system

Autoamputation of the tongue in a patient who was taking flupentixol has been explained as being secondary to an acute, atypical, neuroleptic drug-induced orolingual dyskinesia (1).

- A 21-year-old man, who had been mentally retarded from birth, was given intramuscular flupentixol 50 mg/day and oral diazepam 7.5 mg/day for disruptive and inappropriate behavior, and 84 hours later began chewing his tongue repetitively, with resultant edema and bleeding from multiple lacerations. The biting abated within 2 hours after intravenous biperiden 5 mg and diazepam 5 mg; ampicillin, cloxacillin, and metronidazole were also given. One week after admission to hospital, he had an autoamputation of the distal one-third of his tongue.

Reference

1. Pantanowitz L, Berk M. Auto-amputation of the tongue associated with flupenthixol induced extrapyramidal symptoms. Int Clin Psychopharmacol 1999;14(2):129–31.

Fluphenazine

General Information

Fluphenazine is a phenothiazine neuroleptic drug.

In a double-blind comparison of a group of stabilized outpatients taking a low dosage of fluphenazine enanthate (1.25–5 mg every 2 weeks) with a group taking a standard dosage (12.5–50 mg every 2 weeks), relapse rates were higher in the low-dose group (56%) than in the standard-dose group (7%) (1). However, patients in the low-dose group had a better outcome in terms of some measures of psychosocial adjustment and family satisfaction. Patients in the low-dose group had fewer signs of tardive dyskinesia, and relapses led less often to re-admission to hospital; they also responded more readily to treatment with temporary increases in medication than patients treated with standard doses.

In a comparison of a low dose of fluphenazine decanoate (5 mg) with a standard dose (25 mg) every 2 weeks, there was no significant difference in relapse at 1 year (2), nor was there a difference in survival at 1 year, but at 2 years, survival was significantly better with the 25 mg dose (64%) than with the 5 mg dose (31%) (3).

Organs and Systems

Gastrointestinal

Neuroleptic drugs, particularly phenothiazine derivatives, have been reported to cause colitis (SED-14, 150) (4) (SEDA-20, 43). The mechanism was thought to be an anticholinergic effect. Two other cases have been reported (5,6), one with positive rechallenge.

- A 41-year-old man developed acute abdominal pain with profuse diarrhea and fever (39°C) while receiving intramuscular fluphenazine decanoate 125 mg once every 3 weeks. During the previous 3 months he had also taken oral alimemazine 50 mg/day, levomepromazine 50 mg/day, and amitriptyline 100 mg/day. Colonoscopy showed necrotic ulcers in the mucosa of the sigmoid and descending colon. After three weeks of parenteral nutrition, there was a marked reduction in the colonic lesions and he recovered. Levomepromazine 50 mg/day and fluphenazine decanoate 100 mg/day were reintroduced. Two days later he complained again of abdominal pain, and tomodensitometry confirmed distension.

Second-Generation Effects

Fetotoxicity

Severe rhinorrhea and respiratory distress occurred in a neonate exposed to fluphenazine hydrochloride prenatally (7).

Drug–Drug Interactions

Vitamin C

In one patient, ascorbic acid reduced serum fluphenazine concentrations, perhaps by liver enzyme induction or by reduced absorption (8).

References

1. Kane JM, Rifkin A, Woerner M, Reardon G, Sarantakos S, Schiebel D, Ramos-Lorenzi J. Low-dose neuroleptic treatment of outpatient schizophrenics. I. Preliminary results for relapse rates. Arch Gen Psychiatry 1983;40(8):893–6.
2. Marder SR, Van Putten T, Mintz J, McKenzie J, Lebell M, Faltico G, May PR. Costs and benefits of two doses of fluphenazine. Arch Gen Psychiatry 1984;41(11):1025–9.
3. Marder SR, Van Putten T, Mintz J, Lebell M, McKenzie J, May PR. Low- and conventional-dose maintenance therapy with fluphenazine decanoate. Two-year outcome. Arch Gen Psychiatry 1987;44(6):518–21.
4. Larrey D, Lainey E, Blanc P, Diaz D, David R, Biaggi A, Barneon G, Bottai T, Potet F, Michel H. Acute colitis associated with prolonged administration of neuroleptics. J Clin Gastroenterol 1992;14(1):64–7.

5. Capron M, Lafitte B, Benedit M, Camard CN, Nicolas F, Beligon C, Baillet C. Colite nécrosante chez un homme de 29 ans sous forte dose de neuroleptiques. [Necrotizing colitis in a 29-year-old man following high doses of neuroleptics.] Reanim Urgences 1999;8:701–4.
6. Filloux MC, Marechal K, Bagheri H, Morales J, Nouvel A, Laurencin G, Montastruc JL. Phenothiazine-induced acute colitis: a positive rechallenge case report. Clin Neuropharmacol 1999;22(4):244–5.
7. Nath SP, Miller DA, Muraskas JK. Severe rhinorrhea and respiratory distress in a neonate exposed to fluphenazine hydrochloride prenatally. Ann Pharmacother 1996;30(1):35–7.
8. Dysken MW, Cumming RJ, Channon RA, Davis JM. Drug interaction between ascorbic acid and fluphenazine. JAMA 1979;241(19):2008.

Haloperidol

General Information

Haloperidol is a butyrophenone neuroleptic drug. The enzymes involved in its biotransformation include oxidative cytochrome P450 isozymes, carbonyl reductase, and uridine diphosphoglucose glucuronosyltransferase (1). It is mainly cleared by glucuronidation.

There has been a randomized, placebo-controlled comparison of haloperidol (mean dose 1.8 mg/day), trazodone (200 mg/day), and behavior management techniques in 149 patients with Alzheimer's disease (2). Although 34% of the subjects improved relative to baseline, there were no significant differences in outcomes among the four arms; there were significantly fewer cases of bradykinesia and parkinsonian gait in those given behavioral therapy. These results suggest that other treatments for agitation in dementia need to be considered and evaluated; likewise, they are consistent with the results of a meta-analysis (3) and a clinical trial (4).

Haloperidol 2–3 mg/day and 0.50–0.75 mg/day have been compared in 71 outpatients with Alzheimer's disease (5). After 12 weeks, there was a favorable therapeutic effect of haloperidol 2–3 mg/day, although 25% of the patients developed moderate to severe extrapyramidal signs.

A thorough review of the pharmacokinetics of haloperidol, with special emphasis on interactions, has been published (1).

Organs and Systems

Cardiovascular

Intravenous haloperidol is often prescribed to treat agitation, and torsade de pointes has on occasions occurred (SEDA-20, 36). In a cross-sectional cohort study QT_c intervals were measured before the intravenous administration of haloperidol plus flunitrazepam, and continuous electrocardiographic monitoring was performed for at least 8 hours after ($n = 34$) (6); patients who received only flunitrazepam served as controls. The mean QT_c interval after 8 hours in those who were given haloperidol was longer than in those who were given flunitrazepam alone; four patients given haloperidol had a QT_c interval of more than 500 ms after 8 hours. However, none developed ventricular tachydysrhythmias.

In a case-control study, haloperidol-induced QT_c prolongation was associated with torsade de pointes (7). The odds ratio of developing torsade de pointes in a patient with QT_c prolongation to over 550 ms compared with those with QT_c intervals shorter than 550 ms was 33 (95% CI = 6, 195). The sample consisted of all critically ill adult patients in medical, cardiac, and surgical intensive care units at a tertiary hospital who received intravenous haloperidol and had no metabolic, pharmacological, or neurological risk factors known to cause torsade de pointes, or if the dysrhythmia developed more than 24 hours after intravenous haloperidol. Of 223 patients who fulfilled the inclusion criteria, eight developed torsade de pointes. A group of 41 patients, randomly selected from the 215 without torsade de pointes, served as controls. The length of hospital stay after the development of haloperidol-associated torsade de pointes was significantly longer than that after the maximum dose of intravenous haloperidol in the control group. The overall incidence of torsade de pointes was 3.6% and 11% in patients who received intravenous haloperidol 35 mg or more over 24 hours.

Several cases of torsade de pointes have been reported with intravenous haloperidol used with low-dose oral haloperidol (8).

The effects of haloperidol dose and plasma concentration and CYP2D6 activity on the QT_c interval have been studied in 27 Caucasian patients taking oral haloperidol (aged 23–77 years, dosages 1.5–30 mg/day) (9). Three patients had a QT_c interval longer than 456 ms, which can be considered as the cut-off value for a risk of cardiac dysrhythmias. There was no correlation between QT_c interval and haloperidol dosage or plasma concentrations or CYP2D6 activity.

Asystolic cardiac arrest has been reported after intravenous haloperidol (10).

In one case, the use of carbamazepine and haloperidol led to prolongation of the QT_c interval and cardiac complications (11).

- A 75-year-old man developed ventricular fibrillation and cardiac arrest after intravenous haloperidol (12). His past history included coronary bypass surgery and coronary angioplasty. As he continued to have severe chest pain, emergency angioplasty was performed. On day 3 he received haloperidol by infusion 2 mg/hour, with 2 mg increments every 10 minutes (up to 20 mg in 6 hours) as needed for relief of agitation. Before haloperidol, his QT_c interval was normal; after haloperidol it increased to 570 ms. The next day he developed ventricular fibrillation. Subsequent electrocardiograms showed prolonged QT_c intervals of 579 and 615 ms, and haloperidol was withdrawn; the QT_c returned to normal.

- A 76-year-old man developed torsade de pointes while taking tiapride 300 mg/day; the QT_c interval 1 day after starting treatment was 600 ms; the dysrhythmia resolved when tiapride was withdrawn (13).

- A 39-year-old man died suddenly 1 hour after taking a single oral dose of haloperidol 5 mg (14). He had myasthenia, alcoholic hepatitis, and electrolyte abnormalities due to inadequate nutritional state. His electrocardiogram showed prolongation of the QT_c interval (460 ms). Autopsy showed a cardiomyopathy but no explanation for sudden death.

However, malignant dysrhythmias can occur without changes in the QT interval (SEDA-24, 54).

- A 64-year-old woman underwent coronary artery bypass surgery and was given intravenous haloperidol for agitation and to avoid postoperative delirium; she developed torsade de pointes (15).
- Asystolic cardiac arrest occurred in a 49-year-old woman after she had received haloperidol 10 mg intramuscularly for 2 days; no previous QT_c prolongation had been observed (16).

Respiratory

A single case of fatal pulmonary edema was reported in 1982 in association with haloperidol (17).

Nervous system

Three cases of radial nerve palsy were reported in demented elderly patients confined to wheelchairs who were treated with haloperidol. The combination of extrapyramidal and sedative adverse effects, added to wheelchair confinement, may have resulted in pressure on the upper arm with subsequent neuropathy (18).

Extrapyramidal effects

- A 28-year-old woman simultaneously developed four types of tardive extrapyramidal symptoms (dystonia, dyskinesia, choreoathetotic movements, and myoclonus) while taking haloperidol; the symptoms were subsequently relieved by the use of low-dose risperidone (3 mg/day) (19).

Akathisia

A nocturnal eating/drinking syndrome secondary to neuroleptic drug-induced restless legs syndrome has been attributed to low-dose haloperidol in a 51-year-old schizophrenic woman (20).

Neuroleptic malignant syndrome

Neuroleptic malignant syndrome has been reported with haloperidol (21).

- A 21-year-old Turkish man with succinic semialdehyde dehydrogenase deficiency and mental retardation developed neuroleptic malignant syndrome after a single dose of haloperidol 10 mg for anxiety and agitation, having never received a neuroleptic drug before.

Several reports have suggested a higher incidence and severity of extrapyramidal symptoms during haloperidol treatment in congenitally poor metabolizers of substrates

of CYP2D6 and a patient with a poor metabolizer polymorphism of the CYP2D6 gene developed neuroleptic malignant syndrome after receiving haloperidol (22). All frequent polymorphisms of CYP2D6 were therefore investigated in the second patient, who was a carrier of the wild-type genotype CYP2D6 *1/ *1, which is common in subjects of Caucasian origin. The authors concluded that a genetic defect of haloperidol metabolism via CYP2D6 was unlikely as a reason for the neuroleptic malignant syndrome in this case.

Tardive Tourette's syndrome

Tardive Tourette's syndrome is characterized by multiple motor and vocal tics.

- A 48-year-old woman, who had been taking haloperidol 10 mg/day for 8 years, suddenly stopped taking it and about 2 weeks later developed a Tourette-like syndrome (23). Her symptoms did not respond to increased doses of typical or atypical neuroleptic drugs, but she derived significant sustained improvement from clonazepam 3 mg/day. At review 6 months later, her symptoms remained controlled, with just occasional facial grimaces, and there was no recurrence of her psychotic symptoms.

Endocrine

The prevalence of hyperprolactinemia in patients with chronic schizophrenia taking long-term haloperidol has been studied in 60 patients in Korea (28 women; illness mean duration, 15.5 years) (24). There was hyperprolactinemia, defined as a serum prolactin concentration over 20 ng/ml in men and 24 ng/ml in women, in 40; the prevalence of hyperprolactinemia in women (93%) was significantly higher than in men (47%). There was also a significant correlation between haloperidol dose and serum prolactin concentration in women, but not in men.

The relation of prolactin concentrations and certain adverse events has been explored in large randomized, double-blind studies. In 813 women and 1912 men, haloperidol produced dose-related increases in plasma prolactin concentrations in men and women, but they were not correlated with adverse events such as amenorrhea, galactorrhea, or reduced libido in women or with erectile dysfunction, ejaculatory dysfunction, gynecomastia, or reduced libido in men (25).

No further rise in plasma prolactin concentration was observed with dosages of haloperidol over 100 mg/day, which was explained as being related to saturation of the pituitary dopamine receptors by a modest amount of haloperidol (26).

The time-course of the prolactin increase has been examined in 17 subjects whose prolactin concentrations rose during the first 6–9 days of treatment with haloperidol (27). The increase was followed by a plateau that persisted, with minor fluctuations, throughout the 18 days of observation. Patients whose prolactin concentrations increased above 77 ng/ml ($n = 2$) had hypothyroidism, and it is known that TRH (thyrotropin) stimulates the release of prolactin (28).

The effects of haloperidol and quetiapine on serum prolactin concentrations have been compared in 35 patients with schizophrenia during a drug-free period for at least 2 weeks in a randomized study (29). There was no significant difference in prolactin concentration between the groups at the start of the study; control prolactin concentrations were significantly lower with quetiapine than with haloperidol. Two patients taking haloperidol had galactorrhea related to hyperprolactinemia.

Metabolism

Glucose metabolism has been studied in 10 patients taking haloperidol; none had impaired glucose tolerance and only one had a glycemic peak delay (30).

Liver

Occasional reports of cholestatic jaundice with haloperidol have been published (31,32).

Urinary tract

Retroperitoneal fibrosis has been attributed to haloperidol; since this condition affects the kidney, it should be differentiated from other causes of obstructive uropathy (33).

Skin

Haloperidol has been reported to have caused skin rashes.

- A 41-year-old man who had been taking weekly methotrexate 15 mg for 10 months for psoriasis started to take haloperidol 1.5 mg bd for a psychotic illness, and 2 weeks later developed sudden redness and swelling of the face and hands accompanied by redness and watering of both eyes (34). He had diffuse erythema, edema, scaling, and erosions over his face, the anterior aspect of the neck, and the backs of both hands. A skin biopsy showed parakeratosis, acanthosis, spongiosis, focal epidermal cell degeneration, and dermal edema, accompanied by a moderate lymphomononuclear infiltrate, consistent with subacute dermatitis. A diagnosis of pellagra-like photosensitivity dermatitis, caused by combined deficiency of niacin, riboflavin, and other water-soluble vitamins, probably precipitated by haloperidol, was therefore considered. Haloperidol was withdrawn, vitamins were administered, and the condition resolved in the next 5 days.
- A 55-year-old man who had been treated for a manic-depressive disorder for about 5 years developed toxic epidermal necrolysis after being given carbamazepine and haloperidol (35).

Sexual function

Sexual dysfunction can occur with neuroleptic drugs (SED-14, 149; SEDA-22, 54; 36,37).

- A 49-year-old man with bipolar disorder had erectile dysfunction shortly after starting to take haloperidol 50 mg/day and lithium 1500 mg/day (38). Before this he had had normal sexual function. After 2 months, the dosage of haloperidol was reduced to 20 mg/day, but the sexual dysfunction persisted and did not improve with sildenafil. He was then switched to olanzapine 10 mg/day and lithium 1200 mg/day and 1 week later his sexual dysfunction had disappeared.

Second-Generation Effects

Pregnancy

Although studies have failed to identify an increased rate of malformations with haloperidol, isolated cases have been reported (SED-14, 152) (39) (SEDA-22, 54) (40,41).

Teratogenicity

Several instances of limb reduction after the use of haloperidol during pregnancy suggest that it would be prudent not to use this drug during the first trimester, the period of limb development (39). There is also reason to argue that prenatally administered drugs of this class influence the offspring after the drug has been eliminated, and can produce behavioral teratogenicity; since neurotransmitter systems continue to develop long after birth, such drugs might influence behavior in an adverse way over a very long time (42).

Fetotoxicity

- A neonate had severe hypothermia after antenatal exposure to haloperidol (43). He weighed 3710 g at birth and did not need resuscitation; his axillary temperature was 35°C and he had severe generalized hypotonia. His temperature rose to 36.5°C after 6 hours of rewarming with an overhead radiant heater.

Drug Administration

Drug additives

General allergic reactions have also been reported following injections of parabens-containing formulations of lidocaine and hydrocortisone and after oral use of barium sulfate contrast suspension, haloperidol syrup, and an antitussive syrup, all of which contained parabens (44).

Drug–Drug Interactions

General

Based on published data in humans, concomitant medications were classified as potential inhibitors (cimetidine, fluoxetine, levopromazine, paroxetine, and thioridazine) or inducers of haloperidol metabolism (carbamazepine, phenobarbital, and phenytoin).

Alcohol

Haloperidol increased blood alcohol concentrations (45).

Anticoagulants

Haloperidol has been reported to lower anticoagulant effectiveness through enzyme induction (46).

Antiparkinsonian drugs

In a study of the interindividual variation in serum haloperidol concentrations, Japanese patients taking concomitant antiparkinsonian drugs (n = 145) had a mean haloperidol concentration:dose ratio that was 25% higher than in patients who were not taking antiparkinsonian drugs (n = 95) (47).

Buspirone

Buspirone increases haloperidol concentrations in some patients (SEDA-21, 39).

Carbamazepine

Haloperidol inhibits the metabolism of carbamazepine. In Japanese patients with schizophrenia, serum concentrations of carbamazepine were about 40% lower in the absence of haloperidol (48).

Chlorpromazine

- A 40-year-old man with schizophrenia developed a raised plasma concentration of haloperidol in combination with chlorpromazine and during overlap treatment with clozapine (49). Like haloperidol, chlorpromazine is a competitive inhibitor of CYP2D6; however, clozapine appears to be largely metabolized by CYP1A2.

Fluvoxamine

The addition of low-dose fluvoxamine (50–100 mg/day) to neuroleptic drug treatment may improve the negative symptoms in patients with schizophrenia, but involves a risk of a drug interaction. In 12 in-patients with schizophrenia receiving 6 mg/day of haloperidol, incremental doses of fluvoxamine (25, 75, and 150 mg/day for 2 weeks each) respectively increased haloperidol plasma concentrations by 120%, 139%, and 160% of those before fluvoxamine co-administration; in spite of the increase, there were no particular adverse effects (50).

Itraconazole

Adverse effects can result from increased plasma concentrations of haloperidol during itraconazole treatment. This has been observed in 13 schizophrenic patients treated with haloperidol 12 or 24 mg/day who took itraconazole 200 mg/day for 7 days (51). Plasma concentrations of haloperidol were significantly increased and neurological adverse effects were more common. Itraconazole is a potent inhibitor of CYP3A4.

The effects of itraconazole 200 mg/day for 7 days on the steady-state plasma concentrations of haloperidol and its reduced metabolite have been investigated in schizophrenic patients receiving haloperidol 12 or 24 mg/day (51). Itraconazole significantly increased trough plasma concentrations of both haloperidol and reduced haloperidol (17 versus 13 ng/ml and 6.1 versus 4.9 ng/ml respectively). There was no change in clinical symptoms, but neurological adverse effects of haloperidol were significantly increased during itraconazole co-administration.

Similar findings in a similar study were reported for bromperidol and its reduced metabolite, although there were no differences in clinical symptoms or neurological adverse effects during concomitant itraconazole therapy (52).

Ketamine

The interaction of the dopamine antagonist haloperidol 5 mg orally with subanesthetic doses of ketamine has been studied in a placebo-controlled study in 20 healthy volunteers over 4 days (53). Haloperidol pretreatment reduced impairment of executive cognitive functions produced by ketamine and reduced the anxiogenic effects of ketamine. However, it failed to block the ability of ketamine to produce psychosis, perceptual changes, negative symptoms, or euphoria, and it increased the sedative and prolactin responses to ketamine. These results imply that ketamine may impair executive cognitive functions via dopamine receptor activation in the frontal cortex, but that the psychoactive effects of ketamine are not mediated via dopamine receptors, but rather via NMDA receptor antagonism.

Lithium

Persistent dysarthria with apraxia has been reported with a combination of lithium carbonate and haloperidol (54).

The neuroleptic malignant syndrome has been described when lithium was added to haloperidol (55).

- A pharmacodynamic drug interaction could not be excluded when a 60-year-old man developed delirium at a serum lithium concentration of 0.97 mmol/l when taking lithium and haloperidol (56).

Lorazepam

Several cases of torsade de pointes have been reported with intravenous haloperidol used with lorazepam to treat delirium (SEDA-18, 30) (SEDA-18, 47). Acid mucopolysaccharide deposition may be associated with neuroleptic drug treatment as a possible mechanism contributing to rare cardiovascular adverse events (57).

A thorough review of the pharmacokinetics of haloperidol, with special emphasis on interactions, has been published (58). The interactions include one with lorazepam.

Methyldopa

Dementia occurred when methyldopa was combined with haloperidol (59).

Nefazodone

Hepatic enzyme inhibition by nefazodone can significantly raise concentrations and toxicity of haloperidol (1).

Rifamycins

Blood concentrations of haloperidol are reduced during rifampicin administration, owing to shortening of the half-life of haloperidol (SEDA-12, 258) (60).

Valproate

An interaction between valproate and haloperidol has been reported (61).

- A 30-year-old man with bipolar affective disorder was given intravenous sodium valproate 20 mg/kg/day without any major problem. Later, he was given oral haloperidol 10 mg bd for persistent aggressive behavior and manic symptoms. The next day he developed a sense of imbalance and started swaying. He was drowsy and had cerebellar ataxia. Haloperidol was withdrawn and within 1 day his ataxia had disappeared completely in spite of continuing valproate.

Food–Drug Interactions

Grapefruit inhibits CYP3A4, but in one study there was no interaction with haloperidol (62).

Smoking

The extent to which smoking, the genotype for CYP2D6, and the concomitant use of enzyme inducers or inhibitors can explain variations in steady-state plasma concentrations of haloperidol has been evaluated in 92 patients (63). Smokers were treated with higher doses of haloperidol than non-smokers, supporting the view that smokers require larger doses of neuroleptic drugs than non-smokers to achieve therapeutic effects. Poor metabolizers had higher concentrations of haloperidol metabolites, but not of haloperidol, than extensive metabolizers. Altogether, the patients took about 150 different drugs; compared with the rest of the group, there was a tendency for patients taking inducers to be treated with slightly higher doses of haloperidol on average and to have lower dose-normalized plasma concentrations of haloperidol, although the differences were not significant.

References

1. Kudo S, Ishizaki T. Pharmacokinetics of haloperidol: an update. Clin Pharmacokinet 1999;37(6):435–56.
2. Teri L, Logsdon RG, Peskind E, Raskind M, Weiner MF, Tractenberg RE, Foster NL, Schneider LS, Sano M, Whitehouse P, Tariot P, Mellow AM, Auchus AP, Grundman M, Thomas RG, Schafer K, Thal LJ. Alzheimer's Disease Cooperative Study. Treatment of agitation in AD: a randomized, placebo-controlled clinical trial. Neurology 2000;55(9):1271–8.
3. Schneider LS, Pollock VE, Lyness SA. A metaanalysis of controlled trials of neuroleptic treatment in dementia. J Am Geriatr Soc 1990;38(5):553–63.
4. Sultzer DL, Gray KF, Gunay I, Berisford MA, Mahler ME. A double-blind comparison of trazodone and haloperidol for treatment of agitation in patients with dementia. Am J Geriatr Psychiatry 1997;5(1):60–9.
5. Devanand DP, Marder K, Michaels KS, Sackeim HA, Bell K, Sullivan MA, Cooper TB, Pelton GH, Mayeux R. A randomized, placebo-controlled dose-comparison trial of haloperidol for psychosis and disruptive behaviors in Alzheimer's disease. Am J Psychiatry 1998;155(11):1512–20.
6. Hatta K, Takahashi T, Nakamura H, Yamashiro H, Asukai N, Matsuzaki I, Yonezawa Y. The association between intravenous haloperidol and prolonged QT interval. J Clin Psychopharmacol 2001;21(3):257–61.
7. Sharma ND, Rosman HS, Padhi ID, Tisdale JE. Torsades de pointes associated with intravenous haloperidol in critically ill patients. Am J Cardiol 1998;81(2):238–40.
8. Jackson T, Ditmanson L, Phibbs B. Torsade de pointes and low-dose oral haloperidol. Arch Intern Med 1997;157(17):2013–5.
9. LLerena A, Berecz R, de la Rubia A, Dorado P. QTc interval lengthening and debrisoquine metabolic ratio in psychiatric patients treated with oral haloperidol monotherapy. Eur J Clin Pharmacol 2002;58(3):223–4.
10. Huyse F, van Schijndel RS. Haloperidol and cardiac arrest. Lancet 1988;2(8610):568–9.
11. Iwahashi K. Significantly higher plasma haloperidol level during cotreatment with carbamazepine may herald cardiac change. Clin Neuropharmacol 1996;19(3):267–70.
12. Douglas PH, Block PC. Corrected QT interval prolongation associated with intravenous haloperidol in acute coronary syndromes. Catheter Cardiovasc Interv 2000;50(3):352–5.
13. Iglesias E, Esteban E, Zabala S, Gascon A. Tiapride-induced torsade de pointes. Am J Med 2000;109(6):509.
14. Remijnse PL, Eeckhout AM, van Guldener C. Plotseling overlijden na eenmalige orale toediening van haloperidol. [Sudden death following a single oral administration of haloperidol.] Ned Tijdschr Geneeskd 2002;146(16):768–71.
15. Perrault LP, Denault AY, Carrier M, Cartier R, Belisle S. Torsades de pointes secondary to intravenous haloperidol after coronary bypass grafting surgery. Can J Anaesth 2000;47(3):251–4.
16. Johri S, Rashid H, Daniel PJ, Soni A. Cardiopulmonary arrest secondary to haloperidol. Am J Emerg Med 2000;18(7):839.
17. Mahutte CK, Nakasato SK, Light RW. Haloperidol and sudden death due to pulmonary edema. Arch Intern Med 1982;142(10):1951–2.
18. Sloane PD, McLeod MM. Radial nerve palsy in nursing home patients: association with immobility and haloperidol. J Am Geriatr Soc 1987;35(5):465–6.
19. Suenaga T, Tawara Y, Goto S, Kouhata SI, Kagaya A, Horiguchi J, Yamanaka Y, Yamawaki S. Risperidone treatment of neuroleptic-induced tardive extrapyramidal symptoms. Int J Psychiatry Clin Pract 2000;4:241–3.
20. Horiguchi J, Yamashita H, Mizuno S, Kuramoto Y, Kagaya A, Yamawaki S, Inami Y. Nocturnal eating/drinking syndrome and neuroleptic-induced restless legs syndrome. Int Clin Psychopharmacol 1999;14(1):33–6.
21. Neu P, Seyfert S, Brockmoller J, Dettling M, Marx P. Neuroleptic malignant syndrome in a patient with succinic semialdehyde dehydrogenase deficiency. Pharmacopsychiatry 2002;35(1):26–8.
22. Mihara K, Suzuki A, Kondo T, Yasui N, Furukori H, Nagashima U, Otani K, Kaneko S, Inoue Y. Effects of the CYP2D6*10 allele on the steady-state plasma concentrations of haloperidol and reduced haloperidol in Japanese patients with schizophrenia. Clin Pharmacol Ther 1999;65(3):291–4.
23. Reid SD. Neuroleptic-induced tardive Tourette treated with clonazepam. Clin Neuropharmacol 2004;27:101–4.
24. Jung DU, Seo YS, Park JH, Jeong CY, Conley RR, Kelly DL, Shim JC. The prevalence of hyperprolactinemia after long-term haloperidol use in patients with chronic schizophrenia. J Clin Psychopharmacol 2005;25:613–5.

25. Kleinberg DL, Davis JM, de Coster R, Van Baelen B, Brecher M. Prolactin levels and adverse events in patients treated with risperidone. J Clin Psychopharmacol 1999;19(1):57–61.

26. Zarifian E, Scatton B, Bianchetti G, Cuche H, Loo H, Morselli PL. High doses of haloperidol in schizophrenia. A clinical, biochemical, and pharmacokinetic study. Arch Gen Psychiatry 1982;39(2):212–5.

27. Spitzer M, Sajjad R, Benjamin F. Pattern of development of hyperprolactinemia after initiation of haloperidol therapy. Obstet Gynecol 1998;91(5 Pt 1):693–5.

28. Feek CM, Sawers JS, Brown NS, Seth J, Irvine WJ, Toft AD. Influence of thyroid status on dopaminergic inhibition of thyrotropin and prolactin secretion: evidence for an additional feedback mechanism in the control of thyroid hormone secretion. J Clin Endocrinol Metab 1980;51(3):585–9.

29. Atmaca M, Kuloglu M, Tezcan E, Canatan H, Gecici O. Quetiapine is not associated with increase in prolactin secretion in contrast to haloperidol. Arch Med Res 2002;33(6):562–5.

30. Chae BJ, Kang BJ. The effect of clozapine on blood glucose metabolism. Hum Psychopharmacol 2001;16(3):265–71.

31. Ishak KG, Irey NS. Hepatic injury associated with the phenothiazines. Clinicopathologic and follow-up study of 36 patients. Arch Pathol 1972;93(4):283–304.

32. Dincsoy HP, Saelinger DA. Haloperidol-induced chronic cholestatic liver disease. Gastroenterology 1982;83(3):694–700.

33. Jeffries JJ, Lyall WA, Bezchlibnyk K, Papoff PM, Newman F. Retroperitoneal fibrosis and haloperidol. Am J Psychiatry 1982;139(11):1524–5.

34. Thami GP, Kaur S, Kanwar AJ. Delayed reactivation of haloperidol induced photosensitive dermatitis by methotrexate. Postgrad Med J 2002;78(916):116–7.

35. Arima Y, Iwata K, Hamasaki Y, Katayama I. A case of toxic epidermal necrolysis associated with two drugs. Nishinihon J Dermatol 2001;63:63–5.

36. Weiner DM, Lowe FC. Psychotropic drug-induced priapism: Incidence, mechanism and management. CNS Drugs 1998;9:371–9.

37. Michael A, Calloway SP. Priapism in twins. Pharmacopsychiatry 1999;32(4):157.

38. Tsai SJ, Hong CJ. Haloperidol-induced impotence improved by switching to olanzapine. Gen Hosp Psychiatry 2000;22(5):391–2.

39. Kopelman AE, McCullar FW, Heggeness L. Limb malformations following maternal use of haloperidol. JAMA 1975;231(1):62–4.

40. Austin MP, Mitchell PB. Psychotropic medications in pregnant women: treatment dilemmas. Med J Aust 1998;169(8):428–31.

41. Cohen LS, Rosenbaum JF. Psychotropic drug use during pregnancy: weighing the risks. J Clin Psychiatry 1998;59(Suppl 2):18–28.

42. Coyle I, Wayner MJ, Singer G. Behavioral teratogenesis: a critical evaluation. Pharmacol Biochem Behav 1976;4(2):191–200.

43. Mohan MS, Patole SK, Whitehall JS. Severe hypothermia in a neonate following antenatal exposure to haloperidol. J Paediatr Child Health 2000;36(4):412–3.

44. Kaminer Y, Apter A, Tyano S, Livni E, Wijsenbeek H. Delayed hypersensitivity reaction to orally administered methylparaben. Clin Pharm 1982;1(5):469–70.

45. Morselli PL. Further observations on the interaction between ethanol and psychotropic drugs. Arzneimittelforschung 1971;2:20.

46. Risch SC, Groom GP, Janowsky DS. The effects of psychotropic drugs on the cardiovascular system. J Clin Psychiatry 1982;43(5 Pt 2):16–31.

47. Yukawa E, Ichimaru R, Maki T, Matsunaga K, Anai M, Yukawa M, Higuchi S, Goto Y. Interindividual variation of serum haloperidol concentrations in Japanese patients–clinical considerations on steady-state serum level-dose ratios. J Clin Pharmacol Ther 2003;28:97–101.

48. Iwahashi K, Miyatake R, Suwaki H, Hosokawa K, Ichikawa Y. The drug–drug interaction effects of haloperidol on plasma carbamazepine levels. Clin Neuropharmacol 1995;18(3):233–6.

49. Allen SA. Effect of chlorpromazine and clozapine on plasma concentrations of haloperidol in a patient with schizophrenia. J Clin Pharmacol 2000;40(11):1296–7.

50. Yasui-Furukori N, Kondo T, Mihara K, Inoue Y, Kaneko S. Fluvoxamine dose-dependent interaction with haloperidol and the effects on negative symptoms in schizophrenia. Psychopharmacology 2004;171:223–7.

51. Yasui N, Kondo T, Otani K, Furukori H, Mihara K, Suzuki A, Kaneko S, Inoue Y. Effects of itraconazole on the steady-state plasma concentrations of haloperidol and its reduced metabolite in schizophrenic patients: in vivo evidence of the involvement of CYP3A4 for haloperidol metabolism. J Clin Psychopharmacol 1999;19(2):149–54.

52. Furukori H, Kondo T, Yasui N, Otani K, Tokinaga N, Nagashima U, Kaneko S, Inoue Y. Effects of itraconazole on the steady-state plasma concentrations of bromperidol and reduced bromperidol in schizophrenic patients. Psychopharmacology (Berl) 1999;145(2):189–92.

53. Krystal JH, D'Souza DC, Karper LP, Bennett A, Abi-Dargham A, Abi-Saab D, Cassello K, Bowers MB Jr, Vegso S, Heninger GR, Charney DS. Interactive effects of subanesthetic ketamine and haloperidol in healthy humans. Psychopharmacology (Berl) 1999;145(2):193–204.

54. Bond WS, Carvalho M, Foulks EF. Persistent dysarthria with apraxia associated with a combination of lithium carbonate and haloperidol. J Clin Psychiatry 1982;43(6):256–7.

55. Lin PY, Wu CK, Sun TF. Concomitant neuroleptic malignant syndrome and lithium intoxication in a patient with bipolar I disorder: case report. Chang Gung Med J 2000;23(10):624–9.

56. Omata N, Murata T, Omori M, Wada Y. A patient with lithium intoxication developing at therapeutic serum lithium levels and persistent delirium after discontinuation of its administration. Gen Hosp Psychiatry 2003;25(1):53–5.

57. Ellman JP. Sudden death. Can J Psychiatry 1982;27(4):331–3.

58. Kudo S, Ishizaki T. Pharmacokinetics of haloperidol: an update. Clin Pharmacokinet 1999;37(6):435–56.

59. Thornton WE. Dementia induced by methyldopa with haloperidol. N Engl J Med 1976;294(22):1222.

60. Takeda M, Nishinuma K, Yamashita S, Matsubayashi T, Tanino S, Nishimura T. Serum haloperidol levels of schizophrenics receiving treatment for tuberculosis. Clin Neuropharmacol 1986;9(4):386–97.

61. Ranjan S, Jagadheesan K, Nizamie SH. Cerebellar ataxia with intravenous valproate and haloperidol. Aust NZ J Psychiatry 2002;36(2):268.

62. Yasui N, Kondo T, Suzuki A, Otani K, Mihara K, Furukori H, Kaneko S, Inoue Y. Lack of significant pharmacokinetic interaction between haloperidol and grapefruit juice. Int Clin Psychopharmacol 1999;14(2):113–8.

63. Pan L, Vander Stichele R, Rosseel MT, Berlo JA, De Schepper N, Belpaire FM. Effects of smoking, CYP2D6 genotype, and concomitant drug intake on the steady state plasma concentrations of haloperidol and reduced haloperidol in schizophrenic inpatients. Ther Drug Monit 1999;21(5):489–97.

Loxapine

General Information

Loxapine is a dibenzoxazepine neuroleptic drug.

After 6 and 12 weeks of treatment with loxapine, patients with schizophrenia (n = 24; aged 18–70 years) showed both lymphocyte D_2 dopamine-like and $5HT_{2A}$ platelet receptor binding down-regulation, which suggests that both receptors are involved in the mechanism of action of the drug, as well as its possible extrapyramidal adverse effects (1).

Organs and Systems

Hematologic

Three cases of leukocytosis attributed to loxapine have been reported; all resolved when loxapine was withdrawn (2).

Drug Administration

Drug overdose

A review of acute loxapine overdosage suggested that it has a high potential for causing serious neurological problems and cardiotoxicity (3).

In a study of overdosages of different neuroleptic drugs, loxapine had the highest fatality index (the number of deaths divided by the number of prescriptions) (4).

References

1. Singh AN, Barlas C, Saeedi H, Mishra RK. Effect of loxapine on peripheral dopamine-like and serotonin receptors in patients with schizophrenia. J Psychiatry Neurosci 2003;28:39–47.
2. Haffen E, Vandel P, Vandel S, Bonin B, Kantelip JP. Loxapine et hyperleucocytose: à propos de 3 cas. [Loxapine side-effects: 3 case reports of hyperleukocytosis.] Therapie 2001;56(1):61–3.
3. Peterson CD. Seizures induced by acute loxapine overdosage. Vet Hum Toxicol 1980;22(Suppl 2):52.
4. Buckley N, McManus P. Fatal toxicity of drugs used in the treatment of psychotic illnesses. Br J Psychiatry 1998;172:461–4.

Olanzapine

General Information

Several reviews have echoed the efficacy and tolerability of olanzapine (1–3). In several reviews it has been emphasized that the most common adverse effects are somnolence and weight gain, since about 40% of patients gain weight (especially if on a high starting dose and if they were underweight before treatment); sexual dysfunction can also be a problem for many patients (4–6).

Clinical trials

A thorough and extensive review has summarized five pivotal clinical trials comprising 3252 patients (7). In some studies olanzapine has failed to show efficacy in treatment-resistant schizophrenia (8,9).

In addition to clozapine (10,11), olanzapine has also been reported to improve tardive dyskinesia (12–14).

● An elderly woman who developed moderately severe neuroleptic drug-induced dyskinetic movements responded to olanzapine 10 mg/day (15).

Observational studies

Olanzapine, mean dose 5.4 mg/day, has been given to 21 patients with apathy in the absence of depression after long-term treatment with selective serotonin reuptake inhibitors for non-psychotic depression in an open, flexible-dose study (16). The more frequent adverse effects were sedation (n = 12), increased appetite (n = 8), stiffness (n = 7), edema (n = 6), and dry mouth (n = 5).

The efficacy and safety of switching 108 Asian patients from their regimen of neuroleptic medications to olanzapine (initial dose 10 mg/day) for 6 weeks has been studied in an open, multicenter, randomized study (17). They were randomly assigned to one of two groups: the direct switch group (n = 54) received only olanzapine, while the start-taper switch group (n = 54) received olanzapine and their usual neuroleptic drug in decreasing doses for the first 2 weeks. There were statistically significant improvements from baseline to end-point in both switch groups in the Clinical Global Impressions—Severity of Illness Scale score and the Positive and Negative Syndrome Scale total score. Nevertheless, there were no significant differences between the switch groups in any measure of efficacy. Weight gain occurred in both switch groups and both showed statistically significant improvement from baseline to end-point on the Simpson–Angus Scale and Barnes Akathisia Scale.

In 19 patients previously treated with clozapine, olanzapine was used instead (9). Eight were considered to be responders and the rest decompensated, seven of them enough to require hospitalization. Overall Brief Psychiatric Rating Scale (BPRS) scores increased significantly from baseline to final assessment.

In an open trial to determine the efficacy of olanzapine in the treatment of bipolar mixed state (n = 9), the results showed improvement in acute symptoms and the drug was well tolerated (18).

In a study supported by Eli Lilly and Company it was concluded that olanzapine may be effective in a significant number of neuroleptic drug-resistant schizophrenic patients (19). However, of 25 patients who entered an open trial for 6 months, 14 discontinued olanzapine, one because of an adverse effect (depression), two because of lack of compliance, and 11 because of lack of efficacy.

Of 14 consecutive patients with bipolar I disorder, who were inadequately responsive to standard psychotropic agents and who were given olanzapine, 8 improved (20). The most common adverse effects were sedation, tremor, dry mouth, and increased appetite with weight gain.

The response to olanzapine in 150 consecutive patients has been assessed by reviewing their records (21). Patients with a moderate-to-marked response to olanzapine were more likely to be younger, to be female, and to have a diagnosis of bipolar disorder. No information on adverse effects was provided.

The efficacy of olanzapine in treatment-refractory childhood-onset schizophrenia has been examined in eight patients (mean age 15 years) over 8 weeks (22). There was a 17% improvement in the BPRS total score. Olanzapine was moderately well tolerated. The most common adverse events were increased appetite ($n = 6$), constipation ($n = 5$), nausea/vomiting ($n = 6$), headache ($n = 6$), somnolence ($n = 6$), insomnia ($n = 7$), difficulty in concentrating ($n = 5$), sustained tachycardia ($n = 6$), transient rises in liver transaminases ($n = 7$), and increased agitation ($n = 6$).

In contrast, olanzapine was discontinued in five preadolescent children (aged 6–11 years) within the first 6 weeks because of adverse effects or lack of therapeutic response (23). The adverse effects included sedation ($n = 3$), weight gain ($n = 3$), and akathisia ($n = 2$).

Substance abuse is a major complication in the treatment of schizophrenia. Of 60 patients with schizophrenia in an open 7-week trial of olanzapine up to 25 mg/day, there was substance abuse in 23 (24). There were no differences in response between the substance-abusing patients and the others. Baseline rates of extrapyramidal adverse effects did not differ between the groups. Overall the patients did not improve significantly.

The medical records of 151 hospitalized elderly psychiatric patients (mean age 71 years) have been analysed (25). Of 37 patients treated with olanzapine (mean duration of treatment 20 days, mean dose 10 mg/day), 75% responded. Adverse effects were reported in six patients: sedation in four patients and extrapyramidal signs and postural hypotension in one each.

In 16 patients with treatment-resistant schizophrenia (defined as non-responsiveness to at least three antipsychotic drugs from at least two different chemical classes), olanzapine was effective treatment in a significant proportion; no serious adverse events were associated with maintenance doses of 10–40 mg/day, and no subjects dropped out (26).

In five consecutive cases, olanzapine was effective for treatment-refractory psychosis in patients previously responsive to, but intolerant of, clozapine (27).

Olanzapine was effective in an open trial in 10 patients with obsessive-compulsive disorder refractory to selective serotonin reuptake inhibitors, who were given additional olanzapine; they had minimal adverse effects, primarily sedation (27).

In 23 patients with obsessive-compulsive disorder who had not responded to a 6-month course of fluvoxamine (300 mg/day), olanzapine (5 mg/day) was added in an open comparison (28). There was a significant reduction in the mean score on the Yale–Brown Obsessive–Compulsive Scale; concomitant schizotypal personality disorder was the only factor significantly associated with a response. The most common adverse effects were mild to moderate weight gain and sedation.

The use of olanzapine for the psychiatric manifestations of hereditary coproporphyria has been studied (29).

In an open study, eight patients with Lewy body dementia associated with psychotic and behavioral difficulties were given olanzapine 2.5–7.5 mg (30). Three of the eight could not tolerate olanzapine, even at the lowest dose, two had a clear improvement in psychotic and behavioral symptoms, and three tolerated olanzapine but gained only minimal benefit. The authors concluded that olanzapine in the doses used conferred little advantage over conventional neuroleptic drugs and should be given only with great caution to patients with Lewy body dementia.

In a prospective open 12-week study in eight children, adolescents, and adults with developmental disorders, the most significant adverse effects of olanzapine were increased appetite and weight gain in six patients and sedation in three (31).

The efficacy and adverse effects of high doses of olanzapine (up to 30 mg/day) have been assessed in seven patients with schizophrenia and schizoaffective disorder (32). None discontinued olanzapine because of adverse effects; the only adverse effects were weight gain and self-limited diarrhea. However, poor design and small sample size precluded definitive conclusions.

In an observational prospective study sponsored by Lilly, 2128 patients were treated with olanzapine as monotherapy (mean dose 13 mg/day) or combined with other drugs, and 821 were treated with other neuroleptic drugs as monotherapy or combined with other drugs (control group) (33). Olanzapine was well tolerated and effective, and the overall incidence of adverse events was significantly lower than in controls, although weight gain (6.9%) was significantly more frequent. Dropouts were statistically similar in the two groups.

To evaluate anticholinergic effects, differences among patients taking olanzapine ($n = 12$; average dose, 15 mg/day) or clozapine ($n = 12$; average dose, 444 mg/day) for at least 8 weeks were measured in an unblinded study (34). Altered salivation was significantly more common with clozapine (increased salivation) and olanzapine (decreased salivation), whereas constipation, urinary disturbances, and tachycardia/palpitation were significantly more common with clozapine. There were no global cognitive problems in either group.

Olanzapine has fewer extrapyramidal effects than typical neuroleptic drugs such as haloperidol (SEDA-24, 66). Even some cases of tardive dystonia have been successfully managed with atypical neuroleptic drugs. The antidystonic efficacy of olanzapine has been studied in an open video-blinded study in four patients with tardive cervical dystonia after several years of neuroleptic drug treatment (35). There was moderate to marked improvement in dystonia in all of them, and no serious adverse effects at the maximum dosage reached (7.5 mg/day).

Two patients, one with treatment-resistant rapid cycling bipolar disorder (36) and the other with severe mood changes during corticosteroid therapy (37), improved with olanzapine.

Data from 904 patients with schizophrenia have been collected in a prospective, comparative, non-randomized, open, observational study funded by Eli Lilly (38). In all, 483 patients took olanzapine and 421 took typical antipsychotic drugs. Somnolence was reported in 3.3% of cases in the olanzapine group and in 2.9% in the typical antipsychotic drug group; weight gain was more common in the olanzapine group than in the typical antipsychotic group (1.7% versus 0.2%), although there was no systematic recording of weights and participating clinicians reported weight changes as adverse events when they deemed it appropriate. The incidence of extrapyramidal symptoms was significantly lower in the olanzapine group, and the incidences of dystonia, hypertonia, hypokinesia, tremor, akathisia, and dyskinesia with olanzapine were also significantly lower; the open and non-randomized design of this study limited its conclusions.

In a retrospective study in 499 in-patients with schizophrenia or schizoaffective disorder (39); 259 subjects were taking olanzapine and 240 risperidone. Treatment was considered effective in most cases (74% with olanzapine and 78% with risperidone). There were adverse effects in 19% of the patients taking olanzapine and 22% of those taking risperidone; they were mainly somnolence (n = 15 and n = 17 respectively) and extrapyramidal symptoms (n = 9 and n = 6 respectively); there were also three cases of weight gain with olanzapine.

The efficacy of olanzapine monotherapy for the acute hypomania or mania has also been studied in a consecutive series of 15 patients entering an open, uncontrolled, 8-week trial of olanzapine monotherapy (40). Most of the patients had significant reductions in mania rating and more limited improvement in depression rating, but more reported adverse events, consistent with other studies The most commonly reported adverse effects were moderate to severe dry mouth (80%), weight gain (77%), mild dizziness (60%), edema (53%), mild to moderate drowsiness (53%), and constipation (47%).

The efficacy and safety of olanzapine in particular disorders or particular groups of patients have been previously studied (SEDA-24, 67), and new studies have been published. For example, in a 4-week open trial, 94 elderly psychiatric in-patients (aged 65 years or older) were treated with a mean daily dose of olanzapine of 10 mg (range 2.2–20). The most common adverse effects were somnolence (18%), dizziness (18%), and weakness of the legs or bradykinesia (16%); body weight and fasting triglyceride and glucose concentrations were significantly increased (41).

In a review of the charts of 98 oncology patients who were given olanzapine, 28 took it specifically for prevention of delayed emesis (42). The authors claimed that olanzapine was well tolerated and might reduce the incidence of delayed emesis in patients receiving emetogenic chemotherapy.

In 18 patients with trichotillomania given olanzapine (maximum dose 10 mg/day) in a 3-month open study the most common adverse effects were sedation and weight gain (43). The authors claimed that olanzapine may be effective monotherapy for trichotillomania.

Three patients with chronic pain were successfully treated with olanzapine as an adjunct therapy; two had fatigue associated with olanzapine (44).

Two patients with refractory panic attacks (45) and five patients with long-lasting and untreatable nightmares and insomnia (46) benefited from olanzapine with no adverse effects after several months of treatment.

In a recent non-randomized, open study in 37 patients (21 women, 16 men) with essential tremor, the most common movement disorder, olanzapine significantly reduced the median tremor score from 3.3 to 1.1 (47); seven of the patients reported adverse effects such as sedation, which disappeared in about 7 days, and three complained of weight gain.

In a 7-month open study 39 patients (mean age 52 years; 26% men) taking olanzapine (mean dose 15 mg/day) had limited non-significant improvement in the Wisconsin Card Sorting Test for cognitive assessment (48). The following adverse events were considered to have been treatment-related: weight gain (n = 6), extrapyramidal disorders (n = 2), and increased appetite, weakness, and confusion (n = 1 each).

Comparative studies

Olanzapine has been compared with other typical and atypical antipsychotic drugs in a 3-year prospective observational investigation, the European Schizophrenia Outpatient Health Outcomes (SOHO) study, sponsored by the market authorization holder of olanzapine, in 10 European countries (49). From the initial sample of 10 972 patients who started or changed antipsychotic drug medication for schizophrenia, 8400 were given only one antipsychotic drug at the baseline visit and composed the reference sample (olanzapine 4636; risperidone 1671; quetiapine 651; amisulpride 278; clozapine 291; oral typical 499; and depot typical 374). There were significantly fewer patients with extrapyramidal symptoms in all treatment groups after 6 months, the largest improvements being observed in patients taking olanzapine or clozapine. Conversely, there were increases in weight and body mass index (BMI) in all treatment groups; these were significantly greater in patients taking olanzapine or clozapine (mean body weight increases 2.4 and 2.3 kg respectively; mean BMI changes 0.9 and 0.8 kg/m^2 respectively). In all groups the patients with a lower BMI at baseline had a greater increase in weight and BMI than patients with a higher BMI at baseline; there were no sex-related differences in weight or BMI. Mean doses of all the antipsychotic drugs increased from baseline to 6 months and most patients maintained treatment for 6 months with the single antipsychotic drug that was prescribed at baseline. Reasons for drop-outs (12%) were not given.

Olanzapine versus amisulpride

Two atypical antipsychotic drugs, olanzapine and amisulpride, which have different receptor occupancy profiles, have been compared in a double-blind, randomized, 6-month study in 72 different centers in 11 European and African countries (50). Schizophrenic patients, mainly with acute psychotic symptoms, were assigned to olanzapine (mean dose 13 mg/day; mean age 37 years, 64% men; n = 188) or amisulpride (mean dose 504 mg/day; mean age 38 years, 66% men; n = 189). The incidence and reasons for withdrawal were similar in both groups (63 patients in the olanzapine group and 72 in the amisulpride group left the study prematurely). The two drugs produced similar symptomatic improvement and similar adverse event frequencies (56% with olanzapine and 57% with amisulpride); patients taking olanzapine had more weight gain (50 versus 30 patients; mean weight gain 3.9 versus 1.6 kg), but less amenorrhea (0 versus 4), extrapyramidal symptoms (1 versus 11), insomnia (10 versus 14), and constipation (5 versus 10); there was somnolence in 12 patients in each group and serum transaminase activities were raised in 32 patients taking olanzapine and seven taking amisulpride. Three patients (two taking amisulpride and one taking olanzapine) left the study because of extrapyramidal symptoms. Diabetes was aggravated in one patient taking olanzapine. Two patients died during the study, one from cerebral hemorrhage (taking amisulpride) and one because of suicide (taking olanzapine).

There were moderate but significant improvements in neurocognition (including executive function, working memory, and declarative memory) in a randomized, double-blind, 8-week study in 52 patients with schizophrenia assigned either to olanzapine (10–20 mg/day; n = 18) or amisulpride (400–800 mg/day; n = 18) (51). Of 16 dropouts, six were due to adverse events: olanzapine—sedation (n = 2) and increased transaminases (n = 1); amisulpride—rash, extrapyramidal symptoms, and galactorrhea (n = 1 each).

Olanzapine versus clozapine

Compared with clozapine, considered the gold standard for treatment of patients with refractory schizophrenia, olanzapine had the same level of efficacy (according to the Positive and Negative Syndrome Scale (PANSS), and the Brief Psychiatric Rating Scale (BPRS)) in 150 patients (mean age, 38 years; 60% men) who had failed to respond to conventional neuroleptic drugs because of either insufficient effectiveness or intolerable adverse effects (52). Of these, 147 patients, 52 from Hungary and 95 from South Africa, were randomized to olanzapine (n = 75) or clozapine (n = 72); there were no statistically significant differences in demographic or baseline illness characteristics; the final mean doses were 17 mg/day of olanzapine and 108 mg/day of clozapine. Adverse events occurred in 9.2% of those taking olanzapine and 9.5% of those taking clozapine. There were two cases of leukopenia in those taking clozapine. Patients taking olanzapine had significantly more back pain than those taking clozapine (4 versus 0), but also significantly less somnolence (2 versus 11) and dizziness (1 versus 6); there was similar weight gain with the two drugs (olanzapine 3.3 kg; clozapine 4.1 kg, n = 7 in both groups). Akathisia, parkinsonism, and dyskinesia (measured by specific scales) did not change significantly from baseline to the study endpoint in either group; there were clinically insignificant changes in laboratory measures (urinary pH and phosphate).

Olanzapine versus divalproex

The efficacy and safety of olanzapine in disorders other than schizophrenia have been studied (SEDA-24, 67; SEDA-25, 64; SEDA-26, 61). In a 3-week, randomized, double-blind trial, the effects of a flexible dose of olanzapine (5–20 mg/day) and divalproex (500–2500 mg/day) for the treatment of patients with acute bipolar manic or mixed episodes have been compared (53). The olanzapine treatment group (n = 125) had significantly greater mean improvements in mania rating and a significantly greater proportion of patients achieved protocol-defined remission. With olanzapine there was significantly more weight gain (12 versus 7.9%), dry mouth (34 versus 6.3%), increased appetite (12 versus 2.4%), and somnolence (39 versus 21%), while more cases of nausea (29 versus 10%) were reported with divalproex.

Olanzapine versus fluphenazine

Weight gain and fewer extrapyramidal symptoms were the major adverse events in 60 patients who were randomly assigned to olanzapine (n = 30) or fluphenazine (n = 30) (54). In this 22-week study, final doses were 15 mg/day of olanzapine and 12 mg/day of fluphenazine; adverse events were more frequent in patients taking fluphenazine (n = 23) than in those taking olanzapine (n = 15); adverse effects in the fluphenazine group included akathisia (n = 9), insomnia (n = 6), hypertonia (n = 3), and tremor (n = 2); and in the olanzapine group, weight gain (n = 5), akathisia (n = 3), tremor (n = 3), and hypertonia (n = 1). Patient satisfaction, measured with the Drug Attitude Inventory, was better with olanzapine than with fluphenazine, perhaps also because of less daytime sedation.

Olanzapine versus haloperidol

In a multicenter, double-blind, randomized, controlled trial in patients with schizophrenia or schizoaffective disorder, 159 received olanzapine (mean age 47 years) and 150 received haloperidol (mean age 46 years); they were almost exclusively men (97% in the olanzapine group and 96 in the haloperidol group) (55). During the first 6 weeks of the 12-month trial, mean doses were 11.4 mg/day for olanzapine and 11.2 mg/day for haloperidol; for the rest of the first 6 months they were 14.7 and 13.5 mg/day, and during the last 6 months, 15.8 and 14.3 mg/day. Prophylactic benzatropine 1–4 mg/day was added in those taking haloperidol. There were no significant differences in efficacy between the two groups when considering positive, negative, or total symptoms of schizophrenia or quality of life. Several reasons accounted for dropouts (olanzapine n = 86; haloperidol n = 91). They included: adverse effects (10% haloperidol and 4% olanzapine); lack of efficacy or worsening of symptoms; loss to follow-up, missed appointments, or moving out; consent withdrawal or unhappiness with blinded treatment. There were no significant differences between the two groups in reasons for treatment withdrawal. Olanzapine was associated with significantly overall lower scores on the Barnes scale for akathisia, but not on the AIMS measure of tardive dyskinesia or the Simpson–Angus scale for extrapyramidal symptoms. Motor functioning and memory significantly improved in patients taking olanzapine. However, weight gain was significantly more common with olanzapine than haloperidol at 3 months (17% versus 28%), 6 months (13% versus 33%), and 12 months (8.3% versus 25%). Medication costs were 4–5 times higher with olanzapine; several measures of service use and health costs were also significantly higher in patients taking olanzapine.

Accelerated dose titration of olanzapine with adjunctive lorazepam has been used in acutely agitated patients with schizophrenia and other related disorders in a double-blind study, in which newly admitted acutely agitated patients were randomized to olanzapine (10 mg/day; n = 52) or haloperidol (10 mg/day; n = 48) (56). Baseline characteristics were similar in the two groups (olanzapine, age 39 years, 71% men; haloperidol, age 40 years, 71% men). The same drug titration regimen was followed with the two drugs: on the first day the dose was 10 mg/day, on the second day an increase of 5 mg was allowed, and the maximum dose of 20 mg/day was reached on day 3; lorazepam was permitted only if it was considered clinically necessary; mean daily doses throughout the whole study were 2.6 mg in the olanzapine group and 2.9 mg in the haloperidol group. Both drugs significantly reduced agitation from baseline as early as 1 hour after the first dose; however, haloperidol doses were relatively high. Of the 57 patients who completed the study, there were more taking olanzapine than haloperidol (67% versus 46%); adverse events were the only reasons for withdrawal, with significantly different rates between the treatment groups (17% with haloperidol versus 1.9% with olanzapine).

Dystonia, hypertonia, and increased salivation were reported significantly more often in patients taking haloperidol (8.3% versus 0% for the three effects); somnolence (25%), anxiety (12%), and headache (12%) were the most frequent events in patients taking olanzapine.

There was a different adverse effects profile in a 16-week, double-blind study in 63 out-patients with schizophrenia, who had previously been receiving fluphenazine, when comparing olanzapine (n = 29; mean age 42 years; 22 men) and haloperidol (n = 34; mean age 46 years; 24 men) (57). Patients taking olanzapine had significantly fewer extrapyramidal symptoms than those taking haloperidol. However, they had significantly higher systolic, but not diastolic, blood pressure. Weight gain was also greater in the patients taking olanzapine.

Olanzapine versus lithium
In a comparison of olanzapine and lithium in 87 patients with bipolar disorder in a 12-month, randomized, double-blind trial olanzapine was equally effective in preventing recurrence of depression and more effective than lithium in preventing recurrence of manic and mixed episodes (58). Weight gain was significantly greater with olanzapine (mean 1.8 kg). One patient committed suicide during the initial open phase and two taking lithium died during the double-blind period.

Olanzapine versus risperidone
In an 8-week study, pre-school-age children with bipolar disorder (aged 4–6 years) took either olanzapine (n = 15; mean age 5.0 years; 10 boys; mean dose 6.3 mg/day) or risperidone (n = 16; mean age 5.3 years; 12 boys; mean dose 1.4 mg/day) (59). There were significantly more dropouts with olanzapine (6 versus 1), including one patient who withdrew because of adverse events (increased appetite and hand tremor). The main adverse events, found with both treatments, were significant increases in prolactin concentrations and weight gain. With both treatments, increased appetite, flu-like symptoms, headaches, and sedation were the most commonly reported adverse effects.

Olanzapine versus valproate
Although mania has been associated with olanzapine (SEDA-24, 68; SEDA-25, 68; SEDA-26, 62), it has also been used in the treatment of acute mania. In a 12-week, double-blind, double-dummy, randomized trial, 120 patients with bipolar disorder type I hospitalized for an acute manic episode were randomly assigned to either sodium valproate (n = 63) or olanzapine (n = 57) and were followed in hospital for up to 21 days (60). Valproate and olanzapine had similar short-term effects on clinical or health-related quality of life outcomes in bipolar disorder; adverse effects that occurred in a higher percentage of olanzapine-treated than valproate-treated patients included somnolence (47% versus 29%), weight gain (25% versus 10%), rhinitis (14% versus 3%), edema (14% versus 0%), and slurred speech (7% versus 0%); no adverse events occurred significantly more often with valproate.

In a recent pooled analysis, three previously published trials have been reviewed to compare the efficacy, safety, and tolerability of oral-loaded valproate with standard-titration valproate, lithium, olanzapine, or placebo in patients with acute mania associated with bipolar I disorder (61). Valproate loading was as well tolerated as the other active treatment or better tolerated, as measured by adverse events and changes in laboratory parameters, and was of better efficacy than placebo; however, there were no efficacy differences between valproate loading and olanzapine.

Olanzapine versus ziprasidone

In a 28-week multicenter, randomized, double-blind, parallel-group study sponsored by Eli Lilly, the market authorization holder of olanzapine, in 548 patients with schizophrenia, 277 were assigned to olanzapine (mean 15.3 mg/day) and 271 to ziprasidone (mean 116 mg/day) (62). The proportions of men in the two groups were similar (65 and 64% respectively) and the mean age was slightly higher in the olanzapine group (40 versus 38 years). Significantly more patients withdrew with ziprasidone than olanzapine (58 versus 41%). However, the patients who took olanzapine had significantly greater weight gain (13% versus 1.8%; mean changes 3.1 and – 1.1 kg respectively) and increased appetite (7.2 versus 2.6%). There were significant increases in total cholesterol and low-density lipoprotein cholesterol in those who took olanzapine. On the other hand, those who took ziprasidone had more insomnia (22% versus 6.9%), anorexia (2.6 versus 0.4%), dystonias (2.2 versus 0%), and hypotension (1.8 versus 0%).

Patients taking olanzapine (mean dose 12.6 mg/day; n = 71) had significantly more *weight gain* than those taking ziprasidone (mean dose 135.2 mg/day; n = 55) in a 6-month, randomized, double-blind, multicenter study of 126 patients (63c). The mean changes in body weight and body mass index were 5.0 kg and 1.3 kg/m^2 respectively with olanzapine and –0.8 kg and –0.6 kg/m^2 with ziprasidone.

Combination studies

Combination with valproate or lithium

The efficacy of adding olanzapine to either valproate or lithium alone in acute manic or mixed bipolar episodes has been studied in a 6-week, double-blind, randomized, placebo-controlled trial (64). Compared with valproate or lithium alone, the addition of olanzapine provided better efficacy. Olanzapine was associated with somnolence, dry mouth, weight gain, increased appetite, tremor, and slurred speech.

In another study, 30 patients with mania were randomized to olanzapine or lithium in a double-blind 4-week study (65). Olanzapine did not differ from lithium in terms of efficacy or of extrapyramidal adverse effects, as measured by the Simpson–Angus Scale.

Placebo-controlled studies

A multicenter, double-blind, placebo-controlled study in patients with Alzheimer's disease and symptoms of agitation/aggression and/or psychosis but few or no psychotic symptoms at baseline, and data from a subgroup of patients have been analysed (66). Three subsets of patients were identified on the basis of their symptoms at baseline: those with no clinically significant hallucinations, those with no clinically significant delusions, and those with no clinically significant delusions or hallucinations. Of the patients without hallucinations or delusions

at baseline (n = 75), the placebo-treated patients had significantly more of these symptoms than olanzapine-treated patients. Similarly, among the patients without baseline hallucinations (n = 153), the placebo-treated patients had higher hallucination scores than olanzapine-treated patients, whereas patients without baseline delusions (n = 87) had no significant treatment effects. Abnormal gait, a term comprising leaning, limp, stooped posture, and unsteady gait, occurred in a higher incidence among olanzapine-treated patients (14/72) than placebo-treated patients (0/33). Somnolence was also reported at higher rates with olanzapine (27/72) than placebo (3/33).

Olanzapine has been proposed as a treatment for acute mania. In a randomized, double-blind, placebo-controlled study for 3 weeks, patients were assigned to either olanzapine 10 mg/day (n = 70) or placebo (n = 69) (67). Significantly more olanzapine-treated patients responded (49%) compared with those assigned to placebo (24%). Somnolence, dizziness, dry mouth, and weight gain occurred significantly more often with olanzapine, but there were no statistically significant differences with respect to measures of parkinsonism, akathisia, and dyskinesia.

Promoted by Eli-Lilly, the market authorization holder of olanzapine, three randomized, double-blind, placebo- and active medication-controlled trials in agitated patients have been reanalyzed looking for a calming effect (68); the studies were conducted in patients with schizophrenia (n = 311), bipolar mania (n = 201), or dementia (n = 206) to compare intramuscular olanzapine with intramuscular haloperidol, lorazepam, or placebo. There were no significant between-group differences in Agitation–Calmness Evaluation Scale scores. The incidence of adverse events was not significantly more reduced with olanzapine versus comparators; haloperidol-treated patients with schizophrenia had more acute dystonias and akathisia than olanzapine-treated patients, but no parkinsonism, and no treatment-related adverse effect had an incidence of 10% or more among olanzapine-treated patients, although one had two serious adverse effects, overdose and psychosis.

Olanzapine has been compared with placebo in different mental disorders, including the non-approved indications of psychotic depression, Alzheimer's disease, obsessive-compulsive disorder, and alcohol dependence.

A combination of olanzapine and fluoxetine was used in two randomized, double-blind simultaneous 8-week trials in 249 patients with major depression with psychotic features (trial 1: n = 124, mean age 41 years, 52% women; trial 2: n = 125, mean age, 41 years, 50% women), which have been jointly published (69). This multicenter study was completed by 51 subjects in trial 1 (41%) and 59 subjects in trial 2 (47%). Altogether, there were no significant differences in the rates of discontinuation due to adverse events among the different treatment groups: placebo (n = 100), monotherapy with olanzapine 5–20 mg/day (n = 101), and olanzapine 5–20 mg/day plus fluoxetine 20–80 mg/day (n = 48). Dropout percentages were 59% in trial 1 (similarly distributed in the three groups) and 53% in trial 2 (ranging from 40% of dropouts in the combined therapy group to 59% in the placebo group); lack of efficacy accounted for discontinuation in 8–25% of patients in all subgroups; other reasons were adverse events in patients taking combined therapy, the subject's own decision, loss to follow-up in those taking olanzapine monotherapy, and loss to follow-up in those taking placebo. There was no significant improvement in depressive symptoms in patients taking olanzapine mono-therapy compared with those taking placebo; however, those taking olanzapine + fluoxetine had greater improvement than those taking placebo. Mean weight gain at the endpoint was significantly higher with both olanzapine (3.79 kg) and olanzapine + fluoxetine (2.74 kg) than placebo (0.39 kg). Dry mouth was signifi-cantly more common with the drugs than placebo. Gamma-glutamyl transferase activity increased signifi-cantly with olanzapine + fluoxetine compared with olan-zapine monotherapy or placebo. Prolactin concentrations increased significantly with both olanzapine + fluoxetine and olanzapine monotherapy. There was a mean increase in cholesterol of 0.35 mmol/l (13.5 mg/dl) with olanzapine + fluoxetine (placebo –0.12 mmol/l; olanza-pine +0.19 mmol/l). There were no significant differences in usual laboratory tests or extrapyramidal symptoms.

In a 6-week double-blind study, subjects with obsessive-compulsive disorder who had failed to respond comple-tely to fluoxetine alone (43 adults and one adolescent, mean age, 37 years, 26 women) received either fluoxetine + olanzapine (n = 22), or fluoxetine + placebo (n = 22) (70). Olanzapine was begun at 5 mg/day and titrated upwards to a maximum of 10 mg/day by as early as the second week. There were seven dropouts because of adverse events —five with olanzapine + fluoxetine (weight gain and shakiness) and two with fluoxetine + placebo (increased anxiety and emotional numbing). Mean weight gain was 2.8 kg with olanzapine + fluoxe-tine and 0.5 kg with fluoxetine + placebo.

In a multicenter study, 652 patients with Alzheimer's disease and psychotic symptoms from 61 different centers in Europe, Australia, and Africa (mean age, 77 years; 75% women) were randomly assigned to 10 weeks of double-blind treatment with either olanzapine (1 mg, n = 129; 2.5 mg, n = 134; 5 mg, n = 125; 7.5 mg, n = 132) or placebo (n = 129) (71). Overall, 49% of all the patients had at least one adverse event during the study, with no significant differences in either prevalence or severity between the groups. Weight gain, anorexia, and urinary incontinence were more common with olanzapine. Prolactin concentrations increased significantly more with the highest dose of olanzapine, 7.5 mg. No other individual events, including extrapyramidal symptoms, cognition, and the usual laboratory measures, differed significantly from placebo. There were 17 deaths during the treatment period or in the next 30 days; two taking placebo, four taking olanzapine 1 mg, three taking 2.5 mg, five taking 5 mg, and three taking 7.5 mg; overall mortality was 2.9% of those taking olanzapine and 1.5% of those taking placebo.

In a placebo-controlled study of the addition of olanza-pine for 6 weeks to lithium or valproate in 344 manic patients, of whom 85 had dysphoric mania, those who took olanzapine had significantly greater improvement than those who took placebo; adverse effects were not mentioned (72).

In a 12-week, double-blind study 60 patients with bor-derline personality disorders were randomized to either olanzapine 5–20 mg/day (n = 30; mean age 26 years; five men) or placebo (n = 30; mean age 26 years; three men) (73). Those who took olanzapine had significantly greater *weight gain* (mean increase with olanzapine 2.7 kg; range –9 to 7 kg) than those who took placebo (mean increase –0.05 kg; range –8 to 3 kg). Olanzapine was also associated with a small increase in cholesterol concentration (olan-zapine 8 μmol/l, placebo 3 μmol/l).

Organs and Systems

Cardiovascular

An episode of asystole (at which time olanzapine was withdrawn), followed 6 days later by a brain stem stroke, occurred during a double-blind parallel study for 2 weeks in 39 demented patients with agitation (mean age 83 years) (74); olanzapine (n = 20, mean daily dose 6.65 mg, modal dose 10 mg) or risperidone (n = 19, mean daily dose 1.47 mg, modal dose 2 mg) were given once a day at bedtime.

QT interval prolongation

It has been suggested that olanzapine often causes QT interval prolongation (SEDA-25, 64). However, in an analysis of electrocardiograms obtained as part of the safety assessment of olanzapine in four controlled rando-mized trials ($n = 2700$) the incidence of maximum QT_c prolongation beyond 450 ms during treatment was approximately equal to the incidence of prolongation of the QT_c beyond 450 ms at baseline (75). The authors therefore suggested that olanzapine does not contribute to QT_c prolongation. This has been supported by the report of a patient in whom QT_c prolongation while he was taking clozapine reversed when he switched to olan-zapine (76).

There are multiple susceptibility factors for olanzapine-induced prolongation of the QT interval: female sex, old age, concomitant medications, and underlying cardiac conduction disorder.

- A 66-year-old woman taking chlorpromazine and que-tiapine had QT_c interval prolongation, which improved when these drugs were withdrawn (77). However, pro-longation later recurred while she was taking high-dosage olanzapine (60 mg/day), which had not occurred with a smaller dosage (40 mg/day).
- QT_c prolongation occurred in a 28-year-old woman while she was taking olanzapine 40 mg/day; after olan-zapine withdrawal, the QT_c interval returned to nor-mal (78).
- A 61-year-old woman with Wolff–Parkinson–White syndrome, who had previously had QT_c interval pro-longation with both sulpiride 1200 mg/day and

clozapine 50 mg/day, had a QT_c interval of 390 ms, which increased to 466 ms when she took olanzapine 5 mg/day and returned to 395 ms in 2 days when olanzapine was withdrawn (79). When she was given olanzapine again in the same dose, the QT_c interval increased to 473 ms in 2 weeks and returned to baseline when olanzapine was withdrawn. She was also taking daily valproate 1500 mg/day, lithium 300 mg/day, lorazepam 2 mg/day, and propranolol 40 mg/day.

Two cases of light-headedness or "fainting" in patients taking olanzapine have been reported (80). Electrocardiograms showed first-degree heart block and AV conduction delay, which normalized after dosage reduction.

- A 44-year-old man took clozapine 10 mg/day and the dose was increased to 50 mg/day, after which he complained of episodes of light-headedness. An electrocardiogram showed prolongation of the PR interval to 227 ms. The dosage was reduced to 30 mg and the PR interval returned to 187 ms within 2 days.
- A 36-year-old man took clozapine 20 mg/day and within 2 weeks began to complain of intermittent, unpredictable "fainting" attacks. An electrocardiogram showed a prolonged PR interval at 230 ms. The dose of olanzapine was reduced to 17.5 mg/day, and a repeat electrocardiogram 1 week later was normal.

Hypertension

Subclinical cases of increased blood pressure related to olanzapine have previously been reported (SEDA-25, 64).

- A 29-year-old man developed transient rises in systolic and diastolic blood pressures (160/90; previous blood pressure 130/84 mmHg) with raised transaminases and dependent pitting edema of both feet (81).

Nervous system

Extrapyramidal effects

To ascertain to what extent olanzapine occupies $5-HT_2$ and dopamine D_2 receptors, a positron emission tomography study has been conducted in 12 patients with schizophrenia randomly assigned to 5, 10, 15, or 20 mg/day of olanzapine (82). Olanzapine is a potent $5-HT_2$ antagonist and has higher $5-HT_2$ than D_2 occupancy at all doses. Its D_2 receptor occupancy is higher than that of clozapine and similar to that of risperidone. At the usual clinical dose range of 10–20 mg/day, D_2 receptor occupancy is 71–80%, and this restricted range may explain its freedom from extrapyramidal adverse effects. However, doses of 30 mg/day and higher are associated with more than 80% D_2 receptor occupancy and may have a higher likelihood of extrapyramidal adverse effects.

The relation between extrapyramidal adverse effects and the negative symptoms of schizophrenia has been studied (83). Correlation analysis after 6 weeks of treatment showed that extrapyramidal symptoms correlated significantly in patients taking haloperidol ($n = 10$) but not in those taking olanzapine ($n = 13$). The results of

multiple regression analysis suggested that ratings of negative signs were confounded by extrapyramidal symptoms in patients treated with haloperidol. This confusion occurred to a lesser extent with olanzapine.

Acute dystonia

Patients taking olanzapine reported a low incidence of dystonias, which may be about 0.3% (SEDA-22, 56). In the light of two new cases of acute dystonia associated with olanzapine in patients with previous history of dystonia or parkinsonism related to antipsychotic treatment, comparative figures have been reported. Acute dystonia occurred in 1.4% of patients who took olanzapine, compared with 5.0–6.3% of those taking haloperidol (84).

Two patients developed acute dystonias while taking the lowest therapeutic dose of olanzapine (5 mg/day) (85), and tardive dyskinesia and tardive dystonia have also been reported.

- A 29-year-old woman with paranoid schizophrenia took olanzapine 20 mg/day for 3 months, stopped taking it, restarted it within 2 months, and 2.5 months later developed involuntary intermittent choreoathetoid knee movements and foot squirming in both legs; the movements persisted for more than 4 weeks (86). She denied any subjective sense of restlessness; other reasons for symptoms were excluded and tardive dyskinesia was diagnosed. The movements ceased after 4 months of treatment with quetiapine 300 mg/day later increasing to 600 mg/day. Some months later, her psychotic symptoms increased and she was given additional risperidone 1 mg/day. This resulted in writhing lower limb movements, which disappeared with risperidone withdrawal.

As a possible mechanism of these effects, the authors proposed dopamine D_2 receptor hypersensitivity.

Akathisia

Akathisia has been reported in 16% of patients taking olanzapine (SEDA-21, 56). Three patients developed severe akathisia during treatment with olanzapine (20–25 mg/day) (87). In two, the akathisia resolved after withdrawal of olanzapine and in one of those olanzapine was well tolerated when reintroduced in combination with lorazepam. In the third patient, the akathisia was controlled by dosage reduction. A 33-year-old man with AIDS and a prior history of extrapyramidal symptoms with both typical antipsychotic drugs and risperidone developed dose-dependent akathisia with olanzapine 15–19 mg/day; the akathisia responded to dosage reduction and beta-blockade (88).

- A 30-year-old woman with obsessive-compulsive disorder, who took olanzapine 10 mg/day titrated to 15 mg/day, developed severe akathisia after 3 weeks; the condition resolved on withdrawal and reappeared on rechallenge (89).

Of 10 patients with refractory panic disorder (mean age 35 years; three men) who were given olanzapine (mean dose 12.3 mg/day) one (no age and sex data given) developed significant akathisia at week 4 while taking 17.5 mg/day; it did not resolve at a lower dose of 12.5 mg/day (90[A]).

Restless legs syndrome shares clinical features with akathisia.

- A 41-year-old man complained of paresthesia in both legs at rest, much worse at night, and was relieved by walking around (91). The reaction occurred during the sixth week of treatment, 36 hours after an increase in dose from 10 to 20 mg/day; no other drug was administered.

Neuroleptic malignant syndrome

Several reports of neuroleptic malignant syndrome associated with olanzapine have been published (92–98).

- A 67-year-old man with bipolar disorder became confused, delirious, and manic (99). His only medications were olanzapine 10 mg/day and divalproex sodium 500 mg bd. On day 6, typical neuroleptic malignant syndrome developed. He had a fever (39.9°C), obtundation, rigidity, tremor, sweating, fluctuating pupillary diameter, labile tachycardia and hypertension, hypernatremia, and raised serum creatine kinase. Olanzapine was withdrawn and the syndrome resolved by day 12.

This patient had all of the major manifestations of this condition and there was no other likely explanation for his illness; he had taken no other drug likely to be associated with the syndrome.

- Another case of possible neuroleptic malignant syndrome associated with olanzapine has been reported in a patient who had taken clozapine for 3 years without incident (100). Symptoms suggestive of neuroleptic malignant syndrome appeared 19 days after the addition of olanzapine.

This severe reaction could have resulted from clozapine alone or from an interaction of clozapine with olanzapine.

- A 16-year-old boy developed fever, generalized rigidity, leukocytosis, and increased serum transaminase and creatine kinase activities while taking olanzapine and lithium; when both drugs were withdrawn, his fever and rigidity subsided and the biochemical tests returned to normal, without any complications (101).
- A 75-year-old man developed typical neuroleptic malignant syndrome while taking olanzapine; he had previously had haloperidol-associated neuroleptic malignant syndrome (102).
- A 39-year-old woman with no previous psychiatric history, who was given cefuroxime and gentamicin intravenously for a suspected pneumonia, became manic and delirious; she was given olanzapine 15 mg/day and then 30 mg/day and oxazepam 45 mg/day (103). During the next month she developed a fluctuating fever up to 40.9°C, including a 10-day normothermic period; she also developed autonomic instability, with a labile blood pressure and slight rigidity.

However, serum creatine kinase activity did not increase. The symptoms disappeared when olanzapine was withdrawn.

- A 23-year-old woman had some of the features of the serotonin syndrome (mental status changes, sweating, tremor, and fever); however, the large rise in creatine kinase activity, extreme lead-pipe rigidity, and the abrupt onset suggested neuroleptic malignant syndrome rather than the serotonin syndrome (104).
- An 85-year-old man developed fever, muscle rigidity, and changes in mental state, but his serum creatine kinase activity was not increased (105).
- A 42-year-old man took olanzapine for 3 weeks and developed hyperpyrexia, tremor, labile blood pressure, and mental changes; he had a metabolic acidosis and an escalating creatine kinase activity (106).
- A 78-year-old woman developed fulminant neuroleptic malignant syndrome complicated by pneumonia while taking olanzapine and levomepromazine. When the neuroleptic drugs were withdrawn she recovered; however, when the combination was restarted later, because of severe agitation and hallucinations, the symptoms of neuroleptic malignant syndrome recurred (107).
- A 30-year-old man who had taken olanzapine 20 mg/day for 10 days developed typical neuroleptic malignant syndrome with raised body temperature (39.7°C), obtundation, tremor, rigidity, sweating, fluctuating pupillary diameter, tachycardia, labile hypertension, raised serum creatine kinase activity, and severe hypernatremia (190 mmol/l) (108).
- A 53-year-old man developed atypical neuroleptic malignant syndrome, with fever, altered mental status, and autonomic dysfunction, but without rigidity (a usual feature of this condition) while taking olanzapine (109).

The authors of the last report pointed out that such atypical cases may support either a spectrum concept of neuroleptic malignant syndrome or the theory that neuroleptic malignant syndrome secondary to atypical neuroleptic drugs differs from that caused by conventional neuroleptic drugs. They suggested that more flexible diagnostic criteria than currently mandated in DSM-IV may be warranted.

Following a case of olanzapine-associated weight gain, hyperglycemia, and neuroleptic malignant syndrome in a 64-year-old woman the authors suggested that neuroleptic malignant syndrome might occur more often in patients aged 60 years or older (110). However, this patient had previously had well-documented neuroleptic malignant syndrome secondary to haloperidol, and on another occasion fluphenazine decanoate.

In contrast to these cases, a 34-year-old man, who had had clozapine-induced neuroleptic malignant syndrome, was successfully treated with olanzapine (111).

Parkinsonism

Whether treatment with olanzapine is useful in psychotic patients with neuroleptic drug-induced parkinsonism, or even in dopaminergic psychosis in Parkinson's disease, remains unclear. In a retrospective study of 19 patients

with parkinsonism, 10 had some worsening of their parkinsonism after taking olanzapine, 3 had some motor
benefit, and 7 had improvement in their psychosis (112).

Olanzapine was given to five patients with idiopathic
Parkinson's disease and hallucinosis, and after initial
treatment for 9 days, the frequency of hallucinations was
significantly reduced; during this early phase of treatment,
parkinsonian motor disability increased, which resulted in
medication discontinuation in two of the patients (113).

Worsening parkinsonism was observed in two patients
after treatment with olanzapine 5 mg/day (114). In contrast, coarse tremors induced by fluphenazine or haloperidol disappeared in three patients within days of the start
of treatment with olanzapine (10 mg/day), without discontinuation or reduction in the dosage of fluphenazine
or haloperidol (115). Olanzapine is active at muscarinic
cholinergic receptors, which may account for the
observed suppression of neuroleptic drug-induced tremor;
however, two of the three patients had been taking benzatropine, an antagonist at muscarinic acetylcholine
receptors, with little tremor relief, suggesting that olanzapine could suppress tremor by means of an action other
than muscarinic blockade.

Whether treatment with olanzapine is useful in psychotic symptoms in patients with Parkinson's disease is
unclear (SEDA-23, 66). Nine patients of twelve with
drug-induced psychosis in Parkinson's disease had worsening of motor function while taking olanzapine (2.5 mg/
day and increased in 2.5 mg increments as needed); this
worsening was considered dramatic in six of these
patients, and only one was still taking olanzapine at the
time of the analysis (12 months) (116). In addition, two
elderly patients, one with and one without pre-existing
parkinsonism, had marked rigidity induced by olanzapine
5 mg/day (117).

Tardive dyskinesias

Anecdotal reports of olanzapine-induced tardive dyskinesia have been described as occurring in 1% of treated
patients (SEDA-22, 65). Data from three pivotal trials
of olanzapine (SEDA-21, 56) have been reanalysed by
investigators from the manufacturers, Eli Lilly, with a
focus on tardive dyskinesia (118). In a previous study,
they had presented the percentage of patients developing
dyskinesia without regard to the time to development:
tardive dyskinesia occurred in 1.0% and 4.6% with olanzapine and haloperidol respectively (SEDA-22, 65). They
subsequently calculated the number of cases per patient
exposure time. When patients were stratified by abnormal
involuntary movement scale (AIMS) at baseline, the incidence rate ratios for those with AIMS = 0 was 5.67
(CI = 2.45, 13.1) and for those with AIMS > 0 it was
2.55 (CI = 1.15, 5.68), both favorable to olanzapine.
However, since only the highest dosages of olanzapine
were as effective as haloperidol in the trials, choosing all
the patients treated with olanzapine without regard to the
dosages yielded an underestimation of the risk. Since
olanzapine has a better profile of extrapyramidal adverse
effects, it has been proposed as an alternative in the long-
term treatment of schizophrenia (119).

Two cases of olanzapine-induced tardive dyskinesia
have been described (120).

- A 55-year-old man with Huntington disease took olanzapine 5 mg/day and rapidly developed tardive dyskinesia (121). After 2 months, his chorea improved
  markedly, but orofacial dyskinesia, which he had
  never had in the past, appeared. Olanzapine was withdrawn, and 1 week later the orofacial dyskinesia
  improved while the chorea worsened.

However, olanzapine may also improve pre-existing tardive dyskinesia (SEDA-22, 65) (SEDA-23, 66) (14,122). A
case of remission of tardive dyskinesia after changing
from flupenthixol to olanzapine has been reported (124).

- A 59-year-old man with a long history of neuroleptic
  drug exposure improved markedly when given olanzapine (2.5 mg/day, titrated within 5 days to 17.5 mg/
  day) (14).
- Respiratory dyskinesia, which was conceptualized as a
  form of tardive dyskinesia, including grunting, tachypnea, and impaired vocalization, improved after the
  patient started to take olanzapine 2.5 mg/day (122).

Tardive dystonia

- A 40-year-old woman with bipolar disorder developed
  tardive dystonia after taking olanzapine 10 mg at bedtime for 7 months (124).
- A 30-year-old woman with a schizophrenic disorder had a
  recurrence of tardive dystonia while taking olanzapine and
  was successfully treated with clozapine 150 mg/day (125).

Seizures

The incidence of seizures with olanzapine, which has been
estimated at 0.9% of treated patients, is probably comparable to that with other antipsychotic drugs.

- A 32-year-old woman with genetically confirmed
  Huntington's disease of 6 year's duration, who was
  treated with increasing doses of olanzapine and
  responded well to 30 mg/day, had a seizure (126).

The author pointed out that seizures are common in
juvenile-onset Huntington's disease but rare in adult-
onset Huntington's disease.

Myoclonic status can be triggered in susceptible
patients (127).

- A 54-year-old woman, with probable Alzheimer's disease, who was taking citalopram 20 mg/day and donepezil 5 mg/day and had paranoid ideas and agitation, was
  also given olanzapine 5 mg/day. During the next 48 hours
  she developed spontaneous and action-induced myoclonus in the trunk and limbs, which responded to clonazepam 1 mg/day and stopped when olanzapine was
  withdrawn. Electroencephalography showed slower
  background activity, with high-amplitude generalized
  spikes and continuous spike–wave and polyspike–wave
  complexes. She remained seizure free after withdrawal of
  clonazepam and when 9 months later she was given
  haloperidol 3 mg/day for new neuropsychiatric symptoms.

Fatal status epilepticus occurred in a patient taking olanzapine with no known underlying cause or predisposing factor for seizures.

- A 41-year-old woman developed seizures that progressed to status epilepticus, and died from secondary rhabdomyolysis and disseminated intravascular coagulation (128). She had been taking olanzapine 10 mg/day for 5 months. No other toxic, metabolic, or anatomical abnormalities were identified pre- or postmortem to explain the seizures. However, her medications also included levothyroxine 0.15 mg/day, clonazepam 1.0 mg qds, and propranolol 20 mg tds.

Non-fatal seizures have also been reported.

- A 31-year-old woman with multiple psychiatric and medical disorders, including a generalized seizure disorder (a probable confounding factor in this case), developed seizures when she switched from haloperidol to olanzapine (129).
- A 27-year-old woman had a seizure while taking a stable dosage of olanzapine 15 mg/day 1 day after the introduction of quetiapine 100 mg in the evening (130). She suddenly fell to the ground and had generalized shaking and inarticulate vocalization for about 30–60 seconds.

Seizures associated with olanzapine in premarketing studies have been estimated to occur in 0.88% of patients, a reported incidence probably comparable to that found with many conventional agents.

Speech dysfunction

Four patients taking olanzapine developed speech dysfunction (131). The authors suggested that the incidence of speech abnormalities may be higher than listed in the package insert of olanzapine (impairment of articulation 2% and voice alteration less than 1%).

Sleep-related eating disorder consists of partial arousal from sleep followed by rapid ingestion of food, commonly with at least partial amnesia for the episode on the next day; this disorder has been reported, purportedly for the first time, in association with an atypical neuroleptic drug (132).

- A 52-year-old man with bipolar I disorder and a family history of sleepwalking took olanzapine 10 mg/day and after several days had episodes of sleep-related eating disorder, witnessed by his wife; he had no memory of these episodes. After olanzapine withdrawal, the episodes disappeared rapidly.

Sensory systems

Esotropia (an inward squint) has been reported in a patient taking olanzapine and fluoxetine for psychosis; it resolved promptly on withdrawal of olanzapine (133).

- A 14-year-old girl with psychotic depression took fluoxetine 40 mg/day and olanzapine 5 mg/day, and 6 months later developed a severe headache, menorrhagia, diplopia, and eye irritation; her mother had also noted a "lazy eye." She had corrected visual acuity of 20/20 in both eyes, and intermittent esotropia of 14–16 diopters when fixing at 6 meters, and 8 diopters when fixing at 1/3 meters. A month later, her deviation had increased to 20 and 14 diopters respectively. Olanzapine was withdrawn and within 1 week the diplopia and headaches cleared, and the esotropia resolved.
- A 37-year-old man who had taken olanzapine 10 mg/day at bedtime for 1 week developed an excessive whitish discharge from the eyes (134).

The authors suggested that in some patients olanzapine may cause heavy mucus secretion in any of the mucus-secreting surfaces of the body.

Psychological, psychiatric

Hypofrontal symptoms, with problematic discourtesy and socially inconsiderate conduct, occurred in a 31-year-old man shortly after the introduction of olanzapine 20 mg/day; he had taken typical neuroleptic drugs for 13 years without such symptoms (135).

Anxiety and panic attacks

Anxiety has been reported as a common adverse effect (in 36% of patients) associated with olanzapine (SEDA-22, 65).

- A 36-year-old woman taking olanzapine 5 mg tds had new-onset panic attacks, with feelings of severe anxiety, shortness of breath, trembling, palpitation, and sweating, and felt that she was going to die (136). She had no previous history of panic attacks, which responded to the addition of alprazolam. The authors hypothesized that the 5-HT$_2$ antagonist action of olanzapine had triggered the onset of this condition.

Mania

Mania has rarely been associated with classic neuroleptic drugs, but has been described in patients treated with new antipsychotic drugs, especially risperidone (SEDA-22, 69). Several cases of mania, presumably associated with olanzapine, have been reported (99,137–144). The mechanism is not clear. It has been suggested that olanzapine could lead to manic symptoms in patients with schizophrenia because of its potent anti-5HT$_{2A}$ action.

- An 85-year-old woman with a 3-year history of delusional disorder developed florid manic symptoms, which resolved 2 weeks after withdrawal (145).
- A 36-year-old woman with paranoid schizophrenia was given olanzapine 15 mg/day and 3 weeks later she had a hypomanic episode (Young Mania Rating Scale, YMRS, score 30). The dosage of olanzapine was reduced to 10 mg/day and then to 5 mg/day, with minimal improvement of her manic symptoms (YMRS score 27). Olanzapine was withdrawn and a conventional neuroleptic drug given instead. The manic episode resolved over the next 4 days (YMRS score 5).
- Mania and hypomania occurred in four men and one woman aged 18–36 years taking olanzapine 10–20 mg/day (146).

Obsessive-compulsive disorder

Olanzapine-induced de novo obsessive-compulsive disorder has been reported in two cases (147). One of the patients developed de novo obsessive-compulsive symptoms with the introduction of olanzapine, whereas in the other case the patient had undisturbing obsessive symptoms before olanzapine treatment.

Olanzapine-induced worsening of obsessive-compulsive symptoms has also been reported.

- Obsessive-compulsive symptoms worsened in a 35-year-old man who had had a similar response to both clozapine and risperidone in previous trials (148).
- A 35-year-old woman developed obsessive-compulsive symptoms after starting to take olanzapine (149). Her symptoms remitted after withdrawal and recurred on rechallenge, which provides evidence for de novo provocation of obsessive-compulsive symptoms by olanzapine.
- A 42-year-old woman with bipolar II disorder who had taken olanzapine 5 mg/day for 5 weeks began to have severe enduring panic-like anxiety and serious obsessive-compulsive symptoms (150).

In three other cases, olanzapine caused significant exacerbation of obsessive-compulsive symptoms in schizophrenia (two cases) and obsessive-compulsive disorder (one case) (151).

The aims of a prospective study in 113 consecutively hospitalized young patients (mean age 22 years) were to determine whether the severity of obsessive-compulsive symptoms differs during treatment with olanzapine or risperidone and to establish whether the duration of neuroleptic treatment is related to the severity of obsessive-compulsive symptoms (152). At baseline and week 6 assessments, obsessive-compulsive symptoms were found in 32 of 106 evaluable cases and 16 met DSM-IV criteria for obsessive-compulsive disorder, but there were no differences in patients taking olanzapine or risperidone. However, the severity of obsessive-compulsive symptoms was associated with the duration of treatment with olanzapine.

Paranoia

Paranoia and agitation occurred in two patients, a 34-year-old man and a 30-year-old woman, who had taken olanzapine for a few weeks (153). Their symptoms improved when olanzapine was withdrawn.

Stuttering

Stuttering, which has previously been described with several neuroleptic drugs, has also been described with olanzapine. After they had seen one case, the authors of a recent study searched for this particular adverse effect in both out-patients and in-patients who had attended their clinic over the previous 3 years (154). Of 2100 new patients per year, 600 were taking neuroleptic drugs and there were seven patients with drug-induced stuttering, six taking olanzapine and one clozapine. The stuttering occurred on average 2–21 days after the start of treatment and ceased 2–5 days after withdrawal. Pre-existing brain pathology or concomitant antidepressants may have

contributed to this effect; only one of the seven patients had a history of stuttering.

Endocrine

Prolactin

Olanzapine can cause increased serum prolactin concentrations and galactorrhea, but probably to a lesser extent than haloperidol (SEDA-22, 65).

- A depressed 27-year-old woman taking mirtazapine developed hyperprolactinemia and galactorrhea after taking olanzapine 10 mg/day for 5 weeks (155).
- A 19-year-old woman with mild mental retardation and a history of birth anoxia who took olanzapine 15 mg/day for 3 weeks developed euprolactinemic galactorrhea (156). Laboratory tests were all normal, including a normal thyroid-stimulating hormone concentration and a prolactin concentration of 13 ng/ml (reference range 3–30 ng/ml). She had mild to moderate akathisia. Both the galactorrhea and the akathisia abated after substitution by quetiapine.

The authors stated that the reason why galactorrhea occurred is unclear but suggested that it may have been due to structural damage and greater sensitivity to prolactin resulting from the patient's anoxia at birth.

However, paradoxical cases of improvement in galactorrhea have also been observed (SEDA-23, 67).

Two women with neuroleptic-drug induced hyperprolactinemia, menstrual dysfunction, and galactorrhea had improvement in these adverse effects during treatment with olanzapine (157).

- A 35-year-old woman developed hyperprolactinemia after 2 months of risperidone treatment; the effects persisted after she switched to olanzapine, mean dose 2.5 mg/day (158).
- A 34-year-old woman, who developed amenorrhea while taking risperidone, regained her normal menstrual pattern along with a marked fall in serum prolactin concentration 8 weeks after being switched to olanzapine, whereas amantadine had failed to normalize the menses and had apparently reactivated the psychotic symptoms (159).

The authors suggested that olanzapine may offer advantages for selected patients in whom hyperprolactinemia occurs during treatment with other antipsychotic drugs.

Improvement in galactorrhea has also been observed in a case of trichotillomania refractory to a selective serotonin reuptake inhibitor (160). The patient only had a positive response with risperidone in combination with fluoxetine, but developed hyperprolactinemia and an intolerable galactorrhea. Olanzapine in combination with fluoxetine was started, with significant clinical improvement and without symptoms of galactorrhea; however, the patient had undesired weight gain of 3.6 kg after 22 weeks.

Metabolism

Diabetes mellitus

Hyperglycemia associated with olanzapine has a frequency of 1/100 to 1/1000. Hyperglycemia and diabetes have been associated with olanzapine, and published cases have suggested that these adverse effects may be caused by a mechanism related to weight gain (SEDA-23, 67; SEDA-24, 69; SEDA-25, 65). However, several cases of de novo onset or exacerbation of existing diabetes mellitus in patients treated with neuroleptic drugs have been reported and were not significantly related to weight gain. These included patients taking clozapine (161,162) or olanzapine (162,163).

- A 51-year-old woman with an 18-year history of type II diabetes mellitus developed glucose dysregulation with persistent hyperglycemia within 3 weeks of starting treatment with olanzapine, in the absence of weight gain (164).

The author suggested that olanzapine can cause glucose dysregulation by a mechanism other than weight gain.

- A 32-year-old African-American man with no prior history of diabetes mellitus or glucose intolerance had a raised blood glucose concentration after 6 weeks of olanzapine therapy, and required insulin (165). Olanzapine was withdrawn and blood glucose concentrations returned to normal about 2 weeks later. At rechallenge hyperglycemia occurred again.
- A 50-year-old man developed acute ketoacidosis with de novo diabetes mellitus after 8 months of adjunctive olanzapine (166). His dosage was then gradually titrated to 30 mg/day over 6 months, and after withdrawal of olanzapine his diabetes mellitus disappeared completely.
- A 31-year-old man taking olanzapine 10 mg/day, who had no family history of diabetes and had never had any laboratory evidence of diabetes or glucose intolerance, developed diabetic ketoacidosis; obesity was the only predisposing factor, his BMI being 40 kg/m^2 (167).
- A 45-year-old man with a 4-year history of diet-controlled diabetes had hyperglycemia with polyuria, polydipsia, and blurred vision associated with the use of olanzapine 10 mg/day. Within 1 week after olanzapine was withdrawn his blood glucose returned to normal and insulin was discontinued.
- A 54-year-old woman developed severe glucose dysregulation with exacerbation of type 2 diabetes 12 days after starting to take olanzapine 10 mg/day; she also gained 13 kg (168).
- A 39-year-old woman developed hyperglycemia after the dose of olanzapine was increased from 10 to 15 mg/day (169).

Six patients with new-onset diabetes mellitus in patients taking olanzapine (from 10 mg for 2 months to 25 mg for 22 months) were switched to quetiapine (170). Five of the six had known risk factors for diabetes mellitus (positive family history, obesity, race, and hyperlipidemia); only one gained significant body weight with olanzapine. There was a close temporal relation between the onset of therapy and the appearance of diabetes in three patients. The authors made recommendations about the detection and management of this effect in patients taking

neuroleptic drugs; they suggested that consideration should be given to testing for diabetes mellitus 3–7 months after starting neuroleptic drug treatment and that screening is ideally carried out by measuring the fasting plasma glucose concentration.

The effects of olanzapine on glucose–insulin homeostasis have been studied in 14 patients in an attempt to elucidate the possible mechanisms of olanzapine-associated weight gain (171). Olanzapine caused weight gain of 1–10 kg in 12 patients and raised concentrations of insulin, leptin, and blood lipids, as well as insulin resistance; three patients developed diabetes mellitus. The authors concluded that both increased insulin secretion and hyperleptinemia may be mechanisms behind olanzapine-induced weight gain. They also suggested that the metabolite N-desmethylolanzapine has a normalizing effect on the metabolic abnormalities.

In an open study in seven men and four women taking olanzapine (mean daily dose 12 mg and mean treatment duration 23 months), although the mean fasting triglyceride concentrations and mean fasting plasma glucose concentrations were similar to those found in the previous study, the mean fasting insulin concentrations were lower (143 pmol/l versus 228 pmol/l), and four of the subjects had hyperinsulinemia, compared with ten in the other study (172). However, the small sample sizes precluded any clear conclusions.

Susceptibility factors, such as family history, obesity, and concomitant medications can predispose an individual taking olanzapine to diabetes mellitus. New-onset diabetes mellitus developed after olanzapine was given to a 31-year-old man and a 44-year-old man (173). In both cases, the family history included diabetes mellitus, type unknown. The patients were taking various psychotropic drugs. In the first case body weight increased by about 12 kg (BMI 32 kg/m^2) 6 weeks after starting olanzapine, when his diabetes mellitus started; in the second case (BMI 26 kg/m^2, weight 81 kg) the previous weight was unknown.

Olanzapine-induced non-ketotic hyperglycemia has also been reported in the absence of obesity (174).

- A non-obese 51-year-old man without a history of diabetes mellitus had a serum glucose concentration of 89 mmol/l and was non-ketotic. Treatment with olanzapine had been started less than 6 months before; about 2 months before the event, his blood glucose concentration was 6.0 mmol/l, and 8 days after withdrawal the glucose concentration returned to normal; he no longer required insulin nor any other hypoglycemic drug.

The authors suggested that olanzapine can cause hyperglycemia by a mechanism other than weight gain.

Glucose concentrations have been studied in 47 patients with non-responsive schizophrenia taking olanzapine (175). Three of them, who had taken olanzapine for 3–6 months, had persistently high blood glucose concentrations. However, this is similar to what would be expected on the basis of the prevalence of diabetes mellitus in US adults.

Death from olanzapine-induced hyperglycemia has been reported (176).

- A 31-year-old man took olanzapine 10 mg/day for 1 week, and his fasting blood glucose rose to 11 mmol/l (200 mg/dl). For more aggressive treatment of his psychosis, the dosage of olanzapine was increased to 20 mg/day, and his fasting blood glucose rose to 16 mmol/l (280 mg/dl). He became progressively weaker and developed polydipsia and polyuria and died 3 weeks after starting to take olanzapine. He had no personal or family history of diabetes mellitus and was taking no other drugs at the time of his death. Death was attributed to hyperosmolar non-ketotic hyperglycemia.

The authors recommended including vitreous glucose and gammahydroxybutyrate analysis as part of postmortem toxicology work-up when the drug screen reveals either olanzapine or clozapine.

On the other hand, seven cases of asymptomatic lowered blood glucose concentrations have also been reported in patients with Tourette's syndrome who were taking olanzapine (mean dose 12 mg/day) during an 8-week, open-label trial (177). The mean serum glucose concentration was 4.8 mmol/l at baseline and 4.1 mmol/l during the study; the average weight gain was 4.5 kg. The authors suggested that increased insulin release may have been responsible for the changes observed; however, non-insulin mechanisms, such as a low carbohydrate intake, may also have played a role.

Weight gain

Significant weight gain occurs more often with olanzapine than with either haloperidol or risperidone (SEDA-22, 64).

Weight gain of 38.5 kg occurred in a 15-year-old adolescent who had taken olanzapine 5–10 mg/day for 14 months (178).

The medical records of ten patients with a DSM-IV diagnosis of cluster B personality disorder (narcissistic personality disorder) who had taken olanzapine 2.5–20 mg/day for 8 weeks have been reviewed (179). The mean Social Dysfunction and Aggression Scale score was 29 for the 8 weeks before olanzapine therapy and improved to 14 after 8 weeks of treatment. Five of the ten patients developed severe weight gain.

In 15 patients with excessive weight gain associated with olanzapine the mean weight gain was 11 (range 3.6–25) kg and the mean duration of treatment was 7 (range 2–11) months (180).

In a retrospective study of 16 patients (7 men, 9 women) taking olanzapine (mean dose 14 mg/day, range 10–30) there was weight gain of over 7% in 15 of them; no change in diet, access to food, or change in exercise pattern had occurred (181).

Weight gain was observed in nine patients (seven men, two women, mean age 41 years) treated with olanzapine (mean dose 19, range 10–30, mg/day) (182). The patients had a mean weight gain of 9.9 kg and triglyceride concentrations (reference range 0.3–2.3 mmol/l) increased from a mean of 1.9 mmol/l (range 0.8–4.3) to a mean of 2.7 mmol/l (range 1.5–4.2).

Average weight gain was 8 kg in patients with refractory schizophrenia (n = 8) who were taking olanzapine in high doses (20–40 mg over an average of 40 weeks) (183).

In a retrospective chart review, 20 consecutive patients, who requested a switch from their previous neuroleptic drug therapy to olanzapine, were monitored over 12 months to note changes in weight. After 12 months of olanzapine treatment, 13 had a mean weight gain of 7.3 kg and three had no significant change in weight (184). Paradoxically, four patients lost weight when taking olanzapine, and the authors claimed that this is the first report of patients who had weight loss with olanzapine, although in these cases it is difficult to determine which factors contributed to the weight loss.

Long-term olanzapine has been assessed in 27 outpatients with schizophrenia or schizoaffective disorders (mean age 40 years; 13 men) (185). At entry to the study the mean dose of olanzapine was 8.52 mg/day and the mean body mass index (BMI) 25 (range 19–35). At the end (mean treatment duration 22 months, range, 6–42) the mean dose of olanzapine was 8.70 mg/day and the mean BMI was 29 (range 20–40). Weight gain was more pronounced in the first year than in the second (7.7 versus 1.7 kg), especially during the first 3 months. Weight gain per month was significantly higher in patients with lower BMIs, but the greatest weight gain was in the most obese patient (BMI 35, weight gain 29 kg).

Weight gain has also been reported in 26 patients with bipolar affective disorder who were followed for 1 year while taking the combination of topiramate + olanzapine (mean modal doses 271 and 9.9 mg/day respectively). Although most of them gained weight during the first month of combined therapy (mean weight gain 0.7 kg), there was slight weight loss after 12 months (0.5 kg); weight loss was more pronounced in the obese patients (n = 5), and no patient with a low BMI (n = 3) lost weight (186). Patients who were switched to quetiapine from olanzapine lost weight after 10 weeks (mean weight loss 2.02 kg; n = 16 (187); although most of the patients lost weight, four gained about 2.6 kg.

In a randomized double-blind study there was weight gain in nine of 29 patients with alcohol dependence treated with olanzapine versus four of 31 taking placebo (188).

There was a mean weight gain of 7.9 kg (range 0–25 kg) between baseline and end-point weights in eight adolescents with psychoses (age range 12–18 years) who took olanzapine for 17.5 (range 4–26) weeks (189).

In contrast, in a study sponsored by Eli Lilly, there was improvement in 21 hospitalized elderly patients with schizophrenia or schizoaffective disorder who were taking olanzapine, mean final dose 13 mg/day. There was no significant weight gain (mean baseline weight 73.6 kg, mean final weight, 72.8 kg; mean treatment duration around 10 months) (190). However, the propensity for studies that are sponsored by pharmaceutical companies to be more favorable to their drug is well known (191).

In a study supported by the market authorization holder risperidone data from two double-blind trials in

552 adult and elderly patients with schizophrenia or schizoaffective disorders, weight gain after 8 weeks was higher with olanzapine than risperidone (mean doses not stated) in the adult and elderly patients and in smokers and non-smokers (192). For example, among the elderly patients, weight gain with olanzapine was 1.18 kg on average in smokers (n = 27) and 1.30 kg in non-smokers (n = 35); with risperidone weight gain was 0.08 kg in smokers (n = 20) and 1.06 kg in non-smokers (n = 31).

Famotidine did not prevent or attenuate weight gain in 14 patients taking olanzapine who were randomly assigned to either famotidine 40 mg/day (n = 7) or placebo (n = 7) in addition to olanzapine 10 mg/day for 6 weeks (mean weight gain 4.8 kg versus 4.9 kg respectively) (193).

There was significant weight gain in 12 drug-naive patients with a first-episode of psychosis who took olanzapine for 3–4 months (mean dose 10.7 mg/day) compared with a control group of four healthy volunteers (8.8 kg versus 1.2 kg) (194).

In 55 subjects randomized to olanzapine 10 mg/day, risperidone 4 mg/day, or placebo for 2 weeks, there were significant increases in weight with olanzapine (2.25 kg) and risperidone (1.05 kg) (195).

Taking advantage of this effect on weight, olanzapine 10 mg has been used in an open trial in 20 patients with anorexia nervosa (196). Of the 14 patients who completed the 10-week study, 10 gained an average of 3.9 kg and three of these attained their ideal body weight. The other four patients who completed the study lost a mean of 1.0 kg. The most common adverse effects were sedation ($n = 13$), headache ($n = 5$), fatigue ($n = 4$), and hypoglycemia ($n = 4$).

The genetic basis of some reactions associated with neuroleptic drugs is of particular interest. Different polymorphisms have been studied in connection with atypical neuroleptic drug-induced weight gain (SEDA-26, 57). The relation between the CYP2D6 genotype and weight gain in patients taking atypical neuroleptic drugs has been addressed in a study in 11 Caucasian patients taking a fixed dose of olanzapine 7.5–20 mg/day for up to 47 weeks (197). They had their DNA analysed for the CYP2D6*1, CYP2D6*3, and CYP2D6*4 alleles; six had two *1 alleles and the other five had either *1/*3 or *1/*4. Subjects with a heterozygous genotype (*1/*3 or *1/*4) had a statistically significantly larger percentage change in body mass index than those who were homozygous for the *1 allele.

Histamine H_2 receptor antagonists, like nizatidine, can control appetite in overweight patients (198).

- A 23-year-old man, who had had repeated episodes of weight gain during olanzapine treatment, had good control and subsequent weight reduction after 4–5 weeks of therapy with nizatidine.

In a double-blind trial, the efficacy of nizatidine in limiting weight gain has been evaluated in 175 patients with schizophrenia and related disorders who took olanzapine 5–20 mg, nizatidine 150 mg or 300 mg, or placebo for up to 16 weeks (199). There was significantly less weight gain

on average at weeks 3 and 4 with olanzapine plus nizatidine 300 mg compared with olanzapine plus placebo, but the difference was not statistically significant at 16 weeks.

Electrolyte balance

Hyponatremia has been reported in patients treated with olanzapine (200).

Fluid balance

Peripheral edema might be more frequent than expected in patients taking olanzapine. In a recent open, non-randomized study in 49 subjects taking olanzapine, 28 reported edema, which was severe in five (201). There were no significant differences regarding sex, dose of olanzapine or duration of treatment, concomitant diagnoses, or other psychotropic drugs, but there was a tendency toward greater frequency of thyroid abnormalities and older age in those with edema, in whom there was a positive correlation between age and severity.

Hematologic

Neutropenia and agranulocytosis
Although hemotoxicity was not observed during premarketing studies of olanzapine (202), cases of neutropenia (neutrophil count below 1.5×10^9/l) (203) or agranulocytosis (neutrophil count below 0.5×10^9/l) (204) have been published.

Six cases of hematological reactions, including leukopenia, granulocytopenia, and neutropenia, were reported in Canada in the 2 years after the approval of the drug in 1996 (205).

- A 20-year-old man with schizophrenia took olanzapine 5 mg/day increasing to 10 mg/day after 3 days (206). His total leukocyte count fell to 3.9×10^9/l on the fifth day, with an absolute neutrophil count of 1.52×10^9/l. Olanzapine was withdrawn and the leukocyte count normalized within 72 hours. A few days later rechallenge was positive. He was given lithium 900 mg/day and continued to take olanzapine 20 mg/day.
- A 17-year-old woman who was taking valproate 1000 mg/day and olanzapine 20 mg/day was changed to olanzapine monotherapy 25 mg/day; after 12 days her white blood cell count fell to 3.9×10^9/l and the absolute neutrophil count to 1.2×10^9/l; the counts returned to baseline when olanzapine was withdrawn (207).
- Neutropenia occurred in a 21-year-old man taking olanzapine 20 mg/day, haloperidol 10 mg/day, and lorazepam 2 mg/day (208). The white cell count was 2.6×10^9/l, with a 22% differential neutrophil count.

The authors of the second case stated that, in spite of decades of experience with haloperidol, an associated literature search has revealed only very rare instances of hemotoxicity, and since lorazepam is not known to cause leukopenia, olanzapine may have been the more likely offender; the simultaneous effect of both antipsychotic drugs, olanzapine and haloperidol, is unknown.

Mechanism

Some explanations of the lower risk of agranulocytosis have been advanced after an in vitro cytotoxicity study (209). Like clozapine, olanzapine is oxidized to a reactive nitrenium ion by HOCl, the major oxidant produced in activated neutrophils. However, the olanzapine-reactive metabolite has a lower propensity to cause toxicity to human neutrophils, monocytes, and HL-60 cells than the reactive clozapine nitrenium ion. The lower toxic potential of the olanzapine reactive metabolite, in conjunction with the lower therapeutic plasma concentrations of olanzapine compared with clozapine, may help to explain this difference between the drugs.

Dose relation

In three cases of leukopenia with olanzapine, a reduction in dose was followed by normalization of the leukocyte count and allowed continued treatment (210). Two of the patients had a previous history of neutropenia and agranulocytosis associated with typical neuroleptic drugs and the third developed neutropenia for the first time while taking olanzapine.

Time course

Of two cases of olanzapine-induced neutropenia, one occurred 17 days after the first dose of olanzapine and the other more than 5 months after the first dose (203); in the second patient, re-exposure to olanzapine caused the neutrophil count to fall again. In neither case was there evidence of infection, and the white blood cell counts increased immediately to the reference ranges after withdrawal of olanzapine, no special treatment being necessary. In addition, two cases of reversible leukopenia during treatment with olanzapine have also been reported (211).

Susceptibility factors

Genetic

Susceptibility factors for neutropenia and agranulocytosis have been investigated in two patients, one of whom had neutropenia after taking other drugs and one of whom was also taking amitriptyline (212). The risk of clozapine-induced agranulocytosis was significantly higher in patients with the HLA haplotypes DQB*01, DQB*05, and DQB*02. One of the patients who had neutropenia had HLA-DQB*05 and both patients had haplotype HLA-A2.

If there is an HLA association for olanzapine, it seems to be different from the HLA antigens incriminated for clozapine (SEDA-25, 66).

- In a 46-year-old man taking olanzapine 10 mg/day, leukopenia and neutropenia were associated with HLA types A1 24, B7, B35, DRB1*15, DRB1*11, DRB3*01–03, DRB5*01–02, a haplotype distinct from that previously observed in clozapine-induced hemotoxicity (213).

Cross-reaction with clozapine

Olanzapine-induced blood disorders have been reviewed and compared with clozapine-induced agranulocytosis (214). Whether clozapine-induced granulocytopenia or leukopenia predisposes a patient to hematological adverse effects during treatment with olanzapine is unclear: three patients who had previously stopped clozapine owing to hematologic adverse effects had improvement with olanzapine that equated to a 16–31 point reduction in rating scale scores during 1 year of follow-up without any hematological abnormalities (215). In contrast, one case associated with clozapine worsened when olanzapine was given (216).

- In a 31-year-old woman clozapine monotherapy 75 mg/day caused neutropenia (neutrophil count $1.1 \times 10^9/l$); 5 days after clozapine withdrawal the neutrophil count normalized ($2.6 \times 10^9/l$). Olanzapine was then introduced at 5 mg/day and the next day increased to 10 mg/day. After a week the neutrophil count fell to $0.9 \times 10^9/l$ and olanzapine was withdrawn. The neutrophil count was normal 4 weeks after olanzapine withdrawal.

In another case, olanzapine caused agranulocytosis after clozapine had had no effect on the white cell count (204).

- A 27-year-old man taking clozapine, benperidol, and risperidone developed severe extrapyramidal adverse effects and had insufficient antipsychotic effects and was given olanzapine, rapidly increasing to 40 mg/day. After 9 days of treatment with olanzapine, his total white cell count fell from $5.8 \times 10^9/l$ to $3.4 \times 10^9/l$, and olanzapine was withdrawn; 3 days later his neutrophil count was $0.39 \times 10^9/l$. He developed a fever of 39.5°C and was treated with granulocyte colony-stimulating factor.

Prolonged granulocytopenia due to olanzapine occurred in a 39-year-old woman after clozapine withdrawal (217). In contrast, two patients with severe clozapine-induced granulocytopenia and agranulocytosis were successfully treated with olanzapine in a dose greater than 25 mg/day (218). Furthermore, a 65-year-old man who had previously developed leukopenia and neutropenia, first with clozapine and then also with risperidone, took olanzapine (20 mg/day for 2 years with only a transient reduction in leukocyte and neutrophil (but not erythrocyte or platelet) counts) during a flu-like illness (219).

Seven patients, after developing eosinophilia ($n = 1$), agranulocytosis ($n = 2$), or neutropenia ($n = 4$) during neuroleptic drug therapy, were given olanzapine, with improvement and without evidence of blood dyscrasias (220,221). In a review it was suggested that this might reflect the fact that olanzapine shares many pharmacological properties with clozapine, which might make it more likely to prolong the clozapine-induced effect on white cells (222).

Thrombocytopenia

Thrombocytopenia, with a platelet count of $20 \times 10^9/l$, possibly associated with olanzapine and subsequently with benzatropine mesylate, has been reported in a 38-year-old woman (223).

A fatal case has also been reported (224).

- There was a fall in platelet count to 4000 x 10^9/l in a 78-year-old man who had taken olanzapine 10 mg/day for the last 3 weeks. He had massive spontaneous nasal and gingival bleeding and died from progressive bleeding complications. His plasma olanzapine concentration was 230 ng/ml (about 10 times higher than the usual target concentrations. He had had idiopathic thrombocytopenic purpura 13 years before.

Pancytopenia

Pancytopenia associated with exacerbation of motor disability induced by olanzapine (10 mg/day) in a patient with Parkinson's disease has been reported (225). Olanzapine withdrawal increased the neutrophil, total granulocyte, erythrocyte, and platelet counts to normal in 3 weeks, while the motor disability improved only moderately.

Eosinophilia

Eosinophilia occurred in a 30-year-old man taking olanzapine 22.5 mg/day (226). After 5 weeks, the white blood cell count increased to 15×10^9/l, of which 7.9×10^9/l were eosinophils. Olanzapine was withdrawn, and 5 weeks later the white blood cell count was 8.5×10^9/l with 14% eosinophils.

Salivary glands

While hypersalivation is a common adverse effect of clozapine (SEDA-22, 60), dry mouth is associated with olanzapine (SEDA-21, 57; SEDA-22, 64). Drooling as an adverse effect of olanzapine has been reported in a 20-year-old woman (227). Excessive, chalky-white, frothy, sticky salivation has been reported in a 27-year-old man who had taken olanzapine 10 mg/day at bedtime for 4 days (134).

Gastrointestinal

Olanzapine-induced fecal incontinence has been reported (228).

- A 65-year-old man who had had primary insomnia for 20 years, was given olanzapine 2.5 mg/day at night-time because of lack of response to various anxiolytics; he developed fecal incontinence during the 20 days of olanzapine treatment in combination with two anxiolytic drugs. The frequency of incontinence varied from 1 to 3 times a day, and withdrawal of olanzapine resulted in complete recovery.

Liver

Initial trials and other observational studies with olanzapine detected transient rises in liver transaminases (SEDA-23, 65). Several other cases have further illustrated this.

- There were significantly raised transaminases, up to 5 times the reference range, in a 37-year-old woman taking olanzapine 10 mg/day and in a 62-year-old woman taking 15 mg/day (229).

- A 29-year-old man had raised transaminases, pitting edema of both feet, and transient rises in systolic and diastolic blood pressures after taking olanzapine 20 mg/day; 14 days after olanzapine withdrawal, the transaminases returned to baseline, the edema cleared completely, and the blood pressure returned to normal (Farooque 203).
- Olanzapine caused increased transaminases in a 38-year-old man with hereditary coproporphyria; the enzyme changes were not associated with symptoms or evidence of either acute liver failure or exacerbation of his porphyria (230).

Olanzapine-induced acute hepatitis has been reported, supposedly for the first time (231).

- A 78-year-old woman with no history of liver disease, alcohol intake, intravenous drug use, or blood transfusions, took olanzapine 10 mg/day; her only other medications were calcium + vitamin D, multivitamins, and occasional paracetamol. After 13 days she developed symptoms of acute hepatitis with abnormal liver function tests; hepatobiliary ultrasonography showed several gallstones without dilatation of the biliary ducts or changes of acute cholecystitis. The plasma paracetamol concentration was less than 7 μmol/l, and serological tests for hepatitis A, B, and C viruses, cytomegalovirus, Epstein–Barr virus, and antimitochondrial and antinuclear antibodies were negative.

Pancreas

Acute pancreatitis has been reported in a patient who did not take alcohol and who had undergone a cholecystectomy in the past; other medical causes of pancreatitis were ruled out (232).

Acute pancreatitis was subsequently discussed (233,234). The main points of debate were the interpretation of attributability and the use of the Naranjo probability scale in a case in which the patient was taking another drug that could not be completely excluded as at least a partial contributor to the acute pancreatitis.

Urinary tract

Urinary incontinence has been attributed to olanzapine.

- A 61-year-old man with bipolar disorder treated with olanzapine and lithium developed urinary incontinence 4 days later (235). He was successfully treated with ephedrine 25 mg/day.

The authors suggested that the alpha-blocking effect of olanzapine was involved and that an alpha-adrenoceptor agonist (such as ephedrine) would reduce urinary incontinence in such cases.

Sexual function

Priapism has been attributed to olanzapine.

- A 68-year-old taking olanzapine 5 mg at bedtime developed priapism and required emergency surgery (236).

- In a 46-year-old man taking olanzapine 15 mg/day, priapism remitted after withdrawal (237).
- A 27-year-old man had taken olanzapine 15 mg at bedtime for 12 days for hallucinations and then developed priapism (238). Partial detumescence was followed by recurrence, and he required an operation to insert a glandular shunt. He subsequently required mechanical support to achieve an erection.
- Recurrent priapism during clozapine and then olanzapine therapy occurred in a 43-year-old man (239).
- A 26-year-old man, who had previously taken a variety of psychotropic medications, including typical neuroleptic drugs and risperidone, without sexual adverse effects, took olanzapine 10 mg at bedtime (240). Soon after, he developed priapism; 24 hours after withdrawal of olanzapine, the adverse effect disappeared.
- A 51-year-old man developed priapism 16 hours after attempting suicide by taking olanzapine 100 mg and gabapentin 1500 mg (241). Detumescence was produced by two injections of lidocaine and 8 hours later an intracorporeal shunt.

Irreversible priapism has been reported during treatment with olanzapine and lithium.

- A 30-year-old man with bipolar disorder who had previously taken haloperidol and lithium was also given trihexyphenidyl (benzhexol) because of rigidity, tremor, and akathisia, but persistent symptoms led to his being given olanzapine in place of haloperidol; 6 days later he developed priapism (242).

Olanzapine, was the likely causal agent; lithium was not thought to be causative in this instance.

Immunologic

Hypersensitivity syndrome, defined as a drug-induced complex consisting of fever, rash, and internal organ involvement, has been associated with olanzapine (243).

- A 34-year-old man took clozapine for several months, but developed a cardiomyopathy. Clozapine was withdrawn and olanzapine 20 mg/day was given instead; 60 days later he developed a recurrent high fever, rash, and pruritus. There was bilateral periorbital edema and generalized erythroderma without target lesions or bullae and no mucosal involvement. He also had an eosinophilia and hepatitis.

A Guillain–Barré-like syndrome has been associated with olanzapine hypersensitivity in a patient who already had an immunological disorder (244).

- A 58-year-old man with Vogt–Koyanagi–Harada syndrome, alopecia, vitiligo, and poliosis started to take olanzapine 5 mg/day for hypomania; he was also taking clonazepam 6 mg/day, valproic acid 600 mg/day, and prednisone 5 mg/day. Three weeks later he developed rapidly progressive numbness and weakness culminating in paresis of all four limbs, a generalized erythematous macular rash on the trunk and limbs, and hepatic dysfunction. He improved after withdrawal of olanzapine and 5 courses of plasma exchange over 10

days, and 6 months later his strength had returned to normal.

Body temperature

Neuroleptic drugs can cause a reduction in body temperature, and several cases of hypothermia, defined as a body temperature lower than 35°C, have been reported (SEDA-26, 58). Now hypothermia related to olanzapine has been published (245).

- A 54-year-old man with end-stage renal disease on hemodialysis took olanzapine 2.5 mg/day for 21 days because of night-time delirium, including visual hallucinations and abnormal behaviors; the delirium disappeared completely and his body temperature returned to normal. However, the delirium reappeared 7 days later and he took olanzapine again for 10 days; his body temperature suddenly fell to below 34° C after the first dose.

Susceptibility factors

Genetic

Gilbert's syndrome has been reported to increase susceptibility to the adverse effects of olanzapine.

- A 19-year-old man with Gilbert's syndrome took olanzapine 2.5 mg/day for 2 days, then increased the dosage to 5 mg/day on day 3 (246). On the day 4, because of a suicide attempt and extreme agitation, the patient was admitted to a psychiatric center. He was given oral olanzapine 10 mg and lorazepam 5 mg. On day 6, he was conscious but did not respond to verbal stimuli, and his symptoms of mutism persisted over the next few days. Communication was only possible by monosyllables on day 8. On day 10 he was bradypsychic, oriented, and able to articulate short sentences with great effort. Speech returned to normal on day 12. He described his experience as a sensation of not being able to find the words in his head. He had not previously had speech alterations, nor did they occur later.

Since mutism with olanzapine has been reported in cases of overdose, and detoxification of bilirubin by conjugation with glucuronic acid, the pathway olanzapine uses, is altered in Gilbert's syndrome, which affects 10% of the population, the authors claimed that we should keep in mind idiopathic unconjugated hyperbilirubinemia when prescribing olanzapine.

The role of the −759C/T polymorphism in the $5HT_{2C}$ receptor gene, located at q24 of the X chromosome, in weight gain from olanzapine has been examined in 42 subjects (age data not provided; 34 men) with schizophrenia who took olanzapine 7.5–20 mg/day for 4 weeks (mean endpoint serum concentration 24 ng/ml) (247). There was no difference in mean olanzapine dose between patients with the alleles T or C. Of the 42 patients, 15 gained more than 10% of their body weight and there were no T alleles in those subjects; of the other

27 patients without a 10% weight gain, 11 had a T allele (41%). Conversely, subjects with a C allele gained a mean of 12% over their initial body weight compared to those with a T allele, who gained a mean of 4.7%. These significant differences suggest a possible protective effect of the T allele on weight gain associated with olanzapine, although other confounding variables (such as diet and exercise) were not recorded in this study.

Smoking

Plasma concentrations of olanzapine are lower in smokers than in non-smokers, mainly because of induction of cytochrome CYP1A2 (SEDA-27, 62). In addition, in a recent study it has been investigated whether the smoking-inducible cytochrome CYP1A2 and the polymorphic CYP2D6 play significant roles in the metabolism of olanzapine and its clinical effects at steady-state; caffeine and debrisoquine were used as measures of CYP1A2 and CYP2D6 respectively (248). Psychiatric patients, smokers (n = 8) and non-smokers (n = 9), took oral olanzapine for 15 days. The mean urinary caffeine indexes of non-smokers and smokers indicated that smoking had induced a six-fold higher activity of CYP1A2; likewise, the dose-corrected plasma olanzapine concentration was about five-fold lower in smokers than in non-smokers. The authors suggested that a simple caffeine test might assist in individualization of the dosage of olanzapine.

Drug Administration

Drug overdose

Fatal overdoses in which novel antipsychotic drugs were the sole ingestant have been reported with olanzapine (249). In two cases blood olanzapine concentrations were 237 ng/ml in one and 675 ng/ml in the other (250). The usual target range for plasma olanzapine concentrations is 9–23 ng/ml.

- A 59-year-old woman died after taking an unknown quantity of olanzapine; the blood concentration was 4900 ng/ml (249).
- A 43-year-old man (251) died within hours of ingestion of 600 mg of olanzapine; the drug concentration in a postmortem analysis of the blood was 12 400 ng/ml.
- A 24-year-old man took a presumed cumulative dose of 420 mg of olanzapine and 10 mg of alprazolam (252). He had a cardiac arrest with asystole, from which he was initially resuscitated. Recurrent cardiac arrest, probably caused by hyperkalemia, occurred in the intensive care unit and he died within 1 hour.

In a fatal case of olanzapine overdose a cardiac dysrhythmia, non-convulsive status epilepticus, and persistent choreoathetosis occurred consecutively (253).

- A 62-year-old man with bipolar disorder taking olanzapine 30 mg/day and lithium 1200 mg/day attempted suicide with an estimated 750 mg of olanzapine. He developed delirium, a ventricular tachycardia, and cardiac asystole, which responded to resuscitation. On day

3 electroencephalography showed generalized frequent small amplitude spike and wave complexes with no limb movements, which resolved with intravenous fosphenytoin. His consciousness improved over the next 2 weeks, but he then lapsed progressively into coma and had choreoathetosis and dystonia in his head and all limbs. He died on day 57 from congestive heart failure and pneumonia. Post-mortem findings showed bilateral bronchopneumonia, mild cortical atrophy, and enlargement of the lateral ventricles without other major lesions; at microscopy, there was reactive astrocytosis in the striatum, globus pallidus, and thalamus bilaterally.

Several non-fatal cases have also been published.

- Profound central nervous system depression and tachycardia without other dysrhythmias occurred within 2 hours of an olanzapine overdose (800 mg; the highest serum concentration was 991 ng/ml) (254). The patient recovered after intubation, gut decontamination, and supportive care.
- A patient took an overdose of 1110 mg of olanzapine and had tachypnea, sinus tachycardia, fluctuating blood pressure, and brief hypoxemia; respiratory and cardiovascular function returned to normal within 16 hours of ingestion with minimal interventions (255).
- A patient attempted suicide by taking 120 mg of olanzapine, but had no serious clinical adverse effects, and only fatigue, dizziness, and headache (256).
- A 22-year-old man took about 800 mg of olanzapine; his olanzapine serum concentration reached a maximum of 200 ng/ml (257). His vital signs were stable at all times, but he started to become progressively somnolent, with short periods of aggressive agitation. Gastric lavage was performed and after 10 hours he was alert and oriented.

In four cases of overdose with up to 1000 mg there was significant central nervous system depression and miosis after acute overdosage of olanzapine (258). Olanzapine concentrations were over 250, 59, 54, and 151 ng/ml. All four patients recovered with supportive care and two required advanced airway support. The authors pointed out that all the patients had marked miosis and depressed mental status, findings that are usually associated with intoxication with opioids or alpha$_2$-adrenoceptor agonists.

Overdose of olanzapine has reportedly caused diabetes insipidus.

- A 17-year-old who took olanzapine 75 mg and prazepam 7.5 mg in a suicide attempt developed polyuria (5400 ml/24 hours), reduced urine osmolality (166 mosmol/kg H$_2$O), normal plasma osmolality, and an increasing serum sodium concentration, consistent with a diagnosis of diabetes insipidus (259).

Several cases of olanzapine overdose have been reported in children.

- An 11-kg, 1-year-old child developed agitation followed by prolonged lethargy after accidental ingestion of an unknown amount of olanzapine (260).

- An 18-month-old boy took 30–40 mg of olanzapine and had respiratory distress and mental status changes (261).
- Prolonged central nervous system depression occurred after a medication error in a 17-kg, 6-year-old girl who was given Zypresa (olanzapine 10 mg) instead of Zyrtec (cetirizine 10 mg) (262).
- In a dispensing error a 14-year-old boy was given olanzapine 80 mg/day, eight times the recommended dose in adults (263).
- A 2.5-year-old boy took one or two tablets of 7.5 mg and exhibited agitation, aggressive behavior, miosis, hypersalivation, tachycardia, and ataxia (264). A similar dose was well tolerated by an 8-year-old autistic boy (265).

Drug–Drug Interactions

General

Pharmacokinetic and pharmacodynamic profiles of olanzapine have been extensively reviewed (266). Olanzapine does not inhibit CYP isozymes, and no clinically significant metabolic interactions were found of olanzapine with aminophylline, biperiden, diazepam, ethanol, fluoxetine, imipramine, lithium, or R/S-warfarin.

Carbamazepine

Patients co-medicated with carbamazepine had a median dose-corrected plasma olanzapine concentration 36% lower than those taking monotherapy (267).

The pharmacokinetics of two single therapeutic doses of olanzapine have been determined in 11 healthy volunteers, before and after carbamazepine (268). The dose of olanzapine given after pretreatment with carbamazepine for weeks was cleared more rapidly than olanzapine given alone. Olanzapine C_{max} and AUC were significantly lower after the second dose, the half-life was significantly faster, and the clearance and volume of distribution were significantly increased. This interaction may be attributed to induction of CYP1A2 by carbamazepine, leading to increased first-pass and systemic metabolism of olanzapine.

The effect of carbamazepine on the glucuronidation of olanzapine has been studied in 30 patients taking olanzapine monotherapy (dosage 2.5–30, median 15, mg/day) and in 15 patients being co-medicated with carbamazepine (dosage 5–50, median 20, mg/day) (269). The median ratio of unbound olanzapine concentration to daily dose in the carbamazepine group was 38% lower than in the monotherapy group, confirming that carbamazepine accelerates the metabolism of olanzapine. Furthermore, in the carbamazepine group, the median glucuronidated olanzapine fraction was 79% of the unbound fraction, compared with 43% in the monotherapy group, which suggests that an increased rate of olanzapine glucuronidation contributes to the increased rate of metabolism of olanzapine induced by carbamazepine.

Ciprofloxacin

The plasma concentration of olanzapine doubled in a patient who also took ciprofloxacin, a potent inhibitor of CYP1A2 (270). The magnitude of the interaction was surprising, because available data suggest that CYP1A2-mediated oxidation of olanzapine accounts for only a small portion of the biotransformation of olanzapine relative to glucuronidation.

Diazepam

There were orthostatic changes when olanzapine and diazepam were co-administered (266).

Ethanol

There were orthostatic changes when olanzapine and ethanol were co-administered (266).

Inhibitors of CYP2D6

Of 56 patients, 22 of whom took olanzapine monotherapy and the rest took other psychotropic drugs, those co-medicated with inhibitors of CYP2D6 and other drugs had a median dose-corrected concentration about 40% higher than those taking monotherapy.

Insulin

Olanzapine has been reported to have precipitated diabetes (271).

- A 31-year-old man with a treatment-refractory psychiatric disorder without prior diabetes was given olanzapine 10 mg/day. After 3 months he developed hyperglycemia and an acidosis (pH 7.11). After treatment he needed at least 64 U/day of insulin, but 15 days after stopping olanzapine his insulin requirements fell and 15 days later insulin was withdrawn.

Lithium

The neuroleptic malignant syndrome occurred when lithium was added to olanzapine (272).

- A 13-year-old boy with rhabdomyolysis ascribed to olanzapine was also taking lithium, so that a drug interaction could not be excluded (273).
- A 16-year-old boy developed the neuroleptic malignant syndrome when his olanzapine dose was increased (274).

In a double-blind study of 344 patients inadequately responsive to lithium or valproate who were randomized to olanzapine or lithium for 6 weeks, 21% gained weight on lithium plus olanzapine compared with 4.9% taking lithium and placebo (275). Whether lithium contributed to weight gain in the olanzapine group is unclear.

Selective serotonin re-uptake inhibitors (SSRIs)

Fluoxetine
Pharmacokinetic and pharmacodynamic profiles of olanzapine have been extensively reviewed (266). Olanzapine

does not inhibit CYP isozymes, and no clinically significant metabolic interaction was found with fluoxetine.

Fluvoxamine

The atypical antipsychotic drug olanzapine is also metabolized by CYP1A2.

- A 21-year-old woman with schizophrenia and depression, who had been taking fluvoxamine (150 mg/day) and olanzapine (15 mg/day) for several months, developed an extrapyramidal movement disorder, including rigidity and tremor (276). The plasma fluvoxamine concentration was 70 ng/ml (usual target range is 20–500 ng/ml), while that of olanzapine was 120 ng/ml (usual target range is 9–25 ng/ml). The dosage of olanzapine was reduced to 5 mg/day and the plasma olanzapine concentration fell to 38 ng/ml, with resolution of the tremor and rigidity. When fluvoxamine was replaced with paroxetine (20 mg/day) the olanzapine concentration fell further to 22 ng/ml.

Of the SSRIs, fluvoxamine is the most potent inhibitor of CYP1A2 and is therefore likely to increase plasma olanzapine concentrations. The extrapyramidal effects in this case were presumably due to excessive blockade of dopamine D_2 receptors by raised olanzapine concentrations.

Fluvoxamine under steady-state conditions increases the systemic availability of olanzapine and inhibits the metabolism of clozapine, as shown in 21 male non-smoking Chinese volunteers (mean age 27 years) (277). This could be related to the different metabolic pathways and secretion rates of the two drugs; it would be advisable to reduce the dosage of olanzapine and to extend the dosing interval of clozapine when they are combined with fluvoxamine.

Similar results and recommendations have been published in a study of 71 patients with schizophrenia (mean age 33 years, range 18–63, 40 men who were given olanzapine monotherapy or olanzapine + flupentixol, benperidol, carbamazepine, and other drugs (278).

To determine whether a subtherapeutic dose of fluvoxamine, a potent inhibitor of CYP1A2, could affect the metabolism of olanzapine, male smokers with stable psychotic illnesses taking olanzapine (mean dose 17.5 mg/day) were switched to a mean dose of 13.0 mg/day and were given fluvoxamine 25 mg/day (279). At 2, 4, and 6 weeks there were no significant changes in olanzapine plasma concentration, antipsychotic response, or metabolic indices (for example serum glucose and lipids). The ratio of 4'-N-desmethylolanzapine:olanzapine fell from 0.45 at baseline to 0.25 at week 6, suggesting inhibition of CYP1A2-mediated olanzapine 4'-N-demethylation by fluvoxamine. In conclusion, these results suggested that a 26% reduction in olanzapine therapeutic dose requirement may be achieved by co-administration of a subtherapeutic oral dose of fluvoxamine.

Paroxetine

There have been reports that selective serotonin reuptake inhibitors, which inhibit CYP1A2, increase plasma olanzapine concentrations (SEDA-24, 71; SEDA-26, 63). In a recent open add-on trial, 21 patients with obsessive-compulsive disorder unresponsive to treatment with paroxetine 60 mg/day for at least 12 weeks, took additional olanzapine 10 mg/day (280). Steady-state plasma concentrations of paroxetine were not changed, and 7 patients were rated as responders at final evaluation. Sedation (n = 12), weight gain up to 3 kg (n = 8), dry mouth (n = 6), and constipation (n = 3) were the most frequent adverse effects.

Tricyclic antidepressants

Clomipramine and olanzapine are both metabolized by CYP1A2 and CYP2D6, and it is therefore possible that raised concentrations of both compounds can result from co-administration.

- Seizures in a 34-year-old man were attributed to an interaction of olanzapine with clomipramine (281).

A pharmacodynamic interaction occurred between olanzapine and imipramine (266).

Smoking

Plasma concentrations of olanzapine and clozapine are lower in smokers than in non-smokers, mainly because of induction of CYP1A2. Two patients who smoked tobacco and cannabis have been reported (282). One took olanzapine and had extrapyramidal motor symptoms after reducing his consumption of tobacco. The second, who was taking clozapine, developed confusion after he stopped smoking tobacco and cannabis, related to increased clozapine concentrations. The authors recommended that when patients stop smoking, appropriate dosage adjustments should be made to ensure that the plasma neuroleptic drug concentrations remain within the usual target range.

Management of adverse drug reactions

Various therapies have been used to avoid or to control weight gain during olanzapine treatment, including nizatidine (SEDA-25, 65), famotidine (SEDA-29, 74), and behavioral therapy (SEDA-26, 58). Topiramate has also been used to treat weight gain in 43 women who had taken olanzapine for at least 3 months in a 10-week, double-blind, placebo-controlled study (283). Those who took topiramate lost on average 5.6 kg (95% CI= 3.0, 8.5).

Sibutramine up to 15 mg/day has also been assessed in 37 subjects with schizophrenia or schizoaffective disorder and olanzapine-associated weight gain in a 12-week, double-blind, randomized study (284). Those who took sibutramine had significantly greater mean weight loss than those who took placebo (3.76 kg versus 0.82 kg). However, the patients who took sibutramine had a mean increase in systolic blood pressure of 2.1 mmHg, anticholinergic adverse effects, and sleep disturbances.

Monitoring therapy

In an open 2-week study in 54 in-patients with schizophrenia (aged 18–75 years; 38 men), olanzapine had a beneficial effect at a plasma concentration of 20–50 ng/ml (285); the authors suggested that olanzapine plasma concentration measurement may be useful in optimizing acute treatment in some patients.

References

1. Wood A. Clinical experience with olanzapine, a new atypical antipsychotic. Int Clin Psychopharmacol 1998;13(Suppl 1):S59–62.
2. Kasper S. Risperidone and olanzapine: optimal dosing for efficacy and tolerability in patients with schizophrenia. Int Clin Psychopharmacol 1998;13(6):253–62.
3. Gray R. Olanzapine: efficacy in treating the positive and negative symptoms of schizophrenia. Ment Health Care 1998;1(6):193–4.
4. Tollefson GD, Kuntz AJ. Review of recent clinical studies with olanzapine. Br J Psychiatry Suppl 1999;37:30–5.
5. Stephenson CM, Pilowsky LS. Psychopharmacology of olanzapine. A review. Br J Psychiatry Suppl 1999;38:52–8.
6. Green B. Focus on olanzapine. Curr Med Res Opin 1999;15(2):79–85.
7. Bever KA, Perry PJ. Olanzapine: a serotonin-dopamine-receptor antagonist for antipsychotic therapy. Am J Health Syst Pharm 1998;55(10):1003–16.
8. Conley RR, Tamminga CA, Bartko JJ, Richardson C, Peszke M, Lingle J, Hegerty J, Love R, Gounaris C, Zaremba S. Olanzapine compared with chlorpromazine in treatment-resistant schizophrenia. Am J Psychiatry 1998;155(7):914–20.
9. Henderson DC, Nasrallah RA, Goff DC. Switching from clozapine to olanzapine in treatment-refractory schizophrenia: safety, clinical efficacy, and predictors of response. J Clin Psychiatry 1998;59(11):585–8.
10. Simpson GM, Varga E. Clozapine—a new antipsychotic agent. Curr Ther Res Clin Exp 1974;16(7):679–86.
11. Pi EH, Simpson GM. Atypical neuroleptics: clozapine and the benzamides in the prevention and treatment of tardive dyskinesia. Mod Probl Pharmacopsychiatry 1983;21:80–6.
12. Littrell KH, Johnson CG, Littrell S, Peabody CD. Marked reduction of tardive dyskinesia with olanzapine. Arch Gen Psychiatry 1998;55(3):279–80.
13. O'Brien J, Barber R. Marked improvement in tardive dyskinesia following treatment with olanzapine in an elderly subject. Br J Psychiatry 1998;172:186.
14. Lykouras L, Malliori M, Christodoulou GN. Improvement of tardive dyskinesia following treatment with olanzapine. Eur Neuropsychopharmacol 1999;9(4):367–8.
15. Almeida OP. Olanzapine for the treatment of tardive dyskinesia. J Clin Psychiatry 1998;59(7):380–1.
16. Marangell LB, Johnson CR, Kertz B, Zboyan HA, Martinez JM. Olanzapine in the treatment of apathy in previously depressed participants maintained with selective serotonin reuptake inhibitors: an open-label, flexible-dose study. J Clin Psychiatry 2002;63(5):391–5.
17. Lee CT, Conde BJ, Mazlan M, Visanuyothin T, Wang A, Wong MM, Walker DJ, Roychowdhury SM, Wang H, Tran PV. Switching to olanzapine from previous antipsychotics: a regional collaborative multicenter trial assessing 2 switching techniques in Asia Pacific. J Clin Psychiatry 2002;63(7):569–76.
18. Sharma V, Pistor L. Treatment of bipolar mixed state with olanzapine. J Psychiatry Neurosci 1999;24(1):40–4.
19. Sacristan JA, Gomez JC, Martin J, Garcia-Bernardo E, Peralta V, Alvarez E, Gurpegui M, Mateo I, Morinigo A, Noval D, Soler R, Palomo T, Cuesta M, Perez-Blanco F, Massip C. Pharmacoeconomic assessment of olanzapine in the treatment of refractory schizophrenia based on a pilot clinical study. Clin Drug Invest 1998;15:29–35.
20. McElroy SL, Frye M, Denicoff K, Altshuler L, Nolen W, Kupka R, Suppes T, Keck PE Jr, Leverich GS, Kmetz GF, Post RM. Olanzapine in treatment-resistant bipolar disorder. J Affect Disord 1998;49(2):119–22.
21. Zarate CA Jr, Narendran R, Tohen M, Greaney JJ, Berman A, Pike S, Madrid A. Clinical predictors of acute response with olanzapine in psychotic mood disorders. J Clin Psychiatry 1998;59(1):24–8.
22. Kumra S, Jacobsen LK, Lenane M, Karp BI, Frazier JA, Smith AK, Bedwell J, Lee P, Malanga CJ, Hamburger S, Rapoport JL. Childhood-onset schizophrenia: an open-label study of olanzapine in adolescents. J Am Acad Child Adolesc Psychiatry 1998;37(4):377–85.
23. Krishnamoorthy J, King BH. Open-label olanzapine treatment in five preadolescent children. J Child Adolesc Psychopharmacol 1998;8(2):107–13.
24. Conley RR, Kelly DL, Gale EA. Olanzapine response in treatment-refractory schizophrenic patients with a history of substance abuse. Schizophr Res 1998;33(1–2):95–101.
25. Madhusoodanan S, Suresh P, Brenner R, Pillai R. Experience with the atypical antipsychotics—risperidone and olanzapine in the elderly. Ann Clin Psychiatry 1999;11(3):113–8.
26. Dursun SM, Gardner DM, Bird DC, Flinn J. Olanzapine for patients with treatment-resistant schizophrenia: a naturalistic case-series outcome study. Can J Psychiatry 1999;44(7):701–4.
27. Weiss EL, Longhurst JG, Bowers MB Jr, Mazure CM. Olanzapine for treatment-refractory psychosis in patients responsive to, but intolerant of, clozapine. J Clin Psychopharmacol 1999;19(4):378–80.
28. Bogetto F, Bellino S, Vaschetto P, Ziero S. Olanzapine augmentation of fluvoxamine-refractory obsessive-compulsive disorder (OCD): a 12-week open trial. Psychiatry Res 2000;96(2):91–8.
29. Strauss J, DiMartini A. Use of olanzapine in hereditary coproporphyria. Psychosomatics 1999;40(5):444–5.
30. Walker Z, Grace J, Overshot R, Satarasinghe S, Swan A, Katona CL, McKeith IG. Olanzapine in dementia with Lewy bodies: a clinical study. Int J Geriatr Psychiatry 1999;14(6):459–66.
31. Potenza MN, Holmes JP, Kanes SJ, McDougle CJ. Olanzapine treatment of children, adolescents, and adults with pervasive developmental disorders: an open-label pilot study. J Clin Psychopharmacol 1999;19(1):37–44.
32. Fanous A, Lindenmayer JP. Schizophrenia and schizoaffective disorder treated with high doses of olanzapine. J Clin Psychopharmacol 1999;19(3):275–6.
33. Gomez JC, Sacristan JA, Hernandez J, Breier A, Ruiz Carrasco P, Anton Saiz C, Fontova Carbonell E. The safety of olanzapine compared with other antipsychotic drugs: results of an observational prospective study in

patients with schizophrenia (EFESO Study). Pharmacoepidemiologic Study of Olanzapine in Schizophrenia. J Clin Psychiatry 2000;61(5):335–43.

34. Chengappa KN, Pollock BG, Parepally H, Levine J, Kirshner MA, Brar JS, Zoretich RA. Anticholinergic differences among patients receiving standard clinical doses of olanzapine or clozapine. J Clin Psychopharmacol 2000;20(3):311–6.

35. Lucetti C, Bellini G, Nuti A, Bernardini S, Dell'Agnello G, Piccinni A, Maggi L, Manca L, Bonuccelli U. Treatment of patients with tardive dystonia with olanzapine. Clin Neuropharmacol 2002;25(2):71–4.

36. Ananth J. Is olanzapine a mood stabilizer? Can J Psychiatry 1999;44(9):927–8.

37. Brown ES, Khan DA, Suppes T. Treatment of corticosteroid-induced mood changes with olanzapine. Am J Psychiatry 1999;156(6):968.

38. Alvarez E, Bobes J, Gómez J-C, Sacristán JA, Cañas F, Carrasco JL, Gascón J, Gubert J, Gutiérrez M; EUROPA Study Group. Safety of olanzapine versus conventional antipsychotics in the treatment of patients with acute schizophrenia. A naturalistic study. Eur Neuropsychopharmacol 2003;13:39-48.

39. Taylor DM, Wright T, Libretto SE, for the Risperidone Olanzapine Drug Outcomes Studies on Schizophrenia (RODOS) UK Investigator Group. Risperidone compared with olanzapine in a naturalistic clinical study: a cost analysis. J Clin Psychiatry 2003;64:589-97.

40. Dennehy EB, Doyle K, Suppes T. The efficacy of olanzapine monotherapy for acute hypomania or mania in an outpatient setting. Int Clin Psychopharmacol 2003;18:143-5.

41. Hwang JP, Yang CH, Lee TW, Tsai SJ. The efficacy and safety of olanzapine for the treatment of geriatric psychosis. J Clin Psychopharmacol 2003;23:113-8.

42. Passik SD, Kirsh KL, Theobald DE, Dickerson P, Trowbridge R, Gray D, Beaver M, Comparet J, Brown J. A retrospective chart review of the use of olanzapine for the prevention of delayed emesis in cancer patients. J Pain Symptom Manage 2003;25:485-9.

43. Stewart RS, Nejtek VA. An open-label, flexible-dose study of olanzapine in the treatment of trichotillomania. J Clin Psychiatry 2003;64:49-52.

44. Gorski ED, Willis KC. Report of three case studies with olanzapine for chronic pain. J Pain 2003;4:166-8.

45. Khaldi S, Kornreich C, Dan B, Pelc I. Usefulness of olanzapine in refractory panic attacks. J Clin Psychopharmacol 2003;23:100-1.

46. Jakovljević, Šagud M, Mihaljević-Peleš A. Olanzapine in the treatment-resistant, combat-related PTSD–a series of case reports. Acta Psychiatr Scand 2003;107:394-6.

47. Yetimalar Y, Irtman G, Gürgör N, Başoğlu M. Olanzapine efficacy in the treatment of essential tremor. Eur J Neurol 2003;10:79-82.

48. Stratta P, Donda P, Rossi A, Rossi A. Executive function assessment of patients with schizophrenic disorder residual type in olanzapine treatment: an open study. Hum Psychopharmacol 2005;20:401-8.

49. Lambert M, Haro JM, Novick D, Edgell ET, Kennedy L, Ratcliffe M, Naber D. Olanzapine vs. other antipsychotics in actual out-patient settings: six months tolerability results from the European Schizophrenia Out-patient Health Outcomes study. Acta Psychiatr Scand 2005;111:232-43.

50. Mortimer A, Martin S, Lôo H, Peuskens J, for the SOLIANOL Study Group. A double-blind, randomized comparative trial of amisulpride versus olanzapine for 6 months in the treatment of schizophrenia. Int Clin Psychopharmacol 2004;19:63-9.

51. Wagner M, Quednow BB, Westheide J, Schlaepfer TE, Maier W, Kuhn KU. Cognitive improvement in schizophrenic patients does not require a serotonergic mechanism: randomized controlled trial of olanzapine vs amisulpride. Neuropsychopharmacology 2005;30:381-90.

52. Bitter I, Dossenbach MR, Brook S, Feldman PD, Metcalfe S, Gagiano CA, Furedi J, Bartko G, Janka Z, Banki CM, Kovacs G, Breier A; Olanzapine HGCK Study Group. Olanzapine versus clozapine in treatment-resistant or treatment-intolerant schizophrenia. Prog Neuro-psychopharmacol Biol Psychiatry 2004;28:173-80.

53. Tohen M, Baker RW, Altshuler LL, Zarate CA, Suppes T, Ketter TA, Milton DR, Risser R, Gilmore JA, Breier A, Tollefson GA. Olanzapine versus divalproex in the treatment of acute mania. Am J Psychiatry 2002;159(6):1011–7.

54. Dossenbach MRK, Folnegovic-Smalc V, Hotujac L, Uglesic B, Tollefson GD, Grundy SL, Friedel P, Jakovljevic M, and Olanzapine HGCH Study Group. Double-blind, randomized comparison of olanzapine versus fluphenazine in the long-term treatment of schizophrenia. Prog Neuropsychopharmacol Biol Psychiatry 2004;28:311-8.

55. Rosenheck R, Perlick D, Bingham S, Liu-Mares W, Collins J, Warren S, Leslie D, Allan E, Campbell EC, Caroff S, Corwin J, Davis L, Douyon R, Dunn L, Evans D, Frecska E, Grabowski J, Graeber D, Herz L, Kwon K, Lawson W, Mena F, Sheikh J, Smelson D, Smith-Gamble V, for the Department of Veterans Affairs Cooperative Study Group on the Cost-Effectiveness of Olanzapine. Effectiveness and cost of olanzapine and haloperidol in the treatment of schizophrenia. A randomized controlled trial. JAMA 2003;290:2693-702.

56. Kinon BJ, Ahl J, Rotelli MD, McMullen E. Efficacy of accelerated dose titration of olanzapine with adjunctive lorazepam to treat acute agitation in schizophrenia. Am J Emerg Med 2004;22:181-6.

57. Buchanan RW, Ball MP, Weiner E, Kirkpatrick B, Gold JM, McMahon RP, Carpenter WT Jr. Olanzapine treatment of residual positive and negative symptoms. Am J Psychiatry 2005;162:124-9.

58. Tohen M, Greil W, Calabrese JR, Sachs GS, Yatham LN, Oerlinghausen BM, Koukopoulos A, Cassano GB, Grunze H, Licht RW, Dell'Osso L, Evans AR, Risser R, Baker RW, Crane H, Dossenbach MR, Bowden CL. Olanzapine versus lithium in the maintenance treatment of bipolar disorder: a 12-month, randomized, double-blind, controlled clinical trial. Am J Psychiatry 2005;162:1281-90.

59. Biederman J, Mick E, Hammerness P, Harpold T, Aleardi M, Dougherty M, Wozniak J. Open-label, 8-week trial of olanzapine and risperidone for the treatment of bipolar disorder in preschool-age children. Biol Psychiatry 2005;58:589-94.

60. Revicki DA, Paramore LC, Sommerville KW, Swann AC, Zajecka JM. Divalproex sodium versus olanzapine in the treatment of acute mania in bipolar disorder: health-related quality of life and medical cost outcomes. J Clin Psychiatry 2003;64:288-94.

61. Hirschfeld RMA, Baker JD, Wozniak P, Tracy K, Sommerville KW. The safety and early efficacy of oral-loaded divalproex versus standard-titration divalproex, lithium, olanzapine, and placebo in the treatment of acute mania associated with bipolar disorder. J Clin Psychiatry 2003;64:841-6.

62. Breier A, Berg PH, Thakore JH, Naber D, Gattaz WF, Cavazzoni P, Walker DJ, Roychowdhury SM, Kane JM. Olanzapine versus ziprasidone: results of a 28-week double-blind study in patients with schizophrenia. Am J Psychiatry 2005;162:1879-87.

63. Simpson GM, Weiden P, Pigott T, Murray S, Siu CO, Romano SJ. Six-month, blinded, multicenter continuation study of ziprasidone versus olanzapine in schizophrenia. Am J Psychiatry 2005;162:1535-8.

64. Tohen M, Chengappa KN, Suppes T, Zarate CA Jr, Calabrese JR, Bowden CL, Sachs GS, Kupfer DJ, Baker RW, Risser RC, Keeter EL, Feldman PD, Tollefson GD, Breier A. Efficacy of olanzapine in combination with valproate or lithium in the treatment of mania in patients partially nonresponsive to valproate or lithium monotherapy. Arch Gen Psychiatry 2002;59(1):62–9.

65. Berk M, Ichim L, Brook S. Olanzapine compared to lithium in mania: a double-blind randomized controlled trial. Int Clin Psychopharmacol 1999;14(6):339–43.

66. Clark WS, Street JS, Feldman PD, Breier A. The effects of olanzapine in reducing the emergence of psychosis among nursing home patients with Alzheimer's disease. J Clin Psychiatry 2001;62(1):34–40.

67. Tohen M, Sanger TM, McElroy SL, Tollefson GD, Chengappa KN, Daniel DG, Petty F, Centorrino F, Wang R, Grundy SL, Greaney MG, Jacobs TG, David SR, Toma VOlanzapine HGEH Study Group. Olanzapine versus placebo in the treatment of acute mania. Am J Psychiatry 1999;156(5):702–9.

68. Battaglia J, Lindborg SR, Alaka K, Meehan K, Wright P. Calming versus sedative effects of intramuscular olanzapine in agitated patients. Am J Emerg Med 2003;21:192-8.

69. Rothschild AJ, Williamson DJ, Tohen MF, Schatzberg A, Andersen SW, Van Campen LE, Sanger TM, Tollefson GD. A double-blind, randomized study of olanzapine and olanzapine/fluoxetine combination for major depression with psychotic features. J Clin Psychopharmacol 2004;19:365-73.

70. Shapira NA, Ward HE, Mandoki M, Murphy TK, Yang MCK, Blier P, Goodman WK. A double-blind, placebo-controlled trial of olanzapine addition in fluoxetine-refractory obsessive-compulsive disorder. Biol Psychiatry 2004;55:553-5.

71. De Deyn PP, Martín M, Deberdt W, Jeandel C, Hay DP, Feldman PD, Young CA, Lehman DL, Breier A. Olanzapine versus placebo in the treatment of psychosis with or without associated behavioural disturbances in patients with Alzheimer's disease. Int J Geriatr Psychiatry 2004;19:115-26.

72. Baker RW, Brown E, Akiskal HS, Calabrese JR, Ketter TA, Schuh LM, Trzepacz P, Watkin JG, Tohen M. Efficacy of olanzapine combined with valproate or lithium in the treatment of dysphoric mania. Br J Psychiatry 2004;185:472-8.

73. Soler J, Pascual JC, Campins J, Barrachina J, Puigdemont D, Alvarez E, Perez V. Double-blind, placebo-controlled study of dialectical behavior therapy plus olanzapine for borderline personality disorder. Am J Psychiatry 2005;162:1221-4.

74. Fontaine CS, Hynan LS, Koch K, Martin-Cook K, Svetlik D, Weiner MF. A double-blind comparison of olanzapine versus risperidone in the acute treatment of dementia-related behavioral disturbances in extended care facilities. J Clin Psychiatry 2003;64:726-30.

75. Czekalla J, Beasley CM Jr, Dellva MA, Berg PH, Grundy S. Analysis of the QT_c interval during olanzapine

76. Cohen H, Loewenthal U, Matar MA, Kotler M. Reversal of pathologic cardiac parameters after transition from clozapine to olanzapine treatment: a case report. Clin Neuropharmacol 2001;24(2):106–8.

77. Gurovich I, Vempaty A, Lippmann S. QTc prolongation: chlorpromazine and high-dosage olanzapine. Can J Psychiatry 2003;48:348.

78. Dineen S, Withrow K, Voronovitch L, Munshi F, Nawbary MW, Lippmann S. QTc prolongation and high-dose olanzapine. Psychosomatics 2003;44:174-5.

79. Su K-P, Lane H-Y, Chuang C-L, Chen K-P, Shen WW. Olanzapine-induced QTc prolongation in a patient with Wolff–Parkinson–White syndrome. Schizophr Res 2004;66:191-2.

80. Kosky N. A possible association between high normal and high dose olanzapine and prolongation of the PR interval. J Psychopharmacol 2002;16(2):181–2.

81. Farooque R. Uncommon side effects associated with olanzapine. Pharmacopsychiatry 2003;36:83.

82. Kapur S, Zipursky RB, Remington G, Jones C, DaSilva J, Wilson AA, Houle S. 5-HT_2 and D_2 receptor occupancy of olanzapine in schizophrenia: a PET investigation. Am J Psychiatry 1998;155(7):921–8.

83. Allan ER, Sison CE, Alpert M, Connolly B, Crichton J. The relationship between negative symptoms of schizophrenia and extrapyramidal side effects with haloperidol and olanzapine. Psychopharmacol Bull 1998;34(1):71–4.

84. Landry P, Cournoyer J. Acute dystonia with olanzapine. J Clin Psychiatry 1998;59(7):384.

85. Alevizos B, Papageorgiou C, Christodoulou GN. Acute dystonia caused by low dosage of olanzapine. J Neuropsychiatry Clin Neurosci 2003;15:241.

86. Bressan RA, Jones HM, Pilowsky LS. Atypical antipsychotic drugs and tardive dyskinesia: relevance of D2 receptor affinity. J Psychopharmacol 2004;18:124-7.

87. Jauss M, Schroder J, Pantel J, Bachmann S, Gerdsen I, Mundt C. Severe akathisia during olanzapine treatment of acute schizophrenia. Pharmacopsychiatry 1998;31(4):146–8.

88. Meyer JM, Marsh J, Simpson G. Differential sensitivities to risperidone and olanzapine in a human immunodeficiency virus patient. Biol Psychiatry 1998;44(8):791–4.

89. Kirrane RM. Olanzapine-induced akathisia in OCD. Ir J Psychol Med 1999;16:118.

90. Hollifield M, Thompson PM, Ruiz JE, Uhlenhuth EH. Potential effectiveness and safety of olanzapine in refractory panic disorder. Depress Anxiety 2005;21:33-40.

91. Kraus T, Schuld A, Pollmacher T. Periodic leg movements in sleep and restless legs syndrome probably caused by olanzapine. J Clin Psychopharmacol 1999;19(5):478–9.

92. Margolese HC, Chouinard G. Olanzapine-induced neuroleptic malignant syndrome with mental retardation. Am J Psychiatry 1999;156(7):1115–6.

93. Levenson JL. Neuroleptic malignant syndrome after the initiation of olanzapine. J Clin Psychopharmacol 1999;19(5):477–8.

94. Apple JE, Van Hauer G. Neuroleptic malignant syndrome associated with olanzapine therapy. Psychosomatics 1999;40(3):267–8.

95. Burkhard PR, Vingerhoets FJ, Alberque C, Landis T. Olanzapine-induced neuroleptic malignant syndrome. Arch Gen Psychiatry 1999;56(1):101–2.

96. Gheorghiu S, Knobler HY, Drumer D. Recurrence of neuroleptic malignant syndrome with olanzapine treatment. Am J Psychiatry 1999;156(11):1836.

97. Hickey C, Stewart C, Lippmann S. Olanzapine and NMS. Psychiatr Serv 1999;50(6):836-7.

98. Marcus EL, Vass A, Zislin J. Marked elevation of serum creatine kinase associated with olanzapine therapy. Ann Pharmacother 1999;33(6):697-700.

99. Filice GA, McDougall BC, Ercan-Fang N, Billington CJ. Neuroleptic malignant syndrome associated with olanzapine. Ann Pharmacother 1998;32(11):1158-9.

100. Moltz DA, Coeytaux RR. Case report: possible neuroleptic malignant syndrome associated with olanzapine. J Clin Psychopharmacol 1998;18(6):485-6.

101. Berry N, Pradhan S, Sagar R, Gupta SK. Neuroleptic malignant syndrome in an adolescent receiving olanzapine-lithium combination therapy. Pharmacotherapy 2003;23:255-9.

102. Goveas JS, Hermida A. Olanzapine induced "typical" neuroleptic malignant syndrome. J Clin Psychopharmacol 2003;23:101-2.

103. Nielsen J, Bruhn AM. Atypical neuroleptic malignant syndrome caused by olanzapine. Acta Psychiatr Scand. 2005;112:238-40.

104. Sierra-Biddle D, Herran A, Diez-Aja S, Gonzalez-Mata JM, Vidal E, Diez-Manrique F, Vazquez-Barquero JL. Neuroleptic malignant syndrome and olanzapine. J Clin Psychopharmacol 2000;20(6):704-5.

105. Nyfort-Hansen K, Alderman CP. Possible neuroleptic malignant syndrome associated with olanzapine. Ann Pharmacother 2000;34(5):667.

106. Stanfield SC, Privette T. Neuroleptic malignant syndrome associated with olanzapine therapy: a case report. J Emerg Med 2000;19(4):355-7.

107. Jarventausta K, Leinonen E. Neuroleptic malignant syndrome during olanzapine and levomepromazine treatment. Acta Psychiatr Scand 2000;102(3):231-3.

108. Arnaout MS, Antun FP, Ashkar K. Neuroleptic malignant syndrome with olanzapine associated with severe hypernatremia. Hum Psychopharmacol 2001;16(3):279-81.

109. Reeves RR, Torres RA, Liberto V, Hart RH. Atypical neuroleptic malignant syndrome associated with olanzapine. Pharmacotherapy 2002;22(5):641-4.

110. Malyuk R, Gibson B, Procyshyn RM, Kang N. Olanzapine associated weight gain, hyperglycemia and neuroleptic malignant syndrome: case report. Int J Geriatr Psychiatry 2002;17(4):326-8.

111. Nemets B, Geller V, Grisaru N, Belmaker RH. Olanzapine treatment of clozapine-induced NMS. Hum Psychopharmacol 2000;15(2):77-8.

112. Friedman J. Olanzapine in the treatment of dopaminomimetic psychosis in patients with Parkinson's disease. Neurology 1998;50(4):1195-6.

113. Graham JM, Sussman JD, Ford KS, Sagar HJ. Olanzapine in the treatment of hallucinosis in idiopathic Parkinson's disease: a cautionary note. J Neurol Neurosurg Psychiatry 1998;65(5):774-7.

114. Jimenez-Jimenez FJ, Tallon-Barranco A, Orti-Pareja M, Zurdo M, Porta J, Molina JA. Olanzapine can worsen parkinsonism. Neurology 1998;50(4):1183-4.

115. Strauss AJ, Bailey RK, Dralle PW, Eschmann AJ, Wagner RB. Conventional psychotropic-induced tremor extinguished by olanzapine. Am J Psychiatry 1998;155(8):1132.

116. Molho ES, Factor SA. Possible tardive dystonia resulting from clozapine therapy. Mov Disord 1999;14(5):873-4.

117. Granger AS, Hanger HC. Olanzapine: extrapyramidal side effects in the elderly. Aust NZ J Medical 1999;29(3):371-2.

118. Beasley CM, Dellva MA, Tamura RN, Morgenstern H, Glazer WM, Ferguson K, Tollefson GD. Randomised double-blind comparison of the incidence of tardive dyskinesia in patients with schizophrenia during long-term treatment with olanzapine or haloperidol. Br J Psychiatry 1999;174:23-30.

119. Kane J. Olanzapine in the long-term treatment of schizophrenia. Br J Psychiatry Suppl 1999;37:26-9.

120. Herran A, Vazquez-Barquero JL. Tardive dyskinesia associated with olanzapine. Ann Intern Med 1999;131(1):72.

121. Benazzi F. Rapid onset of tardive dyskinesia in Huntington disease with olanzapine. J Clin Psychopharmacol 2002;22(4):438-9.

122. Gotto J. Treatment of respiratory dyskinesia with olanzapine. Psychosomatics 1999;40(3):257-9.

123. Haberfellner EM. Remission of tardive dyskinesia after changing from flupenthixol to olanzapine. Eur Psychiatry 2000;15(5):338-9.

124. Dunayevich E, Strakowski SM. Olanzapine-induced tardive dystonia. Am J Psychiatry 1999;156(10):1662.

125. García-Lado I, García-Caballero A, Recimil MJ, Area R, Ozaita G, Lamas S. Reappearance of tardive dystonia with olanzapine treated with clozapine. Schizophr Res 2005;76:357-8.

126. Bonelli RM. Olanzapine-associated seizure. Ann Pharmacother 2003;37:149-50.

127. Camacho A, Garcia-Navarro M, Martinez B, Villarejo A, Pomares E. Olanzapine-induced myoclonic status. Clin Neuropharmacol 2005;28:145-7.

128. Wyderski RJ, Starrett WG, Abou-Saif A. Fatal status epilepticus associated with olanzapine therapy. Ann Pharmacother 1999;33(7-8):787-9.

129. Lee JW, Crismon ML, Dorson PG. Seizure associated with olanzapine. Ann Pharmacother 1999;33(5):554-6.

130. Hedges DW, Jeppson KG. New-onset seizure associated with quetiapine and olanzapine. Ann Pharmacother 2002;36(3):437-9.

131. Gaile S, Noviasky JA. Speech disturbance and marked decrease in function seen in several older patients on olanzapine. J Am Geriatr Soc 1998;46(10):1330-1.

132. Paquet V, Strul J, Servais L, Pelc I, Fossion P. Sleep-related eating disorder induced by olanzapine. J Clin Psychiatry 2002;63(7):597.

133. Singh HK, Markowitz GD, Myers G. Esotropia associated with olanzapine. J Clin Psychopharmacol 2000;20(4):488.

134. Wahid Z, Ali S. Side effects of olanzapine. Am J Psychiatry 1999;156(5):800-1.

135. Swartz C, Walder M. Hypofrontal symptoms from olanzapine: a case report. Ann Clin Psychiatry 1999;11(1):17-9.

136. Mandalos GE, Szarek BL. New-onset panic attacks in a patient treated with olanzapine. J Clin Psychopharmacol 1999;19(2):191.

137. Pozo P, Alcantara AG. Mania-like syndrome in a patient with chronic schizophrenia during olanzapine treatment. J Psychiatry Neurosci 1998;23(5):309-10.

138. Benazzi F, Rossi E. Mania induced by olanzapine. Hum Psychopharmacol 1998;13:585-6.

139. Reeves RR, McBride WA, Brannon GE. Olanzapine-induced mania. J Am Osteopath Assoc 1998;98(10):549-50.

140. London JA. Mania associated with olanzapine. J Am Acad Child Adolesc Psychiatry 1998;37(2):135-6.

141. Lindenmayer JP, Klebanov R. Olanzapine-induced manic-like syndrome. J Clin Psychiatry 1998;59(6):318-9.

142. Benazzi F. Olanzapine-induced psychotic mania in bipolas schizoaffective disorder. Can J Psychiatry 1999;44(6):607-8.

143. Fitz-Gerald MJ, Pinkofsky HB, Brannon G, Dandridge E, Calhoun A. Olanzapine-induced mania. Am J Psychiatry 1999;156(7):1114.

144. Simon AE, Aubry JM, Malky L, Bertschy G. Hypomania-like syndrome induced by olanzapine. Int Clin Psychopharmacol 1999;14(6):377–8.

145. Narayan G, Puranik A. Olanzapine-induced mania. Int J Psychiatry Clin Pract 2000;4:333–4.

146. Akdemir A, Türkçapar H, Örsel S. Mania and hypomania with olanzapine use. Eur Psychiatry 2004;19:175-6.

147. Mottard JP, de la Sablonniere JF. Olanzapine-induced obsessive-compulsive disorder. Am J Psychiatry 1999;156(5):799–800.

148. Morrison D, Clark D, Goldfarb E, McCoy L. Worsening of obsessive-compulsive symptoms following treatment with olanzapine. Am J Psychiatry 1998;155(6):855.

149. al-Mulhim A, Atwal S, Coupland NJ. Provocation of obsessive-compulsive behaviour and tremor by olanzapine. Can J Psychiatry 1998;43(6):645.

150. Jonkers F, De Haan L. Olanzapine-induced obsessive-compulsive symptoms in a patient with bipolar II disorder. Psychopharmacology (Berl) 2002;162(1):87–8.

151. Lykouras L, Zervas IM, Gournellis R, Malliori M, Rabavilas A. Olanzapine and obsessive-compulsive symptoms. Eur Neuropsychopharmacol 2000;10(5):385–7.

152. de Haan L, Beuk N, Hoogenboom B, Dingemans P, Linszen D. Obsessive-compulsive symptoms during treatment with olanzapine and risperidone: a prospective study of 113 patients with recent-onset schizophrenia or related disorders. J Clin Psychiatry 2002;63(2):104–7.

153. al Jeshi A. Paranoia and agitation associated with olanzapine treatment. Can J Psychiatry 1998;43(2):195.

154. Bär KJ, Häger F, Sauer H. Olanzapine and clozapine induced stuttering. Pharmacopsychiatry 2004;37:131-4.

155. Licht RW, Arngrim T, Cristensen H. Olanzapine-induced galactorrhea. Psychopharmacology (Berl) 2002;162(1):94–5.

156. Kingsbury SJ, Castelo C, Abulseoud O. Quetiapine for olanzapine-induced galactorrhea. Am J Psychiatry 2002;159(6):1061.

157. Canuso CM, Hanau M, Jhamb KK, Green AI. Olanzapine use in women with antipsychotic-induced hyperprolactinemia. Am J Psychiatry 1998;155(10):1458.

158. Mendhekar DN, Jiloha RC, Srivastava PK. Effect of risperidone on prolactinoma. A case report. Pharmacopsychiatry 2004;37:41-2.

159. Gazzola LR, Opler LA. Return of menstruation after switching from risperidone to olanzapine. J Clin Psychopharmacol 1998;18(6):486–7.

160. Potenza MN, Wasylink S, Epperson CN, McDougle CJ. Olanzapine augmentation of fluoxetine in the treatment of trichotillomania. Am J Psychiatry 1998;155(9):1299–300.

161. Popli AP, Konicki PE, Jurjus GJ, Fuller MA, Jaskiw GE. Clozapine and associated diabetes mellitus. J Clin Psychiatry 1997;58(3):108–11.

162. Wirshing DA, Spellberg BJ, Erhart SM, Marder SR, Wirshing WC. Novel antipsychotics and new onset diabetes. Biol Psychiatry 1998;44(8):778–83.

163. Ober SK, Hudak R, Rusterholtz A. Hyperglycemia and olanzapine. Am J Psychiatry 1999;156(6):970.

164. Ramankutty G. Olanzapine-induced destabilization of diabetes in the absence of weight gain. Acta Psychiatr Scand 2002;105(3):235–6.

165. Fertig MK, Brooks VG, Shelton PS, English CW. Hyperglycemia associated with olanzapine. J Clin Psychiatry 1998;59(12):687–9.

166. Lindenmayer JP, Patel R. Olanzapine-induced ketoacidosis with diabetes mellitus. Am J Psychiatry 1999;156(9):1471.

167. Gatta B, Rigalleau V, Gin H. Diabetic ketoacidosis with olanzapine treatment. Diabetes Care 1999;22(6):1002–3.

168. Bettinger TL, Mendelson SC, Dorson PG, Crismon ML. Olanzapine-induced glucose dysregulation. Ann Pharmacother 2000;34(7–8):865–7.

169. Folnegovic-Smalc V, Jukić V, Kozumplik O, Mimica N, Uzun S. Olanzapine use in a patient with schizophrenia and the risk of diabetes. Eur Psychiatry 2004;19:62-4.

170. Ashim S, Warrington S, Anderson IM. Management of diabetes mellitus occurring during treatment with olanzapine: report of six cases and clinical implications. J Psychopharmacol 2004;18:128-32.

171. Melkersson KI, Hulting AL, Brismar KE. Elevated levels of insulin, leptin, and blood lipids in olanzapine-treated patients with schizophrenia or related psychoses. J Clin Psychiatry 2000;61(10):742–9.

172. Cohn TA, Remington G, Kameh H. Hyperinsulinemia in psychiatric patients treated with olanzapine. J Clin Psychiatry 2002;63(1):75–6.

173. Bonanno DG, Davydov L, Botts SR. Olanzapine-induced diabetes mellitus. Ann Pharmacother 2001;35(5):563–5.

174. Roefaro J, Mukherjee SM. Olanzapine-induced hyperglycemic nonketotic coma. Ann Pharmacother 2001;35(3):300–2.

175. Lindenmayer JP, Smith RC, Singh A, Parker B, Chou E, Kotsaftis A. Hyperglycemia in patients with schizophrenia who are treated with olanzapine. J Clin Psychopharmacol 2001;21(3):351–3.

176. Meatherall R, Younes J. Fatality from olanzapine induced hyperglycemia. J Forensic Sci 2002;47(4):893–6.

177. Budman CL, Gayer AI. Low blood glucose and olanzapine. Am J Psychiatry 2001;158(3):500–1.

178. Bryden KE, Kopala LC. Body mass index increase of 58% associated with olanzapine. Am J Psychiatry 1999;156(11):1835–6.

179. Zullino DF, Quinche P, Hafliger T, Stigler M. Olanzapine improves social dysfunction in cluster B personality disorder. Hum Psychopharmacol 2002;17(5):247–51.

180. Gupta S, Droney T, Al-Samarrai S, Keller P, Frank B. Olanzapine-induced weight gain. Ann Clin Psychiatry 1998;10(1):39.

181. Gupta S, Droney T, Al-Samarrai S, Keller P, Frank B. Olanzapine: weight gain and therapeutic efficacy. J Clin Psychopharmacol 1999;19(3):273–5.

182. Sheitman BB, Bird PM, Binz W, Akinli L, Sanchez C. Olanzapine-induced elevation of plasma triglyceride levels. Am J Psychiatry 1999;156(9):1471–2.

183. Bronson BD, Lindenmayer JP. Adverse effects of high-dose olanzapine in treatment-refractory schizophrenia. J Clin Psychopharmacol 2000;20(3):382–4.

184. Littrell KH, Petty RG, Hilligoss NM, Peabody CD, Johnson CG. Weight loss associated with olanzapine treatment. J Clin Psychopharmacol 2002;22(4):436–7.

185. Haberfellner EM, Rittmannsberger H. Weight gain during long-term treatment with olanzapine: a case series. Int Clin Psychopharmacol 2004;19:251-3.

186. Vieta E, Sánchez-Moreno J, Goikolea JM, Colom F, Martínez-Arán A, Benabarre A, Corbella B, Torrent C, Comes M, Reinares M, Brugue E. Effects of weight and outcome of long-term olanzapine–topiramate combination treatment in bipolar disorder. J Clin Psychopharmacol 2004;24:374-8.

187. Gupta S, Masand PS, Virk S, Schwartz T, Hameed A, Frank BL, Lockwood K. Weight decline in patients switching from olanzapine to quetiapine. Schizophr Res 2004;70:57-62.

188. Guardia J, Segura L, Gonzalvo B, Iglesias L, Roncero C, Cardús M, Casas M. A double-blind, placebo-controlled study of olanzapine in the treatment of alcohol-dependence disorder. Alcohol Clin Exp Res 2004;28:736-45.

189. Ercan ES, Kutlu A, Varan A, Çikoğlu S, Coşkunol H, Bayraktar E. Olanzapine treatment of eight adolescent patients with psychosis. Hum Psychopharmacol Clin Exp 2004;19:53-6.

190. Barak Y, Shamir E, Mirecki I, Weizman R, Aizenberg D. Switching elderly chronic psychotic patients to olanzapine. Int J Neuropsychopharmacol 2004;7:165-9.

191. Lexchin J, Bero LA, Djulbegovic B, Clark O. Pharmaceutical industry sponsorship and research outcome and quality: systematic review. BMJ 2003;326(7400):1167-70.

192. Lasser RA, Mao L, Gharabawi G. Smokers and nonsmokers equally affected by olanzapine-induced weight gain: metabolic implications. Schizophr Res 2004;66:163-7.

193. Poyurovsky M, Tal V, Maayan R, Gil-Ad I, Fuchs C, Weizman A. The effect of famotidine addition on olanzapine-induced weight gain in first-episode schizophrenia patients: a double-blind placebo-controlled pilot study. Eur Neuropsychopharmacol 2004;14:332-6.

194. Sengupta SM, Klink R, Stip E, Baptista T, Malla A, Joober R. Weight gain and lipid metabolic abnormalities induced by olanzapine in first-episode, drug-naive patients with psychotic disorders. Schizophr Res 2005;80:131-3.

195. Roerig JL, Mitchell JE, de Zwaan M, Crosby RD, Gosnell BA, Steffen KJ, Wonderlich SA. A comparison of the effects of olanzapine and risperidone versus placebo on eating behaviors. J Clin Psychopharmacol 2005;25:413-8.

196. Powers PS, Santana CA, Bannon YS. Olanzapine in the treatment of anorexia nervosa: an open label trial. Int J Eat Disord 2002;32(2):146–54.

197. Ellingrod VL, Miller D, Schultz SK, Wehring H, Arndt S. CYP2D6 polymorphisms and atypical antipsychotic weight gain. Psychiatr Genet 2002;12(1):55–8.

198. Sacchetti E, Guarneri L, Bravi D. H(2) antagonist nizatidine may control olanzapine-associated weight gain in schizophrenic patients. Biol Psychiatry 2000;48(2):167–8.

199. Cavazzoni P, Tanaka Y, Roychowdhury SM, Breier A, Allison DB. Nizatidine for prevention of weight gain olanzapine: a double-blind placebo-controlled trial. Eur Neuropsychopharmacol 2003;13:81-5.

200. Littrell KH, Johnson CG, Littrell SH, Peabody CD. Effects of olanzapine on polydipsia and intermittent hyponatremia. J Clin Psychiatry 1997;58(12):549.

201. Ng B, Postlethwaite A, Rollnik J. Peripheral oedema in patients taking olanzapine. Int Clin Psychopharmacol 2003;18:57-9.

202. Beasley CM Jr, Tollefson GD, Tran PV. Safety of olanzapine. J Clin Psychiatry 1997;58(Suppl 10):13-7.

203. Steinwachs A, Grohmann R, Pedrosa F, Ruther E, Schwerdtner I. Two cases of olanzapine-induced reversible neutropenia. Pharmacopsychiatry 1999;32(4):154–6.

204. Naumann R, Felber W, Heilemann H, Reuster T. Olanzapine-induced agranulocytosis. Lancet 1999;354(9178):566–7.

205. Anonymous. Olanzapine: hematological reactions. CMAJ 1998;159(1):81–285–6.

206. Gajwani P, Tesar GE. Olanzapine-induced neutropenia. Psychosomatics 2000;41(2):150–1.

207. Duggal HS, Gates C, Pathak PC. Olanzapine-induced neutropenia: mechanism and treatment. J Clin Psychopharmacol 2004;24:234-5.

208. Abdullah N, Voronovitch L, Taylor S, Lippmann S. Olanzapine and haloperidol: potential for neutropenia? Psychosomatics 2003;44:83-4.

209. Gardner I, Zahid N, MacCrimmon D, Uetrecht JP. A comparison of the oxidation of clozapine and olanzapine to reactive metabolites and the toxicity of these metabolites to human leukocytes. Mol Pharmacol 1998;53(6):991–8.

210. Kodesh A, Finkel B, Lerner AG, Kretzmer G, Sigal M. Dose-dependent olanzapine-associated leukopenia: three case reports. Int Clin Psychopharmacol 2001;16(2):117–9.

211. Meissner W, Schmidt T, Kupsch A, Trottenberg T, Lempert T. Reversible leucopenia related to olanzapine. Mov Disord 1999;14(5):872–3.

212. Dettling M, Cascorbi I, Hellweg R, Deicke U, Weise L, Muller-Oerlinghausen B. Genetic determinants of drug-induced agranulocytosis: potential risk of olanzapine? Pharmacopsychiatry 1999;32(3):110–2.

213. Buchman N, Strous RD, Ulman AM, Lerner M, Kotler M. Olanzapine-induced leukopenia with human leukocyte antigen profiling. Int Clin Psychopharmacol 2001;16(1):55–7.

214. Felber W, Naumann R, Schuler U, Fulle M, Reuster T, Garcia K, Heilemann H. Are there genetic determinants of olanzapine-induced agranulocytosis? Pharmacopsychiatry 2000;33(5):197–9.

215. Swartz JR, Ananth J, Smith MW, Burgoyne KS, Gadasally R, Arai Y. Olanzapine treatment after clozapine-induced granulocytopenia in 3 patients. J Clin Psychiatry 1999;60(2):119–21.

216. Benedetti F, Cavallaro R, Smeraldi E. Olanzapine-induced neutropenia after clozapine-induced neutropenia. Lancet 1999;354(9178):567.

217. Konakanchi R, Grace JJ, Szarowicz R, Pato MT. Olanzapine prolongation of granulocytopenia after clozapine discontinuation. J Clin Psychopharmacol 2000;20(6):703–4.

218. Oyewumi LK, Al-Semaan Y. Olanzapine: safe during clozapine-induced agranulocytosis. J Clin Psychopharmacol 2000;20(2):279–81.

219. Dernovsek MZ, Tavcar R. Olanzapine appears haematologically safe in patients who developed blood dyscrasia on clozapine and risperidone. Int Clin Psychopharmacol 2000;15(4):237–8.

220. Chatterton R. Experiences with clozapine and olanzapine. Aust NZ J Psychiatry 1998;32(3):463.

221. Finkel B, Lerner A, Oyffe I, Rudinski D, Sigal M, Weizman A. Olanzapine treatment in patients with typical and atypical neuroleptic-associated agranulocytosis. Int Clin Psychopharmacol 1998;13(3):133–5.

222. Lambert T. Olanzapine after clozapine: the rare case of prolongation of granulocytopaenia. Aust NZ J Psychiatry 1998;32(4):591–2.

223. Bogunovic O, Viswanathan R. Thrombocytopenia possibly associated with olanzapine and subsequently with benztropine mesylate. Psychosomatics 2000;41(3):277–8.

224. Carrillo JA, González JA, Gervasini G, López R, Fernández MA, Martín G. Thrombocytopenia and fatality associated with olanzapine. Eur J Clin Pharmacol 2004;60:295-6.

225. Onofrj M, Thomas A. One further case of pancytopenia induced by olanzapine in a Parkinson's disease patient. Eur Neurol 2001;45(1):56–7.

226. Mathias S, Schaaf LW, Sonntag A. Eosinophilia associated with olanzapine. J Clin Psychiatry 2002;63(3):246–7.

227. Perkins DO, McClure RK. Hypersalivation coincident with olanzapine treatment. Am J Psychiatry 1998;155(7):993–4.

228. Mendhekar DN, Srivastav PK, Sarin SK, Jiloha RC. A case report of olanzapine-induced fecal incontinence. J Clin Psychiatry 2003;64:339.

229. Kolpe M, Ravasia S. Effect of olanzapine on the liver transaminases. Can J Psychiatry 2003;48:210.

230. Horgan P, Jones H. Olanzapine use in acute porphyria. Int J Psychiatry Clin Pract 2003;7:67-9.

231. Jadallah KA, Limauro DI, Colatrella AM. Acute hepato-cellular-cholestatic liver injury after olanzapine therapy. Ann Intern Med 2003;138:357-8.

232. Doucette DE, Grenier JP, Robertson PS. Olanzapine-induced acute pancreatitis. Ann Pharmacother 2000;34(10):1128–31.

233. Woodall BS, DiGregorio RV. Comment: olanzapine-induced acute pancreatitis. Ann Pharmacother 2001;35(4):506–8.

234. Doucette DE, Robertson PS. Comment: olanzapine-induced acute pancreatitis. Ann Pharmacother 2001;35:508.

235. Vernon LT, Fuller MA, Hattab H, Varnes KM. Olanzapine-induced urinary incontinence: treatment with ephedrine. J Clin Psychiatry 2000;61(8):601–2.

236. Heckers S, Anick D, Boverman JF, Stern TA. Priapism following olanzapine administration in a patient with multiple sclerosis. Psychosomatics 1998;39(3):288–90.

237. Deirmenjian JM, Erhart SM, Wirshing DA, Spellberg BJ, Wirshing WC. Olanzapine-induced reversible priaprism: a case report. J Clin Psychopharmacol 1998;18(4):351–3.

238. Gordon M, de Groot CM. Olanzapine-associated priapism. J Clin Psychopharmacol 1999;19(2):192.

239. Compton MT, Saldivia A, Berry SA. Recurrent priapism during treatment with clozapine and olanzapine. Am J Psychiatry 2000;157(4):659.

240. Kuperman JR, Asher I, Modai I. Olanzapine-associated priapism. J Clin Psychopharmacol 2001;21(2):247.

241. Matthews SC, Dimsdale JE. Priapism after a suicide attempt by ingestion of olanzapine and gabapentin. Psychosomatics 2001;42(3):280–1.

242. Jagadheesan K, Thakur A, Akhtar S. Irreversible priapism during olanzapine and lithium therapy. Aust NZ J Psychiatry 2004;38:381.

243. Raz A, Bergman R, Eilam O, Yungerman T, Hayek T. A case report of olanzapine-induced hypersensitivity syndrome. Am J Med Sci 2001;321(2):156–8.

244. Benito-Leon J, Mitchell AJ. Guillain–Barré-like syndrome associated with olanzapine hypersensitivity reaction. Clin Neuropharmacol 2005;28:150-1.

245. Fukunishi I, Sato Y, Kino K, Shirai T, Kitaoka T. Hypothermia in a hemodialysis patient treated with olanzapine monotherapy. J Clin Psychopharmacol 2003;23:314.

246. Dueñas-Laita A, Pérez-Castrillón JL, Herreros-Fernández V. Olanzapine toxicity in unconjugated hyperbilirubinaemia (Gilbert's syndrome). Br J Psychiatry 2003;182:267.

247. Ellingrod VL, Perry PJ, Ringold JC, Lund BC, Bever-Stille K, Fleming F, Holman TL, Miller D. Weight gain associated with the -759C/T polymorphism of the 5HT2C receptor and olanzapine. Am J Med Genet B Neuropsychiatr Genet 2005;134:76-8.

248. Carrillo JA, Herráiz AG, Ramos SA, Gervasini G, Vizcaíno S, Benítez J. Role of the smoking-induced cytochrome P450 (CYP)1A2 and polymorphic CYP2D6 in steady-state concentration of olanzapine. J Clin Psychopharmacol 2003;23:119-27.

249. Elian AA. Fatal overdose of olanzepine. Forensic Sci Int 1998;91(3):231–5.

250. Gerber JE, Cawthon B. Overdose and death with olanzapine: two case reports. Am J Forensic Med Pathol 2000;21(3):249–51.

251. Stephens BG, Coleman DE, Baselt RC. Olanzapine-related fatality. J Forensic Sci 1998;43(6):1252–3.

252. Favier JC, Da Conceicao M, Peyrefitte C, Aussedat M, Pitti R. Intoxication mortelle à l'olanzapine. [Fatal intoxication with olanzapine.] Cah Anesthesiol 2002;50:29–31.

253. Davis LE, Becher MW, Tlomak W, Benson BE, Lee RR, Fisher EC. Persistent choreoathetosis in a fatal olanzapine overdose: drug kinetics, neuroimaging, and neuropathology. Am J Psychiatry 2005;162:28-33.

254. Cohen LG, Fatalo A, Thompson BT, Di Centes Bergeron G, Flood JG, Poupolo PR. Olanzapine overdose with serum concentrations. Ann Emerg Med 1999;34(2):275–8.

255. Gardner DM, Milliken J, Dursun SM. Olanzapine overdose. Am J Psychiatry 1999;156(7):1118–9.

256. Dobrusin M, Lokshin P, Belmaker RH. Acute olanzapine overdose. Hum Psychopharmacol 1999;14:355–6.

257. Bosch RF, Baumbach A, Bitzer M, Erley CM. Intoxication with olanzapine. Am J Psychiatry 2000;157(2):304–5.

258. O'Malley GF, Seifert S, Heard K, Daly F, Dart RC. Olanzapine overdose mimicking opioid intoxication. Ann Emerg Med 1999;34(2):279–81.

259. Etienne L, Wittebole X, Liolios, A, Hantson P. Polyuria after olanzapine overdose. Am J Psychiatry 2004;161:1130.

260. Bonin MM, Burkhart KK. Olanzapine overdose in a 1-year-old male. Pediatr Emerg Care 1999;15(4):266–7.

261. Catalano G, Cooper DS, Catalano MC, Butera AS. Olanzapine overdose in an 18-month-old child. J Child Adolesc Psychopharmacol 1999;9(4):267–71.

262. Bond GR, Thompson JD. Olanzapine pediatric overdose. Ann Emerg Med 1999;34(2):292–3.

263. Heimann SW. High-dose olanzapine in an adolescent. J Am Acad Child Adolesc Psychiatry 1999;38(5):496–8.

264. Yip L, Dart RC, Graham K. Olanzapine toxicity in a toddler. Pediatrics 1998;102(6):1494.

265. Malek-Ahmadi P, Simonds JF. Olanzapine for autistic disorder with hyperactivity. J Am Acad Child Adolesc Psychiatry 1998;37(9):902.

266. Callaghan JT, Bergstrom RF, Ptak LR, Beasley CM. Olanzapine. Pharmacokinetic and pharmacodynamic profile. Clin Pharmacokinet 1999;37(3):177–93.

267. Olesen OV, Linnet K. Olanzapine serum concentrations in psychiatric patients given standard doses: the influence of comedication. Ther Drug Monit 1999;21(1):87–90.

268. Lucas RA, Gilfillan DJ, Bergstrom RF. A pharmacokinetic interaction between carbamazepine and olanzapine: observations on possible mechanism. Eur J Clin Pharmacol 1998;54(8):639–43.

269. Linnet K, Olesen OV. Free and glucuronidated olanzapine serum concentrations in psychiatric patients: influence of carbamazepine comedication. Ther Drug Monit 2002;24(4):512–7.

270. Markowitz JS, DeVane CL. Suspected ciprofloxacin inhibition of olanzapine resulting in increased plasma concentration. J Clin Psychopharmacol 1999;19(3):289–91.

271. Gatta B, Rigalleau V, Gin H. Diabetic ketoacidosis with olanzapine treatment. Diabetes Care 1999;22(6):1002–3.

272. Sierra-Biddle D, Herran A, Diez-Aja S, Gonzalez-Mata JM, Vidal E, Diez-Manrique F, Vazquez-

Barquero JL. Neuropletic malignant syndrome and olanzapine. J Clin Psychopharmacol 2000;20(6):704–5.

273. Rosebraugh CJ, Flockhart DA, Yasuda SU, Woosley RL. Olanzapine-induced rhabdomyolysis. Ann Pharmacother 2001;35(9):1020–3.

274. Berry N, Pradhan S, Sagar R, Gupta SK. Neuroleptic malignant syndrome in an adolescent receiving olanzapine–lithium combination therapy. Pharmacotherapy 2003;23(2):255–9.

275. Tohen M, Chengappa KN, Suppes T, Zarate CA Jr, Calabrese JR, Bowden CL, Sachs GS, Kupfer DJ, Baker RW, Risser RC, Keeter EL, Feldman PD, Tollefson GD, Breier A. Efficacy of olanzapine in combination with valproate or lithium in the treatment of mania in patients partially nonresponsive to valproate or lithium monotherapy. Arch Gen Psychiatry 2002;59(1):62–9.

276. de Jong J, Hoogenboom B, van Troostwijk LD, de Haan L. Interaction of olanzapine with fluvoxamine. Psychopharmacology (Berl) 2001;155(2):219–20.

277. Wang CY, Zhang ZJ, Li WB, Zhai YM, Cai ZJ, Weng YZ, Zhu RH, Zhao JP, Zhou HH. The differential effects of steady-state fluvoxamine on the pharmacokinetics of olanzapine and clozapine in healthy volunteers. J Clin Pharmacol 2004;44:785-92.

278. Bergemann N, Frick A, Parzer P, Kopitz J. Olanzapine plasma concentration, average daily dose, and interaction with co-medication in schizophrenic patients. Pharmacopsychiatry 2004;37:63-8.

279. Albers LJ, Ozdemir V, Marder SR, Raggi MA, Aravagiri M, Endrenyi L, Reist C. Low-dose fluvoxamine as an adjunct to reduce olanzapine therapeutic dose requirements: a prospective dose-adjusted drug interaction strategy. J Clin Psychopharmacol 2005;25:170-4.

280. D'Amico G, Cedro C, Muscatello MR, Pandolfo G, Di Rosa AE, Zoccali R, La Torre D, D'Arrigo C, Spina E. Olanzapine augmentation of paroxetine-refractory obsessive-compulsive disorder. Prog Neuropsychopharmacol Biol Psychiatry 2003;27:619-23.

281. Zullino DF, Delessert D, Eap CB, Preisig M, Baumann P. Tobacco and cannabis smoking cessation can lead to intoxication with clozapine or olanzapine. Int Clin Psychopharmacol 2002;17(3):141–3.

282. Deshauer D, Albuquerque J, Alda M, Grof P. Seizures caused by possible interaction between olanzapine and clomipramine. J Clin Psychopharmacol 2000;20(2):283–4.

283. Nickel MK, Nickel C, Muehlbacher M, Leiberich PK, Kaplan P, Lahmann C, Tritt K, Krawczyk J, Kettler C, Egger C, Rother WK, Loew TH. Influence of topiramate on olanzapine-related adiposity in women: a random, double-blind, placebo-controlled study. J Clin Psychopharmacol 2005;25:211-7.

284. Henderson DC, Copeland PM, Daley TB, Borba CP, Cather C, Nguyen DD, Louie PM, Evins AE, Freudenreich O, Hayden D, Goff DC. A double-blind, placebo-controlled trial of sibutramine for olanzapine-associated weight gain. Am J Psychiatry 2005;162:954-62.

285. Mauri MC, Steinhilber CP, Marino R, Invernizzi E, Fiorentini A, Cerveri G, Baldi ML, Barale F. Clinical outcome and olanzapine plasma levels in acute schizophrenia. Eur Psychiatry 2005;20:55-60.

Perphenazine

General Information

Perphenazine is a phenothiazine neuroleptic drug.

Placebo-controlled studies

Following remission of manic symptoms in 37 patients who had taken perphenazine and a mood stabilizer (lithium, carbamazepine, or valproate), treatment was randomly assigned double-blind to perphenazine or placebo for 6 months, while continuing the mood stabilizer (1). Those who took perphenazine had worse outcomes than those who took placebo, in that they were more likely to have a shorter time to a depressive relapse, were more likely to discontinue treatment, or were more likely to have depression or extrapyramidal symptoms. The authors tentatively concluded that perphenazine may not be beneficial in maintenance treatment for bipolar I patients.

Organs and Systems

Hematologic

Aplastic anemia, defined by the presence of pancytopenia and a hypocellular bone marrow in the absence of any abnormal blood cells, is a serious reaction that has been attributed to perphenazine in a single case (2).

- A 23-year-old man with schizophrenia taking perphenazine 4 mg bd, benzatropine mesylate 2 mg/day, lithium carbonate 600 mg each morning and 900 mg at bedtime, and famotidine 40 mg/day developed fatigue, shortness of breath, dizziness, light-headedness, and general debility. He had a pancytopenia, which persisted in spite of blood transfusions. Bone-marrow aspiration showed hypocellularity, absent megakaryocytes, reduced erythropoiesis and myelopoiesis, increased iron storage, and a relative excess of lymphoid cells. All medications were withdrawn and he was given lorazepam 2 mg bd. He recovered after 8 months.

References

1. Oyewumi LK. Acquired aplastic anemia secondary to perphenazine. Can J Clin Pharmacol 1999;6(3):169–71.

2. Zarate C, Tohen M. Double-blind comparison of the continued use of antipsychotic treatment versus its discontinuation in manic patients. Am J Psychiatry 2004;161:169–71.

Pimozide

General Information

Pimozide is a diphenylbutylpiperidine neuroleptic drug, structurally similar to the butyrophenones.

Organs and Systems

Cardiovascular

Pimozide can cause QT interval prolongation and torsade de pointes; a total of 40 reports (16 deaths) of serious cardiac reactions, mainly dysrhythmias, were reported to the Committee on Safety of Medicines in the UK from 1971 to 1995 (1).

Drug Administration

Drug overdose

In a study of overdosages of different neuroleptic drugs, pimozide had the lowest fatality index (the number of deaths divided by the number of prescriptions) (2).

Drug–Drug Interactions

Clarithromycin

In 12 healthy volunteers given oral pimozide 6 mg after 5 days of treatment with clarithromycin (500 mg bd) or placebo, pimozide significantly prolonged the QT_c interval in the first 20 hours in both groups (maximum changes in QT_c 16 and 13 ms respectively) (3).

Clarithromycin inhibits the metabolism of pimozide, pimozide plasma concentrations increase, and there is an increased risk of cardiotoxicity through prolongation of the QT interval and fatal ventricular dysrhythmias (3).

References

1. Committee on Safety of Medicines—Medicines Control Agency. Cardiac arrhythmias with pimozide (Orap). Curr Probl Pharmacovigilance 1995;21:1.
2. Buckley N, McManus P. Fatal toxicity of drugs used in the treatment of psychotic illnesses. Br J Psychiatry 1998;172:461–4.
3. Desta Z, Kerbusch T, Flockhart DA. Effect of clarithromycin on the pharmacokinetics and pharmacodynamics of pimozide in healthy poor and extensive metabolizers of cytochrome P450 2D6 (CYP2D6). Clin Pharmacol Ther 1999;65(1):10–20.

Prochlorperazine

General Information

Prochlorperazine is a phenothiazine derivative.

Organs and Systems

Nervous system

In 192 consecutive patients attending an emergency department for nausea/vomiting or headache, akathisia occurred in 16% of those treated with prochlorperazine (5–10 mg intravenously or intramuscularly); 4% (all of them women) developed dystonias (1).

Slow infusion of prochlorperazine has been used to try to minimize the risk of akathisia in 160 patients randomly assigned to two groups; akathisia developed in 31 of 84 who were given a 2-minute infusion and in 18 of 76 patients who were given a 15-minute infusion (2).

Liver

Phenothiazines can cause cholestatic jaundice.

- Cholestasis occurred in a woman with alpha-1 antitrypsin deficiency (phenotype PiZZ) who had taken prochlorperazine 5-10 mg qds for 27 months (3). She developed jaundice and ascites. Liver biopsy confirmed diffuse advanced chronic cholestasis, moderate portal and periportal inflammation, and bridging necrosis. Her liver function tests normalized within days of withdrawal of prochlorperazine.

References

1. Olsen JC, Keng JA, Clark JA. Frequency of adverse reactions to prochlorperazine in the ED. Am J Emerg Med 2000;18(5):609–11.
2. Vinson DR, Migala AF, Quesenberry CP Jr. Slow infusion for the prevention of akathisia induced by prochlorperazine: a randomized controlled trial. J Emerg Med 2001;20(2):113–9.
3. Mindikoglu A, Anantharaju A, Hartman G, Li S, Villanueva J, Thiel D. Prochlorperazine induced cholestasis in a patient with alpha 1 antitrypsin deficiency. Hepato-Gastroenterology 2003;50:1338–40.

Quetiapine

General Information

The pharmacology, efficacy, and safety of quetiapine, an atypical neuroleptic drug, have been extensively reviewed (1). Quetiapine interacts with a broad range of neurotransmitter receptors and has a higher affinity for serotonin

(5-HT$_{2A}$) receptors than dopamine (D$_2$) receptors in the brain. In a meta-analysis, quetiapine was as effective as haloperidol with fewer extrapyramidal adverse effects (2).

Data from short-term clinical trials (6 weeks) suggest that quetiapine may be useful for the management of psychotic disorders in patients who do not tolerate the adverse effects of the typical antipsychotic drugs or clozapine (3). The most common adverse effects of quetiapine were dizziness, hypotension, somnolence, and weight gain. Raised hepatic enzymes have also been reported. In addition, two patients with idiopathic Parkinson's disease and psychosis were treated with quetiapine for 52 weeks (4). Psychotic symptoms were successfully controlled without worsening of motor disability.

Observational studies

A post-hoc analysis of the Spectrum trial, an international open non-comparative study, sponsored by AstraZeneca Pharmaceutical, has recently been published; this study was purportedly carried out to evaluate improvements in efficacy and tolerability gained by switching to quetiapine in patients who had previously taken haloperidol (n = 43), olanzapine (n = 66), or risperidone (n = 55) (5). Switching to quetiapine produced improvements from baseline in Positive and Negative Syndrome Scale and in Calgary Depression Scale for Schizophrenia scores. There were significant reductions in extrapyramidal adverse effects on the Simpson–Angus scale and Barnes Akathisia scale. Patients who switched to quetiapine from haloperidol had a mean weight gain of 2 kg, while those who switched from olanzapine had a mean loss of 1 kg and those who switched from risperidone had a mean gain of 0.7 kg.

In seven patients with refractory schizophrenia taking high-dose quetiapine 1200–2400 mg/day, there were mild to marked improvements in positive symptoms, violent behavior, behavioral disturbances, and sociability (6). Sedation, orthostasis, dysphagia, and a nocturnal startle reaction were reported and were responsive to dosage reduction. Weight gain was 4.1 kg and there were no significant electrocardiogram abnormalities.

Placebo-controlled studies

Quetiapine has been approved by the FDA as monotherapy for the treatment of acute mania. It has been evaluated in combination with lithium or divalproex in 191 patients who had recently been manic (7). After treatment with quetiapine plus lithium or divalproex for 7–28 days, the patients were randomized to either additional quetiapine or placebo and followed for 3 weeks more. Early discontinuation was more frequent with placebo than with quetiapine. The intention-to-treat population included 81 taking quetiapine and 89 taking placebo. The mean dose during the last week in patients taking quetiapine was 504 mg/day. Patients taking quetiapine had a greater improvement in the Young Mania Rating Scale score (YMRS) than patients taking placebo. The response rate (50% or greater improvement from baseline using the YMRS) was significantly higher in the group with added quetiapine than added placebo.

Common adverse events included somnolence, dry mouth, weakness, and postural hypotension.

In 542 outpatients with bipolar I or II disorder experiencing a major depressive episode, who were randomly assigned to 8 weeks of quetiapine (300 or 600 mg/day; n = 181 and n = 180 respectively) or placebo (n = 181), both doses produced statistically significant improvement in Montgomery–Asberg Depression Rating Scale (MADRS) total scores compared with placebo from week 1 onward (8). There were extrapyramidal symptoms in 8.9% of those who took 600 mg/day, 6.7% of those who took 300 mg/day, and 2.2% of those who took placebo. Patients who took quetiapine 600 mg/day had a mean weight gain of 1.6 kg, compared with 1.0 kg in those who took 300 mg/kg and 0.2 kg in those who took placebo.

Organs and Systems

Nervous system

Although quetiapine seems to cause a lower incidence of extrapyramidal symptoms, a case of neuroleptic malignant syndrome has been described (9).

- A 40-year-old man with chronic schizophrenia and borderline intelligence presented with acute psychotic decompensation. He had previously taken several different neuroleptic drugs and had had significant extrapyramidal symptoms but never neuroleptic malignant syndrome. He was given quetiapine 25 mg bd, increasing to 250 mg bd by day 13. He then had increasing symptoms of restlessness, agitation, and episodic sweating. Loxapine 25 mg and lorazepam 2 mg were added. On day 14 he developed confusion, lead-pipe muscle rigidity, a temperature of 38.2°C, a labile blood pressure, tachypnea, and tachycardia; creatine kinase activity was 18 354 IU/l and he had myoglobinuria. Quetiapine was withdrawn and supportive treatment was instituted. He recovered 5 days after withdrawal.

Two patients with schizophrenia who developed focal tardive dystonia with atypical antipsychotic drugs (risperidone and olanzapine) had marked sustained improvement when quetiapine was gradually introduced and the other antipsychotic drugs were withdrawn; there was no loss of control of psychotic symptoms (10).

Psychiatric

Two patients developed resistant auditory hallucinations when taking quetiapine (11). One of these patients, a 39-year-old woman who took 600 mg/day, developed mild sedation, which resolved spontaneously by week 4.

Endocrine

The effects of haloperidol and quetiapine on serum prolactin concentrations have been compared in 35 patients with schizophrenia during a drug-free period of at least 2 weeks in a randomized study (12). There was no significant difference in prolactin concentration between the groups at the start of the study, and control prolactin concentrations were significantly lower with quetiapine than haloperidol. No patients taking quetiapine had galactorrhea.

Dose-dependent decreases in total T3 and T4 and free T4, without an increase in TSH, have been reported (13,14). Such changes have not been observed with other neuroleptic drugs.

Long-Term Effects

Drug abuse

An unusual case of quetiapine abuse has been reported (15).

- A 34-year-old woman with a history of polysubstance dependence (alcohol, cannabis, and cocaine), depressive episodes associated with multiple suicide attempts, and borderline personality disorder, who had been incarcerated after conviction on charges of physical assault and possession of controlled substances, complained of difficulty in sleeping, poor impulse control, irritability, and depressed mood. She was given oral quetiapine 600 mg/day. On one occasion, she crushed two 300-mg tablets, dissolved them in water, boiled them, drew the solution through a cotton swab, and injected the solution intravenously. Apart from having "the best sleep I ever had" she described no dysphoric, euphoric, or other effects. She admitted to previous intranasal abuse of crushed quetiapine tablets.

Drug withdrawal

Incapacitating quetiapine withdrawal has been reported (16).

- A 36-year-old woman with rapid-cycling bipolar II disorder and premenstrual mood exacerbation was treated as an out-patient with lamotrigine 400 mg/day, clonazepam 0.5 mg tds, and quetiapine 100 mg/day. She gained 9 kg in 6 months and was advised to reduce the dose of quetiapine to 50 mg/day. After 1 day, she reported nausea, dizziness, headache, and anxiety severe enough to preclude normal daily activities. She was instructed to take quetiapine 75 mg/day, but her symptoms continued and only resolved when she took 100 mg/day. Slower reduction in the dose of quetiapine (by 12.5 mg/day every 5 days) with an antiemetic, ondansetron, also failed. On a third attempt, prochlorperazine successfully reduced her withdrawal symptoms, although moderate nausea persisted for 2 days after complete withdrawal.

No other medications were changed, so quetiapine withdrawal was the most likely explanation for the symptoms in this case.

Drug-Drug Interactions

Selective serotonin reuptake inhibitors (SSRIs)

Escitalopram

Among the atypical antipsychotic drugs available in Europe, quetiapine seems to be associated with less weight gain. However, severe weight gain can occur and has been reported in a supposed interaction with escitalopram (17).

- A 16-year-old adolescent girl was given quetiapine for a first psychotic episode for 5 months and had no weight gain. Because her depressive symptoms were pronounced after remission of her psychotic symptoms, she was given escitalopram and had a dramatic increase in weight (8 kg over 1 month). Quetiapine was replaced by amisulpride and there was a transient fall in weight. However, severe psychotic symptoms led to reintroduction of quetiapine, and her weight again rose dramatically (8 kg over 1 month). Withdrawal of escitalopram and replacement by topiramate was followed by weight stabilization.

The authors did not propose a mechanism for this supposed interaction.

Fluvoxamine

Concomitant administration of quetiapine and fluvoxamine reportedly caused neuroleptic malignant syndrome (18).

- A 57-year-old man with major depression took fluvoxamine 150 mg/day for 1 year with no adverse effects and in remission stopped taking it. He later developed agitation and was given risperidone, and then, because of extrapyramidal symptoms, quetiapine 150 mg/day. However, 2 months later he again became depressed and fluvoxamine 100 mg/day was added. After 10 days of he stopped eating and drinking and developed muscle rigidity. On day 13 he had a fever, severe extrapyramidal symptoms, a high blood pressure, and a tachycardia, and was becoming stuporose. He had a raised creatine kinase activity (7500 IU/l) and leukocyte count (13 x 10^9/l). All psychotropic drugs were stopped and he was given dantrolene by intravenous infusion. His symptoms gradually improved.

The authors suggested that since the doses of quetiapine and fluvoxamine were relatively low and since they are metabolized by different CYP isozymes, this was probably not a pharmacokinetic interaction. Instead, they suggested that it may have been caused by dopamine–serotonin disequilibrium.

References

1. Goldstein JM. Quetiapine fumarate (Seroquel): a new atypical antipsychotic. Drugs Today (Barc) 1999;35(3):193–210.
2. Leucht S, Pitschel-Walz G, Abraham D, Kissling W. Efficacy and extrapyramidal side-effects of the new antipsychotics olanzapine, quetiapine, risperidone, and sertindole compared to conventional antipsychotics and placebo. A meta-analysis of randomized controlled trials. Schizophr Res 1999;35(1):51–68.
3. Misra LK, Erpenbach JE, Hamlyn H, Fuller WC. Quetiapine: a new atypical antipsychotic. S D J Med 1998;51(6):189–93.

4. Parsa MA, Bastani B. Quetiapine (Seroquel) in the treatment of psychosis in patients with Parkinson's disease. J Neuropsychiatry Clin Neurosci 1998;10(2):216–9.

5. Larmo I, de Nayer A, Windhager E, Lindenbauer B, Rittmannsberger H, Platz T, Jones AM, Altman C; Spectrum Study Group. Efficacy and tolerability of quetiapine in patients with schizophrenia who switched from haloperidol, olanzapine or risperidone. Hum Psychopharmacol 2005;20;573-81.

6. Pierre JM, Wirshing DA, Wirshing WC, Rivard JM, Marks R, Mendenhall J, Sheppard K, Saunders DG. High-dose quetiapine in treatment refractory schizophrenia. Schizophr Res 2005;73;373-5.

7. Sachs G , Chengappa KNR, Suppes T, Mullen JA, Brecher M, Devine NA, Sweitzer DE. Quetiapine with lithium or divalproex for the treatment of bipolar mania: a randomized, double-blind, placebo-controlled study. Bipolar Disord 2004;6;213-23.

8. Calabrese JR, Keck PE Jr, Macfadden W, Minkwitz M, Ketter TA, Weisler RH, Cutler AJ, McCoy R, Wilson E, Mullen J. A randomized, double-blind, placebo-controlled trial of quetiapine in the treatment of bipolar I or II depression. Am J Psychiatry 2005;162;1351-60.

9. al-Waneen R. Neuroleptic malignant syndrome associated with quetiapine. Can J Psychiatry 2000;45(8):764–5.

10. Gourzis P, Polychronopoulos P, Papapetropoulos S, Assimakopoulos K, Argyriou AA, Beratis S. Quetiapine in the treatment of focal tardive dystonia induced by other atypical antipsychotics: a report of 2 cases. Clin Neuropharmacol 2005;28;195-6.

11. Mosolov SN, Kabanov S. Quetiapine in the treatment of patients with resistant auditory hallucinations: two case reports with long-term cognitive assessment. Eur Psychiatry 2005;20;430.

12. Atmaca M, Kuloglu M, Tezcan E, Canatan H, Gecici O. Quetiapine is not associated with increase in prolactin secretion in contrast to haloperidol. Arch Med Res 2002;33(6):562–5.

13. Henderson DC, Nasrallah RA, Goff DC. Switching from clozapine to olanzapine in treatment-refractory schizophrenia: safety, clinical efficacy, and predictors of response. J Clin Psychiatry 1998;59(11):585–8.

14. Sacristan JA, Gomez JC, Martin J, Garcia-Bernardo E, Peralta V, Alvarez E, Gurpegui M, Mateo I, Morinigo A, Noval D, Soler R, Palomo T, Cuesta M, Perez-Blanco F, Massip C. Pharmacoeconomic assessment of olanzapine in the treatment of refractory schizophrenia based on a pilot clinical study. Clin Drug Invest 1998;15:29–35.

15. Hussain MZ, Waheed W, Hussain S. Intravenous quetiapine abuse. Am J Psychiatry 2005;162;1755-56.

16. Kim DR, Staab JP. Quetiapine discontinuation syndrome. Am J Psychiatry 2005;162;1020.

17. Holzer L, Paiva G, Halfon O. Quetiapine-induced weight gain and escitalopram. Am J Psychiatry 2005;162;192-193.

18. Matsumoto R, Kitabayashi Y, Nakatomi Y, Tsuchida H, Fukui K. Neuroleptic malignant syndrome induced by quetiapine and fluvoxamine. Am J Psychiatry 2005;162;812.

Risperidone

General Information

Risperidone is an atypical benzisoxazole neuroleptic drug. It is a dopamine D_2 receptor antagonist, with antagonistic actions at $5\text{-}HT_2$ receptors, alpha-adrenoceptors, and histamine H_1 receptors.

Observational studies

In a Prescription Event Monitoring study of the safety of risperidone in 7684 patients treated in general practice, information on risperidone prescriptions issued to patients in England was gathered between July 1993 and April 1996 (1). After 6 months, 76% of the patients for whom data were available were still taking risperidone. Drowsiness/sedation was the most frequent reason for stopping risperidone and the most frequently reported event (4.6 cases per 1000 patient-months). Extrapyramidal symptoms were rarely reported, the incidence being 3.2 per 1000 patient-months; they were more frequent in elderly patients (7.8 per 1000 patient-months). There were only four reports of dyskinesias and one report of tardive dyskinesia, which resulted in withdrawal of risperidone. Eight overdoses of risperidone alone were reported, with no serious clinical sequelae. Nine patients took risperidone during ten pregnancies, with seven live births and three early therapeutic terminations. There were no abnormalities among the live births.

Long-term data on the efficacy and tolerability of risperidone are scant, as most of the clinical trials have been of short duration (no longer than 12 weeks). However, some additional data from open studies have emerged. In one study, 386 patients with chronic schizophrenia took risperidone 2–16 mg/day for up to 57 weeks; 247 patients were treated for at least 1 year (2). All but 48 patients (88%) had been treated with antipsychotic drugs before entering the study. At the end of the study, 64% of the patients were rated as having improved on the Clinical Global Impression change scale, and extrapyramidal symptoms (scored on the Extrapyramidal Symptom Rating Scale, ESRS) tended to be lower in severity or remained unchanged over the course of risperidone treatment; 27% of the patients required antiparkinsonian medication during the study, and 6.5% discontinued treatment prematurely because of adverse events. One or more adverse events were reported by 221 patients (57%) during risperidone treatment. Extrapyramidal symptoms occurred in 23%. Insomnia and anxiety were reported by 13% and 12% of patients. Two patients died during the 1-year study: one patient drowned and another committed suicide by hanging after 3.5 months. At the end of the study the mean increase in body weight was 1.8 kg.

In a retrospective study of 97 patients taking risperidone, under 30% of the patients were still taking risperidone after a mean period of follow-up of 102 (range 13–163) weeks (3). Reasons for discontinuation included not achieving the desired therapeutic effect ($n = 39$), noncompliance ($n = 22$), adverse effects ($n = 26$), the patient's not liking the drug and requesting a change to a different medication ($n = 17$), and symptom remission ($n = 6$). The authors stated that in routine clinical practice, the use of risperidone is plagued by many of the same problems of older antipsychotic drugs.

The efficacy and safety of risperidone have been examined in special groups of patients, such as those with psychotic depression (4), autistic disorders (41), bipolar disorder (5), mental retardation (6), and children and adolescents (7).

Patients with bipolar disorders may benefit from risperidone. This has been observed in an open trial of ten patients with rapid cycling bipolar disorder who were refractory to lithium carbonate, carbamazepine, and valproate; eight improved after 6 months of treatment. One patient dropped out through non-adherence to therapy and one because of adverse effects (agitation, anxiety, insomnia, and headache) (5). There was a similar beneficial effect in eight adults with moderate to profound mental retardation (6). Risperidone was associated with a significant reduction in aggression and self-injurious behavior, whereas adverse effects were primarily those of sedation and restlessness.

Eight of a heterogeneous group of eleven children and adolescents (mean age 9.8, range 5.5–16 years) with mood disorders and aggressive behavior, improved with a low dose of risperidone (0.75–2.5 mg/day) (7). Treatment was stopped in two children because of drowsiness; the most bothersome adverse effect of risperidone was weight gain in two cases (mean increase 4 kg).

The medical records of 151 hospitalized elderly psychiatric patients (mean age 71 years) have been analysed (8). Of 114 patients treated with risperidone (mean duration of treatment 17 days; mean dose 3 mg/day), 78% responded. Adverse events were reported in 20 patients, including new-onset extrapyramidal effects in four; tremor in four; sedation in three; hypotension in three; diarrhea in two; tardive dyskinesia in two; and chest pain, anxiety, restlessness, itching, insomnia, and falls in one each.

A review of both core information and new findings concerning risperidone concluded that it can be associated with transient drowsiness (probably no greater than that observed with haloperidol), postural hypotension (which is avoided by dose titration), weight gain, reduced sexual interest, and erectile dysfunction. Risperidone can also cause dose-dependent hyperprolactinemia (which can cause amenorrhea, sexual dysfunction, and galactorrhea). Risperidone tends not to cause extrapyramidal signs at therapeutic doses (the optimal dose being 6 mg/day) but they do occur dose-dependently (9).

In an 8-week open prospective study of risperidone in 20 patients, mean age 34 (range 19–53) years, adverse effects included giddiness ($n = 3$), headache ($n = 2$), and agitation ($n = 2$); one woman reported galactorrhea and another developed obsessive-compulsive symptoms; 16 of 20 patients were taking antiparkinsonian drugs before the study, compared with 12 patients at the end (10).

Objections have been raised on the ways in which some clinical trials are interpreted (11). Furthermore, it has been stated that a major problem with the risperidone literature is that the original data can be very difficult to decipher or even obtain. Huston and Moher have described their frustration in trying to perform a meta-analysis of the effects of risperidone; they found "obvious redundancy in the results of a single-center trial being published twice," as well as problems with "changing authorship, lack of transparency in reporting, and frequent citation of abstracts and unpublished reports" (12). These concerns were reiterated in an editorial (13).

In a prospective, 6-week open trial in 31 Chinese patients with acute exacerbation of schizophrenia, risperidone doses were titrated to 6 mg/day (if tolerated) over 3 days, but were reduced thereafter if adverse effects occurred (14). Efficacy and adverse effects were assessed on days 0, 4, 14, 28, and 42. End-point steady-state plasma concentrations of risperidone and 9-hydroxyrisperidone were analysed. Of the 30 patients who completed the trial, 17 tolerated the 6 mg dose well, while the other 13 received lower final doses (mean 3.6 mg) for curtailing adverse effects. At end-point, 92% of the 13 low-dose patients had responded to treatment (a 20% or more reduction in the total score on the Positive and Negative Syndrome Scale), compared with 53% of the 17 high-dose subjects. There were no significant between-group differences in other minor efficacy measures. End-point plasma concentrations of the active moieties (risperidone plus 9-hydroxyrisperidone) were 40 ng/ml in the low-dose group and 50 ng/ml in the high-dose group; this difference was not significant, suggesting that the different responses were pharmacokinetic in origin. The results of this preliminary trial suggest that up to 6 mg/day of risperidone is efficacious in treating patients with an acute exacerbation of schizophrenia. Nearly 60% of the patients tolerated 6 mg/day; in the others, reducing the dosage to relieve adverse effects still yielded efficacy.

The efficacy and safety of long-term risperidone have been assessed in children and adolescents ($n = 11$) in a prospective study (15). Subjects with autism or pervasive developmental disorder not otherwise specified took risperidone for 6 months, after which their parents were given the option of continuing for a further 6 months. Weight gain was common, although the rate of increase abated with time. After 6 months two patients developed facial dystonias, which disappeared after reducing the dosage in one case and after withdrawal in the other. Amenorrhea was also observed, but there were no changes in liver function, blood tests, or electroencephalography. The authors concluded that risperidone may be effective and relatively safe in the long-term treatment of behavioral disruption in autistic children and adolescents.

Youths with behavioral disorders (16,17), and patients with disturbing neuroleptic drug-induced extrapyramidal symptoms (18) have been studied. Ten youths (aged 6–14 years) were randomly assigned to receive placebo and 10 to receive risperidone (0.75–1.50 mg/day) for conduct disorder in a preliminary study lasting 10 weeks (16). Of those assigned to risperidone, six completed the course; three completed placebo. Those who took risperidone were significantly less aggressive during the last weeks of the study than those who did not. Eight youths who took risperidone and four who took placebo had at least one adverse effect, including increased appetite ($n = 3$ for risperidone), sedation ($n = 3$ for risperidone; $n = 2$ for placebo), headache ($n = 1$ for risperidone; $n = 1$ for placebo), insomnia ($n = 1$ for risperidone), restlessness ($n = 1$ for risperidone), irritability ($n = 1$ for risperidone), enuresis ($n = 1$ for placebo), and nausea/vomiting ($n = 1$ for risperidone; $n = 1$ for placebo). These adverse effects were mild and transient.

There was marked reduction in aggression in 14 of 26 subjects (10–18 years old) in an open study of risperidone (0.5–4 mg/day) for 2–12 months (17). Two subjects had marked weight gain (8 and 10 kg) in the first 8 weeks; another participant who took lithium (1400 mg/day, serum concentration 0.9 mmol/l) presented with moderate akathisia and hand tremor; in seven, tiredness and sedation occurred after week 8.

Observational studies

A prospective open study of the effects of risperidone (mean dose 1.8 mg/day) has been carried out in 21 sites across Canada in 108 patients (mean age 44 years) with bipolar I disorder, with manic or mixed episodes (19). The usual mood stabilizers and antidepressants, but no other antipsychotic drugs or newer generation anticonvulsants, were permitted during the 12-week study. There were significant reductions in manic and depressive symptoms, as measured by conventional scales (Young Mania Rating Scale). The antimanic effect was observed from week 1 on: mean baseline score 28 ($n = 107$); mean change from baseline on week 1, –11 ($n = 103$); week 3, –18 ($n = 92$); week 12, –23 ($n = 77$). There were 15 serious unspecified adverse events in nine patients, resulting in five dropouts; adverse events that occurred in at least 10% of the patients included headache (24%), depression (15%), fatigue (13%), nausea (12%), constipation (10%), and diarrhea (10%); 27% developed at least one of the following extrapyramidal symptoms: dyskinesia, dystonia, hyperkinesia, involuntary muscle contractions, and tremor; none developed tardive dyskinesia. At week 12 there was weight gain of 2.4 kg among completers; and 21 had a weight gain of at least 7% during the study.

Risperidone has also been used in combination with topiramate in a Spanish multicenter study in 58 patients (28 men and 30 women; mean age 41 years) with bipolar I disorder, with manic but not mixed episodes (20). Risperidone (mean dose 2.7 mg/day) and topiramate (mean dose 236 mg/day) were started with a maximum 48-hour time difference; risperidone was used for acute manic symptoms and topiramate for longer-term stabilization and prevention of relapse. The incidence of any adverse event was 64%, mostly somnolence, paresthesia, dizziness, tremor, weight loss ($n = 27$; mean change – 1.1 kg), extrapyramidal disorders, gastrointestinal effects, and cognitive disturbances. One patient developed tardive dyskinesia during the study and there were five dropouts because of adverse effects; adverse effects that required withdrawal of risperidone but not topiramate were amenorrhea ($n = 3$) and sexual dysfunction ($n = 1$).

In 21 Turkish children and adolescents with conduct disorder (17 boys and 4 girls; mean age 11 years), the mean dose of risperidone at the 8-week endpoint was 1.27 mg/day (21). There were significant improvements in several symptoms, including inattention and hyperactivity/impulsivity. There was mild and transient sedation at the beginning of the study in all patients, and a mean increase in sleep duration of 0.9 hours (range 0–3 hours). There were no extrapyramidal symptoms or other severe adverse events.

Several symptoms of dementia can be improved by risperidone. In 18 patients with Alzheimer's disease (no sex or age data reported), delusions of theft, hallucinations, and agitation/aggression improved significantly after 12 weeks of treatment (22). The modal optimal dosage was 1 mg/day, the same already suggested for this pathology (SEDA-26, 64). There were mild extrapyramidal symptoms at some point during the trial in one patient.

Risperidone has been used in children with severe disruptive behavior disorders (SEDA-27, 62; SEDA-28, 74). In 107 children, aged 5–12 years (86 boys) who entered a 48-week open study and received risperidone (mean dose 1.5 mg/day) there were significant improvements in conduct at 4 weeks. The most common adverse events were somnolence ($n = 35$), headache ($n = 35$), rhinitis ($n = 30$), and weight gain ($n = 22$; mean increase from baseline 5.5 kg, half attributable to developmentally expected growth) (23). Adverse events led to dropouts in 11 patients; they were mainly weight gain ($n = 4$), depression ($n = 3$), and suicide attempts ($n = 2$); other reasons for dropouts were loss to follow-up ($n = 15$), consent withdrawal ($n = 12$), insufficient response ($n = 11$), and non-adherence to therapy ($n = 7$). Although uncontrolled extrapyramidal symptoms were an exclusion criterion, there were no significant adverse extrapyramidal effects, and vital signs, cognition, and laboratory analyses were not found except for transient increases in prolactin concentrations at week 4 (maximum of 28 ng/ml in boys and 24 ng/ml in girls).

In a 6-week open study in 146 Chinese in-patients, risperidone was titrated to 6 mg/day within 7 days; 29 patients withdrew prematurely (24). Higher risperidone doses did not appear to have greater efficacy in treating acute schizophrenia and were related to more severe adverse effects.

In another study only 38 patients out of 79 with chronic schizophrenia completed 52 weeks of treatment with risperidone; the most common dose was 6 mg/day (25).

In a two-phase long-term multicenter study in 74 patients (mean age 30 years; 56 men), the starting dose of risperidone was 1 mg/day, increased to 2 mg after 3 days and adjusted to 8 mg/day maximum; treatment duration was 2 weeks for phase 1 (mean dose at the end 3.8 mg/day) and 1 year for phase 2 (mean dose at the end 3.5 mg/day) (26). Of 12 patients who did not complete phase 1, two had adverse events (agitation, somnolence, and self-harm); there were nine dropouts during phase 2, but none was apparently due to adverse events.

In a review based on the Medline and PsycINFO databases from 1999 to 2002, six articles focusing on the issue of adherence to therapy in schizophrenia and two meta-analyses of depot typical antipsychotic agents in schizophrenia were identified (27). The usual adverse effects were often reasons for non-adherence. Clinicians' fear of managing adverse effects such as acute dystonia or neuroleptic malignant syndrome, which can be prolonged with long-acting agents, were also related to non-adherence. However, according to this review, with proper dosing these agents do not increase the occurrence of adverse events compared with their oral counterparts. Extrapyramidal symptoms were the most significant complications of typical neuroleptic drug therapy; parkinsonism was a generally irreversible long-term complication. Some studies have suggested that, among other factors, patients who receive depot neuroleptic drugs are at increased risk of extrapyramidal symptoms, with Parkinsonism over the short term being indistinguishable from negative symptoms and tardive dyskinesia being usually irreversible.

In a 6-month multicenter study in 96 manic patients (mean age 41 years; 50% women) who took risperidone monotherapy 4.2 mg/day, 16 withdrew from the study, in four cases because of adverse events: akathisia, impotence, drowsiness, and weight gain (28).

Comparative studies

In a 1-year randomized comparison of risperidone (mean age 39 years; 236 men; n = 349) and conventional neuroleptic drugs (mean age 38 years; 231 men; n = 326), risperidone produced statistically superior scores on the Positive and Negative Syndrome Scale for Schizophrenia, Barnes Akathisia Scale, and 36-Item Short Form Health Survey scale; however, there was no statistically significant difference in resource utilization between the two groups (29). This study, the Risperidone Outcome Study of Effectiveness (ROSE), was supported by Janssen, the market authorization holder of risperidone.

Risperidone versus haloperidol

In a randomized comparison of risperidone (mean dose 4.4, range 1–12, mg/day; n = 21) and haloperidol (mean dose 11, range 2–20, mg/day; n = 20) in patients with schizophrenia, tremor and rigidity were the most frequent adverse effects in both groups (risperidone—tremor, 8; rigidity, 6; haloperidol—tremor, 13; rigidity, 10) (30). There was akathisia in two patients taking haloperidol and in none taking risperidone. The higher incidence of extrapyramidal symptoms in patients receiving haloperidol might have been partly explained by the relatively higher doses that were used. Serum prolactin concentrations increased significantly in both groups after 8 weeks: 42 ng/ml with risperidone and 41 ng/ml with haloperidol.

In a double-blind, randomized, controlled, flexible-dose trial of the Early Psychosis Global Working Group, 555 patients (mean age 25 years; 71% men; median treatment length 206 days) with first-episode psychosis were assigned either to haloperidol (n = 277; mean modal dose 2.9 mg/day) or risperidone (n = 278; mean modal dose 3.3 mg/day). Haloperidol caused significantly more frequent and severe extrapyramidal symptoms (mainly emergent dyskinesia, parkinsonism, parkinsonian dystonia, and akathisia) associated with more concomitant medication (mainly anticholinergic drugs and benzodiazepines) (31). In contrast, risperidone caused significantly more weight gain than haloperidol at month 3 (4.6 versus 3.5 kg) but not at the end point (7.5 versus 6.5 kg); risperidone caused higher prolactin concentrations (women, n = 73, mean 74 ng/ml; men, n = 185, mean 34 ng/ml) than haloperidol (women, n = 71, mean 48 ng/ml; men, n = 178, mean 22 ng/ml) (reference prolactin concentrations in women < 25 ng/ml, in men < 18 ng/ml); haloperidol caused hyperprolactinemia in one patient and risperidone caused it in 14; risperidone also caused galactorrhea (n = 6), gynecomastia (n = 3), and moderate hyperglycemia (n = 1).

Similarly, risperidone caused extrapyramidal symptoms in fewer patients (24%) than haloperidol did (43%) in a two-phase study in patients with acute bipolar mania (phase I, 3 weeks, patients receiving either risperidone 1–6 mg/day, haloperidol 2–12 mg/day, or placebo (32). Plasma prolactin concentration was higher with risperidone (no data provided); prolactin-related adverse events included non-puerperal lactation, breast pain, dysmenorrhea, and reduced libido or sexual dysfunction; these effects occurred in six patients on risperidone (4%) and in two on haloperidol (1.3%).

Risperidone versus olanzapine

Several adverse events, including one death (no further information provided), have been described with risperidone in a study in which elderly patients with schizophrenia (mean age 70 years) were randomly assigned to risperidone (n = 32) or olanzapine (n = 34) for 4 weeks (33).

Placebo-controlled studies

Patients with bipolar disorder may benefit from risperidone, but this conclusion has mostly come from open studies with small sample sizes (SEDA-23, 69). Experience with risperidone has been reviewed with data from Canadian studies (34).

In a 6-week open study of risperidone (mean dosage 4.7 mg/day) in combination with mood-stabilizing treatments (usually lithium, carbamazepine, or valproate) for the treatment of schizoaffective disorder in 102 patients, 95 of whom completed the trial, at week 4 most patients had improved symptom severity and 9.3% were completely symptom-free (35). There were no statistically significant differences between baseline and week 4 in the severity of extrapyramidal symptoms, as measured by the UKU Side-Effect Rating Scale subscale for neurological adverse effects; other adverse effects included depressive symptoms (n = 13), exacerbation of mania (n = 5), drowsiness (n = 3), and impotence (n = 2).

The efficacy and safety of risperidone (mean dose 2.9, range 1.5–4 mg/day) have been examined in a double-blind, randomized, parallel-group, 6-week study of the treatment of aggression in 35 adolescents with a primary diagnosis of disruptive behavior disorder and subaverage intelligence (36). Risperidone significantly improved aggression, and extrapyramidal symptoms were absent or very mild; there was transient tiredness in 11 of the 19 drug-treated subjects compared with one of the 16 placebo-treated subjects; other adverse effects were sialorrhea ($n = 4$ for risperidone, $n = 0$ for placebo), nausea ($n = 3$ for risperidone, $n = 0$ for placebo), and weight gain (mean of 3.5% of body weight with risperidone). There were no clinically important changes in laboratory parameters, electrocardiography, heart rate, or blood pressure.

In a double-blind, placebo-controlled study of the addition of low-dose risperidone (mean dosage 2.2 mg/day) to a 5-HT re-uptake inhibitor in refractory obsessive-compulsive disorder in 70 adults, 18 of 20 risperidone-treated patients had at least one adverse effect (37). The adverse effects in both groups included sedation ($n = 17$ for risperidone, $n = 8$ for placebo), increased appetite (6 and 3), restlessness (6 and 6), and dry mouth (5 and 5).

The results of two major controlled clinical trials of risperidone in patients with dementia were not conclusive, but risperidone was more effective than placebo in agitated and demented patients (SEDA-25, 68). The long-term data have been reviewed and they suggest that although in the two pivotal previous comparisons of risperidone with placebo the risk of adverse events was similar in the two groups when risperidone was given in the optimal dosage (1 mg/day), during a 12-month open extension of these studies, the incidence of de novo tardive dyskinesia was very low, and there were no clinically important adverse events or changes in vital signs or laboratory signs (38).

Risperidone has been used in bipolar disorder, dementia, disruptive behavior disorder with subaverage intelligence, Tourette's syndrome, and autism (SEDA-23, 69) (SEDA-25, 67) (SEDA-26, 64). The efficacy and safety of low doses of risperidone in the treatment of autism and serious behavioral problems have been studied in 101 children aged 5–17 years with autistic disorder accompanied by severe tantrums, aggression, or self-injurious behavior, who were randomly assigned to risperidone for 8 weeks ($n = 49$; dosage 0.5–3.5 mg/day) or placebo ($n = 52$) (39). Risperidone produced a 57% reduction in the Irritability Score, compared with a 14% reduction in the placebo group; all other parameters were also significantly improved. Risperidone was associated with an average weight gain of 2.7 kg, compared with 0.8 kg with placebo; increased appetite, fatigue, drowsiness, dizziness, and drooling were more common with risperidone. In two-thirds of the children with a positive response to risperidone at 8 weeks, the benefit was maintained at 6 months.

Risperidone was also effective and well tolerated in 118 children aged 5–12 years with subaverage intelligence and severely disruptive behavior in a 6-week, multicenter, double-blind, randomized trial (40). Risperidone produced significantly greater improvement than placebo on the conduct problem subscale of the Nisonger Child Behavior Rating Form from week 1 (respective reductions in score of 15 and 6). The most common adverse effects of risperidone (mean dose at end-point 1.16 mg/day) were headache and somnolence; the extrapyramidal symptom profile of risperidone was comparable to that of placebo, and there were respective mean weight increases of 2.2 and 0.9 kg.

Adults with autistic disorder ($n = 17$) or pervasive developmental disorder not otherwise specified ($n = 14$) participated in a randomized, 12-week, double-blind, placebo-controlled trial of risperidone (41). Among those who completed the study, risperidone ($n = 14$) was superior to placebo ($n = 16$) in reducing the symptoms of autism, and the most prominent adverse effect was mild transient sedation during the initial phase of drug administration. Abnormal gait was reported in one patient taking risperidone.

According to a thorough review, risperidone is more effective than placebo in patients with dementia (42). Nevertheless, the results of the two major controlled clinical trials have not been conclusive (see Table 1). There was no statistical difference between the treatments in the first study, although there was in the second; extrapyramidal symptoms were notably more common with risperidone. In both studies, there was a high rate of placebo response, which implies that these patients responded favorably to the increased care that is given during a clinical trial. Long-term data in patients with dementia are lacking, but they are crucial in identifying tardive dyskinesia.

In a 9-week double-blind, randomized, placebo-controlled study, subjects were assigned to either risperidone ($n = 14$, 13 men and 1 woman; mean age 42 years; dose titrated upwards up to 2 mg/day) or to placebo ($n = 9$, 6 men and 3 women; mean age 39 years) (43). Adverse effects were reported in seven subjects who took risperidone, including dry mouth, tiredness, weakness, reduced sexual arousal and delayed ejaculation, and a mild dystonic reaction. However, five placebo-treated subjects also reported adverse effects, which might have been due to the strong tendency of these patients to become somatically preoccupied. There were no group differences in dropout rates due to adverse effects.

Table 1 The results of two 12-week, randomized, double-blind studies of risperidone in elderly patients with behavioral symptoms associated with dementia (28,29)

Mean age (years)	Number	Dose (mg/day)	Response (%)	Extrapyramidal symptoms (%)
81	344	Risperidone (1.1)	54	15
		Haloperidol (1.2)	63	22
		Placebo	47	11
83	625	Risperidone (1)	45	13
		Risperidone (2)	50	21
		Placebo	33	7.4

Patients with post-traumatic stress disorder may also benefit from risperidone. In a randomized study in 40 male Vietnam combat veterans, 37 completed at least 1 week of treatment, of whom 19 took risperidone and 18 took placebo; their mean ages were 51 and 54 years respectively (44). The dose of risperidone was individually adjusted (maximum 6 mg/day); the mean dose at endpoint was 2.5 mg/day. Risperidone was generally well tolerated; adverse effects included only mild akathisia in one patient taking 3 mg/day and gastrointestinal effects in one patient taking 4 mg/day; no patient withdrew because of adverse effects.

Irritable aggression and intrusive thoughts in post-traumatic stress disorder are reduced by low-dose risperidone as adjunctive therapy, according to the results of a double-blind, randomized trial in 16 male combat veterans, who took either risperidone (n = 7; mean age 49 years) or placebo (n = 8; mean age 54 years) for 6 weeks; one subject taking risperidone dropped out because of urinary retention (45). Concurrent antidepressant medication and anxiolytic drugs were allowed in both groups.

Risperidone also appears to be safe and effective in the short-term treatment of tics in children or adults with Tourette syndrome, according to the results of a randomized, double-blind study in 34 subjects (mean age 20, range 6–62 years), of whom 26 were children (25 boys and 1 girl; mean age 11, range 6–18 years) (46). After 8 weeks of treatment (mean dose 2.5 mg/day; maximum 4 mg/day for adults and 3 mg/day for children), those who took risperidone had significant improvements in tic severity, as evidenced by a 32% drop in the Yale Global Tic Severity Scale total score. Furthermore, 12 children randomized to risperidone had a 36% reduction in tic symptoms compared with an 11% reduction in the 14 children who took placebo. The most frequent adverse effects with risperidone were: increased appetite (n = 7), fatigue (n = 6), sedation (n = 3), foggy thinking (n = 2), blurred vision (n = 2). In and acute social phobia (which might be also diagnosed as panic attack) (n = 2); the last of these was managed by dosage reduction in one subject but required treatment withdrawal in the other. Two men had sexual adverse effects (either erectile difficulties or reduced libido); both improved with dosage reduction. There was a mean weight gain of 2.8 kg.

In a multicenter, randomized, double-blind, 12-week trial in Australia and New Zealand, 384 patients with dementia, mainly Alzheimer's disease, were initially enrolled and received at least one dose of risperidone (n = 167; 71% women; mean age 83 years; modal dose 0.99 mg/day) or placebo (n = 170; 72% women; mean age 83 years) (47). Clinical improvement in aggression and psychotic symptoms was evidenced by means of specific scales; 45 subjects taking risperidone and 56 taking placebo did not complete the trial, mainly because of insufficient responses and adverse effects. In the whole sample there was a high prevalence and variety of adverse events (94% of those taking risperidone and 92% of those taking placebo), mainly injuries, falls, somnolence, and urinary tract infections; however, only the last two were more common in those taking risperidone than in those taking placebo. A total of 39 patients taking risperidone group (23%) and 36 (21%) taking placebo had at least one severe adverse effect, including cerebrovascular events (9% with risperidone, n = 15; 1.8% with placebo, n = 3). Ten patients (3.6% with risperidone, n = 6; 2.4% with placebo, n = 4) died during the course of the trial, pneumonia and stroke being the most frequent causes (three due to pneumonia and two to stroke in the risperidone group; one due to pneumonia in the placebo group). Nevertheless, the investigators considered that relations between risperidone and adverse events that led to death were doubtful or non-existent.

In a multicenter, double-blind, 3-week study, 259 patients with an acute episode of mania were randomly assigned to risperidone (mean age 38 years; 71 men; n = 134; mean modal dose 4.1 mg/day) or placebo (mean age 40 years; 76 men; n = 125) (48). Improvement in the mean Young Mania Rating Scale total score (adjusted for covariates) was significantly greater with risperidone than placebo at the end-point. The rates of withdrawal due to adverse events were similar in the two groups (risperidone n = 10, placebo n = 7). The most common serious adverse events were manic reactions (risperidone n = 10, placebo n = 6), agitation (risperidone n = 3, placebo n = 0), and two accidental deaths in the placebo group. Adverse events that occurred in more than 10% of patients were somnolence (risperidone n = 38, placebo n = 9), hyperkinesia (risperidone n = 21, placebo n = 6), headache (n = 19 in both groups), dizziness (risperidone n = 15, placebo n = 11), dyspepsia (risperidone n = 15, placebo n = 8), and nausea (risperidone n = 15, placebo n = 3). Risperidone significantly increased the total score in the Extrapyramidal Symptom Rating Scale from baseline to end-point (baseline 1.2; change 0.6). Risperidone increased plasma prolactin concentrations in both men and women and five patients had at least a 7% weight gain.

In a 12-week, double-blind, randomized, placebo-controlled study in 40 patients with treatment-resistant schizophrenia (funded by Johnson & Johnson Pharmaceutical Research & Development), the addition of risperidone to clozapine improved overall symptoms and positive and negative symptoms (49). The adverse events profile of clozapine + risperidone was similar to that of clozapine + placebo. Clozapine + risperidone did not cause additional weight gain, agranulocytosis, or seizures compared with clozapine + placebo. All the patients completed 12 weeks of treatment; however, the small sample size precluded definitive conclusions.

From an initial sample of 36 patients with autism spectrum disorder (aged 5–17 years) who started an 8-week open study with risperidone, responders continued treatment for another 16 weeks (n = 26); two children withdrew because of unacceptable weight gain and the other 24 entered a double-blind withdrawal phase and were assigned to either risperidone (n = 12) or placebo (n = 12). Increased appetite and weight gain were the most important adverse events (mean increase 5.7 kg during 6 months, expected developmental weight gain 2.4 kg) (50).

In a two-phase placebo-controlled study with an initial sample of 45 patients, 39 of whom completed the study, the addition of low doses of risperidone (0.5 mg/day) appeared to improve symptoms in patients with obsessive–compulsive disorder taking fluvoxamine monotherapy (51). The main adverse events included transient sedation and mildly increased appetite.

Of 65 men (mean age 52 years) with post-traumatic stress disorder randomized to risperidone (n = 33) or placebo (n = 32) in a 4-month double-blind study supported by the Janssen Research Foundation, 22 and 26 patients respectively completed the study (52). There were no significant differences in weight.

The success of masking procedures in double-blind studies has recently been studied by the Research Units on Pediatric Psychopharmacology Autism Network following an 8-week placebo-controlled trial of risperidone in 101 autistic children (aged 5–17 years) (53). Clinicians attributed improvement to risperidone and lack of improvement to placebo, whether the child was taking risperidone or placebo; in contrast, the parents attributed improvement to risperidone only in the placebo group. Although the parents reported that adverse events influenced their guesses, the presence of adverse events was not associated with correct guesses. Adverse events therefore did not threaten the blindness of the study.

Organs and Systems

Cardiovascular

Cardiac arrest was attributed to risperidone in a patient with no history of cardiac disease (56).

The possible association of risperidone with QT_c interval prolongation is controversial (SEDA-24, 54). There were no significant changes in 73 patients with schizophrenia (mean age 34 years; 59% men) who took risperidone for 42 days (mean dose 3.7, range 4–6 mg/day) (57).

One of two children aged 29 and 23 months with autistic disorder developed a persistent tachycardia and dose-related QT_c interval prolongation while taking risperidone (58).

Cardiotoxicity of risperidone has recently been discussed (59) in the light of a death in a patient taking a therapeutic dose (SEDA-22, 68).

Acute massive pulmonary thromboembolism has been attributed to olanzapine in two patients (60):

- a 64-year-old woman taking bromperidol chronically, who also took risperidone (last dose 6 mg) for 40 days before the event;
- a 48-year-old woman who took risperidone for 6 days (last dose 2 mg).

The first patient died and the second survived. In the same series from a Japanese Emergency Center, seven patients (two men and five women, aged 23–70) who took chlorpromazine, levomepromazine, or propericiazine also had thromboembolic disorders; none of them had any known risk factor for thrombotic disease. Suggested mechanisms are: reduced movement at night as a consequence of the sedative effect of neuroleptic drugs (symptoms developed in all cases in the early morning); an effect of anticardiolipin antibodies, which can occur in some patients taking phenothiazines; or increased $5HT_{2A}$-induced platelet aggregation.

The debate about the need to restrict drug therapy in relation to the risk of cardiovascular events is open, and some authors have already expressed agreement (61) or disagreement about it (62), as well as pointing to the need for individual patient meta-analyses and significant changes to clinical trial methods in order to better assess the effectiveness, in contrast to the efficacy, of risperidone and other drugs in dementia (63).

Hypotension has been associated with risperidone (see Drug formulations below).

Nervous system

Seizures

Stuttering, a rare adverse effect of neuroleptic drugs, is thought to be a harbinger of seizures. Stuttering without seizures has been attributed to risperidone (64).

- A 32-year-old Korean man with delusions took oral risperidone 1.0 mg/day and lorazepam 0.5 mg bd. The dose of risperidone was increased to 4 mg/day over 4 days. On day 5, he began to stutter and could not articulate what he wanted to say without stuttering. The dose of risperidone was increased to 8 mg/day and the stuttering became more pronounced. However, after 1 year of continuous treatment, he did not stutter any more.

The authors suggested that since the patient had a history of stuttering risperidone had reactivated the speech pattern.

Extrapyramidal effects

Risperidone produces dose-related extrapyramidal adverse effects, but at concentrations lower than those of conventional antipsychotic drugs (SEDA-21, 58); at equieffective doses it has fewer such effects than haloperidol (SEDA-21, 57) (SEDA-22, 67) (SEDA-23, 68). The relation between the degree of receptor occupancy and the presence of extrapyramidal symptoms is not clear, as observed in a SPECT study in 20 patients (65). The frequency of risperidone-induced extrapyramidal signs, on the other hand, is intermediate between clozapine and conventional antipsychotic drugs, according to an open study in patients treated for at least 3 months with clozapine (n = 41; mean dose 426 mg/day), risperidone (n = 23; 4.7 mg/day), or conventional antipsychotic drugs (n = 42; 477 mg/day chlorpromazine equivalents) (66). The point prevalence of akathisia was 7.3% in those who took clozapine, 13% in those who took risperidone, and 24% in those who took conventional antipsychotic drugs; the point prevalences of rigidity and cogwheeling were 4.9% and 2.4% respectively with clozapine, 17% and 17% with risperidone, and 36% and 26% with conventional antipsychotic drugs.

A combined analysis of 12 double-blind studies in schizophrenic patients comparing the use of risperidone

($n = 2074$) with placebo ($n = 140$) or haloperidol ($n = 517$) has further stressed the dose-related extrapyramidal effects associated with risperidone (67). After covariance analysis to adjust for baseline ESRS (Extrapyramidal Symptom Rating Scale) scores, sex, race, age, height, duration of symptoms, age at first hospitalization, hospitalization status, and diagnosis, the effects of the maximum dose of risperidone on the mean shift to worse ESRS total scores were 1.4 (CI = 0.73, 2.03) at 1–4 mg/day ($n = 319$); 2.1 (CI = 1.65, 2.50) at 4–8 mg/day ($n = 932$); 3.3 (CI = 2.61, 3.89) at 8–12.5 mg/day ($n = 439$); and 3.8 (CI = 2.99, 4.55) at 13 mg/day or more ($n = 361$). The results also showed a significant dose-dependent increase in the use of antiparkinsonian drugs; the percentages of those requiring antiparkinsonian drugs were 14% at 1–4 mg/day ($n = 319$); 25% at 4–8 mg/day ($n = 900$); 27% at 8–12.5 mg/day ($n = 407$); and 31% at 13 mg/day or more ($n = 335$). Of 882 patients who took risperidone for at least 12 weeks, based on unsolicited reports, two developed tardive dyskinesia.

In an international, multicenter, double-blind study in 183 patients with a first psychotic episode treated with flexible doses of risperidone or haloperidol for 6 weeks, the severity of extrapyramidal symptoms and the use of antiparkinsonian drugs were significantly lower in patients taking low doses (up to 6 mg/day) than high doses (over 6 mg/day) of risperidone or haloperidol (68). These findings are consistent with the suggestion that patients with a first psychotic episode may require low doses of antipsychotic drugs. Furthermore, the severity of extrapyramidal symptoms was significantly lower in the risperidone-treated patients. Also, risperidone-treated subjects were significantly less likely than haloperidol-treated subjects to require concomitant anticholinergic drugs after 4 weeks (20 versus 63%); they had significantly less observable akathisia (24 versus 53%) and significantly less severe tardive dyskinesia. This was observed in a randomized double-blind comparison of risperidone 6 mg/day ($n = 34$) and haloperidol 15 mg/day ($n = 33$) (69).

In a 9-week open study of risperidone for agitated behavior in 15 patients with dementia (modal dose 0.5 mg/day), extrapyramidal symptoms developed at some point during the trial in 8 patients, and cognitive skills were impaired in 3 patients (70). Similarly, in 22 patients with dementia and behavioral disturbances, treated with risperidone 1.5 mg/day (range 0.5 mg qds to 3 mg bd), 50% had significant improvement, but 50% had some extrapyramidal symptoms (71). A further case of a severe extrapyramidal reaction in an old patient with dementia further illustrated these susceptibility factors (72).

Extrapyramidal symptoms have been reported within 24 hours of an injection of depot risperidone in patients with schizophrenia (73).

- A 32-year-old man who had been taking olanzapine 15 mg/day for 4 months was switched to injectable risperidone 25 mg because of poor adherence. On the next day he had an oculogyric crisis, dysarthria, torticollis, dysphagia, tremor, and rigidity, which resolved with procyclidine.

- A 36-year-old man had worse extrapyramidal symptoms.
- A 28-year-old man developed akathisia.

In order to limit the risk of extrapyramidal symptoms, the authors suggested that the dose of the oral neuroleptic drug should be reduced or omitted in the days after the injection, and that attention should be paid to the half-life of any other depot drug that has previously been given.

Although several cases of sensitivity to risperidone with extrapyramidal signs in Lewy body dementia have been published (SEDA-20, 52), a case of successful treatment without extrapyramidal adverse effects has also been reported (74).

- A 74-year-old man with Lewy body dementia treated with a combination of donepezil (5 mg in the evening) and risperidone (0.25 mg/day) had significant improvement, objectively and subjectively, within 2 weeks.

Akathisia

Restless legs syndrome has been reported in association with risperidone (75).

- A 31-year-old woman with schizoaffective disorder taking risperidone 6 mg/day complained of uncomfortable tingling and tearing sensations deep inside the calves and less severe sensations in her arms after 5 days; the symptoms vanished after replacement by quetiapine 400 mg/day.

Parkinsonism

Intolerable exacerbation of parkinsonism with risperidone has been reported (SEDA-20, 52), and even a 12-year-old boy reportedly developed parkinsonian tremor while taking risperidone (76). However, in contrast, eight patients (five women, three men) with advanced Parkinson's disease, motor fluctuations, and levodopa-induced dyskinesia took part in an open study with a low dosage of risperidone (mean 0.187 mg/day); after an average of 11 months all the patients had moderate to pronounced reductions in levodopa-induced dyskinesias (77).

Whether risperidone should be used in patients with Parkinson's disease is a subject of debate (78,79).

The efficacy and safety of risperidone have been evaluated in 44 patients (25 women and 19 men) with Parkinson's disease (80). There was either complete or near-complete resolution of hallucinations in 23, but an unsatisfactory response ($n = 6$) or worsening of parkinsonism ($n = 6$) in 12. Excluding patients with diffuse Lewy body disease, there was no significant worsening of scores on the Unified Parkinson's Disease Rating Scale after either 3 or 6 months of treatment, and the presence of dementia did not predict the response to treatment.

The long-term effect of risperidone on basal ganglia volume, measured by MRI scanning, has been studied in 30 patients with a first episode of schizophrenia who took risperidone, 12 patients taking long-term typical neuroleptic drugs, and 23 healthy controls (81). Treatment with risperidone for 1 year (mean dosage 2.7, range 1–6 mg/

day) did not alter basal ganglia volume, although there were movement disorders in both groups of treated patients, suggesting effects of both illness and medications.

Tardive dyskinesia

Tardive dyskinesia has occasionally been reported with risperidone (SEDA-20, 53; SEDA-21, 59; SEDA-22, 68) (82–87).

The incidence of tardive dyskinesia in patients with chronic schizophrenia taking risperidone is said to be 0.34% per year. Advanced age and dementia may be contributing factors (SEDA-21, 59; SEDA-22, 68; SEDA-23, 70).

Tardive dyskinesia/dystonia developed in four patients treated with risperidone at an early intervention facility for young people with psychosis (88). Other cases that have emerged were in:

- a 13-year-old young girl treated with risperidone 6 mg/day (89);
- a 21-year-old woman without previous exposure to other neuroleptic drugs or systemic illnesses that affected the central nervous system (90);
- a 58-year-old man with chronic alcoholism who developed tardive dyskinesia after exposure to risperidone that was aggravated by olanzapine (91).
- a 16-year-old girl who developed buccolingual masticatory tardive dyskinesia after taking risperidone 6 mg/day (92); when she restarted risperidone 2 mg/day, increasing to 6 mg/day later on, the dyskinesia improved.
- a 74-year-old woman who developed persistent tardive dyskinesia following a short trial (3 weeks) of a low dose (0.5 mg bd) of risperidone (93).

Tardive dyskinesia was studied in 330 elderly patients with dementia (mean age 83 years) (94). They were enrolled in a 1-year open study, in which the modal risperidone dose was 0.96 mg/day and the median duration of use was 273 days. The 1-year cumulative incidence of persistent tardive dyskinesia among the 225 patients without dyskinesia at baseline was 2.6%, and patients with dyskinesia at baseline had significant reductions in severity.

Attention deficit hyperactivity disorder (ADHD) may be a susceptibility factor for risperidone-induced tardive dyskinesia and withdrawal dyskinesia. Both conditions have occurred in patients with a past or recent history of attention deficit hyperactivity disorder.

- A 34-year-old woman developed dyskinesia after starting risperidone, with a marked increase in prolactin concentrations (95).
- A 13-year-old boy developed mild mouth movements, neck twisting, and intermittent upward gaze approximately 2 weeks after withdrawal of risperidone (1.5 mg and then 0.5 mg) (96).

Tardive dystonia

Dystonia and tardive dystonia have been attributed to risperidone.

- A 23-year-old man taking risperidone 8 mg/day developed blepharospasm (97).
- A 25-year-old man, who had never taken any other psychotropic medication, developed tardive dyskinesia with severe blepharospasm and tardive dystonia 2 months after withdrawal of risperidone (98).
- Possible risperidone-induced tardive dystonia has been reported in a 47-year-old man (99).

Marked improvement of tardive dystonia after replacing haloperidol with risperidone in a schizophrenic patient has been reported (100).

Pisa syndrome, a tardive axial dystonia with flexion of the trunk towards one side, is a rare reaction that occurs during treatment with neuroleptic drugs (SED-14, 146; SEDA-24, 57). Two cases related to risperidone have been published.

- A 24-year-old woman with mental retardation and an unspecified psychosis took risperidone 2 mg/day and trihexyphenidyl 2 mg/day and after 2 weeks developed symptoms that included tilting of her body backwards and to the left and tremors and cogwheel rigidity of the limbs (101). Risperidone was withdrawn and olanzapine 5 mg/day started; after 4 weeks there was no improvement and she was then lost to follow-up.
- A 25-year-old man with auditory hallucinations took risperidone 7 mg/day for 15 months plus biperiden 6 mg/day (102). He then complained of leaning to the right and being unable to straighten up. He had tonic flexion of the trunk to the right with slight backward axial rotation. After 5 months of risperidone withdrawal, the condition had not resolved; it later improved with co-beneldopa 400/100 mg/day and then after the addition of cabergoline 0.75 mg/day.

Rabbit syndrome

Rabbit syndrome has been reported in patients taking risperidone.

- A 27-year-old man took risperidone 6 mg/day and after 4 months the dosage was reduced to 4 mg/day; 7 months after the start of treatment he developed fine rapid pouting and puckering of the lips (103). These movements were accompanied by a strange, irritating, involuntary popping sound. The dosage of risperidone was reduced to 2 mg/day and an anticholinergic drug was added. Within days, there was symptomatic improvement, but a trial withdrawal of the anticholinergic drug resulted in worsening of the symptoms and treatment was renewed.
- A 38-year-old man with major depressive disorder and psychotic features developed rabbit syndrome after taking risperidone 4 mg/day and paroxetine 40 mg/day for 4 months; he was also taking simvastatin 10 mg/day, thiamine 100 mg/day, and folic acid 1 mg/day (104).

Co-administration of lithium and risperidone has been associated with the rabbit syndrome (105). This reaction was probably caused by the risperidone, and the role of lithium was not clear.

Neuroleptic malignant syndrome

Neuroleptic malignant syndrome has been reported in patients taking risperidone (106,107); most cases occurred within the first months, and even as early as 12 hours (SEDA-22, 68) (SEDA-23, 70).

Delayed risperidone-induced neuroleptic malignant syndrome has been reported in a 27-year-old man after 21 months (108), and in a 17-year-old girl who took risperidone 0.5 mg bd (109). Risperidone-induced neuroleptic malignant syndrome has also been reported in a 63-year-old woman with probable Lewy body dementia, who had previously had an episode of neuroleptic malignant syndrome with trifluoperazine (110).

A patient with schizoaffective disorder, who developed risperidone-related neuroleptic malignant syndrome, responded satisfactorily to supportive management and vitamin E plus vitamin B6 (111).

- A 30-year-old man with a history of bipolar disorder had substantial weight gain with olanzapine and was switched to risperidone (dose unknown) and lithium carbonate (450 mg bd) (112). A few days later he developed confusion, mild muscle rigidity, a raised temperature, and increased creatine kinase activity. The medications were withdrawn and he responded to supportive therapy.

Catatonia

Catatonia has been reported in relation to risperidone (SEDA-23, 72) (113,114).

- Catatonia occurred in a 61-year-old woman who was taking risperidone 5 mg/day for prominent paranoid delusions after a post-frontal lobotomy some 35 years ago. The catatonic disorder was dose-dependent and resolved immediately after changing to clozapine.

Pseudotumor cerebri

Two patients with hydrocephalus and learning difficulties developed headache, nausea, vomiting, drowsiness, lethargy, and episodes of collapse after starting to take risperidone for aggressive outbursts; the condition mimicked increased intracranial pressure (115). Withdrawal of the drug resulted in complete resolution of all the symptoms within 72 hours. The authors pointed out the striking degree of overlap between the adverse effects profile of risperidone and the symptoms of raised intracranial pressure due to shunt malfunction, which has not been previously highlighted.

Stroke in patients with dementia

Stroke is a matter of increasing concern with some antipsychotic drugs, and the Canadian Medicine Agency has issued a warning that there is a risk of stroke with risperidone (116). On April 2005, the FDA issued an alert and asked the company to add the following information to both the oral and depot formulations: "The FDA has found that older patients treated with atypical neuroleptic drugs for dementia had a higher chance for death than patients who did not take the medicine. This is not an approved use" (117). In some countries, the use of risperidone in demented patients has been restricted (118). It is important to remember, as the company itself, Janssen Pharmaceuticals, stated in a letter of April 2003, that risperidone is not approved for the treatment of dementia (119).

Based on data from four placebo-controlled trials (n = 1230), Janssen warned about cerebrovascular adverse events, including stroke, in elderly patients with dementia: "Cerebrovascular adverse events (e.g. stroke, transient ischemic attack), including fatalities, were reported in patients (mean age 85 years; range 73–97) in trials of risperidone in elderly patients with dementia-related psychosis. In placebo-controlled trials, there was a significantly higher incidence of cerebrovascular adverse events in patients treated with risperidone compared to patients treated with placebo. (Risperidone) has not been shown to be safe or effective in the treatment of patients with dementia-related psychosis". The higher risk of stroke with risperidone or olanzapine compared with placebo has been addressed by the World Health Organization (120), the European Agency for the Evaluation of Medicinal Products (EMEA), and other national agencies, which have also warned of this possible association (121,122,123). Furthermore, the Medicines and Healthcare products Regulatory Agency (MHRA) of the UK has recommended avoiding risperidone and olanzapine in elderly patients with dementia (124). In the UK, the Committee on Safety of Medicines (CSM) has also advised in a message of 9 March 2004 that there is clear evidence of an increased risk of stroke in elderly patients with dementia who take risperidone or olanzapine (125): "...the magnitude of this risk is sufficient to outweigh likely benefits in the treatment of behavioral disturbances associated with dementia and is a cause for concern in any patient with a high baseline risk of stroke".

Surprisingly, there was no statistically significant increased risk of stroke with either risperidone or olanzapine in a retrospective population-based study of 11 400 patients over the age of 66 years, in which three cohorts were identified: users of typical antipsychotic drugs (n = 1015), risperidone (n = 6964), or olanzapine (n = 3421) (126). During 13 318 person-years of follow-up, there were 92 admissions for stroke, distributed as follows: typical antipsychotic drug users (n = 10), risperidone users (n = 58), and olanzapine users (n = 24); the crude stroke rate per 1000 person-years did not differ significantly among the patients taking typical antipsychotic drugs (5.7), risperidone (7.8), or olanzapine (5.7). Relative to typical antipsychotic drug users, model-based estimates adjusted for covariates showed risk ratios for stroke of 1.1 (95% CI = 0.5, 2.3) with olanzapine and 1.4 (95% CI = 0.7, 2.8) with risperidone. Relative to olanzapine, users of risperidone were not at significantly increased risk of stroke-related hospital admission (adjusted risk ratio = 1.3, 95% CI = 0.8, 2.2). Finally, the authors of this report suggested that the possibility of a type II error to detect small differences in stroke-related outcomes between groups could not be excluded; assuming that the relative risk of stroke in risperidone-treated patients was 1.4, this would translate into about two extra strokes per 1000 person-years. Once more, experimental and observational studies have not yielded consistent results. Further studies to assess the specific risks of atypical antipsychotic drugs in elderly patients with dementia are required.

Psychological, psychiatric

Anxiety and behavioral stimulation, characterized by anxiety, insomnia, and restlessness, during risperidone treatment have been reported (SEDA-20, 53) (SEDA-21, 59) (SEDA-23, 71) (127). Six of thirteen outpatients who took part in a 10-week open trial of risperidone had a good initial response, followed by intolerable effects, including feelings of agitation and depression and periods of crying and insomnia (128). The patients who developed this syndrome had a significantly higher mean baseline rating on the Brief Psychiatric Rating Scale anxiety subscale.

Visual distortion with generalized anxiety and panic attacks has been attributed to risperidone (SEDA-22, 69). Visual disturbance resembling hallucinogen persistent perception disorder occurred after each of three consecutive risperidone dosage increases in a 55-year-old woman; there was absence of substance abuse (129).

It has been speculated that the antiserotonergic properties of risperidone could lead to obsessive and depressive symptoms, since a patient taking risperidone 4 mg/day developed major depression and obsessions, which resolved with fluoxetine 29 mg/day, relapsing when fluoxetine was withdrawn (130).

Psychomotor performance and cognitive function have been explored in a double-blind, randomized, four-way, crossover trial in 12 healthy volunteers (six men, six women), aged 66–77 years (mean age 69), who took single doses of risperidone 0.25 or 0.5 mg, lorazepam 1 mg, or placebo(131). Compared with placebo those who took risperidone had minor impairment of motor activity (reduced finger tapping at 6 hours after dosing in both the 0.25 and 0.5 mg groups), postural instability (displacement from the center of gravity at 3 hours after dosing), and impaired information processing (impaired digit symbol substitution at 2.5 hours after dosing in the 0.25 mg group). There were no significant effects on speed of reaction, vigilance and sustained attention, working and long-term memory, cortical arousal, or electroencephalography. The same study showed detrimental effects of lorazepam on cognitive function and cortical arousal.

Mania

Mania has rarely been associated with typical neuroleptic drugs, but has been described in patients treated with new antipsychotic drugs, especially risperidone (SEDA-22, 69) (SEDA-23, 71) (132). Risperidone-induced mania occurred in a 23-year-old man and a 21-year-old woman, who developed acute mania with euphoria, psychomotor agitation, and hypersexuality, at dosages of 4–8 mg/day (133).

Nevertheless, risperidone has been used in the treatment of mania in combination with mood stabilizers (SEDA-23, 69) (SEDA-26, 64).

Obsessive-compulsive disorder

Obsessive-compulsive symptoms associated with risperidone have been reported (SEDA-22, 69).

- Obsessive-compulsive symptoms developed in a 26-year-old Chinese woman taking risperidone for a chronic schizophrenic illness (134). She had no history of obsessive-compulsive symptoms. Risperidone 2 mg/day, benzhexol 2 mg/day, and diazepam 10 mg at night had been prescribed after she had had adverse effects with other antipsychotic drugs.
- A man who had taken risperidone 4 mg/day for 18 months developed an obsessional image of a person's face that repeatedly appeared in his mind as he went about his activities; the recurrent images disappeared after the dosage of risperidone was reduced to 3 mg/day (135).

In one case, reintroduction of risperidone did not cause obsessive-compulsive symptoms to re-emerge in a 29-year-old man who had previously developed obsessive-compulsive features when first treated with risperidone (136).

Endocrine

Prolactin

In one study, the prevalence of hyperprolactinemia among women taking risperidone was 88% (n = 42) versus 48% (n = 105) in those taking conventional antipsychotic drugs; 48% of these women of reproductive age taking risperidone had abnormal menstrual cycles (137). In the whole sample (147 women and 255 men) there were trends towards low concentrations of reproductive hormones associated with rises in prolactin; patients taking concomitant medications known to increase prolactin had been excluded. Raised prolactin concentrations were also observed in 13 (9 women and 4 men) of 20 patients (13 women and 7 men; mean age 36 years) (138). In premenopausal women there was a good correlation between prolactin concentrations and age, but there was no clear correlation between duration of treatment, dose, prolactin concentration, and prolactin-related adverse effects.

There was a significant rise in baseline serum prolactin concentration in 10 patients after they had taken risperidone for a mean of 12 weeks compared with 10 patients who were tested after a neuroleptic drug-free wash-out period of at least 2 weeks (139). A non-significant increase in serum prolactin has also been observed in an open comparison of risperidone with other neuroleptic drugs in 28 patients (140). However, in a meta-analysis of two independent studies (n = 404), prolactin was greatly increased by risperidone (mean change 45–80 ng/ml), a larger effect than with olanzapine and haloperidol (141).

Five patients (four women and one man, aged 30–45 years), who were evaluated for risperidone-induced hyperprolactinemia, had significant hyperprolactinemia, with prolactin concentrations of 66–209 µg/l (142). All but one had manifestations of hypogonadism, and in these four patients, risperidone was continued and a dopamine receptor agonist (bromocriptine or cabergoline) was added; in three patients this reduced the prolactin concentration and alleviated the hypogonadism.

The relation of prolactin concentrations and certain adverse events has been explored by using data from two large randomized, double-blind studies ($n = 2725$; 813 women, 1912 men) (143). Both risperidone and haloperidol produced dose-related increases in plasma prolactin concentrations in men and women, but they were not correlated with adverse events such as amenorrhea, galactorrhea, or reduced libido in women or with erectile dysfunction, ejaculatory dysfunction, gynecomastia, or reduced libido in men. Nevertheless, in five patients risperidone (1–8 mg/day) caused amenorrhea in association with raised serum prolactin concentrations (mean 122 ng/ml, range 61–230 ng/ml; reference range 2.7–20 ng/ml) (144).

In 41 schizophrenia patients who took either risperidone (11 men, 9 women; mean dose 4 mg/day) or perospirone (10 men, 11 women; mean dose 24 mg/day) for at least 4 weeks, prolactin concentrations increased only in those taking risperidone (5.3-fold in women and 4.2-fold in men) (145).

Hyperprolactinemia was found after about 30 months in 12 premenopausal women with schizophrenia or schizoaffective disorder (aged 15–55 years) taking risperidone but not in those taking olanzapine ($n = 14$) (146). Prolactin concentrations were significantly higher in the first group than in the second (123 ng/ml versus 26 ng/ml).

There was a high correlation between prolactin concentrations and risperidone in 14 men (mean age 53 years) (147). In 47 men (mean age 23 years) receiving acute treatment with several neuroleptic drugs, including risperidone (mean dose 3.6, range 1–6, mg; mean duration 45 days; risperidone monotherapy in 35 subjects), there was hyperprolactinemia, defined as a prolactin concentration greater than 636 mU/l (male reference range 45–375 mIU/l), in 27 of the 37 patients who could be assessed (148). There was neither gynecomastia nor galactorrhea. One year later, 38 of the 47 patients were reassessed (mean dose 2.3, range 1–6, mg); prolactin concentrations could be determined in 20 of these patients, and there was hyperprolactinemia in six of them. In the 35 patients taking risperidone monotherapy, two had high prolactin concentrations at baseline, but this rose to 21 of 29 patients at the end of the acute phase and fell to four of 16 after 1 year. According to the authors, the lower concentrations at 1 year might have been due to the development of tolerance to prolactin.

Risperidone-induced galactorrhea associated with a raised prolactin has been reported (149,150,151), as have amenorrhea and sexual dysfunction (9). It is suggested that this condition can occur after many weeks of risperidone treatment, with small dosages (2–4 mg/day), and at times even after drug withdrawal.

- Galactorrhea associated with a rise in prolactin occurred after a few weeks of treatment with risperidone in two women aged 24 and 39 (149). One of them was switched to thioridazine, with an improvement in the galactorrhea, and the other continued to take

risperidone owing to a robust response; her galactorrhea was partially treated with bromocriptine.

- A 34-year-old woman, who developed amenorrhea while taking risperidone, regained her normal menstrual pattern along with a marked fall in serum prolactin concentration 8 weeks after being switched to olanzapine, whereas amantadine had failed to normalize the menses and had apparently reactivated the psychotic symptoms (152).

The authors suggested that olanzapine may offer advantages for selected patients in whom hyperprolactinemia occurs during treatment with other antipsychotic drugs.

- Galactorrhea and gynecomastia occurred in a 38-year-old hypothyroid man who took risperidone for 14 days (153).

The authors suggested that men with primary hypothyroidism may be particularly sensitive to neuroleptic drug-induced increases in prolactin concentrations.

- A 17-year-old man developed galactorrhea and breast tenderness within weeks of starting to take risperidone.

The authors suggested that patients who have galactorrhea, amenorrhea, or both while taking risperidone should be gradually switched to olanzapine, quetiapine, or clozapine (154). Indeed, when 20 women with schizophrenia who were taking risperidone and had menstrual disturbances, galactorrhea, and sexual dysfunction (SEDA-24, 72) (SEDA-26, 65) were switched from risperidone to olanzapine over 2 weeks and then took olanzapine 5–20 mg/day for 8 further weeks, serum prolactin concentrations fell significantly (155). Scores on the Positive and Negative Syndrome Scale, Abnormal Involuntary Movement Scale, and Simpson–Angus Scale for extrapyramidal symptoms at the end-point were also significantly reduced. There were improvements in menstrual functioning and patients' perceptions of sexual adverse effects.

There were no significant adverse effects on growth or sexual maturation in a retrospective study based on a sample of 700 children aged 5–15 years who had been taking risperidone (0.02–0.06 mg/kg/day) for 11 or 12 months because of disruptive behavior disorders (156).

Risperidone-induced hyperprolactinemia has been reported to resolve with quetiapine, a low-potency dopamine D_2 receptor antagonist (157).

In a randomized, double-blind, 12-week study in 78 inpatients with schizophrenia assigned to either risperidone 6 mg/day (73% men; $n = 41$) or haloperidol 20 mg/day (81% men; $n = 37$), prolactin concentrations increased significantly in men in both groups (158). Adjusted for haloperidol dose equivalents (risperidone 6 mg/day equivalent to haloperidol 12 mg/day), risperidone caused a significantly larger rise in prolactin than haloperidol. The study was limited by the small number of women in the sample, which allowed the comparison of prolactin concentrations by sex but without consideration of treatment; the women had a significantly larger rise in prolactin than the men.

Amenorrhea presumed to have been induced by risperidone has been successfully treated with Shakuyaku-kanzo-to, a Japanese herbal medicine that contains *Peoniae radix* and *Glycyrrhizae radix* (159).

The risk of prolactinoma in patients taking risperidone and other neuroleptic drugs, accompanied by hyperprolactinemia, amenorrhea, and galactorrhea has been discussed in the light of a case of hyperprolactinemia (160).

- A 35-year-old woman who had taken lithium carbonate 800 mg/day for 2 years was also given risperidone 6 mg/day for a manic relapse. She missed two menstrual periods and had galactorrhea. A head CT scan showed a pituitary microadenoma and the prolactin concentration was 125 µg/l (reference range up to 20 µg/l). Risperidone withdrawal resulted in disappearance of the prolactinoma. Her other symptoms persisted and did not change with olanzapine 2.5 mg/day; however, bromocriptine 12.5 mg/day for 2 weeks relieved her symptoms and the prolactin concentration normalized.

Metabolism

Diabetes mellitus

There has been a report of diabetic ketoacidosis in a 42-year-old man, without a prior history of diabetes mellitus, who took risperidone (2 mg bd) (161). The authors pointed out that in premarketing studies of risperidone, diabetes mellitus occurred in 0.01–1% of patients.

Weight gain

Pathological weight gain has been increasingly identified as a problem when atypical neuroleptic drugs are given to children (SEDA-21, 57) (SEDA-22, 69). In one case, unremitting weight gain, triggered by risperidone, was eventually curbed through the use of a diet containing slowly absorbed carbohydrates and a careful balance of carbohydrates, proteins, and fats (162).

- A 9-year-old boy with autism and overactivity was unresponsive to several drugs. Risperidone 0.5 mg bd was effective, reducing his Aberrant Behavior Checklist score from 103 to 57 by the end of the first week. Four weeks later his weight had risen from 34.6 to 37 kg. This rate of weight gain (0.6 kg/week) continued over the next 12 weeks. His weight was then contained by the use of the "Zone" diet, with an emphasis on slowly absorbed carbohydrates (examples include apples, oatmeal, kidney beans, whole-grain pasta, and sweet potatoes) in a calorie-reduced diet containing 30% proteins and 30% fats.

In 37 children and adolescent inpatients treated with risperidone for 6 months, compared with 33 psychiatric inpatients who had not taken atypical neuroleptic drugs, risperidone was associated with significant weight gain in 78% of the treated children and adolescents compared with 24% of those in the comparison group (163). Risperidone dosage, concomitant medicaments, and other demographic characteristics (such as age, sex, pubertal status, and baseline weight and body mass index)

were not associated with an increased risk of morbid weight gain.

Weight gain has been studied in 146 Chinese patients with schizophrenia (85 men and 61 women; mean age 33 years) who took risperidone in a maximum dose of 6 mg/day (164). Mean body weight rose gradually from 61 kg at baseline to 62 kg on day 14, 63 kg on day 28, and 64 kg on day 42; the mean dose at 6 weeks was 4.3 mg/day. Weight gain was associated with a lower baseline body weight, younger age, undifferentiated subtype, a higher dosage of risperidone, and treatment response (for positive, negative, and cognitive symptoms and social functioning). However, a possible ethnic difference might contribute to the marked weight increase found in these Chinese patients, since non-white patients reported more weight gain than white patients.

Risperidone-induced weight gain varies throughout the age span, young people being the most sensitive; although preadolescents and adolescents take substantially lower daily doses and lower mg/kg doses than adults, they gain as much or more weight (corrected for age-expected growth) (165). This effect is hardly, or not at all, experienced by those aged over 65. Furthermore, young people also had a greater percentage of drug-induced weight gain relative to baseline body weight, and the percentage increases in body mass index during risperidone treatment were consistently greater in the young.

There was weight gain of 17% (mean 5.6 kg) after risperidone treatment for 6 months in 63 autistic children and adolescents (mean age 8.6 years), taken from an initial sample of 101 outpatients; this gain exceeded the developmentally expected norms and decelerated over time (166). Body mass index increased by 10.6% (mean 2.0 kg/m^2). Changes in serum leptin did not reliably predict risperidone-associated weight gain.

- Long-term weight gain has been described in a 35-year-old man with schizophrenia taking risperidone 4 mg/day; weight changed from 94 to 121 kg and there was also "carbohydrate craving" and persistent hunger (167).

The underlying mechanism may involve the adipose tissue hormone leptin or immune modulators such as tumor necrosis factor α.

In contrast, in a multicenter, open study in 127 elderly psychotic patients (median age 72, range 54–89 years) taking risperidone (mean dose 3.7 mg/day) there was no significant weight gain after 12 months (168).

Risperidone-induced obesity can cause sleep apnea (169).

- A 50-year-old man with schizophrenia gained 29 kg over 31 months and developed diabetes while taking risperidone 6 mg/day. He reported difficulty in sleeping and frequent daytime napping that left him unsatisfied, and his wife reported prominent snoring and apnea at night.

Nasal continuous positive airway pressure produced improvement.

Electrolyte balance

Polydipsia with hyponatremia has been reported in patients taking risperidone (170,171).

- A 28-year-old man complained of unbearable thirst 2 weeks after starting risperidone 8 mg/day, and would drink 4–5 liters of water within a variable period of a few minutes to 8 hours; he did not develop hyponatremia (172). The condition lasted about 2 years and remitted after withdrawal of risperidone. After a drug-free interval of 2 weeks, clozapine was started and the condition had not recurred after 6 months.

The mechanism of this effect was unclear. The authors thought that SIADH was unlikely, considering the features of polyuria, a low urine osmolality (172 mosmol/kg), and absence of hyponatremia during polydipsia.

Hematologic

Agranulocytosis, leukopenia, neutropenia, lymphopenia, and thrombocytopenia have been reported in patients taking risperidone (173–176).

- A 40-year-old woman developed agranulocytosis after taking risperidone for 2 weeks (175). She had also developed agranulocytosis after treatment with several other antipsychotic drugs (chlorpromazine, haloperidol, and zuclopenthixol).
- A 90-year-old man with vascular dementia, delusions, and hallucinations developed a fever of 39.5°C (177). During the previous 6 months he had been taking risperidone 2 mg/day for the psychotic symptoms; he was also given unidentified antibiotics for pneumonia. His white blood cell count was $1.4 \times 10^9/l$ (no neutrophils). Risperidone was withdrawn and the white cell count rose to $2.2 \times 10^9/l$ (30% neutrophils) on day 4, $2.5 \times 10^9/l$ (33% neutrophils) on day 14, and $5.4 \times 10^9/l$ (59% neutrophils) on day 18.

It is not clear whether this event could have been due to an interaction with other drugs that the patient received.

Two cases of nose bleeding have been described in association with risperidone, probably for the first time (178).

- A 57-year-old woman with no history of hypertension and not taking other medicines began having profuse nose bleeds and headaches immediately after starting to take risperidone 1 mg/day; risperidone was withdrawn 4 days later and the nose bleeds stopped.
- A 42-year-old man began to have spontaneous nose bleeds while taking risperidone; coagulation tests were normal.

The authors suggested that the mechanisms of these bleeding episodes might have been thrombocytopenia, a recognized adverse effect of certain neuroleptic drugs, and/or reduced platelet aggregation due to an antagonist effect on $5HT_{2A}$ receptors.

The WHO International Drug Monitoring database contains 54 reports of nose bleeds associated with risperidone, 37 with an established cause-and-effect relation (179).

Mouth

Hypersalivation or sialorrhea has been reported with all neuroleptic drugs, and has been associated with risperidone as one of the most frequently mentioned adverse effects in patients with disturbing extrapyramidal symptoms during previous neuroleptic drug treatment (SEDA-25, 68). Hypersalivation is a troublesome adverse effect that can contribute to non-adherence to therapy, but it can be treated with clonidine.

- Hypersalivation in a 22-year-old man was rapidly and markedly reduced by clonidine 0.1 mg bd over 3 days (180).

Liver

Several cases of hepatotoxicity have been reported in adults (181) and boys (SEDA-22, 69) taking long-term risperidone.

- Hepatotoxicity occurred in an 81-year-old man who took only two doses of risperidone 0.5 mg (182).
- In a 25-year-old woman, liver function tests were 2.6–7.4 times higher than the upper limit of the reference range during risperidone therapy (183).
- A 13-year-old girl developed liver enzyme rises and fatty liver infiltration in the context of pre-existing obesity after taking risperidone 0.5 mg/day for 3 days (184).
- Transient increases in liver enzymes induced by risperidone occurred in two men aged 19 and 22 (185).
- Cholestatic jaundice occurred in a 37-year-old man taking risperidone (186).

Urinary tract

Hemorrhagic cystitis has been associated with risperidone (187).

- An 11-year-old boy developed acute dysuria and increased frequency accompanied by gross hematuria. He was taking fluoxetine, valproic acid, benzatropine, haloperidol, clonidine, trazodone, and nasal desmopressin. One week before presentation, risperidone had been introduced instead of haloperidol to improve behavioral control. The risperidone was discontinued and haloperidol resumed, and his symptoms resolved during the following week.

Several cases of urinary incontinence have been associated with risperidone, and the manufacturers report that this adverse effect occurs in up to 1% of patients (SEDA-22, 70). The authors of a report of two patients, in both of whom the adverse effect was clearly temporally related to the drug, stated that at least 28% of patients developed transient urinary incontinence after starting risperidone (188).

Skin

Photosensitivity has been attributed to risperidone.

- A 69-year-old woman taking risperidone developed an erythematous rash with areas of blistering and early desquamation. It was most pronounced in exposed areas, although there was some spread beyond (189).

Musculoskeletal

In premenopausal women with schizophrenia or schizoaffective disorder aged 15–55 years, there was a significantly lower speed of sound transmission in the radius and the phalanges of the hand in patients taking risperidone (n = 12) compared with those taking olanzapine (n = 14) (van Os 229). Furthermore, the speed of sound in bone in this group correlated inversely with urinary deoxypyridinoline excretion, indicating a high bone turnover rate in this subset of patients, which is usually considered a risk factor for fragility fracture, independent of bone mineral density. Hyperprolactinemia is considered to be a risk factor for osteoporosis and those taking risperidone had significantly higher prolactin concentrations than those taking olanzapine (123 ng/ml versus 26 ng/ml) (see also Endocrine above). The relative risks for fragility fracture in the women taking risperidone compared with those taking olanzapine were 1.78 and 1.23 (speed of sound in bone measured at the phalanges and the radius respectively).

Sexual function

The prevalence of sexual dysfunction in patients taking psychotropic medications, based on spontaneous reports, has been underestimated on the basis of a randomized open study of the sexual effects of risperidone (n = 24; mean age 25 years) and quetiapine (n = 25; mean age 27 years) as assessed by a semistructured interview, the Antipsychotics and Sexual Functioning Questionnaire (190). Sexual dysfunction was reported by 12 patients taking risperidone (mean dose 3.2 mg/day) and four taking quetiapine (mean dose 580 mg/day), mainly reduced libido and/or impaired orgasm. Furthermore, there were significantly higher concentrations of prolactin in men taking risperidone (47 µg/l, n = 15) than in those taking quetiapine (12 µg/l, n = 19).

Prolonged erection has been reported with risperidone (SEDA-24, 71; 196,192,193).

- Priapism associated with risperidone occurred in a 19-year-old man who had taken 2 mg/day for 4 days (194).
- A 32-year-old man who had taken flupentixol (up to 3 mg/day) over 5 years was switched to risperidone 4 mg/day because of recurrent psychotic episodes, and 2 months later, in view of a partial response, the dose was increased to 5 mg/day (195). He developed a persistent, painful penile erection 2 weeks later and was treated; a second episode occurred on the next day, and episodes of rigid erection that subsided without surgical intervention were detected during the next 4 days. The results of investigations, a complete hemogram, a sickling test, a penile Doppler study, and a penile biopsy, were all normal.

- A 26-year-old man who had taken risperidone 3 mg/day and sodium valproate 1500 mg/day for 1 year developed a persistent erection, dysuria, and urinary incontinence, which did not respond to irrigation of the corpora cavernosa on two occasions and required surgical treatment (196). Prolonged priapism also resulted in penile fibrosis, associated with a high risk of permanent erectile dysfunction.

In one case priapism followed the use of first risperidone and then ziprasidone YET TO start HERE(197). In two other cases, presumed to be due to risperidone (198,199), penile irrigation with isotonic saline and phenylephrine injection resulted in detumescence. Risperidone has a high affinity for alpha-1 adrenoceptors, and alpha-1 blockade leads to direct arteriolar dilatation, which results in increased blood inflow and reduced outflow secondary to effacement and subsequent obstruction of emissary veins.

Ejaculatory dysfunction has also been associated with risperidone in three cases.

- A 21-year-old patient with bipolar schizoaffective disorder developed absent ejaculation with normal orgasm 3 weeks after starting to take risperidone (200).
- A 37-year-old man with paranoid schizophrenia had ejaculatory difficulty during sexual intercourse with his wife, compatible with retrograde ejaculation, 1–2 weeks after starting to take risperidone (201). He reported complete failure to emit semen but a normal desire, erection, and sense of orgasm. Semen was seen in postcoital urine. The dosage of risperidone was reduced to 3 mg/day and anterograde ejaculation was partially restored.
- A 17-year-old man, with a history of paranoid schizophrenia and irregular abuse of cannabis and alcohol, took risperidone (202). Starting at 0.5 mg/day, the dose was titrated up to 4 mg/day over 3 weeks. In a few weeks, he stopped taking risperidone because of a recurrent inability to ejaculate, despite normal libido, erection, and sense of orgasm; he also had difficulty in urinating. These effects disappeared after drug withdrawal. However, risperidone was restarted because of relapse of schizophrenia, and the ejaculatory disturbance recurred within a few days.

Risperidone may have caused retrograde ejaculation by altering sympathetic tone and allowing semen to pass retrogradely into the bladder during ejaculation.

Immunologic

Risperidone has been rarely associated with allergic reactions (SEDA-22, 70).

Long-Term Effects

Drug withdrawal

Serious withdrawal effects have occasionally been reported with risperidone (SEDA-21, 60; SEDA-22, 70). Both manic and psychotic symptoms have been described in a patient with chronic schizophrenia after risperidone withdrawal (203).

- A 38-year-old Chinese man responded to risperidone monotherapy for 2 weeks after 19 years of resistance to typical neuroleptic drugs. Three days later he lost his medicine and 2 days later his auditory hallucinations and persecutory delusions recurred. Meanwhile, vivid manic symptoms (such as heightened mood, irritability, reduced need for sleep, hyperactivity, pressured speech, flight of ideas, and grandiosity) emerged for the first time throughout the history of his illness.
- Dyskinesia occurred for 5 days in an 82-year-old woman after withdrawal of risperidone and citalopram, which she had taken for about 3 months (204).

It is not clear whether this was a withdrawal effect or a tardive dyskinesia in response to risperidone.

Tardive dyskinesia at multiple sites, including respiratory dyskinesia, has been associated with withdrawal of risperidone (205).

- An 84-year-old Japanese woman with mixed dementia taking bromperidol and biperiden was switched to risperidone 2 mg/day. After several weeks she began to have limb and orofacial dyskinesia and staggered while walking. The risperidone was abruptly withdrawn. During the following days the previous abnormal movements increased and extended to the trunk. There was respiratory dyskinesia with dyspnea. The symptoms resolved completely in risperidone withdrawal and treatment with haloperidol and biperiden.

Second-Generation Effects

Teratogenicity

All neuroleptic drugs cross the placenta and reach the fetus in potentially significant amounts; the best recommendation is to avoid any drug during the first trimester and only to use drugs thereafter if the benefits to the mother and fetus outweigh any possibility of risk (SED-14, 152) (SEDA-22, 54). Risperidone has been used in two women before and throughout pregnancy without developmental abnormalities in the children after 9 months and 1 year; the dosages started at 2 mg/day and were increased to 4–6 mg/day (206). In a large postmarketing study of 7684 patients who took risperidone, nine women took it during ten pregnancies; there were seven live births and three therapeutic terminations of pregnancy (SEDA-23, 69) (207).

Lactation

The distribution and excretion of risperidone and 9-hydroxyrisperidone into the breast milk of a young woman with puerperal psychosis, who was treated with risperidone, has been reported (208).

- A 21-year-old woman with a 2-year history of bipolar disorder stopped all of her medication when she discovered that she was pregnant; she was given risperidone 2.5 months after childbirth, gradually increasing

to a steady-state dosage of 6 mg/day. Risperidone and 9-hydroxyrisperidone concentrations in plasma and breast milk were measured, and calculations indicated that a suckling infant would receive only 0.84% of the maternal dose as risperidone and 3.46% as 9-hydroxyrisperidone.

The transfer to milk of risperidone and its active metabolite 9-hydroxyrisperidone has been examined in two breast-feeding women and in one woman with risperidone-induced galactorrhea (209). The milk:plasma concentration ratio was under 0.5 for both compounds; the calculated relative infant doses were 2.3%, 2.8%, and 4.7% of that of women's weight-adjusted doses; neither compound was detected in the plasma of the two babies, who achieved their developmental milestones satisfactorily and did not have any adverse effect attributable to risperidone. The authors concluded that maternal risperidone therapy is unlikely to pose a significant hazard to the breast-fed infant in the short term, and recommended an individual benefit–harm analysis to take decisions about this issue.

Susceptibility Factors

Genetic factors

Impaired metabolism of risperidone can increase the risk of adverse effects.

- A homozygous non-functional genotype, CYP2D6*4, was found in a 17-year-old patient with schizophrenia who developed severe akathisia, parkinsonism, and drowsiness after taking risperidone 6 mg/day for 3 months; he had high plasma concentrations of risperidone and an active metabolite (210).

The effect of the Ser9Gly polymorphism of the dopamine D_3 receptor (DRD3) gene in response to risperidone has been studied in 123 Han Chinese with acute schizophrenia treated for up to 42 days (211). Compared with patients with the Gly9Gly genotype, after adjusting, those with Ser9Gly had significant better performance on negative symptoms and better functioning but not better positive symptoms.

Age

Elderly people

Elderly patients with dementia have been said to be at particular risk of developing extrapyramidal adverse effects, even with very low doses.

Delirium occurs in 1.6% of elderly patients newly treated with risperidone (SEDA-22, 70), and cases have been reported in patients of advanced age (212,213). It is suggested that in these patients, treatment should begin with low dosages (0.25–0.5 mg/day) and that the dosage be gradually increased over several days, with close monitoring.

The safety, tolerability, and efficacy of risperidone have been assessed in 103 patients with schizophrenia (52 men and 51 women) aged 65 years or older in an open, multicenter, 12-week study (214). The mean risperidone dose at end-point was 2.4 mg/day. Adverse events occurred in 91 patients and included dizziness (*n* = 23), insomnia (*n* = 17), agitation (*n* = 15), somnolence (*n* = 15), injury (*n* = 12), constipation (*n* = 11), and extrapyramidal disorders (*n* = 10); 11 patients withdrew because of adverse events. Among the 91 patients with normal baseline QT_c intervals (below 450 ms), 9 had a prolonged QT_c interval during the study (range 450–516 ms).

Adverse effects leading to dosage reduction or discontinuation were also observed in a retrospective study in 57 patients mean age 84 (range 66–97) years, who took risperidone (doses 0.5–4 mg/day) for more than 1 year (average 2 years) for dementia-related behavioral disturbances (215). Adverse effects included hypotension (*n* = 4), agitation (*n* = 6), and sedation (*n* = 5), and six patients developed a new movement disorder.

In a retrospective study, a substantial proportion of patients who needed antiparkinsonian medication while taking risperidone (mean daily dose 4.4 mg) were identified. Twelve of fifty-five elderly inpatients (aged over 65 years) taking risperidone received antiparkinsonian drugs (216).

A large proportion of new or worsened extrapyramidal adverse effects (32%) was observed in a review of the charts of 41 patients with dementia (mean age 75 years) treated with risperidone (mean 1.8 mg/day) (217).

In 129 patients (aged 18–93 years) patients over 40 years had higher risperidone total plasma concentrations (estimated as 35% per decade in patients over 42 years) (218). According to the authors, this might lead to an increased incidence of adverse events in elderly subjects.

Children

There was a high incidence (44%) of new movement disorders in 36 children and adolescents who were treated with risperidone (the highest dose of risperidone was 6 mg/day and the average maintenance dose was 4 mg/day) (219). The higher incidence reported in the latter series has been countered by a contrasting report that the most common adverse effect was excessive weight gain (*n* = 10) in a series of 30 children and adolescents (aged 6–21 years) taking risperidone (0.5–6 mg/day) for attention-deficit disorder (220). Vomiting and drowsiness each occurred in one patient; one patient had withdrawal dyskinesia, but the reintroduction of risperidone and slower withdrawal produced no recurrence.

Risperidone has been assessed in children with autistic disorder and disruptive behavior in a multicenter, two-part, open study (221). Part one consisted of a 4-month open phase in 63 children (aged 5–17 years; 49 boys) taking risperidone 2.0 mg/day; of the 12 dropouts, five were due to loss of efficacy and one to an adverse event (constipation); there was a mean weight gain of 5.1 kg, which was significantly greater than expected from developmental norms. Part two was a randomized, double-blind study in which risperidone was either substituted by placebo (n = 16) or continued (n = 16); there were more relapses in those who took placebo (n = 10) than in

those who continued to take risperidone (n = 2). One of six dropouts in the first part of the study was due to an adverse event (constipation); over the 6 months of the study there was a mean weight gain of 5.1 kg.

Other features of the patient

Hypothermia associated with hypothalamic and thermoregulatory dysfunction has been reported in a patient with Prader–Willi syndrome taking risperidone and olanzapine (222). Hypothermia in response to these drugs is said to result from $5-HT_2$ receptor blockade, and it is recommended that patients with hypothalamic dysfunction should be carefully monitored if risperidone or olanzapine are used.

Drug Administration

Drug formulations

Depot formulations of neuroleptic drugs are specifically recommended over oral formulations for long-term treatment in patients with schizophrenia with a history of poor adherence to therapy; they also improve systemic availability, maintain stable plasma concentrations, and reduce the risk of overdose (223). A long-acting injection of risperidone, a combination of extended-release microspheres and a diluent for parenteral use, has recently been introduced in several countries, including the USA, for the treatment of schizophrenia. Long-acting depot risperidone has been reviewed (Harrison 113). According to the market authorization holder, the recommended dose is 25 mg intramuscularly (maximum 50 mg) every 2 weeks, the half-life of absorption being 3–6 days, and oral risperidone should be given with the first injection and continued for 3 weeks to establish tolerability and ensure adequate plasma concentrations (224). The most common adverse events were insomnia, agitation, headache, psychosis, rhinitis, pain, dizziness, anxiety, depression, and hyperkinesia; there were six deaths (suicide, n = 4; cardiac failure, n = 2) (Harrison 113).

Generally, the same rules and warnings as for oral risperidone must be followed; elderly patients and those predisposed to orthostatic hypotension should be instructed in non-pharmacological interventions to reduce hypotension (for example, rising slowly to the seated position and sitting on the edge of the bed for several minutes before trying to stand up in the morning), and they should avoid circumstances that accentuate hypotension, including sodium depletion and dehydration (FDA). Patients with renal impairment may be less able to eliminate risperidone and those with impaired hepatic function may have an increase in the unbound fraction of the drug, possibly resulting in an enhanced effect. Women should notify their doctor if they become or intend to become pregnant during therapy and for at least 12 weeks after the last injection; they should also be advised not to breast-feed an infant during treatment and for at least 12 weeks after the last injection. Patients taking risperidone who start to take carbamazepine or other hepatic enzyme inducers should be monitored closely during the first 4–8 weeks and the dose of risperidone should perhaps be increased (FDA).

Long-acting risperidone has been used in a multicenter, open study in 725 patients with schizophrenia or schizoaffective disorders; however, only the data from 46 stable patients who were initially taking conventional oral neuroleptic monotherapy were presented (225). During a 2-week run-in period, neuroleptic drugs other than risperidone were withdrawn, and patients who were not currently taking risperidone received flexible doses of oral risperidone 1–6 mg/day. At the start of the 12-month treatment phase, they were assigned, on the judgement of the investigator, to intramuscular long-acting risperidone 25, 50, or 75 mg every 2 weeks. In all, 18 patients did not complete the study, for several reasons; two of them committed suicide. Patients improved during the study; however, since there was no control group it is not possible to know whether the improvement was due to the treatment.

Conventional depot neuroleptic drugs were changed to long-acting injectable risperidone in 196 patients, of whom 166 (mean age 43 years; 67% men) completed the run-in period, consisting of two cycles of their current depot neuroleptic (226). Of these patients, 152 completed the 12-week treatment period, although only 62% of the 166 patients received the six intended injections. The modal doses of long-acting risperidone were 25 mg in 86% of the patients and 37.5 mg in 14%; the mean duration of treatment was 68 days. The patients also received oral risperidone (27%), neuroleptic drugs or sedative/hypnotics (51%), antidepressants (12%), and antiparkinsonian drugs (33%). At least one adverse effect was reported in 96 patients (58%), most of them mild or moderate in intensity. They were hyperkinesia (n = 3), mild tardive dyskinesia (n = 1), and another extrapyramidal disorder (n = 1). In 14 patients there was a serious adverse effect, mainly psychosis. Two of the dropouts were due to adverse effects: one attempted suicide and the other had a severe psychosis. There was hyperprolactinemia in 18 patients; one had ejaculatory failure throughout the study and another had mild impotence that resolved at day 31. Adverse effects potentially related to prolactin (other than hyperprolactinemia) were reported in four patients, two with reduced libido. Body weight (1.0 kg) and body mass index (0.3 kg/m^2) increased.

In a multicenter trial across 22 European countries, a total of 715 stable patients (mean age 40 years; 63% men), most with schizophrenia, who had already participated in a run-in 6-month trial with the same medication, entered the 12-month extension phase (227). The subjects received risperidone injections every 2 weeks; the initial dose was 25 mg in most cases (84%) and the rest were given 37.5 mg (9%) or 50 mg (7%); the end point dose was 25 mg in 39% of patients, 37.5 mg in 23%, and 50 mg in 37%. There were 207 dropouts (29% of the whole sample), adverse events being the reason for withdrawal in 20 patients (2.8%)—exacerbation of the disease (n = 3), delirium, relapse, and weight gain (all n = 2); one patient died with pneumonia. There was weight gain in 54 patients (8%), with significant increases in body weight (79.1 versus 80.4 kg) and body mass index (27.0 versus

27.5 kg/m^2); at endpoint, 20% of patients had an increase in body weight of 7% or more. One patient developed new-onset diabetes mellitus and one hyperglycemia. Other observed adverse events were anxiety (n = 83, 12%), insomnia (n = 73, 10%), depression (n = 52, 7%), headache (n = 38, 5%), and sexual dysfunction (n = 12, 1.7%).

A similar design was used in 249 patients with schizoaffective disorder who received injectable risperidone for 6 months (initial dose 25 mg in 82% of patients, end-point doses ranging from 25 mg in 49% of patients to 75 mg in one); oral risperidone supplementation was needed in 19% (mean modal dose 3 mg/day) (228). Three patients died during the study with heart attack, stroke, and gastrointestinal bleeding; other important adverse events were increases in body weight and body mass index (mean increases 1.4 kg and 0.5 kg/m^2), sexual dysfunction (4%), and new-onset diabetes mellitus (0.4%).

In a third study, a 6-month multicenter open trial using risperidone injections in 382 patients in an early illness stage (mean age 29 years; 69% men), mainly with schizophrenia; non-adherence was the reason for medication change in 42% of the patients (229). Of the whole sample, 88% received a starting dose of 25 mg every 2 weeks; at 6 months, 45% were still being treated with this dose, while 27% and 28% received larger parenteral doses (37.5 and 50 mg respectively), and oral risperidone was added in 19% of patients (mean modal dose 2.7 mg/day). A total of 278 subjects (78%) completed the study. Adverse events accounted for dropouts in 21 cases (6% of the whole sample). Treatment-related adverse events were reported by 217 patients (57%), the most frequent being insomnia (7%), exacerbation of disease (6%), depression (5%), anxiety (5%), weight increase (4%), and relapse and headache (3% each); other observed events were injection site pain (2%), sexual dysfunction (2%), and new-onset diabetes mellitus, gynecomastia, hyperprolactinemia, and non-puerperal lactation (0.3% each); body weight and body mass index increased by 1.8 and 0.6 kg/m^2 respectively.

In a 12-week, double-blind study, 193 Caucasian, 174 African–American, and 72 subjects from other races, all of them with schizophrenia or schizoaffective disorder, were randomized to long-acting injectable risperidone (25, 50, or 75 mg every 2 weeks) or placebo (230). Race influenced neither efficacy nor extrapyramidal symptoms nor body weight.

Finally, of 192 patients who had remained symptomatically stable for at least 1 month with another major second-generation antipsychotic drug, olanzapine (mean age 38 years; 63% men), 70% completed a 6-month study with injectable risperidone; treatment-related adverse events were reported by 121 patients (63%), mostly anxiety (12%), exacerbation of disease (10%), insomnia (9%), depression (6%), and akathisia (5%) (231).

Drug dosage regimens

The safety and tolerability of a rapid oral loading regimen for risperidone, developed to achieve therapeutic doses within 24 hours, have been evaluated (232). Risperidone was begun in a dose of 1 mg, increasing by 1 mg every 6–8 hours up to 3 mg. Dose increases were contingent on the tolerance of the last administered dose. Of 11 consecutive inpatients who were treated with this protocol, seven tolerated the most rapid titration, achieving a dose of 3 mg bd in 16 hours; three required slightly slower titration and achieved the target dose in 24 hours; one could not tolerate the 3 mg dose but tolerated 2 mg tds; no patient had serious extrapyramidal adverse effects, sedation, or any other adverse event during the rapid titration, and in no case did risperidone have to be withdrawn. The authors concluded that aggressive dosing with risperidone is well tolerated in most psychiatric inpatients.

A randomized double-blind comparison of two dosage regimens of risperidone, 8 mg od and 4 mg bd for 6 weeks, in 211 patients has provided further information (233). Neither efficacy nor ESRS scores differed significantly. At least one adverse event was reported in 72 of the patients taking once-daily therapy and 87 of the patients taking twice-daily therapy. The most frequently reported were insomnia, anxiety, extrapyramidal symptoms, agitation, and headache. The only statistically significant difference between the groups was in the incidence of anxiety, which was reported by 31% of those taking twice-daily therapy and 17% of those taking once-daily therapy.

The optimal dose of risperidone in first-episode schizophrenia has been studied in 17 drug-naive patients (12 women, 5 men; mean age 29 years) (234). The mean optimal dosage of risperidone was 2.70 mg/day. All the patients reached the optimal dose before developing extrapyramidal adverse effects; four developed parkinsonism and one developed akathisia at a mean dosage of 5.20 mg/day. In contrast, acute exacerbations of schizophrenia may require a higher dose.

Of 82 subjects, those who initially took higher oral doses of risperidone were more likely to have extrapyramidal adverse effects (mean dose 4.3 mg/day) (235).

Drug overdose

During 13 months, a regional poisons center gathered information by telephone on 31 patients with reported risperidone overdose (236). Risperidone was the sole ingestant in 15 cases (1–180 mg). The major effects in this group included lethargy ($n = 7$), spasm/dystonia ($n = 3$), hypotension ($n = 2$), tachycardia ($n = 6$), and dysrhythmias ($n = 1$). One patient who co-ingested imipramine died of medical complications, but symptoms resolved within 24 hours in most of the others; all the patients were asymptomatic at 72 hours after ingestion.

- A 41-year-old man who took risperidone 270 mg developed a prolonged QT_c interval (480 ms) and sinus bradycardia (44/minute), without hemodynamic compromise. After 9 hours, he had episodes of asymptomatic supraventricular tachycardia with a maximum frequency of 150/minute. After 30 hours he was in sinus rhythm with a normal QT_c interval (360 ms).

He was discharged 72 hours after admission, asymptomatic and with a normal electrocardiogram.

- A 15-year-old who took 110 mg of risperidone in a suicide attempt developed only transient lethargy, hypotension, and tachycardia without any other significant effects (237).

Delayed complications, including respiratory depression, can occur after risperidone overdose (238).

- A 26-year-old woman with schizophrenia who had taken risperidone 6 mg/day for 3 months was found unconscious. She later reported having taken 30 mg of risperidone in a suicidal attempt. On day 3 she developed transient respiratory distress, with lethargy, tongue protrusion, and pharyngeal-laryngeal muscle spasm.

Drug–Drug Interactions

Anesthetics

Interactions with anesthetics should be considered in patients taking α-adrenoceptor antagonists, such as risperidone (239).

- A 32-year-old woman, who had had a previous uncomplicated vaginal delivery with epidural anesthesia at age 21 and a cesarean delivery with a spinal anesthetic at 28, was at week 39 of her third gestation on treatment with risperidone 2 mg/day and lithium 1200 mg/day, which she had taken throughout pregnancy. In preparation for cesarean section she was given a spinal anesthetic with hyperbaric 0.75% bupivacaine 12 mg, fentanyl 10 micrograms, and preservative-free morphine 0.2 mg. Two minutes later, her blood pressure fell from 120/50 to 70/30 mmHg and did not resolve with 50 mg of ephedrine or 2 liters of Ringer's lactate solution, but did rise with phenylephrine 600 micrograms.

Carbamazepine

Carbamazepine induces CYP3A, and the metabolism of risperidone, which mainly involves CYP2D6, may also involve CYP3A. Carbamazepine can therefore reduce risperidone plasma concentrations (SEDA-22, 71). However, since carbamazepine alters the biotransformation of many agents, non-specific enzyme induction has been suggested for the risperidone and carbamazepine interaction (240).

Plasma concentrations of risperidone and 9-hydroxyrisperidone were measured in 44 patients (aged 26–63 years) treated with risperidone alone ($n = 23$) or co-medicated with carbamazepine ($n = 11$) (241). Carbamazepine markedly reduced the plasma concentrations of risperidone and 9-hydroxyrisperidone.

Mean plasma concentrations of risperidone and 9-hydroxyrisperidone (5 ng/ml and 35 ng/ml) fell significantly during carbamazepine co-administration (2.5 ng/ml and 19 ng/ml) in 11 schizophrenic patients taking risperidone 6 mg/day and then carbamazepine 400 mg/day for 1 week; the changes in risperidone concentrations correlated positively with the concentration ratio of risperidone/9-hydroxyrisperidone, which was closely associated with CYP2D6 genotype (242).

Conversely, carbamazepine concentrations can increase when risperidone is added; when risperidone 1 mg/day was added in eight patients taking carbamazepine (mean dose 625 mg/day) carbamazepine plasma concentrations increased from 6.7 µg/ml at baseline to 8.0 µg/ml 2 weeks later (243).

- A 23-year-old man had raised 9-hydroxyrisperidone concentrations in association with carbamazepine dosage reduction and concomitant fluvoxamine therapy (244).
- In a 50-year-old man with deficient CYP2D6 activity, the addition of carbamazepine to pre-existing risperidone therapy resulted in a marked reduction in the plasma concentrations of risperidone and 9-hydroxyrisperidone and an acute exacerbation of his psychosis (245).

Clozapine

Risperidone increases plasma clozapine concentrations (246). The effects of risperidone 3.25 mg/day on cytochrome P450 isozymes have therefore been assessed in eight patients by determination of the metabolism of caffeine (for CYP1A2), dextromethorphan (for CYP2D6), and mephenytoin (for CYP2C19) (246). The results suggested that risperidone is a weak in vivo inhibitor of CYP2D6, CYP2C19, and CYP1A2. The authors concluded that inhibition by risperidone of those isozymes is an unlikely mechanism to explain increased clozapine concentrations.

Donepezil

Possible interactions between donepezil and risperidone, which are both metabolized by CYP2D6 and CYP3A4, have been studied (SEDA-26, 65) (247). Of 24 healthy men (mean age 40 years) were assigned to risperidone 1 mg/day, donepezil 5 mg/day, or both, 20 reported at least one adverse event, mostly headache, nervousness, or somnolence. However, measures of pharmacokinetics showed no interaction.

HIV protease inhibitors

An interaction of risperidone with ritonavir and indinavir has been reported (248).

- A 34-year-old man with AIDS took risperidone 4 mg/day for a Tourette-like tic disorder. Ritonavir and indinavir were added, and 1 week later he developed significantly impaired swallowing, speaking, and breathing, and worsening of his existing tremors.

The authors hypothesized that inhibition of CYP2D6 and CYP3A4 by ritonavir and indinavir may have resulted in accumulation of the active moiety of risperidone.

Lithium

Dystonia occurred in an 81-year-old man who took lithium in addition to risperidone 1 mg/day, valproic acid 2250 mg/day, and benzatropine 4 mg/day (249).

There were no changes in lithium pharmacokinetics when risperidone was substituted open-label for another neuroleptic drug in 13 patients (250). On the other hand, an 81-year-old man had an acute dystonic reaction 4 days after lithium was added to a regimen of risperidone, valproic acid, and benzatropine (251).

- A 17-year-old man had taken risperidone for 2 years without adverse effects, but 12 weeks after lithium was added, he reported prolonged erections (lasting 1–3 hours) 2–5 times daily; risperidone was tapered and withdrawn and the problem resolved (252).

In an in vitro study, there was no visible precipitate formation when lithium citrate syrup was mixed with risperidone solution (253).

Macrolide antibiotics

Drugs that inhibit CYP3A, such as the macrolides, can significantly alter risperidone concentrations, especially in patients with CYP2D6 deficiency (254).

Maprotiline

In three patients, maprotiline plasma concentrations increased when risperidone was added (255). The data suggest that risperidone is a modest inhibitor of CYP2D6 and should therefore be used with caution in combination with drugs such as maprotiline, where increased plasma concentrations have the potential to cause serious toxicity.

Opioids

Two patients who were hospitalized with a diagnosis of opioid dependence received concomitant treatment with methadone 50 mg/day in one case, and levorphanol 14 mg/day in the other, each in association with risperidone. After several days, both had symptoms of opioid withdrawal despite having no change in their opioid doses (256). The withdrawal symptoms resolved soon after risperidone was withdrawn. According to the authors, this finding suggests that risperidone may precipitate opioid withdrawal in opioid-dependent patients.

Phenytoin

An interaction between risperidone and phenytoin resulted in extrapyramidal symptoms (257).

Selective serotonin reuptake inhibitors (SSRIs)

The possibility of a pharmacodynamic interaction between risperidone and serotonin re-uptake inhibitors has been discussed (258–260). Published cases of amelioration and deterioration have perpetuated the debate. Amelioration was observed in four patients with depression that had responded inadequately to selective serotonin re-uptake inhibitors by the addition of risperidone 1 mg bd ($n = 2$) or 0.5 mg at night ($n = 2$) (261).

Citalopram and escitalopram

The effect of citalopram on plasma concentrations of risperidone has been prospectively studied in 15 patients with schizophrenia (262). The addition of citalopram did not alter plasma risperidone concentrations.

Fluoxetine

A pharmacokinetic interaction of risperidone with fluoxetine has been reported (SEDA-22, 71). When 10 schizophrenic patients stabilized on risperidone 4–6 mg/day took fluoxetine 20 mg/day for concomitant depression the mean plasma risperidone concentration increased from 12 to 56 ng/ml at week 4; the concentration of 9-hydroxyrisperidone was not significantly affected (263). One patient dropped out after 1 week because of akathisia associated with a markedly increased plasma risperidone concentration.

In an open, 30-day trial, the pharmacokinetics, safety, and tolerability of a combination of risperidone 4 or 6 mg/day with fluoxetine 20 mg/day were evaluated in 11 psychotic inpatients (264). CYP2D6 genotyping showed that three were poor metabolizers and eight were extensive metabolizers. The mean AUC of risperidone increased from 83 and 398 h.ng/ml to 341 and 514 h.ng/ml when risperidone was co-administered with fluoxetine in extensive and poor metabolizers respectively. However, despite this pharmacokinetic interaction, the severity and incidence of extrapyramidal symptoms and adverse events did not increase significantly when fluoxetine was added; 10 of the 11 patients improved clinically.

- Catastrophic deterioration, with the severity of obsessive-compulsive symptoms returning to pretreatment levels, was observed in a 21-year-old man when risperidone was added to fluoxetine in a dosage that was stepped up to 3 mg/day (265).

Combined treatment with atypical neuroleptic drugs and SSRIs is common and case reports have suggested that SSRIs can increase risperidone concentrations and increase the risk of extrapyramidal disorders (SEDA-23, 18).

- Severe parkinsonism with urinary retention occurred when fluoxetine 20 mg/day was added to risperidone 2 mg/day in a 46-year-old man with schizophrenia. Risperidone had been prescribed at this dose for 1 month without any adverse effects, and the authors considered that a pharmacokinetic interaction between fluoxetine and risperidone was the most likely mechanism (266).

In a systematic open study in 11 hospitalized patients taking a steady dose of risperidone (4–6 mg/day), fluoxetine (20 mg/day) increased plasma concentrations of active antipsychotic medication (combined concentrations of risperidone and 9-hydroxyrisperidone) by 50% after treatment for 25 days (264). Despite this, the treatment was well tolerated and there were improvements in rating scales of psychosis and depressed mood. Whether this was due to the introduction of fluoxetine or the higher plasma concentrations of risperidone is not clear.

Fluoxetine and norfluoxetine would require at least a further 2 weeks to reach steady state, so additional increases in risperidone concentrations might be anticipated over this time.

Fluvoxamine

- A 24-year-old woman with auditory hallucinations taking risperidone 6 mg/day developed neuroleptic malignant syndrome after adding fluvoxamine 50 mg/day (267).

Paroxetine

The serotonin syndrome has been reported during treatment with paroxetine and risperidone (268). A case of edema in a patient taking risperidone and paroxetine has also been reported (269).

The effects of paroxetine 20 mg/day for 4 weeks on steady-state plasma concentrations of risperidone and its active metabolite 9-hydroxyrisperidone have been studied in 10 patients taking risperidone 4–8 mg/day (270). During paroxetine administration, mean plasma risperidone concentrations increased significantly, while 9-hydroxyrisperidone concentrations fell slightly but not significantly; after 4 weeks, the sum of the risperidone and 9-hydroxyrisperidone concentrations increased significantly by 45% over baseline, and the mean plasma risperidone/9-hydroxyrisperidone concentration ratio was also significantly changed. However, the drug combination was generally well tolerated, with the exception of one patient who developed parkinsonian symptoms during the second week, and whose total plasma risperidone and 9-hydroxyrisperidone concentrations increased by 62%.

The addition of paroxetine 20 mg/day to risperidone 4–8 mg/day in 10 patients with schizophrenia produced a 45% increase in plasma concentrations of risperidone and its active metabolite, 9-hydroxyrisperidone (270). One of the patients developed signs of drug-induced Parkinson's disease following the addition of paroxetine.

The serotonin syndrome occurred in a patient taking paroxetine plus an atypical antipsychotic drug, risperidone (268).

- A 53-year-old man took paroxetine 40 mg/day and risperidone 6 mg/day, having previously taken lower doses of both. Within 2 hours he developed ataxia, shivering, and tremor. He had profound sweating but was apyrexial, and was confused, with involuntary jerking movements of his limbs. He recovered without specific treatment over the next 2 days.

This was the first report of the serotonin syndrome in a patient taking an SSRI and an atypical antipsychotic drug. The reaction was unexpected because risperidone, in addition to being a potent dopamine receptor antagonist, is also a 5-HT$_2$-receptor antagonist. Subsequent animal studies suggested that 5-HT$_2$-receptor antagonists increase the firing of serotonergic neurons, perhaps through a postsynaptic feedback loop. This could account for potentiation of the effects of SSRIs by 5-HT$_2$-receptor antagonists, such as risperidone.

- A 78-year-old woman with hypertension, angina, diabetes, depression, and dementia took venlafaxine 37.5 mg bd for depression and risperidone 0.25 mg at bedtime for agitation for more than 1 year, as well as isosorbide dinitrate, lisinopril, and glibenclamide (271). Because of periods of extreme agitation, her medication was changed to paroxetine 20 mg and risperidone 0.5 mg at bedtime. Her agitation worsened, and after 3 days the dose of risperidone was increased to 0.5 mg bd. Two days later she did not respond to verbal or tactile stimuli and developed a tremor, dizziness, and muscle incoordination. The risperidone and paroxetine were withdrawn and her agitation was managed with low doses of lorazepam. She recovered and 6 months later was stable, taking paroxetine 40 mg/day and risperidone 0.25 mg at bedtime.

Venlafaxine

In eight patients with major depressive disorder without psychotic features, who did not respond to serotonin re-uptake inhibitors therapy when risperidone was added, all improved within 1 week. Furthermore, risperidone also seemed to have beneficial effects on sleep disturbance and sexual dysfunction (272). In an open study in 30 healthy subjects who took risperidone 1 mg orally before and after venlafaxine dosing to steady state, the oral clearance of risperidone fell by 38% and the volume of distribution by 17%, resulting in a 32% increase in AUC; renal clearance of 9-hydroxyrisperidone also fell by 20% (273). The authors concluded that these small effects were consistent with the fact that venlafaxine is unlikely to alter the clearance of risperidone, which is mainly by CYP2D6.

Unlike SSRIs, venlafaxine only weakly inhibits CYP2D6. Venlafaxine (up to 150 mg/day for 10 days) modestly but significantly increased the plasma AUC of a single dose of risperidone (1 mg) in 38 healthy volunteers in an open longitudinal design, but did not affect its active metabolite, 9-hydroxyrisperidone (273). The authors suggested that their study had shown a small potentiating effect of venlafaxine on risperidone availability, probably through weak inhibition by venlafaxine of CYP2D6. Venlafaxine is therefore less likely to increase risperidone concentrations than are fluoxetine and paroxetine.

Tetracycline

- A possible interaction of risperidone with tetracycline has been reported in a 15-year-old adolescent with Asperger's syndrome, Tourette's syndrome, and obsessive-compulsive disorder (274). Acute exacerbation of motor and vocal tics occurred when tetracycline 250 mg bd was introduced for acne; withdrawal of tetracycline resulted in an improvement in the tics.

Thioridazine

A 23-year-old man had high risperidone plasma concentrations secondary to concurrent thioridazine use (244).

Thioridazine 25 mg bd significantly increased the steady-state plasma concentration of risperidone (3 mg bd) and reduced 9-hydroxyrisperidone in 12 patients with schizophrenia (275). Since risperidone mainly undergoes 9-hydroxylation yielding an active metabolite, 9-hydroxyrisperidone, metabolic inhibition could account for this increase. Moreover, there was a greater rise in risperidone concentrations in subjects with more capacity for CYP2D6-dependent 9-hydroxylation of risperidone.

Valproate

Plasma concentrations of risperidone and 9-hydroxyrisperidone were measured in 44 patients (aged 26–63 years) taking risperidone alone ($n = 23$) or co-medicated with sodium valproate ($n = 10$) (154). Valproate had no major effect on plasma risperidone concentrations.

However, an anecdotal report has suggested that some individuals may be susceptible to an interaction.

- Catatonia occurred in a 42-year-old woman taking valproic acid, sertraline, and risperidone (79). The catatonic features evolved for the first time after a single dose of valproate and were alleviated by lorazepam; the same catatonic signs recurred after a second dose of valproate and again remitted after lorazepam.

The authors considered that this was a possible interaction, since catatonia has not been reported with valproate alone.

The addition of risperidone 10 mg/day over 2 months to valproate and clonazepam in a 40-year-old woman provoked marked edema in the legs and moderate edema in the arms (276). The authors considered that this was a possible interaction, since edema has not been reported with either of these drugs separately.

In contrast, a beneficial interaction has been observed when valproic acid was added to risperidone; the previous addition of valproic acid to treatment with chlorpromazine had no effect on psychotic symptoms (277).

Diagnosis of Adverse Drug Reactions

A therapeutic target range for serum risperidone concentrations has not been established, but in 20 of 22 patients taking 6 mg/day, which is considered the optimum dosage for most patients, risperidone serum concentrations were 50–150 nmol/l (278). Steady-state serum concentrations of risperidone and 9-hydroxyrisperidone, the active moiety, were also measured in 42 patients; there was no correlation between the serum concentration of the active moiety and adverse effects.

References

1. Mackay FJ, Wilton LV, Pearce GL, Freemantle SN, Mann RD. The safety of risperidone: a post-marketing study on 7684 patients. Hum Psychopharmacol 1998;13:413–8.
2. Moller HJ, Gagiano CA, Addington DE, Von Knorring L, Torres-Plank JF, Gaussares C. Long-term treatment of chronic schizophrenia with risperidone: an open-label,

multicenter study of 386 patients. Int Clin Psychopharmacol 1998;13(3):99–106.

3. Binder RL, McNiel DE, Sandberg DA. A naturalistic study of clinical use of risperidone. Psychiatr Serv 1998;49(4):524–6.

4. Muller-Siecheneder F, Muller MJ, Hillert A, Szegedi A, Wetzel H, Benkert O. Risperidone versus haloperidol and amitriptyline in the treatment of patients with a combined psychotic and depressive syndrome. J Clin Psychopharmacol 1998;18(2):111–20.

5. Vieta E, Gasto C, Colom F, Martinez A, Otero A, Vallejo J. Treatment of refractory rapid cycling bipolar disorder with risperidone. J Clin Psychopharmacol 1998;18(2):172–4.

6. Cohen SA, Ihrig K, Lott RS, Kerrick JM. Risperidone for aggression and self-injurious behavior in adults with mental retardation. J Autism Dev Disord 1998;28(3):229–33.

7. Schreier HA. Risperidone for young children with mood disorders and aggressive behavior. J Child Adolesc Psychopharmacol 1998;8(1):49–59.

8. Madhusoodanan S, Suresh P, Brenner R, Pillai R. Experience with the atypical antipsychotics—risperidone and olanzapine in the elderly. Ann Clin Psychiatry 1999;11(3):113–8.

9. Keks NA, Culhane C. Risperidone (Risperdal): clinical experience with a new antipsychosis drug. Expert Opin Investig Drugs 1999;8(4):443–52.

10. Chong SA, Yap HL, Low BL, Choo CH, Chan AO, Wong KE, Mahendran R, Chee KT. Clinical evaluation of risperidone in Asian patients with schizophrenia in Singapore. Singapore Med J 1999;40(1):41–3.

11. Meibach RC, Mazurek MF, Rosebush P. Neurologic side effects in neuroleptic-naive patients treated with haloperidol or risperidone. Neurology 2000;55(7):1069.

12. Huston P, Moher D. Redundancy, disaggregation, and the integrity of medical research. Lancet 1996;347(9007):1024–6.

13. Rennie D. Fair conduct and fair reporting of clinical trials. JAMA 1999;282(18):1766–8.

14. Lane HY, Chiu WC, Chou JC, Wu ST, Su MH, Chang WH. Risperidone in acutely exacerbated schizophrenia: dosing strategies and plasma levels. J Clin Psychiatry 2000;61(3):209–14.

15. Zuddas A, Di Martino A, Muglia P, Cianchetti C. Long-term risperidone for pervasive developmental disorder: efficacy, tolerability, and discontinuation. J Child Adolesc Psychopharmacol 2000;10(2):79–90.

16. Findling RL, McNamara NK, Branicky LA, Schluchter MD, Lemon E, Blumer JL. A double-blind pilot study of risperidone in the treatment of conduct disorder. J Am Acad Child Adolesc Psychiatry 2000;39(4):509–16.

17. Buitelaar JK. Open-label treatment with risperidone of 26 psychiatrically-hospitalized children and adolescents with mixed diagnoses and aggressive behavior. J Child Adolesc Psychopharmacol 2000;10(1):19–26.

18. Heck AH, Haffmans PM, de Groot IW, Hoencamp E. Risperidone versus haloperidol in psychotic patients with disturbing neuroleptic-induced extrapyramidal symptoms: a double-blind, multi-center trial. Schizophr Res 2000;46(2–3):97–105.

19. Yatham LN, Binder C, Riccardelli R, Leblanc J, Connolly M, Kusumakar V, on behalf of the RIS-CAN 25 Study Group. Risperidone in acute and continuation treatment of mania. Int Clin Psychopharmacol 2003;18:227–35.

20. Vieta E, Goikolea JM, Olivares JM, González-Pinto A, Rodríguez A, Colom F, Comes M, Torrent C, Sánchez-Moreno J. 1-year follow-up of patients treated with risperidone and topiramate for a manic episode. J Clin Psychiatry 2003;64:834–9.

21. Ercan ES, Kutlu A, Cikoglu S, Veznedaroglu B, Erermis S, Varan A. Risperidone in children and adolescents with conduct disorder: a single-center, open-label study. Curr Ther Res 2003;64:55–64.

22. Shigenobu K, Ikeda M, Fukuhara R, Komori K, Tanabe H. A structured, open trial of risperidone therapy for delusions of theft in Alzheimer Disease. Am J Geriatr Psychiatry 2003;11:256–7.

23. Findling RL, Aman MG, Eerdekens M, Derivan A, Lyons B (The Risperidone Disruptive Behavior Study Group). Long-term, open-label study of risperidone in children with severe disruptive behaviors and below-average IQ. Am J Psychiatry 2004;161:677–84.

24. Lane H-Y, Chang Y-C, Chiu C-C, Lee S-H, Lin C-Y, Chang W-H. Fine-tuning risperidone dosage for acutely exacerbated schizophrenia: clinical determinants. Psychopharmacology 2004;172:393–9.

25. Reveley MA, Libretto SE for the RIS-GBR-31 investigators. Treatment outcome in patients with chronic schizophrenia during long-term administration with risperidone. J Clin Psychopharmacol 2004;24:260–7.

26. Huq Z-U, on behalf of the RIS-GBR-31 investigators. A trial of low doses of risperidone in the treatment of patients with first-episode schizophrenia, schizophreniform disorder, or schizoaffective disorder. J Clin Psychopharmacol 2004;24:220–4.

27. Bhanji NH, Chouinard G, Margolese HC. A review of compliance, depot intramuscular antipsychotics and the new long-acting injectable atypical antipsychotic risperidone in schizophrenia. Eur Neuropsychopharmacol 2004;14:87–92.

28. Vieta E, Brugué E, Goikolea JM, Sánchez-Moreno J, Reinares M, Comes M, Colom F, Martínez-Arán A, Benabarre A, Torrent C. Acute and continuation risperidone monotherapy in mania. Hum Psychopharmacol Clin Exp 2004;19:41–5.

29. Mahmoud RA, Engelhart LM, Janagap CC, Oster G, Ollendorf D. Risperidone versus conventional antipsychotics for schizophrenia and schizoaffective disorder. Symptoms, quality of life and resource use under customary clinical care. Clin Drug Invest 2004;24:275–86.

30. Yen Y-C, Lung F-W, Chong M-Y. Adverse effects of risperidone and haloperidol treatment in schizophrenia. Prog Neuropsychopharmacol Biol Psychiatry 2004;28:285–90.

31. Schooler N, Rabinowitz J, Davidson M, Emsley R, Harvey PD, Kopala L, McGorry PD, Van Hove I, Eerdekens M, Swyzen W, De Smedt G; Early Psychosis Global Working Group. Risperidone and haloperidol in first-episode psychosis: a long-term randomized trial. Am J Psychiatry 2005;162:947–53.

32. Smulevich AB, Khanna S, Eerdekens M, Karcher K, Kramer M, Grossman F. Acute and continuation risperidone monotherapy in bipolar mania: a 3-week placebo-controlled trial followed by a 9-week double-blind trial of risperidone and haloperidol. Eur Neuropsychopharmacol 2005;15:75–84.

33. Ritchie CW, Chiu E, Harrigan S, Hall K, Hassett A, Macfarlane S, Mastwyk M, O'Connor DW, Opie J, Ames D. The impact upon extra-pyramidal side effects, clinical symptoms and quality of life of a switch from conventional

34. Iskedjian M, Hux M, Remington GJ. The Canadian experience with risperidone for the treatment of schizophrenia: an overview. J Psychiatry Neurosci 1998;23(4):229–39.

35. Vieta E, Herraiz M, Fernandez A, Gasto C, Benabarre A, Colom F, Martinez-Aran A, Reinares M. Efficacy and safety of risperidone in the treatment of schizoaffective disorder: initial results from a large, multicenter surveillance study. Group for the Study of Risperidone in Affective Disorders (GSRAD). J Clin Psychiatry 2001;62(8):623–30.

36. Buitelaar JK, van der Gaag RJ, Cohen-Kettenis P, Melman CT. A randomized controlled trial of risperidone in the treatment of aggression in hospitalized adolescents with subaverage cognitive abilities. J Clin Psychiatry 2001;62(4):239–48.

37. McDougle CJ, Epperson CN, Pelton GH, Wasylink S, Price LH. A double-blind, placebo-controlled study of risperidone addition in serotonin reuptake inhibitor-refractory obsessive-compulsive disorder. Arch Gen Psychiatry 2000;57(8):794–801.

38. Davidson M. Long-term safety of risperidone. J Clin Psychiatry 2001;62(Suppl. 21):26–8.

39. McCracken JT, McGough J, Shah B, Cronin P, Hong D, Aman MG, Arnold LE, Lindsay R, Nash P, Hollway J, McDougle CJ, Posey D, Swiezy N, Kohn A, Scahill L, Martin A, Koenig K, Volkmar F, Carroll D, Lancor A, Tierney E, Ghuman J, Gonzalez NM, Grados M, Vitiello B, Ritz L, Davies M, Robinson J, McMahon D. Research Units on Pediatric Psychopharmacology Autism Network. Risperidone in children with autism and serious behavioral problems. N Engl J Med 2002;347(5):314–21.

40. Aman MG, De Smedt G, Derivan A, Lyons B, Findling RLRisperidone Disruptive Behavior Study Group. Double-blind, placebo-controlled study of risperidone for the treatment of disruptive behaviors in children with subaverage intelligence. Am J Psychiatry 2002;159(8):1337–46.

41. McDougle CJ, Holmes JP, Carlson DC, Pelton GH, Cohen DJ, Price LH. A double-blind, placebo-controlled study of risperidone in adults with autistic disorder and other pervasive developmental disorders. Arch Gen Psychiatry 1998;55(7):633–41.

42. Bhana N, Spencer CM. Risperidone: a review of its use in the management of the behavioural and psychological symptoms of dementia. Drugs Aging 2000;16(6):451–71.

43. Koenigsberg HW, Reynolds D, Goodman M, New AS, Mitropoulou V, Trestman RL, Silverman J, Siever LJ. Risperidone in the treatment of schizotypal personality disorder. J Clin Psychiatry 2003;64:628–34.

44. Hamner MB, Faldowski RA, Ulmer HG, Frueh BC, Huber MG, Arana GW. Adjunctive risperidone treatment in post-traumatic stress disorder: a preliminary controlled trial of effects on comorbid psychotic symptoms. Int Clin Psychopharmacol 2003;18:1–8.

45. Monnelly EP, Ciraulo DA, Knapp C, Keane T. Low-dose risperidone as adjunctive therapy for irritable aggression in posttraumatic stress disorder. J Clin Psychopharmacol 2003;23:193–6.

46. Scahill L, Leckman JF, Schultz RT, Katsovich L, Peterson BS. A placebo-controlled trial of risperidone in Tourette syndrome. Neurology 2003;60:1130–5.

47. Brodaty H, Ames D, Snowdon J, Woodward M, Kirwan J, Clarnette R, Lee E, Lyons B, Grossman F. A randomized placebo-controlled trial of risperidone for the treatment of aggression, agitation, and psychosis of dementia. J Clin Psychiatry 2003;64:134–43.

48. Hirschfeld RMA, Keck PE, Kramer M, Karcher K, Canuso C, Eerdekens M, Grossman F. Rapid antimanic effect of risperidone monotherapy: a 3-week multicenter, double-blind, placebo-controlled trial. Am J Psychiatry 2004;161:1057–65.

49. Josiassen RC, Joseph A, Kohegyi E, Stokes S, Dadvand M, Paing WW, Shaughnessy RA. Clozapine augmented with risperidone in the treatment of schizophrenia: a randomized, double-blind, placebo-controlled trial. Am J Psychiatry 2005;162:130–6.

50. Troost PW, Lahuis BE, Steenhuis MP, Ketelaars CE, Buitelaar JK, van Engeland H, Scahill L, Minderaa RB, Hoekstra PJ. Long-term effects of risperidone in children with autism spectrum disorders: a placebo discontinuation study. J Am Acad Child Adolesc Psychiatry 2005;44:1137–44.

51. Erzegovesi S, Guglielmo E, Siliprandi F, Bellodi L. Low-dose risperidone augmentation of fluvoxamine treatment in obsessive-compulsive disorder: a double-blind, placebo-controlled study. Eur Neuropsychopharmacol 2005;15:69–74.

52. Bartzokis G, Lu PH, Turner J, Mintz J, Saunders CS. Adjunctive risperidone in the treatment of chronic combat-related posttraumatic stress disorder. Biol Psychiatry 2005;57:474–9.

53. Vitiello B, Davies M, Arnold LE, McDougle CJ, Aman M, McCracken JT, Scahill L, Tierney E, Posey DJ, Swiezy NB, Koenig K. Assessment of the integrity of study blindness in a pediatric clinical trial of risperidone. J Clin Psychopharmacol 2005;25:565–9.

54. De Deyn PP, Rabheru K, Rasmussen A, Bocksberger JP, Dautzenberg PL, Eriksson S, Lawlor BA. A randomized trial of risperidone, placebo, and haloperidol for behavioral symptoms of dementia. Neurology 1999;53(5):946–55.

55. Katz IR, Jeste DV, Mintzer JE, Clyde C, Napolitano J, Brecher MRisperidone Study Group. Comparison of risperidone and placebo for psychosis and behavioral disturbances associated with dementia: a randomized, double-blind trial. J Clin Psychiatry 1999;60(2):107–15.

56. Ravin DS, Levenson JW. Fatal cardiac event following initiation of risperidone therapy. Ann Pharmacother 1997;31(7–8):867–70.

57. Chiu CC, Chang WH, Huang MC, Chiu YW, Lane HY. Regular-dose risperidone on QTc intervals. J Clin Psychopharmacol 2005;25:391–3.

58. Posey DJ, Walsh KH, Wilson GA, McDougle CJ. Risperidone in the treatment of two very young children with autism. J Child Adolesc Psychopharmacol 1999;9(4):273–6.

59. Henretig FM. Risperidone toxicity acknowledged. J Toxicol Clin Toxicol 1999;37:893–4.

60. Kamijo Y, Soma K, Nagai T, Kurihara K, Ohwada T. Acute massive pulmonary thromboembolism associated with risperidone and conventional phenothiazines. Circ J 2003;67:46–8.

61. Qureshi N. Atypical antipsychotics and dementia: some reflections! http: //bmj.bmjjournals.com/cgi/eletters/328/7450/1262-b.

62. Mowat D, Fowlie D, MacEwan T. CSM warning on atypical antipsychotics and stroke may be detrimental for dementia. http: //bmj.bmjjournals.com/cgi/eletters/328/7450/1262-b.

63. Schneider L, Dagerman K. Meta-analysis of atypical antipsychotics for dementia patients: balancing efficacy and adverse events. http://ipa.confex.com/ipa/11congress/tech-program/paper_4201.htm.

64. Lee HJ, Lee HS, Kim L, Lee MS, Suh KY, Kwak DI. A case of risperidone-induced stuttering. J Clin Psychopharmacol 2001;21(1):115–6.

65. Dresel S, Tatsch K, Dahne I, Mager T, Scherer J, Hahn K. Iodine-123-iodobenzamide SPECT assessment of dopamine D2 receptor occupancy in risperidone-treated schizophrenic patients. J Nucl Med 1998;39(7):1138–42.

66. Miller CH, Mohr F, Umbricht D, Woerner M, Fleischhacker WW, Lieberman JA. The prevalence of acute extrapyramidal signs and symptoms in patients treated with clozapine, risperidone, and conventional antipsychotics. J Clin Psychiatry 1998;59(2):69–75.

67. Lemmens P, Brecher M, Van Baelen B. A combined analysis of double-blind studies with risperidone vs. placebo and other antipsychotic agents: factors associated with extrapyramidal symptoms. Acta Psychiatr Scand 1999;99(3):160–70.

68. Emsley RARisperidone Working Group. Risperidone in the treatment of first-episode psychotic patients: a double-blind multicenter study. Schizophr Bull 1999;25(4):721–9.

69. Wirshing DA, Marshall BD Jr, Green MF, Mintz J, Marder SR, Wirshing WC. Risperidone in treatment-refractory schizophrenia. Am J Psychiatry 1999;156(9):1374–9.

70. Lavretsky H, Sultzer D. A structured trial of risperidone for the treatment of agitation in dementia. Am J Geriatr Psychiatry 1998;6(2):127–35.

71. Herrmann N, Rivard MF, Flynn M, Ward C, Rabheru K, Campbell B. Risperidone for the treatment of behavioral disturbances in dementia: a case series. J Neuropsychiatry Clin Neurosci 1998;10(2):220–3.

72. Hong R, Matsuyama E, Nur K. Cardiomyopathy associated with the smoking of crystal methamphetamine. JAMA 1991;265(9):1152–4.

73. Adamou MA, Hale AS. Extrapyramidal syndrome and long-acting injectable risperidone. Am J Psychiatry 2004;161:756–7.

74. Geizer M, Ancill RJ. Combination of risperidone and donepezil in Lewy body dementia. Can J Psychiatry 1998;43(4):421–2.

75. Wetter TC, Brunner J, Bronisch T. Restless legs syndrome probably induced by risperidone treatment. Pharmacopsychiatry 2002;35(3):109–11.

76. Roberts MD. Risperdal and parkinsonian tremor. J Am Acad Child Adolesc Psychiatry 1999;38(3):230.

77. Meco G, Fabrizio E, Alessandri A, Vanacore N, Bonifati V. Risperidone in levodopa induced dyskinesiae. J Neurol Neurosurg Psychiatry 1998;64(1):135.

78. Friedman JH, Ott BR. Should risperidone be used in Parkinson's disease? J Neuropsychiatry Clin Neurosci 1998;10(4):473–5.

79. Workman RH. In reply. J Neuropsychiatry Clin Neurosci 1998;10:474–5.

80. Leopold NA. Risperidone treatment of drug-related psychosis in patients with parkinsonism. Mov Disord 2000;15(2):301–4.

81. Lang DJ, Kopala LC, Vandorpe RA, Rui Q, Smith GN, Goghari VM, Honer WG. An MRI study of basal ganglia volumes in first-episode schizophrenia patients treated with risperidone. Am J Psychiatry 2001;158(4):625–31.

82. Haberfellner EM. Tardive dyskinesia during treatment with risperidone. Pharmacopsychiatry 1997;30(6):271.

83. Saran BM. Risperidone-induced tardive dyskinesia. J Clin Psychiatry 1998;59(1):29–30.

84. Silberbauer C. Risperidone-induced tardive dyskinesia. Pharmacopsychiatry 1998;31(2):68–9.

85. Friedman JH. Rapid onset tardive dyskinesia ("fly catcher tongue") in a neuroleptically naive patient induced by risperidone. Med Health R I 1998;81(8):271–2.

86. Sakkas P, Liappas J, Christodoulou GN. Tardive dyskinesia due to risperidone. Eur Psychiatry 1998;13:107–8.

87. Fischer P, Tauscher J, Kufferle B. Risperidone and tardive dyskinesia in organic psychosis. Pharmacopsychiatry 1998;31(2):70–1.

88. Campbell M. Risperidone-induced tardive dyskinesia in first-episode psychotic patients. J Clin Psychopharmacol 1999;19(3):276–7.

89. Carroll NB, Boehm KE, Strickland RT. Chorea and tardive dyskinesia in a patient taking risperidone. J Clin Psychiatry 1999;60(7):485–7.

90. Hong KS, Cheong SS, Woo JM, Kim E. Risperidone-induced tardive dyskinesia. Am J Psychiatry 1999;156(8):1290.

91. Snoddgrass PL, Labbate LA. Tardive dyskinesia from risperidone and olanzapine in an alcoholic man. Can J Psychiatry 1999;44(9):921.

92. Kumar S, Malone DM. Risperidone implicated in the onset of tardive dyskinesia in a young woman. Postgrad Med J 2000;76(895):316–7.

93. Spivak M, Smart M. Tardive dyskinesia from low-dose risperidone. Can J Psychiatry 2000;45(2):202.

94. Jeste DV, Okamoto A, Napolitano J, Kane JM, Martinez RA. Low incidence of persistent tardive dyskinesia in elderly patients with dementia treated with risperidone. Am J Psychiatry 2000;157(7):1150–5.

95. Silver H, Aharon N, Schwartz M. Attention deficit-hyperactivity disorder may be a risk factor for treatment-emergent tardive dyskinesia induced by risperidone. J Clin Psychopharmacol 2000;20(1):112–4.

96. Lore C. Risperidone and withdrawal dyskinesia. J Am Acad Child Adolesc Psychiatry 2000;39(8):941.

97. Mullen A, Cullen M. Risperidone and tardive dyskinesia: a case of blepharospasm. Aust NZ J Psychiatry 2000;34(5):879–80.

98. Bassitt DP, de Souza Lobo Garcia L. Risperidone-induced tardive dyskinesia. Pharmacopsychiatry 2000;33(4):155–6.

99. Narendran R, Young CM, Pato MT. Possible risperidone-induced tardive dystonia. Ann Pharmacother 2000;34(12):1487–8.

100. Yoshida K, Higuchi H, Hishikawa Y. Marked improvement of tardive dystonia after replacing haloperidol with risperidone in a schizophrenic patient. Clin Neuropharmacol 1998;21(1):68–9.

101. Jagadheesan K, Nizamie SH. Risperidone-induced Pisa syndrome. Aust N Z J Psychiatry 2002;36(1):144.

102. Harada K, Sasaki N, Ikeda H, Nakano N, Ozawa H, Saito T. Risperidone-induced Pisa syndrome. J Clin Psychiatry 2002;63(2):166.

103. Levin T, Heresco-Levy U. Risperidone-induced rabbit syndrome: an unusual movement disorder caused by an atypical antipsychotic. Eur Neuropsychopharmacol 1999;9(1–2):137–9.

104. Hoy JS, Alexander B. Rabbit syndrome secondary to risperidone. Pharmacotherapy 2002;22(4):513–5.

105. Mendhekar DN. Rabbit syndrome induced by combined lithium and risperidone. Can J Psychiatry 2005;50:369.

106. Rohrbach P, Collinot JP, Vallet G. Syndrome malin des neuroleptiques induit par la rispéridone. [Neuroleptic

malignant syndrome induced by risperidone.] Ann Fr Anesth Reanim 1998;17(1):85–6.

107. Aguirre C, Garcia Monco JC, Mendibil B. Síndrome neuroléptico maligno asociado a risperidona. [Neuroleptic malignant syndrome associated with risperidone.] Med Clin 1998;110:239.

108. Lee MS, Lee HJ, Kim L. A case of delayed NMS induced by risperidone. Psychiatr Serv 2000;51(2):254–5.

109. Robb AS, Chang W, Lee HK, Cook MS. Case study. Risperidone-induced neuroleptic malignant syndrome in an adolescent. J Child Adolesc Psychopharmacol 2000;10(4):327–30.

110. Sechi G, Agnetti V, Masuri R, Deiana GA, Pugliatti M, Paulus KS, Rosati G. Risperidone, neuroleptic malignant syndrome and probable dementia with Lewy bodies. Prog Neuropsychopharmacol Biol Psychiatry 2000;24(6):1043–51.

111. Dursun SM, Oluboka OJ, Devarajan S, Kutcher SP. High-dose vitamin E plus vitamin B6 treatment of risperidone-related neuroleptic malignant syndrome. J Psychopharmacol 1998;12(2):220–1.

112. Bourgeois JA, Kahn DR. Neuroleptic malignant syndrome following administration of risperidone and lithium. J Clin Psychopharmacol 2003;23:315–17.

113. Lauterbach EC. Catatonia-like events after valproic acid with risperidone and sertraline. Neuropsychiatry Neuropsychol Behav Neurol 1998;11(3):157–63.

114. Bahro M, Kampf C, Strnad J. Catatonia under medication with risperidone in a 61-year-old patient. Acta Psychiatr Scand 1999;99(3):223–6.

115. Edwards RJ, Pople IK. Side-effects of risperidone therapy mimicking cerebrospinal fluid shunt malfunction: implications for clinical monitoring and management. J Psychopharmacol 2002;16(2):177–9.

116. Health Canada. www.hc-sc.gc.ca.

117. Food and Drug Administration. Patient Information Sheet. Risperidone Tablets (marketed as Risperdal). http://www.fda.gov/cder/drug/InfoSheets/patient/risperidonePIS.htm.

118. Agemed. Agencia Española de Medicamentos y Productos Sanitarios. Nota informativa sobre risperidona. http://www.agemed.es/documentos/notasPrensa/csmh/2004/cont_risperidone.htm (COMPROBAR).

119. Janssen Pharmaceutica Inc. 2003 safety alert–Risperdal. http://www.fda.gov/medwatch/SAFETY/2003/risperdal.htm.

120. World Health Organization. Regulatory matters–olanzapine, risperidone. WHO Pharm Newslett 2004;2:1–2.

121. EMEA Public Statement, EMEA/CPMP/856/04 Final, 9 March 2004. http://www.emea.eu.int.

122. Agemed. Agencia Española de Medicamentos y Productos Sanitarios. Nota informativa sobre olanzapina y risperidona. http://www.agemed.es/documentos/notasPrensa/csmh/2004/cont_olanzapina.htm.

123. Wooltorton E. Risperidone (Risperdal): increased rate of cerebrovascular events in dementia trials. Can Med Assoc J 2002;167:1269–70.

124. Medicines and Healthcare products Regulatory Agency (MHRA). Atypical antipsychotic drugs: questions and answers, 9 March 2004. http://www.mca.gov.uk/aboutagency/regframework/com/csmhome.htm.

125. Committee on Safety of Medicines. Atypical antipsychotic drugs and stroke. http://www.mca.gov.uk/aboutagency/regframework/com/csmhome.htm.

126. Hermann N, Mamdani M, Lanctôt KL. Atypical antipsychotics and risk of cerebrovascular accidents. Am J Psychiatry 2004;161:1113–5.

127. Hori M, Shiraishi H. Risperidone-induced anxiety might also develop "awakening" phenomenon. Psychiatry Clin Neurosci 1999;53(6):682.

128. Ashleigh EA, Larsen PD. A syndrome of increased affect in response to risperidone among patients with schizophrenia. Psychiatr Serv 1998;49(4):526–8.

129. Lauterbach EC, Abdelhamid A, Annandale JB. Posthallucinogen-like visual illusions (palinopsia) with risperidone in a patient without previous hallucinogen exposure: possible relation to serotonin $5HT_{2a}$ receptor blockade. Pharmacopsychiatry 2000;33(1):38–41.

130. Bakaras P, Georgoussi M, Liakos A. Development of obsessive and depressive symptoms during risperidone treatment. Br J Psychiatry 1999;174:559.

131. Allain H, Tessier C, Bentué-Ferrer D, Tarral A, Le Breton S, Gandon HM, Bouhours P. Effects of risperidone on psychometric and cognitive functions. Psychopharmacology 2003;165:419–29.

132. Zolezzi M, Badr MG. Risperidone-induced mania. Ann Pharmacother 1999;33(3):380–1.

133. Guzelcan Y, de Haan L, Scholte WF. Risperidone may induce mania. Psychopharmacology (Berl) 2002;162(1):85–6.

134. Mahendran R. Obsessional symptoms associated with risperidone treatment. Aust NZ J Psychiatry 1998;32(2):299–301.

135. Mahendran R, Andrade C, Saxena S. Obsessive-compulsive symptoms with risperidone. J Clin Psychiatry 1999;60(4):261–3.

136. Sinha BN, Duggal HS, Nizamie SH. Risperidone-induced obsessive-compulsive symptoms: a reappraisal. Can J Psychiatry 2000;45(4):397–8.

137. Kinon BJ, Gilmore JA, Liu H, Halbreich UM. Prevalence of hyperprolactinemia in schizophrenic patients treated with conventional antipsychotic medications or risperidone. Psychoneuroendocrinology 2003;28:55–68.

138. Brunelleschi S, Zeppegno P, Risso F, Cattaneo CI, Torre E. Risperidone-associated hyperprolactinemia: evaluation in twenty psychiatric outpatients. Pharmacol Res 2003;48:405–9.

139. Jones H, Curtis VA, Wright PA, Lucey JV. Risperidone is associated with blunting of D-fenfluramine evoked serotonergic responses in schizophrenia. Int Clin Psychopharmacol 1998;13(5):199–203.

140. Shiwach RS, Carmody TJ. Prolactogenic effects of risperidone in male patients—a preliminary study. Acta Psychiatr Scand 1998;98(1):81–3.

141. David SR, Taylor CC, Kinon BJ, Breier A. The effects of olanzapine, risperidone, and haloperidol on plasma prolactin levels in patients with schizophrenia. Clin Ther 2000;22(9):1085–96.

142. Tollin SR. Use of the dopamine agonists bromocriptine and cabergoline in the management of risperidone-induced hyperprolactinemia in patients with psychotic disorders. J Endocrinol Invest 2000;23(11):765–70.

143. Kleinberg DL, Davis JM, de Coster R, Van Baelen B, Brecher M. Prolactin levels and adverse events in patients treated with risperidone. J Clin Psychopharmacol 1999;19(1):57–61.

144. Kim YK, Kim L, Lee MS. Risperidone and associated amenorrhea: a report of 5 cases. J Clin Psychiatry 1999;60(5):315–7.

145. Togo T, Iseki E, Shoji M, Oyama I, Kase A, Uchikado H, Katsuse O, Kosaka K. Prolactin levels in schizophrenic patients receiving perospirone in comparison to risperidone. J Pharmacol Sci 2003;91:259–62.

146. Becker D, Liver O, Mester R, Rapoport M, Weizman A, Weiss M. Risperidone, but not olanzapine, decreases bone mineral density in female premenopausal schizophrenia patients. J Clin Psychiatry 2003;64:761–6.

147. Spollen JJ, Wooten RG, Cargile C, Bartztokis G. Prolactin levels and erectile function in patients treated with risperidone. J Clin Psychopharmacol 2004;24:161–6.

148. Češková E, Přikryl R, Kašpárek T, Ondrušová M. Prolactin levels in risperidone treatment of first-episode schizophrenia. Int J Psych Clin Pract 2004;8:31–6.

149. Popli A, Gupta S, Rangwani SR. Risperidone-induced galactorrhea associated with a prolactin elevation. Ann Clin Psychiatry 1998;10(1):31–3.

150. Schreiber S, Segman RH. Risperidone-induced galactorrhea. Psychopharmacology (Berl) 1997;130(3):300–1.

151. Gupta SC, Jagadheesan K, Basu S, Paul SE. Risperidone-induced galactorrhoea: a case series. Can J Psychiatry 2003;48(2):130–1.

152. Gazzola LR, Opler LA. Return of menstruation after switching from risperidone to olanzapine. J Clin Psychopharmacol 1998;18(6):486–7.

153. Mabini R, Wergowske G, Baker FM. Galactorrhea and gynecomastia in a hypothyroid male being treated with risperidone. Psychiatr Serv 2000;51(8):983–5.

154. Gupta S, Frank B, Madhusoodanan S. Risperidone-associated galactorrhea in a male teenager. J Am Acad Child Adolesc Psychiatry 2001;40(5):504–5.

155. Kim KS, Pae CU, Chae JH, Bahk WM, Jun TY, Kim DJ, Dickson RA. Effects of olanzapine on prolactin levels of female patients with schizophrenia treated with risperidone. J Clin Psychiatry 2002;63(5):408–13.

156. Dunbar F, Kusumakar V, Daneman D, Schulz M. Growth and sexual maturation during long-term treatment with risperidone. Am J Psychiatry 2004;161:918–20.

157. Kunwar AR, Megna JL. Resolution of risperidone-induced hyperprolactinemia with substitution of quetiapine. Ann Pharmacother 2003;37:206–8.

158. Zhang XY, Zhou DF, Cao LY, Zhang PY, Wu GY, Shen YC. Prolactin levels in male schizophrenic patients treated with risperidone and haloperidol: a double-blind and randomized study. Psychopharmacology 2005;178:35–40.

159. Yamada K, Kanba S, Yagi G, Asai M. Herbal medicine (Shakuyaku-kanzo-to) in the treatment of risperidone-induced amenorrhea. J Clin Psychopharmacol 1999;19(4):380–1.

160. Mendhekar DN, Jiloha RC, Srivastava PK. Effect of risperidone on prolactinoma. A case report. Pharmacopsychiatry 2004;37:41–2.

161. Croarkin PE, Jacobs KM, Bain BK. Diabetic ketoacidosis associated with risperidone treatment? Psychosomatics 2000;41(4):369–70.

162. Horrigan JP, Sikich L. Diet and the atypical neuroleptics. J Am Acad Child Adolesc Psychiatry 1998;37(11):1126–7.

163. Martin A, Landau J, Leebens P, Ulizio K, Cicchetti D, Scahill L, Leckman JF. Risperidone-associated weight gain in children and adolescents: a retrospective chart review. J Child Adolesc Psychopharmacol 2000;10(4):259–68.

164. Lane H-Y, Chang Y-C, Cheng Y-C, Liu G-C, Lin X-R, Chang W-H. Effects of patient demographics, risperidone dosage, and clinical outcome on body weight in acutely exacerbated schizophrenia. J Clin Psychiatry 2003;64:316–20.

165. Safer DJ. A comparison of risperidone-induced weight gain across the age span. J Clin Psychopharmacol 2004;24:429–36.

166. Martin A, Scahill L, Anderson GM, Aman M, Arnold LE, McCracken J, McDougle CJ, Tierney E, Chuang S, Vitiello B (The Research Units on Pediatric Psychopharmacology Autism Network). Weight and leptin changes among risperidone-treated youths with autism: 6-month prospective data. Am J Psychiatry 2004;161:1125–7.

167. Ziegenbein M, Kropp S. Risperidone-induced long-term weight gain in a patient with schizophrenia. Aust NZ J Psychiatry 2004;38:175–6.

168. Barak Y. No weight gain among elderly schizophrenia patients after 1 year of risperidone treatment. J Clin Psychiatry 2002;63(2):117–9.

169. Wirshing DA, Pierre JM, Wirshing WC. Sleep apnea associated with antipsychotic-induced obesity. J Clin Psychiatry 2002;63(4):369–70.

170. Whitten JR, Ruehter VL. Risperidone and hyponatremia: a case report. Ann Clin Psychiatry 1997;9(3):181–3.

171. Kern RS, Marshall BD, Kuehnel TG, Mintz J, Hayden JL, Robertson MJ, Green MF. Effects of risperidone on polydipsia in chronic schizophrenia patients. J Clin Psychopharmacol 1997;17(5):432–5.

172. Kar N, Sharma PS, Tolar P, Pai K, Balasubramanian R. Polydipsia and risperidone. Aust NZ J Psychiatry 2002;36(2):268–70.

173. Edleman RJ. Risperidone side effects. J Am Acad Child Adolesc Psychiatry 1996;35(1):4–5.

174. Dernovsek Z, Tavcar R. Risperidone-induced leucopenia and neutropenia. Br J Psychiatry 1997;171:393–4.

175. Finkel B, Lerner AG, Oyffe I, Sigal M. Risperidone-associated agranulocytosis. Am J Psychiatry 1998;155(6):855–6.

176. Assion HJ, Kolbinger HM, Rao ML, Laux G. Lymphocytopenia and thrombocytopenia during treatment with risperidone or clozapine. Pharmacopsychiatry 1996;29(6):227–8.

177. Sluys M, Güzelcan Y, Casteelen G, de Haan L. Risperidone-induced leucopenia and neutropenia: a case report. Eur Psychiatry 2004;19:117.

178. Harrison-Woolrych M, Clark DW. Nose bleeds associated with use of risperidone. BMJ 2004;328:1416.

179. Srinivas VR. Postmarketing surveillance is needed. BMJ 2004;329:51.

180. Gajwani P, Franco-Bronson K, Tesar GE. Risperidone-induced sialorrhea. Psychosomatics 2001;42(3):276.

181. Fuller MA, Simon MR, Freedman L. Risperidone-associated hepatotoxicity. J Clin Psychopharmacol 1996;16(1):84–5.

182. Phillips EJ, Liu BA, Knowles SR. Rapid onset of risperidone-induced hepatotoxicity. Ann Pharmacother 1998;32(7–8):843.

183. Benazzi F. Risperidone-induced hepatotoxicity. Pharmacopsychiatry 1998;31(6):241.

184. Landau J, Martin A. Is liver function monitoring warranted during risperidone treatment? J Am Acad Child Adolesc Psychiatry 1998;37(10):1007–8.

185. Whitworth AB, Liensberger D, Fleischhacker WW. Transient increase of liver enzymes induced by risperidone: two case reports. J Clin Psychopharmacol 1999;19(5):475–6.

186. Krebs S, Dormann H, Muth-Selbach U, Hahn EG, Brune K, Schneider HT. Risperidone-induced cholestatic hepatitis. Eur J Gastroenterol Hepatol 2001;13(1):67–9.

187. Hudson RG, Cain MP. Risperidone associated hemorrhagic cystitis. J Urol 1998;160(1):159.

188. Agarwal V. Urinary incontinence with risperidone. J Clin Psychiatry 2000;61(3):219.

189. Almond DS, Rhodes LE, Pirmohamed M. Risperidone-induced photosensitivity. Postgrad Med J 1998;74(870):252–3.

190. Knegtering R, Castelein S, Bous H, van der Linde J, Bruggeman R, Kluiter H, van den Bosch RJ. A randomized open-label study of the impact of quetiapine versus risperidone on sexual functioning. J Clin Psychopharmacol 2004;24:56–61.

191. Tekell JL, Smith EA, Silva JA. Prolonged erection associated with risperidone treatment. Am J Psychiatry 1995;152(7):1097.

192. Slauson SD, LoVecchio F. Risperidone-induced priapism with rechallenge. J Emerg Med 2004;27:88–9.

193. Loh C, Leckband SG, Meyer JM, Turner E. Risperidone-induced retrograde ejaculation: case report and review of the literature. Int Clin Psychopharmacol 2004;19:111–2.

194. Sirota P, Bogdanov I. Priapism associated with risperidone treatment. Int J Psychiatry Clin Pract 2000;4:237–9.

195. Relan P, Gupta N, Mattoo S. A case of risperidone-induced priapism. J Clin Psychiatry 2003;64:482–3.

196. Bourgeois JA, Mundh H. Priapism associated with risperidone: a case report. J Clin Psychiatry 2003;64:218–9.

197. Reeves RR, Mack JE. Priapism associated with two atypical antipsychotic agents. Pharmacotherapy 2002;22(8):1070–3.

198. Ankem MK, Ferlise VJ, Han KR, Gazi MA, Koppisch AR, Weiss RE. Risperidone-induced priapism. Scand J Urol Nephrol 2002;36(1):91–2.

199. Freudenreich O. Exacerbation of idiopathic priapism with risperidone–citalopram combination. J Clin Psychiatry 2002;63(3):249–50.

200. Kaneda Y. Risperidone-induced ejaculatory dysfunction: a case report. Eur Psychiatry 2001;16(2):134–5.

201. Shiloh R, Weizman A, Weizer N, Dorfman-Etrog P, Munitz H. Risperidone-induced retrograde ejaculation. Am J Psychiatry 2001;158(4):650.

202. Holtmann M, Gerstner S, Schmidt MH. Risperidone-associated ejaculatory and urinary dysfunction in male adolescents. J Child Adolesc Psychopharmacol 2003;13:107–9.

203. Lane HY, Chang WH. Manic and psychotic symptoms following risperidone withdrawal in a schizophrenic patient. J Clin Psychiatry 1998;59(11):620–1.

204. Miller LJ. Withdrawal-emergent dyskinesia in a patient taking risperidone/citalopram. Ann Pharmacother 2000;34(2):269.

205. Komatsu S, Kirino E, Inoue Y, Arai H. Risperidone withdrawal-related respiratory dyskinesia: a case diagnosed by spirography and fibroscopy. Clin Neuropharmacol 2005;28:90–3.

206. Ratnayake T, Libretto SE. No complications with risperidone treatment before and throughout pregnancy and during the nursing period. J Clin Psychiatry 2002;63(1):76–7.

207. Mackay FJ, Wilton GL, Pearce SN, Freemantle SN, Mann RD. The safety of risperidone a postmarketing study on 7684 patients. Hum Psychopharmacol 1998;13:413–8.

208. Hill RC, McIvor RJ, Wojnar-Horton R, Hackett LP, Ilett KF. Risperidone distribution and excretion into human milk: report and estimated infant exposure during breast-feeding. J Clin Psychopharmocol 2000;20:285–6.

209. Ilett KF, Hackett P, Kristensen JH, Vaddadi KS, Gardiner SJ, Begg EJ. Transfer of risperidone and 9-hydroxyrisperidone into human milk. Ann Pharmacother 2004;38:273–6.

210. Kohnke MD, Griese EU, Stosser D, Gaertner I, Barth G. Cytochrome P450 2D6 deficiency and its clinical relevance in a patient treated with risperidone. Pharmacopsychiatry 2002;35(3):116–8.

211. Lane HY, Hsu SK, Liu YC, Chang YC, Huang CH, Chang WH. Dopamine D3 receptor Ser9Gly polymorphism and risperidone response. J Clin Psychopharmacol 2005;25:6–11.

212. Ravona-Springer R, Dolberg OT, Hirschmann S, Grunhaus L. Delirium in elderly patients treated with risperidone: a report of three cases. J Clin Psychopharmacol 1998;18(2):171–2.

213. Tavcar R, Dernovsek MZ. Risperidone-induced delirium. Can J Psychiatry 1998;43(2):194.

214. Madhusoodanan S, Brecher M, Brenner R, Kasckow J, Kunik M, Negron AE, Pomara N. Risperidone in the treatment of elderly patients with psychotic disorders. Am J Geriatr Psychiatry 1999;7(2):132–8.

215. Goldberg RJ. Long-term use of risperidone for the treatment of dementia-related behavioral disturbances in a nursing home population. Int J Geriatr Psychopharmacol 1999;2:1–4.

216. Cates M, Collins R, Woolley T. Antiparkinsonian drug prescribing in elderly inpatients receiving risperidone therapy. Am J Health Syst Pharm 1999;56(20):2139–40.

217. Irizarry MC, Ghaemi SN, Lee-Cherry ER, Gomez-Isla T, Binetti G, Hyman BT, Growdon JH. Risperidone treatment of behavioral disturbances in outpatients with dementia. J Neuropsychiatry Clin Neurosci 1999;11(3):336–42.

218. Aichhorn W, Weiss U, Marksteiner J, Kemmler G, Walch T, Zernig G, Stelzig-Schoeler R, Stuppaeck C, Geretsegger C. Influence of age and gender on risperidone plasma concentrations. J Psychopharmacol 2005;19:395–401.

219. Demb HB, Nguyen KT. Movement disorders in children with developmental disabilities taking risperidone. J Am Acad Child Adolesc Psychiatry 1999;38(1):5–6.

220. Kewley GD. Risperidone in comorbid ADHD and ODD/CD. J Am Acad Child Adolesc Psychiatry 1999;38(11):1327–8.

221. Research Units on Pediatric Psychopharmacology Autism Network. Risperidone treatment of autistic disorder: longer-term benefits and blinded discontinuation after 6 months. Am J Psychiatry 2005;162:1361–9.

222. Phan TG, Yu RY, Hersch MI. Hypothermia induced by risperidone and olanzapine in a patient with Prader–Willi syndrome. Med J Aust 1998;169(4):230–1.

223. Harrison TS, Goa KL. Long-acting risperidone. A review of its use in schizophrenia. CNS Drugs 2004;18:113–32.

224. Food and Drug Administration. Risperdal Consta. Long-acting injection. Summary of product characteristics. http://www.fda.gov/medwatch/safety/2005/aug_PI/Risperdal_%20Consta_PI.

225. van Os J, Bossie CA, Lasser RA. Improvements in stable patients with psychotic disorders switched from oral conventional antipsychotics therapy to long-acting risperidone. Int Clin Psychopharmacol 2004;19:229–32.

226. Turner M, Eerdekens E, Jacko M, Eerdekens M. Long-acting injectable risperidone: safety and efficacy in stable patients switched from conventional depot antipsychotics. Int Clin Psychopharmacol 2004;19:241–9.

227. Kissling W, Heres S, Lloyd K, Sacchetti E, Bouhours P, Medori R, Llorca PM. Direct transition to long-acting risperidone—analysis of long-term efficacy. J Psychopharmacol 2005;19 (5 Suppl):15–21.

228. Mohl A, Westlye K, Opjordsmoen S, Lex A, Schreiner A, Benoit M, Bräunig P, Medori R. Long-acting risperidone in stable patients with schizoaffective disorder. J Psychopharmacol 2005;19(5 Suppl):22–31.

229. Parellada E, Andrezina R, Milanova V, Glue P, Masiak M, Turner MS, Medori R, Gaebel W. Patients in the early phases of schizophrenia and schizoaffective disorders effectively treated with risperidone long-acting injectable. J Psychopharmacol 2005;19(5 Suppl):5–14.

230. Ciliberto N, Bossie CA, Urioste R, Lasser RA. Lack of impact of race on the efficacy and safety of long-acting risperidone versus placebo in patients with schizophrenia or schizoaffective disorder. Int Clin Psychopharmacol 2005;20:207–12.

231. Gastpar M, Masiak M, Latif MA, Frazzingaro S, Medori R, Lombertie ER. Sustained improvement of clinical outcome with risperidone long-acting injectable in psychotic patients previously treated with olanzapine. J Psychopharmacol 2005;19(5 Suppl):32–8.

232. Feifel D, Moutier CY, Perry W. Safety and tolerability of a rapidly escalating dose-loading regimen for risperidone. J Clin Psychiatry 2000;61(12):909–11.

233. Nair NP, Reiter-Schmitt B, Ronovsky K, Vyssoki D, Baeke J, Desseilles M, Kindts P, Mesotten F, Peuskens J, Addington D, et alThe Risperidone Study Group. Therapeutic equivalence of risperidone given once daily and twice daily in patients with schizophrenia. J Clin Psychopharmacol 1998;18(2):103–10.

234. Kontaxakis VP, Havaki-Kontaxaki BJ, Stamouli SS, Christodoulou GN. Optimal risperidone dose in drug-naive, first-episode schizophrenia. Am J Psychiatry 2000;157(7):1178–9.

235. Riedel M, Schwarz MJ, Strassnig M, Spellmann I, Muller-Arends A, Weber K, Zach J, Muller N, Moller HJ. Risperidone plasma levels, clinical response and side-effects. Eur Arch Psychiatry Clin Neurosci 2005;255:261–8.

236. Acri AA, Henretig FM. Effects of risperidone in overdose. Am J Emerg Med 1998;16(5):498–501.

237. Catalano G, Catalano MC, Nunez CY, Walker SC. Atypical antipsychotic overdose in the pediatric population. J Child Adolesc Psychopharmacol 2001;11(4):425–34.

238. Akyol A, Senel AC, Ulusoy H, Karip F, Erciyes N. Delayed respiratory depression after risperidone overdose. Anesth Analg 2005;101:1490–1.

239. Williams JH, Hepner DL. Risperidone and exaggerated hypotension during a spinal anesthetic. Anesth Analg 2004;98:240–1.

240. Lane HY, Chang WH. Risperidone–carbamazepine interactions: is cytochrome P450 3A involved? J Clin Psychiatry 1998;59(8):430–1.

241. Spina E, Avenoso A, Facciola G, Salemi M, Scordo MG, Giacobello T, Madia AG, Perucca E. Plasma concentrations of risperidone and 9-hydroxyrisperidone: effect of comedication with carbamazepine or valproate. Ther Drug Monit 2000;22(4):481–5.

242. Ono S, Mihara K, Suzuki A, Kondo T, Yasui-Furukori N, Furukori H, de Vries R, Kaneko S. Significant pharmacokinetic interaction between risperidone and carbamazepine: its relationship with CYP2D6 genotypes. Psychopharmacology (Berl) 2002;162(1):50–4.

243. Mula M, Monaco F. Carbamazepine–risperidone interactions in patients with epilepsy. Clin Neuropharmacol 2002;25(2):97–100.

244. Alfaro CL, Nicolson R, Lenane M, Rapoport JL. Carbamazepine and/or fluvoxamine drug interaction with risperidone in a patient on multiple psychotropic medications. Ann Pharmacother 2000;34(1):122–3.

245. Spina E, Scordo MG, Avenoso A, Perucca E. Adverse drug interaction between risperidone and carbamazepine in a patient with chronic schizophrenia and deficient CYP2D6 activity. J Clin Psychopharmacol 2001;21(1):108–9.

246. Eap CB, Bondolfi G, Zullino D, Bryois C, Fuciec M, Savary L, Jonzier-Perey M, Baumann P. Pharmacokinetic drug interaction potential of risperidone with cytochrome P450 isozymes as assessed by the dextromethorphan, the caffeine, and the mephenytoin test. Ther Drug Monit 2001;23(3):228–31.

247. Zhao Q, Xie C, Pesco-Koplowitz L, Jia X, Parier J-L. Pharmacokinetic and safety assessments of concurrent administration of risperidone and donepezil. J Clin Pharmacol 2003;43:180–6.

248. Kelly DV, Beique LC, Bowmer MI. Extrapyramidal symptoms with ritonavir/indinavir plus risperidone. Ann Pharmacother 2002;36(5):827–30.

249. Durrenberger S, de Leon J. Acute dystonic reaction to lithium and risperidone. J Neuropsychiatry Clin Neurosci 1999;11(4):518–9.

250. Demling J, Huang ML, De Smedt G. Pharmacokinetics and safety of combination therapy with lithium and risperidone in adult patients with psychosis. Int J Neuropsychopharmacol 1999;2:S63.

251. Durrenberger S, de Leon J. Acute dystonic reaction to lithium and risperidone. J Neuropsychiatry Clin Neurosci 1999;11(4):518–9.

252. Owley T, Leventhal B, Cook EH Jr. Risperidone-induced prolonged erections following the addition of lithium. J Child Adolesc Psychopharmacol 2001;11(4):441–2.

253. Park SH, Gill MA, Dopheide JA. Visual compatibility of risperidone solution and lithium citrate syrup. Am J Health Syst Pharm 2003;60(6):612–3.

254. Bork JA, Rogers T, Wedlund PJ, de Leon J. A pilot study on risperidone metabolism: the role of cytochromes P450 2D6 and 3A. J Clin Psychiatry 1999;60(7):469–76.

255. Normann C, Lieb K, Walden J. Increased plasma concentration of maprotiline by coadministration of risperidone. J Clin Psychopharmacol 2002;22(1):92–4.

256. Wines JD Jr, Weiss RD. Opioid withdrawal during risperidone treatment. J Clin Psychopharmacol 1999;19(3):265–7.

257. Sanderson DR. Drug interaction between risperidone and phenytoin resulting in extrapyramidal symptoms. J Clin Psychiatry 1996;57(4):177.

258. Caley CF. Extrapyramidal reactions from concurrent SSRI and atypical antipsychotic use. Can J Psychiatry 1998;43(3):307–8.

259. Baker RW. Possible dose–response relationship for risperidone in obsessive-compulsive disorder. J Clin Psychiatry 1998;59(3):134.

260. Stein DJ, Hawkridge S, Bouwer C, Emsley RA. Dr Stein and colleagues reply. J Clin Psychiatry 1998;59:134.

261. O'Connor M, Silver H. Adding risperidone to selective serotonin reuptake inhibitor improves chronic depression. J Clin Psychopharmacol 1998;18(1):89–91.

262. Avenoso A, Facciola G, Scordo MG, Gitto C, Ferrante GD. No effect of citalopram on plasma levels of clozapine, risperidone and their active metabolites in patients with chronic schizophrenia. Clin Drug Invest 1998;16:393–8.

263. Spina E, Avenoso A, Scordo MG, Ancione M, Madia A, Gatti G, Perucca E. Inhibition of risperidone metabolism by fluoxetine in patients with schizophrenia: a clinically

relevant pharmacokinetic drug interaction. J Clin Psychopharmacol 2002;22(4):419–23.

264. Bondolfi G, Eap CB, Bertschy G, Zullino D, Vermeulen A, Baumann P. The effect of fluoxetine on the pharmacokinetics and safety of risperidone in psychiatric patients. Pharmacopsychiatry 2002;35(2):50–6.

265. Andrade C. Risperidone may worsen fluoxetine-treated OCD. J Clin Psychiatry 1998;59(5):255–6.

266. Bozikas V, Petrikis P, Karavatos A. Urinary retention caused after fluoxetine–risperidone combination. J Psychopharmacol 2001;15(2):142–3.

267. Reeves RR, Mack JE, Beddingfield JJ. Neurotoxic syndrome associated with risperidone and fluvoxamine. Ann Pharmacother 2002;36(3):440–3.

268. Hamilton S, Malone K. Serotonin syndrome during treatment with paroxetine and risperidone. J Clin Psychopharmacol 2000;20(1):103–5.

269. Masson M, Elayli R, Verdoux H. Rispéridone et edème: à propos d'un cas. [Risperidone and edema: apropos of a case.] Encephale 2000;26(3):91–2.

270. Spina E, Avenoso A, Facciola G, Scordo MG, Ancione M, Madia A. Plasma concentrations of risperidone and 9-hydroxyrisperidone during combined treatment with paroxetine. Ther Drug Monit 2001;23(3):223–7.

271. Karki SD, Masood GR. Combination risperidone and SSRI-induced serotonin syndrome. Ann Pharmacother 2003;37:388–90.

272. Ostroff RB, Nelson JC. Risperidone augmentation of selective serotonin reuptake inhibitors in major depression. J Clin Psychiatry 1999;60(4):256–9.

273. Amchin J, Zarycranski W, Taylor KP, Albano D, Klockowski PM. Effect of venlafaxine on the pharmacokinetics of risperidone. J Clin Pharmacol 1999;39(3):297–309.

274. Steele M, Couturier J. A possible tetracycline–risperidone–sertraline interaction in an adolescent. Can J Clin Pharmacol 1999;6(1):15–7.

275. Nakagami T, Yasui-Furukori N, Saito M, Mihara K, De Vries R, Kondo T, Kaneko S. Thioridazine inhibits risperidone metabolism: a clinically relevant drug interaction. J Clin Psychopharmacol 2005;25:89–91.

276. Sanders RD, Lehrer DS. Edema associated with addition of risperidone to valproate treatment. J Clin Psychiatry 1998;59(12):689–90.

277. Chong SA, Tan CH, Lee EL, Liow PH. Augmentation of risperidone with valproic acid. J Clin Psychiatry 1998;59(8):430.

278. Olesen OV, Licht RW, Thomsen E, Bruun T, Viftrup JE, Linnet K. Serum concentrations and side effects in psychiatric patients during risperidone therapy. Ther Drug Monit 1998;20(4):380–4.

Sertindole

General Information

Sertindole is an antipsychotic drug that was approved in several countries in 1997, but was suspended in January 2000, following concerns about reports of cardiac dysrhythmias and sudden death (SEDA-22, 71) (SEDA-25, 71) (1). To gather further information, data collected for prescription-event-monitoring (PEM) studies have been analysed (2). Patients taking sertindole ($n = 462$; 5482 months of observation) were compared with patients taking risperidone or olanzapine ($n = 16\,542$; 139 987 months of observation). There were seven deaths in the sertindole group and 415 in the other, and the death rates were not significantly different, although the confidence intervals were wide because of the relatively small number of patients in the sertindole group. There were six cases of prolonged QT_c interval in the patients taking sertindole and one in the controls. The authors concluded that the sertindole group was too small to rule out an association between sertindole and cardiovascular deaths.

On 28 June 2001 an ad hoc expert committee was convened to review the available clinical and preclinical data related to the cardiovascular activity of sertindole and to consider whether such data supported the then current marketing authorization status of sertindole (3). Based on a re-evaluation of all the available data, including additional data submitted by the Marketing Authorization Holder, it was concluded that the re-introduction of sertindole could be supported by further clinical safety data, strong safeguards (including extensive contraindications and warnings for patients at risk of cardiac dysrhythmias), a reduction in the recommended maximum dose from 24 mg to 20 mg in all but exceptional cases, and extensive electrocardiographic monitoring before and during treatment. The Committee for Proprietary Medicinal Products (CPMP) of The European Agency for the Evaluation of Medicinal Products later recommended lifting the ban on sertindole-containing medicinal products on the basis of additional data provided by the marketing authorization holders (see monthly report from October, 2001: www.emea.eu.int).

Organs and Systems

Cardiovascular

Sertindole was associated with 27 deaths (16 cardiac) in 2194 patients who were enrolled in premarketing studies; further fatal cases have been collected, and the Committee on Safety of Medicines has described reports of 36 deaths (including some sudden cardiac deaths) and 13 serious but non-fatal dysrhythmias also associated with sertindole (4).

Metabolism

Four patients developed tardive dyskinesia while taking conventional antipsychotic drugs and were switched to sertindole (5). Three apparently recovered from the movement disorder. In the other patient sertindole monotherapy was not sufficient to reduce the movement effects, but combination treatment with tetrabenazine resulted in a greater reduction in extrapyramidal symptoms. There was no evidence of QT_c prolongation in these patients, but one patient gained 8 kg in weight.

References

1. Rawlins M. Suspension of availability of Serdolect (sertindole). Media Release 1998;2:3–7.
2. Wilton LV, Heeley EL, Pickering RM, Shakir SA. Comparative study of mortality rates and cardiac dysrhythmias in post-marketing surveillance studies of sertindole and two other atypical antipsychotic drugs, risperidone and olanzapine. J Psychopharmacol 2001;15(2):120–6.
3. EMEA. The European Agency for the Evaluation of Medicinal Products. http://www.emea.eu.int/pdfs/human/referral/Sertindole/285202en.pdf
4. Committee on Safety of Medicines-Medicines Control Agency. Cardiac arrhythmias with pomizode (Orap). Curr Probl Pharmacovigilance 1995;21:1.
5. Perquin LN. Treatment with the new antipsychotic sertindole for late-occurring undesirable movement effects. Int Clin Psychopharmacol 2005;20:335–8.

Thioridazine

General Information

Thioridazine is a phenothiazine neuroleptic drug.

Organs and Systems

Cardiovascular

Cardiac dysrhythmias

Thioridazine has been associated with QT interval prolongation (1,2) and several cases of torsade de pointes have been reported (3,4).

Two types of T wave changes have also been described after treatment with thioridazine: type I (with rounded, flat, or notched T waves) and type II (with biphasic T waves) (5).

On July 7, 2000, doctors and pharmacists in the USA were notified about the addition of extensive new safety warnings, including a boxed warning to the professional product label for the neuroleptic drug thioridazine (Melleril, Novartis Pharmaceuticals). The text of the new boxed warning read: "Melleril (thioridazine) has been shown to prolong the QT_c interval in a dose-related manner, and drugs with this potential, including Melleril, have been associated with torsade de pointes-type arrhythmias and sudden death. Due to its potential for significant, possibly life-threatening, prodysrhythmic effects, Melleril should be reserved for use in the treatment of patients with schizophrenia who fail to show an acceptable response to adequate courses of treatment with other neuroleptic drugs, either because of insufficient effectiveness or the inability to achieve an effective dose due to intolerable adverse effects from those drugs."

The new labelling changes were based primarily on the FDA's review of three published studies. The first of these showed increased blood concentrations of thioridazine in patients with a genetic defect, resulting in the slow inactivation of debrisoquine (6). In this study, 19 healthy subjects (six poor and 13 extensive metabolizers of debrisoquine) took a single oral dose of thioridazine 25 mg. The poor metabolizers reached higher blood concentrations of thioridazine 2.4 times more quickly than the extensive metabolizers. There was a 4.5-fold increase in the systemic availability of the drug in the poor metabolizers, in whom thioridazine remained in the blood twice as long. The second study showed dose-related prolongation of the QT_c interval from 388 (range 370–406) to 411 (range 397–425) ms 4 hours after an oral dose of thioridazine 50 mg (7). The average maximal increase was 23 ms. This change was statistically greater than that for either placebo or thioridazine 10 mg. In the third study the effect of the selective serotonin re-uptake inhibitor (SSRI) fluvoxamine, 25 mg bd for 1 week, on thioridazine blood concentrations was evaluated in 10 hospitalized men with schizophrenia (8). The concentrations of thioridazine and its two active breakdown products, mesoridazine and sulforidazine, increased three-fold after the administration of fluvoxamine.

The possibility that some of the cardiac effects of thioridazine may be related to these metabolites as well as the parent compound has been explored (5,9) but needs further investigation.

Several regulatory measures (www.mca.gov.uk; www.medsafe.govt.nz; www.imb.ie; www.hc-sc.gc.ca; www.bpfk.org; www.fda.gov; www.who.int/medicines) have been adopted in different countries with regard to thioridazine.

The role of cardiac effects of neuroleptic drugs in sudden death is controversial (SEDA-18, 47; SEDA-20, 26; SEDA-20, 36). There may be multiple non-cardiac causes, including asphyxia, convulsions, or hyperpyrexia. However, some cases of sudden death in apparently young healthy individuals may be directly attributable to cardiac dysrhythmias after treatment with thioridazine or chlorpromazine (10), and from time to time, cases of sudden death are reported (SEDA-20, 36; SEDA-22, 46), including four cases in which thioridazine in standard doses was implicated as the cause of death or as a contributing factor (11).

- A 68-year-old man with a 5-year history of Alzheimer's disease was treated with thioridazine 25 mg tds because of violent outbursts (12). His other drugs, temazepam 10–30 mg at night, carbamazepine 100 mg bd for neuropathic pain, and droperidol 5–10 mg as required, were unaltered. Five days later, he was found dead, having been in his usual condition 2 hours before. Post-mortem examination showed stenosis of the coronary arteries, but no coronary thrombosis, myocardial infarction, or other significant pathology. The certified cause of death was cardiac dysrhythmia due to ischemic heart disease. Thioridazine was considered as a possible contributing factor.

Hypotension

Hypotension is the most commonly observed cardiovascular adverse effect of neuroleptic drugs, particularly after administration of those that are also potent alpha-adrenoceptor antagonists, such as thioridazine. A central mechanism involving the vasomotor regulatory center may also contribute to the lowering of blood pressure.

Severe orthostatic hypotension has been observed in older volunteers ($n = 14$; aged 65–77 years) who participated in a randomized, double-blind, three-period, crossover study, in which they took single oral doses of thioridazine (25 mg), remoxipride (50 mg), or placebo (13). Compared with placebo, there were falls in supine and erect systolic and diastolic blood pressures after thioridazine, but not remoxipride. Standing systolic blood pressures fell by a maximum of 26 mmHg. There were similar falls in blood pressure in young volunteers.

Hypotension has been reported in two other elderly patients who were taking thioridazine (14). The patients, men aged 68 and 70 years, had traumatic brain injury and were taking oral thioridazine 25 mg/day for agitation. A few days later they developed mild hypotension (100/50 and 100/60 mmHg respectively).

Ear, nose, throat

Epistaxis has been reported in three patients with hypertension taking thioridazine (15).

Nervous system

In one study, there was no significant difference in serum concentrations of thioridazine, its metabolites, or radioreceptor activity between patients with and without tardive dyskinesia (16).

Sensory systems

Acute toxic effects of thioridazine on the eyes include nyctalopia, blurred vision, and dyschromatopsia, which typically become evident after 2–8 weeks of dosages over 800 mg/day.

Lens and corneal deposits
Deposits in the lens or cornea probably result from melanin-drug complex deposition and are best detected by slit-lamp examination. These deposits are probably dose- and time-related, since they generally occur only after years of treatment. Fortunately, they are in large part reversible, but if undetected they may progress to interfere with vision. They are most often reported with chlorpromazine or thioridazine and can occur in association with pigmentary changes in the skin.

Pigmentary retinopathy
Pigmentary retinopathy, which can seriously impair vision, is specifically associated with thioridazine, and has occurred more often with high and prolonged dosage (for example 1200–1800 mg/day for weeks to months) (17), although in one case the daily dose was only 700 mg (18). Large-scale surveys have confirmed the relative safety of dosages up to 800 mg/day (19); at any dosage, however, any complaint of brownish discoloration of vision or impaired dark adaptation requires immediate evaluation.

- A 28-year-old woman with a long history of psychiatric problems was taking fluoxetine, diazepam, methylphenidate, and thioridazine 800 mg qds (17). Fluorescein angiography showed confluent areas of punctate hyperfluorescence, consistent with diffuse retinal pigment epithelial alteration secondary to acute thioridazine toxic effects.
- A 51-year-old woman with a long history of psychiatric problems and no family history of hereditary retinal degeneration had reduced vision in both eyes for several years while she was taking thioridazine 300 mg/day and chlorpromazine 600 mg/day (20). She had large patches of atrophy outside the arcades and within the macula, with sparing of foveal pigmentation; there were diffuse increases in hyperpigmentation in the peripheries of both eyes, and hypopigmented or unpigmented retinal epithelium.

The authors thought that the pigmentary changes were probably due to thioridazine, as adverse effects on the retina are rare with chlorpromazine.

Fluid balance

Water intoxication has been reported during treatment with thioridazine and may be due to its pronounced anticholinergic properties and/or direct stimulation of the hypothalamic thirst center (21).

Hematologic

The Committee on Safety of Medicines and the erstwhile Medicines Control Agency in the UK received 999 reports of hemopoietic disorders related to neuroleptic drugs between 1963 and 1996; there were 65 deaths (22). There were 182 reports of agranulocytosis; chlorpromazine and thioridazine were associated with the highest number of deaths—27 of 56 and nine of 24 respectively.

Skin

Photosensitivity reactions have been observed in patients taking thioridazine (23).

- A 72-year-old woman developed well-delimited hyperpigmented papular lesions on exposed areas after 4 years of treatment.
- A 63-year-old woman developed maculopapular pigmented skin lesions on the face, neck, and forearms after several years; histology showed a lichenoid dermatitis.
- A 72-year-old woman developed papular pigmented skin lesions on the face, neck, and the backs of the hands; a skin biopsy showed a lichenoid reaction.

In all three cases the reaction disappeared after withdrawal of thioridazine.

Sexual function

Thioridazine and other highly anticholinergic drugs can cause male sexual dysfunction, including impotence and retrograde ejaculation 24,25. A case of thioridazine-induced inhibition of female orgasm has also been reported (26).

Death

A retrospective case-control study conducted by the same group who first identified a relation effect of pimozide and thioridazine on the QT interval (SEDA-24, 55) has been published (27). The study was carried out in five large psychiatric hospitals in England and included all inpatients with sudden unexplained death over a period of 12 years (1984–1995) and two controls for each case from the same hospital matched for age, sex, and duration of inpatient stay, one of whom was also matched for primary psychiatric diagnosis. The patients were aged 18–74 years, and there were 69 cases and 132 controls (63 matched for diagnosis). Since the presence of an organic disorder was significantly associated with sudden unexpected death, this was adjusted for. Sudden death was associated with hypertension, ischemic heart disease, and thioridazine therapy (OR = 5.3; 95% CI = 1.7, 15). Among the limitations of this study was the incompleteness of the records, which often lacked information about important risk factors, including underlying cardiac disease, smoking, or the use of alcohol or illicit drugs; furthermore, the low rate of post-mortem examination raises the possibility that some of the deaths were from causes other than cardiac dysrhythmias, although other causes were found in only three of 30 cases when a post-mortem examination was done.

Long-Term Effects

Drug withdrawal

The consequences of restricting the indications for thioridazine in patients with learning disabilities have been a matter of reflection (28). Of 155 psychiatric patients, 18 were regularly taking thioridazine at the time of the directive; all stopped taking it and seven had moderate or severe difficulties during the following 3 months and one developed probable neuroleptic malignant syndrome 4 weeks after switching to an alternative drug. According to the authors, regulatory authorities should take account of the adverse consequences of drug changes when making judgements about the benefit-to-harm balance, especially in vulnerable patients, such as those with learning disabilities and psychiatric illnesses.

Other consequences of banning thioridazine have been reported in a rural general practice in Ireland, in which 29 of 40 GPs responded to a questionnaire and 17 reported management problems and adverse reactions (29). There was increased service demand, as 44% of the GPs described up to a 50% increase in referrals to the mental health service; although most of the GPs (67%) reported satisfaction with alternative agents, 37% described adverse effects associated with the alternative agents. It seems reasonable that directives should incorporate the flexibility required to accommodate the needs of patients who are already successfully stabilized on these drugs.

Susceptibility Factors

Genetic factors

The higher serum concentrations in poor CYP2D6 metabolizers are associated with an increased risk of thioridazine toxicity (11).

Age

Age increases the risk of thioridazine toxicity (11).

Sex

Female sex increases the risk of thioridazine toxicity (11).

Other features of the patient

Several factors increase the risk of thioridazine toxicity: pre-existing cardiac disease, hypokalemia, a glucose load, alcohol, exercise, and concomitant therapy with tricyclic antidepressants, erythromycin, co-trimoxazole, cisapride, risperidone, hydroxyzine, and drugs that inhibit CYP2D6 (some SSRIs, fluphenazine, and perphenazine) (11).

Drug Administration

Drug overdose

Brugada-like electrocardiographic abnormalities have occurred after thioridazine overdose (30).

- A previously healthy 58-year-old woman whose blood concentration of thioridazine was 1480 µg/l (usual therapeutic concentrations are up to 200 µg/l) became comatose and had muscular rigidity. An electrocardiogram showed sinus rhythm with significant QT prolongation and 1 day later evolved into a Brugada-like pattern. Over the next 72 hours, both the electrocardiogram and the clinical abnormalities resolved.

Drug–Drug Interactions

Beta-adrenoceptor antagonists

Lipophilic beta-adrenoceptor antagonists are metabolized to varying degrees by oxidation by liver microsomal cytochrome P450 (for example propranolol by CYP1A2 and CYP2D6 and metoprolol by CYP2D6). They can therefore reduce the clearance and increase the steady-state plasma concentrations of other drugs that undergo similar metabolism, potentiating their effects. Drugs that interact in this way include thioridazine (31).

The combination of propranolol or pindolol with thioridazine should be avoided if possible (32).

Diuretics

Diuretic-induced hypokalemia can potentiate thioridazine-induced cardiotoxicity (10).

Fluvoxamine

Fluvoxamine increased thioridazine concentrations three-fold (8) in 10 patients with schizophrenia taking steady-state thioridazine who were given fluvoxamine (25 mg bd) for 1 week. Fluvoxamine interferes with the metabolism of thioridazine, probably via CYP2C19 and/or CYP1A2.

Fluvoxamine inhibits CYP1A2, and a low dosage (50 mg/day) produced a 225% increase in plasma thioridazine concentrations in 10 patients with schizophrenia (8). This was not reflected in an increased incidence of clinical adverse events; however, thioridazine prolongs the QT_c interval, which was not measured.

Lithium

Several cases of neurotoxicity in patients taking lithium and thioridazine have been reported. The cause of this interaction has not been resolved, but lithium seems compatible with all neuroleptic drugs, although patients should be carefully monitored (33–35).

Methylphenidate

Multiple involuntary movements, consisting of jaw grinding, oral dyskinesias, bilateral hand rolling, vermiform tongue movements, and bilateral choreiform movements of the digits, have been described in an 11-year-old boy taking thioridazine 150 mg/day and methylphenidate 10 mg bd (36). The methylphenidate was discontinued and within 4 weeks his movement disorder had completely disappeared.

Quinidine

Concurrent administration of quinidine with neuroleptic drugs, particularly thioridazine, can cause myocardial depression (10).

Risperidone

A 23-year-old man had high risperidone plasma concentrations secondary to concurrent thioridazine use (37).

Thioridazine 25 mg bd significantly increased the steady-state plasma concentration of risperidone (3 mg bd) and reduced 9-hydroxyrisperidone in 12 patients with schizophrenia (38). Since risperidone mainly undergoes 9-hydroxylation yielding an active metabolite, 9-hydroxyrisperidone, metabolic inhibition could account for this increase. Moreover, there was a greater rise in risperidone concentrations in subjects with more capacity for CYP2D6-dependent 9-hydroxylation of risperidone.

Tricyclic antidepressants

The interaction of thioridazine with a tricyclic antidepressant is particularly dangerous, since both cause cardiac toxicity. One report concerned two young patients who developed ventricular dysrhythmias, from which they recovered (39).

References

1. Warner JP, Barnes TR, Henry JA. Electrocardiographic changes in patients receiving neuroleptic medication. Acta Psychiatr Scand 1996;93(4):311–3.

2. Reilly JG, Ayis SA, Ferrier IN, Jones SJ, Thomas SH. QT$_c$-interval abnormalities and psychotropic drug therapy in psychiatric patients. Lancet 2000;355(9209):1048–52.

3. Kiriike N, Maeda Y, Nishiwaki S, Izumiya Y, Katahara S, Mui K, Kawakita Y, Nishikimi T, Takeuchi K, Takeda T. Iatrogenic torsade de pointes induced by thioridazine. Biol Psychiatry 1987;22(1):99–103.

4. Connolly MJ, Evemy KL, Snow MH. Torsade de pointes ventricular tachycardia in association with thioridazine therapy: report of two cases. New Trends Arrhythmias 1985;1:157.

5. Axelsson R, Aspenstrom G. Electrocardiographic changes and serum concentrations in thioridazine-treated patients. J Clin Psychiatry 1982;43(8):332–5.

6. von Bahr C, Movin G, Nordin C, Liden A, Hammarlund-Udenaes M, Hedberg A, Ring H, Sjoqvist F. Plasma levels of thioridazine and metabolites are influenced by the debrisoquin hydroxylation phenotype. Clin Pharmacol Ther 1991;49(3):234–40.

7. Hartigan-Go K, Bateman DN, Nyberg G, Martensson E, Thomas SH. Concentration-related pharmacodynamic effects of thioridazine and its metabolites in humans. Clin Pharmacol Ther 1996;60(5):543–53.

8. Carrillo JA, Ramos SI, Herraiz AG, Llerena A, Agundez JA, Berecz R, Duran M, Benitez J. Pharmacokinetic interaction of fluvoxamine and thioridazine in schizophrenic patients. J Clin Psychopharmacol 1999;19(6):494–9.

9. Dahl SG. Active metabolites of neuroleptic drugs: possible contribution to therapeutic and toxic effects. Ther Drug Monit 1982;4(1):33–40.

10. Risch SC, Groom GP, Janowsky DS. The effects of psychotropic drugs on the cardiovascular system. J Clin Psychiatry 1982;43(5 Pt 2):16–31.

11. Timell AM. Thioridazine: re-evaluating the risk/benefit equation. Ann Clin Psychiatry 2000;12(3):147–51.

12. Thomas SH, Cooper PN. Sudden death in a patient taking antipsychotic drugs. Postgrad Med J 1998;74(873):445–6.

13. Swift CG, Lee DR, Maskrey VL, Yisak W, Jackson SH, Tiplady B. Single dose pharmacodynamics of thioridazine and remoxipride in healthy younger and older volunteers. J Psychopharmacol 1999;13(2):159–65.

14. Rampello L, Raffaele R, Vecchio I, Pistone G, Brunetto MB, Malaguarnera M. Behavioural changes and hypotensive effects of thioridazine in two elderly patients with traumatic brain injury: post-traumatic syndrome and thioridazine. Gaz Med Ital Arch Sci Med 2000;159:121–3.

15. Idupuganti S. Epistaxis in hypertensive patients taking thioridazine. Am J Psychiatry 1982;139(8):1083–4.

16. Widerlov E, Haggstrom JE, Kilts CD, Andersson U, Breese GR, Mailman RB. Serum concentrations of thioridazine, its major metabolites and serum neuroleptic-like activities in schizophrenics with and without tardive dyskinesia. Acta Psychiatr Scand 1982;66(4):294–305.

17. Shah GK, Auerbach DB, Augsburger JJ, Savino PJ. Acute thioridazine retinopathy. Arch Ophthalmol 1998;116(6):826–7.

18. Meredith TA, Aaberg TM, Willerson WD. Progressive chorioretinopathy after receiving thioridazine. Arch Ophthalmol 1978;96(7):1172–6.

19. Simpson GM, Pi EH, Sramek JJ Jr. Adverse effects of antipsychotic agents. Drugs 1981;21(2):138–51.

20. Borodoker N, Del Priore LV, De A Carvalho C, Yannuzzi LA. Retinopathy as a result of long-term use of thioridazine. Arch Ophthalmol 2002;120(7):994–5.

21. Rao KJ, Miller M, Moses A. Water intoxication and thioridazine (Mellaril). Ann Intern Med 1975;82(1):61.

22. King DJ, Wager E. Haematological safety of antipsychotic drugs. J Psychopharmacol 1998;12(3):283–8.

23. Llambrich A, Lecha M. Photoinduced lichenoid reaction by thioridazine. Photodermatol Photoimmunol Photomed 2004;20:108-9.

24. Mitchell JE, Popkin MK. Antipsychotic drug therapy and sexual dysfunction in men. Am J Psychiatry 1982;139(5):633–7.

25. Siris SG, Siris ES, van Kammen DP, Docherty JP, Alexander PE, Bunney WE Jr. Effects of dopamine blockade on gonadotropins and testosterone in men. Am J Psychiatry 1980;137(2):211–4.

26. Shen WW, Park S. Thioridazine-induced inhibition of female orgasm. Psychiatr J Univ Ott 1982;7(4):249–51.

27. Reilly JG, Ayis SA, Ferrier IN, Jones SJ, Thomas SH. Thioridazine and sudden unexplained death in psychiatric in-patients. Br J Psychiatry 2002;180:515–22.

28. Davies SJ, Cooke LB, Moore AG, Potokar J. Discontinuation of thioridazine in patients with learning disabilities: balancing cardiovascular toxicity with adverse consequences of changing drugs. BMJ 2002;324(7352):1519–21.

29. Bailey P, Russell V. Restricting the use of thioridazine. Br J General Pract 2002;52(479):499–500.

30. Copetti R, Proclemer A, Pillinini PP. Brugada-like ECG abnormalities during thioridazine overdose. Br J Clin Pharmacol 2005;59:608.

31. Greendyke RM, Kanter DR. Plasma propranolol levels and their effect on plasma thioridazine and haloperidol concentrations. J Clin Psychopharmacol 1987;7(3):178–82.

32. Markowitz JS, Wells BG, Carson WH. Interactions between antipsychotic and antihypertensive drugs. Ann Pharmacother 1995;29(6):603–9.

33. Jefferson JW, Greist JH, Baudhuin M. Lithium: interactions with other drugs. J Clin Psychopharmacol 1981;1(3):124–34.

34. Waddington JL. Some pharmacological aspects relating to the issue of possible neurotoxic interactions during combined lithium-neuroleptic therapy. Hum Psychopharmacol 1990;5:293–7.

35. Batchelor DH, Lowe MR. Reported neurotoxicity with the lithium/haloperidol combination and other neuroleptics. A literature review. Hum Psychopharmacol 1990;5:275–80.

36. Connor DF. Stimulants and neuroleptic withdrawal dyskinesia. J Am Acad Child Adolesc Psychiatry 1998;37(3):247–8.

37. Alfaro CL, Nicolson R, Lenane M, Rapoport JL. Carbamazepine and/or fluvoxamine drug interaction with risperidone in a patient on multiple psychotropic medications. Ann Pharmacother 2000;34(1):122–3.

38. Nakagami T, Yasui-Furukori N, Saito M, Mihara K, De Vries R, Kondo T, Kaneko S. Thioridazine inhibits risperidone metabolism: a clinically relevant drug interaction. J Clin Psychopharmacol 2005;25:89-91.

39. Heiman EM. Cardiac toxicity with thioridazine–tricyclic antidepressant combination. J Nerv Ment Dis 1977;165(2):139–43.

Tiapride

General Information

Tiapride is a substituted benzamide related to sulpiride, a selective dopamine D_2 and D_3 receptor antagonist with little propensity for causing catalepsy and sedation. It has antidyskinetic activity, reflecting its antidopaminergic action, and also anxiolytic activity mediated by mechanisms that are poorly understood. Unlike the benzodiazepines, tiapride does not affect vigilance and has a low potential for interaction with ethanol, and possibly for abuse. From a clinical point of view, it can be considered as an atypical neuroleptic drug. It has reported efficacy in neuroleptic-drug induced tardive dyskinesia, levodopa-induced dyskinesias, psychomotor agitation in elderly patients, and chorea (1).

Tiapride appears to be useful in alcohol withdrawal as an alternative to the benzodiazepines (2). It facilitates the management of ethanol withdrawal, but its use in patients at risk of severe reactions in acute withdrawal should be accompanied by adjunctive therapy for hallucinosis and seizures. Since it may prove difficult to identify such patients and since there is also a small risk of the neuroleptic malignant syndrome (particularly with parenteral administration), the usefulness of tiapride in this setting is likely to be limited. The potential risk of tardive dyskinesia at the dosage used in alcoholic patients following detoxification (300 mg/day) requires evaluation and necessitates medical supervision. It is unlikely to produce problems of dependence or abuse.

Tiapride has been assessed for the treatment of agitation and aggressiveness in elderly patients with cognitive impairment in a multicenter, double-blind study, in which patients were randomly allocated to tiapride 100 mg/day ($n = 102$), haloperidol 2 mg/day ($n = 101$), or placebo ($n = 103$) (3). The percentage of responders after 21 days, according to the Multidimensional Observation Scale for Elderly Subjects (MOSES) irritability/aggressiveness subscale, was significantly greater in both of the active treatment groups (haloperidol 63%, tiapride 69%) than in the placebo group (49%). There were 10 dropouts in the tiapride group, 21 in the haloperidol group, and 16 in the placebo group. The number of patients with at least one extrapyramidal symptom was significantly smaller with tiapride (16%) than with haloperidol (34%) and identical to that with placebo (17%); there was no significant difference across the groups in the numbers of patients with endocrinological adverse events. Four deaths were reported: one with placebo (stroke), one with tiapride (pneumonia), and two with haloperidol (stroke and heart failure).

Organs and Systems

Cardiovascular

Torsade de pointes has been attributed to tiapride.

- A 76-year-old man developed torsade de pointes while taking tiapride 300 mg/day; the QT_c interval 1 day after starting treatment was 600 ms; the dysrhythmia resolved when tiapride was withdrawn (4).

References

1. Dose M, Lange HW. The benzamide tiapride: treatment of extrapyramidal motor and other clinical syndromes. Pharmacopsychiatry 2000;33(1):19–27.

2. Peters DH, Faulds D. Tiapride. A review of its pharmacology and therapeutic potential in the management of alcohol dependence syndrome. Drugs 1994;47(6):1010–32.
3. Allain H, Dautzenberg PH, Maurer K, Schuck S, Bonhomme D, Gerard D. Double blind study of tiapride versus haloperidol and placebo in agitation and aggressiveness in elderly patients with cognitive impairment. Psychopharmacology (Berl) 2000;148(4):361–6.
4. Iglesias E, Esteban E, Zabala S, Gascon A. Tiapride-induced torsade de pointes. Am J Med 2000;109(6):509.

Tiotixene

General Information

Tiotixene is a thioxanthene neuroleptic drug.

Organs and Systems

Nervous system

Neuroleptic malignant syndrome has been reported as a severe complication of tiotixene (SEDA-21, 49).

- A 42-year-old man with a history of paranoid schizophrenia developed confusion and emesis. His medications included buspirone and tiotixene. He also had a generalized seizure that resolved spontaneously. His serum sodium concentration was 114 mmol/l, which was thought to be secondary to psychogenic polydipsia and which was corrected. The next day his temperature rose to 40°C and the day after that he became lethargic and non-verbal and developed generalized muscle rigidity. A diagnosis of probable neuroleptic malignant syndrome was made, and he was given dantrolene sodium 1 mg/kg intravenously; neuroleptic drugs were withdrawn. The serum creatine kinase activity was over 234 500 U/l. Urinalysis showed more than 3 g/l of protein, a trace of ketones, and hemoglobin and leukocytes. By the following day, his muscle rigidity and fever had resolved and his mental status was markedly improved.

Skin

Cutaneous lesions consisting of telangiectatic macules have been reported with tiotixene (1).

Reference

1. Matsuoka LY. Thiothixene drug sensitivity. J Am Acad Dermatol 1982;7(3):405.

Ziprasidone

General Information

Ziprasidone is a dibenzotheolylpiperazine compound with a receptor binding profile similar to that of other atypical antipsychotic drugs, and a high affinity for serotonin ($5-HT_{2A}$) receptors and a lower affinity for dopamine (D_2) receptors; it also has high affinities for $5-HT_{1A}$, $5-HT_{1D}$ and $5-HT_{2C}$ receptors and inhibits serotonin and noradrenaline reuptake. Two extensive reviews of the clinical pharmacology of ziprasidone have appeared (1,2).

On February 5, 2000, the FDA approved ziprasidone for the treatment of schizophrenia (www.fda.gov). However, the FDA has been concerned with the possibility that ziprasidone and a number of other drugs might increase the risk of the specific potentially fatal cardiac dysrhythmia, torsade de pointes. The FDA did not approve ziprasidone in 1998, because of evidence that it can cause prolongation of the QT interval, and they asked that specific safety data be gathered. The safety data were submitted in 1999. Although QT prolongation is still a theoretical concern, over 4000 patients have been treated in clinical trials without evidence of torsade de pointes. In addition, overall mortality in the trials was similar to that seen with placebo and other neuroleptic drugs. The FDA labelling does not include a so-called "black-box warning" and does not require an electrocardiogram before or during treatment. However, the labelling does warn physicians and patients about QT interval prolongation and the possible risk of sudden death. The labelling suggests that doctors use their best judgment, based on the health status of the individual, as to whether to use ziprasidone as first-line treatment or only after other available drugs have failed. There is no requirement that patients have regular heart check-ups while taking this drug.

Observational studies

In a PET study in the brain region of interest, intended to delineate receptor occupancy, 16 patients with schizophrenia or schizoaffective disorders were randomly assigned to receive the recommended dosage of ziprasidone: 40, 80, 120, or 160 mg/day (19). Ziprasidone was more like risperidone and olanzapine in its receptor occupancy profile than clozapine and quetiapine; the optimal effective dose of ziprasidone is closer to 120 mg/day than to the lower doses suggested by previous PET studies. One subject taking 120 mg/day withdrew because of severe somnolence, and another developed oculogyric crises. Given concerns about prolongation of the QT_c interval with ziprasidone, this was reported in some detail. The number of individuals who had shortening of the QT_c interval was the same as the number who had prolongation (n = 8). In subjects in whom the interval was prolonged the ranges were as follows: 0–25 msec (n = 4), 26–50 msec (n = 2), and over 50 msec (n = 2); the two individuals with increases of over 50 msec were taking 120 mg/day. Only one subject had a QT_c interval greater than 450 msec; (due to prolongation by 51 msec).

Ziprasidone is apparently well tolerated, with a limited potential to cause extrapyramidal adverse effects or weight gain (4). Out-patients who partly respond to conventional antipsychotic drugs, risperidone, or olanzapine may have improved control of psychotic symptoms after switching to ziprasidone, according to the results of a re-analysis of 6-week, multicenter, randomized, open, parallel-group studies in patients with schizophrenia who had previously taken conventional antipsychotic drugs (n = 108), olanzapine (n = 104), or risperidone (n = 58); these results have been published in two different journals (5, 6).

In an open 12-week study of ziprasidone in 12 patients with Parkinson's disease and psychosis, two withdrew because of adverse effects; one had increased diurnal sedation on day 5 and the other had deterioration of gait at 1 week (7^c). The other 10 patients reported significant improvement in psychiatric symptoms and no deterioration in motor symptoms. The small sample size and lack of a control group precluded definitive conclusions.

In an open study nine patients with treatment-resistant schizophrenia took clozapine + ziprasidone for 6 months (8). Mental state improved in seven, with significant reductions in the mean Brief Psychiatric Rating Scale score. The combination allowed an 18% reduction in the daily dose of clozapine. All had some adverse effects before combination treatment (each had been unsuccessfully treated with at least two first-generation antipsychotic drugs and/or two second-generation drugs as monotherapy) and during combination treatment, but the co-administration of ziprasidone did not result in a corresponding increase in adverse effects; for example, there was neither further weight gain (n = 5) nor weight loss in patients with previous weight gain (n = 2). However, the small sample size and the lack of a control group precluded definitive conclusions.

Published and unpublished studies from 1995 to 2004, in which intramuscular ziprasidone was assessed, have been reviewed (9). The most common adverse events in the 921 patients were nausea, headache, dizziness, anxiety, somnolence, insomnia, and injection-site pain; 1.1 to 6.1% withdrew because of treatment-related adverse events.

Comparative studies

Ziprasidone 20 mg (*n* = 30) has been compared with diazepam 10 mg (*n* = 30) and placebo (*n* = 30) in a randomized, parallel-group, double-blind study in non-psychotic subjects who were anxious before undergoing minor dental surgery (10). The peak anxiolytic effect of ziprasidone compared with placebo was similar to that of diazepam but had a later onset. However, at 3 hours after the dose, the anxiolytic effect of ziprasidone was significantly greater than that of placebo and somewhat greater than that of diazepam. The sedative effect of ziprasidone was never greater than that of placebo, whereas diazepam was significantly more sedative than placebo 1–1.5 hours after the

dose. Ziprasidone was generally well tolerated; only one patient reported treatment-related adverse events (nausea and vomiting) and, unlike diazepam, ziprasidone did not reduce the blood pressure. Dystonia, extrapyramidal syndrome, akathisia, and postural hypotension were not seen with ziprasidone. Ziprasidone may therefore have anxiolytic effects in addition to its neuroleptic properties.

Placebo controlled studies

Ziprasidone has been used in 28 children and adolescents (aged 7–17 years) with Tourette's syndrome in an 8-week pilot study (11). They were randomly assigned to ziprasidone (5–40 mg/day; *n* = 16) or placebo (*n* = 12). Ziprasidone significantly reduced tic frequency. There was one case each of somnolence and akathisia, both with the highest dose of ziprasidone; these were considered to be severe but did not necessitate withdrawal.

Ziprasidone is the first atypical antipsychotic drug to be available in both intramuscular and oral formulations in the USA. These two formulations have been compared with haloperidol in a 6-week, multicenter, parallel-group, flexibly dosed study in patients with an acute exacerbation of schizophrenia or schizoaffective disorder (12). They were randomized to ziprasidone (n = 427; intramuscularly for 3 days, then orally 40–80 mg bd) or haloperidol (n = 138; intramuscularly for 3 days then orally 5–20 mg/day). At the end of the intramuscular phase the patients who had receiving ziprasidone had significantly better Brief Psychiatric Rating Scale Total scores than those who had received haloperidol, but at end point there were no significant between-group differences. However, ziprasidone produced significantly greater improvement in negative subscale scores, both at the end of the intramuscular phase and at the end of the study. Haloperidol caused more extrapyramidal effects, including akathisia and movement disorders. At end of the study, mean weight change in the ziprasidone group was 0.25 kg compared with –0.15 kg in the haloperidol group; mean QT_c changes from baseline were +3.2 ms versus –3.5 ms respectively.

In a double-blind study, 144 agitated patients who required emergency sedation were randomized to ziprasidone 20 mg (n = 46), droperidol 5 mg (n = 50), or midazolam 5 mg (n = 48) (13). Those who were sedated with droperidol or ziprasidone required rescue medications to achieve adequate sedation less often than those who were sedated with midazolam; more remained agitated at 15 minutes after ziprasidone. There was akathisia in one patient who received droperidol, one who received midazolam (and subsequently droperidol rescue sedation), and four who received ziprasidone. There were no other adverse events.

Organs and Systems

Cardiovascular

Three extensive reviews of ziprasidone have devoted particular attention to the possibility of QT interval prolongation (14–16). Ziprasidone up to 160 mg/day prolongs the QT_c interval on average 5.9–9.7 ms (data from 4571 patients); a QT_c interval of over 500 ms was seen in two of 2988 ziprasidone recipients and in one of 440 placebo recipients. In an open study in 31 patients with schizophrenia, ziprasidone given for 21–29 days prolonged the QT_c interval by 20 ms (95% CI = 14, 26).

- A 38-year-old woman with a psychosis who took 4020 mg of ziprasidone had borderline intraventricular conduction delay (QRS duration 111 ms); the QT_c interval was 445 ms (17). She oscillated between being drowsy and calm, and alert and agitated; her blood pressure fell from 129/81 to 99/34 mmHg 4 hours later. She also had diarrhea and urinary retention.

The question of the effect of ziprasidone on the QT_c interval has been analysed in an extensive review (18). It is generally accepted that 440 ms is the upper limit of normality, and the authors concluded that ziprasidone clearly prolongs the QT_c interval, but that the clinical consequences of this effect are uncertain, and that so far no direct association with torsade de pointes, sudden death, or increased cardiac mortality has been observed. However, they provided recommendations about its use.

1. Before starting treatment, conditions that might predispose to a higher risk of QT_c interval prolongation or torsade de pointes (either cardiac, metabolic, or others) should be ruled out; a careful medical history is recommended.
2. In people with stress, shock, and extreme or prolonged physical exertion already taking ziprasidone, therapy can be continued, but electrocardiographic monitoring is advised when episodes are severe or prolonged.
3. Although no special precautions are suggested when ziprasidone is co-prescribed with metabolic inhibitors, drugs that do not inhibit hepatic enzymes should be preferred.
4. Ziprasidone should be avoided in the following cases:
   a. when there is evidence of long QT syndrome, a history of myocardial infarction or ischemic heart disease, and persistent or recurrent bradycardia; a cardiologist should be consulted if uncertain.
   b. in conditions often associated with electrolyte disturbance, including anorexia or bulimia; electrolyte disturbance should be always ruled out by means of a blood sample, and low concentrations of calcium, potassium, or magnesium should be corrected before treatment.
   c. in patients taking other drugs that prolong the QT interval, including certain antidysrhythmic drugs, antidepressants, antihistamines, antimicrobial drugs, and others; if co-prescription is unavoidable, electrocardiographic monitoring by a specialist is advised.
   d. in patients taking thioridazine, droperidol, sertindole, or pimozide; these drugs should be always withdrawn before starting ziprasidone.
   e. in patients taking other antipsychotic drugs; this is permissible only when cross-tapering and excessive doses of antipsychotic drugs should not be used in combination.

Ziprasidone is probably the only antipsychotic drug *not* to be unequivocally associated with weight gain, which might result in a relatively reduced effect on cardiac mortality compared with some other antipsychotic drugs. However, reports of cardiovascular adverse events led the FDA to include a "black box" warning in the official labeling of the product. Furthermore, in a warning letter the FDA stated that the manufacturers, Pfizer Inc, had promoted the product in a misleading manner, because they had minimized the greater capacity of ziprasidone to cause QT prolongation, as well as its potential to cause torsade de pointes and sudden death (19). The Indications and Usage section of the manufacturers' approved product labelling stated that ziprasidone has a "greater capacity to prolong the QT/QT_c interval compared with several other antipsychotic drugs" and that this effect "is associated in some other drugs with the ability to cause torsade de pointes-type arrhythmia, a potentially fatal polymorphic ventricular tachycardia, and sudden death. Whether ziprasidone will cause torsade de pointes or increase the rate of sudden death is not yet known." Although torsade de pointes was not observed in pre-marketing studies, experience is too limited to rule out an increased risk. Furthermore, the FDA has received several spontaneous reports of QT interval prolongation greater than 500 ms, all suggestive of a potential risk of this dysrhythmia. There are also reports of sudden death of unknown cause, which could have been due to unrecognized torsade de pointes.

Nervous system

In a randomized, Phase III, double-blind study, ziprasidone 80 mg/day and 160 mg was more effective than placebo in patients with acute exacerbations of schizophrenia or schizoaffective disorders (n = 302) (20). After 6 weeks, somnolence (19%) and akathisia (13%) were more frequent with ziprasidone 160 mg than with placebo (5 and 7% each). Benzatropine was required at some time during the study by 20% of the patients taking ziprasidone 80 mg/day, 25% of those taking ziprasidone 160 mg/day, and 13% of those taking placebo. The long-term safety of ziprasidone is unknown.

Tardive dyskinesia has been associated with ziprasidone in a 49-year-old man with bipolar disorder (21) and in a 70-year-old woman within 2 months of starting ziprasidone treatment (22).

Neuroleptic malignant syndrome has been reported in a 52-year-old woman with Parkinson's disease and psychotic symptoms who took ziprasidone (23).

Psychiatric

Three cases of hypomania in patients with depression have been reported (24).

In four other cases the symptoms occurred within 7 days of the start of ziprasidone therapy and improved substantially when the drug dosage was lowered or the drug was withdrawn (25).

- An association between mania and ziprasidone has been suggested in a 20-year-old man taking 160 mg at bedtime who suddenly had increased energy, elated mood, and agitation (26). The symptoms resolved after 48 hours when ziprasidone was substituted by olanzapine.

Metabolism

Ziprasidone is said to be associated with less weight gain than the other atypical neuroleptic drugs and than most typical ones (1,2).

However, significant weight gain can occur.

- A 12-year-old boy had significant weight gain within 3 months of starting to take ziprasidone, from 63 kg (BMI = 26.5) before treatment to 69 kg (BMI = 28.2) after treatment (27).

Musculoskeletal

Rhabdomyolysis with pancreatitis and hyperglycemia has been reported in a middle-aged woman with schizoaffective disorder (28).

- A 50-year-old man who was treated with ziprasidone 40 mg bd for 3 weeks had a substantial rise in creatine kinase activity without any evidence of muscle trauma, stiffness, or swelling or any signs of neuroleptic malignant syndrome (29). There was no renal insufficiency or compartment syndrome.

The authors suggested that ziprasidone may enhance muscle cell permeability leading to rhabdomyolysis.

Sexual function

Priapism has been attributed to ziprasidone.

- A 32-year-old patient with schizophrenia taking ziprasidone 40 mg bd developed several spontaneous involuntary erections, which lasted about 20–30 minutes and did not resolve with ejaculation (30). No physical or laboratory abnormalities were found.

A possible explanation of this event would be that novel antipsychotic drugs are antagonists at α_1-adrenoceptors with very high affinities.

Drug administration

Drug overdose

Various reports of overdose of ziprasidone have been published. In a retrospective chart review of isolated cases reported to a Poisons Center during 2001 to 2003, 30 patients met the criteria among about 150 000 exposures that were reviewed (31). Eight patients accidentally took ziprasidone and 22 intentionally. The average dose of ziprasidone was 205 mg (recommended dose 80–160 mg/day) and only one required significant medical intervention.

- A 17-year-old man with a 5-year history of severe depression took 120 tablets of ziprasidone 20 mg (2400 mg), 15–20 tablets of bupropion SR 150 mg (2250–3000 mg), 15 tablets of clonazepam 0.5 mg, and 4 tablets of lorazepam 0.5 mg (32). He was somnolent but responded to vocal commands. Over the next 45

minutes he became lethargic and was intubated. The initial electrocardiogram 1 hour after ingestion was normal, with a QT_c interval of less than 440 ms, but 2.5 hours after ingestion he developed a widened QRS interval (200 ms), which resolved with intravenous lidocaine 75 mg. A subsequent 12-lead electrocardiogram showed a QT_c interval of 480 ms and a QRS interval of 120 ms. He was alkalinized to maintain a pH of 7.45–7.50. The QT_c interval varied from 420 to 480 ms. About 40 hours after ingestion, the QT_c interval finally stabilized at around 440 ms.

- In a 37-year-old woman, ingestion of 1200 mg of ziprasidone was associated with a QT_c interval of 459 ms (33).
- In a 17-month-old girl who was given ziprasidone 400 mg, the QT_c interval was 480 ms (Bryant 81).
- In a 50-year-old woman, 1760 mg of ziprasidone and 1000 of quetiapine caused an increase in the QT_c interval to 638 ms (Bryant 81).

All the last three patients had somnolence, which resolved completely in a few days.

Drug–Drug Interactions

Pfizer, the marketing authorization holder of ziprasidone, has promoted several pharmacokinetic studies. Oral contraceptives (ethinylestradiol 30 µg/day plus levonorgestrel 150 µg/day) (34), lithium 900 mg/day (35), ketoconazole 400 mg/day (36), and carbamazepine (100–400 mg/day) (37) had no effects on the pharmacokinetics of ziprasidone (40 mg/day).

Ketoconazole

Ziprasidone is oxidatively metabolized by CYP3A4, but it does not inhibit CYP3A4 or other isoenzymes at clinically relevant concentrations. The effect of ketoconazole 400 mg qds for 6 days on the single-dose pharmacokinetics of ziprasidone 40 mg has been evaluated in an open, placebo-controlled, crossover study in healthy volunteers (38). Ketoconazole caused a modest increase in the mean AUC (33%) and the mean C_{max} (34%) of ziprasidone. This effect was not considered clinically relevant and suggests that other inhibitors of CYP3A4 are unlikely to affect the pharmacokinetics of ziprasidone significantly. Most of the reported adverse events were mild. The adverse events that were most commonly reported in subjects who took the drugs concomitantly were dizziness, weakness, and somnolence. There were no treatment-related laboratory abnormalities or abnormal vital signs during the study and at the 6-day follow-up evaluation.

Lithium

In 34 healthy men, ziprasidone did not alter serum lithium concentrations or renal lithium clearance (39).

In a placebo-controlled, open-label study in 25 healthy subjects there were no changes in serum lithium concentration or renal lithium clearance when ziprasidone (40–80 mg/day) was added for 7 days (39).

References

1. Buckley PF. Ziprasidone: pharmacology, clinical progress and therapeutic promise. Drugs Today 2000;36:583–9.

2. Daniel DG, Copeland LF. Ziprasidone: comprehensive overview and clinical use of a novel antipsychotic. Expert Opin Investig Drugs 2000;9(4):819–28.

3. Mamo D, Kapur S, Shammi C, Papatheodorou G, Mann S, Therrien F, Remington G. A pet study of dopamine D2 and serotonin 5-HT2 receptor occupancy in patients with schizophrenia treated with therapeutic doses of ziprasidone. Am J Psychiatry 2004;161:818–25.

4. Editorial. Ziprasidone: zealous in the treatment of schizophrenia and schizoaffective disorder. Drug Ther Perspect 2003;19:1–4.

5. Weiden PJ, Simpson GM, Potkin SG, O'Sullivan RL. Effectiveness of switching to ziprasidone for stable but symptomatic outpatients with schizophrenia. J Clin Psychiatry 2003;64:580–8.

6. Weiden PJ, Daniel DG, Simpson G, Romano SJ. Improvement in indices of health status in outpatients with schizophrenia switched to ziprasidone. J Clin Psychopharmacol 2003;23:595–600.

7. Gómez-Esteban JC, Zarranz JJ, Velasco F, Lezcano E, Lachen MC, Rouco I, Barcena J, Boyero S, Ciordia R, Allue I. Use of ziprasidone in parkinsonian patients with psychosis. Clin Neuropharmacol 2005;28:111–4.

8. Ziegenbein M, Kropp S, Kuenzel HE. Combination of clozapine and ziprasidone in treatment-resistant schizophrenia: an open clinical study. Clin Neuropharmacol 2005;28:220–4.

9. Preskorn SH. Pharmacokinetics and therapeutics of acute intramuscular ziprasidone. Clin Pharmacokinet 2005;44:1117–33.

10. Wilner KD, Anziano RJ, Johnson AC, Miceli JJ, Fricke JR, Titus CK. The anxiolytic effect of the novel antipsychotic ziprasidone compared with diazepam in subjects anxious before dental surgery. J Clin Psychopharmacol 2002;22(2):206–10.

11. Sallee FR, Kurlan R, Goetz CG, Singer H, Scahill L, Law G, Dittman VM, Chappell PB. Ziprasidone treatment of children and adolescents with Tourette's syndrome: a pilot study. J Am Acad Child Adolesc Psychiatry 2000;39(3):292–9.

12. Brook S, Walden J, Benattia I, Siu CO, Romano SJ. Ziprasidone and haloperidol in the treatment of acute exacerbation of schizophrenia and schizoaffective disorder: comparison of intramuscular and oral formulations in a 6-week, randomized, blinded-assessment study. Psychopharmacology 2005;178:514–23.

13. Martel M, Sterzinger A, Miner J, Clinton J, Biros M. Management of acute undifferentiated agitation in the emergency department: a randomized double-blind trial of droperidol, ziprasidone, and midazolam. Acad Emerg Med 2005;12:1167–72.

14. Stimmel GL, Gutierrez MA, Lee V. Ziprasidone: an atypical antipsychotic drug for the treatment of schizophrenia. Clin Ther 2002;24(1):21–37.

15. Gunasekara NS, Spencer CM, Keating GM. Ziprasidone: a review of its use in schizophrenia and schizoaffective disorder. Drugs 2002;62(8):1217–51.

16. Caley CF, Cooper CK. Ziprasidone: the fifth atypical antipsychotic. Ann Pharmacother 2002;36(5):839–51.

17. House M. Overdose of ziprasidone. Am J Psychiatry 2002;159(6):1061–2.

18. Taylor D. Ziprasidone in the management of schizophrenia. The QT interval issue in context. CNS Drugs 2003;17:423–30.

19. FDA warning letter. Pfizer Inc. Geodon (ziprasidone HCl). http://www.pharmcast.com/WarningLetters/Yr2002.

20. Daniel DG, Zimbroff DL, Potkin SG, Reeves KR, Harrigan EP, Lakshminarayanan M. Ziprasidone 80 mg/day and 160 mg/day in the acute exacerbation of schizophrenia and schizoaffective disorder: a 6-week placebo-controlled trial Neuropsychopharmacol 1999;20:491–505.

21. Rosenquist KJ, Walker SS, Ghaemi SN. Tardive dyskinesia and ziprasidone. Am J Psychiatry 2002;159(8):1436.

22. Keck M, Müller M, Binder E, Sonntag A, Holsboer F. Ziprasidone-related tardive dyskinesia. Am J Psychiatry 2004;161:175–6.

23. Gray N. Ziprasidone-related neuroleptic malignant syndrome in a patient with parkinson's disease: a diagnostic challenge. Hum Psychopharmacol 2004;19:205–7.

24. Davis R, Risch SC. Ziprasidone induction of hypomania in depression? Am J Psychiatry 2002;159(4):673–4.

25. Baldassano CF, Ballas C, Datto SM, Kim D, Littman L, O'Reardon J, Rynn MA. Ziprasidone-associated mania: a case series and review of the mechanism. Bipolar Disord 2003;3:72–5.

26. Nolan BP, Schulte JJ. Mania associated with initiation of ziprasidone. J Clin Psychiatry 2003;64:336.

27. Jaworowski S, Hauser S, Mergui J, Hirsch H. Ziprasidone and weight gain. Clin Neuropharmacol 2004;27:99–100.

28. Yang SH, McNeely MJ. Rhabdomyolysis, pancreatitis, and hyperglycemia with ziprasidone. Am J Psychiatry 2002;159(8):1435.

29. Zaidi AN. Rhabdomyolysis after correction of hyponatremia in psychogenic polydipsia possibly complicated by ziprasidone. Ann Pharmacother 2005;39:1726–31.

30. Reeves RR, Kimble R. Prolonged erections associated with ziprasidone treatment: a case report. J Clin Psychiatry 2003;64:97–8.

31. LoVecchio F, Watts D, Eckholdt P. Three-year experience with ziprasidone exposures. Am J Emerg Med 2005;23:586–7.

32. Biswas AK, Zabrocki LA, Mayes KL, Morris-Kukoski CL. Cardiotoxicity associated with intentional ziprasidone and bupropion overdose. J Toxicol Clin Toxicol 2003;41:79–82.

33. Bryant SM, Zilberstein J, Cumpston KL, Magdziarz DD, Costerisan D. A case series of ziprasidone overdoses. Vet Hum Toxicol 2003;45:81–2.

34. Muirhead GJ, Harness J, Holt PR, Oliver S, Anziano RJ. Ziprasidone and the pharmacokinetics of a combined oral contraceptive. Br J Clin Pharmacol 2000;49(Suppl. 1):S49–56.

35. Apseloff G, Mullet D, Wilner KD, Anziano RJ, Tensfeldt TG, Pelletier SM, Gerber N. The effects of ziprasidone on steady-state lithium levels and renal clearance of lithium. Br J Clin Pharmacol 2000;49(Suppl. 1):S61–4.

36. Miceli JJ, Smith M, Robarge L, Morse T, Laurent A. The effects of ketoconazole on ziprasidone pharmacokinetics—a placebo-controlled crossover study in healthy volunteers. Br J Clin Pharmacol 2000;49(Suppl. 1):71S–6S.

37. Miceli JJ, Anziano RJ, Robarge L, Hansen RA, Laurent A. The effect of carbamazepine on the steady-state pharmacokinetics of ziprasidone in healthy volunteers. Br J Clin Pharmacol 2000;49(Suppl. 1):65S–70S.

38. Miceli JJ, Smith M, Robarge L, Morse T, Laurent A. The effects of ketoconazole on ziprasidone pharmacokinetics—a placebo-controlled crossover study in healthy volunteers. Br J Clin Pharmacol 2000;49(Suppl 1):S71–6.

39. Apseloff G, Mullet D, Wilner KD, Anziano RJ, Tensfeldt TG, Pelletier SM, Gerber N. The effects of

ziprasidone on steady-state lithium levels and renal clearance of lithium. Br J Clin Pharmacol 2000;49(Suppl 1):61S–4S.

Zotepine

General Information

Zotepine is a dibenzothiepine neuroleptic drug, an antagonist at D_1 dopamine receptors and at $5\text{-}HT_1$ and $5\text{-}HT_2$ receptors (1). It rarely causes extrapyramidal disturbances, and when they do occur they are usually mild.

The efficacy and safety of zotepine have been explored in a 1-year open study in 253 patients with schizophrenia (mean age 38 years, range 18–65) who took zotepine 75–450 mg/day (2). The mean total BPRS score was reduced from 52 at baseline to 41. Since concomitant treatment was allowed, 173 patients reported 205 ongoing and 448 new neuroleptic medicaments during the study. A total of 826 adverse events were reported by 220 patients; 50 had serious adverse events and 5 died during the study, 2 taking zotepine; one death was a suicide and the other was due to a ventricular dysrhythmia. In all, 138 patients (55%) withdrew from the study; 60 withdrawals were due to adverse events. The most frequently reported adverse events were weight gain (28%; mean weight gain 4.3 kg), somnolence (15%), and weakness (13%). There were adverse events that could be related to extrapyramidal effects in 5%. In 14 patients with normal baseline electrocardiography there were abnormalities at the end of the study, most commonly sinus tachycardia; there were no reports of torsade de pointes. No clinically important hematological abnormalities have been reported to date.

References

1. Ackenheil M. Das biochemische Wirkprofil von Zotepin im Vergleich zu anderen Neuroleptika. [The biochemical effect profile of zotepine in comparison with other neuroleptics.] Fortschr Neurol Psychiatr 1991;59(Suppl. 1):2–9.
2. Palmgren K, Wighton A, Reynolds CW, Butler A, Tweed JA, Raniwalla J, Welch CP, Bratty JR. The safety and efficacy of zotepine in the treatment of schizophrenia: results of a one-year naturalistic clinical trial. Int J Psychiatry Clin Pract 2000;4:299–306.

Zuclopenthixol

General Information

Zuclopenthixol is a thioxanthene neuroleptic drug.

Organs and Systems

Hematologic

Neutropenia has been associated with zuclopenthixol (1).

- A 66-year-old man with schizophrenia took zuclopenthixol 10 mg tds for 18 days and developed a mild leukopenia $(2.9 \times 10^9/l)$ and thrombocytopenia $(109 \times 10^9/l)$. He was asymptomatic, with no evidence of infection or a bleeding tendency. Zuclopenthixol was withdrawn, without any change in the rest of his drug therapy (glibenclamide 5 mg tds, biperiden 2 mg bd, oxazepam 10 mg tds, dipyridamole 75 mg tds, and ranitidine 150 mg/day). The leukocyte and platelet counts rose over the next 5 days.

Sexual function

Priapism, although infrequent, can occur during treatment with neuroleptic drugs and necessitates prompt urological consultation and sometimes even surgical intervention (SEDA-14, 149). A case has been associated with zuclopenthixol (2).

- A 31-year-old man developed priapism after taking zuclopenthixol 30 mg/day for 8 days, the dose having been increased to 75 mg the day before, while he was still taking oral carbamazepine 600 mg/day and clorazepate dipotassium 30 mg/day. He had a history of perinatal anoxic encephalopathy with severe motor sequelae and dyslalia, alcohol dependence, and a personality disorder. On the day before the priapism occurred, he had been physically restrained and given an extra dose of intramuscular clorazepate dipotassium 50 mg. When priapism occurred, all drugs except clorazepate were withdrawn and about 6 hours later the corpora cavernosa were washed and infused with noradrenaline in glucose (8 doses of 40 µg), after which the priapism resolved.

References

1. Coutinho E, Fenton M, Adams C, Campbell C. Zuclopenthixol acetate in psychiatric emergencies: looking for evidence from clinical trials. Schizophr Res 2000;46(2–3):111–8.
2. Salado J, Blazquez A, Diaz-Simon R, Lopez-Munoz F, Alamo C, Rubio G. Priapism associated with zuclopenthixol. Ann Pharmacother 2002;36(6):1016–8.

HYPNOSEDATIVES

BENZODIAZEPINES

General Information

The benzodiazepines typically share hypnotic, anxiolytic, myorelaxant, and anticonvulsant activity. Because their efficacy and tolerability are generally good, especially in the short term, they have been used extensively and are likely to continue to be used for many years to come. However, their less specific use in the medically or psychiatrically ill, and in healthy individuals experiencing the stresses of life or non-specific symptoms has often been inappropriate and sometimes dangerous (SEDA-18, 43). The pharmacoepidemiology of benzodiazepine use has been carefully studied in various countries (1), including the USA (2) and France, where 7% of the adult population (17% of those over 65 years) are regular users (3). In Italy, consumption of benzodiazepines remained stable (50 defined daily doses per 1000 population) from 1995 to 2003, while expenditure increased by 43% to €565M per annum (4). The need to limit spending on pharmaceutical products, as well as the very real likelihood of inducing iatrogenic disease (for example cognitive impairment, accidents, drug dependence, withdrawal syndromes), has prompted many reviews and policy statements aimed at discouraging inappropriate use. Despite this, the available evidence suggests that there continues to be expensive and inappropriate use in several countries (5).

A comprehensive review of manufacture, distribution, and use has described the rather marked international variation in use of the drugs and the role of the International Narcotics Control Board, a United Nations agency, in the restriction of these drugs (1). Most countries are signatories to the UN Convention on Psychotropic Substances 1971, and are thus obliged to implement controls on the international trade in abusable drugs, including benzodiazepines. Some countries, such as Australia and New Zealand, have imposed further stringent controls on certain drugs, such as flunitrazepam, which are thought to have particular abuse liability (6). A wide-ranging discussion of benzodiazepine regulation has pointed out both the potential merits of the approach and the fact that some restrictions in the past have turned out to be counterproductive (7).

A comprehensive review of benzodiazepine-induced adverse effects and liability to abuse and dependence, in which it was concluded that most benzodiazepine use is both appropriate and helpful (2), has been challenged (8–10). Balanced clinical reviews of benzodiazepine use (11,12) include sets of recommendations on appropriate prescribing and avoiding adverse effects, including tolerance/dependence. Similarly, guidelines for the management of insomnia and the judicious use of hypnotics have been reviewed (13,14). Benzodiazepines are often over-prescribed in hospital (SEDA-17, 42), and their continued prescription after discharge constitutes a significant source of long-term users. Anxiety symptoms and insomnia are common in the medically ill population and can be due to specific physical causes, a reaction to illness, or a co-morbid psychiatric illness, such as depression. Moreover, caffeine, alcohol, nicotine, and a variety of medications can cause insomnia (15). Accordingly, the systematic assessment of such patients allows remediable causes to be identified and the use of hypnosedatives to be minimized. Elderly people and medically ill patients are susceptible to the adverse effects of benzodiazepines, and alternatives are worth considering (11,16), particularly given evidence that behavioral therapies can be more effective and more durable than drug therapy (17).

The important advantages of the benzodiazepines over their predecessors are that they cause relatively less psychomotor impairment, drowsiness, and respiratory inhibition, and are consequently relatively safe in overdose. However, it must be emphasized that these advantages are relative, and that the low toxicity potential does not apply when they are combined with other agents, particularly alcohol (18) and opioids (19).

As well as the added toxicity seen in co-administration with other CNS depressants, benzodiazepines facilitate self-injurious behavior by disinhibiting reckless or suicidal impulses (20). Benzodiazepines are commonly used in both attempted and completed suicide (21). A German study has suggested that hypnosedatives are the commonest drugs used in self-poisoning, that most are prescribed by physicians, and that in nearly half of those taking them chronically, adverse effects were considered to be a possible cause of self-poisoning (22). Before prescribing any drugs of this class, clinicians are exhorted to assess both suicidality and alcohol problems; there is a quick screen for the latter, the Alcohol Use Disorders Identification Test (AUDIT), which consists of a 10-item questionnaire and an 8-item clinical procedure (23).

Hypersensitivity reactions are rare. A few cases of anaphylaxis have been described, although usually these have been with the injectable forms and may have involved the stabilizing agents (24). Serious skin reactions to clobazam (SEDA-21, 38) and tetrazepam (25) have been reported. Lesser reactions have also been reported with diazepam, clorazepate (via N-methyldiazepam) (26), and midazolam (SEDA-17, 44, 45).

Tumor-inducing effects have been observed in animals (SEDA-6, 39), but human reports are essentially negative.

First-trimester exposure appears to confer a small but definite increased risk (from a baseline of 0.06% up to 0.7%) of oral cleft in infants (27). However, second-generation effects are infrequent and usually reversible (28), although some doubt remains about the extent of developmental delay in children who have been exposed in utero (27). A review has emphasized that concerns about second-generation effects are mainly theoretical, and has concluded that some agents (for example chlordiazepoxide) are probably safe during pregnancy and lactation and that others (for example alprazolam) are best avoided (29).

Pharmacokinetics

As far as is currently known, benzodiazepines and similar drugs (zopiclone, zolpidem) act by a single mechanism, interacting at the GABA receptor complex to enhance the ability of GABA to open a chloride ion channel and thereby hyperpolarize the neuronal membrane. It is usual, therefore, to classify benzodiazepines, and recommend their clinical use, on the basis of their duration of action or their half-life. While this is without doubt a useful classification, it is simplistic and does not take into account other important pharmacokinetic factors.

The first factor that is considered significant is the metabolism of benzodiazepines to pharmacologically active metabolites. Many newer benzodiazepines intended for use as long-acting anxiolytic or sedative agents were in fact intended to be so metabolized to ensure stable blood concentrations over prolonged periods. Drugs with long durations of effect, attributable at least in part to the formation of active metabolites, are listed in Table 1. Individuals vary considerably in their metabolism of benzodiazepines, and interpatient variation in concentrations of the parent compounds and of (generally active) metabolites is usual. In addition, ethnicity plays a major role in determining the frequency of poor and extensive metabolizers, with notable differences between Caucasians and East Asians (30).

Another, often neglected, aspect of the pharmacokinetics of benzodiazepines is their rate of onset of action, since their properties and therapeutic benefits depend to a considerable degree on the rapidity of onset of their perceived effects. Within a given drug class, the more rapidly the hypnotic effect occurs, the greater the abuse potential. For most drugs of abuse, it is the affective and behavioral changes associated with a rapid rise in drug blood concentration that is sought, whether the drug is abused by intravenous injection, nasal or bronchial absorption, or (as with alcohol) rapid oral absorption from an empty stomach (31). Diazepam and flunitrazepam are effective hypnotics because they are rapidly absorbed and there is a quick rise in blood concentrations, even though after tissue redistribution and loss of their immediate effects they have long half-lives. It also explains the preference, and so the increased liability for abuse, for drugs like diazepam (31) and flunitrazepam, especially when the latter is snorted (32). In general, polar molecules, such as lorazepam, oxazepam, and temazepam (all of which have a hydroxyl group), gain access to the CNS more slowly than their more lipophilic cousins. Since temazepam is much more quickly absorbed from a soft gelatin liquid-containing capsule than from a hard capsule or tablet, it is the preferred form for both hypnotic use (Table 2) and recreational use (and for this reason is restricted in some countries). Kinetic differences between drugs and between formulations partially explain why comparing equipotent doses of benzodiazepines is difficult.

The route of metabolism can also be significant, particularly in those with liver disease or who are taking concomitant hepatic enzyme inhibitors, such as erythromycin (SEDA-20, 31). The complex interaction between hepatic dysfunction and benzodiazepines has been reviewed (33); these drugs more readily affect liver function in individuals with liver disease and may also directly contribute to hepatic encephalopathy, as shown by the ability of benzodiazepine antagonists to reverse coma transiently in such patients (33). Elderly people appear to be at increased risk only if they are physically unwell, and particularly if they are taking many medications.

Rapid absorption, often followed by rapid redistribution to tissue stores with consequent falls in brain and blood drug concentrations, plays a significant role in the quick onset and cessation of perceived effects, but long-term actions, for example mild sedative and antianxiety effects, are a consequence of slow hepatic clearance, either by hydroxylation and subsequent conjugation to a glucuronide or by microsomal metabolism to other possibly pharmacologically active metabolites. Agents that are subject to microsomal metabolism and/or oxidation accumulate more rapidly in patients with reduced liver function (for example frail elderly people); only the metabolism of drugs such as oxazepam, lorazepam, and temazepam, which predominantly undergo glucuronidation, is not affected by liver function (Table 3) nor suffer from interference by drugs, such as cimetidine, estrogens, or erythromycin, which compete for the enzyme pathways (see Drug–Drug Interactions).

Pharmacodynamics

The use of techniques of molecular biology to clone the benzodiazepine receptor and the other components of the GABA receptor/chloride channel complex has shown that there are likely to be many variant forms of the receptor, owing to the multiplicity of protein subunits that constitute it. This has given rise to the hope that more selective agonist drugs, for example the "Z drugs" (zaleplon, zolpidem, and zopiclone), may produce fewer adverse effects (35,36); however, this hope appears to have been over-optimistic (10). There are also many ways in which different drugs interact with receptor sites to produce their effects, including agonism, partial agonism, antagonism, inverse agonism (contragonism), and even partial inverse agonism; this increases the complexity considerably. The suggestion that partial agonists (such as alpidem and abecarnil) have greater anxiolytic than sedative potency (37), or that they will be less likely to give rise to abuse (35) or dependence (38), is yet to be established.

Table 1 Benzodiazepines with active metabolites that have long half-lives; metabolism and predominant metabolite half-lives

To desmethyldiazepam*	$t_{1/2}$ (hours)	To other metabolites	$t_{1/2}$ (hours)
Lorazepate	40–100	Chlordiazepoxide	40–100
Diazepam	36	Clobazam	30–150
Halazepam	20	Flurazepam	40–120
Medazepam	2	Quazepam	40–75
Prazepam	120+		

*Half-life about 60 hours.

Table 2 Rates of absorption and half-lives of benzodiazepines

	t_{max} (hours)	$t_{1/2}$ (hours)
Slow absorption		
Clonazepam	2–4	20–40
Loprazolam	2–5	5–15
Lorazepam	2	10–20
Oxazepam	2	5–15
Temazepam (hard capsules)	3	8–20
Intermediate absorption and elimination		
Alprazolam	1–2	12–15
Bromazepam	1–4	10–25
Chlordiazepoxide	1–2	10–25
Intermediate absorption, slow elimination (with active metabolites)		
Flurazepam	1.5	40–120
Clobazam	1–2	20–40
Chlorazepate	1	40–100
Quazepam	1.5	15–35
Rapid absorption, slow elimination, but rapid redistribution		
Diazepam	1	20–70
Flunitrazepam	1	10–40
Nitrazepam	1	20–30
Rapid absorption, rapid elimination, rapid redistribution		
Lormetazepam (soft capsules)	1	8–20
Temazepam (soft capsules)	1	8–20
Rapid absorption, rapid elimination		
Brotizolam	1	4–7
Zolpidem	1.5	2–5
Zopiclone	1.5	5–8
Rapid absorption, very rapid elimination		
Midazolam	0.3	1–4
Triazolam	1	2–5

Table 3 Predominant metabolic pathways for benzodiazepines and related agonists

Via CYP3A oxidation	*Via glucuronidation*
Alprazolam	Lorazepam
Anidazolam	Oxazepam
Bromazepam	Temazepam
Brotizolam	
Chlordiazepoxide	
Clobazam (also CYP2C19)	
Clonazepam	
Chlorazepate	
Diazepam (also CYP2C19)	
Estazolam	
Flunitrazepam	
Flurazepam	
Halazepam	
Loprazolam	
Lormetazepam	
Medazepam	
Midazolam	
Nitrazepam	
Prazepam	
Quazepam	
Triazolam	
Zaleplon	
Zolpidem	
Zopiclone	

Medicolegal considerations

Medicolegal problems, especially with the use of triazolam, have been discussed (SEDA-13, 33); debate continues on the interpretation of evidence that points to an increased incidence of adverse behavioral effects with triazolam (39), flunitrazepam, and other short-acting high-potency agents (12). A review has highlighted a substantial rate (0.3–0.7%) of aggressive reactions to benzodiazepines, and the fact that a majority so affected may have intended a disinhibitory effect, with clear forensic implications (40). High rates of benzodiazepine consumption, much of it illicit, continue in prison populations.

Efforts to restrict benzodiazepines in New York State (2,7) and to ban triazolam (Halcion) in the UK (7,41) and the Netherlands (SEDA-4,v; 42) have likewise been fraught with controversy. For example, the requirement to use triplicate prescription forms in New York has been effective in reducing prescription volumes, including arguably necessary and appropriate prescriptions (43). A 1979 suspension of triazolam availability in the Netherlands was overturned in 1990, while a 1993 formal ban in the UK has remained in force (44). Two extensive reports have included recommendations for resolving the special problems posed by the Halcion controversy (44,45).

General adverse effects

Benzodiazepines have a high therapeutic index of safety, with little effect on most systems (other than the CNS) in high doses. However, their toxicity increases markedly when they are combined with other CNS depressant drugs, such as alcohol or opioid analgesics. Medically ill and brain injured patients are particularly susceptible to adverse neurological or behavioral effects (SEDA-18, 43; SEDA-20, 30; 46).

The most frequent adverse effect which occurs in at least one-third of patients is drowsiness, often accompanied by incoordination or ataxia. Problems with driving, operating machinery, or falls can result, particularly in the elderly, and can be an important source of morbidity, loss of physical function, and mortality (47,48). Memory impairment, loss of insight, and transient euphoria are common; "paradoxical" reactions of irritability or aggressive behavior have been well documented (11) and appear to occur more often in individuals with a history of impulsiveness or a personality disorder (40), and in the context of interpersonal stress and frustration (49). Tolerance to the sedative and hypnotic effects generally occurs more rapidly than to the anxiolytic or amnestic effects (1).

Physical dependence on benzodiazepines is recognized as a major problem, and occurs after relatively short periods of treatment (50,51), particularly in patients with a history of benzodiazepine or alcohol problems. Abrupt withdrawal can cause severe anxiety, perceptual changes, convulsions, or delirium. It can masquerade as a return of the original symptoms in a more severe form (rebound), or present with additional features (SEDA-17, 42; 11). Up to 90% of regular benzodiazepine users have adverse symptoms on withdrawal. The differences between rebound, withdrawal syndrome, and recurrence have been reviewed in detail (3).

Rebound insomnia or heightened daytime anxiety can occur, particularly after short-acting benzodiazepine hypnotics (12,52,53), and constitute a major reason for continuing or resuming drug use (11).

The use of intravenous benzodiazepines administered by paramedics for the treatment of out-of-hospital status epilepticus has been evaluated in a double-blind, randomized trial in 205 adults (54). The patients presented either with seizures lasting 5 minutes or more or with repetitive generalized convulsive seizures, and were randomized to receive intravenous diazepam 5 mg, lorazepam 2 mg, or placebo. Status epilepticus was controlled on arrival at the hospital in significantly more patients taking benzodiazepines than placebo (lorazepam 59%, diazepam 43%, placebo 21%). The rates of respiratory or circulatory complications related to drug treatment were 11% with lorazepam, 10% with diazepam, and 23% with placebo, but these differences were not significant.

Intranasal midazolam 0.2 mg/kg and intravenous diazepam 0.3 mg/kg have been compared in a prospective randomized study in 47 children (aged 6 months to 5 years) with prolonged (over 10 minutes) febrile seizures (55). Intranasal midazolam controlled seizures significantly earlier than intravenous diazepam. None of the children had respiratory distress, bradycardia, or other adverse effects. Electrocardiography, blood pressure, and pulse oximetry were normal in all children during seizure activity and after cessation of seizures.

Organs and Systems

Cardiovascular

Hypotension follows the intravenous injection of benzodiazepines, but is usually mild and transient (SED-11, 92) (56), except in neonates who are particularly sensitive to this effect (57). Local reactions to injected diazepam are quite common and can progress to compartment syndrome (SEDA-17, 44). In one study (58), two-thirds of the patients had some problem, and most eventually progressed to thrombophlebitis. Flunitrazepam is similar to diazepam in this regard (59). Altering the formulation by changing the solvent or using an emulsion did not greatly affect the outcome (60). Midazolam, being water-soluble, might be expected to produce fewer problems; in five separate studies there were no cases of thrombophlebitis, and in two others the incidence was 8–10%, less than with diazepam but similar to thiopental and saline (61).

Respiratory

Respiratory depression has been reported as the commonest adverse effect of intravenous diazepam (56), especially at the extremes of age. Midazolam has similar effects (62). All benzodiazepines can cause respiratory depression, particularly in bronchitic patients, through drowsiness and reduction in exercise tolerance (63). Rectal administration of, for example, diazepam can offer advantages in unconscious or uncooperative patients, and is less likely than parenteral administration to produce respiratory depression.

A previous report that rectal and intravenous diazepam can cause respiratory depression in children with seizures (SEDA-24, 84) has been challenged (64,65). The authors of the second comment stated that this complication does not occur when rectal diazepam gel is used without other benzodiazepines; they also recommended that during long-term therapy families should be instructed not to give rectal diazepam more than once every 5 days or five times in 1 month.

In a prospective study of children admitted to an accident and emergency department because of seizures, there were 122 episodes in which diazepam was administered rectally and/or intravenously; there was respiratory depression in 11 children, of whom 8 required ventilation (66). The authors questioned the use of rectal or intravenous diazepam as first-line therapy for children with acute seizures.

Nervous system

Falls

The role of different types of benzodiazepines in the risk of falls in a hospitalized geriatric population has been examined in a prospective study of 7908 patients, consecutively admitted to 58 clinical centers during 8 months

(67). Over 70% of the patients were older than 65 years, 50% were women, and 24% had a benzodiazepine prescription during the hospital stay. The findings suggested that the use of benzodiazepines with short and very short half-lives is an important and independent risk factor for falls. Their prescription for older hospitalized patients should be carefully evaluated.

In a case-control study using the Systematic Assessment of Geriatric Drug Use via Epidemiology (SAGE) database, the records of 9752 patients hospitalized for fracture of the femur during the period 1992–1996 were extracted and matched by age, sex, state, and index date to the records of 38 564 control patients (68). Among older individuals, the use of benzodiazepines slightly increased the risk of fracture of the femur. Overall, non-oxidative benzodiazepines do not seem to confer a lower risk than oxidative agents. However, the latter may be more dangerous among very old individuals (85 years of age or older), especially if used in high dosages.

In a similar case-control study, 245 elderly patients were matched with 817 controls (69). Benzodiazepines as a group were not associated with a higher risk of hip fracture, but patients who used lorazepam or two or more benzodiazepines had a significantly higher risk.

Effects on performance

All benzodiazepines can cause drowsiness and sedation, and can affect motor and mental performance. Driving is one motor and mental task that is particularly likely to be impaired (SEDA-7, 46), with dangerous consequences; hypnosedatives, like alcohol, impair both actual driving performance (70) and laboratory psychomotor tests (35), and are over-represented in blood samples from delinquent drivers (71). As with alcohol, the maximal impairment occurs while the drug blood concentrations are rising (72), rather than when they have peaked, are stable, or are falling. Somewhat surprisingly, zopiclone 7.5 mg, but not triazolam 0.25 mg, produced deficits in simulated aircraft flight performance 2 and 3 hours after the dose (73). The motor and mental performance reductions induced by hypnotics, especially in elderly people (74), result in an increased incidence of falls (SEDA-21, 38) which can cause hip fractures (75). Agents with short half-lives, including the "Z drugs", were previously thought to carry a reduced risk or even none, but earlier reassuring data have been supplanted by convincing evidence of harm, particularly during the first 2 weeks of prescription (48,76).

Fit young subjects had no impairment of their exercise ability after temazepam or nitrazepam, although nitrazepam caused a subjective feeling of hangover (77).

Seizures

Benzodiazepines can provoke seizures and occasionally precipitate status epilepticus.

- A 28-year-old man with complex partial status, which lasted for 2 months, had a paradoxical worsening of seizure activity in response to diazepam and midazolam (78).

Of 63 neonates receiving lorazepam, diazepam, or both in an intensive care unit, 10 had serious adverse events, including 6 with seizures (57).

Psychological, psychiatric

Cognition

The amnestic effects of benzodiazepines are pervasive and appear to derive from disruption of the consolidation of short-term into long-term memory (79). Amnesia appears to underlie the tendency of regular hypnotic users to overestimate their time asleep, because they simply forget the wakeful intervals (80); in contrast, the same patients underestimate their time spent asleep when drug-free. This amnestic property (SEDA-17, 42) (SEDA-19, 33) has been used to advantage in minor surgery, particularly with midazolam and other short-acting compounds (although male doctors and dentists are advised to have a chaperone present when performing benzodiazepine-assisted procedures with female patients). However, unwanted amnesia can occur, particularly with triazolam, when used as a hypnotic or as an aid for travelers (81,82). The combination of a short half-life and high potency, especially when it was used in the higher doses that were recommended when the drug was initially launched, makes triazolam particularly likely to cause this problem. Studies of low-dose lorazepam (1 mg) in healthy young adults have shown specific deficits in episodic memory (SEDA-21, 38) (SEDA-19, 35). Flurazepam and temazepam have initiated relatively few reports of adverse effects on memory, although flurazepam did cause daytime sedation. Temazepam was uncommonly mentioned in adverse reaction reports, but was also reported more often as being without adequate hypnotic effect. Ironically, temazepam produces more, and oxazepam less, sedation than other benzodiazepines in overdose (83).

The role of benzodiazepines in brain damage has been reviewed (SEDA-14, 36). Cognitive impairment in long-term users can be detected in up to half of the subjects, compared with 16% of controls, but the issue of reversibility with prolonged abstinence is unresolved. Cognitive toxicity is more common with benzodiazepines than other anticonvulsants, with the possible exception of phenobarbital (84).

Patients often have memory deficits after taking benzodiazepines and alcohol. In a study of hippocampal presynaptic glutamate transmission in conjunction with memory deficits induced by benzodiazepines and ethanol, reductions in hippocampal glutamate transmission closely correlated with the extent of impairment of spatial memory performance. The results strongly suggested that presynaptic dysfunction in dorsal hippocampal glutamatergic neurons would be critical for spatial memory deficits induced by benzodiazepines and ethanol (85).

When the relation between benzodiazepine use and cognitive function was evaluated in a prospective study of 2765 elderly subjects, the authors concluded that current benzodiazepine use, especially in recommended or higher dosages, is associated with worse memory among community-dwelling elderly people (86).

In a prospective study, 1389 people aged 60–70 years were recruited from the electoral rolls of the city of Nantes, France (Epidemiology of Vascular Aging Study) (87). A range of symptoms was examined, including cognitive functioning and symptoms of depressive anxiety, and data were also collected on psychotropic and other drugs, as well as tobacco use and alcohol consumption at baseline and thereafter at 2 and 4 years. Users of benzodiazepines were divided into episodic users, recurrent users, and chronic users. Chronic users of benzodiazepines had a significantly higher risk of cognitive decline in the global cognitive test and two attention tests than non-users. Overall, episodic and recurrent users had lower cognitive scores compared with non-users, but the differences were not statistically significant. These findings suggest that long-term use of benzodiazepines is a risk factor for increased cognitive decline in elderly people.

A detailed review has confirmed a relation between impaired memory and benzodiazepine use (88). Different drugs had a similar profile in relation to memory impairment and this was independent of sedation. The benzodiazepines produced anterograde amnesia but not retrograde amnesia, and retrieval processes remained intact.

Delirium

Excessive anxiety and tremulousness, hyperexcitability, confusion, and hallucinations were all reported more often with triazolam than with temazepam or flurazepam, when spontaneous reporting was analysed (81). Whether this is dose-related, and perhaps related to the rapid changes in blood concentration with triazolam, is not clear. Delirium is common, particularly in elderly people, who may have impaired drug clearance, and must always be regarded as possibly drug-induced. Of considerable relevance to hospital practice is the finding of a three-fold increased risk of postoperative delirium in patients given a benzodiazepine (89). Dose- and age-related increases in adverse cognitive and other central nervous effects from benzodiazepines (82) are well documented. The use of these drugs in elderly people has been reviewed, with recommendations about maximizing the benefit-to-harm balance in this group of individuals who are susceptible to cognitive and other adverse effects (90,91).

Sleep

The benzodiazepines typically suppress REM sleep, with consequent rebound dreaming and restlessness on withdrawal, leading to poorer sleep patterns (SEDA-12, 42) (92).

The use of benzodiazepines, particularly the short-acting compounds such as triazolam, for the induction of sleep has provoked much discussion (SEDA-17, 42). The debate rages over the risks and benefits of short-acting compounds, in inducing bizarre behavior or rapid withdrawal with daytime anxiety, compared with the possibility of hangover sedation and performance deficits with longer-acting compounds (93). The treatment of sleep disorders is multifaceted, because of the complex nature of sleep and the variety of factors that can give rise to sleep disorders (82). Consequently, such treatment should be selected and proffered carefully, with due regard for all the factors, not treated cavalierly with the latest flavor-of-the-month benzodiazepine receptor agonist. Non-drug treatments are effective (17) and should be considered first; pharmacological treatment should take into consideration any pre-existing factors, for example anxiety, depression, the duration and nature of medical problems (including any painful condition), concomitant medications, and other substance use (13).

Psychoses

Depression is commonly seen (94), either during benzodiazepine treatment or as a complication of withdrawal (SEDA-17, 42). Relief of anxiety symptoms can uncover pre-existing depression, rather than causing depression per se. In addition to their euphoriant effects in some individuals, benzodiazepines can directly increase irritability and depression and, less commonly, lead to full-blown manic episodes (95,96).

Review of a Canadian adverse drug reactions database showed several cases of previously unreported benzodiazepine-induced adverse effects, including hallucinations and encephalopathy (97), although whether benzodiazepines alone were responsible is difficult to confirm. Visual hallucinations have also been reported in association with zolpidem (98).

Benzodiazepine withdrawal, like alcohol withdrawal, can cause schizophreniform auditory hallucinations (99).

Behavior

While they are generally regarded as being tranquillizers, benzodiazepines and related hypnosedatives can release aggression and induce antisocial behavior (100), particularly in combination with alcohol (101) and in the presence of frustration (49). Aggression can occur during benzodiazepine intoxication and withdrawal (100). Non-medical use of flunitrazepam (102) seems particularly likely to reveal paradoxical rage and aggression, with consequent forensic problems. The combination of abnormal disinhibited behavior and amnesia produced by benzodiazepines can be singularly dangerous. Anecdotal cases suggest that hypnosedatives can also disinhibit violent behavior in individuals taking antidepressants. A literature review of behavioral adverse effects associated with benzodiazepines (clonazepam, diazepam, and lorazepam) has shown that 11–25% of patients with mental retardation have these adverse effects (103). In two controlled studies, lorazepam was more likely to provoke aggression than oxazepam (104,105).

Gastrointestinal

Nausea due to benzodiazepines has been reported as being commoner in children, but the incidence does not usually greatly exceed that found with placebo. Gastrointestinal disturbances are more common with the newer non-benzodiazepine agents, for example zopiclone and buspirone (106,107).

Liver

Jaundice has been reported after benzodiazepines, although in only a few cases have they been the only drugs involved (SED-12, 97).

Sexual function

Female orgasm is inhibited by some central depressant and psychotropic drugs, including antipsychotic drugs, antidepressants, and anxiolytic benzodiazepines (108). A survey of patients with bipolar affective disorder taking lithium showed that the co-administration of benzodiazepines was associated with a significantly increased risk (49%) of sexual dysfunction in both men and women (109). Reduced libido is uncommonly reported, as is sexual inhibition, but the actual incidences of these complications may be considerably higher. Hypnosedatives, particularly flunitrazepam and gammahydroxybutyrate, are implicated in sexual assault and "date rape" (110).

Long-Term Effects

Drug tolerance

Animal studies have suggested a possible mechanism for tolerance, in that chronic treatment of rats with triazolam reduced the mRNA coding for certain GABA receptor proteins (111).

Drug dependence

The likelihood and possible severity of dependence on benzodiazepines has been discussed (50).

In 1048 consecutive patients attending 20 primary care health centers in the Canary Islands, who had taken benzodiazepines for 1 month or more, 47% developed dependence (112). Benzodiazepine dependence was more prevalent among women who were middle-aged, separated, of low educational background, unemployed, or housewives. Benzodiazepine dependence was closely related only to three of the variables considered: the dose, the duration of use, and the concomitant use of antidepressants.

Drug withdrawal

The likelihood and possible severity of withdrawal from benzodiazepines has been discussed, especially with regard to the newer short-acting compounds (50).

Withdrawal symptoms occur in at least one-third of long-term users (over 1 year), even if the dose is gradually tapered (113). Symptoms come on within 2–3 days of withdrawal of a short-acting or medium-acting benzodiazepine, or 7–10 days after a long-acting drug; short-acting benzodiazepines tend to produce a more marked withdrawal syndrome (35). Lorazepam and alprazolam are particularly difficult to quit. Symptoms usually last 1–6 weeks, but can persist for many months, leaving the patient in a vulnerable state, with likely recurrence of the original disorder and of self-medication. Withdrawal symptoms can occur within 4–6 weeks of daily long-acting

benzodiazepine use (114), and possibly earlier in susceptible individuals.

Symptoms on withdrawal are variable in nature and degree. Rebound insomnia can occur one or two nights after withdrawal of short-acting drugs. Anxiety is common, with both psychological and physical manifestations, including apprehension, panic, insomnia, palpitation, sweating, tremor, and gastrointestinal disturbances. Irritability and aggression also occur, notably after triazolam. Depression has been reported after benzodiazepine withdrawal (115). There may be increased or distorted sensory perceptions, such as photophobia, altered (metallic) taste, and hypersensitivity to touch and pain. Flu-like muscle aches and spasms, unsteadiness, and clumsiness are common. Perceptual distortions include burning or creeping of the skin and apparent movement or changes in objects or self (113). General malaise with loss of appetite can occur. As with alcohol, paranoid psychosis, delirium, and epileptic fits are possible on withdrawal (SED-12, 97). With careful handling, often involving psychological and sometimes adjunctive pharmacological support, motivated patients who depend on benzodiazepines can usually be successfully withdrawn. In particular, the combination of gradual dose-tapering and cognitive behavioral therapy can be helpful (116). Guidelines on the management of such patients have been concisely presented (51).

Second-Generation Effects

Teratogenicity

Benzodiazepines readily pass from the mother to fetus through the placenta (117). There may be a risk of congenital malformations, particularly oral cleft, if a pregnant woman takes a benzodiazepine during the first trimester, but the data are inconsistent across drugs (alprazolam having the most clearly defined risk), and any overall effect is probably small (27,28). The risk of benzodiazepine-induced birth defects thus remains uncertain (118), despite two cases of fetal-alcohol syndrome reported after benzodiazepine exposure alone (119).

The occurrence of congenital abnormalities associated with the use of benzodiazepines (alprazolam, clonazepam, medazepam, nitrazepam, and tofisopam) during pregnancy has been analysed in a matched case-control study (120). The cases and controls were drawn from the Hungarian Case-Control Surveillance of Congenital Abnormalities from 1980 to 1996. Of the 38 151 pregnant women who delivered babies without congenital anomalies, 75 had taken benzodiazepines during pregnancy, compared with 57 of 22 865 who delivered offspring with anomalies. Thus, treatment with these benzodiazepines during pregnancy did not cause a detectable teratogenic risk. However, the true relevance of these findings needs to be supported by prospective case ascertainment.

Fetotoxicity

Benzodiazepines readily pass from the mother to fetus through the placenta (117). There is a further concern about cognitive development after in utero exposure to benzodiazepines. It now appears that the slowed intellectual progress seen in some children exposed in utero will "catch up" in most cases by age 4 (28). Unfortunately, the impact of sedative-hypnotic use during pregnancy is often complicated by the abuse of multiple agents and poor maternal nutrition and antenatal care, and may be further confounded by social and environmental deprivation, which the infant often faces after birth (28,121). More definite but short-lived problems occur with benzodiazepines given in late pregnancy and during labor; here floppiness, apnea, and withdrawal in the infant can pose problems (28,122) but usually resolve uneventfully (SEDA-21, 38). Pregnant women should avoid benzodiazepines if possible, especially during late pregnancy and labor; if required, chlordiazepoxide appears to have the best established record (29). On the other hand, alprazolam should be avoided, and temazepam plus diphenhydramine appears to be a particularly toxic combination in late pregnancy, based on animal research and one case report of fetal activation followed by stillbirth (123).

Lactation

Benzodiazepines are secreted into the milk in relatively small amounts (28). During lactation, longer-acting agents are relatively contraindicated, particularly with continued administration beyond 3–5 days, owing to the likelihood of infant sedation (122,124). Short-acting benzodiazepines and zopiclone are probably safe, especially if restricted to single doses or for short courses of therapy (28,125). Zopiclone and midazolam, for example, become undetectable in breast milk 4–5 hours after a dose (126).

Susceptibility Factors

Age

The safety of benzodiazepines in neonates has been assessed in a retrospective chart review of 63 infants who received benzodiazepines (lorazepam and/or midazolam) as sedatives or anticonvulsants (57). Five infants had hypotension and three had respiratory depression. In all cases of respiratory depression, ventilatory support was initiated or increased. Significant hypotension was treated with positive inotropic drugs in two cases. Thus, respiratory depression and hypotension are relatively common when benzodiazepines are prescribed in these patients. However, both depression and hypotension could also have been due to the severe underlying illnesses and concomitant medications. Matched controls were not studied.

Other features of the patient

Benzodiazepines are more likely to cause adverse effects in patients with HIV infection and other causes of organic brain syndrome (46).

Drug Administration

Drug overdose

Overdosage of benzodiazepines alone is generally thought to be safe, but deaths have occasionally been reported (127–129). In 204 consecutive suicides seen by the San Diego County Coroner during 1981–1982, drugs were detected in 68%, and anxiolytics and hypnotics in 11% and 12% respectively; although benzodiazepines were found in under 10% of the group as a whole, they were found in one-third of those who died by overdose (130). In one series of 2827 intentional cases of poisoning, in which there were ten deaths, three were associated with benzodiazepines; death was related to a delay between ingestion and medical intervention (131), and advanced age has also been described as a risk factor (132). In other cases death has been attributed to combined overdose with other drugs, such as alcohol (128), oxycodone (133,134), tramadol (135), and amitriptyline (136).

Concomitant benzodiazepine overdose has also been reported to be an independent risk factor in the development of hepatic encephalopathy (OR = 1.91; CI = 1.00, 3.65) and renal dysfunction (OR = 1.81; CI = 1.00, 3.22) in patients who take a paracetamol overdose (137).

Drug–Drug Interactions

Antibiotics

Antibiotics (erythromycin, chloramphenicol, isoniazid) compete for hepatic oxidative pathways that metabolize most benzodiazepines, as well as zolpidem, zopiclone, and buspirone (SEDA-22, 39) (SEDA-22, 41).

Macrolides cause increases in the serum concentrations, AUCs, and half-lives and reductions in the clearance of triazolam and midazolam (138–140). These changes can result in clinical effects, such as prolonged psychomotor impairment, amnesia, or loss of consciousness (141). Erythromycin can increase concentrations of midazolam and triazolam by inhibition of CYP3A4, and dosage reductions of 50% have been proposed if concomitant therapy is unavoidable (142).

Antifungal imidazoles

Antifungal imidazoles (ketoconazole, itraconazole, and analogues) compete for hepatic oxidative pathways that metabolize most benzodiazepines, as well as zolpidem, zopiclone, and buspirone (SEDA-22, 39, 41–43).

Antihistamines

The potentiation of sedative effects from benzodiazepines when combined with centrally acting drugs with antihistamine properties (for example first-generation antihistamines, tricyclic antidepressants, and neuroleptic drugs) can pose problems (143). Antihistamines that do not have central actions do not interact with benzodiazepines as in the case of mizolastine and lorazepam (144), ebastine and diazepam (145), and terfenadine and diazepam (143).

Calcium channel blockers

Diltiazem and verapamil compete for hepatic oxidative pathways that metabolize most benzodiazepines, as well as zolpidem, zopiclone, and buspirone (SEDA-22, 39) (SEDA-22, 41).

Central stimulants

Caffeine and other central stimulants can reverse daytime sedation from benzodiazepine use. There was a positive effect of caffeine (250 mg) on early-morning performance after both placebo and flurazepam (30 mg) given the night before (146,147), particularly in terms of subjective assessments of mood and sleepiness. However, one cannot assume that the alerting effect of caffeine necessarily reverses the amnestic, disinhibiting, or insight-impairing effects of benzodiazepines. Indeed, caffeine can actually worsen learning and performance already impaired by lorazepam (148). Other drugs with direct or indirect CNS stimulant activity (theophylline, ephedrine, amphetamine, and their analogues) have similar effects and can counteract the effects of benzodiazepines, at least subjectively. Another worrying feature of stimulant use, particularly in drug misusers, is that it commonly increases the perceived need for hypnosedatives.

Cimetidine

Cimetidine can impair benzodiazepine metabolism and lead to adverse effects (SEDA-18, 43). In contrast, a few benzodiazepines are metabolized exclusively by glucuronide conjugation (lorazepam, oxazepam, temazepam), and are therefore unaffected by concomitant therapy with cimetidine and other oxidation inhibitors (123).

Clozapine

Caution has been recommended when starting clozapine in patients taking benzodiazepines (SEDA-19, 55). Three cases of delirium associated with clozapine and benzodiazepines (149) have been reported. There have been several reports of synergistic reactions, resulting in increased sedation and ataxia, when lorazepam was begun in patients already taking clozapine (150).

- Syncope and electrocardiographic changes (sinus bradycardia of 40/minute with deep anteroseptal inverted T waves and minor ST changes in other leads) have been observed with the concurrent administration of clozapine (after the dosage was increased to 300 mg/day) and diazepam (30 mg/day) in a 50-year-old man (151).

CNS depressants

The interactions of benzodiazepines with other nervous system depressants, especially alcohol and other GABA-ergic drugs, have been reviewed (152). Other drugs with nervous system depressant effects (opioids, anticonvulsants, general anesthetics) also can add to, and complicate, the depressant action of benzodiazepines.

Phenothiazines and butyrophenones can counteract intoxication from lysergic acid diethylamide (LSD); benzodiazepines can inhibit this useful effect of antipsychotic drugs (153).

Disulfiram

Disulfiram competes for hepatic oxidative pathways that metabolize most benzodiazepines, as well as zolpidem, zopiclone, and buspirone (SEDA-22, 39) (SEDA-22, 41).

Enzyme inducers

Enzyme induction can be problematic with co-administration of benzodiazepines and rifampicin or certain anticonvulsants (phenobarbital, phenytoin, carbamazepine). However, despite enzyme stimulation, the net effect of adding these anticonvulsants can be augmentation of benzodiazepine-induced sedation.

Rifampicin, and presumably other enzyme inducers, reduces concentrations of zolpidem, zopiclone, and buspirone (SEDA-22, 42). Drugs that are solely glucuronidated (lorazepam, oxazepam, and temazepam) are not affected.

HIV protease inhibitors

Some protease inhibitors (saquinavir) compete for hepatic oxidative pathways that metabolize most benzodiazepines, as well as zolpidem, zopiclone, and buspirone (SEDA-22, 39) (SEDA-22, 41).

Levodopa

The question of whether starting a benzodiazepine in patients taking levodopa is followed by a faster increase in antiparkinsonian drug requirements has been studied using drug dispensing data for all the residents in six Dutch cities (154). All were 55 years old or older and had used levodopa for at least 360 days. There were 45 benzodiazepine starters and 169 controls. Antiparkinsonian drug doses increased faster in the benzodiazepine group, but the difference was not significant (RR = 1.44; 95% CI = 0.89, 2.59).

Lithium

In 18 patients treated with benzodiazepines and/or antipsychotic drugs there were increased chromosomal aberrations and increased sister chromatid exchange, but there were no significant differences between this group and another group of 18 patients taking lithium in addition to benzodiazepines and/or antipsychotic drugs (155).

Moxonidine

Moxonidine can potentiate the effect of benzodiazepines (156).

Muscle relaxants

Laboratory investigations have shown that some benzodiazepines can produce biphasic effects on the actions of neuromuscular blocking agents (157,158), higher doses potentiating the effects (157,159); however, several human investigations have failed to show a significant effect (160–162). It has been suggested that agents that are added to commercial formulations of some benzodiazepines to render them more water-soluble may mask the benzodiazepine effect (162).

Nevertheless, some interactions of benzodiazepines with muscle relaxants used in anesthesia have been described. Diazepam has been reported to potentiate the effects of tubocurare (163) and gallamine (164) and to reduce the effects of suxamethonium (164). However, in 113 patients undergoing general anesthesia, intravenous diazepam 20 mg, lorazepam 5 mg, and lormetazepam 2 mg did not potentiate the neuromuscular blocking effects of vecuronium or atracurium (162).

In 113 patients undergoing general anesthesia, intravenous midazolam 15 mg slowed recovery of the twitch height after vecuronium and atracurium compared with diazepam. The recovery index was not altered (162). However, in another study in 20 patients, midazolam 0.3 mg/kg did not affect the duration of blockade, recovery time, intensity of fasciculations, or adequacy of relaxation for tracheal intubation produced by suxamethonium 1 mg/kg, nor the duration of blockade and adequacy of relaxation for tracheal intubation produced by pancuronium 0.025 mg/kg in incremental doses until 99% depression of muscle-twitch tension was obtained (161). Furthermore, in 60 patients undergoing maintenance anesthesia randomly assigned to one of six regimens (etomidate, fentanyl, midazolam, propofol, thiopental plus nitrous oxide, or isoflurane plus nitrous oxide), midazolam did not alter rocuronium dosage requirements (165).

Neuroleptic drugs

Because of the frequency of co-administration of benzodiazepines with neuroleptic drugs, it is important to consider possible adverse effects that can result from such combinations. In a brief review, emphasis has been placed on pharmacokinetic interactions between neuroleptic drugs and benzodiazepines, as much information on their metabolic pathways is emerging (166). Thus, the enzyme CYP3A4, which plays a dominant role in the metabolism of benzodiazepines, also contributes to the metabolism of clozapine, haloperidol, and quetiapine, and neuroleptic drug plasma concentrations can rise. Intramuscular levomepromazine in combination with an intravenous benzodiazepine has been said to increase the risk of airways obstruction, on the basis of five cases of respiratory impairment; the doses of levomepromazine were higher in the five cases that had accompanying airways obstruction than in another 95 patients who did not (167).

Omeprazole

Omeprazole can impair benzodiazepine metabolism and lead to adverse effects (SEDA-18, 43).

Opioids

Fentanyl competes for hepatic oxidative pathways that metabolize most benzodiazepines, as well as zolpidem, zopiclone, and buspirone (SEDA-22, 39) (SEDA-22, 41).

Oral contraceptives

Oral contraceptives alter the metabolism of some benzodiazepines that undergo oxidation (alprazolam, chlordiazepoxide, diazepam) or nitroreduction (nitrazepam) (168). For these drugs, oral contraceptives inhibit enzyme activity and reduce clearance. There is nevertheless no evidence that this interaction is of clinical importance. It should be noted that for other benzodiazepines that undergo oxidative metabolism, such as bromazepam or clotiazepam, no change has ever been found in oral contraceptive users. Some other benzodiazepines, such as lorazepam, oxazepam, and temazepam, are metabolized by glucuronic acid conjugation. The clearance of temazepam was increased when oral contraceptives were co-administered, but the clearances of lorazepam and oxazepam were not (169). Again, it is unlikely that this is an interaction of clinical importance.

Selective serotonin reuptake inhibitors (SSRIs)

Some SSRIs (notably fluvoxamine and to a lesser extent fluoxetine) and their metabolites inhibit hepatic oxidative enzymes, particularly CYP2C19 and CYP3A, that metabolize most benzodiazepines, as well as zaleplon, zolpidem, zopiclone, and buspirone (SEDA-22, 39) (SEDA-22, 41) (170,171). Apart from fluvoxamine, SSRIs do not generally have a clinically prominent effect on hyposedative effects; studies vary from those that have found that fluoxetine has a moderate but functionally unimportant impact on diazepam concentrations (172) to results that suggest significant aggravation of the cognitive effects of alprazolam when co-prescribed with the SSRI (173).

Tricyclic antidepressants

Four patients developed adverse effects attributable to combinations of benzodiazepines with tricyclic antidepressants, including exacerbations of delusional disorder (174).

Nefazodone competes for hepatic oxidative pathways that metabolize most benzodiazepines, as well as zolpidem, zopiclone, and buspirone (SEDA-22, 39) (SEDA-22, 41).

References

1. Fraser AD. Use and abuse of the benzodiazepines. Ther Drug Monit 1998;20(5):481–9.
2. Woods JH, Winger G. Current benzodiazepine issues. Psychopharmacology (Berl) 1995;118(2):107–15.
3. Pelissolo A, Bisserbe JC. Dependance aux benzodiazepines. Aspects clinique et biologiques. [Dependence on benzodiazepines. Clinical and biological aspects.] Encephale 1994;20(2):147–57.
4. Ciuna A, Andretta M, Corbari L, Levi D, Mirandola M, Sorio A, Barbui C. Are we going to increase the use of antidepressants up to that of benzodiazepines? Eur J Clin Pharmacol 2004;60(9):629–34.
5. Anonymous. What's wrong with prescribing hypnotics? Drug Ther Bull 2004;42(12):89–93.

6. Judd F. Flunitrazepam—schedule 8 drug. Australas Psychiatry 1998;6:265.

7. Woods JH. Problems and opportunities in regulation of benzodiazepines. J Clin Pharmacol 1998;38(9):773–82.

8. Griffiths RR. Commentary on review by Woods and Winger. Benzodiazepines: long-term use among patients is a concern and abuse among polydrug abusers is not trivial. Psychopharmacology (Berl) 1995;118(2):116–7.

9. Lader M. Commentary on review by Woods and Winger. Psychopharmacology (Berl) 1995;118:118.

10. Holbrook AM. Treating insomnia. BMJ 2004;329(7476):1198–9.

11. Lader M. Psychiatric disorders. In: Speight T, Holford N, editors. Avery's Drug Treatment. 4th ed.. Auckland: ADIS Internation Press, 1997:1437.

12. Ashton H. Guidelines for the rational use of benzodiaze-pines. When and what to use. Drugs 1994;48(1):25–40.

13. Pagel JF. Treatment of insomnia. Am Fam Physician 1994;49(6):1417–211423–4.

14. Mendelson WB, Jain B. An assessment of short-acting hypnotics. Drug Saf 1995;13(4):257–70.

15. Mellinger GD, Balter MB, Uhlenhuth EH. Insomnia and its treatment. Prevalence and correlates. Arch Gen Psychiatry 1985;42(3):225–32.

16. Wise MG, Griffies WS. A combined treatment approach to anxiety in the medically ill. J Clin Psychiatry 1995;56(Suppl 2):14–9.

17. Morin CM, Colecchi C, Stone J, Sood R, Brink D. Behavioral and pharmacological therapies for late-life insomnia: a ran-domized controlled trial. JAMA 1999;281(11):991–9.

18. Gaudreault P, Guay J, Thivierge RL, Verdy I. Benzodiazepine poisoning. Clinical and pharmacological considerations and treatment. Drug Saf 1991;6(4):247–65.

19. Megarbane B, Gueye P, Baud F. Interactions entre benzo-diazepines et produits opioides. [Interactions between ben-zodiazepines and opioids.] Ann Med Interne (Paris) 2003;154(Spec No 2):S64–72.

20. Taiminen TJ. Effect of psychopharmacotherapy on suicide risk in psychiatric inpatients. Acta Psychiatr Scand 1993;87(1):45–7.

21. Michel K, Waeber V, Valach L, Arestegui G, Spuhler T. A comparison of the drugs taken in fatal and nonfatal self-poisoning. Acta Psychiatr Scand 1994;90(3):184–9.

22. Schwarz UI, Ruder S, Krappweis J, Israel M, Kirch W. Epidemiologie medikamentöser Parasuizide. Eine Erhebung aus dem Universitätsklinikum Dresden. [Epidemiology of attempted suicide using drugs. An inquiry from the Dresden University Clinic.] Dtsch Med Wochenschr 2004;129(31–32):1669–73.

23. Bohn MJ, Babor TF, Kranzler HR. The Alcohol Use Disorders Identification Test (AUDIT): validation of a screening instrument for use in medical settings. J Stud Alcohol 1995;56(4):423–32.

24. Deardon DJ, Bird GL. Acute (type 1) hypersensitivity to i.v. Diazemuls Br J Anaesth 1987;59(3):391.

25. Pirker C, Misic A, Brinkmeier T, Frosch PJ. Tetrazepam drug sensitivity—usefulness of the patch test. Contact Dermatitis 2002;47(3):135–8.

26. Sachs B, Erdmann S, Al-Masaoudi T, Merk HF. In vitro drug allergy detection system incorporating human liver microsomes in chlorazepate-induced skin rash: drug-speci-fic proliferation associated with interleukin-5 secretion. Br J Dermatol 2001;144(2):316–20.

27. Altshuler LL, Cohen L, Szuba MP, Burt VK, Gitlin M, Mintz J. Pharmacologic management of psychiatric illness during pregnancy: dilemmas and guidelines. Am J Psychiatry 1996;153(5):592–606.

28. McElhatton PR. The effects of benzodiazepine use during pregnancy and lactation. Reprod Toxicol 1994;8(6):461–75.

29. Iqbal MM, Sobhan T, Ryals T. Effects of commonly used benzodiazepines on the fetus, the neonate, and the nursing infant. Psychiatr Serv 2002;53(1):39–49.

30. Kim K, Johnson JA, Derendorf H. Differences in drug pharmacokinetics between East Asians and Caucasians and the role of genetic polymorphisms. J Clin Pharmacol 2004;44(10):1083–105.

31. Griffiths RR, McLeod DR, Bigelow GE, Liebson IA, Roache JD, Nowowieski P. Comparison of diazepam and oxazepam: preference, liking and extent of abuse. J Pharmacol Exp Ther 1984;229(2):501–8.

32. Bond A, Seijas D, Dawling S, Lader M. Systemic absorp-tion and abuse liability of snorted flunitrazepam. Addiction 1994;89(7):821–30.

33. Ananth J, Swartz R, Burgoyne K, Gadasally R. Hepatic disease and psychiatric illness: relationships and treatment. Psychother Psychosom 1994;62(3–4):146–59.

34. Hesse LM, von Moltke LL, Greenblatt DJ. Clinically important drug interactions with zopiclone, zolpidem and zaleplon. CNS Drugs 2003;17(7):513–32.

35. Lader M. Clin pharmacology of anxiolytic drugs: Past, present and future. In: Biggio G, Sanna E, Costa E, editors. GABA-A Receptors and Anxiety. From Neurobiology to Treatment. New York: Raven Press, 1995:135.

36. Anonymous. Zopiclone, zolpidem and zaleplon. Get your "zzz's" without affecting performance the next day. Drugs Ther Perspect 2004;20(2):16–8.

37. Haefely W, Martin JR, Schoch P. Novel anxiolytics that act as partial agonists at benzodiazepine receptors. Trends Pharmacol Sci 1990;11(11):452–6.

38. Rickels K, DeMartinis N, Aufdembrinke B. A double-blind, placebo-controlled trial of abecarnil and diazepam in the treatment of patients with generalized anxiety dis-order. J Clin Psychopharmacol 2000;20(1):12–8.

39. O'Donovan MC, McGuffin P. Short acting benzodiaze-pines. BMJ 1993;306(6883):945–6.

40. Michel L, Lang JP. Benzodiazepines et passage à l'acte criminel. [Benzodiazepines and forensic aspects.] Encephale 2003;29(6):479–85.

41. Dyer C. Halcion edges its way back into Britain in low doses. BMJ 1993;306:1085.

42. Te Lintelo J, Pieters T. Halcion: de lotgevallen van de "Dutch Hysteria". Pharm Wkbl 2003;138(46):1600–5.

43. Wagner AK, Soumerai SB, Zhang F, Mah C, Simoni-Wastila L, Cosler L, Fanning T, Gallagher P, Ross-Degnan D. Effects of state surveillance on new post-hos-pitalization benzodiazepine use. Int J Qual Health Care 2003;15(5):423–31.

44. Abraham J. Transnational industrial power, the medical profession and the regulatory state: adverse drug reactions and the crisis over the safety of Halcion in the Netherlands and the UK. Soc Sci Med 2002;55(9):1671–90.

45. Klein DF. The report by the Institute of Medicine and postmarketing surveillance. Arch Gen Psychiatry 1999;56(4):353–4.

46. Ayuso JL. Use of psychotropic drugs in patients with HIV infection. Drugs 1994;47(4):599–610.

47. Gray SL, LaCroix AZ, Blough D, Wagner EH, Koepsell TD, Buchner D. Is the use of benzodiazepines

associated with incident disability? J Am Geriatr Soc 2002;50(6):1012–8.

48. Wagner AK, Zhang F, Soumerai SB, Walker AM, Gurwitz JH, Glynn RJ, Ross-Degnan D. Benzodiazepine use and hip fractures in the elderly: who is at greatest risk? Arch Intern Med 2004;164(14):1567–72.

49. Salzman C, Kochansky GE, Shader RI, Porrino LJ, Harmatz JS, Swett CP Jr. Chlordiazepoxide-induced hostility in a small group setting. Arch Gen Psychiatry 1974;31(3):401–5.

50. Woods JH, Katz JL, Winger G. Abuse liability of benzodiazepines. Pharmacol Rev 1987;39(4):251–413.

51. Ashton H. The treatment of benzodiazepine dependence. Addiction 1994;89(11):1535–41.

52. Adam K, Oswald I. Can a rapidly-eliminated hypnotic cause daytime anxiety? Pharmacopsychiatry 1989;22(3):115–9.

53. Kales A, Manfredi RL, Vgontzas AN, Bixler EO, Vela-Bueno A, Fee EC. Rebound insomnia after only brief and intermittent use of rapidly eliminated benzodiazepines. Clin Pharmacol Ther 1991;49(4):468–76.

54. Alldredge BK, Gelb AM, Isaacs SM, Corry MD, Allen F, Ulrich S, Gottwald MD, O'Neil N, Neuhaus JM, Segal MR, Lowenstein DH. A comparison of lorazepam, diazepam, and placebo for the treatment of out-of-hospital status epilepticus. N Engl J Med 2001;345(9):631–7.

55. Wassner E, Morris B, Fernando L, Rao M, Whitehouse WP. Intranasal midazolam for treating febrile seizures in children. Buccal midazolam for childhood seizures at home preferred to rectal diazepam. BMJ 2001;322(7278):108.

56. Donaldson D, Gibson G. Systemic complications with intravenous diazepam. Oral Surg Oral Med Oral Pathol 1980;49(2):126–30.

57. Ng E, Klinger G, Shah V, Taddio A. Safety of benzodiazepines in newborns. Ann Pharmacother 2002;36(7–8):1150–5.

58. Glaser JW, Blanton PL, Thrash WJ. Incidence and extent of venous sequelae with intravenous diazepam utilizing a standardized conscious sedation technique. J Periodontol 1982;53(11):700–3.

59. Mikkelsen H, Hoel TM, Bryne H, Krohn CD. Local reactions after i.v. injections of diazepam, flunitrazepam and isotonic saline Br J Anaesth 1980;52(8):817–9.

60. Jensen S, Huttel MS, Schou Olesen A. Venous complications after i.v. administration of Diazemuls (diazepam) and Dormicum (midazolam) Br J Anaesth 1981;53(10):1083–5.

61. Reves JG, Fragen RJ, Vinik HR, Greenblatt DJ. Midazolam: pharmacology and uses. Anesthesiology 1985;62(3):310–24.

62. Dundee JW, Halliday NJ, Harper KW, Brogden RN. Midazolam. A review of its pharmacological properties and therapeutic use. Drugs 1984;28(6):519–43.

63. Woodcock AA, Gross ER, Geddes DM. Drug treatment of breathlessness: contrasting effects of diazepam and promethazine in pink puffers. BMJ (Clin Res Ed) 1981;283(6287):343–6.

64. Mackereth S. Use of rectal diazepam in the community. Dev Med Child Neurol 2000;42(11):785.

65. Kriel RL, Cloyd JC, Pellock JM. Respiratory depression in children receiving diazepam for acute seizures: a prospective study. Dev Med Child Neurol 2000;42(6):429–30.

66. Norris E, Marzouk O, Nunn A, McIntyre J, Choonara I. Respiratory depression in children receiving diazepam for acute seizures: a prospective study. Dev Med Child Neurol 1999;41(5):340–3.

67. Passaro A, Volpato S, Romagnoni F, Manzoli N, Zuliani G, Fellin R. Benzodiazepines with different half-life and falling in a hospitalized population. The GIFA study. Gruppo Italiano di Farmacovigilanza nell'Anziano. J Clin Epidemiol 2000;53(12):1222–9.

68. Sgadari A, Lapane KL, Mor V, Landi F, Bernabei R, Gambassi G. Oxidative and nonoxidative benzodiazepines and the risk of femur fracture. The Systematic Assessment of Geriatric Drug Use Via Epidemiology Study Group. J Clin Psychopharmacol 2000;20(2):234–9.

69. Pierfitte C, Macouillard G, Thicoipe M, Chaslerie A, Pehourcq F, Aissou M, Martinez B, Lagnaoui R, Fourrier A, Begaud B, Dangoumau J, Moore N. Benzodiazepines and hip fractures in elderly people: case-control study. BMJ 2001;322(7288):704–8.

70. O'Hanlon JF, Volkerts ER. Hypnotics and actual driving performance. Acta Psychiatr Scand 1986;332(Suppl):95–104.

71. Heinemann A, Grellner W, Preu J, Kratochwil M, Cordes O, Lignitz E, Wilske J, Puschel K. Zur Straenverkehrsdelinquenz durch psychotrope Substanzen bei Senioren in drei Regionen Deutschlands. Teil I: Medikamente und Betäubungsmittel. Blutalkohol 2004;41:117–27.

72. Ellinwood EH Jr, Linnoila M, Easler ME, Molter DW. Onset of peak impairment after diazepam and after alcohol. Clin Pharmacol Ther 1981;30(4):534–8.

73. Jing BS, Zhan H, Li YF, Zhou YJ, Guo H. [Effects of short-action hypnotics triazolam and zopiclone on simulated flight performance.]Space Med Med Eng (Beijing) 2003;16(5):329–31.

74. Kruse WH. Problems and pitfalls in the use of benzodiazepines in the elderly. Drug Saf 1990;5(5):328–44.

75. Ray WA, Griffin MR, Schaffner W, Baugh DK, Melton LJ 3rd. Psychotropic drug use and the risk of hip fracture. N Engl J Med 1987;316(7):363–9.

76. Vermeeren A. Residual effects of hypnotics: epidemiology and clinical implications. CNS Drugs 2004;18(5):297–328.

77. Charles RB, Kirkham AJ, Guyatt AR, Parker SP. Psychomotor, pulmonary and exercise responses to sleep medication. Br J Clin Pharmacol 1987;24(2):191–7.

78. Al Tahan A. Paradoxic response to diazepam in complex partial status epilepticus. Arch Med Res 2000;31(1):101–4.

79. Ghoneim MM, Mewaldt SP. Benzodiazepines and human memory: a review. Anesthesiology 1990;72(5):926–38.

80. Schneider-Helmert D. Why low-dose benzodiazepine-dependent insomniacs can't escape their sleeping pills. Acta Psychiatr Scand 1988;78(6):706–11.

81. Bixler EO, Kales A, Brubaker BH, Kales JD. Adverse reactions to benzodiazepine hypnotics: spontaneous reporting system. Pharmacology 1987;35(5):286–300.

82. Gillin JC, Byerley WF. Drug therapy: the diagnosis and management of insomnia. N Engl J Med 1990;322(4):239–48.

83. Buckley NA, Dawson AH, Whyte IM, O'Connell DL. Relative toxicity of benzodiazepines in overdose. BMJ 1995;310(6974):219–21.

84. Meador KJ. Cognitive side effects of antiepileptic drugs. Can J Neurol Sci 1994;21(3):S12–6.

85. Shimizu K, Matsubara K, Uezono T, Kimura K, Shiono H. Reduced dorsal hippocampal glutamate release significantly correlates with the spatial memory deficits produced by benzodiazepines and ethanol. Neuroscience 1998;83(3):701–6.

86. Hanlon JT, Horner RD, Schmader KE, Fillenbaum GG, Lewis IK, Wall WE Jr, Landerman LR, Pieper CF, Blazer DG, Cohen HJ. Benzodiazepine use and cognitive

function among community-dwelling elderly. Clin Pharmacol Ther 1998;64(6):684–92.

87. Paterniti S, Dufouil C, Alperovitch A. Long-term benzodiazepine use and cognitive decline in the elderly: the epidemiology of vascular aging study. J Clin Psychopharmacol 2002;22(3):285–93.

88. Ghoneim MM. Drugs and human memory (Part 2). Clinical, theoretical and methodological issues. Anaesthesiology, 2004;100:1277-97.

89. Marcantonio ER, Juarez G, Goldman L, Mangione CM, Ludwig LE, Lind L, Katz N, Cook EF, Orav EJ, Lee TH. The relationship of postoperative delirium with psychoactive medications. JAMA 1994;272(19):1518–22.

90. Shorr RI, Robin DW. Rational use of benzodiazepines in the elderly. Drugs Aging 1994;4(1):9–20.

91. Madhusoodanan S, Bogunovic OJ. Safety of benzodiazepines in the geriatric population. Expert Opin Drug Saf 2004;3(5):485–93.

92. Gillin JC, Spinweber CL, Johnson LC. Rebound insomnia: a critical review. J Clin Psychopharmacol 1989;9(3):161–72.

93. McClure DJ, Walsh J, Chang H, Olah A, Wilson R, Pecknold JC. Comparison of lorazepam and flurazepam as hypnotic agents in chronic insomniacs. J Clin Pharmacol 1988;28(1):52–63.

94. Patten SB, Williams JV, Love EJ. Self-reported depressive symptoms following treatment with corticosteroids and sedative-hypnotics. Int J Psychiatry Med 1996;26(1):15–24.

95. Strahan A, Rosenthal J, Kaswan M, Winston A. Three case reports of acute paroxysmal excitement associated with alprazolam treatment. Am J Psychiatry 1985;142(7):859–61.

96. Rigby J, Harvey M, Davies DR. Mania precipitated by benzodiazepine withdrawal. Acta Psychiatr Scand 1989;79(4):406–7.

97. Patten SB, Love EJ. Neuropsychiatric adverse drug reactions: passive reports to Health and Welfare Canada's adverse drug reaction database (1965–present). Int J Psychiatry Med 1994;24(1):45–62.

98. Tsai MJ, Huang YB, Wu PC. A novel clinical pattern of visual hallucination after zolpidem use. J Toxicol Clin Toxicol 2003;41(6):869–72.

99. Roberts K, Vass N. Schneiderian first-rank symptoms caused by benzodiazepine withdrawal. Br J Psychiatry 1986;148:593–4.

100. Bond AJ. Drug-induced behavioural disinhibition. CNS Drugs 1998;9:41–57.

101. Brahams D. Iatrogenic crime: criminal behaviour in patients receiving drug treatment. Lancet 1987;1(8537):874–5.

102. Dobson J. Sedatives/hypnotics for abuse. NZ Med J 1989;102(881):651.

103. Kalachnik JE, Hanzel TE, Sevenich R, Harder SR. Benzodiazepine behavioral side effects: review and implications for individuals with mental retardation. Am J Ment Retard 2002;107(5):376–410.

104. Kochansky GE, Salzman C, Shader RI, Harmatz JS, Ogeltree AM. The differential effects of chlordiazepoxide and oxazepam on hostility in a small group setting. Am J Psychiatry 1975;132(8):861–3.

105. Bond A, Lader M. Differential effects of oxazepam and lorazepam on aggressive responding. Psychopharmacology (Berl) 1988;95(3):369–73.

106. Monchesky TC, Billings BJ, Phillips R. Zopiclone: a new nonbenzodiazepine hypnotic used in general practice. Clin Ther 1986;8(3):283–91.

107. Newton RE, Marunycz JD, Alderdice MT, Napoliello MJ. Review of the side-effect profile of buspirone. Am J Med 1986;80(3B):17–21.

108. Shen WW, Sata LS. Inhibited female orgasm resulting from psychotropic drugs. A five-year, updated, clinical review. J Reprod Med 1990;35(1):11–4.

109. Ghadirian AM, Annable L, Belanger MC. Lithium, benzodiazepines, and sexual function in bipolar patients. Am J Psychiatry 1992;149(6):801–5.

110. Smith KM, Larive LL, Romanelli F. Club drugs: methylenedioxymethamphetamine, flunitrazepam, ketamine hydrochloride, and gamma-hydroxybutyrate. Am J Health Syst Pharm 2002;59(11):1067–76.

111. Ramsey-Williams VA, Carter DB. Chronic triazolam and its withdrawal alters GABA$_A$ receptor subunit mRNA levels: an in situ hybridization study. Brain Res Mol Brain Res 1996;43(1–2):132–40.

112. De las Cuevas C, Sanz E, De la Fuente J. Benzodiazepines: more "behavioural" than addiction dependence. Psychopharmacology 2003;167:297-303.

113. Lader M, Morton S. Benzodiazepine problems. Br J Addict 1991;86(7):823–8.

114. Miller NS, Gold MS. Benzodiazepines: tolerance, dependence, abuse, and addiction. J Psychoactive Drugs 1990;22(1):23–33.

115. Olajide D, Lader M. Depression following withdrawal from long-term benzodiazepine use: a report of four cases. Psychol Med 1984;14(4):937–40.

116. Baillargeon L, Landreville P, Verreault R, Beauchemin JP, Gregoire JP, Morin CM. Discontinuation of benzodiazepines among older insomniac adults treated with cognitive-behavioural therapy combined with gradual tapering: a randomized trial. CMAJ 2003;169(10):1015–20.

117. Ashton H. Disorders of the foetus and infant. In: Davies DM, editor. Textbook of Adverse Drug Reactions. 3rd ed.. Oxford: Oxford University Press, 1985:77.

118. Rosenberg L, Mitchell AA, Parsells JL, Pashayan H, Louik C, Shapiro S. Lack of relation of oral clefts to diazepam use during pregnancy. N Engl J Med 1983;309(21):1282–5.

119. Laegreid L, Olegard R, Wahlstrom J, Conradi N. Abnormalities in children exposed to benzodiazepines in utero. Lancet 1987;1(8524):108–9.

120. Eros E, Czeizel AE, Rockenbauer M, Sorensen HT, Olsen J. A population-based case-control teratologic study of nitrazepam, medazepam, tofisopam, alprazolum and clonazepam treatment during pregnancy. Eur J Obstet Gynecol Reprod Biol 2002;101(2):147–54.

121. Thadani PV. Biological mechanisms and perinatal exposure to abused drugs. Synapse 1995;19(3):228–32.

122. Boutroy MJ. Drug-induced apnea. Biol Neonate 1994;65(3–4):252–7.

123. Anonymous. Benzodiazepines: general statement. AHFS Drug Information 1998;1934.

124. Spigset O. Anaesthetic agents and excretion in breast milk. Acta Anaesthesiol Scand 1994;38(2):94–103.

125. Pons G, Rey E, Matheson I. Excretion of psychoactive drugs into breast milk. Pharmacokinetic principles and recommendations. Clin Pharmacokinet 1994;27(4):270–89.

126. Matheson I, Lunde PK, Bredesen JE. Midazolam and nitrazepam in the maternity ward: milk concentrations and clinical effects. Br J Clin Pharmacol 1990;30(6):787–93.

127. Michalodimitrakis M, Christodoulou P, Tsatsakis AM, Askoxilakis I, Stiakakis I, Mouzas I. Death related to midazolam overdose during endoscopic retrograde cholangiopancreatography. Am J Forensic Med Pathol 1999;20(1):93–7.

128. Drummer OH, Syrjanen ML, Cordner SM. Deaths involving the benzodiazepine flunitrazepam. Am J Forensic Med Pathol 1993;14(3):238–43.

129. Aderjan R, Mattern R. Eine tödlich verlaufene Monointoxikation mit Flurazepam (Dalmadorm). Probleme bei der toxikologischen Beurteilung. [A fatal monointoxication by flurazepam (Dalmadorm). Problems of the toxicological interpretation.] Arch Toxicol 1979;43(1):69–75.

130. Mendelson WB, Rich CL. Sedatives and suicide: the San Diego study. Acta Psychiatr Scand 1993;88(5):337–41.

131. Bruyndonckx RB, Meulemans AI, Sabbe MB, Kumar AA, Delooz HH. Fatal intentional poisoning cases admitted to the University Hospitals of Leuven, Belgium from 1993 to 1996. Eur J Emerg Med 2002;9(3):238–43.

132. Shah R, Uren Z, Baker A, Majeed A. Trends in suicide from drug overdose in the elderly in England and Wales, 1993–1999. Int J Geriatr Psychiatry 2002;17(5):416–21.

133. Burrows DL, Hagardorn AN, Harlan GC, Wallen ED, Ferslew KE. A fatal drug interaction between oxycodone and clonazepam. J Forensic Sci 2003;48(3):683–6.

134. Drummer OH, Syrjanen ML, Phelan M, Cordner SM. A study of deaths involving oxycodone. J Forensic Sci 1994;39(4):1069–75.

135. Michaud K, Augsburger M, Romain N, Giroud C, Mangin P. Fatal overdose of tramadol and alprazolam. Forensic Sci Int 1999;105(3):185–9.

136. Kudo K, Imamura T, Jitsufuchi N, Zhang XX, Tokunaga H, Nagata T. Death attributed to the toxic interaction of triazolam, amitriptyline and other psychotropic drugs. Forensic Sci Int 1997;86(1–2):35–41.

137. Schmidt LE, Dalhoff K. Concomitant overdosing of other drugs in patients with paracetamol poisoning. Br J Clin Pharmacol 2002;53(5):535–41.

138. Warot D, Bergougnan L, Lamiable D, Berlin I, Bensimon G, Danjou P, Puech AJ. Troleandomycin–triazolam interaction in healthy volunteers: pharmacokinetic and psychometric evaluation. Eur J Clin Pharmacol 1987;32(4):389–93.

139. Phillips JP, Antal EJ, Smith RB. A pharmacokinetic drug interaction between erythromycin and triazolam. J Clin Psychopharmacol 1986;6(5):297–9.

140. Gascon MP, Dayer P, Waldvogel F. Les interactions médicamenteuses du midazolam. [Drug interactions of midazolam.] Schweiz Med Wochenschr 1989;119(50):1834–6.

141. Hiller A, Olkkola KT, Isohanni P, Saarnivaara L. Unconsciousness associated with midazolam and erythromycin. Br J Anaesth 1990;65(6):826–8.

142. Amsden GW. Macrolides versus azalides: a drug interaction update. Ann Pharmacother 1995;29(9):906–17.

143. Moser L, Huther KJ, Koch-Weser J, Lundt PV. Effects of terfenadine and diphenhydramine alone or in combination with diazepam or alcohol on psychomotor performance and subjective feelings. Eur J Clin Pharmacol 1978;14(6):417–23.

144. Patat A, Perault MC, Vandel B, Ulliac N, Zieleniuk I, Rosenzweig P. Lack of interaction between a new antihistamine, mizolastine, and lorazepam on psychomotor performance and memory in healthy volunteers. Br J Clin Pharmacol 1995;39(1):31–8.

145. Mattila MJ, Aranko K, Kuitunen T. Diazepam effects on the performance of healthy subjects are not enhanced by treatment with the antihistamine ebastine. Br J Clin Pharmacol 1993;35(3):272–7.

146. Johnson LC, Spinweber CL, Gomez SA, Matteson LT. Daytime sleepiness, performance, mood, nocturnal sleep: the effect of benzodiazepine and caffeine on their relationship. Sleep 1990;13(2):121–35.

147. Johnson LC, Spinweber CL, Gomez SA. Benzodiazepines and caffeine: effect on daytime sleepiness, performance, and mood. Psychopharmacology (Berl) 1990;101(2):160–7.

148. Rush CR, Higgins ST, Bickel WK, Hughes JR. Acute behavioral effects of lorazepam and caffeine, alone and in combination, in humans. Behav Pharmacol 1994;5(3):245–54.

149. Jackson CW, Markowitz JS, Brewerton TD. Delirium associated with clozapine and benzodiazepine combinations. Ann Clin Psychiatry 1995;7(3):139–41.

150. Cobb CD, Anderson CB, Seidel DR. Possible interaction between clozapine and lorazepam. Am J Psychiatry 1991;148(11):1606–7.

151. Tupala E, Niskanen L, Tiihonen J. Transient syncope and ECG changes associated with the concurrent administration of clozapine and diazepam. J Clin Psychiatry 1999;60(9):619–20.

152. Hollister LE. Interactions between alcohol and benzodiazepines. Recent Dev Alcohol 1990;8:233–9.

153. Vardy MM, Kay SR. LSD psychosis or LSD-induced schizophrenia? A multimethod inquiry. Arch Gen Psychiatry 1983;40(8):877–83.

154. van de Vijver DA, Roos RA, Jansen PA, Porsius AJ, de Boer A. Influence of benzodiazepines on antiparkinsonian drug treatment in levodopa users. Acta Neurol Scand 2002;105(1):8–12.

155. Bigatti MP, Corona D, Munizza C. Increased sister chromatid exchange and chromosomal aberration frequencies in psychiatric patients receiving psychopharmacological therapy. Mutat Res 1998;413(2):169–75.

156. Wesnes K, Simpson PM, Jansson B, Grahnen A, Weimann HJ, Kuppers H. Moxonidine and cognitive function: interactions with moclobemide and lorazepam. Eur J Clin Pharmacol 1997;52(5):351–8.

157. Driessen JJ, Vree TB, van Egmond J, Booij LH, Crul JF. In vitro interaction of diazepam and oxazepam with pancuronium and suxamethonium. Br J Anaesth 1984;56(10):1131–8.

158. Wali FA. Myorelaxant effect of diazepam. Interactions with neuromuscular blocking agents and cholinergic drugs. Acta Anaesthesiol Scand 1985;29(8):785–9.

159. Driessen JJ, Vree TB, van Egmond J, Booij LH, Crul JF. Interaction of midazolam with two non-depolarizing neuromuscular blocking drugs in the rat in vivo sciatic nerve–tibialis anterior muscle preparation. Br J Anaesth 1985;57(11):1089–94.

160. Asbury AJ, Henderson PD, Brown BH, Turner DJ, Linkens DA. Effect of diazepam on pancuronium-induced neuromuscular blockade maintained by a feedback system. Br J Anaesth 1981;53(8):859–63.

161. Cronnelly R, Morris RB, Miller RD. Comparison of thiopental and midazolam on the neuromuscular responses to succinylcholine or pancuronium in humans. Anesth Analg 1983;62(1):75–7.

162. Driessen JJ, Crul JF, Vree TB, van Egmond J, Booij LH. Benzodiazepines and neuromuscular blocking drugs in patients. Acta Anaesthesiol Scand 1986;30(8):642–6.

163. Feldman SA, Crawley BE. Diazepam and muscle relaxants. BMJ 1970;1(697):691.

164. Feldman SA, Crawley BE. Interaction of diazepam with the muscle-relaxant drugs. BMJ 1970;1(5705):336–8.

165. Olkkola KT, Tammisto T. Quantifying the interaction of rocuronium (Org 9426) with etomidate, fentanyl, midazolam, propofol, thiopental, and isoflurane using closed-loop

feedback control of rocuronium infusion. Anesth Analg 1994;78(4):691–6.

166. Bourin M, Baker GB. Therapeutic and adverse effect considerations when using combinations of neuroleptics and benzodiazepines. Saudi Pharm J 1998;3–4:262–5.

167. Hatta K, Takahashi T, Nakamura H, Yamashiro H, Endo H, Kito K, Saeki T, Masui K, Yonezawa Y. A risk for obstruction of the airways in the parenteral use of levomepromazine with benzodiazepine. Pharmacopsychiatry 1998;31(4):126–30.

168. Jochemsen R, van der Graaff M, Boeijinga JK, Breimer DD. Influence of sex, menstrual cycle and oral contraception on the disposition of nitrazepam. Br J Clin Pharmacol 1982;13(3):319–24.

169. Patwardhan RV, Mitchell MC, Johnson RF, Schenker S. Differential effects of oral contraceptive steroids on the metabolism of benzodiazepines. Hepatology 1983;3(2):248–53.

170. Nemeroff CB, DeVane CL, Pollock BG. Newer antidepressants and the cytochrome P450 system. Am J Psychiatry 1996;153(3):311–20.

171. Dresser GK, Spence JD, Bailey DG. Pharmacokinetic–pharmacodynamic consequences and clinical relevance of cytochrome P450 3A4 inhibition. Clin Pharmacokinet 2000;38(1):41–57.

172. Lemberger L, Rowe H, Bosomworth JC, Tenbarge JB, Bergstrom RF. The effect of fluoxetine on the pharmacokinetics and psychomotor responses of diazepam. Clin Pharmacol Ther 1988;43(4):412–9.

173. Lasher TA, Fleishaker JC, Steenwyk RC, Antal EJ. Pharmacokinetic pharmacodynamic evaluation of the combined administration of alprazolam and fluoxetine. Psychopharmacology (Berl) 1991;104(3):323–7.

174. Beresford TP, Feinsilver DL, Hall RC. Adverse reactions to benzodiazepine–tricyclic antidepressant compound. J Clin Psychopharmacol 1981;1(6):392–4.

Abecarnil

General Information

Abecarnil is a partial agonist at the benzodiazepine-GABA receptor complex, and is used in generalized anxiety disorder. Its pharmacology suggests that it may be less likely to produce sedation and tolerance, but data thus far have not shown clear differences in its adverse effects from those of classical benzodiazepines, such as alprazolam, diazepam, and lorazepam. As expected, both acute adverse effects and tolerance are dose-related.

In a multicenter, double-blind trial, abecarnil (mean daily dose 12 mg), diazepam (mean daily dose 22 mg), or placebo were given in divided doses for 6 weeks to 310 patients with generalized anxiety disorder (1). Those who had improved at 6 weeks could volunteer to continue double-blind treatment for a total of 24 weeks. Slightly more patients who took diazepam (77%) and placebo (75%) completed the 6-week study than those who took abecarnil (66%). The major adverse events during abecarnil therapy were similar to those of diazepam, namely drowsiness, dizziness, fatigue, and difficulty in

coordination. Abecarnil and diazepam both produced statistically significantly more symptom relief than placebo at 1 week, but at 6 weeks only diazepam was superior to placebo. In contrast to diazepam, abecarnil did not cause withdrawal symptoms. The absence of a placebo control makes it difficult to interpret the results of another study of the use of abecarnil and diazepam in alcohol withdrawal, which appeared to show comparable efficacy and adverse effects of the two drugs (2).

References

1. Rickels K, DeMartinis N, Aufdembrinke B. A double-blind, placebo-controlled trial of abecarnil and diazepam in the treatment of patients with generalized anxiety disorder. J Clin Psychopharmacol 2000;20(1):12–8.

2. Anton RF, Kranzler HR, McEvoy JP, Moak DH, Bianca R. A double-blind comparison of abecarnil and diazepam in the treatment of uncomplicated alcohol withdrawal. Psychopharmacology (Berl) 1997;131(2):123–9.

Alprazolam

General Information

Alprazolam, a triazolobenzodiazepine, has been marketed as an anxiolytic with additional antidepressant properties; an analogue, adinazolam, also has partial antidepressant activity (1) and is useful in panic disorder. Like other benzodiazepines, alprazolam is effective in acute and generalized anxiety; its efficacy in panic disorder (2,3), premenstrual syndrome (4), and chronic pain (5) is complicated by high rates of adverse effects (6). On the other hand, low-dose alprazolam (1.4 mg/day) is useful and well tolerated in the treatment of anxiety associated with schizophrenia (SEDA-19, 34).

The value of the Saskatchewan data files in an acute adverse event signalling scheme has been evaluated using two benzodiazepines (7). The first 20 000 patients taking lorazepam and the first 8525 patients taking alprazolam were followed for 12 months after the initial prescription. The most frequent adverse drug reactions associated with these benzodiazepines were drowsiness, depression, impaired intellectual function and memory, lethargy, impaired coordination, dizziness, nausea and/or vomiting, skin rashes, and respiratory disturbance. Sleep disorders, depression, dizziness and/or vertigo, respiratory depression, gastrointestinal disorders, and inflammatory skin conditions occurred significantly more often during the first 30 days after the initial prescription than during the next 6 months.

Comparative studies

In a randomized, crossover, open study of the control of nausea and vomiting in 19 patients with operable breast cancer, granisetron alone was compared with granisetron plus alprazolam (8). Alprazolam increased the efficacy of granisetron. The addition of alprazolam did not increase the incidence of adverse reactions to granisetron, but neither the adverse effects nor their frequencies were specified.

Placebo-controlled studies

Alprazolam 0.25 mg or 1 mg has been evaluated in 47 otherwise healthy subjects, selected for a moderate to high degree of anxiety before oral surgery in a three-arm, parallel design, double-blind, randomized, placebo-controlled study (9). There were 27 adverse events in the interval between dosing and surgery: 2, 10, and 15 events each with placebo, alprazolam 0.25 mg and alprazolam 1 mg respectively. The most common events were drowsiness with placebo and drowsiness, dizziness, lightheadedness, and nausea with alprazolam. There were also single reports of trembling, feeling cold, anxiety, panic attacks, a desire to smoke, increased appetite, sleepiness, and dry mouth with alprazolam. There were no serious adverse events and no subject withdrew from the study because of adverse events.

Organs and Systems

Cardiovascular

Alprazolam has been associated with hypotension (10).

- A 76-year-old woman, who had a history of hypertension, valvular heart disease (mitral regurgitation) with chronic atrial fibrillation, chronic obstructive airways disease, diverticular disease of the sigmoid colon, and generalized anxiety disorder, developed severe hypotension with a tachycardia after taking alprazolam for 7 days. She also had severe weakness, depressed mood, and impaired gait and balance, without clinical features of neuromuscular disease.

Psychological, psychiatric

Rapid and sometimes serious mood swings to mania or depression, and other adverse effects, including enuresis, aggression, impaired memory, sedation, and ataxia, can occur in patients with panic disorder treated with alprazolam (SEDA-19, 34; SEDA-20, 31).

Disinhibition has been reported as a major problem with alprazolam, particularly in patients with borderline personality disorder (11). Several case reports have suggested that alprazolam can cause behavioral disinhibition (12), in common with other benzodiazepines that are occasionally used for recreational or criminal purposes (13). In one study, covering the period January 1989 to June 1990, the medical records of 323 psychiatric inpatients treated with alprazolam, clonazepam, or no benzodiazepine were reviewed (14). The frequencies of behavioral disturbances were not significantly different in the different groups, suggesting that alprazolam does not have unique disinhibitory activity and that disinhibition with benzodiazepines may not be an important

clinical problem in all psychiatric populations. The study design did not allow the establishment of a relation between the prescription of the benzodiazepine and worsening behaviors, and the findings need to be interpreted conservatively, because it was a retrospective review of a heterogeneous population.

Agoraphobia/panic disorder occurred in 31 patients, 15 of whom had originally been treated with alprazolam and 16 with placebo, had been previously followed during an 8-week treatment period, and had alprazolam-induced memory impairment (15). These patients were reviewed 3.5 years after treatment to determine whether the memory impairment persisted. Those who had used alprazolam performed as well as those who had taken placebo on the memory task and other objective tests. The performances in both groups were similar to pretreatment values. However, there were differences in subjective ratings: those who had used alprazolam rated themselves as less attentive and clear-headed and more incompetent and clumsy. Memory impairment found while patients were taking alprazolam did not persist 3.5 years later.

Abrupt withdrawal of alprazolam after prolonged treatment of panic disorder is associated with panic attacks.

- A 77-year-old married woman with panic attacks did not experience them while she took alprazolam 0.5 mg bd for 5 months; however, the attacks recurred after an increase in dose to 0.5 mg qds (16).

The authors suggested that the duration of action of alprazolam is too brief to prevent rebound anxiety with administration four times a day, but this explanation is highly speculative. This case illustrates the potential severity of alprazolam rebound and how its long-term use can exacerbate the symptoms for which it was originally administered.

In a placebo-controlled, within-subject, repeated-measures study of the effects of alprazolam on human risk-taking behavior, 16 adults were given placebo or alprazolam 0.5, 1.0, and 2.0 mg (17). Alprazolam produced dose-related changes in subjective effects and response rates and dose-dependently increased selection of the risky response option. At a dose of 2.0 mg there was an increased probability of making consecutive risky responses following a gain on the risky response option. Thus, alprazolam increased risk-taking under laboratory conditions. In agreement with previous studies, the observed shift in trial-by-trial response probabilities suggested that sensitivity to consequences (for example oversensitivity to recent rewards) may be an important mechanism in the psychopharmacology of risky decision making. Additionally, risk-seeking personality traits may predict the acute effects of drugs on risk-taking behavior.

In a double-blind, crossover, placebo-controlled study in 12 healthy men of the impact of alprazolam 0.25 and 1.00 mg on aspects of action monitoring, i.e. the monitoring of response conflict and the detection and correction of errors by means of neurophysiological measures, alprazolam significantly reduced the amplitude of the error-related negativity (ERN) and therefore affected brain correlates of error detection (18). It increased reaction time and the latencies of lateralized readiness potentials (LRP), thereby affecting motor preparation. It had no effect on amplitude differences in the N2 amplitude component between congruent and incongruent trials, and therefore did not disturb conflict monitoring on correct trials. Alprazolam did not disturb post-error adjustments of behavior.

The cognitive effects of a single dose of alprazolam 0.5 or 1 mg on measures of psychomotor function, visual attention, working memory, planning, and learning have been assessed in 36 healthy adults in a double-blind, parallel-group study (19). Alprazolam 0.5 mg reduced only the speed of attentional performance, although the magnitude of this reduction was large (d = 0.8). At a dose of 1.0 mg, there was impairment of psychomotor function, equivalent to that seen for attentional function at the lower dose. In addition, there was moderate impairment (d approx = 0.5) in working memory and learning. These results suggest that low-dose alprazolam primarily alters visual attentional function. At the higher dose psychomotor functions also became impaired, and it is likely that a combination of these led to the observed moderate impairment of higher-level executive and memory processes.

In a double-blind, placebo-controlled, repeated-measures design 16 healthy volunteers took alprazolam 1 mg, L-theanine 200 mg, or placebo (20). The acute effects of alprazolam and L-theanine were assessed under relaxed conditions and in experimentally induced anxiety. Subjective self-reports of anxiety were obtained during both task conditions before and after drug treatment. The results showed some evidence for a relaxing effect of L-theanine during the baseline condition. Alprazolam did not have any anxiolytic effects compared with placebo on any of the measures during the relaxed state. Neither L-theanine nor alprazolam had any significant anxiolytic effects during experimentally induced anxiety. Adverse events were not reported.

Endocrine

Alprazolam can alter dehydroepiandrosterone and cortisol concentrations. Of 38 healthy volunteers who received a single intravenous dose of alprazolam 2 mg over 2 minutes (phase I), 15 of 25 young men (aged 22–35 years) and all 13 elderly men (aged 65–75 years) responded to alprazolam and agreed to participate in a crossover study of placebo and alprazolam infusion to plateau for 9 hours (21). Plasma samples at 0, 1, 4, and 7 hours were assayed for steroid concentrations. Alprazolam produced:

(a) significant increases in dehydroepiandrosterone concentrations at 7 hours in both the young and elderly men;
(b) significant reductions in cortisol concentrations;
(c) no change in dehydroepiandrosterone-S concentrations.

These results suggest that alprazolam modulates peripheral concentrations of dehydroepiandrosterone and that

dehydroepiandrosterone and/or dehydroepiandrosterone-S may have an in vivo role in modulating GABA receptor-mediated responses.

The acute and chronic effects (3 weeks) of alprazolam and lorazepam on plasma cortisol have been examined in 68 subjects (aged 60–83 years), who took oral alprazolam 0.25 or 0.50 mg bd, or lorazepam 0.50 or 1.0 mg bd, or placebo according to a randomized, double-blind, placebo-controlled, parallel design (22). Plasma cortisol concentrations were significantly affected compared with placebo, but only by the 0.5 mg dose of alprazolam. During the first and last days of treatment, there was a significant fall in cortisol at 2.5 hours after alprazolam compared with placebo. The predose cortisol concentrations increased significantly during chronic alprazolam treatment, and there were correlations between the cortisol changes and changes in depression, anxiety, and memory scores. These findings suggest that even a short period of chronic treatment with alprazolam, but not lorazepam, may result in interdose activation of the hypothalamic–pituitary–adrenal axis in the elderly, consistent with drug withdrawal. If confirmed, this effect may contribute to an increased risk for drug escalation and dependence during chronic alprazolam treatment.

In a parallel, double-blind, placebo-controlled study in 13 elderly women and 12 elderly men, alprazolam 0.5 mg bd for 3 weeks caused significant rises in inter-dose morning plasma cortisol concentrations in the women but not in the men (23). In addition, higher morning plasma cortisol concentrations were significantly associated with better cognitive performance. The authors concluded that elderly women had greater inter-dose activation of the hypothalamic–pituitary–adrenal axis during treatment with therapeutic doses of alprazolam than men, but they stated that this could have been related to drug withdrawal.

In a double-blind, crossover, placebo-controlled study of the effects of alprazolam 5 mg and dehydroepiandrosterone (DHEA) 100 mg/day, alone and in combination, on hypothalamic–pituitary–adrenal axis activity in 15 men (aged 20–45 years; body mass index 20–25 kg/m^2), alprazolam significantly increased basal growth hormone and blunted the responses to exercise of plasma cortisol, ACTH, AVP, and DHEA (24). DHEA and alprazolam in combination significantly increased the growth hormone response to exercise. The authors concluded that DHEA and alprazolam up-regulate growth hormone during exercise, perhaps by blunting a suppressive (HPA axis) system and potentiating an excitatory (glutamate receptor) system.

Skin

Alprazolam, which is lipid-soluble, can cause photosensitivity after a long duration of administration.

- A 65-year-old man developed pruritic erythema on sun-exposed areas (photosensitivity) due to alprazolam (25). A photopatch test was negative, but an oral photochallenge test with UVA irradiation was positive after he had taken alprazolam for 17 days.

Long-Term Effects

Drug dependence

Dependence on alprazolam and withdrawal symptoms appear to present greater problems than with other benzodiazepines (SED-12, 98).

Drug withdrawal

Withdrawal symptoms have been described with alprazolam.

- A woman with paranoid schizophrenia developed catatonia 5 days after the abrupt withdrawal of olanzapine and alprazolam (26). The catatonic symptoms included mutism, prostration, waxy flexibility, oculogyric movements, and an inability to swallow. Her symptoms disappeared after administration of alprazolam and haloperidol, and there was no recurrence.
- A 39-year-old woman had withdrawal symptoms after her dose of alprazolam was reduced (27). Cognitive symptoms made it almost impossible for her to stop taking alprazolam or to continue psychotherapeutic treatment. The medication was stopped by means of a behavioral experiment, in which both patient and therapist were unaware of the way in which the medication was reduced, after which continuation of treatment became possible.

Pharmacological strategies for withdrawing alprazolam, by switching to a longer-acting agent, have been proposed (28).

The potential interaction of paroxetine 20 mg/day and alprazolam 1 mg/day for 15 days on polysomnographic sleep and subjective sleep and awakening quality has been evaluated in a randomized, double-blind, double-dummy, placebo-controlled, repeated-dose, four-period, crossover study in 22 young subjects with no history of sleep disturbances (29). There were subjective withdrawal symptoms after abrupt discontinuation of alprazolam, including increased subjective sleep latency and reduced subjective sleep efficiency.

Drug Administration

Drug overdose

The effects of alprazolam overdose have been reported (30,31).

- A 28-year-old African-American man took alprazolam 12 mg. He denied using alcohol, other prescription medications, over-the-counter medications, or illicit drugs. He denied any suicidal intent. He stated that he had taken this large dose because his usual dose of 1–2 mg had failed to relieve his anxiety. He was drowsy and his heart rate was 58/minute. He had

marked first-degree atrioventricular block, with a PR interval of 500 ms.

- A 30-year-old woman, with a history of depression, was found dead after taking an unknown quantity of alprazolam, tramadol, and alcohol. At autopsy, only slight decomposition and diffuse visceral congestion were observed. Blood concentrations of alprazolam, alcohol, and tramadol were 0.21 mg/l, 1.29 g/kg, and 38 mg/l respectively.

The relative toxicity of alprazolam compared with other benzodiazepines has been assessed from a database of consecutive poisoning admissions to a regional toxicology service (32). There were 2065 admissions for single benzodiazepine overdose: alprazolam 131 overdoses, diazepam 823 overdoses and other benzodiazepine 1109 overdoses. The median length of stay for alprazolam overdoses was 19 hours, which was 1.27 times longer than for other benzodiazepines. Of patients with alprazolam overdoses, 22% were admitted to ICU, which was 2.06 times more likely than with other benzodiazepines. Flumazenil was given to 14% of alprazolam patients and 16% were ventilated, which was significantly more than for other benzodiazepine overdoses (8% and 11% respectively). Of those with alprazolam overdoses 12% had a Glasgow Coma Scale score under 9, compared with 10% for other benzodiazepines. The authors concluded that alprazolam was significantly more toxic after overdose than other benzodiazepines.

Drug–Drug Interactions

Alcohol

In common with other benzodiazepines, alprazolam produces additional impairment of performance when it is taken together with alcohol (33). The combination can also produce behavioral disturbance and aggression (13,34).

Alosetron

In an open, randomized, crossover study in 12 healthy men and women, alosetron 1 mg bd did not affect the pharmacokinetics of a single oral dose of alprazolam 1 mg (35).

Amfetamine

Six healthy volunteers learned to recognize the effects of oral D-amfetamine 15 mg and then the effects of a range of doses of D-amfetamine (0, 2.5, 5, 10, and 15 mg), alone and after pre-treatment with alprazolam (0 and 0.5 mg), were assessed (36). Amfetamine alone functioned as a discriminative stimulus and produced stimulant-like self-reported drug effects related to dose. Alprazolam alone did not have amfetamine-like discriminative stimulus effects, nor did it increase ratings of sedation or impair performance. Alprazolam pre-treatment significantly attenuated the discriminative stimulus effects of amfetamine, and some of the self-reported drug effects. The authors suggested that future human laboratory-based studies should compare the behavioral effects of amfetamine alone and after pre-treatment with alprazolam, using other behavioral arrangements, such as drug self-administration. They also suggested that benzodiazepines with lower abuse potential (for example oxazepam) might also attenuate the behavioral effects of amfetamine.

Dextropropoxyphene

Inhibition of alprazolam metabolism by dextropropoxyphene has been reported (37).

Grapefruit juice

There have been two studies of the effects of repeated ingestion of grapefruit juice on the pharmacokinetics and pharmacodynamics of both single and multiple oral doses of alprazolam in a total of 19 subjects (38). Grapefruit juice altered neither the steady-state plasma concentration of alprazolam nor its clinical effects.

Ketoconazole

In a double-blind, crossover, pharmacokinetic and pharmacodynamic study of the interaction of ketoconazole with alprazolam and triazolam, two CYP3A4 substrate drugs with different kinetic profiles, impaired clearance by ketoconazole had more profound clinical consequences for triazolam than for alprazolam (39).

Miocamycin

Hydroxylation of miocamycin metabolites is mainly performed by CYP3A4. Some macrolide antibiotics cause drug interactions that result in altered metabolism of concomitantly administered drugs by the formation of a metabolic intermediate complex with CYP450 or competitive inhibition of CYP450 (40). The resulting interactions can cause rhabdomyolysis (associated with the coadministration of some statins, for example lovastatin or simvastatin), hypoprothrombinemia (associated with warfarin), excessive sedation (associated with certain benzodiazepines, for example alprazolam, diazepam, midazolam, or triazolam), ataxia (associated with carbamazepine), and ergotism (associated with ergotamine).

Moclobemide

- A 44-year-old man developed the serotonin syndrome after taking moclobemide and alprazolam for 1 year (41). The symptoms developed after 4 days of extreme heat, which was thought to have contributed.

Nefazodone

Nefazodone is a weak inhibitor of CYP2D6 but a potent inhibitor of CYP3A4 and it increases plasma concentrations of drugs that are substrates of CYP3A4, such as alprazolam, astemizole, carbamazepine, ciclosporin, cisapride, terfenadine, and triazolam.

Oral contraceptives

Oral contraceptives alter the metabolism of some benzodiazepines that undergo oxidation (alprazolam, chlordiazepoxide, and diazepam) or nitroreduction (nitrazepam) (42). Oral contraceptives inhibit enzyme activity and reduce the clearances of these drugs. There is nevertheless no evidence that this interaction is of clinical importance. It should be noted that for other benzodiazepines that undergo oxidative metabolism, such as bromazepam or clotiazepam, no change has ever been found in oral contraceptive users. Some other benzodiazepines are metabolized by glucuronic acid conjugation. The clearance of temazepam was increased when oral contraceptives were administered concomitantly, but the clearances of lorazepam and oxazepam were not (43). Again, it is unlikely that this is an interaction of clinical importance.

Ritonavir

The inhibitory effect of ritonavir (a viral protease inhibitor) on the metabolism of alprazolam, a CYP3A-mediated reaction, has been investigated in a double-blind study (44). Ten subjects took alprazolam 1.0 mg plus either low-dose ritonavir (four doses of 200 mg) or placebo. Ritonavir reduced alprazolam clearance by 60%, prolonged its half-life, and magnified its benzodiazepine agonist effects, such as sedation and impairment of performance.

Selective serotonin reuptake inhibitors (SSRIs)

Fluoxetine

A within-subject, double-blind, placebo-controlled, parallel design has been used to measure the effects of citalopram (20 mg/day) and fluoxetine (20 mg/day) on the pharmacokinetics and pharmacodynamics of alprazolam (1 mg/day) (45). Fluoxetine significantly impaired the metabolism of a single oral dose of alprazolam 1 mg, leading to prolongation of the half-life and an increased AUC, whereas citalopram did not. Neither SSRI significantly affected the pharmacodynamic effects of alprazolam. This experiment suggests differential effects of citalopram and fluoxetine on alprazolam kinetics.

Fluvoxamine

The effects of co-administration of fluvoxamine on plasma concentrations of alprazolam have been studied in 23 Japanese outpatients (46). All patients were taking fluvoxamine (25–100 mg/day) either before or after monotherapy with alprazolam (0.4–1.6 mg/day). Co-administration with fluvoxamine produced on average a 58% increase in plasma alprazolam concentrations. However, there were wide variations in the plasma concentrations of alprazolam. The interaction was attributed to the CYP2C19 genotype.

Paroxetine

In a double-blind, double-dummy, placebo-controlled, repeated-dose (15 days), 4-period crossover study, each of 25 young adult volunteers received each of four treatment sequences (paroxetine + alprazolam placebo, alprazolam + paroxetine placebo, paroxetine + alprazolam, and paroxetine placebo + alprazolam placebo) in randomized order (47). There was no pharmacodynamic interaction at steady state. The most commonly reported adverse event was drowsiness, with a higher incidence when alprazolam was used, both alone and in combination with paroxetine.

Sertraline

Sertraline (50–150 mg/day) had no effects on alprazolam metabolism in a randomized, double-blind, placebo-controlled study in 10 healthy volunteers (48).

Troleandomycin

Troleandomycin inhibits the metabolism of ecabapide and alprazolam by inhibition of CYP3A4 (49,50).

Venlafaxine

The effects of venlafaxine on the pharmacokinetics of alprazolam have been investigated in 16 healthy volunteers. Steady-state venlafaxine 75 mg bd did not inhibit CYP3A4 metabolism of a single dose of alprazolam 2 mg (51).

References

1. Ansseau M, Devoitille JM, Papart P, Vanbrabant E, Mantanus H, Timsit-Berthier M. Comparison of adinazolam, amitriptyline, and diazepam in endogenous depressive inpatients exhibiting DST nonsuppression or abnormal contingent negative variation. J Clin Psychopharmacol 1991;11(3):160–5.
2. Andersch S, Rosenberg NK, Kullingsjo H, Ottosson JO, Bech P, Bruun-Hansen J, Hanson L, Lorentzen K, Mellergard M, Rasmussen S, et al. Efficacy and safety of alprazolam, imipramine and placebo in treating panic disorder. A Scandinavian multicenter study. Acta Psychiatr Scand 1991;365(Suppl):18–27.
3. O'Sullivan GH, Noshirvani H, Basoglu M, Marks IM, Swinson R, Kuch K, Kirby M. Safety and side-effects of alprazolam. Controlled study in agoraphobia with panic disorder. Br J Psychiatry 1994;165(2):79–86.
4. Mortola JF. A risk-benefit appraisal of drugs used in the management of premenstrual syndrome. Drug Saf 1994;10(2):160–9.
5. Reddy S, Patt RB. The benzodiazepines as adjuvant analgesics. J Pain Symptom Manage 1994;9(8):510–4.
6. Verster JC, Volkerts ER. Clinical pharmacology, clinical efficacy, and behavioral toxicity of alprazolam: a review of the literature. CNS Drug Rev 2004;10(1):45–76.
7. Rawson NS, Rawson MJ. Acute adverse event signalling scheme using the Saskatchewan Administrative health care utilization datafiles: results for two benzodiazepines. Can J Clin Pharmacol 1999;6(3):159–66.
8. Abali H, Oyan B, Guler N. Alprazolam significantly improves the efficacy of granisetron in the prophylaxis of emesis secondary to moderately emetogenic chemotherapy in patients with breast cancer. Chemotherapy 2005;51(5):280–5.
9. Wolf DL, Desjardins PJ, Black PM, Francom SR, Mohanlal RW, Fleishaker JC. Anticipatory anxiety in moderately to highly-anxious oral surgery patients as a screening model for anxiolytics: evaluation of alprazolam. J Clin Psychopharmacol 2003;23:51–7.
10. Ranieri P, Franzoni S, Trabucchi M. Alprazolam and hypotension. Int J Geriatr Psychiatry 1999;14(5):401–2.
11. Gardner DL, Cowdry RW. Alprazolam-induced dyscontrol in borderline personality disorder. Am J Psychiatry 1985;142(1):98–100.
12. Cowdry RW, Gardner DL. Pharmacotherapy of borderline personality disorder. Alprazolam, carbamazepine, trifluoperazine, and tranylcypromine. Arch Gen Psychiatry 1988;45(2):111–9.
13. Michel L, Lang JP. Benzodiazepines et passage à l'acte criminel. [Benzodiazepines and forensic aspects.] Encephale 2003;29(6):479–85.
14. Rothschild AJ, Shindul-Rothschild, Viguera A, Murray M, Brewster S. Comparison of the frequency of behavioral disinhibition on alprazolam, clonazepam, or no benzodiazepine in hospitalized psychiatric patients. J Clin Psychopharmacol 2000;20(1):7–11.
15. Kilic C, Curran HV, Noshirvani H, Marks IM, Basoglu M. Long-term effects of alprazolam on memory: a 3.5 year follow-up of agoraphobia/panic patients Psychol Med 1999;29(1):225–31.
16. Bashir A, Swartz C. Alprazolam-induced panic disorder. J Am Board Fam Pract 2002;15(1):69–72.
17. Lane SD, Tcheremissine OV, Lieving LM, Nouvion S, Cherek DR. Acute effects of alprazolam on risky decision making in humans. Psychopharmacology 2005;181(2):364–73.
18. Riba J, Rodriguez-Fornells A, Munte TF, Barbanoj MJ. A neurophysiological study of the detrimental effects of alprazolam on human action monitoring. Cognitive Brain Res 2005;25(2):554–65.
19. Snyder PJ, Werth J, Giordani B, Caveney AF, Feltner D, Maruff P. A method for determining the magnitude of change across different cognitive functions in clinical trials: the effects of acute administration of two different doses alprazolam. Hum Psychopharmacol 2005;20(4):263–73.
20. Lu K, Gray MA, Oliver C, Liley DT, Harrison BJ, Bartholomeusz CF, Phan KL, Nathan PJ. The acute effects of L-theanine in comparison with alprazolam on anticipatory anxiety in humans. Hum Psychopharmacol 2004;19:457–65.
21. Kroboth PD, Salek FS, Stone RA, Bertz RJ, Kroboth FJ 3rd. Alprazolam increases dehydroepiandrosterone concentrations. J Clin Psychopharmacol 1999;19(2):114–24.
22. Pomara N, Willoughby LM, Ritchie JC, Sidtis JJ, Greenblatt DJ, Nemeroff CB. Interdose elevation in plasma cortisol during chronic treatment with alprazolam but not lorazepam in the elderly. Neuropsychopharmacology 2004;29:605–11.
23. Pomara N, Willoughby LM, Ritchie LC, Sidtis JJ, Greenblatt DJ, Nemeroff CB. Sex-related elevation in cortisol during chronic treatment with alprazolam associated with enhanced cognitive performance. Psychopharmacology 2005;182(3):414–9.
24. Deuster PA, Faraday MM, Chrousos GP, Poth MA. Effects of dehydroepiandrosterone and alprazolam on hypothalamic-pituitary responses to exercise. J Clin Endocrinol Metab 2005;90(8):4777–83.
25. Watanabe Y, Kawada A, Ohnishi Y, Tajima S, Ishibashi A. Photosensitivity due to alprazolam with positive oral photo-challenge test after 17 days administration. J Am Acad Dermatol 1999;40(5 Pt 2):832–3.
26. Roberge C, Mosquet B, Hamel F, Crete P, Starace J. Catatonie après sevrage d'un traitement par olanzapine et alprazolam. [Catatonia after olanzapine and alprazolam withdrawal.] J Clin Pharm 2001;20:163–5.
27. Meesters Y, Van Velzen CJM, Horwitz EH. Een cognitief aspect bij de afbouw van alprazolum. [A cognitive aspect

related to the withdrawal of alprazolam; a case study.] Tijdschr Psychiatrie 2002;44:199–203.

28. Rosenbaum JF. Switching patients from alprazolam to clonazepam. Hosp Community Psychiatry 1990;41(12):1302.

29. Barbanoj MJ, Clos S, Romero S, Morte A, Giménez S, Lorenzo JL, Luque A, Dal-Ré R. Sleep laboratory study on single and repeated dose effects of paroxetine, alprazolam and their combination in healthy young volunteers. Neuropsychobiology 2005;51(3):134–47.

30. Mullins ME. First-degree atrioventricular block in alprazolam overdose reversed by flumazenil. J Pharm Pharmacol 1999;51(3):367–70.

31. Michaud K, Augsburger M, Romain N, Giroud C, Mangin P. Fatal overdose of tramadol and alprazolam. Forensic Sci Int 1999;105(3):185–9.

32. Isbister GK, O'Regan L, Sibbritt D, Whyte IA. Alprazolam is relatively more toxic than other benzodiazepines in overdose. Br J Clin Pharmacol 2004;58(1):88–95.

33. Linnoila M, Stapleton JM, Lister R, Moss H, Lane E, Granger A, Eckardt MJ. Effects of single doses of alprazolam and diazepam, alone and in combination with ethanol, on psychomotor and cognitive performance and on autonomic nervous system reactivity in healthy volunteers. Eur J Clin Pharmacol 1990;39(1):21–8.

34. Bond AJ, Silveira JC. The combination of alprazolam and alcohol on behavioral aggression. J Stud Alcohol 1993;11(Suppl):30–9.

35. D'Souza DL, Levasseur LM, Nezamis J, Robbins DK, Simms L, Koch KM. Effect of alosetron on the pharmacokinetics of alprazolam. J Clin Pharmacol 2001;41(4):452–4.

36. Rush CR, Stoops WW, Wagner FP, Hays LR, Glaser PEA. Alprazolam attenuates the behavioral effects of d-amphetamine in humans. J Clin Psychopharmacol 2004;24(4):410–20.

37. Hansen BS, Dam M, Brandt J, Hvidberg EF, Angelo H, Christensen JM, Lous P. Influence of dextropropoxyphene on steady state serum levels and protein binding of three anti-epileptic drugs in man. Acta Neurol Scand 1980;61(6):357–67.

38. Yasui N, Kondo T, Furukori H, Kaneko S, Ohkubo T, Uno T, Osanai T, Sugawara K, Otani K. Effects of repeated ingestion of grapefruit juice on the single and multiple oral-dose pharmacokinetics and pharmacodynamics of alprazolam. Psychopharmacology (Berl) 2000;150(2):185–90.

39. Greenblatt DJ, Wright CE, von Moltke LL, Harmatz JS, Ehrenberg BL, Harrel LM, Corbett K, Counihan M, Tobias S, Shader RI. Ketoconazole inhibition of triazolam and alprazolam clearance: differential kinetic and dynamic consequences. Clin Pharmacol Ther 1998;64(3):237–47.

40. Rubinstein E. Comparative safety of the different macrolides. Int J Antimicrob Agents 2001;18(Suppl 1):S71–6.

41. Butzkueven H. A case of serotonin syndrome induced by moclobemide during an extreme heatwave. Aust NZ J Med 1997;27(5):603–4.

42. Jochemsen R, van der Graaff M, Boeijinga JK, Breimer DD. Influence of sex, menstrual cycle and oral contraception on the disposition of nitrazepam. Br J Clin Pharmacol 1982;13(3):319–24.

43. Patwardhan RV, Mitchell MC, Johnson RF, Schenker S. Differential effects of oral contraceptive steroids on the metabolism of benzodiazepines. Hepatology 1983;3(2):248–53.

44. Greenblatt DJ, von Moltke LL, Harmatz JS, Durol AL, Daily JP, Graf JA, Mertzanis P, Hoffman JL, Shader RI. Alprazolam-ritonavir interaction: implications for product labeling. Clin Pharmacol Ther 2000;67(4):335–41.

45. Hall J, Naranjo CA, Sproule BA, Herrmann N. Pharmacokinetic and pharmacodynamic evaluation of the inhibition of alprazolam by citalopram and fluoxetine. J Clin Psychopharmacol 2003;23:349–57.

46. Suzuki Y, Shioiri T, Muratake T, Kawashima Y, Sato S, Hagiwara M, Inoue Y, Shimoda K, Someya T. Effects of concomitant fluvoxamine on the metabolism of alprazolam in Japanese psychiatric patients: interaction with CYP2C19 mutated alleles. Eur J Clin Pharmacol 2003;58:829–33.

47. Calvo G, Garcia-Gea C, Luque A, Morte A, Dal-Ré, Barbanoj M. Lack of pharmacologic interaction between paroxetine and alprazolam at steady state in health volunteers. J Clin Psychopharmacol 2004;24(3):268–76.

48. Hassan PC, Sproule BA, Naranjo CA, Herrmann N. Dose-response evaluation of the interaction between sertraline and alprazolam in vivo. J Clin Psychopharmacol 2000;20(2):150–8.

49. Juurlink DN. Ito S. Comment: clarithromycin–digoxin interaction. Ann Pharmacother 1999;33(12):1375–6.

50. Piquette RK. Torsade de pointes induced by cisapride/clarithromycin interaction. Ann Pharmacother 1999;33(1):22–6.

51. Amchin J, Zarycranski W, Taylor KP, Albano D, Klockowski PM. Effect of venlafaxine on the pharmacokinetics of alprazolam. Psychopharmacol Bull 1998;34(2):211–9.

Bentazepam

General Information

Bentazepam is a benzodiazepine with properties similar to those of diazepam.

Organs and Systems

Liver

In three cases chronic hepatocellular injury developed with oral bentazepam (1).

Reference

1. Andrade RJ, Lucena MI, Aguilar J, Lazo MD, Camargo R, Moreno P, Garcia-Escano MD, Marquez A, Alcantara R, Alcain G. Chronic liver injury related to use of bentazepam: an unusual instance of benzodiazepine hepatotoxicity. Dig Dis Sci 2000;45(7):1400–4.

Bromazepam

General Information

Bromazepam, a moderately short-acting benzodiazepine (half-life about 12 hours), has been used in the treatment of anxiety states and has the usual effects of benzodiazepines, for example amnesia and depressed psychomotor performance, although it causes less depression than lorazepam (SED-12, 98).

Organs and Systems

Nervous system

The mechanism of rare extrapyramidal effects with bromazepam is unexplained (SEDA-18, 44).

Long-Term Effects

Drug withdrawal

Catatonia has been described as an effect of bromazepam withdrawal (1).

- A 51-year-old man who had been taking bromazepam for 9 years in a dosage that he had gradually increased to 18 mg/day abruptly stopped taking it. He initially developed psychotic symptoms and on day 5 after withdrawal became mute and was posturing. A provisional diagnosis of a catatonic syndrome was made, and he was given oral lorazepam 3 mg and risperidone 4 mg. It was later concluded that the catatonia had most probably been due to benzodiazepine withdrawal; the risperidone was withdrawn, and diazepam was substituted for lorazepam and slowly tapered. He had no recurrence of catatonic or psychotic symptoms and his fatigue improved.

Withdrawal-induced seizures have been described in a woman taking various benzodiazepines (2).

- A 43-year-old woman had had insomnia since she was a child. At the age of 15, benzodiazepine therapy improved her sleeping, but when she gradually stopped taking benzodiazepines the insomnia returned after a few days. At the age of 26 she was abusing several benzodiazepines, including diazepam and flunitrazepam. At the age of 30 she was taking high doses of bromazepam every evening before going to sleep. After 1 month, she abruptly stopped taking bromazepam and during withdrawal had an epileptic seizure. During the next few years, she had periods of relative well-being, but also two further periods of benzodiazepine abuse, both resulting in seizures after withdrawal. She later had withdrawal seizures with zolpidem.

Drug Administration

Drug overdose

A fatal overdose of bromazepam has been reported (3).

- A 42-year-old woman with a history of depression was found unconscious, lying near her car. The lower part of her body was undressed and there were multiple purple spots and excoriations on her body, suggesting a sexual assault. She was hypothermic (core temperature 28.4°C). Her plasma bromazepam concentration

was 7.7 mg/l, the highest concentration reported in a case of fatal intoxication.

Drug–Drug Interactions

Fluconazole

The interaction of bromazepam with fluconazole has been studied in 12 healthy men in a randomized, double-blind, four-way crossover study (4). The subjects took a single oral or rectal dose of bromazepam (3 mg) after pretreatment for 4 days with oral fluconazole 100 mg/day or placebo. Pharmacodynamic effects of bromazepam were assessed using self-rated drowsiness, the continuous number addition test, and electroencephalography. After rectal administration there was a higher AUC (1.7-fold) and a higher C_{max} (1.6-fold) than after oral administration; there were electroencephalographic effects and subjective drowsiness after rectal bromazepam, and the electroencephalographic effects correlated closely with mean plasma bromazepam concentrations. However, fluconazole caused no significant changes in the pharmacokinetics or pharmacodynamics of oral or rectal bromazepam.

References

1. Deuschle M, Lederbogen F. Benzodiazepine withdrawal-induced catatonia. Pharmacopsychiatry 2001;34(1):41–2.
2. Aragona M. Abuse, dependence, and epileptic seizures after zolpidem withdrawal. Review and case report. Clin Neuropharmacol 2000;23(5):281–3.
3. Michaud K, Romain N, Giroud C, Brandt C, Mangin P. Hypothermia and undressing associated with non-fatal bromazepam intoxication. Forensic Sci Int 2001;124(2–3):112–4.
4. Ohtani Y, Kotegawa T, Tsutsumi K, Morimoto T, Hirose Y, Nakano S. Effect of fluconazole on the pharmacokinetics and pharmacodynamics of oral and rectal bromazepam: an application of electroencephalography as the pharmacodynamic method. J Clin Pharmacol 2002;42(2):183–91.

Brotizolam

General Information

Brotizolam is a triazolothienodiazepine used in the treatment of insomnia and also has anticonvulsant, antianxiety and muscle relaxant properties. It reduces latency to sleep, reduces the number of awakenings and waking time during sleep, and increases total sleep time (1). It may delay the onset of REM sleep but has no effect on slow-wave sleep (2). It has an intermediate half-life of about 5 hours, and is said to cause no early-morning rebound insomnia and minimal morning drowsiness (3). Its most common adverse effects are drowsiness, headache, and dizziness; mild rebound insomnia can occur after withdrawal.

Long-Term Effects

Drug withdrawal

In 63 out-patients with chronic insomnia who were given either brotizolam 0.25 mg or 0.5 mg at night for 3 weeks (n = 29) or placebo (n = 34), those who switched abruptly from brotizolam to placebo had rebound insomnia, which was most marked on the first post-brotizolam placebo night (4).

Drug–Drug Interactions

The metabolism of brotizolam is altered by drugs that inhibit CYP3A4, such as erythromycin and itraconazole.

Erythromycin

Erythromycin 1200 mg/day or placebo for 7 days was given to 14 healthy men in a double-blind, randomized, crossover design (5). On the sixth day they took a single oral dose of brotizolam 0.5 mg and blood samples were taken for 24 hours. Erythromycin significantly increased the C_{max}, AUC, and half-life of brotizolam.

Itraconazole

In a randomized, double-blind, placebo-controlled, cross-over study in 10 healthy men, itraconazole 200 mg/day for 4 days significantly altered the pharmacokinetics of a single dose of brotizolam 0.5 mg; it reduced the apparent oral clearance four-fold, increased the $AUC_{0 \rightarrow 24h}$ more than two-fold, and prolonged the half-life five-fold (6). The digit symbol substitution test and sleepiness were significantly altered as a result.

References

1. Roehrs T, Zorick F, Koshorek GL, Wittig R, Roth T. Effects of acute administration of brotizolam in subjects with disturbed sleep. Br J Clin Pharmacol 1983;16 Suppl 2:371S–376S.
2. Nicholson AN. Brotizolam: studies of effects on sleep and on performance in young adulthood and in middle age. Br J Clin Pharmacol 1983;16 Suppl 2:365S–369S.
3. Langley MS, Clissold SP. Brotizolam. A review of its pharmacodynamic and pharmacokinetic properties, and therapeutic efficacy as an hypnotic. Drugs 1988;35(2):104–22.
4. Rickels K, Morris RJ, Mauriello R, Rosenfeld H, Chung HR, Newman HM, Case WG. Brotizolam, a triazolothienodiazepine, in insomnia. Clin Pharmacol Ther 1986;40(3):293–9.
5. Tokairin T, Fukasawa T, Yasui-Furukori N, Aoshima T, Suzuki A, Inoue Y, Tateishi T, Otani K. Inhibition of the metabolism of brotizolam by erythromycin in humans: in vivo evidence for the involvement of CYP3A4 in brotizolam metabolism. Br J Clin Pharmacol 2005;60(2):172–5.
6. Osanai T, Ohkubo T, Yasui N, Kondo T, Kaneko S. Effect of itraconazole on the pharmacokinetics and pharmacodynamics of a single oral dose of brotizolam. Br J Clin Pharmacol 2004;58(5):476–81.

Chlordiazepoxide

General Information

Chlordiazepoxide, which has a long duration of action ($t_{1/2}$ = 10–25 hours), is useful for the management of alcohol withdrawal and is arguably better tolerated than other benzodiazepines when used for this indication. As with diazepam, loading doses are possible and simplify clinical management.

Organs and Systems

Metabolism

A San Francisco woman with a history of diabetes and high blood pressure was hospitalized in January 2001 with a life-threatening low blood sugar concentration after she consumed Anso Comfort capsules (1). The authors conjectured that hospitalization may have been necessitated by a drug interaction of chlordiazepoxide with medications that she was taking for other medical conditions.

Hematologic

- A 68-year-old man, who had been taking lorazepam, perphenazine, and amitriptyline for many years, developed acute thrombocytopenic purpura after combination therapy of chlordiazepoxide 5 mg and clidinium 2.5 mg tds for irritable bowel syndrome (2). His disease improved after withdrawal of chlordiazepoxide and clidinium and treatment with intravenous prednisolone.

In this case it was not clear which of the two compounds caused the purpura; it is possible that it was due to the combination.

Second-Generation Effects

Teratogenicity

The teratogenic potential of oral chlordiazepoxide has been studied by comparing 22 865 cases with congenital abnormalities and 38 151 matched healthy controls (3). Chlordiazepoxide had been used during pregnancy in 201 cases (0.88%) and 268 controls (0.70%). There was a significantly higher odds ratio for chlordiazepoxide use during the second and third months of gestation in those with congenital cardiovascular malformations. However, this association was found when exposure data were based mainly on maternal self-reported chlordiazepoxide use. There was no increase in the rate of any specific congenital cardiovascular malformation type. In conclusion, therapeutic doses of chlordiazepoxide during pregnancy are unlikely to pose a substantial teratogenic risk to the human fetus, although a somewhat higher rate of congenital cardiovascular malformations cannot be excluded.

Drug Administration

Drug formulations

The California State Health Director has warned consumers to stop using the herbal product Anso Comfort capsules immediately, because the product contains the undeclared prescription drug chlordiazepoxide (1). Anso Comfort capsules, available by mail or telephone order from the distributor in 60-capsule bottles, are clear with dark green powder inside. The label is yellow with green English printing and a picture of a plant. An investigation by the California Department of Health Services Food and Drug Branch and Food and Drug Laboratory showed that the product contains chlordiazepoxide. The ingredients for the product were imported from China and the capsules were manufactured in California. Advertising for the product claims that the capsules are useful for the treatment of a wide variety of illnesses, including high blood pressure and high cholesterol, in addition to claims that it is a natural herbal dietary supplement. The advertising also claims that the product contains only Chinese herbal ingredients and that consumers may reduce or stop their need for prescribed medicines. No clear medical evidence supports any of these claims. The distributor, NuMeridian (formerly known as Top Line Project), has voluntarily recalled the product.

Drug–Drug Interactions

Heparin

In normal non-fasting subjects, 100–1000 IU of heparin given intravenously caused a rapid increase in the free fractions of diazepam, chlordiazepoxide, and oxazepam (4,5), but no change in the case of lorazepam (4). The clinical implications of this finding are not known.

Oral contraceptives

In 7 healthy young women who had taken oral contraceptives for more than 6 months the protein binding of chlordiazepoxide was reduced and its volume of distribution increased (6). The clearance of chlordiazepoxide is also reportedly reduced by oral contraceptives (7).

Mid-cycle spotting occurred in a large proportion of 72 women taking oral contraceptives and chlordiazepoxide; however, there were no pregnancies (8).

References

1. Anonymous. Herbal medicine. Warning: found to contain chlordiazepoxide. WHO Pharm Newslett 2001;1:2–3.
2. Alexopoulou A, Michael A, Dourakis SP. Acute thrombocytopenic purpura in a patient treated with chlordiazepoxide and clidinium. Arch Intern Med 2001;161(14):1778.
3. Czeizel AE, Rockenbauer M, Sorenson, Olsen J. A population-based case-control study of oral chlordiazepoxide use during pregnancy and risk of congenital abnormalities. Neurotoxicol Teratol 2004;26(4):593-8.
4. Desmond PV, Roberts RK, Wood AJ, Dunn GD, Wilkinson GR, Schenker S. Effect of heparin administration on plasma binding of benzodiazepines. Br J Clin Pharmacol 1980;9(2):171–5.
5. Routledge PA, Kitchell BB, Bjornsson TD, Skinner T, Linnoila M, Shand DG. Diazepam and N-desmethyldiazepam redistribution after heparin. Clin Pharmacol Ther 1980;27(4):528–32.
6. Roberts RK, Desmond PV, Wilkinson GR, Schenker S. Disposition of chlordiazepoxide: sex differences and effects of oral contraceptives. Clin Pharmacol Ther 1979;25(6):826–31.
7. Back DJ, Orme ML. Pharmacokinetic drug interactions with oral contraceptives. Clin Pharmacokinet 1990;18(6):472–84.
8. Somos P. Interaction between certain psychopharmaca and low-dose oral contraceptives. Ther Hung 1990;38(1):37–40.

Clobazam

General Information

Clobazam, a 1,5-benzodiazepine, differs in its chemical structure from most other benzodiazepines. It has been claimed to have less sedative effects for its effective anticonvulsant and anti-anxiety effects (SED-12, 98). Whether because of tolerance or not, clobazam tends to be less sedative than clonazepam. Both the therapeutic and adverse effects of clobazam have been related to its major metabolite N-desmethylclobazam, the formation of which depends on CYP2C19 activity. Mutant alleles that confer high CYP2C19 activity, and are therefore associated with high concentrations of the metabolite, are particularly common (30–40%) in Asian populations (1).

Observational studies

In an open study 25 patients with new-onset focal and primary generalized epilepsy were treated with clobazam at a single centre (2). After a mean follow-up of 16 months (range 7–24), 16 patients were seizure free, while five had more than a 50% reduction in seizure frequency. Sedation was the most common adverse event, reported by four patients; however it was always mild and did not require withdrawal of clobazam. Other adverse effects, reported in one patient each, were weight gain, ataxia, loss of short-term memory, and breakthrough seizures.

Comparative studies

In a randomized double-blind comparison of clobazam, carbamazepine, and phenytoin monotherapy in children with epilepsy, there were no differences in tests of intelligence, memory, attention, psychomotor speed, and impulsivity between clobazam and the other drugs after 6 and 12 months of therapy, suggesting that the adverse effects of clobazam on cognition and behavior may be less common than generally thought (3). However, the authors did not discuss a trend for some scores, particularly items in the Wechsler Intelligence Scale for Children–Revised, to improve significantly only in children taking non-benzodiazepine anticonvulsants. Moreover, many children withdrew from the study before completion of the follow-up, resulting in potential bias.

In a prospective multicenter double-blind comparison of clobazam with phenytoin or carbamazepine monotherapy in children with partial or generalized tonic-clonic seizures, the retention rate after one year did not differ, but exit due to inefficacy tended to be more common with clobazam (19 versus 11% for the other drugs combined), while exit due to adverse effects tended to be more common with carbamazepine or phenytoin (15 versus 4% for clobazam) (4). Although all treatments were claimed to have similar efficacy, detailed descriptions of the changes in seizure frequency and the proportion of patients who gradually achieved seizure control in each treatment group were not given. Behavioral and mood problems tended to be more common with clobazam than with the other drugs (38/119 versus 29/116). Drooling was more common with clobazam (7/119 versus 2/116), whereas rash or vomiting were more common with the other treatments (9/116 versus 4/119 and 10/116 versus 4/119 respectively). Tolerance was reported in 7.5% of patients taking clobazam, in 4.2% of those taking carbamazepine, and in 6.7% of those taking phenytoin; however, the definition of tolerance (no seizures for 3–6 months, followed by seizures sufficiently numerous to require a switch to another drug) was questionable, and no information was given about patients with seizure relapses who required an increase in dosage. Although these results suggest that clobazam is a valuable alternative to phenytoin and carbamazepine in childhood epilepsy, more precise characterization of responses would have been desirable.

Organs and Systems

Respiratory

Patients receiving intravenous benzodiazepines must be monitored for respiratory depression, and may need artificial ventilation during intensive treatment.

Nervous system

Clobazam is better tolerated than other benzodiazepines used in epilepsy (5). Its most common adverse effects are mild and transient drowsiness, dizziness, or fatigue; rather less common are muscle weakness, restlessness, aggressiveness, weight increase, ataxia, mood disorders, psychotic and behavioral disturbances, vertigo, hypotonia, hypersalivation, and edema (SED-13, 152). There may be a loss of therapeutic response over time.

Akathisia has been rarely seen with benzodiazepines (SED-13, 152).

Psychological, psychiatric

Of 63 children with refractory epilepsy given add-on clobazam (mean dosage 0.8 mg/kg/day) and followed for 15–64 months, 15 (24%) had to discontinue treatment owing to adverse effects, which included severe aggressive outbursts, hyperactivity, insomnia, and depression with suicidal ideation (6). Likewise, there were behavioral or mood problems in 38 of 119 children taking clobazam

monotherapy over one year of follow-up, while drooling was reported in seven children (4). In another study, 7 of 63 children treated with clobazam developed aggressive agitation, self-injurious behavior, insomnia, and incessant motor activity. All the affected children were relatively young (mean age 6 years) and mentally disabled (7).

In controlled trials with clonazepam, adverse events were recorded in 60–90% of cases, and led to withdrawal rates as high as 36% (8). The most common effects were drowsiness, ataxia, and behavioral and personality changes. Other problems were hypersalivation, tolerance, and sometimes a paradoxical increase in seizure frequency.

Skin

Toxic epidermal necrolysis has been associated with clobazam (SEDA-21, 48). Bullae with sweat gland necrosis rarely complicate coma, but have recently been reported in association with clobazam, used as adjunctive therapy for resistant epilepsy in a 4-year-old girl (9).

Long-Term Effects

Drug tolerance

Clobazam has similar effects on anxiety to other benzodiazepines, but may be better tolerated (SEDA-20, 31). Used as an anticonvulsant, clobazam is generally well tolerated in epileptic patients, many showing little evidence of tolerance (5). On the other hand, children with epilepsy appear unusually prone to adverse behavioral reactions when taking clobazam (SEDA-19, 34).

Drug withdrawal

Withdrawal effects can be troublesome. Of 13 patients taken off clonazepam 0.01–0.5 mg/day because of adverse effects, 8 had withdrawal seizures and 5 had other withdrawal symptoms (SEDA-19, 63). Choreoathetosis was described in one patient completing withdrawal of clonazepam (SED-13, 152).

Epileptic-negative myoclonus status (almost continuous lapses in muscle tone associated with epileptiform discharges and interfering with postural control and motor coordination) rarely occur after rapid withdrawal of clobazam or valproate (10).

Susceptibility factors

Genetic

Measurement of clobazam and N-desmethylclobazam plasma concentrations and genetic analysis may be useful when unusual adverse effects occur (11).

- A 10-year-old girl had two epileptic seizures and a subcontinuous spike and wave pattern during sleep. She was given clobazam and developed severe somnolence, weight gain, and severe enuresis. She had a high plasma concentration of N-desmethylclobazam, the major metabolite of clobazam. She and her parents underwent molecular analysis of the CYP2C19 gene, which is implicated in the metabolism of this drug. She had one copy of the most common mutation (CYP2C19*2), as did her mother, and probably another rare mutation.

Drug administration

Drug overdose

Clobazam toxicity can cause respiratory depression.

- A 49-year-old woman, a chronic alcoholic who had been undergoing psychiatric treatment, was found dead at home (12). Autopsy findings were unremarkable. Further detailed analysis showed that the clobazam concentration found in post-mortem blood was 3.9 µg/ml higher than the usual target concentration (0.1–0.4 µg/ml). All the available information suggested that death had resulted from respiratory depression due to clobazam toxicity.

Drug–Drug Interactions

The specific cytochrome P450 isoforms that mediate the biotransformation of clobazam and of its metabolites N-desmethylclobazam and 4'-hydroxyclobazam have been identified using cDNA-expressed P450 and P450-specific chemical inhibitors in vitro (13). The results of this study showed that:

- clobazam is mainly demethylated by CYP3A4, CYP2C19, and CYP2B6;
- clobazam is 4'-hydroxylated by CYP2C19 and CYP2C18;
- N-desmethylclobazam is 4'-hydroxylated by CYP2C19 and CYP2C18;
- The formation of N-desmethylclobazam is mediated by CYP3A4, CYP2C19, and CYP2B6;
- N-desmethylclobazam is hydroxylated to 4'-hydroxy-desmethylclobazam by CYP2C19.

These findings explain some pharmacokinetic interactions of clobazam with ketoconazole (which inhibits the demethylation of clobazam by 70%) and omeprazole (which inhibits the hydroxylation of N-desmethylclobazam by 26%). In addition, in 22 patients with epilepsy who were genotyped for CYP2C19, there was a higher plasma metabolic ratio of N-desmethylclobazam:clobazam in patients with one CYP2C19*2 mutated allele than in those with the wild-type genotype.

Antiepileptic drugs

Several metabolic interactions between clobazam and other antiepileptic drugs have been reported, in particular phenytoin intoxication after the addition of clobazam (SED-12, 98).

Carbamazepine

Negative myoclonus and more typical signs of carbamazepine intoxication (fatigue, ataxia, clumsiness) occurred in a 66-year-old man after he took add-on clobazam (10 mg/day) for 4 weeks (14). Plasma concentrations of carbamazepine (58 µmol/l) and carbamazepine-10,11-epoxide (19 µmol/l) were higher than before clobazam therapy, and his symptoms resolved quickly when carbamazepine dosage was reduced and clobazam was withdrawn. The interaction was confirmed on rechallenge.

This interaction does not occur in most patients.

Stiripentol

After the addition of stiripentol (50 mg/kg) in 20 children treated with clobazam, mean serum clobazam concentrations increased about twofold and norclobazam concentrations increased about threefold; a mean 25% reduction in clobazam dose was required because of adverse effects (15). Serum concentrations of concomitantly administered valproic acid rose by about 20%. These findings are in agreement with evidence that stiripentol is a potent metabolic inhibitor.

References

1. Kosaki K, Tamura K, Sato R, Samejima H, Tanigawara Y, Takahashi T. A major influence of CYP2C19 genotype on the steady-state concentration of N-desmethylclobazam. Brain Dev 2004;26(8):530–4.
2. Mehndiratta MM, Krishnamurthy M, Rajesh KN, Singh G. Clobazam monotherapy in drug naive adult patients with epilepsy. Seizure 2003;12:226–8.
3. Bawden HN, Camfield CS, Camfield PR, Cunningham C, Darwish H, Dooley JM, Gordon K, Ronen G, Stewart J, van Mastrigt RCanadian Study Group for Childhood Epilepsy. The cognitive and behavioural effects of clobazam and standard monotherapy are comparable. Epilepsy Res 1999;33(2–3):133–43.
4. Canadian Study Group for Childhood Epilepsy. Clobazam has equivalent efficacy to carbamazepine and phenytoin as monotherapy for childhood epilepsy. Epilepsia 1998;39(9):952–9.
5. Remy C. Clobazam in the treatment of epilepsy: a review of the literature. Epilepsia 1994;35(Suppl 5):S88–91.
6. Sheth RD, Ronen GM, Goulden KJ, Penney S, Bodensteiner JB. Clobazam for intractable pediatric epilepsy. J Child Neurol 1995;10(3):205–8.
7. Sheth RD, Goulden KJ, Ronen GM. Aggression in children treated with clobazam for epilepsy. Clin Neuropharmacol 1994;17(4):332–7.
8. Sato S. Clonazepam. In: Levy R, Mattson R, Meldrum B, Penry J, Dreifuss F, editors. Antiepileptic Drugs. 3rd ed.. New York: Raven Press, 1989:65.
9. Setterfield JF, Robinson R, MacDonald D, Calonje E. Coma-induced bullae and sweat gland necrosis following clobazam. Clin Exp Dermatol 2000;25(3):215–8.
10. Gambardella A, Aguglia U, Oliveri RL, Russo C, Zappia M, Quattrone A. Negative myoclonic status due to antiepileptic drug tapering: report of three cases. Epilepsia 1997;38(7):819–23.

11. Parmeggiani A, Posar A, Sangiorgi S, Giovanardi-Rossi P. Unusual side-effects due to clobazam: a case report with genetic study of CYP2C19. Brain Dev 2004;26:63–6.
12. Proença P, Teixeira H, Pinheiro J, Marques EP, Vieira DN. Forensic intoxication with clobazam: HPLC/DAD/MSD analysis. Forensic Sci Int 2004;143:205–9.
13. Giraud C, Tran A, Rey E, Vincent J, Treluyer JM, Pons G. In vitro characterization of clobazam metabolism by recombinant cytochrome P450 enzymes: importance of CYP2C19. Drug Metab Dispos 2004;32(11):1279–86.
14. Genton P, Nguyen VH, Mesdjian E. Carbamazepine intoxication with negative myoclonus after the addition of clobazam. Epilepsia 1998;39(10):1115–8.
15. Rey E, Tran A, D'Athis P, Chiron C, Dulac O, Vincent J, Pons G. Stiripentol potentiates clobazam in childhood epilepsy: a pharmacological study. Epilepsia 1999;40(Suppl 7):112–3.

Clonazepam

General Information

Clonazepam is a benzodiazepine that is used predominantly in epilepsy, panic disorder, and mania, and also appears to be effective in relieving antipsychotic drug-induced akathisia (1). The use of clonazepam in psychiatric disorders is complicated by significant drowsiness in a majority of patients, and additional behavioral problems in children (SEDA-19, 34).

Although they are commonly used in the adjunctive management of chronic pain, benzodiazepines are generally not analgesic per se. One exception may be the stabbing/lancinating neuropathic pain that often responds to anticonvulsants, including clonazepam. Nevertheless, the use of benzodiazepines in pain syndromes is generally contraindicated, and clonazepam, although often effective, should be used with caution (SEDA-17, 42). This is because of the availability of other agents with comparable or superior efficacy and the significant incidence of adverse effects of clonazepam, including depression, self-poisoning, cognitive impairment, and dependence (2), as well as the potential for diversion.

Placebo-controlled studies

In a randomized, double-blind, placebo-controlled, three-arm study in 60 patients with panic disorder, paroxetine alone (40 mg/day) was compared with paroxetine co-administered with clonazepam (2 mg/day) followed either by a tapered benzodiazepine withdrawal phase or continuing combination treatment (3). The outcomes in the three groups were similar. Most of the patients had at least one adverse effect: 68% of patients given paroxetine alone, 85% in those given the two drugs, and 94% in those given the two drugs followed by withdrawal. The most common adverse effects of combined treatment were sedation, sexual dysfunction, and jitteriness; jitteriness and gastrointestinal symptoms were most common with

monotherapy. Sedation and sleep disturbances were the most common adverse effects that made patients withdraw from the study.

Clonazepam is widely used for the treatment of sleep disturbances related to post-traumatic stress disorder, despite very limited published data supporting its use for this indication. In a randomized, single-blind, placebo-controlled, crossover trial of clonazepam 1 mg at bedtime for 1 week followed by 2 mg at bedtime for 1 week in six patients with combat-related post-traumatic stress disorder there were no statistically significant differences between clonazepam and placebo (4). Adverse effects of clonazepam were generally mild and essentially indiscernible from those attributed to placebo. Only one patient elected to continue taking clonazepam at the end of the trial. The small sample size was a significant limitation of the study.

Organs and Systems

Nervous system

In controlled trials with clonazepam, adverse events were recorded in 60–90% of cases and led to withdrawal rates as high as 36% (5). The most common effects were drowsiness, ataxia, and behavioral and personality changes. Other problems were hypersalivation, tolerance, and sometimes a paradoxical increase in seizure frequency.

Psychological, psychiatric

Behavioral adverse events associated with clonazepam include agitation, aggression, hyperactivity, irritability, property destruction, and temper tantrums. These adverse effects can be inadvertently confused with other behavioral or psychiatric conditions, especially if exacerbation of existing challenging behavior occurs.

- A 49-year-old man with severe mental retardation was given clonazepam 2 mg/day to treat aggression, self-injurious behavior, property destruction, and screaming, which got worse instead of better (6). When clonazepam was withdrawn, the behavior improved.

The addition of clonazepam to clomipramine has been reported to have caused acute mania (7).

- A 48-year-old Japanese man with a history of bipolar affective disorder became depressed again. He was already taking lithium carbonate 800 mg/day and carbamazepine 800 mg/day. Clomipramine was added, and the dose was increased to 225 mg/day over 2 months and then maintained for 2 months. Because clomipramine had little effect, clonazepam 3 mg/day was added. On the first day after he took clonazepam, symptoms of hyperthymia, haughtiness, talkativeness, and flight of ideas suddenly appeared once more. Drug-induced delirium was excluded, because orientation was not disrupted and the symptoms did not fluctuate over time. Clomipramine and clonazepam were withdrawn. Because the symptoms were similar to the previous manic episode, the same prescription was reinstated, with the addition of sodium valproate 800 mg/day. After 3 months, he was discharged in remission and had no recurrence. He had not taken other benzodiazepines throughout the treatment.

This report suggests that clonazepam induced a switch to mania, possibly in combination with an effect of clomipramine.

Metabolism

In a randomized, double-blind, placebo-controlled, crossover study in 15 men (mean age 22 years), diazepam 10 mg and clonazepam 1 mg infused over 30 minutes both reduced insulin sensitivity and increased plasma glucose, but the effect of clonazepam was significantly greater (8).

Mouth and teeth

A sensation of a burning mouth has been attributed to clonazepam (9).

- A 52-year-old white woman developed a burning mouth. She had previously taken alprazolam for anxiety, but this was changed to clonazepam because of increased anxiety and panic. Clonazepam relieved her symptoms, but after 4 weeks of therapy she continued to have a constant, mild, oral burning sensation. Examination of the mouth was normal and laboratory tests were unremarkable. The dose of clonazepam was reduced and her symptoms abated but remained intolerable. Clonazepam was withdrawn and her symptoms completely resolved. Since no other medications relieved her anxiety and panic she took clonazepam again, but again developed an intolerable burning mouth. Clonazepam was again withdrawn and her symptoms resolved.

Urinary tract

Urinary retention has been attributed to clonazepam.

- A 2-year old girl with epilepsy and dyskinetic cerebral palsy due to kernicterus, who was taking carbamazepine and valproate, was also given clonazepam 0.05 mg/day and 3 days later developed urinary retention, which did not improve with antibiotic treatment (10). A urine sample obtained by catheterization was sterile. Urinary retention persisted for 10 days, requiring repeated catheterization, but resolved after clonazepam was withdrawn. She was symptom free for the next 6 months.

Long-Term Effects

Drug withdrawal

Withdrawal effects can be troublesome. Of 13 patients taken off clonazepam at a rate of 0.016–0.5 mg/day because of adverse effects, 8 had withdrawal seizures and five had other withdrawal symptoms (SEDA-19, 63). Choreoathetosis was described in one patient completing withdrawal of clonazepam (SED-13, 152).

- A 43-year-old man underwent an incomplete transcranial removal of a pituitary growth-hormone-secreting macroadenoma (11). His daily insulin dose was reduced from more than 300–104 U/day and he was given hydrocortisone and levothyroxine replacement therapy, together with lanreotide injections. A month after discharge, he was given high-dose clonazepam. Three months later, the clonazepam was withdrawn abruptly and he developed hypoglycemic coma.

The author concluded that interruption of benzodiazepine treatment had caused reduced growth hormone secretion and insulin requirements.

Second-Generation Effects

Teratogenicity

The medical records of 28 565 infants were surveyed as part of a hospital-based malformation surveillance program to identify those who had been exposed prenatally to an anticonvulsant, including clonazepam (12). During 32 months, 166 anticonvulsant-exposed infants were identified; 52 had been exposed to clonazepam, 43 as monotherapy, 33 of those during the first trimester. One infant had dysmorphic features, growth retardation, and a heart malformation. There was no increase in major malformations in births exposed to clonazepam monotherapy. However, the study was not large enough to have adequate power to determine whether or not the rate of major malformations is increased in clonazepam-exposed pregnancies.

Drug–Drug Interactions

Carbamazepine

The mutual interaction of clonazepam and carbamazepine has been investigated in 183 children and adults with epilepsy during routine clinical care (13). Carbamazepine increased the clearance of clonazepam by 22% and clonazepam reduced the clearance of carbamazepine by 21%.

The effects of concomitant carbamazepine, phenytoin, sodium valproate, and zonisamide on the steady-state serum concentrations of clonazepam have been investigated in 51 epileptic in-patients under 20 years of age (14). Serum concentrations of clonazepam correlated positively with the dose of clonazepam and negatively with the doses of carbamazepine and valproic acid, but not with phenytoin or zonisamide. These results confirm that as the oral doses of carbamazepine and sodium valproate increase, the serum concentration of clonazepam falls, but there is no interaction with either phenytoin or zonisamide. In the case of carbamazepine the mechanism of action is thought to be enzyme induction, increasing the metabolism of clonazepam. It is not known what the mechanism is with sodium valproate. In patients with epilepsy, the co-administration of either sodium valproate or carbamazepine will reduce the serum concentration of clonazepam and increase the risk of a seizure. When clonazepam is used in the treatment of epilepsy, sodium valproate and carbamazepine should be avoided; phenytoin and zonisamide would be safer alternatives.

Oxycodone

A fatal drug interaction was caused by the ingestion of oxycodone and clonazepam.

- A 38-year-old white woman was found dead (15). She had physical evidence of previous drug abuse and positive hepatitis B and C serology. Her plasma clonazepam concentration was 1.41 µg/ml and her plasma oxycodone concentration was 0.60 µg/ml.

Postmortem findings suggested severe nervous system and respiratory depression produced by high concentrations of clonazepam and oxycodone, including collapsed lungs, aspirated mucus, and heart failure.

Phenelzine

A flushing reaction has been associated with an interaction of phenelzine with clonazepam (SEDA-17, 17).

Sertraline

In a randomized, double-blind, placebo-controlled, crossover study in 13 subjects, sertraline did not affect the pharmacokinetics or pharmacodynamics of clonazepam (16).

Valproate

The effects of concomitant carbamazepine, phenytoin, sodium valproate, and zonisamide on the steady-state serum concentrations of clonazepam have been investigated in 51 epileptic in-patients under 20 years of age (14). Serum concentrations of clonazepam correlated positively with the dose of clonazepam and negatively with the doses of carbamazepine and valproic acid, but not with phenytoin or zonisamide. These results confirm that as the oral doses of carbamazepine and sodium valproate increase, the serum concentration of clonazepam falls, but there is no interaction with either phenytoin or zonisamide. In the case of carbamazepine the mechanism of action is thought to be enzyme induction, increasing the metabolism of clonazepam. It is not known what the mechanism is with sodium valproate. In patients with epilepsy, the co-administration of either sodium valproate or carbamazepine will reduce the serum concentration of clonazepam and increase the risk of a seizure. When clonazepam is used in the treatment of epilepsy, sodium valproate and carbamazepine should be avoided; phenytoin and zonisamide would be safer alternatives.

References

1. Lima AR, Weiser KV, Bacaltchuk J, Barnes TR. Anticholinergics for neuroleptic-induced acute akathisia. Cochrane Database Syst Rev 2004;(1):CD003727.
2. Reddy S, Patt RB. The benzodiazepines as adjuvant analgesics. J Pain Symptom Manage 1994;9(8):510–4.
3. Pollack MH, Simon NM, Worthington JJ, Doyle AL, Peters P, Toshkov F, Otto MW. Combined paroxetine and clonazepam treatment strategies compared to paroxetine monotherapy for panic disorder. J Psychopharmacol 2003;17:276–82.
4. Cates ME, Bishop MH, Davis LL, Lowe JS, Woolley TW. Clonazepam for treatment of sleep disturbances associated with combat-related posttraumatic stress disorder. Ann Pharmacother 2004;38:1395–9.
5. Sato S. Clonazepam. In: Levy R, Mattson R, Meldrum B, Penry J, Dreifuss F, editors. Antiepileptic Drugs. 3rd ed.. New York: Raven Press, 1989:765.
6. Kalachnik JE, Hanzel TE, Sevenich R, Harder SR. Brief report: clonazepam behavioral side effects with an individual with mental retardation. J Autism Development Dis 2003;33:349–54.
7. Ikeda M, Fujikawa T, Yanai I, Horiguchi J, Yamawaki S. Clonazepam-induced maniacal reaction in a patient with bipolar disorder. Int Clin Psychopharmacol 1998;13(4):189–90.
8. Chevassus H, Mourand I, Molinier N, Lacarelle B, Brun JF, Petit P. Assessment of single-dose benzodiazepines on insulin secretion, insulin sensitivity and glucose effectiveness in healthy volunteers: a double-blind, placebo-controlled, randomised, cross-over trial. BMC Clin Pharmacol 2004;4(3):1–10.
9. Culhane NS, Hodle AD. Burning mouth syndrome after taking clonazepam. Ann Pharmacother 2001;35(7–8):874–6.
10. Caksen H, Odabas D. Urinary retention due to clonazepam in a child with dyskinetic cerebral palsy. J Emerg Med 2004;26(2):244.
11. Shuster J. Benzodiazepines and glucose control; mycophenolate mofetil-induced dyshidrotic eczema; concomitant use of bupropion and amantadine causes neurotoxicity; intra-articular steroids and acute adrenal crisis. Hosp Pharm 2000;35:489–91.
12. Lin AE, Peller AJ, Westgte MN, Houde K, Franz A, Holmes LB. Clonazepam use in pregnancy and the risk of malformations. Birth Defects Res Clin Mol Teratol 2004;70(Part A):534–6.
13. Yukawa E, Nonaka T, Yukawa M, Ohdo S, Higuchi S, Kuroda T, Goto Y. Pharmacoepidemiologic investigation of a clonazepam–carbamazepine interaction by mixed effect modeling using routine clinical pharmacokinetic data in Japanese patients. J Clin Psychopharmacol 2001;21(6):588–93.
14. Ikawa K, Eshima N, Morikawa N, Kawashima H, Izumi T, Takeyama M. Influence of concomitant anticonvulsants on serum concentrations of clonazepam in epileptic subjects: an age- and dose-effect linear regression model analysis. Pharm Pharmacol Commun 1999;5:307–10.
15. Burrows DL, Hagardorn AN, Harlan GC, Wallen EDB, Ferslew KE. A fatal drug interaction between oxycodone and clonazepam. J Forensic Sci 2003;48:683–6.
16. Bonate PL, Kroboth PD, Smith RB, Suarez E, Oo C. Clonazepam and sertraline: absence of drug interaction in a multiple-dose study. J Clin Psychopharmacol 2000;20(1):19–27.

Diazepam

General Information

Diazepam produces less sedation in cigarette smokers, and higher (not lower, as stated in SEDA-20) doses may be required for the same sedative or anxiolytic effect. Owing in part to its continued widespread use, several unusual adverse effects of diazepam continue to be reported. These include cases of urinary retention and compartment syndrome, which are not explicable by its pharmacology. On the other hand, accumulation of diazepam and attendant complications of obtundation and respiratory depression may be understood in terms of its long half-life, particularly in elderly people and medically ill patients. Caution about the intravenous use of diazepam comes from a study that showed cardiac dysrhythmias (mainly ventricular extra beats) in a quarter of oral surgery patients; midazolam and lorazepam were much safer (1).

Placebo-controlled studies

In a double-blind, placebo-controlled, randomized trial in 120 children with spastic cerebral palsy received a bedtime dose of diazepam or placebo (2). A bedtime dose of diazepam to reduce hypertonia and muscle spasm, with passive stretching exercises, significantly improved behavior. The diazepam relaxed the muscles making the passive stretching easy, and the movements sustained muscle relaxation during the day. There was a significant improvement in wellbeing, improved activities of daily living, and reduced family burden of caring. There were fewer unwarranted crying spells during the day and less wakefulness during the night. There was no daytime sedation.

Organs and Systems

Cardiovascular

Cases of inadvertent intra-arterial injection of diazepam have been reported.

- A 51-year-old woman with an acute claustrophobic anxiety attack developed gangrene of the fingers after she was inadvertently given diazepam 10 mg intra-arterially (3).
- Inadvertent intra-arterial injection of diazepam (2.5 mg in 0.5 ml) has been reported in an 8-year-old girl (4). Gangrene resulted and amputation of the 4th and 5th fingers was required.

Gangrene has been previously reported with intra-arterial injection of diazepam and is also well known with other classes of drugs, such as barbiturates and phenothiazines. It appears to be caused by the drug rather than the solvent used in the intravenous formulations.

Respiratory

Respiratory difficulties are a major potential adverse effect of rectal diazepam (5).

Of 94 children who presented with seizures, 11 had respiratory depression after intravenous or rectal diazepam (6). However, this finding was challenged (7,8). The authors of the second comment stated that this complication does not occur when rectal diazepam gel is used without other benzodiazepines; they also recommended that during long-term therapy families should be instructed not to give rectal diazepam more than once every 5 days or five times in 1 month.

Patients receiving intravenous benzodiazepines must be monitored for respiratory depression, which may demand artificial ventilation during intensive treatment. Diazepam may cause more respiratory depression than lorazepam at equieffective dosages (SEDA-20, 59) and is contraindicated in neonates for this reason and because it produces unacceptably prolonged sedation (9).

Nervous system

In an open study in 104 patients with acute stroke, diazepam 10 mg bd for 3 days was well tolerated (10).

In a multicenter, double-blind study, 310 patients with generalized anxiety disorder were treated for 6 weeks with abecarnil (mean daily dose 12 mg), diazepam (mean daily dose 22 mg), or placebo in divided doses for 6 weeks (11). Those who had improved at 6 weeks could volunteer to continue double-blind treatment for a total of 24 weeks. Slightly more patients who took diazepam (77%) and placebo (75%) completed the 6-week study than those who took abecarnil (66%). The major adverse events during abecarnil therapy were similar to those of diazepam, namely drowsiness, dizziness, fatigue, and difficulty in coordination. Abecarnil and diazepam both produced statistically significantly more symptom relief than placebo at 1 week, but at 6 weeks only diazepam was superior to placebo. In contrast to diazepam, abecarnil did not cause withdrawal symptoms. The absence of a placebo control makes it difficult to interpret the results of another study of the use of abecarnil and diazepam in alcohol withdrawal, which appeared to show comparable efficacy and adverse effects of the two drugs (12).

The mechanism of rare extrapyramidal effects with diazepam is unexplained (SEDA-18, 44).

Seizures

In a randomized trial, seizures occurred in 14 (16%) of 86 patients with gliomas undergoing contrast CT examinations; however, in 83 other patients with gliomas receiving diazepam prophylaxis, seizures occurred in only two patients (2.2%).

Six patients had untoward effects from excessive rectal diazepam (13). In three cases, seizures reappeared and were interrupted by rectal diazepam, followed by sedation and gradual awakening; the intervals were about 4 days. The other three patients had variable and complex symptoms, with serial seizures and alternating states of tension, apathy, and sleepiness. The plasma concentrations of diazepam and desmethyldiazepam showed rapid fluctuations.

- A 20-year-old man with complex partial seizures presented with exacerbation of his disease (14). He was taking phenytoin and sodium valproate, with plasma concentrations in the target ranges. During a video electroencephalogram recording he was given diazepam 10 mg, and the partial seizures developed into frequent generalized seizures. The same response was seen on a subsequent occasion.

The authors commented that although paradoxical reactions to benzodiazepines are rare, they should be considered in cases of refractory epilepsy.

- A 28-year old man with complex partial status, which lasted for 2 months, had a paradoxical worsening of seizure activity in response to diazepam and midazolam (14).

Neuromuscular function

Muscle rigidity after high-dose opioids can be reduced by the benzodiazepines midazolam and diazepam (SEDA-19, 82).

Psychological, psychiatric

A fugue-like state with retrograde amnesia has been associated with diazepam (15).

- A 23-year-old military officer on active duty took diazepam 5 mg tds and ibuprofen for back spasms. Three days later he was found sitting in a church, having assumed a previous role from his past life. He identified the date as 14 months before and his memory before that time was intact. However, he had no memory of events during the previous 14 months. There were no symptoms suggesting a schizophrenic disorder and his mental function was normal. His symptoms resolved within 24 hours of withdrawal of diazepam, except for amnesia of the event. He assumed his correct identity and was aware of the correct date. He had taken ibuprofen in the past with no adverse effects and this was his first exposure to a benzodiazepine. No other medications were involved and a full medical review found no cause for his symptoms other than diazepam use.

Healthy men and women (n = 46) were randomly assigned to placebo or diazepam 5 or 10 mg in a double-blind, between-groups design to examine the effect of diazepam on self-aggressive behavior under controlled laboratory conditions (16). The participants were then provided with the opportunity to self-administer electric shocks during a competitive reaction time task. Self-aggression was defined by the intensity of shock chosen. Diazepam 10 mg was associated with higher average shocks than placebo and a greater likelihood of attempting to self-administer a shock that they were led to believe was severe and painful. There were sedative effects, but diazepam did not impair memory, attention, concentration, pain threshold, or reaction-time performance. The authors concluded that clinically relevant doses of diazepam may be associated with self-aggressive behavior without significantly impairing basic cognitive processes or psychomotor performance.

Capgras syndrome has been attributed to diazepam.

- A 78-year-old man with a long history of generalized anxiety disorder had benefited from diazepam for at least 30 years (17). During the 6 months before evaluation, he developed a fixed delusion that his sister-in-law had disguised herself as his wife and had replaced her at home. His anxiety symptoms remained at baseline and cognitive function was unimpaired on detailed testing. Medications included diazepam 5 mg bd, paroxetine 40 mg/day, levothyroxine, rabeprazole, ranitidine, and finasteride. The dose of diazepam was tapered and withdrawn and risperidone 0.5 mg qds was started. Within 10 days, the Capgras delusion had completely resolved and he readily recognized his wife during visits.

Endocrine

Gynecomastia, with raised estradiol, has been reported in men taking diazepam (18).

Acute diazepam administration causes a reduction in plasma cortisol concentrations, consistent with reduced activity of the hypothalamic–pituitary–adrenal axis, especially in individuals experiencing stress. However, the effects of chronic diazepam treatment on cortisol have been less well studied, and the relation to age, anxiety, duration of treatment, and dose are poorly understood. In a double-blind, placebo-controlled, crossover study, young (19–35 years, n = 52) and elderly (60–79 years, n = 31) individuals with and without generalized anxiety disorder took diazepam 2.5 or 10 mg for 3 weeks (19). The elderly had significant reductions in plasma cortisol concentrations compared with placebo, both after the first dose and during chronic treatment, but the younger subjects did not. A final challenge with the same dose did not produce any significant cortisol effects in either group and the cortisol response in the elderly was significantly reduced compared with the initial challenge. These results are consistent with the development of tolerance to the cortisol-reducing effects of diazepam. The effect was more apparent in the elderly, was not modulated by generalized anxiety disorder or dosage, and was not related to drug effects on performance and on self-ratings of sedation and tension.

Skin

Cutaneous adverse effects of diazepam are rare. The incidence was 0.4 per 1000 in the Boston Collaborative Surveillance Program (20).

- A 50-year-old woman was referred to hospital for chronic depression with alcohol dependence (21). There was no history of drug allergy. She was given oral thioridazine 100 mg/day and diazepam 10 mg qds. After 2 days she noticed an erythematous eruption on her ankles. Thioridazine was withdrawn, but the eruption became more erythematous and affected both extremities and flanks within a few hours. She was given methylprednisolone 80 mg/day. The next day the eruption became bullous and she became pyrexial (39.4° C). Urea and creatinine concentrations were normal. Blood cultures were negative. A skin biopsy showed bullous vasculitis with numerous eosinophils in the dermis. Diazepam was then withdrawn, the pyrexia resolved, and the skin lesions healed, although post-inflammatory ulcers persisted on both ankles for 2 months. A lymphocyte blast transformation test was positive for diazepam.

Sweet's syndrome has been attributed to diazepam (22).

- A 70-year-old white man, with no significant preceding medical history, developed an acute painful rash, a fever (38.4° C), and severe arthralgia 5 days after starting to take diazepam 10 mg bd for lumbar muscular contracture due to hard physical exercise. He had taken no other medications. There were well-defined purple-red skin plaques, surmounted by vesicular and hemorrhagic blisters. He had a leukocytosis. Sweet's syndrome was confirmed by punch biopsy of a lesion. Diazepam was withdrawn, and prednisolone 30 mg/day was given for 2 weeks and then tapered. The patient improved quickly and the eruption cleared in 10 days.

Immunologic

Hypersensitivity reactions after diazepam are very rare and usually mild. However, some severe reactions have been reported.

- A 50-year-old woman with chronic depression or dysthymic disorder and alcohol dependence was given oral thioridazine 100 mg/day and diazepam 10 mg qds (23). She had no history of drug allergy. Two days later she noticed an erythematous eruption on her ankles. Thioridazine was withdrawn, but the eruption became more widespread over a few hours. She was given methylprednisolone 80 mg/day, but the following day the eruption progressively became bullous and her condition worsened. She developed a fever of 39.4° C, felt ill, and had a neutrophilia, but blood cultures were sterile and her renal function was normal. A skin biopsy showed bullous vasculitis with numerous eosinophils in the dermis. Diazepam was then withdrawn, which led to resolution of pyrexia and gradual healing of the skin lesions over the next 2 months. The lymphocyte blast transformation test was positive for diazepam.
- A 28-year-old nurse had generalized urticaria and collapsed while she was undergoing a gastroscopy for suspected *Helicobacter pylori* infection (24). Before the start of the procedure she was given lidocaine oral spray and intravenous diazepam 10 mg, and at the end intravenous flumazenil 1 mg. Skin prick tests and intradermal tests with diazepam 5 mg/ml produced a weal-and-flare reaction; flumazenil 0.1 mg/ml and lidocaine 2% had no effect.

Although in the second case, for safety reasons, a challenge test was not performed, it was suggested that the reaction had been IgE-mediated.

Long-Term Effects

Genotoxicity

The possible genotoxic effects of propofol and diazepam have been investigated in 45 patients undergoing open heart surgery (25). Peripheral blood samples were collected before and at the end of anesthesia the anesthesia with either diazepam 0.2 mg/kg + fentanyl 10 micrograms/kg (n = 24) or propofol 1 mg/kg + fentanyl 10 micrograms/kg (n = 21). Anesthesia was maintained by pancuronium and fentanyl plus diazepam 5 mg/kg or propofol 2–4 mg/kg/hour. The mean frequencies of chromosomal aberrations, before and at the end of the anesthesia, were not significantly different. Age, smoking, and sex were not confounding factors.

Second-Generation Effects

Teratogenicity

To study the possible teratogenicity of short-term (about 3 weeks) oral diazepam during pregnancy, a matched case-population control pair analysis was conducted in the population-based data set of Hungarian Case-Control Surveillance of Congenital Abnormalities (26). The investigators compared the "total" (maternal self-reported plus medically recorded) and "medically recorded" diazepam treatments, and compared cases and controls from 1980 to 1996. Among 38 101 neonates without any congenital abnormality 4130 (11%) were exposed to diazepam, compared to 2746 of 22 865 (12%) neonates or fetuses with congenital abnormalities, and 97 of 812 (12%) neonates or fetuses with Down's syndrome. Based on maternal self-reported and medically recorded information, the matched case-population control pair analysis showed a higher rate of limb deficiencies, rectal-anal atresia/stenosis, cardiovascular malformations, and multiple congenital abnormalities after diazepam use during the second and third months of gestation. However, the evaluation of only medically recorded diazepam use did not show a higher use of diazepam in any congenital abnormality group. The authors suggested that the higher occurrence of diazepam treatment among cases in the primary analysis may have been due to a lower proportion of mothers who recalled having take diazepam during their pregnancy in the population control group, i.e. a recall bias. They concluded that short-term diazepam treatment in usual therapeutic doses during pregnancy did not cause detectable teratogenicity.

Fetotoxicity

Postnatal, longitudinal, somatic, neurological, mental, and behavioral development has been studied at birth and at 8, 15, and 24 months of life in children whose mothers had been treated during pregnancy with diazepam (n = 126) or promethazine (n = 127) and in children whose mothers had not been exposed (27). The children in the diazepam group weighed less at birth but not at 8 months or subsequently.

Diazepam has been reported to cause inappropriate ADH secretion in a neonate (28).

- A female infant was delivered vaginally at 41 weeks. Her 30-year-old mother had taken diazepam for epilepsy and hysterical attacks throughout the pregnancy. The pregnancy and delivery were uneventful. The baby was admitted to the neonatal ward in anticipation of neonatal drug withdrawal syndrome. On the first day of life, milk feeding was stopped because of poor sucking, vomiting, and increased gastric aspirate volume. On the same day oliguria was reported and the urine osmolality was increased. Secretion of antidiuretic hormone was suspected as the cause of the oliguria, and so fluid intake was restricted and a diuretic was given. Subsequently the urine output increased and the urine osmolality gradually fell. The baby's condition became stable and she was discharged on day 16.

Susceptibility Factors

Age

Five neonates who suffered an unexpected long period of respiratory failure, muscular hypotonia, and drowsiness were retrospectively investigated (9). Unusually high doses of diazepam had been given by intravenous bolus injection and serum concentrations of diazepam and its active metabolites were high. The authors emphasized the persistence of the very long-acting N-desmethyldiazepam, particularly in neonates and even more exaggeratedly in premature infants, owing to reduced capacity of hepatic uridine diphosphate glucuronyl transferase activity (9).

Drug Administration

Drug administration route

Although intravenous diazepam is the preferred route, the undiluted intravenous solution of diazepam can be given rectally, and is effective in the emergency management of seizures in children (29). Rectal gel is an alternative, and can be given by non-medical personnel (30). Adverse effects of rectal diazepam are rare and mild. Animal studies and clinical experience have not shown damage to the rectal mucosa.

The safety of rectal diazepam gel (Dyastat) has been reviewed (31). Sedation and somnolence were the most common adverse events, ranging from 13 to 51%, but the real incidence was difficult to determine, because it was not always possible to distinguish between drug-related and postictal sedation. Neurocognitive effects were similar to those reported with intravenous diazepam, but with a slightly delayed onset and a longer duration. When neuropsychological testing was performed, test scores returned to baseline within 4 hours after administration. Hyperactivity was rarely appreciated in children treated with rectal diazepam gel. A few adults (<1%) reported agitation, euphoria, nervousness, and hyperkinesia. The incidence of respiratory depression or apnea was much

lower than with intravenous administration: no respiratory depression occurred in 200 children included in controlled clinical studies of rectal diazepam gel, while only two children had respiratory depression in a study of 246 doses and there were two more cases of hypoventilation in another study of 1578 doses. Whether these instances were related to the medication or to seizures could not be definitely ascertained. In a controlled study, mean and minimum respiratory rates 15 minutes to 4 hours after treatment were similar in patients who received rectal diazepam compared with those who received placebo. Despite the high level of concern, an out-of-hospital study showed a 23% complication rate (hypotension, cardiac dysrhythmias, and respiratory depression) among patients treated with placebo, compared with 11% for intravenous diazepam and 10% for intravenous lorazepam, suggesting that the respiratory complications of some seizures are significant and the very small risk of respiratory depression with rectal diazepam is outweighed by the much greater risk of delaying treatment. Other adverse effects were pruritus and rash (3.1–4.4%) and rectal irritation (3.6–4.4%, compared with 3.1–3.4% in the placebo group).

Rectal diazepam is not intended for use more than five times per month (one dose every 5 days), because repeated administration can exacerbate seizures and cause withdrawal seizures. There were at least six cases of seizure worsening, some with a clear cyclic pattern, suggesting a relation to intermittent use of diazepam. All improved after rectal diazepam was restricted or withdrawn. However, in some cases intermittent diazepam had to be replaced by long-term low-dose treatment with oral benzodiazepines.

As of February 2003 a total of 40 reports describing 63 adverse events attributed to rectal diazepam gel were filed at the FDA's MedWatch program. These included convulsions or ineffective treatment in eight patients; gastrointestinal symptoms (vomiting, constipation, diarrhea, or abnormal stools) in seven; emotional and cognitive adverse effects (disorientation, confusion, memory impairment, stupor, nervousness, emotional lability) in six; respiratory depression and dyspnea in six; somnolence in four; hypotension or vasodilatation, pain, and skin rash or anaphylactic reactions in two each; infection, a positive drug screen, an unevaluable reaction, rectal hemorrhage, and abnormal vision in one each. There were three deaths, whose causes were not conclusively determined.

Drug overdose

- A 54-year-old man took 2 g of laboratory-grade diazepam and was treated with activated charcoal, diuresis, and flumazenil infusion (32). He wakened, but had drowsiness, dysarthria, diplopia, and dizziness for 9 days. Blood concentrations of diazepam and its main metabolite, N-desmethyldiazepam, remained high for over 4 weeks.

Of 149 patients, 10 received an overdose of rectal diazepam indicated for acute repetitive seizures (51 overdoses in total) (33). There were no untoward events in 40 cases, and the adverse events were most often not drug-related. No patient had bradypnea or apnea.

In 40 of 51 instances of overdose of rectal diazepam there were no adverse events (Pellock). However, in 11 cases adverse events included vomiting (n = 3), otitis media (n = 3), and bronchitis, convulsion, cough, fever and somnolence (n = 1 each). There was no cardiac or respiratory depression in any case of overdose, and all events resolved without incident.

Drug–Drug Interactions

Antihistamines

Drugs that depress the CNS, such as the benzodiazepines, have their effects increased by interaction with the classic antihistamines. However, the second-generation antihistamines have not yet been proven to interact with CNS depressants such as alcohol or diazepam (34–37).

Beta-adrenoceptor antagonists

Lipophilic beta-adrenoceptor antagonists are metabolized to varying degrees by oxidation by liver microsomal cytochrome P450 (for example propranolol by CYP1A2 and CYP2D6 and metoprolol by CYP2D6). They can therefore reduce the clearance and increase the steady-state plasma concentrations of other drugs that undergo similar metabolism, potentiating their effects. Drugs that interact in this way include diazepam (38).

Bupivacaine

Animal studies have shown that diazepam can prolong the half-life of bupivacaine (39).

Caffeine

At least 40 drugs interact with caffeine, including benzodiazepines (for example diazepam, whose sedative effect is counteracted by caffeine) (40).

Cholinesterase inhibitors

Although diazepam does not have anticholinergic properties, it is possible to reverse diazepam-induced delirium by the use of cholinesterase inhibitors, such as physostigmine; however, physostigmine can on occasion induce severe arterial hypertension, especially if the dose exceeds 2 mg intravenously. In healthy volunteers sedated with diazepam, an increase in awareness was established with the use of physostigmine, but there was also a reduction in ventilatory drive (SEDA-10, 119).

Cisapride

Cisapride increases the absorption of diazepam (41).

Clozapine

Hypotension, collapse, and respiratory arrest occurred when low doses (12.5–25 mg) of clozapine were added to a pre-existing diazepam regimen (SEDA-22, 41).

- A 50-year-old man with symptoms of chronic paranoid schizophrenia resistant to typical neuroleptic drugs had a brief syncopal attack with significant electrocardiographic changes (sinus bradycardia and deep antero-septal inverted T waves and minor ST segment changes) after the dosage of clozapine was increased to 300 mg/day while he was taking diazepam 30 mg/day (42).

The mechanism of this presumed interaction is unknown.

Disulfiram

Disulfiram inhibits hepatic drug metabolism and can prolong the effects of substances that are normally metabolized in the liver. This has been studied for various benzodiazepines. The clearances of chlordiazepoxide and diazepam were significantly reduced and their half-lives prolonged by disulfiram (43).

Heparin

In healthy non-fasting subjects, 100–1000 IU of heparin given intravenously caused a rapid increase in the unbound fractions of chlordiazepoxide, diazepam, and oxazepam (44,45), but no change in the case of lorazepam (44). The clinical implications of this finding are not known.

Ibuprofen

The effect of diazepam on the pharmacokinetics of ibuprofen has been studied in eight healthy subjects, who took ibuprofen or ibuprofen plus diazepam at 10.00 or 22.00 hours in a randomized, crossover study (46). Diazepam significantly prolonged the half-life of ibuprofen at 22.00 hours but not at 10.00 hours. The mean clearance of ibuprofen was therefore reduced by diazepam at night. This time-dependent effect of diazepam on the pharmacokinetics of ibuprofen may be due to circadian variation in the pattern of protein production in the liver and/or competitive protein binding of the two drugs during the night.

Lithium

There has been a well-documented case of profound hypothermia in a patient taking lithium and diazepam; it did not occur with either drug alone (47). Otherwise, benzodiazepines and lithium are compatible.

Naltrexone

The effects of naltrexone on diazepam intoxication were investigated in 26 non-drug-abusing subjects who received either naltrexone 50 mg or placebo and 90 minutes later oral diazepam 10 mg in a double-blind, crossover trial (48). Naltrexone was significantly associated with negative mood states, such as sedation, fatigue, and anxiety, compared with placebo, while positive states (friendliness, vigor, liking the effects of diazepam, feeling high from diazepam) were significantly more common with placebo. Naltrexone significantly delayed the time to peak diazepam concentrations (135 minutes) compared with placebo

(75 minutes), but there were no significant differences in the concentrations of nordiazepam, the main metabolite of diazepam, at any stage in the study.

Olanzapine

There were orthostatic changes when olanzapine and diazepam were co-administered (49).

Omeprazole

In human liver microsomes, the metabolism of diazepam was mainly to 3-hydroxydiazepam (90%); omeprazole inhibited this conversion (50).

In a double blind, placebo-controlled, crossover study in eight white and seven Chinese men who were extensive metabolizers of debrisoquine and mephenytoin, omeprazole 40 mg/day reduced the oral clearance of diazepam by 38% and increased desmethyldiazepam AUC by 42%. In contrast, in the Chinese men the oral clearance of diazepam fell by only 21% and desmethyldiazepam AUC by 25%. The authors concluded that the extent of the inhibitory effect of omeprazole on diazepam metabolism depends on ethnicity (51). Differences between Caucasians and Asians may account for such effects.

Penicillamine

In one patient the use of penicillamine led to exacerbation of phlebitis that had been caused by intravenous diazepam (52).

Phenytoin

Neurological abnormalities have been attributed to phenytoin toxicity caused by an interaction with diazepam (53).

- A 44-year-old man with a long-standing seizure disorder developed headache, nystagmus, diplopia, and ataxia. His antiepileptic drug regimen of phenytoin, phenobarbital, and lamotrigine had been unchanged for almost 5 months. Two days before admission he had been given amoxicillin and diazepam. The serum phenytoin concentration was 148 μmol/l, having been 32 μmol/l 2 weeks before. Both phenytoin and diazepam were withdrawn, and his symptoms resolved.

Rifamycins

The interaction of rifampicin with diazepam has been reviewed (54). The mean half-life of diazepam fell from 58 to 14 hours in seven patients who took isoniazid, rifampicin, and ethambutol (55).

References

1. Roelofse JA, van der Bijl P. Cardiac dysrhythmias associated with intravenous lorazepam, diazepam, and midazolam during oral surgery. J Oral Maxillofac Surg 1994;52(3):247–50.

2. Mathew A, Mathew MC. Bedtime diazepam enhances well being in children with spastic cerebral palsy. Pediatr Rehabil 2005;8(1):63–6.

3. Joist A, Tibesku CO, Neuber M, Frerichmann U, Joosten U. Fingergangrän nach akziden teller intraarterieller Injektion von Diazepam. [Gangrene of the fingers caused by accidental intra-arterial injection of diazepam.] Dtsch Med Wochenschr 1999;124(24):755–8.

4. Derakshan MR. Amputation due to inadvertent intra-arterial diazepam injection. Iran J Med Sci 2000;25:84–6.

5. Dooley JM. Rectal use of benzodiazepines. Epilepsia 1998;39(Suppl 1):S24–7.

6. Norris E, Marzouk O, Nunn A, McIntyre J, Choonara I. Respiratory depression in children receiving diazepam for acute seizures: a prospective study. Dev Med Child Neurol 1999;41(5):340–3.

7. Mackereth S. Use of rectal diazepam in the community. Dev Med Child Neurol 2000;42(11):785.

8. Kriel RL, Cloyd JC, Pellock JM. Respiratory depression in children receiving diazepam for acute seizures: a prospective study. Dev Med Child Neurol 2000;42(6):429–30.

9. Peinemann F, Daldrup T. Severe and prolonged sedation in five neonates due to persistence of active diazepam metabolites. Eur J Pediatr 2001;160(6):378–81.

10. Lodder J, Luijckx G, van Raak L, Kessels F. Diazepam treatment to increase the cerebral GABAergic activity in acute stroke: a feasibility study in 104 patients. Cerebrovasc Dis 2000;10(6):437–40.

11. Rickels K, DeMartinis N, Aufdembrinke B. A double-blind, placebo-controlled trial of abecarnil and diazepam in the treatment of patients with generalized anxiety disorder. J Clin Psychopharmacol 2000;20(1):12–8.

12. Anton RF, Kranzler HR, McEvoy JP, Moak DH, Bianca R. A double-blind comparison of abecarnil and diazepam in the treatment of uncomplicated alcohol withdrawal. Psychopharmacology (Berl) 1997;131(2):123–9.

13. Brodtkorb E, Aamo T, Henriksen O, Lossius R. Rectal diazepam: pitfalls of excessive use in refractory epilepsy. Epilepsy Res 1999;35(2):123–33.

14. Al Tahan A. Paradoxic response to diazepam in complex partial status epilepticus. Arch Med Res 2000;31(1):101–4.

15. Simmer ED. A fugue-like state associated with diazepam use. Mil Med 1999;164(6):442–3.

16. Berman ME, Jones GD, McCloskey MS. The effects of diazepam on human self-aggressive behaviour. Psychopharmacology 2005;178:100–6.

17. Stewart JT. Capgras syndrome related to diazepam treatment. Southern Med J 2004;97(1):65–6.

18. Bergman D, Futterweit W, Segal R, Sirota D. Increased oestradiol in diazepam-related gynaecomastia. Lancet 1981;2(8257):1225–6.

19. Pomara N, Willoughby LM, Sidtis J, Cooper TB, Greenblatt DJ. Cortisol response to diazepam: its relationship to age, dose, duration of treatment and presence of generalised anxiety disorder. Psychopharmacology 2005;178:1–8.

20. Bigby M, Jick S, Jick H, Arndt K. Drug-induced cutaneous reactions. A report from the Boston Collaborative Drug Surveillance Program on 15,438 consecutive inpatients, 1975 to 1982. JAMA 1986;256(24):3358–63.

21. Olcina GM, Simonart T. Severe vasculitis after therapy with diazepam. Am J Psychiatry 1999;156(6):972–3.

22. Guimera FJ, Garcia-Bustinduy M, Noda A, Saez M, Dorta S, Sanchez R, Martin-Herrera A, Garcia-Montelongo R. Diazepam-associated Sweet's syndrome. Int J Dermatol 2000;39(10):795–8.

23. Olcina GM, Simonart T. Severe vasculitis after therapy with diazepam. Am J Psychiatry 1999;156(6):972–3.

24. Asero R. Hypersensitivity to diazepam. Allergy 2002;57(12):1209.

25. Karahalil B, Yagar S, Bahadir G, Durak P, Sardas S. Diazepam and propofol used as anaesthetics during open-heart surgery do not cause chromosomal aberrations in peripheral blood lymphocyctes. Mutat Res 2005;581:181–6.

26. Czeizel AE, Erös E, Rockenbauer M, Sorensen HT, Olsen J. Short-term oral diazepam treatment during pregnancy. A population-based teratological case-control study. Clin Drug Invest 2003;23:451–62.

27. Czeizel AE, Szegal BA, Joffe JM, Racz J. The effect of diazepam and promethazine treatment during pregnancy on the somatic development of human offspring. Neurotoxicol Teratol 1999;21(2):157–67.

28. Nako Y, Tachibana A, Harigaya A, Tomomasa T, Morikawa A. Syndrome of inappropriate secretion of anti-diuretic hormone complicating neonatal diazepam withdrawal. Acta Paediatr 2000;89(4):488–9.

29. Seigler RS. The administration of rectal diazepam for acute management of seizures. J Emerg Med 1990;8(2):155–9.

30. Shafer PO. New therapies in the management of acute or cluster seizures and seizure emergencies. J Neurosci Nurs 1999;31(4):224–30.

31. Pellock JM. Safety of Diastat, a rectal gel formulation of diazepam for acute seizure treatment. Drug Saf 2004;27(6):383–92.

32. de Haro L, Valli M, Bourdon JH, Iliadis A, Hayek-Lanthois M, Arditti J. Diazepam poisoning with one-month monitoring of diazepam and nordiazepam blood levels. Vet Hum Toxicol 2001;43(3):174–5.

33. Brown L, Bergen DC, Kotagal P, Groves L, Carson D. Safety of Diastat when given at larger-than-recommended doses for acute repetitive seizures. Neurology 2001;56(8):1112.

34. Bhatti JZ, Hindmarch I. The effects of terfenadine with and without alcohol on an aspect of car driving performance. Clin Exp Allergy 1989;19(6):609–11.

35. Moser L, Huther KJ, Koch-Weser J, Lundt PV. Effects of terfenadine and diphenhydramine alone or in combination with diazepam or alcohol on psychomotor performance and subjective feelings. Eur J Clin Pharmacol 1978;14(6):417–23.

36. Rombaut N, Heykants J, Vanden Bussche G. Potential of interaction between the H_1-antagonist astemizole and other drugs. Ann Allergy 1986;57(5):321–4.

37. Doms M, Vanhulle G, Baelde Y, Coulie P, Dupont P, Rihoux JP. Lack of potentiation by cetirizine of alcohol-induced psychomotor disturbances. Eur J Clin Pharmacol 1988;34(6):619–23.

38. Ochs HR, Greenblatt DJ, Verburg-Ochs B. Propranolol interactions with diazepam, lorazepam, and alprazolam. Clin Pharmacol Ther 1984;36(4):451–5.

39. Yan AC, Newman RD. Bupivacaine-induced seizures and ventricular fibrillation in a 13-year-old girl undergoing wound debridement. Pediatr Emerg Care 1998;14(5):354–5.

40. Mattila MJ, Nuotto E. Caffeine and theophylline counteract diazepam effects in man. Med Biol 1983;61(6):337–43.

41. Bateman DN. The action of cisapride on gastric emptying and the pharmacodynamics and pharmacokinetics of oral diazepam. Eur J Clin Pharmacol 1986;30(2):205–8.

42. Tupala E, Niskanen L, Tiihonen J. Transient syncope and ECG changes associated with the concurrent administration of clozapine and diazepam. J Clin Psychiatry 1999;60(9):619–20.

43. MacLeod SM, Sellers EM, Giles HG, Billings BJ, Martin PR, Greenblatt DJ, Marshman JA. Interaction of disulfiram with benzodiazepines. Clin Pharmacol Ther 1978;24(5):583–9.

44. Desmond PV, Roberts RK, Wood AJ, Dunn GD, Wilkinson GR, Schenker S. Effect of heparin administration on plasma binding of benzodiazepines. Br J Clin Pharmacol 1980;9(2):171–5.

45. Routledge PA, Kitchell BB, Bjornsson TD, Skinner T, Linnoila M, Shand DG. Diazepam and N-desmethyldiaze-pam redistribution after heparin. Clin Pharmacol Ther 1980;27(4):528–32.

46. Bapuji AT, Rambhau D, Srinivasu P, Rao BR, Apte SS. Time dependent influence of diazepam on the pharmacokinetics of ibuprofen in man. Drug Metabol Drug Interact 1999;15(1):71–81.

47. Naylor GJ, McHarg A. Profound hypothermia on combined lithium carbonate and diazepam treatment. BMJ 1977;2(6078):22.

48. Swift R, Davidson D, Rosen S, Fitz E, Camara P. Naltrexone effects on diazepam intoxication and pharmacokinetics in humans. Psychopharmacology (Berl) 1998;135(3):256–62.

49. Callaghan JT, Bergstrom RF, Ptak LR, Beasley CM. Olanzapine. Pharmacokinetic and pharmacodynamic profile. Clin Pharmacokinet 1999;37(3):177–93.

50. Zomorodi K, Houston JB. Diazepam–omeprazole inhibition interaction: an in vitro investigation using human liver microsomes. Br J Clin Pharmacol 1996;42(2):157–62.

51. Caraco Y, Tateishi T, Wood AJ. Interethnic difference in omeprazole's inhibition of diazepam metabolism. Clin Pharmacol Ther 1995;58(1):62–72.

52. Brandstetter RD, Gotz VP, Mar DD, Sachs D. Exacerbation of diazepam-induced phlebitis by oral penicillamine. BMJ (Clin Res Ed) 1981;283(6290):525.

53. Murphy A, Wilbur K. Phenytoin-diazepam interaction. Ann Pharmacother 2003;37:659–63.

54. Baciewicz AM, Self TH. Rifampin drug interactions. Arch Intern Med 1984;144(8):1667–71.

55. Ochs HR, Greenblatt DJ, Roberts GM, Dengler HJ. Diazepam interaction with antituberculosis drugs. Clin Pharmacol Ther 1981;29(5):671–8.

Flumazenil

General Information

Flumazenil is used as a benzodiazepine antagonist in the treatment of poisoning or the reversal of benzodiazepine effects in anesthesia 1,2) or in neonates (3). Guidelines for its use have been summarized (4). The problems in its use are those of dose adjustment, the risks of panic anxiety, seizures, or other signs of excessively rapid benzodiazepine withdrawal, and pharmacokinetic problems due to the short half-life of flumazenil (about 1 hour) compared with the longer half-lives of most benzodiazepines (5). Its use is also commonly associated with vomiting and headache, and rarely with psychosis or sudden cardiac death (SEDA-17,

46), especially in mixed overdoses. Flumazenil was not effective in reversing the amnesic effects of midazolam (6), but it may be useful in hepatic coma, regardless of cause (7). In patients with a history of seizures or chronic benzodiazepine dependence, or after mixed drug overdose, flumazenil can trigger convulsions, which are occasionally fatal. It is not generally helpful to measure benzodiazepine plasma concentrations, but they can assist in the diagnosis of overdose and thus guide the use of antagonists (8).

Midazolam can cause paradoxical reactions, including increased agitation and poor cooperation (9,10). Often other drugs are required to continue the procedure successfully. Reversal of these reactions by flumazenil has been reported. In 58 patients undergoing surgery under spinal or epidural anesthesia, flumazenil 0.1 mg over 10 seconds abolished the agitation without reversing sedation (total dose range 0.1–0.5 mg) (11). In 30 patients who had been given midazolam, flumazenil 0.15–0.5 mg resulted in cessation of the agitation without reversal of sedation (9). Adverse effects of flumazenil were not reported in these studies.

The usefulness and relative safety of midazolam in children have been reviewed (12). Myoclonic-like movements associated with midazolam in three full-term newborns were reversed by flumazenil (13). However, care must be taken when considering the use of flumazenil for reversal of midazolam-induced agitation, as no controlled trials have been published.

Organs and Systems

Nervous system

A case of opisthotonos after flumazenil has been reported (14).

- A healthy 17-year-old man received an interscalene brachial plexus block using mepivacaine 600 mg and bupivacaine 150 mg. He became disorientated and showed signs of local anesthetic toxicity, for which he was given midazolam 5 mg. Flumazenil 0.5 mg was given 23 minutes after the end of the procedure, causing opisthotonos.

Similar reports have appeared in the past in patients with seizure disorders. It is recommended that flumazenil not be used in patients predisposed to seizures.

Endocrine

The effects of flumazenil and midazolam on adrenocorticotrophic hormone and cortisol responses to a corticotrophin-releasing hormone challenge have been assessed in eight healthy men (15). Flumazenil significantly reduced adrenocorticotrophic responses compared with midazolam or placebo, but had no effects on cortisol secretion. The authors suggested that this agonist effect of flumazenil on the pituitary–adrenal axis might account for the anxiolytic activity of flumazenil, which has been observed during simulated stress.

Long-Term Effects

Drug withdrawal

Flumazenil can provoke acute withdrawal reactions and extreme anxiety (16). Its duration of action (less than 1 hour) is generally shorter than that of the original benzodiazepine, whose effects can therefore return while the patient is unobserved.

References

1. Gaudreault P, Guay J, Thivierge RL, Verdy I. Benzodiazepine poisoning. Clinical and pharmacological considerations and treatment. Drug Saf 1991;6(4):247–65.
2. Brogden RN, Goa KL. Flumazenil. A reappraisal of its pharmacological properties and therapeutic efficacy as a benzodiazepine antagonist. Drugs 1991;42(6):1061–89.
3. Richard P, Autret E, Bardol J, Soyez C, Barbier P, Jonville AP, Ramponi N. The use of flumazenil in a neonate. J Toxicol Clin Toxicol 1991;29(1):137–40.
4. Cone AM, Stott SA. Flumazenil. Br J Hosp Med 1994;51(7):346–8.
5. Geller E, Halpern P. Benzodiazepine antagonists and inverse agonists. Curr Opin Anaesthesiol 1990;3:568.
6. Curran HV, Birch B. Differentiating the sedative, psychomotor and amnesic effects of benzodiazepines: a study with midazolam and the benzodiazepine antagonist, flumazenil. Psychopharmacology (Berl) 1991;103(4):519–23.
7. Ananth J, Swartz R, Burgoyne K, Gadasally R. Hepatic disease and psychiatric illness. relationships and treatment. Psychother Psychosom 1994;62(3–4):146–59.
8. Nishikawa T, Suzuki S, Ohtani H, Eizawa NW, Sugiyama T, Kawaguchi T, Miura S. Benzodiazepine concentrations in sera determined by radioreceptor assay for therapeutic-dose recipients. Am J Clin Pathol 1994;102(5):605–10.
9. Fulton SA, Mullen KD. Completion of upper endoscopic procedures despite paradoxical reaction to midazolam: a role for flumazenil? Am J Gastroenterol 2000;95(3):809–11.
10. Saltik IN, Ozen H. Role of flumazenil for paradoxical reaction to midazolam during endoscopic procedures in children. Am J Gastroenterol 2000;95(10):3011–2.
11. Weinbroum AA, Szold O, Ogorek D, Flaishon R. The midazolam-induced paradox phenomenon is reversible by flumazenil. Epidemiology, patient characteristics and review of the literature. Eur J Anaesthesiol 2001;18(12):789–97.
12. Aviram EE, Ben-Abraham R, Weinbroum AA. Flumazenil use in children. Paediatr Perinatal Drug Ther 2003;5(4):202–9.
13. Zaw W, Knoppert DC, da Silva O. Flumazenil's reversal of myoclonic-like movements associated with midazolam in term newborns. Pharmacotherapy 2001;21(5):642–6.
14. Watanabe S, Satumae T, Takeshima R, Taguchi N. Opisthotonos after flumazenil administered to antagonize midazolam previously administered to treat developing local anesthetic toxicity. Anesth Analg 1998;86(3):677–8.
15. Strohle A, Wiedemann K. Flumazenil attenuates the pituitary response to CRH in healthy males. Eur Neuropsychopharmacol 1996;6(4):323–5.
16. Lopez A, Rebollo J. Benzodiazepine withdrawal syndrome after a benzodiazepine antagonist. Crit Care Med 1990;18(12):1480–1.

Flunitrazepam

General Information

Flunitrazepam has acquired a reputation for toxicity, abuse potential (1) and associated forensic problems (2), including being implicated in sexual assault ("date rape") (3). It has been withdrawn from general availability in various countries, including the USA, Australia, and New Zealand, and it is considered to be a narcotic in various European countries. It has a rapid onset of action but a long half-life. The earlier recommended hypnotic dose (1–2 mg) is excessive, and like some other benzodiazepines, such as triazolam, flunitrazepam is safer in a smaller dose (4). Like triazolam, it has been associated with nocturnal binge eating (5). Although its outpatient use in individuals who are susceptible to abuse is hazardous, intravenous flunitrazepam is useful for alcohol withdrawal delirium, but assisted ventilation should be available.

Long-Term Effects

Drug withdrawal

Withdrawal syndrome and delirium has been attributed to flunitrazepam (6).

- A 69-year-old man developed acute benzodiazepine withdrawal delirium following a short course of flunitrazepam after an acute exacerbation of chronic obstructive pulmonary disease. He was not an alcohol- or drug-abuser and he had not previously taken benzodiazepines. Six days after withdrawal of flunitrazepam he became agitated and confused, and had visual hallucinations, disorganized thinking, insomnia, increased psychomotor activity, disorientation in time and place, and memory impairment. Tachycardia and significant anxiety were also noted. He fulfilled the DSM IV criteria for withdrawal syndrome and delirium, and had spontaneous remission of symptoms within 48 hours.

The authors commented that physicians should be more aware of drug withdrawal syndromes, even after limited periods of administration of sedative drugs.

References

1. Bond A, Seijas D, Dawling S, Lader M. Systemic absorption and abuse liability of snorted flunitrazepam. Addiction 1994;89(7):821–30.
2. Michel L, Lang JP. Benzodiazepines et passage à l'acte criminel. [Benzodiazepines and forensic aspects.] Encephale 2003;29(6):479–85.
3. Smith KM, Larive LL, Romanelli F. Club drugs: methylenedioxymethamphetamine, flunitrazepam, ketamine hydrochloride, and gamma-hydroxybutyrate. Am J Health Syst Pharm 2002;59(11):1067–76.
4. Grahnen A, Wennerlund P, Dahlstrom B, Eckernas SA. Inter- and intraindividual variability in the concentration-effect (sedation) relationship of flunitrazepam. Br J Clin Pharmacol 1991;31(1):89–92.
5. Lowenstein W, LeJeunne C, Fadlallah JP, Hughes FC, Haas C, Durand H. Binge eating and flunitrazepam. Eur J Intern Med 1994;5:57.
6. Diehl JL, Guillibert E, Guerot E, Kimounn E, Labrousse J. Acute benzodiazepine withdrawal delirium after a short course of flunitrazepam in an intensive care patient. Ann Med Interne (Paris) 2000;151(Suppl. A):A44–6.

Lorazepam

General Information

Lorazepam is a benzodiazepine with CNS, depressant, anxiolytic, and sedative properties, used as a hypnotic, sedative, and anxiolytic drug.

Comparative studies

In a multicenter, randomized, double-blind comparison of diazepam (0.15 mg/kg followed by phenytoin 18 mg/kg), lorazepam (0.1 mg/kg), phenobarbital (15 mg/kg), and phenytoin (18 mg/kg) in 518 patients with generalized convulsive status epilepticus, lorazepam was more effective than phenytoin and at least as effective as phenobarbital or diazepam plus phenytoin (1). Drug-related adverse effects did not differ significantly among the treatments and included hypoventilation (up to 17%), hypotension (up to 59%), and cardiac rhythm disturbances (up to 9%).

Intramuscular lorazepam 4 mg has been compared with the combination of intramuscular haloperidol 10 mg + promethazine 50 mg in 200 emergency psychiatric patients with agitation, aggression, or violence (2). The treatments were comparably effective and well tolerated overall, but two patients who took lorazepam had moderate adverse effects: one had worse bronchial asthma and one had nausea and dizziness.

Placebo-controlled studies

The use of intravenous benzodiazepines administered by paramedics for the treatment of out-of-hospital status epilepticus has been evaluated in a double-blind, randomized trial in 205 adults (3). The patients presented either with seizures lasting 5 minutes or more or with repetitive generalized convulsive seizures and were randomized to receive intravenous diazepam 5 mg, lorazepam 2 mg, or placebo. Status epilepticus was controlled on arrival at the hospital in significantly more patients taking benzodiazepines than placebo (lorazepam 59%, diazepam 43%, placebo 21%). The rates of respiratory or circulatory complications related to drug treatment were 11% with lorazepam, 10% with diazepam, and 23% with placebo, but these differences were not significant.

Organs and Systems

Respiratory

Patients receiving intravenous benzodiazepines must be monitored for respiratory depression, which may demand artificial ventilation during intensive treatment. Lorazepam may cause less respiratory depression than diazepam at equieffective dosages (SEDA-20, 59).

Nervous system

Lorazepam is often used to manage anxiety, presurgically, and as a sedative. Common adverse effects include sedation, dizziness, weakness, unsteadiness, and disorientation, and lorazepam can cause significant impairment of driving ability. All positive lorazepam drug-impaired driving cases submitted to the Washington State Toxicology Laboratory between January 1998 and December 2003 were reviewed (4). The mean whole lorazepam blood concentration in these drivers (n=170, 56% male, mean age 40 years) was 48 ng/ml, but 86% of drivers tested positive for other drugs, which may have contributed to their impairment. In 23 cases lorazepam was the only drug detected, at a mean blood concentration of 51 ng/ml. In ten of the other cases (in which no drugs other than lorazepam were present) Drug Recognition Expert reports were obtained, containing details of events surrounding arrests and performances on field sobriety tests; lorazepam concentrations in these cases averaged 50 ng/ml. These results suggested that lorazepam can cause significant impairment of driving and psychomotor abilities, independent of the blood concentration.

In a placebo-controlled, double-blind, parallel-group comparison of single oral doses of lorazepam (2 mg) and flunitrazepam (1.2 mg), 36 young, healthy subjects completed a test battery before and after treatment including classic behavioral tests and visual and auditory event-related potentials (5). Differences in the impairment profile between equipotent doses of lorazepam and flunitrazepam suggested that lorazepam causes atypical central visual processing changes.

In a comparison of interventions commonly used for controlling agitation or violence in people with serious psychiatric disorders, 200 people were randomized to intramuscular lorazepam 4 mg or intramuscular haloperidol 10 mg + promethazine 25–50 mg (6). The haloperidol + promethazine combination produced a faster onset of tranquillization/sedation and more clinical improvement over the first 2 hours. The intervention did not differ in the need for additional interventions or physical restraints, numbers absconding, or adverse effects. Adverse effects were uncommon in both groups, but were only very briefly mentioned; they included respiratory difficulty in one patient given lorazepam and nausea and dizziness in another. Both interventions are effective in controlling violent/agitated behavior. If speed of sedation is required, the haloperidol + promethazine combination has advantages over lorazepam.

Psychological, psychiatric

Lorazepam causes some rare adverse effects, including a manic-like reaction on withdrawal, delirium, and paradoxical precipitation of tonic seizures or myoclonus in children (SEDA-19, 35). It can both relieve and worsen behavioral disturbances in demented elderly patients (SEDA-20, 32).

The effects of lorazepam on three neuropsychiatric measures of attention and psychomotor performance have been investigated in 40 patients, 20 of whom were given placebo, 10 were given lorazepam 1 mg, and 10 were given lorazepam 2.5 mg (7). Performance on digit cancellation, digit-symbol substitution, and the Paced Auditory Serial Addition Task was significantly impaired by lorazepam (2.5 mg) and this was significantly worse in the middle-aged subjects compared with the younger. These results suggest that older people are more susceptible to these adverse effects.

Lorazepam disinhibits aggression more than its chemically and kinetically similar analogue oxazepam (8).

Cognition

The psychomotor and amnesic effects of single oral doses of lorazepam 2 mg were studied in 48 healthy subjects in a double-blind, placebo-controlled, randomized, parallel-group study (9). The effects were assessed by a battery of subjective and objective tests that explored mood and vigilance, attention, psychomotor performance, and memory. Vigilance, psychomotor performance, and free recall were significantly impaired by lorazepam.

Lorazepam shares with other benzodiazepines the ability to impair explicit memory, but has a distinct further effect on implicit memory as well (10). However, it impairs memory more than its chemically and kinetically similar analogue, oxazepam (11). The effects of lorazepam 2.5 mg and diazepam 0.3 mg/kg on explicit and implicit memory tasks have been examined in 24 men and 24 women randomly allocated to lorazepam, diazepam, or placebo (10). An implicit word-stem completion task and explicit memory tasks of immediate and delayed word recall and word recognition were administered 90 minutes after drug administration. Both diazepam and lorazepam significantly impaired performance on explicit memory measures. Only lorazepam significantly impaired performance on the implicit memory task.

In a separate study, pharmacokinetic–pharmacodynamic modeling of the psychomotor and amnesic effects of a single oral dose of lorazepam 2 mg was investigated in 12 healthy volunteers in a randomized, double-blind, placebo-controlled, two-way, crossover study using the following tasks: choice reaction time, immediate and delayed cued recall of paired words, and immediate and delayed free recall and recognition of pictures (12). The delayed recall trials were more impaired than the immediate recall trials; similar observations were made with the recognition versus recall tasks.

The effects of lorazepam and diazepam on false memories and related states of awareness have been investigated in 36 healthy volunteers, randomly assigned to one of three groups (placebo, diazepam 0.3 mg/kg, lorazepam

0.038 mg/kg) (13). The results suggested that diazepam and lorazepam cause impaired conscious recollection, associated with true, but not false, memories.

In a double-blind, placebo-controlled study of sex differences in the effects of lorazepam in trained social drinkers, lorazepam substituted for alcohol equally in both sexes and increased associated scores for light-headedness (14). The women had much greater performance impairment in a digital symbol substitution test after lorazepam than the men. These results suggest that the stimulus and cognitive effects of benzodiazepine receptor agonists are modulated by different brain mechanisms.

The effects of lorazepam on the allocation of study time, memory, and judgement of learning have been investigated in a cognitive task in which the repetition of word presentation was manipulated (15). The study was placebo-controlled in 30 healthy volunteers. In a measure of the accuracy of delayed judgement of learning, all the participants benefited from word repetition; although lorazepam reduced overall performance, accuracy was not affected. All the participants then benefited from repetition of learning, although performance in those who took lorazepam remained lower than in those who took placebo. Repetition of learning had an effect on judgement of learning in both groups. Finally, study time fell significantly with the frequency of presentation an effect that was prevented by lorazepam. These findings suggest that lorazepam has a differential effect on the monitoring and the control processes involved in learning.

Metabolism

Metabolic acidosis and hyperlactatemia have been attributed to lorazepam (16).

- A 34-year-old woman with a history of renal insufficiency induced by long-term use of cocaine developed respiratory failure and was intubated and sedated with intravenous lorazepam (65 mg, 313 mg, and 305 mg on 3 consecutive days). After 2 days she had a metabolic acidosis, with hyperlactatemia and hyperosmolality. Propylene glycol, a component of the lorazepam intravenous formulation, was considered as a potential source of the acidosis, as she had received more than 40 times the recommended amount over 72 hours. Withdrawal of lorazepam produced major improvements in lactic acid and serum osmolality.

Long-Term Effects

Drug withdrawal

Lorazepam has considerable abuse potential, and poses particular difficulties in withdrawal (17). On the other hand, a sizeable sample ($n = 97$) of chronic users who wanted to discontinue were generally able to use stable or decreasing doses on an as-needed basis (18).

The Omnibus Budget Reconciliation Act of 1987 (OBRA '87) regulations (www.elderlibrary.org) specify when antipsychotic drugs can and cannot be used to treat behavioral disturbances in nursing home residents in the USA. Accordingly, antipsychotic drugs can be used in patients with delirium or dementia only if there are psychotic or agitated features that present a danger to the patient or others. Preventable causes of agitation must be excluded and the nature and frequency of these behaviors must be documented. Non-dangerous agitation, uncooperativeness, wandering, restlessness, insomnia, and impaired memory are insufficient in isolation to justify the use of antipsychotic drugs. With this in mind, the effects of withdrawing haloperidol, thioridazine, and lorazepam have been examined in a double-blind, crossover study in 58 nursing home residents (43 women and 15 men, mean age 86 years), half of whom continued to take the psychotropic drugs that had been prescribed, while the other half were tapered to placebo (19). After 6 weeks, the drugs were tapered to the reverse schedule for another 6 weeks. There were no differences between drug and placebo in functioning, adverse effects, and clinical global impression. Cognitive functioning improved during placebo. The authors concluded that gradual dosage reductions of psychoactive medications must be attempted, unless clinically contraindicated, in an effort to withdraw these drugs. Similar conclusions have been reached in other studies (SEDA-22, 54).

Drug Administration

Drug administration route

The pharmacokinetics of intranasal lorazepam compared with oral administration have been evaluated in 11 volunteers in a randomized, crossover study (20). Lorazepam had favorable pharmacokinetics for intranasal administration compared with standard methods. Intranasal delivery could provide an alternative non-invasive delivery route for lorazepam.

Drug–Drug Interactions

Caffeine

Caffeine aggravates, rather than attenuates, lorazepam-induced impairment in learning and performance (21). This contradicts popular wisdom that stimulants are useful in perking up patients taking benzodiazepines, and invites research into what may be a very dangerous practice in benzodiazepine users, that of taking caffeine before driving or operating machinery (see also General Introduction in the monograph on Benzodiazepines).

Clozapine

Caution has been recommended when starting clozapine in patients taking benzodiazepines (SEDA-19, 55). Three cases of delirium associated with clozapine and benzodiazepines (22) have been reported. There have been several reports of synergistic reactions, resulting in increased sedation and ataxia, when lorazepam was begun in patients already taking clozapine (23).

Haloperidol

A thorough review of the pharmacokinetics of haloperidol, with special emphasis on interactions, has been published (24). The interactions include one with lorazepam.

Several cases of torsade de pointes have been reported with intravenous haloperidol used with lorazepam to treat delirium (SEDA-18, 30) (SEDA-18, 47). Acid mucopolysaccharide deposition may be associated with neuroleptic drug treatment as a possible mechanism contributing to rare cardiovascular adverse events (25).

Moxonidine

When co-administered with lorazepam 0.4 mg, moxonidine increased impairment of attentional tasks (choice, simple reaction time and digit vigilance performance, memory tasks, immediate word recall, delayed word recall accuracy, and visual tracking). These effects should be considered when moxonidine is coadministered with lorazepam, although they were smaller than would have been produced by a single dose of lorazepam 2 mg alone (26).

References

1. Treiman DM, Meyers PD, Walton NY, Collins JF, Colling C, Rowan AJ, Handforth A, Faught E, Calabrese VP, Uthman BM, Ramsay RE, Mamdani MBVeterans Affairs Status Epilepticus Cooperative Study Group. A comparison of four treatments for generalized convulsive status epilepticus. N Engl J Med 1998;339(12):792–8.
2. Alexander J, Tharyan P, Adams C, John T, Mol C, Philip J. Rapid tranquillisation of violent or agitated patients in a psychiatric emergency setting. Pragmatic randomised trial of intramuscular lorazepam v. haloperidol plus promethazine. Br J Psychiatry 2004;185:63–9.
3. Alldredge BK, Gelb AM, Isaacs SM, Corry MD, Allen F, Ulrich S, Gottwald MD, O'Neil N, Neuhaus JM, Segal MR, Lowenstein DH. A comparison of lorazepam, diazepam, and placebo for the treatment of out-of-hospital status epilepticus. N Engl J Med 2001;345(9):631–7.
4. Clarkson JE, Gordon AM, Logan BK. Lorazepam and driving impairment. J Anal Toxicol 2004;28:475–80.
5. Pompéia S, Manzano GM, Galduroz JCF, Tufik S, Bueno OFA. Lorazepam induces an atypical dissociation of visual and auditory event-related potentials. J Psychopharmacol 2003;17:31–40.
6. Alexander J, Tharyan P, Adams C, John T, Mol C, Philip J. Rapid tranquillisation of violent or agitated patients in a psychiatric emergency setting. Br J Psychiatry 2004;185:63–9.
7. Fluck E, Fernandes C, File SE. Are lorazepam-induced deficits in attention similar to those resulting from aging? J Clin Psychopharmacol 2001;21(2):126–30.
8. Bond A, Lader M. Differential effects of oxazepam and lorazepam on aggressive responding. Psychopharmacology (Berl) 1988;95(3):369–73.
9. Micallef J, Soubrouillard C, Guet F, Le Guern ME, Alquier C, Bruguerolle B, Blin O. A double blind parallel group placebo controlled comparison of sedative and mnesic effects of etifoxine and lorazepam in healthy subjects. Fundam Clin Pharmacol 2001;15(3):209–16.
10. Le Roi S, Kirby KC, Montgomery IM, Daniels BA. Differential effects of lorazepam and diazepam on explicit and implicit memory. Aust J Psychopharmacol 1999;9:48–54.
11. Curran HV, Schiwy W, Lader M. Differential amnesic properties of benzodiazepines: a dose-response comparison of two drugs with similar elimination half-lives. Psychopharmacology (Berl) 1987;92(3):358–64.
12. Blin O, Jacquet A, Callamand S, Jouve E, Habib M, Gayraud D, Durand A, Bruguerolle B, Pisano P. Pharmacokinetic–pharmacodynamic analysis of amnesic effects of lorazepam in healthy volunteers. Br J Clin Pharmacol 1999;48(4):510–2.
13. Huron C, Servais C, Danion JM. Lorazepam and diazepam impair true, but not false, recognition in healthy volunteers. Psychopharmacology (Berl) 2001;155(2):204–9.
14. Jackson A, Stephens D, Duka T. Gender differences in response to lorazepam in a human drug discrimination study. J Psychopharmacol 2005;19(6):614–9.
15. Izaute M, Bacon E. Specific effects of an amnesic drug: effect of lorazepam on study time allocation and judgement on learning. Neuropsychopharmacology 2005;30:196–204.
16. Cawley MJ. Short-term lorazepam infusion and concern for propylene glycol toxicity: case report and review. Pharmacotherapy 2001;21(9):1140–4.
17. Lader M. Clin pharmacology of anxiolytic drugs: Past, present and future. In: Biggio G, Sanna E, Costa E, editors. GABA-A Receptors and Anxiety. From Neurobiology to Treatment. New York: Raven Press, 1995:135.
18. Romach M, Busto U, Somer G, Kaplan HL, Sellers E. Clinical aspects of chronic use of alprazolam and lorazepam. Am J Psychiatry 1995;152(8):1161–7.
19. Cohen-Mansfield J, Lipson S, Werner P, Billig N, Taylor L, Woosley R. Withdrawal of haloperidol, thioridazine, and lorazepam in the nursing home: a controlled, double-blind study. Arch Intern Med 1999;159(15):1733–40.
20. Wermeling DP, Miller JL, Archer SM, Manaligod JM, Rudy AC. Bioavailability and pharmacokinetics of lorazepam after intranasal, intravenous, and intramuscular administration. J Clin Pharmacol 2001;41(11):1225–31.
21. Rush CR, Higgins ST, Bickel WK, Hughes JR. Acute behavioral effects of lorazepam and caffeine, alone and in combination, in humans. Behav Pharmacol 1994;5(3):245–54.
22. Jackson CW, Markowitz JS, Brewerton TD. Delirium associated with clozapine and benzodiazepine combinations. Ann Clin Psychiatry 1995;7(3):139–41.
23. Cobb CD, Anderson CB, Seidel DR. Possible interaction between clozapine and lorazepam. Am J Psychiatry 1991;148(11):1606–7.
24. Kudo S, Ishizaki T. Pharmacokinetics of haloperidol: an update. Clin Pharmacokinet 1999;37(6):435–56.
25. Ellman JP. Sudden death. Can J Psychiatry 1982;27(4):331–3.
26. Wesnes K, Simpson PM, Jansson B, Grahnen A, Weimann HJ, Kuppers H. Moxonidine and cognitive function: interactions with moclobemide and lorazepam. Eur. J Clin Pharmacol 1997;52(5):351–8.

Lormetazepam

General Information

Lormetazepam is a short-acting benzodiazepine with effects similar to those of diazepam.

Organs and Systems

Psychological, psychiatric

Auditory hallucinations have been attributed to lormetazepam (1).

- A 45-year-old woman with moderate depression and anxiety took lormetazepam 4 mg/day. After a few days, she noticed musical auditory hallucinations like children's songs. There were no neurological, otological, or psychiatric causes for the hallucinations. When the dose of lormetazepam was reduced to 2 mg the auditory experience changed to "classic tinnitus." The lormetazepam was eventually withdrawn, but a slight degree of tinnitus persisted.

In a randomized, double-blind, placebo-controlled, crossover study in 18 young adults (mean age, 27 years), a single dose of lormetazepam 1 mg had no significant effect on either visual simple reaction time or visual choice reaction time (2). Lormetazepam caused mild dizziness in two subjects.

Drug Administration

Drug formulations

A new oral solution formulation of lormetazepam has been compared with lormetazepam tablets in an open, randomized, parallel group study in 108 out-patients with insomnia. They were given 0.5 mg on the first night and were allowed to increase their dosage by 0.25 mg each day (for the oral solution) or 0.5 mg every 2 days (for the tablets). The mean daily dose of lormetazepam was lower in those who used the oral solution than in those who used the tablets, and the cumulative dose of lormetazepam was lower with the oral solution (3). There were no significant differences between the groups in sleep characteristics or adverse effects. The occurrence of adverse effects did not differ between the two groups. Adverse events related to the lormetazepam tablet included mouth ulcers (n = 1), drowsiness (3), headache (2), frequent dreams (1), insomnia (3), night sweats (1). Adverse events associated with the lormetazepam oral solution included bad taste (n = 2), bitterness (2), dry mouth (1), drowsiness (5), hypotension (1), nausea/constipation (2), insomnia (3), and headache (1). These results suggest that using an oral solution of lormetazepam allows easier determination of the minimal individual effective dose.

References

1. Curtin F, Redmund C. Musical hallucinations during a treatment with benzodiazepine. Can J Psychiatry 2002;47(8):789–90.
2. Fabbrini M, Fritelli C, Bonanni E, Maestri M, Mance ML, Iudice A. Psychomotor performance in healthy young adult volunteers receiving lormetazepam and placebo: A single-dose, randomised, double-blind, crossover trial. Clin Ther 2005;27(1):78-83.
3. Ancolio C, Tardieu S, Soubrouillard C, Alquier C, Pradel V, Micallef J, Blin O. A randomized clinical trial comparing doses and efficacy of lormetazepam tablets or oral solution for insomnia in a general practice setting. Hum Psychopharmacol 2004;19:129-34.

Mexazolam

General Information

Mexazolam is a benzodiazepine with effects similar to those of diazepam.

The anxiolytic effects of mexazolam have been compared with those of alprazolam in 64 outpatients with generalized anxiety disorder in a multicenter, double-blind, parallel-group, randomized trial (1). Five mexazolam and nine alprazolam recipients reported mild adverse events: drowsiness in three patients in each group; dizziness in one taking mexazolam and two taking alprazolam; blurred vision in one patient taking mexazolam; and weight gain, nausea, and insomnia in one patient taking alprazolam. Both drugs were effective anxiolytics and both were well tolerated.

Reference

1. Vaz-Serra A, Figueira ML, Bessa-Peixoto A, Firmino H, Albuquerque R, Paz C, Dolgner A, Vaz-Silva M, Almeida L. Mexazolam and alprazolam in the treatment of generalised anxiety disorder. Clin Drug Invest 2001;21:257–63.

Midazolam

General Information

Midazolam is used mainly in parenteral form in anesthesia, as a sedative adjunct to medical and dental procedures, and in status epilepticus (1). Its pharmacology and therapeutics have been extensively reviewed (2). It produces greater amnesia than diazepam, useful in terms of its anesthetic use, but carries a risk of cardiorespiratory depression and death (SED-12, 99), particularly at the extremes of age and when combined with the opioid fentanyl (SEDA-22, 42), which can cause accumulation of midazolam (see Drug–Drug Interactions in this monograph). Vomiting has been reported in 10% of children having midazolam sedation before radiology (SEDA-22, 41). Behavioral disinhibition (SEDA-18, 44), acute withdrawal, and hiccups appear to be relatively common (SEDA-22, 41); hallucinations, flumazenil-reversible dystonia, and hypersensitivity have all been observed (SED-12, 99) (SEDA-17, 44).

Clinical electrophysiological procedures can be very complex and prolonged, requiring safe and effective conscious sedation. A study in 700 patients has shown that intermittent midazolam plus fentanyl in electrophysiological procedures is safe and efficacious (3). All the staff were ACLS-certified and had successfully completed conscious sedation training courses, but none was an anesthetist; one team member was dedicated to monitoring conscious sedation and providing rescue defibrillation if required.

The pharmacology and adverse effects of midazolam in infants and children have been reviewed (4).

Midazolam has been carefully evaluated for adverse effects when used in critically ill infants and children; several difficulties, including prolonged obtundation and paradoxical behavioral and withdrawal reactions, have been noted (SEDA-19, 35) (5). Midazolam by the buccal route has been evaluated in children with persistent seizures; it was both effective and well tolerated (6), offering obvious practical advantages to rectal or parenteral administration. The availability of flumazenil, a specific benzodiazepine antagonist, to correct any adverse or overdose effects from injected midazolam should not encourage laxity in its use. Recent reports have highlighted kinetic interactions between midazolam and a variety of other drugs (see Drug–Drug Interactions in this monograph), and its effects can be magnified or prolonged in patients with hepatic or renal insufficiency (SEDA-20, 32).

Midazolam was used in a wide range of doses (0.03–0.6 mg/kg) in 91 children undergoing diagnostic or minor operative procedures with intravenous midazolam sedation (7). Opioids were co-administered in 84% and oxygen desaturation occurred in 32%, most of whom had received high doses of opioids in addition to the midazolam. Other adverse events included airway obstruction ($n = 3$) and vomiting ($n = 1$). The presence of independent appropriate trained personnel not directly involved in performing the procedure, appropriate resuscitation equipment, and monitoring were recommended whenever midazolam and opioids are co-administered for intravenous sedation.

A cherry-flavored midazolam syrup was evaluated for premedication in 85 children requiring general anesthesia (8). The patients received a randomly assigned dose of 0.25, 0.5, or 1 mg/kg. All clinicians and observers were blinded to the treatment group. There was satisfactory dose-related sedation in 81%, and 83% had satisfactory non-dose-related anxiolysis at separation from parents and at anesthetic induction. One or more adverse events occurred in 36%, but only 31% of these were judged as possibly related to midazolam (hiccups 6%, hypoxemia 6%, vomiting 5%, hallucinations 4%, drooling 4%, agitation 2%, coughing 2%, diplopia 2%, dizziness 2%, and hypotension 2%). The authors suggested that although adverse effects were common, they were minor.

Intranasal midazolam 0.2 mg/kg and intravenous diazepam 0.3 mg/kg have been compared in a prospective randomized study in 47 children (aged 6 months to 5 years) with febrile seizures that lasted over 10 minutes (9). Intranasal midazolam controlled seizures significantly earlier than intravenous diazepam. None of the children had respiratory distress, bradycardia, or other adverse effects. Electrocardiography, blood pressure, and pulse oximetry were normal in all children during seizure activity and after cessation of seizures.

In a Canadian multicenter, open, randomized trial in 156 patients to determine whether sedation with propofol would lead to shorter times to tracheal extubation and length of stay in ICU than sedation with midazolam, the patients who received propofol spent longer at the target sedation level than those who received midazolam (60 versus 44% respectively) (10). Propofol allowed clinically significantly earlier tracheal extubation than midazolam (6.7 versus 25 hours). However, this did not result in earlier discharge from the ICU.

In a double-blind, randomized, placebo-controlled study during coronary angiography in 90 patients, midazolam with or without fentanyl and local anesthesia provided better hemodynamic stability than placebo (11).

In a randomized study in 301 agitated or aggressive patients, intramuscular midazolam was more rapidly sedating than a mixture of haloperidol + promethazine (12). There was only one important adverse event, transient respiratory depression, in one of the 151 patients who were given midazolam.

Observational studies

In a 1-year retrospective survey of the use of intramuscular midazolam in a 30-bed acute inpatient general adult unit in Sydney, Australia, 212 doses of intramuscular midazolam were given, predominantly 5 mg (48%) or 10 mg (50%) (13). An antipsychotic drug was co-administered in 2.4%. Adverse effects were documented in eight episodes (3.8%), seven cases of excess sedation and one of urinary incontinence. None of the adverse effects required medical intervention.

In 27 children with refractory generalized convulsive status epilepticus, midazolam 0.2 mg/kg as a bolus followed by 1–5 (mean 3.1) micrograms/kg/minute as a continuous infusion achieved complete control of seizures in 26 children within 65 minutes (14). There were no adverse effects, such as hypotension, bradycardia, or respiratory depression. In one patient with acute meningoencephalitis, status epilepticus could not be controlled. Five patients died of the primary disorders, one with progressive encephalopathy.

Comparative studies

In a randomized trial of intramuscular midazolam 15 mg (n=151) or intramuscular haloperidol 10 mg plus promethazine 50 mg (n=150) in agitated patients in three psychiatric emergency rooms, both treatments were effective (15). Midazolam was more rapidly sedating than haloperidol + promethazine, reducing the time people were exposed to aggression. One important adverse event occurred in each group; a patient given midazolam had transient respiratory depression, and one given haloperidol + promethazine had a generalized tonic-clonic seizure.

In a comparison of intranasal midazolam 0.2 mg/kg and intravenous diazepam 0.2 mg/kg in the treatment of acute childhood seizures in 70 children aged 2 months to 15 years with acute seizures (febrile or afebrile), the two drugs were equally effective and there were no significant adverse effects in either group (16). Although intranasal midazolam was as safe and effective as diazepam, seizures were controlled more quickly with intravenous diazepam.

In a multicenter, randomized controlled comparison of buccal midazolam and rectal diazepam for emergency-room treatment of 219 separate episodes of active seizures in 177 children aged 6 months and older with and without intravenous access, the dose varied with age, from 2.5 to 10 mg (17). The primary end point was therapeutic success—cessation of seizures within 10 minutes and for at least 1 hour without respiratory depression requiring intervention. The therapeutic response was 56% (61 of 109) for buccal midazolam and 27% (30 of 110) for rectal diazepam. When center, age, known diagnosis of epilepsy, use of antiepileptic drugs, prior treatment, and length of seizure before treatment were taken into account by logistic regression, buccal midazolam was more effective than rectal diazepam. The rates of respiratory depression did not differ.

Placebo-controlled studies

In a double-blind, randomized, placebo-controlled trial 130 patients were randomized to either midazolam 7.5 mg of orally (n = 65) or a placebo (n = 65) as pre-medication before upper gastrointestinal endoscopy (18). The median anxiety score during the procedure was significantly reduced by midazolam. Significantly more of those who took midazolam graded overall tolerance as "excellent or good" and reported a partial to complete amnesia response. Those who took midazolam were more willing to repeat the procedure if necessary. Midazolam significantly prolonged the median recovery time. There were no significant effects on satisfaction score or hemodynamic changes.

Organs and Systems

Cardiovascular

The incidence of hypotension with the use of midazolam for pre-hospital rapid-sequence intubation of the trachea has been assessed in a retrospective chart review of two aeromedical crews (19). The rapid-sequence protocols were identical, except for the dose of midazolam. Both crews used 0.1 mg/kg, but one crew had a maximum dose of 5 mg imposed. This meant that patients over 50 kg received lower doses of midazolam; they also had a higher incidence of hypotension. This relation was also present in patients with traumatic brain injury, implying that cerebral perfusion could be compromised at a critical time in those without dosage restriction.

Midazolam depresses both cardiovascular and respiratory function, especially in elderly patients (20). As little as 0.01 mg/kg can obtund the response to hypoxia and hypercapnia (21). The simultaneous use of opiates (such as fentanyl) commonly produces hypoxia (22).

Hypertension and tachycardia during coronary angiography can cause significant problems. In a double-blind,

randomized, placebo-controlled study during coronary angiography in 90 patients, midazolam with or without fentanyl under local anesthesia provided better hemodynamic stability than placebo (11).

Midazolam is often used for conscious sedation during transesophageal echocardiography. In a prospective study of the effects of midazolam or no sedation in addition to pharyngeal local anesthesia with lidocaine on the cardiorespiratory effects of transesophageal echocardiography in patients in sinus rhythm midazolam (median dose 3.3, range 1–5 mg) caused significantly higher heart rates and significantly lower blood pressures and oxygen saturations (23).

Respiratory

Intranasal midazolam is a successful route of administration for sedating children. However, it can cause nasal burning, irritation, and lacrimation (24). In a study of an alternative route of administration, namely inhalation via a nebulizer, bronchospasm developed in two of the 10 patients studied. This formulation of midazolam has a pH of 3.0, and this was thought to be the reason it caused bronchospasm.

Nervous system

Two premature neonates, who already had epileptic manifestations related to severe hypoxic ischemic encephalopathy, developed seizures (one tonic and the other tonic-clonic) within a few seconds of receiving intravenous midazolam (0.15 µg/kg) for sedation (25). In one, the seizure recurred after rechallenge on the same day. Benzodiazepines occasionally cause tonic seizures, especially after intravenous administration to children with Lennox–Gastaut syndrome. This seems to be the first report related to midazolam in newborns.

Involuntary epileptiform movements have been described in three of six premature infants given midazolam for sedation (26). The infants had been born at 24–26 weeks gestation, were aged 23–32 days, and weighed on average 671 g. They were given midazolam 100 µg/kg by slow bolus injection for sedation and then had accentuated myoclonic jerks resembling clonic seizures within 5 minutes. In all cases it was the first dose of midazolam and the cardiorespiratory parameters remained normal. The abnormal movements resolved within 5–10 minutes. There has also been a report of convulsions caused by midazolam in two preterm infants (27). They were of 26 and 28 weeks gestation and were given midazolam 100 and 150 µg/kg intravenously before tracheal intubation. Both were successfully treated with flumazenil 10 µg/kg.

The safety of midazolam in very low birth-weight neonates is being questioned. In 200 children weighing 3–15 kg premedicated with rectal midazolam 0.5 or 1.0 mg/kg before minor surgery, the incidence of hiccups was 22% and 26% respectively (28). The mean age of children with hiccups was 6 months and of children without hiccups 20 months. Intranasal ethyl chloride spray was 100% successful in treating the hiccups. The incidence of hiccups was related to age but not dose. The effectiveness of ethyl chloride was postulated to be via cold nasopharyngeal stimulation.

The effects of sevoflurane and halothane anesthesia, including the effect of midazolam 0.5 mg/kg premedication, on recovery characteristics have been studied in 100 children aged 6 months to 6 years undergoing myringotomy (29). Children who received sevoflurane had about 50% faster recovery times and discharge-home times than those who received halothane. However, sevoflurane without midazolam premedication was associated with 67% postoperative agitation compared with 40% in the premedicated group. Midazolam delayed early recovery times by about 5 minutes, but had no effect on discharge times. In view of the very high incidence of postoperative agitation in the control group midazolam was seen as an effective premedicant.

Midazolam can cause paradoxical reactions, including increased agitation and poor cooperation (30,31). Often other drugs are required to continue the procedure successfully. Reversal of these reactions by flumazenil, a benzodiazepine antagonist, has been reported.

Sevoflurane often causes postoperative delirium and agitation in children, and this may be severe. The effect of intravenous clonidine 2 μg/kg on the incidence and severity of postoperative agitation has been assessed in a double-blind, randomized, placebo-controlled trial in 40 boys who had anesthetic induction with sevoflurane after oral midazolam premedication (32). There was agitation in 16 of those who received placebo and two of those who received clonidine; the agitation was severe in six of those given placebo and none of those given clonidine.

The effect of a single bolus dose of midazolam before the end of sevoflurane anesthesia has been investigated in a double-blind, randomized, placebo-controlled trial in 40 children aged 2–7 years (33). Midazolam significantly reduced the incidence of delirium after anesthesia. However, when it was used for severe agitation, midazolam only reduced the severity of agitation without abolishing it.

In the presence of acute neurological injuries, midazolam produces a high risk of raised intracranial pressure (34), and the risk of airway obstruction (35) is a further concern.

Recovery after propofol or midazolam has been compared in two studies (36,37). Memory was significantly impaired by midazolam, an effect that was reminiscent of the problems experienced with short-acting oral benzodiazepine hypnotics, such as triazolam.

Forty anxious day-case patients undergoing extraction of third molar teeth under local anesthesia with sedation, were studied in a randomized, double-blind, controlled trial (38). A target-controlled infusion of propofol was compared with patient-controlled propofol for sedation, combined with a small dose of intravenous midazolam (0.03 mg/kg) to improve amnesia. Five patients became over-sedated in the target-controlled group compared with none in the patient-controlled group.

Sedation, cognition, and mood during midazolam infusion in 20 volunteers with red hair and 19 with non-red (blond or brown) hair were studied in a randomized, placebo-controlled, crossover design, to test the hypothesis that patients with red hair may require more drug to attain desired degrees of sedation (39). The red-haired volunteers had significantly greater alertness and lower drowsiness scores than non-red-haired subjects during midazolam infusion. Visuospatial scores were significantly higher in the subjects with red hair than in those with non-red hair during both placebo and midazolam trials. Delayed memory scores were significantly higher during midazolam infusion in subjects with red hair than in those with non-red hair. Midazolam appears to cause significantly less sedation and cognitive impairment in red-haired subjects.

The potential of intrathecal midazolam to produce symptoms suggestive of neurological damage has been investigated in a comparison of patients (n = 1100) who received intrathecal anesthesia with or without intrathecal midazolam 2 mg (40). Eighteen risk factors were evaluated with respect to symptoms representing potential neurological complications. Intrathecal midazolam was not associated with an increased risk of neurological symptoms. In contrast, neurological symptoms were increased in patients aged over 70 years and in those with a blood-stained spinal tap.

Psychological, psychiatric

Midazolam can cause an unpleasant state of dysphoria if surgical stimuli are applied to the patient, who may nevertheless appear calm and untroubled. Amnesia is also routinely produced and can be beneficial. Delayed recovery of cognitive function occurs after the use of benzodiazepines as premedication (41).

There was a variety of significant nervous system adverse effects in six of 104 patients who underwent transesophageal echocardiography, including aggression, euphoria, depression, and intense hiccups (42). These effects occurred despite careful titration and relatively low doses of intravenous midazolam (mean 4.8 mg), and were generally reversible with intravenous flumazenil 0.25–0.5 mg.

In a placebo-controlled study of the effects of midazolam 0.5 mg/kg as a premedicant in 40 children aged 4–6 years having myringotomy, midazolam caused significant *amnesia* on a cued recall task (43). In addition, free recall for post-drug events was also impaired by midazolam, suggesting that benzodiazepine-induced amnesia occurs even for highly salient information.

Hematologic

Fat emulsions affect coagulation and fibrinolysis (44). In 36 patients undergoing aortocoronary bypass operations with midazolam/fentanyl- or propofol/alfentanil-based anesthesia, factor XIIa concentrations and kallikrein-like activity were about 30% higher in the propofol group. The authors suggested that there had been stronger activation of the contact phase at the start of recirculation and stronger fibrinolysis in the propofol group. They also found more hypotension in the propofol group, which they assumed to be due to release of kallikrein, resulting in release of bradykinin. Propofol has not been proven to cause increased perioperative bleeding.

In a 40-month-old boy a withdrawal syndrome with neurological symptoms was accompanied by thrombocytosis, which peaked at $1230 \times 10^9/l$ (45). Recovery from the withdrawal syndrome was accompanied by normalization of the platelet count. The relevance of this change in platelet count was not clear. The boy had also been given fentanyl, and the authors suggested that the combination of midazolam with fentanyl should be used with caution.

Urinary tract

The conjugates of the main metabolite of midazolam, α-hydroxymidazolam, accumulate in renal insufficiency. In five patients with severe renal insufficiency (creatinine clearance 7 ml/minute or less), in whom prolonged sedation after midazolam was immediately reversed by flumazenil, there were high serum concentrations of glucuronidated α-hydroxymidazolam, even at times when the concentrations of the unconjugated metabolite and midazolam itself were low (46). Glucuronidated α-hydroxymidazolam is about one-tenth as potent as midazolam and unconjugated α-hydroxymidazolam, and accumulation of the conjugated metabolite in renal insufficiency may be important.

Body temperature

Hypothermia is common during anesthesia, and adversely affects outcome. It primarily results from internal redistribution of body heat from the core to the periphery. Premedication with sedative agents can affect perioperative heat loss by altering core-to-peripheral heat distribution. This has been analysed in a prospective randomized study in 45 patients undergoing arthroscopic knee ligament reconstruction surgery (47). Heavy premedication caused initial hypothermia. Moderate premedication reduced perioperative heat loss. No premedication was associated with significantly lower intraoperative core temperatures than in sedated patients.

Midazolam premedication caused exaggerated perioperative hypothermia in 15 elderly surgical patients compared with 15 young patients (48). The same group also showed that atropine prevents midazolam-induced core hypothermia in 40 elderly patients (49). The thermoregulatory effects of benzodiazepine agonists and cholinergic inhibitors oppose each other, and the combination leaves core temperature unchanged.

Long-Term Effects

Drug withdrawal

The withdrawal of an infusion of midazolam, used as sedation in intensive care units, is associated with occasional severe and bizarre behavioral disturbances, particularly in children (50). These are similar in nature to the withdrawal effects seen with other short-acting benzodiazepines.

Susceptibility Factors

Age

The pharmacology and adverse effects of midazolam in infants and children have been reviewed (4). The optimal dose of intramuscular midazolam for preoperative sedation has been studied in a double-blind prospective study of 600 patients who were age-stratified (51). The patients received intramuscular atropine 0.6 mg and one of five doses of midazolam 15 minutes before induction of anesthesia. For the age groups 20–39, 40–59, and 60–79 years, the optimal sedative and amnesic effects of midazolam were 0.10, 0.08, and 0.04 mg/kg respectively. The frequency with which the undesirable adverse effects of reduced blood pressure, oxygen desaturation, oversedation, loss of eyelash reflex, and tongue root depression occurred increased with age, and optimal doses for a low incidence of adverse effects were 0.08, 0.06, and 0.04 mg/kg in the same age groups respectively.

Drug Administration

Drug administration route

Midazolam nasal spray 2 mg/kg has been compared with a citric acid placebo for conscious sedation in children undergoing painful procedures (52). Citric acid was added to the placebo, so that the sensation of nasal burning caused by midazolam did not unblind the observers. Parents and nurses judged the procedure to be more comfortable with midazolam, but the children rated the discomfort of the procedure similar in the two groups. Anxiety was significantly reduced by midazolam. There was nasal discomfort in 43% of the midazolam group. The authors concluded that midazolam intranasal spray effectively reduces anxiety, but that its use may be limited by nasal discomfort.

Drug–Drug Interactions

Antifungal azole derivatives

The pharmacokinetics and pharmacodynamics of midazolam are significantly affected by antifungal azoles. The effects of fluconazole (400 mg loading dose followed by 200 mg/day) on the kinetics of midazolam have been studied in 10 mechanically ventilated adults receiving a stable infusion of midazolam (53). Concentrations of midazolam were increased up to fourfold after the start of fluconazole therapy; these changes were most marked in patients with renal insufficiency. During the study, the ratio of α-hydroxymidazolam to midazolam progressively fell. The authors concluded that in ICU patients receiving fluconazole, reduction of the dose of midazolam should be considered if the degree of sedation is increasing.

In a study of the effects of itraconazole 200 mg/day and rifampicin 600 mg/day on the pharmacokinetics and pharmacodynamics of oral midazolam 7.5–15 mg during and 4 days after the end of the treatment, switching from inhibition to induction of metabolism caused an up to 400-fold change in the AUC of oral midazolam (54).

In an in vitro study of midazolam biotransformation using human liver microsomes, midazolam metabolism was competitively inhibited by the antifungal azoles ketoconazole, itraconazole, and fluconazole, and the antidepressant fluoxetine and its metabolite norfluoxetine (55). The degree of inhibition was consistent with the inhibition reported in pharmacokinetic studies, and suggests that in vitro assay is useful for predicting significant interactions.

Atorvastatin

Midazolam is metabolized by CYP3A4, as is atorvastatin. In a matched-pair study the effects of long-term atorvastatin on the pharmacokinetics of midazolam 0.15 mg/kg intravenously as a single dose were studied in 14 patients undergoing general anesthesia for elective surgery (56). Atorvastatin significantly reduced the clearance of midazolam by 33% and increased the AUC by 40%.

Diltiazem

Diltiazem caused a 43% mean increase in the half-life of midazolam in 30 patients who underwent coronary artery bypass grafting (57). Similar effects were observed with alfentanil. The proposed mechanism was diltiazem-induced inhibition of benzodiazepine metabolism by CYP3A. Patients taking diltiazem had delayed early postoperative recovery as a result.

Drugs that influence CYP3A

Midazolam is selectively metabolized by CYP3A4, with which several drugs interact, influencing its pharmacokinetics and pharmacodynamics.

Itraconazole, an inhibitor of CYP3A, and rifampicin, an inducer of CYP3A, altered the pharmacokinetics and pharmacodynamics of oral midazolam in nine healthy volunteers (54). The half-life was prolonged from 2.7 to 7.6 hours by itraconazole and reduced to 1.0 hour by rifampicin. These effects were still present, although less marked, at 4 days after withdrawal of itraconazole and rifampicin. Similarly, after acute administration, the period of drowsiness was increased from 76 to 201 minutes with itraconazole and fell to 35 minutes with rifampicin; the effects were again less marked 4 days after withdrawal.

The protease inhibitor saquinavir, propofol, and fluconazole (53,58,59) increased the systemic availability and peak plasma concentrations and prolonged the half-life of midazolam, thus increasing its sedative effects. The dosage of midazolam should be reduced in patients taking these drugs.

Grapefruit juice reduced the metabolism of midazolam; prolonged sedation can be expected in some circumstances (60).

Chronic administration of glucocorticoids, which induce CYP3A4, reduces the sedative effect of midazolam by increasing its clearance; higher doses are therefore required for sedation (61).

Fentanyl

Several adverse effects have been reported with the combined use of fentanyl and midazolam, including chest wall rigidity, making ventilation with a bag and mask impossible (SEDA-16, 79). In neonates, hypotension can occur (SEDA-16, 80), and respiratory arrest in a child and sudden cardiac arrest have been reported (SEDA-16, 80). However, in one study there were no cardiac electrophysiological effects of midazolam combined with fentanyl in subjects undergoing cardiac electrophysiological studies (SEDA-18, 80).

Glucocorticoids

Chronic administration of glucocorticoids, which induce CYP3A4, reduces the sedative effect of midazolam by increasing its clearance; higher doses are therefore required for sedation (61).

Grapefruit juice

Grapefruit juice reduced the metabolism of midazolam; prolonged sedation can be expected in some circumstances (60).

Halothane

Midazolam produced marked reduction of the MAC of halothane in humans at lower serum concentrations than required to cause sleep (62).

Isoflurane

The effects of the combination of midazolam and isoflurane on memory were studied in a randomized, double-blind study in 28 volunteers (63). Midazolam 0.03 mg/kg or 0.06 mg/kg combined with isoflurane 0.2% almost completely abolished explicit and implicit memory, but there were more variable effects on the level of sedation. The duration of the deficit averaged 45 minutes. The study was remarkable for the very low doses required to abolish memory, owing to synergy of the combination of midazolam and isoflurane and abolition of memory at subhypnotic doses with this combination. However, the subjects did not undergo surgery, so caution must be exercised in extrapolating the result to surgical patients, because painful stimuli increase the dosage required to abolish memory.

Ketoconazole and nefazodone

Pharmacokinetic and pharmacodynamic interactions of midazolam with fluoxetine, fluvoxamine, nefazodone, and ketoconazole have been investigated in 40 healthy subjects (64). The mean AUC of midazolam was increased 772% by ketoconazole and 444% by nefazodone. However, fluoxetine and fluvoxamine had no significant effects. Nefazodone and ketoconazole caused significant increases in midazolam-related cognitive impairment, reflecting changed midazolam clearance.

Macrolides

Interactions of macrolide antibiotics with midazolam are clinically important. Increases in serum concentration, AUC, and half-life, and a reduction in clearance have been documented (65). These changes can result in clinical effects, such as prolonged psychomotor impairment, amnesia, or loss of consciousness (66).

Erythromycin potentiates the effects of oral midazolam (SEDA-17, 125). Either midazolam should be avoided in patients taking erythromycin or the dose should be reduced by 50–75% (67,68).

Azithromycin

In an open, randomized, crossover, pharmacokinetic and pharmacodynamic study in 12 healthy volunteers who took clarithromycin 250 mg bd for 5 days, azithromycin 500 mg/day for 3 days, or no pretreatment, followed by a single dose of midazolam (15 mg), clarithromycin increased the AUC of midazolam by over 3.5 times and the mean duration of sleep from 135 to 281 minutes (69). In contrast, there was no change with azithromycin, suggesting that it is much safer for co-administration with midazolam.

Clarithromycin

In an open, randomized, crossover, pharmacokinetic and pharmacodynamic study in 12 healthy volunteers who took clarithromycin 250 mg bd for 5 days, azithromycin 500 mg/day for 3 days, or no pretreatment, followed by a single dose of midazolam (15 mg), clarithromycin increased the AUC of midazolam by over 3.5 times and the mean duration of sleep from 135 to 281 minutes (70). In contrast, there was no change with azithromycin, suggesting that it is much safer for co-administration with midazolam.

Nitrous oxide

The combination of midazolam with nitrous oxide produced retrograde amnesia in 21 women undergoing elective cesarean section (71). All had spinal anesthesia. After delivery the patients received intravenous midazolam, average dose 94 µg/kg, and inhaled nitrous oxide 50%. At the end of surgery, flumazenil was given in 0.1 mg increments until the patient awoke. Another nine women were given only nitrous oxide inhalation after delivery. Of the women who received midazolam and nitrous oxide, 33% could not recall their baby's face, while all of the women not given midazolam could. The results suggest that midazolam plus nitrous oxide can produce retrograde amnesia not reversed by flumazenil.

Pethidine

Rectal pethidine is not advised in children, owing to enormous variability in systemic availability (SEDA-18, 82). When used for sedation in children undergoing esophagogastroduodenoscopy, hypoxia with dysrhythmias was more likely to occur with a combination of pethidine and diazepam than with pethidine and midazolam (SEDA-18, 81).

Rifampicin

In a study of the effects of itraconazole 200 mg/day and rifampicin 600 mg/day on the pharmacokinetics and pharmacodynamics of oral midazolam 7.5–15 mg during and 4 days after the end of the treatment, switching from inhibition to induction of metabolism caused an up to 400-fold change in the AUC of oral midazolam (54).

Saquinavir

Saquinavir substantially potentiates the effects of midazolam by raising its blood concentrations, and this suggests that a parallel interaction could occur with other protease inhibitors (and no doubt certain other benzodiazepines) in similar combinations (59).

References

1. Parent JM, Lowenstein DH. Treatment of refractory generalized status epilepticus with continuous infusion of midazolam. Neurology 1994;44(10):1837–40.
2. Lauven PM. Pharmacology of drugs for conscious sedation. Scand J Gastroenterol Suppl 1990;179:1–6.
3. Pachulski RT, Adkins DC, Mirza H. Conscious sedation with intermittent midazolam and fentanyl in electrophysiology procedures. J Interv Cardiol 2001;14(2):143–6.
4. Blumer JL. Clinical pharmacology of midazolam in infants and children. Clin Pharmacokinet 1998;35(1):37–47.
5. Hughes J, Gill A, Leach HJ, Nunn AJ, Billingham I, Ratcliffe J, Thornington R, Choonara I. A prospective study of the adverse effects of midazolam on withdrawal in critically ill children. Acta Paediatr 1994;83(11):1194–9.
6. Scott RC, Besag FM, Neville BG. Buccal midazolam and rectal diazepam for treatment of prolonged seizures in childhood and adolescence: a randomised trial. Lancet 1999;353(9153):623–6.
7. Karl HW, Cote CJ, McCubbin MM, Kelley M, Liebelt E, Kaufman S, Burkhart K, Albers G, Wasserman G. Intravenous midazolam for sedation of children undergoing procedures: an analysis of age- and procedure-related factors. Pediatr Emerg Care 1999;15(3):167–72.
8. Marshall J, Rodarte A, Blumer J, Khoo KC, Akbari B, Kearns G. Pediatric pharmacodynamics of midazolam oral syrup. Pediatric Pharmacology Research Unit Network. J Clin Pharmacol 2000;40(6):578–89.
9. Wassner E, Morris B, Fernando L, Rao M, Whitehouse WP. Intranasal midazolam for treating febrile seizures in children. Buccal midazolam for childhood seizures at home preferred to rectal diazepam. BMJ 2001;322(7278):108.
10. Hall RI, Sandham D, Cardinal P, Tweeddale M, Moher D, Wang X, Anis AH. Propofol vs midazolam for ICU sedation: a Canadian multicenter randomized trial. Chest 2001;119(4):1151–9.
11. Baris S, Karakaya D, Aykent R, Kirdar K, Sagkan O, Tur A. Comparison of midazolam with or without fentanyl for conscious sedation and hemodynamics in coronary angiography. Can J Cardiol 2001;17(3):277–81.
12. TREC Collaborative Group. Rapid tranquillisation for agitated patients in emergency psychiatric rooms: a randomised trial of midazolam versus haloperidol plus promethazine. BMJ 2003;327(7417):708–13.
13. Bradley N, Malesu RR. The use of intramuscular midazolam in an acute psychiatric unit. Aust NZ J Psychiatry 2003;37:111–2.

14. Ozdemir D, Gulez P, Uran N, Yendur G, Kavakli T, Aydin A. Efficacy of continuous midazolam infusion and mortality in childhood refractory generalised convulsive status epilepticus. Seizure 2005;14:129–32.

15. Huf G, Coutinho ESF, Adams CE, Borges RVS, Ferreira MAV, Silva FJF, Pereira AJCR, Abreu FM, Lugao SM, Santos MPCP, Gewandsznajder M, Mercadante VRP, Lange W Jr, Dias CI. Rapid tranquillisation for agitated patients in emergency psychiatric rooms: a randomised trial of midazolam versus haloperidol plus promethazine. Br Med J 2003;327:708–11.

16. Mahmoudian T, Zadeh MM. Comparison intranasal midazolam with intravenous diazepam for treating acute seizures in children. Epilepsy Behav 2004;5(2):253–5.

17. McIntyre J, Robertson S, Norris E, Appleton R, Whitehouse WP, Phillips B, Martland T, Berry K, Collier J, Smith S, Choonara I. Safety and efficacy of buccal midazolam versus rectal diazepam for emergency treatment of seizures in children: a randomised controlled trial. Lancet 2005;366:205–10.

18. Mui L, Teoh AYB, Enders KWN, Lee Y, Au Yeung ACM, Chan Y, Lau, JYW, Chung SCS. Premedication with orally administered midazolam in adults undergoing diagnostic upper endoscopy: a double-blind, placebo-controlled randomised trial. Gastrointest Endosc 2005;61(2):195–200.

19. Davis DP, Kimbro TA, Vilke GM. The use of midazolam for prehospital rapid-sequence intubation may be associated with a dose-related increase in hypotension. Prehosp Emerg Care 2001;5(2):163–8.

20. Anonymous. Midazolam—is antagonism justified? Lancet 1988;2(8603):140–2.

21. Alexander CM, Gross JB. Sedative doses of midazolam depress hypoxic ventilatory responses in humans. Anesth Analg 1988;67(4):377–82.

22. Bailey PL, Pace NL, Ashburn MA, Moll JW, East KA, Stanley TH. Frequent hypoxemia and apnea after sedation with midazolam and fentanyl. Anesthesiology 1990;73(5):826–30.

23. Blondheim DS, Levi D, Marmor AT. Mild sedation before transesophageal echo induces significant hemodynamic and respiratory depression. Echocardiography 2004;21(3):241–5.

24. McCormick AS, Thomas VL. Bronchospasm during inhalation of nebulized midazolam. Br J Anaesth 1998;80(4):564–5.

25. Montenegro MA, Guerreiro CAM, Guerreiro MM, Moura-Ribeiro MV. Midazolam-induced seizures in the neonatal period. Epilepsia 1999;40(Suppl 7):92.

26. Waisman D, Weintraub Z, Rotschild A, Bental Y. Myoclonic movements in very low birth weight premature infants associated with midazolam intravenous bolus administration. Pediatrics 1999;104(3 Part 1):579.

27. Birrell VL, Wyllie JP, Pagan J. Midazolam causing convulsions. Br J Intensive Care 1999;9:197.

28. Marhofer P, Glaser C, Krenn CG, Grabner CM, Semsroth M. Incidence and therapy of midazolam induced hiccups in paediatric anaesthesia. Paediatr Anaesth 1999;9(4):295–8.

29. Lapin SL, Auden SM, Goldsmith LJ, Reynolds AM. Effects of sevoflurane anaesthesia on recovery in children: a comparison with halothane. Paediatr Anaesth 1999;9(4):299–304.

30. Fulton SA, Mullen KD. Completion of upper endoscopic procedures despite paradoxical reaction to midazolam: a role for flumazenil? Am J Gastroenterol 2000;95(3):809–11.

31. Saltik IN, Ozen H. Role of flumazenil for paradoxical reaction to midazolam during endoscopic procedures in children. Am J Gastroenterol 2000;95(10):3011–2.

32. Kulka PJ, Bressem M, Tryba M. Clonidine prevents sevoflurane-induced agitation in children. Anesth Analg 2001;93(2):335–8.

33. Kulka PJ, Bressem M, Wiebalck A, Tryba M. Prophylaxe des "Postsevoflurandelirs" un Midazolam. [Prevention of "post-sevoflurane delirium" with midazolam.] Anaesthesist 2001;50(6):401–5.

34. Eldridge PR, Punt JA. Risks associated with giving benzodiazepines to patients with acute neurological injuries. BMJ 1990;300(6733):1189–90.

35. Montravers PH, Dureuil B, Desmonts JM. Effects of midazolam on upper airway resistances. Anesthesiology 1988;69:A824.

36. Atanassoff PG, Alon E, Pasch T. Recovery after propofol, midazolam, and methohexitone as an adjunct to epidural anaesthesia for lower abdominal surgery. Eur J Anaesthesiol 1993;10(4):313–8.

37. Crawford M, Pollock J, Anderson K, Glavin RJ, MacIntyre D, Vernon D. Comparison of midazolam with propofol for sedation in outpatient bronchoscopy. Br J Anaesth 1993;70(4):419–22.

38. Burns R, McCrae AF, Tiplady B. A comparison of target-controlled with patient-controlled administration of propofol combined with midazolam for sedation during dental surgery. Anaesthesia 2003;58:170–6.

39. Chua MV, Tsueda K, Doufas AG. Midazolam causes less sedation in volunteers with red hair. Can J Anaesth 2004;51:25–30.

40. Tucker AP, Lai C, Nadeson R, Goodchild CS. Intrathecal midazolam I: A cohort study investigating safety. Anesth Analg 2004;98:1512–20.

41. Gast PH, Fisher A, Sear JW. Intensive care sedation now. Lancet 1984;2(8407):863–4.

42. Wenzel RR, Bartel T, Eggebrecht H, Philipp T, Erbel R. Central-nervous side effects of midazolam during transesophageal echocardiography. J Am Soc Echocardiogr 2002;15(10 Part 2):1297–300.

43. Buffett-Jerrott SE, Stewart SH, Finley GA, Loughlan HL. Effects of benzodiazepines on explicit memory in a paediatric surgery setting. Psychopharmacology 2003;168:377–86.

44. Schulze HJ, Wendel HP, Kleinhans M, Oehmichen S, Heller W, Elert O. Effects of the propofol combination anesthesia on the intrinsic blood-clotting system. Immunopharmacology 1999;43(2–3):141–4.

45. Ducharme MP, Munzenberger P. Severe withdrawal syndrome possibly associated with cessation of a midazolam and fentanyl infusion. Pharmacotherapy 1995;15(5):665–8.

46. Bauer TM, Ritz R, Haberthur C, Ha HR, Hunkeler W, Sleight AJ, Scollo-Lavizzari G, Haefeli WE. Prolonged sedation due to accumulation of conjugated metabolites of midazolam. Lancet 1995;346(8968):145–7.

47. Toyota K, Sakura S, Saito Y, Ozasa H, Uchida H. The effect of pre-operative administration of midazolam on the development of intra-operative hypothermia. Anaesthesia 2004;59:116–21.

48. Matsukawa T, Ozaki M, Nishiyama T, Imamura M, Kumazawa T. Exaggerated perioperative hypothermia in elderly surgical patients. Anesth Resusc 2001;37:53–7.

49. Matsukawa T, Ozaki M, Nishiyama T, Imamura M, Iwamoto R, Iijima T, Kumazawa T. Atropine prevents midazolam-induced core hypothermia in elderly patients. J Clin Anesth 2001;13(7):504–8.

50. Conway EE Jr, Singer LP. Acute benzodiazepine withdrawal after midazolam in children. Crit Care Med 1990;18(4):461.

51. Nishiyama T, Matsukawa T, Hanaoka K. The effects of age and gender on the optimal premedication dose of intramuscular midazolam. Anesth Analg 1998;86(5):1103–8.

52. Ljungman G, Kreuger A, Andreasson S, Gordh T, Sorensen S. Midazolam nasal spray reduces procedural anxiety in children. Pediatrics 2000;105(1 Part 1):73–8.

53. Ahonen J, Olkkola KT, Takala A, Neuvonen PJ. Interaction between fluconazole and midazolam in intensive care patients. Acta Anaesthesiol Scand 1999;43(5):509–14.

54. Backman JT, Kivisto KT, Olkkola KT, Neuvonen PJ. The area under the plasma concentration-time curve for oral midazolam is 400-fold larger during treatment with itraconazole than with rifampicin. Eur J Clin Pharmacol 1998;54(1):53–8.

55. von Moltke LL, Greenblatt DJ, Schmider J, Duan SX, Wright CE, Harmatz JS, Shader RI. Midazolam hydroxylation by human liver microsomes in vitro: inhibition by fluoxetine, norfluoxetine, and by azole antifungal agents. J Clin Pharmacol 1996;36(9):783–91.

56. McDonnell CG, Harte S, O'Driscoll J, O'Loughlin C, Van Pelt FD, Shorten GD. The effects of concurrent atorvastatin therapy on the pharmacokinetics of intravenous midazolam. Anaesthesia 2003;58:899–904.

57. Ahonen J, Olkkola KT, Salmenpera M, Hynynen M, Neuvonen PJ. Effect of diltiazem on midazolam and alfentanil disposition in patients undergoing coronary artery bypass grafting. Anesthesiology 1996;85(6):1246–52.

58. Hamaoka N, Oda Y, Hase I, Mizutani K, Nakamoto T, Ishizaki T, Asada A. Propofol decreases the clearance of midazolam by inhibiting CYP3A4: an in vivo and in vitro study. Clin Pharmacol Ther 1999;66(2):110–7.

59. Palkama VJ, Ahonen J, Neuvonen PJ, Olkkola KT. Effect of saquinavir on the pharmacokinetics and pharmacodynamics of oral and intravenous midazolam. Clin Pharmacol Ther 1999;66(1):33–9.

60. D'Arcy PF. Grapefruit juice, midazolam and triazolam. Int Pharm J 1996;10:223.

61. Nakajima M, Suzuki T, Sasaki T, Yokoi T, Hosoyamada A, Yamamoto T, Kuroiwa Y. Effects of chronic administration of glucocorticoid on midazolam pharmacokinetics in humans. Ther Drug Monit 1999;21(5):507–13.

62. Inagaki Y, Sumikawa K, Yoshiya I. Anesthetic interaction between midazolam and halothane in humans. Anesth Analg 1993;76(3):613–7.

63. Ghoneim MM, Block RI, Dhanaraj VJ. Interaction of a subanaesthetic concentration of isoflurane with midazolam: effects on responsiveness, learning and memory. Br J Anaesth 1998;80(5):581–7.

64. Lam YWF, Alfaro CL, Ereshefsky L, Miller M. Pharmacokinetic and pharmacodynamic interactions of oral midazolam with ketoconazole, fluoxetine, fluvoxamine, and nefazodone. J Clin Pharmacol 2003;43:1274–82.

65. Gascon MP, Dayer P, Waldvogel F. Les interactions médicamenteuses du midazolam. [Drug interactions of midazolam.] Schweiz Med Wochenschr 1989;119(50):1834–6.

66. Hiller A, Olkkola KT, Isohanni P, Saarnivaara L. Unconsciousness associated with midazolam and erythromycin. Br J Anaesth 1990;65(6):826–8.

67. Olkkola KT, Aranko K, Luurila H, Hiller A, Saarnivaara L, Himberg JJ, Neuvonen PJ. A potentially hazardous interaction between erythromycin and midazolam. Clin Pharmacol Ther 1993;53(3):298–305.

68. Amsden GW. Macrolides versus azalides: a drug interaction update. Ann Pharmacother 1995;29(9):906–17.

69. Amacher DE, Schomaker SJ, Retsema JA. Comparison of the effects of the new azalide antibiotic, azithromycin, and erythromycin estolate on rat liver cytochrome P-450. Antimicrob Agents Chemother 1991;35(6):1186–90.

70. Yeates RA, Laufen H, Zimmermann T. Interaction between midazolam and clarithromycin: comparison with azithromycin. Int J Clin Pharmacol Ther 1996;34(9):400–5.

71. Takano M, Takano Y, Sato I. [The effect of midazolam on the memory during cesarean section and the modulation by flumazenil.]Masui 1999;48(1):73–5.

Nitrazepam

General Information

Nitrazepam is a long-acting benzodiazepine used primarily for insomnia. However, it has a long half-life (about 24 hours) and poses a substantial risk of residual daytime effects, including sedation, psychomotor and cognitive impairment, and accidental injury (1).

Organs and Systems

Death

Between January 1983 and March 1994, 302 patients with intractable epilepsy were entered into a nitrazepam compassionate plea protocol (2). Nitrazepam increased the risk of death, especially in young patients with intractable epilepsy. The authors suggested that nitrazepam should be used with caution in young children with intractable epilepsy, especially if they have difficulties in swallowing, aspiration, pneumonia, gastroesophageal reflux, or a combination of these.

References

1. Vermeeren A. Residual effects of hypnotics: epidemiology and clinical implications. CNS Drugs 2004;18(5):297–328.

2. Rintahaka PJ, Nakagawa JA, Shewmon DA, Kyyronen P, Shields WD. Incidence of death in patients with intractable epilepsy during nitrazepam treatment. Epilepsia 1999;40(4):492–6.

Oxazepam

General Information

Oxazepam is a relatively short-acting benzodiazepine with a half-life of about 9 hours. As well as being used as a drug in its own right, it is a metabolite of temazepam and desmethyldiazepam.

Passiflora incarnata extract has been compared with oxazepam in a double-blind, randomized trial in 26 outpatients with DSM-IV generalized anxiety disorder (1). Both were effective, but oxazepam had a more rapid onset of action and caused significantly more problems relating to work performance.

Organs and Systems

Psychological, psychiatric

An important behavioral adverse effect of benzodiazepines, hostility and aggression in response to provocation, is less common with oxazepam than with its closely related analogue lorazepam (2). In a study of the effects of oxazepam on implicit versus explicit memory processes, as a function of time-course the effects of oxazepam (30 mg) or placebo on directly comparable tests of implicit memory and explicit memory were examined at three times in 60 healthy volunteers. Before the plasma concentration had peaked, oxazepam impaired cued recall performance relative to placebo but did not impair priming. At the time of the peak, oxazepam impaired performance in both memory tasks. After the peak, cued recall performance in the oxazepam group remained significantly impaired relative to placebo. However, oxazepam-induced impairments in priming were only marginal, suggesting that oxazepam-induced impairments in implicit memory processes begin to wane after theoretical peak drug concentrations. The results support the hypothesis that benzodiazepines cause impaired implicit memory processes time-dependently (3).

In 30 subjects who were given an acute dose of oxazepam 30 mg, lorazepam 2 mg, or placebo, both drugs impaired explicit memory relative to placebo. Also, both oxazepam and lorazepam impaired priming performance. The results suggested that episodic memory is time-dependently impaired by both benzodiazepines (4).

Long-Term Effects

Drug tolerance

Additional doses of benzodiazepines in long-term users are commonly needed, but it is unknown whether these additional doses have any effect. The effects of an additional 20 mg dose of oxazepam has been assessed in a double-blind, balanced-order, crossover, randomized study in 16 long-term users of oxazepam and 18 benzodiazepine-naïve controls (5). The effects of oxazepam 10 and 30 mg were assessed on: (a) saccadic eye movements as a proxy for the sedative effect; (b) the acoustic startle response as a proxy for the anxiolytic effects; (c) memory; (d) reaction time tasks; (e) subjective measurements. There were dose-related effects on the peak velocity of saccadic eye movement and response probability and on the peak amplitude of the acoustic startle response. Comparison with the controls suggested that the sedative effects might be confounded with the suppression of sedative withdrawal symptoms, whereas the patients were as sensitive as the controls to the effects of an additional dose of oxazepam on the acoustic startle response.

Neither 10 mg nor 30 mg of oxazepam affected the reaction time tasks in the patients, whereas the controls had dose-related impairment. The memory impairing effects did not differ significantly. In contrast to the controls, the patients could not discriminate between a 10 mg and a 30 mg dose, as assessed by visual analogue scales and the Spielberger State-Trait Anxiety Inventory Version 1, which might indicate a placebo effect of 10 mg in patients. The authors concluded that additional doses of oxazepam during long-term treatment have pronounced effects, even after daily use for more than 10 years.

References

1. Akhondzadeh S, Naghavi HR, Vazirian M, Shayeganpour A, Rashidi H, Khani M. Passionflower in the treatment of generalized anxiety: a pilot double-blind randomized controlled trial with oxazepam. J Clin Pharm Ther 2001;26(5):363–7.
2. Bond A. Drug induced behavioural disinhibition. CNS Drugs 1998;9:41–57.
3. Buffett-Jerrott SE, Stewart SH, Bird S, Teehan MD. An examination of differences in the time course of oxazepam's effects on implicit vs explicit memory. J Psychopharmacol 1998;12(4):338–47.
4. Buffett-Jerrott SE, Stewart SH, Teehan MD. A further examination of the time-dependent effects of oxazepam and lorazepam on implicit and explicit memory. Psychopharmacology (Berl) 1998;138(3–4):344–53.
5. Oude Voshaar RC, Verkes R-J, van Luijtelaar GLJM, Edelbroek PM, Zitman FG. Effects of additional oxazepam in long-term users of oxazepam. J Clin Psychopharmacol 2005;25(1):42-50.

Quazepam

General Information

Quazepam is a long-acting benzodiazepine with effects similar to those of diazepam.

Organs and Systems

Nervous system

The hangover effects of night-time administration of triazolam 0.25 mg, flunitrazepam 1 mg, and quazepam 15 mg were compared in 15 healthy subjects, who were given one of the three hypnotics at each session in a single-blind, crossover fashion (1). There were no significant between-drug differences in relation to psychomotor performance. Subjective hangover effects in the morning were prominent with flunitrazepam and quazepam relative to triazolam, whereas objective indices suggested a marked hangover effect of quazepam compared with the other two compounds.

Drug-drug interactions

Hypericum perforatum (St John's wort)

The effect of St John's wort 900 mg/day on a single dose of quazepam15 mg have been examined in a randomized, placebo-controlled, crossover study in 13 healthy subjects (2). Quazepam, but not St John's wort, produced sedative-like effects on a visual analogue test. St John's wort reduced the plasma quazepam concentration but did not affect the pharmacodynamics of quazepam. St John's wort, but not quazepam, impaired psychomotor performance in a digit symbol substitution test. The results suggested that St John's wort reduces plasma quazepam concentrations, probably by increasing CYP3A4 activity, but does not influence the pharmacodynamic effects of the drug.

Itraconazole

The effects of itraconazole 100 mg/day for 14 days on the pharmacokinetics of a single oral dose of quazepam 20 mg and its two active metabolites have been studied in 10 healthy men in a double-blind, randomized, placebo-controlled crossover study (3). Itraconazole did not change the pharmacokinetics of quazepam but significantly reduced the peak plasma concentration and AUC of 2-oxoquazepam and N-desalkyl-2-oxoquazepam. Itraconazole did not affect psychomotor function. The results suggested that CYP3A4 is partly involved in the metabolism of quazepam.

Food–Drug Interactions

The disposition of quazepam 20 mg, in the fasting state and 30 minutes after the consumption of meals containing different amounts of dietary fat, has been studied in a three-arm, randomized, crossover study in nine healthy men (4). Plasma concentrations of quazepam and its metabolite, 2-oxoquazepam, were measured for up to 48 hours after dosing. The peak concentrations of quazepam 30 minutes after low-fat and high-fat meals were 243% and 272%, respectively, of those in the fasted state. The AUCs of quazepam from 0 to 8 hours and from 0 to 48 hours were increased by both the low-fat and high-fat meals by 1.4–2 times. Quazepam was well tolerated, with no significant difference in the Stanford Sleepiness Scale between fasted and fed conditions. Thus, food significantly increased the absorption of quazepam but did not prolong the half-life. Quazepam is lipophilic; it would therefore be more highly dissolved in a fatty meal and more available for absorption.

References

1. Takahashi T, Okajima Y, Otsubo T, Shinoda J, Mimura M, Nakagome K, Kamijima K. Comparison of hangover effects among triazolam, flunitrazepam and quazepam in healthy subjects: a preliminary report. Psychiatry Clin Neurosci 2003;57:303–9.
2. Kawaguchi A, Ohmori M, Tsuruoka S, Nishiki K, Harada K, Miyamori I, Yano R, Nakamura T, Masada M, Fujimura A. Drug interaction between St John's wort and quazepam. Br J Clin Pharmacol 2004;58:403–10.
3. Kato K, Yasui-Furukori N, Fukasawa T, Aoshima T, Suzuki A, Kanno M, Otani K. Effects of itraconazole on the plasma kinetics of quazepam and its two active metabolites after a single oral dose of the drug. Ther Drug Monit 2003;25:473–7.
4. Yasui-Furukori N, Kondo T, Takahata T, Mihara K, Ono S, Kaneko S, Tateishi T. Effect of dietary fat content in meals on pharmacokinetics of quazepam. J Clin Pharmacol 2002;42(12):1335–40.

Temazepam

General Information

Temazepam is a benzodiazepine that was first discovered as a metabolite of diazepam. It is relatively short-acting, having a half-life of about 10 hours.

Temazepam has favorable kinetics for use as a hypnotic (1) and is widely prescribed for this purpose. Rather surprisingly, both 10 and 20 mg caused discernible and comparable psychomotor impairment the following morning (SEDA-20, 32), at a time when the subjective effects had virtually disappeared. This result emphasizes the often subtle but pervasive insight-impairing effects of benzodiazepines. Temazepam in soft capsules is more rapidly absorbed than from tablets. However, drug users have abused temazepam capsules by heating them to liquefaction and injecting the resultant fluid intravenously, often together with other substances; the resulting medical problems are varied and often serious.

The incidence of adverse effects of temazepam and whether the addition of cognitive therapy was associated with a reduction in drug and fewer adverse effects have been studied in 60 patients aged 55 years or older (mean 65) with chronic and primary insomnia (2). They were randomized to placebo (n=20), temazepam (mean dose 20 mg/day) (n=20), or temazepam plus cognitive therapy (n=20). Adverse effects were infrequent: placebo (11%), temazepam 7.8%, and temazepam plus cognitive therapy 8.3%. The adverse events were mild and reduced in severity over the course of treatment. The patients who received cognitive therapy used less temazepam with about the same incidence of adverse effects.

Organs and Systems

Cardiovascular

The effect of hypnotics during spaceflight on the high incidence of post-flight orthostatic hypotension has been studied in astronauts who took no treatment (n=20), temazepam 15 or 30 mg (n=9), or zolpidem 5 or 10 mg (n=8) (3). Temazepam and zolpidem were only taken the night before landing. On the day of landing, systolic pressure fell significantly and heart rate increased significantly in the temazepam group, but not in the control group or in the zolpidem group. Temazepam may aggravate orthostatic hypotension after spaceflight, when astronauts are hemodynamically compromised. It should not be the initial choice as a sleeping aid for astronauts; zolpidem may be a better choice.

Nervous system

In a double-blind, randomized, placebo-controlled, cross-over study of the comparative pharmacodynamics of single doses of temazepam (15 and 30 mg), diphenhydramine (50 and 75 mg), and valerian (400 and 800 mg) in 14 healthy elderly volunteers, temazepam had dose-dependent effects on sedation and psychomotor ability with a distinct time course (4). Temazepam 30 mg and both doses of diphenhydramine elicited significantly greater sedation than placebo, and temazepam had the greatest effect. There were no differences in sedation scores between 50 and 75 mg of diphenhydramine. Psychomotor impairment was evident after diphenhydramine 75 mg compared with placebo on the manual tracking test; this was less than the impairment with temazepam 30 mg but similar to that with temazepam 15 mg. There was no psychomotor impairment with diphenhydramine 50 mg. Valerian was not different from placebo on any measure of psychomotor performance or sedation.

Psychological, psychiatric

Temazepam produces a variety of adverse psychological effects, including restlessness, agitation, irritability, aggression, rage, and psychosis. Perhaps because it is likely to be used recreationally, temazepam has also been used to facilitate crime, particularly in the UK (5).

The effects of oral temazepam 5, 10, and 30 mg on memory have been studied in healthy volunteers subjected to a battery of cognitive tests and analogue mood ratings (6). The lowest dose had no effect and 10 mg significantly increased only self-ratings of well-being. Temazepam 30 mg significantly improved recall of items learned before drug administration, but impaired recall and recognition of word lists acquired after drug administration. There was no impairment of retrieval, suggesting that automatic information processing was unaffected. Temazepam 30 mg significantly reduced self-ratings of anxiety and increased self-ratings of sedation. It also significantly impaired performance in symbol copying, digit-symbol substitution, and number cancellation tasks. It is striking that at a dose that was sedative and impaired many aspects of performance, temazepam nevertheless improved retrieval of items learned before drug administration. In a randomized, cross-over study, psychomotor performance and memory were tested and mood assessed for 3 hours after single doses of placebo, temazepam 20 mg, or temazepam 30 mg in six healthy women aged 21–23 years (7). Psychomotor speed and explicit memory showed dose-dependent slowing and impairment, but there was no change in short-term memory.

In a five-period crossover study in 15 healthy subjects aged 18–25 years, placebo, temazepam (15 and 30 mg), and clonidine (150 and 300 micrograms) were given orally in counterbalanced order in sessions at least 4 days apart (8). Performance on a range of tasks aimed at assessing the role of central noradrenergic mechanisms in cognitive function was significantly impaired in a dose-related fashion, and both drugs caused subjective *sedation*. Clonidine did not affect the formation of new long-term memories, in contrast to temazepam, but did impair measures of working memory.

Drug Administration

Drug dosage regimens

Benzodiazepine prescribing for sleep induction in an elderly medical inpatient population has been examined, to determine if hospital prescribing increases the use of benzodiazepines after discharge (9). The secondary objectives included monitoring for adverse effects and assessment of the quality of sleep in hospital compared with the quality of sleep at home. Inpatient and outpatient prescribing of benzodiazepines used for insomnia was recorded over 3 months. Benzodiazepines were prescribed for 20% of patients, and 94% of the prescriptions were for temazepam. Of the 57 patients who were given benzodiazepines during a hospital admission, 57% had not taken a benzodiazepine at home before admission. Benzodiazepines were effective in the short-term for inducing sleep in hospital, with little evidence of adverse effects.

Drug administration route

Inadvertent intra-arterial injection of temazepam can cause tissue damage.

- A 29-year-old unemployed man developed pain and swelling of the right hand following inadvertent intra-arterial injection of temazepam capsules (10). Over 10 days, necrotic areas involving the index, middle, and little fingers developed and the fingers had to be amputated. He reported episodes of intravenous drug use during the previous 3 days—a single heroin dose followed by temazepam four times (10 mg gel capsules dissolved in hot water). He had injected the drugs into a superficial blood vessel on the back of the right hand.

Drug overdose

In a retrospective analysis of 352 consecutive cases of fatal overdose, temazepam accounted for 65% of all deaths from benzodiazepine overdose (11). Acute rhabdomyolysis, usually associated with intravascular injection, has been seen after an oral overdose of temazepam (SEDA-17, 45).

- An 83-year-old woman developed coma, respiratory depression, hypotonia with generalized hyporeflexia, bilateral extensor plantar responses, and bullous eruptions containing serous fluid over the medial aspects of the knees after taking seven temazepam tablets (12).

Drug–Drug Interactions

Oral contraceptives

The clearance of temazepam was increased when oral contraceptives were co-administered (13). It is unlikely that this is an interaction of clinical importance.

References

1. Ashton H. Guidelines for the rational use of benzodiazepines. When and what to use. Drugs 1994;48(1):25–40.
2. Morin CM, Bastien, CH, Brink D, Brown TR. Adverse effects of temazepam in older adults with chronic insomnia. Hum Psychopharmacol Clin Exp 2003;18:75-82.
3. Shi SJ, Garcia KM, Keck, JV. Temazepam, but not zolpidem, causes orthostatic hypotension in astronauts after spaceflight. J Cardiovasc Pharmacol 2003;41:31-9.
4. Glass JR, Sproule BA, Herrmann N, Streiner D, Busto UE. Acute pharmacological effects of temazepam, diphenhydramine, and valerian in healthy elderly subjects. J Clin Psychopharmacol 2003;23:260-8.
5. Michel L, Lang JP. Benzodiazepine et passage à l'acte criminel. [Benzodiazepines and forensic aspects.] Encephale 2003;29(6):479–85.
6. File SE, Joyce EM, Fluck E, De Bruin E, Bazari F, Nandha H, Fitton L, Adhiya S. Limited memory impairment after temazepam. Hum Psychopharmacol 1998;13:127–33.
7. Wise MG, Griffies WS. A combined treatment approach to anxiety in the medically ill. J Clin Psychiatry 1995;56(Suppl 2):14–9.
8. Tiplady B, Bowness E, Stien L, Drummond G. Selective effects of clonidine and temazepam on attention and memory. J Psychopharmacol 2005;19(3) 259-65.
9. Ramesh M, Roberts G. Use of night-time benzodiazepines in an elderly inpatient population. J Clin Pharm Ther 2002;27(2):93–7.
10. Feeney GF, Gibbs HH. Digit loss following misuse of temazepam. Med J Aust 2002;176(8):380.
11. Obafunwa JO, Busuttil A. Deaths from substance overdose in the Lothian and Borders region of Scotland (1983–91). Hum Exp Toxicol 1994;13(6):401–6.
12. Verghese J, Merino J. Temazepam overdose associated with bullous eruptions. Acad Emerg Med 1999;6(10):1071.
13. Patwardhan RV, Mitchell MC, Johnson RF, Schenker S. Differential effects of oral contraceptive steroids on the metabolism of benzodiazepines. Hepatology 1983;3(2):248–53.

Triazolam

General Information

The history of the introduction of triazolam has been reviewed (SEDA-4, v). Triazolam continues to be controversial (1) and much of the controversy has been reviewed (2–5).

Much of the literature on the adverse effects of triazolam is based on doses of 0.25 mg and above; the reduction in recommended dose to 0.125 mg may improve therapeutic safety, at the cost of compromising hypnotic efficacy in most patients. A low therapeutic index is also suggested by the possibility of amnesia, even after a single dose (6), and triazolam appears to be associated with a greater incidence of behavioral disturbances than other hypnotic benzodiazepine agonists, such as temazepam and zopiclone.

A case of recurrent iatrogenic Kleine–Levin syndrome with irritability, hyperphagia, and amnesia (7) has provided a further example of the bizarre behavior not infrequently seen with triazolam in a high dose (0.5 mg); see also the monograph on flunitrazepam. At this dose, triazolam produces too many adverse effects on sleep, memory, and judgement to be useful to the military or to commercial airline pilots; elderly patients are particularly sensitive, even to lower doses, owing in part to reduced drug clearance (SED-12, 99). Amnesia and confusion, which can occur even after single doses of triazolam, are accentuated by continued use, and frank psychopathology occurred in psychiatric patients during 2 weeks' use and during withdrawal (8).

The demonstrated ability of triazolam to impair acetylcholine release (SEDA-19, 36) suggests a mechanism for its amnestic effects, and should further caution prescribers about its use in elderly people, particularly those with pre-existing cognitive impairment. The extent to which this anticholinergic effect is shared by other benzodiazepines is yet to be established. The ability of triazolobenzodiazepines, such as triazolam and alprazolam, to inhibit the noradrenergic neurons of the locus ceruleus is unlikely to explain the prominent adverse effects of these drugs (SEDA-21, 37). For one thing, all GABA enhancers share this property; for another, a far more specific and selective inhibitor in the locus ceruleus (clonidine) has an adverse effects profile that is dramatically different from that of the benzodiazepines, whose adverse effects probably relate to their actions at a far wider range of GABA-ergic synapses.

Organs and Systems

Nervous system

In a study of the effects of d-amfetamine (20 mg/70 kg) on the sedative and memory-impairing effects of triazolam (0.25 mg/70 kg) in 20 healthy adults, the results suggested that benzodiazepines have specific effects on memory that are not merely a by-product of their sedative effects, and that the degree to which sedative effects contribute to the amnestic effects may vary as a function of the particular memory process being assessed (9). In addition to enhancing the understanding of the mechanisms underlying benzodiazepine-induced amnesia, these results may also contribute to better understanding of the complex relation between specific memory processes and level of arousal.

Psychological, psychiatric

The effect of triazolam on muscarinic acetylcholine receptor binding has been investigated in living brain slices by the use of a positron-based imaging technique (10). Stimulation of GABA/benzodiazepine binding sites lowered the affinity of the muscarinic acetylcholine cholinergic receptor for its ligand, which may underlie benzodiazepine-induced amnesia, a serious clinical adverse effect of benzodiazepines.

Ten healthy men participated in a randomized, double-blind, crossover study involving placebo and triazolam 0.125 mg (11). Resting electroencephalography and event-related potentials under an oddball paradigm were recorded before drug administration, and at 1, 2, 4, 6, and 8 hours after. P300 waveforms were analysed by peak amplitudes. Triazolam can cause cognitive dysfunction without general sedation or apparent sleepiness, and this effect appeared up to 6 hours after administration of a low dose of triazolam.

The subjective and behavioral effects of triazolam were investigated in 20 healthy women, who took oral triazolam 0.25 mg or placebo at the follicular, periovulatory, and luteal phases of their menstrual cycle in a within-subject design (12). After triazolam most of them reported the expected increases in fatigue and decreases in arousal and psychomotor performance. Neither plasma concentrations of triazolam nor mood and performance differed across the three phases. This study shows that the effects of triazolam are highly stable across the menstrual cycle.

Liver

Liver damage has been attributed to triazolam (13).

- A 64-year-old woman developed anorexia, fatigue, and jaundice. She had occasionally taken triazolam 0.25 mg for insomnia, and her liver function tests had been normal 5 months before. There was no history of liver disease or alcohol misuse or other serious medical history. She was jaundiced, with spider angiomata, and gradually deteriorated over the next 16 days, finally losing consciousness. Liver histology was consistent with submassive necrosis, extensive coagulation necrosis, and marked inflammation secondary to triazolam. She had a liver transplant 31 days after the initial presentation and 2 years later was in good health.

Drug Administration

Drug overdose

Death has been attributed to an overdose of triazolam (14).

- A 77-year-old woman was found dead in her bathtub. She had a history of depression, liver disease, spinal stenosis, and diabetes mellitus. An empty bottle of triazolam was found in the bin. At autopsy there was no injury or evidence of drowning. There was triazolam 0.12 mg/l in the heart blood.
- A 57-year-old man was found dead in a bamboo thicket (15). His blood and urine contained triazolam and hydroxytriazolam. Blood concentrations of triazolam and unbound hydroxytriazolam were 62–251 and 10–66 ng/ml respectively. There was a substantial amount of triazolam in his bile.

This man probably died of postural asphyxia caused by triazolam poisoning.

Drug–Drug Interactions

Fluconazole

Fluconazole increases blood concentrations of triazolam (16).

Grapefruit juice

In a randomized, four-phase, crossover study, the effect of grapefruit juice on the metabolism of triazolam interaction was investigated. Even one glass of grapefruit juice increased plasma triazolam concentrations, and chronic consumption produced a significantly greater increase. The half-life of triazolam is prolonged by repeated consumption of grapefruit juice, probably due to inhibition of hepatic CYP3A4 activity (17).

Isoniazid

Isoniazid is a potent hepatic enzyme inhibitor and interferes with the metabolism of many drugs (SEDA-8, 287) (SEDA-11, 271). In healthy volunteers the half-life of triazolam was prolonged from 2.5 to 3.3 hours when it was given with isoniazid, whereas isoniazid did not affect the kinetics of oxazepam (SEDA-9, 267).

Ketoconazole

Ketoconazole can increase the concentrations of triazolam through inhibition of CYP3A4 (18).

If erythromycin, also a CYP3A4 inhibitor, is given in combination with ketoconazole, there is an even more dramatic effect on triazolam concentrations.

In a double-blind, crossover kinetic and dynamic study of the interaction of ketoconazole with alprazolam and triazolam, two CYP3A4 substrate drugs with different kinetic profiles, impaired clearance by ketoconazole had more profound clinical consequences for triazolam than for alprazolam (19).

Macrolide antibiotics

Interactions of macrolides with triazolam are clinically relevant. Increases in serum concentration, AUC, and half-life, and a reduction in clearance have been documented (20–22). These changes can result in clinical effects, such as prolonged psychomotor impairment, amnesia, or loss of consciousness.

In a randomized, double-blind, pharmacokinetic–pharmacodynamic study, 12 volunteers took placebo or triazolam 0.125 mg orally, together with placebo, azithromycin, erythromycin, or clarithromycin. The apparent oral clearance of triazolam was significantly reduced by erythromycin and clarithromycin. The peak plasma concentration was correspondingly increased, and the half-life was prolonged. The effects of triazolam on dynamic measures were nearly identical when triazolam was given with placebo or azithromycin, but benzodiazepine agonist effects were enhanced by erythromycin and clarithromycin (23).

Erythromycin

Erythromycin can increase concentrations of triazolam by inhibition of CYP3A4, and dosage reductions of 50% have been proposed if concomitant therapy is unavoidable (24).

Miocamycin

CYP3A4 is mainly responsible for catalyzing the hydroxylation of miocamycin metabolites, which can alter the metabolism of concomitantly administered drugs by the formation of a metabolic intermediate complex with CYP450 or by competitive inhibition of CYP450 (25). This can cause excessive sedation with benzodiazepines such as triazolam.

Nefazodone

Nefazodone is a weak inhibitor of CYP2D6, but a potent inhibitor of CYP3A4, and increases plasma concentrations of drugs that are substrates of CYP3A4, such as triazolam (26,27).

Ritonavir

The inhibitory effect of ritonavir on the biotransformation of triazolam and zolpidem has been investigated (28). Short-term low-dose ritonavir produced a large and significant impairment of triazolam clearance and enhancement of its clinical effects. In contrast, ritonavir produced small and clinically unimportant reductions in zolpidem clearance. The findings are consistent with the complete dependence of triazolam clearance on CYP3A activity, compared with the partial dependence of zolpidem clearance on CYP3A.

References

1. O'Donovan MC, McGuffin P. Short acting benzodiazepines. BMJ 1993;306(6883):945–6.
2. Ashton H. Guidelines for the rational use of benzodiazepines. When and what to use. Drugs 1994;48(1):25–40.
3. Schneider PJ, Perry PJ. Triazolam—an "abused drug" by the lay press? DICP 1990;24(4):389–92.
4. Abraham J. Transnational industrial power, the medical profession and the regulatory state: adverse drug reactions and the crisis over the safety of Halcion in the Netherlands and the UK. Soc Sci Med 2002;55(9):1671–90.
5. Te Lintelo J, Pieters T. Halcion: de lotgevallen van de "Dutch Hysteria". Historie biedt lessen voor farmaceutische patiëntenzorg. Pharm Wkblad 2003;138(46):1600–5.
6. Bixler EO, Kales A, Manfredi RL, Vgontzas AN, Tyson KL, Kales JD. Next-day memory impairment with triazolam use. Lancet 1991;337(8745):827–31.
7. Menkes DB. Triazolam-induced nocturnal bingeing with amnesia. Aust NZ J Psychiatry 1992;26(2):320–1.
8. Soldatos CR, Sakkas PN, Bergiannaki JD, Stefanis CN. Behavioral side effects of triazolam in psychiatric inpatients: report of five cases. Drug Intell Clin Pharm 1986;20(4):294–7.
9. Mintzer MZ, Griffiths RR. Triazolam-amphetamine interaction; dissociation of effects on memory versus arousal. J Psychopharmacol 2003;17:17–29.
10. Mendelson WB, Jain B. An assessment of short-acting hypnotics. Drug Saf 1995;13(4):257–70.
11. Hayakawa T, Uchiyama M, Urata J, Enomoto T, Okubo J, Okawa M. Effects of a small dose of triazolam on P300. Psychiatry Clin Neurosci 1999;53(2):185–7.
12. Rukstalis M, de Wit H. Effects of triazolam at three phases of the menstrual cycle. J Clin Psychopharmacol 1999;19(5):450–8.
13. Kanda T, Yokosuka O, Fujiwara K, Saisho H, Shiga H, Oda K, Okuda K, Sugawara Y, Makuuchi M, Hirasawa H. Fulminant hepatic failure associated with triazolam. Dig Dis Sci 2002;47(5):1111–4.
14. Levine B, Grieshaber A, Pestaner J, Moore KA, Smialek JE. Distribution of triazolam and alpha-hydroxytriazolam in a fatal intoxication case. J Anal Toxicol 2002;26(1):52–4.
15. Moriya F, Hashimoto Y. A case of fatal triazolam overdose. Legal Med 2003;5:591–5.
16. Varhe A, Olkkola KT, Neuvonen PJ. Effect of fluconazole dose on the extent of fluconazole–triazolam interaction. Br J Clin Pharmacol 1996;42(4):465–70.
17. Lilja JJ, Kivisto KT, Backman JT, Neuvonen PJ. Effect of grapefruit juice dose on grapefruit juice–triazolam interaction: repeated consumption prolongs triazolam half-life. Eur J Clin Pharmacol 2000;56(5):411–5.
18. Bickers DR. Antifungal therapy: potential interactions with other classes of drugs. J Am Acad Dermatol 1994;31(3 Pt 2):S87–90.
19. Greenblatt DJ, Wright CE, von Moltke LL, Harmatz JS, Ehrenberg BL, Harrel LM, Corbett K, Counihan M, Tobias S, Shader RI. Ketoconazole inhibition of triazolam and alprazolam clearance: differential kinetic and dynamic consequences. Clin Pharmacol Ther 1998;64(3):237–47.
20. Warot D, Bergougnan L, Lamiable D, Berlin I, Bensimon G, Danjou P, Puech AJ. Troleandomycin–triazolam interaction in healthy volunteers: pharmacokinetic and psychometric evaluation. Eur J Clin Pharmacol 1987;32(4):389–93.
21. Phillips JP, Antal EJ, Smith RB. A pharmacokinetic drug interaction between erythromycin and triazolam. J Clin Psychopharmacol 1986;6(5):297–9.
22. Gascon MP, Dayer P, Waldvogel F. Les interactions médicamenteuses du midazolam. [Drug interactions of midazolam.] Schweiz Med Wochenschr 1989;119(50):1834–6.
23. Laux G. Aktueller stand der Behand lung mit Benzodiazepines. [Current status of treatment with benzodiazepines.] Nervenarzt 1995;66(5):311–22.
24. Amsden GW. Macrolides versus azalides: a drug interaction update. Ann Pharmacother 1995;29(9):906–17.
25. Rubinstein E. Comparative safety of the different macrolides. Int J Antimicrob Agents 2001;18(Suppl 1):S71–6.
26. Ereshefsky L, Riesenman C, Lam YW. Serotonin selective reuptake inhibitor drug interactions and the cytochrome P450 system. J Clin Psychiatry 1996;57(Suppl 8):17–25.
27. Nemeroff CB, DeVane CL, Pollock BG. Newer antidepressants and the cytochrome P450 system. Am J Psychiatry 1996;153(3):311–20.
28. Greenblatt DJ, von Moltke LL, Harmatz JS, Durol AL, Daily JP, Graf JA, Mertzanis P, Hoffman JL, Shader RI. Differential impairment of triazolam and zolpidem clearance by ritonavir. J Acquir Immune Defic Syndr 2000;24(2):129–36.

NON-BENZODIAZEPINE HYPNOSEDATIVES

Alpidem

General Information

Alpidem, like zolpidem, is an imidazopyridine, chemically distinct from the benzodiazepines. It binds selectively to a subset of benzodiazepine receptors (1), which may account for its apparently milder withdrawal effects and a relative dominance of anxiolytic over sedative and cognitive effects (2). Its effects are reversed by flumazenil (3). Alpidem was withdrawn in 1953 in France, the only country in which it was marketed, because of hepatotoxicity (4).

References

1. Lader M. Psychiatric disorders. In: Speight T, Holford N, editors. Avery's Drug Treatment. 4th ed.. Auckland: ADIS International Press, 1997:1437.
2. Lader M. Clinical pharmacology of anxiolytic drugs: past, present and future. In: Biggio G, Sanna E, Costa E, editors. GABA-A Receptors and AnxietyFrom Neurobiology to Treatment. New York: Raven Press, 1995:135.
3. Zivkovic B, Morel E, Joly D, Perrault G, Sanger DJ, Lloyd KG. Pharmacological and behavioral profile of alpidem as an anxiolytic. Pharmacopsychiatry 1990;23(Suppl 3):108–13.
4. Cassano GB, Petracca A, Borghi C, Chiroli S, Didoni G, Garreau M. A randomized, double-blind study of alpidem vs placebo in the prevention and treatment of benzodiazepine withdrawal syndrome. Eur Psychiatry 1996;11(2):93–9.

Buspirone

General Information

Buspirone, an azapirone drug, similar to its analogues ipsaperone and tandospirone, is chemically and pharmacologically dissimilar to the benzodiazepines and is useful in both generalized anxiety and depression (1). In contrast to benzodiazepines, buspirone has an antidepressant-like therapeutic latency (2), and for this reason it needs to be given with considerable education and encouragement. It is effective in children with anxiety disorders (SEDA-19, 36) and as an antidepressant adjuvant in refractory depression with (3) and without (4) obsessive-compulsive symptoms. Buspirone is also effective in treating anxious alcoholics, and in patients with agitation or aggression associated with organic brain disease (5).

Originally thought to act through a dopaminergic mechanism, buspirone is now regarded as acting as a 5-HT_{1A} receptor (partial) agonist (6) and produces dose-related adverse effects similar to those of the SSRIs (nausea, headache, insomnia, dizziness, and sexual dysfunction) (SEDA-22, 39) (7). Like them, it interacts with monoamine oxidase inhibitors and can produce the serotonin syndrome (8). An uncommon association with extrapyramidal movement disorders (SEDA-17, 46; SEDA-18, 45) may reflect its structural relation to some dopamine receptor antagonists.

Compared with benzodiazepines, buspirone causes less sedation and memory impairment (9), has little interaction with alcohol (2,10), and is essentially devoid of problems of abuse, disinhibition, tolerance, and dependence.

These advantages are important in treating particular groups of patients, including patients with brain injury, elderly people, forensic populations, and the medically ill (11,12). On the other hand, over 50% of patients with cerebellar ataxia reported significant adverse effects, as described above (SEDA-21, 39).

Buspirone may also have advantages in treating individuals who are susceptible to problems with alcohol (13) or other hypnosedatives (10), although habitual users of benzodiazepines are often unwilling to persist with buspirone long enough for it to have a therapeutic effect (2).

Unlike carbamazepine or tricyclic antidepressants, buspirone is not helpful in the management of benzodiazepine withdrawal (14).

Observational studies

Buspirone (40 mg/day for 4 weeks) has been used as an alternative treatment in eight psychiatric outpatients with attention deficit disorder (15). The results suggested that buspirone might be useful in attention deficit disorders, reducing hyperactive behavior and enabling greater attention with few adverse effects. There were no clinically significant changes in blood pressure. There was mild transient tiredness, which lasted about 2 weeks. In one case, there was an initial trend toward a reduced appetite, but that generally stabilized. Headaches and stomach aches occurred in two patients during the first 2 weeks and one of them continued to have mild episodes.

The effects of hydroxyzine 50 mg/day, buspirone 20 mg/day, and placebo have been studied in 244 patients with generalized anxiety disorder in a double-blind placebo-controlled study (16). Hydroxyzine ($n = 81$) was considerably better than placebo ($n = 81$), and buspirone ($n = 82$) was intermediate. The main adverse effects were headache and migraine with buspirone (6.1 versus 4.9% with hydroxyzine and 1.2% with placebo). Somnolence occurred in 9.9% with hydroxyzine, 4.9% with buspirone, and none with placebo. Dizziness occurred in 6.1% with buspirone, none with hydroxyzine, and 2.5% with placebo.

In a 21-day, open, multicenter, dose-escalation study, 13 children and 12 adolescents with anxiety disorder and 14 healthy adults took buspirone 5–30 mg bd titrated over 3 weeks (17). Buspirone was generally safe and well-tolerated at doses up to 30 mg in adolescents and adults and in most of the children. The most common adverse events in children and adolescents were light-headedness (68%), headache (48%), and dyspepsia (20%); two children withdrew from the study at the higher doses (15 mg and 30 mg bd) owing to adverse effects. In adults, the most common adverse event was somnolence (21%); mild light-headedness, nausea, and diarrhea were also reported.

Placebo-controlled studies

Symptoms of anxiety are common among opioid-dependent individuals. Although buspirone has been used successfully for the treatment of anxiety in alcoholic patients, its efficacy in opioid-dependent patients had not been previously examined. In a 12-week, randomized, placebo-controlled trial of buspirone in 36 subjects receiving methadone maintenance treatment who presented with symptoms of anxiety, buspirone did not significantly reduce anxiety symptoms (18). However, buspirone was associated with trends toward reduction in depression scale scores and a slower return to substance use.

Systematic reviews

The safety results from two comparisons of buspirone 15 mg bd and buspirone 10 mg tds in patients with persistent anxiety have been subjected to meta-analysis (19). The incidences of adverse events were similar, except for a significantly higher incidence of bouts of palpitation in patients taking buspirone bd (5%) compared with tds (1%). The most frequently reported adverse events with both regimens were dizziness, headache, and nausea. There were no appreciable differences between treatments in vital signs, physical examination, electrocardiography, or clinical laboratory results. A change to twice-daily dosing with buspirone may offer convenience and possibly greater adherence to therapy in patients with persistent anxiety, without compromising the safety and tolerability profile of the drug.

Organs and Systems

Respiratory

Because of its virtual absence of respiratory depressant effects compared with benzodiazepines (20), buspirone may be especially useful in treating anxious patients with lung disease (21).

Nervous system

The efficacy and safety of buspirone have been evaluated in the management of anxiety and irritability in 22 children with pervasive developmental disorders. One child developed abnormal involuntary movements of the mouth, cheeks, and tongue after having taken buspirone 20 mg/day for 10 months. No other drugs were prescribed. The abnormal movements disappeared completely within 2 weeks of withdrawal of buspirone. Other adverse effects in other children were minimal and included initial sedation, slight agitation, and initial nausea (22).

Long-Term Effects

Drug abuse

Despite continuing scrutiny, buspirone has shown little, if any, abuse potential (7).

Drug Administration

Drug formulations

Because of their short half-lives, buspirone and its analogues are problematic to administer (6). A single dose of buspirone ER (extended-release) 30 mg has been compared with two doses of buspirone IR (immediate-release) 15 mg given 12 hours apart to assess differences in tolerability in an open, crossover, randomized study in 18 healthy men (23). Blood samples were obtained at 22 times. Seven subjects reported a total of 13 adverse events during the study, but none of the events recorded was unexpected. All were mild and resolved by the end of the study without medical intervention. Three adverse events (rhinitis, headache, and light-headedness) were categorized as unrelated to the study drug. The other 10 adverse events included drowsiness, dizziness, depression, tinnitus, and increased blood pressure. There were no significant differences between the two formulations.

Drug–Drug Interactions

Calcium channel blockers

In a randomized placebo-controlled trial, the possible interactions of buspirone with verapamil and diltiazem were investigated. Both verapamil and diltiazem considerably increased plasma buspirone concentrations, probably by inhibiting CYP3A4. Thus, enhanced effects and adverse effects of buspirone are possible when it is used with verapamil, diltiazem, or other inhibitors of CYP3A4 (24).

Erythromycin

Erythromycin, an inhibitor of CYP3A, can increase buspirone concentrations (SEDA-22, 39).

Co-administration of erythromycin with the anxiolytic drug buspirone increased the plasma concentration of buspirone (25).

Grapefruit juice

In a randomized, two-phase crossover study, the effects of grapefruit juice on the pharmacokinetics and pharmacodynamics of oral buspirone were investigated in 10 healthy volunteers. Grapefruit juice increased the mean peak plasma concentration of buspirone 4.3-fold (26). Large amounts of grapefruit juice should be avoided in patients taking buspirone.

Haloperidol

Buspirone increases haloperidol concentrations in some patients (SEDA-21, 39).

Itraconazole

Itraconazole, an inhibitor of CYP3A, can increase buspirone concentrations (SEDA-22, 39).

The interaction of itraconazole with the active 1-(2-pyrimidinyl)-piperazine metabolite of buspirone has been studied after a single oral dose of buspirone 10 mg (27). Itraconazole reduced the mean AUC of the metabolite by 50% and the C_{max} by 57%, whereas the mean AUC and C_{max} of the parent drug were increased 14.5-fold and 10.5-fold respectively. Thus, itraconazole caused relatively minor changes in the plasma concentrations of the active piperazine metabolite of buspirone, although it had major effects on the concentrations of buspirone after a single oral dose.

Monoamine oxidase inhibitors

Buspirone interacts with monoamine inhibitors and can cause the serotonin syndrome (8).

Rifampicin

The effects of rifampicin on the pharmacokinetics and pharmacodynamics of buspirone were investigated in 10 young healthy volunteers. There was a significant reduction in the effects of buspirone in three of the six psychomotor tests used after rifampicin pretreatment. The interaction between rifampicin and buspirone is probably mostly due to increased CYP3A4 activity. Buspirone will most likely have a greatly reduced anxiolytic effect when it is used together with rifampicin or other potent inducers of CYP3A4, such as phenytoin and carbamazepine (28).

Selective serotonin re-uptake inhibitors (SSRIs)

The serotonin syndrome can occur with therapeutic doses of SSRIs (see above), but it occurs most commonly when SSRIs are co-administered with other drugs that also potentiate serotonin function. Recent case reports have suggested that there is a risk of the serotonin syndrome when SSRIs are combined with buspirone (29).

Fluvoxamine

The effects of fluvoxamine on the pharmacokinetics and pharmacodynamics of buspirone have been investigated in 10 healthy volunteers. Fluvoxamine moderately increased plasma buspirone concentrations and reduced the production of the active metabolite of buspirone. The mechanism of this interaction is probably inhibition of CYP3A4. However, this pharmacokinetic interaction was not associated with impaired psychomotor performance and is probably of limited clinical significance (30).

In 10 healthy volunteers, fluvoxamine (100 mg/day for 5 days) significantly increased the peak concentrations of the anxiolytic drug buspirone. Concentrations of the active metabolite, 1-(2-pyrimidinyl)-piperazine, were reduced (30). These effects were probably mediated through inhibition of CYP3A4 by fluvoxamine.

Sertraline

Serotonin syndrome has been attributed to the combination of buspirone and sertraline.

- A 49-year-old man had major adverse effects 11 days after taking a combination of sertraline, buspirone, and loxapine (31). The adverse effects were characteristic of the serotonin syndrome, which is characterized by a constellation of symptoms, including hypomania, agitation, seizures, confusion, restlessness, hyper-reflexia, tremor, myoclonus, ataxia, incoordination, anxiety, double vision, fever, shivering, variable effects on blood pressure, nausea and vomiting, sweating, and diarrhea.

St John's wort

Clinicians should inform their patients about the risks associated with taking St John's wort if they are taking psychotropic drugs that can cause the serotonin syndrome.

- A 27-year-old married woman developed symptoms of generalized anxiety disorder and was given buspirone 30 mg/day (32). During treatment she felt depressed and decided to take St John's wort. Two months later she started to have nervousness, aggressiveness, hyperactivity, insomnia, blurred vision, and very short periods of confusion and disorientation. The symptoms were consistent with serotonin syndrome. St John's wort was withdrawn and her symptoms resolved after 1 week.

Buspirone is a partial agonist at $5HT_{1A}$ receptors; St John's wort is a non-selective inhibitor of 5HT reuptake and also upregulates postsynaptic $5HT_{1A}$ and $5HT_{2A}$ receptors; it therefore causes overstimulation of $5HT_{1A}$ receptors, leading to the serotonin syndrome.

References

1. Rickels K, Amsterdam J, Clary C, Hassman J, London J, Puzzuoli G, Schweizer E. Buspirone in depressed outpatients: a controlled study. Psychopharmacol Bull 1990;26(2):163–7.
2. Lader M. Psychiatric disorders. In: Speight T, Holford N, editors. Avery's Drug Treatment. 4th ed.. Auckland: ADIS International Press, 1997:1437.
3. Menkes DB. Buspirone augmentation of sertraline. Br J Psychiatry 1995;166(6):823–4.
4. Joffe RT, Schuller DR. An open study of buspirone augmentation of serotonin reuptake inhibitors in refractory depression. J Clin Psychiatry 1993;54(7):269–71.
5. Stanislav SW, Fabre T, Crismon ML, Childs A. Buspirone's efficacy in organic-induced aggression. J Clin Psychopharmacol 1994;14(2):126–30.
6. Blier P, Ward NM. Is there a role for 5-HT$_{1A}$ agonists in the treatment of depression? Biol Psychiatry 2003;53(3):193–203.
7. Lader M. Clin pharmacology of anxiolytic drugs: past, present and future. In: Biggio G, Sanna E, Costa E, editors.

GABA-A Receptors and Anxiety: From Neurobiology to Treatment. New York: Raven Press, 1995:135.

8. Napoliello MJ, Domantay AG. Buspirone: a worldwide update. Br J Psychiatry 1991;159(Suppl 12):40–4.

9. Alford C, Bhatti JZ, Curran S, McKay G, Hindmarch I. Pharmacodynamic effects of buspirone and clobazam. Br J Clin Pharmacol 1991;32(1):91–7.

10. Escande M, Frexinos M, Fabre S. Un nouvelle generation de tranquillisants. [A new generation of tranquilizing agents.] Rev Prat 1994;44(17):2316–9.

11. Steinberg JR. Anxiety in elderly patients. A comparison of azapirones and benzodiazepines. Drugs Aging 1994;5(5):335–45.

12. Weiss KJ. Management of anxiety and depression syndromes in the elderly. J Clin Psychiatry 1994;55(Suppl):5–12.

13. Cornelius JR, Bukstein O, Salloum I, Clark D. Alcohol and psychiatric comorbidity. Recent Dev Alcohol 2003;16:361–74.

14. Pelissolo A, Bisserbe JC. Dependance aux benzodiazepine. Aspects clinique et biologiques. [Dependence on benzodiazepines. Clinical and biological aspects.] Encephale 1994;20(2):147–57.

15. Niederhofer H. An open trial of buspirone in the treatment of attention-deficit disorder. Hum Psychopharmacol Clin Exp 2003;18:489–92.

16. Lader M. Anxiolytic effect of hydroxyzine: a double-blind trial versus placebo buspirone. Hum Psychopharmacol 1999;14(Suppl 1):94–102.

17. Salazar DE, Frackiewicz EJ, Dockens R, Kollia G, Fulmor IE, Tigel PD, Uderman HD, Shiovitz TM, Sramek JJ, Cutler NR. Pharmacokinetics and tolerability of buspirone during oral administration to children and adolescents with anxiety disorder and normal healthy adults. J Clin Pharmacol 2001;41(12):1351–8.

18. McRae AL, Sonne SC, Brady KT, Durkalski V, Palesch Y. A randomized, placebo-controlled trial of buspirone for the treatment of anxiety in opioid-dependent individuals. Am J Addict 2004;13:53–63.

19. Sramek JJ, Hong WW, Hamid S, Nape B, Cutler NR. Meta-analysis of the safety and tolerability of two dose regimens of buspirone in patients with persistent anxiety. Depress Anxiety 1999;9(3):131–4.

20. Rapoport DM, Greenberg HE, Goldring RM. Differing effects of the anxiolytic agents buspirone and diazepam on control of breathing. Clin Pharmacol Ther 1991;49(4):394–401.

21. Craven J, Sutherland A. Buspirone for anxiety disorders in patients with severe lung disease. Lancet 1991;338(8761):249.

22. Buitelaar JK, van der Gaag RJ, van der Hoeven J. Buspirone in the management of anxiety and irritability in children with pervasive developmental disorders: results of an open-label study. J Clin Psychiatry 1998;59(2):56–9.

23. Sakr A, Andheria M. Pharmacokinetics of buspirone extended-release tablets: a single-dose study. J Clin Pharmacol 2001;41(7):783–9.

24. Lamberg TS, Kivisto KT, Neuvonen PJ. Effects of verapamil and diltiazem on the pharmacokinetics and pharmacodynamics of buspirone. Clin Pharmacol Ther 1998;63(6):640–5.

25. Mahmood I, Sahajwalla C. Clinical pharmacokinetics and pharmacodynamics of buspirone, an anxiolytic drug. Clin Pharmacokinet 1999;36(4):277–87.

26. Lilja JJ, Kivisto KT, Backman JT, Lamberg TS, Neuvonen PJ. Grapefruit juice substantially increases plasma concentrations of buspirone. Clin Pharmacol Ther 1998;64(6):655–60.

27. Kivisto KT, Lamberg TS, Neuvonen PJ. Interactions of buspirone with itraconazole and rifampicin: effects on the pharmacokinetics of the active 1-(2-pyrimidinyl)-piperazine metabolite of buspirone. Pharmacol Toxicol 1999;84(2):94–7.

28. Lamberg TS, Kivisto KT, Neuvonen PJ. Concentrations and effects of buspirone are considerably reduced by rifampicin. Br J Clin Pharmacol 1998;45(4):381–5.

29. Manos GH. Possible serotonin syndrome associated with buspirone added to fluoxetine. Ann Pharmacother 2000;34(7–8):871–4.

30. Lamberg TS, Kivisto KT, Laitila J, Martensson K, Neuvonen PJ. The effect of fluvoxamine on the pharmacokinetics and pharmacodynamics of buspirone. Eur J Clin Pharmacol 1998;54(9–10):761–6.

31. Bonin B, Vandel P, Vandel S, Sechter D, Bizouard P. Serotonin syndrome after sertraline, buspirone and loxapine? Thérapie 1999;54(2):269–71.

32. Dannawi M. Possible serotonin syndrome after combination of buspirone and St John's wort. J Psychopharmacol 2002;16(4):401.

Chloral hydrate

General Information

Chloral hydrate, which was synthesized by Justus Liebig in 1832, continues to be used for sleep disorders and for sedation before surgery or radiology, especially in children (1). As an alternative to the benzodiazepines, it has essentially similar properties and problems, but is associated with perhaps rather more frequent gastrointestinal disturbances, up to 7% in one large sample of children. Chloral hydrate had a reputation for safety, which has been challenged by data that suggest that it and short-acting barbiturates are particularly lethal when taken in overdose (2). Liver damage and cardiac toxicity, consistent with the chemical similarities between trichloroethanol, chloroform, and alcohol, can occur.

Organs and Systems

Cardiovascular

A single dose (1.2 g) in a 4-year-old girl gave rise to a reversible ventricular dysrhythmia (SEDA-21, 39).

Nervous system

Two generalized seizures occurred in a 2-year-old boy after he was sedated before echocardiography (SEDA-22, 42).

Gastrointestinal

A 78-year-old woman and a 45-year-old man developed pneumatosis cystoides coli. Both were taking chloral hydrate (3). It was speculated that chloral hydrate had caused the abdominal symptoms in these patients.

- A 69-year-old man had daily loose stools with mucous discharge. Repeated stool cultures were negative. Histology confirmed the appearances of pneumatosis cystoides coli. Various treatments had little effect on his symptoms. He had taken chloral hydrate for insomnia for 10 years. Subjectively, his symptoms were worse while taking the drug and resolved after withdrawal. With his consent, an unblinded controlled challenge with chloral hydrate was performed. The results, recorded meticulously in a symptoms diary, were dramatic. Before chloral hydrate, his bowels opened 18 times per week and only two motions contained mucus. After one tablet of chloral hydrate on days 1, 2, 5, and 7 he passed 49 stools in one week, 35 containing mucus. Two weeks later his bowel habit had returned to normal, with 15 movements per week, and only one stool contained mucus. Subsequently his bowel frequency fell to 8–10 times per week with considerable colonoscopic improvement.

Liver

Hyperbilirubinemia in neonates associated with chloral hydrate (4) may be due to the more prolonged half-life of trichloroethanol in newborns compared with adults (37 versus 14 hours).

Skin

Allergic skin reactions continue to be reported (5).

Immunologic

Several reports have highlighted the importance of gelatin allergy in young children, with some deaths due to anaphylaxis.

- A 2-year-old boy and a 4-year-old boy developed anaphylactic symptoms after being given a chloral hydrate suppository, which contained gelatin, for sedation before electroencephalography (6).

Chloral hydrate suppositories are often used to sedate children during various examinations and the authors suggested using gelatin-free formulations.

Long-Term Effects

Drug dependence

Chloral hydrate dependence can occur (7), the most famous case being that of Anna O. (Bertha Pappenheim), who was treated by Breuer from 1880 to 1882, and whose pathology was discussed by him and Freud in 1895 (8).

Tumorigenicity

Concerns about carcinogenicity are not supported by its persisting and generally successful therapeutic use.

Susceptibility factors

Preterm infants

The degree of sedation (COMFORT), feeding behavior, and cardiorespiratory events (bradycardia, apnea) before and after oral chloral hydrate 30 mg/kg have been prospectively evaluated in 26 former preterm infants during procedural sedation at term post-conception age (9). There was a significant increase in sedation up to 12 hours after administration and a minor but significant reduction in oral intake. There was a significant increase in the number of episodes of bradycardia and in the duration of the most severe. The study was therefore stopped when 26 neonates had been recruited. Infants who had severe bradycardia after chloral hydrate had a lower gestational age at birth but no difference in post-conceptional age at time of inclusion. Chloral hydrate was associated with an increase in unintended adverse effects in former preterm infants, probably reflecting differences in the pharmacodynamics and pharmacokinetics of chloral hydrate.

Drug Administration

Drug formulations

Several reports have highlighted the importance of gelatin allergy in young children, with some deaths due to anaphylaxis. Some formulations of chloral hydrate contain gelatin.

- A 2-year-old boy and a 4-year-old boy developed anaphylactic symptoms after being given a chloral hydrate suppository, which contained gelatin, for sedation before electroencephalography (6).

Chloral hydrate suppositories are often used to sedate children during various examinations and the authors suggested using gelatin-free formulations.

Drug overdose

Chloral hydrate toxicity can cause a number of dysrhythmias, including supraventricular tachycardia, ventricular tachycardia, ventricular fibrillation, and torsade de pointes.

- A 27-year-old man with a history of psychiatric illness became unconscious after taking about 20 g of chloral hydrate, 1000 mg of loxapine, and 180 mg of fluoxetine (10). He became unresponsive and hypotensive and had intermittent episodes of ventricular tachycardia, which was successfully treated with intravenous propranolol.

References

1. Fox BE, O'Brien CO, Kangas KJ, Murphree AL, Wright KW. Use of high dose chloral hydrate for ophthalmic exams in children: a retrospective review of 302 cases. J Pediatr Ophthalmol Strabismus 1990;27(5):242–4.

2. Buckley NA, Whyte IM, Dawson AH, McManus PR, Ferguson NW. Correlations between prescriptions and drugs taken in self-poisoning. Implications for prescribers and drug regulation. Med J Aust 1995;162(4):194–7.
3. Marigold JH. Pneumatosis cystoides coli and chloral hydrate. Gut 1998;42(6):899–900.
4. Lambert GH, Muraskas J, Anderson CL, Myers TF. Direct hyperbilirubinemia associated with chloral hydrate administration in the newborn. Pediatrics 1990;86(2):277–81.
5. Lindner K, Prater E, Schubert H, Siegmund S. Arzneimittel exanthem auf chlorhydrat. [Drug exanthema due to chloral hydrate.] Dermatol Monatsschr 1990;176(8):483–5.
6. Yamada A, Ohshima Y, Tsukahara H, Hiraoka M, Kimura I, Kawamitsu T, Kimura K, Mayumi M. Two cases of anaphylactic reaction to gelatin induced by a chloral hydrate suppository. Pediatr Int 2002;44(1):87–9.
7. Stone CB, Okun R. Chloral hydrate dependence: report of a case. Clin Toxicol 1978;12(3):377–80.
8. de Paula Ramos S. Revisiting Anna O: a case of chemical dependence. Hist Psychol 2003;6(3):239–50.
9. Allegaert K, Daniels H, Naulaers G, Tibboel D, Devlieger H. Pharmacodynamics of chloral hydrate in former preterm infants. Eur J Pediatr 2005;164:403–7.
10. Zahedi A, Grant MH, Wong DT. Successful treatment of chloral hydrate cardiac toxicity with propranolol. Am J Emerg Med 1999;17(5):490–1.

Clomethiazole

General Information

Clomethiazole is a sedative-hypnotic that has been used extensively in the treatment of alcohol withdrawal, as well as for inducing sedation and sleep in the elderly. In addition to GABA enhancement, which it shares with the benzodiazepines, clomethiazole also enhances the activity of another inhibitory amino acid, glycine. Whether this property is clinically important is uncertain. As well as the expected effects of sedation and memory impairment, it produces nasal irritation, especially in younger patients, in whom it has a shorter half-life. Its use in alcohol withdrawal is becoming less common, possibly owing to the demonstrated safety and efficacy of longer-acting benzodiazepines, such as chlordiazepoxide, and alternatives such as carbamazepine.

Hypotension, phlebitis, and respiratory depression can occur after intravenous use. While effects in the elderly may be increased, the incidence of adverse effects is similar to that seen in younger subjects (1).

Clomethiazole, like the benzodiazepines, has an additive effect with other CNS depressants, and can cause profound bradycardia when combined with beta-adrenoceptor antagonists.

Organs and Systems

Nervous system

In a double-blind, double-dummy, placebo-controlled comparison of clomethiazole and gammahydroxybutyrate in ameliorating the symptoms of alcohol withdrawal, alcohol-dependent patients were randomized to receive either clomethiazole 1000 mg or gammahydroxybutyrate 50 mg/kg (2). There was no difference between the three treatments in ratings of alcohol withdrawal symptoms or requests for additional medication. After tapering the active medication, there was no increase in withdrawal symptoms, suggesting that physical tolerance did not develop to either clomethiazole or gammahydroxybutyrate during the 5-day treatment period. The most frequently reported adverse effect of gammahydroxybutyrate was transient vertigo, particularly after the evening double dose.

The effect of clomethiazole on cerebral outcome in patients undergoing coronary artery bypass surgery has been investigated in 245 patients, who were randomized double-blind to placebo or clomethiazole (1800 mg over 45 minutes followed by 800 mg/hour until the end of surgery) (3). A battery of eight neuropsychological tests was administered preoperatively and repeated 4–7 weeks after surgery. There were no differences between the clomethiazole and placebo groups in postoperative neuropsychological tests scores. Thus, clomethiazole did not improve or worsen cerebral outcome after coronary artery bypass surgery.

The efficacy and safety of clomethiazole (75 mg/kg by intravenous infusion over 24 hours), as a neuroprotective drug, were studied in a double-blind, placebo-controlled trial (CLASS, the Clomethiazole Acute Stroke Study) in 1360 patients with acute hemispheric stroke (4). Clomethiazole was generally well tolerated and safe. Sedation was the most common adverse event, leading to treatment withdrawal in 16% of patients compared with 4.2% of placebo-treated patients.

In a small subset of CLASS (95 patients) mortality at 90 days was 19% in the clomethiazole group and 23% in the placebo group (5). Sedation was the most common adverse event (clomethiazole 53%, placebo 17%), followed by rhinitis and coughing. The incidence and pattern of serious adverse events was similar between the groups.

A report has illustrated the dangers of driving a motor vehicle whilst taking clomethiazole (6).

- A 53-year-old male car driver was followed by the police for about 2 km, while he drove in an unsafe manner, until the car crashed into the owner's garage. Analysis of his blood showed a clomethiazole concentration of 3.3 µg/ml. Neither alcohol nor any other drug could be detected. One day later he committed suicide by swallowing at least 60 capsules of clomethiazole.

Liver

Reversible cholestatic jaundice has been attributed to clomethiazole (SEDA-21, 39).

Long-Term Effects

Drug abuse

As with other GABA enhancers, clomethiazole is highly abusable in susceptible individuals and should not be used for outpatient detoxification (7).

Drug dependence

Clomethiazole maintains its efficacy with apparently less dependence than temazepam during prolonged use (7).

Drug–Drug Interactions

Alteplase

The safety of the thrombolytic drug alteplase (tPA) plus clomethiazole in patients with acute ischemic stroke has been assessed in a randomized, double-blind study (8). All received alteplase 0.9 mg/kg, beginning within 3 hours of stroke onset, and then either intravenous clomethiazole 68 mg/kg ($n = 97$) over 24 hours, or placebo ($n = 93$) beginning within 12 hours of stroke onset. During follow-up for 90 days the number of serious adverse event reports was 47 in the clomethiazole group and 48 in the placebo group. There were 15 deaths in those given clomethiazole and nine in those given placebo, but this was not significantly different. Sedation was also greater with clomethiazole (42%) than with placebo (13%).

Cimetidine

Cimetidine can reduce clomethiazole clearance by 60–70% (9).

References

1. Fagan D, Lamont M, Jostell KG, Tiplady B, Scott DB. A study of the psychometric effects of chlormethiazole in healthy young and elderly subjects. Age Ageing 1990;19(6):395–402.
2. Nimmerrichter AA, Walter H, Gutierrez-Lobos KE, Lesch OM. Double-blind controlled trial of gamma-hydroxybutyrate and clomethiazole in the treatment of alcohol withdrawal. Alcohol Alcohol 2002;37(1):67–73.
3. Kong RS, Butterworth J, Aveling W, Stump DA, Harrison MJ, Hammon J, Stygall J, Rorie KD, Newman SP. Clinical trial of the neuroprotectant clomethiazole in coronary artery bypass graft surgery: a randomized controlled trial. Anesthesiology 2002;97(3):585–91.
4. Wahlgren NG, Ranasinha KW, Rosolacci T, Franke CL, van Erven PM, Ashwood T, Claesson L. Clomethiazole acute stroke study (CLASS): results of a randomized, controlled trial of clomethiazole versus placebo in 1360 acute stroke patients. Stroke 1999;30(1):21–8.
5. Wahlgren NG, Diez-Tejedor E, Teitelbaum J, Arboix A, Leys D, Ashwood T. Grossman E. Results in 95 hemorrhagic stroke patients included in CLASS, a controlled trial of clomethiazole versus placebo in acute stroke patients. Stroke 2000;31(1):82–5.
6. Logemann E. Risks for driving under the influence of clomethiazole. Probl Forens Sci 2000;XLIII:144–7.
7. Bayer AJ, Bayer EM, Pathy MS, Stoker MJ. A double-blind controlled study of chlormethiazole and triazolam as hypnotics in the elderly. Acta Psychiatr Scand 1986;329(Suppl):104–11.
8. Lyden P, Jacoby M, Schim J, Albers G, Mazzeo P, Ashwood T, Nordlund A, Odergren T. The Clomethiazole Acute Stroke Study in Tissue-type Plasminogen Activator-treated Stroke (CLASS-T): final results. Neurology 2001;57(7):1199–205.
9. Shaw G, Bury RW, Mashford ML, Breen KJ, Desmond PV. Cimetidine impairs the elimination of chlormethiazole. Eur J Clin Pharmacol 1981;21(1):83–5.

Etifoxine

General Information

Etifoxine is a non-benzodiazepine drug, licensed in France for psychosomatic manifestations of anxiety.

Organs and Systems

Psychological, psychiatric

The psychomotor and amnesic effects of single oral doses of etifoxine (50 mg and 100 mg) and lorazepam (2 mg) were studied in 48 healthy subjects in a double-blind, placebo-controlled, randomized, parallel-group study (1). Its effects were assessed by a battery of subjective and objective tests that explored mood and vigilance, attention, psychomotor performance, and memory. Whereas vigilance, psychomotor performance, and free recall were significantly impaired by lorazepam, neither dose of etifoxine (50 mg and 100 mg) produced such effects. The results suggested that 50 mg and 100 mg single doses of etifoxine do not induce amnesia and sedation compared with lorazepam.

Reference

1. Micallef J, Soubrouillard C, Guet F, Le Guern ME, Alquier C, Bruguerolle B, Blin O. A double blind parallel group placebo controlled comparison of sedative and amnesic effects of etifoxine and lorazepam in healthy subjects. Fundam Clin Pharmacol 2001;15(3):209–16.

Lesopitron

General Information

Lesopitron is a non-benzodiazepine anxiolytic drug. Its structure is similar to that of buspirone, and it is an agonist at central serotonin (5-HT$_{1A}$) receptors.

A 6-week, double-blind, randomized, parallel, phase II, single-center, outpatient study has been performed to study the efficacy and safety of lesopitron 40–80 mg/day compared with lorazepam 2–4 mg/day and placebo in 161 patients with generalized anxiety disorder (1). The most common adverse events associated with lesopitron were somnolence, headache, and dyspepsia, compared with headache, somnolence, and insomnia with lorazepam. Patients treated with placebo mainly experienced headache, somnolence, and pharyngitis.

Reference

1. Fresquet A, Sust M, Lloret A, Murphy MF, Carter FJ, Campbell GM, Marion-Landais G. Efficacy and safety of lesopitron in outpatients with generalized anxiety disorder. Ann Pharmacother 2000;34(2):147–53.

Ritanserin

General Information

Ritanserin, a selective $5\text{-}HT_2$ receptor antagonist, increases slow-wave sleep in healthy volunteers (1). It improved sleep in middle-aged poor sleepers (2), but on withdrawal after 20 days treatment there was rebound sleep impairment, which was at its worst 3 nights after withdrawal, consistent with its long half-life (40 hours). Ritanserin is thus not a useful hypnotic, but analogues with more appropriate kinetics will be of interest.

References

1. Idzikowski C, Mills FJ, James RJ. A dose-response study examining the effects of ritanserin on human slow wave sleep. Br J Clin Pharmacol 1991;31(2):193–6.
2. Adam K, Oswald I. Effects of repeated ritanserin on middle-aged poor sleepers. Psychopharmacology (Berl) 1989;99(2):219–21.

Suriclone

General Information

Suriclone, a cyclopyrrolone analogue of zopiclone, has similar pharmacology to the benzodiazepines, binding close to the same site of the GABA receptor–chloride channel complex. It is effective as an anxiolytic and has the notable advantages of minimal sedation and cognitive toxicity, and milder withdrawal effects than those of diazepam or lorazepam (1). Its withdrawal from further development is a mystery.

Reference

1. Lader M. Clin pharmacology of anxiolytic drugs: Past, present and future. In: Biggio G, Sanna E, Costa E, editors. GABA-A Receptors and AnxietyFrom Neurobiology to Treatment. New York: Raven Press, 1995:135.

Zaleplon

General Information

Zaleplon is a non-benzodiazepine that induces sleep comparable to other hypnotics but with significantly fewer residual effects (1), related at least in part to its short half-life. It is a pyrazolopyrimidine hypnotic that binds selectively to the $GABA_{A1A}$ receptor, previously known as the benzodiazepine type 1 (BDZ_1) receptor. Whereas such agonist selectivity was hoped to confer advantages in terms of the risk of adverse effects, in practice zaleplon is similar to the older non-selective benzodiazepines in terms of both efficacy and safety (2–4). The so-called "Z drugs," including zaleplon, are significantly more expensive than benzodiazepines and are therefore likely to be less cost-effective (5).

Hypnotics need to be prescribed appropriately, and guidance has been published (6). In particular, treatable causes for insomnia, such as psychiatric disorders and physical illnesses, need to be identified and treated before prescribing hypnotics.

Pharmacokinetics

After oral administration zaleplon is well absorbed (71%) and peak concentrations are reached in about 60 minutes. However, it undergoes presystemic elimination and has a systemic availability of about 30%. Its adverse effects include anterograde amnesia, depression, paradoxical reactions (for example restlessness, agitation), dependence, and withdrawal symptoms (related to the dose and duration of treatment). Although the data are limited, it is thought to be relatively safe in overdose, unless it is combined with other CNS depressants.

The pharmacokinetics and absolute oral systemic availability of zaleplon have been assessed in a partially randomized, single-dose, crossover study in 23 healthy subjects, who received intravenous infusions of zaleplon 1 and 2.5 mg during the first and second periods and were then randomly assigned to receive an oral dose of 5 mg or an intravenous infusion of 5 mg in a crossover design (7). The oral and intravenous doses of zaleplon were well tolerated. Somnolence, abnormal vision, diplopia, and dizziness were the most commonly reported adverse events.

Pharmacological effects

Initial pharmacodynamic data suggested that sleep latency is improved by zaleplon and there is no significant next-day psychomotor impairment or memory impairment (8), and evaluations at zaleplon peak plasma concentrations show much less impairment than with other hypnotics, suggesting an improved benefit-to-harm balance for zaleplon compared with older agents. However, outcome data are mainly from industry-sponsored trials and are often difficult to compare. Differences between zaleplon and benzodiazepine hypnotics may have been exaggerated (3,4).

Zaleplon can be used to treat symptoms of insomnia with little next-day psychomotor or memory impairment. However, further research is needed.

Observational studies

The results of a 1-year open extension of two randomized, double-blind studies of zaleplon have been reported (9). In 316 older patients who took zaleplon nightly from 6 to 12 months and were then followed through a 7-day single-blind, placebo-controlled, run-out period, the safety profile was similar to that observed in a short-term trial in an equivalent population. The data also suggested that therapy for up to 12 months produced and maintained statistically significant improvement in time to persistent sleep onset, duration of sleep, and the number of nocturnal wakenings. Withdrawal was not associated with rebound insomnia. The authors concluded that placebo-controlled, double-blind trials are needed to confirm these results.

Comparative studies

A review of published studies of zaleplon has shown that it has a quick onset of action and undergoes rapid elimination, which results in an arguably better safety profile than previously available agents (10). In addition, rebound insomnia and other withdrawal effects have not been demonstrated with zaleplon, and it is well tolerated in both young and older patients. These characteristics may be advantageous for patients who should not receive benzodiazepines.

Placebo-controlled studies

Three doses of zaleplon have been compared with placebo in outpatients with insomnia in a 4-week study (11). During week 1, sleep latency was significantly shorter with zaleplon 5, 10, and 20 mg than with placebo. The significant reduction in sleep latency persisted to week 3 with zaleplon 10 mg and to week 4 with zaleplon 20 mg. Compared with placebo, zaleplon 10 mg and 20 mg also had significant positive effects on sleep duration, number of awakenings, and sleep quality. Pharmacological tolerance did not develop with zaleplon and there were no indications of rebound insomnia or withdrawal symptoms after discontinuation. There was no significant difference in the frequency of adverse events with zaleplon compared with placebo. The authors concluded that zaleplon provides effective treatment of insomnia with a favorable safety profile.

Zaleplon versus triazolam

Zaleplon and triazolam have been compared in two concurrent multicenter, randomized, double-blind, placebo-controlled crossover studies in chronic insomniacs (12). Study 1 compared zaleplon (10 and 40 mg) with triazolam (0.25 mg) and placebo; study 2 compared zaleplon (20 and 60 mg) with triazolam (0.25 mg) and placebo. All doses of zaleplon produced significant reductions in sleep latency, and triazolam 0.25 mg reduced sleep latency comparable with zaleplon 10 mg. Only triazolam and zaleplon 60 mg produced significant increases in total sleep time compared with placebo. Zaleplon 40 and 60 mg and triazolam also reduced the percentage of REM sleep compared with placebo. There was no evidence of residual daytime impairment with zaleplon, but triazolam produced significant impairment in performance on a digit copying test. There were more adverse events with zaleplon 60 mg compared with triazolam 0.25 mg and placebo. The most frequently reported adverse events with all treatments included headache, dizziness, and somnolence.

Zaleplon versus zolpidem

Zaleplon and zolpidem have been compared in two concurrent multicenter, randomized, double-blind, placebo-controlled crossover studies in chronic insomniacs (12). In study 1, zaleplon 10 mg, zolpidem 10 mg, or placebo were given double-blind to 36 healthy subjects under standardized conditions in a six-period, incomplete-block, crossover study (13). The subjects were gently awakened and given the medication at predetermined times, 5, 4, 3, or 2 hours before morning awakening, which occurred 8 hours after bedtime. When they awoke in the morning, subjective and objective assessments of residual effects of hypnotics were administered. There were no serious adverse experiences during the study; all adverse events were mild to moderate. The most commonly reported adverse events associated with zaleplon were weakness and somnolence. Weakness, depersonalization, dizziness, and somnolence were the most frequent nervous system adverse events associated with zolpidem.

Zaleplon has been compared with zolpidem 10 mg and placebo in 615 adult outpatients with insomnia (14). After a 7-night placebo (baseline) period, the patients were randomly assigned to receive one of five treatments in double-blind fashion for 28 nights (zaleplon 5, 10, or 20 mg; zolpidem 10 mg; or placebo), followed by placebo for 3 nights. Sleep latency, sleep maintenance, and sleep quality were determined from sleep questionnaires each morning. Rebound insomnia and withdrawal effects on withdrawal were also assessed. There was no evidence of rebound insomnia or withdrawal symptoms on withdrawal of zaleplon after 4 weeks. The frequency of adverse events in the active treatment groups did not differ significantly from that in the placebo group.

The pharmacokinetics and pharmacodynamics of zaleplon (10 or 20 mg) and zolpidem (10 or 20 mg) have been investigated in a randomized, double-blind, crossover, placebo-controlled study in 10 healthy volunteers with no history of sleep disorder (15). The half-life of zaleplon was significantly shorter than that of zolpidem. Zaleplon produced less sedation than zolpidem at the two doses studied, and the sedation scores in the zaleplon groups returned to baseline sooner than in the zolpidem groups. Zaleplon had no effect on recent or remote recall, whereas zolpidem had a significant effect on both measures.

Organs and Systems

Nervous system

Zaleplon has been reported to cause sleepwalking.

- A 14-year-old boy with major depressive disorder responded to paroxetine 20 mg/day with full remission of depressive symptoms except insomnia (16). Diphenhydramine and trazodone did not improve his sleep and caused excessive daytime drowsiness. He then responded well to zaleplon 10 mg, but when he took two extra tablets 3 weeks later he developed complex behavior and sleepwalking. He had slurred speech, was slow in responding to questions, was moderately confused, and was uncoordinated and moved slowly. Physical examination, routine laboratory investigations, and an electrocardiogram were all normal. He remained in hospital for 8 hours and awakened without any memory of his activities. His mental state at 1 week and 1 month were both normal.

In a double-blind, placebo-controlled, repeated-measures study of the effects of zaleplon 10 mg on performance after a short period (1 hour) of daytime sleep in 16 volunteers (eight men and eight women) zaleplon had a statistically significant negative impact on *balance* through the first 2 hours after the dose compared with placebo (17). In addition, symptoms related to "drowsiness" were statistically more prevalent with zaleplon during the first 3 hours. Zaleplon also had a significantly negative impact on memory at 1 hour and 4 hours.

Psychological, psychiatric

- A 25-year-old unmarried Asian woman with high intellectual functioning and psychiatric history developed illusions and hallucinations and a feeling of superior psychosocial adjustment, with no current or prior medical or depersonalization within several minutes of taking zaleplon 10 mg (18). The illusions and visual hallucinations resolved after 15 minutes, but she continued to have light-headedness and fatigue, which gradually resolved by the next day.

Drug Administration

Drug overdose

An overdose of zaleplon can cause psychomotor impairment.

- A 20-year-old man became unsteady on his feet after a road traffic accident (19). He had slow movements and reactions, poor co-ordination, lack of balance, and poor attention. He admitted to having inhaled three crushed zaleplon 10 mg tablets and ingesting three zaleplon 10 mg tablets. His blood zaleplon concentration was 0.13 µg/ml.

Blood concentrations consistent with doses exceeding therapeutic concentrations of zaleplon can impair level of consciousness and driving ability.

Drug–Drug Interactions

Alcohol

The addition of alcohol to the Z drugs, zaleplon, zolpidem, and zopiclone, produces additive sedative effects without altering their pharmacokinetics (20).

The effects of alcohol combined with either zaleplon or triazolam have been studied in 18 healthy volunteers (21). Triazolam, with and without ethanol, impaired digit symbol substitution, symbol copying, simple and complex reaction times, and divided attention performance compared with placebo. Zaleplon without ethanol impaired only digit symbol substitution and divided attention tracking, but when it was combined with ethanol all measures were impaired. However, zaleplon without ethanol was consistently better than triazolam alone. Zaleplon produced less performance impairment and a shorter period of ethanol potentiation than triazolam.

Cimetidine

Cimetidine increases plasma concentrations of the Z drugs and increases their sedative effects (20); this occurs to a lesser extent than the similar effect on benzodiazepines that are exclusively metabolized by CYP3A4.

Digoxin

The interaction of zaleplon with digoxin has been investigated in 20 subjects (22). There were one or more adverse effects in 18% of those who took digoxin alone and 35% of those who took digoxin plus zaleplon, but these were all mild and resolved quickly. Zaleplon had no significant effects on selected pharmacokinetic and pharmacodynamic properties of digoxin.

Erythromycin

Erythromycin increases plasma concentrations of the Z drugs and increases their sedative effects (20); this occurs to a lesser extent than the similar effect on benzodiazepines that are exclusively metabolized by CYP3A4.

Ibuprofen

The interaction of zaleplon with ibuprofen has been investigated in 17 subjects (23). Healthy adult volunteers were given zaleplon 10 mg alone, ibuprofen 600 mg alone, or zaleplon 10 mg plus ibuprofen 600 mg in an open, randomized, crossover study. The adverse effects were mild and resolved without intervention. The authors concluded that there was no evidence of a significant interaction between zaleplon and ibuprofen.

Ketoconazole

Ketoconazole increases plasma concentrations of the Z drugs and increases their sedative effects (20); this occurs to a lesser extent than the similar effect on benzodiazepines that are exclusively metabolized by CYP3A4.

Rifampicin

Rifampicin significantly induces the metabolism of zaleplon and reduces its sedative action (20), although the effect is less than the effect of rifampicin on triazolam or midazolam, probably because they are more exclusively metabolized by CYP3A4.

References

1. Mangano RM. Efficacy and safety of zaleplon at peak plasma levels. Int J Clin Pract Suppl 2001;116:9–13.
2. Landolt HP, Gillin JC. GABA(A1a) receptors: involvement in sleep regulation and potential of selective agonists in the treatment of insomnia. CNS Drugs 2000;13(3):185–99.
3. Anonymous. What's wrong with prescribing hypnotics? Drug Ther Bull 2004;42(12):89–93.
4. Holbrook AM. Treating insomnia. BMJ 2004;329(7476):1198–9.
5. Dundar Y, Boland A, Strobl J, Dodd S, Haycox A, Bagust A, Bogg J, Dickson R, Walley T. Newer hypnotic drugs for the short-term management of insomnia: a systematic review and economic evaluation. Health Technol Assess 2004;8(24):iii–x1–125.
6. National Institute for Health and Clinical Excellence. Zaleplon, zolpidem and zopiclone for the short-term management of insomnia. Technology Appraisal Guidance 2004:77. www.nice.org.uk
7. Rosen AS, Fournie P, Darwish M, Danjou P, Troy SM. Zaleplon pharmacokinetics and absolute bioavailability. Biopharm Drug Dispos 1999;20(3):171–5.
8. Anonymous. Does zaleplon help you sleep and wake refreshed? Drug Ther Perspect 1999;14:1–4.
9. Ancoli-Israel S, Richardson GS, Mangano RM, Jenkins L, Hall P, Jones WS. Long-term use of sedative hypnotics in older patients with insomnia. Sleep Med 2005; 6: 107–13.
10. Israel AG, Kramer JA. Safety of zaleplon in the treatment of insomnia. Ann Pharmacother 2002;36(5):852–9.
11. Fry J, Scharf M, Mangano R, Fujimori MZaleplon Clinical Study Group. Zaleplon improves sleep without producing rebound effects in outpatients with insomnia. Int Clin Psychopharmacol 2000;15(3):141–52.
12. Drake CL, Roehrs TA, Mangano RM, Roth T. Dose–response effects of zaleplon as compared with triazolam (0.25 mg) and placebo in chronic primary insomnia Hum Psychopharmacol 2000;15(8):595–604.
13. Danjou P, Paty I, Fruncillo R, Worthington P, Unruh M, Cevallos W, Martin P. A comparison of the residual effects of zaleplon and zolpidem following administration 5 to 2 h before awakening Br J Clin Pharmacol 1999;48(3):367–74.
14. Elie R, Ruther E, Farr I, Emilien G, Salinas EZaleplon Clinical Study Group. Sleep latency is shortened during 4 weeks of treatment with zaleplon, a novel nonbenzodiazepine hypnotic. J Clin Psychiatry 1999;60(8):536–44.
15. Drover D, Lemmens H, Naidu S, Cevallos W, Darwish M, Stanski D. Pharmacokinetics, pharmacodynamics, and relative pharmacokinetic/pharmacodynamic profiles of zaleplon and zolpidem. Clin Ther 2000;22(12):1443–61.
16. Liskow B, Pikalov A. Zaleplon overdose associated with sleepwalking and complex behaviour. J Am Acad Child Adolesc Psychiatry 2004;43:927–8.
17. Whitmore JN, Fischer JR, Barton EC, Storm WF. Performance following a sudden awakening from daytime nap induced by zaleplon. Aviation Space Environ Med 2004;75(1):29–36.
18. Bhatia SC, Arora M, Bhatia SK. Perceptual disturbances with zaleplon. Psychiatr Serv 2001;52(1):109–10.
19. Stillwell ME. Zaleplon and driving impairment. J Forensic Sci 2003;48:677–9.
20. Hesse LM, von Moltke LL, Greenblatt DJ. Clinically important drug interactions with zopiclone, zolpidem and zaleplon. CNS Drugs 2003;17(7):513–32.
21. Roehrs T, Rosenthal L, Koshorek G, Mangano RM, Roth T. Effects of zaleplon or triazolam with or without ethanol on human performance. Sleep Med 2001;2(4):323–32.
22. Sanchez Garcia P, Paty I, Leister CA, Guerra P, Frias J, Garcia Perez LE, Darwish M. Effect of zaleplon on digoxin pharmacokinetics and pharmacodynamics. Am J Health Syst Pharm 2000;57(24):2267–70.
23. Sanchez Garcia P, Carcas A, Zapater P, Rosendo J, Paty I, Leister CA, Troy SM. Absence of an interaction between ibuprofen and zaleplon. Am J Health Syst Pharm 2000;57(12):1137–41.

Zolpidem

General Information

Zolpidem and alpidem are imidazopyridines, chemically distinct from the benzodiazepines and appear to bind selectively to a subset of benzodiazepine receptors (1). This property may account for their apparently milder withdrawal effects and, in the case of alpidem, a relative dominance of anxiolytic over sedative and cognitive effects (2). Zolpidem has hypnotic efficacy comparable to short- and medium-acting benzodiazepines, and has similar or possibly fewer adverse effects at therapeutic doses (3–5), except for gastrointestinal disturbances, which appear to be more common, and visual hallucinations, especially in women (SEDA-17, 46; SEDA-18, 45; SEDA-21, 40). On the other hand, zolpidem unexpectedly causes as much memory impairment as triazolam, if not more, at least in young healthy adults (SEDA-19, 33). Also surprising is the same author's later contention that zolpidem causes less memory impairment than do the benzodiazepine hypnotics (SEDA-21, 40), presumably including triazolam. Zolpidem may also be relatively toxic in overdose, owing to respiratory depression (6), but later observations have suggested that it may not be any more toxic than the benzodiazepines (SEDA-21, 40). At normal doses zolpidem is said to be at least as safe as standard benzodiazepines in patients with respiratory compromise (SEDA-21, 40). The effects of zolpidem are reversed by flumazenil (7).

Hypnotics need to be prescribed appropriately, and guidance has been published (8). In particular, treatable causes for insomnia, such as psychiatric disorders and physical illnesses, need to be identified and treated before prescribing hypnotics.

Subjective responses to treatment with zolpidem were assessed in 16 944 outpatients with insomnia. Nausea, dizziness, malaise, nightmares, agitation, and headache were the most common adverse events reported. There was one serious adverse reaction in a 48-year-old woman, who developed paranoid symptoms during the documentation phase. There were no life-threatening adverse events (9).

The safety and tolerability of zolpidem have been investigated in two multicenter studies, in which 8.9% and 7.5% of the patients reported an adverse event (10). The most frequent events were related to the central nervous system (somnolence, headache, confusion, vertigo), but gastrointestinal and cutaneous symptoms were also frequently reported.

Improvement in social and occupational function has been attributed to zolpidem (11).

- A recovering 60-year-old alcoholic woman developed reduced cognitive function, including considerable memory loss, praxis disorders, and an inability to join in conversation. A CT scan showed non-specific cerebral atrophy. She was given zolpidem 10 mg for insomnia, which she took at first at 2200 hours, but then earlier, at 1900 hours. After starting to take it at the earlier time she talked more easily and could wash the dishes and do the housework, things that she had lost the ability to do. This beneficial effect was detectable 45–60 minutes after the dose of zolpidem, lasted for 3 hours after each administration, and then abated. The subjective improvement in cognitive function was confirmed several times by her general practitioner.

Three patients had improvements in dystonia and parkinsonism after taking zolpidem 10 mg (12). The improvement in dystonia began at 15–45 minutes and optimal benefits were observed after 1–2 hours. The mean duration of action was 4.5 hours initially, falling to 2–3 hours with chronic use. This is similar to that reported in patients with progressive supranuclear palsy, and corresponds to the drug's half-life (2.5 hours). Sleepiness was noted at doses over 10 mg bd.

Observational studies

In 80 elderly subjects (aged over 70 years) with disorders of sleep and co-morbidities, including diabetes and arterial hypertension, zolpidem was effective and well tolerated (13).

Comparative studies

Of 53 patients with insomnia randomly given zolpidem 10 mg or zaleplon 10 mg on 2 consecutive nights, 62% preferred zolpidem and 38% preferred zaleplon (14). The quality of sleep (for example the ease of getting to sleep) was significantly more improved after zolpidem. Insomniac patients tended to prefer zolpidem to zaleplon on both nocturnal and diurnal assessments. Seven adverse events occurred during the study, three of them with zolpidem (sluggish tongue, impaired concentration, leg complaints) and four with zaleplon (cephalalgia requiring an analgesic, abdominal fullness, headache, vertigo). All the adverse events were mild or moderate in intensity.

Placebo-controlled studies

In a double-blind, placebo-controlled, crossover study, 10 patients with "probable progressive supranuclear palsy" took single oral doses of zolpidem (5 and 10 mg), co-careldopa (levodopa 250 mg plus carbidopa 25 mg), or placebo in four separate trials in random order (15). Zolpidem, unlike levodopa or placebo, reduced voluntary saccadic eye movements, and the 5 mg dose produced a statistically significant improvement in motor function. The adverse effects of zolpidem included drowsiness and increased postural instability and were more marked after a dose of 10 mg.

Zolpidem has been investigated in a multicenter, double-blind, placebo-controlled, parallel-group, randomized study in 138 adults, who were experienced air travelers (16). They were randomized to zolpidem 10 mg or placebo for three (or optionally four) consecutive nights, starting with the first night-time sleep after travel. Sleep was assessed with daily questionnaires. Compared with placebo, zolpidem was associated with significantly improved sleep, longer total sleep time, reduced numbers of awakenings, and improved sleep quality. It was not associated with improvement in sleep latency. No unexpected or serious adverse events were reported and the most common adverse event was headache in both groups.

Dose-comparing studies

Zolpidem 10 mg/day and zopiclone 7.5 mg/day, given at night, have been compared in a 14-day, double-blind study in 479 chronic primary insomniacs (17). With zolpidem 68% of the patients were rated at least "moderately improved," versus 62% with zopiclone. However, with zolpidem sleep-onset latency improved in significantly more patients (86 versus 78%). In addition, significantly fewer patients who took zolpidem had drug-related adverse events (31 versus 45%); bitter taste accounted for 5.8% of such complaints with zolpidem compared with 40% with zopiclone. In conclusion, zolpidem was at least as effective as zopiclone but showed significantly less rebound on withdrawal; overall it was better tolerated.

Organs and Systems

Nervous system

Zolpidem 10 mg, temazepam 15 mg, and placebo have been compared in 630 healthy adults in a multicenter study (18). They were given 15 minutes before lights out, with polysomnographic monitoring for 7.5 hours. Subjective questionnaires and performance tests, including digit symbol substitution and symbol copying, were administered before and after sleep. Neither drug significantly reduced objective sleep latency, but zolpidem reduced awakenings compared with temazepam. Both improved sleep efficiency and most subjective sleep measures, and zolpidem was superior to temazepam in five of six subjective outcome measures. Symbol copying, morning sleepiness, and morning concentration were not altered. Zolpidem 10 mg provided greater subjective hypnotic efficacy than temazepam 15 mg in this model of transient insomnia, with reduced polysomnographic awakenings.

Zolpidem was identified in the blood of 29 subjects arrested for impaired driving (19). In those in whom zolpidem was present with other drugs and/or alcohol, the symptoms reported were generally those of nervous system depression and included slow movements and reactions, slow and slurred speech, poor coordination, lack of balance, flaccid muscle tone, and horizontal and vertical gaze nystagmus. In five subjects in whom zolpidem was the only drug detected, signs of impairment included slow and slurred speech, slow reflexes, disorientation, and lack of balance and coordination. Therapeutic doses of zolpidem can affect driving adversely, and concentrations above the target range further impair both driving ability and consciousness.

Zolpidem 5 mg, zopiclone 3.75 mg, lormetazepam 1 mg, or placebo were given at night to 48 healthy volunteers aged 65 years or over (20). The study included four treatment periods separated by wash-out periods of at least 1 week. The results suggested that compared with placebo, the active drugs increased body sway; however, this effect disappeared after 5 hours with zolpidem, while it disappeared only after 8 hours with lormetazepam and zopiclone. None of the three drugs affected attention. In learning tasks, there was impaired memory with lormetazepam relative to both zolpidem and placebo. Zolpidem had no significant effect on memory.

In in-patients aged 50 years or older who received zolpidem as a hypnotic, 23 patients had respectively 16 and 10 adverse drug reactions possibly and probably related to zolpidem use (19% incidence) (21). Of the total 26 adverse drug reactions, 21 affected the nervous system and occurred with both zolpidem 5 mg and 10 mg. Zolpidem was withdrawn in 39% of those who had a nervous system adverse drug reaction.

- A 28-year-old man sustained anoxic brain damage following an aborted cardiac arrest and subsequently developed severe muscular rigidity and spasticity involving all the limbs (22). The spasticity was refractory to the standard regimens used for spastic hypertonia. Zolpidem dramatically inhibited muscular rigidity, spasticity, and dystonic posturing in a dose-dependent manner, resulting in a sustained improvement in global performance over 4 years. The only adverse effects were mild drowsiness and mild postural instability.

The authors postulated a central mechanism of action by selective inhibition of GABAergic inhibitory neurons.

In a double-blind, placebo-controlled study in eight healthy men, zolpidem 10 mg produced statistically significant postural sway in the tandem stance test, and triazolam 0.25 mg was statistically significant only as defined by the polygonal area of foot pressure center (23). Zolpidem, which has a minimal muscle-relaxant effect, produced more imbalance than triazolam, which is known for its muscle relaxant effect. The authors suggested that in the use of hypnotics, sway derives from suppression of the central nervous system relevant to awakening rather than from muscle relaxation.

Psychological, psychiatric

Like the benzodiazepines, zolpidem can produce a variety of paradoxical effects, including disturbances of mood, perception, and behavior. In 192 surgical patients zolpidem 8 mg but not 16 mg caused such effects, particularly anxiety, 1 hour after administration (24).

The acute effects of zolpidem and triazolam have been compared in 10 non-drug-abusing subjects using a Digit-Enter-and-Recall task with varying delay intervals (9, 10, and 20 seconds) (25). Zolpidem and triazolam impaired performance as a function of dose after all intervals. However, zolpidem produced significantly less impairment than triazolam after the longest delay (20 seconds). Zolpidem and triazolam produced comparable dose-related impairment of the digit symbol substitution, circular lights, and picture recall/recognition tasks. The results suggested that zolpidem may have less potential than triazolam to impair recall, which may be due to differences between these compounds in terms of their benzodiazepine-receptor binding profiles.

Delirium has been attributed to zolpidem (26,27).

- A 26-year-old woman was treated at a psychiatric inpatient unit for psychotic depression. She had neither formal thought disorder nor perceptual disturbances and was cognitively intact. Ten days later she was stabilized on fluoxetine, risperidone, and benzatropine. She then developed flu-like symptoms, and 3 days later took zolpidem 10 mg to help her to sleep. After 30 minutes she was found agitated, confused, and rambling, and wanted to go to the beach. Her speech was disorganized and she had visual hallucinations. Her gait was ataxic. There were no signs of meningeal irritation. Her temperature was 37.3°C, her pulse 114/minute, and her blood pressure 116/78 mmHg. When she was evaluated the next morning, her delirium had cleared and she made a full recovery.

- An 86-year-old white woman with headaches and diplopia took zolpidem 5 mg and about 2 hours later became restless, disoriented, and physically agitated. She was given haloperidol and needed restraining for her own safety. Her symptoms resolved by day 5 and she had no recollection of the incident. Rechallenge was not attempted.

Following 8 hours of undisturbed night-time sleep, 80 subjects (50 men, 30 women) took oral zolpidem 0, 5, 10, or 20 mg at 1000 h (20 per group) and then oral melatonin 0 or 5 mg at 1030 h (thus, 10 subjects per drug combination) (28). They napped from 1000 h to 1130 h, after which they were given cognitive tests (Restricted Reminding, Paired-Associates, and Psychomotor Vigilance). They were tested again after a second nap from 1245 to 1600 h. Melatonin 5 mg plus zolpidem 0 mg enhanced daytime sleep, with no memory or performance impairment. Zolpidem 20 mg plus melatonin 0 mg also enhanced daytime sleep (non-significantly), but memory and vigilance were impaired. The authors concluded that there were no advantages to using melatonin plus zolpidem. Functional coupling of sleep-inducing and memory-impairing effects may be specific to benzodiazepine-receptor agonists such as zolpidem, suggesting potential advantages to using melatonin in the operational environment.

In a double blind, placebo-controlled study in 36 young healthy volunteers, zolpidem 5 or 10 mg increased N2 and P3 latencies and reduced N2 and P3 amplitudes (29). However, contrary to expectations, there was no change in N1 while P2 amplitude was increased by the higher dose. The effects on N2 and P3 amplitudes and latencies were similar to those of other hypnosedatives. However, zolpidem unexpectedly increased P2 amplitude, which the authors attributed to its selective receptor binding profile.

In a double-blind, crossover, randomized, placebo-controlled study in 22 healthy volunteers, single doses of zolpidem 10 mg and triazolam 0.25 mg had no effect on the enhanced non-word recall observed after sleep or on improvement in the performance of a digit symbol substitution test in subjects who slept (30). The authors concluded that the hypnotics did not interfere with nocturnal sleep-induced improvement in memory.

Visual hallucinations have been attributed to zolpidem.

- A 50-year-old woman, who had taken one conjugated estrogen tablet daily for hormone replacement therapy and tricalcium phosphate twice daily for osteoporosis, took zolpidem 10 mg for insomnia and paracetamol 500 mg for a headache at bedtime (31). She began to have visual hallucinations within 20 minutes, lasting for about 30 minutes, and then her vision returned to normal. She only partially recalled the event. She had never taken zolpidem before and had not had any such disturbances in the past.

A paranoid syndrome has been attributed to zolpidem.

- A 67-year-old former teacher with no significant psychiatric history developed progressively worsening confusion, agitation, rapid speech, poor sleep, limited appetite, flight of ideas, and increasingly disorganized and paranoid thoughts (32). Her only medication was zolpidem, 10 mg at bedtime, which had been started 6 weeks earlier. Routine laboratory tests and a CT head scan were normal. The outcome was not reported.

Liver

Liver damage has been attributed to zolpidem (33).

- A 53-year-old woman first took zolpidem for insomnia in July 1996. In September 1996 she again took zolpidem 20 mg/day and 2 days later had developed sudden epigastric pain associated with pale stools and dark urine but no fever; she stopped taking zolpidem and the abdominal pain resolved spontaneously within 12 hours. In April 1997, she had another episode of abdominal pain after taking zolpidem. Eleven days later her serum alanine transaminase and gamma-glutamyltranspeptidase activities were 50 IU/l and 89 IU/l respectively. In June 1997 ultrasound showed a normal biliary tract. Viral hepatitis and concurrent infections with Epstein-Barr virus and cytomegalovirus were excluded.

Long-Term Effects

Drug abuse

There has been a literature review of the abuse potential of zolpidem (34). There were 15 published cases of abuse or dependence. In six patients the abuse was secondary to other forms of abuse or dependence. The authors concluded that the abuse potential of zolpidem is much less than with other hypnotics and that it is also safer than conventional hypnotics. Patients with a history of other substance abuse may be considered as being at risk of later abuse of zolpidem.

However, there is a risk of abuse and dependence from chronic use of zolpidem in high doses (35).

- A 67-year-old Caucasian woman, who had previously been treated for depression, anxiety, and insomnia, as well as alcohol, barbiturate, and benzodiazepine dependence, was given zolpidem 10 mg at bedtime for insomnia. She increased the dose without the knowledge of her physicians, using up to 100 mg/day for 1.5 years, alternating it with various benzodiazepines obtained from multiple physicians when zolpidem was unobtainable. She developed severe generalized tremor, psychomotor agitation, facial flushing, and anxiety, despite taking chlordiazepoxide 300 mg in divided doses during the first 24 hours of detoxification. A tapering dose of zolpidem was initiated and the chlordiazepoxide was tapered. Her symptoms completely subsided within 30 minutes of a single dose of zolpidem 15 mg.

Drug dependence

Four cases of former drug or alcohol abusers with personality disorders have been described; all developed dependence while taking high doses of zolpidem (36).

A rare case of physical and psychological addiction to an excessive dose of zolpidem and subsequent detoxification using diazepam has been reported.

- A 46-year-old white man with a history of polysubstance abuse took for zolpidem for 2 years and gradually increased the total dosage to about 400 mg/day in divided doses; he was detoxified using a standard benzodiazepine 7-day diazepam tapering regimen (37).

Drug withdrawal

Zaleplon has been compared with zolpidem 10 mg and placebo in 615 adult outpatients with insomnia (38). After a 7-night placebo (baseline) period, the patients were randomly assigned to receive one of five treatments in double-blind fashion for 28 nights (zaleplon 5, 10, or 20 mg; zolpidem 10 mg; or placebo), followed by placebo for 3 nights. Sleep latency, sleep maintenance, and sleep quality were determined from sleep questionnaires each morning. Rebound insomnia and withdrawal effects on withdrawal were also assessed. After withdrawal of zolpidem, the incidence of withdrawal symptoms was significantly greater than after withdrawal of placebo, and there was suggestion of significant rebound insomnia in some patients who had taken zolpidem compared with placebo. The frequency of adverse events in the active treatment groups did not differ significantly from that in the placebo group.

Withdrawal-induced seizures have been described in a woman taking various benzodiazepines and zolpidem (39).

- A 43-year-old woman had had insomnia since she was a child. At the age of 15 benzodiazepine therapy improved her sleeping, but when she gradually stopped taking benzodiazepines the insomnia returned after a few days. At the age of 26 she was abusing several benzodiazepines, including diazepam and flunitrazepam. At the age of 30 she was taking high doses of bromazepam every evening before going to sleep. After 1 month, she abruptly stopped taking bromazepam and during withdrawal had an epileptic seizure. During the next few years, she had periods of relative well-being, but also two further periods of benzodiazepine abuse, both resulting in seizures after withdrawal. Finally, a physician prescribed zolpidem. Two months later she increased the dose to 450–600 mg/day. After another month of abuse, she was forced by an unexpected event to discontinue the zolpidem and 4 hours later had an epileptic seizure, similar to the previous ones. She started taking zolpidem again, the drug abuse continued, and her fits settled. Six months later she underwent a planned program of zolpidem withdrawal.

Susceptibility Factors

Hepatic disease

The adverse effects of zolpidem can be enhanced in hepatic cirrhosis (40).

- A 41-year-old white man developed postoperative complications 11 months after liver transplantation. His mental status began to deteriorate secondary to hepatic encephalopathy. One day before admission he was given zolpidem 5 mg for sleep and 1 hour later he awoke in a stupor and was not oriented to place or time. He became increasingly incoherent and verbally abusive.

Although this patient's worsening mental state could have been explained by the natural history of the encephalopathy, it is possible that zolpidem exacerbated his decline.

Drug Administration

Drug dosage regimens

A review of six studies in over 4000 patients has suggested that non-nightly administration of zolpidem is effective and does not appear to be associated with withdrawal symptoms or dose escalation (41). On the other hand, a case report has suggested that stopping and restarting zolpidem can trigger visual hallucinations; the same phenomenon was observed three times in a healthy 23-year-old Chinese woman (42).

Drug overdose

Two deaths due to acute intentional zolpidem overdose have been reported (43).

- A 36-year-old woman with a history of psychiatric illness, including paranoid disorder, depression with panic episodes, and stress disorder, was found dead in bed. Caffeine, risperidone, and zolpidem were found in her urine.

- A 58-year-old woman with a history of hypertension and mental illness (manic depression and schizophrenia) was found dead in bed, with white foam around her mouth. Zolpidem and carbamazepine were found in her urine.

The cause of death in both cases was thought to have been acute zolpidem overdose, but it is not clear how risperidone and carbamazepine could have been excluded as possible contributors.

- A 44-year-old white man, who had had major depression and anxiety disorder for 25 years, became drowsy after swallowing 20 tablets (10 mg) of zolpidem (44). He was not taking any other medications at the time. A few hours later he became unresponsive and comatose and developed respiratory depression with hypoxia and mild hypercapnia. He subsequently made a full recovery after appropriate medical support.

Drug–Drug Interactions

Antifungal imidazoles

Potential interactions of zolpidem with three commonly prescribed azole derivatives (ketoconazole, itraconazole, and fluconazole) have been evaluated in a controlled clinical study. Co-administration of zolpidem with ketoconazole impaired zolpidem clearance and enhanced its benzodiazepine-like agonist pharmacodynamic effects. Itraconazole and fluconazole had a small effect on zolpidem kinetics and dynamics. The findings were consistent with in vitro studies of differentially impaired zolpidem metabolism by azole derivatives (45).

Similarly, itraconazole 200 mg did not alter the pharmacokinetics and pharmacodynamics of zolpidem 10 mg in 10 healthy volunteers (46). Therefore, unlike triazolam, zolpidem may be used in normal or nearly normal doses together with itraconazole.

Drugs that compete for hepatic oxidative pathways

Omeprazole, like cimetidine, can impair benzodiazepine metabolism and lead to adverse effects (SEDA-18, 43). Other drugs, including antibiotics (erythromycin, chloramphenicol, isoniazid), antifungal drugs (ketoconazole, itraconazole, and analogues), some SSRIs (fluoxetine, paroxetine), other antidepressants (nefazodone), protease inhibitors (saquinavir), opioids (fentanyl), calcium channel blockers (diltiazem, verapamil), and disulfiram also compete for hepatic oxidative pathways that metabolize most benzodiazepines, as well as zolpidem, zopiclone, and buspirone (SEDA-22, 39) (SEDA-22, 41).

Fluoxetine

The possible pharmacokinetic and pharmacodynamic interactions of repeated nightly zolpidem dosing with fluoxetine were evaluated in 29 healthy women. There were no clinically significant pharmacokinetic or pharmacodynamic interactions (47).

Rifampicin

Rifampicin, and presumably other CYP3A inducers, reduces concentrations of zolpidem (SEDA-22, 42).

Ritonavir

The inhibitory effect of ritonavir on the biotransformation of triazolam and zolpidem has been investigated (48). Short-term, low-dose ritonavir produced a large and significant impairment of triazolam clearance and enhancement of its clinical effects. In contrast, ritonavir produced small and clinically unimportant reductions in zolpidem clearance. The findings are consistent with the complete dependence of triazolam clearance on CYP3A activity, compared with the partial dependence of zolpidem clearance on CYP3A.

Sertraline

Interactions between zolpidem and sertraline have been studied in 28 healthy women, who took a single dose of zolpidem alone and five consecutive doses of zolpidem 10 mg while taking chronic doses of sertraline 50 mg (49). Co-administration of sertraline 50 mg and zolpidem 10 mg was safe but could result in a shortened onset of action and increased effect of zolpidem.

Valproic acid

An interaction of zolpidem with valproic acid has been reported.

- A 47-year-old white man with a history of bipolar disorder, who was taking citalopram 40 mg/day and zolpidem 5 mg at bedtime, developed manic symptoms and was given valproic acid (50). Soon after this, he had episodes of somnambulism, which stopped when valproic acid was withdrawn. On rechallenge with valproic acid, the somnambulism returned.

This appears to be the first report in which the interaction of valproic acid with zolpidem led to somnambulism.

References

1. Lader M. Psychiatric disorders. In: Speight T, Holford N, editors. Avery's Drug Treatment. 4th ed.. Auckland: ADIS International Press, 1997:1437.
2. Lader M. Clin pharmacology of anxiolytic drugs: Past, present and future. In: Biggio G, Sanna E, Costa E, editors. GABA-A Receptors and Anxiety. From Neurobiology To Treatment. New York: Raven Press, 1995:135.
3. Langtry HD, Benfield P. Zolpidem. A review of its pharmacodynamic and pharmacokinetic properties and therapeutic potential. Drugs 1990;40(2):291–313.
4. Declerck AC. Is "poor sleep" too vague a concept for rational treatment? J Int Med Res 1994;22(1):1–16.
5. Rosenberg J, Ahlstrom F. Randomized, double blind trial of zolpidem 10 mg versus triazolam 0.25 mg for treatment of insomnia in general practice Scand J Prim Health Care 1994;12(2):88–92.
6. Lheureux P, Debailleul G, De Witte O, Askenasi R. Zolpidem intoxication mimicking narcotic overdose: response to flumazenil. Hum Exp Toxicol 1990;9(2):105–7.
7. Zivkovic B, Morel E, Joly D, Perrault G, Sanger DJ, Lloyd KG. Pharmacological and behavioral profile of alpidem as an anxiolytic. Pharmacopsychiatry 1990;23(Suppl. 3):108–13.
8. National Institute for Health and Clinical Excellence. Zaleplon, zolpidem and zopiclone for the short-term management of insomnia. Technology Appraisal Guidance 2004:77. www.nice.org.uk.
9. Hajak G, Bandelow B. Safety and tolerance of zolpidem in the treatment of disturbed sleep: a post-marketing surveillance of 16 944 cases. Int Clin Psychopharmacol 1998;13(4):157–67.
10. Ganzoni E, Gugger M. Sicherheitsprofil von zolpidem: zwei studien mit 3805 patienten bei schweizer praktikem. [Safety profile of zolpidem: two studies of 3805 patients by Swiss practitioners.] Schweiz Rundsch Med Prax 1999;88(25–26):1120–7.
11. Jarry C, Fontenas JP, Jonville-Bera AP, Autret-Leca E. Beneficial effect of zolpidem for dementia. Ann Pharmacother 2002;36(11):1808.
12. Evidente VG. Zolpidem improves dystonia in "Lubag" or X-linked dystonia–parkinsonism syndrome. Neurology 2002;58(4):662–3.
13. Cotroneo A, Gareri P, Lacava R, Cabodi S. Use of zolpidem in over 75-year-old patients with sleep disorders and comorbidities. Arch Gerontol Geriatr 2004;38:93–6.
14. Allain, H, Bentué-Ferrer D, Le Breton S, Polard E, Gandon J-M. The preference of insomnia patients between a single dose of zolpidem 10 mg versus zaleplon 10 mg. Hum Psychopharmacol Clin Exp 2003;1:369–74.
15. Daniele A, Moro E, Bentivoglio AR. Zolpidem in progressive supranuclear palsy. N Engl J Med 1999;341(7):543–4.
16. Jamieson AO, Zammit GK, Rosenberg RS, Davis JR, Walsh JK. Zolpidem reduces the sleep disturbance of jet lag. Sleep Med 2001;2(5):423–30.
17. Tsutsui SZolpidem Study Group. A double-blind comparative study of zolpidem versus zopiclone in the treatment of chronic primary insomnia. J Int Med Res 2001;29(3):163–77.
18. Erman MK, Erwin CW, Gengo FM, Jamieson AO, Lemmi H, Mahowald MW, Regestein QR, Roth T, Roth-Schechter B, Scharf MB, Vogel GW, Walsh JK, Ware JC. Comparative efficacy of zolpidem and temazepam in transient insomnia. Hum Psychopharmacol 2001;16(2):169–76.
19. Logan BK, Couper FJ. Zolpidem and driving impairment. J Forensic Sci 2001;46(1):105–10.
20. Allain H, Bentué-Ferrer D, Tarral A, Gandon J-M. Effects on postural oscillation and memory functions of a single dose of zolpidem 5 mg, zopiclone 3.75 mg and lormetazepam 1 mg in elderly healthy subjects. A randomised, crossover, double-blind study versus placebo. Eur J Clin Pharmacol 2003;59:179–88.
21. Mahoney JE, Webb MJ, Gray, SL. Zolpidem prescribing and adverse drug reactions in hospitalized general medicine patients at a Veterans Affairs Hospital. Am J Geriatr Pharmacother 2004;2(1):66–74.
22. Shadan FF, Poceta JS, Kline LE. Zolpidem for postanoxic spasticity. Southern Med J 2004;98:791–2.
23. Nakamura M, Ishii M, Niwa Y, Yamazaki M, Ito H. Temporal changes in postural sway caused by ultrashort-acting hypnotics: triazolam and zolpidem. ORL 2005;67(2):106–12.
24. Uhlig T, Huppe M, Brand K, Heinze J, Schmucker P. Zolpidem and promethazine in pre-anaesthetic medication. A pharmacopsychological approach. Neuropsychobiology 2000;42(3):139–48.
25. Rush CR, Baker RW. Zolpidem and triazolam interact differentially with a delay interval on a digit-enter-and-recall task. Hum Psychopharmacol 2001;16(2):147–57.

26. Freudenreich O, Menza M. Zolpidem-related delirium: a case report. J Clin Psychiatry 2000;61(6):449–50.

27. Brodeur MR, Stirling AL. Delirium associated with zolpidem. Ann Pharmacother 2001;35(12):1562–4.

28. Wesensten NJ, Balkin TJ, Reichardt RM, Kautz MA, Saviolakis GA, Belenky G. Daytime sleep and performance following a zolpidem and melatonin cocktail. Sleep. 2005;28(1):93–103.

29. Lucchesi LM, Braga NI, Manzano GM, Pompeia S, Tufik S. Acute neurophysiological effects of the hypnotic zolpidem in healthy volunteers. Prog Neuropsychopharmacol Biol Psychiatry 2005;29(4):557–64.

30. Melendez, J, Galli I, Boric K, Ortega A, Zuniga L, Henriquez-Roldan CF, Cardenas AM. Zolpidem and triazolam do not affect the nocturnal sleep-induced memory improvement. Psychopharmacology 2005;181:21–26.

31. Huang, C-L, Chang, C-J, Hung C-F, Lin H-Y. Zolpidem-induced distortion in visual perception. Ann Pharmacother 2003;37:683–6.

32. Hill KP, Oberstar JV, Dunn ER. Zolpidem-induced delirium with mania in an elderly woman. Psychosomatics 2004;45:88–9.

33. Karsenti D, Blanc P, Bacq Y, Metman EH. Hepatotoxicity associated with zolpidem treatment. BMJ 1999;318(7192):1179.

34. Soyka M, Bottlender R, Moller HJ. Epidemiological evidence for a low abuse potential of zolpidem. Pharmacopsychiatry 2000;33(4):138–41.

35. Madrak LN, Rosenberg M. Zolpidem abuse. Am J Psychiatry 2001;158(8):1330–1.

36. Vartzopoulos D, Bozikas V, Phocas C, Karavatos A, Kaprinis G. Dependence on zolpidem in high dose. Int Clin Psychopharmacol 2000;15(3):181–2.

37. Rappa L, Larose-Pierre M, Payne DR, Eraikhuemen NE, Lanes DM, Kearson ML. Detoxification from high-dose zolpidem using diazepam. Ann Pharmacother 2004;38:590–4.

38. Elie R, Ruther E, Farr I, Emilien G, Salinas EZaleplon Clinical Study Group. Sleep latency is shortened during 4 weeks of treatment with zaleplon, a novel nonbenzodiazepine hypnotic. J Clin Psychiatry 1999;60(8):536–44.

39. Aragona M. Abuse, dependence, and epileptic seizures after zolpidem withdrawal: review and case report. Clin Neuropharmacol 2000;23(5):281–3.

40. Clark A. Worsening hepatic encephalopathy secondary to zolpidem. J Pharm Technol 1999;15:139–41.

41. Hajak G, Cluydts R, Allain H, Estivill E, Parrino L, Terzano MG, Walsh JK. The challenge of chronic insomnia: is non-nightly hypnotic treatment a feasible alternative? Eur Psychiatry 2003;18(5):201–8.

42. Tsai MJ, Huang YB, Wu PC. A novel clinical pattern of visual hallucination after zolpidem use. J Toxicol Clin Toxicol 2003;41(6):869–72.

43. Gock SB, Wong SH, Nuwayhid N, Venuti SE, Kelley PD, Teggatz JR, Jentzen JM. Acute zolpidem overdose—report of two cases. J Anal Toxicol 1999;23(6):559–62.

44. Hamad A, Sharma N. Acute zolpidem overdose leading to coma and respiratory failure. Intensive Care Med 2001;27(7):1239.

45. Greenblatt DJ, von Moltke LL, Harmatz JS, Mertzanis P, Graf JA, Durol AL, Counihan M, Roth-Schechter B, Shader RI. Kinetic and dynamic interaction study of zolpidem with ketoconazole, itraconazole, and fluconazole. Clin Pharmacol Ther 1998;64(6):661–71.

46. Luurila H, Kivisto KT, Neuvonen PJ. Effect of itraconazole on the pharmacokinetics and pharmacodynamics of zolpidem. Eur J Clin Pharmacol 1998;54(2):163–6.

47. Allard S, Sainati S, Roth-Schechter B, MacIntyre J. Minimal interaction between fluoxetine and multiple-dose zolpidem in healthy women. Drug Metab Dispos 1998;26(7):617–22.

48. Greenblatt DJ, von Moltke LL, Harmatz JS, Durol AL, Daily JP, Graf JA, Mertzanis P, Hoffman JL, Shader RI. Differential impairment of triazolam and zolpidem clearance by ritonavir. J Acquir Immune Defic Syndr 2000;24(2):129–36.

49. Allard S, Sainati SM, Roth-Schechter BF. Coadministration of short-term zolpidem with sertraline in healthy women. J Clin Pharmacol 1999;39(2):184–91.

50. Sattar SP, Ramaswamy S, Bhatia SC, Petty F. Somnambulism due to probable interaction of valproic acid and zolpidem. Ann Pharmacother 2003;37:1429–33.

Zopiclone

General Information

Although it is chemically distinct to the benzodiazepines, zopiclone, a cyclopyrrolone, has similar pharmacology, binding close to the same site of the GABA receptor–chloride channel complex.

Zopiclone has been widely used as a hypnotic, comparable to estazolam (1), and appears to be relatively safe in overdose. It has no adverse effects that would not be expected from its pharmacological and pharmacokinetic properties, with three exceptions: bitter taste (in 3.6% of 20 513 patients) and, like zolpidem, increased risks of gastrointestinal disturbances and visual hallucinations (2,3). Subchronic zopiclone produces minimal changes in the sleep electroencephalogram (4), a potential advantage over benzodiazepines, and in direct comparisons was at least as effective as triazolam or flunitrazepam, with no rebound insomnia after 1 month (SEDA-19, 37). On the other hand, longer administration of zopiclone can cause physical dependence (5), emphasizing the importance of restricting treatment duration. Likewise, its potential for abuse and release of aggression (6) is similar to that of the benzodiazepines (SEDA-17, 47).

Hypnotics need to be prescribed appropriately, and guidance has been published (7). In particular, treatable causes for insomnia, such as psychiatric disorders and physical illnesses, need to be identified and treated before prescribing hypnotics.

Organs and Systems

Nervous system

In a two-part, placebo-controlled, crossover comparison of the effects of zopiclone and zaleplon on car driving, memory, and psychomotor performance, zaleplon 10 mg had no residual effect on driving when taken at bedtime, 10 hours before driving (8). In contrast, zopiclone 7.5 mg caused marked residual impairment. Patients should be advised to avoid driving the morning after taking zopiclone.

Further concern that zolpidem can impair performance comes from a Chinese study (9). Zopiclone 7.5 mg, but not triazolam 2.5 mg, impaired simulated flight performance 2 hours and 3 hours after a dose at midday and sleeping for 1 hour; performance recovered after 4 hours.

Endocrine

The syndrome of inappropriate secretion of antidiuretic hormone has been attributed to zopiclone (10).

- A woman with a 2-week history of insomnia took zopiclone 7.5 mg nightly and over the next 9 days became confused, lethargic, and depressed, culminating in an overdose of six zopiclone tablets. Her previous medical history included hypertension and two episodes of diuretic-induced SIADH. Her serum sodium was 129 mmol/l and 4 days later fell to 113 mmol/l. Her serum osmolality was low (240 mmol/kg) and her urine sodium was 20 mmol/l. The serum sodium returned to normal 12 days after withdrawal of zopiclone.

The rapid resolution of symptoms and correction of the hyponatremia after withdrawal was consistent with an effect of zopiclone.

Long-Term Effects

Drug withdrawal

The acute polysomnographic effects of withdrawal of standard doses of zopiclone ($n = 11$), zolpidem ($n = 11$), triazolam ($n = 10$), and placebo ($n = 7$) have been studied in healthy men (11). They took zopiclone 7.5 mg, zolpidem 10 mg, triazolam 0.25 mg, or placebo for 4 weeks in double-blind, randomized order. Sleep EEG was performed. Total sleep time and sleep efficiency were lower in the first night after withdrawal of triazolam. After withdrawal from zopiclone or zolpidem there were slight but not significant rebound effects. Self-rating scales showed minimal rebound insomnia after withdrawal of all three hypnotics. In the placebo group there were no changes in sleep. These results suggest that the risks of tolerance and dependency are low after short-term zopiclone or zolpidem in the recommended doses.

Drug Administration

Drug overdose

- A 72-year-old with respiratory debilitation due to bronchogenic carcinoma died after taking zopiclone about 200–350 mg (12).

Drug–Drug Interactions

Nefazodone

An 86-year-old white woman taking nefazodone for depression started to take zopiclone for insomnia, but subsequently had morning drowsiness (13). The plasma concentration of zopiclone was measured 8 hours after administration on two occasions, during and after nefazodone therapy. After withdrawal of nefazodone, the plasma concentration of the *S*-enantiomer of zopiclone fell from 107 to 17 ng/ml, while the plasma concentration of the *R*-enantiomer fell from 21 to 1.5 ng/ml.

The substantial fall in plasma zopiclone concentration after withdrawal of nefazodone probably reflected a drug interaction due to inhibition of CYP3A4 by nefazodone (14). Despite the normally short half-life of zopiclone, the residual sedation initially observed in this case suggests that the interaction had clinical significance.

References

1. Li S, Wang C. A comparative study of imovane and estazolam treatment on sleep disturbances. Chin Med Sci J 1995;10(1):56–8.
2. Goa KL, Heel RC. Zopiclone. A review of its pharmacodynamic and pharmacokinetic properties and therapeutic efficacy as an hypnotic. Drugs 1986;32(1):48–65.
3. Mahendran R, Chee KT, Peh LH, Wong KE, Lim L. A postmarketing surveillance study of zopiclone in insomnia. Singapore Med J 1994;35(4):390–3.
4. Roschke J, Mann K, Aldenhoff JB, Benkert O. Functional properties of the brain during sleep under subchronic zopiclone administration in man. Eur Neuropsychopharmacol 1994;4(1):21–30.
5. Jones IR, Sullivan G. Physical dependence on zopiclone: case reports. BMJ 1998;316(7125):117.
6. Shaw SC, Fletcher AP. Aggression as an adverse drug reaction. Adverse Drug React Toxicol Rev 2000;19(1):35–45.
7. National Institute for Health and Clinical Excellence. Zaleplon, zolpidem and zopiclone for the short-term management of insomnia. Technology Appraisal Guidance 2004:77. www.nice.org.uk
8. Vermeeren A, Riedel WJ, van Boxtel MP, Darwish M, Paty I, Patat A. Differential residual effects of zaleplon and zopiclone on actual driving: a comparison with a low dose of alcohol. Sleep 2002;25(2):224–31.
9. Jing BS, Zhan H, Li YF, Zhou YJ, Guo H. [Effects of short-action hypnotics triazolam and zopiclone on simulated flight performance.]Space Med Med Eng (Beijing) 2003;16(5):329–31.
10. Cubbin SA, Ali IM. Inappropriate antidiuretic hormone secretion associated with zopiclone. Psychiatr Bull 1999;23:306–7.
11. Voderholzer U, Riemann D, Hornyak M, Backhaus J, Feige B, Berger M, Hohagen F. A double-blind, randomized and placebo-controlled study on the polysomnographic withdrawal effects of zopiclone, zolpidem and triazolam in healthy subjects. Eur Arch Psychiatry Clin Neurosci 2001;251(3):117–23.
12. Bramness JG, Arnestad M, Karinen R, Hilberg T. Fatal overdose of zopiclone in an elderly woman with bronchogenic carcinoma. J Forensic Sci 2001;46(5):1247–9.
13. Alderman CP, Gebauer MG, Gilbert AL, Condon JT. Possible interaction of zopiclone and nefazodone. Ann Pharmacother 2001;35(11):1378–80.
14. Nemeroff CB, DeVane CL, Pollock BG. Newer antidepressants and the cytochrome P450 system. Am J Psychiatry 1996;153(3):311–20.

DRUGS OF ABUSE

AMPHETAMINES

See also Dexamfetamine,
Methylenedioxymetamfetamine

General Information

Note on spelling

In International Non-proprietary Names the digraph -ph- is usually replaced by -f-, although usage is not consistent, and -ph- is used at the beginnings of some drug names (for example, compare fenfluramine and phentermine) or when a name that beings with a ph- is modified by a prefix (for example, chlorphentermine). For the amphetamines we have used the following spellings: amfetamine, benzfetamine, dexamfetamine, metamfetamine (methylamphetamine), and methylenedioxymetamfetamine (ecstasy).

Dexamphetamine, metamfetamine, and methylenedioxymetamfetamine (MDMA, ecstasy) are covered in separate monographs.

Pharmacology and general adverse effects

Amfetamine is a sympathomimetic compound derived from phenylethylamine. However, the word amphetamines has become generic for the entire group of related substances, including benzfetamine, dexamfetamine, metamfetamine, and methylenedioxymetamfetamine (MDMA, ecstasy). Metamfetamine, a popular drug of abuse, is also known as "speed," "meth," "chalk," "crank," "ice," "crystal," or "glass." Other amfetamine-like drugs include fenfluramine (used as an appetite suppressant) and methylphenidate (used in narcolepsy and attention deficit hyperactivity disorder (ADHD)). When it was first introduced, one of the most frequent uses of amfetamine was as an anorexigenic agent in the treatment of obesity. A number of anorectic agents, many of them related to amfetamine, have since been manufactured. Most are stimulants of the central nervous system; in descending order of approximate stimulatory potency, they are dexamfetamine, phentermine, chlorphentermine, mazindol, amfepramone (diethylpropion), and fenfluramine.

The amfetamine epidemic of the 1960s and early 1970s has now been superseded by cocaine abuse in the USA and many other Western countries. Realization of the risk of abuse and of dependence has led to the present attitude that there may be only a restricted place for amphetamines in medicine. Perhaps low-dose, short-term use in combating fatigue and altering depressed mood could be justified, but only for specific indications and under continuous medical supervision. However, in the USA there has been a resurgence of metamfetamine abuse on the West coast and in the Southwest and Midwest. This geographical distribution is thought to reflect the traffic from Mexico of ephedrine, a precursor for the synthesis of metamfetamine in the quick-bake method. Because primitive labs can be established in trailers, the spread to adjacent locals has been very rapid, resulting in escalation of migrating epidemics.

Adverse effects of "catecholaminergic stimulants," such as amfetamine and cocaine, fall into several categories, based on dose, time after dose, chronicity of use, and pattern of use/abuse (for example 4–5 day bingeing episodes). Adverse effects include not only responses during the period of use but also intermediate and long-term residual effects after withdrawal. For example, in some abusers once an amfetamine psychosis has developed with chronic abuse, only one or two moderate doses are required to induce the full-blown psychosis in its original form, even long after withdrawal (1). This is also evidenced by the precipitous slide to severe re-addiction in former abusers who are re-introduced to stimulants.

Even with therapeutic use of stimulants, usually in moderate doses, careful monitoring for emergent psychosis, agitation, and abuse is important. Periodic checks for monodelusional syndromes are important with doses in the mid-to-high range (2).

The use of amfetamine-type stimulants for depression, fatigue, and psychasthenia has fallen into disfavor since the early 1970s, because of the potential for abuse and the low rates of success, especially after tolerance is established. However, there have been reports and reviews of successes in carefully selected groups of patients (3–5). The underlying symptoms of patients who respond to stimulants are mild anhedonia, lack of mental and physical energy, easy fatiguability, and low self-esteem, but in the absence of the marked depressed mood disturbance, guilt, and hopelessness that are associated with major depression. Examples include patients with dysthymic disorders, medically ill patients (especially after a stroke), depressed patients, hospitalized cancer patients, and patients with significant cardiovascular disorders, all of whom can have anergia and easy fatiguability. HIV-related neuropsychiatric symptoms, including depression, respond to psychostimulants (4–6). Withdrawn apathetic geriatric patients without major dementia have positive responses (7). General adverse effects, such as tachycardia and agitation, are relatively mild and all reverse on withdrawal (SEDA-17, 1). The combination of stimulants with monoamine oxidase inhibitors in treatment-resistant depressed patients has been reported (8). However, this use should be restricted to patients in whom there is careful monitoring by specialists, because of the potential for hypertensive crisis.

The relative reinforcing effects or abuse potential of these drugs is thought to be related to their potency in releasing dopamine from nerve terminals, compared with serotonin release. Amfetamine, metamfetamine, and phenmetrazine are potent dopamine releasers with high euphoriant and stimulant properties, whereas the compounds with halide substitution in the phenol ring, for example chlorphentermine, are more potent releasers of serotonin and have greater sedative action in anorectic doses. Thus, in summary, those drugs with relatively strong serotonergic to dopaminergic releasing properties seem to provide anorectic effects without euphoria, except at high doses, and might be considered first in any patient who has potential for abuse (2,9,10).

In a study of extended treatment (15 months) of ADHD, amfetamine was clearly superior to placebo in reducing inattention, hyperactivity, and other disruptive behavioral problems. The treatment failure rate was considerably lower and the time to treatment failure was longer in the treated group; adverse effects were few and relatively mild (11).

There is an association between the illicit use of metamfetamine and traumatic accidents. A retrospective review of trauma patients in California showed that metamfetamine rates doubled between 1989 and 1994, while cocaine showed a minimal increase and alcohol a fall. Metamfetamine-positive patients were most likely to be Caucasian or Hispanic and were most commonly injured in motor vehicle collisions. The authors recommended intervention strategies, similar to those used for preventing alcohol consumption and driving, in order to minimize morbidity and mortality (12).

Traumatic shock can be complicated by metamfetamine intoxication (13). Identifying the cause of shock is a key step in the management of patients with severe injuries. This is a challenge, because shock is occasionally caused by more than one mechanism; among the many causes, metabolic derangement attributable to drug abuse should be considered, and masked metabolic acidosis may be a clue to metamfetamine intoxication (14). With the increased emergence of metamfetamine abuse, clinicians should consider it in the differential diagnosis of any patient exhibiting violence, psychosis, seizures, or cardiovascular abnormalities.

Organs and Systems

Cardiovascular

Tachycardia, dysrhythmias, and a rise in blood pressure have been described after the administration of centrally acting sympathomimetic amines. Amfetamine acutely administered to men with a history of amfetamine abuse enhanced the pressor effects of tyramine and noradrenaline, while continuous amfetamine led to tolerance of the pressor response to tyramine. As with intravenous amfetamines, cardiomyopathy, cardiomegaly, and pulmonary edema have been reported with smoking of crystal metamfetamine (15–17).

The cardiovascular response to an oral dose of d-amfetamine 0.5 mg/kg has been determined in 81 subjects with schizophrenia, 8 healthy controls who took amfetamine, and 7 subjects with schizophrenia who took a placebo (18). Blood pressure increased in both amfetamine groups, whereas placebo had no effect. However, pulse rate did not change in the schizophrenic group and only increased after 3 hours in the controls. Intramuscular haloperidol 5 mg produced a more rapid fall in systolic blood pressure in six subjects, compared with 12 subjects who did not receive haloperidol. The authors concluded that increased blood pressure due to amfetamine may have a dopaminergic component. They also suggested that haloperidol may be beneficial in the treatment of hypertensive crises caused by high doses of amfetamine or metamfetamine.

Two cases of myocardial infarction after the use of amfetamine have been reported (19,20).

- A 34-year-old man who smoked a pack of cigarettes a day took amfetamine for mild obesity. He developed an acute myocardial infarction 1 week later. Echocardiography showed inferior left ventricular hypokinesia and a left ventricular ejection fraction of 50%. Coronary cineangiography showed normal coronary arteries but confirmed the inferior left ventricular hypokinesia. Blood and urine toxicology were positive only for amfetamine.

- A 31-year-old man developed generalized discomfort after injecting four doses of amfetamine and metamfetamine over 48 hours, but no chest pain or tightness or shortness of breath. Electrocardiography showed inverted T-waves and left bundle branch block. Echocardiography showed reduced anterior wall motion.

The authors reviewed other reported cases of myocardial infarction associated with amphetamines. The patients were in their mid-thirties and most were men. The interval from the use of amphetamines to the onset of symptoms varied from a few minutes to years. No specific myocardial site was implicated. Coronary angiography in most cases showed non-occlusion. The cause of myocardial ischemia in these cases was uncertain, even though coronary artery spasm followed by thrombus formation was considered the most likely underlying mechanism. Some have suggested that electrocardiographic and biochemical cardiac marker testing should be considered in every patient, with or without symptoms suggesting acute coronary syndrome, after the use of amphetamines. Others have suggested that calcium channel blockers may play an important role in the treatment of myocardial infarction due to amfetamine use or abuse. In one patient, administration of beta-blockers caused anginal pain, suggesting that they should be avoided. All the patients except one had a good outcome.

Coronary artery rupture has been associated with amfetamine abuse (21).

- A 31-year-old woman suddenly developed central chest pain, with a normal electrocardiogram. Changes in troponin and creatine kinase MB were consistent with acute myocardial infarction. Drug screening was positive for amphetamines and barbiturates. Coronary angiography showed an aneurysm with 99% occlusion of the proximal left circumflex coronary artery and extravasation of contrast material. A stent was inserted percutaneously and antegrade flow was achieved without residual stenosis.

An uncommon presentation of amfetamine-related acute myocardial infarction due to coronary artery spasm has been reported (22).

- A 24-year-old man developed an acute myocardial infarction involving the anterior and inferior walls within 3 hours of taking intravenous amfetamine. A coronary angiogram showed plaques in the mid-portion of the left anterior descending artery, which developed spasm after the administration of intracoronary ergonovine. He was discharged after treatment with verapamil, isosorbide mononitrate, and aspirin. He subsequently developed early morning chest tightness 2 weeks, 1 month, 2 months, and 9 months after discharge. On each occasion he left against medical advice.

These findings suggest that coronary artery plaques played a role in endothelial dysfunction resulting from amfetamine use, and that induction of coronary artery spasm, a finding not reported before, was the likely mechanism of amfetamine-related acute myocardial infarction.

During short-term treatment with a modified-release formulation of mixed amfetamine salts in children with ADHD, changes in blood pressure, pulse, and QT_c interval were not statistically significantly different from the changes that were seen in children with ADHD taking placebo (23). Short-term cardiovascular effects were assessed during a 4-week, double-blind, randomized, placebo-controlled, forced-dose titration study with once-daily mixed amfetamine salts 10, 20, and 30 mg (n = 580). Long-term cardiovascular effects were assessed in 568 subjects during a 2-year, open extension study of mixed amfetamine salts 10–30 mg/day. The mean increases in blood pressure after 2 years of treatment (systolic 3.5 mmHg, diastolic 2.6 mmHg) and pulse (3.4/minute) were clinically insignificant. These findings differ from previously reported linear dose-response relations with blood pressure and pulse with immediate-release methylphenidate during short-term treatment (24). These differences may be attributable to differences in timing between dosing and cardiovascular measurements or to differences in formulations. Both amphetamine and methylphenidate have sympathomimetic effects that can lead to increases in systolic blood pressure and diastolic blood pressure at therapeutic doses, although the sizes of the effects on blood pressure may differ (25).

Vertebral artery dissection has been described in a previously healthy man with a 3-year history of daily oral amfetamine abuse (26).

- A healthy 40-year-oldhanded man presented with a 3-day history of an occipital headache and imbalance. He had a 3-year history of daily oral amfetamine abuse with escalating quantities, the last occasion being 12 hours before the onset of the symptoms. He had a history of "speed" abuse and a 20-pack-year history of tobacco use. He had mild right arm dysmetria without ataxia. His brain CT scan without contrast was normal. He then developed nausea, vomiting, visual loss, and progressive obtundation. He had

hypertension (160/90 mmHg), bilateral complete visual loss, right lower facial weakness, mild dysarthria without tongue deviation, divergent gaze attenuated by arousal, bilateral truncal and appendicular dysmetria with inability to stand and walk, and generalized symmetrical hyper-reflexia with extensor plantar reflexes. His urine screen was positive for metamfetamine. A brain MRI scan showed infarction of both medial temporal lobes, the left posteromedial thalamus, and the right superior and left inferior cerebellum. Magnetic resonance angiography and fat saturation MRI showed reduced flow in the left vertebral artery and a ring of increased signal within its lumen, consistent with hematoma and dissection. He was treated with anticoagulants and made a partial recovery.

Since this patient had no known risk factors for vertebral artery dissection and had abused amfetamine daily for 3 years with escalating amounts, an association between metamfetamine and vertebral artery dissection cannot be excluded. The local and systemic vascular impacts of amfetamine could have contributed to initial changes (along with smoking), resulting in dissection.

Of the other central stimulants, aminorex, doxapram, fenfluramine, and fenfluramine plus phentermine can cause chronic pulmonary hypertension, as can chlorphentermine, phentermine, phenmetrazine, and D-norpseudoephedrine (SED-9, 8). A genetic predisposition may be involved (SED-9, 8). Pulmonary hypertension may develop or be diagnosed a long time after the drug has been withdrawn.

Nervous system

Metamfetamine toxicity in infants can mimic scorpion (*Centruroides sculpturatus*) envenomation (27,28). However, the neurotoxic effect of envenomation can be distinguished from amfetamine-induced toxicity by the presence of cholinergic stimulation in scorpion envenomation, producing hypersalivation, bronchospasm, fasciculation of the tongue, purposeless motor agitation, involuntary and conjugate slow and roving eye movements, and often extraocular muscle dysfunction (29). Failure of the antivenin would bring scorpion neurotoxicity into great question (30).

Concentrations of metamfetamine and its metabolite amfetamine were measured in autopsied brain regions of 14 human metamfetamine abusers (31). There was no evidence of variation in the regional distribution of amphetamines in the brain. Post-mortem redistribution of metamfetamine in the heart and lung has been reported before, although peripheral blood concentrations appear to remain constant (32,33).

Stereotyped behavior

A type of automatic behavior, which can continue for hours, has been observed in addicts who inject large doses of central nervous system stimulants. Dyskinesias can occur, with strange facial and tongue movements or jerky motions of the arms and legs and a never-ending repetition of certain actions. Such stereotyped activity is induced in laboratory animals with high doses of amfetamine.

Amphetamines and brain damage

The question of whether amphetamines in large doses can cause permanent brain damage has repeatedly been raised by animal studies (34,35), but definitive studies in man have not been performed. Vasculitis of large elastic vessels, found in chronic animal studies, has been reported to involve the internal carotid artery in man (36); intravenous administration is secondarily implicated. In man and animals, behavioral changes continue for several months after withdrawal of amphetamines; chronic residual changes have been reported mainly in monoaminergic neurons or terminals, either as structural changes or as residual depletion of monoamines and synthesizing enzymes (37). In post-mortem studies (38,39) chronic metamfetamine abusers had significantly lower concentrations of dopamine, tyrosine hydroxylase, and dopamine transporters in the caudate and putamen. It has been suggested that the reduced dopamine concentrations (up to 50% of control), even if not indicative of neurotoxicity, are consistent with amotivational changes reported by metamfetamine abusers after withdrawal (38).

Metamfetamine-induced neurotoxicity in animals, especially involving effects on the mitochondrial membrane potential and electron transport chain and subsequent apoptotic cascade, has been comprehensively reviewed (40). Metamfetamine increases the activity of dopamine, mainly by inhibiting the dopamine transporter. However, this does not explain why psychosis persists even when the metamfetamine is no longer present in the body (41). Chronic metamfetamine use has been reported to reduce dopamine transporter density in the caudate/putamen and nucleus accumbens. However, previous studies have been criticized for not controlling for other drug use.

Dopamine transporter density in the brain has been investigated during a period of abstinence in 11 metamfetamine monodrug users and nine healthy subjects, all men (42). The dopamine transporter density of metamfetamine users was significantly lower in the caudate/putamen, nucleus accumbens, and prefrontal cortex than in the controls. The severity of psychiatric symptoms correlated with the duration of metamfetamine use. The reduction in dopamine transporter density in the caudate/putamen and nucleus accumbens was significantly associated with the duration of metamfetamine use and closely related to the severity of persistent psychiatric symptoms. The reduction in dopamine transporters may be long lasting, even if metamfetamine is withdrawn.

Only some metamfetamine users develop psychosis, not all (41). In laboratory animals, metamfetamine is toxic to dopamine terminals. In 15 subjects (six men and nine women, mean age 32 years), who met the criteria for metamfetamine abuse, and 18 healthy volunteers (12 men and six women), there was a significant reduction in the number of dopamine transporters in detoxified metamfetamine abusers compared with controls (mean values of 28% in the caudate and 21% in the putamen) (43). This was associated with poor motor and memory performance. The reductions in dopamine transporters in the metamfetamine abusers were smaller than those found in patients with Parkinson's disease and occurred in subjects who had been abstinent for 11 months. Since significant reductions in dopamine transporters occur with both age and metamfetamine use, it is possible that metamfetamine will be associated with a higher risk of parkinsonian symptoms in abusers later in life.

Glucose metabolism in the brain has been studied using positron emission tomography after administration of ^{18}F-fluorodeoxyglucose, to look for evidence of functional changes in regions other than those innervated by dopamine neurons in 15 detoxified metamfetamine abusers and 21 controls (44). Whole-brain metabolism in the metamfetamine abusers was 14% higher than in the controls. The difference was largest in the parietal cortex (20%), but there was significantly lower metabolism in the thalamus (17%) and striatum (12% caudate and 6% putamen). The authors suggested that metamfetamine, in doses abused by humans, causes long-lasting metabolic changes in brain regions connected with dopamine pathways, but also in areas that are not innervated by dopamine.

The effects of protracted abstinence on loss of dopamine transporters in the striatum in five metamfetamine abusers have been evaluated during short-term abstinence and then retested during protracted abstinence (12–17 months) (45). The dopamine transporters increased in number, providing hope for effective treatment; however, this regeneration was not sufficient to provide complete functional recovery, as measured by neuropsychological tests.

Chronic amfetamine abusers, chronic opiate abusers, and patients with focal lesions of the orbital prefrontal cortex or dorsal lateral/medial prefrontal cortex were subjected to a computerized decision-making task, in order to compare their capacity for making decisions (46). Chronic amfetamine abusers made suboptimal decisions (correlated with years of abuse) and deliberated significantly longer before making their choices. The opiate abusers had only the second of these behavioral changes. Both the suboptimal choices and the increased deliberation times were also evident in patients with damage to the orbital frontal prefrontal cortex but not other areas. These data are consistent with the hypothesis that chronic amfetamine abusers have similar decision-making deficits to those seen after focal damage to the orbital frontal prefrontal cortex.

The use of proton magnetic resonance scanning (^{1}H MRS) in detecting long-term cerebral metabolite abnormalities in abstinent metamfetamine users has been studied in 26 subjects (13 men) with a history of metamfetamine dependence (mean age 33 years) and 24 healthy subjects with no history of drug dependence (47). The neuronal marker N-acetylaspartate was reduced by 6% in the frontal white matter and by 5% in the basal ganglia of the abstinent metamfetamine users. N-acetylaspartate is a marker for mature neurons, and reduced N-acetylaspartate is thought to indicate reduced neuronal density or neuronal content. According to the authors, these findings suggest neuronal loss or persistent neuronal damage in the absence of significant brain atrophy in metamfetamine users. They speculated that these abnormalities may underlie the persistent abnormal forms of behavior, such as violence, psychosis, and personality changes, seen in some individuals months or even years after their last drug use. Metamfetamine users in the study also had increased concentrations of choline-containing compounds and myoinositol in the frontal gray matter. Myoinositol is a glial cell marker, while the increase in choline-containing compounds reflects increased cell membrane turnover. Thus, these increases in the frontal cortex in drug users may have reflected glial proliferation (astrocytosis). The authors suggested that the finding of reduced N-acetylaspartate accompanied by increased myoinositol, which has been observed in many active brain disorders, indicated glial proliferation in response to neuronal injury. However, they noted that neurotoxicity may not be present in subjects who use amounts of the drugs that are much lower than the amounts used by the chronic abusers they studied. They suggested that future studies should observe whether treatments or long periods of abstinence could reverse these abnormalities.

These findings have given further support to an earlier observation of long-term neurotoxicity associated with MDMA (ecstasy) in animals (SEDA-14, 3). However, it is uncertain whether the reported abnormalities suggestive of neuronal damage are reversible despite continued treatment or beyond 21 months of abstinence.

Dyskinesias

Although controversial, there is a growing consensus that stimulants can provoke, cause, or exacerbate Gilles de la Tourette's syndrome (SEDA-7, 10), based on the observation that stimulants such as the amphetamines, methylphenidate, and pemoline facilitate dopamine retention in the synaptic cleft. There is much evidence that in children with ADHD vulnerable to Tourette's syndrome, stimulants exacerbate motor and phonic tics (48). These studies suggest that Tourette's syndrome and a family history of dyskinesias should be contraindications to stimulant use. However, there is virtually no evidence that stimulants in clinically appropriate doses provoke Tourette's syndrome, and it has been suggested that dyskinesias are a function of high doses (49). Nevertheless, patients taking stimulants should be carefully examined periodically for dyskinesias. It is not known whether structural changes in the central nervous system accompany stimulant-induced dyskinesias.

Stroke

Intracerebral hemorrhage associated with amfetamine has been reported for more than five decades. Eight cases were associated with amfetamine over a period of 3.5 years (50). All had undergone head CT scans and cerebral digital subtraction angiography. Seven had a parenchymal hematoma, three in the frontal lobe and one each in the parietal lobe, frontoparietal region, temporal lobe, and brain stem. One patient had a subarachnoid hemorrhage. The time from exposure to onset of

symptoms ranged from less than 10 minutes to about 2 months (median 1 day). The authors reviewed the literature and found 37 other cases. They observed that young people, mean age 28 years, were at high risk. While most were repeat abusers, one-third claimed to be first time or infrequent users. Intracerebral hemorrhage was seen with all routes of drug use, 57% from oral use, 34% from intravenous use, and 5% after inhalation. Of those who had a CT scan, 84% had a proven intracerebral hemorrhage, three had a subarachnoid hemorrhage, and one had a brainstem hemorrhage. In one patient, with a negative CT scan, the diagnosis of subarachnoid hemorrhage was confirmed by lumbar puncture. In 35 patients who had angiography, 20 were normal or showed only mass effect from a hematoma, 16 had vasculitic beading, and 1 had an arteriovenous malformation. Seven patients died and only 14 had a good recovery.

- A previously healthy 16-year-old schoolboy had mesencephalic ischemia, most probably caused by vasospasm, after combined abuse of amfetamine and cocaine (51). There was a close temporal relation between intake of the drug and the onset of symptoms. Thus, combining these drugs, even in small amounts, may be harmful.

Chorea

Chorea has been attributed to amphetamines.

- A 22-year-old man who had had ADHD since the age of 8 years took methylphenidate, and had an adequate response for 14 years (52). However, his symptoms worsened and he switched from methylphenidate to mixed amfetamine salts 20 mg bd. A month later he continued to have difficulty in focusing on tasks, and the dosage was eventually increased to 45 mg tds over several weeks, with symptomatic improvement. However, 5 days later, he awoke feeling nauseated and agitated and had choreiform movements of his face, trunk, and limbs. He had also taken escitalopram 10 mg/day for anxiety and depression for 2 months before any changes in his ADHD medications. He was treated with intravenous diphenhydramine, lorazepam, and diazepam without improvement in the chorea. Amfetamine was withdrawn and 3 days later his chorea abated. He restarted methylphenidate and the movement disorders did not recur.
- Choreoathetosis worsened in an 8-year-old boy with learning disabilities when he was treated with dexamfetamine, recurred on rechallenge with the same dose, and immediately resolved with diphenhydramine (53).

The authors of the first report speculated that long-term therapy with methylphenidate could have desensitized the patient to the effects of amphetamines, since these drugs act in similar ways. It is also possible that amphetamine therapy interacted with the escitalopram. For this reason, they suggested caution when treating ADHD patients with amphetamines when they are also taking an SSRI.

Psychological, psychiatric

Amphetamines release monoamines from the brain and thereby stimulate noradrenergic, serotonergic, and particularly dopaminergic receptors. Under certain circumstances this leads to psychosis and compulsive behavior, as well as auditory hallucinations similar to those experienced in paranoid schizophrenia. In addition, amphetamines cause an acute toxic psychosis with visual hallucinations, usually after one or two extremely large doses (54).

When an amfetamine is taken, even in a therapeutic dose, most people experience a sensation of enhanced energy or vitality, which, with repetitive administration, follows different patterns. Most often euphoria will develop, usually with a sense of heightened function or perception, and occasionally compulsive behavior as well as hallucinogenic delusions. Dysphoria occurs in some (especially older) individuals. The euphoric effect may enhance craving for amfetamine, and repeated reinforcement can lead to conditioned drug responses, which may facilitate dependence. Progression to severe dependence depends highly on individual vulnerability, the circumstances, the setting, the pattern of use, and especially escalation to high-dose patterns of use. Although most people probably use amphetamines for the original reason they were prescribed, and do not escalate the dosage, a significant proportion do, highlighting the abuse potential. The amphetamines are sometimes used recreationally for years in moderate doses. However, once inhalation and intravenous administration or higher doses are used, a "high-dose transition" into abuse usually occurs, and the capacity for low-dose occasional use is lost, presumably, forever (see the sections on Drug abuse and Drug dependence in this monograph).

Attention

Deficits in attention and motor skills persisted after 1 year of abstinence from stimulant abuse in 50 twin pairs in which only one member had heavy stimulant abuse with cocaine and/or amphetamines (55). Stimulant abusers performed significantly worse on tests of motor skills and attention, and significantly better on one test of visual vigilance. These findings provide evidence of long-term residual effects of stimulant abuse.

Koro

A koro-like syndrome has been related to amfetamine abuse (56).

- A 17-year-old man who had been abusing amfetamine and cannabis for 2 years took amfetamine 1 g orally over the course of an evening and suddenly felt an uncomfortable sensation in his groin and thought that his penis was being sucked into his abdomen. Physical examination was normal. The serum prolactin and bilirubin concentrations were raised. He had normal sexual function, and was able to attain and sustain an erection. He described the phenomenon of penile shrinkage as "Whizz-Dick" and stated that all the amfetamine users with whom he was in contact were

aware of the phenomenon. He was treated with reassurance and supportive counseling.

Reports of koro-like fears of penile shrinkage with amphetamines (57) and cannabis (58,59) are rare. There are no published reports that provide objective evidence that penile shrinkage results from abuse of amphetamines, and the fear is more likely due to altered perception and a poor body image. The authors suggested that it may be an example of an urban myth, a lurid story, or an anecdote based on hearsay and widely circulated as true (Bloor 77).

Memory

Working memory performance may be improved or impaired by amfetamine, depending on dosage and baseline working memory capacity. There was an inverted U-shaped relation between the dose of D-amfetamine and working memory efficiency in 18 healthy people (mean age 24 years, 6 women) who were randomized single-blind to either amfetamine (n = 12) or placebo (n = 6) (60). The primary outcome measures were self-administered questionnaires and blood-oxygenation-level-dependent (BOLD) functional magnetic resonance imaging. Given the overlap between neurochemical systems affected by amfetamine and those disordered in schizophrenia, the effect of amfetamine on working memory in healthy individuals may provide insight into the memory deficits that occur in schizophrenia.

Personality degeneration

In a double-blind, placebo-controlled, short-term study there was significant deterioration of personality in five of 26 children treated with dexamfetamine (61).

Phobias

Social phobia has been attributed to amfetamine (62).

- A 26-year-old woman reported flushing, sweating, palpitation, and shortness of breath, in a range of social situations. She was described as a confident and extroverted woman, with no history of psychiatric problems. She reported daily oral consumption of street amfetamine 1.6 g. At the time of assessment, she had given up her work. Initially, she felt good while taking the drug, but more recently she had been using it to "get going"; there were no symptoms of psychosis or affective disorder.

The authors speculated that dopaminergic dysfunction, reported by some to underlie social phobia, could have resulted in this case from chronic amfetamine-related striatal dopamine depletion.

Psychoses

Psychotic reactions in people taking amphetamines were first reported many years ago and the question was posed whether it was due to the amphetamines or to co-existing and exacerbated paranoid schizophrenia. In one study, most of the psychotic symptoms remitted before the excretion of amines had fallen to its normal basal value

(SED-9, 8). The psychotic syndrome was indistinguishable from paranoid schizophrenia, with short periods of disorientation, and could occur after a single dose (many had taken the equivalent of some 500 mg of amfetamine or metamfetamine orally) with or without simultaneous alcohol, and was most pronounced in addicts (SED-9, 9). Amfetamine psychosis was also seen in 14 people in Australia (1); the predominant hallucinations were visual, which is unusual for schizophrenia (SED-8, 11). Similarly, in contrast to schizophrenia, vision was the primary sensory mode in thinking disorders and body schema distortions in 25 amfetamine addicts (63).

In other studies, volunteers previously dependent on amphetamines were dosed to a level at which amfetamine psychosis was produced, in order to examine the mechanism of action and pharmacokinetics of amfetamine and its possible relation to schizophrenia (64,65). Psychosis was induced by moderately high doses of amfetamine and the psychotic symptoms were often a replication of the chronic amfetamine psychosis, raising the question of whether the establishment of chronic stimulant psychosis leaves residual vulnerability to psychosis precipitated by stimulants. The mechanism might be similar to that which operates in the reverse tolerance that has been seen in experimental animals (66). In some cases an underlying psychosis can be precipitated; an increase in schizophrenic symptoms (SED-8, 12) was observed in 17 actively ill schizophrenic patients after a single injection of amfetamine.

Amfetamine psychosis is relatively rare in children, even in hyperactive children taking large doses of amfetamine; amfetamine psychosis has been reported in an 8-year-old child with a hyperkinetic syndrome (SED-8, 12). Large doses of amfetamine can cause disruption of thinking, but amfetamine psychosis is not usually accompanied by the degree of disorganization normally seen in schizophrenia (SED-9, 8).

Increased sensitivity to stress may be related to spontaneous recurrence of metamfetamine psychosis, triggering flashbacks. Stressful experiences, together with metamfetamine use, induce sensitization to stress associated with noradrenergic hyperactivity, involving increased dopamine release (67,68). This hypothesis has been investigated by determining plasma noradrenaline metabolite concentrations in 26 flashbackers (patients with spontaneous recurrence of metamfetamine psychosis) (11 taking neuroleptic drugs before and during the study and 15 during the course of the study), 18 non-flashbackers with a history of metamfetamine psychosis, 8 with persistent metamfetamine psychosis, and 34 controls (23 metamfetamine users and 11 non-users). The 26 flashbackers had had stressful events and/or metamfetamine-induced, fear-related, psychotic symptoms during previous metamfetamine use. Mild psychosocial stressors then triggered flashbacks. During flashbacks, plasma noradrenaline concentrations increased markedly. Flashbackers with a history of stressful events, whether or not they had had fear-related symptoms, had a further increase in 3-methoxytyramine concentrations. Thus, robust noradrenergic hyperactivity, involving increased dopamine release in

response to mild stress, may predispose to further episodes of flashbacks. The authors pointed out the limitations of their study: (a) plasma noradrenaline concentrations do not accurately reflect central monoamine neurotransmitter function; (b) raised noradrenaline concentrations may reflect heightened autonomic arousal secondary to stress or anxiety; (c) the neuroleptic drugs used may have altered the concentrations of noradrenaline and 3-methoxytyramine; and (d) the study was retrospective and carried out in women in prison.

A paranoid hallucinatory state similar to schizophrenia has been reported in women with a history of metamfetamine abuse in a study of flashbacks in 81 female inmates in Japan (69). Details of symptoms of initial metamfetamine psychosis, stressful experiences, and patterns of abuse were obtained. Plasma monoamine concentrations were also measured during flashback states and in control abusers who had never experienced them. The researchers reported that concreteness of abstract thought and impaired goal-directed thought characteristic of schizophrenia was not usually seen in metamfetamine-induced psychosis. Moreover, it was the use of metamfetamine and not a severe stressor that caused the initial psychotic state, but the flashbacks appeared to be due to mild environmental stressors. The authors described this pattern as "spontaneous psychosis due to previous metamfetamine psychosis." They also observed that plasma concentrations of noradrenaline were significantly higher in women with flashbacks both during flashbacks and during remissions. This suggests a possible role of noradrenergic hyperactivity in sensitivity to mild stress and susceptibility to flashbacks. Furthermore, these noradrenergic findings could be used to predict relapse to a paranoid hallucinatory state in schizophrenia.

Chlorpromazine has been used to treat amfetamine psychosis due to acute poisoning in children who did not respond to barbiturates (70).

Management

Reports have suggested that atypical antipsychotic drugs, such as risperidone (71–73) and olanzapine (74), can be effective in the treatment of acute and residual metamfetamine-induced psychosis. Moreover, adherence to olanzapine for about 8 weeks also effectively controlled cravings for metamfetamine. Rigorous controlled studies are needed to establish the therapeutic efficacy of atypical antipsychotic drugs in the treatment of the psychosis and cravings of metamfetamine addiction.

- A 76-year-old woman, who had taken dexamfetamine since the age of 28 years for narcolepsy, developed an acute schizophreniform psychosis with paranoid delusions and auditory hallucinations. She was initially treated with sulpiride while continuing to take dexamfetamine. Five months later, sulpiride was withdrawn, and her psychotic symptoms recurred. She was given risperidone 3 mg/day and continued to take dexamfetamine 15 mg/day.
- A 24-year-old man, with a history of intravenous metamfetamine abuse since the age of 19 years,

developed psychotic symptoms characterized by auditory hallucinations and persecutory delusions. He had no insight and was given haloperidol and levomepromazine. His symptoms disappeared after 4 months of treatment and he then stopped using metamfetamine. A year and a half later, "odd ideas" recurred, and he became anxious but had insight. He was treated with bromperidol 9 mg/day for 6 months, but the odd ideas persisted. On referral, he was found to fulfil the criteria for obsessive-compulsive disorder according to DSM-IV. No abused substances, including metamfetamine, were identified in his urine. Risperidone 2 mg/day was started and then increased to 5 mg/day. After 3 weeks, the intrusive thoughts and symptoms of "anxious-restless state" gradually subsided and eventually disappeared. His symptoms recurred within a week of stopping risperidone and resolved on reintroduction.

Metabolism

Amphetamines can cause retardation of growth (height and weight) in hyperactive children (SED-9, 9).

Hematologic

Acute myeloblastic leukemia occurred in a 24-year-old man who had taken massive doses of amfetamine for more than 2 years (SED-8, 13).

Teeth

Dental wear has been evaluated prospectively in metamfetamine users at an urban university hospital (75). Information was collected from 43 patients (26 men, 40 tobacco smokers), mean age 39 years, who admitted to having used metamfetamine for more than 1 year. Patients who regularly snorted metamfetamine had higher "tooth-wear" scores for anterior maxillary teeth than patients who injected, smoked, or ingested metamfetamine. The authors suggested that the anatomy of the blood supply to this area possibly explained the association of the regional differences in tooth wear with snorted metamfetamine. The anterior maxillary teeth and the nasal mucosa have a common blood supply. Thus, snorting may cause vasoconstriction, impairing the blood supply both to the nasal mucosa as well as the teeth.

Urinary tract

Acute transient urinary retention associated with metamfetamine and ecstasy (3,4 methylenedioxymetamfetamine, MDMA) in an 18-year-old man has been described (76). Analysis by gas chromatography–mass spectrometry confirmed the presence of metamfetamine (>25 µg/ml), MDMA (> 5 µg/ml), amfetamine (1.4 µg/ml), and methylenedioxyamfetamine (3.7 µg/ml) in the urine. Bladder dysfunction resulting from alpha-adrenergic stimulation of the bladder neck may have explained the observed effect.

Skin

The severe form of erythema multiforme known as toxic epidermal necrolysis has been attributed to a mixture of dexamfetamine and ephedrine (77).

- A 27-year-old woman developed peripheral target plaques, papules, blisters, and lip erosions, consistent with erythema multiforme, 9 days after using "speed" (dexamfetamine and ephedrine), and 3 days later developed widespread lesions with large areas of blistering affecting 40% of her body surface area. She was given intravenous ciclosporin and improved within 24 hours.

Musculoskeletal

There may be an association between metamfetamine abuse and rhabdomyolysis. In a retrospective review of 367 patients with rhabdomyolysis, 166 were positive for metamfetamine (78). They had higher mean initial and lower mean peak activities of creatine phosphokinase. There was no significant difference in the incidence of acute renal insufficiency. The authors suggested screening all patients with rhabdomyolysis of unclear cause for metamfetamine and measuring creatine phosphokinase activity.

Sexual function

Reports of the effects of amfetamine on sexual behavior refer variously to unchanged, reduced, mixed, and heightened sexual performance, but long-term abusers often have sexual dysfunction (SED-9, 9).

Immunologic

An anaphylactic reaction after the injection of crushed tablets equivalent to 45 mg of amfetamine occurred in a young woman; in others injected with the same solution and at the same time there were no adverse effects (SED-9, 8). The reaction may have involved amfetamine or excipients. Scleroderma is a potential consequence of various stimulants used for appetite control (79).

Infection risk

- A 34-year-old woman who had taken intranasal metamfetamine weekly for 15 years developed osteomyelitis of the frontal bone and a subperiosteal abscess. The authors proposed that this was due to chronic abuse of metamfetamine (80).

A rare case of Pott's puffy tumor, anterior extension of a frontal sinus infection that results in frontal bone osteomyelitis and subperiosteal abscess, has been associated with metamfetamine use (81).

- A 34-year-old woman presented with a 9-day history of fever, chills, photophobia, and neck pain. Nine months earlier, she had developed a swelling on her forehead, which enlarged and spontaneously drained pus. Over the next weeks, a fistula developed at the site of the swelling, accompanied by an intermittent bloody purulent drainage that lasted for about 9 months. She had either inhaled metamfetamine or had used it intranasally weekly for 15 years and reported continued use immediately before the development of the forehead lesion. She had a sinocutaneous fistula in the midline of the forehead, with seropurulent discharge but no local erythema or tenderness. A CT scan of the head showed complete opacification of all sinuses, with a 1 cm connection between the anterior frontal sinus and the skin. Cultures grew *Streptococcus milleri* and *Candida albicans*. She responded to extensive medical and surgical treatment.

The authors proposed that intranasal metamfetamine had contributed to chronic sinus inflammation and subsequent complications. Furthermore, the vasoconstriction induced by metamfetamine in the mucosal vessels may have resulted in ischemic injury to the sinus mucosa, providing an environment conducive to bacterial growth.

Death

There has been a retrospective investigation of metamfetamine-related fatalities during a 5-year period (1994–1998) in Southern Osaka city in Japan (82). Among 646 autopsy cases, methamphetamine was detected in 15, most of whom were men in their late thirties. The cause and manner of death were methamphetamine poisoning ($n = 4$), homicide ($n = 4$), accidental falls and aspiration from drug abuse ($n = 4$), death in an accidental fire, myocardial infarction, and cerebral hemorrhage (one each). Blood metamfetamine concentrations were 23–170 μmol/l in fatal poisoning, 4.4–38 μmol/l in deaths from other extrinsic causes, and 14–22 μmol/l in cardiovascular and cerebrovascular accidents. The common complications were cardiomyopathy, cerebral perivasculitis, and liver cirrhosis/interstitial hepatitis. The general profile of patients reported in this series compares with that of a previous study from Taiwan (SEDA-24, 2).

Long-Term Effects

Drug abuse

The most important problem encountered with amphetamines is abuse and the development of dependence. The most rapid amfetamine epidemic occurred in Japan after World War II, where there had been little or no previous abuse (83). Although a high proportion of amfetamine users probably already have emotional and social difficulties, sustained abuse can result in serious psychiatric complications, ranging from severe personality disorders to chronic psychoses (84,85). Whereas signs of intense physical dependence are not thought to occur (SED-9, 9), withdrawal may be associated with intense depression (SED-9, 9) (86), and relapses in psychiatric disorders have often been noted. Some countries in which the problem became widespread banned amphetamines, and Australia restricted their use to narcolepsy and behavioral disorders in children. Amfetamine dependence developed into a serious problem in the USA (and to a lesser extent in the UK), where it followed the typical pattern of drug dependence (SED-9, 7, 10).

Continuing critical re-assessment of the usefulness versus the harmfulness of amphetamines has led to further restrictions in their use (SED-9, 10). They have been subjected to rigid legislative control in many countries, accompanied by recommendations that they should not be prescribed. The World Health Organization and the United Nations have also stressed the need for strict control of amphetamines (SED-9, 10) (87).

There is a high prevalence of the use of amfetamine-like drugs in Brazil, particularly among women, owing to the "culture of slimness as a symbol of beauty" (88). Of 2370 subjects in São Paulo and Brasilia, 72% had already undergone from one to more than 10 previous courses of treatment, usually with amfetamine-like anorectic drugs. Over half of them had taken amfetamine-like drugs in compound formulations containing four or more substances, such as benzodiazepines and/or laxatives, diuretics, and thyroid hormones. There were adverse reactions to the amfetamine-like drugs in 86% and 37% sought medical advice; 3.9% required hospitalization. The authors argued the need for more rigorous legislation and enforcement strategies to stop such misuse of drugs.

Further evidence concerning increased metamfetamine abuse has come from Taiwan (89). Between 1991 and 1996, of 3958 deaths with autopsies, 244 were related to metamfetamine (mean age 31 years, 73% men). The manner of death was natural (13%), accidental (59%), suicidal (11%), homicidal (14%), or uncertain (3%). Owing to the endemic problem and public hazard created by illicit metamfetamine abuse, the authors urged stronger anti-drug programs.

There was a high frequency of amphetamine abuse and withdrawal among patients from the Thai–Myanmar border area admitted to hospital with *Plasmodium falciparum* malaria (90). This co-morbidity can cause diagnostic confusion, alter malaria pathophysiology, and lead to drug interactions. Considering the potential neuropsychiatric adverse effects of mefloquine, an important component of current antimalarial treatment in Southeast Asia, it should be avoided in patients who abuse amphetamines.

Drug dependence

The role of dopamine in the addictive process has been explored (91). The authors raised the possibility that the orbitoprefrontal cortex is linked to compulsive drug abuse. They recruited 15 metamfetamine users and 20 non-drug user controls. The metamfetamine abusers had significantly fewer dopamine D_2 receptors than the controls. There was an association between lower numbers of dopamine D_2 receptors and metabolism in the orbitofrontal cortex in the metamfetamine users. These findings are similar to those observed in cocaine, alcohol, and heroin users. The authors suggested that D_2 receptor-mediated dysregulation of the orbitofrontal cortex could be a common mechanism underlying loss of control and compulsive abuse of drugs.

Second-Generation Effects

Pregnancy

In pregnant women who reported for prenatal care between 1959 and 1966 there was an excess of oral clefts in the offspring of mothers who had taken amphetamines in the first 56 days from their last menstrual period, but this was considered to be either a chance finding or one element in a multifactorial situation (SED-9, 9).

Teratogenicity

The possible neurotoxic effect of prenatal metamfetamine exposure on the developing brain has been studied using ^{1}H magnetic resonance spectroscopy in 12 metamfetamine-exposed children and 14 age-matched unexposed controls (92). There was an increased creatinine concentration in the striatum, with relatively normal concentrations of *N*-acetyl-containing compounds in children exposed to metamfetamine. These findings suggest that exposure to metamfetamine in utero causes abnormal energy metabolism in the brain of children. However, there were no differences in reported behavioral problems among metamfetamine-exposed children compared with controls.

Susceptibility Factors

Genetic

A study in 93 unrelated metamfetamine-dependent subjects and 131 controls did not prove any association between metamfetamine dependence in Caucasians of Czech origin and TaqI A polymorphism of the DRD2 gene, I/D polymorphism of the ACE gene, or M235T polymorphism of the AGT gene (93).

A genetic explanation for individual differences in the response of the brain to amphetamines has been suggested. There are numerous proteins involved in regulating synaptic dopamine activity; catechol-O-methyl transferase, which inactivates released dopamine, appears to play a unique role in regulating dopamine flux in the prefrontal cortex because of the low abundance and minimal role of dopamine transporters (94,95,96). Neuroimaging and genetic analysis in 27 healthy volunteers showed that individuals with the met/met catechol-o-methyl transferase genotype were at risk of an adverse response (related to prefrontal cortex information processing) to amfetamine (97). In populations of European ancestry, individuals with met/met genotypes constitute about 15–20% of the population (98).

Age

Special care should be taken when using amphetamines in elderly patients, in view of the likelihood of stimulation of adrenoceptors and in particular of cardiovascular and respiratory function. Periodic users especially need to be wary of acute use under circumstances of exercise and environmental heat, owing to the risk of heat stroke.

Other features of the patient

The existence of a previous neurological disorder may be a risk factor for treatment-resistant psychosis in metamfetamine abusers (99). It is of particular interest that most of these patients sustained their disorder during childhood. It is not uncommon for metamfetamine patients to continue with psychotic symptoms despite extended periods of abstinence (100). These patients often are labeled as being schizophrenic. This study has shown the importance of considering a history of neurological disorders, especially during childhood, in such patients.

Drug Administration

Drug administration route

Injection as a method of delivery of illicit drugs carries its own special risks. Metamfetamine-dependent subjects ($n = 427$) participated in a study to detect differences between injecting metamfetamine users (13%) and non-injecting users (87%) (101). The patients entered treatment at a center in California between 1988 and 1995. Injectors reported significantly more years of heavy use. Psychological problems were more common in the injectors, more of whom reported depression, suicidal ideation, hallucinations, and episodes of feeling that their body parts "disconnect and leave." Moreover, injectors reported more problems concerning sexual functioning and more episodes of loss of consciousness. The injectors were more commonly HIV-positive and they had more felony convictions and were on parole more often than other users. Although individuals who inject metamfetamine use it more often than non-injectors, the number of grams used per week did not differ between the groups. Thus, injectors use a smaller amount of drug per dose than non-injectors. Eighty percent of the injectors were unemployed, possibly reflecting the extent of impairment related to addiction in this group. The injectors, who had more psychiatric and medical morbidity, warrant special attention and carefully designed treatment plans.

Drug overdose

Overdosage of amphetamines can cause restlessness, dizziness, tremor, increased reflexes, talkativeness, tenseness, irritability, and insomnia; less common effects include euphoria, confusion, anxiety, delirium, hallucinations, panic states, suicidal and homicidal tendencies, excessive sweating, dry mouth, metallic taste, anorexia, nausea, vomiting, diarrhea, and abdominal cramps. Fatal poisoning is usually associated with hyperpyrexia, convulsions, coma, or cerebral hemorrhage. In addition, peripheral excitation of smooth muscle or blood vessels supplying skeletal muscle has been described. Excitatory actions can cause increased heart rate, palpitation, dysrhythmias, and metabolic effects, such as glycogenolysis in liver and adipose tissue.

The problem of fatal overdose is central to the problem of frequently repeated intravenous high-dose abuse of stimulants of unknown quality and quantity. Fatal overdose is less frequent among experienced chronic users than in naive or episodic high-dose users (102–104), in part because of the establishment of tolerance to hyperpyrexia and hypertension. Fatal hypertension can be potentiated by high ambient temperatures and vigorous exercise, as in the use of these drugs by athletes. Rare individuals have used up to 1–3 g/day of oral amfetamine for many years without problems of overdosing, yet acute toxic overdoses have been reported at 100–200 mg (66). Hyperpyrexia, seizures, hypertensive cerebrovascular hemorrhage, ventricular fibrillation, left ventricular failure, and complications of intravenous drug abuse have all been reported as causes of death (105–106). An autopsy study of amfetamine abusers in San Francisco showed that 54% died of drug toxicity, 10% of accidental trauma, 12% by suicide, and 10% by homicide (107).

Two deaths from metamfetamine overdose in drug dealers have been reported from Thailand, which has experienced an increase in metamfetamine abuse (108).

- A 43-year-old male drug dealer swallowed a number of metamfetamine tablets at the time of his arrest. When seen in the emergency room, he was comatose with reactive pupils. He died 6 hours after consuming the tablets. The autopsy findings were non-specific.
- Another 33-year-old female drug dealer, while at the police station, swallowed a number of metamfetamine pills that had been hidden in her undergarments. At the hospital, a gastric lavage was done but she died 10 hours after ingestion.

As described in these cases, there may be an increased risk of death in drug dealers who, in attempting to prevent arrest, may consume toxic amounts without anticipating the consequences.

Drug–Drug Interactions

Adrenergic neuron blocking drugs

Amphetamines and other stimulatory anorectic agents, apart from fenfluramine, would be expected to impair the hypotensive effects of adrenergic neuron blocking drugs such as guanethidine. Not only do they release noradrenaline from stores in adrenergic neurons and block the reuptake of released noradrenaline into the neuron, but they also impair re-entry of the antihypertensive drugs (109).

Alcohol

Alcohol increases blood concentrations of amphetamines (SED-9, 9).

Barbiturates

Barbiturates can enhance amfetamine hyperactivity (SED-8, 9).

Benzodiazepines

Benzodiazepines can enhance amfetamine hyperactivity (SED-8, 9).

Estradiol

Preclinical studies (as well as anecdotal clinical reports) have shown that estrogens, through effects on the central nervous system, can influence behavioral responses to psychoactive drugs. In an unusual crossover study, the subjective and physiological effects of oral D-amfetamine 10 mg were assessed after pretreatment with estradiol (110). One group of healthy young women used estradiol patches (Estraderm TTS, total dose 0.8 mg), which raised plasma estradiol concentrations to about 750 pg/ml, and a control group used placebo patches. Most of the subjective and physiological effects of amfetamine were not affected by acute estradiol treatment, but the estrogen did increase the magnitude of the effect of amfetamine on subjective ratings of "pleasant stimulation" and reduced ratings of "want more." Estradiol also produced some subjective effects when used alone, raising ratings of "feel drug," "energy and intellectual efficiency," and "pleasant stimulation." Some limitations of the study were:

(a) plasma amfetamine concentrations were not measured, so an effect of estradiol on the pharmacokinetics of amfetamine cannot be ruled out;
(b) only single doses of amfetamine and estradiol were tested;
(c) the dose of amfetamine was relatively low and that of estradiol relatively high, maximizing the chances of detecting estradiol-dependent increases in two subjective effects of amfetamine.

Monoamine oxidase inhibitors

The amphetamines should not be used together with or within 14 days of any monoamine oxidase inhibitors; severe hypertensive reactions and on occasion confusional states (for example with fenfluramine) can occur (SED-9, 9).

Mood-stabilizing drugs

Amfetamine reduces regional brain activation during the performance of several cognitive tasks (111). The results of a double-blind, placebo-controlled study in healthy volunteers suggested that both lithium and valproate can significantly attenuate dexamfetamine-induced changes in brain activity in a task-dependent and region-specific manner (112). There is also good evidence that dexamfetamine stimulates the phosphatidylinositol (PI) cycle in vivo (113) and in vitro (114), and this may be the mechanism responsible for its effect on brain activation; both lithium and valproate can attenuate the PI cycle, probably through different mechanisms (115).

Ritonavir

A fatal interaction between ritonavir and metamfetamine has been described (116).

- A 49-year-old HIV-positive Caucasian man had taken ritonavir (400 mg bd), saquinavir (400 mg bd), and stavudine (40 mg bd) for 4 months. His CD4 cell count was 617×10^6 cells/l and HIV-1 RNA less than 400 copies/ml. He had previously taken zidovudine for 7 months. He self-injected twice with metamfetamine and sniffed amyl nitrite and was found dead a few hours later. At autopsy, there was no obvious cause of death. Metamfetamine was detected in the bile (0.5 mg/l) and cannabinoids and traces of benzodiazepines were detected in the blood.

Nitric oxide formed from amyl nitrite inhibits cytochrome P450 (117) and ritonavir inhibits CYP2D6 (118), which has a major role in metamfetamine detoxification (119). This interaction could have led to fatal plasma concentrations of metamfetamine. It is therefore suggested that patients who take protease inhibitors are made aware of the potential risk of using any form of recreational drugs metabolized by CYP2D6, particularly metamfetamine.

SSRIs

A man taking long-term dexamfetamine had two episodes of serotonin syndrome while taking first venlafaxine and later citalopram (120).

- A 32-year-old man, who was taking dexamfetamine 5 mg tds for adult ADHD, developed marked agitation, anxiety, shivering, and tremor after taking venlafaxine for 2 weeks (75 mg/day increased after a week to 150 mg/day). His heart rate was 140/minute, blood pressure 142/93 mmHg, and temperature 37.3°C. His pupils were dilated but reactive. There was generalized hypertonia, hyper-reflexia, and frequent myoclonic jerking. Dexamfetamine and venlafaxine were withdrawn and cyproheptadine (in doses of 8 mg up to a total of 32 mg over 3 hours) was given. His symptoms completely resolved within a few hours.

Dexamfetamine was restarted 3 days later and citalopram was started a few days later. Two weeks later he reported similar symptoms and stopped taking citalopram. He was successfully treated again with cyproheptadine.

Triazolam

In 20 healthy adults who received a placebo, triazolam 0.25 mg/70 kg, amfetamine sulfate 20 mg/70 kg, and a combination of triazolam and amfetamine in a double-blind, crossover study the results supported the conclusion that triazolam-induced impairment of free recall is related to its sedative effects, whereas recognition, memory, and recall differ with respect to the contribution of sedation to the amnesic effect of triazolam (121). Thus, benzodiazepines have specific effects on memory that are not merely a by-product of their sedative effects, and the degree to which sedative effects contribute to their amnesic effects can vary as a function of the particular memory process being assessed. It is important to note that the generalizability of the conclusions of this study is limited by use of a single dose design for both drugs.

Tricyclic antidepressants

Tricyclic antidepressants increase blood concentrations of amfetamine (122,123).

Management of adverse effects

Potential benefit of amfebutamone

It has been suggested, based on few case reports (124), that amfebutamone may be of help in weaning people from amfetamine abuse.

- A 53-year-old woman with a 30-year history of amfetamine abuse gave herself amfebutamone (diethylpropion); this resulted in rapid and successful cessation of amfetamine abuse (125).

Given the importance of craving, withdrawal symptoms, and maintenance treatment in the withdrawal process, the effect of amfebutamone on these processes needs to be systematically evaluated.

References

1. Bell DS. The experimental reproduction of amphetamine psychosis. Arch Gen Psychiatry 1973;29(1):35–40.
2. Ellinwood EH Jr. Emergency treatment of acute adverse reactions to CNS stimulants. In: Bourne P, editor. Acute Drug Abuse Emergencies: A Treatment Manual. New York: Academic Press, 1976:115.
3. Fawcett JF, Busch KA. Stimulants in psychiatry. In: Schatzberg AF, Nemeroff CB, editors. The American Psychiatric Press Textbook of Psychopharmacology. Washington, DC: American Psychiatric Press, 1995:417.
4. Angrist B, d'Hollosy M, Sanfilipo M, Satriano J, Diamond G, Simberkoff M, Weinreb H. Central nervous system stimulants as symptomatic treatments for AIDS-related neuropsychiatric impairment. J Clin Psychopharmacol 1992;12(4):268–72.
5. Satel SL, Nelson JC. Stimulants in the treatment of depression: a critical overview. J Clin Psychiatry 1989;50(7):241–9.
6. Holmes VF, Fernandez F, Levy JK. Psychostimulant response in AIDS-related complex patients. J Clin Psychiatry 1989;50(1):5–8.
7. Chiarello RJ, Cole JO. The use of psychostimulants in general psychiatry. A reconsideration. Arch Gen Psychiatry 1987;44(3):286–95.
8. Fawcett J, Kravitz HM, Zajecka JM, Schaff MR. CNS stimulant potentiation of monoamine oxidase inhibitors in treatment-refractory depression. J Clin Psychopharmacol 1991;11(2):127–32.
9. Jonsson S, O'Meara M, Young JB. Acute cocaine poisoning. Importance of treating seizures and acidosis. Am J Med 1983;75(6):1061–4.
10. Barinerd H, Krupp M, Chatton J, et al. Current Medical Diagnosis and TreatmentLos Altos CA: Lange Medical Publishers;. 1970.
11. Gillberg C, Melander H, von Knorring AL, Janols LO, Thernlund G, Hagglof B, Eidevall-Wallin L,

Gustafsson P, Kopp S. Long-term stimulant treatment of children with attention-deficit hyperactivity disorder symptoms. A randomized, double-blind, placebo-controlled trial. Arch Gen Psychiatry 1997;54(9):857–64.

12. Pacifici R, Zuccaro P, Farre M, Pichini S, Di Carlo S, Roset PN, Ortuno J, Segura J, de la Torre R. Immunomodulating properties of MDMA alone and in combination with alcohol: a pilot study. Life Sci 1999;65(26):PL309–16.

13. Schneider HJ, Jha S, Burnand KG. Progressive arteritis associated with cannabis use. Eur J Vasc Endovasc Surg 1999;18(4):366–7.

14. Stracciari A, Guarino M, Crespi C, Pazzaglia P. Transient amnesia triggered by acute marijuana intoxication. Eur J Neurol 1999;6(4):521–3.

15. Karch SB, Billingham ME. The pathology and etiology of cocaine-induced heart disease. Arch Pathol Lab Med 1988;112(3):225–30.

16. Ellenhorn DJ, Barceloux DG. Amphetamines. In: Medical Toxicology: Diagnosis and Treatment of Human Poisoning. New York: Elsevier Science Publishers, 1988:625.

17. Call TD, Hartneck J, Dickinson WA, Hartman CW, Bartel AG. Acute cardiomyopathy secondary to intravenous amphetamine abuse. Ann Intern Med 1982;97(4):559–60.

18. Angrist B, Sanfilipo M, Wolkin A. Cardiovascular effects of 0.5 milligrams per kilogram oral d-amphetamine and possible attenuation by haloperidol Clin Neuropharmacol 2001;24(3):139–44.

19. Waksman J, Taylor RN Jr, Bodor GS, Daly FF, Jolliff HA, Dart RC. Acute myocardial infarction associated with amphetamine use. Mayo Clin Proc 2001;76(3):323–6.

20. Costa GM, Pizzi C, Bresciani B, Tumscitz C, Gentile M, Bugiardini R. Acute myocardial infarction caused by amphetamines: a case report and review of the literature. Ital Heart J 2001;2(6):478–80.

21. Brennan K, Shurmur S, Elhendy A. Coronary artery rupture associated with amphetamine abuse. Cardiol Rev 2004;12:282-3.

22. Hung MJ, Kuo LT, Cherng WJ. Amphetamine-related acute myocardial infarction due to coronary artery spasm. Int J Clin Pract 2003;57:62-4.

23. Findling RL, Biederman J, Wilens TE, Spencer TJ, McGrough JJ, Lopez FA, Tulloch SJ, on behalf of the SL1381.301 and .302 Study Groups. Short- and long-term cardiovascular effects of mixed amphetamine salts extended release in children. J Pediatr 2005;147:348-54.

24. Findling RL, Short EJ, Manos MJ. Short-term cardiovascular effects of methylphenidate and adderall. J Am Acad Child Adolescent Psychiatry 2001;40:525-9.

25. Gutgesell H, Atkins D, Barst R, Buck M, Franklin W, Humes R, Ringel R, Shaddy R, Taubert KA. AHA scientific statement. Cardiovascular monitoring of children and adolescents receiving psychotropic drugs. J Am Acad Child Adolesc Psychiatry 1999;38:1047-50.

26. Zaidat OO, Frank J. Vertebral artery dissection with amphetamine abuse. J Stroke Cerebrovasc Dis 2001;10:27–9.

27. Sewell RA, Cozzi NV. More about parkinsonism after taking ecstasy. N Engl J Med 1999;341(18):1400.

28. Baggott M, Mendelson J, Jones R. More about parkinsonism after taking ecstasy. N Engl J Med 1999;341(18):1400–1.

29. Mintzer S, Hickenbottom S, Gilman S. More about parkinsonism after taking ecstasy. N Engl J Med 1999;341:1401.

30. Borg GJ. More about parkinsonism after taking ecstasy. N Engl J Med 1999;341(18):1400.

31. Kalasinsky KS, Bosy TZ, Schmunk GA, Reiber G, Anthony RM, Furukawa Y, Guttman M, Kish SJ. Regional distribution of methamphetamine in autopsied brain of chronic human methamphetamine users. Forensic Sci Int 2001;116(2–3):163–9.

32. Barnhart FE, Fogacci JR, Reed DW. Methamphetamine—a study of postmortem redistribution. J Anal Toxicol 1999;23(1):69–70.

33. Moriya F, Hashimoto Y. Redistribution of methamphetamine in the early postmortem period. J Anal Toxicol 2000;24(2):153–5.

34. Escalante OD, Ellinwood EH Jr. Central nervous system cytopathological changes in cats with chronic methedrine intoxication. Brain Res 1970;21(1):151–5.

35. Wagner GC, Ricaurte GA, Seiden LS, Wagner GC, Ricaurte GA, Seiden LS, Schuster CR, Miller RJ, Westley J. Long-lasting depletions of striatal dopamine and loss of dopamine uptake sites following repeated administration of methamphetamine. Brain Res 1980;181(1):151–60.

36. Bostwick DG. Amphetamine-induced cerebral vasculitis. Hum Pathol 1981;12(11):1031–3.

37. Napiorkowski B, Lester BM, Freier MC, Brunner S, Dietz L, Nadra A, Oh W. Effects of in utero substance exposure on infant neurobehavior. Pediatrics 1996;98(1):71–5.

38. Wilson JM, Kalasinsky KS, Levey AI, Bergeron C, Reiber G, Anthony RM, Schmunk GA, Shannak K, Haycock JW, Kish SJ. Striatal dopamine nerve terminal markers in human, chronic methamphetamine users. Nat Med 1996;2(6):699–703.

39. McCann UD, Wong DF, Yokoi F, Villemagne V, Dannals RF, Ricaurte GA. Reduced striatal dopamine transporter density in abstinent methamphetamine and methcathinone users: evidence from positron emission tomography studies with (^{11}C)WIN-35,428. J Neurosci 1998;18(20):8417–22.

40. Davidson C, Gow AJ, Lee TH, Ellinwood EH. Methamphetamine neurotoxicity: necrotic and apoptotic mechanisms and relevance to human abuse and treatment. Brain Res Brain Res Rev 2001;36(1):1–22.

41. Volkow ND. Drug abuse and mental illness: progress in understanding comorbidity. Am J Psychiatry 2001;158(8):1181–3.

42. Sekine Y, Iyo M, Ouchi Y, Matsunaga T, Tsukada H, Okada H, Yoshikawa E, Futatsubashi M, Takei N, Mori N. Methamphetamine-related psychiatric symptoms and reduced brain dopamine transporters studied with PET. Am J Psychiatry 2001;158(8):1206–14.

43. Volkow ND, Chang L, Wang GJ, Fowler JS, Leonido-Yee M, Franceschi D, Sedler MJ, Gatley SJ, Hitzemann R, Ding YS, Logan J, Wong C, Miller EN. Association of dopamine transporter reduction with psychomotor impairment in methamphetamine abusers. Am J Psychiatry 2001;158(3):377–82.

44. Volkow ND, Chang L, Wang GJ, Fowler JS, Franceschi D, Sedler MJ, Gatley SJ, Hitzemann R, Ding YS, Wong C, Logan J. Higher cortical and lower subcortical metabolism in detoxified methamphetamine abusers. Am J Psychiatry 2001;158(3):383–9.

45. Volkow ND, Chang L, Wang GJ, Fowler JS, Franceschi D, Sedler M, Gatley SJ, Miller E, Hitzemann R, Ding YS, Logan J. Loss of dopamine transporters in methamphetamine abusers recovers with protracted abstinence. J Neurosci 2001;21(23):9414–8.

46. Rogers RD, Everitt BJ, Baldacchino A, Blackshaw AJ, Swainson R, Wynne K, Baker NB, Hunter J, Carthy T, Booker E, London M, Deakin JF, Sahakian BJ, Robbins TW. Dissociable deficits in the decision-making cognition of chronic amphetamine abusers, opiate abusers, patients with focal damage to prefrontal cortex, and tryptophan-depleted normal volunteers: evidence for monoaminergic mechanisms. Neuropsychopharmacology 1999;20(4):322–39.

47. Ernst T, Chang L, Leonido-Yee M, Speck O. Evidence for long-term neurotoxicity associated with methamphetamine abuse: a ^{1}H MRS study. Neurology 2000;54(6):1344–9.

48. Lowe TL, Cohen DJ, Detlor J, Kremenitzer MW, Shaywitz BA. Stimulant medications precipitate Tourette's syndrome. JAMA 1982;247(12):1729–31.

49. Shapiro AK, Shapiro E. Do stimulants provoke, cause, or exacerbate tics and Tourette syndrome? Compr Psychiatry 1981;22(3):265–73.

50. Buxton N, McConachie NS. Amphetamine abuse and intracranial haemorrhage. J R Soc Med 2000;93(9):472–7.

51. Strupp M, Hamann GF, Brandt T. Combined amphetamine and cocaine abuse caused mesencephalic ischemia in a 16-year-old boy—due to vasospasm? Eur Neurol 2000;43(3):181–2.

52. Morgan JC, Christopher-Winter W, Wooten GF. Amphetamine-induced chorea in attention-deficit hyperactivity disorder. Mov Disord 2004;19:840-2.

53. Mattson RH, Calverley JR. Dextroamphetamine-sulfate-induced dyskinesias. JAMA 1986;204:108-10.

54. Kramer JC, Fischman VS, Littlefield DC. Amphetamine abuse. Pattern and effects of high doses taken intravenously. JAMA 1967;201(5):305–9.

55. Toomey R, Lyons MJ, Eise SA, Xian H, Chantarujikapong S, Seidman LJ, Faraone SV, Tsuang MT. A twin study of the neuropsychological consequences of stimulant abuse. Arch Gen Psychiatry 2003;60:303-10.

56. Bloor RN. Whizz-Dick: side effect, urban myth or amphetamine-related koro-like syndrome. Int J Clin Pract 2004;58:717-9.

57. Yapp P, Koro A. Culture-bound depersonalization syndrome. Br J Psychiatry 1965;111:43-50.

58. Chowdhury AN, Bera NK. Koro following cannabis smoking: two case reports. Addiction 1994;89:1017-20.

59. Earleywine M. Cannabis-induced koro in Americans. Addiction 2001;96:1663-6.

60. Tipper CM, Cairo TA, Woodward TS, Phillips AG, Liddle PF, Nagan ETC. Processing efficiency of a verbal working memory system is modulated by amphetamine: an fMRI investigation. Psychopharmacology 2005;180:634-43.

61. Greenberg LM, McMahon SA, Deem MA. Side effects of dextroamphetamine therapy of hyperactive children. West J Med 1974;120:105.

62. Williams K, Argyropoulos S, Nutt DJ. Amphetamine misuse and social phobia. Am J Psychiatry 2000;157(5):834–5.

63. Ellinwood EH Jr. Amphetamine psychosis. 1. Description of the individuals and process. J Nerv Ment Dis 1967;144:273.

64. Griffith JD, Cavanaugh JH, Held J, et al. Experimental psychosis induced by the administration of d-amphetamine. In: Costa E, Garattini S, editors. Amphetamines and Related Compounds. New York: Raven Press, 1970:897.

65. Griffith JD, Cavanaugh J, Held J, Oates JA. Dextroamphetamine. Evaluation of psychomimetic properties in man. Arch Gen Psychiatry 1972;26(2):97–100.

66. Ellinwood EH Jr, Kilbey MM. Fundamental mechanisms underlying altered behavior following chronic administration of psychomotor stimulants. Biol Psychiatry 1980;15(5):749–57.

67. Yui K, Goto K, Ikemoto S, Ishiguro T. Stress induced spontaneous recurrence of methamphetamine psychosis: the relation between stressful experiences and sensitivity to stress. Drug Alcohol Depend 2000;58(1–2):67–75.

68. Yui K, Ishiguro T, Goto K, Ikemoto S. Susceptibility to subsequent episodes in spontaneous recurrence of methamphetamine psychosis. Ann NY Acad Sci 2000;914:292–302.

69. Yui K, Ikemoto S, Goto K, Nishijima K, Yoshino T, Ishiguro T. Spontaneous recurrence of methamphetamine-induced paranoid-hallucinatory states in female subjects: susceptibility to psychotic states and implications for relapse of schizophrenia. Pharmacopsychiatry 2002;35(2):62–71.

70. Espelin DE, Done AK. Amphetamine poisoning. Effectiveness chlorpromazine N Engl J Med 1968;278(25):1361–5.

71. Bertram M, Egelhoff T, Schwarz S, Schwab S. Toxic leukencephalopathy following "ecstasy" ingestion. J Neurol 1999;246(7):617–8.

72. Semple DM, Ebmeier KP, Glabus MF, O'Carroll RE, Johnstone EC. Reduced in vivo binding to the serotonin transporter in the cerebral cortex of MDMA ("ecstasy") users. Br J Psychiatry 1999;175:63–9.

73. Misra L, Kofoed L, Oesterheld JR, Richards GA. Risperidone treatment of methamphetamine psychosis. Am J Psychiatry 1997;154(8):1170.

74. Misra LK, Kofoed L, Oesterheld JR, Richards GA. Olanzapine treatment of methamphetamine psychosis. J Clin Psychopharmacol 2000;20(3):393–4.

75. Richards JR, Brofeldt BT. Patterns of tooth wear associated with methamphetamine use. J Periodontol 2000;71(8):1371–4.

76. Delgado JH, Caruso MJ, Waksman JC, Hanigman B, Stillman D. Acute transient urinary retention from combined ecstasy and methamphetamine use. J Emerg Med 2004;26:173-5.

77. Yung A, Agnew K, Snow J, Oliver F. Two unusual cases of toxic epidermal necrolysis. Australas J Dermatol 2002;43(1):35–8.

78. O'Connor A, Cluroe A, Couch R, Galler L, Lawrence J, Synek B. Death from hyponatraemia-induced cerebral oedema associated with MDMA ("ecstasy") use. NZ Med J 1999;112(1091):255–6.

79. Aeschlimann A, de Truchis P, Kahn MF. Scleroderma after therapy with appetite suppressants. Scand J Rheumatol 1990;19(1):87–90.

80. Hall W, Lynskey M, Degenhardt L. Trends in opiate-related deaths in the United Kingdom and Australia, 1985–1995. Drug Alcohol Depend 2000;57(3):247–54.

81. Banooni P, Rickman LS, Ward DM. Pott puffy tumor associated with intranasal methamphetamine. JAMA 2000;283(10):1293.

82. Heinemann A, Iwersen-Bergmann S, Stein S, Schmoldt A, Puschel K. Methadone-related fatalities in Hamburg 1990–1999: implications for quality standards in maintenance treatment? Forensic Sci Int 2000;113(1–3):449–55.

83. Masaki T. The amphetamine problem in Japan: annex to Sixth Report of Expert Committee on Drugs Liable to Produce Addiction. World Health Organ Tech Rep Series 1956;102:14.

84. Unwin JR. Illicit drug use among Canadian youth. I. Can Med Assoc J 1968;98(8):402–7.

85. Kosman ME, Unna DR. Effects of chronic administration of the amphetamines and other stimulants on behavior. Clin Pharmacol Ther 1968;9(2):240–54.

86. Ellinwood EH Jr, Petrie WM. Drug induced psychoses. In: Pickens RW, Heston LL, editors. Psychiatric Factors in Drug Abuse. New York: Grune and Stratton, 1979:301.

87. Ellinwood EH Jr. Assault and homicide associated with amphetamine abuse. Am J Psychiatry 1979;3:25.

88. Niki Y, Watanabe S, Yoshida K, Miyashita N, Nakajima M, Matsushima T. Effect of pazufloxacin mesilate on the serum concentration of theophylline. J Infect Chemother 2002;8(1):33–6.

89. Ashton CH. Adverse effects of cannabis and cannabinoids. Br J Anaesth 1999;83(4):637–49.

90. Newton P, Chierakul W, Ruangveerayuth R, Abhigantaphand D, Looareesuwan S, White NJ. Malaria and amphetamine "horse tablet" in Thailand. Trop Med Intl Health 2003;80:17-18.

91. Volkow ND, Chang L, Wang GJ, Fowler JS, Ding YS, Sedler M, Logan J, Franceschi D, Gatley J, Hitzemann R, Gifford A, Wong C, Pappas N. Low level of brain dopamine D_2 receptors in methamphetamine abusers: association with metabolism in the orbitofrontal cortex. Am J Psychiatry 2001;158(12):2015–21.

92. Smith LM, Chang L, Yonekura ML, Grob C, Osborn D, Ernst T. Brain proton magnetic resonance spectroscopy in children exposed to methamphetamine in utero. Neurology 2001;57(2):255–60.

93. Sery O, Vojtova V, Zvolsky P. The association study of DRD2, ACE and AGT gene polymorphisms and metamphetamine dependence. Physiol Res 2001;50(1):43–50.

94. Sekine Y, Iyo M, Ouchi Y, Matsunaga T, Tsukada H, Okada H, Yoshikawa E, Futatsubashi M, Takei N, Mori N. Metamphetamine-related psychiatric symptoms and reduced brain dopamine transporters studied with PET. Am J Psychiatry 2001, 158:1206-14.

95. Mazei MS, Pluto CP, Kirkbride B, Pehek EA. Effect of catecholamine uptake blockers in the caudate-putamen and sub-regions of the medical prefrontal cortex of the rat Brain Res 2002;936:58-67.

96. Moron JA, Brockington A, Wise RA, Rocha BA, Hope BT. Dopamine uptake through the norepinephrine transporter in brain regions with low levels of the dopamine transporter: evidence from knock-out mouse lines. J Neurosci 2002;22:389-95.

97. Mattay VS, Goldberg TE, Fera F, Hariri AR, Tessitore A, Egan MF, Kolachana N, Callicot JH, Weinberger DR. Catechol-o-methyl transferase Val 158 – met genotype and individual variation in the brain response to amphetamine. Proc Natl Acad Sci 2003;100:6186-91.

98. Palmatier MA, Kang AM, Kidd KK. Global variation in the frequency of functionally different catechol-o-methyl transferase alleles. Biol Psychiatry 1999;46:557-67.

99. Fujii D. Risk factors for treatment-resistive methamphetamine psychosis. J Neuropsychiatry Clin Neurosci 2002;14(2):239-40.

100. Iwanami A, Sugiyama A, Kuroki N, Toda S, Kato N, Nakatani Y, Horita N, Kaneko T. Patients with methamphetamine psychosis admitted to a psychiatric hospital in Japan. A preliminary report. Acta Psychiatr Scand 1994;89(6):428-32.

101. Domier CP, Simon SL, Rawson RA, Huber A, Ling W. A comparison of injecting and noninjecting methamphetamine users. J Psychoactive Drugs 2000;32(2):229-32.

102. Ellinwood EH Jr. Emergency treatment of acute reactions to CNS stimulants. J Psychedelic Drugs 1972;5:147.

103. Nausieda PA. Central stimulant toxicity. In: Vinken PJ, Bruyn GW, editors. Handbook of Clinical Neurology. Intoxications of the Nervous System. Part 11. Amsterdam: Elsevier/North-Holland Biomedical Press, 1979:223.

104. Kalant H, Kalant OJ. Death in amphetamine users: causes and rates. Can Med Assoc J 1975;112(3):299–304.

105. Delaney P, Estes M. Intracranial hemorrhage with amphetamine abuse. Neurology 1980;30(10):1125–8.

106. Olsen ER. Intracranial hemorrhage and amphetamine usage. Review of the effects of amphetamines on the central nervous system. Angiology 1977;28(7):464–71.

107. Karch SB, Stephens BG, Ho CH. Methamphetamine-related deaths in San Francisco: demographic, pathologic, and toxicologic profiles. J Forensic Sci 1999;44(2):359–68.

108. Sribanditmongkol P, Chokjamsai M, Thampitak S. Methamphetamine overdose and fatality: 2 case reports. J Med Assoc Thai 2000;83(9):1120–3.

109. Simpson FO. Antihypertensive drug therapy. Drugs 1973;6(5):333–63.

110. Justice AJ, de Wit H. Acute effects of estradiol pretreatment on the response to d-amphetamine in women. Neuroendocrinology 2000;71(1):51–9.

111. Willson MC, Wilman AH, Bell EC, Asghar SJ, Silverstone PH. Dextroamphetamine causes a change in regional brain activity in vivo during cognitive tasks: an fMRI study utilizing BOLD. Biol Psychiatry 2004;56:284-91.

112. Bell EC, Willson MC, Wilman AH, Dave S, Asghar SJ, Silverstone PH. Lithium and valproate attenuate dextroamphetamine-induced changes in brain activation. Hum Psychopharmacol 2005;20:87-96.

113. Silverstone PH, O'Donnell T, Ulrich M, Asghar S, Hanstock CC. Dextroamphetamine increases phosphoinositol cycle activity in volunteers: an MRS study. Hum Psychopharmacol Clin Exp 2002;17:425-9.

114. Yu M-F, Lin W-W, Li L-T, Yin H-S. Activation of metabotropic glutamate receptor 5 is associated with effect of amphetamine on brain neurons. Synapse 2003;50:333-44.

115. Gurvich N, Klein PS. Lithium and valproic acid: parallels and contrasts in diverse signaling contexts. Pharmacol Ther 2002;96:45-66.

116. Cullen W, Bury G, Langton D. Experience of heroin overdose among drug users attending general practice. Br J Gen Pract 2000;50(456):546–9.

117. Christie B. Gangrene bug killed 35 heroin users. West J Med 2000;173(2):82–3.

118. Dettmeyer R, Schmidt P, Musshoff F, Dreisvogt C, Madea B. Pulmonary edema in fatal heroin overdose: immunohistological investigations with IgE, collagen IV and laminin—no increase of defects of alveolar-capillary membranes. Forensic Sci Int 2000;110(2):87–96.

119. McCreary M, Emerman C, Hanna J, Simon J. Acute myelopathy following intranasal insufflation of heroin: a case report. Neurology 2000;55(2):316–7.

120. Prior FH, Isbister GK, Dawson AH, Whyte IM. Serotonin toxicity with therapeutic doses of dexamphetamine and venlafaxine. Med J Aust 2002;176(5):240–1.

121. Mintzer MZ, Griffiths RR. Triazolam-amphetamine interaction: dissociation of effects on memory versus arousal. J Psychopharmacol 2003;17:17-29.

122. Wharton RN, Perel JM, Dayton PG, Malitz S. A potential clinical use for methylphenidate with tricyclic antidepressants. Am J Psychiatry 1971;127(12):1619–25.

123. Cooper TB, Simpson GM. Concomitant imipramine and methylphenidate administration: a case report. Am J Psychiatry 1973;130(6):721.

124. Chan-Ob T, Kuntawogse N, Boonyanaruthee V. Bupropion for amphetamine withdrawal syndrome. J Med Assoc Thai 2001;84:1763-5.

125. Tardieu S, Poirier Y, Micallef J, Blin O. Amphetamine-like stimulant cessation in an abusing patient treated with bupropion. Acta Psychiatr Scand 2004;109:75-8.

CANNABINOIDS

General Information

Cannabis is the abbreviated name for the hemp plant *Cannabis sativa*. The common names for cannabis include marijuana, grass, and weed. Other names for cannabis refer to particular strains; they include bhang and ganja. The most potent forms of cannabis come from the flowering tops of the plants or from the dried resinous exudate of the leaves, and are referred to as hashish or hash.

Cannabis is one of the oldest and most widely used drugs in the world. In different Western countries the possible therapeutic use of cannabinoids as antiemetics in patients with cancer or in patients with multiple sclerosis has become an issue, because of the prohibition of cannabis, and has polarized opinion about the seriousness of its adverse effects (1,2).

The long history of marijuana use both as a recreational drug and as an herbal medicine for centuries has been reviewed (3). *Cannabis sativa* contains more than 450 substances and only a few of the main active cannabinoids have been evaluated. Cannabis is the most commonly used illicit drug. In 2001, 83 million Americans and 37% of those aged 12 and older had tried marijuana (4).

Pharmacology

The primary active component of cannabis is Δ9-tetrahydrocannabinol (THC), which is responsible for the greater part of the pharmacological effects of the cannabis complex. Δ8-THC is also active. However, the cannabis plant contains more than 400 chemicals, of which some 60 are chemically related to Δ9-THC, and it is evident that the exact proportions in which these are present can vary considerably, depending on the way in which the material has been harvested and prepared. In man, Δ9-THC is rapidly converted to 11-hydroxy-Δ9-THC (5), a metabolite that is active in the central nervous system. A specific receptor for the cannabinols has been identified; it is a member of the G-protein-linked family of receptors (6). The cannabinoid receptor is linked to the inhibitory G-protein, which is linked to adenyl cyclase in an inhibitory fashion (7). The cannabinoid receptor is found in highest concentrations in the basal ganglia, the hippocampus, and the cerebellum, with lower concentrations in the cerebral cortex.

When cannabis is smoked, usually in a cigarette with tobacco, the euphoric and relaxant effects occur within minutes, reach a maximum in about 30 minutes, and last up to 4 hours. Some of the motor and cognitive effects can persist for 5–12 hours. Cannabis can also be taken orally, in foods such as cakes (for example "space cake") or sweetmeats (for example hashish fudge) (8).

Many variables affect the psychoactive properties of cannabis, including the potency of the cannabis used, the route of administration, the smoking technique, the dose, the setting, the user's past experience, the user's expectations, and the user's biological vulnerability to the effects of the drug.

Animal and in vitro toxicology

Δ9-tetrahydrocannabinol, the active component in herbal cannabis, is very safe. Laboratory animals (rats, mice, dogs, monkeys) can tolerate doses of up to 1000 mg/kg, equivalent to some 5000 times the human intoxicant dose. Despite the widespread illicit use of cannabis, there are very few, if any, instances of deaths from overdose (9).

Long-term toxicology studies with THC were carried out by the National Institute of Mental Health in the late 1960s (10). These included a 90-day study with a 30-day recovery period in both rats and monkeys and involved not only Δ9-THC but also Δ8-THC and a crude extract of marijuana. Doses of cannabis or cannabinoids in the range 50–500 mg/kg caused reduced food intake and lower body weight. All three substances initially depressed behavior, but later the animals became more active and were irritable or aggressive. At the end of the study the weights of the ovaries, uterus, prostate, and spleen were reduced and the weight of the adrenal glands was increased. The behavioral and organ changes were similar in monkeys, but less severe than those seen in rats. Further studies were carried out to assess the damage that might be done to the developing fetus by exposure to cannabis or cannabinoids during pregnancy. Treatment of pregnant rabbits with THC at doses up to 5 mg/kg had no effect on birth weight and did not cause any abnormalities in the offspring (10).

A similarly detailed toxicology study was carried out with THC by the National Institute of Environmental Health Sciences in the USA, in response to a request from the National Cancer Institute (11). Rats and mice were given THC up to 500 mg/kg five times a week for 13 weeks; some were followed for a period of recovery over 9 weeks. By the end of the study more than half of the rats treated with the highest dose (500 mg/kg) had died, but all of the remaining animals appeared to be healthy, although in both species the higher doses caused lethargy and increased aggressiveness. The THC-treated animals ate less food and their body weights were consequently significantly lower than those of untreated controls at the end of the treatment period, but returned to normal during recovery. During this period the animals were sensitive to touch and some had convulsions. There was a trend towards reduced uterine and testicular weights.

In further studies rats were treated with doses of THC up to 50 mg/kg and mice with up to 500 mg/kg 5 times a week for 2 years in a standard carcinogenicity test (11). After 2 years, more treated animals had survived than controls, probably because the treated animals ate less and had lower body weights. The treated animals also had a significantly lower incidence of the various cancers normally seen in aged rodents in testes, pancreas, pituitary gland, mammary glands, liver, and uterus. Although there was an increased incidence of precancerous changes in the thyroid gland in both species and in the mouse ovary after one dose (125 mg/kg), these changes were not dose-related. The conclusion was that there was "no evidence of carcinogenic activity of THC at doses up to 50 mg/kg." This was also supported by the failure to detect any genetic toxicity in other tests designed to identify drugs capable of causing chromosomal damage. For example, THC was negative in the so-called "Ames test," in which bacteria are exposed to very high concentrations of a drug to see whether it causes mutations. In another test, hamster ovary cells were exposed to high concentrations of the drug in tissue culture; there were no effects on cell division that might suggest chromosomal damage.

By any standards, THC must be considered to be very safe, both acutely and during long-term exposure. This probably partly reflects the fact that cannabinoid receptors are virtually absent from those regions at the base of the brain that are responsible for such vital functions as breathing and blood pressure control. The available animal data are more than adequate to justify its approval as a human medicine, and indeed it has been approved by the FDA for certain limited therapeutic indications (generic name = dronabinol) (9).

Respiratory

There have been several attempts to address this question by exposing laboratory animals to cannabis smoke. After such exposure on a daily basis for periods of up to 30 months, extensive damage has been observed in the lungs of rats (12), dogs (13), and monkeys (14), but it is very difficult to extrapolate these findings to man, as it is difficult or impossible to imitate human exposure to cannabis smoke in any animal model.

Nervous system

Animal studies on neurotoxicity have yielded conflicting results. Treatment of rats with high doses of THC given orally for 3 months (15) or subcutaneously for 8 months (16) produced neural damage in the hippocampal CA3 zone, with shrunken neurons, reduced synaptic density, and loss of cells. But in perhaps the most severe test of all, rats and mice treated on 5 days each week for 2 years had no histopathological changes in the brain, even after 50 mg/kg/day (rats) or 250 mg/kg/day (mice) (11). Although claims were made that exposure of a small number of rhesus monkeys to cannabis smoke led to ultrastructural changes in the septum and hippocampus (17,18), subsequent larger-scale studies failed to show any cannabis-induced histopathology in monkey brain (19).

Studies of the effects of cannabinoids on neurons in vitro have also yielded inconsistent results. Exposure of rat cortical neurons to THC shortened their survival: twice as many cells were dead after exposure to THC 5 µmol/l for 2 hours than in control cultures (20). Concentrations of THC as low as 0.1 µmol/l had a significant effect. The effects of THC were accompanied by release of cytochrome c, activation of caspase-3, and DNA fragmentation, suggesting an apoptotic mechanism.

All of the effects of THC could be blocked by the antagonist AM-251 or by pertussis toxin, suggesting that they were mediated through CB1 receptors. Toxic effects of THC have also been reported in hippocampal neurons in culture, with 50% cell death after exposure to THC 10 µmol/l for 2 hours or 1 µmol/l for 5 days (21). The antagonist rimonabant blocked these effects, but pertussis toxin did not. The authors proposed a toxic mechanism involving arachidonic acid release and the formation of free radicals. On the other hand, other authors have failed to observe any damage in rat cortical neurons exposed for up to 15 days to THC 1 mmol/l, although they found that this concentration killed rat C6 glioma cells, human astrocytoma U373MG cells, and mouse neuroblastoma N18TG12 cells (22). In a remarkable study, injection of THC into solid tumors of C6 glioma in rodent brain led to increased survival times, and there was complete eradication of the tumors in 20–35% of the treated animals (23). A stable analogue of anandamide also produced a drastic reduction in the tumor volume of a rat thyroid epithelial cell line transformed by K-ras oncogene, implanted in nude mice (24). The antiproliferative effect of cannabinoids has suggested a potential use for such drugs in cancer treatment (25).

Some authors have reported neuroprotective actions of cannabinoids. WIN55,212-2 reduced cerebral damage in rat hippocampus or cerebral cortex after global ischemia or focal ischemia in vivo (26). The endocannabinoid 2AG protected against damage elicited by closed head injury in mouse brain, and the protective effects were blocked by rimonabant (27). THC had a similar effect in vivo in protecting against damage elicited by ouabain (28). Rat hippocampal neurons in tissue culture were protected against glutamate-mediated damage by low concentrations of WIN55,212-2 or CP-55940, and these effects were mediated through CB1 receptors (29). But not all of these effects seem to require mediation by cannabinoid receptors. The protective effects of WIN55,212-2 did not require either CB1 or CB2 cannabinoid receptors in cortical neurons exposed to hypoxia (26), and there were similar findings for the protective actions of anandamide and 2-AG in cortical neuronal cultures (30). Both THC and cannabidiol, which is not active at cannabinoid receptors, protected rat cortical neurons against glutamate toxicity (31) and these effects were also independent of CB1 receptors. The authors suggested that the protective effects of THC might be due to the antioxidant properties of these polyphenolic molecules, which have redox potentials higher than those of known antioxidants (for example ascorbic acid).

Pregnancy

In animals, THC can cause spontaneous abortion, low birth weight, and physical deformities (32). However, these were only seen after treatment with extremely high doses of THC (50–150 times higher than human doses), and only in rodents and not in monkeys.

Tolerance and dependence

Many animal studies have shown that tolerance develops to most of the behavioral and physiological effects of THC

(33). Dependence on cannabinoids in animals is clearly observable, because of the availability of CB_1 receptor antagonists, which can be used to precipitate withdrawal. Thus, a behavioral withdrawal syndrome was precipitated by rimonabant in rats treated for only 4 days with THC in doses as low as 0.5–4.0 mg/kg/day (34). The syndrome included scratching, face rubbing, licking, wet dog shakes, arched back, and ptosis, many of the signs that are seen in rats undergoing opiate withdrawal. Similar withdrawal signs occurred when rats treated chronically with the synthetic cannabinoid CP-55940 were given rimonabant (35). Rimonabant-induced withdrawal after 2 weeks of treatment of rats with the cannabinoid HU-120 was accompanied by a marked increase in release of the stress-related neuropeptide corticotropin-releasing factor in the amygdala, a result that also occurred in animals undergoing heroin withdrawal (36). An electrophysiological study showed that precipitated withdrawal was also associated with reduced firing of dopamine neurons in the ventral tegmental area of rat brain (37).

These data clearly show that chronic administration of cannabinoids leads to adaptive changes in the brain, some of which are similar to those seen with other drugs of dependence. The ability of THC to cause selective release of dopamine from the nucleus accumbens (38) also suggests some similarity between THC and other drugs in this category.

Furthermore, although many earlier attempts to obtain reliable self-administration behavior with THC were unsuccessful (33), some success has been obtained recently. Squirrel monkeys were trained to self-administer low doses of THC (2 µg/kg per injection), but only after the animals had first been trained to self-administer cocaine (39). THC is difficult to administer intravenously, but these authors succeeded, perhaps in part because they used doses comparable to those to which human cannabis users are exposed, and because the potent synthetic cannabinoids are far more water-soluble than THC, which makes intravenous administration easier. Mice could be trained to self-administer intravenous WIN55212-2, but CB_1 receptor knockout animals could not (40).

Another way of demonstrating the rewarding effects of drugs in animals is the conditioned place preference paradigm, in which an animal learns to approach an environment in which it has previously received a rewarding stimulus. Rats had a positive THC place preference after doses as low as 1 mg/kg (41).

Some studies have suggested that there may be links between the development of dependence to cannabinoids and to opiates (42). Some of the behavioral signs of rimonabant-induced withdrawal in THC-treated rats can be mimicked by the opiate antagonist naloxone (43). Conversely, the withdrawal syndrome precipitated by naloxone in morphine-dependent mice can be partly relieved by THC (44) or endocannabinoids (45). Rats treated chronically with the cannabinoid WIN55212-2 became sensitized to the behavioral effects of heroin (46). Such interactions can also be demonstrated acutely. Synergy between cannabinoids and opiate analgesics has been described above. THC also facilitated the antinociceptive effects of RB 101, an inhibitor of enkephalin inactivation, and acute administration of THC caused increased release of Met-enkephalin into microdialysis probes placed into the rat nucleus accumbens (47).

The availability of receptor knockout animals has also helped to illustrate cannabinoid–opioid interactions. CB_1 receptor knockout mice had greatly reduced morphine self-administration behavior and less severe naloxone-induced withdrawal signs than wild type animals, although the antinociceptive actions of morphine were unaffected in the knockout animals (40). The rimonabant–precipitated withdrawal syndrome in THC-treated mice was significantly attenuated in animals with knock-out of the pro-enkephalin gene (48). Knockout of the µ opioid (OP_3) receptor also reduced rimonabant-induced withdrawal signs in THC-treated mice, and there was an attenuated naloxone withdrawal syndrome in morphine-dependent CB_1 knockout mice (49,50).

These findings clearly point to interactions between the endogenous cannabinoid and opioid systems in the CNS, although the neuronal circuitry involved is unknown. Whether this is relevant to the so-called "gateway" theory is unclear. In the US National Household Survey of Drug Abuse, respondents aged 22 years or over who had started to use cannabis before the age of 21 years were 24 times more likely than non-cannabis users to begin using hard drugs (51). However, in the same survey the proportion of cannabis users who progressed to heroin or cocaine use was very small (2% or less). Mathematical modeling using the Monte Carlo method suggested that the association between cannabis use and hard drug use need not be causal, but could relate to some common predisposing factor, for example "drug-use propensity" (52).

Tumorigenicity

THC does not appear to be carcinogenic, but there is plenty of evidence that the tar derived from cannabis smoke is. Bacteria exposed to cannabis tar develop mutations in the standard Ames test for carcinogenicity (53), and hamster lung cells in tissue culture develop accelerated malignant transformations within 3–6 months of exposure to tobacco or cannabis smoke (54).

Observational studies

In an open trial the safety, tolerability, dose range, and efficacy of the whole-plant extracts of *Cannabis sativa* were evaluated in 15 patients with advanced multiple sclerosis and refractory lower urinary tract symptoms (55). The patients took extracts containing delta-9-tetrahydrocannabinol (THC) and cannabidiol (CBD; 2.5 mg per spray) for 8 weeks followed by THC only for a further 8 weeks. Urinary urgency, the number and volume of incontinence episodes, frequency, and nocturia all reduced significantly after treatment with both extracts. Patients' self assessments of pain, spasticity, and quality of sleep improved significantly, and the improvement in pain continued for up to a median of 35 weeks. Most of the patients had symptoms of intoxication, such as mild drowsiness, disorientation, and altered time perception, during the dose titration period. Three had single short-lived hallucinations that did not occur when the dose was reduced. All complained of a worsening of dry mouth that was already present from other treatments and two complained of mouth soreness at the site of drug administration.

Of 220 patients with multiple sclerosis in Halifax, Canada 72 (36%) reported ever having used cannabis (56). Ever use of cannabis for medicinal purposes was associated with male sex, the use of tobacco, and recreational use of cannabis. Of the 34 medicinal cannabis users, 10 reported mild, eight moderate and one strong adverse effects; none reported severe adverse effects. The most common adverse effects were feeling "high" (n = 24), drowsiness (20), dry mouth (14), paranoia (3), anxiety (3), and palpitation (3).

Placebo-controlled studies

Cannabis has been used to treat many medical conditions, especially those involving pain and inflammation. Many studies with improved designs and larger sample sizes are providing preliminary data of efficacy and safety in conditions such as multiple sclerosis and chronic pain syndromes.

In a parallel group, double-blind, randomized, placebo-controlled study undertaken at three sites in 160 patients with multiple sclerosis a cannabis-based medicinal extract containing equal amounts of delta-9-tetrahydrocannabinol (THC) and cannabidiol (CBD) at doses of 2.5–120 mg of each daily in divided doses for 6 weeks, spasticity scores were significantly improved by cannabis (57). However, when the changes in symptoms were measured using the Primary Symptoms Scale, there were no significant differences between cannabis and placebo. The main adverse events were dizziness (33%), local discomfort at the site of application (26%), fatigue (15%), disturbance in attention (8.8%), disorientation (7.5%), a feeling of intoxication (5%), and mouth ulcers (5%).

In a randomized, double-blind, placebo-controlled, crossover trial the effect of the synthetic delta-9-tetrahydrocannabinol dronabinol on central neuropathic pain was evaluated in 24 patients with multiple sclerosis (58). Oral dronabinol reduced central pain. Adverse events were reported by 96% of the patients compared with 46% during placebo treatment. They were more common during the first week of treatment. The most common adverse events during dronabinol treatment were dizziness (58%), tiredness (42%), headache (25%), myalgia (25%), and muscle weakness (13%). There was increased tolerance to the adverse effects over the course of treatment and with dosage adjustments.

Three cannabis-based medicinal extracts in sublingual form recently became available for use against pain. In a randomized, double-blind, placebo-controlled, crossover study for 12 weeks in 34 patients with chronic neuropathic pain THC extracts were effective in symptom control (59). Drowsiness and euphoria/dysphoria were common in the first 2 weeks. Dizziness was less of a problem. Anxiety and panic were infrequent but occurred during the run-in period. Dry mouth was the most common complaint.

General adverse effects

A review has summarized the evidence related to the adverse effects of acute and chronic use of cannabis (60). The effects of acute usage include anxiety, impaired attention, and increased risk of psychotic symptoms. Probable risks of chronic cannabis consumption include bronchitis and subtle impairments of attention and memory.

The adverse effects of cannabis can be considered under two main headings, reflecting psychoactive and autonomic effects, in addition to which there are direct toxic effects. The most frequently reported psychoactive effects include enhanced sensory perception (for example a heightened appreciation of color and sound). Cannabis intoxication commonly heightens the user's sensitivity to other external stimuli as well, but subjectively slows the appreciation of time. In high doses, users may also experience depersonalization and derealization. Various forms of psychomotor performance, including driving, are significantly impaired for 8–12 hours after using cannabis. The most serious possible consequence of cannabis use is a road accident if a user drives while intoxicated.

Adverse reactions have been reported at relatively low doses and principally affect the psyche, leading to anxiety states, panic reactions, restlessness, hallucinations, fear, confusion, and rarely toxic psychosis. These effects appear to be reversible (61). Ingestion of cake with cannabis by people who seldom use or have never used cannabis before can result in mental changes, including confusion, anxiety, loss of logical thinking, fits of laughter, hallucinations, hypertension, and/or paranoid psychosis, which can last as long as 8 hours.

The autonomic effects of cannabis lead to tachycardia, peripheral vasodilatation, conjunctival congestion, hyperthermia, bronchodilatation, dry mouth, nystagmus, tremor, ataxia, hypotension, nausea, and vomiting, that is a spectrum of effects that closely resembles the consequences of overdosage with anticholinergic agents. Some individuals have sleep disturbances. Increased appetite and dry mouth are other common effects of cannabis intoxication.

Hypersensitivity reactions are rare, but a few have been reported after inhalation. Delayed hypersensitivity reactions, particularly affecting vascular tissue, have been recorded with chronic systemic administration. Tumor-inducing effects are difficult to attribute to cannabis alone. Animal studies have shown neoplastic pulmonary lesions superimposed on chronic inflammation, but such pathology may be primarily associated with the "tar" produced by burning marijuana. The most serious potential adverse effects of cannabis use come from the inhalation of the same carcinogenic hydrocarbons that are present in tobacco, and some data suggest that heavy cannabis users are at risk of chronic respiratory diseases and lung cancer.

The effects of oral cannabinoids (dronabinol or *Cannabis sativa* plant extract) in relieving pain and muscle spasticity have been studied in 16 patients with multiple sclerosis (mean age 46 years, mean duration of disease 15 years) in a double-blind, placebo-controlled, crossover study (62). The initial dose was 2.5 mg bd, increasing to 5 mg bd after 2 weeks if the dose was well tolerated. The plant extract was more likely to cause adverse events; five patients had increased spasticity and one rated an adverse event of acute psychosis as severe. All physical measures were in the reference ranges. There were no significant differences in any measure of efficacy score that would indicate a therapeutic benefit of cannabinoids. This study is the largest and longest of its kind, but the authors acknowledged some possible shortcomings. The route of administration could affect subjective ratings, since the gastrointestinal tract is a much slower and more inefficient route than the lungs. Another possibility is that the dose was too small to have the desired therapeutic effects.

Organs and Systems

Cardiovascular

Marijuana has several effects on the cardiovascular system, and can increase resting heart rate and supine blood pressure and cause postural hypotension. It is associated with an increase in myocardial oxygen demand and a decrease in oxygen supply. Peripheral vasodilatation, with increased blood flow, orthostatic hypotension, and tachycardia, can occur with normal recreational doses of cannabis. High doses of THC taken intravenously have often been associated with ventricular extra beats, a shortened PR interval, and reduced T wave amplitude, to which tolerance readily develops and which are reversible on withdrawal. While the other cardiovascular effects tend to decrease in chronic smokers, the degree of tachycardia continues to be exaggerated with exercise, as shown by bicycle ergometry.

Marijuana use is most popular among young adults (18–25 years old). However, with a generation of post-1960s smokers growing older, the use of marijuana in the age group that is prone to coronary artery disease has increased. The cardiovascular effects may present a risk to those with cardiovascular disorders, but in adults with normal cardiovascular function there is no evidence of permanent damage associated with marijuana (61,63,64), and it is not known whether marijuana can precipitate myocardial infarction, although mixed use of tobacco and cannabis make the evaluation of the effects of cannabis very difficult.

Ischemic heart disease

Investigators in the Determinants of Myocardial Infarction Onset Study recently reported that smoking marijuana is a rare trigger of acute myocardial infarction (65). Interviews of 3882 patients (1258 women) were conducted on an average of 4 days after infarction. Reported use of marijuana in the hour preceding the first symptoms of myocardial infarction was compared with use in matched controls. Among the patients, 124 reported smoking marijuana in the previous year, 37 within 24 hours, and 9 within 1 hour of cardiac symptoms. The risk of myocardial infarction was increased 4.8 times over baseline in the 60 minutes after marijuana use and then fell rapidly. The authors emphasized that in a majority of cases, the mechanism that triggered the onset of myocardial infarction involved a ruptured atherosclerotic plaque secondary to hemodynamic stress. It was not clear whether marijuana has direct or indirect hemodynamic effects sufficient to cause plaque rupture.

Two cases of coronary artery disease have been reported (66).

- A 48-year-old man, a chronic user of cannabis who had had coronary artery bypass grafting 10 years before and recurrent angina over the past 18 months, developed chest pain. An electrocardiogram showed intermittent resting ST segment changes and coronary angiography showed that of the three previous grafts, only one was still patent. There was also sub-total occlusion of a stent in the left main stem. After 24 hours he had a cardiac arrest while smoking cannabis and had multiple episodes of ventricular fibrillation, requiring both electrical and pharmacological cardioversion. He then underwent urgent percutaneous coronary intervention which involved stenting of his left main stem. He eventually stabilized and recovered for discharge 11 weeks later.
- A 22-year-old man had two episodes of tight central chest pain with shortness of breath after smoking cannabis. He had been a regular marijuana smoker since his mid-teens and had used more potent and larger amounts during the previous 2 weeks. An electrocardiogram showed ST segment elevation in leads V1-5, with reciprocal ST segment depression in the inferior limb leads. A provisional diagnosis of acute myocardial infarction was made. Thrombolysis was performed, but the electrocardiographic changes continued to evolve. Angiography showed an atheromatous plaque in the left anterior descending artery which was dilated and stented. There was early diffuse disease in the cardiac vessels.

The authors suggested that in the first case ventricular fibrillation had been caused by increased myocardial oxygen demand in the presence of long-standing coronary artery disease. In the second case, they speculated that chronic cannabis use may have contributed to the unexpectedly severe coronary artery disease in a young patient with few risk factors.

- Two young men, aged 18 and 30 years, developed retrosternal pain with shortness of breath, attributed to acute coronary syndrome (67). Each had smoked marijuana and tobacco and admitted to intravenous drug use. Urine toxicology was positive for tetrahydrocannabinol. Aspartate transaminase and creatine kinase activities and troponin-I and C-reactive protein concentrations were raised. Echocardiography in the first patient showed hypokinesia of the posterior and inferior walls and in the second hypokinesia of the basal segment of the anterolateral wall. Coronary angiography showed normal coronary anatomy with coronary artery spasm. Genetic testing for three common genetic polymorphisms predisposing to acute coronary syndrome was negative.

The authors suggested that marijuana had increased the blood carboxyhemoglobin concentration, leading to reduced oxygen transport capacity, increased oxygen demand, and reduced oxygen supply.

Cardiac dysrhythmias

In terms of its potential for inducing cardiac dysrhythmias, cannabis is most likely to cause palpitation due to a dose-related sinus tachycardia. Other reported dysrhythmias include sinus bradycardia, second-degree atrioventricular block, and atrial fibrillation. Also reported are ventricular extra beats and other reversible electrocardiographic changes.

Paroxysmal atrial fibrillation has been reported in two cases after marijuana use (68).

- A healthy 32-year-old doctor, who smoked marijuana 1–2 times a month, had paroxysmal tachycardia for several months. An electrocardiogram was normal and a Holter recording showed sinus rhythm with isolated supraventricular extra beats. He was treated with propranolol. He later secretly smoked marijuana while undergoing another Holter recording, which showed numerous episodes of paroxysmal atrial tachycardia and atrial fibrillation lasting up to 2 minutes. He abstained from marijuana for 12 months and maintained stable sinus rhythm.
- A 24-year-old woman briefly lost consciousness and had nausea and vomiting several minutes after smoking marijuana. She had hyporeflexia, atrial fibrillation (maximum 140/minute with a pulse deficit), and a blood pressure of 130/80 mmHg. Echocardiography was unremarkable. Within 12 hours, after metoprolol, propafenone, and intravenous hydration with electrolytes, sinus rhythm was restored.

The authors discussed the possibility that Δ9-THC, the active ingredient of marijuana, can cause intra-atrial reentry by several mechanisms and thereby precipitate atrial fibrillation.

Sustained atrial fibrillation has also been attributed to marijuana (69).

- A 14-year-old African-American man with no cardiac history had palpitation and dizziness, resulting in a fall, within 1 hour of smoking marijuana. After vomiting several times he had a new sensation of skipped heartbeats. The only remarkable finding was a flow murmur. The electrocardiogram showed atrial fibrillation. Echocardiography was normal. Serum and urine toxicology showed cannabis. He was given digoxin, and about 12 hours later his cardiac rhythm converted to sinus rhythm. Digoxin was withdrawn. He abstained from marijuana over the next year and was symptom free.

The authors noted that marijuana's catecholaminergic properties can affect autonomic control, vasomotor reflexes, and conduction-enhancement of perinodal fibers in cardiac muscle, and thus lead to an event such as this.

Supraventricular tachycardia after the use of cannabis has been reported (70).

- A 35-year-old woman with a 1-month history of headaches was found to be hypertensive, with a blood pressure of 179/119 mmHg. She smoked 20 cigarettes a day and used cannabis infrequently. Her family history included hypertension. Electrocardiography suggested left ventricular hypertrophy but echocardiography was unremarkable. She was given amlodipine 10 mg/day and the blood pressure improved. While in the hospital, she smoked marijuana and about 30 minutes later developed palpitation, chest pain, and shortness of breath. The blood pressure was 233/120 mmHg and the pulse rate 150/minute. Electrocardiography showed atrial flutter with 2:1 atrioventricular block. Cardiac troponin was normal at 12 hours. Urine toxicology was positive for cannabis only. Two weeks later, while she was taking amlodipine 10 mg/day and atenolol 25 mg/day, her blood pressure was 117/85 mmHg.

The authors reviewed the biphasic effect of marijuana on the autonomic nervous system. At low to moderate doses it causes increased sympathetic activity, producing a tachycardia and increase in cardiac output; blood pressure therefore increases. At high doses it causes increased parasympathetic activity, leading to bradycardia and hypotension. They thought that this patient most probably had adrenergic atrial flutter.

Coronary no-flow and ventricular tachycardia after habitual marijuana use has been reported (71).

- A 34-year-old man developed palpitation, shortness of breath, and chest pain. He had smoked a quarter to a half an ounce of marijuana per week and had taken it 3 hours before the incident. He had ventricular tachycardia at a rate of 200/minute with a right bundle branch block pattern. Electrical cardioversion restored sinus rhythm. Angiography showed a significant reduction in left anterior descending coronary artery flow rate, which was normalized by intra-arterial verapamil 200 micrograms.

The authors thought that marijuana may have enhanced triggered activity in the Purkinje fibers along with a reduction in coronary blood flow, perhaps through coronary spasm.

Syncope

Postural syncope after marijuana use has been studied in 29 marijuana-experienced volunteers, using transcranial Doppler to measure cerebral blood velocity in the middle cerebral artery in response to postural changes (72). They were required to abstain from marijuana and other drugs for 2 weeks before the assessment, as confirmed by urine drug screening. They were then given marijuana, tetrahydrocannabinol, or placebo and lying and standing measurements were made. When marijuana or tetrahydrocannabinol was administered, 48% reported a dizziness rating of three or four and had significant falls in standing cerebral blood velocity, mean arterial blood pressure, and systolic blood pressure. Eight subjects were so dizzy that they had to be supported. The authors suggested that marijuana interferes with the protective mechanisms that maintain standing blood pressure and cerebral blood velocity. All but one of the subjects who took marijuana or tetrahydrocannabinol reported some degree of dizziness. Women tended to be dizzier. As the postural dizziness was significant and unrelated to plasma concentrations of tetrahydrocannabinol or other indices, the authors raised concerns about marijuana use in those who are medically compromised or elderly.

Arteritis

A case of progressive arteritis associated with cannabis use has been reported (73).

- A 38-year-old Afro-Caribbean man was admitted after 3 months of severe constant ischemic pain and numbness affecting the right foot. The pain was worse at night. He also had intermittent claudication after walking 100 yards. He had a chronic history of smoking cannabis about 1 ounce/day, mixed with tobacco in the early years of usage. However, at the time of admission, he had not used tobacco in any form for over 10 years. He had patchy necrosis and ulceration of the toes and impalpable pulses in the right foot. The serum cotinine concentrations were consistent with those found in non-smokers of tobacco. Angiography of his leg was highly suggestive of Buerger's disease (thromboangiitis obliterans).

Remarkably, this patient, despite having abstained from tobacco for more than 10 years, developed a progressive arteritis leading to ischemic changes. While arterial pathology with cannabis has been reported before, it has been difficult to dissociate the effects of other drugs.

Popliteal artery entrapment occurred in a patient with distal necrosis and cannabis-related arteritis, two rare or exceptional disorders that have never been described in association (74).

- A 19-year-old man developed necrosis in the distal third right toe, with loss of the popliteal and foot pulses. Arteriography showed posterior popliteal artery compression in the right leg and unusually poor distal vascularization in both legs. An MRI scan did not show a cyst and failed to identify the type of compression and the causal agent. Surgery showed that the patient had type III entrapment. Surprisingly, the pain failed to regress and the loss of distal pulses persisted despite a perfect result on the postoperative MRA scan. The patient then admitted consuming cannabis 10 times a day for 4 years, which suggested a Buerger-type arteritis related to cannabis consumption. A 21-day course of intravenous vasodilators caused the leg pain to disappear and the toe necrosis to regress. An MRA scan confirmed permanent occlusion of three arteries on the right side of the leg and the peroneal artery on the left side. Capillaroscopy excluded Buerger's disease.

The authors suggested that popliteal artery entrapment in a young patient with non-specific symptoms should raise the suspicion of a cannabis-related lesion. Their review of literature suggested that this condition affects young patient and that complications secondary to popliteal artery entrapment did not occur in those who were under 38 years age.

Respiratory

Acute inhalation of marijuana or THC causes bronchodilatation, but with chronic use resistance in the bronchioles increases (75,76). Prolonged use of cannabis by inhalation can cause chronic inflammatory changes in the bronchial tree, in part related to the inhalants that accompany the smoke. In some cases attacks of asthma and glottal and uvular angioedema can occur. Reduced respiratory gas exchange has been reported in long-term smokers, and under experimental conditions THC can depress respiratory function slightly and act as a respiratory irritant. In fact, chronic marijuana cigarette smoking and chronic tobacco cigarette smoking produce very similar changes, but these occur after smoking fewer cigarettes when marijuana is smoked, compared with tobacco-smoking. With marijuana inhalation, when a filter is never used, inhalation is deeper and the smoke is held in the lungs for longer than when smoking commercially produced tobacco-based cigarettes (77). There is therefore a greater build-up of carbon monoxide, reduction in carboxyhemoglobin saturation, and alveolar cellular irritation with depression of macrophages (SEDA-13, 25). Pneumothorax, pneumopericardium, and pneumomediastinum have been reported when positive pulmonary pressure is applied or a Valsalva maneuver used, as often happens (78,79).

A possible role of marijuana use in the formation of large lung bullae has been discussed (80). Four men, who smoked both tobacco and marijuana, developed large, multiple, bilateral, peripheral bullae at their lung apices, with normal parenchymal tissue elsewhere. Three patients with large bullae in the upper lung lobe have been reported (81). All had been heavy marijuana smokers over 10–24 years. However, they all had at least nine pack-years of cigarette exposure and so marijuana may not have been the only cause of their lung bullae. Nevertheless, the authors recommended that all those who present with upper lung bullae should be screened for cannabis use.

While Δ9-THC may not contribute directly to lung bullae, it is possible that the respiratory dynamics of smoking the drug explains it. Typically, a draw on a marijuana joint has, on average, a depth of inspiration that is one-third greater, a volume two-thirds greater, and a breath-holding time four times longer than a draw on a cigarette. The marijuana joint lacks a filter tip, and the practice of smoking "leads to a fourfold greater delivery of tar and a five times greater increase in carboxyhemoglobin per cigarette smoked" (75). Smoking three to four joints of marijuana per day is reported to produce a symptom profile and damage to the respiratory airways similar to that caused by smoking 20 tobacco cigarettes daily.

Cannabis smoking can cause pneumothorax (82).

- A 23-year-old man who had smoked cannabis regularly for about 10 years developed severe respiratory distress. He had bilateral pneumothoraces with complete collapse of the left lung.

No obvious reason for the problem was found and the authors suggested that coughing while breath-holding during cannabis inhalation had caused the problem.

The term "Bong Lung" is used to refer to a histological change that occurs in the lungs of chronic cannabis smokers (83). Patients with cannabis-induced recurrent *pneumothorax* often undergo resection of bullae. In Australia, the histopathology of resected lung was examined in 10 cannabis smokers, 5 heavy tobacco smokers, and 5 non-smokers. All marijuana smokers had irregular emphysema with cystic blebs and bullae in the lung apices. There was also massive accumulation of intra-alveolar pigmented histiocytes or "smoker's macrophages" throughout the pulmonary parenchyma, but sparing of the peribronchioles, similar to desquamative interstitial pneumonia.

Nervous system

Marijuana can interact with the neurotransmitter dopamine, and the effects of marijuana on the brain in schizophrenia have been studied by single photon emission computerized tomography (SPECT) (84).

A 38-year-old man with schizophrenia secretively smoked marijuana during a neuroimaging study. A comparison of two sets of images, before and after marijuana inhalation, showed a 20% reduction in the striatal dopamine D_2 receptor binding ratio, suggestive of increased synaptic dopaminergic activity.

On the basis of this in vivo SPECT study, the authors speculated that marijuana may interact with dopaminergic systems in brain reward pathways.

A review of the evidence has suggested that, particularly with high doses, cannabis users are 3–7 times more likely to cause motor accidents than non-drug users (85).

Electroencephalography

Long-term marijuana alters the electroencephalogram during abstinence (86). In 29 individuals who met DSM-III R criteria for marijuana dependence or abuse and 21 drug-free controls, electroencephalograms were recorded for 3 minutes (87). Marijuana abusers had significantly lower log power for the theta and alpha1 bands during abstinence compared with controls. The authors also observed increased cerebrovascular resistance using transcranial Doppler sonography in an overlapping sample of marijuana abusers. They proposed that this combination of electroencephalographic findings and changes in cerebral blood flow may explain cognitive deficits reported in chronic marijuana users.

Extrapyramidal effects

Extrapyramidal effects have been reported in a patient taking neuroleptic drugs who smoked cannabis (88).

- A 20-year-old man with no previous movement disorders, who had smoked marijuana for 4 years was given risperidone 9 mg/day and clorazepate 10–20 mg/day for paranoid schizophrenia. After 4 weeks he started

using marijuana again and had least two episodes of cervical and jaw dystonia with oculogyric crises, for which intramuscular biperiden was effective. He acknowledged heavy marijuana use before each episode. He was then given oral biperiden 2–4 mg/day and risperidone was replaced by olanzapine 30 mg/day. He again started smoking marijuana and had similar episodes of dystonia and oculogyric crises. No other causes of secondary extrapyramidal disorders were found.

The authors suggested a causal association between use of marijuana and extrapyramidal disorders. The research literature contains evidence that the endogenous cannabinoid system plays a role in basal ganglia transmission circuitry, possibly by interfering with dopamine reuptake. Furthermore, central cannabinoid receptors are located in two areas that regulate motor activity, the lateral globus pallidus and substantia nigra (89).

Seizures

Propriospinal myoclonus has been reported after cannabis use (90).

- A 25-year-old woman developed spinal myoclonus 18 months after having experienced acute-onset repetitive involuntary flexion and extension spasms of her trunk immediately after smoking cannabis. The jerks, which lasted 2–5 seconds, involved the trunk, neck, and to a lesser extent the limbs. The attacks occurred in clusters lasting up to 2 weeks and she was asymptomatic for 2-3 months between clusters. The myoclonus was not present during sleep. During a bout of jerks, myoclonus would occur every few minutes and continue for up to 9 hours, with associated fatigue and back pain. Neurological examination showed repetitive flexion jerks of the trunk with no other abnormal signs. An electroencephalogram, an MRI scan of the head and spine, and a full-length myelogram were all normal. Multi-channel surface electromyography with parallel frontal electroencephalography showed propriospinal myoclonus of mid-thoracic origin.

There have been no previous reports of propriospinal myoclonus precipitated by marijuana. The etiology was not clear but may have involved cannabinoid receptors located in the brain and spinal cord as well as the peripheral nervous system.

Strokes

A cannabis smoker had recurrent transient ischemic attacks.

- A 50-year-old male cigarette smoker with hypertension had episodes of transient right-sided hemisensory loss lasting only a few minutes (91). Coughing while smoking marijuana was the apparent trigger. Electroencephalography was negative. An MRI scan of the brain showed chain-like low-flow infarctions in the white matter of the left parietal subcortex. Duplex sonography and digital subtraction angiography showed a subocclusive stenosis of the left internal carotid artery. Blood flow to the left middle cerebral artery was reduced and delayed and there was steal by the right middle cerebral artery. Endarterectomy of the left internal carotid resolved the symptoms.

The authors suggested that marijuana may increase the risk of reduced cerebral perfusion. Coughing may have contributed by reduced flow velocity within arteries supplying the brain and by a sudden increase in intracranial pressure.

Occipital stroke has been reported after cannabis use (92).

- A 37-year-old Albanian man had an uneventful medical history except that he smoked 20 cigarettes/day and marijuana joints regularly for 10 years. In the previous 6 months he had increased his marijuana smoking to 2-3 joints/week from 1-2 joints/month. He suddenly developed left-sided hemiparesis, left-sided hemihypesthesia, and recurrent double vision 15 minutes after having smoked a joint containing about 250 mg marijuana. Most of the symptoms disappeared within 1 hour after onset. An MRI scan showed an area of impaired diffusion, 2 cm in diameter, in the right occipital area subcortically. He responded well to acetylsalicylic acid with dipyridamole and atorvastatin and was discharged 3 days later with blurred vision when looking to the left. There was no cardiac source of embolism. Other causes of stroke were carefully excluded, and it could only be attributed to the use of marijuana.

The authors, based on previous reports of the vasogenic effects of marijuana, suggested that this event may have been related to increased concentrations of catecholamine and carboxyhemoglobin, and diminishing cerebral autoregulatory capacity.

- A 36-year-old man developed acute aphasia followed by a convulsive seizure a few hours later after heavy consumption of hashish and 3–4 alcoholic beverages at a party (93). He had no previous vascular risk factors. His blood pressure was 120/80 mmHg. An MRI scan showed two ischemic infarcts, one in the left temporal lobe and one in the right parietal lobe. Magnetic resonance angiography of the head and neck showed narrowing of the distal temporal branches of the left middle cerebral artery without involvement of its proximal segment. There was no evidence of diffuse atherosclerotic disease. There was tetrahydrocannabinol in the urine. Electroencephalography and transesophageal echocardiography were normal. He was given ticlopidine. A year later, he had a second episode of aphasia and right hemiparesis immediately after smoking marijuana. His blood pressure was 140/80 mmHg. An MRI scan showed acute left and right frontal cortical infarctions. He had a third stroke 18 months later, when he developed auditory agnosia after heavy use of hashish and 3–4 drinks of alcohol. On this occasion he was normotensive. An MRI scan showed acute infarcts in the right posterior temporal lobe and lower parietal lobe.

The authors discussed the importance of the close temporal relation between the use of cannabis and alcohol and the episodes of stroke. The mechanism was unclear. However, they speculated that a vasculopathy, either toxic or immune inflammatory, was the most likely mechanism.

Transient global amnesia
Transient global amnesia, an amnesia of sudden onset regarding events in the present and recent past, typically occurs in elderly people. Transient global amnesia following accidental marijuana ingestion has been reported in a young boy (94).

- A 6-year-old boy accidentally became intoxicated with marijuana after eating cookies laced with marijuana. He developed retentive memory deficits of sudden onset, later diagnosed as transient global amnesia. He was anxious and had a tachycardia, fine tremors in the upper and lower limbs, and an ataxic gait. His CSF was unremarkable. He had cannabinoids in his urine. His memory returned to normal after 14 hours. His mother admitted baking marijuana cookies and leaving them out on the kitchen table. Up to 12 months later he had no memory of the episode.

This is the first case of transient global amnesia from marijuana in a 6-year-old. With increased use of marijuana in society, children can sometimes be exposed to marijuana inadvertently.

Sensory systems

Eyes
No consistent effects of cannabinoids on the eyes have been reported, apart from a reduction in intraocular pressure (95). The initial reduction in intraocular pressure is followed by a rebound increase associated with increased prostaglandin concentrations.

In contrast, bilateral angle-closure glaucoma has been reported after combined consumption of ecstasy and cannabis (96).

- A 29-year-old woman developed severe headaches, blurred vision, and malaise. Her visual acuity was <20/400 in the right eye and 20/40 in the left eye. Intraocular pressures were raised at 38 and 40 mmHg. Slit lamp examination showed bilateral conjunctival hyperemia, corneal edema, and shallow anterior chambers. Gonioscopy showed bilateral circular closed angles. The pupils were mid-dilated and non-reactive to light. The optic nerve heads in both eyes had slightly enlarged cups. She admitted to recreational use of ecstasy and marijuana before this ophthalmic crisis and also 2 years earlier, when she had had an episode of ophthalmic migraine with headache and transient blurred vision. Ophthalmic examination showed a narrow anterior chamber angle in both eyes.

The authors suggested that the bilateral angle-closure glaucoma had been precipitated by a combined mydriatic effect of ecstasy and cannabis.

Ears

The effect of THC, 7.5 mg and 15 mg, on auditory functioning has been investigated in eight men in a double-blind, randomized, placebo-controlled, crossover trial (97). Blood concentrations of THC were measured for up to 48 hours after ingestion, and audiometric tests were carried out at 2 hours. There were no significant differences across treatments, suggesting that cannabis does not affect the basic unit of auditory perception.

Psychological, psychiatric

The psychological effects of cannabis vary with personal and social factors. However, some guidance to the essential effects of the drug can be derived from investigations with THC and marijuana in non-user volunteers. Blood concentrations of THC over 75 µg/ml under these conditions are associated with euphoria, and somewhat higher concentrations with dissociation of events and memory and impairment of psychomotor tasks lasting over 24 hours (61).

Through random urine testing of draftees to the Italian army, 133 marijuana users were identified, tested, and interviewed (98). Among these marijuana users, 83% of those with cannabis dependence, 46% with cannabis abuse, and 29% of occasional users had at least one DSM-IIIR psychiatric diagnosis. With greater cannabis use, the risk of associated psychiatric disabilities tended to increase progressively.

Occasional and regular users can suffer panic attacks, paranoia, hallucinations, or feelings of unreality (depersonalization and derealization).

Behavior

In a critical English-language literature review of the cannabis research done during the 10 years from 1994 to 2004 the relation between the rate of cannabis use, behavioral problems, and mental disorders in young people was explored (Rey 1194). Although there are shortcomings in the studies done in this area, the data suggest that early and heavy use of cannabis has negative effects on psychosocial functioning and psychopathology. Although infrequent use causes few mental health or behavioral problems, cannabis is not necessarily harmless. Accumulating evidence suggests that regular marijuana use during adolescence may have effects, whether biological, psychological, or social, that are different from those in later life. Most recent data challenge the notion that marijuana relieves psychotic or depressive symptoms.

The use of marijuana is related to risky behaviors that may result in other drug use, high-risk sexual activity, risky car driving, traffic accidents, and crime. The acute effects of marijuana on human risk taking has been investigated in a laboratory setting in 10 adults who were given three doses of active marijuana cigarettes (half placebo and half 1.77%, 1.77%, and 3.58 % tetrahydrocannabinol) and placebo cigarettes (99). There were measurable changes in risky decisions after marijuana. Tetrahydrocannabinol 3.58% increased selection of the risky response option and also caused shifts in trial-by-trial response probabilities, suggesting altered sensitivity

to both reinforced and losing risky outcomes. The authors suggested that the effect on risk taking was possibly seen only at the 3.58% dose because it created a requisite level of impairment to disrupt inhibitory processes in the meso-limbic-prefrontal cortical network.

Cognitive effects

Long-term heavy use of cannabis impairs mental performance, causes defects in memory (especially short-term memory), and leads to impairment of memory, attention, and organization and integration of complex information (100). Adolescents with pre-existing disabilities in learning and cognition have experienced serious aggravation of their problem from regular use of cannabis (101).

The effects of chronic marijuana smoking on human brain function and cognition have been further investigated (75). Normalized regional brain blood flow and regional brain metabolism, measured using PET scanning with ^{15}O, were compared in 17 frequent marijuana users and 12 non-users. Testing was performed after at least 26 hours of monitored abstinence in all subjects. Marijuana users had hypoactivity or reduced brain blood flow in a large region of the posterior cerebellum compared with controls. This is consistent with what was reported in the only previous PET study of chronic marijuana use (102). The cerebellum is hypothesized to have input to aspects of cognition, specifically timing, the processing of sensory information, and attention and prediction of real-time events. Users often report that marijuana smoking is followed by alterations in the sense of time and less efficient cognitive processing.

The safety and possible benefits of long-term marijuana use have been studied in four seriously ill patients in the Missoula Chronic Clinical Cannabis Use Study with a quality-controlled sample of marijuana (103). They were evaluated using an extensive neurocognitive battery.

- A 62-year-old woman with congenital cataracts smoked marijuana illicitly for 12 years (current use 3–4 g/day smoked and 3–4 g/day orally). She had mild-to-moderate difficulty with attention and concentration and minimal-to-mild difficulty with acquisition and storage of very complex new verbal material. Her executive functioning was not affected.

- A 50-year-old man with hereditary osteo-onychodysplasia had smoked marijuana since 1974 to alleviate muscle spasms and pain (current use 7 g/day of 3.75% THC). He had mild-to-moderate impairment of attention and concentration and reduced ability to acquire new verbal material. He scored poorly on the California Verbal Learning Test (CVLT), a measure of short-term memory recall, and had difficulty with motor tasks.

- A 48-year-old man with multiple congenital cartilaginous exostoses had smoked marijuana since the late 1970s (current use 9 g/day of 2.7% THC). His neurocognitive scores suggest mild difficulty in sustaining attention and a minimal-to-mild deficit in the acquisition of new verbal material.

- A 45-year-old woman with multiple sclerosis had smoked cannabis since 1990 to control pain and muscle spasms (current use marijuana cigarettes containing 3.5% THC 10/day). She had impairment of concentration, learning, and memory efficiency. Her ability to acquire new verbal information was also impaired.

The authors attributed these cognitive deficits not to marijuana use but rather to the patients' illnesses, arguing that it is difficult for patients with painful debilitating diseases to concentrate on neurocognitive tasks. Any abnormalities in MRI imaging and electroencephalography were attributed to age-related brain deterioration. There were no significant abnormalities of respiratory function, apart from a "slight downward trend in FEV_1 and FEV_1/FVC ratios, and perhaps an increase in FVC" in three patients, interpretation of these findings being complicated by concomitant tobacco smoking. One patient had mild polycythemia and a raised white cell count. None had abnormal endocrine tests. This was a comprehensive study of the long-term effects of cannabis, but concomitant illnesses and use of tobacco made the results difficult to interpret.

Concerns have been raised about the possible adverse effects of acute as well as chronic medicinal and recreational use of cannabis on cognition and the body (104). The author, while acknowledging the therapeutic role of cannabinoids in the management of pain and other conditions, expressed concern that in recent years the prevalence of recreational cannabis use (especially in the young) and the potency of the available products have markedly increased in the UK.

An unusual account of transient amnesia after marijuana use has been reported from Europe (105).

- A 40-year-old healthy man with a long history of cannabis use was hospitalized with an acute memory disturbance after smoking for several hours a strong type of marijuana called "superskunk." After smoking, he had difficulty recollecting recent events and would ask the same questions repeatedly. While his routine laboratory results were within the reference ranges, his urine and blood toxic screens had very high concentrations of cannabinoids (and no other drugs). He was alert and oriented to his name, address, date, and place of birth, but could not recall his marital status, whether he had children, or the nature of his job. He was disoriented in time. He performed normally in tests of general cognitive functioning (for example Raven's matrices, word fluency, Rey's complex figurecopy) and short-term memory (for example digit span, verbal cues), but showed severe impairment in verbal and non-verbal long-term components of anterograde memory tests. He had a severe retrograde memory defect mainly affecting autobiographical memory, with a temporal gradient such that remote facts were preserved. These memory impairments lasted 4 days and then rapidly improved, leaving amnesia for the acute episode. Electroencephalography during the amnestic episode was normal, except for brief trains of irregular slow activity in the frontal areas bilaterally. A SPECT scan of his brain was normal. A week later, repeat neuropsychological examination showed normal memory and a normal electroencephalogram and MRI scan of the brain with enhancement. One year later, he had stopped using marijuana and had no further amnestic episodes.

The authors found similarities between the memory disorder seen here and transient global amnesia, which consists of anterograde amnesia and a variably graded retrograde amnesia. Such amnesia has been previously reported with a number of substances and medications, but not with marijuana. The authors stated that although memory impairment has been reported with marijuana before, it has never involved retrieval of already learned material. They wondered if the memory impairment was due to marijuana-induced changes in cerebral blood flow and ischemia through vasospasm. However, their SPECT data did not support this theory. They considered the possibility that cannabinoid receptors, which are dense in the hippocampus, could have been occupied by marijuana, resulting in such memory loss. They cautioned that the effects of marijuana on memory may be more severe than previously thought.

The neurocognitive effects of marijuana have been studied in 113 young adults (106). Marijuana users, identified by self-reporting and urinalysis, were categorized as light users (<5 joints per week) or heavy users (5 or more joints per week) and current users or former users, the latter having used the drug regularly in the past (1 or more joint per week) but not for at least 3 months. IQ, memory, processing speed, vocabulary, attention, and abstract reasoning were assessed. Current regular heavy users performed significantly worse than non-users beyond the acute intoxication period. Memory, both immediate and delayed, was most strongly affected. However, after 3 months abstinence, there were no residual effects of marijuana, even among those who had formerly had heavy use.

Delta-9-tetrahydrocannabinol (THC) activates cannabinoid receptors in frontal cortex and hippocampus. Electroencephalograms were obtained from 10 subjects who performed cognitive tasks before and after smoking marijuana or a placebo, to examine the effects on performance and neurophysiology signals of cognitive functions (107). Marijuana increased heart rate and reduced global theta band electroencephalographic power, consistent with increased autonomic arousal. Responses in working memory tasks were slower and less accurate after smoking marijuana, and were accompanied by reduced alphaband electroencephalographic reactivity in response to increased task difficulty. Marijuana disrupted both sustained and transient attention processes, resulting in impaired memory task performance. In the episodic memory task, marijuana use was associated with an increased tendency to identify distracter words erroneously as having been previously studied. In both tasks, marijuana attenuated stimulus-locked event-related potentials (ERP). In subjects most affected by marijuana, a pronounced ERP difference between previously studied words and new distracter words was also reduced, suggesting disruption of neural mechanisms underlying memory for recent episodes.

In a Vietnamese study of 54 monozygotic male twin pairs who were discordant for regular marijuana use and who had not used any other illicit drug regularly the marijuana users significantly differed from the non-users on the general intelligence domain; however, within that domain only the performance of the block design subtests of Wechsler Adult Intelligence Scale-Revised reached statistical significance (108). The marijuana users had not used it for at least 1 year, and a mean of almost 20 years had passed since the last time marijuana had been used regularly. There were no marked long-term residual effects of marijuana use on cognitive abilities.

Marijuana impairs memory, although it is not clear which components of memory are affected. Memory involves two components: an initial delay-independent discrimination or "encoding" and a second delay-dependent discrimination or "recall" of information. In five subjects tetrahydrocannabinol acutely impaired delay-dependent discrimination but not delay-independent discrimination (109). In other words, smoking marijuana increased the rates of forgetting but did not alter initial discriminability.

Psychosis

The causal relation between cannabis abuse and schizophrenia is controversial. Cannabis abuse, and particularly heavy abuse, can exacerbate symptoms of schizophrenia and can be considered as a risk factor eliciting relapse in schizophrenia (110). Chronic cannabis use can precipitate schizophrenia in vulnerable individuals (111).

Four cases in which psychosis developed after relatively small amounts of marijuana were smoked for the first time have been reported (112). All required hospitalization and neuroleptic drug treatment. Each had a mother with manic disorder and two had psychotic features. The authors noted that marijuana is a dopamine receptor agonist, and mania may be associated with excessive dopaminergic neurotransmission. The use of marijuana may precipitate psychosis or mania in subjects who are genetically vulnerable to major mental illness.

Marijuana abuse and its possible associated risks in reinforcing further use, causing dependence, and producing withdrawal symptoms among adolescents with conduct symptoms and substance use disorders has been investigated in 165 men and 64 women selected and then interviewed from a group of 255 consecutive admissions to a university-based adolescent substance abuse treatment program (113). All had DSM-IIIR substance dependence, 82% had conduct disorder, 18% had major depression, and 15% had attention-deficit/hyperactivity disorder. Most (79%) met the criteria for cannabis dependence. Two-thirds of the cannabis-dependent individuals admitted serious drug-related problems and reported associated drug withdrawal symptoms according to the Comprehensive Addiction Severity Index in adolescents (CASI). For the majority, progression from first to regular cannabis use was as rapid as tobacco progression and more rapid than that of alcohol.

Schizophrenia

Neurotrophins, such as nerve growth factor and brain-derived neurotrophic factor (BDNF), are implicated in neuronal development, growth, plasticity, and maintenance of function. Neurodevelopment is impaired in schizophrenia and vulnerable schizophrenic brains may be more sensitive to toxic influences. Thus, cannabis may be more neurotoxic to schizophrenic brains than to non-schizophrenic brains when used chronically. In 157 drug-naïve first-episode schizophrenic patients there were significantly raised BDNF serum concentrations by up to 34% in patients with chronic cannabis abuse or multiple substance abuse before the onset of the disease (114). Thus, raised BDNF serum concentrations are not related to schizophrenia and /or substance abuse itself but may reflect cannabis-related idiosyncratic damage to the schizophrenic brain. Disease onset was 5.2 years earlier in the cannabis-consuming group.

Cannabis is commonly used in by people with schizophrenia, and there is a possible association between the drug and the presentation, symptoms, and course of the illness. The effects of marijuana on psychotic symptoms and cognitive deficits in schizophrenia have been studied in 13 medicated stable patients and 13 healthy subjects in a double-blind, placebo-controlled randomized study using 2.5 and 5 mg of intravenous delta-9-tetrahydrocannabinol (THC) (115). Tetrahydrocannabinol transiently worsened cognitive deficits, perceptual alterations, and a range of positive and negative symptoms in those with schizophrenia. There were no positive effects. These results suggest a role for cannabinoid receptors in the pathophysiology of schizophrenia.

Endocrine

In animals (particularly monkeys), cannabis depresses ovarian and testicular function. In man, chronic use has been associated with reduced serum FSH and LH concentrations in a few people, often accompanied by reduced serum testosterone, oligospermia, reduced sperm motility, and gynecomastia (116). There is no evidence of impairment of male fertility; no studies have been carried out on female fertility. There is evidence of slightly shortened gestation periods in chronic users (117). There are variable non-specific effects on serum prolactin and growth hormone and a rise in plasma cortisol concentrations has been recorded in one study.

Hematologic

Of the hematological changes very occasionally noted, polycythemia appears to be secondary to reduced pulmonary oxygen exchange (see the Respiratory section in this monograph).

Gastrointestinal

Although cannabis has been used as an antiemetic, in 19 patients it was associated with cyclical hyperemesis, and in seven cases withdrawal was followed by the disappearance of symptoms; in three cases there was a positive re-challenge (118).

Immunologic

Tetrahydrocannabinol depresses lymphocyte and macrophage activity in cell cultures, while in rats in vivo it directly suppresses natural killer cell activity and impairs T lymphocyte transformation by phytohemagglutinin in concentrations of cannabinoids achievable with the usual doses (119). Variable results have been obtained in man in tests of circulating T cells and hormonal immunity (120).

In animals and man, chronic use often suppresses the immune system's response to inhaled bacterial or fungal material. In this connection it is relevant to note that a contaminant mould (*Aspergillus*) found in cannabis can predispose immunocompromised cannabis smokers to infection. It has been suggested that baking the cannabis (at 300°F for 15 minutes) before smoking will kill the fungus and reduce the potential risk (121).

The effects of marijuana on immune function have been reviewed (122). The studies suggest that marijuana affects immune cell function of T and B lymphocytes, natural killer cells, and macrophages. In addition, cannabis appears to modulate host resistance, especially the secondary immune response to various infectious agents, both viral and bacterial. Lastly, marijuana may also affect the cytokine network, influencing the production and function of acute-phase and immune cytokines and modulating network cells, such as macrophages and T helper cells. Under some conditions, marijuana may be immunomodulatory and promote disease.

The effects of cannabinoids on the immune system have been examined in two separate studies. In the first of these the effects of oral cannabinoids on immune functioning were studied in 16 patients with multiple sclerosis in a crossover study of dronabinol, *Cannabis sativa* plant extract, or placebo for 4 weeks (123). There was a modest increase in pro-inflammatory cytokine tumor necrosis factor alfa during cannabis plant extract treatment in all the subjects. Those with high adverse event scores (n=7) had significant increases in pro-inflammatory plasma cytokine IL-12p40 while taking the plant extract; this was not the case with tetrahydrocannabinol. Other cytokines were not affected. Tumor necrosis factor alfa and IL-12p40 are known to worsen the course of multiple sclerosis (124). These results are interesting because they suggest immunoactivation by cannabinoids in patients with multiple sclerosis, rather than immunosuppression, as previously reported with the plant extract (125). More studies are needed, because these pro-inflammatory effects could have negative influence on the course of the disease.

A severe allergic reaction after intravenous marijuana has been reported (126).

- A 25-year-old man with intermittent metamfetamine use developed facial edema, pruritus, and dyspnea 45 minutes after injecting a mixture of crushed marijuana leaves and heated water. He was anxious, and had tachypnea, respiratory stridor, wheezing, edema of the face and oral mucosa, and truncal urticaria. There was mild pre-renal uremia and urine toxicology was positive for metamfetamine and marijuana. Skin testing was not done. With appropriate medical intervention there was resolution of symptoms within a day.

The authors noted that marijuana may have contaminants, including *Aspergillus*, *Salmonella*, herbicides, and mercury, which can trigger allergic reactions.

Long-Term Effects

Drug tolerance

Tolerance develops with heavy chronic use in individuals who report problems in controlling their use and who continue to use cannabis despite adverse personal consequences (60).

Drug withdrawal

Withdrawal symptoms occur after chronic heavy use (60). Abrupt withdrawal of high-level use of cannabinoids causes irritability, restlessness, and insomnia, with a rebound increase in REM sleep, tremor, and anorexia lasting up to a week (127–129). Occasional use does not appear to be associated with major consequences.

Tumorigenicity

Three different associations of cannabinoids with cancer have been discussed (130). Firstly, there is a possible direct carcinogenic effect. In in vitro studies and in mice tetrahydrocannabinol alone does not seem to be carcinogenic or mutagenic. However, cannabis smoke is both carcinogenic and mutagenic and contains similar carcinogens to those in tobacco smoke. Cannabis is possibly linked to digestive and respiratory system cancers. Case reports support this association but epidemiological cohort studies and case-control studies have provided conflicting evidence. Secondly, there is conflicting evidence on the beneficial effects of tetrahydrocannabinol and other cannabinoids in patients with cancer. In some in vitro and in vivo studies, tetrahydrocannabinol and synthetic cannabinoids had antineoplastic effects, but in others tetrahydrocannabinol had a negative effect on the immune system. No anticancer effects of tetrahydrocannabinol in humans have so far been reported. Thirdly, cannabis may palliate some of the symptoms and adverse effects of cancer. Cannabis may improve appetite, reduce nausea and vomiting, and alleviate moderate neuropathic pain in patients with cancer. The authors defined the challenge for the medical use of cannabinoids as the development of safe, effective, and therapeutic methods of using it that are devoid of the adverse psychoactive effects. Lastly, they discussed the possible associations between cannabis smoking and tumors of the prostate and brain, noting the need for larger, controlled studies.

Second-Generation Effects

Pregnancy

Behavioral anomalies have been identified in the offspring of monkeys and women exposed to cannabis during pregnancy (131,132). These include reduced visual responses, increased auditory responses, and reduced quietude. Most of the effects resolved within 4–5 weeks postpartum and there were no abnormalities at 1 year.

Teratogenicity

In animals, THC crosses the placenta and is excreted in breast milk. There is conflicting evidence concerning teratogenicity in animals, but no definitive evidence in man. However, there have been many anecdotal reports of abnormalities. Although these were without consistent characteristics, the descriptions would readily fit the fetal alcohol syndrome (133–136) and clinical evaluation of the use of cannabis during pregnancy is complicated by the frequent concomitant use of alcohol and tobacco.

Fetotoxicity

The effect of maternal and prenatal marijuana exposure on offspring from birth to adolescence is being investigated (137). The Ottawa Prenatal Prospective Study (OPPS), a longitudinal project begun in 1978, recently reported its findings in 146 low-risk, middle-class children aged 9–12 years. Their performances on neurobehavioral tasks that focus on visuoperceptual abilities (ranging from basic skills to those requiring integration and cognitive manipulation of such skills) were analysed. Performance outcomes were different in children with prenatal exposure to cigarette smoking and those with prenatal exposure to marijuana. Maternal cigarette smoking affected fundamental visuoperceptual functioning. Prenatal marijuana use had a negative effect on performance in visual problem-solving, which requires integration, analysis, and synthesis. In a second prospective study, the effects of prenatal marijuana exposure and child behavior problems were studied in 763 subjects aged 10 years (138). Prenatal maternal marijuana exposure was associated with increased hyperactivity, impulsivity, and inattention in the children. There was also increased delinquency and externalizing problems. The authors suggested a possible pathway between prenatal marijuana exposure and delinquency, which may be mediated by the effects of marijuana exposure on symptoms of inattention.

The effects of prenatal marijuana exposure on cognitive functioning have been examined in 145 children aged 13-16 years (139). The age breakdown was 45 13-year-olds, 36 14-year-olds, 51 15-year-olds, and 13 16-year-olds. These groups were further classified by maternal marijuana use: less than six joints per week (n = 120) and six or more joints per week (n = 25). A standard neuro-cognitive test battery was administered, and two of the tests differed significantly between non-users/light users and heavy users. On the Abstract Designs test the children of heavy users had significantly slower response

times. Children in the heavy user group also scored significantly lower on the Peabody Spelling test. These results suggest a dose-related effect of prenatal marijuana exposure on cognition. The two tests that differed across groups depend, to a lesser degree, on cognitive manipulation or comprehension and, to a greater degree, on visual memory, analysis, and integration. Unlike cigarette exposure, marijuana does not seem to affect overall intelligence. It is therefore possible that heavy marijuana use during pregnancy causes subtle deficits in visual analysis.

Cannabis is the illicit drug that is most commonly used by young women and they are not likely to withdraw until the early stages of pregnancy. The effects of early maternal marijuana use on fetal growth have been reported in pregnant women who elected voluntary saline-induced abortion at mid-gestation (weeks 17–22) (140). Marijuana (n = 44) and non-marijuana exposed fetuses (n = 95) were compared and adjusted for maternal alcohol and cigarette use. Both fetal foot length and body weight were significantly reduced by marijuana. Fetal growth impairment was greatest in the group with moderate, regular exposure to about 3–6 joints/week and not in those with heavy maternal marijuana use. There was no significant effect on fetal body length and head circumference due to prenatal marijuana exposure.

Susceptibility Factors

Age

Children

In young children, accidental ingestion leads to the rapid onset of drowsiness, hypotonia, dilated pupils, and coma. Fortunately, gradual recovery occurs spontaneously, barring accidents. Passive inhalation of marijuana in infants can have serious consequences.

- A 9-month-old girl presented with extreme lethargy and a modified Glasgow coma scale of 10, after having been exposed to cigarette and cannabis smoke at the home of her teenage sister's friend (141). The physical examination and laboratory results were unremarkable. Cannabinoids were detected in a urine screen.

While chronic adult users can display apathy and impaired concentration, these effects are possibly in part associated with other factors. No permanent organic brain damage has been demonstrated (141,142).

HIV infection

The use of cannabinoids has been studied in 62 patients with HIV-1 infection (143). Cannabinoids and HIV are of interest because there is the chance of an interaction between tetrahydrocannabinol and antiretroviral therapy. Tetrahydrocannabinol inhibits the metabolism of other drugs (144,145) and cannabinoids are broken down by the same cytochrome P-450 enzymes that metabolize HIV protease inhibitors. The subjects were randomly assigned to marijuana, dronabinol (synthetic delta-9-tetrahydrocannabinol), or placebo, given three times a day, 1 hour before meals. The amounts of HIV RNA in the blood did not increase significantly over the course of the study and there were no significant effects on CD4+ or CD8+ cell counts. However, there was significant weight gain in both cannabinoid groups compared with placebo. Although this study was of very short duration, the results suggested that either oral or smoked marijuana may be safe for individuals with HIV-1.

Other features of the patient

People with pre-existing coronary artery disease may have an increased incidence of attacks of angina (146). In individuals who are vulnerable to schizophrenia, cannabis can precipitate psychoses or aggravate schizophrenia. The control of epilepsy may be impaired. Users undergoing anesthesia may react unexpectedly and may have enhanced nervous system depression. Because of impairment of judgement and psychomotor performance, users should not drive or operate machinery for at least 24 hours after administration.

Drug–Drug Interactions

Alcohol

Additive psychoactive effects sought by users may be achieved by combinations of cannabis and alcohol, but at the same time the ability of THC to induce microsomal enzymes will increase the rate of metabolism of alcohol and so reduce the additive effects (127).

Anticholinergic drugs

The anticholinergic effects of cannabis (127) may result in interactions with other drugs with anticholinergic effects, such as some antidysrhythmic drugs.

Barbiturates, short-acting

Additive psychoactive effects sought by users may be achieved by combinations of cannabis and short-acting barbiturates, but at the same time the ability of THC to induce microsomal enzymes will increase the rate of metabolism of barbiturates and so reduce the additive effects (127).

Disulfiram

Concurrent administration with disulfiram is associated with hypomania (147).

Indinavir

The effects of smoked marijuana (3.95% tetrahydrocannabinol; up to three cigarettes per day) and oral dronabinol (2.5 mg tds) on the pharmacokinetics of indinavir 800 mg 8-hourly ($n = 28$) have been evaluated in a randomized, placebo-controlled study in HIV-infected patients (148). On day 14, marijuana reduced the 8-hour AUC of indinavir by 15%, the C_{max} by 14%, and the C_{min}

by 34%. However, only the change in C_{max} was significant. Dronabinol had no effects.

Lysergic acid diethylamide

"Flashbacks," or the return of hallucinogenic effects, occur in almost a quarter of those who have used LSD, particularly if they have also used other CNS stimulants, such as alcohol or marijuana. They can experience distortions of perception of objects, space, or time, which intrude without warning into reality, resulting in delusions, panic, and unusual images. A "trailing phenomenon" has also been reported, in which the visual perception of objects is reduced to a series of interrupted pictures rather than a constant view. The frequency of these events may slowly abate over several years, but in a significant number their incidence later increases (149,150).

Nelfinavir

The effects of smoked marijuana (3.95% tetrahydrocannabinol; up to three cigarettes per day) and oral dronabinol (2.5 mg tds) on the pharmacokinetics of nelfinavir 750 mg tds ($n = 34$) have been evaluated in a randomised, placebo-controlled study in HIV-infected patients (151). At day 14, marijuana reduced the 8-hour AUC of nelfinavir by 10%, the C_{max} by 17%, and the C_{min} by 12%. However, none of these changes was significant. Dronabinol had no effects.

Psychotropic drugs

Cannabis alters the effects of psychotropic drugs, such as opioids, anticholinergic drugs, and antidepressants, although variably and unpredictably (127).

Sildenafil

Myocardial infarction has been attributed to the combination of cannabis and sildenafil.

- A 41-year-old man developed chest tightness radiating down both arms (152). He had taken sildenafil and cannabis recreationally the night before. His vital signs were normal and he had no signs of heart failure. However, electrocardiography showed an inferior evolving non-Q-wave myocardial infarct and his creatine kinase activity was raised (431 U/l).

Cannabis inhibits CYP3A4, which is primarily responsible for the metabolism of sildenafil, increased concentrations of which may have caused this cardiac event.

References

1. Caswell A. Marijuana as medicine. Med J Aust 1992;156(7):497–8.
2. Voelker R. Medical marijuana: a trial of science and politics. JAMA 1994;271(21):1645–8.
3. Goodin D. Marijuana and multiple sclerosis. Lancet Neurol 2004;3(2):79–80.

4. Rey JM, Martin A, Krabman P. Is the party over? Cannabis and juvenile psychiatric disorder: the past 10 years. J Am Acad Child Adolesc Psychiatry 2004;43(10):1194–205.

5. Woody GE, MacFadden W. Cannabis related disorders. In: Kaplan HI, Sadock B, editors. 6th ed.Comprehensive Textbook of Psychiatry vol. 1. Baltimore: Williams & Wilkins, 1995:810–7.

6. Herkenham M. Cannabinoid receptor localization in brain: relationship to motor and reward systems. Ann NY Acad Sci 1992;654:19–32.

7. Childers SR, Fleming L, Konkoy C, Marckel D, Pacheco M, Sexton T, Ward S. Opioid and cannabinoid receptor inhibition of adenylyl cyclase in brain. Ann NY Acad Sci 1992;654:33–51.

8. Aronson J. When I use a word···: Sloe gin. BMJ 1997;314:1106.

9. Iversen LL. The Science of MarijuanaNew York: Oxford University Press;. 2000.

10. Braude MC. Toxicology of cannabinoids. In: Paton WM, Crown J, editors. Cannabis and its Derivatives. Oxford: Oxford University Press, 1972:89–99.

11. Chan PC, Sills RC, Braun AG, Haseman JK, Bucher JR. Toxicity and carcinogenicity of delta 9-tetrahydrocannabinol in Fischer rats and B6C3F1 mice. Fundam Appl Toxicol 1996;30(1):109–17.

12. Fleischman RW, Baker JR, Rosenkrantz H. Pulmonary pathologic changes in rats exposed to marihuana smoke for 1 year. Toxicol Appl Pharmacol 1979;47(3):557–66.

13. Roy PE, Magnan-Lapointe F, Huy ND, Boutet M. Chronic inhalation of marijuana and tobacco in dogs: pulmonary pathology. Res Commun Chem Pathol Pharmacol 1976;14(2):305–17.

14. Fligiel SE, Beals TF, Tashkin DP, Paule MG, Scallet AC, Ali SF, Bailey JR, Slikker W Jr. Marijuana exposure and pulmonary alterations in primates. Pharmacol Biochem Behav 1991;40(3):637–42.

15. Scallet AC, Uemura E, Andrews A, Ali SF, McMillan DE, Paule MG, Brown RM, Slikker W Jr. Morphometric studies of the rat hippocampus following chronic delta-9-tetrahydrocannabinol (THC). Brain Res 1987;436(1):193–8.

16. Landfield PW, Cadwallader LB, Vinsant S. Quantitative changes in hippocampal structure following long-term exposure to delta 9-tetrahydrocannabinol: possible mediation by glucocorticoid systems. Brain Res 1988;443(1–2):47–62.

17. Harper JW, Heath RG, Myers WA. Effects of *Cannabis sativa* on ultrastructure of the synapse in monkey brain. J Neurosci Res 1977;3(2):87–93.

18. Heath RG, Fitzjarrell AT, Fontana CJ, Garey RE. *Cannabis sativa*: effects on brain function and ultrastructure in rhesus monkeys. Biol Psychiatry 1980;15(5):657–90.

19. Scallet AC. Neurotoxicology of cannabis and THC: a review of chronic exposure studies in animals. Pharmacol Biochem Behav 1991;40(3):671–6.

20. Downer E, Boland B, Fogarty M, Campbell V. Delta 9-tetrahydrocannabinol induces the apoptotic pathway in cultured cortical neurones via activation of the CB_1 receptor. Neuroreport 2001;12(18):3973–8.

21. Chan GC, Hinds TR, Impey S, Storm DR. Hippocampal neurotoxicity of Δ9-tetrahydrocannabinol. J Neurosci 1998;18(14):5322–32.

22. Sanchez C, Galve-Roperh I, Canova C, Brachet P, Guzman M. Δ9-tetrahydrocannabinol induces apoptosis in C6 glioma cells. FEBS Lett 1998;436(1):6–10.

23. Galve-Roperh I, Sanchez C, Cortes ML, del Pulgar TG, Izquierdo M, Guzman M. Anti-tumoral action of cannabinoids: involvement of sustained ceramide accumulation and extracellular signal-regulated kinase activation. Nat Med 2000;6(3):313–9.

24. Bifulco M, Laezza C, Portella G, Vitale M, Orlando P, De Petrocellis L, Di Marzo V. Control by the endogenous cannabinoid system of ras oncogene-dependent tumor growth. FASEB J 2001;15(14):2745–7.

25. Guzman M, Sanchez C, Galve-Roperh I. Control of the cell survival/death decision by cannabinoids. J Mol Med 2001;78(11):613–25.

26. Nagayama T, Sinor AD, Simon RP, Chen J, Graham SH, Jin K, Greenberg DA. Cannabinoids and neuroprotection in global and focal cerebral ischemia and in neuronal cultures. J Neurosci 1999;19(8):2987–95.

27. Panikashvili D, Simeonidou C, Ben-Shabat S, Hanus L, Breuer A, Mechoulam R, Shohami E. An endogenous cannabinoid (2-AG) is neuroprotective after brain injury. Nature 2001;413(6855):527–31.

28. van der Stelt M, Veldhuis WB, Bar PR, Veldink GA, Vliegenthart JF, Nicolay K. Neuroprotection by Δ9-tetrahydrocannabinol, the main active compound in marijuana, against ouabain-induced in vivo excitotoxicity. J Neurosci 2001;21(17):6475–9.

29. Shen M, Thayer SA. Cannabinoid receptor agonists protect cultured rat hippocampal neurons from excitotoxicity. Mol Pharmacol 1998;54(3):459–62.

30. Sinor AD, Irvin SM, Greenberg DA. Endocannabinoids protect cerebral cortical neurons from in vitro ischemia in rats. Neurosci Lett 2000;278(3):157–60.

31. Hampson AJ, Grimaldi M, Axelrod J, Wink D. Cannabidiol and (-)Δ9-tetrahydrocannabinol are neuroprotective antioxidants. Proc Natl Acad Sci USA 1998;95(14):8268–73.

32. Zimmer L, Morgan JP. Marijuana Myths, Marijuana FactsNew York: Lindesmith Centre;. 1997.

33. Pertwee RG. Tolerance to and dependence on psychotropic cannabinoids. In: Pratt J, editor. The Biological Basis of Drug Tolerance. London: Academic Press, 1991:232–65.

34. Aceto MD, Scates SM, Lowe JA, Martin BR. Dependence on delta 9-tetrahydrocannabinol: studies on precipitated and abrupt withdrawal. J Pharmacol Exp Ther 1996;278(3):1290–5.

35. Rubino T, Patrini G, Massi P, Fuzio D, Vigano D, Giagnoni G, Parolaro D. Cannabinoid-precipitated withdrawal: a time-course study of the behavioral aspect and its correlation with cannabinoid receptors and G protein expression. J Pharmacol Exp Ther 1998;285(2):813–9.

36. Rodriguez de Fonseca F, Carrera MR, Navarro M, Koob GF, Weiss F. Activation of corticotropin-releasing factor in the limbic system during cannabinoid withdrawal. Science 1997;276(5321):2050–4.

37. Diana M, Melis M, Muntoni AL, Gessa GL. Mesolimbic dopaminergic decline after cannabinoid withdrawal. Proc Natl Acad Sci USA 1998;95(17):10269–73.

38. Tanda G, Pontieri FE, Di Chiara G. Cannabinoid and heroin activation of mesolimbic dopamine transmission by a common mu1 opioid receptor mechanism. Science 1997;276(5321):2048–50.

39. Tanda G, Munzar P, Goldberg SR. Self-administration behavior is maintained by the psychoactive ingredient of marijuana in squirrel monkeys. Nat Neurosci 2000;3(11):1073–4.

40. Ledent C, Valverde O, Cossu G, Petitet F, Aubert JF, Beslot F, Bohme GA, Imperato A, Pedrazzini T,

Roques BP, Vassart G, Fratta W, Parmentier M. Unresponsiveness to cannabinoids and reduced addictive effects of opiates in CB_1 receptor knockout mice. Science 1999;283(5400):401–4.

41. Lepore M, Vorel SR, Lowinson J, Gardner EL. Conditioned place preference induced by delta 9-tetrahydrocannabinol: comparison with cocaine, morphine, and food reward. Life Sci 1995;56(23–24):2073–80.

42. Manzanares J, Corchero J, Romero J, Fernandez-Ruiz JJ, Ramos JA, Fuentes JA. Pharmacological and biochemical interactions between opioids and cannabinoids. Trends Pharmacol Sci 1999;20(7):287–94.

43. Kaymakcalan S, Ayhan IH, Tulunay FC. Naloxone-induced or postwithdrawal abstinence signs in Δ9-tetrahydrocannabinol-tolerant rats. Psychopharmacology (Berl) 1977;55(3):243–9.

44. Hine B, Friedman E, Torrelio M, Gershon S. Morphine-dependent rats: blockade of precipitated abstinence by tetrahydrocannabinol. Science 1975;187(4175):443–5.

45. Yamaguchi T, Hagiwara Y, Tanaka H, Sugiura T, Waku K, Shoyama Y, Watanabe S, Yamamoto T. Endogenous cannabinoid, 2-arachidonoylglycerol, attenuates naloxone-precipitated withdrawal signs in morphine-dependent mice. Brain Res 2001;909(1–2):121–6.

46. Pontieri FE, Monnazzi P, Scontrini A, Buttarelli FR, Patacchioli FR. Behavioral sensitization to heroin by cannabinoid pretreatment in the rat. Eur J Pharmacol 2001;421(3):R1–3.

47. Valverde O, Maldonado R, Valjent E, Zimmer AM, Zimmer A. Cannabinoid withdrawal syndrome is reduced in pre-proenkephalin knock-out mice. J Neurosci 2000;20(24):9284–9.

48. Valverde O, Noble F, Beslot F, Dauge V, Fournie-Zaluski MC, Roques BP. Δ9-tetrahydrocannabinol releases and facilitates the effects of endogenous enkephalins: reduction in morphine withdrawal syndrome without change in rewarding effect. Eur J Neurosci 2001;13(9):1816–24.

49. Lichtman AH, Fisher J, Martin BR. Precipitated cannabinoid withdrawal is reversed by Δ(9)-tetrahydrocannabinol or clonidine. Pharmacol Biochem Behav 2001;69(1–2):181–8.

50. Lichtman AH, Sheikh SM, Loh HH, Martin BR. Opioid and cannabinoid modulation of precipitated withdrawal in Δ(9)-tetrahydrocannabinol and morphine-dependent mice. J Pharmacol Exp Ther 2001;298(3):1007–14.

51. US Department of Health and Human Services. National Household Survey on Drug Abuse, 1982–94. Computer Files (ICPSR Version)Ann Arbor, MI: Inter-University Consortium for Political Social;. 1999.

52. Morral AR, McCaffrey DF, Paddock SM. Reassessing the marijuana gateway effect. Addiction 2002;97(12):1493–504.

53. Wehner FC, van Rensburg SJ, Thiel PG. Mutagenicity of marijuana and Transkei tobacco smoke condensates in the Salmonella/microsome assay. Mutat Res 1980;77(2):135–42.

54. Leuchtenberger C, Leuchtenberger R. Cytological and cytochemical studies of the effects of fresh marihuana smoke on growth and DNA metabolism of animal and human lung cultures. In: Braude MC, Szara S, editors. The Pharmacology of Marijuana. New York: Raven Press, 1976:595–612.

55. Brady CM, DasGupta R, Dalton C, Wiseman OJ, Berkley KJ, Fowler CJ. An open-label pilot study of cannabis-based extracts for bladder dysfunction in advanced multiple sclerosis. Mult Scler 2004;10(4):425–33.

56. Clark AJ, Ware MA, Yazer E, Murray TJ, Lynch ME. Patterns of cannabis use among patients with multiple sclerosis. Neurology 2004;62(11):2098–100.

57. Wade DT, Makela P, Robson P, House H, Bateman C. Do cannabis-based medicinal extracts have general or specific effects on symptoms in multiple sclerosis? A double-blind, randomized, placebo-controlled study on 160 patients. Mult Scler 2004;10(4):434–41.

58. Svendsen KB, Jensen TS, Bach FW. Does the cannabinoid dronabinol reduce central pain in multiple sclerosis? Randomised double blind placebo controlled crossover trial. BMJ 2004;329(7460):253.

59. Notcutt W, Price M, Miller R, Newport S, Phillips C, Simmons S, Sansom C. Initial experiences with medicinal extracts of cannabis for chronic pain: results from 34 ′N of 1′ studies. Anaesthesia 2004;59(5):440–52.

60. Hall W, Solowij N. Adverse effects of cannabis. Lancet 1998;352(9140):1611–6.

61. Institute of Medicine. Marijuana and HealthWashington, DC: National Academy Press;. 1982.

62. Killestein J, Hoogervorst EL, Reif M, Kalkers NF, Van Loenen AC, Staats PG, Gorter RW, Uitdehaag BM, Polman CH. Safety, tolerability, and efficacy of orally administered cannabinoids in MS. Neurology 2002;58(9):1404–7.

63. Avakian EV, Horvath SM, Michael ED, Jacobs S. Effect of marihuana on cardiorespiratory responses to submaximal exercise. Clin Pharmacol Ther 1979;26(6):777–81.

64. Relman AS. Marijuana and health. N Engl J Med 1982;306(10):603–5.

65. Mittleman MA, Lewis RA, Maclure M, Sherwood JB, Muller JE. Triggering myocardial infarction by marijuana. Circulation 2001;103(23):2805–9.

66. Lindsay AC, Foale RA, Warren O, Henry JA. Cannabis as a precipitant of cardiovascular emergencies. Int J Cardiol 2005;104:230–2.

67. Papp E, Czopf L, Habon T, Halmosi R, Horvath B, Marton Z, Tahin T, Komocsi A, Horvath I, Melegh, B, Toth K. Drug-induced myocardial infarction in young patients. Int J Cardiol 2005;98:169–70.

68. Kosior DA, Filipiak KJ, Stolarz P, Opolski G. Paroxysmal atrial fibrillation following marijuana intoxication: a two-case report of possible association. Int J Cardiol 2001;78(2):183–4.

69. Singh GK. Atrial fibrillation associated with marijuana use. Pediatr Cardiol 2000;21(3):284.

70. Fisher BAC, Ghuran A, Vadamalai V, Antonios TF. Cardiovascular complications induced by cannabis smoking: a case report and review of the literature. Emerg Med J 2005;22:679–80.

71. Rezkalla SH, Sharma P, Kloner RA. Coronary no-flow and ventricular tachycardia associated with habitual marijuana use. Ann Emerg Med 2003;42:365–9.

72. Mathew RJ, Wilson WH, Davis R. Postural syncope after marijuana: a transcranial Doppler study of the hemodynamics. Pharmacol Biochem Behav 2003;75:309–18.

73. Schneider HJ, Jha S, Burnand KG. Progressive arteritis associated with cannabis use. Eur J Vasc Endovasc Surg 1999;18(4):366–7.

74. Ducasse E, Chevalier J, Dasnoy D, Speziale F, Fiorani P, Puppinck P. Popliteal artery entrapment associated with cannabis arteritis. Eur J Vasc Endovasc Surg 2004;27(3):327–32.

75. Wu TC, Tashkin DP, Djahed B, Rose JE. Pulmonary hazards of smoking marijuana as compared with tobacco. N Engl J Med 1988;318(6):347–51.

76. Goodyear K, Laws D, Turner J. Bilateral spontaneous pneumothorax in a cannabis smoker. J Roy Soc Med 2004;97:435–6.

77. Gill A. Bong lung: regular smokers of cannabis show relatively distinctive histologic changes that predispose to pneumothorax. Am J Surg Pathol 2005;29:980–1.

78. Tashkin DP, Calvarese BM, Simmons MS, Shapiro BJ. Respiratory status of seventy-four habitual marijuana smokers. Chest 1980;78(5):699–706.

79. Tashkin DP. Pulmonary complications of smoked substance abuse. West J Med 1990;152(5):525–30.

80. Douglass RE, Levison MA. Pneumothorax in drug abusers. An urban epidemic? Am Surg 1986;52(7):377–80.

81. Tashkin DP, Coulson AH, Clark VA, Simmons M, Bourque LB, Duann S, Spivey GH, Gong H. Respiratory symptoms and lung function in habitual heavy smokers of marijuana alone, smokers of marijuana and tobacco, smokers of tobacco alone, and nonsmokers. Am Rev Respir Dis 1987;135(1):209–16.

82. Johnson MK, Smith RP, Morrison D, Laszlo G, White RJ. Large lung bullae in marijuana smokers. Thorax 2000;55(4):340–2.

83. Thompson CS, White RJ. Lung bullae and marijuana. Thorax 2002;57(6):563.

84. Voruganti LN, Slomka P, Zabel P, Mattar A, Awad AG. Cannabis induced dopamine release: an in-vivo SPECT study. Psychiatry Res 2001;107(3):173–7.

85. Ramaekers JG, Berghaus G, van Laar M, Drummer OH. Dose related risk of motor vehicle crashes after cannabis use. Drug Alcohol Depend 2004;73:109–19.

86. Struve FA, Straumanis JJ, Patrick G, Leavitt J, Manno JE, Manno BR. Topographic quantitative EEG sequelae of chronic marihuana use: a replication using medically and psychiatrically screened normal subjects. Drug Alcohol Depend 1999;56:167–79.

87. Herning RI, Better W, Tate K, Cadet JL. EEG deficits in chronic marijuana abusers during monitored abstinence: preliminary findings. Ann N Y Acad Sci 2003;993:75-8; discussion 79–81.

88. Altable CR, Urrutia AR, Martinez MIC. Cannabis-induced extrapyramidalism in a patient on neuroleptic treatment. J Clin Psychopharmacol 2005;25:91–2.

89. Arsenault L, Cannon M, Witton J. Causal association between cannabis and psychosis: examination of the evidence. Br J Psychiatry 2004;184:110–7.

90. Lozsadi DA, Forster A, Fletcher NA. Cannabis-induced propriospinal myoclonus. Mov Disord 2004;19(6):708–9.

91. Haubrich , Diehl R, Donges M, Schiefer J, Loos M, Kosinski C. Recurrent transient ischemic attacks in a cannabis smoker. J Neurol 2005;252:369–70.

92. Finsterer J, Christian P, Wolfgang K. Occipital stroke shortly after cannabis consumption. Clin Neurol Neurosurg 2004;106(4):305–8.

93. Mateo I, Pinedo A, Gomez-Beldarrain M, Basterretxea JM, Garcia-Monco JC. Recurrent stroke associated with cannabis use. J Neurol Neurosurg Psychiatry 2005;76:435–7.

94. Shukla PC, Moore UB. Marijuana-induced transient global amnesia. South Med J 2004;97(8):782–4.

95. Dawson WW, Jimenez-Antillon CF, Perez JM, Zeskind JA. Marijuana and vision—after ten years' use in Costa Rica. Invest Ophthalmol Vis Sci 1977;16(8):689–99.

96. Trittibach P, Frueh BE, Goldblum D. Bilateral angle-closure glaucoma after combined consumption of "ecstasy" and marijuana. Am J Emerg Med 2005;23:813–4.

97. Mulheran M, Middleton P, Henry JA. The acute effects of tetrahydrocannabinol on auditory threshold and frequency resolution in human subjects. Hum Exp Toxicol 2002;21(6):289–92.

98. Troisi A, Pasini A, Saracco M, Spalletta G. Psychiatric symptoms in male cannabis users not using other illicit drugs. Addiction 1998;93(4):487–92.

99. Lane SD, Cherek DR, Tcheremissine OV, Lieving LM, Pietras CJ. Acute marijuana effects on human risk taking. Neuropsychopharmacology 2005;30:800–9.

100. Solowij N. Cannabis and Cognitive FunctioningCambridge: Cambridge University Press;. 1998.

101. Schwartz RH, Gruenewald PJ, Klitzner M, Fedio P. Short-term memory impairment in cannabis-dependent adolescents. Am J Dis Child 1989;143(10):1214–9.

102. Volkow ND, Gillespie H, Mullani N, Tancredi L, Grant C, Valentine A, Hollister L. Brain glucose metabolism in chronic marijuana users at baseline and during marijuana intoxication. Psychiatry Res 1996;67(1):29–38.

103. Russo E, Mathre ML, Byrne A, Velin R, Bach PJ, Sanchez-Ramos J, Kirlin KA. Chronic cannabis use in the compassionate investigational new drug program: an examination of benefits and adverse effects of legal clinical cannabis. J Cannabis Ther 2002;2:3–57.

104. Ashton CH. Adverse effects of cannabis and cannabinoids. Br J Anaesth 1999;83(4):637–49.

105. Stracciari A, Guarino M, Crespi C, Pazzaglia P. Transient amnesia triggered by acute marijuana intoxication. Eur J Neurol 1999;6(4):521–3.

106. Fried PA, Watkinson B, Gray R. Neurocognitive consequences of marihuana—comparison with pre-drug performance. Neurotoxicol Teratol 2005;27:231–9.

107. Ilan AB, Smith ME, Gevins A. Effects of marijuana on neurophysiological signals of working and episodic memory. Psychopharmacology(Berl) 2004;176(2):214–22.

108. Lyons MJ, Bar JL, Panizzon MS, Toomey R, Eisen S, Xian H, Tsuang MT. Neuropsychological consequences of regular marijuana use: a twin study. Psychol Med 2004;34(7):1239–50.

109. Lane SD, Cherek DR, Lieving LM, Tcheremissine OV. Marijuana effects on human forgetting functions. J Exp Analysis Behav 2005;83:67–83.

110. Linszen DH, Dingemans PM, Lenior ME. Cannabis abuse and the course of recent-onset schizophrenic disorders. Arch Gen Psychiatry 1994;51(4):273–9.

111. Andreasson S, Allebeck P, Engstrom A, Rydberg U. Cannabis and schizophrenia. A longitudinal study of Swedish conscripts. Lancet 1987;2(8574):1483–6.

112. Bowers MB Jr. Family history and early psychotogenic response to marijuana. J Clin Psychiatry 1998;59(4):198–9.

113. Crowley TJ, Macdonald MJ, Whitmore EA, Mikulich SK. Cannabis dependence, withdrawal, and reinforcing effects among adolescents with conduct symptoms and substance use disorders. Drug Alcohol Depend 1998;50(1):27–37.

114. Jockers-Scherubl MC, Danker-Hopfe H, Mahlberg R, Selig F, Rentzsch J, Schurer F, Lang UE, Hellweg R. Brain-derived neurotrophic factor serum concentrations are increased in drug-naive schizophrenic patients with chronic cannabis abuse and multiple substance abuse. Neurosci Lett 2004;371(1):79–83.

115. D'Souza DC, Abi-Saab WM, Madonick S, Forselius-Bielen K, Doersch A, Braley G, Gueorguieva R, Cooper TB, Krystal JH. Delta-9-tetrahydrocannabinol effects in

schizophrenia: implications for cognition, psychosis and addiction. Biol Psychiatry 2005;57:594–608.

116. Kolodny RC, Masters WH, Kolodner RM, Toro G. Depression of plasma testosterone levels after chronic intensive marihuana use. N Engl J Med 1974;290(16):872–4.

117. Fried PA, Watkinson B, Willan A. Marijuana use during pregnancy and decreased length of gestation. Am J Obstet Gynecol 1984;150(1):23–7.

118. Allen JH, de Moore GM, Heddle R, Twartz JC. Cannabinoid hyperemesis: cyclical hyperemesis in association with chronic cannabis abuse. Gut 2004;53:1566–70.

119. Klein TW, Newton C, Friedman H. Inhibition of natural killer cell function by marijuana components. J Toxicol Environ Health 1987;20(4):321–32.

120. Pillai R, Nair BS, Watson RR. AIDS, drugs of abuse and the immune system: a complex immunotoxicological network. Arch Toxicol 1991;65(8):609–17.

121. Levitz SM, Diamond RD. Aspergillosis and marijuana. Ann Intern Med 1991;115(7):578–9.

122. Klein TW, Friedman H, Specter S. Marijuana, immunity and infection. J Neuroimmunol 1998;83(1–2):102–15.

123. Killestein J, Hoogervorst EL, Reif M, Blauw B, Smits M, Uitdehaag BM, Nagelkerken L, Polman CH. Immunomodulatory effects of orally administered cannabinoids in multiple sclerosis. J Neuroimmunol 2003;137:140–3.

124. Huang YM, Liu X, Steffensen K, Sanna A, Arru G, Fois ML, Rosati G, Sotgiu S, Link H. Immunological heterogeneity of multiple sclerosis in Sardinia and Sweden. Mult Scler 2005;11:16–23.

125. Van Boxel-Dezaire AH, Hoff SC, Van Ooosten BW, Verweij CL, Drager AM, Ader HJ, Van Houwelingen JC, Barkhof F, Polman CH, Nagelkerken L. Decreased interleukin-10 and increased interleukin-12p40 mRNA are associated with disease activity and characterize different disease stages in multiple sclerosis. Ann Neurol 1999;45:695–703.

126. Perez JA Jr. Allergic reaction associated with intravenous marijuana use. J Emerg Med 2000;18(2):260–1.

127. Jones RT. Cannabis and health. Annu Rev Med 1983;34:247–58.

128. Carney MW, Bacelle L, Robinson B. Psychosis after cannabis abuse. BMJ (Clin Res Ed) 1984;288(6423):1047.

129. Liakos A, Boulougouris JC, Stefanis C. Psychophysiologic effects of acute cannabis smoking in long-term users. Ann NY Acad Sci 1976;282:375–86.

130. Hall W, Christie M, Currow D. Cannabinoids and cancer: causation, remediation and palliation. Lancet Oncol 2005;6:35–42.

131. Fried PA. Marihuana use by pregnant women: neurobehavioral effects in neonates. Drug Alcohol Depend 1980;6(6):415–24.

132. Abel EL. Prenatal exposure to cannabis: a critical review of effects on growth, development, and behavior. Behav Neural Biol 1980;29(2):137–56.

133. Qazi QH, Mariano E, Milman DH, Beller E, Crombleholme W. Abnormalities in offspring associated with prenatal marihuana exposure. Dev Pharmacol Ther 1985;8(2):141–8.

134. Fried PA. Marihuana use by pregnant women and effects on offspring: an update. Neurobehav Toxicol Teratol 1982;4(4):451–4.

135. Greenland S, Staisch KJ, Brown N, Gross SJ. Effects of marijuana on human pregnancy, labor, and delivery. Neurobehav Toxicol Teratol 1982;4(4):447–50.

136. Nahas G, Frick HC. Developmental effects of cannabis. Neurotoxicology 1986;7(2):381–95.

137. Fried PA, Watkinson B. Visuoperceptual functioning differs in 9- to 12-year olds prenatally exposed to cigarettes and marihuana. Neurotoxicol Teratol 2000;22(1):11–20.

138. Goldschmidt L, Day NL, Richardson GA. Effects of prenatal marijuana exposure on child behavior problems at age 10. Neurotoxicol Teratol 2000;22(3):325–36.

139. Fried PA, Watkinson B, Gray R. Differential effects on cognitive functioning in 13- to 16-year-olds prenatally exposed to cigarettes and marihuana. Neurotoxicol Teratol 2003;25:427–36.

140. Hurd YL, Wang X, Anderson V, Beck O, Minkoff H, Dow-Edwards D. Marijuana impairs growth in mid-gestation fetuses. Neurotoxicol Teratol 2005;27:221–9.

141. Wert RC, Raulin ML. The chronic cerebral effects of cannabis use. II. Psychological findings and conclusions. Int J Addict 1986;21(6):629–42.

142. Wert RC, Raulin ML. The chronic cerebral effects of cannabis use. I. Methodological issues and neurological findings. Int J Addict 1986;21(6):605–28.

143. Abrams DI, Hilton JF, Leiser RJ, Shade SB, Elbeik TA, Aweeka FT, Benowitz NL, Bredt BM, Kosel B, Aberg JA, Deeks SG, Mitchell TF, Mulligan K, Bacchetti P, McCune JM, Schambelan M. Short-term effects of cannabinoids in patients with HIV-1 infection: a randomized, placebo-controlled clinical trial. Ann Intern Med 2003;139:258–66.

144. Benowitz NL, Nguyen TL, Jones RT, Herning RI, Bachman J. Metabolic and psychophysiologic studies of cannabidiol-hexobarbital interaction. Clin Pharmacol Ther 1980;28:115–20.

145. Benowitz NL, Jones RT. Effects of delta-9-tetrahydrocannabinol on drug distribution and metabolism. Antipyrine, pentobarbital, and ethanol. Clin Pharmacol Ther 1977;22:259–68.

146. Aronow WS, Cassidy J. Effect of marihuana and placebo-marihuana smoking on angina pectoris. N Engl J Med 1974;291(2):65–7.

147. Lacoursiere RB, Swatek R. Adverse interaction between disulfiram and marijuana: a case report. Am J Psychiatry 1983;140(2):243–4.

148. Watson SJ. Hallucinogens and other psychotomimetics: biological mechanisms. In: Barchas JD, Berger PA, Cioranello RD, Elliot GR, editors. Psychopharmacology from Theory to Practise. New York: Oxford University Press, 1977:1437.

149. Strassman RJ. Adverse reactions to psychedelic drugs. A review of the literature. J Nerv Ment Dis 1984;172(10):577–95.

150. McLeod AL, McKenna CJ, Northridge DB. Myocardial infarction following the combined recreational use of Viagra and cannabis. Clin Cardiol 2002;25(3):133–4.

151. Kosel BW, Aweeka FT, Benowitz NL, Shade SB, Hilton JF, Lizak PS, Abrams DI. The effects of cannabinoids on the pharmacokinetics of indinavir and nelfinavir. AIDS 2002;16(4):543–50.

152. Damle BD, Mummaneni V, Kaul S, Knupp C. Lack of effect of simultaneously administered didanosine encapsulated enteric bead formulation (Videx EC) on oral absorption of indinavir, ketoconazole, or ciprofloxacin. Antimicrob Agents Chemother 2002;46(2):385–91.

COCAINE

General Information

Cocaine is an alkaloid derived from the plant *Erythroxylon coca* and other *Erythroxylon* species in South America. The leaves contain cocaine as the principal alkaloid, plus a variety of minor alkaloids. Only decocainized coca products are legal in the USA, but some commercially available tea products have been found in the past to contain cocaine in a concentration normally found in coca leaves (about 5 mg of cocaine per 1 g tea-bag). This results in only mild symptoms when package directions to drink a few cups per day are followed, but massive overdosing can result in severe agitation, tachycardia, sweating, and raised blood pressure.

Cocaine, as the leaf or as an extracted substance in different forms, has been used for different purposes. Andean Indians have long chewed leaves of the coca plant to reduce hunger and increase stamina. Pure cocaine was first extracted from coca in the 19th century; it was used to treat exhaustion, depression, and morphine addiction and was available in many patent medicines, tonics, and soft drinks. In the USA, after the Harrison Narcotic Act of 1914 and the Narcotic Drugs Import and Export Act of 1922, the use of cocaine fell, and the National Commission on Marihuana and Drug Abuse in 1973 reported that cocaine was little used. Since then, however, use has grown and there is now an epidemic in many countries. There may be about 5 million regular cocaine users in the USA, and users who have significant difficulties, including serious as well as fatal medical complications, continue to be reported frequently (1,2). Deaths occur not only from overdosage but also from drug-induced mental states, which can lead to serious injuries (3).

Cocaine was also the first aminoester local anesthetic, and its adverse effects differ from those of other local anesthetics. Owing to its rapid absorption by mucous membranes, cocaine applied topically can cause systemic toxic effects. There is a wide variation in the rate and amount of cocaine that is systemically absorbed. This variability can be affected by the type and concentration of vasoconstrictor used with cocaine and also accounts for the differences in cocaine pharmacokinetics in cocaine abusers (SEDA-20, 128).

As a recreational drug cocaine can be snorted (sniffed), swallowed, injected, or smoked. The street drug comes in the form of a white powder, cocaine hydrochloride. The hydrochloride salt and the cutting agents are removed to create the free base, which is smoked. The inexpensive widely available crack formulation is prepared by alkalinizing cocaine hydrochloride and precipitating the resultant alkaloidal free-base cocaine, which, unlike the hydrochloride, is not destroyed by heat when smoked. Smoking crack provides a rapid effect, comparable to that of intravenous injection. Intense euphoria, followed within minutes by dysphoria, leads to frequent dosing and

a greater potential for rapid addiction (4). As with amphetamines, the euphoric effect can enhance craving, and repeated reinforcement can lead to conditioned drug responses, which facilitate dependence. Facilitated conditioned effects with cocaine may be due to its rapid elimination and the development of acute tolerance. Frequent repeated dosing becomes necessary to sustain euphoria, thereby promoting a tight temporal juxtaposition of euphoria with recent drug-taking (5).

Rapid intravenous or inhalational administration of cocaine can cause very high concentrations in areas of high vascular perfusion, for example the heart and brain, before eventual distribution to other tissues. Under these conditions there is a catecholaminergic storm in the heart and a local anesthetic effect, with prolongation of conduction. Once beyond the immediate period of vulnerability, accumulation (for example through frequent overdosing or accidents from body packing of condom-filled stimulants to avoid detection) leads to a different cascade of events over a period of hours, leading to death. This cascade includes a catecholaminergic hypermetabolic state, with hyperpyrexia and acidosis, anorexia, and repeated seizures, usually ending in cardiac collapse (1,6). On the other hand, chronic dosing can cause catecholaminergic cardiomyopathy, for example contraction bands, cardiomegaly (7,8), and repeated vasospastic insults to cerebral and coronary arteries (9). Whether these chronic effects predispose to increased sensitivity to acute toxicity has not been systematically explored, but autopsy studies suggest that they do.

In addition to other chronic changes in abusers, personality deterioration carries a significant association with high-risk behaviors, which are a source of physical and psychiatric morbidity and mortality. These include suicide, violent trauma and aggressive behavior, high-risk methods of drug use (for example needle sharing), and high-risk sexual behavior, with increased risks of HIV, hepatitis B, and other infections.

The stimulant properties of cocaine are similar to those of amphetamines, although the differences are notable, in part because of the very short half-life of cocaine. However, cocaine has the same problem of abuse potential as other stimulants, and at high doses causes stimulant psychosis (10). In addition, even when it is used as a local nasopharyngeal anesthetic, it has toxic, even fatal, effects in high doses.

Death from cocaine often occurs within 2–3 minutes, suggesting direct cardiac toxicity, fatal dysrhythmias, and depression of medullary respiratory centers as common causes of death (11,12). Thus, cocaine's local anesthetic properties can contribute additional hazards when high doses are used, reminiscent of deaths reported in the era when it was used as a mucous membrane paste for nasopharyngeal surgery (13).

Periods of increased cocaine use, especially intravenous administration, inhalation of the free base, and high-dose use, are associated with cocaine-related deaths. For example, according to the Drug Abuse Warning Network, there was a three-fold increase in such deaths from 195 to 580 per year in the USA between 1981 and 1985. Despite the importance of these mortality data, relatively little is known of the types of pathophysiological sequences involved in the cascade of events leading to death. More important, there is a paucity of guidelines to appropriate diagnostic and treatment strategies for the various prefatal conditions.

Patients with brain death due to acute intoxication of drugs, including cocaine, can be organ donors. Satisfactory outcomes of graft functions were achieved in orthotopic liver transplants (14,15), kidney transplants (15), and lung transplants (16). The use of grafts from donors with brain death due to cocaine overdosage may be a valid option to expand the donor pool and help maximize the availability of scarce donor organs.

General adverse effects

Cocaine has a spectrum of pharmacological effects. It initially causes excitement and euphoria; later, with higher doses, lower centers become involved, producing reduced coordination, tremors, hyper-reflexia, increased respiratory rate, and at times nausea, vomiting, and convulsions. These symptoms are eventually followed by CNS depression.

Cardiovascular effects include tachycardia, hypertension, and increased cardiac irritability; large intravenous doses can cause cardiac failure. Cardiac dysrhythmias have been ascribed to a direct toxic effect of cocaine and a secondary sensitization of ventricular tissue to catecholamines (17), along with slowed cardiac conduction secondary to local anesthetic effects. Myocardial infarction has increased as a complication of cocaine abuse (7,8). Dilated cardiomyopathies, with subsequent recurrent myocardial infarction, have been associated with long-term use of cocaine, raising the possibility of chronic effects on the heart (18). Many victims have evidence of pre-existing fixed coronary artery disease precipitated by cocaine (SEDA-9, 35) (19–21). However, myocardial infarction has been noted even in young intranasal users with no evidence of coronary disease (22), defined by autopsy or angiography (23,24). If applied to mucous membranes, cocaine causes local vasoconstriction, and, with chronic use, necrosis.

As a general rule, mortality is higher when cocaine is used intravenously or as smoked free base than if taken nasally or orally (25). The symptoms of acute cocaine poisoning include agitation, sweating, tachycardia, tonic-clonic seizures, severe respiratory and metabolic acidosis, apnea, and ventricular dysrhythmias. Seizures occur at high doses, and may be a major determinant of fatal outcomes; their control with sedatives is important to reduce lethality (26). Associated hyperthermia can contribute as a primary cause in cases of fatal hyperpyrexia, and can potentiate the hypoxic cardiovascular events in cardiac deaths in those who survive the initial acute dose (27,28). A study of a very large number of cocaine deaths showed that the morbidity rate increased by four times on days on which the ambient temperature rose above 31.1°C (29). The final agonal events in cocaine deaths involve the combination of sympathomimetic myocardial responses and/or cardiac conduction slowing,

secondary to cocaine's local anesthetic effect, leading to dysrhythmias (30). In reported fatal overdoses, convulsions and death have usually occurred within minutes. Most patients who have survived for the first 3 hours after an initial acute overdose have been likely to recover. Treatment includes respiratory and cardiovascular resuscitative measures. Short-acting barbiturates, benzodiazepines, beta-blockers, and phentolamine have all been used with some success (23,31). Because of a possible risk of coronary vasodilatation with the use of propranolol to manage dysrhythmias in cocaine overdose, the use of labetalol for this indication is recommended, if a beta-blocker is required (32,33). In one study of 60 cocaine-related deaths, autopsy findings were non-specific but typical of those found in respiratory depression of central origin (34).

Organs and Systems

Cardiovascular

Myocardial infarction

> DoTS classification (BMJ 2003; 327: 1222-5):
> Dose-relation: Toxic
> Time-course: Time independent
> Susceptibility factors: Not known

Cocaine abuse is a risk factor for myocardial ischemia, infarction, and dysrhythmias, as well as pulmonary edema, ruptured aortic aneurysm, infectious endocarditis, vascular thrombosis, myocarditis, and dilated cardiomyopathy (35).

By one estimate, since the first report in 1982, over 250 cases of myocardial infarction due to cocaine have been reported, mostly in the USA. The first report from the UK was published in 1999 (63). Acutely, cocaine suppresses myocardial contractility, reduces coronary caliber and coronary blood flow, induces electrical abnormalities in the heart, and increases heart rate and blood pressure. These effects can lead to myocardial ischemia (36,37). However, intranasal cocaine in doses used medicinally or recreationally does not have a deleterious effect on intracardiac pressures or left ventricular performance (38).

As cocaine use has become more widespread, the number of cocaine-related cardiovascular events has increased (39). Myocardial ischemia and infarction associated with cocaine are unrelated to the route of administration, the amount taken, and the frequency of use. The risk of acute myocardial infarction is increased after acute use of cocaine and it can occur in individuals with normal coronary arteries at angiography. The patients are typically young men and smokers and do not have other risk factors for atherosclerosis.

Tachycardia and vasoconstriction from cocaine can exacerbate coronary insufficiency, complicated by dysrhythmias and hypertensive and vascular hemorrhage (1). Sudden deaths have been reported in patients with angina (40). Chronic dosing includes cardiomyopathy and cardiomegaly; other chronic conditions include endocarditis and thrombophlebitis. Crack smoking has led to pneumopericardium (41).

The cardiac effects of intracoronary infusion of cocaine have been studied in dogs and humans (42). The procedure can be performed safely and does not alter coronary arterial blood flow. The effects of direct intracoronary infusion of cocaine on left ventricle systolic and diastolic performance have been studied in 20 patients referred for cardiac catheterization for evaluation of chest pain. They were given saline or cocaine hydrochloride (1 mg/minute) in 15-minute intracoronary infusions, and cardiac measurements were made during the final 2–3 minutes of each infusion. The blood cocaine concentration obtained from the coronary sinus was 3.0 µg/ml, which is similar in magnitude to the blood–cocaine concentration reported in abusers who die of cocaine intoxication. Minimal systemic effects were produced. The overall results were that cocaine caused measurable deterioration of left ventricular systolic and diastolic performance.

Possible predictors of cardiovascular responses to smoked cocaine have been studied in 62 crack cocaine users (24 women and 38 men, aged 20–45 years) who used a single dose of smoked cocaine 0.4 mg/kg (43). Physiological responses to smoked cocaine, such as changes in heart rate and blood pressure, were monitored. The findings suggested that higher baseline blood pressure and heart rate, a greater amount and frequency of current cocaine use, and current cocaine snorting predicted a reduced cardiovascular response to cocaine. By contrast, factors such as male sex, African-American race, higher body weight, and current marijuana use were associated with a greater cardiovascular response.

Anecdotal reports

- A 26-year-old man reported smoking 10 cigarettes/day and using cocaine by inhalation at weekends (44). His electrocardiogram showed raised anterolateral ST segments. He had raised creatine kinase MB activity and troponin I concentration. He was normotensive with signs of pulmonary congestion. Ventriculography showed anterolateral and apical hypokinesia and an ejection fraction of 21%. Angiography showed massive thrombosis of the left anterior descending coronary artery. He was given recombinant tissue plasminogen activator by intravenous infusion. Angiography 4 days later showed thrombus resolution, and ventriculography showed an improved ejection fraction. He was symptom-free 6 months later.
- A 50-year-old man had 12 hours of chest pain and shortness of breath after a cocaine binge. His history included hyperlipidemia, cigarette smoking, and cocaine use (45). The troponin concentration was increased, at 25 ng/ml. An electrocardiogram showed anteroseptal ST segment elevation and inferolateral depression with T wave inversion. An angiogram showed multi-vessel occlusion of the left anterior descending artery, right coronary artery, and left circumflex artery. Ventriculography showed severe anteroapical and inferior wall hypokinesis with an ejection fraction of 25%. Angioplasty was performed urgently and final angiography showed no residual stenosis in the treated vessels.

- In a cocaine abuser, resolution of intracoronary thrombosis with direct thrombin has been successfully attempted (46). Medical treatment was started with tirofiban and low molecular weight heparin, and 48 hours later they were replaced with bivalirudin, a direct thrombin inhibitor, in an initial bolus dose of 0.1 mg/kg followed by an infusion of 0.25 mg/kg/hour. Repeat angiography 48 hours after bivalirudin showed near total thrombus dissolution with resolution of the electrocardiographic abnormalities.

Accelerated extensive atherosclerosis secondary to chronic cocaine abuse has also been reported (47).

- A 32-year-old man, who was a cigarette smoker (10 cigarettes/day) and had been a frequent cocaine user for 16 years, developed acute chest pain 2 hours after heavy cocaine use. He had electrocardiographic abnormalities consistent with an acute myocardial infarction, confirmed by serial enzyme measurements. There was no family history of atherosclerosis. Echocardiography showed a large akinetic anteroapical segment. Serum lipoprotein concentrations were normal. Despite management with aspirin, glyceryl trinitrate, heparin, and morphine, he required emergency cardiac catheterization because of prolonged chest pain. Coronary angiography showed severe atherosclerosis of the middle and distal segments of the left anterior descending coronary artery and a large diagonal branch. Intracoronary glyceryl trinitrate and nitroprusside ruled out coronary spasm. Intravascular ultrasound localized the lesion to the middle of the left anterior descending artery and showed diffuse plaques along the entire artery, with variable composition, including fibrocalcific changes consistent with a chronic process. Two overlapping stents were inserted.

The authors acknowledged that the mechanisms for accelerated atherosclerotic process are not known. However, they cited research that suggests that cocaine has pro-atherogenic effects in blood vessels.

Fiberoptic bronchoscopy is often done after intratracheal injection of 2.5% cocaine solution and lidocaine spray. Acute myocardial infarction after fiberoptic bronchoscopy with intratracheal cocaine has been reported (48).

- A 73-year-old man with a history of breathlessness, cough, and weight loss had some ill-defined peripheral shadow in the upper zones of a chest X-ray. He had fiberoptic bronchoscopy with cocaine and lidocaine and 5 minutes later became distressed, with dyspnea, chest pain, and tachycardia. Electrocardiography showed an evolving anterior myocardial infarction. Coronary angiography showed a stenosis of less than 25% in the proximal left anterior descending artery with coronary artery spasm. He made an uneventful recovery.

The authors suggested that the principal cardiac effects of cocaine can be attributed to or are mediated by the following mechanisms: increased myocardial oxygen demand due to an acute rise in systemic blood pressure and heart rate; coronary vasoconstriction caused by alpha-adrenergic effects and calcium-dependent direct vasoconstriction; and promotion of arteriosclerosis and endothelial dysfunction, which predisposes to vasoconstriction and thrombosis.

A case of cocaine-induced myocardial ischemia in pregnancy mistaken for Wolff–Parkinson–White syndrome has been reported (49).

- A 22-year-old pregnant woman at 38 weeks' gestation developed chest pain, palpitation, and shortness of breath. Her blood pressure was 148/97 mmHg, heart rate 105/minute, and respiratory rate 22/minute. An electrocardiogram showed a short PR interval and a broad QRS complex with slurred upstrokes of the R waves in leads V2 and V3. Other laboratory studies were normal. The fetal heart rate was 160/minute. The differential diagnosis included a new onset dysrhythmia, Wolff–Parkinson–White syndrome, and myocardial ischemia/infarction. With supportive treatment and monitoring the dysrhythmia resolved. Serial creatine kinase measurements were normal. However, there were cocaine metabolites in the urine. The patient admitted to having used cocaine 3–4 hours before the episode.

The use of cocaine among women of reproductive age is increasing and pregnancy also enhances the potential cardiovascular toxicity of cocaine.

Fatal pulmonary edema developed in a 36-year-old man shortly after injecting free-base cocaine intravenously (50).

Symptoms of cardiac ischemia

Cocaine users often present with complaints suggestive of acute cardiac ischemia (chest pain, dyspnea, syncope, dizziness, and palpitation). Two studies have shown that the risk of actual acute cardiac ischemia among cocaine users with such symptoms was low. The first study reviewed the clinical database from the Acute Cardiac Ischemia–Time Insensitive Predictive Instrument Clinical Trial, a multicenter prospective clinical trial conducted in the USA in 1993 (51). Among 10 689 enrolled patients, 293 (2.7%) had cocaine-associated complaints. This rate varied from 0.3% to 8.4% in the 10 participating hospitals. Only six of these patients had a diagnosis of acute cardiac ischemia (2.0%), four with unstable angina and two with acute myocardial infarction. The cocaine users were admitted to the coronary care unit as often as other study participants (14 versus 18%), but were much less likely to have confirmed unstable angina (1.4 versus 9.3%). A second study also suggested that cocaine users who present with chest pain have a very low risk of adverse cardiac events (52). Emergency departments have instituted centers for the evaluation and treatment of patients with chest pain who are at low to moderate risk of acute coronary syndromes. In this particular study, patients with a history of coronary artery disease or presentations that included hemodynamic instability, electrocardiographic changes consistent with ischemia, or

clinically unstable angina were directly admitted to hospital. In a retrospective study of 179 patients with reliable 30-day follow-up in chest pain centers, there was one cardiac complication due to cocaine use.

Frequency

During the hour after cocaine is used, the risk of myocardial infarction is 24 times the baseline risk (53). Cocaine users have a lifetime risk of non-fatal myocardial infarction that is seven times the risk in non-users, and cocaine use accounts for up to 25% of cases of acute myocardial infarction in patients aged 18–45 years (54). In 2000, there were 175 000 cocaine-related visits to emergency departments in the USA (55), with chest discomfort in 40% of the patients (56), 57% of whom were admitted to the hospital and had an admission lasting an average of 3 days (57), involving huge costs (58).

Based on a retrospective study of 344 patients with cocaine-associated chest pain, it has been suggested that patients who do not have evidence of ischemia or cardiovascular complications over 9–12 hours in a chest-pain observation unit have a very low risk of death or myocardial infarction during the 30 days after discharge (59). Nevertheless, patients with cocaine-associated chest pain should be evaluated for potential acute coronary syndromes; those who do not have recurrent symptoms, increased concentrations of markers of myocardial necrosis, or dysrhythmias can be safely discharged after 9–12 hours of observation. A protocol of this sort should incorporate strategies for treating substance abuse, since there is an increased likelihood of non-fatal myocardial infarction in patients who continue to use cocaine.

The incidence of acute myocardial infarction in cocaine-associated chest pain is small but significant (60). The electrocardiogram has a higher false-positive rate in these patients. A normal electrocardiogram reduces the likelihood of myocardial injury but does not exclude it.

Cocaine use may account for up to 25% of acute myocardial infarctions among patients aged 18–45 years. The safety of a 12-hour observation period in a chest pain unit followed by discharge in individuals with cocaine-associated chest discomfort who are at low risk of cardiovascular events has been evaluated in 302 consecutive patients aged 18 years or older (66% men, 70% black, 84% tobacco users) who developed chest pain within 1 week of cocaine use or who tested positive for cocaine (59). Cocaine use was self-reported by 247 of the 302 subjects and rest had urine positive for cocaine; 203 had used crack cocaine, 51 reported snorting, and 10 had used it intravenously. Of the 247 who reported cocaine use, 237 (96%) said they had used it in the week before presentation and 169 (68%) within 24 hours before presentation. Follow-up information was obtained for 300 subjects. There were no deaths from cardiovascular causes. Four patients had a non-fatal myocardial infarction during the 30-day period; all four had continued to use cocaine. Of the 42 who were directly admitted to hospital, 20 had acute coronary syndrome. The authors suggested that in this group of subjects, observation for 9–12 hours with follow-up is appropriate.

Pathophysiology

The hypothesis that cocaine users have increased coronary microvascular resistance, even in the absence of recent myocardial infarction, coronary artery disease, or spasm, has been assessed in 59 consecutive cocaine users without acute or recent myocardial infarction or angiographically significant epicardial stenosis or spasm (61). Microvascular resistance was significantly increased by 26–54% in cocaine users. There was an abnormally high resistance in the left anterior descending artery in 61% of the patients, in the left circumflex artery in 69%, and in the right coronary artery in47%. Increased microvascular resistance may explain many important cardiovascular effects of cocaine and has therapeutic implications. For example, slow coronary filling in diagnostic tests may suggest the possibility of cocaine use in patients in whom it was not otherwise suspected. There was increased microvascular resistance in the coronary bed even after the acute effects of a dose of cocaine would have worn off, suggesting that cocaine may have long-lasting effects on coronary microvasculature. This implies that medical therapy of vasoconstriction should be continued for extended periods. Moreover, heightened microvascular resistance in cocaine users may explain the development of chest pain and myocardial ischemia in patients who do not have epicardial stenosis due to coronary artery disease or spasm. The authors suggested that in the absence of coronary artery disease, the small vessel effects of cocaine may be more important. As the process is diffuse rather than confined to one vascular territory, electrocardiographic findings may not be localizing.

In a report of cocaine-associated chest pain, the authors studied the incidence and predictors of underlying significant coronary disease in 90 patients with and without myocardial infarction (62). Patients with 50% or more stenosis of coronary arteries or major branches or bypass graft were included and 50% of them had significant disease: one-vessel disease in 32%, two-vessel disease in 10%, three-vessel disease in 6%, and significant graft stenosis in 3%. There was significant disease in 77% of patients with myocardial infarction or a raised troponin I concentration, compared with 35% of patients without myonecrosis. Predictors of significant coronary disease included myocardial damage, a prior history of coronary disease, and a raised cholesterol. Only seven of the 39 patients without myonecrosis or a history of coronary disease had significant angiographic disease. The authors concluded that significant disease is found in most patients with cocaine-associated myocardial damage. In contrast, only a minority of those without myonecrosis have significant coronary disease.

In a review of 114 cases, coronary anatomy, defined either by angiography or autopsy, was normal in 38% of chronic cocaine users who had had a myocardial infarction (23). The authors of another review concluded that "the vast majority of patients dying with cocaine toxicity, either have no pathological changes in the heart, or only minimal changes" (24). There can be a delay between the use of cocaine and the development of chest pain (64). The results of a study of 101 consecutive patients admitted with acute chest pain related to cocaine suggested that it commonly causes chest pain that may not be secondary to myocardial ischemia (65). The use of intranasal cocaine for therapeutic purposes (to treat epistaxis) was associated with myocardial infarction in a 57-year-old man with hypertension and stable angina (66).

In a review of the literature, 91 patients with cocaine-induced myocardial infarction were identified (67). Myocardial infarction occurred in 44 patients after intranasal use, in 27 after smoking, and in 19 after intravenous use. Almost half had a prior episode of chest pain. Two-thirds had their myocardial infarction within 3 hours of use. There were acute complications related to the myocardial infarction in 18 patients. Of 24 patients followed up, 58% had subsequent cardiac complications. Two-phase myocardial imaging with ^{99m}Tc-sestamibi can be helpful in the definitive diagnosis of cocaine-induced myocardial infarction in patients with a history of cocaine use, chest pain, and a non-diagnostic electrocardiogram (68). Damage to the myocardium associated with cocaine can be unrecognized by the abuser (69). Major electrocardiographic findings (including myocardial infarction, myocardial ischemia, and bundle branch block) were recorded during a review of 99 electrocardiograms of known cocaine abusers. None of the 11 patients with major electrocardiographic changes had a past history of cardiac disease or had complained of chest pain. The mechanism by which cocaine causes acute myocardial damage is unclear. Of 20 healthy cocaine abusers given intravenous cocaine, in doses commonly self-administered, or placebo, none developed myocardial ischemia or ventricular dysfunction on two-dimensional echocardiography during the test (70).

Myocardial infarction has been documented in 6% of patients who present to emergency departments with cocaine-associated chest pain (71,72). Treatment of cocaine-associated myocardial infarction has previously generally been conservative, using benzodiazepines, aspirin, glyceryl trinitrate, calcium channel blockers, and thrombolytic drugs. In the context of 10 patients with cocaine-associated myocardial infarction, who were treated with percutaneous interventions, including angioplasty, stenting, and AngioJet mechanical extraction of thrombus, the authors suggested that percutaneous intervention can be performed in such patients safely and with a high degree of procedural success (73). Patients with cocaine-associated chest pain and electrocardiographic ST-segment elevation should first undergo coronary angiography, if available, followed by percutaneous intervention. Alternatively, thrombolytic drugs can be used. However, the relative safety and efficacy of thrombolytic drugs compared with percutaneous intervention is undefined in patients with cocaine-associated myocardial infarction.

Management

Although percutaneous revascularization for cocaine-associated myocardial infarction is the preferred method of treatment (74), the feasibility and safety of multivessel primary angioplasty has been demonstrated. In patients with persistent myocardial ischemia despite medical therapy, or evidence of cardiogenic shock, aggressive early intervention is particularly beneficial. Beta-blockers should be avoided in patients who have recently used cocaine; they fail to control the heart rate, enhance cocaine-induced vasoconstriction, increase the likelihood of seizures, and reduce survival (75).

In a randomized, controlled trial in 36 patients with cocaine-associated chest pain, the early use of lorazepam together with glyceryl trinitrate was more efficacious than glyceryl trinitrate alone in relieving the chest pain (76). These findings contrast with those of an earlier study in which there was no evidence of additional benefit from diazepam in managing cocaine-related chest pain (77). However, the Advanced Cardiovascular Life Support (ACLS) guidelines state that in cocaine-associated acute coronary syndromes a nitrate should be first-line therapy together with a benzodiazepine (78).

Coronary artery dissection

Spontaneous acute coronary dissection after cocaine abuse has been reported (79).

- A 34-year-old woman developed chest pain suggestive of acute coronary syndrome, having inhaled cocaine 30 minutes before. Her blood pressure was 180/100 mmHg and an electrocardiogram showed sinus rhythm with anterolateral ischemia, ST segment depression, and T wave inversion. An echocardiogram showed a large hypokinetic area, including the middle and apical segments of the anterior septum and the anterior and lateral walls, and mild reduction in the left ventricular ejection fraction (45%). Troponin I and creatine kinase were slightly raised but the MB fraction was normal. Unstable angina was diagnosed but despite full medical therapy, the chest pain and ischemic changes did not resolve. Immediate catheterization showed a dissection flap within the left main trunk extending to the proximal portion of the descending anterior and circumflex arteries. There was no atherosclerosis in the coronary vessels. The flap resulted in a 90% stenosis of the proximal left anterior descending artery. Urgent coronary artery bypass surgery was successful.

Spontaneous coronary artery dissection is an unusual cause of acute coronary syndrome. Only three other cases secondary to cocaine use have been described.

Infarction in other organs

Aortic thrombus and renal infarction has been reported in a patient who used nasal cocaine (80).

- A 52-year-old woman with a history of hypertension for 15 years developed acute left flank pain, nausea, and vomiting. On a previous similar occasion 2 weeks before she had had a trace of proteinuria and microscopic hematuria. A contrast-enhanced CT scan of the abdomen had not shown stones, hydronephrosis, or morphological abnormalities. She had had no rash. Her urine contained cocaine. Creatine kinase and lactate dehydrogenase activities were raised and there

was a leukocytosis. A second abdominal CT scan with contrast showed a segmental infarct of the left kidney. A transesophageal echocardiogram showed a 2x2 cm mobile mass, consistent with a thrombus, attached to the aortic arch, distal to the left subclavian artery. There was no evidence of atherosclerosis. She was given anticoagulants and aggressive fluid therapy for rhabdomyolysis.

The authors speculated that cocaine may have caused aortic inflammation by assuming that cocaine-related increased sympathetic tone along with possible cocaine-related aortic inflammation (similar to reported cases of cocaine-related vascular injury) may have led to enhance aggregation of platelets at the inflamed area, which would act as a nidus to form a thrombus. The thrombus resolved with anticoagulation within 13 days and similar results have been reported before (81). The patient had very high creatine kinase activity, suggesting rhabdomyolysis, which could have been caused by intense cocaine-related vasoconstriction.

Other vascular disease

Dissection of the aorta has been reported during cocaine use (82,83). The authors of these two reports noted that all six cases of this rare complication reported in the past 5 years were in men with pre-existing essential hypertension. In a review of emergency visits to a hospital during a 20-year period, 14 of 38 cases of acute aortic dissection involved cocaine use; 6 were of type A and 8 of type B (84). Crack cocaine had been smoked in 13 cases and powder cocaine had been snorted in one case. The mean time of onset of chest pain was 12 hours after cocaine use. The chronicity of cocaine use was not known in most of the cases. The cocaine users were typically younger than the non-cocaine users. Chronic untreated hypertension and cigarette smoking were often present.

- A 43-year-old man with untreated hypertension developed transient mild chest pressure followed by shortness of breath for 4 hours (85). He had long used tobacco, alcohol, and cocaine and admitted to having used cocaine within the last 12 hours. He had a tachycardia with a pansystolic murmur suggesting mitral regurgitation. Urine drug screen was positive for cocaine metabolites. A chest X-ray showed mild cardiomegaly and prominent upper lobe vasculature. An electrocardiogram showed atrial flutter at a rate of 130/minute and non-specific T wave changes. The diagnoses were myocardial infarction due to cocaine, with mild congestive heart failure, mitral regurgitation and atrial flutter. However, transesophageal echocardiography showed severe aortic insufficiency and a dissection flap in the ascending aorta. He underwent emergency repair of the aortic root and resuspension of the aortic valve.

Intramural hematoma of the ascending aorta has been reported in a cocaine user (86).

- A healthy 39-year-old man developed retrosternal chest pain radiating to the back with nausea and sweating. About 10–15 minutes before, he had inhaled cocaine for 2 hours and then smoked crack cocaine. He had an aortic dissection, which was repaired surgically.

The authors identified hypertension secondary to the use of cocaine as the risk factor for this complication.

Coronary artery dissection associated with cocaine is rare. The first case was reported in 1994 (87) and two other cases have been reported (88,89).

- A healthy 33-year-old man with prior cocaine use had a small myocardial infarction and, 36 hours later, having inhaled cocaine, developed a dissection of the left main coronary artery, extending distally to the left anterior descending and circumflex arteries. There was marked anterolateral and apical hypokinesis.
- A 23-year-old man with a history of intravenous drug abuse and hepatitis C was found unconscious, hypoxic, and hypotensive. A urine drug screen was positive for cocaine metabolites, benzodiazepines, and opiates. An electrocardiogram suggested a myocardial infarction, verified by raised troponin I and the MB fraction of creatine kinase. He had severe hypokinesia with a left ventricular ejection fraction of 10%, falling to less than 5%. He became septic, developed multiorgan system failure, and died. The postmortem findings included dissection of the left anterior descending artery with complete occlusion of the true lumen and thrombosis of the false lumen. The left ventricle showed extensive transmural myocardial necrosis with adjacent contraction band necrosis. He also had deep vein thromboses in veins in the neck and abdomen and multiple pulmonary infarctions.

Peripheral vascular disease in the fingers has been attributed to cocaine.

- A 48-year-old man who smoked cigarettes and used cocaine developed ischemia of the right index finger due to occlusion of the distal ulnar artery (90). He had a history of recurrent deep vein thrombosis. A venous bypass graft was performed. Two years later he had non-healing gangrene of the left index finger. His blood pressure was normal in both arms. Urine toxicology was positive for cocaine. Angiography of the left arm showed small-vessel vasculitis.
- A healthy 36-year-old man, who had used intranasal crack cocaine daily in increasing doses for 2 weeks, developed pain, numbness, swelling, and cyanosis of the fingers and toes aggravated by cold and an ulcer on one finger (91). Ultrasound Doppler of the hand confirmed ischemic finger necrosis. He was treated unsuccessfully with aspirin, diltiazem, and heparin, but responded to intravenous infusions of iloprost for 5 days.

Another less common complication of cocaine use is cerebral vasculitis (92), and benign cocaine-induced cerebral angiopathy has been reported (93). (Stroke is discussed under the section Nervous system in this monograph).

Cardiac dysrhythmias

Cocaine and its metabolite block sodium and potassium channels and its use is also associated with QT interval prolongation.

- A 37-year-old man developed severe chest pain and shortness of breath after smoking 200+ rocks of crack cocaine in 72 hours while attempting to walk several miles in order to get more drug money (94). His symptoms resolved before the ambulance arrived. His medical history included a prior episode of chest pain after a 3-day crack cocaine binge. He also occasionally used alcohol and marijuana. He was anxious but had a normal heart rate and blood pressure. His drug screen showed cocaine. Cardiac enzymes were within the reference ranges. An initial electrocardiogram showed a QT_c interval of 621 ms which shortened to 605 ms after 2 hours, 530 ms after 7 hours, and 543 ms after 15 hours. One month later the QT_c interval was 453 ms.

Brugada syndrome has been attributed to cocaine.

- A 36-year-old man became comatose 14 hours after inhaling an unspecified amount of heroin for an unspecified duration (95). He had been taking lithium, chlorpromazine, and diazepam for a chronic psychosis. His family reported recreational use of drugs of abuse. His Glasgow coma score was 4 without focal deficits. A toxicology screen was positive for cocaine and remained positive for 4 days after admission. His electrocardiogram showed prominent coved ST elevation and J wave amplitude of at least 2 mm in leads V1–V3, followed by negative T waves with no isoelectric separation, associated with right incomplete bundle branch block indicative of type 1 Brugada syndrome. His serum potassium concentration was 5.9 mmol/l. He was immediately treated with 42% sodium bicarbonate intravenously. The Brugada pattern completely resolved in 24 hours and his creatinine improved significantly in 48 hours. Transthoracic echocardiography was normal.

The authors of this report thought that the Brugada syndrome was probably not due to chlorpromazine or lithium in this patient, and it has not been previously described with heroin. It may have been due to hyperkalemia (as the Brugada pattern normalized when the serum potassium concentration normalized), perhaps facilitated by cocaine. Another case of Brugada syndrome is described under "Drug overdose".

Dysrhythmias seem to be the most likely cause of sudden death from cocaine, but cardiac conduction disorders are more common in patients with acute cocaine toxicity. Severe cocaine toxicity also causes acidemia and cardiac dysfunction (96). Four patients developed seizures, psychomotor agitation, and cardiopulmonary arrest; two of these are briefly summarized here.

- A 43-year-old man injected a large dose of cocaine in a suicide attempt and had a seizure and cardiopulmonary arrest, from which he was resuscitated. His arterial blood pH was 6.72 and his electrocardiogram showed a wide complex tachycardia. An infusion of sodium bicarbonate maintained the blood pH at 7.50 and the electrocardiogram became normal. The bicarbonate infusion was discontinued after 12 hours.

- A 25-year-old man had a cardiac arrest after taking one "knot" or sealed bag of crack cocaine (2.5 g) and was resuscitated. His arterial blood pH was 6.92 and an electrocardiogram showed sinus rhythm, QRS axis 300°, and terminal 40 msec of the QRS axis 285°. After an infusion of sodium bicarbonate, his blood pH was 7.30, his QRS axis 15°, and the terminal 40 msec QRS axis 30°. He passed the bag of cocaine rectally within 12 hours of admission.

These patients' initial laboratory values showed acidosis, prolongation of the QRS complex and QT_c interval, and right axis deviation. Appropriate treatment included hyperventilation, sedation, active cooling, and sodium bicarbonate, which led to correction of the blood pH and of the cardiac conduction disorders. The authors suggested that when intracellular pH is lowered, myocardial contractility is depressed as a result of reduced calcium availability. During acidosis, there are abnormalities of repolarization and depolarization, which potentiate dysrhythmias.

The electrocardiographic PR interval increased with increasing abstinence from crack cocaine in a study of 441 chronic cocaine users who had smoked at least 10 g of cocaine in the 3 months before enrollment (97). The authors suggested that this may have reflected the normalization of a depolarization defect. Chronic cocaine users have shortened PR intervals, indicative of rapid cardiac depolarization.

Pregnancy increases the incidence of dysrhythmias in patients with Wolff–Parkinson–White syndrome. This association may relate to an effect of estrogen, increased plasma volume, or increased maternal stress or anxiety.

Cardiovascular effects of cocaine as a local anesthetic

Cardiovascular effects due to enhanced sympathetic activity include tachycardia, increased cardiac output, vasoconstriction, and increased arterial pressure. Myocardial infarction is the most common adverse cardiac effect (43), and there is an increased risk of myocardial depression when amide-type local anesthetics, such as bupivacaine, levobupivacaine, lidocaine, or ropivacaine are administered with antidysrhythmic drugs.

- A woman who inappropriately used cocaine on the nasal mucosa to treat epistaxis had a myocardial infarction (69).
- A patient who was treated with intranasal cocaine and phenylephrine during a general anesthetic had a myocardial infarction and a cardiac arrest due to ventricular fibrillation (SEDA-20, 128).
- Myocardial ischemia was reported in a fit 29-year-old patient after the nasal application of cocaine for surgery. No relief was gained from vasodilators or intracoronary verapamil, and there were no other signs of cocaine toxicity. Although coronary vasoconstriction and platelet activation are systemic effects of cocaine, pre-existing thrombus may also have played a part (SEDA-22, 142).

Previous cocaine abuse has also been implicated in increasing the risk of myocardial ischemia when other local anesthetics are used.

Cardiac dysrhythmias have also been described in patients after the use of topical cocaine for nasal surgery (SEDA-20, 128).

- A patient who was treated with intranasal cocaine and submucosal lidocaine during general anesthesia developed ventricular fibrillation (SEDA-17, 142).

These events do not appear to have been related to the concomitant use of a vasoconstrictor, but more to excessive doses of cocaine.

Substantial systemic absorption of cocaine can cause severe cardiovascular complications (98).

- An 18-year-old man had both nasal cavities prepared with a pack soaked in 3–5 ml of Brompton solution (3% cocaine, about 3 mg/kg, plus adrenaline 1:4000) 2 hours preoperatively. In the anesthetic room he was anxious and withdrawn, with a mild tachycardia. Ten minutes later the nasal pack was removed and polypectomy was begun, with immediate sinus tachycardia and marked ST depression on lead II of the electrocardiogram. Increasing the depth of anesthesia and giving fentanyl had little effect, and the procedure was terminated. After extubation a further electrocardiogram showed T wave flattening in leads II, III, aVF, and aVL. Further cardiac investigations ruled out a myocardial infarction, an anatomical defect, or other pathological or metabolic processes. On day 4 a stress electrocardiogram showed no ischemic changes.

Absorption of cocaine from the nasal mucosa in eight patients using cotton pledglets soaked in 4 ml of 4% cocaine and applied for 10 or 20 minutes resulted in an absorption rate four times higher than expected, but was not associated with any cardiovascular disturbance; however, one of four patients who received 4 ml of 10% cocaine for 20 minutes developed intraoperative hypertension and another transient ventricular tachycardia (99). The authors advised against topical use of 10% cocaine.

Respiratory

The respiratory effects of cocaine are well known (100). In healthy crack cocaine users, there is evidence of cocaine-related injury to the pulmonary microcirculation, from fiberoptic bronchoscopy and examination of the bronchoalveolar fluid in 10 cocaine-only smokers, six cocaine-plus-tobacco smokers, 10 tobacco smokers, and 10 non-smokers, all with normal respiratory function (101). The percentages of hemosiderin-positive alveolar macrophages (a marker of recent alveolar hemorrhage) were markedly increased in the cocaine smokers compared with the others. Furthermore, the concentrations of endothelin (ET-1), an indicator of cell damage, were significantly raised in the cocaine smokers and to a lesser extent in the cocaine-and-tobacco smokers. These findings suggest that many asymptomatic healthy crack users have chronic alveolar hemorrhage that is not clinically evident.

In 177 heavy cocaine users (compared with 75 non-cocaine users), some of whom were also tobacco or marijuana users, cocaine use was associated with a higher prevalence of acute respiratory symptoms, including black sputum and chest pain. However, chronic respiratory symptoms occurred at similar frequencies in both groups. In cocaine-only smokers, mild impairment in carbon monoxide diffusing capacity suggested pulmonary capillary membrane damage, and abnormal airway conductance suggested injury to the upper airway or the large intrathoracic bronchi. Reported pulmonary complications of crack cocaine range from acute symptoms (coughing, chest pain, and palpitation) to acute syndromes (end-stage lung disease, eosinophilic infiltrates of the lung, and pulmonary infarction). The single-breath carbon monoxide diffusing capacity after the use of crack cocaine was reduced in three of six reports (100). If confirmed, a reduced carbon monoxide diffusing capacity after crack may signify damage to the alveolar capillary membrane or the pulmonary vasculature. It has been suggested that a well-designed controlled study to investigate the true impact of crack on the lung is necessary, since several confounding factors may account for the discrepancy in results (102).

Reports of acute pulmonary syndromes after the inhalation of cocaine have long been familiar.

- A 32-year-old woman rapidly developed progressive deterioration of respiratory function leading to end-stage lung disease (103). An open lung biopsy showed an inflammatory process with extensive accumulation of free silica.

The authors cautioned that some cocaine may contain silica, which could lead to severe pulmonary complications after smoking.

- A 27-year-old man twice developed inflammatory lung disease (with a predominance of eosinophils) after inhaling crack cocaine (104). Glucocorticoid treatment led to prompt resolution on both occasions.
- A 23-year-old woman developed pulmonary infarction associated with the use of crack (105).

Passive inhalation of free-base cocaine in small children can lead to serious consequences.

- A previously healthy 3-week-old boy developed pulmonary edema and autonomic manifestations of cocaine exposure from passive use (106). The urinary drug screen was positive for benzoylecgonine, a cocaine metabolite.
- Fatal pulmonary edema developed in a 36-year-old man shortly after he injected free-base cocaine intravenously (63).

A case of severe bullous emphysema in a cocaine smoker has been described (107).

- A 40-year-old man with cough, shortness of breath, and fever progressed to respiratory failure. He had smoked cocaine for the previous 17 years. His tobacco history was not known. His medical history included recurrent respiratory tract infections. A chest X-ray and CT scan showed findings consistent with bilateral bullous emphysema with a right lung abscess. He was ventilated and given antibiotics but died from respiratory failure secondary to pneumonia. Sputum cultures were positive for *Enterobacter cloacae* and *Streptococcus* species. Alpha-1 antitrypsin deficiency was ruled out.

Pneumothorax and pneumomediastinum
Spontaneous pneumomediastinum and pneumothorax have been reported (108).

- A 20-year-old obese Hispanic man awoke with severe, continuous retrosternal chest pain radiating to the neck and back (109). The pain was aggravated by deep breathing and local chest pressure. He denied substance abuse and gave a history of a flu-like illness 2 months before. His respiratory rate was 19/minute. He had a two-component pericardial rub. Laboratory blood testing ruled out myocardial infarction. His arterial blood gases and pH, electrocardiogram, chest X-ray, and echocardiogram were unremarkable. A later chest X-ray showed air in the mediastinum and chest CT confirmed the diagnosis of pneumomediastinum. Urine toxicology was positive for cocaine and cannabinoids. On further questioning, he admitted to substance use and performing a Valsalva maneuver during inhalation.
- A 22-year-old previously healthy man presented with acute sore throat awakening him from sleep (110). He had palpable crepitation due to extensive cervical subcutaneous emphysema. A chest X-ray showed a pneumomediastinum and bilateral apical pneumothoraces. CT scan did not show any underlying lung disease. Bronchoscopy was unremarkable except for a swollen nasal mucosa and acute bronchitis. Bronchoalveolar lavage was normal without evidence of infection. The patient reported repeated cocaine consumption. No tube drainage was necessary and the air collections resolved spontaneously within days.

Crack, the heat stable form of cocaine, when smoked and followed by deep inhalation plus a Valsalva maneuver to increase uptake, and cough triggered by the sniffed substance can cause pulmonary barotrauma. The increased intra-alveolar pressure can cause alveolar rupture, with consequent air dissection through the peribronchial connective tissue in the mediastinum, pleural space, pericardium, peritoneum, or subcutaneous soft tissues.

Two cases of spontaneous pneumothorax in intranasal cocaine users have been reported from Italy (112).

- A 30-year-old man, a cocaine sniffer, who had used cocaine more than five times a month for 4 years, complained of shortness of breath and acute chest pain. He had episodic cough and bloody sputum. A chest X-ray showed an 80% pneumothorax on the left side. On thoracoscopy the entire lung visceral pleura seemed to be covered by fibrinous exudate. After yttrium aluminium garnet (YAG) laser pleurodesis surgery, which abrades the pleura, he made a full recovery within 4 days.
- A 24-year-old man who had been inhaling cocaine nasally 4–5 times a month for a year developed

respiratory distress and chest pain 2 days after the last use, because of asided pneumothorax. He underwent video-assisted thoracoscopic surgery with laser pleurodesis and responded rapidly.

In both cases, pneumothorax occurred with a delay after cocaine inhalation. The authors suggested that it was therefore unlikely that these cases of pneumothorax were due to direct traumatic effects of the drug powder inhaled, to barotrauma due to exaggerated inspiration, or to a Valsalva maneuver. Histological examination in both cases showed small foreign body granulomas with polarized material in the subpleural parenchyma. The authors proposed that the pleural damage could have been directly caused by a filler substance known as mannite (a fine white powder comprised of insoluble cellulose fibers).

Spontaneous pneumomediastinum has been reported with the inhalation of free-base cocaine (108).

Asthma

Cocaine can cause exacerbation of asthma. All adult visits to an urban emergency room for an asthma attack during a 7-month period were reviewed (113). Of 163 patients (aged 18–55 years), 116 agreed to participate in a facilitated questionnaire and 103 provided urine samples for drug screening. African-Americans made up 89% of the group and 35% were cigarette smokers. Urine toxicology was positive for cocaine in 13% and for opiates in 5.8%. The severity of the exacerbation of asthma was greatest in the cocaine-positive group, 38% of whom were admitted to hospital (compared with 20% of the non-cocaine users). The length of stay was significantly longer in the cocaine-positive patients. Most of the patients did not use inhaled corticosteroids according to the treatment guidelines.

Other respiratory effects

Empyema-like eosinophilic pleural effusion following the use of smoked crack cocaine has been reported (114).

- A 33-year-old man developed a fever of 104°F, sweats, and a productive cough 1 week after using crack cocaine. He had diffuse wheezes at the lung apices and apical infiltrates and pleural effusions on CT scan. He had a raised white cell count at $18 \times 10^9/l$ with 45% eosinophils. He was negative for acid-fast bacilli, HIV, and fungi. There was no history of travel or exposure to ill contacts or other medications. Pus-like pleural fluid 500 ml, drained from a left pneumothorax, was an exudate with large number of white cells with 80% eosinophils. Cultures of blood, sputum, bronchoalveolar lavage fluid, and pleural fluid were negative for bacteria and fungi. The bronchoalveolar lavage fluid contained many eosinophils and transbronchial biopsy showed an acute inflammatory infiltrate, with many eosinophils, edema, and no fibrosis, consistent with eosinophilic pneumonitis. The symptoms improved after chest drainage and glucocorticoid treatment. His illness resolved 1 month after discharge.

The authors speculated that a leak mediator, vascular endothelial growth factor, present in eosinophils, and cytokines implicated in the effects of cocaine and eosinophil activation may have contributed to the pleural effusion. They found the highest concentration of vascular endothelial growth factor ever reported at 20 ng/ml and increased pleural concentrations of IL-5, IL-6, and IL-8, suggesting potential roles for these cytokines in eosinophilic lung disease due to cocaine. They recommended that a pleural effusion that appears grossly to be pus in the setting of cocaine abuse should not be drained until an eosinophil predominant effusion is ruled out. If infection is excluded, an eosinophilic empyema in the setting of inhaled cocaine abuse should be treated with glucocorticoids and may not require drainage.

Inhalation of crack cocaine can mimic pulmonary embolism (115).

- A patient who was a known cocaine inhaler developed respiratory distress. The initial ventilation–perfusion lung scan was highly suggestive of pulmonary embolism, based on multiple segmental and subsegmental perfusion defects with normal ventilation. However, pulmonary angiography was normal. The symptoms resolved rapidly without anticoagulation and angiography was negative 2 weeks later.

The authors concluded that intense pulmonary artery vasospasm secondary to cocaine inhalation may have caused this syndrome.

Handling cocaine pipes can cause thermal injury to the fingertips. The presence of bilateral thumb burns raised suspicion of crack lung in a young woman with suspected community-acquired pneumonia (111). When confronted with urine toxicology positive for cocaine, she admitted to having smoked large quantities of free-base cocaine only a few hours before the onset of symptoms.

Ear, nose, throat

Intranasal use, a common method of cocaine abuse, can damage the sinonasal tract, causing acute and chronic inflammation, necrosis, and osteocartilaginous erosion (SEDA-17, 36). These conditions occur secondary to the combined effects of direct trauma from instrumentation, vasoconstriction of small blood vessels with resultant ischemic necrosis, and chemical irritation from adulterants. Intranasal cocaine users can develop septal perforation, saddle-nose deformities, and sinonasal structural damage.

- A 43-year-old woman with a past history of chronic heavy cocaine use and osteomyelitis of the hard palate and nasal cavity 10 years before had required continuous follow-up for recurrent ethmoid and sphenoid sinusitis (116). Endoscopy showed an absent nasal septum, middle turbinates, anterior two-thirds of the inferior turbinates, and lateral nasal wall.
- Pott's puffy tumor, a subperiosteal infection of the frontal bone, has been described in a 34-year-old man with a history of chronic intranasal cocaine use (117).

This rare complication of frontal sinusitis appeared to develop secondary to the insertion of foreign bodies into the nose to facilitate inhalation of cocaine. Local trauma plus cocaine-induced vasoconstriction may have led to the complication.

Midline nasal and hard palate destruction have been reported in two chronic users of intranasal cocaine (118). The pathophysiology of these lesions is multifactorial, including ischemia secondary to vasoconstriction, chemical irritation from adulterants, impaired mucociliary transport, reduced immunity, and infection secondary to trauma.

In another case there was progression of septal perforation to secondary bone infection in a chronic cocaine user (119).

- A 56-year-old chronic intranasal cocaine abuser with a visible nasal defect presented with a hole in the roof of his mouth. He had been reportedly drug free for 2 weeks. He had an oronasal fistula with adjacent black necrotic areas and erosive destruction of the nasal septum, turbinates, and antrum, with mucoperiosteal thickening of the sphenoid and maxillary sinuses. Treatment included antibiotics and a prosthesis plate construction to cover the defect. Two years later, having continued to inhale cocaine, he had progressive destruction of his sinonasal tract, a fistula between his oral and nasal cavity, a saddle-nose deformity with total cartilage loss, and a complete palatal defect. Biopsy of the nasal septum showed acute osteomyelitis and extensive bacterial overgrowth (including anerobic *Actinomyces*-like organisms). He was given intravenous antibiotics for 6 weeks followed by long-term oral antibiotics.

Cocaine-related erosion of the external nasal structures has been described (120).

- A 43-year-old woman with a T12 paraplegia due to a car accident 24 years earlier and a sublabial abscess 2 years before developed progressive erosion of both the internal and external portions of her nose over 6 months, with nasal crusting and nose bleeds. Several antibiotics were unhelpful. There was partial destruction of the external nasal structure and two oronasal fistulae in the upper gingival sulcus. Intranasal biopsy showed acute and chronic inflammation. On two occasions urine drug screening was positive for cocaine, although she denied using cocaine.

A 30-year-old man developed destructive rhinitis due to cocaine abuse after initially presenting with Henoch–Schönlein purpura (121). Cocaine use can mimic vasculitis and is often accompanied by positive ANCAs. Cocaine-induced midline destructive lesions are characterized by mucosal damage and ischemic necrosis of the nasal septum. Histopathological similarity to leukocytoclastic vasculitis and the presence of PR3-ANCA can lead to confusion between Wegener's granulomatosis and cocaine-induced midline destructive lesions.

From January 1991 to September 2001, 25 patients with cocaine-induced midline destructive lesions were observed at the Department of Otorhinolaryngology of the University of Brescia (122). There was septal perforation in all 25, and 16 also had partial destruction of the inferior turbinate. There was hard palate resorption in six patients and 14 were positive for antineutrophil cytoplasmic antibodies (ANCA). The need to consider substance abuse in the differential diagnosis of destructive lesions of the nasal cavity, even in the presence of antineutrophilic cytoplasmic antibodies, has been emphasized. There was constant progression of the ulceronecrotic process in 17 patients, and in three patients who stopped cocaine use the mucosa slowly normalized. Management consisted of periodic debridement of necrotic tissue and crusts, local and systemic antimicrobial drug therapy based on culture results, the use of saline douches for moistening the nasal mucosa, and surgical correction in severe cases.

All reports of cocaine-induced midline destructive lesions have been reviewed, and retrospective data involving 25 cases have been reported (123). All but three subjects admitted to cocaine abuse during the initial evaluation. There were 15 men and 10 women with a mean age of 36 years (range 22–66 years) and they had abused cocaine for 2–30 years, at a dose of 1–180 g/week. At rhinoscopy, all had necrotizing ulcerative lesions, extensive crusting, and septal perforation. The destructive process extended to the inferior (68%), middle (44%), and superior turbinates (16%). There were hard and/or soft palate perforations in six patients (24%); the lateral wall of the nose was entirely reabsorbed in five. The lesions caused both dysphagia and nasal reflux. In two subjects there was ulceration at the base of the columella at diagnosis. During follow up, two patients developed a nasocutaneous fistula. At diagnosis or during follow-up, none had any laboratory finding suggestive of a systemic disease, but 22 had nasal swabs positive for *Staphylococcus aureus*. Fungi were not grown. During the course of disease, two patients with severe diffuse destructive lesions had acute orbital symptoms and signs (diplopia, pain, and proptosis), caused by infection. One patient had a secretory otitis media. All had septal erosion. In 15 there was altered olfaction. The authors speculated that the possible mechanism included either direct damage to the neuroepithelium from cocaine or its adulterants or obstruction of the olfactory cleft by inflammation and edema of the nasal mucosa. Vascular abnormalities mimicking vasculitis were present in 23 subjects. The authors observed a constant progression of the ulcerative process in 17 patients. Three subjects who stopped using cocaine had slow normalization of the mucosa. The authors suggested that any sinonasal inflammatory condition involving the midline structures characterized by symptoms such as nasal obstruction and crusting that persists or remains refractory to treatment may be the first manifestation of a potentially lethal drug addiction.

Most surgeons in the UK use local anesthesia with cocaine for nasal operations because of the superior operative field it provides and because they consider it to be safe, even with adrenaline. The incidence of adverse reactions to cocaine given in this way is reportedly low, and serious complications are less common than with general anesthesia. These conclusions were based on a postal survey of all British Associations of Otolaryngologists and Head and Neck Surgeons. Only

11% of surgeons had experienced cocaine toxicity in their patients, and there had been only one recorded death (124).

Nervous system

Research in 21 cocaine users and 13 non-drug-using, age-matched controls has suggested that chronic cocaine use may be associated with specific neurochemical changes in the brains of habitual users (125). All the subjects underwent a spectral brain scan in a proton MRS scanner. The significant finding was that the concentration of *N*-acetylaspartate in the left thalamus (but not in the basal ganglia) was significantly lower (17%) in the chronic cocaine users than in the controls. *N*-acetylaspartate is found in adult neurons and not in glia; it is often used as a marker of neuronal viability; a reduction suggests neuronal damage and/or loss.

Assessment of smooth pursuit eye movements has been used in the study of neurophysiological effects of a variety of clinical and subclinical disorders. In 126 patients who met DSM-IIIR criteria for dependence on alcohol, cocaine, or heroin, or dual alcohol and cocaine abuse, there was a significant reduction in tracking accuracy in the heroin-dependent and the dually-dependent subjects relative to controls (126). However, eye movement dysfunction in the drug-dependent groups was not detectable when the effects of antisocial personality disorder were statistically removed. The magnitude of the dysfunction correlated significantly with several antisocial personality-related features, including an increased number of criminal charges, months of incarceration, increased problems associated with drug abuse, and lower intellectual functioning. The findings suggested that there may be an association between premorbid personality traits and eye movement impairment.

Cocaine use has been associated with a reduced inhibitory response of the P50 auditory evoked response, attributed to increased catecholaminergic and reduced cholinergic neurotransmission secondary to cocaine (127). In a double-blind, placebo-controlled study, 11 cocaine users in the first and third weeks of detoxification had electrophysiological testing 10 minutes before and 30 minutes after taking nicotine gum 6 mg. Nicotine briefly reversed the inhibitory deficit.

Chronic substance use has been associated with long-lasting changes in brain function (128). Five brain regions that may be affected (the orbitofrontal gyrus, rectal gyrus, anterior cingulate gyrus, basal ganglia, and thalamus) were selected for analyses of cerebral glucose metabolism by positron emission scanning in controls, cocaine users, and alcoholics (17 in each group), who performed the Stroop test, which assesses cognitive interference and response inhibition. In controls, higher brain glucose metabolism in the orbitofrontal gyrus correlated with poorer performance. In contrast, in substance users, higher brain glucose metabolism was associated with better performance. Chronic abuse appears to be associated with altered function of the orbitofrontal gyrus.

Cerebellar dysfunction

Cocaine has been associated with movement disorders, such as acute dystonias, choreoathetosis, and akathisia. Chronic pancerebellar dysfunction occurred in a cocaine user with schizophrenia treated with risperidone (129).

- A 38-year-old man was found comatose in a crack house. The ambient temperature was 13°C. He had earlier abused cocaine. His temperature was 43°C, heart rate 115, blood pressure 144/89 mmHg, and oxygen saturation 97% on air. His general muscle tone was flaccid. He had a mild leukocytosis and hypophosphatemia. Urine toxicology was positive for benzoylecgonine, a cocaine metabolite. He was mechanically ventilated, cooled, and given intravenous fluids. His temperature fell to 38°C, but he later developed acute disseminated intravascular coagulation and rhabdomyolysis. After 5 days he developed nystagmus, intention tremor, truncal ataxia, dysarthria, ocular dysmetria, and dysmetria of the arms and legs. There were no sensory or motor deficits. Finger-to-nose and heel-to-knee tests were slowed and uncoordinated. He could not stand. Brain imaging studies (CT and MRI scans) were unremarkable. He was given thiamine, propranolol, clonazepam, primidone, and baclofen every 8 hours without improvement. After 1 year he still had nystagmus, intention tremor, ataxic gait, and dysmetria. He could walk short distances slowly.

Cocaine and neuroleptic drugs can both cause hyperthermia and the authors proposed that the combination of cocaine and risperidone may have caused this problem.

Migraine

About 60–75% of chronic cocaine abusers report severe headaches (130), which can resemble migraine; migraine-like symptoms can include auras, visual field changes, and paraphasia (131). About 60–75% of chronic cocaine abusers report severe headaches (130,131). Of 21 patients who were admitted to hospital from January 1985 to December 1988 for acute headache associated with cocaine intoxication 15 had headaches with migrainous features in the absence of neurological or systemic complications (130). None had a history of cocaine-unrelated headaches or a family history of migraine, and all had a favourable outcome. The authors discussed three possible mechanisms of cocaine-related vascular headaches, depending on the interval between cocaine ingestion and development of the headache. They postulated that acute headaches after cocaine use may relate to the sympathomimetic or vasoconstrictive effects of cocaine, while headaches after cocaine withdrawal or exacerbated during a cocaine binge may relate to cocaine-induced effects on the serotoninergic system.

Movement disorders

The effects of cocaine or its withdrawal on neurotransmitter activity have been evaluated in several studies. Although changes in dopaminergic activity appear to be associated with early cocaine abstinence, extrapyramidal symptoms (due to alterations in dopamine functioning) have only infrequently been reported in cocaine users.

However, in a recent report, extrapyramidal symptoms of classic muscle stiffness and cogwheel rigidity at the elbow occurred during the "crash" phase of a 40-year-old man's cocaine withdrawal (132).

Cocaine can cause other movement disorders (133). Motor and vocal tics, pre-existing tremors, and generalized dystonia were induced or exacerbated by drug use, and continued even after cocaine use ended. Lastly, dopamine dysregulation (that is reduced dopamine receptor availability in association with reduced frontal metabolism) was demonstrated by positive emission tomography in 20 chronic cocaine abusers (134). The findings were prominent in the orbitofrontal, cingulate, and prefrontal cortex 3–4 months after detoxification. The author hypothesized that dysregulation of these brain areas may result in compulsive drug-taking behavior.

Dystonia has been described shortly after the use of cocaine (SEDA-16, 24). In a retrospective study of 116 patients taking neuroleptic drugs, 42% of cocaine users versus 14% of non-users developed dystonia (135). This suggests that the use of cocaine may be a major risk factor for acute dystonic reactions secondary to the use of neuroleptic drugs.

Cocaine-induced chronic tics have been reported (136), and cocaine has reportedly exacerbated Gilles de la Tourette syndrome (138).

Choreoathetoid movements have been evaluated in samples of cocaine-dependent men (n = 71), amphetamine-dependent men (n = 9), and 56 controls (137). The cocaine-dependent men had significantly worse non-facial (limbs plus body) choreoathetosis scores, and the differences between groups were most marked in the younger age groups. The facial scores were increased only in those under 32 years of age, as they were in the younger amphetamine-dependent subjects. The authors suggested that the absence of choreoathetoid movements in the older cocaine-dependent men may represent an age-related self-selection effect.

Myasthenia gravis

A 24-year-old woman developed myasthenia gravis during cocaine use (139). The authors speculated that while cocaine did not cause impairment of motor axons for neuromuscular transmission, a reduction in the number of acetylcholine receptors per neuromuscular junction (as occurs in myasthenia gravis) could have increased the susceptibility to an effect of cocaine.

A rare case of polysubstance abuse, which unmasked myasthenia and caused complete external ophthalmoplegia, has been reported (140).

- A 29-year-old woman, who used cocaine 2 g/day, heroin 1 g/day, and methadone 40 mg/day, developed abscesses caused by drug injection. She had had generalized weakness, difficulty in swallowing, and lagging eyelids for 1 week. There was bilateral ptosis, and a diagnosis of myasthenia was made. Edrophonium 10 mg relieved the ptosis and improved ocular movements.

Parkinson's disease

Alpha-synuclein is a presynaptic protein that has been implicated as a possible causative agent in the pathogenesis of Parkinson's disease. In a study of post-mortem neuropathological specimens from cocaine users (n = 21) and age-matched drug-free controls (n = 13), the concentrations of alpha-synuclein in dopamine-containing cells of the substantia nigra and ventral tegmental area were increased threefold in chronic cocaine users compared with controls with changes in the expression of alpha-synuclein mRNA (141). Although alpha-synuclein is prominent in the hippocampus, there was no increase in this area. Alpha-synuclein concentrations were increased in the ventral tegmental but not the substantia nigra area in victims of excited cocaine delirium who had paranoia, marked agitation, and hyperthermia before death. The authors speculated that increased alpha-synuclein may be a protective response to changes in dopamine and increased oxidative stress resulting from cocaine abuse. On the other hand, this accumulation of alpha-synuclein with long-term cocaine abuse may put addicts at increased risk of the motor abnormalities of Parkinson's disease.

Seizures

Cocaine lowers the seizure threshold and may therefore be dangerous to patients already at risk of seizures (50). Exacerbation of generalized tonic-clonic seizures (which occurred initially during the use of crack but later continued independently of the drug) has been described, with progressive electroencephalographic abnormalities (142). In this case, the development of the seizures suggested that cocaine can stimulate kindling, with progressive intensification of after-discharges and the eventual emergence of seizure activity.

Cerebrovascular vasoconstriction and a sudden increase in blood pressure probably underlie the many reports of cocaine-induced strokes, CNS hemorrhage, and migraine (SEDA-17, 1). Three cases of generalized seizures occurring shortly after the intravenous use of cocaine (143).

Sleep disturbance

The disruption of normal sleeping patterns during cocaine withdrawal may be related to effects on the cholinergic system. In a study of nine patients undergoing cocaine withdrawal, rapid eye movement (REM) latency was markedly shortened, REM sleep percentage was increased, REM density was very high, and the total sleep period was long during the first week (144). Changes in REM sleep are thought to be related to changes in cholinergic activity. At week 3, characteristic chronic insomnia was observed.

Stroke

Cocaine has been associated with cerebrovascular events, such as transient ischemic attacks (SEDA-24, 24), cerebral hemorrhage (145,146), and cerebral infarction (SEDA-22, 23) (SEDA-20, 26) (SEDA-20, 21).

Regional cerebral blood flow was assessed using single photon emission computed tomography (SPECT) and tracer HMPAO in 10 cocaine abusers within 72 hours of last cocaine use and then after 21 days of abstinence (145). Compared with controls, recent cocaine abusers had significantly reduced cerebral blood flow in 11 of 14 brain regions, with the largest reductions in the frontal cortex and parietal cortex and greater cerebral blood flow in the brain stem. These perfusion defects appeared to be primarily due to combined abuse of alcohol and cocaine. Frontal but not parietal defects appeared to resolve partially during 21 days of abstinence.

Cocaine has been associated with significant reductions in cerebral blood flow, thought to be secondary to its vasoconstrictor effects (146). In 13 chronic cocaine abusing men (mean age 38, range 28–45) and 10 healthy aged-matched male subjects, cerebrovascular pathology was assessed with functional magnetic resonance imaging (fMRI) to compare the abnormal blood oxygenation level dependent (BOLD) responses to photic visual stimulation. Cocaine abusers had a significantly enhanced positive BOLD response to photic stimulation compared with controls. The authors proposed that the enhanced activation in the cocaine abusers could have resulted from low resting cerebral blood flow secondary to increased vasoconstriction and/or from low oxidative metabolism during activation. Alternatively, the larger signal intensity in the cocaine abusers could have resulted from inefficient neuronal processing, as has been reported in other conditions of cerebral pathology.

In a review of ischemic stroke in young American adults (aged 15–44 years) admitted to 46 regional hospitals between 1988 and 1991, illicit drug use was noted in 12% and was the probable cause of stroke in 4.7% (147). Multidrug use was common among users: 73% used cocaine, 29% used heroin, and 14% used phencyclidine. Drug-associated stroke in these young adults appeared to be related to vascular mechanisms (such as large and small vessel occlusive disease) rather than to hypertension or to diabetes. Cerebral infarction is significantly more common among users of cocaine alkaloid (crack) than cocaine hydrochloride (148). Ischemic stroke has been attributed to combined use of cocaine and amphetamine (149).

- A previously healthy 16-year-old man developed an unsteady gait and double vision. His symptoms began 5 minutes after intranasal "amfetamine" (actually amfetamine cut with cocaine). He had a left-sided internuclear ophthalmoplegia, an incomplete fascicular paresis of the left oculomotor nerve, and saccadic vertical smooth pursuit. Cranial MRI showed a left-sided hyperintense lesion near the midline of the mesencephalon. A repeat MRI scan 9 days later showed that the lesion was much smaller. He made a full recovery within 3 weeks.

The cause of cocaine-related stroke and transient ischemic attacks has been studied by transcranial Doppler sonography, a continuous measure of cerebral blood flow velocity, to monitor the course of cerebral hemodynamic changes during acute intravenous injection of placebo, and of cocaine 10, 25, and 50 mg in seven cocaine abusers (150). There was a significant increase in mean and systolic velocity (lasting about 2 minutes) with all doses of cocaine but not with placebo. Cocaine produced an immediate brief period of vasoconstriction (as demonstrated by an increase in systolic velocity) in the large arteries of the brain.

A bilateral hippocampal stroke, which is rare, has been attributed to cocaine (151).

- An unresponsive 25-year-old white woman was found to be pulseless and was resuscitated. She had used cocaine the previous night and had had numerous psychiatric admissions for mood disorder and drug overdoses in the past. Her pupils measured 4 mm bilaterally and were non-responsive to light. She had roving eye movements and positive corneal responses. She did not respond to verbal or noxious stimuli. Her reflexes were brisk throughout, with clonus in the legs. There were cocaine metabolites in her urine. CT and MRI scans of the brain showed bilateral hippocampal and basal ganglionic hypodensities. She improved and was discharged to a nursing home 3 weeks later but continued to have severe short-term memory difficulties, problems with praxis, and mild quadrispasticity.

The authors cited a combination of ischemia and excitotoxicity due to cocaine exposure as the possible cause of the brain injury. Oxidative stress and free radicals associated with cerebral hypoxia contribute to cell damage and death.

Analysis of intracranial hematomas for detecting drugs with short half-lives in individuals who survive several hours after intracranial bleeding has been recommended for establishing the cause of death. Cocaine metabolites were found in an intracerebral hematoma in a patient who apparently died due to cocaine (152). The hematoma contained ethanol 0.05% w/v and benzoylecgonine 0.43 mg/l.

"Spontaneous" acute subdural hematoma related to cocaine abuse has been described (153).

- A 38-year-old man with a 10-year history of cocaine use became comatose. He had had an acute severe headache and progressive deterioration after abusing cocaine. His Glasgow Coma Scale was 3 and his pupils were dilated but reactive to light. He had hypertension and bradycardia. Routine toxicology was positive for cocaine. Blood tests, including coagulation profile, were normal. A CT scan of the brain showed a left acute subdural hematoma with midline shift and obliteration of the basal cisterns. During emergency craniotomy the source of the bleed was identified as a pinhole rupture of a parietal cortical artery. The patient had no history of head injury and there were no intraoperative findings of head injury. He died 24 hours later without evidence of clot reaccumulation. An autopsy was not performed.

Hypertensive encephalopathy can follow the use of cocaine (154).

Brown–Séquard syndrome after esophageal sclerotherapy and recent crack cocaine abuse has been reported (155).

- A 44-year-old man with hepatitis C and cirrhosis, esophageal varices, and poorly controlled hypertension, who was also a chronic alcoholic and crack cocaine abuser, had his varices injected at endoscopy and developedsided weakness and numbness up to T4. There was flaccid right leg weakness, right T4-6 hypalgesia, left leg hypalgesia up to L1, and reduced sweating on the left up to T4-6. An MRI scan of the thoracic spine showed a lesion at T4-6 involving the anterior and central portions of the spinal cord. A urine screen was positive for cocaine. He recovered spontaneously 2 weeks later.

Cocaine-induced ischemia was the most likely cause of this adverse outcome.

Subarachnoid hemorrhage

There is an association between cocaine use and aneurysmal subarachnoid hemorrhage (SEDA-18, 36; SEDA-21, 26) (156). Subarachnoid hemorrhage was temporally related to cocaine abuse in 12 young adult abusers who had underlying cerebral aneurysms; hypertension was a probable contributing factor (157) and cocaine is a risk factor for cerebral vasospasm after aneurysmal subarachnoid hemorrhage (158). In a retrospective analysis of the medical records of 440 patients who presented to a neurosurgery unit between 1992 and 1999 with aneurysmal subarachnoid hemorrhage, 27 patients (6.1%) had either a positive urine screen for cocaine metabolites ($n = 20$) or a history of cocaine use within 72 hours of subarachnoid hemorrhage ($n = 7$). Cocaine users were more likely to have cerebral vasospasm from 3 to 16 days after subarachnoid hemorrhage than non-exposed patients (63 versus 30%). They were also more likely to be younger and to have aneurysms of the anterior circulation than the control group (97 versus 84%).

The adverse effect of cocaine on the clinical course in subarachnoid hemorrhage has been studied in a retrospective review of the medical records of 151 patients with intracranial aneurysms treated at a Taiwanese hospital between January 1996 and December 2001 (159). Of 108 patients who had subarachnoid hemorrhage, 36 had used cocaine within the previous 24 hours and 20 of them had subarachnoid hemorrhages of greater severity (Hunt and Hess grade of IV or V) compared with eight of 72 non-cocaine users. There was significant angiographically confirmed vasospasm in 28 of the cocaine users and 20 of the non-users. There was a 2.8-fold greater risk of vasospasm associated with cocaine use. Cocaine has vasoactive properties that both influence and increase the occurrence of cerebral vasospasm, the main cause of morbidity and mortality in patients with subarachnoid hemorrhage who survive the initial event.

Sensory systems

Eyes

Ophthalmic effects associated with cocaine can occur during both active drug use and early abstinence. Cocaine abuse has been associated with ophthalmic complications, including ulceration of the cornea, vasoconstrictor effects on the retinal vasculature, irregularities in oculomotor performance, and secondary optic neuropathy.

Corneal ulceration secondary to smoking crack cocaine has been reported in a 27-year-old woman (160).

Cases of the "crack eye syndrome" continue to be reported. Of 14 crack cocaine users with corneal problems, 10 had corneal ulcers infected with both bacterial and fungal organisms; 4 had corneal epithelial defects (161). All were actively smoking crack daily. The authors suggested that crack smoking predisposes users, through an unknown mechanism, to corneal epithelial changes, infection, and perforation. Typical presentations include loss of vision with or without pain.

- A 29-year-old woman with a painful corneal ulcer related to cocaine abuse was found to be putting cocaine powder directly into the affected eye to reduce the pain (162). Her history included prior corneal perforations.

Such topical use of cocaine may aggravate the condition.

Acute iritis has been reported after intranasal use (163).

There has been a report of retinal changes in 60 users of crack (164). Microtalc retinopathy and retinal nerve fiber layer "rake" or "slit" defects were detected by threshold visual field testing and fundus photography.

Orbital infarction has been described after cocaine use (165).

- A 36-year-old woman drank alcohol and snorted cocaine and heroin at a party. She lost consciousness, with her head positioned down with her left face pressed against a desk. She awoke 3 hours later with severe left orbital pain. Her right eye was normal but there was complete visual loss in the left eye and a nearly complete left ptosis. The left pupil did not react to light but reacted consensually. Movements in the right eye were full, but movements in the left eye were severely limited in all directions. In the left fundus there was retinal edema and retinal pigment epithelium disruption. An orbital MRI scan showed diffuse swelling of all the extraocular muscles in the left orbit. A week later the pain had abated and there was mild improvement in the eye movements and ptosis, but no change in vision. She was instructed to wear protective polycarbonate lenses at all times.

Central retinal artery occlusion has previously been reported in cases of intravenous and intranasal cocaine abuse and has also been reported in a man who smoked crack cocaine (166).

- A 42-year-old man smoked crack and developed sudden painless loss of vision in his right eye for 9 hours. He had smoked cigarettes for 20 years and crack cocaine twice a week for the previous 4 years. Visual acuity in the right eye was counting fingers at one meter, and in the left eye 6/4. There was a right relative afferent papillary defect. In the right fundus there was evidence of central retinal artery occlusion and the left fundus was unremarkable. Treatment included intravenous acetazolamide, intermittent ocular massage, and rebreathing into a paper bag. He was found to have sickle cell trait, a risk factor for central retinal artery occlusion.

Ears

Cochleovestibular deficit has been attributed to intralabyrinthine hemorrhage after cocaine consumption (167).

- A 43-year-old man taking maintenance methadone used intranasal cocaine and drank alcohol and suddenly developed tinnitus in the left ear and progressive giddiness. He had left deafness, right spontaneous horizontal nystagmus, and left areflexia on caloric stimulation. A diagnosis of cochleovestibular deficit was made. A week later the right spontaneous horizontal nystagmus had resolved but was elicitable by a head-shaking maneuver. An MRI scan showed blood in the left labyrinth. Neurological and ophthalmological examinations were normal. The giddiness resolved after several weeks but his hearing did not recover.

The authors presumed that this disorder was due to cocaine-induced vascular effects. As cocaine prevents reuptake of amines by presynaptic receptors, increased concentrations of amines may lead to abruptly altered circulation.

Psychological, psychiatric

Single photon emission computerized tomography (SPECT) has suggested that some psychiatric symptoms in cocaine users are associated with changes in blood flow (168). Multiple scalloped areas of reduced cerebral blood flow (especially periventricular regions and deep portions of the brain) have been seen. Hypoperfusion has also been noted in the frontal lobes of cocaine users with mania.

Cognitive function

The effects of cocaine on cognitive functions have been measured in controlled studies. The preliminary results of a study of 20 heavy cocaine abusers and a group of matched controls showed impaired function on neuropsychological tests in 50% of the abusers compared with 15% of the controls (169). There were problems with concentration, memory, problem-solving, and abstract thinking in the cocaine users. The heavy users had the greatest loss of memory. Recent cocaine use was associated with poorer oral fluency and arithmetic scores.

The effect of cocaine on cognitive functioning has been studied in 20 crack users, 37 crack and alcohol users, and 29 controls at 6 weeks and 6 months of abstinence (170). The two substance-dependent groups had significant cognitive impairment in a range of neuropsychological tests compared with the controls at both times. Drug dose was strongly associated with the extent of impairment. Abstinent substance users were more depressed than controls during the test period, but depression had only a slight effect on neuropsychological performance.

Neuropsychological performance was examined in 355 incarcerated adult male felons, who were classified by DSM-IV criteria into four subgroups: alcohol dependence or abuse ($n = 101$), cocaine dependence or abuse ($n = 60$), multisubstance dependence or abuse ($n = 56$), and no history of drug abuse ($n = 138$) (171). The cocaine and control groups had similar neuropsychological test scores. However, both the multisubstance and alcohol groups performed significantly worse on nearly all measures. The multisubstance group had worse short-term memory, long-term memory, and visuomotor ability. Correlations between neuropsychological performance and length of abstinence from drug use showed that after abstinence the alcohol group had the greatest improvement on tests. Although the cocaine group had the least amount of improvement with abstinence, their overall performance was not significantly different from controls.

Even after 4 weeks of abstinence, chronic heavy cocaine users have poor cognitive functioning compared with non-drug-using controls (172). A battery of neuropsychological tests was administered to 30 abstinent chronic cocaine abusers and 21 non-drug-using matched controls. Decrements in areas such as executive functioning, visuoperception, psychomotor speed, and manual dexterity were associated with heavier use of cocaine. Neither frequency nor duration of cocaine use was a strong predictor of performance.

Neurolinguistic functioning has been assessed in six African-American male cocaine abusers undergoing drug rehabilitation (173). A test battery to assess language, cognition, and memory skills was administered at 1 week and 1 month of cocaine abstinence. Participants' performances were compared with the normative data for each test. There was reduced ability for general language knowledge, memory, and verbal learning ability during the period of early abstinence. However, the sample size was small and the duration of study short.

Mood disorders

The rate of co-morbidity has been studied in 208 female African-American crack cocaine users; 148 were in treatment, 54 were active crack users, and 61% reported a history of sexual abuse (174). Many had co-morbid depression (48%) and eating disorders (11%).

Delirium

Cocaine-induced delirium with severe acidosis has been reported (175).

- A 25-year-old man with agitation and paranoia who had consumed a lot of alcohol with cocaine the night before had a clonic seizure lasting 1 minute. In the emergency room, he responded to pain and made incomprehensible sounds. His pulse rate was 116/minute, blood pressure 100/40 mmHg, respiratory rate 28/minute, and temperature 38.3°C. He was acidotic (pH 6.53), with a $PaCO_2$ of 13.1 kPa, a base deficit of 36 mmol/l, a serum potassium concentration of 7 mmol/l, and sodium 153 mmol/l. He was hyperventilated and given sodium bicarbonate, dantrolene, and passive cooling. His acidosis quickly corrected and his temperature fell to 37.6°C within 1 hour.

Obsessive-compulsive disorder

Suspected risk factors for obsessive-compulsive disorder were investigated in a prospective epidemiological study, using data from the Epidemiologic Catchment Area surveys (1980–1984) (176). Users of both cocaine and marijuana were at increased risk of obsessive-compulsive disorder compared with non-users of illicit drugs, but cocaine use alone was not associated with an increased risk, within the limited sample size.

Panic disorder

As with several other drugs, for example marijuana, PCP, and LSD, cocaine can precipitate panic disorder, which continues long after drug withdrawal (177). Among 280 patients in a methadone maintenance clinic, the prevalence of panic disorder increased from 1% to 6% over a decade (178). A marked rise in the frequency of cocaine abuse coincided with this outbreak. The authors suggested that episodes of panic occurring in cocaine users can result in hospitalization for either psychiatric or medical illnesses.

Paranoid psychoses

Of 55 individuals with cocaine dependence, 53% reported transient cocaine-induced psychotic symptoms (179). Paranoid delusions (related to drug use) and auditory hallucinations were often reported. In addition, almost one-third (all of whom also described psychotic symptoms) reported transient behavioral stereotypes.

Paranoid psychosis has also been described in a 64-year-old man who had first begun to use crack cocaine 6 months before. The paranoid symptoms continued for 3 weeks after he stopped using crack.

The author suggested that the man's age may have made him particularly sensitive to the psychiatric effects of cocaine (180).

The possible genetic basis of cocaine-induced paranoia has been studied in 45 European Americans with cocaine dependency (181). Low activity of the enzyme dopamine β-hydroxylase (the enzyme that catalyses the conversion of dopamine to noradrenaline) in the serum or cerebrospinal fluid was positively associated with the occurrence of positive psychotic symptoms in several psychiatric disorders. The activity of dopamine β-hydroxylase is a stable, genetically determined trait that is regulated by genes located at the DBH locus. The haplotype associated with low dopamine β-hydroxylase activity, Del-a, occurred more often in 29 subjects with cocaine-induced paranoia than in 16 without. These findings may have implications for the pharmacological treatment of cocaine dependence.

Endocrine

The association of cocaine withdrawal with hypothalamic–pituitary–adrenal axis dysregulation has previously been reported and may be important in understanding vulnerability to stress response and relapse (182). The hypothesis that withdrawn cocaine-dependent patients would have higher cerebrospinal fluid concentrations of corticotropin-releasing hormone than healthy controls has been tested in 29 cocaine-dependent men (mean age 40 years) who were abstinent for a minimum of 8 days (mean 29 days) and 66 healthy controls. The subjects were 21 African Americans, two Hispanics, and six Caucasians. There were no significant differences in cerebrospinal fluid concentrations of corticotropin-releasing hormone between the subjects and the controls. There was no correlation between the number of days of cocaine withdrawal and the corticotropin-releasing hormone concentrations. This negative study reflected the fact that the hypothalamic–pituitary–adrenal axis in cocaine abstinence is no longer dysregulated.

Cocaine and nicotine share many similarities, including a strong potential for addiction. In a comparison of the acute effects of cocaine and cigarette smoking on luteinizing hormone, testosterone, and prolactin, 24 men who met criteria for cocaine abuse or nicotine dependence were given intravenous cocaine (0.4 mg/kg) or placebo cocaine, or smoked a low-nicotine or high-nicotine cigarette (183). Placebo-cocaine and low-nicotine cigarette smoking did not change luteinizing hormone, testosterone, or prolactin. Luteinizing hormone increased significantly after both intravenous cocaine and high-nicotine cigarette smoking and correlated significantly with increases in cocaine and nicotine plasma concentrations. However, high-nicotine cigarette smoking stimulated significantly greater increases in luteinizing hormone release than intravenous cocaine. On the other hand, testosterone concentrations did not change significantly after either cocaine or high-nicotine cigarette smoking. Prolactin concentrations fell significantly and remained below baseline after intravenous cocaine. However, after high-nicotine cigarette smoking, prolactin increased to hyperprolactinemic concentrations within 6 minutes and remained significantly above baseline for 42 minutes. The increases in luteinizing hormone were temporally related to behavioral and physiological measures of sexual arousal. The authors commented that the rapid increases in luteinizing hormone and reports of subjective high after both intravenous cocaine and high-nicotine cigarette smoking illustrate the similarities between these drugs and they suggested a possible contribution of luteinizing hormone to their abuse-related effects.

Blood sugar concentrations can become labile in people with diabetes mellitus who use cocaine, not only because their diet changes, but also because adrenaline concentrations affect the mobilization of glucose (36).

In a prospective study, endocrine responses to hyperthermic stress were assessed in 10 male cocaine users after 4 weeks of abstinence and again after 1 year of abstinence (184). They sat in a sauna for 30 minutes at a temperature of 90°F and a relative humidity of 10%. At the end of the sauna, they rested for another 30 minutes at room temperature. Sublingual temperature, pulse rate, and blood pressure were recorded just before and immediately after the sauna and 30 minutes after the period at room temperature. Venous β-erythropoietin, ACTH, metenkephalin, prolactin, and cortisol were also measured. There were no significant differences between the two groups in heart rate and blood pressure. At baseline and after 1 year of abstinence, plasma prolactin concentrations were higher in the cocaine users than in the controls. Moreover, the hormonal responses in the cocaine users were different from those in controls. Concentrations of all the hormones, except for metenkephalin, were significantly lower in the cocaine users than in the controls at the end of the sauna; the cocaine users did not have significant hormonal changes to hyperthermia after either 4 weeks or 1 year of abstinence. The authors concluded that cocaine abuse produces alterations in the hypothalamic-pituitary axis, which persist during abstinence.

Electrolyte balance

Cocaine-induced periodic paralysis with hypokalemia has been reported (185).

- A healthy 33-year-old man suddenly developed generalized weakness and became unable to walk or lift his limbs; he also had mild chest pain. He had had similar episodes 10 days and 5 years before, with spontaneous resolution. He had no spontaneous motor activity and his strength was 2/5 in all major muscle groups with a very mild left upper limb predominance. Cardiac enzymes and neuroimaging of the brain and spinal cord were normal. Creatinine kinase was raised (395 IU/l). Acetylcholine receptor antibodies were in the reference range. His serum potassium concentration was 1.9 mmol/l. Urine toxicology screen showed cocaine, cannabinoids, and benzodiazepines and he admitted to cocaine binge use the previous night and also before the previous two episodes. With potassium supplements his strength gradually improved.

Such severe generalized weakness and hypokalemia may be due to intracellular shift of potassium due to adrenergic stimulation by cocaine or a direct effect on potassium channels.

Hematologic

Erythrocytosis has been implicated as one of the factors underlying cocaine-associated cardiac complications. In a prospective study, differences in mean hemoglobin concentration, hematocrit, and reticulocyte count were measured in 79 consecutive cocaine-exposed and cocaine-unexposed patients who developed chest pain (186). The authors hypothesized that the contribution of the bone marrow to cocaine-induced erythrocytosis is negligible. Acute cocaine exposure was of less than 3 hours duration. Hemoglobin and hematocrit levels were significantly higher in cocaine-using subjects than in controls (13.5 versus 12.6 g/dl and 40% versus 38%). There was no corresponding increase in reticulocyte count, suggesting that the bone marrow does not contribute to transient erythrocytosis. Men with chest pain were more likely to be exposed to cocaine then women. Moreover, all relative increases in hemoglobin concentration in the cocaine-exposed group were attributable to sex. Amongst other variables, only a history of diabetes mellitus was significantly associated with an increased reticulocyte count.

The authors therefore concluded that acute cocaine exposure is not associated with erythrocytosis in younger patients with chest pain.

In an attempt to replicate the conclusions of an earlier study, there was no association between cocaine use during pregnancy and acute thrombocytopenia in 326 patients (187). There were similar prevalences of thrombocytopenia in cocaine-using women (13/160) and non-using women (11/160) during pregnancy. Thrombocytopenia occurred more often in the third trimester in both groups.

Splenic infarction related to cocaine abuse is extremely rare and has been reported as a possible complication in patients with sickle cell hemoglobinopathies.

- A 17-year-old cocaine abuser was found dead in bed. Autopsy showed signs of sepsis, splenic infarctions of different ages, and splenic necrosis with abscesses. The splenic abscesses and microabscesses in various other organs showed mixed bacterial infections (188).

These findings show that in cocaine abusers pain and fever can be an expression of severe cocaine-associated complications.

Gastrointestinal

Gastrointestinal symptoms, especially diarrhea, occur after cocaine use. Cases of more severe abdominal distress have required surgical intervention and have been due to bowel infarction (1,189) or pneumoperitoneum (190).

Esophageal damage

The esophagus can undergo thermal injury, in which the inner esophageal wall has a "candy-cane" appearance (alternating pink and white linear bands), when boiling-hot liquids are consumed. This reversible condition is associated with chest pain, difficulty in swallowing, odynophagia, and abdominal pain. Candy-cane esophagus secondary to smoking crack cocaine has been reported (191).

- A 55-year-old man accidentally sucked into his mouth and swallowed a portion of boiling water during his last smoke of free base cocaine and 2 days later developed sudden constant pain in the left shoulder and arm accompanied by sweating. He had melena, a hematocrit of 30%, a blood urea nitrogen of 35 mmol/l, and a serum creatinine of 1.1 mmol/l. The initial electrocardiogram showed sinus rhythm, left atrial enlargement, and borderline left ventricular hypertrophy. However, 2 hours later he started to sweat and became hypotensive (blood pressure 85/50 mmHg). An electrocardiogram showed new biphasic T waves and T wave inversion. Urgent cardiac catheterization showed patent coronary arteries. Esophagogastroduodenoscopy within the hour showed a candy-cane appearance in the distal esophagus, patchy erythema and erosions in the gastric antrum, and an ulcer in the base of the duodenal bulb. Biopsies of the esophagus tissue showed parakeratosis, squamous hyperplasia with regeneration, and minimal inflammation. Biopsies of the stomach showed chronic gastritis and bacteria consistent with *Helicobacter pylori*. Later that day, the electrocardiogram normalized.

The most likely cause of the chest pain was microvascular spasm of the epicardial coronary arteries, due to either thermal injury to the esophagus or a direct effect of cocaine.

Peptic ulceration

There is a higher incidence of gastric ulcers in cocaine users, both perforated ulcers and giant gastroduodenal ulcers, thought to be due to localized ischemia secondary to vasoconstriction (192,193). Two reports have afforded data on gastrointestinal ulcers and cocaine. In one study the authors observed that since the advent of crack cocaine they had seen more than 70 cases of crack-related perforated ulcers (194). They suggested that an ischemic process rather than an acid-producing mechanism was to blame. They described three patients, all of whom had laparoscopic omental patches for ulcers, with good results. In a longitudinal assessment of patients with endoscopically diagnosed gastric ulcers ($n = 98$) or duodenal ulcers ($n = 116$) users of cocaine or metamfetamine were nearly 10 times more likely to have giant gastric or duodenal ulcers (over 2.5 cm) compared with non-users (195). The authors speculated that cocaine and amfetamine-induced catecholamine stimulation of α-adrenoceptors may cause intense vasoconstriction and thus a reduced blood supply to an ulcer, resulting in a giant ulcer.

Five cases of gastric perforation (rather than the more common duodenal perforation) have been reported in young male smokers of crack, all of whom had only brief histories of prodromal symptoms and none of whom had long-standing peptic ulcer disease (196).

Intestinal ischemia

Two women developed chronic mesenteric ischemia, successfully managed by revascularization (197). The authors concluded that in both cases chronic mesenteric ischemia had been caused by intravenous cocaine abuse.

Ischemia of the small bowel and colon after the use of cocaine has been reported (SEDA-22, 33).

- A 38-year-old man presented with a 2-day history of severe abdominal pain and bloody stools after smoking cocaine 48 hours earlier (198). He had abdominal pain, guarding, rebound tenderness, and high-pitched, hypoactive bowel sounds. His white blood cell count was $31 \times 10^9/l$. Radiography showed thumb-printing in the transverse colon. Endoscopy showed friable edematous mucosa with submucosal hemorrhage and patches of yellowish fibrinous material. He recovered fully with intravenous nutrition and supportive measures after 30 days.
- A 36-year-old man who had injected cocaine the day before admission and who occasionally sniffed, smoked, or injected cocaine, presented with a 3-day history of severe abdominal pain and bloody diarrhea (199). His mid-abdomen was very tender, with rebound

tenderness and guarding; bowel sounds were absent. Radiography showed a dilated transverse colon and dilated small bowel loops. He underwent emergency surgery and an edematous dilated transverse colon was removed. The pathology was consistent with ischemic colitis. The blood vessels were dilated but showed no structural abnormalities or thrombosis.

Pneumoperitoneum is a surgical emergency associated with the use of crack cocaine (200).

- A 42-year-old man was intoxicated after an alcohol binge and recreational crack cocaine smoking. He was semi-conscious and in acute respiratory distress. He had subcutaneous emphysema in the face, neck, chest, abdomen, and legs, and tenderness in the left lower back. He had a temperature of 38^0C, a heart rate of 117/minute, and a respiratory rate of 35/minute. There was hypoxia (PaO$_2$ 7.3) and a metabolic acidosis (pH 7.28). A chest X-ray showed free gas under the right hemidiaphragm. At laparotomy no organ perforation was found.

In all likelihood, a prolonged Valsalva maneuver while smoking crack cocaine causes of pneumoperitoneum. Repeated, deep, vigorous inhalation can cause an abrupt imbalance of pressures in the pulmonary alveolar–capillary complex. Free air from ruptured alveoli then dissects along fascial planes and can collect in the mediastinum and retroperitoneum.

Liver

Cocaine has been associated with liver toxicity (SEDA-14, 32; SEDA-13, 27).

- A 23-year-old man became unresponsive and had a seizure after taking cocaine and alcohol (201). Severe liver necrosis developed and hepatocellular damage was documented with ^{99m}Tc-PYP imaging.

Acute hepatitis induced by intranasal cocaine, with transient increases in liver enzymes, has been reported in three HIV-positive patients (202). All had non-active chronic viral hepatitis with normal immunological status; one was seropositive for hepatitis B virus and two were positive for hepatitis C virus. A few days after intranasal cocaine use, serum transaminases rose to high values, and two of the patients had fever, stiffness, sweats, and hepatomegaly. Alcohol and hepatotoxic agents were ruled out. Within a few days, the clinical and laboratory signs of hepatitis improved in all three cases.

Hepatotoxicity secondary to cocaine exposure can be associated with hyperthermia, ischemia, or a direct cocaine effect. Another possible mechanism may be an acquired mitochondrial defect (203).

- A 41-year-old man jumped through a ground-floor window after cocaine abuse. His Glasgow coma score was 7/15. His temperature was 40°C, heart rate 140/minute, and respiratory rate 40/minute. He had a tonic–clonic seizure and was intubated and ventilated. Blood chemistry showed a profound metabolic acidosis (pH 7.19), a raised creatinine at 232 μmol/l, a raised creatine kinase at 14 500 IU/l. Rhabdomyolysis was confirmed by the finding of myoglobinuria. Both urine and plasma toxic screens showed traces of cocaine and benzodiazepines. He was treated with dialysis, vasopressors, and then liver transplantation. Postoperative complications included hypotension, hypothermia, anuria, resistant hyperkalemia, and compartment syndrome, for which he had bilateral forearm fasciotomy. He died 12 hours later and post-mortem examination of the liver showed extensive hepatic necrosis in the acinar zones with macrovesicular steatosis, a pattern seen in toxic liver damage. A muscle biopsy showed a few necrotic fibers, glycogen depletion, and increased lipid staining.

The authors suggested that this patient had a defect in lipid metabolism, based on the muscle biopsy. Muscle mitochondria are a principle site for beta-oxidation of fatty acids. Microvesicular steatosis can progress to liver failure with severe and prolonged impairment of beta-oxidation. This metabolic defect may have exacerbated the direct toxic effects of cocaine.

In one case acute hepatitis and thrombotic microangiopathy occurred simultaneously (204).

- A 22-year-old woman, with a 3-year history of alcohol and intravenous cocaine abuse and chronic hepatitis C virus infection, developed fever, jaundice, vomiting, and weakness. She was jaundiced, with markedly raised transaminases (aspartate transaminase 1264 U/l, alanine transaminase 1305 U/l) increased alkaline phosphatase, and normal gamma-glutamyltransferase and serum total bilirubin. Partial thromboplastin time was 33 seconds. A complete blood count and renal function were normal. Urine screen was positive for cocaine. Over the next 48 hours she developed acute respiratory and renal failure, hypotension, and tonic-clonic seizures. She also developed severe thrombocytopenia (platelet cell count 21 x 10^9/l), impaired coagulation, and a hemolytic anemia with 6–7 schistocytes per microscopic field. A chest X-ray showed bilateral alveolar infiltrates compatible with acute respiratory distress syndrome. An electrocardiogram showed an acute inferior myocardial infarction. Liver biopsy showed multifocal hepatic necrosis and microvesicular steatosis, consistent with toxic hepatitis. She recovered after plasma exchange, blood transfusions, and hemodialysis.

The authors noted that thrombotic microangiopathy due to cocaine is fairly rare. Its pathogenesis is unclear, possible mechanisms being an immune reaction or direct damage to the vascular endothelium. Cocaine-induced acute hepatitis has been linked to several toxic metabolites, including norcocaine and N-hydroxynorcocaine, which are produced by cytochrome P450 enzymes.

Urinary tract

Cocaine can cause acute renal insufficiency (SEDA-21,19) (SEDA-24, 38). Acute renal insufficiency, with malignant hypertension, apparently precipitated by cocaine-induced vasoconstriction, has been described in a 33-year-old woman who had pre-existing scleroderma and normal renal function (205). She was successfully treated with hemodialysis.

Cocaine can also cause chronic renal insufficiency (206). Of hemodialysis patients from an urban center in California, 55 who reported a history of significant cocaine use were compared with 138 non-users. A diagnosis of hypertension-related end-stage renal disease was reported in 49 of the 55 cocaine users (89%) and 64 of the 138 non-users (46%). Of 113 patients with end-stage renal disease, 49 had a history of cocaine use. The patients who had used cocaine had hypertension for a shorter duration (5.3 versus 12.7 years). They were also younger (41 versus 54 years). The authors proposed that this outcome had been caused by several mechanisms: renal vasoconstriction or stenosis, resulting in ischemic nephropathy and secondary hypertension, direct renal damage with progressive renal insufficiency, and recurrent episodes of accelerated hypertension, vasculitis, acute tubular necrosis, and rhabdomyolysis.

Chronic cocaine use can exacerbate pre-existing hypertension, when renal blood vessels narrow secondary to cocaine-induced intimal fibrosis. Acute renal insufficiency with concomitant rhabdomyolysis after cocaine use has been reported (207,208), but it can also occur, albeit rarely, in the absence of rhabdomyolysis (209).

- A healthy 31-year-old man developed acute renal insufficiency 18 hours after inhaling cocaine 5 g. His blood pressure was 150/100 mmHg, his serum creatinine 177 µmol/l, creatine phosphokinase activity 107 U/l, and serum potassium concentration 3.8 mmol/l. The urinary sodium concentration was 30 mmol/l and there was a trace of protein and 1–2 red blood cells per high-power field. Immunological studies were unremarkable. Ultrasound showed kidneys of normal size with hyperechogenity of the right kidney. Over the next 10 days he recovered spontaneously.

The authors suggested that intense cocaine-induced renal vasoconstriction had been the likely underlying mechanism.

Renal infarction is an uncommon adverse effect of cocaine (210).

- After using intranasal cocaine, a 25-year-old African man developed fever and progressive right flank pain over 4 days. He had a temperature of 38.3°C, a blood pressure of 106/54 mmHg, and severe tenderness in the right flank and right lower quadrant of the abdomen. His urine contained cocaine. A CT scan showed reduced uptake in the lower pole of the kidney, confirming renal infarction. Other causes were ruled out.

Musculoskeletal

Non-traumatic rhabdomyolysis is often secondary to alcohol, cocaine, amphetamines, heroin, etc, and is characterized by laboratory features that reflect the release of muscle cell contents into the plasma. Early detection can prevent progression to acute renal insufficiency (211).

- A 37-year-old man sustained multiple fractures without loss of consciousness after an assault. He had a history of smoking crack cocaine and drinking alcohol. He was afebrile and had stable vital signs. His urine analysis was positive for cocaine and blood, thought to be secondary to trauma from a Foley catheter. His temperature rose to 101°F and surgery was postponed. He subsequently developed uremia with a creatinine concentration of 389 µmol/l, a raised uric acid, and a creatine kinase activity of 3055 IU/l, with an MB isoenzyme activity of 12 ng/ml (reference range 0 to 6); serum phosphorus and magnesium were also increased. Ultrasonography of the kidneys was normal and he did not have HIV. His fever resolved spontaneously and his renal function recovered.

After an extensive literature search, the authors reported that the main pathophysiological mechanism underlying cocaine associated rhabdomyolysis is unknown. They suggested that cocaine blocks the reuptake of noradrenaline and dopamine, resulting in increased sympathetic activity. This, coupled with potent vasoconstriction by the cocaine metabolite benzoylecgonine, can lead to skeletal muscle ischemia and injury and result in rhabdomyolysis.

Sexual function

The practice of trading sex for drugs in places where there is a high prevalence of cocaine abuse has been noted in both metropolitan areas and smaller communities along major interstate highways. In Baltimore, there was a 97% increase in the number of primary and secondary cases of syphilis from 1993 to 1995 (212).

Priapism has been reported in men who have used cocaine by inhalation or applied it topically to the glans penis or intraurethrally (SEDA-19, 26).

- Three men developed priapism and delayed seeking treatment (213). Cocaine use within the previous 24 hours was the singular contributing factor in all three cases. Two of the men had had previous episodes of priapism, which had resolved spontaneously. Initial duplex ultrasonography confirmed low penile blood flow. Manual aspiration and irrigation failed in all three cases. Surgical shunting failed in the first two cases. One man required partial penile amputation for infected, gangrenous, distal penile tissue; one responded to angiographic embolization; and one had only a partial response to angiographic embolization, but then refused further intervention; his erection resolved during the next 24 hours.

The authors suggested that acute sexual excitement during cocaine intoxication can cause penile erection, with impaired detumescence. Cocaine can inhibit the reuptake of noradrenaline (by blocking transport in presynaptic sympathetic neurons), thus preventing sinusoidal contraction and the efflux of penile blood.

- A 44-year-old black man developed priapism 2 hours after having overdosed on 30–40 trazodone tablets 50 mg and 10 Tylenol No. 3 (paracetamol plus codeine) tablets (214). Toxicology analysis was positive for cocaine and opiates. The priapism required detumescence twice, on initial presentation and then 6 hours later, and 8 hours after presentation he again developed painless priapism, which resolved spontaneously after 1.5 hours.

Trazodone-induced priapism may be mediated by alpha-adrenoceptor antagonism. While the mechanism for cocaine-induced priapism is unclear, it may result from vasospasm, venous pooling, and sludging of blood in the penis. The authors proposed that the two drugs may act in an additive or synergistic manner, posing a greater hazard than either alone.

Priapism associated with intracavernosal injection of cocaine has also been reported.

- A 43-year-old man developed persistent painful erection after intracavernosal injection of cocaine (215). He had previously administered cocaine in this way to prolong erections. Cavernosal aspiration resulted in partial detumescence, but the condition recurred. Urine screen was positive for cocaine. Aspiration and irrigation fully alleviated the condition.

During penile erection, nitric oxide is released from the endothelium of the cavernous spaces and from nerve endings (non-adrenergic and non-cholinergic). Nitric oxide stimulates guanylate cyclase, which is involved in the conversion of guanosine triphosphate to cyclic guanosine monophosphate (cGMP); the latter relaxes the smooth muscle in the corpora cavernosa, allowing influx of blood for erection. The authors suggested that cocaine directly applied to the cavernosal endothelium can cause nitric oxide production.

Three patients developed priapism after taking cocaine or non-prescription weight loss formulations containing ephedrine (216). Intracavernous injection of phenylephrine and irrigation with heparinized saline, followed by an Al-Ghorab shunt procedure, was effective.

Some believe that topical application of cocaine to the glans penis enhances sexual performance. However, such use can cause complications, including superficial penile necrosis (217).

- A 32-year-old white heterosexual man developed widespread painful, blackened lesions on the penis after applying cocaine as an aphrodisiac. Screens for sexually transmitted infections were negative. He was given a 5-day course of antibiotics and the lesions healed completely.

The authors thought that cocaine applied to the glans penis may have been well absorbed through the thinner keratinized squamous epithelium. They attributed the superficial necrosis to intense skin vasoconstriction caused by cocaine.

Immunologic

The prevalence of infection with the human immunodeficiency virus (HIV) among drug abusers, including cocaine users, is increasing (218). Two separate reports have suggested that cocaine may compromise immunological function. In one study, human mononuclear cells were stimulated in vitro with mitogens in the presence and absence of cocaine; cocaine inhibited the proliferation of the mononuclear cells (219). In a second study, cocaine amplified HIV-1 replication in co-cultures containing cytomegalovirus-activated peripheral blood mononuclear cells (220).

Two cases of connective tissue disease have been reported (221).

A case of urticarial vasculitis, a type III hypersensitivity reaction, has been reported after cocaine use (222).

- A 24-year-old man with acute malaise and fever had a pruritic rash with multiple erythematous circumscribed weals on the trunk, arms, legs, neck, and scalp. He admitted to using intranasal cocaine 6 months, 4 days, and 1 day before the onset of the symptoms. His temperature was 39°C. His erythrocyte sedimentation rate was 80 mm in the first hour, C-reactive protein was 283 mg/l (reference range below 10), and the white blood cell count was 12.4×10^9/l with 89% neutrophils. A biopsy of an urticarial lesion showed a perivascular inflammatory infiltrate in the upper and middle dermis. Bed rest, oral prednisone, oral hydroxyzine, and topical polidocanol led to improvement within 24 hours.

Two cases of cocaine-induced type I hypersensitivity reactions, have been reported (223).

- A 23-year-old woman developed tongue swelling and difficulty in breathing immediately after having sniffed cocaine. The anterior half of her tongue was edematous with bleeding lesions caused by her fingernails. There were cocaine metabolites in the urine. The diagnosis was angioedema of the tongue induced by cocaine or its contaminants, and it resolved with subcutaneous adrenaline, H_1 receptor antihistamines, and intravenous glucocorticoids.
- A 19-year-old man developed generalized urticaria, intense pruritus, and mild bronchospasm 30 minutes after injecting cocaine for the third time. He had weals on the face, neck, arms, and chest, and scattered wheezing in the lungs. Urine toxicology screen was positive for cocaine metabolites. His symptoms resolved several hours after the administration of H_1 receptor antihistamines and intravenous glucocorticoids.

Body temperature

Cocaine can cause hyperthermia, primarily in hot weather, perhaps through a hypermetabolic state; impaired heat dissipation may be another contributing factor. Seven healthy, cocaine-naïve subjects participated in tests of progressive passive heat stress, during which each received intranasal cocaine or lidocaine as placebo (224). Esophageal temperature, skin blood flow, sweat rate, and perceived thermal sensation were measured. Cocaine augmented the temperature increase during heat stress and also increased the temperature threshold for the onset of both cutaneous vasodilatation and sweating. It also impaired the perception of heat. This study

elicited commentary, in which it was pointed out that measured effects of small doses of cocaine may not be reflective of true cocaine poisoning (225). Also, the subjects in the study did not have psychomotor agitation, which is often prominent in cocaine toxicity and which improves with sedatives.

Death

Poisoning can occur with doses of cocaine as low as 20 mg (10 drops of cocaine 4%). Victims generally collapse and die after associated cardiovascular abnormalities, dysrhythmias, and respiratory failure. Signs and symptoms of intoxication include excitement, restlessness, headache, nausea, vomiting, abdominal pain, convulsions, and delirium.

In a retrospective study of 48 men who suffered cocaine-related deaths and a control group of 51 male cocaine users who died of trauma, the blood cocaine concentrations measured in the two groups were similar (226). However, concentrations of the cocaine metabolite benzoylecgonine were higher in those with cocaine-related deaths. This group also had a significantly lower body mass index, with larger hearts and heavier lungs, livers, and spleens than the control subjects. Reduced body weight, an adverse effect of long-term cocaine use, is probably related to its effects on the serotonergic system and therefore appetite. Cardiomegaly is thought to result from chronic cocaine-induced excessive catecholamine stimulation, with circulatory overload, and increased organ weight is a result of passive visceral congestion in cocaine-induced heart failure. Cardiac alterations may explain why similar blood cocaine concentrations can be lethal in some cases but benign in others. This study shows that isolated measurements of postmortem cocaine and benzoylecgonine blood concentrations cannot be used to assess or predict cocaine toxicity.

The increasing prevalence of multisubstance abuse can influence morbidity and mortality (227).

- An 18-year-old man experienced sudden and severe chest pain while drinking alcohol. He vomited, collapsed, and died. On postmortem examination, thrombosis of the left coronary artery, dilated cardiomyopathy with congestive heart failure, and pulmonary embolism were noted. Blood analysis showed raised cocaine and marijuana concentrations and a trace of alcohol.

The author's opinion was that although multidrug use had played a part, the high blood concentration of cocaine had been the main cause of death. He also noted that marijuana can interact with cocaine to produce pronounced sympathomimetic effects.

The smuggling of illicit drugs by the technique called "body packing" carries medical risks. Drug packages can rupture and digestive secretions can seep into packets and allow drug absorption. Consequently, drug intoxication, intestinal obstruction, peritonitis, and death can occur. A man carrying 99 cocaine powder packages weighing 10 g died as a result (228), and the death of a drug dealer has been reported (229).

- A 17-year-old man swallowed a small plastic bag of cocaine in order to avoid arrest. After 1 hour he complained of a headache and 30 minutes later developed palpitation and agitation and collapsed. Histological examination of his heart showed myocardial necrosis. The blood concentration of cocaine was high at 98 µg/ml. The tissue concentrations of cocaine and its metabolites in various organs and fluids were recorded; the highest concentrations of cocaine and metabolites were detected in the liver, lungs, brain, and blood (in descending order).

The cause of sudden death in this case was probably a cardiac dysrhythmia.

Death occurred shortly after intravenous cocaine in a 26-year-old who had also used heroin and methadone 24 hours before (230). Death was attributed to thoracic aortic dissection after the use of crack cocaine. Histological findings showed connective tissue abnormalities, including focal microcystic medial necrosis and fragmentation of the elastic fibers in the arterial wall. Aortic dissection among cocaine users is thought to be related to weakening of the media of the aorta, and to sheering forces that result from sudden and profound hypertension that accompanies cocaine use (231). Whether or not other susceptibility factors, such as Marfan's syndrome and hypertension, can contribute to aortic aneurysm in cocaine abusers is uncertain. The presence of concentric left ventricular hypertrophy is suggestive of chronic hypertension. Transesophageal echocardiography is helpful in diagnosing aortic dissection in cocaine abusers (232,233).

The criteria for the interpretation of cocaine concentrations in biological samples and their relation to the cause of death has been comprehensively reviewed (234). The importance of scene investigation, forensic autopsy, and forensic sampling for drug analysis has been discussed, with particular emphasis on the need to use appropriate blood preservatives and interpretation of the half-life and concentrations of cocaine and its metabolites, benzylecgonine and ethylcocaine, in combined cocaine + alcohol abuse.

About 1–2% of people in Western countries are regular consumers of cocaine and 10% are sporadic users. This proportion increases considerably in the age groups in which organ donors are most often found. Cocaine use is often associated with death, creating opportunities for organ donation (235).

- A 30-year-old woman with a history of bronchial asthma and cocaine abuse had a cardiorespiratory arrest preceded by sudden dyspnea 1 hour after cocaine inhalation. Direct laryngoscopy showed edema of the glottis. After extended cardiopulmonary resuscitation, she went into a deep coma with dilated non-reactive pupils. Toxicological analysis showed cocaine and amphetamines in her urine. She was brain dead 11 hours later and her organs were used for transplantation. Her liver was given to a 14-year-

old boy with acute hepatocellular failure caused by isoniazid; 5 years later he had normal liver function. Her kidneys were given to a 52-year-old woman and a 56-year-old man, both with polycystic kidney disease; in both cases renal function was normal after 5 years.

The grafts took in all three cases. Myoglobinuric acute renal insufficiency in the donor did not affect immediate, short-term, or long-term graft function. In their review of the literature, the authors found one report of eight transplants from three donors. There were no effects attributable to cocaine in any of the recipients in the immediate post-transplantation period. They concluded that organ donation is safe after brain death caused by cocaine toxicity, probably because of the characteristics of the cocaine, such as a short half life.

Long-Term Effects

Drug tolerance

Chronic cocaine exposure and long-term adaptation at the molecular level have been investigated; changes in transcription factor gene expression may be involved (236). NURR1 is a key factor that regulates transcription of the gene that encodes the cocaine-sensitive dopamine transporter and functions in the development of dopamine neurons. In a recent study, postmortem human midbrain specimens from cocaine users and controls underwent various analyses. Human NURR1 gene expression was markedly reduced in dopamine neurons in the cocaine users and normal in the controls. NURR1-deficient cocaine abusers also had dopamine neurons with markedly reduced dopamine transporter gene expression. NURR1 appears to have a critical role in the brain's adaptation to repeated cocaine exposure and in maintenance of dopamine neurons.

Drug dependence

There has been considerable interest in evaluating drugs such as amfepramone (diethylpropion) for attenuating the negative emotional state induced by craving for cocaine, in the hope of finding a drug for long-term treatment of cocaine dependence. However, in 50 cocaine-dependent patients amfepramone was ineffective and caused significant adverse effects (237). Of the patients who took amfepramone 25–75 mg/day, 12% were withdrawn from the study: one developed coronary vasospasm and another atrial fibrillation. These poor results are comparable to those of earlier studies with methylphenidate in cocaine addicts (238,239).

Drug withdrawal

There have been reports of the effects of cocaine withdrawal on cognition. There was impairment of memory, visuospatial abilities, and concentration in 16 cocaine abusers during the first 2 weeks of abstinence (55). Measured deficits were independent of withdrawal-related depression.

Tumorigenicity

Chronic cocaine use, which is associated with immuno-suppression, may be carcinogenic. The possible association between chronic cocaine exposure and pancreatic adenocarcinoma has been investigated (240,241). A study of hospital records in Brazil for the years 1986–1998 showed that of 198 patients with pancreatic adeno-carcinoma, 13 (6.5%) were younger than 40 years; of these, five had a history of chronic cocaine inhalation and one had abused marijuana.

Second-Generation Effects

Pregnancy

Some of the risks run by the pregnant cocaine-using mother and her child (such as preterm labor and premature delivery) have been reported (242). However, the literature on maternal cocaine use and its possible outcomes is problematic, because many studies have been methodologically flawed (243).

A retrospective study of data from a large perinatal registry showed that there was an increased risk of placenta previa among women who used cocaine, compared with those who did not use drugs or alcohol (244).

Another apparent obstetric risk of cocaine use is rupture of a uterine scar (from a previous cesarean section).

- There was extensive laceration of the maternal urinary bladder after a vaginal birth in a 34-year-old woman whose urine tested positive for cocaine (245).

The authors postulated that the injury may have resulted from a cocaine-augmented contractile response of the pregnant uterus.

A "pre-eclampsia–like" syndrome has been described, characterized by acute hypertension and a low platelet count, in a 33-year-old cocaine user; her 20-week-old fetus died (246).

One of the serious medical conditions linked with cocaine use during pregnancy is premature delivery, with an incidence in cocaine users of 17–27%. The mechanism of the effect of cocaine on both spontaneous and agonist-induced contractility of pregnant human myometrium has been evaluated (247). Myometrium samples from 42 women who were undergoing cesarean section at term were examined after exposure to various pharmacological probes in combination with cocaine. The results suggested that cocaine augments the contractility of uterine tissue by both adrenergic and non-adrenergic mechanisms. Cocaine increased spontaneous myometrial contractility over three-fold. Prazosin, an alpha-adrenoceptor antagonist, blocked this effect, but only for the first 35 minutes. Cocaine increased both the sensitivity and maximal tissue response to the alpha-adrenoceptor agonist methoxamine. The maximal response to oxytocin, but not sensitivity, was increased by cocaine; prazosin did not inhibit this effect.

Cocaine has been associated with both preterm delivery and premature rupture of the membranes (248). Among 85 of 604 expectant mothers with premature rupture of the membranes with documented cocaine exposure compared with women with no drug exposure for six conditions of major neonatal morbidity, cocaine users were older and of higher parity. The non-cocaine users had more morbidity, in particular neonatal infection and sepsis. The authors proposed that the mechanism of premature rupture of the membranes in the presence of cocaine may not be related to infection. Instead, cocaine may have a direct effect on the myometrium, stimulating uterine contractility.

The Maternal Lifestyle Study has reported that the prevalence of adverse perinatal complications associated with the use of cocaine or opiates during pregnancy was lower than has been previously reported (249). In 11 811 mother–infant pairs followed prospectively, 11% of the exposed and non-exposed groups were hospitalized at least once. However, violence was a factor (20%) in admissions among the cocaine-exposed women.

Considering the increased prevalence of recreational cocaine abuse among young women (250), the diverse clinical manifestations of acute cocaine intake combined with physiological changes of pregnancy and the pathophysiology of co-existing pregnancy-specific diseases can result in life-threatening complications during anesthesia. When regional anesthesia is selected, combative behaviour, altered pain perception, and ephedrine-resistant hypotension can occur. Cardiac dysrhythmias, hypertension, and myocardial ischemia can occur during general anesthesia. Individualizing anesthesia management in patients presenting with diverse clinical manifestations of acute cocaine intake is important.

- A 26-year-old woman with a history of multiple substance abuse required emergency caesarean section at 30 weeks of gestation as a result of crack cocaine-induced placental abruption and fetal distress (251). Her admission blood pressure was 145/95 mmHg, heart rate 95/minute and respiratory rate 20/minute. The fetal heart rate was 130/minute and non-reactive, with late and variable decelerations and no response to maternal oxygen administration. Spinal block with bupivacaine, fentanyl, and morphine was performed with the patient in a sitting position. No maternal or neonatal postoperative complications were reported.

Teratogenicity

In the USA 100 000 crack cocaine babies are born each year, and an increasing number of anomalies is being linked to maternal cocaine abuse (252, 253). However, it is unclear by what mechanism cocaine affects the fetus. Interruption of the intrauterine blood supply, with subsequent destruction of fetal structures, may account for some of its effects (254).

There are frequent reports of intrauterine growth retardation, neurobehavioral abnormalities, cerebral injury, and cardiac anomalies in "coke babies" (SEDA-14, 15; SEDA-21, 4; SEDA-21, 129) (255,256). Brain hemorrhages (257) and asymmetrical growth retardation (258) associated with maternal cocaine abuse have been discussed.

In 500 neonates ankyloglossia, a defect in the attachment of the tongue within the mouth, was 3.5 times more common in cocaine-exposed neonates than in others (259). Other facial, vertebral, and cardiovascular defects have been described (SEDA-17, 4; 19).

Two other clinical syndromes involving anomalies of multiple organ systems in fetuses of cocaine-abusing women have been described.

- An infant with Pena–Shokeir phenotype (including facial, musculoskeletal, pulmonary, and cardiac malformations accompanied by extensive brain damage) was born to a cocaine-abusing mother (260). The infant died shortly after birth.
- An infant exposed to cocaine in utero had a combination of facial, ear, eye, and vertebral anomalies, accompanied by cardiac, central nervous system, and other malformations (261).

In contrast, in 34 light-to-moderate cocaine users compared with 600 non-users attending a public prenatal care clinic, pronounced untoward effects on the fetus were reportedly less common than in other studies (262). In all cases the cocaine had been taken intranasally, and the majority of users reduced their intake during pregnancy; none had been referred for drug abuse counseling and none was taking drug treatment. There was no significant difference in obstetric complications among these mild cocaine users compared with non-users, and no significant differences in infant growth, morphology, or behavior. However, the cocaine users had histories of more fetal losses and during pregnancy they suffered more infectious diseases, such as hepatitis, *Herpes simplex*, and gonorrhea.

Fetal growth

Among the adverse outcomes of prenatal cocaine exposure, low birth weight and reduced length and head circumference have been reported. In a prospective study in New York City, 386 pairs of cocaine- and crack-using mothers and their infants and 130 matched control pairs were followed during the course of pregnancy and delivery (263). The neonates were assessed by physical and neurological examination, the Brazelton Neonatal Behavioral Assessment Scale (BNBAS), and the Neonatal Stress Scale during the first 48 hours of life. The results corroborated earlier findings of reduced fetal growth in cocaine-exposed infants. Significantly more (17%) of the babies of cocaine users had a head circumference less than the tenth percentile compared with the controls (3%). They performed less well on the BNBA Scale and had higher measures on the Neonatal Stress Scale. They had clinically significant neurological impairment, with jitteriness, increased tone, and an exaggerated Moro reflex. However, some of these findings may have reflected a direct neurotoxic effect of cocaine, since testing was done during the first 48 hours of life. The authors observed that crack had a more adverse outcome than cocaine. They concluded that the most important predictor of neonatal outcome may be the frequency, quantity, and type of cocaine used.

The effect of maternal cocaine use on infant outcome has been prospectively assessed in 224 women, of whom 105 were cocaine users and 119 were controls (264). The infants were of gestational age 34 weeks or more and were not asphyxiated. The infants exposed to cocaine were more likely to be admitted to the newborn intensive care unit, to be treated for congenital syphilis or presumed sepsis, to have a greater length of stay, to have lower birth weight and head circumference, and to be discharged to the care of someone other than the mother. However, the two groups were similar in the incidence of abnormal cranial and renal ultrasonography and abnormal pneumocardiography. Moreover, when controlled for cigarette use and other confounders, there were no significant differences in the groups on growth retardation factors.

Cardiovascular

Intrauterine cocaine exposure is associated with neonatal cardiovascular dysfunction and malformations. The long-term effects of cocaine on the neonate's cardiovascular system and development are unknown. The effect of cocaine on the infant's autonomic function and subsequent development has been reported (265). Heart rate variability, a non-invasive test of autonomic function, was evaluated in 77 prenatally cocaine-exposed infants, 77 healthy controls, and 89 infants who had been exposed prenatally to drugs other than cocaine (alcohol, marijuana, and/or nicotine). Within the first 72 hours of life, the cocaine-exposed infants were asymptomatic but had lower heart rate variability and lower vagal tone than the two comparison groups. At follow-up, the cocaine-exposed infants had recovered at 2–6 months of age and now had higher heart rate variability and vagal tone than the two non-exposed groups. Most of the increase in heart rate variability and vagal tone was seen in the infants who had had light cocaine exposure and was not apparent in those who had had heavy exposure.

The same researchers have published two reports on the cardiovascular effects of intrauterine cocaine. In the first study, 82 healthy neonates with intrauterine cocaine exposure, 108 exposed to drugs other than cocaine, and 87 healthy controls were evaluated for global and segmental systolic and diastolic cardiac function (266). During the first 48 hours of life, the neonates with intrauterine cocaine exposure had significant left ventricular diastolic segmental abnormalities. They had a higher index of asynchrony and global and segmental fractional area changes in contrast to the other two groups. The degree of abnormality in the index of asynchrony was greater in the neonates with heavier cocaine exposure. In a second study at 2–6 months of age, 56 cocaine-exposed infants were compared with 72 who had been exposed to drugs other than cocaine and 60 healthy controls (267). The cocaine-exposed infants had recovered left ventricular diastolic function. Only in infants with heavy cocaine exposure was there an alteration in septal wall diastolic filling.

- A 6-month-old fetus who had been exposed to cocaine had a single-ventricle heart; the authors suggested that coronary spasm, resulting in infarction, may have destroyed the right ventricle (268).

Double aortic arch anomaly has been linked to maternal cocaine abuse (269).

- A girl born to a cocaine-abusing mother had cocaine withdrawal symptoms and at 2 months developed respiratory dysfunction and died. At autopsy, the heart and lung were normal, but there was a double aortic arch anomaly of right persistent dominant arch type. The aorta encircled the trachea and esophagus. The right common carotid, right vertebral, and right subclavian arteries arose from the right aortic arch and the left carotid and left subclavian arteries originated from the left aortic arch.

Cocaine may directly affect the fetal cardiovascular system or do so by increasing the concentrations of circulating catecholamines and activating the sympathetic nervous system.

Nervous system
In two reports it was suggested that brain hemorrhage (257) and asymmetrical growth retardation (258,270) can occur.

Cognitive effects
The influence of exposure to cocaine in utero on the developing human nervous system is not yet clearly understood (SEDA-22, 21). Prenatal cocaine exposure has been associated with neurobehavioral effects in infancy, ranging from no effect to effects on arousal and state regulation, as well as on neurophysiological and neurological functions. A prospective controlled study in 154 cocaine-using pregnant mothers and suitable matched controls from a rural community produced neurobehavioral effects that supported those from previous controlled studies (271). The mothers underwent drug testing and medical examination during each trimester. Their infants were assessed as near to 40 weeks after conception as possible, using the BNBA Scale. When controlled for the effects of marijuana, alcohol, and tobacco use, the use of cocaine in the third trimester was negatively related to state regulation, attention, and responsiveness among the exposed infants. Twice as many cocaine-exposed infants as controls failed to come to and maintain the quiet alert state required for orientation testing.

It is unclear whether abnormalities in early infancy are associated with neurodevelopmental impairment at a later age. Several studies have suggested that the findings are limited to early childhood. The possible effects of prenatal cocaine exposure on later cognitive functioning and difficulties have been reported in three studies. In the first, 236 infants at 8 and 18 months of age were evaluated; 37 had heavy exposure to cocaine in utero, 30 had light exposure, and 169 had no exposure (272). Cognitive functioning was assessed with the Bayley Scales of Infant Development. Information processing was tested with an infant-controlled habituation procedure. At 8 months, cocaine-exposed infants and controls had no differences in cognitive functioning. Their abilities to process information indexed by habituation and response to novelty were comparable. However, at 18 months the infants with high cocaine exposure performed poorly on the Mental Development Index (MDI). The 18-month index covers a wider range of cognitive tasks requiring integrated learning, responsiveness to environmental cues, and memory than the 8-month index. These results suggest that the effects of cocaine are more likely to show up when more challenging measures are used. Infants raised in high-risk environments, with stressors and low support, scored lower at both 8 and 18 months.

In a second study, intellectual functioning at 6–9 years was measured in 88 cocaine-exposed children and 96 unexposed children in New York City (273). The participants were interviewed and underwent medical and neurological examination and psychological assessment. Child intelligence was measured with the Wechsler Intelligence Scale for Children-III (WISC-III). Intelligence quotient scores did not differ between the two groups of children, even when adjustments for co-variables were made.

In a third study the Robert Wood Johnson database of published literature on prenatal cocaine exposure and child outcome was examined (274). Only 8 of 101 studies focused on school-age children. Intelligence quotient (IQ), receptive language, and expressive language were measured. This meta-analysis showed an average difference of 3.12 IQ points between cocaine-exposed and control groups. When the IQ distribution is shifted downwards by this amount, there is a 1.6-fold increase in the number of children with IQs under 70. The authors noted that the calculated decrement in IQ in exposed children is subtle and does not include the possible effect of the drug on domains of function such as language abilities.

Research on the relation between prenatal cocaine exposure and childhood behavior also continues. In a pilot study, 27 children exposed to cocaine in utero and 75 control children were assessed (275). The children had a mean age of 80 months and most were first-grade students. The child's first-grade teacher (blinded to exposure status and study design) rated the children's behavior with the Conners' Teacher Rating Scale (CTRS) and the Problem Behavior Scale (PROBS 14), an investigator-developed scale that measures behaviors associated with cocaine exposure. The drug-exposed children had higher CTRS scores (that is more problematic behavior), but the difference was not significant. On subscales of the PROBS 14, the drug-exposed group had significantly more problematic behavior. These results appear to substantiate teachers' reports of problematic behavior in children with prenatal cocaine exposure.

The effects of prenatal cocaine exposure on information processing and developmental assessment have been studied in 108 infants aged 3 months, 61 of whom had been exposed to cocaine, and 47 controls using an infant-control habituation and novelty responsiveness procedure in a developmental assessment using the Bayley Scales of Infant Development (276). Infants exposed to cocaine prenatally were significantly more likely than controls to fail to start the habituation procedure, and those who did were significantly more likely than controls to react with irritability early in the procedure. Cocaine-exposed infants had a comparatively depressed performance on the motor but not the mental Bayley scales. This information was obtained by raters blind to the history and was controlled for both perinatal and sociodemographic factors. Most of the infants in both groups reached the habituation criteria, and among those who did there were no significant differences between cocaine-exposed and non-exposed infants in habituation or in recovery to a novel stimulus. Thus, differences in reactivity to novelty, but not information processing, between cocaine-exposed and non-cocaine-exposed infants suggested that the effects of prenatal cocaine exposure may be on arousal and attention regulation, rather than on early cognitive processes.

Developmental correlates have been assessed in three groups of children aged 4–6 years (277). In 18 children there had been prenatal exposure to cocaine and the mothers had continued to use crack. Another 28 children had had no prenatal exposure but their mothers had used crack after the children were born. The control group were 28 children whose mothers had never used cocaine. Prenatally exposed children performed significantly worse than the others in tests of receptive language and visual motor drawing. Prenatal crack exposure was associated with poor visual motor performance, even after controlling for intrauterine alcohol and marijuana exposure, age, birth weight, and duration of maternal crack use.

There have been two studies of the neurodevelopmental effects of cocaine during the first 48 hours of life. In the first, 23 cocaine-exposed and 29 non-exposed infants were prospectively assessed within the first 48 hours of life; infant meconium was used to detect cocaine and the BNBA Scale was used for clinical assessment (278). One-third of the cocaine-exposed neonates were born to women who denied cocaine use. In six of the seven clusters assessed, cocaine-exposed infants fared badly compared with control infants. The cocaine-exposed infants had poor autonomic stability and there was a dose–response relation between meconium cocaine concentration and poor performance in relation to orientation and so-called "regulation of state," which refers to how the infant responds when aroused. The authors concluded that cocaine exposure is independently related to poor behavioral performance in areas that are central to optimal infant development. They emphasized the value of the identification and quantification of cocaine in infants.

In another blinded study, neurodevelopmental and neurobehavioral performance were prospectively assessed in 131 neonates (mean age 43 hours) exposed in utero to cocaine, with or without other drugs (279). Cocaine-exposed neonates were developmentally at risk in the tests compared with infants exposed to other drugs alone or in combination. As in the previous study, larger amounts of cocaine were associated with higher neurobehavioral risk scores.

In a study of immediate and late dose–response effects of cocaine exposure in utero on neurobehavioral performance, 251 full-term urban neonates were examined by blinded raters at 2 and 17 days (280). The babies were classified as having been heavily exposed, lightly exposed, or not exposed to cocaine. After controlling for covariates, in contrast to the studies mentioned above (275,280), there were no neurobehavioral effects of exposure at 2 days of age. However, at 17 days there was a significant dose-related effect: heavily exposed infants had poorer state regulation and greater excitability, implying impairment of their ability to modulate arousal. The authors postulated that these late effects might be expected if cocaine exposure in utero is associated with evolving neuroanatomical damage or disruption of the monoaminergic neurotransmitter systems. These effects did not appear to be related to intrauterine growth retardation, as has been suggested by others.

Arousal and attention have been investigated in 180 healthy nursery infants before hospital discharge and at 1 month of age (281). Cocaine-exposed infants showed a lack of arousal-modulated attention and preferred faster frequencies of stimulation, regardless of arousal condition compared with non-exposed infants. There were similar differences 1 month after birth, showing that these effects persisted beyond the period of presence of cocaine or its metabolites at birth. These effects were independent of absence of prenatal care, alcohol use, minority status, or sex, suggesting a direct and even chronic effect of intrauterine cocaine exposure on arousal-modulated attention and presumably on the developing nervous system of the infants.

The behavioral and hormonal responses in 30 preterm cocaine-exposed infants were compared with the responses in 30 non-cocaine-exposed infants of similar gestational age (282). The mothers of cocaine-exposed infants were more often single, had higher parity and more obstetric complications, and were less likely to visit, touch, hold, and feed their infants than the other mothers. Cocaine-exposed infants had smaller head circumferences at birth, spent more time in the neonatal intensive care unit, and had a greater incidence of periventricular or intraventricular hemorrhages. They also had poorer state regulation and difficulty in maintaining alert states and in regulating their own behavior. They spent more time in indeterminate sleep (suggesting nervous system immaturity), with reduced periods of quiet sleep and increased amounts of agitation, tremulousness, mouthing, multiple limb movements, and clenched fists. There were higher urinary noradrenaline, dopamine, and cortisol concentrations and lower plasma insulin concentrations in the cocaine-exposed infants, suggesting that they may have experienced a high degree of stress in the perinatal period.

In a study of 464 inner-city black infants, whose mothers were recruited prenatally based on alcohol and cocaine use during pregnancy, gestational age of less than 38 weeks was significantly correlated with cocaine use in the mothers (283). The infants were tested at 6.5, 12, and 13 months of age; the cocaine-exposed infants were more excitable, preferred faster frequencies of stimulation, had more difficulty habituating, were more reactive, and showed a greater startle response to noise. Moreover, these effects of cocaine on cognitive function were documented beyond the neonatal period, thus eliminating effects from acute cocaine exposure or withdrawal. The authors suggested that two separate mechanisms may underlie the effects of cocaine on gestational age and cognition. The nervous system deficits, poorer cognitive performance, and faster reactivity are probably mediated by a direct action of cocaine (requiring heavy exposure) on neurotransmitters, whereas shortened gestation may be mediated by vasoconstriction, which occurs at lower exposure. Timing of exposure during pregnancy may also play a critical role in determining the type of deficits.

The effects of prenatal maternal cocaine use on neurobehavior have been reported in 2-week old infants (284). The BNBA Scale was administered to the infants of mothers who had a reported high frequency of cocaine use during pregnancy: ($n = 23$, >75th percentile reported days of use) or a low frequency ($n = 32$, <75th percentile). Infants with high intrauterine exposure had higher scores on the BNBAS excitability cluster than infants with low exposure. Infants with a high BNBAS excitability score had poorer tone and motor movement, were more irritable and hard to console, and had difficulties in self-quieting.

In a meta-analysis of 18 published reports (1985–1988, 13 of which failed to meet the inclusion criteria and were excluded) on the effect of in utero exposure to cocaine on infant neurobehavioral outcome, cocaine-exposed infants were compared to non-exposed infants on BNBAS cluster scales at birth and at 3–4 weeks of age (285). Although the sample size was large enough to detect statistical significance in most of the tests of difference between the two groups, the magnitude of all the effects was small. This was true for differences in the motor performance and abnormal reflex clusters in the infant groups (with a slight trend toward increasing standard differences over time) and in the orientation and autonomic regular clusters (with a trend toward a reduced effect size over time). However, the main finding of the study was the small magnitude of the neurobehavior dysfunction at both times. The authors cautioned that these data may not be generalizable, since polydrug exposure, the amount of cocaine exposure, and other variables could have confounded the data.

In 158 cocaine-exposed (82 heavily and 76 lightly exposed) and 161 non-cocaine exposed infants, neurobehavioral function was assessed at 43 weeks after conception (286). Mediating factors (the timing and amount of drug exposure) and maternal psychological distress as a confounding factor were considered in the design and statistical analysis. The infants with heavy cocaine exposure had significantly more jitteriness and attentional difficulties. They were also more likely to be identified with an abnormality and less likely to cooperate with testing procedures than infants in the other groups. Higher concentrations of cocaine metabolites, cocaethylene and benzoylecgonine, were associated with a higher incidence of movement and tone abnormalities, jitteriness, and the presence of any abnormality. Higher cocaethylene concentrations were associated with attentional abnormalities; higher concentrations of metahydroxybenzoylecgonine were associated with jitteriness.

Cognitive, motor, and behavior development, as measured by the Mullen Scales of Early Learning and the Bayley Scales of Infant Development-II, were compared in 56 prenatally cocaine-exposed infants and toddlers (aged 1–3 years) and 56 non-exposed matched controls (287). There were developmental problems in expressive and receptive language areas in those who had been exposed prenatally.

The effect of intrauterine cocaine exposure on visual attention, cognition, and behavior has been investigated in 14 cocaine-exposed children and 20 controls aged 14–60 months (288). The cocaine-exposed children were slower in tests of disengagement and sustained attention. They also had greater difficulties in behavioral regulation.

Research on the effects of prenatal cocaine exposure on development in the first 2 years of life has been reported in 203 full-term infants (289). The infants, who were defined as having had no cocaine exposure, light exposure, or heavy exposure, were tested with the Bayley Scales of Infant Development at 6, 12, and 24 months. Assays of neonatal meconium for cocaine metabolites along with mothers' self-reports were used to evaluate the dose–response relation. There were no significant adverse effects due to cocaine exposure on scores in the major tests up to 24 months of age. Cocaine-exposed infants with the lowest 10th percentile birth weight and those placed with kinship caregivers had less optimal development. Cocaine-exposed infants who participated in child-focused early intervention programs scored higher than the others.

The behavioral effects of prenatal cocaine exposure at age 5 years have been studied in 140 children exposed to cocaine, 61 exposed to alcohol, tobacco, and/or marijuana, and 120 not exposed to any drugs (290). They were evaluated with the Achenbach Child Behavior Checklist. There was no association between behavior and intrauterine cocaine exposure. However, the current behavioral health of the mother, including recent drug use and psychological functioning, did affect the child's internalizing and externalizing behavior.

Gastrointestinal

Cocaine exposure in utero can affect various fetal organs. Gastrointestinal disorders, including ten cases of necrotizing enterocolitis (291), one of intestinal atresia, and one of spontaneous colonic perforation, have been reported (292).

Amelia and humeral "bifurcation" due to humeroradial synostosis are both very rare limb abnormalities and occur in less than 1 in 50 000 births.

- A 29-year-old Canadian Aboriginal woman, who had used cocaine intermittently during the first 8 weeks of pregnancy, gave birth to a boy at 38 weeks. Amniocentesis was normal and labor was unremarkable (293). The boy's left arm was absent and he had right-sided phocomelia with a three-fingered hand. Radiography showed hypoplasia of the left clavicle and scapula, and absence of the left pedicle at T5. There was ulnar aplasia, with a short radius fused to the humerus on the right side. The family history was unremarkable.

This was one of the most severely affected children reported. In most cases the defect is unilateral. When both arms are involved, oligodactyly is often asymmetrical. Such cocaine related defects have also been reported in animal models (294,295,296).

- A 23-year-old woman who was taking methadone 34 mg/day and lorazepam 4 mg/day admitted to irregular alcohol abuse of up to 3 liters of beer per day and regular intravenous use of cocaine and heroin over the previous 3 months at dosages and frequencies that could not be accurately ascertained (297). She was positive for hepatitis B and C. At 12 weeks her urine contained cocaine, cannabis, codeine, and morphine. At 17 weeks, ultrasonography showed ahydramnios, and color Doppler confirmed bilateral renal agenesis. After termination an autopsy confirmed the renal findings but also showed a right upper limb reduction defect and single umbilical artery. The right arm skeleton consisted of a round rudiment of the humerus and a near normal ulna, with no evidence of a radius. The hand lacked the thumb and two fingers.

The authors reviewed the literature and reported that cocaine abuse in pregnancy results in congenital defects in 15–20% of cases, primarily involving the brain, heart, genitourinary tract, and limbs. They reported that at their hospital, since 1996, 35 cases of cocaine abuse in pregnancy had been identified. From this group another four cases of congenital defects, including talipes, optic nerve atrophy, acromelia of the left hand and right fingers, and an isolated single umbilical artery, were identified. The overall defect rate at their center was 14%. They quoted other human and animal studies suggesting an increased risk of vascular disruption defects after in utero cocaine exposure and supported this by the findings from their case of a single umbilical artery. They further suggested that some of the findings are similar to the teratogenic effects of thalidomide, a well known inhibitor of angiogenesis, supporting the notion that these findings are due to cocaine-induced vascular disruption syndrome in pregnancy.

Reproductive system
An increased incidence of genital malformations has often been noted (298).

Fetotoxicity

The prevalence rate of cocaine use during pregnancy is 10–45% in some centers in North America. As cocaine use is increasing and widespread, information on the possible adverse effects secondary to fetal cocaine exposure continues to amass in case reports and studies.

However, in a prospective, large-scale, longitudinal study there was no association between prenatal cocaine exposure and congenital anomalies in 272 offspring of 154 cocaine-using mothers and 154 non-using matched controls (299). The cocaine-exposed group had significantly more premature infants, who were significantly smaller in birth weight, length, and head circumference than the control infants. However, there were no differences in the type or number of abnormalities.

The impact of prenatal exposure to cocaine on fetal growth and fetal head circumference has been studied in 476 African-American neonates, including 253 full-term infants prenatally exposed to cocaine (with or without alcohol, tobacco, or marijuana) and 223 non-cocaine exposed infants (147 drug-free, 76 exposed to alcohol, tobacco, or marijuana) (300). The cocaine-associated deficit in fetal growth was 0.63 standard deviations and for gestational age 0.33 standard deviations. There were also cocaine-associated deficits in birth weight and length, but no evidence of a disproportionate effect on head circumference.

The relation between prenatal cocaine exposure and early childhood outcome has been reviewed (301). Prospective longitudinal studies of perinatal cocaine exposure and associated outcomes were studied in a survey of 36 of 74 reports, published from 1984 to October 2000, in which the examiners were blinded to cocaine exposure. Prenatal cocaine exposure did not alter physical growth, developmental test scores, or receptive and expressive language among children aged 6 years or less. The authors concluded that there is no convincing evidence that prenatal cocaine exposure is associated with effects on a child's physical or behavioral development, and that many findings once thought to be specific effects of in utero cocaine exposure instead correlated with factors such as the quality of the child's environment and prenatal exposure to tobacco, marijuana, or alcohol.

This review generated responses from other authorities in the field. Some commented that the conclusions may be premature, given the age of the subjects, and drew attention to several studies that have shown subtle but consistent deficits in cognitive and attentional processes in 6- and 7-year-old children (302). These effects may become more prominent as development continues and may persist into adulthood. Others criticized the attempt to isolate cocaine exposure from all other associated risk factors; from a public health perspective, prenatal cocaine exposure clusters with other risk factors, such as poor caregiving, child maltreatment, domestic violence, and prenatal exposure to other substances (303). Furthermore, the selection criteria narrowed the total articles reviewed to under half of the 74 articles found. Others suggested that the study had been misinterpreted (304).

The effects of prenatal cocaine exposure have been assessed prospectively in 217 infants, 95 (44%) of whom had benzoylecgonine, a cocaine metabolite, in their meconium (278). Among these infants, benzoylecgonine concentration was inversely related to fetal growth (birth weight, length, and head circumference), whereas maternal self-report of days of cocaine use did not correlate with either fetal growth or meconium benzoylecgonine concentration. The report suggested a dose–response relation between the magnitude of prenatal cocaine exposure and impaired fetal growth.

In 39 cocaine-exposed infants and 39 control infants aged 35 weeks or older, head size was smaller and birth weight tended to be lower in the cocaine-exposed infants (305). Moreover, the head circumference of the cocaine-exposed infants was significantly smaller at any given birth weight than in the control infants. The behavioral scores were significantly higher (on days 1 and 2) in the cocaine-exposed infants; the higher scores were most frequently attributed to increased jitteriness, a hyperactive Moro response, and excessive sucking. Lastly, cocaine-exposed infants had an increase in flow velocity in the anterior cerebral arteries between days 1 and 2; however, there was no increased propensity to ischemic and/or hemorrhagic cerebral injury in the infants exposed to cocaine. The blood flow changes on the second day may have reflected falling infant cocaine concentrations after birth.

In a longitudinal evaluation of 28 infants exposed to cocaine in utero and 22 unexposed controls for 15 months, the cocaine-exposed infants weighed significantly less at birth than the control infants, but not subsequently (306). Compared with controls, motor development was compromised in the cocaine-exposed infants at 4 and 7 months, but not at 1 and 15 months, suggesting that compromised motor performance in the exposed group normalized for later milestones, probably through a self-righting process. A disturbing aspect of this study was the extremely poor performance in all-motor assessments at every age by every infant (including the controls). The investigators postulated that in an inner-city population (such as that studied here), once an infant accumulates three or more risk points (as most infants in the study did), additional risk factors (including exposure to cocaine) have little further negative impact on their development.

The complex interplay between the relative effects of prenatal cocaine exposure and the perinatal and environmental factors on development has been evaluated, using a structural model to describe the direct and indirect effects of prenatal drug exposure on developmental outcome from birth to age 6 months (307). Key variables included prenatal drug exposure, perinatal medical characteristics, maternal/caregiver/family characteristics, the home environment, and neurobehavioral outcomes. The study was based on 154 predominantly crack-using women and 154 control subjects matched for pregnancy risk, parity, race, and socioeconomic status. Prior exclusion criteria included age under 18 years, a major illness diagnosed before pregnancy, chronic use of legal drugs, and any use of illicit drugs other than cocaine and marijuana. Urine specimens were collected at two unanticipated times and positive serum samples were confirmed by gas chromatography/mass spectroscopy. Measures analysed by blinded evaluators included medical assessment at birth and developmental assessments at birth, 1 month, and 6 months, as well as caregiver characteristics and environmental factors at birth and 1 month.

Exposure to cocaine affected development at birth. Increasing exposure was significantly related to poor developmental outcomes, as measured by the Brazelton qualifier scores. Although no direct effects of cocaine were found at either 1 month or 6 months analysed separately, time-dependent analysis showed an effect on development at 6 months. The indirect effects of cocaine exposure were mediated through maternal psychosocial well-being at delivery and birth head circumference. In addition, indirect effects of prenatal cocaine exposure were also related to concomitant alcohol and tobacco use and the birth head circumference. Neither maternal nor caregiver factors at 1 month was directly related to developmental outcome at any time. These findings support previous findings (308–310) that suggest that cocaine is a mild teratogen with regard to neurodevelopmental outcome.

The presence of cocaine during the prenatal period disrupts the development of neural systems involved in mediating visual attention. Of 14 cocaine-exposed children and 20 control children aged 14–60 months, whose visual attention, cognition, and behavior were assessed, the cocaine-exposed children had slower reaction times, supporting the hypothesis that impairment in disengagement and sustained attention are associated with prenatal cocaine exposure (311). There was a trend to slower reaction times to targets presented in the right visual field, but not the left visual field. Cocaine-exposed children also had greater difficulties in behavioral regulation, especially related to an ability to cope with heightened levels of positive and negative emotions (312).

An association between prenatal cocaine exposure and deficits in total language functioning was found in 236 cocaine-exposed and 207 non-cocaine exposed full-term children (313). The link between prenatal cocaine exposure and language deficits during early childhood was not related to cocaine-associated deficits in birth weight, length, or head circumference. Three different but potentially interacting mechanisms whereby maternal cocaine use might affect early language development have been proposed (314):

- subtle dysregulation of attentional systems, with potential for disrupting an infant's ability to extract and process available linguistic information;
- disruptions in parent–child linguistic interactions due to cocaine and other drug use;
- impoverished, unstable, and endangering social and care-giving environments.

Early detection of language deficits allowed ameliorative intervention aimed at improving academic performance and social adaptation in preschool and school-aged children.

The Maternal Lifestyle Study, a prospective randomized study, has followed 717 infants exposed only to cocaine and 7442 non-exposed infants from birth to hospital discharge at 12 network sites (315,316). Cocaine-exposed infants were younger (1.2 weeks), weighed less (536 g), and were smaller (2.6 cm shorter and head circumference 1.5 cm smaller). Congenital anomalies were not increased. Acute subtle changes in central and autonomic nervous system function, such as irritability, jitteriness, tremors, high pitched crying, and excessive suck, were more common in the cocaine-exposed group.

There was a significantly increased prevalence of infection in the infants of cocaine-using mothers. Hepatitis was 42 times more common, syphilis was 15 times as common, and human immunodeficiency virus positivity was 16 times more common (although the overall prevalence was only 0.1%). Exposure to cocaine increased the likelihood of involvement with social services such as child protection agencies.

The preliminary results of the Meconium Project, a study done in Barcelona between October 2002 and February 2004, have been reported (317). The findings were based on the first 830 meconium samples and 549 mother–infant pairs. Overall drug use was 7.9%, and drug screens detected 6-monoacetymorphine and cocaine. There was under-reporting of drug abuse. The self-reporting rates in interviews were opiates 1.3%, cocaine 1.8%, and both drugs 1.3%; meconium analysis showed higher rates (8.7. 4.4, and 2.2% respectively). One declared case of ecstasy consumption was confirmed. Arecoline, the main alkaloid in areca nuts, was found in meconium samples from four Asian neonates whose mothers had consumed betel nuts. The use of opiates and cocaine during pregnancy was associated with active use of cannabis, tobacco smoking, and a higher number of cigarettes smoked. Lower birth weight in newborns was associated with mothers who used cocaine only and both cocaine and opiates. One of the four infants exposed to arecoline had a low birth weight, hypotonia, and hyporeflexia.

Cardiovascular

In a retrospective review of all dysrhythmias in children with prenatal cocaine exposure, 18 cases were detected in 554 infants who had positive urine screens for cocaine (318). In 13 neonates the dysrhythmia occurred beyond the period of direct cocaine exposure and six of the children had dysrhythmias after the neonatal period. Most of the dysrhythmias were supraventricular extra beats. Overall, the rate of consultations for dysrhythmias was higher among cocaine-exposed neonates than expected. Some cocaine-exposed children had symptomatic dysrhythmias that were persistent or recurrent and required treatment to maintain cardiac output and restore normal cardiac rhythm. Children who were exposed prenatally to cocaine appeared to be at increased risk of abnormal responses to stress, manifested by symptomatic dysrhythmias beyond the period of cocaine exposure.

- A child presented at 12 months of age with status epilepticus, sustained ventricular tachycardia, and a positive urine screen for cocaine. At 22 months he returned with a cardiac arrest, a history of a fall, a head injury, and a positive test for cocaine in the urine. He died soon after.

There has been a recent report of a myocardial infarct in a full-term infant born to a 28-year-old woman who had used cocaine 2–3 times per week and methadone 40 mg/day (319).

Respiratory

Respiratory rates in the 3-week-old babies of mothers who had used cocaine during pregnancy were higher than expected; in addition newborn babies who had been exposed prenatally to both cocaine and narcotic analgesics had abnormal control of breathing (following hypercapnia challenge) during the first few months of life (320). The preliminary results of another study are also of some interest; this prospective study of maternal drug abuse showed a reduced incidence of respiratory distress syndrome among premature infants prenatally exposed to cocaine (321). The authors noted that while this finding needs to be confirmed, it may suggest that fetal lung maturation can be accelerated by exposure to cocaine.

Prenatal cocaine exposure appears to have short-term effects on respiratory function in very low birth weight infants. In a retrospective study of 149 such infants, 48 cocaine-exposed and 101 non-exposed, the cocaine-exposed infants had transiently improved respiratory status at time of delivery; they needed surfactant treatment in lower doses and at a lower frequency and intubation less often (322). At 24 and 48 hours there was no significant difference between the treatment requirements in the two groups. The development of bronchopulmonary dysplasia was also similar. The authors suggested that prenatal cocaine exposure affects the fetus by two mechanisms: indirectly through reduced uterine blood flow with placental insufficiency and directly through an adrenergic effect on the fetus. The fetus may experience cocaine as a stressor that leads to accelerated fetal lung maturity.

Nervous system

Fetal microcephaly has been attributed to cocaine abuse during pregnancy (323). Urine toxicology confirmed the presence of morphine, benzoylecgonine, barbiturates, paracetamol, and propoxyphene. Analyses of amniotic fluid, placenta, and fetal serum and urine were also positive for these substances. The authors suggested that vascular disruption was the likely major mechanism of anomalies, both behavioral and malformative, due to prolonged exposure to cocaine in utero.

- An infant born at 37 weeks gestation to a mother who had engaged in discontinuous cocaine abuse during the first and second trimesters of pregnancy had microcrania (below the 10th percentile), a closed anterior fontanelle, and overlapping of all sutures (324). The infant was of low birth weight (2290 g; 25th percentile). There were deep scalp rugae, a prominent occipital bone, and normal hair pattern. An MRI scan of the brain showed enlargement of the lateral ventricles and pericerebral spaces, with severe reduction of the cerebral and cerebellar parenchyma and white matter abnormalities.

These findings are part of the recognizable pattern of defects in the rare condition termed fetal brain disruption sequence. The presence of a normal hair pattern suggests normal brain development during the first 18 weeks of gestation. At a later stage partial destruction of the brain results in reduced intracranial pressure and subsequent collapse of the fetal skull.

Prenatal cocaine exposure has been associated with subependymal hemorrhage and subependymal cyst formation in term neonates and more recently in preterm neonates (<36 weeks of gestation) (325). Medical records and cranial sonograms obtained during 1 year on 122 premature infants showed an increased incidence of subependymal cysts in preterm cocaine-exposed infants (8 of 18) compared with non-exposed infants (8 of 99). There was no increase in the incidence of major structural abnormalities. All subependymal cysts resolved by 4 months of age. The authors noted that the neurodevelopmental implications of such cyst formation are unknown.

An unusual congenital malformation, the cloverleaf skull, has been associated with cocaine exposure in utero (326). In this condition, the cranium is trilobed, with severe brain deformity and hydrocephalus, because of premature fusion of the coronal and lambdoid sutures.

- A girl born by cesarean section at 38 weeks gestation weighed 3515 g and measured 54 cm in length and needed cardiopulmonary resuscitation. She had feeding and respiratory problems. Cranial sonography on day 11 showed a trilobed cranial mass with ventricular enlargement. She was discharged on day 35. The mother, a 24-year-old cocaine user, had engaged in active drug use for the 2 years before and during the first 2 months of pregnancy; she had also used alcohol (three units per day) and smoked marijuana (1–2 joints per day) during the first 5 months, and she had smoked 10 joints per day throughout the entire pregnancy. The father was also a marijuana smoker. The infant failed to thrive (body weight at 6 months 3120 g, height 57 cm), developed sepsis, and died. Autopsy showed adrenal infarction secondary to systemic infection.

Motor development

A second report from the Maternal Lifestyle Study focused on motor development in 392 children prenatally exposed to cocaine and 776 non-exposed control infants who were identified by meconium assay and mothers' self-reporting (327). Motor skills were assessed at 1 month with the NICU Network Neurobehavioral Scale (NNNS), at 4 months with the posture and fine motor assessment of infants (PFMAI), at 12 months with the Bayley Scales of Infant Development-2nd edition (BSID-II), and at 18 months with the Peabody Developmental Motor Scales (PDMS). The infants with prenatal cocaine exposure had motor skill deficits at 1 month, but normal function at 18 months. Heavy cocaine use was associated with poorer motor performance. Both lower and higher nicotine exposures related to poorer motor performance.

Neurodevelopment

Neurodevelopmental and cognitive outcomes among prenatally cocaine exposed children have been documented by measuring school performance in 62 cocaine-exposed and 73 control children who were students at an American inner-city school (328). The children were followed prospectively from birth to the end of the 4th grade. Their report cards, standardized test results, teacher and parent reports, and birth and early childhood data were

studied. Both groups had poor grade progression from grades 1 to 4 (71% versus 84%), low Grade Point averages, reading skills below grade level (30% versus 28%), and below-average standardized test scores. The children with higher Full Scale Intelligence Quotients and better home environments, regardless of drug status, had successful progression.

Prenatal exposure to cocaine can alter the typical developmental trajectory of functional asymmetries. Twenty infants who were prenatally exposed to cocaine performed a grasping task with their right hand for significantly shorter durations and were less likely to show a dominant hand preference at 1 month of age (329). It is unclear from this study whether the absence of side biases in motor functions seen in cocaine-exposed, 1-month-old infants continues beyond the neonatal period. Considering how much plasticity there is in infants' brains, any early nervous system disorganization or damage may not predict later outcomes. Therefore, the possibility that absence of motor asymmetries seen in these infants may reflect a transient effect of prenatal exposure to cocaine must be considered.

The development of motor asymmetries was assessed in 20 infants who were exposed to cocaine prenatally and 23 infants who were not (330). Asymmetries in stepping, grasping, and head orientation were assessed at 1 month of age. As expected, based on the findings of previous research on high-risk infants, infants who were exposed to cocaine prenatally preformed a grasping task with their right hand for significantly shorter durations than the control infants, who were more likely to have a side bias for head orientation and stepping. There was a dose-response relation between maternal substance use during pregnancy and motor asymmetries in the infants. These findings suggest that prenatal exposure to cocaine alters the typical developmental trajectory of functional asymmetries and may have important implications for long-term developmental outcomes. However, longer term studies with larger samples are needed to explore these issues.

Sensory systems

Eyes

Neither cocaine (nor cigarette smoke) exposure in utero is associated with poor acuity or visual abnormalities early in infancy (331). At six weeks of age, infants' visual acuity was measured with the Teller acuity card procedure, and a neurological examination was carried out, including examination of the visual system, with assessment of eyelid edema, visual attention, and gaze ability in 96 infants. This does not exclude the possibility that cocaine exposure has an effect on the developing eye, but suggests that these abnormalities may occur later in life or very rarely, thus necessitating longer follow-up.

The long-term consequences of prenatal cocaine exposure in school-age children on intelligence, visuomotor skills, and motor abilities have been studied by comparing 101 children exposed perinatally to cocaine with 130 unexposed children at age 7 years (332). The children who were exposed prenatally to cocaine continued to display deficits in tasks of verbal, visuomotor integration, and fine motor skills. However, the effect of cocaine was rendered non-significant by inclusion of sociodemographic and environmental variables, especially the care-giver's vocabulary and the child's home environment. These results suggest that prenatal cocaine exposure per se may overlap with other cognitive risk factors experienced by children of low-income families. This study had several limitations: although cocaine exposure was established through a careful review of hospital records, including urine screens and maternal reports, misclassification of drug exposure was possible. The groups were initially recruited at different ages, ranging from birth to 2 years old, and there was a possibility that accuracy of maternal recall of cocaine abuse may have been less reliable the further removed the mother was from the experience of pregnancy. Moreover, both groups often used other drugs, especially alcohol and nicotine. The dichotomous classification (exposed/non-exposed) may have masked important differences in the groups. Alternatively, during the course of 7 years after the time of prenatal cocaine exposure, the postnatal environmental variables may have had a strong effect on development, masking the subtle effects of prenatal cocaine exposure.

Ears

The auditory brainstem responses (ABR) in neonates who were exposed prenatally to cocaine showed prolonged absolute peak latencies compared with non-exposed neonates and may indicate compromise of the auditory system from gestational exposure to cocaine (333). Among 58 infants studied, 21 (36%) were positive by meconium analysis for cocaine, and five (8.5%) were also positive for cannabinoids. There were significant differences in mean maternal age, gravidity, parity, birth weight, and head circumference among cocaine-exposed infants.

Animal studies have shown that cocaine has a direct toxic effect on the organ of Corti or its embryonic precursor, the otic placode (334). Human studies have shown that cocaine has an acute toxic effect on brainstem auditory neurons, causing impaired synaptic efficiency and prolongation of interpeak latencies (335,336). Conversely, no apparent effect of maternal cocaine use during pregnancy on the developing auditory system in otherwise healthy term infants has been noted (337). In addition, abnormalities in the fetal auditory system may also be secondary to the effects of cocaine on the maternal circulation. Impaired fetal growth and undernutrition can also affect brainstem maturity, as a result of delayed myelination, and can result in an abnormal auditory brainstem response. It is well established that maternal cocaine abuse is associated with low birth weight and intra-uterine growth restriction (338,339). Thus, a multiplicity of factors may explain the variations in reports of the effects of cocaine on hearing in neonates. Other factors, such as the timing of cocaine exposure during pregnancy, the amount of exposure, the effect of other abused substances, and the susceptibility of the sensorineural

organs of hearing at various stages of gestation, singly or additively, can determine the effects of cocaine on the auditory system (Tan-Laxa 357).

Psychological

A study in 105 African-American infants suggested that infants exposed prenatally to cocaine are at a high risk of significant problems in arousal and attention (340). The 8-week-old infants had their heart rates recorded when presented with a series of stimuli. The order of the stimuli was as follows: auditory (rattle), visual (red ring), and social (examiner's face and voice). There were four groups of infants: preterm drug-exposed ($n = 25$), full-term drug-exposed ($n = 32$), preterm non-exposed ($n = 22$), and full-term non-exposed ($n = 26$). Preterm infants' ages were corrected to match those of the full-term infants. There were significant differences in the responses to social stimuli. Drug-exposed infants had an accelerated heart rate (indicating distress or arousal), whereas non-drug exposed infants had a slowed response (indicating focused attention). However, there were no heart rate differences among the groups in the auditory or visual conditions.

There have been two studies of the adverse effect of prenatal cocaine on behavior of the offspring. In the first, 31 cocaine-exposed, very low birth weight infants and matched very low birth weight controls followed longitudinally were assessed at 3 years (341). The cocaine-exposed children had delayed cognitive, motor, and language development compared with the controls. Of the exposed children 45% scored in the range of mental retardation compared with 16% of the controls. Infants in the exposed group during the neonatal period were less responsive in their interactions and their mothers were less nurturing and less emotionally available.

In contrast, in a second study there were few differences in interactive behaviors between prenatally cocaine-exposed and non-exposed 12-month-old infants and their mothers (342). Videotapes recorded African-American infants and their mothers engaged in interactions (49 cocaine-exposed, 63 non-exposed). Children who were prenatally exposed to cocaine ignored their mother's departure during separation significantly more often than controls. Mothers who abused cocaine used more verbal behavior with their children than non-abusers.

Psychomotor development

Although in utero exposure to cocaine has been suggested to produce substantial deficits in multiple areas of functioning, this view has been supplanted by reports suggesting that the impact of exposure is subtle (343,344). Some reports using broad measures of developmentally appropriate performance have suggested that prenatal cocaine exposure is associated with modest reductions on the Bayley Mental Development Index (345), whereas other reports have not shown a significant effect (346). Using standardized assessments, including the Bayley Psychomotor Development Index, most studies have shown no deficits in motor functions (Jacobson 581,347,348). Direct effects of prenatal cocaine exposure on mental, motor, and behavioral outcomes have been evaluated longitudinally in 1-, 2-, and 3-year-old infants (349). The cocaine exposed infants (n = 474) scored 1.6 Mental Development Index points below infants who were not exposed to cocaine (n = 655). The effect of cocaine remained significant after controlling for co-variates, including birth weight, socioeconomic status, maternal education, vocabulary size, race, and psychopathology, as well as prenatal exposure to alcohol, cigarettes, and marijuana. Cocaine exposure in this study was not high compared with another large-scale study (Frank 1143).

It has been argued that cocaine exposure effects will be most evident in the domains of attention and affective regulation (Frank 1613,350). It is therefore plausible that the effects of prenatal cocaine exposure may nevertheless become more evident during the development of more advanced motor (351), cognitive (352), language (353), and behavioral skills (354).

Behavior

Sex influenced the impact of prenatal cocaine exposure on child outcome (355). Boys but not girls with persistent prenatal cocaine exposure were most likely to have adverse behavioral outcomes in central processing, abstract thought, and motor skills, with moderate to large effect sizes. In addition, and in contrast to the effect of some exposure, boys with persistent cocaine exposure were rated as more positive to their environment compared with either the groups with some or no exposure. The primary innovation of this study was that it encompassed five critical design elements: (1) analyses stratified by sex; (2) ordinal measures of cocaine exposure; (3) extension of the observation period into early school age; (4) appropriate controlling for confounders (other prenatal drug exposures, socioeconomic status, and postnatal family characteristics, including drug exposure); (5) a sensitive tool specifically designed to index behaviors associated with prenatal cocaine exposure.

The impact of prenatal cocaine exposure on children's physiological and behavioral functioning has been examined. Responses to emotion-inducing stimuli were studied in 27 children aged 3–6 years who were prenatally exposed to cocaine compared with 27 non-exposed controls (356). The children's affect during infant crying, a mildly frustrating task, and simulated maternal distress were observed and electroencephalographic activity was monitored. The drug-exposed group had greater right frontal electroencephalographic asymmetry, a pattern that may relate to negative emotional regulation and greater overall electroencephalographic activation. These children also showed less empathic behavior and an inability to complete mildly frustrating tasks.

It has been argued that a failure to find enduring behavioral effects of prenatal cocaine exposure in some studies may have been related to the use of broad, relatively insensitive measures (357,358). Furthermore, to judge the effect of prenatal cocaine exposure based on a dichotomous variable (yes/no) may not be an appropriate study design (359,360).

In one study, drug-exposed infants did not differ from non-exposed infants on Neonatal Behavioral Assessment Scale (NBAS) clusters or on birth characteristics (361). Infants (n = 137) born to three groups of low-income mothers—cocaine and poly-drug-using mothers in a treatment group (n = 76), users in a treatment rejecter group (n = 18), and non-users (n = 43)—were examined at 2 days and 2–4 weeks. The motor cluster improved and regulation of state worsened from the first to the second examination. There were no interactions of group by time. Regression analyses found no competing variables to explain the group differences. Power analysis showed that sample size was sufficient to have detected differences. The literature on prenatal cocaine exposure on infants is mired with contradictory perspectives, ranging from detrimental findings to the absence of any effects. The studies done thus far suffer from methodological problems or confounding variables, such as the use of other drugs and nicotine and nutritional status, making it harder to generalize.

The degree to which sex-specific effects can be identified in relation to prenatal cocaine exposure was the focus of two studies by the same group of researchers. In the first study 499 subjects, who had been prospectively followed since pregnancy, were evaluated in a laboratory at 7 years (362). The findings support sex- and alcohol-moderated effects on prenatal cocaine exposure. Among boys with prenatal alcohol exposure, those with significant cocaine exposure had significantly higher levels of Delinquent Behavior than the boys with no cocaine exposure. Prenatal cocaine exposure in the alcohol-exposed boys doubled the likelihood of clinically significant Externalizing Behavior scores compared with controls. In the absence of prenatal alcohol exposure, prenatal cocaine exposure alone in boys had no significant effect. In contrast, among girls with no prenatal alcohol exposure, the presence of prenatal cocaine exposure correlated with significant Externalizing Behaviors and Aggressive Behaviors compared with controls. Clinically significant Externalizing Behavior scores were almost five times as likely in this group. In contrast, among girls with prenatal alcohol exposure, no association was found. Prenatal cocaine exposure in the alcohol-exposed group of boys doubled the likelihood of clinically significant Externalizing Behavior scores compared with controls. In contrast, among girls with prenatal alcohol exposure, there was no association between prenatal cocaine exposure and behavior.

In the second study the same group evaluated the effects of prenatal cocaine exposure on child behavior in 506 African–American mother-child pairs (363). The mothers were identified as cocaine users and non-users during the initial prenatal visits with urine screen confirmation. Offspring behavior was assessed 6–7 years later using caregiver reports with the Achenbach Child Behavior Checklist (CBCL). Analyses stratified by sex and prenatal alcohol exposure showed that behaviors in girls without prenatal alcohol exposure but with prenatal cocaine exposure were adverse: 6.5% of the unique variance in behavior was related to prenatal cocaine exposure. Among these girls, the likelihood of scoring in the abnormal range for Aggression was 17 times control.

Cognition

Although data from animal research has also confirmed a sex-specific effect of prenatal cocaine exposure (364), studies in humans have not consistently supported this (365). Nevertheless, there is other evidence that prenatal cocaine exposure has a sex-related effect on cognition at 4 years of age (366).

Language

There have been two new reports on language abilities and prenatal cocaine exposure. Both studies used the Clinical Evaluation of Language Fundamentals (CELF-P). One documented language outcomes in 4-year-old children (189 cocaine-exposed and 185 non-exposed children) (367). There were more mild receptive language delays in the cocaine-exposed group. These children were also less likely to have higher expressive abilities. Of the cocaine-exposed group, children who were in adoptive or foster care performed with higher language skills compared with children who remained in the original household.

In the second study language outcomes were assessed in 3-year-olds (368). CELF-P was administered to 424 children (226 cocaine-exposed and 198 non-cocaine exposed). Structural equation modelling was done on the data for expressive and receptive language functioning. There was a relation between increased level of prenatal cocaine exposure and reduced expressive language functioning. For receptive language functioning, there was no statistically significant association with prenatal cocaine exposure.

The Miami Prenatal Cocaine Study is a prospective study starting at birth. The effect of prenatal cocaine exposure on language functioning in 476 full-term African–American infants has been evaluated longitudinally at six times, from 4 months to 3 years of age (369). The children were categorized as cocaine-exposed (n = 253) or non-cocaine exposed (n = 223) by maternal self report and bioassays using maternal/infant urine and meconium. The Bayley Scale of Infant Development was administered at 4, 8, 12, 18, and 24 months and the Clinical Evaluation of Language Fundamentals-Preschool at 36 months; 464 children received at least one language assessment. In longitudinal analyses, using Generalized Estimating Equations, cocaine-exposed children had lower overall language skills than non-cocaine-exposed children. The findings were stable after evaluation of potential confounding effects, including exposure to other prenatal substances and sociodemographic factors. Preliminary evidence also suggested possible mediation through an intermediary effect involving cocaine-associated defects in fetal growth. Cocaine-exposed-children scored on an average 15% of a standard deviation lower on measures of global language ability. These results were strongest at 18 and 36 months. There was no relation between head circumference and language function, suggesting that language development was affected by a more generalized defect, perhaps associated with low birth weight and related susceptibility factors, as opposed to a specific defect resulting from smaller head circumference. The major limitation of this study was the lack of inclusion of infants with major congenital malformations or HIV or other infections. The authors hypothesized that this phenomenon may be underpinned by impairment of neurobehavioral arousal and attention processes that are essential to processing linguistic cues and information, disruption of specific parent-child interactions critical to language development as a result of parental drug use, and the effect of negative social environments typically associated with parental use of cocaine and other drugs.

The effects of prenatal cocaine exposure on later learning abilities, including language, have been further investigated in 265 infants aged 1 year (134 cocaine-exposed and 131 matched non-exposed), who were tested using the Preschool Language Scale-3 (PLS-3) by blinded examiners (370). The infants were assigned to three cocaine exposure groups (as defined by maternal self-report and infant meconium assay): non-exposure (n = 131), heavier exposure (n = 66), and lighter exposure (n = 68). Fetal cocaine exposure was associated with deficits in developmental precursors of speech/language skills. At 1 year of age, more heavily exposed infants had poorer auditory comprehension than the non-exposed infants and worse total language performance than lighter and non-exposed infants. The more heavily exposed infants were also more likely to be classified as mildly delayed than non-exposed infants. Moreover, the degree of cocaine exposure had an inverse relation to auditory comprehension.

Drug Administration

Drug administration route

With the use of cocaine eye-drops, poisoning can occur with doses as small as 20 mg (10 drops of cocaine 4%). Victims generally collapse and die after associated cardiovascular abnormalities, dysrhythmias, and respiratory failure. Signs and symptoms of intoxication include excitement, restlessness, headache, nausea, vomiting, abdominal pain, convulsions, and delirium.

Children who live in homes in which there is active drug use may be at risk of passive environmental drug exposure (371).

- A 6-year-old boy developed general malaise and mild agitation, tachycardia, hypertension, and dilated pupils. His mother was a cocaine addict. Cocaine and benzodiazepines were found in significant amounts in the child's urine and hair: the urine contained 109 ng/ml of cocaine and 145 ng/ml of benzodiazepines. Hair samples contained 16 ng/ml of cocaine and 0.6 ng/ml of benzodiazepines.

The authors emphasized that passive environmental exposure may be dangerous for children.

Drug overdose

Body packing, the act of swallowing packets holding illegal drugs in order to hide the evidence from legal authorities, can cause symptoms of drug intoxication or overdose (SEDA-22, 44; 372). In an analysis of all cases of cocaine body packers reported to a metropolitan poisons control center from January 1993 to May 1994, 34 of 46 individuals were symptom-free. Eight had mild symptoms (hypertension and tachycardia) that resolved with decontamination (activated charcoal or whole body irrigation) or tranquilizers (one received benzodiazepines). Two had severe symptoms, including seizures and cardiac dysrhythmias, and both died.

An increase in the number of deaths of all body packers in New York has been associated with an increase in deaths among opiate body packers: of 50 deaths among body packers from 1990 to 2001, 42 were due to opiates (373). Four were related to cocaine and four to both opiates and cocaine. In 37 cases open or leaking drug packets in the gastrointestinal tract resulted in acute intoxication and death. Five cases involved intestinal obstruction or perforation, one a gunshot wound, one an intracerebral hemorrhage due to hypertensive disease, and one was undetermined. The number of packets recovered was 1–111 (average 46).

- A 49-year-old man became ill during a plane flight (374). He admitted to having swallowed 102 latex packages of cocaine 5 g each and 20 tablets of activated charcoal 125 mg. After stabilization in an emergency room, he suffered a seizure. After restabilization he had not defecated and was given a laxative of a mineral oil liquid paraffin. During the next 24 hours his condition worsened. His serum cocaine concentration increased from 1.95 to 2.2 µg/ml. During preparation for surgery he developed an untreatable dysrhythmia and died. Autopsy showed cocaine packages in the gut, 71 ruptured and 95 intact.

The reported lethal oral dose of cocaine is 1–3 g. In this case, paraffin may have contributed to rupture of the packages by dissolving the latex.

- A 28-year-old man aboard a flight reported feeling ill, became confused, and had a generalized seizure (375). After landing he had two further generalized seizures and a cardiac arrest. With medical management, his dysrhythmia, a wide complex tachycardia with features of right bundle branch block (Brugada syndrome) and ST segment elevation in the anteroseptal leads, was converted to sinus rhythm. He had raised troponin and creatine kinase MB. In the next 24 hours, he developed acute renal insufficiency and was treated with intermittent hemodialysis. A contrast CT showed generalized edema of the brain and features of coning, subarachnoid hemorrhage, and acute hydrocephalus. A plain abdominal X-ray showed multiple radio-opaque shadows. A CT scan of the abdomen confirmed the presence of multiple radio-opaque densities consistent with foreign bodies. There was free gas in the

abdomen, consistent with bowel perforation. He died 96 hours after admission. Autopsy showed 120 condoms containing cocaine in the gastrointestinal tract; one had ruptured.

Successful management of a cocaine body packer after endoscopic removal has been reported (376).

- A 55-year-old man was found unconscious at Al Doha airport in Qatar later had a generalized seizure. He was unconscious, afebrile, and tachypneic. His blood pressure was 150/90 mmHg and his pulse rate 105/minute. His pupils were fixed and dilated. A brain CT scan was unremarkable. A plain X-ray of the abdomen showed two packets in the stomach and one in the rectum. The packets, each of which weighed about 80 g, were removed by endoscopy, and he recovered consciousness after five days of conservative treatment.

In this case cocaine was found in the urine and the packets contents were also identified as cocaine.

In another cocaine body packer, non-surgical management was followed by the development of a giant gastric ulcer (377).

- A 35-year-old man presented to the emergency room 5 days after swallowing 35 latex-wrapped packages of cocaine. He was asymptomatic but concerned that only 10 of the 35 packets had passed in his stools. He was treated with laxatives and passed only 8 packages during the next 8 days. Radiography showed that 10–15 foreign bodies remained clustered in the stomach 14 days after ingestion. Several fragments of latex wrapping were found in his stools. On exploratory laparotomy, 15 latex packages were found impacted in the antrum just proximal to the pylorus. Beneath the packages there was a giant gastric ulcer, 2.5 cm in diameter. He had an uneventful postoperative course.

Fatal crack cocaine ingestion has been reported in an infant (378).

- A 10-month-old girl developed apnea, ventricular fibrillation, and a metabolic acidosis, and died shortly afterwards. Her 2-year-old brother had fed her crack cocaine. At autopsy the brain had a thinned corpus callosum, ranging in thickness from 0.2 to 0.5 cm. There were two pieces of crack cocaine in the duodenum and high concentrations of cocaine in the blood and other tissues.

The authors noted that the thinned corpus callosum suggested that the infant had been exposed to cocaine in utero or during the early postnatal period.

Aspects of non-fatal cocaine overdose among cocaine users have been studied in Australia in 200 current cocaine users (120 injecting users and 80 non-injecting users), who volunteered for a structured interview; 13% had overdosed on cocaine, 7% in the preceding 12 months (379). Those who had overdosed were more likely to inject cocaine, to be female, to have longer cocaine habits, to have used more cocaine in the preceding month and preceding 6 months, and to have had higher degrees of cocaine dependence and more polydrug use.

Drug–Drug Interactions

Alcohol

Cocaine abusers have reported that alcohol prolongs the euphoriant properties of cocaine, while ameliorating the acutely unpleasant physical and psychological sequelae, primarily paranoia and agitation. It may also lessen the dysphoria associated with acute cocaine abstinence. It has also been proposed that concurrent alcohol abuse may be an integral part of cocaine abuse. The combination of cocaine with alcohol can cause enhanced hepatotoxicity and enhanced cardiotoxicity (380). Trauma in patients who use cocaine plus alcohol has been reported (381). Those who use cocaine plus alcohol are 3–5 times more likely to have homicidal ideation and plans; this is particularly prominent in patients with antisocial personality disorder (382). A large high school survey by the Centers for Disease Control and Prevention showed that illicit substance abuse, prevalence of weapon carrying, and physical fighting were higher among the adolescents who reported recent use of cocaine, marijuana, alcohol, and corticosteroids. Among 215 female homicide offenders, 70% had been regular drug users at some time before imprisonment. Alcohol, crack, and powdered cocaine were the drugs most likely to be related to these homicides (383).

In a double-blind study, subjects meeting DSM-IV criteria for cocaine dependence and alcohol abuse participated in three drug administration sessions, involving intranasal cocaine with oral alcohol, cocaine with oral placebo alcohol, and cocaine placebo with oral alcohol (384). Cocaine plus alcohol produced greater euphoria and increased perception of well-being than cocaine alone. Heart rate was significantly higher with cocaine plus alcohol than with either alone. Cocaine concentrations were higher after cocaine plus alcohol than after cocaine alone. Metabolism of cocaine to cocaethylene was observed only during administration of cocaine plus alcohol. The authors concluded that enhanced psychological effects during abuse of cocaine plus alcohol may encourage the ingestion of larger amounts of these substances, placing users at increased risk of toxicity than with either drug alone.

The adverse effects of the combined use of alcohol and cocaine have been reviewed (385). There is little evidence that this combination acts synergistically or that either drug enhances the negative effects of the other. However, the combination leads to the formation of cocaethylene, which may potentiate cardiotoxic effects and the combination has a greater than additive effect on heart rate. Lastly, cocaine antagonizes the learning and psychomotor performance deficits and driving impairment caused by alcohol.

Reports of liver complications after cocaine use are infrequent. However, fulminant hepatitis with acute renal insufficiency requiring liver transplantation occurred after the use of cocaine and alcohol (386).

- A 33-year-old chronic alcoholic with hepatitis C developed acute liver and renal insufficiency with grade III encephalopathy. Hemodialysis was begun and emergency liver transplantation was performed. The explanted liver showed marked diffuse macrovesicular steatosis with massive coagulative-type necrosis. The postoperative course included a persistently raised gamma-glutamyltransferase, but he recovered fully after 60 days.

Macrovesicular steatosis can be attributed to alcohol or cocaine, but massive liver necrosis is more probably due to cocaine. The mechanisms of cocaine hepatotoxicity, such as increased lipid peroxidation, free radical activity, and impaired calcium sequestration, may be potentiated by alcohol.

Atropine

Anticholinergic poisoning involving adulterated cocaine has been reported (387).

- A 39-year-old man who was a recreational user of alcohol and cocaine presented with agitation, hallucinations, and delirium. He had a dry flushed skin, tachycardia, dilated, minimally reactive pupils, urinary retention, and absent bowel sounds. He was treated with intravenous fluids and a sedative. There were cocaine metabolites in the urine. Reanalysis of a urine sample by thin layer chromatography confirmed the presence of the anticholinergic drug atropine.

Clozapine

An interaction of clozapine with cocaine has been reported (388). Eight male cocaine addicts underwent four oral challenges with increasing doses of clozapine (12.5, 25, and 50 mg) and placebo, followed 2 hours later by cocaine 2 mg/kg intranasally (389). Subjective and physiological responses, and serum cocaine concentrations were measured over 4 hours. Clozapine pretreatment increased cocaine concentrations during the study and significantly increased the peak serum cocaine concentrations dose-dependently. Despite this rise in blood concentrations, clozapine pretreatment significantly reduced subjective responses to cocaine, including "expected high," "high," and "rush" effects, notably at the 50 mg dose. There were also significant effects on "sleepiness," "paranoia" and "nervousness." Clozapine caused a significant near-syncopal episode in one subject, requiring withdrawal. Clozapine had no significant effect on baseline pulse rate or systolic blood pressure, but it attenuated the significant pressor effects of a single dose of intranasal cocaine. These data suggested a possible therapeutic role for clozapine in the treatment of cocaine addiction in humans, but also suggest caution due to the near-syncopal event and the increase in serum cocaine concentrations.

Ephedrine

Chronic cocaine use sensitizes coronary arterial α-adrenoceptors to agonists (SEDA-22, 154).

Indometacin

Cocaine combined with indometacin in a 23-year-old pregnant woman at 34 weeks gestation may have caused fetal anuria and neonatal gastric hemorrhage (390).

Monoamine oxidase inhibitors

The combination of monoamine oxidase inhibitors with cocaine can cause hyperpyrexia (391).

Neuroleptic drugs

Cocaine-abusing psychiatric patients significantly more often develop neuroleptic drug-induced acute dystonia according to a 2-year study carried out on the island of Curaçao, Antilles, where cocaine and cannabis are often abused (392). The sample consisted of 29 men with neuroleptic drug-induced acute dystonia aged 17–45 years who had received high potency neuroleptic drugs in the month before admission; nine were cocaine users and 20 non-users. Cocaine use was a major risk factor for neuroleptic drug-induced acute dystonia and should be added to the list of well-known risk factors, such as male sex, younger age, neuroleptic drug dosage and potency, and a history of neuroleptic drug-induced acute dystonia. The authors suggested that high-risk cocaine-using psychiatric patients who start to take neuroleptic drugs should be provided with an anticholinergic drug as a prophylactic measure to prevent neuroleptic drug-induced acute dystonia.

Nimodipine

It has been suggested that calcium channel blockers can be used to treat cocaine dependence, and some studies have shown reductions in cocaine-induced subjective and cardiovascular responses with nifedipine and diltiazem. The cardiovascular and subjective responses to cocaine have been evaluated in a double-blind, placebo-controlled, crossover study in five subjects pretreated with two dosage of nimodipine (393). Nimodipine 60 mg attenuated the rise in systolic, but not diastolic, blood pressure after cocaine. In three subjects nimodipine 90 mg produced greater attenuation than 60 mg. The subjective effects of cocaine were not altered by either dose of nimodipine.

Opiates

The combination of opiates and cocaine in drug overdose deaths has been examined retrospectively in a review of all accidental drug overdoses that resulted in death in New York City from 1990 to 1998 (394). There were 7451 accidental overdose deaths, of which opiates played a role in 71%, cocaine in 70%, and alcohol in 40%; one of these drugs was identified in 98% of all overdose deaths. As evidence of higher polydrug mortality, 58% of the deaths were caused by more than one drug. The most common combination was opiates plus cocaine, which caused a peak death rate in 1994 of 44 per million person-years, compared with rates of 21 and 31 deaths per million person-years for opiates and cocaine alone respectively. Although it was retrospective, this study has elaborated and highlighted the dangers of polydrug abuse and has shown that the risk of overdose is significantly increased when opiates and cocaine are used together.

Rhabdomyolysis and ventricular fibrillation has been attributed to cocaine plus diamorphine (heroin) ingestion (395).

- A 28-year-old man went into cardiorespiratory arrest after using intravenous cocaine and diamorphine. He was intubated and ventilated and given adrenaline, naloxone, and sodium bicarbonate. During a thoracotomy he developed ventricular fibrillation and was electrically converted to sinus rhythm. He had hyperkalemia and myoglobinuria. He developed acute renal insufficiency, disseminated intravascular coagulopathy, and a right leg compartment syndrome. There were cocaine metabolites and opioids in his urine. Hemodialysis and fasciotomy were performed, but he died 2 months later with a complicating bronchopneumonia.

The authors discussed the possibility that naloxone, an effective opioid antidote, may have been harmful in this case.

Selegiline

In a single-blind, placebo-controlled, within-subject, non-randomized, crossover study in 12 cocaine-dependent subjects, intravenous cocaine (20 mg and 40 mg) was evaluated before and during transdermal selegiline (20 mg), a selective inhibitor of monoamine oxidase B (396). Selegiline attenuated the effects of cocaine on systolic blood pressure and heart rate and many of its subjective effects, including the desire to use cocaine. Selegiline did not alter the pharmacokinetics of cocaine or cocaine-induced changes in prolactin and growth hormone. These results provide further evidence that selegiline may be useful in the treatment of cocaine abuse and dependence and provide safety documentation important in justifying the need for larger-scale therapeutic trials.

Suxamethonium

Procaine and cocaine are esters that are hydrolysed by plasma cholinesterase and may therefore competitively enhance the action of suxamethonium (succinylcholine) (397). Chloroprocaine may have a similar action. Lidocaine also interacts, although the mechanism is not clear unless very high doses are used (398).

References

1. Stein R, Ellinwood EH Jr. Medical complication of cocaine abuse. Drug Ther 1990;10:40.

2. Rowbotham MC. Neurologic aspects of cocaine abuse. West J Med 1988;149(4):442–8.

3. Marzuk PM, Tardiff K, Leon AC, Hirsch CS, Stajic M, Portera L, Hartwell N, Iqbal MI. Fatal injuries after cocaine use as a leading cause of death among young adults in New York City. N Engl J Med 1995;332(26):1753–7.

4. Gawin FH, Ellinwood EH Jr. Cocaine and other stimulants. Actions, Abuse, Treatment. N Engl J Med 1988;318(18):1173–82.

5. Clayton RR. Cocaine use in the United States: in a blizzard or just being snowed? NIDA Res Monogr 1985;61:8–34.

6. Lathers CM, Tyau LS, Spino MM, Agarwal I. Cocaine-induced seizures, arrhythmias and sudden death. J Clin Pharmacol 1988;28(7):584–93.

7. Karch SB, Billingham ME. The pathology and etiology of cocaine-induced heart disease. Arch Pathol Lab Med 1988;112(3):225–30.

8. Jiang JP, Downing SE. Catecholamine cardiomyopathy: review and analysis of pathogenetic mechanisms. Yale J Biol Med 1990;63(6):581–91.

9. Hong R, Matsuyama E, Nur K. Cardiomyopathy associated with the smoking of crystal methamphetamine. JAMA 1991;265(9):1152–4.

10. Ellinwood EH Jr, Petrie WM. Dependence on amphetamine, cocaine, and other stimulants. In: Pradhan SN, editor. Drug Abuse: Clinical and Basic Aspects. New York: CV: Mosby, 1977:248.

11. Barinerd H, Krupp M, Chatton J, et al. Current Medical Diagnosis and TreatmentLos Altos CA: Lange. Medical Publishers;. 1970.

12. Moe GK, Akildskov JA. Antiarrhythmic drugs. In: Gilman AG, Goodman LS, editors. The Pharmacological Basis of Therapeutics. New York: MacMillan, 1970:1437.

13. Ellinwood EH Jr, Petrie WM. Drug induced psychoses. In: Pickens RW, Heston LL, editors. Psychiatric Factors in Drug Abuse. New York: Grune & Stratton, 1979:301.

14. Komokata T, Nishida S, Ganz S, Suzuki T, Olson L, Tzakis AG. The impact of donor chemical overdose on the outcome of liver transplantation. Transplantation 2003;76:705–8.

15. Caballero F, Lopez-Navidad A, Gomez M, Sola R. Successful transplantation of organs from a donor who died from acute cocaine intoxication. Clin Transplant 2003;17:89–92.

16. Bhorade SM, Vigneswaran W, McCabe MA, Garrity ER. Liberalization of donor criteria may expand the donor pool without adverse consequence in lung transplantation. J Heart Lung Transplant 2000;19:1199–204.

17. Nanji AA, Filipenko JD. Asystole and ventricular fibrillation associated with cocaine intoxication. Chest 1984;85(1):132–3.

18. Wiener RS, Lockhart JT, Schwartz RG. Dilated cardiomyopathy and cocaine abuse. Report of two cases. Am J Med 1986;81(4):699–701.

19. Wodarz N, Boning J. "Ecstasy"-induziertes psychotisches Depersonalisationssyndrom. ["Ecstasy"-induced psychotic depersonalization syndrome.] Nervenarzt 1993;64(7):478–80.

20. McCann UD, Ricaurte GA. MDMA ("ecstasy") and panic disorder: induction by a single dose. Biol Psychiatry 1992;32(10):950–3.

21. Williams H, Meagher D, Galligan P. MDMA. ("Ecstasy"); a case of possible drug-induced psychosis. Ir J Med Sci 1993;162(2):43–4.

22. Isner JM, Estes NA 3rd, Thompson PD, Costanzo-Nordin MR, Subramanian R, Miller G, Katsas G, Sweeney K, Sturner WQ. Acute cardiac events temporally related to cocaine abuse. N Engl J Med 1986;315(23):1438–43.

23. Minor RL Jr, Scott BD, Brown DD, Winniford MD. Cocaine-induced myocardial infarction in patients with normal coronary arteries. Ann Intern Med 1991;115(10):797–806.

24. Virmani R. Cocaine-associated cardiovascular disease: clinical and pathological aspects. NIDA Res Monogr 1991;108:220–9.

25. Stark TW, Pruet CW, Stark DU. Cocaine toxicity. Ear Nose Throat J 1983;62(3):155–8.

26. Jonsson S, O'Meara M, Young JB. Acute cocaine poisoning. Importance of treating seizures and acidosis. Am J Med 1983;75(6):1061–4.

27. Catravas JD, Waters IW. Acute cocaine intoxication in the conscious dog: studies on the mechanism of lethality. J Pharmacol Exp Ther 1981;217(2):350–6.

28. Covino BG, Vasalla HG. In: Local Anesthetics: Mechanism of Action and Clinical Use. New York: Grune and Stratton, 1976:127.

29. Marzuk PM, Tardiff K, Leon AC, Hirsch CS, Portera L, Iqbal MI, Nock MK, Hartwell N. Ambient temperature and mortality from unintentional cocaine overdose. JAMA 1998;279(22):1795–800.

30. Jaffe JH. Drug Addiction and Drug Abuse. In: Gilman AG, Goodman LS, Rall TW, Murad F, editors. The Pharmacological Basis of Therapeutics. 7th edn.. New York: McMillan, 1985:54.

31. Hollander JE, Carter WA, Hoffman RS. Use of phentolamine for cocaine-induced myocardial ischemia. N Engl J Med 1992;327(5):361.

32. Lange RA, Cigarroa RG, Flores ED, McBride W, Kim AS, Wells PJ, Bedotto JB, Danziger RS, Hillis LD. Potentiation of cocaine-induced coronary vasoconstriction by beta-adrenergic blockade. Ann Intern Med 1990;112(12):897–903.

33. Boehrer JD, Moliterno DJ, Willard JE, Hillis LD, Lange RA. Influence of labetalol on cocaine-induced coronary vasoconstriction in humans. Am J Med 1993;94(6):608–10.

34. Mittleman RE, Wetli CV. Death caused by recreational cocaine use. An update. JAMA 1984;252(14):1889–93.

35. Cregler LL. Cocaine: the newest risk factor for cardiovascular disease. Clin Cardiol 1991;14(6):449–56.

36. Kloner RA, Hale S, Alker K, Rezkalla S. The effects of acute and chronic cocaine use on the heart. Circulation 1992;85(2):407–19.

37. Thadani P. Cardiovascular toxicity of cocaine: underlying mechanisms. NIDA Res Monogr 1991;108:1–238.

38. Boehrer JD, Moliterno DJ, Willard JE, Snyder RW 2nd, Horton RP, Glamann DB, Lange RA, Hillis LD. Hemodynamic effects of intranasal cocaine in humans. J Am Coll Cardiol 1992;20(1):90–3.

39. Lange RA, Hillis RD. Cardiovascular complications of cocaine use. New Engl J Med 2001;345:351–8.

40. Cohen S. Reinforcement and rapid delivery systems: understanding adverse consequences of cocaine. NIDA Res Monogr 1985;61:151–7.

41. Cregler LL, Mark H. Medical complications of cocaine abuse. N Engl J Med 1986;315(23):1495–500.

42. Pitts WR, Vongpatanasin W, Cigarroa JE, Hillis LD, Lange RA. Effects of the intracoronary infusion of cocaine on left ventricular systolic and diastolic function in humans. Circulation 1998;97(13):1270–3.

43. Sofuoglu M, Nelson D, Dudish-Poulsen S, Lexau B, Pentel PR, Hatsukami DK. Predictors of cardiovascular

response to smoked cocaine in humans. Drug Alcohol Depend 2000;57(3):239–45.

44. Villota JN, Rubio LF, Flores JS, Peris VB, Burguera EP, Gonzalez VB, Banuls MP, Escorihuela AL. Cocaine-induced coronary thrombosis and acute myocardial infarction. Int J Cardiol 2004;96:481–2.

45. Meltser H, Bhakta D, Kalaria VG. Multivessel coronary thrombosis secondary to cocaine use successfully treated with multivessel primary angioplasty. Int J Cardiovasc Interv 2004;1:39–42.

46. Doshi SN, Marmur JD. Resolution of intracoronary thrombus with direct thrombin inhibition in a cocaine abuser. Heart 2004;90:501.

47. Erwin MB, Hoyle JR, Smith CH, Deliargyris EN. Cocaine and accelerated atherosclerosis: insights from intravascular ultrasound. Int J Cardiol 2004;93:301–3.

48. Osula S, Stockton P, Abdelaziz MM, Walshaw MJ. Intratracheal cocaine induced myocardial infarction: an unusual complication of fibreoptic bronchoscopy. Thorax 2003;58:733–4.

49. Kuczkowski KM. Chest pain and dysrhythmias in a healthy parturient: Wolff–Parkinson–White syndrome vs cocaine-induced myocardial ischemia? Anaesth Intens Care 2004;32:143–4.

50. Allred RJ, Ewer S. Fatal pulmonary edema following intravenous "freebase" cocaine use. Ann Emerg Med 1981;10(8):441–2.

51. Feldman JA, Fish SS, Beshansky JR, Griffith JL, Woolard RH, Selker HP. Acute cardiac ischemia in patients with cocaine-associated complaints: results of a multicenter trial. Ann Emerg Med 2000;36(5):469–76.

52. Kushman SO, Storrow AB, Liu T, Gibler WB. Cocaine-associated chest pain in a chest pain center. Am J Cardiol 2000;85(3):394–6.

53. Mittleman MA, Mintzer D, Maclure M, Tofler GH, Sherwood JB, Muller JE. Triggering of myocardial infarction by cocaine. Circulation 1999;99:2737–41.

54. Qureshi AI, Suri MF, Guterman LR, Hopkins LN. Cocaine use and the likelihood of nonfetal myocardial infarction and stroke: data from the Third National Health and Nutrition Examination Survey. Circulation 2001;103:502–6.

55. Office of Applied Statistics. Year end 2000 emergency department data from the Drug Abuse Warning Network. DAWN Series D-18, Rockville, Md: Substance Abuse and Mental Health Services Administration, 2001 (DHHS Publication No. 01-03532).

56. Brody SC, Slovis CM, Wren KD. Cocaine-related medical problems: consecutive series of 233 patients. Am J Med 1990;88:325–31.

57. Hollander JE. The management of cocaine associated myocardial ischemia. New Engl J Med 1995;333:1267–72.

58. Hockstra JW, Gibler WB, Levy RC, Sayre M, Naber W, Chandra A, Kacich R, Magorien R, Walsh R. Emergency department diagnosis of acute myocardial infarction and ischemia: a cost analysis of two diagnostic protocols. Acad Emerg Med 1994;1:103–10.

59. Weber JE, Shofer FS, Larkin L, Kalaria AS, Hollander JE. Validation of brief observation period for patients with cocaine-associated chest pain. New Engl J Med 2003;348:510–7.

60. Carley S, Ali B. Towards evidence based emergency medicine: best BETs from the Manchester Royal Infirmary. Acute myocardial infarction in cocaine induced chest pain presenting as an emergency. Emerg Med J 2003;20:174–5.

61. Kelly RF, Sompalli V, Sattar P, Khankari K. Increased TIMI frame counts in cocaine users: a case for increased microvascular resistance in the absence of epicardial coronary disease or spasm. Clin Cardiol 2003;26:319–22.

62. Kontos MC, Jesse RL, Tatum JL, Ornato J. Coronary angiographic findings in patients with cocaine-associated chest pain. J Emerg Med 2003;24:9–13.

63. Inyang VA, Cooper AJ, Hodgkinson DW. Cocaine induced myocardial infarction. J Accid Emerg Med 1999;16(5):374–5.

64. Amin M, Gabelman G, Karpel J, Buttrick P. Acute myocardial infarction and chest pain syndromes after cocaine use. Am J Cardiol 1990;66(20):1434–7.

65. Sharkey SW, Glitter MJ, Goldsmith SR. How serious is cocaine-associated acute chest pain syndromes after cocaine use. Cardiol Board Rev 1992;9:58–66.

66. Ross GS, Bell J. Myocardial infarction associated with inappropriate use of topical cocaine as treatment for epistaxis. Am J Emerg Med 1992;10(3):219–22.

67. Hollander JE, Hoffman RS. Cocaine-induced myocardial infarction: an analysis and review of the literature. J Emerg Med 1992;10(2):169–77.

68. Yuen-Green MS, Yen CK, Lim AD, Lull RJ. Tc-99m sestamibi myocardial imaging at rest for evaluation of cocaine-induced myocardial ischemia and infarction. Clin Nucl Med 1992;17(12):923–5.

69. Tanenbaum JH, Miller F. Electrocardiographic evidence of myocardial injury in psychiatrically hospitalized cocaine abusers. Gen Hosp Psychiatry 1992;14(3):201–3.

70. Eisenberg MJ, Mendelson J, Evans GT Jr, Jue J, Jones RT, Schiller NB. Left ventricular function immediately after intravenous cocaine: a quantitative two-dimensional echocardiographic study. J Am Coll Cardiol 1993;22(6):1581–6.

71. Rejali D, Glen P, Odom N. Pneumomediastinum following Ecstasy (methylenedioxymetamphetamine, MDMA) ingestion in two people at the same "rave". J Laryngol Otol 2002;116(1):75–6.

72. Morgan MJ, McFie L, Fleetwood H, Robinson JA. Ecstasy (MDMA): are the psychological problems associated with its use reversed by prolonged abstinence? Psychopharmacology (Berl) 2002;159(3):294–303.

73. Fox HC, McLean A, Turner JJ, Parrott AC, Rogers R, Sahakian BJ. Neuropsychological evidence of a relatively selective profile of temporal dysfunction in drug-free MDMA ("ecstasy") polydrug users. Psychopharmacology (Berl) 2002;162(2):203–14.

74. Shrma AK, Hamwi SM, Garg N, Castagna MT, Suddath W, Ellahham S, Lindsay J. Percutaneous intervention in patients with cocaine-associated myocardial infarction: a case series and review. Catheter Cardiovasc Intervention 2002;56:346–52.

75. Hollander JE. The management of cocaine associated myocardial ischemia. New Engl J Med 1995;333:1267–72.

76. Honderick T, Williams D, Seaberg D, Wears R. A prospective, randomized, controlled trial of benzodiazepines and nitroglycerine or nitroglycerine alone in the treatment of cocaine-associated acute coronary syndromes. Am J Emerg Med 2003;21:39–42.

77. Baumann BM, Perrone J, Hornig SE, Shofer FS, Hollander JE. Randomized, double-blind, placebo-controlled trial of diazepam, nitroglycerin, or both for treatment of patients with potential cocaine-associated acute coronary syndromes. Acad Emerg Med 2000;7:878–86.

78. Anonymous. Second American Heart Associations International Evidence Evaluation Conference, Part 6. Advanced cardiovascular life support: Section 1.

Introduction to ACLS from the Guidelines 2000 Conference. Circulation 2000;102 Suppl 1:186–9.

79. Bizzarri F, Mondillo S, Guerrini F, Barbati R, Frati G, Davoli G. Spontaneous acute coronary dissection after cocaine abuse in a young woman. Can J Cardiol 2003;19:297–9.

80. Mochizuki Y, Zhang M, Golestaneh L, Thananart S, Coco M. Acute aortic thrombosis and renal infarction in acute cocaine intoxication: a case report and review of literature. Clin Nephrol 2003;60:130–3.

81. Stollberger C, Kopsa W, Finsterer J. Resolution of an aortic thrombus under anticoagulant therapy. Eur J Cardiothorac Surg, 2001. 20:880–2.

82. Cohle SD, Lie JT. Dissection of the aorta and coronary arteries associated with acute cocaine intoxication. Arch Pathol Lab Med 1992;116(11):1239–41.

83. Berry J, van Gorp WG, Herzberg DS, Hinkin C, Boone K, Steinman L, Wilkins JN. Neuropsychological deficits in abstinent cocaine abusers: preliminary findings after two weeks of abstinence. Drug Alcohol Depend 1993;32(3):231–7.

84. Hsue PY, Salinas CL, Bolger AF, Benowitz NL, Waters DD. Acute aortic dissection related to crack cocaine. Circulation 2002;105(13):1592–5.

85. Riaz K, Forker AD, Garg M, McCullough PA. Atypical presentation of cocaine-induced type A aortic dissection: a diagnosis made by transesophageal echocardiography. J Investig Med 2002;50(2):140–2.

86. Neri E, Toscano T, Massetti M, Capannini G, Frati G, Sassi C. Cocaine-induced intramural hematoma of the ascending aorta. Tex Heart Inst J 2001;28(3):218–9.

87. Jaffe BD, Broderick TM, Leier CV. Cocaine-induced coronary-artery dissection. N Engl J Med 1994;330(7):510–1.

88. Eskander KE, Brass NS, Gelfand ET. Cocaine abuse and coronary artery dissection. Ann Thorac Surg 2001;71(1):340–1.

89. Steinhauer JR, Caulfield JB. Spontaneous coronary artery dissection associated with cocaine use: a case report and brief review. Cardiovasc Pathol 2001;10(3):141–5.

90. Kumar PD, Smith HR. Cocaine-related vasculitis causing upper-limb peripheral vascular disease. Ann Intern Med 2000;133(11):923–4.

91. Balbir-Gurman A, Braun-Moscovici Y, Nahir AM. Cocaine-induced Raynaud's phenomenon and ischaemic finger necrosis. Clin Rheumatol 2001;20(5):376–8.

92. Morrow PL, McQuillen JB. Cerebral vasculitis associated with cocaine abuse. J Forensic Sci 1993;38(3):732–8.

93. Martin K, Rogers T, Kavanaugh A. Central nervous system angiopathy associated with cocaine abuse. J Rheumatol 1995;22(4):780–2.

94. Taylor D Parish D, Thompson L, Cavaliere M. Cocaine induced prolongation of the QT interval. Emerg Med J 2004;21:252–3.

95. Silvain J, Maury E, Qureshi T, Baudel JL, Offenstadt G. A puzzling electrocardiogram. Intensive Care Med 2004;30:340.

96. Wang RY. pH-dependent cocaine-induced cardiotoxicity. Am J Emerg Med 1999;17(4):364–9.

97. Kajdasz DK, Moore JW, Donepudi H, Cochrane CE, Malcolm RJ. Cardiac and mood-related changes during short-term abstinence from crack cocaine: the identification of possible withdrawal phenomena. Am J Drug Alcohol Abuse 1999;25(4):629–37.

98. Laffey JG, Neligan P, Ormonde G. Prolonged perioperative myocardial ischemia in a young male: due to topical intranasal cocaine? J Clin Anesth 1999;11(5):419–24.

99. Liao BS, Hilsinger RL Jr, Rasgon BM, Matsuoka K, Adour KK. A preliminary study of cocaine absorption from the nasal mucosa. Laryngoscope 1999;109(1):98–102.

100. Tashkin DP, Gorelick D, Khalsa ME, Simmons M, Chang P. Respiratory effects of cocaine freebasing among habitual cocaine users. J Addict Dis 1992;11(4):59–70.

101. Baldwin GC, Choi R, Roth MD, Shay AH, Kleerup EC, Simmons MS, Tashkin DP. Evidence of chronic damage to the pulmonary microcirculation in habitual users of alkaloidal ("crack") cocaine. Chest 2002;121(4):1231–8.

102. Ettinger NA, Albin RJ. A review of the respiratory effects of smoking cocaine. Am J Med 1989;87(6):664–8.

103. O'Donnell AE, Mappin FG, Sebo TJ, Tazelaar H. Interstitial pneumonitis associated with "crack" cocaine abuse. Chest 1991;100(4):1155–7.

104. Oh PI, Balter MS. Cocaine induced eosinophilic lung disease. Thorax 1992;47(6):478–9.

105. Delaney K, Hoffman RS. Pulmonary infarction associated with crack cocaine use in a previously healthy 23-year-old woman. Am J Med 1991;91(1):92–4.

106. Batlle MA, Wilcox WD. Pulmonary edema in an infant following passive inhalation of free-base ("crack") cocaine. Clin Pediatr (Phila) 1993;32(2):105–6.

107. van der Klooster JM, Grootendorst AF. Severe bullous emphysema associated with cocaine smoking. Thorax 2001;56(12):982–3.

108. Hunter JG, Loy HC, Markovitz L, Kim US. Spontaneous pneumomediastinum following inhalation of alkaloidal cocaine and emesis: case report and review. Mt Sinai J Med 1986;53(6):491–3.

109. Goel R, Flaker GC. Cardiovascular complications of cocaine use. N Engl J Med 2001;345(21):1575–6.

110. Maeder M, Ullmer E. Pneumomediastinum and bilateral pneumothorax as a complication of cocaine smoking. Respiration 2003;70:407.

111. Gatof D, Albert RK. Bilateral thumb burns leading to the diagnosis of crack lung. Chest 2002;121(1):289–91.

112. Torre M, Barberis M. Spontaneous pneumothorax in cocaine sniffers. Am J Emerg Med 1998;16(5):546–9.

113. Rome LA, Lippmann ML, Dalsey WC, Taggart P, Pomerantz S. Prevalence of cocaine use and its impact on asthma exacerbation in an urban population. Chest 2000;117(5):1324–9.

114. Strong DH, Westcott JY, Biller JA, Morrison JL, Effros RM, Maloney J. Eosinophilic "empyema" associated with crack cocaine use. Thorax 2003;58:823–4.

115. Ramachandaran S, Khan AU, Dadaparvar S, Sherman MS. Inhalation of crack cocaine can mimic pulmonary embolism. Clin Nucl Med 2004;29:756–7.

116. Gupta A, Hawrych A, Wilson WR. Cocaine-induced sinonasal destruction. Otolaryngol Head Neck Surg 2001;124(4):480.

117. Noskin GA, Kalish SB. Pott's puffy tumor: a complication of intranasal cocaine abuse. Rev Infect Dis 1991;13(4):606–8.

118. Smith JC, Kacker A, Anand VK. Midline nasal and hard palate destruction in cocaine abusers and cocaine's role in rhinologic practice. Ear Nose Throat J 2002;81(3):172–7.

119. Talbott JF, Gorti GK, Koch RJ. Midfacial osteomyelitis in a chronic cocaine abuser: a case report. Ear Nose Throat J 2001;80(10):738–43.

120. Vilela RJ, Langford C, McCullagh L, Kass ES. Cocaine-induced oronasal fistulas with external nasal erosion but without palate involvement. Ear Nose Throat J 2002;81(8):562–3.

121. Rowshani A, Schot LJ, ten Berge JM. C-ANCA as a serological pitfall. Lancet 2004;363:782.

122. Trimarchi M, Nicolai P, Lombardi D, Facchetti F, Morassi ML, Maroldi R, Gregorini G, Specks U. Sinonasal

osteocartilaginous necrosis in cocaine abusers: experience in 25 patients. Am J Rhinol 2003;17:33–43.

123. Trimarchi M, Nicolai P, Lombardi D, Facchetti F, Morassi ML, Maroldi R, Gregorini G, Specks U. Sinonasal osteocartilaginous necrosis in cocaine abusers: experience in 25 patients. Am J Rhinol 2003;17:33–43.

124. De R, Uppal HS, Shehab ZP, Hilger AW, Wilson PS, Courteney R. Current practices of cocaine administration by UK otorhinolaryngologists. J Laryngol Otol 2003;117:109–12.

125. Li SJ, Wang Y, Pankiewicz J, Stein EA. Neurochemical adaptation to cocaine abuse: reduction of N-acetyl aspartate in thalamus of human cocaine abusers. Biol Psychiatry 1999;45(11):1481–7.

126. Costa L, Bauer LO. Smooth pursuit eye movement dysfunction in substance-dependent patients: mediating effects of antisocial personality disorder. Neuropsychobiology 1998;37(3):117–23.

127. Adler LE, Olincy A, Cawthra E, Hoffer M, Nagamoto HT, Amass L, Freedman R. Reversal of diminished inhibitory sensory gating in cocaine addicts by a nicotinic cholinergic mechanism. Neuropsychopharmacology 2001;24(6):671–9.

128. Goldstein RZ, Volkow ND, Wang GJ, Fowler JS, Rajaram S. Addiction changes orbitofrontal gyrus function: involvement in response inhibition. Neuroreport 2001;12(11):2595–9.

129. Tanvetyanon T, Dissin J, Selcer UM. Hyperthermia and chronic pancerebellar syndrome after cocaine abuse. Arch Intern Med 2001;161(4):608–10.

130. Dhuna A, Pascual-Leone A, Belgrade M. Cocaine-related vascular headaches. J Neurol Neurosurg Psychiatry 1991;54(9):803–6.

131. Mossman SS, Goadsby PJ. Cocaine abuse simulating the aura of migraine. J Neurol Neurosurg Psychiatry 1992;55(7):628.

132. Satel SL, Swann AC. Extrapyramidal symptoms and cocaine abuse. Am J Psychiatry 1993;150(2):347.

133. Cardoso FE, Jankovic J. Cocaine-related movement disorders. Mov Disord 1993;8(2):175–8.

134. Volkow ND, Fowler JS, Wang GJ, Hitzemann R, Logan J, Schlyer DJ, Dewey SL, Wolf AP. Decreased dopamine D_2 receptor availability is associated with reduced frontal metabolism in cocaine abusers. Synapse 1993;14(2):169–77.

135. Hegarty AM, Lipton RB, Merriam AE, Freeman K. Cocaine as a risk factor for acute dystonic reactions. Neurology 1991;41(10):1670–2.

136. Attig E, Amyot R, Botez T. Cocaine induced chronic tics. J Neurol Neurosurg Psychiatry 1994;57(9):1143–4.

137. Bartzokis G, Beckson M, Wirshing DA, Lu PH, Foster JA, Mintz J. Choreothetoid movements in cocaine dependence. Biol Psychiatry 1999;45:1630–5.

138. Mesulam MM. Cocaine and Tourette's syndrome. N Engl J Med 1986;315(6):398.

139. Berciano J, Oterino A, Rebollo M, Pascual J. Myasthenia gravis unmasked by cocaine abuse. N Engl J Med 1991;325(12):892.

140. Valmaggia C, Gottlob IM. Cocaine abuse, generalized myasthenia, complete external ophthalmoplegia, and pseudotonic pupil. Strabismus 2001;9(1):9–12.

141. Mash DC, Ouyang Q, Pablo J, Basile M, Izenwasser S, Lieberman A, Perrin RJ. Cocaine abusers have an overexpression of alpha-synuclein in dopamine neurons. J Neurosci 2003;23:2564–71.

142. Dhuna A, Pascual-Leone A, Langendorf F. Chronic, habitual cocaine abuse and kindling-induced epilepsy: a case report. Epilepsia 1991;32(6):890–4.

143. Myers JA, Earnest MP. Generalized seizures and cocaine abuse. Neurology 1984;34(5):675–6.

144. Kowatch RA, Schnoll SS, Knisely JS, Green D, Elswick RK. Electroencephalographic sleep and mood during cocaine withdrawal. J Addict Dis 1992;11(4):21–45.

145. Yapor WY, Gutierrez FA. Cocaine-induced intratumoral hemorrhage: case report and review of the literature. Neurosurgery 1992;30(2):288–91.

146. Ramadan NM, Levine SR, Welch KM. Pontine hemorrhage following "crack" cocaine use. Neurology 1991;41(6):946–7.

147. Kosten TR, Cheeves C, Palumbo J, Seibyl JP, Price LH, Woods SW. Regional cerebral blood flow during acute and chronic abstinence from combined cocaine–alcohol abuse. Drug Alcohol Depend 1998;50(3):187–95.

148. Lee JH, Telang FW, Springer CS Jr, Volkow ND. Abnormal brain activation to visual stimulation in cocaine abusers. Life Sci 2003;73:1953–61.

149. Sloan MA, Kittner SJ, Feeser BR, Gardner J, Epstein A, Wozniak MA, Wityk RJ, Stern BJ, Price TR, Macko RF, Johnson CJ, Earley CJ, Buchholz D. Illicit drug-associated ischemic stroke in the Baltimore-Washington Young Stroke Study. Neurology 1998;50(6):1688–93.

150. Levine SR, Brust JC, Futrell N, Brass LM, Blake D, Fayad P, Schultz LR, Millikan CH, Ho KL, Welch KM. A comparative study of the cerebrovascular complications of cocaine: alkaloidal versus hydrochloride—a review. Neurology 1991;41(8):1173–7.

151. Strupp M, Hamann GF, Brandt T. Combined amphetamine and cocaine abuse caused mesencephalic ischemia in a 16-year-old boy—due to vasospasm? Eur Neurol 2000;43(3):181–2.

152. Herning RI, Better W, Nelson R, Gorelick D, Cadet JL. The regulation of cerebral blood flow during intravenous cocaine administration in cocaine abusers. Ann NY Acad Sci 1999;890:489–94.

153. Bolouri MR, Small GA. Neuroimaging of hypoxia and cocaine-induced hippocampal stroke. J Neuroimag 2004;14:290–1.

154. McIntyre IM, Hamm CE, Sherrad JL, Gary RD, Riley AC, Lucas JR. The analysis of an intracerebral haematoma for drugs of abuse. J Forensic Sci 2003;48:680–2.

155. Alves OL, Gomes O. Cocaine-related acute subdural hematoma: an emergent cause of cerebrovascular accident. Acta Neurochir (Wien) 2000;142(7):819–21.

156. Grewal RP, Miller BL. Cocaine induced hypertensive encephalopathy. Acta Neurol (Napoli) 1991;13(3):279–81.

157. Mueller D, Gilden DH. Brown–Sequard syndrome after esophageal sclerotherapy and crack cocaine abuse. Neurology 2002;58(7):1129–30.

158. Chadan N, Thierry A, Sautreaux JL, Gras P, Martin D, Giroud M. Rupture anéurysmale et toxicomanie à la cocaïne. [Aneurysm rupture and cocaine addiction.] Neurochirurgie 1991;37(6):403–5.

159. Oyesiku NM, Colohan AR, Barrow DL, Reisner A. Cocaine-induced aneurysmal rupture: an emergent negative factor in the natural history of intracranial aneurysms? Neurosurgery 1993;32(4):518–26.

160. Conway JE, Tamargo RJ. Cocaine use is an independent risk factor for cerebral vasospasm after aneurysmal subarachnoid hemorrhage. Stroke 2001;32(10):2338–43.

161. Tang BH. Cocaine and subarachnoid hemorrhage. J Neurosurg 2005;102:961–2.

162. Zagelbaum BM, Tannenbaum MH, Hersh PS. *Candida albicans* corneal ulcer associated with crack cocaine. Am J Ophthalmol 1991;111(2):248–9.

163. Sachs R, Zagelbaum BM, Hersh PS. Corneal complications associated with the use of crack cocaine. Ophthalmology 1993;100(2):187–91.

164. Zagelbaum BM, Donnenfeld ED, Perry HD, Buxton J, Buxton D, Hersh PS. Corneal ulcer caused by combined intravenous and anesthetic abuse of cocaine. Am J Ophthalmol 1993;116(2):241–2.

165. Wang ES. Cocaine-induced iritis. Ann Emerg Med 1991;20(2):192–3.

166. Rofsky JE, Townsend JC, Ilsen PF, Bright DC. Retinal nerve fiber layer defects and microtalc retinopathy secondary to free-basing "crack" cocaine. J Am Optom Assoc 1995;66(11):712–20.

167. Van Stavern GP, Gorman M. Orbital infarction after cocaine use. Neurology 2002;59(4):642–3.

168. Michaelides M, Larkin G. Cocaine-associated central retinal artery occlusion in a young man. Eye 2002;16(6):790–2.

169. Nicoucar K, Sakbani K, Vukanovic S, Guyot J-P. Intralabyrinthine haemorrhage following cocaine consumption. Acta Oto-Laryngol 2005;125:899–901.

170. Miller BL, Mena I, Giombetti R, Villanueva-Meyer J, Djenderedjian AH. Neuropsychiatric effects of cocaine: SPECT measurements. J Addict Dis 1992;11(4):47–58.

171. O'Malley S, Adamse M, Heaton RK, Gawin FH. Neuropsychological impairment in chronic cocaine abusers. Am J Drug Alcohol Abuse 1992;18(2):131–44.

172. Di Sclafani V, Tolou-Shams M, Price LJ, Fein G. Neuropsychological performance of individuals dependent on crack-cocaine, or crack-cocaine and alcohol, at 6 weeks and 6 months of abstinence. Drug Alcohol Depend 2002;66(2):161–71.

173. Selby MJ, Azrin RL. Neuropsychological functioning in drug abusers. Drug Alcohol Depend 1998;50(1):39–45.

174. Bolla KI, Rothman R, Cadet JL. Dose-related neurobehavioral effects of chronic cocaine use. J Neuropsychiatry Clin Neurosci 1999;11(3):361–9.

175. Butler LF, Frank EM. Neurolinguistic function and cocaine abuse. J Med Speech-Lang Path 2000;8:199–212.

176. Ross-Durow PL, Boyd CJ. Sexual abuse, depression, and eating disorders in African American women who smoke cocaine. J Subst Abuse Treat 2000;18(1):79–81.

177. Allam S, Noble JS. Cocaine-excited delirium and severe acidosis. Anaesthesia 2001;56(4):385–6.

178. Crum RM, Anthony JC. Cocaine use and other suspected risk factors for obsessive-compulsive disorder: a prospective study with data from the Epidemiologic Catchment Area surveys. Drug Alcohol Depend 1993;31(3):281–95.

179. Aronson TA, Craig TJ. Cocaine precipitation of panic disorder. Am J Psychiatry 1986;143(5):643–5.

180. Rosen MI, Kosten T. Cocaine-associated panic attacks in methadone-maintained patients. Am J Drug Alcohol Abuse 1992;18(1):57–62.

181. Brady KT, Lydiard RB, Malcolm R, Ballenger JC. Cocaine-induced psychosis. J Clin Psychiatry 1991;52(12):509–12.

182. Nambudiri DE, Young RC. A case of late-onset crack dependence and subsequent psychosis in the elderly. J Subst Abuse Treat 1991;8(4):253–5.

183. Cubells JF, Kranzler HR, McCance-Katz E, Anderson GM, Malison RT, Price LH, Gelernter J. A haplotype at the DBH locus, associated with low plasma dopamine beta-hydroxylase activity, also associates with cocaine-induced paranoia. Mol Psychiatry 2000;5(1):56–63.

184. Roy A, Bissette G, Williams R, Berman J, Gonzalez B. CSF CRH in abstinent cocaine-dependent patients. Psychiatry Res 2003;117:277–80.

185. Mendelson JH, Sholar MB, Mutschler NH, Jaszyna-Gasior M, Goletiani NV, Siegel AJ, Mello NK. Effects of intravenous cocaine and cigarette smoking on luteinizing hormone, testosterone, and prolactin in men. J Pharmacol Exp Ther 2003;307:339–48.

186. Vescovi PP. Cardiovascular and hormonal responses to hyperthermic stress in cocaine addicts after a long period of abstinence. Addict Biol 2000;5:91–5.

187. Lajara-Nanson WA. Cocaine induced hypokalaemic periodic paralysis. J Neurol Neurosurg Psychiatry 2002;73(1):92.

188. Weber JE, Larkin GL, Boe CT, Fras A, Kalaria AS, Maio RF, Luchessi B, Ensign L, Sweeney B, Hollander JE. Effect of cocaine use on bone marrow-mediated erythropoiesis. Acad Emerg Med 2003;10:705–8.

189. Miller JM Jr, Nolan TE. Case-control study of antenatal cocaine use and platelet levels. Am J Obstet Gynecol 2001;184(3):434–7.

190. Dettmeyer R, Schlamann M, Madea B. Cocaine-associated abscesses with lethal sepsis after splenic infarction in a 17-year-old woman. Forensic Sci Int 2004;140:210–3.

191. Nalbandian H, Sheth N, Dietrich R, Georgiou J. Intestinal ischemia caused by cocaine ingestion: report of two cases. Surgery 1985;97(3):374–6.

192. Chan YC, Camprodon RA, Kane PA, Scott-Coombes DM. Abdominal complications from crack cocaine. Ann R Coll Surg Engl 2005;87(1):72–3.

193. Cohen ME, Kegel JG. Candy cocaine esophagus. Chest 2002;121(5):1701–3.

194. Arrillaga A, Sosa JL, Najjar R. Laparoscopic patching of crack cocaine-induced perforated ulcers. Am Surg 1996;62(12):1007–9.

195. Pecha RE, Prindiville T, Pecha BS, Camp R, Carroll M, Trudeau W. Association of cocaine and methamphetamine use with giant gastroduodenal ulcers. Am J Gastroenterol 1996;91(12):2523–7.

196. Abramson DL, Gertler JP, Lewis T, Kral JG. Crack-related perforated gastropyloric ulcer. J Clin Gastroenterol 1991;13(1):17–9.

197. Myers SI, Clagett GP, Valentine RJ, Hansen ME, Anand A, Chervu A. Chronic intestinal ischemia caused by intravenous cocaine use: report of two cases and review of the literature. J Vasc Surg 1996;23(4):724–9.

198. Simmers TA, Vidakovic-Vukic M, Van Meyel JJ. Cocaine-induced ischemic colitis. Endoscopy 1998;30(1):S8–9.

199. Papi C, Candia S, Masci P, Ciaco A, Montanti S, Capurso L. Acute ischaemic colitis following intravenous cocaine use. Ital J Gastroenterol Hepatol 1999;31(4):305–7.

200. Chan YC, Camprodon RAM Kane PA, Scott-Coombes DM. Abdominal complications from crack cocaine. Ann R Coll Surg Engl 2004;86:47–50.

201. Whitten CG, Luke BA. Liver uptake of Tc-99m PYP. Clin Nucl Med 1991;16(7):492–4.

202. Peyriere H, Mauboussin JM. Cocaine-induced acute cytologic hepatitis in HIV-infected patients with nonactive viral hepatitis. Ann Intern Med 2000;132(12):1010–1.

203. Rahman TM, Wadsworth C, Anson G, Wendon J. Cocaine-induced hepatotoxicity or an acquired mitochondrial defect? Clin Int Care 2004;15:61–3.

204. Balaguer F, Fernandez J. Cocaine-induced acute hepatitis and thrombotic microangiopathy. JAMA 2005;293:797–8.

205. Lam M, Ballou SP. Reversible scleroderma renal crisis after cocaine use. N Engl J Med 1992;326(21):1435.

206. Norris KC, Thornhill-Joynes M, Robinson C, Strickland T, Alperson BL, Witana SC, Ward HJ. Cocaine use, hypertension, and end-stage renal disease. Am J Kidney Dis 2001;38(3):523–8.

207. Singhal P, Horowitz B, Quinones MC, Sommer M, Faulkner M, Grosser M. Acute renal failure following cocaine abuse. Nephron 1989;52(1):76–8.

208. Roth D, Alarcon FJ, Fernandez JA, Preston RA, Bourgoignie JJ. Acute rhabdomyolysis associated with cocaine intoxication. N Engl J Med 1988;319(11):673–7.

209. Amoedo ML, Craver L, Marco MP, Fernandez E. Cocaine-induced acute renal failure without rhabdomyolysis. Nephrol Dial Transplant 1999;14(12):2970–1.

210. Saleem TM, Singh M, Murtaza M, Singh A, Kasubhai M, Gnanasekaran I. Renal infarction: a rare complication of cocaine abuse. Am J Emerg Med 2001;19(6):528–9.

211. Doctora JS, Williams CW, Bennett CR, Howlett BK. Rhabdomyolysis in the acutely cocaine-intoxicated patient sustaining maxillofacial trauma: report of a case and review of the literature. J Oral Maxillofac Surg 2003;61:964–7.

212. Anonymous. Outbreak of primary and secondary syphilis—Baltimore City, Maryland, 1995. MMWR Morb Mortal Wkly Rep 1996;45(8):166–9.

213. Altman AL, Seftel AD, Brown SL, Hampel N. Cocaine associated priapism. J Urol 1999;161(6):1817–8.

214. Myrick H, Markowitz JS, Henderson S. Priapism following trazodone overdose with cocaine use. Ann Clin Psychiatry 1998;10(2):81–3.

215. Mireku-Boateng AO, Tasie B. Priapism associated with intracavernosal injection of cocaine. Urol Int 2001;67(1):109–10.

216. Munarriz R, Hwang J, Goldstein I, Traish AM, Kim NN. Cocaine and ephedrine-induced priapism: case reports and investigation of potential adrenergic mechanisms. Urology 2003;62:187–92.

217. Carey F, Dinsmore WW. Cocaine-induced penile necrosis. Int J STD AIDS 2004;15:424–5.

218. Robinson AJ, Gazzard BG. Rising rates of HIV infection. BMJ 2005;330(7487):320–1.

219. Delafuente JC, DeVane CL. Immunologic effects of cocaine and related alkaloids. Immunopharmacol Immunotoxicol 1991;13(1–2):11–23.

220. Peterson PK, Gekker G, Chao CC, Schut R, Verhoef J, Edelman CK, Erice A, Balfour HH Jr. Cocaine amplifies HIV-1 replication in cytomegalovirus-stimulated peripheral blood mononuclear cell cocultures. J Immunol 1992;149(2):676–80.

221. Trozak DJ, Gould WM. Cocaine abuse and connective tissue disease. J Am Acad Dermatol 1984;10(3):525.

222. Hofbauer GF, Hafner J, Trueb RM. Urticarial vasculitis following cocaine use. Br J Dermatol 1999;141(3):600–1.

223. Castro-Villamor MA, de las Heras P, Armentia A, Duenas-Laita A. Cocaine-induced severe angioedema and urticaria. Ann Emerg Med 1999;34(2):296–7.

224. Crandall CG, Vongpatanasin W, Victor RG. Mechanism of cocaine-induced hyperthermia in humans. Ann Intern Med 2002;136(11):785–91.

225. Schier JG, Hoffman RS, Nelson LS. Cocaine and body temperature regulation. Ann Intern Med 2002;137(10):855–6.

226. Karch SB, Stephens B, Ho CH. Relating cocaine blood concentrations to toxicity—an autopsy study of 99 cases. J Forensic Sci 1998;43(1):41–5.

227. Daisley H, Jones-Le Cointe A, Hutchinson G, Simmons V. Fatal cardiac toxicity temporally related to poly-drug abuse. Vet Hum Toxicol 1998;40(1):21–2.

228. Furnari C, Ottaviano V, Sacchetti G, Mancini M. A fatal case of cocaine poisoning in a body packer. J Forensic Sci 2002;47(1):208–10.

229. Fineschi V, Centini F, Monciotti F, Turillazzi E. The cocaine "body stuffer" syndrome: a fatal case. Forensic Sci Int 2002;126(1):7–10.

230. Palmiere C, Burkhardt S, Staub C, Hallenbarter M, Pizzolato GP, Dettmeyer R, La Harpe R. Thoracic aortic dissection associated with cocaine abuse. Forensic Sci Int 2004;141:137–42.

231. Eagle KA, Isselbacher EM, DeSanctis RW; International Registry for Aortic Dissection (IRAD) Investigators. Cocaine related aortic dissection in perspective. Circulation 2002;105:1529–30.

232. Om A, Porter T, Mohanty PK. Transesophageal echocardiographic diagnosis of acute aortic dissection. Complication of cocaine abuse. Am Heart J 1992;123:532–4.

233. Riaz K, Forker AD, Garg M, McCullough PA. Atypical presentation of cocaine-induces type-A aortic dissection: a diagnosis made by transesophageal echocardiography. J Invest Med 2002;50:140–2.

234. Stephens BG, Jentzen JM, Karch S, Mash DC, Wetli CV. Criteria for the interpretation of cocaine levels in human biological samples and their relation to the cause of death. Am J Forensic Med Pathol 2004;25:1–10.

235. Caballero F, Lopez-Navidad A, Gomez M, Sola R. Successful transplantation of organs from a donor who died from acute cocaine intoxication. Clin Transplant 2003;17:89–92.

236. Bannon MJ, Pruetz B, Manning-Bog AB, Whitty CJ, Michelhaugh SK, Sacchetti P, Granneman JG, Mash DC, Schmidt CJ. Decreased expression of the transcription factor NURR1 in dopamine neurons of cocaine abusers. Proc Natl Acad Sci USA 2002;99(9):6382–5.

237. Alim TN, Rosse RB, Vocci FJ Jr, Lindquist T, Deutsch SI. Diethylpropion pharmacotherapeutic adjuvant therapy for inpatient treatment of cocaine dependence: a test of the cocaine-agonist hypothesis. Clin Neuropharmacol 1995;18(2):183–95.

238. Grabowski J, Roache JD, Schmitz JM, Rhoades H, Creson D, Korszun A. Replacement medication for cocaine dependence: methylphenidate. J Clin Psychopharmacol 1997;17(6):485–8.

239. Gawin F, Riordan C, Kleber H. Methylphenidate treatment of cocaine abusers without attention deficit disorder: a negative report. Am J Drug Alcohol Abuse 1985;11(3–4):193–7.

240. Duarte JG, do Nascimento AF, Pantoja JG, Chaves CP. Chronic inhaled cocaine abuse may predispose to the development of pancreatic adenocarcinoma. Am J Surg 1999;178(5):426–7.

241. Nahrwold DL. Editorial comment. Am J Surg 1999;178:427.

242. Spence MR, Williams R, DiGregorio GJ, Kirby-McDonnell A, Polansky M. The relationship between recent cocaine use and pregnancy outcome. Obstet Gynecol 1991;78(3 Pt 1):326–9.

243. Chasnoff IJ. Methodological issues in studying cocaine use in pregnancy: a problem of definitions. NIDA Res Monogr 1991;114:55–65.

244. Handler A, Kistin N, Davis F, Ferre C. Cocaine use during pregnancy: perinatal outcomes. Am J Epidemiol 1991;133(8):818–25.

245. Hsu CD, Chen S, Feng TI, Johnson TR. Rupture of uterine scar with extensive maternal bladder laceration after cocaine abuse. Am J Obstet Gynecol 1992;167(1):129–30.

246. Abramowicz JS, Sherer DM, Woods JR Jr. Acute transient thrombocytopenia associated with cocaine abuse in pregnancy. Obstet Gynecol 1991;78(3 Pt 2):499–501.

247. Hurd WW, Betz AL, Dombrowski MP, Fomin VP. Cocaine augments contractility of the pregnant human uterus by both adrenergic and nonadrenergic mechanisms. Am J Obstet Gynecol 1998;178(5):1077–81.

248. Refuerzo JS, Sokol RJ, Blackwell SC, Berry SM, Janisse JJ, Sorokin Y. Cocaine use and preterm premature rupture of membranes: improvement in neonatal outcome. Am J Obstet Gynecol 2002;186(6):1150–4.

249. Bauer CR, Shankaran S, Bada HS, Lester B, Wright LL, Krause-Steinrauf H, Smeriglio VL, Finnegan LP, Maza PL, Verter J. The Maternal Lifestyle Study: drug exposure during pregnancy and short-term maternal outcomes. Am J Obstet Gynecol 2002;186(3):487–95.

250. Kuczkowski KM. Cocaine abuse in pregnancy—anaesthetic implications. Intl J Obstet Anaesth 2002;11:204–10.

251. Kuczkowski KM. Caesarian section in cocaine-intoxicated parturient: regional vs. general anaesthesia? Anaesthesia 2003;58:1042–143.

252. Askin DF, Diehl-Jones B. Cocaine: effects of in utero exposure on the fetus and neonate. J Perinat Neonatal Nurs 2001;14:83–102.

253. Chasnoff IJ, Burns WJ, Schnoll SH, Burns KA. Cocaine use in pregnancy. N Engl J Med 1985;313(11):666–9.

254. Jones KL. Developmental pathogenesis of defects associated with prenatal cocaine exposure: fetal vascular disruption. Clin Perinatol 1991;18(1):139–46.

255. Napiorkowski B, Lester BM, Freier MC, Brunner S, Dietz L, Nadra A, Oh W. Effects of in utero substance exposure on infant neurobehavior. Pediatrics 1996;98(1):71–5.

256. Tsay CH, Partridge JC, Villarreal SF, Good WV, Ferriero DM. Neurologic and ophthalmologic findings in children exposed to cocaine in utero. J Child Neurol 1996;11(1):25–30.

257. Kapur RP, Shaw CM, Shepard TH. Brain hemorrhages in cocaine-exposed human fetuses. Teratology 1991;44(1):11–8.

258. Little BB, Snell LM. Brain growth among fetuses exposed to cocaine in utero: asymmetrical growth retardation. Obstet Gynecol 1991;77(3):361–4.

259. Harris EF, Friend GW, Tolley EA. Enhanced prevalence of ankyloglossia with maternal cocaine use. Cleft Palate Craniofac J 1992;29(1):72–6.

260. Lavi E, Montone KT, Rorke LB, Kliman HJ. Fetal akinesia deformation sequence (Pena–Shokeir phenotype) associated with acquired intrauterine brain damage. Neurology 1991;41(9):1467–8.

261. Lessick M, Vasa R, Israel J. Severe manifestations of oculoauriculovertebral spectrum in a cocaine exposed infant. J Med Genet 1991;28(11):803–4.

262. Richardson GA, Day NL. Maternal and neonatal effects of moderate cocaine use during pregnancy. Neurotoxicol Teratol 1991;13(4):455–60.

263. Datta-Bhutada S, Johnson HL, Rosen TS. Intrauterine cocaine and crack exposure: neonatal outcome. J Perinatol 1998;18(3):183–8.

264. Hurt H, Brodsky NL, Braitman LE, Giannetta J. Natal status of infants of cocaine users and control subjects: a prospective comparison. J Perinatol 1995;15(4):297–304.

265. Mehta SK, Super DM, Connuck D, Kirchner HL, Salvator A, Singer L, Fradley LG, Kaufman ES. Autonomic alterations in cocaine-exposed infants. Am Heart J 2002;144(6):1109–15.

266. Mehta SK, Super DM, Salvator A, Singer L, Connuck D, Fradley LG, Harcar-Sevcik RA, Thomas JD, Sun JP. Diastolic filling abnormalities by color kinesis in newborns exposed to intrauterine cocaine. J Am Soc Echocardiogr 2002;15(5):447–53.

267. Mehta SK, Super DM, Connuck D, Kirchner HL, Salvator A, Singer L, Fradley LG, Thomas JD, Sun JP. Diastolic alterations in infants exposed to intrauterine cocaine: a follow-up study by color kinesis. J Am Soc Echocardiogr 2002;15(11):1361–6.

268. Shepard TH, Fantel AG, Kapur RP. Fetal coronary thrombosis as a cause of single ventricular heart. Teratology 1991;43(2):113–7.

269. Afzal MN, Fatal double aortic arch anomaly and maternal cocaine abuse. J Coll Physicians Surg Pak 2003;13:166–7.

270. Kosofsky BE. The effect of cocaine on developing human brain. NIDA Res Monogr 1991;114:128–43.

271. Eyler FD, Behnke M, Conlon M, Woods NS, Wobie K. Birth outcome from a prospective, matched study of prenatal crack/cocaine use: II. Interactive and dose effects on neurobehavioral assessment. Pediatrics 1998;101(2):237–41.

272. Alessandri SM, Bendersky M, Lewis M. Cognitive functioning in 8- to 18-month-old drug-exposed infants. Dev Psychol 1998;34(3):565–73.

273. Wasserman GA, Kline JK, Bateman DA, Chiriboga C, Lumey LH, Friedlander H, Melton L, Heagarty MC. Prenatal cocaine exposure and school-age intelligence. Drug Alcohol Depend 1998;50(3):203–10.

274. Lester BM, LaGasse LL, Seifer R. Cocaine exposure and children: the meaning of subtle effects. Science 1998;282(5389):633–4.

275. Delaney-Black V, Covington C, Templin T, Ager J, Martier S, Sokol R. Prenatal cocaine exposure and child behavior. Pediatrics 1998;102(4 Pt 1):945–50.

276. Mayes LC, Bornstein MH, Chawarska K, Granger RH. Information processing and developmental assessments in 3-month-old infants exposed prenatally to cocaine. Pediatrics 1995;95(4):39–45.

277. Bender SL, Word CO, Di Clemente RJ, Crittenden MR, Persaud NA, Ponton LE. The developmental implications of prenatal and/or postnatal crack cocaine exposure in preschool children: a preliminary report. J Dev Behav Pediatr 1995;16(6):418–24.

278. Mirochnick M, Frank DA, Cabral H, Turner A, Zuckerman B. Relation between meconium concentration of the cocaine metabolite benzoylecgonine and fetal growth. J Pediatr 1995;126(4):636–8.

279. Martin JC, Barr HM, Martin DC, Streissguth AP. Neonatal neurobehavioral outcome following prenatal exposure to cocaine. Neurotoxicol Teratol 1996;18(6):617–25.

280. Tronick EZ, Frank DA, Cabral H, Mirochnick M, Zuckerman B. Late dose-response effects of prenatal cocaine exposure on newborn neurobehavioral performance. Pediatrics 1996;98(1):76–83.

281. Karmel BZ, Gardner JM, Freedland RL. Arousal-modulated attention at four months as a function of intrauterine cocaine exposure and central nervous system injury. J Pediatr Psychol 1996;21(6):821–32.

282. Scafidi FA, Field TM, Wheeden A, Schanberg S, Kuhn C, Symanski R, Zimmerman E, Bandstra ES. Cocaine-

exposed preterm neonates show behavioral and hormonal differences. Pediatrics 1996;97(6 Pt 1):851–5.

283. Jacobson SW, Jacobson JL, Sokol RJ, Martier SS, Chiodo LM. New evidence for neurobehavioral effects of in utero cocaine exposure. J Pediatr 1996;129(4):581–90.

284. Schuler ME, Nair P. Brief report: frequency of maternal cocaine use during pregnancy and infant neurobehavioral outcome. J Pediatr Psychol 1999;24(6):511–4.

285. Held JR, Riggs ML, Dorman C. The effect of prenatal cocaine exposure on neurobehavioral outcome: a meta-analysis. Neurotoxicol Teratol 1999;21(6):619–25.

286. Singer LT, Arendt R, Minnes S, Farkas K, Salvator A. Neurobehavioral outcomes of cocaine-exposed infants. Neurotoxicol Teratol 2000;22(5):653–66.

287. Chapman JK. Developmental outcomes in two groups of infants and toddlers: prenatal cocaine exposed and noncocaine exposed part 1. Infant-Toddler Interv 2000;10:19–36.

288. Heffelfinger AK, Craft S, White DA, Shyken J. Visual attention in preschool children prenatally exposed to cocaine: implications for behavioral regulation. J Int Neuropsychol Soc 2002;8(1):12–21.

289. Frank DA, Jacobs RR, Beeghly M, Augustyn M, Bellinger D, Cabral H, Heeren T. Level of prenatal cocaine exposure and scores on the Bayley Scales of Infant Development: modifying effects of caregiver, early intervention, and birth weight. Pediatrics 2002;110(6):1143–52.

290. Accornero VH, Morrow CE, Bandstra ES, Johnson AL, Anthony JC. Behavioral outcome of preschoolers exposed prenatally to cocaine: role of maternal behavioral health. J Pediatr Psychol 2002;27(3):259–69.

291. Downing GJ, Horner SR, Kilbride HW. Characteristics of perinatal cocaine-exposed infants with necrotizing enterocolitis. Am J Dis Child 1991;145(1):26–7.

292. Spinazzola R, Kenigsberg K, Usmani SS, Harper RG. Neonatal gastrointestinal complications of maternal cocaine abuse. NY State J Med 1992;92(1):22–3.

293. Marles SL, Reed M, Evans JA. Humeroradial synostosis, ulnar aplasia and oligodactyly, with contralateral amelia, in a child with prenatal cocaine exposure. Am J Med Genet 2003;116A:85–9.

294. Fawcett LB, Buck SJ, Brent RL. Limb reduction defects in the A/J mouse strain associated with maternal blood loss. Teratology 1998;58:183–9.

295. Church MW, Holmes PA, Tilak JP, Hotra JW. Prenatal cocaine exposure influences the growth and life span of laboratory rats. Neurotoxicol Teratol 2004;26:429–41.

296. Church MW. Does cocaine cause birth defects? Neurotoxicol Teratol 1993;15:289; discussion 311–12.

297. Kashiwagi M, Chaoui R, Stallmach T, Hurlimann S, Lauper U, Hebisch G. Fetal bilateral renal agenesis, phocomelia, and single umbilical artery associated with cocaine abuse in early pregnancy. Birth Defects Res Part A Clin Mol Teratol 2003;67:951–2.

298. Greenfield SP, Rutigliano E, Steinhardt G, Elder JS. Genitourinary tract malformations and maternal cocaine abuse. Urology 1991;37(5):455–9.

299. Behnke M, Eyler FD, Garvan CW, Wobie K. The search for congenital malformations in newborns with fetal cocaine exposure. Pediatrics 2001;107(5):E74.

300. Bandstra ES, Morrow CE, Anthony JC, Churchill SS, Chitwood DC, Steele BW, Ofir AY, Xue L. Intrauterine growth of full-term infants: impact of prenatal cocaine exposure. Pediatrics 2001;108(6):1309–19.

301. Frank DA, Augustyn M, Knight WG, Pell T, Zuckerman B. Growth, development, and behavior in early childhood following prenatal cocaine exposure: a systematic review. JAMA 2001;285(12):1613–25.

302. Stanwood GD, Levitt P. Prenatal cocaine exposure as a risk factor for later developmental outcomes. JAMA 2001;286(1):45.

303. Singer LT, Arendt RE. Prenatal cocaine exposure as a risk factor for later developmental outcomes. JAMA 2001;286(1):45–6.

304. Delaney-Black V, Covington CY, Nordstrom-Klee B, Sokol RJ. Prenatal cocaine exposure as a risk factor for later developmental outcomes. JAMA 2001;286(1):46–7.

305. King TA, Perlman JM, Laptook AR, Rollins N, Jackson G, Little B. Neurologic manifestations of in utero cocaine exposure in near-term and term infants. Pediatrics 1995;96(2 Pt 1):259–64.

306. Fetters L, Tronick EZ. Neuromotor development of cocaine-exposed and control infants from birth through 15 months: poor and poorer performance. Pediatrics 1996;98(5):938–43.

307. Behnke M, Eyler FD, Garvan CW, Wobie K, Hou W. Cocaine exposure and developmental outcome from birth to 6 months. Neurotoxicol Teratol 2002;24(3):283–5.

308. Bauer CR. Perinatal effects of prenatal drug exposure. Neonatal aspects. Clin Perinatol 1999;26(1):87–106.

309. Lutiger B, Graham K, Einarson TR, Koren G. Relationship between gestational cocaine use and pregnancy outcome: a meta-analysis. Teratology 1991;44940:405–14.

310. Singer LT, Arendt R, Minnes S, Farkas K, Salvator A. Neurobehavioural outcomes of cocaine-exposed infants. Neurotoxicol Teratol 2000;22(5):653–66.

311. Lidow MS. Prenatal cocaine exposure adversely affects development of the primate cerebral cortex. Synapse 1995;21(4):332–41.

312. Heffelfinger AK, Craft S, White DA, Shyken J. Visual attention in preschool children prenatally exposed to cocaine: implications for behavioural regulation. J Int Neuropsychol Soc 2002;8:12–21.

313. Bandstra ES, Morrow CE, Vogel AL, Fifer RC, Ofir AY, Dausa AT, Xue L, Anthony JC. Longitudinal influence of prenatal cocaine exposure on child language functioning. Neurotoxicol Teratol 2002;24(3):297–308.

314. Malakoff ME, Mayes LC, Schottenfeld R, Howell S. Language production in 24-month-old inner-city children of cocaine- and other drug-using mothers. J Appl Dev Psychol 1999;20:159–80.

315. Bauer CR, Shankaran S, Bada HS, Lester B, Wright LL, Krause-Steinrauf H, Smeriglio VL, Finnegan LP, Maza PL, Verter J. The Maternal Lifestyle Study: drug exposure during pregnancy and short-term maternal outcomes. Am J Obstet Gynecol 2002;186:487–95.

316. Bauer CR, Langer JC, Shankaran S, Bada H, Lester B, Wright LL, Krause-Stenrauf H, Smeriglio VL, Finnegan LP, Maza PL, Verter J. Acute neonatal effects of cocaine exposure during pregnancy. Arch Pediatr Adolesc Med 2005;159:824–34.

317. Pichini S, Puig C, Zuccaro P, Marchei E, Pellegrini M, Murillo J, Vall O, Pacifici R, Garcia-Algar O. Assessment of exposure to opiates and cocaine during pregnancy in a Mediterranean city: preliminary results of the "Meconium Project". Forensic Sci Int 2205;153:59–65.

318. Frassica JJ, Orav EJ, Walsh EP, Lipshultz SE. Arrhythmias in children prenatally exposed to cocaine. Arch Pediatr Adolesc Med 1994;148(11):1163–9.

319. Bulbul ZR, Rosenthal DN, Kleinman CS. Myocardial infarction in the perinatal period secondary to maternal

cocaine abuse. A case report and literature review. Arch Pediatr Adolesc Med 1994;148(10):1092–6.

320. McCann EM, Lewis K. Control of breathing in babies of narcotic- and cocaine-abusing mothers. Early Hum Dev 1991;27(3):175–86.

321. Zuckerman B, Maynard EC, Cabral H. A preliminary report of prenatal cocaine exposure and respiratory distress syndrome in premature infants. Am J Dis Child 1991;145(6):696–8.

322. Hand IL, Noble L, McVeigh TJ, Kim M, Yoon JJ. The effects of intrauterine cocaine exposure on the respiratory status of the very low birth weight infant. J Perinatol 2001;21(6):372–5.

323. Kesrouani A, Fallet C, Vuillard E, Jacqz-Aigrain E, Sibony O, Oury JF, Blot P, Luton D. Pathologic and laboratory correlation in microcephaly associated with prenatal cocaine exposure. Early Hum Dev 2001;63(2):79–81.

324. Bellini C, Massocco D, Serra G. Prenatal cocaine exposure and the expanding spectrum of brain malformations. Arch Intern Med 2000;160(15):2393.

325. Smith LM, Qureshi N, Renslo R, Sinow RM. Prenatal cocaine exposure and cranial sonographic findings in preterm infants. J Clin Ultrasound 2001;29(2):72–7.

326. Esmer MC, Rodriguez-Soto G, Carrasco-Daza D, Iracheta ML, Del Castillo V. Cloverleaf skull and multiple congenital anomalies in a girl exposed to cocaine in utero: case report and review of the literature. Childs Nerv Syst 2000;16(3):176–80.

327. Miller-Loncar C, Lester BM, Seifer R, Lagasse LL, Bauer CR, Shankaran S, Bada HS, Wright LL, Smeriglio VL, Bigsby R, Liu J. Predictors of motor development in children prenatally exposed to cocaine. Neurotoxicol Teratol 2005;27:213–20.

328. Hurt H, Brodsky N, Roth H, Malmud E, Giannetta J. School performance of children with gestational cocaine exposure. Neurotoxicol Teratol 2005;27(2):203–11.

329. Schuetze P, Croff SL, Eiden RD. The development of motor asymmetries in 1-month-old infants who were prenatally exposed to cocaine. Laterality 2003;8:79–93.

330. Schuetze P, Croff S, Das Eiden R. The development of motor asymmetries in 1-month-old infants who were prenatally exposed to cocaine. Laterality 2003;8:79–93.

331. Hajnal BL, Ferriero DM, Partridge JC, Dempsey DA, Good WV. Is exposure to cocaine or ciagarette smoke during pregnancy associated with infant visual abnormalities? Dev Med Child Neurol 2004;46:520–5.

332. Arendt RE, Short EJ, Singer LT, Minnes S, Hewitt J, Flynn S, Carlson L, Min MO, Klein N, Flannery D. Children prenatally exposed to cocaine: developmental outcomes and environmental risks at seven years of age. J Dev Behav Ped 2004;25:83–90.

333. Tan-Laxa MA, Sison-Switala C, Rintelman W, Ostrea Jr EM. Abnormal auditory brainstem response among infants with prenatal cocaine exposure. Pediatrics 2004;113:357–60.

334. Church MW, Overbeck GW. Prenatal cocaine exposure in the Long-Evans rat III. Developmental effects on the brainstem-auditory evoked potential. Neurotoxicol Teratol 1990;12:345–51.

335. Salamy A, Eldredge L, Anderson MA, Bull D. Brainstem transmission time in infants exposed to cocaine in utero. J Pediatr 1990;117:627–9.

336. Lester BM, Lagasse L, Seifer R, Tronick EZ, Bauer CR, Shankaran S, Bada HS, Wright LL, Smeriglio VL, Liu J, Finnegan LP, Maza PL. The Maternal Lifestyle Study (MLS). Effects of prenatal cocaine and or opiate exposure on auditory brain response at one month. J Pediatr 2003;142:279–85.

337. Carzoli R, Murphy S, Kinsley J, Houy J. Evaluation of auditory brain-stem response in full-term infants of cocaine-abusing mothers. Am J Dis Child 1991;145:1013–6.

338. Hulse GK, English DR, Milne E, Holman CD, Bower CI. Maternal cocaine use and low birth weight newborns: a meta-analysis. Addiction 1997;92:1561–70.

339. Bauer CR, Shankaran S, Bada HS, Lester B, Wright LL, Kause-Steinrauf H, Smeriglio VL, Finnegan LP, Maza PL, Verter J. The Maternal Lifestyle Study. Drug exposure during pregnancy and short-term maternal outcomes. Am J Obstet Gynecol 2002;186:487–95.

340. Coles CD, Bard KA, Platzman KA, Lynch ME. Attentional response at eight weeks in prenatally drug-exposed and preterm infants. Neurotoxicol Teratol 1999;21(5):527–37.

341. Singer LT, Hawkins S, Huang J, Davillier M, Baley J. Developmental outcomes and environmental correlates of very low birthweight, cocaine-exposed infants. Early Hum Dev 2001;64(2):91–103.

342. Ukeje I, Bendersky M, Lewis M. Mother–infant interaction at 12 months in prenatally cocaine-exposed children. Am J Drug Alcohol Abuse 2001;27(2):203–24.

343. Lester BM, LaGasse LL, Seifer R. Cocaine exposure and children: the meaning of subtle effects. Science 1998;282:633–4.

344. Frank DA, Augustyn M, Knight WG, Pell T, Zuckerman B. Growth, development and behavior in early childhood following prenatal cocaine exposure: a systematic review. JAMA 2001;285:1613–25.

345. Singer L, Arendt R, Farkas K, Minnes S, Huang J, Yamashita T. Relationship of prenatal cocaine exposure and maternal postpartum psychological distress to child development outcome. Dev Psychopathol 1997;9:473–89.

346. Jacobson SW, Jacobson JL, Sokol RJ, Martier SS, Chiodo LM. New evidence for neurobehavioral effects of in utero cocaine exposure. J Pediatr 1996;129:581–90.

347. Singer LT, Arendt R, Minnes S, Farkas K, Salvator A, Kirchner HL, Kliegman R. Cognitive and motor outcomes of cocaine exposed infants. JAMA 2002;287:1952–60.

348. Frank DA, Jacobs RR, Beeghly M, Augustyn M, Bellinger D, Cabral H, Heeren T. Levels of prenatal cocaine exposure and scores on the Bayley Scale of Infant Development: modifying effects of caregiver, early intervention, and birth weight. Pediatrics 2002;110:1143–52.

349. Messinger DS, Bauer CR, Das A, Seifer R, Lester BM, Lagasse, LL, Wright LL, Shankaran S, Bada HS, Smeriglio VL, Langer JC, Beeghly M, Poole WK. The Maternal Lifestyle Study: cognitive, motor and behavioral outcomes of cocaine-exposed and opiate-exposed infants through the 3 years of age. Pediatrics 2004;113:1677–85.

350. Lester BM, LaGasse L, Freier K, Brunner S. Studies of cocaine exposed infants. NIDA Monogr 1996;164:175–210.

351. Arendt R, Angelopoulos J, Salvator A, Singer L. Motor development of cocaine exposed children at age two years. Pediatrics 1999;103:86–92.

352. Kilbride H, Castor C, Hoffman E, Fuger KL. Thirty-six month outcome of prenatal cocaine exposure for term or near-term infants: impact of early case management. J Dev Behav Pediatr 2000;21:19–26.

353. Singer LT, Arendt R, Minnes S, Salvator A, Siegel AC, Lewis BA. Developing language skills of cocaine-exposed infants. Pediatrics 2001;107:1057–64.

354. Delaney-Black V, Covington C, Templin T, Ager J, Martier S, Sokol R. Prenatal cocaine exposure and child behavior. Pediatrics 1998;102:945–50.

355. Delaney-Black V, Covington C, Nordstrom B, Ager J, Janisse J, Hannigan J, Chiodo L, Sokol R. Prenatal cocaine: quantity of exposure and gender moderation. J Dev Behav Ped 2004;25(4):254–63.

356. Jones NA, Field T, Davalos M, Hart S. Greater right frontal EEG asymmetry and nonempathic behavior are observed in children prenatally exposed to cocaine. Int J Neurosci 2004;114:459–80.

357. Zuckerman B, Frank DA, Mayes L. Cocaine exposed infants and developmental outcomes. "Crack-kids revisited". JAMA 2002;287:1990–1.

358. Tronick EZ, Beeghley M. Prenatal cocaine exposure, child development, and the compromising effects of cumulative risk. Clin Perinatol 1999;26:151–71.

359. Richardson GA, Hamel SC, Goldschmidt L, Day NL. The effects of perinatal cocaine use on neonatal neurobehavioral status. Neurotoxicol Teratol 1996;18:519–28.

360. Bandstra ES, Morrow CE, Anthony JC, Accornero VH, Fried PA. Longitudinal investigation of task persistence and sustained attention in children with prenatal cocaine exposure. Neurotoxicol Teratol 2001;32:545–59.

361. Myers BJ, Dawson KS, Britt GC, Lodder DE, Meloy LD, Saunders MK, Meadows SL, Elswick RK. Prenatal cocaine exposure and infant performance on the Brazelton Neonatal Behavioral Assessment Scale. Subst Use Misuse 2003;38:2065–96.

362. Nordstrom Bailey B, Sood BG, Sokol RJ, Ager J, Janisse J, Hannigan JH, Covington C, Delaney-Black V. Gender and alcohol moderate prenatal cocaine effects on teacher-report of child behavior. Neurotoxicol Teratol 2005;27:181–9.

363. Sood BG, Nordstrom Bailey B, Covington C, Sokol RJ, Ager J, Janisse J, Hannigan JH, Delaney-Black V. Gender and alcohol moderate caregiver reported child behavior after prenatal cocaine. Neurotoxicol Teratol 2005;27:191–201.

364. Markowski VP, Cox C, Wei SS. Prenatal cocaine exposure produced gender-specific motor effects in aged rats. Neurotoxicol Teratol 1998;20:43–53.

365. Field TM, Scaffidi F, Pickens J, Prodromidis M, Pelaez-Nogueras M, Torquati J, Wilcox H, Malphurs J, Schanberg S, Kuhn C. Polydrug-using adolescent mother and their infants receiving early intervention. Adolescence 1998;33:117–43.

366. Bennett DS, Bendersky M, Lewis M. Children's intellectual and emotional behavioral adjustment at 4 years as a function of cocaine exposure, maternal characteristics, and environmental risk. Dev Psychol 2002;38:648–58.

367. Lewis B, Singer LT, Short EJ, Minnes S, Arendt R, Weishampel P, Klein N, Min MO. Four-year language outcomes of children exposed to cocaine in utero. Neurotox Teratol 2004;26:617–27.

368. Morrow CO, Vogel AL, Anthony JC, Ofir AY, Dausa AT, Bandstra E. Expressive and receptive language functioning in preschool children with prenatal cocaine exposure. J Pediatr Psychol 2004;29(7):543–54.

369. Morrow CE, Bandstra ES, Anthony JC, Ofir AY, Xue L, Reyes MB. Influence of prenatal cocaine exposure on early language development: longitudinal findings from four months to three years of age. J Dev Behav Pediatr 2003;24:39–50.

370. Singer LT, Arendt R, Minnes S, Salvator A, Siegel AC, Lewis BA. Developing language skills of cocaine-exposed infants. Pediatrics 2001;107(5):1057–64.

371. De Giorgio F, Rossi SS, Rainio J, Chiarotti M. Cocaine found in a child's hair due to environmental exposure? Int J Legal Med 2004;118:310–2.

372. June R, Aks SE, Keys N, Wahl M. Medical outcome of cocaine bodystuffers. J Emerg Med 2000;18(2):221–4.

373. Gill JR, Graham SM. Ten years of "body packers" in New York City: 50 deaths. J Forensic Sci 2002;47(4):843–6.

374. Visser L, Stricker B, Hoogendoorn M, Vinks A. Do not give paraffin to packers. Lancet 1998;352(9137):1352.

375. Grigorov V, Goldberg L, Foccard JP. Cardiovascular complications of acute cocaine poisoning: a clinical case report. Cardiol J S Africa 2004;15:139–42.

376. Khan FY. The cocaine 'body-packer' syndrome: diagnosis and treatment. Indian J Med Sci 2005;59:457–8.

377. Miller JS, Hendren SK, Liscum KR. Giant gastric ulcer in a body packer. J Trauma 1998;45(3):617–9.

378. Havlik DM, Nolte KB. Fatal "crack" cocaine ingestion in an infant. Am J Forensic Med Pathol 2000;21(3):245–8.

379. Kaye S, Darke S. Non-fatal cocaine overdose among injecting and non-injecting cocaine users in Sydney, Australia. Addiction 2004;99:1315–22.

380. Sands BF, Ciraulo DA. Cocaine drug-drug interactions. J Clin Psychopharmacol 1992;12(1):49–55.

381. Signs SA, Dickey-White HI, Vanek VW, Perch S, Schechter MD, Kulics AT. The formation of cocaethylene and clinical presentation of ED patients testing positive for the use of cocaine and ethanol. Am J Emerg Med 1996;14(7):665–70.

382. Salloum IM, Daley DC, Cornelius JR, Kirisci L, Thase ME. Disproportionate lethality in psychiatric patients with concurrent alcohol and cocaine abuse. Am J Psychiatry 1996;153(7):953–5.

383. Spunt B, Brownstein HH, Crimmins SM, Langley S. Drugs and homicide by women. Subst Use Misuse 1996;31(7):825–45.

384. McCance-Katz EF, Kosten TR, Jatlow P. Concurrent use of cocaine and alcohol is more potent and potentially more toxic than use of either alone—a multiple-dose study. Biol Psychiatry 1998;44(4):250–9.

385. Pennings EJ, Leccese AP, Wolff FA. Effects of concurrent use of alcohol and cocaine. Addiction 2002;97(7):773–83.

386. Hurtova M, Duclos-Vallee JC, Saliba F, Emile JF, Bemelmans M, Castaing D, Samuel D. Liver transplantation for fulminant hepatic failure due to cocaine intoxication in an alcoholic hepatitis C virus-infected patient. Transplantation 2002;73(1):157–8.

387. Weiner AL, Bayer MJ, McKay CA Jr, DeMeo M, Starr E. Anticholinergic poisoning with adulterated intranasal cocaine. Am J Emerg Med 1998;16(5):517–20.

388. Hameedi FA, Sernyak MJ, Navui SA, Kosten TR. Near syncope associated with concomitant clozapine and cocaine use. J Clin Psychiatry 1996;57(8):371–2.

389. Farren CK, Hameedi FA, Rosen MA, Woods S, Jatlow P, Kosten TR. Significant interaction between clozapine and cocaine in cocaine addicts. Drug Alcohol Depend 2000;59(2):153–63.

390. Carlan SJ, Stromquist C, Angel JL, Harris M, O'Brien WF. Cocaine and indomethacin: fetal anuria, neonatal edema, and gastrointestinal bleeding. Obstet Gynecol 1991;78(3 Pt 2):501–3.

391. Tordoff SG, Stubbing JF, Linter SP. Delayed excitatory reaction following interaction of cocaine and monoamine oxidase inhibitor (phenelzine). Br J Anaesth 1991;66(4):516–8.

392. van Harten PN, van Trier JC, Horwitz EH, Matroos GE, Hoek HW. Cocaine as a risk factor for neuroleptic-induced acute dystonia. J Clin Psychiatry 1998;59(3):128–30.

393. Kosten TR, Woods SW, Rosen MI, Pearsall HR. Interactions of cocaine with nimodipine: a brief report. Am J Addict 1999;8(1):77–81.

394. Coffin PO, Galea S, Ahern J, Leon AC, Vlahov D, Tardiff K. Opiates, cocaine and alcohol combinations in accidental drug overdose deaths in New York City, 1990-98. Addiction 2003;98:739–47.

395. Cann B, Hunter R, McCann J. Cocaine/heroin induced rhabdomyolysis and ventricular fibrillation. Emerg Med J 2002;19(3):264–5.

396. Houtsmuller EJ, Notes LD, Newton T, Van Sluis N, Chiang N, Elkashef A, Bigelow GE. Transdermal selegiline and intravenous cocaine: safety and interactions. Psychopharmacology 2004;172:31–40.

397. Matsuo S, Rao DB, Chaudry I, Foldes FF. Interaction of muscle relaxants and local anesthetics at the neuromuscular junction. Anesth Analg 1978;57(5):580–7.

398. Usubiaga JE, Wikinski JA, Morales RL, Usubiaga LE. Interaction of intravenously administered procaine, lidocaine and succinylcholine in anesthetized subjects. Anesth Analg 1967;46(1):39–45.

DEXAMFETAMINE

See also Amphetamines

General Information

Dexamfetamine or (+)-amfetamine is significantly more potent than (−)-amfetamine. The use of dexamfetamine as an appetite suppressant has rapidly declined, because of appreciation of its potential for abuse and addiction. These arise mainly from euphoria, which may be followed by depression as the effect of the drug wears off. Stimulant effects were reported in 23% of 347 patients using dexamfetamine as an appetite suppressant (SED-9, 10).

Dexamfetamine is extremely variable in its effects, and can even produce drowsiness in a small proportion of subjects. Postmenopausal women are more prone to drowsiness, anger, and sadness than euphoria (1). Adverse effects due to sympathetic overactivity are fairly common but not usually serious. However, in view of dexamfetamine's addiction potential, other anorectic drugs should be considered first.

When it was first introduced, one of the most frequent uses of amfetamine was as an anorexigenic agent in the treatment of obesity. A number of anorectic agents, many of them related to amfetamine, have since been manufactured. Most are stimulants of the central nervous system. In descending order of approximate stimulatory potency, they are dexamfetamine, phentermine, chlorphentermine, mazindol, diethylpropion, and fenfluramine. The last of these has a stimulatory effect only in overdosage. One of the problems that has concerned clinicians over the use of anorectic drugs for the treatment of weight reduction is that despite 6 weeks to 3 months of weight reduction efficacy, the effect begins to wear off and on withdrawal weight gain rebounds.

The anorectic agents produce adverse effects mainly of the central nervous system sympathomimetic type. Therapy should therefore only be allowed under strict medical supervision, to ensure the earliest possible detection of any signs of drug abuse. Long-term drug treatment of obesity should be avoided altogether.

Organs and Systems

Metabolism

Hyperinsulinemia secondary to chronic administration of dexamfetamine, with a fall in fasting blood sugar after a few weeks of use, has been described (SED-9, 10).

Drug Administration

Drug formulations

Introduction of modified-release formulations has provided some improvement in the use of anorectic drugs. Steady release of the drug permits a constant concentration in the blood throughout the entire day. Thus, a sudden excess of physiological hunger is prevented, and adverse effects involving the central nervous system are diminished.

Reference

1. Halbreich U, Asnis G, Ross D, Endicott J. Amphetamine-induced dysphoria in postmenopausal women. Br J Psychiatry 1981;138:470–3.

DIAMORPHINE (HEROIN)

General Information

Heroin is a potent opioid that offers no substantial advantages over morphine. In the UK it is the preferred parenteral opioid for subcutaneous administration to cachectic cancer patients, because of its high solubility. The adverse of other opioid analgesis are covered in SED-15.

The prevalence of heroin as a drug of abuse has been reviewed (1). There are over 1 million heroin addicts in the USA. The lifetime prevalence among those aged 12–25 years continues to increase gradually.

Comparative studies

High-dose diamorphine has been compared with morphine in a double-blind, crossover, randomized study in 39 intravenous opioid users who were allocated to either morphine 3% solution or diamorphine 2% solution, gradually increasing up to an individual maintenance dose adjusted to meet the patient's subjective needs (2). Those who started with diamorphine and subsequently switched to morphine terminated prematurely owing to excessive histamine reactions, all of which occurred during crossover to morphine. Symptoms included severe pruritus, flushing, swelling, urticaria, severe headaches, nausea, general malaise, hypotension, and tachycardia. Only 44% of the original cohort finished the 6-week study (14 getting diamorphine at the end and three getting morphine). Average daily doses were 491 mg for diamorphine and 597 mg for morphine. These results suggest that diamorphine produces fewer adverse effects than morphine and may be preferable for high-dose maintenance prescription. However, the study was very small and the subject selection was biased, as were the variables used to determine a successful outcome. The result was contrary to all the well-established pharmacological facts, and the authors did not mention the risks associated with high doses of short-acting opioids.

In a randomized, double-blind study, 64 patients undergoing total knee arthroplasty received either intrathecal morphine 0.3 mg or intrathecal diamorphine, 0.3 mg in 0.3 ml, with 2–2.5 ml of 0.5% heavy spinal bupivacaine (3). The patients given morphine had significantly greater analgesia at 4, 8, and 12 hours postoperatively. The incidence of opioid-related adverse effects was not significantly different between the groups.

In a single-blind, randomized, controlled study, 70 patients scheduled for elective cesarean section under spinal anesthesia using hyperbaric bupivacaine 0.5% received intrathecal fentanyl 20 μg, intrathecal diamorphine 300 μg, or 0.9% saline (4). Significantly less intraoperative and postoperative "analgesic control" was required in the opioid groups, especially in those given diamorphine. Diamorphine produced longer-lasting analgesia than fentanyl (12 hours versus 1 hour).

Nausea, vomiting, and pruritus occurred relatively infrequently, with no differences between the groups; sedation was more frequent with fentanyl.

In two separate open, randomized, controlled but unblinded trials, 549 patients were divided into five treatment groups. In the first study there were 375 patients in three groups:

- methadone alone for 12 months (controls);
- methadone plus inhaled heroin for 12 months;
- methadone alone for 6 months followed by methadone plus inhaled heroin for 6 months.

In the second study there were 174 patients in two similar experimental groups in whom injectable rather than inhaled heroin was used (5). A response to treatment was defined as at least a 40% improvement in physical, mental, or social domains of quality of life, if not accompanied by a substantial (over 20%) increase in the use of another illicit drug, such as cocaine or amphetamines. After 12 months those who took methadone and heroin (smoked or injected) had significantly better outcomes. The incidences of adverse effects (constipation and drowsiness) were similar in all the groups. However, owing to the limitations of the study and the complex nature of drug dependence, the therapeutic outcomes could not be justifiably and solely attributed to the specific drug(s).

Placebo-controlled studies

In a randomized, double-blind study, 14 patients who underwent elective surgery for correction of bilateral arthritic deformities of the feet received 15 ml of 0.9% saline containing diamorphine 2.5 mg into the cannula in one foot and 15 ml of saline into the other foot (6). Intravenous regional diamorphine did not improve postoperative pain relief or secondary hyperalgesia. There were no significant adverse effects.

Organs and Systems

Respiratory

"Chasing the dragon," or inhaling heroin vapor through a straw, is a technique by which heroin users avoid the risks of injection. In Amsterdam, 85% of heroin users smoke or chase the drug. Pulmonary function can be affected by heroin inhalation. It can depress the respiratory center, release histamine (which can trigger asthma), result in septic emboli, and increase susceptibility to infectious diseases, such as tuberculosis and pneumonia. In 100 methadone maintenance users, lung function and shortness of breath were evaluated using spirometry and clinical history (7). Impaired lung function and shortness of breath correlated with chronic heroin smoking.

Heroin insufflation has been identified as a trigger for asthma (8). Of 23 patients aged 50 years and younger who were admitted to Cook County Hospital during 6 months with a primary diagnosis of asthma exacerbation, 13 reported heroin use and identified it as an asthma trigger, four reported heroin use but did not associate it with asthma exacerbation, and six had no recent history of heroin use. The patients with heroin-triggered symptoms stated that asthma exacerbation had not occurred on first use but had developed over time. Five of the seven patients whose asthma developed in adulthood reported that their heroin use had predated their asthma.

The same group of researchers then conducted a retrospective case-control study of all asthma admissions in patients under 50 over a period of 2 years. The charts of patients admitted with diabetic ketoacidosis during the same period were used as controls. Drug histories and urine drug screens were used to identify heroin users. Of 104 admissions (84 patients), 38 acknowledged a history of heroin use and 34 returned a urine drug screen that was positive for opiates. Both a history of heroin use and a positive urine drug screen were significantly more common in those with asthma than in those with diabetic ketoacidosis.

These two studies taken together suggest that heroin insufflation commonly triggers asthma, and that heroin use is more prevalent in patients with asthma. It is possible that heroin causes bronchospasm or pulmonary mast cell degranulation; alternatively, since heroin becomes a trigger over a long period of time, allergic sensitization could be involved. Another possibility is that the cutting agents and contaminants in street heroin act as irritants to the airways.

In a retrospective study of the case notes of patients who had been admitted to hospital with acute attacks of asthma, there was a high prevalence of heroin use—15% had used only heroin and another 16% had used both heroin and cocaine (9). Heroin users had been intubated more often than non-drug users (17% versus 2.3%). Similarly, more heroin users were admitted to ICU than non-users (21% versus 12%). However, they did not spend more time receiving mechanical ventilation or being in hospital. These findings suggest that heroin induced some degree of bronchoconstriction and respiratory depression, which worsened the initial presentation of asthma.

Heroin-induced pulmonary edema, or "heroin lung" (SEDA-19, 29; SEDA-25, 39), is a serious complication, which may be due to release of histamine, with increased pulmonary lymph flow and capillary permeability. There have been 27 reports of non-fatal heroin overdose associated with non-cardiogenic pulmonary edema (10). In a retrospective case-control study there were 23 heroin fatalities and 12 controls with sudden cardiac deaths (11). The authors tried to verify that defects of the alveolar capillary membranes and/or an acute anaphylactic reaction can lead to pulmonary congestion, edema, and hemorrhages. There were defects of the epithelial and endothelial basal laminae of the alveoli in both groups. There was an insignificant increase in IgE-positive cells in the heroin group. The findings suggested that heroin-associated lung edema is generally not caused by an anaphylactic reaction.

Bilateral pulmonary edema associated with heroin abuse has been reported several times (12). Bronchospasm has been noted following the use of street heroin, perhaps due to contaminants (13).

Nervous system

Delayed onset oculogyric crisis and generalized dystonia occurred in a 19-year-old man after intranasal heroin use, possibly due to bilateral hypoxic infarction of the pallidum and pallidothalamic tracts (14).

There has been one report of mixed transcortical aphasia attributed to heroin (15).

Myoclonic spasm has been reported about 24 hours after withdrawal of an epidural infusion of diamorphine (SEDA-16, 81).

Demyelination has been attributed to diamorphine (16).

- A 41-year-old chronic diamorphine user developed an unsteady gait and dysarthria over 2 weeks, followed by severe cerebellar ataxia and moderate dysmetria of the arms and legs. An MRI scan suggested myelin damage, with symmetrical involvement of the cerebellar hemispheres and decussation of the superior cerebellar peduncles, the corticospinal tracts, and the centrum semiovale, suggesting spongiform leukoencephalopathy. Two years later having taken no more diamorphine he was improved, with minor regression of the MRI lesions, especially the white matter lesions.

Myelopathy has been reported after intranasal insufflation of diamorphine (17).

- A 52-year-old man with a history of diamorphine abuse presented with sudden paraplegia a few hours after intranasal insufflation. He had flaccid paralysis of both legs, acute urinary retention, and reduced rectal tone. Deep tendon reflexes were absent and plantar responses were extensor. An MRI scan of the spine and an immunoglobulin profile supported the conclusion that this was a case of acute myelopathy with an immunopathological cause, involving a protein specific to spinal cord parenchyma, triggering acute local inflammation, ischemia, and tissue damage. Seven weeks later he recovered normal neurological function.

This case of heroin myelopathy is similar to other reported cases, except that this case occurred with intranasal rather than intravenous use. The MRI findings were consistent with a transverse myelitis. The authors suggested that hypersensitivity and an immune-mediated attack on the spinal cord was the likely mechanism of injury.

Reflex sympathetic dystrophy, in which there is an excessive or abnormal sympathetic nervous system response in a limb, has been associated with rhabdomyolysis secondary to heroin abuse (18).

- A 37-year-old male heroin smoker developed tea-colored urine and pain, swelling, and tenderness in both feet. He had acute renal insufficiency and rhabdomyolysis and was treated with hemodialysis. Urine toxicology was negative. He also had persistent, burning pain in both feet, with cool, pale, thin skin on both legs, a mild reduction in sensation on the lateral aspects of the lower legs and diminished bilateral knee and ankle reflexes. Walking was restricted, with limited range of movement owing to the severe pain. His feet would swell and redden after a 5-meter walk, suggesting loss of sympathetic regulation. Nerve conduction velocity studies of the tibial, peroneal, and sural nerves were abnormal. Radiographs showed mildly reduced bone mineralization in the legs. Three-phase bone scintigraphy showed diffusely increased radiotracer accumulation over both feet in all three phases, as found in reflex sympathetic dystrophy. The diagnosis was confirmed by local anesthetic sympathetic blockade. Nasal calcitonin spray led to pain relief 2 months later. A follow-up three-phase bone scintigram showed less radiotracer uptake, consistent with a good response to calcitonin therapy.

Progressive spongiform leukoencephalopathy

A rare consequence of inhaling heated heroin ("chasing the dragon") is a progressive spongiform leukoencephalopathy. The first cases of leukoencephalopathy due to inhalation of heroin pyrolysate were described in the Netherlands in 1982. The first three cases in the USA were reported in 1996 (SEDA-22, 35).

The condition has three stages, progressing from cerebellar signs and motor restlessness to pyramidal and pseudobulbar signs and, in a minority of patients (about 25%), to a terminal stage characterized by spasms, hypotonic paresis, and ultimately death. Symmetrical spongiform degeneration occurs, particularly in the cerebral and cerebellar white matter and in corticospinal and solitary tracts. Involvement of the cerebellum and the posterior limb of the internal capsule, with sparing of the anterior limb, appear to be characteristic, helping to distinguish this condition from other causes of leukoencephalopathy. Neuroimaging techniques contribute significantly to the management of such patients, as is evident from recent cases.

- A 16-year-old man was found at home drowsy 36 hours after smoking heroin for the first time (19). He had a gaze paresis and sensorimotor hemiplegia on the left side. A CT scan showed bilateral hypodense lesions in the globus pallidus. An MRI scan 5 days later showed symmetrical hyperintense signals in the T2 weighted images, along with a massive diffusion disorder in the diffusion weighted images. This was predominantly in the deep white matter of the parieto-occipital cortex. There were also bilateral hyperintense T_2 signals in the ventral globi pallidorum. MR spectroscopy suggested combined hypoxic and mitochondrial damage, resulting in axonal injury without demyelination. He was discharged 1 month later with no hemiplegia. A follow-up MRI scan 6 months later showed improvement. There were no signs of brain atrophy.

According to the authors, this is first reported MRI follow-up study of acute leukoencephalopathy after a single inhaled dose of heroin. It most likely involved a complex mechanism triggered by heroin, causing mitochondrial and hypoxic injury limited to specific areas of the white matter of the brain.

- A 21-year-old man became unconscious after smoking heroin at a rave party and developed an aspiration pneumonia, bilateral pyramidal signs, and spastic paresis (20). He was given clonidine and midazolam, but continued to have daily episodes of stretching spasms, hyperventilation, profuse sweating, pyrexia, and mydriasis. He was given coenzyme Q 300 mg qds on day 14. A CT scan showed diffuse hypodensities in the white matter indicative of spongiform leukoencephalopathy. On day 30, he began to improve and an MRI scan showed diffuse abnormal lesions in the white matter. He gradually recovered and an MRI scan on day 59 showed less extensive lesions. At 7 months he had regained all functions but had a residual gait ataxia. A follow-up MRI scan showed recovery of the white matter, but some necrosis.

Another report has provided more details from physical assessments and laboratory and radiological (MRI and MRS) data, and more information about the course of this heroin-related effect (21). The three cases showed raised concentrations of intracerebral lactate (reflecting mitochondrial dysfunction), which suggests a conversion of aerobic to anaerobic metabolism seen in hypoxic-ischemic conditions, including stroke. One patient recovered quite well after antioxidant therapy, supporting a metabolic effect of the heroin-related toxin; a similar response to co-enzyme Q has been found in other mitochondrial disorders with a high CSF lactate. Thus, the authors recommended that although the role of antioxidant therapy in this condition is unclear, it may be prudent to administer oral co-enzyme Q supplemented with vitamins C and E to patients with this syndrome.

Intravenous administration of pure heroin did not cause a leukoencephalopathy in a patient in whom inhalation had caused it (22), and toxicity in these cases may have been due to the heating of the heroin. This might have implications for young heroin users who, because of the known increased risk of HIV infection, prefer to "chase" (smoke) the drug, rather than to inject it intravenously.

- A 23-year-old pregnant woman at 39 weeks of gestation developed tonic-clonic seizures and hypothermia after taking excessive heroin intravenously (23). She developed Cheyne-Stokes respiration needing intubation and a cesarean section was performed, after which she developed inappropriate secretion of antidiuretic hormone and acute renal insufficiency. She made a complete recovery.

The mechanism of heroin-induced spongiform leukoencephalopathy is not known and the disease is not reproducible in animals. Suspicions have been raised about potential contaminants, such as strychnine, caffeine, phenacetin, quinine, and procaine. This man started inhaling heroin 2 weeks before the onset of the disease, but he used it in large amounts, raising the possibility that the severity of the disease is dose-related. In the two cases discussed above, the onset was abrupt, which is not usual. Antioxidant therapy, such as the use coenzyme Q, is supported by its efficacy in previous cases and also because mitochondrial swelling has been reported in patients with such lesions. The authors stressed that there was a favorable outcome with prolonged and supportive care and the use of coenzyme Q, especially in a patient who developed stretching spasms, a feature that has been associated with a very high mortality rate.

Because cases occur in clusters, even though many others who inhale heroin do not get this effect, there is suspicion about the possible role of contaminants of small batches of drug by an unknown substance. In addition, some have suggested that heating could be important, since leukoencephalopathy has not been reported with other means of heroin use (until this year, as reported in the next case). The authors of the report postulated that there might be a relation between the amount of heroin inhaled and the severity of the illness. Once symptoms develop, progression continues, usually for 2–3 weeks, but in some individuals it progresses for up to 6 months after exposure. "Coasting," or the phenomenon of symptom progression after cessation of exposure to toxins, has been observed with other toxins, and it is proposed to result from the storage of toxin in lipid-rich neural or non-neural body tissues, with subsequent release into the bloodstream. Although the "toxin" involved in this condition is unknown, progression could be due to "coasting." Alternatively, oxidative damage could be initiated by the toxin and produce persistent metabolic changes in the affected white matter. Whatever the mechanism, the illness is extremely grave, with no known treatment and with progression to akinetic mutism and death in about 20% of reported cases.

A cluster of five cases of toxic leukoencephalopathy over a span of 5 weeks after inhalation of heroin vapor has been reported, with details of three (24).

- A 27-year-old man was found at home in a low state of consciousness after smoking heroin. He developed aphasia and spastic quadriparesis. A CT scan showed symmetrical white matter hypodensity in the cerebellum and symmetrical white matter hypoattenuation in the posterior limb of the internal capsule and optic radiations. An MRI scan showed white matter hypointensity in the cerebellum, with sparing of the cortex and dentate nuclei. There were also hyperintensities in the medial lemnisci and spinothalamic tracts. He died after a seizure 6 weeks later. An autopsy showed spongiform degeneration of the white matter that coincided with the MRI scans.
- A 39-year-old man developed bradykinesia and ataxia after smoking heroin. A brain MRI scan showed symmetrically increased signals in the white matter of the cerebellum, the peduncles, and the pons, with sparing of the dentate nuclei. There were abnormal signals in the posterior limb of the internal capsule and optic radiations. The signal abnormalities in this patient

were not as pronounced as the previous one, who ultimately died. This patient was also taking methadone in a maintenance program.
- A 32-year-old man in a methadone program developed dysarthria, bradykinesia, and ataxia after smoking heroin. CT scans showed symmetric hypodensities in the cerebellum with sparing of the dentate nuclei. There were abnormal signals in the internal capsule and optic radiations., although not as extensive as in those who died.

These cases were typical of toxic leukoencephalopathy secondary to heroin inhalation. Moreover, they had symptoms involving both cerebellum and non-cerebellar structures.

In two reports from China 10 cases (nine men and one woman) of heroin-induced spongiform leukoencephalopathy have been reviewed. The first report discussed six cases with cerebellar signs and symptoms, with symmetrical lesions on neuroimaging in the white matter of the cerebellum, basal ganglia, posterior crus of the internal capsule, and the semi-oval center (25). The second report, described four cases that occurred during the abstinence period and showed improvement after 4 weeks of comprehensive treatment (26).

Two cases of possible toxic leukoencephalopathy following probable inhalation of heroin vapor have been reported (27).

- A 55-year-old man developed confusion, behavioral change, aggression, poor attention, disorientation in time, and impaired short-term memory. He had full ocular movements with no nystagmus, brisk deep tendon reflexes, and bilateral extensor plantar responses. He became progressively drowsy with myoclonic jerks and died 2 weeks later.
- A 36-year-old man with a history of substance abuse became unresponsive, with his eyes in mid-position gaze, with pinpoint pupils, brisk deep tendon reflexes, and bilateral extensor plantar responses. On day 9 he spontaneously opened his eyes. However, he died 1 month later with persistent pyrexia from methicillin resistant Staphylococcus aureus.

Neuroimaging and neuropathology in both cases showed diffuse symmetrical degeneration of white matter, with sparing of subcortical U fibers, cerebellum, and brain stem. Toxicology was negative, but was done some time after the report of substance use. In these case reports heroin use could not be confirmed. Although the findings suggested the possibility of heroin toxicity due to inhalation, sparing of the cerebellum and brain stem, frontal predominance of degeneration, and the more prominent axonal involvement are not typical of heroin toxicity, throwing speculation on an unidentified impurity.

- A 37-year-old male cocaine abuser was admitted with intoxication, mutism, and substupor (28). His toxicological screening was positive for heroin and cocaine. There was spasticity of all limbs and Babinski reflexes. The CSF contained some erythrocytes. The electroencephalogram showed generalized slowing, and a CT scan showed bifrontal confluent hypodensities in the deep white matter. The cranial MRI scan showed diffuse bihemispheric white matter lesions dominantly in the frontal lobe on T2-weighted images. There were abnormal hyperintense lesions in the pyramidal tracts and the corpus callosum. He gradually improved and made a complete recovery within 6 months, as confirmed by neurological and neuropsychological examination.

The findings of toxic leukoencephalopathy in this patient's brain-imaging studies were similar to those reported in patients who have inhaled impure heroin. However, he had used intravenous heroin and cocaine. This is therefore the first case report of leukoencephalopathy after intravenous use of these drugs. However, it should be noted that the authors did not indicate how the route of drug use was confirmed. They noted that lipophilic substances, such as hexachlorophene or triethylthin, were likely impurities in the abused substances.

- A 53-year-old man with a 7-year history of heroin abuse presented with confused speech and unsteady gait (29). A CT scan showed low attenuation in the white matter tracts and an MRI scan showed increased signal intensity in the white matter tracts extending from the centrum semiovale, corpus callosum, corona radiata, posterior limbs of the internal capsules, cerebral and cerebellar peduncles, and pyramidal tracts, suggestive of spongiform demyelination. He became bed-bound and tetraplegic and died of a chest infection.

- A 37-year-old man, with a short history of heroin and cocaine use, presented with spasticity of all limbs, a confusional state, and mutism (28). The electroencephalogram, CT scan, and MRI scan showed predominantly positive frontal pathology, with other lesions in the pyramidal tracts and corpus callosum. The diagnosis was leukoencephalopathy. The symptoms gradually abated after 4 weeks, and repeat tests after 6 months shown to be normal.

Three cases of toxic and progressive spongiform leukoencephalopathy have also been reported as a result of vapor inhalation of heroin (21). There were generalized white matter abnormalities and pathology in the cerebellum, internal capsule, corpus callosum, and brain stem.

Heroin-induced leuckoencephalopathy has been misdiagnosed as psychiatric illness (30).

- A 47-year-old woman, with a history of amphetamine abuse, depression, and paranoia, smoked heroin for 4 weeks after stopping amphetamines, and 10 days later became drowsy and confused with increased paranoia and depression. She was disoriented and restless. Her speech was garbled. She had frequent non-purposeful movements and an unsteady gait. A CT brain scan was normal. She was given chlorpromazine, doxepin, and diazepam, but her ataxia and incontinence worsened, her speech and all her movements slowed, with increased tone in all limbs and cogwheel rigidity. Her power and sensation were normal. Truncal ataxia impaired walking. An MRI brain scan showed diffuse high-intensity signals in both cerebral hemispheres, and review of the CT scan showed hypodensities in the same regions. She was treated with co-enzyme Q and regained mobility and continence, but with no improvement in cognitive impairment.

Progressive myelopathy
Another form of brain injury, progressive myelopathy, has been reported after inhalation of heroin vapor (31).

- A 46-year-old man chronically abused heroin through inhalation and developed a gait disorder with paresthesia of the legs, incontinence, and impotence. MRI scans showed bilateral subcortical lesions and bilateral signal abnormalities in the corticospinal tract and posterior columns. Motor evoked potentials were slow, with prolonged F wave latency, which is evidence of peripheral nerve disease. Multivitamins and high doses of prednisone did not produce benefit.

This condition was diagnosed as progressive myelopathy affecting only the corticospinal tract and posterior columns, the characteristics of which differ from acute leukoencephalopathy significantly, although both are serious consequences of heroin inhalation. Progressive myelopathy has been reported after heroin insufflation but never after inhalation of vapors. The authors favored an immune mechanism.

Cerebellar lesions
Symmetrical deep cerebellar lesions have been associated with heroin smoking (32).

- A 40-year-old man presented to the emergency room after smoking heroin on 2 consecutive days. He complained of unsteadiness and clumsiness. While speaking to a nurse, his condition dramatically worsened, and he became severely ataxic, dysarthric, and clumsy and was unable to stand. An MRI scan showed symmetric bilateral "C-shaped" lesions in the white matter and dentate nuclei of the cerebellum. The diagnosis was toxic leukoencephalopathy secondary to inhaled heroin. Six months later, his cerebellar dysfunction was unchanged.

Because of clustering of cases, it was thought that an additive in the heroin had brought on the symptoms.

Parkinsonism
Reversible Parkinsonism from heroin vapor inhalation has been reported (33).

- A 38-year-old Caucasian woman, who had taken heroin vapor 2 weeks before, developed confusion, ataxia, and urinary incontinence. She was afebrile and spoke with a slow low-volume monotonous voice. She had marked truncal ataxia with absent postural reflexes, cogwheel rigidity in her limbs, and marked bradykinesia. She had used non-prescription amfetamine up to 4 g/day intermittently for 8 years. After a bout of depression, she came under the care of a psychiatrist and stopped taking amfetamine after she was given chlorpromazine and diazepam. She then started using heroin. In the CSF, protein was raised with normal glucose and lactate; homovanillic acid, a dopamine metabolite, was markedly reduced and the serotonin metabolite 5-hydroxyindoleacetic acid (5-HIAA) was mildly reduced. Tetrahydrobiopterin was just detectable. She was given co-enzyme Q 30 mg tds and became more alert and began to move and talk. She had recovered fully 1 month later.

According to the authors tetrahydrobiopterin has a range of co-factor roles, including being required for the activity of tyrosine and tryptophan hydroxylase, enzymes that are essential for dopamine and serotonin synthesis. They speculated that something present in the heroin pyrolysate, a product formed when heroin is heated to 250°C, inhaled by the patient, acted as a reversible inhibitor of tetrahydrobiopterin metabolism, providing a biochemical explanation for impairment of dopamine metabolism and Parkinsonism in this case.

Sensory systems

Profound reversible deafness with vestibular dysfunction has been attributed to heroin abuse (34).

● A 47-year-old intravenous opiate user, after a period of abstinence, injected about 0.25 g of illicit diamorphine during a period of 24 hours and developed bilateral symmetrical sensorineural hearing loss, ear fullness, and loud tinnitus 20 minutes later. His symptoms gradually subsided with no sequelae after 3 weeks.

The authors pointed out that bilateral deafness after heroin relapse after prolonged abstinence had been reported in previous two cases, suggesting resensitization of a tolerized opioid system or prolonged hypersensitization of a system in withdrawal.

Seven cases of acute strabismus related to opiate abuse in Switzerland between 1993 and 2001 have been reported (35). In five cases the symptoms coincided with heroin withdrawal and acute esotropia occurred a few days after heroin was stopped. The other two patients developed acute exotropia that was related to opiate abuse. All the symptoms disappeared spontaneously. It is likely that changes in the blood opioid concentration disrupted the oculomotor system and affected binocular vision, although the exact mechanism underlying this phenomenon is not known.

Psychiatric

A group of 210 young individuals, aged 18–24 years, participating in the Australian Treatment Outcome Study, was studied for longitudinal treatment outcomes of heroin dependency (36). There was a high rate of psychiatric co-morbidity, including post-traumatic stress disorder (37%), current major depression (23%), antisocial personality disorder (75%), and borderline personality disorder (51%). While 17% had attempted suicide in the preceding year, 41% had overdosed at some time, 24% in the previous year. First heroin use occurred at age 16.8, first monthly use at age 17.2, and first intravenous use at age 17.4 years. The authors concluded that young heroin users moved to problematic drug use more quickly than did older users. Thus, there is a very limited window during which early intervention can be applied before young heroin users progress to problematic use.

Endocrine

The syndrome of inappropriate secretion of antidiuretic hormone (SIADH) has been attributed to heroin (23).

● A 23-year-old pregnant woman developed antepartum bleeding at 35 weeks and a tonic-clonic convulsion and hypothermia at 39 weeks, having used heroin 4 hours before. She had further tonic-clonic seizures, became obtunded, and required intubation. She had occasional runs of ventricular bigeminy. A cesarean section was performed. The neonate had poor respiratory effort and required ventilation. Blood chemistry suggested inappropriate secretion of antidiuretic hormone, acute renal insufficiency, and acute pancreatitis. She and the baby recovered after 2 weeks.

Nutrition

In a case-control study in 106 heroin-dependent individuals undergoing an opioid detoxification program (n = 19) or a methadone maintenance treatment program (n = 87) there were large significant differences in the mean values of some vitamins and minerals between the heroin-dependent individuals and the healthy, non-dependent controls (37). Dependent individuals had higher white cell counts and transaminases and lower erythrocyte counts and cholesterol, albumin, tocopherol, folic acid, sodium, selenium, and copper concentrations.

Gastrointestinal

In a small randomized study, 40 women undergoing elective cesarean section received either diamorphine 300 micrograms or 0.9% saline as part of a standard spinal anesthesia (38). Intrathecal diamorphine may contribute to the delay in gastric emptying that occurs immediately after elective spinal cesarean section. This is relevant within the context of other possible compounding variables that might delay the reintroduction of a solid diet postoperatively.

Urinary tract

Following an observation that many patients develop acute renal insufficiency after using heroin, the authors identified 27 patients (mostly men, average age 29 years) who developed renal insufficiency after intravenous heroin use (39). Rhabdomyolysis was the likely cause of renal insufficiency in all cases. Twelve had a history of polydrug abuse and all had a history of intravenous diamorphine use in the 24 hours before presentation. Eight patients required renal dialysis for an average of 14 days. Patients who required dialysis had a higher admission creatine kinase, a higher peak creatine kinase, and a lower urine output in the initial 24 hours. They also had a longer hospital stay. Some had positive tests for hepatitis B (10%), hepatitis C (74%) and HIV (5%); viral infections can compound rhabdomyolysis and subsequent renal impairment through glomerulonephritis. No patient died and all patients recovered normal renal function. Rhabdomyolysis is a recognized cause of renal insufficiency, but its pathogenesis after heroin use is not fully understood.

In most of 19 renal specimens from autopsies of intravenous diamorphine users there was severe lymphomonocytic glomerulonephritis as a result of activation of the classical pathway of the complement binding system (40). This could have been a result of diamorphine itself, adulterants, or active hepatitis B and/or C infection.

Heroin was presumed to be the cause of reversible nephrotic syndrome in patients dependent on heroin (41).

Renal amyloidosis can be a late effect (41).

Skin

A traumatic skin lesion with blisters and sweat gland necrosis was described in a 24-year-old man who was comatose as a result of heroin overdose; immunofluorescence showed deposits of immunoglobulin and C3 in dermal vessels (42).

Musculoskeletal

Focal myopathy has been reported after intramuscular diamorphine (43).

- A 36-year-old man developed progressive, painless stiffness of both knee joints over 3 months. It had started 4 weeks after he began to give himself heroin injections two to three times a day in alternate thigh muscles. He had a broad-based stiff gait, and he walked without bending his knees. Because of contractures of the quadriceps muscles, which were indurated, active and passive knee flexion was limited to an angle of 5–10 degrees. Electromyography of the right quadriceps muscle showed firm fibrous resistance to needling without insertional activity. Ultrasound showed a preserved but enlarged muscle structure and thickening of the connective tissue. A muscle biopsy showed variation in fiber size with scattered collection of atrophic fibers and perivascular and endomysial infiltrates comprised chiefly of lymphocytes and macrophages. The serum creatine kinase activity was normal. After 7 weeks of physiotherapy, MRI of the thighs showed severe fibrosis of the muscle, suggesting a possible inflammatory component. Following treatment with prednisone and D-penicillamine, he was entirely normal, except for slightly limited knee flexion on both sides.

This patient's main symptom was progressive stiffness, due to contractures of the quadriceps muscles induced by chronic heroin injections. The findings made it very likely that heroin caused a primarily vascular lesion leading to non-specific inflammatory changes and subsequent fibrosis. Clinically, weakness was minimal and there was painless contracture. This presumably reflects the predominantly fibrotic process within muscle tissue. Combination therapy with prednisone and D-penicillamine led to significant improvement. The regenerating

process was confirmed by the second muscle biopsy, and electromyography showed reinnervation. The second biopsy did not show inflammatory cells, indicating absence of the inflammatory component. Thus, this case suggests that heroin-induced fibrotic myopathy is reversible.

Infection risk

Injection of heroin can cause local infection.

- A 32-year-old man with pyomyositis developed abdominal pain and vomiting, fresh rectal bleeding, hematuria, and a swelling on his lower back (44). Two weeks before, he had accidentally given himself an extravascular injection of heroin into his left groin. A CT scan showed a large left sided gluteal abscess communicating through the left sacroiliac joint with the retroperitoneal space. He needed catheterization and multiple open drainage of the abscess and was discharged after 2 months.

Necrotizing fasciitis, a rapidly progressive soft tissue infection with high morbidity and mortality, is most commonly associated with drug abuse b y injection. On the West Coast of the USA it is associated with the use of black tar heroin, which is a dense, gummy, coal-colored substance that is produced from opium grown in northern Mexico. Crude processing leads to contamination. In a retrospective review of cases of necrotizing fasciitis occurring in the year 2000 there were 20 cases associated with black tar heroin; 17 were men (45). The mean age was 44 years and there were nine African Americans, six Caucasians, and five Hispanics. Eleven had involvement of the torso. The hospital mortality was 50%. Four patients had fever, 16 had tachycardia, four had blisters/bullae on their skin, and eight had brawny/woody skin. The mean white cell count was $37 \times 10^9/l$, but survivors had lower counts than non-survivors ($25 \times 10^9/l$ versus $49 \times 10^9/l$). Blood cultures were positive in seven and surgical cultures in 15. The white cell count was the most consistent abnormal finding, and the degree of increase was associated with mortality. Hyponatremia was also common (47%). Lactate concentrations were raised but did not predict survival or mortality. Soft tissue gas on X-ray was present in one-third of the subjects. There was no common or predominant responsible organism, arguing against a single contaminated batch of black tar heroin.

Death

Heroin contributes significantly towards mortality worldwide and heroin users are at significantly greater risk of premature death than their non-heroin-using peers (46). Men are typically over-represented, up to 80% in some instances. The mean age at the time of death is late twenties or early thirties and it typically occurs in those who are regular users rather than novices. Death usually occurs 1–3 hours after heroin use. It is not instantaneous and tends to occur in the company of other people. It is also likely to involve the use of other nervous system depressants taken in conjunction with heroin. Medical help is often sought too late; in only 14% of cases was an ambulance called as a first action. Multiple drug use is common in heroin-related deaths, with alcohol or benzodiazepine as significant co-abused substances. Blood morphine concentrations are generally not higher than in controls. Intravenous use accounts for most fatal overdoses. The validity of the term "heroin overdose" for these deaths has been questioned, as there is frequent co-abuse of another drug and post-mortem morphine concentrations are not in excess of individual tolerance in most instances. Snorting or smoking heroin still carries a considerable risk of a lethal outcome, owing to a combination of polydrug abuse, variability in morphine blood concentration, and reduced tolerance because of periods of reduced or sporadic use. Clinicians should take into account the complexity of multiple drug dependence during assessments and in designing treatment. There should also be raised awareness among drug users of the different potential risk of overdose and subsequent drug-related death (47).

Doctor shopping is a common form of drug-seeking behavior in which there is a fraudulent presentation of disease to multiple doctors and pharmacies in order to procure prescription drugs. The authors of a study from Australia explored drug-seeking behavior patterns among young people who subsequently died of heroin overdose (48). They examined 202 consecutive Coroner's Court cases (149 men and 53 women) from 1994 to 1999, of all heroin-related deaths in individuals aged under 25 years, excluding probable suicides. Their mean age was 21 years (range 14–24). Heroin or its derivatives were detected in all but two subjects. Other drugs were present in 90% of cases. Prescription and pharmacy drugs were present in 80% of the cases, benzodiazepines being most common and especially in women. Benzodiazepines, opiates, and psychotropic drugs accounted for 43%, 11%, and 8% of all prescriptions respectively. They reviewed the use of prescriptions from two different insurance groups. In the Pharmaceutical Benefits Scheme, the average number of prescriptions per person increased in the 4 years before death from four per person per year in the fourth year before death to a peak of 17 in the year before death. Although all prescriptions increased, the rates for benzodiazepines and other opioids increased more than prescriptions for other drugs. In the Medicare Beneficiary Scheme, 22% saw an average of more than 15 medical practitioners per year and 40% averaged more than 30 services per year. Women were disproportionately represented.

In this study there were high levels of polydrug and prescription drug use in heroin-related deaths. The authors suggested that their data provided circumstantial evidence of drug-seeking behavior before death, including increasing visits to multiple doctors and disproportionate increases in prescriptions for drugs likely to be misused. They reiterated that among younger heroin users, such drug-seeking behavior, besides indicating an increased risk of fatal overdose, may also provide an opportunity to intervene. "Doctor shoppers" are defined as those individuals who have seen more than 15 different GPs and have had 30 or more consultations in 1 year.

However, in this study only 22% saw more than 15 GPs and 40% averaged more than 30 medical services, suggesting that most did not meet the official doctor shopping criteria, even when they increased their use of medical services. The authors suggested that an increase in doctor shopping in the years before heroin-related death while certainly a fiscal issue, may represent an opportunity to intervene and reduce mortality.

In a review of 239 cases of heroin-related drug deaths between 1997 and 2000, 18 deaths were associated with non-intravenous administration (49). The median morphine concentration in these non-injectors was 0.05 mg/g and this was significantly lower than in injectors (2.3 mg/g). There was concurrent use of alcohol, other illicit drugs, and/or pharmaceutical formulations in 17 of the 18 cases.

A review of 139 methadone-related deaths between 1998 and 2002 in Palm Beach County supported those of previous investigations and suggested that it is not possible to establish a definitive lethal methadone blood concentration range. Methadone-related death is usually associated with the use of other drugs and toxicological analysis in such cases should be contextualized by the clinical circumstances surrounding the event and even a few months before the incident occurred (50).

The role of methadone and opiates in accidental overdose deaths in New York City has been investigated using data from the Office of Chief Medical Examiner of all accidental drug overdose deaths between 1990 and 1998 (51). There were 7451 overdose deaths in all during this period, of which 1024 were methadone-induced, 4627 were heroin-induced, and 408 were attributable to both. Thus, 70% of the deaths from accidental overdose were due to opiates. Co-variates significantly associated with methadone-induced deaths were female sex, older age, and absence of cocaine, heroin, cannabis, and alcohol in toxic screens. Co-variates associated with heroin overdose were male sex, Caucasian or Hispanic ethnicity, younger age, and the absence of cocaine and methadone and the presence of cannabis and alcohol in toxic screens.

The proportion of accidental deaths in New York City from methadone overdose of 13–16% of total overdose mortality did not change appreciably, although the proportion of overdose deaths attributed to heroin increased significantly (from 54 to 64%) during the study. Deaths from accidental overdose were 3–6 times more common among heroin users than methadone users during the same time. This absence of an increase in deaths from methadone use is noteworthy, because the number of people taking methadone during this period rose from 25 795 to 33 666 (a 31% increase). Thus, both heroin-induced overdose mortality and prescriptions of methadone increased during the same interval. Methadone-induced death risk was highest for those aged 35–44. The risk of heroin-induced overdose death fell with age. The lower likelihood of methadone-induced overdose death in those who had positive toxicology for cocaine, heroin, or alcohol suggests that individuals with methadone-induced deaths were less likely to be using other drugs concurrently. This also suggests that methadone deaths occurred in those who were using medicinal methadone rather than from the street. There was no association of these accidental deaths with the weekend use of these drugs.

Long-Term Effects

Drug abuse

Smoking heroin by heating the free base over tin foil and inhaling the vapors is known as "chasing the dragon," a method that probably originated in Southeast Asia. Some are using this to reverse the stimulant effects of ecstasy. In 102 patients (55 men and 47 women, mean age 21 years) interviewed at four clinics in Dublin, Ireland, three subgroups of opiate users were identified: (a) those who had ever used opiates to come off ecstasy, who were compared with those who had never used opiates for this purpose, (b) those whose first use of opiates had been to come off ecstasy, and (c) those who had started opiates by "chasing" and then did or did not move to injecting (52). Of the 102 patients, 92 reported having taken ecstasy, 68 of whom reported having taken opiates to come off it, and the remaining 24 of whom had not. The 68 patients who reported taking opiates to come off ecstasy had significantly heavier ecstasy use, in terms of the number of nights per week and number of tablets taken per night. Of 36 who reported that their first ever experience of using opiates was in the context of "chasing" to come off ecstasy, 28 reported this as their main reason for starting to use opiates and the other eight reported that they would probably have tried opiates independent of their ecstasy use. Of the 86 patients whose initial route of using heroin was "chasing," 61 reported changing to injecting, 23 continued to smoke heroin, and two switched to an oral formulation of methadone or morphine. When those who came to inject heroin were compared with those who did not (61 versus 23), the injectors had begun illicit drug use earlier, had started heroin at a younger age, were younger at the time of interview, and had been more likely to have a history of ecstasy use. Despite the younger age of onset of illicit drug use in those who came to inject, they had not been using illicit drugs for longer at the time of interview. This study confirmed the authors' previous findings that heroin smoking was associated with ecstasy use.

The Swiss government has developed a program called PROVE, which provides prescriptions of injectable opioids for the treatment of heroin-dependent patients. This program has generally been viewed to be successful in terms of retention in treatment, morbidity and mortality, legal behavior, and cost-effectiveness. However, during the 26-month observation period, epileptic seizures occurred in 11% of the 186 patients treated. This finding, along with previous reports of reduced regional blood supply to the brains of opioid users, led the authors to study cerebral deoxygenation after intravenous opioid administration in ten opioid-dependent subjects and to compare it with intravenous saline in ten matched controls using Near Infrared Spectroscopy (NIRS) (53).

Heroin and methadone produced a rapid and dramatic reduction in both respiratory rate and cortical hemoglobin oxygenation, while saline had no effects. The authors suggested that opioid-induced acute deoxygenation of cortical hemoglobin was probably associated with respiratory depression. In one in three subjects, oxygen saturation after intravenous heroin fell rapidly, a finding that has not previously been described in humans.

The authors suggested two possible mechanisms for this phenomenon. They discounted the possibility that opioids increase the utilization of oxygenated hemoglobin in the CNS, because PET data from other studies suggest decreased utilization of glucose (indicating reduced brain activity) following opioids. They believed that it was more likely that the increase in cortical-reduced hemoglobin was related to opioid-induced respiratory depression, with carbon dioxide retention and resulting vasodilatation. Although preliminary, these data have potential implications for treatment programs involving intravenous opioid maintenance. The authors suggested that intravenous opioids may produce both systemic and cerebral hypoxia, which may at least in part account for the hyperexcitability (as measured by electroencephalography) found after intravenous opioids in other studies. Furthermore, hypoxia may mediate or contribute to the rush sensation, similar to that seen with high altitude or in cases of asphyxiophilia.

Drug withdrawal

The relation between the severity of opiate withdrawal and the dose, duration, and route of administration of heroin has been assessed in a retrospective analysis of heroin withdrawal in 22 patients (54). Abrupt withdrawal from opiates resulted in increased symptom severity, peaking on day 2 and then abating after that until day 7. Both the dose and the route of administration were related to the withdrawal score: intravenous heroin was linked to greater total and maximum withdrawal severity than smoking heroin. The authors speculated that the effect of the route of administration may have been due to lower systemic availability of smoked compared with injected heroin. Their data suggested that even the duration of the withdrawal symptoms seemed to increase with higher doses and intravenous use. However, there were several limitations to this study. It was retrospective and the period of observation lasted only 7 days although the withdrawal symptoms lasted much longer. In addition, many subjects took doxepin and benzodiazepines during the observation phase, which may have reduced or suppressed their withdrawal symptoms.

Opioid withdrawal symptoms should be included as part of the differential diagnosis in young people with restless legs syndrome.

- Restless legs syndrome was a feature of opioid withdrawal on days 3–4 in two heroin-dependent individuals (55). They were treated with levodopa and clonidine.

Second-Generation Effects

Fetotoxicity

A report of death associated with intravenous heroin use has provided insights about the distribution of heroin and its metabolites in the fetus (56).

- A 17-year-old girl with a history of heavy drug abuse for 2 years was found dead in a public restroom, with fresh needle puncture marks. She was 18–20 weeks pregnant with a male fetus, and had massive brain and lung edema from acute intoxication. Analysis of her hair showed that she had used heroin over the previous few months. Drug screening of body fluids showed only opiates (high concentrations of 6-mono-acetyl-morphine and morphine) in the maternal and fetal circulation at the time of death. Unexpectedly high amounts of morphine, 6-monoacetyl-morphine, and morphine-3-glucuronide were also found in the amniotic fluid. Only morphine-3-glucuronide was detected in the fetus, whereas both morphine-3-glucuronide and morphine-6-glucuronide were detected in the mother, in the body fluids, and in all investigated organ tissues except the brain.

The authors noted that heroin is considered a "pro-drug," with 6-monoacetyl-morphine, morphine, and morphine-6-glucuronide accounting for most of its narcotic activity. In blood, diamorphine is rapidly converted to the active metabolite 6-monoacetyl-morphine, which is presumably converted to morphine in the liver. The majority of the morphine is converted to morphine-3-glucuronide and small amounts are converted to morphine-6-glucuronide, which is pharmacologically active. They concluded that morphine-3-glucuronide can cross the placenta and that high concentrations of heroin and its metabolites can be found in fetal compartments during heroin abuse by the mother. Lastly, heroin and its pharmacologically active metabolites appear to be present in the fetal central nervous system for much longer than in the maternal circulation, because of low fetal drug-metabolizing capacity as well as minimal drug elimination from the amniotic fluid in advancing pregnancy.

In nine prospective, longitudinal, multicenter studies, 1227 infants who were exposed in utero to cocaine (n = 474), opiates (n = 50), cocaine + opiates (n = 48), or neither (n = 655) were followed for 1–3 years after birth. Prenatal exposure to cocaine and/or opiates was not associated with mental, motor, or behavioral defects after controlling for birth weight and environmental risks. This result should be treated with caution, since the effects of prenatal opiates or cocaine exposure may become more evident as more advanced motor, cognitive, language, and behavioral skills develop (57).

Drug Administration

Drug contamination

Atypical reactions after the use of heroin have been attributed to contamination with clenbuterol (58).

- Four patients, aged 21–43 years, all developed chest pain, palpitation, and shortness of breath after inhaling heroin. They had tachycardias, low blood pressure, hypokalemia, and hyperglycemia.

In all the cases the heroin had been adulterated with veterinary clenbuterol.

Drug dosage regimens

High-dose intrathecal diamorphine for analgesia after elective cesarean section has been studied in 40 women who were randomized to diamorphine 0.5 or 1 mg (59). All also received diclofenac 100 mg at the end of the cesarean section and morphine via a patient-controlled analgesia system. Postoperative analgesia was more prolonged and reliable in those who were given diamorphine 1 mg, who needed significantly less morphine. There was postoperative nausea in just under half of the patients in each group, and most of the patients (93%) had mild to moderate pruritus. There were no cases of excessive sedation or oxygen desaturation.

Drug administration route

Chasing the dragon

The term "chasing the dragon" originated in East Asia, where typically heroin powder is placed on aluminum foil and heated from beneath with a lighter or matches, causing heroin to liquefy into a reddish brown blob (heroin pyrolysate), which moves around on the foil and emits a white vapor (SEDA-24, 40). The blob or "dragon" is "chased" with the lighter and the vapor is sucked through a straw or pipe. This method of administration avoids the negative consequences of intravenous use. It is also known as "Chinese blowing". This became a popular method of heroin administration when the drug was cheap but impure.

Intranasal

The pharmacokinetic profile of intranasal diamorphine has been studied in adults. Diamorphine is rapidly absorbed as a dry powder and has similar pharmacokinetic properties to intramuscular diamorphine, with similar physiological responses (for example reduced pupil diameter, respiration, and temperature) and behavioral measures (for example euphoria, sedation, and dysphoria).

Intranasal diamorphine has been evaluated in a multi-center, randomized, controlled trial as an alternative to intramuscular morphine in 404 patients aged 3–16 years with suspected limb fractures (i.e. in acute pain of moderate to severe intensity) (60). They were randomized to either intramuscular morphine sulfate 0.2 mg/kg (n = 200) or intranasal diamorphine hydrochloride 0.1 mg/kg (n = 204). Intranasal diamorphine was significantly better tolerated: 80% had no obvious discomfort compared with only 9% of those given morphine. There were no serious adverse effects of diamorphine, but the lack of blinding may have introduced bias.

The intranasal route of administration is also not without problems when abused. Some non-fatal complications include neurological, acute myelopathy, oculogyric crisis and generalized dystonia, hypersensitivity reactions, pneumonitis, pemphigus, and pancreatitis (60).

Intranasal diamorphine is as effective as intramuscular morphine and is much better tolerated by children, with no apparent increased risk of adverse effects (61,62). In a multicenter, randomized, controlled study, 404 children aged 3–16 years with a fracture of an arm or leg were given either nasal diamorphine 0.1 mg/kg or intramuscular morphine 0.2 mg/kg. The onset of pain relief was faster with nasal diamorphine, and there were no serious adverse effects. The frequencies of opioid-related mild adverse effects were similar in the two groups.

Epidural

In a randomized, placebo-controlled, double-blind study of the relative efficacies of patient-controlled analgesia (PCA) regimens (63), 60 patients undergoing elective total hip or knee replacement were randomly allocated to receive epidural diamorphine 2.5 mg followed by a PCA bolus 1 mg with a 20-minute lockout (group 1), subcutaneous diamorphine 2.5 mg followed by a PCA bolus 1 mg with a 10-minute lockout (group 2), or epidural diamorphine 2.5 mg in 4 ml of 0.125% bupivacaine followed by a PCA bolus of 1 mg diamorphine in 4 ml 0.125% bupivacaine with a 20-minute lockout (group 3). Diamorphine demands were significantly higher in group 2 in the first postoperative 24-hour period, but pain scores were only significantly higher in group 2 in the first 3 hours postoperatively compared with group 3 and group 1. There were also fewer opioid-related adverse effects in group 2, and group 3 reported higher incidences of various adverse effects. The conclusion was that PCA diamorphine given with or without bupivacaine provides analgesia of similar efficacy once adequate pain relief has been achieved. Taking the incidences of adverse effect profiles into account, diamorphine subcutaneous PCA was a simple and effective method of providing analgesia.

Intrathecal

In 62 women who asked for regional analgesia in labor and who were randomized to an intrathecal injection of either bupivacaine 2.5 mg with fentanyl 25 μg or bupivacaine 2.5 mg with diamorphine 250 μg, the diamorphine provided longer analgesia (64). There were significant differences in adverse effects between the groups. There were no instances of nausea or vomiting, but pruritus was more common in those who received fentanyl. The dose of diamorphine was deliberately low, and more studies are needed to confirm these findings.

Drug overdose

The main life-threatening complications of heroin intoxication include acute pulmonary edema and delayed respiratory depression with coma after successful naloxone treatment. In a prospective study of the management of 160 heroin and heroin mixture intoxication cases treated in an emergency room in Switzerland between 1991 and 1992, there were no rehospitalizations after discharge

from the emergency room and there was only one death outside the hospital due to pulmonary edema, which occurred at between 2.25 and 8.25 hours after intoxication (65). A literature review found only two reported cases of delayed pulmonary edema, which occurred 4 and 6 hours after hospitalization. The authors recommended surveillance of a heroin user for at least 8 hours after successful opiate antagonist treatment.

In a retrospective chart review (66) of patients with heroin overdose between July 1996 and July 1999 at a Medical Center, 13 of 125 charts indicated the presence of non-cardiogenic pulmonary edema and all the patients were men. The mean field respiratory rate for individuals with non-cardiogenic pulmonary edema was 5.9 compared to 9.7 in those without. All the patients with non-cardiogenic pulmonary edema and 68% without received naloxone. The mean duration of heroin use was less in patients with non-cardiogenic pulmonary edema (2.9 years) than in patients without (14 years). The results of this chart review are limited, but it seems that in a small percentage of cases, heroin overdose results in non-cardiogenic pulmonary edema. These individuals typically have a respiratory rate close to six and need naloxone. The authors postulated that there may be a higher risk of non-cardiogenic pulmonary edema in men who are relatively inexperienced heroin users.

Among heroin users, the annual rate of mortality is 1–4%; overdose and HIV infection being the leading causes. The effect of the frequency and route of heroin administration on the occurrence of non-fatal heroin overdose has been studied (67). Among 2556 subjects with heroin dependence, 10% had taken overdoses requiring emergency care in the prior 12 months. The cumulative risk of overdose increased as the frequency of heroin use fell. Among daily heroin users, the risk was greater with increased frequency of heroin injection, but not among non-daily users. The risk of overdose was greater with injection than with other routes of administration.

Strategies for preventing heroin overdose have been discussed (68). Heroin overdose is a leading cause of morbidity and mortality among active heroin injectors. Education, family support groups, and motivational interviews after overdose have been proposed as complementary strategies to reduce morbidity and mortality. The author concluded that methadone maintenance is the most effective method of reducing mortality from overdose, and that home treatment with naloxone by a significant other is a possible future strategy. However, the use of naloxone is regarded as controversial by some health professionals.

Drug–Drug Interactions

Alcohol

Many heroin users use heroin and alcohol together. There has been an evaluation of the pharmacokinetic interaction between heroin and alcohol and the role of that interaction in the cause of 39 heroin-related deaths that were attributed to either heroin or heroin + ethanol (69). The cases were arbitrarily divided into two groups according to blood ethanol concentration (low-ethanol group, under 1000 µg/ml, and high ethanol group, over 1000 µg/ml. The high-ethanol group was associated with reduced hydrolysis of 6-acetylmorphine to morphine, and there was an inverse correlation between blood ethanol concentration and hydrolysis of 6-acetylmorphine to morphine. The concentration of total morphine was lower in the high-ethanol group. High blood ethanol concentrations were also associated with an increased ratio of unbound to total morphine and with reduced excretion of unbound and total morphine. The relative concentrations of conjugated heroin metabolites were reduced in the presence of a high blood ethanol concentration. The authors hypothesized that alcohol inhibits the glucuronidation of morphine, resulting in less conjugated morphine in the blood. Thus, in patients with high blood ethanol concentrations the additional depressant effects of unconjugated heroin metabolites may contribute to a more acute death.

Anticholinergic drugs

Combining opiates with anticholinergic drugs is a common practice in recreational abuse (SEDA-21, 34). Heroin mixed with hyoscine (scopolamine) is nicknamed "polo" and "point on point." Mixed drug toxicity, with atypical signs and symptoms of opiate abuse, has been reported (70).

- A 41-year-old woman who had taken 11 alprazolam tablets and heroin mixed with an unknown substance developed slurred speech and a staggering gait. She was also taking paroxetine. Her pupils were dilated, her skin warm and dry. Electrocardiography showed a sinus bradycardia. She was given intravenous naloxone 2.0 mg and became acutely agitated and combative. She was delirious, agitated, and disoriented, and was given an intravenous sedative and intubated. Her urine contained codeine, morphine, and atropine.

This case exemplifies the difficulty in identifying anticholinergic drugs such as atropine, and unfortunately the finding of dilated pupils did not raise the suspicion of mixed drug toxicity. The use of naloxone uncovered florid agitation due to anticholinergic drug toxicity.

Interference with Diagnostic Tests

Blood glucose

Diamorphine flattens the glucose tolerance curve and increases glycosylation of HbA_1 (71).

Antithrombin III

Diamorphine depresses the biological activity of antithrombin III (72).

References

1. Kreek MJ, Bart G, Lilly C, Laforge KS, Nielson DA. Pharmacogenetics and human molecular genetics of opiate and cocaine addictions and their treatments. Pharmacol Rev 2005;57(1):1–26.
2. Haemmig RB, Tschacher W. Effects of high-dose heroin versus morphine in intravenous drug users: a randomised double-blind crossover study. J Psychoactive Drugs 2001;33(2):105–10.
3. Riad T, Williams B, Musson J, Wheatley B. Intrathecal morphine compared with diamorphine for postoperative analgesia following unilateral knee arthroplasty. Acute Pain 2002;4:5–8.
4. Cowan CM, Kendall JB, Barclay PM, Wilkes RG. Comparison of intrathecal fentanyl and diamorphine in addition to bupivacaine for caesarean section under spinal anaesthesia. Br J Anaesth 2002;89(3):452–8.
5. Van den Brink W, Hendricks VM, Blanken P, Koeter WJM, Van Zureten JB, Van Ree JM. Medical prescription of heroin to treatment resistant heroin addicts: two randomized controlled trials. Br Med J 2003;327:310–2.
6. Serpell MG, Anderson E, Wilson D, Dawson N. I.v. regional diamorphine for analgesia after foot surgery Br J Anaesth 2000;84(1):95–6.
7. Buster M, Rook L, van Brussel GH, van Ree J, van den Brink W. Chasing the dragon, related to the impaired lung function among heroin users. Drug Alcohol Depend 2002;68(2):221–8.
8. Krantz AJ, Hershow RC, Prachand N, Hayden DM, Franklin C, Hryhorczuk DO. Heroin insufflation as a trigger for patients with life-threatening asthma. Chest 2003;123:510–7.
9. Levine M, Iliescu ME, Margellos-Anast H, Estarziau M, Ansell DA. The effects of cocaine and heroin use on intubation rates and hospital utilization in patients with acute asthma exacerbations. Chest 2005;128(4):1951–7.
10. Servin FS, Raeder JC, Merle JC, Wattwil M, Hanson AL, Lauwers MH, Aitkenhead A, Marty J, Reite K, Martisson S, Wostyn L. Remifentanil sedation compared with propofol during regional anaesthesia. Acta Anaesthesiol Scand 2002;46(3):309–15.
11. Dettmeyer R, Schmidt P, Musshoff F, Dreisvogt C, Madea B. Pulmonary edema in fatal heroin overdose: immunohistological investigations with IgE, collagen IV and laminin— no increase of defects of alveolar-capillary membranes. Forensic Sci Int 2000;110(2):87–96.
12. Reynes AN, Pujol JA, Baixeras RP, Fernandez B. Edema agudo de pulmon unilateral en paciente con sobredosis do heroina y tratado con naloxona intravenosa. Med Clin (Barc) 1990;94:637.
13. Anderson K. Bronchospasm and intravenous street heroin. Lancet 1986;1(8491):1208.
14. Schoser BG, Groden C. Subacute onset of oculogyric crises and generalized dystonia following intranasal administration of heroin. Addiction 1999;94(3):431–4.
15. Chenery HJ, Murdoch BE. A case of mixed transcortical aphasia following drug overdose. Br J Disord Commun 1986;21(3):381–91.
16. Koussa S, Tamraz J, Nasnas R. Leucoencephalopathy after heroin inhalation. A case with partial regression of MRI lesions. J Neuroradiol 2001;28(4):268–71.
17. McCreary M, Emerman C, Hanna J, Simon J. Acute myelopathy following intranasal insufflation of heroin: a case report. Neurology 2000;55(2):316–7.

18. Lee BF, Chiu NT, Chen WH, Liu GC, Yu HS. Heroin-induced rhabdomyolysis as a cause of reflex sympathetic dystrophy. Clin Nucl Med 2001;26(4):289–92.

19. Vella S, Kreis R, Lovblad KO, Steinlin M. Acute leukoencephalopathy after inhalation of a single dose of heroin. Neuropediatrics 2003;34:100–4.

20. Gacouin A, Lavoue S, Signouret T, Person A, Dinard MD, Shpak N, Thomas R. Reversible spongiform leucoencephalopathy after inhalation of heated heroin. Intensive Care Med 2003;29:1012–5.

21. Kriegstein AR, Shungu DC, Millar WS, Armitage BA, Brust JC, Chillrud S, Goldman J, Lynch T. Leukoencephalopathy and raised brain lactate from heroin vapor inhalation ("chasing the dragon"). Neurology 1999;53(8):1765–73.

22. Wolters EC, van Wijngaarden GK, Stam FC, Rengelink H, Lousberg RJ, Schipper ME, Verbeeten B. Leucoencephalopathy after inhaling "heroin" pyrolysate. Lancet 1982;2(8310):1233–7.

23. Cooley S, Lalchandani S, Keane D. Heroin overdose in pregnancy: an unusual case report. J Obstet Gynaecol 2002;22(2):219–20.

24. Keogh CF, Andrews GT, Spacey SD, Forkheim KE, Graeb DA. Neuroimaging features of heroin inhalation toxicity: "chasing the dragon". Am J Roentgenol 2003;180:847–50.

25. Liao Y, Liu SX, Yi WZ, Zhou J, Tang JX, Wu B. Clinic and image features of heroin-induced spongiform leukoencephalopathy. Chin J Clin Rehab 2004;8(25):5448–9.

26. Xiao XC, Chu XF, Gu KY, Dong JZ, Su XL, Zhou CX, Suo P, Zhao HW, Chen ZW. Pathological changes in heroin-induced spongiform leukoencephalopathy and therapeutic effect of comprehensive rehabilitation: a follow-up study. Chin J Clin Rehab 2004;8(7):1339–41.

27. Ryan A, Molloy FM, Farrell MA, Hutchinson M. Fatal toxic leukoencephalopathy: clinical, radiological, and necropsy findings in two patients. J Neurol Neurosurg Psychiatry 2005;76(7):1014–6.

28. Maschke M, Fehlings T, Kastrup O, Wilhelm HW, Leonhardt G. Toxic leukoencephalopathy after intravenous consumption of heroin and cocaine with unexpected clinical recovery. J Neurol 1999;246(9):850–1.

29. Au-Yeung K, Lai C. Toxic leucoencephalopathy after heroin inhalation. Australas Radiol 2002;46(3):306–8.

30. Sayers GM, Green MC, Shaffer RE. Heroin-induced leucoencephalopathy misdiagnosed as psychiatric illness. Int J Psychiatry Clin Pract 2002;6:53–5.

31. Nyffeler T, Stabba A, Sturzenegger M. Progressive myelopathy with selective involvement of the lateral and posterior columns after inhalation of heroin vapour. J Neurol 2003;250:496–8.

32. Ropper AH, Blair R. Symmetric deep cerebellar lesions after smoking heroin. Arch Neurol 2003;60:1605–6.

33. Heales S, Crawley F, Rudge P. Reversible parkinsonism following heroin pyrolysate inhalation is associated with tetrahydrobiopterin deficiency. Mov Disord 2004;19(10):1248–51.

34. Ishiyama A, Ishiyama G, Baloh RW, Evans CJ. Heroin-induced reversible profound deafness and vestibular dysfunction. Addiction 2001;96(9):1363–4.

35. Sutter FK, Landau K. Heroin and strabismus. Swiss Med Wkly 2003;133:293–4.

36. Mills KL, Teesson M, Darke S, Ross J, Lynskey M. Young people with heroin dependence: findings from the Australian Treatment Outcome Study(ATOS). J Subst Abuse Treat 2004;27(1):67–73.

37. Estévez Diaz-Flores JF, Estévez Diaz-Flores F, Calzadillo Hernández C, Rodrigrez Rodrigrez EM, Romero Diáz C, Serra-Majem L. Application of linear discriminant analysis to the biochemical and haematological differentiation of opiate addicts from healthy subjects: a case-controlled study. Eur J Clin Nutr 2004;58:449–55.

38. King H, Barclay P. The effects of intrathecal diamorphine on gastric emptying after elective Caesarian section. Anaesthesia 2004;59:565–9.

39. Rice EK, Isbel NM, Becker GJ, Atkins RC, McMahon LP. Heroin overdose and myoglobinuric acute renal failure. Clin Nephrol 2000;54(6):449–54.

40. Dettmeyer R, Stojanovski G, Madea B. Pathogenesis of heroin-associated glomerulonephritis. Correlation between the inflammatory activity and renal deposits of immunoglobulin and complement? Forensic Sci Int 2000;113(1–3):227–31.

41. Llach F, Descoeudres C, Massry SG. Heroin associated nephropathy: clinical and histological studies in 19 patients. Clin Nephrol 1979;11(1):7–12.

42. Rocamora A, Matarredona J, Sendagorta E, Ledo A. Sweat gland necrosis in drug-induced coma: a light and direct immunofluorescence study. J Dermatol 1986;13(1):49–53.

43. Weber M, Diener HC, Voit T, Neuen-Jacob E. Focal myopathy induced by chronic heroin injection is reversible. Muscle Nerve 2000;23(2):274–7.

44. Crossley A. Temperature pyomyositis in an injecting drug misuser. A difficult diagnosis in a difficult patient. J Accid Emerg Med 2003;20:299–300.

45. Lonergan S, Rodriguez RM, Schaulis M, Navaran P. A case series of patients with black tar heroin-associated necrotizing fasciitis. J Emerg Med 2004;26(1):47–50.

46. Mirakbari SM. Heroin overdose as cause of death: truth or myth. Aust J Forensic Sci 2004;36:73–8.

47. Man L-H, Best D, Gossop M, Stillwell G, Strang J. Relationship between prescribing and risk of opiate overdose among drug users in and out of maintenance treatment. Eur Addiction Res 2004;10:35–40.

48. Martyres RF, Clode D, Burns JM. Seeking drugs or seeking help? Escalating "doctor shopping" by young heroin users before fatal overdose. Med J Aust 2004;180(5):211–4.

49. Thiblin I, Eksborg S, Petersson A, Fugelstad A, Rajs J. Fatal intoxication as a consequence of intranasal administration (snorting) or pulmonary inhalation (smoking) of heroin. Forensic Sci Int 2004;139:241–7.

50. Wolf BC, Lavezzi WA, Sullivan LM, Flannagan LM. Methadone-related deaths in Palm Beach County. J Forensic Sci 2004;49:375–8.

51. Bryant WK, Galea S, Tracy M, Markham Piper T, Tardiff KJ, Vlahov D. Overdose deaths attributed to methadone and heroin in New York City, 1990–1998. Addiction 2004;99(7):846–54.

52. Gervin M, Hughes R, Bamford L, Smyth BP, Keenan E. Heroin smoking by "chasing the dragon" in young opiate users in Ireland: stability and associations with use to "come down" off "Ecstasy". J Subst Abuse Treat 2001;20(4):297–300.

53. Stohler R, Dursteler KM, Stormer R, Seifritz E, Hug I, Sattler-Mayr J, Muller-Spahn F, Ladewig D, Hock C. Rapid cortical hemoglobin deoxygenation after heroin and methadone injection in humans: a preliminary report. Drug Alcohol Depend 1999;57(1):23–8.

54. Smolka M, Schmidt LG. The influence of heroin dose and route of administration on the severity of the opiate withdrawal syndrome. Addiction 1999;94(8):1191–8.

55. Scherbaum N, Stüper B, Bonnet U, Gastpar M. Transient restless-leg-like syndrome as a complication of opiate withdrawal. Pharmacophysciatry 2003;36:70–2.

56. Potsch L, Skopp G, Emmerich TP, Becker J, Ogbuhui S. Report on intrauterine drug exposure during second trimester of pregnancy in a heroin-associated death. Ther Drug Monit 1999;21(6):593–7.

57. Messenger DS, Baver CR, Das A, Seifer R, Lester BM, Lagasse LL, Wright LL, Shankaran S, Bada HS, Smerglio VL, Langer JC, Beeghly M, Poole WK. The Maternal Lifestyle Study: cognitive, motor and behavioural outcomes of cocaine-exposed and opiate-exposed infants through three years of age. Paediatrics 2004;113:1677–85.

58. Hoffman RS, Nelson LS, Chan GM, Halcomb SE, Bouchard NC, Ginsberg BY, Cone J, Jea-Francois Y, Voit S, Marcus S, Ford M, Sanford C, Michels JE, Richardson WH, Bertous LM, Johnson-Arbor K, Thomas J, Belson M, Patel M, Schier J, Wolkin A, Rubin C, Duprey Z. Atypical reactions associated with heroin use—Five States, January–April 2005. JAMA 2005;294(19):2424–7.

59. Stacey R, Jones R, Kar G, Poon A. High-dose intrathecal diamorphine for analgesia after Caesarean section. Anaesthesia 2001;56(1):54–60.

60. Kendall JM, Latter VS. Intranasal diamorphine as an alternative to intramuscular morphine. Clin Pharmacokinet 2003;42:501–3.

61. Davies M, Crawford I. Towards evidence based emergency medicine: best BETs from the Manchester Royal Infirmary. Nasal diamorphine for acute pain relief in children. Emerg Med J 2001;18(4):271.

62. Kendall JM, Reeves BC, Latter VSNasal Diamorphine Trial Group. Multicentre randomised controlled trial of nasal diamorphine for analgesia in children and teenagers with clinical fractures BMJ 2001;322(7281): 261–5.

63. Gopinathan C, Sockalingham I, Fung MA, Peat S, Hanna MH. A comparative study of patient-controlled epidural diamorphine, subcutaneous diamorphine and an epidural diamorphine/bupivacaine combination for postoperative pain. Eur J Anaesthesiol 2000;17(3):189–96.

64. Vaughan DJ, Ahmad N, Lillywhite NK, Lewis N, Thomas D, Robinson PN. Choice of opioid for initiation of combined spinal epidural analgesia in labour—fentanyl or diamorphine. Br J Anaesth 2001;86(4):567–9.

65. Osterwalder JJ. Patients intoxicated with heroin or heroin mixtures: how long should they be monitored? Eur J Emerg Med 1995;2(2):97–101.

66. Sterrett C, Brownfield J, Korn CS, Hollinger M, Henderson SO. Patterns of presentation in heroin overdose resulting in pulmonary edema. Am J Emerg Med 2003;21:32–4.

67. Brugal MT, Barrio G, De LF, Regidor E, Royuela L, Suelves JM. Factors associated with non-fatal heroin overdose: assessing the effect of frequency and route of heroin administration. Addiction 2002;97(3):319–27.

68. Sporer KA. Strategies for preventing heroin overdose. BMJ 2003;326:442–4.

69. Polettini A, Groppi A, Montagna M. The role of alcohol abuse in the etiology of heroin-related deaths. Evidence for pharmacokinetic interactions between heroin and alcohol. J Anal Toxicol 1999;23(7):570–6.

70. Wang HE. Street drug toxicity resulting from opiates combined with anticholinergics. Prehosp Emerg Care 2002;6(3):351–4.

71. Ceriello A, Giugliano D, Dello Russo P, Sgambato S, D'Onofrio F. Increased glycosylated haemoglobin A_1 in opiate addicts: evidence for a hyperglycaemic effect of morphine. Diabetologia 1982;22(5):379.

72. Ceriello A, Dello Russo P, Curcio F, Tirelli A, Giugliano D. Depressed antithrombin III biological activity in opiate addicts. J Clin Pathol 1984;37(9):1040–2.

DIMETHYLTRYPTAMINE

General Information

Dimethyltryptamine is a psychedelic drug (1,2). It can be smoked, inhaled, injected, or used as an enema. It is inactive by mouth, except when there is inhibition of gut monoamine oxidase (MAO type A). Its effects are similar to those of lysergide (3,4). The derivative 5-methoxy-N,N-dimethyltryptamine has similar actions. Both compounds are found in various plants (5,6).

Indigenous peoples of the Amazon have used dimethyltryptamine for spiritual and medicinal purposes for thousands of years (7). South American powdered snuffs are usually prepared from the seeds of *Anadenanthera peregrina* or the bark of *Virola* trees. Ayahuasca is the name given by the Quechua to any of various psychoactive infusions or decoctions prepared from plants that are found in the Amazon Rainforest. Ayahuasca prepared from *Banisteriopsis caapi* vines and *Psychotria viridis* contains reversible beta-carboline alkaloids (harmine, tetrahydroharmine, and harmaline, from the former), which inhibit monoamine oxidase type A, thus enabling the oral absorption of dimethyltryptamine (from the leaves of the latter) (8); this gives hallucinogenic effects for 3–4 hours (9). In some parts of South America *Diplopterys cabrerana* is used instead of *Psychotria viridis* (10). In the USA dimethyltryptamine is found in common plants such as *Phalaris arundinacea*, *Phalaris tuberosa*, or *Phalaris aquatica* (canary grass) (11).

Animal studies of the median lethal dose of dimethyltryptamine and of several harmala alkaloids suggested that the lethal dose of these substances in humans is probably greater than 20 times the typical ceremonial dose (12).

Observational studies

In 15 healthy men, ayahuasca caused pupillary dilatation from 3.7 to 4.7 mm at 40–240 minutes after ingestion (10). There was a small increase in body temperature, from 37°C to a maximum of 37.3°C at 240 minutes. Heart rate increased from 72/minute at baseline to a maximum of 79/minute by 20 minutes, fell to 65 bpm by 120 minutes, and then gradually returned to baseline by 240 minutes. There was a concomitant increase in blood pressure from 126/83 mmHg to a maximum of 137/92 mmHg at 40 minutes, with return to baseline by 180 minutes. Plasma prolactin, cortisol, and growth hormone all increased within 60–120 minutes.

Organs and Systems

Psychological

The effects of intramuscular N,N-dimethyltryptamine 250 micrograms/kg have been assessed in 38 volunteers, of whom 12 received placebo (13). Altered states of consciousness and somatic changes were measured by questionnaires administered after the experiment. In spite of the low dose compared with other studies there were significant differences from placebo in the following syndromes of altered states of consciousness: visual hallucinations, impairment of memory and attention, changes in body image, depersonalization syndrome, derealization syndrome, euphoric state, anxious–depressive state, and delusion. The somatic adverse effects were mainly subjective respiratory problems, dizziness, and weakness.

The effects of dimethyltryptamine have been studied in 15 healthy men who used ritualistic ayahuasca once every 2 weeks on average and 15 age-matched men who had never used ayahuasca (10). On the Tridimensional Personality Questionnaire, ayahuasca users had greater stoic rigidity versus exploratory excitability, greater regimentation versus disorderliness, and a trend toward greater reflection versus impulsivity. On the Harm Reduction Scale, users had significantly greater confidence versus fear of uncertainty and trends toward greater gregariousness versus shyness and greater optimism versus anticipatory worry. On the WHO–UCLA Auditory Learning Verbal Memory Test, the users performed significantly better than controls on word recall tests.

Long-Term effects

Drug abuse

There is no evidence that ayahuasca use leads to substantial or persistent abuse potential (12).

Drug Administration

Drug overdose

- A 25-year-old white man was found dead the morning after he had consumed herbal extracts containing beta-carbolines and hallucinogenic tryptamines (14). No anatomical cause of death was found at post-mortem examination. N,N-dimethyltryptamine (0.02 mg/l), 5-methoxy-N,N-dimethyltryptamine (1.88 mg/l), tetrahydroharmine (0.38 mg/l), harmaline (0.07 mg/l), and harmine (0.17 mg/l) were found in post-mortem blood.

The coroner ruled that the cause of death was hallucinogenic amine intoxication.

References

1. Rubin DR. Dimethyltryptamine, a do-it-yourself hallucinogenic drug. JAMA 1967;201(2):143.
2. Szára S. DMT at fifty. Neuropsychopharmacol Hung 2007;9(4):201–5.

3. Brimblecombe RW. Psychotomimetic drugs: biochemistry and pharmacology. Adv Drug Res 1973;7:165–206.

4. Barker SA, Monti JA, Christian ST. N,N-dimethyltryptamine: an endogenous hallucinogen. Int Rev Neurobiol 1981;22:83–110.

5. Halpern JH. Hallucinogens and dissociative agents naturally growing in the United States. Pharmacol Ther 2004;102(2):131–8.

6. Callaway JC, Grob CS, McKenna DJ, Nichols DE, Shulgin A, Tupper KW. A demand for clarity regarding a case report on the ingestion of 5-methoxy-N, N-dimethyltryptamine (5-MeO-DMT) in an Ayahuasca preparation J Anal Toxicol. 2006;30(6):406–7.

7. Schultes RE, Hofmann A. *Plants of the Gods. Their Sacred, Healing and Hallucinogenic Powers.* Rochester, VT: Healing Arts Press, 1992.

8. Ott J. Pharmahuasca: human pharmacology of oral DMT plus harmine. J Psychoactive Drugs. 1999;31(2):171–7.

9. Riba J, Rodriguez-Fornells A, Urbano G, Morte A, Antonijoan R, Montero M, Callaway JC, Barbanoj MJ. Subjective effects and tolerability of the South American psychoactive beverage ayahuasca in healthy volunteers. Psychopharmacology (Berl) 2001;154:85–95.

10. McKenna DJ. Clinical investigations of the therapeutic potential of ayahuasca: rationale and regulatory challenges. Pharmacol Ther 2004;102(2):111–29.

11. US Department of Agraiculture, Natural Resources Conservation Service. The Plants Database. Baton Rouge, LA: National Plant Data Center, 2004.

12. Gable RS. Risk assessment of ritual use of oral dimethyltryptamine (DMT) and harmala alkaloids. Addiction 2007;102(1):24–34.

13. Bickel P, Dittrich A, Schoepf J. [Altered states of consciousness induced by N,N-dimethyltryptamine (DMT).] Pharmakopsychiatr Neuropsychopharmakol 1976;9(5):220–5.

14. Sklerov J, Levine B, Moore KA, King T, Fowler D. A fatal intoxication following the ingestion of 5-methoxy-N,N-dimethyltryptamine in an ayahuasca preparation. J Anal Toxicol 2005;29(8):838–41.

GAMMAHYDROXYBUTYRATE

General Information

Sodium gammahydroxybutyrate (GHB or sodium oxybate) is an endogenous compound, a precursor of GABA, which increases the release of dopamine and acetylcholine in the brain. It was first synthesized in 1960 as a potential anesthetic, and was popular in the late 1980s and early 1990s as a dietary supplement (as a replacement for L-tryptophan after it had been recalled from the market), an aid to sleep, and a bodybuilding agent. It was subsequently used in the treatment of narcolepsy. However, it has also been used as a party drug, since it causes alcohol-like effects and aroused sexuality. In large doses it can cause disorientation, nausea and vomiting, and muscle spasms. It is also known as BDO, Blue Nitro, Enliven, GBH, Liquid ecstasy, Midnight Blue, RenewTrient, Reviarent, Serenity, and SomatoPro. An analogue, gammavalerolactone (GVL), is a precursor and has also been used recreationally, but it is very expensive.

In 1990 the FDA limited the availability of gammahydroxybutyrate, and in March 2000, the Drug Enforcement Agency made it a Schedule I drug. In 2001 the Expert Advisory Committee on Drugs (EACD) in New Zealand advised the New Zealand Medicines and Medical Devices Safety Authority to schedule gammahydroxybutyrate under the Misuse of Drugs Act 1975 (1). In June 2003 gammahydroxybutyrate was categorized as a Class C drug in the UK.

Ingestion of 0.5–3 teaspoons of sodium gammahydroxybutyrate can produce vomiting, drowsiness, hypotonia, and/or vertigo; loss of consciousness, irregular respiration, tremors, or myoclonus can follow. Seizure-like activity, bradycardia, hypotension, and/or respiratory arrest have also been reported. The severity and duration of symptoms depend on the dose and on the presence of other nervous system depressants, such as alcohol.

In a double-blind, randomized, placebo-controlled, crossover study in 24 patients with narcolepsy, gammahydroxybutyrate 60 mg/kg in a single night-dose for 4 weeks reduced the daily number of hypnagogic hallucinations, daytime sleep attacks, and the severity of subjective daytime sleepiness, and tended to reduce the number of daily attacks of cataplexy (2). It reduced the percentage of wakefulness during REM sleep and the number of awakenings out of REM sleep, and tended to increase slow wave sleep. Adverse events were few and mild.

Organs and Systems

Nervous system

In a double-blind, double-dummy comparison of clomethiazole and gammahydroxybutyrate in ameliorating the symptoms of alcohol withdrawal, alcohol-dependent patients were randomized to receive either clomethiazole 1000 mg or gammahydroxybutyrate 50 mg/kg (3). There was no difference between the treatments in ratings of alcohol withdrawal symptoms or requests for additional medications. After tapering the active medication, there was no increase in withdrawal symptoms, suggesting that physical tolerance did not develop to either clomethiazole or gammahydroxybutyrate during the 5-day treatment period. The most frequently reported adverse effect of gammahydroxybutyrate was transient vertigo, particularly after the evening double dose.

Drug Administration

Drug overdose

Some who have taken gammahydroxybutyrate without knowing what it was have subsequently awakened in hospital having been deeply comatose for a few hours, without after-effects (4).

References

1. Anonymous. Gamma hydroxybutyrate. Fantasy drugs to be classified. WHO Pharm Newslett 2001;2:5–6.
2. Lammers GJ, Arends J, Declerck AC, Ferrari MD, Schouwink G, Troost J. Gammahydroxybutyrate and narcolepsy: a double-blind placebo-controlled study. Sleep 1993;16(3):216–20.
3. Nimmerrichter AA, Walter H, Gutierrez-Lobos KE, Lesch OM. Double-blind controlled trial of gamma-hydroxybutyrate and clomethiazole in the treatment of alcohol withdrawal. Alcohol Alcohol 2002;37(1):67–73.
4. Strange DG, Jensen D. Gammahydroxybutyrat, et nyt rusmiddel. [Gamma-hydroxybutyrate, a new central nervous system stimulant.] Ugeskr Laeger 1999;161(50):6934–6.

KHAT

General information

Khat, or qat, is a stimulant commonly used in East Africa, Yemen, and Southern Saudi Arabia. Khat leaves from the evergreen bush Catha edulis are typically chewed while fresh, but can also be smoked, brewed in tea, or sprinkled on food. Its use is culturally based.

The major effects of khat on the central nervous system can be attributed to cathinone (S-(−)-alpha-aminopropriophenone) in fresh khat leaves and cathine (norpseudoephedrine) in dried khat leaves and stems. Cathinone, a phenylalkylamine, is the major active component and is structurally similar to amfetamine. It degrades to norpseudoephedrine and norephedrine within days of leaf picking. Cathinone increases dopamine release and reduces dopamine re-uptake (1).

Khat is often used in social gatherings called "sessions", which can last 3–4 hours. They are generally attended by men, although khat use among women is growing. Men are also more likely to be daily users. Users pick leaves from the khat branch, chew them on one side of the mouth, swallowing only the juice, and adding fresh leaves periodically. The khat chewer may experience increased alertness and euphoria. About 100–300 grams of khat may be chewed during each session, and 100 grams of khat typically contains 36 mg of cathinone.

Khat has been recognized as a substance of abuse with increasing popularity. It is estimated that 10 million people chew khat worldwide, and it is used by up to 80% of adults in Somalia and Yemen. It now extends to immigrant African communities in the UK and USA. It is banned in Saudi Arabia, Egypt, Morocco, Sudan, and Kuwait. It is also banned in the USA and European countries. However, in Australia, its importation is controlled by a licence issued by the Therapeutic Goods Administration, which allows up to 5 kg of khat per month per individual for personal use.

There have been several review articles describing khat and its growth. The World Health Organization Advisory Group's 1980 report reviewed the pharmacological effects of khat in animals and humans (2). The societal context of khat use has also been reviewed (3).

Organs and Systems

Cardiovascular

Khat is a sympathomimetic amine and increases blood pressure and heart rate. Limited evidence suggests that khat increases the risk of acute myocardial infarction. In Yemen 100 patients admitted to an intensive care unit with an acute myocardial infarction were compared with 100 sex- and age-matched controls recruited from an ambulatory clinic (4). They completed a questionnaire on personal habits, such as khat use and cigarette smoking, past medical history, and a family history of myocardial infarction. Use of khat was an independent risk factor for acute myocardial infarction, with an odds ratio of 5.0 (95% CI = 1.9, 13). The relation was dose-related: "heavy" khat users were at higher risk than "moderate" users, although the extent of use and the potency of khat used were estimated, being hard to quantify. To explain the increased risk of acute myocardial infarction, the authors suggested that it may have been related to increased blood pressure and heart rate, with a resultant increase in myocardial oxygen demand. They also suggested that khat could have acted via the mechanisms proposed to explain acute myocardial infarction after the use of amphetamines, such as catecholamine-induced platelet aggregation and coronary vasospasm.

Nervous system

The prevalence and health effects of headache in Africa have been reviewed (5). Of 66 khat users, 25% reported headaches (6). Among people with migraine, 12% reported using khat (7).

Psychiatric

In addition to the acute stimulant effects of euphoria and alertness caused by khat, there is the question of whether continued khat use alters mood, behavior, and mental health.

- A 33-year-old unemployed Somali man with a 10-year history of khat chewing, who had lived in Western Australia for 4 years, wand who was socially isolated, started to sleep badly, and had weight loss and persecutory delusions (8). His mental state deteriorated over 2–3 months and he thought that his relatives were poisoning him and that he was being followed by criminals. He had taken rifampicin and ethambutol for pulmonary tuberculosis for 1 year but became noncompliant for 2 months before presentation. He had reportedly chewed increasing amounts of khat daily from his backyard for last 2 years. There was no history of other drug use and his urine drug screen was negative. He responded well to olanzapine 20 mg/day and was discharged after 4 weeks, as his psychosis was gradually improving.

The impact of khat use on psychological symptoms was one of several factors considered in a study in which 180 Somali refugees were interviewed about psychiatric symptoms and about migration-related experiences and traumas (9). Suicidal thinking was more common among those who used khat (41 of 180) after migration compared with those who did not (21 of 180). However, a causal relationship cannot be deduced from these data.

The authors raised the concern that khat psychosis could be increasing in Australia because of a growing number of African refugees. Furthermore, factors related to immigration, such as social displacement and unemployment, may predispose to abuse, especially as khat is easily available in Australia.

In a cross-sectional survey of Yemeni adults the self-reported frequency of khat use and psychological symptoms was assessed using face-to-face interviews with members from a random sample of urban and rural households (10). Of 800 adults surveyed, 82% of men and 43% of women had used khat at least once. There was no association between khat and negative adverse psychological symptoms, and khat users had less phobic anxiety (56%) than non-users (38%). The authors were surprised by these results and offered several explanations: that the form of khat used in Yemen is less potent than in other locations; that prior reports of khat-related psychosis occurred in users in unfamiliar environments; that the sampling procedure may have under-represented heavier khat users; and that their measurement tool was not sensitive enough to detect psychological symptoms.

In Hargeisa, Somalia, trained local interviewers screened 4854 individuals for disability due to severe psychiatric problems and identified 169 cases (137 men and 32 women) (11). A subset of 52 positive screening cases was randomly selected for interview and each was matched for age, sex, and education with a control. In all, 8.4% of men screened positive and 83% of those who screened positive had severe psychotic symptoms. Khat chewing and the use of greater amounts of khat were more common in this group. Khat users were also more likely to have had active war experience. Only 1.9% of women had positive screening. Khat use starting at an earlier age and in larger amounts (in "bundles" per day) correlated positively with psychotic symptoms.

Gastrointestinal

Constipation is a common physical complaint among khat users (3). The World Health Organization Advisory Group attributed this adverse effect to the tannins and norpseudoephedrine found in khat (12).

Anorexia due to khat is probably due to norephedrine (12).

The tannins in khat have been associated with delayed intestinal absorption, stomatitis, gastritis, and esophagitis observed with khat use (12).

Khat chewing may be a risk factor for duodenal ulcer. In a case-control study 175 patients with duodenal ulceration (all diagnosed by endoscopy) and 150 controls completed a questionnaire about their health habits (13). Khat use, defined as chewing khat at least 14 hours/week, was significantly more common among the cases (76 versus 35%). Potential confounding variables, including smoking, use of alcohol or NSAIDs, a family history, and chronic hepatic and renal disease, were not significantly different between the two groups. The authors postulated several mechanisms, including a physiological reaction to the stress response to cathine or exposure to chemicals, such as pesticides.

Long-Term Effects

Drug dependence

Khat use can lead to dependence, which is more important because of its social consequences than because of the effects of physical withdrawal. Khat users may devote significant amounts of time to acquiring and using khat, to the detriment of work and social responsibilities. The physical effects of early khat withdrawal are generally mild. Chronic users may experience craving, lethargy, and a feeling of warmth during early khat abstinence.

Tumorigenicity

Three studies of the association of khat with head and neck cancers have been reviewed (14) The studies showed a trend towards an increased risk of oral cancer and head and neck cancer with the use of khat, but there were too few data for a definitive conclusion. Tobacco use, which is common among khat users, and alcohol use were confounding factors.

A possible association of cancer of the oral cavity with khat has been studied in exfoliated buccal and bladder cells from healthy male khat users; the cells were examined for micronuclei, a marker of genotoxic effects (15). Of 30 individuals who did not use cigarettes or alcohol, 10 were non-users of khat and the other 20 used 10–60 g/day. The 10 individuals who used more than 100 g/day had an eight-fold increase in the frequency of micronuclei compared with non-users. There was a statistically significant dose-response relation between khat use and the number of micronuclei in oral mucosal cells but not urothelial cells. In a separate set of samples taken from khat users and non-users who also used cigarettes and alcohol, alcohol and cigarette use caused a 4.5-fold increase in the frequency of micronuclei and use of khat in addition to alcohol and cigarettes further doubled the frequency. In buccal mucosal cells from four individuals who ingested 100 g/day for 3 days, the maximum frequency of micronuclei occurred at 27 days after chewing and returned to baseline after 54 days. These data together suggest that khat, especially in combination with alcohol and smoking, may contribute to or cause oral malignancy.

However, in a small study of biopsies taken from the oral mucosa of 40 Yemeni khat users and 10 non-users there were no histopathological changes consistent with malignancy (16). In the khat users, there were changes such as acanthosis, orthokeratosis, epithelial dysplasia, and intracellular edema on both the chewing and non-chewing sides of the oral mucosa. However, none of these lesions was malignant or pre-malignant. The authors thought that these changes were most probably due to mechanical friction or possibly due to chemical components of the khat or pesticides that had been used on the plants.

Second-Generation Effects

Fertility

Reduced sperm count, semen volume, and sperm motility have been associated with khat dependence (17).

Fetotoxicity

Reduced infant birth weight has been reported in khat-using mothers (1).

Lactation

Reduced lactation has been reported in khat-using mothers (1).

Drug Administration

Drug contamination

Some khat leaves are grown with chemical pesticides. In 114 male khat users in two different mountainous areas of Yemen, users of khat that had been produced in fields in which chemical pesticides were used regularly had more acute gastrointestinal adverse effects (nausea and abdominal pain) and chronic body weakness and nasal problems (18). The authors suggested that organic chemical pesticides such as dimethoatecide can cause such adverse effects.

Drug overdose Acute toxicity requiring emergency medical treatment is rare. When it occurs there is a typical sympathomimetic syndrome, which should be treated with fluids, control of hyperthermia, bed rest, and, if necessary, sedation with benzodiazepines (1).

References

1. Haroz R, Greenberg MI. Emerging drugs of abuse. Med Clin N Am 2005;89:1259–76.
2. World Health Organization Advisory Group. Review of the pharmacology of khat. Bull Narcotics 1980;32:83–93.
3. Al-Motarreb A, Baker K, Broadley KJ. Khat: pharmacological and medical aspects and its social use in Yemen. Phytother Res 2002;16:403–13.
4. Al-Motarreb A, Briancon S, Al-Jaber N, Al-Adhi B, Al-Jailani F, Salek MS, Broadley KJ. Khat chewing is a risk factor for acute myocardial infarction: a case-control study. Br J Clin Pharmacol 2005;59:574–81.
5. Tekle Haimanot R. Burden of headache in Africa. J Headache Pain 2003;4:S47-S54.
6. Mekasha A. The clinical effects of khat. In: The International Symposium on Khat, Ethiopia, 1984;77–81.
7. Tekle Haimanot R, Seraw B, Forsgren L, Ekbom K, Ekstedt J. Migraine, chronic tension-type headache, and cluster headache in an Ethiopian rural community. Cephalagia 1995;15:482–8.
8. Stefan J, Mathew B. Khat chewing: an emerging drug concern in Australia? Aust N Z J Psychiatry 2005;39:842–3.
9. Bhui K, Abdi A, Abdi M, Pereira S, Dualeh M. Traumatic events, migration characteristics and psychiatric symptoms among Somali refugees—preliminary communication. Soc Psychiatry Psychiatr Epidemiol 2003;38:35–43.
10. Numan N. Exploration of adverse psychological symptoms in Yemeni khat users by the Symptoms Checklist-90 [SCL-90]. Addiction 2004;99:61–5.
11. Odenwald M, Neuner F, Schauer M. Khat use as risk factor for psychotic disorders: a cross-sectional and case-control study in Somalia. BMC Med 2005;3:5.
12. World Health Organization Advisory Group. Review of the pharmacology of khat. Bull Narcotics 1980;32:83–93.
13. Raja'a YA, Noma T, Warafi TA. Khat chewing is a risk factor for duodenal ulcer. Saudi Med J 2000;21:887–8.
14. Goldenberg D, Lee J, Koch WM, Kim MM, Trink B, Sidransky D, Moon CS. Habitual risk factors for head and neck cancer. Otolaryngol Head Neck Surg 2004;131:986–93.
15. Kassie F, Darroudi F, Kundi M, Schulte-Hermann R, Knasmuller S. Khat (*Catha edulis*) consumption causes genotoxic effects in humans. Int J Cancer 2001;92:329–32.
16. Ali AA, Al-Sharabi AK, Aguirre JM. Histopathological changes in oral mucosa due to takhzeen al-qat: a study of 70 biopsies. J Oral Pathol Med 2006;35:81–5.
17. Al-Motarreb A, Baker K, Broadley KJ. Khat: pharmacological and medical aspects and its social use in Yemen. Phytother Res 2002;16:403–13.
18. Date J, Tanida N, Hobara T. Qat chewing and pesticides: a study of adverse health effects in people in the mountainous areas of Yemen. Int J Environ Health Res 2004;6:405–14.

LYSERGIDE

General Information

Lysergic acid diethylamide (LSD) is a hallucinogen that is usually taken orally. Its initial effects, anticholinergic and sympathomimetic in type, occur within about half an hour and include tachycardia, hyperthermia, mydriasis, piloerection, hypertension, and occasionally nausea and vomiting. The more important psychoactive effects develop 1–2 hours later and can last 24–48 hours. They are principally changes in perception, mood, and behavior, leading in many cases to acute panic, hallucinations, delusions, and in some cases a classical psychosis. Perception may be strikingly heightened and distorted, and initial perceptions may mask and overshadow later sensory perception. Users often grossly exaggerate their mental and emotional capacities, attributing to themselves extraordinary powers. Visual hallucinations, loss of appreciation of time and space, and instability of mood are common. Meaningfulness and a sense of universal union often predominate. A significant problem in street purchase is the uncertain quality and likely impurity of the material obtained. Doses of 25 mg and more are sufficient to cause its psychophysiological effects, which are generally dose-related up to 500 mg. The half-life is about 3 hours, but the effects last considerably longer (1).

Acute panic attacks and hallucinogen-induced psychotic disorder often occur when people with pre-existing personality disorder or pre-psychotic personalities use hallucinogens. Suicide and self-injury have been reported. Prolonged psychotic disorders can occur, but psychiatric opinion is divided as to whether these occur only in people with pre-existing disorders or in healthy individuals as well. "Flashbacks" occur particularly when there has been prolonged heavy use, but eventually disappear (SED-11, 83) (1,2). Self-injury and suicide can result (3).

Hypersensitivity reactions are exceedingly rare and no reports have been validated. Tumor-inducing effects are possible, as a consequence of various reports from animal and human studies that chromosomal abnormalities may be associated with exposure to LSD. However, its mutagenic effects in practice are questionable, and no useful evidence for or against its potential for carcinogenicity has been produced (4).

Organs and Systems

Cardiovascular

Vasoconstriction, affecting both cerebral and peripheral circulations, has been associated with LSD (5), but it is not usually significant at ordinary doses in people with a normal circulatory system.

Nervous system

Hallucinogen-induced mood disorder is associated with changes in affect, varying from euphoria to manic-like symptoms, panic/fear, and depression, often occurring within minutes and often varying in the same individual on different occasions. Changes in sensory perception, with a loss of ability to distinguish temporal or spatial reality and sensory hallucinations, particularly visual and tactile, are frequent, with a tendency to assume godlike attributes. These features often merge in a psychosis, particularly with repeated use. Whether chronic psychosis after LSD is the result of the drug or of a combination of the drug and predisposing factors is currently unanswerable (6–8).

The repeated use of LSD is associated not only with psychoses, but also with more specific neurological signs and symptoms, including ataxia, incoordination, dysphasia, paresthesia, and tremor. Convulsions have been reported. "Flashback," or the return of hallucinogenic effects, occurs in almost a quarter of those who have used LSD, particularly if they have also used other CNS stimulants, such as alcohol or marijuana (2,9). They can experience distortions of perception of objects, space, or time, which intrude without warning into reality, resulting in delusions, panic, and unusual images. A "trailing phenomenon" has also been reported, in which the visual perception of objects is reduced to a series of interrupted pictures rather than a constant view (10). The frequency of these events may slowly abate over several years, but in a significant number their incidence later increases (1,3).

Sensory systems

Apart from visual hallucinations (discussed under the section Nervous system in this monograph), diplopia, blurred vision, mydriasis, and other visual disturbances occur (11). Pupillary dilatation, combined with altered sensory appreciation, has led to a number of instances of retinal damage after continued direct exposure to the sun (12).

Hematologic

The only hematological effect reported has been an increased rate of blood clotting associated with severe hyperthermia (13).

Gastrointestinal

Retroperitoneal fibrosis has been reported, not unexpectedly, in view of the structural similarity between LSD and methysergide (3).

Body temperature

Hyperthermia can occur but does so very seldom with usual doses (14). It has been produced experimentally with high doses.

Second-Generation Effects

Teratogenicity

Animal studies have not shown that LSD is teratogenic. Various publications have referred to a high incidence of abortions and congenital abnormalities associated with LSD. However, none of these has shown any consistent pattern of abnormality, and neither has there been an acceptable control group (4).

Susceptibility Factors

People who are predisposed to psychosis, schizophrenia, or a family history thereof, may be at considerable risk of developing LSD psychosis (3,8). People with epilepsy may be more prone to convulsions, with a reduction in convulsant threshold.

Drug–Drug Interactions

Monoamine oxidase inhibitors

Chronic administration of a monoamine oxidase inhibitor causes a subjective reduction in the effects of LSD, perhaps due to differential changes in central serotonin and dopamine receptor systems (15).

Phenothiazines and butyrophenones

Phenothiazines and butyrophenones can counteract the psychoactive effects of LSD and benzodiazepines can depress their effects (16).

Reserpine

Reserpine can accentuate the effects of LSD (17).

Interference with Diagnostic Tests

Ambroxol

A urine sample from a patient with a severe head injury tested positive for LSD (18). The test was carried out using a homogeneous immunoassay CEDIA DAU LSD (Boehringer, Mannheim). Although LSD has a half-life of 3 hours, the patient's urine tested positive for several days. Analysis of the same urine samples using high-performance liquid chromatography (HPLC) failed to detect LSD. LSD screening was then performed in urine samples obtained from ten other patients in the same ward. All samples tested positive for LSD by the CEDIA DAU LSD assay but negative using HPLC. All of the patients were taking ambroxol. Ambroxol was detected in the urine by HPLC. In a volunteer, LSD was not detected in a fasting urine sample. After a dose of ambroxol 15 mg a urine sample obtained 90 minutes later tested positive for LSD. The addition of 50 µl of Mucosolvan juice (which contains ambroxol) to the negative fasting urine sample resulted in a positive test for LSD. The authors concluded that ambroxol should be excluded when LSD screening is performed using the CEDIA DAU LSD test.

References

1. Watson SJ. Hallucinogens and other psychotomimetics: biological mechanisms. In: Barchas JD, Berger PA, Cioranello RD, Elliot GR, editors. Psychopharmacology from Theory to Practise. New York: Oxford University Press, 1977:1437.
2. Alarcon RD, Dickinson WA, Dohn HH. Flashback phenomena. Clinical and diagnostic dilemmas. J Nerv Ment Dis 1982;170(4):217–23.
3. Strassman RJ. Adverse reactions to psychedelic drugs. A review of the literature. J Nerv Ment Dis 1984;172(10):577–95.
4. Tuchmann-Duplessis H. Drug effects on the fetus. In: Avery GS, editor. Monographs on Drugs vol. 2. London: ADIS, 1975:158.
5. Lieberman AN, Bloom W, Kishore PS, Lin JP. Carotid artery occlusion following ingestion of LSD. Stroke 1974;5(2):213–5.
6. Sarwer-Foner GJ. Some clinical and social aspects of lysergic acid diethylamidel. II Psychosomatics 1972;13(5):309–16.
7. McWilliams SA, Tuttle RJ. Long-term psychological effects of LSD. Psychol Bull 1973;79(6):341–51.
8. Bowers MB Jr. Acute psychosis induced by psychotomimetic drug abuse. I. Clinical findings. Arch Gen Psychiatry 1972;27(4):437–40.
9. Tec L. Phenothiazine and biperiden in LSD reactions. JAMA 1971;215(6):980.
10. Asher H. "Trailing" phenomenon—a long lasting LSD side effect. Am J Psychiatry 1971;127:1233.
11. Abraham HD. Visual phenomenology of the LSD flashback. Arch Gen Psychiatry 1983;40(8):884–9.
12. Schatz H, Mendelblatt F. Solar retinopathy from sun-gazing under the influence of LSD. Br J Ophthalmol 1973;57(4):270–3.
13. Klock JC, Boerner U, Becker CE. Coma, hyperthermia and bleeding associated with massive LSD overdose. A report of eight cases. West J Med 1974;120(3):183–8.
14. Friedman SA, Hirsch SE. Extreme hyperthermia after LSD ingestion. JAMA 1971;217(11):1549–50.
15. Bonson KR, Murphy DL. Alterations in responses to LSD in humans associated with chronic administration of tricyclic antidepressants, monoamine oxidase inhibitors or lithium. Behav Brain Res 1996;73(1–2):229–33.
16. Vardy MM, Kay SR. LSD psychosis or LSD-induced schizophrenia? A multimethod inquiry. Arch Gen Psychiatry 1983;40(8):877–83.
17. Resnick O, Krus DM, Raskin M. Accentuation of the psychological effects of LSD-25 in normal subjects treated with reserpine. Life Sci 1965;4(14):1433–7.
18. Rohrich J, Zorntlein S, Lotz J, Becker J, Kern T, Rittner C. False-positive LSD testing in urine samples from intensive care patients. J Anal Toxicol 1998;22(5):393–5.

MESCALINE

General Information

Mescaline is one of eight hallucinogenic alkaloids derived from the peyote cactus, slices of which ("peyote buttons") have been used in religious rites by North and South American Indian tribes. Mescaline itself is only one of the alkaloids present in peyote, but it produces the same effects as the crude preparation. Chemically, it is related to amfetamine. In doses of some 300–500 mg it depresses nervous system activity and produces visual and occasionally auditory hallucinations, illusions, depersonalization, and depressive symptoms [1]. The total picture can closely resemble that caused by lysergic acid diethylamide. Its physical effects include nausea, tremor, and sweating.

Organs and Systems

Psychological, psychiatric

The effects of mescaline have been investigated in a psychiatric research study [2]. Psychosis induced during the experiment was measured with the Brief Psychiatric Rating Scale and the Paranoid Depression Scale. During use of mescaline neuropsychological measures showed reduced functioning of the right hemisphere. Single photon emission computed tomography (SPECT) studies showed a hyperfrontal pattern, with an emphasis on the right hemisphere. The authors discussed the possible educational value of experimentally induced psychosis in understanding the psychotic state.

References

1. Huxley A. *The Doors of Perception*. London: Chatto and Windus, 1954.
2. Hermle L, Funfgeld M, Oepen G, Botsch H, Borchardt D, Gouzoulis E, Fehrenbach RA, Spitzer M. Mescaline-induced psychopathological, neuropsychological, and neurometabolic effects in normal subjects: experimental psychosis as a tool for psychiatric research. Biol Psychiatry 1992;32(11):976–91.

METAMFETAMINE

See also Amphetamines

General Information

Metamfetamine has been abused for more than 80 years. It is easily synthesized in home laboratories and has a low street price, more prolonged effects, and a high potential for abuse/dependency. It is therefore not surprising that there has been a worldwide surge in its use in recent years. It is often sold as "crank", "speed", "shabu", "meth", "chalk", "crystal", "glass", or "ice."

Organs and Systems

Cardiovascular

A retrospective chart review was conducted to explore metamfetamine-associated acute coronary syndromes in patients who presented to the emergency room at a University Center between 1994 and 1996 (1). There were 36 admissions, three of which were repeat patients. Nine of these patients had acute coronary syndrome. Of these, one had an acute anterior Q wave myocardial infarction with cardiac arrest, seven had non-Q wave myocardial infarctions, and one had unstable angina. There were potentially life-threatening cardiac complications in three subjects (8%). The authors suggested that acute coronary syndromes and life-threatening complications associated with the use of metamfetamine are not uncommon, as evidenced by their experience in this study.

The frequency of acute coronary syndrome in patients who developed chest pain after using metamfetamine has been described. In 33 patients (25 men, 8 women, mean age 40 years) metamfetamine abuse was confirmed by urine screening. Acute coronary syndrome was diagnosed in nine patients. Three patients (two of whom had acute coronary syndrome) had cardiac dysrhythmias. The authors concluded that acute coronary syndrome is common in patients with chest pain after metamfetamine use, and that the frequency of other potentially life-threatening cardiac complications is not negligible. A normal electrocardiogram reduces the likelihood of acute coronary syndrome, but an abnormal electrocardiogram is not helpful in distinguishing patients with or without acute coronary syndrome (2). These conclusions were based on a small population that included hospitalized patients, not all of whom underwent non-invasive cardiac stress testing or coronary angiography. Thus the generalizality of these findings is limited.

- A 44-year-old female metamfetamine abuser died unexpectedly due to right-sided infective endocarditis (3).
- A 27-year-old man who had used intravenous amfetamine had an acute myocardial infarction (4).

The authors suggested that coronary angiography with an ergometrine provocation test might be necessary in patients who use amfetamine and who develop acute coronary syndrome.

The cardiovascular effects of metamfetamine in 11 patients with Parkinson's disease have been described and compared with six healthy controls (5). All were tested twice, once with intravenous saline and once with intravenous metamfetamine 0.3 mg/kg. Cardiovascular measurements were taken for 15 minutes before drug administration and for 103 minutes after. Both groups had significant increases in blood pressure after metamfetamine, but the patients with Parkinson's disease had a shorter duration of increased blood pressure and a lower increase from baseline than the controls. The authors proposed that these results suggested cardiac and vascular hyposensitivity to metamfetamine in Parkinson's disease due to impaired metamfetamine-induced catecholamine release.

Nervous system

Subcortical hemorrhage after metamfetamine abuse has been reported (6).

- A 32-year-old woman was found comatose with left hemiplegia. A brain CT scan showed a subcortical hemorrhage in the right parietal lobe with a midline shift of more than 10 mm. Cerebral angiography did not show any vascular anomalies. A craniotomy was performed immediately to remove the hematoma. The serum metamfetamine concentration was very high at 120 ng/ml. She admitted using intravenous metamfetamine before she became unconscious. She was discharged 40 days later with a residual left hemiparesis.

The authors strongly recommended testing for drugs in young, non-hypertensive patients with angiographically negative intracranial hemorrhage.

Intracerebral hemorrhage after metamfetamine abuse has been reported (7).

- A 31-year-old man was found dead 9.5 hours after smoking and snorting metamfetamine. He developed a headache with nausea and vomiting 15 minutes after taking the drug and decided to sleep it off. His friends checked on him throughout the night and noted that he had slurred speech and was falling off his seat. He also complained of left-sided numbness. He was later found dead, with only the left side of his body clothed. Autopsy showed a subarachnoid hemorrhage and an intracerebral hemorrhage lateral to the basal ganglia. His blood metamfetamine concentration was 300 ng/ml.

According to the authors, the cause of death was a spontaneous intracerebral hemorrhage and subarachnoid hemorrhage without vasculitis.

Chronic cerebral vasculitis and delayed ischemic stroke due to metamfetamine has been reported (8).

- A 19-year-old woman used metamfetamine intravenously four times over 2 months and had a headache on each occasion, except the second. She stopped using it, but the headache continued, and about 3 months later she developed severe right-sided headache. She also noticed blurred vision on the left side and numbness of the left arm and leg. She denied any other drug use, including oral contraceptives. Her blood pressure was 110/70 mmHg. Magnetic resonance angiography showed bleeding of the right posterior cerebral artery and characteristic features of vasculitis. Her symptoms gradually improved.

Delayed-onset stroke due to metamfetamine is rare, but is associated with the use of other sympathomimetics, such as ephedrine and cocaine.

Electroencephalographic abnormalities were examined in 11 recently abstinent metamfetamine abusers and 11 non-drug-using controls (9). The metamfetamine-dependent subjects were hospitalized for 4 days to ensure abstinence during the study. The abstinent metamfetamine users had increased power in their electroencephalograms for lower frequency bands, with robust effects in all regions. This effect did not appear to be due to drowsiness. There was also a higher rate of generalized electroencephalographic slowing than in controls. Two of the 11 controls and seven of the 11 metamfetamine abusers had abnormal electroencephalograms. The authors contended that the patterns seen here are indicative of the development of an encephalopathy, suggesting that the neurotoxic effects of metamfetamine may contribute to electroencephalographic abnormalities. The authors suggested that these abnormalities may be associated with other cognitive dysfunction.

Psychological

The effects of metamfetamine abuse on cognition and on brain structures and function

The effects of metamfetamine on cognition have been studied in 25 subjects who remained abstinent after treatment for metamfetamine abuse, 25 who relapsed, and 25 who continued to use it throughout treatment (10). The cognitive test battery covered six domains: manipulation of information and perceptual speed; ability to ignore irrelevant information; executive function, mental flexibility, and logical thinking; learning; episodic memory; and working memory. There were significant differences between groups in measures of selective reminding and the repeated memory test. There were no differences between the abstinent and relapse groups. On the selective reminding task the continuing use group had a significantly higher number of correct responses than the abstinent group. For the episodic memory tasks, the relapse group performed significantly worse than continuous use group. These results suggest that there is cognitive impairment after a subject stops using metamfetamine. There were no significant differences between those who relapsed and those who remained abstinent, although the relapse group performed better on most tasks. This study was limited, because the relapse group included individuals who relapsed after varying times of abstinence. It is possible that differing periods of abstinence could have distinctive effects on cognition. However, these results do suggest that relapse has a deleterious effect on episodic memory compared with abstinence in ex-users and continued use.

The association of metamfetamine dependence and cognitive function in the initial stages of abstinence has been examined in 27 metamfetamine-dependent subjects and 18 non-drug using controls (11). The subjects were required to stop using metamfetamine and to produce a negative urine sample on the day of cognitive testing with a standard neurocognitive battery. Metamfetamine-dependent subjects had significantly impaired measures of attention, verbal learning, and memory, and fluency-based measures of executive function compared with control subjects. These results suggest that metamfetamine abuse may be associated with cognitive impairment across a number of domains. The authors claimed this to be the first evidence that metamfetamine dependence is associated with a broad range of cognitive deficits, the degree of which is substantial and greater than in cocaine dependence. However, causality cannot be inferred, owing to the study design and the sample size, and it is possible that withdrawal symptoms were related to poor performance in those using metamfetamine.

Structural changes in the brain and cognitive impairment

Since metamfetamine interacts with brain mechanisms, it may affect brain structures. The cortical surface and subcortical structures have been compared in 22 metamfetamine abusers (15 men, 7 women) and 21 controls (10 men, 11 women) (12). In addition to MRI scans, the subjects were also administered a neuropsychological battery to assess episodic memory and gave self-ratings on depressive symptoms and anxiety. There was a significant gray matter deficit in metamfetamine abusers in the cingulate gyrus, subgenual cortex, and paralimbic belts. The most severe deficits were in the cingulate regions, where gray matter volumes were 11% below control. In addition, hippocampal volume in metamfetamine abusers was significantly lower than in the controls by 7.8%. This effect occurred across the left and right hemispheres and remained significant even after controlling for extensive marijuana use in the metamfetamine group. Furthermore, hippocampal volume correlated positively with episodic memory performance. Metamfetamine abusers also had significantly more white matter in the temporal and occipital regions. These findings are similar to the structural abnormalities seen in schizophrenia or dementia. It is not clear how these abnormalities occur and whether they are progressive or reversible.

In another study PET was used to investigate metabolic abnormalities in 20 recently abstinent metamfetamine abusers and 22 controls (13). The mood disturbances in these individuals were also compared with their cerebral metabolism. Metamfetamine abusers were tested after 4–7 days of abstinence. All gave self-ratings on depression, anxiety, and craving for metamfetamine. PET images were acquired

50 minutes after administration of FDG (fluorodeoxyglucose) for 30 minutes. Metamfetamine abusers had significantly higher scores on the Beck Depression Inventory, which correlated with recent metamfetamine use. There was higher cerebral glucose metabolism in a portion of the brain extending from the central to the posterior region of the cingulate gyrus in abstinent metamfetamine abusers. Glucose metabolism was significantly lower in metamfetamine users than in controls in the following regions: infragenual accumbens; left perigenual accumbens; right insula. Metamfetamine abusers had significantly higher activity in the following regions: lateral orbitofrontal cortex; right, middle, and posterior cingulate; amygdala; ventral striatum; cerebellar vermis. In addition, depression scores in metamfetamine abusers correlated positively with activity in the right perigenual accumbens. Depression scores also correlated with activity in the amygdala in metamfetamine abusers. Since most of the systems examined in this study have dopaminergic connections, it is possible that these abnormalities are caused by changes in the dopamine system; a decrease in dopamine input to a specific region could cause a metabolic deficit. It is unknown whether the effects seen in this study reflected abstinence, chronic metamfetamine abuse, or factors that predated metamfetamine abuse.

Magnetic resonance imaging (MRI) and new computerized brain-mapping techniques have been used to evaluate the pattern of structural brain alterations associated with chronic metamfetamine abuse and how these changes are related to deficits in cognitive impairment (14,15). In 22 human subjects who abused metamfetamine compared with 21 age-matched, healthy controls, cortical maps showed severe gray-matter deficits in cingulate, limbic, and paralimbic cortices of metamfetamine abusers, averaging 11% below control. Hippocampal deficits, measured as volume, and white-matter hypertrophy were correlated with memory performance on a word-recall test. MRI-based cortical maps suggested that chronic metamfetamine abuse causes a selective pattern of cerebral deterioration that contributes to impaired memory performance. Metamfetamine abuse can also selectively damage the medial temporal lobe and, consistent with metabolic studies, the cingulate-limbic cortex, including neuroadaptation, neuropil[1] reduction, or cell death.

Prominent white-matter hypertrophy has been suggested to result from altered myelination and adaptive glial changes, including gliosis secondary to neuronal damage (Thompson6028). A potential limitation of this study relates to matching of the two groups: most of the metamfetamine abusers were smokers, compared with only two of the control subjects.

In smokers structural changes in the brain have been reported (16), although these were more restricted than in metamfetamine abusers.

Metamfetamine current abusers show evidence of deficits in performance on tests of memory, perceptual motor speed, inhibition, problem solving, manipulation of information, abstract thinking, and mental flexibility (17,18). Studies on those who have abused metamfetamine for periods of 5 days to several years have shown that these performance deficits continue during abstinence. These individuals have impaired performance on tests of the ability to inhibit irrelevant information (19,20), decision making (21), memory (22,23), and spatial processing and learning (Kalechstein215,24). Studies of other cognitive functions, such as tests of verbal memory, have yielded inconsistent results, with reports of deficits (Simon 61,Kalechstein 215) or normal performance (Chang 65).

Effects of relapse

Data from 75 participants in a longitudinal study of metamfetamine abusers has attempted to differentiate the cognitive performance of those who remained abstinent, relapsed, or continued to use metamfetamine during treatment (25). Relapse of metamfetamine abuse can affect episodic memory differently than it affects other cognitive functions. This highlights the fact that individuals who are either abstinent or relapsing may have more problems with treatment that requires their attention, understanding, and memory for compliance than those who continue to use metamfetamine.

Susceptibility factors for impaired cognition

Metamfetamine use is high in individuals with HIV infection, who already have neurocognitive impairment (26). Furthermore, metamfetamine and HIV both have influences on the striatal and striatal–cortical pathways, so there could be additive effects when the two are combined. This has been studied in 200 individuals in four groups: HIV-positive and metamfetamine dependent (n = 43); HIV-negative and metamfetamine dependent (n = 47); HIV-positive and metamfetamine non-user (n = 50); and HIV-negative and metamfetamine non-user (n = 60). The neurocognitive test battery consisted of tasks dependent on seven different domains: speed of information processing; learning; recall; abstraction/executive functioning; verbal fluency; attention/working memory; and motor skills. In addition, a Global Deficit Score (GDS) was calculated, displaying the severity of impairment across all tests in the battery (calculated by adding deficit ratings for each test and dividing by the total number of tests). Each of the three groups with risk factors for cognitive impairment had higher GDS scores than controls. Among individual domains, HIV-negative metamfetamine abusers had more impairment in tests of attention and recall than controls. HIV-positive metamfetamine abusers performed significantly worse than controls on tests of learning and motor skills. In addition, the percentage global impairment was greatest for HIV-positive metamfetamine abusers than for any other group. These results confirm the separate neurocognitive deficits of HIV infection and metamfetamine abuse, and also show that the combination may have increased deleterious effects on cognition. Since HIV's mechanism of brain injury is not completely understood, it cannot be said with certainty how these two susceptibility factors interact.

---

[1] The The brain parenchyma is composed of neurons supported by a framework of glial cells (astrocytes, oligodendrocytes, and ependyma), blood vessels, and microglia. The processes of these cells combine to form a delicate fibrillar background termed the 'neuropil'.

Given the increased risk of brain injury among metamfetamine abusers, it has been questioned whether concomitant hepatitis C (HCV) infection, which produces deficits in cognitive functions (27,28), could have a further detrimental effect. In six hepatitis C-positive metamfetamine abusers, 10 hepatitis C-negative abusers, and 10 controls, MR spectroscopy showed that hepatitis C infection may worsen metamfetamine-associated neuronal injury in white-matter, as measured with N-acetylaspartate; a reduction in the concentration of this marker correlated with worse global neuropsychological deficits (29). Without a longitudinal study, it is not possible to determine whether the suggested neuronal injury is reversible. However, the findings of this study vary from those of a previous report (30).

Psychiatric

The toxic effects of methamphetamine on brain neurotransmitters have been investigated (31). Loss of the dopamine transporter in the brain due to metamfetamine abuse was studied, in order to determine its correlation with psychiatric symptoms, in 11 abstinent metamfetamine abusers and nine healthy controls. In the metamfetamine group, the mean duration of use was 4.8 years, and the mean length of abstinence was 5.6 months. Dopamine transporter density was measured by analysing PET scans of the orbitofrontal cortex, dorsolateral prefrontal cortex, and amygdala. Dopamine transporter density was significantly lower in metamfetamine abusers in all three regions. The reductions in dopamine transporter densities in the orbitofrontal and dorsolateral prefrontal cortices were significantly associated with the duration of metamfetamine use and with the severity of persistent psychiatric symptoms.

Two young women with metamfetamine abstinence developed mania after taking fluvoxamine, which was prescribed for persistent depressive symptoms (32). Both had abused methamphetamine for several years. Two weeks after starting to take fluvoxamine 100 mg/day and brotizolam 0.5 mg/day they became manic, with elevated mood, talkativeness, and increased activity. When fluvoxamine was withdrawn the manic state gradually abated and they were discharged from hospital 3 months after admission. It is not known whether a manic switch in metamfetamine users with depression is specific to fluvoxamine, or can occur with other SSRIs.

Once metamfetamine psychosis has developed, patients in remission are liable to spontaneous relapse without again consuming metamfetamine (33). It has been postulated that a sensitization phenomenon induced by repeated consumption of metamfetamine develops in the brain in patients with metamfetamine psychosis and is the neural basis for increased susceptibility to relapse (34).

Metabolism

Antioxidant systems were generally preserved in the post-mortem brains of 20 metamfetamine abusers. In the subgroup of abusers with very low dopamine concentrations,

there were changes in several antioxidant systems in the caudate. These data suggest that metamfetamine may cause dopamine-related oxidative stress in the brain (35). Animal studies also support this view (36). A massive reduction in post-mortem striatal dopamine has also been shown in metamfetamine abusers (37).

Metamfetamine abusers during early abstinence have dysfunction in the limbic and paralimbic regions, which has been linked with negative affective states (London 73). In a comparison of 17 metamfetamine abusers who abstained for 4–7 days and 18 controls, measuring mood and cerebral glucose metabolism during the performance of vigilance tasks, earlier reports of long-term neurotoxic effects of metamfetamine assessed by impaired cerebral metabolism (38,39) and reduced numbers of dopamine transporters (40) were supported. This study had limitations, including the fact that although all of the regions tested influence mood, they also contribute to other behavioral states not addressed in this study. While the groups were similar in most categories, most of the metamfetamine abusers, but none of the control subjects, were tobacco smokers; a potential confounding variable was the craving associated with abstinence in nicotine-dependent smokers. Notwithstanding these caveats, these results suggest that the long term neurotoxic effects of metamfetamine are associated with discrete changes in cerebral metabolism. It is not clear how these deficits emerge over time. Whether they are progressive, and to what extent therapy or abstinence can reverse them, is not known.

Urinary tract

Acute transient urinary retention associated with metamfetamine and ecstasy (3,4 methylenedioxymetamfetamine, MDMA) in an 18-year-old man has been described (41). Analysis by gas chromatography–mass spectrometry confirmed the presence of metamfetamine (>25 µg/ml), MDMA (> 5 µg/ml), amfetamine (1.4 µg/ml), and methylenedioxyamfetamine (3.7 µg/ml) in the urine. Bladder dysfunction resulting from alpha-adrenergic stimulation of the bladder neck may have explained the observed effect.

Death

Metamfetamine-related deaths have been reported (42).

- A 22-year-old man was found dead in a field. At autopsy, all organs were severely congested. Concentrations of metamfetamine in the heart, urine, and stomach were 0.8, 17, and 6.2 µg/ml respectively. Immunohistochemical tests on skeletal muscle showed lower than normal immunoreactivity of myoglobin, and 70-kDa heat shock protein was positive in the kidney. Because the concentration of metamfetamine in the blood was not lethal, acute intoxication was not deemed to have been the cause of death. Rather, based on the immunohistochemical findings, it was suggested that the patient had died of hyperthermia and metabolic acidosis. The patient's muscular hyperactivity had led to hyperthermia and metabolic acidosis.

• An 18-year-old woman died after taking a single dose of oral metamfetamine. The autopsy showed severe congestion and edema in the lungs. The metamfetamine concentration in blood from the heart was 17 µg/ml and metamfetamine was also found in the urine and stomach contents. This patient also had reduced concentrations of myoglobin in skeletal muscle, but there was no evidence of 70-kDa heat shock protein in the kidney, in contrast to the previous case. As the concentration of metamfetamine in the patient's heart was above the lethal concentration, she was reported to have died of acute metamfetamine intoxication.

The authors proposed that immunohistochemical staining can be useful in the diagnosis of metamfetamine poisoning.

An unexpected death due to right-sided infective endocarditis has been reported in a metamfetamine abuser (43).

• A 44-year-old female metamfetamine abuser was brought to hospital after losing consciousness. She had a low blood pressure, anuria, a high fever, and a significant increase in white blood cell count. She had many scars from previous intravenous injections and old scars of cuts from previous suicide attempts. She died 6 hours later. The postmortem g/ml. An autopsy showed mycoticμmetamfetamine blood concentration was 0.6 emboli in the pulmonary artery and primary infective endocarditis in the tricuspid valve. The emboli had also been disseminated to other organs in the body. There was no evidence of acute infection at the sites of scars or intravenous injections. She was said to have died from septicemic shock due to right-sided infective endocarditis.

According to the authors, right-sided infective endocarditis, often a consequence of intravenous drug abuse, is rare and has a better prognosis than left-sided infective endocarditis. Among drug abusers, the injected particulate matter can damage the tricuspid valve through continuous bombardment of the endothelial surface. The authors recommended a high degree of suspicion of drug abuse in right-sided infective endocarditis, even when an injection site is not visible.

Metamfetamine has been implicated in another death that was attributed to impaired performance due to nervous system effects (44).

• A 44-year-old man was found dead after his Cessna airplane crashed in mountainous terrain. A gas chromatography/mass spectrometer was used to conduct stereochemical analyses of metamfetamine and amfetamine present in the pilot's body. At autopsy the total concentration of amphetamines in the urine was 8.0 µg/ml and the blood concentration of metamfetamine was 1.13 µg/ml, enough to produce toxic effects that could have impaired his performance.

The authors concluded that he had taken oral metamfetamine in a dose high enough to impair nervous system function. A combination of poor weather conditions and metamfetamine intoxication probably caused him to crash.

Long-Term Effects

Drug abuse

Concerns have been raised in an editorial about the effects on children, through abuse and neglect, of metamfetamine abuse in parents (45). The author provocatively suggested that the drug users are not victims but rather violators of the law and thus criminals who needed to be confined when necessary and rehabilitated to protect society and families. Indeed, in some regions in the USA more than 50% of the inmates are being held on metamfetamine-related crimes. One rural county of 11 000 people in Colorado had more than 4 dozen children placed in foster care in just 1 year because of metamfetamine-related abuse and neglect. The health-care costs for dentistry, psychiatry, and social services are substantial and increasing in this group, and violence is also on the increase. This epidemic has led the US congress to form a Meth Caucus to consider legislation to chart an effective course of action.

Drug dependence

Metamfetamine dependence was associated with impairment across a range of neurocognitive domains in users whose abstinence was continually monitored with urine screening (46). The authors compared 27 metamfetamine-dependent individuals who achieved abstinence for 5–14 days and 18 control subjects and evaluated neurocognitive measures sensitive to psychomotor speed, measures of verbal learning and memory, and executive systems measures sensitive to fluency. The differential performance across the test and control groups was not attributable to demographic factors, estimated premorbid IQ, or self-reported depression. It is unlikely that the observed neurocognitive deficits resulted from residual symptoms of withdrawal, because symptoms of metamfetamine withdrawal resolved to minimal levels by the fifth day of abstinence (47). Moreover, metamfetamine-dependent individuals had poor test performances even after 12 months of abstinence, suggesting that impairment observed during the early phase of abstinence was relatively stable (48). It is possible that neurocognitive impairment from amphetamines is associated with worse functional outcomes, including poorer vocational functioning during the rehabilitation program, thus undermining the effectiveness of psychosocial treatment for metamfetamine dependence.

Genotoxicity

The genetic toxicity of metamfetamine in human abusers has been studied (49). Previous research has suggested that amfetamine can act as a mutagen, so the researchers sought to understand the genotoxicity of metamfetamine. In order to obtain samples for analysis, they recruited 76 current metamfetamine abusers from a treatment facility. The control group comprised 98 healthy volunteers. Blood samples were taken from all participants, added to cultures, and incubated for 72 hours. Assays were then run for sister chromatid exchange and micronucleus, indicators of genotoxicity. The results suggested that metamfetamine abuse significantly increases the incidence of sister chromatid exchange and micronucleus compared with healthy controls. In addition, the amount of metamfetamine consumption correlated positively with the degree of increase. These results suggest that metamfetamine is a genotoxic agent whose effects are dose-related. The authors speculated that long-term metamfetamine abuse produces high concentrations of reactive oxygen species, which in turn consume natural antioxidants and scavengers in the body. This reduction could make a person more vulnerable to genotoxicity.

Second-Generation Effects

Fetotoxicity

The effects of prenatal metamfetamine exposure on fetal growth and drug withdrawal symptoms in infants born at term have been described (50). In 294 mother–infant pairs of methamphetamine-exposed and non-exposed pairs, withdrawal symptoms requiring pharmacological intervention were observed in 4% of methamphetamine-exposed infants. There was growth restriction in methamphetamine-exposed infants and a confounding effect of nicotine. There was also growth restriction in infants who had been exposed to metamfetamine during all three trimesters of pregnancy. Since this study included only infants seen at term, the effect of metamfetamine on premature delivery was not addressed. Since toxicology screens were not conducted on control infants, it is possible that some of the neonates in the control group had also been exposed to drugs.

The effects of in utero exposure to metamfetamine on fetal growth and withdrawal symptoms have been studied (51). Consenting pregnant women with a positive urine toxicology screen were interviewed and followed until delivery. A control group of unexposed infants was obtained through newborn logbooks during the same time. A total of 294 mothers took part in the study, 134 in the exposed group and 160 in the unexposed group. There were no significant differences found between exposed and unexposed infants for any of the outcome measures. The percentage of infants who were small for gestational age was higher in the exposed group, but this did not reach significance. However, there were significant differences in the metamfetamine-exposed group. For instance, infants who were exposed to metamfetamine during all three trimesters had reduced weight and head circumference compared with infants who were exposed for only a portion of the pregnancy. Another important finding was that concurrent use of metamfetamine and nicotine resulted in significantly reduced growth parameters compared with infants who were exposed to metamfetamine alone. These results suggest that the frequency of metamfetamine use and its combination with nicotine could negatively affect the growth of a developing fetus. Withdrawal symptoms requiring pharmacological intervention were observed in 4% of metamfetamine-exposed infants.

Susceptibility factors

Genetic

Glutathione S-transferases play an important role in the defence against oxidative stress due to metamfetamine. The genes encoding glutathione S-transferases, and specifically the GSTP1 gene, may have a role in inducing genetic vulnerability (52). Genotyping in 189 metamfetamine abusers and 199 controls showed that a functional polymorphism on exon 5 of the GSTP1 gene, especially the G allele of the GSTP1 polymorphism, may contribute to a vulnerability to psychosis associated with metamfetamine abuse in the Japanese population. Specifically, variant GSTP1 genes may lead to an excess of metabolic products of the oxidative process induced by metamfetamine and may lead to metamfetamine-induced neurotoxicity, including damage to dopamine neurons.

Deletion of glutathione S-transferase M1 (GSTM1) has been examined in metamfetamine abusers (53). As GSTM1 plays a role in the antioxidant systems of the brain that help prevent neurotoxicity, the authors sought to explore the presence of the alleles of GSTM1 in 157 metamfetamine abusers and 200 controls. Female abusers had a higher frequency of the deletion allele than female controls but the effect did not reach statistical significance. Also, the frequency of carrying the deletion allele was significantly higher in female abusers than male abusers. These results suggest that deletion of the GSTM1 gene may contribute to the development of metamfetamine abuse in women.

Genetic variation in dopaminergic function may contribute to the risk of becoming a metamfetamine abuser. In a case-control study in 416 metamfetamine abusers and 435 healthy controls, two polymorphisms in genes encoding proteins of the dopaminergic system, the Val 158 Met polymorphism in the catechol-O-methyl transferase (COMT) gene and the 120-bp VNTR polymorphism in the promoter of the dopamine D4 receptor gene, were investigated for their association with metamfetamine abuse (54). All the subjects were Han Chinese from Taiwan. There was an excess of the high-activity Val158 allele in the metamfetamine abusers, consistent with previous reports of an association of this allele with drug abuse. The 120-bp VNTR polymorphism in the promoter of the dopamine D4 receptor gene itself did not show a significant association with metamfetamine abuse.

However, analysis of the 120-bp VNTR polymorphism and the exon 3 VNTR in the dopamine D4 receptor as a haplotype showed a significant association. There were interactive effects between polymorphisms in the COMT and dopamine D4 genes. However, an earlier similar study did not find such interactive effects (55). It is possible that substance abuse is a complex polygenic trait, and a number of genes may act as susceptibility factors, each gene having a weak or moderate effect on risk. The effects of the genes may be masked by the effects of other susceptibility genes, so that epistatic gene–gene interactions are more significant in determining the association (56).

Metamfetamine is reported to be the most popular drug among young abusers in Japan, where genetic factors have been studied and may contribute to vulnerability to the effects of metamfetamine. Methamphetamine-associated psychosis resembles paranoid schizophrenia. Oxidative stresses in dopaminergic pathways are postulated to underlie this neurotoxicity, and polymorphisms in the quinone oxidoreductase (NQO2) gene may contribute (57). In 191 Japanese subjects polymorphisms in the NQO1 and NQO2 genes were determined. The genotype and allele frequencies for the polymorphism (Pro 187 Ser) of the NQO1 gene did not differ across the subgroups of patients and controls. In contrast, the genotype frequency for the insertion/deletion polymorphism was significantly different in patients with prolonged-type metamfetamine psychosis. If this is confirmed, the insertion/deletion polymorphism in the promoter region of the NQO2 gene would be a specific mechanism by which genetic variation leads to a risk of metamfetamine-induced psychosis. There is evidence for dopamine–quinone mediated metamfetamine psychosis (58), and detoxification of dopamine–quinones is catalysed by the quinoreductases, NQO1 and NQO2 (59) (60).

The hDAT1 gene (SLC6A3) encodes the human dopamine transporter. A possible genetic influence of hDAT1 gene variants on the development of metamfetamine dependence or psychosis has been investigated. Analysis of four exonic polymorphisms of the hDAT gene—242 C/T (exon 2), 1342 A/G (exon 9), 2319 G/A (3'UTR of exon 15), and VNTR (3'UTR of exon 15)—in 124 Japanese patients with metamfetamine dependence or psychosis showed that the presence of nine or fewer repeat alleles in the hDAT1 gene is a strong risk factor for a worse prognosis of metamfetamine psychosis (61).

Six single nucleotide polymorphisms (SNPs) in the GABA$_A$ receptor γ2 subunit gene (GABRG2), three of which are new, have been identified (62). Two of these SNPs, 315C>T and 1128+99C>A, were used as representatives of the linkage disequilibrium blocks for further case-control association analysis. No associations were found in either allelic or genotype frequencies. There was a haplotypic association in GABRG2 with metamfetamine use disorder. These findings suggest that GABRG2 may be one of the susceptibility genes for metamfetamine use disorder.

Drug-drug interactions

Amfebutamone (bupropion)

When bupropion and metamfetamine were co-administered to 26 subjects, 20 of whom completed the protocol, there was no evidence of additive cardiovascular effects (63). The subjects received metamfetamine 0, 15, and 30 mg intravenously before and after randomization to bupropion 150 mg bd in a modified-release formulation or matched placebo. There was a non-significant trend for bupropion to reduce metamfetamine-associated increases in blood pressure and a significant reduction in the metamfetamine-associated increase in heart rate. Bupropion reduced the plasma clearance of metamfetamine and the appearance of amfetamine in the plasma. Metamfetamine did not alter the peak and trough concentrations of bupropion or its metabolites. These findings are relevant to the potential use of bupropion in ameliorating acute abstinence in metamfetamine users. However, the risk of seizures during bupropion treatment for metamfetamine abuse has not been estimated.

Morphine

There have been two new reports of the fatal combination of metamfetamine with morphine (64).

- A 43-year-old man was found dead in bed after injecting metamfetamine and morphine the night before. An autopsy showed mild edema of the brain and lungs, fatty liver, and active HCV hepatitis. The postmortem findings suggested hyperthermia. The blood metamfetamine concentration was 550 ng/ml, and the blood morphine concentration was 760 ng/ml.
- A 21-year-old man was found dead in bed after injecting metamfetamine and morphine the previous evening. He had severe edema of the brain and lungs, swollen tonsils, and enlarged deep cervical lymph nodes. The postmortem findings suggested hyperthermia. The blood metamfetamine concentration was 2640 ng/ml, and the blood morphine concentration was 500 ng/ml.

In both cases, the blood metamfetamine concentration was less than the lethal concentration of 4.5 µg/ml. Morphine concentrations were higher than the non-toxic concentration of 0.3 µg/ml. It is unlikely that morphine was the cause of death, because it would have caused hypothermia instead of hyperthermia. It is more likely that morphine interacted with metamfetamine, increasing the hyperthermic effect that is typical of metamfetamine overdose. This would explain why hyperthermia caused death, despite a non-lethal blood concentration of metamfetamine.

Management of adverse reactions

Many have raised concerns about the public health problem created by the use of metamfetamine, especially in urban gay and bisexual men. Some communities have prevalences of metamfetamine use 20 times that in the general population. High-risk sexual behaviors facilitated by metamfetamine has been consistently associated with a high rate of HIV infection. Cognitive behavioral therapy and contingency management techniques have been used in the treatment of cocaine dependence and are now being tried for metamfetamine dependence. In a randomized controlled trial of four behavioral treatments (cognitive behavioral therapy, contingency management, cognitive behavioral therapy + contingency management, and gay-specific cognitive behavioral therapy) for 16 weeks with follow-up for 1 year in 162 treatment-seeking, metamfetamine-dependent gay and bisexual men, 61% of whom were HIV-positive, treatments that included contingency management produced maximum suppression of metamfetamine use (65). The contingency management therapy included an operant reinforcement schedule that provided increasingly valuable incentives delivered in the form of vouchers for consecutive urine samples that documented abstinence; these vouchers could be exchanged for goods or services promoting an addiction-free lifestyle. Heavy users were excluded. Maximum reductions in unprotected receptive anal intercourse resulted from the gay-specific cognitive behavioral therapy, which also produced the fastest rate of reduction in reported unprotected receptive anal intercourse. Cognitive behavioral therapy + contingency management significantly reduced metamfetamine use (measured using urine drug screens) and increased attendance at therapy sessions over standard cognitive behavioral therapy. Extent of drug use and psychiatric problems reduced during the treatment period across conditions, with maintenance of improvements up to 1 year after randomization, suggesting that the specific treatments delivered to gay and bisexual men seeking treatment for metamfetamine dependence are less important than that they receive significant exposure to some treatment. The structural effects of regular clinic visits, urine screens, etc may contribute towards maintenance. Thus, drug abuse treatments merit consideration as a primary strategy for preventing HIV infection in this population.

References

1. Turnipseed SD, Richards JR, Kirk JD, Diercks DB, Amsterdam EA. Frequency of acute coronary syndrome in patients presenting to the emergency department with chest pain after methamphetamine use. J Emerg Med 2003;24:369–73.
2. Turnipseed SD, Richards JR, Kirk D, Diercks DB, Amsterdam EA. Frequency of acute coronary syndrome in patients presenting to the emergency department with chest pain after methamphetamine use. J Emerg Med 2003;24:369–73.
3. Takasaki T, Nishida N, Esaki R, Ikeda N. Unexpected death due to right-sided infective endocarditis in a methamphetamine abuser. Legal Med 2003;5:65–8.
4. Hung M-J, Kuo L-T, Cherng W-J. Amphetamine-related acute myocardial infarction due to coronary artery spasm. Intl J Clin Pract 2003;57:62–4.

5. Pavese N, Rimoldi O, Gerhard A, Brooks D, Piccini P. Cardiovascular effects of methamphetamine in Parkinson's disease patients. Movement Disord 2004;19 (3):298–30.
6. Inamasu J, Nakamura Y, Saito R, Kuroshima Y, Mayanagi K, Ohba S, Ichikizaki K. Subcortical hemorrhage caused by methamphetamine abuse: efficacy of the triage system in the differential diagnosis—case report. Neurol Med Chir (Tokyo) 2003;43:82–4.
7. McGee SM, McGee DN, McGee MB. Spontaneous intracerebral hemorrhage related to methamphetamine abuse. Am J Forensic Med Pathol 2004;25 (4):334–7.
8. Ohta K, Mori M, Yoritaka A, Okamoto K, Kishida S. Delayed ischaemic stroke associated with methamphetamine use. J Emerg Med 2005;28:165–7.
9. Newton TF, Cook IA, Kalechstein AD, Duran S, Monroy F, Ling W, Leuchter AF. Quantitative EEG abnormalities in recently abstinent methamphetamine dependent individuals. Clin Neurophysiol 2003;114:410–15.
10. Simon SL, Dacey J, Glynn S, Rawson R, Ling W. The effect of relapse on cognition in abstinent methamphetamine abusers. J Subst Abuse Treat 2004;27:59–66.
11. Kalechstein AD, Newton TF, Green M. Methamphetamine dependence is associated with neurocognitive impairment in the initial phases of abstinence. J Neuropsychiatry Clin Neurosci 2003;15:215–20.
12. Thompson PM, Hayashi KM, Simon SL, Geaga JA, Hong MS, Sui Y, Lee JY, Toga AW, Ling W, London ED. Structural abnormalities in the brains of human subjects who use methamphetamine. J Neurosci 2004;24 (26):6028–36.
13. London ED, Simon SL, Berman SM, Mandelkern MA, Lichtman AM, Bramen J, Shinn AK, Miotto K, Learn J, Dong Y, Matochik JA, Kurian V, Newton T, Woods R, Rawson R, Ling W. Mood disturbances and regional cerebral metabolic abnormalities in recently abstinent methamphetamine abusers. Arch Gen Psychiatry 2004;61 (1):73–84.
14. Thompson PM, Hayashi KM, Simon SL, Geaga JA, Hong MS, Sui Y, Lee JY, Toga AW, Ling W, London ED. Structural abnormalities in the brain of human subjects who use methamphetamine. J Neurosci 2004;24:6028–36.
15. London ED, Simon SL, Berman SM, Manderlkern MA, Lichtman AM, Bramen J, Shinn AK, Miotto K, Learn J, Dong Y, Matochik JA, Kurian V, Newton T, Woods RP, Rawson R, Ling W. Regional cerebral dysfunction associated with mood disturbances in abstinent methamphetamine abusers. Arch Gen Psychiatry 2004;61:73–84.
16. Brody AL, Mendelkern MA, Jarvik ME, Lee GS, Smith EC, Huang JC, Bota RG, Bartzokis G, London ED. Differences between smokers and non-smokers in regional gray matter volumes and densities. Biol Psychiatry 2004;55:77–84.
17. Simon SL, Dornier C, Carnell J, Brethen P, Rawson R, Ling W. Cognitive impairment in individuals currently using methamphetamine. Am J Addict 2004;9:222–31.
18. Simon SL, Dornier C, Sim T, Richardson K, Rawson R, Ling W. Cognitive performance of current methamphetamine and cocaine abusers. J Addict Dis 2002;21:61–74.
19. Kalechstein AD, Newton TF, Green M. Methamphetamine dependence is associated with neurocognitive impairment in the initial phases of abstinence. J Neuropsychiatry Clin Neurosci 2003;15:215–20.
20. Salo R, Nordahl TE, Possin K, Leamon M, Gibson DR, Galloway GP, Flynn NM, Henik A, Pfefferbaum A, Sullivan EV. Preliminary evidence of reduced cognitive inhibition in methamphetamine-dependent individuals. Psychiatry Res 2002;111:65–74.

21. Paulus MP, Hozack NE, Zauscher BE, Frank L, Brown GG, Braff DL, Schukit MA. Behavioral and functional neuroimaging evidence for prefrontal dysfunction in methamphetamine-dependent subjects. Neuropsychopharmacology 2001;26:53–63.

22. Chang L, Ernst T, Speck O, Patel H, DeSilva M, Leonido-Yee M, Miller EN. Perfusion MRI and computerized cognitive test abnormalities in abstinent methamphetamine users. Psychiatry Res 2002;114:65–79.

23. McKetin R, Mattick RP. Attention and memory in illicit amphetamine users: comparison with nondrug-using controls. Drug Alcohol Depend 1998;50:181–4.

24. Ornstein TJ, Iddon JL, Baldacchino AM, Sahakian BJ, London M, Everitt BJ, Robbins TW. Profiles of cognitive dysfunction in chronic amphetamine and heroin abusers. Neuropsychopharmacology 2000;23:113–26.

25. Simon SL, Dacey J, Glynn S, Rawson R, Long W. The effect of relapse on cognition in abstinent methamphetamine abusers. J Subst Abuse Treat 2004;27:59–69.

26. Rippeth JD, Heaton RK, Carey CL, Marcotte TD, Moore DJ, Gonzalez R, Wolfson T, Grant I. Methamphetamine dependence increases risk of neuropsychological impairment in HIV infected persons. J Int Neuropsychol Soc 2004;10:1–14.

27. Forton DM, Allsop JM, Main J, Foster GR, Thomas HC, Taylor-Robinson SD. Evidence of cerebral effect of the hepatitis C virus. Lancet 2001;358:38–9.

28. Hilsabeck RC, Perry W, Hassanein TI, Neuropsychological impairments in patients with chronic hepatitis C. Hepatology 2002;35:440–6.

29. Taylor MJ, Letendre SL, Schweinsburg BC, Al Hassoon OM, Brown GG, Gongwatana A, Grant I, and the HNRC. Hepatitis C virus infection is associated with reduced white matter N-acetylaspartate in abstinent methylamphetamine users. J Int Neuropsychol Soc 2004;10:110–3.

30. Forton DM, Thomas HC, Murphy CA, Allsop JM, Foster GR, Main J, Wesnes KA, Taylor-Robinson SD. Hepatitis C and cognitive impairment in a cohort of patients with mild liver disease. Hepatology 2002;35:433–9.

31. Sekine Y, Minabe Y, Ouchi Y, Takei N, Iyo M, Nakamura K, Suzuki K, Tsukada H, Okada H, Yoshikawa E, Futatsubashi M, Mori N. Association of dopamine transporter loss in the orbitofrontal and dorsolateral prefrontal cortices with methamphetamine-related psychiatric symptoms. Am J Psychiatry 2003;160:1699–701.

32. Won M, Minabe Y, Sekine Y, Takei N, Kondo N, Mori N. Manic switch induced by fluvoxamine in abstinent pure methamphetamine abusers. Rev Psychiatr Neurosci 2003;28:134–5.

33. Sato M, Numachi Y, Hamamura T. Relapse of paranoid psychotic state in methamphetamine model of schizophrenia. Schizophr Bull 1992;18:115–22.

34. Ujike H. Stimulant-induced psychosis and schizophrenia : the role of sensitization. Curr Psychiatry Rep 2002;4:177–84.

35. Mirecki A, Fitzmaurice P, ang L, Kalasinsky KS, Peretti FJ, Aiken SS, Wickham DJ, Sherwin A, Nobrega JN, Forman HJ, Kish SJ. Brain anti-oxidant systems in human methamphetamine users. J Neurochem 2004;89:1396–408.

36. Cadet JL, Brannock C. Free radicals and pathobiology of brain dopamine systems. Neurochem Int 1998;32:117–31.

37. Moszczynska A, Fitzmaurice P, Ang L, Kalasinsky KS, Schmunk GA, Peretti FJ, Aiken SS, Wickham DJ, Kish SJ. Why is Parkinsonism not a feature of human methamphetamine users? Brain 2004;127:363–70.

38. Ernst ND, Chang L, Leonido-Yee M, Speck O. Evidence of long-term neurotoxicity associated with methamphetamine abuse: a ^{1}H MRS study. Neurology 2000;54:1344–9.

39. Volkow ND, Chang L, Wang GJ, Fowler JS, Franceschi D, Sedler MJ, Gatley SJ, Hitzeman R, Ding YS, Wong C, Logan J. Higher cortical and lower subcortical metabolism in detoxified methamphetamine abusers. Am J Psychiatry 2001;158:383–9.

40. McCann UD, Wong DF, Yokoi F, Viltemagne V, Dannals RF, Ricaurte GA. Reduced striatal dopamine transporter density in abstinent methamphetamine and methcathinone users: evidence from positron emission tomography studies with [^{11}C]WIN- 35,428. J Neurosci 1998;18:8417–22.

41. Delgado JH, Caruso MJ, Waksman JC, Hanigman B, Stillman D. Acute transient urinary retention from combined ecstasy and methamphetamine use. J Emerg Med 2004;26:173–5.

42. Ishigami A, Kubo S, Gotohda T, Tokunaga I. The application of immunohistochemical findings in the diagnosis in methamphetamine-related death-two forensic autopsy cases. J Med Invest 2003;50:112–6.

43. Takasaki T, Nishida N, Esaki R, Ikeda N, Unexpected death due to right-sided infective endocarditis in a methamphetamine abuser. Leg Med (Tokyo) 2003;5:65–8.

44. Chaturvedi AK, Cardona PS, Soper JW, Canfield DV. Distribution and optical purity of methamphetamine found in toxic concentration in a civil aviation accident pilot fatality. J Forensic Sci 2004;49 (4):832–6.

45. Assael LA. Methamphetamine: an epidemic of oral health neglect, loss of access to care, abuse and violence. J Oral Maxillofac Surg 2005;63:1253–4.

46. Kalechstein AD, Newton TF, Green M. Methamphetamine dependence is associated with neurocognitive impairment in the initial phases of abstinence. J Neuropsychiatry Clin Neurosci 2003;15:215–20.

47. Srisurapanont M, Jarusuraisin N, Jittiwutikan J. Amphetamine withdrawal: I. Reliability, validity and factor structure of a measure. Aust NZ J Psychiatry 1999;33:89–93.

48. Volkow ND, Chang L, Wang GJ, Fowler JS, Ding YS, Sedler M, Lagan J, Franceschi D, Gatley J, Hitzemann R, Gifford A. Association of dopamine transporter reduction with psychomotor impairment in methamphetamine abusers. Am J Psychiatry 2001;158:377–88.

49. Li JH, Hu HC, Chen WB, Lin SK. Genetic toxicity of methamphetamine in vitro and in human abusers. Environ Mol Mutagen 2003;42:233–42.

50. Smith L, Yonikura ML, Wallace T, Berman N, Kuo J, Berkowitz C. Effects of prenatal methamphetamine exposure on fetal growth and drug withdrawal symptoms in infants born at term. Dev Beh Pediatrics 2003;24:17–23.

51. Smith L, Yonekura ML, Wallace T, Berman N, Kuo J, Berkowitz C. Effects of prenatal methamphetamine exposure on fetal growth and drug withdrawal symptoms in infants born at term. J Dev Behav Pediatr 2003;24:17–23.

52. Hashimoto T, Hashimoto K, Matsuzawa D, Shimizu E, Sekine Y, Inada T, Ozaki N, Iwata N, Harano M, Komiyama T, Yamada M, Sora I, Ujike H, Iyo M. A functional glutathione S-transferase P1 gene polymorphism is associated with methamphetamine-induced psychosis in Japanese population. Am J Med Gen Part B (Neuropsych Genet) 2005;135B:5–9.

53. Koizumi H, Hashimoto K, Kumakiri C, Shimizu E, Sekine Y, Ozaki N, Inada T, Harano M, Komiyama T, Yamada M, Sora I, Ujike H, Takei N, Iyo M. Association between the

glutathione S-transferase M1 gene deletion and female methamphetamine abusers. Am J Med Genet 2004;126B (Part B):43–5.

54. Li T, Chen C, Hu X, Ball D, Lin S, Chen W, Sham PC, Loh E, Murray RM, Collier DA. Association analysis of the DRDH and COMT genes in methamphetamine abusers. Am J Med Genet 2004;129B:120–4.

55. Vandenberg DJ, Rodriguez LA, Hivert E, Schiller JH, Villareal G, Pugh EW, Lachman H, Uhl GR. Long forms of the dopamine receptor (DRD4) gene VNTR are more prevalent in substance abusers: no interaction with functional alleles of the catchol-O-methyl transferase (COMT) gene. Am J Med Genet 2000;96:678–83.

56. Martinetti M, Dugoujon JM, Tinelli C, Cipriani A, Cortelazzo A, Salvaneschi L, Casali L, Semenzato G, Cuccia M, Luisetti M. HLA-Gm/kappam interaction in sarcoidosis. Suggestions for a complex genetic structure. Eur Resp J 2000;16:74–80.

57. Ohgake S, Hashimoto K, Shimizu E, Koizumi H, Okamura N, Koike K, Matsuzawa D, Sekine Y, Inada T, Ozaki N, Yamada M, Sora I, Ujike H, Shirayama Y, Iyo M. Functional polymorphism of the NQO2 gene is associated with methamphetamine psychosis. Addict Biol 2005;10:145–8.

58. Hashimoto S, Tsukada H, Nishiyama S, Fukumoto D, Kakiuchi T, Shimizu E, Iyo M. Protective effects of N-acetylcysteine on the reduction of dopamine transporters in the striatum of the monkeys treated with methamphetamine. Neuropsychopharmacology 2004;29:2018–13.

59. Ross D, Kepa JK, Winski SL, Beall HD, Anwar A, Siegel D. NAD(P)H: Quinone oxidoreductase 1 (NQO1): chemoprotection, bioactivation, gene regulation and genetic polymorphism. Chem Biol Interact 2000;129:77–97.

60. Long DJ II, Jaiswal AK. NRH: Quinone oxidoreductase 2 (NQO2). Chem Biol Interact 2000;129:99–112.

61. Ujike H, Harano M, Inada T, Yamada M, Komiyama T, Sekine Y, Sora I, Iyo M, Katsu T, Nomura A, Nakata K, Oyaki N. Nine or fewer repeat alleles in VNTR polymorphism of the dopamine transporter gene is a strong risk factor for prolonged methamphetamine psychosis. Pharmacogenomics J 2003;3:242–7.

62. Nishiyama T, Ikeda M, Iwata N, Suzuki T, Kitajima T, Yamanouchi Y, Sekine Y, Iyo M, Harano M, Komiyama T, Yamada M, Sora I, Ujike H, Inada T, Furukawa T, Ozaki N. Haplotype association between GABAA receptor γ2 subunit gene (GABRG2) and methamphetamine use disorder. Pharmacogenomics J 2005;5:89–95.

63. Newton TF, Roache JD, De la Garza R II, Fong T, Wallace CL, Li S-H, Elkashef A, Chiang N, Kahn R. Safety of intravenous methamphetamine administration during treatment with bupropion. Psychopharmacology 2005;182:426–35.

64. Uemura K, Sorimachi Y, Yashiki M, Yoshida K. Two fatal cases involving concurrent use of methamphetamine and morphine. J Forensic Sci 2003;48:1179–81.

65. Shoptaw S, Reback CJ, Peck JA, Yang X, Rotheram-Fuller E, Larkins S, Veniegas RC, Freese TE, Hucks-Ortiz C. Behavioral treatment approaches for methamphetamine dependence and HIV-related sexual risk behaviors among urban gay and bisexual men. Drug Alcohol Depend 2005;78:125–34.

METHADONE

General Information

Methadone is an MOR(OP$_3$, μ) receptor agonist with pharmacological properties similar to those of morphine. It is an attractive alternative MOR opioid receptor analgesic, because of its lack of neuroactive metabolites, a clearance that is independent of renal function, good oral systemic availability, a longer half-life with fewer doses needed per day, and extremely low cost. It is mainly metabolized by CYP3A4.

Drug studies

Pain relief

Experience with methadone in cancer pain is limited. Its long half-life tends to produce delayed toxicity, especially in older patients, but in chronic renal insufficiency and stable liver disease methadone is safe, unlike morphine.

Methadone is being used increasingly for treating chronic pain and cancer pain (neuropathic and somatic) that is non-responsive or has lost responsiveness because of tolerance to high-dose MOR opioid receptor agonists (for example morphine, fentanyl, oxycodone) (1). There are several protocols for converting from morphine to methadone and for initiating and stabilizing maintenance dosages.

In a prospective uncontrolled study, 45 patients with advanced cancer were given 0.1% methadone 2–3 times a day as required (2). Ten had nausea and vomiting, none had drowsiness, and 17 had constipation. In another study, nine of 29 patients in a tertiary level cancer pain clinic could not take opioid analgesics owing to uncomfortable adverse effects: nausea, vomiting, and drowsiness in four and other adverse effects in five (3). The average daily dose of methadone at the end of the titration phase (range 1–79 days) was 208 (range 15–1520) mg. Twenty patients had methadone toxicity during titration. In 12 patients mild drowsiness was a problem, six patients had nausea, and one patient each had confusion and severe headaches. In a third cross-sectional prospective study, 24 patients with advanced cancer pain were rapidly switched form oral morphine to oral methadone using a fixed ratio of 1:5 (4). There was a significant reduction in pain intensity and adverse effects intensity within 24 hours of substitution, although five patients required alternative treatments.

In a prospective, open, uncontrolled study 50 patients with a history of cancer taking daily oral morphine (90–800 mg) but with uncontrolled pain with or without severe opioid adverse effects were switched to oral 8-hourly methadone in a dose ratio of 1:4 for patients receiving less than 90 mg of morphine daily, 1:8 for patients receiving 90–300 mg daily, and 1:12 for patients receiving more than 300 mg daily (5). Methadone was effective in 80% of the patients when comparing analgesic response with opioid-related adverse effects. Ten patients were switched because of uncontrolled pain, eight because of moderate or severe adverse effects in the presence of acceptable pain control, and 32 because of uncontrolled pain with morphine-related adverse effects. In the last 32 there were significant improvements in pain intensity, nausea and vomiting, constipation, and drowsiness, with a 20% increase in methadone dose over and above the recommended starting dose.

In a prospective uncontrolled study of intrathecal methadone in 24 patients with a history of intractable chronic non-malignant pain, methadone was a better analgesic than morphine, with improved quality of life and no adverse effects in 13 patients (6). The final rates of methadone infusion were 20% higher than the preceding morphine rates.

Opioid dependence

An analysis of the balance of benefit to harm during methadone maintenance treatment for diamorphine dependence has shown lower mortality and morbidity with improvement in quality of life (7). The risks of methadone treatment include an increased risk of opiate overdosage during induction into treatment, and adverse effects of methadone in some patients. However, with careful management the benefits of prescribing methadone outweigh the risks.

The validity of self-reported opiate and cocaine use has been studied in 175 veterans enrolled in a methadone treatment program (8). Urine analysis showed higher rates of substance use than the patients themselves reported. The authors encouraged the development of more objective measures for assessing patient progress and the performance of the methadone program.

Restless legs syndrome

Methadone 16 mg/day was effective in 29 patients with restless legs syndrome that had not responded to dopamine receptor agonists (9). Most (n = 17) were still taking methadone at follow-up and reported a 75% reduction in symptoms. Of 27 patients, 17 reported at least one adverse event while taking methadone, including constipation (n = 11), fatigue (n = 2), and insomnia, sedation, rash, reduced libido, confusion, and hypertension (one each). Five patients stopped treatment because of adverse events.

Organs and Systems

Cardiovascular

A variety of complications following parenteral self-administration of oral methadone were noted, including regional thrombosis, often associated with shock and multiorgan failure (10).

The use of methadone/dihydrocodeine has been linked to an acute myocardial infarction (11).

- A 22-year-old man with a 6-year history of intravenous heroin use was maintained on methadone 60 mg/day and dihydrocodeine 0.5 g/day. He had an extensive anterior myocardial infarction as a result of occlusion of the left anterior descending coronary artery, which was reopened by percutaneous transluminal coronary angioplasty.

This case presents circumstantial evidence only, and the association was probably not a true one.

QT interval prolongation

Opioids block the cardiac human ether-a-go-go-related gene (HERG) potassium current in susceptible patients without any apparent heart disease and can thus prolong the QT interval (12). Two cases of QT_c interval prolongation and torsade de pointes have been reported in patients taking methadone (13).

- A patient was found unconscious with plasma concentrations of bromazepam 277 µg/ml and methadone 3500 µg/ml, both of which were above the toxic threshold. The QT_c interval was 688 ms. After initial improvement, torsade de pointes occurred and the patient was treated with DC shock 200 J, isoprenaline, magnesium, and potassium. The QT_c interval improved to 440 ms after 3 days.
- The second patient was admitted to the hospital in a comatose state. Sinus bradycardia was present with a QT_c interval of 736 ms. The plasma concentration of methadone was 1740 µg/ml. Ventricular bigeminy was followed by torsade de pointes. The patient was treated with DC shock 200 J, lidocaine, and magnesium. By the fifth day after the episode, the QT_c interval had improved to 502 ms.

This potentially life-threatening dysrhythmia has been reported previously in association with methadone and is probably under-recognized in this population. The authors did not provide details about the sex of the patients and reasons why methadone concentrations were high.

Following reports similar to those mentioned above, changes in the QT_c interval were studied in 132 heroin-dependent patients as they were starting treatment with methadone (14). After baseline electrocardiography methadone 30–150 mg/day was given and a second electrocardiogram was obtained 2 months later. Across all doses of methadone, the QT_c interval increased significantly by a mean of 11 ms over the first 2 months of treatment. No episodes of torsade de pointes were reported. Male sex and methadone doses over 110 mg/day were associated with the greatest prolongation. The average follow-up QT_c interval was 428 ms. Clinical significance is generally attributed to an increase in QT_c interval of 40 ms or greater or a value above 500 ms.

None of these patients had an increase that was above this threshold. While these results were statistically significant, the authors were not sure of their clinical significance.

The synthetic opioid levacetylmethadol, a metabolite of methadone, can also cause torsade de pointes, and its use requires electrocardiographic screening before treatment and during titration (Krantz 1615).

There have been another 11 cases showing a direct link between QT interval prolongation and oral methadone maintenance treatment at doses of 14–360 micrograms/day (15) (16). QT interval prolongation can lead to arrhythmias such as torsade de pointes, especially when high doses of methadone are given intravenously and associated with concomitant use of cocaine and/or medications that inhibit the hepatic clearance of methadone (e.g. antidepressants and antihistamines).

Methadone-related torsade de pointes has been reported in a patient with chronic bone and vaso-occlusive pain due to sickle cell disease (17).

- A 40-year-old man with sickle cell disease, hypertension, congestive heart failure, and a past history of cocaine and marihuana abuse, was given a large dose of oral methadone 560 mg/day, following hydromorphone 170 mg intravenously and by PCA for progressive back and leg pain. On day 2, he developed asymptomatic bradycardia and QT_c prolongation (454–522 msec). On day 3, he developed profuse sweating and non-sustained polymorphous ventricular tachycardia consistent with torsade de pointes. He had hypokalemia and hypocalcaemia. Echocardiography showed normal bilateral ventricular function, mild pulmonary hypertension, and trivial four-valve regurgitation. Methadone was replaced by modified-release morphine and a continuous epidural infusion of hydromorphone + bupivacaine. Daily electrocardiography showed a heart rate of 50–69/minute, a QT_c interval of 375–463 msec, and no further dysrhythmias.

This case highlights the importance of very careful monitoring especially when prescribing such large doses of methadone. The effects of methadone on cardiac function are potentially fatal.

Another report has highlighted the potential risks of combining prodysrhythmic drugs on cardiovascular function (18).

- A 39-year-old man had recurrent episodes of sinus tachycardia at 115/minute, with no other abnormalities. He was taking methadone 120 mg/day for opioid dependency and doxepin 100 mg/day for anxiety, and was given metoprolol 50 mg/day. During the next few weeks he had episodes of recurrent syncope with sinus bradycardia (47/minute) and prolongation of the QT interval (542 ms). The QT interval and heart rate normalized after withdrawal of all treatment.

In this case it is likely that the myocardial repolarization potential of methadone and doxepin may have been influenced or triggered by bradycardia induced by metoprolol. This shows the importance of cardiac monitoring in patients receiving combination therapy with potential adverse cardiac effects. Patients with co-morbidities are at high risk.

The association between methadone treatment and QT_c interval prolongation, QRS widening, and bradycardia has been explored prospectively in 160 patients with at least a 1-year history of opioid misuse (19). The QT_c interval increased significantly from baseline at 6 months (n = 149) and 12 months (n = 108). The QRS duration and heart rate did not change. There were no cases of torsade de pointes, cardiac dysrhythmias, syncope, or sudden death. There was a positive correlation between methadone concentration and the QT_c interval.

There has been a report of five cases of episodes of syncope and an electrocardiogram showing ventricular tachydysrhythmias with prolonged QT intervals and episodes of torsade de pointes; all the patients were taking high doses of methadone (270–660 mg/day) with no previous history of cardiac disease (20). Torsade de pointes also occurred when high doses (3 mg/kg) of the long-acting methadone derivative, levomethadyl acetate HCl (LAAM), were given to a 41-year-old woman with a history of heroin dependence (21). She was also taking fluoxetine and intravenous cocaine, which can prolong the QT interval, and fluoxetine and marijuana, which inhibit the activity of CYP3A4, which is responsible for the metabolism of LAAM and its active metabolite.

In a retrospective case study in methadone maintenance treatment programs in the USA and a pain management center in Canada, 17 methadone-treated patients developed torsade de pointes during 5 years (22). The dose of methadone was 65–1000 mg/day. Six patients had had an increase in methadone dose in the months just before the onset of torsade de pointes. One patient had taken nelfinavir, a potent inhibitor of CYP3A4, begun just before the development of torsade de pointes. The above two risk factors (increased drug dosage and drug interactions) are important when eliciting the cause of torsade de pointes in patients taking methadone.

Respiratory

In 10 stable patients maintained on methadone (50–120 mg/day) and nine healthy subjects assessed using polysomnography, the methadone-maintained patients had more abnormalities of sleep architecture, with a higher prevalence of central sleep apnea (23). Methadone depresses respiration, probably by acting on μ opioid receptors in the ventral surface of the medulla and possibly on other receptor sites in the lung and spinal cord. All the patients taking methadone also used benzodiazepines and cannabis, which may have influenced the above findings.

Nervous system

Reversible choreic movements of the upper limbs, torso, and speech mechanism developed in a 25-year-old man taking methadone as a heroin substitute (24).

Spastic paraparesis has been attributed to methadone (25).

- A 43-year-old patient taking methadone for pain secondary to a squamous cell carcinoma of the larynx, which progressed despite surgery and radiation therapy, developed reversible spastic paraparesis with prominent extensor spasms in the legs while receiving an infusion of high-dose intravenous methadone 100 mg/hour. On the second day, after 5 hours on 100 mg/hour, he noted weakness in both legs, uncontrollable trembling, bilateral tinnitus, and generalized anxiety. Dexamethasone 6 mg intravenously every 6 hours was started and the methadone was reduced to 60 mg/hour. Dexamethasone was withdrawn when an MRI scan confirmed the absence of metastases in the thoracic and cervical spinal cord. Because of persistent spastic paraparesis, methadone was switched to levorphanol 40 mg/hour intravenously, and there was complete resolution of symptoms 24 hours later.

Methadone can cause movement disorders characterized by tremor, choreiform movements, and a gait abnormality (26).

- A 41-year-old woman with a 15-year history of chronic neuropathic pain was given methadone 5 mg tds and then qds. One month after the final increase she had bilateral tremor spreading from her arm up to her neck, followed by choreiform movements of the torso, a broad-based gait, and staccato-like speech. She was switched from methadone to modified-release oxycodone 60 mg/day, with complete resolution after 3 weeks.

Coprolalia is a typical symptom of Tourette's syndrome that can take on a malignant quality in response to pharmacological agents. Malignant coprolalia in association with heroin abuse has been reported (27).

- A 37-year-old woman developed Tourette's syndrome at 9 years of age, with motor and phonic tics. At 25 she began to smoke heroin weekly. After 3 months, her motor tics became uncontrollable and she began to have coprolalia for the first time at a rate of about 10 words per minute. Heroin was withdrawn over 6 months but her motor tics and coprolalia did not improve, despite various drug treatments. Six months later she smoked heroin again and was readmitted with violent motor tics and constant coprolalia. She was sedated and her condition improved slightly, after which she was given sulpiride 600 mg/day and clonazepam 4 mg/day. She made a partial recovery with inadequate control of motor tics.

Since the mechanisms that underlie heroin abuse and Tourette's syndrome overlap, specifically involving dopaminergic innervation, it is likely that this condition was caused by the effects of heroin in the ventral tegmental area and its projections.

Psychological, psychiatric

In a randomized, double-blind, crossover study of 20 patients on a stable methadone regimen, a single dose of methadone caused episodic memory deficits

(28). This was significant in patients with a history of diamorphine use averaging more than 10 years duration. Such deficits can be avoided by giving methadone in divided doses.

Psychomotor and cognitive performance has been studied in 18 opioid-dependent methadone maintenance patients and 21 non-substance abusers (29). Abstinence from heroin and cocaine for the previous 24 hours was verified by urine testing. The methadone maintenance patients had a wide range of impaired functions, including psychomotor speed, working memory, decision making, and metamemory. There was also possible impairment of inhibitory mechanisms. In the areas of time estimation, conceptual flexibility, and long-term memory, the groups performed similarly.

The combinations of methadone + carbamazepine and buprenorphine + carbamazepine have been compared in the treatment of mood disturbances during the detoxification of 26 patients with co-morbidities (30). The buprenorphine combination had more of an effect. More patients taking the methadone combination dropped out of the study (58% versus 36%). However, both regimens were considered safe and without unexpected adverse effects. The results of this study need to be interpreted with caution because of the small sample size.

Endocrine

Prolonged therapy with methadone causes increases in serum thyroid hormone-binding globulin, triiodothyronine, and thyroxine, as well as albumin, globulin, and prolactin, and these must be monitored (SEDA-15, 71; SEDA-17, 81).

Fluid balance

Edema has been reported after methadone treatment (SEDA-17, 81; 31).

- A 31-year-old white man with depression, hepatitis C, and cirrhosis of the liver was hospitalized for alcohol detoxification. He had taken methadone 50 mg bd for opium dependence for 6 months. He developed bilateral pedal edema and 27 kg weight gain. There was no ascites, portal hypertension, or congestive heart failure. Most of his laboratory tests were within the reference ranges, except for reduced prothrombin time and platelet count. After stopping alcohol, his methadone dose was reduced to 60 mg/day; his edema resolved 15 days later. When the dose of methadone was increased to 70 mg/day there was a progressive increase in the edema. When methadone was withdrawn his edema completely resolved and he lost 8 kg in 2 weeks.

The exact frequency of fluid retention from methadone is not known. Based on a review of previous case reports, the authors suggested that the usual time necessary to develop edema is 3–6 months, but it can take several years. Marked fluid retention occurs mostly at high doses of methadone and the resultant edema is refractory to diuretics alone. Edema is reversible after withdrawal of methadone and recurs with re-challenge. The exact

mechanism is not clear, but it has been speculated to be related to increased secretion of antidiuretic hormone, abnormalities in the globin fraction of total serum proteins, orthostatic circulatory congestion, or release of histamines from mast cells or basophils causing increased venular permeability leading to angioedema.

Gastrointestinal

In a randomized, double-blind, placebo-controlled trial of the efficacy of intravenous methylnaltrexone (0.015–0.095 mg/kg) in treating chronic methadone-induced constipation in 22 patients attending a methadone maintenance program (oral methadone linctus 30–100 mg/day), methylnaltrexone induced immediate bowel movements in all subjects (32). There were no opioid withdrawal symptoms or significant adverse effects.

Skin

Subcutaneous administration can cause skin erythema and induration at the injection site (SEDA-16, 81).

Parenteral self-administration of oral methadone can cause cellulitis, abscess formation, and necrosis of the skin and deeper tissues (10).

Immunologic

The immunotoxic potential of methadone has been studied in rats that were given methadone 20 or 40 mg/kg/day for 6 weeks (33). The higher dose increased serum IgG concentrations but had no effect on functioning of the immune system. This suggests that methadone is not associated with immunotoxicity, even at dosages that were very high compared with usual clinical doses. The author advised caution in extrapolating animal data to humans.

Death

There has been a cross-sectional survey of 238 patients in New South Wales who died during a methadone maintenance program in a 5-year period (34). There were 50 deaths (21%) in the first week of methadone maintenance treatment, 88% of which were drug-related. These findings reinforce the importance of a thorough drug and alcohol assessment of people seeking methadone maintenance treatment, cautious prescribing of methadone, frequent clinical review of patients, and tolerance to methadone during stabilization.

In a retrospective study of cases from the Jefferson County Coroner/Medical Examiners Office, Alabama, USA between January 1982 and December 2000 there were 101 deaths in patients in whom methadone was detected in the blood (35). Methadone was the sole intoxicant in 15 cases, with a mean concentration of 0.27 µg/ml. A benzodiazepine was the most frequently detected co-intoxicant in 60 of the 101 cases and the only co-intoxicant in another 30 cases. In 26 cases methadone had been taken with a range of non-benzodiazepine substances, including antidepressants, antipsychotic drugs, antiepileptic drugs, and cocaine. The high incidence of benzodiazepine + methadone related deaths can be explained by synergistic respiratory depression. Higher concentrations of methadone can occur with chronic abuse of methadone plus benzodiazepines, because over time benzodiazepines inhibit the hepatic enzymes that metabolize methadone. This might explain why the mean methadone concentration in the 30 deaths attributed to methadone plus a benzodiazepine was only 0.6 μg/ml.

Long-Term Effects

Drug abuse

Since its first use as a treatment for opioid dependence, methadone has been the subject of much debate. Since it is itself an opioid, there is a potential for abuse. In one study all unexpected deaths positive for methadone in the Strathclyde Police region of Scotland from 1991 to 2001 were identified and were classified as being "methadone-only", "methadone-related", or "not methadone-related" (36). Of 352 cases, 82 were considered not methadone-related; the other 270 were thought to be caused by methadone alone or in combination with other drugs. Of these drug-related deaths, methadone was identified as the sole cause in 56, while methadone in combination with other drugs was responsible for 140 deaths. The other 74 cases were positive for methadone, but the concentration was not high enough to be considered contributory. When methadone-only deaths and methadone-related deaths were compared, there was a significant difference in blood methadone concentrations, which were higher on average in methadone-only cases (800 versus 400 ng/ml). This does not mean that methadone is unsafe, but rather that its use in methadone maintenance programs needs to be closely monitored.

Drug tolerance

Several methadone studies have focused on opioid-dependent or opioid-abusing subjects. For example, six opioid-dependent individuals maintained on methadone subsequently developed cancer and continued to use methadone, but in a higher dose as an analgesic (25,37). The first five were partly refractory to the analgesic effects of opioids other than methadone, but all six achieved adequate analgesia without sedation or respiratory depression from aggressive upward intravenous methadone titration using an infusion of 100 mg/hour. Methadone was given in divided doses every 6–12 hours rather than once daily, as is customary in maintenance therapy for opioid dependence. The reasons for increasing the methadone dosage and frequency of administration are cross-tolerance to other opioids and the presence in methadone-maintained individuals of hyperalgesia to pain (a low pain tolerance to pain detection ratio) (38). These issues are also relevant to determining whether other drugs are more effective than morphine in managing acute pain in these patients.

Drug withdrawal

Four patients with methadone withdrawal psychosis have been described (SEDA-20, 79).

Drug withdrawal

The neonatal abstinence syndrome occurs in 30–80% of infants whose mothers have taken opiates during pregnancy. The incidence is higher in those whose mothers have a history of opioid dependence and are taking methadone maintenance than in those who are taking methadone for chronic pain (39). The methadone blood concentration may be a useful predictor of the likelihood of severe withdrawal requiring treatment, but clinical assessment by a standardized scoring system is still required to determine the need to treat the neonatal abstinence syndrome (40).

Second-Generation Effects

Pregnancy and fetotoxicity

Methadone is extensively used in opioid withdrawal and maintenance programs (see Drug tolerance in this monograph), and has been safely used for this purpose in pregnancy, with only mild effects on the offspring (41). However, fetal exposure to methadone in utero can cause a neonatal abstinence syndrome after delivery.

The outcomes in 100 chronic opiate-dependent pregnant women who received levomethadone substitution treatment have been reported (42). The average gestational age at delivery was 38 weeks and the mean birth weight was 2869 g. The rate of premature labor was 19% and the risk of premature delivery 11%. There were withdrawal symptoms in 74% of the neonates at a mean of 39 hours and all responded well to levomethadone.

A newborn girl born of an HIV-positive mother who took antiretroviral drugs and methadone during pregnancy developed a methadone abstinence syndrome at day 7 (43). She was HIV-negative and was treated symptomatically for 15 days with chlorpromazine. The platelet count was 1049×10^9/l on day 17 and fell progressively to 290×10^9/l at 8 weeks. The authors suggested that the thrombocytosis had been secondary to intrauterine methadone exposure.

In a randomized controlled trial in 18 pregnant women in the second trimester, a change from short-acting morphine to methadone or buprenorphine was explored (44). The transition was accomplished without any adverse events in mother or fetus and with minimal withdrawal discomfort.

In 42 methadone-maintained women methadone had profound effects on fetal neurobehavioral functioning, implying a disruption of or threat to fetal neural development (45). At peak concentrations the fetuses had slower heart rates, less heart rate variability, fewer heart rate accelerations, reduced duration of movements, reduced motor activity, and a lower degree of coupling between fetal movement and fetal heart rate. The long-term effects of such daily changes in the fetus are not known. There were very few effects on maternal physiology.

Susceptibility Factors

From a literature search and subsequent analysis of data on the relation between methadone prescribing and mortality, it was concluded that (46):

(a) 69% of deaths attributed to methadone occurred in subjects who had not previously received methadone;
(b) 51% of deaths attributed to methadone occurred during the dose-stabilizing period of methadone maintenance treatment;
(c) the dose of illicit methadone exceeded that prescribed for methadone maintenance therapy;
(d) deaths were attributed to discharge from prison and immediate intravenous injection of methadone in people who had lost their tolerance to high doses of methadone when incarcerated.

Subsequent advice related to the above identifiable susceptibility factors included:

(a) restriction of take-home prescriptions with daily supervised consumption of methadone in pharmacy premises;
(b) meticulous evaluation of substance abuse history;
(c) slowing down of increases and tolerance testing during the stabilization period of methadone maintenance; enhanced psychosocial assistance during the first months out of prison;
(d) use of naloxone as an adjunct to methadone syrup.

Drug Administration

Drug dosage regimens

The role of opioid rotation in cancer pain management has been described, highlighting the limitations of equianalgesic tablets and the need for monitoring and individualization of dose. This is particularly important when methadone is used as the opioid for conversion. The authors referred to a greater than expected potency of methadone, with excessive sedation and opioid-related adverse effects, if the switch is done on a one-to-one basis. They suggested that the calculated equianalgesic dose of methadone should be reduced by 75–90% and the dose then titrated upwards if necessary (47,48).

Drug administration route

Methadone has been used for intrathecal administration. Although this route can provide prolonged analgesia, the adverse effects have been reported to be unacceptable (SEDA-16, 81).

Of 90 patients undergoing abdominal or lower limb surgery randomly assigned double-blind to two groups, 60 received racemic methadone in initial doses

of 3–6 mg followed by 6–12 mg by continuous infusion over 24 hours, and 30 received repeated boluses of 3–6 mg every 8 hours (49). In both groups the highest visual analogue score occurred 2 hours after surgery. From then on the pain diminished gradually and significantly at each recording. Opioid-related adverse effects were not different between the two groups, except for miosis, which was significantly more common in the bolus group. The results suggested that both epidural methadone protocols used in this study provide effective and safe postoperative analgesia. However, the infusion method should be preferred, as the doses of methadone can be reduced after the first day of treatment.

Drug–Drug Interactions

Antiretroviral drugs

Methadone is often used for opioid replacement therapy in intravenous drug abusers. The incidence of HIV infection is significantly higher in this population than in the general public, and interactions with drugs used for the treatment of AIDS are therefore important.

Methadone is predominantly metabolized by CYP3A4. Antiretroviral therapy with a non-nucleoside reverse transcriptase inhibitor (for example efavirenz, abacavir, and nevirapine) and/or a protease inhibitor (for example amprenavir) will induce the metabolism of methadone. This therapeutic combination is becoming increasingly common in HIV-positive substance misusers. Two studies have explicitly shown a significant reduction of methadone concentration by 28–87%. In the first study, 11 patients taking methadone maintenance therapy were given efavirenz and had a mean increase in methadone dosage requirement of 22% (50). In the second study, five methadone-maintained opioid-dependent individuals were given a combination of abacavir and amprenavir; the methadone concentration fell to 35% of the original concentration within 14 days (51).

In a prospective study of 54 patients taking antiretroviral drugs who also took methadone and a further 154 patients who did not take methadone there were similar clinical, virological, and immunological outcomes after 12 months (52). These results support the usefulness of methadone in the management of intravenous drug users with HIV infection.

Protease inhibitors

In an in vitro study of the effects of the HIV-1 protease inhibitors, ritonavir, indinavir, and saquinavir, which are metabolized by the liver CYP3A4, all three protease inhibitors inhibited methadone demethylation and buprenorphine dealkylation in rank order of potency ritonavir > indinavir > saquinavir (53). Clinical studies are required to establish the further relevance of these observations.

Zidovudine

The metabolism of the antiviral nucleoside zidovudine to the inactive glucuronide form in vitro was inhibited by methadone (54). The concentration of methadone required for 50% inhibition was over 8 µg/ml, a supratherapeutic concentration, thus raising questions about the clinical significance of the effect. However, in eight recently detoxified heroin addicts, acute methadone treatment increased the AUC of oral zidovudine by 41% and of intravenous zidovudine by 19%, following the start of oral methadone (50 mg/day) (55). These effects resulted primarily from inhibition of zidovudine glucuronidation, but also from reduced renal clearance of zidovudine, and methadone concentrations remained in the target range throughout. It is recommended that increased toxicity surveillance, and possibly reduction in zidovudine dose, are indicated when the two drugs are co-administered.

Cimetidine

Cimetidine increases the effects of methadone, probably by inhibition of methadone metabolism (56).

Drugs that affect CYP3A

In a randomized four-way crossover study in healthy subjects, the effects of intravenous and oral methadone were measured after pre-treatment with rifampicin (hepatic/intestinal CYP3A induction), troleandomycin (hepatic/intestinal CYP3A inhibition), grapefruit juice (selective intestinal CYP3A inhibition), or nothing (57). Intestinal and hepatic CYP3A activity affected methadone N-demethylation only slightly and had no significant effects on methadone concentrations, clearance, or clinical effects. There was a significant correlation between methadone oral availability and intestinal availability, since only rifampicin altered oral methadone availability. This suggests a role of intestinal metabolism and in first-pass extraction of methadone. This study used a single, relatively low dose of methadone (15 micrograms) rather than a therapeutic dose at steady state (80–100 micrograms/day), when tolerance will be taken into consideration.

Enzyme inducers

Enzyme-inducing drugs, such as carbamazepine, phenobarbital, phenytoin, and rifampicin, enhance the metabolism of methadone, leading to lower serum methadone concentrations (58).

Fluconazole

In a randomized, double-blind, placebo-controlled trial, oral fluconazole increased the serum methadone AUC by 35% (59). Although renal clearance was not significantly affected, mean serum methadone peak and trough concentrations rose significantly, while renal clearance was not significantly altered.

Grapefruit juice

In an unblinded study, the effect of grapefruit juice on the steady-state pharmacokinetics of methadone was evaluated for 5 days in eight patients taking methadone (mean dose 107, range 63–150, micrograms/day) (60). Grapefruit juice was associated with a modest increase in methadone availability that would not normally enhance its adverse effects. Only 6–8 glasses of grapefruit juice per day can lead to inhibition of hepatic CYP3A. Further studies need to be done to clarify to what extent intestinal and/or hepatic metabolic activities play a part in methadone availability and the subsequent risk of overdosage in individuals taking high maintenance doses of methadone. Since the therapeutic effect of methadone is mainly mediated by the R-enantiomer, monitoring plasma concentrations of R-methadone could be recommended, but it is an imprecise indicator of therapeutic activity (61).

Phenytoin

Phenytoin enhances the metabolism of methadone (59).

Rifampicin

Enzyme-inducing drugs, such as rifampicin, enhance the metabolism of methadone, leading to lower serum methadone concentrations (59). This interaction is thought to have caused acute methadone withdrawal symptoms in two patients with AIDS (SEDA-16, 81).

Selective serotonin re-uptake inhibitors

Fluvoxamine

Fluvoxamine increases the effects of methadone, probably by inhibition of methadone metabolism (62).

Paroxetine

Paroxetine 20 mg/day, a selective CYP2D6 inhibitor, was given for 12 days to 10 patients on methadone maintenance (63). Eight were genotyped as CYP2D6 homozygous extensive metabolizers and two as poor metabolizers. Paroxetine increased the steady-state concentrations of R-methadone and S-methadone, especially in the extensive metabolizers.

Use of methadone in opioid withdrawal

A widely used technique for opioid detoxification, pioneered by Isbell and Vogel (64), involves the substitution of methadone for the illicit opioid, followed by a gradual reduction in the amount of methadone taken.

Methadone maintenance treatment was established in 1964 in New York City by Vincent Dole and Marie Nyswander. In the initial studies, subjects who were heavily addicted to heroin were evaluated and stabilized on daily methadone doses as inpatients before transfer to an outpatient clinic for continued treatment. With further experience, it was feasible to drop the inpatient phase (65).

Methadone is used to substitute for a variety of opioid drugs. It is well absorbed after oral ingestion, with peak blood concentrations after about 4 hours. Steady-state concentrations are reached after about 5 days. By virtue of its long duration of action (the half-life with regular dosing is about 22 hours), methadone suppresses opioid withdrawal symptoms for 24–36 hours. In the early stages of treatment patients may report problems such as drowsiness, insomnia, nausea, euphoria, difficulty in micturition, and excessive sweating. With the exception of chronic constipation and excessive sweating, these effects do not generally persist.

Methadone maintenance treatment is considered to be a medically safe treatment with relatively few and minimal adverse effects. However, the danger of serious adverse effects and death with the increasing use of methadone as maintenance therapy in drug addicts has been highlighted. It must be emphasized that a daily maintenance dose of 50–100 mg is toxic in a non-tolerant adult and as little as 10 mg can be fatal in a child. There is an increasing number of reports of the deaths of children of mothers on maintenance therapy from inadvertent ingestion.

British studies have shown that, using methadone, about 80% of inpatients, but only 17% of outpatients, were successfully withdrawn (66,67). However, the technique is not without problems, one being that the methadone reduces but does not eliminate withdrawal symptoms. The withdrawal response has been described as being akin to a mild case of influenza, objectively mild but subjectively severe (68). The fear of withdrawal symptoms expressed by those dependent on drugs should not be underestimated: these factors are associated with the subsequent severity of withdrawal symptoms, and they are more closely related to symptom severity than drug dosage (69). Methadone substitution can result in a protracted withdrawal response, with patients still experiencing significantly more symptoms than controls 2 weeks after withdrawal (70).

In a study of methadone withdrawal, patients who were withdrawn over 10 days had a withdrawal syndrome that began to increase in severity from day 3, with peak severity of symptoms on day 13; in those who were withdrawn over 21 days, symptoms began to increase about day 10 with a peak on day 20 and abated thereafter, although some patients did not recover fully until 40 days after starting withdrawal (71). Thus, the duration of the withdrawal syndrome is much the same for both treatments in terms of symptom severity. It is possible that an exponential rather than linear reduction in dosage may improve the withdrawal response. These results may be of clinical significance, in that patients may feel it important that they recover from withdrawal as quickly as possible, in order to participate fully in other aspects of drug withdrawal programs. However, although there was no difference between the 10-day and 21-day programs regarding completion rates for detoxification (70 and 79% respectively), the dropout rates after detoxification were significantly different. During the 10 days after the last dose of methadone, the dropout rate in the 21-day group was 18% compared with 30% in the 10-day group. These results may also have financial implications in respect of the number of subjects who can be admitted to treatment programs.

In some treatment programs, total abstinence is not considered to be a practical objective and treatment may involve the use of drugs such as methadone as maintenance therapy with the expectation of reducing illicit drug consumption (72). Well-organized methadone maintenance treatment can reduce the intake of illicit opioids in many injecting drug users (73,74).

Outcome studies of methadone maintenance treatment have reported favorable results. High rates of patient retention, reduced criminality, and improved social rehabilitation are reported. Despite its proved effectiveness, it remains a controversial approach among substance abuse treatment providers, public officials, policy makers, the medical profession, and the public at large. Nevertheless, almost every nation with a significant narcotic addiction problem has established a methadone maintenance treatment program.

For patients entering treatment from an institution where they have been drug-free, initial daily methadone doses should be no more than 20 mg. Otherwise initial daily doses of 30–40 mg should be sufficient to obtain the necessary balance between withdrawal and narcotic symptoms. Thereafter, stabilization is achieved by gradually increasing the dose. When methadone is given in adequate oral doses (usually 60 mg/day or more), a single dose in a stabilized patient lasts 24–36 hours, without creating euphoria and sedation. Tolerance to methadone seems to remain steady, and patients can be maintained on the same dose, in some cases for more than 20 years. The methadone dose must be determined individually, owing to individual variability in pharmacokinetics and pharmacodynamics. Maintenance of appropriate methadone blood concentrations is recommended.

Tolerance to the narcotic properties of methadone develops within 4–6 weeks, but tolerance to the autonomic effects (for example constipation and sweating) develops more slowly.

The major adverse effects during treatment occur during the initial stabilization phase. In addition to constipation and sweating, the most frequently reported adverse effects are transient skin rash, weight gain, and fluid retention. Since the main metabolic pathway of methadone is CYP3A4, numerous drug interactions can be expected. Drugs that interact with methadone are listed in the table in the monograph on opioids.

References

1. Ayonrinde OT, Bridge DT. The rediscovery of methadone for cancer pain management. Med J Aust 2000;173(10):536–40.
2. Mercadante S, Casuccio A, Agnello A, Barresi L. Methadone response in advanced cancer patients with pain followed at home. J Pain Symptom Manage 1999;18(3):188–92.
3. Hagen NA, Wasylenko E. Methadone: outpatient titration and monitoring strategies in cancer patients. J Pain Symptom Manage 1999;18(5):369–75.
4. Mercadante S, Casuccio A, Calderone L. Rapid switching from morphine to methadone in cancer patients with poor response to morphine. J Clin Oncol 1999;17(10):3307–12.
5. Mercadante S, Casuccio A, Fulfaro F, Groff L, Boffi R, Villari P, Gebbia V, Ripamonti C. Switching from morphine to methadone to improve analgesia and tolerability in cancer patients: a prospective study. J Clin Oncol 2001; 19(11):2898–904.

6. Mironer YE, Tollison CD. Methadone in the intrathecal treatment of chronic nonmalignant pain resistant to other neuraxial agents: the first experience. Neuromodulation 2001;4:25–31.

7. Bell J, Zador D. A risk-benefit analysis of methadone maintenance treatment. Drug Saf 2000;22(3):179–90.

8. Chermack ST, Roll J, Reilly M, Davis L, Kilaru U, Grabowski J. Comparison of patient self-reports and urinalysis results obtained under naturalistic methadone treatment conditions. Drug Alcohol Depend 2000;59(1):43–9.

9. Ondo WG. Methadone for refractory restless legs syndrome. Movement Disord 2005;20(3):345–8.

10. Nathan HJ. Narcotics and myocardial performance in patients with coronary artery disease. Can J Anaesth 1988; 35(3 Pt 1):209–13.

11. Backmund M, Meyer K, Zwehl W, Nagengast O, Eichenlaub D. Myocardial Infarction associated with methadone and/or dihydrocodeine. Eur Addict Res 2001; 7(1):37–9.

12. Krantz MJ, Mehler PS. Synthetic opioids and QT prolongation. Arch Intern Med 2003;163:1615; author reply 1615.

13. De Bels D, Staroukine M, Devriendt J. Torsades de pointes due to methadone. Ann Intern Med 2003;139:E156.

14. Martell BA, Arnsten JH, Ray B, Gourevitch MN. The impact of methadone induction on cardiac conduction in opiate users. Ann Intern Med 2003;139:154–5.

15. Piquet V, Desmeules J, Enret G, Stoller R, Dayer P. QT interval prolongation in patients on methadone with concomitant drugs. J Clin Psychopharmacol 2004;24:446–8.

16. Decerf JA, Gressens B, Brohet C, Liolios A, Hantson P. Can methadone prolong the QT interval? Intensive Care Med 2004;30:1690–1.

17. Porter BO, Coyn PJ, Smith WR. Methadone-related torsade de pointes in a sickle cell patient treated for chronic pain. Am J Hematol 2005;78(4):316–7.

18. Rademacher S, Dietz R, Haverkamp W. QT prolongation and syncope with methadone, doxepin, and a beta-blocker. Ann Pharmacother 2005;39(10):1762–3.

19. Martell BA, Arnsten JH, Krantz MJ, Gourevitch MN. Impact of methadone treatment on cardiac repolarisation and conduction in opioid users. Am J Cardiol 2005;95(7):915–8.

20. Hays H, Woodroffe MA. High dosing methadone and a possible relationship to serious cardia arrhythmias. Pain Res Manag 2001;6(2):64.

21. Deamer RL, Wilson DR, Clark DS, Prichard JG. Torsades de pointes associated with high dose levomethadyl acetate (ORLAAM). J Addict Dis 2001;20(4):7–14.

22. Krantz MJ, Lewkowiez L, Hays H, Woodroffe MA, Robertson AD, Mehler PS. Torsade de pointes associated with very-high-dose methadone. Ann Intern Med 2002;137(6):501–4.

23. Teichtahl H, Prodromidis A, Miller B, Cherry G, Kronborg I. Sleep-disordered breathing in stable methadone programme patients: a pilot study. Addiction 2001; 96(3):395–403.

24. Wasserman S, Yahr MD. Choreic movements induced by the use of methadone. Arch Neurol 1980;37(11):727–8.

25. Manfredi PL, Gonzales GR, Payne R. Reversible spastic paraparesis induced by high-dose intravenous methadone. J Pain 2001;2(1):77–9.

26. Clark JD, Elliott J. A case of a methadone-induced movement disorder. Clin J Pain 2001;17(4):375–7.

27. Berthier ML, Campos VM, Kulisevsky J, Valero JA. Heroin and malignant coprolalia in Tourette's syndrome. J Neuropsychiatry Clin Neurosci 2003;15:116–7.

28. Curran HV, Kleckham J, Bearn J, Strang J, Wanigaratne S. Effects of methadone on cognition, mood and craving in detoxifying opiate addicts: a dose-response study. Psychopharmacology (Berl) 2001;154(2):153–60.

29. Mintzer MZ, Stitzer ML. Cognitive impairment in methadone maintenance patients. Drug Alcohol Depend 2002;67(1):41–51.

30. Seifert J, Metzner C, Paetzold W, Borsutzky M, Ohlmeier M, Passie T, Hauser U, Becker H, Wiese B, Emrich HM, Schneider U. Mood and affect during detoxification of opiate addicts: a comparison of buprenorphine versus methadone. Addiction Biol 2005;10:157–64.

31. Mahe I, Chassany O, Grenard AS, Caulin C, Bergmann JF. Methadone and edema: a case-report and literature review. Eur J Clin Pharmacol 2004;59(12):923–4.

32. Yuan CS, Foss JF, O'Connor M, Osinski J, Karrison T, Moss J, Roizen MF. Methylnaltrexone for reversal of constipation due to chronic methadone use: a randomized controlled trial. JAMA 2000;283(3):367–72.

33. Ryle PR. Justification for routine screening of pharmaceutical products in immune function tests: a review of the recommendations of Putman et al.(2003). Fundam Clin Pharmacol 2005;19:317–22.

34. Zador D, Sunjic S. Deaths in methadone maintenance treatment in New South Wales, Australia 1990–1995. Addiction 2000;95(1):77–84.

35. Mikolaenko I, Robinson CA Jr, Davis GG. A review of methadone deaths in Jefferson County, Alabama. Am J Forensic Med Pathol 2002;23(3):299–304.

36. Seymour A, Black M, Jay J, Cooper G, Weir C, Oliver J. The role of methadone in drug-related deaths in the west of Scotland. Addiction 2003;98:995–1002.

37. Manfredi PL, Gonzales GR, Cheville AL, Kornick C, Payne R. Methadone analgesia in cancer pain patients on chronic methadone maintenance therapy. J Pain Symptom Manage 2001;21(2):169–74.

38. Doverty M, Somogyi AA, White JM, Bochner F, Beare CH, Menelaou A, Ling W. Methadone maintenance patients are cross-tolerant to the antinociceptive effects of morphine. Pain 2001;93(2):155–63.

39. Sharpe C, Kuschel C. Outcomes of infants born to mothers receiving methadone for pain management in pregnancy. Arch Dis Child Fetal Neonatal Ed 2004;89:F33–F36.

40. Kuschel CA, Austerberry L, Cornwell M, Couch R, Rowley RSH. Can methadone concentrations predict the severity of withdrawal in infants at risk of neonatal abstinence syndrome? Arch Dis Child Fetal Neonatal Ed 2004;89:F390–F393.

41. Pinto F, Torrioli MG, Casella G, Tempesta E, Fundaro C. Sleep in babies born to chronically heroin addicted mothers. A follow up study Drug Alcohol Depend 1988;21(1):43–7.

42. Kastner R, Hartl K, Lieber A, Hahlweg BC, Knobbe A, Grubert T. Substitutionsbehandlung von opiatabhängigen schwangeren'—Analyse der Behandlungverläufe an der 1. Ufk München. [Maintenance therapy in opiate-dependent pregnant patients—analysis of the course of therapy at the clinic of the University of Munich.] Geburtschilfe Frauenheilkd 2002;62:32–6.

43. Garcia-Algar O, Brichs LF, Garcia ES, Fabrega DM, Torne EE, Sierra AM. Methadone and neonatal thrombocytosis. Pediatr Hematol Oncol 2002;19(3):193–5.

44. Jones HE, Johnson RE, Jasinski DR, Milio L. Randomized controlled study transitioning opioid-dependent pregnant women from short-acting morphine to buprenorphine or methadone. Drug Alcohol Depend 2005;78, 33–8.

45. Jannson Lm, DiPietro, J, Elko A. Fetal response to maternal methadone administration. Am J Obstet Gynaecol 2005;193(3):611–7.

46. Vormfelde SV, Poser W. Death attributed to methadone. Pharmacopsychiatry 2001;34(6):217–22.

47. Indelicato RA, Portenoy RK. Opioid rotation in the management of refractory cancer pain. J Clin Oncol 2002;20(1):348–52.

48. Watanabe S, Tarumi Y, Oneschuk D, Lawlor P. Opioid rotation to methadone: proceed with caution. J Clin Oncol 2002;20(9):2409–10.

49. Prieto-Alvarez P, Tello-Galindo I, Cuenca-Pena J, Rull-Bartomeu M, Gomar-Sancho C. Continuous epidural infusion of racemic methadone results in effective postoperative analgesia and low plasma concentrations. Can J Anaesth 2002;49(1):25–31.

50. Bart PA, Rizzardi PG, Gallant S, Golay KP, Baumann P, Pantaleo G, Eap CB. Methadone blood concentrations are decreased by the administration of abacavir plus amprenavir. Ther Drug Monit 2001;23(5):553–5.

51. Clarke SM, Mulcahy FM, Tjia J, Reynolds HE, Gibbons SE, Barry MG, Back DJ. The pharmacokinetics of methadone in HIV-positive patients receiving the non-nucleoside reverse transcriptase inhibitor efavirenz. Br J Clin Pharmacol 2001;51(3):213–7.

52. Moreno A, Perez-Elias MJ, Casado JL, Munoz V, Antela A, Dronda F, Navas E, Moreno S. Long-term outcomes of protease inhibitor-based therapy in antiretroviral treatment-naive HIV-infected injection drug users on methadone maintenance programmes. AIDS 2001;15(8):1068–70.

53. Iribarne C, Berthou F, Carlhant D, Dreano Y, Picart D, Lohezic F, Riche C. Inhibition of methadone and buprenorphine N-dealkylations by three HIV-1 protease inhibitors. Drug Metab Dispos 1998;26(3):257–60.

54. Trapnell CB, Klecker RW, Jamis-Dow C, Collins JM. Glucuronidation of 3′-azido-3′-deoxythymidine (zidovudine) by human liver microsomes: relevance to clinical pharmacokinetic interactions with atovaquone, fluconazole, methadone, and valproic acid. Antimicrob Agents Chemother 1998;42(7):1592–6.

55. McCance-Katz EF, Rainey PM, Jatlow P, Friedland G. Methadone effects on zidovudine disposition (AIDS Clinical Trials Group 262). J Acquir Immune Defic Syndr Hum Retrovirol 1998;18(5):435–43.

56. Dawson GW, Vestal RE. Cimetidine inhibits the in vitro N-demethylation of methadone. Res Commun Chem Pathol Pharmacol 1984;46(2):301–4.

57. Kharasch ED, Hoffer C, Whittington D, Sheffels P. Role of hepatic and intestinal cytochrome P450 3A and 2B6 in the metabolism, disposition and mioitic effects of methadone. Clin Pharmacol Ther 2004;76:250–69.

58. Finelli PF. Letter: Phenytoin and methadone tolerance. N Engl J Med 1976;294(4):227.

59. Cobb MN, Desai J, Brown LS Jr, Zannikos PN, Rainey PM. The effect of fluconazole on the clinical pharmacokinetics of methadone. Clin Pharmacol Ther 1998;63(6):655–62.

60. Benmerbarek M, Devaud C, Gex-Fabry M, Powell Golay K, Brogli C, Baumann P, Gravier B, Eap C B. Effects of grapefruit juice on the pharmacokinetics of the enantiomers of methadone. Clin Pharmacol Ther 2004;76:55–63.

61. Esteban J, de la Cruz Pellin M, Gimeno C, Barril J, Mora E, Gimenez J, Vilenova E. Detection of clinical interactions between methadone and anti-retroviral compounds using an enantioselective capillary electrophoresis for methadone analysis. Toxicol Lett 2004;151:243–9.

62. Iribarne C, Picart D, Dreano Y, Berthou F. In vitro interactions between fluoxetine or fluvoxamine and methadone or buprenorphine. Fundam Clin Pharmacol 1998;12(2):194–9.

63. Begre S, von Bardeleben U, Ladewig D, Jaquet-Rochat S, Cosendai-Savary L, Golay KP, Kosel M, Baumann P, Eap CB. Paroxetine increases steady-state concentrations of (R)-methadone in CYP2D6 extensive but not poor metabolizers. J Clin Psychopharmacol 2002;22(2):211–5.

64. Isbell H, Vogel VH, Chapman KW. Present status of narcotic addiction with particular reference to medical indications and comparative addiction liability of the newer and older analgesic drugs. JAMA 1948;138:1019.

65. Dole VP, Nyswander M. A medical treatment for diacetyl-morphine (heroin) addiction. A clinical trial with methadone hydrochloride. JAMA 1965;193:646–50.

66. Glossop M, Johns A, Green L. Opiate withdrawal: in-patient vs out-patient programmes and preferred vs random assignment to treatment. BMJ (Clin Res Ed) 1986;293:103.

67. Gossop M, Green L, Phillips G, Bradley B. What happens to opiate addicts immediately after treatment: a prospective follow up study. BMJ (Clin Res Ed) 1987;294(6584):1377–80.

68. Kleber HD. Detoxification from narcotics. In: Lowinson L, Ruiz P, editors. Substance Abuse. Baltimore: Williams and Wilkins, 1981:317.

69. Phillips GT, Gossop M, Bradley B. The influence of psychological factors on the opiate withdrawal syndrome. Br J Psychiatry 1986;149:235–8.

70. Gossop M, Bradley B, Phillips GT. An investigation of withdrawal symptoms shown by opiate addicts during and subsequent to a 21-day in-patient methadone detoxification procedure. Addict Behav 1987;12(1):1–6.

71. Gossop M, Griffiths P, Bradley B, Strang J. Opiate withdrawal symptoms in response to 10-day and 21-day methadone withdrawal programmes. Br J Psychiatry 1989;154:360–3.

72. Newman RG, Whitehill WB. Double-blind comparison of methadone and placebo maintenance treatments of narcotic addicts in Hong Kong. Lancet 1979;2(8141):485–8.

73. Lowinson JH, Marion IJ, Joseph H, Dole VP. Methadone maintenance. In: Lowinson JH, Ruiz P, Millman RB, editors. Substance Abuse. A Comprehensive Textbook. 2nd ed. Baltimore: Williams and Wilkins, 1992:550.

74. Ball JC, Ross A. The Effectiveness of Methadone Maintenance TreatmentNew York: Springer-Verlag;. 1991.

METHYLENEDIOXYMETAMFETAMINE (MDMA, ECSTASY)

See also Amphetamines

General Information

Methylenedioxymetamfetamine (MDMA), commonly known by names such as "ecstasy", "XTC", "E", or "Love Drug", was synthesized in 1914 for use in chemical warfare, but has more recently become a popular drug of abuse among young people, especially at "raves". It is relatively easy to obtain and is erroneously regarded as a safe drug. However, it has actions like those of amphetamine.

It is the usual and expected constituent of the tablets that are known as ecstasy, although adulteration with other substances is not uncommon (see the section on Drug contamination in this monograph). MDMA and other drugs, such as its *N*-demethylated derivative (MDA), 3,4-methylenedioxyamfetamine (MDA), 3,4-methylenedioxyethylamfetamine (MDEA), *N*-methyl-benzodioxazolylbutamine (MBDB), and 4-bromo-2,5-dimethoxyphenylethylamine (2-CB or Nexus), are often grouped together as "ecstasy". Some have used the term "enactogen", meaning "touching within", to describe ecstasy.

Clinical effects

Ecstasy that is sold on the streets is a heterogeneous substance, with enormous variations in its main active ingredients. It most often contains derivatives of MDMA and 3,4-methylenedioxy-*N*-ethylamfetamine (MDEA). Other amphetamine derivatives that it can contain include 3,4-MDA, *N*-methyl-1-(1,3-benzodioxol-5-yl)-2-butanamine (MBDB), and 2,5-dimethoxy-4-bromamfetamine (DOB). The amount of active ingredient in street ecstasy ranges from none to very high. In addition, other amphetamines or hallucinogens can be mixed in low doses MDMA produces a pleasant altered state of mind, with enhanced emotional closeness, but it is also used in high doses and settings in which toxicity is often reported (1). In the UK, considerable adverse effects have been reported from its use at rave dances. In the 1970s and 1980s its mind-altering effect caused some clinicians to advocate its use as an adjunct to psychotherapy (2).

Concern has been raised about the increasing use of ecstasy in Europe (3), particularly the UK and the Netherlands. The patterns and trends of substance use among college students have been evaluated over a 30-year period (4). Alcohol use remained stable, but illicit drug use peaked in 1978 and fell sharply over the next 20 years. Ecstasy was the exception: its use rose from 4.1% in 1989 to 10% in 1999. Ecstasy was the second most frequently tried illicit drug after marijuana.

Associated with increased physical activity and altered thermoregulation, ecstasy has been reported to cause unconsciousness, seizures, hyperthermia, tachycardia, hypotension, disseminated intravascular coagulation, and acute renal insufficiency, as well as death.

Ecstasy has a mild stimulant effect and is modestly hallucinogenic. The results of one study suggested that tolerance to its effects develops, but that adverse effects can increase with continued use (5). Recent reports have highlighted some disturbing effects, particularly when it is used while dancing vigorously at rave parties (6). In this setting, with increased physical activity, ecstasy can cause unconsciousness, seizures, hyperthermia, tachycardia, hypotension, disseminated intravascular coagulation, rhabdomyolysis, and acute renal insufficiency (7). Severe complications are also linked to uncontrolled fluid intake, hemodilution, and salt-losing syndromes. Deaths after the use of ecstasy in such settings have been described (8). It has been suggested that severe toxicity from ecstasy can result from altered thermoregulation in the face of excessive activity in warm environments (9).

The toxic effects of ecstasy have been reported in seven individuals who took it in a nightclub and developed varying degrees of toxicity (10). Three collapsed in or around the nightclub and arrived in an ambulance. Four came in themselves.

- A 20-year-old man collapsed at the nightclub. He had tachycardia, hypotension, hyperglycemia, hyperthermia, and significant hyperkalemia. He was ventilated but died 1 hour later.
- A 22-year-old man collapsed after falling 15 feet through a glass roof into a stairwell. He was comatose and had hypoglycemia, tachycardia, hypotension, hyperthermia, raised liver enzymes, significantly raised creatine kinase activity, and hyperkalemia. Although he was treated vigorously, metabolic acidosis persisted. He developed significant myoglobinuria and his creatine kinase activity peaked at 215 000 IU/l. His liver, kidney, respiratory, and cardiovascular function started to fail and he died 58 hours after admission.
- An 18-year-old man was found collapsed outside the nightclub. He had taken five ecstasy tablets and some "powder" that was later confirmed as ecstasy. He was vomiting and agitated, had a tachycardia and hyperthermia, and needed mechanical ventilation. He later developed rhabdomyolysis and renal impairment with raised liver enzymes. He went on to develop pneumonia and a urinary tract infection. He was discharged after 32 days with a mildly ataxic gait and dysphonia secondary to vocal cord damage.
- A 23-year-old man took two tablets of ecstasy and developed a tachycardia and a fever, which responded to treatment.

- An 18-year-old man took four tablets of ecstasy and became anxious but showed no clinical signs of MDMA toxicity.
- An 18-year-old woman took two tablets and had no clinical signs of toxicity except a tachycardia.
- A 17-year-old man took one ecstasy tablet and was well without any symptoms.

No other drugs were detected in the serum samples from any of these patients. Detailed analyses of the tablets obtained from the patients showed ecstasy and no contaminants. The authors commented on the unpredictable nature of toxicity with ecstasy, especially when death can occur with one tablet in some cases, while others survive even after they have consumed large quantities. In this series there was no correlation between the amount of ecstasy taken and the resulting serum MDMA concentration in most patients. However, high serum concentrations correlated with the severity of symptoms, including death. The local news headlines prominently implied that "poisoned" ecstasy had led to death when in fact there was no contamination. The authors suggested that such headlines lead people to believe erroneously that ecstasy use is safe except when it is contaminated.

Deaths related to ecstasy and MDEA in seven young white men, two of whom had hyperthermia, have been reviewed (11). In all cases, autopsy showed striking liver damage with necrosis; five patients had heart damage (contraction band necrosis and cell necrosis with inflammation), and others had brain damage, including focal bleeding, gross edema, and hypoxic changes. In one patient, who died of acute water intoxication, the pituitary gland was necrotic and there was accompanying cerebral edema. The authors proposed that the spectrum of pathological findings suggested more than one mechanism of damage, injury being caused by hyperthermia in some cases and a toxic effect (directly accountable for damage to liver and other organs) in others.

In a retrospective review of all violent deaths from 1992 to 1997 in South Australia, six deaths were associated with ecstasy abuse; all occurred after September 1995. Three victims had documented hyperthermia and there was evidence of hyperthermia in another. The authors suggested that individual susceptibility to MDMA may be caused by impaired metabolism by CYP2D6 or through genetically poor metabolism (seen in 5–9% of Caucasians). One woman, who died with a cerebral hemorrhage, had fluoxetine (a CYP2D6 inhibitor) present in her blood. Furthermore, toxicology identified paramethoxyamfetamine (PMA) in all the cases, amfetamine/metamfetamine in four cases, and MDMA in only two cases. PMA, which is sold as an MDMA substitute or is present as a contaminant, is associated with a high rate of lethal complications (12).

- A 35-year-old male criminal died under suspicious circumstances (13). The police had seen him alive about 1.5 hours before the alleged time of death during a patrol visit to his home. Evaluation of the corpse showed an obvious head injury and the body was in an advanced stage of rigor mortis, despite the fact that the alleged time of death had been less than 4 hours earlier. The body temperature was significantly raised (42°C). A witness testified that the deceased had taken ecstasy at various times during the night, after which he had been groaning, before taking off his clothes and thrashing on the floor while hitting his head and bumping into things. When resuscitation had been attempted, his jaw had been locked. Toxicology detected amfetamine, metamfetamine, and PMA in the blood.

The authors suggested that in a subgroup of amfetamine abusers, a triad of amfetamine use, prolonged exertion, and hyperthermia can be potentially lethal. Any temperature above 42°C requires active cooling (to below 38.5°C) and carries a poor prognosis. In this case rigor mortis may have started almost at the time of death. An ecstasy tablet that was allegedly from the same batch contained 50 mg of PMA.

Epidemiology of the use of ecstasy

"Club drugs", which are used primarily by young adults at all night "raves", dance parties, including ecstasy, have been frequently reviewed (14) (15) (16). Ecstasy is classified as an empathogen or enactogen, as the subjective experience has been described by users as intensely emotional and as creating a perception that one can experience the emotions of others.

A survey of 3021 young adults (14–24 years old) in Germany showed that regular use of ecstasy by itself is uncommon (2.6%). Among lifetime users, 97% have also used cannabinoids, 59% cocaine, 48% other substances, 46% hallucinogens, and 26% opiates. However, the interviews revealed that the use of ecstasy and hallucinogens is increasing, especially in young people. The authors observed that a large number of first-time users are at risk of regular use (17).

Recent US data suggest a decline in the use of ecstasy. Programs sponsored by the US Drug Enforcement Administration have systematically collected results from toxicological analyses conducted by state and local forensic laboratories on substances seized by law enforcement operations. The numbers of seizures of ecstasy in 2003 were down from a peak in 2001. These data are concordant with other surveys done in the USA, including a decline in the prevalence of use by students in secondary schools, which appears to be related to reduced availability and perceptions of the risks associated with its use. The percentage of 12th graders who said that there was a great risk of harm associated with using ecstasy increased from 38% in 2000 to 58% in 2004. The perceived availability of ecstasy by students fell from 62% in 2001 to 48% on 2004. Moreover, the disapproval of people who tried ecstasy once or twice increased from 81% in 2001 to 88% in 2004.

Surveys conducted by the Office of Applied Studies of the Substance Abuse and Mental Health Services Administration (SAMHSA) on the prevalence, patterns, and consequences of the use and abuse of alcohol, tobacco, and illegal drugs in the general US civilian non-institutionalized population in 2003 showed that 4.6% of the US population aged 12 and above had ever used ecstasy, 0.9% had used it in the past year, and 0.2% had used it in the past month. About 2.4% of those aged 12–17 years had ever used ecstasy compared with 15% of those aged 18–25 years and 3.1% of those aged 26 and over. However, the trend was different in Australia, where a national survey showed an increase in lifetime use of ecstasy from 2.4% in 1998 to 6.1% in 2001, with past year use rising from 0.9% to 2.4%. Furthermore, 10% of teenage ecstasy users and 6.9% of users in their twenties used ecstasy daily or weekly. Lifetime use of ecstasy among 10th grade students in Turkey rose from 2.7% in 1998 to 3.3% in 2001.

New data suggest that use of ecstasy occurs in a variety of settings and is not any more restricted to raves. Amongst US university undergraduate students, the number of sexual partners increased the likelihood of ecstasy use, as did self-reported sexual identity. Gay, lesbian, and bisexual students were more than two times as likely to have used ecstasy in the past year. Poly-substance use amongst ecstasy users was the norm (Maxwell).

In an epidemiological study of 2000randomly selected individuals in West Germany and 1000 in East Germany in 2001, the lifetime prevalences of ecstasy and amfetamine use were about 4% and 3% respectively. The percentage of people who reported that "one should never try (ecstasy) at all" increased from 72% in 1993 to 87% in 2001. In another epidemiological study in Germany involving 8139 randomly selected adults aged 18–59, 46% responded. In West Germany, 1.5% reported having tried ecstasy at least once. The younger subjects were more experienced—5.2% and 5.7% of the 18–20 and 21–24 year-olds respectively reported using it; 1.8% and 3.7% of the 18–20 year-olds and 3.7% of the 21-24 year-olds had used ecstasy within the last 12 months. During the 30 days before the interview, 0.3% of the interviewees reported having used ecstasy. However, in East Germany, ecstasy was the only substance with a higher prevalence, which, when extrapolated, suggested that 1.2 million of the 18–59 year-olds reported having taken ecstasy within the 12-months before the interview.

The Early Developmental Stages of Psychopathology Study recruited a representative sample of 14–24 year olds from metropolitan Munich and its surrounds; the longitudinal analyses involved 2446 of the 3021 interviews done at baseline. The findings suggested that men use ecstasy more often than women. While the onset of use was unlikely before the age of 14, it was followed by a sudden surge in use, which appeared to stagnate at age 24 for women and 26 for men. The increase in use was in the younger cohort. Ecstasy users, compared with non-users, had a higher probability of using other illicit drugs. Use of ecstasy appeared to

be a transient phenomenon—88% of the occasional users were non-users at follow-up and only 0.8% of the sample fulfilled the diagnostic criteria for a lifetime ecstasy-related substance use disorder. Moreover, increased proportions of the subjects in the group of lifetime ecstasy users had at least one mental disorder besides abuse or dependency. In most of the cases (88%), the onset of mental disorders occurred before the first use of ecstasy or a related substance. There were no significant differences in the incidence of mental disorders prospectively between the baseline ecstasy abstainers and ecstasy users in the follow-up period.

In the 1990s, a new dance and music culture called "techno" emerged in many European countries. Techno music was often associated with raves, which were also known as Techno raves. Around the same time, in Berlin, the Love Parade movement, ardent techno fans, started and attracted 1.5 million people in 1999. This caught the attention of the leisure industry, and magazines, music, and other paraphernalia were aimed at the adolescents and young adults involved in the techno scene. The use of club drugs is closely associated with this culture. In a study of 1664 adolescents and young adults designed to investigate the phenomenon of ecstasy use in the "techno party scene" in 1996, cannabis was the most commonly used drug in this group. About 50% of the group had used ecstasy once in their life and 44% had used speed; the 12-month and last-month usage rates were 40% and 28% respectively for ecstasy and 46% and 35% for speed. Ecstasy users were likely to have used combination drugs, the most prevalent sequences being "cannabis–ecstasy–speed" (10%) followed by "cannabis–ecstasy–speed–hallucinogens" (8%). The authors reported that the exclusive use of ecstasy in the techno party scene was almost non-existent. In the "Techno Study", in which 1412 persons were interviewed, the variability and stability of ecstasy use behavior in the techno party scene were explored. This study found even more drug use in 1998–9. Again cannabis was the more commonly used drug. The lifetime, 12-month, and last-month prevalences were respectively 40, 30, and 20% for ecstasy and 41, 31, and 21% for speed.

During 19 in-depth interviews, with a response rate of 46%, exploring the reasons for ecstasy use, the three main explanations identified were (a) affiliation, (b) stimulation, and (c) relaxation (Soellner). Ecstasy was perceived as being extremely helpful in stimulating interactions with people and stimulating effects were mentioned, especially in combination with music. "Feeling good" or "a desire to feel good" were the most commonly cited reasons for using ecstasy. Fear of reduced efficiency (75%) and fear of damage to health (62%) and developing addiction (36%) were common reasons for quitting. However, 44% stated that "ecstasy did nothing to me". The author concluded that ecstasy is second only to cannabis in illegal drug preferences among adolescents and young adults in Germany. Increased rates of ecstasy use in East Germany paralleled increases in ecstasy-related crimes by 60% in that area. Moreover, ecstasy users were more often polydrug users with increase degrees of psychopathology; the existence of mental illness increased the likelihood of future use of ecstasy. According to the author, ecstasy does not seem to determine substance-specific drug-use-related behavior. It rather seems to be yet another substance in the youth-related illicit substance carousel. Most people who begin to use a variety of illicit substances for whatever reasons, stop on their own without treatment of any sort.

Many large metropolitan areas saw a surge in the use of ecstasy among its young adults during the late 1990s. In the USA, Seattle reported high levels of club drug use compared with the national average and increases in problematic behavior and morbidity/mortality associated with the use of these drugs. Seattle also had a history of providing permits for large rave dance parties. Those who had ever used ecstasy were much more likely to use almost all other drugs; 26% of ecstasy users at rave parties mentioned ever using "research chemicals" (a broad array of poorly studied psychoactive chemicals used to explore new psychedelic experiences) compared with 1% of non-ecstasy users. At raves, behaviors associated with the use of ecstasy in the prior 6 months included unprotected sex by nearly one-third and driving under the influence of ecstasy by 38%. The proportion who reported having "overdosed, passed out, or had a bad experience" caused by ecstasy was 15%; 6% agreed that their use had been out of control in the previous 6 months. About 26% said that they worried that they might later have problems with health and memory. The use of adulterated ecstasy was reported by respondents in all groups, from 29% to 59% at rave parties. About 12–18% of those who had ever used ecstasy reported that they had used antidepressants to control its effects; 16% took vitamins and 11% took 5-hydroxytryptophan to protect against the depressant effects of ecstasy. Lifetime ecstasy use was 11% among 12th graders nationally, compared with 16% in Seattle; usage during the prior 30 days was 2% nationally and 6% locally. Use of ecstasy was second only to use of cannabis among illicit drugs used by school children. About 50% of all those surveyed who had ever used ecstasy had not used it in the past 6 months and most of those who had used it reported using it less than monthly. Not only were ecstasy users less likely to be active users than active drinkers, but those who did use it did so much less frequently.

Among men who have sex with men, drug use and sexual activity were inextricably connected for all participants who also reported strong expectations that ecstasy and other drugs lowered inhibitions and increased courage in finding sex partners, trying new sexual experiences, or going to gay venues they might not otherwise visit. All except African–Americans described ecstasy as a commonly used drug. Those who had ever used ecstasy were more likely to report unprotected anal sex and were more likely to have a sexually transmitted disease than those who had never used ecstasy.

Of 13 deaths in which ecstasy was identified, three were determined to have resulted at least partly from ecstasy. Nine out of 10 cases in which ecstasy was present but was ruled as not having directly caused the death were due to gun shot wounds or motor vehicle accidents. There was one case of overdose with ecstasy and metamfetamine. In four of 10 cases, ecstasy concentrations were above the concentration associated with recreational use, but death was ruled to be caused by physical trauma. There were three suicides with gun shot wounds to the head, of whom two were positive for ecstasy.

Community surveys have shown that Latinos report significantly lower levels of lifetime use of ecstasy, while STD clinic data point to significantly lower ecstasy use among African–Americans. Thus, ecstasy users were primarily Caucasians and men in their late teens or twenties (18). Most people who have ever used ecstasy either no longer continue to use it or use it infrequently, so opportunities for acute negative consequences are infrequent compared with alcohol. Acute toxicity due to moderate doses of ecstasy appears to be low, based on self-reporting and low mortality rates in the community. Rates of emergency department admissions and student use of ecstasy were higher in Seattle than the national average. Among men who have sex with men, ecstasy use was higher than among heterosexual men and was associated with higher levels of both unprotected sex and sexually transmitted diseases.

In a report from Korea, hair and urine samples were collected from 791 subjects aged 20–62 years suspected of drug use (19). Ecstasy and/or MDA (its main metabolite) were found in 5.6%. Only four subjects were positive for MDA alone. However, only 9 subjects had both ecstasy and MDA in the urine. Abuse of ecstasy or MDA was found principally among young adults, of whom 73% were aged 20–29 and 27% aged 30–39; in this group 88% of men and 18% of women had positive hair samples. The concentrations of ecstasy and MDA in the hair of male abusers were higher than in female abusers. The authors speculated that it may be difficult to detect ecstasy or MDA in urine samples from occasional abusers. They also observed that polydrug use was not common among the Korean users of ecstasy.

Pharmacokinetics

The pharmacokinetic effects of ecstasy have been studied in healthy volunteers (20). In the pilot phase, two subjects each took ecstasy 50, 100, and 150 mg. In the second phase, eight subjects took ecstasy 75 and 125 mg. All were CYP2D6 extensive metabolizers. The ecstasy plasma concentrations were not proportional to dose, probably indicating non-linear kinetics in the dosage range usually taken recreationally. While the results were not conclusive (owing to problems in the study design) and require further exploration, the finding that relatively small increases in the dose of ecstasy ingested can translate to disproportionate rises in ecstasy plasma concentrations, if confirmed, would be important.

Organs and Systems

Cardiovascular

Cardiotoxicity following ecstasy use has been reported (21).

- A 16-year-old boy took three tablets of ecstasy and amfetamine 0.3 g and several hours later had convulsions and a temperature of 40.9°C. His heart rate was 210/minute and his blood pressure 100/75 mmHg. His creatine kinase activity was raised and he had myoglobinuria, renal impairment, hyperkalemia, and hypocalcemia. An electrocardiogram showed ventricular and supraventricular tachycardias but no myocardial ischemia. A diagnosis of serotonin syndrome due to ecstasy ingestion with associated hyperpyrexia and rhabdomyolysis was made. Following active treatment, his condition stabilized, with restoration of sinus rhythm and normal urine output. However, 12 hours later he developed jaundice, raised liver enzymes, and coagulopathy, suggesting acute liver failure due to ecstasy. With supportive treatment, his liver function improved. However, another 12 hours later, he developed shortness of breath associated with-sided chest signs and X-ray changes compatible with aspiration pneumonia, and required emergency intubation 4 days later. He developed pulmonary edema, his pulmonary artery was occluded, and an echocardiogram showed globally impaired left ventricular function with an ejection fraction of 30–35%; there was electrocardiographic T wave inversion. Primary myocardial damage causing cardiac dysfunction was investigated using serial creatine kinase and troponin measurements. He recovered completely with treatment and an echocardiogram showed an ejection fraction of 60%.

The authors reported that this was the first case report of clinical, radiological, biochemical, and echocardiographic evidence of myocardial damage and cardiac dysfunction following ecstasy and amfetamine use.

Transient myocardial ischemia associated with ecstasy has been reported (22).

- A 25-year-old man, a regular alcohol drinker with a history of asthma, had been out drinking 8 pints of lager and 4 gins. His last drink was spiked with a tablet that was presumably ecstasy. He scooped the tablet out, and even though some of the tablet may have dissolved, he finished the drink. He awoke 3 hours later with restlessness, nausea, and abdominal cramps. In the emergency room, his temperature was 37.2°C, and he was sweating. His heart rate was 120/minute and his blood pressure 130/70 mmHg. He had some abdominal discomfort. A diagnosis of ecstasy ingestion was made, although urine MDMA concentrations were not measured. His electrocardiogram on admission showed sinus tachycardia, with T wave inversion in leads I, aVL, and V4–6 and the voltage criteria for left ventricular hypertrophy. The next day the electrocardiogram had returned to normal. An echocardiogram was within normal limits. He was well on discharge.

The authors reasoned that the myocardial ischemia did not proceed to necrosis or a dysrhythmia because the amount of drug exposure was low.

Coronary spasm has been attributed to ecstasy (23).

- A previously healthy 20-year-old Caucasian woman developed intermittent left-sided chest tightness associated with palpitation but no dyspnea, hemoptysis, calf pain, or swelling. She had been out the night before with her friends to a club, since when she had become more restless, anxious, and unable to sleep. She was very talkative, but cold and sweaty and had dilated pupils. She had normal heart sounds and sinus rhythm. An electrocardiogram showed ischemia, with wide spread ST segment depression and T wave inversion, consistent with coronary artery spasm. Her creatine kinase and troponin T concentrations were not raised and an echocardiogram was normal. She made a complete recovery with standard intravenous doses of nitrates. Amphetamine derivatives, presumed to have been from a drink laced with ecstasy, were found in her urine.

The authors speculated that the mechanism of action was endothelial dysfunction, similar to that postulated for cocaine-induced coronary spasm.

Atrial fibrillation after use of ecstasy has been reported (24).

- A 17-year-old previously healthy man had a generalized tonic-clonic seizure. He denied drug abuse, but his urine drug screen was positive for ecstasy. He had an irregular heart rhythm with a normal blood pressure. An electrocardiogram showed atrial fibrillation with a ventricular rate of 102/minute. His routine laboratory investigations, a brain CT scan, and an electroencephalogram were normal. There were no underlying cardiac lesions. He was stabilized with medical treatment and was doing well 6 months later.

Since atrial fibrillation is unusual in young people, the authors speculated that ecstasy may have contributed in this case even though this adverse effect has not been reported before.

Of 50 participants, randomly recruited on four different nights from a popular night club, users of ecstasy, alcohol, cannabis, and other psychostimulants did not have significant differences in their body temperature despite increases in environmental temperature and period of dancing (25). However, increased blood pressure was predicted by the number of ecstasy tablets ingested and the amount of cocaine used. There were significant differences in both heart rate and blood pressure between polysubstance users compared with the alcohol and cannabis users. The authors suggested that polysubstance users may be at higher risk of cardiovascular toxicity. However, more data are needed to confirm these observations.

Cardiovascular autonomic functioning during MDMA use has been investigated in 12 MDMA users and a matched group of non-users (26). Resting heart rate variability (an index of parasympathetic tone) and heart rate response to the Valsalva maneuver (Valsalva ratio, an index of overall autonomic responsiveness) were both reduced in the drug users. Thus, seemingly healthy users of MDMA had autonomic dysregulation, comparable to that seen in diabetes mellitus. In several users there was a total absence of post-Valsalva release bradycardia, a sign of parasympathetic dysfunction. Since no cardiac data were available for these patients before their use of ecstasy, and since all were multidrug users, the findings must be interpreted with caution.

Extensive aortic dissection with cardiac tamponade and mesenteric ischemia has been attributed to ecstasy (27).

- A 29-year-old man who took ecstasy and alcohol at a rave had no immediate adverse effects, slept well later on, and was in good health until he suddenly collapsed to the floor about 2 hours after waking. When seen 36 hours after the last dose of ecstasy he was short of breath and had abdominal pain, diarrhea, and vomiting. He had a loose bloody bowel movement but refused further investigations. He was discharged with a diagnosis of gastroenteritis, only to be readmitted 8 hours later after sudden deterioration and hypertension. Despite extensive efforts, his condition deteriorated and he died 5 hours later. At autopsy, there was a type I aortic dissection, starting at the root and spreading to the bifurcation, which had resulted in cardiac tamponade. The dissection had involved the mesenteric arteries, resulting in bowel ischemia.

Since this condition is rare in young adults, diagnosis can be difficult. The authors believed that this was the first case report of aortic dissection secondary to ecstasy.

Ecstasy has been associated with sudden death and cardiovascular complications. Eight healthy self-reported ecstasy users participated in a four-session, ascending-dose, double-blind, placebo-controlled comparison of the echocardiographic effects of ecstasy and those of dobutamine (28). Ecstasy 1.5 mg/kg increased the mean heart rate by 28 beats/minute, systolic blood pressure by 25 mmHg, diastolic blood pressure by 7 mmHg, and cardiac output by 2 l/minute. The effects of ecstasy were similar to those produced by dobutamine (40 µg/kg/minute), except that ecstasy had no measurable inotropic effects. Thus, ecstasy increases systolic and diastolic blood pressures in the absence of a significant change in cardiac contractility and end-systolic wall thickness. The resulting increase in the tension of the ventricular wall leads to disproportionately higher myocardial oxygen consumption than would be expected from the observed changes in the heart rate and blood pressure. The authors commented that the behavioral and environmental factors accompanying the use of ecstasy—sustained exercise from dancing, often in crowded nightclubs with high ambient temperature and humidity—could further potentiate toxicity. They recommended a combination of beta-blockers and vasodilators for the emergency treatment of ecstasy-associated vascular instability.

Respiratory

Pneumomediastinum after MDMA has been reported.

- Spontaneous pneumomediastinum occurred in a 17-year-old man who presented with chest pain and vomiting after taking two tablets of ecstasy (29).
- A 16-year-old boy took six ecstasy tablets over 2 hours (30). A period of vomiting ensued and he developed chest pain and swelling in the neck. A chest X-ray showed pneumomediastinum. He was managed conservatively and his emphysema settled uneventfully within 2 days.

The authors of the second case reported that five such cases have been reported in the past 4 years in the British literature; the respiratory complications were thought to be due to severe physical exercise or secondary to vomiting. It is therefore possible that this complication may not be due to a direct effect of MDMA, but rather a consequence of repeated Valsalva maneuvers associated with the dance habits of ecstasy users or vomiting induced by the drug.

Two cases of pneumomediastinum occurred in people who took ecstasy at the same rave party (31).

- A 23-year-old woman developed chest, back, and neck pain, and surgical emphysema over the chest and neck 7 hours after taking ecstasy. She had surgical emphysema in the mediastinum and neck and was given intravenous fluids and antibiotics.
- A 22-year-old man developed anterior chest and neck pain and surgical emphysema over the neck and chest 8 hours after taking four ecstasy tablets. He had surgical emphysema in the mediastinum and neck and contrast swallow showed a leak at the posterolateral aspect of the mid-esophagus. He was given intravenous fluids and antibiotics.

These symptoms were possibly related to the use of ecstasy, but it is also possible that a corrosive additive in the ecstasy was responsible. Ecstasy can cause gastrointestinal dysmotility, which could have resulted in esophageal rupture.

Adverse effects can develop after amphetamines are abused in combination (32).

- A 28-year-old healthy woman had left-sided pleuritic chest pain for 18 hours, having taken one tablet of ecstasy and one tablet of speed (metamfetamine hydrochloride) 4 hours earlier. There was surgical emphysema in her neck. On auscultation there was a crunching cardiac systolic sound. Chest X-ray showed pneumopericardium and pneumomediastinum. She was given analgesics and monitored. The chest pain subsided after 4 days.

Her respiratory problems could have been due to alveolar rupture, caused by an increase in intra-alveolar pressure, due to the exertion while she had been dancing strenuously. Alternatively, it could have been secondary to her use of positive ventilatory pressure after taking the drugs; this is done by a partner, either by direct mouth-to-mouth contact or through a cardboard cylinder, to enhance the user's experience of the stimulant's effects.

Hemopneumothorax has been reported in association with MDMA abuse (33).

- A 33-year-old man developed shortness of breath and right-sided chest pain 12 hours after taking two tablets of MDMA. His blood pressure was 110/60 mmHg and his pulse 120/minute. A chest X-ray showed a complete right-sided pneumothorax with left shift of the mediastinal structures. Needle decompression was completed and 200 ml of blood was drained over 1 hour. A CT scan showed apical bullae with hemopneumothorax. Another 1500 ml of blood was drained over the next 2 days.

Spontaneous hemopneumothorax has not been previously reported in association with MDMA. The authors suggested that ruptured bullae with torn apical vascular adhesions were responsible.

Nervous system

Two previously reported cases of cerebral edema associated with the syndrome of inappropriate secretion of antidiuretic hormone (SIADH) after ingestion of ecstasy elicited strong responses (34). The writers questioned the certainty of the clinical diagnosis on the basis of the biochemical details, specifically the plasma and urine osmolalities (35). They pointed out that cerebral edema can occur secondary to severe water consumption without SIADH and suggested that the use of intravenous fluids or water consumption after exposure to MDMA requires close monitoring. The original authors replied that they strongly agreed with the potential risks of unlimited fluid intake in MDMA users.

Dry mouth, with possible increased risk of enamel erosion and dental caries, has also been described (36).

Since the toxic effects of MDMA resemble certain features of both the serotonin syndrome and the neuroleptic malignant syndrome, some have suggested that MDMA may have combined actions on both the dopamine and serotonin systems (37). However, dantrolene (which may be helpful in cases of neuroleptic malignant syndrome) does not appear to be useful in dealing with MDMA toxicity (38).

A report linking ecstasy use and parkinsonism generated considerable discussion (39).

- A 29-year-old man developed difficulty in walking and lost the ability to write or drive 4 weeks after developing clumsiness of his extremities. He was unable to work or live independently. His electroencephalogram, lumbar puncture, and MRI and positron emission tomography (PET) scans were normal. Eleven weeks after the onset of symptoms he had disturbed gait and fine motor coordination; his condition deteriorated and 8 weeks later he had bradykinesia of the face and limbs, absence of blinking, hypokinesia in relation to speech, postural instability, and a markedly parkinsonian gait. However, his cognitive function was intact.

This patient had taken ecstasy 10 times during the year before, the last time about 3 months before the onset of symptoms. Apart from marijuana, he denied using other substances. He was treated with maximal tolerable doses of levodopa and pramipexole, without improvement. The authors reported that they had no explanation for this patient's symptoms, other than the use of ecstasy. They felt that the parkinsonian symptoms most closely resembled MPTP-induced parkinsonism. They further postulated that this could be a delayed neurotoxic effect of ecstasy in the substantia nigra and striatum and could have occurred as a result of neuronal damage by free radicals.

This report brought many responses. The first criticized the authors' conclusion that the neurotoxic effects in their patient were similar to those seen with MPTP (40). They argued that the PET scan was normal, and that the patient's condition did not respond to the use of antiparkinsonian drugs, unlike MPTP-induced damage, which does respond. They further argued that the extensive animal and human experience involving MDMA has not shown damage to the dopaminergic system. Furthermore, MDMA and MPTP are chemically unrelated and share neither precursors nor metabolites. Although the purity of street drugs is suspected and a contaminant could have been responsible, they seriously doubted that MPTP was one of the contaminants ingested.

Another group expressed the concern that although other types of amfetamine can cause dopaminergic neurotoxicity in animals and are sometimes sold as MDMA, they have not been associated with parkinsonism, despite more than 60 years of worldwide therapeutic and illicit use (41). Furthermore, they observed that the pattern in this case, in which symptoms began 8 weeks after drug exposure, did not fit the expected pattern of drug toxicity. They proposed that the problem had either been caused by contaminants or by a highly idiosyncratic reaction. However, if the toxicity had been from a contaminant, a cluster of cases would have been expected, as was seen with MPTP. On the other hand, if this had been an idiosyncratic case, the symptoms would have emerged soon after drug use. They were also puzzled by the absence of hair analysis to confirm MDMA use.

The authors of the original report responded by citing evidence that MDMA can be toxic to dopaminergic neurons and can cause lasting changes in neuronal responses to dopamine (42). They further stated that although parkinsonism has not been reported after MDMA, the possibility that such an association exists cannot be excluded. They raised the possibility that clinical evidence of parkinsonism can be missed, particularly when the disorder is mild and, as in their patient, when tremor is absent. They pointed out that with valproate the first full report of parkinsonism did not appear until 18 years after its introduction into the USA, even though those who commonly prescribe it (neurologists and psychiatrists) should have noticed it earlier. The authors clarified that they had mentioned MPTP as an example of a substance that may have a delayed neurotoxic effect on monoaminergic neurons, without intending to suggest that MDMA acts chemically in the same manner as MPTP. They agreed that a contaminant could have played a role, and raised a concern that contaminants also put MDMA users at risks of adverse effects. They countered the argument that all idiosyncratic responses have to be immediate, by quoting examples from the literature to suggest that the course of a reaction depends on the underlying mechanism.

The patient discussed in the case report also responded, alleging that his permission had not been obtained before publication (43). Contradicting the original published report, he stated that he had never used marijuana; the authors defended their data by stating that when the patient's friend, who had accompanied him to the clinic, mentioned cannabis use the patient did not deny it. He also reported using creatine, ephedrine, caffeine, and aspirin, none of which, the authors noted, has previously been associated with parkinsonian symptoms.

- A 38-year-old man developed Parkinsonism that progressed to Hoehn and Yahr stage 5 within 4 years of onset (44). Treatment with ropinirole resulted in further deterioration, levodopa was not tolerated, and subthalamic nucleus stimulation provided only partial relief. The patient reluctantly reported heavy use of ecstasy through most of his twenties and thirties. He had a family history of Parkinsonism, but other investigations to determine the underlying cause, such as urine copper and heavy metals, were non-contributory.

The authors cautioned against making a strong link between ecstasy use and parkinsonian symptoms, because there are other potential toxins that could have caused it and the use of ecstasy may have been coincidental.

Impaired cerebral blood flow after ecstasy intoxication has been reported (45).

- A 19-year-old woman had a grand mal seizure 4 hours after taking 10 ecstasy tablets and developed coma, hyperthermia, tachycardia, tachypnea, raised liver enzymes, and renal insufficiency. After wakening, she reported hallucinations, helplessness, panic attacks, and amnesia, though she was oriented. Her brain CT and MRI scans were normal. Her urine toxicology screen was positive for ecstasy and opiates. A SPECT study 20 days after intoxication showed reduced non-homogeneous supratentorial tracer uptake bilaterally, suggesting hypoperfusion. Electroencephalography showed diffuse slowing and occasionally generalized sharp waves. After treatment with valproic acid, she had slight amnesia, her neuropsychological deficits disappeared, and the SPECT scan normalized 29 days later, followed by normalization of the electroencephalogram.

The authors thought that the abnormal SPECT scan was best explained by reduced cerebral blood flow due to vasoconstriction, which could have been due to ecstasy-induced reduction in serotonin.

Two cases of non-aneurysmal subarachnoid hemorrhage associated with ecstasy have been reported (46).

- A 29-year-old woman had a sudden severe occipital headache and photophobia with nausea and vomiting. She had marked nuchal rigidity. Her head CT scan was normal, but the CSF was blood-stained with spectro-photometric evidence of xanthochromia. Cerebral angiography showed beading of vessels in the posterior circulation. However, MRI/MRA did not show any abnormality and the ESR was normal. The patient admitted to having used ecstasy for the first time. Cerebral angiography 2 months later showed that the abnormalities had completely resolved.
- A 24-year-old man smoked marijuana and then consumed one and a half ecstasy tablets at a party, and several hours later had a sudden severe retro-orbital headache and felt unwell. Shortly after, he had a

generalized tonic-clinic seizure, after which he was alert and oriented with an unremarkable neurological examination. He then had another seizure and a brain CT scan showed a subarachnoid hemorrhage, localized to several parasaggital sulci towards the vertex. Cerebral angiography showed focal beading of peripheral branches to the right anterior cerebral cortex, consistent with arteritis.

In both cases, angiography suggested vasculitis, when there was subarachnoid hemorrhage in the absence of a vascular malformation, probably associated with the use of ecstasy. The authors recommended that an accurate drug history is essential in any young person with a subarachnoid hemorrhage. The precise etiology of ecstasy-associated subarachnoid hemorrhage is not known.

Other neurological adverse effects have been reported in a first-time ecstasy user (47).

- A 19-year-old man, an occasional user of benzodiazepines and heroin, developed dizziness and loss of consciousness the morning after taking ecstasy for the first time. He was intubated and ventilated. He had a monocular hematoma, pulmonary edema, and cutaneous emphysema, without focal neurological deficits. There was rhabdomyolysis with renal insufficiency but no rise in temperature. Body fluid toxicological analysis showed only benzodiazepines and amphetamines. After he was weaned from the respirator, although awake, he did not react to noxious stimuli and showed no spontaneous movements. A CT scan of the brain had been normal on admission, but 2 and 3 weeks later it showed multiple hypodense lesions in the white matter. One month after admission, he was in a vegetative state, with marked spasticity, intermittent non-specific reactions to noise, and sustained generalized myoclonus. An electroencephalogram showed generalized slowing with intermittent theta wave activity. An MRI scan showed bilateral severe white matter damage.

Brain damage in this patient was restricted to the white matter. Although hypoxia was present at the time of admission, the authors did not believe that it was the only causative factor. By a process of elimination, they concluded that ecstasy might have caused the toxic encephalopathy, although it was not detected by toxicological analysis. They suggested that myelin damage might be an indirect sequel of MDMA-related metabolic oxidative stress and multiorgan failure due to individual susceptibility. They also entertained the idea of a lipophilic toxic contaminant rather than MDMA as the causative toxic agent.

An unusual case of bilateral sixth nerve palsy associated with ecstasy has been reported (48).

- A 17-year-old man developed horizontal diplopia in all directions of gaze while using ecstasy tablets every 5–7 days for 2 months. A diagnosis of bilateral sixth nerve palsy was confirmed. Ocular motility returned to normal within 5 days without treatment. There was no evidence of inflammation or degenerative disease of the central nervous system.

The authors speculated that the most likely cause of the lesion was either an interaction of ecstasy with serotonergic neurons or cerebral edema (albeit not detected by MRI) secondary to ecstasy.

Sensory systems

Ten heavy ecstasy users and 10 age-matched controls underwent single-pulse transcranial magnetic stimulation (TMS) of the occipital cortex to evaluate the hypothesis that use of ecstasy can increase excitability in the visual cortex, which can result in visual hallucinations (49). Transcranial magnetic stimulation can elicit conscious subjective light sensations (phosphenes) in the absence of visual stimuli. The minimum intensity that evokes phosphenes, also known as the phosphene threshold is thought to be a useful measure of cortical excitability. The phosphene threshold was significantly reduced in ecstasy users compared with controls and correlated negatively with the frequency of ecstasy consumption but not the duration of ecstasy use. The phosphene threshold of subjects with hallucinations was lower than that of subjects without hallucinations. However, the presence of hallucinations correlated directly with the frequency of ecstasy use. The authors suggested that this may represent neurotoxicity of ecstasy linked to massive serotonin release, followed by serotonin depletion in this cortical area.

Psychological, psychiatric

Concerns have been raised about the long-term nervous system effects of MDMA use from animal data, which suggest that the dose of MDMA used for recreational purposes by humans can cause toxic effects in non-human primates, especially involving the serotonin system. In a case-control study, 10 long-term MDMA users were compared with 10 controls who had not used MDMA but were drug abusers and were matched for age, sex, education, and premorbid intellect (50). All participated in a single photon emission computed tomography (SPECT) study with a serotonin transporter (SERT) ligand. Dopamine transporter binding was determined from scans acquired 23 hours after injection of the tracer. Hair analysis, which covers about the last 4 weeks of drug use, generally confirmed the drug use history, although two ecstasy users tested negative for both amfetamine and metamfetamine. On neuropsychological testing, both groups showed comparable performances in tests of verbal and spatial memory, psychomotor speed, attention, and executive function. Larger lifetime doses of ecstasy were associated with reduced verbal memory performance on the California Verbal Learning Test and more errors in the Spatial Working Memory Test. Ecstasy users showed a cortical reduction of SERT binding prominently in primary sensorimotor cortex, with normal dopamine transporter binding in the lenticular nuclei. There was also an association between the length

of the MDMA-free period (mean 18 days) before the scan and ligand binding in the cingulate. Spatial working memory performance and SERT binding in the left calcarine cortex were negatively correlated with the estimated lifetime dose of MDMA in the ecstasy users. While these results could be coincidental, or reduced SERT binding could be the imaging correlate of a psychobiological predisposition to heavy use of ecstasy, the authors suggested that this study provided evidence for specific albeit temporary serotonergic neurotoxicity of MDMA in humans. If confirmed, these effects on the serotonergic system could underlie psychiatric morbidity in ecstasy users.

The neurotransmitter systems involved in the psychological and information processing effects of ecstasy have been studied in 16 ecstasy-naïve subjects (51). Ecstasy produced a state of enhanced mood, well-being, increased emotional sensitiveness, little anxiety, moderate thought disturbances, but no hallucinations or panic reactions. It caused thought disturbances, such as difficulty in concentrating, thought blocking, and difficulty in reaching decisions. Women had greater subjective responses to ecstasy than men. For instance, they scored higher on ratings of anxiety, adverse effects, and thought disturbances, suggesting differences in the metabolism of ecstasy. Prepulse inhibition of the startle response was used to measure physiological changes. Prepulse inhibition and startle habituation have been used as operational measures of sensorimotor gating and habituation functions, respectively, in investigations of attention. Moreover, some studies have associated prepulse inhibition with endogenous serotonin release. Ecstasy 1.7 mg/kg increased prepulse inhibition, suggesting that the overflow of endogenous serotonin caused by ecstasy may interfere with the startle response and cause a breakdown of cognitive integrity. This effect further suggests serotonin dysregulation in response to ecstasy.

Reports of adverse neuropsychiatric effects of MDMA have included descriptions of flashbacks, anxiety, insomnia (52), panic attacks (35,53), and psychosis (54). Subacute adverse effects that have been reported following MDMA use include drowsiness, depression, anxiety, and irritability (55). Prolonged or chronic effects have also been reported, including panic disorder (56,57), psychosis (54,58,59), flashbacks (54), major depressive disorder (35,60), and memory disturbance (35). The observation that only certain individuals develop neuropsychiatric disturbances after using ecstasy suggests that certain predisposing psychiatric factors (or high-dose regimens) may make some individuals more vulnerable to these untoward effects.

Two cases of ecstasy abuse with unusual neuropsychiatric complications have been reported (61).

- A 24-year-old man had perinatal asphyxia and subsequently a slight delay in psychomotor development. At age 3–4 years, he had problems in school because of ADHD and had difficulty with relationships, persisting into adulthood. From age 21 he abused ecstasy and marijuana. He suffered a "horror trip" with a panic attack after using LSD at age 23. Although he abstained from LSD, he used more ecstasy and cannabis. Three months later, he developed a psychosis. He was anxious and paranoid and reported bizarre delusions and auditory and visual hallucinations. He had a normal brain CT scan and no medical, neurological, or laboratory abnormalities. Electroencephalography showed normal basal activity, with paroxysmal discharges in both temporal regions but no evidence of seizures. The diagnosis was paranoid schizophrenia and he was treated with antipsychotic drugs, with progressive improvement over 6 weeks. After discharge, he stopped taking his medications and resumed sporadic ecstasy and cannabis abuse. Drug intake was often followed by short-lasting prepsychotic decompensation with increased impulsivity and hyperactivity. However, he recovered after such episodes within 2–3 days without any antipsychotic medication.

- A 23-year-old woman who had been heroin dependent for several years and was taking methadone maintenance and who had abused cannabis and benzodiazepine in the past, used ecstasy for the first time and within 3 hours developed a wide range of sympathomimetic symptoms, including tachycardia, tremor, mydriasis, and headache. She felt extremely anxious, had psychomotor agitation, and was incoherent in thinking and disoriented in time and place. Her urine drug screen was positive for benzodiazepines, cannabis, and ecstasy. Electroencephalography suggested temporal lobe epilepsy with a normal CT scan. She then had two series of four and five complex-partial seizures, two with secondary tonic-clonic generalization. Although her psychotic symptoms persisted for a while, even with treatment, her seizures responded well to anticonvulsant drugs (phenobarbital 100 mg/day, clonazepam 3 mg/day, and diazepam as required) and her electroencephalogram normalized.

While acknowledging that only a minority of ecstasy users develop psychosis and even fewer develop seizures, the authors argued that symptoms suggestive of toxic psychosis were present in both cases and that the electroencephalographic abnormalities showed seizure activity. They suggested that individuals with potential genetic vulnerability to psychosis and seizures may be at higher risk when they use ecstasy, especially in combination with cannabis.

With increasing use of ecstasy, there is considerable interest in its effects on psychopathology and cognition. The residual effects of ecstasy on psychopathology and cognition have been examined in 18 current regular users, 15 ex-users with an average abstinence of 2 years, 16 ecstasy-naïve polydrug users, and 15 non-drug users (62). Both current and previous users had significantly worse psychopathology than the controls and polydrug users. Ecstasy users had higher scores on the Symptoms Check List SCL-90-R Global Severity Index and the Positive Symptom Distress Index. They also scored significantly higher on eight specific factors on the SCL-90-R: somatization, obsessive-compulsive disorder, anxiety, phobic anxiety, interpersonal sensitivity, depression, paranoid ideation, and altered appetite/restless sleep.

Current users had higher scores across more categories, but the values were not significantly different from ex-users. Current and ex-users had lower working memory and verbal learning abilities than the controls and polydrug users, but there were no significant differences between the two groups. Using regression analyses, the investigators found that the best indicator of increased psychopathology was a high consumption of cannabis, while cognitive deficits were best predicted by the amount of previous ecstasy use. Thus, the investigators suggested that ecstasy-induced cognitive impairment may not be reversible over time and with abstinence, suggesting that ecstasy is a potent neurotoxin.

The neuropsychological effects of ecstasy have been studied in 20 polydrug users (63). Various functions and processes throughout the brain, such as verbal fluency, spatial working memory, and attention, were assessed. Ecstasy users showed a significant deficit in complex visual pattern recognition and spatial working memory compared with the polydrug abusers who were not taking ecstasy. Since these tests are sensitive to temporal lobe functioning, the authors suggested that there is a selective temporal lobe deficit in ecstasy users, with relative sparing of executive functioning. Based on previous studies, they postulated that serotonin dysfunction in the temporal lobe may be associated with impaired visuospatial working memory. However, they did not study a drug-free control group and the ecstasy group used higher amounts of drugs in general. In addition, the sample size was small.

Some reports have suggested that the relation between the dose of ecstasy and its complications may not be straightforward (64). Some people have problems with very small amounts.

- A 21-year-old woman developed a protracted psychotic depersonalization disorder with suicidal tendency after taking two tablets of ecstasy for the first time (65).
- A 23-year-old man with no prior psychiatric history developed panic disorder after taking a single dose of ecstasy (56).
- A psychosis occurred in an 18-year-old man who had used ecstasy on an occasional recreational basis (66).

The acute and short-term effects of a recreational dose of MDMA (1.7 mg/kg) given to 13 MDMA-naïve healthy volunteers in a double-blind, placebo-controlled study have been reported (67). MDMA produced a state of enhanced mood, well-being, and enhanced emotional responsiveness, with mild depersonalization, derealization, thought disorder, and anxiety. The subjects also had changes in their sense of space and time, heightened sensory awareness, and increased psychomotor drive. MDMA increased blood pressure moderately, except in one case of a transient hypertensive reaction. The most frequent somatic adverse effects were jaw clenching, poor appetite, restlessness, and insomnia. Lack of energy, difficulty in concentrating, fatigue, and feelings of restlessness during the next day were also described. The authors suggested that MDMA produces a psychological profile different from classic hallucinogens or psychostimulants.

The potential risk of hypertensive effects of recreational dosages of ecstasy should also be considered in the safety profile.

Using PET with a radioligand that selectively labels the serotonin (5-HT) transporter, 14 MDMA users who were currently abstaining (for 3 weeks) from use and 15 MDMA-naïve controls were studied (68). MDMA users showed a reduction in global and regional brain 5-HT transporter binding, a measure of the number of 5-HT neurons, compared with controls. Deficiency in SERT correlated positively with the extent of previous drug use. The authors suggested that ecstasy users are susceptible to MDMA-induced serotonin neural injury.

The effect of MDMA on densities of the serotonin transporter has been studied with PET imaging in 30 current users, 29 former users, 29 users of other drugs, and 29 drug-naïve controls (69). Former users were required to be abstinent for 20 weeks before the study. On the day of PET scanning, hair samples were taken for drug analysis. Serotonin transporter volumes were reduced in current MDMA users in the posterior cingulate gyrus, left caudate, thalamus, occipital cortex, medial temporal lobes, hippocampus, and brainstem. Among current users men and women had different deficit profiles: women had lower serotonin binding in all areas except the left caudate and auditory cortex; men had deficits in the occipital cortex and medial temporal lobes. There were no significant differences in the other three groups. In addition, there was a negative correlation between the usual dose of MDMA and the serotonin density in the occipital lobe and the left precentral sulcus in current users. The fact that serotonin transporter numbers were higher in female than male current users suggests that MDMA may be more damaging to the female brain. These results suggest that MDMA causes changes in the serotonin transporter system, but that the effect is also reversible after abstention. One possible confounding factor was that a period of 3 days without drugs was required before PET imaging. However, it is possible that MDMA remained in the brain tissue, a factor that would certainly have affected the results of the study.

Although alterations in serotonergic systems after the use of ecstasy have been documented, it is unclear whether these neurotoxic effects are reversible. In 117 subjects who underwent PET studies using a serotonin transporter (SERT) ligand, 30 subjects were actual ecstasy users, 29 were former ecstasy users, 29 were drug-naïve controls, and 29 were using drugs other than ecstasy. The distribution volume ratios in ecstasy users were significantly reduced in the mesencephalon and the thalamus (70). The distribution volume ratio in former ecstasy users was very close to that in drug-naïve controls in all brain regions. The distribution volume ratio in polydrug users was slightly higher than that in the naïve group. The authors therefore concluded that ecstasy causes protracted but fully reversible alterations in the serotonin transporter. However, this does not imply full reversibility of the neurotoxic effects.

Panic disorder

A case of ecstasy-related panic disorder has been reported (71).

- A 21-year-old man used increasing dosages of ecstasy (seven tablets or 400–500 mg/day) over a period of 5 months. After taking six tablets on one day, he developed palpitation, chest pain, sweating, vertigo, and a fear of dying. He responded well to alprazolam and was discharged. Several days later, the panic attacks recurred spontaneously. Five weeks later, he had a complete medical work-up that was unrevealing. Since his panic attacks persisted in the absence of any drugs, he was given the SSRI paroxetine 20 mg/day plus alprazolam; his panic attacks gradually abated and stopped 3 months later. The paroxetine was gradually tapered over the next 3 months and he continued to be symptom-free at 6 months.

The authors suggested that this case may have exemplified dose-dependent serotonergic neurotoxicity from ecstasy abuse.

Acute psychoses

Acute psychoses have been reported following the use of ecstasy.

- A 26-year-old man without previous psychopathology (except social phobia) unknowingly consumed ecstasy with alcohol and developed an acute psychosis 12 hours later (58,72).

The authors reported that 12 cases of acute psychosis have previously been reported after the use of ecstasy once or twice. Based on a previously suggested hypothesis, they proposed that the psychosis was probably due to the indirect effects of MDMA on the dopaminergic system, secondary to serotonergic deregulation. Most patients who went on to develop a chronic psychosis were either chronic ecstasy users or multiple substance users. However, in the case mentioned here, after 6 months the patient still had symptoms of psychosis. The authors further suggested that genetically slow metabolizers of ecstasy are probably more vulnerable to this adverse effect, even with a single exposure.

Depression

Past reports have suggested that ecstasy use is associated with increased scores on self-report measures of depression. The long-term effects of ecstasy consumption on depression have been examined in 29 individuals who had consumed large quantities of the drug in the past, but were now leading relatively drug-free lives (73). They had taken an average of 1.5 ecstasy tablets in the last month, 8.4 in the last 6 months, and 23.3 in the last 12 months. None had taken it in the last 14 days. The former chronic ecstasy users had not taken ecstasy for an average of 26 weeks. The female former users had taken ecstasy more recently than the men (15 versus 31 weeks respectively). The levels of depression, as measured by Beck's depression inventory (BDI), were significantly increased compared with a matched non-drug using control group. The depression scores were independent of alcohol and cannabis use. Peak usage (maximum ecstasy usage in 12 hours) and current levels of perceived stress together significantly predicted depression scores. Thus, former chronic ecstasy users may be at a high risk of developing more severe depression.

In 430 regular ecstasy users, a semi-structured interview was administered to evaluate the psychological effects of different patterns of ecstasy use in men and women (74). Factor analysis established three main categories of acute effects of ecstasy—namely positive and negative effects on mental health and physical effects. In terms of suba-cute effects, 83% reported low mood and 80% reported impaired concentration between ecstasy-taking sessions. Susceptibility factors influencing these effects included age, sex, extent of ecstasy use, and concomitant use of cocaine or amfetamine. Specifically, the individuals who had mid-week lows were older than those who did not, and in men length of use was also a significant factor. Women who reported impaired concentration between ecstasy-taking sessions were older than those who did not. Surprisingly, men who had mid-week lows consumed ecstasy less often. This may be because they were aware, based on their previous experience, that higher consump-tion leads to an increase in mid-week symptoms. The most common long-term effects included tolerance to ecstasy (59%), impaired ability to concentrate (38%), depression (37%), and feeling more open towards people (31%). In terms of what might persuade abusers to stop using ecstasy, their most prominent concern was the drug's long-term effects on mental health. Those who took cocaine with ecstasy had higher scores on the nega-tive effect factor.

Serotonin deficits and mood disorders in MDMA users have been studied (75). This topic is of interest because mood disorders and the effects of MDMA may be modu-lated through the serotonin system. The researchers com-pared the prevalence of mood disorders to serotonin transporter densities in short-term, long-term, and absti-nent MDMA users, in the following subgroups: moderate users (n = 15), heavy users (n = 23), former heavy users (n = 16), and MDMA-naïve controls (n = 15). Former MDMA users had had to be abstinent from the drug for at least 12 months. The subjects agreed to abstain from any drug use for 3 weeks before the study, and this condition was confirmed by a urine drug screen. Mood disorders were evaluated using the Composite International Diagnostic Interview (CIDI) and the Beck Depression Inventory (BDI). Serotonin transporter densities were calculated using SPECT imaging techniques. Across all groups, CIDI scores did not differ for current or lifetime mood disorders. BDI scores differed significantly between former users and controls, former users scoring higher. Additionally, BDI scores correlated positively with the total number of MDMA tablets taken. Serotonin transporter densities were lower in heavy MDMA users than in the other groups, and lower in women than in men. There were no relations between

serotonin transporter densities and mood disorder scores. An interesting component of the results is that serotonin transporter densities appear to recover after heavy MDMA use stops, whereas the prevalence of mood dis-orders does not improve. However, the results of this study were limited by the small sample size.

National drug information strategies convey the mes-sage that the use of ecstasy is associated with an increase in both the incidence and severity of major affective disorder. However, very little research supports this. Many reviewers have felt that the use of ecstasy results in a depressed mood. In a meta-analysis of 22 studies of self-reported depressive symptoms in ecstasy users there was a significant but small effect size of 0.31 (95% CI = 0.17, 0.37) (76). Estimated lifetime ecstasy use, but not duration of use, usual dose per episode, or abstention, predicted the effect size. Although the effect size was small, it contrasted with most published literature, which has generally shown no association between the use of ecstasy and depressive symptoms. It is possible that individual studies were not statistically powerful enough to detect effects that only emerged through meta-analysis. However, the authors noticed that smaller studies produced larger effect sizes. Furthermore, they questioned the quality of the primary data sources. Only nine of the 23 studies controlled for use of cannabis between the groups, and only eight of these recorded use parameters. This meant that it was not possible to differ-entiate the effects of ecstasy accurately from those of cannabis. Very few studies reported wider drug histories. The authors cited the example of a study that showed that although most of the volunteers self-reported a pre-ference for using ecstasy, their actual behavior was dis-crepant and they valued intoxication over the effects of individual drugs. A posteriori classification of most ecstasy-using populations would more accurately define them as having a preference for alcohol and cannabis. There was no association between effect size and absten-tion from ecstasy. The authors suggested that although use of ecstasy may have been postponed or stopped, other substance misuse had continued, preventing a lin-ear relation between abstention and effect size. They mentioned problems with the measuring scales used and heterogeneity amongst the ecstasy users. Specifically, self-reported scales may be heavily confounded by the somatic effects of substance misuse. Interpretation is further complicated by the high prevalence of psychiatric co-morbidity amongst drug users. The observed effects may be common to polysubstance misuse in general rather than to the use of ecstasy per se. The authors recommended that drug users should be informed that substance misuse has both acute and subacute effects that may have negative effects on health, work, and relation-ships.

Cognitive effects

Several studies have focused on the cognitive effects of ecstasy, and a review of the adverse effects of MDMA in animals has raised concerns about potential toxicity in humans (77). MDMA can selectively damage brain serotonin (5-HT) neurons in animal brains, even at doses within the range of those typically used recreationally. In many studies, lasting reductions in various brain 5-HT markers have been reported in MDMA-treated animals. Axons of 5-HT neurons were damaged in MDMA-treated animals, including monkeys, and neurotoxic effects of MDMA have been observed in every animal species studied so far. In non-human primates, the toxic dose of MDMA closely approaches the dose used by humans.

In 30 regular ecstasy users (drug use 10 or more times per month) and 31 ecstasy-free controls, prospective memory—the process of remembering to do things at some future time—was assessed using a self-report questionnaire (78). Ecstasy users reported global impairment in prospective memory, even after controlling for the use of other drugs. The authors observed that their finding of impairment of cognitive functions, especially memory, in regular ecstasy users is similar to findings in other recent studies. They postulated that serotonergic toxicity secondary to chronic ecstasy exposure can cause cognitive impairment, as serotonergic pathways are involved in memory function.

More support for a relation between memory impairment, serotonin neurotoxicity, and chronic ecstasy use has come from a study of cognitive performance and serotonin function in two groups of 21 men with moderate or heavy ecstasy use and a control group of 20 men who had not used ecstasy (79). Ecstasy users had a broad spectrum of statistically significant but clinically subtle impairment of memory and prolonged reaction times. Heavy users had larger effects than moderate users. Serotonergic function was assessed in a double-blind crossover challenge with dexfenfluramine 30 mg or placebo. Release of cortisol, but not prolactin, after dexfenfluramine was significantly reduced in both groups of ecstasy users compared with controls. According to the authors, these neuroendocrine findings are similar to those observed in animals and humans. Recent exposure to ecstasy, psychosocial profiles, and the use of other drugs did not explain the differences.

The density of serotonin transporters (SERTs) in humans can be measured by neuroimaging techniques, such as SPECT and PET. Two groups of ecstasy users (22 recent but abstinent and 16 ex-users) were compared with ecstasy-naïve controls (80). In subjects who had stopped using ecstasy more than 1 year before the study, cortical densities of SERTs did not differ from those of controls. However, recent ecstasy users had global reductions in SERTs. In addition, individuals who had stopped using ecstasy had a deficit in verbal memory similar to that in current ecstasy users. Higher lifetime doses of ecstasy were associated with greater impairment of immediate verbal memory. The authors suggested that the absence of reductions in SERT densities in ex-ecstasy users suggested reversibility of ecstasy-induced changes in brain SERTs. Thus, ecstasy use can lead to neurotoxic changes in human cortical 5-HT brain neurons, but these changes may be reversible. However, functional consequences of ecstasy on cortical brain 5-HT neurons may

not be reversible, as seen by impaired memory function in this group, similar to that in current ecstasy users.

The amino acid N-acetylaspartate is a robust, non-specific marker of neuronal loss or brain dysfunction. It is detectable by proton magnetic resonance spectroscopy (MRS). The ratio of creatine to phosphocreatine, another marker that remains stable in many brain diseases, can also be assessed using MRS. In a recent study, the ratio of N-acetylaspartate to creatine/phosphocreatine was used to determine whether there are memory deficits in ecstasy users and whether there are changes in specific brain regions (81). Eight men with a history of ecstasy use and seven ecstasy-naïve male controls took part. The findings included a significant difference in delayed recall between ecstasy users and ecstasy-naïve controls. There was a strong association between impaired memory function in ecstasy users and neuronal pathology, specific to the prefrontal cortex. Poorer performance on verbal memory testing was associated with greater neuronal loss or dysfunction in ecstasy users. These findings implicate ecstasy as a cause of serotonergic neuronal damage and memory impairment.

Another group has assessed cognitive function in 80 subjects who were non-users, novice users, regular users, or currently abstinent users of ecstasy (82). Compared with the non-users, all three groups of ecstasy users had significantly poorer verbal fluency and immediate and delayed prose recall. Days since last use and total lifetime consumption of ecstasy made separate contributions to the variance in recall scores. Novice and currently abstinent users likewise had significantly poorer immediate recall than non-users. The novice users had milder impairment than the regular users, while those who were currently abstinent performed at a similar level to regular users. These deficits were not attributable either to differences in general reasoning ability or to impairment of working memory. The authors expressed concern that increased use of ecstasy in large numbers of young people and its enduring effects on memory are under-appreciated consequences.

However, cognitive impairment and ecstasy use is a complex association to study, with difficult confounding variables. For example, ecstasy users may consume cannabis to relieve the negative experience that occurs when ecstasy-related euphoria diminishes. In a recent study the concurrent use of cannabis was controlled by the recruitment of 31 drug-naïve controls, 11 ecstasy/cannabis users, and 18 cannabis users (83). Users were instructed to abstain from drug use for 48 hours before testing. The ecstasy/cannabis user group had deficits in learning, memory, verbal word fluency, speed of processing, and manual dexterity compared with the healthy controls. The authors suggested that the deficits in the drug group were not related to ecstasy. They observed that the study group did not perform poorly compared with those who used cannabis only. Moreover, the finding that the ecstasy/cannabis group was no different from the cannabis group suggested that ecstasy does not cause significant cognitive deficits. The poorer performance of the ecstasy/cannabis users was very little affected by the frequency of use or total ecstasy consumption. However, co-varying for indices of cannabis consumption removed most of the significant differences between the groups. On the other hand, ecstasy did affect the measures of working memory, such as forward and backward digit span. The authors raised the question of whether previous evaluations of the effects of ecstasy on cognition had been confounded by the concomitant use of cannabis.

In one study, investigators used a working memory task and functional magnetic resonance imaging (fMRI) to investigate cerebral activation in 11 previously heavy but currently abstinent ecstasy users and two equal-sized groups of moderate users and non-users (84). Surprisingly, there were no significant group differences in working memory and no differences in cortical activation patterns for a conservative level of significance. However, ecstasy users had stronger activation in the right parietal cortex than controls. Furthermore, heavy users had a weaker blood oxygenation level-dependent (BOLD) response than moderate users and controls. Although these results suggest subtly altered brain functioning associated with prior use of ecstasy, the authors cautioned that an alternative interpretation of the group differences must be considered. The ecstasy users in this study had also used other amphetamines and cannabis, further confounding the data.

A meta-analysis of 10 studies that met a priori inclusion/exclusion criteria evaluated the possible functional neurotoxic effects of ecstasy use in humans on verbal short-term memory, verbal long-term memory, processing speed, and percent errors (attention) (85). The mean effect sizes were significant. Ecstasy users had poorer verbal short-term memory and long-term memory, reacted more slowly, and made more errors. However, the meta-regressions of effect sizes against total lifetime ecstasy consumption were not significant, and the effect sizes for long-term memory became insignificant, suggesting that ecstasy does not impair long-term memory. Although the conclusions were based on a very low number of studies, the results supported the idea that chronic ecstasy use impairs short-term memory. There was no support for a link between lifetime ecstasy consumption and functional neurotoxicity. The author suggested that there may be a stepwise relation rather than a continuous one. Moreover, it is possible that the damage threshold is reached after the first few doses of ecstasy, so that lifetime use does not matter. Furthermore, apparent neurotoxic effects of ecstasy may actually be due to the effects of concomitant drugs used, such as alcohol.

The effects of MDMA use on visuospatial memory have been reported in 25 current MDMA users (11 men, 14 women), 10 former users who were abstinent for at least six months (6 men, 4 women), and 18 non-using controls (6 men, 12 women) (86). Both user groups performed more poorly than the controls, even controlled for age, years of education, intelligence, and alcohol and tobacco use. However, when cannabis use was considered, the main effect became non-significant. These results suggest that current and former MDMA users have deficits in visuospatial memory compared with controls. Since the tasks in question here involve executive functioning in addition to visuospatial memory, it is possible that these deficits are mediated through the prefrontal cortex. However, the authors did not propose a mechanism for this interaction.

The effects of MDMA on verbal working memory (reading span and computation span) have been studied in 42 current users (22 men, 20 women), 17 previous users (9 men, 8 women), and 31 non-users (12 men, 19 women) (87). Previous users were required to have been abstinent for at least 6 months and users were asked to abstain for 7 days before the test. Both MDMA groups performed significantly worse on computation span; reading span was not significantly affected. After controlling for other types of drug use, all the effects remained. The fact that there was impairments in current and former users of MDMA suggests that there are long-term effects of MDMA abuse on working memory. It is not known whether this impairment is caused by changes in executive function or a more specific effect on working memory processes. One potential problem with the current study was that there was no guarantee that subjects had not used drugs before testing.

Executive functioning (cognitive flexibility, working memory, and response inhibition) has been investigated in relation to MDMA use in 59 participants (26 MDMA users, 33 non-users) (88). Male users had a specific set of deficits compared with controls: they performed poorly on tasks that required set shifting (cognitive flexibility) and complex executive function tasks. These functions are related to the ability to adapt quickly to changes in the environment. Other processes remained intact. These effects were noted in moderate MDMA users (a mean of 1.7 MDMA tablets per month). The fact that these deficits were not seen in female users could be explained by the fact that men had significantly more MDMA consumption than women (mean use of 54 tablets versus 39 tablets). It is not known whether these deficits are severe enough to affect behavior in real life.

In an analysis of the nature of cognitive deficits in ecstasy users, 60 currently abstinent ecstasy users and 30 non-users were given memory tests (89). Heavy ecstasy users (n=30; lifetime dose at least 80 ecstasy tablets) had poorer memory than both non-users and moderate users (n=30). However, there were no group differences in central executive function, working memory, planning ability, and cognitive impulsivity between ecstasy users and controls. Poorer memory was associated with a heavier pattern of ecstasy use. Poor memory performance did not predict poor working memory, planning ability, central executive control, or high cognitive impulsivity. Thus, the authors concluded that primary memory dysfunction in heavy ecstasy users (lifetime consumption of about 500 ecstasy tablets) may be related to a particularly high vulnerability of the hippocampus to the neurotoxic effects of ecstasy. Hippocampal dysfunction after the use of ecstasy may be a susceptibility factor for earlier onset and/or more severe age-related memory impairment in later years.

Concerns have been raised about the effect of long-term ecstasy use on cognition. Visual evoked potentials have been studied during digit discrimination in eight heavy ecstasy users, 8 moderate ecstasy users, and 18 drug-free controls (90^c). Only one subject had used ecstasy in the previous 6 months. The heavy users made significantly more errors than the other two groups. They had reduced amplitude but not latency of visual evoked potentials at both the occipital (Oz) and frontal (Fz) leads during the task. The effect in the occipital leads was present in P200 for heavy users only and in P300 components for both groups of users. In the frontal leads the amplitude effect was present in N250 for heavy users only and in P300 components for both groups. These results suggest that ecstasy consumption affects cortical activity, reflected in both exogenous and endogenous aspects of information processing, especially those components that affect attentional demands. Long-term exposure to ecstasy causes long-term electroencephalographic deficits suggesting altered cortical activity.

Ecstasy-related deficits in cognitive functioning, such as shifting, inhibition, updating, and access to long-term memory elements of the central executive functions, have been evaluated in two studies, one of which focused on updating and access to long-term memory, and the other on switching and inhibition (91). The first study supported an ecstasy-related deficit in memory updating and access to long-term memory that was not related to sex, intelligence, amphetamine use, or sleep quality. Access to long-term memory, as indexed by word fluency scores, was more related to aspects of cocaine use than to equivalent indices of ecstasy use. Contrary to the investigators' expectations, updating executive component process was influenced by cannabis. In the second study, there was no evidence of any ecstasy-related deficit on inhibition and switching measures. Thus, ecstasy/polydrug users reached a lower level on the computation span task and recalled fewer letters correctly on the letter-updating task. Ecstasy/polydrug users scored higher on an intelligence test and significantly higher on a sleep questionnaire, and the main effect remained significant after controlling for these co-variates. Contrary to expectations, ecstasy/polydrug users actually performed better than controls on the random letter generation task used to measure inhibition. There were no significant ecstasy/polydrug-related effects on the tasks used to measure switching. Thus, both studies provide support for ecstasy/polydrug-related deficits in memory updating and access to semantic memory, but not shifting or inhibition. The authors questioned the unanticipated effects of ecstasy in producing more letters and caution in over-interpreting these findings, saying that on other similar measures these differences did not exist. As ecstasy users were frequently poly-drug users, and since the impact of concurrently used drugs in ecstasy users appears to be significantly large, the contribution of other drug use cannot be minimized. Effects of ecstasy/polydrug use on executive functions are not uniform—ecstasy/polydrug users perform worse on updating and access tasks, but not shifting and inhibition tasks, which appear to be relatively unaffected.

Susceptibility factors

Sex differences in subjective experiences of ecstasy use have been reported. Three previously published controlled studies of the acute effects of ecstasy in healthy subjects with no or single previous ecstasy experience have been summarized (92). There were 74 subjects (54 men and 20 women), aged 20–49 (mean 27) years, of whom 69 were ecstasy-naïve and five had used it on one or two occasions before. All had been screened using a semi-structured interview, and anyone with a personal or family history of mood disorders, schizophrenia, or other psychiatric disorders was excluded. The analysis included psychological and physiological effects of ecstasy in controlled settings. Generally, subjective effects of ecstasy were more intense in women than men. Women had especially higher scores for ecstasy-induced perceptual changes, anxiety, and adverse effects. In contrast, men were slightly activated by ecstasy compared with women and had significantly higher increases in systolic blood pressure. The authors suggested that women may be more sensitive to the effects of ecstasy. Many previous studies have suggested sex differences in markers for serotonergic activity in ecstasy users, with suggestions that serotonergic function is relatively impaired in women compared with men. The authors, based on their findings coupled with data from previous reports, raised the question of whether women might be more susceptible to ecstasy-induced depletion of serotonin.

Nutrition

In a study of plasma concentrations of 33 amino acids, 159 subjects were recruited, of whom 107 were ecstasy users (93). The subjects were grouped according to cumulative lifetime use: under 100 tablets (n = 34), 100–499 tablets (n = 42), 500–2500 tablets (n = 30), abstinent subjects (n = 11), and never users (n = 41). All were ecstasy free for at least 3 days, as verified by toxicological analysis. In 49% of the users, the time to the last use of ecstasy was 1 month or less. There were significant reductions in the serum concentrations of phosphoserine, glutamate, citrulline, methionine, tyrosine, and histidine. Based on findings from other studies, the authors speculated that the reductions in serine and methionine may underlie psychosis associated with the use of ecstasy. Reduced glutamate may also add to the burden of psychiatric symptoms in ecstasy users.

Electrolyte balance

Hyponatremia secondary to the syndrome of inappropriate antidiuretic hormone secretion (SIADH) has been reported (SEDA-23, 36; SEDA-25, 37).

- A previously healthy 18-year-old woman had an altered mental state after using ecstasy, followed by excessive thirst and consumption of a lot of water within a few hours (94). She became anxious, remorseful, and mildly agitated, with visual hallucinations. She vomited several times, became lethargic and unresponsive, and had pronounced bruxism. Her pupils were dilated with a sluggish response, she was hypothermic, and she had a serum sodium concentration of 124 mmol/l; serum and urine osmolalities suggested SIADH and serum ADH concentration was inappropriately high. A urine drug screen was only positive for amphetamines. She was given isotonic saline, and her sodium concentration fell further to 114 mmol/l. She was given hypertonic saline and recovered completely after medical treatment.

The combination of SIADH and excessive fluid intake after the use of ecstasy played a significant role in causing water intoxication, which was worsened by isotonic saline in the early stages of treatment. The authors reviewed the literature, and found that 17 of the reported 18 cases were young women aged 15–30 and the majority admitted taking only one tablet along with large quantities of fluid. The interval between the consumption of ecstasy and the onset of symptoms was 4–24 hours, with serum sodium concentrations of 101–130 mmol/l. There were three deaths and all were women. Hyperthermia was extremely rare (one case). Initial treatment appears to have played an important role in reducing mortality. When treatment other than hypertonic saline was used initially, there was a higher risk of death. The authors strongly recommended a high degree of suspicion of water intoxication and aggressive treatment with hypertonic saline instead of diuretics and water restriction.

A death from hyponatremia after ecstasy use has been reported from New Zealand, where it is a commonly used recreational drug (95).

- A 27-year-old woman was admitted to the intensive care unit with coma and respiratory arrest. She had been seen taking two ecstasy tablets while dancing at a nightclub about 5 hours before admission. She had drunk copious amounts of water to cool herself. When she reached the hospital, her Glasgow score was 3 and she was in respiratory arrest with fixed dilated pupils. Her serum sodium concentration was 124 mmol/l with a low serum osmolarity (267 mmol/l) and normal renal function. Her toxicology screen was positive for ethanol, and MDMA was found in her serum. The electrocardiogram showed a sinus tachycardia without ischemia, and a chest X-ray showed gross pulmonary edema. A CT scan of the brain showed generalized cerebral edema, diffuse cerebral swelling, and significantly increased intracranial pressure; the basal cisterns and extracerebral spaces were completely effaced. Brain death was confirmed by cerebral angiography, which showed no intracranial flow in the carotid or vertebral arteries. The post-mortem report confirmed marked cerebral edema with signs of cerebellar tonsillar herniation.

The authors reviewed the possible explanations, which included excess water ingestion, SIADH, and the serotonin syndrome. While many ecstasy users drink water to counter its hyperthermic effects, others do it because of "health advice" given at such parties to prevent toxic effects. However, death can occur due to water intoxication, albeit infrequently.

In a retrospective analysis of all cases of hyponatremia associated with ecstasy (SEDA-25, 37) at the London Centre of the National Poisons Information Service from December 1993 to March 1996, 17 patients were identified with a serum sodium concentration under 130 (range 107–128 mmol/l) (96). In 10, ecstasy was identified analytically, and six of them had SIADH. The clinical presentation was very consistent, with initial vomiting and delirium, and 11 had seizures. There was complete recovery in 14, but two died of cerebral edema 5 hours after ingestion.

- A 20-year-old woman attended a "rave", where she took ecstasy and drank large amounts of water (often suggested to prevent dehydration and other life-threatening consequences) (97). She felt drowsy and had a headache. After lying still for 3 hours, she had a tonic-clonic seizure and was brought to hospital, where her plasma sodium concentration was 112 mmol/l and she had cerebral edema. She was treated with hypertonic saline and recovered fully.

These reports show that hyponatremia due to ecstasy occurs primarily due to inappropriate secretion of anti-diuretic hormone and relative excess liquid intake, usually in the setting of a warm environment and possible hyperthermic effects of ecstasy. Excess water ingestion should be discouraged in ecstasy users.

Two deaths in patients with hyponatremia after ecstasy intoxication have been reported (98).

- A 15-year-old woman developed impaired consciousness and psychomotor agitation and 10 minutes later reactive bilateral mydriasis and decerebration. A CT scan showed diffuse brain edema and subarachnoid hemorrhage. She had hyponatremia (119 mmol/l) and was considered brain dead 32 hours later.
- A 19-year-old woman became disoriented and had hyponatremia (131 mmol/l) and a serum ethanol concentration of 2.4 mmol/l. Her serum sodium had fallen to 120 mmol/l 10 hours later, and 5 hours later she suddenly deteriorated and had a respiratory arrest, coma, and sinus tachycardia of 180/minute with bigeminy. A CT scan showed diffuse edema in the posterior fossa. She was considered brain dead 23 hours after taking the ecstasy.

Hyperkalemia has been reported in association with fatal MDMA toxicity (99).

- A 19-year-old man took 12 tablets of MDMA and danced in a hot environment throughout the night. His heart rate rose to 180/minute and his blood pressure fell to 70/50 mmHg. He had a cardiac arrest and was resuscitated and mechanically ventilated. His potassium blood concentration was 6.3 mmol/l. He remained comatose until a second cardiac arrest proved fatal 2.5 hours later. The plasma MDMA concentration was 3.65 µg/ml.

The authors suggested that the patient's hyperkalemia was precipitated by a hypermetabolic state in which there was reduced glomerular filtration and the beginning of rhabdomyolysis.

Mineral balance

SIADH has been associated with MDMA (100,101).

- A healthy 19-year-old woman complained of nausea and vomited 8 hours after taking unknown quantities of MDMA and beer; 3 hours later, she suddenly clenched her jaw, had tonic contractions of all four limbs, and collapsed. She was obtunded, with occasional moaning and non-purposeful movements of the limbs. Head CT scan showed mild cerebral edema. Her serum electrolytes, including a sodium of 115 mmol/l and a corresponding urine osmolality of 522 mosm/kg, suggested SIADH. Despite treatment, the serum sodium concentration 10 hours later was 116 mmol/l, but 18 hours after treatment, it rose to 125 mmol/l. She became progressively more responsive, with normalization of her sodium concentration, and after 48 hours was awake and alert, with a serum sodium concentration of 136 mmol/l.

The authors reviewed nine other reported patients with MDMA-related SIADH, all of whom were women. They concluded that MDMA-associated SIADH is multifactorial and that MDMA may stimulate vasopressin secretion in susceptible individuals. They further suggested that hyponatremia can also occur secondary to voluntary increases in fluid or water intake aimed at preventing the adverse effects of MDMA. With appropriate treatment, full recovery is possible in almost all cases of this life-threatening condition.

- An 18-year-old woman developed impaired consciousness, psychomotor shaking, hallucinations, tics, and delirium. Her serum sodium concentration was low at 120 mmol/l with a plasma osmolality of 242 mosm/kg and a urine osmolality of 562 mosm/kg, suggesting SIADH. Most other blood tests were within the reference ranges, except for a raised creatine kinase. Urine toxic screen was positive for amphetamines. Treatment with hypertonic saline brought about resolution of symptoms. The patient recalled taking three ecstasy tablets over 6 hours.

Hematologic

Transient anemia has been associated with ecstasy (102).

- A 36-year-old HIV-positive man taking zidovudine, lamivudine, and indinavir was noticed to have a mild asymptomatic anemia (hemoglobin 10.2 g/l, white cell count 10.4×10^9/l with a normal differential count, and platelets 237×10^9/l). The blood film suggested hemolysis. He had taken ecstasy 2 weeks earlier for the first time. The anemia was reportedly "secondary to oxidative stress, probably due to drug toxicity." Three weeks later his hemoglobin returned to normal.

In the absence of other explanations, the authors concluded that this syndrome had probably been caused by ecstasy toxicity.

Mouth

Oral complications of topical application of ecstasy have not been described previously (103).

- A 15-year-old boy developed an atraumatic painful swelling of the upper lip, fever, and malaise. He had good oral hygiene and no pathological periodontal pockets, but there was a swelling of the maxillary labial vestibule in relation to the upper central incisors. Both maxillary central incisors had grade II mobility and were tender to percussion. The dentoalveolar abscess was incised and drained and culture yielded commensal oral flora. His white cell count was normal with raised eosinophils; the ESR was slightly raised at 23 mm/hour and he had a mild rise in hepatic enzymes. He had used ecstasy 1 day before the onset of symptoms and had stored the drug in the upper anterior labial vestibule adjacent to the site of periodontal destruction. He denied previous use of ecstasy or other recreational drugs.

The authors diagnosed local drug-induced necrotizing gingivitis and said that although similar lesions have been reported with local cocaine, none has been reported with ecstasy.

Teeth

An unusual adverse effect of MDMA, tooth wear, has been reported (104).

- A healthy 17-year-old boy, who complained of dental sensitivity that occurred only when he consumed fizzy drinks, presented with marked tooth wear. All the teeth were involved, especially the premolars and permanent molars. He ate a poorly balanced diet, with 500 ml carbonated drinks twice daily. He subsequently admitted to frequent MDMA abuse.

Current health promotions advise that ecstasy users should frequently consume "sports" type or fruit drinks to counteract dehydration and avoid ion imbalances. These particular drinks may be erosive to the teeth. Bruxism and trismus associated with ecstasy use can also contribute to tooth wear.

Liver

Cases of ecstasy-associated hepatotoxicity have been reported.

- A 27-year-old man developed jaundice without fever (105). He used ecstasy regularly and had recently increased his consumption. All other possible causes of acute hepatitis were ruled out. After withdrawal of ecstasy, the condition resolved completely.
- A 17-year-old girl developed progressive jaundice and weight loss (106). Four months before, she had had malaise, anorexia, a sore throat, and tender cervical lymph nodes. Two months later she reported eating

"hallucinogenic mushrooms." Five weeks before admission she observed blood clots in her stool. Three weeks later, she ate more mushrooms. She then developed progressive jaundice, with vomiting, dark urine, light-colored stools, and raised aminotransferases. She drank alcohol in binges and reported marijuana use since the age of 15. She had used ecstasy on several occasions during the previous 2 months. She had a minimally tender enlarged liver. Her viral hepatitis screen was negative. Genetic, metabolic, and autoimmune disorders were also ruled out. A urine screen for drugs was positive only for cannabinoids. Abdominal ultrasound showed periportal edema and contraction of the gall bladder, but no gallstones or dilated bile ducts. A CT scan showed moderate amounts of free fluid in the abdomen and pelvis and periportal hepatic edema. Percutaneous liver biopsy showed acute cholestatic hepatitis with cholangitis, eosinophils, and histiocytes, strongly suggesting a hypersensitivity reaction. There was no evidence of chronic liver disease. Within 24 hours after the liver biopsy, her liver enzymes and coagulopathy had begun to improve.

- A 21-year-old soldier had nausea, vomiting, abdominal pain, and fever, having taken 7–8 ecstasy tablets a week (107). He was treated with intravenous fluids and paracetamol (650 mg/day) for 3 days. His symptoms did not improve and he developed jaundice. He had increased transaminase activities and total bilirubin concentrations and all tests for viral hepatitis were negative. He was considered for liver transplantation, but over the next 2 weeks improved spontaneously.

The authors of the last report postulated a hyperthermic effect of ecstasy on the liver or a direct toxic effect of ecstasy on hepatocytes as possible mechanisms.

Cases of successful liver transplantation after ecstasy-induced hepatotoxicity have been reported (98,108,109).

- A 19-year-old man with no significant past history of medical problems took $1^{1/2}$ tablets of ecstasy and some alcohol. Within 2–4 days, he developed tiredness, nausea, malaise, and vomiting and on the fifth day weakness and anorexia; 12 days later, he was admitted to hospital with marked jaundice and weight loss. He deteriorated and developed hepatomegaly and an abnormal prothrombin time; his total bilirubin was 571 (reference range 5–17) µmol/l, aspartate transaminase 213 (14–50) U/l, and alanine transaminase 336 (11–60) U/l. By day 20, his condition had deteriorated, with increased jaundice, somnolence, mild disorientation, and altered glucose metabolism. His total bilirubin was 654 µmol/l, aspartate transaminase 1290 U/l, and alanine transaminase 1932 U/l. A liver biopsy showed swollen hepatocytes, patches of necrosis, and patchy cholestasis. Two days later, he developed a grade III encephalopathy and further disturbances of coagulation. Abdominal ultrasonography showed hepatic atrophy, and a liver biopsy showed massive necrosis. Serological tests for viral infections were negative. The probable diagnosis was toxic hepatitis secondary to MDMA. On day 31, he had a right

auxiliary liver transplantation, but there was no clinical and laboratory improvement during the next 48 hours. Histopathology of the transplanted liver showed massive liver necrosis consistent with a "diagnosis of primary non-function". A second liver transplant on day 33 was successful.

- A 17-year-old girl with no history of drug abuse took two tablets of ecstasy at a disco 10 days apart, and reported malaise, constipation, and icterus 6 days after taking the second. She had severe acute hepatic failure. Viral hepatitis was ruled out. Following rapid clinical and neurological deterioration (encephalopathy grade I–II), she had an urgent liver transplant and made a good recovery. In the affected liver, there was submassive centrilobular hemorrhagic necrosis (75–80%) with massive periportal and lobular lymphocytic infiltration (CD 8+) and moderate fatty changes.
- A 25-year-old woman developed abdominal pain, jaundice, and vomiting 5 days after consuming ecstasy. She had hepatocellular failure with a prothrombin ratio of 6.5, cytolysis, cholestasis, renal insufficiency, and encephalopathy. She had a liver transplant 2 days later and recovered fully.
- A 17-year-old man developed toxic subacute hepatitis with grade II encephalopathy and coagulation disorders a few days after consuming ecstasy. Liver transplantation was performed and he recovered fully.
- A 16-year-old woman consumed ecstasy and developed jaundice, hepatic failure, a prothrombin ratio of 9, and grade I encephalopathy. She had a liver transplant and was asymptomatic at 11 months.

Many cases of hepatotoxicity and some deaths have occurred in drug-naïve subjects after the ingestion of relatively small amounts of ecstasy. The authors reported that more than 70 deaths have been reported worldwide between 1990 and 1998 after the onset of severe hepatic damage from ecstasy. Although many patients died after liver transplantation, a significant number recovered. The authors suggested that in emergency care the use of ecstasy should be suspected in young people who present with unexplained jaundice, hepatomegaly, or altered liver function, in the absence of other known substance exposure. Early referral for liver transplantation may be significantly beneficial. The prognosis may be better with grade I–II encephalopathy than grade III–IV, which is usually associated with rapid deterioration and a poor prognosis.

Urinary tract

Transient proximal tubular renal injury following ecstasy has been reported (110).

- An 18-year-old woman presented with new onset seizures and polydipsia. She had a hyponatremia of 117 mmol/l, polyuria for several hours, renal glycosuria with urine glucose of over 55 mmol/l, a blood glucose of 6.6 mmol/l, and solute diuresis. She had low tubular reabsorption of phosphorus, with an appropriate trans-tubular potassium gradient of 3.0 and a serum potassium of 3.7 mmol/l. After medical treatment and gradual correction of her hyponatremia, her tubular reabsorption of phosphorus normalized and her glyco-suria resolved. An extensive drug screen was positive for ecstasy.

The authors thought that this was the first report of acute transient proximal tubular injury with ecstasy. In contrast to SIADH, there was a high urine output in the presence of hyponatremia and solute diuresis.

Urinary retention has been reported (111).

- A 17-year-old man developed abdominal pain and urinary retention after taking ecstasy the previous evening. He had a tachycardia, mydriasis, and a tender suprapubic mass, which resolved with catheterization. He was discharged the next day.

The authors suggested that ecstasy had caused release of noradrenaline, which had caused urinary retention through alpha-adrenoceptor stimulation.

Acute transient urinary retention associated with metamfetamine and ecstasy (3,4 methylenedioxymetamfetamine, MDMA) in an 18-year-old man has been described (112). Analysis by gas chromatography–mass spectrometry confirmed the presence of metamfetamine (>25 µg/ml), MDMA (> 5 µg/ml), amfetamine (1.4 µg/ml), and methylenedioxyamfetamine (3.7 µg/ml) in the urine. Bladder dysfunction resulting from alpha-adrenergic stimulation of the bladder neck may have explained the observed effect.

Skin

Two cases of facial papules have been reported (113).

- A 20-year-old woman developed diarrhea and a pruritic yellowish skin 7 days after she had taken half a tablet of ecstasy. Her liver was enlarged and tender. Her urine was positive for MDMA. A diagnosis of acute hepatotoxicity after MDMA was made. She rapidly developed reddish papules over the face with a perioral and acne-like distribution.
- A 21-year-old-man developed similar skin lesions after using ecstasy, without hepatotoxicity.

Both patients responded to a low fat diet and 1% metronidazole ointment. The authors suggested that serotonin indirectly affects the nerve endings of the eccrine glands via other peptides, and that the interaction of MDMA with serotonin may have caused the rapid development of pimples in these abusers.

Guttate psoriasis has been reported after the use of MDMA (114).

- A 23-year-old man reported to the hospital 5 days after MDMA ingestion with a pruritic generalized eruption. The rash, small papules of 10 mm in a guttate pattern, covered his legs, arms, trunk, and face. There were parakeratosis, elongated rete ridges, epidermal spongiosis, hypogranulosis, and superficial mononuclear cell infiltration. The rash responded completely to glucocorticoids and narrow band UVB at a maximum dose of 0.85 J/cm2.

The authors described this as a case of a lichenoid drug eruption caused by MDMA or a contaminant in the tablet. The patient denied using any other illicit medications.

Immunologic

Four healthy male MDMA users volunteered for a randomized, double-blind, double-dummy, crossover pilot study in which they took single oral doses of MDMA 75 mg ($n = 2$) or 100 mg ($n = 2$), alcohol (0.8 mg/kg), MDMA plus alcohol, or placebo to study the effects on their immune system. The doses of MDMA were compatible with those used for recreational use (115). The baseline immunological parameters were within the reference ranges. Acute MDMA use produced time-dependent immune dysfunction, which paralleled MDMA plasma concentrations and MDMA-induced cortisol stimulation kinetics. The changes in the immune system after MDMA peaked at 1–2 hours. Although the total leukocyte count remained unchanged, there was a fall in the ratio of CD4 to CD8 T cells and in the percentage of mature T lymphocytes (CD3 cells), probably because of a fall in both the percentage and the absolute number of T helper cells. The fall in CD4 cell count and in the functional responsiveness of lymphocytes to mitogenic stimulation with phytohemagglutinin A was MDMA dose-dependent. Alcohol produced a decrease in T helper cells, B lymphocytes, and mitogen-induced lymphocyte proliferation. Combined MDMA and alcohol use produced the greatest suppressive effect on CD4 cell count and mitogen-stimulated lymphoproliferation. Immune function was partially restored at 24 hours. According to the authors, these results provided the first evidence that recreational use of MDMA alone or in combination with alcohol alters immunological status. The reaction of the immune system to MDMA appears to be an alteration of physiological homeostasis, similar to that seen in volunteers exposed to acute physiologic stress, suggesting that MDMA could be a "chemical stressor". Moreover, combined MDMA and alcohol use produced additive effects. This is an important finding, since in the general population alcohol and MDMA are commonly taken together.

Immunosuppression has been attributed to ecstasy (116).

- A 24-year-old African–Americans man developed a vesicular rash on his left forehead and eyelids consistent with herpes zoster ophthalmicus. He reported safe sexual practices and no intravenous drug use, but had used ecstasy three times a day for 4 days before the onset of symptoms. He slowly improved with intravenous aciclovir.

Varicella zoster reactivation is a potential complication of immunosuppression and is uncommon under the age of 50. After ruling out all the potential causes for the infection, the authors speculated that ecstasy had caused immunosuppression, leading to the infection. In support of this they invoked previous reports of immunosuppression with ecstasy.

Because other drugs of abuse can cause immune dysfunction in regular users, the effects of acute administration of ecstasy on the immune system have been studied in both controlled and natural settings (117). In the controlled study, 18 male ecstasy users were given two doses of ecstasy 100 mg at intervals of 4 or 24 hours. There were significant reductions in CD4 T helper cells (30%) and the lymphoproliferative response to phytohemagglutinin mitogenic stimulation (68%) 1.5 hours after the first dose, and a 103% increase in the number of natural killer cells. At 4 hours, CD4 T helper cells and lymphocyte proliferative responses were reduced by 40% and 87%, but natural killer cell numbers increased to 141%. At 24 hours, the second dose augmented the alterations in the numbers of CD4 T helper cells and natural killer cells about threefold. The authors suggested that this large effect after repeated administration of ecstasy increases the interval during which the immune response is compromised, leading to a higher risk of illness and infection in ecstasy abusers.

In the uncontrolled study, 30 recreational users of ecstasy (mean age 24 years) were observed for 2 years and had lymphocyte counts at yearly intervals. The ecstasy users tended to have lower white blood cell counts over time. Lymphocyte counts were significantly lower than in healthy controls by year 1 and significantly lower than that the following year. CD4 and CD19 cell numbers fell significantly from basal to year 1 and from year 1 to year 2. Natural killer cell numbers were always lower than in healthy controls but did not fall with time. The authors extended these results to suggest a possible role of serotonin dysregulation caused by ecstasy in compromising immune function. These findings suggest that ecstasy abusers may be at a significantly higher risk of infectious diseases.

Body temperature

Abuse of ecstasy occurs in a variety of settings. Six cases of severe hyperthermia over 2 months have been reported (118).

- A 20-year-old woman was found unresponsive at a rave. She was hot to the touch and on the way to hospital had a tonic–clonic seizure and became pulseless and apneic. Aggressive resuscitation was unsuccessful. An autopsy showed gross pulmonary congestion and edema. She had acute neuronal ischemia and mild hepatic steatosis, without any evidence of myocardial damage. Her blood ecstasy concentration was 1.21 mg/l. Death was reported to have been secondary to ecstasy toxicity.
- A 20-year-old man, a previously healthy student, became agitated within 4 hours of ingesting two tablets of ecstasy at the same rave. His friends reported previous uneventful use of similar amounts. He was comatose and had minimal purposeful movements and a temperature of 41.5°C. His heart rate was 200/minute and his BP 123/57 mmHg. He had hot dry skin and diffuse intermittent myoclonus. Urine toxicology was positive for ecstasy only. His glucose was 2.5 mmol/l (42 mg/dl), platelets 80 x 10^9/l, sodium 148 mmol/l, creatinine 186 µmol/l (2.1 mg/dl), and total bilirubin 34 µmol/l (2.0 mg/d; peak creatine kinase activity was 17 000 iu/l. He recovered with cooling.
- An 18-year-old man at the same rave developed an altered mental state after taking two "tulips" (slang

for ecstasy tablets, which contained pure ecstasy). He was alert but disoriented in person, place, and time. His rectal temperature was 40.7°C. His heart rate was 177/minute, sodium 144 mmol/l, creatine 175 µmol/l (2.3 mg/dl), blood urea nitrogen 6.4 mmol/l (18 mg/dl), and peak creatine kinase activity 4964 iu/l. His urine was positive for ecstasy only. He recovered as his core temperature fell with treatment.

- A 20-year-old man became unconscious and had generalized tonic–clonic seizures 2 hours after taking one ecstasy tablet. He had a rectal temperature of 40.6°C, a heart rate of 170/minute, and a blood pressure of 92/21 mmHg. He had roving eye movements and myoclonus. His blood pH was 6.92, the serum sodium was 144 mmol/l, and his urine was positive for ecstasy. He recovered with cooling.
- A 22-month-old boy took a pill from a "Tic-Tac" container that had been filled with ecstasy tablets by his parents. Two hours later he was responsive to only painful stimuli and had roving eye movements. He had a rectal temperature of 39.3°C and a heart rate of 190/minute. He had bruxism, sweating, and rigid limbs with intermittent writhing. His urine was positive for ecstasy only. He gradually improved with cooling.
- A 27-year-old quadriplegic man became unconscious 2 hours after taking one tablet of ecstasy for recreational purposes. He had a rectal temperature of 41.3°C, a heart rate of 130/minute, and a blood pressure of 120/58 mmHg. He had roving eye movements with 5 mm equal reactive pupils. His urine was positive for ecstasy and cannabinoids. He recovered with cooling.

The authors observed that three of the six patients had been exposed to ecstasy outside of a rave and had hyperthermia despite not having vigorous muscle activity, thereby questioning previous assumptions about pathogenesis. There were no contaminants in the tablets, based on laboratory analyses. The authors speculated that another cause of hyperthermia and acute toxicity could be a genetic predisposition to defective metabolism of ecstasy, especially CYP2D6 deficiency. They postulated that hyperthermia in ecstasy users could be heterogeneous and was more likely to be due to a combination of causes.

A high-grade fever occurred in a young woman after the use of MDMA (119).

- A 20-year-old woman became unresponsive after taking two tablets of MDMA. She had a heart rate of 172/minute, a blood pressure of 164/118 mmHg, a respiratory rate of 30/minute, and a temperature of 41.8°C. She was placed under a cooling blanket, with ice packs and fans and 30 minutes later her temperature was 40.8°C, with a subsequent fall to 37.9°C 20 minutes after that. She also had a deep vein thrombosis in the arm. She continued to improve and there were no sequelae.

Death

MDMA and MDEA, or "eve," have rapidly become popular drugs of abuse in recent years in Europe and to some extent in other developed countries as well. Over the years, reports of deaths following their use at "rave" parties have generated considerable public concern.

In a study of all ecstasy-positive deaths (22 of 19 366 deaths) in New York City from January 1997 to June 2000, 18 were men, average age 27 years (120). The deaths fell into three categories: acute drug intoxication ($n = 13$), mechanical injury ($n = 7$), and a combination of natural causes and acute drug intoxication ($n = 2$). Only two of the deaths due to acute drug intoxication were caused by ecstasy alone, and one death was caused by a combination of ecstasy and coronary artery disease. Seven deaths were caused by a combination of cocaine/opiates and ecstasy. Acute ecstasy poisoning includes symptoms such as hypertension, hyperthermia, and delirium, and can progress to intracranial hemorrhage, status epilepticus, and death. Based on reports from harm-reduction organizations and previous studies, the authors discussed the lack of standardization of ecstasy tablets and the danger of harmful additives that could react with other drugs to cause life-threatening symptoms.

A fatal outcome after the use of ecstasy in a private setting rather than at a rave has been reported (121).

- A 19-year-old woman consumed beer, cannabis, and four ecstasy tablets with friends in her apartment. Two hours later, she suddenly jumped up and made some uncontrolled movements. She had cramps, was unable to speak, and needed support to walk. Following bouts of trembling and sweating she seemed to fall asleep, and her colleagues left her in her bedroom. An hour later they found that she had vomited and was trembling and barely conscious. She could not be resuscitated. At autopsy, there was cerebral edema and massive pulmonary edema and her distal bronchi were filled with gastric contents. The right ventricle was dilated and filled with blood. Histologically, there was generalized vascular congestion, perivascular edema of the brain, and pulmonary edema with focal atelectasis. There were no signs of disseminated intravascular coagulation. There was fatty degeneration of the liver. These findings were consistent with asphyxiation, hypoxia, shock, and cardiopulmonary failure. No ethanol was detected in the blood or cerebrospinal fluid but the urine concentration was 100 mg/l. There were cannabinoids and amphetamines in the urine and blood.

The authors bemoaned the fact that ecstasy is still falsely considered to be harmless by most users. They emphasized that the classic user profile has changed over recent years, with non-rave party use increasing, as ecstasy is more readily available. They also observed that a quieter environment, which is supposed to reduce the risk of adverse effects, did not prevent this woman's death. According to her friends, who had ingested similar amounts of ecstasy from the same supply without any untoward effects, this had been her first experience with ecstasy. Her blood ecstasy concentration of 3.8 mg/l was highly toxic. As her friends did not have any untoward reactions, the authors proposed that different people have

different sensitivity. Ecstasy probably had a direct toxic effect on vital centers in the brain stem, leading to central nervous impairment after a short initial state of excitation. Cannabinoid concentrations did not suggest a likely contribution to death.

Epidemiology

In a report from the UK it was noted that the annual death rate from ecstasy use is uncertain (122). A meta-analysis of surveys of illegal drug use by 15–24-year-old individuals in the UK in 1996 showed that 7% had used ecstasy in the previous year and 3% in the previous month, a rate 44% higher than in 1993–1995. But the actual numbers of ecstasy-related deaths are more difficult to ascertain. In Scotland, ecstasy-related deaths are defined as "ecstasy found in the body"; in England, on the other hand, it is defined as "ecstasy written on the death certificate." Using these disparate definitions, 11 deaths were reported in Scotland and 18 in England in the 15–24-year-old category. Based on all available data, the author of the report suggested a 26-fold range (0.2–5.3) of possible ecstasy-related deaths per 10 000 within this age group. To make death estimates more reliable, the author recommended that the definition of "ecstasy-related-death" should be unified, and that surveys should ask directly about regular, sporadic, and first-time drug use, so as to determine which group is at most risk. The author expressed concern that these data (poor as they are) may be used by the courts and policy makers as if they were more valid, citing the example of Switzerland, which has recently liberalized the guidelines on sentencing people who supply ecstasy tablets, in contrast to suppliers of heroin, based on their interpretation of the data that the death rate in ecstasy users is at the lower end of the range.

This report generated two responses. One group felt that estimating such death rates is even more complex than noted (123). They asked whether, if someone purchased ecstasy but instead got a contaminated or a very different compound, causing death, the death would be classified as ecstasy-related or as something else? Furthermore, they wondered if the definition of ecstasy-related death is based on the detection of MDMA (or its metabolites) in the blood or on detection of other compounds (consumed under the presumption of its being ecstasy). Citing their own experience, they suggested that criteria for inclusion should include: "Would the deceased still be alive if (s)he had not abused the substance?" Thus, even deaths from causes such as traffic accidents, when the death was secondary to intoxication with the compound, should be included.

The second group, while agreeing with the original report, raised the concern that surveys that are more frequent would be economically prohibitive (124). They noted that in 1997 The National Program on Substance Abuse Deaths (NPSAD) was established to monitor drug-related deaths in the UK. It uses a "cause of death ratio," which looks at the number of drug-related deaths in specific categories (for example 15–24-year-olds) and the drugs implicated in these deaths. According to the authors, this method allows rapid surveillance of the pattern of deaths over time.

Deaths related to ecstasy either alone or in combination with other drugs have been reported from England and Wales using the National Programme on Substance Abuse Deaths (np-SAD) database (125). This database receives information about all drug-related deaths from coroners. A total of 202 ecstasy related deaths occurred in the period from 1996 to 2002. There was a steady increase in the number of deaths each year. The male to female ratio was 4:1 and 75% of the victims were under 29 years. In 17% of cases, ecstasy was the sole drug implicated in death, and in the other cases a number of other drugs (mainly alcohol, cocaine, amfetamine, and opiates) were found. Based on toxicology reports, MDMA accounted for 86% of the cases and 3,4-methylenedioxyamfetamine (MDA) for 13% of the cases. There was one death each associated with 3, 4-methylenedioxy-N-ethylamfetamine (MDEA) and paramethoxyamfetamine (PMA). The authors reported that this was the largest sample of ecstasy-related deaths on record, with a death-rate of 3.4 per month; 31% of the subjects were living independently and 25% were with parents; 47% were employed and 10% were students; 49% died at home and 39% in hospitals. The most frequent verdict by the coroner was accident/misadventure at 49%.

The authors commented on the possible reasons for the increase in the number of deaths from ecstasy. Specifically, the UK has possibly the greatest availability of ecstasy among all European Union countries. Unfortunately, with increased use and supply, the purchase price of ecstasy tablets has fallen significantly over the years to nearly half its original price. Moreover, it is not uncommon for more lethal forms of amphetamines to be passed off as ecstasy. With increased publicity and awareness, coroners may have improved and thereby increased their reporting of such deaths. There is a particular concern that deaths from ecstasy tend to occur in younger individuals. The authors speculated that multiple compounds were probably consumed, in order to boost the effects of the single compound or to overcome its untoward effects. In such cases, death possibly occurred from the cumulative effects of all the drugs used.

In a retrospective chart review of the prevalence of MDMA-related deaths in the USA from 1999 to 2001 102 cases were identified in which MDMA was detected post-mortem: 10% were from 1999, 38% from 2000, and 52% from 2001 (126). The mean age was 25 years. Cases were coded as drug-related (drug toxicity contributed to death) or drug-unrelated (drug found but not related to the cause of death). Of the 102 reports 71 were considered to have been drug-related. The mean time from onset of symptoms to contacting emergency services was 6 hours and 42 minutes. There was no difference in MDMA concentration between drug-related and drug-unrelated deaths. There was also a high rate of polysubstance use (73%). The authors emphasized that there was a 400% increase in MDMA-related deaths from 1999 to 2001. Furthermore, the long delay between the onset of symptoms and medical intervention suggests that many of these deaths could have been prevented with timely attention. Because only 8% of the medical examiners who were contacted for information responded, non-response bias was possible. Furthermore, a disproportionate number of deaths (77%) were reported by one state (Florida), suggesting possible geographic reporting bias.

Pathology

Three deaths have been reported after the ingestion of MDMA/MDEA, in which immunohistological studies further elucidated the causes of the deaths (127). One death was caused by MDMA intoxication, one was from MDEA, and the third was from combined intoxication. One case had been reported previously (SEDA-22, 31).

- A 19-year-old man consumed ecstasy for the entire duration of a discotheque party (lasting until morning) and had respiratory difficulty, uncoordinated movements, generalized hypertonia, and hyperpyrexia (40.6°C). The diagnosis was disseminated intravascular coagulation and was treated with heparin but suffered severe blood loss from the oral cavity and injection wounds. He subsequently had a cardiac arrest.
- A 20-year-old man took many ecstasy tablets at a discotheque, returned home complaining of feeling feverish (axillary temperature 40°C), went to bed, and was found dead 10 hours later, his pillow soaked with blood.
- A 19-year-old man was found unconscious near a discotheque. His course progressively worsened, with the appearance of diffuse subcutaneous petechiae. While being treated for hypotension, he developed sustained convulsions and uremia and later died.

In all cases, amphetamines were detected in the urine. The pathological findings showed diffuse subserous petechiae and polyvisceral stasis. The brains showed massive edema and signs of neuronal hypoxia. In two cases, the heart showed coagulative myocytolysis, while one had areas of subendocardial hemorrhage. The lungs showed subpleural and intra-alveolar hemorrhages with severe edema, and in two cases microthrombotic formations inside lung capillaries. In two cases, the liver showed evidence of microvesicular steatosis and in one case centrilobular necrosis. Liver cells in the central zones showed coagulation necrosis with precipitation of fibrin in the whole area affected by necrosis. In the kidney, fibrin thrombi in the renal glomeruli were observed in two of the cases, while acute tubular necrosis was observed in one.

The findings in these three cases were similar to those seen in deaths from hyperthermia and disseminated intravascular coagulation. The myocardial necrosis (coagulative myocytolysis) without infarct necrosis suggested adrenergic overdrive. In two cases the muscle showed the typical pathological changes observed in deaths due to malignant hyperpyrexia. Myoglobin was detected in the proximal tubules in all three cases. The authors interpreted the clinical, histopathological, and toxicological data as indicative of an idiosyncratic response to ecstasy.

Long-Term Effects

Drug tolerance

Ecstasy can release large amounts of serotonin in the synaptic cleft, with an 80% loss of brain serotonin and its metabolites within 4 hours of an injection of ecstasy. However, ecstasy inhibits monoamine oxidases A and B, leading to an increase in serotonin and other catecholamines (128). Chronic tolerance to ecstasy in humans is a robust empirical phenomenon, which leads to dosage escalation and bingeing. Tolerance to ecstasy develops rapidly; many users describe their first experience as their best, and with repeated use its positive effects subside rapidly. Many predicted that ecstasy would not become a drug of abuse, specifically because of the early reduction in subjective efficacy. This is opposite to cocaine. Many recreational users increase the dosage; however, taking a double dose does not double the supposedly beneficial effects but increases the negative effects. Chronic bingeing is statistically linked to higher rates of drug-related psychobiological problems. Thus, the authors suggested that regular ecstasy users should balance their desire for an optimal on-drug experience with the need to minimize the post-drug consequences. Some individuals intensify their use to maximize the on-drug experience, while others use it sparingly to minimize the long-term consequences. The rate of development of chronic tolerance can be variable, even with regular ecstasy use. The authors postulated that serotonergic mechanisms may partly explain the unusual pattern of chronic tolerance with ecstasy.

Drug dependence

There are few preclinical or clinical data to suggest that repeated use of MDMA is associated with increased tolerance or dependence. However, anecdotal reports suggest that in some individuals, increasing amounts of MDMA are used in order to achieve the same reinforcing psychoactive effects (58,129).

Second-Generation Effects

Pregnancy

Workers in Canada have tried to characterize women who reported gestational exposure to ecstasy (130). The Motherisk Program is a large Teratogen Information Service based in Toronto and receives over 150 calls daily about exposure to various agents during pregnancy. The authors reviewed the data from 1998 to 2000. The study group consisted of all pregnant women who had been exposed to ecstasy. The control group was randomly selected pregnant women who visited the Motherisk Clinic during the same week as the subject who called about ecstasy. The 132 ecstasy-exposed women were significantly younger, earlier in gestational age, and weighed less than the non-exposed controls. The ecstasy users had had significantly fewer pregnancies and live births and had a higher rate of therapeutic abortions but not spontaneous abortions. Significantly more ecstasy users reported

unplanned pregnancies and were more likely to be single and white. The ecstasy users were more likely than controls to have had alcohol exposure in pregnancy, and significantly more drank heavily. The ecstasy-exposed women were more likely to binge-drink and smoke cigarettes during pregnancy, and more were significantly heavy smokers. Ecstasy users also had a greater tendency to use marijuana, cocaine, amphetamines, ketamine, gamma-hydroxybutyrate, and psilocybin.

Of the 132 women who reported ecstasy use, 129 had used it during pregnancy, of whom 101 reported previous use of ecstasy before their pregnancy, but had discontinued it. The mean gestational age of last ecstasy exposure was 5.0 weeks (range 1–24 weeks). All but three used tablets: two snorted and one used a liquid formulation. The mean dose taken on one occasion was 1.24 tablets. Of the 122 patients whose data were included in the analysis, most (57%) had only one exposure to ecstasy during pregnancy; 10 women had more than five exposures. One 15-year-old girl did not realize she was pregnant until 24 weeks gestation and had used two tablets of ecstasy four times a day. Only seven of the ecstasy group reported exposure to ecstasy alone. Ecstasy-associated adverse events were reported by 33 of 77 respondents. The physical adverse effect most commonly reported was vomiting (23%). The authors raised significant concerns about the potential teratogenic effects in these women, due to clustering of risk factors.

Susceptibility factors

Age

Infants
The effects of ingestion, accidental or otherwise, of ecstasy in infants continue to be reported (131).

- An 11-month-old, breast-fed boy of Argentinian–Italian origin developed generalized seizures, depressed consciousness (Glasgow coma scale 12), repetitive and jerky movements of limbs with associated hyper-reflexia and muscle rigidity, dilated non-reactive pupils, and perioral cyanosis. He had a sinus tachycardia of 190/minute and pulse oximetry of 83%. There was ecstasy 11.7 mg/l and its metabolite MDA 1.2 mg/l in the urine. Cocaine 1.3 mg/l and benzoylecgonine 0.4 mg/l were found in a proximal hair sample and cocaine 4.6 mg/l and benzoylecgonine 0.5 mg/l in a distal segment. His mother's hair was negative. He recovered completely within 24 hours.

This infant appeared to have accidentally ingested ecstasy. The extremely high urinary concentration of the parent drug and metabolites excluded the possibility of intoxication through breast-feeding. The urine concentration of ecstasy was in the range encountered in acute intoxication and adult deaths. The child's symptoms were consistent with cardiovascular and autonomic effects and acute toxic effects of ecstasy. The hair analysis showed chronic exposure to cocaine, which could have been through passive inhalation of cocaine, ingestion of powder residues, or intentional administration/exposure through breast milk. This case highlighted the importance of drug testing in unusual cases, even in infants.

- A 15-month-old infant developed seizures (132). Her heart rate was 200/minute, here peripheries were cyanosed, and she had labored breathing, dilated pupils reacting sluggishly to light, and a temperature of 37.3°C. There was a rash over the right iliac fossa. Her neck and back were arched. Her liver was enlarged. There was no response to rectal diazepam, rectal paraldehyde, or intravenous lorazepam, but she responded to phenytoin and anesthesia with thiopental and suxamethonium. Amphetamine was detected in her urine.

Discussion with parents revealed that accidental ingestion of ecstasy may have occurred after a teenaged uncle's party.

Drug Administration

Drug formulations

In New Zealand, "herbal ecstasy" is a term used for many different herbal formulations, none of which contains ecstasy. Some of the names for these herbs (which can be sold in stores) include "The Bomb", "Reds", and "Sublime". Analysis of "The Bomb" showed substantial amounts of ephedrine; the Ministry of Health in New Zealand removed it from the market. Some symptoms associated with herbal ecstasy include headache, dizziness, palpitation, tachycardia, and raised blood pressure. Thus, in countries where the term "herbal ecstasy" is commonly used, it is important that those who see patients who have taken herbal ecstasy should not confuse it with ecstasy, as toxicity and medical management may be quite different (133).

Drug contamination

It is not uncommon for people to be deceived into consuming other substances, believing them to be MDMA. At dance parties and "raves" for young adults, compounds passed as ecstasy may be more lethal than MDMA, either because they contain more potent amphetamines than MDMA or because of adulteration with other substances. In three fatal cases reported in the USA, the victims (two men aged 19 and 24 and a woman aged 18) believed that they were using MDMA but had in fact taken PMA, a more potent central stimulant with structural and pharmacological similarities to MDMA (134). They became agitated and developed bruxism, severe hyperthermia, convulsions, and hemorrhages. The presence of PMA was confirmed by enzyme immunoassay, and MDMA was not detected. PMA is not a contaminant of MDMA.

Drug adulteration

A potential confounding variable in studies of MDMA is that the subjects may not actually be taking MDMA, but rather other substances, such as amfetamine or metamfetamine. In 21 subjects who claimed to have taken only MDMA and no other drugs (135) a hair sample showed that 19 had MDMA present, while seven had concentrations of 3,4-methylenedioxyamfetamine (MDA) similar to or greater than those of MDMA. Eight subjects also tested positive for of amfetamine or metamfetamine. At a follow-up interview with those who tested positive for drugs other than MDMA, none admitted knowledge of taking MDA, amfetamine, or metamfetamine. These results suggest that not all street ecstasy tablets contain pure MDMA. Often, MDA, amfetamine, or metamfetamine is disguised as MDMA. It is unknown whether the combination of MDMA with these drugs poses a greater health risk to abusers. The main limitation of this study was that it relied on the subjects' own reports. The authors suggested that hair testing be implemented in all MDMA research trials to ensure that the study sample is accurate.

Drug overdose

Accidental MDMA overdose has been reported in a baby (136).

- A 14-month-old boy began convulsing 40 minutes after taking an unknown medication. The convulsions lasted for 20 minutes. He became cyanotic with a heart rate of 130/minute and a temperature of 38°C. He was treated with oxygen, intravenous benzodiazepines, and dipyrone, but continued to have isolated ventricular extra beats, hypertension, and tachycardia. The serum and urine concentrations of MDMA 8 hours after ingestion were 0.591 and 1477 mg/l respectively. After 12 hours, the tachycardia and hypertension resolved and the child was discharged 9 days later with no residual symptoms.

Drug–Drug Interactions

Alcohol

The pharmacokinetic and pharmacodynamic interactions of single doses of ecstasy 100 mg and alcohol 0.8 g/kg have been investigated in nine healthy men (mean age 23 years) in a double-blind, double-dummy, randomized, placebo-controlled crossover design (137). Each underwent four 10-hour experimental sessions, including blood sampling, with 1 week between each. For the task used to test the recognition and recording of visual information, the conditions involving ethanol yielded significantly more errors and fewer responses than ecstasy alone or placebo alone. The combination of ecstasy with ethanol reversed the subjective effect of sedation caused by alcohol alone. In addition, the combination extended the sense of euphoria caused by ecstasy to 5.25 hours. The addition of ethanol caused plasma ecstasy concentrations to rise by 13%. These results show that the combination of ecstasy with alcohol potentiates the euphoria of ecstasy and reduces perceived sedation. However, psychomotor impairment of visual processing caused by alcohol is not reversed. This is a concern for road safety, as people who take both drugs would feel sober, but their driving would still be compromised, although the extent of driving impairment under these conditions is not known. The increase in plasma concentrations of ecstasy caused by alcohol could exacerbate the adverse effects of ecstasy.

Moclobemide

There have been four deaths after interactions of moclobemide, a monoamine oxidase A inhibitor, with ecstasy (138).

- An 18-year-old woman took an unknown quantity of ecstasy and the next day became confused and had seizures, loss of consciousness, and respiratory arrest. At autopsy, no cause of death was found. There were no specific findings on histology except visceral hyperemia. MDMA, moclobemide, and some alcohol were found in the blood. There was no history of mental illness or evidence of prescription drugs.
- A 23-year-old man was found dead in his apartment, having previously been confused. At autopsy there was no clear cause of death. His urine was positive for amphetamines, opiates, moclobemide, dextromethorphan, cyclizine, oxazepam, diazepam, and temazepam. The coroner concluded that the death was from a combination of ecstasy and moclobemide, with a possible contribution from dextromethorphan.
- An 18-year-old man with a long history of amfetamine abuse became confused, was rolling about on the ground, and then collapsed and died. He had ecstasy and moclobemide in his possession. Autopsy did not show a clear cause of death. His urine was positive for amphetamines and marijuana. Histology showed congestion in the brain, liver, kidney, and lungs and signs of aspiration. The cause of death was recorded as accidental poisoning from moclobemide and ecstasy.
- A 19-year-old man took 10 ecstasy tablets and became unconscious, had problems breathing, and later died. At autopsy, his lungs were edematous and there was general visceral congestion. His blood contained MDMA, MDA, cannabis, and moclobemide. The cause of death was recorded as having been due to the combination of moclobemide and ecstasy.

The authors suggested that the serotonin syndrome possibly occurred as a result of the combination of moclobemide and ecstasy, although it was not clear why the combination had been used.

Ritonavir

An important consideration in the use of all HIV-1 protease inhibitors, but of ritonavir in particular, is their potential for drug interactions through their effects on cytochrome P450 isozymes. The various interactions of ritonavir with other antiretroviral drugs have been reviewed (139). Ritonavir, which inhibits CYP2D6, the principal pathway by which MDMA is metabolized, can also produce clinically relevant interactions with recreational drugs (140).

- A 32-year-old HIV-positive man, who added ritonavir 600 mg bd to his antiretroviral regimen of zidovudine and lamivudine, became unwell within hours after having taken two and a half tablets of ecstasy, estimated to contain 180 mg of MDMA. He was hypertonic, sweating profusely, tachypneic, tachycardic, and cyanosed. Shortly after he had a tonic-clonic seizure and a cardiorespiratory arrest. Attempts at resuscitation were unsuccessful. Blood concentrations obtained post-mortem showed an MDMA concentration of 4.56 mg/l, in the range of that reported in a patient with a life-threatening illness and symptoms similar to this patient after an overdose of 18 tablets of MDMA.

A patient infected with HIV-1 who was taking ritonavir and saquinavir had a prolonged effect from a small dose of MDMA and a near-fatal reaction from a small dose of gamma-hydroxybutyrate (141).

Selective serotonin re-uptake inhibitors (SSRIs)

An MDMA-related psychiatric adverse effect may have been enhanced by an SSRI (142).

- A 52-year-old prisoner, who was taking the SSRI citalopram 60 mg/day, suddenly became aggressive, agitated, and grandiose after using ecstasy. He carried out peculiar compulsive movements and had extreme motor restlessness, but no fever or rigidity. He was given chlordiazepoxide and 2 days later was asymptomatic. Citalopram was reintroduced, and 2 days later he reported visual hallucinations of little bugs in the cell. Promazine was substituted for citalopram and his condition improved 2 days later.

The authors suggested that SSRIs such as citalopram can potentiate the neurochemical and behavioral effects of MDMA.

References

1. Solowij N, Hall W, Lee N. Recreational MDMA use in Sydney: a profile of "Ecstacy" users and their experiences with the drug. Br J Addict 1992;87(8):1161–72.
2. Liester MB, Grob CS, Bravo GL, Walsh RN. Phenomenology and sequelae of 3,4-methylenedioxymethamphetamine use. J Nerv Ment Dis 1992;180(6):345–52.
3. Vaiva G, Boss V, Bailly D, Thomas P, Lestavel P, Goudemand M. An "accidental" acute psychosis with ecstasy use. J Psychoactive Drugs 2001;33(1):95–8.
4. Pope HG Jr, Ionescu-Pioggia M, Pope KW. Drug use and life style among college undergraduates: a 30-year longitudinal study. Am J Psychiatry 2001;158(9):1519–21.
5. Lessick M, Vasa R, Israel J. Severe manifestations of oculoauriculovertebral spectrum in a cocaine exposed infant. J Med Genet 1991;28(11):803–4.
6. Henry JA. Ecstasy and the dance of death. BMJ 1992;305(6844):5–6.
7. Gelenberg AJ. One man's ecstasy. Biol Ther Psychiatry Newslett 1992;15:45–7.
8. Screaton GR, Singer M, Cairns HS, Thrasher A, Sarner M, Cohen SL. Hyperpyrexia and rhabdomyolysis after MDMA ("ecstasy") abuse. Lancet 1992;339(8794):677–8.

9. Henry JA, Jeffreys KJ, Dawling S. Toxicity and deaths from 3,4-methylenedioxymethamphetamine ("ecstasy"). Lancet 1992;340(8816):384–7.

10. Greene SL, Dargan PI, O'Connor N, Jones AL, Kerins M. Multiple toxicity from 3,4-methylenedioxymethamphetamine ("ecstasy"). Am J Emerg Med 2003;21:121–4.

11. Campkin NT, Davies UM. Another death from Ecstacy. J R Soc Med 1992;85(1):61.

12. Byard RW, Gilbert J, James R, Lokan RJ. Amphetamine derivative fatalities in South Australia-is "Ecstasy" the culprit? Am J Forensic Med Pathol 1998;19(3):261–5.

13. James RA, Dinan A. Hyperpyrexia associated with fatal paramethoxyamphetamine (PMA) abuse. Med Sci Law 1998;38(1):83–5.

14. Britt GC, McCance-Katz EF. A brief overview of the clinical pharmacology of "club drugs". Subst Use Misuse 2005;40:1189–201.

15. Maxwell JC. Party drugs: properties, prevalence, patterns and problems. Subst Use Misuse 2005;40:1203–40.

16. Soellner R. Club drug use in Germany. Subst Use Misuse 2005;40:1279–93.

17. Schuster P, Lieb R, Lamertz C, Wittchen HU. Is the use of ecstasy and hallucinogens increasing? Results from a community study. Eur Addict Res 1998;4(1-2):75–82.

18. Banta-Green C, Goldbaum G, Kingston S, Golden M, Harruff R, Logan BK. Epidemiology of MDMA and associated club drugs in the Seattle area. Subst Use Misuse 2005;40:1295–315.

19. Han E, Yang W, Lee J, Park Y, Kim E, Lim M, Chung H. The prevalence of MDMA/MDA in both hair and urine in drug users. Forensic Sci Int 2005;152:73–7.

20. de la Torre R, Farre M, Ortuno J, Mas M, Brenneisen R, Roset PN, Segura J, Cami J. Non-linear pharmacokinetics of MDMA ("ecstasy") in humans. Br J Clin Pharmacol 2000;49(2):104–9.

21. Barrett PJ, Taylor GT. "Ecstasy" ingestion: a case report of severe complications. J R Soc Med 1993;86(4):233–4.

22. McCann UD, Ricaurte GA. Lasting neuropsychiatric sequelae of (+-)methylenedioxymethamphetamine ("ecstasy") in recreational users. J Clin Psychopharmacol 1991;11(5):302–5.

23. Bassi S, Ritto D. Ecstasy and chest pain due to coronary artery spasm. Int J Cardiol 2005;99:485–7.

24. Madhok A, Boxer R, Chowdhury D. Atrial fibrillation in an adolescent—the agony of ecstasy. Pediatr Emerg Care 2003;19:348–9.

25. Cole JC, Sumnall HR, Smith GW, Rostami-Hodjegan A. Preliminary evidence of cardiovascular effects of polysubstance misuse in nightclubs. J Psychopharmacol 2005;19(1):67–70.

26. Duxbury AJ. Ecstasy—dental implications. Br Dent J 1993;175(1):38.

27. Ames D, Wirshing WC. Ecstasy, the serotonin syndrome, and neuroleptic malignant syndrome—a possible link? JAMA 1993;269(7):869–70.

28. Campkin NJ, Davies UM. Treatment of "ecstasy" overdose with dantrolene. Anaesthesia 1993;48(1):82–3.

29. Mintzer S, Hickenbottom S, Gilman S. Parkinsonism after taking ecstasy. N Engl J Med 1999;340(18):1443.

30. Sewell RA, Cozzi NV. More about parkinsonism after taking ecstasy. N Engl J Med 1999;341(18):1400.

31. Baggott M, Mendelson J, Jones R. More about parkinsonism after taking ecstasy. N Engl J Med 1999;341(18):1400–1.

32. Mintzer S, Hickenbottom S, Gilman S. More about parkinsonism after taking Ecstasy. N Engl J Med 1999;341:1401.

33. Ng CP, Chau LF, Chung CH. Massive spontaneous haemopneumothorax and ecstasy abuse. Hong Kong J Emerg Med 2004;11:94–7.

34. Borg GJ. More about parkinsonism after taking ecstasy. N Engl J Med 1999;341(18):1400.

35. Bertram M, Egelhoff T, Schwarz S, Schwab S. Toxic leukencephalopathy following "ecstasy" ingestion. J Neurol 1999;246(7):617–8.

36. Schroeder B, Brieden S. Bilateral sixth nerve palsy associated with MDMA ("ecstasy") abuse. Am J Ophthalmol 2000;129(3):408–9.

37. Semple DM, Ebmeier KP, Glabus MF, O'Carroll RE, Johnstone EC. Reduced in vivo binding to the serotonin transporter in the cerebral cortex of MDMA ("ecstasy") users. Br J Psychiatry 1999;175:63–9.

38. Vollenweider FX, Liechti ME, Gamma A, Greer G, Geyer M. Acute psychological and neurophysiological effects of MDMA in humans. J Psychoactive Drugs 2002;34(2):171–84.

39. Greer G, Strassman RJ. Information on "Ecstasy". Am J Psychiatry 1985;142(11):1391.

40. Whitaker-Azmitia PM, Aronson TA. "Ecstasy" (MDMA)-induced panic. Am J Psychiatry 1989;146(1):119.

41. Creighton FJ, Black DL, Hyde CE. "Ecstasy" psychosis and flashbacks. Br J Psychiatry 1991;159:713–5.

42. Peroutka SJ, Newman H, Harris H. Subjective effects of 3,4-methylenedioxymethamphetamine in recreational users. Neuropsychopharmacology 1988;1(4):273–7.

43. McCann UD, Ricaurte GA. MDMA ("ecstasy") and panic disorder: induction by a single dose. Biol Psychiatry 1992;32(10):950–3.

44. O'Suilleabhain P, Giller C. Rapidly progressive parkinsonism in a self-reported user of ecstasy and other drugs. Mov Disord 2003;18:1378–81.

45. Finsterer J, Stollberger C, Steger C, Kroiss A. Long lasting impaired cerebral blood flow after ecstasy intoxication. Psychiatry Clin Neurosci 2003;57:221–5.

46. Yin Foo Lee G, Wooi Kee Gong G, Vrodos N, Patrick Brophy B. 'Ecstasy'-induced subarachnoid haemorrhage: an under-reported neurological complication? J Clin Neurosci 2003;10:705–7.

47. Pallanti S, Mazzi D. MDMA (Ecstasy) precipitation of panic disorder. Biol Psychiatry 1992;32(1):91–5.

48. McGuire P, Fahy T. Chronic paranoid psychosis after misuse of MDMA ("ecstasy"). BMJ 1991;302(6778):697.

49. Oliveri M, Calvo G. Increased visual cortical excitability in ecstasy users: a transcranial magnetic stimulation study. J Neurol Neurosurg Psychiatry 2003;74:1136–8.

50. Schifano F. Chronic atypical psychosis associated with MDMA ("ecstasy") abuse. Lancet 1991;338(8778):1335.

51. Benazzi F, Mazzoli M. Psychiatric illness associated with "ecstasy". Lancet 1991;338(8781):1520.

52. Morgan MJ, McFie L, Fleetwood H, Robinson JA. Ecstasy (MDMA): are the psychological problems associated with its use reversed by prolonged abstinence? Psychopharmacology (Berl) 2002;159(3):294–303.

53. Fox HC, McLean A, Turner JJ, Parrott AC, Rogers R, Sahakian BJ. Neuropsychological evidence of a relatively selective profile of temporal dysfunction in drug-free MDMA ("ecstasy") polydrug users. Psychopharmacology (Berl) 2002;162(2):203–14.

54. Barrett PJ. "Ecstasy" misuse—overdose or normal dose? Anaesthesia 1993;48(1):83.

55. Wodarz N, Boning J. "Ecstasy"-induziertes psychotisches Depersonalisationssyndrom. ["Ecstasy"-induced psychotic depersonalization syndrome.] Nervenarzt 1993;64(7):478–80.

56. Williams H, Meagher D, Galligan P. M.D.M.A. ("Ecstasy"); a case of possible drug-induced psychosis Ir J Med Sci 1993;162(2):43–4.

57. Vollenweider FX, Gamma A, Liechti M, Huber T. Psychological and cardiovascular effects and short-term sequelae of MDMA ("ecstasy") in MDMA-naive healthy volunteers. Neuropsychopharmacology 1998;19(4):241–51.

58. McCann UD, Szabo Z, Scheffel U, Dannals RF, Ricaurte GA. Positron emission tomographic evidence of toxic effect of MDMA ("Ecstasy") on brain serotonin neurons in human beings. Lancet 1998;352(9138):1433–7.

59. Windhaber J, Maierhofer D, Dantendorfer K. Panic disorder induced by large doses of 3,4-methylenedioxymethamphetamine resolved by paroxetine. J Clin Psychopharmacol 1998;18(1):95–6.

60. Winstock AR. Chronic paranoid psychosis after misuse of MDMA. BMJ 1991;302(6785):1150–1.

61. Vecellio M, Schopper C, Modestin J. Neuropsychiatric consequences (atypical psychosis and complex-partial seizures) of ecstasy use: possible evidence for toxicity-vulnerability predictors and implications for preventative and clinical care. J Psychopharmacol 2003;17:342–5.

62. MacInnes N, Handley SL, Harding GF. Former chronic methylenedioxymethamphetamine (MDMA or ecstasy) users report mild depressive symptoms. J Psychopharmacol 2001;15(3):181–6.

63. Ricaurte GA, McCann UD. Experimental studies on 3,4-methylenedioxymethamphetamine (MDA, "ecstasy") and its potential to damage brain serotonin neurons. Neurotox Res 2001;3(1):85–99.

64. Heffernan TM, Ling J, Scholey AB. Subjective ratings of prospective memory deficits in MDMA ("ecstasy") users. Hum Psychopharmacol 2001;16(4):339–44.

65. Verkes RJ, Gijsman HJ, Pieters MS, Schoemaker RC, de Visser S, Kuijpers M, Pennings EJ, de Bruin D, Van de Wijngaart G, Van Gerven JM, Cohen AF. Cognitive performance and serotonergic function in users of ecstasy. Psychopharmacology (Berl) 2001;153(2):196–202.

66. Reneman L, Lavalaye J, Schmand B, de Wolff FA, van den Brink W, den Heeten GJ, Booij J. Cortical serotonin transporter density and verbal memory in individuals who stopped using 3,4-methylenedioxymethamphetamine (MDMA or "ecstasy"): preliminary findings. Arch Gen Psychiatry 2001;58(10):901–6.

67. Reneman L, Majoie CB, Schmand B, van den Brink W, den Heeten GJ. Prefrontal N-acetylaspartate is strongly associated with memory performance in (abstinent) ecstasy users: preliminary report. Biol Psychiatry 2001;50(7):550–4.

68. Bhattachary S, Powell JH. Recreational use of 3,4-methylenedioxymethamphetamine (MDMA) or "ecstasy": evidence for cognitive impairment. Psychol Med 2001;31(4):647–58.

69. Buchert R, Thomasius R, Wilke F, Peterson K, Nebeling B, Obrocki J, Schulze O, Schmidt U, Clausen M. A voxel-based PET investigation of the long-term effects of 'ecstasy' consumption on brain serotonin transporters. Am J Psychiatry 2004;161:1181–9.

70. Buchert R, Thomasius R, Nebeling B, Petersen K, Obrocki J, Jenicke L, Wilke F, Wartberg L, Zapletalova P, Clausen M. Long-term effects of "ecstasy" use on serotonin transporters of the brain investigated by PET. J Nucl Med 2003;44:375–84.

71. Croft RJ, Mackay AJ, Mills AT, Gruzelier JG. The relative contributions of ecstasy and cannabis to cognitive impairment. Psychopharmacology (Berl) 2001;153(3):373–9.

72. Liechti ME, Gamma A, Vollenweider FX. Gender differences in the subjective effects of MDMA. Psychopharmacology (Berl) 2001;154(2):161–8.

73. Harry RA, Sherwood R, Wendon J. Detection of myocardial damage and cardiac dysfunction following ecstasy ingestion. Clin Intensive Care 2001;12:85–7.

74. Verheyden SL, Henry JA, Curran HV. Acute, sub-acute and long-term subjective consequences of 'ecstasy' (MDMA) consumption in 430 regular users. Hum Psychopharmacol 2003;18:507–17.

75. De Win MML, Reneman L, Reitsma JB, den Heeten GJ, Booij J, van den Brink W. Mood disorders and serotonin transporter density in ecstasy users- the influence of long-term abstention, dose, and gender. Psychopharmacology 2004;173:376–82.

76. Sumnall HR, Cole JC. Self-reported depressive symptomatology in community samples of polysubstance misusers who report ecstasy use: a meta-analysis. J Psychopharmacol 2005;19(1):84–92.

77. D'Costa DF. Transient myocardial ischaemia associated with accidental Ecstasy ingestion. Br J Cardiol 1998;5:290–1.

78. Brody S, Krause C, Veit R, Rau H. Cardiovascular autonomic dysregulation in users of MDMA ("Ecstasy"). Psychopharmacology (Berl) 1998;136(4):390–3.

79. Duflou J, Mark A. Aortic dissection after ingestion of "ecstasy" (MDMA). Am J Forensic Med Pathol 2000;21(3):261–3.

80. Lester SJ, Baggott M, Welm S, Schiller NB, Jones RT, Foster E, Mendelson J. Cardiovascular effects of 3,4-methylenedioxymethamphetamine. A double-blind, placebo-controlled trial. Ann Intern Med 2000;133(12):969–73.

81. Levine AJ, Drew S, Rees GM. "Ecstasy" induced pneumomediastinum. J R Soc Med 1993;86(4):232–3.

82. Ryan J, Banerjee A, Bong A. Pneumomediastinum in association with MDMA ingestion. J Emerg Med 2001;20(3):305–6.

83. Rejali D, Glen P, Odom N. Pneumomediastinum following Ecstasy (methylenedioxymetamphetamine, MDMA) ingestion in two people at the same 'rave'. J Laryngol Otol 2002;116(1):75–6.

84. Daumann J, Fimm B, Willmes K, Thron A, Gouzoulis-Mayfrank E. Cerebral activation in abstinent ecstasy (MDMA) users during a working memory task: a functional magnetic resonance imaging (fMRI) study. Brain Res Cogn Brain Res 2003;16:479–87.

85. Verbaten MN. Specific memory deficits in ecstasy users? The results of a meta-analysis. Hum Psychopharmacol 2003;18:281–90.

86. Wareing M, Murphy PN, Fisk JE. Visuospatial memory impairments in users of MDMA ('ecstasy'). Psychopharmacology 2004;173:391–7.

87. Wareing M, Fisk JE, Murphy P, Montgomery C. Verbal working memory deficits in current and previous users of MDMA. Hum Psychopharmacol Clin Exp 2004;19:225–34.

88. Von Geusau NA, Stalenhoef P, Huizinga M, Snel J, Ridderinkhof KR. Impaired executive function in male MDMA ('ecstasy') users. Psychopharmacology 2004;175:331–41.

89. Gouzoulis-Mayfrank E, Thimm B, Rezk M, Hensen G, Daumann J. Memory impairment suggests hippocampal dysfunction in abstinent ecstasy users. Prog Neuropsychopharmacol Biol Psychiatry 2003;27:819–27.

90. Casco C, Forcella MC, Beretta G, Grieco A, Campana G. Long-term effects of MDMA (ecstasy) on the human

central nervous system revealed by visual evoked potentials. Addict Biol 2005;10:187–95.

91. Montgomery C, Fisk JE, Newcombe R, Murphy PN. The differential effects of ecstasy/polydrug use on executive components: shifting, inhibition, updating and access to semantic memory. Psychopharmacology 2005;182:262–76.

92. Ahmed JM, Salame MY, Oakley GD. Chest pain in a young girl. Postgrad Med J 1998;74(868):115–6.

93. Stuerenburg HJ, Petersen K, Buhmann C, Rosenkranz M, Baeumer T, Thomasius R. Plasma amino acids in ecstasy users. Neuro Endocrinol Lett 2003;24:348–9.

94. Budisavljevic MN, Stewart L, Sahn SA, Ploth DW. Hyponatremia associated with 3,4-methylenedioxymethylamphetamine ("ecstasy") abuse. Am J Med Sci 2003;326:89–93.

95. Ajaelo I, Koenig K, Snoey E. Severe hyponatremia and inappropriate antidiuretic hormone secretion following ecstasy use. Acad Emerg Med 1998;5(8):839–40.

96. Gomez-Balaguer M, Pena H, Morillas C, Hernandez A. Syndrome of inappropriate antidiuretic hormone secretion and "designer drugs" (ecstasy). J Pediatr Endocrinol Metab 2000;13(4):437–8.

97. Roques V, Perney P, Beaufort P, Hanslik B, Ramos J, Durand L, Le Bricquir Y, Blanc F. Hepatite aiguë a l'ecstasy. [Acute hepatitis due to ecstasy.] Presse Méd 1998;27(10):468–70.

98. Jonas MM. Case records of the Massachusetts General Hospital. Weekly clinicopathological exercises. Case 6-2001. A 17-year-old girl with marked jaundice and weight loss. N Engl J Med 2001;344(8):591–9.

99. Ravina P, Quiroga JM, Ravina T. Hyperkalemia in fatal MDMA ('ecstasy') toxicity. Int J Cardiol 2004;93:307–8.

100. Hwang I, Daniels AM, Holtzmuller KC. "Ecstasy"-induced hepatitis in an active duty soldier. Mil Med 2002;167(2):155–6.

101. Garbino J, Henry JA, Mentha G, Romand JA. Ecstasy ingestion and fulminant hepatic failure: liver transplantation to be considered as a last therapeutic option. Vet Hum Toxicol 2001;43(2):99–102.

102. De Carlis L, De Gasperi A, Slim AO, Giacomoni A, Corti A, Mazza E, Di Benedetto F, Lauterio A, Arcieri K, Maione G, Rondinara GF, Forti D. Liver transplantation for ecstasy-induced fulminant hepatic failure. Transplant Proc 2001;33(5):2743–4.

103. Brazier WJ, Dhariwal DK, Patton DW, Bishop K. Ecstasy related periodontitis and mucosal ulceration—a case report. Br Dent J 2003;194:197–9.

104. Caballero F, Lopez-Navidad A, Cotorruelo J, Txoperena G. Ecstasy-induced brain death and acute hepatocellular failure: multiorgan donor and liver transplantation. Transplantation 2002;74(4):532–7.

105. Wollina U, Kammler HJ, Hesselbarth N, Mock B, Bosseckert H. Ecstasy pimples—a new facial dermatosis. Dermatology 1998;197(2):171–3.

106. Gill JR, Hayes JA, deSouza IS, Marker E, Stajic M. Ecstasy (MDMA) deaths in New York City: a case series and review of the literature. J Forensic Sci 2002;47(1):121–6.

107. Gore SM. Fatal uncertainty: death-rate from use of ecstasy or heroin. Lancet 1999;354(9186):1265–6.

108. Ramsey JD, Johnston A, Holt DW. Death rate from use of ecstasy or heroin. Lancet 1999;354(9196):2166.

109. Lind J, Oyefeso A, Pollard M, Baldacchino A, Ghodse H. Death rate from use of ecstasy or heroin. Lancet 1999;354(9196):2167.

110. Kwon C, Zaritsky A, Dharnidharka VR. Transient proximal tubular renal injury following ecstasy ingestion. Pediatr Nephrol 2003;18:820–2.

111. Inman DS, Greene D. The agony and the ecstasy: acute urinary retention after MDMA abuse. BJU Int 2003;91:123.

112. Delgado JH, Caruso MJ, Waksman JC, Hanigman B, Stillman D. Acute transient urinary retention from combined ecstasy and methamphetamine use. J Emerg Med 2004;26:173–5.

113. Fineschi V, Centini F, Mazzeo E, Turillazzi E. Adam (MDMA) and Eve (MDEA) misuse: an immunohistochemical study on three fatal cases. Forensic Sci Int 1999;104(1):65–74.

114. Tan B, Foley P. Guttate psoriasis following ecstasy ingestion. Australas J Dermatol 2004;45:167–9.

115. O'Connor A, Cluroe A, Couch R, Galler L, Lawrence J, Synek B. Death from hyponatraemia-induced cerebral oedema associated with MDMA ("Ecstasy") use. NZ Med J 1999;112(1091):255–6.

116. Zwick OM, Fischer DH, Flanagan JC. "Ecstasy" induced immunosuppression and herpes zoster ophthalmicus. Br J Ophthalmol 2005;89:923–4.

117. Hartung TK, Schofield E, Short AI, Parr MJ, Henry JA. Hyponatraemic states following 3,4-methylenedioxymethamphetamine (MDMA, "ecstasy") ingestion. Quart J Med 2002;95(7):431–7.

118. Patel MM, Belson MG, Longwater AB, Olson KR, Miller MA. MDMA (ecstasy)-related hyperthermia. J Emerg Med 2005;29(4):451–4.

119. Bordo DJ, Dorfman MA. Ecstasy overdose: rapid cooling leads to successful outcome. Am J Emerg Med 2004;22(4):326–7.

120. Cherney DZ, Davids MR, Halperin ML. Acute hyponatraemia and 'ecstasy': insights from a quantitative and integrative analysis. Quart J Med 2002;95(7):475–83.

121. Libiseller K, Pavlic M, Grubwieser, Rabl W. Ecstasy—deadly risk even outside rave parties. Forensic Sci Int 2005;153:227–30.

122. Pacifici R, Zuccaro P, Farre M, Pichini S, Di Carlo S, Roset PN, Ortuno J, Segura J, de la Torre R. Immunomodulating properties of MDMA alone and in combination with alcohol: a pilot study. Life Sci 1999;65(26):PL309–16.

123. Pacifici R, Zuccaro P, Farre M, Pichini S, Di Carlo S, Roset PN, Palmi I, Ortuno J, Menoyo E, Segura J, de la Torre R. Cell-mediated immune response in MDMA users after repeated dose administration: studies in controlled versus noncontrolled settings. Ann NY Acad Sci 2002;965:421–33.

124. Murray MO, Wilson NH. Ecstasy related tooth wear. Br Dent J 1998;185(6):264.

125. Schifano F, Oyefeso A, Corkery J, Cobain K, Jambert-Gray R, Martinotti G, Ghodse AH. Death rates from ecstasy (MDMA, MDA) and polydrug use in England and Wales 1996-2002. Hum Psychopharmacol 2003;18:519–24.

126. Patel MM, Wright DW, Ratcliff JJ, Miller MA. Shedding new light on the "safe" club drug: methylenedioxymethamphetamine (ecstasy)-related fatalities. Acad Emerg Med 2004;11(2):208–9.

127. Goorney BP, Scholes P. Transient haemolytic anaemia due to ecstasy in a patient on HAART. Int J STD AIDS 2002;13(9):651.

128. Parrott AC. Chronic tolerance to recreational MDMA (3,4-methlenedioxymethamphetamine) or ecstasy. J Psychopharmacol 2005;19(1):71–83.

129. McCann UD, Ricaurte GA. Major metabolites of (+/-)3,4-methylenedioxyamphetamine (MDA) do not mediate its toxic effects on brain serotonin neurons. Brain Res 1991;545(1–2):279–82.

130. Ho E, Karimi-Tabesh L, Koren G. Characteristics of pregnant women who use ecstasy (3, 4-methylenedi-oxymethamphetamine). Neurotoxicol Teratol 2001;23(6):561–7.

131. Garcia-Algar O, Lopez N, Bonet M, Pellegrini M, Marchei E, Pichini S. 3,4-Methylenedioxymethamphetamine (MDMA) intoxication in an infant chronically exposed to cocaine. Ther Drug Monit 2005;27(4):409–11.

132. Campbell S, Qureshi T. Taking ecstasy··· it's child's play! Pediatr Anesth 2005;15:256–60.

133. Yates KM, O'Connor A, Horsley CA. "Herbal Ecstasy": a case series of adverse reactions. NZ Med J 2000;113(1114):315–7.

134. Kraner JC, McCoy DJ, Evans MA, Evans LE, Sweeney BJ. Fatalities caused by the MDMA-related drug paramethoxyamphetamine (PMA). J Anal Toxicol 2001;25(7):645–8.

135. Kalasinsky KS, Hugel J, Kish SJ. Use of MDA (the 'love drug') and methamphetamine in Toronto by unsuspecting users of ecstasy (MDMA). J Forensic Sci 2004;49(5):1106–12.

136. Melian AM, Burillo-Putze G, Campo CD, Padron AG, Ramos CO. Accidental ecstasy poisoning in a toddler. Pediatr Emerg Care 2004;20(8):534–5.

137. Hsu A, Granneman GR, Bertz RJ. Ritonavir. Clinical pharmacokinetics and interactions with other anti-HIV agents. Clin Pharmacokinet 1998;35(4):275–91.

138. Henry JA, Hill IR. Fatal interaction between ritonavir and MDMA. Lancet 1998;352(9142):1751–2.

139. Harrington RD, Woodward JA, Hooton TM, Horn JR. Life-threatening interactions between HIV-1 protease inhibitors and the illicit drugs MDMA and gamma-hydroxybutyrate Arch Intern Med 1999;159(18):2221–4.

140. Vuori E, Henry JA, Ojanpera I, Nieminen R, Savolainen T, Wahlsten P, Jantti M. Death following ingestion of MDMA (ecstasy) and moclobemide. Addiction 2003;98:365–8.

141. Lauerma H, Wuorela M, Halme M. Interaction of seroto-nin reuptake inhibitor and 3,4-methylenedioxymetham-phetamine? Biol Psychiatry 1998;43(12):929.

142. Hernandez-Lopez C, Farre M, Roset PN, Menoyo E, Pizarro N, Ortuno J, Torrens M, Cami J, de La Torre R. 3,4-Methylenedioxymethamphetamine (ecstasy) and alco-hol interactions in humans: psychomotor performance, subjective effects, and pharmacokinetics. J Pharmacol Exp Ther 2002;300(1):236–44.

ORGANIC SOLVENTS

General Information

Lighter fuels, benzene, toluene, cleaning fluids (carbon tetrachloride), petrol, paraffin, and even the fluorocarbon propellants found in various household sprays and medications have all been used, particularly by children, to produce changes in consciousness. They are all inhaled, often with the aid of a plastic bag, and, since they are lipid-soluble, they are readily concentrated in brain tissue. As with many anesthetics there is an early period of hyperactivity, excitement, and intoxication, followed by sedation and confusion. Prolonged or regular use can cause serious toxicity, with bone-marrow depression, cardiac dysrhythmias, peripheral neuropathy, cerebral damage, and liver and kidney disorders (1).

Many organic solvents are used in pharmaceutical products. These include propylene glycol, polyethylene glycols, ethanol, dimethyl sulfoxide, N-methyl-2-pyrrolidone, glycofurol, Solketal, glycerol formal, and acetone; evidence of harm from such solvents is scant (2).

Much of the information about the harmful effects of organic solvents comes from studies of industrial exposure, although toluene abuse through sniffing of glues and other household sources of solvents (acrylic paints, adhesive cements, aerosol paints, lacquer thinners, shoe polish, typewriter correction fluids, varnishes, and fuels) has also been widely reported.

Epidemiology of inhalant abuse

The epidemiology of inhalant abuse has been widely studied in the USA (3). There is no overall sex difference in solvent abuse, although girls are more likely to be users in younger age groups and boys in older ones. Inhalant abuse is more common among school dropouts. Native Americans and Hispanic Americans are over-represented and blacks under-represented.

In the UK 3.5–10% of children under 13 years have abused volatile substances, and 0.5–1% are long-term users (3). In 1980 24 cases of solvent abuse were reported in Singapore, but by 1984 the number had increased to 763 and from 1987–91, 1781 glue sniffers were identified. In 2004, it was reported that street children in India were abusing typewriter eraser fluid, which contains toluene. In low-income families in Sao Paulo, Brazil, 24% of children had inhaled a volatile substance at some time and 4.9% had inhaled within the last month.

Sniffing petrol (which contains 13% toluene) has been reported among Native Americans in Canada and Aborigines in Australia. In Japan, there is an increased likelihood of illicit drug use, including toluene abuse, among children with attention deficit hyperactivity disorder.

General effects

The acute and chronic effects of solvent sniffing have been thoroughly reviewed (4). Solvents can be sniffed through the nose or huffed through the mouth, direct from the container, or from a filled bag, can, or bottle, or from a soaked rag. The acute effects include euphoria, exhilaration, dizziness, visual and auditory hallucinations, disinhibited behavior, relaxation, and sleep. Tolerance to these effects occurs but there are no withdrawal symptoms. Coughing, sneezing, salivation, flushing of the skin, nausea, vomiting, photophobia, disorientation, tinnitus, diplopia, headache, ataxia, slurred speech, depressed reflexes, nystagmus, and unconsciousness can occur.

Observational studies

In 25 adults, aged 18–40 years, three different patterns of symptoms led to hospitalization: muscle weakness (n = 9), gastrointestinal complaints, including abdominal pain and hematemesis (n = 6), and neuropsychiatric disorders, including altered mental status, cerebellar abnormalities, and peripheral neuropathy (n = 10) (5). Hypokalemia (n = 13), hypophosphatemia (n = 10), hyperchloremia (n = 22), and hypobicarbonatemia (n = 23) were common. There was rhabdomyolysis in 10 patients. The muscle weakness and gastrointestinal syndromes resolved within 1–3 days of abstinence from sniffing and repletion of fluid and electrolytes.

Organs and Systems

Cardiovascular

Myocardial infarctions have been attributed to toluene abuse (6,7).

Heart failure with a dilated cardiomyopathy occurred in a 15-year-old boy with a 2-year history of intermittent solvent abuse (8).

There were myocardial degenerative changes in the heart of a 14-year-old boy who died suddenly after abuse of typewriter correction fluid (9).

Respiratory

In 42 solvent inhalers aged 11–31, residual volumes were significantly higher than in 20 controls aged 10–26; lung tissue obtained at autopsy from three inhalers contained microscopic abnormalities similar to those seen in experimental panlobular emphysema (10).

Goodpasture's syndrome has been reported after exposure to toluene and other organic solvents (11,12,13).

Nervous system

Chronic moderate- to high-level exposure to some organic solvents, such as carbon disulfide, n-hexane and methyl n-butyl ketone, can cause encephalopathies (14), which have been variously described as "organic solvent syndrome", "painters' syndrome", "psycho-organic syndrome", and "chronic solvent encephalopathy" (15).

Some, such as bromopropane, may also cause polyneuropathies (3,4). There is some evidence of marginal atrophic abnormalities in the brain or impaired nerve conduction velocity in solvent-exposed workers (16).

It has been suggested that the adverse effects of organic solvents on the nervous system may be, at least in part, mediated by effects on dopaminergic function (17), perhaps via active metabolites (18). The proposed metabolites are shown in Table 1.

Cerebellar damage is a common feature of chronic neurological toxicity due to solvents.

- A 24-year-old man, who had sniffed toluene-containing compounds since he was 17, developed a syndrome that included psychiatric impairment, signs of bilateral pyramidal tract damage, altered cerebellar and sensory functions, and a peripheral neuropathy (19).
- An 18-year-old girl, who had inhaled pure toluene for 6 years, developed neurological symptoms, with a broad-based ataxic gait, incoordination of the arms and legs, unsteadiness, dysarthria, downbeat nystagmus, bilateral positive Babinski signs, and poor concentration and abstracting abilities (20). After withdrawal of toluene her symptoms improved and at 8 months had disappeared.

Of 24 solvent abusers (mean age 23 years), who had used substances containing a mean of 425 mg of toluene per day for an average of 6.3 years, 16 had marked impairment in neurological and neuropsychological tests; cerebellar symptoms were particularly prominent (21). The impairment correlated significantly with CT scan measurements of the cerebellum, ventricles, and cortical sulci, all of which were abnormal compared with age-matched controls.

Sensory systems

Eyes
The effects of occupational exposure to organic solvents on color discrimination have been reviewed (22). Workers who have been exposed to styrene have subtle impairment of color discrimination compared with age-matched controls. The impairment generally tends to be of the tritan (blue–yellow) type, although some cases of

protan (red–green) impairment have also been found. Toluene has an acute effect on color discrimination, even with relatively high exposure, but its chronic effects are not known. Tetrachloroethylene may cause slight impairment.

Ears
Organic solvents are ototoxic (23) and can cause vestibular damage and extracochlear high-frequency hearing loss (24). Ototoxic solvents include carbon disulfide, n-hexane, styrene, toluene, and trichloroethylene, and xylene (25,26).

- A 27-year-old glue sniffing woman developed sensorineural hearing loss, optic atrophy, and global brain damage (27).

Psychological

The neurobehavioral effects of long-term occupational exposure to organic solvents have been reviewed (28,29).

Psychiatric

Chronic toluene abuse produces a leukoencephalopathy, characterized chiefly by dementia (30).

Electrolyte balance

Hypokalemia is a common accompaniment of the metabolic acidosis that toluene inhalation can cause (see below). However, severe hypokalemia has also been reported in the absence of acidosis (31).

Acid-base balance

There have been many anecdotal reports of metabolic acidosis in glue sniffers, attributed to renal tubular acidosis. However, it has been suggested that it is in fact due to overproduction of hippuric acid resulting from the metabolism of toluene, with or without a reduced rate of urinary ammonium ion excretion (32).

Hematologic

Exposure to organic solvents is associated with an increased risk of hematological malignancies (see below, under Tumorigenicity).

Liver

- A 19-year-old boy developed hepatorenal failure after having sniffed a proprietary brand of liquid cleaner from a rag for 6 hours (33).

Urinary tract

Organic solvents can cause nephropathies.

- A 38-year-old man developed acute oliguric renal failure after repeated glue sniffing for about 8 hours (34). He also had severe liver damage, mild muscle necrosis, and bone marrow depression. He made a complete recovery.
- An 18-year-old man, who had been a regular glue sniffer (Evostik, which at that time contained toluene)

Table 1 Possible toxic metabolites of organic solvents (19)

Solvent	Suggested active metabolite
Aromatic monocyclic hydrocarbons Ethylbenzene Styrene Vinyltoluene	Phenyglyoxylic acid
Chlorinated hydrocarbons Trichloroethylene Perchloroethylene Trichloroethane	Chloral (trichloroacetaldehyde)
Glycols	Glyoxylic acid

for 5 years (one large tin a week), developed mesangiocapillary glomerulonephritis with heavy proteinuria and immune complex deposition (35). He continued to sniff glue occasionally and later developed nephrotic features and hypertension.

- A 16-year-old girl, with a history of heavy smoking and sniffing of Pattex glue, developed rapidly progressive glomerulonephritis and renal insufficiency, with high serum titers of antiglomerular basement membrane antibodies and linear deposits of immunoglobulin G and diffuse epithelial crescents on renal biopsy (36). Repeated plasmapheresis and immunosuppression caused the autoantibodies to disappear from the serum but renal function was not affected.

Of 63 adults with advanced renal insufficiency, those with biopsy-proven primary proliferative glomerulonephritis and those whose clinical presentation was consistent with glomerulonephritis had significantly greater exposure to organic solvents than patients with a variety of other renal diseases (37).

In a case-control study of 50 patients with biopsy-proven glomerulonephritis and 100 sex- and age-matched controls (50 patients each with non-glomerular renal disease or acute appendicitis), half of the patients with glomerulonephritis reported more than slight exposure to organic solvents compared with 20% of the controls (38).

Fanconi's syndrome has also been reported (39), as have severe tubulo-interstitial nephritis (40), and acute toxic tubular necrosis (41).

Three patients who regularly sniffed a lacquer thinner, whose major component was toluene, or a cement containing xylene and cyclohexane, for 2–6 years developed urolithiasis (42), as did a young man who was a persistent sniffer of toluene (43).

Immunologic

Although there are reports of associations between exposure to organic solvents and various connective tissue diseases, such as systemic sclerosis, scleroderma, undifferentiated connective tissue disease, systemic lupus erythematosus, and rheumatoid arthritis, the evidence of a causal association is weak (44).

Death

Death from solvent sniffing can occur either indirectly, through asphyxiation or aspiration of vomit (4). Direct causes include reflex vagal inhibition, cardiac dysrhythmias, respiratory depression, and anoxia.

Long-Term Effects

Tumorigenicity

There is evidence of varying quality of increased risks of cancer following exposure to the following solvents (45):

- benzene (leukemias and lung and nasopharyngeal cancers);
- carbon tetrachloride (hematological malignancies);
- methylene chloride (liver and biliary tract cancers);
- styrene (leukemias);
- tetrachloroethylene (esophageal and cervical cancers and non-Hodgkin's lymphoma);
- toluene (gastrointestinal and lung cancers);
- trichloroethane (multiple myeloma);
- trichloroethylene (tumors of the liver and biliary tract and non-Hodgkin's lymphomas);
- xylene (hematological malignancies).

Exposure to organic solvents is particularly associated with an increased risk of hematological malignancies, including leukemias (46) and non-Hodgkin's lymphomas (47). In one study there was a seven-fold increased risk of acute lymphoblastic leukemia in workers exposed to solvents (48). In another study there was an increased risk of non-Hodgkin's lymphoma in those who had been exposed to organic solvents daily for at least 1 year (OR = 3.3; 95% CI = 1.9, 5.8) (49). The risk of Hodgkin's disease may also be increased. Exposure to some solvents may increase the risk of multiple myeloma (50,51).

Exposure to benzene carries a risk of acute leukemias, usually myeloid or monocytic and preceded by a preleukemic phase, with anemia, leukopenia, and thrombocytopenia. Inhalation of toluene has occasionally been associated with aplastic anemia and acute leukemias.

Second-Generation Effects

Pregnancy

The effects of exposure to organic solvents in pregnancy have been reviewed (52). There is evidence of a moderate increase in the risks of spontaneous abortion and congenital malformations, especially facial clefts.

Teratogenicity and fetotoxicity

Exposure to toluene by glue sniffing during pregnancy can cause intrauterine growth retardation, premature delivery, congenital malformations, and postnatal developmental retardation (53).

A syndrome resembling that of the fetal alcohol syndrome has been described in infants who had been exposed to toluene in utero (54). Of all toluene-exposed infants, 39% were born prematurely and 9% died during the perinatal period; 54% were small for gestational age and 52% had continued postnatal growth deficiency; 33% had prenatal microcephaly, 67% postnatal microcephaly, and 80% developmental delay. In 83% there were craniofacial features similar to the fetal alcohol syndrome, and 89% of these children had other minor anomalies. The authors suggested that toluene and alcohol have a common mechanism of craniofacial teratogenesis, namely deficiency of craniofacial neuroepithelium and mesodermal components owing to increased embryonic cell death.

References

1. Errebo-Knudsen EO, Olsen F. Organic solvents and presenile dementia (the painters' syndrome). A critical review of the Danish literature. Sci Total Environ 1986;48(1-2):45–67.

2. Mottu F, Laurent A, Rufenacht DA, Doelker E. Organic solvents for pharmaceutical parenterals and embolic liquids: a review of toxicity data. PDA J Pharm Sci Technol 2000;54(6):456–69.

3. Martin KA. Toxicity, toluene. http://www.emedicine.com/EMERG/topic594.htm.

4. Meadows R, Verghese A. Medical complications of glue sniffing. South Med J 1996;89(5):455–62.

5. Streicher HZ, Gabow PA, Moss AH, Kono D, Kaehny WD. Syndromes of toluene sniffing in adults. Ann Intern Med 1981;94(6):758–62.

6. Hussain TF, Heidenreich PA, Benowitz N. Recurrent non-Q-wave myocardial infarction associated with toluene abuse. Am Heart J 1996;131(3):615–6.

7. Carder JR, Fuerst RS. Myocardial infarction after toluene inhalation. Pediatr Emerg Care 1997;13(2):117–9.

8. Wiseman MN, Banim S. "Glue sniffer's" heart? Br Med J (Clin Res Ed) 1987;294(6574):739.

9. Banathy LJ, Chan LT. Fatality caused by inhalation of "liquid paper" correction fluid. Med J Aust 1983;2(12):606.

10. Schikler KN, Lane EE, Seitz K, Collins WM. Solvent abuse associated pulmonary abnormalities. Adv Alcohol Subst Abuse 1984;3(3):75–81.

11. Beirne GJ. Goodpasture's syndrome and exposure to solvents. JAMA 1972;222(12):1555.

12. Robert R, Touchard G, Meurice JC, Pourrat O, Yver L. Severe Goodpasture's syndrome after glue sniffing. Nephrol Dial Transplant 1988;3(4):483–4.

13. Nathan AW, Toseland PA. Goodpasture's syndrome and trichloroethane intoxication. Br J Clin Pharmacol 1979;8(3):284–6.

14. Triebig G, Grobe T, Dietz MC. Polyneuropathie und Enzephalopathie durch organische Lösungsmittel und Lösungsmittelgemische. Arbeitsmedizinische und neurologische Aspekte zur neuen Berufskrankheit. [Polyneuropathy and encephalopathy caused by organic solvents and mixed solvent solutions. Occupational medicine and neurologic aspects of a new occupational disease.] Nervenarzt 1999;70(4):306–14.

15. Matsuoka M. [Neurotoxicity of organic solvents—recent findings.] Brain Nerve 2007;59(6):591–6.

16. Ridgway P, Nixon TE, Leach JP. Occupational exposure to organic solvents and long-term nervous system damage detectable by brain imaging, neurophysiology or histopathology. Food Chem Toxicol 2003;41(2):153–87.

17. Gralewicz S, Dyzma M. Organic solvents and the dopaminergic system. Int J Occup Med Environ Health 2005;18(2):103–13.

18. Mutti A, Franchini I. Toxicity of metabolites to dopaminergic systems and the behavioural effects of organic solvents. Br J Ind Med 1987;44(11):721–3.

19. Ferreiro JL, Isern Longares JA. [Chronic toluene poisoning.] Neurologia 1990;5(6):205–7.

20. Malm G, Lying-Tunell U. Cerebellar dysfunction related to toluene sniffing. Acta Neurol Scand 1980;62(3):188–90.

21. Fornazzari L, Wilkinson DA, Kapur BM, Carlen PL. Cerebellar, cortical and functional impairment in toluene abusers. Acta Neurol Scand 1983;67(6):319–29.

22. Lomax RB, Ridgway P, Meldrum M. Does occupational exposure to organic solvents affect colour discrimination? Toxicol Rev 2004;23(2):91–121.

23. Fuente A, McPherson B. Organic solvents and hearing loss: the challenge for audiology. Int J Audiol 2006;45(7):367–81.

24. Bazydło-Golińska G. [The effect of organic solvents on the inner ear.] Med Pr 1993;44(1):69–78.

25. Morata TC, Dunn DE, Sieber WK. Occupational exposure to noise and ototoxic organic solvents. Arch Environ Health 1994;49(5):359–65.

26. Bilski B. Wpływ rozpuszczalników organicznych na narzad słuchu. [Effect of organic solvents on hearing organ.] Med Pr 2001;52(2):111–8.

27. Williams DM. Hearing loss in a glue sniffer. J Otolaryngol 1988;17(6):321–4.

28. Iregren A. Behavioral methods and organic solvents: questions and consequences. Environ Health Perspect 1996;104 Suppl 2:361–6.

29. Lees-Haley PR, Williams CW. Neurotoxicity of chronic low-dose exposure to organic solvents: a skeptical review. J Clin Psychol 1997;53(7):699–712.

30. Filley CM, Halliday W, Kleinschmidt-DeMasters BK. The effects of toluene on the central nervous system. J Neuropathol Exp Neurol 2004;63(1):1–12.

31. Baskerville JR, Tichenor GA, Rosen PB. Toluene induced hypokalaemia: case report and literature review. Emerg Med J 2001;18(6):514–6.

32. Carlisle EJ, Donnelly SM, Vasuvattakul S, Kamel KS, Tobe S, Halperin ML. Glue-sniffing and distal renal tubular acidosis: sticking to the facts. J Am Soc Nephrol 1991;1(8):1019–27.

33. O'Brien ET, Yeoman WB, Hobby JA. Hepatorenal damage from toluene in a "glue sniffer". Br Med J 1971;2(5752):29–30.

34. Gupta RK, van der Meulen J, Johny KV. Oliguric acute renal failure due to glue-sniffing. Case report. Scand J Urol Nephrol 1991;25(3):247–50.

35. Venkataraman G. Renal damage and glue sniffing. Br Med J (Clin Res Ed) 1981;283(6304):1467.

36. Bonzel KE, Müller-Wiefel DE, Ruder H, Wingen AM, Waldherr R, Weber M. Anti-glomerular basement membrane antibody-mediated glomerulonephritis due to glue sniffing. Eur J Pediatr 1987;146(3):296–300.

37. Zimmerman SW, Groehler K, Beirne GJ. Hydrocarbon exposure and chronic glomerulonephritis. Lancet 1975;2(7927):199–201.

38. Ravnskov U, Forsberg B, Skerfving S. Glomerulonephritis and exposure to organic solvents. A case-control study. Acta Med Scand 1979;205(7):575–9.

39. Moss AH, Gabow PA, Kaehny WD, Goodman SI, Haut LL. Fanconi's syndrome and distal renal tubular acidosis after glue sniffing. Ann Intern Med 1980;92(1):69–70.

40. Taverner D, Harrison DJ, Bell GM. Acute renal failure due to interstitial nephritis induced by 'glue-sniffing' with subsequent recovery. Scott Med J 1988;33(2):246–7.

41. Gupta RK, van der Meulen J, Johny KV. Oliguric acute renal failure due to glue-sniffing. Case report. Scand J Urol Nephrol 1991;25(3):247–50.

42. Kaneko T, Koizumi T, Takezaki T, Sato A. Urinary calculi associated with solvent abuse. J Urol 1992;147(5):1365–6.

43. Kroeger RM, Moore RJ, Lehman TH, Giesy JD, Skeeters CE. Recurrent urinary calculi associated with toluene sniffing. J Urol 1980;123(1):89–91.

44. Garabrant DH, Dumas C. Epidemiology of organic solvents and connective tissue disease. Arthritis Res 2000;2(1):5–15.

45. Lynge E, Anttila A, Hemminki K. Organic solvents and cancer. Cancer Causes Control 1997;8(3):406–19.

46. Brandt L. Exposure to organic solvents and risk of haematological malignancies. Leuk Res 1992;16(1):67–70.

47. Rêgo MA. Non-Hodgkin's lymphoma risk derived from exposure to organic solvents: a review of epidemiologic studies. Cad Saude Publica 1998;14 Suppl 3:41–66.

48. McMichael AJ, Spirtas R, Kupper LL, Gamble JF. Solvent exposure and leukemia among rubber workers: an epidemiologic study. J Occup Med 1975;17:234.

49. Olsson H, Brandt L. Risk of non-Hodgkin's lymphoma among men occupationally exposed to organic solvents. Scand J Work Environ Health 1988;14:246.

50. Greene MH, Hoover RN, Eck RL, Fraumeni JF. Cancer mortality among printing plant workers. Environ Res 1979;20:66.

51. Gallagher RP, Threlfall WJ. Cancer mortality in metal workers. CMAJ 1983;129:191.

52. Saillenfait AM, Robert E. Exposition professionnelle aux solvants et grossesse. Etat des connaissances épidémiologiques. [Occupational exposure to organic solvents and pregnancy. Review of current epidemiologic knowledge.] Rev Epidémiol Santé Publique 2000;48(4):374–88.

53. Donald JM, Hooper K, Hopenhayn-Rich C. Reproductive and developmental toxicity of toluene: a review. Environ Health Perspect 1991;94:237–44.

54. Pearson MA, Hoyme HE, Seaver LH, Rimsza ME. Toluene embryopathy: delineation of the phenotype and comparison with fetal alcohol syndrome. Pediatrics 1994;93(2):211–5.

PHENCYCLIDINE

General Information

Phencyclidine or 1-(1-phenylcyclohexy-1) piperidine (known as PCP, "angel dust", and many other names) was originally developed as an anesthetic, but was abused as an illicit drug from the late 1960s onwards. It is an antagonist at the N-methyl-d-aspartate (NMDA) subtype of glutamate receptors and a dopamine receptor agonist. It has anticholinergic properties through blockade of ion channels in acetylcholine receptors. It is still used in some countries as an anti-parkinsonian agent (SED-11, 86; 1–3).

The psychoactive effects of phencyclidine are stimulant and similar to the effects of hallucinogens. Hallucinations are often bizarre, frightening, and challenging. Aggressive behavior, usually with amnesia, is common. Self-destructive actions are also seen. Overdosage is associated with paresthesia, slurred speech, ataxia, and later catatonia, dilated pupils, and coma, with tachycardia, hypertension, and dysrhythmias. Seizures and deaths have occurred (SED-11, 86; 4).

Although the effects of phencyclidine are usually short-lived, it can cause prolonged and severe behavioral disturbances, exaggeration of pre-existing thoughts, and serious medical complications (5).

Observational studies

Among 107 consecutive patients with phencyclidine intoxication, the diagnosis was confirmed by positive urine assay in 27 (6). The most common abnormalities were mental/behavioral (89%) and nystagmus (85%). There were also increases in blood pressure, temperature, and heart rate. The most common serious medical complication requiring hospitalization was rhabdomyolysis, which occurred in three patients, two of whom developed acute renal insufficiency.

Four major and five minor clinical patterns of acute phencyclidine intoxication have been described in 1000 patients (7). Major patterns were acute brain syndrome (24.8%), toxic psychosis (16.6%), catatonic syndrome (11.7%), and coma (10.6%). Minor patterns included lethargy or stupor (3.8%) and combinations of bizarre behavior, violence, agitation, and euphoria in patients who were alert and oriented (32.5%). Patients with major patterns of toxicity usually required hospitalization and had most of the complications. Patients with minor patterns generally had mild intoxication and did not require hospitalization, except for treatment of injuries or autonomic effects of phencyclidine. There were various types of injuries in 16%, and aspiration pneumonia in 1.0%. There were 22 cases of rhabdomyolysis (2.2%), and three patients required dialysis for renal insufficiency. One patient who had been comatose died suddenly with a pulmonary embolism.

Of 68 users of phencyclidine (37 men, 31 women; aged 14–38 years), 42 used it daily, and 14 used it intravenously; 25 considered themselves to be addicted to it (8). The effects that they reported are listed in Table 1, the withdrawal effects in Table 2, and the unwanted behaviors in Table 3.

Organs and Systems

Cardiovascular

In seven cases of poisoning with phencyclidine, death followed a hypertensive crisis (9). In one case an acute episode of hypertension resulted in coma and blindness (10).

Nervous system

The electroencephalographic effects of phencyclidine intoxication have been reported in a patient who was comatose, with nystagmus and waxy rigidity of the limbs (11). The electroencephalogram showed widespread sinusoidal theta rhythm, interrupted every few seconds by periodic slow-wave complexes, similar to that seen in deep ketamine anesthesia.

In another patient, a 23-year old man, the electroencephalogram also showed monomorphic non-reactive generalized theta rhythm, interrupted by periodic bilaterally synchronous high voltage slow paroxysms similar to those described in subacute sclerosing panencephalitis (12). The authors suggested that this unusual finding supported the hypothesis that phencyclidine acts by reversible deafferentation of cortical neurons.

Table 1 Self-reported effects of phencyclidine in 68 users

Effects	%	Effects	%
Decreased appetite	61.8	Increased sex drive	29.4
Confused thoughts	60.3	Visual hallucinations	27.9
Loss of memory	58.8	Paranoia	26.5
Increased strength	57.4	Ringing in ears	23.5
Feel "speedy"	55.9	Headaches	22.1
Euphoria	54.4	Decreased sex drive	20.6
Drowsiness	51.5	Fatigue	20.6
Bad trips	44.1	Auditory hallucinations	20.6
Insomnia	41.2	Feel hot	19.1
Depression	35.3	Decreased strength	17.6
Dizziness	33.8	Vomiting	13.2
Increased anger	33.8	Feeling of detachment	13.2
Increased anxiety	33.8	Increased appetite	7.4
Increased violence	32.4	Decreased anxiety	7.4

Table 2 Self-reported symptoms after withdrawal of chronic phencyclidine in 68 users

Experiences	%	Experiences	%
Craving for PCP	51.5	Headaches	16.2
Increased need for sleep	48.5	Insomnia	14.7
Poor memory	45.6	None	14.7
Depression	44.1	Recurring tastes	13.2
Laziness	44.1	Panic	11.8
Increased appetite	38.2	Decreased appetite	8.8
Confused thoughts	35.3	Decreased need for sleep	4.4
Flashbacks	32.4	Feeling "speedy"	1.5
Irritable	30.9	Feeling hot	1.5
Feeling weak	30.9	Ringing in ears	1.5
Increased anxiety	22.1		

Table 3 Self-reported unwanted behaviors in 68 users of phencyclidine

Behavior	%	Behavior	%
Lost money	48.5	Hurt someone else	23.5
Got lost	39.7	Unwanted sexual encounter	19.1
Took drugs	35.3	Attempted suicide	13.2
Got into fight	30.9	Committed crime	11.8
Hurt yourself	26.5	Had car accident	10.3

Sensory systems

Amaurosis fugax has been attributed to phencyclidine (13).

Psychiatric

After ingesting street drugs sold as "PCP", "THC", and "methadone", three young men developed schizophreniform psychoses, analgesia, anesthesia, and amnesia (14). Except for the unusually long duration (2–4 weeks), these reactions resembled phencyclidine-induced psychoses.

Phencyclidine use can cause a psychosis lasting several weeks in a small number of users, who have premorbid personalities similar top those who develop psychoses after using lysergide or cannabis (15). Nine patients with phencyclidine-induced psychoses had hostility, agitation, tangentiality, and delusions of influence and religious grandiosity; six reported auditory hallucinations; four were disoriented in at least one sphere. Despite treatment with antipsychotic drugs, the episodes often persisted for more than 30 days (16).

Mechanism

Phencyclidine-induced psychosis is probably mediated via dopamine D_2 receptors. This is consistent with the results of studies with different dopamine receptor antagonists in the treatment of this psychosis. In one study haloperidol (a predominantly D_2 antagonist with noradrenergic effects) and pimozide (a predominantly D_2 antagonist with no noradrenergic activity) were equally effective and both were superior to chlorpromazine (a non-selective D_1 and D_2 antagonist with noradrenergic effects) (17).

Musculoskeletal

Two young patients with phencyclidine toxicity developed acute rhabdomyolysis and myoglobinuria, probably secondary to skeletal muscle injury due to an acute dystonic motor reaction (18) and 30 reported cases of rhabdomyolysis associated with acute renal insufficiency have been reviewed (19).

Body temperature

Three men with phencyclidine intoxication developed severe malignant hyperthermia, respiratory failure, and coma (20). Two days later serum transaminase activity and bilirubin concentration rose markedly with prolongation of the prothrombin time. Liver biopsies showed marked perivenular necrosis and collapse. One patient died.

Multiorgan failure

- A 42 year old woman developed confusion and a bradycardia of 20/minute. She was disoriented and only partly responsive to verbal stimuli (21). Her pupils were dilated and partly responsive to light, with circular nystagmus. An electrocardiogram showed wide QRS complexes, peaked T waves, and no P waves. She had a severe metabolic and respiratory acidosis, acute renal insufficiency, and markedly raised liver enzymes. She had a prolonged prothrombin time and a reduced concentration of clotting factor V. There were opiates and phencyclidine in the urine. She was intubated and mechanically ventilated but later had a cardiac arrest. She stabilized after cardiopulmonary resuscitation and underwent urgent hemodialysis. Her temperature rose to 39.5°C and she died on day 3 in multiorgan failure.

Death

Of 19 deaths associated exclusively with phencyclidine intoxication, 13 were due to asphyxia by drowning or trauma (22). In two cases the probable cause of death was primary respiratory depression accompanied by seizure activity. A secondary drug effect or concurrent disease may have contributed to the deaths of the remaining four individuals.

Second-Generation Effects

Fetotoxicity

The neurodevelopmental consequences of exposure to phencyclidine during pregnancy have been reviewed (23).

In a retrospective, case-control study of the use of phencyclidine in 23 of 13 653 pregnant women, the phencyclidine users had smaller infants (2698 versus 3011 g), which may have been partly accounted for by a reduction in gestational age (37.3 versus 38.3 weeks) (24). The users were more likely to have used tobacco, alcohol, or marijuana and had a higher incidence of syphilis and diabetes mellitus.

Two neonates whose mothers had used phencyclidine during pregnancy had jitteriness, hypertonicity, vomiting, and in one case diarrhea (25). Both had phencyclidine in the urine. Both remained jittery and slightly hypertonic despite treatment with phenobarbital. One was microcephalic.

When 94 neonates born to mothers who had taken phencyclidine during pregnancy were compared with 94 controls the former had poor attention, hypertonia, and depressed neonatal reflexes (26).

Drug administration

Drug overdose

Two phencyclidine-intoxicated patients had bizarre combinations of disorientation, hallucination, agitation, and dyskinetic motor activity (27).

In nine cases of phencyclidine hydrochloride poisoning, early signs of overdose included drowsiness, nystagmus, miotic pupils, raised blood pressure, increased deep tendon reflexes, ataxia, anxiety, and agitation. In more severe cases, seizures, spasticity, and opisthotonos were seen, in addition to deep coma and respiratory depression (28).

The management of acute phencyclidine intoxication has been reviewed (29,30,31,32). Acidification of the urine enhances its renal clearance (33). All of the following have been used to achieve this: ammonium chloride 1 g qds with sufficient water or cranberry juice; lysine dihydrochloride 2 g tds with sufficient water or cranberry juice; lysine hydrochloride 2 g qds with water or cranberry juice; cranberry juice 18 or more oz/day alone or plus lysine, ammonium chloride, or ascorbic acid (34).

References

1. Balster RL, Wessinger WD. Central nervous system depressant effects of phencyclidine. In: Kameka JM, Domino EF, Geneste P, editors. Phencyclidine and Related Amylcyclohexylamines. Ann Arbor, MI: NPP Books, 1983:291–309 Preset and Future Applications.
2. McCarron MM, Schulze BW, Thompson GA, Conder MC, Goetz WA. Acute phencyclidine intoxication: incidence of clinical findings in 1,000 cases. Ann Emerg Med 1981;10(5):237–42.
3. Peterson RC, Stillman RC. Phencyclidine. A review Rockville, MD: National Institute on Drug Abuse;. 1978.
4. Garey RE. PCP (Phencyclidine): an update. J Psychedelic Drugs 1979;11(14):265–75.
5. Showalter CV, Thornton WE. Clinical pharmacology of phencyclidine toxicity. Am J Psychiatry 1977;134(11):1234–8.
6. Barton CH, Sterling ML, Vaziri ND. Phencyclidine intoxication: clinical experience in 27 cases confirmed by urine assay. Ann Emerg Med 1981;10(5):243–6.
7. McCarron MM, Schulze BW, Thompson GA, Conder MC, Goetz WA. Acute phencyclidine intoxication: clinical patterns, complications, and treatment. Ann Emerg Med 1981;10(6):290–7.
8. Rawson RA, Tennant FS Jr, McCann MA. Characteristics of 68 chronic phencyclidine abusers who sought treatment. Drug Alcohol Depend 1981;8(3):223–7.
9. Eastman JW, Cohen SN. Hypertensive crisis and death associated with phencyclidine poisoning. JAMA 1975;231(12):1270–1.
10. Stratton MA, Witherspoon JM, Kirtley T. Hypertensive crisis and phencyclidine abuse. Va Med 1978;105(8):569–72.
11. Stockard JJ, Werner SS, Aalbers JA, Chiappa KH. Electroencephalographic findings in phencyclidine intoxication. Arch Neurol 1976;33(3):200–3.
12. Fariello RG, Black JA. Pseudoperiodic bilateral EEG paroxysms in a case of phencyclidine intoxication. J Clin Psychiatry 1978;39(6):579–81.
13. Ubogu E. Amaurosis fugax associated with phencyclidine inhalation. Eur Neurol 2001;46(2):98–9.
14. Rainey JM Jr, Crowder MK. Prolonged psychosis attributed to phencyclidine: report of three cases. Am J Psychiatry 1975;132(10):1076–8.
15. Fauman B, Aldinger G, Fauman M, Rosen P. Psychiatric sequelae of phencyclidine abuse. Clin Toxicol 1976;9(4):529–38.
16. Allen RM, Young SJ. Phencyclidine-induced psychosis. Am J Psychiatry 1978;135(9):1081–4.
17. Giannini AJ, Nageotte C, Loiselle RH, Malone DA, Price WA. Comparison of chlorpromazine, haloperidol and pimozide in the treatment of phencyclidine psychosis: DA-2 receptor specificity. J Toxicol Clin Toxicol 1984–1985;22(6):573–9.
18. Cogen FC, Rigg G, Simmons JL, Domino EF. Phencyclidine-associated acute rhabdomyolysis. Ann Intern Med 1978;88(2):210–2.
19. Patel R, Connor G. A review of thirty cases of rhabdomyolysis-associated acute renal failure among phencyclidine users. J Toxicol Clin Toxicol 1985–1986;23(7-8):547–56.
20. Armen R, Kanel G, Reynolds T. Phencyclidine-induced malignant hyperthermia causing submassive liver necrosis. Am J Med 1984;77(1):167–72.
21. Stein GY, Fradin Z, Ori Y, Singer P, Korobko Y, Zeidman A. Phencyclidine-induced multi-organ failure. Isr Med Assoc J 2005;7(8):535–7.
22. Burns RS, Lerner SE. Phencyclidine deaths. JACEP 1978;7(4):135–41.
23. Deutsch SI, Mastropaolo J, Rosse RB. Neurodevelopmental consequences of early exposure to phencyclidine and related drugs. Clin Neuropharmacol 1998;21(6):320–32.
24. Mvula MM, Miller JM Jr, Ragan FA. Relationship of phencyclidine and pregnancy outcome. J Reprod Med 1999;44(12):1021–4.
25. Strauss AA, Modaniou HD, Bosu SK. Neonatal manifestations of maternal phencyclidine (PCP) abuse. Pediatrics 1981;68(4):550–2.
26. Golden NL, Kuhnert BR, Sokol RJ, Martier S, Williams T. Neonatal manifestations of maternal phencyclidine exposure. J Perinat Med 1987;15(2):185–91.
27. Tong TG, Benowitz NL, Becker CE, Forni PJ, Boerner U. Phencyclidine poisoning. JAMA 1975;234(5):512–3.
28. Liden CB, Lovejoy FH Jr, Costello CE. Phencyclidine. Nine cases of poisoning. JAMA 1975;234(5):513–6.
29. Dorand RD. Phencyclidine ingestion: therapy review. South Med J 1977;70(1):117–9.
30. Aronow R, Done AK. Phencyclidine overdose: an emerging concept of management. JACEP 1978;7(2):56–9.
31. Sioris LJ, Krenzelok EP. Phencyclidine intoxication: a literature review. Am J Hosp Pharm 1978;35(11):1362–7.
32. Rappolt RT, Gay GR, Farris RD. Emergency management of acute phencyclidine intoxication. JACEP 1979;8(2):68–76.
33. Domino EF, Wilson AE. Effects of urine acidification on plasma and urine phencyclidine levels in overdosage. Clin Pharmacol Ther 1977;22(4):421–4.
34. Simpson GM, Khajawall AM. Urinary acidifiers in phencyclidine detoxification. Hillside J Clin Psychiatry 1983;5(2):161–8.

PSILOCYBIN

General Information

Psilocybin (4-phosphoryloxy-N,N-dimethyltryptamine) is a psychedelic indole, as is its active metabolite, psilocin. It is found in about 200 species of fungi, including those of the genus *Psilocybe*, such as *Psilocybe semilanceata* (the liberty cap) and *Psilocybe cubensis* (golden top or golden cap), also called "magic mushrooms" or simply "shrooms" (1).

Psychedelic drugs (which include lysergide, mescalin, and psilocybin) act primarily by sertonergic mechanisms and indirect effects on dopamine function (2,3). Psilocin is an agonist at $5HT_{1A}$ and $5HT_{2A}$ receptors (4).

The effects of psilocybin last for a few hours. Its physical effects include pupillary dilatation, flushing, nausea, tremor, pyrexia, hyper-reflexia, tachycardia, weakness, and dizziness (5). The psychedelic effects include visual and other types of hallucinations, including enhanced perception of colours, synesthesia, time slowing, and a sensation of ego fragmentation. Psilocybin also causes illusions of motion in otherwise stationary objects (6). Psychotic reactions ("bad trips") can occur, accompanied by fear, panic, and dangerous behavior, especially when psilocybin is used in combination with other drugs and alcohol or by psychiatrically unstable patients. During such an attack, self-mutilation can occur (7).

In a survey of the use of magic mushrooms in the UK in 2004, the 174 self-selecting respondents were predominantly in their 20s, white, British, and in education or employed; 64% were men (8). They reported infrequent but intense consumption (47% used the mushrooms 4–12 times/year and the average consumption in one sitting was 12 g). They did so to obtain laughter, hallucinations, altering perspective (41–74%), and feelings of being closer to nature (49%). Negative effects included paranoia (35%) and anxiety (32%).

Organs and Systems

Cardiovascular

In a double-blind study nine patients with obsessive-compulsive disorder were given four single-doses of psilocybin 25-300 micrograms/kg (9). One had transient hypertension unrelated to anxiety or somatic symptoms, but there were no other significant adverse effects.

Psychological

In a double-blind study in 12 healthy volunteers, psilocybin 115 and 250 micrograms/kg significantly impaired their ability to reproduce interval durations longer than 2.5 seconds and to synchronize to inter-beat intervals longer than 2 seconds; it also caused them to be slower in their preferred tapping rate (10). These objective effects on timing performance were accompanied by deficits in working memory and subjective changes in conscious state (depersonalization, derealization, and disturbances of subjective time sense).

Psychiatric

In a placebo-controlled study using PET scanning in seven healthy volunteers, oral psilocybin 250 micrograms/kg produced changes in mood, disturbed thinking, illusions, elementary and complex visual hallucinations, and impaired ego-functioning (11). Psilocybin significantly reduced (^{11}C)-raclopride receptor binding potential bilaterally in the caudate nucleus (19%) and putamen (20%), consistent with an increase in endogenous dopamine. Changes in (^{11}C)-raclopride binding in the ventral striatum correlated with depersonalization associated with euphoria. The authors concluded that stimulation of both $5HT_{1A}$ and $5HT_{2A}$ receptors may be important for modulation of striatal dopamine release in acute psychoses.

Five Japanese patients who ate an indigenous mushroom, *Psilocybe argentipes* (hikageshibiretake), had various effects (12). One became stuporose with complete amnesia; one had a psychedelic state with dreamy consciousness, and three had acute psychotic reactions with vivid visual hallucinations. All had acute anxiety and panic reactions.

Hallucinogen persisting perception disorder (HPPD) has been reported after the use of psilocybin (13).

- An 18-year-old student, who had used cannabis moderately for many years developed perceptual impairment and dysphoric mood, which lasted for 8 months. The perceptual disturbances initially appeared after he took 40 hallucinogenic mushrooms in an infusion. He re-experienced the symptoms the following day after a cannabis snort. He reported visual disturbances such as distortion of objects, auditory disturbance with a sensation of resonance, depersonalization, derealization, changes in the perception of body weight, spatiotemporal disturbances, and inability to distinguish illusion from reality. These symptoms were similar to those experienced after initial intoxication with the mushrooms. Flashbacks occurred daily and got worse in the dark. Because these symptoms were distressing he stopped using cannabis 2 months later. The symptoms abated but then increased again 4 months later. An MRI scan, electroencephalography, and blood tests were normal. He was depressed and had a social phobia tendency but no thought disorder or hallucinations. He was initially treated with amisulpiride and then with olanzapine, to provide more sedation. However, this exacerbated his symptoms and was replaced by risperidone 2 mg/day. He was given sertraline for persistent dysphoric mood and anxiety. After 6 months, the flashbacks disappeared and his mood and social interactions improved.

This case suggests that HPPD can result from co-intox-ication by psilocybin and cannabis and can persist after drug consumption has stopped.

Endocrine

In a placebo-controlled study in eight healthy subjects, psilocybin 45–315 micrograms/kg had no effect on plasma concentrations of thyroid-stimulating hormone (TSH), prolactin, or cortisol; plasma concentrations of corticotropin (ACTH) were increased by doses of 215 and 315 micrograms/kg (14).

References

1. Letcher A. Shrooms. *A Cultural History of the Magic Mushroom.* London: Faber & Faber Limited, 2006.
2. Vollenweider FX, Vollenweider-Scherpenhuyzen MF, Bäbler A, Vogel H, Hell D. Psilocybin induces schizophrenia-like psychosis in humans via a serotonin-2 agonist action. Neuroreport 1998;9(17):3897–902.
3. Vollenweider FX. Advances and pathophysiological models of hallucinogenic drug actions in humans: a preamble to schizophrenia research. Pharmacopsychiatry 1998;31 Suppl 2:92–103.
4. Vollenweider FX, Csomor PA, Knappe B, Geyer MA, Quednow BB. The effects of the preferential 5-HT$_{2A}$ agonist psilocybin on prepulse inhibition of startle in healthy human volunteers depend on interstimulus interval. Neuropsychopharmacology 2007;32(9):1876–87.
5. Martin WR, Sloan JW. Relationship of CNS tryptaminergic processes and the action of LSD-like hallucinogens. Pharmacol Biochem Behav 1986;24(2):393–9.
6. Carter OL, Pettigrew JD, Burr DC, Alais D, Hasler F, Vollenweider FX. Psilocybin impairs high-level but not low-level motion perception. Neuroreport 2004;15(12):1947–51.
7. Attema-de Jonge ME, Portier CB, Franssen EJ. Automutilatie na gebruik van hallucinogene paddenstoelen. [Automutilation after consumption of hallucinogenic mushrooms.] Ned Tijdschr Geneeskd 2007;151(52):2869–72.
8. Riley SC, Blackman G. Between prohibitions: patterns and meanings of magic mushroom use in the UK. Subst Use Misuse 2008 ;43(1):55–71.
9. Moreno FA, Wiegand CB, Taitano EK, Delgado PL. Safety, tolerability, and efficacy of psilocybin in 9 patients with obsessive-compulsive disorder. J Clin Psychiatry 2006;67(11):1735–40.
10. Wittmann M, Carter O, Hasler F, Cahn BR, Grimberg U, Spring P, Hell D, Flohr H, Vollenweider FX. Effects of psilocybin on time perception and temporal control of behaviour in humans. J Psychopharmacol 2007;21(1):50–64.
11. Vollenweider FX, Vontobel P, Hell D, Leenders KL. 5-HT modulation of dopamine release in basal ganglia in psilocybin-induced psychosis in man—a PET study with [^{11}C]raclopride. Neuropsychopharmacology 1999;20(5):424–33.
12. Musha M, Ishii A, Tanaka F, Kusano G. Poisoning by hallucinogenic mushroom hikageshibiretake (*Psilocybe argentipes K. Yokoyama*) indigenous to Japan. Tohoku J Exp Med 1986;148(1):73–8.
13. Espiard ML, Lecardeur L, Abadie P, Halbecq I, Dollfus S. Hallucinogen persisting perception disorder after psilocybin consumption: a case study. Eur Psychiatry 2005;20:458–6.
14. Hasler F, Grimberg U, Benz MA, Huber T, Vollenweider FX. Acute psychological and physiological effects of psilocybin in healthy humans: a double-blind, placebo-controlled dose-effect study. Psychopharmacology (Berl) 2004;172(2):145–56.

DRUGS USED IN ALZHEIMER'S DISEASE

DRUGS USED IN ALZHEIMER'S DISEASE

Donepezil

General Information

The approval of donepezil in several countries in America and Europe has been hailed as a major milestone, because it has met regulatory guidelines for the approval of anti-dementia drugs (1). Donepezil belongs to a piperidine class of reversible acetylcholinesterase inhibitors, chemically unrelated to either tacrine or physostigmine. It is highly specific for acetylcholinesterase and does not inhibit butyrylcholinesterase. The incidence of adverse effects with donepezil is comparable to that of placebo in controlled trials, and unlike tacrine, liver enzyme monitoring is not required. The long-term effectiveness of donepezil in large populations is yet to be established. Moreover, while it improves cognitive symptoms, it does not alter the course of the disease. Based on a limited number of studies, support for the use of donepezil in Alzheimer's disease has emerged (2–5).

Clinical trials of donepezil were funded by the manufacturer and appear to have been methodologically sound, although the absence of caregiver quality-of-life measures and outcomes related to activities of daily living is difficult to reconcile. An earlier 12-week study showed no improvement in caregiver quality of life with donepezil (4). As with most phase 3 studies, extrapolation of these results to routine practice is hampered by the fact that study populations are likely to be healthier than patients seen in routine clinical practice. Whether the results would be any different in a more heterogeneous population remains to be seen (6).

The launch of donepezil has attracted intense interest among both the scientific community and the public. The debate regarding "lessons for health care policy" (7) has been summarized:

(1) Licensing trials in highly selected patients may provide insignificant information on which to base clinical decisions, especially when the effect sizes are small and comorbidity is common.
(2) All trial evidence should be published before new drugs are marketed, and medical journals should not carry advertisements referring to unpublished data.
(3) Communication of benefits and risks should emphasize clinical effect sizes rather than statistical significance.
(4) Claims about effects on populations or services should be based on evidence.
(5) Secrecy surrounding licensing should be ended, and data from trials should be available for independent analysis.
(6) Overvaluation of new technology could threaten funding for vital but more mundane care.

There was intense debate in response to the above publication, as reflected by several letters to the Editor of the *British Medical Journal* (8–13).

Although once-daily donepezil is effective and well tolerated for the symptoms of mild-to-moderate Alzheimer's disease (14), two practical questions arise:

(1) Which patients are too severely affected to be treated with donepezil?
(2) At what point should donepezil be discontinued if the patient continues to deteriorate?

The current status of donepezil in the management of Alzheimer's disease has been comprehensively reviewed (15). Several recent studies have confirmed the efficacy and tolerability of donepezil using different doses, study designs, and durations of treatment (16–20). Some relevant conclusions were the following:

(1) Younger patients should be targeted for assessment and treatment (21).
(2) Of the patients 6% discontinued medication owing to adverse events (20).
(3) Sleep disturbances were more common in trials with bedtime dosing of donepezil (19).
(4) Long-term safety and realistic improvement were observed over a period of up to 4.9 years (22).
(5) The presence of the apolipoprotein E4 allele did not predict donepezil failure (17).

The adverse effects of donepezil in general practice have been evaluated in a post-marketing pharmacovigilance study in 1762 patients in the UK (23). This observational cohort study used the technique of prescription-event monitoring for a minimum period of 6 months. The commonest adverse events were nausea, diarrhea, malaise, dizziness, and insomnia. Aggression, agitation, and abnormal dreams were uncommonly associated with the drug. There were no causally associated cardiac rhythm disturbances or liver disorders. The authors suggested that the abnormal dreams and psychiatric disturbances were possible adverse drug reactions that require further confirmation.

Observational studies

Seven elderly patients with psychotic or non-psychotic behavioral symptoms in Lewy body dementia had some benefit from donepezil (24). Donepezil was withdrawn prematurely in three patients owing to poor response and/or adverse events. The adverse events were sedation, somnolence, worsening of chronic obstructive pulmonary disease, syncope, sweating, and bradycardia. These results have to be confirmed in controlled trials.

In two patients with Alzheimer's disease, donepezil provided some benefit in relieving cognitive symptoms, but there were increased behavioral problems, such as anxiety, agitation, irritability, and lack of impulse control; these were then successfully controlled by adding gabapentin (25).

Further evidence that donepezil is effective and well tolerated in treating symptoms of mild to moderately severe Alzheimer's disease has emerged from a multinational trial (26). Common adverse effects were nausea, vomiting, diarrhea, anorexia, dizziness, and confusion, consistent with previous findings.

The beneficial effect of donepezil on global ratings of dementia symptoms, cognition, and activities of daily living has been confirmed (27,28). Donepezil was to be well tolerated for periods up to 1 year, and adverse events were usually mild and transient, lasted only an initial few days, and typically resolved without the need for dosage modification. It has been suggested that patients with Alzheimer's disease do best while taking donepezil 10 mg/day and when the dosage is maintained at that level without interruption. Donepezil treatment effects that are lost after prolonged withdrawal do not fully recover when the drug is restarted (29).

In an open study patients with Alzheimer's disease and psychotic symptoms benefited from the addition of donepezil to neuroleptic drugs (30).

The AD 2000 West Midlands Donepezil Trial was based on 565 community-resident patients with mild to moderate Alzheimer's disease who took donepezil 5 or 10 mg/day for 12 weeks(31). The results suggested that donepezil is not cost-effective, with benefits below minimally relevant thresholds. Common adverse events and those contributing to withdrawal of medication could have major impact on quality of life and costs, but were not measured or reported.

Following the publication of the findings of AD 2000, several correspondents criticized the design of the study and its conclusions (32,33,34,35). Furthermore, the need for comparisons of cholinesterase inhibitors, alone and in combination, with other drugs has been acknowledged (36).

The efficacy of cholinesterase inhibitors in vascular dementia (37,38) and in dementia with Lewy bodies (39) is uncertain.

Sertraline augmentation in the treatment of the behavioral manifestations of Alzheimer's disease in outpatients treated with donepezil has been evaluated (40). There were no significant differences on primary endpoints. Diarrhea was more common with donepezil + sertraline group compared with donepezil + placebo.

There was partial improvement in cognitive measurements with donepezil after 6–8 months in three patients with Wernicke–Korsakoff syndrome (41). Previous reports have suggested promising results (42,43,44). However, the extent of benefit in each case is hard to interpret, because of a variety of confounding factors, including short treatment periods, lack of reporting of sequential cognitive testing, and variables such as the spontaneous partial recovery that occurs in the first few months after diagnosis of Wernicke–Korsakoff syndrome

and treatment with thiamine. No clear conclusions can therefore be drawn from these case reports.

Comparative studies

In a double-blind study 769 subjects with an amnestic subtype of mild cognitive impairment, a transitional state between the cognitive changes of normal aging and early Alzheimer's disease, were randomly assigned to receive vitamin E 2000 IU/day, donepezil 10 mg/day, or placebo for 3 years (45). Possible or probable Alzheimer's disease developed in 211 subjects. Compared with placebo, there were no significant differences in the probability of progression to Alzheimer's disease in the vitamin E group (hazard ratio = 1.02; 95% CI = 0.74, 1.41) or in the donepezil group (hazard ratio = 0.80; 95% CI = 0.57, 1.13) during the 3 years of treatment. However, during the first 12 months donepezil reduced the likelihood of progression to Alzheimer's disease, a finding that was supported by secondary outcome measures. Among carriers of one or more Apo-E ε4 alleles, the benefit of donepezil was evident throughout the 3-year follow up. There were no significant differences in the rates of progression to Alzheimer's disease between the vitamin E and placebo groups at any time, either among all patients or among Apo-E ε4 carriers. Vitamin E produced no benefit in patients with mild cognitive impairment. Although donepezil was associated with a lower rate of progression to Alzheimer's disease during the first 12 months of treatment, the rate of progression to Alzheimer's disease after 3 years was not lower among patients who took donepezil than among those who took placebo.

Placebo-controlled studies

In a 12-week double-blind, placebo-controlled, parallel-group study, aimed at establishing the efficacy and safety of donepezil in patients with mild to moderately severe Alzheimer's disease, donepezil (5 and 10 mg od) was well tolerated and efficacious (46). Adverse events significantly more common with donepezil were nausea, insomnia, and diarrhea, which appeared to be dose related and did not require treatment. Seven patients treated with placebo and six in each of the donepezil groups had serious adverse events during the trial. Three had events that were considered possibly related to donepezil. These included gastric ulceration with hemorrhage, syncope and a transient ischemic attack, nausea, aphakia, tremor, and sweating. Both groups of patients treated with donepezil had falls in mean heart rate that were larger than with placebo. Two patients treated with donepezil had electrocardiographic changes: one developed an intraventricular conduction defect and ventricular extra beats, while the other had sinus arrhythmia, left axis deviation, and increased QRS voltage, possibly secondary to left ventricular enlargement. Neither reported cardiovascular adverse events. Two patients taking placebo also had electrocardiographic abnormalities: one with bundle branch block, the other with sinus bradycardia and ventricular extra beats.

In a double-blind, randomized, placebo-controlled trial, 69 patients with multiple sclerosis and cognitive impairment were treated with donepezil (10 mg/day) or placebo for 24 weeks (47). Of those treated with donepezil, 65% had significant improvement on a test of verbal learning and memory, compared with 50% of those given placebo. The patients and clinicians judged that there was significantly greater memory improvement in those who took donepezil. Unusual dreams occurred more frequently with donepezil (34%) than placebo (8.8%). Although these results are encouraging, the therapeutic efficacy of donepezil was not particularly impressive. The study did not include assessment by care givers, who would have provided more reliable information on patients' cognitive status. Moreover, since the evaluating physician was the treating physician, bias may have been introduced. The sample size was small and randomization resulted in unequal representation of patients for disease course and disability. The reported adverse effects, such as abnormal dreams (34%), diarrhea (26%), nausea (26%), spasticity (17%), and numbness (17%), were more common than in trials in Alzheimer's disease. In patients with multiple sclerosis and cognitive impairment there is little justification for off-label use of donepezil since the benefit to harm balance is unfavorable (48).

Systematic reviews

A meta-analysis of various drugs approved for treating Alzheimer's disease in the USA and Canada has suggested that donepezil can delay cognitive impairment and deterioration in global health for at least 6 months in patients with mild-to-moderate Alzheimer's disease (49). Patients taking active treatment will have more favorable Alzheimer's Disease Assessment Scale cognitive subscale (ADAS-cog) scores for at least 6 months, after which their scores will begin to converge with those who are taking placebo. The cost-effectiveness data were inconclusive.

A meta-analysis of randomized, double-blind, placebo-controlled, multicenter, multinational trials of donepezil (5 and 10 mg/day) in patients with probable Alzheimer's disease (n = 2376) or probable/ possible vascular dementia (n = 1219) according to NINCDS-ADRDA criteria and NINDS-AIREN criteria respectively (10 studies lasting 12–24 weeks) has been published (50). In both conditions the percentage of withdrawals from the donepezil 10 mg/day group was higher than from the donepezil 5 mg/day and placebo groups. Withdrawals because of adverse events were higher in all those with vascular dementia than in those with Alzheimer's disease, perhaps because of inappropriate polypharmacy (patients with vascular dementia took on average eight other medications). Cardiovascular events were more common in those with vascular dementia but were not increased by donepezil. In both Alzheimer's disease and vascular dementia, donepezil produced significant benefits compared with placebo on measures of cognition and global function. Placebo-treated patients with Alzheimer's disease had reduced cognition and global function, whereas placebo-treated patients with vascular dementia remained stable,

suggesting that the effects of donepezil in vascular dementia were driven by improvement rather than stabilization or reduced decline.

Organs and Systems

Cardiovascular

Symptomatic sinus bradycardia is a possible adverse effect of treatment with donepezil in Alzheimer's disease (51).

- An 84-year-old patient with hypertensive cardiomyopathy developed bradycardia, fainting, and left-sided heart failure 3 weeks after starting treatment with donepezil. When donepezil was withdrawn, the sinus bradycardia disappeared; 24-hour electrocardiography showed no signs of sinus node disease, and no episodes of this type recurred during the next 6 months.

It is important to emphasize that disorders of cardiac rhythm associated with the use of donepezil are extremely unusual.

Soon after the start of donepezil treatment three patients with Alzheimer's disease developed cardiac syncope (52). In two cases, a bradydysrhythmia was documented and pacemaker implantation was considered justified rather than donepezil withdrawal.

Exaggeration of hypotension during donepezil treatment, due to interference with autonomic control, has been described (53).

The causes of syncope in patients with Alzheimer's disease treated with donepezil have been reported in 16 consecutive patients (12 women, 4 men) with Alzheimer's disease, mean age 80 years, who underwent staged evaluation, ranging from physical examination to electrophysiological testing (54). The mean dose of donepezil was 7.8 mg/day and the mean duration of donepezil treatment at the time of syncope was 12 months. Among the causes of syncope, carotid sinus syndrome (n = 3), complete atrioventricular block (n = 2), sinus node dysfunction (n = 2), and paroxysmal atrial fibrillation (n = 1) were diagnosed. No cause of syncope was found in six patients. Non-invasive evaluation is recommended before withdrawing cholinesterase inhibitors in patients with Alzheimer's disease and unexplained syncope.

Nervous system

Convulsions have been reported during treatment with donepezil (55).

- A patient with mild Alzheimer's disease taking donepezil, 5 mg/day for 2 weeks and then 10 mg/day for 23 days, was admitted with convulsions. His only other medication was aspirin 100 mg/day. Blood analysis was normal, and a computerized tomographic (CT) scan showed a mild degree of cortical atrophy with no structural lesions. Donepezil was withdrawn, and no other drug treatment was given. Six weeks later, donepezil 5 mg/day was restarted. On day 52, he developed loss of consciousness and convulsions, necessitating withdrawal of donepezil.

Convulsions in Alzheimer's disease are very rare until late in the illness, and the authors attributed this patient's convulsions to donepezil.

Restless legs, mumbling, and stuttering have been reported in a patient taking donepezil (56). According to the Naranjo probability scale, the causality was probable, since rechallenge was positive.

Extrapyramidal effects have been reported in three patients taking donepezil; in two cases, the effects disappeared when donepezil was withdrawn (57).

Increased rates of hypnotic drug use among patients taking donepezil has been linked to sleep problems (58).

Kinematic analysis of handwriting movements in patients with Alzheimer's disease taking donepezil did not show deterioration (59,60). Indeed there was a non-significant trend towards smoother movement in those taking donepezil. This is consistent with the observation of improved handwriting in a subject with dementia with Lewy bodies during donepezil treatment (61).

Psychological, psychiatric

Behavioral worsening in seven patients with Alzheimer's disease after the start of donepezil therapy has been described (62). Their mean age was 76 years, and their mean score on the Mini-Mental State Examination was 18. Five patients had had dementia-related delusions and irritability before taking donepezil, one had had a history of major depression, and another had had a history of somatization disorder. At the start of treatment with donepezil, four were taking sertraline, one paroxetine, one venlafaxine, and four risperidone. All took donepezil 5 mg/day, and after 4–6 weeks the dosage in five patients was increased to 10 mg/day. In the other two cases, donepezil was discontinued after 5 weeks: in one case because of gastrointestinal symptoms and in the other because of increasing agitation. After an average of 7.3 (range 1–13) weeks after starting donepezil, all seven patients had a recurrence of previous behavioral problems. Five became agitated, one became depressed, and the other became more anxious and somatically preoccupied. The pattern of behavioral change involves regression to an earlier behavioral problem.

Violent behavior has been described with donepezil (63).

- A 76-year-old man who was taking oxybutynin 3 mg tds for bladder instability took donepezil 5 mg/day for presumed Alzheimer's disease and 5 days later became very paranoid, believing that his wife had been stealing his money. He beat her and held her hostage in their house with a knife until their daughter intervened. He was given haloperidol 0.5 mg bd, and donepezil and oxybutynin were withdrawn. His paranoid ideation resolved within a few days and did not recur despite withdrawal of haloperidol.

Although a causal relation between this violent incident and donepezil cannot be proved, the temporal relation was suggestive.

Endocrine

In healthy men aged 61–70 years, donepezil 5 mg/day (n = 12) or placebo (n = 12) for 4 weeks, followed by donepezil 10 mg/day for another 4 weeks reversed age-related down-regulation of the growth hormone/insulin-like growth factor-1 (IGF-1) axis (64). In view of this, it would be important to investigate whether donepezil or other cholinesterase inhibitors, such as rivastigmine or galantamine, can restore the senile decline of growth hormone secretion in the long term, and to evaluate the benefit to harm balance as an intervention for the somatopause.

Liver

Donepezil has not been associated with hepatotoxic effects, which is a distinct advantage over tacrine. Donepezil treatment benefit persisted over 98 weeks, with no evidence of hepatotoxicity (65).

Urinary tract

Urinary incontinence may often be disregarded as a manifestation of dementia, but it has also been attributed to donepezil.

Of 94 patients with mild-to-moderate disease treated with recommended dosages of donepezil (3 mg/day during the first week and then 5 mg/day), 7 developed urinary incontinence (66). In five of them, the incontinence was transient, and there was no need to change the prescription. Incontinence occurred in both sexes, in relatively young and old patients, and in those with very mild-to-moderate dementia. In six patients the incontinence occurred at the higher dosage of 5 mg/day, in one patient it disappeared when donepezil was withdrawn, and in another an increase in dosage caused the reappearance of incontinence, suggesting a likely causal, dose-dependent relation between donepezil and urinary incontinence.

Urinary incontinence in patients with Down syndrome treated with donepezil has been described before (67). Urinary incontinence can be a major concern and a source of distress, not only for patients but also for caregivers. Clinicians should be alert to the possibility of urinary incontinence when prescribing donepezil for individuals with Alzheimer's disease. The authors emphasized that the incontinence may often be transient and not serious, but it could limit a patient's activities and quality of life and could also affect therapeutic adherence.

Clinicians often encounter patients with dementia and urge incontinence who might benefit from both anticholinergic medication and a cholinesterase inhibitor (68), a paradoxical combination.

- A 76-year-old woman with Alzheimer's disease and urge incontinence had been taking donepezil 10 mg at bedtime and tolterodine 2 mg in the morning and 4 mg at night. Her dementia deteriorated, and there was concern that tolterodine may have contributed. However, multiple episodes of nocturia, poor sleep, and worsening of behavior followed tolterodine dosage reduction to 2 mg bd. These symptoms were relieved when the prior tolterodine dosage was resumed, and her agitation resolved.

This pharmacodynamic interaction suggests that donepezil should be avoided in patients with Alzheimer's disease who are taking anticholinergic drugs.

Skin

A purpuric rash associated with donepezil has been reported (69).

- An 82-year-old woman with hypertension, taking long-term atenolol and doxazosin, developed moderate cognitive impairment attributed to Alzheimer's disease. She was given donepezil 5 mg/day and, after 4 days, developed diarrhea, vomiting, and a purpuric rash on her trunk, arms, and legs. Platelet counts were 119–157 × 10^9/l. Donepezil was withdrawn, with resolution of the gastrointestinal symptoms.

Donepezil was the probable cause of this rash, because of the temporal association with treatment and its recurrence on rechallenge.

Long-Term Effects

Drug withdrawal

In two patients taking donepezil, withdrawal symptoms developed (70). The symptoms, severe agitation, difficulty in concentrating and sleeping, and rapid mood changes, typically developed 5–6 days after withdrawal of donepezil and disappeared by around days 9–10.

Susceptibility Factors

Down's syndrome

Down's syndrome was associated with higher plasma donepezil concentrations than in healthy volunteers, and patients with higher concentrations developed adverse reactions more often (71). In 14 patients (9 men) aged 15–37 years and six healthy controls aged 21–27 years, the mean plasma donepezil concentrations were 18 and 28 mg/ml at doses of 3 and 5 mg/day respectively in those with Down's syndrome and 7.8 and 18 mg/ml respectively in healthy volunteers taking 2 and 5 mg/day. Although slightly different dosages were used in the two groups, making a comparison of low dosages inappropriate, nevertheless at a dose of 5 mg/day there was a clear difference in plasma concentrations. Considering the potential use of donepezil in peoples with Down's syndrome (72,73) and the serious adverse effects of donepezil in these patients (74), this difference in pharmacokinetics of donepezil is important. The authors proposed that the usual maintenance dose of 10 mg/day in the USA and EU is probably too much for patients with Down's syndrome; they recommended 3–5 mg/day instead.

Three patients with Alzheimer's disease associated with Down syndrome were treated with donepezil (67). One became agitated and aggressive; the other two developed urinary incontinence. In all three cases donepezil was withdrawn. These results are important, because many individuals with Down syndrome

develop clinical and neuropathological evidence of Alzheimer's disease after the age of 40 years. Also, patients with Down syndrome were excluded from donepezil clinical trials. Therefore, lesser data on the efficacy or safety of donepezil are available for this population.

Drug Administration

Drug overdose

Two cases of donepezil overdose have been reported (75,76).

- A 79-year-old nursing home patient was given donepezil 50 mg in error. She developed nausea, vomiting, and persistent bradycardia—typical cholinergic adverse effects. She was treated with atropine, 0.2 mg as needed, for bradycardia (total dose 3 mg over 18 hours) and was discharged on the second day.
- A 74-year-old woman with a history of stroke, myocardial infarction, hypothyroidism, and probable multi-infarct dementia took nine donepezil tablets (a total dose of 45 mg). She developed nausea and vomiting 2 hours later. She fell asleep for 4–5 hours but remained rousable. About 9 hours after ingestion, she became flushed and had a bout of diarrhea. Donepezil was withdrawn for 3 days, and there were no adverse effects when it was reintroduced.

Drug–Drug Interactions

Ginkgo biloba

Gingko supplementation had no significant effect on the pharmacokinetics and pharmacodynamics of donepezil in 14 patients with Alzheimer's disease (77).

Maprotiline

Neuroleptic malignant syndrome has been attributed to a combination of donepezil plus maprotiline.

- Concomitant treatment with donepezil and maprotiline in a 73-year-old patient with Alzheimer's disease and stroke produced a syndrome resembling the neuroleptic malignant syndrome (78). The patient responded to withdrawal of maprotiline and donepezil and intravenous fluids, and did not require dantrolene.

Since the patient had been taking maprotiline for almost 2 months before this episode, the authors suggested that both these drugs may cause an imbalance in acetylcholine/dopamine in the striatum.

Memantine

Pharmacokinetic and pharmacodynamic data from an open study in 24 healthy subjects suggested that there is no interaction between memantine (an N-methyl-D-aspartate receptor antagonist approved for treatment of

Alzheimer's disease) and donepezil (79). These findings support the potential for combining memantine and cholinesterase inhibitors in patients with Alzheimer's disease.

Neostigmine

An additive inhibitory effect of donepezil and neostigmine on acetylcholinesterase has been proposed to explain prolonged neuromuscular blockade during anesthesia in an 85-year-old woman taking donepezil (80).

Neuromuscular blocking drugs

The potential for interactions between donepezil and neuromuscular blocking agents has major implications for the anesthetic care of people taking donepezil (81,82,83). Prolonged paralysis resulting from an interaction of donepezil with suxamethonium has been reported (84,85).

Paroxetine

A possible interaction between donepezil and paroxetine has been described (86).

- Two elderly patients with Alzheimer's disease and a mood disorder were treated with donepezil 5 mg/day and paroxetine 20 mg/day. One of them became agitated, confused, and aggressive, and donepezil was withdrawn after 8 days. On reintroduction of donepezil she again became rapidly confused, irritable, and verbally aggressive. In the other case, while the patient was taking paroxetine, donepezil 5 mg/day resulted in severe diarrhea, flatulence, and insomnia. The dosage of donepezil was reduced to 5 mg on alternate days, but the diarrhea and flatulence persisted. The symptoms resolved when donepezil was stopped.

Donepezil is metabolized in the liver by CYP2D6 and CYP3A4. Selective serotonin re-uptake inhibitors (SSRIs), such as paroxetine, are potent inhibitors of CYP2D6. Mood disorders are common in patients with Alzheimer's disease, and SSRIs are used in these patients and can increase the plasma concentration of donepezil, increasing the risk of severe adverse reactions.

Risperidone

Extrapyramidal effects occurred in a patient who took donepezil and risperidone concurrently (87). Although risperidone is less likely than conventional antipsychotic drugs to cause extrapyramidal effects, and is therefore particularly useful for older patients who are very susceptible to developing extrapyramidal disturbances, an increase in brain acetylcholine resulting from donepezil, along with dopamine receptor blockade by risperidone, would have led to an imbalance between cholinergic and dopaminergic systems. Although a clinically significant interaction between donepezil and risperidone

seems to be rare, clinicians should be alert to such a possibility.

An open comparison of donepezil and risperidone, alone or in combination, in 24 healthy men showed no significant pharmacokinetic differences (88). Adverse events such as headache, nervousness, and somnolence were minor and comparable in all groups. These results suggest that no clinically significant interactions occur between risperidone and donepezil at steady state. However, whether these conclusions can be extrapolated to the elderly patients with dementia, who may eliminate both donepezil and risperidone slowly, is uncertain.

Tiapride

Parkinsonism has been reported in a patient concurrently taking donepezil and tiapride, probably through a pharmacodynamic interaction (89).

Monitoring therapy

P_{300}, one of the cognitive event-related potentials of the cerebral cortex may serve as a marker for measuring the course of Alzheimer's disease during treatment with donepezil. There was reduced P_{300} latency associated with parallel improvement of ADAS-J Cog scores after administration of donepezil 5 mg/kg in 13 patients with Alzheimer's disease (8 women and 5 men, aged 70–88 year) (90).

References

1. Whitehouse PJ. Donepezil. Drugs Today (Barc) 1998;34(4):321–6.
2. Barner EL, Gray SL. Donepezil use in Alzheimer disease. Ann Pharmacother 1998;32(1):70–7.
3. Peruche B, Schulz M. Donepezil—a new agent against Alzheimer's disease. Pharm Ztg 1998;143:38–42.
4. Rogers SL, Friedhoff LT, Apter JT, Richter RW, Hartford JT, Walshe TM, Baumel B, Linden RD, Kinney FC, Doody RS, Borison RL, Ahem GL. The efficacy and safety of donepezil in patients with Alzheimer's disease: results of a US multicentre, randomized, double-blind, placebo-controlled trial. The Donepezil Study Group. Dementia 1996;7(6):293–303.
5. Rogers SL, Farlow MR, Doody RS, Mohs R, Friedhoff LT. A 24-week, double-blind, placebo-controlled trial of donepezil in patients with Alzheimer's disease. Donepezil Study Group. Neurology 1998;50(1):136–45.
6. Warner JP. Commentary on donepezil. Evid Based Med 1998;3:155.
7. Melzer D. New drug treatment for Alzheimer's disease: lessons for healthcare policy. BMJ 1998;316(7133):762–4.
8. Dening T, Lawton C. New drug treatment for Alzheimer's disease. Doctors want to offer more than sympathy. BMJ 1998;317(7163):945.
9. Levy R. New drug treatment for Alzheimer's disease. Effects of drugs can be variable. BMJ 1998;317(7163):945.
10. Evans M. New drug treatment for Alzheimer's disease. Drugs should not need to show cost effectiveness to justify their prescription. BMJ 1998;317(7163):945–6.

11. Johnstone P. New drug treatment for Alzheimer's disease. Information from unpublished trials should be made available. BMJ 1998;317(7163):946.

12. Zamar AC, Wise ME, Watson JP. New drug treatment for Alzheimer's disease. Treatment with metrifonate warrants multicentre trials. BMJ 1998;317(7163):946.

13. Baxter T, Black D, Prempeh H. New drug treatment for Alzheimer's disease. SMAC's advice on use of donepezil is contradictory. BMJ 1998;317(7163):946.

14. Doody RS. Clinical profile of donepezil in the treatment of Alzheimer's disease. Gerontology 1999;45(Suppl 1):23–32.

15. Dooley M, Lamb HM. Donepezil: a review of its use in Alzheimer's disease. Drugs Aging 2000;16(3):199–226.

16. Cameron I, Curran S, Newton P, Petty D, Wattis J. Use of donepezil for the treatment of mild–moderate Alzheimer's disease: an audit of the assessment and treatment of patients in routine clinical practice. Int J Geriatr Psychiatry 2000;15(10):887–91.

17. Greenberg SM, Tennis MK, Brown LB, Gomez-Isla T, Hayden DL, Schoenfeld DA, Walsh KL, Corwin C, Daffner KR, Friedman P, Meadows ME, Sperling RA, Growdon JH. Donepezil therapy in clinical practice: a randomized crossover study. Arch Neurol 2000;57(1):94–9.

18. Homma A, Takeda M, Imai Y, Udaka F, Hasegawa K, Kameyama M, Nishimura T. Clinical efficacy and safety of donepezil on cognitive and global function in patients with Alzheimer's disease. A 24-week, multicenter, double-blind, placebo-controlled study in Japan. E2020 Study Group. Dement Geriatr Cogn Disord 2000;11(6):299–313.

19. Knopman DS. Management of cognition and function: new results from the clinical trials programme of Aricept® (donepezil HCl). Int J Neuropsychopharmacol 2000;3(7):13–20.

20. Matthews HP, Korbey J, Wilkinson DG, Rowden J. Donepezil in Alzheimer's disease: eighteen month results from Southampton Memory Clinic. Int J Geriatr Psychiatry 2000;15(8):713–20.

21. Evans M, Ellis A, Watson D, Chowdhury T. Sustained cognitive improvement following treatment of Alzheimer's disease with donepezil. Int J Geriatr Psychiatry 2000;15(1):50–3.

22. Rogers SL, Doody RS, Pratt RD, Ieni JR. Long-term efficacy and safety of donepezil in the treatment of Alzheimer's disease: final analysis of a US multicentre open-label study. Eur Neuropsychopharmacol 2000;10(3):195–203.

23. Dunn NR, Pearce GL, Shakir SA. Adverse effects associated with the use of donepezil in general practice in England. J Psychopharmacol 2000;14(4):406–8.

24. Lanctot KL, Herrmann N. Donepezil for behavioural disorders associated with Lewy bodies: a case series. Int J Geriatr Psychiatry 2000;15(4):338–45.

25. Dallocchio C, Buffa C, Mazzarello P. Combination of donepezil and gabapentin for behavioral disorders in Alzheimer's disease. J Clin Psychiatry 2000;61(1):64.

26. Burns A, Rossor M, Hecker J, Gauthier S, Petit H, Moller HJ, Rogers SL, Friedhoff LT. The effects of donepezil in Alzheimer's disease—results from a multinational trial. Dement Geriatr Cogn Disord 1999;10(3):237–44.

27. Mohs RC, Doody RS, Morris JC, Ieni JR, Rogers SL, Perdomo CA, Pratt RD; "312" Study Group. A 1-year, placebo-controlled preservation of function survival study of donepezil in AD patients. Neurology 2001;57(3):481–8.

28. Winblad B, Engedal K, Soininen H, Verhey F, Waldemar G, Wimo A, Wetterholm AL, Zhang R, Haglund A, Subbiah P; Donepezil Nordic Study Group. A 1-year, randomized, placebo-controlled study of donepezil in patients with mild to moderate AD. Neurology 2001;57(3):489–95.

29. Doody RS, Geldmacher DS, Gordon B, Perdomo CA, Pratt RD; Donepezil Nordic Study Group. Open-label, multicenter, phase 3 extension study of the safety and efficacy of donepezil in patients with Alzheimer disease. Arch Neurol 2001;58(3):427–33.

30. Bergman J, Brettholz I, Shneidman M, Lerner V. Donepezil as add-on treatment of psychotic symptoms in patients with dementia of the Alzheimer's type. Clin Neuropharmacol 2003;26:88–92.

31. Anonymous. AD 2000 Collaborative Group. Long-term donepezil treatment in 565 patients with Alzheimer's disease (AD 2000): randomized double blind trial. Lancet 2004;363:2105–15.

32. Holmes C, Burns A, Passmore P, Forsyth D, Wilkinson D. AD 2000: design and conclusions. Lancet 2004;364:1213–14.

33. Akinatade L, Zaiac M, Leni JR, McRae T. AD 2000: design and conclusions. Lancet 2004;364:1214.

34. Howe I. AD 2000: design and conclusions. Lancet 2004;364:1215.

35. Clarke N. AD 2000: design and conclusions. Lancet 2004;364:1216.

36. Gray R, Bentham P, Hills R, on behalf of the Collaborative Group. AD 2000: design and conclusions. Authors' reply. Lancet 2004;364:1216.

37. Erkinjuntti T, Roman G, Gauthier S. Treatment of vascular dementia – evidence from clinical trials with cholinesterase inhibitors. J Neurol Sci 2004;226:63–6.

38. Black S, Roman GC, Geldmacher DS, Salloway S, Hecker J, Burns A, Perdomo C, Kumar D, Pratt R; Donepezil 307 Vascular Dementia Study Group. Efficacy and tolerability of donepezil in vascular dementia: positive results of a 24-week, multicenter, international, randomized, placebo-controlled clinical trial. Stroke 2003;34(10):2323–30.

39. Kaufer DI. Pharmacologic treatment expectations in the management of dementia with Lewy bodies. Dementia Geriatr Cogn Dis 2004;17 Suppl 1:32–9.

40. Finkel SI, Mintzer JE, Dysken M, Krishnan KRR, Burt T, McRae T. A randomized placebo-controlled study of the efficacy and safety of sertraline in the treatment of the behavioral manifestations of Alzheimer's disease in outpatients treated with donepezil. Int J Geriatr Psychiatry 2004;19:9–18.

41. Cochrane M, Cochrane A, Jauhar P, Ashton E. Acetylcholinesterase inhibitors for the treatment of Wernicke-Korsakoff syndrome – three further cases show response to donapezil. Alcohol Alcohol 2005;40:151–4.

42. Angunawela II, Barker A. Anticholinesterase drugs for alcoholic Korsakoff syndrome. Int J Geriatr Psychiatry 2001;16:338–9.

43. Casadevall CT, Pascual MLF, Fernandez TT, Escalza CI, Navas VI, Fanlo MC, Morales AF. Pharmacological treatment of Korsakoff psychosis: a review of the literature and experience in two cases. Rev Neurol 2002;35:341–5.

44. Sahin HA, Gurvit IH, Bilgic B, Hanagasi HA, Emre M. Therapeutic effects of an acetylcholinesterase inhibitor (donepezil) on memory in Wernicke–Korsakoff's disease. Clin Neuropharmacol 2002;25:16–20.

45. Peterson RC, Thomas RG, Grundman M, Bennett D, Doody R, Ferris S, Galasko D, Jis S, Kaye J, Levey A, Pfeiffer E, Sano M, Van Dyck CH, Thal LJ, for the Alzheimer's Disease Cooperative Study Group. Vitamin E

and donepezil for the treatment of mild cognitive impairment. N Engl J Med 2005;352:2379–88.

46 Rogers SL, Doody RS, Mohs RC, Friedhoff LT. Donepezil improves cognition and global function in Alzheimer disease: a 15-week, double-blind, placebo-controlled study. Donepezil Study Group. Arch Intern Med 1998;158(9):1021–31.

47. Krup LB, Christodoulou C, Melville RN, Scherl WF, McAllister WS, Elkin LE. Donepezil improved memory in multiple sclerosis in a randomized clinical trial. Neurology 2004;63:1579–85.

48. Amato MP. Donepezil for memory impairment in multiple sclerosiss. Lancet Neurol 2005;4:72–3.

49. Wolfson C, Oremus M, Shukla V, Momoli F, Demers L, Perrault A, Moride Y. Donepezil and rivastigmine in the treatment of Alzheimer's disease: a best-evidence synthesis of the published data on their efficacy and cost-effectiveness. Clin Ther 2002;24(6):862–86.

50. Passmore AP, Bayer AJ, Steinhagen-Thiessen E. Cognitive, global, and functional benefits of donepezil in Alzheimer's disease and vascular dementia: results from large-scale clinical trials. J Neurol Sci 2005;229-230:1241–6.

51. Calvo-Romero JM, Ramos-Salado JL. Bradycardia sinusal sintomatica associada a donepecilo. [Symptomatic sinus bradycardia associated with donepezil.] Rev Neurol 1999;28(11):1070–2.

52. Bordier P, Garrigue S, Barold SS, Bressolles N, Lanusse S, Clementy J. Significance of syncope in patients with Alzheimer's disease treated with cholinesterase inhibitors. Europace 2003;5:429–31.

53. McLaren AT, Allen J, Murray A, Ballard CG, Kenny RA. Cardiovascular effects of donepezil in patients with dementia. Dement Geriatr Cogn Disord 2003;15:183–8.

54. Bordier P, Lanusse S, Garrigue S, Reynard C, Robert F, Gencel L, Lafitte A. Causes of syncope in patients with Alzheimer's disease treated with donepezil. Drugs Aging 2005;22:687–94.

55. Babic T, Zurak N. Convulsions induced by donepezil. J Neurol Neurosurg Psychiatry 1999;66(3):410.

56. Amouyal-Barkate K, Bagheri-Charabiani H, Montastruc JL, Moulias S, Vellas B. Abnormal movements with donepezil in Alzheimer disease. Ann Pharmacother 2000;34(11):1347.

57. Carcenac D, Martin-Hunyadi C, Kiesmann M, Demuynck-Roegel C, Alt M, Kuntzmann F. Syndrome extrapyramidal sous donepezil. [Extra-pyramidal syndrome induced by donepezil.] Presse Méd 2000;29(18):992–3.

58. Stahl SM, Markowitz JS, Gutterman EM, Papadopoulos G. Co-use of donepezil and hypnotics among Alzheimer's disease patients living in the community. J Clin Psychiatry 2003;64:466–72.

59. Hegerl U, Mergl R, Henkel V, Gallinat J, Koffer G, Muller-Siecheneder F, Pogarell O, Juckel G, Schroter A, Bahra R, Emir B, Laux G, Moller HJ. Kinematic analysis of the effects of donepezil hydrochloride on hand motor function in patients with Alzheimer's dementia. J Clin Psychopharmacol 2003;23:214–6.

60. Bohnen N, Kaufer D, Hendrickson R, Ivanco L, Moore R, Dekosky ST. Effects of donepezil on motor function in patients with Alzheimer's disease. J Clin Psychopharmacol 2004;24:354–6.

61. Kaufer DI, Catt KE, Lopez OL, DeKosky ST. Dementia with Lewy bodies: response to delirium-like features to donepezil. Neurology 1998;51:1512.

62. Wengel SP, Roccaforte WH, Burke WJ, Bayer BL, McNeilly DP, Knop D. Behavioral complications associated with donepezil. Am J Psychiatry 1998;155(11):1632–3.

63. Bouman WP, Pinner G. Violent behavior associated with donepezil. Am J Psychiatry 1998;155(11):1626–7.

64. Obermayr RP, Mayerhofer L, Knechtelsorfer M, Mersich N, Huber ER, Geyer G, Trgl K-H. The age-related down-regulation of the growth hormone/insulin-like growth factor-1 axis in the elderly male is reversed considerably by donepezil, a drug for Alzheimer's disease. Exp Gerontol 2005;40:157–63.

65. Rogers SL, Friedhoff LT. Long-term efficacy and safety of donepezil in the treatment of Alzheimer's disease: an interim analysis of the results of a US multicentre open label extension study. Eur Neuropsychopharmacol 1998;8(1):67–75.

66. Hashimoto M, Imamura T, Tanimukai S, Kazui H, Mori E. Urinary incontinence: an unrecognised adverse effect with donepezil. Lancet 2000;356(9229):568.

67. Hemingway-Eltomey JM, Lerner AJ. Adverse effects of donepezil in treating Alzheimer's disease associated with Down's syndrome. Am J Psychiatry 1999;156(9):1470.

68. Siegler EL, Reidenberg M. Treatment of urinary incontinence with anti-cholinergics in patients taking cholinesterase inhibitors for dementia. Clin Pharmacol Ther 2004;75:484–8.

69. Bryant CA, Ouldred E, Jackson SH, Kinirons MT. Purpuric rash with donepezil treatment. BMJ 1998;317(7161):787.

70. Singh S, Dudley C. Discontinuation syndrome following donepezil cessation. Intl J Geriat Psychiatry 2003;18:282–4.

71. Kondoh T, Nakashima M. Pharmacokinetics of donepezil in Down's syndrome. Ann Pharmacother 2005;39:572.

72. Krishnani PS, Sullivan JA, Walter BK, Spiridiqliozzi GA, Doraiswami PM, Krishnan KR. Cholinergic therapy for Down's syndrome. Lancet 1999;353:1064–5.

73. Cipriani G, Bianchetti A, Trabucchi M. Donepezil use in the treatment of dementia associated with Down's syndrome. Arch Neurol 2003;60:292.

74. Hemigway-Eltomey JM, Lerner AJ. Adverse effects of donepezil in treating Alzheimer's disease associated with Down's syndrome. Am J Psychiatry 1999;156:1470.

75. Shepherd G, Klein-Schwartz W, Edwards R. Donepezil overdose: a tenfold dosing error. Ann Pharmacother 1999;33(7–8):812–15.

76. Greene YM, Noviasky J, Tariot PN. Donepezil overdose. J Clin Psychiatry 1999;60(1):56–7.

77. Yasui-Furukori N, Furukori H, Kaneda A, Kaneko S, Tateishi T. The effects of Ginkgo biloba extracts on the pharmacokinetics and pharmacodynamics of donepezil. J Clin Pharmacol 2004;44:538–42.

78. Ohkoshi N, Satoh D, Nishi M, Shoji S. Neuroleptic malignant-like syndrome due to donepezil and maprotiline. Neurology 2003;60:1051.

79. Periclou AP, Ventura D, Sherman T, Rao N, Abramowitz WT. Lack of pharmacokinetic or pharmacodynamic interaction between memantine and donepezil. Ann Pharmacother 2004;38(9):1389–94.

80. Sprung J, Castellani WJ, Srinivasan V, Udayashankar S. The effects of donepezil and neostigmine in a patient with unusual pseudocholinesterase activity. Anesth Analg 1998;87(5):1203–5.

81. Heath ML. Donepezil, Alzheimer's disease and suxamethonium. Anaesthesia 1997;52:1018.

82. Heath ML. Donepezil and succinylcholine. Anaesthesia 2003;58:202.

83. Walker C, Perks D. Do you know about donepezil and succinylcholine? Anaesthesia 2002;57:1041.

84. Crowe S, Collins L. Suxamethonium and donepezil: a cause of prolonged paralysis. Anaesthesiology 2003;98:574–5.

85. Morillo S, Ferrari DA, Lopez RTA. Interaction of donepezil and muscular blockers in Alzheimer's disease. Rev Esp Anestesiol Reanim 2003;50:97–100.

86. Carrier L. Donepezil and paroxetine: possible drug interaction. J Am Geriatr Soc 1999;47(8):1037.

87. Magnuson TM, Keller BK, Burke WJ. Extrapyramidal side effects in a patient treated with risperidone plus donepezil. Am J Psychiatry 1998;155(10):1458–9.

88. Zao Q, Xie C, Pesco-Koplowitz L, Jia X, Parier J-L. Pharmacokinetic and safety assessment of concurrent administration of risperidone and donepezil. J Clin Pharmacol 2003;43:180–6.

89. Arai M. Parkinsonism onset in a patient concurrently using tiapride and donepezil. Intern Med 2000;39(10):863.

90. Katsada E, Sato K, Sawaki A, Dohi Y, Ueda R, Ojika K. Long term effects of donepezil on P300 auditory event-related potentials in patients with Alzheimer's disease. J Geriat Psychiatry Neurol 2003;16:39–43.

Memantine

Memantine is an amantadine derivative that is a non-competitive antagonist at the N-methyl-D-aspartate (NMDA) glutamate receptor. It significantly attenuates the progression of Alzheimer's disease from moderate to severe (1,2) and has an additive effect when co-administered with donepezil (3,4). It has also been studied in patients with Parkinson's disease, and considering that it can provide protection against the neurotoxic effect of the HIVgp120 protein, it is being evaluated for the treatment of HIV-associated dementia (5). Its adverse effects include agitation, restlessness, insomnia, pronounced delirious states, and muscular hypotonia. All were reversible after dosage reduction or withdrawal.

Systematic reviews

In a meta-analysis of trials of memantine in people with moderate to severe Alzheimer's disease, an analysis of the change from baseline at 28 weeks showed statistically significant results in favor of memantine 20 mg/day on cognition, activities of daily living, and in the global clinical impression of change measured by the CIBIC-Plus (6). There were no significant differences between memantine and placebo in the number of drop-outs and the total numbers of adverse effects, but a significant difference in favor of memantine in the number who had agitation.

Organs and Systems

Psychiatric

A toxic psychosis has been reported in two patients taking memantine (7).

References

1. Reisberg B, Doody R, Stoffler A, Schmitt F, Ferris S, Mobius HJ, Memantine Study Group. Memantine in moderate-to-severe Alzheimer's disease. N Engl J Med 2003;348:1333–41.

2. Bullock R. Efficacy and safety of memantine in moderate-to-severe Alzheimer disease: the evidence to date. Alzheimer Dis Assoc Disord 2006;20(1):23–9.

3. Tarriot PN, Farlow MR, Grossberg GT, Graham SM, McDonald S, Gergel I, Memantine Study Group. Memantine treatment in patients with moderate to severe Alzheimer's disease already receiving donepezil: a randomized controlled trial. JAMA 2004;291:317–24.

4. Rossom R, Adityanjee, Dysken M. Efficacy and tolerability of memantine in the treatment of dementia. Am J Geriatr Pharmacother 2004;2(4):303–12.

5. Alisky JM. Could cholinesterase inhibitors and memantine alleviate HIV dementia? J Acquir Immune Defic Syndr 2004;38:1–3.

6. Areosa SA, Sherriff F. Memantine for dementia. Cochrane Database Syst Rev 2003;(3):CD003154.

7. Riederer P, Lange KW, Kornhuber J, Danielczyk W. Pharmacotoxic psychosis after memantine in Parkinson's disease. Lancet 1991;338(8773):1022–3.

Metrifonate

General Information

Metrifonate has been used for the treatment of schistosomiasis for almost 40 years. Its identification as a cholinesterase inhibitor, together with recognition of the cholinergic deficit in Alzheimer's disease, has led to its use in Alzheimer's disease.

The pharmacology and pharmacokinetics of metrifonate and experience with its use in Alzheimer's disease have been reviewed (1).

Two reviews (2,3) of the use of metrifonate in Alzheimer's disease have been published. Both reported a positive effect of metrifonate, with generally mild and usually transient adverse effects, consisting of gastrointestinal symptoms (such as abdominal pain, diarrhea, flatulence, and nausea, probably reflecting cholinergic overactivation) and leg cramps, possibly caused by the overstimulation of nicotinic receptors at the neuromuscular junction. No laboratory abnormalities were reported.

In a randomized, double-blind, placebo-controlled trial in 408 patients with Alzheimer's disease, metrifonate (20 mg/kg/day for 2 weeks followed by 0.65 mg/kg/day for 24 weeks) significantly improved several mental performance scales (4). Of the 273 patients treated with metrifonate 12% discontinued treatment because of adverse effects, compared with 4% of the 134 patients treated with placebo. The adverse effects that led to withdrawal were mainly gastrointestinal in nature. Diarrhea occurred in 18% of the patients treated with metrifonate (leading to withdrawal in 3%) and in 8% of the patients treated with placebo; 2% of the patients treated with metrifonate discontinued treatment because of nausea and vomiting and 1% because of dyspepsia. Nausea occurred in 12% of the patients treated with metrifonate and in 10% of the patients treated with placebo. Vomiting occurred in 7% and 4% respectively.

In a second randomized, double-blind, placebo-controlled trial (5) 480 patients were randomized to receive placebo ($n = 120$), a low dose of metrifonate (0.5 mg/kg/day for 2 weeks followed by 0.2 mg/kg/day for 10 weeks) ($n = 121$), a moderate dose of metrifonate (0.9 mg/kg/day for 2 weeks followed by 0.3 mg/kg/day for 10 weeks) ($n = 121$), or a high dose (2.0 mg/kg/day for 2 weeks followed by 0.65 mg/kg/day for 10 weeks) ($n = 118$). These doses were selected to achieve steady-state erythrocyte acetylcholinesterase inhibition of 30%, 50%, and 70% respectively. There was a significant dose-related improvement in several mental performance scales with metrifonate. Most of the adverse events were mild and transient. Adverse events that occurred more often in the patients treated with metrifonate were abdominal pain (placebo 4%, low dose 3%, moderate dose 11%, high dose 12%), diarrhea (8%, 9%, 11%, 19%, respectively), flatulence (9%, 2%, 8%, 16%), and leg cramps (1%, 1%, 3%, 8%). Bradycardia, presumably related to the vagotonic effect of acetylcholinesterase inhibition, led to withdrawal of treatment in three patients with asymptomatic bradycardia. All three were in the loading-dose phase of the highest metrifonate dosage regimen.

In 16 patients with Alzheimer's disease, adverse effects were dose-related (6). In eight patients who took 2.5 mg/kg/day for 14 days, 4 mg/kg/day for 3 days, and then 2.0 mg/kg/day for 14 days (acetylcholinesterase inhibition 88–94%) treatment had to be withdrawn because of moderate to severe adverse effects in 6 patients on day 28 of the planned 31. In eight patients treated with 2.5 mg/kg/day for 14 days followed by 1.5 mg/kg/day for 35 days (acetylcholinesterase inhibition 89–91%), the frequency of more severe adverse effects was considerably lower despite similar acetylcholinesterase inhibition. The most frequent adverse events in the high-dose group were muscle cramps and abdominal discomfort during the loading-dose phase, followed by increasing gastrointestinal symptoms in the second loading phase, accompanied by headache and muscle aches. The adverse events profile initially improved during the maintenance phase. However, after 11 days of maintenance treatment, six patients complained of generalized moderate to severe muscle cramps, weakness, inability to resume daily activities, and difficulties with coordinations. The adverse events profile was much more favorable with the lower dose, with the same, but much less severe, range of adverse effects. Again, the most frequent adverse effects were gastrointestinal disturbances, muscle cramps, and light-headedness. One patient had increased sweating, dizziness, and palpitation on day 29 and another developed severe abdominal tenderness on day 31, which led to termination of treatment. There were no abnormalities in laboratory parameters. The authors proposed a maximum tolerated dose of 1.5 mg/kg/day of metrifonate for maintenance therapy in patients with Alzheimer's disease.

The Metrifonate Alzheimer's Trial (MALT) was designed to evaluate a wide range of symptoms and efficacy measures in four main clinical domains of Alzheimer's disease: cognition, psychiatric and behavioral features, activities of daily living, and global functioning (7). These are considered key targets for antidementia drugs. This prospective, multicenter, randomized, double-blind, parallel group study was conducted at 71 independent study centers, and 605 patients were randomized to placebo ($n = 208$), metrifonate 40 or 50 mg/day ($n = 200$), and metrifonate 60 or 80 mg/day ($n = 197$); within each treatment group the dose was determined according to body weight, and treatment lasted for 24 weeks. Metrifonate improved a wide range of symptoms across all four clinical domains of Alzheimer's disease in a dose-dependent manner, and was well tolerated at both doses. One patient withdrew from the study during the high-dose loading phase owing to cholinergic effects, resulting in muscle weakness. Metrifonate was also associated with a slight gradual lowering of hemoglobin, hematocrit, and erythrocyte count in both groups during the first 12 weeks; these were nonprogressive, stabilized over the 6-month treatment period, and hence were not considered to be clinically relevant. Metrifonate was associated with a higher incidence of bradycardia (under 50/minute); serious adverse events possibly related to bradycardia (hypotension, postural hypotension, syncope, dizziness, malaise, and accidental injury) occurred in three patients taking placebo, eight taking low-dose metrifonate (three of whom were withdrawn from the study), and two taking high-dose metrifonate.

In patients with mild to moderate Alzheimer's disease metrifonate significantly improved behavior as well as cognition, function in activities of daily living, and global functional status, as shown by a pooled analysis of four prospective, multicenter, randomized, double-blind, parallel-group, placebo-controlled trials, meeting FDA guidelines (8).

The safety and tolerability of once-daily oral metrifonate has been evaluated in patients with probable mild to moderate Alzheimer's disease in a randomized, double-blind, placebo-controlled, parallel-group study (9). Metrifonate was given to 29 patients as a loading dose (2.5 mg/kg) for 2 weeks, followed by maintenance dose (1 mg/kg) for 4 weeks; 10 patients received placebo. The proportion of patients who had at least one adverse event was comparable in the two groups: metrifonate 76%, placebo 80%. Selected adverse events, defined as those for which the incidence in the metrifonate and placebo group differed by at least 10%, were diarrhea, nausea, leg cramps, and accidental injury. The adverse events were predominantly mild and transient. Those who took metrifonate had a significantly lower heart rate. Metrifonate had no clinically important effect on laboratory tests, such as liver function tests, and did not affect exercise tolerance or pulmonary function.

Organs and Systems

Nervous system

Patients with Alzheimer's disease taking long-term treatment with metrifonate suffered seizures after abrupt withdrawal of antimuscarinic agents (10).

- A 58-year-old woman was given hyoscyamine for abdominal cramps by her local physician without the knowledge of her neurologists. She had a generalized seizure 36 hours after stopping the drug.
- A 66-year-old woman was treated for a skin allergy with doxepin cream in large doses. Withdrawal of this treatment led to two complex partial seizures.

The authors speculated that the antimuscarinic drugs impaired the cholinergic receptor down-regulation that would normally occur in the presence of the increased concentrations of acetylcholine caused by acetylcholinesterase inhibition. Withdrawal of the antagonist therefore abruptly exposed the receptors to high concentrations of the neurotransmitter, leading to seizures.

Second-Generation Effects

Teratogenicity

The birth of a hydrocephalic infant with a large meningomyelocele to a mother who had been treated with metrifonate for an infection with *Schistosoma hematobium* during the second month of pregnancy has been reported, but the report is not recent and the association was probably coincidental (SED-11, 598).

Drug Administration

Drug dosage regimens

The effects of a loading dose on the adverse events of metrifonate 50 mg/day have been studied in Alzheimer's disease. The regimen without a loading dose was better tolerated during the 4-week study period (11).

Drug–Drug Interactions

Organophosphorus insecticides

Theoretically, suxamethonium will be potentiated after administration of metrifonate and, in view of its ability to suppress enzyme activity, metrifonate should be used with great caution in areas where organophosphorus insecticides are used, since they have a similar effect.

References

1. Cummings JL, Ringman JM. Metrifonate (Trichlorfon): a review of the pharmacology, pharmacokinetics and clinical experience with a new acetylcholinesterase inhibitor for Alzheimer's disease. Expert Opin Investig Drugs 1999;8(4):463–71.
2. Cummings JL. Metrifonate: overview of safety and efficacy. Pharmacotherapy 1998;18(2 Part 2):43–6.
3. Mucke HAM. Metrifonate: treatment of Alzheimer's disease, acetylcholinesterase-inhibitor. Drugs Future 1998;23:491–7.
4. Morris JC, Cyrus PA, Orazem J, Mas J, Bieber F, Ruzicka BB, Gulanski B. Metrifonate benefits cognitive, behavioral, and global function in patients with Alzheimer's disease. Neurology 1998;50(5):1222–30.
5. Cummings JL, Cyrus PA, Bieber F, Mas J, Orazem J, Gulanski B. Metrifonate treatment of the cognitive deficits of Alzheimer's disease. Metrifonate Study Group. Neurology 1998;50(5):1214–21.
6. Cutler NR, Jhee SS, Cyrus P, Bieber F, TanPiengco P, Sramek JJ, Gulanski B. Safety and tolerability of metrifonate in patients with Alzheimer's disease: results of a maximum tolerated dose study. Life Sci 1998;62(16):1433–41.
7. Dubois B, McKeith I, Orgogozo JM, Collins O, Meulien D. A multicentre, randomized, double-blind, placebo-controlled study to evaluate the efficacy, tolerability and safety of two doses of metrifonate in patients with mild-to-moderate Alzheimer's disease: the MALT study. Int J Geriatr Psychiatry 1999;14(11):973–82.
8. Farlow MR, Cyrus PA. Metrifonate therapy in Alzheimer's disease: a pooled analysis of four randomized, double-blind, placebo-controlled trials. Dement Geriatr Cogn Disord 2000;11(4):202–11.
9. Blass JP, Cyrus PA, Bieber F, Gulanski B. Randomized, double-blind, placebo-controlled, multicenter study to evaluate the safety and tolerability of metrifonate in patients with probable Alzheimer disease. The Metrifonate Study Group. Alzheimer Dis Assoc Disord 2000;14(1):39–45.
10. Piecoro LT, Wermeling DP, Schmitt FA, Ashford JW. Seizures in patients receiving concomitant antimuscarinics and acetylcholinesterase inhibitor. Pharmacotherapy 1998;18(5):1129–32.
11. Jann MW, Cyrus PA, Eisner LS, Margolin DI, Griffin T, Gulanski B. Efficacy and safety of a loading-dose regimen versus a no-loading-dose regimen of metrifonate in the symptomatic treatment of Alzheimer's disease: a randomized, double-masked, placebo-controlled trial. Metrifonate Study Group. Clin Ther 1999;21(1):88–102.

Piracetam

General Information

Piracetam is a so-called "nootropic" drug, one of a class of drugs that affect mental function (1). In healthy volunteers it improves the higher functions of the brain involved in cognitive processes, such as learning and memory. Its mechanisms of action are not known but may include increased cholinergic neurotransmission.

Piracetam has been used to treat fetal distress during labor, but there is insufficient evidence to assess its efficacy (2).

A meta-analysis of 19 double-blind, placebo-controlled trials in elderly patients with dementia or cognitive impairment showed significant improvement with piracetam (3).

Piracetam has been used to treat Alzheimer's disease, sometimes in combination with lecithin, but without major beneficial effets (4,5,6,7,8), beyond a slight increase in alertness (9).

Systematic reviews

In a systematic review of trials of piracetam in Alzheimer's disease the evidence of effects on cognition and other measures was inconclusive (10). The reviewers concluded that the evidence available from the published literature does not support the use of piracetam in the treatment of people with dementia or cognitive impairment.

Organs and Systems

Nervous system

Piracetam 80–100 mg/day for 4 months has been evaluated in a randomized, double-blind, placebo-controlled, crossover study in 25 children with Down syndrome (11). Piracetam did not enhance cognition or behavior but was associated with adverse events: 18 children completed the study, 4 withdrew, and 3 were excluded at baseline. The adverse events were related to the nervous system and included aggression ($n = 4$), agitation/irritability ($n = 2$), sexual arousal ($n = 2$), poor sleep ($n = 1$), and reduced appetite ($n = 1$).

References

1. Vernon MW, Sorkin EM. Piracetam. An overview of its pharmacological properties and a review of its therapeutic use in senile cognitive disorders. Drugs Aging 1991;1(1):17–35.
2. Hofmeyr GJ, Kulier R. Piracetam for fetal distress in labour. Cochrane Database Syst Rev 2002;(1):CD001064.
3. Waegemans T, Wilsher CR, Danniau A, Ferris SH, Kurz A, Winblad B. Clinical efficacy of piracetam in cognitive impairment: a meta-analysis. Dement Geriatr Cogn Disord 2002;13(4):217–24.
4. Smith RC, Vroulis G, Johnson R, Morgan R. Comparison of therapeutic response to long-term treatment with lecithin versus piracetam plus lecithin in patients with Alzheimer's disease. Psychopharmacol Bull 1984;20(3):542–5.
5. Samorajski T, Vroulis GA, Smith RC. Piracetam plus lecithin trials in senile dementia of the Alzheimer type. Ann N Y Acad Sci 1985;444:478–81.
6. Growdon JH, Corkin S, Huff FJ, Rosen TJ. Piracetam combined with lecithin in the treatment of Alzheimer's disease. Neurobiol Aging 1986;7(4):269–76.
7. Davidson M, Mohs RC, Hollander E, Zemishlany Z, Powchik P, Ryan T, Davis KL. Lecithin and piracetam in Alzheimer's disease. Biol Psychiatry 1987;22(1):112–4.
8. Croisile B, Trillet M, Fondarai J, Laurent B, Mauguière F, Billardon M. Long-term and high-dose piracetam treatment of Alzheimer's disease. Neurology 1993;43(2):301–5.
9. Pierlovisi-Lavaivre M, Michel B, Sebban C, Tesolin B, Chave B, Sambuc R, Melac M, Gastaut JL, Poitrenaud J, Millet Y. [The significance of quantified EEG in Alzheimer's disease. Changes induced by piracetam.] Neurophysiol Clin 1991;21(5-6):411–23.
10. Flicker L, Grimley Evans G. Piracetam for dementia or cognitive impairment. Cochrane Database Syst Rev 2001;(2):CD001011.
11. Lobaugh NJ, Karaskov V, Rombough V, Rovet J, Bryson S, Greenbaum R, Haslam RH, Koren G. Piracetam therapy does not enhance cognitive functioning in children with Down syndrome. Arch Pediatr Adolesc Med 2001;155(4):442–8.

Rivastigmine

General Information

Rivastigmine (1) was the second drug after donepezil in a class of second-generation acetylcholinesterase inhibitors to become commercially available. It is now marketed in over 60 countries worldwide, including those in Europe and South America and the United Kingdom. It has central selectivity, suggesting fewer peripheral adverse effects. These include nausea, vomiting, abdominal pain, and anorexia (2,3). Daily doses up to 12 mg were tolerated and produced improvement in patients with Alzheimer's disease (4).

Although therapy with newer as well as older cholinesterase inhibitors can prevent cognitive decline, psychiatric and behavioral disturbances, and impaired ability to perform basic activities of daily living, and improve global state of patients with mild-to-moderately-severe Alzheimer's disease, more potent therapies are needed, particularly for modifying the rate of disease progression. Future research efforts need to focus on identifying predictors and better measures of response, timing of treatment, optimum dosage regimens, longer-term follow-up, and establishing how and when acetylcholinesterase inhibitors should be stopped (5,6).

Comparative studies

Dual cholinesterase inhibitors, such as rivastigmine, may be neuroprotective in Alzheimer's disease (7). Brain grey matter density changes were measured using voxel-based morphometry in 26 patients with minimal to mild Alzheimer's disease treated with donepezil, rivastigmine, or galantamine for 20 weeks. Patients whose drug treatment inhibited both acetylcholinesterase and butyrylcholinesterase, i.e. rivastigmine, did not have the widespread cortical atrophic changes in parietotemporal regions that are invariably reported in untreated patients, and that were detectable in those who took selective cholinesterase inhibitors, such as donepezil and galantamine. A strength of this study was its combined clinical, neuropsychological, and neuroimaging characterization of small groups of carefully screened patients, a powerful means of assessing specific drug effects in neuropsychiatric disorders. However, residual neurobiological and genetic heterogeneity in small groups of patients, despite careful selection, might have influenced the findings. None of the patients had the wild type butyrylcholinesterase variant, Apo-E status was not a significant co-variate in any analysis, and there were no interactions between Apo-E status and drug groups.

In two studies of rivastigmine, 38% of users suffered from nausea, 23% from vomiting, and 24% from vertigo (8). Over 6 months there was a dropout rate of 25%. The gastrointestinal effects led to significant loss of weight. The positive response rate was only 10%, compared with a 6% response to placebo.

The efficacy and safety of rivastigmine have been investigated in several subsequent studies (9–12). Neurologists generally titrated this drug more slowly than recommended in the prescribing information (9). In daily practice, over 50% of the patients were unable to tolerate rivastigmine, mainly because of cholinergic gastrointestinal adverse effects (10), whereas others have found that the adverse events were most frequently mild and transient (11,12). It has been suggested that early treatment with rivastigmine (6–12 mg/day) is associated with sustained long-term (up to 52 weeks) cognitive benefit in patients with Alzheimer's disease (12).

Placebo-controlled studies

Using the data from two pooled open extensions of four 6-month, randomized, placebo-controlled trials, projections of decline, had the patients not been treated, were made using a baseline-dependent mathematical model (13). MMSE data were available for 1998 rivastigmine-treated patients with Alzheimer's disease and 657, 298, and 83 were still taking it at 3, 4, and 5 years respectively. According to the global deterioration scale (GDS), the severity of dementia was as follows: very mild 2%, mild 27%, moderate 38%, moderately severe 30%, severe 4%. Projected mean scores in untreated patients with Alzheimer's disease fell below 10 points in the mini-mental state examination (MMSE) at about 3 years, while the mean MMSE score of patients who continued to take rivastigmine stayed above 10 points for 5 years. The most common adverse event was nausea, at least once during the open extensions in 865 patients (40%), followed by vomiting (n = 560; 28%), agitation (n = 504; 25%), accidental trauma (n = 422; 21%), dizziness (n = 421; 21%), and diarrhea (n = 379; 19%). In most cases, these events were mild or moderate, although in 451 patients (22%) withdrawal was required. These adverse events during long-term exposure to rivastigmine are similar in type and severity to those observed with rivastigmine and other cholinesterase inhibitors during short-term use. The incidence of new adverse events and the rate of withdrawal due to adverse events fell with treatment duration.

The neural correlates of the effects of rivastigmine given as add-on therapy to patients with schizophrenia taking antipsychotic drugs who displayed moderate cognitive impairment on visual sustained attention have been studied in a double-blind, placebo-controlled, longitudinal design using BOLD functional magnetic resonance imaging (14). Rivastigmine improved attention and was associated with increased activity in the neural regions that are involved in visual processing and attention systems, namely the occipital/fusiform gyrus and the frontal regions. The 12-week study initially involved 36 patients, but only 20 completed it. Unwillingness to continue with the study was not because of adverse effects of rivastigmine. The dosage of rivastigmine was 1.5 mg bd orally for 2 weeks, 3 mg bd for 2 weeks, 4.5 mg bd for 2 weeks, and 6 mg bd thereafter. In line with previous data (15,16), the results of this study suggested that rivastigmine increases cerebellar activity and influences, albeit non-significantly, attentional processes in schizophrenia. There were no differences between the groups in tardive dyskinesia, other adverse effects, or symptoms.

Evaluation of short-latency afferent inhibition (SAI) may be useful in identifying patients with Alzheimer's disease who are likely to respond to cholinesterase inhibitors (17). In 14 patients with Alzheimer's disease, pathologically reduced short-latency afferent inhibition was increased by a single oral dose of rivastigmine, and the baseline value and the change were associated with a response to long-term treatment. About two-thirds of the patients had an abnormal baseline value. In contrast, normal baseline short-latency afferent inhibition, or an abnormal value that was not greatly increased by a single dose of rivastigmine, was invariably associated with a poor response to long-term treatment.

Systematic reviews

A meta-analysis of various drugs approved for the treatment of Alzheimer's disease in the USA and Canada has suggested that rivastigmine can delay cognitive impairment and deterioration in global health for at least 6 months in patients with mild-to-moderate Alzheimer's disease (18). Patients taking active treatment will have more favorable ADAS-Cog (Alzheimer's Disease Assessment Scale-Cognitive) scores for at least 6 months, after which their scores will begin to converge with those who are taking placebo. The cost-effectiveness data were inconclusive.

Organs and Systems

Cardiovascular

Based on an electrocardiographic analysis of pooled data from four 26-week, phase III, multicenter, double-blind, placebo-controlled trials of rivastigmine (n = 2791), there were no adverse effects on cardiac function (19). Rivastigmine can therefore be safely given to patients with Alzheimer's disease, without the need for cardiac monitoring.

Psychological, psychiatric

Three patients with dementia, with no prior psychiatric history, deteriorated while taking rivastigmine (20). The time course suggested an association with rivastigmine, and each improved after withdrawal.

Gastrointestinal

Potentially fatal rupture of the esophagus has been associated with untitrated use of rivastigmine tablets in a patient with Alzheimer's disease (21).

- A 67-year-old Caucasian woman had a 2-year history of progressive memory loss. She had arterial hypertension successfully controlled with lisinopril and no history of ethanol abuse. A diagnosis of probable Alzheimer's disease was made, and she was given rivastigmine 1.5 mg/day, increasing to 9 mg/day by weekly increments of 1.5 mg. During the titration period, there were no significant adverse effects. After 13 weeks, weight loss was observed and rivastigmine was withdrawn. After 8 weeks she developed marked cognitive deterioration, and she and her carer were advised to restart rivastigmine 1.5 mg/day. However, she mistakenly took one tablet of 4.5 mg. About 30 minutes later she started to vomit several times. Nearly 2 hours later, she complained of severe chest pain, followed by high-grade fever. A chest X-ray showed mediastinal and soft-tissue emphysema, and a contrast X-ray showed rupture of the distal part of the esophagus. Emergency surgery was performed and she recovered.

Rivastigmine and other acetylcholinesterase inhibitors can produce chest pain because of increased esophageal contractions, although consequent rupture of the esophagus has not previously been reported. In this case, the failure to titrate the dosage of rivastigmine could have resulted in rupture secondary to severe vomiting. This confirms the need for careful titration of the dose of rivastigmine, even when restarting treatment.

The transient use of centrally acting antiemetics during rivastigmine titration may be useful in combination with dose-withholding and titration strategies to allow more patients to reach higher doses, resulting in a rapid and robust therapeutic effect. Patients who were treated with trihexyphenidyl and trimethobenzamide were more likely to be able to maintain or to increase their dose of rivastigmine than patients who were treated with glycopyrrolate or ondansetron (22).

Drug–Drug Interactions

General

Rivastigmine did not interact significantly with a wide range of concomitant medications prescribed for elderly patients with Alzheimer's disease, based on an analysis of 2459 patients (rivastigmine 1696, placebo 763) from four randomized placebo-controlled studies (23). However, the Breslow–Day analysis used in this study detected only differences in the odds ratios of rivastigmine versus placebo among patients taking concomitant medications. Thus, these results have to be cautiously interpreted.

References

1. Gottwald MD, Rozanski RI. Rivastigmine, a brain-region selective acetylcholinesterase inhibitor for treating Alzheimer's disease: review and current status. Expert Opin Invest Drugs 1999;8(10):1673–82.

2. Corey-Bloom J, Anand R, Veach J, for the ENA 713 B352 Study Group. A randomized trial evaluating efficacy and safety of ENA-713 (rivastigmine tartrate), a new acetylcholinesterase inhibitor, in patients with mild to moderately severe Alzheimer's disease. Int J Geriatr Psychopharmacol 1998;1:55–65.

3. Rosler M, Anand R, Cicin-Sain A, Gauthier S, Agid Y, Dal-Bianco P, Stahelin HB, Hartman R, Gharabawi M. Efficacy and safety of rivastigmine in patients with Alzheimer's disease: international randomised controlled trial. BMJ 1999;318(7184):633–8.

4. Forette F, Anand R, Gharabawi G. A phase II study in patients with Alzheimer's disease to assess the preliminary efficacy and maximum tolerated dose of rivastigmine (Exelon). Eur J Neurol 1999;6(4):423–9.

5. Knopman DS. Metrifonate for Alzheimer's disease: is the next cholinesterase inhibitor better? Neurology 1998;50(5):1203–5.

6. Bayer T. Another piece of the Alzheimer's jigsaw. BMJ 1999;318(7184):639.

7. Venneri A, McGeown WJ, Shanks MF. Empirical evidence of neuroprotection by dual cholinesterase inhibition in Alzheimer's disease. Neuropharmacol Neurotoxicol 2005;16:107–10.

8. Van Gool WA. Het effect van rivastigmine bij de ziekte van Alzheimer. Geneesmiddelenbulletin 2000;34:17–22.

9. Schmidt R, Lechner A, Petrovic K. Rivastigmine in outpatient services: experience of 114 neurologists in Austria. Int Clin Psychopharmacol 2002;17(2):81–5.

10. Richard E, Walstra GJ, van Campen J, Vissers E, van Gool WA. [Rivastigmine for Alzheimer disease; evaluation of preliminary results and of structured assessment of efficacy.] Ned Tijdschr Geneeskd 2002;146(1):24–7.

11. Bilikiewicz A, Opala G, Podemski R, Puzynski S, Lapin J, Soltys K, Ochudlo S, Barcikowska M, Pfeffer A, Bilinska M, Paradowski B, Parnowski T, Gabryelewicz T. An open-label study to evaluate the safety, tolerability and efficacy of rivastigmine in patients with mild to moderate probable Alzheimer's disease in the community setting. Med Sci Monit 2002;8(2):PI9–15.

12. Doraiswamy PM, Krishnan KR, Anand R, Sohn H, Danyluk J, Hartman RD, Veach J. Long-term effects of rivastigmine in moderately severe Alzheimer's disease: does early initiation of therapy offer sustained benefits? Prog Neuropsychopharmacol Biol Psychiatry 2002;26(4):705–12.

13. Small GW, Kaufer D, Mendiondo MS, Quarg P, Spiegel R. Cognitive performance in Alzheimer's disease patients receiving rivastigmine for up to 5 years. Int J Clin Pract 2005;59:473–7.

14. Aasen I, Kumari V, Sharma T. Effects of rivastigmine on sustained attention in schizophrenia. An fMRI study. J Clin Psychopharmacol 2005;25:311–7.

15. Schmahmann JD. The role of the cerebellum in affect and psychosis. J Neurolinguist 2000;13:189–214.

16. Friedman JI. Cholinergic targets for cognitive enhancement in schizophrenia: focus on cholinesterase inhibitors and muscarinic agonists. Psychopharmacology 2004;174:45–53.

17. Di Lazzaro V, Oliviero A, Pilato F, Saturno E, Dileone M, Marra C, Ghirlanda S, Ranieri F, Gainotti G, Tonali P. Neurophysiological predictors of long term response to AChE inhibitors in AD patients. J Neurosurg Psychiatry 2005;76:1064–9.

18. Wolfson C, Oremus M, Shukla V, Momoli F, Demers L, Perrault A, Moride Y. Donepezil and rivastigmine in the treatment of Alzheimer's disease: a best-evidence synthesis

of the published data on their efficacy and cost-effectiveness. Clin Ther 2002;24(6):862–86.

19. Weber JE, Chudnofsky CR, Boczar M, Boyer EW, Wilkerson MD, Hollander JE. Cocaine-associated chest pain: how common is myocardial infarction? Acad Emerg Med 2000;7(8):873–7.

20. Smith DJ, Yukhnevich S. Adverse reactions to rivastigmine in three cases of dementia. Aust N Z J Psychiatry 2001;35(5):694–5.

21. Simon T, Becquemont L, Mary-Krause M, de Waziers I, Beaune P, Funck-Brentano C, Jaillon P. Combined glutathione-S-transferase M1 and T1 genetic polymorphism and tacrine hepatotoxicity. Clin Pharmacol Ther 2000;67(4):432–7.

22. Jhee SS, Shiovitz T, Hartman RD, Messina J, Anand R, Sramek J, Cutler NR. Centrally acting antiemetics mitigate nausea and vomiting in patients with Alzheimer's disease who receive rivastigmine. Clin Neuropharmacol 2002;25(2):122–3.

23. Grossberg GT, Stahelin HB, Messina JC, Anand R, Veach J. Lack of adverse pharmacodynamic drug interactions with rivastigmine and twenty-two classes of medications. Int J Geriatr Psychiatry 2000;15(3):242–7.

Tacrine

General Information

Tacrine (tetrahydroaminoacridine) was one of the first drugs to be widely marketed for the loss of memory and intellectual decline in Alzheimer's disease. However, its efficacy is controversial. A Cochrane review of the use of tacrine in Alzheimer's disease produced results that were compatible with improvement, no change, or even harm (1). For measures of overall clinical improvement, the intention-to-treat analyses did not detect any difference between tacrine and placebo (OR = 0.87; 95%CI = 0.61, 1.23). There was no effect on behavioral disturbance (SMD = 0.04; 95%CI = –0.52, 0.43) or cognitive function (SMD = 0.14; 95%CI = –0.02, 0.30). The odds ratio for withdrawal due to an adverse event was significantly different from 1, the control group experiencing fewer events (OR = 0.7; 95%CI = 4.1, 7.9). Raised serum liver enzymes caused the most withdrawals. Gastrointestinal adverse effects (diarrhea, anorexia, dyspepsia, and abdominal pain) were the other major adverse events and the odds ratio for withdrawal was also in favor of the control group (OR = 3.8; 95%CI = 2.8, 5.1). No deaths were reported in any of the studies, which lasted up to 6 months.

In a prospective randomized, double blind, parallel-group, multicenter comparison of idebenone 360 mg/day (*n* = 104) and tacrine up to 160 mg/day (*n* = 99) for 60 weeks in 203 patients with mild-to-moderate dementia of the Alzheimer's type, idebenone produced more benefit than tacrine (2). The benefit : harm balance was favorable for idebenone compared with tacrine. More patients taking tacrine (65%) reported adverse events than those taking idebenone (50%). The dropout rate in those taking tacrine was also higher. There were statistically significant differences between tacrine and idebenone for gastrointestinal adverse events (nausea, vomiting) and hepatobiliary toxicity (raised serum transaminase activity).

Organs and Systems

Nervous system

Myoclonus has been attributed to tacrine (3).

- A 68-year-old woman, who had had dementia of probable Alzheimer's type for 4 years, was given tacrine 40 mg/day and 24 hours later progressively developed generalized uncontrolled abnormal movements affecting all her limbs and her mouth, suggestive of myoclonus and controlled by clonazepam. The myoclonus disappeared 24 hours after tacrine was withdrawn. Causation was established a few months later by rechallenge: myoclonus occurred during the next 48 hours.

Liver

Hepatic failure associated with tacrine occurred outside the usual time of onset (within 9 months) and resulted in the death of a 75-year-old woman with Alzheimer's disease who had taken tacrine for 14 months (4). Thus, the potential for delayed, life-threatening hepatotoxicity with tacrine, although unusual, should not be overlooked.

The presence of the combined alleles M1 and T1, which mark deficiencies in glutathione-*S*-transferase genes, increases susceptibility to tacrine hepatotoxicity (5). It would be interesting to use this molecular epidemiological approach to identify the role of combinations of glutathione-*S*-transferase genotypes in other adverse drug reactions.

Tacrine-induced hepatotoxicity was reduced by ursodeoxycholic acid (13 mg/kg/day for 105 days) in a pilot study in 14 patients with Alzheimer's disease (6). Serum activity of alanine transaminase in 100 patients taking ursodeoxycholic acid was normal in 93% of cases, compared with 69% of patients who had taken tacrine alone.

Authors evaluating the adverse effects of tacrine in Alzheimer's disease recommend regular monitoring for hepatotoxicity (SEDA-15, 136).

References

1. Qizilbash N, Birks J, Lopez-Arrieta J, Lewington S, Szeto S. Tacrine for Alzheimer's disease. Cochrane Database Syst Rev 2000;(2):CD000202.

2. Gutzmann H, Kuhl KP, Hadler D, Rapp MA. Safety and efficacy of idebenone versus tacrine in patients with Alzheimer's disease: results of a randomized, double-blind, parallel-group multicenter study. Pharmacopsychiatry 2002;35(1):12–18.

3. Abilleira S, Viguera ML, Miquel F. Myoclonus induced by tacrine. J Neurol Neurosurg Psychiatry 1998;64(2):281.

4. Blackard WG Jr, Sood GK, Crowe DR, Fallon MB. Tacrine. A cause of fatal hepatotoxicity? J Clin Gastroenterol 1998;26(1):57–9.

5. Simon T, Becquemont L, Mary-Krause M, de Waziers I, Beaune P, Funck-Brentano C, Jaillon P. Combined glutathione-S-transferase M1 and T1 genetic polymorphism and tacrine hepatotoxicity. Clin Pharmacol Ther 2000;67(4):432–7.

6. Salmon L, Montet JC, Oddoze C, Montet AM, Portugal H, Michel BF. Acide ursodesoxycholique et prévention de l'hépatotoxicité de la tacrine: une étude pilote. [Ursodeoxycholic acid and prevention of tacrine-induced hepatotoxicity: a pilot study.] Therapie 2001;56(1):29–34.

PSYCHOLOGICAL AND PSYCHIATRIC ADVERSE EFFECTS OF NON-PSYCHOACTIVE DRUGS

Aciclovir

Reversible psychiatric adverse effects have been described in three dialysis patients receiving intravenous aciclovir (8–10 mg/kg/day) (1).

High-dose aciclovir can cause disturbed cognition, changes in the level of consciousness, tremor, asterixis, hallucinations, and psychiatric syndromes. Often, symptoms of aciclovir toxicity are difficult to differentiate from neurological deterioration caused by Herpes encephalitis.

- A 67-year-old man developed severe neurotoxicity after a 7-day course of high-dose aciclovir (800 mg/day) for *Herpes zoster* (2). He had severe disturbance of consciousness, stupor, and loss of spontaneous movement but regained consciousness after only two sessions of hemodialysis.

The authors concluded that hemodialysis is highly effective in eliminating aciclovir and that it can help to differentiate aciclovir-induced neurotoxicity from neurological deterioration due to the underlying encephalitis.

In a retrospective analysis of samples sent for analysis of aciclovir concentrations, neuropsychiatric syndromes correlated with concentrations of 9-carboxymethoxymethylguanine (CM MG), the main metabolite of aciclovir (3). Based on a retrospective chart analysis, one psychiatrist unaware of the drug concentrations divided patients in one group with neuropsychological symptoms (n = 49) and one group without (n = 44). By ROC analysis, CM MG was the strongest predictor of neuropsychiatric symptoms. Symptoms included agitation, confusion, pronounced tiredness, lethargy, coma, dysarthria, myoclonus, and hallucinations and started within 1–2 days of aciclovir administration. Using a cut-off value of 11 µmol/l of CM MG, sensitivity and specificity were 91 and 93% respectively. CM MG was by far the best predictor of neuropsychiatric disorders, compared with aciclovir drug concentrations or exposure, serum creatinine or creatinine clearance. Thus, CM MG concentrations might be a useful alternative tool to differentiate conditions associated with Herpes infection from adverse effects of aciclovir. This group also reported the effect of hemodialysis. Among 39 patients who recovered from their condition, 12 received hemodialysis and recovered shortly thereafter. Thus, this study supports the preliminary findings reported in the case report above.

Aldesleukin

Aldesleukin can cause moderate impairment of cognitive function, with disorientation, confusion, hallucinations, sleep disturbances, and sometimes severe behavioral changes requiring transient neuroleptic drug administration (4–6). Some of the cognitive deficits mimicked those observed in dementias, such as Alzheimer's disease. Several studies have also shown increased latency and reduced amplitude of event-related evoked potentials in patients with cognitive impairment (6,7). Other infrequent adverse effects included paranoid delusions, hallucinations, loss of interest, sleep disturbances or drowsiness, reduced energy, fatigue, anorexia, and malaise. Coma and seizures

were exceptionally noted. Symptoms occurred within 1 week of treatment and complete recovery was usually noted after aldesleukin withdrawal.

In 10 patients with advanced tumors, low-dose subcutaneous aldesleukin produced significant psychological changes; increased depression scores, psychasthenia, and conversion hysteria were the most common findings (8).

The short-term occurrence of depressive symptoms has been investigated by using the Montgomery and Asberg Depression Rating Scale (MADRS) before and after 3 and 5 days of treatment in 48 patients without a previous psychiatric history and treated for renal cell carcinoma or melanoma with aldesleukin alone (n = 20), aldesleukin plus interferon alfa-2b (n = 6), or interferon alfa-2b alone (n = 22) (9). On day 5, patients in the aldesleukin groups had significantly higher MADRS scores, whereas there were no significant changes in the patients who received interferon alfa-2b alone. Eight of 26 patients given aldesleukin and only three of 22 given interferon alfa-2b alone had severe depressive symptoms. Depressive symptoms occurred as early as the second day of aldesleukin treatment and were more severe in the patients who received both cytokines. Early detection of mood changes can be useful in pinpointing patients at risk of subsequent severe neuropsychiatric complications.

Neuropsychiatric symptoms are less frequent with subcutaneous aldesleukin (10). No predictive or predisposing factors have been clearly identified. Whether a direct effect of aldesleukin on neuronal tissues, an increased vascular brain permeability with a subsequent increased brain water content, or an aldesleukin-induced release of neuroendocrine hormones (beta-endorphin, ACTH, or cortisol), accounted for these effects, is unknown. A possible immune-mediated cerebral vasculitis has also been reported in one patient (SEDA-20, 334).

Amantadine

The risk of mental complications seems to increase substantially if doses of 200 mg or more are given (11). Amantadine can cause mania and is contraindicated in patients with bipolar affective disorder (12).

- While taking amantadine, an elderly man developed the Othello syndrome, a severe delusion of marital infidelity, as described in Shakespeare's plays *Othello* and *A Winter's Tale*; it abated with drug withdrawal (SEDA-17, 170).
- Amantadine and phenylpropanolamine may have caused intense recurrent déjà vu experiences in a healthy 39-year-old within 24 hours of starting both drugs for influenza (13).

Amiodarone

Delirium has rarely been reported with amiodarone (SEDA-13, 140).

- A 54-year-old man with no previous psychiatric history took amiodarone 400 mg bd (14). After a few days he became depressed and paranoid, suffered from insomnia, and had rambling speech. The dosage of amiodarone was reduced to 200 mg bd and he improved. However, 3 days later he became confused, with tangential thinking, labile effect, and a macular rash on the limbs. His serum sodium was reduced at 127 mmol/l and his blood urea nitrogen was raised. A CT scan of the head was normal. Amiodarone was withdrawn and 4 days later he was alert and oriented. About a week later he started taking amiodarone again and within 4 days became increasingly agitated, confused, and paranoid. He once more recovered after withdrawal of amiodarone.

- Depression has been attributed to amiodarone in a 65-year-old woman who was taking amiodarone (dosage not stated) (15). Because the mode of presentation was atypical in onset, course, duration, and its response to antidepressant drugs, amiodarone was withdrawn, and she improved rapidly. There was no evidence of thyroid disease.

Anagrelide

There has been a single case report of visual hallucinations in a patient taking anagrelide, with recurrence on re-challenge (16).

Androgens and anabolic steroids

In the late 1980s various reports seemed to show that the use of anabolic steroids was linked to aggressive behavior and mood changes, even to the extent of inducing or potentiating violent crime (17,18). During the decade that followed, a series of other papers similarly linked high circulating concentrations of testosterone to increased degrees of aggression and related changes in mood. Undoubtedly, some of these findings are well-founded, but one must always be alert to the fallacy that individuals with particular pre-existent personality traits might be more susceptible than others to become body-builders, to use anabolic steroids, or to take testosterone. This possibility remains open after the completion of a thorough study of weightlifters at various American academic centers. In 20 male weightlifters, 10 of whom were taking anabolic steroids (metandrostenolone, testosterone, and nandrolone), supranormal testosterone concentrations were associated with increased aggression (19). Users tended to have a greater degree of depression, agitation, psychic or somatic anxiety, hypochondriasis, and hopelessness on the Hamilton Depression Scale; on the Modified Manic State Rating Scale, users showed more talkativeness, restlessness, threatening language, irritability, and sexual preoccupation. However, the results of the Personality Disorder Questionnaire suggested that this finding, while valid, was to some extent confounded by the personality disorder profile of the steroid users. The latter showed "cluster B" personality disorder traits for antisocial, borderline, and histrionic personality disorder, significantly differing in this respect from non-users.

In a useful review there is particular reference to the contested evidence on the behavioral effects of the androgenic anabolic steroids (20). The authors observed that certain of these complications, in particular hypomania and increased aggressiveness, have been confirmed in some, but not all, randomized controlled studies. Epidemiological attempts to determine whether anabolic steroids trigger violent behavior have failed, primarily because of high rates of non-participation. Studies of the use of anabolic steroids in different populations typically report a prevalence of repeated use of 1–5% among adolescents. The symptoms and signs of the use of anabolic steroids seem to be often overlooked by health-care professionals, and the number of cases of complications is virtually unknown. The authors suggested that future epidemiological research in this area should focus on retrospective case-control studies and perhaps also on prospective cohort studies of populations selected for a high prevalence of anabolic steroid use, rather than large-scale population-based studies.

Apart from the physical effects of exogenous anabolics and androgens, they can also have behavioral effects, including promotion of sexual behavior (which may or may not be regarded as an unwanted effect) and perhaps enhanced aggressiveness. Men who use androgenic anabolic steroids to enhance their sporting achievements seem to be more likely to have cyclic depression (21), but young men who have stopped using anabolic steroids can also develop depression and fatigue as withdrawal effects (22).

Androgens can rarely cause psychotic mania.

- A 28-year-old man with AIDS and a history of bipolar disorder was given a testosterone patch to counter progressive weight loss and developed worsening mania with an elevated mood, racing thoughts, grandiose delusions, and auditory hallucinations (23). His condition improved in hospital after removal of the patch and the administration of antipsychotic drugs. No cause for the psychosis, other than the use of testosterone, was found.

Suicide—or attempted suicide—in eight users of anabolic steroids has been described in Germany; the cases were related variously to hypomanic states during use of anabolic steroids or depression after withdrawal (24). Some of the users had committed acts of violence while using the drugs. In all cases, there were risk factors for suicidality and the drugs may simply have triggered the suicidal decision.

Anticholinergic drugs

See also individual compounds

Anticholinergic drugs can cause vivid and sometimes exotic hallucinations and this has led to their misuse. Plants containing atropine and related substances were used in witches' brews in the Middle Ages to conjure up the devil, but even synthetic tertiary amines given in eye-drops and depot plasters containing atropine (SEDA-13, 114) have caused hallucinations. Postoperative confusion in elderly patients has been clearly correlated with drugs that have anticholinergic properties (SEDA-13, 114), and the use of anticholinergic drugs for the treatment of Parkinson's disease has long been associated with neuropsychiatric adverse effects. Cyclopentolate is a short-acting cycloplegic with a rapid onset (and considerable intensity) of action, which particularly in children has been reported to cause hallucinations and psychotic episodes (25). It has been suggested that a partial structural affinity of the side-chain to some hallucinogens aggravates the problems associated with cyclopentolate.

Anticholinergic drugs can impair short-term memory. The effects in non-demented patients are reversible, receding within a few weeks of withdrawing treatment (26). Comparisons of dopaminergic drugs with anticholinergic drugs in healthy volunteers have shown that anticholinergic drugs caused significant impairment of memory function, more confusion, and dysphoria (SEDA-13, 115).

Anticoagulant proteins

In a phase III trial, hallucinations occurred in 1.1% of patients treated with drotrecogin alfa versus 0.1% in the placebo group (27).

Antiepileptic drugs

See also individual compounds

Cognitive effects

Antiepileptic drugs can cause cognitive impairment. Several factors affect cognition, including the cause of the epilepsy, the effect of seizures, and the antiepileptic medication, including the rate of titration, the dosage, the number of drugs taken, co-medications, and individual sensitivity. Experimental design flaws in assessing the cognitive adverse effects of antiepileptic drugs have been identified in a Medline review, cross-referencing terms related to cognition and anticonvulsants (28). The authors observed that there is incomplete information about newer antiepileptic drugs, but some of them may have favorable cognitive profiles.

In a comparison of the adverse cognitive effects and sedation caused by topiramate, lamotrigine, and phenobarbital, phenobarbital had 0.6–4.8 times the efficacy of topiramate and 1.1–10 times the efficacy of lamotrigine (29). Sedation rates were similar with topiramate and phenobarbital but were 2–4 times worse than with lamotrigine. The estimated monthly cost of treatment with phenobarbital was $10, compared with $177 for topiramate and $137 for lamotrigine. In this systematic review only studies on medically refractory focal epilepsy were included; this provides only a partial view of these drugs, since phenobarbital, lamotrigine, and topiramate differ in efficacy among different epilepsy syndromes (primary generalized epilepsy, Lennox–Gastaut syndrome) and in

less severe forms of epilepsy. In addition, other adverse effects and drug interactions, which also differ among the three drugs, were not considered.

Behavioral effects

A comparison of social skills in individuals with intellectual disability and epilepsy treated with carbamazepine, valproic acid, and phenytoin showed less positive social skills in those who received phenytoin (30).

Behavioral and psychiatric disturbances are not uncommon (31). Although epilepsy is itself associated with an increased risk of such disturbances, drugs play an important role. Phenobarbital-induced behavioral disturbances, especially hyperkinesia, are especially common in children, with an incidence of 20–50% and need for drug withdrawal in 20–30% of cases, whereas it is unclear whether and to what extent adults are affected.

Among older drugs, valproic acid and carbamazepine are least likely to cause adverse psychiatric effects, though valproate rarely causes encephalopathy and reversible pseudodementia. Phenytoin has been implicated in psychiatric adverse effects with or without other signs of toxicity, and at serum concentrations above or below the upper limit of the target range, but the actual incidence of these reactions is unknown. Benzodiazepines can cause paradoxical excitation, particularly in children and in anxious patients, and several other psychiatric symptoms can complicate the benzodiazepine withdrawal syndrome. Psychiatric or behavioral disorders have been reported with ethosuximide, but the lack of systematic studies prevents assessment of incidence and cause-and-effect relation. Among newer drugs, vigabatrin has been implicated most commonly in psychiatric adverse effects. With gabapentin, lamotrigine, and levetiracetam aggressiveness or hyperactivity can occur, especially in patients with previous behavioral problems or learning disability. Adverse psychiatric reactions to lamotrigine are uncommon, whereas with topiramate, felbamate, and other new drugs information is still insufficient.

Overall, the problem of drug-induced psychiatric disorders can be minimized by avoiding unnecessarily large dosages and drug combinations and by careful monitoring of the clinical response. In patients with a previous history of psychiatric disorders, carbamazepine and valproate are the first-line drugs, and are least likely to cause behavioral disturbances. The ideal management of such disturbances is withdrawal of the offending agent. When continuation of treatment is necessary for seizure control, psychosocial intervention and psychotropic medication can be useful.

In a retrospective study of 89 patients who developed psychiatric symptoms during treatment with tiagabine, topiramate, or vigabatrin, the psychiatric problem was either an affective or a psychotic disorder (not including affective psychoses) (32). All but one of the patients had complex partial seizures with or without secondary generalization. More than half were taking polytherapy. Nearly two-thirds had a previous psychiatric history, and there was a strong association between the type of previous psychiatric illness and the type of emerging psychiatric problem. Patients taking vigabatrin had an earlier onset of epilepsy and more neurological abnormalities than those taking topiramate.

Patients with chronic epilepsy have a higher likelihood of psychosis than the healthy population (33,34). Psychosis is especially frequent in patients with temporal lobe epilepsy (35). Antiepileptic drugs have been reported to precipitate psychosis, although the literature is confounded by the inclusion of affective and confusional psychoses in this category. Moreover, the purported association has mostly been made through isolated case reports or small non-controlled case series. In fact, most antiepileptic drugs have been associated with psychosis: phenytoin and phenobarbital (36), carbamazepine and valproate (37), felbamate (38), gabapentin (39), levetiracetam (40), topiramate (41), vigabatrin (42), and zonisamide (43). There have been no reports of psychosis associated with lamotrigine.

A retrospective chart review of 44 consecutive patients with epilepsy who had psychotic symptoms with clear consciousness has shown the difficulties in associating psychosis with drug effects (44). These patients were divided into two groups based on the presence or absence of changes in their drug regimen before the onset of the first episode of psychosis. In 27 patients the first episode of psychosis was unrelated to changes in their antiepileptic drug regimen, and in 23 of them the psychosis was temporally related to changes in seizure frequency. In 17 patients the first episode of psychosis developed in association with changes in their antiepileptic drug treatment, and in 12 of them the psychosis was temporally related to seizure attenuation or aggravation. This study therefore highlights the fact that psychosis can occur in relation to changes in seizure frequency, sometimes due to lack of effect of the new medication or to concomitant withdrawal of an efficacious medication.

Withdrawal of anticonvulsants with favorable mood stabilization properties, such as carbamazepine, has often been associated with acute psychosis (45,46). Moreover, the phenomenon of "forced normalization," by which complete seizure freedom in a patient with previous refractory epilepsy can lead to a psychotic state, may also contribute to the apparent association between drugs and psychosis (47).

Information from double-blind studies of psychosis as an adverse event is relatively scarce. A double-blind, randomized, add-on, placebo-controlled trial with carbamazepine showed that there was no increase in chronic psychotic symptoms in patients with suspected temporal lobe seizures (48).

The relation between psychosis and tiagabine has been assessed in an analysis of data from two multicenter, double-blind, randomized, placebo-controlled trials of add-on tiagabine therapy (32 or 56 mg/day) in 554 adolescents and adults with complex partial seizures during 8–12 weeks (49). There were psychotic symptoms (hallucinations) in 3 (0.8%) of 356 patients taking tiagabine and none of the 198 taking placebo, a non-significant difference. Thus, it appears that tiagabine does not increase the risk of psychosis, but the result is inconclusive.

An analysis of double-blind, placebo-controlled trials of vigabatrin as add-on therapy for treatment-refractory partial epilepsy showed that compared with placebo patients taking vigabatrin had a significantly higher incidence of events coded as psychosis (2.5% versus 0.3%) (50). There were no significant differences between treatment groups for aggressive reaction, manic symptoms, agitation, emotional lability, anxiety, or suicide attempts. In an open trial of topiramate, psychosis was seen in 30/1001 (3%) of the patients, and was severe enough to require withdrawal in eight (51).

Should certain antiepileptic drugs be contraindicated in patients with active psychosis? Unfortunately there is not enough solid information to answer this question. Undoubtedly, anticonvulsants that are less likely to cause psychosis (lamotrigine, carbamazepine, oxcarbazepine, valproate) should be preferred (52,53). However, patients with psychoses have been successfully treated even with drugs that are believed to be associated with psychosis, such as vigabatrin. For example, in a prospective study in 10 patients with psychosis and epilepsy to whom vigabatrin was added, there was no aggravation of the psychiatric disorder (54).

The appropriate methods and timing in assessing cognitive and behavioral adverse events during drug development programs have been thoroughly reviewed (55).

The authors of a critical review focusing on pediatric data concluded that adverse effects on learning and behavior may have been over-rated (56). Because of methodological flaws, many early studies could not discriminate between effects of drugs and the influence of heredity, brain damage, seizures, and psychosocial factors. In fact, the majority of children taking antiepileptic drugs do not experience major cognitive or behavioral effects from these medications. In some patients, however, drugs do produce detrimental effects, barbiturates and benzodiazepines being among those most commonly implicated (57). At least with some agents, such as gabapentin, behavioral adverse effects occur mainly in children with pre-existing learning disability. As to phenytoin, carbamazepine, and valproate which are the drugs most commonly recommended for first-line use, recent investigations have failed to show major differences in cognitive effects between these agents, although in some studies, patients taking phenytoin tended to have lower motor and information processing speeds (SED-13, 140) (58,59) (SEDA-19, 72) (SEDA-20, 64).

Antihistamines

See also individual compounds

Cognitive effects

In 42 healthy naval aviation personnel in a double-blind, randomized, placebo-controlled, crossover study, subjective drowsiness, cognitive performance, and vigilance were measured after three treatments: fexofenadine 180 mg, diphenhydramine 50 mg as a positive control, or placebo. Diphenhydramine significantly impaired cognitive performance, while fexofenadine had similar effects on complex cognitive skills to placebo. The authors concluded that their findings provided additional support for the safe use of fexofenadine by aviation personnel (60).

In a double-blind, placebo-controlled, crossover study of the acute effects of single doses of fexofenadine 120 mg, olopatadine 10 mg, and chlorphenamine 4 mg on cognitive and psychomotor performance in 11 healthy Japanese volunteers, both chlorphenamine and olopatadine reduced behavioral activity while fexofenadine was similar to placebo (61).

In a randomized, double blind, six-way, crossover study of the cognitive effects of fexofenadine 180 mg, both alone and in combination with alcohol, fexofenadine had no disruptive effects on objective measures related to driving a car and aspects of psychomotor and cognitive function, even when combined with a dose of alcohol equivalent to 0.3 g/kg (62).

The effect of fexofenadine on cognitive performance was assessed using the test of variables of attention (TOVA) in a double-blind, placebo-controlled, randomized, crossover design in 42 healthy subjects (63). Each subject rated their subjective feelings of drowsiness on a visual analogue scale and then completed four separate TOVA tests: at baseline and after the administration of placebo, diphenhydramine 50 mg, or fexofenadine 180 mg. Diphenhydramine caused significant increases in omission errors and response time on the TOVA and increases in self-reported drowsiness compared with placebo, while fexofenadine had no significant effects. All of these findings suggest that fexofenadine has a favorable nervous system adverse effects profile.

In healthy volunteers promethazine caused impaired cognitive function and psychomotor performance (64). The test battery consisted of critical flicker fusion, choice reaction time, compensatory tracking task, and assessment of subjective sedation. Cetirizine and loratadine at all doses tested were not significantly different from placebo in any of the tests used.

School performance in 63 children aged 8–10 years was not impaired by short-term diphenhydramine or loratadine (65).

The effects of ebastine 10 mg on cognitive impairment have been assessed in 20 healthy volunteers who performed six types of attention-demanding cognitive tasks, together with objective measurements of reaction times and accuracy (66). Ebastine was compared with placebo and a positive control, chlorphenamine (chlorphenamine, 2 mg and 6 mg). Compared with placebo, ebastine had no effect on any objective cognitive test nor any effect on subjective sleepiness. In contrast, chlorphenamine significantly increased reaction times, decreased accuracy in cognitive tasks, and increased subjective sleepiness. The effect of chlorphenamine increased with plasma concentration.

In a double-blind, placebo-controlled, randomized trial of the effects of levocetirizine 5 mg and diphenhydramine 50 mg on objective measurements (a word-learning test, the Sternberg Memory Scanning Test, a tracking test, and a divided attention test that measured both tracking and memory scanning simultaneously) in 48 healthy

volunteers (24 men and 24 women). Levocetirizine had no effect, while diphenhydramine significantly affected divided attention and tracking after acute administration (67). However, on day 4 the effects of diphenhydramine did not reach significance, suggesting a degree of tolerance to this first-generation drug.

The effects of levocetirizine on cognitive function have been assessed in two comprehensive and well-controlled studies. The first analysed the effects of single and multiple doses of levocetirizine on measures of nervous system activity, using integrated measures of cognitive and psychometric performance. In a three-way crossover design, 19 healthy men took either levocetirizine 5 mg, diphenhydramine 50 mg (positive control), or placebo once-daily on five consecutive days. Critical flicker fusion tests were performed on days 1 and 5 at baseline and up to 24 hours after drug administration. The primary outcome was that, in contrast to diphenhydramine, levocetirizine did not have any deleterious effect on any cognitive or psychometric function compared with placebo (68). In a double-blind, crossover study levocetirizine 5 mg once-daily for 4 days was compared with cetirizine 10 mg, loratadine 10 mg, promethazine 30 mg, and placebo in terms of CNS inhibitory effects in 20 healthy volunteers (69). With the exception of promethazine none of the drugs had disruptive or sedative effects on objective measurements in a comprehensive battery of psychomotor and cognitive tests.

Memory

In a double-blind, placebo-controlled, randomized study (70) of levocetirizine 5 mg/day, diphenhydramine 50 mg/day, or placebo in 48 healthy volunteers (24 men and 24 women) levocetirizine did not impair performance or cause memory deficits after acute and subchronic administration while diphenhydramine significantly affected divided attention and tracking after acute administration (71). Levocetirizine had no deleterious effect on cognitive and psychometric functions compared with placebo, as assessed by a comprehensive battery of psychometric tests in healthy men.

Driving performance

In a study of the effect of fexofenadine on driving and psychomotor behavior there were no differences between fexofenadine and placebo on reaction times, decision-making, or driver behavior (72). However, one criticism of this study was the failure to include a positive control, such as diphenhydramine.

The effect of levocetirizine 5 mg/day on actual driving performance during normal traffic has been compared with the effect of the first-generation antihistamine diphenhydramine 50 mg/day in 48 healthy volunteers in a double-blind, placebo-controlled, randomized trial (73). Treatments were given on days 1, 2, 3, and 4, at 90 minutes before the start of a standardized driving test on days 1 and 4. In contrast to diphenhydramine, driving performance was not significantly affected by levocetirizine 5 mg/day.

Delirium

A 56-year-old Caucasian developed acute delirium having taken diphenhydramine 300 mg/day for 2 days to treat a pruritic rash. He subsequently developed visual and auditory hallucinations with erratic aggressive behavior. The author concluded that the drug-induced delirium was associated with the combination of treatment for an infected wound with linezolid with diphenhydramine given for secondary drug-induced rash (74).

Hallucinations

Auditory hallucinations have been reported in a 42-year-old man taking a combination of meclozine 25 mg bd and metaxalone 800 mg bd (75). The hallucinations ceased when the medications were withdrawn.

Apomorphine

Four men with Parkinson's disease underwent long-term treatment with apomorphine and developed dose-related psychosexual disorders (76).

Aprotinin

Psychotic reactions, including delirium, hallucinations, and confusion, have also been reported in patients given aprotinin, but it is possible that the symptoms were due to underlying pancreatitis (77).

Araliaceae

A woman with prior episodes of depression had a manic episode several days after starting to take *P. ginseng* (78).

Artificial sweeteners

Aspartame has been associated with mood disturbances, but only anecdotally (79).

Atropine

Atropine can cause slight memory impairment, detectable if special studies of mental function are performed (SEDA-13, 115).

Azathioprine

Azathioprine has been newly associated with psychiatric adverse events (80).

- A 13-year-old boy with Wegener's granulomatosis developed incapacitating obsessive–compulsive symptoms and severe panic attacks 4 weeks after switching from cyclophosphamide to azathioprine. He had obsessions about dying, committing suicide, and harming others, obsessive negative thoughts about himself and others, compulsive behavior, severe panic attacks more than once a day, and sleep disturbances. He was given fluvoxamine 100 mg/day, but 18 months later the

symptoms suddenly disappeared, 3 weeks after he switched from azathioprine to methotrexate. In the next 4 years, he had no relapse.

Psychiatric adverse effects have not previously been reported with azathioprine. Neither does the database of the WHO Uppsala Monitoring Centre mention obsessive–compulsive symptoms or panic attacks as a possible adverse effect of azathioprine. However, the time course in this case and the absence of symptoms before and after azathioprine therapy suggest a causal relation. It is possible that the combination of subtle cerebral dysfunction as a result of the vasculitis and the use of azathioprine may have caused the symptoms in this patient.

Azithromycin

Azithromycin can cause delirium (81).

Baclofen

Euphoria or depression can occur, and mania has been reported in a patient with schizophrenia (82).

Benzatropine and etybenzatropine

Benzatropine can cause slight memory impairment, detectable if special studies of mental function are performed (SEDA-13, 115) (SEDA-15, 137).

Beta-adrenoceptor antagonists

Beta-adrenoceptor antagonists impair performance in psychomotor tests after single doses. These include effects of atenolol, oxprenolol, and propranolol on pursuit rotor and reaction times (83,84). However, other studies with the same drugs have failed to show significant effects (85–89), and the issue has remained controversial. A report that sotalol improved psychomotor performance in 12 healthy individuals in a dose of 320 mg/day but impaired performance at 960 mg/day (90) has been interpreted to indicate that the water-soluble beta-adrenoceptor antagonists would be less likely than the fat-soluble drugs to produce nervous system effects. Both atenolol and propranolol alter the electroencephalogram; atenolol affects body sway and alertness and propranolol impairs short-term memory and the ability to concentrate (91,92). These results suggest that both lipophilic and hydrophilic beta-adrenoceptor antagonists can affect the central nervous system, although the effects may be subtle and difficult to demonstrate.

In 27 hypertensive patients aged 65 years or more, randomized to continue atenolol treatment for 20 weeks or to discontinue atenolol and start cilazapril, there was a significant improvement in the choice reaction time in the patients randomized to cilazapril (93). This study has confirmed previous reports that chronic beta-blockade can determine adverse effects on cognition in elderly patients. Withdrawal of beta-blockers should be considered in any elderly patient who has signs of mental impairment.

In a placebo-controlled trial of propranolol in 312 patients with diastolic hypertension, 13 tests of cognitive function were assessed at baseline, 3 months, and 12 months (94). Propranolol had no significant effects on 11 of the 13 tests. Compared with placebo, patients taking propranolol had fewer correct responses at 3 months and made more errors of commission.

Bipolar depression affects 1% of the general population, and treatment resistance is a significant problem. The addition of pindolol can lead to significant improvement in depressed patients who are resistant to antidepressant drugs, such as selective serotonin reuptake inhibitors or phenelzine. Of 17 patients with refractory bipolar depression, in whom pindolol was added to augment the effect of antidepressant drugs, eight responded favorably (95). However, two developed transient hypomania, and one of these became psychotic after the resolution of hypomanic symptoms. In both cases transient hypomanic symptoms resolved without any other intervention, while psychosis required pindolol withdrawal.

Anxiety and depression have been reported after the use of nadolol, which is hydrophilic (96). In a study of the co-prescribing of antidepressants in 3218 new users of beta-blockers (97), 6.4% had prescriptions for antidepressant drugs within 34 days, compared with 2.8% in a control population. Propranolol had the highest rate of co-prescribing (9.5%), followed by other lipophilic beta-blockers (3.9%) and hydrophilic beta-blockers (2.5%). In propranolol users, the risk of antidepressant use was 4.8 times greater than the control group, and was highest in those aged 20–39 (RR = 17; 95% CI = 14, 22).

The development of a severe organic brain syndrome has been reported in several patients taking beta-adrenoceptor antagonists regularly without a previous history of psychiatric illness (98–100). A similar phenomenon was seen in a young healthy woman who took propranolol 160 mg/day (101). The psychosis can follow initial therapy or dosage increases during long-term therapy (102). The symptoms, which include agitation, confusion, disorientation, anxiety, and hallucinations, may not respond to treatment with neuroleptic drugs, but subsides rapidly when the beta-blockers are withdrawn. Symptoms are also ameliorated by changing from propranolol to atenolol (103).

A schizophrenia-like illness has also been seen in close relation to the initiation of propranolol therapy (104).

Beta-lactamase inhibitors

Behavioral changes occurred in four children aged 1.5–10.5 years, taking co-amoxiclav (105).

Bisphosphonates

Bisphosphonates, regardless of route of administration, have also been associated with hallucinations (auditory and olfactory) (106) and visual disturbances (107).

- A 79-year-old Caucasian woman who had been taking alendronate 10 mg/day for over 2 years to prevent osteoporosis reported hearing "voices in her head" along with red-colored visual disturbances (108). These disturbances began shortly after her regimen had been changed from alendronate 10 mg/day to 70 mg once/week. Assessment of causality revealed "probable" and "highly probable" relationships respectively between the adverse events and the switch from daily to weekly alendronate therapy.

Other bisphosphonates, such as etidronate and pamidronate, have caused both reversible and irreversible auditory, visual, and olfactory hallucinations beginning 2 hours to 1 week after drug administration. The mechanism of these adverse effects is unknown but is thought to be independent of calcium homeostasis.

Bromocriptine

Up to 10% of patients have to be withdrawn from treatment because of psychiatric symptoms. Bromocriptine-induced psychosis is well known and particular caution is warranted in patients with a family history of mental disorders (109). Even very low doses of bromocriptine can cause psychotic reactions (SEDA-9, 126) (SEDA-10, 117), and well-recognized problems include confusion, hallucinations, delusions, and paranoia.

Buflomedil

Depression has been reported in a few frail elderly women taking buflomedil (SEDA-13, 169).

Cabergoline

When cabergoline is used in patients with Parkinson's disease, the same spectrum of dyskinesias and psychiatric complications as with bromocriptine is observed (110).

The successful use of ergot derivatives to shrink macroprolactinomas can have unwanted neurological consequences, and this has been described in two case reports.

Three Italian men aged 39, 42, and 53 years with invasive prolactinomas took cabergoline 1.0–3.0 mg/week and all developed CSF rhinorrhea after 2–7 months (111). This was clearly a consequence of loss of the "stopper" effect of the tumor, owing to shrinkage, and in each case was successfully treated by endoscopic trans-sphenoidal surgery.

- A 42-year-old Spanish man took cabergoline (up to 3 mg/day) for a large prolactinoma causing hypopituitarism and symptomatic chiasmal compression (112). After 18 months there was only a minimal tumor remnant on the floor of the sella turcica, but there was chiasmal herniation. However, there were no clinical effects of this, and in particular the visual fields were normal.

Caffeine

Red Bull, a mixture of caffeine, taurine, and inositol, a widely consumed "power drink," affects mental performance and mood.

- A 36-year-old man with bipolar-I disorder had a second manic episode, after having been in remission for 5 years while taking lithium to maintain a serum lithium concentration of 0.8–1.1 mmol/l (113). One week before this episode, he drank three cans of Red Bull at night, as he needed less sleep; 3 days later he drank three more cans. After 4 days he was feeling euphoric, hyperactive, and insomniac. He gradually became more hyperactive and had increased libido and irritability. He took no more Red Bull and improved within 7 days.

Based on this report, the authors suggested that stimulant beverages that contain caffeine might cause cognitive and behavioral changes, especially in vulnerable patients with bipolar illness.

Calcium channel blockers

A patient taking diltiazem developed the signs and symptoms of mania (114) and another developed mania with psychotic features (115). There have also been reports that nifedipine can cause agitation, tremor, belligerence, and depression (116), and that verapamil can cause toxic delirium (117). Nightmares and visual hallucinations have been associated with nifedipine (118). Depression has been reported as a possible adverse effect of nifedipine (119).

Some reports have suggested that calcium channel blockers may be associated with an increased incidence of depression or suicide. However, there is a paucity of evidence from large-scale studies. A study of the rates of depression with calcium channel blockers, using data from prescription event monitoring, involved gathering information on symptoms or events in large cohorts of patients after the prescription of lisinopril, enalapril, nicardipine, and diltiazem by general practitioners (120). The crude overall rates of depression during treatment were 1.89, 1.92, and 1.62 per 1000 patient-months for the ACE inhibitors, diltiazem, and nicardipine respectively. Using the ACE inhibitors as the reference group, the rate ratios for depression were 1.07 (95% CI = 0.82, 1.40) and 0.86 (0.69, 1.08) for diltiazem and nicardipine respectively. This study does not support the hypothesis that calcium channel blockers are associated with depression.

Carbamazepine

Behavioral and psychiatric disturbances are less common with carbamazepine than with other anticonvulsants.

- A 9-year-old boy with seizures developed intermittent complex visual hallucinations during therapy with fosphenytoin and, on a separate occasion, carbamazepine (121).

In 10 children with rolandic epilepsy, carbamazepine impaired memory and possibly visual search tasks (122). Evaluation of individual data suggested that some children were especially vulnerable to the adverse effects of carbamazepine on cognition. The authors did not comment on the fact that rolandic epilepsy is regarded as a syndrome for which treatment in most cases is not indicated.

The cognitive effects of carbamazepine and gabapentin have been compared in a double-blind, crossover, randomized study in 34 healthy elderly adults, of whom 19 withdrew (15 while taking carbamazepine, probably because of excessively rapid dosage titration) (123). The primary outcome measures were standardized neuropsychological and mood state tests, yielding 17 variables. Each subject had cognitive testing at baseline (before drug treatment), at the end of the first drug phase, the end of the second drug phase, and 4 weeks after completion of the second drug phase. Adverse events were frequently reported with both anticonvulsants, although they were more common with carbamazepine. There were significant differences between carbamazepine and gabapentin for only one of 11 cognitive variables, with better attention/vigilance for gabapentin, although the effect was modest. Both carbamazepine and gabapentin can cause mild cognitive deficits in elderly subjects, and gabapentin has a slightly better profile.

Cardiac glycosides

Acute psychosis and delirium can occur in digitalis toxicity, particularly in elderly people (124–126), and can be accompanied by visual or auditory hallucinations (127,128).

- Acute delirium occurred in a 61-year-old man whose serum digitoxin concentration was 44 ng/ml (129).

Digitalis toxicity can occasionally cause depression (130).

- A 77-year-old woman developed extreme fatigue, anorexia, psychomotor retardation, and social withdrawal 1 month after starting to take digoxin 0.5 mg/day for congestive heart failure (131). She did not respond to intravenous clomipramine 25 mg/day for 7 months. Her serum digoxin concentration was 3.2 ng/ml. Digoxin was withdrawn, and 12 days later, when her serum digoxin concentration was 0.5 ng/ml, she had improved, but was left with a memory disturbance, which was attributed to background dementia.

Celastraceae

Khat has amphetamine-like effects and can cause psychoses (132–138), including mania (139) and hypnagogic hallucinations (140). Two men developed relapsing short-lasting psychotic episodes after chewing khat leaves; the psychotic symptoms disappeared without any treatment within 1 week (141).

In 800 Yemeni adults (aged 15–76 years) symptoms that might have been caused by the use of khat were elicited by face-to-face interviews; 90 items covered nine scales of the following domains: somatization, depression, anxiety, phobia, hostility, interpersonal sensitivity, obsessive-compulsive, hostility, interpersonal sensitivity, paranoia, and psychoticism (142). At least one life-time episode of khat use was reported in 82% of men and 43% of women. The incidence of adverse psychological symptoms was not greater in khat users, and there was a negative association between the use of khat and the incidence of phobic symptoms.

In 25 daily khat-chewing flight attendants, 39 occasional khat-chewing flight attendants, and 24 non-khat-chewing aircrew members, memory function test scores were significantly lower in khat chewers than non-chewers and in regular chewers than occasional chewers (143).

Celecoxib

A 78-year-old woman had auditory hallucinations while taking celecoxib for osteoarthritis (144). Her symptoms occurred after she had taken celecoxib 200 mg bd for 48 hours and progressed over the next 8 days. Celecoxib was withdrawn and her hallucinations gradually disappeared over the next 4 days. Rechallenge with a lower dose (100 mg bd) caused recurrence.

There have been two reports of visual hallucinations in patients taking celecoxib.

- A 79-year-old woman presented to her optometrist with a 2-day history of seeing orange spots in both visual fields 2 months after starting to take celecoxib 100 mg/day (145). Physical examination and a CT scan were normal. Celecoxib was withdrawn and her symptoms resolved within 3 days.
- An 81-year-old woman took celecoxib 100 mg/day, and over the next 2 weeks developed delirium and auditory and visual hallucinations (146). Celecoxib was withdrawn and her symptoms resolved over several days. She took a few doses of rofecoxib 12.5 mg/day 6 months later without any problem. She began to take rofecoxib regularly again 2 months later, and after 1 month developed agitation, confusion, and hallucinations. Physical examination suggested no cause of the delirium other than rofecoxib. A CT scan was negative. The rofecoxib was withdrawn, and over the next 2 days her symptoms resolved

Auditory hallucinations have been previously reported in a patient taking celecoxib (SEDA-25, 134) but are probably uncommon.

Cephalosporins

It has long been known that intramuscular procaine penicillin can cause some peculiar psychological adverse reactions, and that other penicillin derivatives, such as amoxicillin, can cause psychiatric reactions, such as hallucinations (SEDA-21, 259). In a report from the Netherlands, neuropsychiatric symptoms occurred in six patients who received cefepime for febrile neutropenia (147). The patients, two men and four women, aged 32–75 years, received 6 g/day ($n = 5$) or 3 g/day ($n = 1$).

The symptoms started 1–5 days after the first dose and varied from nightmares, anxiety, agitation, and visual and auditory hallucinations to coma and seizures. After withdrawal of cefepime, they recovered within 1–5 days. The causality between their neuropsychiatric symptoms and cefepime was considered as probable (WHO criteria) because of the temporal relation, lack of other causal neurological explanations, and positive rechallenge in five patients.

The mechanisms of these adverse effects are unknown, although it might be of value to take a closer look at the theory that drug-induced limbic kindling may be the principal pathogenic factor (SEDA-21, 259).

Cetirizine

In a double-blind, crossover study levocetirizine 5 mg once daily for 4 days was compared with cetirizine 10 mg, loratadine 10 mg, promethazine 30 mg, and placebo in terms of nervous system inhibitory effects in 20 healthy volunteers (148). With the exception of promethazine, none of the drugs had disruptive or sedative effects on objective measurements in a comprehensive battery of psychomotor and cognitive tests.

The Early Treatment of the Atopic Child (ETAC) Study has provided evidence that cetirizine may be able to halt the progression to asthma in high-risk groups of young children and infants with atopic dermatitis (149). However, the study involved giving relatively high doses of cetirizine to very young children (aged 1–2 years at study entry) over a long period of time. Therefore, the impact of prolonged use of high-dose cetirizine (0.25 mg/kg bd over 18 months) on behavior and cognitive ability in young children and infants has been assessed in a double-blind, randomized, placebo-controlled study (150). Well-validated and standardized assessments of behavior or cognition were used, and the ages at which psychomotor milestones were attained were established. The authors concluded that, compared with placebo, cetirizine had no significant effects on behavior, cognition, or psychomotor milestones in young children with atopic dermatitis.

Chlorphenamine maleate

Functional neuroimaging of cognition was impaired by chlorphenamine (151).

Chloroquine and hydroxychloroquine

Many mental changes attributed to chloroquine have been described, notably agitation, aggressiveness, confusion, personality changes, psychotic symptoms, and depression. Acute mania has also been recognized (SEDA-18, 287). The mental changes can develop slowly and insidiously. Subtle symptoms, such as fluctuating impairment of thought, memory, and perception, can be early signs, but may also be the only signs. The symptoms may be connected with the long half-life of chloroquine and its accumulation, leading to high tissue concentrations (SEDA-11, 583). Chloroquine also inhibits glutamate dehydrogenase activity and can reduce concentrations of the inhibitory transmitter GABA.

When severe psychosis following treatment with chloroquine and hydroxychloroquine occurs it is usually during treatment for malaria, but it can follow treatment for connective tissue disorders. Hallucinations have been reported after hydroxychloroquine treatment for erosive lichen planus (152).

- A 75-year-old woman was given hydroxychloroquine 400 mg/day for erosive lichen planus in conjunction with topical glucocorticoids and a short course of oral methylprednisolone 0.5 mg/kg/day. After 10 days she became disoriented in time and place, followed by feelings of depersonalization and kinesthetic hallucinations, preceded by nightmares. She stopped taking hydroxychloroquine 1 week later and the hallucinations progressively disappeared. She recovered her normal mental state within 1 month and had not relapsed 2 years later.

In some cases with psychosis after the administration of recommended doses, symptoms developed after the patients had taken a total of 1.0–10.5 g of the drug, the time of onset of behavioral changes varying from 2 hours to 40 days. Most cases occurred during the first week and lasted from 2 days to 8 weeks (SEDA-11, 583).

Transient global amnesia occurred in a healthy 62-year-old man, 3 hours after he took 300 mg chloroquine. Recovery was spontaneous after some hours (SEDA-16, 302).

In one center, toxic psychosis was reported in four children over a period of 18 months (SEDA-16, 302). The children presented with acute delirium, marked restlessness, outbursts of increased motor activity, mental inaccessibility, and insomnia. One child seemed to have visual hallucinations. In each case, chloroquine had been administered intramuscularly because of fever. The dosages were not recorded. The children returned to normal within 2 weeks.

The potential for severe psychiatric adverse events must always be considered in patients taking long-term chloroquine and hydroxychloroquinine. The onset may be a few hours to many days after the start of therapy and it can occur in a patient without a preceding history of mental illness. The mechanism is unknown. Recovery is rapid and occurs within days of stopping treatment.

Probable suicide after quinine treatment for chloroquine-resistant malaria has been reported (153).

- A 27-year-old man with falciparum malaria was given an infusion of quinine 600 mg in 5% dextrose tds until his vomiting stopped. Five hours later he was found dead by hanging using his turban. There was no adverse social history and the patient was in general good health.

The cause of this man's suicide is not known and it was probably not related to quinine.

Cholera vaccine

Severe complications connected with cholera (or combined) immunization are extremely rare and the causal relation is always doubtful. However, when they do occur they constitute a contraindication to further administration. There are occasional reports of neurological and psychiatric reactions (SED-8, 706) (SEDA-1, 246), Guillain–Barré syndrome (SEDA-1, 246), myocarditis (154,155), myocardial infarction (SEDA-3, 261), a syndrome similar to immune complex disease (156), acute renal insufficiency accompanied by hepatitis (157), and pancreatitis (158).

Choline

Occasionally, doses of choline up to 9 g/day have been found to produce severe depression, presumably by altering the adrenaline/acetylcholine balance (159).

Chromium

There have been three reports of the efficacy of chromium in depression, with adverse effects that included dizzy spells and vivid dreams (160).

- A 50-year-old man developed bipolar II disorder with the onset of a major depressive episode in his late twenties. His mood stabilized with lithium, but he continued to have periods of irritability and breakthrough depression. He started to take chromium picolinate 400 micrograms/day and within 2 days felt more relaxed and stable than he had since the onset of his disorder. He stopped taking lithium. Several months later he forgot to take his chromium, and within a few days his symptoms returned. In order to catch up he took 800 micrograms/day and developed sweating each morning and a mild hand tremor. After reducing the dosage to 600 micrograms/day he again went into complete remission. After more than 1 year of chromium treatment he developed uric acid kidney stones. One year after switching to a different chromium salt (the polynicotinate), there was no recurrence of kidney stones.
- A 38-year-old man with bipolar II disorder took chromium polynicotinate 400 micrograms/day. Shortly after the first dose his mood started to improve, but he had unusually vivid intense dreams. The dose of chromium was increased to 600 micrograms/day. He then developed intermittent brief dizzy spells due to orthostatic hypotension. After switching to chromium picolinate his dizzy spells did not recur.
- A 47-year-old man with a dysthymic disorder and intermittent panic attacks and rage outbursts took chromium 400 micrograms/day, and after 1 day had strikingly vivid dreams. Over the next several days there was a dramatic improvement in his mood and behavior. The efficacy of chromium was later confirmed by a double-blind, placebo-controlled, n-of-1 trial.

Cimetidine

Cimetidine crosses the blood–brain barrier and can cause confusion, particularly in elderly or sick individuals with compromised hepatic or renal function, and especially after intravenous treatment. Very rarely an acute confusional psychosis has been seen in a younger person (161). Delirium has been thought to be a particular problem with intravenous use, but this is more likely to be a reflection of patient selection. Depression has occasionally been attributed to cimetidine (162,163).

Ciprofloxacin

The administration of ciprofloxacin has been associated with psychosis (164,165) and hypoactive delirium (166,167).

- A 27-year-old woman developed an acute psychotic reaction following the use of ciprofloxacin eye-drops (1 drop hourly to each eye) (168).
- Psychosis occurred in a 32-year-old woman who was taking ciprofloxacin for multidrug resistant tuberculosis; the symptoms resolved within 48 hours after the ciprofloxacin was withdrawn (169).

Cisapride

In 16% of children given cisapride for intestinal pseudo-obstruction, treatment was followed by mild irritability and hyperactivity (SEDA-18, 370). In one adult, aggressive behavior seemed to be a direct complication of treatment (170).

Clarithromycin

Two patients, a man aged 74 and a woman aged 56 years, developed delirium after taking clarithromycin (171).

Two patients (aged 21 and 33 years) with late-stage AIDS had acute psychoses shortly after taking clarithromycin (2 g/day) for MAC bacteremia (172). In both cases the psychosis resolved on withdrawal but recurred on rechallenge. In one case treatment with azithromycin was well tolerated.

Of cases of mania attributed to antibiotics and reported to the WHO, 28% were due to clarithromycin (173).

- A 77-year-old man who was HIV-negative developed mania after 6 days treatment with clarithromycin 1 g/day for a soft tissue infection; his mental state resolved on withdrawal (174).
- A 53-year-old Canadian lawyer taking long-term fluoxetine and nitrazepam developed a frank psychosis 1–3 days after starting to take clarithromycin 500 mg/day for a chest infection (175). His symptoms resolved on withdrawal of all three drugs, and did not recur with erythromycin or when fluoxetine and nitrazepam were restarted in the absence of antibiotics.

The symptoms may have been due to a direct effect of clarithromycin or else inhibition of hepatic cytochrome P450 metabolism, leading to fluoxetine toxicity.

Clarithromycin occasionally causes hallucinations.

- Visual hallucinations with marked anxiety and nervousness occurred after the second dose of oral clarithromycin 500 mg in a 32-year-old woman (176). Clarithromycin was withdrawn and the symptoms disappeared a few hours later.
- Visual hallucinations developed in a 56-year-old man with chronic renal insufficiency and underlying aluminium intoxication maintained on peritoneal dialysis 24 hours after he started to take clarithromycin 500 mg bd for a chest infection, and resolved completely 3 days after withdrawal (177).

Clonidine and apraclonidine

Clonidine has been used to treat hypertension and migraine. It ameliorates the opioid withdrawal syndrome by reducing central noradrenergic activity. Its role in the treatment of psychiatric disorders has been the subject of an extensive review, but without new information on its safety (178). Clonidine is also used epidurally, in combination with opioids, neostigmine, and anesthetic and analgesic agents, to produce segmental analgesia, particularly for postoperative relief of pain after obstetrical and surgical procedures.

Clusiaceae

Delirium has been attributed to St. John's wort (*Hypericum perfurtum*).

- A 76-year-old woman began taking an extract of St. John's wort (75 mg/day) and developed delirium and psychosis 3 weeks later (179). She had no relevant medical history and did not take any other medications. She was given risperidone and donepezil hydrochloride, and her paranoid delusions and visual hallucinations improved.

The final diagnosis was acute psychotic delirium associated with St. John's wort in a woman with underlying Alzheimer's dementia.

Hypomania has been reported with St. John's wort (180).

- A 47-year-old woman with an 8-year history of nocturnal panic attacks and a recent history of major depression had a poor response to SSRIs and instead took a 0.1% tincture of St. John's wort. After 10 days she noted racing and distorted thoughts, increased irritability, hostility, aggressive behavior, and a reduced need for sleep. After discontinuing the herbal treatment, her symptoms resolved within 2 days.

The author suggested that St. John's wort had caused this episode of hypomania.

Two cases of mania have been associated with the use of St. John's wort (181). The authors pointed out that

St. John's wort, like all antidepressants, can precipitate hypomania, mania, or increased cycling of mood states, particularly in patients with occult bipolar disorder. Alternatively, the mania experienced by these patients could simply be the expression of the natural cause of their psychiatric illness.

Codeine

The same investigators further analysed the data according to codeine abuse/dependent and non-dependent status (182). There was codeine abuse/dependence in 41% of the subjects. The most common psychological and physical problems attributed to codeine use in the dependent/abuse group ($n = 124$) were depression (23%), anxiety (22%), gastrointestinal disturbances (15%), constipation (6%), and headache/migraine (5%). The codeine-dependent subjects were younger than those who were not dependent. The mean age of the respondents when they first started using codeine was 26 (range 2–78) years. A total of 563 codeine products had been used on a regular basis; the most common combination involved codeine with paracetamol (70%). Codeine was used for headaches (41%), back pain (22%), and other types of pain (25%). Those in the dependent group were more likely to have used codeine initially for other reasons, for example for pleasure and to relax or reduce stress. Most subjects said that they had obtained their codeine from one physician (66%) or by purchasing it over the counter (54%). Subjects in the dependent group also obtained codeine from friends (32%), from family (11%), "off the street" (19%), and through prescriptions from more than one physician (11%). Overall, more subjects in the dependent group considered themselves to have problems with a larger number of substances (3.3: range 0–15) compared with those in the non-dependent group (1.2: range 0–8). In the dependent group, more subjects said that their physical or mental health problems interfered with normal social activities at least "quite a bit." Significantly more subjects in the dependent group had sought help for a mental health problem, had had an inpatient psychiatric admission, or had sought help for a substance-use disorder, especially alcohol and stimulants. A larger number of subjects in the dependent group identified at least one family member (usually male) with substance-use problems compared with those in the non-dependent group.

Corticosteroids—glucocorticoids

The psychostimulant effects of the glucocorticoids are well known (183), and their dose dependency is recognized (SED-11, 817); they may amount to little more than euphoria or comprise severe mental derangement, for example mania in an adult with no previous psychiatric history (SEDA-17, 446) or catatonic stupor demanding electroconvulsive therapy (184). In their mildest form, and especially in children, the mental changes may be detectable only by specific tests of mental function (185). Mental effects can occur in patients treated with fairly low doses; they can also occur after withdrawal or omission of treatment, apparently because of adrenal suppression (186,187).

- A 32-year-old woman developed irritability, anger, and insomnia after taking oral prednisone (60 mg/day) for a relapse of ileal Crohn's disease (188). The prednisone was withdrawn and replaced by budesonide (9 mg/day), and the psychiatric adverse effects were relieved after 3 days. A good clinical response was maintained, with no relapse after 2 months of budesonide therapy.

Dexamethasone has been used in ventilator-dependent preterm infants to reduce the risk and severity of chronic lung disease. Usually it is given in a tapering course over a long period (42 days). The effects of dexamethasone on developmental outcome at 1 year of age has been evaluated in 118 infants of very low birthweights (47 boys and 71 girls, aged 15–25 days), who were not weaning from assisted ventilation (189). They were randomly assigned double-blind to receive placebo or dexamethasone (initial dose 0.25 mg/kg) tapered over 42 days. A neurological examination, including ultrasonography, was done at 1 year of age. Survival was 88% with dexamethasone and 74% with placebo. Both groups obtained similar scores in mental and psychomotor developmental indexes. More dexamethasone-treated infants had major intracranial abnormalities (21 versus 11%), cerebral palsy (25 versus 7%; OR = 5.3; CI = 1.3, 21), and unspecified neurological abnormalities (45 versus 16%; OR = 3.6; CI = 1.2, 11). Although the authors suggested an adverse effect, they added other possible explanations for these increased risks (improved survival in those with neurological injuries or at increased risk of such injuries).

Children have marked increases in behavioral problems during treatment with high-dose prednisone for relapse of nephrotic syndrome, according to the results of a study conducted in the USA (190). Ten children aged 2.9–15 years (mean 8.2 years) received prednisone 2 mg/kg/day, tapering at the time of remission, which was at week 2 in seven patients. At baseline, eight children had normal behavioral patterns and two had anxious/depressed and aggressive behavior using the Child Behaviour Checklist (CBCL). During high-dose prednisone therapy, five of the eight children with normal baseline scores had CBCL scores for anxiety, depression, and aggressive behavior above the 95th percentile for age. The two children with high baseline CBCL scores had worsening behavioral problems during high-dose prednisone. Behavioral problems occurred almost exclusively in the children who received over 1 mg/kg every 48 hours. Regression analysis showed that prednisone dosage was a strong predictor of increased aggressive behaviour.

Intravenous methylprednisolone was associated with a spectrum of adverse reactions, most frequently behavioral disorders, in 213 children with rheumatic disease, according to the results of a US study (191). However, intravenous methylprednisolone was generally well tolerated. The children received their first dose of intravenous methylprednisolone 30 mg/kg over at least 60 minutes, and if the first dose was well tolerated they were given further infusions at home under the supervision of a nurse. There was at least one adverse reaction in 46 children (22%) of whom 18 had an adverse reaction within the first three doses. The most commonly reported adverse reactions were behavioral disorders (21 children), including mood changes, hyperactivity, hallucinations, disorientation, and sleep disorders. Several children had serious acute reactions, which were readily controlled. Most of them were able to continue methylprednisolone therapy with premedication or were given an alternative glucocorticoid. The researchers emphasized the need to monitor treatment closely and to have appropriate drugs readily available to treat adverse reactions.

Large doses are most likely to cause the more serious behavioral and personality changes, ranging from extreme nervousness, severe insomnia, or mood swings to psychotic episodes, which can include both manic and depressive states, paranoid states, and acute toxic psychoses. A history of emotional disorders does not necessarily preclude glucocorticoid treatment, but existing emotional instability or psychotic tendencies can be aggravated by glucocorticoids. Such patients as these should be carefully and continuously observed for signs of mental changes, including alterations in the sleep pattern. Aggravation of psychiatric symptoms can occur not only during high-dose oral treatment, but also after any increase in dosage during long-term maintenance therapy; it can also occur with inhalation therapy (192). The psychomotor stimulant effect is said to be most pronounced with dexamethasone and to be much less with methylprednisolone, but this concept of a differential psychotropic effect still has to be confirmed.

The effects of prednisone on memory have been assessed (SEDA-21, 413) (193). Glucocorticoid-treated patients performed worse than controls in tests of explicit memory. Pulsed intravenous methylprednisolone (2.5 g over 5 days, 5 g over 7 days, or 10 g over 5 days) caused impaired memory in patients with relapsing-remitting multiple sclerosis, but this effect is reversible, according to the results of an Italian study (194). Compared with ten control patients, there was marked selective impairment of explicit memory in 14 patients with relapsing-remitting multiple sclerosis treated with pulsed intravenous methylprednisolone. However, this memory impairment completely resolved 60 days after methylprednisolone treatment.

Glucocorticoids can regulate hippocampal metabolism, physiological functions, and memory. Despite evidence of memory loss during glucocorticoid treatment (SEDA-23, 428), and correlations between memory and cortisol concentrations in certain diseases, it is unclear whether exposure to the endogenous glucocorticoid cortisol in amounts seen during physical and psychological stress in humans can inhibit memory performance in otherwise healthy individuals. In an elegant experiment on the effect of cortisol on memory, 51 young healthy volunteers (24 men and 27 women) participated in a double-blind, randomized, crossover, placebo-controlled trial of cortisol 40 mg/day or 160 mg/day for 4 days (195). The lower dose of cortisol was equivalent to the cortisol delivered during a mild stress and the higher dose to major stress.

Cognitive performance and plasma cortisol were evaluated before and until 10 days after drug administration. Cortisol produced a dose-related reversible reduction in verbal declarative memory without effects on nonverbal memory, sustained or selective attention, or executive function. Exposure to cortisol at doses and plasma concentrations associated with physical and psychological stress in humans can reversibly reduce some elements of memory performance.

- Psychosis occurred in a 32-year-old woman who was taking ciprofloxacin for multidrug resistant tuberculosis; the symptoms resolved within 48 hours after the ciprofloxacin was withdrawn (169).

Prednisone, 10 mg/day for 1 year, has been evaluated in 136 patients with probable Alzheimer's disease in a double-blind, randomized, placebo-controlled trial (196). There were no differences in the primary measures of efficacy (cognitive subscale of the Alzheimer Disease Assessment Scale), but those treated with prednisone had significantly greater memory impairment (Clinical Dementia sum of boxes), and agitation and hostility/suspicion (Brief Psychiatric Rating Scale). Other adverse effects in those who took prednisone were reduced bone density and a small rise in intraocular pressure.

In healthy individuals undergoing acute stress, there was specifically impaired retrieval of declarative long-term memory for a word list, suggesting that cortisol-induced impairment of retrieval may add significantly to the memory deficits caused by prolonged treatment (197).

The effects of acute systemic dexamethasone administration on sleep structure have been investigated. Dexamethasone caused significant increases in REM latency, the percentage time spent awake, and the percentage time spent in slow-wave sleep. There were also significant reductions in the percentage time spent in REM sleep and the number of REM periods (SEDA-21, 413) (198).

Chronic glucocorticoid exposure is associated with reduced size of the hippocampus, resulting in impaired declarative memory. In 52 renal transplant recipients (mean age 45 years, 34 men and 18 women) taking prednisone (100 mg/day for 3 days followed by 10 mg/day for as long as needed; mean dose 11 mg/day) there was a major reduction in immediate recall but not delayed recall (199). However, there was a significant correlation between mean prednisone dose and delayed recall.

In animals, phenytoin pretreatment blocks the effects of stress on memory and hippocampal histology. In a double blind, randomized, placebo-controlled trial 39 patients (mean age 44 years, 8 men) with allergies or pulmonary or rheumatological illnesses who were taking prednisone (mean dose 40 mg/day) were randomized to either phenytoin (300 mg/day) or placebo for 7 days (200). Those who took phenytoin had significantly smaller increases in a mania self-report scale. There was no effect on memory. Thus, phenytoin blocked the hypomanic effects of prednisone, but not the effects on declarative memory.

Seventeen patients taking long-term glucocorticoid therapy (16 women, mean age 47 years, mean prednisone dose 16 mg, mean length of current treatment 92 months) and 15 matched controls were assessed with magnetic resonance imaging and proton magnetic resonance spectroscopy, neurocognitive tests (including the Rey Auditory Verbal Learning Test, Stroop Colour Word Test, Trail Making Test, and estimated overall intelligent quotient), and psychiatric scales (including the Hamilton Rating Scale for Depression, Young Mania Rating Scale, and Brief Psychiatric Rating Scale) (201). Glucocorticoid-treated patients had smaller hippocampal volumes and lower N-acetylaspartate ratios than controls. They had lower scores on the Rey Auditory Verbal Learning Test and Stroop Colour Word Test (declarative memory deficit) and higher scores on the Hamilton Rating Scale for Depression and the Brief Psychiatric Rating Scale (depression). These findings support the idea that chronic glucocorticoid exposure is associated with changes in hippocampal structure and function.

Mania has been attributed to glucocorticoids (202).

- A 46-year-old man, with an 8-year history of cluster headaches and some episodes of endogenous depression, took glucocorticoids 120 mg/day for a week and then a tapering dosage at the start of his latest cluster episode. His headaches stopped but then recurred after 10 days. He was treated prophylactically with verapamil, but a few days later, while the dose of glucocorticoid was being tapered, he developed symptoms of mania. The glucocorticoids were withdrawn, he was given valproic acid, and his mania resolved after 10 days. Verapamil prophylaxis was restarted and he had no more cluster headaches.

The authors commented that the manic symptoms had probably been caused by glucocorticoids or glucocorticoid withdrawal. They concluded that patients with cluster headache and a history of affective disorder should not be treated with glucocorticoids, but with valproate or lithium, which are effective in both conditions. Lamotrigine, an anticonvulsive drug with mood-stabilizing effects, may prevent glucocorticoid-induced mania in patients for whom valproate or lithium are not possible (203).

Glucocorticoids can cause neuropsychiatric adverse effects that dictate a reduction in dose and sometimes withdrawal of treatment. Of 32 patients with asthma (mean age 47 years) who took prednisone in a mean dosage of 42 mg/day for a mean duration of 5 days, those with past or current symptoms of depression had a significant reduction in depressive symptoms during prednisone therapy compared with those without depression (204). After 3–7 days of therapy there was a significant increase in the risk of mania, with return to baseline after withdrawal.

Of 92 patients with systemic lupus erythematosus (78 women, mean age 34 years) followed between 1999 and 2000, psychiatric events occurred in six of those who were treated with glucocorticoids for the first time or who received an augmented dose, an overall 4.8% incidence

(205). The psychiatric events were mood disorders with manic features (delusions of grandiosity) (n = 3) and psychosis (auditory hallucinations, paranoid delusions, and persecutory ideas) (n = 3). Three patients were first time users (daily prednisone dose 30–45 mg/day) and three had had mean increases in daily prednisone dose from baseline of 26 (range 15–33) mg. All were hypoalbuminemic and none had neuropsychiatric symptoms before glucocorticoid treatment. All the events occurred within 3 weeks of glucocorticoid administration. In five of the six episodes, the symptoms resolved completely after dosage reduction (from 40 mg to 18 mg) but in one patient an additional 8-week course of a phenothiazine was given. In a multivariate regression analysis, only hypoalbuminemia was an independent predictor of psychiatric events (HR = 0.8, 95% CI = 0.60, 0.97).

Although mood changes are common during short-term, high-dose, glucocorticoid therapy, there are virtually no data on the mood effects of long-term glucocorticoid therapy. Mood has been evaluated in 20 outpatients (2 men, 18 women), aged 18–65 years taking at least 7.5 mg/day of prednisone for 6 months (mean current dose 19 mg/day; mean duration of current prednisone treatment 129 months) and 14 age-matched controls (1 man, 13 women), using standard clinician-rated measures of mania (Young Mania Rating Scale, YMRS), depression (Hamilton Rating Scale for Depression, HRSD), and global psychiatric symptoms (Brief Psychiatric Rating Scale, BPRS, and the patient-rated Internal State Scale, ISS) (206). Syndromal diagnoses were evaluated using a structured clinical interview. The results showed that symptoms and disorders are common in glucocorticoid-dependent patients. Unlike short-term prednisone therapy, long-term therapy is more associated with depressive than manic symptoms, based on the clinician-rated assessments. The Internal State Scale may be more sensitive to mood symptoms than clinician-rated scales.

Two women developed secondary bipolar disorder associated with glucocorticoid treatment and deteriorated to depressive–catatonic states without overt hallucinations and delusions (207).

- A 21-year-old woman, who had taken prednisolone 60 mg/day for dermatomyositis for 1 year developed a depressed mood, pessimistic thought, irritability, poor concentration, diminished interest, and insomnia. Although the dose of prednisolone was tapered and she was treated with sulpiride, a benzamide with mild antidepressant action, she never completely recovered. After 5 months she had an exacerbation of her dermatomyositis and received two courses of methylprednisolone pulse therapy. Two weeks after the second course, while taking prednisolone 50 mg/day, she became hypomanic and euphoric. She improved substantially with neuroleptic medication and continued to take prednisolone 5 mg/day. About 9 months later she developed depressive stupor without any significant psychological stressor or changes in prednisolone dosage. She had mutism, reduction in contact and reactivity, immobility, and depressed mood. Manic or mixed state and psychotic symptoms were not observed. She was initially treated with intravenous clomipramine 25 mg/day followed by oral clomipramine and lithium carbonate. She improved markedly within 2 weeks with a combination of clomipramine 100 mg/day and lithium carbonate 300 mg/day. Prednisolone was maintained at 5 mg/day.

- A 23-year-old woman with ulcerative colitis and no previous psychiatric disorders developed emotional lability, euphoria, persecutory delusions, irritability, and increased motor and verbal activity 3 weeks after starting to take betamethasone 4 mg/day. She improved within a few weeks with bromperidol 3 mg/day. After 10 months she became unable to speak and eat, was mute, depressive, and sorrowful, and responded poorly to questions. There were no neurological signs and betamethasone had been withdrawn 10 months before. She was treated with intravenous clomipramine 25 mg/day and became able to speak. Intravenous clomipramine caused dizziness due to hypotension, and amoxapine 150 mg/day was substituted after 6 days. All of her symptoms improved within 10 days. Risperidone was added for mood lability and mild persecutory ideation.

There have been sporadic case reports of adverse psychiatric effects in patients using inhaled glucocorticoids. Of 60 preschool children with a recent diagnosis of asthma taking inhaled budesonide 100–200 micrograms/day, nine had suspected neuropsychological adverse events after 18 months, according to their parents (208). The symptoms reported were irritability, depression, aggressiveness, excitability, and hyperactivity. These adverse events disappeared when the medication was terminated or reduced and recurred when budesonide was restarted at higher doses. Most of the symptoms occurred within 2 days from starting the high dose (200 micrograms 2–4 times a day).

The management of a psychotic reaction in an Addisonian patient taking a glucocorticoid needs special care (SED-8, 820). Psychotic reactions that do not abate promptly when the glucocorticoid dosage is reduced to the lowest effective value (or withdrawn) may need to be treated with neuroleptic drugs; occasionally these fail and antidepressants are needed (SEDA-18, 387). However, in other cases, antidepressants appear to aggravate the symptoms.

- Two patients with prednisolone-induced psychosis improved on giving the drug in three divided daily doses. Recurrence was avoided by switching to enteric-coated tablets.

This suggests that in susceptible patients the margin of safety may be quite narrow (SED-12, 982). It is possible that reduced absorption accounted for the improvement in this case, but attention should perhaps be focused on peak plasma concentrations rather than average steady-state concentrations.

In one case, glucocorticoid-induced catatonic psychosis unexpectedly responded to etomidate (209).

- A 27-year-old woman with myasthenia gravis taking prednisolone 100 mg/day became unresponsive and had respiratory difficulties. She was given etomidate 20 mg intravenously to facilitate endotracheal intubation. One minute later she became alert and oriented, with normal muscle strength, and became very emotional. Eight hours later she again became catatonic and had a similar response to etomidate 10 mg. Glucocorticoid-induced catatonia was diagnosed, her glucocorticoid dosage was reduced, and she left hospital uneventfully 4 days later.

The effect of etomidate on catatonia, similar to that of amobarbital, was thought to be due to enhanced GABA receptor function in patients with an overactive reticular system.

A case report has suggested that risperidone, an atypical neuroleptic drug, can be useful in treating adolescents with glucocorticoid-induced psychosis and may hasten its resolution (210).

- A 14-year-old African-American girl with acute lymphocytic leukemia was treated with dexamethasone 24 mg/day for 25 days. Four days after starting to taper the dose she had a psychotic reaction with visual hallucinations, disorientation, agitation, and attempts to leave the floor. Her mother refused treatment with haloperidol. Steroids were withdrawn and lorazepam was given as needed. Nine days later the symptoms had not improved. She was given risperidone 1 mg/day; within 3 days the psychotic reaction began to improve and by 3 weeks the symptoms had completely resolved.

Obsessive-compulsive behavior after oral cortisone has been described (211).

- A 75-year-old white man, without a history of psychiatric disorders, took cortisone 50 mg/day for 6 weeks for pulmonary fibrosis and developed severe obsessive-compulsive behavior without affective or psychotic symptoms. He was given risperidone without any beneficial effect. The dose of cortisone was tapered over 18 days. An MRI scan showed no signs of organic brain disease and an electroencephalogram was normal. His symptoms improved 16 days after withdrawal and resolved completely after 24 days. Risperidone was withdrawn without recurrence.

Corticotrophins (corticotropin and tetracosactide)

Mood changes continue to be reported in association with corticotropin (212). Emotional instability or psychotic tendencies can be aggravated, while euphoria, insomnia, and personality changes such as hypomania and depression can be precipitated, sometimes even with psychotic manifestations. Although it seems reasonable to assume that one is dealing here mainly with an effect of the glucocorticoids secreted in response to corticotropin, it should be recalled that segments of the corticotropin molecule themselves have effects on brain function and could conceivably play a role.

Co-trimoxazole

Delirium and psychosis have been rarely reported with co-trimoxazole, but are more likely in elderly people (213).

COX-2 inhibitors

Psychiatric effects have been previously reported with celecoxib (SEDA-26, 123), and may represent a class effect of the COX-2 inhibitors, according to the Australian Adverse Reactions Advisory Committee (ADRAC), which has received 142 reports of acute neuropsychiatric reactions attributed to celecoxib and 49 to rofecoxib (214). The most common reactions associated with celecoxib were: confusion (n = 23) somnolence (n = 22), and insomnia (n = 21), while those associated with rofecoxib were confusion (n = 18) and hallucinations (n = 11). In many cases the onset of the reaction occurred within 24 hours of the first dose of the drug.

Cyclobenzaprine

Rarely, manic psychosis can be activated in patients with bipolar affective disorders (215).

First-onset paranoid psychosis has also been reported (216).

- A 36-year-old woman with no past psychiatric problems took 23 tablets of cyclobenzaprine (10 mg each) over 6 weeks to ease back pain resulting from a back injury. She developed insomnia, reduced appetite, poor concentration, irritability, disorganized thoughts, persecutory delusions, and auditory hallucinations. Cyclobenzaprine was withdrawn and a course of loxapine was started, leading to rapid and complete resolution of her agitation and psychotic symptoms within 72 hours. Loxapine was subsequently quickly withdrawn with no ill effects and she recovered fully.

The authors thought that this psychotic episode was related to cyclobenzaprine, in view of the temporal relation of the symptoms to the administration of cyclobenzaprine and their rapid resolution after withdrawal.

Cycloserine

Cycloserine can cause altered mood, cognitive deterioration, dysarthria, confusion, and even psychotic crises (217).

Cyproheptadine

A central anticholinergic syndrome with psychotic symptoms in a 9-year-old boy on therapeutic doses has also been described (218).

Cyproheptadine (6 mg/day) was considered to be the most likely cause of aggressive behaviour in a 5-year-old boy (219).

Dapsone

Dapsone-induced psychosis has rarely been reported (SEDA-15, 331) (220).

Desloratadine

The results of several studies suggest that desloratadine has minimal or no effects on cognitive functions and psychomotor performance (221–223).

Dextromethorphan

Dextromethorphan-induced psychotic and/or manic-like symptoms have been reported.

- A 2-year-old child developed hyperirritability, incoherent babbling, and ataxia after being over-medicated with a pseudoephedrine/dextromethorphan over-the-counter combination cough formulation for upper respiratory symptoms (224). The symptoms abated after withdrawal of the product.

In another three cases (girls aged 10, 13, and 15 years) severe acute psychosis was associated with the use of an over-the-counter formulation containing ephedrine or pseudoephedrine and dextromethorphan combined with other compounds (225). The psychopathology included agitation, depressed mood, flat affect, pressure of speech, visual and auditory hallucinations, and paranoia. All three improved dramatically, with residual symptoms of irritability, 2–4 days after withdrawal of the mixture and treatment with risperidone 0.5–2.0 mg/day.

- An 18-year-old student had dissociative phenomenon, nihilistic and paranoid delusions, vivid visual hallucinations, thought insertion, and broadcasting after having consumed 1–2 bottles of cough syrup (dextromethorphan 711 mg per bottle) every day for several days (226). The psychotic symptoms remitted completely without any treatment 4 days after withdrawal of dextromethorphan. He was hospitalized twice more over the next 2 months with similar symptoms; each time he had consumed large doses of dextromethorphan.

Cautious use of over-the-counter formulations is recommended in patients with a predisposition to affective illness (SEDA-21, 87).

Cognitive deterioration has been reported from prolonged use of dextromethorphan (227).

Diethyltoluamide

Acute manic psychosis has been attributed to percutaneous absorption of diethyltoluamide (SEDA-12, 138).

Diphenhydramine

Use of diphenhydramine, mainly as a sleeping aid, has been associated with cognitive impairment in elderly people without dementia (228).

Diphenhydramine has been associated with acute delirium in elderly patients with mild dementia, even in single doses of 25 mg (SEDA-19, 173).

Children and adolescents who are given diphenhydramine as premedication, often intravenously as a bolus, to prevent the adverse effects of blood transfusion, can develop drug-seeking behavior. It is recommended that in these circumstances antihistamines should be given orally or infused slowly (229).

Disopyramide

Acute psychosis has been attributed to disopyramide (230,231).

Disulfiram

Disulfiram can cause distressing neuropsychiatric effects including paranoia, impaired memory and concentration, ataxia, dysarthria, and frontal release signs (signs that can be indicative of permanent structural damage or temporary metabolic or infectious changes to the brain's frontal lobes), such as snout and grasp reflexes (232,233). The mechanisms are not properly understood, but adverse effects develop more frequently in subjects with low plasma dopamine beta-hydroxylase activity. In one study, a research subject with low dopamine beta-hydroxylase activity developed a schizophrenic reaction to disulfiram; it would be useful to know whether determining blood dopamine beta-hydroxylase activities predicts the risk of adverse reactions to disulfiram (234).

Of 52 patients with alcohol dependence/abuse who were given disulfiram 250 mg bd after food, six developed psychotic symptoms; all had a mood disorder but no thought disorder (235). The psychotic symptoms remitted completely after withdrawal and a short course of antipsychotic therapy, except in one patient who had to be given lithium.

- A 47-year-old man with alcohol abuse took disulfiram (236). He developed a psychosis while taking it and for 2 weeks after. He stated that he had not taken alcohol. He was successfully treated with antipsychotic drugs. Afterwards it was discovered that his family history was positive for schizophrenia; it is therefore possible that he was more vulnerable to develop psychosis due to disulfiram.

Prolonged toxic delirium related to disulfiram and alcohol intake has been reported (237). The predominant presenting feature was neuropsychiatric rather than autonomic symptoms.

- A 50-year-old woman with a history of bipolar disorder type I and alcohol dependence taking disulfiram had a 4-day history of a change in mental status, including visual hallucinations and deficits in orientation and concentration. Other features included a tachycardia and non-focal neurological signs. Extensive metabolic, infectious, and neurological investigations revealed no abnormalities that alone could explain her acute confusional state. It was subsequently discovered that she had drunk alcohol on at least two separate occasions while taking disulfiram before her change in mental status, and that a similar, although shorter, episode had occurred previously.

Donepezil

Behavioral worsening in seven patients with Alzheimer's disease after the start of donepezil therapy has been described (238). Their mean age was 76 years, and their mean score on the Mini-Mental State Examination was 18. Five patients had had dementia-related delusions and irritability before taking donepezil, one had had a history of major depression, and another had had a history of somatization disorder. At the start of treatment with donepezil, four were taking sertraline, one paroxetine, one venlafaxine, and four risperidone. All took donepezil 5 mg/day, and after 4–6 weeks the dosage in five patients was increased to 10 mg/day. In the other two cases, donepezil was discontinued after 5 weeks: in one case because of gastrointestinal symptoms and in the other because of increasing agitation. After an average of 7.3 (range 1–13) weeks after starting donepezil, all seven patients had a recurrence of previous behavioral problems. Five became agitated, one became depressed, and the other became more anxious and somatically preoccupied. The pattern of behavioral change involves regression to an earlier behavioral problem.

Violent behavior has been described with donepezil (239).

- A 76-year-old man who was taking oxybutynin 3 mg tds for bladder instability took donepezil 5 mg/day for presumed Alzheimer's disease and 5 days later became very paranoid, believing that his wife had been stealing his money. He beat her and held her hostage in their house with a knife until their daughter intervened. He was given haloperidol 0.5 mg bd, and donepezil and oxybutynin were withdrawn. His paranoid ideation resolved within a few days and did not recur despite withdrawal of haloperidol.

Although a causal relation between this violent incident and donepezil cannot be proved, the temporal relation was suggestive.

Dopamine receptor agonists

In seven patients with Parkinson's disease (age 55–66, 5 men), none of whom was demented, intravenous infusion of levodopa produced some reduction in self-reported learning performance and a reduction in activation of the ipsilateral occipital association area (240). The authors suggested that levodopa may have some hitherto undetected but subtle effects on cognitive performance.

Complex behavioral changes can apparently be caused by medication in Parkinson's disease. High-dose dopamine receptor agonist therapy has been associated with excessive gambling, which can be regarded as pathological (241). In a review of the records of 1884 patients, of whom 529 were taking pramipexole, 421 ropinirole, and 331 pergolide there were nine patients (7 men, 2 women) whose gambling behavior had led to financial hardship, in two cases with losses in excess of $60 000. None of these patients was taking levodopa alone and eight were taking pramipexole. The mean age of the affected patients was 57 years and the mean duration of their illness was 12 years. None was demented, but it is noteworthy that four had a previous history of depression and one of panic attacks. Overall the incidence of gambling behavior in the whole patient population was 0.5% (surely not 0.05% as stated) and 1.5% in those taking pramipexole, although as the authors pointed out this is close to some estimates of this problem in the general population, so a causal association was by no means established and the striking preponderance of one drug may have been coincidental.

Doxazosin

Doxazosin, 16 mg/day, has been reported to have caused an acute psychosis.

- A 71-year-old woman with type II diabetes and hypertension began to hear voices and to have auditory hallucinations. Doxazosin was progressively withdrawn over the next 14 days and by the time the dosage had been reduced to 8 mg a day the psychosis was much less severe; it disappeared completely after withdrawal (242).

Ebastine

The effects of ebastine 10 mg on cognitive impairment have been assessed in 20 healthy volunteers who performed six types of attention-demanding cognitive tasks, together with objective measurements of reaction times and accuracy (243). Ebastine was compared with placebo and a positive control, chlorphenamine 2 mg and 6 mg. Compared with placebo, ebastine had no effect on any objective cognitive test nor any effect on subjective sleepiness. In contrast, chlorphenamine significantly increased reaction times, reduced accuracy in cognitive tasks, and increased subjective sleepiness. The effect of chlorphenamine increased with plasma concentration.

Efavirenz

Efavirenz has been associated with psychiatric problems, such as anxiety, depression, and confusion (244,245). In a retrospective study of 1897 patients, dementia and depression were significantly associated with efavirenz compared with other drugs; the respective odds ratios were 4.0 (95% CI = 1.2, 14) and 1.7 (1.0, 3.0) (246). However, those who were given efavirenz were perhaps more ill than those who were not, judging by CD4 counts and opportunistic infections.

Most clinicians tend to avoid efavirenz in patients with a psychiatric history. However, it is important to remember that efavirenz can precipitate sudden and severe psychiatric symptoms in patients with no such history. Three patients developed sudden irritability; excitability with anxiety; and insomnia, confusion, and amnesia (247).

The psychiatric adverse effects of efavirenz correlate with its plasma concentrations. In 130 HIV-infected patients, toxicity was three times more common in patients who had an efavirenz concentration over 4000 ng/ml (248).

In a questionnaire survey of 152 patients who stopped taking efavirenz 82 did so because of neuropsychiatric symptoms; a history of multiple episodes of depression was associated with efavirenz discontinuation (249).

Enflurane

There was a reduced capacity for learning and decision-making in healthy volunteers after exposure to subanesthetic concentrations of enflurane (250,251).

Ephedra, **ephedrine, and pseudoephedrine**

Ma huang, which contains ephedrine, can cause psychiatric complications, which can last several weeks. These have been reviewed in the context of two cases of psychotic reactions (252).

- A 27-year-old US Marine presented with depressed affect, irritability, and poor concentration and eventually admitted 2 years self-medication with Ma huang to improve workout performance.
- A 27-year-old US Marine developed a frank psychosis with ideas of reference and some paranoid ideation. He had been taking two preparations containing Ma huang, although the duration of use was unclear.

After discontinuing the drug both made a full recovery. The authors emphasized the importance of recognizing possible abuse of such "natural" medications, widely perceived to be harmless despite warnings and attempted restrictions by regulatory authorities. Treatment is supportive while awaiting spontaneous recovery after drug withdrawal.

Pseudoephedrine is often used by scuba divers to avoid ear barotrauma. The psychometric and cardiac effects of pseudoephedrine have been evaluated at 1 atmosphere (100 kPa, sea level) and 3 atmospheres (30 kPa, 20 m) in a double-blind, placebo-controlled, crossover study in 30 active divers in a hyperbaric chamber (253). Pseudoephedrine did not cause significant alterations in psychometric performance at 3 atmospheres.

A young child suffered from visual hallucinations caused by high doses of pseudoephedrine (254).

Ergot derivatives

Unlike the dopamine-mimetic ergolines, ergotamine is not usually regarded as a drug with major effects on the brain. However, this may not always be true.

- A 75-year-old woman in Arizona developed progressive confusion, auditory hallucinations, and aggressive behavior (255). Her systolic pressure was raised (194 mmHg). She had a history of migraine and many other problems, including hypertension. Her headaches had become more severe over the weeks before admission and she had been gradually increasing her intake of a formulation containing ergotamine 1 mg and caffeine 100 mg: just before the emergence of her psychiatric symptoms she was taking 14 tablets per week. After withdrawal of the ergotamine/caffeine, her mental state and blood pressure returned to normal, without antihypertensive medication, and remained so 12 months later.

Although there has been extensive discussion of the phenomenon of ergotism in medieval and early modern times, its features do not closely resemble this case, perhaps because of the co-administration of caffeine. However, regardless of that, there does seem to be a strong case for a drug effect here.

A syndrome resembling reversible dementia has been described in chronic ergotamine intoxication (SEDA-3, 121).

Erythromycin

Erythromycin has been associated with complications such as confusion, paranoia, visual hallucinations, fear, lack of control, and nightmares. These suspected psychiatric adverse effects were seen within 12–48 hours of starting therapy with conventional doses. Such complications may even be under-reported (256–259).

Epoetin

Visual hallucinations have been reported in patients receiving epoetin (260).

Estrogens

The effects of estrogens on mood tend to be positive, and improved performance in intellectual tests has been described (SEDA-20, 382) (261); this is in parallel with the known effects of endogenous estrogens. During the menopause some women become depressed and irritable, and the ability of estrogens to correct this has been delineated in various studies, including work with estradiol given transdermally (262). Some workers also claim increased vigilance, and have concluded that this is reflected in encephalographic changes. There is even some evidence of an improvement in mental balance and self-control when estrogens are given to demented and aggressive old people of both sexes (263). However, all of these effects of estrogens on mood or mental

performance are only likely to last for as long as the treatment does, and the effects on mood may occur only at the start of treatment; altered mood can follow acute withdrawal.

Ethionamide and protionamide

Psychotic reactions have been described in patients taking ethionamide (264,265) and may be exacerbated by alcohol (266).

Fenfluramine

The effect of fenfluramine is usually not one of stimulation but of calmness or drowsiness. However, in predisposed subjects, it can precipitate psychotic illness. Several published cases illustrate this (267). It is wise to avoid fenfluramine in patients prone to endogenous depression or psychosis.

Fentanyl

Mood alteration during patient-controlled epidural anesthesia with either morphine or fentanyl was compared in a randomized, double-blind study of 52 patients undergoing elective hip or knee joint arthroplasty under general anesthesia (268). Mood was assessed preoperatively and at 24, 48, and 72 hours, using the bipolar version of the Profile of Mood States. Pain intensity postoperatively did not vary with morphine or fentanyl and, as expected, both fentanyl and morphine users had significant somnolence, pruritus, and nausea compared with baseline. With morphine, the mean score for measures of composure/anxiety, elation/depression and clear-headedness/confusion increased, indicating a change toward the more positive pole, but there were negative changes for the fentanyl users' scores for five of the six components of the Profile of Mood States. The difference in test scores between morphine and fentanyl was significant at 48 hours of patient-controlled anesthesia and 24 hours after withdrawal. There was no correlation between mood scores and pain scores, and mood scores with fentanyl fell with increasing plasma concentrations. Previous investigations have shown transient positive feelings with intravenous fentanyl, followed by more negative feelings in the longer term. The authors suggested that the differences in mood between the two groups may have been explained by differences in the lipid solubility and pharmacokinetics of epidural morphine and fentanyl.

Fexofenadine

Even in a high dose (360 mg bd versus the recommended dose of 60 mg bd) fexofenadine had no disruptive effects on psychomotor performance and cognitive function in healthy volunteers (269).

In one study fexofenadine did not alter driving and psychomotor performance when taken in the recommended dosage of 60 mg bd (270).

Fluorouracil

Confusion and cerebral cognitive defects have been attributed to fluorouracil (271,272).

Fluoroquinolones

Psychiatric adverse effects in patients taking fluoroquinolones occur at a rate of 2–4%, causing headache (2–4% of patients), dizziness (2–3%), and other symptoms (under 1%), including confusion, agitation, insomnia, depression, somnolence, vertigo, light-headedness, and tremors. Seizures are rare. Isolated cases of depression (273) and psychosis have been described (274).

Some quinolones displace GABA or compete with GABA binding at receptor sites in the nervous system. Substitution of compounds containing 7-piperazinyl or 7-pyrrolidinyl, such as gatifloxacin, gemifloxacin, and moxifloxacin, is associated with reduced seizure-causing potential. Administration of non-steroidal anti-inflammatory drugs concurrently with certain quinolones has been linked to an increase in the possibility of seizures (275).

- Delirium occurred in a 69-year-old white man with a history of depression, non-insulin-dependent diabetes mellitus, hypertension, and atherosclerotic disease who was treated with intravenous gatifloxacin 400 mg/day (276). After the first dose of gatifloxacin he had numerous hallucinations and the symptoms got worse after each dose. After withdrawal no further hallucinations occurred.

Ofloxacin can cause serious psychiatric adverse effects, particularly in those with a past psychiatric history (277).

In a retrospective study the data on fluoroquinolones and other antibacterial drugs, rufloxacin was associated with a reporting rate of 221 reports/daily defined dose/1000 inhabitants/day, and the most frequent were psychiatric disorders (278).

Flurotyl

Flurotyl caused toxic delirium in two of 135 patients with schizophrenia (SED-13, 12).

Gabapentin

Gabapentin has rarely been implicated in psychiatric reactions. Two cases, one of mania (279) and one of catatonia (280), have been reported.

- A 35-year-old woman with epilepsy without a history of psychiatric disorders developed elevated mood after being stabilized on gabapentin monotherapy (3200 mg/day). After 5 months she developed a manic episode, which remitted when gabapentin was withdrawn.
- Catatonia was reported in a 48-year-old woman with bipolar disorder within 48 hours of withdrawal of gabapentin, 500 mg/day. The condition lasted for several days and disappeared after treatment with a benzodiazepine.

In the first case the presentation was strongly suggestive for a causative role of the drug. In the second it was speculated that the condition might have been provoked by altered GABAergic transmission after gabapentin withdrawal.

- Two women, aged 37 and 38 years, took gabapentin and after a few days developed behavioral changes associated with euphoria (281). In one case the symptoms were transient and in the other they resolved after withdrawal. The behavioral changes were not related to seizure activity.

In a double-blind, crossover comparison of the cognitive effects of carbamazepine (mean dose 731 mg/day) and gabapentin (2400 mg/day), each given for 5 weeks, in 35 healthy volunteers performance on gabapentin was better than on carbamazepine for 8 of 31 variables (visual serial addition test, choice reaction times at initiation and total, memory paragraph I and delayed recall, Stroop Word and Interference, Vigor measure), whereas carbamazepine was better than gabapentin on none (282). Although these data suggest that gabapentin produces fewer cognitive effects than carbamazepine, the applicability of the findings to long-term therapy in epileptic patients is uncertain.

Behavioral disturbances, including aggressiveness, irritability, hyperactivity, and/or dysphoric changes, occur in up to 22% of patients. Children and patients with mental retardation or a history of similar disorders may be at special risk (SEDA-20, 61). In a recent study, hyperactivity, irritability, and agitation occurred in 15 of 32 mostly mentally retarded children, and required drug withdrawal in four (283).

The cognitive effects of carbamazepine and gabapentin have been compared in a double-blind, crossover, randomized study in 34 healthy elderly adults, of whom 19 subjects withdrew (15 while taking carbamazepine, probably due to excessively rapid dosage titration) (284). The primary outcome measures were standardized neuropsychological and mood state tests, yielding 17 variables. Each subject had cognitive testing at baseline (pre-drug), the end of the first drug phase, the end of the second drug phase, and 4 weeks after completion of the second drug phase. Adverse events were frequently reported with both antiepileptic drugs, although they were more common with carbamazepine. There were significant differences between carbamazepine and gabapentin for only 1 of 11 cognitive variables, with better attention/vigilance for gabapentin, although the effect was modest. Both carbamazepine and gabapentin can cause mild cognitive deficits in elderly subjects, and gabapentin has a slightly better profile.

Gabapentin has been associated with hostility, especially in children. Two cases of aggression in adults taking gabapentin for bipolar disorder have been reported (285). Neither patient had a history of aggression. In one the symptoms appeared after 3 days of treatment (1200 mg/day), and in the other after 48 hours (600 mg/day). In the second case, aggression was associated with auditory hallucinations. It is hard to associate new antiepileptic drugs with psychiatric adverse effects in patients with severe psychiatric disorders, and rechallenge with the offending drug should ideally have been tried before postulating a causal relation.

Gentamicin

There are several case reports of acute toxic psychoses due to gentamicin (286).

Gonadorelin

Depressed mood and emotional lability occur in up to 75% of gonadorelin recipients, and there are rare reports of more severe mood disturbances (287). Defects of verbal memory have been described and may be reversed by "add-back" estrogen treatment (287) and sertraline (288).

- A 32-year-old woman had psychotic symptoms of persecutory delusions, agitation, and auditory hallucinations a few days after her second injection of triptorelin (289). Her symptoms recurred after a pregnancy, suggesting that they were due to the rapid fall in estrogen in both instances.

During a 6-month, randomized trial, men randomized to gonadorelin agonists had reduced attention and memory test scores, compared with men who were not given gonadorelin agonists but were closely monitored, in whom there was no change (290).

Griseofulvin

The psychiatric effects of griseofulvin can be very disturbing and are aggravated by alcohol (SED-12, 676) (291).

Halothane

Slight depression of mood, lasting up to 30 days, along with a non-specific slowing of the electroencephalogram for 1–2 weeks, was observed after halothane. In 16 healthy young men, halothane anesthesia had negative effects on postoperative mood and intellectual function, the changes being greatest 2 days after anesthesia, with restoration of function after 8 days (292). In seven subjects, serial electroencephalography, serum bromide determinations, and psychological tests before and after halothane anesthesia showed that there was significant psychological impairment 2 days after anesthesia (293).

The effect of a single preoperative dose of the opioid oxycodone on emergence behavior has been studied in a randomized trial in 130 children (294). Oxycodone prophylaxis, compared with no premedication, significantly reduced the incidence of post-halothane agitation.

HMG coenzyme-A reductase inhibitors

Animal and cross-sectional studies have suggested that serum lipid concentrations can cause altered cognitive function, mood, and behavior (295).

HMG-CoA reductase inhibitors

Emerging data associate statins with a reduced risk of Alzheimer's disease; however, two women had significant cognitive impairment temporally related to statin therapy (296). One took atorvastatin, and the other first took atorvastatin then simvastatin. Cognitive impairment and dementia as potential adverse effects associated with statins has been reviewed (297).

The risk of Alzheimer's disease has previously been mentioned (SEDA-28, 535). In 308 hypercholesterolemic adults aged 35–70 years, daily treatment with placebo, simvastatin 10 mg, or simvastatin 40 mg for 6 months was associated with decremental effects of simvastatin on tests previously observed to be sensitive to statins and on tests not previously administered, but not on tests previously observed to be insensitive to statins (298). For the three tests specifically affected by simvastatin, effects on cognitive performance were small, manifesting only as a failure to improve during the 6 months of treatment, and were confounded by baseline differences on one test. This study provides partial support for minor decrements in cognitive functioning with statins. Whether such effects have any long-term sequelae or occur with other cholesterol-lowering interventions is not known.

When the MedWatch drug surveillance system of the Food and Drug Administration (FDA) from November 1997 to February 2002 was searched for reports of statin-associated memory loss, 60 patients were identified; 36 had taken simvastatin, 23 atorvastatin, and one pravastatin (Wagstaff 871). About a half of the patients noted cognitive adverse effects within 2 months of therapy and 14 of 25 patients noted improvement when the statin was withdrawn. Memory loss recurred in four patients who were rechallenged. The current literature is conflicting with regard to the effects of statins on memory loss. Experimental studies support links between cholesterol intake and amyloid synthesis; however, observational studies suggest that patients taking statins have a reduced risk of dementia. However, available prospective studies show no cognitive or antiamyloid benefits of any statin.

Some studies have shown increased risks of violent death and depression in subjects with reduced serum cholesterol concentrations. Serum and membrane cholesterol concentrations, the microviscosity of erythrocyte membranes, and platelet serotonin uptake have been determined in 17 patients with hypercholesterolemia (299). There was a significant increase in serotonin transporter activity only during the first month of simvastatin therapy. This suggests that within this period some patients could be vulnerable to depression, violence, or suicide. This is an important paper, in that it explains why mood disorders are not regularly seen in clinical trials with statins, as has been summarized in a recent review (300).

Hormonal contraceptives—oral

Psychiatric symptoms have been described in women taking oral contraceptives in isolated case reports (301,302), probably reflecting non-specific effects in susceptible individuals. As to psychological effects, many physicians have found that certain women react to oral contraceptives by becoming morose or unhappy (303), but this does not necessarily mean that they meet the clinical criteria of true depression, the incidence of which has not been found to be increased (304). Several possible biological mechanisms for mood changes have been suggested; however, when nervousness and depression among combined oral contraceptive users are carefully evaluated over time, the pattern is so inconsistent that it is difficult to study. If one looks for anything like consistent depression one is unlikely to find it (305).

Many women who change to an oral contraceptive after unsatisfactory experience with other forms of contraception find greater sexual satisfaction (304) because of relief from worry about pregnancy.

Hormonal contraceptives—progestogen implants

Two cases of major depression and panic disorder, developing soon after insertion of Norplant and resolving after removal, have been reported, but a causal association was not proven (306).

Five women using the Norplant system developed major depression, two of whom also developed obsessive-compulsive disorder and one of whom also developed agoraphobia (307). They had no prior psychiatric history but developed major depression within 1–3 months after insertion of Norplant. The depression worsened over time and in all cases resolved within 1–2 months after removal of Norplant. There was no recurrence of depression after 7–8 months in four cases available for follow-up. In addition to major depression, obsessive-compulsive disorder developed in two women and symptoms of agoraphobia developed in one woman during Norplant treatment, which resolved after removal.

Hormone replacement therapy

The mechanisms underlying the mood changes that are often associated with menstruation, the menopause, and hormonal therapy are not understood, but there is now some evidence that they have an association with the response to the neuroactive steroid pregnanolone and that progestogens and estrogens might alter this response. In a randomized, double-blind, crossover study 26 postmenopausal women with climacteric symptoms took oral estradiol 2 mg/day continuously for two cycles and either vaginal progesterone 800 mg or a placebo during the last 14 days of each cycle (308). Before treatment and again at the end of each treatment cycle pregnanolone was administered intravenously, and its effects on saccadic eye velocity, saccade acceleration, saccade latency, and self-rated sedation were examined. During treatment with either estradiol alone or with added progesterone the effect of pregnanolone on saccadic eye movements and self-rated sedation was increased. Saccadic eye velocity, saccade acceleration, and sedation responses to pregnanolone were also increased in women who usually experienced cyclicity of mood during HRT treatment, but not in those with no history of mood cyclicity.

Hyoscine

In one study hyoscine premedication had detrimental effects on memory and on motor tasks compared with placebo, while atropine did not (309), although the difference is unlikely to be absolute. In view of certain of these effects, hyoscine hydrobromide is not a suitable antiemetic for those likely to drive vehicles before the effect has worn off, for example air passengers.

Hypericum perforatum (St John's wort)

The clinical evidence associating *Hypericum* with psychotic events has been summarized in a systematic review of 17 case reports (310). In 12 instances the diagnosis was *mania or hypomania*. In most of these cases, causality between the herbal remedy and the adverse effect was rated as possible; in no instance was there a positive rechallenge.

Insulin

Using evidence from auditory-evoked brain potentials and hypoglycemic clamps, it has been argued that antecedent hypoglycemia not only reduces awareness, but also that several aspects of cognitive function are attenuated during subsequent hypoglycemia 18–24 hours later (311). However, there were no effects of repeated hypoglycemia on cognitive function in patients included in the DCCT, a large American study that included more than 1400 patients, which showed that normalization of blood glucose prevents or delays the development of secondary (microvascular) complications in type 1 diabetes (312).

Nevertheless, long periods of hypoglycemia can cause permanent brain damage. There is concern that frequent attacks of hypoglycemia impair brain function but there are few hard data.

Hypoglycemic coma due to insulin with extensive mental changes has been reported, including a review of six comparable cases in patients aged 37–56 years, whose coma lasted from 36 hours to 31 days (313).

- A 37-year-old man could not be wakened in the morning. He had injected insulin without eating. His blood glucose was 1.5 mmol/l and he did not improve with intravenous glucose 16 g. In hospital he remained unconscious with a blood glucose of 12.2 mmol/l. There was no alcohol in the blood, his pH was 7.35, and he had a normal anion gap (18 mmol/l). His serum creatinine concentration was 288 μmol/l and his creatine kinase activity was high, suggesting rhabdomyolysis. A brain CT scan was normal and repeated electroencephalography showed slow waves with reduced voltages but no focal changes or irritation. He gradually recovered and was discharged after 6 days. Because of the dissociation between physical and mental improvement he was checked after 6 months and still had antegrade memory loss and problems with memory, complaining that he needed reminders on paper, and had less vitality and reduced emotionality.

Of the six reviewed patients, two died in coma; the other four had neuropsychological problems that did not improve after 6 months and up to 2 years. They had comparable electroencephalographic changes. During coma there was hypokalemia and hypocalcemia combined with increased lipolysis; this may have accounted for the permanent cerebral changes.

Of 20 patients with severe hypoglycemic coma and 20 with no or light coma, those with hypoglycemia had chronic depression and anxiety and performed persistently more poorly in several cognitive tests (314).

In 42 patients with at least two episodes of severe hypoglycemia in the previous 2 years and 51 patients with no episodes, low blood glucose, hypoglycemia-impaired ability to do mental subtractions, and awareness of neuroglycopenia and hypoglycemia predicted future severe attacks of hypoglycemia (315). In another study, blood glucose awareness training increased adrenaline responses to hypoglycemia (316). However, in a reanalysis of data from the Diabetes Control and Complications Study, a large study relating the development of secondary complications to less strict control of blood glucose (317), there was no effect of repeated hypoglycemia (312).

The effect of hypoglycemia on cognitive function has been investigated in 142 children aged 6–15 years with type 1 diabetes intensively treated for 18 months; 58 had 111 periods of treatment. There were no effects on cognitive functions (318). In 29 prepubertal children, with diabetes for at least 12 months and using twice-daily mixed insulin, observed for two nights, asymptomatic hypoglycemia occurred in 13 children on the first night and in 11 children on the second night; cognitive performance was not altered, but mood was reduced (319).

In healthy volunteers hypoglycemia caused significant deterioration in short-term attention, whereas sustained attention and intelligence scores did not deteriorate (320).

Interferon alfa

Neuropsychiatric complications of interferon alfa were recognized in the early 1980s and represent one of the most disturbing adverse effects of interferon alfa (SED-13, 1091; SEDA-20, 327; SEDA-22, 400). Reviews have provided comprehensive analysis of the large amount of experimental and clinical data that have accumulated since 1979 (321,322).

Clinical features

Within a large spectrum of symptoms, complications are classified as acute, subacute, or chronic.

Acute neuropsychological disturbances are usually associated with the flu-like syndrome and include headache, fatigue, and weakness, drowsiness, somnolence, subtle impairment of memory or concentration, and lack of initiative (323). This pattern of cognitive impairment is similar to changes observed during influenza and has also been described in healthy patients who have received a single dose of interferon alfa (324). More severe acute manifestations (for example, marked somnolence or lethargy, frank encephalopathy with visual hallucinations, dementia or delirium, and sometimes coma) have been almost exclusively described in patients receiving more than 20–50 MU (323); vertigo, cramps, apraxia, tremor and dizziness were also reported.

The subacute or chronic neuropsychiatric effects of long-term therapy are usually non-specific, with cognitive impairment (for example visuospatial disorientation, attentional deficits, memory disturbances, slurred speech, difficulties in reading and writing), changes in emotion, mood, and behavior (for example psychomotor slowing, hypersomnia, loss of interest, affective disorders, irritability, agitation, delirium, paranoia, aggressiveness, and murderous impulses). Post-traumatic stress symptoms have also been reported (SEDA-22, 400). As a result, severe psychic distress can be observed during long-lasting treatment or in patients who are otherwise not severely affected (323,325). The most severe psychiatric complications of interferon alfa include rare cases of homicidal ideation, suicidal ideation, and attempted suicide (326).

Behavioral effects

Patients with a wide variety of medical illnesses have behavioral alterations, including depression, at rates 5–10 times higher than in the general population. Recent theories have proposed that inflammatory mediators, notably cytokines, are involved. Interferon alfa causes behavioral symptoms, including depression, fatigue, and cognitive dysfunction. The psychiatric effects of low-dose interferon alfa on brain activity have been assessed using functional MRI during a task of visuospatial attention in patients infected with hepatitis C virus (327). Despite having symptoms of impaired concentration and fatigue, the 10 patients who received interferon alfa had similar task performance and activation of parietal and occipital brain regions to 11 control subjects infected with hepatitis C. However, in contrast to the controls, the patients who received interferon alfa had significant activation in the dorsal part of the anterior cingulate cortex, which correlated highly with the number of task-related errors; there was no such correlation in the controls. Consistent with the role of the anterior cingulate cortex in conflict monitoring, this activation of the anterior cingulate cortex suggests that interferon alfa might increase the processing of conflict or reduce the threshold for conflict detection, thereby signalling the need for mental effort to maintain performance.

Cognitive effects

Neurocognitive performance has been studied in 70 patients receiving interferon alfa 2b (pegylated or conventional) and ribavirin, because impairment of concentration is common during antiviral therapy of chronic hepatitis C (328). Repeated computer-based testing showed significantly increased reaction times. Accuracy measures, reflected by the number of false reactions, were affected only for the working-memory task. Cognitive performance returned to pre-treatment values after the end of therapy. Cognitive impairment was not significantly correlated with the degree of concomitant depression.

A number of mild to moderate frontal–subcortical brain symptoms, including cognitive and behavioral slowing, apathy, impaired executive function, and reduced memory have also been attributed to interferon alfa. These symptoms may alter the quality of life. Of 30 adults treated with interferon-alfa alone (n = 13) or in combination with chemotherapy (low-dose cytarabine arabinoside in 15 or hydroxycarbamide in two), 16 had a significant reduction in one or more cognitive tests compared with baseline (329). The combination with chemotherapy was associated with greater risks of impaired cognitive performance, but these patients also took higher cumulative doses of interferon alfa. Although one-third of patients had depressive symptoms, there was no significant correlation between new depressive symptoms and most of the cognitive decline, suggesting that depression alone did not account for cognitive dysfunction. Very similar marked cognitive impairment, assessed by a battery of computer-assisted psychological tests, was also found in 70 patients with chronic hepatitis C treated with interferon alfa and ribavirin; there was no difference in the quality or intensity of the cognitive changes when comparing standard and pegylated interferon alfa, and no correlation with age or education (330). An important finding was that cognitive performances returned to pre-treatment values after the end of treatment.

Psychiatric effects

The clinical features of mania have been described in four patients with malignant melanoma, with a detailed review of nine other published cases (331). Although seven suffered from depression during treatment, the onset of mania or hypomania was often associated with interferon alfa dosage fluctuation (withdrawal or dose reduction) or introduction of an antidepressant for interferon alfa-induced depression. In these patients, the risk of mood fluctuations persisted for several months after interferon alfa withdrawal, and low-dose gabapentin was considered useful in treating manic disorders and in preventing mood fluctuations. Interferon alfa was suggested as a possible cause of persistent manic-depressive illness for more than 4 years in a 40-year-old man (332). Although the manic episodes may have been coincidental, the negative history and the age of onset are in keeping with a possible role of interferon alfa treatment.

The clinical features, management, and prognosis of psychiatric symptoms in patients with chronic hepatitis C have been reviewed using data from 943 patients treated with interferon alfa (85%) or interferon beta (15%) for 24 weeks (333). Interferon-induced psychiatric symptoms were identified in 40 patients (4.2%) of those referred for psychiatric examination. They were classified in three groups according to the clinical profile: 13 cases of generalized anxiety disorder (group A), 21 cases of

mood disorders with depressive features (group B), and six cases of other psychiatric disorders, including psychotic disorders with delusions/hallucinations (n = 4), mood disorders with manic features (n = 1), and delirium (n = 1) (group C). The time to onset of the symptoms differed significantly between the three groups: 2 weeks in group A, 5 weeks in group B, and 11 weeks in group C. Women were more often affected than men. There was no difference in the incidence or nature of the disorder according to the type of interferon used. Whereas most patients who required psychotropic drugs were able to complete treatment, 10 had to discontinue interferon treatment because of severe psychiatric symptoms, 5 from group B and five from group C. Twelve patients still required psychiatric treatment for more than 6 months after interferon withdrawal. In addition, residual symptoms (anxiety, insomnia, and mild hypothymia) were still present at the end of the survey in seven patients. Delayed recovery was mostly observed in patients in group C and in patients treated with interferon beta. Although several patients with a previous history of psychiatric disorders are sometimes successfully treated with interferon alfa, severe decompensation with persistent psychosis should be regarded as a major possible complication (334).

Evaluation

The neuropsychological adverse effects of long-term treatment have been assessed in 14 patients with myeloproliferative disorders using a battery of psychometric and electroneurological tests before and after 3, 6, 9, and 12 months of treatment (median dose 25 MU/week) (335). In contrast to several previous studies, there was no significant impairment of neurological function, and attention and short-term memory improved during treatment. Despite the small number of patients, these results suggest that prolonged interferon alfa treatment did not cause severe cognitive dysfunction, at least in patients with cancer.

Electroencephalographic (EEG) findings show reversible cerebral changes with slowing of dominant alpha wave activity, and occasional appearance of one and two activity in the frontal lobes, suggesting a direct effect on fronto-subcortical functions. Marked electroencephalographic abnormalities are sometimes observed in asymptomatic patients. The pattern of changes is identical whatever the dose, but the severity of symptoms is dose- and schedule-related. Most patients improve or recover after dosage reduction or withdrawal, and protracted toxicity, with impaired memory, deficits in motor coordination, persistent frontal lobe executive functions, Parkinson-like tremor, and mild dementia, have been occasionally reported (336).

In a study of 67 patients with chronic viral hepatitis, the self-administered Minnesota Multiphasic Personality Inventory (MMPI), which determines the patient's psychological profile, significantly correlated with the clinical evaluation and was a sensitive and reliable tool for identifying patients at risk of depressive symptoms before the start of interferon alfa therapy (337). It was also successfully used to monitor patients during treatment.

Incidence

Psychiatric symptoms have been prospectively examined in 104 patients with chronic hepatitis C, of whom 84 received interferon alfa-2b 15 MU/week and 20 were not treated (338). The incidence of clinically relevant scores for depression, anxiety, or anger/hostility increased from 23% of patients before interferon alfa to 58% of patients during treatment, and returned to 30% and 19% of patients 4 weeks and 6 months after withdrawal respectively. In contrast, there were no significant changes in the reference group. There were also significantly higher scores in the 40 patients who took concomitant ribavirin. Six patients successfully received antidepressant therapy, but withdrawal because of untreatable psychiatric symptoms was needed in 8.3% of patients, i.e. about half of the patients who had interferon-induced major depressive disorders.

The most typical psychiatric symptoms reported by patients taking interferon alfa are depressive symptoms, at rates of 10–40% in most studies (339–342). In four clinical studies in a total of 210 patients with chronic hepatitis C, the rate of major depressive disorders during interferon alfa treatment was 23–41% (343–346).

Suicidal ideation or suicidal attempts have been reported in 1.3–1.4% of patients during interferon alfa treatment for chronic viral hepatitis or even within the 6 months after withdrawal (347,348), but the excess risk related to interferon alfa is not known.

Predictive factors

As regular psychiatric assessment is not always possible, the identification of easily detectable predictive factors of severe psychiatric disorders may help select which patients should undergo close psychiatric assessment. In 71 patients treated with interferon alfa alone or combined with ribavirin for chronic hepatitis C, female sex, scores on the MADRS at 4 months of treatment, sleep disorders, and prior antidepressant use were independent risk factors of suicidal behavior or depression (349). This study also suggested that prolonged follow-up is required, as 8% of patients still had suicidal behavior 6 months after the end of treatment.

Time-course

Subacute or chronic neuropsychiatric manifestations are more typically identified after several weeks of treatment and are among the most frequent treatment-limiting adverse effects (325,341,350–352). The onset can be insidious in patients treated with low doses, or subacute in those who receive high doses. Most patients develop severe depressive symptoms within the first 3 months of treatment (343–346).

Although psychiatric manifestations usually appear during interferon alfa therapy, delayed reactions can occur.

- A 37-year-old man without a previous psychiatric history developed major depression with severe psychotic features within days after the discontinuation of a 1-year course of interferon alfa-2b (353).

In 10 patients with melanoma and no previous psychiatric disorders, depression scores measured on the Montgomery-Asberg Depression Rating Scale were significantly increased after 4 weeks of high-dose interferon alfa (354). Patients whose scores were higher before treatment developed the worst symptoms of depression during treatment. This positive correlation provides striking evidence that baseline and regular assessment of mood and cognitive functions are necessary to detect disorders as early as possible.

Pathophysiology

The pathophysiology of neuropsychiatric adverse effects of interferon alfa has been reviewed. Interactions of interferon alfa with the central opioid, serotonin, dopamine, and glutamate neurotransmitter systems are probably involved (355). The possible predominant role of the serotonergic system has again been emphasized in two studies, with evidence of early and significant changes in tryptophan metabolism resulting in persistently low tryptophan concentrations and increased kynurenine concentrations and kynurenine/tryptophan ratio during interferon alfa treatment compared with values obtained before treatment (356) and in an untreated control group (357). In addition, the fall in tryptophan concentrations correlated with the development and intensity of mood and cognitive symptoms during interferon alfa treatment (Capuron 906).

Mechanisms

Several mechanisms involved in the pathogenesis of interferon alfa-induced psychiatric adverse effects have been hypothesized (Loftis 175), but the mechanisms by which interferon alfa enters the brain to produce neurotransmitter changes are unclear. In a prospective study in 48 patients who received adjuvant interferon alfa for malignant melanoma there was a positive correlation between the increase in serum concentration of soluble ICAM-1 and depression scores; the authors suggested that induction of soluble ICAM-1 by interferon alfa increased the permeability of the blood–brain barrier, allowing the drug to enter the brain more readily (358).

The relation between interferon alfa-induced depressive disorders and the viral response to treatment has been examined in 39 patients with hepatitis C infection and no history of active psychiatric disease (359). After treatment with interferon alfa-2b and ribavirin for 6–12 months, 13 developed major depressive disorders requiring treatment with citalopram. The end-of-treatment response rates and sustained viral response rates were significantly greater in patients who developed a major depressive disorder than in those who did not (62% versus 27% and 39% versus 12% respectively). Despite the small sample, these results suggest that interferon alfa-induced depression occurs at doses or concentrations that are associated with a therapeutic effect (i.e. is a collateral adverse reaction) and that antidepressant therapy could allow patients to take an optimal dose.

Although very few studies have specifically investigated the role of the underlying disease, the findings of significant neuropsychiatric deterioration during interferon alfa treatment compared with placebo or no treatment in chronic hepatitis C, chronic myelogenous leukemia, or amyotrophic lateral disease strongly suggested a causal role of interferon alfa (360–362).

The mechanism by which the systemic administration of interferon alfa produces neurotoxicity is unclear, and might result from a complex of direct and indirect effects involving the brain vasculature, neuroendocrine system, neurotransmitters and the secondary cytokine cascade with cytokines which exert effects on the nervous system, for example interleukin-1, interleukin-2, or tumor necrosis factor alfa (363). Whether a clinical effect is directly mediated through the action of a given cytokine or results from a secondary pathway through the induction of other cytokines or second messengers is difficult to determine.

A study in 18 patients treated with interferon alfa for chronic hepatitis C has given insights into the possible pathophysiological mechanism of depression (364). Depression rating scales, plasma tryptophan concentrations, and serum kynurenine and serotonin concentrations were measured at baseline and after 2, 4, 16, and 24 weeks of treatment with interferon alfa 3–6 MU 3–6 times weekly. During treatment, tryptophan and serotonin concentrations fell significantly, while kynurenine concentrations rose significantly. Depression rating scales also rose from baseline after the first month of treatment, with continued increases thereafter. In addition, there was a relation between increased scores of depression and changes in serum kynurenine and serotonin concentrations. These changes suggested a predominant role of the serotonergic system in the pathophysiological mechanisms of interferon alfa-associated depression. Accordingly, 35 of 42 patients included in three open trials of antidepressant treatment responded to a selective serotonin reuptake inhibitor drug, such as citalopram or paroxetine, and were able to complete interferon treatment (345,365,366).

Susceptibility factors

Various possible susceptibility factors have been analysed in several studies (344–346,367). Sex, the dose or type of interferon alfa (natural or recombinant), a prior personal history of psychiatric disease, substance abuse, the extent of education, the duration and severity of the underlying chronic hepatitis, and scores of depression before interferon alfa treatment were not significantly different between patients with and without interferon alfa-induced depression. Advanced age was suggested to be a risk factor in only one study (346). Although a worsening of psychiatric symptoms was noted during treatment in 11 patients receiving psychiatric treatment before starting interferon, only one was unable to complete the expected 6-month course of interferon alfa and ribavirin therapy (344).

Of 91 patients treated with interferon alfa-2b and low-dose cytarabine for chronic myelogenous leukemia, 22 developed severe neuropsychiatric toxicity (368). Their symptoms consisted mostly of severe depression or psychotic behavior, which resolved on withdrawal in all patients. The time to toxicity ranged from as early as 2 weeks to as long as 184 weeks after the start of treatment. Five of six patients had recurrent or worse symptoms after re-administration of both drugs. Several baseline factors were analysed, but only a pretreatment history of neurological or psychiatric disorders was considered to be a reliable risk factor. Severe neuropsychiatric toxicity developed during treatment in 63% of patients with previous neuropsychiatric disorders compared with 10% in patients without. It is unlikely that the combination of interferon alfa-2b with low-dose cytarabine potentiated the neuropsychiatric adverse effects of interferon alfa in this study. Indeed, previous experience with this combination, but after exclusion of patients with a psychiatric history, was not associated with such a high incidence of neuropsychiatric toxicity or any significant difference in toxicity between interferon alfa alone and interferon alfa plus low-dose cytarabine.

Patients receiving high doses of interferon alfa or long-term treatment are more likely to develop pronounced symptoms (351). A previous history of psychiatric disorders, organic brain injury, or addictive behavior are among potential susceptibility factors, but worsening of an underlying psychiatric disease is not the rule, provided that strict psychiatric surveillance and continuation of psychotropic drugs are maintained (369). Other putative susceptibility factors include the intraventricular administration of interferon alfa, previous or concomitant cranial irradiation, asymptomatic brain metastases, and pre-existing intracerebral ateriosclerosis (SED-13, 1092) (341,370–374). Despite early findings, co-infection with HIV has not been confirmed to be a susceptibility factor (SEDA-20, 327).

The most frequent susceptibility factor for interferon alfa-associated depression is a history of mood and anxiety symptoms before treatment. The roles of a previous history of depression, female sex, and high dose or long duration of interferon alfa treatment have been discussed (Raison 105). High doses undoubtedly play a role in the development of depression, but the duration of treatment is not a consistently replicated susceptibility factor. Although it was initially found that a past history of depression might be a susceptibility factor, this has not been convincingly associated with an increased risk of interferon alfa-induced depression in further studies. Similar conclusions were reached in patients with a past history of drug or alcohol abuse, provided that patients remain abstinent during treatment. Female sex and age have also not emerged as consistent susceptibility factors.

Although depressive symptoms are usually ascribed to interferon alfa alone, ribavirin could also contribute to their occurrence. Among 162 patients with chronic hepatitis C treated with peginterferon alfa-2b (1.5 micrograms/kg once weekly) plus weight-based or standard-dose ribavirin and prospectively evaluated for the incidence and severity of depression, 39% developed moderate to severe depressive symptoms and six patients withdrew because of behavioral symptoms, such as depression, anxiety, or fatigue (375). Depressive symptoms developed after 4 weeks of treatment and maximal depression scores were reached after 24 weeks. Baseline depression scores, a history of major depressive disorders, and higher doses of ribavirin were significant predictive factors of depression during treatment, in contrast to sex, age, or a history of substance abuse.

In 33 patients with chronic hepatitis C, of whom 10 developed major depressive disorders during interferon alfa treatment, there was no relation between changes in thyroid function and the development of depression (376). No patients with depression had clinical hypothyroidism and the sole patient with overt hypothyroidism had no depressive symptoms.

A number of other factors, including genetic or biological factors, that are possibly associated with the development of psychiatric disorders, have been explored. In a retrospective study of 110 patients with chronic hepatitis C, those with the apolipoprotein E ε4 allele, the inheritance of which may be associated with several neuropsychiatric outcomes, were more likely than those without this allele to have psychiatric referrals and neuropsychiatric symptoms, in particular irritability, anger, anxiety, or other mood symptoms, during interferon alfa treatment (377). In addition, patients with this allele had neuropsychiatric events sooner than patients without it. In another study in 14 patients with chronic hepatitis C, lower baseline serum activities of propylendopeptidase and dipeptyl peptidase IV, which both play a role in the pathophysiology of major depression, were possible predictors of higher depressive and anxiety scores during interferon alfa treatment (378). Whether these factors help to predict susceptibility to interferon alfa-induced neuropsychiatric disorders in clinical practice needs to be further investigated.

The occurrence of psychiatric disorders has been prospectively investigated in 63 patients who received a 6-month course of interferon alfa (9 MU/week) for hepatitis C (379). All were assessed at baseline with the Structured Clinical Interview for DSM-III-R (SCID) and monitored monthly with the Hopkins Symptoms Checklist (SCL-90). Most had a history of alcohol or polysubstance dependence, and 12 had a lifetime diagnosis of major depression. There were no significant changes in the SCL-90 scores during the 6-month period of survey in the 49 patients who completed the study, even in those who had a lifetime history of major depression. At 6 months, there was probable minor depression in eight patients and major depression in one; none had attempted suicide.

In a prospective study, 50 patients with chronic hepatitis B or C who received 18–30 MU/week of natural or recombinant interferon alfa were followed for 12 months (380). The SCID before starting interferon alfa identified 16 patients with a current psychiatric diagnosis and eight with a previous psychiatric disorder; 26 patients free of any psychiatric history constituted the control group. Psychiatric manifestations during treatment occurred in

11 patients (five from the control group), major depression in five, depressive disorders in three, severe dysphoria in two, and generalized anxiety disorder in one. Most of them were successfully treated with psychological support and drug therapy. Overall, 20 patients interrupted interferon alfa (10 in each group), including three for psychiatric adverse effects, but patients with a pre-existing or recent psychiatric diagnosis were no more likely to withdraw from treatment than the controls.

Of 33 patients with chronic hepatitis C treated with interferon alfa, 9 MU/week for 3–12 months, prospectively evaluated using the Montgomery-Asberg Depression Rating Scale (MADRS) before and after 12 weeks of treatment, eight developed depressive symptoms, of whom four had major depression without a previous psychiatric history (381). All four recovered after treatment with antidepressants. This study confirmed that a high baseline MADRS is significantly associated with the occurrence of depressive symptoms.

A prospective study in 81 patients with chronic hepatitis C taking interferon alfa-2a and ribavirin compared adherence to treatment, adverse effects, response rates, and dropout rates in 23 patients without any psychiatric history or drug addiction (control group), 16 patients with psychiatric disorders, 21 patients in a methadone substitution programme, and 21 patients with former intravenous drug addiction (382). There were no significant differences between groups as regards the frequency and severity of new depressive episodes during treatment, but significantly more patients in the psychiatric groups had to take antidepressants before and during interferon alfa treatment. The rate of dropouts was significantly higher in the drug addiction group (43%) compared with the other groups (13–18%). The cause was psychiatric or somatic adverse effects in half of these patients. No patients in the psychiatric or methadone groups had to stop treatment because of psychiatric adverse effects. Although the limited number of patients and heterogeneity in psychiatric diagnoses precluded any definitive conclusion, these results suggested that treatment with interferon alfa and special psychiatric care is possible in at-risk psychiatric groups.

These studies have confirmed that previous psychiatric disorders are not necessarily a contraindication to a potentially effective treatment. However, patients with depressive symptoms immediately before treatment are still regarded at risk of severe psychiatric deterioration with treatment (339).

Management

Cognitive defects, depression, and mania can occur in patients taking interferon alfa. Interferon alfa-induced psychiatric adverse effects and their recommended management have been comprehensively reviewed (383,384,385).

Various pharmacological and non-pharmacological interventions have been discussed (386), and prompt intervention should be carefully considered in every patient who develops significant neuropsychiatric adverse effects while receiving interferon alfa. Depending on the clinical manifestations, proposed treatment options include antidepressants, psychostimulants, or antipsychotic drugs.

Based on a possible reduction in central dopaminergic activity mediated by the binding of interferon alfa to opioid receptors, naltrexone has been proposed as a means of improving cognitive dysfunction (351).

Selective serotonin re-uptake inhibitors have been advocated as the drugs of choice to allow completion of interferon alfa treatment (339), but that was based on very limited experience and the unproven assumption that SSRIs are safe in patients with underlying liver disease. The preliminary results of a double-blind, placebo-controlled study showed that 2 weeks of pretreatment with paroxetine significantly reduced the occurrence of major depression in 16 patients on high-dose interferon alfa for malignant melanoma (387). In a placebo-controlled trial, the preventive effects of paroxetine (mean maximal dose of 31 mg) were studied in 40 patients with high-risk malignant melanoma and interferon alfa-induced depression (388). Treatment started 2 weeks before adjuvant high-dose interferon alfa. Paroxetine significantly reduced the incidence of major depression (45% in the placebo group and 11% in the paroxetine group) and the rate of interferon alfa withdrawal (35 versus 5%). Although the number of patients was small and the duration of the survey short (12 weeks), this suggests that paroxetine effectively prevents the risk of depressive disorders in patients eligible for high-dose interferon alfa. However, these results are limited, because patients with melanoma who receive adjuvant high-dose interferon alfa are particularly likely to develop depression. The safety of prophylaxis with paroxetine also requires additional data, because three patients taking paroxetine developed retinal hemorrhages, including one with irreversible loss of vision.

- In contrast to this study, a 31-year-old woman with major depressive disorder, which responded to paroxetine and trazodone, had progressive recurrence of mood disorders after the introduction of interferon alfa for essential thrombocythemia (389).

This suggests that interferon alfa can also reverse the response to antidepressants.

The successful management of psychiatric symptoms during interferon alfa treatment has been detailed in 60 patients with chronic viral hepatitis, treated for 1 year (390). A new psychiatric disease developed during treatment in 18 patients (depression in 12, predominant irritability in four, and anxiety in two), and five others had major depression at baseline. Based on the potential mechanisms of interferon alfa-induced psychiatric disorders, and depending on the clinical features, a variety of psychopharmacological treatments were used in 12 of these patients, including selective serotonin reuptake inhibitors in four cases, low-dose pre-synaptic D2 dopamine receptor antagonists (sulpiride and amisulpride) in three, neuroleptic drugs in two, and benzodiazepines or related drugs in three. Only four patients failed to

respond to treatment, and one had to be withdrawn from the study because of persistent irritability. During the study, the 12 patients taking psychiatric drugs had significantly less severe psychiatric symptoms than the 11 untreated patients.

Interferon beta

There have been reports of depression, suicidal ideation, and attempted suicide in patients receiving interferon beta (391,392,393). The lifetime risk of depression in patients with multiple sclerosis is high, and there has been a lively debate about whether interferon beta causes or exacerbates depression in such patients. Impressions of a possibly raised incidence of depression among patients treated with interferon beta for multiple sclerosis should be interpreted in the light of the spontaneous tendency to depressive disorders and suicidal ideation, which is encountered even in patients with untreated multiple sclerosis. Moreover, no raised incidence of these complications has been recorded in some studies (394,395). A critical review of the methodological limitations in studies that assessed mood disorders in patients on disease-modifying drugs for multiple sclerosis may help explain the widely divergent results from one study to another (396). Some results have argued against a specific role of interferon beta in the risk of depressive disorders.

A multicenter comparison of 44 and 22 micrograms of interferon beta-1a and placebo in 365 patients showed no significant differences in depression scores between the groups over a 3-year period of follow-up (397). In 106 patients with relapsing–remitting multiple sclerosis, depression status was evaluated before and after 12 months of interferon beta-1a treatment (398). According to the Beck Depression Inventory II scale, most of the patients had minimum (53%) or mild (32%) depression at baseline, and depression scores were not significantly increased after 1 year of treatment. There were no cases of suicidal ideation. In another study of 42 patients treated with interferon beta-1b, major depression at baseline was found in 21% of patients and was associated with a past history of psychiatric illness in most cases (399). Major depression was not considered as an exclusion criterion for interferon beta treatment when patients were on antidepressant therapy. There was a three-fold reduction in the prevalence of depression over the 1-year course of interferon treatment, suggesting a possible beneficial effect of treatment on mood. Finally, a single subcutaneous injection of interferon beta-1b did not alter cognitive performance and mood states in eight healthy volunteers (400).

The emotional state of 90 patients with relapsing–remitting multiple sclerosis has been carefully assessed with a battery of psychological tests at baseline and after 1 and 2 years of treatment with interferon beta-1b (401). In contrast to what was expected, and despite the lack of controls, there was significant improvement in emotional state, as shown by significant reductions in scores of anxiety and depression over time. In addition, there was no effect of low-dose oral glucocorticoids in a subgroup of 46 patients.

Depression has been quantified by telephone interview in 56 patients with relapsing multiple sclerosis 2 weeks before treatment, at the start of treatment, and after 8 weeks of treatment (402). Patients with a high depressive score 2 weeks before treatment significantly improved on starting treatment and returned to baseline within 8 weeks, whereas the depression score in non-depressed patients remained essentially unchanged. The investigators therefore suggested that patients' expectations had temporarily resulted in improvement of depression, and that increased depression during treatment is more likely to reflect pretreatment depression.

The clinical features, management, and prognosis of psychiatric symptoms in patients with chronic hepatitis C have been reviewed using data from 943 patients treated with interferon alfa (85%) or interferon beta (15%) for 24 weeks (403). Interferon-induced psychiatric symptoms were identified in 40 patients (4.2%) of those referred for psychiatric examination. They were classified in three groups according to the clinical profile: 13 cases of generalized anxiety disorder (group A), 21 cases of mood disorders with depressive features (group B), and six cases of other psychiatric disorders, including psychotic disorders with delusions/hallucinations ($n = 4$), mood disorders with manic features ($n = 1$), and delirium ($n = 1$) (group C). The time to onset of the symptoms differed significantly between the three groups: 2 weeks in group A, 5 weeks in group B, and 11 weeks in group C. Women were more often affected than men. There was no difference in the incidence or nature of the disorder according to the type of interferon used. Whereas most patients who required psychotropic drugs were able to complete treatment, 10 had to discontinue interferon treatment because of severe psychiatric symptoms, five from group B and five from group C. Twelve patients still required psychiatric treatment for more than 6 months after interferon withdrawal. In addition, residual symptoms (anxiety, insomnia, and mild hypothymia) were still present at the end of the survey in seven patients. Delayed recovery was mostly observed in patients in group C and in patients treated with interferon beta.

Although isolated reports of psychotic delusional symptoms and depression continue to be published (404), recent controlled trials or longitudinal studies have not provided evidence of an increase in depression scores or in the rate of depression in patients treated with interferon beta (SEDA-27, 389). In a meta-analysis of seven trials in 1215 patients with relapsing remitting multiple sclerosis, the incidence of depression was 16% and did not differ between interferon-beta and controls, but the scales used to assess depression were specified in only three trials (405). Using a public reimbursement database for multiple sclerosis, the prevalence and incidence of depression and depression scores were not different in 163 patients treated with interferon beta or glatiramer, but the study was poorly controlled for potential biases (406). Overall, the current data suggest that interferon-beta is not substantially associated with depression.

According to data obtained from six controlled and 17 non-controlled studies sponsored by the manufacturer of interferon beta-1a, depression was more frequently reported as an adverse event in the treated patients (407). However, this was not associated with an increase in depression scores in the only study that used appropriate rating scales. There was also no significant difference in the rate of suicide attempts or suicide in patients taking interferon compared with placebo. There were 42 cases of serious and non-serious reports of depression and suicide attempts, including only five cases of suicide, in postmarketing experience covering more than 161 000 patient-years. Overall these data suggest that interferon beta-1a does not produce typical depression.

One debatable case of visual pseudo-hallucinations occurred only, but not reproducibly, within 30–60 minutes after interferon beta-1a injection in a 37-year-old woman with disseminated encephalomyelitis (408).

Interferon gamma

Neuropsychiatric disturbances have not been consistently found in patients receiving interferon gamma, despite electroencephalographic monitoring and psychometric tests (409). However, careful examination led to the impression that interferon gamma can cause neurophysiological changes similar to those of interferon alfa (410), and data from the manufacturers also point to rare cases of nervous system adverse effects in patients treated with high-dose interferon gamma (411).

Interleukin 2

According to a study in 32 patients with cancers, the presence of emotional symptoms and sleep disturbance before treatment can predict the early development of severe depressive symptoms after subsequent therapy with IL-2 and/or interferon alfa (412). Although the number of patients was small, the authors suggested that this potentially high-risk population might benefit from prophylactic antidepressant treatment.

Iodinated contrast media

To reduce the incidence of generalized reactions to contrast media in high-risk patients some authors have advocated the prophylactic administration of glucocorticoids (prednisolone 30 mg orally or methylprednisolone 32 mg orally, 12 and 2 hours before contrast injections). In one case an acute psychosis complicated glucocorticoid premedication to reduce the risk of contrast reactions (413).

- A 13-year-old girl with bipolar disorder and a history of adverse reactions to contrast media was given methylprednisolone (32 mg/day) and ranitidine (300 mg/day) before a CT scan of the head with intravenous contrast enhancement. One day after, she developed psychiatric symptoms, which were more severe than her initial symptoms, including extreme agitation and mental confusion. All medications were withdrawn and her symptoms resolved within 2 weeks.

The authors suggested that the recurrence of the manic symptoms could have been due to premedication with prednisolone. Exacerbation of manic symptoms after the use of glucocorticoids has been documented before, but never in a case of short-term premedication before contrast-enhanced radiographic examination. This report shows that even a short-term course of glucocorticoids can have significant adverse effects in patients with a history of mood disorders.

In 2500 cases of cervical myelography with metrizamide, there were transient mental reactions in 25, including 13 cases of confusion or disorientation, four of depression, two of hallucinations, two of psychosis, and one each of anxiety, drowsiness, dysphasia, and nightmares (414).

Of 18 German patients undergoing lumbar myelography with metrizamide, six had an organic psychosis, characterized by impaired memory and depression, but it was demonstrable only by psychometric tests and disappeared within 5 days (415). In four of the 18 patients there was hyporeflexia or areflexia and in three there were electroencephalographic changes; there was no correlation between these various types of effect.

Visual hallucinations are very rare adverse effects of contrast media, with isolated reports after vertebral angiography or myelography. The mechanism of this adverse reaction could be similar to that reported in transient cortical blindness after infusion of contrast agents. However, other possibilities include a toxic effect of contrast media on the optic nerve, transient impairment of cerebral blood flow, which could be mediated through the release of the potent vasoconstrictor endothelin, or the formation of microclots. Two cases of visual hallucinations after coronary angiography have been reported (416).

- A 70-year-old woman with a history of mastectomy developed syncope which lasted a few seconds. She had taken tamoxifen 10 mg bd for 10 years and had no history of allergic reactions. Doppler ultrasound showed aortic stenosis and coronary angiography was performed using 150 ml of iopromide (a non-ionic contrast medium, iodine 370 mg/ml). She had visual hallucinations (spiders on the wall, moving curtains) 30 minutes after the injection of iopromide. The symptoms resolved 72 hours later without any specific treatment. Neurological and psychiatric examinations were normal, as were brain MRI and Doppler ultrasound of the carotid and vertebral arteries.
- A 64-year-old man with a history of ischemic heart disease underwent coronary angiography with 150 ml of iopromide (iodine 370 mg/ml). One hour later he had visual hallucinations (moving objects, pictures of familiar persons), which resolved about 40 hours later without any treatment. He had taken the following drugs for a year: nifedipine 10 mg tds, metoprolol 50 mg bd, and aspirin 325 mg/day. His serum creatinine concentration was in the reference range and there was no history of allergies or previous exposure to contrast media.

Myelography with either iopamidol or metrizamide can cause transitory deterioration in memory as determined by psychological tests, but the effect is less with iopamidol (417).

Many of the psychiatric complications of cerebral angiography may be due to arterial trauma rather than to the toxic effect of the contrast agent. If the investigation is undertaken under general anesthesia, the use of a volatile anesthetic can in itself cause an increase in intracranial pressure and thereby constitute an aggravating factor (418). Focal electroencephalographic changes can occur on the side of the injection, and if these are prolonged they can be followed by evidence of neurological involvement. Transient global amnesia and confusional states have been reported after cerebral angiography, even with non-ionic media (SEDA-9, 410) (419).

Ipratropium bromide

Ipratropium produces no significant impairment of cognitive or psychomotor function in elderly patients with chronic airway obstruction (SEDA-21, 187).

Isoflurane

Memory function and its relation to depth of hypnotic state has been prospectively evaluated in anesthetized and non-anesthetized subjects, using the Bispectral Index during general anesthesia and an auditory word stem completion test and process dissociation procedure after anesthesia (420). Isoflurane was used in 47 patients and propofol in one. There was evidence of memory for words presented during light anesthesia (Bispectral Index score 61–80) and adequate anesthesia (score 41–60) but not during deep anesthesia (score 21–40). The process dissociation procedure showed a significant implicit memory contribution but not reliable explicit memory contribution. Memory performance was better in non-anesthetized subjects than in anesthetized patients, with a higher contribution from explicit memory and a comparable contribution from implicit memory. The authors concluded that during general anesthesia for elective surgery, implicit memory persists, even in adequate hypnotic states, to a comparable degree as in non-anesthetized subjects.

Isoniazid

Isoniazid can cause neuropsychiatric syndromes, including euphoria, transient impairment of memory, separation of ideas and reality, loss of self-control, psychoses (421), and obsessive-compulsive neurosis (422). Isoniazid should be used with caution in patients with pre-existing psychoses, as it can cause relapse of paranoid schizophrenia (423). Patients on chronic dialysis appear to be vulnerable to neurological adverse drug reactions, because of abnormal metabolism of uremic toxins. It is therefore recommended that a higher than usual dose of pyridoxine be given to patients on dialysis taking isoniazid (424,425).

Itraconazole

Visual hallucinations with confusion have been reported in a 75-year-old woman, occurring on three separate occasions, each time about 2 hours after a 200 mg dose of itraconazole. Her symptoms abated spontaneously over about 8 hours (426).

Ketamine

In a randomized, double-blind, crossover study of cognitive impairment in 24 volunteers who received *S*-ketamine 0.25 mg/kg, racemic ketamine 0.5 mg/kg, or *R*-ketamine 1.0 mg/kg, the ketamine isomers caused less tiredness and cognitive impairment than equianalgesic doses of racemic ketamine (427). In addition, *S*-ketamine caused less reduction in concentration capacity and primary memory.

A placebo-controlled study of low-dose ketamine infusion in ten volunteers showed formal thought disorder and impairments in working and semantic memory (428). The degree of thought disorder correlated with the impairment in working memory.

The psychotomimetic effects of ketamine, apart from encouraging illicit use, can lead to distressing psychic disturbances, particularly in children (429); there can be nightmares, delirium, and hallucinations (430). Oral ketamine is an effective analgesic in patients with chronic pain. In 21 patients with central and peripheral chronic neuropathic pain treated with oral ketamine, the starting dose was ketamine 100 mg/day, titrated upward by 40 mg/day increments every 2 days until a satisfactory effect was achieved, or until adverse effects became limiting (431). Nine patients discontinued ketamine because of intolerable adverse effects, including psychotomimetic symptoms, such as "elevator" effect or dissociative feelings, somnolence or insomnia, and sensory changes such as taste disturbance and somatic sensations.

The pharmacological effects of the *R*- and *S*-enantiomers of ketamine have been compared in 11 subjects who received *R*-ketamine 0.5 mg and then *S*-ketamine 0.15 mg, separated by 1 week (432). Before and after each drug administration they were subjected to a painful stimulus using a nerve stimulator applied to the right central incisor tooth. Pain suppression was equal with the two drugs. The subjects reported more unpleasant psychotomimetic effects with *S*-ketamine and more pleasant effects with *R*-ketamine. Seven of eleven subjects preferred *R*-ketamine, while none preferred *S*-ketamine. These results suggest that the neuropsychiatric effect of ketamine may be predominantly due to the *S*-enantiomer, and that *R*-ketamine may be a better alternative. This study is in direct distinction to earlier work suggesting that *R*-ketamine is responsible for most of the undesirable neuropsychiatric side effects of ketamine.

A placebo-controlled study in 10 healthy young men showed a linear relation between ketamine plasma concentrations of 50–200 ng/ml and the severity of psychotomimetic effects (433). The psychedelic effects were also similar to those observed in a previous study of dimethyltryptamine, an illicit LSD-25 type of drug, and were a function of plasma concentration rather than simply an emergence phenomenon. Clinically useful analgesia was obtained at plasma concentrations of 100–200 ng/ml. At plasma concentrations of 200 ng/ml, all subjects had lateral nystagmus. When ketamine is given in large doses, patients rapidly become unresponsive, and so the effects described in this study are usually only observed during the recovery phase.

The effects of ketamine 50 or 100 ng/ml on memory have been investigated in a double-blind, placebo-controlled, randomized, within-subject study in 12 healthy volunteers (434). Deleterious effects of ketamine on episodic memory were primarily attributable to its effects on encoding, rather than retrieval. The authors suggested that the effects they observed were similar to the memory deficits seen in schizophrenia and thus provide some support for the ketamine model of the disease.

Subanesthetic low-dose ketamine is thought to cause delirium and disturbing dreaming. A systematic review of NMDA receptor antagonists in preventive analgesia has shown that only one of 20 studies documented adverse psychotomimetic effects attributable to ketamine (435). In that study, ketamine was given by the epidural route in a relatively high dose.

Ketamine often causes emergence delirium and disturbing dreaming. Benzodiazepines are often co-administered to attempt to manage this. The optimal dose of diazepam to add to ketamine–fentanyl field anaesthesia has been assessed in a randomized double-blind study in 400 patients from Vanuatu; the optimal dose was 0.1 mg/kg (436).

There have been several attempts to understand the pathophysiology of schizophrenia using subanesthetic doses of ketamine to probe glutaminergic function in healthy and schizophrenic volunteers; no long-term adverse consequences were attributable to ketamine (437).

There have been several attempts to attenuate the unpleasant psychological adverse effects that occur after sedation with ketamine.

Prior use of benzodiazepines or opiates limits the psychotomimetic effects of ketamine. There has been a double-blind, placebo-controlled study of the role of lorazepam in reducing these effects after subanesthetic doses of ketamine in 23 volunteers who received lorazepam 2 mg or placebo, 2 hours before either a bolus dose of ketamine 0.26 mg/kg followed by an infusion of 0.65 mg/kg/hour or a placebo infusion (438). The ability of lorazepam to block the undesirable effects of ketamine was limited to just some effects. It reduced the ketamine-associated emotional distress and perceptual alterations, but exacerbated the sedative, attention-impairing, and amnesic effects of ketamine. However, it failed to reduce many of the cognitive and behavioral effects of ketamine. There were no pharmacokinetic interactions between subanesthetic doses of ketamine and lorazepam.

The effect of intravenous midazolam 0.05 mg/kg on emergence phenomena after ketamine 1.5 mg/kg intravenously for painful procedures has been assessed in a randomized, double-blind, placebo-controlled study in 104 children (439). Midazolam was given 2 minutes after the ketamine. There was no significant difference between the two groups in levels of agitation. The overall rate of agitation was low, but probably high enough to detect any significant differences between the groups.

The neuropsychiatric effects of ketamine were modulated by lamotrigine, a glutamate release inhibitor, in 16 healthy volunteers (440). Lamotrigine 300 mg was given 2 hours before ketamine 0.26 and 0.65 mg/kg on two separate days. There were fewer ketamine-induced perceptual abnormalities, fewer schizophreniform symptoms, and less learning and memory impairments. Mood-elevating effects were increased with lamotrigine. The authors commented that the results were experimental and that further studies are needed to confirm the potential benefits in a larger group of patients.

The hypothesis that the unpleasant emergence phenomena that often accompany the use of ketamine, including odd behavior, vacant stare, and abnormal affect, would be reduced by the use of a selected recorded tape played during the perioperative period has been tested in 28 adults (441). The incidence of dreams was higher when the recorded tape was connected. This report emphasizes the current recommendations that a quiet room with minimal stimuli is best for reducing emergence phenomena after ketamine sedation.

Ketoconazole

In one patient taking a high dosage for prostate cancer, weakness was associated with mental disturbances, notably confabulation and disorientation in time and space (SEDA-13, 233).

Lamotrigine

The cognitive and behavioral effects of lamotrigine 150 mg/day have been compared with those of carbamazepine 696 mg/day in 25 healthy adults in a double-blind, crossover, randomized study with two 10-week treatment periods (442). Lamotrigine had significantly fewer cognitive and behavioral adverse effects than carbamazepine, and 48% of the variables favored lamotrigine. The differences encompassed cognitive speed, memory, graphomotor coding, neurotoxic symptoms, mood, sedation, perception of cognitive performance, and other quality-of-life perceptions. The cognitive and behavioral changes favored lamotrigine over carbamazepine, but the magnitude of the observed effects was modest, although it could be relevant in some patients.

Lamotrigine caused hypomania in a 23-year-old woman with major depression and no history of bipolar illness; it resolved 2 weeks after withdrawal (443).

A 23-year-old woman with schizophrenia had worsening of psychiatric symptoms after exposure to lamotrigine (444). This patient was taking quetiapine 400 mg/day and divalproex sodium 1500 mg/day. Because of excessive somnolence valproate was withdrawn and lamotrigine was started at 12.5 mg bd. On the second day the patient had worsening of psychotic symptoms, which the authors attributed to lamotrigine.

Two patients taking lamotrigine 225 mg/day and 200 mg/day developed forced normalization (445). Lamotrigine led to seizure control and disappearance of interictal epileptiform discharges from the electroencephalogram. However, simultaneously they had de novo psychopathology and disturbed behavior. Reduction of the dose of lamotrigine led to disappearance of the symptoms and reappearance of the spikes but not the seizures.

Levamisole

Anxiety and depression have been associated with levamisole (446).

Psychosis has also been reported (447).

- A 28-year-old man, without a psychiatric history, developed a paranoid psychosis. He had been taking levamisole twice a week in an unspecified dose for 2 years for a stage 4 melanoma and metastatic lymph nodes in the axilla. Physical examination, a CT scan, an electroencephalogram, and standard laboratory tests were all normal. He was treated with perphenazine, with partial success, but after tapering of the dose his symptoms reappeared. It was thought likely that the psychosis had been caused by levamisole, which was discontinued. Three weeks later he had recovered completely. Levamisole was not reintroduced.

Levetiracetam

Psychiatric adverse events, most often hostility and irritability, have been reported in children taking levetiracetam. Four cases of acute psychosis have also been reported (448).

- A 5-year-old girl with refractory epilepsy treated with a ketogenic diet was given levetiracetam 250 mg bd (25 mg/kg/day). She had a history of mild mental retardation and was receiving special education. Two weeks later she started to have visual hallucinations, became agitated, bit relatives, and could not sleep. Levetiracetam was withdrawn and her symptoms resolved within 24 hours and did not recur.

In a retrospective analysis of 118 patients with epilepsy with intellectual disability taking levetiracetam, 15 had psychiatric adverse events, including affective disorder in two, aggressive behavior in nine, emotional lability in two, and other personality changes (agitation, hostility, anger) in two (449) . There was a significant correlation with a previous history of status epilepticus and a previous psychiatric history, but no relation to the titration schedule of levetiracetam, electroencephalography and MRI, and the type and duration of epilepsy.

In six patients with Lennox–Gastaut syndrome levetiracetam improved myoclonic, atonic, and generalized tonic–clonic seizures (450). The most common adverse effect was irritability during the start of treatment in two patients, requiring withdrawal in one.

In a retrospective analysis of 14 patients (aged over 60 years) levetiracetam 500–3000 mg/day was generally well tolerated (451). There were no major adverse events. One patient reported dizziness but was able to continue taking levetiracetam while the symptoms improved.

The prevalence and clinical features of psychiatric adverse events in patients taking levetiracetam for a mean duration of 8.3 months have been investigated prospectively in 517 patients, 50% with symptomatic partial epilepsy, 27% with cryptogenic partial epilepsy, 8.7% with idiopathic generalized epilepsy, 6.8% undetermined, and 5% with other syndromes (Lennox–Gastaut syndrome, West syndrome, and symptomatic generalized epilepsy) (452). There were psychiatric adverse events in 53 patients (10%): 19 (3.5%) developed aggressive behavior, 13 (2.5%) affective disorders, 12 (2.3%) emotional lability, 6 (1.2%) psychosis, and 3 (0.6%) other behavioral abnormalities, such as agitation, anger, or hostile behavior, and personality changes: four (0.7%) reported suicidal ideation. There was a significant association with a history of febrile seizures, status epilepticus, and a previous psychiatric history. In contrast, lamotrigine co-therapy had a protective effect, while the levetiracetam titration regimen had no effect on the occurrence of psychiatric adverse events.

Psychiatric adverse events and seizure exacerbation in patients taking levetiracetam have been reviewed, including randomized clinical trials in which levetiracetam was used in combination with other antiepileptic drugs, withdrawal to monotherapy studies, long-term extension studies, and postmarketing surveillance of adverse events in programs such as MEDWATCH (453). In contrast to previous reviews of safety data, the frequency of adverse events in patients taking levetiracetam were compared with those taking placebo in controlled trials: 15% of patients taking levetiracetam discontinued therapy or had their dosage reduced because of adverse events, compared with 12% of those who took placebo. The adverse events most commonly associated with withdrawal or dosage reduction were somnolence (4.4% versus 1.6% with placebo), convulsions (3.0% versus 3.4%), dizziness (1.4% versus 0%), weakness (1.3% versus 0.7%), and rashes (0% versus 1.1%). Most of the adverse events were mild or moderate in intensity. Overall, 113 patients (15%) taking levetiracetam and 49 (11%) of those who took placebo had severe adverse events. Seizure exacerbation in placebo-controlled trials (defined as a greater than 25% increase in seizure frequency) occurred in 14% of patients taking levetiracetam, compared with 26% of those taking placebo; seizure aggravation in these studies could therefore have represented spontaneous fluctuation in seizure frequency. The authors cited the Postmarketing Antiepileptic Drug Survey (454), a prospective registry that polls information from 16 epilepsy centers, behavioral adverse events were reported in 288 patients taking levetiracetam, and led to withdrawal in 27 patients (8.5%). Behavioral adverse events included depression

(n = 14, 4.9%), anxiety (n = 8, 2.8%), aggression (n = 5, 1.7%), irritability (n = 4, 1.4%), mood swings (n = 3, 1.0%) and unspecified behavioral changes (n = 5, 1.7%). There was a previous history of psychiatric disease, behavioral problems, or mental retardation in 51% of patients with behavioral adverse events, compared with 50% of all the patients taking levetiracetam. As the authors observed, it is difficult to ascertain from this information how much levetiracetam contributes to behavioral and psychiatric symptoms, since these can be related to the epilepsy, the direct effect of seizures, drug interactions, or fluctuations in seizure frequency.

Levocetirizine

The effects of levocetirizine on cognitive function have been assessed in two comprehensive well-controlled studies. The first analysed the effects of single and multiple doses of levocetirizine on measures of nervous system activity, using integrated measures of cognitive and psychometric performance. In a three-way crossover design, 19 healthy men took either levocetirizine 5 mg, diphenhydramine 50 mg (positive control), or placebo once daily on 5 consecutive days. Critical flicker fusion tests were performed on days 1 and 5 at baseline and up to 24 hours after drug administration. The primary outcome was that, in contrast to diphenhydramine, levocetirizine did not have any deleterious effect on any cognitive or psychometric function compared with placebo (455).

In a double-blind, crossover study levocetirizine 5 mg once daily for 4 days was compared with cetirizine 10 mg, loratadine 10 mg, promethazine 30 mg, and placebo in terms of nervous system inhibitory effects in 20 healthy volunteers (456). With the exception of promethazine none of the drugs had disruptive or sedative effects on objective measurements in a comprehensive battery of psychomotor and cognitive tests. These studies suggest that levocetirizine has minimal sedative effects in healthy individuals when given in its recommended dose.

Levodopa and dopa decarboxylase inhibitors

Toxic psychoses and toxic delirium can occur, particularly in individuals with a history of postencephalitic parkinsonism or psychiatric disease (457). Nearly half of all patients with Parkinson's disease are demented or have significant cognitive impairment (458). These patients are particularly susceptible to delirium induced by levodopa and other antiparkinsonian drugs. The author recommended dosage reduction or elimination of drugs that may be responsible, starting with anticholinergic drugs but leaving levodopa unchanged if possible. He also recommended the use of the atypical neuroleptic drugs clozapine or olanzapine in severely disturbed patients in whom drug withdrawal is not feasible.

Much more common are such symptoms as confusion (13% of cases), depression (9% of cases, often requiring antidepressant drugs), and sleep disturbances (20% of cases). Vivid hallucinations are common. Psychotic symptoms appear to be more common when enzyme inhibitors are used or anticholinergic drugs given. Panic attacks seem to occur in some 20% of cases during the "off" phases of an "off/on" cycle (459).

Libido has been found to increase, at least for some time, in a proportion of patients (460); whilst perhaps in part reflecting the improved mobility and sense of well-being, a causative relation with prolactin inhibition by levodopa has been suggested.

There may be subtle neuropsychiatric adverse effects of dopaminergic medication. The syndrome of hedonistic homeostatic dysregulation, generally associated with substance abuse, has been described in 15 patients with Parkinson's disease (461). Four patients were described in detail. The authors noted that 12 of the 15 patients were men (three of the four described in the paper) and had early onset of Parkinson's disease: those described were aged only 36–42 years. Characteristically, these patients were taking large and increasing doses of levodopa (or other dopamine agonists, including apomorphine) despite worsening dyskinesias; they had impaired social functioning, including absence from work, belligerent behavior, and hypersexuality; they had a hypomanic or bipolar affect; they underwent a withdrawal reaction on reducing levodopa dosage, with depression, dysphoria, and anxiety; and they had a disorder with the above features lasting at least 6 months. All of the four patients described in detail had taken apomorphine at some time (75–170 mg/day) and tended to use higher than prescribed doses by intermittent injection. All were also taking levodopa (maximum dose 1875–5500 mg/day). The authors noted that the long-term management of this condition can be very difficult. They suggested management of acute psychosis with atypical neuroleptic drugs, such as olanzapine or risperidone, the use of antidepressants if needed, and careful supervision of patient self-medication. They conceded that this may be extremely difficult.

In a prospective study of 89 patients with Parkinson's disease, of whom 60 were free of hallucinations at entry, though most of these had disturbed sleep patterns, after 4 years 50% of the original non-hallucinators were experiencing hallucinations, while only 14% of those with hallucinations at entry were no longer affected by them (462). Those classified as having severe hallucinations increased from 10% at entry to 35% after 4 years. The development or worsening of hallucinations was not associated with levodopa dosage but was strongly correlated with the use of dopamine receptor agonists in combination with levodopa: some 59% of patients with hallucinations were taking agonists as against 33% of non-hallucinators.

It is a pharmacological truism that exogenous dopamine is not a central dopamine receptor agonist, because it cannot cross the blood–brain barrier. However, the possibility that it might, and could thus contribute to the development of delirium, has been examined (463). Of over 21 000 inpatients in Stanford University Hospital, 1164 were given intravenous dopamine. Although the authors conceded that the methods that they used to

assess the presence of delirium were imprecise, they nevertheless calculated that intravenous administration of dopamine nearly triples the likelihood of the need for subsequent antipsychotic medication. This raises at least the possibility that in some pathological states significant amounts of exogenous dopamine can enter the brain. A prospective study is needed to clarify this. Meanwhile the authors have proposed that it might be useful to give dopamine receptor antagonists to patients who are about to receive dopamine infusions. However, most clinicians would probably be reluctant to do this on the existing evidence.

Local anesthetics

- A 59-year-old woman, grade ASA I, had psychiatric effects associated with local anesthetic toxicity after receiving bupivacaine 50 mg and mepivacaine 75 mg for an axillary plexus block. She complained of dizziness and a "near death experience" (464).

Lovastatin

In a double-blind study, 209 healthy adults were randomized to placebo or lovastatin 20 mg/day for 6 months (465). Placebo-treated subjects improved between baseline and post-treatment periods on neuropsychological tests in all performance domains (neuropsychological performance, depression, hostility, and quality of life), consistent with the effects of practice on test performance, whereas those treated with lovastatin improved only on tests of memory recall. Comparisons of the changes in performance between placebo and lovastatin showed small but significant differences for tests of attention and psychomotor speed, and were consistent with greater improvement with placebo. Psychological well-being was not affected by lovastatin. The authors concluded that treatment of hypercholesterolemia with lovastatin did not cause psychological distress or substantially alter cognitive function. However, treatment did result in slight impairment of performance in neuropsychological tests of attention and psychomotor speed, the clinical importance of which is uncertain.

Macrolide antibiotics

See also individual compounds

Although there is no evidence that neuropsychiatric complications of macrolides develop more readily in uremic patients, several factors may predispose toward these adverse effects, such as reduced drug clearance, altered plasma protein binding, different penetration of drug across the blood–brain barrier, and an increased propensity for drug interactions.

Two women, aged 49 and 50 years, developed altered mental status a few days after starting to take clarithromycin for eradication of *H. pylori* (466). There was incoherent speech with perseveration, inability to sustain attention, impaired ability to comprehend, coprolalia, euphoria, restlessness, visual hallucinations, anxiety, and inappropriate affect. Similarly, in three cases, a 46-year-old man, a 39-year-old woman, and a 4-year-old boy, treatment with clarithromycin was followed by nervous system and psychiatric symptoms that included euphoria, insomnia, aggressive behavior, hyperactivity, and emotional lability (467).

Azithromycin caused delirium in two elderly patients who took 500 mg initially followed by 250 mg/day (468).

Mannitol

Acute mania has been reported in a patient who was given intravenous mannitol (469).

- A previously well 75-year-old woman without a personal or family history of mental disorders developed severe major depression and was given nortriptyline 50 mg/day. After 10 days a diagnosis of bilateral acute angle-closure glaucoma was made and she was given a continuous intravenous infusion of mannitol 20%, oral acetazolamide 500 mg, and topical pilocarpine 2%, timolol 0.5%, and dexamethasone 0.1% (each 1 drop every 15 minutes). She became euphoric 30 minutes later and remained overactive, overly affectionate, and talkative, telling jokes, with pressured speech and flight of ideas. Her manic state remitted 1 hour after the end of the mannitol infusion and her severe depression recurred dramatically.

Mannitol can cause both acute expansion of extracellular fluid volume and a rapid reduction of intracellular fluid volume with retention of brain electrolytes; these may have been the mechanisms in this case.

Measles, mumps, and rubella vaccines

The suggestion that measles/MMR immunization can cause autism has not been confirmed. Autism is characterized by absorption in self-centered subjective mental activities (such as day-dreams, fantasies, hallucinations), especially when accompanied by marked withdrawal from reality.

The scientific and public response to the 1998 publication of Wakefield and colleagues (470), and Wakefield's subsequent press conference, in which he suggested that immunization with MMR might be associated with Crohn's disease and autism was enormous and controversial. For example, Black and colleagues (471) stated that the publicity generated by this paper was out of proportion to the strength of the evidence it contained; Beale (472) suggested that the Lancet would bear a heavy responsibility for acting against the public health interest that the journal usually aims to promote; O'Brien and colleagues (473) considered that the substantial amount of evidence that contradicts the findings of Wakefield and colleagues did not achieve the same prominence in the popular press.

In replying to these letters, Wakefield (474) defended the clinician's duty to his patients and the researcher's

obligation to test hypotheses. For his part, the editor of the Lancet pointed out that the paper had been presented with a commissioned commentary in the same issue; peer review had confirmed that the paper merited publication, with suitable revisions and editing, as an early report; finally, he considered that the press had presented the information in a balanced way (475). However, it was subsequently discovered that the work had been sponsored by a pharmaceutical company which had not been declared at the time (476). Furthermore, the children who were included in the study had been selected other than at random. Subsequently, 10 of Wakefield's colleagues withdrew the findings that they had initially reported and one other could not be traced; only Wakefield did not retract (477).

The Lancet then reviewed the developments in the discussion (478), with emphasis on evidence that contradicts the alleged association and new data presented by Wakefield. The last part of the editorial read as follows:

In a new twist, Wakefield's crusade fuelled further anxiety among parents when he and John O'Leary, director of pathology at Coombe Women's Hospital, Dublin, Ireland, presented unpublished data to the US Senate's congressional oversight committee in Washington on April 6 [2000]. The hearing was called by the chairman, Dan Burton, an Indiana Republican, whose grandson has autism and visited the Royal Free Hospital in November last year. At the hearing, six parents of children with autism gave moving testimonies of their children's illness. ··· Scientific evidence was presented by six chosen 'experts'. According to Wakefield's testimony, he has now studied more than 150 children with 'autistic enterocolitis'—an unproven association had become a disease—and a detailed analysis of the first 60 cases is to be published in the American Journal of Gastroenterology later this year.

Wakefield presented uninterpretable fragments of results only and concentrated on refuting studies that had contradicted his findings. His conclusions were surprisingly non-committal: 'the virological data indicate that this may be measles virus in some children'; he added that it would be imprudent to interpret the temporal relationship with MMR as a chance finding, in the absence of thorough investigation. O'Leary explained that gut-biopsy material from 24 of 25 children with autism was positive for measles virus compared with one of 15 controls, and that this material was presented to him by Wakefield using 'blinded protocols'. Since the controls are not further described and the details of these findings remain unpublished, this evidence raises far more questions than it answers.

"Autism is a poorly understood neurodevelopmental-disease spectrum with a heart-breaking personal story behind every case. But parents of such children have not been served well by these latest claims made well beyond the publically [sic] available evidence. A congressional hearing, like a press conference, is no place to make controversial scientific assessments. And if scientists question the safety of vaccines without making their evidence fully transparent, harm will be done to many more children than they purport to protect."

Wakefield and Montgomery subsequently raised doubts about the adequacy of the evidence that secured the license for MMR vaccine (479). Particularly in view of the immunosuppressive properties of the measles virus, they suggested that there is a potential for adverse interactions between the component live viruses. They therefore proposed that spaced monovalent measles, mumps, and rubella immunization should replace the use of the combined MMR vaccine. The continuing publications of Wakefield led to reduced MMR coverage in some parts of the UK and to well-publicized concerns about the potential for measles outbreaks among primary school entrants. In an editorial in the British Medical Journal, Elliman and Bedford replied to Wakefield's paper (480). They considered that the current concerns were idiosyncratic and presented reviews confirming the vaccine's safety. The Medicines Control Agency and the Department of Health in the UK rejected any suggestion by Wakefield and colleagues that combined MMR vaccines were licensed prematurely. A review of the licensing of MMR vaccines led to the assurance that the licensing procedure was normal and was based on robust studies (481). This position was shared by the Committee on Safety of Medicines and the Joint Committee on Vaccination and Immunisation. At the end of November 2001, Wakefield left his post at the Royal Free and University College Medical School in London. The college said: "Dr Wakefield's research was no longer in line with the Department of Medicine's research strategy and he left the university by mutual agreement" (482). The WHO strongly endorsed the use of MMR vaccine. The combination vaccine was recommended rather than monovalent presentations. There was no evidence to suggest impaired safety of MMR (483).

In April 2001, the Institute of Medicine's Immunization Safety Review Committee released its report "MMR vaccine and autism". Although scientists generally agreed that most cases of autism result from events that occur in the prenatal period or shortly after birth, there was concern because the symptoms of autism typically do not emerge until the child's second year, and this is the same time at which MMR vaccine is first administered in most developed countries. The committee took also into consideration the papers published by Wakefield and other groups and scientists suggesting evidence of a link between MMR vaccine and Crohn's disease and autism. Following review of the numerous research efforts on the MMR–autism hypothesis the committee concluded in its report "that the evidence favors rejection of a causal relationship at the population level between MMR vaccine and autistic spectrum disorders". Epidemiological evidence showed no association between MMR vaccine and autism, and the committee did not find a proven biological mechanism that would explain such a relation. Therefore, the committee did not recommend a policy review at this time of the licensure of MMR vaccine or of the current schedules and recommendations for MMR administration (484).

A conference of the American Academy of Pediatrics on "new challenges in childhood immunizations" was convened in Oak Brook, IL, on 12–13 June 2000 and reviewed data on what is known about the pathogenesis, epidemiology, and genetics of autism and the available data on hypothesized associations with Crohn's disease, measles, and MMR vaccine. The participants concluded that the available evidence did not support the hypothesis that MMR vaccine causes either Crohn's disease or autism or associated disorders. They recommended continued scientific efforts directed to the identification of the causes of autism (485).

The response to a newly published adverse event due to immunization must be rapid. If the reported association is correct, urgent re-evaluation of the immunization program is necessary. Otherwise, if the reported association is false, a credible counter-message is necessary to minimize the negative impact on the immunization program (486). The rapidity of response to the 1998 publication of Wakefield and colleagues, including the convening of an independent review panel in the UK, was very useful.

Two studies suggested a link between measles/MMR immunization and autism. Fudenberg reported that 15 of 40 patients with infantile autism developed symptoms within a week after MMR immunization (487). Wakefield and colleagues evaluated 12 children with chronic enterocolitis and regressive developmental disorders (470). The onset of behavioral symptoms was associated with MMR immunization in eight cases, as reported by the parents. Both reports were non-comparative and anecdotal. By chance alone some cases of autism will occur shortly after immunization, and most children in developed countries receive their first measles or MMR vaccination in the second year of life, when autism typically manifests. The imprecision of the interval between immunization and the onset of behavioral symptoms in the study by Wakefield and colleagues made these data suspect, even before their retraction.

Inaccuracies in the study of Fudenberg, for example referring to hepatitis B vaccine as a live vaccine, cast some doubts on the carefulness of the entire report. Developmental delay is likely to be detected by a gradual awareness over a period of time, not on a particular day. Epidemiological studies in various countries (UK, Sweden, Finland), comparing the introduction and use of vaccines and the incidence of autism, have not supported a relation between measles/MMR vaccine and autism (488–491). Wing reviewed 16 studies in Europe, North America, and Japan and found no increase in autism with increasing use of measles or MMR vaccines (492). An analysis of two large European datasets produced similar results (493). In early 1998, experts in various medical disciplines reviewed the work of the Inflammatory Disease Study Group of the Royal Free Hospital in detail and concluded that there is no evidence for a link between measles/MMR vaccine and either Crohn's disease or autism (494).

Data from an earlier study have been reanalysed to test the hypothesis that MMR vaccine might cause autism but that the induction interval needs to be short (495). Evidence for an increased incidence was sought using the case-series method. The study used data on all

MMR vaccines, including booster doses. The results of this study, combined with results obtained earlier by the same authors, provided powerful evidence against the hypothesis that MMR vaccine causes autism at any time after immunization.

However, the controversy, particularly in the USA, has not ended. A survey commissioned by "Generation Rescue" (a parent-founded, parent-funded, and parent-led organization of more than 350 families) was published in June 2007. Data were gathered by Survey USA, a national market-research company, which carried out a telephone survey of the parents of more than 17 000 children, aged 4–17 years, in five counties in California (San Diego, Sonoma, Orange, Sacramento, and Marin) and four counties in Oregon (Multnomah, Marion, Jackson, and Lane) (496). They asked parents whether their child had been immunized, and whether that child had one or more of the following diagnoses: attention deficit disorder (ADD), attention deficit hyperactivity disorder (ADHD), Asperger's syndrome, pervasive development disorder not otherwise specified, or autism. Among more than 9000 boys aged 4–17 years, they found that immunized boys were 155% more likely to have neurological disorders compared with their non-immunized peers. Immunized boys were 224% more likely to have ADHD, and 61% more likely to have autism. Older immunized boys in the 11–17 age bracket were 158% more likely to have a neurological disorder, 317% more likely to have ADHD, and 112% more likely to have autism. It is open for discussion whether these results, obtained by this method, can really make a substantial contribution to a scientific question.

Meclozine

Loss of memory, confusion, and disorientation has been ascribed to meclozine in an 85-year-old woman (SEDA-13, 132).

Medroxyprogesterone

Mental or personality changes, a typical glucocorticoid effect, have been reported to be more severe and more frequent with combined aminoglutethimide plus medroxyprogesterone acetate treatment than with monotherapy in patients with bone metastases from breast cancer. The increased frequency of depressive syndromes on the two-drug therapy could not be attributed to the physical adverse effects of the combination, since the mental disorders appeared only during the first weeks of treatment, whereas the Cushingoid features did not become apparent until some 6–8 weeks of treatment had been given (497).

Mefloquine

At first thought to occur only after therapeutic doses of mefloquine, it is now clear that neuropsychiatric reactions occur after prophylactic use as well. The incidence is estimated at about one in 13 000 with prophylactic use, but as high as one in 215 with therapeutic use (SEDA-17, 329). Combination with other drugs that affect the nervous system can result in unpredictable reactions. The symptoms vary in type and severity: non-cooperation, disorientation, mental confusion, hallucinations, agitation, and impaired consciousness. An acute psychiatric syndrome with attempted suicide was reported in one case. A single dose can be all that is needed to evoke a mental reaction. Convulsions have been reported, with or without psychiatric symptoms; it seems that mefloquine can aggravate and perhaps even provoke latent epilepsy (SEDA-13, 809; SEDA-16, 307; SEDA-17, 329) (498).

- A severe psychiatric and neurological syndrome, with agitation, progressive delirium, and generalized rigors, was seen in a 47-year-old man after he had taken mefloquine 1500 mg over 24 hours (499).
- A 7-year-old Indian boy was diagnosed as having "cerebral malaria" and received quinine followed by mefloquine (dose not given) (500). He developed hallucinations and removed his clothes and danced. His symptoms resolved within 24 hours of stopping mefloquine. This case highlights the fact that mefloquine should not be given after quinine in cases of severe malaria.
- A 42-year-old man with no previous psychiatric history suddenly developed visual symptoms after the third dose (total dose 750 mg) of prophylactic mefloquine (501). The symptoms consisted of an impression of focusing on two different planes and of perceiving his surroundings as very far from him. They were associated with slurred speech and altered comprehension. They occurred daily, lasting up to an hour, for 6 months. He had previously taken a course of mefloquine for 7 weeks without any adverse events.
- A 52-year-old woman with no psychiatric history developed anxiety, paranoia, visual hallucinations, confusion, and depressive symptoms after 3 doses of prophylactic mefloquine (250 mg/week) (502). She had previously taken mefloquine prophylaxis intermittently for 4 years with no adverse events.

These case reports illustrate important neuropsychiatric adverse effects of mefloquine in individuals who had previously taken mefloquine safely and had no psychiatric history.

- A 48-year-old woman developed anxiety, tremor, depression, dry mouth, nausea, and marked weight loss (503). Physical examination, electrocardiography, chest X-ray, CT scan, and laboratory investigations were unremarkable. The Hamilton D score was 44 for 17 items. She had taken mefloquine 250 mg/week for 8 weeks for malaria prophylaxis, and after 2 weeks had started to feel unwell, with dysphoria, depression, and weakness. She was given fluoxetine 20 mg/day and alprazolam 1.5 mg/day. Her condition continued to deteriorate. The dose of fluoxetine was increased to 40 mg/day and flunitrazepam was added. She was later instead given milnacipran, a serotonin and noradrenaline reuptake inhibitor. Five months after the first course of mefloquine she had recovered sufficiently to return to work. However, she relapsed and she was eventually stabilized on venlafaxine 75 mg/day.

A postal survey of 5446 returning Danish travellers examined the adverse effects of unstated doses of mefloquine, chloroquine, and chloroquine plus proguanil for malaria prophylaxis (504). There were 4158 responses (76%); 1223 travellers took chloroquine, 1827 took chloroquine plus proguanil, and 809 took mefloquine. Overall, although chloroquine and chloroquine plus proguanil were associated with a large number of mild (mainly gastrointestinal) adverse effects, 30–50% had diarrhea and about 20% had nausea or abdominal pain. There was a significantly larger number of reported "unacceptable symptoms" (not defined) with mefloquine: 2.7%, 1.0%, and 0.6% for mefloquine, chloroquine, and chloroquine plus proguanil respectively. Most of the more serious adverse events were in those who took mefloquine. Compared with chloroquine alone the relative risk (95% CI) of "depression," experiencing "strange thoughts," or having altered spatial perception were 5.1 (2.7, 9.5), 6.4 (2.5, 16.1), and 3.0 (1.4, 6.2) respectively. There was also a higher incidence of depression in women than in men. The relative risk of hospital admission or early termination of travel possibly related to prophylaxis was higher with mefloquine than with either chloroquine or chloroquine plus proguanil; the relative risks (95% CI) were 162 (69, 498), 612 (169, 5054), and 261 (127, 649) respectively.

A postal survey of the incidence of psychiatric disturbances in 2500 returning Israeli travellers (505) showed that travellers with this class of adverse effects were more likely to have taken mefloquine than other antimalarial drugs. Of 117 travellers with psychiatric adverse effects, 115 had taken mefloquine compared with 948/1340 for the entire cohort. This was a retrospective postal study with a response rate of 54% (1340 out of 2500), and of those who responded 71% had taken mefloquine, 5% had taken chloroquine, and 24% had taken no prophylaxis. In this study 11% (117) of the respondents reported psychiatric disturbances, mainly sleep disturbance, fatigue, vivid dreams, or "lack of mood." Only 16 of the respondents had symptoms lasting 2 months or more. Those who had had a psychiatric disturbance were also more likely to have been female and to have taken recreational drug use.

Although the above studies were limited by retrospective design, their results are in broad agreement with the results of other studies over the past few years that indicate that women have a higher incidence of psychiatric adverse effects from mefloquine than men (506–509).

In a prospective, double-blind, randomized, placebo-controlled study in 119 healthy volunteers (mean age 35 years), who took either atovaquone 250 mg/day + chloroguanide 100 mg/day or mefloquine 250 mg/week, depression, anger, and fatigue occurred during the use of mefloquine but not atovaquone + chloroguanide (510).

Mefloquine can also cause psychosis.

- After 4 weeks of malaria prophylaxis with mefloquine 250 mg/week, a 25-year-old woman developed bizarre paranoid delusions with auditory and visual hallucinations (511). MRI and MRA scans of the brain were unremarkable, but electroencephalography showed diffuse cerebral dysfunction. A malaria smear was negative. Mefloquine was withdrawn and the psychotic symptoms resolved over 6 days with temporary risperidone treatment. The symptoms did not recur during a 2-year follow-up period.

In a review of 10 trials (n = 2750 non-immune adult travelers) (512) the effects of mefloquine in adult travellers were compared with the effects of other regimens in relation to episodes of malaria, withdrawal from prophylaxis, and adverse effects. Five trials were field studies of male soldiers. One comparison of mefloquine with placebo showed that mefloquine was effective in an area of drug resistance (OR = 0.04; 95% CI = 0.02, 0.08) and withdrawals in the mefloquine group were consistently higher in four placebo-controlled trials (OR = 3.56; 95% CI = 1.67, 7.60).

In five comparisons of mefloquine with other chemoprophylaxis regimens, there was no difference in tolerability. The only consistent adverse effects consistently specific to mefloquine in the controlled trials were insomnia and fatigue, but there were also 516 reports of adverse effects of mefloquine, 63% of which were in tourists and travellers. Four deaths were attributed to mefloquine.

In another major review the risk of depression, psychosis, a panic attack, or a suicide attempt during current or previous use of mefloquine was compared with the risk during the use of proguanil and/or chloroquine or doxycycline (513). The study population (n = 35 370) was aged 17–79 years (45% men). There was no evidence that the risk of depression was increased during or after the use of mefloquine, but psychoses and panic attacks were more frequent in current users of mefloquine than in those using other antimalarial drugs.

Airline pilots should not take routine prophylaxis with mefloquine because of the small risk of neuropsychiatric reactions.

Melatonin

A severely depressed woman developed a mixed affective state after taking melatonin for 7 days in a clinical trial (514). Confusion, hallucinations, and paranoia temporally related to melatonin have also been described (515).

Memantine

A toxic psychosis was reported in two patients taking memantine (516).

Mepacrine

Acute psychosis has been seen with the use of large parenteral doses of mepacrine in the treatment of malignancies (SEDA-11, 592).

- An acute transient psychosis (screaming, kicking, and hallucinations) was seen in an 11-year-old boy after 5 days' treatment with quinacrine 100 mg tds for *Giardia lamblia* infestation. Recovery occurred after withdrawal (SEDA-16, 309).

Mercury and mercurial salts

The safety of mercury in vaccines has also been discussed in relation to the possibility of neurocognitive damage (517).

There is continuing debate that there is an association between autism and thiomersal-containing vaccines. Some authors believe that review of the literature supports the hypothesis that mercury in vaccines may be a factor in the pathogenesis of autism (518). The World Health Organization's Global Advisory Committee on Vaccine Safety (GACvS) has also kept this issue under review and concluded in November 2002 that there is no evidence of toxicity in infants, children, or adults who have been exposed to thiomersal in vaccines. The CSM is also keeping the issue under close review and studies of the possible toxicology of thiomersal continue to appear (519).

Two studies, a population-based cohort study from Denmark (520), and a two-phase retrospective cohort study from the USA (521), have both suggested that there is no causal relation between autism and childhood vaccination with thiomersal-containing vaccines. A third epidemiological study using data from the US Vaccine Adverse Event Reporting System showed an association with increasing thiomersal from vaccines with neurodevelopmental disorders (522). Following calls from regulatory agencies for thiomersal to be withdrawn from vaccines, new childhood vaccines have been developed and in many countries are replacing those that contain thiomersal.

The usual dose of ethyl mercury in pediatric vaccines is small (about 12.5–25 micrograms of mercury). However, the metabolism of ethyl mercury in infants who receive vaccines containing thiomersal is unknown. The mean doses of mercury in 40 full-term infants exposed to thiomersal-containing vaccines were 46 (range 38–63) micrograms in 2-month-old children and 111 (range 88–175) micrograms in 6-month-old children. Blood mercury concentrations in the thiomersal-exposed 2-month-old children ranged from less than 3.8 to 21 nmol/l; in 6-month-old children all the concentrations were below 7.5 nmol/l. Only one of 15 blood samples from 21 controls contained measurable concentrations of mercury. Urine concentrations of mercury were low after vaccination, but stool concentrations were high in thiomersal-exposed 2-month-old children (mean 82 ng/g dry weight). The mean half life of ethyl mercury was 7 days.

This study was not designed as a formal assessment of the pharmacokinetics of mercury. However, it showed that the administration of vaccines containing thiomersal did not seem to raise blood concentrations of mercury above safe values in infants. Ethyl mercury seems to be eliminated from the blood rapidly via the stools. The

authors concluded that the thiomersal in routine vaccines poses very little risk in full-term infants.

- Previously well, developmentally normal 20-month-old twin girls presented with weakness, anorexia, a papular rash, and increasingly swollen, red, painful hands and feet of 1 month's duration (523). They had no history of fever, conjunctivitis, lymphadenopathy, or oral changes characteristic of Kawasaki disease. They were irritable and unwell and were sweating but afebrile. Both had a tachycardia, and one had a raised blood pressure of 130/90 mmHg (95th centile for age 108/62 mmHg). Both had reduced muscle power and diminished reflexes. Their palms and soles were erythematous and indurated, with desquamation, judged to be acrodynia. The infants had been given a mercury-containing teething powder from India once or twice a week over the 4 preceding months. Their blood mercury concentrations were 176 and 209 (reference range below 18) μmol/l. Chelation therapy with 2,3-dimercaptosuccinic acid was administered by nasogastric tube. Before admission the twins had regressed developmentally and were unable to feed, sit, or walk. Over the next 8 weeks they had some minor neurocognitive improvements, but their long-term prognosis was uncertain.

Methadone

In a randomized, double-blind, crossover study of 20 patients on a stable methadone regimen, a single dose of methadone caused episodic memory deficits (524). This was significant in patients with a history of diamorphine use averaging more than 10 years duration. Such deficits can be avoided by giving methadone in divided doses.

Psychomotor and cognitive performance has been studied in 18 opioid-dependent methadone maintenance patients and 21 non-substance abusers (525). Abstinence from heroin and cocaine for the previous 24 hours was verified by urine testing. The methadone maintenance patients had a wide range of impaired functions, including psychomotor speed, working memory, decision making, and metamemory. There was also possible impairment of inhibitory mechanisms. In the areas of time estimation, conceptual flexibility, and long-term memory, the groups performed similarly.

Methotrexate

There was a significantly higher risk of late cognitive impairment (concentration and memory) in patients (*n* = 39) taking adjuvant cyclophosphamide, fluorouracil, and methotrexate than in controls matched for age, disease, surgery, and radiation dose (526).

In studies of the neurotoxic effects of low-dose methotrexate treatment, dizziness, headache, visual disturbances or hallucinations, lack of concentration, cognitive dysfunction, and depression-like symptoms were detected in 1–35% of patients (527,528). Advanced age and mild renal insufficiency were possible susceptibility factors (529).

Methylphenidate

Methylphenidate-induced obsessive-compulsive symptoms are very rare (530); a second case has been reported (531).

- An 8-year-old boy had unusual difficulties in completing a trial of methylphenidate for ADHD. Methylphenidate 10 mg/day initially improved his school performance, but by the end of the second week of treatment severe obsessive-compulsive behavior began to emerge, requiring drug withdrawal. He eventually refused to eat because he suspected that his food was being poisoned. His parents reported that he became more sensitive and easily upset, appearing increasingly tense and fearful. He continued to have intrusive distressing contamination worries. He had repetitive tic-like movements of his head and neck, frequently rubbed his face with his shirtsleeves, and avoided touching his face directly with his fingers for fear of contamination. His symptoms gradually and completely dissipated over 2–3 months without specific intervention. While very mild symptoms of ADHD persisted, he remained entirely free of symptoms at 1 year. Medication was not restarted.

This case lends further support to existing anecdotal evidence associating methylphenidate with obsessive-compulsive disorder of acute onset. However, rechallenge was not attempted, and causality should be viewed as being uncertain.

In a small percentage of hyperactive children, there were gross behavioral changes with hallucinations after brief administration of modest doses of methylphenidate; the reaction subsided after withdrawal (532).

Methysergide

An LSD-like reaction has been observed with as little as 2 mg (533). There is a risk of dependence and the drug is better avoided in patients with a history of mental disorders.

Metirosine

In 14 drug-free subjects with a history of depression metirosine 5 g caused marked, transient increases in core depressive and anxiety symptoms (534).

When tested in 21 healthy subjects, metirosine produced panic attacks on several occasions in three of them (535).

Metoclopramide

Supersensitivity psychosis has been reported in two men, aged 74 and 65 years, who had taken metoclopramide 5 mg qds for 6 and 3 months respectively (536). Hallucinations and delusions developed 12 hours after the drug was withdrawn in one patient and 3 days after withdrawal in the other. Both recovered after treatment with risperidone.

Metronidazole

Metronidazole can cause delirium.

- A 75-year-old man took oral metronidazole 500 mg tds for Clostridium difficile colitis and 48 hours after the start of therapy he became withdrawn and less responsive; during the next 24 hours he developed hallucinations and confusion (537). Metronidazole was switched to oral vancomycin, and his symptoms resolved within 24 hours. One month later, he was re-challenged with metronidazole for recurrent Clostridium difficile diarrhea without knowledge of his prior adverse drug reaction. Soon after taking the first dose of metronidazole, he again developed hallucinations, which resolved after switching to vancomycin, confirming that metronidazole was the cause of his mental confusion.

Miconazole

Acute toxic psychosis is a rare consequence of miconazole (SED-12, 679; 538).

Minocycline

Transient depersonalization symptoms have been attributed to by minocycline (539) and may have been caused by increased intracranial pressure.

Mizolastine

Mizolastine in doses up to 10 mg does not differ from placebo with regard to cognitive function and psychomotor performance, but in higher doses, there is dose-dependent impairment (SEDA-19, 175).

Morphine

Hallucinations have been described after the use of morphine in various dosage forms; in one series, patients experienced adequate pain relief and no further hallucinations or nightmares when changed to oxycodone (540). Delusions and hallucinations have been reported in a patient who was also taking dosulepin (541). Restlessness, vomiting, and disorientation were described in two male patients over 60 years of age taking modified-release morphine for relief of pain in advanced cancer (542).

In a double-blind, crossover, randomized, placebo-controlled study of the acute effects of immediate-release morphine on everyday cognitive functioning in 14 patients who were also taking modified-release opioids, immediate-release morphine on top of modified-release morphine produced discrete impairment of cognitive functioning (543). Both immediate and delayed memory recall were affected. Impairment in delayed recall was more pronounced. Retrograde amnesia suggested that morphine produces additional difficulties in retrieval of information. Simple tracking tasks were enhanced in the morphine group; however, more demanding tasks and set shifting were impaired. There was no effect on backwards digit span. These findings suggest that morphine produces discrete impairment that is likely to affect quality of life.

Myristica fragrans (nutmeg)

Nutmeg is the dried kernel of the evergreen tree *Myristica fragrans*. At sufficiently high doses it has psychoactive effects.

- A 23-year-old man developed delusions and hallucinations (544). His medical history was unremarkable. For several months he had been regularly taking nutmeg 5 g/day. He improved after withdrawal of nutmeg and was treated with olanzapine. After 6 months he was still taking antipsychotic medication but could resume work.

Nalbuphine

- A 53-year-old man with no past psychiatric history was found by the police walking aimlessly, unclothed, and responding to auditory hallucinations (545). He had slurred speech and generalized tremors. Lumbar puncture, a CT head scan, and urine drug screen were all normal. He responded to risperidone 1 mg bd with dramatic improvement after 2 days. He reported chronic use of nalbuphine, with recent increased use.

Nalidixic acid

Disturbances of body perception, hallucinations, confusion, confabulation, and depression can rarely occur with nalidixic acid (546).

Naloxone

Behavioral effects have been noted after high doses of naloxone (2 and 4 mg/kg) in volunteers. Most subjects experienced an initial rush or buzz, tingling or numbness in the extremities, dizziness, a heavy head, and reluctance to move or initiate activities, which usually subsided within 15–30 minutes of administration. Transient sweating and yawning occurred later. Nausea and stomachache were often experienced and in two subjects persisted throughout the day, although in mild form. There were no pupillary changes or sleepiness. Increasing doses of naloxone were associated with increasingly impaired cognitive performance. Violent behavior has been reported after the use of naloxone to reverse sedation (SEDA-17, 88).

Naltrexone

Panic attacks precipitated by naltrexone have been reported (547).

- A 29-year-old woman with bulimia nervosa and a family history of anxiety was enrolled in a trial of naltrexone (100 mg/day). She had no history of opioid use. Within hours of her first dose she experienced alarm, anxiety, chest discomfort, shortness of breath, a fear of dying, sweating, nausea, and derealization. She was unable to remain at home or to go out alone. For 3 days she continued to take naltrexone, with an increasing frequency of panic attacks. On day 4 she was treated with alprazolam (0.5 mg) but relapsed after further naltrexone. Withdrawal of naltrexone led to complete remission of symptoms.

This effect was attributed to an action of naltrexone in removing the endogenous opioid effect at OP_3 (μ) opioid receptors in the locus ceruleus and thus resulting in unchecked noradrenergic hyperactivity.

Acute psychosis secondary to the use of naltrexone has been reported (548).

- A 44-year-old physically healthy woman developed auditory and visual hallucinations and persecutory delusions. She had been given naltrexone hydrochloride 50 mg/day 3 days before the acute psychotic incident in order to prevent relapse of alcohol dependence. Her psychotic symptoms completely resolved 48 hours after withdrawal of naltrexone.

In studies of its use in treating alcohol, opioid, and nicotine dependence, naltrexone has not been reported to cause depression or dysphoria. Patients who complain of naltrexone-associated dysphoria often have co-morbid depressive disorders or depression resulting from opioid or alcohol withdrawal states (549). Co-morbid depression is not a contraindication to naltrexone. Small pilot studies have supported the use of naltrexone in combination with antidepressants for the treatment of patients with co-morbid depression. The risk of non-fatal overdose is significantly increased after naltrexone treatment, as a result of reduced tolerance, compared with patients taking substitution methadone (550).

Naproxen

Abnormalities in cognitive capacity and altered behavior have been described in elderly patients taking naproxen (SEDA-8, 107; 551).

Nevirapine

Various psychiatric abnormalities (delirium, an affective state, and a psychosis) have been described in three patients who took nevirapine for 10–14 days (552):

- cognitive impairment, clouding of consciousness, and a paranoid episode in a 35-year-old man
- delusions of persecutory and depressive thoughts in a 42-year-old woman
- delusions of persecution and infestation and hallucinations in a 36-year-old woman

All three responded to withdrawal of nevirapine.

Nicotine replacement therapy

A woman developed hallucinations and cerebral arterial narrowing after using nicotine patches (553). In another case delusions and hallucinations occurred when a woman who had been using nicotine to try to stop smoking took a cigarette (554).

Nitrates, organic

In an elderly woman, isosorbide dinitrate caused visual hallucinations and subsequent suicidal ideation, thought to be due to hypotension and cerebral ischemia (555).

Nitrofurantoin

Rarely, concomitant dysphoric, euphoric, or even psychotic reactions have been reported in patients taking nitrofurantoin (556).

Non-steroidal anti-inflammatory drugs (NSAIDs)

See also COX-2 inhibitors

Behavioral changes have been reported with indometacin (557).

- A 92-year-old man with a history of senile dementia of the Alzheimer type, glaucoma, and constipation took indometacin 25 mg for pseudogout. After six doses he became very agitated and confused and was physically and verbally aggressive. Indometacin was withdrawn and he recovered over 10 days with the help of haloperidol 0.5 mg/day (558).

NSAIDs should probably be included as one of the many groups of drugs that can cause confusion in the elderly.

Cognitive function in elderly out-patients was assessed in a large retrospective study by a questionnaire; there were no significant differences in the total scores of users and non-users of NSAIDs (SEDA-17, 106). Another study showed that performance in sensorimotor coordination and short-term memory tests can improve in healthy elderly volunteers who take indometacin (559). NSAIDs have also been associated with a reduced risk of Alzheimer's disease (560).

Norfloxacin

There have been reports of hallucinations with norfloxacin (561).

Omeprazole

- Delirious psychosis has been attributed to intravenous omeprazole 40 mg bd in a 77-year-old woman with Guillain–Barré syndrome, who also had *Helicobacter pylori*-associated gastritis (562). Delirium developed 2 days after omeprazole was begun and resolved completely after drug withdrawal.

Opioid analgesics

See also individual compounds

Psychomotor symptoms have been noted subsequent to epidural buprenorphine (563).

Oxolamine

Evidence has accumulated that oxolamine can cause hallucinations in children, especially those under 10 years of age (564,565); some cases have been confirmed by re-exposure to the drug.

Oxybutynin

Neuropsychiatric adverse effects seem to pose a greater problem with oxybutynin than other anticholinergic drugs. A group of pharmacologists involved in pharmacovigilance have described five spontaneous reports of night terrors associated with oxybutinin (566). Four were children, aged 5–8 years, being treated for enuresis, and the fifth was a 77-year-old woman with incontinence. Total daily doses were 8–12 mg in the children and 10 mg in the elderly woman: these are all toward the upper end of the recommended ranges.

The effects of oxybutynin on cognitive function have been studied in 25 children (aged 5–17 years, 14 girls) with daytime enuresis (567). Ten were treated with behavior modification and 15 with behavior modification plus oxybutynin 7.5–20 mg/day, depending on body weight. Patient allocation was not random but according to parental choice. Neuropsychological testing was done at baseline (after 4 weeks of behavior modification) and at the end of the treatment period. There was no cognitive impairment in the oxybutynin-treated children, but surprisingly baseline function was lower in those who were allocated to oxybutynin. This may have introduced selection bias, making it difficult to interpret any possible drug effect.

Pentagastrin

In a double-blind placebo-controlled study intravenous pentagastrin led to panic attacks in nine of 19 patients with social phobia, seven of 11 with panic disorder, and two of 19 healthy controls (568). Anxiety, blood pressure, and pulse increased in all three groups.

In a double-blind, placebo-controlled study in seven patients with obsessive-compulsive disorder and seven healthy controls, intravenous pentagastrin 0.6 microgram/kg produced panic-like reactions in six (86%) of the seven patients and only two (29%) of the controls (569).

The anxiety caused by pentagastrin is prevented by inhibiting cholecystokinin receptors (570).

Pentamidine

Confusion and hallucinations have occasionally been reported with pentamidine. Magnesium deficiency may affect mental function; a flat affect, slow speech, and mental withdrawal are some of the typical effects (SEDA-13, 825) (571). The symptoms of hypomagnesemia can be ill defined; unexplained symptoms, despite improvement of the *P. jiroveci* infection, demand measurement of the serum magnesium concentration.

Pentazocine

Perceptual disturbances are generally thought to occur more often with pentazocine than with other opioids. Objective definition of such phenomena is difficult, but in a study of postoperative dreaming after the use of pentazocine and morphine as premedicants there was no statistically significant difference between the two drugs (SED-11, 148) (572).

Pentoxifylline

Rare cases of hallucinations in elderly people have been ascribed to a stimulant effect of pentoxifylline on the central nervous system (SEDA-18, 219).

Phenobarbital

Phenobarbital-induced behavioral disturbances, especially hyperkinesia, are especially common in children, with an incidence of 20–50%; drug withdrawal is required in 20–30% of cases. It is unclear whether and to what extent adults are affected in this way.

Phenoxybenzamine

Panic attacks have been described in one case, a week after withdrawal of the drug; the causal association was not certain (SEDA-17, 163).

Phentermine

A schizophreniform-like psychotic disorder was attributed to phentermine in a young woman without a personal or family history of psychiatric disorders (573).

Phenylbutazone

Psychomotor reactions to phenylbutazone when driving have been reported (574).

Phenylpropanolamine (norephedrine)

Psychosis has been attributed to phenylpropanolamine (575,576). Risk factors include symptoms or a history of mood spectrum disorder, a history of psychosis, female sex, and a family history of psychiatric disorder.

Although irritability and insomnia are frequent in adults, behavioral disturbances (restlessness, irritability, aggressiveness, and sleep disturbances), seizures, and delirium with hallucinations have been most often observed in children (SEDA-11, 1) (577).

Phenytoin and fosphenytoin

Phenytoin has been implicated in psychiatric adverse effects with or without other signs of toxicity, and at serum concentrations above or below the upper limit of the target range, but the actual incidence of these reactions is unknown (578).

- A 9-year-old boy with seizures developed intermittent complex visual hallucinations during therapy with fosphenytoin and, on a separate occasion, carbamazepine (579).

Five patients with Down's syndrome and dementia, aged 44–67 years, taking phenytoin had progressive cognitive decline (580). This resolved once the drug was withdrawn. Cognitive decline was not related to high serum concentrations. Older patients with Down's syndrome might be especially sensitive to the effects of phenytoin.

Pramipexole

Hallucinations occur in about 10% of patients with Parkinson's disease, regardless of the stage of disease (581).

Primaquine

Severe mental depression and confusion was reported in one patient who had been treated with chloroquine beforehand; all the symptoms disappeared on withdrawal (SEDA-11, 588).

Procainamide

Acute psychosis has been attributed to procainamide (582).

- A 45-year-old woman developed an acute psychosis within 72 hours of starting to take procainamide 75 mg intravenously, followed by a continuous infusion of 2 mg/minute for atrial fibrillation. The plasma procainamide concentration was 8.2 µg/ml and the plasma concentration of the main acetylated metabolite, acecainide, was 4.6 µg/ml. She was then given oral procainamide 500 mg qds, and 2 days later her trough concentrations of procainamide and acecainide were 4.5 and 4.9 µg/ml respectively. The following day she had visual hallucinations and was later found wandering the hospital asking about the babies under her bed. She had no previous history of psychiatric illness and she recovered completely 24 hours after withdrawal of procainamide.

There have been a few previous reports of similar adverse effects with procainamide in therapeutic dosages, and in most cases the plasma concentrations of procainamide and acecainide have been within the usual target ranges, as in this case.

Progesterone

Progesterone, the physiological progestogen, has long been overshadowed in medical treatment by its synthetic analogues, because of their superior oral absorption. However, it can be administered in other ways, particularly as a vaginal suppository, and the hypothesis has long existed that it might be more effective or better tolerated than the synthetic agents. In a double-blind, placebo-controlled, randomized study a Swedish group examined the mental effects of progesterone in 36 women with climacteric symptoms (583). All received estradiol 2 mg/day during three 28-day cycles. Vaginal progesterone suppositories (400 or 800 mg/day) or placebo were added sequentially for 14 days per cycle. Women without a history of premenstrual syndrome had cyclical variations in both negative mood and physical symptoms while taking progesterone 400 mg/day, but not the higher dose or the placebo. Women without a history of premenstrual syndrome had more physical symptoms on progesterone treatment compared with placebo. Women with prior premenstrual syndrome reported no progesterone-induced symptom cyclicity. The authors concluded that in women without prior premenstrual syndrome natural progesterone caused negative mood effects similar to those induced by synthetic progestogens, but the dose-effect relation was complex.

Prolintane

Prolintane can cause visual hallucinations (584).

Propofol

The association of propofol with a range of excitatory events is well recognized. Behavioral disturbances with repeated propofol sedation have been reported in a 30-month-old child (585). Propofol was well tolerated initially, but the child then became increasing irritable, aggressive, and uncooperative during awakening from subsequent sedations, including screaming, kicking, hitting, and biting. The next two sedations were performed using methohexital and were not followed by any behavioral disturbances.

Prolonged delirium after emergence from propofol anesthesia has also been reported (586).

A psychotic reaction has been reported (587).

- A 37-year-old man who had abused metamfetamine, paint thinner, psychotomimetic drugs, and alcohol for 20 years was given chlorpromazine, haloperidol, and flunitrazepam just before surgery. After spinal anesthesia he was given propofol 5 mg/kg/hour intravenously. However, euphoria and excitement occurred 10 minutes after the start of the infusion and he had excitement, hallucinations, and delirium. His symptoms were suppressed by intravenous haloperidol 5 mg.

The authors speculated that propofol may produce psychotic symptoms when it is used in patients with a history of drug abuse.

Quinapril

Depression has been attributed to quinapril (588).

- A 90-year-old man with a history of peripheral arterial disease and mild heart failure presented with reduced appetite, insomnia, anhedonia, reduced energy, and suicidal ideation. His symptoms had started a month before, when he had begun to take quinapril 10 mg/day. He was also taking furosemide 20 mg/day and digoxin. He was alert, coherent, and cognitively intact, but depressed with a flat affect. He had no psychotic symptoms. Quinapril was withdrawn. He improved within 48 hours and recovered fully in 5 days.

The authors cited several reports of depression, mania, and psychosis with various ACE inhibitors.

Acute psychosis has been attributed to quinapril (589).

- A 93-year-old woman with heart failure was given quinapril 2.5 mg bd. Two hours after the first dose she became confused, disoriented, and anxious. During the night, she made many frantic telephone calls to her daughter and other family members complaining that she was being assaulted. Her anxiety, disorientation, and visual hallucinations continued for 5 days. She had had an episode of hallucinations 2 years before, while taking a beta-blocker. Quinapril was withdrawn. She recovered over the next day.

The authors commented on three previous cases of visual hallucination with other ACE inhibitors.

Ranitidine

Depression has been estimated to occur in 1–5% of patients taking ranitidine (SEDA-17, 418) (590).

Occasional confusion and behavioral disturbances have been attributed to ranitidine (591–593).

Remifentanil

In 201 patients scheduled for day-case surgery, those who received remifentanil had significantly fewer responses to surgical stimulation and had better psychomotor and psychometric function during the recovery period, although there was no significant difference in time to recovery room or hospital discharge (594).

Reserpine

Depression is a very common adverse effect of reserpine (SED-9, 328) (595).

- A 66-year-old woman was admitted to hospital suffering from an agitated depressive psychosis. This settled with standard antipsychotic therapy. It was subsequently found that she had been taking reserpine in an over-the-counter formulation and had stopped this a week before her admission (596). Her doctor felt that the syndrome had occurred as a result of nervous system hypersensitivity after reserpine withdrawal.

Rivastigmine

Three patients with dementia, with no prior psychiatric history, deteriorated while taking rivastigmine (597). The time course suggested an association with rivastigmine, and each improved after withdrawal.

Saquinavir

Saquinavir can occasionally be associated with acute paranoid psychotic reactions (598).

- A 41-year-old woman took zidovudine and didanosine for HIV-1 infection after an acute seroconversion illness. Zidovudine and didanosine had to be withdrawn because of neutropenia and nausea, so she was given stavudine plus lamivudine without any adverse effects. Saquinavir (600 mg tds) was added 12 months later because of weight loss and a falling CD4 cell count. Within 24 hours of starting to take saquinavir she developed agitated depression with paranoid ideation. After drug withdrawal her mental health returned to normal over 5 days. Over the next 6 weeks stavudine and lamivudine were reintroduced without any adverse effects and continued for a further 11 months. Saquinavir was then reintroduced at the previous dosage, and within 2 days she again became extremely mentally agitated with paranoid ideation. Saquinavir was withdrawn and she recovered within 7 days. She was later given indinavir without problems. Her mother had a history of major depressive illness.

Sevoflurane

Delirium during emergence from sevoflurane anesthesia has often been documented. Four patients, an adult and three children aged 3–8 years, who were able to recount the experience, have been reported (599). They had full recall of postoperative events, were terrified, agitated, and distressed, and hence presented with acute organic mental state dysfunction which was short-lived. Two were disoriented and had paranoid ideation. They were not in any pain or were not distressed by pain if it was present. The authors hypothesized that misperception of environmental stimuli associated with sevoflurane's particular mode of action may have been the underlying cause of this phenomenon. Anxiolytic premedication and effective analgesia did not necessarily prevent the problem.

The effect of intravenous clonidine 2 micrograms/kg on the incidence and severity of postoperative agitation has been assessed in a double-blind, randomized, placebo-controlled trial in 40 boys who had anesthetic induction

with sevoflurane after oral midazolam premedication (600). There was agitation in 16 of those who received placebo and two of those who received clonidine; the agitation was severe in six of those given placebo and none of those given clonidine.

The effects of intravenous and caudal epidural clonidine on the incidence and severity of postoperative agitation have been assessed in a randomized, double-blind study in 80 children, all of whom received sevoflurane as the sole general anesthetic for induction and maintenance (601). A caudal epidural block was performed before surgery for analgesia with 0.175% bupivacaine 1 ml/kg. The children were assigned randomly to four groups: (I) clonidine 1 microgram/kg added to the caudal bupivacaine; (II) clonidine 3 micrograms/kg added to the caudal bupivacaine; (III) clonidine 3 micrograms/kg intravenously; and (IV) no clonidine. The incidences of agitation were 22, 0, 5, and 39% in the four groups respectively. Thus, clonidine 3 micrograms/kg effectively prevented agitation after sevoflurane anesthesia independent of the route of administration.

The effect of a single preoperative dose of the opioid oxycodone on emergence behavior has been studied in a randomized trial in 130 children (602). Oxycodone prophylaxis had no effect on post-sevoflurane delirium.

The effect of a single bolus dose of midazolam before the end of sevoflurane anesthesia has been investigated in a double-blind, randomized, placebo-controlled trial in 40 children aged 2–7 years (603). Midazolam significantly reduced the incidence of delirium after anesthesia. However, when it was used for severe agitation midazolam only reduced the severity without abolishing agitation. The authors concluded that midazolam attenuates, but does not abolish, agitation after sevoflurane anesthesia.

Silver salts and derivatives

In one case there was a rapid decline in mental function at silver concentrations of 191 ng/ml, but the patient was seriously ill in other respects and although high concentrations of silver were found in the brain postmortem, it is not at all clear that the silver was responsible for the mental state (604).

Simvastatin

In the 4S study there were five suicides in the simvastatin group and four in controls (605).

Sulfonamides

Headache, drowsiness, lowered mental acuity, and other psychiatric effects can be caused by sulfonamides (606). However, these adverse effects are rare, and the causative role of the drug is usually not clearly established.

Sulindac

Psychiatric symptoms, bizarre behavior, and paranoia have been attributed to sulindac (607).

Sultiame

- A 9-year-old boy with rolandic epilepsy developed impaired vigilance, depressed mood, fatigue, loss of drive, and listlessness when given sultiame 5 mg/kg/day (608). All his symptoms disappeared within a few days of withdrawal.

Tacrolimus

After kidney transplantation, 15 of 20 children tolerated tacrolimus after switching from ciclosporin for immunological reasons or adverse effects (609). The most frequent adverse effects were neuropsychological and behavioral symptoms in three children, ranging from anorexia nervosa-like symptoms, with weight loss, amenorrhea, depression, and school problems, to severe insomnia and in one child aggressive and anxious behavior. Only the last child was exposed to toxic tacrolimus blood concentrations. All the adverse effects were fully reversible after withdrawal.

- A 5 year old girl developed speech arrest, agitation, tremor, ataxia, and deviation of downward gaze 13 days after liver transplantation (610). The tacrolimus trough concentration was 44 ng/ml. Three days after dosage reduction, she could say a few words and her extra-ocular movements returned to normal. Four months later, she continued to have reduced fluency, dysarthria, and ataxia. One year later, the reduced fluency and mild ataxia persisted.

Rapid identification of speech loss linked to tacrolimus may be important, because dosage reduction or withdrawal may be associated with reverse of speech loss.

Tetrachloroethylene

Reversible neuropsychiatric symptoms, readily resembling alcoholic intoxication, have occurred after a single dose of 5 ml. Sleep apnea causing neuropsychiatric abnormalities has been attributed to exposure to high concentrations of tetrachloroethylene and *N*-butanol vapors; however, the patient was obese and the association with the solvents was not clear (611).

Thalidomide

Mood changes are common with thalidomide (612). Three cases of paranoid reactions have been reported in patients with multiple myeloma taking thalidomide; however, causality was not proved (613).

- Reversible dementia occurred in a patient with multiple myeloma taking thalidomide 200 mg/day plus dexamethasone (614). Memory deficit and mania occurred

about 2 months after the start of therapy and did not resolve on withdrawal of dexamethasone and administration of risperidone. The memory loss worsened and proceeded to disorientation, apraxia, and tremor 4 months after the start of therapy. The dementia resolved completely within 48 hours after withdrawal of thalidomide.

Theophylline

The concern that cognitive function may be impaired in children taking theophylline is still not resolved (SEDA-14, 1) (SEDA-15, 1); a review has suggested that detrimental effects of theophylline on various measures of cognitive function may be measure-specific (615).

Thyroid hormones

Several cases of mania have been reported even after dosages of levothyroxine that are usually considered safe (616).

Previous case reports have suggested that psychosis and mania can be the result of starting thyroid hormone replacement at too high a dosage (617). Two further cases of mania associated with levothyroxine have been reported (618,619), suggesting that caution should be exercised when prescribing levothyroxine, especially in elderly people.

Tiagabine

Although psychosis has been described occasionally (620), in three double-blind trials the incidence of psychotic symptoms was comparable to placebo (SEDA-21, 74). However, patients with a past history of psychiatric disturbances are often excluded from trials, and caution is recommended when using tiagabine in these patients.

Timolol

Depressive symptoms were reported in 17 of 165 patients after the administration of timolol over two decades (621). Depression accounted for 17% of 369 central nervous system reactions to timolol reported to a National Registry of Drug-induced Ocular Side Effects during 7 years: of these, 20 cases were of acute suicidal depression.

Tobramycin

As with gentamicin, toxic psychoses can occur with tobramycin (622).

Tolterodine

Short-term memory loss has been attributed to tolterodine.

- A 46-year-old woman took tolterodine 4 mg/day for stress incontinence (623). Her urinary symptoms were successfully controlled, but she complained of deteriorating short-term memory, so that she had to keep notes to remind herself of simple tasks. In fact the deterioration had started some 2 years earlier, but had become more noticeable during the 3 months of treatment with tolterodine. Psychometric testing showed quite severe impairment of verbal learning, although other aspects of memory seemed unimpaired. One month after stopping the drug her verbal learning had improved very significantly: she was now above the 75th percentile in the Hopkins Verbal Learning Test, having been at the 1st percentile at the initial test (albeit with different material).

This patient probably had some memory deficit before drug administration, but this appears to have been very modest. She had an exceptionally strong family history of dementia, Alzheimer's disease having affected her mother, several maternal aunts and uncles, and her maternal grandmother. It is therefore possible that this apparent sensitivity to anticholinergic drugs may have ominous implications for this particular patient. However, the authors also pointed out that she might have a genetic polymorphism causing a deficiency of the CYP2D6 isoenzyme that metabolizes tolterodine, a trait that is found in about 7% of the population. Of course both circumstances can co-exist.

Topiramate

Depression and psychosis have been observed in respectively 15 and 3% of patients taking topiramate (SEDA-22, 91) (624). Three patients taking topiramate for bipolar disorder developed substantial depression (625). The symptoms began or increased within 1 week of topiramate treatment (25 mg/day) or with an increase in dosage to 50 mg/day. All had significant relief from depression 1–2 weeks after withdrawal of topiramate. The close association with the onset of the most severe depression these patients had ever experienced suggests an adverse effect of topiramate. However, all these patients had bipolar disorder, so the onset of depression could have been coincidental. Moreover, their depression might also have been due to a synergistic interaction between topiramate and their other medications.

Two adult patients with refractory partial seizures without a previous history of psychosis had an acute psychotic episode with hallucinations and psychomotor agitation after taking topiramate 200–300 mg/day (626).

A single case report of new-onset panic attacks has been described.

- A 24-year-old woman with a history of bipolar disorder and binge eating had a history of "isolated" panic attacks 8 years before and the attacks subsided 14 days after topiramate withdrawal (627).

Because she had a history of psychiatric diseases and panic attacks, the relation of these symptoms to topiramate was doubtful.

All anticonvulsants have been associated with adverse cognitive events, and many, sometimes contradictory, studies of classical anticonvulsants have been published (628). Cognitive adverse effects are large for phenobarbital, and possibly larger for phenytoin than for carbamazepine or valproic acid (628). Most often cognitive adverse events result in mild general psychomotor slowing. Although the severity of cognitive adverse effects is considered mild to moderate for most anticonvulsants, their impact may be substantial in some patients, especially in those with pre-existing impaired cognitive function. There is relatively little reliable information on cognitive adverse events of new anticonvulsants. Most of the published studies are on polytherapy and there is little information about healthy volunteers (629).

Topiramate can cause cognitive adverse effects in some patients. Those affected often have impaired verbal learning and fluency. Slow titration reduces the likelihood of cognitive adverse events. Two patients had neuropsychological deficits during topiramate treatment and cognitive improvement after withdrawal (630). One patient was assessed during and after topiramate withdrawal and the other before, during, and after.

The cognitive adverse effects of gabapentin, lamotrigine, and topiramate in healthy volunteers have been compared in a randomized, single-blind, parallel-group study (629). Neurobehavioral testing was conducted at baseline, during the acute oral dosing period 3 hours after medication administration, and at 2 and 4 weeks during chronic dosing. Acutely, those who took topiramate (2.8 mg/kg) performed significantly worse than those who took gabapentin (17 mg/kg) or lamotrigine (3.5 mg/kg) on tasks of letter and category word fluency and visual attention; cognitive effects were not different between those who took gabapentin or lamotrigine. The doses were then increased to topiramate 5.7 mg/kg/day, lamotrigine 7.1 mg/kg/day, and gabapentin 35 mg/kg/day. At 2 and 4 weeks, those taking topiramate had significant verbal memory deficit and slow psychomotor speed compared with baseline; those taking gabapentin and lamotrigine did not. However, the clinical impact of the trial was limited, owing to the small sample size (17 patients) and the very rapid topiramate titration, much faster than is currently recommended (631).

There has been a retrospective analysis of neuropsychological scores before and after the use of topiramate 125–600 mg/day for at least 3 months in 18 patients (632). Topiramate was associated with significant deterioration in verbal IQ, learning, and fluency. Withdrawal or dosage reduction was associated with significant improvement. There was no correlation between individual topiramate dose and the change in test score. This study was retrospective and the patients were selected because they had cognitive problems; the resultant bias makes it difficult to generalize these results to wider populations.

A group of 14 US epilepsy centers has published the results of a post-marketing surveillance study of 701 patients taking topiramate (633). Although 41% of the patients reported cognitive adverse events at any time during treatment, only 5.8% of them discontinued for that reason. Cognitive effects were the most frequent reason for withdrawal because of adverse events (41/170 or 24% of those who discontinued). The central nervous system-related adverse effects profile in these patients included psychomotor slowing, fatigue, slurred speech, irritability, behavioral changes, confusion, inappropriate laughter, and hallucinations. Only 2.4% of the patients were taking monotherapy, and the mean dose of topiramate at 6 months was 385 mg/day, with a mean weekly dose during titration of 36 mg/day. Risk factors for discontinuation were evaluated. A slow titration rate (slower than 25 mg/week), but not the total dose at discontinuation, was significantly associated with a lower discontinuation rate. There was no specific population, dose titration, or concomitant antiepileptic drug that increased the risk of treatment discontinuation because of cognitive complaints. Psychomotor slowing was the most common complaint, but most patients elected to continue treatment because of improved seizure control.

In a retrospective survey, behavioral changes occurred in nine of 69 children (median age 12 years) taking topiramate (634). Dosages in the affected children were 2.4–8.5 mg/kg, and three children were taking monotherapy. Manifestations included school difficulties (635), mood swings (636), and aggression, irritability, and hallucinations (one case each). There was no apparent relation to dosage or titration rate.

The cognitive effects of topiramate and valproate as adjunctive therapy to carbamazepine have been compared in 53 patients (637). Topiramate was given in an initial dose of 25 mg and increased weekly by 25 mg/day increments to a minimum of 200 mg/day. Cognition was significantly worsened by topiramate and improved by valproate. Gradual introduction of topiramate reduced the extent of cognitive impairment.

The effects of topiramate and valproate, when added to carbamazepine, on cognitive status in adults have been compared in a randomized, observer-blinded, parallel-group study (637). Topiramate was introduced slowly at a starting dose of 25 mg/day and increased weekly by 25 mg/day increments over 8 weeks to a minimum dosage of 200 mg/day. The target dosages were 200–400 mg/day for topiramate and 1800 mg/day for valproate. There were significant differences between topiramate and valproate in short-term verbal memory—worsening with topiramate and improvement with valproate—but the differences were small. There were no effects on mood disorders, psychiatric symptoms, or motor and mental speed and language tests. These results suggest that if the dose of topiramate is slowly titrated cognitive adverse events can be minimized. However, although most patients tolerated topiramate if properly titrated and dosed, a subset of patients had clinically significant deficits, possibly as an idiosyncratic reaction. In another multicenter, parallel-group, randomized study similar results were obtained when topiramate or valproate was added to carbamazepine (638,639).

The cognitive adverse effects of topiramate have been studied in 62 patients taking carbamazepine who had add-on therapy with either topiramate, valproic acid, or placebo in a multicenter, randomized, double-blind study (640). A neuropsychological test battery was administered before the start of treatment with topiramate, valproic acid, or placebo, at the end of an 8-week titration period, and at the end of a 12-week maintenance period. Slightly more patients taking topiramate dropped out of the study; 62 patients completed the study: 27 in the topiramate group, 25 at the valproic acid group, and 10 on placebo. At the end of maintenance therapy, the effects of topiramate and valproic acid were comparable, except for two variables, the Symbol Digit Modalities Test and the Controlled Oral Word Association Test, in which topiramate had greater negative effects than valproic acid. The statistical differences were due in large part to a small subset of patients who were more negatively affected by topiramate. Cognitive adverse effects of topiramate compared with valproic acid were greater at the end of titration than at the end of maintenance.

Changes in neuropsychological testing have been investigated in two groups of patients: 22 patients who were taking topiramate in addition to other antiepileptic drugs, in whom topiramate was withdrawn during hospitalization, and 16 patients before and after using topiramate (641). Since each group used a different order of testing, this open design allowed comparison of the same patient on and off the drug, without respect to which condition was tested first. Topiramate was associated with impairment of fluency, attention/concentration, processing speed, language skills, and perception; working memory was affected but not retention.

Of 431 consecutive patients taking topiramate, 31 (7.2%) developed word-finding difficulties (642). This adverse event was not associated with a rapid titration schedule, but it correlated with the presence of simple partial seizures and left temporal epileptiform activity on the electroencephalogram, suggesting that there is a subgroup of patients with a specific susceptibility.

A significant improvement in neuropsychological functions associated with the frontal lobe after withdrawal of topiramate in epilepsy patients was reported in an open controlled study (643). The results suggested that verbal fluency and working memory were very sensitive to topiramate. Improvement in this function was observed after withdrawal of topiramate but not in the control group.

The effects of topiramate on tests of intellect and other cognitive processes have been studied in 18 patients (632). Repeat assessments in those taking topiramate were associated with a significant deterioration in many domains, which were not seen in controls. The greatest changes were for verbal IQ, verbal fluency, and verbal learning. There were improvements in verbal fluency, verbal learning, and digit span in patients who had topiramate withdrawn or reduced.

The factors associated with behavioral and cognitive abnormalities in children taking topiramate have been studied retrospectively (644). There were behavioral or cognitive abnormalities in 11 of 75 children at 2–4 months after the start of therapy. The mean dosage (4.6 mg/kg/day) at which these abnormalities were observed was similar to the mean final dose (5.8 mg/kg/day) in children without abnormalities. Five of the eleven children with behavioral or cognitive abnormalities had a previous history of behavioral or cognitive abnormalities, but only nine of the 64 children without abnormalities had a previous history of behavioral or cognitive abnormalities.

Angelman's syndrome, a genetic disorder that involves a defect in the DNA coding for subunits of the GABA-A receptor, is often associated with intractable epilepsy. Topiramate was effective in five children with Angelman's syndrome and epilepsy (645). One patient had transient insomnia and one had akathisia and insomnia that persisted until topiramate was withdrawn.

Slowed mental function is a relatively common adverse effect of topiramate, especially at high dosages, and has been confirmed in a randomized single-blind, parallel-group study in healthy volunteers. After single doses, topiramate (2.8 mg/kg) reduced performance in attention and word fluency tests, whereas lamotrigine (3.5 mg/kg) and gabapentin (17 mg/kg) had no effects (629). After 4 weeks of multiple dosing, topiramate (5.7 mg/kg) was still associated with impairment in verbal memory and psychomotor speed, while there was no impairment with lamotrigine (7.1 mg/kg) or gabapentin (35 mg/kg). While these data suggests that topiramate can cause greater cognitive dysfunction in the short-term than lamotrigine and gabapentin, these findings should be interpreted cautiously. First, the design was less than ideal: the use of a double-blind, crossover design and inclusion of placebo would have been preferable. Secondly, the speed of topiramate titration was much faster than currently recommended, and it is known that neurotoxicity can be reduced by slow dose escalation. Finally, the dosage of topiramate was very high, in view of the fact that no enzyme-inducing co-medication was present.

Word-finding difficulties are a known adverse effect of topiramate. Language disturbances (anomia or impaired verbal expression) without other complaints suggestive of impaired cognition were reported by four of 51 patients; five others had language problems associated with cognitive dysfunction (646). One severely retarded patient with a limited vocabulary became mute, although otherwise alert. Anomia or dysnomia were also seen in two of eight patients treated with zonisamide, which shares with topiramate a sulfa structure and carbonic anhydrase inhibitory properties. Psychotic symptoms are a significant adverse effect in occasional patients. In a retrospective survey, five of 80 patients developed psychosis 2–46 days after starting to take topiramate; manifestations included paranoid delusions in four and auditory hallucinations in three (647). The dosage at the time of onset of symptoms was 50–500 mg/day, and recovery occurred rapidly after drug withdrawal (three patients), dosage reduction (one patient), or neuroleptic drug treatment (one patient). Three of the patients had no significant psychiatric history.

The prevalence of psychiatric adverse events in relation to topiramate has been studied in 431 consecutive patients (648). There were psychiatric adverse events in 103 patients (24%), including affective disorder in 46 (11%),

psychotic disorder in 16 (3.7%), aggressive behavior in 24 (5.6%), and other behavioral abnormalities (agitation, anger/hostility, and anxiety) in 17 (3.9%). Psychiatric disorders were more prevalent with a high starting dose, rapid titration, a family history of a psychiatric disorder, a family history of epilepsy, a personal history of febrile seizures, and the presence of tonic–atonic seizures. The seizure frequency before starting topiramate and lamotrigine co-administration protected against psychiatric adverse events.

- A 41-year-old woman with a diagnosis of bipolar disorder developed severe suicidality when topiramate was added to her previous treatment (carbamazepine and levothyroxine) for mood stabilization (649).

Topiramate has been associated with several psychiatric adverse events, such as nervousness, depression, behavioral problems, mood lability, and psychosis. Obsessive–compulsive disorder has also been reported (650).

- A 19-year-old man with focal epilepsy took carbamazepine 1000 mg/day and lamotrigine 300 mg/day. Because his seizures persisted topiramate was added up to 200 mg/day and the dose of carbamazepine was reduced to 300 mg/day. Behavioral problems started within a week and worsened over the following months. He finally developed obsessive-compulsive disorder. Citalopram was given in doses up to 60 mg/day and topiramate was tapered within 2 weeks. The symptoms improved.

The authors discussed the hypothesis of a coincidental finding as well as the development of an alternative psychosis but finally attributed the patient's psychiatric disorder to topiramate.

Three patients with hypomania associated with topiramate have been reported. All had chronic mood disorders and all developed symptoms of hypomania soon after starting to take topiramate titrated up to 100 mg/day in addition to pre-existing psychotropic drugs (651). The authors speculated that the antidepressant effects of topiramate had caused hypomania by release of monoamine neurotransmitters.

Worsening of psychosis could have been an adverse event of topiramate in a 50-year-old woman with a long history of a schizoaffective disorder treated with risperidone 4 mg/day, venlafaxine 300 mg/day, and topiramate 200 mg/day (652). Her behavior worsened after topiramate was introduced. After it was withdrawn her mood, spontaneous speech, and psychomotor activity improved.

Daytime vigilance has been assessed in 14 newly diagnosed never medicated adults with focal epilepsy at baseline and 2 months after slow titration of topiramate to 200 mg/day (653). Multiple Sleep Latency Test (MSLT), visual simple and choice reaction time (VRT), and self-rating with the Epworth Sleepiness Scale were used for quantification of sleepiness, and compared with 14 healthy volunteers. At baseline MSLT scores were comparable. Two months after topiramate monotherapy, MSLT, VRT, and self-rating did not change significantly.

The authors concluded that a short course of topiramate monotherapy dose not impair vigilance in this population.

Tramadol

Hallucinations have been attributed to tramadol, both visual (654) and auditory (655).

- A 66-year-old tetraparetic man developed hallucinations while taking tramadol, paroxetine, and dosulepin for chronic pain (654).
- A 74-year-old man with lung cancer took tramadol 200 mg/day for chest pain (655). Soon afterwards (time not specified) he had vivid auditory hallucinations in the form of "two voices singing accompanied by an accordion and a banjo." They resolved 48 hours after withdrawal.

Trifluoromethane

In a phase 1, dose-ranging study, trifluoromethane 10–60% impaired neuropsychological function in five healthy men. Symptoms that occurred more often at higher doses included tiredness, difficulty in concentrating, tingling, dry mouth, and loss of appetite (656).

Trihexyphenidyl

Although it has been thought that high doses of trihexyphenidyl might impair learning in children, a careful study of this question suggested that there is in practice little interference (657).

As little as 8 mg/day combined with levodopa has caused an acute toxic confusional state in some patients.

- A 75-year-old man with a 10-year history of parkinsonism developed fever and acute delirium after taking levodopa plus trihexyphenidyl (658). He had visual hallucinations, was disoriented in time and place, and could respond only to simple questions. He was unable to stand or walk and had paratonic rigidity in all limbs and marked bradykinesia. There were occasional myoclonic jerks in the arms and legs. Deep reflexes were reduced bilaterally, and the plantar reflexes were absent.

Impairment of memory was attributed to trihexyphenidyl in healthy volunteers (SEDA-13, 115).

Trimethoprim and co-trimoxazole

- Delirium occurred after treatment with co-trimoxazole in a patient with AIDS; the episode completely resolved within 72 hours of drug withdrawal (659).

Triptorelin

In 20 women (mean age 67 years) who received intramuscular triptorelin 3.75 mg 6-weekly for endometriosis, nine developed mood disturbance after the second injection and appeared to be cumulative, since the symptoms only started after the second injection and worsened with successive injections (660). One woman withdrew after the third injection because of severe irritability.

Tumor necrosis factor alfa

Evaluation of cognitive function in patients receiving tumor necrosis factor alfa alone or combined with interleukin-2 showed reversible attentional deficits, memory disorders, deficits in motor coordination and frontal lobe executive functions (661). There was reversible hypoperfusion in the frontal lobes.

Valaciclovir

At high doses (8 g/day) hallucinations and confusion were a significant concern (662,663), but similar symptoms have also occurred at lower doses and in patients with renal insufficiency.

- Ocular and auditory hallucinations have been reported in a 60-year-old female patient on CAPD (664).
- A 58-year-old man with chronic renal insufficiency, who was hemodialysed twice a week, was treated with valaciclovir (1 g tds) for *Herpes zoster* (665). Two days later he became disoriented, dizzy, dysarthric, and experienced hallucinations. The serum aciclovir concentration was 21 µg/ml. Treatment was discontinued and he was treated with hemodialysis for 6 hours, resulting in marked clinical improvement. The next day his symptoms of dysarthria recurred, but immediately and completely resolved after a second hemodialysis.

Valproic acid

Psychotic reactions had been reported to the WHO monitoring system on 16 occasions by 1993, and other occasional cases have been published (SED-13, 149) (666).

Behavioral problems are less frequent with valproate than with other anticonvulsants and include sedation, hyperactivity, irritability, and aggression (667).

There has been a randomized, double-blind, single crossover study of the effects of sodium valproate on cognitive performance and behavior in eight children with learning and behavioral problems associated with electroencephalographic epileptiform discharges but without clinical seizures (668). The children became more distractable, had increased delay in response time, and had lower memory scores while taking valproate. Their parents reported higher internalizing scores on the Child Behavior Checklists.

Although valproate has little effect on cognitive function compared with other anticonvulsants, it can occasionally cause reversible dementia, sometimes associated with MRI evidence of pseudoatrophy of the brain (SEDA-20, 67) (SEDA-21, 76). The presentation can be striking, with progressive motor and intellectual deterioration and apathy appearing shortly after the introduction of valproate, or sometimes after as long as 2 years. Many affected patients have relatively high serum drug concentrations (100–120 mg/l), other signs of toxicity, such as tremor and ataxia, and normal electroencephalographic background activity. The presentation differs from valproate-induced stupor, in which consciousness is impaired, and there may be clinical or electroencephalographic signs of increased seizure activity. In all cases, clinical and imaging signs improved rapidly after drug withdrawal. The incidence of this syndrome is probably very low, although there is concern about possible under-diagnosis.

Valsartan

Two different reports have pointed to nightmares and depression with valsartan.

- Nightmares from valsartan for hypertension improved when it was withdrawn and recurred when it was restarted in a 64-year-old woman (669).
- A 43-year-old woman reported depression and attempted suicide after taking valsartan in combination with hydrochlorothiazide for hypertension (670). The patient was also taking atenolol, but the time course and complete resolution after withdrawal of valsartan/hydrochlorothiazide suggested that this was a drug-induced episode of depression.

Vigabatrin

Although abnormal behavior, psychosis, and depression occur with increased frequency (around 2–4% or even higher) in patients with refractory epilepsy, the risk is increased during treatment with vigabatrin or on its abrupt withdrawal (SEDA-18, 71) (SEDA-19, 76) (SEDA-20, 70) (SEDA-21, 77) (671). The risk depends on the patient's characteristics, the dosage, and the rate of dose escalation. In the manufacturer's safety database, the incidence of psychosis was reported as being 1.1% or 0.64% based on postmarketing Prescription Event Monitoring (672). However, in adults with refractory partial epilepsy, severe behavioral or mood disturbances can be anticipated in 26% of treated cases (SEDA-18, 72), although higher figures have also been presented (SEDA-20, 70). In one study depression occurred with a frequency of about 9% with vigabatrin and 1% with placebo (673).

Psychiatric complications occur in both children and adults. While there is some evidence that a positive psychiatric history may increase the risk (and may represent a relative contraindication), behavioral and mood disturbances can occur in patients who had no similar episodes in the past. A retrospective survey of 50 patients with vigabatrin-associated psychosis ($n = 28$) or depression ($n = 22$) suggested that psychosis tends to occur in patients with more severe epilepsy and may be related to asided EEG focus or suppression of seizures (64% of patients were free of seizures). On the other hand,

patients with depression tended to have a past history of depressive illness and to have had little or no change in seizure frequency (674). To minimize risks, vigabatrin should be introduced in low dosages and increased slowly, and it should not be withdrawn abruptly. Psychiatric symptoms usually remit with drug withdrawal, although if treatment needs to be continued, neuroleptic medication may be useful.

That vigabatrin can cause psychosis and depression has been confirmed in an analysis of data from double-blind, controlled, add-on trials (675). Compared with placebo, vigabatrin-exposed patients had a significantly higher incidence of events coded as depression (12 versus 3.5%) and psychosis (2.5 versus 0.3%). Although these events occurred during the first 3 months, most of the studies lasted 12–18 weeks and therefore definite conclusions could not be reached about timing.

Vincamine

Vincamine has minor psychoactive effects (676).

Vitamin A: Carotenoids

Daily doses of 1 000 000 µg RE of vitamin A can cause acute symptomatic psychosis, with striking disturbances in the electroencephalogram and pathological changes in the cerebrospinal fluid (677). After withdrawal of vitamin A and administration of low doses of neuroleptic drugs the psychotic symptoms disappeared within a week.

Neuropsychiatric changes were reported to be the earliest dose-limiting symptomatic toxicity in patients with cancer taking high doses of retinol (678).

Vitamin A: Retinoids

Since 1982, according to the FDA, isotretinoin has been linked to over 400 cases of psychiatric illness, including depression, suicidal ideation, suicide attempts, and completed suicide (679). In the light of questionable psychiatric safety, evidence examining the link between isotretinoin and psychiatric illness in adolescents and young adults has been assessed (680). The data sources were primary literature located via MEDLINE (1966–2002). The key terms were isotretinoin, depression, psychosis, suicide, and adolescents. The authors concluded that although there may be a causal relation between isotretinoin and psychiatric illness in adolescents and young adults, this is not demonstrated by the literature. They identified some limitations in the available evidence, including the low numbers of adolescents in published studies, as well as methodological errors in many of the studies analysed. In addition, it is clear that acne alone can predispose patients to psychiatric illness, and that therefore isotretinoin is not the only contributing factor.

A series of cases suggesting an association between exposure to isotretinoin and manic psychosis has been reported (681). Five young adults developed manic psychosis during 1 year in association with isotretinoin treatment, resulting in suicidality and progression to long-standing psychosis. Associated risk factors were a family and personal history of psychiatric morbidity. The cases were drawn from 500 soldiers who had been evaluated in a military specialist dermatology clinic for severe acne.

In 1998 depression, psychosis, and suicidal ideation, suicide attempts, and suicide were added to the product label of isotretinoin. Since then the FDA has received increasing number of reports of these problems (682). From the time that isotretinoin was marketed in 1982 up to May 2000 the FDA received 37 reports of patients taking isotretinoin who committed suicide, 110 reports of patients who were hospitalized for depression, suicidal ideation, or suicide attempts, and 284 reports of patients with depression who did not need hospitalization (683). In 62% of the suicide cases a psychiatric history or possible contributing factors were identified, and 69% of patients hospitalized for depression had either a previous psychiatric history of possible contributing factors. Drug withdrawal led to improvement in about one-third of the patients, while in 29% depression persisted after withdrawal. In 24 cases dechallenge and rechallenge were positive. However, since this was a series of spontaneous reports, and since there are no good data on the incidence of depression and suicide among adolescents with acne, a causal relation cannot be concluded.

A change in dreaming pattern has been reported in two patients, occurring within 2–3 weeks after the start of treatment with isotretinoin 40 mg/day for cystic acne (684). One patient also reported increased irritability and bouts of depression. In both patients all the symptoms abated after 4–5 weeks without a change in isotretinoin dosage.

Voriconazole

Voriconazole can be associated with visual and other hallucinations.

- A 78-year-old man began to have auditory hallucinations, specifically of Christmas music, on the second day of voriconazole therapy (685). Psychiatric evaluation was otherwise unremarkable. After withdrawal of voriconazole the hallucinations reduced in intensity by 2 days and ceased altogether by the third day.

An extensive literature search, including Pfizer drug trial safety data, yielded no other reports of auditory hallucinations with voriconazole. Several other cases of musical hallucinations secondary to a variety of causes have been reported. They tend to occur secondary to temporal lobe insults and often have religious or patriotic themes.

Xenon

The subjective, psychomotor, and physiological properties of subanesthetic concentrations of xenon have been studied in 10 volunteers (686). Xenon sedation was well

tolerated and was not associated with any adverse physiological effects. In particular, there was no nausea or vomiting. It was preferred to sedation with nitrous oxide and was subjectively dissimilar (xenon was more pleasant).

Yohimbine

Yohimbine commonly causes anxiety; in eight patients with panic disorder this effect was reduced by fluvoxamine (687).

Manic symptoms have been attributed to yohimbine (688).

In a placebo-controlled study in 18 combat veterans with post-traumatic stress disorder and 11 healthy controls, intravenous yohimbine 0.4 mg/kg significantly increased the amplitude, magnitude, and probability of the acoustic startle reflex (used as a model to investigate the neurochemical basis of anxiety and fear states) in the veterans with post-traumatic stress disorder but not in the controls (689).

Zonisamide

Behavioral problems and acute mania have been described (SEDA-16, 75; SEDA-19, 76).

Of 74 epileptic patients who had taken zonisamide 14 had psychotic episodes, diagnosed retrospectively (690). The authors estimated that the incidence of psychotic episodes during zonisamide treatment was several times higher than the previously reported prevalence of epileptic psychosis, and that the risk was higher in young patients. In 13 patients, psychotic episodes occurred within a few years of starting zonisamide. In children, obsessive-compulsive symptoms were related to psychotic episodes.

A unique form of paramnesia has been attributed to zonisamide (691).

- After an episode of zonisamide-induced psychosis, a 28-year-old man with epilepsy consistently mistook people who were unknown to him, such as hospital staff, for people whom he had met long ago. However, he did not misidentify their names or other attributes, such as their occupations.

The authors could not fit this extraordinary form of misidentification into any known subcategory of misidentification syndromes, but rather thought that it fitted Kraepelin's description of "assoziierende Erinnerungsfälschungen".

Complex visual hallucinations occurred in three patients taking zonisamide for different syndromes and types of seizures (Landau–Kleffner syndrome in a 7-year-old girl, myoclonic and generalized tonic seizures in a 21-year-old woman, and partial epilepsy in a 13-year-old girl) (692). None of the patients had visual hallucinations before zonisamide was started, and the symptoms resolved after withdrawal.

- A 5-year-old girl presented behavioral problems after her seizures had been controlled with zonisamide

(693). This was interpreted as a case of forced normalization, and her behavioral difficulties disappeared when zonisamide was withdrawn and her seizures reappeared.

References

1. Tomson CR, Goodship TH, Rodger RS. Psychiatric side-effects of acyclovir in patients with chronic renal failure. Lancet 1985;2(8451):385–6.

2. Chiang C-K, Fang C-C, Hsu W-D, Chu T-S, Tsai T-J. Hemodialysis reverses acyclovir-induced nephrotoxicity and neurotoxicity. Dial Transplant 2003;32:624.

3. Hellden A, Odar-Cederlof I, Diener P, Barkholt L, Medin C, Svensson JO, Sawe J, Stahle L. High serum concentrations of the acyclovir main metabolite 9-carboxymethoxy-methylguanine in renal failure patients with acyclovir-related neuropsychiatric side effects: an observational study. Nephrol Dial Transplant 2003;18:1135–41.

4. Denicoff KD, Rubinow DR, Papa MZ, Simpson C, Seipp CA, Lotze MT, Chang AE, Rosenstein D, Rosenberg SA. The neuropsychiatric effects of treatment with interleukin-2 and lymphokine-activated killer cells. Ann Intern Med 1987;107(3):293–300.

5. Caraceni A, Martini C, Belli F, Mascheroni L, Rivoltini L, Arienti F, Cascinelli N. Neuropsychological and neurophysiological assessment of the central effects of interleukin-2 administration. Eur J Cancer 1993;29A(9):1266–9.

6. Walker LG, Wesnes KP, Heys SD, Walker MB, Lolley J, Eremin O. The cognitive effects of recombinant interleukin-2 (rIL-2) therapy: a controlled clinical trial using computerised assessments. Eur J Cancer 1996;32A(13):2275–83.

7. Pace A, Pietrangeli A, Bove L, Rosselli M, Lopez M, Jandolo B. Neurotoxicity of antitumoral IL-2 therapy: evoked cognitive potentials and brain mapping. Ital J Neurol Sci 1994;15(7):341–6.

8. Pizzi C, Caraglia M, Cianciulli M, Fabbrocini A, Libroia A, Matano E, Contegiacomo A, Del Prete S, Abbruzzese A, Martignetti A, Tagliaferri P, Bianco AR. Low-dose recombinant IL-2 induces psychological changes: monitoring by Minnesota Multiphasic Personality Inventory (MMPI). Anticancer Res 2002;22(2A):727–32.

9. Capuron L, Ravaud A, Dantzer R. Early depressive symptoms in cancer patients receiving interleukin 2 and/or interferon alfa-2b therapy. J Clin Oncol 2000;18(10):2143–51.

10. Vial T, Descotes J. Clinical toxicity of interleukin-2. Drug Saf 1992;7(6):417–33.

11. Hunter KR. Treatment of parkinsonsim. Prescr J 1976;16:101.

12. Rego MD, Giller EL Jr. Mania secondary to amantadine treatment of neuroleptic-induced hyperprolactinemia. J Clin Psychiatry 1989;50(4):143–4.

13. Taiminen T, Jaaskelainen SK. Intense and recurrent deja vu experiences related to amantadine and phenylpropanolamine in a healthy male. J Clin Neurosci 2001;8(5):460–2.

14. Barry JJ, Franklin K. Amiodarone-induced delirium. Am J Psychiatry 1999;156(7):1119.

15. Ambrose A, Salib E. Amiodarone-induced depression. Br J Psychiatry 1999;174:366–7.

16. Swords R, Fay M, O'Donnell R, Murphy PT. Anagrelide-induced visual hallucinations in a patient with essential thrombocythemia. Eur J Haematol 2004;73(3):223–4.

17. Pope HG Jr, Katz DL Affective and psychotic symptoms associated with anabolic steroid use. Am J Psychiatry 1988;145:487–90.

18. Conacher GN, Workman DG. Violent crime possibly associated with anabolic steroid use. Am J Psychiatry 1989;146:679.

19. Perry PJ, Kutscher EC, Lund BC, Yates WR, Holman TL, Demers L. Measures of aggression and mood changes in male weightlifters with and without androgenic anabolic steroid use. J Forensic Sci 2003;48:646–51.

20. Thiblin I, Petersson A. Pharmacoepidemiology of anabolic androgenic steroids: a review. Fundam Clin Pharmacol 2005;19:27–44.

21. Copeland J, Peters R, Dillon P. Anabolic–androgenic steroid use disorders among a sample of Australian competitive and recreational users. Drug Alcohol Depend 2000;60(1):91–6.

22. Christiansen K. Behavioural effects of androgen in men and women. J Endocrinol 2001;170(1):39–48.

23. Weiss EL, Bowers MB Jr, Mazure CM. Testosterone-patch-induced psychotic mania. Am J Psychiatry 1999;156(6):969.

24. Thiblin I, Runeson B, Rajs J. Anabolic androgenic steroids and suicide. Ann Clin Psychiatry 1999;11(4):223–31.

25. Khurana AK, Ahluwalia BK, Rajan C, Vohra AK. Acute psychosis associated with topical cyclopentolate hydrochloride. Am J Ophthalmol 1988;105(1):91.

26. van Herwaarden G, Berger HJ, Horstink MW. Short-term memory in Parkinson's disease after withdrawal of long-term anticholinergic therapy. Clin Neuropharmacol 1993;16(5):438–43.

27. Olsen KM, Martin SJ. Pharmacokinetics and clinical use of drotrecogin alfa (activated) in patients with severe sepsis. Pharmacotherapy 2002;22(12 Pt 2):S196–205.

28. Meador KJ. Newer anticonvulsants: dosing strategies and cognition in treating patients with mood disorders and epilepsy. J Clin Psychiatry 2003;64 Suppl 8:s30–4.

29. Lathers CM, Schraeder PL, Claycamp HG. Clinical pharmacology of topiramate versus lamotrigine versus phenobarbital: comparison of efficacy and side effects using odds ratios. J Clin Pharmacol 2003;43:491–503.

30. Matson JL, Luke MA, Mayville SB. The effects of antiepileptic medications on the social skills of individuals with mental retardation. Res Dev Disabil 2004;25(2):219–28.

31. Wong I, Tavernor S, Tavernor R. Psychiatric adverse effects of anticonvulsant drugs: incidence and therapeutic implications. CNS Drugs 1997;8:492–509.

32. Trimble MR, Rusch N, Betts T, Crawford PM. Psychiatric symptoms after therapy with new antiepileptic drugs: psychopathological and seizure related variables. Seizure 2000;9(4):249–54.

33. Adachi N, Onuma T, Hara T, Matsuura M, Okubo Y, Kato M, Oana Y. Frequency and age-related variables in interictal psychoses in localization-related epilepsies. Epilepsy Res 2002;48(1–2):25–31.

34. Bredkjaer SR, Mortensen PB, Parnas J. Epilepsy and non-organic non-affective psychosis. National epidemiologic study. Br J Psychiatry 1998;172:235–8.

35. Kanemoto K, Tsuji T, Kawasaki J. Reexamination of interictal psychoses based on DSM IV psychosis classification and international epilepsy classification. Epilepsia 2001;42(1):98–103.

36. Iivanainen M, Savolainen H. Side effects of phenobarbital and phenytoin during long-term treatment of epilepsy. Acta Neurol Scand Suppl 1983;97:49–67.

37. McKee RJ, Larkin JG, Brodie MJ. Acute psychosis with carbamazepine and sodium valproate. Lancet 1989;1(8630):167.

38. McConnell H, Snyder PJ, Duffy JD, Weilburg J, Valeriano J, Brillman J, Cress K, Cavalier J. Neuropsychiatric side effects related to treatment with felbamate. J Neuropsychiatry Clin Neurosci 1996;8(3):341–6.

39. Jablonowski K, Margolese HC, Chouinard G. Gabapentin-induced paradoxical exacerbation of psychosis in a patient with schizophrenia. Can J Psychiatry 2002;47(10):975–6.

40. Kossoff EH, Bergey GK, Freeman JM, Vining EP. Levetiracetam psychosis in children with epilepsy. Epilepsia 2001;42(12):1611–3.

41. Stella F, Caetano D, Cendes F, Guerreiro CA. Acute psychotic disorders induced by topiramate: report of two cases. Arq Neuropsiquiatr 2002;60(2-A):285–7.

42. Sander JW, Hart YM, Trimble MR, Shorvon SD. Vigabatrin and psychosis. J Neurol Neurosurg Psychiatry 1991;54(5):435–9.

43. Miyamoto T, Kohsaka M, Koyama T. Psychotic episodes during zonisamide treatment. Seizure 2000;9(1):65–70.

44. Matsuura M. Epileptic psychoses and anticonvulsant drug treatment. J Neurol Neurosurg Psychiatry 1999;67(2):231–3.

45. Darbar D, Connachie AM, Jones AM, Newton RW. Acute psychosis associated with abrupt withdrawal of carbamazepine following intoxication. Br J Clin Pract 1996;50(6):350–1.

46. Heh CW, Sramek J, Herrera J, Costa J. Exacerbation of psychosis after discontinuation of carbamazepine treatment. Am J Psychiatry 1988;145(7):878–9.

47. Wolf P. Acute behavioral symptomatology at disappearance of epileptiform EEG abnormality. Paradoxical or "forced" normalization. Adv Neurol 1991;55:127–42.

48. Neppe VM. Carbamazepine as adjunctive treatment in nonepileptic chronic inpatients with EEG temporal lobe abnormalities. J Clin Psychiatry 1983;44(9):326–31.

49. Sackellares JC, Krauss G, Sommerville KW, Deaton R. Occurrence of psychosis in patients with epilepsy randomized to tiagabine or placebo treatment. Epilepsia 2002;43(4):394–8.

50. Somerville ER. Aggravation of partial seizures by antiepileptic drugs: is there evidence from clinical trials? Neurology 2002;59(1):79–83.

51. Shorvon SD. Safety of topiramate: adverse events and relationships to dosing. Epilepsia 1996;37(Suppl 2):S18–22.

52. Dietrich DE, Kropp S, Emrich HM. Oxcarbazepine in affective and schizoaffective disorders. Pharmacopsychiatry 2001;34(6):242–50.

53. Besag FM. Behavioural effects of the new anticonvulsants. Drug Saf 2001;24(7):513–36.

54. Veggiotti P, De Agostini G, Muzio C, Termine C, Baldi PL, Ferrari Ginevra O, Lanzi G. Vigabatrin use in psychotic epileptic patients: report of a prospective pilot study. Acta Neurol Scand 1999;99(3):142–6.

55. Aldenkamp AP. Cognitive and behavioural assessment in clinical trials: when should they be done? Epilepsy Res 2001;45(1–3):155–7.

56. Bourgeois BF. Antiepileptic drugs, learning, and behavior in childhood epilepsy. Epilepsia 1998;39(9):913–21.

57. Smith DB, Mattson RH, Cramer JA, Collins JF, Novelly RA, Craft B. Results of a nationwide Veterans Administration Cooperative Study comparing the efficacy and toxicity of carbamazepine, phenobarbital, phenytoin, and primidone. Epilepsia 1987;28(Suppl 3):S50–8.

58. Aldenkamp AP, Alpherts WC, Diepman L, van't Slot B, Overweg J, Vermeulen J. Cognitive side-effects of phenytoin compared with carbamazepine in patients with localization-related epilepsy. Epilepsy Res 1994;19(1):37–43.

59. Pulliainen V, Jokelainen M. Comparing the cognitive effects of phenytoin and carbamazepine in long-term monotherapy: a two-year follow-up. Epilepsia 1995;36(12):1195–202.

60. Bower EA, Moore JL, Moss M, Selby KA, Austin M, Meeves S. The effects of single-dose fexofenadine, diphenhydramine, and placebo on cognitive performance in flight personnel. Aviat Space Environ Med. 2003;74:145–52.

61. Kamei H, Noda Y, Ishikawa K, Senzaki K, Muraoka I, Hasegawa Y, Hindmarch I, Nabeshima T. Comparative study of acute effects of single doses of fexofenadine, olopatadine, d-chlorpheniramine and placebo on psychomotor function in healthy volunteers. Hum Psychopharmacol 2003;18:611–8.

62. Ridout F, Shamsi Z, Meadows R, Johnson S, Hindmarch I. A single-center, randomised, double-blind, placebo-controlled, crossover investigation of the effects of fexofenadine hydrochloride 180 mg alone and with alcohol, with hydroxyzine hydrochloride 50 mg as a positive internal control, on aspects of cognitive and psychomotor function related to driving a car. Clin Ther 2003;25:1518–38.

63. Mansfield L, Mendoza C, Flores J, Meeves SG. Effects of fexofenadine, diphenhydramine, and placebo on performance of the test of variables of attention (TOVA). Ann Allergy Asthma Immunol. 2003;90:554–9.

64. Shamsi Z, Kimber S, Hindmarch I. An investigation into the effects of cetirizine on cognitive function and psychomotor performance in healthy volunteers. Eur J Clin Pharmacol 2001;56(12):865–71.

65. Bender BG, McCormick DR, Milgrom H. Children's school performance is not impaired by short-term administration of diphenhydramine or loratadine. J Pediatr 2001;138(5):656–60.

66. Tagawa M, Kano M, Okamura N, Higuchi M, Matsuda M, Mizuki Y, Arai H, Fujii T, Komemushi S, Itoh M, Sasaki H, Watanabe T, Yanai K. Differential cognitive effects of ebastine and (+)-chlorpheniramine in healthy subjects: correlation between cognitive impairment and plasma drug concentration. Br J Clin Pharmacol 2002;53(3):296–304.

67. Verster JC, Volkerts ER, van Oosterwijck AW, Aarab M, Bijtjes SI, De Weert AM, Eijken EJ, Verbaten MN. Acute and subchronic effects of levocetirizine and diphenhydramine on memory functioning, psychomotor performance, and mood. J Allergy Clin Immunol 2003;111(3):623–7.

68. Gandon JM, Allain H. Lack of effect of single and repeated doses of levocetirizine, a new antihistamine drug, on cognitive and psychomotor functions in healthy volunteers. Br J Clin Pharmacol 2002;54(1):51–8.

69. Hindmarch I, Johnson S, Meadows R, Kirkpatrick T, Shamsi Z. The acute and sub-chronic effects of levocetirizine, cetirizine, loratadine, promethazine and placebo on cognitive function, psychomotor performance, and weal and flare. Curr Med Res Opin 2001;17(4):241–55.

70. Verster JC, Volkerts ER, Van Oosterwijck AW, Aarab M, Bijtjes SI, De Weert AM, Eijken EJ, Verbaten MN. Acute and subchronic effects of levocetirizine and diphenhydramine on memory functioning, psychomotor performance, and mood. J Allergy Clin Immunol 2003;111:623–7.

71. Gandon JM, Allain H. Lack of effect of single and repeated doses of levocetirizine, a new antihistamine

drug, on cognitive and psychomotor functions in healthy volunteers. Br J Clin Pharmacol 2002;54:51–8.

72. Potter PC, Schepers JM, Van Niekerk CH. The effects of fexofenadine on reaction time, decision-making, and driver behavior. Ann Allergy Asthma Immunol 2003;91:177–81.

73. Verster JC, De Weert AM, Bijtjes SI, Aarab M, Van Oosterwijck AW, Eijken EJ, Verbaten MN, Volkerts ER. Driving ability after acute and sub-chronic administration of levocetirizine and diphenhydramine: a randomized, double-blind, placebo-controlled trial. Psychopharmacol (Berl) 2003;169:84–90.

74. Serio RN. Acute delirium associated with combined diphenhydramine and linezolid use. Ann Pharmacother 2004;38(1):62–5.

75. Kuykendall JR, Rhodes RS. Auditory hallucinations elicited by combined meclizine and metaxalone use at bedtime. Ann Pharmacother 2004;38:1968–9.

76. Courty E, Durif F, Zenut M, Courty P, Lavarenne J. Psychiatric and sexual disorders induced by apomorphine in Parkinson's disease. Clin Neuropharmacol 1997;20(2):140–7.

77. Vonk J. Ervaringen met Trasylol bij de behandeling van acute pancreatitis. Ned Tijdschr Geneeskd 1965;109:1510.

78. Gonzalez-Seijo JC, Ramos YM, Lastra I. Manic episode and ginseng: report of a possible case. J Clin Psychopharmacol 1995;15(6):447–8.

79. Koehler SM, Glaros A. The effect of aspartame on migraine headache. Headache 1988;28(1):10–4.

80. van der HJ, Duyx J, de Langen JJ, van Royen A. Probable psychiatric side effects of azathioprine. Psychosom Med 2005;67(3):508.

81. Sirois F. Delirium associé à l'azithromycine. [Delirium associated with azithromycin administration.] Can J Psychiatry 2002;47(6):585–6.

82. Wolf ME, Almy G, Toll M, Mosnaim AD. Mania associated with the use of baclofen. Biol Psychiatry 1982;17(6):757–9.

83. Bryan PC, Efiong DO, Stewart-Jones J, Turner P. Propranolol on tests of visual function and central nervous activity. Br J Clin Pharmacol 1974;1:82.

84. Glaister DH, Harrison MH, Allnutt MF. Environmental influences on cardiac activity. In: Burley DM, Frier JH, Rondel RK, Taylor SH, editors. New Perspectives in Beta-blockade. Horsham, UK: Ciba Laboratories, 1973:241.

85. Landauer AA, Pocock DA, Prott FW. Effects of atenolol and propranolol on human performance and subjective feelings. Psychopharmacology (Berl) 1979;60(2):211–5.

86. Salem SA, McDevitt DG. Central effects of beta-adrenoceptor antagonists. Clin Pharmacol Ther 1983;33(1):52–7.

87. Ogle CW, Turner P, Markomihelakis H. The effects of high doses of oxprenolol and of propranolol on pursuit rotor performance, reaction time and critical flicker frequency. Psychopharmacologia 1976;46(3):295–9.

88. Turner P, Hedges A. An investigation of the central effects of oxprenolol. In: Burley DM, Frier JH, Rondel RK, Taylor SH, editors. New Perspectives in Beta-blockade. Horsham, UK: Ciba Laboratories, 1973:269.

89. Tyrer PJ, Lader MH. Response to propranolol and diazepam in somatic and psychic anxiety. BMJ 1974;2(909):14–6.

90. Greil W. Central nervous system effects. Curr Ther Res 1980;28:106.

91. Currie D, Lewis RV, McDevitt DG, Nicholson AN, Wright NA. Central effects of beta-adrenoceptor antagonists. I—Performance and subjective assessments of mood. Br J Clin Pharmacol 1988;26(2):121–8.

92. Nicholson AN, Wright NA, Zetlein MB, Currie D, McDevitt DG. Central effects of beta-adrenoceptor antagonists. II—Electroencephalogram and body sway. Br J Clin Pharmacol 1988;26(2):129–41.

93. Hearing SD, Wesnes KA, Bowman CE. Beta blockers and cognitive function in elderly hypertensive patients: withdrawal and consequences of ACE inhibitor substitution. Int J Geriatr Psychopharmacol 1999;2:13–7.

94. Perez-Stable EJ, Halliday R, Gardiner PS, Baron RB, Hauck WW, Acree M, Coates TJ. The effects of propranolol on cognitive function and quality of life: a randomized trial among patients with diastolic hypertension. Am J Med 2000;108(5):359–65.

95. Yatham LN, Lint D, Lam RW, Zis AP. Adverse effects of pindolol augmentation in patients with bipolar depression. J Clin Psychopharmacol 1999;19(4):383–4.

96. Russell JW, Schuckit NA. Anxiety and depression in patient on nadolol. Lancet 1982;2(8310):1286–7.

97. Thiessen BQ, Wallace SM, Blackburn JL, Wilson TW, Bergman U. Increased prescribing of antidepressants subsequent to beta-blocker therapy. Arch Intern Med 1990;150(11):2286–90.

98. Topliss D, Bond R. Acute brain syndrome after propranolol treatment. Lancet 1977;2(8048):1133–4.

99. Helson L, Duque L. Acute brain syndrome after propranolol. Lancet 1978;1(8055):98.

100. Kurland ML. Organic brain syndrome with propranolol. N Engl J Med 1979;300(7):366.

101. Gershon ES, Goldstein RE, Moss AJ, van Kammen DP. Psychosis with ordinary doses of propranolol. Ann Intern Med 1979;90(6):938–9.

102. Kuhr BM. Prolonged delirium with propanolol. J Clin Psychiatry 1979;40(4):198–9.

103. McGahan DJ, Wojslaw A, Prasad V, Blankenship S. Propranolol-induced psychosis. Drug Intell Clin Pharm 1984;18(7–8):601–3.

104. Steinhert J, Pugh CR. Two patients with schizophrenic-like psychosis after treatment with beta-adrenergic blockers. BMJ 1979;1(6166):790.

105. Macknin ML. Behavioral changes after amoxicillin–clavulanate. Pediatr Infect Dis J 1987;6(9):873–4.

106. Burnet SP, Petrie JP. "Wake up and smell the roses"—a drug reaction to etidronate. Aust NZ J Med 1999;29(1):93.

107. Foley-Nolan D, Daly MJ, Williams D, Wasti A, Martin M. Pamidronate-associated hallucinations. Ann Rheum Dis 1992;51:927–8.

108. Coleman CI, Parkerson KA, Lewis A. Alendronate-induced auditory hallucinations and visual disturbances. Pharmacotherapy. 2004;24:799–802.

109. Le Feuvre CM, Isaacs AJ, Frank OS. Bromocriptine-induced psychosis in acromegaly. BMJ (Clin Res Ed) 1982;285(6351):1315.

110. Clarke CE, Deane KD. Cabergoline versus bromocriptine for levodopa-induced complications in Parkinson's disease. Cochrane Database Syst Rev 2001;(1):CD001519.

111. Cappabianca P, Lodrini S, Felisati G, Peca C, Cozzi R, Di Sarno A, Cavallo LM, Giombini S, Colao A. Cabergoline-induced CSF rhinorrhea in patients with macroprolactinoma. Report of three cases. J Endocrinol Invest 2001;24(3):183–7.

112. Marcos L, De Luis DA, Botella I, Hurtado A. Tumour shrinkage and chiasmal herniation after successful cabergoline treatment for a macroprolactinoma. Clin Endocrinol (Oxf) 2001;54(1):126–7.

113. Machado-Vieira R, Viale CI, Kapczinski F. Mania associated with an energy drink: the possible role of caffeine, taurine, and inositol. Can J Psychiatry 2001;46(5):454–5.

114. Brink DD. Diltiazem and hyperactivity. Ann Intern Med 1984;100(3):459–60.

115. Ahmad S. Nifedipine-induced acute psychosis. J Am Geriatr Soc 1984;32(5):408.

116. Palat GK, Hooker EA, Movahed A. Secondary mania associated with diltiazem. Clin Cardiol 1984;7(11):611–2.

117. Jacobsen FM, Sack DA, James SP. Delirium induced by verapamil. Am J Psychiatry 1987;144(2):248.

118. Pitlik S, Manor RS, Lipshitz I, Perry G, Rosenfeld J. Transient retinal ischaemia induced by nifedipine. BMJ (Clin Res Ed) 1983;287(6408):1845–6.

119. Eccleston D, Cole AJ. Calcium-channel blockade and depressive illness. Br J Psychiatry 1990;156:889–91.

120. Dunn NR, Freemantle SN, Mann RD. Cohort study on calcium channel blockers, other cardiovascular agents, and the prevalence of depression. Br J Clin Pharmacol 1999;48(2):230–3.

121. Benatar MG, Sahin M, Davis RG. Antiepileptic drug-induced visual hallucinations in a child. Pediatr Neurol 2000;23(5):439–41.

122. Seidel WT, Mitchell WG. Cognitive and behavioral effects of carbamazepine in children: data from benign rolandic epilepsy. J Child Neurol 1999;14(11):716–23.

123. Martin R, Meador K, Turrentine L, Faught E, Sinclair K, Kuzniecky R, Gilliam F. Comparative cognitive effects of carbamazepine and gabapentin in healthy senior adults. Epilepsia 2001;42(6):764–71.

124. Singh RB, Singh VP, Somani PN. Psychosis: a rare manifestation of digoxin intoxication. J Indian Med Assoc 1977;69(3):62–3.

125. Shear MK, Sacks MH. Digitalis delirium: report of two cases. Am J Psychiatry 1978;135(1):109–10Digitalis delirium: psychiatric considerations. Int J Psychiatry Med 1977–78;8(4):371–81.

126. Portnoi VA. Digitalis delirium in elderly patients. J Clin Pharmacol 1979;19(11–12):747–50.

127. Gorelick DA, Kussin SZ, Kahn I. Paranoid delusions and auditory hallucinations associated with digoxin intoxication. J Nerv Ment Dis 1978;166(11):817–9.

128. Volpe BT, Soave R. Formed visual hallucinations as digitalis toxicity. Ann Intern Med 1979;91(6):865–6.

129. Kardels B, Beine KH. Acute delirium as a result of digitalis intoxication. Notf Med 2001;27:542–5.

130. Wamboldt FS, Jefferson JW, Wamboldt MZ. Digitalis intoxication misdiagnosed as depression by primary care physicians. Am J Psychiatry 1986;143(2):219–21.

131. Song YH, Terao T, Shiraishi Y, Nakamura J. Digitalis intoxication misdiagnosed as depression—revisited. Psychosomatics 2001;42(4):369–70.

132. Alem A, Shibre T. Khat induced psychosis and its medico-legal implication: a case report. Ethiop Med J 1997;35(2):137–9.

133. Jager AD, Sireling L. Natural history of khat psychosis. Aust NZ J Psychiatry 1994;28(2):331–2.

134. Pantelis C, Hindler CG, Taylor JC. Use and abuse of khat (Catha edulis): a review of the distribution, pharmacology, side effects and a description of psychosis attributed to khat chewing. Psychol Med 1989;19(3):657–68.

135. Maitai CK, Dhadphale M. khat-induced paranoid psychosis. Br J Psychiatry 1988;152:294.

136. McLaren P. Khat psychosis. Br J Psychiatry 1987;150:712–713.

137. Kalix P. Amphetamine psychosis due to khat leaves. Lancet 1984;1(8367):46.

138. Dhadphale M, Mengech A, Chege SW. Miraa (Catha edulis) as a cause of psychosis. East Afr Med J 1981;58(2):130–5.

139. Giannini AJ, Castellani S. A manic-like psychosis due to khat (Catha edulis Forsk.) J Toxicol Clin Toxicol 1982;19(5):455–9.

140. Granek M, Shalev A, Weingarten AM. Khat-induced hypnagogic hallucinations. Acta Psychiatr Scand 1988;78(4):458–61.

141. Nielen RJ, van der Heijden FM, Tuinier S, Verhoeven WM. Khat and mushrooms associated with psychosis. World J Biol Psychiatry 2004;5(1):49–53.

142. Numan N. Exploration of adverse psychological symptoms in Yemeni khat users by the Symptoms Checklist-90 (SCL-90). Addiction 2004;99(1):61–5.

143. Khattab NY, Amer G. Undetected neuropsychophysiological sequelae of khat chewing in standard aviation medical examination. Aviat Space Environ Med 1995;66(8):739–44.

144. Lantz MS, Giambanco V. Acute onset of auditory hallucinations after initiation of celecoxib therapy. Am J Psychiatry 2000;157(6):1022–3.

145. Lund BC, Neiman RF. Visual disturbance associated with celecoxib. Pharmacotherapy 2001;21(1):114–5.

146. Macknight C, Rojas-Fernandez CH. Celecoxib- and rofecoxib-induced delirium. J Neuropsychiatry Clin Neurosci 2001;13(2):305–6.

147. Diemont W, MacKenzie M, Schaap N, Goverde G, Van Heereveld H, Hekster Y, Van Grootheest K. Neuropsychiatric symptoms during cefepime treatment. Pharm World Sci 2001;23:36.

148. Hindmarch I, Johnson S, Meadows R, Kirkpatrick T, Shamsi Z. The acute and sub-chronic effects of levocetirizine, cetirizine, loratadine, promethazine and placebo on cognitive function, psychomotor performance, and weal and flare. Curr Med Res Opin 2001;17(4):241–55.

149. Early Treatment of the Atopic Child. Allergic factors associated with the development of asthma and the influence of cetirizine in a double-blind, randomised, placebo-controlled trial: first results of ETAC. Pediatr Allergy Immunol 1998;9(3):116–24.

150. Stevenson J, Cornah D, Evrard P, Vanderheyden V, Billard C, Bax M, Van Hout AETAC Study Group. Long-term evaluation of the impact of the H_1-receptor antagonist cetirizine on the behavioral, cognitive, and psychomotor development of very young children with atopic dermatitis. Pediatr Res 2002;52(2):251–7.

151. Okamura N, Yanai K, Higuchi M, Sakai J, Iwata R, Ido T, Sasaki H, Watanabe T, Itoh M. Functional neuroimaging of cognition impaired by a classical antihistamine, d-chlorpheniramine. Br J Pharmacol 2000;129(1):115–23.

152. Ferraro V, Mantoux F, Denis K, Lay-Macagno M-A, Ortonne J-P, Lacour J-P. Hallucinations au cours d'un traitement par hydroxychloroquine. Ann Dermatol Venereol 2004;131:471–3.

153. Adam I, Elbashir MI. Suicide after treatment of chloroquine resistant falciparum malaria with quinine. Saudi Med J 2004;25 (2):248–9.

154. Gavrilesco S, Streian C, Constantinesco L. Tachycardie ventriculaire et fibrillation auriculaire associées après vaccination anticholériques. [Associated ventricular tachycardia and auricular fibrillation after anticholera vaccination.] Acta Cardiol 1973;28(1):89–94.

155. Driehorst J, Laubenthal F. Akute Myocarditis nach Choleraschutzimpfung. [Acute myocarditis after cholera vaccination.] Dtsch Med Wochenschr 1984;109(5):197–8.

156. Mall T, Gyr K. Episode resembling immune complex disease after cholera vaccination. Trans R Soc Trop Med Hyg 1984;78(1):106–7.

157. Eisinger AJ, Smith JG. Acute renal failure after TAB and cholera vaccination. BMJ 1979;1(6160):381–2.

158. Gatt DT. Pancreatitis following monovalent typhoid and cholera vaccinations. Br J Clin Pract 1986;40(7):300–1.

159. Tamminga C, Smith RC, Chang S, Haraszti JS, Davis JM. Depression associated with oral choline. Lancet 1976;2(7991):905.

160. McLeod MN, Golden RN. Chromium treatment of depression. Int J Neuropsychopharmacol 2000;3(4):311–4.

161. Bhatia MS, Agrawal P, Khastgir U, Malik SC. Cimetidine induced psychosis. Indian Pediatr 1989;26(10):1061–2.

162. Pierce JR Jr. Cimetidine-associated depression and loss of libido in a woman. Am J Med Sci 1983;286(3):31–4.

163. Billings RF, Tang SW, Rakoff VM. Depression associated with cimetidine. Can J Psychiatry 1981;26(4):260–1.

164. Zabala S, Gascon A, Bartolome C, Castiella J, Juyol M. Ciprofloxacino y psicosis aguda. [Ciprofloxacin and acute psychosis.] Enferm Infecc Microbiol Clin 1998;16(1):42.

165. James EA, Demian AZ. Acute psychosis in a trauma patient due to ciprofloxacin. Postgrad Med J 1998;74(869):189–90.

166. Grassi L, Biancosino B, Pavanati M, Agostini M, Manfredini R. Depression or hypoactive delirium? A report of ciprofloxacin-induced mental disorder in a patient with chronic obstructive pulmonary disease. Psychother Psychosom 2001;70(1):58–9.

167. Imani K, Druart F, Glibert A, Morin T. Troubles neuro-psychiatriques induits par la ciprofloxacine. [Neuropsychiatric disorders induced by ciprofloxacin.] Presse Méd 2001;30(27):1356.

168. Tripathi A, Chen SI, O'Sullivan S. Acute psychosis following the use of topical ciprofloxacin. Arch Ophthalmol 2002;120(5):665–6.

169. Norra C, Skobel E, Breuer C, Haase G, Hanrath P, Hoff P. Ciprofloxacin-induced acute psychosis in a patient with multidrug-resistant tuberculosis. Eur Psychiatry 2003;18:262–3.

170. Anonymous. Cisapride-aggressive behaviour. Bull Swed Adverse Drug React Advisory Comm 1991;57:1.

171. Pijlman AH, Kuck EM, van Puijenbroek EP, Hoekstra JB. Acuut delies, waarschijulijk uitgelokt door clarithromycine. [Acute delirium, probably precipitated by clarithromycin.] Ned Tijdschr Geneeskd 2001;145(5):225–8.

172. Nightingale SD, Koster FT, Mertz GJ, Loss SD. Clarithromycin-induced mania in two patients with AIDS. Clin Infect Dis 1995;20(6):1563–4.

173. Abouesh A, Stone C, Hobbs WR. Antimicrobial-induced mania (antibiomania): a review of spontaneous reports. J Clin Psychopharmacol 2002;22(1):71–81.

174. Cone LA, Sneider RA, Nazemi R, Dietrich EJ. Mania due to clarithromycin therapy in a patient who was not infected with human immunodeficiency virus. Clin Infect Dis 1996;22(3):595–6.

175. Pollak PT, Sketris IS, MacKenzie SL, Hewlett TJ. Delirium probably induced by clarithromycin in a patient receiving fluoxetine. Ann Pharmacother 1995;29(5):486–8.

176. Jimenez-Pulido SB, Navarro-Ruiz A, Sendra P, Martinez-Ramirez M, Garcia-Motos C, Montesinos-Ros A. Hallucinations with therapeutic doses of clarithromycin. Int J Clin Pharmacol Ther 2002;40(1):20–2.

177. Steinman MA, Steinman TI. Clarithromycin-associated visual hallucinations in a patient with chronic renal failure on continuous ambulatory peritoneal dialysis. Am J Kidney Dis 1996;27(1):143–6.

178. Ahmed I, Takeshita J. Clonidine: a critical review of its role in the treatment of psychiatric disorders. Drug Ther 1996;6:53–70.

179. Laird RD, Webb M. Psychotic episode during use of St. John's wort. J Herbal Pharmacother 2001;1:81–7.

180. Schneck C. St. John's wort and hypomania. J Clin Psychiatry 1998;59(12):689.

181. Nierenberg AA, Burt T, Matthews J, Weiss AP. Mania associated with St. John's wort. Biol Psychiatry 1999;46(12):1707–8.

182. Sproule BA, Busto UE, Somer G, Romach MK, Sellers EM. Characteristics of dependent and nondependent regular users of codeine. J Clin Psychopharmacol 1999;19(4):367–72.

183. Klein JF. Adverse psychiatric effects of systemic glucocorticoid therapy. Am Fam Physician 1992;46(5):1469–74.

184. Doherty M, Garstin I, McClelland RJ, Rowlands BJ, Collins BJ. A steroid stupor in a surgical ward. Br J Psychiatry 1991;158:125–7.

185. Satel SL. Mental status changes in children receiving glucocorticoids. Review of the literature. Clin Pediatr (Phila) 1990;29(7):383–8.

186. Alpert E, Seigerman C. Steroid withdrawal psychosis in a patient with closed head injury. Arch Phys Med Rehabil 1986;67(10):766–9.

187. Hassanyeh F, Murray RB, Rodgers H. Adrenocortical suppression presenting with agitated depression, morbid jealousy, and a dementia-like state. Br J Psychiatry 1991;159:870–2.

188. Nahon S, Pisanté L, Delas N. A successful switch from prednisone to budesonide for neuropsychiatric adverse effects in a patient with ileal Crohn's disease. Am J Gastroenterol 2001;96(1):1953–4.

189. O'Shea TM, Kothadia JM, Klinepeter KL, Goldstein DJ, Jackson BG, Weaver RG III, Dillard RG. Randomized placebo-controlled trial of a 42-day tapering course of dexamethasone to reduce the duration of ventilator dependency in very low birth weight infants: outcome of study participants at 1-year adjusted age. Pediatrics 1999;104(1 Part 1):15–21.

190. Soliday E, Grey S, Lande MB. Behavioral effects of corticosteroids in steroid–sensitive nephrotic syndrome. Pediatrics 1999;104(4):e51.

191. Klein-Gitelman MS, Pachman LM. Intravenous corticosteroids: adverse reactions are more variable than expected in children. J Rheumatol 1998;25(10):1995–2002.

192. Kaiser H. Psychische Storungen nach Beclomethasondipropionat-Inhalation?. [Mental disorders following beclomethasone dipropionate inhalation?.] Med Klin 1978;73(38):1334.

193. Keenan PA, Jacobson MW, Soleymani RM, Mayes MD, Stress ME, Yaldoo DT. The effect on memory of chronic prednisone treatment in patients with systemic disease. Neurology 1996;47(6):1396–402.

194. Oliveri RL, Sibilia G, Valentino P, Russo C, Romeo N, Quattrone A. Pulsed methylprednisolone induces a reversible impairment of memory in patients with relapsing-remitting multiple sclerosis. Acta Neurol Scand 1998;97(6):366–9.

195. Newcomer JW, Selke G, Melson AK, Hershey I, Craft S, Richards K, Alderson AL. Decreased memory performance in healthy humans induced by stress-level cortisol treatment. Arch Gen Psychiatry 1999;56(6):527–33.

196. Aisen PS, Davis KL, Berg JD, Schafer K, Campbell K, Thomas RG, Weiner MF, Farlow MR, Sano M, Grundman M, Thal LJ. A randomized controlled trial of prednisone in Alzheimer's disease. Alzheimer's Dis Cooperative Study. Neurology 2000;54(3):588–93.

197. de Quervain DJ, Roozendaal B, Nitsch RM, McGaugh JL, Hock C. Acute cortisone administration impairs retrieval of long-term declarative memory in humans. Nat Neurosci 2000;3(4):313–4.

198. Moser NJ, Phillips BA, Guthrie G, Barnett G. Effects of dexamethasone on sleep. Pharmacol Toxicol 1996;79(2):100–2.

199. Bermond B, Surachno S, Lok A, ten Berge IJ, Plasmans B, Kox C, Schuller E, Schellekens PT, Hamel R. Memory functions in prednisone-treated kidney transplant patients. Clin Transplant 2005;19(4):512–7.

200. Brown ES, Stuard G, Liggin JD, Hukovic N, Frol A, Dhanani N, Khan DA, Jeffress J, Larkin GL, McEwen BS, Rosenblatt R, Mageto Y, Hanczyc M, Cullum CM. Effect of phenytoin on mood and declarative memory during prescription corticosteroid therapy. Biol Psychiatry 2005;57(5):543–8.

201. Brown ES, J Woolston D, Frol A, Bobadilla L, Khan DA, Hanczyc M, Rush AJ, Fleckenstein J, Babcock E, Cullum CM. Hippocampal volume, spectroscopy, cognition, and mood in patients receiving corticosteroid therapy. Biol Psychiatry 2004;55:538–45.

202. Preda A, Fazeli A, McKay BG, Bowers MB Jr, Mazure CM. Lamotrigine as prophylaxis against steroid-induced mania. J Clin Psychiatry 1999;60(10):708–9.

203. Preda A, Fazeli A, McKay BG, Bowers MB Jr, Mazure CM. Lamotrigine of prophylaxis against steroid-induced mania. J. Clin Psychiatry 1999;60(10):708–9.

204. Brown ES, Suppes T, Khan DA, Carmody TJ 3rd. Mood changes during prednisone bursts in outpatients with asthma. J Clin Psychopharmacol 2002;22(1):55–61.

205. Chau SY, Mok CC. Factors predictive of corticosteroid psychosis in patients with systemic lupus erythematosus. Neurology 2003;61:104–7.

206. Bolanos SH, Khan DA, Hanczyc M, Bauer MS, Dhanani N, Brown ES. Assessment of mood states in patients receiving long-term corticosteroid therapy and in controls with patient-rated and clinician-rated scales. Ann Allergy Asthma Immunol 2004;92:500–5.

207. Wada K, Suzuki H, Taira T, Akiyama K, Kuroda S. Successful use of intravenous clomipramine in depressive–catatonic state associated with corticosteroid treatment. Int J Psych Clin Pract 2004;8:131–3.

208. Hederos CA. Neuropsychologic changes and inhaled corticosteroids. J Allergy Clin Immunol 2004;114:451–2.

209. Ilbeigi MS, Davidson ML, Yarmush JM. An unexpected arousal effect of etomidate in a patient on high-dose steroids. Anesthesiology 1998;89(6):1587–9.

210. Kramer TM, Cottingham EM. Risperidone in the treatment of steroid-induced psychosis. J Child Adolesc Psychopharmacol 1999;9(4):315–6.

211. Scheschonka A, Bleich S, Buchwald AB, Ruther E, Wiltfang J. Development of obsessive-compulsive behaviour following cortisone treatment. Pharmacopsychiatry 2002;35(2):72–4.

212. Minden SL, Orav J, Schildkraut JJ. Hypomanic reactions to ACTH and prednisone treatment for multiple sclerosis. Neurology 1988;38(10):1631–4.

213. Masters PA, O'Bryan TA, Zurlo J, Miller DQ, Joshi N. Trimethoprim–sulfamethoxazole revisited. Arch Intern Med 2003;163:402–10.

214. ADRAC. Acute neuropsychiatric events with celecoxib and rofecoxib. Aust Adv Drug React Bull 2003;22:3.

215. Beeber AR, Manring JM Jr. Psychosis following cyclobenzaprine use. J Clin Psychiatry 1983;44(4):151–2.

216. O'Neil BA, Knudson GA, Bhaskara SM. First episode psychosis following cyclobenzaprine use. Can J Psychiatry 2000;45(8):763–4.

217. Mitchell RS, Lester W. Clinical experience with cycloserine in the treatment of tuberculosis. Scand J Respir Dis Suppl 1970;71:94–108.

218. Watemberg NM, Roth KS, Alehan FK, Epstein CE. Central anticholinergic syndrome on therapeutic doses of cyproheptadine. Pediatrics 1999;103(1):158–60.

219. Strayhorn JM Jr. Case study: cyproheptadine and aggression in a five-year-old boy. J Am Acad Child Adolesc Psychiatry 1998;37(6):668–70.

220. Balkrishna, Bhatia MS. Dapsone induced psychosis. J Indian Med Assoc 1989;87(5):120–1.

221. Wilken JA, Kane RL, Ellis AK, Rafeiro E, Briscoe MP, Sullivan CL, Day JH. A comparison of the effect of diphenhydramine and desloratadine on vigilance and cognitive function during treatment of ragweed-induced allergic rhinitis. Ann Allergy Asthma Immunol 2003;91(4):375–85.

222. Nicholson AN, Handford AD, Turner C, Stone BM. Studies on performance and sleepiness with the H_1-antihistamine, desloratadine. Aviat Space Environ Med 2003;74(8):809–15.

223. Valk PJ, Van Roon DB, Simons RM, Rikken G. Desloratadine shows no effect on performance during 6 h at 8,000 ft simulated cabin altitude Aviat Space Environ Med 2004;75(5):433–8.

224. Roberge RJ, Hirani KH, Rowland PL 3rd, Berkeley R, Krenzelok EP. Dextromethorphan- and pseudoephedrine-induced agitated psychosis and ataxia: case report. J Emerg Med 1999;17(2):285–8.

225. Soutullo CA, Cottingham EM, Keck PE Jr. Psychosis associated with pseudoephedrine and dextromethorphan. J Am Acad Child Adolesc Psychiatry 1999;38(12):1471–2.

226. Price LH, Lebel J. Dextromethorphan-induced psychosis. Am J Psychiatry 2000;157(2):304.

227. Hinsberger A, Sharma V, Mazmanian D. Cognitive deterioration from long-term abuse of dextromethorphan: a case report. J Psychiatry Neurosci 1994;19(5):375–7.

228. Basu R, Dodge H, Stoehr GP, Ganguli M. Sedative-hypnotic use of diphenhydramine in a rural, older adult, community-based cohort: effects on cognition. Am J Geriatr Psychiatry 2003;11(2):205–13.

229. Dinndorf PA, McCabe MA, Frierdich S. Risk of abuse of diphenhydramine in children and adolescents with chronic illnesses. J Pediatr 1998;133(2):293–5.

230. Falk RH, Nisbet PA, Gray TJ. Mental distress in patient on disopyramide. Lancet 1977;1(8016):858–9.

231. Padfield PL, Smith DA, Fitzsimons EJ, McCruden DC. Disopyramide and acute psychosis. Lancet 1977;1(8022):1152.

232. Peachey JE, Brien JF, Roach CA, Loomis CW. A comparative review of the pharmacological and toxicological properties of disulfiram and calcium carbimide. J Clin Psychopharmacol 1981;1(1):21–6.

233. Gostout CJ. Patient assessment and resuscitation. Gastrointest Endosc Clin N Am 1999;9(2):175–87.

234. Ewing JA, Rouse BA, Mueller RA, Silver D. Can dopamine beta-hydroxylase levels predict adverse reactions to disulfiram? Alcohol Clin Exp Res 1978;2(1):93–4.

235. Murthy KK. Psychosis during disulfiram therapy for alcoholism. J Indian Med Assoc 1997;95(3):80–1.

236. Verbon H, de Jong CA. Psychose tijdens en na disulfiramgebruik. [Psychosis during and after disulfiram use.] Ned Tijdschr Geneeskd 2002;146(12):571–3.

237. Park CW, Riggio S. Disulfiram–ethanol induced delirium. Ann Pharmacother 2001;35(1):32–5.

238. Wengel SP, Roccaforte WH, Burke WJ, Bayer BL, McNeilly DP, Knop D. Behavioral complications

associated with donepezil. Am J Psychiatry 1998;155(11):1632–3.

239. Bouman WP, Pinner G. Violent behavior associated with donepezil. Am J Psychiatry 1998;155(11):1626–7.

240. Feigin A, Ghilardi MF, Carbon M, Edwards C, Fukuda M, Dhawan V, Margouleff C, Ghez C, Eidelberg D. Effects of levodopa on motor sequence learning in Parkinson's disease. Neurology 2003;60:1744–9.

241. Driver-Dunkley E, Samanta J, Stacy M. Pathological gambling associated with dopamine agonist therapy in Parkinson's disease. Neurology 2003;61:422–3.

242. Evans M, Perera PW, Donoghue J. Drug induced psychosis with doxazosin. BMJ 1997;314(7098):1869.

243. Tagawa M, Kano M, Okamura N, Higuchi M, Matsuda M, Mizuki Y, Arai H, Fujii T, Komemushi S, Itoh M, Sasaki H, Watanabe T, Yanai K. Differential cognitive effects of ebastine and (+)-chlorpheniramine in healthy subjects: correlation between cognitive impairment and plasma drug concentration. Br J Clin Pharmacol 2002;53(3):296–304.

244. Morales-Ramirez J, Tashima K, Hardy D, et al. In: A phase III, multicenter randomized open-label study to compare the antiretroviral activity and tolerability of efavirenz (EFV) + indinavir (IDV), versus EFV + zidovudine (ZDV) + lamivudine (3TC), versus IDV + ZDV + 3TC at 36 weeks (DMP 266-006)38th Interscience Conference on Antimicrobial Agents and Chemotherapy (San Diego). Washington DC: American Society for Microbiology, 1998:I-103.

245. Mayers D, Jemesk J, Eyster E, et al. In: A double-blind, placebo-controlled study to assess the safety, tolerability and antiretroviral activity of efavirenz (EFV DMP 266) in combination with open-label zidovudine (ZDV) and lamivudine (3TC) in HIV infected patients (DMP 266–004)38th Interscience Conference on Antimicrobial Agents and Chemotherapy (San Diego). Washington DC: American Society for Microbiology, 1998:22340.

246. Welch KJ, Morse A. Association between efavirenz and selected psychiatric and neurological conditions. J Infect Dis 2002;185(2):268–9.

247. Peyriere H, Mauboussin JM, Rouanet I, Fabre J, Reynes J, Hillaire-Buys D. Management of sudden psychiatric disorders related to efavirenz. AIDS 2001;15(10):1323–4.

248. Marzolini C, Telenti A, Decosterd LA, Greub G, Biollaz J, Buclin T. Efavirenz plasma levels can predict treatment failure and central nervous system side effects in HIV-1-infected patients. AIDS 2001;15(1):71–5.

249. Spire B, Carrieri P, Garzot MA, L'henaff M, Obadia Y. Factors associated with efavirenz discontinuation in a large community-based sample of patients. AIDS Care 2004;16(5):558–64.

250. Bentin S, Collins GI, Adam N. Decision-making behaviour during inhalation of subanaesthetic concentrations of enflurane. Br J Anaesth 1978;50(12):1173–8.

251. Bentin S, Collins GI, Adam N. Effects of low concentrations of enflurane on probability learning. Br J Anaesth 1978;50(12):1179–83.

252. Jacobs KM, Kirsch KA. Psychiatric complications of Mahuang. Psychosomatics 2000;41:58–62.

253. Taylor D, O'Toole K, Auble T, Ryan C, Sherman D. The psychometric and cardiac effects of pseudoephedrine and antihistamine in the hyperbaric environment. S Pac Underw Med Soc J 2001;31:50–7.

254. Sauder KL, Brady WJ Jr, Hennes H. Visual hallucinations in a toddler: accidental ingestion of a sympathomimetic over-the-counter nasal decongestant. Am J Emerg Med 1997;15(5):521–6.

255. Gulbranson SH, Mock RE, Wolfrey JD. Possible ergotamine–caffeine-associated delirium. Pharmacotherapy 2002;22(1):126–9.

256. Umstead GS, Neumann KH. Erythromycin ototoxicity and acute psychotic reaction in cancer patients with hepatic dysfunction. Arch Intern Med 1986;146(5):897–9.

257. Black RJ, Dawson TA. Erythromycin and nightmares. BMJ (Clin Res Ed) 1988;296(6628):1070.

258. Williams NR. Erythromycin: a case of nightmares. BMJ (Clin Res Ed) 1988;296(6616):214.

259. Murdoch J. Psychiatric complications of erythromycin and clindamycin. Can J Hosp Pharm 1988;41:277.

260. van den Bent MJ, Bos GM, Sillevis Smitt PA, Cornelissen JJ. Erythropoietin induced visual hallucinations after bone marrow transplantation. J Neurol 1999;246(7):614–6.

261. Kimura D. Estrogen replacement therapy may protect against intellectual decline in postmenopausal women. Horm Behav 1995;29(3):312–21.

262. Saletu B, Brandstatter N, Metka M, Stamenkovic M, Anderer P, Semlitsch HV, Heytmanek G, Huber J, Grunberger J, Linzmayer L, Kurz CH, Decker K, Binder G, Knogler W, Koll B. Double-blind, placebo-controlled, hormonal, syndromal and EEG mapping studies with transdermal oestradiol therapy in menopausal depression. Psychopharmacology (Berl) 1995;122(4):321–9.

263. Kay PAJ, Yurkow J, Forman LJ, Chopra A, Cavalieri T. Transdermal estradiol in the management of aggressive behavior in male patients with dementia. Clin Gerontol 1995;15:54–8.

264. Narang RK. Acute psychotic reaction probably caused by ethionamide. Tubercle 1972;53(2):137–8.

265. Sharma GS, Gupta PK, Jain NK, Shanker A, Nanawati V. Toxic psychosis to isoniazid and ethionamide in a patient with pulmonary tuberculosis. Tubercle 1979;60(3):171–2.

266. Lansdown FS, Beran M, Litwak T. Psychotoxic reaction during ethionamide therapy. Am Rev Respir Dis 1967;95(6):1053–5.

267. Shannon PJ, Leonard D, Kidson MA. Letter: fenfluramine and psychosis. BMJ 1974;3(5930):576.

268. Tsueda K, Mosca PJ, Heine MF, Loyd GE, Durkis DA, Malkani AL, Hurst HE. Mood during epidural patient-controlled analgesia with morphine or fentanyl. Anesthesiology 1998;88(4):885–91.

269. Hindmarch I, Shamsi Z, Kimber S. An evaluation of the effects of high-dose fexofenadine on the central nervous system: a double-blind, placebo-controlled study in healthy volunteers. Clin Exp Allergy 2002;32(1):133–9.

270. Vermeeren A, O'Hanlon JF. Fexofenadine's effects, alone and with alcohol, on actual driving and psychomotor performance. J Allergy Clin Immunol 1998;101(3):306.

271. Tuxen MK, Hansen SW. Neurotoxicity secondary to antineoplastic drugs. Cancer Treat Rev 1994;20(2):191–214.

272. Lynch HT, Droszcz CP, Albano WA, Lynch JF. "Organic brain syndrome" secondary to 5-fluorouracil toxicity. Dis Colon Rectum 1981;24(2):130–1.

273. McCarty JM, Richard G, Huck W, Tucker RM, Tosiello RL, Shan M, Heyd A, Echols RM. A randomized trial of short-course ciprofloxacin, ofloxacin, or trimethoprim/sulfamethoxazole for the treatment of acute urinary tract infection in women. Ciprofloxacin Urinary Tract Infection Group. Am J Med 1999;106(3):292–9.

274. Reeves RR. Ciprofloxacin-induced psychosis. Ann Pharmacother 1992;26(7–8):930–1.

275. Saravolatz LD, Leggett J. Gatifloxacin, gemifloxacin, and moxifloxacin: the role of 3 newer fluoroquinolones. Clin Infect Dis 2003;37:1210–5.

276. Sumner CL, Elliott RL. Delirium associated with gatifloxacin. Psychosomatics 2003;44:85–6.

277. Hall CE, Keegan H, Rogstad KE. Psychiatric side effects of ofloxacin used in the treatment of pelvic inflammatory disease. Int J STD AIDS 2003;14:636–7.

278. Leone R, Venegoni M, Motola D, Moretti U, Piazzetta V, Cocci A, Resi D, Mozzo F, Velo G, Burzilleri L, Montanaro N, Conforti A. Adverse drug reactions related to the use of fluoroquinolone antimicrobials: an analysis of spontaneous reports and fluoroquinolone consumption data from three Italian regions. Drug Saf 2003;26:109–20.

279. Leweke FM, Bauer J, Elger CE. Manic episode due to gabapentin treatment. Br J Psychiatry 1999;175:291.

280. Rosebush PI, MacQueen GM, Mazurek MF. Catatonia following gabapentin withdrawal. J Clin Psychopharmacol 1999;19(2):188–9.

281. Trinka E, Niedermuller U, Thaler C, Doering S, Moroder T, Ladurner G, Bauer G. Gabapentin-induced mood changes with hypomanic features in adults. Seizure 2000;9(7):505–8.

282. Meador KJ, Loring DW, Ray PG, Murro AM, King DW, Nichols ME, Deer EM, Goff WT. Differential cognitive effects of carbamazepine and gabapentin. Epilepsia 1999;40(9):1279–85.

283. Khurana DS, Riviello J, Helmers S, Holmes G, Anderson J, Mikati MA. Efficacy of gabapentin therapy in children with refractory partial seizures. J Pediatr 1996;128(6):829–33.

284. Mujica R, Weiden P. Neuroleptic malignant syndrome after addition of haloperidol to atypical antipsychotic. Am J Psychiatry 2001;158(4):650–1.

285. Pinninti NR, Mahajan DS. Gabapentin-associated aggression. J Neuropsychiatry Clin Neurosci 2001;13(3):424.

286. Kane FJ Jr, Byrd G. Acute toxic psychosis associated with gentamicin therapy. South Med J 1975;68(10):1283–5.

287. Warnock JK, Bundren JC, Morris DW. Depressive symptoms associated with gonadotropin-releasing hormone agonists. Depress Anxiety 1998;7(4):171–7.

288. Moghissi KS, Schlaff WD, Olive DL, Skinner MA, Yin H. Goserelin acetate (Zoladex) with or without hormone replacement therapy for the treatment of endometriosis. Fertil Steril 1998;69(6):1056–62.

289. Mahe V, Nartowski J, Montagnon F, Dumaine A, Gluck N. Psychosis associated with gonadorelin agonist administration. Br J Psychiatry 1999;175:290–1.

290. Green HJ, Pakenham KI, Headley BC, Yaxley J, Nicol DL, Mactaggart PN, Swanson C, Watson RB, Gardiner RA. Altered cognitive function in men treated for prostate cancer with luteinizing hormone-releasing hormone analogues and cyproterone acetate: a randomized controlled trial. BJU Int 2002;90(4):427–32.

291. Gotz H, Reichenberger M. Ergebnisse einer Fragebogenaktion bei 1670 Dermatologen der Bundesrepublik Deutschland uber Nebenwirkungen bei der Griseofulvin Therapie. [Results of questionnaires of 1670 dermatologists in West Germany concerning the side effects of griseofulvin therapy.] Hautarzt 1972;23(11):485–92.

292. Davison LA, Steinhelber JC, Eger EI 2nd, Stevens WC. Psychological effects of halothane and isoflurane anesthesia. Anesthesiology 1975;43(3):313–24.

293. Bruchiel KJ, Stockard JJ, Calverley RK, Smith NT, Scholl ML, Mazze RI. Electroencephalographic abnormalities following halothane anesthesia. Anesth Analg 1978;57(2):244–51.

294. Neunteufl T, Berger R, Pacher R. Endothelin receptor antagonists in cardiology clinical trials. Expert Opin Investig Drugs 2002;11(3):431–43.

295. Farmer JA, Torre-Amione G. Comparative tolerability of the HMG-CoA reductase inhibitors. Drug Saf 2000;23(3):197–213.

296. King DS. Cognitive impairment associated with atorvastatin and simvastatin. Pharmacotherapy 2003;23:1663–7.

297. Wagstaff LR, Mitton MW, Arvik BM, Doraiswamy PM. Statin-associated memory loss: analysis of 60 case reports and review of the literature. Pharmacotherapy 1920;23:871–80.

298. Muldoon MF, Ryan CM, Sereika SM, Flory JD, Manuck SB. Randomized trial of the effects of simvastatin on cognitive functioning in hypercholesterolemic adults. Am J Med 2004;117(11):823–9.

299. Vevera J, Fisar Z, Kvasnicka T, Zdenek H, Starkova L, Ceska R, Papezova H. Cholesterol-lowering therapy evokes time-limited changes in serotonergic transmission. Psychiatry Res 2005;133(2-3):197–203.

300. Waters DD. Safety of high-dose atorvastatin therapy. Am J Cardiol 2005;96(5A Suppl):69F-75F.

301. Calanchini C. Die Auslösung eines Zwangssyndroms durch Ovulationshemmer. [Development of a compulsive syndrome by ovulation inhibitors.] Schweiz Arch Neurol Psychiatr 1986;137(4):25–31.

302. Van Winter JT, Miller KA. Breakthrough bleeding in a bulimic adolescent receiving oral contraceptives. Pediat Adolesc Gynecol 1986;4:39.

303. Chang AM, Chick P, Milburn S. Mood changes as reported by women taking the oral contraceptive pill. Aust NZ J Obstet Gynaecol 1982;22(2):78–83.

304. Fleming O, Seager CP. Incidence of depressive symptoms in users of the oral contraceptive. Br J Psychiatry 1978;132:431–40.

305. Goldzieher JW, Moses LE, Averkin E, Scheel C, Taber BZ. A placebo-controlled double-blind crossover investigation of the side effects attributed to oral contraceptives. Fertil Steril 1971;22(9):609–23.

306. Wagner KD, Berenson AB. Norplant-associated major depression and panic disorder. J Clin Psychiatry 1994;55(11):478–80.

307. Wagner KD. Major depression and anxiety disorders associated with Norplant. J Clin Psychiatry 1996;57(4):152–7.

308. Wihlback A-C, Nyberg S, Backstrom T, Bixo M, Sundstrom-Poromaa I. Estradiol and the addition of progesterone increase the sensitivity to a neurosteroid in postmenopausal women. Psychoneuroendocrinology 2005;30:38–50.

309. Anderson S, McGuire R, McKeown D. Comparison of the cognitive effects of premedication with hyoscine and atropine. Br J Anaesth 1985;57(2):169–73.

310. Stevinson C, Ernst E. Can St John's wort trigger psychoses? Int J Clin Pharmacol Ther 2004;42:473–80.

311. Fruehwald-Schultes B, Born J, Kern W, Peters A, Fehm HL. Adaptation of cognitive function to hypoglycemia in healthy men. Diabetes Care 2000;23(8):1059–66.

312. Austin EJ, Deary IJ. Effects of repeated hypoglycemia on cognitive function: a psychometrically validated reanalysis of the Diabetes Control and Complications Trial data. Diabetes Care 1999;22(8):1273–7.

313. Berger A, Croisier M, Jacot E, Kehtari R. Coma hypoglycémique de longue durée. [Hypoglycemic coma of long duration.] Rev Med Suisse Romande 1999;119(1):49–53.

314. Strachan MW, Deary IJ, Ewing FM, Frier BM. Recovery of cognitive function and mood after severe hypoglycemia

in adults with insulin-treated diabetes. Diabetes Care 2000;23(3):305–12.

315. Cox DJ, Gonder-Frederick LA, Kovatchev BP, Young-Hyman DL, Donner TW, Julian DM, Clarke WL. Biopsychobehavioral model of severe hypoglycemia. II. Understanding the risk of severe hypoglycemia. Diabetes Care 1999;22(12):2018–25.

316. Kinsley BT, Weinger K, Bajaj M, Levy CJ, Simonson DC, Quigley M, Cox DJ, Jacobson AM. Blood glucose awareness training and epinephrine responses to hypoglycemia during intensive treatment in type 1 diabetes. Diabetes Care 1999;22(7):1022–8.

317. The Diabetes Control and Complications Trial Research Group. The effect of intensive treatment of diabetes on the development and progression of long-term complications in insulin-dependent diabetes mellitus. N Engl J Med 1993;329(14):977–86.

318. Wysocki T, Harris MA, Mauras N, Fox L, Taylor A, Jackson SC, White NH. Absence of adverse effects of severe hypoglycemia on cognitive function in school-aged children with diabetes over 18 months. Diabetes Care 2003;26(4):1100–5.

319. Matyka KA, Wigg L, Pramming S, Stores G, Dunger DB. Cognitive function and mood after profound nocturnal hypoglycaemia in prepubertal children with conventional insulin treatment for diabetes. Arch Dis Child 1999;81(2):138–42.

320. McAulay V, Deary IJ, Ferguson SC, Frier BM. Acute hypoglycemia in humans causes attentional dysfunction while nonverbal intelligence is preserved. Diabetes Care 2001;24(10):1745–50.

321. Schaefer M, Engelbrecht MA, Gut O, Fiebich BL, Bauer J, Schmidt F, Grunze H, Lieb K. Interferon alpha (IFNalpha) and psychiatric syndromes: a review. Prog Neuropsychopharmacol Biol Psychiatry 2002;26(4):731–46.

322. Van Gool AR, Kruit WH, Engels FK, Stoter G, Bannink M, Eggermont AM. Neuropsychiatric side effects of interferon-alfa therapy. Pharm World Sci 2003;25(1):11–20.

323. Vial T, Descotes J. Clinical toxicity of the interferons. Drug Saf 1994;10(2):115–50.

324. Smith A, Tyrrell D, Coyle K, Higgins P. Effects of interferon alpha on performance in man: a preliminary report. Psychopharmacology (Berl) 1988;96(3):414–6.

325. Meyers CA, Valentine AD. Neurological and psychiatric adverse effects of immunological therapy. CNS Drugs 1995;3:56–68.

326. James CW, Savini CJ. Homicidal ideation secondary to interferon. Ann Pharmacother 2001;35(7–8):962–3.

327. Capuron L, Pagnoni G, Demetrashvili M, Woolwine BJ, Nemeroff CB, Berns GS, Miller AH. Anterior cingulate activation and error processing during interferon-alpha treatment. Biol Psychiatry 2005;58(3):190–6.

328. Kraus MR, Schafer A, Wissmann S, Reimer P, Scheurlen M. Neurocognitive changes in patients with hepatitis C receiving interferon alfa-2b and ribavirin. Clin Pharmacol Ther 2005;77(1):90–100.

329. Scheibel RS, Valentine AD, O'Brien S, Meyers CA. Cognitive dysfunction and depression during treatment with interferon-alpha and chemotherapy. J Neuropsychiatry Clin Neurosci 2004;16:185–91.

330. Kraus MR, Schafer A, Wissmann S, Reimer P, Scheurlen M. Neurocognitive changes in patients with hepatitis C receiving interferon alfa-2b and ribavirin. Clin Pharmacol Ther 2005;77:90–100.

331. Greenberg DB, Jonasch E, Gadd MA, Ryan BF, Everett JR, Sober AJ, Mihm MA, Tanabe KK, Ott M,

Haluska FG. Adjuvant therapy of melanoma with interferon-alpha-2b is associated with mania and bipolar syndromes. Cancer 2000;89(2):356–62.

332. Monji A, Yoshida I, Tashiro K, Hayashi Y, Tashiro N. A case of persistent manic depressive illness induced by interferon-alfa in the treatment of chronic hepatitis C. Psychosomatics 1998;39(6):562–4.

333. Hosoda S, Takimura H, Shibayama M, Kanamura H, Ikeda K, Kumada H. Psychiatric symptoms related to interferon therapy for chronic hepatitis C: clinical features and prognosis. Psychiatry Clin Neurosci 2000;54(5):565–72.

334. Schafer M, Boetsch T, Laakmann G. Psychosis in a methadone-substituted patient during interferon-alpha treatment of hepatitis C. Addiction 2000;95(7):1101–4.

335. Mayr N, Zeitlhofer J, Deecke L, Fritz E, Ludwig H, Gisslinger H. Neurological function during long-term therapy with recombinant interferon alpha. J Neuropsychiatry Clin Neurosci 1999;11(3):343–8.

336. Rohatiner AZ, Prior PF, Burton AC, Smith AT, Balkwill FR, Lister TA. Central nervous system toxicity of interferon. Br J Cancer 1983;47(3):419–22.

337. Scalori A, Apale P, Panizzuti F, Mascoli N, Pioltelli P, Pozzi M, Redaelli A, Roffi L, Mancia G. Depression during interferon therapy for chronic viral hepatitis: early identification of patients at risk by means of a computerized test. Eur J Gastroenterol Hepatol 2000;12(5):505–9.

338. Kraus MR, Schafer A, Faller H, Csef H, Scheurlen M. Psychiatric symptoms in patients with chronic hepatitis C receiving interferon alfa-2b therapy. J Clin Psychiatry 2003;64:708–14.

339. Dieperink E, Willenbring M, Ho SB. Neuropsychiatric symptoms associated with hepatitis C and interferon alpha: A review. Am J Psychiatry 2000;157(6):867–76.

340. Zdilar D, Franco-Bronson K, Buchler N, Locala JA, Younossi ZM. Hepatitis C, interferon alfa, and depression. Hepatology 2000;31(6):1207–11.

341. Renault PF, Hoofnagle JH, Park Y, Mullen KD, Peters M, Jones DB, Rustgi V, Jones EA. Psychiatric complications of long-term interferon alfa therapy. Arch Intern Med 1987;147(9):1577–80.

342. Prasad S, Waters B, Hill PB, et al. Psychiatric side effects of interferon alpha-2b in patients treated for hepatitis C. Clin Res 1992;40:840A.

343. Bonaccorso S, Marino V, Biondi M, Grimaldi F, Ippoliti F, Maes M. Depression induced by treatment with interferon-alpha in patients affected by hepatitis C virus. J Affect Disord 2002;72(3):237–41.

344. Dieperink E, Ho SB, Thuras P, Willenbring ML. A prospective study of neuropsychiatric symptoms associated with interferon-alpha-2b and ribavirin therapy for patients with chronic hepatitis C. Psychosomatics 2003;44(2):104–12.

345. Hauser P, Khosla J, Aurora H, Laurin J, Kling MA, Hill J, Gulati M, Thornton AJ, Schultz RL, Valentine AD, Meyers CA, Howell CD. A prospective study of the incidence and open-label treatment of interferon-induced major depressive disorder in patients with hepatitis C. Mol Psychiatry 2002;7(9):942–7.

346. Horikawa N, Yamazaki T, Izumi N, Uchihara M. Incidence and clinical course of major depression in patients with chronic hepatitis type C undergoing interferon-alpha therapy: a prospective study. Gen Hosp Psychiatry 2003;25(1):34–8.

347. Janssen HL, Brouwer JT, van der Mast RC, Schalm SW. Suicide associated with alfa-interferon therapy for chronic viral hepatitis. J Hepatol 1994;21(2):241–3.

348. Rifflet H, Vuillemin E, Oberti F, Duverger P, Laine P, Garre JB, Cales P. Pulsions suicidaires chez des malades atteints d'hépatite chronique C au cours ou au décours du traitement par l'interféron alpha. [Suicidal impulses in patients with chronic viral hepatitis C during or after therapy with interferon alpha.] Gastroenterol Clin Biol 1998;22(3):353–7.

349. Gohier B, Goeb JL, Rannou-Dubas K, Fouchard I, Cales P, Garre JB. Hepatitis C, alfa interferon, anxiety and depression disorders: a prospective study of 71 patients. World J Biol Psychiatry 2003;4:115–8.

350. Valentine AD, Meyers CA, Kling MA, Richelson E, Hauser P. Mood and cognitive side effects of interferon-alpha therapy. Semin Oncol 1998;25(1 Suppl 1):39–47.

351. Adams F, Quesada JR, Gutterman JU. Neuropsychiatric manifestations of human leukocyte interferon therapy in patients with cancer. JAMA 1984;252(7):938–41.

352. Bocci V. Central nervous system toxicity of interferons and other cytokines. J Biol Regul Homeost Agents 1988;2(3):107–18.

353. Prior TI, Chue PS. Psychotic depression occurring after stopping interferon-alpha. J Clin Psychopharmacol 1999;19(4):385–6.

354. Capuron L, Ravaud A. Prediction of the depressive effects of interferon alfa therapy by the patient's initial affective state. N Engl J Med 1999;340(17):1370.

355. Schaefer M, Schwaiger M, Pich M, Lieb K, Heinz A. Neurotransmitter changes by interferon-alfa and therapeutic implications. Pharmacopsychiatry 2003;36 Suppl 3:S203–6.

356. Capuron L, Neurauter G, Musselman DL, Lawson DH, Nemeroff CB, Fuchs D, Miller AH. Interferon-alfa-induced changes in tryptophan metabolism. Relationship to depression and paroxetine treatment. Biol Psychiatry 2003;54:906–14.

357. Wihlback A-C, Nyberg S, Backstrom T, Bixo M, Sundstrom-Poromaa I. Estradiol and the addition of progesterone increase the sensitivity to a neurosteroid in postmenopausal women. Psychoneuroendocrinology 2005;30:38–50.

358. Schaefer M, Horn M, Schmidt F, Schmid-Wendtner MH, Volkenandt M, Ackenheil M, Mueller N, Schwarz MJ. Correlation between sICAM-1 and depressive symptoms during adjuvant treatment of melanoma with interferon-alpha. Brain Behav Immun 2004;18:555–62.

359. Loftis JM, Socherman RE, Howell CD, Whitehead AJ, Hill JA, Dominitz JA, Hauser P. Association of interferon-alpha-induced depression and improved treatment response in patients with hepatitis C. Neurosci Lett 2004;365:87–91.

360. McDonald EM, Mann AH, Thomas HC. Interferons as mediators of psychiatric morbidity. An investigation in a trial of recombinant alpha-interferon in hepatitis-B carriers. Lancet 1987;2(8569):1175–8.

361. Pavol MA, Meyers CA, Rexer JL, Valentine AD, Mattis PJ, Talpaz M. Pattern of neurobehavioral deficits associated with interferon alfa therapy for leukemia. Neurology 1995;45(5):947–50.

362. Poutiainen E, Hokkanen L, Niemi ML, Farkkila M. Reversible cognitive decline during high-dose alpha-interferon treatment. Pharmacol Biochem Behav 1994;47(4):901–5.

363. Licinio J, Kling MA, Hauser P. Cytokines and brain function: relevance to interferon-alpha-induced mood and cognitive changes. Semin Oncol 1998;25(1 Suppl 1):30–8.

364. Bonaccorso S, Marino V, Puzella A, Pasquini M, Biondi M, Artini M, Almerighi C, Verkerk R, Meltzer H, Maes M. Increased depressive ratings in patients with hepatitis C receiving interferon-alpha-based immunotherapy are related to interferon-alpha-induced changes in the serotonergic system. J Clin Psychopharmacol 2002;22(1):86–90.

365. Gleason OC, Yates WR, Isbell MD, Philipsen MA. An open-label trial of citalopram for major depression in patients with hepatitis C. J Clin Psychiatry 2002;63(3):194–8.

366. Kraus MR, Schafer A, Faller H, Csef H, Scheurlen M. Paroxetine for the treatment of interferon-alpha-induced depression in chronic hepatitis C. Aliment Pharmacol Ther 2002;16(6):1091–9.

367. Pariante CM, Landau S, Carpiniello BCagliari Group. Interferon alfa-induced adverse effects in patients with a psychiatric diagnosis. N Engl J Med 2002;347(2):148–9.

368. Hensley ML, Peterson B, Silver RT, Larson RA, Schiffer CA, Szatrowski TP. Risk factors for severe neuropsychiatric toxicity in patients receiving interferon alfa-2b and low-dose cytarabine for chronic myelogenous leukemia: analysis of Cancer and Leukemia Group B 9013. J Clin Oncol 2000;18(6):1301–8.

369. Van Thiel DH, Friedlander L, Molloy PJ, Fagiuoli S, Kania RJ, Caraceni P. Interferon-alpha can be used successfully in patients with hepatitis C virus-positive chronic hepatitis who have a psychiatric illness. Eur J Gastroenterol Hepatol 1995;7(2):165–8.

370. Adams F, Fernandez F, Mavligit G. Interferon-induced organic mental disorders associated with unsuspected pre-existing neurologic abnormalities. J Neurooncol 1988;6(4):355–9.

371. Meyers CA, Obbens EA, Scheibel RS, Moser RP. Neurotoxicity of intraventricularly administered alpha-interferon for leptomeningeal disease. Cancer 1991;68(1):88–92.

372. Hagberg H, Blomkvist E, Ponten U, Persson L, Muhr C, Eriksson B, Oberg K, Olsson Y, Lilja A. Does alpha-interferon in conjunction with radiotherapy increase the risk of complications in the central nervous system? Ann Oncol 1990;1(6):449.

373. Laaksonen R, Niiranen A, Iivanainen M, Mattson K, Holsti L, Farkkila M, Cantell K. Dementia-like, largely reversible syndrome after cranial irradiation and prolonged interferon treatment. Ann Clin Res 1988;20(3):201–3.

374. Mitsuyama Y, Hashiguchi H, Murayama T, Koono M, Nishi S. An autopsied case of interferon encephalopathy. Jpn J Psychiatry Neurol 1992;46(3):741–8.

375. Raison CL, Borisov AS, Broadwell SD, Capuron L, Woolwine BJ, Jacobson IM, Nemeroff CB, Miller AH. Depression during pegylated interferon-alpha plus ribavirin therapy: prevalence and prediction. J Clin Psychiatry 2005;66:41–8.

376. Loftis JM, Wall JM, Linardatos E, Benvenga S, Hauser P. A quantitative assessment of depression and thyroid dysfunction secondary to interferon-alpha therapy in patients with hepatitis C. J Endocrinol Invest 2004;27:RC16–20.

377. Gochee PA, Powell EE, Purdie DM, Pandeya N, Kelemen L, Shorthouse C, Jonsson JR, Kelly B. Association between apolipoprotein E epsilon4 and neuropsychiatric symptoms during interferon alpha treatment for chronic hepatitis C. Psychosomatics 2004;45:49–57.

378. Maes M, Bonaccorso S. Lower activities of serum peptidases predict higher depressive and anxiety levels following interferon-alpha-based immunotherapy in patients with hepatitis C. Acta Psychiatr Scand 2004;109:126–31.

379. Mulder RT, Ang M, Chapman B, Ross A, Stevens IF, Edgar C. Interferon treatment is not associated with a worsening of psychiatric symptoms in patients with hepatitis C. J Gastroenterol Hepatol 2000;15(3):300–3.

380. Pariante CM, Orru MG, Baita A, Farci MG, Carpiniello B. Treatment with interferon-alpha in patients with chronic hepatitis and mood or anxiety disorders. Lancet 1999;354(9173):131–2.

381. Castera L, Zigante F, Bastie A, Buffet C, Dhumeaux D, Hardy P. Incidence of interferon alfa-induced depression in patients with chronic hepatitis C. Hepatology 2002;35(4):978–9.

382. Schaefer M, Schmidt F, Folwaczny C, Lorenz R, Martin G, Schindlbeck N, Heldwein W, Soyka M, Grunze H, Koenig A, Loeschke K. Adherence and mental side effects during hepatitis C treatment with interferon alfa and ribavirin in psychiatric risk groups. Hepatology 2003;37:443–51.

383. Hauser P. Neuropsychiatric side effects of HCV therapy and their treatment: focus on IFN alpha-induced depression. Gastroenterol Clin North Am 2004;33 (1 Suppl):S35–50.

384. Loftis JM, Hauser P. The phenomenology and treatment of interferon-induced depression. J Affect Disord 2004;82:175–90.

385. Raison CL, Demetrashvili M, Capuron L, Miller AH. Neuropsychiatric adverse effects of interferon-alpha: recognition and management. CNS Drugs 2005;19:105–23.

386. Valentine AD. Managing the neuropsychiatric adverse effects of interferon treatment. BioDrugs 1999;11:229–37.

387. Miller A, Musselman D, Pena S, Su C, Pearce B, Nemeroff C. Pretreatment with the antidepressant paroxetine, prevents cytokine-induced depression during IFN-alpha therapy for malignant melanoma. Neuroimmunomodulation 1999;6:237.

388. Musselman DL, Lawson DH, Gumnick JF, Manatunga AK, Penna S, Goodkin RS, Greiner K, Nemeroff CB, Miller AH. Paroxetine for the prevention of depression induced by high-dose interferon alfa. N Engl J Med 2001;344(13):961–6.

389. McAllister-Williams RH, Young AH, Menkes DB. Antidepressant response reversed by interferon. Br J Psychiatry 2000;176:93.

390. Maddock C, Baita A, Orru MG, Sitzia R, Costa A, Muntoni E, Farci MG, Carpiniello B, Pariante CM. Psychopharmacological treatment of depression, anxiety, irritability and insomnia in patients receiving interferon-alpha: a prospective case series and a discussion of biological mechanisms. J Psychopharmacol 2004;18:41–6.

391. The IFNB Multiple Sclerosis Study GroupThe University of British Columbia MS/MRI Analysis Group. Interferon beta-1b in the treatment of multiple sclerosis: final outcome of the randomized controlled trial. Neurology 1995;45(7):1277–85.

392. Neilley LK, Goodin DS, Goodkin DE, Hauser SL. Side effect profile of interferon beta-1b in MS: results of an open label trial. Neurology 1996;46(2):552–4.

393. Lublin FD, Whitaker JN, Eidelman BH, Miller AE, Arnason BG, Burks JS. Management of patients receiving interferon beta-1b for multiple sclerosis: report of a consensus conference. Neurology 1996;46(1):12–8.

394. Ebers GC, Hommes O, Hughes RAC, et alPRISMS (Prevention of Relapses and Disability by Interferon beta-1a Subcutaneously in Multiple Sclerosis) Study Group. Randomised double-blind placebo-controlled study of interferon beta-1a in relapsing/remitting multiple sclerosis. Lancet 1998;352(9139):1498–504.

395. European Study Group on Interferon Beta-1b in Secondary Progressive MS. Placebo-controlled multicentre randomised trial of interferon beta-1b in treatment of secondary progressive multiple sclerosis. Lancet 1998;352(9139):1491–7.

396. Feinstein A. Multiple sclerosis, disease modifying treatments and depression: a critical methodological review. Mult Scler 2000;6(5):343–8.

397. Patten SB, Metz LMSPECTRIMS Study Group. Interferon beta1a and depression in secondary progressive MS: data from the SPECTRIMS Trial. Neurology 2002;59(5):744–6.

398. Zephir H, De Seze J, Stojkovic T, Delisse B, Ferriby D, Cabaret M, Vermersch P. Multiple sclerosis and depression: influence of interferon beta therapy. Mult Scler 2003;9(3):284–8.

399. Feinstein A, O'Connor P, Feinstein K. Multiple sclerosis, interferon beta-1b and depression. A prospective investigation. J Neurol 2002;249(7):815–20.

400. Exton MS, Baase J, Pithan V, Goebel MU, Limmroth V, Schedlowski M. Neuropsychological performance and mood states following acute interferon-beta-1b administration in healthy males. Neuropsychobiology 2002;45(4):199–204.

401. Borras C, Rio J, Porcel J, Barrios M, Tintore M, Montalban X. Emotional state of patients with relapsing-remitting MS treated with interferon beta-1b. Neurology 1999;52(8):1636–9.

402. Mohr DC, Likosky W, Dwyer P, Van Der Wende J, Boudewyn AC, Goodkin DE. Course of depression during the initiation of interferon beta-1a treatment for multiple sclerosis. Arch Neurol 1999;56(10):1263–5.

403. Hosoda S, Takimura H, Shibayama M, Kanamura H, Ikeda K, Kumada H. Psychiatric symptoms related to interferon therapy for chronic hepatitis C: clinical features and prognosis. Psychiatry Clin Neurosci 2000;54(5):565–72.

404. Goeb JL, Cailleau A, Lainé P, Etcharry-Bouyx F, Maugin D, Duverger P, Gohier B, Rannou-Dubas K, Dubas F, Garre JB. Acute delirium, delusion, and depression during IFN-beta-1a therapy for multiple sclerosis: a case report. Clin Neuropharmacol 2003;26:5–7.

405. Filippini G, Munari L, Incorvaia B, Ebers GC, Polman C, D'Amico R, Rice GP. Interferons in relapsing remitting multiple sclerosis: a systematic review. Lancet 2003;361:545–52.

406. Patten SB, Fridhandler S, Beck CA, Metz LM. Depressive symptoms in a treated multiple sclerosis cohort. Mult Scler 2003;9:616–20.

407. Patten SB, Francis G, Metz LM, Lopez-Bresnahan M, Chang P, Curtin F. The relationship between depression and interferon beta-1a therapy in patients with multiple sclerosis. Mult Scler 2005;11:175–81.

408. Moor CC, Berwanger C, Welter FL. Visual pseudo-hallucinations in interferon-beta 1a therapy. Akt Neurol 2002;29:355–7.

409. Mattson K, Niiranen A, Pyrhonen S, Farkkila M, Cantell K. Recombinant interferon gamma treatment in non-small cell lung cancer. Antitumour effect and cardiotoxicity. Acta Oncol 1991;30(5):607–10.

410. Born J, Spath-Schwalbe E, Pietrowsky R, Porzsolt F, Fehm HL. Neurophysiological effects of recombinant interferon-gamma and -alpha in man. Clin Physiol Biochem 1989;7(3–4):119–27.

411. Todd PA, Goa KL. Interferon gamma-1b. A review of its pharmacology and therapeutic potential in chronic granulomatous disease. Drugs 1992;43(1):111–22.

412. Capuron L, Ravaud A, Miller AH, Dantzer R.Baseline mood and psychosocial characteristics of patients developing depressive symptoms during interleukin-2 and/or interferon-alfa cancer therapy. Brain Behav Immun 2004;18:205–13.

413. Mesurolle B, Ariche M, Cohen D. Premedication before i.v. contrast-enhanced CT resulting in steroid-induced psychosis Am J Roentgenol 2002;178(3):766–7.

414. Nyegaard and Co. Summarising notes from the Amipaque Symposium 1977;.

415. Richert S, Sartor K, Holl B. Subclinical organic psychosyndromes on intrathecal injection of metrizamide for lumbar myelography. Neuroradiology 1979;18(4):177–84.

416. Iliopoulou A, Giannakopoulos G, Goutou P, Pagou H, Stamatelopoulos S. Visual hallucinations due to radiocontrast media. Report of two cases and review of the literature. Br J Clin Pharmacol 1999;47(2):226–7.

417. Hammeke TA, Haughton VM, Grogan JP, et al. A preliminary study of cognitive and affective alterations following intrathecal administration of iopamidol or metrizamide. Invest Radiol 1984;19(Suppl):S268.

418. Jennett WB, Barker J, Fitch W, McDowall DG. Effect of anaesthesia on intracranial pressure in patients withoccupying lesions. Lancet 1969;1(7585):61–4.

419. Brady AP, Hough DM, Lo R, Gill G. Transient global amnesia after cerebral angiography with iohexol. Can Assoc Radiol J 1993;44(6):450–2.

420. Iselin-Chaves IA, Willems SJ, Jermann FC, Forster A, Adam SR, Van der Linden M. Investigation of implicit memory during isoflurane anaesthesia for elective surgery using the process dissociation procedure. Anesthesiology. 2005;103(5):925–33.

421. Mandell GL, Sande MA. Antimicrobial agents: drugs used in the chemotherapy of tuberculosis and leprosy. In: Goodman Gilman A, Rall TW, Nies AS, Taylor P, editors. Goodman and Gilman's The Pharmacological Basis of Therapeutics. 8th ed.. New York: Pergamon Press, 1990:241 Chapter 49.

422. Bhatia MS. Isoniazid-induced obsessive compulsive neurosis. J Clin Psychiatry 1990;51(9):387.

423. Bernardo M, Gatell JM, Parellada E. Acute exacerbation of chronic schizophrenia in a patient treated with antituberculosis drugs. Am J Psychiatry 1991;148(10):1402.

424. Siskind MS, Thienemann D, Kirlin L. Isoniazid-induced neurotoxicity in chronic dialysis patients: report of three cases and a review of the literature. Nephron 1993;64(2):303–6.

425. Cheung WC, Lo CY, Lo WK, Ip M, Cheng IK. Isoniazid induced encephalopathy in dialysis patients. Tuber Lung Dis 1993;74(2):136–9.

426. Cleveland KO, Campbell JW. Hallucinations associated with itraconazole therapy. Clin Infect Dis 1995;21(2):456.

427. Pfenninger EG, Durieux ME, Himmelseher S. Cognitive impairment after small-dose ketamine isomers in comparison to equianalgesic racemic ketamine in human volunteers. Anesthesiology 2002;96(2):357–66.

428. Adler CM, Goldberg TE, Malhotra AK, Pickar D, Breier A. Effects of ketamine on thought disorder, working memory, and semantic memory in healthy volunteers. Biol Psychiatry 1998;43(11):811–6.

429. Tarnow J, Hess W. Pulmonale Hypertonie und Lungenödem nach Ketamin. [Pulmonary hypertension and pulmonary edema caused by intravenous ketamine.] Anaesthesist 1978;27(10):486–7.

430. Klausen NO, Wiberg-Jorgensen F, Chraemmer-Jorgensen B. Psychomimetic reactions after low-dose ketamine infusion. Comparison with neuroleptanaesthesia. Br J Anaesth 1983;55(4):297–301.

431. Enarson MC, Hays H, Woodroffe MA. Clinical experience with oral ketamine. J Pain Symptom Manage 1999;17(5):384–6.

432. Rabben T. Effects of the NMDA receptor antagonist ketamine in electrically induced A delta-fiber pain. Methods Find Exp Clin Pharmacol 2000;22(3):185–9.

433. Bowdle TA, Radant AD, Cowley DS, Kharasch ED, Strassman RJ, Roy-Byrne PP. Psychedelic effects of ketamine in healthy volunteers: relationship to steady-state plasma concentrations. Anesthesiology 1998;88(1):82–8.

434. Honey GD, Honey RA, Sharar SR, Turner DC, Pomarol-Clotet E, Kumaran D, Simons JS, Hu X, Rugg MD, Bullmore ET, Fletcher PC. Impairment of specific episodic memory processes by sub-psychotic doses of ketamine: the effects of levels of processing at encoding and of the subsequent retrieval task. Psychopharmacology (Berl) 2005;181(3):445–57.

435. McCartney C, Sinha A, Katz J. A qualitative systematic review of N-methyl-D-aspartate receptor antagonists in preventive analgesia. Anesth Analg 2004;98:1385–400.

436. Grace RF. The effect of variable dose diazepam on dreaming and emergence phenomena in 400 cases of ketamine–fentanyl anaesthesia. Anaesthesia 2003;58:904–10.

437. Lahti AC, Warfel D, Michaelidis T, Weiler MA, Frey K, Tamminga CA. Long-term outcome of patients who receive ketamine during research. Biol Psychiatry 2001;49(10):869–75.

438. Krystal JH, Karper LP, Bennett A, D'Souza DC, Abi-Dargham A, Morrissey K, Abi-Saab D, Bremner JD, Bowers MB Jr, Suckow RF, Stetson P, Heninger GR, Charney DS. Interactive effects of subanesthetic ketamine and subhypnotic lorazepam in humans. Psychopharmacology (Berl) 1998;135(3):213–29.

439. Sherwin TS, Green SM, Khan A, Chapman DS, Dannenberg B. Does adjunctive midazolam reduce recovery agitation after ketamine sedation for pediatric procedures? A randomized, double-blind, placebo-controlled trial. Ann Emerg Med 2000;35(3):229–38.

440. Anand A, Charney DS, Oren DA, Berman RM, Hu XS, Cappiello A, Krystal JH. Attenuation of the neuropsychiatric effects of ketamine with lamotrigine: support for hyperglutamatergic effects of N-methyl-D-aspartate receptor antagonists. Arch Gen Psychiatry 2000;57(3):270–6.

441. Lauretti GR, Ramos MBP, De Mattos AL, De Oliveira AC. Emergence phenomena after ketamine anesthesia: influence of a selected recorded tape and midazolam. Rev Bras Anesthesiol 1996;46:329–34.

442. Meador KJ, Loring DW, Ray PG, Murro AM, King DW, Perrine KR, Vazquez BR, Kiolbasa T. Differential cognitive and behavioral effects of carbamazepine and lamotrigine. Neurology 2001;56(9):1177–82.

443. Margolese HC, Beauclair L, Szkrumelak N, Chouinard G. Hypomania induced by adjunctive lamotrigine. Am J Psychiatry 2003;160:183–4.

444. Chan Y-C, Miller KM, Shaheen N, Votolato NA, Hankins MB. Worsening of psychotic symptoms in schizophrenia with addition of lamotrigine: a case report. Schizophr Res 2005;78:343–5.

445. Clemens B. Forced normalisation precipitated by lamotrigine. Seizure 2005;14:485–9.

446. Hsu WH. Toxicity and drug interactions of levamisole. J Am Vet Med Assoc 1980;176(10 Spec No):1166–9.

447. Jeffries JJ, Cammisuli S. Psychosis secondary to long-term levamisole therapy. Ann Pharmacother 1998;32(1):134–5.

448. Kossoff EH, Bergey GK, Freeman JM, Vining EP. Levetiracetam psychosis in children with epilepsy. Epilepsia 2001;42(12):1611–3.

449. Mula M, Trimble MR, Sander JW. Psychiatric adverse events in patients with epilepsy and learning disabilities taking levetiracetam. Seizure 2004;13(1):55–7.

450. De Los Reyes EC, Sharp GB, Williams JP, Hale SE. Levetiracetam in the treatment of Lennox–Gastaut syndrome. Pediatr Neurol 2004;30(4):254–6.

451. Alsaadi TM, Koopmans S, Apperson M, Farias S. Levetiracetam monotherapy for elderly patients with epilepsy. Seizure 2004;13(1):58–60.

452. Mula M, Trimble MR, Yuen A, Liu RS, Sander JW. Psychiatric adverse events during levetiracetam therapy. Neurology 2003;61:704–6.

453. Arroyo S, Crawford P. Safety profile of levetiracetam. Epileptic Disord 2003;5 Suppl 1:S57–63.

454. Sadek AH, Fix A, French J. Levetiracetam related adverse events: a post-marketing study. Epilepsia 2002;43 Suppl 8:151.

455. Gandon JM, Allain H. Lack of effect of single and repeated doses of levocetirizine, a new antihistamine drug, on cognitive and psychomotor functions in healthy volunteers. Br J Clin Pharmacol 2002;54(1):51–8.

456. Hindmarch I, Johnson S, Meadows R, Kirkpatrick T, Shamsi Z. The acute and sub-chronic effects of levocetirizine, cetirizine, loratadine, promethazine and placebo on cognitive function, psychomotor performance, and weal and flare. Curr Med Res Opin 2001;17(4):241–55.

457. Winkelman AC, DiPalma JR. Drug treatment of parkinsonism. Semin Drug Treatm 1972;1:10.

458. Soliman IE, Park TS, Berkelhamer MC. Transient paralysis after intrathecal bolus of baclofen for the treatment of post-selective dorsal rhizotomy pain in children. Anesth Analg 1999;89(5):1233–5.

459. Vazquez A, Jimenez-Jimenez FJ, Garcia-Ruiz P, Garcia-Urra D. "Panic attacks" in Parkinson's disease. A long-term complication of levodopa therapy. Acta Neurol Scand 1993;87(1):14–8.

460. Presthus J, Holmsen R. Appraisal of long-term levodopa treatment of parkinsonism with special reference to therapy limiting factors. Acta Neurol Scand 1974;50(6):774–90.

461. Giovannoni G, O'Sullivan JD, Turner K, Manson AJ, Lees AJ. Hedonistic homeostatic dysregulation in patients with Parkinson's disease on dopamine replacement therapies. J Neurol Neurosurg Psychiatry 2000;68(4):423–8.

462. Goetz CG, Leurgans S, Pappert EJ, Raman R, Stemer AB. Prospective longitudinal assessment of hallucinations in Parkinson's disease. Neurology 2001;57(11):2078–82.

463. Sommer BR, Wise LC, Kraemer HC. Is dopamine administration possibly a risk factor for delirium? Crit Care Med 2002;30(7):1508–11.

464. Marsch SC, Schaefer HG, Castelli I. Unusual psychological manifestation of systemic local anesthetic toxicity. Anesthesiology 1998;88(2):531–3.

465. Muldoon MF, Barger SD, Ryan CM, Flory JD, Lehoczky JP, Matthews KA, Manuck SB. Effects of lovastatin on cognitive function and psychological well-being. Am J Med 2000;108(7):538–46.

466. Gomez-Gil E, Garcia F, Pintor L, Martinez JA, Mensa J, de Pablo J. Clarithromycin-induced acute psychoses in peptic ulcer disease. Eur J Clin Microbiol Infect Dis 1999;18(1):70–1.

467. Geiderman JM. Central nervous system disturbances following clarithromycin ingestion. Clin Infect Dis 1999;29(2):464–5.

468. Cone LA, Padilla L, Potts BE. Delirium in the elderly resulting from azithromycin therapy. Surg Neurol 2003;59:509–11.

469. Navarro V, Vieta E, Gasto C. Mannitol-induced acute manic state. J Clin Psychiatry 2001;62(2):126.

470. Wakefield AJ, Murch SH, Anthony A, Linnell J, Casson DM, Malik M, Berelowitz M, Dhillon AP, Thomson MA, Harvey P, Valentine A, Davies SE, Walker-Smith JA. Ileal-lymphoid-nodular hyperplasia, non-specific colitis, and pervasive developmental disorder in children. Lancet 1998;351(9103):637–41.

471. Black D, Prempeh H, Baxter T. Autism, inflammatory bowel disease, and MMR vaccine. Lancet 1998;351(9106):905–6.

472. Beale AJ. Autism, inflammatory bowel disease, and MMR vaccine. Lancet 1998;351(9106):906.

473. O'Brien SJ, Jones IG, Christie P. Autism, inflammatory bowel disease, and MMR vaccine. Lancet 1998;351(9106):906–7.

474. Author's reply Wakefield AJ. Lancet 1998;351(9106):908.

475. Editor's reply Horton R. Lancet 1998;351(9106):908.

476. Horton R. The lessons of MMR. Lancet 2004;363(9411):747–9.

477. Wakefield AJ, Harvey P, Linnell J. MMR—responding to retraction. Lancet 1998;351(9112):1356.

478. Anonymous. Measles, MMR, and autism: the confusion continues. Lancet 2000;355(9213):1379.

479. Wakefield AJ, Montgomery SM. Measles, mumps, rubella vaccine: through a glass, darkly. Adverse Drug React Toxicol Rev 2000;19(4):265–83.

480. Elliman D, Bedford H. MMR vaccine: the continuing saga. BMJ 2001;322(7280):183–4.

481. Letter from the Chief Medical Officer, the Chief Nursing Officer, and the Chief Pharmaceutical Officer, Department of Health, London. Current vaccine and immunisation issues. 1. MMR vaccine. http://www.doh.gov.uk/cmo/cmoh.htm.

482. BBC News. MMR research doctor resigns. http://news.bbc.co.uk/1/hi/health/1687967.stm.

483. WHO statement on MMR vaccine, 25 January 2001.

484. Immunization Safety Review Committee. Immunization safety review: measles–mumps–rubella vaccine and autism. In: Stratton K, Gable A, Shetty P, McCormick M, editors. Institute of MedicineNational Academy of Sciences, 2001:241.

485. Halsey NA, Hyman SLConference Writing Panel. Measles–mumps–rubella vaccine and autistic spectrum disorder: report from the New Challenges in Childhood Immunizations Conference convened in Oak Brook. Illinois, June 12–13, 2000Pediatrics 2001;107(5):E84.

486. Duclos P, Ward BJ. Measles vaccines: a review of adverse events. Drug Saf 1998;19(6):435–54.

487. Fudenberg HH. Dialysable lymphocyte extract (DLyE) in infantile onset autism: a pilot study. Biotherapy 1996;9(1–3):143–7.

488. Lee JW, Melgaard B, Clements CJ, Kane M, Mulholland EK, Olive JM. Autism, inflammatory bowel disease, and MMR vaccine. Lancet 1998;351(9106):905.

489. Immunizations Practices Advisory Committee (ACIP). Recommendations on BCG vaccines. MMWR Morb Mortal Wkly Rep 1979;28:241.

490. Global Advisory Group of the Expanded Programme on Immunization (EPI). Report on the meeting 13–17 October, 1986. New Delhi. Unedited document. WHO/87/1, 1986.

491. Corti Ortiz D, Rivera Garay P, Aviles Jasse J, Hidalgo Carmona F, MacMillan Soto G, Coz Canas LF, Vargas Delaunoy R, Susaeta Saenz de San Pedro R. Profilaxis del cancer vesical superficial con 1 mg de BCG endovesical: comparacion con otras dosis. [Prophylaxis of superficial

bladder cancer with 1 mg of intravesical BCG: comparison with other doses.] Actas Urol Esp 1993;17(4):239–42.

492. Wing L. Autistic spectrum disorders. BMJ 1996;312(7027):327–8.

493. Fombonne E. Inflammatory bowel disease and autism. Lancet 1998;351(9107):955.

494. Department of Health, England and Wales. MMR vaccine is not linked to Crohn's disease or autism: conclusion of an expert scientific seminar. London: Press release 98/109, 24 March 1998.

495. Farrington CP, Miller E, Taylor B. MMR and autism: further evidence against a causal association. Vaccine 2001;19(27):3632–5.

496. The California–Oregon Unvaccinated Children Survey. <www.GenerationRescue.org>.

497. Wander HE, Nagel GA, Blossey HC, Kleeberg U. Aminoglutethimide and medroxyprogesterone acetate in the treatment of patients with advanced breast cancer. A phase II study of the Association of Medical Oncology of the German Cancer Society (AIO). Cancer 1986;58(9):1985–9.

498. Chamberland M, Duperval R, Marcoux JA, Dube P, Pigeon N. Severe falciparum malaria in nonendemic areas: an unrecognized medical emergency. CMAJ 1991;144(4):455–8.

499. Speich R, Haller A. Central anticholinergic syndrome with the antimalarial drug mefloquine. N Engl J Med 1994;331(1):57–8.

500. Havaldar PV, Mogale KD. Mefloquine-induced psychosis. Pediatr Infect Dis J 2000;19(2):166–7.

501. Borruat FX, Nater B, Robyn L, Genton B. Prolonged visual illusions induced by mefloquine (Lariam): a case report. J Travel Med 2001;8(3):148–9.

502. Javorsky DJ, Tremont G, Keitner GI, Parmentier AH. Cognitive and neuropsychiatric side effects of mefloquine. J Neuropsychiatry Clin Neurosci 2001;13(2):302.

503. Whitworth AB, Aichhorn W. First time diagnosis of depression. Induced by Mefloquine? J Clin Pyschopharmacol 2005;25(4):399–400.

504. Petersen E, Ronne T, Ronn A, Bygbjerg I, Larsen SO. Reported side effects to chloroquine, chloroquine plus proguanil, and mefloquine as chemoprophylaxis against malaria in Danish travelers. J Travel Med 2000;7(2):79–84.

505. Potasman I, Beny A, Seligmann H. Neuropsychiatric problems in 2,500 long-term young travelers to the tropics. J Travel Med 2000;7(1):5–9.

506. Schwartz E, Potasman I, Rotenberg M, Almog S, Sadetzki S. Serious adverse events of mefloquine in relation to blood level and gender. Am J Trop Med Hyg 2001;65(3):189–92.

507. Huzly D, Schonfeld C, Beuerle W, Bienzle U. Malaria chemoprophylaxis in German tourists: a prospective study on compliance and adverse reactions. J Travel Med 1996;3(3):148–55.

508. Phillips MA, Kass RB. User acceptability patterns for mefloquine and doxycycline malaria chemoprophylaxis. J Travel Med 1996;3(1):40–5.

509. Schlagenhauf P, Steffen R, Lobel H, Johnson R, Letz R, Tschopp A, Vranjes N, Bergqvist Y, Ericsson O, Hellgren U, Rombo L, Mannino S, Handschin J, Sturchler D. Mefloquine tolerability during chemoprophylaxis: focus on adverse event assessments, stereochemistry and compliance. Trop Med Int Health 1996;1(4):485–94.

510. van Riemsdijk MM, Sturkenboom MC, Ditters JM, Ligthelm RJ, Overbosch D, Stricker BH. Atovaquone

plus chloroguanide versus mefloquine for malaria prophylaxis: a focus on neuropsychiatric adverse events. Clin Pharmacol Ther 2002;72(3):294–301.

511. Kukoyi O, Carney CP. Curses, madness, and mefloquine. Psychosomatics 2003;44:339–41.

512. Cayley WE Jr. Mefloquine for preventing malaria in non-immune adult travelers. Am Fam Phys 2004;69 (3):521–2.

513. Meier CR, Wilcock K, Jick SS. The risk of severe depression, psychosis or panic attacks with prophylactic antimalarials. Drug Saf 2004;27 (3):203–13.

514. Dalton EJ, Rotondi D, Levitan RD, Kennedy SH, Brown GM. Use of slow-release melatonin in treatment-resistant depression. J Psychiatry Neurosci 2000;25(1):48–52.

515. Herxheimer A, Petrie KJ. Melatonin for preventing and treating jet lag. Cochrane Database Syst Rev 2001;(1):CD001520.

516. Riederer P, Lange KW, Kornhuber J, Danielczyk W. Pharmacotoxic psychosis after memantine in Parkinson's disease. Lancet 1991;338(8773):1022–3.

517. Hessel L. Mercury in vaccines. Bull Acad Natl Med 2003;187:1501–10.

518. Novak M, Klezlova V. [Allergy to merthiolate (Thimerosalum) in a set of standard epicutaneous tests in patients with eczematous diseases and leg ulcer during three periods between 1979 and 1999.]Cesko-Slov Dermatol 2000;75:3–10.

519. Bigazzi PE. Metals and kidney autoimmunity. Environ Health Perspect 1999;107(Suppl 5):753–65.

520. Hviid A, Stellfeld M, Wohfahrts J, Melbye M. Association between thiomersal-containing vaccine and autism. J Am Med Assoc 2003;13:1763–6.

521. Verstraetan T, Davis RL, DeStefano F, Lieu TA, Rhodes PH, Black SB, Shinefield H, Chen RT. Safety of thiomersal containing vaccines: a two phased study of computerised health maintenance organisation data bases. Pediatrics 2003;112:1039–48.

522. Geier MR, Geier DA. Neurodevelopmental disorders after thiomersal containing vaccines: a brief communication. Exp Biol Med 2003;228:660–4.

523. Weinstein M, Bernstein S. Pink ladies: mercury poisoning in twin girls. Can Med Assoc J 2003;168:201.

524. Curran HV, Kleckham J, Bearn J, Strang J, Wanigaratne S. Effects of methadone on cognition, mood and craving in detoxifying opiate addicts: a dose-response study. Psychopharmacology (Berl) 2001;154(2):153–60.

525. Mintzer MZ, Stitzer ML. Cognitive impairment in methadone maintenance patients. Drug Alcohol Depend 2002;67(1):41–51.

526. Schagen SB, van Dam FS, Muller MJ, Boogerd W, Lindeboom J, Bruning PF. Cognitive deficits after postoperative adjuvant chemotherapy for breast carcinoma. Cancer 1999;85(3):640–50.

527. Rau R, Schleusser B, Herborn G, Karger T. Longterm combination therapy of refractory and destructive rheumatoid arthritis with methotrexate (MTX) and intramuscular gold or other disease modifying antirheumatic drugs compared to MTX monotherapy. J Rheumatol 1998;25(8):1485–92.

528. Wernick R, Smith DL. Central nervous system toxicity associated with weekly low-dose methotrexate treatment. Arthritis Rheum 1989;32(6):770–5.

529. Dhondt JL, Farriaux JP, Millot F, Taret S, Hayte JM, Mazingue F. Methotrexate a haute close et hyperphenylalaninennie. [High-dose methotrexate and hyperphenylalaninemia.] Arch Fr Pediatr 1991;48(4):249–51.

530. Koizumi HM. Obsessive-compulsive symptoms following stimulants. Biol Psychiatry 1985;20(12):1332–3.

531. Kouris S. Methylphenidate-induced obsessive-compulsiveness. J Am Acad Child Adolesc Psychiatry 1998;37(2):135.

532. Janowsky DS, el-Yousel MK, Davis JM, Sekerke HJ. Provocation of schizophrenic symptoms by intravenous administration of methylphenidate. Arch Gen Psychiatry 1973;28(2):185–91.

533. Persyko I. Psychiatric adverse reactions to methysergide. J Nerv Ment Dis 1972;154(4):299–301.

534. Berman RM, Narasimhan M, Miller HL, Anand A, Cappiello A, Oren DA, Heninger GR, Charney DS. Transient depressive relapse induced by catecholamine depletion: potential phenotypic vulnerability marker? Arch Gen Psychiatry 1999;56(5):395–403.

535. McCann UD, Penetar DM, Belenky G. Panic attacks in healthy volunteers treated with a catecholamine synthesis inhibitor. Biol Psychiatry 1991;30(4):413–6.

536. Lu ML, Pan JJ, Teng HW, Su KP, Shen WW. Metoclopramide-induced supersensitivity psychosis. Ann Pharmacother 2002;36(9):1387–90.

537. Mahl TC, Ummadi S. Metronidazole and mental confusion. J Clin Gastroenterol 2003;36:373–4.

538. Cohen J. Antifungal chemotherapy. Lancet 1982;2(8297):532–7.

539. Cohen PR. Medication-associated depersonalizing symptoms: report of transient depersonalization symptoms induced by minocycline. South Med J 2004;97:70–3.

540. Kalso E, Vainio A. Hallucinations during morphine but not during oxycodone treatment. Lancet 1988;2(8616):912.

541. D'Souza M. Unusual reaction to morphine. Lancet 1987;2(8550):98.

542. Jellema JG. Hallucination during sustained-release morphine and methadone administration. Lancet 1987;2(8555):392.

543. Kamboj SK, Tookman A, Jones L, Curran HV. The effects of immediate-release morphine on cognitive functioning in patients receiving chronic opioid therapy in palliative care. Pain 2005;117(3):388–95.

544. Kelly BD, Gavin BE, Clarke M, Lane A, Larkin C. Nutmeg and psychosis. Schizophr Res 2003;60:95–6.

545. Camacho A, Matthews SC, Dimsdale JE. "Invisible" synthetic opiates and acute psychosis. N Engl J Med 2001;345(6):469.

546. Mougeot G, Hugues FC. Bouffées délirantes confuso–oniriques provoquées par l'acide nalidixique. [Acute confusional–hallucinatory states caused by nalidixic acid.] Nouv Presse Méd 1980;9(7):455.

547. Maremmani I, Marini G, Fornai F. Naltrexone-induced panic attacks. Am J Psychiatry 1998;155(3):447.

548. Amraoui A, Burgos V, Baron P, Alexandre JY. Psychose delirante aiguë au cours d'un traitement par chlorhydrate de naltrexone. [Acute delirium psychosis induced by naltrexone chlorhydrate.] Presse Méd 1999;28(25):1361–2.

549. Miotto K, McCann M, Basch J, Rawson R, Ling W. Naltrexone and dysphoria: fact or myth? Am J Addict 2002;11(2):151–60.

550. Ritter AJ. Naltrexone in the treatment of heroin dependence: relationship with depression and risk of overdose. Aust NZ J Psychiatry 2002;36(2):224–8.

551. Wysenbeek AJ, Klein Z, Nakar S, Mane R. Assessment of cognitive function in elderly patients treated with naproxen. A prospective study. Clin Exp Rheumatol 1988;6(4):399–400.

552. Wise ME, Mistry K, Reid S. Drug points: Neuropsychiatric complications of nevirapine treatment. BMJ 2002;324(7342):879.

553. Jackson M. Cerebral arterial narrowing with nicotine patch. Lancet 1993;342(8865):236–7.

554. Foulds J, Toone B. A case of nicotine psychosis? Addiction 1995;90(3):435–7.

555. Rosenthal R. Visual hallucinations and suicidal ideation attributed to isosorbide dinitrate. Psychosomatics 1987;28(10):555–6.

556. D'Arcy PF. Nitrofurantoin. Drug Intell Clin Pharm 1985;19(7–8):540–7.

557. Carney MW. Paranoid psychosis with indomethacin. BMJ 1977;2(6093):994–5.

558. Mallet L, Kuyumjian J. Indomethacin-induced behavioral changes in an elderly patient with dementia. Ann Pharmacother 1998;32(2):201–3.

559. Bruce-Jones PN, Crome P, Kalra L. Indomethacin and cognitive function in healthy elderly volunteers. Br J Clin Pharmacol 1994;38(1):45–51.

560. Stewart WF, Kawas C, Corrada M, Metter EJ. Risk of Alzheimer's disease and duration of NSAID use. Neurology 1997;48(3):626–32.

561. Kundu AK. Norfloxacin-induced hallucination—an unusual CNS toxicity of 4-fluoroquinolones. J Assoc Physicians India 2000;48(9):944.

562. Heckmann JG, Birklein F, Neundorfer B. Omeprazole-induced delirium. J Neurol 2000;247(1):56–7.

563. MacEvilly M, Carroll CO. Hallucination repression after epidural buprenorphine. BMJ 1989;298:928.

564. Anonymous. Hallucinaties door oxolamine (Bredon). Bull Bijwerk Geneesmd 1986;2:10–1.

565. McEwen J, Meyboom RH, Thijs I. Hallucinations in children caused by oxolamine citrate. Med J Aust 1989;150(8):449–50.

566. Valsecia ME, Malgor LA, Espindola JH, Carauni DH. New adverse effect of oxybutynin: "night terror". Ann Pharmacother 1998;32(4):506.

567. Sommer BR, O'Hara R, Askari N, Kraemer HC, Kennedy WA. The effect of oxybutynin treatment on cognition in children with diurnal incontinence. J Urol 2005;173:2125–7.

568. McCann UD, Slate SO, Geraci M, Roscow-Terrill D, Uhde TW. A comparison of the effects of intravenous pentagastrin on patients with social phobia, panic disorder and healthy controls. Neuropsychopharmacology 1997;16(3):229–37.

569. de Leeuw AS, Den Boer JA, Slaap BR, Westenberg HG. Pentagastrin has panic-inducing properties in obsessive compulsive disorder. Psychopharmacology (Berl) 1996;126(4):339–44.

570. Lines C, Challenor J, Traub M. Cholecystokinin and anxiety in normal volunteers: an investigation of the anxiogenic properties of pentagastrin and reversal by the cholecystokinin receptor subtype B antagonist L-365,260. Br J Clin Pharmacol 1995;39(3):235–42.

571. Gradon JD, Fricchione L, Sepkowitz D. Severe hypomagnesemia associated with pentamidine therapy. Rev Infect Dis 1991;13(3):511–2.

572. Heaney RM, Gotlieb N. Granulocytopenia after intravenous abuse of pentazocine and tripelennamine ("Ts and blues"). South Med J 1983;76(5):654–6.

573. Lee SH, Liu CY, Yang YY. Schizophreniform-like psychotic disorder induced by phentermine: a case report Zhonghua Yi Xue Za Zhi . (Taipei) 1998;61(1):44–7.

574. Linnoila M, Seppala M, Mattila MJ. Acute effect of antipyretic analgesics alone or in combination with alcohol on human psychomotor skills related to driving. Br J Clin Pharmacol 1974;1:477.

575. Cornelius JR, Soloff PH, Reynolds CF 3rd. Paranoia, homicidal behavior, and seizures associated with phenylpropanolamine. Am J Psychiatry 1984;141(1):120–1.

576. Marshall RD, Douglas CJ. Phenylpropanolamine-induced psychosis: Potential predisposing factors. Gen Hosp Psychiatry 1994;16(5):358–60.

577. Dupuis L, Spielberg S. Oral decongestants: facts and fiction. On Contin Pract 1985;12:22.

578. Wong I, Tavernor S, Tavernor R. Psychiatric adverse effects of anticonvulsant drugs: incidence and therapeutic implications. CNS Drugs 1997;8:492–509.

579. Nousiainen I, Kalviainen R, Mantyjarvi M. Color vision in epilepsy patients treated with vigabatrin or carbamazepine monotherapy. Ophthalmology 2000;107(5):884–8.

580. Tsiouris JA, Patti PJ, Tipu O, Raguthu S. Adverse effects of phenytoin given for late-onset seizures in adults with Down syndrome. Neurology 2002;59(5):779–80.

581. Weiner WJ, Factor SA, Jankovic J, Hauser RA, Tetrud JW, Waters CH, Shulman LM, Glassman PM, Beck B, Paume D, Doyle C. The long-term safety and efficacy of pramipexole in advanced Parkinson's disease. Parkinsonism Relat Disord 2001;7(2):115–20.

582. Bizjak ED, Nolan PE Jr, Brody EA, Galloway JM. Procainamide-induced psychosis: a case report and review of the literature. Ann Pharmacother 1999;33(9):948–51.

583. Andreen L, Bixo M, Nyberg S, Sundstrom-Poromaa I, Backstrom T. Progesterone effects during sequential hormone replacement therapy. Eur J Endocrinol 2003;148:571–7.

584. Paya B, Guisado JA, Vaz FJ, Crespo-Facorro B. Visual hallucinations induced by the combination of prolintane and diphenhydramine. Pharmacopsychiatry 2002;35(1):24–5.

585. Gozal D, Gozal Y. Behavior disturbances with repeated propofol sedation in a child. J Clin Anesth 1999;11(6):499.

586. Seppelt IM. Neurotoxicity from overuse of nitrous oxide. Med J Aust 1995;163(5):280.

587. Yamaguchi S, Mishio M, Okuda Y, Kitajima T. [A patient with drug abuse who developed multiple psychotic symptoms during sedation with propofol.]Masui 1998;47(5):589–92.

588. Gunduz H, Georges JL, Fleishman S. Quinapril and depression. Am J Psychiatry 1999;156(7):1114–5.

589. Tarlow MM, Sakaris A, Scoyni R, Wolf-Klein G. Quinapril-associated acute psychosis in an older woman. J Am Geriatr Soc 2000;48(11):1533.

590. Stocky A. Ranitidine and depression. Aust NZ J Psychiatry 1991;25(3):415–8.

591. Delerue O, Muller JP, Destee A, Warot P. Mania-like episodes associated with ranitidine. Am J Psychiatry 1988;145(2):271.

592. Tamarin F, Brandstetter RD, Price W. Mental confusion and ranitidine. Crit Care Med 1988;16(8):819.

593. Sonnenblick M, Yinnon A. Mental confusion as a side effect of ranitidine. Am J Psychiatry 1986;143(2):257.

594. Cartwright DP, Kvalsvik O, Cassuto J, Jansen JP, Wall C, Remy B, Knape JT, Noronha D, Upadhyaya BK. A randomized, blind comparison of remifentanil and alfentanil during anesthesia for outpatient surgery. Anesth Analg 1997;85(5):1014–9.

595. Riddiough MA. Preventing, detecting and managing adverse reactions of antihypertensive agents in the ambulant patient with essential hypertension. Am J Hosp Pharm 1977;34(5):465–79.

596. Samuels AH, Taylor AJ. Reserpine withdrawal psychosis. Aust N Z J Psychiatry 1989;23(1):129–30.

597. Smith DJ, Yukhnevich S. Adverse reactions to rivastigmine in three cases of dementia. Aust N Z J Psychiatry 2001;35(5):694–5.

598. Finlayson JA, Laing RB. Acute paranoid reaction to saquinavir. Am J Health Syst Pharm 1998;55(19):2016–7.

599. Wells LT, Rasch DK. Emergence "delirium" after sevoflurane anesthesia: a paranoid delusion? Anesth Analg 1999;88(6):1308–10.

600. Kulka PJ, Bressem M, Tryba M. Clonidine prevents sevoflurane-induced agitation in children. Anesth Analg 2001;93(2):335–8.

601. Tabak F, Mert A, Ozaras R, Biyikli M, Ozturk R, Ozbay G, Senturk H, Aktuglu Y. Losartan-induced hepatic injury. J Clin Gastroenterol 2002;34(5):585–6.

602. Neunteufl T, Berger R, Pacher R. Endothelin receptor antagonists in cardiology clinical trials. Expert Opin Investig Drugs 2002;11(3):431–43.

603. Kulka PJ, Bressem M, Wiebalck A, Tryba M. Prophylaxe des "Postsevoflurandelirs" mit Midazolam. [Prevention of "post-sevoflurane delirium" with midazolam.] Anaesthesist 2001;50(6):401–5.

604. Iwasaki S, Yoshimura A, Ideura T, Koshikawa S, Sudo M. Elimination study of silver in a hemodialyzed burn patient treated with silver sulfadiazine cream. Am J Kidney Dis 1997;30(2):287–90.

605. Pedersen TR, Tobert JA. Benefits and risks of HMG-CoA reductase inhibitors in the prevention of coronary heart disease: a reappraisal. Drug Saf 1996;14(1):11–24.

606. Wade A, Reynolds JE. In: Sulfonamides. London: The Pharmaceutical Press, 1982:1457.

607. Kruis R, Barger R. Paranoid psychosis with sulindac. JAMA 1980;243(14):1420.

608. Weglage J, Pietsch M, Sprinz A, Feldmann R, Denecke J, Kurlemann G. A previously unpublished side effect of sulthiame in a patient with Rolandic epilepsy. Neuropediatrics 1999;30(1):50.

609. Kemper MJ, Sparta G, Laube GF, Miozzari M, Neuhaus TJ. Neuropsychologic side-effects of tacrolimus in pediatric renal transplantation. Clin Transplant 2003;17:130–4.

610. Sokol DK, Molleston JP, Filo RS, Van Valer J, Edwards-Brown M. Tacrolimus (FK506)-induced mutism after liver transplant. Pediatr Neurol 2003;28:156–8.

611. Muttray A, Randerath W, Ruhle KH, Gajsar H, Gerhardt P, Greulich W, Konietzko J. Obstruktives Schlafapnoesyndrom durch eine berufliche Losungsmittelexposition. [Obstructive sleep apnea syndrome caused by occupational exposure to solvents.] Dtsch Med Wochenschr 1999;124(10):279–81.

612. Tseng S, Pak G, Washenik K, Pomeranz MK, Shupack JL. Rediscovering thalidomide: a review of its mechanism of action, side effects, and potential uses. J Am Acad Dermatol 1996;35(6):969–79.

613. Clark TE, Edom N, Larson J, Lindsey LJ. Thalomid (thalidomide) capsules: a review of the first 18 months of spontaneous postmarketing adverse event surveillance, including off-label prescribing. Drug Saf 2001;24(2):87–117.

614. Morgan AE, Smith WK, Levenson JL. Reversible dementia due to thalidomide therapy for multiple myeloma. N Engl J Med 2003;348(18):1821–2.

615. International Coffee Organization. United States of America: coffee drinking studyLondon: International Coffee Organization;. 1989.

616. Evans DL, Strawn SK, Haggerty JJ Jr, Garbutt JC, Burnett GB, Pedersen CA. Appearance of mania in drug-resistant bipolar depressed patients after treatment

with L-triiodothyronine. J Clin Psychiatry 1986;47(10):521–2.

617. Josephson AM, Mackenzie TB. Thyroid-induced mania in hypothyroid patients. Br J Psychiatry 1980;137:222–8.

618. El Kaissi S, Kotowicz MA, Berk M, Wall JR. Acute delirium in the setting of primary hypothyroidism: the role of thyroid hormone replacement therapy. Thyroid 2005;15(9):1099–101.

619. Goldstein BI, Levitt AJ. Thyroxine-associated hypomania. J Am Acad Child Adolesc Psychiatry 2005;44(3):211.

620. Trimble MR, O'Donoghue M, Sander L, Duncan J. Psychosis with tiagabine. Epilepsia 1997;38(Suppl 3):40.

621. Schweitzer I, Maguire K, Tuckwell V. Antiglaucoma medication and clinical depression. Aust NZ J Psychiatry 2001;35(5):569–71.

622. McCartney CF, Hatley LH, Kessler JM. Possible tobramycin delirium. JAMA 1982;247(9):1319.

623. Womack KB, Heilman KM. Tolterodine and memory. Arch Neurol 2003;60:771–3.

624. Shorvon SD. Safety of topiramate: adverse events and relationships to dosing. Epilepsia 1996;37(Suppl 2):S18–22.

625. Klufas A, Thompson D. Topiramate-induced depression. Am J Psychiatry 2001;158(10):1736.

626. Stella F, Caetano D, Cendes F, Guerreiro CA. Acute psychotic disorders induced by topiramate: report of two cases. Arq Neuropsiquiatr 2002;60(2-A):285–7.

627. Goldberg JF. Panic attacks associated with the use of topiramate. J Clin Psychopharmacol 2001;21(4):461–2.

628. Aldenkamp AP. Effects of antiepileptic drugs on cognition. Epilepsia 2001;42(Suppl 1):46–9.

629. Martin R, Kuzniecky R, Ho S, Hetherington H, Pan J, Sinclair K, Gilliam F, Faught E. Cognitive effects of topiramate, gabapentin, and lamotrigine in healthy young adults. Neurology 1999;52(2):321–7.

630. Rorsman I, Kallen K. Recovery of cognitive and emotional functioning following withdrawal of topiramate maintenance therapy. Seizure 2001;10(8):592–5.

631. Biton V, Edwards KR, Montouris GD, Sackellares JC, Harden CL, Kamin MTopiramate TPS-TR Study Group. Topiramate titration and tolerability. Ann Pharmacother 2001;35(2):173–9.

632. Thompson PJ, Baxendale SA, Duncan JS, Sander JW. Effects of topiramate on cognitive function. J Neurol Neurosurg Psychiatry 2000;69(5):636–41.

633. Tatum WO IV, French JA, Faught E, Morris GL III, Liporace J, Kanner A, Goff SL, Winters L, Fix APADS Investigators. Post-marketing antiepileptic drug survey. Postmarketing experience with topiramate and cognition. Epilepsia 2001;42(9):1134–40.

634. Hamiwka LD, Gerber PE, Connolly MB, Farrell K. Topiramate-associated behavioural changes in children. Epilepsia 1999;40(Suppl 7):116.

635. Mohamed K, Appleton R, Rosenbloom L. Efficacy and tolerability of topiramate in childhood and adolescent epilepsy: a clinical experience. Seizure 2000;9(2):137–41.

636. Galvez-Jimenez N, Hargreave M. Topiramate and essential tremor. Ann Neurol 2000;47(6):837–8.

637. Aldenkamp AP, Baker G, Mulder OG, Chadwick D, Cooper P, Doelman J, Duncan R, Gassmann-Mayer C, de Haan GJ, Hughson C, Hulsman J, Overweg J, Pledger G, Rentmeester TW, Riaz H, Wroe S. A multicenter, randomized clinical study to evaluate the effect on cognitive function of topiramate compared with valproate as add-on therapy to carbamazepine in patients with partial-onset seizures. Epilepsia 2000;41(9):1167–78.

638. Meador KJ, Hulihan JF, Karim R. Cognitive function in adults with epilepsy; effect of topiramate and valproate added to carbamazepine. Epilepsia 2001;42(Suppl 2):75.

639. Meador KJ. Effects of topiramate on cognition. J Neurol Neurosurg Psychiatry 2001;71(1):134–5.

640. Meador KJ, Loring DW, Hulihan JF, Kamin M, Karim R; CAPSS-027 Study Group. Differential cognitive and behavioral effects of topiramate and valproate. Neurology 2003;60:1483–8.

641. Lee S, Sziklas V, Andermann F, Farnham S, Risse G, Gustafson M, Gates J, Penovich P, Al-Asmi A, Dubeau F, Jones-Gotman M. The effects of adjunctive topiramate on cognitive function in patients with epilepsy. Epilepsia 2003;44:339–47.

642. Mula M, Trimble MR, Thompson P, Sander JW. Topiramate and word-finding difficulties in patients with epilepsy. Neurology 2003;60:1104–7.

643. Kockelmann E, Elger CE, Helmstaedter C. Significant improvement in frontal lobe associated neuropsychological functions after withdrawal of topiramate in epilepsy patients. Epilepsy Res 2003;54:171–8.

644. Gerber PE, Hamiwka L, Connolly MB, Farrell K. Factors associated with behavioral and cognitive abnormalities in children receiving topiramate. Pediatr Neurol 2000;22(3):200–3.

645. Franz DN, Glauser TA, Tudor C, Williams S. Topiramate therapy of epilepsy associated with Angelman's syndrome. Neurology 2000;54(5):1185–8.

646. Ojemann LM, Crawford CA, Dodrill CB, Holmes MD, Kutsy R, Wilensky AJ. Language disturbances as a side effect of topiramate and zonisamide therapy. Epilepsia 1999;40(Suppl 7):66.

647. Khan A, Faught E, Gilliam F, Kuznliecky R. Acute psychotic symptoms induced by topiramate. Seizure 1999;8(4):235–7.

648. Mula M, Trimble MR, Lhatoo SD, Sander JW. Topiramate and psychiatric adverse events in patients with epilepsy. Epilepsia 2003;44:659–63.

649. Litman LC. Sexual sadism with lust-murder proclivities in a female? Can J Psychiatry 2003;48:127.

650. Ozkara C, Ozmen M, Erdogan A, Yalug I. Topiramate related obsessive–compulsive disorder. Eur Psychiatry 2005;20:78–9.

651. Kaplan M. Hypomania with topiramate. J Clin Psychopharmacol 2005;25:196–7.

652. Duggal HS, Singh I. Worsening of psychosis or topiramate-induced adverse event? Gen Hosp Psychiatry 2004;26(3):245–7.

653. Bonanni E, Galli R, Maestri M, Pizzanelli C, Fabbrini M, Manca ML, Iudice A, Murri L. Daytime sleepiness in epilepsy patients receiving topiramate monotherapy. Epilepsia 2004;45(4):333–7.

654. Devulder J, De Laat M, Dumoulin K, Renson A, Rolly G. Nightmares and hallucinations after long-term intake of tramadol combined with antidepressants. Acta Clin Belg 1996;51(3):184–6.

655. Keeley PW, Foster G, Whitelaw L. Hear my song: auditory hallucinations with tramadol hydrochloride. BMJ 2000;321(7276):1608.

656. Rahill AA, Brown GG, Fagan SC, Ewing JR, Branch CA, Balakrishnan G. Neuropsychological dose effects of a freon, trifluoromethane (FC-23), compared to N_2O. Neurotoxicol Teratol 1998;20(6):617–26.

657. Marsden CD, Marion MH, Quinn N. The treatment of severe dystonia in children and adults. J Neurol Neurosurg Psychiatry 1984;47(11):1166–73.

658. Tanabe K, Yokochi F, Hirai S, Mori H, Suda K, Kondo T, Mizuno Y. [A 75-year-old man with parkinsonism and delirium.]No To Shinkei 1994;46(1):85–92.

659. Salkind AR. Acute delirium induced by intravenous trimethoprim–sulfamethoxazole therapy in a patient with the acquired immunodeficiency syndrome. Hum Exp Toxicol 2000;19(2):149–51.

660. Wong AYK, Tang L. An open and randomised study comparing the efficacy of standard danazol and modified triptorelin regimens for postoperative disease management of moderate to severe endometriosis. Fertil Steril 2004;81:1522–7.

661. Meyers CA, Valentine AD, Wong FC, Leeds NE. Reversible neurotoxicity of interleukin-2 and tumor necrosis factor: correlation of SPECT with neuropsychological testing. J Neuropsychiatry Clin Neurosci 1994;6(3):285–8.

662. Feinberg JE, Hurwitz S, Cooper D, Sattler FR, MacGregor RR, Powderly W, Holland GN, Griffiths PD, Pollard RB, Youle M, Gill MJ, Holland FJ, Power ME, Owens S, Coakley D, Fry J, Jacobson MA. A randomized, double-blind trial of valaciclovir prophylaxis for cytomegalovirus disease in patients with advanced human immunodeficiency virus infection. AIDS Clinical Trials Group Protocol 204/Glaxo Wellcome 123-014 International CMV Prophylaxis Study Group. J Infect Dis 1998;177(1):48–56.

663. Lowance D, Neumayer HH, Legendre CM, Squifflet JP, Kovarik J, Brennan PJ, Norman D, Mendez R, Keating MR, Coggon GL, Crisp A, Lee ICInternational Valacyclovir Cytomegalovirus Prophylaxis Transplantation Study Group. Valacyclovir for the prevention of cytomegalovirus disease after renal transplantation. N Engl J Med 1999;340(19):1462–70.

664. Izzedine H, Launay-Vacher V, Aymard G, Legrand M, Deray G. Pharmacokinetic of nevirapine in haemodialysis. Nephrol Dial Transplant 2001;16(1):192–3.

665. Linssen-Schuurmans CD, van Kan EJ, Feith GW, Uges DR. Neurotoxicity caused by valacyclovir in a patient on hemodialysis. Ther Drug Monit 1998;20(4):385–6.

666. Chadwick DW, Cumming WJ, Livingstone I, Cartlidge NE. Acute intoxication with sodium valproate. Ann Neurol 1979;6(6):552–3.

667. American Academy of Pediatrics. Committee on Drugs. Behavioral and cognitive effects of anticonvulsant therapy. Pediatrics 1985;76(4):644–7.

668. Ronen GM, Richards JE, Cunningham C, Secord M, Rosenbloom D. Can sodium valproate improve learning in children with epileptiform bursts but without clinical seizures? Dev Med Child Neurol 2000;42(11):751–5.

669. Kastalli K, Aïdli S, Klouz A, Sraïri S. Nightmares induced by valsartan. Pharmacoepidemiol Drug Saf 2003;12(Suppl 2):236.

670. Ullrich H, Passenberg P, Agelink MW. Episodes of depression with attempted suicide after taking valsartan with hydrochlorothiazide. Dtsch Med Wochenschr 2003;128(48):2534–6.

671. Ferrie CD, Robinson RO, Panayiotopoulos CP. Psychotic and severe behavioural reactions with vigabatrin: a review. Acta Neurol Scand 1996;93(1):1–8.

672. Levinson D, Mumford J. Vigabatrin and psychosis. Epilepsia 1995;36(Suppl. 4):32.

673. Beran RG, Berkovic SF, Buchanan N, Danta G, Mackenzie R, Schapel G, Sheean G, Vajda F. A double-blind, placebo-controlled crossover study of vigabatrin 2 g/day and 3 g/day in uncontrolled partial seizures Seizure 1996;5(4):259–65.

674. Guberman A, Andermann F, McLachlan R, Savard G, Manchanda RThe Vigabatrin Behavioral Effects Study Group. Severe behavioral/psychiatric reactions on vigabatrin: a Canadian study. Epilepsia 1996;37(Suppl. 5):170.

675. Levinson DF, Devinsky O. Psychiatric adverse events during vigabatrin therapy. Neurology 1999;53(7):1503–11.

676. Crispi G, Di Lorenzo RS, Gentile A, Florino A, Pannone B, Sciorio G. Azione psicoattiva della vincamina in un gruppo di soggetti affetti da "sindrome depressiva recidivante". Nota preventiva. [Psychoactive effect of vincamine in a group of subjects affected by recurrent depressive syndrome. Preliminary note.] Minerva Med 1975;66(70):3683–5.

677. Haupt R. Akute symptomatische Psychose bei Vitamin A-Intoxikation. [Acute symptomatic psychosis in vitamin A intoxication.] Nervenarzt 1977;48(2):91–5.

678. Goodman GE, Alberts DS, Earnst DL, Meyskens FL. Phase I trial of retinol in cancer patients. J Clin Oncol 1983;1(6):394–9.

679. Wysowski DK, Pitts M, Beitz J. Depression and suicide in patients treated with isotretinoin. New Engl J Med 2001;344:460.

680. Enders SY, Enders JM. Isotretinoin and psychiatric illness in adolescents and young adults. Ann Pharmacother 2003;37:1124–7.

681. Barak YAC, Wohl YBC, Greenberg YA, Dayan YBB, Friedman TB, Shoval GA, Knobler HAD. Affective psychosis following Accutane (isotretinoin) treatment. Int Clin Psychopharmacol 2005;20:39–41.

682. Wysowski DK, Pitts M, Beitz J. An analysis of reports of depression and suicide in patients treated with isotretinoin. J Am Acad Dermatol 2001;45(4):515–9.

683. Wysowski DK, Pitts M, Beitz J. Depression and suicide in patients treated with isotretinoin. N Engl J Med 2001;344(6):460.

684. Gupta MA, Gupta AK. Isotretinoin use and reports of sustained dreaming. Br J Dermatol 2001;144(4):919–20.

685. Agrawal AK, Sherman LK. Voriconazole-induced musical hallucinations. Infection 2004;32:293–5.

686. Bedi A, McCarroll C, Murray JM, Stevenson MA, Fee JP. The effects of subanaesthetic concentrations of xenon volunteers. Anaesthesia 2002;57(3):233–41.

687. Goddard AW, Woods SW, Sholomskas DE, Goodman WK, Charney DS, Heninger GR. Effects of the serotonin reuptake inhibitor fluvoxamine on yohimbine-induced anxiety in panic disorder. Psychiatry Res 1993;48(2):119–33.

688. Price LH, Charney DS, Heninger GR. Three cases of manic symptoms following yohimbine administration. Am J Psychiatry 1984;141(10):1267–8.

689. Morgan CA 3rd, Grillon C, Southwick SM, Nagy LM, Davis M, Krystal JH, Charney DS. Yohimbine facilitated acoustic startle in combat veterans with post-traumatic stress disorder. Psychopharmacology (Berl) 1995;117(4):466–71.

690. Miyamoto T, Kohsaka M, Koyama T. Psychotic episodes during zonisamide treatment. Seizure 2000;9(1):65–70.

691. Murai T, Kubota Y, Sengoku A. Unknown people believed to be known: the "assoziierende Erinnerungs falschungen" by Kraepelin. Psychopathology 2000;33(1):52–4.

692. Akman CI, Goodkin HP, Rogers DP, Riviello JJ Jr. Visual hallucinations associated with zonisamide. Pharmacotherapy 2003;23:93–6.

693. Hirose M, Yokoyama H, Haginoya K, Iinuma K. [A five-year-old girl with epilepsy showing forced normalization due to zonisamide] No To Hattatsu 2003;35:259–63.

Index of drug names

Note: The letter '*t*' with the locater refers to tables.

Printed and bound by CPI Group (UK) Ltd, Croydon, CR0 4YY

03/10/2024

01040330-0020